POWER SYSTEM PROTECTION

IEEE Press
445 Hoes Lane, P.O. Box 1331
Piscataway, NJ 08855-1331

POWER SYSTEM PROTECTION

P. M. Anderson
Power Math Associates, Inc.

IEEE Power Engineering Society, *Sponsor*

IEEE Press Power Engineering Series
P. M. Anderson, *Series Editor*

WILEY-INTERSCIENCE
A JOHN WILEY & SONS, INC., PUBLICATION

IEEE PRESS

The Institute of Electrical
and Electronics Engineers, Inc.,
New York

For general information on our other products and services please contact our Customer Care Department within the U.S. at 877-762-2974, outside the U.S. at 317-572-3993 or fax 317-572-4002.

Library of Congress Cataloging-in-Publication Data

Anderson, P.M. (Paul M.), 1926—
Power system protection / by P.M. Anderson.
p. cm. -- (IEEE Press power engineering series)
Includes bibliographical references and index.
ISBN 0-7803-3427-2
1. Electric power systems—Protection. I. Title. II Series.
TL1010.A53 1998
621.31'7—dc21 98-28659
CIP

Printed in the United States of America
10 9 8 7 6 5 4 3

To Ginny

BOOKS IN THE IEEE PRESS POWER ENGINEERING SERIES

ELECTRIC POWER APPLICATIONS OF FUZZY SYSTEMS
Mohamed E. El-Hawary, *Dalhousie University*
1998 Hardcover 384 pp IEEE Order No. PC5666 ISBN 0-7803-1197-3

RATING OF ELECTRIC POWER CABLES: Ampacity Computations for Transmission, Distribution, and Industrial Applications
George J. Anders, *Ontario Hydro Technologies*
1997 Hardcover 464 pp IEEE Order No. PC5647 ISBN 0-7803-1177-9

ANALYSIS OF FAULTED POWER SYSTEMS, Revised Printing
Paul M. Anderson, *Power Math Associates, Inc.*
1995 Hardcover 536 pp IEEE Order No. PC5616 ISBN 0-7803-1145-0

ELECTRIC POWER SYSTEMS: Design and Analysis, Revised Printing
Mohamed E. El-Hawary, *Dalhousie University*
1995 Hardcover 808 pp IEEE Order No. PC5606 ISBN 0-7803-1140-X

POWER SYSTEM STABILITY, VOLUMES I, II, III
An IEEE Press Classic Reissue Set
Edward Wilson Kimbark, *Iowa State University*
1995 Softcover 1008 pp IEEE Order No. PC5600 ISBN 0-7803-1135-3

ANALYSIS OF ELECTRIC MACHINERY
Paul C. Krause and Oleg Wasynczuk, *Purdue University*
Scott D. Sudhoff, *University of Missouri at Rolla*
1994 Hardcover 584 pp IEEE Order No. PC4556 ISBN 0-7803-1101-9

POWER SYSTEM CONTROL AND STABILITY, Revised Printing
Paul M. Anderson, *Power Math Associates, Inc.*
A. A. Fouad, *Iowa State University*
1993 Hardcover 480 pp IEEE Order No. PC3789 ISBN 0-7803-1029-2

SUBSYNCHRONOUS RESONANCE IN POWER SYSTEMS
P. M. Anderson, *Power Math Associates, Inc.*
B. L. Agrawal, *Arizona Public Service Company*
J. E. Van Ness, *Northwestern University*
1990 Softcover 282 pp IEEE Order No. PP2477 ISBN 0-7803-5350-1

Contents

Preface

This is an academic book on power system protection. It was prepared to serve as an introductory text in power system protection for graduate students in Electric Power Systems. The present edition is the outgrowth of previous editions that have been used by several universities for 25 years.

There is a fundamental problem in developing a graduate course in power system protection at a university. Most faculty members are not experts at relaying, and very few practice this art on a regular basis. Without consistent contact with the field, it is difficult to keep up-to-date, particularly with respect to new relay hardware. But there is a more fundamental problem in this regard. One has to question whether the teaching of protection hardware and relay settings is appropriate material for academic study. Therefore, I chose to concentrate on the analytical techniques that are useful in system protection, with very little emphasis on the hardware that is used to implement a protective scheme. Analysis is one subject that the university professor can teach with authority, and is one that many practicing engineers need to understand more thoroughly. Therefore, it is hoped that this book will be of value to both the graduate student and the practicing protection engineer.

In writing a book such as this, the writer stands on the shoulders of giants. I was greatly influenced in the preparation of this work by the excellent texts that have been available for some time. The fine book by Mason stands as a classic in this field and is a necessary reference. The excellent books of Warrington are recommended as references, particularly because of their thorough coverage of relay hardware, a subject that is purposely omitted here. The fine book by Atabekov is recommended for its excellent analytical treatment. The excellent work of Horowitz and Phadke provides an authoritative treatment of the subject that is recommended. Finally, the manufacturers all publish very good material on relays and relay applications. Several of these documents are referenced throughout the book.

The treatment of the subject is presented at the graduate level; that is, it is assumed that the reader has a bachelor's degree in electrical engineering. Moreover, it is assumed that the reader has mastered those power system fundamentals that are covered in the excellent introductory books of Stevenson and Grainger, Gross, Neuenswander, Gönen, and El-Hawary.

These fundamentals are not reviewed here. It is also presumed that the reader is familiar with symmetrical components and with the mathematical principles of functions of a complex variable. We also assume that the reader is familiar with computers and computer programming, such that the computer can be used as a tool in problem solving.

This book has been under development for at least 25 years. The author taught this subject for many years and shared his notes with professors from other universities. The comments were gratifying, and encouraged the development of a full textbook on the subject of system protection. There is no doubt that protection is one of the most demanding technical subjects in power systems, and protection engineers are a breed apart. The author shares an intense interest in the subject with these masters of the art and science of protection. Most of the engineers who practice power system protection learn this craft by practical application—the author included. There is an opportunity, however, for the academic pursuit of protection as a science in its own right. Therefore, I hope this book assists those who make this study their life's work—a noble pursuit and one that I enthusiastically recommend.

A family of computer programs is included as an attachment to the book. These programs are briefly described as follows:

`STUFLT` A student fault program that includes a feature for finding the fault equivalent for any network branch, as described in Chapter 5.

`RELAY` A small computer program used to compute the fault current and apparent impedance for a three-phase fault placed at any location along a series compensated transmission line. See Chapter 15 for examples.

`PWRMAT` A program developed by the power faculty at Iowa State University to perform complex matrix analysis typical of power system problems. A maximum of 30 complex matrices of up to 400 elements can be stored and manipulated.

`FREQ` A program used to compute the frequency trajectory as a function of time following the islanding of a portion of a power system, where the island has an imbalance between generation and load. Problems of this type are described in Chapter 20, and this program was used for the computations of frequency vs. time found in that chapter.

Once data files are constructed as text files and named, the user simply double-clicks on the program icon and identifies the data file by name in order to launch the computation and view the results. Typical data files are provided for all programs so that the reader can duplicate text computations or produce results for similar problems of interest. Brief descriptive manuals are provided for all of the programs.

P. M. Anderson
Power Math Associates, Inc.

Acknowledgments

In preparing the manuscript of a technical book, the author must draw on many different resources. I have tried throughout the book to provide appropriate citations to the source of information. If I have neglected to mention anyone who provided information or advice, I apologize for this lapse in memory and assure the injured party that it was not intentional. I would particularly like to cite the following individuals who made a special effort on my behalf by providing inspiration, advice, data, or other items that have been found useful and helpful.

The data on current transformer saturation in Chapter 2 were provided by Bill Kotheimer, who has made a special study of this phenomenon and has written a useful computer program to perform this type of analysis. The author has made a special effort to obtain data for fuses in the form of mathematical equations, but this was unsuccessful. I was able, however, to obtain digital files representing the common types of ANSI fuses from the S&C Electric Company. Their contribution of these data is acknowledged. Also appearing in Chapter 3 is the description of electromechanical relays in equation form, which were provided by Stan Zocholl of Schweitzer Engineering Laboratories.

The description of pilot protection schemes presented in Chapter 13 was a difficult subject for the author due to the many different schemes and in the details associated with their installation. The author is indebted to Alan Taylor of the Los Angeles Department of Water and Power for his patient guidance and assistance on this subject.

Chapter 22 on HVDC protection was reviewed by Don Martin of ABB, who made a number of valuable comments and suggestions. Chapter 23 on SSR protection was reviewed by R. G. Farmer, and data were provided by Baj Agrawal. Their contributions are greatly appreciated.

The author is also indebted to many academic colleagues throughout the country who offered many suggestions and comments on the book. Special thanks are extended to Dr. James Tudor of the University of Missouri, Dr. Turan Gönen of Sacramento State University, and Dr. Bruce Wollenberg of the University of Minnesota, and to their students, who thoroughly tested draft manuscripts and offered many helpful comments. Thanks also to Professor John Pavlat and Dr. Ken Kruempel of Iowa State University for permitting use of their software.

A special thanks goes to Washington State University for affording the author the opportunity to test the manuscript in an actual classroom situation during the 1996–1997 academic year. During this time the author was appointed as the Schweitzer Visiting Professor to continue a Washington State University program sponsored by the Schweitzer Engineering Laboratory of Pullman, Washington.

The author also thanks his colleagues at Power Math Associates for their special help. Simon Magbuhat helped in system analysis for many simulations, and contributed to the fault tree analysis of Chapter 27. Dr. Sudhir Agarwal prepared the computer program RELAY that is used to examine the effect of fault location on fault measurements. Also, Dr. Mahmood Mirheydar prepared the program FREQ and also modified the fault program to compute fault equivalents for all network branches.

Finally, we acknowledge the expert assistance of the publisher and editors of the book, whose skill and special expertise is acknowledged.

P. M. Anderson
September, 1998

List of Symbols

The general rules used for symbols in this book are as follows. Letter symbols for units of physical quantities (unit symbol abbreviations) are given in Roman or Greek typeface, e.g., A for amperes, VA for voltamperes, or Ω for ohms. Mathematical variables are always given in italic typeface, and may be either Roman or Greek, e.g., v for instantaneous voltage, V for rms phasor voltage magnitude, or λ for failure rate of a component. Complex numbers, such as phasors, are given in bold italic typeface, e.g., $\boldsymbol{I}$ for current and $\boldsymbol{V}$ for voltage. Matrices are specified by bold Roman typeface, e.g., $\mathbf{Z}$ for an impedance matrix. The following tabulations give specific examples of symbols used in this book.

1. CAPITALS

A	ampere; unit symbol abbreviation for current
$\boldsymbol{A}$	complex two-port network transmission parameter
B	$=\mathrm{Im}\,\boldsymbol{Y}$, susceptance
$\boldsymbol{B}$	complex two-port transmission parameter
C	capacitance
CTI	coordination time interval
$\boldsymbol{C}$	complex two-port transmission parameter
D	damping constant
$\boldsymbol{D}$	complex two-port transmission parameter
E	source emf; voltage
E_S	source voltage, relay equivalent circuit
E_U	source voltage, relay equivalent circuit
F	failure probability
F	farad; unit symbol abbreviation for capacitance
G	$=\mathrm{Re}\,\boldsymbol{Y}$, conductance
H	inertia constant
H	henry, unit symbol abbreviation for inductance
Hz	hertz, unit symbol abbreviation for frequency

I	rms current magnitude
$\boldsymbol{I}$	rms phasor current
K	spring constant; controller gain
L	inductance
LL	line-to-line
LN	line-to-neutral
M	$=10^6$, mega, a prefix
M	matrix defining relay voltage and current
MOC	minimum operating current, overcurrent relay
N	newton, unit symbol abbreviation for force
P	average real power
P_n	natural power of a transmission line, *SIL*
Q	average reactive power, unavailability
R	$=\text{Re}\mathbf{Z}$, resistance
S	$=P+jQ$, complex apparent power
S_B	base complex or apparent power, VA
SIL	surge impedance loading of a transmission line
T	time constant
V	rms voltage magnitude
$\boldsymbol{V}$	rms phasor voltage
V	volt, unit symbol abbreviation for voltage
VA	volt-ampere, unit symbol abbreviation for apparent power
VAR	volt-ampere, reactive, unit symbol abbreviation for var
W	number of failures in a specified time period
W	watt, unit symbol abbreviation for active power
X	$=\text{Im}\mathbf{Z}$, reactance
$\boldsymbol{Y}$	$=G+jB$, complex admittance
$\boldsymbol{Y}_C$	characteristic admittance of a transmission line + $1/\mathbf{Z}_C$
Y	admittance magnitude
Y	admittance matrix
$\mathbf{Z}$	$=R+jX$, complex impedance
$\mathbf{Z}_R$	impedance seen at the relay at R
$\mathbf{Z}_E$	external system impedance of a network equivalent
$\mathbf{Z}_S$	source impedance of a network equivalent
$\mathbf{Z}_U$	source impedance of a network equivalent
$\mathbf{Z}_F$	fault impedance
Z	impedance magnitude
Z	impedance matrix

2. LOWERCASE

ac	alternating current
a-b-c	phase designation of three phase currents or voltages
b	$=\omega c$, line susceptance per unit length
c	capacitance per unit length
dc	direct current
det	determinant of a matrix
e	base for natural logarithms
f	frequency
h	distance measurement parameter

i	instantaneous current designation
j	$\sqrt{-1}$, a 90° operator
k	= 10^3, kilo, an mks prefix
k	degree of series compensation
l	inductance per unit length
ℓ	length of a transmission line
ln, log	natural (base e) logarithm, base 10 logarithm
m	= 10^{-3}, milli..., an mks prefix
m	mutual inductance per unit length
pu	per unit
p	instantaneous power; Maxwell's coefficient
r	apparatus resistance
s	distance along a transmission line
t	time
v	instantaneous voltage
var	voltampere reactive, when used as a noun or adjective
w	frequency of failure
x	line reactance per unit length; apparatus reactance
y	line shunt admittance per unit length
z	line impedance per unit length

3. UPPERCASE GREEK

Δ	delta connection; determinant (of a matrix)
Σ	summation notation
Ω	ohm, unit symbol abbreviation for impedance

4. LOWERCASE GREEK

γ	propagation constant of a transmission line
δ	voltage phase angle
ζ	magnitude of impedance
θ	phase angle of voltage or current, angle of impedance
λ	item failure in reliability calculations
μ	= 10^{-6}, micro, an mks prefix
π	pi
ρ	resistivity, magnitude of admittance
τ	time constant
ϕ	phase angle of voltage or current, angle of admittance
ω	radian frequency

5. SUBSCRIPTS

a	phase a
A	phase a
b	phase b
B	phase b
B	base quantity
c	phase c
C	phase c

LN	line-to-neutral
LL	line-to-line
max	maximum
min	minimum
R	receiving end, of a transmission line
S	sending end, of a transmission line
u	per unit, sometimes used for clarity

6. SUPERSCRIPTS AND OVERBARS

$(\ )^t$	transpose (of a matrix)
$(\ )^{-1}$	inverse (of a matrix)
$\tilde{A}$	tilde, distinguishing mark for various quantities
$\hat{A}$	circumflex, distinguishing mark for various quantities
$(\ )^*$	conjugate, of a phasor or matrix

PART I
PROTECTIVE DEVICES AND CONTROLS

1

Introduction

1.1 POWER SYSTEM PROTECTION

The purpose of this book is to present an introductory treatment of power system protection. This includes a study of common types of abnormal occurrences, such as faults, that can lead to power system failure, and the methods for detecting and clearing these abnormalities to restore normal operation. The method of detecting and clearing faults requires the use of special hardware, which is designed for this purpose. This book does not emphasize the hardware, as this requires timely information concerning the products marketed by various manufacturers. Instead, the emphasis will be on analytical techniques that are needed to compute the system conditions at the point of detection and the methods of making detection fast and effective. Most of these techniques are applicable to protective devices of any manufacturer and, therefore, constitute general methods for protective system analysis.

The treatment of the subject of system protection is intended primarily for the engineer or student who is learning the subject. This does not mean, however, that the presentation is elementary. It is presumed that the reader has mastered the usual requirements of the bachelors degree in electrical engineering and is familiar with the use of digital computers. It is also presumed that the reader has a working knowledge of symmetrical components, as presented in [1].[1]

System protection has evolved, over the years, from relatively primitive devices with limited capability, to complex systems that involve extensive hardware. These modern protective systems are more selective in their detection and operation, and often require greater analytical effort in their application. This book is concerned with this analytical effort. We begin with a review of the basic equipment arrangements that provide the protective equipment with its raw data. We then progress to the computation of protective system quantities, beginning with simple circuit arrangements and progressing to more complex situations. This involves a treatment of the Thevenin impedance seen by the protective device at the point of application. This

[1]Numbered references, shown in brackets [ ], are provided at the end of every chapter.

concept is examined in settings designed to protect lines, generators, transformers, and bus structures. Finally, we examine some special topics, including system aspects of protection, and protective system reliability.

1.2 PREVENTION AND CONTROL OF SYSTEM FAILURE

Most of the failure modes in a power system can be controlled to limit damage and thereby enhance reliability. Mechanical failures are controlled by designing the system to withstand all but the unusually severe mechanical loads such as extensive ice buildup, hurricanes, and tornadoes. This is done in a way that tends to minimize the total transport cost of energy, which requires a balance between initial cost and maintenance. Insulation design is coordinated to minimize damage to expensive equipment due to electrical surges. Since it is not economical to design a system to withstand all possible system failures, the alternative is to design a protective system that can quickly detect abnormal conditions and take appropriate action. The type of action taken depends on the protective device and on the environmental condition that is observed by that device. The two basic types of protective systems are defined in this book as follows:

1. *Reactionary devices* These devices are designed to recognize a certain hazard in the power system environment and to take predetermined action to remove that hazard. In most cases, the hazard is related to an abnormal system operating condition that would eventually cause failure of one or more system components. Therefore, the action is usually to isolate that portion of the system experiencing the hazard so that the rest of the system can operate normally.
2. *Safeguard devices* These devices are designed to recognize a certain hazard in the power system environment and to take predetermined action to change that environment to a less hazardous condition.

Each type of device will be discussed briefly.

1.2.1 Reactionary Devices

Reactionary devices are designed to detect a specific system hazard, such as a short circuit, on a system component and to remove that hazard. The power system is not designed to operate continually in the presence of hazards such as short circuits. The protective system action restores the system to the best possible operating condition under the circumstances. In many cases, after testing or inspection, the failed component can often be restored to operation. This may not be true, however, if the failed component were to continue to operate for a long time under the influence of the hazard.

The protection of the system against total failure must be adaptive in the sense that it must operate in an acceptable manner for any kind of abnormal occurrence in a reliable manner. The protective system is required to make decisions quickly, typically in a few milliseconds, and usually on the basis of limited information as to the history and state of the system at a given point of observation.

The condition of the electrical system as it experiences and recovers from a major fault or failure may be characterized in terms of a set of unique system states. Thus, we say that the system is in the *normal state* when all items of equipment that should be in operation are actually working and are operating within normal design limits. When an event occurs that causes the operation of any system component to exceed its normal operating limits, we say that the system has entered an *abnormal state*, which implies that something must be done to relieve the abnormality before a serious failure occurs. Abnormalities can be transient in nature, and, depending on the nature of the condition, it may be prudent to wait a bit before taking any action to see if the abnormality clears itself. If not, we enter an *action state*, in which certain prescribed actions must be taken, usually without further intentional delay. Following this action the system is in an *outage state*, in which the faulted device is removed from service. Since this state is not the desired operating condition, the system is usually caused to enter a *restorative state*, wherein any required inspections or other repair actions are taken in order to again reach the *normal state*. This process is depicted in Figure 1.1.

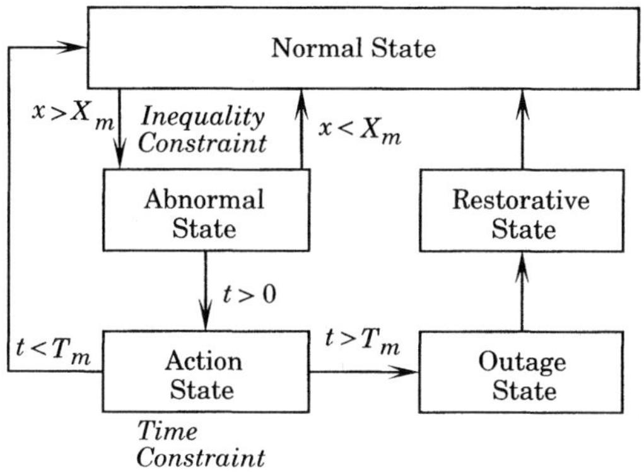

Figure 1.1 Operative states of a protection system.

From Figure 1.1, it is apparent that two conditions must be met in order to remove a component from normal service.

To trip the device:

1. Violate the inequality constraint, $x > X_m$, *and*
2. Violate the time constraint, $t > T_m$.

It is also convenient to think of this process in terms of a decision flow chart, which might be implemented in a computer control system. Such a flow chart is shown in Figure 1.2. Usually the system is in the normal state, and the protective device is set to assume that the normal state prevails at startup. As time is incremented, the protective device checks the observed system variables, represented by x in the flow chart, to determine if any variable exceeds its threshold value. If not, time is incremented to observe the next measured value. If the threshold is exceeded, the time threshold is checked and tripping action is withheld until the time threshold expires. When both the quantity and time thresholds are exceeded, the circuit is tripped. This type of logic is designed to prevent tripping for short, temporary disturbances that might be observed. Such disturbances are often a part of the normal operating condition of the network, and tripping should not be initiated for such events.

In some protective systems, automatic restoration is begun following a preset time delay. This concept has proven valuable since most power system disturbances are temporary, including short circuits. Once the circuit is de-energized, the abnormal condition clears itself, and the circuit can be successfully restored. If this is not a part of the programmed response of the device, the circuit is "locked out" and remains in the outage state until repair personnel can determine the cause of the outage and take appropriate action. This state is shown at the bottom of the flow chart, where the system is in the outage state with no automatic escape.

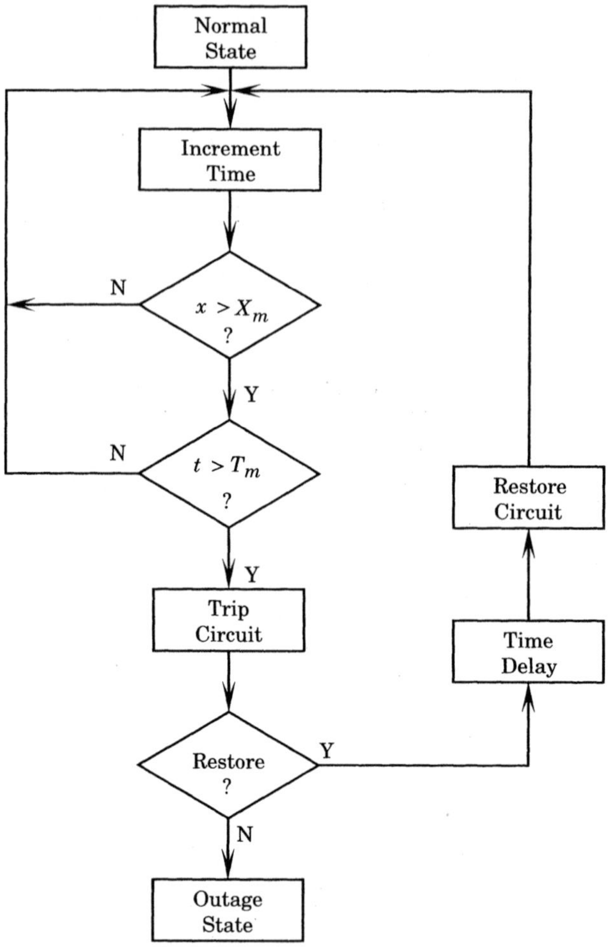

Figure 1.2 Flow chart of the tripping decision process.

1.2.2 Safeguard Devices

Safeguard devices are different from reactionary devices, since safeguards are designed to change the environmental conditions that are the cause of the emergency. Examples of safeguard devices are fire sprinkler systems, apparatus supplementary cooling systems, and detectors that monitor unbalanced currents or voltages in equipment. These systems are designed to change the operating environment when protective sensors detect a specified hazardous condition. Thus, the fire sprinkler puts out the fire and the apparatus supplementary cooling system withdraws additional heat from the operating machine to restore a normal working environment. Some safeguards are designed to alarm the operator, who makes a decision as to the severity of the hazard and may take preemptive action, such as shutting down a facility that is experiencing a hazardous condition. Other safeguards are designed to take action, which may include removing of equipment from service if other remedial actions are not effective.

Some safeguard devices are extremely important in power systems. A good example is the emergency core cooling system of a nuclear reactor, which has the capability of preventing core meltdown following a dangerous reactor failure or accident. Such safeguards must be carefully designed for very high reliability and security because of the high cost of failure.

For the most part, this book deals with reactionary protective systems, although many features of these systems are shared by safeguards as well.

1.2.3 Protective Device Operation

The protective device usually consists of several elements that are arranged to test the system condition, make decisions regarding the normality of observed variables, and take action as required. These elements are depicted graphically in Figure 1.3.

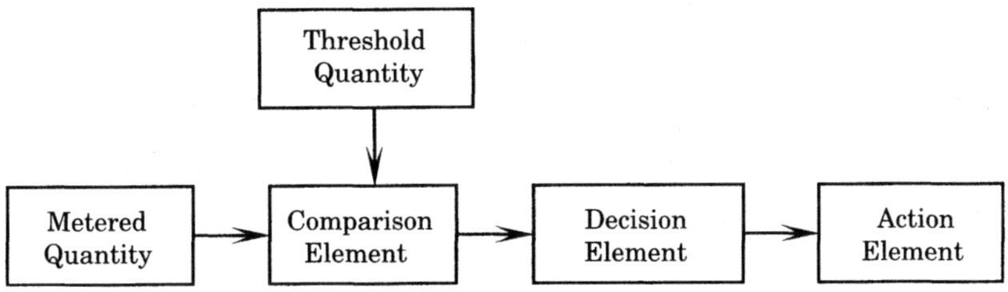

Figure 1.3 Protective device functional elements.

The protective system always measures certain system quantities, such as voltages and currents, and compares these system quantities, or some combination of these quantities, against a threshold setting that is computed by the protection engineer and is set into the device. If this comparison indicates an alert condition, a decision element is triggered. This may involve a timing element, to determine the permanence of the condition, and may require other checks on the system at other points in the network. Finally, if all checks are satisfied, an action element is released to operate, which usually means that circuit breakers are instructed to open and isolate a section of the network.

The time required to take any necessary corrective action is called the *clearing time* and is defined as follows:

$$T_c = T_p + T_d + T_a \tag{1.1}$$

where T_c = clearing time
T_p = comparison time
T_d = decision time
T_a = action time, including circuit breaker operating time.

The clearing time is very important since other protective systems in the network may be time-coordinated with this protective device in order to ensure that only the necessary portions of the network are interrupted. There is an important message here, namely, that many protective devices will observe a given disturbance and many of them will find that disturbance to exceed their threshold settings. Each device should have some kind of restraint to allow those closest to the disturbance to trip first. Time is one kind of restraint that is often used. Other restraint mechanisms will be discussed later.

Clearing time is also important because some disturbances, such as short circuits, must be cleared promptly in order to preserve system stability. This depends on many factors, including the location and type of disturbance. However, it is a general rule that abnormal system conditions must be corrected, and speed of correction is always important.

1.3 PROTECTIVE SYSTEM DESIGN CONSIDERATIONS

A protective system should be designed to recognize certain system abnormalities which, if undetected, could lead to damage of equipment or extended loss of service. The design and specification of the system components is an important part of the protective strategy, and power systems are designed to withstand the usual operating contingencies that accompany load changes and line switching operation. The coordination of insulation is an important design consideration, and protective devices are commonly installed to protect expensive apparatus from damaging over voltages. These problems, although a part of the overall system protection, are outside the scope of this book. Our concern here is the detection, clearing, and restoration of circuits from damaging abnormal conditions, which we usually call *faults*. This requires a knowledge of the types of faults that are likely to be experienced and the kinds of protective devices that can be used to recognize faults and initiate action to clear the fault from the system. The protective system designer must develop a strategy to accomplish this within the framework of available protective equipment while optimizing the restoration of the system to the normal state.

There are several design considerations that must be weighed against cost in devising a protection strategy. We have mentioned an optimum restoration, but this does not necessarily imply the fastest possible restoration. For example, if the protective system intelligence determines that a fault is permanent, there is no point in repeating the reclosing of the circuit before a repair crew has located and repaired the difficulty. Only then can the circuit be restored to its normal state. Thus every occurrence has a unique optimum pattern for returning to normal following a disturbance and this may involve human intervention, such as a physical repair. Normally, there is no human intervention in the protective system action, however, as this would cause the abnormal condition to persist for an extended time. In cases where this is feasible, the protective system issues an alarm, following which the operator can analyze the situation and manually take any action that is required.

The protective system should also be designed for minimum loss-of-load. There is no need, usually, to de-energize the entire system because of an isolated fault. The system should have *selectivity* to isolate the fault such that the minimum interruption occurs. Often this requires automatic reclosing following a circuit opening, since experience has shown that the large majority of faults are temporary and that reclosures are very often successful. Minimum loss-of-load may also require that alternate circuits be available to serve important loads. For example, bulk transmission systems usually have the capability of serving all load with one or more major circuit components out of service.

The protective system should also be designed with due regard for its own unreliability. This means that backup protective systems should be installed to operate in case of primary protective equipment failure so that system damage can be minimized and restoration of normal service can be achieved quickly.

It is also important that the protective system be designed such that the system can perform under normal operating conditions. The protective equipment senses system voltages and currents and from these measurements computes a relaying quantity which is compared to a threshold or trip value. This threshold must not be set too low or the protected circuit may be interrupted unnecessarily. Furthermore, threshold values must be periodically reviewed to make certain that these settings are satisfactory for current system loading conditions. This is an operating as well as a design problem.

The operation of protective equipment must be accurate and fast. Bulk power system reliability standards require that systems survive severe fault conditions without causing a system collapse. This in turn requires fast, reliable protective system operation. Thus, there is a direct dependency upon the protective system to achieve a given level of system reliability. This adds to the challenge of designing an effective protective system and a reliable power delivery system.

1.4 DEFINITIONS USED IN SYSTEM PROTECTION

In this section we define certain terms which are used in system protection engineering. Other, more specific terms, will be introduced later, as required.

Protective relaying is the term used to signify the science as well as the operation of protective devices, within a controlled strategy, to maximize service continuity and minimize damage to property and personnel due to system abnormal behavior. The strategy is not so much that of protecting the equipment from faults, as this is a design function, but rather to protect the normal system and environment from the effect of a system component which has become faulted.

Reliability of a protective system is defined as the probability that the system will function correctly when required to act. This reliability has two aspects: first, the system must operate in the presence of a fault that is within its zone of protection and, second, it must refrain from operating unnecessarily for faults outside its protective zone or in the absence of a fault [2].

Security in protective systems is a term sometimes used to indicate the ability of a system or device to refrain from unnecessary operations. Often we use security as a generic term to indicate that the system is operating correctly, whereas reliability is usually taken to be a quantifiable variable.

Sensitivity in protective systems is the ability of the system to identify an abnormal condition that exceeds a nominal "pickup" or detection threshold value and which initiates protective action when the sensed quantities exceed that threshold.

Selectivity in a protective system refers to the overall design of protective strategy wherein only those protective devices closest to a fault will operate to remove the faulted component. This implies a grading of protective device threshold, timing, or operating characteristics to obtain the desired selective operation. This restricts interruptions to only those components that are faulted.

Protection zones (primary protection zones) are regions of primary sensitivity. Figure 1.4 shows a small segment of a power system with protection zones enclosed by dashed lines.

Coordination of protective devices is the determination of graded settings to achieve selectivity.

Primary relays (primary sensitivity) are relays within a given protection zone that should operate for prescribed abnormalities within that zone. In Figure 1.4, for example, consider a fault on line JK. For this condition, relays supervising breakers J and K should trip before any others and these relays are called primary relays.

Backup relays are relays outside a given primary protection zone, located in an adjacent zone, which are set to operate for prescribed abnormalities within the given primary protection zone and independently of the primary relays. For example, suppose a fault on line JK of Figure 1.4 cannot be cleared by breaker J due to relay or breaker J malfunc-

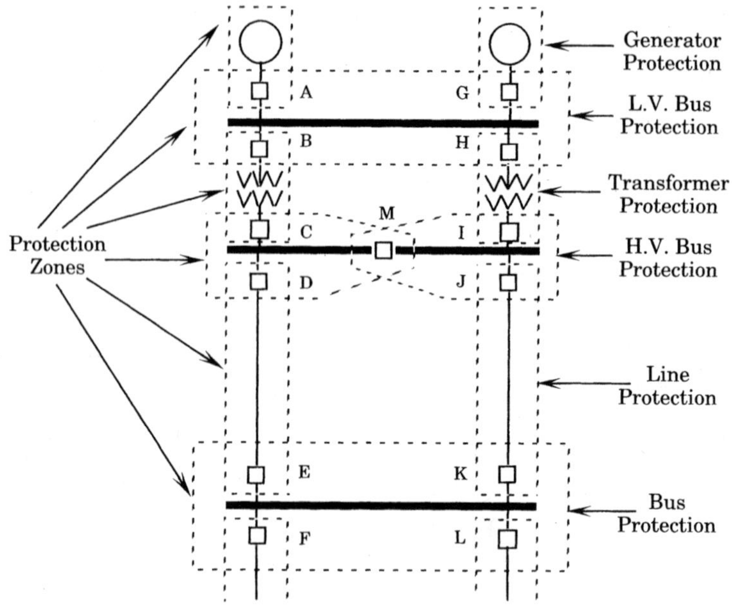

Figure 1.4 One line diagram of a system showing primary protection zones [3].

tion. Assume that breaker K operates normally leaving the fault connected to the bus terminated by breakers IJM. Backup relays at locations I and M should be set to operate for the fault on line JK, but only after a suitable delay that would allow breaker J to open first, if possible.

Local backup relays are an alternate set of relays in a primary protection zone that operate under prescribed conditions in that protection zone. Often such local backup relays are a duplicate set of primary relays set to operate independently for the same conditions as the primary set. This constitutes an OR logic trip scheme and is an effective safeguard against relay failures.

Undesired tripping (false tripping) results when a relay trips unnecessarily for a fault outside its protection zone or when there is no fault at all. This can occur when the protective system is set with too high a sensitivity. Such operation may cause an unnecessary load outage, for example, on a radial circuit, or may cause overloading of adjacent lines of a network. Thus, in some cases, unnecessary tripping is merely an inconvenience, which, although undesirable, may not cause serious damage or overloading. In other cases, where an important line is falsely tripped, it can lead to cascading outages and very serious consequences.

Failure to trip is a protective system malfunction in which the protective system fails to take appropriate action when a condition exists for which action is required. This type of failure may result in extensive damage to the faulted component if not rectified by backup protection.

Other definitions used in system protection are given in [2], which is a standard for such definitions in the United States. Many of these definitions will be introduced as needed for clarity and precision.

A summary of several important terms and definitions is also given in Appendix A.

1.5 SYSTEM DISTURBANCES

The disturbances that occur on electric networks are varied in both magnitude and character. In this section we examine the nature of disturbances and attempt to identify those disturbances for which protective system deployment is feasible and/or necessary.

A disturbance is defined as follows by the IEEE [4].

Disturbance (General). An undesired variable applied to a system that tends to affect adversely the value of a controlled variable.

Clearly, what appears as a disturbance to one kind of apparatus may be of little consequence to another type, irrespective of the magnitude of the disturbance. Our classification is general, to begin with, and from this general classification we shall speculate on which disturbance classes may require special protective system applications.

There are many possible ways to categorize disturbance types and characteristics. One reference divides disturbances into two major categories, *load* disturbances and *event* disturbances [5]. These are defined as follows:

Load disturbances: Small random fluctuations superimposed on slowly varying loads.

Event disturbances:

(a) Faults on transmission lines due to equipment malfunctions or natural phenomena such as lightning.
(b) Cascading events due to protective relay action following severe overloads or violation of operating limits.
(c) Generation outages due to loss of synchronism or malfunction.

As defined in [5], load disturbances are a part of the system normal operating conditions. In an operating power system, frequency and voltage are always in a state of change due to load disturbances. Any departure from normal frequency and voltage, due to a load disturbance, is usually small and requires no explicit power plant or protective system response. Occasionally, however, major load disturbances do occur. These major disturbances are usually caused by important transmission or generation outages, and are characterized by low, high, or widely varying frequency and voltage on the power system.

Small event disturbances are also a part of the normal power system operating environment. Event disturbances, however, imply a need for rapid response by the protective systems and can lead to larger upsets if this action fails or is delayed. Large event disturbances always require fast protective system action and may lead to complete system failure if this action is not correct and fast.

Disturbances, both large and small, may be classified as shown in Table 1.1 [6]. Usually, the system protection engineer is interested in small disturbances from the standpoint that the protective system should not act for this type of disturbance, since such action would usually make the situation worse. Indeed, for this type of disturbance the removal of additional system components is likely to make the situation worse.

The large disturbances, on the other hand, will often require correct response by the protective system. This means that the protective system should act promptly to remove damaged or faulted components, and should not act except in carefully defined conditions. Usually, it is the system disturbances that require protective system action, for example, transmission line faults or other destructive natural phenomena, or random failure of system components. The protective system must be carefully designed to trip only for those load disturbances that

TABLE 1.1 Disturbance Cause and Effect Matrix

Event Magnitude	Type of disturbance			
	Load disturbance		System disturbance	
	Cause	Effect	Cause	Effect
Small	Daily load cycle	Frequency error	Load trips	Slight over Frequency
	Small overload	Voltage deviation	Equipment outage	Possible load Shedding
	Random load fluctuation	Spontaneous oscillation	Large load Variation	Sustained system Oscillations
Large	Generation overload	Frequency error	Network faults	Loss of network elements
	Winter power plant freeze up	Time error	Plant outages	Load shedding
	Inadequate reserves	Low voltage	Line outages	Unit tripping
	Circuit overloads	Loss of plant auxiliaries	Destructive natural events	Cascading/ separation/ islanding
		Plant trips		Instability
		Line trips		Blackouts

would lead to permanent equipment damage and possible sustained outage. The table entries are not exhaustive and do not describe fully the conditions under which system protection must act, or refrain from action. The intent of the table is simply to note that there are many types of disturbances for which normal protective system action is not the proper corrective action. Normal protective action is required in response to system *component failures*, where the prompt removal of the failed element is necessary for continued system operation. Failure of the protective systems, or improper protective system response, may lead to serious system operating conditions. There are other types of systems that are designed to respond to some of the large disturbances mentioned in Table 1.1, but these are not what we would usually call "normal" protections. These schemes are designed to recognize a particular stressed system condition and to take remedial action. Such schemes are sometimes called *special protection schemes*, or *remedial action schemes*, and their function is often to prevent instability or the cascading of outages that may lead to blackout.

1.6 THE BOOK CONTENTS

The contents of this book are divided into logical units of study, and these logical units are designated as "parts" each with a defined objective.

Part 1, *Protective Devices and Controls*, provides very basic information as to the connection and intrinsic operation of system controls that are designed to remove a severe disturbance, such as a short circuit, from the operating power system. This requires the operation of some type of device that will separate the fault from the system in an timely and effective manner. The separation device may be a fuse, a circuit breaker, or other device designed for a particular application and with a given rating. The interruption of short circuits provides a severe test to the interruption device, and this will be the subject of study in this initial part of the book. The final portion of Part 1 investigates the mathematical characteristics of the power system under faulted conditions and provides analytical techniques for the analysis of any type of fault condition.

Part 2, *Protection Concepts*, investigates mathematics of the power system under faulted conditions. Faults on radial feeders, such as those usually found in distribution systems, are presented in Chapter 6. This introduces the problems associated with the coordination of time-current devices, such as fuses or circuit reclosers. These studies are applicable to most distribution protective systems and their study introduces basic concepts regarding the necessity of recognizing faulted conditions, and clearing the fault in a timely manner, based solely on the magnitude of the fault current.

The detection of faults on transmission systems, introduced in Chapter 7, is more complex because the system is usually meshed, as opposed to the radial systems examined in Chapter 6. This means that the protective logic must be more sophisticated as the direction of current flow is dependent on the fault location. Many schemes have been devised for the protection of transmission elements, and some of these concepts are introduced here.

The remainder of Part 2 examines how a faulted condition can be viewed in the impedance or admittance planes, as functions of a complex variable. Some protective devices use measurements of both voltage and current that can be interpreted as loci in the Z or Y planes, with trip zones set as regions of those planes. This is an important concept for certain types of relays.

Part 3, *Transmission Protection*, concentrates on transmission systems, beginning with an analysis of distance protection, which utilizes Z plane loci as a measurement of distance from the relay to the fault. The mutual induction of fault currents flowing in lines parallel to the faulted lines is presented in Chapter 12. This concept can complicate fault detection and clearing involving zero sequence currents.

Pilot protection schemes are commonly used on high-voltage transmission lines to provide fast, dependable operation. Commonly used pilot schemes are described in Chapter 13. Chapter 14 investigates several topics that add complexity to transmission protection, and describes methods of overcoming these complexities. One important form of complexity in transmission is the use of series compensation, which is the subject of Chapter 15.

Part 4, *Apparatus Protection*, is investigated in Chapters 16 through 19. This includes the protection of buses, transformers, generators, and motors. This type of protection is different than line protection since all terminals of the protected device are available to the protection equipment without need for communication. Apparatus protection can also take advantage of the nature of the item being protected and its unique requirements. Also, since repair of items such as large transformers or generators can require weeks or even months to complete, multiple special protective schemes are often used to limit damage of the equipment by fast recognition of a particular hazard.

Part 5, *System Aspects of Protection*, examines disturbance conditions that have widespread effects. This includes underfrequency protection, out-of-step conditions that may lead to instability, HVDC disturbances that inject abnormal effects at all terminals of the dc system, and subsynchronous oscillations that can affect one or more generators on the interconnected system.

Finally, Part 6 of the book examines the *Reliability of Protective Systems*. The subject is introduced in an elementary manner such that all required basic concepts are presented in Chapters 24 and 25 prior to their use in reliability calculations that follow. The application of reliability concepts is demonstrated through elementary examples first, and is then used in the analysis of typical protection equipment. Emphasis is on the fault tree methods and Markov modeling for the study of complex systems. This leads to the reliability modeling of typical protective systems and the opitimization of the scheduling of protection equipment inspections, based on probabilistic techniques.

REFERENCES

[1] Anderson, P. M., *Analysis of Faulted Power Systems*, IEEE Press, Piscataway, NJ, 1992.

[2] American National Standard, C37.90-1989, *IEEE Standard for Relays and Relay Systems Associated with Electric Power Apparatus*, New York, 1989.

[3] Mason, C. R., *The Art and Science of Protective Relaying*, John Wiley & Sons, Inc., New York, 1956.

[4] IEEE Std-100-1992, *IEEE Standard Dictionary of Electrical and Electronics Terms*, IEEE, Piscataway, NJ, 1993.

[5] Debs, A. S., and A. R. Benson, *Security Assessment of Power Systems*, U. S. Energy Research and Development Administration, Washington, D.C., 1975.

[6] Weber, G. A., M. C. Winter, S. M. Follen, and P. M. Anderson, *Nuclear Plant Response to Grid Electrical Disturbances*, EPRI, Palo Alto, 1983.

PROBLEMS

1.1 A transmission line 100 km long is being designed as part of high reliability bulk power transmission system. Suppose that the basic line cost is $80,000/km and you are to evaluate the economics of adding two overhead ground wires which will increase the cost by 20%. However, the overhead ground wires are expected to reduce the incidence of lightning flashovers from 20 per year to 1 per year. Discuss the alternatives involved and decide whether you would recommend the added expense of the ground wires.

1.2 List all the reasonably probable abnormal occurrences you can think of that would lead to protective system operation together with your estimate of the probability of occurrence of each, in connection with the following system components:

(a) transmission line
(b) transformer
(c) generator

1.3 Distinguish among the terms power system reliability, security, and service continuity.

1.4 Consider the system portion shown in Figure P1.4.

(a) Sketch the zones of protection.
(b) Describe a possible backup scheme for failure of breaker 7 for a fault on line BC. Is load 1 interrupted? load 2?
(c) Describe a means of clearing a fault on line BC without a momentary interruption of load 1.
(d) Make any comment you would like about the system shown in Figure P1.4.

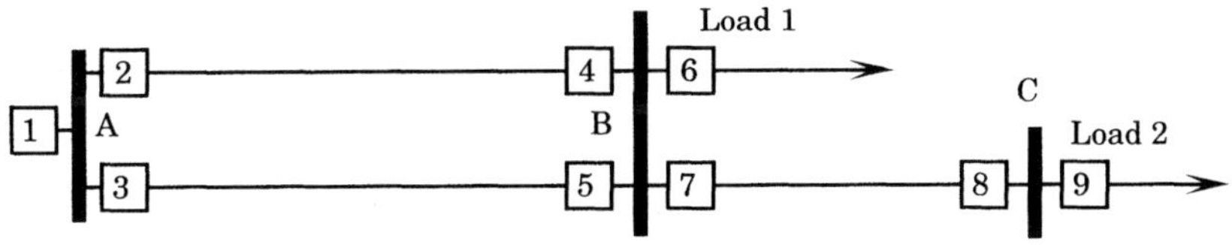

Figure P1.4

1.5 The system of Figure P1.4 is rearranged as shown in Figure P1.5 where the bus arrangement at B has been changed to a ring bus arrangement.

(a) Compare the cost of the two systems

(b) Compare the operation of the two systems for when a fault occurs on line BC with

1. A failure of breaker 7.
2. A failure of breaker 6.

(c) Sketch zones of protection.

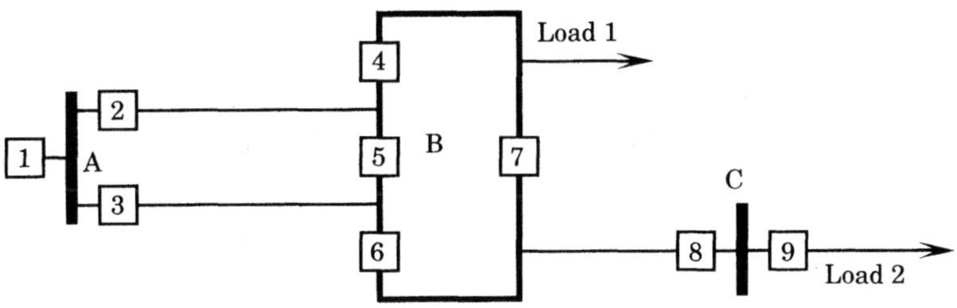

Figure P1.5

1.6 Consider a power system with several optional designs for all of its components, with the more reliable and fault resistant components having higher cost, but lower outage rates. However, the use of the lower cost components will require more investment in protective equipment in order to assure prompt removal of faulted equipment. Is it possible to find an optimum solution of this problem, where the cost of both components and their protection is a minimum?

2

Protection Measurements and Controls

This chapter presents some basic control configurations for protective systems. The method of connecting protective devices into the power system are presented, and some of the problems of making accurate observations of system conditions will be explored. We also investigate the methods by which circuit breakers are controlled, both for manual and automatic operation. Finally, we present some basic information on instrument transformers, which represent the interface between the protective system and the power system.

2.1 GRAPHIC SYMBOLS AND DEVICE IDENTIFICATION

Graphic symbols are important in communicating protective system information. As an introduction to the basic relaying circuits, we review briefly the IEEE standards for graphic symbols that are used in this book.

The symbols commonly used in protection engineering are IEEE standards, and many of them have also been adopted by the International Electrotechnical Commission (IEC) [1]. The symbols most used in protective systems are those shown in Figure 2.1.

The first two symbols show the correct graphical symbol for electrical contacts. The "a" contact is a normally-open contact and is always depicted in drawings in the open position even though, in a particular application, the contact may be nearly always closed. This permits us to distinguish immediately that this contact is one that is open when no current flows in its operating coil. The a contact is sometimes called a "front contact."

The second contact, or "b" contact, is always illustrated in the closed position, since it always returns to this position when there is no current in its operating coil. The b contact is also referred to as a "back contact." The operating coils that are associated with relay contacts and with the circuit breakers are depicted as shown in Figure 2.1 (older diagrams may show operating coils as circles, but this practice is depreciated).

The graphic symbols for current and potential transformers are also illustrated in Figure 2.1. Note that these transformers are shown with polarity dots to clarify the phase rela-

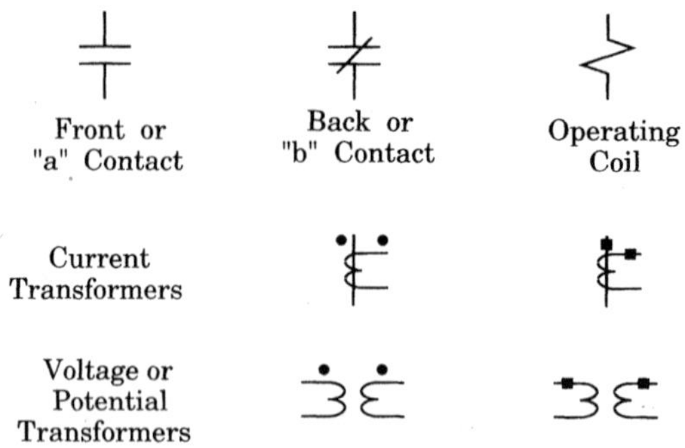

Figure 2.1 Graphic symbols used in protective relaying [1].

tionship of the currents and voltages. The IEEE standards note that the potential transformer may also be referred to as a "voltage transformer." In practice, the more common terms are current transformer, or CT, and potential transformer, or PT (or voltage transformer, VT) [2]. These terms are used in this book.

Protective relays have two circuits or sets of circuits, one for ac and one for dc quantities. The ac circuits are replicas of the ac quantities in the actual power system, which are transformed to suitable magnitudes by current and potential transformers. The dc circuit controls the tripping of the circuit breaker by permitting current to flow through the breaker trip coil under control of one or more relays. The relays provide the control intelligence and a suitable set of contacts to control the flow of current in the dc trip circuit of the circuit breakers.

Protective system control drawings also use a formal system of device function numbering to clearly identify objects that are used in graphic displays. These numbers conform to ANSI/IEEE Standard C37.2, which defines the devices and their function, and gives each device a function number for use in drawings, diagrams, manuals, and other publications. A partial listing of these standard device function numbers is given in Appendix B. The ANSI/IEEE Standard should be consulted for more complete information [3].

Using device numbers, each relay coil or contact may be identified as to the type or function of the device to which that item belongs. This provides a shorthand notation for use on drawings and other media that is brief and is readily understood.

2.2 TYPICAL RELAY CONNECTIONS

As an introduction to relay and circuit breaker connections, consider the system shown in Figure 2.2, where a radial line is protected by time-overcurrent relays (51) in each phase. Only the connection in one phase is illustrated for simplicity. The circuit breaker (52) control circuit consists of a battery that is connected through the circuit breaker auxiliary contacts (52a) and the circuit breaker trip coil (52TC), and finally to the relay contact (51). During normal operation, the circuit breaker is closed and load currents are flowing (downward in the figure). The circuit breaker front contacts (52a) are closed under this condition. Think of the "a" in 52a as meaning in *agreement* with the circuit breaker main contacts.[1] If a fault is detected such that the current in the relay coil exceeds a given preset threshold value, the relay will close its contacts in a measured time, which depends on the magnitude of the current and on the relay characteristics. This will cause current to flow in the circuit breaker trip coil (52TC),

[1] There may be other contacts in a relay control scheme that use the letter "a" in the contact identification, but such contacts are not part of the circuit breaker and their open/close position is independent of the circuit breaker state.

tripping the circuit breaker main contacts and also the auxiliary contacts. Tripping the main contacts removes the fault from the system and allows the relay to reset itself in a short time, which depends on the relay design. Tripping the auxiliary contacts opens the control circuit and interrupts the flow of current in that circuit.

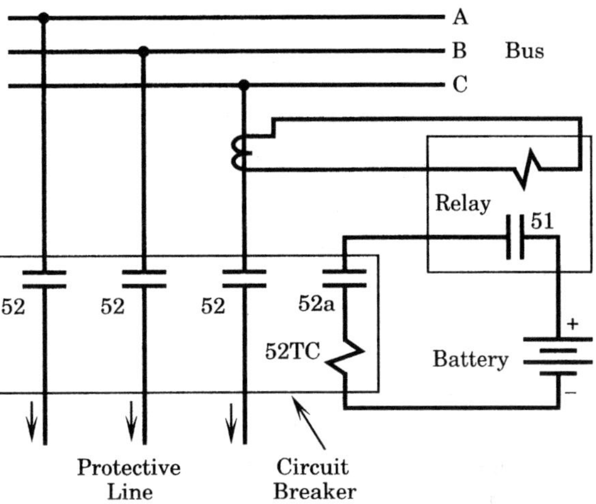

Figure 2.2 A simple relay and circuit breaker control circuit.

The control circuit of Figure 2.2 is simpler than that found in most practical applications. A practical circuit would include a relay for each phase conductor and may have a fourth relay to measure ground currents. Also, since most faults are temporary, it may be desirable to include a means of automatically reclosing the circuit breaker after allowing time for the fault to deionize. These additional features will be introduced later. Finally, we note that this is a special case, being a radial line, making a simple overcurrent relay adequate to provide the necessary selectivity and control required. Most transmission lines are not operated radially and require more elaborate relay protection. Figure 2.3 shows a typical ac connection for a relay that requires both current and potential supplies. The connection of Figure 2.3 is typical of that used for a directional relay. Note that relay element 1 sees current I_a and voltage V_{bc}, and that these quantities are nearly in phase for a transmission line fault, which usually has the current lagging the phase voltage by nearly 90 degrees. One reason the connection of Figure 2.3 has been popular goes back to a principle of electromechanical relays, where having the relay current and voltage in phase on a wattmeter type element produces maximum torque on the relay element. Obviously, other connections of the current and voltage transformers are possible.

The dc circuit of the relay is the circuit breaker tripping circuit, as shown in Figure 2.4, which shows a tripping circuit that could be used with the relay connections of Figure 2.3. This dc trip circuit incorporates a holding coil or "seal-in" relay labeled "S" in the figure. The operation is as follows. If one of the relay elements detects a fault condition, the corresponding relay contact R is closed by the relay logic. Since the breaker auxiliary relay "a" contacts are closed (note that the breaker is still closed) closing R causes current to flow in the circuit breaker trip coil (TC). In many cases, the relay contacts are not designed for the relative severe duty of interrupting the trip circuit, hence the R contacts are paralleled by the seal-in relay contacts S, which remain closed throughout the breaker operation even though the relay contacts may drop out. When the circuit breaker main contacts open, the breaker auxiliary contacts "a" also open, interrupting the current flow in the dc control circuit. This interruption also causes the seal-in relay to drop out and the circuit is ready for reclosing and for tripping the next fault.

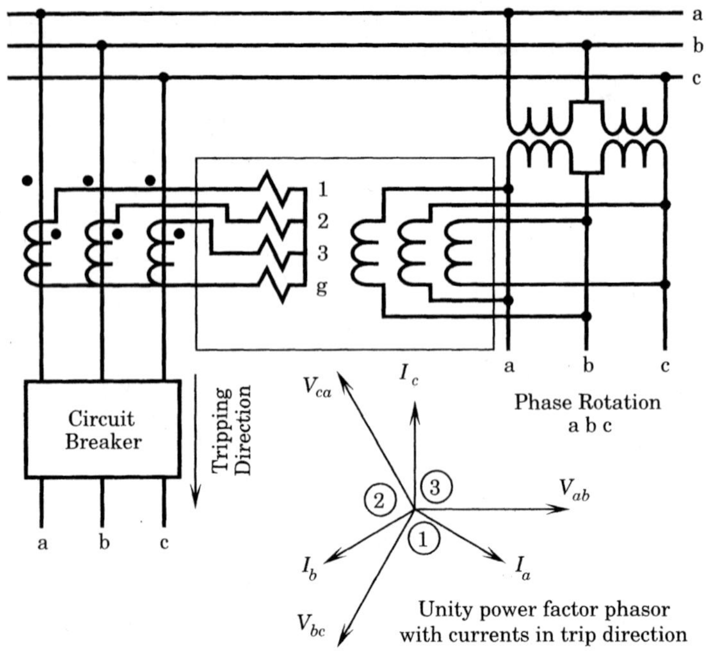

Figure 2.3 Typical ac relay connection showing both the current and voltage supplies (90 degree connection).

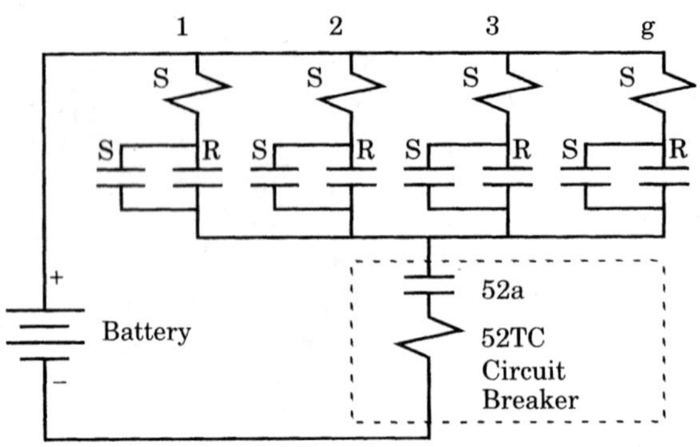

Figure 2.4 Typical dc trip circuit connection.

In small stations, where a battery supply cannot be justified, the battery can be replaced by a capacitor that is kept charged from the ac line by a rectifier. The capacitor is sized to have sufficient energy to trip the circuit breaker.

Another method of arranging the trip circuit is the series trip connection shown in Figure 2.5. Here, the circuit breaker must be equipped with three trip coils, labeled TC, rather than the one coil used in the shunt trip circuit. Series trip is convenient at locations where it is impractical to have a battery supply, such as small remote breaker locations. An arrangement similar in philosophy is used in low-cost, distribution system oil circuit "reclosers," where the actual line current is sometimes used as the tripping current. This saves the expense of current transformers, but requires a trip circuit capable of handling fault current magnitudes. These devices will be discussed further in Chapter 3.

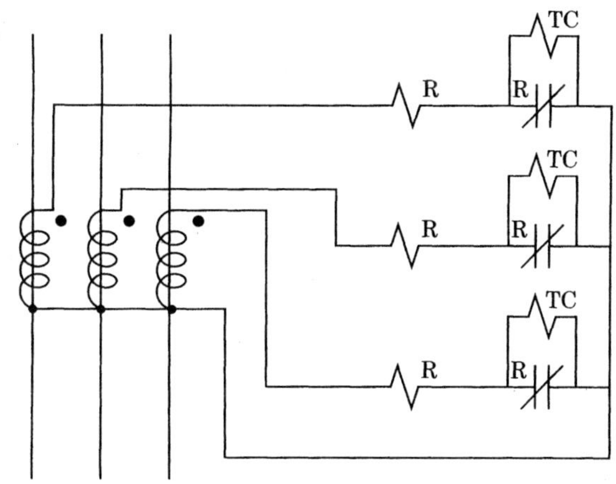

Figure 2.5 A series trip coil arrangement.

2.3 CIRCUIT BREAKER CONTROL CIRCUITS

The circuit breaker control circuits shown in Figures 2.2 and 2.3 are simplified and do not illustrate typical circuits that are found in the industry. One shortcoming of these circuits is that they have no means of manual operation of the breaker, either for opening or closing. Other features are required in a practical system. These features will be discussed in connection with the description of a typical control circuit.

Consider the control circuit shown in Figure 2.6, which is a complete tripping and closing circuit for a circuit breaker. Here, the protective relay contacts are shown as a single contact labeled "R" and this should be understood to include as many contacts as are actually available from the various relays at a given installation.

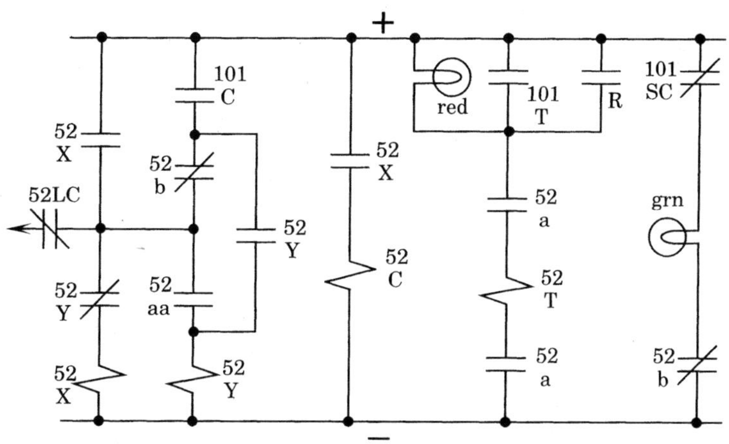

Figure 2.6 Typical circuit breaker control circuit.

This control circuit, which is often called the X-Y control scheme, is designed to provide several unique safeguards, as follows:

1. The control is electrically trip free.
2. The control includes an anti-pumping feature.
3. A provision for reclosing is provided.

First, we examine the general concepts of the control scheme. Then we shall examine the above special features. Contacts in Figure 2.6 labeled 101 are manual control switches. There are three of these switches:

101C: Manual closing contact

101T: Manual tripping contact

101SC: Manual slip contact

First, assume that the breaker is open, and the green light is on, indicating a non-energized breaker. The operator now wishes to manually close the breaker. This is accomplished by manually closing contact 101C. Since the breaker is initially open, contacts 52a and 52aa are both open. Similarly, contacts 52b are closed. Closing 101C momentarily energizes coils 52X since 52Y "b" contacts are closed. Coil 52X picks up its respective contacts in the close circuit causing current to flow through the circuit breaker closing coil 52C, thereby closing the circuit breaker. Another 52X contact (to the left of 101C) seals in the closing contact 101C.

Now, when the breaker main contacts close, the 52 auxiliary contacts change their open/close status. Thus, 52aa closes, which picks up coil 52Y, thereby opening the 52X coil and de-energizing the closing coil 52C. Note that 52b opens, which assures that the 101C circuit remains open. Contact 52Y is used for anti-pumping and is discussed below. Thus, by momentarily depressing 101C, the operator puts in motion a number of control features. The end result is that the breaker is closed, the red light is on and the green light is off. The lamp current flows through 52T, but the current magnitude is much too small to operate the breaker.

To manually trip the breaker, the operator closes contacts 101T, which causes current to flow through the trip coil 52T, thereby tripping the circuit breaker, turning off the red light, and energizing the green light.

Now, suppose the operator manually closes 101C and closes the breaker when there is a permanent fault on the line. Moreover, suppose the operator stubbornly holds the 101C contacts closed. Should this occur, the first reaction after closing will be the pick up of the relay contacts because of the fault, which trips the breaker. However, the initial breaker closure also picks up 52aa. This energizes coil 52Y, the anti-pumping coil, which is held closed by contacts 52Y as long as 101C is depressed. At the same time, coil 52Y also opens the circuit of closing auxiliary coil 52X, preventing further closing of the breaker. Thus the breaker is closed, but opens immediately and remains open, even if the operator holds 101C in the closed position.

The reclosing feature uses contact 52LC, a latching contact, not shown in Figure 2.6. After the breaker is tripped, the mechanical breaker closing mechanism is latched to permit closing. This breaker action closes contacts 52LC. These contacts can be connected to a reclosing relay, which can apply positive potential to coil 52X, initiating the automatic reclosure of the line.

Note that it is essential that the trip circuit be energized from the battery supply, since the ac line or bus voltage may be badly depressed during a fault condition. The breaker closing voltage may be supplied from the ac bus, however. In this case, the control circuit is the same except that 52X, 52Y, and 52C are connected to an ac supply.

2.4 INSTRUMENT TRANSFORMERS

Protective systems for power systems are designed as system control components with the inherent intelligence to perform the required control functions. Most of the relay equipment involved in this function is relatively small and is mounted on low-voltage relay panels in a control building. This makes the relay equipment convenient and safe to work with for calibration and testing. It also requires that the currents and voltages used in the relays themselves must be transformed from transmission levels to appropriate lower voltage levels for safety and convenience of personnel. This transformation is accomplished by means of current transformers (CT's) and potential or voltage transformers (VT's), which are collectively referred to as "instrument transformers." These transformers are insulated for the appropriate primary voltage level of the system and with secondary currents and voltages that match the rated values of the relay apparatus. In North America, these secondary standard ratings are 5 amperes and 120 volts rms at 60 hertz, for CT's and VT's, respectively.

There are two concerns in applying instrument transformers; transformer selection for accuracy and transformer connections.

2.4.1 Instrument Transformer Selection

Many instrument transformers are iron-core transformers that are designed to give secondary currents or voltages that are accurate replicas of the primary quantities. The protection engineer must select the appropriate transformers based on the relays to be used in the protection scheme and the connection (e.g., wye or delta) of the relays and transformers to be used [4].

For current transformers, an important criterion in selecting the correct transformation is the maximum load current. The CT secondary current, under normal conditions, will represent the load on the protected power system circuit and this load current will flow through the relay circuits all the time. The relay is designed for a given maximum load current, and this value must not be exceeded. Most relays are designed for a 5 ampere rated current, hence the CT should be selected to provide about 5 amperes at normal load conditions.

For voltage transformers, the transformation ratio is seldom a problem, since both the secondaries and relays are designed for 120 volt continuous service. In some applications, the VT primaries are connected line-to-line and the secondaries line-to-neutral and this must be taken into account.

Since many instrument transformers are iron core transformers, the quality of the iron and its saturation characteristics are important. This is especially true for current transformers, which might be expected to saturate when carrying fault currents. This may or may not be a problem, depending on the application, since even badly saturated transformers may still give the correct tripping signal to the relays. Generally speaking, the transformers used should be of as high quality as possible, as this tends to reduce problems and to provide better relay accuracy. Transformer accuracy is especially important in differential relaying schemes, where the relay sees the difference in currents.

Saturation of the current transformer can be estimated by any one of three methods [4]:

1. The excitation (saturation) curve method
2. The formula method
3. The computer simulation method

In all cases, we represent the current transformer by the equivalent circuit, shown in Figure 2.7. The primary current is transformed through the ideal transformer with ratio $1 : N$. The equivalent circuit parameters are defined as follows:

$\mathbf{Z}_H$ = primary leakage impedance
$\mathbf{Z}_S$ = secondary leakage impedance
R_M = core loss component of the excitation branch
X_M = excitation component of the excitation branch

The basic equivalent circuit is simplified as shown in Figure 2.7(b). Here the primary leakage impedance and core loss elements are neglected. The exciting current, flowing through the shunt excitation branch is defined as shown.

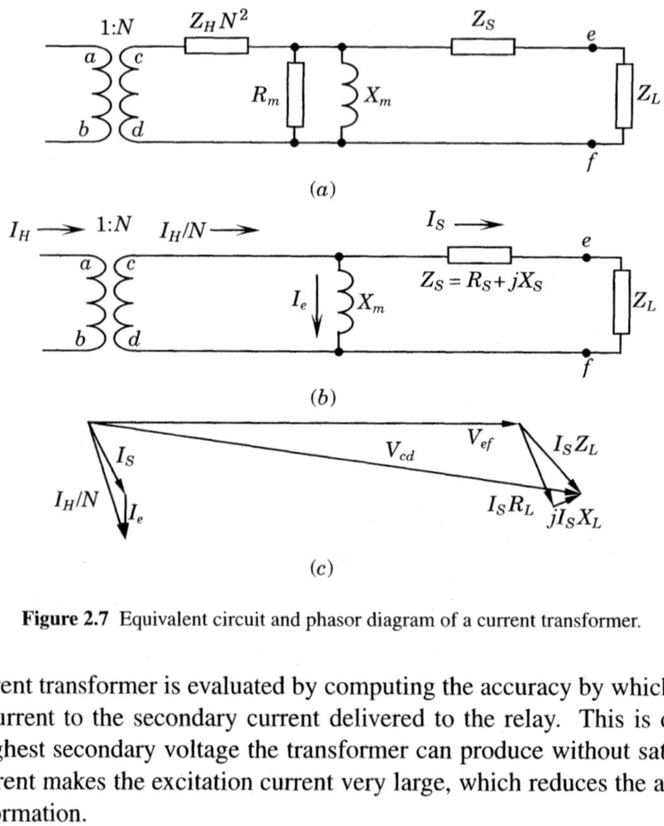

Figure 2.7 Equivalent circuit and phasor diagram of a current transformer.

The current transformer is evaluated by computing the accuracy by which it transforms the primary current to the secondary current delivered to the relay. This is determined by finding the highest secondary voltage the transformer can produce without saturation. High secondary current makes the excitation current very large, which reduces the accuracy of the current transformation.

From Figure 2.7(b), we define the secondary voltage as follows:

$$\mathbf{V}_{cd} = \mathbf{V}_S = \mathbf{I}_S(\mathbf{Z}_S + \mathbf{Z}_L) = \mathbf{I}_S\mathbf{Z}_B \tag{2.1}$$

where $\mathbf{V}_S$ = secondary voltage, (V)
$\mathbf{I}_S$ = maximum secondary current, (A)
$\mathbf{Z}_S$ = secondary leakage impedance, (Ω)
$\mathbf{Z}_L$ = the external impedance or "burden," (Ω)
$\mathbf{Z}_B$ = the secondary burden, (Ω)

In most applications, the maximum secondary current can be estimated by dividing the known fault current by the transformation ratio of the CT.

2.4.1.1 ANSI Standard CT Accuracy Classes. The ANSI relaying accuracy classes are specified in ANSI Standard C57.13-1993 [5]. These standards use a letter designation and voltage rating to define the capability of the current transformer. The letter designation code is given as follows:

Code C—Indicates that the transformer ratio can be calculated

Code T—Indicates that the ratio must be determined by test

The C classification covers most bushing current transformers with uniformly distributed windings and any other transformers whose core leakage flux has negligible effect on the ratio, within the defined limits. The T classification covers most wound-type current transformers and any others whose core leakage flux affects the ratio appreciably.

An ANSI Accuracy Standard Chart for Class C current transformers is shown in Figure 2.8. Here, the transformer secondary voltage capability is plotted as a function of secondary current for various Class C transformers. This chart gives a limit (10%) in the ratio of the CT with a given accuracy class and a given burden. For example, for a burden of 4 ohms, the curves specify that the ratio error for class C400 will not exceed 10% between one and 20 times normal secondary current. This computation is checked as follows:

$$V_s = (4\Omega)(5 \times 20\text{A}) = 400\text{V}$$

The relaying accuracy class of a given current transformer can be obtained from the manufacturer. For T-class current transformers, the manufacturer can supply typical overcurrent ratio curves, such as the one shown in Figure 2.9 [4]. As implied by the class name, data for these curves must be determined by test on the actual transformer.

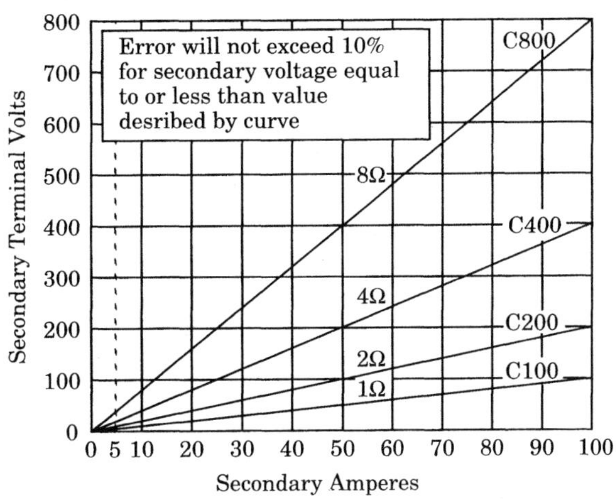

Figure 2.8 ANSI Accuracy Standard Chart for Class C current transformers.

2.4.1.2 Excitation Curve Method. This method requires the use of an excitation curve for the current transformers to be used. Such curves are available from the manufacturers. As a substitute, a typical set of curves could be used, such as the curves shown in [4], which are reproduced here as Figure 2.10. These curves represent data obtained by applying rms sec-

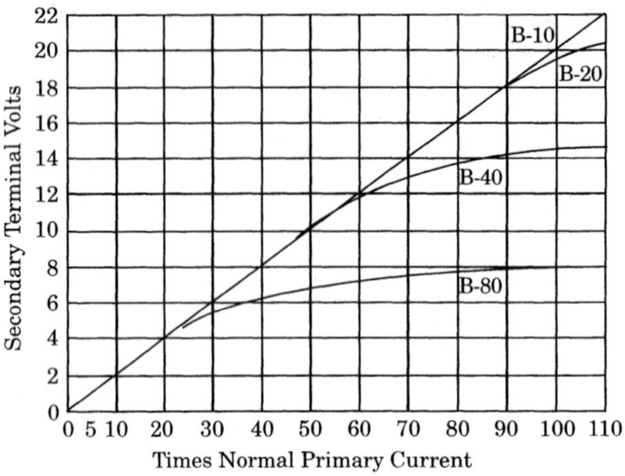

Figure 2.9 Typical overcurrent ratio curves for a T-class current transformer [5].

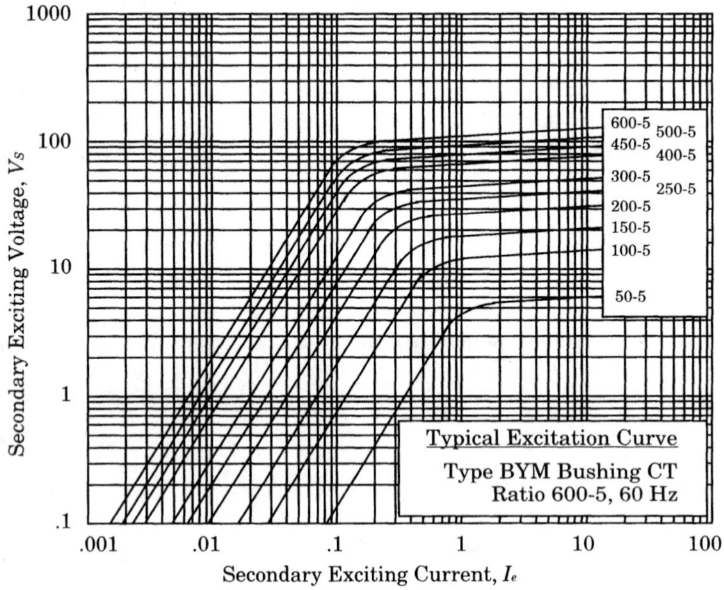

Figure 2.10 Excitation curves for a multiratio bushing CT with an ANSI accuracy classification of C100 [4].

ondary voltages to the current transformer with the primary circuit open, and give approximate exciting current requirements for the CT for a given secondary voltage.

These curves can be used very simply to determine if the CT becomes saturated at any given fault current. From (2.1), given the fault current and CT ratio, one can determine the secondary voltage. From Figure 2.8, for the computed voltage, one can readily see if the operating point is in the saturated region without making any further computation. This method, including several examples, is discussed further in [4].

2.4.1.3 The Formula Method. An excellent method estimating the CT performance is based on a knowledge of CT design principles. Table 2.1 shows the relationship between the standard secondary burden of the C Class of current transformers and the rated secondary voltage. The rated voltage is based on voltage the CT will support across a standard burden with 20 times rated current without exceeding 10% ratio correction.

TABLE 2.1 Standard Burden and Rated Voltage of C Class CT's

C Class	Standard Z_B (1)	Rated Voltage (2)
C100	1 Ω	100 V
C200	2 Ω	200 V
C400	4 Ω	400 V
C800	8 Ω	800 V

(1) Assumed impedance angle of 60°
(2) Computed as 20 × 5A secondary current = A

The secondary voltage is a function of the CT secondary fault current I_F and the total secondary burden Z_B. We may write this voltage as

$$v = N\frac{d\phi}{dt} \quad \text{V} \tag{2.2}$$

where N is the number of secondary turns and ϕ is the core flux in webers. Rearranging, we compute the total flux in terms of the flux density as

$$N\phi = NBA = \int_0^t v\,dt \tag{2.3}$$

For a fully offset voltage this becomes

$$\begin{aligned} N\phi = NBA &= \int_0^t Z_B i_F (e^{-Rt/L} - \cos\omega t)\,dt \\ &= Z_B i_F \left[\frac{L}{R}(1 - e^{-Rt/L}) - \sin\omega t\right] \end{aligned} \tag{2.4}$$

Using the maximum value of the expression in square brackets, we write

$$NBA\omega = Z_B i_F \left(\frac{X}{R} + 1\right) \tag{2.5}$$

Now, the secondary voltage rating of the CT is the voltage that the CT will support across a standard burden with 20 times rated current, without exceeding a 10% ratio error. Thus, we can write

$$Z_B \le \frac{20}{i_F\left(\frac{X}{R} + 1\right)} \tag{2.6}$$

where the burden is in per unit based on the standard CT burden and the fault current is in per unit based on the CT rated current. Since we use an extreme value of the quantity in parentheses, this will yield a conservatively small value of the maximum tolerable secondary burden [6].

For example, for a transmission line with X/R of 12 and a maximum fault current of four times rated current of a C800 CT, saturation will be avoided when Z_B is less than 0.38 per unit of the standard 8 ohm burden, or about 3 ohms.

2.4.1.4 The Simulation Method. The ANSI accuracy charts, such as Figures 2.8 and 2.9, do not provide an accurate insight as to the waveform distortion that occurs when a large primary current drives the current transformer into saturation. This problem has been addressed and results published to show the type of distortion that may occur, especially from fully offset primary currents of large magnitude [7], [8]. These publications show that substantial waveform distortion is likely with high primary currents, especially if the current is fully offset. A computer simulation has been prepared to permit the engineer to examine any case of interest [8].

EXAMPLE 2.1

An example of a current transformer simulation is to be run for a current transformer of the C400 accuracy class and with 40,000 amperes rms primary current, fully offset. Specifications for the current transformer are shown in Table 2.2.

TABLE 2.2 Data for C400 Current Transformer Calculation

CT ratio	150/5
CT relaying accuracy class	C400
Core cross-section area	43.1 in.^{2}
Length of magnetic path	24 in.
Secondary winding resistance	0.1 Ω
Secondary burden	0.1 + j 0 Ω
CT secondary cable resistance	0.1 Ω
Frequency	60 Hz
Primary current	40,000 A rms
Incident angle	0° (fully offset)
Primary current time constant	0.1 sec

Solution

The results of the computer simulation are shown in Figure 2.11, where the primary current is fully offset with a typical decrement time constant.[2] The secondary current has an initial high pulse that persists for less than 4 milliseconds in each half cycle. The performance of conventional overcurrent relays is not specified when confronted with such currents. The relay will be affected by saturation in the armature circuit and will have eddy currents induced due to the fast current rise. Note that this example is determined for a CT that is operating at over 266 times its rating, but such a condition can occur in power systems, depending on the availability of fault currents of high magnitude. The simulation method is flexible since any transformer operating under any specified condition can be studied. ■

Since the performance of relays under the conditions described in the example are not predictable, laboratory testing of the relay is advised to determine the relay behavior [8].

2.4.2 Instrument Transformer Types and Connections

Instrument transformers are available in a number of types and can be connected in a number of different ways to provide the required relay quantities.

2.4.2.1 Current Transformers. Current transformers are available primarily in two types: bushing CT's and wound CT's. Bushing CT's are usually less expensive than wound

[2]The author is indebted to W. C. Kotheimer of Kotheimer Associates for information regarding the saturation of current transformers and for the plot data for Figure 2.11.

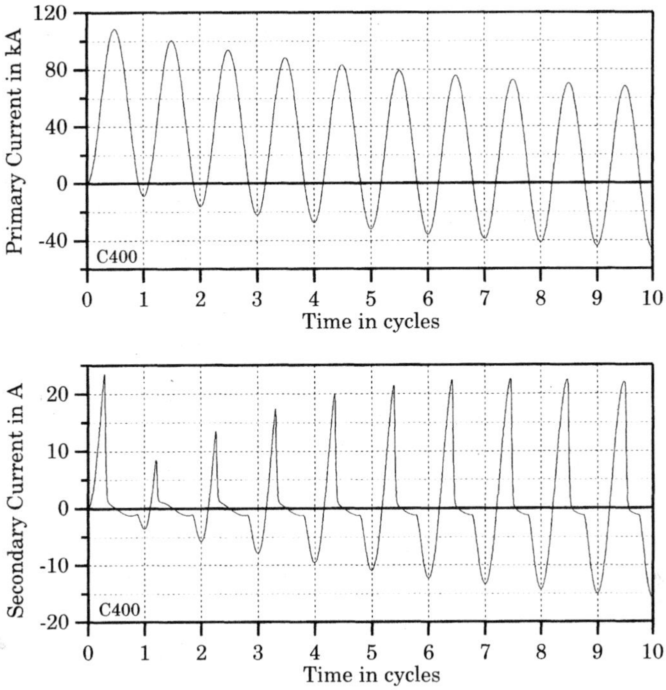

Figure 2.11 Example of CT secondary saturation due to large, fully offset primary current.

CT's, but they have lower accuracy. They are often used for relaying because of their favorable cost and because their accuracy is often adequate for relay applications. Moreover, bushing CT's are conveniently located in the bushings of transformers and circuit breakers, and therefore take up no appreciable space in the substation.

Bushing CT's are designed with a core encircling an insulating bushing, through which the primary current lead of the bushing passes. This means that the diameter of the core is relatively large, giving a large mean magnetic path length compared to other types. The bushing CT also has only one primary turn, namely, the metallic connection through the center of the bushing. To compensate for the long path length and minimum primary turn condition, the cross-sectional area of iron is increased. This has the advantage for relaying that the bushing CT tends to be more accurate than wound CT's at large multiples of secondary current rating. The bushing CT, however, is less accurate at low currents because of its large exciting current. This makes the bushing CT a poor choice for applications, such as metering, which require good accuracy at normal currents.

Current transformers are labeled with terminal markings to ensure correct polarity of a given connection. The markings label the primary winding H and the secondary winding X, each with appropriate subscripts, as shown in Figure 2.12. The usual practice is to indicate

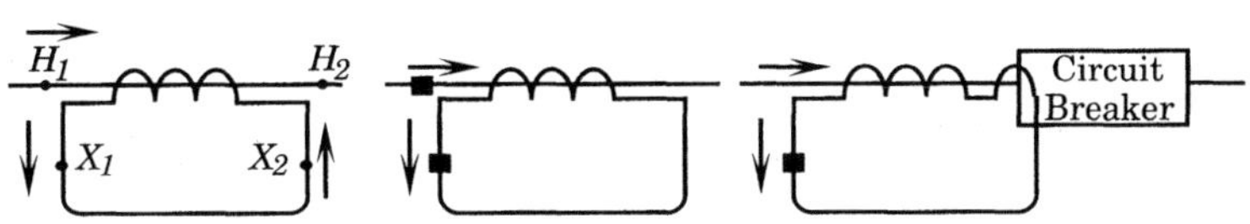

Figure 2.12 Polarity convention for current transformers [9].

polarity by dots, as shown in the two right-hand illustrations in Figure 2.12. Polarity marks are essential where two or more current transformers are connected together so that the resulting current definition can be clearly determined. For the bushing CT on the right in Figure 2.12, the polarity designation can be omitted since the primary current is, by definition, assumed to be flowing toward the breaker from the system.

Figure 2.13 shows a wye connection of current transformers, where the phasor primary and secondary currents in each phase are exactly in phase, but differ by the magnitude of the turns ratio.

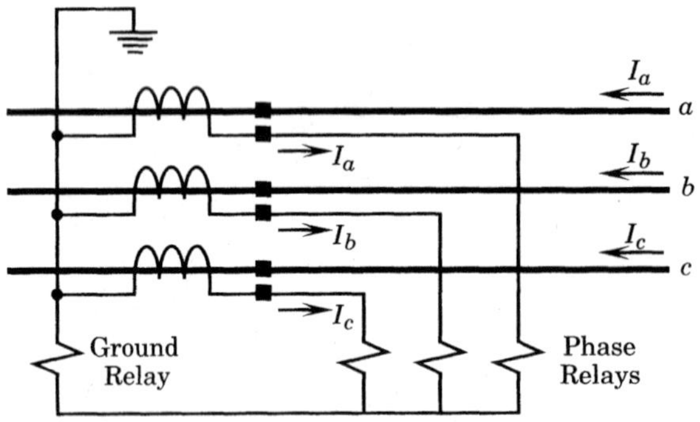

Figure 2.13 Wye connection of current transformers [9].

The delta connection of CT's can be made in two ways, and these are shown in Figure 2.14, together with the resulting phasor diagrams for each connection. It can be easily shown that the output secondary currents for these connections contain no zero sequence component. Note that delta connection B is the reverse of connection A.

The delta connection of current transformers is important for distance relaying. The subject is explored in Chapter 11.

*2.4.2.2 **Voltage (Potential) Transformers**.* Two types of voltage measuring devices are used in protective relaying. These are the instrument potential transformer, which is a two-winding transformer, and the capacitance potential device or coupling capacitor voltage transformer (CCVT), which is a capacitive voltage divider.

The wound potential transformer is much like a conventional transformer except that it is designed for a small constant load and hence cooling is not as important as accuracy.

The capacitance potential devices in common use are of two types: the coupling-capacitor device and the bushing device. These are shown in Figure 2.15. The coupling capacitor device is a series stack of capacitors with the secondary tap taken from the last unit, which is called the auxiliary capacitor. Bushing voltage dividers are constructed from capacitance bushings, where a particular level is tapped as a secondary voltage.

The equivalent circuit of a capacitance potential device is shown in Figure 2.16. The equivalent reactance of this circuit is defined by the equation

$$X_L = \frac{X_{C1}X_{C2}}{X_{C1} + X_{C2}} \tag{2.7}$$

This reactance is adjusted to make the applied voltage and the tapped voltage in phase, in which case the device is called a resonant potential device. Since the bottom capacitor is much larger than the top capacitor

$$X_{C2} << X_{C1} \tag{2.8}$$

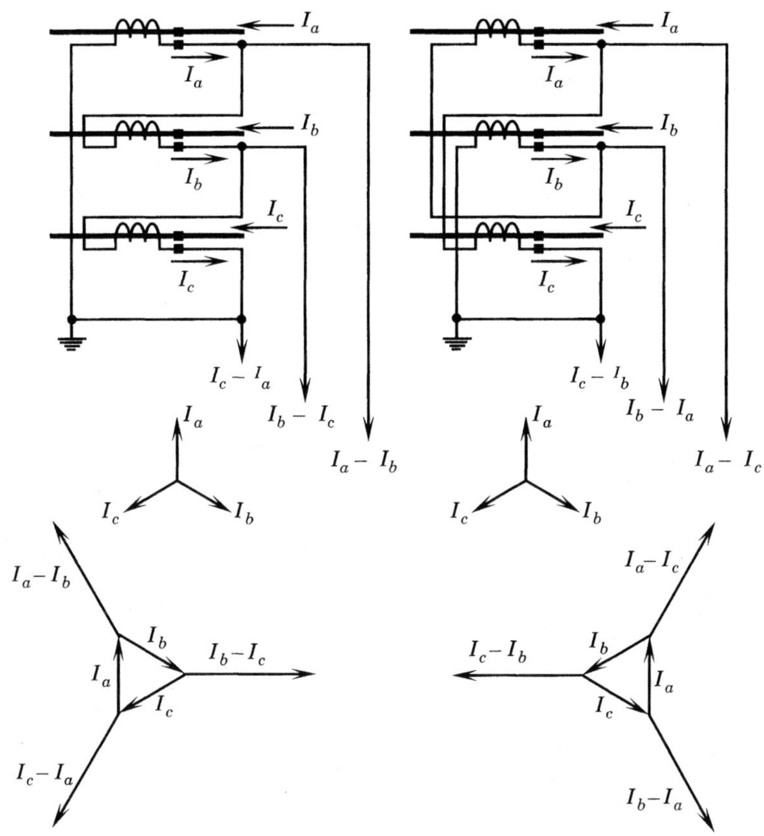

Figure 2.14 Delta connection of current transformers and the phasor diagrams for balanced three-phase currents [9].

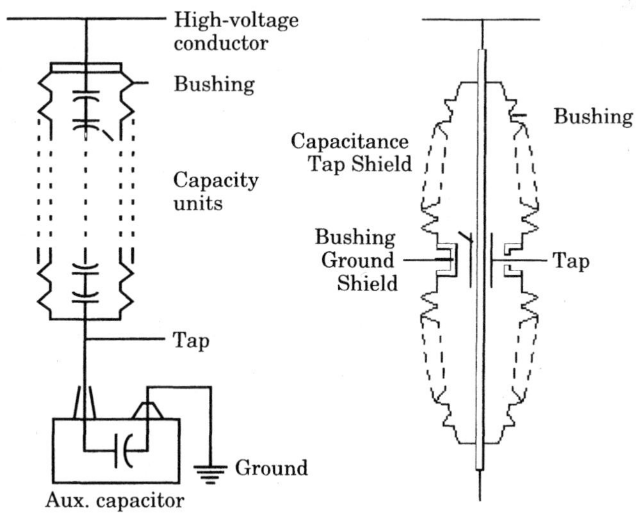

Figure 2.15 Capacitor potential devices [9], [10].

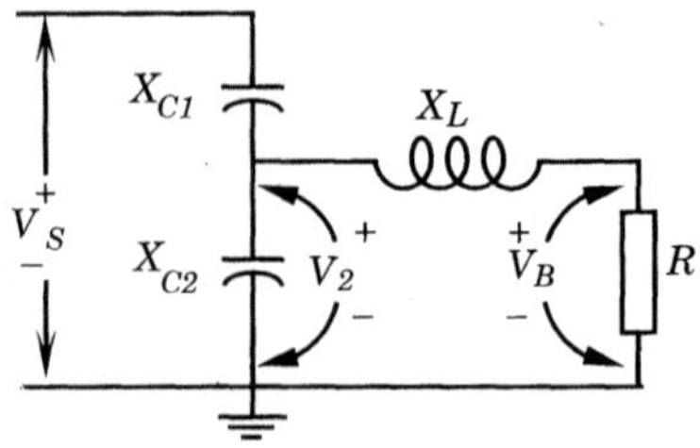

Figure 2.16 Equivalent circuit of a capacitor potential device [9].

which means that, practically

$$X_L \cong X_{C2} \tag{2.9}$$

Potential transformers (or capacitance devices) are connected Y-Y, Δ-Δ, Y-Δ, or Δ-Y, as required for particular applications. In many applications the open delta connection is used so that one potential device can be saved and the three-phase voltages can still be provided.

CCVT's are usually designed to reduce the transmission voltage to a safe metering level by a capacitive voltage divider, although a magnetic core transformer may be needed to further reduce the voltage to relay voltages, usually 67 volts line-to-neutral (115 V line-to-line).

2.4.2.3 Optical Current and Voltage Transducers. The foregoing discussion indicates that there are problems associated with the accurate acquisition of system currents and voltages due to faulted system conditions. This is especially a problem in capturing the transient currents and voltages that are required to correctly analyze the faulted system conditions. Current transformers tend to saturate, and voltage transformers, especially CCVT's, suffer from transient errors, especially for faults causing significant voltage collapse [10–16].

A new type of current and voltage transducer has been introduced, which solves many of the problems cited for ferromagnetic transducers. The principle of optical devices is based on a measurement of the magnetic field in the vicinity of the current-carrying conductor. The measurement is based on optical modulation and demodulation of the Faraday effect [11], [12]. Using this technique it is possible, in principle, to measure even dc current. Some of the advantages of this new method are the following:

1. The signal obtained from the current carrying conductor is transmitted to electronic processing equipment using fiber optic cables, which have the advantage of electrical insulation and rejection of electromagnetic induction noise.
2. The dynamic range of the optical devices are projected to be greatly superior to electromagnetic transformers.
3. The transducers should be compact and lightweight devices.

A simplified view of this type of optical current transducer (OCT) implementation is shown in Figure 2.17. The principal elements of the system are the sensor assembly, where the field measurements in the vicinity of the conductors are made, the fiber-optic cable that transmits the measured signals, and the signal processing unit, which consists of an optical interface and a computer.

Several devices of this type have been introduced [13–18], and others are sure to follow. All proposed systems use fiber optics to isolate the grounded parts from the high-voltage parts of the system, as shown in Figure 2.17.

Considerable effort has been concentrated in producing an optical current transducer (OCT). These devices are not current transformers, but optical electronic measurement systems.

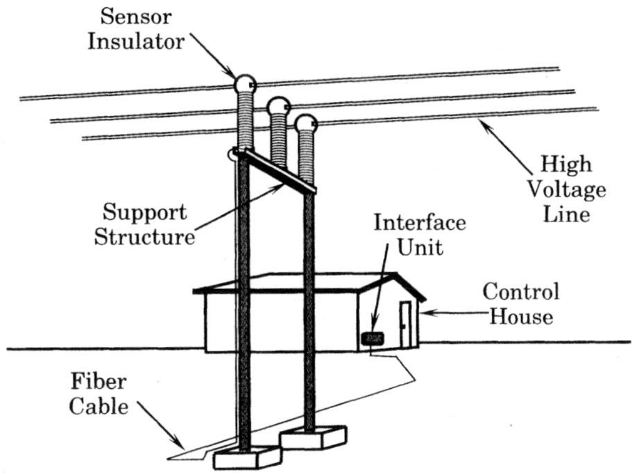

Figure 2.17 Typical arrangement of an optical current transducer [16].

There are several different methods that can be used to design an OCT, and most of the methods explored are not based on transformer principles. The power level of the signal available for ground-based processing is weak, being typically in the microwatt range. This is in contrast to the signal level of ordinary current transformers, which is at a level of several watts. The OCT has several advantages over conventional CT's. The OCT is light in weight, being much lighter than an oil-filled CT of similar rating, which results in savings in installation cost. Optical systems are immune from electrical noise. They provide safety advantages due to the natural insulating quality of the optical transmission fibers. The optical systems are also less likely to fail catastrophically than conventional current transformers.

Optical current measurements have been investigated since the late 1960s, but were not extensively developed until the late 1970s and early 1980s. As a result of extensive research, several different approaches have been explored. Most of the systems under active development employ some technique to measure the magnetic field associated with the current in the conductor of interest. This is an application of Ampere's law, which can be written as

$$I = \oint \mathbf{H} \bullet dl \tag{2.10}$$

where I is the current, **H** is the magnetic field intensity, and dl is the closed path of integration. The path of integration is optional and is accomplished in different ways by the developers, using circular, square, or other path configurations. Other methods can be used, but the conversion from magnetic to optical signals is currently the most common. This type of conversion is usually referred to as the "Faraday effect" or the "magneto-optic effect" in the technical literature. In practice, transparent glasses or crystals are used to construct the Faraday effect devices. These glasses have the property that the value of the refractive index depends on the direction of propagation and the polarization of the light and the refractive index has different values for two mutually orthogonal polarizations of the light wave. The plane of polarization is proportional to the magnetic field in the material and is measured by the rotation of the plane of polarization using various methods.

The physical devices that have been developed can be classified into five different types [18], as shown in Figure 2.18. Type 1 uses an ordinary current transformer with an added insulated optical transducer added. Type 2 uses a magnetic circuit around the conductor and measures the field inside the magnetic core optically in an air gap. Type 3 uses an optical path in a block of optically active material, with the light path enclosing the current in the conductor exactly once, which is an optical implementation of a conventional CT. Type 4 uses an optical path inside a fiber that is wound around the conductor any number of times. Type 5 measures the magnetic field at a point near the conductor, and is therefore not considered a true current transducer. The devices shown in Figure 2.18 represent the state of the art in the mid-1990s. Additional transducers are anticipated as the technology matures and refinements are implemented by developers.

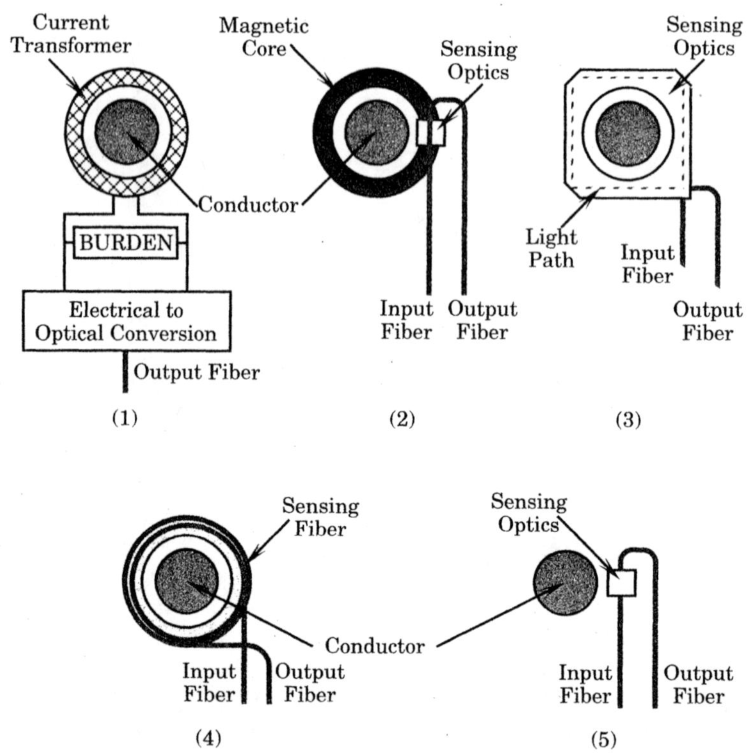

Figure 2.18 Types of optical current transducers [18]. (1) Conventional CT with optical readout. (2) Magnetic concentrator with optical measurement. (3) OCT using bulk optics. (4) Fiber optics based current measurement. (5) Witness sensor.

2.5 RELAY CONTROL CONFIGURATIONS

As a final topic on relay control, we consider the many ways in which system data can be monitored and transmitted to relays. Consider the relay connections in Figure 2.19, which might be considered the maximum practical redundancy in relay connections. Note that there are two independent relay systems, each of which could contain several relays to detect phase and ground faults on the protected line section.

This system is made very reliable by the use of completely independent systems for

- dc power supply (batteries)
- potential supply to each relay system
- current monitoring for each relay system
- dual trip coils in the circuit breaker

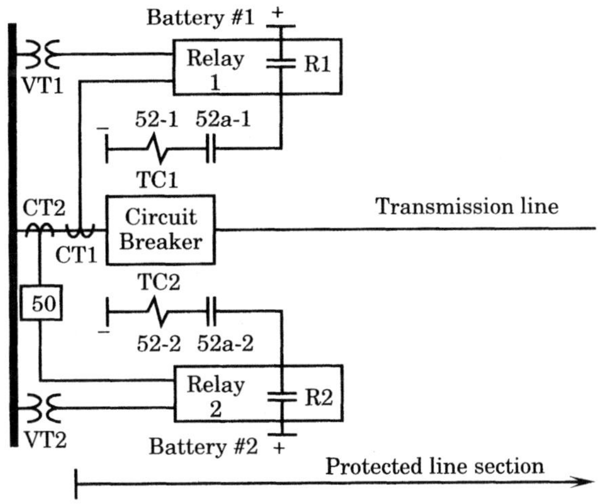

Figure 2.19 Transmission line protection with redundant relay systems and independent system data gathering systems [4].

Dual communications systems to the remote end of the line can also be provided. The two systems are designated the "primary" and the "backup" systems, although they could, in fact, be identical systems. In the illustration, the secondary system is "supervised" by an overcurrent device (50) so that it will function only when an overcurrent is detected. This supervision is optional, but it may be used to ensure that the primary relay operates first, for example, where the primary system has superior selectivity. The overcurrent device (50) is often connected in series with transmission line distance relays to prevent false tripping on loss of potential due to blown fuses or other causes that result in loss of voltage measurement to the relay.

The redundancy in Figure 2.19 is not uncommon, except for the redundant battery, which is seldom specified. Note that all system functions are duplicated except the circuit breaker. It would be possible to place redundant circuit breakers in series and have them controlled by independent relays. This would be very costly, and would probably not even be considered except for a circuit that is considered very important for some reason.

Figure 2.20 shows control configurations that are commonly used in power system protection. Part (1) has redundancy only in the relays and the two relay systems share the same battery and the same instrument transformers.

In part (2), the system is made more reliable by duplicating the instrument transformers, giving each relay its own independent supply. This leaves the circuit breaker trip coil and breaker mechanism as the most vulnerable to failure of the protective system. Part (3) of Figure 2.20 uses duplicate trip coils and has each relay connected to its own trip coil. Since the added trip coil can be obtained at quite a reasonable cost, this is often considered to be prudent, especially for the higher voltage circuits that carry large amounts of power. On these circuits, failure to properly clear a fault can have very high cost, hence greater redundancy is readily justified. Part (4) of Figure 2.20 uses redundancy in all subsystems except the circuit breaker mechanism, which would be very expensive to duplicate. This arrangement would be used at stations where high reliability is very important.

Obviously, other control configurations can be devised. It is not likely that a utility would use the same configuration for all applications, since the protected circuits are not equally important to the integrity of the entire system. Generally, the high voltage bulk power transmission lines will be protected by highly redundant protective systems, since these lines carry large amounts of power and high availability is essential.

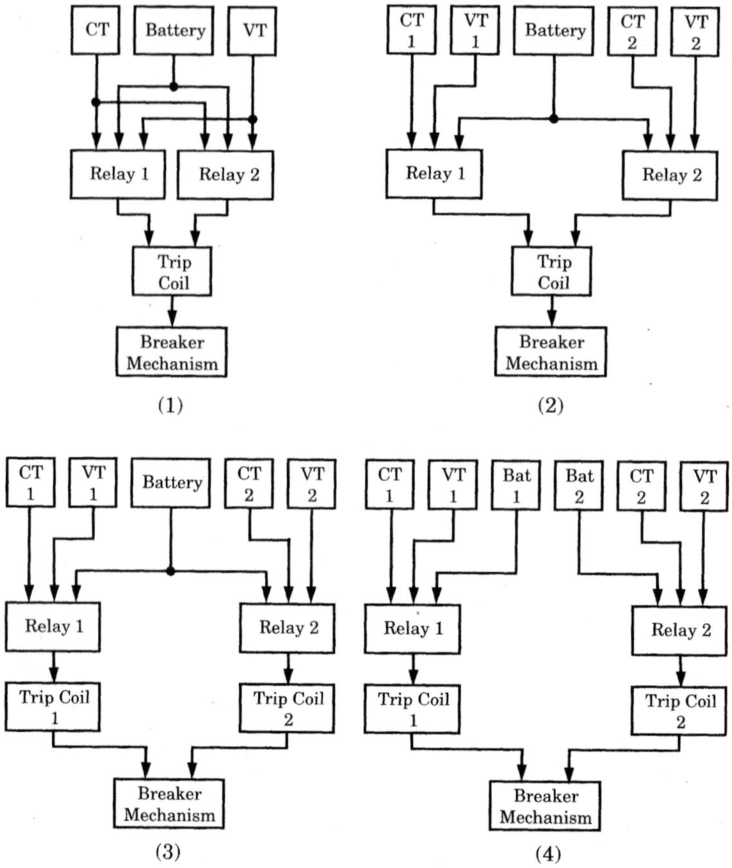

Figure 2.20 Block diagram of typical control configurations. (1) Redundant relays. (2) Redundant instrument transforms and relays. (3) Redundant instrument transforms, relays, and trip coils. (4) Redundant instrument transforms, relays, trip coils, and batteries.

2.6 OPTICAL COMMUNICATIONS

One of the most difficult technologies in power system protection is that of communications. In many types of protection, control, and measurement, the information must be transmitted from one location to another, where the data transmitter and receiver may be a great distance apart. Moreover, both the sending and receiving end of the transmission are often at high-voltage switching stations, where power frequency electromagnetic interference (EMI), radio frequency interference (RFI), switching transients, and even lightning are a part of the operating environment. These environmental problems have plagued protection engineers for years and have often been the source of numerous false trips of transmission lines and other protected components. This problem has become even more difficult with the advent of digital systems, which generate tremendous amounts of data that must be transmitted without error to remote points.

Fortunately, a solution to these communications problems has emerged in the form of an optical waveguide or optical fiber, in which light propagates along the fiber by total internal reflection. The optical fiber consists of a core material that has a refractive index higher than that of the cladding material surrounding the fiber. Transmission with this type of optical waveguide has many advantages over wire communications and makes it possible to transmit large volumes of data from point to point with high reliability and low error rate. The high data rate is possible because of the high bandwidth and low loss of the fiber. Moreover, this medium is immune to outside electromagnetic fields, which pose such a difficult problem for wire communications, especially in the environment of high-voltage substations.

There are three basic types of light guides that are usually identified according to fiber design:

(a) Multimode, stepped refractive index profile
(b) Multimode, graded index
(c) Single mode, stepped index

The basic differences in the three modes are illustrated in Figure 2.21. Multimode, stepped refractive index profile fibers, *(a)* in Figure 2.21, are often used for image transmission and short distance data transmission. The number of rays or modes of light that can be guided by this type of fiber depend on the core size and on the difference in refractive index between the core and the cladding. A transmitted pulse flattens out as it travels down the fiber because the higher angle modes have a greater distance to travel than the low angle modes. This limits the data transmission rate and the distance because it determines how close the pulses can be spaced without creating overlap at the receiving end.

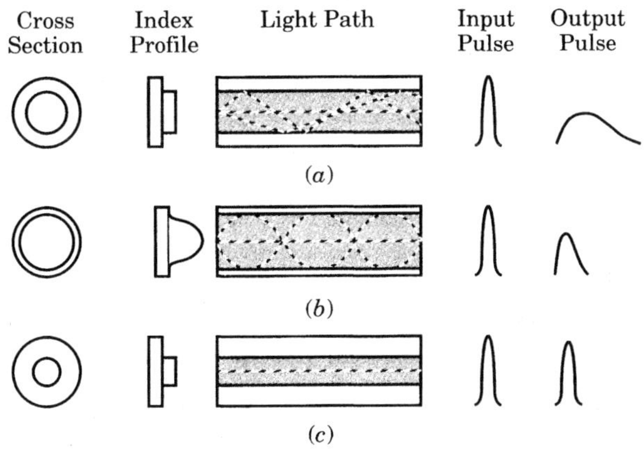

Figure 2.21 Types of optical fiber transmission.

In the graded index multimode fiber, *(b)* in Figure 2.21, the refractive index decreases with radial distance from the center. This tends to minimize pulse broadening due to the mode dispersion since the light rays travel more slowly near the center of the fiber. This type of fiber is used for medium distance, intermediate rate transmission.

High rate transmission systems use the single mode fiber, *(c)* in Figure 2.21. These fibers have low refractive index difference and a small core size, which tends to eliminate pulse dispersion since only one mode is transmitted. This type of fiber is useful for long distance, high data rate transmission systems.

The wavelength of light transmitted is a critical parameter in determining the attenuation of the signal. Experience has shown that the lowest attenuation occurs with infrared frequencies of 850, 1300, and 1500 nanometers (nm). The 850 nm wavelength, commonly used for utility applications, is available from light-emitting diodes (LED) or laser diodes (LD). Use of the higher frequencies is spreading rapidly, however.

Splicing of fiber cables has proven to be rather easy in spite of the small size of the fibers. Almost lossless splices may be fabricated in the field and have proven to be quite reliable. A number of splicing techniques have been developed, some of which are mechanical while others depend on fusing or welding the glass ends together. Splicing is important since some applications link stations that are many kilometers apart. The physical arrangement employs transmission static wires that incorporate fiber-optic cables in several optional configurations. Testing has also been performed in constructing high-voltage phase conductors that incorporate fiber-optic strands for communications [19].

Utility applications for fiber-optic data transmission are growing rapidly. The need for data acquisition and communications for supervision, control, and protection are the primary applications. In the past, these needs were met using a variety of communications media, such as microwave, power line carrier, and hard-wired circuits. Optical systems, however, offer an almost ideal replacement for these media due to three major advantages: immunity to high electromagnetic fields, wide bandwidth, and the nonconducting characteristic of the fiber cables. The fiber systems offer excellent tolerance to vibration, no cross-talk with adjacent cables, immunity to EMI and RFI, no spark or fire hazard, no short-circuit loading, no ringing or echoes, and no contact discontinuity [19]. Moreover, test installations show that the cost is often competitive.

One application that is attractive is interstation communications, control, and protection. Interstation links are typically from a few hundred meters to a few tens of kilometers. Repeater stations are required at about 15 km intervals, but this is a function of the attenuation and will improve with future development.

Intrastation applications are also being tested with excellent results. Here, the fiber-optic alternative is attractive because of its freedom from interference problems and because of the insulation characteristics of the fiber cables themselves. Signals from transducers and circuit breakers are converted to digital form and transmitted to a control room, where they are fed to microcomputers for processing. These microcomputers work with low-voltage input signals and would be damaged by transients. Here, the use of fiber cables is an ideal solution, since the cable isolates the sensitive equipment from the high-voltage equipment and shields the transmitted data from any type of outside interference. Applications in generating stations are also growing rapidly, due to the distributed controls now being used in power plants. Here, fiber optics eliminates such problems as ground loops and interference, and also requires much less space than the older equipment. Fiber-optic systems are even being placed inside light-water nuclear reactors to gather data on the reactor operating conditions.

The actual measurement of power system parameters is critical to any communications and control system. The measurement of voltage, current, temperature, pressure, and other physical parameters is the heart of any control and protection system. The potential of optical devices as sensors provides important new opportunities, as noted in Section 2.4.2.3. The ideal sensor for utility applications should have the following characteristics [19]:

- Nonmagnetic—impervious to external influence
- Passive—self-powered
- Fully dielectric—no need for insulating support structures
- Accurate—capable of precision measurement

New optics-based sensors promise benefits in all of these areas. One type, the "Pockels" voltage sensor, is promising. It is based on a characteristic of lithium niobate ($LiNbO_3$) crystal. When exposed to an electric field, the index of refraction of one of its axes changes in proportion to the strength of the field, a characteristic called *birefringence*. The shift of axis can be analyzed by passing a beam of polarized light through the crystal, permitting accurate calculation of the voltage producing the field. Present sensors are accurate to 0.5%, but 0.1% should be possible [20]. These devices do not need to contact the high-voltage conductor, but can be located a few feet off the ground under the substation bus structure, where the bus electric field can be sensed.

Optical current sensors use a Faraday glass material that rotates its polarization under the influence of a magnetic field. Several such sensors are under development. Sensors for temperature and other physical quantities are also under development.

The development of optical sensors, together with optical data transmission systems, promise to provide new methods of data acquisition and transmission that will eliminate problems that have always been difficult for the protection engineer. Future protection and control systems should have the benefit of cleaner signals, uncorrupted by outside interference. Moreover, the optical concept meshes perfectly with the low-voltage ratings of digital devices that will be the heart of future protective systems.

REFERENCES

[1] IEEE Std 315-1975 (ANSI Y32.2-1975), "IEEE Graphic Symbols for Electrical and Electronics Diagrams," IEEE, New York, 1975.

[2] ANSI/ASME Y1.1-1972, "Abbreviations, For Use on Drawings and in Text," American Society of Mechanical Engineers, New York, 1972.

[3] ANSI/IEEE C37.2-1991, "IEEE Standard Electrical Power System Device Function Numbers," IEEE, New York, 1991.

[4] Blackburn, J. L., Ed., *Applied Protective Relaying*, Westinghouse Electric Corp., Newark, NJ, 1976.

[5] ANSI Std C57.13-1993, "IEEE Standard Requirements for Instrument Transformers," IEEE, New York, 1993.

[6] Zocholl, Stanley E., Jeff Roberts, and Gabriel Benmouyal, "Selecting CTs to Optimize Relay Performance," a paper presented at the 23rd Annual Western Protective Relay Conference, Spokane, Washington, October 15–17, 1996.

[7] IEEE Power System Relaying Committee, *Transient Response of Current Transformers*, IEEE Publication 76 CH 1130-4 PWR, IEEE, New York, January 1976.

[8] Garrett, R., W. C. Kotheimer, and S. E. Zocholl, "Computer Simulation of Current Transformers and Relays for Performance Analysis," a paper presented at the 14th Annual Western Relay Conference, Spokane, Washington, October 20–23, 1987.

[9] Mason, C. Russell, *The Art and Science of Protective Relaying*, John Wiley & Sons, New York, 1956.

[10] Phadke, Arun G. and James S. Thorp, *Computer Relaying for Power Systems*, John Wiley & Sons, Inc., New York, 1988

[11] Poljac, M. and N. Kolibas, "Computation of Current Transformer Transient Performance," paper 88 WM 046-5, presented at the IEEE PES Winter Meeting, New York, 1988.

[12] Zocholl, S. E., W. C. Kotheimer, and F. Y. Tajaddodi, "An Analytical Approach to the Application of Current Transformers for Protective Relaying," a paper presented at the 15th Annual Western Relay Conference, Spokane, October 1988.

[13] Saito, S., J. Hamasaki, Y. Fujii, K. Yokoyama, and Y. Ohno, "Development of the Laser Current Transformer for Extra-High-Voltage Power Transmission Lines," IEEE J. Quant. Elec., QE-3, November 1967, pp. 589–597.

[14] Rogers, A. J., "Method for the Simultaneous Measurement of Current and Voltage on High-voltage Lines using Optical Techniques," Proc. IEE, 123, October 1976, pp. 957–960.

[15] Sawa, T., K. Kurosawa, T. Kaminishi, and T. Yokota, "Development of Optical Instrument Transformers," paper 89 TD 380-7 PWRD, presented at the IEEE PES 1989 Transmission and Distribution Conference, April 2–7, 1989, New Orleans.

[16] Ulmer, Edward A., Jr., "A High-Accuracy Optical Current Transducer for Electric Power Systems," paper 89 TD 382-3 PWRD, presented at the IEEE PES 1989 Transmission and Distribution Conference, April 2–7, 1989, New Orleans.

[17] Aikawa, Eiya, Masami Watanabe, Hisamitsu Takahashi, and Mitsumasa Imataki, "Development of New Concept Optical Zero-Sequence Current/Voltage Transducers for Distribution Systems," IEEE Trans. PWRD (6), January 1991, pp. 414–420.

[18] IEEE Committee Report, "Optical Current Transducers for Power Systems: A Review," IEEE paper 94 WM 241-0 PWRD, presented at the IEEE PES Winter Meeting, New York, January 30–February 3, 1994.

[19] Nagel, Suzanne R., "Optical Fibers," *McGraw-Hill Encyclopedia of Science and Technology*, 5th Ed., McGraw-Hill Book Co., New York, 1982.

[20] Hayes, William C., "Fiber Optics: The future is now," *Electrical World*, February 1984, pp. 51–59.

PROBLEMS

2.1 The 90 degree connection of system ac voltages and currents to a set of phase relays, as shown in Figure 2.3, is sometimes altered by inserting a resistor R in series with each of the potential coils of the relay. This has the effect of making the relay voltage lead the line-to-line voltage applied to the relay and brings the relay voltage and current more nearly in phase. Sketch this connection and its phasor diagram.

2.2 Sketch an ac relay connection similar to the 90 degree connection of Figure 2.3, but one that results in a 30 degree phase relationship between the relay voltage and current at unity power factor.

2.3 What is meant by the term *electrically trip free*? What is meant by the term *mechanically trip free*?

2.4 Distinguish between dropout and reset of a relay. Consult the definitions of Appendix A.

2.5 Calculate the secondary phase and sequence currents flowing in the phase relays and in the ground relay for the wye connection of Figure 2.13 for the following fault conditions:
(a) A three-phase fault
(b) A one-line-to-ground fault
(c) A line-to-line fault

2.6 Calculate the secondary phase currents for the two delta connections of Figure 2.14 for the following conditions:
(a) A three-phase fault
(b) A one-line-to-ground fault
(c) A line-to-line fault

2.7 Devise a current transformer connection scheme that permits only positive sequence currents in the phase relays and only zero sequence currents in the ground relay.

2.8 Derive (2.1).

2.9 Estimate the primary current that will just saturate a current transformer of high permeability silicon steel, with a cross section of 0.0017 square meters. The total secondary burden is 2.7 ohms, and the CT ratio is 2000:5.

2.10 Consider the application of a bushing current transformer with a 1500:5 turns ratio. This CT is to be used in a circuit with a maximum fault current of 25,000 A. The relay burden is 2.0 ohms, including the secondary leakage and the lead impedance. The current transformer iron circuit has a cross section of 0.002 square meters, and it saturates at 1.5 Tesla. Use the formula method to find out if the current transformer will saturate under the given conditions.

2.11 Repeat problem 2.10 using a CT ratio of 1000:5.

2.12 A radial circuit is protected by overcurrent relays that should be adjusted to operate for a fault at the extreme end of the radial line, giving a fault current of 60 A. The circuit breaker has multiple ratio bushing CT's, with ratios as shown in Figure 2.8. The relay has available tap settings by which the minimum relay pickup current may be adjusted. However, each tap setting of the relay results in a different relay burden, which we approximate by the formula

$$Z_B = 10/\text{tap}$$

Taps available are given as 3, 6, 9, and 12 A, where the tap value is taken as the minimum pickup current of the relay. Use the excitation curves of Figure 2.8 to determine a suitable tap and CT ratio, assuming we wish to keep the excitation current to less than 3% of the total primary current.

2.13 Figure P2.13 shows the bus connection at Station X and the lines leading to adjacent stations R, S, T, U, V, and W. Sketch the voltage and current transformer connections for the protection of line XS.

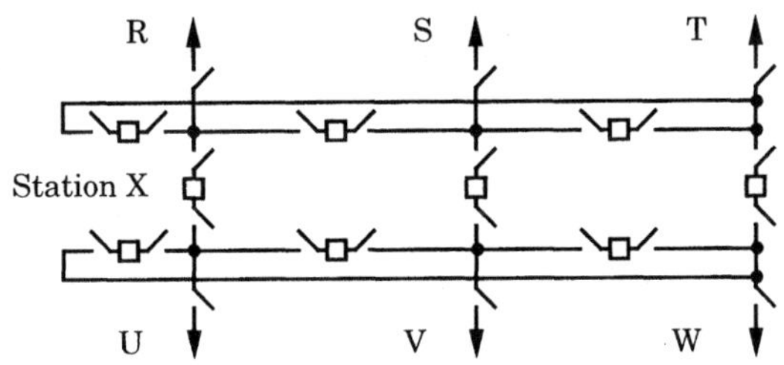

Figure P2.13 Bus arrangement of Station X.

2.14 In computing the total impedance of a given fault on an overhead line, it is often important to estimate the resistance of the arcing fault. Warrington [21] gives the following formula for the arc resistance.

$$R_{\text{arc}} = \frac{8750(s + ut)}{I^{1.4}}$$

where R_{arc} = arc resistance, ohms
s = conductor spacing, ft
u = wind velocity, mile/hr
I = fault current, A rms
t = time, sec

Suppose that the following impedance values are given:
Source: $\mathbf{Z}_S = 0 + j25\Omega$
Line: $\mathbf{Z}_L = 11 + j22\Omega$

(a) Determine the fault impedance, both with and without arc resistance, when the line is radial from the source.
(b) Repeat (a) for a fault at the sending end of the line.
(c) Plot the results of (a) and (b) in the complex R-X plane.

3

Protective Device Characteristics

3.1 INTRODUCTION

In this chapter we introduce a number of different devices that are important in the study of power system protection. The treatment of device characteristics is brief and somewhat superficial. The reason for this is that this book places its emphasis on analysis, rather than hardware. As an academic subject, protection is challenging because of the analytical techniques that are required. These techniques form the core of this text. Still, techniques require the application of hardware items, many of which are complex and interesting in their own right. Our purpose here is to discuss those device characteristics that are important to the understanding of protective system analysis. One who intends to pursue protection professionally will find experienced application engineers and manufacturers who can explain the characteristics of any given device. Indeed, this is the only way to obtain up-to-date information on hardware items that are in a constant state of change and improvement.

Our discussion begins with the characteristics of fuses, which are the simplest and most basic of hardware items used in system protection. This will include a discussion of fuse types and the way in which fuse performance can be specified and displayed for analysis.

The second item for discussion is relays. Relays are available in many types and are designed for many different purposes, as noted in the tabulations of Appendix A. Our approach is to separate the relays into distinct types as a general classification, and to discuss the general characteristics of each type. This will provide an overview of relays and an appreciation of the many options open to the protective system designer.

Power circuit breakers will be discussed only briefly. The purpose of this section is not to treat arc interruption as a scientific study, but to present the definitions and ratings used for power circuit breakers that are important to the study of protective system analysis.

The chapter closes with a discussion of automatic circuit reclosers and sectionalizers. These devices are used on lower voltages, particularly 25 kV and below, for the protection of radial distribution lines. They form a special class of equipment that must be understood in order to coordinate with other protective system equipment.

Any fault protective device must be selected with regard to three different ratings: the voltage rating, the continuous (load) current rating, and the interrupting rating. The voltage rating must be high enough to withstand voltages normally experienced in system operation. The continuous current rating must be adequate for the normal load current that is expected to flow in the circuit of application. This current rating is often chosen to exceed the maximum load current by a margin of 30% or so, at the time of installation, in order to allow for future load growth. The interrupting rating refers to the highest current the device will be called upon to interrupt at rated voltage. This rating is often expressed in MVA. These and other ratings will be discussed in connection with the various protective devices.

3.2 FUSE CHARACTERISTICS

The simplest of all circuit interrupting devices is the fuse. Nearly everyone is familiar with household socket-type fuses and the fact that they are available in different current ratings, which correspond to the maximum allowable continuous current. These small fuses are installed in series with the supply service lines and, if properly installed, carry normal load current without interruption. If the circuit load exceeds the fuse rating or if a fault develops in the wiring or a connected appliance, the fuse melts and creates an open circuit at the service entrance.

A fuse is an "overcurrent protective device with a circuit-opening fusible part that is heated and severed by the passage of overcurrent through it" [1]. For circuits operating at 600 volts and above, fuses are called "power fuses," although the standards make a distinction between a "distribution cutout" and a "power fuse" [2]. Both devices are defined to include an assembly that acts as a fuse support and fuse holder, and may or may not include the fusible link as well. Fuses designed for use at high voltages, in the range from about 2 kV to 200 kV, are of two general types: distribution fuse cutouts and power fuses.

3.2.1 Distribution Fuse Cutouts

Distribution cutouts are primarily for use on power distribution circuits, are designed for typical distribution voltages (say, 35 kV and below), and are often designed for pole or cross-arm mounting that are common for overhead distribution lines. Power fuses are also designed for transmission and subtransmission voltage levels, and for station or substation mounting. Both devices usually include a fuse link, which is a "replaceable part or assembly, comprised entirely or principally of the conducting element, required to be replaced after each circuit interruption to restore the fuse to operating condition" [2]. Fuse links are usually made of tin, lead, or silver in various alloys or combinations, to achieve a desired time-current characteristic.

A typical cutout design is shown in Figure 3.1, which shows both substation and pole-top types. The fuse is contained in a long tube called the fuse link holder, which is usually made of a fiber insulating material. The holder is designed for easy removal by a hot stick. The operator inserts a hook at the end of the hot stick into the hookeye and pulls. This disengages the locked top of the fuse link holder and allows it to swing down, suspended from the lower hinged mechanism. The hot stick can now be used to lift the holder out of the mechanism, completely freeing the holder. The cutout serves both as a fusible element and a manual switch, which allows repair personnel to open a circuit manually. Where the load to be interrupted is great, load-break cutouts can be used that have a special load break mechanism to extinguish the arc created when the cutout is opened.

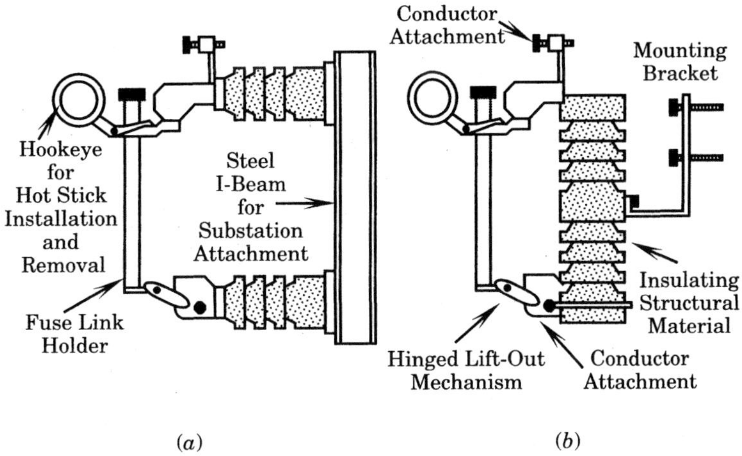

Figure 3.1 Typical cutouts for substation and pole-top installation. (*a*) Substation type. (*b*) Pole-top type.

Cutouts of the type described above are called *open* cutouts. Many cutouts are designed such that the fuse link holder is unlatched at the top when a fault current is interrupted. This makes it easy to locate the blown fuse since the holder of the blown fuse hangs down from the lift out mechanism and is clearly visible from the ground. It is also noted that a circuit can be opened for repair by opening cutouts, and the opened cutouts provide visual proof that the circuit has been de-energized. This is a valuable safety feature for the repair crew.

The fuse replacement elements are designed to fit inside the fuse link holder. A typical fuse link and holder are shown in Figure 3.2. The spent fuse link is removed by unscrewing the screw cap and removing the fuse link assembly. The fuse link contains one or more fusible elements, as shown in Figure 3.2, and may contain a spring to aid in separating the melted elements under low short circuit conditions. The fuse link is inserted in the holder, flexible leader first, and is pulled down until the button is tight against the top of the fuse link holder. The flexible leader is then attached to a bolted connector at the bottom of the holder, so that good metallic contact is made with the hinged mechanism, and the screw cap is replaced to hold the button head in place and form a good electrical connection.

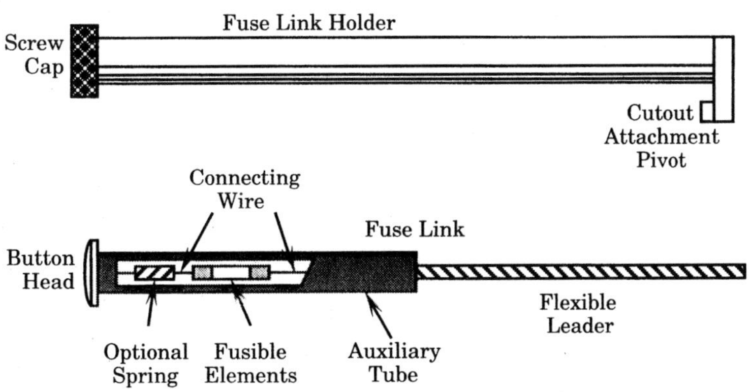

Figure 3.2 A typical fuse link with cutaway of fusible element.

3.2.2 Fuse Types

Fuses are designed for many different applications and with a wide variety of characteristics to meet the requirements both routine and special situations. Only the general types of fuses are described here, but the engineer with either routine or special fuse requirements will find useful application data from the fuse manufacturers.

3.2.2.1 Standard Zero-Current-Clearing Fuses. The most common fuse can be described as a zero-current-clearing device, since the fuse must wait until the current passes through zero before successful clearing is accomplished [3]. No special design of the fuse is required to ensure this type of behavior, since that is the point on the current wave when the current will naturally be interrupted as the fusible element increases its impedance. On a 60 hertz system, a natural zero crossing occurs every 8.33 ms, which represents the maximum time that clearing will be delayed. This type of fuse is very common and finds useful application in many different situations. Examples include distribution transformer primary fusing, distribution branch feeder protection, motor protection, and industrial load protection. Some of these applications may have special requirements, however, which will require special fuse characteristics, and fuses are available to satisfy many different special needs.

Fuse units used in distribution system applications are usually one of two types: expulsion or filled fuse units [2]. An expulsion fuse unit is a vented fuse unit, with provision for escape of arc gases, in which an expulsion effect of gases produced by the arc and the lining of the fuse holder, either alone or aided by a spring, stretches, cools, and eventually extinguishes the arc by blasting the high-pressure gases out of the fuse link holder. The filled fuse unit is one in which the arc is drawn through a filling material, which may be a granular, liquid, or solid material, and the filling aids the arc extinction. Fuses generally operate by melting the fusible material, which creates an arc. The fuse designer uses one of the above principles to lengthen and cool the arc, so that the current can be interrupted safely. Some of the energy created by the arc creates pressure, which is also used to unlatch the top of the fuse holder and cause it to swing down and thereby provide visual evidence of the blown fuse. One of the advantages of the expulsion type of fuse is that the cutout can be reloaded following fault interruption with a relatively inexpensive fuse link. These links also come in a large variety of sizes and ratings, making coordination possible with many different devices.

The oil fuse cutout is another type of expulsion fuse. In this device, the expulsion gases consist of the products produced by the breakdown of oil that surrounds the fusible element.

Another type of fuse is the vacuum fuse. This type is similar in design and operation to the expulsion fuse, and it has a similar time-current characteristic. The major difference is that the vacuum link is a completely sealed unit, which utilizes no expulsion action. Interruption takes place because of the rapid dielectric buildup that occurs in a vacuum after a current zero is reached.

3.2.2.2 Current Limiting Fuses. Another type of fuse sometimes used in power applications is the *current-limiting fuse*. This fuse is defined as "a fuse that, when it is melted by a current within its specified current-limiting range, abruptly introduces a high arc voltage to reduce the current magnitude and duration." This type of fuse is basically different from the current zero waiting types. Here the principle is called current limiting or energy limiting. It does this by introducing a high resistance into the circuit. This limits the current, but it also improves the power factor, making the current more in phase with the voltage. Figure 3.3 shows a fault current that is completely offset, in which case the magnitude is very high and the first current zero is delayed. The current-limiting fuse, however, causes the current to fall to zero as the voltage goes to zero, thereby limiting the maximum current to an amount called the "let-through current."

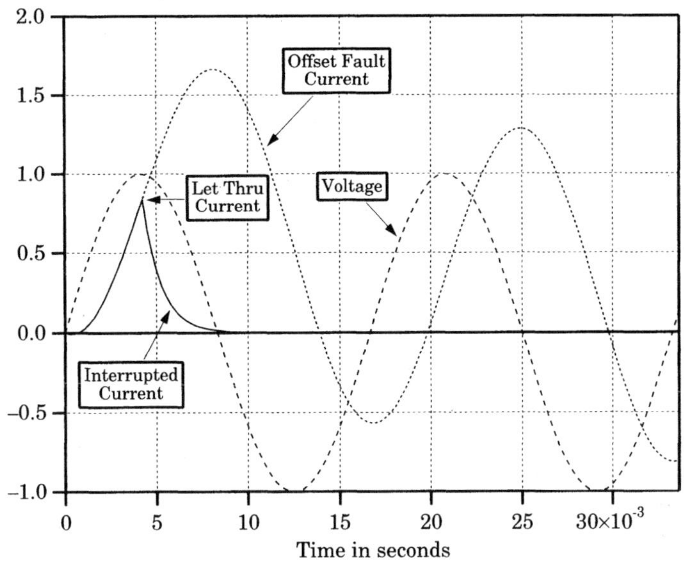

Figure 3.3 Current limiting of an offset fault current.

There are three basic types of current limiting fuses [3]. The first type is the *backup current-limiting fuse*. This fuse is very effective at high fault currents, but is not able to interrupt low currents, hence it must always be used with a conventional overcurrent device to react to low fault currents. The second type is the *general purpose current-limiting fuse* that, according to ANSI standard, is a current-limiting fuse that can interrupt a current that causes the fuse to operate in one hour or less. The third type is a *full range* fuse, which is designed to interrupt any current that causes its fusible element to melt under normal fusing conditions.

The best way to show the effectiveness of the current-limiting fuse is by computing the so-called I^2t factor, which is the current time integral. This factor represents the heating that can occur per increment of resistance. Therefore, it is proportional to the allowable energy in the circuit [3]. There are two parts of the I^2t factor. The first part is the melting I^2t, which can be determined by calculation [3]. The second part of the I^2t factor is that which occurs after arcing begins and continues until full current interruption occurs. This part must be determined by test.

The first requirement of any fuse is to extinguish the arc. The second requirement is that the cutout must be able to withstand the normal full rated voltage across the open fuse link. This usually involves a high frequency transient that rapidly damps out, followed by the normal power frequency voltage, which will be applied across the open cutout. The third requirement of the fuse design is that it must be capable of being coordinated with other fuses or other devices, so that the outage caused by the fault is restricted to the minimum protective zone.

3.2.2.3 Special Fuses. As noted above, special fuses are available to satisfy a variety of different system conditions. A few of these are described below.

One application that sometimes requires special treatment is the fusing of capacitor banks. In some installations, changes in the supply system may cause the available fault current to increase, requiring a change in the capacitor bank fuses. In some cases, it may not be possible to change from a fuse to a more expensive circuit breaker because of space limitations in the existing installation, even though a circuit breaker could be selected with more than adequate interrupting rating. One solution that has been developed for such an application is a special type of current-limiting fuse that has an unusually large interrupting rating [4]. Capacitor bank fusing is also complicated by having a high-frequency inrush current when the bank is energized, which may cause problems in the fuse application.

Another problem situation sometimes encountered is that of the continuous current rating of the fuse for a given application. For satisfactory service, the fuse must have the correct voltage rating, full load current rating, and be able to sustain repeated overloads. In most cases, it is not recommended that fuses be placed in parallel in order to increase current rating, as there is no assurance that the two parallel fuses will melt and clear the circuit at exactly the same time, whether of current-limiting or zero current clearing types. Special parallel fuse cutouts have been developed and are approved for application where increased current ratings are required. It must be emphasized that this is a special fusing device that is designed for this purpose and has been tested and approved for parallel current interruption. Paralleling common fuse cutouts is not recommended.

Another special type of fuse is the "electronic fuse" [5]. This fuse uses integrated electronic circuitry to provide current sensing, the time-current (TC) characteristic, and the control of the fuse performance. The advantage of this fuse is that the time-current characteristic can be provided in several different types, making it possible to select the characteristic that best fits a particular application. The fuse design includes a control module that uses a current transformer for current sensing, for input power to the electronic circuits, and to provide energy to operate the interrupting module. The control module also provides the time-current characteristic that controls circuit interruption. The interrupting module is controlled by the control module to provide current interruption. This is accomplished by signaling a gas-generating power cartridge to move an insulating piston, thereby shunting the fault current to fusible elements. After fault clearing, the interrupting module is replaced.

The special fuses mentioned here are examples of the ingenuity of fuse manufacturers in developing new fuses to meet the varied needs of the protection engineer. The descriptions are by no means complete, and the engineer should consult the manufacturers and suppliers for fuses that may be required for special system protection applications.

3.2.3 Fuse Time-Current Characteristics

Fuse links are manufactured for interchangeability with standard dimensions and with time-current (TC) characteristics designed to meet particular qualifications. One class of power fuses are identified by the letter "E" to signify that their TC characteristics conform to the interchangeability requirements given in Table 3.1.

TABLE 3.1 Melting Time-Current Characteristics of E Rated Links [6]

Line Current Reading	Melting Time	Continuous Current
100 amperes and below	300 sec	200–240% of rating
Above 100 amperes	600 sec	220–264% of rating

Power fuses are available with rated continuous current ratings of 0.5, 1, 2, 3, 5, 7, 10, 15, 20, 30, 40, 50, 65, 80, 100, 125, 150, 200, 250, 300, and 400 amperes. Rated nominal and maximum voltages are specified in Table 3.2.

TABLE 3.2 Rated Voltages for E Rated Power Fuses [6]

Rated Nominal Voltage in kV	Rated Maximum Voltage in kV
2.4*	2.75*
4.8*	5.50*
7.2	8.25
13.8*	15.00*
14.4	15.50
23.0	25.80
34.5	38.00
46.0	48.30
69.0	72.50
92.0	**
115.0	121.00
138.0	145.00
161.0	169.00

* For indoor power fuses only

** Not specified

Power fuses are used at installations where the cost of an oil circuit breaker and all of its associated equipment is not considered economical. There are several factors that influence the decision to use a fuse rather than a more elaborate installation, for example, the anticipated frequency of operation, the required speed of restoration of service following a fault, and the return on the required investment.

Power fuses of the E rated type are applied in many different situations, such as

- Potential transformer protection
- Power transformer protection
- Capacitor bank protection
- High-voltage feeder circuits

Power fuses are available for both indoor and outdoor, and in expulsion or current limiting types. The choice depends on the location and ratings available. For example, expulsion types should be installed outdoors where there is ample room. Indoor installation will often favor current-limiting types.

The E rated fuse is a 100% current fuse; that is, the rating must be equal to or greater than the maximum continuous load current. Overloads that persist for the duration of the melting time may damage the fusible element and change its characteristic. Therefore, in selecting a fuse for installation, consideration must be given to the likelihood of long duration overloads, such as starting currents of large motors. A rule of thumb is to select a fuse with current rating at least as high as the value of current anticipated for 5 seconds. For magnetizing inrush of large power transformers, the exciting current should not exceed 75% of the fuse melting current. Moreover, the power fuse for transformers should be able to carry at least 12 times the rated transformer primary current for one-tenth of a second, as shown by the fuse melting time curve.

At distribution voltages, fuses are often used for feeder circuit protection, especially for relatively short lengths of line supplying a small number of customers. These situations do not usually merit more elaborate protective systems, and fuses have proven quite beneficial in these applications, since proper coordination of fuses restricts the outage due to a fault to a small portion of the affected feeder. For circuit protection, the *coordination* of the fuses is a primary concern. Figure 3.4 shows several applications of fuses on distribution circuits.

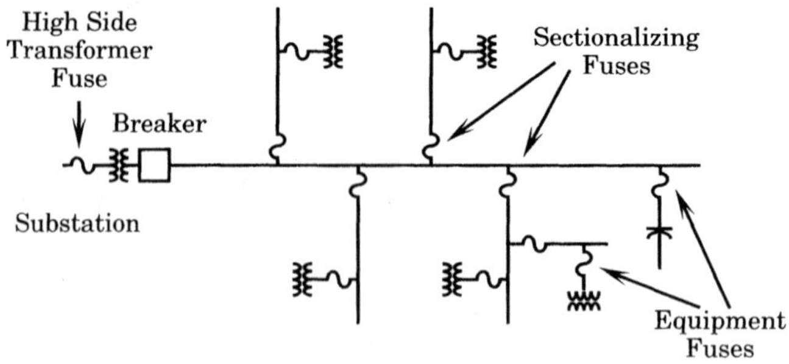

Figure 3.4 Applications of fuses on distribution circuits.

Distribution fuse links are manufactured for interchangeability in two types, designated K and T. The difference in these types is in the relative melting time, which is measured by a variable called the *speed ratio*, and which is defined as follows:

$$\text{Speed ratio} = \frac{\text{Melting current at 0.1 s}}{\text{Melting current at 300 or 600 s}} \tag{3.1}$$

The denominator of (3.1) uses 300 seconds as the time value for fuse links rated 100 amperes or less, and 600 seconds for fuse links rated 140 amperes and 200 amperes. The specified melting currents of Types K and T links are given in Tables 3.3 and 3.4 (see pages 51 and 52), respectively. Note that the speed ratio of the type K links is in the range of 6.0 to 8.1, whereas the type T links have speed ratios of 10.0 to 13.0. The reason for stating certain ratings to be "preferred" is that, by rule of thumb, one can be reasonably sure that any fuse in the series will protect the next higher rating in that series. The "intermediate ratings" fall between the preferred ratings and will also coordinate by rule of thumb with other fuses in the intermediate series.

Fuse TC curves are available from the manufacturer in two forms: minimum melting and total (or maximum) clearing time. Typical examples of log-log plots of time versus current, are shown in Figure 3.5, 3.6, and 3.7. The minimum melting curve is an average melting time measured in low voltage tests where arcing does not occur. Thus, for a given current value, the time for the fuse to open the circuit represents the melting time, which must fall within the tolerances given by the standards of Tables 3.3 and 3.4. A second test, run at several thousand volts (typically 7200 V), provides a measure of the total time to clear the circuit at a given current, including the melting time and the arcing time. These two curves represent the range of clearing times that might be expected for a given fault current. The total clearing curve should be used in coordinating against the minimum melting characteristic of a larger fuse, located toward the power source. Similarly, the minimum melting curve should be used in coordinating with the total clearing of a smaller fuse, located on the load side. Figure 3.5 compares the minimum melting time-current characteristics of the Type K and Type T fuses. Figure 3.6 compares the characteristic shapes of minimum melting and total clearing curves for Type K fuses of the same rating. Much longer times are required to interrupt current for the Type T rating (note the log scale). Figure 3.7 shows an entire family of curves, in this case the Type K minimum melting curves.

Note that the minimum melting current of all links is about twice the fuse nominal current rating. For any current above the minimum pickup, the T link melts more slowly than

TABLE 3.3 Melting Currents for Type K (Fast) Fuse Links [3]

Rated Continuous Current	300 or 600 sec Melting Current* Min	Max	10 sec Melting Current Min	Max	.01 sec Melting Current Min	Max	Speed Ratio
			Preferred Ratings				
6	12.0	14.4	13.5	20.5	72.0	86.0	6.0
10	19.5	23.4	22.5	34.0	128.0	154.0	6.6
15	31.0	37.2	37.0	55.0	215.0	258.0	6.9
25	50.0	60.0	60.0	90.0	350.0	420.0	7.0
40	80.0	96.0	98.0	146.0	565.0	680.0	7.1
65	128.0	153.0	159.0	237.0	918.0	1100.0	7.2
100	200.0	240.0	258.0	388.0	1520.0	1820.0	7.6
140	310.0	372.0	430.0	650.0	2470.0	2970.0	8.0
200	480.0	576.0	760.0	1150.0	3880.0	4650.0	8.1
			Intermediate Ratings				
8	15.0	18.0	18.0	27.0	97.0	116.0	6.5
12	25.0	30.0	29.5	44.0	166.0	199.0	6.6
20	39.0	47.0	48.0	71.0	273.0	328.0	7.0
30	63.0	76.0	77.5	115.0	447.0	546.0	7.1
50	101.0	121.0	126.0	188.0	719.0	862.0	7.1
80	160.0	192.0	205.0	307.0	1180.0	1420.0	7.4
			Ratings Below 6 Amperes				
1	2.0	2.4	†	10.0	†	58.0	—
2	4.0	4.8	†	10.0	†	58.0	—
3	6.0	7.2	†	10.0	†	58.0	—

* 300 s for fuse links rated 100 A or less; 600 s for fuse links rated 140 A and 200 A.

† No minimum value indicated since the standard requirement is that 1, 2, or 3 A ratings shall coordinate with the 6 A rating but not necessarily with each other.

the K link. This difference in fuse characteristic can be used to advantage in a coordination study, with the choice of fuse type used depending on the kind of protective devices with which coordination is required and on the available fault current.

Distribution fuse links are given voltage ratings of 7.2, 14.4, and 17 kV nominal, or 7.8, 15, and 18 kV maximum for use in open-link cutouts. Links for use in enclosed and open cutouts are rated 14.4 and 25 kV nominal or 15 and 27 kV maximum.

3.2.4 Fuse Coordination Charts

Referring again to Figures 3.5 to 3.7, it is apparent that fuses of a given type have a similar shape and coordinate well together, whereas a mixture of T and K fuses may make coordination difficult and the rule of thumb coordination for adjacent sizes would be impossible. It is true that fuses of a given type do coordinate well with other fuses of that same type. However, because of the arcing time, which must be included in the total clearing time to determine the source side coordination, there is a maximum current for safe coordination, even for fuses of

TABLE 3.4 Melting Currents for Type T (Slow) Fuse Links [3]

Rated Continuous Current	300 or 600 sec Melting Current* Min	300 or 600 sec Melting Current* Max	10 sec Melting Current Min	10 sec Melting Current Max	.01 sec Melting Current Min	.01 sec Melting Current Max	Speed Ratio
			Preferred Ratings				
6	12.0	14.4	15.3	23.0	120.0	144.0	10.0
10	19.5	23.4	26.5	40.0	244.0	269.0	11.5
15	31.0	37.2	44.5	67.0	388.0	466.0	12.5
25	50.0	60.0	73.5	109.0	635.0	762.0	12.7
40	80.0	96.0	120.0	178.0	1040.0	1240.0	13.0
65	128.0	153.0	195.0	291.0	1650.0	1975.0	12.9
100	200.0	240.0	319.0	475.0	2620.0	3150.0	13.1
140	310.0	372.0	520.0	775.0	4000.0	4800.0	12.9
200	480.0	576.0	850.0	1275.0	6250.0	7470.0	13.0
			Intermediate Ratings				
8	15.0	18.0	20.5	31.0	166.0	199.0	11.1
12	25.0	30.0	34.5	52.0	296.0	355.0	11.8
20	39.0	47.0	57.0	85.0	496.0	595.0	12.7
30	63.0	76.0	93.0	138.0	812.0	975.0	12.9
50	101.0	121.0	152.0	226.0	1310.0	1570.0	13.0
80	160.0	192.0	248.0	370.0	2080.0	2500.0	13.0
			Ratings Below 6 Amperes				
1	2.0	2.4	†	11.0	†	100.0	—
2	4.0	4.8	†	11.0	†	100.0	—
3	6.0	7.2	†	11.0	†	100.0	—

* 300 s for fuse links rated 100 A or less; 600 s for fuse links rated 140 A and 200 A.

† No minimum value indicated since the standard requirement is that 1, 2, or 3 A ratings shall coordinate with the 6 A rating but not necessarily with each other.

the same type. This coordination limit is a function of the type of cutout used, as this may affect the arcing time slightly. Manufacturers often provide coordination charts similar to those given in Tables 3.5 and 3.6[1] to indicate the recommended coordination limits that are recommended for use with specified cutouts. In these tables, reference is made to the "protected link" and the "protecting link." These terms are clarified in Figure 3.8.

The logic in these designations is as follows. When a short circuit appears on the load side of the downstream protecting links, that link should clear, thereby removing the fault and making it unnecessary for the upstream protected link to clear. The protected link must have a higher rating and be properly coordinated with the protecting link to assure proper operation.

An essential rule of coordination requires that the maximum clearing time of the protecting link shall not be greater than 75% of the minimum melting time of the protected link. This ensures that the protecting link will operate fast enough to prevent damage of the

[1]Tables 3.5 to 3.7 show maximum values of fault current at which EEI-NEMA Type K and T fuse links will coordinate with each other. These tables are based on maximum clearing time curves for protecting links and 75% of minimum melting time curves for protected links. Preferred ratings are shown in bold typeface and nonpreferred ratings in italics. [Tables provided by McGraw-Edison Power Systems, Cooper Industries.]

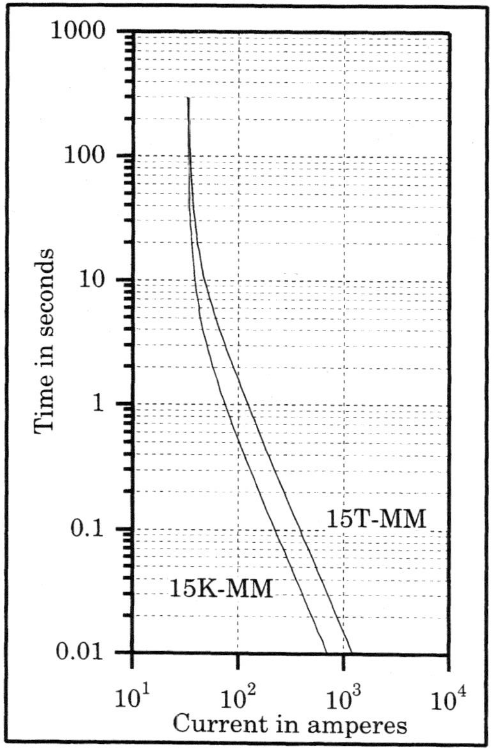

Figure 3.5 Type K (fast) and type T (time delayed) time-current curves of the same current rating.

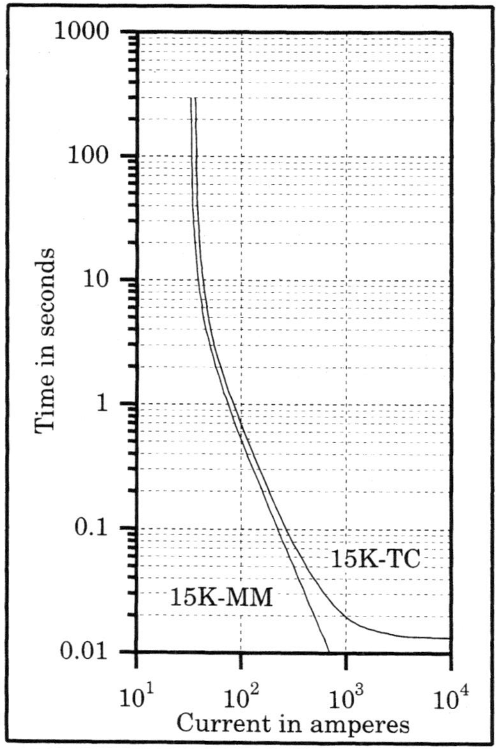

Figure 3.6 Type K (fast) minimum melting and total clearing time-current curves.

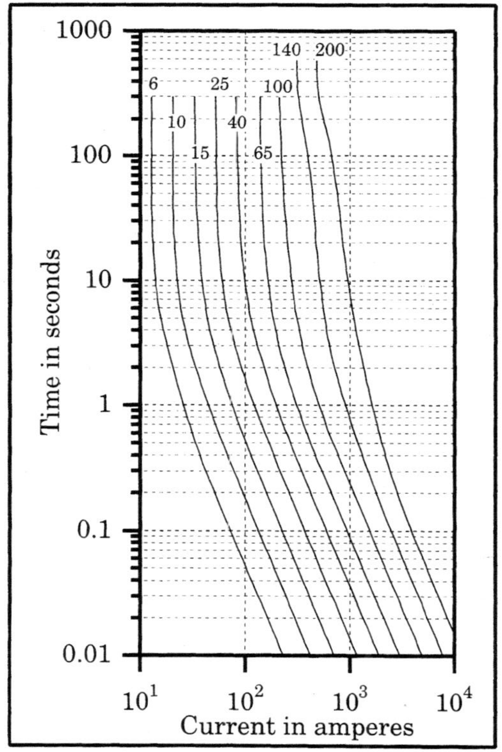

Figure 3.7 The entire set of type K minimum melting time-current fuse characteristics.

TABLE 3.5 Coordination between EEI-NEMA Type K Fuse Links

Protecting Fuse Link Rating, A	Protected link rating (amperes) 8K	10K	12K	15K	20K	25K	30K	40K	50K	65K	80K	100	140	200
	Maximum fault current at which B will protect A (amperes)													
6K		190	350	510	650	840	1060	1340	1700	2200	2800	3900	5800	9200
8K			210	440	650	840	1060	1340	1700	2200	2800	3900	5800	9200
10K				300	540	840	1060	1340	1700	2200	2800	3900	5800	9200
12K					320	710	1050	1340	1700	2200	2800	3900	5800	9200
15K						430	870	1340	1700	2200	2800	3900	5800	9200
20K							500	1100	1700	2200	2800	3900	5800	9200
25K								660	1350	2200	2800	3900	5800	9200
30K									850	1700	2800	3900	5800	9200
40K										1100	2200	3900	5800	9200
50K											1450	3500	5800	9200
65K												2400	5800	9200
80K													4500	9200
100K													2000	9100
140K														4000

TABLE 3.6 Coordination between EEI-NEMA Type T Fuse Links

Protecting Fuse Link Rating, A	Protected link rating (amperes) 8T	10T	12T	15T	20T	25T	30T	40T	50T	65T	80T	100	140	200
	Maximum fault current at which B will protect A (amperes)													
6T		350	600	920	1200	1500	2000	2540	3200	4100	5000	6100	9700	15.2
8T			375	800	1200	1500	2000	2540	3200	4100	5000	6100	9700	15.2
10T				530	1100	1500	2000	2540	3200	4100	5000	6100	9700	15.2
12T					680	1280	2000	2540	3200	4100	5000	6100	9700	15.2
15T						730	1700	2500	3200	4100	5000	6100	9700	15.2
20T							990	2100	3200	4100	5000	6100	9700	15.2
25T								1400	2600	4100	5000	6100	9700	15.2
30T									1500	3100	5000	6100	9700	15.2
40T										1700	3800	6100	9700	15.2
50T											1750	4400	9700	15.2
65T												2200	9700	15.2
80T													7200	15.2
100T													4000	13.8
140T														7.5

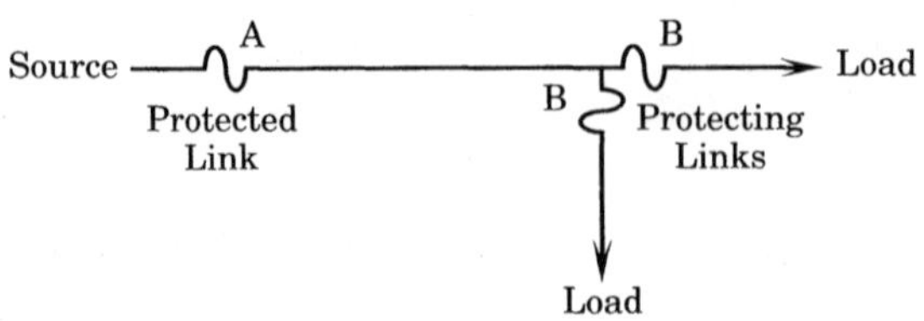

Figure 3.8 Definition of protected and protecting links.

protected link due to partial melting. The 75% factor compensates for operating variables, such as preloading, ambient temperatures, and the like.

In addition to the fuse coordination values, the application engineer also needs to know the continuous current capability of each fuse rating. These values are given in Table 3.7.

TABLE 3.7 Continuous Current-Carrying Capacity of EEI-NEMA Fuse Links

EEI-NEMA K or T Rating	Continuous Current (amperes)	EEI-NEMA K or T Rating	Continuous Current (amperes)	EEI-NEMA K or T Rating	Continuous Current (amperes)
6	9	20	30	65	95
8	12	25	38	80	120†
10	15	30	45	100	150†
12	18	40	60*	140	190
15	23	50	75*	200	200

* Only when used in a 100 or 200 ampere cutout.

† Only when used in a 200 ampere cutout.

Coordination tables, such as Tables 3.5 and 3.6, provide a very convenient method of coordinating fuse links. Use of the tables assumes that the links are operated within their continuous current ratings and are used in a cutout that is designed for the selected link. The tables list the protected link, the protecting link, and the maximum fault current for which proper coordination of the protected and protecting links are assured.

EXAMPLE 3.1

Consider the radial distribution line shown in Figure 3.9, where customers are served all along the length of the feeders. Fuse A is the main feeder protection, and Fuses B and C are installed on lateral feeders to limit the outage due to remote faults, for example, for faults beyond B or C.

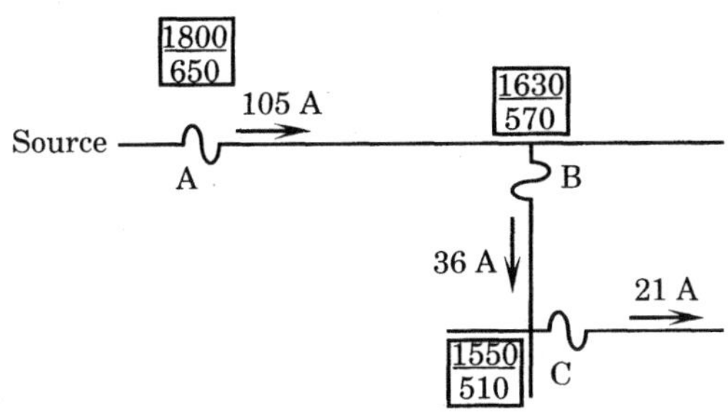

Figure 3.9 Distribution system fuse data.

The maximum and minimum available fault currents, in amperes, at each location are shown in the boxes. Also shown is the normal load current flowing through each fuse. Check the coordination of the fuses. Select fuse ratings for A, B, and C that will coordinate properly.

Solution

As a first trial, let fuse C be a 15T fuse. The load current is 21 A, but the 15T is capable of 23 A, according to Table 3.7. Therefore this fuse is of adequate rating, although there is little room for load growth. From Table 3.6 for T links, we see that the 15T will coordinate with the 25T fuse at location B for currents up to 730 A, but the maximum fault current is 1550 A. Therefore, we select the 30T fuse for location B. The 30T can carry 45 A continuously (OK) and, from Table 3.6, will coordinate with the 15T protecting fuse up to 1700 A. This is a good choice.

The 30T must coordinate with A for fault currents up to 1800 A. To carry the load current at A, we must select the 80T fuse, which can carry 120 A. The 80T will coordinate with the 30T for fault currents up to 5000 A, and this system has only 1800 A available. Thus, a workable solution is 80T at A, 30T at B, and 15T at C. The engineer may wish to allow for a greater load growth at C, depending on the nature of the load served and its likelihood for growth. This would require a larger fuse at C, which will then require that all fuse selections be reconsidered. ■

3.3 RELAY CHARACTERISTICS

This section presents a broad overview of relay characteristics. The intent is to summarize briefly the various characteristics of devices used in power system protection. More detailed treatment of particular devices will be introduced later as required for clarification in a particular application. Relays are available in many different types, serving a host of different purposes and having different design characteristics. The treatment here is a very limited sampling of the many devices that are available commercially. The intent is not to favor any particular type, but simply to describe a limited number of quite different devices, such that the student of relaying will understand that the choice of relaying devices is great, and that there are often many different choices for performing a given protective function.

3.3.1 Relay Types

For the purpose of system protection, we may classify relays according to the following functional descriptions [1]:

Overcurrent relay A relay that operates or picks up when its current exceeds a predetermined value.

Overcurrent relays can be instantaneous, that is, with no intentional time delay [8]. Figure 3.10 shows the general operating characteristic of a time-overcurrent relay. Note the inverse nature of the characteristic, i.e., the higher the current, the smaller the operating time. Overcurrent relays are the simplest of all relaying devices. Only one system variable, the current, needs to be measured to operate an overcurrent relay. However, simple as this seems, there are many different types of relay characteristics that can be developed with only the variables current and time. The choice as to the type of relay characteristic depends on the application, and on the need for coordination with other types of devices, such as fuses or other relays.

Overcurrent relays are not inherently directional. Measuring a current gives no indication of the current direction, which requires some other variable to provide a sense of "polarization," i.e., to provide a reference against which the measured current can be compared. This is sometimes provided by measuring the voltage along with the current using a wattmeter element to determine the direction of power flow.

One application of nondirectional overcurrent relays is on radial distribution systems, where the direction of current flow is always known. In this application the relays must coordinate with a variety of other types of devices, such as fuses and circuit reclosers. The extremely inverse characteristic is not unlike a fuse time-current characteristic, and this type is often used on distribution feeders where the laterals and load centers are protected by fuses. At the other extreme, the definite time-overcurrent relay is useful in cases where the system fault current varies widely between maximum and minimum system conditions. With this relay characteristic, the time is nearly constant over a wide range of conditions. This type of relay

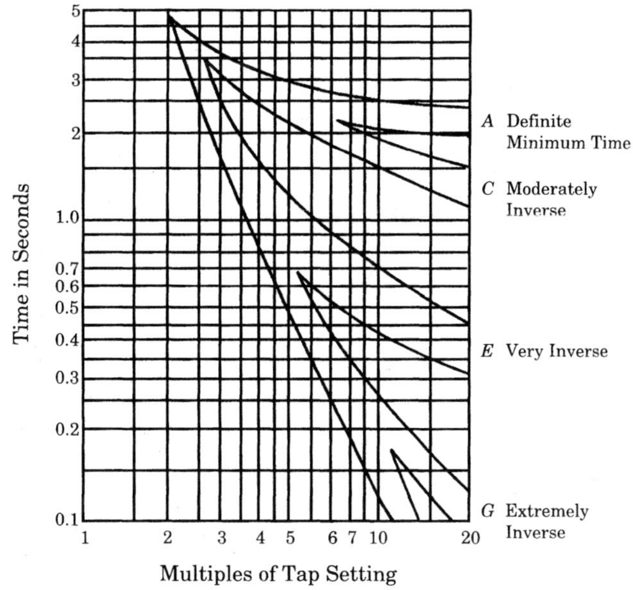

Figure 3.10 General operating characteristic of various inverse time relays [7].

can also be used in combination with another type, such as a very inverse type, to gain higher speed clearing at high fault currents, with the definite time characteristic predominating the lower current ranges.

The inverse characteristic is between the two extremes. This characteristic is suitable in cases where there is a substantial reduction in available fault current as the distance from the source is increased, as on a radial transmission system.

Differential relay A relay that, by its design or application is intended to respond to the difference between incoming and outgoing electrical quantities associated with protective apparatus [8].

Figure 3.11 shows several typical applications of differential protection to buses, generators, and transformers.

Directional relay A relay that responds to the relative phase position of a current with respect to another current or voltage reference [8].

Consider, for example, the high-voltage transmission lines in Figure 3.11. The relays at the ends of these lines could use directional relays since fault currents on these lines can be supplied from generators at both ends of the system. Knowing the direction of current flow helps in designing a protective scheme that is selective and less likely to act incorrectly.

Distance relay A generic term covering those forms of protective relays in which the response to the input quantities is primarily a function of the electrical circuit distance between the relay location and the fault point [8].

Distance relays are designed to respond to current, voltage, and the phase angle between the current and voltage. These quantities can be used to compute the impedance seen by the relay, which is proportional to the distance to the fault.

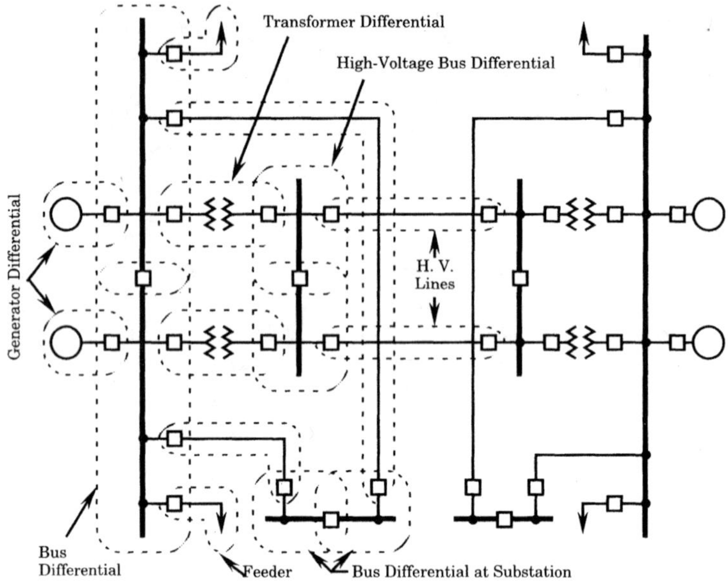

Figure 3.11 A typical power system showing zones of protection.

Distance relays are usually characterized by their impedance plotted in the complex R-X plane, as shown in Figures 3.12 and 3.13. Note that it is possible to design a distance relay that is inherently directional, as shown by the characteristics of Figure 3.13.

Pilot protection A form of line protection that uses a communication channel as a means of comparing electrical quantities at the terminals of the line [8].

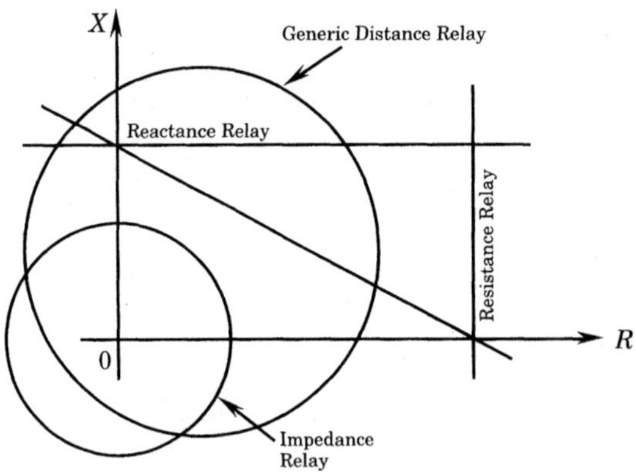

Figure 3.12 General operating characteristics of distance relays.

Pilot relaying is often named for the communication medium used in the application. Thus we speak of wire pilot, carrier pilot, microwave pilot, and fiber-optic pilot protection systems. Figure 3.14 shows the equipment arrangement for a power line carrier (PLC) pilot system, where the communications signal is coupled to the power line as a high-frequency carrier by means of a coupling capacitor or CVT. The coupling capacitor presents a high impedance to the power line frequency, but is practically a short circuit to the communications

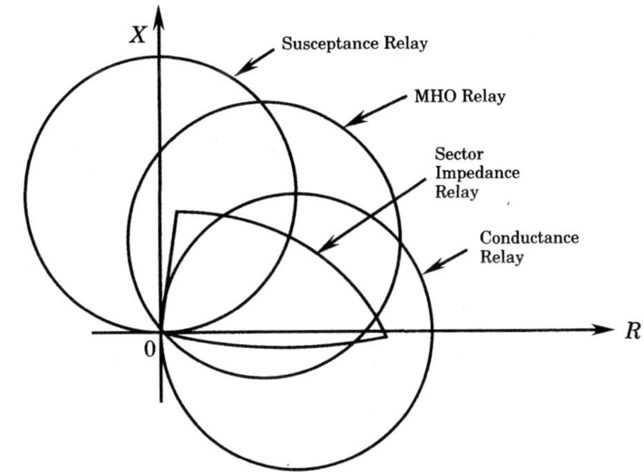

Figure 3.13 General operating characteristics of distance relays that are inherently directional.

frequency carrier. A hybrid coil (HYB) is used to match the impedance of the electronic equipment to that of the CVT. PLC is a common type of pilot scheme for transmission line relaying. Pilot protection provides more information concerning the observed disturbance than systems that are limited only to local information. Information from the two ends of the line are compared to determine the location of the fault. Pilot relaying is discussed in detail in Chapter 13.

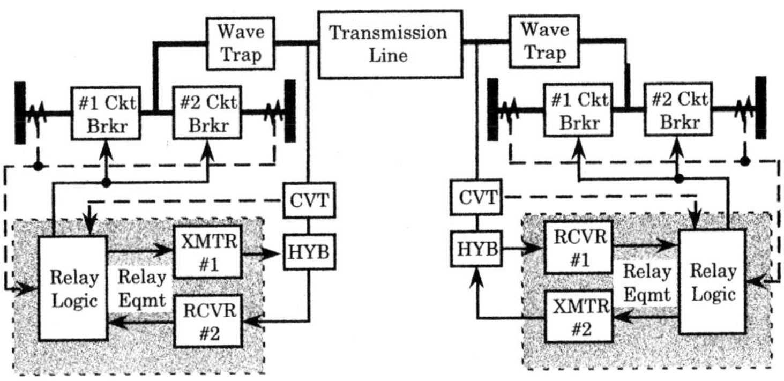

Figure 3.14 Directional comparison carrier-pilot relay system.

Finally, we note that relays are often classified according to the physical technology of their construction. Hence, we speak of electromechanical relays, static relays, and digital relays, for example. Indeed, this classification is convenient in a discussion of relay characteristics and will be used in the following sections.

3.3.2 Electromechanical Relay Characteristics

The characteristics of electromechanical relays have been exhaustively treated in the literature [8]. Excerpts of a rather famous paper [9] are given here to provide a few examples of the wide variety of relay characteristics that can be devised from electromechanical devices. These characteristics are shown in a series of figures. Note that the characteristics of all relay types, described in the previous section, can be identified in these figures.

Electromechanical relays represent a mature technology for protective devices that have been widely used for many years and are still applied for many purposes. These devices have been proven to be sturdy and reliable and are often favored by protection engineers for many applications because of their reliable performance and low cost.

Figure 3.15 illustrates two of simplest forms of electromechanical devices. Relay (1) is a simple solenoid. For current above the threshold I_o the force developed by the solenoid plunger overcomes the force of gravity and closes the open contacts. The solenoid relay is often referred to as an "instantaneous relay," a somewhat ambiguous term that generally is meant to mean "fast," or at least to operate without intentional delay. The speed of this type of relay actually depends on the magnitude of current flowing in the solenoid, and if the current is large the relay will trip in about one cycle.

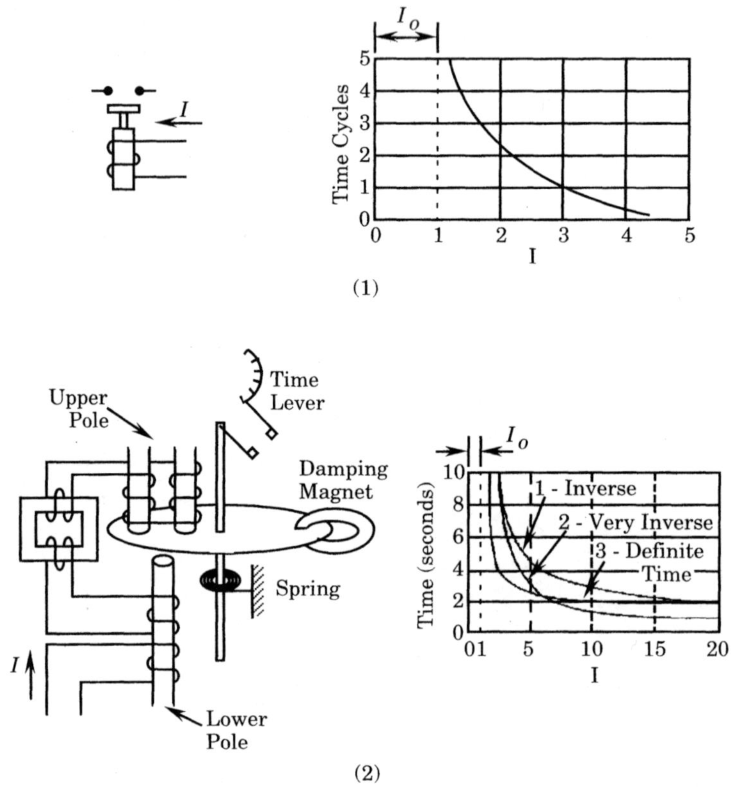

Figure 3.15 Solenoid and induction disk overcurrent relays. (1) Solenoid instantaneous unit. (2) Induction disk inverse time overcurrent unit.

Figure 3.15 (2) illustrates one of the simplest applications of the induction disk in protective relaying. The disk can be caused to rotate due to eddy currents that flow in the disk, the currents being induced due to the fields established by the poles. There are many ingenious forms of induction disk relays. This one measures only the current, but the shape of the relay time-current characteristic can be changed to represent the various generic types described in Figure 3.10. The time required to trip for a given current depends on the angle of rotation required to cause the movable contact to reach the fixed contact. This angle, and hence the time to trip, is adjustable by the "time lever" or dial setting, whereby the fixed contact can be adjusted to a desired angular displacement. This simple feature makes the relay very flexible in its application and provides a valuable characteristic for coordination with other protective devices.

Typical overcurrent relay characteristics for a family of similar relays of the same manufacturer are shown in Figure 3.16, which distinguishes qualitatively between characteristic shapes from definite time (bottom curve) to extremely inverse time (top curve). Most relay manufacturers offer these various characteristics in electromechanical relays of the same basic type of induction disk devices.

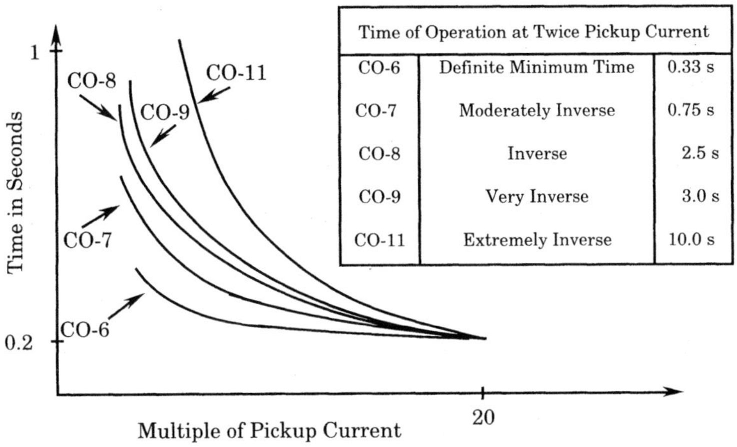

Time of Operation at Twice Pickup Current		
CO-6	Definite Minimum Time	0.33 s
CO-7	Moderately Inverse	0.75 s
CO-8	Inverse	2.5 s
CO-9	Very Inverse	3.0 s
CO-11	Extremely Inverse	10.0 s

Figure 3.16 Type CO curve shape comparison [9].

The induction disk relay can be analyzed by summing the torques acting on the disk. The current flowing in the poles develops a flux that creates eddy currents in the induction disk. These currents interact with the flux to produce torque that tends to rotate the disk. The spring creates a retarding torque. A damping torque is also produced that is proportional to the angular velocity of rotation. We can summarize these torques as follows:

$$T_I - T_S - T_D = 0 \tag{3.2}$$

where T_I = torque due to current in the coil
T_S = retarding torque due to the spring
T_D = damping torque

The driving torque is proportional to the square of the current in the current coil, the spring torque is a constant retarding torque, and the damping torque is proportional to the angular velocity. Therefore, we may write (3.2) as follows [10–13]:

$$k_Q I^2 - T_S = k_D \omega \tag{3.3}$$

where appropriate constants of proportionality have been introduced. We can determine the first constant by means of a simple experiment. If the disk is at rest, the right-hand side of (3.3) is zero. If we slowly increase the current until the disk begins to rotate, this establishes the threshold value of current, which is usually called the "pickup" current. Thus we have the relation

$$k_Q I_p^2 = T_S \tag{3.4}$$

where I_p = Pickup current

Substituting (3.4) into (3.3), we have

$$T_S(M^2 - 1) = k_D\omega = k_D\frac{d\theta}{dt} \tag{3.5}$$

where $M = \dfrac{I}{I_p}$ = Multiple of pickup current

θ = Angle or disk rotation

Then the total angle of disk rotation for a given time is computed as

$$\theta = \theta_o + \frac{T_S}{k_d}\int_0^t (M^2 - 1)\,dt \tag{3.6}$$

where θ_o = Initial (rest) position of disk

It is reasonable to ignore the initial acceleration of the disk, since the disk is very light and accelerated quickly to its final constant velocity. If this simplification is introduced, we approximate (3.5) as

$$\frac{\theta}{t} = \frac{T_S}{k_D}(M^2 - 1) \tag{3.7}$$

for any current greater than pickup. As long as this current continues to flow, the disk rotates at constant velocity until the contacts close. If we designate the angle of travel required to make these contacts as θ_p we can find the time required for pickup. From (3.7), this time is given by

$$t_p = \left(\frac{k_D\theta_p}{T_S}\right)\frac{1}{M^2 - 1} = \frac{T_1}{M^2 - 1} \tag{3.8}$$

where t_p is the time to pickup. Note that the coefficient in the numerator on the right-hand side has the dimensions of seconds, and is therefore recognized as the time constant T_1. This time constant is a relay design parameter and will have different values depending on the shape of relay characteristic curve that is desired.

The foregoing ignores the saturation of the magnetic circuit. Large currents, corresponding to large values of M, cause the electromagnet to saturate. This causes the flux to reach a limiting value, which produces a constant operating time, which we designate as T_2. However, large values of M cause (3.8) to approach zero. Therefore, the effect of saturation is to add a constant T_2 to (3.8). Equation (3.8) also fails to account for the fact that some relay designs require different exponents of the variable M. We can account for these additional concepts as follows. Let

$$t_p = \frac{T_1}{M^p - 1} + T_2 \tag{3.9}$$

where we have added a second time constant to account for saturation and have changed the exponent on M to a variable p that can be changed according to the relay design. Commercially available induction disk relays have values of the exponent p that vary over a rather wide range. This flexibility, in addition to being able to select the two time constants in (3.9) makes it possible to develop relays with many different characteristics.

Table 3.8 shows values of the various design parameters for three common relay designs, where the time to pickup is given for the same multiple M of pickup current.

One might question the reason for expressing the operating characteristic of the induction disk relay by an equation, when the actual characteristic of the device can be measured

TABLE 3.8 Typical Values of Inverse Overcurrent Relay Parameters

Relay Type	M	p	t_p	T_1	T_2	T_1/T_2
Modified inverse	5	0.02	3.00	0.092	0.149	0.46
Very inverse	5	2.00	1.28	18.92	0.492	38.50
Extremely inverse	5	2.00	1.30	28.08	0.130	216.00

experimentally. One reason for interest in the equation is to facilitate the design of digital overcurrent relays that will have similar characteristics to the induction disk, and will coordinate well with induction disk relays that are already in service [14–20].

Equation (3.9) implies that the induction disk relay responds to the true rms value of the input current. In fact, this is not the case, as the induction disk relay has a poor frequency response. This has proven to be a valuable asset, since the induction disk relay does not respond to harmonic frequencies that are often found in industrial applications. Indeed, a relay that would respond to harmonic frequencies may be quite useless in certain industrial applications. This suggests that future digital overcurrent relays will probably have input filters to limit the frequency response, in addition to a time-current relationship similar to (3.9).

Figure 3.17 gives a semilogarithmic plot of TC characteristics for an inverse relay to illustrate the many choices of time dial settings that are usually available on devices of this type. As noted in Figure 3.15, the various time dial or lever settings provide a means of adjusting the angle traveled by the rotating contact in order to reach the fixed contact. The various curves shown in Figure 3.17 correspond to distinct settings of the time dial or lever setting. The relay used for the illustration has an inverse characteristic that corresponds to one of the steeper curves shown in Figure 3.10. The adjective "inverse" is a relative term. Thus we see inverse, moderately inverse, very inverse, and many other names used to distinguish the various characteristics.

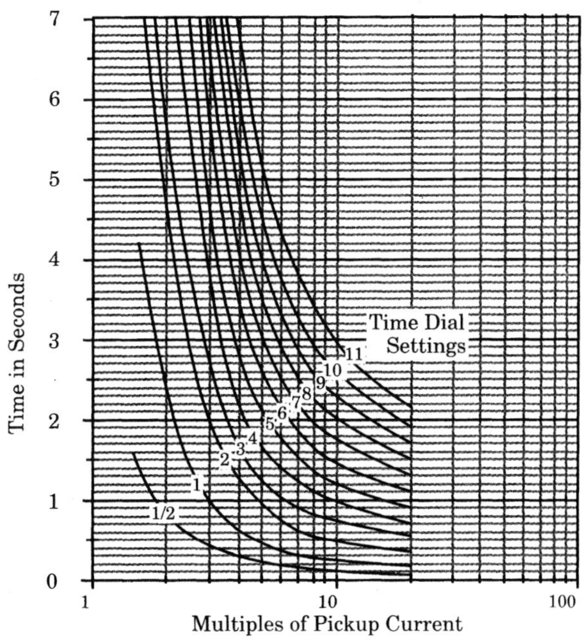

Figure 3.17 Inverse overcurrent time-current characteristic curves.

Figure 3.18 illustrates two other common electromechanical relays that have been widely used. Device (3) is another variation of the induction disk, but this time with electromagnets for both current and voltage. This makes the relay perform as a power measuring device, and hence makes it directional. If the current phasor has a component in the same direction as the voltage phasor, the relay will have a torque in the operate direction. Otherwise the torque is negative and the relay is restrained from operation. The lag loop produces the necessary phase lag to facilitate the rotation of the disk.

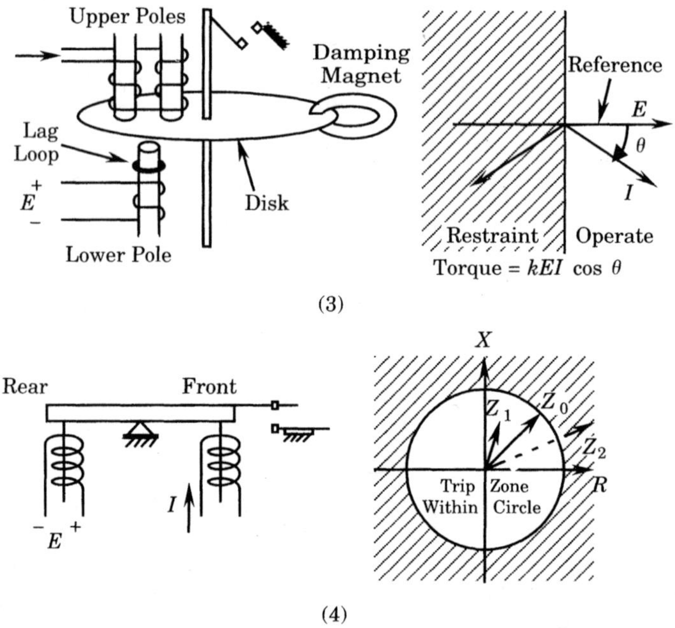

Figure 3.18 Directional overcurrent and balanced beam distance relays. (3) Induction disk directional element. (4) Balanced beam impedance element.

Clearly, the direction sensing capability is very important as it provides the relay with greater selectivity.

Device (4) in Figure 3.18 is called a *balanced beam* relay. In this device, two electromagnets cause forces that tend to tilt the beam in one direction or the other. If the current is large enough, the current force prevails and the contacts are caused to close. The characteristic is best described in the R-X plane, where the trip zone is that area inside a circle with the center at the origin. This relay is very fast, but does not have the directional sensitivity of device (3). Hence, the balanced beam impedance relay is usually used in connection with a directional relay in order to provide the directional selectivity. The relay operating threshold is sometimes called the *balance point*.

Figure 3.19 shows two more electromechanical relays that are based on the principle of the induction cup or cylinder. These devices utilize the principle of the induction motor to produce torque on the rotating element in response to magnetic fields set up by currents and voltages in the stator windings. The rotor is a hollow cylinder and, in some cases, includes rotor windings. These structures have been adapted to many different relay types that have been widely used for many years. Two of these types are illustrated in Figure 3.19.

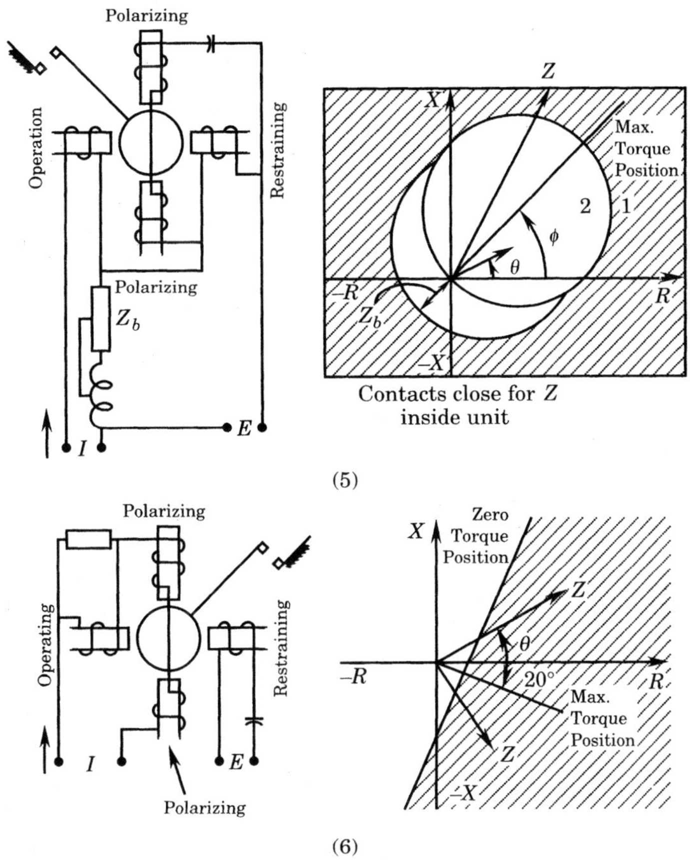

Figure 3.19 Examples of induction cup relays. (5) Induction cylinder mho unit. (6) Induction cylinder directional starting unit (blinder).

Device (5) is an induction cylinder "mho" unit. The relay characteristic is best described in the R-X plane, where it is represented as a circle that may be offset from the origin. The maximum torque line is along a radial in the first quadrant that is represented by the adjustable angle ϕ. Since the circle is offset into the first quadrant, the relay provides directional sensing.

One problem with either the impedance relay of (4) or the mho relay of (5) is that the trip region tends to be much larger than the impedance region of the fault along the line. One way to narrow the trip region is to add relays that narrow the trip zone. These relays are often called *blinders*, and one such device is shown in Figure 3.19(6). Picture this relay characteristic overlaying that of (4) or (5) to narrow the impedance of the trip zone.

The induction cylinder is a product type relay, where the torque is proportional to VI products.[2] The vertical poles provide a vertical polarizing flux, which induces a current in the cylinder perpendicular to the plane of the paper. The current in the operating pole thus acts to produce a torque proportional to I^2, in a direction to close the contacts. The current in the

[2]See the article in [8], "Application of the Ohm and Mho Principles to Protective Relays," by A. R. van C. Warrington, pp. 77–85.

cylinder near the E pole produces a restraining torque, tending to prevent the contacts from closing. Since the torque is the product of the polarizing flux and the fluxes from the I and E poles, the basic equilibrium equation for the device is

$$E[I\cos(\phi - \theta) - K'E] = 0 \tag{3.10}$$

where K' is a winding constant, ϕ is the phase angle of the protected circuit, and θ is the relay characteristic (maximum torque) angle. Dividing (3.10) by E_2 we get the admittance characteristic

$$Y\cos(\phi - \theta) = K' \tag{3.11}$$

which is a straight line in the admittance plane, hence the name "mho" characteristic. Solving for the inverse of this admittance, we have the impedance representation

$$Z = \left(\frac{1}{K'}\right)\cos(\phi - \theta) \tag{3.12}$$

This is a circle in the impedance plane with diameter $1/K'$ and which passes through the origin.

The general form of the torque equation is usually written as

$$KI^2 - K'E^2 + EI\cos(\phi - \theta) = 0 \tag{3.13}$$

Dividing by $K'I^2$ and rearranging, we get

$$Z^2 - \frac{Z\cos(\phi - \theta)}{K'} + \left(\frac{1}{2K'}\right)^2 = \frac{1 + 4KK'}{(2K')^2} \tag{3.14}$$

This is the polar form of the equation of a circle with the following parameters:

$$\begin{aligned} &\text{Radius:} && \frac{\sqrt{1 + 4KK'}}{2K'} \\ &\text{Center:} && R = \frac{1}{2K'}\cos\phi \\ & && X = \frac{1}{2K'}\sin\phi \end{aligned} \tag{3.15}$$

Note that the radius is larger than the distance from the origin to the center of the circle, so that the origin falls within the trip zone. The induction cylinder is capable of being configured to represent several different types of useful relay characteristics, two of which are shown in Figure 3.19.

3.3.3 Static Relay Characteristics

Static relays were introduced in the early 1960s and have been growing in popularity ever since [8]. For the purpose of this discussion, we limit the definition of a static relay to the following [21]:

> **Static relay** A relay or relay unit in which there is no armature or other moving element, the designed response being developed by electronic, solid-state, magnetic, or other components without mechanical motion [21].

The introduction of static relays has resulted in devices that have improved sensitivity, speed, and repeatability over the electromechanical designs. Nearly all static relays are more shockproof than electromechanical designs, and most are very fast in their operation. The reset time is much shorter than most electromechanical designs since there is no mechanical travel required as with parting contacts. This allows closer coordination. The static designs have lower maintenance, lower (VA) burden, and smaller size, all of which are favorable.

A major disadvantage of static relays is their susceptibility to transients of even small magnitude, causing them to require more care for shielding the installation. Also, static designs are more temperature sensitive, and therefore operate over a narrower temperature range than electromagnetic devices.

Functionally, static relays can be designed for almost all relay applications. They are widely used for EHV transmission protection and for generator differential, timing, distance, phase and direction comparison, and overcurrent applications.

The solid-state overcurrent relay of one manufacturer provides, in a single device, all the characteristics usually needed for overcurrent devices, such as the family of curves in Figures 3.20–3.23. Table 3.9 lists the TC characteristics, which are all described by the manufacturer using the following equation [22].

$$t = \frac{aD}{I^n - C} + bD + K \tag{3.16}$$

where t = time to trip, in seconds
a = a constant, in seconds
C = 1 for standard relays
n = an exponent
b = a constant, in seconds
D = time dial setting
K = a constant = 0.01 seconds for standard relays
I = operating current, expressed in Multiples of Pickup (the equation is valid for $I > 1$)

TABLE 3.9 Typical Overcurrent Relay Characteristics [22]

Name	Area of Application	Unique Characteristic of Application
Definite time	Subtransmission lines Distribution feeders	Use if source impedance is subject to wide variations and selectively depends on relay time.
Moderately inverse	Subtransmission lines Distribution feeders	Use if source impedance is subject to moderate variations and for parallel line protection.
Short time	Differential protection Buses and transformers	Use for differential protection of transformers or buses where restraint windings are not used.
Modified inverse	Subtransmission lines Distribution feeders	Use for phase and ground fault protection where a wide range of characteristics is needed to ensure selective operation with forward relays, line reclosers, and sectionalizers.
Inverse	Subtransmission lines Distribution feeders	
Modified very inverse	Subtransmission lines Distribution feeders	Where source impedance is nearly constant, advantage can be taken of a greater degree of inverseness to minimize the tripping time.
Very inverse	Subtransmission lines Distribution feeders	Excellent characteristic for use in ground overcurrent relaying for transmission lines.
Extremely inverse	Subtransmission lines Distribution feeders	Use on distribution feeders where relays must coordinate with main and branch sectionalizing fuses or on feeders with high initial load following extended outages.

Relay parameters a, b, and n determine the curve shape. The two other relay parameters select the pickup current and characteristic curve family. The parameter D selects the member of any curve family which is to be used and is set by adjusting the "Time Dial" setting.

The values of a, b, and n are determined by the type of TC characteristic used. D is the time dial setting and may have any value from 0.5 to 11, as shown in Figures 3.20 through 3.23, which are drawn by solving (3.16). Most of the curves can be obtained with $n = 1$, but the inverse, very inverse, and extremely inverse characteristics require $n = 2$. The exponent n is varied by changing an inexpensive module in the relay. Table 3.10 gives values of a, b, and n for the different characteristic shapes.

TABLE 3.10 Constants for Tripping Time Equation (3.16) [22]

Characteristic Curves	a	b	n
Definite time	0.20	0.180	1
Moderately inverse time	0.55	0.180	1
Short time	0.20	0.015	1
Modified inverse time	1.35	0.055	1
Modified very inverse time	1.35	0.015	1
Inverse time	5.40	0.180	2
Very inverse time	5.40	0.110	2
Extremely inverse time	5.40	0.030	2

Definite time relays have an almost constant time of response for any current exceeding about 10 times minimum pickup. This feature makes this type of relay a good choice where there is a large difference between the fault currents at maximum versus minimum generation. The characteristic shape is greatly different from that of fuses or other devices, but at least the time of pickup will be known for widely varying values of fault currents.

Inverse overcurrent relays coordinate better with fuses than definite time-overcurrent relays. The more inverse characteristic provides a better coordination with fuses than the definite time characteristic. In many cases, this characteristic also provides the same coordination time between adjacent relays but with faster tripping.

Very inverse time-overcurrent relays are a good choice for coordination with fuses and reclosers on distribution lines and are also used in subtransmission circuits where a wide range of characteristics is required to provide for selectivity. This characteristic offers faster tripping than the more definite type, for those cases where the fault current is nearly constant.

Extremely inverse relays are a good choice for providing fast tripping with good time coordination margins between adjacent relays, especially when there is a considerable drop in fault current from one relay location to the next.

The formulas used to compute the relay characteristics are those provided by the manufacturer [22]. These are not the only possible formulas for overcurrent relay characteristics. An IEEE Committee Report [17] provides other formulas that have been tried. One type of expression is an exponential equation similar to (3.16). The other approach is to use a polynomial expression such as

$$t = \left[\sum_{j=1}^{m} \sum_{i=1}^{n} a_{ji} D^j I^i \right]^k \tag{3.17}$$

where a_{ji}, m, n, and k are constants.

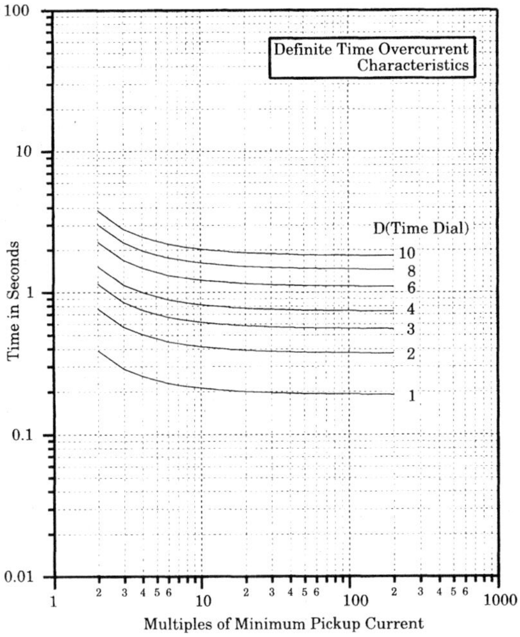

Figure 3.20 Definite time-overcurrent characteristics.

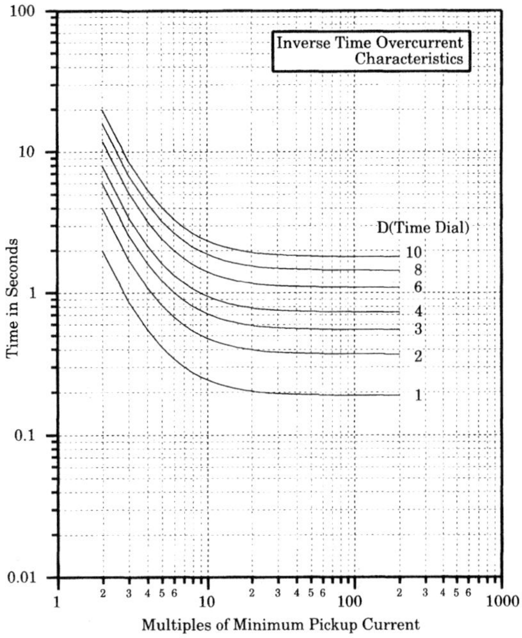

Figure 3.21 Inverse time-overcurrent characteristics.

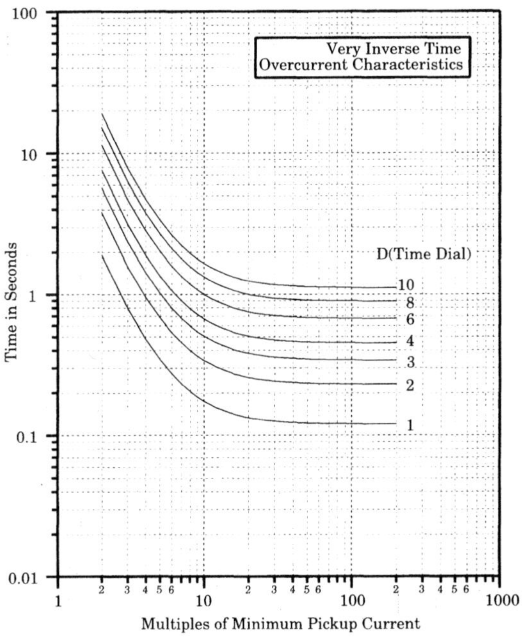

Figure 3.22 Very inverse time-overcurrent characteristics.

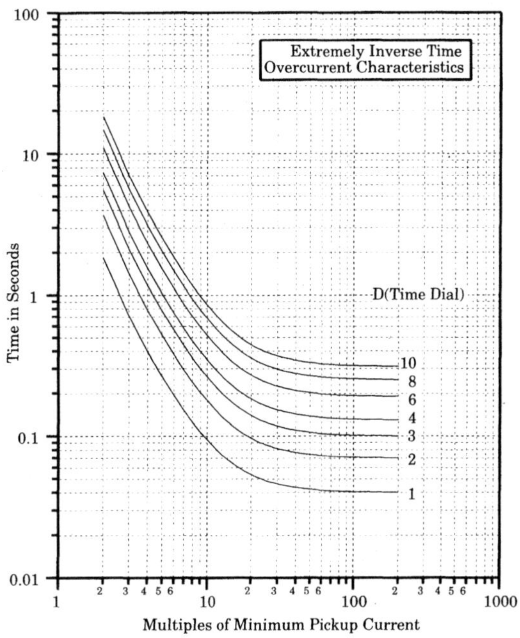

Figure 3.23 Extremely inverse time-overcurrent characteristics.

Other approaches and other equations are also suggested in [17]. None of these approaches seems to be universally accepted as the best method, but research on computer representation of overcurrent relays continues.

3.3.4 Differential Relays

The relays considered in the foregoing sections, with the exception of pilot relays, have been devices that use measurements of system quantities at a particular location in the power system and act according to these measurements. For some relays, the measured quantity is current, in others both current and voltage are used by the relay to determine the requirement for tripping.

Another important type of relay uses measurements at two or more points in the network as a means of determining the presence of a fault in the region defined by the measurement locations. Such relays are called *differential relays*, since they measure the difference between the currents at two different points of measurement. A typical connection is shown in Figure 3.24. Since there is no fault on the protected element, the current entering that element is exactly the same as the current leaving and the difference current, which flows through the relay, is zero. The protected element could be a line or apparatus that has terminals as shown such that the currents entering (or leaving) the device can be measured. The relay, therefore, is a *current differential* device.

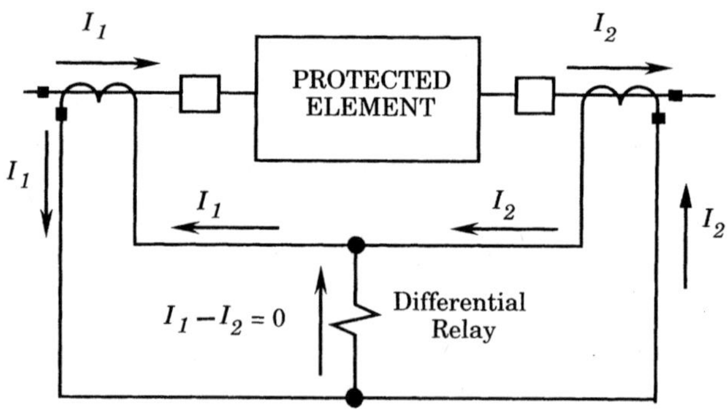

Figure 3.24 A typical differential relay connection.

The principle of the current differential relay is based on Kirchhoff's current law, which states that the sum of currents entering a point must be zero. For the device pictured in Figure 3.24, the currents entering at the two terminals must be zero unless there is an internal fault, in which case a net current must flow to the fault point, as shown in Figure 3.25.

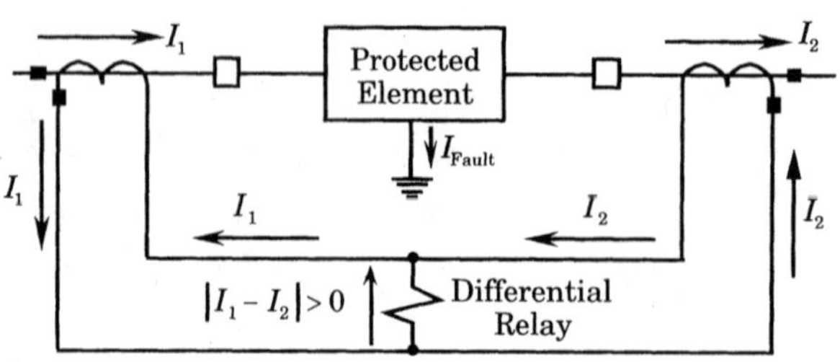

Figure 3.25 Trip condition for the current differential relay.

In many types of equipment, the fault develops as a short circuit to ground, as shown in Figure 3.25. This type of fault is readily detected by the differential relay if there are no CT errors. Practical current transformers are subject to various errors, even if they are of the same type and rating. The errors may be due to manufacturing differences, differences in prefault loading, and differences in saturation, for example. These differences may require that the relay threshold be set higher than zero, and this reduces the relay sensitivity.

The solution to this problem is to design the relay with two types of windings, called the *operating* and *restraining* windings, as shown in Figure 3.26. Currents in the operating winding tend to cause tripping, whereas those in the restraint windings prevent tripping. This is often done in a magnetic circuit, for example, such that the ampere-turns of the operate and restraint windings are arranged to oppose each other. Relays of this type are called *percentage differential relays*.

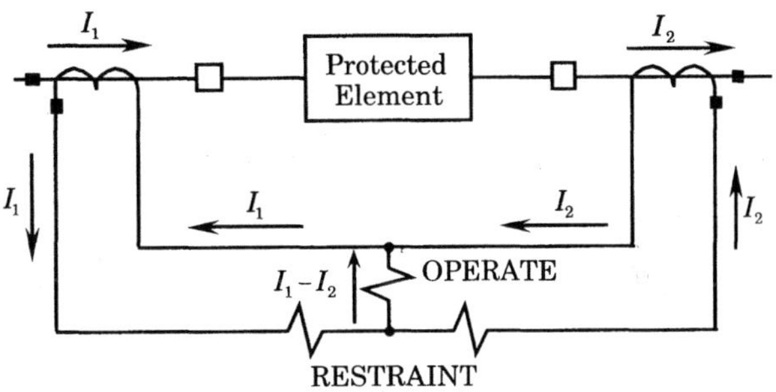

Figure 3.26 Schematic of the percentage differential relay.

The reason for the term *percentage differential* is that the current in the operating coil is a fixed fraction of the total current in the restraining coils. This gives a tripping characteristic as shown in Figure 3.27. With this arrangement, the current required to trip the relay increases as the fault current increases. Usually, the ratings of the percentage differential relays are designed to trip at given values, such as 10% or 25%, with these values expressing the percent unbalance current required to operate the relay, expressed in terms of the smallest current required for operation [17].

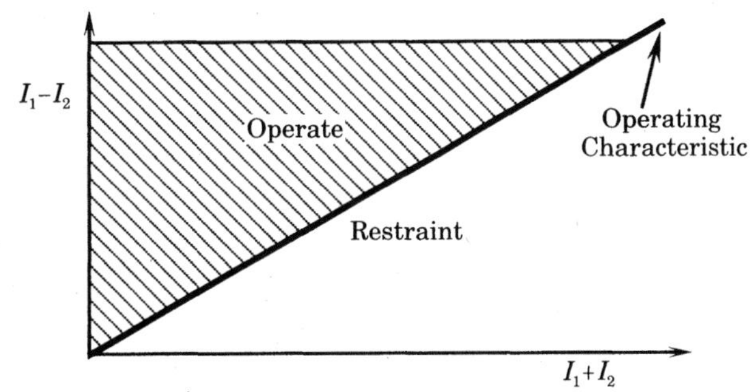

Figure 3.27 Operating characteristic of the percentage differential relay.

3.3.5 Digital Relays

Static relays have been described as the first and second generations of the application of electronic technology to power system protection [19].

Electronic systems were not widely used in protective systems during the vacuum tube era. With the advent of solid-state devices, however, protective relays have increasingly become dependent on solid-state electronic devices rather than the traditional electromechanical designs. The first generation in this trend utilized the transistor in discrete component designs. The second generation introduced the integrated circuits and operational amplifiers into relay designs.

Digital relaying was the next step in the application of electronic technologies to protective relaying [19], [23]. The first development is represented by the early digital devices, many of which were applied as experiments, sometimes utilizing small mainframe computers and small minicomputers. The development of the microprocessor, however, provided significant computational power at low cost that provided a new emphasis on digital protection.

3.3.5.1 Historical Perspective of Digital Relaying. Digital relaying has been the subject of research by the universities and manufacturers for many years. One of the earliest concepts proposed for performing protective functions using digital devices was described in Rockefeller's 1969 technical paper [23]. The approach described in the paper was to provide a digital computer that would provide all of the necessary relay protection for an entire substation. This was probably motivated by the high cost of digital computers at that time, which made it necessary to utilize the computer for many different tasks in order to justify the high investment in the computer. The digital computers of that era were not fast enough to perform all protective functions, so the idea was not extensively pursued. Rockefeller's paper, however, provided a thoughtful description of several protection algorithms that are still studied and discussed. Other investigators soon pursued the concept of digital protection [23–25], which signaled the beginning of a host of other investigations. For the most part, these ideas had to wait for the advent of less expensive computer equipment. The wait was not nearly as long as most thought it would be, as the cost of computers began a remarkable downward spiral in price, coupled with an equally remarkable increase in the speed of performance. The cost of memory, which is required for some types of protection, also followed a steep downward trend. These factors, coupled with the advantages of digital technology, have made the computer relay a viable competitor in power system protection. Many digital relays are now available commercially, and the future will almost certainly be dominated by digital devices. Digital systems are now available as overcurrent relays [18], [26–29] and as full transmission system protection systems [30], and in many other types and configurations.

Digital relaying has several possible advantages over analog protective devices, since the digital computer has the capability of monitoring its own behavior. This is a distinct advantage in a protection environment, where the relay is required to work only on rare occasions, although it continues to monitor the system condition. It would be a great improvement in reliability if the protection engineer could be assured that, although not taking specific action, the relay is indeed functional. The digital computer relay, if properly designed and programmed, has this capability, at least to a limited extent.

The digital devices available are more versatile than other relay designs. One feature is the programmable nature of these devices. For example, the overcurrent relays are capable of being set for any of a wide variety of time-current characteristic curves and these settings are easily changed in the field without changing the physical device. Many of the digital relays are almost insensitive to dc offset of the fault current, which improves the relay selectivity. They also present a high burden to the instrument transformers, which is essential for ground fault relaying in high impedance or ungrounded systems.

The evolution of protective relays has been described by four "generations" of equipment designs, which are defined as follows [30–33]:

1. Electromechanical relays
2. Discrete solid-state relays (static relays)
3. Rack-mounted, integrated solid-state equipment, packaged for multiple protective functions.
4. All digital microprocessor-based relays that measure currents and voltages by sampling the waveforms.

The first generation equipment is the traditional method of relaying, with many different electromechanical devices of widely different designs that perform many different relaying functions. The second generation systems were introduced in the early 1960s and represented a natural step in the evolution of protective equipment. These relays employ discrete solid-state electronic components and are often referred to as "static" relays due to the absence of any mechanical or moving element. The third generation equipment is typical of devices introduced in the 1970s, which utilize integrated circuits. These systems are more complex and often perform many different functions for logic and control.

The fourth generation is represented by equipment generally available in the 1980s and is quite different from the previous technologies in their use of entirely digital equipment. These systems are characterized by devices that are self-checking and programmable. They can perform the usual protective functions and also save the digital records of the event for later analysis by the engineer. There is also the prospect that the digital relays will offer *adaptive* settings that can be changed as system conditions change, probably occurring first as manual intervention and eventually as automatic action of the relay system.

Adaptive relaying is defined as follows [34–38]: "Adaptive Protection is a protection philosophy which permits and seeks to make adjustments to various protection functions in order to make them more attuned to prevailing power system conditions." The important feature is that the protective system is capable of observing changes in the power system caused by changing loads, switching, or other patterns, and making changes in its threshold or other settings in response to these observations.

The future of protective relaying will probably be dominated by digital devices. The application of microprocessors provides hardware that will see greater standardization since the devices are capable of great flexibility; therefore, it will probably not be necessary to have so many different types of devices. Perhaps only the software will require change rather than complete redesign of the hardware system.

The future of digital protection seems to be headed toward the fifth generation: the combination or integration of protection, metering, control, and communications into an integrated control and protection system for a complete substation [31], [39], [40]. This implies the application of not just a few microprocessors, but many processors in a distributed processing environment with central control and surveillance.

Another advancement in the state-of-the-art in power system protection is that of synchronized sampling of the power system through phasor measurements, which can be used for relaying or control functions [41], [50]. Phasor measurements are made by using highly accurate clocks at the points of measurement and noting the time from a waveform zero crossing with respect to a sampling instant or marker. It has been noted that a clock accuracy of a few microseconds is needed for these measurements [44] and accuracies of this order of magnitude are apparently within reach of the current technology [50]. Several uses of phase angle measurement have been discussed, and all applications are digital. An early application was in the measurements for static-state estimation of the power system [48], [49], where the phasor measurements were available from the symmetrical component distance relays (SCDR) [42],

[45]. Phasor measurements on transmission lines have been reported in the literature using radio receivers tuned to WWV or to satellite systems [43]–[47]. The application of phase angle measurements in an advanced adaptive relay has also been investigated [50].

One of the problems in using accurate satellite data is the high cost of receivers. However, if the receiver can be used for several different functions, this choice of time synchronization becomes attractive. It is also possible to link nearby substations using fiber-optic communications links, but these alternatives are also expensive unless required for several functions. It has been noted that accuracies of 1 millisecond are adequate for time-tagging data or meter readings, but this corresponds to an error of 21.6 degrees on a 60 hertz waveform. For phasor measurements, the accuracy of measurements should be in the neighborhood of a few microseconds.

Phasor measurements show a great deal of promise for relaying and for other functions, such as stability enhancement. It is also very important for synchronizing measurements made by monitors that may be located throughout the power system to record large disturbances for later study. The future will almost surely see an increase in the use of phasor meaurements, especially as accurate time measurements become available throughout the power system.

3.3.5.2 Digital Relay Configuration. The digital protective system interfaces with the power system, and to the protected element, in the same way as analog relays. The performance of the protected element is determined by measurements obtained from current and voltage transformers attached directly to the high-voltage conductors. Other input data are obtained by monitoring the status of contacts in the power system. The output of the protective device is also a contact status, as it is in all relay equipment.

A simplified functional block diagram of a digital protective system is shown in Figure 3.28 [32]. The inputs to the relay are analog signals derived from current and voltage transformers. These are primarily 60 hertz signals, but are usually corrupted by harmonics when a fault is present.

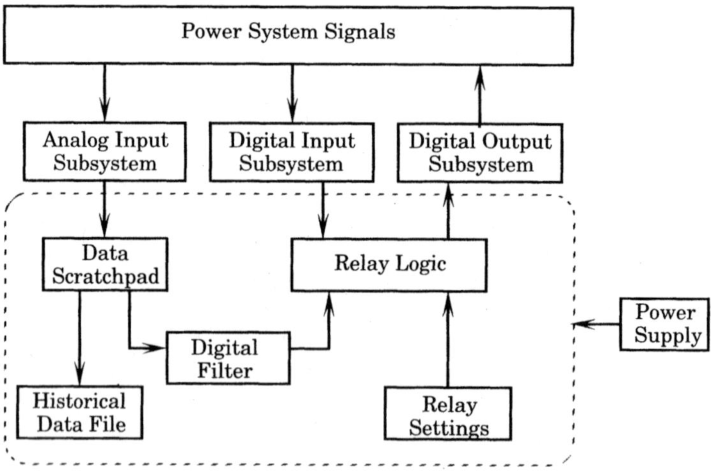

Figure 3.28 Functional block diagram of a digital relay [32].

There may be a large number of analog input signals, for example, a bus protection with five circuits connected to the bus will have 15 current inputs to the relay. In most cases there will be between three and 30 inputs. The raw input signals are measured in kiloamperes and kilovolts, but are reduced to 5 ampere and 67 volt (line-to-neutral) levels. These levels are still too high for digital devices, which require only voltage inputs in the range of ± 10 volts maximum. This means that the current inputs must be fed to low resistance shunts and the voltage signals must be further reduced.

The digital input subsystem provides the relay with any contact position or voltage sensing information that the relay requires. The number of these digital inputs is on the order of five to 10. These signals may also contain transient voltages that must be buffered to protect the digital equipment.

The digital relay output subsystem provides for a maximum of about 10 digital signals, which are low-level signals that must also be buffered.

The input analog data is sampled by an analog-to-digital (A/D) converter. Sampling rates vary between about 240 hertz and 2000 hertz. The samples are entered into computer memory, either through program control or through a direct access channel. These data are retained in the scratch pad random access memory (RAM), and are readily available to the CPU for processing. The raw data are also stored in an historical data file as a record of the transient. Such event snapshots are usually moved to another file after the transient is complete.

The digital filter is required to process the raw data, which is almost always corrupted by noise. Much of the noise is not pertinent to the relay decision-making function and is therefore removed by the digital filter.

The relay logic is a program that runs on the central processor. It emulates the desired protective function by examining the input data and computing secondary quantities. For example, a transmission line distance relay function will compute the impedance seen by the relay, which can be compared against the relay threshold settings, which is usually a region of the R-X plane that is stored as a relay setting.

A more detailed view of a digital relay system is shown in Figure 3.29 [31]. The central processor is the focal point of the system. The CPU executes the protective programs and also performs various communications and maintenance functions. The input sampled data are deposited in RAM, which also acts as a scratch pad during the relaying algorithm execution. The read-only memory (ROM) or programmable read only memory (PROM) is used for permanent program storage, and some programs may execute directly from ROM if it is fast enough. In other cases, the programs are read from ROM to RAM for execution. The erasable ROM (EROM) is required for storing the relay parameters, which may have to be changed occasionally.

A detailed discussion of the design parameters of the digital relay equipment is beyond the scope of this book. Details are available in a number of excellent references [31–34].

3.3.5.3 The Substation Computer Hierarchy. Another concept that has received consideration is that of a hierarchy of digital computers that monitor and control the protective function of the power system. Figure 3.30 shows the basic structure of such a system. The digital computer is ideally suited to such an integrated system, with each level performing certain tasks, and with communications between levels. Such a structure would be difficult to achieve using analog equipment.

The relaying computers and their data acquisition systems are at the lowest end of the hierarchy. These computers communicate with the local switching station through their data acquisition system and their control outputs. Since the relays have control capability over the circuit breakers, they can also be used as links in a control chain where control commands are passed from the central computer requiring the operation of the breakers for reasons other

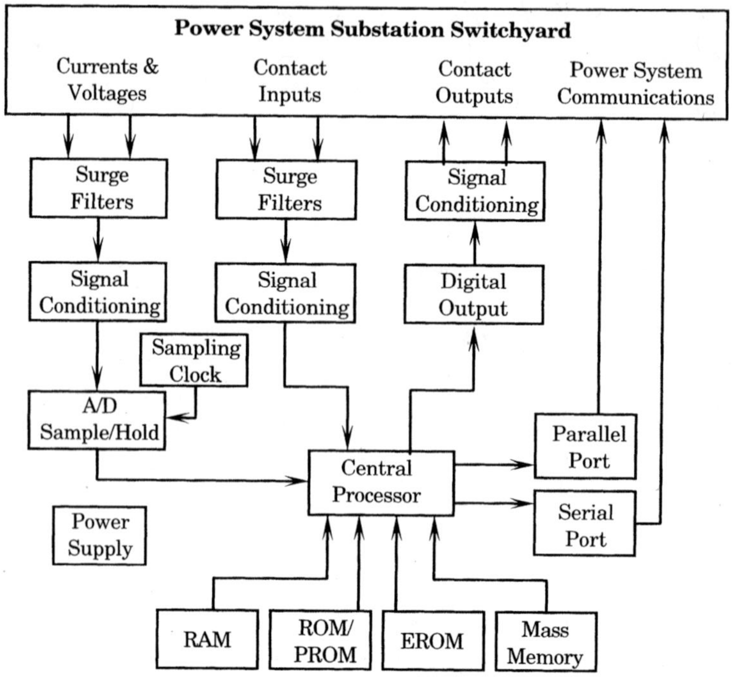

Figure 3.29 The subsystems of a relaying computer.

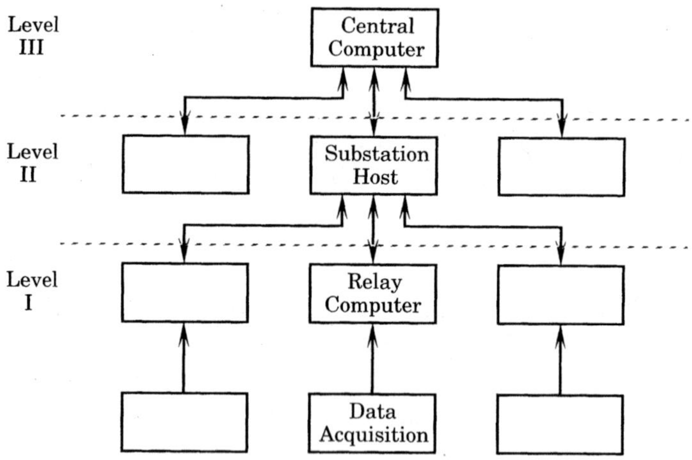

Figure 3.30 A system-wide hierarchical computer system [31].

than fault removal. The communications chain can also be used for passing historical records of faults or other disturbances up to the substation computer and then to the control center, thereby acting as a data accumulation and reporting medium.

The substation computer provides an interface between the control center and the individual relays. The substation relays information regarding the protective system settings, calibration, interrogation, and diagnostic or maintenance functions. The substation computer can also be a storage system to maintain sequence-of-events records for all relays that report to that station.

3.3.5.4 Summary. The digital computer is a very flexible device for control, record keeping, and communications of events. The same relay computer can be used for many different relaying functions, given the availability of input quantities, since the protective function is simply a program. Moreover, as digital hardware evolves, the same program can be passed on from one generation of hardware to another, thus maintaining the reliability of thoroughly tested relay algorithms.

The digital environment is also ideal for the recording of events as they occur on the power system. The older analog methods of starting an oscillograph, were limited to the amount of paper in the oscillograph and the preset scaling of the measured quantities. In digital storage, there may still be a limit in storage, but memory is relatively cheap and long records are clearly possible to store in high-speed RAM, and later download to more permanent and less expensive storage. These records are ideal for the engineer to examine. The scaling of the waveforms can be changed at will, since the data is digital and not waveforms recorded permanently on paper or film. This provides a valuable tool for post-fault analysis that is far superior to the oscillographic methods of the past.

The many advantages of digital systems will probably make these devices the preferred choice for future protective systems. However, there are still many installations of analog equipment that are performing quite adequately. This suggests that there will be a long period in which both systems will be in service, with digital equipment eventually becoming the predominant protective equipment.

3.4 POWER CIRCUIT BREAKERS

The power circuit breaker is the predominant member of the family of devices that is designed to interrupt an electric circuit. A *power circuit breaker* is defined as a device for closing, carrying, and interrupting a circuit by parting separable contacts under either load or fault conditions. For the most part we are only concerned with high-voltage circuit breakers in power system protection engineering. These are defined to be circuit breakers that are rated at least 1000 volts.

Circuit breakers are often identified in terms of the physical mechanism used to aid in the circuit interruption. Thus, we speak of oil circuit breakers, for example, wherein the arc extinction occurs, and is aided by, a surrounding medium of oil. Other common types include air circuit breakers, air blast circuit breakers, sulfur hexaflouride (SF_6) circuit breakers, and vacuum circuit breakers. In this book we are not directly concerned with the physical phenomenon of arc extinction, but we shall be concerned with the interrupting time required for the breaker to complete its function.

3.4.1 Circuit Breaker Definitions

There are many formally defined terms used in the manufacture and application of power circuit breakers. We are concerned with those terms and definitions that are needed for clarity and understanding in protection studies. Other definitions can be found in the standards [54].

The first set of circuit breaker definitions deal with circuit breaker voltages that the device must withstand under various system conditions.

Operating voltage The rms line-to-line voltage of the system on which the circuit breaker is operated.

Recovery voltage The voltage that occurs across the terminals of the ac circuit-interrupting device upon interruption of the current.

Restrike voltage The voltage that occurs at the resumption of current between the contacts of the interrupting device during an opening operation. This voltage is usually specified at a given maximum interval, such as 1/4 cycle, after the establishment of zero current.

The second set of definitions pertains to currents.

Symmetrical component of current The ac component of a short circuit current at normal system frequency, shown in Figure 3.31.

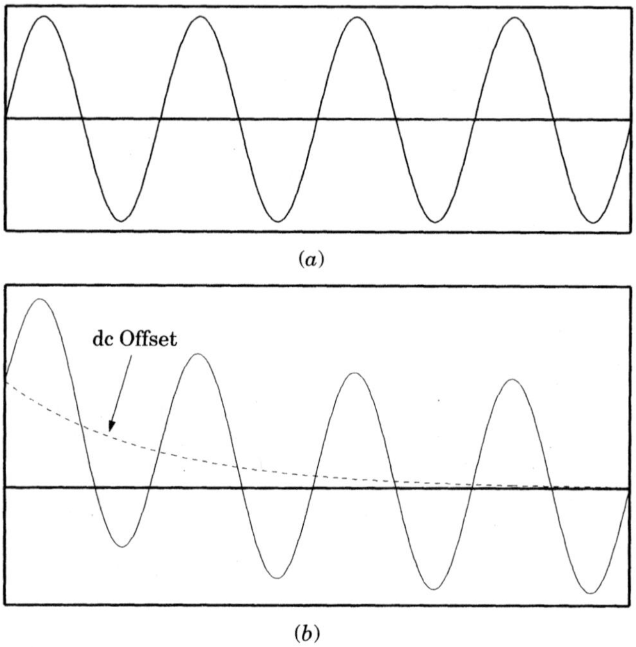

Figure 3.31 Short circuit current waves of one phase. (*a*) Symmetrical waveform. (*b*) Asymmetrical waveform.

DC component of current That component of current in a short circuit that offsets the current wave from its normal zero axis, shown in Figure 3.31.

Asymmetrical current (total current) The combination of the symmetrical and dc components of a short-circuit current, also shown in Figure 3.31.

Making current The rms value of total current, measured from the envelope of the current wave at the time of its first major peak, when a circuit breaker is closed on a short circuit.

Latching current The making current during a closing operation in which the circuit breaker latches.

At any instant of time the total rms current may be computed as

$$I_{\text{total}} = \sqrt{I_{\text{ac}}^2 + I_{\text{dc}}^2} \tag{3.18}$$

It should be apparent that the difference between the maximum symmetrical and asymmetrical current can be large for maximum dc offset, and can approach about 2.8, neglecting the current decay from zero time to the time of the peak.

Another set of definitions refers to various operating characteristics of a circuit breaker.

Reignition The resumption of current between the contacts of a switching device during an opening operation after an interval of zero current of less than 1/4 cycle at normal frequency.

Restrike A resumption of current between the contacts of a switching device during an opening operation after an interval of zero current of 1/4 cycle at normal frequency or longer.

Operating time The operating time of a circuit breaker involves a number of carefully defined time intervals, some of which are shown in Figure 3.32.

Other circuit breaker definitions of interest are given in Appendix A.

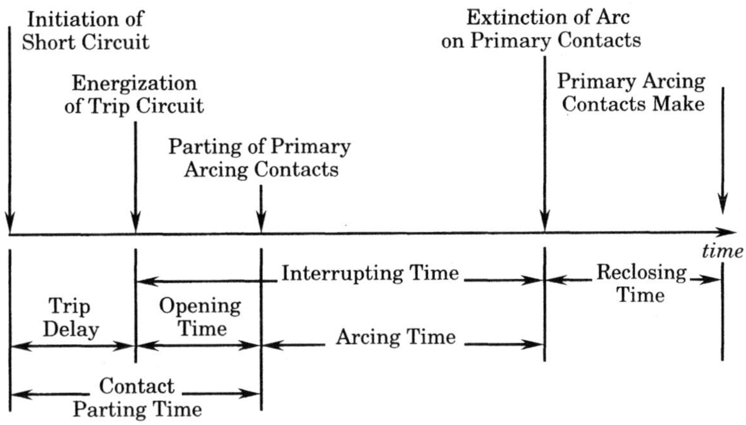

Figure 3.32 Circuit breaker operating times.

3.4.2 Circuit Breaker Ratings

The ratings of circuit breakers are very complex and detailed. Many of the ratings are devised to clarify the performance of the manufactured product and are not of interest here. Some ratings are important to the protection design process and these will be discussed briefly.

Prior to the 1950s, circuit breakers in North America were rated on a total current basis. In 1951, industry committees began the process of changing the rating standards from a total current basis to a symmetrical current basis. Several reasons for doing this were:

1. To simplify application of fast circuit breakers for high-speed relaying applications.
2. To bring the ANSI standards in agreement with the IEC standards, which were on a symmetrical basis.

3. To require that circuit breakers be proven to demonstrate a definite relationship between symmetrical capability and asymmetrical short-circuit ratings.

Industry committees developed standards for the symmetrical ratings and, since about 1960, the ANSI standards are given in terms of both the old (asymmetrical) and the new (symmetrical) bases. Eventually all circuit breakers will be rated on the symmetrical basis and the asymmetrical ratings can then be retired. In the meantime, there exist breakers that were manufactured and rated on the older system, and these must be understood and respected.

Circuit breakers are rated on many operating conditions, all of which are defined in ANSI 37.04. Some of the more important ratings are given in Table 3.11. Most of the quantities listed require no further definition and the need for such ratings is fairly obvious. The voltage range factor does require a more formal definition:

Rated voltage range factor (*K*) The ratio of the rated maximum voltage to the lower limit of the range of operating voltages in which the required symmetrical and asymmetrical interrupting capabilities vary in inverse proportion to the operating voltage.

TABLE 3.11 Quantities by Which Circuit Breakers Are Rated

Maximum voltage	Frequency
Continuous current	Short-circuit current
Standard operating duty	Permissible tripping current
Interrupting time	Reclosing time
Dielectric strength	Control voltage
Transient recovery voltage rate	Load current switching and life
Shunt reactor current switching	Capacitor switching current
Line charging current switching	Excitation current switching
	Voltage range factor (K)

The application of the K factor is shown graphically in Figure 3.33. The rated maximum voltage is the highest voltage, above the nominal system voltage, for which the breaker is designed, and is the upper limit for operation. This maximum voltage, divided by K, defines the voltage at which the circuit breaker must be able to interrupt K times its rated short-circuit current.

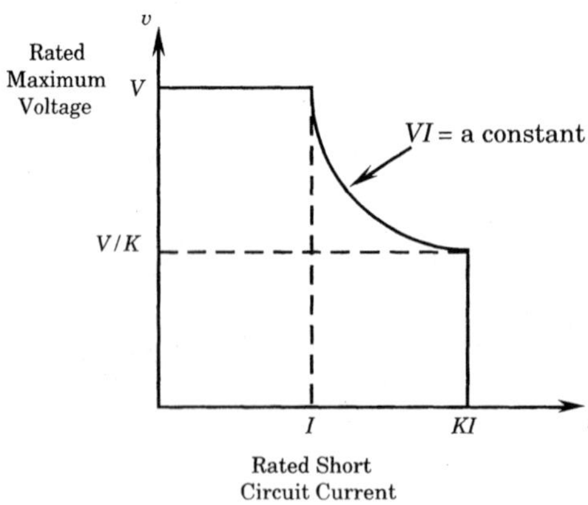

Figure 3.33 Relationship between applied voltage and required symmetrical current for circuit breakers [54].

To compute the asymmetrical capability, the ratings must take into account the time delay between the inception of the fault and the actual time of contact separation, since the current envelope decays as the dc component returns to zero, as shown in Figure 3.34. This is determined by a defined ratio "S" called the asymmetry factor, which is defined as

$$S = \frac{\text{Required asymmetrical current rating}}{\text{Required symmetrical current rating}} \tag{3.19}$$

Figure 3.34 gives the required value of S in terms of the circuit breaker operating time. For example, a three cycle breaker must be able to interrupt at least 1.2 times the symmetrical interrupting rating. Note that the slope of the curve in Figure 3.34 depends on the X/R ratio of the system feeding the fault, which is taken to be approximately 16 to 17 for the purpose of this definition.

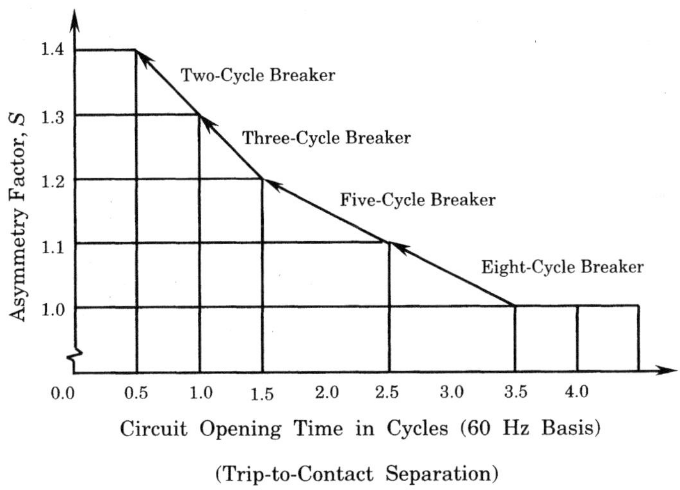

Figure 3.34 Asymmetry factor S as determined by interrupting time rating for circuit breakers rated on a symmetrical basis.

Another important rating of circuit breakers is that of the circuit breaker duty cycle. This is defined as follows:

Standard operating duty The standard operating duty of a circuit breaker is two unit operations with 15 seconds between the operations.

This is often abbreviated CO-15s-CO, where CO is taken to mean "Close-Open."

The total rms short-circuit current through a circuit breaker over any period T is computed, by definition, as

$$I_{rms} = \frac{1}{T}\int_0^T i^2\,dt \tag{3.20}$$

where i = instantaneous current in amperes
t = time in seconds

This rms current enters into the definition of permissible delays in breaker operation.

The time delay between the energizing of the breaker trip coil and the actual parting of the breaker main contacts is another important rated value. Two ratings apply to the tripping delay.

Rated permissible tripping delay (*Y*) The permissible tripping delay of a circuit breaker is designated Y seconds and is the maximum time for which the circuit breaker is required to carry K times rated short-circuit current after closing on this current and before interrupting.

Permissible tripping delay (*T*) Tripping of the circuit breaker may be delayed beyond the rated permissible tripping delay at lower values of current in accordance with the relationship:

$$T = Y\left(\frac{K(\text{rated short-circuit current})}{\text{Short-circuit current through breaker}}\right)^2 \tag{3.21}$$

The total tripping delay on all operations within any 30-minute period must not exceed the time obtained by computing T.

3.4.3 Circuit Breaker Design

The basic function of a circuit breaker is to insert an insulating medium in a circuit to stop the flow of current and maintain a permanent isolation between the two network elements connected to the breaker. This requires that the circuit breaker be able to interrupt even very high currents that are expected to flow during short circuits and that the device be capable of establishing the insulating barrier very quickly. This is accomplished by having contacts in the circuit breaker, at least one of which can be moved quickly away from the other contact, thereby establishing a separation that is adequate to serve as a permanent insulation barrier.

As the contacts are moved apart, an arc is drawn between them that must be extinguished by some means. The contacts are placed in a fluid medium that will aid in extinguishing the arc and also provide the insulation required for permanent isolation of the two breaker contacts and between the breaker contacts and ground. Arc extinction is a science in its own right and is beyond the scope of this book, but is highly recommended for study by the protection engineer. The fluid medium surrounding the contacts is selected to aid the arc extinction. This is assisted by elongating and cooling the arc by some means to hasten the interruption.

The breaker mechanism used to separate the contacts must be fast and reliable. Usually speed is most important in separating the contacts, for example, when interrupting a fault current, but there are occasions when fast closing is more important. Common operating mechanisms include impulse coils, solenoids, spring-motor-pneumatic and hydraulic schemes.

3.4.3.1 Circuit Breaker Fluids. The fluid medium commonly used in circuit breakers depends on the type and rating of the breaker, but will usually be one of the following [55]:

Air at atmospheric pressure

Compressed air

Mineral oil

Ultra-high vacuum

Sulfur hexafluoride (SF_6)

These fluids vary widely in their relative dielectric strengths, which are summarized in Table 3.12.

Oil has the capability of interrupting an arc due to the formation of a hydrogen bubble, which cools the arc and assists in extinction. Therefore, oil does not require the assistance of external energy from the operating mechanism in order to accomplish extinction. Other

TABLE 3.12 Relative Dielectric Strengths of Gases [56]

Gas	Dielectric Constant 20°C, 1 atm	Relative Dielectric Strength, rms Volts†
Air	1.000536	0.95
N_2	1.000547	1.0
CO_2	1.000921	0.9
He	1.000065	0.14
Ar	1.000517	0.28
H_2	1.000254	0.57
SF_6	1.002084	2.3–2.5

† For a 0.1 in. gap

fluids do require outside energy in order to operate. Compressed air and SF_6 operate under high pressure or are required to deliver a blast of gas to elongate and cool the arc during interruption.

Oil has a high dielectric strength, which varies considerably with the state of purity of the oil. If moisture or particulate matter are permitted to enter the oil, the dielectric strength can be greatly reduced. The measured dielectric strength also varies significantly due to the method of its experimental measurement. For example, measured dielectric strength between two needles is given as about 50 kV for a 1 cm spacing between the needles, but increases to about 110 kV between 1 cm spheres and to 160 kV between 25 cm spheres. It has also been shown that the dielectric strength cannot be maintained indefinitely due to oxidation and the interruption of high currents will eventually introduce particulate matter into the oil.

3.4.3.2 Oil Circuit Breakers. There are two types of oil circuit breakers; *dead-tank* and *live-tank* designs. A dead-tank breaker has a steel tank that is partly filled with insulating fluid. Porcelain bushings are mounted on the top of the tank to provide insulation from line voltage to the grounded tank and to permit the connection of the high-voltage elements to the interrupting mechanism, which is submerged in the oil. The contacts are typically bridged by a crosshead that can be moved by a connecting rod, using either linear or rotational motion, to separate the crosshead from the stationary contacts. In some designs the two bushings and one crosshead can be arranged to provide two, four, six, or more breaks per pole, thereby limiting the required separation distance per break. Many oil circuit breakers are of "oil tight" design, which means that the gases generated by the fault can be permitted to escape to the atmosphere, but the oil is not allowed to escape. Escaping oil can be a serious fire hazard. Many modern oil breakers have incorporated arc control devices, using various ingenious design concepts. All oil breakers make use of the oil pressure generated by the gas created by the arc to force fresh oil through the arc. This provides the means of establishing full insulation at the first current zero and prevents restrike of the arc, thereby interrupting the circuit. Some oil breakers use an oil-blast feature to improve the interrupting action.

3.4.3.3 Oilless Circuit Breakers. Oilless circuit breakers are almost universally used for indoor breaker installations and are the only breakers available for the extra high voltage applications. There are three primary types of oilless circuit breakers; magnetic-air, compressed-air, and SF_6 designs. A vacuum circuit breaker is also manufactured for lower voltage applications, but may eventually become available at higher voltages.

The magnetic-air breakers usually employ a stored-energy mechanism to operate the interrupter with the assistance of a strong magnetic field. The magnetic field is used to force the arc into a specially designed arc chute, which cools and lengthens the arc, thereby assisting extinction. Some designs employ a "puffer" to assist in the interruption of small currents, such as load pickup, which do not provide adequate current to create a large magnetic field. The puffer blows a blast of air across the arc, thereby forcing it into the arc chute.

Compressed-air breakers are available for use in high-voltage applications, and are commonly used for indoor installation in the 14.4–34.5 kV range of voltages. Outdoor designs are available in 34.5–765 kV ratings. These breakers use compressed air to operate the breaker as well as to assist in arc extinguishing. A blast of air is directed at the arc, forcing it into a special arc chute to provide the cooling and deionization required for interruption.

Live-tank oilless circuit breakers also employ compressed air for arc interruption, with the air exhausted to the atmosphere. The supporting columns house current transformers and are filled with SF_6 for insulation.

Both dead-tank and live-tank oilless SF_6 breakers are also available. These breakers use SF_6 as an interrupting medium and have been applied at EHV levels of 345, 500, and 765 kV.

3.5 AUTOMATIC CIRCUIT RECLOSERS

Automatic circuit reclosers are defined as "self-controlled devices for automatically interrupting and reclosing an alternating-current circuit with a predetermined sequence of opening and reclosing followed by resetting, hold closed, or lockout" [1]. *Reclosers*, as they are usually called, are popular circuit interrupting devices for distribution systems in which the magnitudes of fault currents are limited. Reclosers have time-current characteristics that make them easy to coordinate with other reclosers, at least of the same manufacture, and with other devices, such as relays and fuses.

The most useful feature of reclosers, from a coordination viewpoint, is their flexibility. Most reclosers are designed for several operations in a fixed sequence, and the TC characteristics of the operations can be selected from a given family of characteristics. These range from instantaneous tripping or fast interrupting operation to various time-delayed tripping choices. Thus, for a temporary fault, one or more fast operations may give the fault adequate time to deionize such that a subsequent reclosure will restore the circuit to normal without any permanent outages to loads served along the feeder. Semipermanent faults may be made to clear, or burn clear, with a time-delayed opening and reclosure. Permanent faults will operate the recloser to lockout or to a hold closed condition, which gives a protected fuse the opportunity to clear.

Figure 3.35 provides a definition of the various operating times associated with a tripping and reclosing operation. The actual times involved depend on the recloser design and must be obtained from the manufacturer. This sequence for detecting a fault and interrupting a circuit, followed by a reclosure, is called a *unit operation*.

Note that the recloser is similar to a circuit breaker although designed for lighter interrupting duty and for use on lower voltage circuits. One type of recloser is a self-contained unit that employs a series trip mechanism and is intended for pole mounting. Such units require no external batteries or other accessories for normal operation. Other reclosers are more like circuit breakers, with parallel tripping and generally higher interrupting ratings for substation applications. All reclosers have a fault interrupting capability, which may employ an oil or vacuum interrupter.

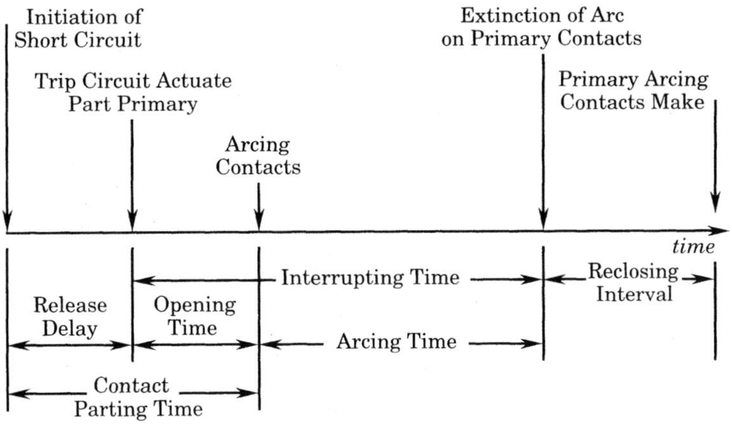

Figure 3.35 Recloser operating time from fault initiation to reclosure (a unit operation).

3.5.1 Recloser Ratings

Reclosers have ratings specified for maximum voltage, frequency, continuous current, minimum tripping current (series trip coil models), symmetrical current interruption, making current, and impulse withstand voltage. Rated frequency is normal power frequency in all cases (60 hertz in North America). Rated maximum voltage is 15 kV on many models, but ranges up to 72 kV. Rated continuous currents can be selected from a wide range, from five to over 1000 amperes. The minimum trip current is always twice the continuous rated current (with a 10% tolerance) for series trip reclosers, but it is variable for shunt trip models. Symmetrical interrupting ratings are generally below 15,000 amperes, and are graded depending on the continuous current rating. Rated making current is the rms current against which the recloser is required to close and is the same as the rated interrupting current in all cases.

3.5.2 Recloser Time-Current Characteristics

Modern automatic circuit reclosers are available in several control configurations that provide a variety of TC characteristics. The simplest control scheme is the series trip coil arrangement shown in Figure 3.36. The series trip coil is connected in series with the recloser main contacts and are rated to carry full line current. When a current greater than the minimum trip current flows, a solenoid plunger is drawn into the trip coil. This plunger trips a latch, which releases the charged opening springs for fast contact separation. Closing energy, plus energy to recharge the opening springs, is supplied by a high voltage closing solenoid connected phase to ground.

Another series trip scheme works in the opposite sense to that just described. In this scheme, the series trip solenoid plunger is directly attached to the main contacts. Opening the contacts also charges a closing spring, which is merely toggled to effect the closing when required.

Typical series trip TC characteristics are shown in Figures 3.37 and 3.38. Note the difference in the characteristics. Actually many other different characteristic shapes are available, as well as many ratings other than those illustrated. In many cases, the recloser is programmed for four operations, which might typically include two fast operations (curve A) and two time-delayed operations (curve B or C). Many other operating sequences are possible.

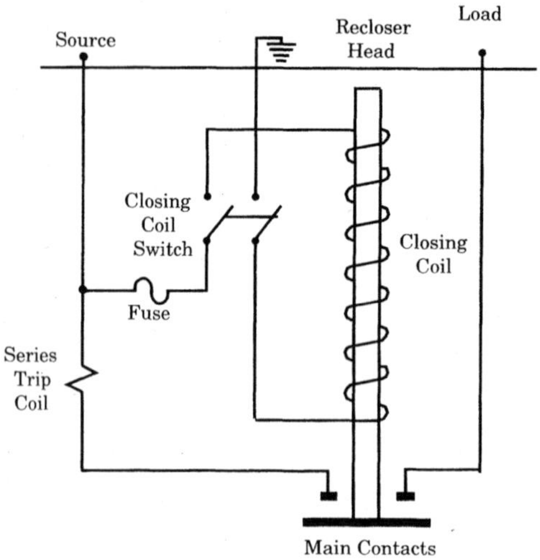

Figure 3.36 Series trip coil arrangement for reclosers. [Courtesy McGraw-Edison Power Systems.]

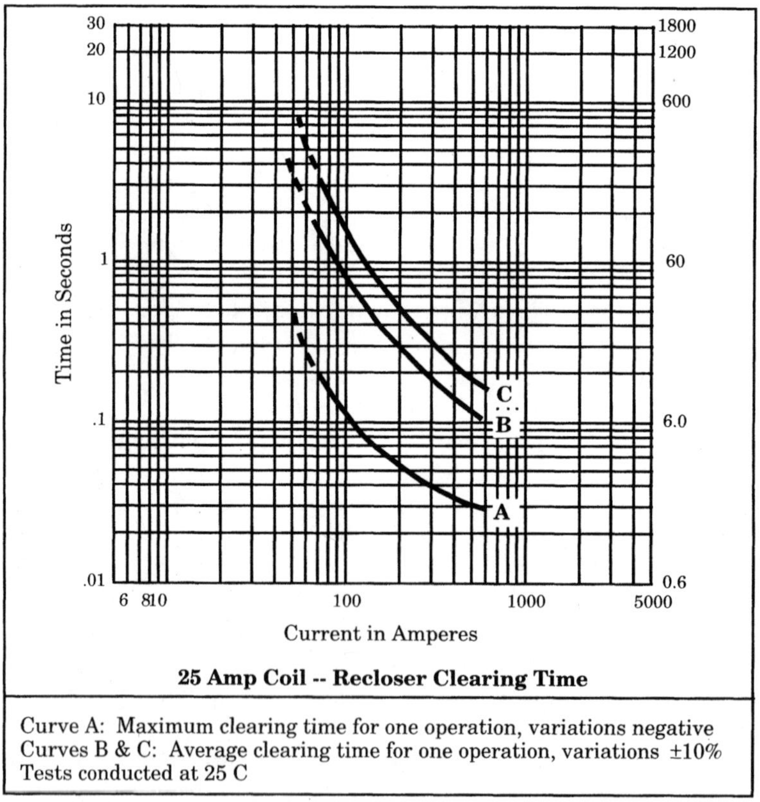

Figure 3.37 Typical TC characteristics for a recloser with hydraulic control. [Courtesy McGraw-Edison.]

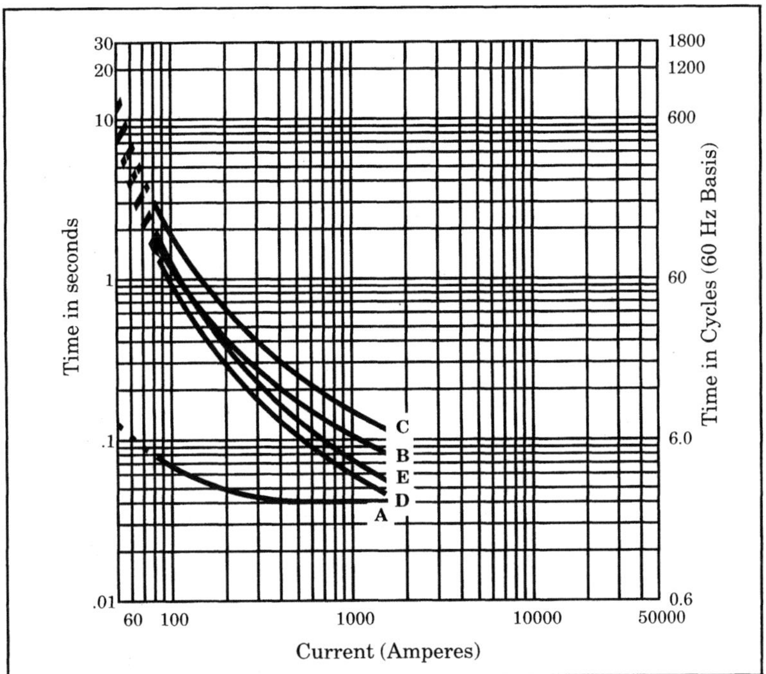

25 Amp Coil -- Recloser Clearing Time					
Curve A: Maximum clearing time for one operation, variations negative	Maximum Interrupting Rating Symmetrical Amperes				
	Phase-to-Phase KV				
Curve B, C, D, & E: Average clearing time for one operation, variations ±10%	Type	4.8	8.32	14.4	34.5
	R	1500	1500	1500	-
Tests conducted at 25C	RV	-	-	-	1500

Figure 3.38 Typical TC characteristics for a recloser with hydraulic control. [Courtesy McGraw-Edison].

Another important type of recloser control uses an electronic control. A block diagram for an electronic control system is shown in Figure 3.39 and is explained as follows. The sensing CT's provide basic ac current level information on each phase. Signals are generated that are proportional to these currents and are presented to a level detection and timing circuit.

Here, any current in excess of the minimum trip is detected and the trip circuit timing circuit is activated. Current magnitude and time then integrate the timing circuit to match tripping with a preselected TC characteristic shape. After timing, the trip signal is amplified and the trip solenoid is energized. Action of the trip solenoid releases the precharged trip springs, opening the main contacts. After tripping, the reclose timing circuits are energized to sequence the reclosing circuit at the appropriate time.

One advantage of the electronic control is the variety of TC characteristic shapes available. The characteristics in Figure 3.40 are an example, but several others are available. Figure 3.41 shows another family of TC characteristics available from another manufacturer.

Electronic controls for reclosers offer several advantages. These controls are simple to adjust and are accurate. Probably the most important advantage to the protection engineer is

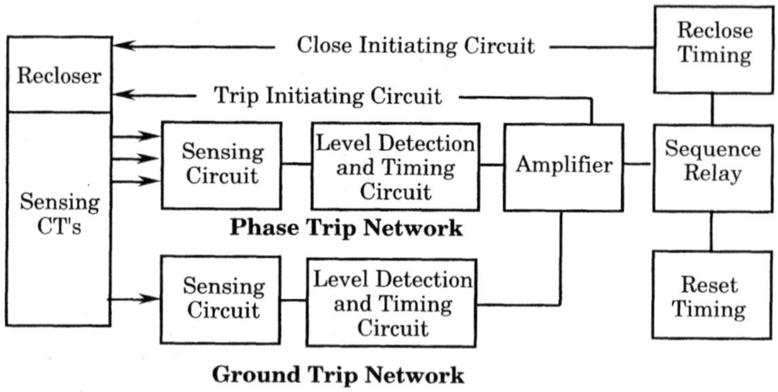

Figure 3.39 Block diagram of an electronic recloser control system. [Courtesy McGraw-Edison.]

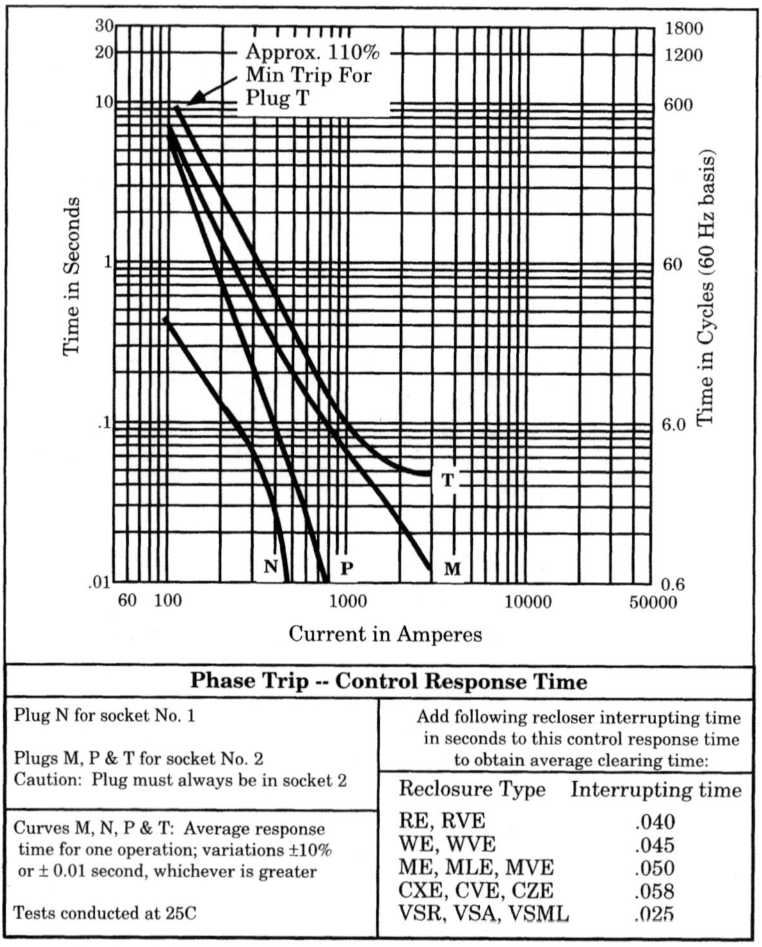

Phase Trip -- Control Response Time

Plug N for socket No. 1 Plugs M, P & T for socket No. 2 Caution: Plug must always be in socket 2	Add following recloser interrupting time in seconds to this control response time to obtain average clearing time:
Curves M, N, P & T: Average response time for one operation; variations ±10% or ± 0.01 second, whichever is greater Tests conducted at 25C	

Reclosure Type	Interrupting time
RE, RVE	.040
WE, WVE	.045
ME, MLE, MVE	.050
CXE, CVE, CZE	.058
VSR, VSA, VSML	.025

Figure 3.40 Example TC characteristics for a recloser with electronic control. [Courtesy McGraw-Edison.]

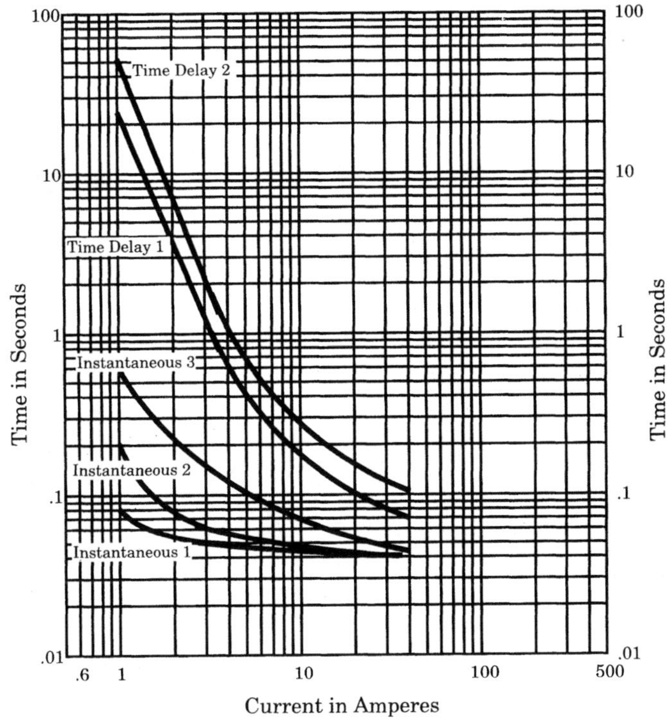

Phase time delay curve is selected by setting the phase time delay curve selector to position 1 or 2

Phase instantaneous curve is selected by setting phase instantaneous curve selector to position 1, 2, or 3

Time delay curves variation from curves ±10% above 0.1 seconds

Figure 3.41 Example TC characteristics for a recloser with electronic control. [Courtesy Westinghouse.]

their flexibility. One manufacturer offers a choice of 18 TC characteristics for phase faults and 14 for ground faults. This simplifies coordinating the recloser with other devices. Moreover, the timing between reclosers is variable, which allows controlling the melting of fuses in the same circuits.

3.6 AUTOMATIC LINE SECTIONALIZERS

Another protective device that is often used in conjunction with reclosers on distribution circuits to optimize the protective scheme is the *automatic line sectionalizer*. The automatic line sectionalizer, which is usually called simply the "sectionalizer," is a self-contained circuit-opening device that opens the main electrical circuit while it is disconnected from the source of power. The circuit opening action takes place after the device has sensed and responded to a predetermined number of successive main-circuit impulses of a predetermined magnitude. Sectionalizers also have a provision for manual operation to interrupt load currents.

Sectionalizers are available in a wide range of ratings and are chosen to work in conjunction with reclosers or circuit breakers. The sectionalizer functions as a counter of the backup recloser or circuit breaker operations. After a predetermined number of circuit interruptions, it can be programmed to open while the recloser has the circuit de-energized. This means that the sectionalizer contacts do not need to interrupt fault currents and can be of simpler, and less expensive, design. The sectionalizer is essentially a fault detector (to determine when to count), a counter, and a switch. Usually, the switch contacts are immersed in oil and are rated to interrupt load current by manual means. This makes the device a convenient manual circuit switcher as well as a sectionalizer.

Sectionalizers provide a convenient method of providing protection to branch lines on a distribution system, thereby protecting the main feeder from branch line faults. Sectionalizers are often installed on poles on overhead distribution circuits.

Two examples of sectionalizer application are shown in Figure 3.42. In (*a*) the recloser R is set for two fast and two retarded operations. For a fault at P, the fuse would ordinarily clear during the third operation of R, i.e., the first retarded operation. If sectionalizer S is set for two counting operations, it will lock out before the third recloser operation, thereby ensuring service continuity to all loads beyond F except for those served through S. In part (*b*), the fuse F will blow on the second recloser operation, but S will not lock out. Therefore, S must be set for three operations.

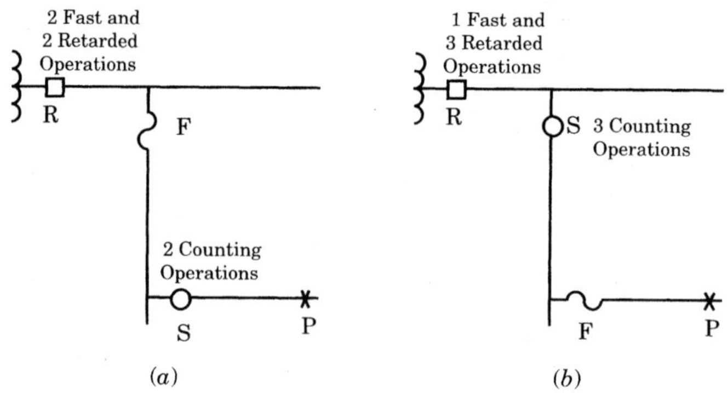

Figure 3.42 Examples of sectionalizer application.

Sectionalizers should not be installed in series between two reclosers. However, two sectionalizers can be used in series downstream from a recloser with the one farthest from the source and set for fewer count operations than the one closer to the source. Since sectionalizers are much cheaper than reclosers, this may be a good investment as the coordination with the backup recloser is more certain than with fuses.

3.7 CIRCUIT SWITCHERS

The circuit switcher is a device that looks and operates much like an air-break switch, but has fault current interruption capability. Therefore, it finds a place in many protective schemes where the cost of a circuit breaker is not warranted, but fault interruption is required. Moreover, the open blade switch has the benefit of visual confirmation of the switch position, i.e., whether open or closed.

A typical circuit switcher consists of an SF_6 interrupter and a motor-driven disconnect switch. The operation of the device is as follows. On opening, the disconnect blade begins to move toward the open position, but before the blade leaves the jaw, a stored- energy spring trips the interrupter to open the circuit. The blade then leaves the jaw, but before reaching its fully open position, the interrupter recloses and the tripping spring energy is restored. Therefore, after the blade is fully open, the interrupter is closed and ready for the next trip operation. The circuit switcher can be equipped with a shunt trip device which provides current interruption in from five to eight cycles [57].

One of the advantages of the circuit switcher is its low cost, compared to a circuit breaker. One report gives the cost as less than 40% of the cost of a circuit breaker and disconnects [58]. The circuit switcher also costs less to install, and it requires less space in the substation. The disadvantage is that the circuit switcher has a lower interrupting rating than a circuit breaker. Thus, the circuit switcher occupies a position between a breaker and fuse-plus-open switch protection, with a cost and space requirement that is attractive. It can be controlled by relays in any desired configuration such as transfer trip schemes to clear other devices on fault detection. Circuit switchers are often favored for the protection of transformers, capacitor banks, shunt reactors, and line or cable switching.

3.8 DIGITAL FAULT RECORDERS

The digital fault recorder is not a protective device, but rather a monitoring and data collection system. However, it plays such an important role in power system protection that it deserves consideration in this chapter, which treats various types of protective equipment.

The traditional method of recording fault currents on the power system has been the cathode ray oscillograph. The oscillograph uses mirrors to focus a beam of light on a moving strip of photo-sensitive paper. The mirror deflection is made proportional to the current or other quantity being recorded so that, with the light-sensitive paper moving at a constant rate, a steady-state alternating current traces a sine wave on the paper. For many years, this has been a common method for capturing fault data on the power system and many substations have been equipped with oscillographs.

There are obvious drawbacks to the use of the cathode ray oscillograph. The devices are costly, so that their installation is usually reserved for only a few very important lines. They require careful calibration and regular cleaning and maintenance. Following the recording of a fault, it is often necessary to install a new paper roll, so that the next fault can be recorded. This means that the device is a problem to maintain at nonattended stations. Moreover, the basic concept of recording the fault information on a strip of paper is not the best choice, since a prolonged period of fault activity causes the device to run out of paper, and after which all fault data is lost. In some cases, this may result in the loss of the most important fault information, depending on the sequence of fault events.

The oscillographic record also presents several problems. The early devices had no means of storing information, so the recorder was triggered after the fault began. This means that the prefault condition was not recorded and, in some cases, the exact sequence of early events was lost. Later devices incorporated a storage mechanism that largely solved this problem. However, the oscillographic record has another serious flaw in the permanent nature of the recording. It is not possible to change the scale of the waveforms, either for magnification or reduction, nor is it possible to change the time axis to examine more closely the exact sequence of events. All of these problems have been solved by the introduction of digital fault recording devices.

A typical digital fault recorder configuration is shown in Figure 3.43. The heart of the substation installation is the central processor unit (CPU). To note the exact time of all events, an antenna is aimed at a navigation satellite, which is backed up by a crystal clock. This gives a nearly exact time synchronization across the power system, which is important for disturbances that cause widespread relay operation and helps the engineer understand the exact sequence of events. Analog events are processed through isolators (Iso) and analog-to-digital converters (A/D) to provide the "channel" inputs to the CPU. Digital events are processed through isolators and signal conditioners (S/C) to provide "event" inputs to the CPU. A recording is started by a "start" signal, which may be due to any of the following conditions:

- Over or under voltage
- Over or under frequency
- Over current
- Negative sequence voltage
- Zero sequence voltage
- Manual trigger

Other connections to the CPU include the keyboard, display, plotter, and data storage through floppy disks and tapes.

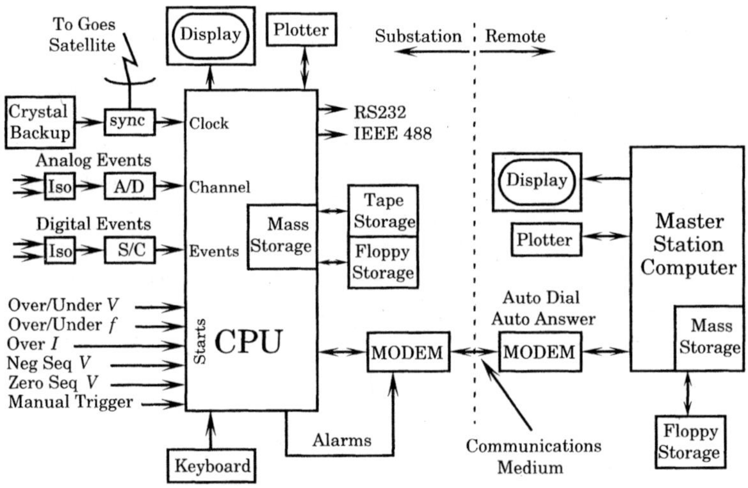

Figure 3.43 Typical digital fault recorder configuration.

Outputs from the substation CPU are fed through a high-speed modulator/demodulator (MODEM) through a communications medium to a similar modem at the Master Station Computer, which is usually located in the protection engineer's office. Fault events are transferred to the Master Station immediately following their occurrence, 24 hours a day. At the Master Station, the engineer can view the time records of the fault on the display and can plot for a permanent record. Magnetic storage is also provided for a digital record of the event. The viewed or plotted record can be scaled as required for convenient and accurate viewing of the event.

An interesting extension of the concept of the digital fault recorder is now provided relay manufacturers. These protective systems are designed for digital line relaying, but also incorporate digital fault recording features. This permits oscillographic viewing of the fault event and remote monitoring of the fault data through a data link from any personal computer, through a suitable communications channel, to the digital relay. This is an example of the flexibility of digitally recording information, which allows the basic data to be transmitted to any location, as required. This concept will probably to emulated in many future protective devices.

REFERENCES

[1] IEEE Std 100-1992, *IEEE Standard Dictionary of Electrical and Electronics Terms*, John Wiley & Sons, Inc., New York, 1992.

[2] ANSI C37.40-1994, "IEEE Standard Service Conditions and Definitions for High Voltage Fuses, Distribution Enclosed Single-Pole Air Switches, Fuse Disconnecting Switches, and Accessories," American National Standards Institute, New York, 1994.

[3] IEEE Tutorial Course, "Application and Coordination of Reclosers, Sectionalizers, and Fuses," IEEE Course Text 80 EHO157-8-PWR, IEEE, New York, 1980.

[4] Wold, K. M., "Current-Limiting Fuses Protect Capacitor Bank Switches," *Transmission and Distribution*, March 1989.

[5] S&C Electric Company, "S&C Fault Fiter® Electronic Fuses," Manufacturer's Descriptive Bulletin 441-30, S&C Electric Company, Chicago, April 1989.

[6] ANSI Std C37.46-1981, "American National Standard Specifications for Power Fuses and Fuse Disconnecting Switches," American National Standards Institute, New York, 1981.

[7] McGraw-Edison Company, "Distribution System Protection and Apparatus Coordination," Bulletin 55020, 1955.

[8] Horowitz, S. H., Ed., *Protective Relaying for Power Systems*, IEEE Press, New York, 1980.

[9] Harder, E. L., and W. E. Marter, "Principles and Practices of Relaying in the United States," *Trans AIEE*, 67, 1948, pp. 1005–1022.

[10] Sonnemann, W. K., and W. E. Glassburn, "Principles of Induction Type Relay Design," *Trans. AIEE*, 72 (3), 1953, pp. 23–27.

[11] Sonnemann, W. K., "A New Inverse Time Overcurrent Relay with Adjustable Characteristics," *Electrical Engineering*, 72, 1953, pp. 607–609.

[12] ANSI/IEEE Standard C37.90-1989, "IEEE Standard Relays and Relay Systems Associated with Electrical Power Apparatus," IEEE, New York, 1989.

[13] IEC Standard, "Single Input Energizing Measuring Relays with Dependent Specified Time," IEC Publication 255-4, 1st Ed., Bureau Central de la Electrotechnique Internationale, Geneva, 1976.

[14] Schweitzer, E. O., and A. Aliaga, "Digital Programmable Time-Parameter Relay Offers Versatility and Accuracy," *IEEE Trans.*, PAS-99 (1), January–February 1980.

[15] Garrett, R., W. C. Kotheimer, and S. E. Zocholl, "Computer Simulation of Current Transformers and Relays for Performance Analysis," 14th Annual Western Relay Conference, Spokane, October 20–23, 1987.

[16] Zocholl, S. E., "Integrated Metering and Protective Relay Systems," Pacific Coast 53rd Annual Engineering and Operations Conf., San Francisco, March 16–17, 1988.

[17] IEEE Committee Report, "Computer Representation of Overcurrent Relay Characteristics," *IEEE Trans.*, PWRD-4 (2), July 1989, pp. 1659–1667.

[18] Kramer, C. E., and W. A. Elmore, "Flexible Inverse Overcurrent Relaying Using a Microprocessor," IEEE paper 89TD378-1, *IEEE Trans.*, PWRD, April 1990, pp. 915–923.

[19] IEEE Committee Report, "Computer Representation of Overcurrent Relay Characteristics," *IEEE Trans.*, PWRD-4 (2), July 1989, pp. 1659–1667.

[20] Benmouyal, G., "Some Aspects of the Digital Implementation of Time Protection," *IEEE Trans.*, PWRD-5 (4), October 1990, p. 1705.

[21] IEEE Committee Report, "Supplement to Recent Practices and Trends in Protective Relaying," *IEEE Trans.*, PAS-83, 1964, pp. 1064–1067.

[22] Hathaway Overcurrent Relays, "Non-Directional Time Overcurrent Relays," Product Bulletin 12-101, Hathaway Instruments, Inc., Denver, November 1972.

[23] Rockefeller, G. D., "Fault Protection with a Digital Computer," *IEEE Trans.*, PAS-88 (4), April 1969, pp. 438–461.

[24] Mann, B. J., and I. F. Morrison, "Digital Calculation of Impedance for Transmission Line Protection," *Ibid.*, PAS-90 (1), January/February 1971, pp. 270–279.

[25] Poncelet, R., "The Use of Digital Computers for Network Protection," CIGRÉ Paper No. 32-08, August 1972.

[26] Redfern, M. A. , "The Application of Microprocessors to Power System Protection Relays," Pennsylvania Electric Association Fall Meeting, September 1984.

[27] Basler Electric, "Time Overcurrent Relays, BE1-51, BE1-51/27C, BE1- 51/27R," Product Bulletin, Basler Electric, Highland, IL, 1983.

[28] ASEA, "RACID "Proper" Overcurrent Relay," ASEA Relays Product Bulletin A 03-2312E, Västerås, Sweden, 1985.

[29] ASEA, "RACIF Three Phase and Ground Overcurrent Relay," ASEA Relays Product Bulletin A 03-2314E, Västerås, Sweden, 1986.

[30] Westinghouse Electric Corp., "MDAR-the first fully digital, micro-processor based transmission line relay," Westinghouse Relay and Telecommunications Division, Product Bulletin B-986, Coral Springs, FL, 1988.

[31] IEEE Committee Report, "Supplement to Recent Practices and Trends in Protective Relaying," *IEEE Trans.*, PAS-83, October 1964, pp. 1064–1067.

[32] Phadke, A. G., and J. S. Thorp, *Computer Relaying for Power Systems*, John Wiley & Sons, Inc., New York, 1988.

[33] IEEE Tutorial Course, "Computer Relaying," IEEE Course Text 79 EH0148-7-PWR, IEEE, New York, 1979.

[34] IEEE Tutorial Course, "Microprocessor Relays and Protection Systems," IEEE Course Text 88EHO269-1-PWR, IEEE, New York, 1988.

[35] Horowitz, S. H., A. G. Phadke, and J. S. Thorp, "Adaptive Transmission System Relaying," *IEEE Trans.*, PWRD-3 (4), October 1988, pp. 1436–1445.

[36] Phadke, A. G., J. S. Thorp, and S. H. Horowitz, "Impact of Adaptive Protection on Power System Control," Proc. 9th PSCC Conf., Lisbon, 1987, pp. 283–290.

[37] Thorp, J. S., S. H. Horowitz, and A. G. Phadke, "The Application of an Adaptive Technology to Power System Protection and Control," *CIGRÉ*, Paris, 1988.

[38] Phadke, A. G., J. S. Thorp, and S. H. Horowitz, "Study of Adaptive Transmission System Protection and Control," Final Report Prepared for Oak Ridge National Laboratory, Virginia Polytechnic Institute, Blacksburg, VA, 1988.

[39] Rockefeller, G. D., C. L. Wagner, J. R. Linders, K. L. Hicks, and D. T. Rizy, "Adaptive Transmission Relaying Concepts for Improved Performance," *IEEE Trans.*, PWRD-3, October 1988, pp. 1446–1458.

[40] Zocholl, S. E., "Integrated Metering and Protective Relay Systems," *IEEE Trans. Industry Applications*, 25 (5), September/October 1989.

[41] Phadke, A. G., M. Ibrahim, and T. Hilbka, "State Estimation with Phasor Measurements," *IEEE Trans.*, PAS-96 (2), March/April 1977, pp. 635–646.

[42] Phadke, A. G., T. Hilbka, M. Adamiak, M. Ibrahim, and J. S. Thorp, "A Micro-Computer Based Ultra-High-Speed Distance Relay - Field Tests," *IEEE Trans.*, PAS-100 (4), April 1981, pp. 2926–2936.

[43] Misout, G., J. Beland, and G. Bedard, "Dynamic Measurement of the Absolute Voltage Angle on Long Transmission Lines," *IEEE Trans.*, PAS-100 (11), November 1981, pp. 4428–4434.

[44] Bonanomi, P., "Phase Angle Measurements with Synchronized Clocks - Principles and Applications," *IEEE Trans.*, PAS-100 (12), December 1981, pp. 5036–5043.

[45] Phadke, A. G., J. S. Thorp, and M. G. Adamiak, "A New Measurement Technique for Tracking Voltage Phasors, Local System Frequency, and Rate of Change of Frequency," *IEEE Trans.*, PAS-102 (5), May 1983, pp. 1026–1038.

[46] Misout, G., J. Beland, G. Bedard, and P. Bussieres, "Study of Time Dissemination Methods Used on an Electric Power System with Particular Reference to Hydro Quebec," *IEEE Trans.*, PAS-103 (4), April 1984, pp. 861–868.

[47] Burnett, R. O., Jr., "Field Experience with Absolute Time Synchronization between Remotely Located Fault Recorders and Sequence of Event Recorders," *IEEE Trans.*, PAS-103 (7), July 1984, pp. 1739–1742.

[48] Thorp, J. S., A. G. Phadke, and K. J. Karimi, "Realtime Voltage Phasor Measurements for Static State Estimation," *IEEE Trans.*, PAS-104 (11), November 1985, pp. 3098–3107.

[49] Phadke, A. G., J. S. Thorp, and K. J. Karimi, "State Estimation with Phasor Measurements," *IEEE Trans.*, PWRS-1 (1), February 1986, pp. 233–251.

[50] Thorp, J. S., A. G. Phadke, S. H. Horowitz, and M. M. Begovic, "Some Applications of Phasor Measurements to Adaptive Protection," IEEE Power Engineering Computer Applications Conference, Montreal, May 18–21, 1987; also included in *Protective Relaying for Power Systems II*, Stanley H. Horowitz, Ed., IEEE Press, 1992, pp. 77–84.

[51] IEEE Committee Report, "Synchronized Sampling and Phasor Measurements for Relaying and Control," *IEEE Trans.*, PWRD-9 (1), January 1994, pp. 442–452.

[52] Thorp, J. S., A. G. Phadke, S. H. Horowitz, and M. M. Begovic, "Some Applications of Phasor Measurements to Adaptive Protection," *Protective Relaying for Power Systems II*, Stanley H. Horowitz, Ed., IEEE Press, 1992.

[53] Dolnic, T. G., and J. A. Kischefsky, "Fault & Operating Data Collection Features of a New Microprocessor Recloser Control," *IEEE Trans.*, PWRD-5 (1), January 1990, pp. 456–459.

[54] ANSI C37.03-1964, "American Standard Definitions for AC High Voltage Circuit Breakers," ANSI, New York, 1964.

[55] Wadha, C. L., *Electrical Power Systems*, 2nd Ed., John Eastern Limited, New Delhi, India, 1991.

[56] Dakin, T. W., "Insulating Gases," contribution to the *Standard Handbook for Electrical Engineers*, 10th Ed., D. G. Fink and J. M. Carroll, Editors, McGraw-Hill Book Co., New York, 1969.

[57] S&C Electric Company Brochure, "The New Mark IV Circuit Switcher," S&C Electric Company, Chicago.

[58] Pettus, L. T., "The Application of Circuit-Switchers for Transformer Protection," a paper presented at the Missouri Valley Electric Association, Engineering Conference, Kansas City, 1967.

PROBLEMS

3.1 What is a fuse link and what is its purpose? How are fuse links used?

3.2 What are the components of a fuse link and what are their purposes?

3.3 Explain what a fuse link is designed to do.

3.4 Explain what is meant by the minimum melting time of a fuse link.

3.5 Explain what is meant by the total clearing time of a fuse link.

3.6 What factors govern the time-current characteristic of a fuse link?

3.7 Where are fuse links applied in the power system?

3.8 Explain what is meant by "fuse coordination."

3.9 Think of yourself as a design engineer for a 15 kV cutout, which is to be marketed for use on distribution systems. Clearly, the continuous current rating of the design will be important, and the maximum fault current that can be interrupted must be determined. What other features should be considered in the design, if it is to meet the requirements of the distribution engineer?

3.10 What kinds of electrical tests can you think of that will ensure proper operation of the design considered in the previous problem.

3.11 In example 3.1, it was noted that the selection of a 15T fuse at location C gives very little room for load growth on the feeder beyond C. Suppose that we select a 20T fuse at C rather than the 15 T. Determine the correct fuses to install at A and B, under these new conditions.

4

Relay Logic

4.1 INTRODUCTION

The purpose of this chapter is to introduce the logic of protective relaying and to illustrate the development of this logic through many advances in the science of system protection. The very concept of protective relaying is based on logic, in particular, that type of logic that can be represented using electrical or electronic circuits. The term *logic* has a specific meaning in this context [1].

> Logic (1) The result of planning a data-processing system or of synthesizing a network of logic elements to perform a specified function. (2) Pertaining to the type or physical realization of logic elements used, for example, diode logic, AND logic.

Clearly, protective relaying is a logic function and the relay itself is a device consisting of logical elements. The protective relay purpose is to recognize a particular abnormal system condition and to initiate a preplanned responsive action. All relays, from the simplest to the most complex, consist of logic elements. In the simplest of relays, this may be only one element, for example, a simple overcurrent device that will recognize a current in excess of a given amount. In complex relays, many logic elements are required to analyze the incoming information regarding the system condition and to determine the appropriate action that is required. This often involves making comparisons of electrical quantities, noting the time duration or repetition of the quantities, and finally making a decision regarding the observed characteristics.

However, relays also have the capability of initiating a particular action in response to the decision of the logic element(s). This action is always a go-no go decision, which is clearly a digital logic action. For example, we could set "trip" be a logical 1 and "no trip" be a logical 0, defining a binary logic in the decision element. All protective relays have this type of digital logic in their decision element. The logic elements themselves can be either analog or digital, but the resulting action is always digital.

This chapter reviews the logic of different types of relays. The purpose is to observe how relays conform to a logic structure, irrespective of the type of electrical or electronic circuits making up the relay hardware. This helps us understand that a relay is a type of computer that has been designed with the capability for analysis and for decision making. In some cases, the logic is simple, but in many cases the logic is very complex.

4.2 ELECTROMECHANICAL RELAY LOGIC

It may seem a bit unusual to think of an electromechanical relay as a logic device. Functionally, however, electromechanical relays are designed to recognize a particular system characteristic and to take appropriate action if that characteristic persists longer than a specified time. There is a logic in this activity that is unmistakable. The relay designer may not have thought of the device in terms of its logic. The specifications were likely spelled out in terms of currents, voltages, and time durations. The logic is built into the relay design, and is the product of the creativity of the inventor. Many electromechanical relays are based on the rotation of the induction disk, if sufficient torque is available to overcome the restraining spring. This design cleverly makes use of the fact that fault currents are usually high currents that will provide high torque if applied to the induction disk. The disk rotation itself is used as the timer, so the fault must exceed a given current for a given time in order to require tripping of the circuit. Some electromechanical relays are capable of performing quite difficult tasks, where selectivity is finely tuned to a specific type of condition.

4.2.1 The Overcurrent Relay

Consider an induction disk overcurrent relay. It has been noted that these relays can be designed with a wide variety of time-current characteristics, such that coordination between the relays and other devices is practical. In terms of the logical operations of an instantaneous overcurrent relay, the following summary is suggested.

1. Current measurement and filtering out high frequencies.
2. Comparison of measured current to a trip threshold value.
3. Timing of overcurrent according to prescribed time-current pickup characteristic.
4. Logic output in form of current through a relay coil, which picks up the relay contacts.

These functions can be performed in many different ways, and various inventors of electromechanical relays found ingenious methods of performing the desired task with relatively simple devices that have proved their value over many years of service.

4.2.2 The Distance Relay

In the electromechanical distance relay, the relationship between the observed voltage and current is measured to determine if the total impedance seen by the relay corresponds to a trip region of the complex R-X plane. A good example of this is illustrated in Figure 3.19 (5), which shows an induction cylinder type of unit, often called an offset mho relay. The interaction between the measured voltage and current are displayed in the R-X plane as a circle that either goes through the origin or is offset from passing through the origin by a prescribed amount. The measured quantities literally compute the impedance seen from the relay to the fault, which constitutes a distance measurement. In terms of a logic operations of the distance relay, the operation can be described as follows:

1. Current measurement and filtering out high frequencies.
2. Computation of impedance from relay to fault.
3. Comparison of computed impedance against relay circular offset mho characteristic to determine if measured impedance is inside the circle.
4. Timed delay of trip action, if fault is in second or third zone.
5. Logic output in form of current through a relay coil, which picks up the relay contacts.

All of these separate logic and decision functions are performed by the one relay hardware system. However, these operations can be performed by other methods, either electromechanical or electronic.

4.3 ELECTRONIC LOGIC CIRCUITS

The early designs for electronic relays were vacuum tube devices, which were introduced in the 1930s. These systems required considerable maintenance and were not as reliable as electromechanical systems, hence their applications were limited [2]. In the early 1950s transistors became available and immediately showed promise for providing the reliable electronic devices required for protective relaying. Still it was not until the mid-1960s that electronic protective equipment began to appear in quantity. These early transistor relays were largely analog devices, except for the digital final stage, which is always digital.

Digital relays were first discussed in 1969 as an application of a digital computer to perform protective functions for an entire substation. An early paper by George Rockefeller is often cited as the first comprehensive proposal for a digital relaying application [3]. This was followed by a large number of new research reports on digital relaying, but little was done in terms of industrial applications. It was not until the microprocessor was introduced that digital relaying began rapid development [4] and because of the many advantages provided by the microprocessor, the future of system protection will probably be largely digital.

The relay hardware for electronic relays consists of both analog and digital devices. The input signals are analog and require, at the very minimum, a conversion to digital form. Therefore, the relays design is often a mixture of analog electronic devices and digital hardware. The goal here is to become familiar with the important electronic modules that are commonly used in electronic relays of all types. The emphasis is on the electronic devices only. The relays may also contain transformers or other components that are also found in electromagnetic and electromechanical protective devices.

4.3.1 Analog Logic Circuits

Certainly the most common of all analog logic devices is the operational amplifier, or *op amp* as it is often called. The op amp is an integrated circuit device that usually contains 20 or more transistors, arranged to provide a stable gain A of 10^4 or more at low frequencies. The op amp is usually depicted simply as an amplifier, shown by the triangular box of Figure 4.1, where the triangle represents the high-gain amplifier [5].

The amplifier can have any number of active states with total gain A, and with an output voltage that is inverted with respect to the input. The resistor R_f provides shunt feedback around the gain element, which stabilizes the gain and lowers the output resistance. The

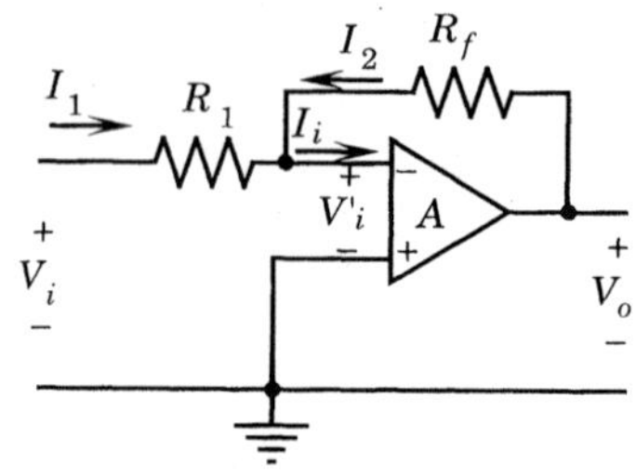

Figure 4.1 The operational amplifier.

internal gain of the op amp is defined as

$$A = \frac{V_o}{V_i'} \tag{4.1}$$

Since the output voltage is not large, say 10 volts or less, the input voltage to the operational amplifier is very small, and we can approximate this voltage as zero, in which case we can also approximate the input current I_i to the amplifier as being approximately zero. This permits us to write the currents shown in Figure 4.1 as follows.

$$I_1 = \frac{V_i}{R_1} \tag{4.2}$$

$$I_2 = \frac{V_o}{R_f} \tag{4.3}$$

But these currents sum to zero. Therefore, equating (4.2) and (4.3), we can compute the output voltage as a function of the input voltage, as follows.

$$V_o = -\frac{R_f}{R_1} V_i \tag{4.4}$$

This equation is the basic gain equation of the op amp. Note that the output voltage is always inverted with respect to the input, as noted by the minus sign. It is also interesting that the gain equation is not a function of the true amplifier gain, A, but does depend on the requirement that A be a very large number. The *effective gain* of the device is noted as

$$A' = -\frac{R_f}{R_1} \tag{4.5}$$

The op amp is used in many different configurations, some of which are reviewed briefly as follows.

4.3.1.1 The Isolator. The first application of the op amp is the *isolator*, which is shown in Figure 4.2(a). This is a unity-gain device that is useful for isolating one circuit from another, for example, where one circuit may have changes in loading that can disturb the operation of the other circuit.

For the isolator, we can write the following equation.

$$V_o = V_i + V_i' \cong V_i \tag{4.6}$$

The isolator output voltage follows the input voltage, without a phase reversal. Also, since the output impedance is low, there is ample current to drive the following circuit. This arrangement is sometimes referred to as a "buffer" amplifier or a noninverting voltage follower [5].

4.3.1.2 The Comparator or Level Detector. In relaying there is often a need to compare a measured quantity against some reference value. This function can be performed by an op amp

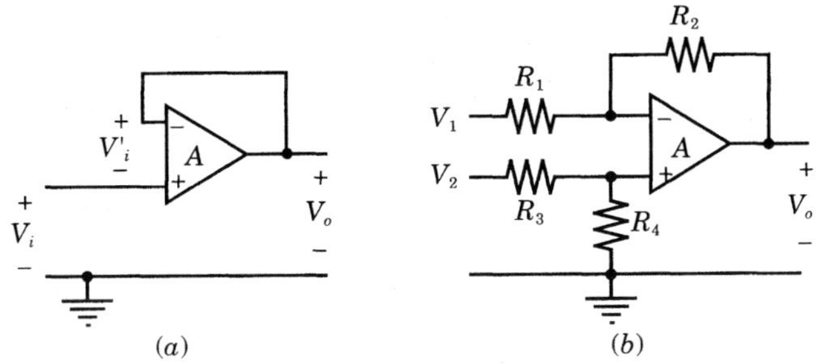

Figure 4.2 Op amp applications: (*a*) The isolator. (*b*) The comparator.

arranged as shown in Figure 4.2(*b*). This is a basic circuit for digital computer applications. When the input voltage is greater than the reference voltage, the output voltage goes into saturation and remains at this value as long as the input voltage is applied. When the input voltage goes below the reference, the output voltage goes into negative saturation. Note that the reference voltage need not be considered a constant, but can be varying. If this is the case, the comparator simply determines the larger of the two voltages.

The IEEE defines a comparator as follows [6]:

1. A circuit for performing amplitude selection between either two variables or between a variable and a constant.
2. A device capable of comparing a measured value with predetermined limits to determine if the value is within these limits.
3. A circuit, having only two logic output states, for comparing the relative amplitudes of two analog variables, or of a variable and a constant, such that the logic signal output of the comparator uniquely determines which variable is the larger at all times.

Our use of the term is most often in the sense of definition (3).

The comparator circuit can be analyzed in a direct way to determine that the output voltage is a function of the two input voltages [7].

$$V_o = K_2 V_2 - K_1 V_1 \tag{4.7}$$

$$\text{where } K_1 = R_2/R_1 \tag{4.8}$$

$$K_2 = \frac{(R_1 + R_2)R_4}{(R_3 + R_4)R_1}$$

It can be shown that, to be physically realizable, we require that $K_1 + 1 > K_2$.

The ideal model of an analog comparator is shown in Figure 4.3, where the output is given by the expression

$$V_o = \begin{cases} V^+ & \text{if } V_i > V_2 \\ V^- & \text{if } V_i < V_2 \end{cases} \tag{4.9}$$

The practical characteristic of the comparator is shown in Figure 4.3(*c*), where it is noted that the difference voltage must exceed some small value before switching to saturation occurs. This minimum difference is called the *resolution* or *sensitivity* of the comparator. This resolution is a function of the *gain*, which is the slope of the transfer characteristic in moving from negative to positive saturation.

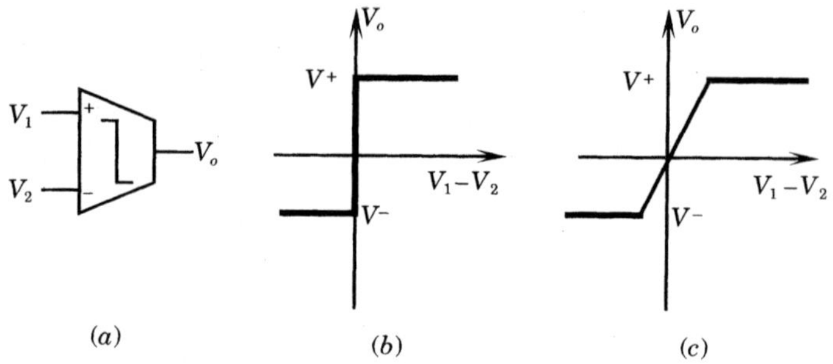

Figure 4.3 Comparator representation and circuit characteristics. (*a*) Symbolic representation. (*b*) Ideal characteristic. (*c*) Practical characteristic.

The comparator is useful in relay circuits. Consider a relay that generates a voltage ramp during the time that a fault condition is observed to exist. If such a ramp is used as the input, the output will remain low until the input ramp reaches the reference level, at which time the output will go high. The reference can be set to correspond to the time or level at which the output should go high, which will cause the protected circuit to trip.

4.3.1.3 The Summer. In many circuits it is necessary to sum several variables, which are represented by voltages. An op amp circuit to perform this type of operation is shown in Figure 4.4.

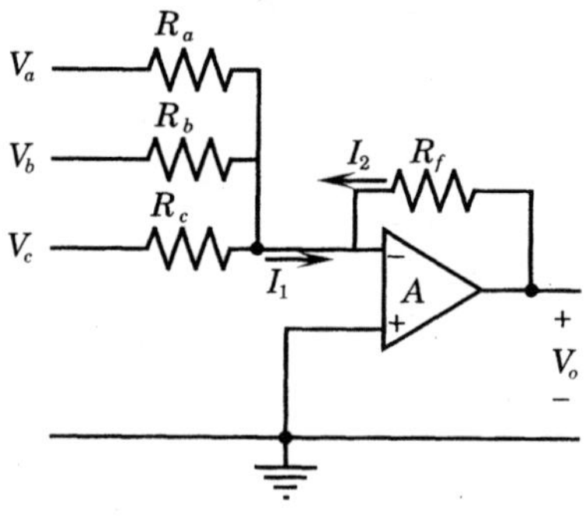

Figure 4.4 The op amp as a summer.

Since the two currents are equal and opposite, we can write

$$\frac{V_a}{R_a} + \frac{V_b}{R_b} + \frac{V_c}{R_c} = -\frac{V_o}{R_f} \tag{4.10}$$

Then

$$V_o = -R_f\left(\frac{V_a}{R_a} + \frac{V_b}{R_b} + \frac{V_c}{R_c}\right) \tag{4.11}$$

The output can be thought of as the weighted sum of the input voltages. However, if the input resistors are set to the same value, the output is simply the sum of the input voltages.

4.3.1.4 The Integrator. The next circuit variation using an op amp is the circuit shown in Figure 4.5(*a*). Here, the feedback resistor has been replaced by a capacitor. The currents

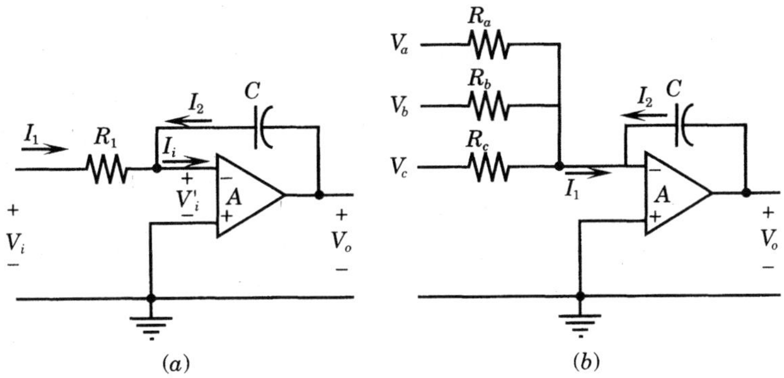

Figure 4.5 The op amp as an integrator. (*a*) The integrator. (*b*) Multiple input integrator.

for this case can be written as follows.

$$I_1 = \frac{V_i}{R_1}$$
$$I_2 = -I_1 = -\frac{V_i}{R_1} \tag{4.12}$$

We can also write the voltage across the capacitor as

$$V_o = \frac{1}{C}\int_0^t I_2\,dt = -\frac{1}{R_1 C}\int_0^t V_i\,dt \tag{4.13}$$

or the output voltage is the integral of the input voltage with a scale factor of $-1/RC$. For example, if we choose $R = 1\text{M}\Omega$ and $C = 1\mu F$, then the scale factor is -1 second.

Figure 4.5(*b*) shows a circuit that is a combination of a summer and an integrator. Here, the output voltage is written as

$$V_o = -\frac{1}{C}\int_0^t \left(\frac{V_a}{R_a} + \frac{V_b}{R_b} + \frac{V_c}{R_c}\right) dt \tag{4.14}$$

If the input resistors are chosen to be of equal value, the integrand is simply the sum of the voltages and the resistance can be factored out.

4.3.1.5 Active Filters. The op amp can also be configured to represent an active filter [8]. The general system configuration is shown in Figure 4.6. The op amp filters can be designed with high Q and adjustable gains and resonant frequencies are readily obtained.

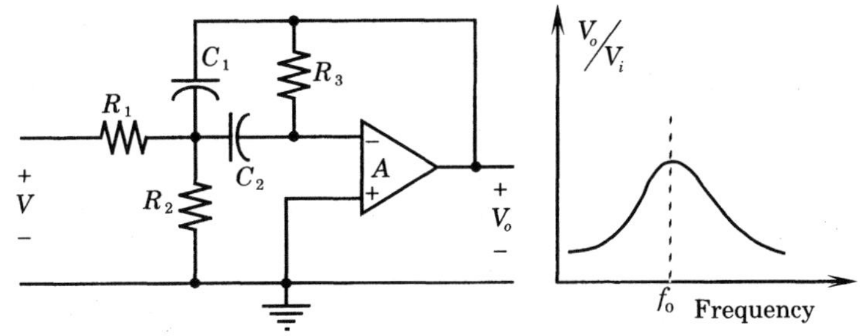

Figure 4.6 A band-pass filter unit.

4.3.1.6 Other Op Amp Applications. There are many other applications that can be designed using op amp circuits, sometimes in cascaded combinations. Examples include zero crossing detectors, light (or dark) detectors, hysteresis comparators, window comparators, non-symmetrical threshold detectors, and many others [9]. Circuits that combine analog and digital devices may also be designed for special purposes. The opportunities are almost without limit.

4.3.2 Digital Logic Circuits

Many of the new protective relays use logic that is primarily or exclusively digital. The use of digital logic circuitry exploded in usage during the 30-year period from 1960 to 1990. The transistorized computer was new in 1960 and flip-flops new devices (see introduction to [10]). In 1962, the first diode-transistor logic (DTL) circuits became available and transistor-transistor logic (TTL) devices were to be introduced a year later. The year 1962 saw the introduction of a new military standard, *MIL-STD 806B*, *Graphic Symbols for Logic Diagrams*, which introduced an entirely new vocabulary and grammar of a symbolic language. In 1973 IEEE published a new standard, *IEEE Std 91-1974*, *Logic Symbols and Diagrams*, which was revised in 1984. In 1986 IEEE published a new standard *IEEE Std 991-1986*, *IEEE Standard for Logic Circuit Diagrams*. Both IEEE 91 and 991 are now adopted as the military standard, and these standards have become the symbolic language of logic devices. The discussion here presents only the very basic symbols and terms of this symbolic language. The interested reader is encouraged to examine the extensive references available on the subject.

The trend to digital protective devices has followed development and availability of logic devices that are cheap and reliable components for the design and construction of decision-making equipment. In the application of digital relays, it is not essential to have the same understanding of the logic components as the digital system designer, but it is, nonetheless, useful to understand the basic diagrams of the equipment and to know how it is intended to work. Indeed, it is useful to be able to compare competitive devices in a meaningful way. This section presents some of the basic logic devices and introduces the symbolic representation that is used in the description of digital relays.

4.3.2.1 Boolean Logic Circuits. Boolean logic circuits and integrated circuit (IC) devices are available in many different types, so the circuit designer can select the appropriate type for a given application. For example, space applications do not have the problem of noise interference that is found in power system protection and many industrial environments. In some applications, such as high-speed digital computers, the speed requirement dictates the use of devices that are free from excessive propagation delays.

There have evolved seven different families of IC logic circuits that are available commercially. These are: resistor-transistor logic (RTL), diode-transistor logic (DTL), high-threshold logic (HTL), transistor-transistor logic (TTL), emitter-coupled logic (ECL), metal-oxide semiconductor (MOS) logic, and complementary metal-oxide (CMOS) logic circuits. Each of these has found applications

The terminal of a logic unit has only two states; *no output*, represented by 0 (zero), and *output*, represented by 1 (one). There are two different logic conventions in use, which indicate the voltage associated with the two logic states. In the *normal* logic, 0 is equivalent to zero voltage, and 1 is equivalent to normal voltage. In the *reverse* logic, 0 is equivalent to normal voltage, and 1 is equivalent to zero voltage. Power system protective devices usually use the *normal* logic convention.

The voltage level is an integral part of the IC design and low voltages are very common, such as 3 to 5 volts. HTL devices employ higher voltages, say in the 15 volt range. Power system protective devices typically employ 20 volts as the normal voltage. This higher voltage is appropriate for the noisy environment of a high voltage substation.

Graphical symbols of logic devices follow two forms, one that employs special graphic symbols for each special type of gate, and another that displays the device as a box with the gate name or symbol printed inside the box [10]. In power systems, both types of symbolic representations are used, with the "box" convention probably being the more common [8].

4.3.2.2 The AND Logic. The AND logic element is shown in Figure 4.7. The circuit shown in Figure 4.7(*a*) as a simple representation of an AND logic. When implemented by resistor and diode elements, the AND gate consists of forward-biased diodes in parallel. If either A or B is low, the output will be held low. Stated another way, both A AND B inputs must have voltages greater than the bias voltage before the output will be high.

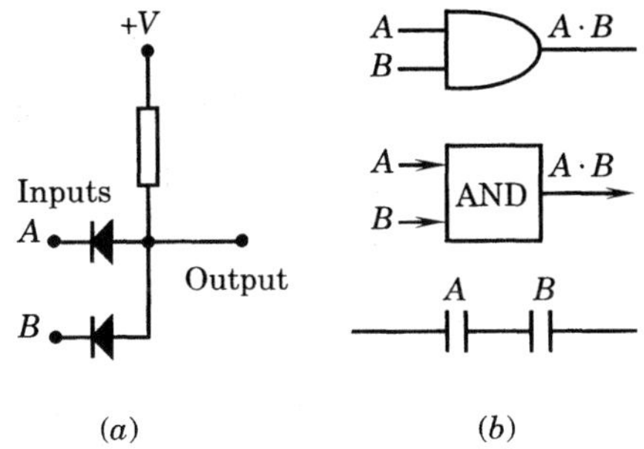

Inputs		Output
A	B	
0	0	0
0	1	0
1	0	0
1	1	1

(*c*)

Figure 4.7 AND logic circuit, symbols, and logic table. (*a*) Circuit. (*b*) Symbols. (*c*) Logic table.

Part (*b*) shows three different symbolic representations of the AND gate. The top figure shows the standard graphic symbol for the AND gate, with rounded sides and a perpendicular straight line at the input. An alternative symbolic representation is a box with the word AND, or & [10], is often used. The electromechanical relay equivalent of the AND logic is two relay contacts in series. Finally, the logic table shows all possible combinations of two input states, taken two at a time, and the corresponding output states. All of these representations may be used to symbolize an AND logic, but the graphic symbol is probably the most common. The logic can be extended to cases with more than two inputs.

There are several variations of the AND logic that are important. Three variations are shown in Figure 4.8, all of which involve negation of one or more signals [8]. Negation is represented in logic diagrams by a small open circle, like (o).

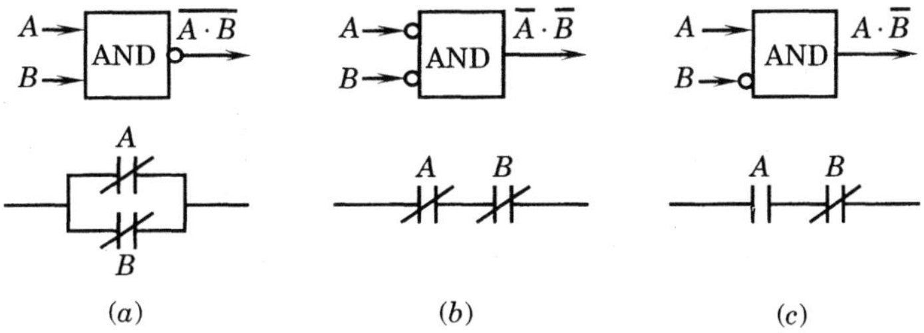

Figure 4.8 Variations of the AND logic. (*a*) Inverse AND. (*b*) Negation AND. (*c*) Mixed AND.

4.3.2.3 The OR logic. The OR logic element is shown in Figure 4.9, which shows a diode-resistor implementation of the OR logic. For this gate, if either A or B or both A and B have a voltage greater than the reference, the output will also have that voltage. Part (*b*) shows the usual graphic symbol for the OR logic, the rectangular box symbol, and the electromechanical relay equivalent. The OR logic table, shown in part (*c*) of the figure, completes the description

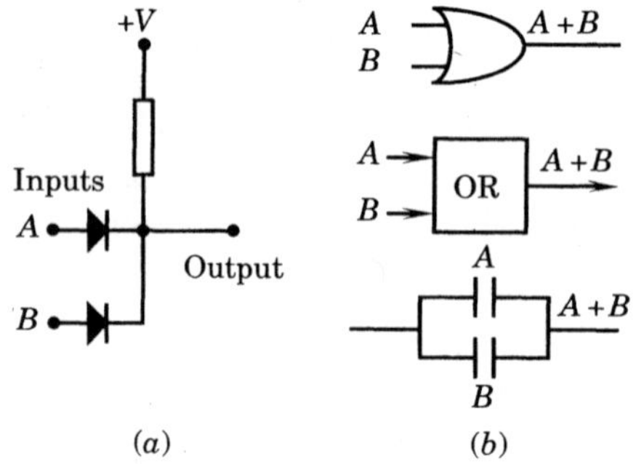

Inputs A	Inputs B	Output
0	0	0
0	1	1
1	0	1
1	1	1

(*c*)

Figure 4.9 OR Logic circuit, symbols, and logic table. (*a*) Circuit. (*b*) Symbols. (*c*) Logic table.

Three variations of the OR logic are shown in Figure 4.10, representing the inverse OR, the negation OR, and a mixed OR logic. The OR logic can be extended to cases where there are more than two input quantities.

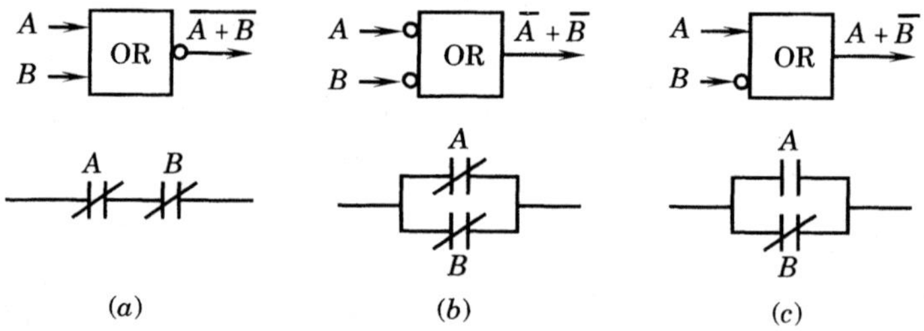

Figure 4.10 Variations of the OR logic. (*a*) Inverse OR. (*b*) Negation OR. (*c*) Mixed OR.

4.3.2.4 The Exclusive OR Logic. An important variation of the OR logic is given a different symbol and has an entirely different meaning. The exclusive OR logic has an output if either of the inputs, but not both inputs, is present. This type of logic is shown in Figure 4.11. Note the slightly different logic symbols from the case shown in Figure 4.9.

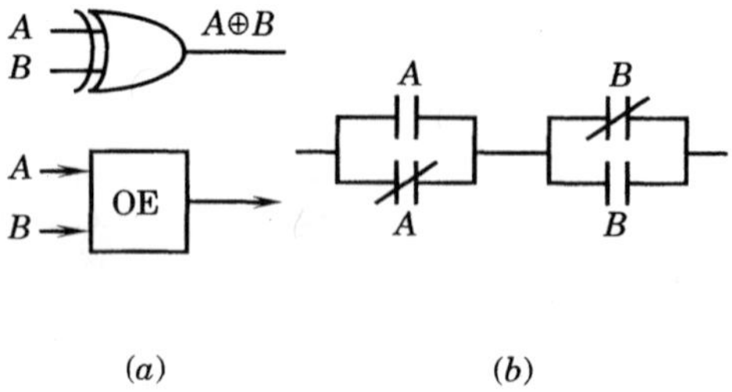

Inputs A	Inputs B	Output
0	0	0
0	1	1
1	0	1
1	1	0

(*c*)

Figure 4.11 Exclusive OR logic symbols and logic table. (*a*) Logic symbols. (*b*) Electromechanical equivalent. (*c*) Logic table.

The exclusive OR is the same as the OR gate, except that it does not provide an output when both A and B inputs are present.

4.3.2.5 The Buffer. Most logic circuits can be thought of as two-port networks, with a port for the input variable and another for the output. If these two-port networks can be designed such that they do not interact, then the circuits can be designed and tested separately from their ultimate utilization in a complex circuit. In some applications, however, the loading effect of one device may affect the neighboring devices. This can often be corrected by inserting a buffer amplifier between the two interacting networks. The buffer amplifier has a high input impedance and a low output impedance, with a gain of unity.

The standard logic symbols for the buffer are shown in Figure 4.12 [10]. The triangle symbol is often used to represent an amplifier, so the buffer symbol needs little explanation. However, the symbols shown in the figure are IEEE standards [10].

(a)

Input	Output
0	0
1	1

(b)

Figure 4.12 The buffer logic symbols. (*a*) Logic symbols. (*b*) Logic table.

4.3.2.6 The NOT or Negation Logic. The NOT or negation logic provides a 0 output if and only if the input is a 1. The negation is often represented by placing a bar over the letter symbol for that variable, as shown in Figure 4.13. The negation logic is often included in the symbol for other types of logic by adding a small circle to the variable that should be negated.

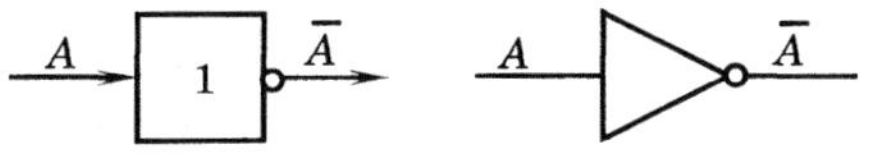

(a)

Input	Output
0	1
1	0

(b)

Figure 4.13 The NOT gate logic symbols and logic table. (*a*) Logic symbols. (*b*) Logic table.

4.3.2.7 The NOR Logic. The NOR logic is equivalent to two normally closed electromechanical contacts in series. The NOR logic is presented in Figure 4.14. The NOR logic standard symbols are not always used by the power system protection community, which tends to favor a symbol with simply the word NOR inside a box. The standard symbol uses an

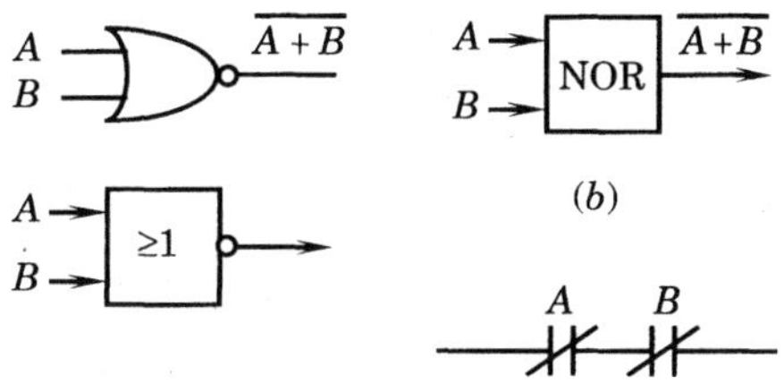

Inputs		Output
A	B	
0	0	1
0	1	0
1	0	0
1	1	0

(a) (c) (d)

Figure 4.14 NOR logic symbols and logic table. (*a*) IEEE standard logic symbols. (*b*) Unofficial symbol in protection. (*c*) Electromechanical equivalent. (*d*) Logic table.

up-slanting triangle on the output, which means that an internal 1 state produces the low (L) or zero level at the output [10]. Note that the output is high if and only if both inputs are low. This logic can be extended to cases with more than two inputs.

4.3.2.8 The NAND Logic. The NAND logic is shown in Figure 4.15. The acronym NAND means "not A and not B" and can be formed from the AND logic by the addition of a NOT circle on the AND symbols. Occasionally, however, the NAND logic is represented by a box with the work NAND, as shown in the figure.

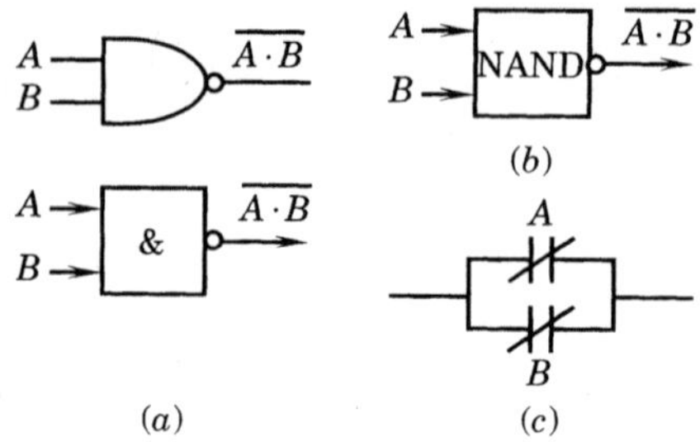

Inputs		Output
A	B	
0	0	1
0	1	0
1	0	0
1	1	0

(d)

Figure 4.15 The NAND logic symbols and logic table. (*a*) IEEE standard logic symbols. (*b*) Unofficial symbol in protection. (*c*) Electromechanical equivalent. (*d*) Logic table.

4.3.2.9 The Time Delay Unit. In protective devices it is often important to carefully time the switching in order to coordinate with other devices or with apparatus characteristics. The IEEE standard logic symbols do not depict a time delay unit, so the protection community has adopted the representation shown in Figure 4.16.

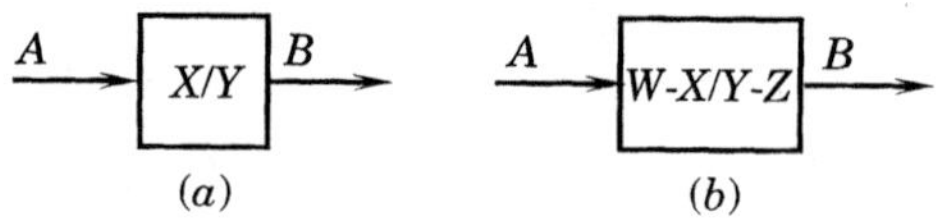

Note: Upper Left Value = On Delay
Lower Right Value = Off Delay

Figure 4.16 Time delay elements. (*a*) Fixed timer. (*b*) Adjustable timer.

For the fixed timer, the X value is the *pickup time* or the time between the input signal first being received until the output signal is generated. The Y value is the *dropout time*, that is, the time following the disappearance of the input that the output signal goes to zero. This is illustrated in Figure 4.17. The adjustable timer can be shown in a similar way. Any of the timer parameters can be set to zero, indicating no intentional delay.

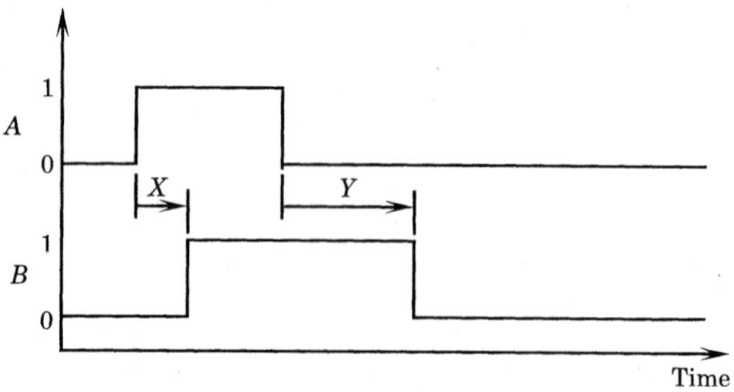

Figure 4.17 Example of time delay control with the timer unit.

4.3.2.10 The Flip-Flop. There are two basic types of components in a sequential digital circuit. These are the logic gates and the binary *storage* or *memory* elements [7]. The most widely used binary storage element is called the *flip-flop*, which is available in many different IC forms. There are other types of memory elements, such as magnetic core, magnetic disk, and IC memory that are used for mass storage. The flip-flop is a different type of memory element.

The flip-flop usually has two or more input terminals and two output terminals, as shown symbolically in Figure 4.18. The logic level of output terminals are the complement of each other, i.e., if one output is 1, the other output is 0. The *state* of the flip-flop is determined by the state of these output terminals. Specifically, the state of the flip-flop is said to be a logical-1 state if the Q output is high (or 1) and is in a logical-0 state if the Q output is low (or 0). Once a given state is established, the flip-flop remains in that state as long as there is no change in the input logic levels.

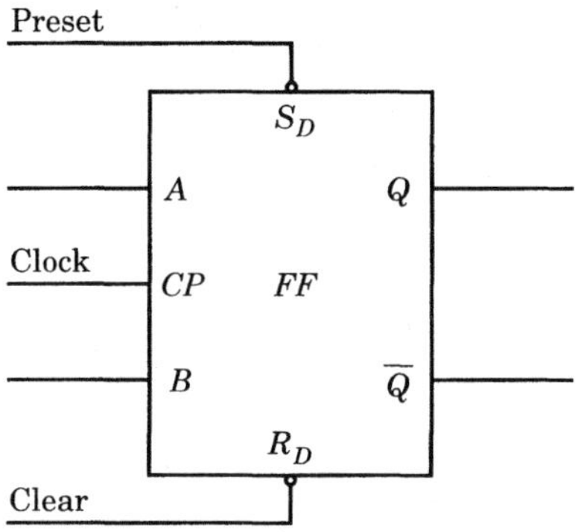

Figure 4.18 Symbolic notation for a typical flip-flop.

The inputs labeled A and B are called *synchronous* or *data inputs* and they work in conjunction with the clock, which is labeled *CP*. The clock input must be at a specific state, such as high, in order for the data inputs to have any effect on the state of the flip-flop.

The *Preset* and *Clear* inputs to the flip-flop are asynchronous and operate independently of the other inputs. These inputs are normally in their high state. A low input at the clear terminal *clears* or *resets* the flip-flop to a logical-0 state, and a low input at the preset terminal *sets* the flip-flop to a logical-1 state, with these operations being independent of the clock. These inputs are often used to preset the flip-flop to a known state, before allowing it to accept data inputs.

There are several types of flip-flops that are used in sequential logic circuits. These are the R-S flip-flop, the J-K flip-flop, the *Toggle* flip-flop, and the *Delay* flip-flop. These devices have different logic transition tables and are used for different tasks. A detailed analysis of these devices is beyond the scope of this book. However, it is noted that flip-flops, or combinations of flip-flops, are useful in the design of many different functional devices, such as storage registers, shift registers, counters, arithmetic units, code converters, and many other devices. The flip-flop is a basic building block of a sequential logic system and is used in conjunction with logic elements of various kinds in order to meet a given specification for a digital system.

4.3.2.11 Sampling of Analog Signals. An electromechanical protective system operates using continuous signals from the power system, obtained through current and voltage transducers at the point of application. These measured quantities are exact replicas of the high-voltage system behavior, except for transducer error. These signals are applied to analog devices that, in the case of electromechanical relays, are tiny machines that respond to the applied signals by producing forces or torques in response to the applied signals. These forces or torques work against preset threshold springs or other restraints to determine if the measured condition warrants tripping of the protected component.

Electronic analog protective systems operate in a different way, but still receive the continuous input signals. These systems perform electronic signal conditioning that is designed to make a binary decision based on the outcome of the signal processing. This processing is usually performed through a series of operational amplifiers and flip-flops to emulate the same type of performance as electromagnetic relays. For example, a comparator can be designed to behave like a distance relay.

Digital protective systems operate in an entirely different manner. The digital system is provided with a series of samples of the measured signal, spaced timewise according to a predetermined sampling rate. The digital protective system then evaluates these sampled quantities using software that is designed to determine the state of the power system condition based on a limited set of discrete observations. This is a fundamentally different type of analysis, and it uses a different mathematical structure, based on discrete-time system theory.

Consider the arbitrary sampled signal shown in Figure 4.19. The signal is represented by a sequence of numbers, represented in the figure by the dots, where each dot represents the measured value of that signal at that instant. The nth number in this sequence is denoted $x(n)$, is formally written as

$$x = \{x(n)\} \qquad -\infty < n < \infty \tag{4.15}$$

and is usually referred to as "the sequence $x(n)$" [11]. It is important to recognize that the sequence (4.15) is only defined for integer values of n. It is not correct to consider that the value of x for nonintegral values of n as being zero; $x(n)$ is simply undefined for nonintegral values of n. Digital relays are designed to perform a particular kind of analysis on numerical sequences and to make a trip or no-trip decision on the basis of the analysis.

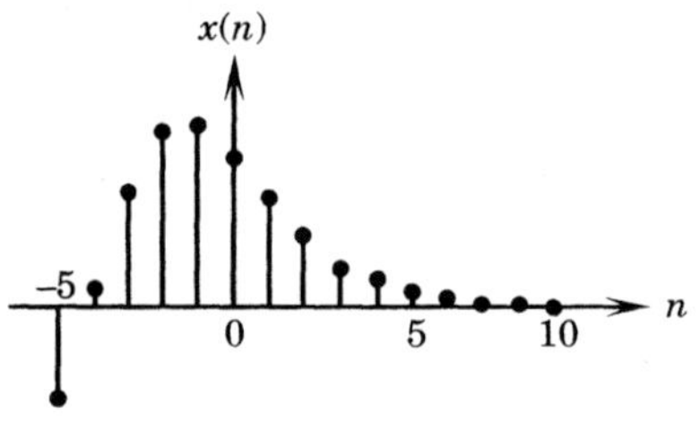

Figure 4.19 An arbitrary sequence.

4.3.2.12 The Analog-to-Digital (A/D) Converter. The logic devices discussed in this section work with signals that can exist in only two states, which we usually call 1 and 0. In this environment, a value is either on or off, and there is no interest in any other values for that signal. All digital devices work with signals having this property.

In many problems, however, there may be a mixture of analog and digital signals. This is certainly true in power system protection, where the raw input quantities are analog signals that come from transducers, which are attached to the high voltage circuit. These analog signals must be converted to a digital representation, similar to Figure 4.19, before being introduced into the digital relay for analysis. This conversion is performed by a special device called the *analog-to-digital converter*, which is often abbreviated as the A/D converter.

A detailed explanation of the A/D converter is beyond the scope of this book, but an appreciation of the process can be gained by considering a very simple approach. Consider the arrangement shown in Figure 4.20, where an analog input V_A is to be converted to digital form. Let us assume that the digital precision required is not great, and that only four discrete levels or values will suffice to characterize the analog voltage.

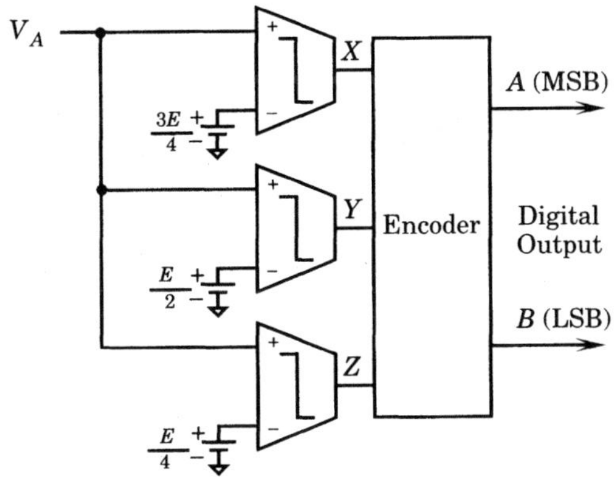

Figure 4.20 A/D converter based on the direct conversion method [7].

In the direct conversion method, the analog input is simultaneously compared against three reference values, which are chosen to be $0.75E$, $0.5E$, and $0.25E$. The states of the comparator output levels, X, Y, and Z depend on the input analog voltage. We can summarize these values as follows.

$$Z = \begin{cases} V^- & \text{if } V_A < E/4 \\ V^+ & \text{if } V_A \geq E/4 \end{cases}$$
$$Y = \begin{cases} V^- & \text{if } V_A < E/2 \\ V^+ & \text{otherwise} \end{cases} \tag{4.16}$$
$$X = \begin{cases} V^- & \text{if } V_A < 3E/4 \\ V^+ & \text{if } V_A \geq 3E/4 \end{cases}$$

The relationship between the comparator outputs and the analog input voltage is shown in Table 4.1.

TABLE 4.1 Comparator Outputs

Range of V_A	X	Y	Z
0.75 E to E	1	1	1
0.5 E to 0.75 E	0	1	1
0.25 E to 0.5 E	0	0	1
0 to 0.25 E	0	0	0

Note that three comparators are required in order to divide the analog input voltage into four distinct levels, which are called *quantum* levels. In this case, the quantum level is $E/4$. In the encoder, two bits is adequate to code these four levels. The encoder must be designed to convert the outputs of the comparators into this binary number.

The encoder outputs are labeled A and B, where A denotes the most significant bit (MSB) and B denotes the least significant bit (LSB). A truth table is constructed to relate A and B to the comparator outputs X, Y, and Z. This is given in Table 4.2.

Mathematically, we can write the encoder output as follows.

$$\begin{aligned} A &= \bar{X}YZ + XYZ = YZ \\ B &= \bar{X}\bar{Y}Z + XYZ \end{aligned} \tag{4.17}$$

Mitra [7] gives a realization of this logic, as shown in Figure 4.21.

TABLE 4.2 A/D Truth Table

X	*Y*	*Z*	*A*	*B*
0	0	0	0	0
0	0	1	0	1
0	1	1	1	0
1	1	1	1	1

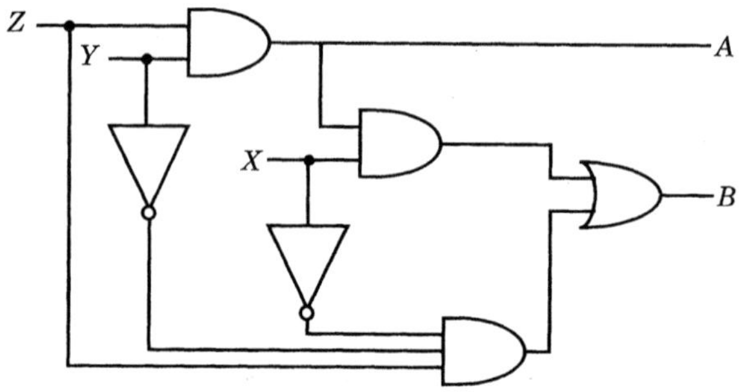

Figure 4.21 A realization of the direct method encoder [7].

The direct method is fast, since the switching time is essentially that of the comparators plus the propagation times of the gates. However, this type of A/D converter usually requires many comparators. For n bits, requiring 2^n levels, this method requires $2^n - 1$ comparators. For example, if $n = 10$, a total of 1,023 comparators would be required. This means that this method is used only where high speed is required and low resolution is acceptable. Commercial A/D converters are available in many different configurations to meet almost any requirement. These are usually packaged as integrated circuits, and the system designer can often find a converter that meets any practical requirement.

4.4 ANALOG RELAY LOGIC

The previous section has presented many of the analog logic functions that are used in the design of protective systems. The purpose of this section is to examine some electronic relay designs that use primarily analog logic.

4.4.1 An Instantaneous Overcurrent Relay

The instantaneous overcurrent relay is a good example of the application of using op amp circuits to practical protective relaying. One analog circuit design for this type of relay is shown in Figure 4.22 [8].

The input sine wave is first filtered to remove any high-frequency components, since the relay is designed to trip only due to fundamental frequency overcurrent conditions. The filtered signal is amplified and then rectified. The integral of the rectified wave is compared to a threshold setting in a level detector, which gives a trip or no-trip output signal.

4.4.2 Phase Comparison Distance Relay

One type of distance relay is designed to pick up when the ratio of voltage and current, or the impedance seen by the relay, falls inside a specified area in the complex R-X plane. Figure 4.23 shows a typical mho relay characteristic. The impedance Z_r represents the diameter

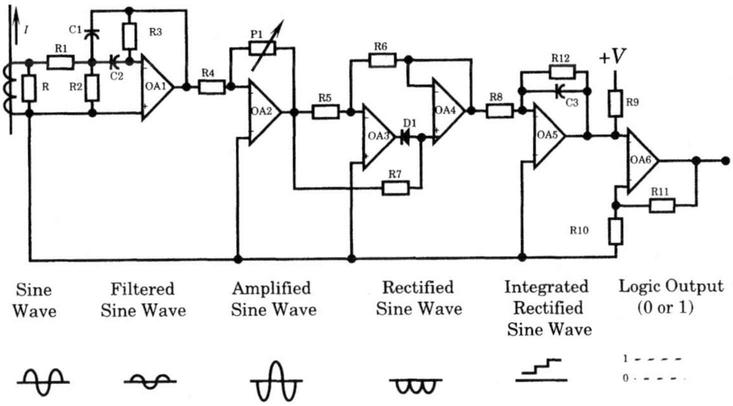

Figure 4.22 An instantaneous overcurrent unit [8].

of the circle in the complex R-X plane. If the impedance seen by the relay is Z, and that impedance is exactly on the relay circular threshold, the angle ψ is exactly 90°. This is illustrated by two different impedances in the figure. Clearly, if the impedance seen by the relay is inside the circle, this angle ψ will be greater than 90°. Therefore, tripping can be determined by comparing the phase angle of the impedance and the difference between this impedance and the diagonal of the circle, that is,

$$\psi \geq \text{ang}Z - \text{ang}(Z - Z_r) \tag{4.18}$$

When the impedance seen by the relay is inside the threshold circle, the angle ψ is greater than 90° and the relay should be caused to trip.

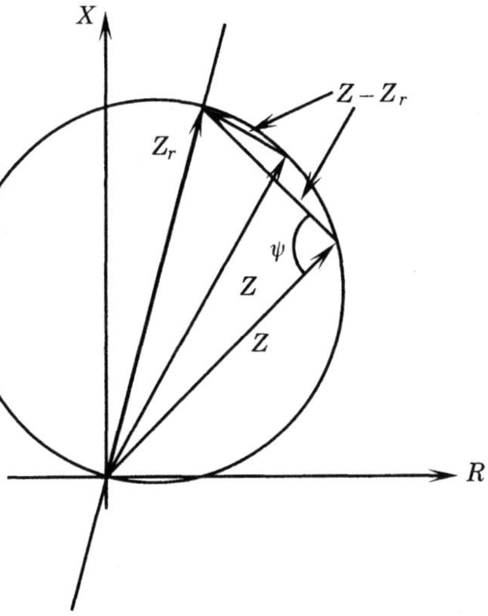

Figure 4.23 The Mho characteristic.

One way of performing this logic is shown in Figure 4.24 [12]. This logic uses op amps in saturation to produce a square wave corresponding to the input sinusoidal values. The square waves of the two inputs are compared in a coincidence detector, which produces a negative signal during the time both square waves agree in polarity, and a positive signal for noncoincidence. This results in a square wave of double frequency, with equal positive and

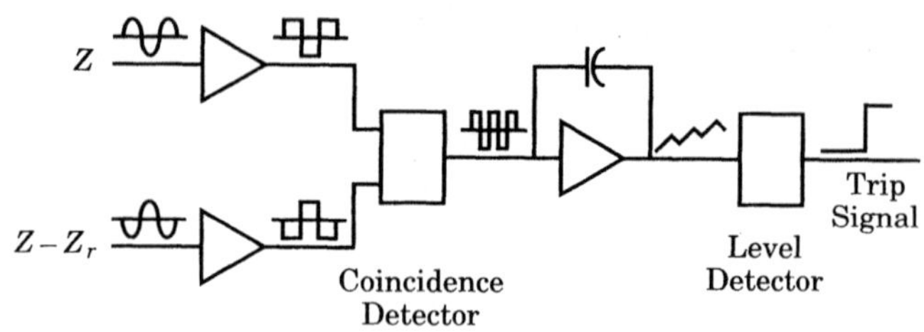

Figure 4.24 Phase comparator to obtain a Mho characteristic.

negative pulses with the input waves have a phase difference of ±90, which corresponds to the relay threshold. As the impedance seen falls inside the threshold circle, the coincidence is biased, such that the integrator produces a growing offset that the level detector will eventually detect, causing a trip.

4.4.3 A Directional Comparison Pilot Relay

Another example of analog logic in relay design is provided by a modern directional comparison relay system that uses pilot communications as an integral part of the trip logic [13]. The relay is a general purpose transmission protection system that uses a directional positive sequence distance relay and a negative sequence directional overcurrent relay, including positive, negative, and zero sequence level detectors, or overcurrent functions. The logic diagram for the positive sequence directional comparison is shown in Figure 4.25.

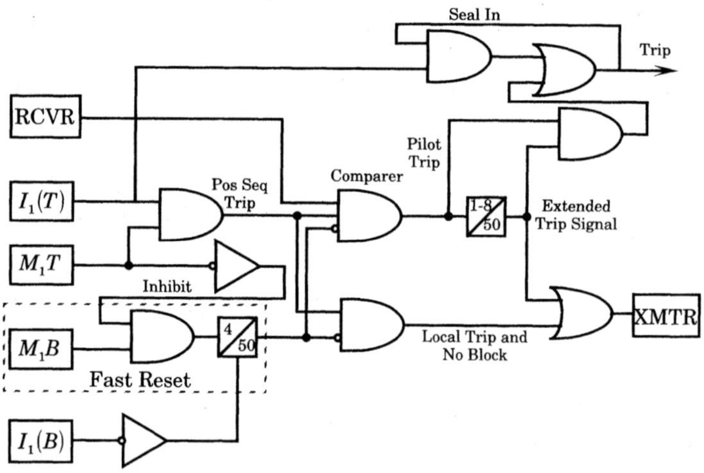

Figure 4.25 Positive sequence directional comparison logic [13].

The positive sequence trip function, M_1T, is supervised by the positive sequence trip current $I_1(T)$. It operates through an AND gate to key the pilot trip channel, and through a second AND gate to produce a local trip signal. Pilot tripping is supervised by a timer for added security, which creates an extended trip signal, which ensures that both the pilot and local trip signals are present. The M_1B is a blocking function, which is timed by the 4/50 timer. It blocks local tripping by inhibiting the local trip AND gate and also blocks transmitter keying through its comparer gate. The M_1B blocking function is supervised by a positive sequence current through a fast reset connection. This means that the 4/50 timer is blocked or reset if the $I_1(B)$ is not present. The M_1B is also supervised, or inhibited, by the M_1T through a NOT gate, which ensures that the M_1B timer is present only if the M_1T has no output.

4.4.4 Conclusions Regarding Solid-State Analog Logic

The circuits described above are typical of the solid-state logic that is used in power system protection. It is not possible to discuss all of the many devices that have been designed, nor is that necessary. The protection engineer develops a skill for reading and understanding these diagrams. The logic symbols become a new language that is both descriptive and powerful, such that complex circuits can be described almost without the need for words.

4.5 DIGITAL RELAY LOGIC

The logic of an analog relay is that of a comparator. The analog relay uses a transformed replica of the actual system voltages and currents to operate a relaying device, which is often like a small motor that generates a small torque or force in response to the system voltages and currents. In many cases, a threshold of operation is set by the adjustment of a mechanical restraint and the relay device pushes or pulls or twists against that restraint to determine whether tripping is to be performed. Electronic analog relays are more sophisticated, but these devices are still analog machines and use the full spectrum of the input currents and voltages, which are processed in various ways to electronically reach a trip decision.

The digital relay also operates as a comparator, but the process is entirely different. First of all, the digital relay does not record the analog signal, but only samples of the signal, which are spaced in time. Therefore, the relay does not deal with continuous signals. Indeed, the mathematics of continuous functions does not apply in this digital device. Instead, the mathematics of discrete signal processing is used [4]. The relay is programmed to apply various forms of digital signal processing algorithms to the observed samples and, based on the results of these computations, the decision to trip is made. In Chapter 3, a discussion of the architecture of a digital protective system is presented. Here, we examine the mathematical concepts that make up the digital logic. Chapter 3 deals with the hardware system, but here we consider the software problems of digital relaying. The presentation is brief and omits many of the details that the serious student of digital relaying must confront. For further study on the subject, the interested reader is referred to the many excellent references, for example, [5], [8], [12], [15].

Before examining the mathematics of digital relaying, a few observations can be made concerning the digital process. There are significant advantages in digital signal processing. One advantage is due to the hardware and the standards that have evolved for digital systems. These make it possible to interconnect digital systems and to exchange data between devices of different manufacturers that are long distances apart, which is often a condition in power systems. Digital communication conforms to a kind of universal language, which can be used to advantage in protective systems. Another advantage is the larger inventory of software for digital signal processing that is useful in relaying as in many other fields. For example, digital filtering programs have been used to extract phasor components from a sampled signal, with the filter rejecting the harmonics. This can be applied to transformer protection, for example, to avoid the false trips that often occur with analog relays due to the transient inrush current [17].

Another advantage of digital systems is the low cost of memory. The stored data provide very useful information to the protection engineer. Whereas the analog relay is unable to remember even its last operation, digital relays can be programmed to display, on command, their last 10 or 20 operations, giving complete information about each event [17].

Digital systems provide a convenient separation between the hardware and the software that is also beneficial. Therefore, identical hardware modules can be programmed to perform different functions, or the same software can be used in different hardware systems, with a substantial savings in development cost.

4.5.1 Digital Signal Processing

In order to design a completely digital protective system, the first stage in the development must address processing of the incoming signals into digital sequences. Previous sections have introduced the analog-to-digital converter, and the concept of representing a signal as a sequence of numbers has also been presented. Now, we carry this concept a step further, to provide a more comprehensive description of the mathematics required to analyze sequences of data.

An extensive mathematics of discrete functions has been developed so that many of the operations that are common in continuous functions are also possible with discrete functions. Functions similar, but distinctly different from continuous functions have been defined that can be used to perform mathematical operations quite similar to those in continuum mathematics. Some of these functions are shown in Figure 4.26, which will be immediately recognized by the engineer or mathematician by their similarity to the unit impulse, unit step, and sine function of continuous functions. The development of these mathematical techniques for discrete variables

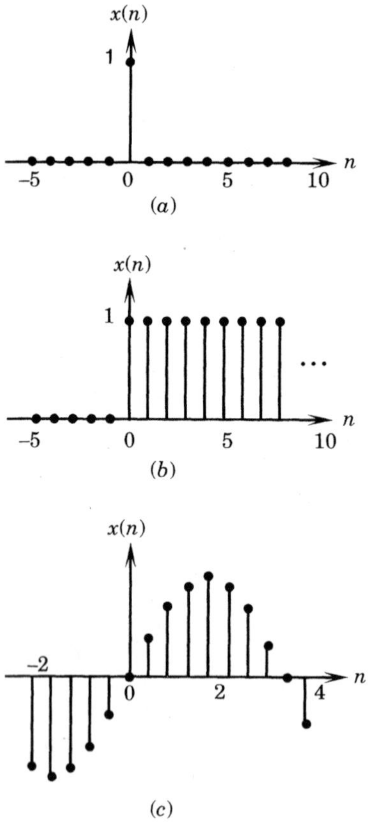

Figure 4.26 Some common discrete sequences. (*a*) A unit sample sequence. (*b*) A unit step sequence. (*c*) A sine sequence.

is beyond the scope of this book, but is highly recommended for those involved in digital signal processing and in digital relaying.

4.5.1.1 Linear Transformations. A *system* is defined as a unique transformation T that maps the input sequence $x(n)$ into an output sequence $y(n)$. This can be expressed mathematically as

$$y(n) = T[x(n)] \tag{4.19}$$

The transformation T may be thought of as a transfer function, which is depicted graphically as shown in Figure 4.27.

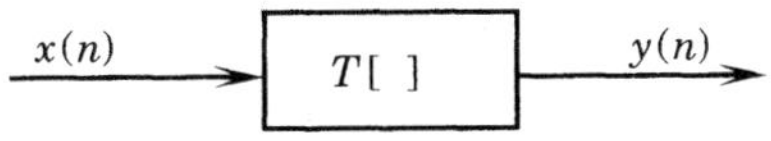

Figure 4.27 The transformation mapping x into y.

One class of functions that is useful and interesting to examine is the class of linear, shift-invariant systems. Linear systems are defined as those that conform to the principal of superposition, that is, if y_1 and y_2 are the responses to inputs x_1 and x_2, respectively, then the system is linear if and only if

$$T[ax_1(n) + bx_2(n)] = ay_1(n) + by_2(n) \tag{4.20}$$

for arbitrary constants a and b.

It is also possible to show that an arbitrary sequence can be represented as the sum of scaled, delayed unit impulse samples, similar to Figure 4.26(a), but scaled appropriately. That is, we can write

$$x(n) = \sum_{k=-\infty}^{\infty} x(k)\delta(n-k) \tag{4.21}$$

Then, the transfer function (4.20) can be expressed as

$$y(n) = \sum_{k=-\infty}^{\infty} x(k)T[\delta(n-k)] = \sum_{k=-\infty}^{\infty} x(k)h_k(n) \tag{4.22}$$

If only linearity is imposed, then $h_k(n)$ will depend on both k and n, which limits the utility of (4.22). However, if the system is also shift-invariant a more useful result is obtained.

A shift-invariant system has the property that if $y(n)$ is the response to $x(n)$, then $y(n-k)$ is the response to $x(n-k)$, where k is a positive or negative integer. If the parameter n refers to time, then shift-invariance corresponds to time invariance in continuous functions. If the system is shift-invariant, this implies that if $h(n)$ is the response to $\delta(n)$, then the response to the input $\delta(n-k)$ is $h(n-k)$. In this case we can write (4.22) as

$$y(n) = \sum_{k=-\infty}^{\infty} x(k)h(n-k) \tag{4.23}$$

Thus, any linear shift-invariant system is completely characterized by its unit sample response. This result is usually called the *convolution sum* and can be written using the familiar convolution notation

$$y(n) = x(n)^*h(n) \tag{4.24}$$

Linear shift-invariant systems can be assembled in cascade to form linear shift-invariant systems with a unit sample response that is the convolution of the two-unit sample responses.

The similarity to linear, time-invariant continuous systems is evident.

4.5.1.2 Frequency Response. A linear shift-invariant system has a fundamental property that the response to a sinusoidal input is a sinusoidal response of the same frequency, with the amplitude and phase determined by the system characteristics. This is a very important characteristic, and it is one that makes the representation in the frequency domain so useful. For a discrete-time system, let the input sequence be the exponential function.

$$x(n) = \sum_{k=-\infty}^{\infty} h(k)e^{j\omega(n-k)} = e^{j\omega n} \sum_{k=-\infty}^{\infty} h(k)e^{-j\omega k} \tag{4.25}$$

Now, we define the *frequency response* function of the system whose unit-sample response is $h(n)$ as follows.

$$H(j\omega) = \sum_{k=-\infty}^{\infty} h(k)e^{-j\omega k} \tag{4.26}$$

Then we can write

$$y(n) = H(j\omega)e^{j\omega n} \tag{4.27}$$

Note that $H(j\omega)$ is a complex function, which can be expressed in terms of its real and imaginary parts. A sinusoid can also be expressed as a combination of complex exponentials, hence the frequency response is related to the system response to a sinusoidal input.

Since $H(j\omega)$ is periodic with frequency ω, it can also be represented by a Fourier series. Then we can write

$$h(n) = \frac{1}{2\pi} \int_{-\pi}^{\pi} H(e^{j\omega})e^{j\omega n}\, d\omega \tag{4.28}$$

where

$$H(e^{j\omega}) = \sum_{n=-\infty}^{\infty} h(n)e^{-j\omega n} \tag{4.29}$$

Equation (4.28) represents the sequence $h(n)$ as the superposition of exponentials whose complex amplitudes are given by (4.29). We call (4.28) and (4.29) a "Fourier transform pair," for the sequence $h(n)$ and this representation exists in a mathematical sense, if the sequence converges. This representation can be applied to any sequence if the series (4.29) converges.

4.5.1.3 Periodic Sequences. Consider a sequence $x(n)$ that is periodic of period N. For this special case we can write

$$x(n) = x(n + kN) \tag{4.30}$$

because of the periodic nature of the function. This sequence can be represented by a Fourier series with frequencies of the terms in the series being integer multiples of the fundamental frequency $2\pi/N$ of the periodic sequence. The Fourier series representation of this periodic sequence has only N complex exponentials of the form

$$x(n) = \frac{1}{N} \sum_{k=0}^{N-1} X(k)e^{j(2\pi/N)nk} \tag{4.31}$$

where the coefficients $X(k)$ are given by

$$X(k) = \sum_{n=0}^{N-1} x(n)e^{-j(2\pi/N)nk} \tag{4.32}$$

The equations (4.31) and (4.32) are considered a Fourier transform pair. The pair is usually called a discrete Fourier series (DFS), which represents a periodic function [11]. For convenience, we simplify the notation, as follows. Let

$$W_N = e^{-j(2\pi/N)} \tag{4.33}$$

Then the DFS transform pair can be written as

$$X(k) = \sum_{n=0}^{N-1} x(n) W_N^{kn} \tag{4.34}$$

$$x(n) = \frac{1}{N} \sum_{k=0}^{N-1} X(k) W_N^{-kn} \tag{4.35}$$

Note the duality of the transform pair. The representation of the periodic sequence in the time and frequency domain are, essentially, equivalent.

The DFS is important because it can be used as a method of representing finite duration sequences. Suppose that a finite duration sequence is obtained from a field measurement. This sequence can be replicated to form a periodic sequence, which can be analyzed using the transform pair (4.34–35). In fact, only the one period is required to do this. The unique frequency characteristics of the finite duration sequence will be obtained in this manner. The Fourier representation obtained in this manner is called the discrete Fourier transform (FDT). One can readily appreciate the importance of this concept. Indeed, there is *never* adequate time to measure a function as an infinite sequence. As a practical matter, every measurement is of finite duration. The properties of the DFT are described in the literature [11] and will not be repeated here. We simply note that the subject has been thoroughly studied by many investigators, and it now forms an important part of digital signal analysis, and in the design of digital signal processing algorithms.

4.5.1.4 The Fast Fourier Transform. Because of the importance of the DFT in processing of discrete sequences, there has been considerable interest in devising methods of computation that are fast and accurate. Most approaches to improving the speed of computation make use of the fact that the complex variable W_N has symmetry of the cosine and sine terms and the fact that the functions are periodic. The most successful methods to date are based on a method introduced in 1965 by Cooley and Tukey [14]. This caused an intense interest in devising new computational methods that are generally called fast Fourier transform (FFT) algorithms. All of these algorithms are based on the concept of decomposing the computation of the DFT of a sequence of length N into successively smaller discrete Fourier transforms. Two general types of decimation are used; decimation-in-time methods and decimation-in-frequency methods. Most of these methods require that N be a power of 2. Many computer programs have been written to perform the FFT signal processing. The methods of programming an FFT algorithm is beyond the scope of this book, but many commercial methods are available for application.

This method of signal analysis is presented here because it is one method of rapidly analyzing finite sequences of information, such as that obtained by a digital relay in its observation of an abnormal signal over a finite time period. Many types of power system disturbances, such as faults, cause a sudden change in the currents and voltages. This means that the current and voltage waveforms immediately after the fault are rich in harmonic content of both supersynchronous and subsynchronous nature. High-frequency components result from the traveling waves on the transmission system, and these waves have been used as the basis for protective relays. Low-frequency components of the signal result from the electromechanical oscillations that result from the disturbance, which are important to detect in certain types of protective systems. The FFT provides a method of extracting the frequency content of a complex wave. This frequency "signature" is often useful in determining the nature of the original disturbance and may be used as a variable in the trip logic.

It is important to note that continuous mathematics is not applicable to the problem at hand, since the information is entirely in the form of sequences of numbers, and is defined only at the instants of recorded measurement.

4.5.2 The Data Window Method

Many of the digital relay methods that have been proposed make use of the concept of a moving data window [15]. The following notation will be used to describe this approach [15].

$y(t)$ = instantaneous value of an ac waveform
y_k = the kth sample value of $y(t)$
ω_o = the fundamental power system radian frequency
Δt = the fixed interval between samples
θ = the fundamental angle between samples, i.e., $\theta = \omega_o \Delta t$

Suppose that we assume the form of the continuous waveform to be

$$y(t) = Y_c \cos \omega_o t + Y_s \sin \omega_o t \tag{4.36}$$

where the coefficients on the right-hand side are real constants. Now assume that three samples are taken at $-\Delta t$, 0, and $+\Delta t$. We can write the values of y_k as

$$\begin{aligned} y_{-1} &= y(-\Delta t) \\ y_0 &= y(0) \\ y_1 &= y(\Delta t) \end{aligned} \tag{4.37}$$

Then the samples can be related to the coefficients by

$$\begin{bmatrix} y_{-1} \\ y_0 \\ y_1 \end{bmatrix} = \begin{bmatrix} \cos\theta & -\sin\theta \\ 1 & 0 \\ \cos\theta & \sin\theta \end{bmatrix} \begin{bmatrix} Y_c \\ Y_s \end{bmatrix} \tag{4.38}$$

Then one looks for a solution of (4.38) for the variables Y_c and Y_s.

The least square solution to (4.38) is given by [15]

$$\hat{Y}_c^{(k)} = \frac{y_{k+1}\cos\theta + y_{k-1}\sin\theta}{1 + 2\cos^2\theta} \tag{4.39}$$

$$\hat{Y}_s^{(k)} = \frac{y_{k+1} - y_{k-1}}{2\sin\theta} \tag{4.40}$$

based only on samples y_{k-1} and y_{k+1}.

The approach of (4.38) uses three terms as a means of filtering out higher frequency harmonics or random perturbations in the solution of (4.36). We say that the algorithm described by this process has a "data window" of three samples. As a new sample becomes available, the oldest of the original sample set is discarded and the new sample is added to the set. In this way, each sample is used in three calculations, once as the parameter y_{k-1}, once as y_k, and once as y_{k+1}.

The moving data window is shown in Figure 4.28. Note that each window contains exactly three samples of data. At the point of the disturbance, window W1 has one prefault sample and two postfault samples.

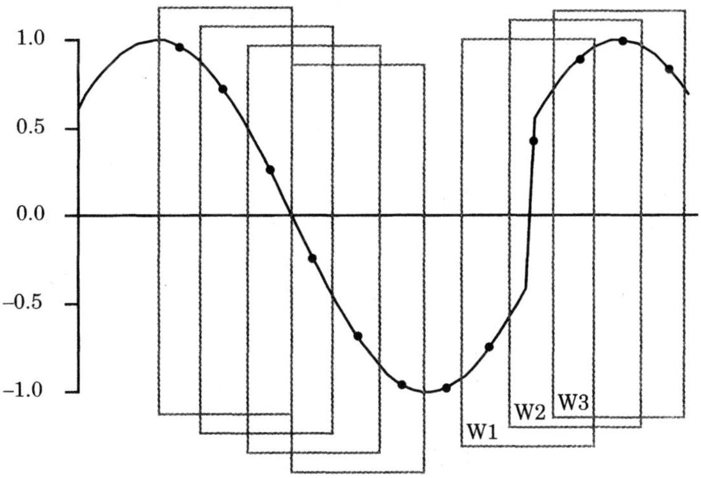

Figure 4.28 A moving three-sample window during a disturbance.

Any window that contains three samples of prefault data will produce the correct phasors using (4.39) and (4.40). Windows, such as W1, however, contain both prefault and postfault data. These windows will not be fitted to a pure sinusoid and the values computed by the process described above have no meaning.

Using the windowing process, the computer must complete the evaluation in the time Δt, which sets a specification for computer speed that limits the complexity of the computation. For example, if there are 12 samples per cycle in a 60 hertz system, then $\Delta t = 1.3889$ ms. Most algorithms use sampling rates of from four to 64 samples per cycle [15]. High sampling rates require very fast computation.

The length of the data window is another important variable in this process. Since there is a problem associated with the windows that contain both pre-and postfault information, some processes attempt to determine when this transition is complete. Longer windows, giving longer sequences of sampled data require a longer time to clear the time period during which windowed sequences contain a mixture of pre-and postfault data. Faster decisions can be made using short windows and short data sequences. This means that there is an inverse relationship between relay speed and accuracy.

4.5.3 The Phasor Method

Another method that has been used for digital relay logic is based on the construction of phasors from the fundamental frequency measurements at the relay [16], [17]. If measurements are taken at a rate corresponding to four measurements per cycle of the fundamental frequency waves, then the sampling rate is $4f_o$, where f_o, is the fundamental frequency of the power system. For 60 hertz systems, this gives a sampling rate of only 240 hertz, which is easily achieved using existing hardware. The key to this method of sampling is to convert these measurements into phasor quantities. This is done by combining the values of successive samples, with the most recent sample being assigned as the real part of the phasor, and the previous sample being the imaginary part. An example is shown in Figure 4.29, where fictitious values of current and voltage are shown immediately after the occurrence of a fault. For the case shown, the voltage is depressed due to the fault and the current is initially quite large, with significant dc offset, then decays exponentially in both the ac and dc components. Sampling

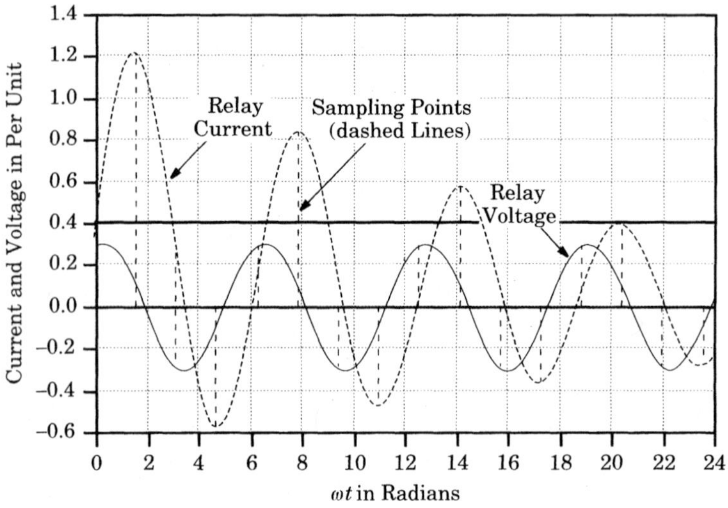

Figure 4.29 Example of quarter-cycle phasor measurements.

instants are shown as the vertical dashed lines, where values of the current and voltage waves are sampled and stored in computer memory.

To illustrate the phasor method, we can tabulate the data as follows. Let V_P be the real part of the phasor and V_Q the imaginary part, with a similar notation for the current. Then we tabulate the following records of quarter-cycle measurements of the voltage and current.

From Table 4.3, we see that the nth sample is always assigned as the real component of the phasor and sample $n - 1$ is always taken as the imaginary part. Since the samples are exactly one-quarter cycle apart, these measurements will be exactly the same as the *instantaneous* phasor values, assuming that the waveform is of constant magnitude and of constant frequency.

TABLE 4.3 Assignment of Measured Quantities to Phasor Components [1]

Sample Number	Real Part of *V*	Imaginary Part of *V*	Real Part of *I*	Imaginary Part of *I*
1	V_{P1}	—	I_{P1}	—
2	V_{P2}	V_{P1}	I_{P2}	I_{P1}
3	V_{P3}	V_{P2}	I_{P3}	I_{P2}
4	V_{P4}	V_{P3}	I_{P4}	I_{P3}
—				

We immediately recognize at least two potential problems with this form of measurement that require investigation; the assumption regarding constant frequency and of constant amplitude. At the onset of a fault, it is known that high-frequency oscillations are usually generated at the time of discontinuity. These high-frequency components usually damp rapidly, but relay designers often provide low-pass filters to remove these high-frequency oscillations from the relay quantities, since it is the fundamental frequency information that is most important. Let us assume that filtering removes all high-frequency components of the wave, thereby satisfying the concern that the observations be those of fundamental frequency wave. A passive, low-pass filter with cutoff frequency of 84 hertz has been shown to work effectively [16].

The second concern is that of constant amplitude of the measured waveform. This will be obtained only if the fault persists long enough to damp any decay in amplitude and any dc offset. Therefore, it is desirable to have these components also removed, which can be accomplished by digital filtering [16]. A simple and effective method of performing the digital filtering uses the addition and subtraction of consecutive samples. If we let X_1, X_2, X_3, and X_4 be the latest four samples of a measured variable P, then the filter is defined as follows:

$$P = X_1 - X_2 - X_3 + X_4 \tag{4.41}$$

If dc offset exists, all four samples will have the same value and the filter output is zero. The filter also eliminates ramps, which can be verified by setting the successive sampled values to 1, 2, 3, and 4, which also results in an output of zero. This eliminates, at least partially, the troublesome exponential decay that is often present in the raw sampled data.

Since the relay computes phasor values of both current and voltage, these values can be combined as desired to provide directional sensing, distance measurements, and even metering during unfaulted conditions. Since the technique is simple, separate relay modules can be constructed to detect all types of phase and ground faults, initiate clearing, and save data for transmission to the relay engineer by data transfer using a modem.

4.5.4 Digital Relaying Applications

Digital relaying is a relatively new concept, and new devices, using different algorithms, are introduced frequently. A few of the developments that have been reported, either as commercial devices or as products of research and development, will be described.

4.5.4.1 Digital Overcurrent Protection. Considerable work has been performed on digital overcurrent protection. It has been noted that most of the electromechanical overcurrent relays have characteristics based on the performance of the induction disk, making the characteristic time-current shape difficult to modify. The modern microprocessor-based relay, on the other hand, can be readily modified to improve coordination for any given condition [18]. Progress has also been made toward developing an analytic standard for the inverse time characteristic, and that can be used with microprocessor relays [19]. An IEEE committee has described the problems of representing overcurrent relay characteristics in digital computer programs and has recommended methods for digital representation [20].

4.5.4.2 Digital Distance Relaying. Distance relaying is widely applied for transmission protection, and many digital devices have been described that use the distance measurement principle, with these descriptions beginning to appear in the literature in the 1960s [21]. Experimenters have reported on distance relays using a variety of different algorithms for high-speed distance protection [22–33]. Algorithms have been proposed that are based on simple arithmetic processing of the sequences, symmetrical component computation, differential equation analysis, least-square error digital filtering, Fourier analysis, Walsh transforms, and general curve fitting. Others will surely be suggested in the future. Digital distance relays are available from several manufacturers, and have been applied in a large number of transmission systems with excellent success.

The trend seems to be developing in the direction of providing new digital systems that can do more than simply recognize and clear faults. For example, some work has been done on using the digital relays as a part of a state estimation scheme. Others have proposed making the digital protective system adaptive.

4.5.4.3 Transformer Protection. Most digital transformer protective schemes use some version of the time-tested percentage differential concept [34–45]. This is a natural choice since all terminals of a transformer are available for measurement and there is no reason not to make use of information obtained from all transformer terminals. However, differential protection of transformers is not without problems, such as the inrush current when the transformer is energized, which appears to the differential relay as an internal fault. Other problems are associated with overvoltage, underfrequency, and the saturation of the transformer iron core.

Digital protection has been shown to be effective in solving some of these problems. Digital filtering is effective in blocking the detection of second harmonic inrush currents and the fifth harmonic present due to overexcitation. Several types of digital harmonic restraint have been used including digital filtering, harmonic identification using DFT, and others. Restraint algorithms have also been devised that are based on voltage rather than current [37], [39], [41], [43].

4.5.4.4 Generator Protection. Generator protection by digital computer has not been pursued to as great a degree as for lines and transformers. However, several promising digital algorithms have been proposed [46–49]. One approach, which differs from the usual practice, is to provide many different types of protection to be programmed in a single digital processing system [49]. This is in contrast to the usual generator protection, which uses different types of relays with each protecting against a particular type of hazard. Digital protection offers the promise of monitoring many different types of problems in a single device. This raises a question of reliability, since the separate relay systems provide a redundancy that is beneficial, but there are advantages in the digital approach.

4.5.4.5 Digital Substation Protection. The concept of integrating the protection of individual station components into a complete digital system has received considerable attention, beginning in the early 1970s [50–63]. The early systems were based on a central computer of the minicomputer class, but later systems have been mostly microcomputer based. Functions such as bus protection can be performed using digital systems, and there are certain advantages offered by the flexibility of software in addressing problems such as current transformer mismatch. However, there are advantages to the integrated control and protection schemes, and these may prevail in the future [60], [62].

4.5.4.6 Other Types of Digital System Protection. Digital protection has been proposed for many other applications, and using new and interesting new methods. Although it is not possible to discuss all of the proposed devices, a few new concepts should be mentioned.

Digital underfrequency protection has been proposed by some. Different forms of digital frequency and rate-of-change of frequency are used to determine whether frequency deviations are large enough to require load shedding [65–67].

Digital directional comparison transmission relaying has been proposed by several investigators [68]. Digital current differential relaying has also been investigated and field tested [69], [70].

A digital transmission line thermal overload relay has been developed that is based on a thermal model that uses direct solar measurements as well as inputs from the line current and temperature [71].

4.5.4.7 Unique Concepts in Digital Protection. Most of the devices described above conform to concepts of system protection that have been used for many years, such as overcurrent, distance measurement, and differential protection. It is not surprising that some totally new concepts have emerged that may change system protection in years to come.

One new concept proposed is the use of Walsh functions for high-speed relaying [15], [72]. Walsh functions are orthogonal functions on the interval [0], [1] that have values of ± 1. They are appealing in applications such as system protection because multiplication by a Walsh function involves only algebraic operations [15]. It is not clear what the role of this type of orthogonal function might be in relaying, but it has been investigated as a research topic.

Kalman filtering is an algorithm for processing discrete measurement data in an optimal manner [73]. Its early application has been in the field of navigation, which is made difficult by noisy observations. This is not unlike the problems in power system protection, where unwanted noise and multiple frequency responses tend to obscure the power frequency response of a disturbance. Applying this powerful technique to power system relaying is a natural development that has been investigated as a research topic, especially by A. A. Girgis [74–81].

The application of statistical decision theory in fault detection has received only limited attention [82]. It is based on the concept of recognizing a discriminant value from measurements of currents or voltages, and basing the trip decision on this discriminant exceeding a threshold value. This bears some similarity to the Kalman filtering method, but the details of the two concepts are different.

Another new concept that looks promising is that of adaptive relaying [15], [83–86]. Adaptive relaying is a protection philosophy that permits and seeks to make adjustments in various protective functions automatically, in order to make them more attuned to prevailing power system conditions [86]. The adaptive concept is probably not possible without digital relays. The functions to be performed are determined through software and the capability to alter the software or the resulting commands, perhaps from a remote control center. Some proposed applications for an adaptive relay are in recognizing load changes, in cold load pickup, end-of-line transmission protection, transformer protection, and automatic reclosing.

4.5.5 Example of a Digital Relay System

A digital transmission line relay provides an excellent example of a relay utilizing digital processing. The block diagram of the relay is shown in Figure 4.30 [86]. The characteristic of the relay is a mho circle in the complex R-X plane, with three zones of protection being represented. Blinders are provided to narrow the trip zone. The relay finds the sum of positive and negative sequence currents of each phase, which is used to identify the type of fault. The relay has capability of using pilot communications with similar relays at the opposite end of the transmission line. Note that none of these characteristics are evident from the block diagram of Figure 4.30, since the unique relay logic is contained in the software, not in the hardware.

The relay block diagram shows the major items of hardware. This is a very general arrangement that could be used for relays having quite different characteristics. The heart of the system is a microprocessor that operates at a clock rate of 10 MHz. Memory is separate, and can be easily replaced, if necessary. Random access memory (RAM) is a volatile read-write memory for working storage. Also included, however, is a nonvolatile RAM, identified as NOVRAM, which stores settings and targets. The input quantities include both voltages and currents, which are converted to digital signals of 15 bits dynamic range. The sampling rate is eight samples per cycle, on a 60 hertz system.

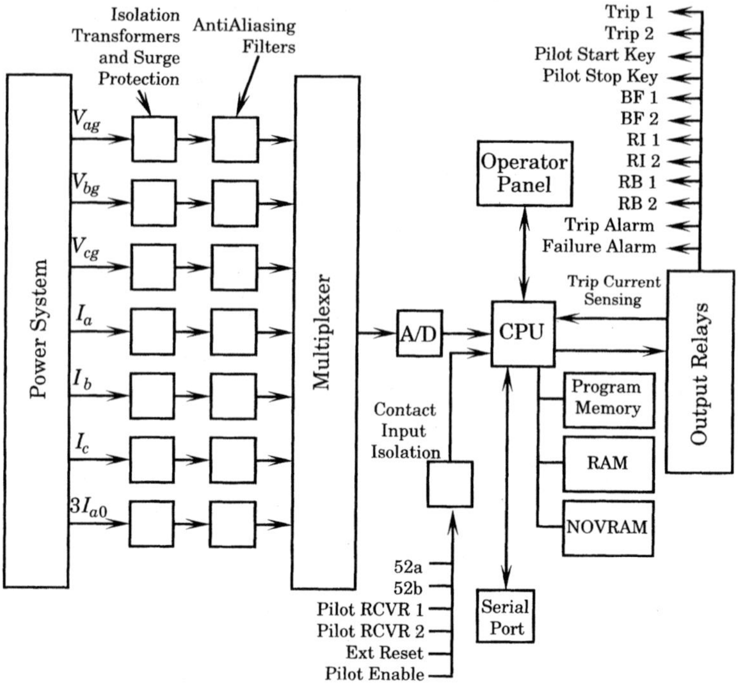

Figure 4.30 Example of a digital transmission relay.

4.6 HYBRID RELAY LOGIC

Another type of relay logic that is being applied is referred to as "hybrid" logic, since it combines some of the features of both analog and digital devices, plus some special devices that are programmable. One of the disadvantages of digital logic is the time required to make the necessary computations, especially if the function to be analyzed is complex. This can be solved by using faster, or more advanced, digital hardware, but this increases the cost. Moreover, in some cases it can be argued that there is nothing wrong with analog logic and it may have certain advantages, for example, in speed of making complex computations.

Protective relay systems have been designed using a type of integrated circuit called an "Application Specific Integrated Circuit" (ASIC) which encompasses a family of devices that include Programmable Logic Devices (PLD). The PLD is a prefabricated integrated circuit that contains many logic gates, which are manufactured in a generic configuration. The device user may then program the device to create a specific protective relay application. The relay systems designed with these devices are sometimes called "hybrid" systems, since they are programmed, but do not necessarily use discrete samples of current and voltage to provide the protective function.

The protective relay designer, therefore, has a host of integrated circuit devices to work with in creating a specific protective relay applications. Thus, we see analog relays, digital relays, and hybrid relays, each having its own special advantages for specific applications.

One portion of nearly all relays is digital, however, and that is the portion dealing with relay monitoring and event recording. The relay monitoring devices perform checks on the hardware system to detect failures and notify the engineer of problems in a timely way. Event recorders log the operation of protective relays and can transmit these records to the engineer's office for analysis. This has become a valuable feature in nearly all modern relay systems. It gives the engineer timely information, provides a method of checking relay settings, and gives prompt notification of any failures.

4.7 RELAYS AS COMPARATORS

The preceding sections of this chapter have examined various types of analog and digital logic that are used in the design of protective systems. We now examine a concept that has been a part of protective relaying for many years, and remains an important method for the development of relays and in understanding their performance. The concept is that of the comparator. This section will examine this idea in greater depth, and show that all relays are, in fact, comparators, but with differences in purpose and design.

4.7.1 Relay Design

Relays may be classified according to the number of input or metered quantities. We shall designate relays with one metered quantity as simplex relays and those with two or more metered quantities as complex relays.

Simplex relays have, by definition, only one metered quantity M, which is usually a current or a voltage. The reference quantity R is usually a physical or electrical constant, which is often adjustable. A few examples of simplex relays are shown in Figure 4.31.

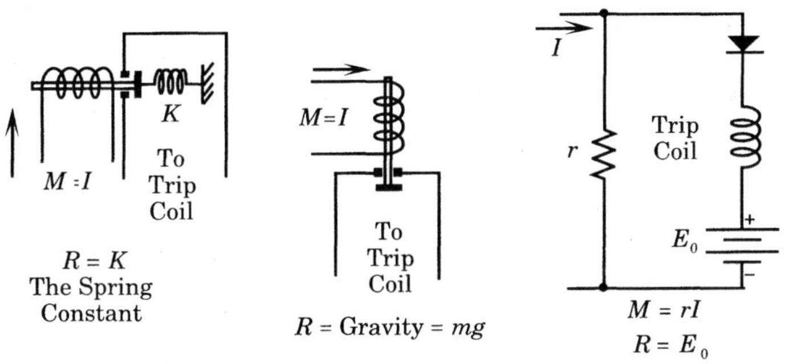

Figure 4.31 Examples of simplex relay design.

Most relays used in power system protection are complex relays and utilize two or more input quantities for improved selectivity and speed. An example is the differential protection relay shown in Figures 3.24 and 3.25, which uses current comparison as a means of obtaining selectivity. Other complex relays use an impedance, admittance, phase comparison, current ratio, or other relay computations to improve speed and selectivity in a given abnormal system situation.

4.7.2 Phase and Amplitude Comparison

Two common relay comparator concepts involve the use of phase and amplitude comparison [87]. To show the relationship between these concepts we define two phasor quantities A and B where[1]

$$\begin{aligned} \boldsymbol{A} &= |\boldsymbol{A}|\, e^{j\theta} \\ \boldsymbol{B} &= |\boldsymbol{B}|\, e^{j\phi} \end{aligned} \tag{4.42}$$

First, we note the following interesting properties of the sum and difference phasors $\boldsymbol{A}+\boldsymbol{B}$ and $\boldsymbol{A}-\boldsymbol{B}$. If the angle between $\boldsymbol{A}$ and $\boldsymbol{B}$ is defined as $(\phi-\theta)$ then we note that, since $|B-A|=|A-B|$ and taking $\bar{A}$ as the reference $(\theta=0)$

$$\begin{aligned} &(i) && \text{when } \phi \ge 90^\circ, \text{ then } |\boldsymbol{A}-\boldsymbol{B}| \ge |\boldsymbol{A}+\boldsymbol{B}| \\ &(ii) && \text{when } \phi = 90^\circ, \text{ then } |\boldsymbol{A}-\boldsymbol{B}| = |\boldsymbol{A}+\boldsymbol{B}| \\ &(iii) && \text{when } \phi \le 90^\circ, \text{ then } |\boldsymbol{A}-\boldsymbol{B}| \le |\boldsymbol{A}+\boldsymbol{B}| \end{aligned} \tag{4.43}$$

These relations are shown graphically in Figure 4.32. If we construct a relay for amplitude comparison a trip signal is sent when

$$|\boldsymbol{A}| \ge |\boldsymbol{B}| \tag{4.44}$$

or vice versa. Such a relay is shown in Figure 4.33 and corresponds to a comparator circuit where both R and M are metered quantities. We can depict the amplitude comparator by the schematic of Figure 4.33.

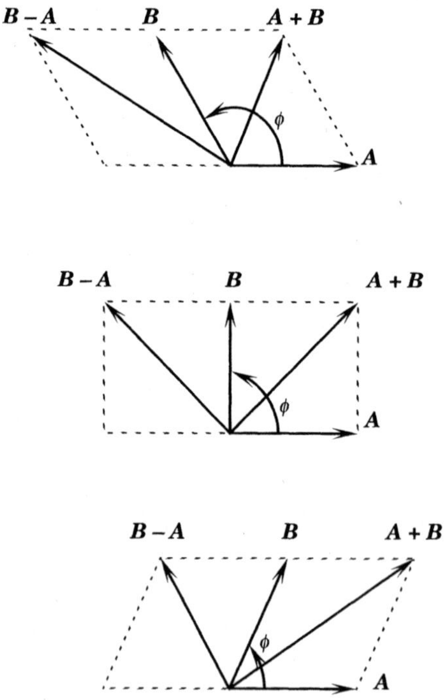

Figure 4.32 Amplitude and phase comparison.

From Figure 4.32 it is apparent that we can determine if the phase difference between $\boldsymbol{A}$ and $\boldsymbol{B}$ is greater than or less than 90 degrees by making an amplitude comparison of $|\boldsymbol{A}+\boldsymbol{B}|$

[1] Note that "a" defined here is not $\exp(j2\pi/3)$, the symmetrical component complex a-operator.

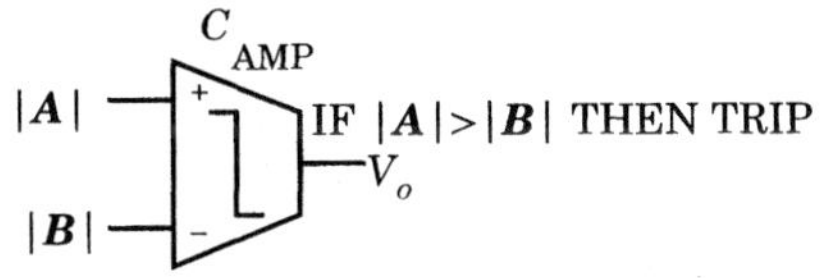

Figure 4.33 Block diagram for an amplitude comparator.

and $|\boldsymbol{A} - \boldsymbol{B}|$. If we supply $|\boldsymbol{A} + \boldsymbol{B}|$ and $|\boldsymbol{A} - \boldsymbol{B}|$ to the amplitude comparator of Figure 4.32 it becomes a phase comparator and determines the phase error with respect to 90 degrees. If we first shift $\boldsymbol{B}$ by +90 degrees and then form $|\boldsymbol{A} + \boldsymbol{B}|$ and $|\boldsymbol{A} - \boldsymbol{B}|$, the phase error is computed with respect to 0 degrees.

The foregoing discussion shows that there is essentially no difference between phase and amplitude comparison and the same comparator device can be used to do both. Signal preparation is necessary, however, to make a phase comparator out of an amplitude comparator. It should also be noted that the quantities $\boldsymbol{A}$ and $\boldsymbol{B}$ are phasor quantities. This implies relay filtering of harmonics so that only the fundamental frequency components are being compared.

4.7.3 The Alpha and Beta Planes

In the previous section, we defined the phasors $\boldsymbol{A}$ and $\boldsymbol{B}$ as follows.

$$\begin{aligned} \boldsymbol{A} &= |\boldsymbol{A}|\, e^{j\theta} = Ae^{j\theta} \\ \boldsymbol{B} &= |\boldsymbol{B}|\, e^{j\phi} = Be^{j\phi} \end{aligned} \tag{4.45}$$

Then we may define a quantity alpha as follows.

$$\vec{\alpha} = \frac{\boldsymbol{A}}{\boldsymbol{B}} = \frac{A}{B}e^{j(\theta-\phi)} = \alpha e^{j(\theta-\phi)} = \alpha_p + j\alpha_q \tag{4.46}$$

Alpha quantities may be plotted in a complex alpha plane.

We may also define complex beta quantities as follows.

$$\vec{\beta} = \frac{\boldsymbol{B}}{\boldsymbol{A}} = \frac{B}{A}e^{j(\phi-\theta)} = \beta e^{j(\phi-\theta)} = \beta_p + j\beta_q \tag{4.47}$$

which may be plotted in the complex beta plane.

Alpha and beta plane diagrams are useful in describing the behavior of certain comparator circuits.

4.7.4 The General Comparator Equations

Consider the relay logic configuration shown in Figure 4.34, which consists of two weighted summers W_{12} and W_{34}, and the amplitude comparator C_{AMP}. If we define scalar constants $k_1, \ldots, k_4$ then we may also define the complex quantities $\boldsymbol{N}_1$ and $\boldsymbol{N}_2$ as follows.

$$\begin{aligned} \boldsymbol{N}_1 &= k_1\boldsymbol{A} + k_2\boldsymbol{B} \\ &= (k_1 A\cos\theta + k_2 B\cos\phi) + j(k_1 A\sin\theta + k_2 B\sin\phi) \end{aligned} \tag{4.48}$$

$$\begin{aligned} \boldsymbol{N}_2 &= k_3\boldsymbol{A} + k_4\boldsymbol{B} \\ &= (k_3 A\cos\theta + k_4 B\cos\phi) + j(k_3 A\sin\theta + k_4 B\sin\phi) \end{aligned} \tag{4.49}$$

A phasor diagram, using arbitrary values of $k_1, \ldots, k_4$, is shown in Figure 4.35. The amplitude comparator of Figure 4.34 compares inputs $\boldsymbol{N}_1$ and $\boldsymbol{N}_2$ for inequality of the moduli; i.e., the locus of the relay tripping characteristic in the complex plane is

$$|\boldsymbol{N}_1| \le |\boldsymbol{N}_2| \tag{4.50}$$

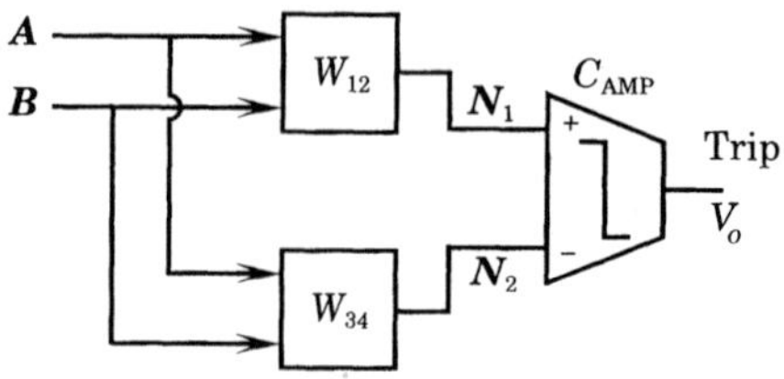

Figure 4.34 A general comparator arrangement.

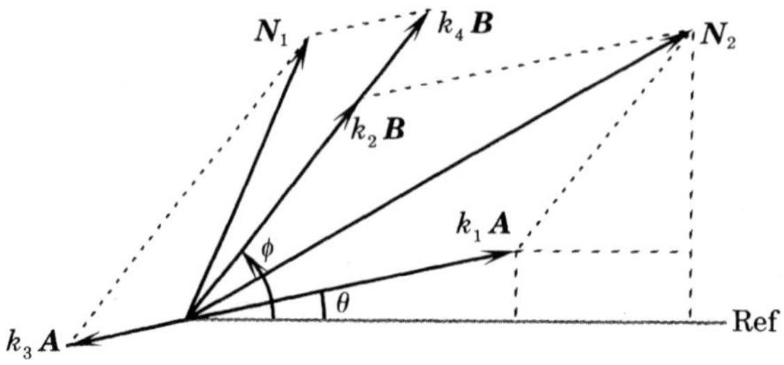

Figure 4.35 Phasor diagram of N_1 and N_2.

Substituting (4.48) and (4.49) into (4.50) we compute

$$\begin{aligned}(k_1 A\cos\theta + k_2 B\cos\phi)^2 + (k_1 A\sin\theta + k_2 B\sin\phi)^2 \\ \le (k_3 A\cos\theta + k_4 B\cos\phi)^2 + (k_3 A\sin\theta + k_4 B\sin\phi)^2\end{aligned} \tag{4.51}$$

Rearranging, we may write (4.51) in the following form.

$$(k_1^2 - k_3^2)^2 A^2 + 2(k_1 k_2 - k_3 k_4) AB\cos(\phi - \theta) + (k_2^2 - k_4^2) B^2 \le 0 \tag{4.52}$$

We now divide through by $(k_2^2 - k_4^2)A^2$ to write (4.52) in terms of the beta magnitude, $\beta = B/A$, with the result

$$\beta^2 - 2\beta_o\beta\cos(\phi - \theta) + \beta_o^2 \le \beta_o^2 - \frac{k_1^2 - k_3^2}{k_2^2 - k_4^2} \tag{4.53}$$

where we have defined

$$\beta_o = \frac{k_3 k_4 - k_1 k_2}{k_2^2 - k_4^2} \tag{4.54}$$

Equation (4.53) is a polar form for the equation of a circle in the complex beta plane. To show this we recall the polar form of a complex variable **Z** which we may write as

$$\mathbf{Z} = \rho e^{j\theta} = \rho\cos\theta + j\rho\sin\theta \tag{4.55}$$

Let **Z** be any point on a circle which has its center at $\mathbf{Z}_o$ where

$$\mathbf{Z}_o = \rho_o e^{j\theta_o} = \rho_o\cos\theta_o + j\rho_o\sin\theta_o \tag{4.56}$$

If the radius of the circle is k, then the equation representing the circle and the area within the circle may be written as

$$|\mathbf{Z} - \mathbf{Z}_o| \le k \tag{4.57}$$

Substituting (4.55) and (4.56) into (4.57) and squaring the result we have

$$(\rho\cos\theta - \rho_o\cos\theta_o)^2 + (\rho\sin\theta - \rho_o\sin\theta_o)^2 \le k^2 \tag{4.58}$$

Simplifying, we may write this result in the standard polar form for a circle of radius k and with its center located a distance ρ_o from the origin along radial making an angle θ_o the horizontal [88].

$$\rho^2 - 2\rho\rho_o\cos(\theta - \theta_o) + \rho_o^2 \le k^2 \tag{4.59}$$

Obviously (4.53) is also the equation of a circle with center located a distance β_o from the origin along a radial making the angle θ_o with the horizontal. Moreover, this circle has radius k where

$$k^2 = \beta_0^2 - \frac{k_1^2 - k_3^2}{k_2^2 - k_4^2} = \left[\frac{k_1k_4 - k_2k_3}{k_2^2 - k_4^2}\right]^2 \tag{4.60}$$

This beta circle is shown in Figure 4.36.

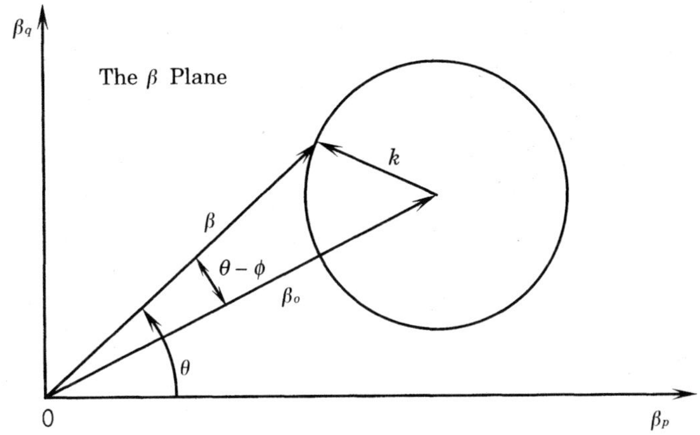

Figure 4.36 The beta circle characteristics.

Referring again to (4.52), we may also divide by $(k_1^2 - k_3^2)B^2$ to write the equation in terms of the alpha magnitude $\alpha = A/B$ with the result

$$\alpha^2 - 2\alpha\alpha_o \cos(\phi - \theta) + \alpha_o^2 \leq k^2 \tag{4.61}$$

where

$$\alpha_o = \frac{k_3k_4 - k_1k_2}{k_1^2 - k_3^2} \tag{4.62}$$

and

$$k = \frac{k_1k_2 - k_3k_4}{k_1^2 - k_3^2} \tag{4.63}$$

It is clear from the foregoing analysis that the relay characteristics of the comparator $|\boldsymbol{N}_1| = |\boldsymbol{N}_2|$ is a circle in either the alpha or the beta plane.

EXAMPLE 4.1

Consider the differential current relay of Figure 3.24 where the threshold equation is given as

$$|\boldsymbol{Q}_O| = |\boldsymbol{Q}_R| \tag{4.64}$$

or

$$|k_O(\boldsymbol{I}_1 - \boldsymbol{I}_2)| = |k_R(\boldsymbol{I}_1 + \boldsymbol{I}_2)| \tag{4.65}$$

where $\boldsymbol{Q}_O$ and $\boldsymbol{Q}_R$ are the operating and restraint quantities, respectively, and k_O and k_R are design parameters. Determine the relay operating characteristic in the alpha plane.

Solution

Since both k_O and k_R are constants we can write (4.65) as

$$|\boldsymbol{I}_1 - \boldsymbol{I}_2| = |k(\boldsymbol{I}_1 + \boldsymbol{I}_2)| \tag{4.66}$$

where $k = k_R/k_O$. Also, from the definition of alpha we can write

$$\vec{\alpha} = |\alpha|e^{j\gamma} = \frac{\boldsymbol{I}_1}{\boldsymbol{I}_2} \tag{4.67}$$

describes a complex number in the α plane and where γ is the angle by which $\boldsymbol{I}_1$ leads $\boldsymbol{I}_2$.

Substituting (4.67) into (4.66) all variables $\boldsymbol{I}_2$ will cancel with the result

$$||\alpha|e^{j\gamma} - 1| = |k(|\alpha|e^{j\gamma} + 1)| \tag{4.68}$$

This result may be written in terms of the magnitude α and angle γ.

$$[(\alpha\cos\gamma - 1) + j\alpha\sin\gamma]^2 = k^2[(\alpha\cos\gamma + 1) + j\alpha\sin\gamma]^2 \tag{4.69}$$

Squaring and simplifying we may write the result as

$$\alpha^2 - 2\alpha\alpha_o\cos\gamma + \alpha_o^2 = r^2 \tag{4.70}$$

where

$$\alpha_o = \frac{1+k^2}{1-k^2} \tag{4.71}$$

and

$$r = \left[\frac{2k}{1-k^2}\right] \tag{4.72}$$

This is a circle with center on the positive real axis at $(\alpha_o, 0)$ and with radius r [89]. This result could also be obtained by writing alpha in rectangular form and solving in rectangular coordinates. This is left as an exercise (see problem 4.11).

It is also apparent that the relay threshold characteristic is a circle in the beta plane too. Verification is left as an exercise. ■

4.7.5 The Amplitude Comparator

Suppose we write the threshold equation of the amplitude comparator as follows [90].

$$\boldsymbol{Q}_O \geq \boldsymbol{Q}_R \tag{4.73}$$

or

$$\left|\frac{\boldsymbol{I}_1 - \boldsymbol{I}_2}{s}\right| \geq \left|\frac{\boldsymbol{I}_1 + \boldsymbol{I}_2}{2}\right| \tag{4.74}$$

where s is a design parameter. Warrington [90] gives values for s in the range of 5% for generator protection and 10–40% for transformer protection.

In terms of the quantities defined in Section 4.7.3, let

$$\begin{aligned} \boldsymbol{I}_1 &= \boldsymbol{A} = ae^{j\theta} = ae^{j0} = I_1e^{j0} \\ \boldsymbol{I}_2 &= \boldsymbol{B} = be^{j\phi} = I_2e^{j\phi} \end{aligned} \tag{4.75}$$

where, since only the difference $\phi - \theta$ is needed, we arbitrarily set $\theta = 0$. Then the comparator inputs are computed from (4.48) and (4.49) to be

$$\begin{aligned} \boldsymbol{Q}_O &= \boldsymbol{N}_1 = k_1\boldsymbol{I}_1 + k_2\boldsymbol{I}_2 = \boldsymbol{I}_1 - \boldsymbol{I}_2 \\ \boldsymbol{Q}_R &= \boldsymbol{N}_2 = k_3\boldsymbol{I}_1 + k_4\boldsymbol{I}_2 = \frac{s}{2}(\boldsymbol{I}_1 + \boldsymbol{I}_2) \end{aligned} \tag{4.76}$$

In terms of the current magnitudes, we write

$$\begin{aligned} \boldsymbol{Q}_O &= k_1I_1 + k_2I_2(\cos\phi + j\sin\phi) \\ &= I_1 - I_2(\cos\phi + j\sin\phi) \end{aligned} \tag{4.77}$$

From this relation we determine two equations

$$\begin{aligned} k_1 I_1 + k_2 I_2 \cos\phi &= I_1 - I_2 \cos\phi \\ k_2 I_2 \sin\phi &= -I_2 \sin\phi \end{aligned} \tag{4.78}$$

Then we can solve for k_1 and k_2 with the result

$$\begin{aligned} k_1 &= 1 \\ k_2 &= -1 \end{aligned} \tag{4.79}$$

In a like manner we can write

$$\begin{aligned} Q_R &= k_3 I_1 + k_4 I_2(\cos\phi + j\sin\phi) \\ &= \frac{sI_1}{2} + \frac{sI_2}{2}(\cos\phi + j\sin\phi) \end{aligned} \tag{4.80}$$

from which we compute

$$k_3 = k_4 = \frac{s}{2} \tag{4.81}$$

Then the relay characteristic in the beta plane is found from (4.53) and (4.60). The center of the threshold circle is at $(\beta_o, 0)$ where

$$\beta_o = \frac{k_1 k_2 - k_3 k_4}{k_2^2 - k_4^2} = \frac{4+s^2}{4-s^2} \tag{4.82}$$

and with radius k where

$$k = \frac{k_1 k_4 - k_2 k_3}{k_2^2 - k_4^2} = \frac{4s}{4-s^2} \tag{4.83}$$

The tripping characteristic is described by (4.53)

$$\beta^2 + 2\left[\frac{4+s^2}{4-s}\right]\beta\cos\phi + \left[\frac{4+s^2}{4-s^2}\right]^2 \geq \left[\frac{4s}{4-s^2}\right] \tag{4.84}$$

4.7.6 The Phase Comparator

The phase comparator circuit is described schematically by Figure 4.37 where the circuit is the same as the amplitude comparator except that the last block C_ϕ compares the phase of $\boldsymbol{N}_1$ and $\boldsymbol{N}_2$ rather than their amplitudes.

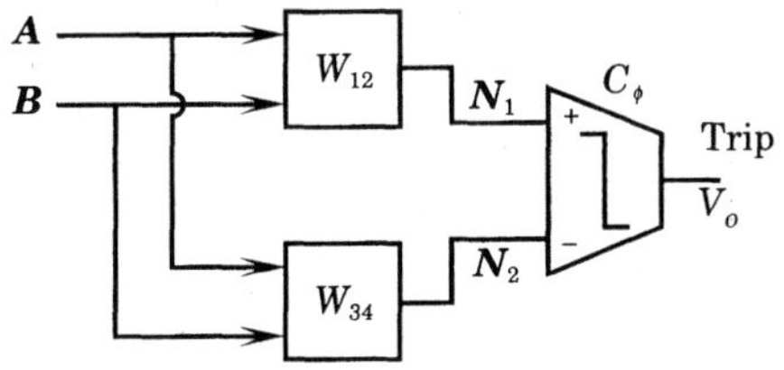

Figure 4.37 A general phase comparator.

Warrington [88] gives the threshold characteristic for the phase comparator to be

$$|\boldsymbol{N}_1 \boldsymbol{N}_2^*| \leq 0 \tag{4.85}$$

which means that $\boldsymbol{N}_1$ has a positive projection on $\boldsymbol{N}_2$. Mathematically (4.85) may be written as

$$\mathrm{Re}(\boldsymbol{N}_1 \boldsymbol{N}_2^*) \leq 0 \tag{4.86}$$

Suppose we let

$$\begin{aligned} \boldsymbol{N}_1 &= N_1 e^{j\sigma_1} \\ \boldsymbol{N}_2 &= N_2 e^{j\sigma_2} \end{aligned} \tag{4.87}$$

then (4.86) becomes

$$\text{Re}(\boldsymbol{N}_1\boldsymbol{N}_2^*) = \text{Re}(N_1 N_2 e^{j(\sigma_1-\sigma_2)}) \leq 0 \tag{4.88}$$

This expression must hold for N_1 and N_2 nonzero or

$$\text{Re}(e^{j(\sigma_1-\sigma_2)}) \leq 0 \tag{4.89}$$

This will be satisfied as long as

$$\sigma_1 - \sigma_2 \leq \pi/2 \tag{4.90}$$

which determines the relay threshold and operating characteristic.

The threshold characteristic (4.90) may be expressed by taking the tangent of terms on both sides. Thus, we compute

$$\tan(\sigma_1 - \sigma_2) \leq \tan(\pi/2) = \pm\infty \tag{4.91}$$

or

$$\frac{\tan\sigma_1 - \tan\sigma_2}{1 + \tan\sigma_1 \tan\sigma_2} \leq \pm\infty \tag{4.92}$$

This requires that

$$1 + \tan\sigma_1 \tan\sigma_2 \geq 0 \tag{4.93}$$

or

$$\tan\sigma_1 \geq \frac{-1}{\tan\sigma_2} \tag{4.94}$$

Thus, we can say that the relay operates when

$$\boldsymbol{Q}_O > \boldsymbol{Q}_R$$

or

$$N_1 \tan\sigma_1 > \frac{-1}{N_2 \tan\sigma_2} \tag{4.95}$$

From the definition of N_1 and N_2 in (4.48) and (4.49), we compute

$$\begin{aligned} \tan\sigma_1 &= \frac{k_1 a \sin\theta + k_2 b \sin\phi}{k_1 a \cos\theta + k_2 b \cos\phi} \\ \tan\sigma_2 &= \frac{k_3 a \sin\theta + k_4 b \sin\phi}{k_3 a \cos\theta + k_4 b \cos\phi} \end{aligned} \tag{4.96}$$

Substituting into (4.95) we get

$$\frac{k_1 a \sin\theta + k_2 b \sin\phi}{k_1 a \cos\theta + k_2 b \cos\phi} \geq -\frac{k_3 a \cos\theta + k_4 b \cos\phi}{k_3 a \sin\theta + k_4 b \sin\phi} \tag{4.97}$$

which can be easily simplified, by cross-multiplying, to the form

$$k_1 k_3 a^2 + k_2 k_4 b^2 + (k_1 k_4 + k_2 k_3) ab \cos(\phi - \theta) \leq 0 \tag{4.98}$$

This expression may be converted to the beta plane by dividing through by $k_2 k_4 a^2$ with the result

$$\beta^2 + \left[\frac{k_1 k_4 + k_2 k_3}{k_2 k_4}\right] \beta \cos(\phi - \theta) + \frac{k_1 k_3}{k_2 k_4} \leq 0 \tag{4.99}$$

where $\beta = b/a$, by definition. Equation (4.99) is easily rearranged into the standard form for the equation of a circle.

$$\beta^2 - \beta_o\beta\cos(\phi - \theta) + \beta_o^2 \le r^2 \tag{4.100}$$

if we define

$$\beta_o = \frac{-(k_1k_4 + k_2k_3)}{2k_2k_4} \tag{4.101}$$

and

$$r^2 = \beta_o^2 - \frac{k_1k_3}{k_2k_4} = \left[\frac{k_1k_4 - k_2k_3}{2k_2k_4}\right]^2 \tag{4.102}$$

and the relay tripping characteristic has a circular threshold in the beta plane.

EXAMPLE 4.2

Find the constants $k_1, \ldots, k_4$ that will give the phase comparator the same characteristic in the beta plane as the differential amplitude relay characteristic given by (4.84).

Solution

To equate the beta plane characteristics we equate the radii and center parameters.

$$\beta_o = \frac{-(k_1k_4 + k_2k_3)}{2k_2k_4} = \frac{-(4+s)^2}{4-s^2} \tag{4.103}$$

$$r = \frac{k_1k_4 - k_2k_3}{2k_2k_4} = \frac{4s}{4-s^2} \tag{4.104}$$

This gives only two nonlinear constraints among the four unknowns. If these constraints are satisfied we should be able to make an arbitrary choice of the remaining two constants.

It is helpful to compute

$$\beta_o + r = \frac{-k_1}{k_2} = \frac{-(s-2)}{s+2} \tag{4.105}$$

$$\beta_o - r = \frac{-k_3}{k_4} = \frac{-(s+2)}{s-2} \tag{4.106}$$

Then let

$$\begin{aligned} k_1 &= k_4 = s - 2 \\ k_2 &= k_3 = s + 2 \end{aligned} \tag{4.107}$$

and the characteristics are determined. ■

4.7.7 Distance Relays as Comparators

For a distance relay we write the threshold equation as follows.

$$|\mathbf{Z} - \mathbf{Z}_o| \le r \tag{4.108}$$

The choice of $\mathbf{Z}_o$ determines the type of characteristic. For example, when $\mathbf{Z}_o = 0$ we refer to the characteristic as an "impedance" relay, but when $|\mathbf{Z}_o| = r$ we usually call it a "mho" relay. These two characteristics are shown in Figure 4.38.

As an amplitude comparator, we write the impedance characteristic (4.108) in rectangular form.

$$|(R - R_o) + j(X - X_o)| \le r \tag{4.109}$$

or

$$(R - R_o)^2 + (X - X_o)^2 \le r^2 \tag{4.110}$$

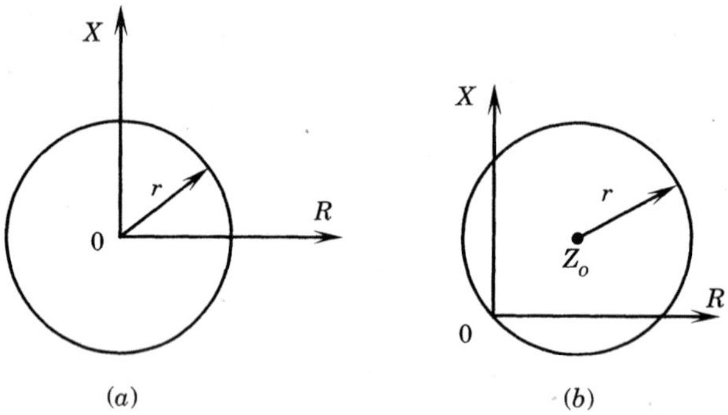

Figure 4.38 Two common distance relay characteristics. (*a*) Impedance characteristic. (*b*) Mho characteristic.

This is the equation of a circle in rectangular form. In polar form we have exactly this same characteristic in (4.57)–(4.59).

Consider the phasor diagram of Figure 4.39 where $\boldsymbol{V}_R$ and $\boldsymbol{I}_R$ are the current and voltage phasors at the relay location. We write these phasor quantities in polar form as follows.

$$\begin{aligned} \boldsymbol{V}_R &= V_R e^{j\theta} \\ \boldsymbol{I}_R &= I_R e^{j\phi} \end{aligned} \tag{4.111}$$

Then, from (4.48) and (4.49) we have

$$\begin{aligned} \boldsymbol{N}_1 &= (k_1 V_R \cos\theta + k_2 I_R \sin\phi) + j(k_1 V_R \sin\theta + k_2 I_R \sin\phi) \\ \boldsymbol{N}_2 &= (k_3 V_R \cos\theta + k_4 I_R \sin\phi) + j(k_3 V_R \sin\theta + k_4 I_R \sin\phi) \end{aligned} \tag{4.112}$$

For an amplitude comparator let the trip threshold be given by

$$|\boldsymbol{N}_1| \le |\boldsymbol{N}_2| \tag{4.113}$$

or

$$N_1^2 \le N_2^2$$

From (4.51) we write

$$(k_1^2 - k_3^2)V_R^2 + (k_2^2 - k_4^2)I_R^2 + 2(k_1 k_2 - k_3 k_4)V_R I_R \cos\delta \le 0 \tag{4.114}$$

and divide by $(k_1^2 - k_3^2)I_R^2$ to write

$$Z_R^2 - 2Z_o Z_R \cos\delta + Z_o^2 \le r^2 \tag{4.115}$$

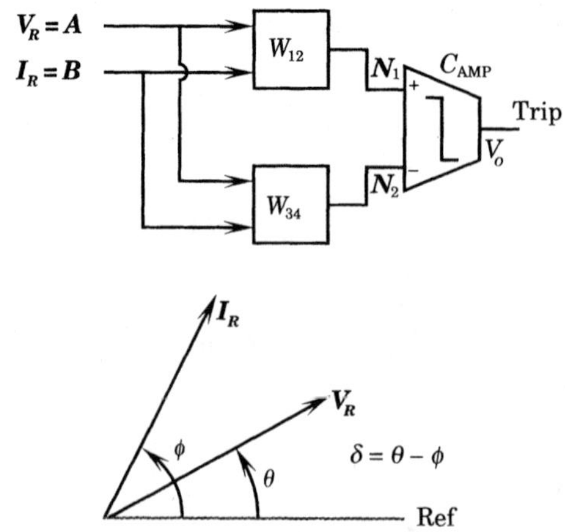

Figure 4.39 The distance relay as an amplitude comparator.

$$\text{where } Z_o = \frac{k_3k_4 - k_1k_2}{k_1^2 - k_3^2}$$
$$r = \frac{k_1k_4 - k_2k_3}{k_1^2 - k_3^2} \qquad (4.116)$$

This is the general circular characteristic of distance type relays as shown in Figure 4.38.

EXAMPLE 4.3

Determine the constants $k_1, \ldots, k_4$ that are required to construct the impedance element of Figure 4.38(a) with a radius $r = s$.

Solution

For the impedance element we have the constraints

$$Z_o = \frac{k_3k_4 - k_1k_2}{k_1^2 - k_3^2} = 0$$
$$r = \frac{k_1k_4 - k_2k_3}{k_1^2 - k_3^2} = s \qquad (4.117)$$

Since $Z_o = 0$ we have the constraint

$$k_1k_2 = k_3k_4 \qquad (4.118)$$

We may also compute

$$s = r + Z_o = \frac{k_1k_4 - k_2k_3 - k_1k_2 + k_3k_4}{k_1^2 - k_3^2} = \frac{k_4 - k_2}{k_1 - k_3} \qquad (4.119)$$

and

$$s = r - Z_o = \frac{k_1k_4 - k_2k_3 + k_1k_2 - k_3k_4}{k_1^2 - k_3^2} = \frac{k_4 + k_2}{k_1 + k_3} \qquad (4.120)$$

Since we have only two constraints, we may select two constants arbitrarily.

Let us arbitrarily choose

$$k_2 = k_3$$

Then

$$s = \frac{k_4}{k_1}$$

and if

$$k_1 = 1$$

then

$$k_4 = s = \text{radius}$$

The only design parameter required for this characteristic is the radius. ■

The impedance characteristic may also be obtained by using a phase comparator instead of the amplitude comparator. This is left as an exercise (see problem 4.19).

4.7.8 General Beta Plane Characteristics

Many of the relay characteristics that are useful in power system protection can be obtained by using a comparator circuit to define the tripping zone. Since phase and amplitude comparators are equivalent, either type of circuit can be used to determine a given desired characteristic.

Tables 4.4 and 4.5 provide a summary of relay characteristics and their design parameters for representation in the beta plane.

TABLE 4.4 Comparator Quantities for Beta Plane Characteristics [88]

Relay Type	Comparator Quantities to be Compared		
Impedance	Amplitude	$\|KI\|$	$\|V\|$
	Phase	$KI - V$	$KI + V$
Mho	Amplitude	$\|KI\|$	$\|-KI + 2V\|$
	Phase	$KI - V$	V
Ohm	Amplitude	$\|2KI - V\|$	$\|V\|$
	Phase	KI	K
Balanced current	Amplitude	$K\|I_1\|$	$\|I_2\|$
	Phase	$KI_1 - I_2$	$KI_1 + I_2$
Differential current	Amplitude	$K\|I_1 - I_2\|$	$\|I_1 + I_2\|$
	Phase	$(K-1)I_1 - (K+1)I_2$	$(K+1)I_1 + (K-1)I_2$
% Differential current (slope 2s)	Amplitude	$\|I_1 - I_2\|$	$s\|I_1 + I_2\|$
	Phase	$(1-s)I_1 - (1+s)I_2$	$-(1+s)I_1 + (1-s)I_2$
Circulating I pilot scheme A	Amplitude	$\|I_A - \gamma I_B\|$	$K\|I_A - \gamma I_B\|$
	Phase	$(1-K)I_A - (1+K)\gamma I_B$	$(1+K)I_A - (1-K)\gamma I_B$
Circulating I pilot scheme B	Amplitude	$\|I_A - \gamma I_B\|$	$K\|I_A\|$
	Phase	$(1-K)I_A - \gamma I_B$	$(1+K)I_A - \gamma I_B$

TABLE 4.5 Design Parameters for Beta Plane Characteristics [88]

Relay Type	Comp	K_1	K_2	K_3	K_4	r	s
Impedance	Amplitude	K	0	0	1	K	0
	Phase	K	−1	K	1	K	0
Mho	Amplitude	K	0	$-K$	2	$K/2$	$K/2$
	Phase	K	−1	0	1	$K/2$	$K/2$
Ohm	Amplitude	$2K$	−1	0	1	∞	∞
	Phase	K	0	K	−1	∞	∞
Balanced current	Amplitude	K	0	0	1	K	0
	Phase	K	−1	K	1	K	0
Differential	Amplitude	K	$-K$	1	1	$\frac{2K}{K^2-1}$	$\frac{K^2+1}{K^2-1}$
current	Phase	$K-1$	$-(K+1)$	$K+1$	$-(K-1)$	$\frac{2K}{K^2-1}$	$\frac{K^2+1}{K^2-1}$
% Diff I	Amplitude	1	−1	s	s	$\frac{s}{1-s^2}$	$\frac{1+s^2}{1-s^2}$
Slope = 2s	Phase	$1-s$	$1+s$	$-(1+s)$	$1-s$	$\frac{s}{1-s^2}$	$\frac{1+s^2}{1-s^2}$
Circ current	Amplitude	1	$-\gamma$	K	$K\gamma$	$\frac{2K}{1-K^2}\left\|\frac{1}{\gamma}\right\|$	$\frac{1-K^2}{1-K^2}\left\|\frac{1}{\gamma}\right\|\angle\gamma$
Pilot scm A	Phase	$1-k$	$-(1+K)\gamma$	$1+K$	$-(1-K)\gamma$	$\frac{2K}{1-K^2}\left\|\frac{1}{\gamma}\right\|$	$\frac{1-K^2}{1-K^2}\left\|\frac{1}{\gamma}\right\|\angle\gamma$
Circ current	Amplitude	1	$-\gamma$	K	0	$K\left\|\frac{1}{\gamma}\right\|$	$K\left\|\frac{1}{\gamma}\right\|\angle-\gamma$
Pilot scm B	Phase	$1-K$	$-\gamma$	$1+K$	$-\gamma$	$K\left\|\frac{1}{\gamma}\right\|$	$K\left\|\frac{1}{\gamma}\right\|\angle-\gamma$

Many useful relay characteristics can be constructed from circles or arcs of circles, which are readily derived from the comparator equations. This forms a powerful methodology for treating the relay mathematically, and also of realizing the device physically.

A thorough discussion and comparison of relay types, developed in terms of both phase and amplitude comparators is given by Warrington [88], [90] and these books are highly recommended for further study.

REFERENCES

[1] Booth, C. J. Ed., *The New IEEE Standard Dictionary of Electrical and Electronics Terms*, IEEE Std. 100-1992, IEEE, Inc. New York, 1992.

[2] Blackburn, J. L., and G. D. Rockefeller, "Solid-State Relaying for Transmission Lines," a paper presented at the Conference on Protective Relaying, Georgia Institute of Technology, Atlanta, GA, May 6–7, 1965, and published by Westinghouse Relay-Instrument Division as Silent Sentinels, RPL 65-4, May 1965.

[3] Rockefeller, G. D., "Fault Protection with a Digital Computer," *IEEE Trans. on Power Apparatus and Systems*, PAS-88, April 1969, pp. 438–464.

[4] Venkata, S. S., M. J. Damborg, A. K Jampala, and R. Ramaswami, "Annotated Bibliography in the report: Computer Aided Relay Protection Coordination," EPRI Final Report EL-6145, December 1988.

[5] Ryder, J. D., and Charles M. Thomson, *Electronic Circuits and Systems*, Prentice-Hall, Inc., Englewood Cliffs, NJ, 1976.

[6] Booth, C. J., Ed., IEEE Std 100-1992, *The New IEEE Standard Dictionary of Electrical and Electronics Terms*, IEEE, New York, 1992.

[7] Mitra, Sanjit, K., *An Introduction to Digital and Analog Integrated Circuits and Applications*, Harper & Row, Publishers, New York, 1980.

[8] Blackburn, J. L., Ed., *Applied Protective Relaying*, Westinghouse Electric Corporation, Newark, NJ, 1976.

[9] Horn, D. T., *The Comparator Book*, Tab Books, Blue Ridge Summit, PA, 1990.

[10] The Institute of Electrical and Electronics Engineers, *Logic Symbols and Diagrams*, ANSI/IEEE Std 91-1984 and ANSI/IEEE Std 991-1986, IEEE, New York, 1987.

[11] Oppenheim, A. V., and R. W. Shafer, *Digital Signal Processing*, Prentice-Hall, Inc., Englewood Cliffs, NJ, 1975.

[12] GEC Measurements, *Protective Relays Application Guide*, The General Electric Company p.l.c. of England, 1983.

[13] Type SLYP-SLCN, Static Directional Comparison Relaying, Description and Application Manual, General Electric Company, GET-6456A.

[14] Cooley, J. W., and J. W. Tukey, "An Algorithm for the Machine Calculation of Complex Fourier Series," *Math. Computation*, 19, 1965, pp. 297–301.

[15] Phadke, A. G., and J. S. Thorp, *Computer Relaying for Power Systems*, Research Studies Press, Ltd., John Wiley & Sons, Inc., New York, 1988, 1990.

[16] Schweitzer, E. O., III, "Unified Protection, Monitoring, and Control of Overhead Transmission Lines Achieves Performance and Economy," a technical paper presented at the 38th Annual Conference for Protective Relay Engineers, Atlanta, GA, April 23–24, 1985.

[17] Schweitzer, E. O., III., "Four New Digital Relays for Overhead Transmission Line Protection," a technical paper presented at the 12th Western Protective Relay Conference, Spokane, WA, October 22–24, 1985.

[18] Kramer, C. A., and W. A. Elmore, "Flexible Inverse Overcurrent Relaying Using a Microprocessor," IEEE paper 89 TD 378-1, PWRD, Conference Record, IEEE PES 1989 Transmission and Distribution Conference, New Orleans, April 2–7, 1989, IEEE, New York, 1989.

[19] Zocholl, S. E., "Developing a Standard for Overcurrent Relay Characteristics," a paper presented at the 16th Western Relay Conference, Spokane, WA, October 1989.

[20] IEEE Committee Report, "Computer Representation of Overcurrent Relay Characteristics," *IEEE Trans. on Power Delivery*, 4, (2), July 1989, pp. 1659–1667.

[21] Mann, B. J., "Real Time Computer Calculation of the Impedance of a Faulted Single-Phase Line," *Electrical Engineering Transactions*, Australia, March 1969, pp. 26–28.

[22] Mann, B. J., and I. F. Morrison, "Digital Calculation of Impedance for Transmission Line Protection," *IEEE Trans. on Power Apparatus & Systems*, PAS-90 (1), January 1971, pp. 270–279.

[23] McInnes, A. D., and I. F. Morrison, "Real Time Calculation of Resistance and Reactance for Transmission Line Protection by Digital Computer," Electrical Engineering, *Trans. Inst. Engineers*, Australia, EE7 (1), January 1971, pp. 16–23.

[24] Gilcrest, G. B., G. D. Rockefeller, and E. A. Udren, "High Speed Distance Relaying Using a Digital Computer, I - System Description," *IEEE Trans. on Power Apparatus & Systems*, PAS-91, (3), May 1972, pp. 1235–1244.

[25] Gilbert, J. G., and R. J. Shovlin, "High Speed Transmission Line Fault Impedance Calculation Using a Dedicated Minicomputer," *IEEE Trans. on Power Apparatus & Systems*, PAS-94 (3), June 1975, pp. 872–883.

[26] Phadke, A. G., M. Ibrahim, and T. Hlibka, "Fundamental Basis for Distance Relaying with Symmetrical Components," *IEEE Trans. on Power Apparatus & Systems*, PAS-96 (2), March/April 1977, pp. 635–646.

[27] Schweitzer, E. O., and A. J. Flechsig, Jr., "Design and Testing of a Microprocessor-Based Directional Distance Relay," IEEE paper A 78 545-6, presented at the IEEE PES Summer Meeting, July 1978, Los Angeles, CA.

[28] Breingan, W. D., M. M. Chen, and T. F. Gallen, "The Laboratory Investigation of a Digital System for the Protection of Transmission Lines," *IEEE Trans. on Power Apparatus & Systems*, PAS-98 (2), March/April 1979.

[29] Phadke, A. G., T. Hlibka, M. Ibrahim, and M. G. Adamiak, "A Microcomputer Based Symmetrical Component Distance Relay," *Proc. PICA Conference*, Cleveland, May 1979.

[30] Sachdev, M. S., and M. A. Baribeau, "A New Algorithm for Digital Impedance Relays," *IEEE Trans. on Power Apparatus & Systems*, PAS-98 (6), November/December 1979, pp. 2232–2240.

[31] Mainka, M. K. Renz, and G. Koch, "Design Aspects of a Fully Microcomputer-Based Feeder Protection System for HV Lines," CIGRÉ paper 34-12, presented at the 1986 Session, 27 August–4 September, Paris.

[32] Adamiak, M. G., and J. P. Jauch, "Field Experience with the AEP Digital Relay," IEEE Trans. on Power Delivery, PWRD-1, October 1986, pp. 91–98.

[33] Phadke, A. G., A. Politis, and J. S. Thorp, "Improved Protection and Control of Power Systems with Digital Computers," CIGRÉ paper 34-09, presented at the 1986 Session, 27 August-4 September, Paris.

[34] Sykes, J. A., and I. F. Morrison, "A Proposed Method of Harmonic Restraint Differential Protection of Transformers by Digital Computer," *IEEE Trans. on Power Apparatus & Systems*, PAS-91 (3), May/June 1972, pp. 1266–1276.

[35] Einvall, C. H., and J. R. Linders, "A Three-Phase Differential Relay for Transformer Protection," *IEEE Trans. on Power Apparatus & Systems*, PAS-94 (6), November/December 1975, pp. 1971–1980.

[36] Malik, O. P., P. K. Dash, and G. S. Hope, "Digital Protection of a Power Transformer," IEEE paper A76 107, presented at the IEEE PES Winter Meeting, New York, 1976.

[37] Larson, R. R., A. J. Flechsig, and E. O. Schweitzer, "An Efficient Inrush Current Detection Algorithm for Digital Relay Protection of Transformers," IEEE paper A77 510-1, presented at the IEEE PES Summer Meeting, Mexico City, 1977.

[38] Larson, R. R., A. J. Flechsig, and E. O. Schweitzer, "The Design and Test of a Digital Relay Transformer Protection," *IEEE Trans. on Power Apparatus & Systems*, PAS-98, 1979, pp. 795–804.

[39] Yacmini, R., and A. Abu-Nassu, "Numerical Calculation of Inrush Current in Single-Phase Transformers, *IEE Proceedings*, 128, Part B (6), November 1981, pp. 327–334.

[40] Rahman, M. A., and P. K. Dash, "Fast Algorithm for Digital Protection of Power Transformers," *IEE Proc.*, Part C, 129 (2), March 1982.

[41] Degens, A. J., "Microprocessor Implemented Digital Filters for Inrush Detection," *Electrical Power and Energy Systems*, 4 (3), July 1982, pp. 196–205.

[42] Thorp, J. S., and A. G. Phadke, "A Microprocessor-Based Three-Phase Transformer Differential Relay," *IEEE Trans. on Power Apparatus & Systems*, PAS-102 (2), February 1982, pp. 426–432.

[43] Phadke, A. R., and J. S. Thorp, "A New Computer Based, Flux Restrained, Current Differential Relay for Power Transformer Protection," *IEEE Trans. on Power Apparatus & Systems*, PAS-102 (11), November 1983, pp. 3624-3629.

[44] Habib, M., and M. A. Marin, "A Comparative Analysis of Digital Relaying Algorithms for the Differential Protection of Three Phase Transformers," *PICA Proc.*, Montreal, May 1987.

[45] Inagaki, K., M. Higaki, Y. Matsui, M. Suzuki, M. Yoshida, and T. Maeda, "Digital Protection Method for Power Transformers Based on an Equivalent Circuit Composed of Inverse Inductance," IEEE Trans. on Power Delivery, PWRD-3, October 1988, pp. 1501–1510.

[46] Sachdev, M. S., and D. W. Wind, "Generator Differential Protection Using a Hybrid Computer," *IEEE Trans. on Power Apparatus & Systems*, PAS-92 (6), November/December 1973, pp. 2063–2072.

[47] Hope, G. S., P. K. Dash, and O. P. Malik, "Digital Differential Protection of a Generation Unit: Scheme and Real-Time Results," *IEEE Trans. on Power Apparatus & Systems*, PAS-96 (2), March/April 1977, pp. 502–512.

[48] Dash, P. K., O. P. Malik, and G. S. Hope, "Fast Generator Protection Against Internal Asymmetrical Faults," *IEEE Trans. on Power Apparatus & Systems*, PAS-96 (5), September/October 1977, pp. 1498–1506.

[49] Lanz, O. E., and W. Fromm, "A New Approach to Digital Protection," CIGRÉ paper 34-12, 1988 Session, 28 August–3 September 1988, Paris.

[50] Walker, L. N., A. D. Ogden, G. E. Ott, and J. R. Tudor, "Special Purpose Digital Computer Requirements for Power System Substation Needs," IEEE paper 70CP 142-PWR, presented at the IEEE PES Winter Meeting, New York, January 1970.

[51] Walker, L. N., G. E. Ott, and J. R. Tudor, "Simulated Power Transmission Substation," SWIEECO Record of Technical Papers, Dallas, April 1970.

[52] Cory, B. J., and J. F. Moont, "Application of Digital Computers to Busbar Protection," IEE Conference on the Application of Computers to Power System Protection and Metering, Bournemouth, England, May 1970.

[53] Couch, G. H., G. C. Dewsnap, A. D. McInnes, and B. J. Mann, "Application of Digital Computers to Protection, Switching, and Metering," IEE Conference on the Application of Computers to Power System Protection and Metering, Bournemouth, England, May 1970.

[54] Rockefeller, G. D., "What are the Prospects for Substation Computer Relaying," *Westinghouse Engineer*, September 1972, pp. 152–156.

[55] Ratti, U., and L. N. Walker, "A Direct Digital Controlled Power Transmission Substation," *Energ. Elect.* (*Italy*), 49 (10), October 1972, pp. 684–689.

[56] Luckett, R. G., P. J. Munday, and B. E. Murray, "A Substation-Based Computer for Control and Protection," IEE (England) Conference Publication 125, Developments in Power System Protection, London, March 1975.

[57] Couch, G. H., and I. F. Morrison, "Computer Control in Substations; Allocation of Zones and Formulation of Switching Strategies for Primary and Backup Protection," *IEEE Trans. on Power Apparatus & Systems*, PAS-94 (2), March/April 1975, pp. 579–590.

[58] Phadke, A. G., M. Ibrahim, and T. Hlibka, "Computer in an EHV Substation: Programming Considerations and Operating Experience," a paper presented at the CIGRÉ Colloquium, October 1975.

[59] Phadke, A. G., T. Hlibka, and M. Ibrahim, "A Digital Computer System for EHV Substations: Analysis and Field Tests," *IEEE Trans. on Power Apparatus & Systems*, PAS-95 (1), January/February 1976, pp. 291–301.

[60] Udren, E. A., and M. Sackin, "Relaying Features of an Integrated Microprocessor-Based Substation Control and Protection System," IEE Conference Publication 185, Developments in Power System Protection, London, June 1980, pp. 88–92.

[61] Barrett, J. P., M. Pavard, P. Bornard, and J. M. Tesserón, "Digital Processing of Control and Protection Functions in EHV Substations," CIGRÉ paper 34-07, 1982 Session, September 1–9, 1982, Paris.

[62] Udren, E. A., "An Integrated, Microprocessor Based System for Relaying and Control of Substations - Design and Testing Program," 12th Annual Western Protective Relaying Conf., Spokane, WA, October 1985.

[63] Russell, B. D., and K. Watson, "Power Substation Automation Using a Knowledge Based System - Justification and Preliminary Field Experiments," *IEEE Trans. on Power Delivery*, PWRD-2 (4), October 1987, pp. 1090-1097.

[64] Widrevitz, B. C., and R. E. Armington, "A Digital Rate-of-Change Underfrequency Protective System for Power Systems," *IEEE Trans. on Power Apparatus & Systems*, PAS-96 (5), September/October 1977, pp. 1707–1718.

[65] Girgis, Adly A., and F. Ham, "A New FFT-Based Digital Frequency Relay for Load Shedding," *IEEE Trans. on Power Apparatus & Systems*, PAS-101 (2), February 1982, pp. 433–439.

[66] Girgis, A. A., and W. Peterson, "Load Shedding: Present Technology and Future Needs," IEEE Proc. 15th Annual Conference on Modeling and Simulation, April 1984, pp. 1703–1706.

[67] Girgis, Adly A., and W. L. Peterson, "Adaptive Estimation of Power System Frequency Deviation and Its Rate of Change for Calculating Sudden Power System Overloads," *IEEE Trans. on Power Delivery*, PWRD-5 (2), April 1990, pp. 585-595.

[68] Gallen, T. F., M. M. Chen, and W. D. Breingan, "A Digital System for Directional Comparison Relaying," *IEEE Trans. on Power Apparatus & Systems*, PAS-99, (3), May/June 1979, pp. 948–956.

[69] Rijanto, H. H. Prutzer, B. Wienhold, and F. Schindele, "HV-Line Differential Protection with Digital Data Transmission Using Light Fibre Optic Transmission Systems," CIGRÉ paper 34-05, 1982 Session, September 1–9, 1982, Paris.

[70] Kwong, W. S., M. J. Clayton, A. Newbould, and J. A. Downes, "A Microprocessor-Based Current Differential Relay For Use With Digital Communication Systems - Its Performance During Field Tests," CIGRÉ paper 34-02, 1986 Session, August 27–September 4, 1982, Paris.

[71] Horton, J. W., "The Use of Walsh Functions for High Speed Digital Relaying," IEEE Pub. 75, CH1034-8 PWR, paper A75 5827, p. 109.

[72] Brown, R. G., *Introduction to Random Signal Analysis and Kalman Filtering*, John Wiley & Sons, Inc., New York, 1983.

[73] Schweitzer, E. O., III., "Four New Digital Relays for Overhead Transmission Line Protection," a paper presented at the 12th Annual Western Protective Relay Conference, Spokane, WA, October 22–24, 1985.

[74] Girgis, A. A., and R. G. Brown, "Application of Kalman Filtering in Computer Relaying," *IEEE Trans. on Power Apparatus & Systems*, PAS-100 (7), July 1981, pp. 3387–3397.

[75] Girgis, A. A., "A New Kalman Filtering Based Digital Distance Relay," *IEEE Trans. on Power Apparatus & Systems*, PAS-101 (9), September 1982, pp. 3471–3480.

[76] Girgis, A. A., and R. G. Brown, "Modelling of Fault-Induced Noise Signals for Computer Relaying Applications," *IEEE Trans. on Power Apparatus & Systems*, PAS-102 (9), 1983, pp. 2831–2841.

[77] Girgis, A. A., and D. Hwang, "Optimal Estimation of Voltage Phasors and Frequency Deviation Using Linear and Non-Linear Kalman Filtering Theory and Limitations," *IEEE Trans. on Power Apparatus & Systems*, PAS-103 (10), 1984, pp. 2943–2952.

[78] Girgis, A. A., and R. G. Brown, "Application of Adaptive Kalman Filtering in Computer Relaying: Fault Classification Using Voltage Models," *IEEE Trans. on Power Apparatus & Systems*, PAS-104 (5), May 1985, pp. 1168-1177.

[79] Girgis, A. A., "Concepts of Kalman Filtering in Power System Protection Using Microprocessors," Proc. 17th Southeastern Symposium on System Theory, April 1985, pp. 16–20.

[80] Girgis, A. A., and E. B. Makram, "Application of Adaptive Kalman Filtering in Fault Classification, Distance Protection, and Fault Location Using Microprocessors," *IEEE Transactions on Power Systems*, PWRS-3, February 1988, pp. 301–309.

[81] Girgis, A. A., and D. G. Hart, "Implementation of Kalman and Adaptive Kalman Filtering Algorithms for Digital Distance Protection on a Vector Signal Processor," *IEEE Trans. on Power Delivery*, PWRD-4, January 1989, pp. 141–156.

[82] Sakaguchi, T., "A Statistical Decision Theoretical Approach to Digital Relaying," *IEEE Trans. on Power Apparatus & Systems*, PAS-99 (5), September/October 1980, pp. 1918–1926.

[83] Phadke, A. G., J. S. Thorp, and S. H. Horowitz, "Study of Adaptive Transmission System Protection and Control," ORNL/SUB/85-2205C, U. S. Department of Energy, Oak Ridge National Laboratory.

[84] Rockefeller, G. D., C. L. Wagner, J. R. Linders, K. L. Hicks, and D. T. Rizy, "Adaptive Power System Transmission Protection" ORNL/SUB/85-2201C, U. S. Department of Energy, Oak Ridge National Laboratory.

[85] Thorp, J. S., S. H. Horowitz, and A. G. Phadke, "The Application of an Adaptive Technology to Power System Protection and Control," CIGRÉ paper 34-03, 1988 Session, August 28–September 3, 1988, Paris.

[86] Phadke, A. G., and S. H. Horowitz, "Adaptive Relaying," *IEEE Computer Applications in Power*, July 1990, pp. 47–51.

[87] Westinghouse Electric Corp., MDAR Relay Descriptive Bulletin B-986, Westinghouse Relay and Telecommunications Division, Coral Springs, FL, 1988.

[88] Warrington, A. R. van C., *Protective Relays: Their Theory and Practice*, 1, John Wiley and Sons, New York, 1962.

[89] Beyer, William H., Ed., *CRC Standard Mathematical Tables*, 26th Ed., CRC Press, Inc., Boca Raton, FL, 1981.

[90] Warrington, A. R. van C., *Protective Relays: Their Theory and Practice*, 2, John Wiley and Sons, New York, 1969.

PROBLEMS

4.1 The network shown in Figure P4.1 is called a composite sequence current network [8] and is used as a filter to isolate sequence currents for relaying needs. The network is capable of performing different functions depending on the position of switches r and s. The network analysis is usually performed assuming that the V_F terminal is open, or that this terminal is

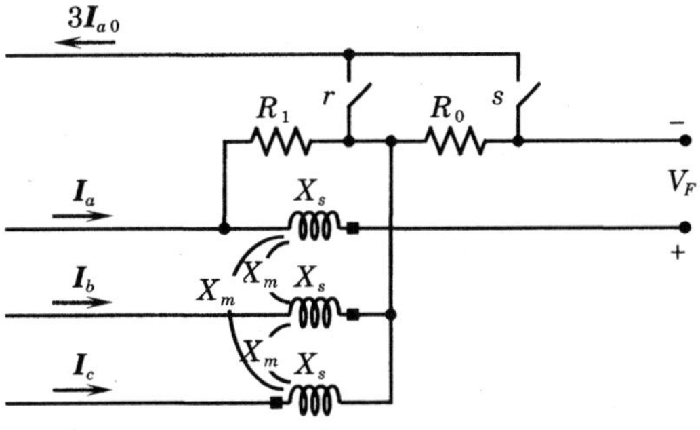

Figure P4.1 A composite sequence network.

loaded into a very high impedance network. The output voltage of the network is a function of the sequence voltages.

(a) Analyze the network assuming switch r is open and switch s is closed. Also, assume that X_s is negligibly small.

(b) Analyze the network assuming that switch r is closed and switch s is open.

4.2 In some cases, the input to an operational amplifier is modified by the addition of a potentiometer, shown in Figure P4.2, so that the input voltage can be adjusted to a desired value. However, the potentiometer has a total resistance that is finite, and this causes an error in the input voltage compared to the potentiometer fractional position, k.

(a) Compute the error in the voltage V_1 that is due to the finite potentiometer resistance.

(b) Find the percent error if $R_1 = 100\text{k}\Omega$, $R_2 = 40\text{k}\Omega$, and $k = 0.4$.

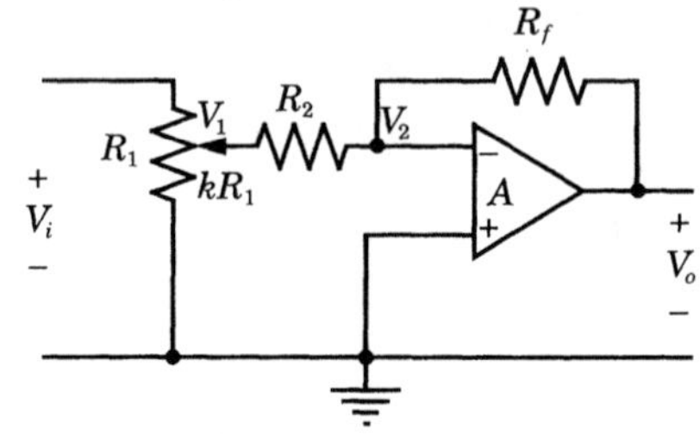

Figure P4.2 Operational amplifier with potentiometer input.

4.3 For the op amp circuit of Figure 4.1, let $R_1 = 8\text{k}\Omega$ and $R_f = 100\text{k}\Omega$.

(a) Find the output voltages corresponding to input voltages of 0.4, 1.0, 2.0, and 2.5 volts.

(b) Assume that the op amp saturates at 15 volts. What limitation does this place on the range of input voltages?

4.4 An equivalent circuit for an op amp has been suggested that is constructed as shown in Figure P4.4. We would like to check the accuracy of the equivalent circuit.

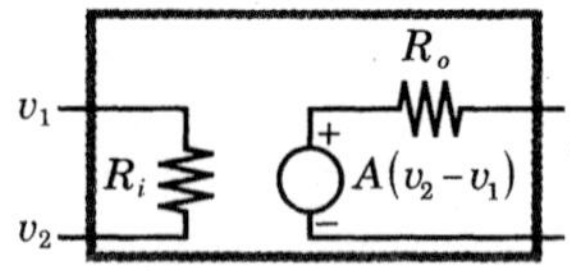

Figure P4.4 Equivalent circuit of an op amp.

(a) Determine the transfer function of the equivalent circuit when connected with input resistance R_1 and feedback resistance R_f as shown in Figure 4.1.

(b) Find the limiting values of the transfer function with the following limits:

(1) $R_o \to 0$

(2) $R_i \to \infty$

(3) $A \to \infty$

4.5 Modify the network of problem 4.4(a) to add a load resistance R_L to the output of the op amp. Write the general solution for this case.

4.6 Consider a digital sequence $y(n)$ where the values of the sequence are obtained by sampling a 60 hertz sinusoidal voltage signal given in the time domain as

$$v(t) = V_{\max} \sin(\omega t + \pi/6)$$

and with the sampling specified to occur every $\pi/6$ radians or 12 samples per cycle of the 60 hertz waveform.

(a) Write out the numerical values of the sampled waveform, in per unit of the peak voltage, for one complete cycle. Let $a(n)$ be the value of the sequence at n.

(b) Determine the value of the specified sequence corresponding to the value $\omega t = 45°$.

4.7 Define new sequences that are defined in terms of the unit sample. Write out the following in terms of the value $a(n)$ and the unit sample.

(a) The sequence $v_1(n)$, which is defined as the first five samples of the sequence defined in problem 4.6.

(b) The sequence $v_2(n)$, which is defined as only those samples of the sequence of problem 4.6 that represent the maximum and zero values of that sequence.

4.8 Consider a system with unit sample sequence

$$h(n) = \begin{cases} e^n, & n \geq 0 \\ 0, & n < 0 \end{cases}$$

Find the response of this system to an input defined in terms of the unit step sequence as

$$x(n) = u(n) - u(n-5)$$

The input sequence is shown in Figure P4.8.

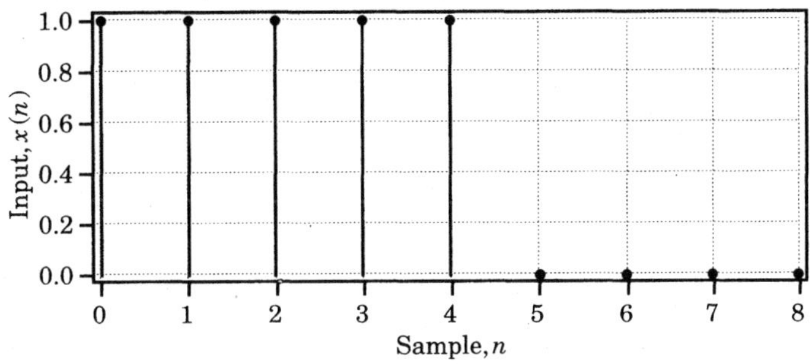

Figure P4.8 Input sequence, $x(n)$.

(a) Plot the exponential system sequence defined in the problem statement.

(b) Plot the response of the system to the input sequence defined in Figure P4.8.

4.9 Given an amplitude comparator that monitors two complex voltages or currents $\boldsymbol{A}$ and $\boldsymbol{B}$, and that is set to respond when $|\boldsymbol{A}| > |\boldsymbol{B}|$.

(a) Define the threshold of operation of the device.

(b) Show how this amplitude comparator can be used as a phase comparator.

4.10 Show that the characteristic of the amplitude comparator with threshold

$$|k_1\boldsymbol{A} + k_2\boldsymbol{B}| = |k_3\boldsymbol{A} + k_4\boldsymbol{B}|$$

where $k_1, \ldots, k_4$ are design constants, is a circle in the α plane, where $\alpha = \boldsymbol{A}/\boldsymbol{B}$.

4.11 Determine the differential current relay characteristic of example 4.1 in rectangular form.

4.12 Verify that the relay characteristic of the differential current relay is a circle in the alpha plane.

4.13 Given a differential current relay that has operate and restraint quantities

$$\boldsymbol{Q}_o = (\boldsymbol{I}_1 - \boldsymbol{I}_2)$$
$$\boldsymbol{Q}_R = \frac{s}{2}(\boldsymbol{I}_1 + \boldsymbol{I}_2)$$

and threshold $\boldsymbol{Q}_o = k\boldsymbol{Q}_R$. Compute the following in terms of k and s if the relay operates as an amplitude comparator:

(a) The polar equation for the threshold characteristic in the α plane.

(b) The polar equation for the threshold characteristic in the β plane.

(c) The rectangular equation for the threshold characteristic in the α plane.

(d) The rectangular equation for the threshold characteristic in the β plane.

4.14 Sketch the threshold characteristic computed in problem 4.5 for $s = 0.1$
(a) In the α plane.
(b) In the β plane,
and verify that the polar and rectangular characteristics give identical results. Let $\theta = 0$.

4.15 Verify (4.90) by writing $e^{j(\sigma_1-\sigma_2)}$ in rectangular form.

4.16 Find an expression for the radius and center for the circular characteristic of a phase comparator
(a) in the α plane
(b) in the β plane

4.17 Find an expression for the operating quantity $\boldsymbol{Q}_o$ and the restraint quantity $\boldsymbol{Q}_R$ for a phase comparator.

4.18 Repeat 4.17 if the system is biased to operate only when one measured quantity $\boldsymbol{N}_2$ is within a certain known angle λ of $\boldsymbol{N}_2$.

4.19 Compute the values of k_1, k_2, k_3, and k_4 that must be used in a phase comparator to obtain exactly the same characteristic in both the alpha and beta planes.

4.20 Compute the necessary coordinates and plot the beta plane characteristic of the phase comparator in example 4.2 where $s = 3$. Compare with the amplitude comparator of problem 4.14.

4.21 Compute a set of constants $k_1, \ldots, k_4$ that may be used to obtain an impedance characteristic using a phase comparator circuit.

4.22 Compute a set of constants $k_1, \ldots, k_4$ that may be used to obtain a mho characteristic in the impedance plane with center at $\boldsymbol{Z}_o$ using
(a) an amplitude comparator
(b) a phase comparator

4.23 Compute a set of constants $k_1, \ldots, k_4$ that may be used to obtain a reactance characteristic in the admittance plane using
(a) an amplitude comparator
(b) a phase comparator

4.24 Verify (4.99).

4.25 Consider the phase comparison logic shown in Figure P4.25, where we define the complex input signals as

$$\boldsymbol{S}_O = \text{The operate signal}$$
$$\boldsymbol{S}_R = \text{The restraint signal}$$

Let these input signals be derived from relay measurements that can be written as

$$\boldsymbol{S}_O = \boldsymbol{k}_1\boldsymbol{V}_R + \boldsymbol{Z}_{k1}\boldsymbol{I}_R = k_1 e^{j\alpha_1} V_R e^{j0} + Z_{k1} e^{j\theta_1} I_R e^{-j\phi_R}$$
$$\boldsymbol{S}_R = \boldsymbol{k}_2\boldsymbol{V}_R + \boldsymbol{Z}_{k2}\boldsymbol{I}_R = k_2 e^{j\alpha_2} V_R e^{j0} + Z_{k2} e^{j\theta_2} I_R e^{-j\phi_R}$$

where we have let the relay voltage be the phasor reference.
(a) Show that the ratio of the operate to the restraint signal corresponds to a circle in the Z plane.
(b) Determine the values of the parameters if the desired circular characteristic is that of an offset mho relay.
(c) Determine the values of the parameters if the desired circular characteristic is that of an mho relay.
(d) Determine the values of the parameters if the desired circular characteristic is that of an impedance relay.

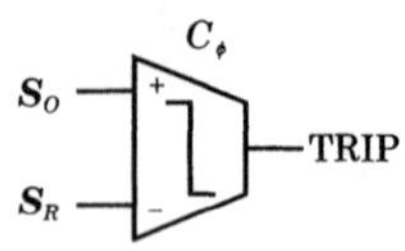

Figure P4.25 A phase comparison logic.

5

System Characteristics

There are several aspects of power system performance that are important in protective system design that are dependent on the active and passive components that make up the system, but are independent of the protective devices themselves. We consider several concepts in this chapter that are a necessary prerequisite to the analysis of protective systems.

We begin with the characteristics of short-circuit currents themselves, and the way in which the currents are affected by the synchronous generators and ac motors on the system. This is a cursory review of this complex subject, and will present only important results that have an immediate bearing on system protection.

The second system aspect is that of switching station arrangements and how these arrangements affect the reliability and security of the power system. These arrangements also affect the protective system design directly. The goal here is to present commonly used switching station arrangements, so that the student of protection will be familiar with the options available for switching and protection in system design.

Finally, we review the characteristics of the passive transmission and distribution circuits and the impedance calculations that are necessary in order to solve for fault currents.

5.1 POWER SYSTEM FAULTS

Short circuits on power systems are usually shunt disturbances of one of the following types (shown along with their common abbreviation):

Three-phase short circuit	(3PH)	
Phase-to-phase short circuit	(L-L)	(5.1)
Two-phase-to-ground short circuit	(2LG)	
One-phase-to-ground short circuit	(1LG)	

The total current flowing to the fault depends on the type of fault and the phase in which the current is measured. It also depends on the location in the system where the fault occurs, since the Thevenin equivalent impedance seen looking back into the system varies with location, with the amount of generation in service at the time, and with the branch switching of the network. This impedance also varies with load since the generators (and motors) are switched on and off in the normal course of scheduled activity on the network. All of these variations can be important, both in determining the exact fault current available at a given place and time, but also in fixing the critical parameters of the protective system. The total maximum fault current is important in determining the interrupting rating of devices. Both the maximum and minimum fault currents and voltages are important for ensuring correct operation of the protective system over all possible operating conditions.

In some cases it is important to recognize system configurations that are not short circuits, but still should be considered "faults." These are often grouped together as longitudinal faults, as opposed to the lateral faults described in (5.1). These faults are described as follows:

One line open	(1LO)	
Two lines open	(2LO)	(5.2)
Three lines open	(3LO)	

The first two of these conditions present unbalanced current flow in the three-phase system, and may require protective response if the unbalance presents a serious threat to equipment. The third fault in (5.2) is nothing more than a line being open, which will not usually require any special protective system action except, perhaps, reclosure.

5.1.1 System Fault Characteristics

Fault currents in a power system depend on the impedance of the system between the generating sources and the fault location, on the prefault current flowing in a given line contributing to the total fault, and on the point on the sine wave of voltage at which the fault occurs. Consider the system shown in Figure 5.1 where a fault is applied at some point in a power system.

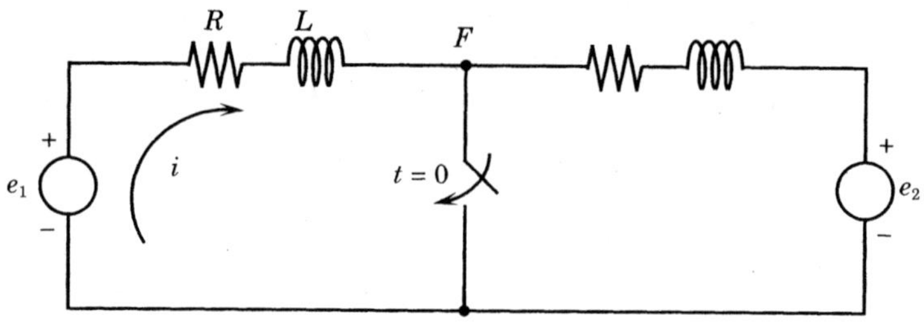

Figure 5.1 Fault applied to a power system.

We shall concentrate only on the contribution to the fault labeled i that is radial from the source e on the left in Figure 5.1. The power system is represented by the resistance R and the inductance L, which are Thevenin equivalent parameters of the entire system to the left of the fault point F. We can write the differential equation of the circuit on the left side as follows.

$$L\frac{di}{dt} + Ri = E_m \sin(\omega t + \beta) \tag{5.3}$$

where E_m is the maximum value of the sinusoidal source voltage and β represents the angle of the supply voltage at which the fault is applied. This representation of the system neglects the shunt susceptances of the transmission lines, but this is often an acceptable assumption in fault studies since the voltages are severely depressed in the vicinity of the fault. This assumption needs to be checked, however, for high-voltage systems.

Solving (5.3) for the fault current, we get

$$\begin{aligned} i(t) &= i_s(t) + i_t(t) \\ &= \frac{E_m}{Z}\sin(\omega t + \beta - \theta) + \left[\frac{E_m}{Z}\sin(\theta - \beta) + i(0^+)\right]e^{-Rt/L} \end{aligned} \tag{5.4}$$

where

$$\begin{aligned} Z &= \sqrt{R^2 + X^2} \\ X &= \omega L \\ \theta &= \tan^{-1}\left(\frac{\omega L}{R}\right) \end{aligned} \tag{5.5}$$

Note that the total fault current has been divided into two components, a steady-state component i_s and a transient component i_t. The steady-state component has the frequency of the applied voltage, but shifted in phase by the angle β and the constant angle of the system impedance, θ, and with a magnitude that is determined by the magnitude of the applied voltage and of the system impedance. The transient component has two parts, one that depends on the angle β on the voltage wave at which the fault is applied. The other component is a function of the prefault current that is flowing at the instant the fault is applied. In many cases, this prefault load current in negligible compared to the larger magnitude of the transient fault current.

The total fault current can be described as a sinusoidal current with a dc offset that decays with a time constant $t = L/R$. We can estimate the value of this time constant as follows.

$$\tau = \frac{L}{R} = \frac{\omega L}{\omega R} = \frac{X/R}{\omega} \tag{5.6}$$

We know that power systems have X/R values of 10 to 20, with the higher ratios being characteristic of the higher voltage (EHV) systems. The value of the system time constant for 50 and 60 Hz systems is shown in Figure 5.2. The transient component of current will be reduced to $1/ce$ of its initial value in c time constants. For example, in one time constant, the transient will be reduced to $1/e$, or about 36.8% of its initial value. On a low-voltage system with a system X/R of 8 and time constant of 0.02 seconds, the transient current reaches this 36.8% value in 0.02 seconds or about 1.2 cycles on an 60 Hz system. On low-voltage systems, the protection may require a few cycles to complete its operation, so the transient will be decayed a few time constants before the breaker operates. However, on a high-voltage system with an X/R of 20, it takes a little over 0.05 seconds for the decay to reach 36.8%, or just over

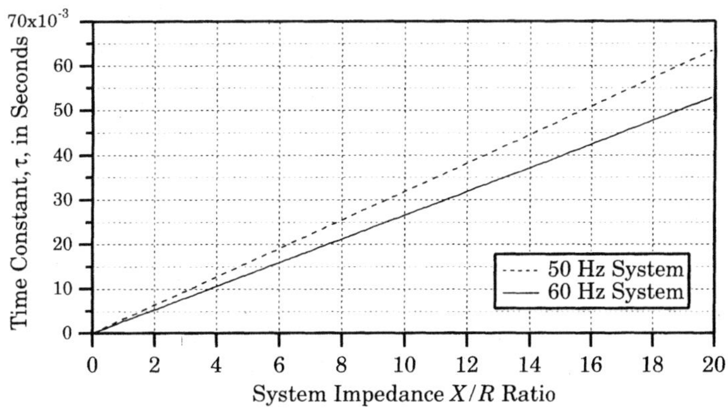

Figure 5.2 System time constant for typical power systems.

three cycles. This is a large difference considering that the protection system on EHV systems should be much faster than three cycles.

This transient component has its maximum value at $t = 0$, and this is the value of greatest interest as it represents the maximum current. The maximum positive current offset occurs under the following conditions:

$$\begin{aligned} \sin(\theta - \beta) &= +1 \\ i(0^+) &> 0 \end{aligned} \tag{5.7}$$

where both conditions apply simultaneously. Ignoring the prefault load current, for the moment, the maximum positive offset occurs when

$$\beta = \theta - \pi/2 \tag{5.8}$$

The maximum negative offset can be similarly computed, but we are not so much concerned by the sign of the offset as its magnitude and rate of decay.

Considering only the maximum condition, but ignoring the prefault load current, we can write the maximum current as

$$\begin{aligned} i_P(t) &= \frac{E_m}{Z}[\sin(\omega t \mp \pi/2) \pm e^{-t/\tau}] \\ &= \frac{E_m}{Z}(\mp \cos \omega t \pm e^{-t/\tau}) \end{aligned} \tag{5.9}$$

where the choice of sign is determined arbitrarily on whether one is interested in a positive or a negative dc offset. The magnitudes are the same in either case. Also note that the current (5.9) is subscripted "P" to indicate that this is the current in the primary of the current transformer of the protective system.

The measurement of the fault current is shown in Figure 5.3, where the secondary current sees a burden Z_B that consists of the relay impedance as well as the impedance of the leads from the CT secondary to the relay. Since the primary fault current is often very large, there is always a concern for the saturation of the current transformer. If we ignore the transformer excitation current, for the moment, we can write the voltage across the burden as

$$v_B = Z_B i_S = \frac{Z_B i_P}{n} \tag{5.10}$$

For an ideal transformer, we can write the voltage in terms of the rate of change of flux linkages.

$$v_B = n\frac{d\phi}{dt} \tag{5.11}$$

where n is the turns ratio of the transformer. The flux is then computed by integrating (5.11)

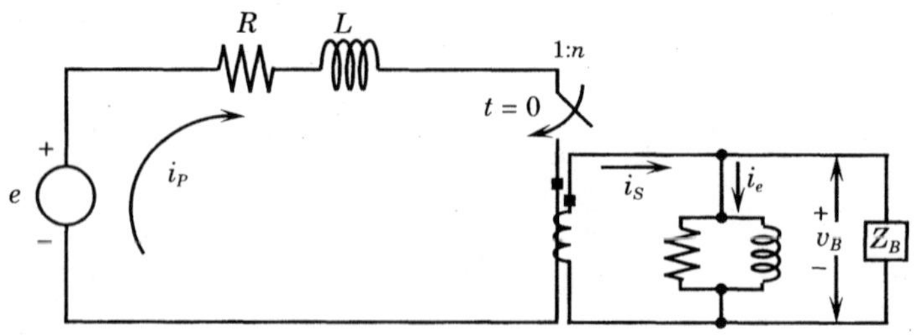

Figure 5.3 Measurement of the fault current.

for a total system impedance of Z [also see (2.4) for review].

$$\begin{aligned}\phi &= \frac{1}{n}\int_0^t v_B\,dt = K\int_0^t (e^{-t/\tau} - \cos\omega t)\,dt \\ &= \frac{K}{a}(1 - e^{-at}) - \frac{K}{\omega^2}\sin\omega t \quad \text{Wb} = \phi_t + \phi_s\end{aligned} \tag{5.12}$$

where

$$K = \frac{Z_B E_m}{Z n^2} \qquad a = \frac{\omega}{X/R}$$

and Z is the total impedance seen by the source voltage.

A plot of the primary current and the flux, with arbitrary vertical scaling, is shown in Figure 5.4. Note that the primary current is completely offset at time zero. The flux also has a dc offset that is relatively large compared to the steady-state component because the time constant multiplier for the transient term is relatively large compared to the $1/\omega$ multiplier of the steady-state component in (5.12). The transient flux plotted in Figure 5.4 assumes an ideal transformer and represents an unrealistic condition. Any physical transformer will saturate with high values of dc flux unless the core has a very large cross section. Therefore, a more realistic picture would be for the flux to level off at some saturated value immediately after zero time. This would provide very small values of secondary voltage due to the small (saturated) value of the flux rate of change. This means that the secondary current will be much smaller than the ideal value shown in the figure until the core saturation reduces to a small value, which may take quite a long time [1], [2].

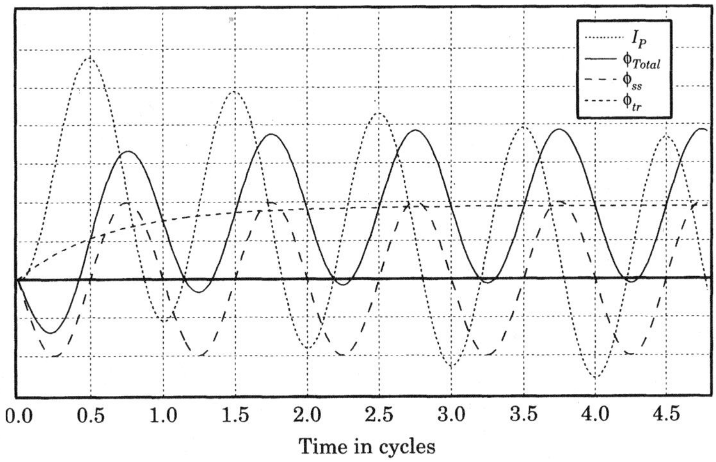

Figure 5.4 Plot of primary current and transient current transformer flux.

One way to measure the relative magnitude of the severity of the saturation is to integrate the steady-state peak flux over 0.5–1.5 cycles and compare this with the transient flux integrated from zero to infinity. If this is done, we can compute the ratio of the two quantities as

$$\frac{\phi_t}{\phi_s} = \frac{X}{R} \tag{5.13}$$

Using this result, we can write the total flux as

$$\phi = \phi_s\left(1 + \frac{X}{R}\right) \tag{5.14}$$

The multiplier in parentheses is called the "transient factor," which gives a measure of the degree of saturation during the current transient. This emphasizes the importance of the system X/R ratio in protective system behavior [2].

5.1.2 Fault Currents Near Synchronous Machines

Another important characteristic that affects certain protective system decisions is the nature of the fault current in close proximity to synchronous generators. This characteristic is largely due to the way in which fault currents decay with time in the stator windings of the machine. At the instant the fault occurs, each phase current has a value that depends on the generator load at that instant. Following the incidence of the fault, the currents in each phase change rapidly to very high values that depend on the type of fault, its location (severity), and the generator subtransient reactance, x_d''. This very high current begins immediately to decay, first at a rapid rate determined by the subtransient time constant t_d'', and later at a slower rate determined by the transient time constant t_d'. Depending on the speed of the relaying and circuit breaker action, the large initial current may have to be interrupted. It is important, therefore, that the protective system designer check the highest possible value that this current can reach, which would occur for a fault at the maximum generation condition and with no impedance from the relay location to the fault location, in other words, a close-in fault. Usually this is done for the three-phase fault with maximum dc current offset in one phase.

It may also be important to estimate the contribution to the fault of nearby motors, particularly large motors. Synchronous motors contribute fault current in exactly the same way as synchronous generators, since they have their own source of excitation. Induction motors, however, experience a rapid decay in air gap flux and their contribution to the fault is usually negligible after two or three cycles. In both cases, the prime mover power for the motor comes from the shaft load. This load will often have sufficient inertia to cause the motor to feed the fault for a few cycles.

To estimate the fault current contribution of the synchronous generator, consider a fault on the terminals of a generator that is initially unloaded. At the instant the fault occurs, the generator is operating with an open circuit terminal voltage E, which is equal to the rated rms voltage. The flux in the air gap is maintained by the field excitation system at a value that will just provide E at the correct value. When the fault strikes, the peak air gap flux will have some value and this flux will have a given angle with respect to the a-phase stator winding, which we consider the reference position. This flux is trapped at its prefault value and currents will flow in the three-phase windings, the field winding, and the damper windings (or solid rotor body) to try and maintain this trapped flux at its prefault magnitude and angle, according to Lenz's law. The energy stored in this magnetic field is quickly absorbed by ohmic losses and the flux begins to decay immediately, and the currents decay as well, finally reaching a new steady-state condition. The actual stator currents have three types of components, a dc offset, a double frequency component, and a rated frequency component. The rated frequency component of five to 10 times normal is initially present in the stator and this current decays to its final value, which depends on the machine synchronous reactance. We are most interested in this rated frequency component and its rms value.

It is convenient to divide the rated frequency current into three components, as follows [3], [4]:

$$I_d = \frac{E}{x_d} \tag{5.15}$$

$$I'_d = \frac{E}{x'_d} \tag{5.16}$$

$$I''_d = \frac{E}{x''_d} \tag{5.17}$$

where E is the rated rms generator phase voltage which is a function of the field current. These equations give the synchronous, transient, and subtransient components of the symmetrical ac stator current at time $t = 0+$, i.e., immediately following the fault. The transient and subtransient currents decay, each with its own time constant. These three currents and their decay pattern are shown in Figure 5.5. Note the different decay time constants of the transient and subtransient components.

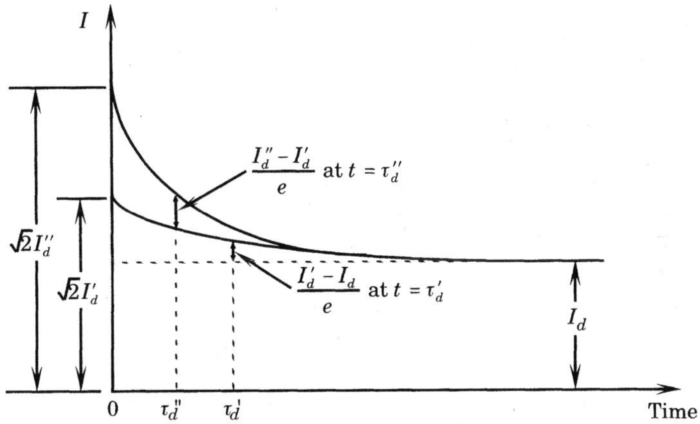

Figure 5.5 Decay patterns of transient and subtransient stator currents following a three-phase fault on an unloaded generator.

Also, the field current can be written as

$$I_F = \frac{E_x}{r_F}(1 - e^{-t/\tau'_{do}}) \tag{5.18}$$

where E_x = exciter voltage
r_F = field winding resistance
I_F = field winding current
t'_{do} = open circuit transient time constant

Table 5.1 provides a set of typical generator reactances and time constants.

TABLE 5.1 Machine Reactances and Time Constants

Item	Units	Turbogenerators	Hydrogenerators
x_d	pu	1.40–1.80	0.90–1.10
x'_d	pu	0.23–0.35	0.25–0.45
x''_d	pu	0.15–0.23	0.20–0.35
τ'_{do}	s	2.80–9.20	1.50–9.50
τ'_d	s	0.40–1.80	0.50–3.30
τ''_d	s	0.02–0.05	0.01–0.05
τ_a	s	0.04–0.35	0.03–0.30

We compute the total ac stator current as the sum of the three ac components, as follows. First we define the following incremental currents.

$$\Delta I' = (I'_d - I_d)e^{-t/\tau'_d} \tag{5.19}$$

$$\Delta I'' = (I''_d - I'_d)e^{-t/\tau''_d} \tag{5.20}$$

where

$$\begin{aligned} \tau'_d &= \frac{x'_d}{x_d}\tau'_{do} \\ \tau''_d &= \frac{x''_d}{x'_d}\tau''_{do} \end{aligned} \tag{5.21}$$

Then the total ac rms symmetrical current is given by

$$I_{\text{ac}} = I_d + \Delta I' + \Delta I'' \tag{5.22}$$

The dc component of stator current can take on any initial value from zero to I''_d, and it decays at its own time constant. This current component is written as

$$I_{\text{dc}} = \sqrt{2}I''_d e^{-t/\tau_a} \tag{5.23}$$

where

$$\tau_a = \frac{L_2}{r_a} \tag{5.24}$$

Note that the time constant depends on the negative sequence inductance of the generator and the armature resistance.

Now the total fault current of interest is the rms value of the total current, which has both an ac and dc component. We write this total rms current as

$$I_{\text{total}} = \sqrt{I_{\text{ac}}^2 + I_{\text{dc}}^2} \tag{5.25}$$

We can compute the total fault current at $t = 0$ from (5.22) with the result

$$I_{\text{max}} = \sqrt{2I''^2_d + I''^2_d} = \sqrt{3}I''_d \tag{5.26}$$

The two components of the total current decay at quite different rates. The ac component decays at a rate determined by the transient and subtransient time constants of the generator. The dc component, however, decays at a rate that depends on the X/R of the system to which the generator is connected. In practical situations, it is the dc component that will dominate the total current decay.

To visualize this decay pattern, consider a simple series R-L circuit with a constant dc initial current I_{dc} flowing. If the R-L portion of this circuit is suddenly shorted, the current will decay as governed by the equation

$$i(t) = I_{\text{dc}}e^{-Rt/L} \tag{5.27}$$

Now, if we think of this as the decaying dc component, whose initial value is

$$I_{\text{dc}} = \sqrt{2}I_{\text{ac}} \tag{5.28}$$

then we may compute the total current as

$$I_{\text{total}} = \sqrt{I_{\text{ac}}^2 + 2I_{\text{ac}}^2 e^{-2Rt/L}} = I_{\text{ac}}\sqrt{1 + 2e^{-2Rt/L}} \tag{5.29}$$

We may write (5.29) as

$$I_{\text{total}} = F_a I_{\text{ac}} \tag{5.30}$$

where we have defined the "asymmetry factor" F_a as

$$F_a = \sqrt{1 + 2e^{-2Rt/L}} = \sqrt{1 + 2\exp\left(\frac{-4\pi f t}{X/R}\right)} \tag{5.31}$$

Figure 5.6 shows a symmetrical fault current for phase a of the generator and for a fault on the generator terminals. This will occur only when the phase angle of the voltage is exactly zero when the fault occurs and with resistance neglected. In this case the fault current is exactly symmetrical and decays as dictated by (5.22). Note that the rate of decay is much greater initially, with the decay at a slower rate after the subtransient current has died away. This case is special because of its symmetry.

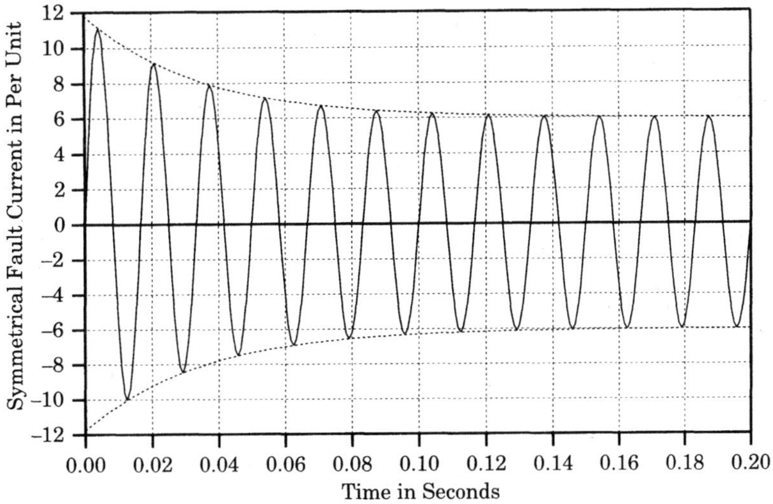

Figure 5.6 Plots of the symmetrical ac component and the total fault current envelope, including the dc component.

Figure 5.7 shows a plot of the three-phase currents of the generator that corresponds to the symmetrical phase a current of Figure 5.6. In this case, the other two-phase currents have dc decrements that are equal and opposite, with the phase b decrement being the negative of the phase c decrement. The actual phase currents are displaced in phase, however, in the normal manner.

The asymmetry factor takes on its maximum value at $t = 1/2$ cycle or 1/120 second. The value of the asymmetry factor at this value of t on a 60 hertz system is given by

$$F_{a(\max)} = \sqrt{1 + 2\exp\left(\frac{-2\pi}{X/R}\right)} \tag{5.32}$$

The dc offset is different in each phase, however, so the average offset is not as great as expressed by (5.31). We can easily compute an average value of the asymmetry factor, which

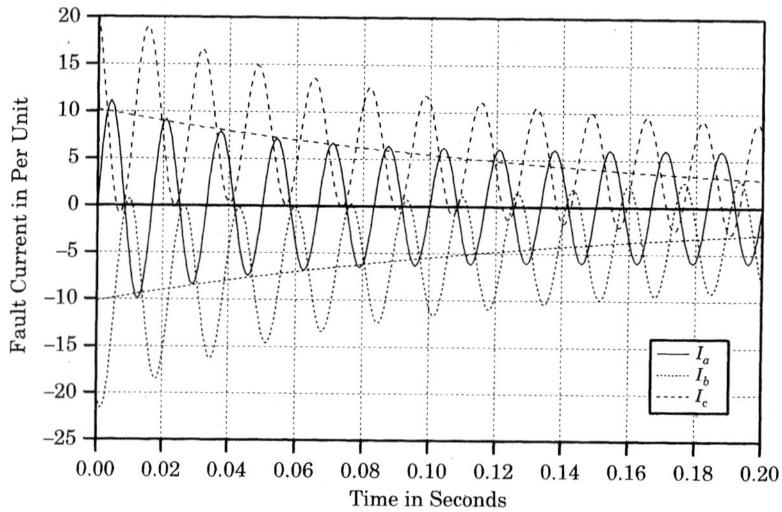

Figure 5.7 Total fault current showing maximum offset.

is given at $t = 1/120$ seconds as

$$F_{a(\text{ave})} = \sqrt{1 + \frac{8}{9}\exp\left(\frac{-2\pi}{X/R}\right)} \tag{5.33}$$

Thus, we see that the dc decay depends on the X/R of the system and values of interest lie between the average and maximum values of asymmetry as given by (5.32) and (5.33), respectively. These values of asymmetry factor are plotted in Figure 5.8.

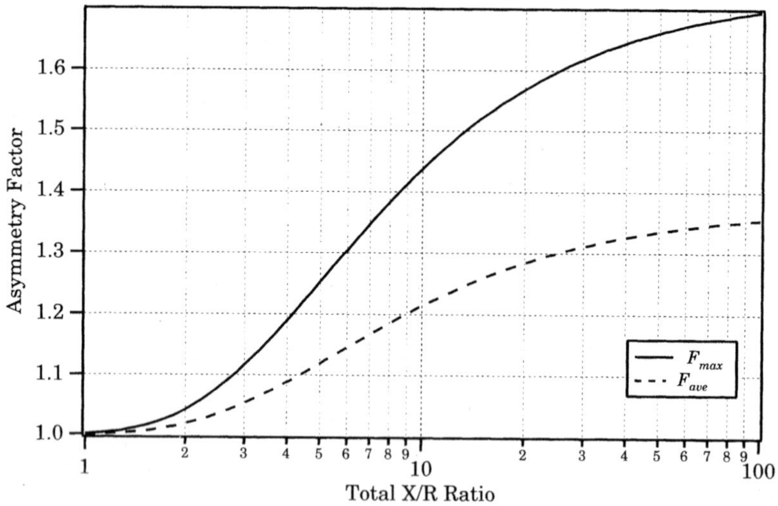

Figure 5.8 Maximum and average asymmetry factors as a function of system X/R.

Using Figure 5.8, one can estimate the maximum asymmetry of the fault current for a particular system condition. This permits an evaluation of the maximum dc component that can exist for that system. The circuit breaker must be able to interrupt the maximum offset current. However, this maximum has a lower probability than the average and this may be an important factor, depending on how much risk one is willing to take and the cost of reducing that risk to zero. Figure 5.9 shows another plot of three-phase fault currents on an unloaded generator. This is the case of maximum dc offset. In this example, the current in phase a fails to cross zero for the first six cycles, due to the subtransient offset. This makes interrupting the current of phase a very difficult, since the moment of current interruption by the circuit breaker is normally at the normal current zero.

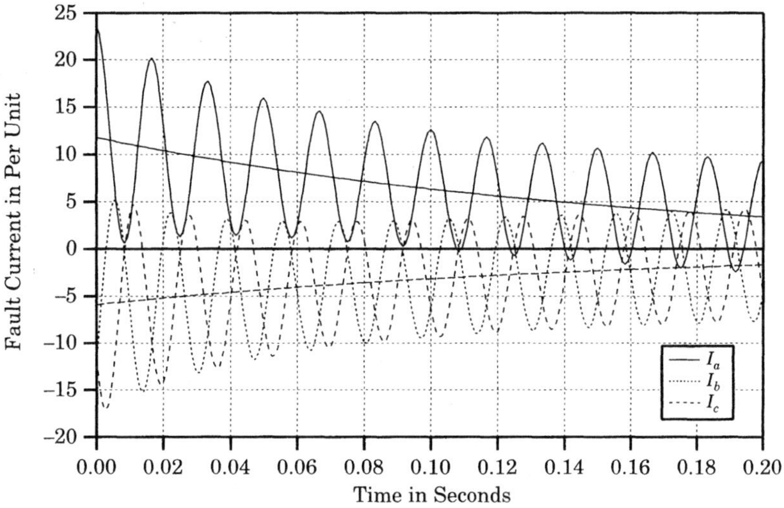

Figure 5.9 Typical three-phase fault on an unloaded generator.

The exact value of the zero crossing depends on the generator parameters and on its prefault loading [5], [6]. The equation for the stator current of the generator can be written from the basic generator equations [7]. For a turbogenerator with effective amortisseur windings in the d and q axes, we write the phase a current following a simultaneous three-phase fault as follows.

$$\begin{aligned}
i_a = -\sqrt{2}V_{qo} &\left\{ \begin{array}{l} \left[\dfrac{1}{x_d} + \left(\dfrac{1}{x'_d} - \dfrac{1}{x_d}\right) e^{-t/\tau'_d} + \left(\dfrac{1}{x''_d} - \dfrac{1}{x'_d}\right) e^{-t/\tau''_d}\right] \cos(\omega t + \alpha) \\ -\dfrac{1}{2}\left(\dfrac{1}{x''_d} + \dfrac{1}{x''_q}\right) e^{-t/\tau_a} \cos\alpha - \dfrac{1}{2}\left(\dfrac{1}{x''_d} - \dfrac{1}{x''_q}\right) e^{-t/\tau_a} \cos(2\omega t + \alpha) \end{array} \right\} \\
-\sqrt{2}V_{do} &\left\{ \begin{array}{l} -\left[\dfrac{1}{x_q} + \left(\dfrac{1}{x''_q} - \dfrac{1}{x'_q}\right) e^{-t/\tau''_q}\right] \sin(\omega t + \alpha) \\ +\dfrac{1}{2}\left(\dfrac{1}{x''_d} + \dfrac{1}{x''_q}\right) e^{-t/\tau_a} \sin\alpha - \dfrac{1}{2}\left(\dfrac{1}{x''_d} - \dfrac{1}{x''_q}\right) e^{-t/\tau_a} \sin s(2\omega t + \alpha) \end{array} \right\} \\
&- \left[\sqrt{2}I_{do}\cos(\omega t + \alpha) + \sqrt{2}I_{qo}\sin(\omega t) + \alpha\right]
\end{aligned} \tag{5.34}$$

where the angle α represents the point on the current wave at which the fault begins.

If there are no current zeros for an extended period, the fault interruption is delayed, perhaps for a considerable period. This will be especially troublesome for generator circuit breakers. In practice, many faults develop sequentially and sequential interruption of the phase currents reduces the severity of the problem. The problem is also less severe when the fault is farther removed from the generator terminals.

It is possible to estimate the time to the first current zero by solving (5.34) for only the fundamental component of current and neglecting the subtransient components and subtransient saliency. This results in the approximate equation for the time to the first zero, as follows [6].

$$t_Z = \frac{\tau_d' \tau_a}{\tau_d' - \tau_a} \ln \left[\frac{x_d' V_o |\cos(\alpha - \delta)|}{x_d''(I_{do} x_d' + V_{qo})} \right] \tag{5.35}$$

where V_o is the prefault voltage. This has a maximum when the fault occurs exactly at the angle $a = \delta$. Clearly, the largest value of (5.35) occurs when the generator is unloaded, when $\delta = 0$, and $V_{qo} = V_o$, which gives

$$t_{Z\max} = \frac{\tau_d' \tau_a}{\tau_d' - \tau_a} \ln \left[\frac{x_d'}{x_d''} \right] \tag{5.36}$$

It is possible to compute the exact dc offset for each phase by use of the simple three-phase phasor diagram shown in Figure 5.10. Here we define

$$\begin{aligned} I_A &= \text{initial dc component of current } I_a \\ I_B &= \text{initial dc component of current } I_b \\ I_C &= \text{initial dc component of current } I_c \end{aligned} \tag{5.37}$$

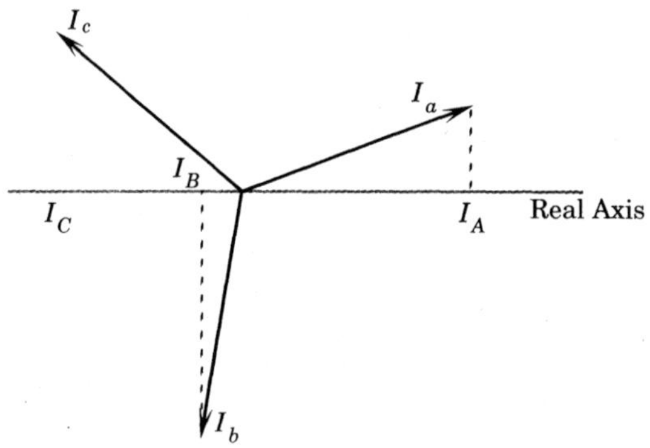

Figure 5.10 Device for computing initial dc offset.

It should be clear from the figure that the initial dc component in each phase is the projection of the phasor current for that phase on the reference real axis. Note that the phasor diagram is usually drawn for rms currents, in which case each of the initial values in (5.37) must be multiplied by $\sqrt{2}$.

For the case of faults removed from the generator terminals by some distance, the impedance between the generator and the fault must be added into the computations. For these cases, we revise the computation as follows. First we define

$$z_e = r_e + jx_e \quad \text{per unit} \tag{5.38}$$

where this impedance in on the generator base. Then we redefine the currents and time constants as follows:

$$I_d = \frac{E}{x_d + x_e} \tag{5.39}$$

$$I_d' = \frac{E}{x_d' + x_e} \tag{5.40}$$

$$I''_d = \frac{E}{x''_d + x_e} \tag{5.41}$$

$$\tau'_d = \frac{x'_d + x_e}{x_d + x_e}\tau'_{do} \tag{5.42}$$

$$\tau''_d = \frac{x''_d + x_e}{x'_d + x_e}\tau''_{do} \tag{5.43}$$

$$\tau_a = \frac{x_2 + x_e}{(r_a + r_e)\omega_B} \tag{5.44}$$

The concepts discussed above are accurate enough for most normal circuit breaker application guidelines and for preliminary relay settings. In some special cases, however, it may be considered necessary to make more accurate calculations. For example, if the machine time constants are unusually long or when one wishes to consider the relay currents a long time after fault inception, more accurate procedures may be desired. For these cases, the manufacturers can furnish accurate decrement curves that can be used for more precise computations. Even these curves, however, are constructed using a set of assumptions regarding the generator prefault loading, the effect of excitation, and other simplifying assumptions. Still, this offers an optional alternate method that may be considered by the protection engineer.

5.1.3 Saturation of Current Transformers

The foregoing sections illustrate the fact that fault currents are likely to have high dc offset in addition to large fundamental frequency components that are likely to cause saturation of current transformers. Relays are usually designed for fundamental frequency sine wave operation and the performance of relay may not be specified for other waveforms or for dc components of current. Therefore, the rating and burden of the current transformer should be specified by the protection engineer in order to ensure undistorted secondary current under maximum fault conditions. The problem is in the interpretation of the CT standards in terms of the limiting value of burden that will assure satisfactory performance. We can determine those limiting values by investigating the flux equation for the current transformer.

From (5.12), we can write the flux due to a fully offset voltage as

$$n\phi = nBA_c = \int_0^t Z_B I_F (e^{-Rt/L} - \cos\omega t) \tag{5.45}$$

or, evaluating the integral we get

$$\begin{aligned} nBA_c\omega &= Z_B I_F \left[\frac{X}{R}(1 - e^{-Rt/L}) - \sin\omega t\right] \\ &= Z_B I_F \left(\frac{X}{R} + 1\right) \le 20V_{Rated} = 20Z_{BR}I_{FR} \end{aligned} \tag{5.46}$$

In the final computation we take the limiting value of the quantity in square brackets. The inequality is due to the ANSI standards, which specify that the C-Rated CT support a voltage across the standard burden Z_{BR} when carrying 20 times rated fault current I_{FR} without exceeding a 10% ratio error. Dividing (5.46) by these ratings, gives the simple expression

$$Z_B \le \frac{20}{I_F\left(\frac{X}{R} + 1\right)} \tag{5.47}$$

where both the burden impedance and the fault current are in per unit on bases determined by the CT rated values [8].

The final relationship provides the relay engineer with an excellent way to determine the maximum value of burden. If a given design results in too great a value of burden, this equation shows how much the impedance must be reduced to avoid exceeding design limits. In many cases, choosing a larger size of control wiring in the substation will reduce the burden enough to satisfy (5.47).

CT saturation problems often occur due to limited space for the installation, requiring the use of low ratio, low accuracy CT's. The larger CT specified by the calculation also adds to the cost of the installation. Undersized CTs result in a burden requirement that cannot be achieved. Therefore, the ideal goals set forth in the above procedure are not always achieved in practice and saturation of the current transformers is possible. This changes the problem to one of trying to minimize saturation and of determining its effect on the quality of the protection. See Section 2.4 for additional information on the saturation of current transformers.

5.2 STATION ARRANGEMENTS

An important consideration in the design and implementation of protective systems is the switching station arrangement. There are many different configurations in use, each of which has its advantages and disadvantages. First, we shall examine some of the more common switching station designs and will study the features of each. Then a few of the less common designs will be presented and some of their unusual features discussed.

5.2.1 Single Bus, Single Breaker Arrangement

The first and simplest of the switching station arrangements is the single bus scheme shown in Figure 5.11(*a*). This arrangement is relatively inexpensive and is simple to construct and operate. It has some obvious disadvantages, however, that limit this design to remote stations where service continuity can be restricted without affecting large numbers of customers. For example, the only way a breaker can be maintained is to remove that terminal from service. This may cause no problems if the load served by that terminal has an alternate source of supply, or can be switched to other circuits temporarily. For high-voltage systems in general, however, it should not be necessary to sacrifice the use of a line or transformer just to perform maintenance on a circuit breaker. Note that this or any station has several types of equipment:

- circuit breakers
- disconnects
- buses
- terminations for lines or transformers (bus sections)

Of all items, the circuit breaker is the only active component, the rest being passive. This means that the circuit breakers are generally less reliable and more subject to failure, due to the existence of moving parts, sometimes with complex mechanisms. The availability of the circuit breakers are kept at a reasonably high level by routine scheduled maintenance. Hence, the removal of circuit breakers on a routine basis is a basic requirement for any switching station. For the single bus, single breaker design of Figure 5.11(*a*), the breaker can be removed for servicing by first opening the breaker, and then manually isolating the breaker by opening the

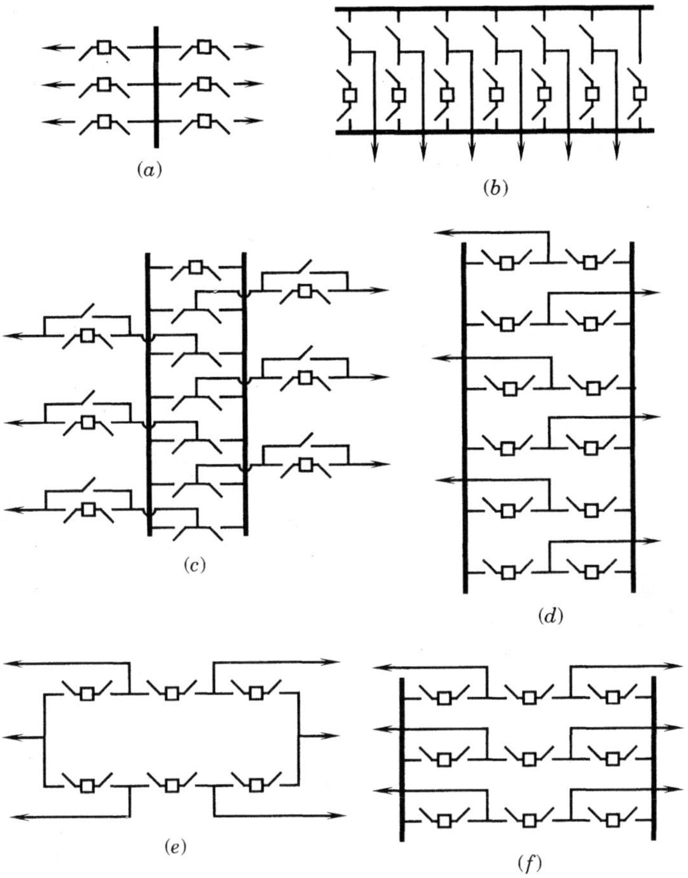

Figure 5.11 Some common station arrangements. (*a*) Single bus, single breaker 6 breakers, 6 lines 1.0 breakers/line. (*b*) Main and transfer 7 breakers, 6 lines 1.167 breakers/line. (*c*) Double bus, single breaker 7 breakers, 6 lines 1.167 breakers/line. (*d*) Double bus, double breaker 12 breakers, 6 lines 2.0 breakers/line. (*e*) Ring bus 6 breakers, 6 lines 1.0 breakers/line. (*f*) Breaker and a half 9 breakers, 6 lines 1.5 breakers/line.

disconnects at both breaker terminals. The disconnects provide maintenance personnel with visual assurance that the breaker is isolated and safe to work on.

Another weakness of this design is that each outgoing line is served by only one breaker. Should the mechanism fail in the open position, there is no way to energize that circuit. Moreover, should the breaker fail to open when there is a fault on the protected branch, then all breakers on the bus that provides a source of fault current must be opened to clear the fault.

Note that the single bus, single breaker arrangement requires only one breaker per connection, which makes it very economical.

5.2.2 Main and Transfer Arrangement

The main and transfer station arrangement is shown in Figure 5.11(*b*). The normal operation of this station would be with each terminal served through its own breaker from the lower bus and with the bus tie breaker open. To perform maintenance on one of the breakers, the bus tie breaker is closed and the disconnect between the breaker to be maintained and the transfer bus is closed. Then the breaker requiring maintenance is opened and its disconnects are opened. During maintenance, there is no outage on any terminal, and the terminal served from the transfer bus is protected through the bus tie breaker.

The advantages of this arrangement are as follows:

1. Low initial cost.
2. Flexible operation.
3. Breaker or line relays can be taken out of service for maintenance with slight modification of line protection, through the bus tie breaker.
4. Potential devices may be used on the main bus for relaying.

The disadvantages of the main and transfer bus scheme are as follows:

1. Requires an extra breaker for the bus tie.
2. Switching is somewhat complicated when removing a breaker for maintenance.
3. The protective relaying for the bus tie breaker may be complicated, since it must be able to substitute for any breaker.
4. Requires a separate bus protection for each bus.
5. Breaker or bus failure takes the entire station out of service until the fault is isolated.

The main and transfer arrangement always requires one more breaker than the total number of connections. For the case illustrated in Figure 5.11, this equates to 1.167 breakers per connection. This is only slightly more costly than the single bus, single breaker arrangement and is often considered a good trade-off, offering reasonable cost and effective breaker maintenance.

5.2.3 Double Bus, Single Breaker Arrangement

The next station arrangement is the single breaker, double bus arrangement shown in Figure 5.11(*c*). This station arrangement also employs two buses with a bus tie breaker connecting them. It also associates a circuit breaker with each terminal connection. The difference is in the connection of the breaker and in the use of a breaker bypass disconnect.

The advantages of this arrangement are:

1. Very flexible arrangement.
2. Breaker or line protective relays can be taken out of service at any time for maintenance with only a slight modification of the protective relaying scheme.
3. Either main bus may be isolated for maintenance.

The disadvantages of this arrangement are:

1. Requires an extra breaker for the bus tie.
2. Five disconnects are required for each terminal.
3. The switching is complicated for isolating a breaker for maintenance.
4. The protective relaying for the bus tie breaker is complicated since it must be capable of substituting for any breaker.
5. Requires a separate and complicated bus protective scheme.
6. Greater exposure to bus faults than previous designs.

7. Line breaker failure takes half of the substation out of service.
8. The failure of the bus tie breaker takes the entire station out of service until the fault is isolated.

It is noted in Figure 5.11(*c*) that this arrangement requires one more breaker than the number of connections served, the same as the main and transfer bus arrangement.

5.2.4 Double Bus, Double Breaker Arrangement

The next switching station arrangement we shall consider is the double bus, double breaker scheme, shown in Figure 5.11(*d*). Note that this scheme uses two breakers for each terminal, hence the designation "double breaker."

The advantages of this switching arrangement are as follows:

1. Flexible operation.
2. High reliability.
3. All switching is done with breakers.
4. Either main bus can be taken out of service at any time for maintenance.
5. Bus failure does not remove any circuit from service.
6. Each connection is served by two breakers, providing greater reliability that a connection can always be made.

The disadvantages of the double bus, double breaker scheme are:

1. Two breakers per circuit raises the cost of the station.
2. The protective relaying must trip two circuit breakers to isolate a faulted line from the station.
3. Clearing a fault on any connection requires the opening of two circuit breakers, which increases the probability of failure to clear the fault successfully.

The double bus, double breaker arrangement requires two breakers per connection, which is the highest cost of any configuration considered so far.

5.2.5 Ring Bus Arrangement

The ring bus station arrangement is shown in Figure 5.11(*e*). This bus arrangement is characterized as having only one breaker per terminal, it serves each connection from two breakers.

The advantages of this scheme are as follows:

1. Low initial and ultimate cost.
2. Flexible operation for breaker maintenance at any time without interrupting load or requiring complex switching.
3. No complicated bus protective scheme is required.
4. Requires only one breaker per connection.
5. Breaker failure removes only two circuits from service under normal operating conditions.
6. Each circuit is fed by two breakers.
7. All switching is performed by breakers.

The disadvantages of the ring bus scheme are:

1. If a fault occurs during the time one line is out of service, the ring can be separated into two sections.
2. Automatic reclosing circuits are rather complex.
3. If a single set of protective relays is used, the transmission line must be taken out of service to maintain the relays.
4. Requires voltage devices on all circuits since there is no definite potential reference point.

The ring bus arrangement is economical in terms of the number of breakers required per connection served. However, the power system is placed at risk when a breaker in a ring bus is being serviced. Whether this risk is acceptable depends on the voltage level, the amount of power transferred through the station, and the frequency of interruptions that could occur during maintenance.

5.2.6 Breaker-and-a-Half Arrangement

The next switching station arrangement to be considered is called the breaker-and-a-half scheme because it uses three breakers for every two circuits.

The advantages of this scheme are as follows:

1. Flexible operation.
2. High reliability.
3. Breaker failure on bus-side breakers removes only one circuit from service.
4. All switching is performed using breakers.
5. Simple operation; no air break disconnect switching required for normal operation.
6. Either main (rack) bus can be taken out of service at any time for maintenance.
7. Bus failure does not remove any circuit from service.

The disadvantages of the breaker-and-a-half scheme are:

1. Requires 1.5 breakers per circuit.
2. Protective relaying is somewhat complicated since the middle breaker must be responsive to either of its associated circuits.
3. Breaker failure of the center breaker will cause the loss of an unfaulted circuit.

The circuit breaker requirement for the breaker-and-a-half arrangement is midway between the very inexpensive arrangements and the expensive, but secure, double bus, double breaker scheme. These factors have made this one of the most favored switching arrangements for high-voltage and extra-high-voltage stations.

5.2.7 Other Switching Arrangements

The foregoing represent the most common switching arrangements and provide for flexibility and reliability of varying amounts. They also represent a wide range of choices for ease of repair, simplicity in relay application, and cost. They do not, however, represent the only choices. Additional station designs are shown in Figure 5.12.

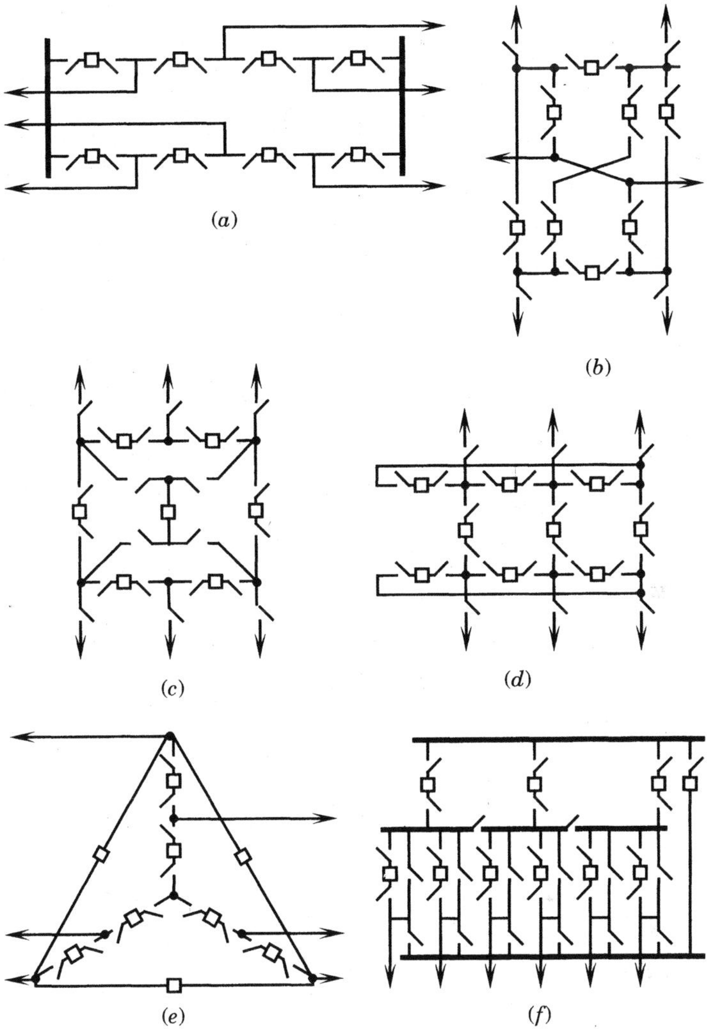

Figure 5.12 Other station arrangements. (*a*) Breaker and a third 8 breakers, 6 lines 1.333 breakers/line. (*b*) Ring tripod 8 breakers, 6 lines 1.333 breakers/line. (*c*) Ring bridge 7 breakers, 6 lines 1.167 breakers/line. (*d*) Crossed ring 9 breakers, 6 lines 1.5 breakers/ line. (*e*) 4×6 network 9 breakers, 6 lines 1.5 breakers/line. (*f*) Pyramid scheme 10 breakers, 6 lines 1.667 breakers/line.

5.2.7.1 Breaker and a Third Arrangement. The breaker-and-a-third arrangement is similar to the breaker-and-a-half topology, but it is more economical in terms of circuit breaker cost. The example shown in Figure 5.12(*a*) is not typical due to the use of only two paths between the rack buses, making this particular example similar to a ring bus. Usually, the breaker-and-a-third would have more than two paths.

This arrangement has the same advantages and disadvantage as the breaker-and-a-half arrangement, but it has one additional disadvantage. A stuck breaker on the center two breakers will always cause the outage of an unfaulted circuit. This makes the design less reliable than the breaker-and-a-half, and probably explains the reason that it seems to be favored less often than the breaker-and-a-half design. The number of breakers per connection for this design is low, making this scheme economical.

5.2.7.2 The Ring Tripod Arrangement. The ring tripod is one several arrangements described in the same technical paper [9]. This arrangement has a favorable index of breakers per connection. It is unusual in that some connections are served by three circuit breakers rather than one or two breakers of other, more common designs. This has the advantage that a breaker that fails to close will have a very low probability of causing outage of a circuit. However, it has the disadvantage that three breakers must operate successfully to clear a fault.

The ring tripod has a difficult crossover of connections at the center of the layout, which may make it difficult to build.

5.2.7.3 The Ring Bridge Arrangement. The ring bridge of Figure 5.12(*c*) has also been described as a ring bus with bridging breaker [9]. The number of breakers per connection is low, making this arrangement economical compared to the breaker-and-a-half. Adding another circuit breaker adds redundancy and makes breaker maintenance of the ring bus breakers less risky. The bridging breaker would normally be in standby, similar to the bus interconnector in the main and transfer scheme.

5.2.7.4 The Crossed Ring Arrangement. The crossed ring of Figure 5.12(*d*) features three circuit breakers for each connection, but because of the topology the cost is the same as the breaker-and-a-half scheme. This has both advantages and disadvantages, as noted above. The crossover of the crossing connections may make this design difficult to build. This station is operated exactly like a ring bus, with the ring connecting breakers normally open. It is also possible to operate the station as two partial rings, with the partial rings either separate or coupled through a breaker.

Arrangements of this type that make three circuit breakers available to each connection may be favored in cases where there is a very high thermal loading and a need to divide the load current between several supply breakers.

5.2.7.5 The 4 × 6 Network Arrangement. The 4×6 network arrangement of Figure 5.12(*e*) is still another variation of station design that attempt to increase the redundancy of supply to outgoing circuits [10]. In this case the connections at the corners are served by three breakers, but the inside connections are served by only two.

5.2.7.6 The Pyramid Station Arrangement. The pyramid station arrangement is quite different from the others in that it provides two breakers in series for each connection. The outgoing circuits are divided into groups, in this case illustrated in Figure 5.12(*f*) having two circuits per group. Each circuit has its own breaker, but is backed up by another breaker. Should the first breaker suffer from a stuck breaker condition, the fault can be cleared by the backup breaker, but at the expense of clearing an unfaulted circuit. A rather complex arrangement, utilizing a separate breaker, is needed to allow for maintenance of any breaker.

In conclusion, there is no one best switching station arrangement and the arrangement that is used for a given station depends on many factors, one of which is the protective system. It is important that the protective system always be available for maintenance and that it not be unduly complex. For the most part, a protective system can be designed for any switching station arrangement. Some may require the tripping of multiple breakers in order to clear certain faults, but this does not create a problem for the relay engineer.

5.3 LINE IMPEDANCES

The final system component that must be described for protection studies is the characteristic of the lines, particularly the line reactance. The methods of computing line parameters is well known and well documented [4], [5], [11], [12]. For protective system analysis it is usually the line series impedance that is required. This reactance, for a transposed three-phase line is computed as

$$\begin{aligned} x_L &= 0.07539 \ln \frac{D_m}{D_s} \quad \Omega/\text{km} \\ &= 0.12134 \ln \frac{D_m}{D_s} \quad \Omega/\text{mile} \end{aligned} \tag{5.48}$$

where D_m = Geometric mean distance between phase conductors
D_s = Geometric mean radius of the phase conductors

(Note that the distances used by North American conductor manufacturers for these calculations are often in English units.) Values of the geometric mean radius are published by the conductor manufacturers and are readily available. Also published are tables of the conductor resistance in ohms per mile. Given these two values we may compute the line impedance as

$$z_L = r_L + jx_L \quad \Omega/\text{unit length} \tag{5.49}$$

The actual computation of the line reactance is simplified by writing (5.48) as

$$x_L = x_a + x_d \quad \Omega/\text{unit length} \tag{5.50}$$

where x_a = reactance of conductor including flux linkages out to a radius of 1 ft
x_d = reactance spacing factor including effect of flux linkages beyond a 1 ft radius

Using this technique, the first term is due almost entirely to the conductor itself and its internal flux linkages, as well as the effect of flux linkages within the constant radius of 1 ft. The second term is a function only of the spacing between conductors.

For distribution circuits it is often more convenient to measure the line length in thousands of feet, or kft, rather than miles. One may then compute [4]

$$x_d = 0.1213 \ln D'_m = 0.2794 \log_{10} D'_m \quad \Omega/\text{mile} \tag{5.51}$$

where D'_m = Geometric mean distance in feet

For distribution circuits it is more convenient to measure this distance in inches, rather than feet and to measure the total reactance in ohms per kft. Thus, we compute

$$x_d = \frac{0.2794}{5.28} \log_{10} \frac{D''_m}{12} \quad \Omega/\text{kft} \tag{5.52}$$

where D''_m = Geometric mean distance in inches

It can easily be shown that (5.52) may be written as

$$x_d = -0.0571 + 0.0529 \log_{10} D''_m \quad \Omega/\text{kft} \tag{5.53}$$

Values of resistance and reactance factors are tabulated in Appendix C for a wide variety of conductor types and spacings.

5.4 COMPUTATION OF AVAILABLE FAULT CURRENT

The first requirement of any protective system coordination study is to know the available fault current at the point of application. This requires that the results of a short-circuit study be available to the protection engineer. Computer programs are available that can quickly make the necessary computation. Here, we briefly review the different types of faults and their computation using the methods of symmetrical components [4].

The maximum fault current condition occurs when the largest number of generators are in service, which usually occurs at peak load. The fault current available at any point in the network depends on the Thevenin equivalent impedance seen looking into the network from the fault point. Maximum generation conditions make the voltage high behind the Thevenin impedance. Also, because of the heavy loading, the Thevenin impedance will be minimum because no lines would be on scheduled outage for this condition. For maximum fault conditions we assume that the fault is a "bolted" fault, with zero fault impedance. This maximum fault current should be computed at every node in the transmission system. Maximum fault currents are required in order to determine the maximum interrupting rating of circuit breakers or other fault-current-interrupting devices.

The minimum fault current occurs at off-peak, when the number of generators in service is small, so that the Thevenin equivalent impedance is high. Often we add fault resistance to further limit the minimum current, since arc resistance usually exists in some amount. These minimum currents should also be computed at all nodes in the network as these conditions are sometimes critical in the coordination of protective equipment. Minimum fault currents are also important because these currents must be compared with maximum load currents to make sure that there is no ambiguity between the two values and that faults can be clearly distinguished from heavy loading conditions.

Four types of faults can be considered; three phase (3PH), double line-to-ground (2LG), one-line-to-ground (1LG), and line-to-line (L-L). Usually, either the 3PH or 1LG is the most severe and the L-L the least severe.

In most cases, the fault currents are computed in per unit, based on an arbitrarily chosen volt-ampere base, such as 100 MVA. A possible source of confusion exists in system computations in knowing whether a given algebraic expression or a numerical result is in system meter-kilogram-second (mks) units (volts, amperes, ohms) or in per unit (pu). We adopt the convention of adding the subscript "u" where necessary to emphasize "per unit." Usually any system equations are exactly the same whether written in mks units or in per unit.

Transmission system fault calculations are nearly always performed in per unit, and the results of these studies are often available in the form of an open circuit driving point and transfer impedance matrix called the *impedance* matrix, or the Z matrix. The Z matrix is defined by the matrix equation

$$\mathbf{V} = \mathbf{ZI} \tag{5.54}$$

where both $\mathbf{V}$ and $\mathbf{I}$ are $n \times 1$ vectors of voltage and current phasors, and the matrix $\mathbf{Z}$ is an $n \times n$ matrix of complex impedances called the driving point and transfer impedances.[1]

[1]Note carefully the notation used for (5.54) and (5.55). In this book, vectors and matrices are displayed in **Roman bold** typeface, as in (5.54). Phasors and complex numbers are displayed in ***bold italic*** typeface, as in (5.55).

Writing this matrix equation in expanded form, we have

$$\begin{bmatrix} V_1 \\ V_2 \\ \dots \\ V_n \end{bmatrix} = \begin{bmatrix} Z_{11} & Z_{12} & \dots & Z_{1n} \\ Z_{21} & Z_{22} & \dots & Z_{2n} \\ \dots & \dots & \dots & \dots \\ Z_{n1} & Z_{n2} & \dots & Z_{nn} \end{bmatrix} \begin{bmatrix} I_1 \\ I_2 \\ \dots \\ I_n \end{bmatrix} \tag{5.55}$$

where the voltages and currents are defined in Figure 5.13 [4]. From (5.55) the impedance seen looking into the network at node i may be written as

$$Z_{ii} = \left.\frac{V_i}{I_i}\right|_{I_k=0, i=1,2,\dots n, k\neq i} \tag{5.56}$$

where we call Z_{ii} the driving point impedance. Clearly, from (5.55), Z_{ii} is the impedance seen looking into node i with all other nodes open. For a fault study we usually assume that load currents are negligible compared to fault currents, or that the load buses are all open. Then Z_{ii} is the Thevenin equivalent impedance seen looking into the transmission network at bus i when bus i is faulted. The Thevenin equivalent voltage is the open circuit voltage at bus i or the normal prefault voltage. Usually we take this voltage to be 1.0 per unit, although the value can be scaled as desired since the network is linear. Impedance Z_{ik} is an impedance relating voltages at a given node i to currents injected at another node ($k \neq i$).

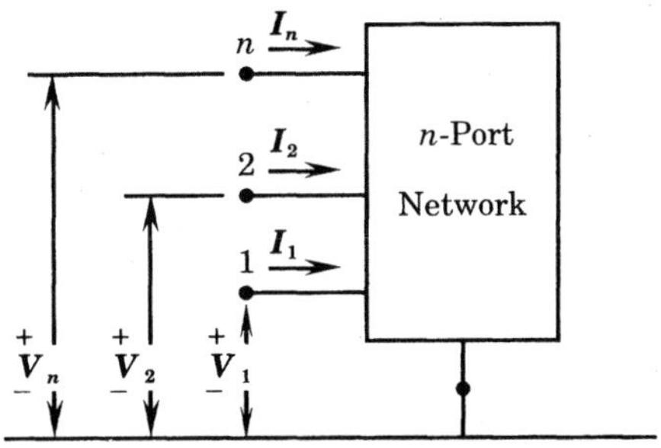

Figure 5.13 An n-port network [4].

The Z matrix of (5.54) or (5.55) is usually available from computer studies for both the positive and zero sequence networks. Usually we assume that the positive and negative sequence networks have equal impedances, that is, ($Z_1 = Z_2$). This is true for transmission lines but not for machines. This assumption usually leads to negligible error except perhaps at or very near a generator bus.

Knowing the driving point or Thevenin impedance at any transmission fault bus we know the sequence network impedances Z_0, Z_1, and $Z_2 = Z_1$ shown in Figure 5.14, where we usually let the Thevenin voltage $V_F = 1.0$.

Knowing these sequence networks, we can solve for any of the desired fault currents at the transmission bus. Figure 5.14 defines the various quantities needed to solve for the fault current.

5.4.1 Three-Phase (3PH) Faults

The arrangement of the sequence networks and the derived sequence and phase currents are shown in Figure 5.15. Obviously there are no negative or zero sequence currents for this type of fault as it is completely balanced. This means that there is no current through the ground impedance and it makes no difference if Z_G is zero or infinite.

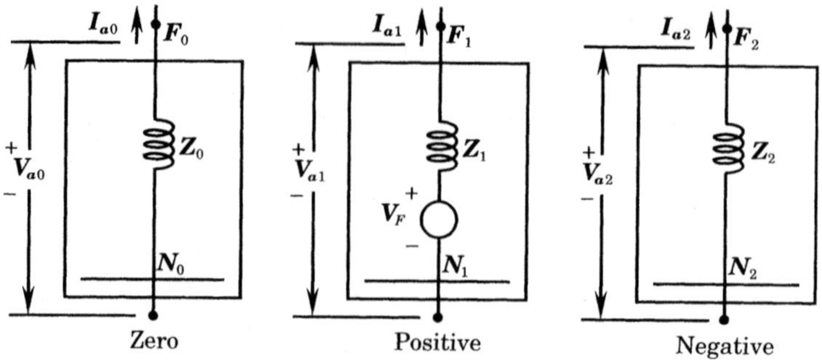

Figure 5.14 Sequence networks with defined sequence quantities [4].

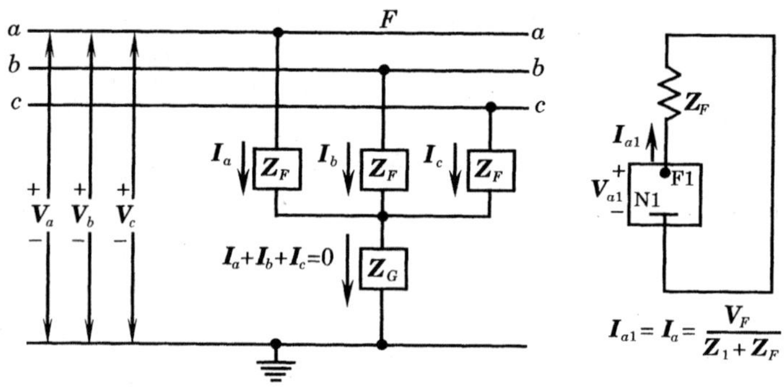

Figure 5.15 Three-phase fault connection and current equations [4].

The Thevenin equivalent voltage V_F completes the fault data necessary for three-phase faults. This voltage will not usually be known precisely and is usually taken to be between 1.0 and 1.1, with 1.05 a good estimated value. The angle is usually taken to be zero, and this voltage then becomes the phasor reference for the calculations.

5.4.2 Double Line-to-Ground (2LG) Faults

The faulted phase designation and sequence network connection for the 2LG fault is shown in Figure 5.16 where is noted that the sequence networks are connected in parallel [4]. The constant "a" in Figure 5.16 is the familiar 120 degree operator, i.e.,

$$\boldsymbol{a} = e^{j2\pi/3} = e^{j120^\circ} = -\frac{1}{2} + j\frac{\sqrt{3}}{2} \tag{5.57}$$

There are three currents of interest for this type of fault, the line currents I_b and I_c, and the ground current $I_b + I_c$. The line currents I_b and I_c are not equal and must be calculated separately.

The positive and negative sequence impedances are exactly equal except for machines, and equating these sequence impedances is usually a very good approximation.

The zero sequence impedance Z_0 is very difficult to determine accurately for multi-grounded systems. Chapter 6 provides a method of estimating this impedance for certain systems.

Fault impedances Z_F and Z_G are chosen arbitrarily, or based on data from typical faults. For simplicity we often let $Z_F = 0$ and give Z_G an estimated value.

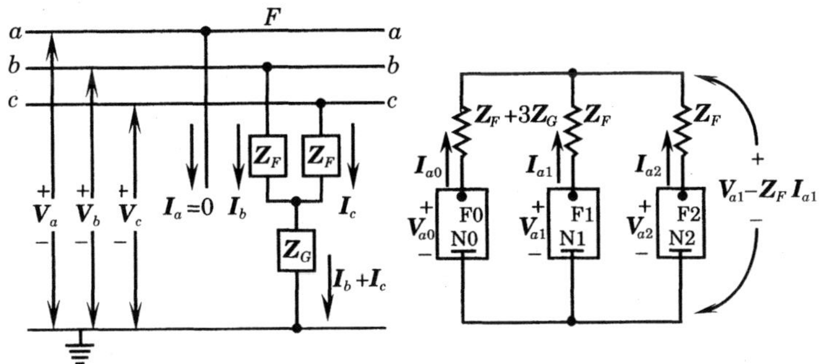

$$I_a = I_{a0} + I_{a1} + I_{a2}$$
$$I_b = I_{a0} + a^2 I_{a1} + a I_{a2}$$
$$I_c = I_{a0} + a I_{a1} + a^2 I_{a2}$$
$$I_b\ I_c = 3I_{a0}$$

$$I_{a1} = \frac{V_F}{Z_1 + Z_F + \dfrac{(Z_2 + Z_F)(Z_0 + Z_F + 3Z_G)}{Z_0 + Z_2 + 2Z_F + 3Z_G}}$$
$$I_{a2} = -\frac{Z_0 + Z_F + 3Z_G}{Z_0 + Z_2 + 2Z_F + 3Z_G} I_{a1}$$
$$I_{a0} = \frac{Z_2 + Z_F}{Z_0 + Z_2 + 2Z_F + 3Z_G} I_{a1}$$

Figure 5.16 2LG fault connection and current equations [4].

5.4.3 Line-to-Line (L-L) Fault

The faulted phase designation and sequence network connections for the line-to-line fault are given in Figure 5.17. Usually, the positive and negative sequence impedances are taken to be equal impedances. No zero sequence current flows for the line-to-line fault since there is no path to ground. Since the fault is unbalanced and we wish to keep phase a as the symmetrical phase, this fault is usually represented as a phase b-c fault, as shown in Figure 5.17.

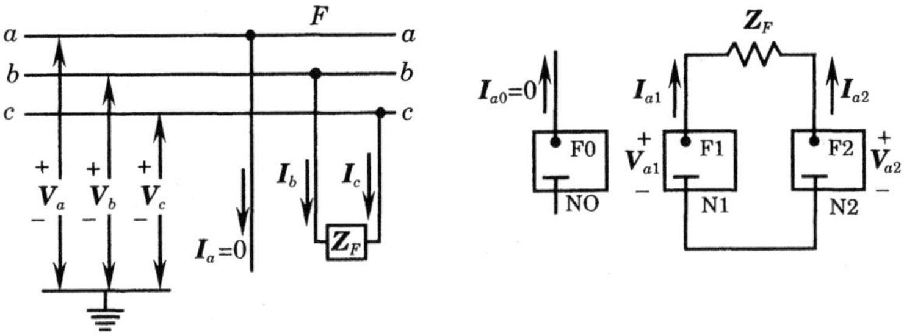

$$I_a = 0 \qquad I_{a1} = \frac{V_F}{Z_1 + Z_2 + Z_F}$$
$$I_b = -I_c = -j\sqrt{3} I_{a1} \qquad I_{a2} = -I_{a1}$$
$$I_{a2} = 0$$

Figure 5.17 Line-to-line fault connections and current equations [4].

5.4.4 One-Line-to-Ground (1LG) Fault

The 1LG fault configuration and sequence network connections are shown in Figure 5.18. The comments above regarding the sequence impedances are applicable here, as well. With phase a as the symmetrical phase, the 1LG fault is usually represented as a fault on phase a. Sequence currents for this type of fault flow in all three sequence networks and are equal.

Synthesis of the phase currents from the sequence currents will show that the currents in phases b and c are zero for this fault type.

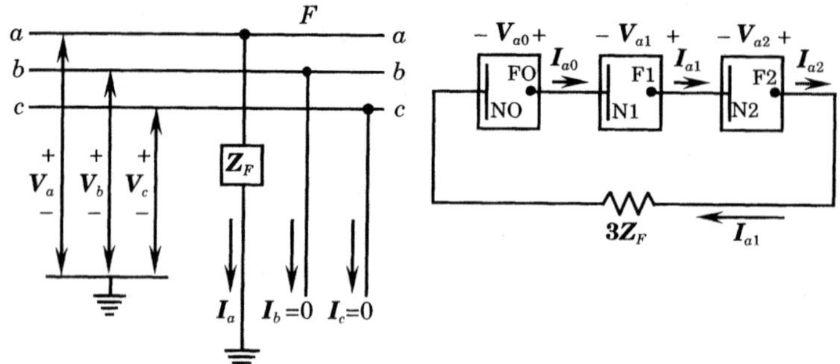

$$I_a = 3I_{a0} \qquad I_b = I_c = 0 \qquad I_{a0} = I_{a1} = I_{a2} = \frac{V_F}{Z_0 + Z_1 + Z_2 + 3Z_F}$$

Figure 5.18 One-line-to-ground fault connections and current equations [4].

5.4.5 Summary of Fault Currents

The foregoing summary provides the basic sequence network connections and the formula used in computing fault currents for common types of fault configurations. Fault current information is needed in all types of protective system coordination studies. Examples of the use of fault currents will be given in Chapter 6 and in other chapters. In all coordination studies it is assumed that these basic fault conditions are known.

5.5 SYSTEM EQUIVALENT FOR PROTECTION STUDIES

Consider a protected transmission line or other longitudinal element that is a component in a large power system, and for which protection studies are required. The network to which the component is connected is an active network, containing sources as well as passive branches, shunt loads, and passive shunt devices such as capacitors and reactors. In many cases, the network is large, and the number of equations required to describe the network is equal to the number of external ports that must be retained. The engineer may require information for any node, requiring the solution of a large number of simultaneous equations to provide the needed information.

The protection engineer, however, is interested in the effect of the power system on only the protected component, and would often prefer a reduced network representation to simplify the calculations for protection studies of that component. Simplifying the network representation may also have the advantage of providing greater insight into system effects that might otherwise be masked by the large number of equations. We will consider the protected element to be a transmission line, although it could be a transformer or other network branch that connects two nodes. The protected line is shown in Figure 5.19, where relays R and Q provide the protection for the line by making current and voltage measurements at the two terminals of the protected component. These currents and voltages are shown in the figure, using appropriate subscripts for the relays at each terminal.

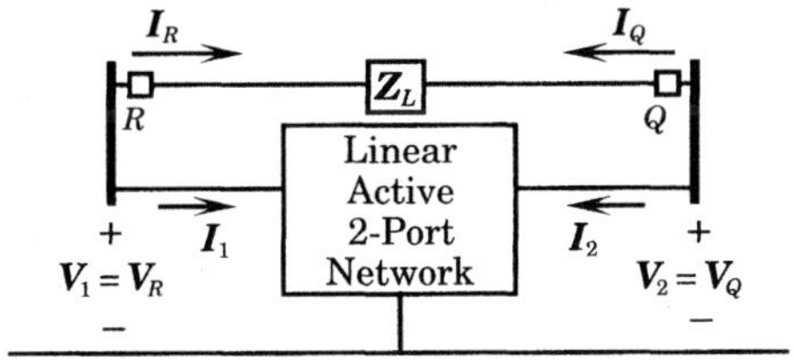

Figure 5.19 The protected element and the active network.

Since the protected component is connected to the network at two points, the required network description is one that measures the network performance as viewed from these two points. This is called a *two-port network description*, the treatment of which is exhaustively treated in the literature [13], [14]. The network described here differs from the ordinary two-port network only in the sense that it is an *active* network, containing internal energy sources. This does not greatly complicate the network representation, however, as the effect of the internal sources can be easily accommodated by a simple modification of the passive two-port network equations.

5.5.1 The Open-Circuit Impedance Matrix

The network shown in Figure 5.19 is to be represented by its open-circuit two-port equations, which may be written as follows [4].

$$\begin{bmatrix} V_1 \\ V_2 \end{bmatrix} = \begin{bmatrix} Z_{11} & Z_{12} \\ Z_{21} & Z_{22} \end{bmatrix} \begin{bmatrix} I_1 \\ I_2 \end{bmatrix} + \begin{bmatrix} V_{S1} \\ V_{S2} \end{bmatrix} \tag{5.58}$$

The impedances in the Z matrix are called "open-circuit parameters" because they are defined with certain network ports open, i.e., with either I_1 or I_2 equal to zero. The voltage vector on the right of (5.58) is the source voltage vector, which represents the effect of the internal sources in the network. In what follows, we shall examine the two-port representation of the large network with the protected line in Figure 5.19 disconnected, that is, with both ports 1 and 2 open. The voltages of the ports are defined as *voltage drops* from nodes 1 and 2 to reference and the port currents are defined as the *currents entering* the network. This is the usual convention for defining two-port parameters [13], [14].

First consider (5.58) when both of the two ports are open. Mathematically, this gives the following results.

$$\begin{bmatrix} V_1 \\ V_2 \end{bmatrix}_{I_1=I_2=0} = \begin{bmatrix} V_{S1} \\ V_{S2} \end{bmatrix} \tag{5.59}$$

Clearly, the source voltage terms in (5.58) represent the open-circuit voltages that appear at the two ports when the external connections are removed. These voltages are due to internal sources, and the effect at the two ports of interest are represented by these open-circuit source voltages. These sources can be thought of as the two-port Thevenin equivalent voltages, since their determination is found according to the rules dictated by Thevenin's theorem, namely, with the two ports open. A power flow solution of the network, with the protected component removed, will provide these voltages, both in magnitude and angle, for any desired network loading condition.

Second, consider (5.58) when all internal sources are properly removed. The term "properly removed" means that all constant voltage sources are replaced by short circuits and all constant current sources are replaced by open circuits, such that there in no energy transferred to the passive network from any internal source. Under this specification, the source terms on the right of (5.58) become zero, and we are left with the familiar passive two-port description of the network. Now we can find the two-port parameters of the network, which is accomplished by injecting a current to the each port in turn. Under this injection condition, the voltages at both ports can be measured or computed, and compared against the injected current, with the following results.

$$\begin{aligned} Z_{11} &= \left.\frac{V_1}{I_1}\right]_{I_2=0} & Z_{12} &= \left.\frac{V_1}{I_2}\right]_{I_1=0} \\ Z_{21} &= \left.\frac{V_2}{I_1}\right]_{I_2=0} & Z_{22} &= \left.\frac{V_2}{I_2}\right]_{I_1=0} \end{aligned} \tag{5.60}$$

This result is nothing more than the definition of the open-circuit impedance matrix parameters, dictated by the defining equation (5.58). The easiest way to determine the impedances is to drive one per unit current into each port, with the two voltages measured for each injection. This gives the two matrix impedances of either the first or second column of the matrix, as noted in (5.60).

The important thing to observe is that the two-port impedance representation can be found irrespective of the size of the power system. Moreover, the fact that the power system contains internal energy sources can be accommodated in a straightforward manner, as noted by (5.58).

5.5.2 Computation of the Two-Port Representation

The two-port parameters can be computed from the equations of the entire system, which is defined as having n nodes [15], [16]. Either the impedance or admittance equations can be used to describe this system, but the admittance equations will be used here. Thus we write

$$\begin{bmatrix} I_1 \\ I_2 \\ \cdots \\ I_n \end{bmatrix} = \begin{bmatrix} Y_{11} & Y_{12} & \cdots & Y_{1n} \\ Y_{21} & Y_{22} & \cdots & Y_{2n} \\ \cdots & \cdots & \cdots & \cdots \\ Y_{n1} & Y_{n2} & \cdots & Y_{nn} \end{bmatrix} \begin{bmatrix} V_1 \\ V_2 \\ \cdots \\ V_n \end{bmatrix} + \begin{bmatrix} I_{s1} \\ I_{s2} \\ \cdots \\ I_{sn} \end{bmatrix} \tag{5.61}$$

where the current source term on the right represents the effect of all internal energy sources in the network flowing into the network nodes when all nodes are shorted, i.e., when all voltages are set to zero. The currents on the left are defined as current injections at each port. All voltages are voltage drops from node to reference. Since we are interested only in the ports 1 and 2, we can set all other currents equal to zero, which requires that no external connections are to be made at these nodes. This reduces (5.61) to the following form.

$$\begin{bmatrix} I_1 \\ I_2 \\ \hline 0 \\ \cdots \\ 0 \end{bmatrix} = \left[\begin{array}{cc|ccc} Y_{11} & Y_{12} & Y_{13} & \cdots & Y_{1n} \\ Y_{21} & Y_{22} & Y_{23} & \cdots & Y_{2n} \\ \hline Y_{31} & Y_{32} & Y_{33} & \cdots & Y_{3n} \\ \cdots & \cdots & \cdots & \cdots & \cdots \\ Y_{n1} & Y_{n2} & Y_{n3} & \cdots & Y_{nn} \end{array}\right] \begin{bmatrix} V_1 \\ V_2 \\ \hline V_3 \\ \cdots \\ V_n \end{bmatrix} + \begin{bmatrix} I_{s1} \\ I_{s2} \\ \hline I_{s3} \\ \cdots \\ I_{sn} \end{bmatrix} \tag{5.62}$$

or, grouping the equations according to the partitions defined in (5.62), we rewrite the matrix equation as

$$\begin{bmatrix}\mathbf{I}\\\mathbf{0}\end{bmatrix}=\begin{bmatrix}\mathbf{Y}_a & \mathbf{Y}_b\\\mathbf{Y}_c & \mathbf{Y}_d\end{bmatrix}\begin{bmatrix}\mathbf{V}_a\\\mathbf{V}_b\end{bmatrix}+\begin{bmatrix}\mathbf{I}_{sa}\\\mathbf{I}_{sb}\end{bmatrix} \tag{5.63}$$

Note that bold Roman typeface characters represent matrices or vectors that are defined by the partitions of (5.62). Equation (5.63) can be solved for the injection currents of the two ports of interest by solving the second equation of (5.63) for the voltage $\mathbf{V}_b$ and substituting this voltage into the first equation. The result can be written as

$$\mathbf{I}=\begin{bmatrix}I_1\\I_2\end{bmatrix}=(\mathbf{Y}_a-\mathbf{Y}_b\mathbf{Y}_d^{-1}\mathbf{Y}_c)\mathbf{V}_a+(\mathbf{I}_{sa}-\mathbf{Y}_d^{-1}\mathbf{I}_{sb})=\hat{\mathbf{Y}}\mathbf{V}_a+\hat{\mathbf{I}}_{sa} \tag{5.64}$$

The current vector $\mathbf{I}$ represents the two-port current phasors of interest, and the voltage vector $\mathbf{V}_a$ represents the port voltage phasors. Now, (5.64) is the two-port equation for the power system in admittance form, but it can be solved for the voltage, with the result being an impedance form, exactly like (5.58). Note that this process can be performed on any network of any size, as long as the network elements are linear. The result is a simple representation of the power system in terms of the network open circuit impedance parameters, given by (5.60).

The two-port impedance parameters for any network can be found using (5.64), but with sources set to zero since the source terms do not affect the matrix quantities. The result is a network reduction, pivoting on the preserved ports of the original n-port network. This is possible since the injected currents are always zero in all ports except for those identifying the protected branch. The resulting admittance matrix for the two-port network is given by

$$\mathbf{Y}_2=(\mathbf{Y}_a-\mathbf{Y}_b\mathbf{Y}_d^{-1}\mathbf{Y}_c) \tag{5.65}$$

This operation is readily performed in small networks where the computational burden is small. For large networks, however, this is a more computationally difficult task.

However, there is an alternative. We explore this by writing, not the admittance, but the impedance matrix representation of the n-port network. We may always write the impedance equation in the following form, where the voltage sources are ignored for simplicity.

$$\begin{bmatrix}V_1\\\cdots\\V_i\\\cdots\\V_k\\\cdots\\V_n\end{bmatrix}=\begin{bmatrix}Z_{11} & \cdots & Z_{1i} & \cdots & Z_{1k} & \cdots & Z_{1n}\\\cdots & \cdots & \cdots & \cdots & \cdots & \cdots & \cdots\\Z_{i1} & \cdots & Z_{ii} & \cdots & Z_{ik} & \cdots & Z_{in}\\\cdots & \cdots & \cdots & \cdots & \cdots & \cdots & \cdots\\Z_{k1} & \cdots & Z_{ki} & \cdots & Z_{kk} & \cdots & Z_{kn}\\\cdots & \cdots & \cdots & \cdots & \cdots & \cdots & \cdots\\Z_{n1} & \cdots & Z_{ni} & \cdots & Z_{nk} & \cdots & Z_{nn}\end{bmatrix}\begin{bmatrix}I_1\\\cdots\\I_i\\\cdots\\I_k\\\cdots\\I_n\end{bmatrix} \tag{5.66}$$

Now, suppose that the protected branch is that branch connecting nodes i and k. Then the currents injected into the network are the currents flowing out of the ends of that branch. These are the currents identified as I_1 and I_2 in Figure 5.19. All other injected currents are zero. If we replace all currents in (5.66) by zero, except for the currents I_i and I_k then we can simplify (5.66) to the form

$$\begin{bmatrix}V_i\\V_k\end{bmatrix}=\begin{bmatrix}Z_{ii} & Z_{ik}\\Z_{ki} & Z_{kk}\end{bmatrix}\begin{bmatrix}I_i\\I_k\end{bmatrix} \tag{5.67}$$

The network with appropriate terminations is shown in Figure 5.20.

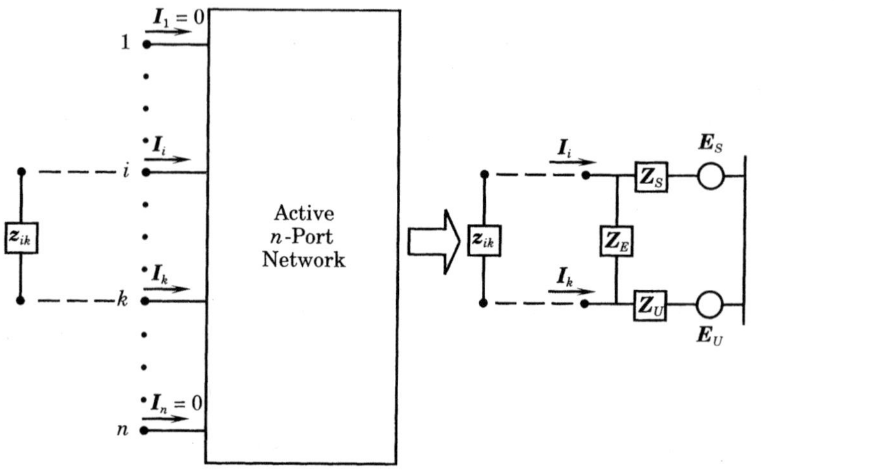

Figure 5.20 Selection of network parameters for the equivalent network.

In the usual two-port notation, we usually write (5.67) as

$$\begin{bmatrix} V_1 \\ V_2 \end{bmatrix} = \begin{bmatrix} Z_{11} & Z_{12} \\ Z_{21} & Z_{22} \end{bmatrix} \begin{bmatrix} I_1 \\ I_2 \end{bmatrix} \tag{5.68}$$

which is the two port representation for the protected branch i-k. Here, we arbitrarily identify the node i as node 1 and node k as node 2. Note that (5.67) must be determined from the impedance matrix computed with the protected branch removed from the network. Since fault current programs are designed to determine the impedance matrix (5.66), it is much simpler to employ (5.67), rather than (5.65), to find the two port parameters of the protected branch. Either method results in exactly the same two-port Z matrix because of the identical assumptions regarding the injected currents. Once the Z matrix with the protected line removed is determined, the desired two-port impedance can simply be read from the appropriate locations in that matrix.

5.5.3 A Simple Two-Port Equivalent

A simple network equivalent is convenient when solving problems associated with a single longitudinal protected element, such as a transmission line. The simplest general equivalent is that shown inside the dashed lines of Figure 5.21. There are three nodes in the two-port equivalent, node 1, node 2, and the reference node. A general equivalent requires

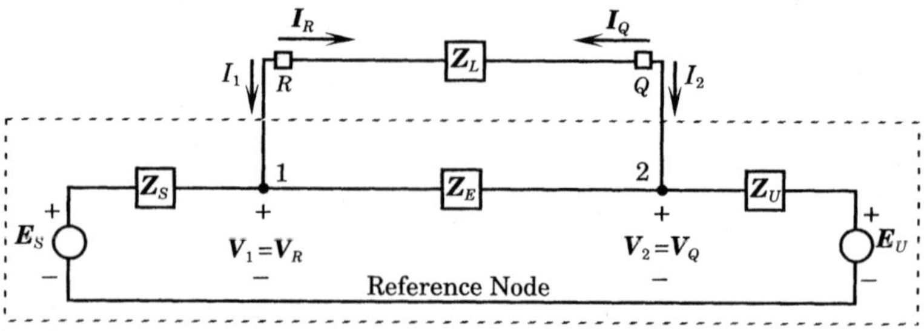

Figure 5.21 A simple two-port network equivalent.

three impedances, which are shown in Figure 5.21. The internal sources are related to the open circuit voltages determined by open-circuit tests of the power system at nodes 1 and 2. The protected component is pictured as a transmission line, with impedance $\mathbf{Z}_L$, with protective devices R and Q located at the line terminals.

5.5.4 Tests of the Equivalent Circuit

The equivalent circuit of the two-port will now be tested to determine its two-port parameters. These tests must be conducted *open circuit*, i.e., with the protected line $\mathbf{Z}_L$ removed.

The first test is the measurement of the voltages at the two nodes under the open-circuit condition. With both ports open, we can compute the current flowing in the clockwise direction, as follows:

$$\mathbf{I} = \frac{\mathbf{E}_S - \mathbf{E}_U}{\mathbf{Z}_S + \mathbf{Z}_E + \mathbf{Z}_U} = \frac{\mathbf{E}_S - \mathbf{E}_U}{\mathbf{Z}_T} \tag{5.69}$$

where, for convenience we have defined

$$\mathbf{Z}_T = \mathbf{Z}_S + \mathbf{Z}_E + \mathbf{Z}_U \tag{5.70}$$

Knowing the current, we readily compute the two-port voltages, which we write in matrix form as follows.

$$\begin{bmatrix} \mathbf{V}_1 \\ \mathbf{V}_2 \end{bmatrix} = \begin{bmatrix} \mathbf{V}_R \\ \mathbf{V}_Q \end{bmatrix} = \frac{1}{\mathbf{Z}_T} \begin{bmatrix} \mathbf{Z}_U + \mathbf{Z}_E & \mathbf{Z}_S \\ \mathbf{Z}_U & \mathbf{Z}_S + \mathbf{Z}_E \end{bmatrix} \begin{bmatrix} \mathbf{E}_S \\ \mathbf{E}_U \end{bmatrix} \tag{5.71}$$

Next, we test the equivalent network with the protected line open and with all internal sources properly removed, which gives the circuit of Figure 5.22.

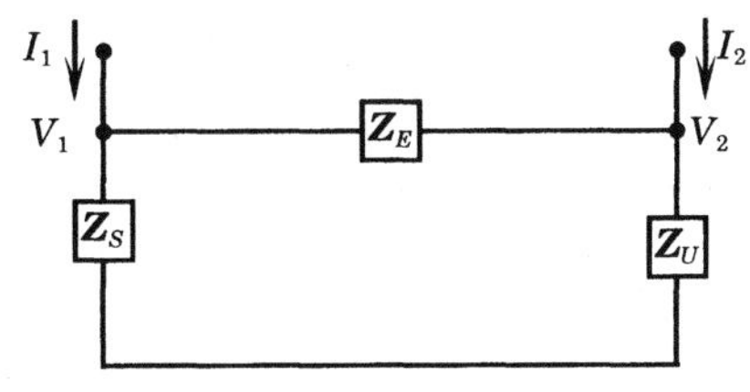

Figure 5.22 The open-circuited equivalent with sources removed.

To determine the two-port parameters, we conduct the following two tests on the equivalent network.

1. Let $I_2 = 0$ and $I_1 = 1.0$ per unit. This determines the first column of the Z matrix.

$$\begin{aligned} \mathbf{Z}_{11} &= \left.\frac{\mathbf{V}_1}{\mathbf{I}_1}\right]_{I_2=0} = \left.\frac{\mathbf{V}_1}{1.0}\right]_{I_2=0} = \frac{\mathbf{Z}_S(\mathbf{Z}_U + \mathbf{Z}_E)}{\mathbf{Z}_T} \\ \mathbf{Z}_{21} &= \left.\frac{\mathbf{V}_2}{\mathbf{I}_1}\right]_{I_2=0} = \left.\frac{\mathbf{V}_2}{1.0}\right]_{I_2=0} = \frac{\mathbf{Z}_S\mathbf{Z}_U}{\mathbf{Z}_T} \end{aligned} \tag{5.72}$$

2. Let $I_1 = 0$ and $I_2 = 1.0$ per unit, to determine the second column of the Z matrix.

$$\begin{aligned} \mathbf{Z}_{12} &= \left.\frac{\mathbf{V}_1}{\mathbf{I}_2}\right]_{I_1=0} = \left.\frac{\mathbf{V}_1}{1.0}\right]_{I_1=0} = \frac{\mathbf{Z}_S\mathbf{Z}_U}{\mathbf{Z}_T} \\ \mathbf{Z}_{22} &= \left.\frac{\mathbf{V}_2}{\mathbf{I}_2}\right]_{I_1=0} = \left.\frac{\mathbf{V}_2}{1.0}\right]_{I_1=0} = \frac{\mathbf{Z}_U(\mathbf{Z}_S + \mathbf{Z}_E)}{\mathbf{Z}_T} \end{aligned} \tag{5.73}$$

The computation of $\boldsymbol{Z}_{12}$ in part 2 gives a partial check on the calculation of part 1, since it is known that the off diagonal terms are equal, which is required of any reciprocal network. Equations (5.71)–(5.73) complete the two-port representation of the equivalent network.

5.5.5 System Equivalent from Two-Port Parameters

The foregoing calculations have determined the two-port parameters of the power system and of the equivalent network. But these results must be equal. The two-port parameters of the power system are determined with the entire network represented and calculations made of the effects observed at the two ports of interest. We can assume that these results are known numerically, as given by (5.68). The equivalent network has two-port parameters too, and these have been computed symbolically, as given by (5.72) and (5.73). However, the numerical values of the equivalent network parameters are not known. To determine the parametric values of the equivalent network requires that we solve for these equivalent parameters in terms of the known two-port parameters. This means that we must solve for $\boldsymbol{E}_S, \boldsymbol{E}_U, \boldsymbol{Z}_S, \boldsymbol{Z}_U$, and $\boldsymbol{Z}_E$ in terms of $\boldsymbol{V}_R, \boldsymbol{V}_Q, \boldsymbol{Z}_{11}, \boldsymbol{Z}_{12}, \boldsymbol{Z}_{21}$, and $\boldsymbol{Z}_{22}$. There are five equations and five unknowns.

Solving (5.72) and (5.73) for the equivalent parameters is difficult because these equations are nonlinear. However, this can be accomplished with diligent effort. The results are as follows.

$$\boldsymbol{Z}_S = \frac{\boldsymbol{Z}_{11}\boldsymbol{Z}_{22} - \boldsymbol{Z}_{12}\boldsymbol{Z}_{21}}{\boldsymbol{Z}_{22} - \boldsymbol{Z}_{21}} \tag{5.74}$$

$$\boldsymbol{Z}_U = \frac{\boldsymbol{Z}_{11}\boldsymbol{Z}_{22} - \boldsymbol{Z}_{12}\boldsymbol{Z}_{21}}{\boldsymbol{Z}_{11} - \boldsymbol{Z}_{12}} \tag{5.75}$$

$$\boldsymbol{Z}_E = \frac{\boldsymbol{Z}_{11}\boldsymbol{Z}_{22} - \boldsymbol{Z}_{12}\boldsymbol{Z}_{21}}{\boldsymbol{Z}_{12}} = \frac{\boldsymbol{Z}_{11}\boldsymbol{Z}_{22} - \boldsymbol{Z}_{12}\boldsymbol{Z}_{21}}{\boldsymbol{Z}_{21}} \tag{5.76}$$

The equivalent source voltages, $\boldsymbol{E}_S$ and $\boldsymbol{E}_U$, can be easily solved in terms of the open-circuit port voltages $\boldsymbol{V}_R$ and $\boldsymbol{V}_Q$ because of the linear relationship given by (5.71). The result is

$$\begin{bmatrix} \boldsymbol{E}_S \\ \boldsymbol{E}_U \end{bmatrix} = \frac{1}{\boldsymbol{Z}_E} \begin{bmatrix} \boldsymbol{Z}_S + \boldsymbol{Z}_E & -\boldsymbol{Z}_S \\ -\boldsymbol{Z}_U & \boldsymbol{Z}_U + \boldsymbol{Z}_E \end{bmatrix} \begin{bmatrix} \boldsymbol{V}_R \\ \boldsymbol{V}_Q \end{bmatrix} \tag{5.77}$$

The equivalent source voltages can be expressed in terms of the power system two-port parameters by substituting (5.74)–(5.76) into (5.77).

5.5.6 Equivalent of a Line with Shunt Faults

The purpose of developing the two-port equivalent is to simplify the examination of the protected element under faulted conditions. The network for this condition is shown in Figure 5.23, where the fault is located a fractional distance h from line terminal R. The representation shown in Figure 5.23 is for the positive sequence network only and applies only to three-phase faults. The methods of symmetrical components can be applied for unbalanced faults, using the negative and zero sequence networks, with the fault point at the same place in all networks [4]. The fault current $\boldsymbol{I}_F$ represents the total fault current flowing toward the reference.

Equation (5.58) applies directly, with the following special constraints on the port currents.

$$\boldsymbol{I}_R = -\boldsymbol{I}_1 = \frac{\boldsymbol{V}_1 - \boldsymbol{V}_F}{h\boldsymbol{Z}_L} \tag{5.78}$$

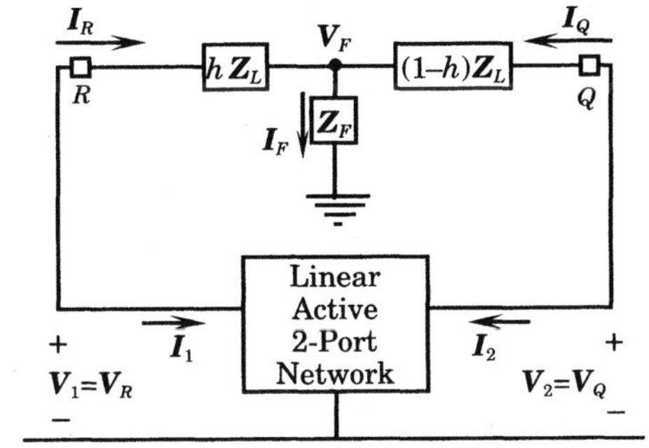

Figure 5.23 The two-port network with faulted line.

$$I_Q = -I_2 = \frac{V_2 - V_F}{(1-h)Z_L} \tag{5.79}$$

$$I_F = \frac{V_F}{Z_F} \tag{5.80}$$

Substituting these values into (5.58), gives the following results, which are stated in terms of the two-port parameters.

$$\begin{bmatrix} V_{S1} \\ V_{S2} \end{bmatrix} = \begin{bmatrix} (Z_{11} + hZ_L + Z_F) & (Z_{12} + Z_F) \\ (Z_{21} + Z_F) & (Z_{22} + (1-h)Z_L + Z_F) \end{bmatrix} \begin{bmatrix} I_R \\ I_Q \end{bmatrix} \tag{5.81}$$

These equations can be solved for the currents seen at the ends of the protected line, with the following results.

$$\begin{bmatrix} I_R \\ I_Q \end{bmatrix} = \frac{1}{Z_\Delta^2} \begin{bmatrix} (Z_{22} + (1-h)Z_L + Z_F) & (Z_{12} + Z_F) \\ (Z_{21} + Z_F) & (Z_{11} + hZ_L + Z_F) \end{bmatrix} \begin{bmatrix} V_{S1} \\ V_{S2} \end{bmatrix} \tag{5.82}$$

where we have defined the determinant as an impedance squared, since it has that dimension.

$$Z_\Delta^2 = (Z_{11} + hZ_L + Z_F)(Z_{22} + (1-h)Z_L + Z_F) - (Z_{12} + Z_F)(Z_{21} + Z_F) \tag{5.83}$$

The result (5.82) gives the relay currents in terms of the open-circuit two-port source voltages. Using (5.71), however, the result can be expressed in terms of the equivalent circuit voltages, if this is desired. The result is given as follows.

$$\begin{bmatrix} I_R \\ I_Q \end{bmatrix} = \frac{Z_L + Z_E}{F_1F_3 - F_2^2} \begin{bmatrix} [(1-h)Z_L(F_2 - F_3) + Z_EF_3] & -[(1-h)Z_L(F_2 - F_1) + Z_EF_2] \\ -[hZ_L(F_2 - F_3) + Z_EF_2] & [hZ_L(F_1 - F_2) + Z_EF_1] \end{bmatrix} \begin{bmatrix} E_S \\ E_U \end{bmatrix} \tag{5.84}$$

where, for convenience, we have defined the following complex functions.

$$\begin{aligned} F_1 &= (Z_S + hZ_L + Z_F)(Z_L + Z_E) - (hZ_L)^2 \\ F_2 &= Z_F(Z_L + Z_E) - h(1-h)Z_L^2 \\ F_3 &= (Z_U + (1-h)Z_L + Z_F)(Z_L + Z_E) - (1-h)^2Z_L^2 \end{aligned} \tag{5.85}$$

Clearly, for shunt faults, the expressions are complex when viewed in this form, much more so than the equations written in terms of the two-port parameters, given by (5.82).

5.5.7 Applications of the Equivalent to Series Faults

For series faults,[2] the network of Figure 5.24 is applicable. For this condition, a very special relationship exists.

$$I_1 = -I_R = I_Q = -I_2 \tag{5.86}$$

[2]Series faults are those where one or more lines are open [2].

From (5.58), we write

$$\begin{aligned}\begin{bmatrix} V_1 \\ V_2 \end{bmatrix} &= \begin{bmatrix} Z_{11} & Z_{12} \\ Z_{21} & Z_{22} \end{bmatrix} \begin{bmatrix} I_1 \\ -I_1 \end{bmatrix} + \begin{bmatrix} V_{S1} \\ V_{S2} \end{bmatrix} \\ &= \begin{bmatrix} Z_{11} & Z_{12} \\ Z_{21} & Z_{22} \end{bmatrix} \begin{bmatrix} -I_R \\ +I_Q \end{bmatrix} + \begin{bmatrix} V_{S1} \\ V_{S2} \end{bmatrix}\end{aligned} \tag{5.87}$$

Solving for I_1 we get

$$I_1 = \frac{V_1 - V_{S1}}{Z_{11} - Z_{12}} = \frac{V_{S2} - V_2}{Z_{22} - Z_{21}} \tag{5.88}$$

An inspection of (5.87) shows that the two ports are completely uncoupled. This means that the series fault condition for the positive sequence network may be represented as shown in Figure 5.24.

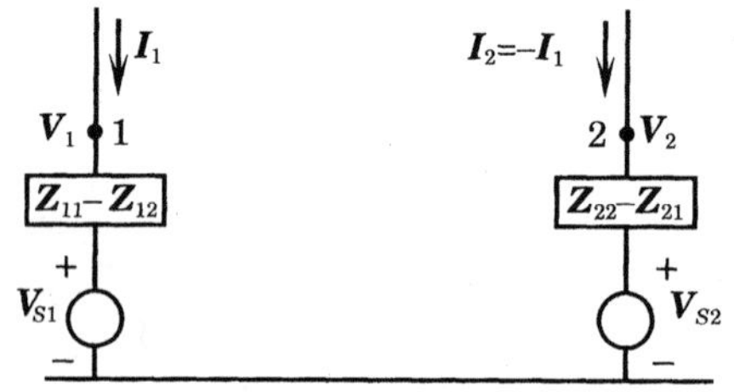

Figure 5.24 Positive sequence network equivalent for series faults.

This simple result is due entirely to the constraint on connections external to the sequence network, which ensures equality of the current entering node 1 to that leaving node 2. In terms of the equivalent circuit parameters, the impedances of Figure 5.24 are

$$Z_{11} - Z_{12} = \frac{Z_S Z_E}{Z_T} \tag{5.89}$$

$$Z_{22} - Z_{21} = \frac{Z_U Z_E}{Z_T} \tag{5.90}$$

By inspection of Figure 5.24, for the special case where only Z_L is connected external to the equivalent of Figure 5.24, the relay current is given by

$$I_R = \frac{V_1 - V_2}{Z_L} \tag{5.91}$$

Substituting (5.71), (5.72), and (5.73) into (5.89), we can find the relay current as a function of the equivalent circuit parameters.

$$I_R = \frac{E_S - E_U}{(Z_S + Z_U)(1 + Z_L/Z_E) + Z_L}. \tag{5.92}$$

This denominator reduces to simply the series impedances between the sources when Z_E becomes large, which is obviously correct.

These equations apply only for those conditions where (5.86) is true. This includes normal load conditions as well as series fault conditions where the currents at each end of the line obey (5.86). For unbalanced series faults, the interpretation of Z_L must be changed to include the impedances of all sequence networks, arranged as required to represent the series unbalance of interest. For a detailed description of the usual cases of interest, see [4].

EXAMPLE 5.1

The foregoing development is applied to a well-known six bus system first introduced in [17] and fully described, including all parameters, in [4]. The system is shown in Figure 5.25.

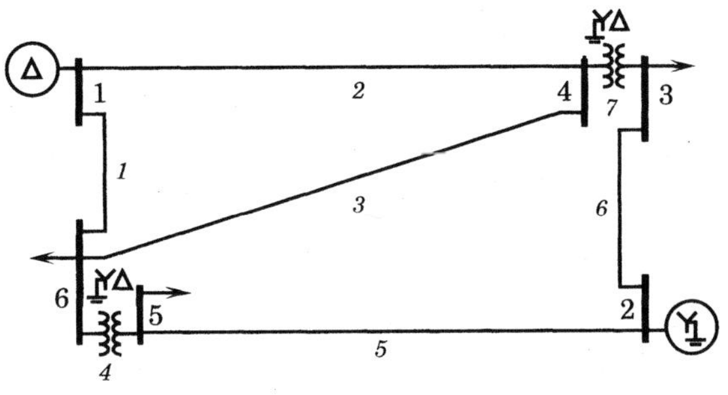

Figure 5.25 A six bus power system.

To determine the two-port parameters for any branch, the impedance matrix of the network is computed with the protected branch open. This yields the two-port parameters, from which the equivalent network parameters for shunt faults may be computed using (5.73)–(5.75) or the parameters for series faults using (5.89)–(5.90). Only the equivalent circuit for shunt faults is investigated here.

Solution

The solution to this problem requires the computation of seven impedance matrices, corresponding to each of the seven branches, but solved with that branch out of service. A computer program is a practical necessity for this type of calculation. The results are shown in Table 5.2.

TABLE 5.2 Computed Shunt Fault Equivalent Circuit Parameters Corresponding to the Seven Branches of the Six Bus Network of Figure 5.25

Line	Z_{11}	Z_{12}	Z_{22}	Z_S	Z_U	Z_E	Z_L
1	0.02188	0.01442	0.22696	0.02158	0.53088	0.36477	0.24600
(1–6)	+*j*0.21980	+*j*0.12107	+*j*0.98368	+*j*0.23338	+*j*2.03427	+*j*1.66840	+*j*1.03600
2	0.02152	0.01674	0.34256	0.02014	0.79700	0.53810	0.16000
(1–4)	+*j*0.22098	+*j*0.11328	+*j*1.12841	+*j*0.23248	+*j*2.16566	+*j*2.10486	+*j*0.74000
3	0.18918	0.01252	0.18519	0.20607	0.19181	1.54585	0.19400
(4–6)	+*j*0.74096	+*j*0.15826	+*j*0.82955	+*j*0.87959	+*j*1.00890	+*j*3.62274	+*j*0.81400
4	0.61099	–0.01303	0.17283	0.59755	0.16425	15.2468	0.00000
(5–6)	+*j*1.70409	+*j*0.06465	+*j*0.78756	+*j*1.85851	+*j*0.81428	+*j*15.9900	+*j*0.60000
5	0.04689	–0.01303	0.17283	0.04389	0.11895	3.81002	0.56400
(2–5)	+*j*0.42409	+*j*0.06465	+*j*1.38756	+*j*0.44184	+*j*1.62326	+*j*8.14622	+*j*1.28000
6	0.03744	–0.00083	0.12992	0.03608	0.13497	1.17768	1.44600
(2–3)	+*j*0.41052	+*j*0.07887	+*j*0.94592	+*j*0.44051	+*j*1.15303	+*j*4.77051	+*j*2.10000
7	1.48344	–0.00083	0.12992	1.59952	0.12417	17.1233	0.00000
(3–4)	+*j*2.51051	+*j*0.07887	+*j*0.67992	+*j*2.85689	+*j*0.69595	*j*18.9404	+*j*0.26600

The six bus system of Figure 5.25 is not necessarily typical of large power systems, although it has characteristics that are found in most transmission networks. The computed results for this small system show that the equivalent impedance Z_E can be of the same order of magnitude as that of the protected line, which is noted by comparing the rightmost two columns of Table 5.2. This suggests that neglecting the parallel equivalent impedance Z_E may sometimes lead to significant error in the calculations of relay settings for the protected component. ■

The test of any equivalent circuit is the accuracy of the solution found using that equivalent. The proposed equivalent is readily checked by forming the Z matrix of the equivalent circuit of Figure 5.20 and comparing this matrix with the appropriate elements from the Z matrix of the complete network. The test shows that the equivalent form of system representation is exact. This means that fault currents derived using the equivalent are just as accurate as those found using the matrix solution for entire network. This test is left as an exercise.

EXAMPLE 5.2

Using the data for the equivalent circuit obtained in Example 5.1, determine the source voltages E_1 and E_2 to complete the specification of the equivalent. Let the protected line be the transmission line from bus 1 to bus 6, for which the equivalent impedances are given in the first row of Table 5.2.

Solution

The solution for the equivalent voltages is given by (5.73). To use this expression, it is necessary to determine the open-circuit port voltages of the network. This requires that the six bus system be solved to satisfy (5.59), that is, with the protected line open. This requires a power flow solution of the network, with the computed port voltages found to be the following:

$$\mathbf{V}_{S1} = 1.05\angle 0^\circ$$
$$\mathbf{V}_{S2} = 0.672\angle - 47.51^\circ$$

This result is substituted into (5.77) to compute the equivalent source voltages.

$$\begin{bmatrix} \mathbf{E}_S \\ \mathbf{E}_U \end{bmatrix} = \frac{1}{\mathbf{Z}_E} \begin{bmatrix} \mathbf{Z}_S + \mathbf{Z}_E & -\mathbf{Z}_S \\ -\mathbf{Z}_U & \mathbf{Z}_U + \mathbf{Z}_E \end{bmatrix} \begin{bmatrix} \mathbf{V}_{S1} \\ \mathbf{V}_{S2} \end{bmatrix}$$

$$= \frac{1}{1.708\angle 77.667^\circ} \begin{bmatrix} 1.941\angle 78.525^\circ & -0.234\angle 84.717 \\ -2.102\angle 75.374^\circ & 3.809\angle 76.402^\circ \end{bmatrix} \begin{bmatrix} 1.05\angle 0^\circ \\ 0.672\angle - 47.51^\circ \end{bmatrix}$$

$$= \begin{bmatrix} 1.125\angle 3.959^\circ \\ 1.118\angle - 105.767 \end{bmatrix}$$

In the application of the equivalent circuit to fault studies, the difference in the two-source voltages, both in terms of magnitude and angle, is often of little interest. The complex source voltage difference determines the load current flowing through the protected branch. In most cases, this load current can be neglected in protection studies. Therefore, the source voltages are often combined into a single source, which will determine the total fault current at the fault point, including contributions from both ends of the protected line. Since the network is linear, this total fault current will be proportional to the magnitude of the source voltage. For the system under study, it would be appropriate to use a source voltage magnitude of about 1.12 per unit. The angle can be set to zero, for convenience. The procedure outlined in this example helps to determine a suitable magnitude of the source voltage. Setting the source voltage to 1.0, for example, would result in fault currents that are about 12 percent low. ■

5.5.8 Conclusions Regarding Two-Port Equivalents

An equivalent circuit is derived, which can be used in the study of the protection of a power system component, such as a transmission line. The equivalent is readily derived from the transmission network equations using active two-port network concepts. Moreover, the derived two-port parameters can be related to a simple physical equivalent circuit to provide a convenient computational framework for the study of the protection of a network component. No assumptions are required regarding the equivalent circuit, other than the requirement for linearity of the network equations, although the proposed equivalent is limited to the network topology assumed in the derivation, and as given in Figure 5.20. The equivalent circuit parameters can easily be found from the results of a power system short-circuit study. Indeed, it can be asserted that the purpose of short-circuit studies is to determine the parameters of the protection equivalent circuit derived above. This enables the protection engineer to study any type of fault on the protected component without the need for considering the entire power system.

5.5.9 Multiport Equivalents

The two-port equivalent provides a method of simplifying the power system for the study of faults on a protected branch of the network. However, there are situations where a protected line has more than two terminals. This results when lines are tapped, but a full switching station is not installed at the tap point, resulting in a three-terminal line. This can also occur when mutually coupled lines must be analyzed. For such cases a multiport equivalent is required.

The general equation for the number of network elements for a multiport network equivalent is given by

$$N = \frac{n(n+1)}{2} \tag{5.93}$$

where n is the number of ports in the protected system and N is the number of required network elements in the supply system. In some cases, some of the network elements can be ignored, for example, if they are extremely large, representing essentially an open circuit between a pair of nodes. For the system of Figure 5.21, the number n is two, therefore requiring three elements in the equivalent of the power system. One element is identified in the figure as the impedance $\mathbf{Z}_E$ and connects the two line terminals. The other two impedances are associated with the Thevenin equivalent source voltages, which are connected to each node supplying the protected system. These sources are arbitrarily identified by the letters S and U, as shown in the figure. The power system equivalent is that portion of the diagram inside the dashed lines. In many cases, it is convenient to ignore the load currents flowing between the nodes, applying the compensation theorem as described in Section 5.6.

Note that no shunt admittance is shown for the protected line, either from line to ground or between lines. This is an acceptable simplification for short transmission lines, but may not be acceptable for long EHV circuits, where the admittances are quite large. Even for long lines, however, the external connections would be the same as that of Figure 5.21, but each line would also have shunt admittance connections to the other line and to the reference node. The network equivalent will also have shunt admittances, in the general case.

5.5.9.1 The Two-Port System Equivalent. The two-port system equivalent has been discussed above, and its structure is repeated here for convenience in comparing the various equivalents. For this case $N = 3$, resulting in the equivalent shown in Figure 5.26, where the system equivalent is that portion inside the dashed box.

Figure 5.26 illustrates an important point. There may be more than one transmission line between the two buses that terminate the system equivalent. In many cases, it is important to study the protection of the two lines together, for example, when the lines are mutually

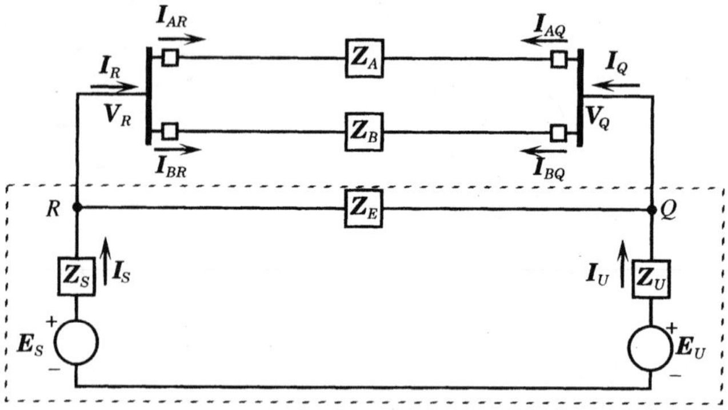

Figure 5.26 The two-port system equivalent.

coupled. However, since the lines terminate at the same two buses, the equivalent is the same whether there is one protected line or more than one line connecting these terminal nodes.

5.5.9.2 The Three-Port System Equivalent. The next class of networks is Type 2, where the transmission lines are bused at only one end. Typical systems having this characteristic are shown in Figure 5.27. A common problem is where a transmission line is tapped at some midline point, resulting in a three-terminal line, as shown in Figure 5.27(*a*). Another common system is the case of a mutually coupled system of lines. This occurs, for example, where two lines leave a switching station and are located along a common right of way for some distance, after which they branch off in different directions.

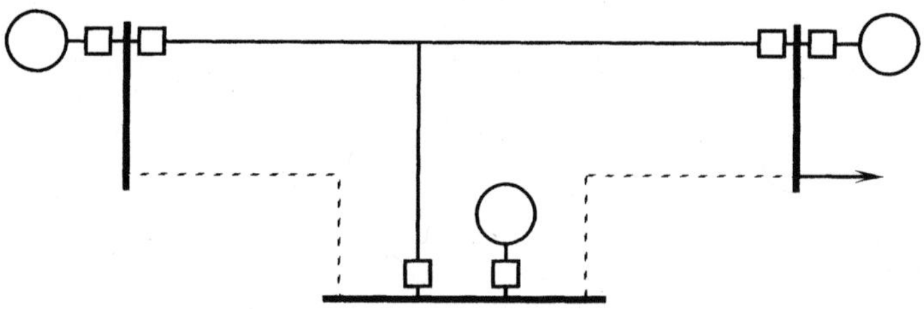

(*a*)

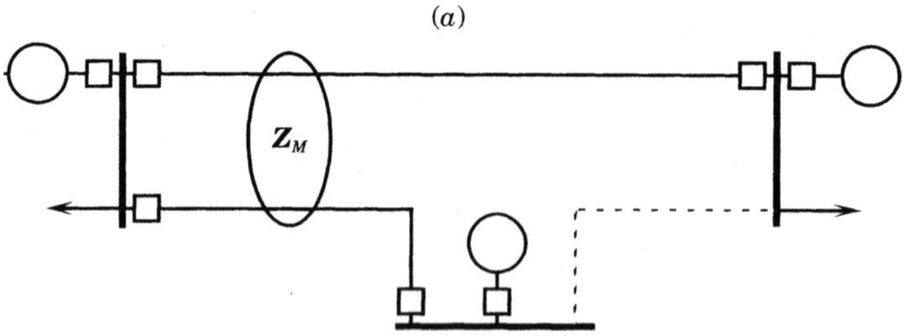

(*b*)

Figure 5.27 Protected lines requiring a three-port system equivalent. (*a*) Tapped transmission line. (*b*) Mutually coupled lines bused with common end bus.

In this case, the number n is three, therefore requiring six network elements connecting the terminal nodes, as shown in Figure 5.28, which illustrates the case where the protected lines are mutually coupled. This requires that both lines be represented as mutually coupled

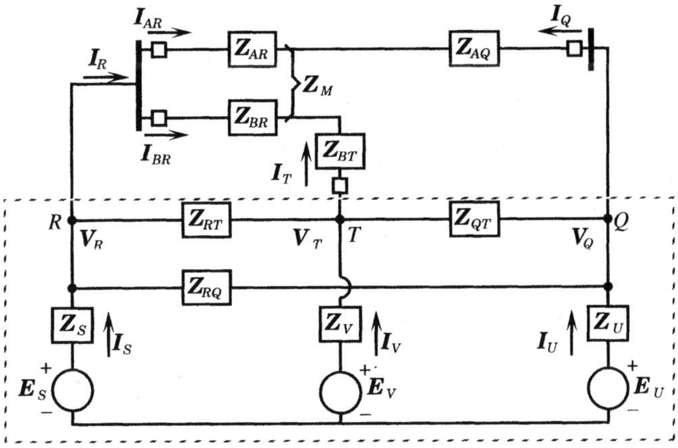

Figure 5.28 Three-port equivalent for two lines bused at one end.

lines for a portion of the total line length where they share the common right of way, but as separate and nonmutually coupled lines for the remainder of their length. Computation of mutual coupling of parallel lines show that the lines can be separated by quite a large distance and still have significant mutual coupling. The best practice is to make the computation and then determine the error that would be introduced by ignoring mutual induction.

The system equivalent results in the delta connection of the nodes R, Q, and T, as shown in the figure, plus the source impedances. Thevenin equivalent sources are also connected to each of the system nodes, and are arbitrarily designated S, U, and V. The entire system equivalent is enclosed by dashed lines in the figure.

5.5.9.3 The Four-Port System Equivalent. There are other configurations of transmission lines where four ports must be considered in protection studies. One such case is that involving two mutually coupled transmission lines, where the lines do not terminate at common nodes, as shown in Figure 5.29.

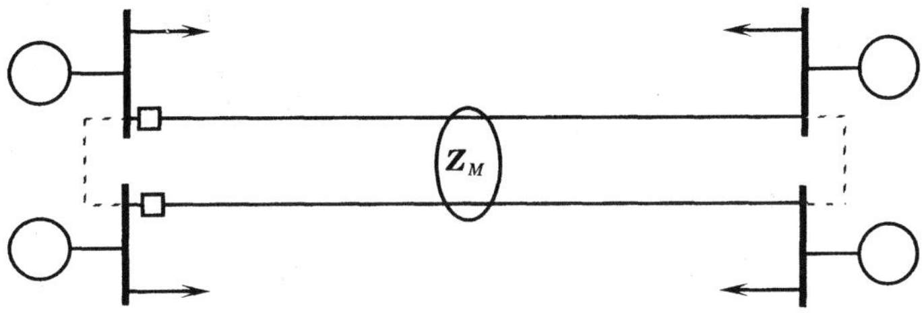

Figure 5.29 Mutually coupled lines with no common terminations.

For this case, the system equivalent network becomes more complex. The result is shown in Figure 5.30, where it is noted that the four-port network requires six impedances in a mesh arrangement, plus the four equivalent sources, for a total of 10 impedances.

In some cases, some of the computed network impedances may be large enough to ignore, which will simplify the equivalent. Also, it is common practice to apply the compensation theorem to the positive sequence network, thereby eliminating three of the four sources. This is discussed in the next section.

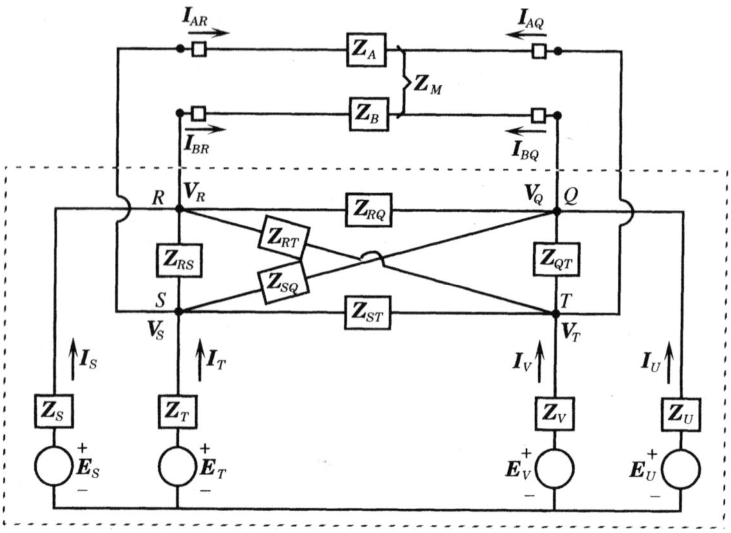

Figure 5.30 A four-port equivalent for two lines not bused at either end.

5.6 THE COMPENSATION THEOREM

The compensation theorem from network theory provides a useful method for examining the change in the currents and voltages of a linear network when a change in impedance occurs [18], [19]. The system is easy to apply for the special case shown in Figure 5.31. The network can be described in terms of either impedance or admittances. When impedances are used, one would normally write voltage equations, but using admittances, current equations are usually used. Either approach is correct, but the admittance approach has certain advantages in this case, as will be noted more clearly once the solutions are determined. Applying the compensation theorem, we first examine the network with an arbitrary admittance, $\boldsymbol{Y}_3$, connected from the point F to the reference, and then, in a second phase of the solution, we solve the same network with a new admittance added to the original admittance. We examine the process in two steps. We identify the solution to the first step as that existing *before* the change in admittance, and the second step that *after* the change in $\boldsymbol{Y}_3$.

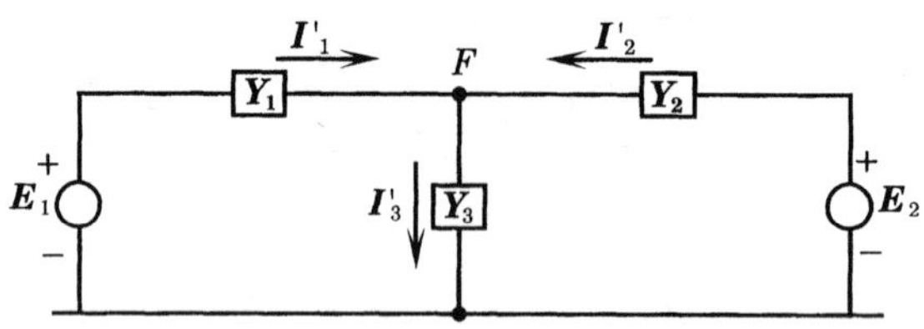

Figure 5.31 Two sources supplying a fault current.

5.6.1 Network Solution *Before* Changing Y_3

For the simple circuit shown in Figure 5.31, the two sources and the network admittances are all constant system parameters. We seek the network currents and voltages in terms of these parameters. The network is easily solved by writing the node voltage equation at node

F, as follows.

$$(Y_1 + Y_2 + Y_3)V'_F = Y_1E_1 + Y_2E_2 \tag{5.94}$$

where we have used the prime notation to indicate that this solution applies *before* we change the network admittances. This matrix equation is easily solved, with the result

$$V'_F = \frac{Y_1E_1 + Y_2E_2}{Y_1 + Y_2 + Y_3} \tag{5.95}$$

The current I_3 flowing through impedance Y_3 can be found from the above voltage, or

$$I'_3 = Y_3V'_F = \frac{Y_3(Y_1E_1 + Y_2E_2)}{Y_1 + Y_2 + Y_3} \tag{5.96}$$

This completes the solution of the network in the *before* state. The source currents can be found from the above results. This is left as an exercise.

5.6.2 Network Solution *After* Changing Y_3

We now consider the effect of a change in the admittance from point F to the neutral bus, where the change is represented by ΔY, as shown in Figure 5.32. The increase in admittance from F to reference will be interpreted as the application of a fault at node F.

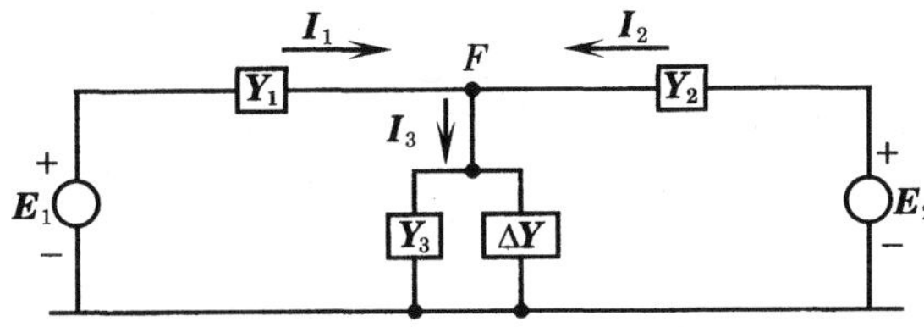

Figure 5.32 The network *after* the change in admittance.

For this new network condition, we again write the system equations as

$$(Y_1 + Y_2 + Y_3 + \Delta Y)V_F = Y_1E_1 + Y_2E_2 \tag{5.97}$$

Solving for the node F voltage, we have

$$V_F = \frac{Y_1E_1 + Y_2E_2}{Y_1 + Y_2 + Y_3 + \Delta Y} \tag{5.98}$$

We can also solve for the new current leaving node F.

$$\begin{aligned} I_3 &= (Y_3 + \Delta Y)V_F \\ &= \frac{(Y_3 + \Delta Y)(Y_1E_1 + Y_2E_2)}{Y_1 + Y_2 + Y_3 + \Delta Y} \end{aligned} \tag{5.99}$$

5.6.3 The Incremental Change in Current and Voltage

The incremental change in current is defined by the equation

$$I''_3 = I_3 - I'_3 \tag{5.100}$$

where the primed notation refers to the prefault case and the double-primed notation denotes the incremental change in current due to the fault. The unprimed current is the total current, consisting of the prefault load current and the incremental current added due to the fault. Substituting the currents from (5.96) and (5.99), we find the incremental change in current as follows.

$$
\begin{aligned}
I_3'' &= I_3 - I_3' \\
&= (Y_1E_1 + Y_2E_2)\left(\frac{(Y_3 + \Delta Y)}{Y_1 + Y_2 + Y_3 + \Delta Y} - \frac{Y_3}{Y_1 + Y_2 + Y_3}\right) \\
&= \frac{\Delta Y(Y_1 + Y_2)(Y_1E_1 + Y_2E_2)}{(Y_1 + Y_2 + Y_3 + \Delta Y)(Y_1 + Y_2 + Y_3)}
\end{aligned}
\tag{5.101}
$$

In a similar fashion, we can find the incremental change in voltage at F.

$$
\begin{aligned}
V_F'' &= V_F - V_F' \\
&= (Y_1E_1 + Y_2E_2)\left(\frac{1}{Y_1 + Y_2 + Y_3 + \Delta Y} - \frac{1}{Y_1 + Y_2 + Y_3}\right) \\
&= \frac{-\Delta Y(Y_1E_1 + Y_2E_2)}{(Y_1 + Y_2 + Y_3 + \Delta Y)(Y_1 + Y_2 + Y_3)}
\end{aligned}
\tag{5.102}
$$

Comparing (5.101) and (5.102) with the solution of the original network in Section 5.6.1, we recognize that the prefault node F voltage (5.95) appears in both of the incremental results. Substituting (5.95) into (5.101) and (5.102), we can write

$$
I_3'' = \frac{\Delta Y(Y_1 + Y_2)V_F'}{Y_1 + Y_2 + Y_3 + \Delta Y}
\tag{5.103}
$$

$$
V_F'' = \frac{-\Delta Y V_F'}{Y_1 + Y_2 + Y_3 + \Delta Y}
\tag{5.104}
$$

Thus, the incremental change in current and voltage can be written in terms of the incremental change in admittance and the prefault voltage. On closer examination of (5.103) and (5.104), we can realize these results in terms of the circuit shown in Figure 5.33, where we define the source voltage as the prefault voltage at F, i.e.,

$$
E_F = V_F'
\tag{5.105}
$$

The compensation theorem shows how the incremental change in currents and voltages at a network node can be found by properly removing[3] all of the original sources and adding a new source, equal to the prefault voltage, at the node where the admittance is changed.

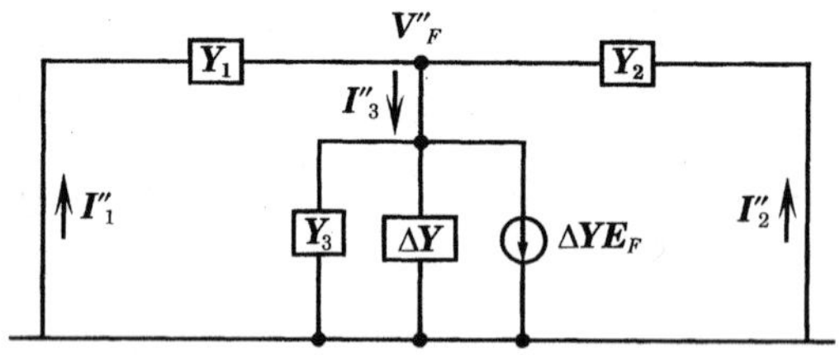

Figure 5.33 Circuit realization for the incremental solution.

If the network is solved in terms of impedances, rather than admittances, then the incremental impedance is added in series with the original impedance. In this case, the incremental solution network is that shown in Figure 5.34. For the impedance method, the prefault voltage is given as

$$
V_F' = \frac{Z_3(Z_2E_1 + Z_1E_2)}{Z_1Z_2 + Z_3(Z_1 + Z_2)} = E_F
\tag{5.106}
$$

[3]Proper source removal should be interpreted as shorting all voltage sources and opening all current sources.

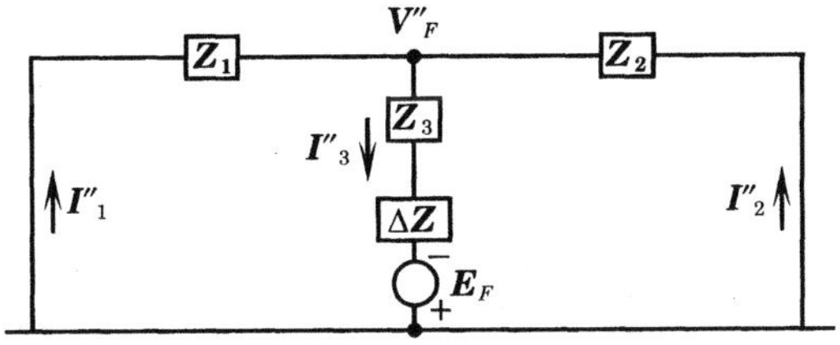

Figure 5.34 Incremental solution using impedances.

The impedance method must be used with caution, as the circuit of Figure 5.31 is not correct if Z_3 is infinite. In this case (5.106) can be taken in the limit as Z_3 becomes large; however, the network diagram of Figure 5.34 is no longer applicable. This special case is left as an exercise.

5.6.4 The Compensation Theorem in Fault Studies

The applications of the compensation theorem in fault studies is direct and useful. The original network of Figure 5.31 can be thought of as the prefault network, where F is the fault point in that network. In some cases the admittance Y_3 might represent a load, a shunt, or other admittance, or it can be set to zero to represent an open circuit at F prior to the fault. Since the network is linear, the change in current in moving from the unfaulted condition to the faulted condition can be solved in terms of the change in admittance ΔY and the prefault voltage V'_F, as given by (5.103) and (5.104). If it is desired to find the load current contribution to the total currents, this corresponds to the unfaulted network currents plus the incremental change in currents, as given by (5.99).

The foregoing results are applicable only for problems in which the network of Figure 5.31 is an accurate equivalent for the power system. In many cases, this is not an adequate equivalent system. The more general case is examined in the next section.

5.7 COMPENSATION APPLICATIONS IN FAULT STUDIES

The general case for applications of the compensation theorem follows from the basic concepts of the previous section with two differences. First, in many cases a more general equivalent circuit of the power system is required and, second, it is usually not necessary to consider a fault impedance to exist prior to the fault.

5.7.1 Prefault Conditions

The equivalent circuit of Figure 5.35 provides a suitable network for consideration of shunt faults at F. Note that the prefault impedance to neutral at F is infinite. This system rep-

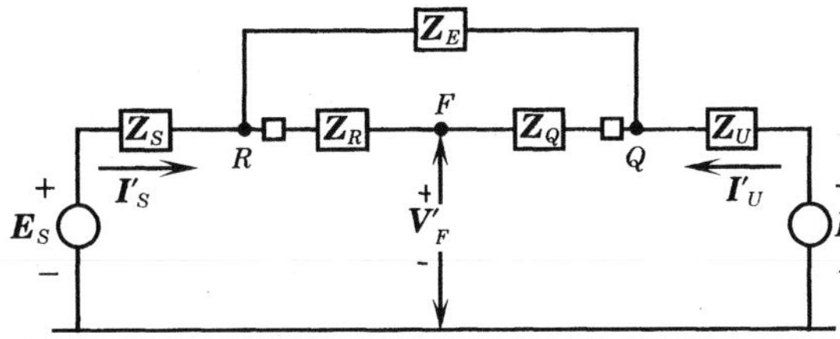

Figure 5.35 The prefault system.

resents the normal system with a prefault power flow existing that depends on the impedances and the source voltages at both ends of the equivalent.

The prefault system is solved using any type of network analysis, such as loop currents or node voltages. The resulting network currents are found to be

$$
\begin{aligned}
\mathbf{I}'_S &= \frac{\mathbf{Z}_K(\mathbf{E}_S - \mathbf{E}_U)}{\mathbf{Z}_K(\mathbf{Z}_S + \mathbf{Z}_U) + \mathbf{Z}_E(\mathbf{Z}_R + \mathbf{Z}_Q)} \\
\mathbf{I}'_E &= \frac{(\mathbf{Z}_R + \mathbf{Z}_Q)(\mathbf{E}_S - \mathbf{E}_U)}{\mathbf{Z}_K(\mathbf{Z}_S + \mathbf{Z}_U) + \mathbf{Z}_E(\mathbf{Z}_R + \mathbf{Z}_Q)} \\
\mathbf{I}'_U &= -\mathbf{I}'_S
\end{aligned}
\tag{5.107}
$$

where we define

$$\mathbf{Z}_K = \mathbf{Z}_R + \mathbf{Z}_Q + \mathbf{Z}_E \tag{5.108}$$

These currents are shown as "primed" quantities to represent the fact that these are prefault currents. The prefault voltage at the fault point F is

$$\mathbf{V}'_F = \frac{(\mathbf{Z}_K\mathbf{Z}_U + \mathbf{Z}_Q\mathbf{Z}_E)\mathbf{E}_S + (\mathbf{Z}_K\mathbf{Z}_S + \mathbf{Z}_R\mathbf{Z}_E)\mathbf{E}_U}{\mathbf{Z}_K(\mathbf{Z}_S + \mathbf{Z}_U) + \mathbf{Z}_E(\mathbf{Z}_R + \mathbf{Z}_Q)} \tag{5.109}$$

For this case, the fault current is zero, so to complete the solution for this system, we write

$$\mathbf{I}'_F = 0 \tag{5.110}$$

The foregoing quantities represent one set of conditions. The next system condition of interest is that of the same system with a shunt fault at the fault point F.

5.7.2 The Faulted Network Condition

When a shunt fault is applied at the fault point, the system is that shown in Figure 5.36, where the fault is assumed to have impedance $\mathbf{Z}_F$. Any type of shunt fault can be applied by appropriate choice of $\mathbf{Z}_F$.

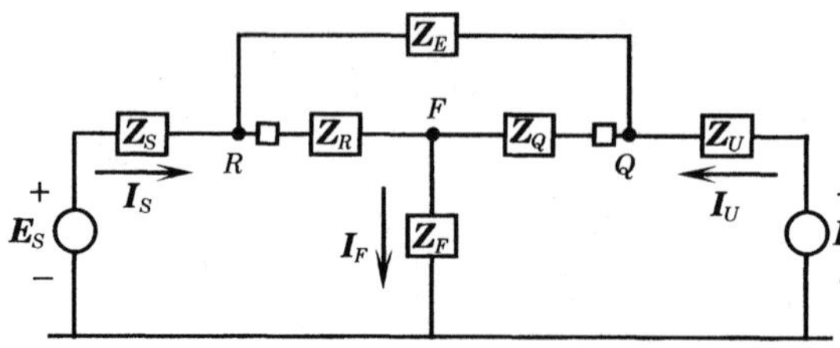

Figure 5.36 The faulted system.

This system is readily solved for the network currents and voltages, to get

$$
\begin{aligned}
\mathbf{I}_S &= \frac{[\mathbf{Z}_K(\mathbf{Z}_U + \mathbf{Z}_F) + \mathbf{Z}_Q(\mathbf{Z}_R + \mathbf{Z}_E)]\mathbf{E}_S - (\mathbf{Z}_K\mathbf{Z}_F + \mathbf{Z}_R\mathbf{Z}_Q)\mathbf{E}_U}{DEN} \\
\mathbf{I}_U &= \frac{[\mathbf{Z}_K(\mathbf{Z}_S + \mathbf{Z}_F) + \mathbf{Z}_R(\mathbf{Z}_Q + \mathbf{Z}_E)]\mathbf{E}_U - (\mathbf{Z}_K\mathbf{Z}_F + \mathbf{Z}_R\mathbf{Z}_Q)\mathbf{E}_S}{DEN}
\end{aligned}
\tag{5.111}
$$

$$
\begin{aligned}
\mathbf{I}_F &= \mathbf{I}_S + \mathbf{I}_U = \frac{(\mathbf{Z}_K\mathbf{Z}_U + \mathbf{Z}_Q\mathbf{Z}_E)\mathbf{E}_S + (\mathbf{Z}_K\mathbf{Z}_S + \mathbf{Z}_R\mathbf{Z}_E)\mathbf{E}_U}{DEN} \\
\mathbf{V}_F &= \mathbf{Z}_F\mathbf{I}_F
\end{aligned}
\tag{5.112}
$$

where the denominator is given by

$$DEN = \frac{\left\{ \begin{array}{c} (\mathbf{Z}_K\mathbf{Z}_F + \mathbf{Z}_R\mathbf{Z}_Q)[\mathbf{Z}_K(\mathbf{Z}_S + \mathbf{Z}_U) + \mathbf{Z}_E(\mathbf{Z}_R + \mathbf{Z}_Q)] \\ + (\mathbf{Z}_K\mathbf{Z}_S + \mathbf{Z}_R\mathbf{Z}_E)(\mathbf{Z}_K\mathbf{Z}_U + \mathbf{Z}_Q\mathbf{Z}_E) \end{array} \right\}}{\mathbf{Z}_K} \tag{5.113}$$

and where $\mathbf{Z}_K$ is defined in (5.108).

The various combinations of impedances that appear in the above equations are so common in these computations that it is convenient to define the following special notation to represent these complex expressions. This allows the equations to be written more compactly and simplifies algebraic computation. Therefore, we define the following:

$$\begin{aligned} \mathbf{Z}_{KS}^2 &= \mathbf{Z}_K\mathbf{Z}_S + \mathbf{Z}_R\mathbf{Z}_E \\ \mathbf{Z}_{KU}^2 &= \mathbf{Z}_K\mathbf{Z}_U + \mathbf{Z}_Q\mathbf{Z}_E \\ \mathbf{Z}_{KF}^2 &= \mathbf{Z}_K\mathbf{Z}_F + \mathbf{Z}_R\mathbf{Z}_Q \\ \mathbf{Z}_{KE}^2 &= \mathbf{Z}_K(\mathbf{Z}_S + \mathbf{Z}_U) + \mathbf{Z}_E(\mathbf{Z}_R + \mathbf{Z}_Q) = \mathbf{Z}_{KS}^2 + \mathbf{Z}_{KU}^2 \end{aligned} \tag{5.114}$$

All of these double subscripted terms have the dimensions of impedance squared. Using these expressions allows us to write the denominator term (5.111) as

$$DEN = \frac{\mathbf{Z}_{KF}^2\mathbf{Z}_{KE}^2 + \mathbf{Z}_{KS}^2\mathbf{Z}_{KU}^2}{\mathbf{Z}_K} \tag{5.115}$$

which can be verified by inspection of (5.113). The *DEN* quantity is noted to have dimensions of impedance-cubed. Then the notation of (5.111) to (5.113) can be simplified to the following form.

$$\mathbf{I}_S = \frac{(\mathbf{Z}_{KF}^2 + \mathbf{Z}_{KU}^2)\mathbf{E}_S - \mathbf{Z}_{KF}^2\mathbf{E}_U}{DEN} \tag{5.116}$$

$$\begin{aligned} \mathbf{I}_U &= \frac{(\mathbf{Z}_{KF}^2 + \mathbf{Z}_{KS}^2)\mathbf{E}_U - \mathbf{Z}_{KF}^2\mathbf{E}_S}{DEN} \\ \mathbf{I}_F &= \frac{\mathbf{Z}_{KU}^2\mathbf{E}_S + \mathbf{Z}_{KS}^2\mathbf{E}_U}{DEN} \\ \mathbf{V}_F &= \frac{\mathbf{Z}_F(\mathbf{Z}_{KU}^2\mathbf{E}_S + \mathbf{Z}_{KS}^2\mathbf{E}_U)}{DEN} \end{aligned} \tag{5.117}$$

The currents given by (5.117) are the total current, representing the load component plus the fault component of currents in the network.

5.7.3 The Fault Conditions Without Load Currents

We can solve for the fault currents and voltages, without the load components of current, using the compensation theorem. The network is shown in Figure 5.37.

The solution of this network gives only the fault component of the network currents, ignoring any prefault load currents that may exist. Similarly, the voltages given by this solution reflect only the effect of the fault, and not of load conditions. The network solution is as follows,

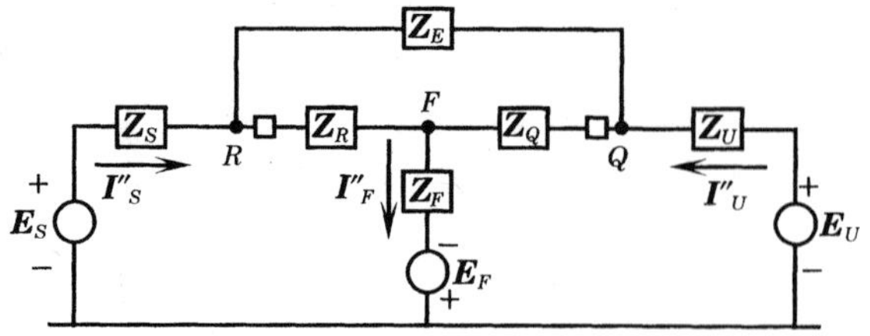

Figure 5.37 Network for the fault-only condition.

using the short-hand notation of (5.114).

$$I''_S = \frac{(Z_K Z_U + Z_Q Z_E)E_F}{DEN} = \frac{Z^2_{KU} E_F}{DEN}$$

$$I''_U = \frac{(Z_K Z_S + Z_R Z_E)E_F}{DEN} = \frac{Z^2_{KS} E_F}{DEN} \tag{5.118}$$

$$I''_F = \frac{[Z_K(Z_S + Z_U) + Z_E(Z_R + Z_Q)]E_F}{DEN} = \frac{Z^2_{KE} E_F}{DEN}$$

$$V''_F = -\frac{(Z_R Z_Q Z^2_{KE} + Z^2_{KS} Z^2_{KU})E_F}{DEN} \tag{5.119}$$

5.7.4 Summary of Load and Fault Conditions

A summary of the network fault and load conditions are shown in Table 5.3 where the results are given in two different ways. The upper portion of the table gives all computed currents and voltages in terms of the phasor source voltages. These are the values discussed in the preceding sections. The results are also given in the lower part of the table in terms of the prefault voltage, which provides a convenient way of expressing the system currents under the faulted system condition. Since, in most cases, the load currents are negligible compared to the fault currents, these (double-primed) currents are the only ones of interest.

Note that the faulted system is assumed to have a driving voltage of only the prefault voltage at F, as shown in the rightmost figure in Table 5.3. The solution of this system leads directly to the currents and voltages of the network expressed in terms of E_F, with the results given in the lower part of the table. However, the pre-fault voltage can be expressed in terms of the two equivalent system source voltages, with this result given in the pre-fault portion of the table. Therefore, the faulted system results can be expressed in terms of these source voltages, with these results given in the top portion of the table under the "faulted" heading.

Note that shorthand notation is used throughout the table. The bottom portion of the table provides the definition of certain simplifying impedance expressions. Other defined quantities of interest are given in (5.114).

For most power systems, the fault currents are much larger than the load currents and the compensation theorem is used to simplify the fault calculations, using the equivalent system on the rightmost column of Table 5.3. Since the load currents are small, this sacrifices little in accuracy. Occasionally, however, a question may arise as to the performance of a relay due to large prefault load currents, in which case the total solution, prefault plus faulted conditions, are desired. This requires the solution of the system on the left of Table 5.3. Note that it is never required that the entire power system be solved, since the fault equivalent provides an exact solution for any fault point F along the faulted component.

TABLE 5.3 Solution of the Fault Equivalent for Total, Pre-fault, and Faulted Conditions

Total = Pre-fault + Faulted	Pre-fault	Faulted
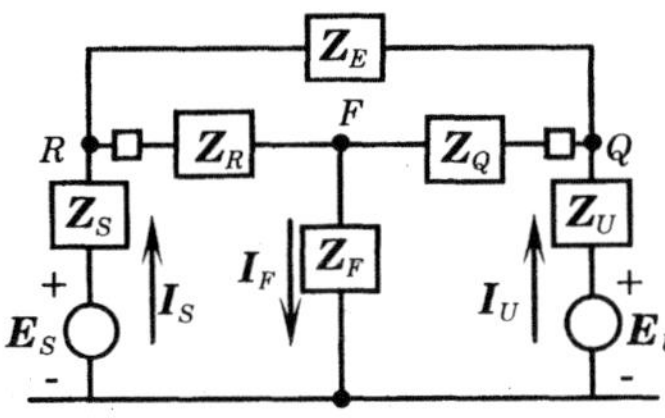	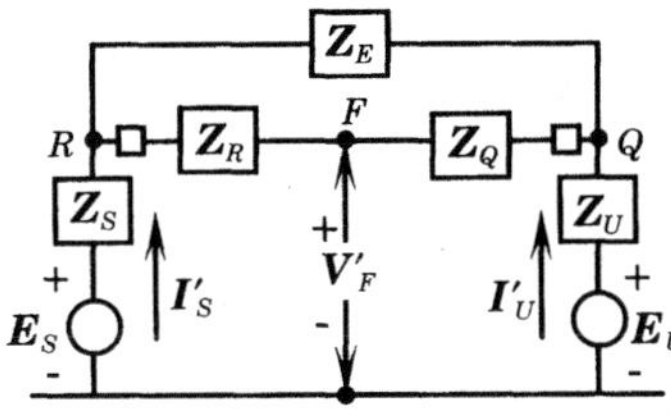	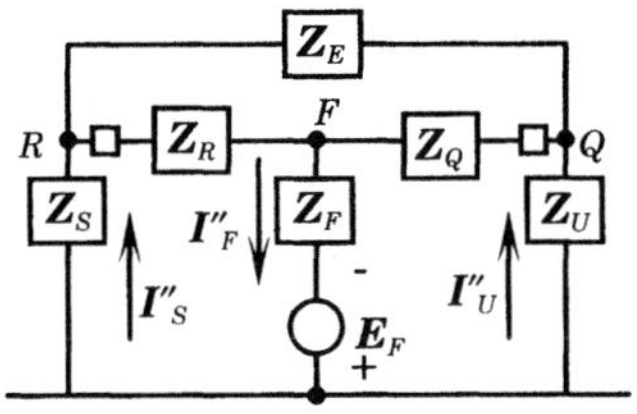
$I_S = \dfrac{(Z_{KF}^2 + Z_{KU}^2)E_S - Z_{KF}^2 E_U}{DEN}$	$I'_S = \dfrac{Z_K(E_S - E_U)}{Z_{KE}^2} = -I'_U$	$I''_S = \dfrac{Z_{KU}^2(Z_{KU}^2 E_S + Z_{KS}^2 E_U)}{Z_{KE}^2 DEN}$
$I_U = \dfrac{(Z_{KF}^2 + Z_{KS}^2)E_U - Z_{KF}^2 E_S}{DEN}$	$I'_U = \dfrac{-Z_K(E_S - E_U)}{Z_{KE}^2} = -I'_S$	$I''_U = \dfrac{Z_{KS}^2(Z_{KU}^2 E_S + Z_{KS}^2 E_U)}{Z_{KE}^2 DEN}$
$I_F = \dfrac{Z_{KU}^2 E_S + Z_{KS}^2 E_U}{DEN}$	$I'_F = 0$	$I''_F = \dfrac{Z_{KU}^2 E_S + Z_{KS}^2 E_U}{DEN}$
$V_F = \dfrac{Z_F(Z_{KU}^2 E_S + Z_{KS}^2 E_U)}{DEN}$	$V'_F = \dfrac{Z_{KU}^2 E_S + Z_{KS}^2 E_U}{Z_{KE}^2}$	$V''_F = \dfrac{-Z_X^4(Z_{KU}^2 E_S + Z_{KS}^2 E_U)}{Z_K Z_{KE}^2 DEN}$
$I_S = \dfrac{Z_{KF}^2 E_S - (Z_{KF}^2 + Z_{KS}^2)E_U + Z_{KE}^2 E_F}{DEN}$	$I'_S = \dfrac{Z_K(E_S - V'_F)}{Z_{KS}^2}$	$I''_S = \dfrac{Z_{KU}^2 E_F}{DEN}$
$I_U = \dfrac{Z_{KF}^2 E_U - (Z_{KF}^2 + Z_{KU}^2)E_S + Z_{KE}^2 E_F}{DEN}$	$I'_U = \dfrac{Z_K(E_U - V'_F)}{Z_{KU}^2}$	$I''_U = \dfrac{Z_{KS}^2 E_F}{DEN}$
$I_F = \dfrac{Z_{KE}^2 E_F}{DEN}$	$I'_F = 0$	$I''_F = \dfrac{Z_{KE}^2 E_F}{DEN}$
$V_F = \dfrac{Z_F Z_{KE}^2 E_F}{DEN}$	$V'_F = E_F$	$V''_F = \dfrac{-Z_X^4 E_F}{Z_K DEN}$
$DEN = (Z_{KF}^2 Z_{KE}^2 + Z_{KS}^2 Z_{KU}^2)/Z_K$	$Z_K = Z_R + Z_Q + Z_E$	$Z_X^4 = Z_R Z_Q Z_{KE}^2 + Z_{KS}^2 Z_{KU}^2$

REFERENCES

[1] Zocholl, S. E., W. C. Kotheimer, and F. Y. Tajadoddi, "An Analytical Approach to the Application of Current Transformers for Protective Relaying," paper 5.3, *Protective Relaying for Power Systems II*, Stanley H. Horowitz, Editor, IEEE Press, NJ, 1992.

[2] GEC Measurements, *Protective Relays Application Guide*, General Electric Company p.l.c. of England, 1975.

[3] McPherson, G., *An Introduction to Electrical Machines and Transformers*, John Wiley & Sons, New York, 1981.

[4] Anderson, P. M., *Analysis of Faulted Power Systems*, IEEE Press, Piscataway, NJ, 1995.

[5] Central Station Engineers, *Electrical Transmission and Distribution Reference Book*, Westinghouse Electric Corporation, 4th Ed., Pittsburgh, 1950.

[6] Lim, L. S., and I. R. Smith, "Turbogenerator Short Circuits With Delayed Current Zeros," Proc. IEE, 124 (12), December 1977, pp. 1163–1169.

[7] Anderson, P. M., and A. A. Fouad, *Power System Control and Stability*, IEEE Press, Piscataway, NJ, 1994.

[8] Zochol, S. E., "Selecting CTs to Optimize Relay Performance." Conference Proc., Western Protective Relay Conference, Spokane, WA, October 15–17, 1997.

[9] Koeppl, G., B. Stepinski, H. Frey, and W. Kolbe, "The Crossed-Ring Arrangement—A New Concept for H.V. Switchgear Installations," *IEEE Trans*, PAS-102 (2), February 1983, pp. 355–363.

[10] Page, R., K. Kishikawara, and F. Stromotich, "The 4×6 Network, A New Power Switching Scheme," *IEEE Trans.*, PWRD-5, October 1989, pp. 2060–2066.

[11] Project EHV, *EHV Transmission Line Reference Book*, General Electric Co., published by Edison Electric Institute, New York, 1968.

[12] Stevenson, W. D., Jr., *Elements of Power System Analysis*, 4th Ed., McGraw-Hill, 1983.

[13] Guilleman, E. A., *Communication Networks, Vol. II*, John Wiley & Sons, Inc., New York, 1935.

[14] Nilsson, J. W., *Electric Circuits*, Addison-Wesley Publishing Co., Reading, MA, 1983.

[15] Anderson, P. M., "Analysis of Simultaneous Faults by Two-Port Network Theory," *IEEE Trans. on Power Apparatus & Systems*, PAS-90 (5), September/October 1971, pp. 2199–2205.

[16] Galicia, D. E., and P. M. Anderson, "Digital Techniques for Finding the Two-Port Parameters of a Large System," a paper presented at the Midwest Power Symposium, University of Michigan, October 21–22, 1971.

[17] Ward, J. B., and H. W. Hale, "Digital Solution of Power-Flow Problems," AIEE Trans., 75, pt. 3, 1956, pp. 398–404.

[18] Boast, W. B., *Principles of Electric and Magnetic Circuits*, 2nd Ed., Harper & Brothers, New York, 1957.

[19] Zaborszky, J., and J. W. Rittenhouse, *Electric Power Transmission*, Ronald Press Company, New York, 1954.

PROBLEMS

5.1 Consider the power system model shown in Figure 5.1 and the equation for the fault current given by (5.3). Make a sketch of the applied voltage showing the angle β measured with respect to the point at which the voltage crosses the horizontal axis with a positive slope. Label the horizontal axis in units of angle as well as time.

5.2 Solve the differential equation (5.3) to verify the solution given by (5.4) and (5.5).

5.3 Determine the condition that leads to maximum positive value of the transient component of current in problem 5.2 and find the angle of the voltage at switch closing that leads to this maximum value.

5.4 Determine the condition that leads to maximum negative value of the transient component of current in problem 5.2 and find the angle of the voltage at switch closing that leads to this minimum value.

5.5 A three-phase 50 MVA, 11 kV, 60 Hz synchronous generator is subjected to several different kinds of short circuits while operating at rated voltage and no load. The sustained short-circuit currents are found to be as follows:
3PH Fault: 2000 A
L-L Fault: 3150 A
1LG Fault: 5240 A
The instantaneous symmetrical three-phase short circuit current is found to be 20 kA. Determine the following generator parameters in per unit:

$$x_d \quad x_d'' \quad x_2 \quad x_0$$

5.6 The dc components of current in the three phases of a short-circuited generator are 1416, 1932, and 516 A. In this test, the short circuit occurred at such an instant as not to produce the maximum asymmetry. Determine the following:

(a) The maximum possible value of the direct-current component.
(b) The peak value of the alternating-current component
(c) The maximum possible rms total fault current

5.7 The following data are given for a synchronous generator that is rated 991 MVA, 26 kV at 90% power factor.

$$x_d = 1.990 \text{ pu} \qquad \tau'_{do} = 4.700 \text{ s}$$
$$x'_d = 0.245 \text{ pu} \qquad \tau''_{do} = 0.031 \text{ s}$$
$$x''_d = 0.200 \text{ pu} \qquad \tau_a = 0.230 \text{ s}$$

Compute the following currents at times $t = 0, 2, 4$, and 6 cycles after fault initiation of a three-phase fault (60 Hz system): I_{ac}, I_{dc}, and I_{total}.

5.8 Compute the total fault MVA that the generator circuit breaker must interrupt at full voltage for each solution of problem 5.7.

5.9 A three-phase fault occurs on a radial 34.5 kV transmission line connected to the generator of problem 5.5. The distance from the generator terminals to the fault is 10 km, and the line impedance is

$$z_{line} = 0.02 + j0.38\Omega/\text{km}$$

The generator step-up transformer has a reactance of 0.07 per unit on the generator base (negligible resistance).
Compute the asymmetry factors for the maximum and average conditions.

5.10 Assume that the line described in problem 5.9 is connected to the generator described in problem 5.5, and that the fault is located exactly as described in problem 5.9. Calculate the following:
(a) The subtransient current
(b) The sustained three-phase fault current
(c) The sustained line-to-line fault current

5.11 Prepare a matrix showing the advantages and disadvantages of the following substation arrangements.
(a) single bus, single breaker
(b) main and transfer bus
(c) double bus, single breaker
(d) double bus, double breaker
(e) ring bus
(f) breaker-and-a-half

5.12 Tabulate the following for each bus arrangement of the previous problem:
(a) number of breakers
(b) number of breakers for each terminal
(c) number of disconnects
(d) number of disconnects for each terminal
(e) number of special bus protective systems

5.13 Add the crossed ring arrangement to the considerations addressed in problems 5.11 and 5.12.

5.14 Add the ring bridge to the considerations addressed in problems 5.11 and 5.12.

5.15 Compute the impedance in ohms per mile of a 34.5 kV distribution circuit constructed with an equivalent spacing of 5.0 ft and using 336, 400 cm ACSR conductor with 26/7 stranding.

5.16 Compute the impedance per mile of the 500 kV transmission line shown in Figure P5.13. The phase conductors are twin bundled 2156 kcm ACSR (Bluebird) with a stranding of 84/19. The shield wires are 7#8 alumoweld conductors. The conductor height above the ground is nominally 90 ft, with tower base extensions used to increase ground clearance where needed.

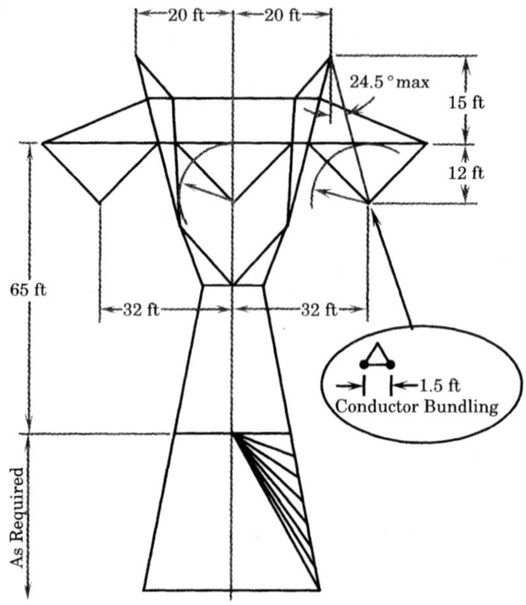

Figure P5.13 Configuration of a 500 kV transmission line for problem 5.16.

5.17 Compute the shunt fault equivalent circuit parameters for the system of Figure 5.25 and using the data of Example 5.1. Compare your results with those given in Table 5.2.

5.18 Replace Line #1 in Figure 5.25 by a double circuit line, where each of the two lines has exactly the same impedance as the original line. Then construct a protection equivalent for one of these parallel lines and make comparison between the equivalent impedance Z_E and that of the protected line Z_L.

5.19 Consider the results of Example 5.1, which are given in Table 5.2. Explain why Z_E is so large in the equivalents for branches 4 and 7, when compared to the protected line impedance.

5.20 Derive the equivalent circuit parameters for series faults on the network of Figure 5.25.

5.21 Show that the protection equivalent of Figure 5.20 gives exact solutions for currents flowing in any of the protected branches of the six bus system of Figure 5.25.

5.22 Verify (5.74), (5.75), and (5.76).

5.23 Verify (5.84).

5.24 Verify (5.88).

5.25 Compute the source currents for the system of Figure 5.31 and check that their sum agrees with (5.96).

5.26 The results of applying the compensation theorem, given by (5.96) and (5.103) are stated in terms of the pre-fault voltage at the fault point. However, these results can also be expressed in terms of the pre-fault current flowing in the impedance Y_3 or the pre-fault load current. Find the expressions for the incremental fault voltage and current, stated in terms of the pre-fault load current at F. Is this result still applicable if the prefault load current is zero?

5.27 Apply the compensation theorem, as described in Section 5.6, but write the impedance equations rather than the admittance equations. Solve these equations for the incremental change in voltage at node F.

5.28 Explain how to determine the parameters of the three-terminal protection equivalent circuit shown in Figure P5.28.

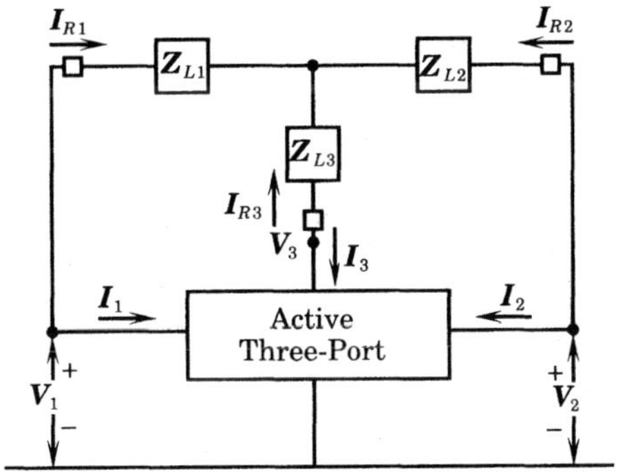

Figure P5.28 A three-terminal line and active three-port network.

5.29 Develop the equations for applying the compensation theorem for the circuit of Figure 5.31 using the impedance method of solution rather than the admittance method. Do this for two conditions:

(a) $\mathbf{Z}_3$ finite, and

(b) $\mathbf{Z}_3$ infinite.

Note that the solution of case (b) can not be determined directly from case (a) without special care. For example, solving case (a) and replacing $\mathbf{Z}_3$ by an infinitely large quantity gives an open circuit for the center leg, which results in zero current for all conditions.

5.30 Verify the solution of the system of Figure 5.36, given in Section 5.7.2.

5.31 Verify the entries in Table 5.3.

5.32 Verify Figure 5.4.

PART II
PROTECTION CONCEPTS

6

Fault Protection of Radial Lines

This chapter presents methods for the study of fault protection of radial lines and is generally applicable to radial distribution feeders and radial transmission lines. Radial lines provide an appropriate introduction to fault protection since there is only one source of supply in a radial system. This permits one to become familiar with the characteristics and applications of protective equipment without the distraction and added complexity of loop or network system configurations. The coordination of fault protective equipment in a radial system is an interesting and useful study, and it provides a good basis for the study of more complex problems.

A general observation regarding fault protection of radial lines is that the protective equipment need only sense current. Directional discrimination is not required, nor is distance measuring equipment. Therefore, radial circuits can always be protected using overcurrent relays, reclosers, sectionalizers, and fuses.

6.1 RADIAL DISTRIBUTION SYSTEMS

Distribution lines are different from transmission and subtransmission lines in that (1) they operate at lower voltages than transmission lines, (2) they are usually radial, and (3) they usually have loads tapped all along the line, not just at the terminals. Devising a protective strategy, then, involves optimizing the service continuity to the maximum number of users at the minimum cost. Usually this means applying a combination of circuit breakers, automatic circuit reclosers, sectionalizers, and fuses to clear temporary faults with fast reclosing, and to isolate permanent faults by an appropriate switching strategy.

Usually we think of a distribution system as a three-phase main line with lateral branches, some of which may be one- or two-phase circuits. Usually the system is wye connected with a ground at the supply transformer and often with multiple additional neutral grounds. Figure 6.1 shows a typical distribution feeder arrangement with a main line and several branches, supplying step-down distribution transformers along both the main line and the branches or laterals.

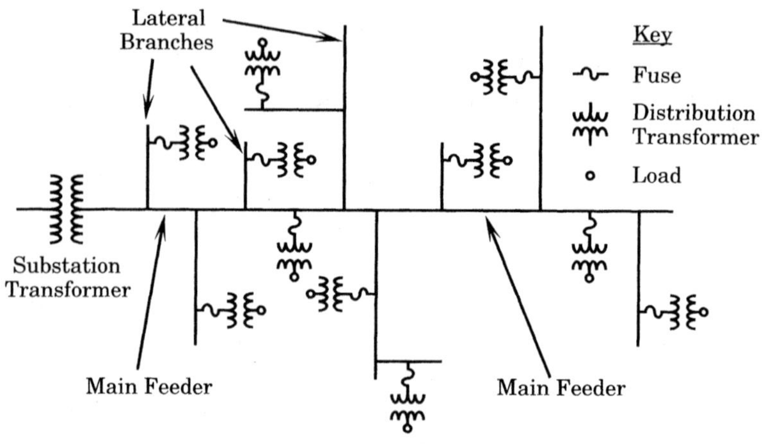

Figure 6.1 One-line diagram of a distribution circuit [1].

The circuit of Figure 6.1 is a small system, but could also represent a larger distribution feeder if we let each distribution transformer symbol in the figure correspond to several transformers or a given kVA per unit length of feeder. Note that each of the distribution transformers has a primary or high side fuse. This could be an external fuse, which is usually located on the high side of the transformer in a fused cutout, or it could be an internal fuse located inside the transformer tank, if the unit is completely-self-protected (CSP) transformer. In either case, the purpose of this fuse is to isolate a faulted transformer or secondary main from the distribution system. Customers served from the secondary main are expected to have their own circuit protection and would not rely on the transformer high-side fuse for this purpose.

To achieve a reasonable service continuity on distribution feeders it is necessary to "sectionalize" the feeder. This means that protective devices should be located at strategic places along the feeder to recognize and clear all faults and to "lockout" or open permanently only for permanent faults. To achieve this coordination it is sometimes necessary to sectionalize the main line as shown in Figure 6.2 where a recloser or breaker with reclosing relays is located at A. This recloser can recognize and clear faults within its protective zone, shown by the dashed arc anchored to A. Faults beyond the protective zone of recloser A are too small in magnitude to be cleared by A so a second recloser B must be added which has a smaller minimum pickup current than A.

It is often advisable to add protection to laterals or branches in addition to the main line reclosers or breakers. The purpose of branch protection is to isolate permanent branch faults and permit restoration of normal service on all but the faulted branch. Thus in Figure 6.2 the lateral branches might be protected by fuses, particularly if the branch is a half-mile or more long. These fuses must be coordinated both with the main line reclosers and the distribution transformer fuses. Branch fuses are relatively inexpensive and provide inexpensive protection for the service continuity of the majority of users supplied by the feeder circuit.

It was pointed out in Chapter 1 that the protective strategy of a system should be devised to ensure maximum protection at minimum total cost. This includes the cost of restoration, the customers' good will, and the energy revenues that must be balanced against the protective equipment cost. Distribution systems often have extensive exposure to faults due to lightning, trees, traffic accidents, and other natural or human-caused incidents. Single-fuse installations will clear faults, but since the number of faults may be large and, by their nature, usually temporary, it may be cheaper in the long term to protect the system against temporary faults

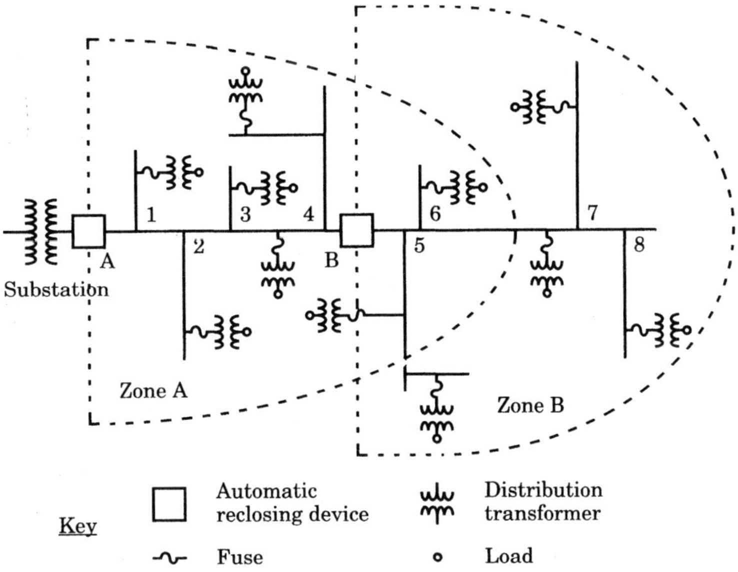

Figure 6.2 Distribution circuit with protective device [2].

using reclosers or circuit breakers with reclosing relays. Actually we must protect against both permanent and temporary faults so we usually use a combination of protective devices.

The suppliers of protective equipment provide excellent technical data to assist the engineer in making fault protection studies. Some of this material is tutorial and is recommended reading for engineers engaged in fault protection work. Of particular value are the application manuals on overcurrent protection by General Electric [1] and McGraw Edison [2]. Numerous articles on the subject also appear in distribution magazines and manufacturer's bulletins periodically, and these are also recommended.

6.2 RADIAL DISTRIBUTION COORDINATION

A radial distribution coordination study requires the use of a considerable amount of information concerning the supply system, the substation, the feeders, the loads, and the protective device characteristics. The purpose of this section is to outline the data requirements and to suggest an orderly procedure for making a study (from [2]).

6.2.1 Supply System Information

One of the first requirements of a coordination study is the computation of the available fault current (or MVA) at every point where a protective device might be located. This is equivalent to computing the Thevenin equivalent impedance at these points. Furthermore, this must be done for both minimum and maximum generation conditions as these may be quite different, depending on the location of the generators and their schedules of operation.

It is convenient to consider two separate problems in determining available fault current. It would be impractical to compute fault data on large systems for all transmission and distribution buses. Usually, protection engineers compute faults for all transmission and subtransmission voltage buses. Thus, for distribution fault studies, we may assume that fault data is available at the high-voltage bus at the distribution substation. These data are taken from computer studies called *fault* studies and are commonly run for both maximum and minimum generation conditions.

For small systems, such as small utilities or small rural systems, computer fault studies may not be available but complete representation of the entire supply system is feasible. Fault studies can be performed on small power systems as well as large ones, and many computer programs are available for personal computers that will perform this task.

Requirements for both large and small systems are outlined below. (Also see [1] and [2].)

Large system data

1. Line-to-line supply side voltage at the substation.
2. Maximum and minimum three-phase and line-to-ground fault current magnitudes at supply voltage.
3. Maximum size or rating of high-side fuses specified by transmission relay engineer.
4. Make and type of high-side fuses, if specified by transmission relay engineer.
5. Type of protection, if other than high-side fuse, specified by transmission relay engineer, e.g.,
   a. type of relays, if used
   b. setting of relays

Small system data

1. Distance between the substation and the power plant.
2. Size and configuration of the circuit between substation and the power plant.
3. Line to line supply voltage.
4. For each generator,
   a. rating in kVA
   b. direct axis transient reactance in percent
   c. direct axis synchronous reactance in percent
   d. negative sequence reactance in percent
   e. type of prime mover
5. Identification of generators normally running during minimum and maximum loads.
6. Maximum size or rating of substation high-side fuses, if specified by a transmission relay engineer.
7. Make and type of substation high-side fuse, if specified by transmission relay engineer.
8. Type of protection if other than high-side fuse, specified by transmission relay engineer, e.g.,
   a. type of relays used
   b. setting of relays

The first task in the study, then, is to assemble the foregoing information. This may require computer studies to obtain fault current data.

6.2.2 Distribution Substation Information

The following data concerning the step down substation should be known.

1. Schematic diagram showing transformer, connections, protective devices on high-voltage and low-voltage sides, and outgoing circuit configuration.

2. Substation transformer capacity, voltage (high and low side) and percent impedance.
3. Substation transformer time-current damage curve. If this curve is not available from the manufacturer, information from the ANSI standard [3] may be used. This standard provides a short time overload (following full load current which is stated as "times rated current" and is based on the equivalent self-cooled rating). This overload data is given in Table 6.1.

Data in Table 6.1 must be used in conjunction with rated full load currents of transformers. These rated values are given in Table 6.2 for single-phase and Table 6.3 for three-phase transformers.

TABLE 6.1 Allowable Short Time Overload Current Following Load for Oil Immersed Transformers [3]

Overload Time (s)	Time Rated Current (per unit)
2.0	25.00
10.0	11.30
30.0	6.70
60.0	4.75
300.0	3.00
1800.0	2.00

6.2.3 Distribution System Information

The following data are required for each distribution feeder for which coordination is to be studied.

1. Circuit diagram (map) of feeder with distance scale.
2. Location of customers or loads for which a lengthy power interruption would be costly or detrimental.
3. Location and size of large power loads.
4. Location of self-protected transformers larger than 25 kVA.
5. Maximum peak metered load current at the substation and at the tap-off point of each heavy branch circuit.

6.2.4 Protective Equipment Information

The following data for protective devices contemplated should be available.

1. High-side substation fuse data: manufacturer, type, time-current characteristic curves, rating.
2. Make and rating of feeder circuit breakers and relays.
3. Automatic circuit reclosers: make, type, table of ratings, and time-current characteristic curves.
4. Line sectionalizing fuses: make, type, and time-current characteristics; both melting and total clearing curves

TABLE 6.2 Full Load Currents of Single-Phase Distribution Transformers [3]

kVA	Circuit Nominal Voltage									
	120	**240**	**480**	**2400**	**4160**	**4800**	**6900**	**7200**	**11500**	**13200**
1.5	12.5	6.3	3.1	0.63	0.36	0.31	0.22	0.21	0.13	0.11
2.5	20.8	10.4	5.2	1.04	0.60	0.52	0.36	0.35	0.22	0.19
3.0	25.0	12.5	6.3	1.25	0.72	0.63	0.43	0.42	0.26	0.23
5.0	41.7	20.8	10.4	2.08	1.20	1.04	0.72	0.69	0.43	0.38
7.5	62.5	31.3	15.6	3.13	1.80	1.56	1.09	1.04	0.65	0.57
10.0	83.3	41.7	20.8	4.17	2.40	2.08	1.45	1.39	0.87	0.76
15.0	125.0	62.5	31.3	6.25	3.61	3.13	2.17	2.08	1.30	1.14
25.0	208.3	104.2	52.1	10.42	6.01	5.21	3.62	3.47	2.17	1.89
37.5	312.5	156.3	78.1	15.63	9.01	7.81	5.43	5.21	3.26	2.84
50.0	416.7	208.3	104.2	20.83	12.02	10.42	7.25	6.94	4.35	3.79
75.0	625.0	312.5	156.3	31.25	18.03	15.63	10.87	10.42	6.52	5.68
100.0	833.3	416.7	208.3	41.67	24.04	20.83	14.49	13.89	8.70	7.58
150.0	1250.0	625.0	312.5	62.50	36.06	31.25	21.74	20.83	13.04	11.36
200.0	1666.7	833.3	416.7	83.33	48.08	41.67	28.99	27.78	17.39	15.15
250.0	2083.3	1041.7	520.8	104.17	60.10	52.08	36.23	34.72	21.74	18.94
333.3	2777.5	1388.8	694.4	138.88	80.12	69.44	48.30	46.19	28.98	25.25
500.0	4166.7	2083.3	1041.7	208.33	120.19	104.17	72.46	69.44	43.48	37.88

TABLE 6.3 Full Load Currents of Three-Phase Distribution Transformers [3]

kVA	Circuit Voltage									
	208	**240**	**480**	**2400**	**4160**	**4800**	**7200**	**12470**	**13200**	**33000**
4.5	12.5	10.8	5.4	1.08	0.62	0.54	0.36	0.21	0.20	0.08
7.5	20.8	18.0	9.0	1.80	1.04	0.90	0.61	0.35	0.33	0.13
9.0	25.0	21.7	10.8	2.17	1.25	1.08	0.73	0.42	0.39	0.16
10.0	27.7	24.1	12.0	2.41	1.39	1.20	0.80	0.46	0.44	0.17
15.0	41.6	36.1	18.0	3.61	2.08	1.80	1.20	0.69	0.66	0.26
22.5	62.5	54.1	27.1	5.41	3.12	2.71	1.80	1.04	0.98	0.39
25.0	69.4	60.1	30.1	6.01	3.47	3.01	2.00	1.16	1.09	0.44
30.0	83.3	72.2	36.1	7.22	4.16	3.61	2.41	1.39	1.31	0.52
37.5	104.1	90.2	45.1	9.02	5.20	4.51	3.01	1.74	1.64	0.66
45.0	124.9	108.3	54.1	10.83	6.25	5.41	3.60	2.08	1.97	0.79
50.0	138.8	120.3	60.1	12.03	6.94	6.01	4.01	2.32	2.19	0.87
75.0	208.2	180.4	90.2	18.04	10.41	9.02	6.01	3.47	3.28	1.31
100.0	277.6	240.6	120.3	24.06	13.88	12.03	8.02	4.64	4.37	1.75
112.5	312.3	270.6	135.3	27.06	15.61	13.53	9.02	5.21	4.92	1.97
150.0	416.4	360.08	180.4	36.08	20.82	18.04	12.03	6.94	6.56	2.62
200.0	555.1	481.1	240.6	48.11	27.76	24.06	16.04	9.27	8.75	3.50
225.0	624.5	541.3	270.6	54.13	31.23	27.06	18.04	10.42	9.84	3.94
300.0	832.7	721.7	360.8	72.17	41.64	36.08	24.06	13.89	13.12	5.25
450.0	1249.1	1082.5	541.3	108.25	62.45	54.13	36.08	20.83	19.68	7.87
500.0	1387.9	1202.8	601.4	120.28	69.39	60.14	40.09	23.15	21.87	8.74
600.0	1665.4	1443.4	721.7	144.34	83.27	72.17	48.11	27.78	26.24	10.50
750.0	2081.8	1804.2	902.1	180.42	104.09	90.21	60.14	34.72	32.80	13.12
1000.0	2775.7	2405.6	1202.8	240.56	138.79	120.28	80.19	46.30	43.74	17.50
1500.0	4163.6	3608.4	1804.2	360.84	208.18	180.42	120.28	69.45	65.61	26.24

5. Distribution transformer external or internal fuses: make, type, and time-current characteristic curves.

The manufacturers of fuses, reclosers, and other protective devices are always helpful in supplying this type of information. The protection engineer should create a library of information on protective equipment used on the systems to be studied. This will be helpful in determining the initial studies, and in later follow-up studies in analyzing faults or other disturbances that occur on the system. Also, as the system changes and loads grow, the same data will be required for redesign and coordination of the protective system.

6.2.5 Step-by-Step Study Procedure

With all the necessary data at hand the coordination study can proceed. The following step-by-step ordering of tasks provides a logical procedure to follow [1].

1. Assemble the necessary data described above.
2. Establish tentative locations of sectionalizing devices.
3. Calculate maximum and minimum values of fault currents at each of the tentative sectionalizing points, and at the end of the main, branch, and lateral circuits. Calculate line-to-ground, three-phase, and line-to-line currents.
4. Select the devices at the substation to give complete and adequate protection to the substation transformer from fault currents on the distribution lines.
5. Coordinate the sectionalizing devices from the substation out, or from the ends of the circuit back to the substation. Revise tentative locations of sectionalizing points if necessary.
6. Check the selected protective devices for current-carrying capacity, interrupting capability, and minimum pickup rating.
7. Prepare a circuit diagram to show circuit configuration, maximum and minimum fault-current values, rating of sectionalizing devices, and other relevant information.

In addition to the above some engineers like to prepare a composite time-current (TC) characteristic curve showing the coordination of all devices, with curves drawn for a common base voltage. Since TC curves of most commonly used fuses and reclosers are readily available, this requires transferring these curves to a separate TC log-log plot (using K & E, number 48 5257 paper). This plot can be filed with the substation records and reviewed as future system changes take place to see if any protective system changes are required.

6.3 RADIAL LINE FAULT CURRENT CALCULATIONS

The calculation of fault currents on the transmission network are described in Chapter 5. These calculations apply to any configuration of the power system, which is invariably a meshed network. For the purpose of this chapter, which considers only radial lines, it is assumed that the fault currents at the radial system source have been computed using the methods of Chapter 5. There are several problems to be addressed, however, for determining the fault current along radials from the meshed network. These problems are discussed below. Many of the general concepts presented in Chapter 5 still apply to the radial line.

6.3.1 General Considerations for Radial Faults

The determination of the sequence network Thevenin impedances of Figure 5.14 provides information for computing the fault current at any transmission bus. For distribution feeders such as Figure 6.1, which are radial from a transmission bus, we simply add incremental impedances to the Thevenin bus impedance as the fault is moved away from the substation along the three-phase main feeder. The fault study of the transmission network gives the fault currents at the high side of the substation transformer. Since the distribution system is radial, it is a relatively simple matter to determine the Thevenin impedance, and therefore the fault currents, at any point along the radial circuit.

Generally there are two types of distribution substation connections of interest; $\Delta - \Delta$ and $\Delta - Y$, and these are shown in Figure 6.3. Other connections are possible but these two are the most common. Connected to the low voltage station bus is a radial three-phase distribution feeder. If the source is delta connected, this feeder consists of three-phase wires which are almost always identical conductor sizes. If the feeder is wye-grounded at the transformer, the neutral wire is nearly always carried along with the phase wires, making a four-wire wye-connected system. In many cases the neutral wire may be one or two wire sizes smaller than the phase wires. This is possible since the load is nearly balanced making the neutral load current small. Also, the neutral is grounded at least at every distribution transformer and sometimes at every pole, in which case the circuit is described as a multigrounded system. This provides a neutral (zero sequence) return path of neutral conductor and the earth in parallel. The impedance of this zero-sequence path in a multigrounded circuit is difficult to determine accurately and empirical results are often used to estimate the earth impedance.

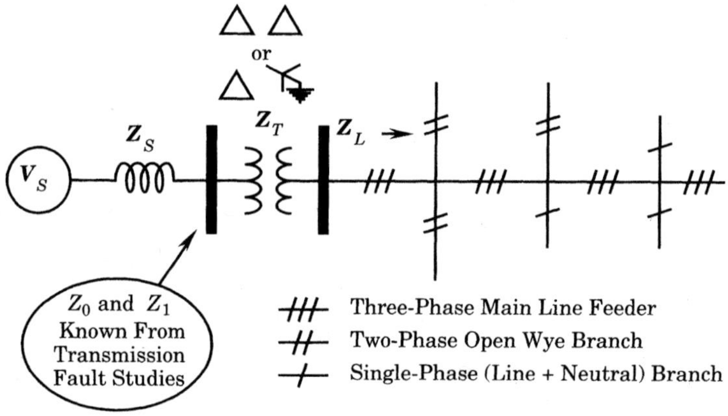

Figure 6.3 Distribution substation and radial feeder.

The branch lines of Figure 6.3 may be three-phase branches but are often single phase. If served from a delta-connected source the branches can only be three- phase or single-phase lines and all distribution transformers are connected line to line. If served from a wye-grounded source the branches may be two-phase, three-wire open wye or single-phase, two-wire (one line plus neutral) with all loads usually connected line to neutral.

6.3.2 Main Line Feeder Faults

By definition, the main feeder is a three-phase circuit. Therefore, faults along the main line feeder may be any of the four common types; 3PH, 2LG, 1LG, or LL.

6.3.2.1 Three-Phase (3PH) Faults. The sequence network arrangements and currents are given in Figure 5.15 for the three-phase fault. Using the distribution system impedances defined in Figure 6.3 we compute

$$\mathbf{Z}_1 = \mathbf{Z}_S + \mathbf{Z}_T + \mathbf{Z}_L \quad \text{pu} \tag{6.1}$$

where, as noted in Figure 6.3, $\mathbf{Z}_L$ is a function of the distance along the feeder to the fault point. The impedances in (6.1) are all positive-sequence impedances. $\mathbf{Z}_L$ is the positive-sequence distribution line impedance, defined approximately by the formula [4]

$$\mathbf{Z}_L = s\mathbf{z}_L = s\left[r_L + j\omega k \ln \frac{D_{eq}}{D_s}\right] \quad \Omega \tag{6.2}$$

where
s = length of line (kft)
r_L = line resistance (Ω/kft)
D_{eq} = equivalent spacing in ft = $(D_{12}D_{23}D_{31})^{1/3}$
D_s = self GMD of conductor (ft) (from tables)
$\omega = 2\pi f = 376.99$ rad/s for $f = 60$ Hz
$\omega k = 0.12134$ for s (miles)
$= 0.02298$ for s (kft)

Equation (6.2) assumes that the line is completely transposed, which usually is not the case for distribution circuits. However, the error in making this assumption is considered acceptable in view of the other variables and uncertainties in these computations.

Dividing (6.2) by the base impedance

$$Z_B = \frac{V_{B-LN}^2}{S_{S-1\phi}} = \frac{V_{B-LL}^2}{S_{B-3\phi}} \tag{6.3}$$

gives the impedance $\mathbf{Z}_L$ in per unit

$$\mathbf{Z}_{Lu} = \frac{\mathbf{Z}_L}{Z_B} \quad \text{pu} \tag{6.4}$$

The fault impedance $\mathbf{Z}_F$ needed to compute the fault current is often taken to be 30 to 40 ohms of resistance for minimum faults and zero for maximum fault conditions. Using the larger value we define, arbitrarily,

$$\mathbf{Z}_F = \begin{cases} 0 & \text{pu for maximum faults} \\ \dfrac{R_{\text{FLT}}}{Z_B} & \text{pu for minimum faults} \end{cases} \tag{6.5}$$

where R_{FLT} is the arcing fault resistance in ohms and the fault current is

$$\mathbf{I}_a = \frac{\mathbf{V}_F}{\mathbf{Z}_1 + \mathbf{Z}_F} = \frac{\mathbf{V}_F}{\mathbf{Z}_S + \mathbf{Z}_T + \mathbf{Z}_L + \mathbf{Z}_F} \tag{6.6}$$

The Thevenin equivalent voltage $\mathbf{V}_F$ completes the fault data necessary for three-phase faults. This voltage will not usually be known precisely and is usually taken to be between 1.0 and 1.1, with 1.05 a good estimated value. The angle is usually taken to be zero, and this voltage then becomes the phasor reference for the calculations to follow. It is important that the protection engineer understand the normalization process, reviewed above.

6.3.2.2 Double Line-to-Ground (2LG) Faults. The faulted phase designation and sequence network connection for the 2LG fault is shown in Figure 5.16. The positive and negative sequence impedances are exactly equal except for machines, and equating these sequence impedances is a very good approximation when the radial feeder is not electrically close to a generator. From (6.1) we know that $\mathbf{Z}_1 = \mathbf{Z}_2$ where these impedances are calculated exactly as given in (6.1)–(6.4) for the 3PH fault.

The zero sequence impedance $\mathbf{Z}_0$ is very difficult to determine accurately for multigrounded neutral systems, but is usually greater than the positive sequence impedance $\mathbf{Z}_L$. If we assume that $\mathbf{Z}_0$ is related to $\mathbf{Z}_L$ by a constant multiplier, we may write

$$\mathbf{Z}_{L0} = k_0 \mathbf{Z}_L \qquad \text{pu} \tag{6.7}$$

where k_0 depends on the size of the neutral wire compared to the phase wires, the ground impedance and the current division between neutral wire and earth. If the earth is a perfect conductor, which may be approached in a system with multiple water-pipe grounds, then $k_0 = 1.0$. If the earth is a very poor conductor then k_0 depends entirely on the neutral wire impedance. Most cases would fall between these two extremes in which case k_0 must be estimated. Morrison [5] gives estimated values of between 3.8 and 4.2 with an average of 4.0. The range of possibilities for k_0 is given in Table 6.4. The average value of 4.0 is suggested where exact conditions are not known. Note that (6.7) gives the magnitude of $\mathbf{Z}_{L0}$ only. We often assume that $\mathbf{Z}_{L0}$ and $\mathbf{Z}_L$ have the same impedance angle, although this is hard to defend. A more rigorous analysis of the impedance in multigrounded lines is given in [6].

TABLE 6.4 Estimated Values of k_0

k_0	Earth and Ground Wire Conditions
1.0	Perfectly conducting earth
3.8–4.2	Finite earth impedance
4.0	Ground wire same size as phase wire
4.6	Ground wire one size smaller than phase wire
4.9	Ground wire two sizes smaller than phase wire

Using (6.7) for the zero sequence line impedance we may write the total zero sequence impedance to the fault point as (see Figure 5.10)

$$\mathbf{Z}_0 = \begin{cases} \infty & \text{delta-connected source} \\ \mathbf{Z}_{L0} + \mathbf{Z}_T & \text{wye-connected source} \end{cases} \tag{6.8}$$

Fault impedances $\mathbf{Z}_F$ and $\mathbf{Z}_G$ are chosen arbitrarily, or based on data from typical faults. Some set $\mathbf{Z}_F = 0$ and give $\mathbf{Z}_G$ the value computed from (6.5).

6.3.2.3 Line-to-Line (LL) Fault. The faulted phase designation and sequence network connections for the LL fault are given in Figure 5.17. The positive and negative sequence impedance are usually considered equal, taking the value of (6.4). The fault impedance is the same as computed previously in (6.5).

6.3.2.4 One-Line-to-Ground (1LG) Fault. The 1LG fault configuration and sequence network connections are shown in Figure 5.18. Here again, the sequence impedances are the same as calculated previously in (6.6) and (6.8).

6.3.2.5 Summary of Main Feeder Faults. Using the equations given in the preceding discussion for the various fault types, a table of main line feeder fault formulas may be constructed. We are interested in determining only the maximum and minimum fault values, i.e., the values that correspond to maximum and minimum generating conditions, respectively. Since the 2LG fault is always intermediate between maximum and minimum it is omitted from the tabulation. Table 6.5 summarizes the previous results for 3PH, LL and 1LG faults. Usually the 3PH fault is the largest, although the 1LG fault, with $\mathbf{Z}_S = \mathbf{Z}_F = 0$ and $k_0 = 1.0$, has exactly the same magnitude. These assumptions are occasionally made.

TABLE 6.5 Fault Currents on a Radial Low-Voltage Feeder

Fault Type	Fault Current Equation	Substation Connection
3PH	$\mathbf{I}_a = \dfrac{\mathbf{V}_F}{\mathbf{Z}_S + \mathbf{Z}_T + \mathbf{Z}_L + \mathbf{Z}_F}$	Delta or wye grounded
LL	$\mathbf{I}_b = -\mathbf{I}_c = \dfrac{-j\sqrt{3}\mathbf{V}_F}{2(\mathbf{Z}_S + \mathbf{Z}_T + \mathbf{Z}_L) + \mathbf{Z}_F}$	Delta or wye grounded
1LG	$\mathbf{I}_a = 0$	Delta
1LG	$\mathbf{I}_a = \dfrac{3\mathbf{V}_F}{2\mathbf{Z}_S + 3\mathbf{Z}_T + (2 + k_0)\mathbf{Z}_L + 3\mathbf{Z}_F}$	Grounded wye

If the distribution substation is supplied from a very large system, then $\mathbf{Z}_S$ will be small. A common assumption, when no exact system data are available, is to set $\mathbf{Z}_S = 0$, which is equivalent to an infinitely large system. This is a good assumption if the substation transformer is small as this makes $\mathbf{Z}_T$ so large as to dominate the impedance, irrespective of the power system strength. Tables of transformer impedances and line impedances are given in the appendices.

EXAMPLE 6.1

Consider the six-node network shown in Figure 6.4, which shows a small transmission network supplying loads at nodes 3, 5, and 6. Data for the network lines are given in Table 6.6 with all values specified in per unit on a 50 MVA base.

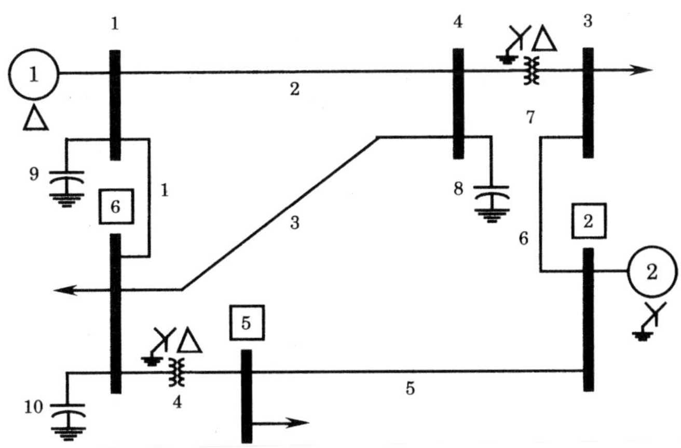

Figure 6.4 A six-node network [4].

TABLE 6.6 Six-Node Network with Per Unit Impedances on a 50 MVA Base

		Self-Impedance		Mutual-Impedance	
Z No.	Z Nodes	$Z_1 = Z_2$	Z_0	Z_M	Branch
1	1–6	0.123 + j0.518	0.492 + j1.042	—	—
2	1–4	0.080 + j0.370	0.400 + j0.925	0.250 + j0.475	3
3	4–6	0.097 + j0.407	0.450 + j1.030	0.250 + j0.475	2
4	5–6	0.000 + j0.300	0.000 + j0.300	...	...
5	2–5	0.282 + j0.640	1.410 + j1.920	...	...
6	2–3	0.723 + j1.050	1.890 + j2.630	...	...
7	3–4	0.000 + j0.133	0.000 + j0.133	...	...
8	0–4	0.000 – j34.10	...	...	...
9	0–1	0.000 – j29.50	...	...	...
10	0–6	0.000 – j28.50	...	...	...
11	Gen 1	0.010 + j0.012	0.000 + j0.032	...	...
12	Gen 2	0.015 + j0.240	0.000 + j0.016	...	...

Using these data the positive- and zero-sequence impedance matrices are computed and these values are given in Table 6.7 on a 100 MVA base.

TABLE 6.7 Positive- and Zero-Sequence Z Matrices on a 100 MVA Base [4]

Positive Sequence, per Unit				Zero Sequence, per Unit			
Bus	Bus	*R*	*X*	Bus	Bus	*R*	*X*
1	1	0.02253	0.21503	1	1	0.36392	1.11336
1	2	–0.00609	0.04974	1	2	0.00000	0.00000
1	3	0.02635	0.16117	1	3	0.00000	0.00000
1	4	0.02254	0.17266	1	4	–0.01120	0.11474
1	5	0.02118	0.12825	1	5	0.00000	0.00000
1	6	0.01831	0.16379	1	6	0.02526	0.34120
2	2	0.04422	0.38094	2	2	0.00000	0.03200
2	3	–0.01594	0.15713	2	3	0.00000	0.03200
2	4	–0.00786	0.13434	2	4	0.00000	0.00000
2	5	–0.00698	0.22306	2	5	0.00000	0.03200
2	6	0.00023	0.15223	2	6	0.00000	0.00000
2	3	0.16244	0.73912	3	3	3.78000	5.29200
3	4	0.14333	0.53368	3	4	0.00000	0.00000
3	5	0.06192	0.27007	3	5	0.00000	0.03200
3	6	0.07295	0.32786	3	6	0.00000	0.00000
4	4	0.13269	0.57694	4	4	0.00756	0.24138
4	5	0.06506	0.27818	4	5	0.00000	0.00000
4	6	0.06881	0.34726	4	6	–0.01706	0.05554
5	5	0.16569	0.80649	5	5	2.82000	3.87200
5	6	0.13256	0.46538	5	6	0.00000	0.00000
6	6	0.13034	0.61119	6	6	0.03849	0.47472

Consider the load on bus 5 and assume that it consists of a 5000 kVA, delta-wye (grounded) transformer bank feeding a 12.47 kV (line-to-line) three-phase, four-wire radial distribution feeder 20,000 ft long, consisting of 1/0 ACSR phase wires and #2 a ACSR neutral wire, with an equivalent three-phase spacing of 50 in. Assume the subtransmission voltage to be 34.5 kV.

Find the 3PH, LL, and 1LG fault currents available at the substation and at the end of the distribution feeder. Let $k_0 = 4.0$.

Solution

First we find the source impedance $\mathbf{Z}_S$ and express it in per unit on a 5000 kVA base (the station transformer base). From Table 6.7

$$\begin{aligned}\mathbf{Z} = \mathbf{Z}_{55} &= (0.16569 + j0.80649)\frac{5000}{100{,}000} \\ &= 0.0083 + j0.0403 \quad \text{pu}\end{aligned}$$

Note that we do not need the zero-sequence source impedance since the zero-sequence network need be represented only to the transformer. From Appendix D, we estimate the transformer impedance to be

$$\mathbf{Z}_T = 0.0 + j0.07 \quad \text{pu}$$

For the 1/0 ACSR distribution line, we have, from Table C.1, with the equivalent spacing $D_{eq} = 50$ in:

$$\begin{aligned}z_L &= r_L + j(x_d + x_a) \\ &= 0.168 + j(0.124 + 0.0528\log_{10}50 - 0.57)\ \Omega/\text{kft}\end{aligned}$$

Then

$$z_L = 0.168 + j0.1567\ \Omega/\text{kft}$$

or

$$\mathbf{Z}_L = 3.36 + j3.134\ \Omega \text{ for 20 kft}$$

and

$$\mathbf{Z}_{L0} = 4\mathbf{Z}_L = 13.44 + j12.536\ \Omega$$

We have set the zero-sequence value arbitrarily to the value noted above.

The base impedance is

$$Z_B = \frac{V_B^2}{S_B} = \frac{(12{,}470)^2}{5\times 10^6} = 31.10\ \Omega$$

Dividing by Z_B we get the per unit values

$$\begin{aligned}\mathbf{Z}_L &= 0.108 + j0.101 \ \text{pu} \\ \mathbf{Z}_{L0} &= 0.432 + j0.403 \ \text{pu}\end{aligned}$$

Finally, we let $\mathbf{Z}_F = 10\Omega$ and 30Ω to learn the effect of fault impedance

$$\mathbf{Z}_F = \frac{R_{\text{FLT}}}{31.1} = \begin{cases}0.322 \ \text{pu} \\ 0.965 \ \text{pu}\end{cases} \quad \text{for } R_{\text{FLT}} = \begin{cases}10\ \Omega \\ 30\ \Omega\end{cases}$$

Then using the formulas of Table 6.5, with $\mathbf{V}_F = 1.0$, we compute the impedances and currents shown in Table 6.8.

A careful study of Table 6.8 is instructive. Note that the largest fault currents occur with 1LG faults. This often the case. It can be explained by comparing the following equations for the fault current magnitude at F.

$$\text{3PH:} \quad \mathbf{I}_F = \frac{\mathbf{V}_F}{\mathbf{Z}_{\text{TOT}}} = \frac{\mathbf{V}_F}{\mathbf{Z}_S + \mathbf{Z}_T + \mathbf{Z}_L + \mathbf{Z}_F} \tag{6.9}$$

$$\text{LL:} \quad \mathbf{I}_F = \frac{\mathbf{V}_F}{\mathbf{Z}_{\text{TOT}}} = \frac{\sqrt{3}\,\mathbf{V}_F}{2(\mathbf{Z}_S + \mathbf{Z}_T + \mathbf{Z}_L) + \mathbf{Z}_F} \tag{6.10}$$

$$\text{1LG:} \quad \mathbf{I}_F = \frac{\mathbf{V}_F}{\mathbf{Z}_{\text{TOT}}} = \frac{\mathbf{V}_F}{2/3\mathbf{Z}_S + \mathbf{Z}_T + 2\mathbf{Z}_L + \mathbf{Z}_F} \tag{6.11}$$

TABLE 6.8 Per Unit Impedance and Fault Current for Common Fault Types

Location>>		At Substation			At Feeder End		
Fault	**Z & I**	$Z_F = 0$	$Z_F = 10$	$Z_F = 30$	$Z_F = 0$	$Z_F = 10$	$Z_F = 30$
3PH	Z_{TOT}	0.0083 +j0.1103	0.3298 +j0.1103	0.9729 +j0.1103	0.1163 +j0.2111	0.4379 +j0.2111	1.0809 +j0.2111
	I_a	9.039 −85.70°	2.875 −18.49°	1.021 −6.47°	4.149 −61.14°	2.057 −25.74°	0.908 −11.05°
LL	Z_{TOT}	0.0166 +j0.2206	0.3381 +j0.2206	0.9812 +j0.2206	0.2326 +j0.4222	0.5542 +j0.4222	1.1973 +j0.4222
	I_b	7.829 −175.70°	4.290 −123.13°	1.722 −102.67°	3.593 −151.14°	2.486 −127.30°	1.364 −109.42°
1LG	Z_{TOT}	0.0166 +j0.2906	0.9812 +j0.2906	2.9104 +j0.2906	0.6648 +j0.8953	1.6294 +j0.8953	3.5587 +j0.8953
	Ia	10.305 −86.74°	2.932 −16.50°	1.026 −5.70°	2.690 −53.40°	1.614 −28.79°	0.818 −14.12°

where we define $\mathbf{I}_F$ to be the fault current and $\mathbf{Z}_{TOT}$ as the total impedance seen by $\mathbf{V}_F$ at the fault point F for any given fault type. Then, for each fault type, it is simply a matter of dividing $\mathbf{V}_F$ by $\mathbf{Z}_{TOT}$ to determine the fault current. The current phase angle can be found, but is not needed for simple overcurrent protection.

For close in faults, with $\mathbf{Z}_L = 0$, a careful inspection of these equations shows that the 1LG fault will always be the highest current fault. For end-of-line faults, the minimum fault current depends on the relative size of $\mathbf{Z}_S$ and $\mathbf{Z}_L$ and depends a great deal on $\mathbf{Z}_F$. For this example, the end-of-line 1LG current decreases in proportion to the other fault types as fault resistance is added, as shown in Figures 6.5 and 6.6.

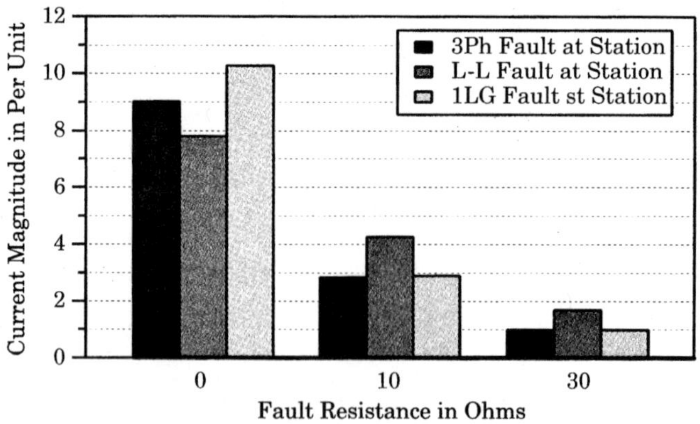

Figure 6.5 Comparison of fault current magnitudes for substation faults.

For close-in faults, the LL fault is the smallest. For zero fault resistance it is always 0.866 of the 3PH fault. This relation does not hold as $\mathbf{Z}_F$ increases since $\mathbf{Z}_F$ is more important in limiting the 3PH fault. Since it is hard to imagine a 3PH fault with equal fault impedance in each phase, we usually compute the 3PH fault without $\mathbf{Z}_F$ to compare against 1LG fault to determine the maximum fault. Then use other fault types with $\mathbf{Z}_F$ to determine the minimum fault. In this example we are given only one value of $\mathbf{Z}_S$, but usually this impedance changes in going from maximum to minimum conditions. In some problems, as in this example, $\mathbf{Z}_F$ is large enough to swamp out any small error in $\mathbf{Z}_S$.

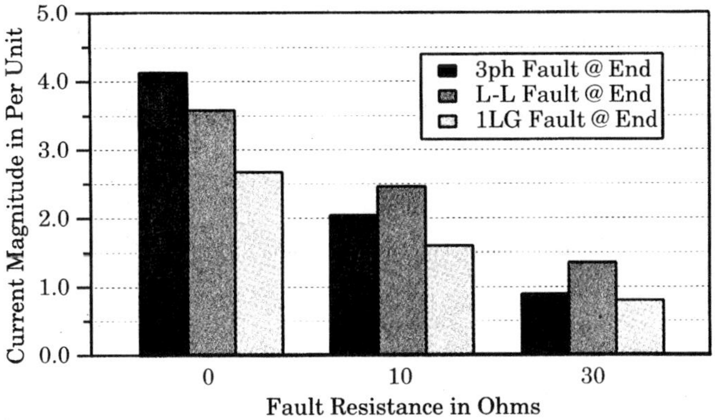

Figure 6.6 Comparison of fault current magnitudes for end-of-line faults.

As a final step in the computation, all per unit currents should be converted to amperes by multiplying by the base current

$$I_B = \frac{S_{B1\phi}}{V_{B-LN}} = \frac{5 \times 10^6/3}{12470/\sqrt{3}} = 231.5 \text{ A}$$

We multiply all per unit currents by this base value to convert them into amperes, with the following results:

The maximum fault is
at substation: $I = 2386A$ (1LG)
at line end: $I = 960A$ (3PH)
The minimum fault with $\mathbf{Z}_F = 30\Omega$
at substation: $I = 236A$ (3PH)
at line end: $I = 189A$ (1LR)

In most problems, the currents are converted back to mks quantities to compare against equipment ratings. ■

6.3.3 Branch Line Faults

Faults on branch lines are easily computed by considering the branch line as a part of the fault impedance $\mathbf{Z}_F$ (or $\mathbf{Z}_G$) with the fault location on the main line feeder. This concept is illustrated in Figure 6.7 where the fault impedance is identified with the branch line. Since

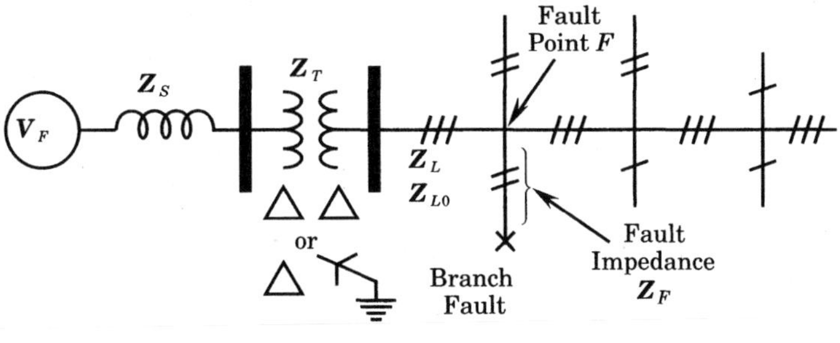

Figure 6.7 Branch line fault configuration.

branch lines are usually single-phase or two-phase (open wye), faults on the branches are limited to unbalanced one- and two-phase fault conditions.

Assigning fault impedance values for branch line unbalanced faults is explained with reference to Figure 6.8 where four common fault configurations are shown and each is identified as a LL, 1LG, or 2LG faults. For an ungrounded single-phase lateral only the LL fault is possible as in 6.8(*a*). A uniground system may have a line-to-neutral fault, in which case $\mathbf{Z}_F$ depends upon whether the neutral is the same size as the phase wire or a reduced size. Values for reduced size cases are average values.

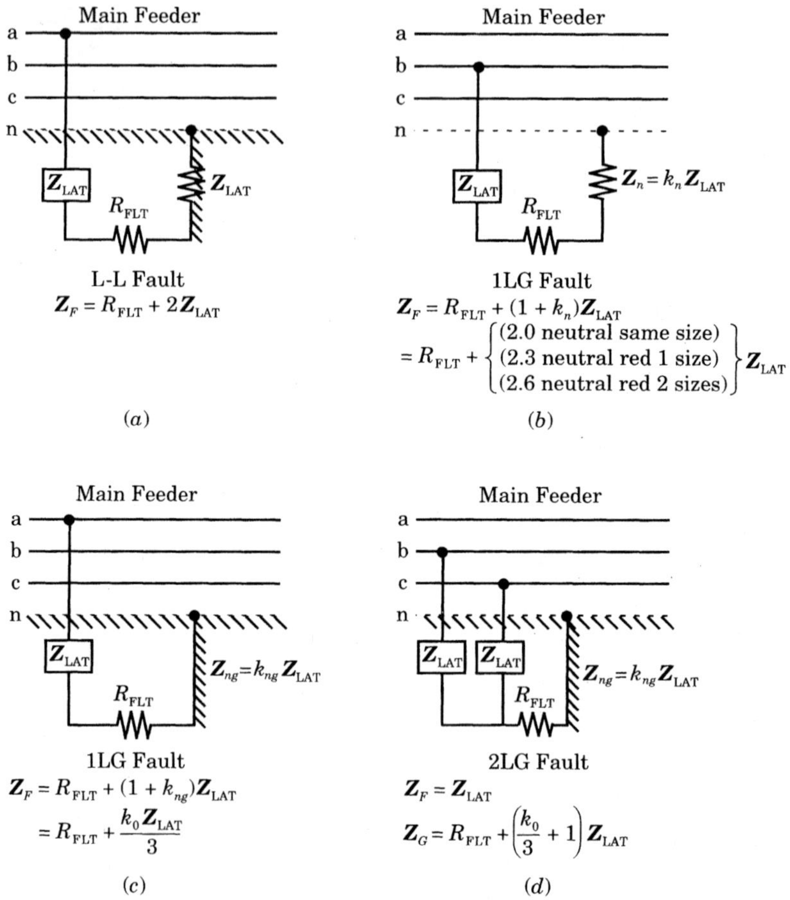

Figure 6.8 Branch line fault configurations. (*a*) 1ϕ - 2 Wire Lateral Ungrounded or Line-to-Line Lateral. (*b*) 1ϕ - 2 Wire Lateral Unigrounded. (*c*) 1ϕ - 2 Wire Lateral Multigrounded. (*d*) 2ϕ - 3 Wire Lateral Multigrounded.

For multigrounded neutral laterals the fault impedance depends on the earth impedance. For the single-phase lateral the only fault possible is the 1LG fault in which case the fault impedance is related to $\mathbf{Z}_0$ and depends on k_0, given in Table 6.4. If a two-phase, open wye system the fault is a 2LG fault with $\mathbf{Z}_F = \mathbf{Z}_L$ of the lateral and $\mathbf{Z}_G$ is the neutral-plus-ground impedance which is related to $\mathbf{Z}_0$.

Since only maximum and minimum fault values are required, it is adequate to compute only the 1LG and LL faults on branches. A suitable fault resistance may also be included as a part of $\mathbf{Z}_F$ or $\mathbf{Z}_G$ in computing minimum fault conditions. Since it is not clear whether the LL or 1LG will be the smaller, both must be checked with and without $\mathbf{Z}_F$.

EXAMPLE 6.2

Consider the fault point F in Figure 6.7 as that fault for which currents of Example 6.1 were computed. Compute the maximum and minimum fault currents at the end of the two-phase, open wye lateral shown in Figure 6.7 if the lateral is 15,000 ft long and is constructed of #2 ACSR with 90 in. spacing. Assume a fault resistance of 30 ohms for minimum conditions. Base quantities are 5 MVA and 12.47 kV.

Solution

Faults for maximum and minimum conditions at the fault point F are already known. We now compute the maximum fault at the end of the lateral, which we define by the notation $\mathbf{Z}_{\text{LAT}}$.

$$\begin{aligned} z_{\text{LAT}} &= r_{\text{LAT}} + j(x_d + x_a) \\ &= 0.267 + j(0.126 + 0.0528\log_{10} 90 - 0.057) \\ &= 0.267 + j0.172\ \Omega/\text{kft} \end{aligned}$$

and for the total 15,000 ft length of the lateral the ohmic value is

$$\mathbf{Z}_{\text{LAT}} = 4.005 + j2.58\ \Omega$$

On the station transformer base, the per unit lateral impedance is

$$\mathbf{Z}_{\text{LAT}} = 0.129 + j0.083 \text{ pu}$$

The sequence impedances at the fault point are computed as follows.

$$\begin{aligned} \mathbf{Z}_1 &= \mathbf{Z}_2 = \mathbf{Z}_S + \mathbf{Z}_T + \mathbf{Z}_L \\ \mathbf{Z}_0 &= \mathbf{Z}_T + \mathbf{Z}_{L0} \end{aligned}$$

1LG Fault: For convenience in finding the 1LG fault currents, we define

$$\begin{aligned} \mathbf{Z}_{012} &= \mathbf{Z}_1 + \mathbf{Z}_2 + \mathbf{Z}_0 = 2\mathbf{Z}_S + 3\mathbf{Z}_T + (2 + k_0)\mathbf{Z}_L \\ &= 2(0.1163 + j0.2113) + (0.432 + j0.403) \\ &= 0.6648 + j0.8953 \qquad \text{pu} \end{aligned}$$

Then with fault resistance assumed to be 30Ω, the arcing fault resistance is

$$R_{\text{FLT}} = 30\Omega = 0.965 \text{ pu}$$

$$\mathbf{Z}_F = R_{\text{FLT}} + \frac{k_0\mathbf{Z}_{\text{LAT}}}{3} = \begin{Bmatrix} 0.1717 + j0.1107 \text{ pu} \\ 1.1363 + j0.1107 \text{ pu} \end{Bmatrix} \quad \text{for} \quad \begin{Bmatrix} R_{\text{FLT}} = 0 \\ R_{\text{FLT}} = 30\ \Omega \end{Bmatrix}$$

Then, for the one-line-to-ground fault we can write

$$\begin{aligned} \mathbf{Z}_{\text{TOT}} &= \mathbf{Z}_{012} + 3\mathbf{Z}_F = 0.6648 + j0.8953 + 3\left\{\begin{pmatrix} 0.1717 \\ 1.1363 \end{pmatrix} + j0.1107\right\} \\ &= \begin{Bmatrix} 1.1799 \\ 4.0738 \end{Bmatrix} + j1.2275 \text{ with } R_{\text{FLT}} = \begin{Bmatrix} 0\ \Omega \\ 30\ \Omega \end{Bmatrix} \end{aligned}$$

Then

$$|\mathbf{I}_a| = \frac{\mathbf{V}_F}{\mathbf{Z}_{\text{TOT}}} = \begin{Bmatrix} 0.5873 \\ 0.2350 \end{Bmatrix} \text{pu} = \begin{Bmatrix} 135.97 \\ 54.41 \end{Bmatrix} \text{A} \quad \text{for} \quad R_{\text{FLT}} = \begin{Bmatrix} 0 \\ 30\ \Omega \end{Bmatrix}$$

LL fault. From Figure 6.8, we see that for the LL fault

$$\mathbf{Z}_F = R_{\text{FLT}} + 2\mathbf{Z}_{\text{LAT}} = \begin{pmatrix} 0.2576 \\ 1.2222 \end{pmatrix} + j0.1661 \text{ pu}$$

The total impedance is (see Table 6.8 for the LL fault)

$$Z_{TOT} = \frac{Z_1 + Z_2 + Z_F}{\sqrt{3}} = \begin{pmatrix} 0.2830 \\ 0.8399 \end{pmatrix} + j0.3396 \text{ pu for } R_{FLT} = \begin{pmatrix} 0 \\ 30\,\Omega \end{pmatrix}$$

Note that the fault resistance in this example makes a very large difference in the resulting total impedance. Then, for the LL fault,

$$|I_F| = \frac{V_F}{Z_{TOT}} = \begin{pmatrix} 2.2619 \\ 1.1037 \end{pmatrix} \text{ pu} = \begin{pmatrix} 523.6 \\ 255.2 \end{pmatrix} \text{ A for } R_{FLT} = \begin{pmatrix} 0 \\ 30\,\Omega \end{pmatrix}$$

Thus, for this lateral the maximum faults are both line-to-line faults and minimum faults are both 1LG faults . A comparison of results for the end-of-branch faults is shown in Figure 6.9.

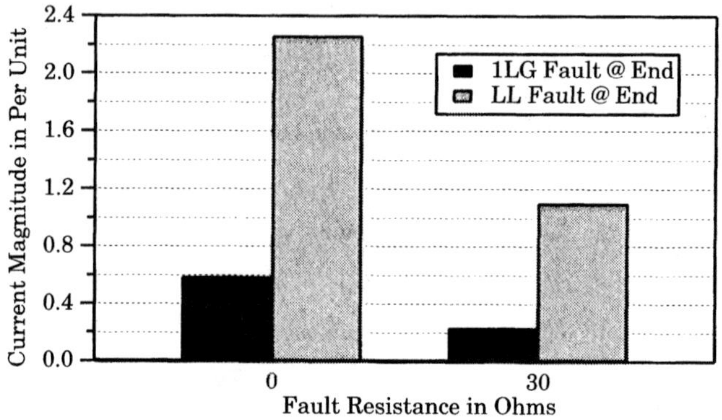

Figure 6.9 Comparison of end-of-branch faults for varying fault resistance. ■

6.4 RADIAL SYSTEM PROTECTIVE STRATEGY

Most of the faults on distribution systems are temporary or transient in nature. Therefore, the first requirement of a protective strategy is to deal effectively with temporary faults by providing a means of fast fault recognition, clearing, and reclosing of the circuit after a brief pause to allow an arcing fault to deionize. To do this requires either circuit breakers with overcurrent and reclosing relays or automatic circuit reclosers. In a great many applications the recloser is adequate but circuit breakers are required in large substations where the available short circuit currents are beyond the rating of reclosers. This basic function, the fast clearing of all faults, is the first requirement of radial system protective strategy. Moreover, if the fault is temporary, a good design will also provide for a method of reclosing, after a brief delay to allow time for the arc to deionize.

The second requirement of radial protection is to isolate permanent faults such that (1) the line section to be isolated is as short as possible and (2) the isolated line is easy to locate. This will restrict the service interruption to a small group of users and will permit the fastest possible location and repair of the trouble.

6.4.1 Clearing Temporary Faults

Referring again to Figure 6.2 we have an example of the usual method for clearing temporary faults. The automatic reclosing device at A will recognize faults along the main line to just short of branch 7, although it will not reach to the extreme end of branch 5. Usually we would set recloser A to trip instantaneously two or three times, using a fast TC curve such as curve A of Figure 3.37. If the fault is temporary, it will have two or three chances to clear and the recloser can reset to normal and await the next fault. Since recloser A is not able to recognize faults at the extreme end of the feeder, a second recloser B is installed as shown in Figure 6.2. This recloser can reach to the ends of all branches on its load side and provides temporary fault clearing for the balance of the circuit. Obviously recloser B must have a smaller minimum pick up than A and it too will be set for two (or three) instantaneous trips followed by two (or three) time delayed openings.

The two reclosers A and B provide satisfactory temporary fault clearing for the main feeder of Figure 6.2. If a permanent fault occurs, say on branch 4, it would operate recloser A through all its instantaneous and time-delayed operations to lock out. This de-energizes the entire circuit until the cause of trouble can be located by patrolling the entire feeder and all branches close to A. This is not an optimum protective scheme as the close in lateral branches provide hazards that can cause long outages of the entire feeder. Clearly, something more needs to be done.

6.4.2 Isolating Permanent Faults

For permanent faults on the main feeder, the reclosers stationed along the feeder provide adequate permanent fault isolation. In Figure 6.2, for example, any permanent fault beyond recloser B would operate B to lockout and A would continue to serve the region between the substation and B. For long feeders, three or more reclosers of graded ratings could be used to confine the main feeder outages to the minimum possible circuit length.

For branch lines it is poor strategy to isolate permanent faults using the main feeder reclosers. As pointed out before, such a permanent branch fault could cause an extended outage of the entire feeder. Adding a recloser, or even a sectionalizer, on every lateral would be costly and unnecessary. Thus, we often protect these branch lines by fuses which are coordinated with the main feeder recloser. Such a protective scheme is shown in Figure 6.10. Here a fused cutout is installed at the source end of every lateral branch and an additional cutout is located in the main feeder at 7 to reduce the area of outage for a permanent fault.

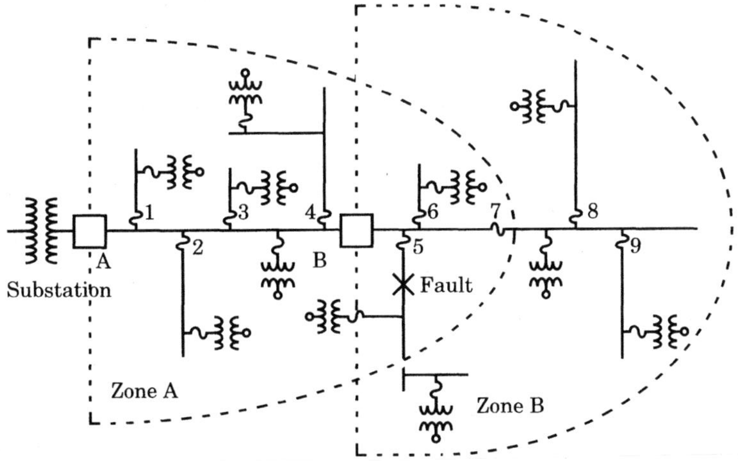

Figure 6.10 Isolation of permanent faults with fuses [1].

Thus a permanent fault X will interrupt only branch 5. Furthermore, the repairman need only inspect the main feeder to find blown fuse 5, and then patrol branch 5 to find and repair the trouble.

The fuse on each branch must be able to carry the load current continuously without melting. Furthermore, it must pick up for minimum faults at the extreme end of the branch. It must have an adequate voltage rating and its interrupting rating must be adequate for close in (maximum) faults.

The fuse must also be coordinated with the recloser that backs it up. For instantaneous recloser operations the fuse should not melt or be damaged, even when there is a temporary fault beyond it. For a permanent fault the fuse should blow before the recloser locks out.

6.5 COORDINATION OF PROTECTIVE DEVICES

One of the important coordination problems on a radial feeder is the proper selection of protective devices for correct sequential operation. There are many different ways of providing protection of transmission and distribution lines. A survey of industry practices in the protection of distribution circuits showed that a large percentage of utilities employ phase and ground overcurrent protection with instantaneous tripping for temporary faults, with time-delayed tripping for permanent faults [7]. Nearly all utilities surveyed also use automatic reclosing of the protective devices, which is a result of the statistics that show most faults to be temporary in nature. This section examines the problems of selecting and coordinating different devices in a coordinated protective scheme.

Most radial protective schemes involve the coordination of fuses, reclosers, and relays. We have already discussed fuse-to-fuse and recloser-to-recloser coordination in Chapter 3. We now consider the coordination of unlike devices. This is more difficult, generally, because fuses, reclosers, and relays have TC characteristics of different shape. This means that the desired coordination is often achieved in only a restricted range of currents.

6.5.1 Recloser-Fuse Coordination

If a recloser with a choice of fast and retarded operations is installed ahead of a fuse as shown in Figure 6.11 we have a common example of the recloser-fuse coordination problem. In this situation we want the recloser fast operations to protect the fuse, or fall "below" the fuse TC characteristic, that is, we want the recloser to clear before the fuse can melt. Curve A does fall below curve C for currents less than the value corresponding to point b, where curves A and C cross. If the fault beyond fuse C is permanent, then we want fuse C to totally clear as the recloser goes through a delayed operation B. This coordination will be correct as long as the current is greater than the crossing point a. Thus we have found approximately the coordination range

$$a < I < b \tag{6.12}$$

This range is approximate because it fails to account for the alternate heating and cooling of the fuse. To account for the fuse heating and cooling we may write equations that closely estimate these effects as exponentials. Thus for heating we write an equation for the temperature of the fuse element, θ.

$$\theta = \theta_f(1 - e^{-t/\tau}) \tag{6.13}$$

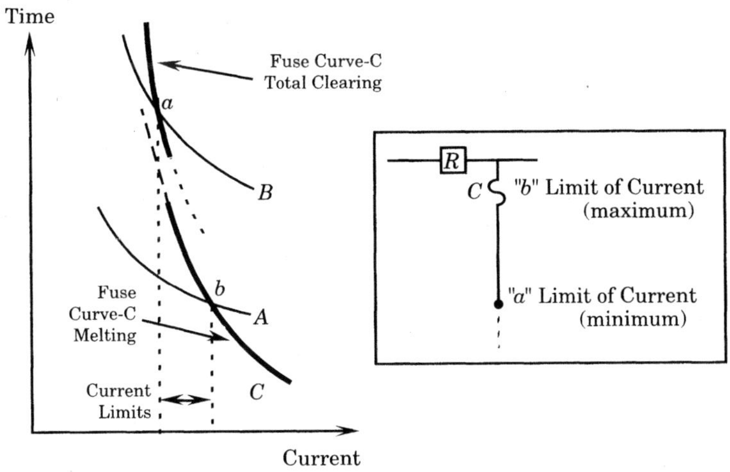

Figure 6.11 Time-current characteristic curves of recloser R superimposed on fuse curve [2].

where θ = temperature (C)

θ_f = final steady-state temperature assuming fuse does not melt (C)

t = time (s)

τ = fuse time constant (s)

At the final steady-state temperature a balance is reached between heat input and heat loss, i.e.,

$$I^2R = \theta_f K \tag{6.14}$$

where K is the heat dissipation constant in watts per degree centigrade (W/°C). If we apply (6.14) for long periods of melting time, corresponding to melting temperatures of θ_m we may write

$$I_m^2 R = \theta_m K \tag{6.15}$$

At the lower end of the TC characteristic, for times in the neighborhood of 0.1 second, very little heat is dissipated. Instead, most of the applied heat energy I^2Rt is used to raise the metal temperature to the melting point. Here we write

$$I^2Rt = C_h\theta_m \quad \text{J} \tag{6.16}$$

where C_h is a median value of heat capacity of the fusible element in joules per degree centigrade (J/°C) and θ_m is the fuse melting temperature.

If we assume that (6.16) applies at 0.1 second we write

$$0.1 I_{0.1}^2 R = C_h\theta_m = C_h \frac{I_m^2 R}{K} \tag{6.17}$$

Then we compute

$$\frac{C_h}{K} = 0.1\left(\frac{I_{0.1}}{I_m}\right)^2 \tag{6.18}$$

We now recognize that $(I_{0.1}/I_m)$ is the fuse speed ratio S and that C_h/K is the fuse time constant. Thus, we write

$$\tau = 0.1S^2 \tag{6.19}$$

as a convenient approximation to the fuse time constant. We may also write a fuse cooling equation

$$\theta = \theta_f^{-t/\tau} \tag{6.20}$$

where we assume that heating and cooling occur with the same time constant. If we normalize (6.13) and (6.20), we may write the per unit temperature equations as follows.

Heating:

$$\frac{\theta}{\theta_f} = \theta_u = 1 - e^{-t/\tau} \quad \text{pu} \tag{6.21}$$

Cooling:

$$\frac{\theta}{\theta_f} = \theta_u = e^{-t/\tau} \quad \text{pu} \tag{6.22}$$

These equations are plotted against normalized time t/τ in Figure 6.12 as curves A and B, respectively. Also plotted is a typical sequence of heating and cooling sequences, shown as curve C, which is pieced together by sections of A and B for appropriate lengths of time. Each time the fuse is subjected to a fault current, heat accumulates in the fuse metal. When the recloser opens the circuit, and the current is interrupted, the fuse cools. Both the heating and cooling action are approximated by exponential function, as shown in Figure 6.12. If enough heat accumulates, the fuse will melt. The fuse melting is a function of the fuse design melting time and the total accumulated time the fuse is heated, less the time it is cooled.

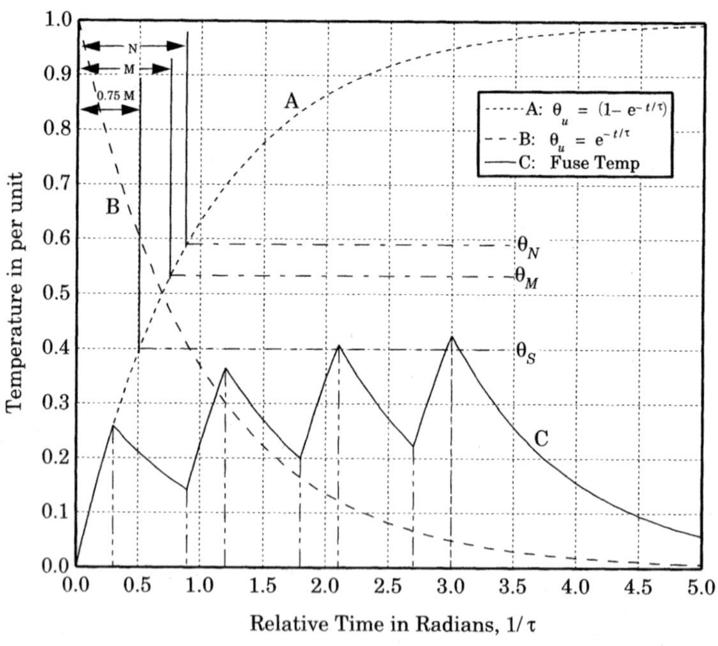

Figure 6.12 Temperature cycle of fuse during reclosure operation [8].

To determine if the fuse will melt during recloser opening and closing cycles, we begin by finding the total fault current. The corresponding fuse minimum melting time M is then taken from the fuse TC characteristic. Opposite the time M in Figure 6.12 is the melting temperature θ_m.

Obviously curve C must remain below θ_m. To ensure a reasonable margin for safe coordination it is convenient to define a safe temperature level θ_s which corresponds to 0.75 M. Then the alternate heating and cooling cycles can be plotted as curve C. The curve C plotted in Figure 6.12 shows the fuse temperature just exceeding this safe temperature level, indicating that the fuse may melt during the four recloser fault cycles illustrated.

One way to check this coordination without plotting a fuse heating-cooling curve is by use of a fuse damage curve which is defined as 75% of the melting time curve. This is illustrated in Figure 6.13 where the 75% curve is compared against curve $A' = 2 \times A$ (in time). Similarly point a' is found from the more conservative curve B' which includes the recloser operating times as shown in Figure 6.13. With experience this extra precaution can be omitted by simply allowing an extra margin in the coordination scheme. This requires experience, however, and there is no guarantee that the extra margin will always be adequate.

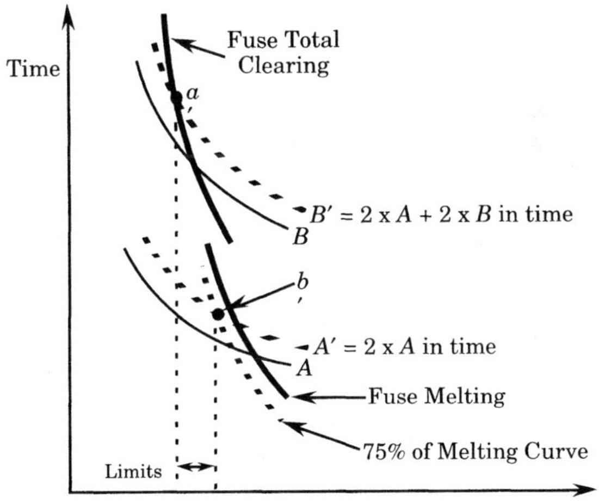

Figure 6.13 Recloser-fuse coordination with the fuse corrected for heating and cooling [1].

6.5.2 Recloser-Relay Coordination

Recloser ratings are usually less than about 400 MVA at rated voltages of 15 kV. Where available fault currents exceed this value, a power circuit breaker must be used so that greater interrupting capacity can be provided. If the substation transformer is supplied from an infinite bus we may compute

$$I_u = \frac{V_F}{X_T} = \frac{1.0}{X_T} \quad \text{per unit} \tag{6.23}$$

or in amperes

$$I_F = I_u I_B = \frac{S_{B3\phi}}{\sqrt{3} V_{BLL} X_T} \quad \text{A} \tag{6.24}$$

and we compute the three-phase fault volt-amperes S_F to be

$$S_F = \sqrt{3} V_{BLL} I_F = \frac{S_{B3\phi}}{X_T} \tag{6.25}$$

If $S_{B3\phi}$ is the substation transformer rating, a 400 MVA fault with a 6% transformer reactance restricts the rating of the three-phase transformer bank to 2400 kVA. Thus for distribution substations larger than about 2400 kVA a power circuit breaker will be required.

Usually the relay used in a radial distribution substation is the simple overcurrent relay, either the traditional electromechanical relay or the newer static or digital devices [9–11]. For a radial system such as Figure 6.2 it is often necessary to coordinate the relay at A with a recloser at B. For a fault beyond B the recloser will operate on its first TC characteristic and open. During this operation the overcurrent relay will also begin to travel in the trip direction, if an electromechanical device. Furthermore the relay may not completely reset before the next tripping sequence begins. With four recloser operations to lockout, the electromechanical relay can accumulate enough angular motion of the induction disk in the trip direction to cause false tripping. This is not a problem with static relays, as they usually reset instantaneously after each recloser operation.

EXAMPLE 6.3

Consider the system of Figure 6.14 where a circuit breaker B is controlled by overcurrent relays that must be set to coordinate with downstream reclosers R rated 35A. The time-current coordination is illustrated in Figure 6.15, where recloser curves A (instantaneous) and B (retarded) must coordinate with overcurrent relay characteristic C. It is usually assumed that the angular velocity of the disk for an induction relay is constant both in the pickup and reset directions. Since the complete pickup time for a given current is known from the characteristic, any lesser time will provide a proportional fraction of pickup angle. The total reset time is a constant which depends on the time-dial (lever) setting and this too can be proportioned according to the reset time available between recloser operations. Referring to Figure 6.15 the relay curve C corresponds to a very inverse overcurrent relay set on the 1.0 time-dial adjustment and a 4 ampere tap (160 ampere primary with 200/5 current transformer). Assume a permanent fault current of 500 amperes located at X in Figure 6.14. We assume the following operating times:

Recloser:	Instantaneous (curve A)	0.036 s
	Time delay (curve B):	0.250 s
Relay:	Pickup (curve C):	0.650 s
	Reset (1/10)(60)	6.000 s

where it is assumed that the reset time for this relay is 60 seconds with a number 10 time-dial setting [1].

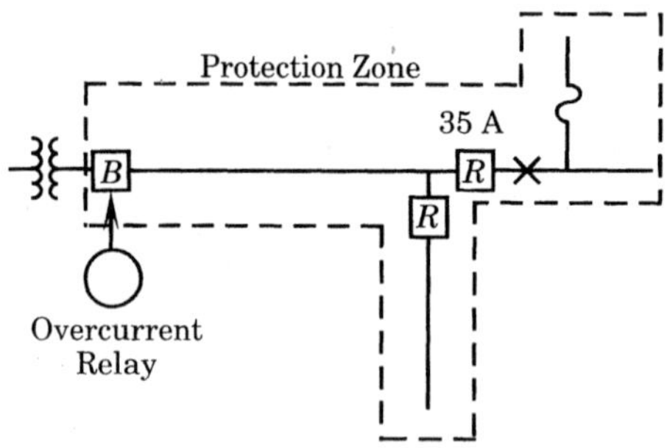

Figure 6.14 Coordination of relay at B with reclosers R.

We compute the relay travel as follows, using (+) for trip direction and (−) for reset direction. For the instantaneous (curve A) operation we compute

$$\text{Closing relay travel} = \frac{0.036}{0.650}(100\%) = 5.5\%$$

If the recloser is open for 1 second,

$$\text{Reset relay travel} = -\frac{1}{6}(100\%) = -16.67\%$$

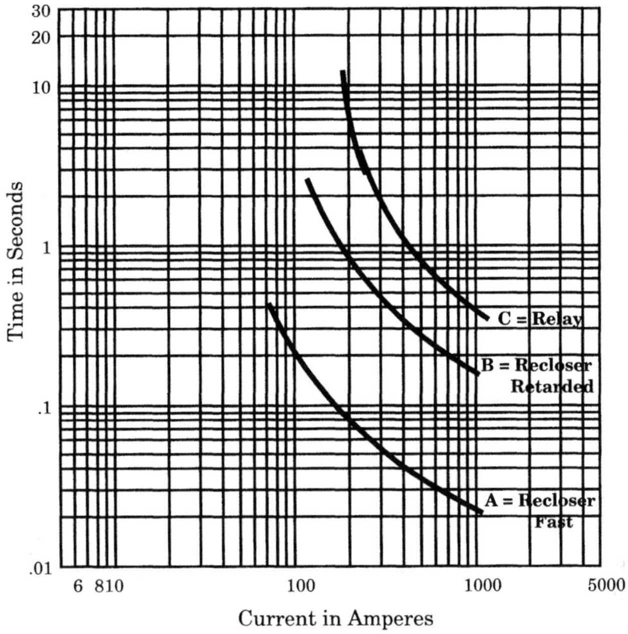

Figure 6.15 Time-current coordination curves.

Obviously the relay will completely reset for this operation. Note that it traveled very little in the pickup direction since the recloser trip time was relatively fast.

For two retarded operations we compute the following travel percentages as follows.

$$\begin{aligned}
\text{First retarded trip travel} &= \frac{0.25}{0.65}(100) = +38.5\% \\
\text{Reset travel for 1 second} &= \frac{-1}{6}(100) = -16.7\% \\
\text{Second retarded trip travel} &= \qquad\quad\ \ +38.5\% \\
\text{Total net travel} &= \qquad\quad\ \ +60.3\%
\end{aligned} \tag{6.26}$$

Since the total relay travel is less than 100% the desired coordination is achieved. Usually a 0.15 to 0.20 second margin is considered desirable to guard against variations in published curves and other errors. ■

6.6 RELAY COORDINATION ON RADIAL LINES

The coordination schemes discussed to this point have been applicable to circuits of limited short circuit capacity. For higher capacity circuits, circuit breakers must be specified to interrupt the greater fault currents. Circuit breakers do not have their own tripping intelligence and must be used in conjunction with relays. It is the relays that provide the intelligence and control logic for tripping the circuit breakers. Since relays are available in a wide variety of designs, this arrangement is very flexible and can be adapted to a wide variety of system configurations.

The system configuration of interest here is a radial line on which a series of relays must be coordinated. Such a system is shown in Figure 6.16. Here, the transmission line is connected radially from a transmission substation at bus G and supplies loads at buses G, H, and R. The line is sectionalized at each load supply point by circuit breakers. Since the line is radial, the relays that control these breakers must be coordinated so that selective system protection can be achieved. The relays can be nondirectional, since there is only one source of power and the current is always flowing from left to right. This may not always be the case, since some large load centers, such as industrial plants, often have generating sources of their own, and these generators can contribute to the total fault current. Thus, the protection engineer must have a knowledge of the load served as well as the transmission source.

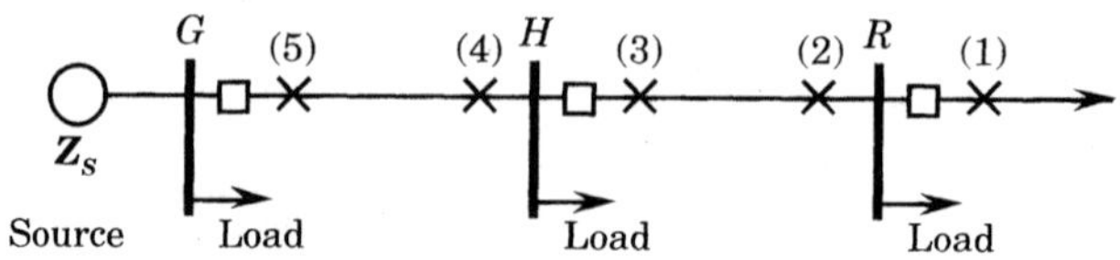

Figure 6.16 A radial transmission line supplying three loads.

Actually, the line shown can be a part of a loop and proper relay selectivity achieved if the backfeed from the right in the figure is small. Usually, it is reasonable to consider the circuit to be radial (from a coordination viewpoint) if the backfeed is less than 25% of the minimum fault current for which the relay must operate [8].

The relays used at each breaker position can be arranged in many different connection arrangements and still achieve the desired coordination and selectivity. Usually, there are both phase and ground relays to coordinate. For simplicity, we will assume a relay configuration exactly like that shown in Figure 2.3, where we have wye-connected phase relays in phases a, b, and c and one ground relay located in the neutral connection. This arrangement has the advantage that any one relay may be taken out of service for maintenance and the circuit still has protection for all types of faults.

6.6.1 Coordination Procedure

Consider the radial circuit shown in Figure 6.16. Faults are shown at different locations, labeled (1) through (5). We now consider the problems of coordinating the relays for proper selectivity for these different fault locations.

The relay at H should be commanded to trip for any fault on line H-R under any system condition, from maximum generation to minimum generation, and with any reasonable value of fault resistance. This coordination is usually achieved without any difficulty.

A more difficult problem is that of ensuring selectivity for a fault at (2), which is near the end of line H-R, and a fault at (1), which is just beyond breaker R. Since the fault currents are almost exactly the same for these two faults, there is no way that selectivity can be achieved on the basis of current magnitude or direction. Even using a more complex relay that could measure the exact impedance from the relay at H to the fault point could not distinguish precisely whether the fault is at (1) or at (2). Clearly, some additional concept must be used for selectivity. The network condition is illustrated by means of an example.

EXAMPLE 6.4

Consider a specific example of the system shown in Figure 6.16. The line is a 34.5 kV subtransmission circuit with characteristics given as follows:

Line Section	*G-H*	*H-R*	*R*-End
Length	10 miles	20 miles	30 miles
ACSR conductor	Linnet	Linnet	Linnet
Equivalent spacing	5.0 ft	5.0 ft	5.0 ft

The fault capability at bus G is known from transmission system fault studies, which provide the following data:

System Fault Capability

Maximum conditions: 200 MVA
Minimum conditions: 100 MVA

Compute the maximum and minimum three-phase fault currents at each fault location (1) through (5), shown in Figure 6.16.

Solution

We begin by computing the line impedance on an ohm per mile basis. From Table C.2, and assuming a conductor operating temperature of 50 C, we read the following.

$$r_a = 0.294 \quad \Omega/\text{mile}$$
$$x_a = 0.4511 \quad \Omega/\text{mile}$$
$$x_d = 0.1953 \quad \Omega/\text{mile}$$

Therefore

$$z = r_a + j(x_a + x_d) = 0.294 + j0.6464 \quad \Omega/\text{mile} \tag{6.27}$$

Now, convert to per unit by selecting, arbitrarily, a 100 MVA base.

$$S_B = 100 \text{ MVA}$$
$$V_B = 34.5 \text{ kV (line-to-line)}$$

Then

$$Z_B = 11.9025 \quad \Omega$$
$$I_B = 1673.4798 \quad \text{A}$$

The per unit impedance is computed by dividing (6.27) by the base impedance.

$$z = 0.02470 + j0.05431 \text{ per unit/mile}$$

Then the per unit line impedances are as follows.

$$\mathbf{Z} = 0.247 + j0.543 \quad \text{per unit} \quad 10 \text{ miles}$$
$$\mathbf{Z} = 0.494 + j1.086 \quad \text{per unit} \quad 20 \text{ miles}$$
$$\mathbf{Z} = 0.741 + j1.629 \quad \text{per unit} \quad 30 \text{ miles}$$

The system impedance may be found as follows:

Maximum condition: 200 MVA:

$$I_{3PH} = 2.0 \text{ pu}$$
$$Z_S = 0 + j0.5 \text{ pu}$$

Minimum condition: 100 MVA:

$$I_{3PH} = 1.0 \text{ pu}$$
$$Z_S = 0 + j1.0 \text{ pu}$$

This completes the determination of all system impedances. We may now construct a table of impedances from the source to all points in the circuit and the three-phase fault currents at each point.

This is shown in Table 6.9. Obviously, the fault currents at 1 and 2 are identical, as are the currents at 3 and 4. Therefore it is impossible to achieve relay selectivity on the basis of current magnitude alone.

TABLE 6.9 System Impedances in Per Unit and Three-Phase Fault Currents in per Unit and in Amperes

Fault Location	Total pu Z to Fault		Three-Phase Fault Current	
	Minimum	Maximum	Minimum	Maximum
1 or 2	0.741+j2.629 |2.732|	0.741+j2.199 |2.255|	|0.366| 612.6 A	|0.444| 742.3 A
3 or 4	0.247+j1.541 |1.561|	0.247+j1.041 |1.070|	|0.641| 1072.3 A	|0.935| 1564.1 A
5	0.000+j1.000 |1.000|	0.000+j0.500 |0.500|	|1.000| 1673.5 A	|2.000| 3347.0 A

■

Two options are available to provide the correct selectivity. The first is to grade the operating time of the relays, making each relay on the source side somewhat slower than the next relay on the load (downstream) side. The second option is to use "pilot relaying," which is a relaying scheme that transmits information from the remote end of the line section to ensure the correct discrimination between faults internal to the line from those external.

Adding time delay is inexpensive but will cause faults close to the supply substation to be tripped more slowly, which is not desirable. Pilot relaying does not have this disadvantage, but it is much more expensive. In many cases time delay is used with quite satisfactory results.

Time delay can be introduced in different ways, depending on the relay time-current characteristic. As noted in Figures 3.20 to 3.23, relays typically are designed with characteristics that vary from definite time (for any current) to extremely inverse. Figure 3.16 shows a specific range of characteristics for one relay manufacturer. Note that the more inverse characteristics cause the relay to operate more slowly for smaller fault currents. This means that close-in faults are tripped quickly, but remote faults are tripped more slowly. A comparison between the definite time and inverse time characteristics for the circuit of interest is shown in Figure 6.17. This figure shows the relay time for relays at R, H, and G if a *coordination time interval* (CTI) is used as the required separation between the adjacent relay time-current characteristic curves. Clearly, the inverse characteristic has the advantage of providing faster operation of relay G for maximum faults, and probably for minimum faults, although the inverse characteristic has an operating time that varies considerably from minimum to maximum conditions. Usually, this difference can be accommodated in relay applications, and the inverse type of relay characteristic is often preferred.

We now consider a procedure that can be used to coordinate the relays in a radial circuit. Obviously, we need to compute the available fault currents at every relay location and this must be done for both minimum and maximum conditions. We also need to grade the settings of the relays, applying an appropriate CTI between adjacent settings. We must also determine the settings of the relay itself so that it will perform as required.

6.6.2 Procedure for Phase and Ground Relays

Relay coordination is often performed graphically, usually using log-log coordinate paper to show several decades of time and current ranges for each device. Computer aided

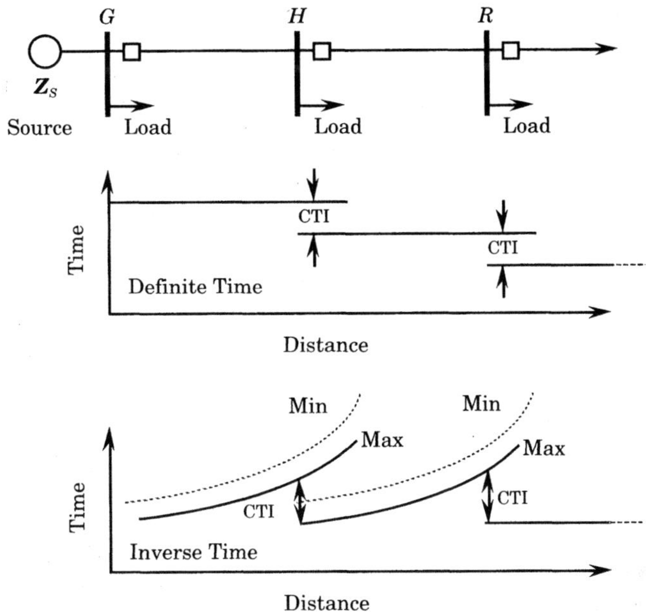

Figure 6.17 Comparison between definite and inverse relays.

procedures have been devised for coordination [12]. The procedure here, however assumes that the engineer is to perform the task manually. In performing the relay coordination, the relays are considered in pairs, usually starting from the load end (sometimes called the "downstream" end) of the circuit and progressing back toward the source (upstream). In performing this task it is helpful to set down an orderly procedure for stepping through the process. Such a procedure is suggested below.

1. Relay MOC Determine the relay minimum operating current (MOC). There are several criteria that should be checked in fixing the MOC. These are different for phase and for ground relays.

Phase Relays

1. Set the MOC at 50% or less than the minimum end-of-line phase-to-phase fault.
2. Check to see that the MOC is 200% or more than the maximum load current.
3. Set the instantaneous trip element at six times the load current or 125% of the maximum three-phase fault current at the first downstream protective device, whichever is greater.

Ground Relays

1. Set the MOC at 33% or less than the phase MOC.
2. Set the MOC at 50% or less than the 1LG end-of-line fault current.
3. Set the instantaneous trip MOC at 150% of the 1LG fault at the first downstream protective device.

2. Relay tap Relays have tap settings that can be used to adjust the minimum current in the relay for which the relay will just pick up. This provides great versatility in the relay application and permits the same relay to be used at many locations. One can think of the relay tap as if it were an autotransformer, as shown in Figure 6.18. We analyze the currents in the system as follows:

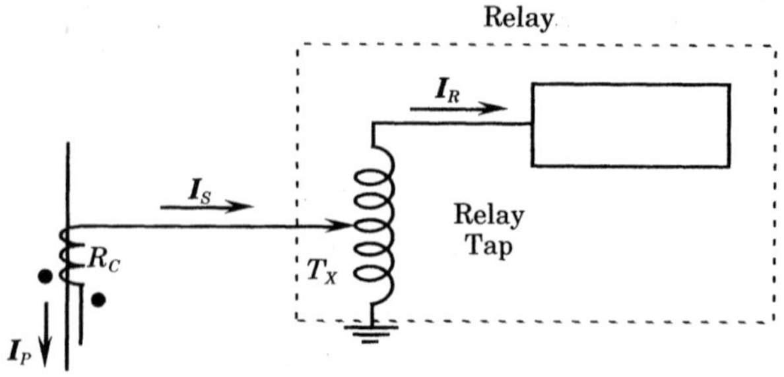

Figure 6.18 Conceptual circuit for the relay tap.

Let

$$\begin{aligned} R_C &= \text{CT ratio} \\ T_X &= \text{Relay tap} \end{aligned} \tag{6.28}$$

Then we may write

$$\begin{aligned} I_S &= T_X I_R \\ I_P &= R_C I_S = R_C T_X I_R \end{aligned} \tag{6.29}$$

Now define the minimum pickup current for the relay to be

$$I_R = 1.0 \tag{6.30}$$

Then (6.30) can be used to give the primary current that will just pick up the relay. The tap that will give pickup at a given primary current, then is

$$T_X = \frac{I_P}{R_C} \tag{6.31}$$

where

$$I_P = \text{primary current in } A \text{ at pickup} = \text{MOC}$$

3. Relay critical fault current Relays are coordinated in pairs, with the upstream relay being adjusted to coordinate properly with the downstream relay. In doing this, there is always one current value that is critical. For inverse relays this is almost always the maximum current at the downstream relay. This is always a good choice of critical current to check first.

Once the critical current is selected, convert this ampere value into a multiple of pickup current, which we shall call *XPU* (for "times pickup"). This *XPU* value corresponds to the horizontal (current) axis of the relay characteristic curves, as shown in Figures 3.20 to 3.23. From (6.29)

$$XPU_{CR} = \frac{I_{CR}}{T_X R_C} \tag{6.32}$$

where I_{CR} is the critical current. It will be necessary, of course, to check later to make sure that the current chosen for this test is actually the critical current. It may be that some other value will be critical, depending on the devices being coordinated and the shape of their characteristics.

4. Relay time dial Relay characteristic curves are always given as a family of curves, as shown in Figure 3.17. Each member of the family corresponds to a different "Time Dial" setting (sometimes called a "Lever Setting"). We now wish to select the correct time dial setting that will provide coordination at the critical current.

We have determined the XPU of the critical current for this relay (the horizontal axis). The other thing we need to know is the time parameter (the vertical axis) that corresponds to the critical current. To select the operating time of this relay, we must find the operating time of the downstream relay at this value of (critical) current, which is the maximum current at the downstream relay. Therefore, we compute this time at the downstream relay (which we designate here by the letter "D"). Referring to the relay characteristic of the downstream relay, we compute the horizontal coordinate as follows.

$$XPU_{D\,\max} = \frac{I_{D\,\max}}{T_{XD}R_{CD}} \tag{6.33}$$

We may then look up the relay operating time of relay D from the published characteristic curves of the D relay type. We designate this operating time as

$$t_{D\,\max} = \text{Operating time of relay } D \text{ at } I_{D\,\max} \tag{6.34}$$

Then, the desired operating time at the relay upstream from D, which is the relay now being set, we find the operating time by adding a *coordination time interval* (*CTI*) to the above time, as shown in Figure 6.19, where

$$\begin{aligned} CTI &= \text{relay detection time} \\ &\quad + \text{relay pickup time} \\ &\quad + \text{margin for error} \\ &\approx 0.2 \text{ to } 0.3 \text{ s} \end{aligned} \tag{6.35}$$

Then

$$t_{CR} \geq t_{D\,\max} + CTI \tag{6.36}$$

This is the vertical axis parameter we need to ensure adequate time coordination. We now enter the relay characteristic curves at the horizontal axis value given by (6.33) and move vertically at that XPU to find the time (6.36). The required time dial setting should be the next setting above this critical time.

This is done rather easily using a graphical technique. The maximum current for the downstream relay should be marked on the plot, and its time coordinate noted. Then the *CTI* is added and a point marked on the log-log coordinate paper. This is the critical coordination point. The time, given by (6.36), must lie above this point.

Some protection engineers like to add a criterion that the time dial be chosen to always ensure tripping in a given time, such as 2 seconds, for example. This is to prevent allowing time margins between adjacent relays that are unnecessarily large and to assure reasonably fast response. This criterion will vary from company to company, but it is a good idea. We shall arbitrarily adopt 2 seconds as our goal.

5. Check relay reset If the downstream protective device utilizes a reclosing scheme, the relay upstream from that device will partially pick up on every closing on the fault and will partially reset on every opening, as discussed in Section 6.5.2. This should be checked if such reclosing is used on the downstream relay device, which may be a relay or other device.

6. Relay maximum fault Find the maximum fault current for this relay and the relay operating time corresponding to this current. This value will be needed in coordinating with the next relay upstream, especially when inverse relays are used. See item #4, above.

7. Repeat the procedure for the next relay upstream The relay that is farthest downstream must be set first so that it picks up for a fault at the farthest end of the radial and still has a pickup current well above maximum load current. Then the next relay upstream is coordinated

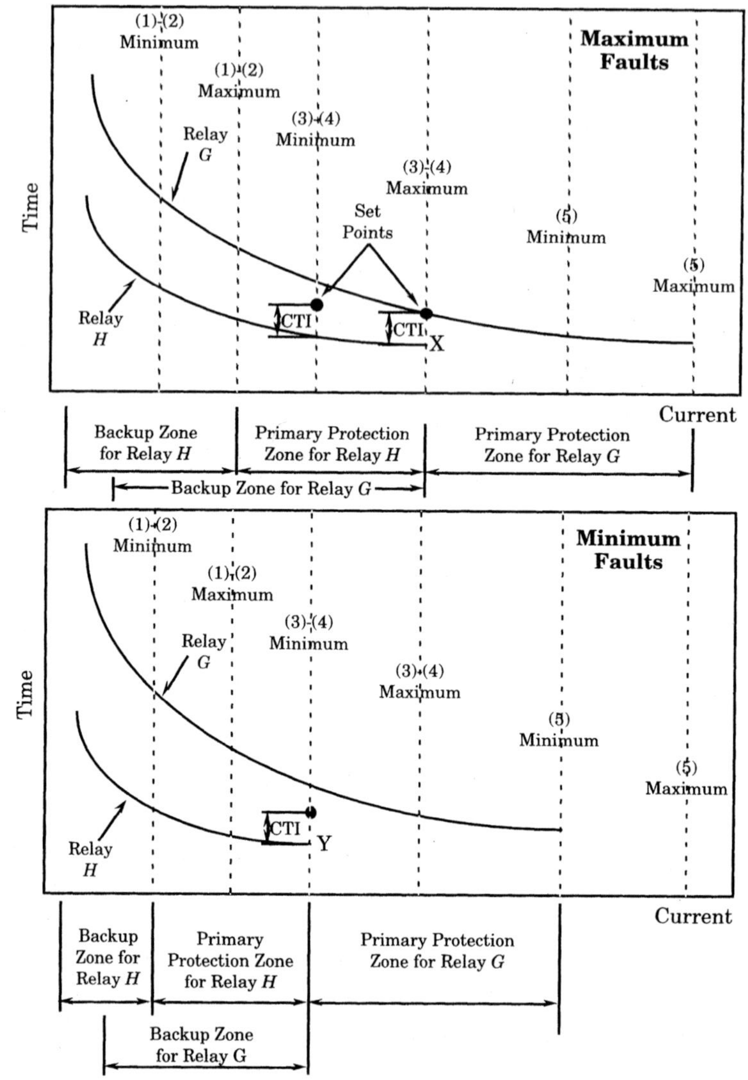

Figure 6.19 Separation of relay characteristics by the CTI.

with this downstream relay and the process repeated, always working with pairs of relays, back toward the source.

EXAMPLE 6.5

Extend Example 6.4 to coordinate relays *G*, *H*, and *R*. The following additional data are given for the system of Figure 6.16.

1. *Loads.* Each of the four loads at buses *G*, *H*, *R*, and the radial line from *R* is a 3 MVA load and is projected to grow to 5 MVA. Each load is served by a 5 MVA step-down transformer that is fused on the 34.5 kV side. The 34.5 kV fuse has the total clearing characteristic shown in Table 6.10.

   It is considered sufficiently accurate to connect these points by straight lines. Relays at *G*, *H*, and *R* all must coordinate with this fuse type. There is no generation at any of the load sites.

TABLE 6.10 Step-Down Transformer Fuse Characteristic

Time (s)	500.0	10.0	1.0	0.1
Current (A)	160	220	520	1600

2. *Relays.* The relays at G, H, and R are all inverse-time relays with characteristics shown in Figure 3.17. It is required that the *MOC* be twice the maximum estimated load current. It is also required that each relay have a coordination time interval *CTI* of 0.3 seconds between its operation and that of the adjacent relay.
3. *Current transformers.* The current transformer ratios are given as

   G: 200/5 H: 200/5 R: 100/5

4. *Backup coordination.* The relay at G must coordinate with relays on the source side of G that are also inverse-time types (Figure 3.17) and with the following settings:

   CT: 200/5 Tap: 8 TD: 4

Completely determine the relay settings so that they coordinate with each other, with the load transformer fuse, and with the transmission source backup relays. Plot your solution on log-log paper showing all required data, including critical currents and coordinating time intervals.

Tabulate the result for each relay giving all pertinent data in a clear and logical manner.

Perform this coordination only for the phase relays.

Solution

We begin by coordinating the two relays that are the farthest downstream, i.e., the relays at R and H.

Relay R Always begin with the relay farthest downstream.

1. *MOC.* The maximum load current is computed based on the maximum estimated load MVA.

$$I_{Ld\,\max} = \frac{5000\text{ kVA}}{\sqrt{3}(34.5\text{ kV})} \cong 80\text{ A}$$

$$MOC_R = 2 \times 80 = 160\text{ A}$$

2. *Relay tap.* The CT ratio is 100:5 or $R_C = 20$

$$T_{XR} = \frac{I_P}{R_C} = \frac{160}{20} = 8$$

3. *Critical fault current.* The critical fault current at relay R is not known since this relay must coordinate with a downstream fuse. This coordination is best done graphically.
4. *Relay time dial setting.* Refer to Figure 3.17 for the characteristics of the inverse time relay. Since this is the last relay in the radial, we must coordinate only with the transformer fuse. This is done graphically to determine that the time dial should be set at 3 to give the required separation from the fuse characteristic.

$$TD_R = 3$$

5. *Check reset.* There is no reclosing device downstream so this requirement is waived for relay R.
6. *Maximum fault current.* The maximum fault current for this relay is found in Table 6.9 to be 742 A.

$$XPU_{R\,\max} = \frac{I_{D\max}}{T_{XD}R_{CD}} = \frac{742}{20 \times 8} = 4.6$$

Then from Figure 3.17, for $TD = 3$ we read approximately

$$t_{R\,\max} = 1.3\text{ s}$$

Relay H

1. *MOC.* The total load at H is 160 A. Therefore

$$MOC_H = 2 \times 160 = 320 \text{ A}$$

2. *Relay tap.* The CT ratio is 200:5 or $R_C = 40$. Then

$$T_{XH} = \frac{320}{40} = 8$$

3. *Critical fault current.* For inverse relays, the critical fault current is almost always the maximum current of the downstream relay. The maximum current R is known from the relay R computations to be 742 A (see Table 6.9).

$$I_{CRH} = 742 \text{ A}$$

Then

$$XPU_H = \frac{742}{8 \times 40} = 2.3$$

4. *Relay time dial setting.* For relay R, we found that the operating time at maximum current was 1.3 seconds.

$$t_{R\max} = 1.3$$
$$CTI = 0.3$$
$$t_{CRH} \geq 1.3 + 0.3 = 1.6 \text{ s}$$

From Figure 3.17, we choose a time dial setting that will give an operating time greater than 1.6 seconds at an $XPU = 2.3$. We select

$$TD_H = 2$$

Actually, it appears from Figure 3.17 that a time dial of about 1.5 or so would be adequate, but we will not try to interpolate. The value selected is generous and ultimately safe.

5. *Check reset.* Relay R has no reclosing capability, hence there is no need to check reset.
6. *Maximum fault current.* The maximum fault current for relay H is taken from Table 6.9.

$$I_{H\max} = 1564 \text{ A}$$

$$XPU_{H\max} = \frac{1564}{8 \times 40} = 4.9$$

$$t_{H\max} = 0.9 \text{ s}$$

Relay G

1. *MOC.*

$$MOC_G = 2 \times 240 = 480 \text{ A}$$

2. *Relay tap.*

$$T_{XG} = \frac{480}{40} = 12$$

3. *Critical fault current.*

$$I_{CRG} = 1564$$

$$XPU_G = \frac{1564}{40 \times 12} = 3.26$$

4. *Relay time dial.*

$$t_{H\max} = 0.9$$
$$t_{CRG} = 0.9 + 0.3 = 1.2 \text{ s}$$

From Figure 3.17, to exceed 1.2 seconds at 3.26 XPU, we require the time dial setting

$$TD_G = 2$$

5. *Check reset.* Not required.
6. *Maximum fault current.* From Table 6.9,

$$I_{G\max} = 3347 \text{ A}$$

$$XPU_G = \frac{3347}{40 \times 12} \cong 7.0$$

$$t_{G\max} = 0.6 \text{ s}$$

Results are summarized in Table 6.11. The relay currents and operating times required for plotting the results are shown in Table 6.12. The pickup multiples and relay times are taken directly from Figure 3.17. The currents are computed from the pickup multiples.

The coordination curves, plotted from Table 6.12, are shown in Figure 6.20.

TABLE 6.11 Summary of Example 6.5 Coordination Parameters

Relay Parameter	Relay *R*	Relay *H*	Relay *G*
Max load current, A	80	160	240
MOC, A	160	320	480
CT ratio	100/5	200/5	200/5
Relay tap	8	8	12
Critical current, A	–	742	1564
Critical *XPU*	–	2.3	3.26
Critical time, s	–	1.6	1.1
Time dial	3	2	2
Maximum fault *I*, A	742	1564	3347
Maximum *XPU*	4.6	4.9	7.0
Maximum *XPU* time, s	1.3	0.8	0.6

TABLE 6.12 Inverse Relay Currents and Times for Example 6.5

Pickup Multiple	Relay *R*		Relay *H*		Relay *G*	
	I in A	*t* in s	*I* in A	*t* in s	*I* in A	*t* in s
2.0	320	6.2	640	4.0	960	4.0
2.5	400	3.6	800	2.4	1200	2.4
3.0	480	2.6	960	1.65	1440	1.65
4.0	640	1.6	1280	1.05	1920	1.05
5.0	800	1.18	1600	0.77	2400	0.77
6.0	960	1.0	1920	0.65	2880	0.65
7.0	—	—	—	—	3360	0.58
8.0	—	—	—	—	3840	0.52

■

6.6.3 Procedure for Instantaneous Relay Settings

A common practice in relaying is to utilize a separate "instantaneous" relay in addition to the usual phase relays to provide fast tripping for close-in faults. Actually, the instantaneous capability may be available as a second element in the phase relay, rather than as a completely

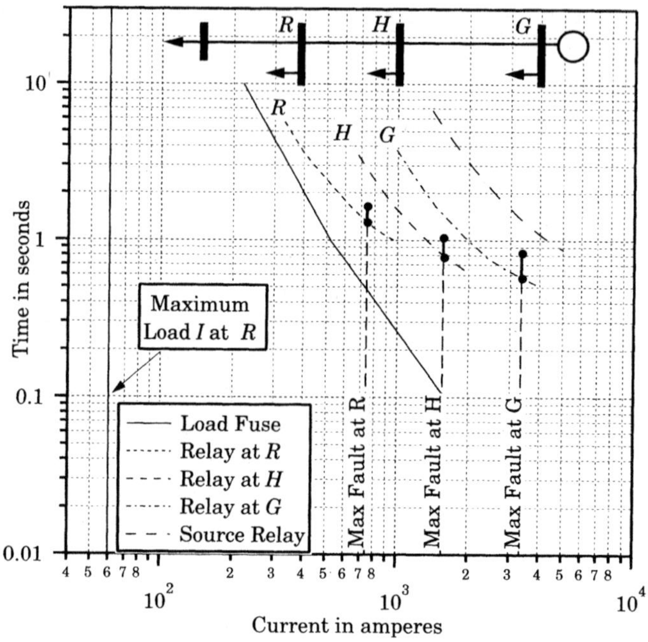

Figure 6.20 Time-current coordination for Example 6.5.

separate relay. The relay logic, then, is an OR logic, which means that the fault will be detected and tripped by either the phase relay or the instantaneous relay element.

Now, define the following currents:

$$\begin{aligned} I_{CI} &= \text{close-in fault current} \\ I_{FE} &= \text{far-end fault current} \end{aligned} \tag{6.37}$$

Then if

$$I_{CI} \geq 1.3 I_{FE} \tag{6.38}$$

the instantaneous trip relay addition will probably be worthwhile. If (6.38) is not satisfied, then there is not enough difference between close-in and remote faults, and it will be difficult to coordinate the relays so that the instantaneous relay can protect a reasonable fraction of the line length.

To probe this idea further, suppose that a relay is protecting a line where we can define the following:

$$\begin{aligned} \mathbf{E}_S &= \text{thevenin equivalent open circuit voltage} \\ \mathbf{Z}_L &= \text{impedance of the protected line} \\ \mathbf{Z}_{SR} &= \text{thevenin source impedance at the relay} \end{aligned} \tag{6.39}$$

Then the close-in and far-end currents are computed as follows.

$$\mathbf{I}_{CI} = \frac{\mathbf{E}_S}{\mathbf{Z}_{SR}} \tag{6.40}$$

$$\mathbf{I}_{FE} = \frac{\mathbf{E}_S}{\mathbf{Z}_{SR} + \mathbf{Z}_L} \tag{6.41}$$

Now define the complex constant

$$\boldsymbol{K}_{SR} = \frac{\boldsymbol{Z}_{SR}}{\boldsymbol{Z}_L} \tag{6.42}$$

Then we may compute

$$\boldsymbol{I}_{FE} = \frac{\boldsymbol{I}_{CI}}{1 + \dfrac{1}{\boldsymbol{K}_{SR}}} \tag{6.43}$$

Now let the extreme end of the reach of the instantaneous relay be defined in terms of the fraction "h" of the total line length to be protected by the instantaneous relay, that is,

$$\text{Extreme end of instantaneous reach} = h\boldsymbol{Z}_L \tag{6.44}$$

The fault current for faults at the end of the reach (ER) is given by

$$\boldsymbol{I}_{ER} = \frac{\boldsymbol{E}_S}{\boldsymbol{Z}_{SR} + h\boldsymbol{Z}_L} \tag{6.45}$$

Now define the ratio

$$K_I = \frac{|\boldsymbol{I}_{ER}|}{|\boldsymbol{I}_{FE}|} = \frac{|\boldsymbol{K}_{SR}| + 1}{|\boldsymbol{K}_{SR}| + h} \tag{6.46}$$

from which we compute the fraction protected as

$$h = \frac{|\boldsymbol{K}_{SR}|(1 - K_I) + 1}{K_I} \tag{6.47}$$

The recommended value of K_I is from 1.15 to 1.3. Usually, it is possible to have instantaneous protection for one-half of the line length, i.e., $h \geq 0.5$, as shown in Figure 6.21.

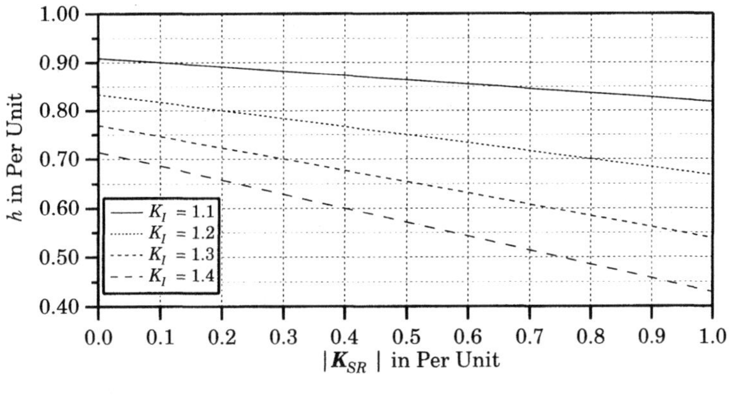

Figure 6.21 Fraction of line protected by instantaneous unit.

The criteria for actually setting the instantaneous relays is usually a function of either the maximum load current "seen" by the relay or the maximum fault current at the next downstream bus. Mathematically, these criteria may be stated as follows. Let

$$\begin{aligned} \boldsymbol{I}_{IPU} &= \text{instantaneous unit pickup current} \\ &\geq 6 \times \text{maximum load current} \\ &\geq 1.25 \times \text{max far end 3PH fault current} \end{aligned} \tag{6.48}$$

To be safe, choose the larger of these quantities. These two criteria are taken from references 8 and 12, respectively. Some utilities develop their own setting criteria, based on their experience. As a general rule, however, the above criteria should be satisfactory.

EXAMPLE 6.6

Consider the system in Figure 6.16 and Example 6.4, where the following impedances are given.

$$\begin{aligned} &\text{Source:} && \mathbf{Z}_S = 0.0 + j0.5 \text{ (max)} \\ & && \quad = 0.0 + j1.0 \text{ (min)} \\ &\text{Line } GH: && \mathbf{Z}_L = 0.247 + j0.543 = 0.597\angle 65.5^\circ \\ &\text{Line } HR: && \mathbf{Z}_L = 0.494 + j1.086 = 1.193\angle 65.5^\circ \end{aligned}$$

Compute the fraction of the line protected by the instantaneous relay element, using 1.15 as the criterion for the current ratio and determine the settings of the instantaneous relays at G and at H.

Solution

The coordination begins at the farthest downstream relay and works back upstream.

For Relay H,

$$\mathbf{Z}_{SR} = \begin{cases} 0.247 + j1.043 = 1.072\angle 76.7^\circ \text{ (max)} \\ 0.247 + j1.543 = 1.563\angle 80.9^\circ \text{ (min)} \end{cases}$$

$$\mathbf{K}_{SR} = \frac{\mathbf{Z}_{SR}}{\mathbf{Z}_L} = \begin{cases} \dfrac{1.072\angle 76.7}{1.193\angle 65.5} = 0.898\angle 11.2^\circ \text{ (max)} \\ \dfrac{1.563\angle 80.9}{1.193\angle 65.5} = 1.279\angle 15.4^\circ \text{ (min)} \end{cases}$$

The above calculations show that the angle of $\mathbf{K}_{SR}$ is usually small and can often be considered to be zero. From the above values we compute

$$h = \frac{1 - 0.15\mathbf{K}_{SR}}{1.15} = \begin{cases} 0.75 \text{ (max)} \\ 0.70 \text{ (min)} \end{cases}$$

Use $h = 0.70$ to be safe. For this value of h we may compute the actual fault current at the far end of the relay reach, as follows.

$$h\mathbf{Z}_L = 0.3458 + j0.7602$$

Then the positive sequence impedance at this value of reach is

$$\mathbf{Z}_1 = \mathbf{Z}_{SR} + h\mathbf{Z}_L = \begin{cases} 0.5928 + j1.8322 \text{ (max)} \\ 0.5928 + j2.3032 \text{ (min)} \end{cases}$$

Then the 3PH fault currents at relay H for this reach are

$$I_{3PH} = \begin{cases} 0.5193 \text{ pu} = 869 \text{ A} \quad \text{max} \\ 0.4205 \text{ pu} = 704 \text{ A} \quad \text{min} \end{cases}$$

Finally, we compute the settings and see how they compare with the computed fault currents. From the criteria of (6.48) we compute

$$\begin{aligned} &I_{Ld\,\max} = 2 \times 80 = 160 \text{ A} \\ &(1) \quad I_{IPU} = 6 \times 160 = 960 \text{ A} \\ &(2) \quad I_{IPU} = 1.25 \times 742 = 927 \text{ A} \end{aligned}$$

where the 742 A maximum fault current is taken from Table 6.9. The setting should be about 950 A, which means that the reach will be just short of $h = 0.7$. Note that the setting criteria do not use the value of computed h, but the solution is quite close.

For Relay G,

$$I_{3PH} = \begin{cases} 0.5193 \text{ pu} = 869 \text{ A} \quad \text{(max)} \\ 0.4205 \text{ pu} = 704 \text{ A} \quad \text{(min)} \end{cases}$$

$$K_{SR} = \begin{cases} 0.839 \quad \text{(max)} \\ 1.687 \quad \text{(min)} \end{cases}$$

Then

$$h = \frac{1 - 0.15\boldsymbol{K}_{SR}}{1.15} = \begin{cases} 0.76 \text{ (max)} \\ 0.65 \text{ (min)} \end{cases}$$

Use a value of $h = 0.65$ to be safe. Then we compute

$$h\boldsymbol{Z}_L = 0.1606 + j0.3530$$

and we readily compute the three phase fault currents as

$$I_{3PH} = \begin{cases} 1.1522 \text{ pu} = 1928 \text{ A max} \\ 0.7340 \text{ pu} = 1228 \text{ A min} \end{cases}$$

$$I_{Ld\,\max} = 4 \times 80 = 320 \text{ A}$$

$$(1) \quad I_{IPU} = 6 \times 320 = 1920 \text{ A}$$

$$(2) \quad I_{IPU} = 1.25 \times 1564 = 1955 \text{ A}$$

Thus, a setting of about 1940 A would give almost exactly a reach of h = 0.65. This completes the instantaneous element settings. ■

The instantaneous element also modifies the coordination of the phase relays slightly, as shown in Figure 6.22. Since the critical coordination is often at the high current end of the time-current characteristic, and the instantaneous element lowers the time at this end, it improves the coordination time interval. In tight situations, this will be very helpful.

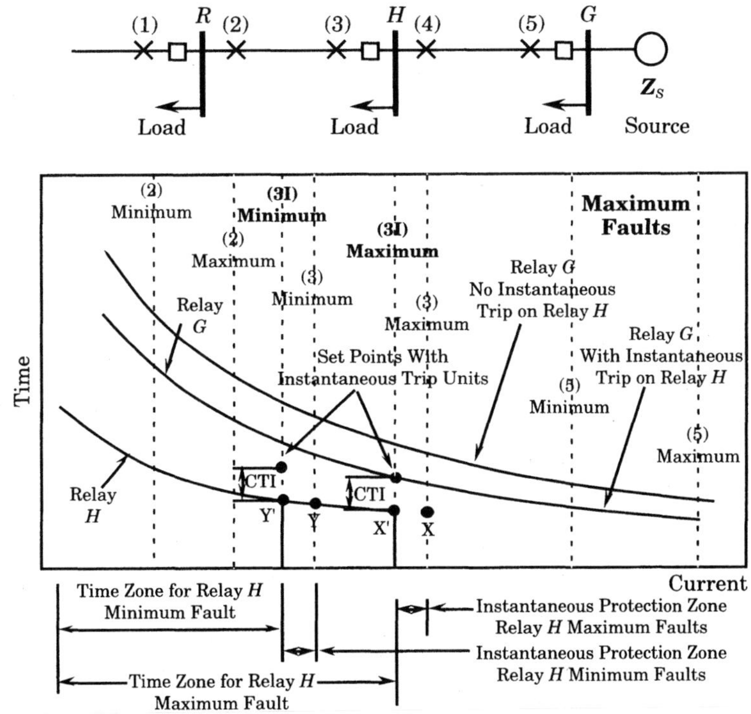

Figure 6.22 Coordination of line section G-H with instantaneous relays at G.

REFERENCES

[1] "Overcurrent Protection for Distribution Systems," Application Manual GET-1751A. General Electric Company, 1962.

[2] "Distribution System Protection and Apparatus Coordination," Line Material Company (now McGraw Edison), Bulletin 55020, 1955.

[3] American National Standard ANSI C57.91-1995. "Guide for loading mineral-oil-immersed power transformers up to and including 100 MVA with 55°C or 65°C average winding rise," American National Standards Institute, New York. 1995.

[4] Anderson, P. M., *Analysis of Faulted Power Systems*, revised ed., IEEE Press, Piscataway, NJ, 1995.

[5] Morrison, C., "A linear approach to the problem of planning new feed points into a distribution system," *IEEE Trans.*, PAS-82, December 1963, pp. 819–32.

[6] McCotter, R. E., H. A. Smollech, S. H. Ramade, and W. H. Kersting, "An Investigation of the Fundamental-Frequency Impedance of a Single-Phase Distribution Lateral," *IEEE Trans.*, PWRD-1 (1), January 1986.

[7] IEEE Committee Report, "Distribution Line Protection Practices: Industry Survey Analysis," *IEEE Trans.*, PAS-102 (10), 1983, pp. 3279–3287.

[8] General Electric Co., "Distribution Data Book," Manual GET-1008L, General Electric Co., Schenectady, NY, 1972.

[9] Westinghouse Electric Corp., "Applied Protective Relaying," Relay Instrument Division, Newark, NJ, 1976.

[10] Hathaway Overcurrent Relays, "Non-directional Time Overcurrent Relays," Prod. Bulletin 12-101, Hathaway Instruments, Inc., Denver, CO, November 1972.

[11] Asea Brown Boveri, "Microprocessor Based Distribution Protection System," Bulletin IB 7.11.1.7-1, Issue C, ABB Power Transmission, Inc., Protective Relay Division, Allentown, PA, 1988.

[12] Venkatta, S. S., M. J. Damborg, A. K. Jampala, and R. Ramaswami, "Computer Aided Relay Protection Coordination," Final Report EL-6145, Electric Power Research Institute, Palo Alto, 1988.

[13] Mason, C. R., *The Art and Science of Protective Relaying*, John Wiley & Sons, Inc., New York, 1956.

PROBLEMS

6.1 Consider the radial distribution system of Figure 6.2. Assume the distribution system is a 7200/12470Y voltage system with equivalent spacing of 50 inches. Let the distance between adjacent lateral taps be 700 ft and assume each lateral to have a length of 2000 ft. Assume the main feeder and laterals are all three-phase lines, with 3/0 ACSR mains and 1/0 ACSR laterals. Compute the total line impedance from the substation to the #8 lateral tap and to the end of that tap in ohms.

6.2 Repeat problem 6.1 but assume a more typically rural arrangement with 1 mile between laterals and each lateral 4 miles long. Let the main feeder be #2 ACSR and each lateral be #4 ACSR.

6.3 Referring to Figure 6.2, devise a sectionalizing strategy using a substation recloser and
(a) sectionalizers on all laterals
(b) sectionalizers on laterals and on main feeder

6.4 Assume that recloser B in Figure 6.2 is the 25 ampere recloser whose TC characteristic is given in Figure 3.37. If all the distribution transformers have primary fuses in the range 3T to 6T, what size fuse could be used on the laterals to coordinate?

6.5 Consider a transmission system that is completely specified by its Z matrix. It is desired to examine the fault protection on a radial line served from a step-down transformer at transmission bus K, which is a Δ-Y connected bank with grounded neutral on the low-voltage side. The transformer serves a three-phase main feeder, with a lateral feeder tapped at the main feeder at a point called F. The objective of the study will be to determine fault currents at the transformer secondary, at the point F on the main feeder, and at the end of the lateral feeder connected radially at point F.

(a) Sketch the entire system, carefully labeling all components of interest.

(b) Sketch the positive, negative, and zero-sequence networks and clearly identify the point F, as well as the zero potential bus, on each sequence network.

6.6 Using the sequence networks of the previous problem for a fault on a radial feeder out of a wye-grounded transformation, sketch the use of the sequence networks to solve a fault on the radial feeder, where the fault is identified as

(a) a line-to-line fault.

(b) a one-line-to-ground fault.

6.7 Assume that the radial distribution system of problem 6.1 is served from bus 6 of the system of Figure 6.4 through a 1000 kVA transformer with 6% reactance. Compute the available three-phase fault current at sectionalizing point #8 and at the extreme end of the #8 lateral. Use the line data from problem 6.1.

6.8 Assume that the radial distribution system of problem 6.2 is served from bus 3 of the system of Figure 6.4 through a 3000 kVA transformer with 7% reactance. Compute the available fault current at every sectionalizing point #8 and at the extreme end of the lateral at point 8.

6.9 A radial distribution system is served from bus 4 of the system of Figure 6.4 through a 10 MVA station transformer with 6% reactance. The station low-voltage rating is 12.47 kV. Compute the three-phase, line-to-line and one-line-to-ground fault currents along the main feeder at a distance of 5 miles from the station if the line conductor is 1/0 ACSR. Assume that the parameter $k_0 = 4$. Let $D_{eq} = 50"$.

6.10 Consider the fault described in problem 6.9 at the end of the 5 mile main feeder as being the fault point F for faults along a 10 kft open wye lateral of 1.0 ACSR conductor. Compute the fault currents at the end of this lateral for conditions of 0 to 30 ohms of fault arc resistance. Both line-to-line and one-line-to-ground faults are needed.

6.11 For the system described in example 6.1 and Figure 6.4, compute the two-line-to-ground fault currents for the following conditions:

(a) at the substation and with $\mathbf{Z}_F = 0$ and $\mathbf{Z}_G = 0$, 10 and 30 Ω.

(b) at the feeder end and with $\mathbf{Z}_F = 0$ and $\mathbf{Z}_G = 0$, 10 and 30 Ω.

6.12 Investigate the constant k_0 used in Table 6.4 and justify the value 4 used as an average. Hint: see Morrison [5].

6.13 Compute the line impedance $\mathbf{Z}_L$ used in calculating positive and zero sequence distribution line impedance for the three configurations of Figure P6.13 if the conductor is 3/0 ACSR and the nominal voltage is

(a) 12.74Y/7.2 kV

(b) 24.94Y/14.4 kV

(c) 34.5Y/19.92 kV

6.14 For the substation arrangement of Figure P6.14, let the following data be given:

$$V_1 = 69000 \text{ volts}$$
$$V_2 = 2400/4160 \text{ Y V}$$
$$V_3 = 7200/12470 \text{ Y V}$$
$$\text{T1:3000 kVA}, \ X = 0.07 \text{ pu}$$
$$\text{T2:1000 kVA}, \ X = 0.06 \text{ pu}$$

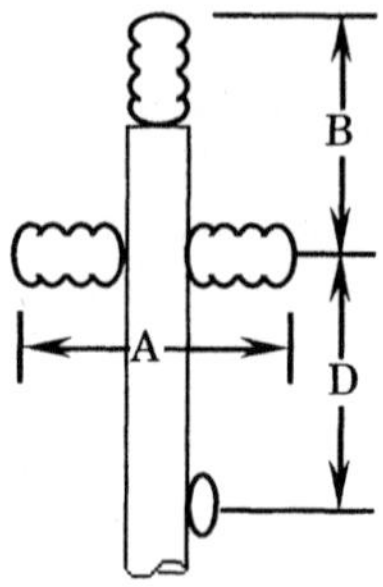

Dimensions Inches	Nominal Voltage - kV 12.47Y/7.2	24.94Y/14.4	34.5Y/19.92
A	32	40	50
B	28	44	51
D	65	72	72
Equivalent Spacing	32.0	45.4	54.4

(*a*)

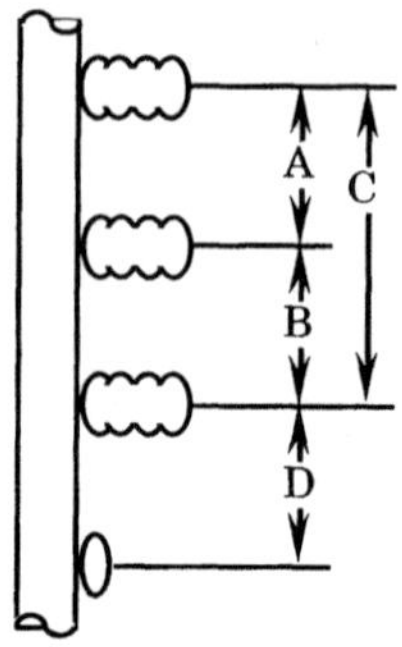

Dimensions Inches	Nominal Voltage - kV 12.47Y/7.2	24.94Y/14.4	34.5Y/19.92
A	24	40	69
B	24	40	69
C	48	80	138
D	24	72	81
Equivalent Spacing	30.2	50.4	87.0

(*b*)

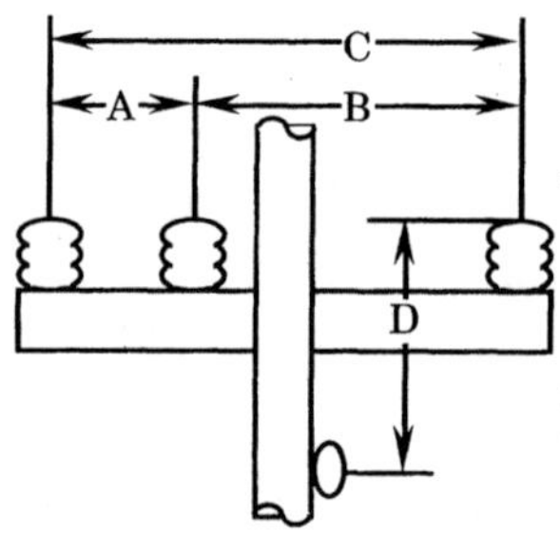

Dimensions Inches	Nominal Voltage - kV 12.47Y/7.2	24.94Y/14.4	34.5Y/19.92
A	29	41	41
B	59	71	71
C	88	112	112
D	44	44	47
Equivalent Spacing	53.2	68.9	68.9

(*c*)

Figure P6.13 Spacing of conductors for use in calculating line reactance of distribution circuits [6]. (*a*) Triangular Phase Spacing. (*b*) Vertical Phase Spacing. (*c*) Horizontal Phase Spacing.

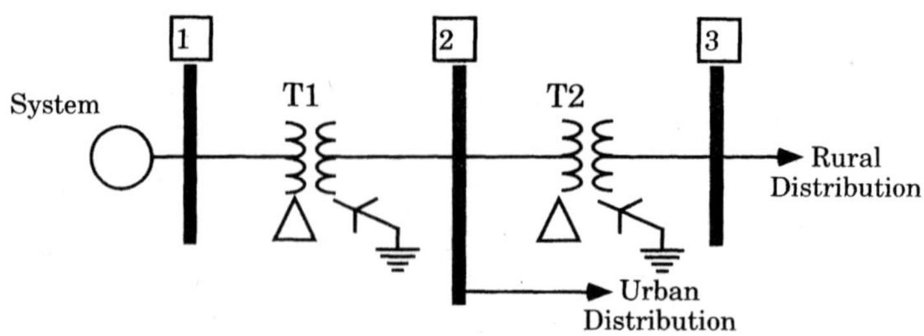

Figure P6.14 Station arrangement for problems 6.14 and 6.15.

System:
Maximum: 500 MVA
Minimum: 300 MVA

Determine the following for maximum and minimum conditions:

(a) Available three-phase fault current in amperes and in MVA at buses 1, 2, and 3.

(b) Ratings of circuit interrupting devices for application to buses 1, 2, and 3. Give a voltage, current, and interrupting rating.

Perform maximum fault calculations with zero fault impedance, but use $Z_F = 0, 5,$ and 10 ohms of arcing fault impedance for minimum conditions. Use a base MVA of 3.0.

6.15 The rural distribution system connected to bus 3 of Figure P6.14 is shown in Figure P6.15. Assume the following conductor sizes.

3ϕ-4W sections: 2/0 ACSR	Fig. P6.13 (c)	$D_{eq} = 53.2''$
2ϕ-3W sections: 1/0 ACSR	Fig. P6.13 (c)	$D_{eq} = 88.0''$
1ϕ-2W sections: 2 ACSR	Fig. P6.13 (a)	$D_{eq} = 93.0''$

Compute maximum and minimum fault conditions at all points shown in Figure P6.15. Assume a multigrounded system with $k_0 = 4$. Assume that the arcing fault resistance is 5 ohms to determine the minimum fault current.

Construct a spreadsheet to compute the following:

(a) The maximum and minimum faults at each sectionalizing point along the main feeder to find the three-phase, line-to-line, and one-line-to-ground fault currents.

(b) The maximum and minimum end-of-lateral one-line-to-ground faults for all seven laterals.

(c) The two-line-to-ground fault currents on lateral #7.

(d) List the one-line-to-ground faults on the following system map:

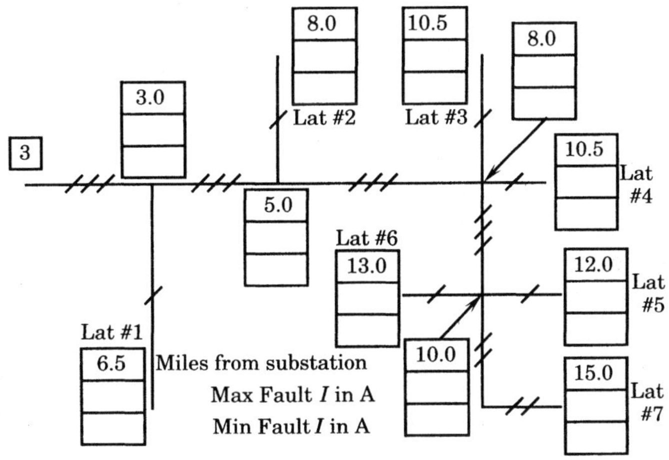

Figure P6.15 System for problem 6.16.

6.16 Repeat problem 6.15 with $Z_F = 10$ ohms for all minimum faults. Note that this should be quite easy to accomplish if the spreadsheet for the previous problem is properly constructed.

6.17 The 4.16 kV urban distribution system connected to bus 2 of Figure P6.14 is shown in Figure P6.17. Let $Z_F = 0$ in all cases. Let all three-phase main feeder sections be 3/0 all aluminum and all branches or laterals 1/0 all aluminum. Line spacings are shown in Figure P6.13(b). Assume a multigrounded system with $k_0 = 4$. Compute the fault currents at the substation and at the end of the longest laterals. Select a recloser from the tabulation of three-phase units in Table P6.17 that can be used at bus 2.

6.18 Consider the system of example 6.4 with the following tabulations of system data (Tables P6.18a and b).

TABLE P6.17 Tabulation of Recloser Ratings for problem 6.17

No.	Maximum Continuous Current Rating (A)	Interrupting Rating (A)
1	50	1250
2	100	2000
3	200	2000
4	400	4000
5	400	6000
6	560	10000
7	560	12000
8	560	16000
9	1120	16000

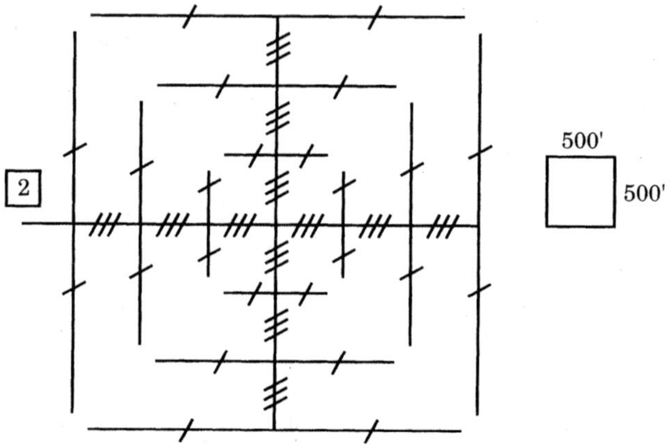

Figure P6.17 System for problem 6.17.

TABLE P6.18a Line and Load Data

Line Lengths	Loads	Line Data
$d_{GH} = 10$ mi	$S_G = 5$ MVA	336,400 ACSR
$d_{HR} = 20$ mi	$S_H = 5$ MVA	$D_{eq} = 5.0$ ft
$d_{Rend} = 5$ mi	$S_{Rend} = 5$ MVA	Temp = 50C

TABLE P6.18b System Data

System Data	Relay Data	CT Data
Max: 200 MVA	Figure 3.17	*G*: 200/5
Min: 100 MVA		*H*: 200/5
		R: 200/5

Determine the maximum and minimum fault capability at each relay location. Plot coordinating curves for the three relays using a coordination time interval (*CTI*) of 0.3 seconds. Use the inverse relay characteristic of Figure 3.17. Coordinate relay *R* with a high-voltage fuse similar to that specified in example 6.4.

6.19 Consider the line and load configuration of problem 6.18.

(a) Bus G is fed by a 10 MVA, 138-34.5 kV transformer with a reactance of 0.07 per unit. The system impedance is as follows:

$$Z_{\text{min}} = j1.0 \text{ pu}$$
$$Z_{\text{max}} = j0.8 \text{ pu}$$

(b) Repeat the calculation of problem 6.18 using the very inverse relay characteristics given in Figure 3.22. Assume that each load is 5 MVA maximum at 90% power factor and 60% load factor, and that the daily peak is 8 hours long.

(c) Repeat the calculation of (b) using the extremely inverse relay characteristic given in Figure 3.23.

6.20 Select a high-voltage power fuse that will remove the step-down transformer of the previous problem in case of a transformer fault, given the following points (Table P6.20) that approximate the minimum melting characteristic of two sizes of power fuses.

TABLE P6.20 Approximate Minimum Melting Time of Fuses

125E time (s)	10.0	5.0	1.5	0.75	0.6
125E current (A)	650	860	1500	2000	2360
100E time (s)	10.0	5.0	2.0	1.0	0.5
100E current (A)	380	500	810	1060	1500

6.21 Select a high-voltage power fuse that could be used to protect the load taps in problem 6.19. Allow 0.1 seconds coordination above minimum melting to obtain total clearing time. Typical minimum melting times for power fuses in ratings corresponding to the load currents are shown in Table P6.21.

TABLE P6.21 Typical Minimum Melting Time of Fuses

50E time (s)	2.0	1.0	0.5	0.3	0.1
50E current A	280	400	570	740	1300
40E time (s)	2.0	1.0	0.5	0.3	0.1
40E current A	210	300	440	530	950
30E time (s)	2.0	1.0	0.5	0.3	0.1
30E current A	140	200	285	375	630
25E time (s)	2.0	1.0	0.5	0.3	0.1
25E current A	120	170	235	300	510

6.22 Extend the computations for the radial transmission system of example 6.3 to include a consideration of ground faults and the setting of the ground relays. Let the zero sequence line impedance be

$$Z_{L0} = 0.0494 + j0.1629 \text{ pu/mile}$$
$$Z_{S0} = 0.0500 + j0.1500 \text{ pu (max and min)}$$

Prepare a table of the total impedances and ground fault currents, similar to Table 6.12. Do this for both the maximum and minimum conditions.

6.23 Using the table of ground fault currents computed in problem 6.22, determine the settings of the ground relays at G, H, and R, assuming that the relays have an over-current characteristic exactly like that of Figure 3.17. Assume a relay connection diagram like that of Figure 2.3. Use a *CTI* of 0.3 seconds.

6.6.4 Time-Current Characteristics for Problem Solving

Time-current characteristic curves for several types of fuses and relays are provided on pages 249 to 250. The characteristics are shown using log-log coordinates, similar to those used by fuse and relay manufacturers. All curves are exactly to scale, but are plotted in a smaller size than usually used in order to fit the restrictions of the book page size. However, these curves are adequate for accurate coordination assessment.

A blank page of log-log graph paper is also provided for the reader to copy in working the exercises. All of the curves provided can be readily copied, using a copy machine, and viewed by holding against a window glass to trace the curves of interest onto the graph paper worksheet.

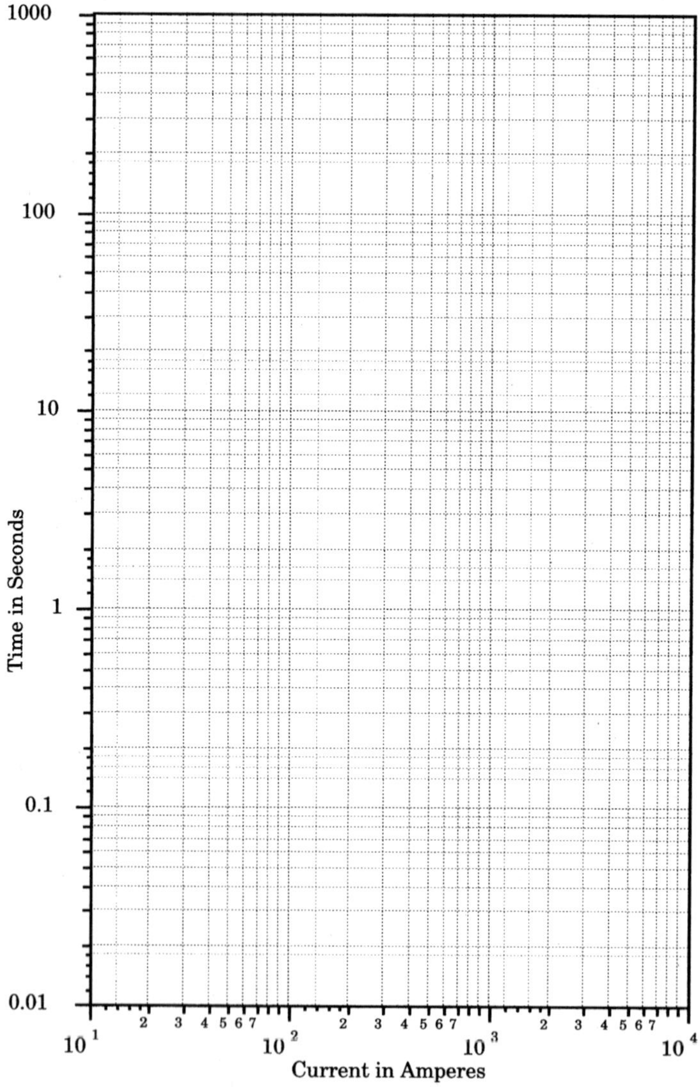

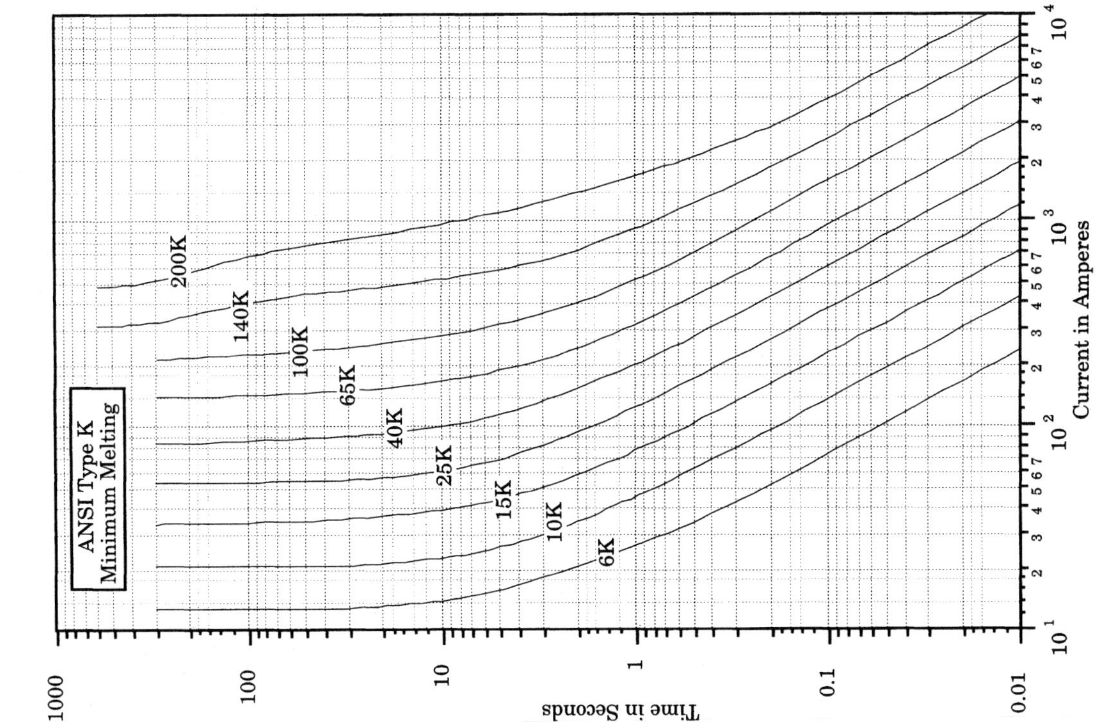
ANSI Type K
Minimum Melting
200K
140K
100K
65K
40K
25K
15K
10K
6K
Time in Seconds
1000
100
10
1
0.1
0.01
Current in Amperes
10^1
10^2
10^3
10^4

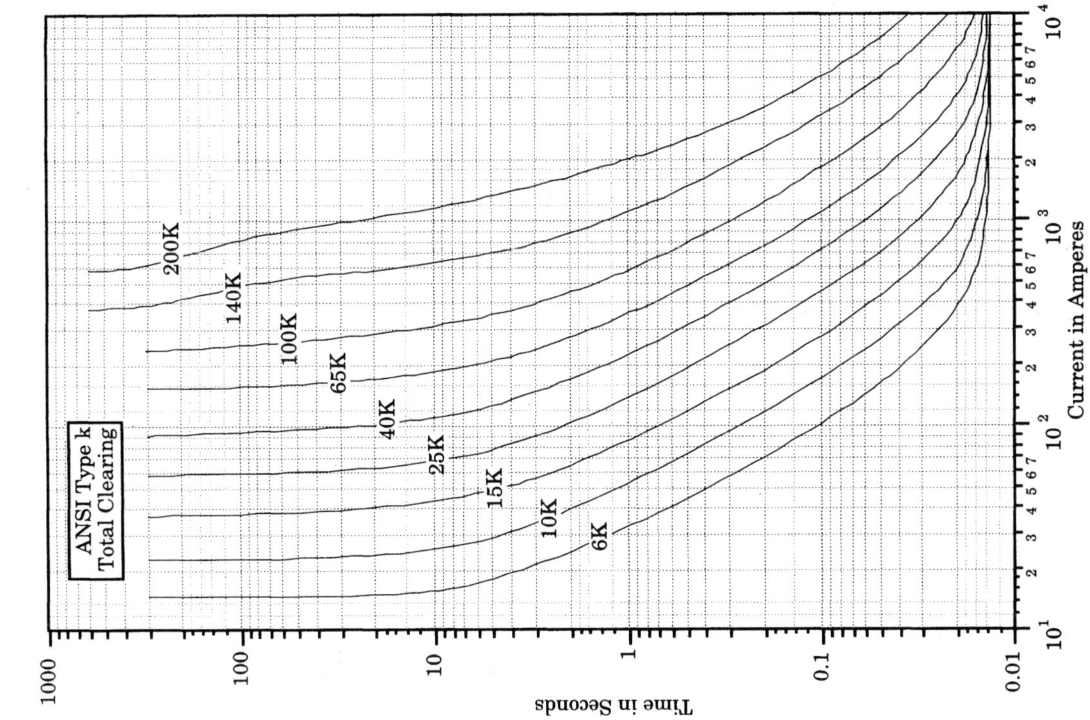
ANSI Type k
Total Clearing
200K
140K
100K
65K
40K
25K
15K
10K
6K
Time in Seconds
1000
100
10
1
0.1
0.01
Current in Amperes
10^1
10^2
10^3
10^4

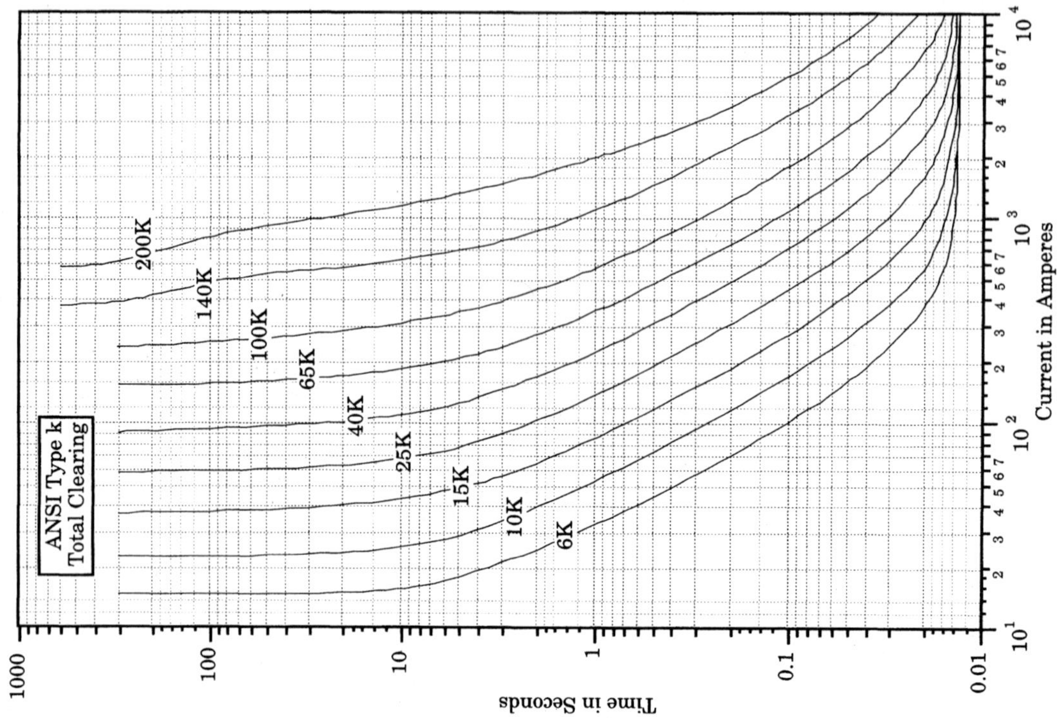
ANSI Type k
Total Clearing
200K
140K
100K
65K
40K
25K
15K
10K
6K
Time in Seconds
1000
100
10
1
0.1
0.01
Current in Amperes
10^1
10^2
10^3
10^4

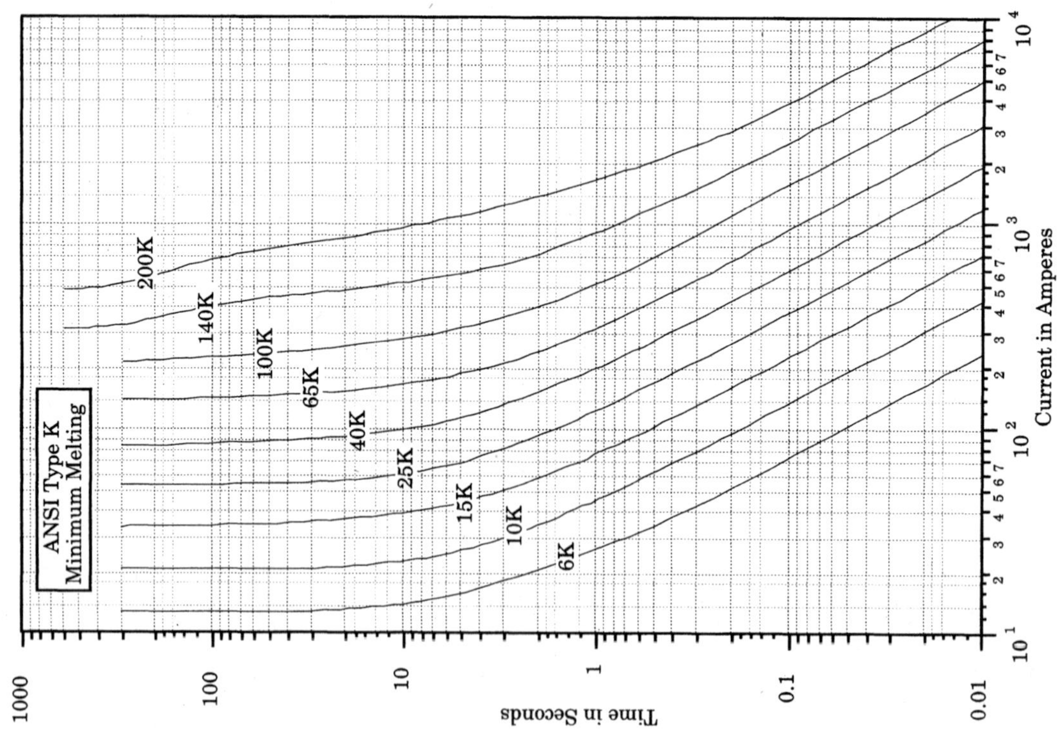
ANSI Type K
Minimum Melting
200K
140K
100K
65K
40K
25K
15K
10K
6K
Time in Seconds
1000
100
10
1
0.1
0.01
Current in Amperes
10^1
10^2
10^3
10^4

7

Introduction to Transmission Protection

This chapter presents a description of several methods that are commonly used for the protection of transmission lines. In Chapter 6, we considered the coordination of relays on radial transmission lines and the coordination between relays and other protective devices on radial systems. This chapter extends the relay coordination and application to looped transmission systems, where one may not assume a radial topology.

7.1 INTRODUCTION

Most high-voltage transmission systems are interconnected in a network system of circuit elements, usually of more than one voltage. The interconnection of many lines presents a new set of conditions on the coordination of protective devices, since the fault currents may flow to the fault point from both ends of any meshed line element. This means that overcurrent relays, which were quite adequate protective devices for radial circuits, are not generally capable of being properly coordinated for meshed transmission systems. Because of this inadequacy of overcurrent relays, other types of relays have been devised that are more selective and that have performance features that make them more applicable to the needs of high voltage transmission circuits. In this chapter, we consider three types of transmission relays, each of which has its own application advantage.

The first relay type to be considered for transmission protection is the directional overcurrent relay. This is the same overcurrent relay that has been treated previously, except it has been improved by the addition of a directional element. This added feature makes the directional overcurrent relay applicable to certain types of transmission protection problems in meshed networks.

The second relay type to be discussed is the distance relay. This relay is immune to some of the inadequacies of the overcurrent relay, and variations of distance measuring devices are widely used for transmission protection.

Third, pilot relays will be discussed and the advantages of this important class of equipment will be explored. Pilot relays add a very important and effective feature to the protective system—communication. Pilot relays at one terminal of a protected element have the advantage of being able to communicate with devices at other terminals. This provides a new dimension of information, and permits much more intelligent decision making regarding the location of a disturbance and whether the observed disturbance is one for which the protection should act.

Of the three types, directional overcurrent relays are the simplest and least expensive, but are the most difficult to apply. Overcurrent relays also have the disadvantage in changing their coordination characteristics as the network and generation sources change. Thus, these relays may require periodic readjustment. Overcurrent relays are widely used for ground protection, however, or at least for backup ground protection. Thus, the overcurrent relay is still an important device that is widely used on power systems.

Distance relays are often a first choice for replacing overcurrent relays when the overcurrent relays are found to be inadequate for an application. Distance relays are very common in applications for phase and ground protection of transmission lines, where a short time delay in clearing end-zone faults can be tolerated. Distance relays are not affected nearly as much as overcurrent devices by changing system conditions. Since distance relays essentially measure the impedance to the fault from the relay location, they are not greatly affected by the amount of generation in service. Distance relays are available in a wide variety of characteristics too, which is an advantage when seeking the best device for a given application.

Finally, pilot relays are considered as the most selective of all kinds of transmission protection. This relay concept is superior because it combines observations on the network from both ends of the protected line to provide greater selectivity and greater assurance of correct tripping. These systems cost more than those with only local intelligence, so they are usually applied on lines where tripping errors are very costly and where delay or sequential tripping is inadequate. Practically speaking, almost all EHV systems are protected by pilot relaying because of the large power transfers commonly carried on these lines, the need for fast relay action for system stability, and because of the relatively high price that is paid for protective system error on these systems. Pilot systems are also used on important lower voltage transmission lines where their application is required to achieve proper selectivity or is considered a good investment because of the importance of the protected circuit.

7.2 PROTECTION WITH OVERCURRENT RELAYS

Overcurrent relays can be applied successfully on transmission networks under certain conditions. Consider the simple network shown in Figure 7.1, where two parallel transmission lines interconnect two systems, each of which contains sources of fault current (the double-ended arrows in the figure indicate that the relays have no directional sensing capability).

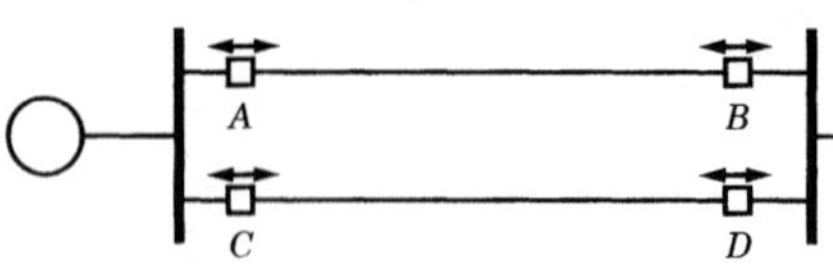

Figure 7.1 Parallel transmission lines interconnecting two systems with generating sources.

In order to coordinate the relays at the four breaker locations, it is necessary to maintain a coordinating time interval between adjacent relays. Thus, for a fault on line A-B, we reason, by inspection, that

$$\begin{aligned} t_A &< t_C \\ t_B &< t_D \end{aligned} \tag{7.1}$$

where t is the relay time for a given fault current magnitude and the subscript identifies the relay location. Similarly, if the fault is on line CD,

$$\begin{aligned} t_D &< t_B \\ t_C &< t_A \end{aligned} \tag{7.2}$$

for proper coordination. Clearly, it is not possible to satisfy both (7.1) and (7.2) simultaneously. The simple overcurrent relay does not have adequate selectivity for this application.

One way of improving the selectivity of the overcurrent relay is the addition of a directional element. This can be done by adding a wattmeter-type element, for example, that senses whether the current is within 90 degrees of being in phase with the voltage, hence establishing the positive direction of power flow. This is not particularly complex, but it does require the addition of voltage transformers in addition to the current transformers, which adds to the cost of the installation.

Overcurrent relays with current directional sensing are usually depicted as shown in Figure 7.2, where we use directional arrows to indicate the tripping direction. Using directional relays will often alleviate the problem described above, simply because the relays have directional discrimination.

Figure 7.2 Directional overcurrent relays applied to a system with fault current sources from both directions.

The criteria for deciding whether directional relays should be required are sometimes determined by the ratio of currents flowing in relays at the two ends of a line [1]. Consider a portion of a large system depicted in Figure 7.3, where we have numbered three fault locations and have defined currents in breaker A corresponding to each fault.

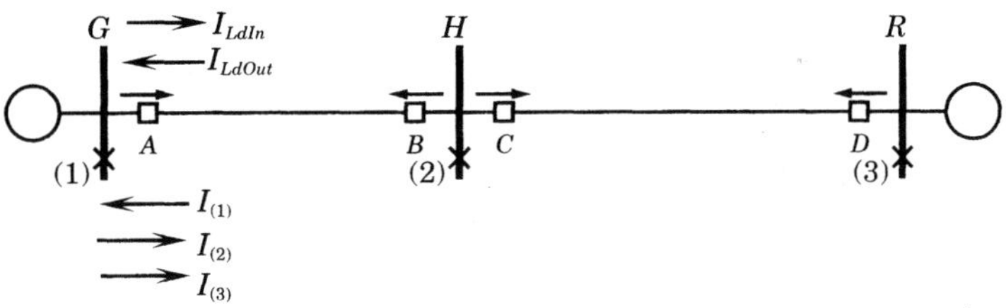

Figure 7.3 Criteria for requiring a directional overcurrent relay.

The criteria that have been developed through practice require that a directional relay be applied at A if any of the following load or fault current conditions exist (numbers in parentheses refer to fault locations) [1]:

$$\begin{aligned} &(1) \qquad I_{(1)\,\max} \geq 0.25 I_{(2)\,\min} \\ &(2) \qquad I_{(1)\,\max} \geq 0.25 I_{(3)\,\min} \\ &(3) \qquad I_{Ld\,\text{Out}} > I_{Ld\,\text{In}} \end{aligned} \tag{7.3}$$

where $I_{Ld\,\text{In}}$ and $I_{Ld\,\text{Out}}$ refer to the maximum load currents flowing in and out of the protected line at A, respectively.

Both load and fault currents are sensed at the bus G side of breaker A. The need for directional capability depends, then, on the ratio of currents seen at the breaker, where the current might be caused to flow in different directions. Fault current for fault (1) flows toward a fault "behind" the relay at A, while the faults at (2) and (3) are "in front" of the relay. If it is possible to have fault or load currents flowing to the left in the figure and those currents exceed the magnitudes noted, then directional relays are required.

If the load current from H to $G(I_{Ld\,\mathrm{Out}})$ is greater than that from G to $H(I_{Ld\,\mathrm{In}})$ the use of directional relays will permit more sensitive settings. It is assumed that relay A is also used for remote backup of relay C for faults on line HR.

We now examine two different cases of applying overcurrent directional relays to power systems, where the protected line or lines are situated such that the fault current can flow in either direction. The first case is that of a loop system, but where there is only one source of current. The second case extends the study to networks with more than one source of fault current.

7.2.1 Loops with One Current Source

First, consider the case of a looped transmission system where there is only one source of fault current, such as the system shown in Figure 7.4. The source of fault current is to the left of bus R. Therefore, it is permissible to use nondirectional overcurrent relays at 1 and 10, since a fault on bus R will not cause currents to flow in the transmission loop in either direction.

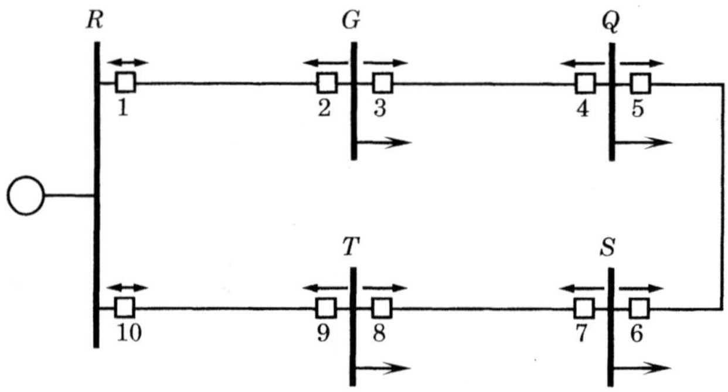

Figure 7.4 A Transmission loop circuit with only one source of fault current.

Consider that all breakers in the looped system are normally closed and that the relays are to be coordinated for this normal condition. We note that, should any breaker be normally open, then we have two radial lines, which can be coordinated as discussed in Chapter 6. We also note that all relays are directional except those at 1 and 10.

For those relays to the right of buses G and T, the normal load current could flow in either direction, depending on the relative size of the loads, and this should be taken into account in setting the minimum operating current of all relays in that area. For load that may be tapped between buses R and G, the load current could flow in either direction in relay 2, and similarly for load on line RT and relay 9. Tapped load is not shown in the figure, but should always be considered as a possible system condition.

Now, consider a fault on line RG. We have already observed that a fault on bus R causes zero fault current in line RG. As the fault location is moved to the right, the fault current in relay 2 increases monotonically until it reaches the value for a fault at bus G. Since the limiting value of current is zero, the relay at 2 can be set very fast, with a very low pickup value that is set just above the tapped load current, if such load exists. This means that the relay at 2 will be an instantaneous device and need have no intentional time delay. Obviously, the same conclusion can be reached for relay 9.

Now, as the fault is moved from R toward G, the fault current at 2 begins to increase from zero. The current in relay 1, however, is at its maximum value for a fault just to the right of the breaker 1 location, and this current decreases as the fault moves toward bus G, as shown in Figure 7.5. Since the fault current at 1 is great, this relay will be set to trip first, thereby opening the loop and creating a counterclockwise radial line from 10 through 2. This actually increases the fault current in relays 2, 4, 6, 8, and 10 (see example 7.1, below). We then coordinate these relays in exactly the same way as a radial line.

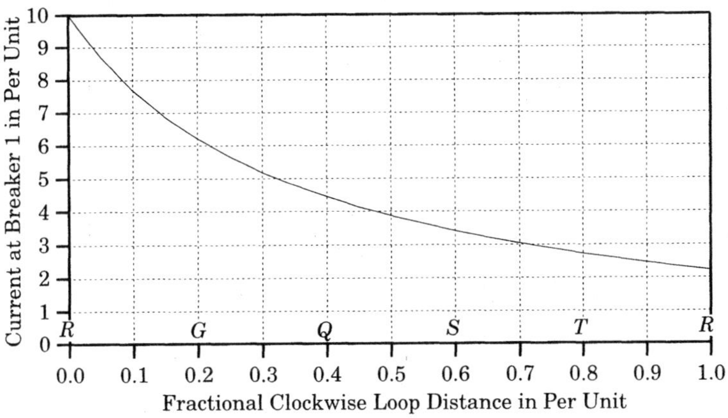

Figure 7.5 Fault current magnitude in breaker 1 at bus R as the fault is moved clockwise around the loop.

EXAMPLE 7.1

To illustrate the division of currents in the loop circuit of Figure 7.4, let us assume that all lines have an impedance of 0.1 per unit (reactance only), with 0.1 also in the generator equivalent. For this system condition, compute the total fault currents at buses G and Q, the line contributions to those faults, and the change in line flows as the nondirectional relays at 1 or 10 trip.

Solution

We easily compute the values in Table 7.1. First, consider a fault at bus G. The current in relay 1 on line RG would increase from 4.444 with the loop closed to 5.000 per unit if line RG were to open (at breaker 3). Also, relay 4 on line GQ would see an increase in current from 1.111 to 2.000 if line RG were to open (at breaker 1). We see, therefore, that both relays 1 and 4 see an increase in fault current after the loop is broken. ■

TABLE 7.1 Computed Fault Currents for Example 7.1

Faulted Bus	Total Fault X	Total Fault I	Loop Closed (Relay)	Line I at (Relay)	Loop Open at Line	Line I at (Relay)
G	0.180	5.555	$RG(1)$	4.444	GQ	$RG(1) = 5.00$
			$QG(4)$	1.111	RG	$QG(4) = 2.00$
Q	0.220	4.545	$GQ(3)$	2.727	SQ	$GQ(3) = 3.33$
			$SQ(6)$	1.818	GQ	$SQ(6) = 2.50$

For inverse-time overcurrent relays, the critical coordination is always at maximum current. Therefore, the criteria for coordination of overcurrent relays on the single-source transmission loop is as follows:

1. Consider maximum conditions.
2. Set the last relay in the loop to trip instantaneously (2 or 9).
3. Open the loop at the extreme end (2 or 9).
4. Compute maximum current at the remote bus (G or T).
5. Coordinate upstream relays as if the line were radial.
6. Repeat, traversing the loop in the opposite direction.

Note that instantaneous trip relays at 2 and 9 cannot overreach, since the fault current for faults at bus R are zero. Therefore, these relays can be set for a very low minimum pickup, even below load current of tapped load.

For establishing the minimum operating current (minimum pickup):

1. Compute maximum load and add a safety factor for cold load pickup and load growth.
2. Establish minimum pickup with the loop open.

Since maximum fault current flows at any relay location with the loop open, we coordinate the relays for this condition, i.e., with the loop open at one end. We do the same for the minimum operating currents. Coordinating at maximum current conditions will ensure the greatest *coordination time interval* (CTI) between adjacent settings. As in a radial system, the relays are coordinated in pairs. Thus, with breaker 1 open, we coordinate 4 with 2, 6 with 4, 8 with 6, and 10 with 8. A similar pattern would be set up in coordinating the loop in the opposite direction.

7.2.2 Loops with Multiple Current Sources

For looped transmission systems with multiple sources, the coordination process is much more complex. Now, all overcurrent relays must be directional and each pair of relays must be coordinated, moving around the loop in both directions. Moreover, at points where external sources are interconnected with the loop, the relays of that interconnection must also be coordinated with those within the loop.

To illustrate the problems of coordinating overcurrent relays in a system with multiple current sources, we again consider the system of Figure 7.4, but with the addition of a new source at bus Q, as shown in Figure 7.6. Considering the same loop impedances as noted previously, the impedance from R to Q through bus G is $j0.2$ and that from R to Q through

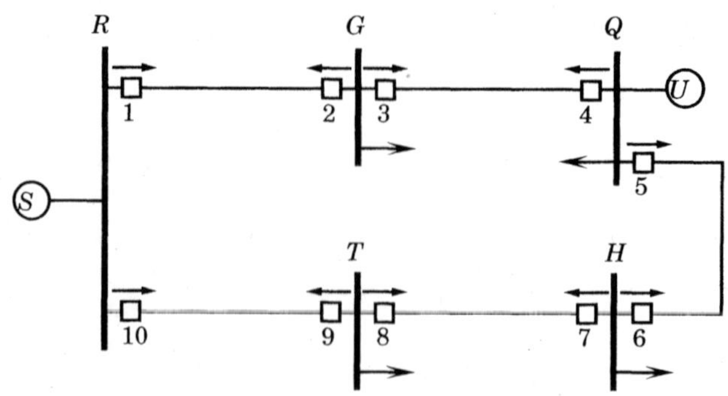

Figure 7.6 A loop system with two sources.

buses T and H is $j0.3$. Consider the currents only on the branch R-G-Q. The sources at S and at U are assumed to be equal, each having an impedance of $j0.1$ per unit. For this system, the fault currents seen at breakers 1 and 4 are shown in Figure 7.7.

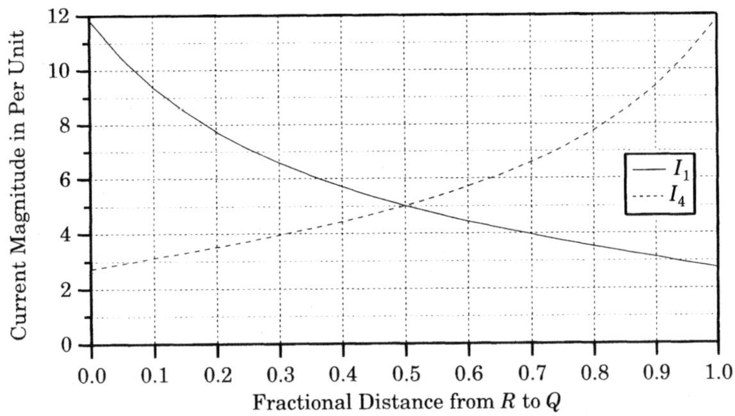

Figure 7.7 Relay currents at R and Q for faults between R and Q.

Figure 7.7 shows the fault current seen at relay locations 1 and 4 as the fault is moved from R to Q. The current magnitudes are perfectly symmetrical since the sources are identical. For faults at or near bus G it is impossible to coordinate relays 4 and 2, or 1 and 3 since they see the same fault current because bus G is equidistant from the two sources. For faults farther removed from bus G coordination is possible, since the relays closest to the fault will trip first, seeing a higher current, thereby making the fault radial from the opposite end.

There is no best starting point for coordinating multiple loop systems. Usually, the process is cut and try, and it may require several attempts to realize complete coordination. A good starting point for the process may be at the point where the largest external current source is connected, if there is one largest source. Computer programs have been prepared for this cut and try process and these programs may be used to advantage [1].

The cut and try procedure described above is time consuming and tedious. Moreover, it is not guaranteed to work at all for some systems and may dictate using distance or pilot relays to achieve correct coordination, or operate the loop with one breaker normally open. This is surely not recommended, since it negates much of the value of providing a two way feed to every load bus.

An excellent example of the coordination of a transmission loop that has two sources of fault current is presented in [1]. This system is shown in Figure 7.8 [1] and is described in example 7.2.

EXAMPLE 7.2

Consider the looped 23 kV system shown in Figure 7.8. Sources of fault current are available from a 69 kV system on the left and on the right, through identical step-down transformers. The 23 kV loop serves identical loads at the top and bottom of the figure. Thus, the system is observed to be perfectly symmetrical. Note also that all overcurrent relays in the loop, labeled A, B, ..., H, are directional. The relays looking into the 69 kV system, at I and J, are also directional.

Solution

Fault locations are noted by the letter "X" and are numbered 1 through 4. The total fault currents flowing for these faults are given in Table 7.2. All faults are three-phase faults with zero fault resistance.

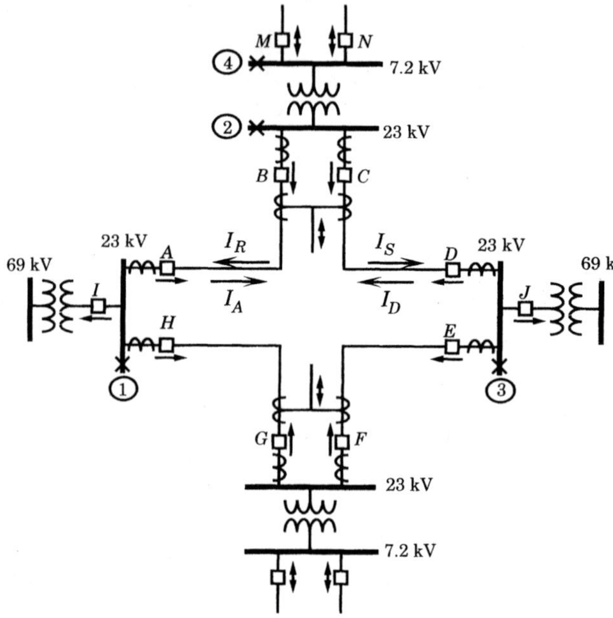

Figure 7.8 Example of relay currents for a multiple-source transmission loop [1].

TABLE 7.2 Bus Fault Current Magnitudes

Fault Location	Max/Min Bus Fault Current, amperes	Max/Min Source Contribution through Breaker *I*	Max/Min Source Contribution through Breaker *J*
1	3000/1620	2500/1200	—
2	2100/1550	—	—
3	3400/1360	—	3000/1000
4	1200/850	—	—

Line current contributions to the various bus faults are given in Table 7.3 for both maximum and minimum system conditions. For example, consider fault 1 at the 23 kV bus on the left. For this fault, breakers A, B, C, and D all see 250 A maximum and 210 A minimum. By symmetry, the same current will flow through breakers H, G, F, and E.

TABLE 7.3 Line Currents for Designated Bus Faults

	Line Currents, Max/Min			
Fault Location	I_A	I_D	I_R(1)	I_S(2)
1	—	250/210	420/380	—
2	1000/750	1100/800	—	—
3	200/180	—	—	320/300
4	550/400	650/450	—	—

(1) Current for a fault at A and Breaker A open.
(2) Current for a fault at D and Breaker D open.

Now, consider fault 1 again, but this time place the fault just to the right of breaker A. Obviously, the total fault current is exactly the same (3000 A). Now, however, breaker A will be set to open quickly because of the 2750 A flowing through that breaker, and thereby breaking the loop. The current distribution will now be different, as shown in Figure 7.8 [1].

Again, looking for fault 1 line currents, we see that the current is different than before. Summarizing, for relays B and D:

Before A opens:	$I_B = I_D = 250$ A max, 210 A min
After A opens:	$I_B = I_D = 420$ A max, 380 A min

Now, consider breaker B. Here, the fault current flowing for a bus fault (fault 2) is greater than the current flowing for a fault at the far end of the protected line (at breaker A). As a rule of thumb [1],

$$\text{if } I_2 \geq 0.25 I_{1A} \tag{7.4}$$

then a directional element is required at breaker B (the subscript numbers refer to fault numbers, the letter "A" means "after breaker A opens"). If this inequality is satisfied, the bus fault (fault 2) would be so small that it could be distinguished entirely by its current magnitude and a nondirectional relay at B would be satisfactory. Since the bus fault current is large, there is no way that proper coordination can be achieved except by the use of directional elements. It would appear, however, that a nondirectional element could be used at D. ■

The procedure for coordinating time-overcurrent relays in a multiple source network is difficult. Often, this suggests that relays be used that have better selectivity. In the remainder of this chapter, we discuss two such relay types; distance protection and unit protection.

7.3 DISTANCE PROTECTION OF LINES

It was noted in the last section that transmission protection using overcurrent relays is difficult and may not be possible, even if the overcurrent relays are directional. In such cases a more selective relay is required. The distance relay is often the preferred solution to this problem.

7.3.1 Distance Relay Characteristics

The distance relay operates on the principle of comparing the voltage and current in some way to obtain a measure of the ratio between these quantities. The method of determining this ratio varies with the relay design and manufacturers of distance relays have created a number of devices that can perform this function. We can analyze the basic idea by considering the relay at station G in Figure 7.9, where the lower line is the protected line of interest. The upper line in the figure is an equivalent impedance representing the other network connections between buses G and H. We would like the relay at A to trip for phase faults within a fractional distance h_S of the total distance between stations G and H. We call this fraction the "reach setting" of the distance relay. This means that the relay will trip only with programmed time delay for faults beyond this reach, but will trip without time delay for faults closer to bus G.

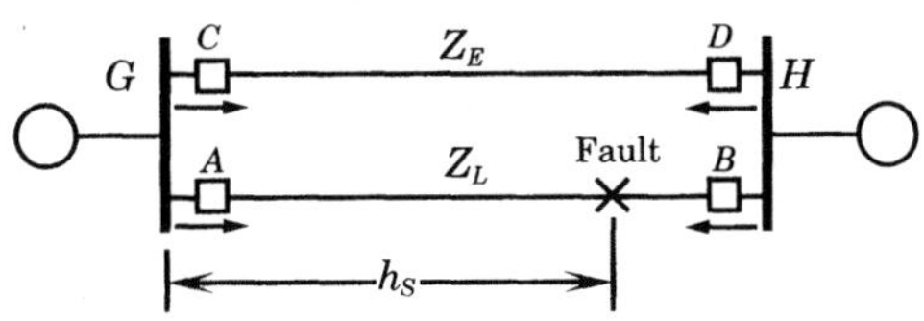

Figure 7.9 Distance protection of a line between G and H.

The reach thus becomes a "balance point" for the relay at A, or a tripping threshold that is expressed in distance units.

Now consider a solid three-phase fault at the balance point. For this condition we compute

$$\boldsymbol{V}_R = h_S \boldsymbol{Z}_L \boldsymbol{I}_{RS} \tag{7.5}$$

where $\boldsymbol{V}_R$ = voltage at the relay
$\boldsymbol{I}_{RS}$ = current at the relay
$\boldsymbol{Z}_L$ = total line impedance from G to H
h_S = fraction of line impedance to balance point

Since the relay measures the ratio of relay voltage to relay current, we compute the relay *apparent impedance*, $\boldsymbol{Z}_{RS}$, which is defined as

$$\frac{\boldsymbol{V}_R}{\boldsymbol{I}_{RS}} = \boldsymbol{Z}_{RS} = h_S \boldsymbol{Z}_L \tag{7.6}$$

For this condition, tripping would be marginal, that is, tripping may or may not occur. However, if the fault is closer to G, the voltage at the relay will be slightly lower or the current slightly higher. Then the measured impedance is

$$\boldsymbol{Z}_R \le h_S \boldsymbol{Z}_L \tag{7.7}$$

Equation 7.7 thus expresses the tripping threshold. Clearly, if the fault is farther from G than the fraction h_S, (7.7) will not be satisfied and tripping will not occur.

This method of fault detection is inherently selective. Moreover, the tripping criterion (7.7) is not dependent on system conditions and is the same for maximum and for minimum system generation and for different system line switching conditions. It is also noted that the tripping criterion requires that measurements be made only at the relay location, which is another advantage. Later, we shall discuss other relay concepts that require the transmission of system information from remote locations. This can make the relay tripping criteria more selective, but it requires more hardware and is more expensive.

The tripping criterion (7.7) describes how the relay will perform but does not give any guidance for setting the relay. Usually, for transmission line protection, the relay is set with a particular value of h_S that is not too close to unity. Clearly, with $h_S = 1$, tripping may occur for a fault just beyond the bus at H, which must be avoided. It is common practice to set h_S to about 0.8 to 0.9, thus protecting 80% to 90% of the line length. For faults beyond this balance point the far-end relay will trip first, thereby making the fault radial from G. This leads to sequential tripping and provides the desired relay coordination and selective fault detection. Sequential tripping is somewhat slower than simultaneous tripping, and this may not be considered adequate for some transmission lines.

Consider the equivalent system of Figure 7.10, where the line Z_L is protected by distance relays at points R and Q. A three-phase fault is located at a fractional distance h along the length of the line, as shown in the figure. The distance relay at R includes a fault locator that determines the fault position by measuring the total impedance seen at R. Since the length of the line is known, the relay can be set to reach a specified fraction of the total line length, say 90%. Faults beyond this fraction should not be tripped immediately, but will be cleared after the Q relay trips the breaker at Q. Faults closer than the design reach of 90% should be picked up by the relay at R. The distance measurement at relay R is a function of all the system parameters, such as the impedances in the network and the prefault angles of the Thevenin equivalent source voltages, since all of these parameters affect the current and voltage at R.

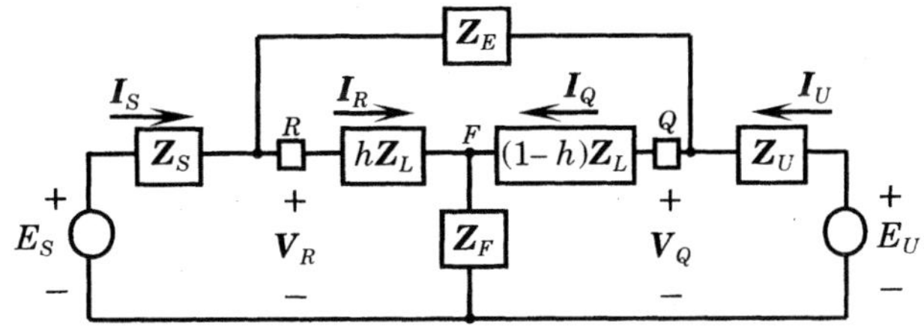

Figure 7.10 System equivalent for a protected line.

It is convenient to display the fault characteristic for protected line in the complex R-X plane. If the origin of the Z plane is taken to be the point R in the network, any impedance measured along the protected line can be represented by a line from the origin to that point. A plot of the impedance seen by the relay at R is shown in Figure 7.11 for several different system conditions. Several observations regarding this figure can be made:

1. The line impedance is shown as a heavy black line stretching from the origin to the point labeled $\mathbf{Z}_L$.
2. A mho relay characteristic passing through the origin is set to give the relay a reach of 90% of the total line length. The diameter of the mho characteristic is plotted as having exactly the same angle as the transmission line. Therefore, the line from the origin to $0.9\mathbf{Z}_L$ is the diameter of the mho circle.
3. The solid black line that lies just to the right of the line impedance is a plot of the impedance seen by the relay at R as the fault is moved from $h = 0$ to $h = 1$. This line corresponds to a fault resistance of 0.01 per unit, which can be visualized at the starting end of the line, just to the right of the origin. It is noted in Figure 7.11 that the measured distance to the fault at the relay threshold is 0.85 rather than the desired 0.9. This is due to several factors including the source impedances and the value of

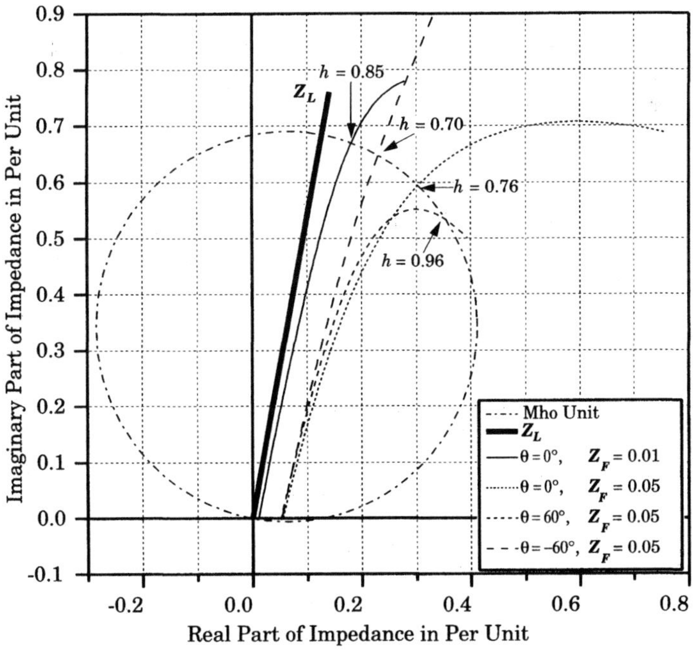

Figure 7.11 Z-Plane representation of faults on line section R-Q.

the external equivalent impedance through which a portion of the total fault current passes. These impedances are discussed below in greater detail. At this point we simply note that a fault at a reach of 0.9 will not be picked up by the relay. The common terminology is to say that the relay *underreaches* the fault.

4. The dotted line that flares far to the right represents the impedance seen by the relay for fault resistance value of 0.05, which is five times greater than that for the solid line. Again, this is easily observed at the $h = 0$ end of this line, just to the right of the origin. This line crosses the relay threshold at $h = 0.76$, which represents a greater error than that observed for the lower value of fault resistance, represented by the solid line.
5. Two other lines are plotted in Figure 7.11, each representing high values of prefault system loading. The first is represented by the dashed line that curves sharply to the right, crossing the relay threshold at $h = 0.96$. This represents a prefault condition where the voltage of the sources on the left E_S is 1.0 per unit and leading the source on the right by 60 degrees. The second dashed line, with the long dashes, represents a case where the source on the right has a voltage of $E_U = 1.0$ per unit and leading the source on the left by 60 degrees. This latter case causes the impedance seen by the relay to cross the relay threshold at a value of $h = 0.70$, which represents the largest error of the cases shown in Figure 7.11.

We conclude that the performance of the distance relay, although largely immune from differences in system conditions, is affected by the system to some extent. Several cases are compared in Table 7.4, including those shown in Figure 7.11. The value in the rightmost column, h_{cr}, is the critical value of h, i.e., the value at the relay threshold. For convenience, the calculations of Table 7.4 assume that the source impedances are purely inductive reactances and the fault impedance is purely resistive. If the fault were other than a three-phase fault, the fault impedance would include parameters from the negative and zero sequence networks.

TABLE 7.4 Distance Relay Performance Under Varying System Conditions

Case	θ_S	X_S	θ_U	X_U	Z_E	Z_F	h_{cr}
a	0	0.1	0	0.1	0.2 + j1.0	0.01 + j0	0.85
b	0	0.1	0	0.1	0.2 + j1.0	0.05 + j0	0.76
c	60°	0.1	0	0.1	0.2 + j1.0	0.05 + j0	0.96
d	0	0.1	60°	0.1	0.2 + j1.0	0.05 + j0	0.70
e	0	1.0	60°	0.1	0.2 + j1.0	0.05 + j0	0.62
f	0	0.1	60°	1.0	0.2 + j1.0	0.05 + j0	0.79
g	0	1.0	60°	0.1	0.4 + j2.0	0.05 + j0	0.60
h	0	1.0	60°	0.1	0.1 + j0.5	0.05 + j0	0.65

From these computed results, we may draw a few conclusions. First, high values of fault resistance tend to reduce the critical value of h, which is partly due to the movement of the line locus to the right as well as to the higher value of resistance. Second, a high source impedance behind the relay tends to make the relay underreach, as noted in case *e*. Third, a high value of the external impedance causes a decrease in the critical value of h, as noted in case *g*. If more than one of these conditions occurs simultaneously, as in case *g*, the error is quite pronounced. The loading of the system is also important, and the error is greater when the phase angle of the source on the right *leads* that of the source on the left. When the source on the left is leading, the relay tends to overreach, as noted by case *c*.

Several types of distance relay characteristics are available from the manufacturers, and names of the various types have come into common use. Some of these are noted in Chapter 3, Figures 3.12 and 3.13, where the characteristic shapes called impedance relay, reactance relay, mho relay, and several others are described.

Since we can represent both the transmission line and the relay characteristic as a plot in the complex Z plane, a natural extension of this concept is to plot both on the same diagram, thereby showing the protected section of line and the relay trip zone together. Examples of this idea are shown in Figure 7.12, where the protected zone of the line from G to H are shown for several different types of relay characteristic. Several comments concerning these characteristics are appropriate here.

1. The impedance relay and the modified impedance relay are inherently nondirectional and must be supervised by a directional element to prevent tripping for faults "behind" the relay, i.e., in the negative Z direction.
2. The mho Relay is inherently directional.
3. The trip zone for impedance and mho relays is the region inside the circles.
4. The reactance relay provides an exact definition of the relay reach, but will allow tripping for faults in the reverse direction. Hence this type of relay requires supervision by another characteristic. Note also that the reactance relay would likely trip for normal load currents without this added supervision.

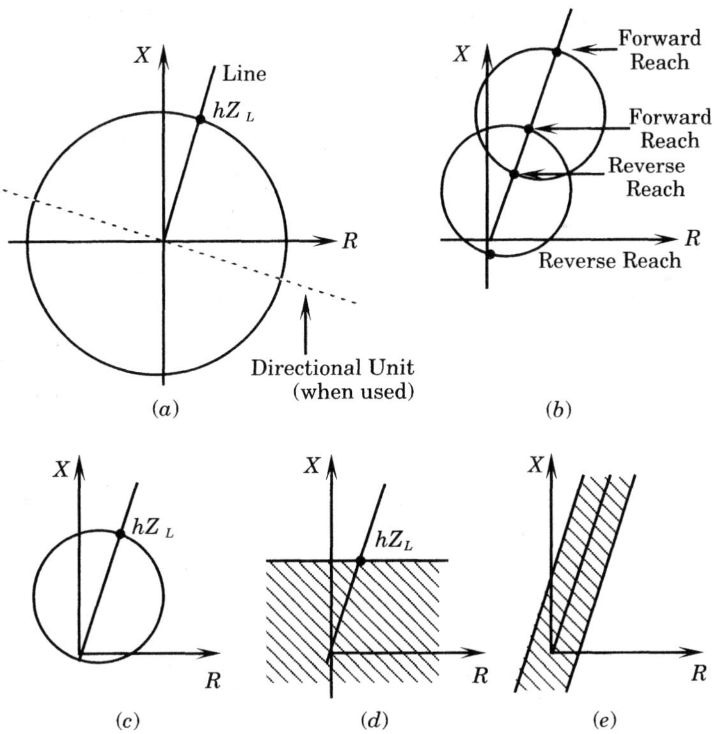

Figure 7.12 Protection of line G-H using different types of distance relay characteristics. (*a*) Impedance. (*b*) Modified impedance or offset mho. (*c*) Mho. (*d*) Reactance. (*e*) Blinders.

5. The blinders, shown in Figure 7.12(*e*), narrow the trip zone very effectively, but would permit tripping well beyond the desired relay balance point. Hence, the blinder characteristic also requires supervision by another characteristic type.

The many distance relay types offer a host of possible solutions for any given transmission protection problem. Note that the calculation of the relay setting is not difficult and that the coordination between relays, which is such a difficult problem for overcurrent relays, is almost eliminated with distance relays. There may be a coordination problem for faults near the end of a line where the fault is beyond the reach of opposite end relays. For this condition, the fault clearing time may be critical and coordination may be necessary with nearby overcurrent devices.

The clearing time of distance relays is usually very fast for any fault within the relay threshold setting. Nominally, we can consider the relay clearing time to be in the neighborhood of one cycle. The operating time characteristic of high-speed impedance relays are shown in Figure 7.13 [2]. The curve in Figure 7.13(*a*) shows a typical operating time as a function of percent of pickup impedance or threshold impedance. The Figure 7.13(*b*) shows an approximation that is often used for simplification.

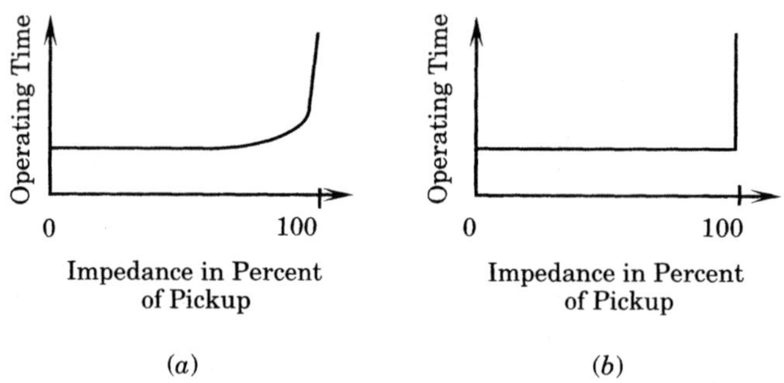

Figure 7.13 Impedance relay operating time versus percent of threshold impedance for a typical high-speed unit. (*a*) Actual characteristic. (*b*) Approximate characteristic.

7.3.2 Zoned Distance Relays

It is noted above that a distance relay will trip a faulted line in a very short time as long as the fault is within the distance threshold. For faults near the far end of the line, just beyond the threshold, the fault must be cleared by some means. This is done by providing more than one distance relaying element within the same relay package and setting the different elements to different thresholds and with different relaying times. Figure 7.14 shows a typical application. In the system illustrated, the distance relays have three zones. Zone 1 is set to protect about 90% of the line length and to operate with no intentional time delay. Zone 2 is set for 100% of the protected line plus about 50% of the shortest adjacent line, and is set to operate with time delay T_2. Zone 3 is set to reach through 100% of the impedance of two lines and 25% of the third line, and to operate the zone 3 element with time delay T_3. The application of the timer, such as T_2, provides a *coordination time interval* (CTI) for the relay operation in the end zone of each line. Similarly, T_3 is set to provide a CTI with zone 2 relays.

Figure 7.15 shows a family of zoned impedance relay characteristics in the complex Z plane. This figure also shows the impedance angle of the protected line (θ) and, perpendicular

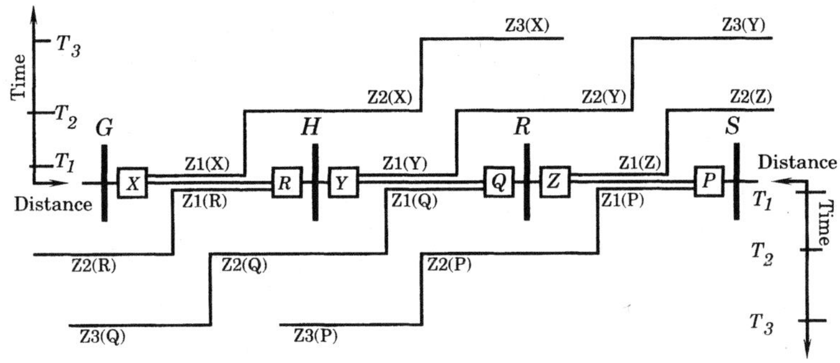

Figure 7.14 Step time and impedance zones for distance relays.

to θ, the tripping characteristic of a directional element (positive torque[1] indicates the trip direction). Such a relay has a set of relay contacts for each zone and, in addition, a set of contacts for the time delay for T_2 and T_3. Note that the time T_1 for zone 1 is set to zero.

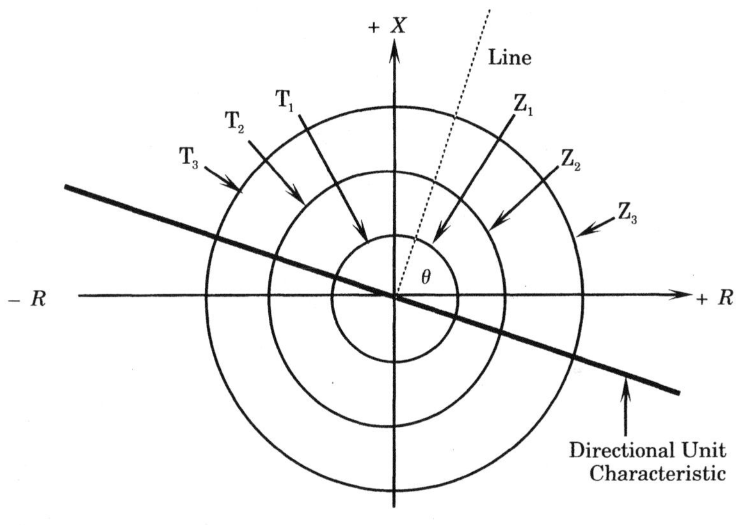

Figure 7.15 Operating characteristics of zoned impedance relays.

The trip circuit connection for a three-zone distance relay is shown in Figure 7.16 [2]. The directional unit supervises all tripping in this circuit. For impedance-type elements, a fault in a given zone will pick up the appropriate Z contacts. Tripping will not occur, however, unless the fault is in the tripping direction. Note that zone 3 has the longest reach so that this element will pick up for any fault from the relay location to the end of the zone 3 reach. This closes the contacts marked Z_3 in Figure 7.16, which starts the timer. If the fault is also within the reach of the zone 2 element, then contact Z_2 also closes, but the zone 2 relay will not trip until the time reaches T_2. If the fault is inside the reach of Z_1, tripping is performed immediately.

[1]The term *torque* refers to the direction perpendicular to the directional unit characteristic. Electromechanical relays are designed to have maximum torque in this direction. The term torque is used for static relays as well, however, although there is no torque developed in the strict sense of the word.

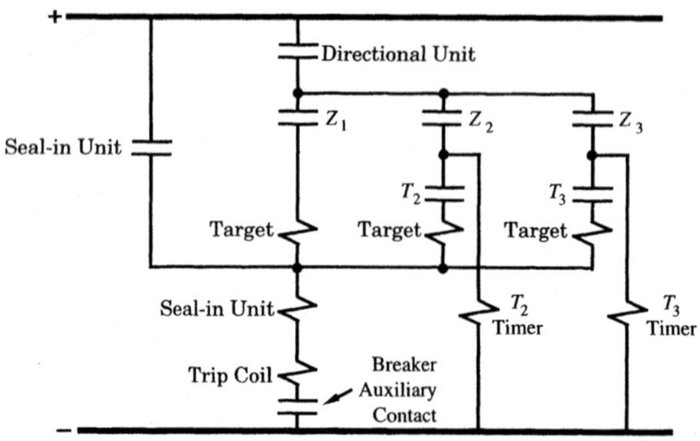

Figure 7.16 Schematic trip circuit connections for the impedance relays of Figure 7.15.

For modified impedance relays the three zones might be arranged as shown in Figure 7.17 [2]. This is basically the same scheme as that discussed previously, except for the location of the circles in the Z plane. Again, note the directional element.

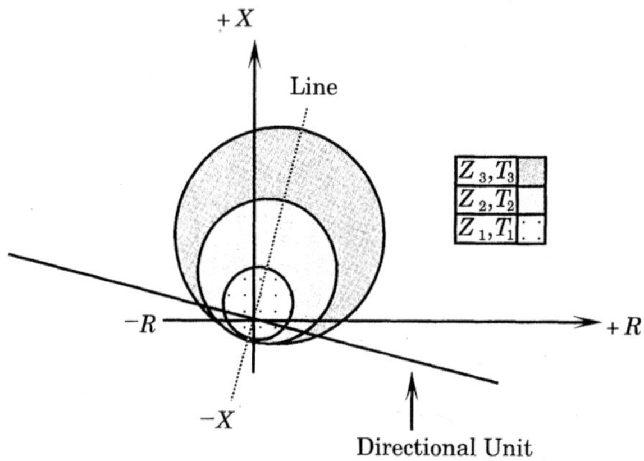

Figure 7.17 Operating characteristics of zoned modified impedance relays.

Finally, for mho-type distance relays, a three-zone relay characteristic is shown in Figure 7.18. This relay is inherently directional, hence no special directional element is required. Note, however, that the trip zone extends far to the right of the X axis, which may come

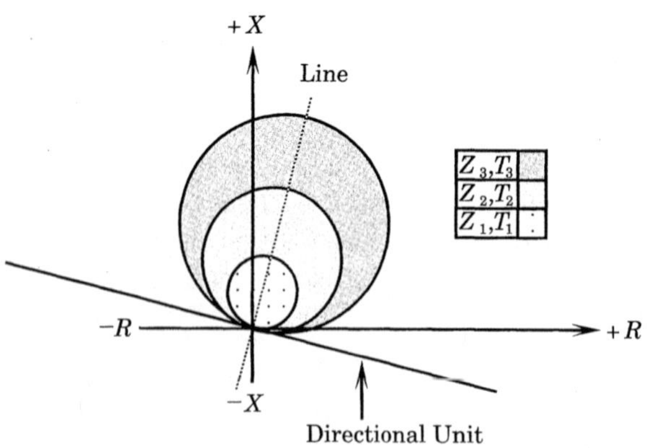

Figure 7.18 Operating characteristics of zoned mho relays.

close to a region representing normal load conditions. If this is the case, the illustrated relay characteristic may have to be supervised by blinders.

7.3.3 Effect of Fault Resistance

The impedance of a typical transmission line is highly inductive. Typical impedance phase angles will be in the range of 65–85 degrees, depending on the conductor size, spacing, and bundling. This means that the line impedance characteristic in the complex Z plane, which is inclined at the impedance angle, will usually be oriented as shown in the previous figures. This impedance, however, does not include the impedance of the fault itself, which is typically an arcing resistance.

Warrington [3] gives the following formula for the resistance of an arcing fault.

$$R_{\text{arc}} = \frac{8750(s + ut)}{I^{1.4}}\ \Omega \tag{7.8}$$

where s = conductor spacing (ft)
u = wind velocity (mi/hr)
t = time (s)
I = rms fault current (A)
R_{arc} = fault arc resistance (Ω)

The effect of the arc resistance on distance relaying is to move the end-of-reach impedance to the right in the Z plane, due to the addition of the arc resistance. This effect is shown in Figure 7.19. Note that the arc resistance is usually more pronounced at the far end of the reach because the current is smaller at the greater distance and current appears in the denominator of (7.8). An example illustrates the point.

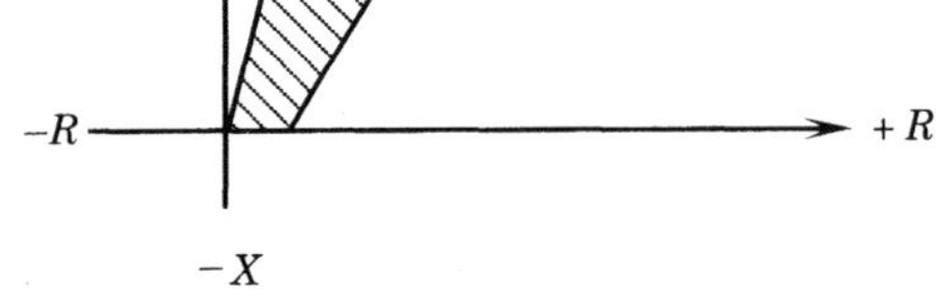

Figure 7.19 Effect of arc resistance on impedance seen by the relay.

EXAMPLE 7.3

Consider a radial 69 kV line 30 miles long with an equivalent spacing of 10 ft. Let the total impedance of this line and the source be

$$Z_L = 11 + j22\ \Omega$$
$$Z_S = 0 + j25\ \Omega$$
$$\mathbf{Z}_{\text{total}} = 11 + j47\ \Omega = 48.27\angle 76.83^\circ$$

1. *End of line fault*

$$I = \frac{69{,}000}{48.3\sqrt{3}} = 825.3 \text{ A}$$

$$R_{\text{arc}} = \frac{8750(10)}{(825.3)^{1.4}} = 7.2\ \Omega$$

2. *Close-in fault*

$$I = \frac{69{,}000}{25.0\sqrt{3}} = 1593.5 \text{ A}$$

$$R_{\text{arc}} = \frac{8750(10)}{(1593.5)^{1.4}} = 2.9\ \Omega$$

In this case the arc resistance is about 2.5 times greater at the far end of the line due to the difference in fault current. For very long lines the difference can be great, especially if the source impedance is small. ■

Clearly, the reach of the distance relay is reduced because of the arc resistance. One might be willing to accept a small reduction in the reach of zone 1, the instantaneous zone. However, it would not be good practice to have zone 2 fail to reach beyond the far end of the protected line, because this would mean that line tripping would be delayed for time T_3, which is considered unacceptable. This should be checked for any given relay application. Warrington suggests the following method of checking this reduction in reach [4]. Figure 7.20 shows a typical mho unit protecting a transmission line of impedance Z with the circle illustrated being that of zone 2. The reach of zone 2 is set as follows.

$$\text{Reach of zone 2} = Z_L + K_Z = d \tag{7.9}$$

This reach corresponds to the diameter, d, of the circle. We may also compute, from Figure 7.20(a),

$$d^2 = a^2 + b^2 \tag{7.10}$$

where, from the law of cosines

$$a^2 = R_{\text{arc}}^2 + K_Z^2 - 2R_{\text{arc}}K_Z\cos\phi \tag{7.11}$$

and, from Figure 7.20

$$b^2 = (Z_L\cos\phi + R_{\text{arc}})^2 + (Z_L\sin\phi)^2 \tag{7.12}$$

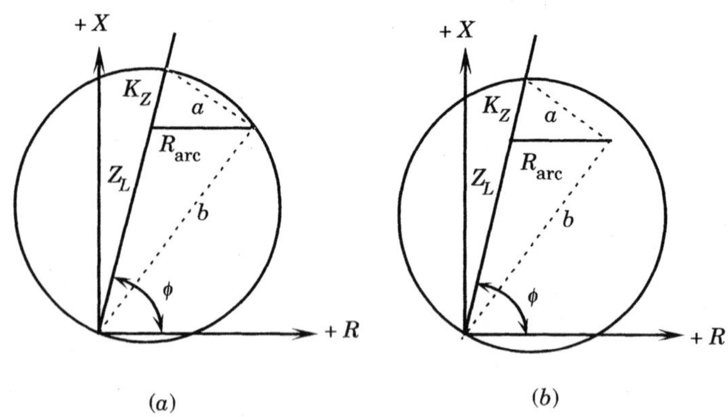

Figure 7.20 Loss of reach due to arc resistance with a mho relay characteristic. (a) Normal orientation. (b) Circle rotated.

Solving (7.10) for R_{arc}, we compute

$$R_{\text{arc}} = \frac{Z_L}{2}\left[\sqrt{\left(1 - \frac{K_Z}{Z_L}\right)^2 \cos^2\phi + \frac{4K_Z^2}{Z_L^2}} \mp \left(1 - \frac{K_Z}{Z_L}\right)\cos\phi\right] \tag{7.13}$$

This value of arc resistance is the maximum value that will ensure fault clearing time T_2 or less for all line faults on the protected line. For example, with

$$K_Z = 0.2 \text{ and } \phi = 60^\circ \tag{7.14}$$

then

$$R_{\text{arc}} = 0.29 Z_L \tag{7.15}$$

or

$$Z_L = 3.45 R_{\text{arc}} \tag{7.16}$$

One method of correcting a difficult situation is to rotate the major diameter of the mho characteristic in the clockwise direction, as shown in Figure 7.20(*b*).

The arc resistance on a line that is supplied from both ends is not as simple to compute as that for a simple radial case. For the more general case, consider a circuit where there is a source to both the left and right of the faulted line. Let the total impedance, including source impedance, be defined as follows

$$\begin{aligned} Z_1 &= \text{total } Z \text{ to the left of the fault} \\ Z_2 &= \text{total } Z \text{ to the right of the fault} \end{aligned} \tag{7.17}$$

For a fault along the protected line, the total fault current can be defined as

$$\text{total fault current} = I_1 + I_2 \tag{7.18}$$

$$\begin{aligned} \text{where } I_1 &= \text{contribution from the left source} \\ I_2 &= \text{contribution from the right source} \end{aligned} \tag{7.19}$$

Then one can compute the total impedance seen by the relay on the left as

$$\mathbf{Z}_R = \frac{\mathbf{V}_R}{\mathbf{I}_1} = \mathbf{Z}_L + \left(\frac{\mathbf{Z}_1}{\mathbf{Z}_2} + 1\right) R_{\text{arc}} \tag{7.20}$$

$$\begin{aligned} \text{where } \mathbf{Z}_L &= \text{actual line impedance from relay to fault} \\ \mathbf{Z}_1 &= \mathbf{Z}_L + \mathbf{Z}_S \end{aligned}$$

Thus, the apparent arc resistance is increased from its actual value to that corresponding to the second term of (7.20). Since the source impedance depends on the generation level, the ratio of impedances in (7.20) will depend on the system conditions, and the apparent arc resistance will be greatest for faults close to the relay. Note also that the apparent arc resistance may have an imaginary part since the impedance ratio of the second term of (7.20) may have a phase angle different from zero.

7.3.4 Summary of Distance Relay Concepts

The introductory analysis of distance relays presented in this section has considered only the three-phase fault case, which is the simplest case. Although this analysis demonstrates the concept of distance measurement, it is readily apparent that the problem is not always as simple as the three-phase case. As noted in (7.4), the impedance seem by the relay for the three-phase fault is

$$Z_R = hZ_L \tag{7.21}$$

Suppose, however, the fault is a line-to-line fault located at the same distance from the relay. Now the relay sees a total impedance of

$$Z_R = 2hZ_L \tag{7.22}$$

Clearly, this is not adequate selectivity. The methods for compensating the relay to correctly discriminate for different types of faults is presented in Chapter 11, which treats the subject of distance relaying in greater detail.

The elementary concepts of distance protection have illustrated the benefit of viewing the relay characteristic in the R-X plane. This useful concept will be developed more fully in the next chapter.

7.4 UNIT PROTECTION

This chapter has illustrated the difficulty of protecting transmission systems using overcurrent relays, even those with directional sensing. These relays are too sensitive to the system condition and the calculation of graded settings is not always possible. The concept is not robust and works only under special conditions.

Distance relaying is a considerable improvement over overcurrent relaying. The distance relay is less sensitive to the system condition and distance measurement is a logical and useful concept in protection. However, this concept also has weaknesses. It is not possible to protect the entire circuit by distance relaying because of the ambiguity of fault location in the neighborhood of the far end of the line. To overcome this deficiency, it is necessary to pull back the reach setting of the protected zone by 10 to 20% of the line length in order to be sure that a fault detected in the trip zone is truly on the protected line and not on some other line. If the protected circuit has a tap at some point along the line, then distance protection becomes quite difficult and additional compromises must be made. Moreover, distance protection for all types of faults makes this technique very difficult and the protection calculations are tedious.

The final method of transmission protection that will be addressed is the concept of "unit protection" of the transmission line [5]. With this concept, the protected line is defined by its protective zone, as shown in Figure 7.21. The zone of protection includes everything inside the current transformers where current measurements are made for the protected component.

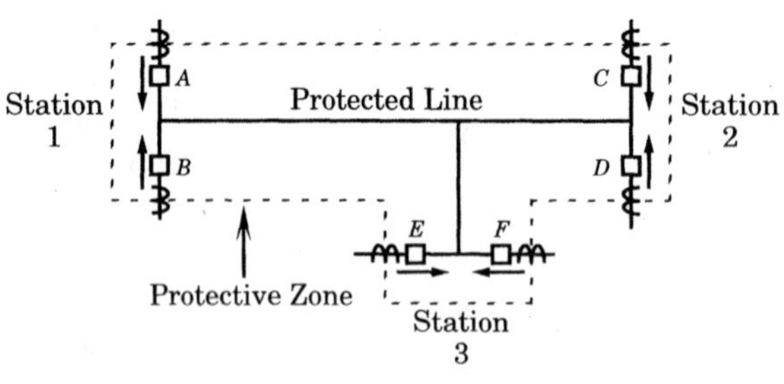

Figure 7.21 A protected line and its protective zone.

For the case illustrated in Figure 7.21, the line has three terminals, but it can have any number of terminals in the most general case. All faults that occur on the protected line are monitored at the current transformers that surround the protected zone. In many cases, measurements of voltage at the terminals may also be used, but it is the current transformers that define the protective zone boundaries.

Figure 7.21 shows each terminal of the protected line to have two circuit breakers. This is a common arrangement, as noted for the station configurations in Chapter 3. Indeed, for high-voltage stations, this is probably the most common arrangement. To clear a fault on the protected line in Figure 7.21, six circuit breakers, A through F, must all clear successfully. If even one breaker fails to open, the fault is still attached to the system and must be cleared by backup relays or breaker failure relays on all buses or circuits connected to the failed breaker.

The system of Figure 7.21 is difficult to protect using distance relays. This subject is explored in Chapter 14. Using overcurrent relays, the task would be almost impossible, although it may work for a special system configuration. Unit protection, however, can easily protect the line of Figure 7.21. Unit protection constructs a replica of the protected line and can readily determine if there is more net current entering the line than leaving it through its terminals. If this is the case, it can only be due to a *ground fault*, i.e., a fault that directs some of the current to ground.

To construct a replica, one must first define the positive direction of current flow. We arbitrarily designate current entering the protected line to be positive, and that leaving to be negative. The positive direction of current flow is shown in Figure 7.21. It is noted that distance relaying is not a unit system, since there is no measurement taken at the remote distance zone boundary.

A replica for the three-terminal line of Figure 7.21 is shown in Figure 7.22. This is a circulating current differential scheme. Note that the secondary CT circuit is an exact replica of the power transmission circuit. For clarity, the power transmission line is shown by a heavy line and the CT secondary by a lighter line. The CT return circuit is shown dashed, so that the lines and connections are easier to follow. There is no significance to the dashed lines except to clarify the current paths.

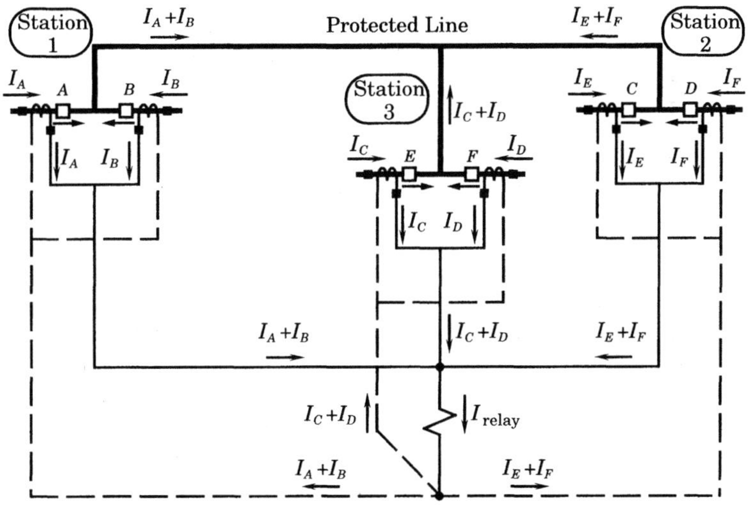

Figure 7.22 Unit tripping scheme for a three-terminal line.

When there is no fault on the transmission lines, the currents in the three terminals add to zero at their connection node. This is also true of the CT secondary circuit. If the primary currents do not add to zero due to a fault, the secondary currents will not add to zero and relay current will flow.

The circulating current replica is only one of several arrangements that can be used. Another arrangement connects the CT secondaries in opposition so that, under normal conditions, no current flows in the replica circuit under normal conditions.

The circulating current unit is a simple concept, but it represents a powerful tool for the protection of circuits where other methods are difficult to coordinate. This type of protection can be applied to a protected element with any number of terminals.

The way in which the circulating current system has been drawn implies a wire pilot network, where pilot wires are connected to each of the terminals. This makes the unit protective system expensive for long lines, but it is often used for important lines in urban areas, where the lines are relatively short. Other means of current comparison are possible. Currents have both magnitude and phase. Measurements of the magnitude and phase can be made at each terminal and this information transmitted to one terminal, where the comparison of the currents is made. It is not a simple matter to transmit all of this information over pilot channels. It is possible to transmit only the magnitudes, or only the phases of the currents. Magnitude comparison alone is usually not considered an ideal method of discrimination, but current direction of phase is adequate, at least on simple two-terminal protected elements.

Pilot systems can make use of any method of transmitting information. The wire pilot system uses a telephone type of wired network. Other systems use power line carrier, microwave, or fiber-optic cables to transmit the information. If the utility does not care to invest in these circuits, they can often be leased from a telephone company or other common carrier.

7.5 GROUND FAULT PROTECTION

An important aspect of transmission line protection is related to the fast detection and clearing of ground faults on transmission systems that have grounded neutrals. In the protection of transmission lines, ground faults are given special treatment. Ground faults are detected using different relays than those used for phase faults, although it is possible that phase relays may detect and properly clear ground faults. Ground relays, however, take advantage of unique features of the power system that make it possible to detect grounded conditions very quickly. These special aspects of ground fault relaying are considered in this section.

7.5.1 Importance of Ground Fault Protection

Most high-voltage and extra-high voltage transmission lines are grounded neutral transmission systems, either solidly grounded or grounded through a resistance or a reactance. It has been estimated that, on these high-voltage systems, over 90% of all transmission line faults are ground faults [6], [7]. It has been observed by one protection engineer that, on 500 kV transmission lines, one-line-to-ground faults "predominate to the extent that on many well designed circuits, no other type of fault has ever occurred, even after years of service" [8]. It may be that some faults involve phase-to-phase as well as ground short circuits, but the ground relays pick up these faults before the phase relays. On the system referenced, the ground relays are applied on the basis of two principles [8]:

1. Install only those relays that are required to properly protect the line.
2. Provide redundancy in the form of two completely independent relay schemes at each line terminal.

The first principle refers to the dependability of the installed systems to properly perform correct detection and tripping to clear the fault, without unnecessary trips, and with *all necessary speed* [8]. This means that the relays are not set to operate at the fastest possible speed, but as fast as is reasonably possible following detection and analysis of the observed system condition. The redundancy principle guards against the unobserved failure of one relay system by having a second system installed that is fully capable of performing the ground relaying function. These principles, or similar ones, are followed by many utilities.

Because of the high incidence of ground faults, it is important that transmission protection include a well-designed ground relaying system that embraces the two basic principles stated above. The design and calibration of ground fault relays is different, in many respects, from phase relays. We now consider some of these differences and discuss ways of taking advantage of the unique characteristics of ground faults on transmission lines.

7.5.2 Unique Characteristics of Ground Faults

It is assumed here that the transmission has multiple grounding points at wye connected transformer neutrals, located throughout the system. When this condition is satisfied, any arcing fault between a phase conductor and the ground will be supplied by zero-sequence currents originating in the neutral connection of the high-voltage transformer banks. We often refer to these neutral connections as the "sources" of ground current, since very little current would flow to the ground fault if there were no grounded neutrals to provide a complete circuit for the fault current. When there are multiple ground sources, the current flowing to the ground may be very large.

Any current flowing to the ground contains zero-sequence components and, under grounded conditions, a zero-sequence voltage will be measured at any nearby relay installation. Negative-sequence currents and voltages will also be observed, and these are sometimes used by the protective system. However, most ground relay systems depend on detecting zero-sequence currents, for this is a sure sign of an abnormal system condition. No significant zero-sequence currents flow during normal operation of the power system, with those that do appear being the result of the unbalance in the operating condition of the three phases. These unbalanced currents are very small compared to fault currents, so it is a good approximation to think of the normal power system as being free of zero-sequence voltages or currents. This is the first principle of ground fault relaying, namely, that a unique type of current exists during a ground fault and the relay needs only to be designed to detect the zero-sequence current in order to make positive identification of a ground fault.

Zero-sequence currents are confronted by zero-sequence impedances that depend on the structure of the power system. This structure does not change based on the loading of the power system, and changes only when switching occurs. Therefore, except for occasional switching, the zero-sequence impedances are almost constants. The zero-sequence impedance is affected by the generation and will change slightly as generators are added or removed. However, the line impedances are more important than the generator impedances for most fault currents. This situation is quite different from positive-sequence currents, which fluctuate with the loadings of the lines as they respond to system load and generation changes. This is the second principle of ground relaying, viz., that the impedance seen by the zero-sequence fault currents are nearly constant from maximum load to minimum load conditions.

Another characteristic of the zero-sequence network is the magnitude of the impedance of the transmission lines. Zero-sequence line impedance is two to six times greater than positive-sequence line impedance. This means that, over the length of a transmission line, there will be a large difference in impedance seen by the fault current as the fault is moved from one end of the line to the other. Figure 7.23 shows the ground fault currents for a 76 mile, 230 kV transmission line [8]. The source behind the high-voltage bus labeled Bonaire is strong, delivering 21,000 amperes to a Bonaire bus fault. However, if the fault is moved to the North Tifton bus, 76 miles away, the current delivered from the Bonaire source is only 700 amperes. A similar comparison of fault currents can be made from the source behind the North Tifton bus, which delivers 19,600 amperes to a bus fault, but only 815 amperes to a fault at the end of the transmission line.

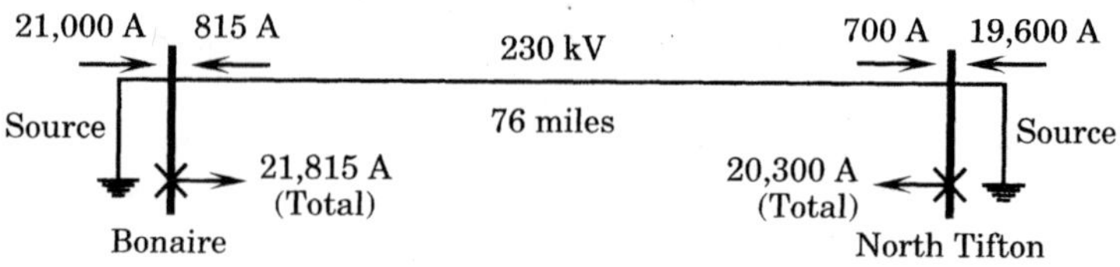

Figure 7.23 230 kV fault currents with strong current sources [8].

For the example cited in Figure 7.23, and many other similar situations, the current supplied to an end-of-line single-line-to-ground fault will be about the same at both terminals. Moreover, this is true for a wide range of system conditions [8]. The reason for this similarity is high zero-sequence impedance of the relatively long line. It should be noted that this may not be true if the line is mutually coupled with another nearby transmission line. There are two important points to observe here. First, there is a large difference in the fault current as the fault is moved from the relay location to the far end of the line. Second, the source impedances are usually small compared to the line impedance, hence the far-end fault currents are about the same at both ends.

Another requirement of ground faults is the need to determine the direction of the fault current. For a radial line, there is no problem in determining the direction of current flow, but this is not true in other parts of a power system. For this reason, many ground relays are directional relays. In order to get a sense of zero-sequence current direction, it is necessary to have a reference current or voltage against which the actual fault current can be compared. This type of comparison is called *polarization* [1], [9], [10]. By means of polarization, it is possible for the ground relay to determine if the fault is ahead or behind the relay location, giving the relay a measurement of the current direction as well as its magnitude. This topic is further explored in the next section.

7.5.3 Polarization of Ground Relays

In order to determine the direction of a fault current from the ground relay location it is necessary to provide the relay with a reference or polarizing quantity against which the zero-sequence line (fault) current can be compared, giving the relay a sense of the current direction. The polarizing quantity can be either a zero-sequence voltage or current. In some cases, negative-sequence quantities may also be used.

7.5.3.1 Voltage Polarization. Consider the bus in an interconnected transmission system with multiple grounded neutrals. The voltage measured at the bus or near the bus, but on any of the connected lines, will be the same voltage. Now consider the fault current flowing away from the bus toward a fault on one of the lines. We say that the current flows in a positive direction toward that fault, or away from the bus.

Now move the fault to another line and the same convention is true for that line, i.e., the current has a positive direction away from the bus for that line. However, for this new fault location, the current direction on the original line is reversed, assuming that the original line provides a contribution to the new fault location. Clearly, the bus voltage can be used as a reference for all currents.

For ground faults, the currents of interest are zero-sequence currents. Therefore, it is necessary to use a zero-sequence voltage as the polarizing quantity. One method of obtaining the zero-sequence polarizing voltage is shown in Figure 7.24. The connection is wye on the line side and open delta on the secondary side. The voltage across this open delta is $3V_{a0}$ and is a suitable quantity for polarization of the ground current relays.

The voltage across the open delta may be written from symmetrical components theory as

$$3V_{a0} = V_a + V_b + V_c \tag{7.23}$$

When the phase voltages are balanced, during normal system conditions, the zero-sequence voltage measured from this circuit is zero, since the three phase voltages are equal in magnitude and 120 degrees apart in phase. However, during unbalanced ground faults, the three voltages will not add to zero and a zero-sequence voltage will appear across the terminals of the circuit shown in Figure 7.24.

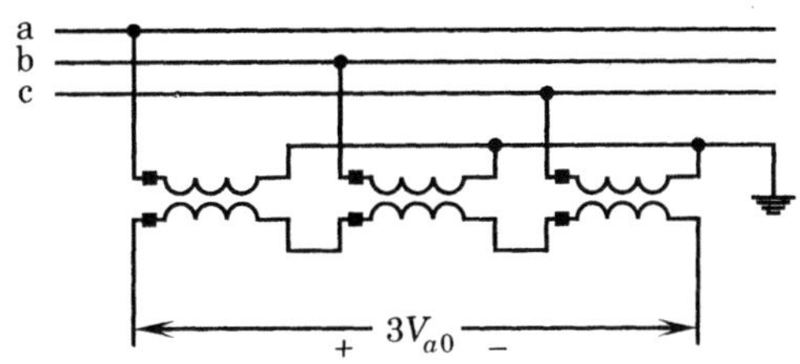

Figure 7.24 Wye to open-delta transformer connection.

7.5.3.2 Current Polarization. Another source of polarization quantity is the current flowing in the neutral of a power transformer. This source of polarization has the advantage that it usually requires only one current transformer, whereas voltage polarization requires three voltage transformers. Current polarization using the neutral of a transformer does not always provide satisfactory results, however, so this method requires careful study to ensure success.

One way of providing current polarization of the relays is by measuring the zero sequence current flowing in the neutral of a two winding transformer bank. The possible arrangements are shown in Figure 7.25. For most ground faults on the wye side of the two winding transformer, the current flows up the neutral and toward the ground fault. Thus, as in Figure 7.25(*a*), the neutral CT measures a current of $3I_{a0}$ for current polarization. The same is true for the zig-zag transformer shown in Figure 7.25(*d*). The wye-wye banks, shown in Figure 7.25(*b*) and (*c*) are not suitable for current polarization. One problem in using the neutral current of the wye-connected transformer as a polarization reference is where there are mutually coupled lines, a situation that requires further study by the protection engineer to make sure the coupling does not cause neutral current reversal [1], [9].

Three winding transformer banks used for polarization are shown in Figure 7.26. In some cases, both wye windings are grounded. When this is the case, it is necessary to connect parallel current transformers in both wye neutrals, with inverse ratios in the two current transformers. For example, if the wye windings are rated 69 kV and 138 kV, the current transformers might be selected as 600/5 and 300/5, respectively.

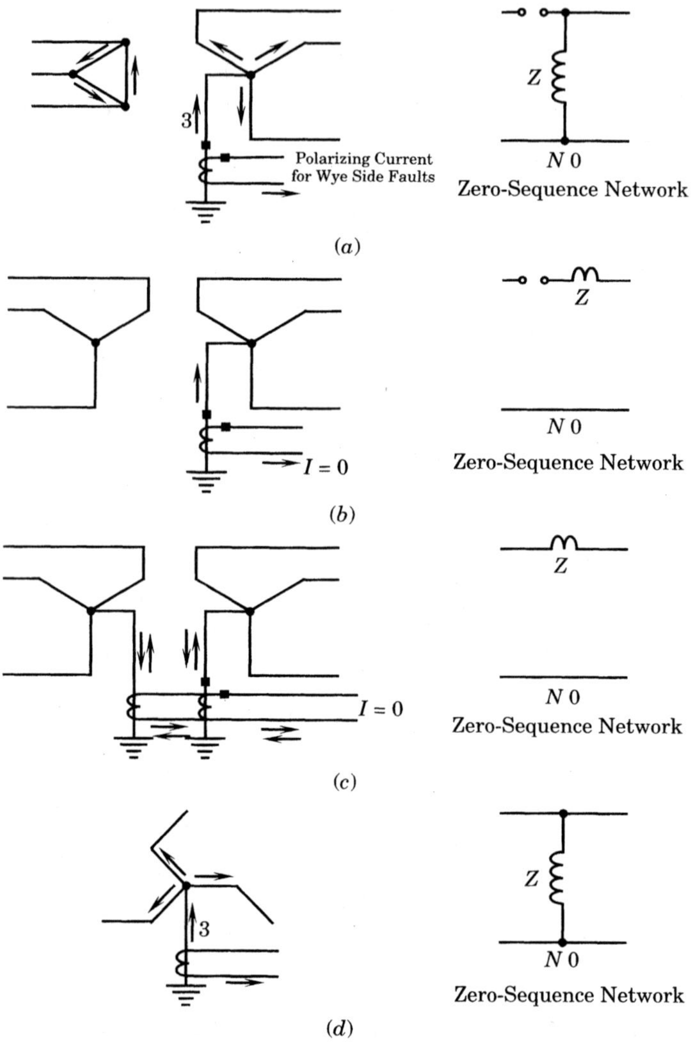

Figure 7.25 Polarization with two-winding transformer banks [1]. (*a*) Delta wye-grounded transformer bank (suitable for polarization). (*b*) Wye wye-grounded transformer bank (not suitable for polarization). (*c*) Grounded wye-wye transformer bank (not suitable for polarization). (*d*) Zig-zag transformer (suitable for polarization).

It should also be noted that the current circulating in the delta can also be used as a polarization current, which may be convenient where it is not possible to measure both wye neutrals. If this is done, only one current transformer is required if there is no load connected to the tertiary. If tertiary load is connected, then three CT's are required, one in each leg of the delta. This will prevent an unbalanced load or fault from giving incorrect polarization. If the three CT's are connected in parallel, they will measure $3I_{a0}$, a connection which cancels the balanced load currents.

The third type of transformer that can be used for polarization in some cases is the wye-grounded autotransformer. However, this application must be examined very carefully, since

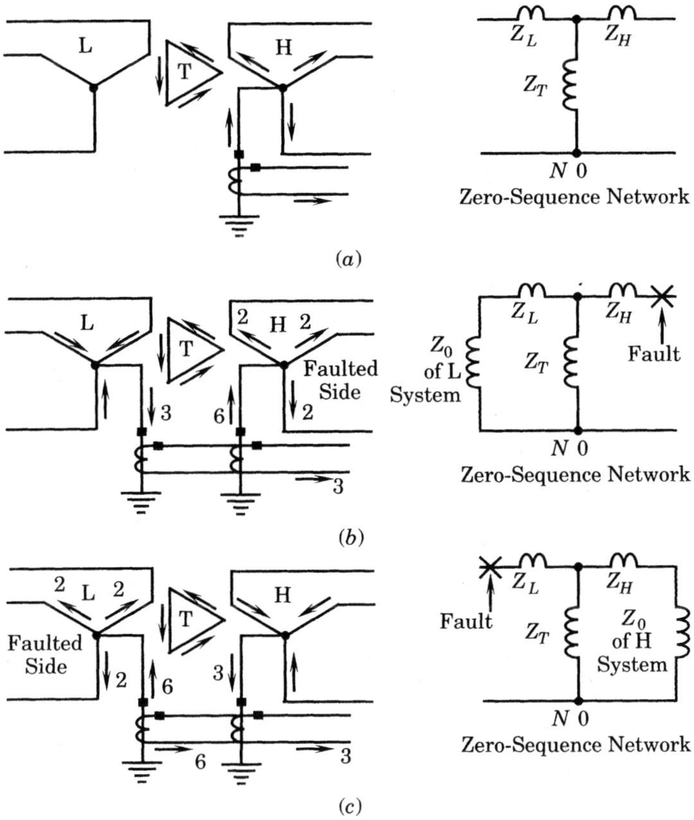

Figure 7.26 Polarization with three-winding transformer banks [1] (suitable for polarization). (*a*) Wye delta wye-grounded. (*b*) Wye-grounded delta wye-grounded. (*c*) Wye-grounded delta wye-grounded.

there is a danger of ambiguity in the current direction in the neutral. The autotransformer circuits are shown in Figure 7.27, where faults are represented first on the high side (*a*), and then on the low side (*b*). Autotransformers that are not grounded or those without a delta tertiary are unsuitable for polarization.

For ground faults on the high side of the autotransformer, the currents are shown in Figure 7.27(*a*). A fraction h_0 of the total fault current flows through the relay location, with the remainder feeding in from the other end of the transmission line. This current is divided between the low-voltage winding and the tertiary, as shown in the figure. From the circuit arrangement, we can compute the neutral current for the high side fault as follows [1].

$$I_N = 3\left(h_0 I_{H0} - s_0 I_{H0}\frac{V_H}{V_L}\right) = 3I_{H0}\left(h_0 - s_0\frac{V_H}{V_L}\right) \tag{7.24}$$

From the zero-sequence network, we determine by inspection that

$$s_0 = h_0\left(\frac{Z_T}{Z_T + Z_H + Z_{HS}}\right) \tag{7.25}$$

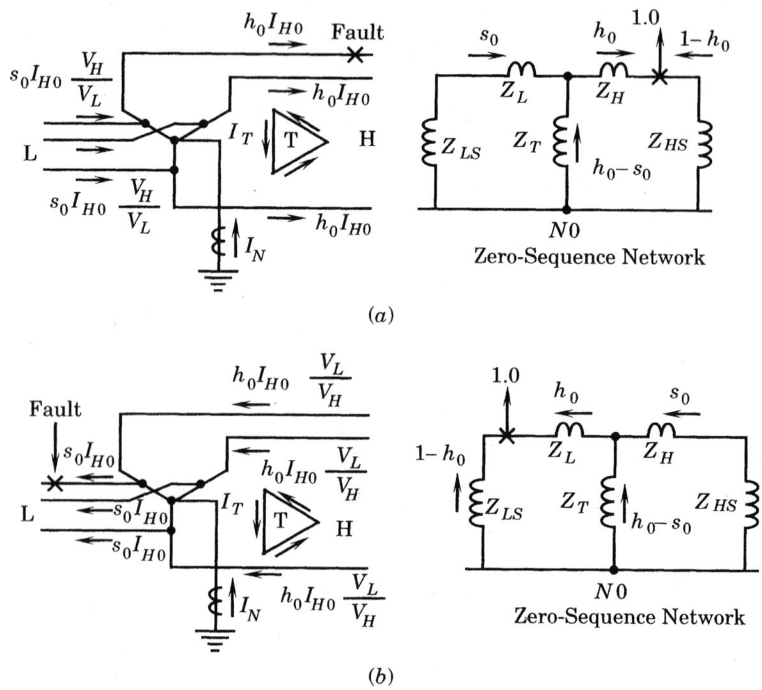

Figure 7.27 Polarization with three-winding autotransformers [1]. (*a*) Ground fault on high side. (*b*) Ground fault on low side [1].

Solving for s_0 and substituting into (7.24), we compute the neutral current.

$$I_N = h_0 I_{H0}\left(1 - \frac{V_H Z_T}{V_L(Z_T + Z_H + Z_{HS})}\right) \tag{7.26}$$

Clearly, there exists the possibility of neutral current reversal, depending on the magnitude of the second term in parentheses.

For the fault represented on the low side, as shown in Figure 7.27(b), a similar situation exists, where the neutral current can be computed as follows.

$$I_N = s_0 I_{L0}\left(1 - \frac{V_L Z_T}{V_H(Z_T + Z_H + Z_{HS})}\right) \tag{7.27}$$

As a result of the uncertainty in neutral current direction, the system installation requires careful study to ensure proper polarizing current direction if the autotransformer neutral current is to be used as a reference for current direction.

7.5.4 Types of Ground Relays

There are three basic types of relays that are used for ground relaying; overcurrent relays, distance relays, and pilot relays. Ground relays are almost always completely independent of phase relays, even though any fault current, including ground fault current, will flow through one or more of the phase relays. The ground relays, however, can be provided with much greater sensitivity to the zero-sequence currents by using higher tap settings. This means that the ground relays will pick up much faster than phase relays for a fault involving the ground.

Directional or nondirectional overcurrent relays are widely used at most voltage levels because of their low cost and reliable service record. Many relay engineers prefer an overcurrent relay with an inverse or very inverse time-current characteristic. This means that the pickup will be very fast for close-in faults and delayed for faults at the end of the transmission line. This delay makes coordination with adjacent lines relatively easy because of the rapid change in fault current for the more remote faults. The ground relay must coordinate with bus differential relays, as well as ground relays on any outgoing lines at the remote end.

In systems with multiple grounds, which is usually the case, the ground overcurrent relays will need to be directional relays, using one of the polarizing methods described in the previous section. The ground relays in a looped system must be coordinated all around the loop in both directions, in exactly the same way that phase relays are coordinated. This is a cut and try process. An excellent example of this type of coordination is given in [8] which is recommended for study.

Instantaneous overcurrent relays are usually applied to supplement the ground fault protection when overcurrent relays are used. Instantaneous ground relays can reduce the fault clearing time to about one cycle in many cases, for faults on a large fraction of the line length.

Directional ground distance relays are responsive to impedance or reactance between the relay and the fault. These relays, although more expensive than overcurrent relays, can provide almost instantaneous protection for most of the line length. For many years, distance relays were not widely used for ground protection due to the inherent problem of measuring zero-sequence impedance or reactance in the presence of a fault. Ground faults usually involve fault resistance of widely varying magnitude. This may prevent the relay from responding to a ground fault. Some relay engineers back up ground distance relays with overcurrent relays to make sure that all faults are recognized in a timely way. Many of the problems associated with ground distance relays have been solved by newer devices, making this a good alternative where overcurrent or directional overcurrent coordination is a problem.

Pilot relaying is used for ground protection in special cases where the other methods are inadequate for reasons of security or dependability. Pilot relays use either directional comparison or phase comparison to determine if the fault is within the protected zone. This might be a good solution for a three-terminal line, for example, where other types of relays are difficult to coordinate. Pilot relaying is sometimes selected on the more important lines because of the high speed and security offered by the pilot schemes. Some engineers argue that pilot relays are not required for ground fault protection, but should be used where stability or other considerations make it necessary to have both terminals of the transmission line tripped at the same time [8].

7.6 SUMMARY

Transmission protection is one of the most important tasks of the protection engineer. Power systems have more transmission lines than any other type of component that requires protection, which means that transmission protection probably occupies most of the engineer's working hours. The task is made more interesting due to the fact that the lines are all different and each presents a new challenge.

It has been noted that transmission protection using nondirectional overcurrent relays is usually inadequate. Even directional overcurrent relays present problems and require considerable checking of different system conditions to make sure that the settings are robust enough for all situations that arise. There is great appeal, however, in the simplicity and reliability of the overcurrent relay. Many have been installed and have provided excellent service for many years.

The next step up in relay performance is the distance relay, which is a very popular relay for transmission line protection at all voltage levels. Many different distance relay characteristics are available, making it possible to mold the relay characteristic to the protected line. Distance relays are fast and will clear faults on 80% or so of the line length very quickly.

Pilot protection has been shown to treat the transmission line as a unit, by measuring the inputs to the line from all terminals. This makes pilot protection valuable for certain multiterminal lines. It is more expensive than other types, but provides fast, secure protection, even in the most difficult situations. Pilot protection is also important in cases where system stability may require fast tripping of both ends of a transmission line. The subject of pilot protection is discussed in greater detail in Chapter 13.

This chapter has also presented an introduction into ground fault relaying. Ground faults comprise the vast majority of transmission line faults and are a very important aspect of transmission protection. Ground relays can be made very sensitive because they make use of the zero-sequence currents that are only available during ground faults. Therefore, most faults on transmission lines are cleared by the ground relays. Ground faults can be recognized by all different types of relays used for phase protection, but the simple overcurrent ground relay is likely the choice of most protection engineers because of its simplicity, reliability, and low cost.

REFERENCES

[1] Blackburn, J. L., Ed., *Applied Protective Relaying*, Westinghouse Electric Corp., Newark, NJ, 1976.

[2] Mason, C. R., *The Art and Science of Protective Relaying*, John Wiley & Sons, New York, 1956.

[3] Warrington, A. R. van C., *Protective Relays: Their Theory and Practice, Vol. 1*, John Wiley and Sons, New York, 1962.

[4] Warrington, A. R. van C., *Protective Relays: Their Theory and Practice, Vol. 2*, John Wiley and Sons, New York, 1969.

[5] Horowitz, S. H., Ed., *Protective Relaying for Power Systems*, "Application of the Ohm and Mho Principles to Protective Relaying," pp. 77–85, IEEE Press, 1980.

[6] GEC Measurements, *Protective Relays Application Guide*, Marketing Department, GEC Measurements, Stafford, England, 1975.

[7] Blackburn, J. L., "Ground Fault Protection of Transmission Lines," *AIEE Transactions, Part III*, 71, August 1952, pp. 685–691.

[8] Griffin, C. H., "Principles of Ground Relaying for High Voltage and Extra High Voltage Transmission Lines," *IEEE Trans. on Power Apparatus & Systems*, PAS-102 (2), February 1983, pp. 420–432.

[9] Blackburn, J. L., "The How and Why of Ground Fault Protection," *Electric Light & Power*, November 1945.

[10] Blackburn, J. L., "Ground Relay Polarization," *AIEE Transactions, Part III*, 71, December 1951, pp. 1088-1093.

PROBLEMS

7.1 Consider the system shown in Figure 7.3. The line impedances and loads are exactly as given in examples 6.3 and 6.4, namely,
G = 5 MVA
H = 5 MVA
R = 10 MVA

all at 95% power factor lagging. The system voltage is 34.5 kV. The generators are set to the following output conditions:

$P_G = 14.4$ MW	$P_R = 5.0$ MW
$Q_G = 0.3$ MVAR	$Q_R = 6.8$ MVAR
$V_G = 1.000$ at 0°	$V_R = 1.000$ at -7.14°

The network solution gives the voltage at H to be 0.984 at -3.23 degrees. Find the line load currents with a system voltage of 34.5 kV.

7.2 Extend Problem 7.1 to determine the fault currents for faults labeled (1), (2), and (3) in Figure 7.3. Let the generator impedances be

$$\begin{aligned} Z_{\text{gen}} &= 0 + j0.1 \text{ per unit-maximum conditions} \\ &= 0 + j0.3 \text{ per unit-minimum conditions} \end{aligned}$$

on a 100 MVA base. Compute the fault contributions from each generator and the line flows, as well as the total fault currents. Assume each generator has an internal voltage of 1.0 per unit and neglect any phase angle difference in these voltages.

7.3 Based on the results of problems 7.1 and 7.2, are directional relays required at locations A through D? Explain.

7.4 Prove that, for a transmission loop with only one source of fault current, the fault current flowing through any relay around the loop is greater with the loop open than with the loop closed.

7.5 Consider the loop transmission system in Figure 7.4. Let the impedances of the various lines be as follows

Per Unit Line Impedance

G-H	$j0.08$
H-R	$j0.10$
R-S	$j0.06$
S-T	$j0.12$
T-G	$j0.10$
Source	$j0.10$

Base Values $S_B = 100$ MVA
$V_B = 34.5$ kV
$I_B = 1673.479$ A

Assume a load of 5 MVA at each bus. Coordinate the loop relays using the inverse relay characteristics given in Figure 3.17. Make your computation assuming available taps of 2, 3, ..., 12 (all integers) and CT ratios of 10, 20, ..., 120 (in increments of 10).

7.6 Consider the single-source transmission loop of Figure 7.4 with a particular loading condition specified. The load conditions for the network are given as follows:

Bus G	5.7 MW	1.9 MVAR
Bus Q	3.8 MW	1.3 MVAR
Bus S	4.8 MW	1.6 MVAR
Bus T	1.9 MW	0.6 MVAR

The line impedances are as follows:

Line R-T	$0.247 + j0.543$
Line R-G	$0.247 + j0.543$
Line G-Q	$0.494 + j1.086$
Line Q-S	$0.494 + j1.086$
Line S-T	$0.494 + j1.086$

The bus voltage at the supply bus is 1.0 per unit. Use a power flow program to compute the line loadings and from this solution determine the find the load current seen by each relay.

7.7 For the system of problem 7.6 let the source impedance be $0 + j0.1$ per unit for all conditions (base MVA = 100, base kV = 34.5). Compute the following fault currents:

(a) The total bus H fault current

1. The contributions in lines G-H and R-H.
2. The contributions with breaker 1 open.

(b) The total bus R fault current

1. The contributions in lines H-R and S-R.
2. The contributions with breaker 3 open.

(c) What can you conclude concerning the magnitude of fault currents at every relay location?

7.8 Coordinate the directional overcurrent relays for the system of problems 7.6 and 7.7. Let the relays all have very inverse characteristics similar to Figure 3.22, assume relay taps of 2, 3, ..., 12 (all integers), and assume multiratio current transformers (see Figure 2.8).

7.9 Complete the relay coordination of example 7.2 for phase time overcurrent relays and instantaneous relays. Assume a relay characteristic exactly like that of Figure 6.18. All lines have 200:5 current transformers, and the full load current through each breaker may be taken to be 5 MVA, which corresponds to 125.5 A primary or 3.13 A secondary. Assume relay taps are available in the range of 2 to 12 amperes. Perform the coordination in the following order:

Coordinate L with M and N. Assume a minimum operating current (MOC) at M or N based on tap 10, and that the overcurrent relays at M and N are set on time dial 2. Assume the instantaneous trip on M and N to be 720 A at 7.2 kV.

7.10 Let the impedance seen at a distance relay location be given as

$$\mathbf{Z}_R = \frac{\mathbf{V}_R}{\mathbf{I}_R}$$

where all quantities are primary, that is, all quantities are in mks units measured at actual system primary voltage and current levels. Show that the relay impedance on the secondary side of the instrument current and voltage transformations is given by

$$\mathbf{Z}_{\text{sec}} = \mathbf{Z}_R \frac{\text{CT ratio}}{\text{VT ratio}}$$

7.11 Consider the single-source loop transmission system described in problem 7.6. Determine the settings, in ohms at the secondary side, for distance relays at each relay location. Assume the relays are all set to protect 90% of the line length.

7.12 Plot the line impedance and relay characteristic in the complex Z plane for relay G on line G-H in problem 7.11. The relay is specified as a mho relay with a maximum torque angle of 60 degrees.

7.13 Repeat the arc resistance calculation of example 7.3 for values of time of

$$t = 0.1, 0.2, 0.3 \text{ s}$$

and for a wind velocity of 20 miles per hour. Plot these arc resistances, together with the line impedance of problem 7.12, in the complex Z plane.

7.14 Verify (7.13).

7.15 Consider the system of problem 7.2, shown in Figure 7.3. An arcing fault at H is to be considered. Compute the actual arc resistance for a 5 ft arc length (the approximate line spacing is 5 ft). Then compute the effective arc resistance as seen by relay A at bus G.

7.16 Consider the protective zone identified for the transmission line of Figure 7.17. Describe the types of failure for which the protection should operate. Look carefully at the protective zone. There are more things within this zone than just the transmission line. Can you identify some of those items? In particular, is the circuit breaker a part of the *protective system* or a part of the *protected system*?

7.17 Carry the question of problem 7.16 a bit further. Identify all the hardware items that should be considered a part of the protective system for a transmission line.

7.18 Consider a transmission line with only two terminals that is protected by a unit protective system. Sketch the unit protection replica circuit for a circulating current system.

7.19 Consider a transmission line with only two terminals that is protected by a unit protective system.

(a) Sketch the unit protective system for the line where the current transformers are connected with their secondary windings opposed, so that no current flows in the replica circuit during normal conditions.

(b) What are the advantage of this type of system?

(c) What are the disadvantages?

(d) This arrangement is sometimes called a balanced voltage system. Explain why this name is appropriate.

7.20 Suppose you set out to design a unit protective system that makes use of high-speed data transmission, rather than telephone wires, for the transfer of information gathered at the terminals of the protected element. Assume that current and voltage are to be measured at each terminal and the protective logic requires data on both the currents and voltages at all terminals.

(a) How can you describe the currents and voltages prior to their transmission to the central processor.

  (i) The current is in a transient during a fault. Do you need to consider just the fundamental frequency current or should the transient dc offset be included? How will you measure the dc offset if it is required?

  (ii) Fault currents often contain high-frequency transients. Is it important to transmit data on these high-frequency current components?

  (iii) How will you measure the phase angle of the currents and voltages? What will you use for a reference to measure the phase angle?

(b) Describe the problems of coding the data for transmission, after it has been suitably processed.

(c) Write a specification for the communications system that will be used to transmit the data.

(d) Write a specification for the speed of response of the system. To do this, assume the following:

  (i) The system will go unstable following a fault if the fault is not cleared in three cycles, without reclosing the line.

  (ii) The circuit breakers are nominally two cycle breakers.

7.21 Verify (7.20). Note carefully any required assumptions.

7.22 Verify (7.25).

8

Complex Loci in the Z and Y Planes

The occurrence of a fault on a power system causes a sudden change in the system currents and voltages. Since the system is largely inductive, the current cannot change instantaneously, but currents and voltages of the network will change as they oscillate toward a new steady-state condition. In the case of a fault, a true steady-state condition will not be reached before fault clearing occurs, adding another step change in system conditions and prolonging the return to a steady state. These changes will include high-frequency transient oscillations of both currents and voltages as the network inductances and capacitances exchange energy during the period of readjustment. It can also involve subsynchronous frequencies as well, if the network includes series capacitors, and will usually include dc offset in all three phases, but in differing amounts [1]. The exact solution of the network under these conditions requires a consideration of the differential equations of the network to find all currents and voltages as functions of time.

The solution of the network differential equations is difficult and costly. As an alternative, engineers have learned to consider the network in a quasi-static sense and solve the network at a particular time as if the system were in the steady state, at least briefly. Since steady-state solutions are easily accomplished in the phasor domain, this provides a time saving method of network solution. Phasor solutions are commonly applied, even to faulted networks, and the solutions have been shown through experience to be sufficiently accurate for protective system design.

Since the phasor domain is so useful in power system analysis, we explore the concepts of impedance and admittance in this chapter as these are phasor domain quantities. We shall investigate the concept of the impedance "seen" by the directional protective device as it "looks" toward its protective zone. For a fault within the protective zone, the relay should compute an impedance that falls within the trip threshold. This concept is useful for any kind of protective device but is essential for a distance relay or fault locator. These devices are very common, especially in transmission line protection. Because of the importance of these concepts in protection engineering, the next three chapters deal with the general subject of describing a relay trip characteristic as a region in the complex plane. Many of the ideas in this chapter come from Atabekov [2] and Churchill [3], both of which are recommended reading on the subject.

8.1 THE INVERSE Z TRANSFORMATION

The quantities impedance and admittance, designated $\boldsymbol{Z}$ and $\boldsymbol{Y}$, respectively, are complex quantities. If the real and imaginary parts of these quantities are variable, then we refer to $\boldsymbol{Z}$ and $\boldsymbol{Y}$ as complex variables. As complex variables, these familiar quantities obey a number of interesting and useful rules, one of which is the important $1/Z$ transformation [3].

We define an impedance in rectangular and polar form by the equations

$$\boldsymbol{Z} = R + jX = \zeta e^{j\theta} \tag{8.1}$$

where, in mks units (ohms or mhos) or in per unit, we define

$$\begin{aligned}
R &= \text{resistance} = \text{Re}(\boldsymbol{Z}) \\
X &= \text{reactance} = \text{Im}(\boldsymbol{Z}) \\
\zeta &= \text{impedance magnitude} = |\boldsymbol{Z}| = Z = \sqrt{R^2 + X^2} \\
\theta &= \text{impedance angle} = \tan^{-1}\left[\frac{X}{R}\right]
\end{aligned}$$

Note that we write the complex number $\boldsymbol{Z}$ in boldface italic typeface, as is done with phasor voltages and currents.

Similarly we define admittance by the relation

$$\boldsymbol{Y} = G + jB = \rho e^{j\phi} \tag{8.2}$$

where

$$\begin{aligned}
G &= \text{conductance} = \text{Re}(\boldsymbol{Y}) \\
B &= \text{susceptance} = \text{Im}(\boldsymbol{Y}) \\
\rho &= \text{admittance magnitude} = |\boldsymbol{Y}| = Y = \sqrt{G^2 + B^2} \\
\phi &= \text{admittance angle} = \tan^{-1}\left[\frac{B}{G}\right]
\end{aligned}$$

The quantities $\boldsymbol{Z}$ and $\boldsymbol{Y}$ are usually plotted in the complex Z and Y planes, respectively, as shown in Figure 8.1 for arbitrary values of $\boldsymbol{Z}$ and $\boldsymbol{Y}$. There is a one-to-one correspondence between the ordered pairs of real numbers (R, X) and (G, B) and the complex numbers $\boldsymbol{Z}$ and $\boldsymbol{Y}$. Since we are accustomed to representing ordered pairs of numbers such as (R, X) and (G, B) as points in a plane, the complex planes of Figure 8.1 are a natural and useful way of representing $\boldsymbol{Z}$ and $\boldsymbol{Y}$.

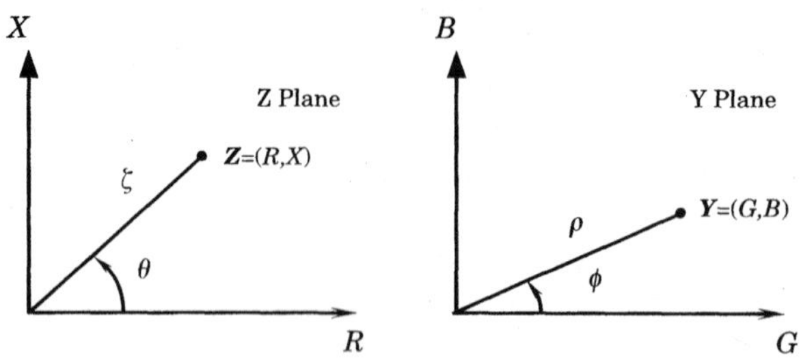

Figure 8.1 Representation of $\boldsymbol{Z}$ and $\boldsymbol{Y}$ in Z plane and Y plane, respectively.

These two-dimensional plots are called *Argand diagrams*, or simply complex planes. Note that the magnitude ζ or ρ of the complex number is the Euclidean distance from the origin to the point (R, X) or (G, B) and that the angles θ or ϕ are always measured in the positive (counterclockwise) sense.

Since admittance is the inverse of impedance, we can compute the admittance $\boldsymbol{Y}$ for a particular value of $\mathbf{Z}$ as follows. Write

$$\boldsymbol{Y} = \frac{1}{\mathbf{Z}} = \left[\frac{1}{\zeta}\right] e^{-j\theta} = \rho e^{j\phi} \tag{8.3}$$

Then

$$\rho = \frac{1}{\zeta} \tag{8.4}$$

and

$$\phi = -\theta \tag{8.5}$$

for this particular transformation. This is an important and useful result. Note that if ζ is outside the unit circle (is greater than unity), as in Figure 8.2, then ρ is always inside the unit circle (less than unity) and the angle θ is always exactly the negative of the angle ϕ.

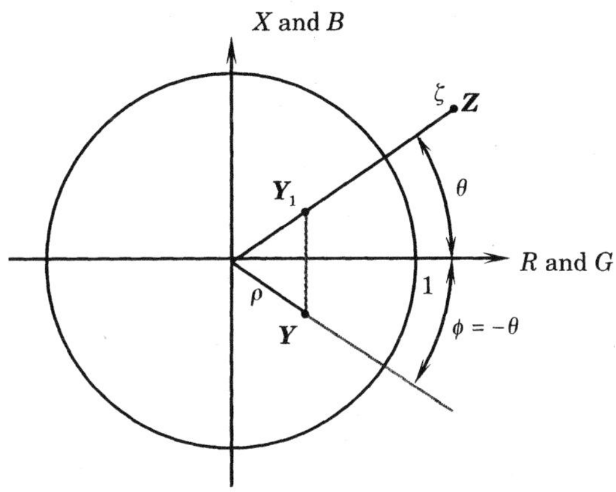

Figure 8.2 $\boldsymbol{Y} = 1/\boldsymbol{Z}$ with respect to the unit circle.

It is convenient to think of the $1/Z$ transformation as two successive transformations. Let

$$\boldsymbol{Y}_1 = [1/\zeta]\, e^{j\theta} \tag{8.6}$$

as shown in Figure 8.2. $\boldsymbol{Y}_1$ is an inversion about the unit circle with $\boldsymbol{Y}_1$ taking the inverse of the magnitude of $\mathbf{Z}$, but with the same angle. Then

$$\boldsymbol{Y} = \boldsymbol{Y}_1^* \tag{8.7}$$

is a reflection of $\boldsymbol{Y}_1$ about the real axis. The symbol (*) denotes conjugation.

In Cartesian coordinates we compute

$$\boldsymbol{Y} = G + jB = \frac{1}{R + jX} = \frac{R - jX}{R^2 + X^2} \tag{8.8}$$

Then

$$G = \frac{R}{R^2 + X^2} \tag{8.9}$$

$$B = \frac{-X}{R^2 + X^2} \tag{8.10}$$

From (8.8) we may also show that

$$R = \frac{G}{G^2 + B^2} \tag{8.11}$$

$$X = \frac{-B}{G^2 + B^2} \tag{8.12}$$

Therefore, any impedance can be converted easily to an equivalent admittance and plotted in the complex admittance plane using the above equations. In many cases it is convenient to consider the impedance seen by a relay, but in other cases it is more convenient to examine the admittance seen by the relay. Therefore, both concepts are useful and it is helpful to quickly visualize changing from impedance to admittance, or vice versa.

8.2 LINE AND CIRCLE MAPPING

The concept of mapping of real variables is familiar to engineers. For example, we define a function $y = f(x)$ of the real variable x and plot the result in a plane with y plotted as a measured distance from the x axis.

In a general way it is also possible to form the complex mapping of a complex variable $\mathbf{w} = f(z)$ from the complex z plane to the complex w plane, where $\mathbf{w}$ is a complex variable with a defined relationship to $\mathbf{Z}$. In doing so each point (R, X) in the Z plane corresponds one-to-one to a point in the w plane.

The $1/Z$ transformation is a special transformation for which a mapping of points from the Z to the Y plane is often desired. Usually we are interested in a certain locus of points in one plane or the other with the line and the circle being loci of particular interest in protection engineering.

A general equation for a line or circle can be written as [3]

$$a(R^2 + X^2) + bR + cX + d = 0 \tag{8.13}$$

where the equation is a circle when $a \neq 0$ and is a line when $a = 0$. Substituting (8.11) and (8.12) into the general circle (8.13) we can perform a $1/Z$ transformation with the result

$$d(G^2 + B^2) + bG - cB + a = 0 \tag{8.14}$$

But (8.14) is also a circle or, if $d = 0$, a line. Since a straight line is the limiting case of a circle, that is, a circle with infinite radius, we say that the $1/Z$ transformation always transforms circles into circles, with lines as the limiting case. A more careful examination of the transformation from (8.13) to (8.14) or vice versa reveals several interesting and useful results.

1. If $a \neq 0$ and $d \neq 0$, both the Z curve (8.13) and the Y curve (8.14) are circles. Furthermore,
   (i) $d \neq 0$ implies $R = X = 0$ (Z origin) is not on the Z curve. If we also have
   (ii) $a \neq 0$, this implies $G = B = 0$ (Y origin) is not on the Y curve so neither circle goes through the origin.

   Hence we conclude: If $a \neq 0$ and $d \neq 0$, then

Z circles not through origin	$\longleftrightarrow$	Y circles not through origin	(8.15)

2. If $a \neq 0$ and $d = 0$, then $R = X = 0$ is a solution to (8.13) and the Z circle goes through the origin in the Z plane. But when $d = 0$, (8.14) is a straight line

$$\begin{aligned} bG - cB + a &= 0 \\ B &= \left(\frac{b}{c}\right) G + \left(\frac{a}{c}\right) \end{aligned} \tag{8.16}$$

where Slope $= b/c$
B intercept $= a/c$.

Stated compactly, if $a \neq 0$ and $d = 0$, then

Z circle through Z origin	$\longleftrightarrow$	Straight line in Y plane	(8.17)

3. If $a = 0$ and $d \neq 0$, the reverse of (8.17) is true, i.e., if $a = 0$ and $d \neq 0$, then

Straight line in Z plane	$\longleftrightarrow$	Y circle through Y origin	(8.18)

4. If $a = d = 0$, we have the special case for lines, i.e., both the Z locus and the Y locus are lines. Furthermore, under this condition, from (8.13)

$$\begin{aligned} bR + cX &= 0 \\ X &= -\left(\frac{b}{c}\right) R \end{aligned} \tag{8.19}$$

and

$$\begin{aligned} bG - cB &= 0 \\ B &= \left(\frac{b}{c}\right) G \end{aligned} \tag{8.20}$$

Thus the lines (8.19) and (8.20) both go through the origin and have the same slope, but of opposite sign. Thus, if $a = d = 0$, then
(i)

Line through Z origin	$\longleftrightarrow$	Line through Y origin	(8.21)

(ii) The Z line slope is the negative of the Y line slope.

There are several special cases of interest in protective systems. Later we shall compute and plot these special cases as an aid to the solution of more complex problems. Remember that all cases considered here are examples of the 1/Z transformation.

8.2.1 The Half Z Plane: $a = c = 0$

Consider the half plane

$$R \geq k_1 \qquad k_1 > 0 \tag{8.22}$$

which is shown by the hatched region of Figure 8.3(a) for k_1 positive. The line $R = k_1$, under the $1/Z$ transformation, corresponds in the Y plane to

$$\frac{G}{G^2 + B^2} = k_1 \tag{8.23}$$

Note that the straight line shown in Figure 8.3(a) is of finite length. The inverse of this line will not be a complete circle in the Y plane, but will have an arc of the circle near the origin missing. For the sake of illustration in Figure 8.3, we imagine that the Z plane line illustrated extends to infinity in both the positive and negative X directions. Such a line of infinite length gives the complete circle shown in Figure 8.3(b). Hereafter, in this book, lines necessarily shown of finite length may need to be interpreted as lines of infinite length, unless otherwise described. Such an infinite line can be thought of as a circle of infinite radius.

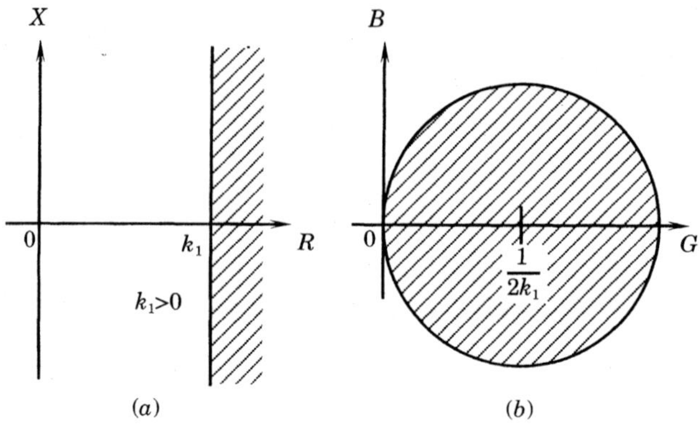

Figure 8.3 $1/Z$ transformation of the half plane $R \leq k_1 > 0$. (a) Z plane. (b) Y plane.

There are two cases of interest—k_1 positive and k_1 negative.

(i) $k_1 > 0$. With $k_1 > 0$ we compute

$$G^2 + B^2 \leq \frac{G}{k_1} \tag{8.24}$$

Rearranging and completing the square, (8.24) may be written as

$$\left(G^2 - \frac{G}{k_1} + \frac{1}{4k_1^2}\right) + B^2 \leq \frac{1}{4k_1^2} \tag{8.25}$$

or

$$\left(G - \frac{1}{2k_1}\right)^2 + B^2 \leq \left(\frac{1}{2k_1}\right)^2 \tag{8.26}$$

This is a circle with center at $G = 1/2k_1$ and radius $1/2k_1$, as shown in Figure 8.3(b).

(ii) $k_1 < 0$. When k_1 is negative, the algebraic manipulation with the inequality (8.24) requires a change in sign. Thus we compute.

$$G^2 + B^2 \geq \frac{G}{k_1} \tag{8.27}$$

which gives the result

$$\left(G - \frac{1}{2k_1}\right)^2 + B^2 \geq \left(\frac{1}{2k_1}\right)^2 \tag{8.28}$$

as shown in Figure 8.4. Note the origin of the Y circle is in the left half plane because the constant k_1 is negative.

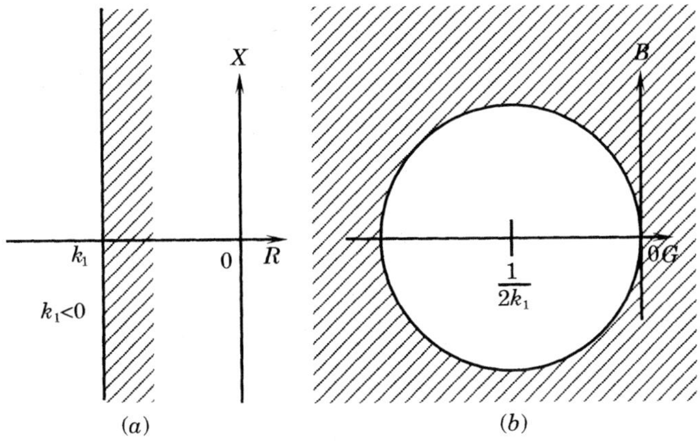

Figure 8.4 $1/Z$ transformation of the half plane $R \leq k_1 < 0$. (*a*) Z plane. (*b*) Y plane.

8.2.2 The Half *Z* Plane: $R \leq -k_2$

The half plane $R \leq -k_2 < 0$ ($k_2 > 0$) is a corollary of the case shown in Figure 8.3. For this case, we compute

$$\frac{G}{G^2 + B^2} \leq -k_2 \qquad k_2 > 0 \tag{8.29}$$

or

$$G^2 + \frac{G}{k_2} + B^2 \leq 0 \tag{8.30}$$

Completing the square, we have

$$\left(G + \frac{1}{k_2}\right)^2 + B^2 \leq \left(\frac{1}{k_2}\right)^2 \tag{8.31}$$

This result is shown in Figure 8.5 The similarity between Figure 8.5 and Figure 8.3 should be apparent.

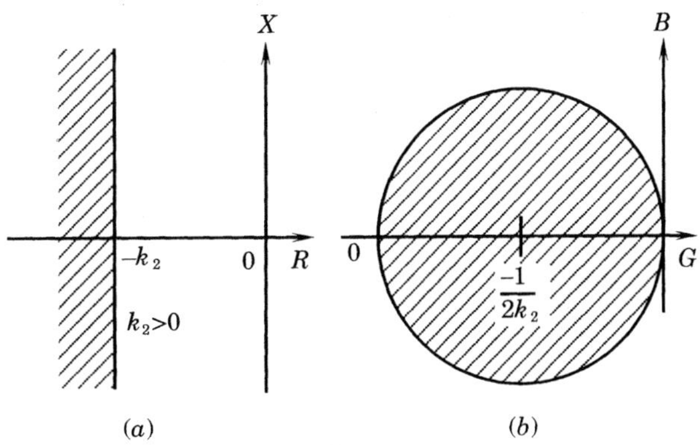

Figure 8.5 $1/Z$ transformation of the half plane $R \leq -k_2 < 0 (k_2 > 0)$. (*a*) Z plane. (*b*) Y plane.

There is one more half plane transformation, namely,

$$R \leq -k_2 > 0 \tag{8.32}$$

where k_2 is a negative quantity. This result is the image of Figure 8.4 and is left as an exercise (see problem 8.8).

8.2.3 The Half Plane: $a = b = 0$

When $a = b = 0$, the half plane is described by

$$X = -\frac{d}{c} \geq -k_3 \tag{8.33}$$

where the constant parameter k_3 can be either positive or negative. This describes a region above a horizontal straight line in the Z plane, and where that line is below the origin for positive values of k_3 and above the origin for negative values. Moreover, in the Y plane, the constraint $a = b = 0$ gives

$$d(G^2 + B^2) - cB = 0 \tag{8.34}$$

or, using only negative values of $k_3(-k_3 > 0)$, we have

$$\frac{B}{G^2 + B^2} = \frac{d}{c} = -k_3 \tag{8.35}$$

Let us consider the case where k_3 is negative. We can rearrange (8.35) to write

$$G^2 + B^2 \leq \frac{B}{k_3} \tag{8.36}$$

Completing the square, this gives the equation for a circle in the form

$$G^2 + \left(B - \frac{1}{2k_3}\right)^2 \leq \frac{1}{4k_3^2} \tag{8.37}$$

This is the region inside the circle with the following parameters:

$$\begin{aligned} &\text{Center:} \quad (0, 1/2k_3) \\ &\text{Radius:} \quad 1/2k_3 \end{aligned} \tag{8.38}$$

The region of the Y plane corresponding to (8.37) is that region *inside* the circle, and this maps into the region *above* the horizontal line (8.33). Figure 8.6 illustrates these regions. The straight line characteristic in the Z plane is typical of protective devices called ohm relays, which are often used as one or more zones of a distance relaying scheme.

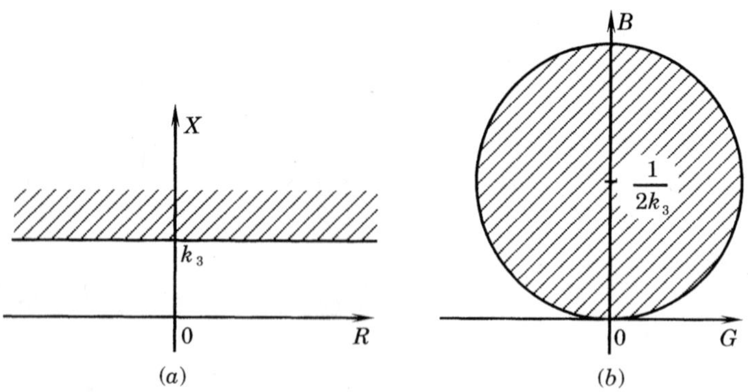

Figure 8.6 $1/Z$ transformation of the half plane $X \geq -k_3 > 0$ and $k_3 < 0$. (*a*) Z plane. (*b*) Y plane.

8.2.4 The Half Plane: $a = 0$

A more general half plane transformation is that in which $a = 0$, which gives a line in the Z plane

$$bR + cX + d = 0 \tag{8.39}$$

Now consider the half plane

$$bR + cX + d \geq 0 \tag{8.40}$$

or

$$\begin{aligned} X &\geq \left(\frac{-b}{c}\right) R + \left(\frac{-d}{c}\right) \\ &\geq mR + k \end{aligned} \tag{8.41}$$

where we define the line parameters

$$\begin{aligned} \text{Slope} &= m = -b/c \\ \text{Intercept} &= k = -d/c \end{aligned} \tag{8.42}$$

Substituting (8.11) and (8.12) into (8.41), we compute

$$\left(\frac{-B}{G^2 + B^2}\right) \geq \left(\frac{mG}{G^2 + B^2}\right) + k \tag{8.43}$$

This result is a circle but the center and radius are not clear from this form of the equation. Rearranging, we compute

$$G^2 + \left(\frac{m}{k}\right) G + B^2 + \left(\frac{1}{k}\right) B \leq 0 \tag{8.44}$$

Completing the square, we have

$$\left(G + \frac{m}{2k}\right)^2 + \left(B + \frac{1}{2k}\right)^2 \leq \left(\frac{m^2 + 1}{4k^2}\right) = S^2 \tag{8.45}$$

where we have defined the radius S as

$$S = \sqrt{\frac{m^2 + 1}{4k^2}} \tag{8.46}$$

and the center is located at $(-m/2k, -1/2k)$. This transformation is sketched in Figure 8.7 for m and k both positive. There are several important observations to be made from Figure 8.7

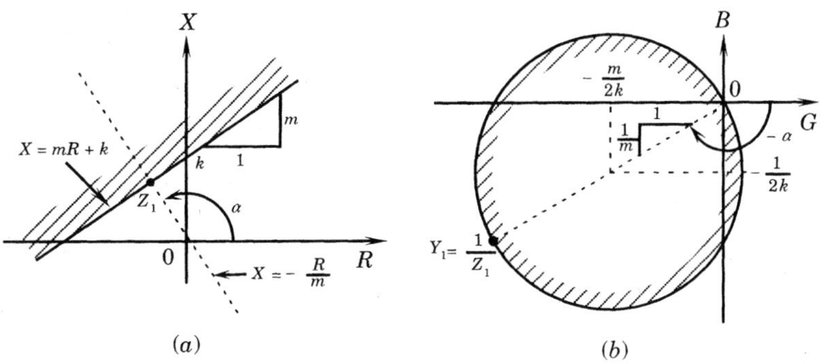

Figure 8.7 $1/Z$ transformation $X \leq mR + k(m, k > 0)$. (a) Z plane. (b) Y plane.

and (8.39)–(8.45). First, for $k \neq 0$ this means that $d \neq 0$, otherwise the line goes through the Z origin and this results in a Y circle of infinite radius, or a line. The G coordinate of the center depends on both the slope m and the intercept k, but the B coordinate depends only on the line intercept k.

Figure 8.7 also shows a simple way of constructing the Y circle from the Z line. First find the point Z_1 located at the intersection of the Z line and a perpendicular through the origin. This is the closest the Z line comes to the Z origin and corresponds to the diameter of the Y circle Y_1. The line 0-Y_1 has slope $1/m$. If the line comes very close to the Z origin then the Y circle will be large and vice versa. The center of the circle lies halfway to the point Y_1, as noted in the figure.

8.2.5 The Half Plane: $d = 0$

A case of particular interest in power system protection is that in which the coefficient d is zero. For this case, we have the Y plane equation

$$bG - cB + a = 0 \tag{8.47}$$

Consider the half plane given by

$$bG - cB + a \geq 0 \tag{8.48}$$

or

$$B \geq \frac{b}{c}G + \frac{a}{c} = mG + k \tag{8.49}$$

where we define

$$\begin{aligned} \text{Slope of line} &= m = b/c \\ B\text{-axis intercept} &= k = a/c \end{aligned} \tag{8.50}$$

Substituting (8.9) and (8.10) into (8.49) we compute

$$R^2 + \frac{m}{k}R + X^2 + \frac{1}{k}X \leq 0 \tag{8.51}$$

Completing the square, we have the equation of a Z plane circle.

$$\left(R + \frac{m}{2k}\right)^2 + \left(X + \frac{1}{2k}\right)^2 = \frac{m^2+1}{4k^2} = S^2 \tag{8.52}$$

This result can also be written in terms of the parameters a, b, and c.

$$\left(R + \frac{b}{2a}\right)^2 + \left(X + \frac{c}{2a}\right)^2 = \frac{b^2+c^2}{4a^2} = S^2 \tag{8.53}$$

Figure 8.8 shows the resulting Y plane line and Z plane circle, where the figure is constructed for $b < 0$ and $c < 0$. This gives a characteristic of a mho relay in the Z plane and its corresponding Y plane mho characteristic line.

EXAMPLE 8.1

A mho relay has a characteristic exactly like that shown in Figure 8.8. The relay is an induction cylinder device with a maximum torque angle of $\theta = 60°$, as shown in the figure, and this angle also defines the direction of the impedance $\mathbf{Z}_1$, which corresponds to the major diameter of the circular mho characteristic. The transmission line that is protected by the relay has an angle ϕ, which is greater than θ. For convenience, we define the impedance $\mathbf{Z}_2$ as the impedance that equals the mho relay threshold, but lies along the angle of the transmission line impedance, as shown in Figure 8.9. Find the magnitude of $\mathbf{Z}_2$ in terms of the magnitude of $\mathbf{Z}_1$ and the two angles θ and ϕ.

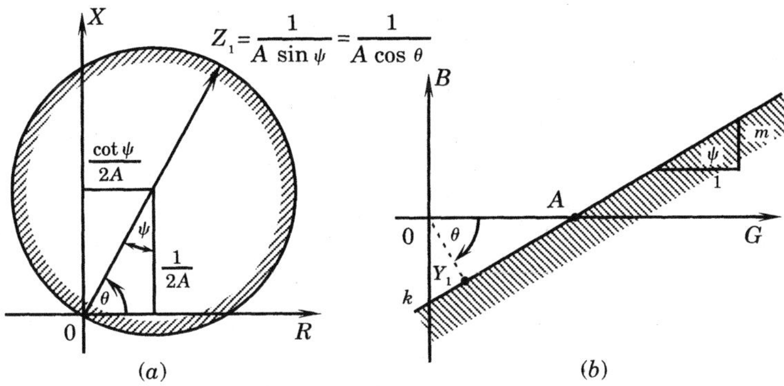

Figure 8.8 $1/Z$ transformation $B \leq mG + k$, sketched for $m > 0$, $k < 0$. (*a*) Z plane. (*b*) Y plane.

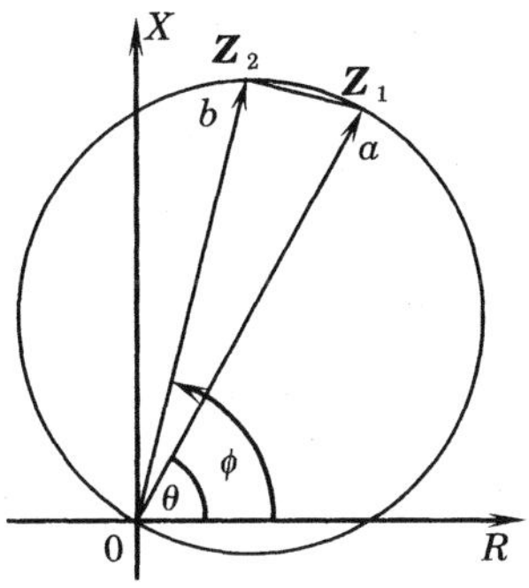

Figure 8.9 Mho characteristic and line impedance at relay threshold.

Solution

The key to solving the problem is to recognize that the angle $a - b - 0$ is a right angle and that impedance magnitude Z_1 is the hypotenuse of the right triangle. Then we can write

$$\cos(\phi - \theta) = \frac{Z_2}{Z_1}$$

and

$$Z_2 = Z_1 \cos(\phi - \theta) \tag{8.54}$$

Using this right triangle relationship, any impedance magnitude Z_2 on the mho relay threshold circle can always be written in terms of the magnitude Z_1 and the angle between Z_1 and the impedance measured from the origin to the threshold. The relationship (8.54) is not restricted to that pictured in Figure 8.9, but can be applied to any point on the mho circle. ■

8.3 THE COMPLEX EQUATION OF A LINE

A general expression that can be modified to represent a variety of loci in the complex plane is given by

$$\mathbf{Z} - \mathbf{Z}_0 = \rho e^{j\phi} \tag{8.55}$$

where $\mathbf{Z} - \mathbf{Z}_0$ is a line of fixed length and direction in the Z plane. However, if we let the parameter ρ be a variable and ϕ be a constant, then (8.55) becomes the equation of a line.

Moreover, if ρ is a constant and ϕ is a variable, the equation becomes that of a circle. Other loci can be traced by letting both ρ and ϕ vary in any desired parametric fashion.

It is useful to have a general expression for a line in the Z plane. Let us define $\mathbf{Z}_1$ and $\mathbf{Z}_2$ as two complex constants, not necessarily equal or collinear. These are shown in Figure 8.10(a). If we add to $\mathbf{Z}_1$ a number $k\mathbf{Z}_2$ where k is a parameter, we have

$$\mathbf{Z} = \mathbf{Z}_1 + k\mathbf{Z}_2 \tag{8.56}$$

where $k\mathbf{Z}_2$ is a straight line in the Z plane with the same slope as $\mathbf{Z}_2$ and which goes through the point $\mathbf{Z}_1$, as shown in Figure 8.10. Then, letting k take on all real values, (8.56) becomes the equation of a straight line in the Z plane having the same slope as $\mathbf{Z}_2$ and displaced from the origin by $\mathbf{Z}_1$.

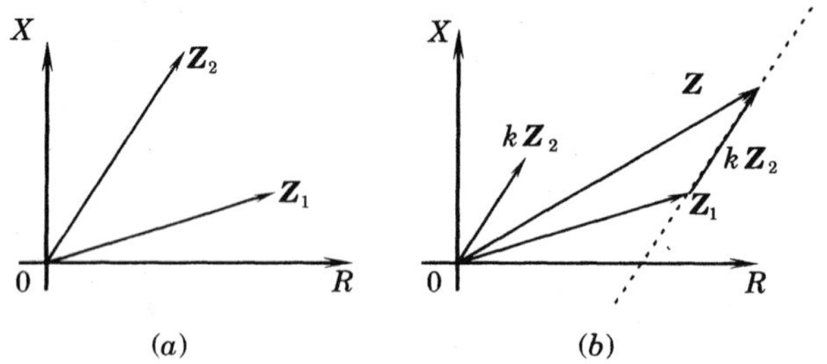

Figure 8.10 Defining a line in the complex Z plane. (a) Two complex constants $\mathbf{Z}_1$ and $\mathbf{Z}_2$. (b) The line $\mathbf{Z} = \mathbf{Z}_1 + k\mathbf{Z}_2$.

8.4 THE COMPLEX EQUATION OF A CIRCLE

There are several ways to define a circle in the complex plane. One method uses the real and imaginary components as variables, as proposed by the general equations (8.13) and (8.14). We can also define a circle in terms of complex quantities $\mathbf{Z}$ or $\mathbf{Y}$ without specifying the components. In some applications, this technique is preferred.

One way of specifying a circle is

$$|\mathbf{Z} - \mathbf{Z}_0| = k \tag{8.57}$$

where $\mathbf{Z}_o$ is a constant and k is the (constant) radius of the circle as shown in Figure 8.11.

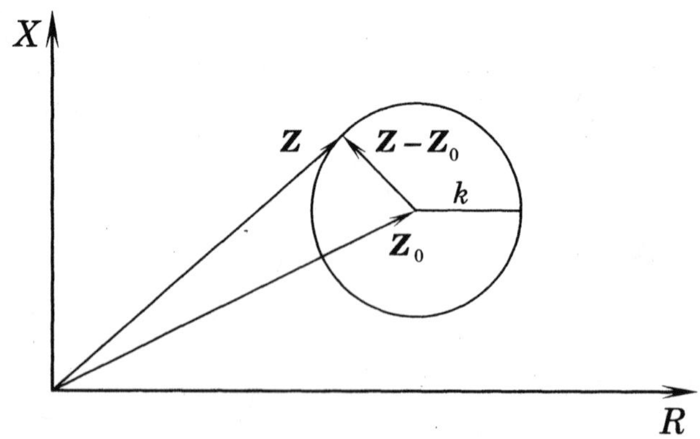

Figure 8.11 The circle $|\mathbf{Z} - \mathbf{Z}_o| = k$.

Another definition of a circle that follows from the definition (8.55) for a line is to let $k = e^{j\phi}$, which always has unit magnitude and variable angle ϕ. We then define a circle as

$$\mathbf{Z} = \mathbf{Z}_1 + \mathbf{Z}_2 e^{j\phi} \tag{8.58}$$

where $\mathbf{Z}_1$ and $\mathbf{Z}_2$ are complex constants and ϕ is a parameter, as shown in Figure 8.12. In (8.18) we found that a straight line in the Z plane transforms, through the $1/Z$ transformation, to a circle through the origin in the Y plane. Figure 8.7 shows a mathematical formulation of this general statement. We now use this concept, stated in (8.18), to find still another mathematical statement of a circle.

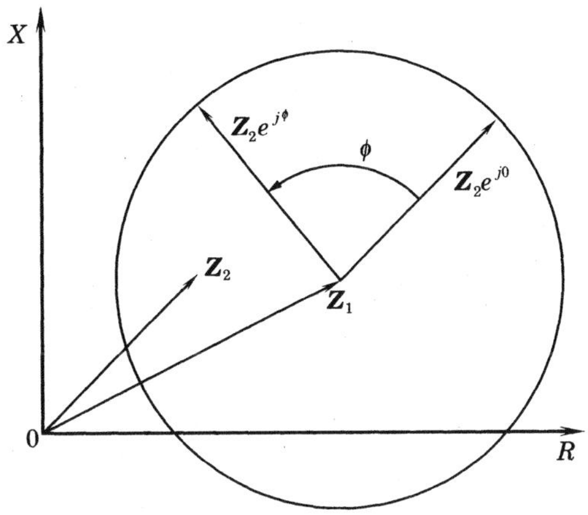

Figure 8.12 The circle $\mathbf{Z} = \mathbf{Z}_1 + \mathbf{Z}_2 e^{j\phi}$.

From (8.56) we have the equation for a line, i.e., $\mathbf{Z} = \mathbf{Z}_1 + k\mathbf{Z}_2$. The inverse of this quantity is, from (8.18), a circle through the origin. We use this fact to show that the complex quantity

$$\mathbf{Z} = \frac{\mathbf{Z}_1 + k\mathbf{Z}_2}{\mathbf{Z}_3 + k\mathbf{Z}_4} \tag{8.59}$$

is a circle, where $\mathbf{Z}_1, \mathbf{Z}_2, \mathbf{Z}_3$, and $\mathbf{Z}_4$ are complex constants and k is a parameter. A straightforward division of (8.59) gives

$$\mathbf{Z} = \frac{\mathbf{Z}_2}{\mathbf{Z}_4} + \frac{\mathbf{Z}_1/\mathbf{Z}_3 - \mathbf{Z}_2/\mathbf{Z}_4}{1 + k\mathbf{Z}_4/\mathbf{Z}_3} \tag{8.60}$$

or

$$\mathbf{Z} = \mathbf{Z}_5 + \frac{\mathbf{Z}_6}{1 + k\mathbf{Z}_7} = \mathbf{Z}_5 + \mathbf{Z}' \tag{8.61}$$

Obviously $\mathbf{Z}_5$ is a constant. From (8.18), $\mathbf{Z}'$ is the inverse of a straight line, or this quantity is a circle, multiplied by a constant $\mathbf{Z}_6$. The circle $\mathbf{Z}'$ goes through the origin in the Z' plane as shown in Figure 8.13.

Atabekov [7] gives a method for constructing the following circle.

$$\mathbf{Z}' = \frac{\mathbf{Z}_6}{1 + k\mathbf{Z}_7} = \frac{\mathbf{Z}_1/\mathbf{Z}_3 - \mathbf{Z}_2/\mathbf{Z}_4}{1 + k\mathbf{Z}_4/\mathbf{Z}_3} \tag{8.62}$$

The construction is shown in Figure 8.14 and is explained as follows. Lay off the complex

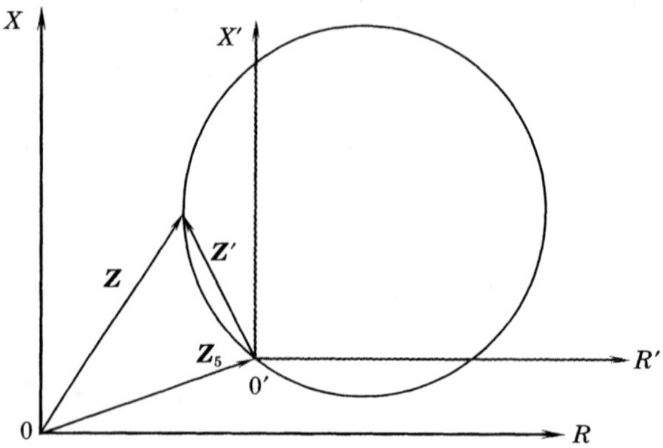

Figure 8.13 The circle $\mathbf{Z} = \mathbf{Z}_5 + \mathbf{Z}'$.

constant $\mathbf{Z}_6 = O'P$ in the Z' plane. If we define

$$\begin{aligned} \mathbf{Z}_3 &= Ce^{j\gamma} \\ \mathbf{Z}_4 &= De^{j\delta} \end{aligned} \tag{8.63}$$

then

$$\frac{\mathbf{Z}_4}{\mathbf{Z}_3} = \frac{D}{C}e^{j(\delta-\gamma)} \tag{8.64}$$

Now, construct the angle $QPL = \delta - \gamma$ with vertex at P. Then the center C' lies at 90 degrees to the tangent PL and along a perpendicular bisector MC' to $O'P$.

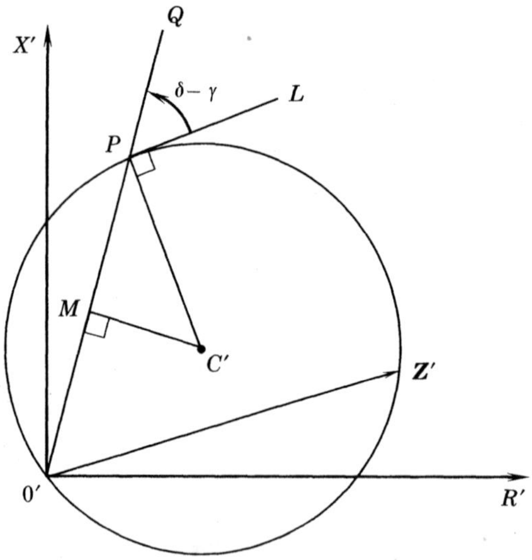

Figure 8.14 Construction of the $\mathbf{Z}'$ circle.

8.5 INVERSION OF AN ARBITRARY ADMITTANCE

There are several special cases where the inversion of an admittance to determine the corresponding impedance is of special interest. Before considering these special cases, however, we first examine the more general problem of the inversion of an admittance, where the admittance is expressed in the form

$$\boldsymbol{Y} = \boldsymbol{Y}_A \pm \boldsymbol{Y}_K = \boldsymbol{Y}_A \pm Y_K e^{j\psi} \tag{8.65}$$

where $\boldsymbol{Y}_A = G_A + jB_A$ = a complex constant admittance
$\boldsymbol{Y}_K = Y_K e^{j\psi}$ = a variable admittance (8.66)

The inverse of (8.65) is an impedance, which can be written as

$$\boldsymbol{Z} = \frac{1}{\boldsymbol{Y}} = \frac{1}{\boldsymbol{Y}_A \pm \boldsymbol{Y}_K} = R + jX \tag{8.67}$$

This expression can be solved for $\boldsymbol{Y}_K$, with the result

$$\pm\boldsymbol{Y}_K = \pm Y_K e^{j\psi} = \frac{(1 - RG_A + XB_A) - j(XG_A + RB_A)}{R + jX} \tag{8.68}$$

This equation can be solved in two ways; (1) with the parameter Y_K held at a given value with the angle ψ taking on all real values, or (2) with the angle ψ held at a given value and the parameter Y_K taking on all real values. Both options are considered below.

8.5.1 Inversion of Y with $|Y_K|$ Constant and ψ Variable

The solution of (8.68) for a given value of Y_K is performed by first rearranging the equation in a standard form of complex variables, as follows.

$$\pm Y_K e^{j\psi} = \frac{(1 - RG_A + XB_A) - j(XG_A + RB_A)}{R + jX} \tag{8.69}$$

Now, compute the magnitude square of (8.69), using either the + or − sign.

$$Y_K^2 = \frac{(1 - RG_A + XB_A)^2 + (XG_A + RB_A)^2}{R^2 + X^2} \tag{8.70}$$

This equation can be rearranged to get a new equation in the R-X plane. Expanding the squared terms and combining coefficients of R and X, we have

$$(Y_K^2 - G_A^2 - B_A^2)R^2 + 2G_A R + (Y_K^2 - G_A^2 - B_A^2)X^2 - 2B_A X = 1 \tag{8.71}$$

Rearranging, we may write

$$R^2 - \frac{2G_A}{Y_A^2 - Y_K^2}R + X^2 + \frac{2B_A}{Y_A^2 - Y_K^2}X = \frac{-1}{Y_A^2 - Y_K^2} \tag{8.72}$$

where we recognize that

$$Y_A^2 = G_A^2 + B_A^2 \tag{8.73}$$

Equation (8.72) is the equation of a circle, but is not in a convenient form. Completing the square on the left-hand side gives the following result.

$$\left(R - \frac{G_A}{Y_A^2 - Y_K^2}\right)^2 + \left(X + \frac{B_A}{Y_A^2 - Y_K^2}\right)^2 = \left(\frac{Y_K}{Y_A^2 - Y_K^2}\right)^2 \tag{8.74}$$

This is the equation of a circle with the following parameters.

$$\begin{aligned} \text{Center:} \quad R &= \frac{G_A}{Y_A^2 - Y_K^2} \\ X &= \frac{-B_A}{Y_A^2 - Y_K^2} \\ \text{Radius:} \quad S &= \frac{Y_K}{Y_A^2 - Y_K^2} \end{aligned} \tag{8.75}$$

Recall that Y_K is a constant parameter. Allowing Y_K to take on different values will result in a family of circles, each of which requires the variable parameter ψ to take on all values, at least from 0 through 360 degrees. This can be verified by writing (8.74) in a more general form, as follows. Write

$$(R-a)^2 + (X-b)^2 = s^2 \tag{8.76}$$

This circle can also be described in parametric form.

$$\begin{aligned} R &= a + s\cos\psi \\ X &= b + s\sin\psi \end{aligned} \tag{8.77}$$

These equations can be shown to satisfy (8.76). If one plots the first of the (8.77) equations against the second equation, the result is a circle where the angle ψ takes on a wide enough range of values to complete the circle.

8.5.2 Inversion of Y with ψ Constant and $|Y_K|$ Variable

The second variation of the original equation (8.68) occurs when a given value is assigned to the parameter ψ and Y_K is allowed to take on all real values. This form of equation is developed by writing (8.68) as follows.

$$\begin{aligned} \pm Y_K &= \pm Y_K e^{j\psi} = \pm(Y_K\cos\psi + jY_K\sin\psi) \\ &= \frac{(1 - RG_A + XB_A) - j(XG_A + RB_A)}{R + jX} \end{aligned} \tag{8.78}$$

Rationalizing the denominator, we have

$$\begin{aligned} \pm(Y_K\cos\psi + jY_K\sin\psi) &= \frac{(1 - RG_A + XB_A) - j(XG_A + RB_A)}{R + jX} \times \frac{R - jX}{R - jX} \\ &= \frac{[R(1 - RG_A + XB_A) - X(XG_A + RB_A)] - j[R(XG_A + RB_A) + X(1 - RG_A + XB_A)]}{R^2 + X^2} \end{aligned} \tag{8.79}$$

This results in two equations, one for the real and one for the imaginary parts of (8.79).

$$\pm Y_K\cos\psi = \frac{R(1 - RG_A + XB_A) - X(XG_A + RB_A)}{R^2 + X^2} \tag{8.80}$$

$$\pm Y_K\sin\psi = -\frac{R(XG_A + RB_A) + X(1 - RG_A + XB_A)}{R^2 + X^2} \tag{8.81}$$

Now, dividing (8.80) by (8.81), we get (using either the + or − sign)

$$\cot\psi = -\frac{R(1 - RG_A + XB_A) - X(XG_A + RB_A)}{R(XG_A + RB_A) + X(1 - RG_A + XB_A)} \tag{8.82}$$

By cross-multiplying and rearranging, this equation can be written as

$$R^2 - \frac{1}{G_A - B_A\cot\psi}R + X^2 - \frac{\cot\psi}{G_A - B_A\cot\psi}X = 0 \tag{8.83}$$

This result shows clearly the characteristics of an equation of the circle. Completing the square, we obtain the following result.

$$\left(R - \frac{1}{2H}\right)^2 + \left(X + \frac{\cot\psi}{2H}\right)^2 = \frac{1}{4H^2} + \frac{\cot^2\psi}{4H^2} = S^2 \tag{8.84}$$

where $H = G_A - B_A \cot\psi$

$$S^2 = \frac{1 + \cot^2\psi}{4H^2} = \left(\frac{1}{2H\sin\psi}\right)^2 \tag{8.85}$$

8.5.3 Summary of Y Inversion Equations

The equations developed above show that the inversion of an admittance of the form (8.65) always leads to a circle in the Z plane. Moreover, there are two different types of circles, depending on which of the independent variables is fixed. In the case of fixed Y_K, the result is a family of circles with k as the independent variable. When ψ is fixed, a different family of circles is obtained. The method of developing the circles will now be applied in special cases of interest. However, it will be noted that the methods of deriving the Z-plane circles is always one of the methods derived above. It will also be shown that the two families of circles are orthogonal.

8.6 INVERSION OF A STRAIGHT LINE THROUGH (1, 0)

There are several examples where it is necessary to consider the inversion of the special line

$$\mathbf{Y} = 1 \pm \mathbf{Y}_K = 1 \pm Y_K e^{j\psi} \tag{8.86}$$

where the line goes through the point (1,0) and is further identified by the slope $\tan\psi =$ a constant. The parameter Y_K takes on all real values to define points along the line $\mathbf{Y}$, as shown in Figure 8.15.

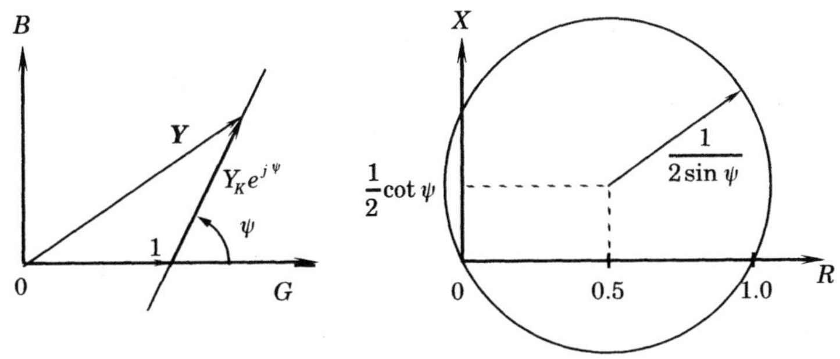

Figure 8.15 Inversion of the line $\mathbf{Y} = 1 \pm Y_K e^{j\psi}$.

The inverse of (8.86) will be a circle through the origin in the Z plane. We are interested in learning more about this circle. From (8.86) we write

$$\mathbf{Z} = \frac{1}{\mathbf{Y}} = \frac{1}{1 \pm \mathbf{Y}_K} \tag{8.87}$$

which we solve for $\pm \boldsymbol{Y}_K$ with the result

$$\pm \boldsymbol{Y}_K = \pm Y_K e^{j\psi} = \frac{1 - \boldsymbol{Z}}{\boldsymbol{Z}} \tag{8.88}$$

Since $\boldsymbol{Z} = R + jX$, we make this substitution in (8.88). The result is, after rationalizing,

$$\pm Y_K e^{j\psi} = \pm(Y_K \cos\psi + jY_K \sin\psi) = \frac{R(1-R) - X^2 - jX}{R^2 + X^2} \tag{8.89}$$

From this equation we can find R and X in terms of the line parameter ψ. By inspection of (8.89), we write

$$\cot\psi = \frac{R(1-R) - X^2}{-X} \tag{8.90}$$

Then we compute directly the quadratic

$$X \cot\psi = R^2 - R + X^2 \tag{8.91}$$

and complete the square to write

$$\left(R - \frac{1}{2}\right)^2 + \left(X - \frac{\cot\psi}{2}\right)^2 = \frac{1}{4} + \frac{1}{4}\cot^2\psi = \left(\frac{1}{2\sin\psi}\right)^2 = S^2 \tag{8.92}$$

This equation is that of a circle in the Z plane with radius $S = 1/(2\sin\psi)$. Note that (0,0) and (1,0) both satisfy (8.92) so these points are always on the circle. The R coordinate of the center is always +0.5 but the X coordinate is a function of ψ as shown in Figure 8.15.

As the line parameter ψ increases from zero, where both the radius and X coordinate of the center are infinite, these parameters become smaller with the radius reaching its minimum value of 0.5 at $\psi = 90$ degrees. If we define

$$\begin{aligned} X_C &= X \text{ coordinate of center} \\ &= 0.5 \cot\psi \end{aligned} \tag{8.93}$$

and plot X_c and S as functions of ψ, the result is shown in Figure 8.16.

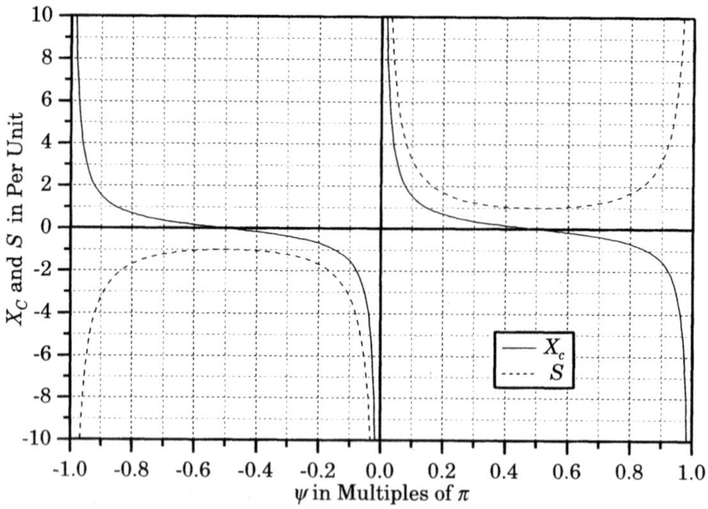

Figure 8.16 X coordinate of circle center and radius versus ψ.

We can also devise a graphical method of performing the inversion of the line $\boldsymbol{Y} = 1 \pm \boldsymbol{Y}_K$, given by (8.86), as follows. Plot the line in the Y plane as shown in Figure 8.17. The perpendicular O-$S_1 < 1$ is the shortest distance from the line to the origin in the Y plane.

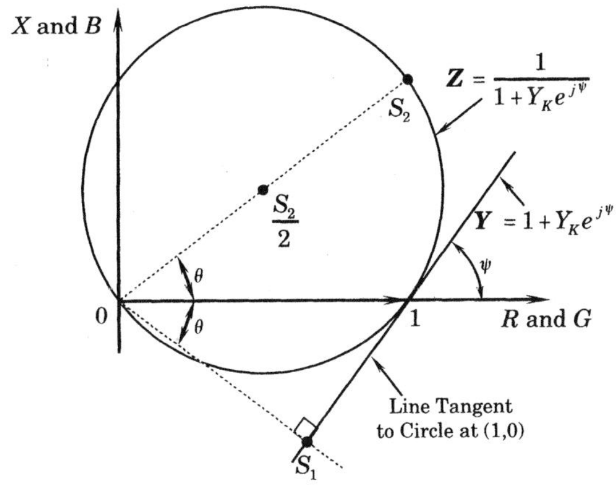

Figure 8.17 Graphical method for inverting the line $Y = 1 + ke^{j\psi}$.

The inverse O-S_1 must lie at the same angle θ reflected about the R axis and corresponds to the furthest point on the circle from the Z origin. Then O-S_2 is the diameter of the circle.

8.7 INVERSION OF AN ARBITRARY STRAIGHT LINE

As an extension of the previous result for the inversion of a line through (1,0), consider the more general case of inverting the line through $(G_A, 0)$, i.e.,

$$Y = G_A \pm Y_K = G_A \pm Y_K e^{j\psi} \tag{8.94}$$

where G_A is restricted to be a real admittance or a constant conductance. The family of straight lines represented by (8.94) is perfectly general except for lines parallel to the horizontal axis. This special case is similar to the half plane of 8.2.1 and need not be examined further.

For (8.94) to be a line we consider ψ to be constant and k to be a variable parameter. Then

$$Z = R + jX = \frac{1}{G_A \pm Y_K} \tag{8.95}$$

Rearranging and rationalizing we have

$$\pm Y_K e^{j\psi} = \pm(Y_K \cos\psi + jY_K \sin\psi) = \frac{[R(1 - G_A R) - G_A X^2] - jX}{R^2 + X^2} \tag{8.96}$$

Then

$$\cot\psi = \frac{R(1 - G_A R) - G_A X^2}{-X} \tag{8.97}$$

which can be rearranged as

$$R^2 - \frac{R}{G_A} + X^2 - \frac{\cot\psi}{G_A} X = 0 \tag{8.98}$$

Completing the square we have, for any G_A,

$$\left(R^2 - \frac{1}{G_A}R + \frac{1}{4G_A^2}\right) + \left(X^2 - \frac{\cot\psi}{G_A}X + \frac{\cot^2\psi}{4G_A^2}\right) = \left(\frac{1 + \cot^2\psi}{4G_A^2}\right) \tag{8.99}$$

which can be simplified to

$$\left(R - \frac{1}{2G_A}\right)^2 + \left(X - \frac{\cot\psi}{2G_A}\right)^2 = \left(\frac{1}{2G_A \sin\psi}\right)^2 \tag{8.100}$$

This is a circle in the Z plane with

$$\text{Center:} \quad \left(\frac{1}{2G_A}, \frac{\cot\psi}{2G_A}\right)$$
$$\text{Radius:} \quad \left(\frac{1}{2G_A \sin\psi}\right) \tag{8.101}$$

Note that the circle location and size are functions only of G_A and ψ. The parameter Y_K gives a correspondence between points along the $\boldsymbol{Y}$ line and the circumference of the $\boldsymbol{Z}$ circle.

8.8 INVERSION OF A CIRCLE WITH CENTER AT (1,0)

We again define the admittance $\boldsymbol{Y}$ in the usual way

$$\boldsymbol{Y} = 1 \pm \boldsymbol{Y}_K = 1 \pm Y_K e^{j\psi} \tag{8.102}$$

as in (8.86) except in this case we let Y_K = a constant and ψ = a variable. Then (8.102) is the equation of a circle with center at (1,0) and radius Y_K, as shown in Figure 8.18. Note that this circle does not go through the origin except for the special case, $Y_K = 1$. Then, from (8.15) the circle (8.102) will map into a Z circle that also will not go through the origin. There is something special about the circle (8.102), however. Its center is always on the G axis. Thus the Z circles will be special too and their centers will be always on the R axis.

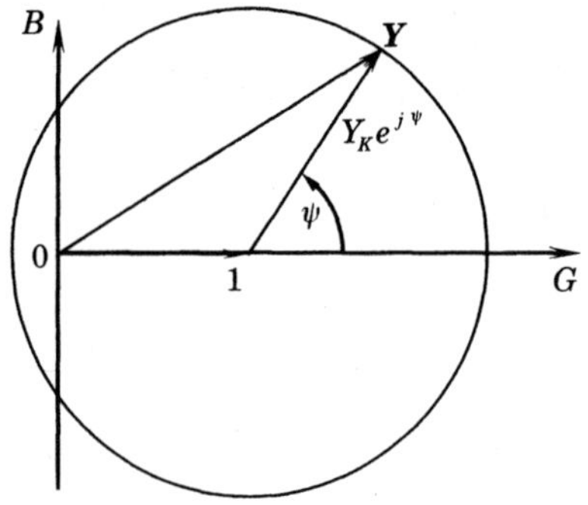

Figure 8.18 The circle $\boldsymbol{Y} = 1 \pm y_K e^{j\psi}$.

To find the inverse of (8.102) we compute

$$\pm\boldsymbol{Y}_K = \pm Y_K e^{j\psi} = \frac{1 - \boldsymbol{Z}}{\boldsymbol{Z}} = \frac{(1 - R) - jX}{R + jX} \tag{8.103}$$

Then the magnitude square of $\boldsymbol{Y}_K$ is computed as

$$Y_K^2 = \frac{(1 - R)^2 + X^2}{R^2 + X^2} \tag{8.104}$$

Rearranging and completing the square we have

$$\left(R - \frac{1}{1 - Y_K^2}\right)^2 + X^2 = \left(\frac{Y_K}{1 - Y_K^2}\right)^2 = S^2 \tag{8.105}$$

which is a circle with

$$\text{Center:} \quad \left(\frac{1}{1-Y_K^2}, 0\right)$$
$$\text{Radius:} \quad S = \frac{Y_K}{1-Y_K^2} \tag{8.106}$$

This circle is shown in Figure 8.19. The R intercepts may be computed from (8.105) by setting $X = 0$ and are found to be $1/(1+Y_K)$ and $1/(1-Y_K)$. Note that when $Y_K = 1$ the circle has infinite radius and becomes a line (Why?).

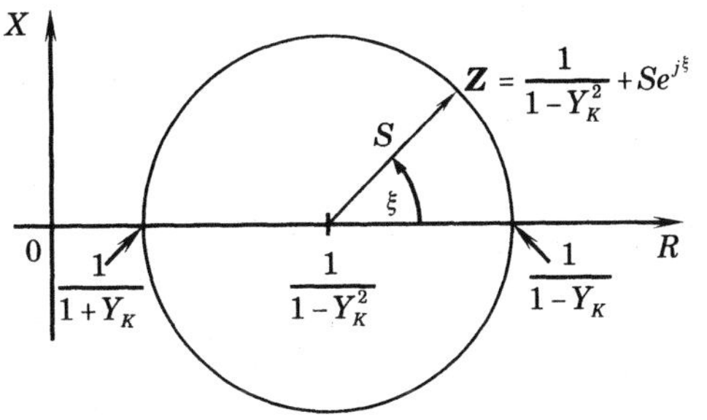

Figure 8.19 Inversion of the circle $Y = 1 \pm ke^{j\psi}$ with k a constant.

If we eliminate $Y_K \pm 1$ from our consideration, there are two ranges of positive Y_K that are of interest; that is, for $Y_K > 0$, we can consider

(i) $0 < Y_K < 1$
(ii) $Y_K > 1$

For (i), we can easily show that there is no encirclement of the origin in either the Y or Z plane. For (ii), the origin is encircled in both planes. These ideas are expressed graphically in Figure 8.20, where the origin encirclements are obtained by examining the intercepts on the horizontal axis in each plane. We also note that circles in the admittance plane always have their center at (1,0), but encircle the origin only for $Y_K > 1$. Impedance circles, on the other hand, encircle (1,0) for $Y_K < 1$, and encircle the origin for $Y_K > 1$.

We may write the equation of the Z circles in rectangular form by (8.105) or, as noted in Figure 8.19, by the complex expression

$$\mathbf{Z} = \frac{1}{1-Y_K^2} + \frac{Y_K}{1-Y_K^2}e^{j\xi} = \frac{1}{1-Y_K^2} + Se^{j\xi} \tag{8.107}$$

But

$$\mathbf{Z} = \frac{1}{\mathbf{Y}} = \frac{1}{1 \pm Y_K e^{j\xi}} \tag{8.108}$$

Equating (8.107) and (8.108) we obtain a relationship between ψ and ξ that can be written in the form

$$\tan\xi = \frac{(1-Y_K^2)\sin\psi}{2Y_K + (1+Y_K^2)\cos\psi} \tag{8.109}$$

The location of the Y and Z circles with respect to each other on the horizontal axes, and their radii, are functions of the admittance Y_K. If both circles are plotted on the same graph and to the same scale it is interesting to note that the Z and Y circles always cross.

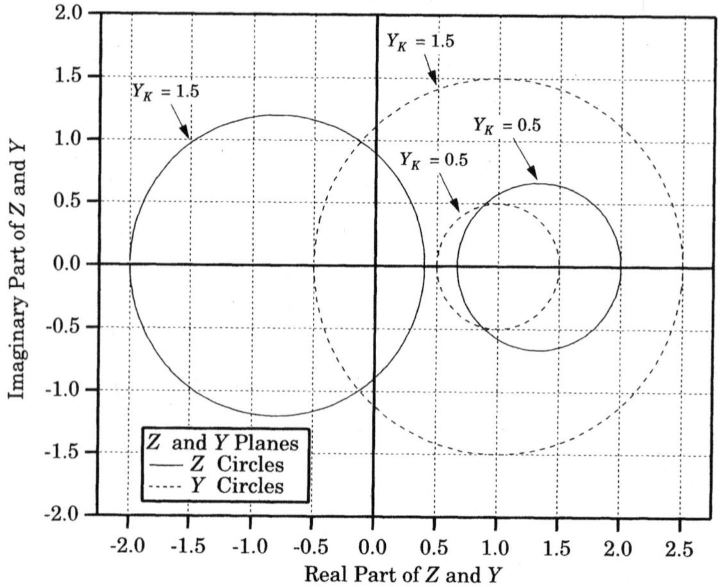

Figure 8.20 Circle inversions for $0 < k < 1$ and $k > 1$.

8.9 INVERSION OF AN ARBITRARY CIRCLE

An arbitrary circle is one that can lie anyplace in the plane and have any desired radius. Let us define an arbitrary circle by the following equation.

$$\boldsymbol{Y} = \boldsymbol{Y}_A \pm \boldsymbol{Y}_K = \boldsymbol{Y}_A \pm Y_K e^{j\psi} \tag{8.110}$$

where $\boldsymbol{Y}_A$ is a complex constant. Equation (8.110) is shown in Figure 8.21 where the complex constant $\boldsymbol{Y}_A$ is defined as

$$\boldsymbol{Y}_A = Y_A e^{j\alpha} = G_A + jB_A \tag{8.111}$$

Then the inverse of $\boldsymbol{Y}$ is

$$\boldsymbol{Z} = \frac{1}{\boldsymbol{Y}} = \frac{1}{\boldsymbol{Y}_A \pm \boldsymbol{Y}_K} = \frac{1}{\boldsymbol{Y}_A \pm Y_K e^{j\psi}} \tag{8.112}$$

Expanding the right side of (8.112) in rectangular form we have

$$\boldsymbol{Z} = R + jX = \frac{1}{G_A + jB_A \pm Y_K \cos\psi \pm jY_K \sin\psi} \tag{8.113}$$

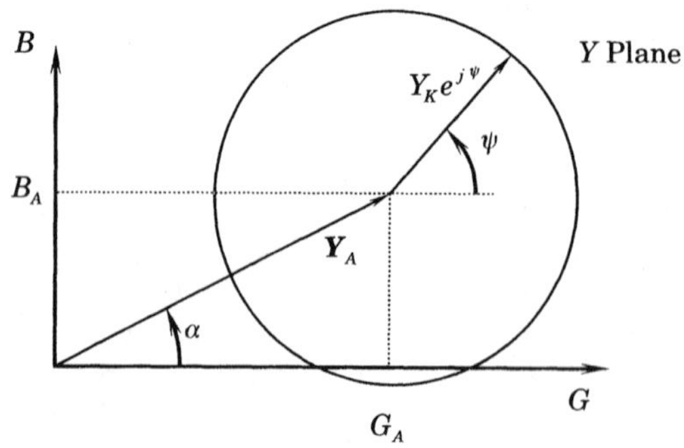

Figure 8.21 The arbitrary Y circle, $\boldsymbol{Y} = \boldsymbol{Y}_A \pm YKe^{j\psi}$.

Cross-multiplying we form two simultaneous equations:

$$\begin{aligned} R(G_A \pm Y_K \cos\psi) \mp X(B_A \pm Y_K \sin\psi) &= 1 \\ R(B_A \pm Y_K \sin\psi) \pm X(G_A \pm Y_K \cos\psi) &= 0 \end{aligned} \tag{8.114}$$

For a given $\boldsymbol{Y}_A$ and Y_K, (8.114) will give a point-by-point mapping in the Z plane corresponding to every value of ψ.

A more useful form of the Z circle equation can be found by solving (8.112) for $\boldsymbol{Y}_K$ with the result

$$\pm\boldsymbol{Y}_K = \pm Y_K e^{j\psi} = \frac{1 - \boldsymbol{Y}_A\boldsymbol{Z}}{\boldsymbol{Z}} = \frac{1 - (G_A + jB_A)(R + jX)}{R + jX} \tag{8.115}$$

Rationalizing, we compute the magnitude squared

$$Y_K^2 = \frac{(1 - G_A R + B_A X)^2 + (G_A X + B_A R)^2}{R^2 + X^2} \tag{8.116}$$

This equation can be rearranged in the form

$$\begin{aligned} -1 = R^2(G_A^2 + B_A^2 - Y_K^2) + R(-2G_A) \\ + X^2(G_A^2 + B_A^2 - Y_K^2) + X(2B_A) \end{aligned} \tag{8.117}$$

Now define

$$N = G_A^2 + B_A^2 - Y_K^2 = Y_A^2 - Y_K^2 \tag{8.118}$$

If we divide (8.117) by N and complete the square of that result we have

$$\left(R - \frac{G_A}{N}\right)^2 + \left(X + \frac{B_A}{N}\right)^2 = \left(\frac{Y_K}{N}\right)^2 \tag{8.119}$$

This is the equation of a circle with the following parameters:

$$\begin{aligned} &\text{Center:} \quad \left(\frac{G_A}{N}, \frac{-B_A}{N}\right) \\ &\text{Radius:} \quad S = \frac{Y_K}{N} \end{aligned} \tag{8.120}$$

The Y and Z circles are both shown in Figure 8.22 on the same set of axes.

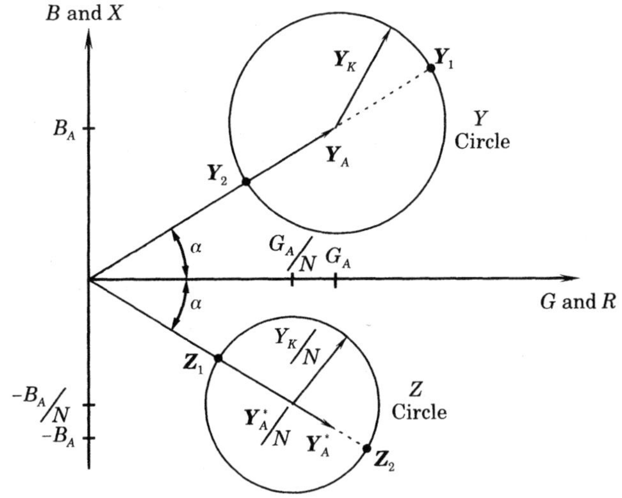

Figure 8.22 The arbitrary Y circle inversion.

A close examination of this figure suggests that the inversion can easily be done graphically once N has been computed. We also observe from Figure 8.22 that we can write the Z circle equation as

$$\boldsymbol{Z} = \frac{\boldsymbol{Y}_A^*}{N} + \left(\frac{Y_K}{N}\right) e^{j\xi} \tag{8.121}$$

Another important fact to be observed from Figure 8.22 is that the point $\boldsymbol{Y}_1$ on the Y circle, which is farthest from the origin, is transformed into $\boldsymbol{Z}_1$ which is the point closest to the Z origin. Just the opposite is true of points $\boldsymbol{Y}_2$ and $\boldsymbol{Z}_2$. Depending on the relationship of these points to the unit circle, we can draw some interesting conclusions, which are illustrated in Figure 8.23. First, if one circle is completely outside the unit circle, then the other is completely inside. Second, if one circle encloses the origin, the inverse circle does likewise. Generally speaking the circles will "cross" unless one of them is completely inside (or outside) the unit circle.

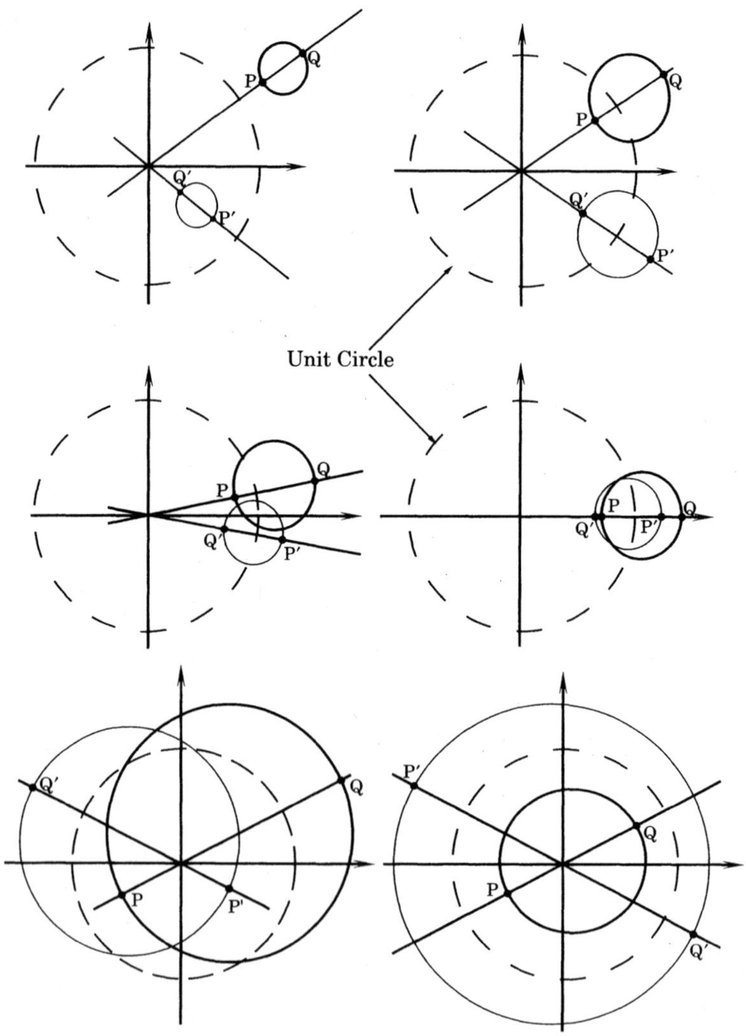

Figure 8.23 Circle inversions with respect to the unit circle.

8.10 SUMMARY OF LINE AND CIRCLE INVERSIONS

Table 8.1 summarizes all findings regarding line and circle inversions. Part 1 of Table 8.1 gives the results of line and circle inversions through the admittance point $\boldsymbol{Y} = 1 + j0$. When $\boldsymbol{Y}$ is a line the Z plane trace is a circle, but when $\boldsymbol{Y}$ is a circle the Z plane trace is also a circle.

TABLE 8.1 Summary of Line and Circle Inversion Formulas

1. Through the Point $Y = 1 + j0$

Y Plane	Z Plane
Line: $Y = 1 \pm Y_K = 1 \pm Y_K e^{j\psi}$, Y_K = variable, ψ = constant	Circle: $\left(R-\frac{1}{2}\right)^2 + \left(X - \frac{\cot\psi}{2}\right)^2 = \left(X - \frac{1}{2\sin\psi}\right)^2 = S^2$
Circle: $Y = 1 \pm Y_K = 1 \pm Y_K e^{j\psi}$, Y_K = constant, ψ = variable	Circle: $\left(R - \frac{1}{1-Y_K^2}\right)^2 + X^2 \left(\frac{Y_K}{1-Y_K^2}\right)^2 = S^2$

2. Through the Point $Y = Y_A$

Y Plane	Z Plane
Line: $Y = G_A \pm Y_K = G_A \pm Y_K e^{j\psi}$, G_A = scalar constant, ψ = variable	Circle: $\left(R - \frac{1}{2G_A}\right)^2 + \left(X - \frac{\cot\psi}{2G_A}\right)^2 = \left(\frac{1}{2G_A \sin\psi}\right)^2$
Circle: $Y = Y_A \pm Y_K = Y_A \pm Y_K e^{j\psi}$, Y_K = constant (complex), ψ = variable	Circle: $\left(R - \frac{G_A}{N}\right)^2 + \left(X + \frac{B_A}{N}\right)^2 = \left(\frac{Y_K}{N}\right)^2$, $N = G_A^2 + B_A^2 - Y_K^2 = Y_A^2 - Y_K^2$

Part 2 of Table 8.1 gives the results of line and circle inversions through the general point $\boldsymbol{Y} = \boldsymbol{Y}_A$. The first case is for $\boldsymbol{Y}_A$ a real number on the G axis with its inversion being a circle. In the second case $\boldsymbol{Y}_A$ is a complex number in the admittance plane, and again the Z plane inverse is a circle.

8.11 ANGLE PRESERVATION IN CONFORMAL MAPPING

As an illustration of angle preservation we begin with an example. Consider the family of Y lines Y_1, Y_2, Y_3, and Y_4 shown in Figure 8.24 where lines Y_1 and Y_2 are parallel as are Y_3 and Y_4. Moreover, the lines Y_1 and Y_2 intersect lines Y_3 and Y_4 at right angles forming the square hatched region. The images of the four lines are circles $Z_1, \ldots, Z_4$, which may be constructed analytically or graphically from the lines. Note that the four circles also intersect at right angles to form the curvilinear hatched square, which is the image of the hatched square formed by the Y plane lines. The fact that the angles are exactly the same in both planes is a general result of mapping any analytic function [3]. Since the impedance function $\boldsymbol{Z} = 1/\boldsymbol{Y}$ is analytic everywhere except at the origin $\boldsymbol{Y} = 0$, the mapping is conformal and angles are preserved exactly. Actually any mapping $\boldsymbol{Y} = f(\boldsymbol{Z})$ is conformal as long as $f(\boldsymbol{Z})$ is analytic and $f'(\boldsymbol{Z})$ is not 0. The transformation $\boldsymbol{Y} = 1/\boldsymbol{Z}$ is a special case of interest in power

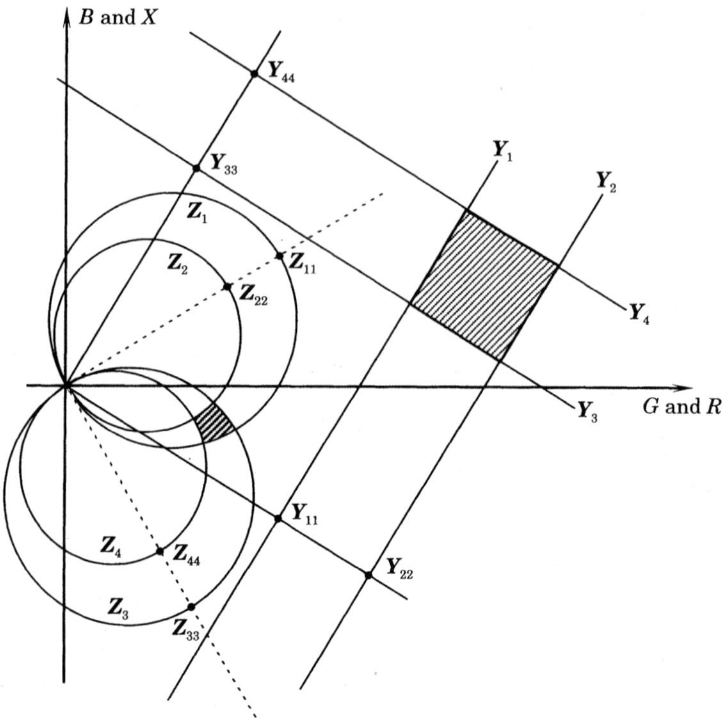

Figure 8.24 An illustration of angle preservation.

system protection. Note also that the right angle between $\boldsymbol{Y}_1$ and $\boldsymbol{Y}_3$ (or $\boldsymbol{Y}_4$) is also a special case and any other angle between Y lines will have a Z image between circles of exactly the same angle.

8.12 ORTHOGONAL TRAJECTORIES

Consider a family of curves in the Z plane described by the equation

$$\frac{dX}{dR} = f(R, X) \tag{8.122}$$

Then a second family of curves in the Z plane is said to be orthogonal to the family (8.122) if the second family has the differential equation

$$\frac{dX}{dR} = \frac{-1}{f(R, X)} \tag{8.123}$$

Simply stated this means that the trajectories all meet at right angles. If these conditions are satisfied the two sets of trajectories are called orthogonal trajectories [3], [4].

In protection studies, we are interested in the Z and Y planes. Consider the transformation

$$Z = R + jX = \frac{1}{Y} = \frac{1}{G + jB} = \frac{G - jB}{G^2 + B^2} \tag{8.124}$$

We can think of (8.124) as the function pair

$$\begin{aligned} R(G, B) &= \frac{G}{G^2 + B^2} \\ X(G, B) &= \frac{-B}{G^2 + B^2} \end{aligned} \tag{8.125}$$

From the definition of the total differential, we compute

$$\begin{aligned} dR &= \frac{\partial R}{\partial G} dG + \frac{\partial R}{\partial B} dB \\ dX &= \frac{\partial X}{\partial G} dG + \frac{\partial X}{\partial B} dB \end{aligned} \tag{8.126}$$

For the condition $R(G, B) = C_1$ the R functions are a family of straight lines parallel to the X axis. Then from (8.126) with $dR = 0$, we compute

$$\left.\frac{dB}{dG}\right|_R = \frac{-\partial R/\partial G}{\partial R/\partial B} \tag{8.127}$$

Similarly, if $X(G, B) = C_2$ represents a family of lines parallel to the X axis we compute, with $dX = 0$ in (8.126)

$$\left.\frac{dB}{dG}\right|_X = \frac{-\partial X/\partial G}{\partial X/\partial B} \tag{8.128}$$

Now we note that the function

$$\mathbf{Y} = f(\mathbf{Z}) = \frac{1}{\mathbf{Z}} \tag{8.129}$$

is an analytic function everywhere except at the Z origin. Therefore (8.129) satisfies the Cauchy-Riemann condition [3]

$$\begin{aligned} \frac{\partial G}{\partial R} &= \frac{\partial B}{\partial X} \\ \frac{\partial G}{\partial X} &= -\frac{\partial B}{\partial R} \end{aligned} \tag{8.130}$$

Substituting (8.130) into (8.127) and (8.128) we can show immediately that the slope of the image families corresponding to $R(G, B) = C_1$ and $X(G, B) = C_2$ are related by

$$\left.\frac{dB}{dG}\right|_R = \frac{-\partial R/\partial G}{\partial R/\partial B} = \frac{-\partial X/\partial B}{-\partial X/\partial G} = \frac{-1}{\left.\frac{dB}{dG}\right|_X} \tag{8.131}$$

and these families of curves are orthogonal in the Y plane as shown in Figure 8.25. If the straight lines are not parallel to the Z axes the orthogonality still holds as observed in Figure 8.24.

We may write the general expression for a line or circle as

$$\begin{aligned} \mathbf{Y} = G + jB &= \mathbf{Y}_A + Y_K e^{j\psi} \\ &= (G_A + Y_K \cos\psi) + j(B_A + Y_K \sin\psi) \end{aligned} \tag{8.132}$$

where $\mathbf{Y}$ is a family of circles when Y_K is constant and is a family of lines when ψ is constant. Then

$$\mathbf{Z} = R + jX = \frac{1}{\mathbf{Y}_A + Y_K e^{j\psi}} \tag{8.133}$$

always is a family of circles for either Y_K or ψ as parameters.

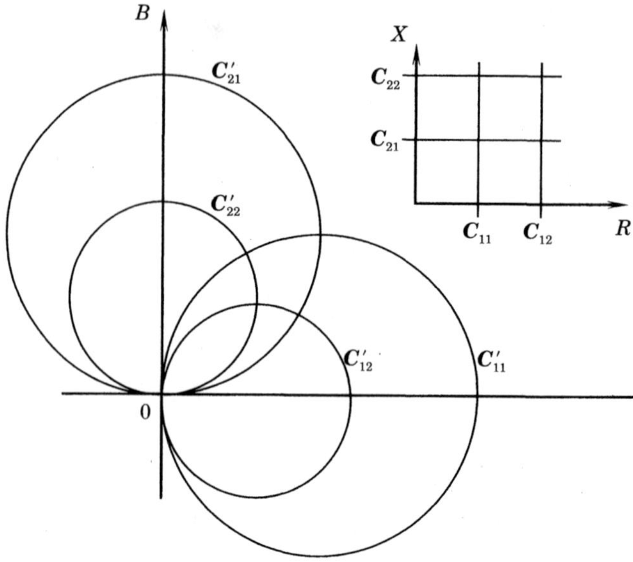

Figure 8.25 Orthogonal trajectories of the families $R = C_{11}, C_{12}$ and $X = C_{21}, C_{22}$ with their images.

We may write (8.133) as two parametric functions G and B and using the parameters Y_K and ψ. Thus

$$\begin{aligned} G(Y_K, \psi) &= G_A + Y_K \cos\psi \\ B(Y_K, \psi) &= B_A + Y_K \sin\psi \end{aligned} \tag{8.134}$$

The slope for constant ψ is given by

$$\left.\frac{dB}{dG}\right|_{\psi} = \frac{\partial B/\partial Y_K}{\partial G/\partial Y_K} = \frac{\sin\psi}{\cos\psi} = \tan\psi \tag{8.135}$$

and for constant Y_K the family of curves has slope

$$\left.\frac{dB}{dG}\right|_{k} = \frac{\partial B/\partial\psi}{\partial G/\partial\psi} = \frac{+Y_K\cos\psi}{-Y_K\sin\psi} = -\cot\psi = \frac{-1}{\left.\frac{dB}{dG}\right|_{\psi}} \tag{8.136}$$

and these families are always orthogonal. This may also be shown by computing

$$\begin{aligned} \frac{\partial \mathbf{Y}}{\partial Y_K} &= e^{j\psi} \\ \frac{\partial \mathbf{Y}}{\partial \psi} &= jY_K e^{j\psi} = jY_K \frac{\partial \mathbf{Y}}{\partial Y_K} \end{aligned} \tag{8.137}$$

which again indicates that the trajectories are orthogonal.

In the Z plane we may also compute

$$\begin{aligned} \frac{\partial \mathbf{Z}}{\partial k} &= \frac{-e^{j\psi}}{(\mathbf{Y}_A + Y_K e^{j\psi})^2} \\ \frac{\partial \mathbf{Z}}{\partial \psi} &= \frac{-jY_K e^{j\psi}}{(\mathbf{Y}_A + Y_K e^{j\psi})^2} = +jY_K \frac{\partial \mathbf{Z}}{\partial Y_K} \end{aligned} \tag{8.138}$$

and these curves are also orthogonal. These results show that families of lines or circles in the Y or Z plane and their images are always orthogonal. This is a useful property in plotting protective device characteristics in the complex plane.

EXAMPLE 8.2

To illustrate the orthogonality of constant Y_K and constant ψ circles, plot the line and circle inversions where $\mathbf{Y}_A = 3 - j4$.

Solution

First we compute the line inversions. We are given that

$$\mathbf{Y}_A = G_A + jB_A = 3 - j4 = 5e^{-53.13°}$$

Then from (8.118)

$$N = G_A^2 + B_A^2 - Y_K^2 = 25 - Y_K^2$$

can be computed for any Y_K. Also, for the image Z circle, from (8.120),

$$\text{Center:} \quad \left(\frac{G_A}{N}, \frac{-B_A}{N}\right) = \left(\frac{3}{N}, \frac{4}{N}\right)$$

$$\text{Radius:} \quad \frac{Y_K}{N}$$

can also be computed for any Y_k. A few values are listed in Table 8.2, and the constant Y_K Z circles are plotted in Figure 8.26.

TABLE 8.2 Computation of Y Circle Inversions

Y_K	N	$a = G_A/N$	$b = B_A/N$	$s = Y_K/N$
0.00	25.00000	0.120000	−0.160000	0.000000
1.00	24.00000	0.125000	−0.166667	0.041667
2.00	21.00000	0.142857	−0.190476	0.095238
3.00	16.00000	0.187500	−0.250000	0.187500
8.34	−44.44444	−0.067500	0.090000	0.187500
12.50	−131.2500	−0.022857	0.030476	−0.095238
25.00	−600.0000	−0.005000	0.006667	−0.041667

The line inversions are computed similarly. The inverse of $\mathbf{Y}_A = 3 - j4$ is a fixed point $\mathbf{Y}_o$ with image $\mathbf{Z}_o$ through which all Z circles must pass.

Also, we know that all Z circles must go through the origin. Then the real parameter G_A must be found for each value of ψ.

$$\mathbf{Y}_o = 3 - j4 = G_A + Y_K e^{j\psi}$$

Then

$$3 = G_A + Y_K \cos\psi$$
$$4 = Y_K \sin\psi$$

or

$$G_A = 3 + 4\cot\psi$$

We also know, from (8.101), that the Z circles are defined by

$$\text{Center:} \quad \left(\frac{1}{2G_A}, \frac{\cot\psi}{2G_A}\right)$$

$$\text{Radius:} \quad \left(\frac{1}{2G_A \sin\psi}\right)$$

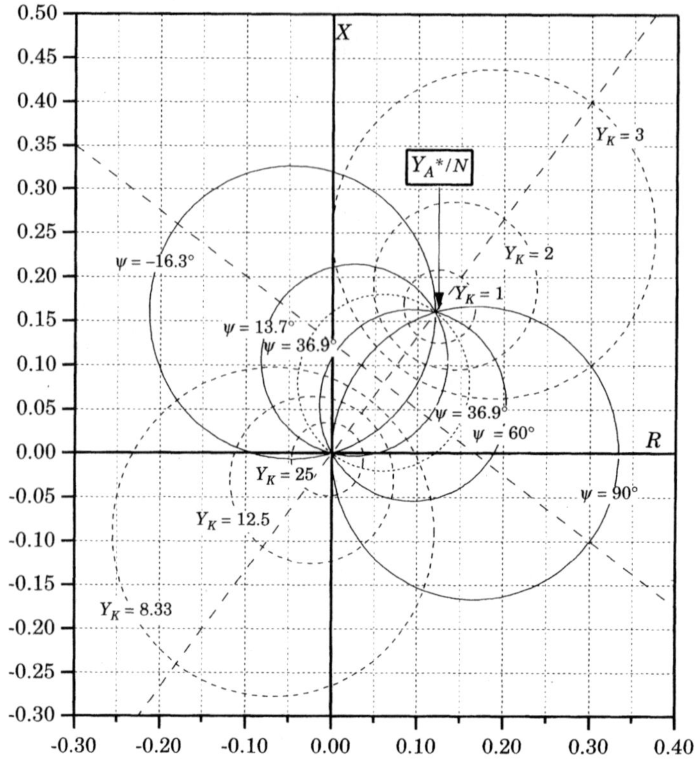

Figure 8.26 Example illustrating orthogonality of k and ψ circles.

The results for several values of ψ are given in Table 8.3 and are plotted in Figure 8.26, where both families of circles are plotted using the parametric (8.77), which defines the plot parameters a, b, and s.

TABLE 8.3 Computation of Y Line Inversions

Ψ	$\cot \Psi$	G_A	$a = \dfrac{1}{2G_A}$	$b = \dfrac{\cot \Psi}{2G_A}$	$s = \dfrac{1}{2G_A \sin \Psi}$
36.86990	1.33333	8.33333	0.06000	0.08000	0.10000
60.00000	0.57735	5.30940	0.09417	0.05371	0.10874
90.00000	0.00000	3.00000	0.16667	0.00000	0.16667
13.74000	4.08975	19.35902	0.02583	0.10563	0.10874
16.26020	−3.42857	10.71429	−0.04667	0.16000	0.16667

■

8.13 IMPEDANCE AT THE RELAY

The concepts of impedance and admittance are useful in relay applications as a means of visualizing what the relay sees under both normal and fault conditions. Consider the general system shown in Figure 8.27, where a protected element is shown with a circuit breaker at each end and an equivalent system for the power system. A complete analysis of this system requires a consideration of both symmetrical and asymmetrical fault conditions.

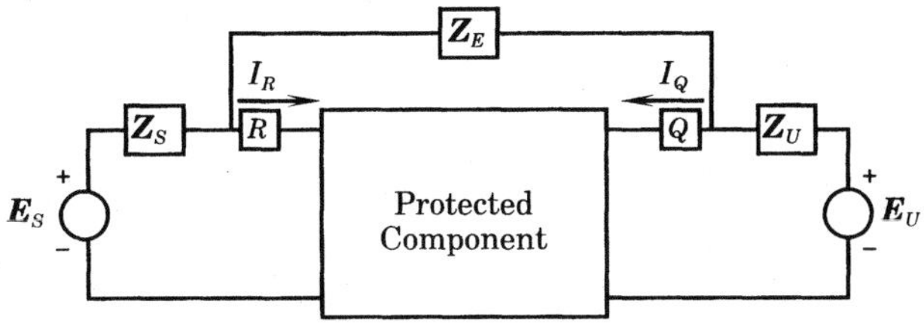

Figure 8.27 Equivalent circuit of a relay protective scheme.

The method of symmetrical components is usually used for asymmetrical conditions. In these unbalanced fault problems we need the Thevenin equivalent circuits of the positive, negative, and zero sequence networks as shown in Figure 8.28. Usually the protected zone can be represented by a π section, as shown in Figure 8.28. The three networks must then be interconnected as required to represent desired unbalanced condition with one or more wires shorted or opened [1].

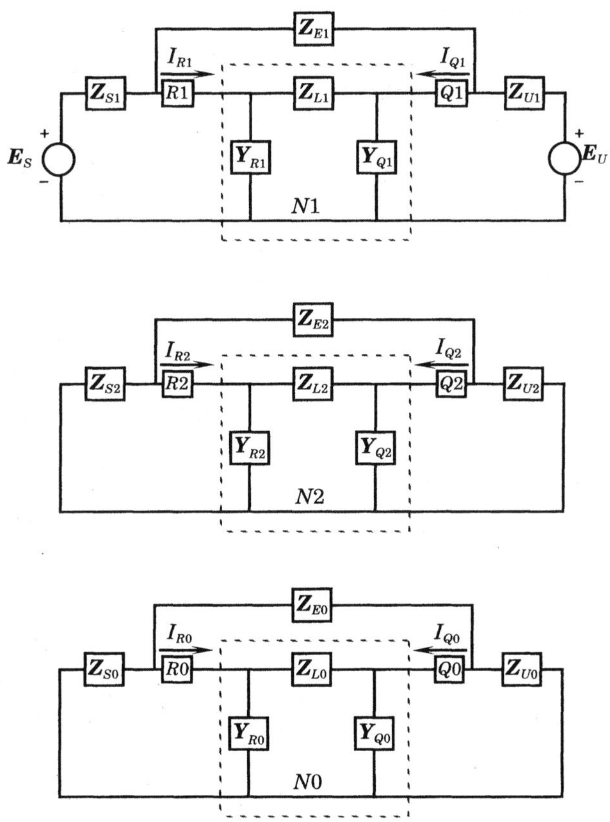

Figure 8.28 Sequence equivalents of a protective scheme.

There are many balanced conditions that require a consideration of only the positive sequence network. These include three-phase faults and swinging out-of-step or other system conditions which follow a disturbance. Viewing the protected zone from the local relay position R is used to determine relay action and to design relay strategies for added selectivity.

The impedance seen by the local relay of Figure 8.27 is a function of both the system operating condition and of the VT and CT connections. Consider a symmetrical system where, at the local relay

V_a = phase-to-neutral voltage, V

I_a = phase a current, A

Then the impedance seen by the local relay, Z_R, is

$$Z_R = \boldsymbol{p}\frac{\boldsymbol{V}_a}{\boldsymbol{I}_a} \tag{8.139}$$

where $\boldsymbol{p}$ is a complex constant depending on the relay ac connection scheme. For example, in the 90 degree connection of Figure 2.3 we have

$$Z_R = \frac{\boldsymbol{V}_{bc}}{\boldsymbol{I}_a} = \boldsymbol{p}\frac{\boldsymbol{V}_a}{\boldsymbol{I}_a} \tag{8.140}$$

and therefore

$$\boldsymbol{p} = \frac{\boldsymbol{V}_{bc}}{\boldsymbol{V}_a} = \sqrt{3}e^{j\pi/2} = -j\sqrt{3} \tag{8.141}$$

The so-called 30 degree connection, which is also used occasionally (see problem 2.2), gives a $\boldsymbol{p}$ value of

$$\boldsymbol{p} = \frac{\boldsymbol{V}_{ac}}{\boldsymbol{V}_a} = \sqrt{3}e^{-j\pi/6} \tag{8.142}$$

For unbalanced faults the value to use for $\boldsymbol{p}$ is not so obvious as it depends on the impedance of all three sequence networks and of the relay ac connection scheme.

For some problems it is more helpful to plot $\boldsymbol{Y}_R = 1/\boldsymbol{Z}_R$ for the local relay conditions. In these problems it is often convenient to be able to sketch the relay characteristics in both planes or to compute these characteristics by hand calculator or computer. The concepts presented in this chapter are helpful in performing these calculations and also for clarifying conceptually the relay behavior.

REFERENCES

[1] Anderson, P. M., *Analysis of Faulted Power Systems*, IEEE Press, Piscataway, NJ, 1995.

[2] Atabekov, G. E., *The Relay Protection of High Voltage Networks*, Pergamon Press, NY, 1960.

[3] Churchill, R. V., *Introduction to Complex Variables and Applications*, McGraw-Hill, NY, 1948.

[4] Reddick, H. W., and F. H. Miller, *Advanced Mathematics for Engineers*, John Wiley and Sons, Inc., NY, 1955.

PROBLEMS

8.1 Review the concept of analytic functions and show that the function $\boldsymbol{Y} = f(\boldsymbol{Z}) = 1/\boldsymbol{Z}$ is analytic.

8.2 What are the Cauchy-Riemann Conditions? Does the function $\boldsymbol{Y} = f(\boldsymbol{Z}) = 1/\boldsymbol{Z}$ satisfy the Cauchy-Riemann conditions?

8.3 What is a conformal mapping?

8.4 Sketch the following transformations in both the Z and Y planes, using the transformation $\mathbf{Z} = 1/\mathbf{Y}$. Note the orthogonality of the resulting curves.

(a) The half plane $R \geq R_o > 0$ (e) The half plane $X \geq X_o > 0$

(b) The half plane $R \geq R_o < 0$ (f) The half plane $X \geq X_o < 0$

(c) The half plane $R \leq R_o > 0$ (g) The half plane $X \leq X_o > 0$

(d) The half plane $R \leq R_o < 0$ (h) The half plane $X \leq X_o < 0$

8.5 Plot the Y image of the infinite strip $0 < X < 1/2$.

8.6 Plot the Y image of the region $R > 2$, $X > 0$.

8.7 Show that the line $X = R - 1$ maps into the circle $G^2 + B^2 - G - B = 0$ and the line $R \geq 0$ maps into $G \leq 0$. Then show that the region between $X \geq R - 1$ and $R \geq 0$ maps into a segment of the Y circle to the left of the B axis.

8.8 Find the Y image of half plane

$$R \leq k, k > 0$$

and plot the result. Compare the result with Figures 8.4 and 8.5.

8.9 Find the inverse of (8.13) when

$$a = 0, b = 2, c = 6, d = -1$$

8.10 Find the Y image of the line passing through the two points (4, 0) and (0, −3) in the Z plane.

8.11 Find the inverse of

$$\mathbf{Z} = 2 + me^{j\pi/4}$$

8.12 Find the inverse of the family of lines passing through

$$R = 3k,\ X = -4k,\ k = 0, 1, 2, \ldots, k \neq 0$$

8.13 Find the inverse of the circle

$$\mathbf{Y} = 1 + 2e^{j\psi}$$

8.14 Find the inverse of the circle

$$\mathbf{Z} = 1 + \frac{1}{2}e^{j\psi}$$

8.15 Find the inverse of the function

$$\mathbf{Z} = \frac{\mathbf{P}}{1 + ke^{j\psi}}$$

where k is a variable, ψ is a constant, and $\mathbf{P} = c + jd$.

8.16 Repeat problem 8.15 if ψ is a variable and k is a constant.

8.17 Plot curves of the R intercept and radius as a function of the constant k when $\mathbf{Y} = 1 + ke^{j\psi}$ and the inversion gives $\mathbf{Z}$ circles similar to those of Figure 8.19.

8.18 Find the Y plane representation of the line $R \geq k_1$, where $k_1 > 0$.

8.19 Find the Y plane representation of the line $X \geq k_1$, where $k_1 > 0$.

8.20 Find the region in the Y plane which corresponds to the semi-infinite Z plane region

$$R > k_1$$

8.21 Find the Y plane image of the half plane $X > k$ under the conditions

(a) $k > 0$

(b) $k = 0$

(c) $k < 0$

8.22 What is an orthogonal trajectory? Review this concept and then compute the trajectories which are orthogonal to the family of circles

$$R^2 + X^2 = cX$$

8.23 Find the trajectories that are orthogonal to the family of hyperbolas

$$R^2 - X^2 = cX$$

8.24 Find the orthogonal k and ψ circles if the constant $\mathbf{A} = 5 - j12$. (See example 8.2.)

8.25 Verify the computation of $\boldsymbol{p}$ for the 30 degree relay connection given by (8.121).

8.26 Consider an impedance circle that passes through the Z origin and let $\mathbf{Z}_1$ be any impedance lying on the circle. Prove that the point $\mathbf{Z}_1$ always forms a right triangle, with the origin being one corner of the triangle, the impedance representing the circle diameter being the corner of the triangle farthest from the origin, and the impedance point $\mathbf{Z}_1$ being the right angle of the triangle. See Figure 8.9 for an example of such a triangle lying inside the circle.

9

Impedance at the Relay

A great deal can be learned about protective system behavior by considering the impedance seen by the relay in the direction of the protected component. We shall refer to this impedance as $\boldsymbol{Z}_R$ and will determine a method of making $\boldsymbol{Z}_R$ plots in the complex Z plane. Since $\boldsymbol{Z}_R$ is a function of the system conditions, we shall make a study of the relay behavior as the system conditions change.

9.1 THE RELAY APPARENT IMPEDANCE, Z_R

To compute the relay impedance $\boldsymbol{Z}_R$ as a function of system conditions in a general form, we begin with the system configuration of Figure 9.1 where we define $\boldsymbol{Z}_R$ as the impedance seen looking toward the protected component.

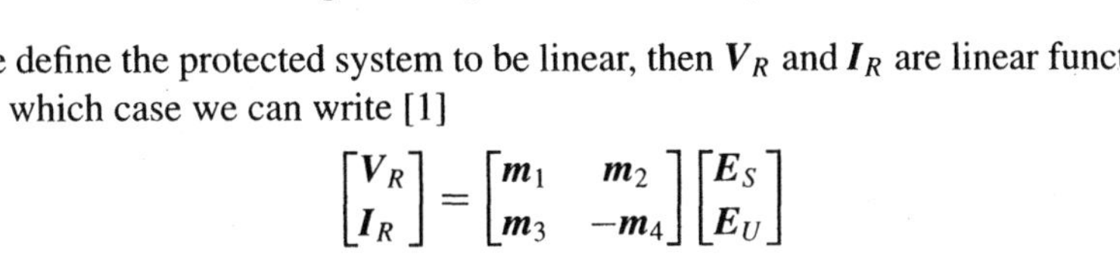

Figure 9.1 Z_R defined for a linear system.

If we define the protected system to be linear, then $\boldsymbol{V}_R$ and $\boldsymbol{I}_R$ are linear functions of $\boldsymbol{E}_S$ and $\boldsymbol{E}_U$, in which case we can write [1]

$$\begin{bmatrix} \boldsymbol{V}_R \\ \boldsymbol{I}_R \end{bmatrix} = \begin{bmatrix} \boldsymbol{m}_1 & \boldsymbol{m}_2 \\ \boldsymbol{m}_3 & -\boldsymbol{m}_4 \end{bmatrix} \begin{bmatrix} \boldsymbol{E}_S \\ \boldsymbol{E}_U \end{bmatrix} \tag{9.1}$$

where the matrix

$$\mathbf{M} = \begin{bmatrix} \boldsymbol{m}_1 & \boldsymbol{m}_2 \\ \boldsymbol{m}_3 & -\boldsymbol{m}_4 \end{bmatrix} \tag{9.2}$$

is a matrix of complex constants, which are functions of the system parameters and the relay connection scheme. Also, from (9.1) we can write

$$\boldsymbol{m}_1 = \left[\frac{\boldsymbol{V}_R}{\boldsymbol{E}_S}\right]_{\boldsymbol{E}_U=0} \qquad \boldsymbol{m}_2 = \left[\frac{\boldsymbol{V}_R}{\boldsymbol{E}_U}\right]_{\boldsymbol{E}_S=0} \tag{9.3}$$

are dimensionless constants, but

$$\boldsymbol{m}_3 = \left[\frac{\boldsymbol{I}_R}{\boldsymbol{E}_S}\right]_{\boldsymbol{E}_U=0} \qquad \boldsymbol{m}_4 = \left[\frac{-\boldsymbol{I}_R}{\boldsymbol{E}_U}\right]_{\boldsymbol{E}_S=0} \tag{9.4}$$

are admittances. All elements of **M** are called short-circuit parameters because they are defined with either $\boldsymbol{E}_S$ or $\boldsymbol{E}_U$ equal to zero, i.e., with one terminal shorted.

Using the definition (9.1), we compute the impedance seen by the relay as

$$\boldsymbol{Z}_R = \frac{\boldsymbol{V}_R}{\boldsymbol{I}_R} = \frac{\boldsymbol{m}_1\boldsymbol{E}_S + \boldsymbol{m}_2\boldsymbol{E}_U}{\boldsymbol{m}_3\boldsymbol{E}_S - \boldsymbol{m}_4\boldsymbol{E}_U} \tag{9.5}$$

Dividing (9.5) by $\boldsymbol{m}_1\boldsymbol{E}_U$, we have

$$\boldsymbol{Z}_R = \frac{(\boldsymbol{E}_S/\boldsymbol{E}_U) + (\boldsymbol{m}_2/\boldsymbol{m}_1)}{(\boldsymbol{m}_3/\boldsymbol{m}_1)(\boldsymbol{E}_S/\boldsymbol{E}_U) - (\boldsymbol{m}_4/\boldsymbol{m}_1)} \tag{9.6}$$

or

$$\boldsymbol{Z}_R = \frac{\boldsymbol{m}_1}{\boldsymbol{m}_3}\left[\frac{(\boldsymbol{E}_S/\boldsymbol{E}_U) + (\boldsymbol{m}_2/\boldsymbol{m}_1)}{(\boldsymbol{E}_S/\boldsymbol{E}_U) - (\boldsymbol{m}_4/\boldsymbol{m}_3)}\right] \tag{9.7}$$

This can be arranged in a more convenient form by performing a long division of the bracketed quantity and simplifying, with the result

$$\boldsymbol{Z}_R = \frac{\boldsymbol{m}_1}{\boldsymbol{m}_3} - \left[\frac{(\boldsymbol{m}_1/\boldsymbol{m}_3) + (\boldsymbol{m}_2/\boldsymbol{m}_4)}{1 - (\boldsymbol{m}_3/\boldsymbol{m}_4)(\boldsymbol{E}_S/\boldsymbol{E}_U)}\right] \tag{9.8}$$

which is recognized to be the difference of an impedance and a Z plane circle.

An alternate form of $\boldsymbol{Z}_R$ is obtained by dividing (9.5) by $\boldsymbol{m}_2\boldsymbol{E}_S$ and following the above procedure to compute

$$\boldsymbol{Z}_R = \frac{-\boldsymbol{m}_2}{\boldsymbol{m}_4} + \left[\frac{(\boldsymbol{m}_1/\boldsymbol{m}_3) + (\boldsymbol{m}_2/\boldsymbol{m}_4)}{1 - (\boldsymbol{m}_4/\boldsymbol{m}_3)(\boldsymbol{E}_U/\boldsymbol{E}_S)}\right] \tag{9.9}$$

We now examine the complex parameters in (9.8) and (9.9). Two of these parameters are impedances, which are defined as follows.

$$\boldsymbol{Z}_X = \frac{\boldsymbol{m}_2}{\boldsymbol{m}_4} \quad \text{ohms} \tag{9.10}$$

$$\boldsymbol{Z}_Z = \frac{\boldsymbol{m}_1}{\boldsymbol{m}_3} \quad \text{ohms} \tag{9.11}$$

The remaining ratio is a dimensionless complex parameter, which we shall call $\boldsymbol{n}$, viz.,

$$\boldsymbol{n} = \frac{\boldsymbol{m}_3}{\boldsymbol{m}_4} \tag{9.12}$$

It is also convenient to define the sum of $\boldsymbol{Z}_X$ and $\boldsymbol{Z}_Z$ as

$$\boldsymbol{Z}_Y = \boldsymbol{Z}_X + \boldsymbol{Z}_Z = \frac{\boldsymbol{m}_1}{\boldsymbol{m}_3} + \frac{\boldsymbol{m}_2}{\boldsymbol{m}_4} \tag{9.13}$$

Then (9.8) and (9.9) can be written, respectively, as

$$\boldsymbol{Z}_R = \boldsymbol{Z}_Z - \frac{\boldsymbol{Z}_Y}{1 - (\boldsymbol{n}\boldsymbol{E}_S/\boldsymbol{E}_U)} \quad \text{ohms} \tag{9.14}$$

or

$$\boldsymbol{Z}_R = -\boldsymbol{Z}_X + \frac{\boldsymbol{Z}_Y}{1 - (\boldsymbol{E}_U/\boldsymbol{n}\boldsymbol{E}_S)} \quad \text{ohms} \tag{9.15}$$

In some cases, it is more convenient to work with the magnitude and angles of the complex quantities separately. Therefore, we define the following.

$$\begin{aligned} \boldsymbol{Z}_X &= Z_X\angle\alpha_x & \boldsymbol{n} &\equiv \boldsymbol{m}_3/\boldsymbol{m}_4 = \nu\angle\alpha_n \\ \boldsymbol{Z}_Y &= Z_Y\angle\alpha_y & \boldsymbol{E} &\equiv \boldsymbol{E}_S/\boldsymbol{E}_U = E\angle\theta \\ \boldsymbol{Z}_Z &= Z_Z\angle\alpha_z \end{aligned} \tag{9.16}$$

Then, using (9.16), (9.14) may be written as

$$\boldsymbol{Z}_R = Z_Z\angle\alpha_z - \frac{Z_Y\angle\alpha_y}{1 - \nu E\angle(\alpha_n + \theta)} \tag{9.17}$$

and (9.15) becomes

$$\boldsymbol{Z}_R = -Z_X\angle\alpha_x - \frac{Z_Y\angle\alpha_y}{1 - \left(\frac{1}{\nu E}\right)\angle - (\alpha_n + \theta)} \tag{9.18}$$

Note that no special conditions of the protected component are assumed in the equations developed above. The protected component can be simple or complex, as there are no restrictions on the derivations.

9.2 PROTECTION EQUIVALENT M PARAMETERS

Consider the protection equivalent shown in Figure 9.2. The protected component is represented by the pi circuit, consisting of impedances $\boldsymbol{Z}_L$, $\boldsymbol{Z}_1$, and $\boldsymbol{Z}_2$. When the protected component is an unfaulted transmission line the shunt impedances $\boldsymbol{Z}_1$ and $\boldsymbol{Z}_2$ may be equal, depending on the line shunt compensation, if any. Other protected components, such as transformers, have pi circuit equivalents where these shunt elements are unequal. The goal is to determine the two-port parameters $\boldsymbol{m}_1$, $\boldsymbol{m}_2$, $\boldsymbol{m}_3$, and $\boldsymbol{m}_4$ for this general protection equivalent, from which we may find the ratios given by (9.10)–(9.12). Note that this can be considered a general case, even though no fault is shown in the figure, since a faulted two-port element can always be represented by an equivalent pi. The results, therefore, will provide insight as to the meaning of the system parameters defined in Section 9.1

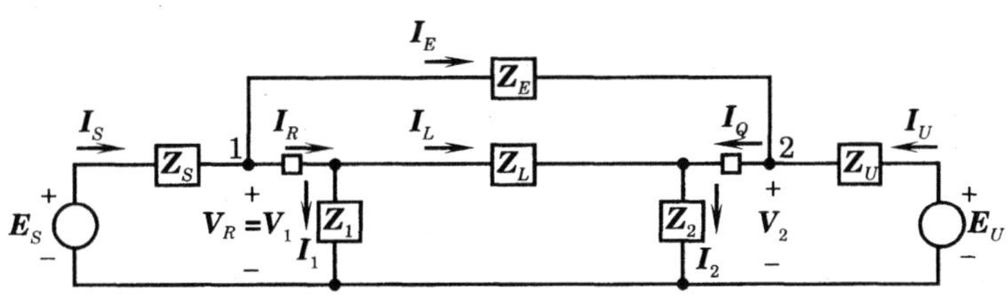

Figure 9.2 Protection equivalent for a pi equivalent protected element.

For the circuit of Figure 9.2, we may write the following general equations.

$$\begin{aligned} \boldsymbol{I}_S &= \boldsymbol{I}_R + \boldsymbol{I}_E & \boldsymbol{I}_U &= \boldsymbol{I}_Q - \boldsymbol{I}_E \\ \boldsymbol{I}_R &= \boldsymbol{I}_1 + \boldsymbol{I}_L = \boldsymbol{I}_S - \boldsymbol{I}_E & \boldsymbol{I}_Q &= \boldsymbol{I}_2 - \boldsymbol{I}_L = \boldsymbol{I}_U + \boldsymbol{I}_E \\ \boldsymbol{E}_S &= \boldsymbol{V}_1 + \boldsymbol{Z}_S\boldsymbol{I}_S & \boldsymbol{E}_U &= \boldsymbol{V}_2 + \boldsymbol{Z}_U\boldsymbol{I}_U \end{aligned} \tag{9.19}$$

We may also write the voltage drop from 1 to 2 as

$$V_1 - V_2 = Z_L I_L = Z_E I_E \tag{9.20}$$

which is the familiar constraint for currents flowing in parallel impedances.

Now, referring to Figure 9.3, we define the new current I_P, representing the total current flowing in the parallel branches Z_L and Z_E.

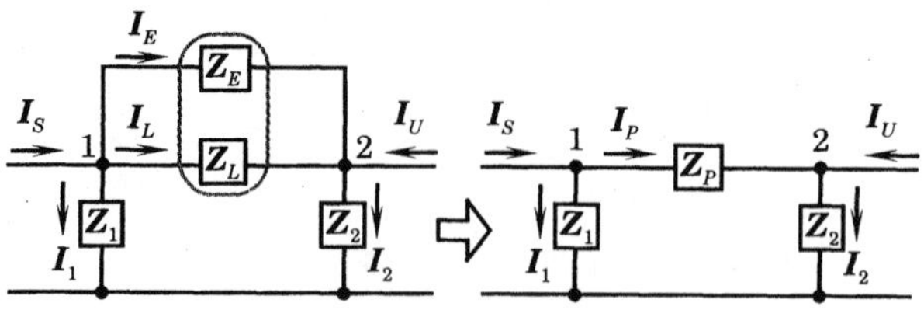

Figure 9.3 Equivalent parallel impedance Z_P.

Then we may write this new current as

$$I_P = I_L + I_E = I_S - I_1 = I_2 - I_U \tag{9.21}$$

and we recognize the new impedance of the parallel combination

$$Z_P = \frac{Z_L Z_E}{Z_L + Z_E} \tag{9.22}$$

Then, using (9.20) and (9.22), we may write

$$I_L = \frac{Z_E}{Z_L} I_E = \frac{Z_P}{Z_L} I_P \tag{9.23}$$

Finally, referring to Figure 9.2, we may write the relay current as

$$I_R = I_1 + I_L = \frac{V_1}{Z_1} + \frac{Z_P}{Z_L} I_P \tag{9.24}$$

The **M** matrix parameters are defined in terms of short-circuit tests on the network. These test are now evaluated.

9.2.1 Network Test with E_U Shorted

The network condition for this test is shown in Figure 9.4, where we have defined impedances observed at key network locations.

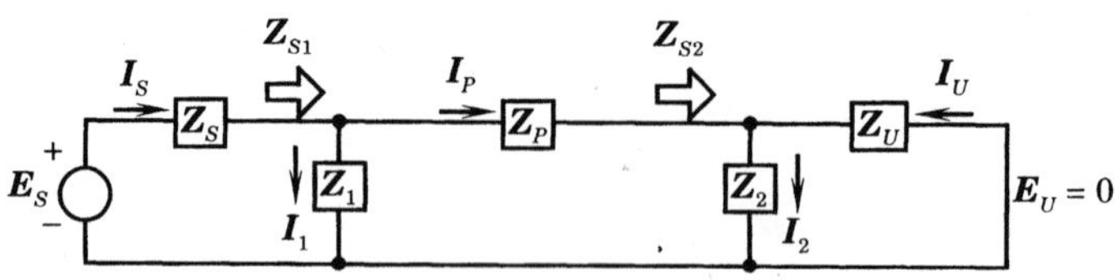

Figure 9.4 Test of the circuit with $E_U = 0$.

From Figure 9.4, we compute the following impedances, seen looking to the right at the defined locations.

$$Z_{S2} = \frac{Z_U Z_2}{Z_U + Z_2} \tag{9.25}$$

$$Z_{S1} = \frac{(Z_P + Z_{S2})Z_1}{Z_P + Z_{S2} + Z_1} = \frac{Z_1[Z_P(Z_U + Z_2) + Z_U Z_2]}{(Z_P + Z_1)(Z_U + Z_2) + Z_U Z_2} \tag{9.26}$$

Then the total impedance seen by the source on the left is

$$\mathbf{Z}_{\text{TOTS}} = \mathbf{Z}_S + \mathbf{Z}_{S1} \tag{9.27}$$

The source current may now be computed as

$$\mathbf{I}_S|_{\mathbf{E}_U=0} = \frac{\mathbf{E}_S}{\mathbf{Z}_{\text{TOTS}}} = \frac{\mathbf{E}_S[(\mathbf{Z}_P + \mathbf{Z}_1)(\mathbf{Z}_U + \mathbf{Z}_2) + \mathbf{Z}_U\mathbf{Z}_2]}{\text{DEN}U} \tag{9.28}$$

where we have defined the denominator term as

$$\text{DEN}U = \mathbf{Z}_S[(\mathbf{Z}_P + \mathbf{Z}_1)(\mathbf{Z}_U + \mathbf{Z}_2) + \mathbf{Z}_U\mathbf{Z}_2] + \mathbf{Z}_1[\mathbf{Z}_P(\mathbf{Z}_U + \mathbf{Z}_2) + \mathbf{Z}_U\mathbf{Z}_2] \tag{9.29}$$

The voltage at node 1 is also the relay voltage, which is computed from (9.19).

$$\mathbf{V}_1|_{\mathbf{E}_U=0} = \mathbf{V}_R|_{\mathbf{E}_U=0} = \mathbf{E}_S - \mathbf{Z}_S\mathbf{I}_S = \frac{\mathbf{E}_S\mathbf{Z}_1[\mathbf{Z}_P(\mathbf{Z}_U + \mathbf{Z}_2) + \mathbf{Z}_U\mathbf{Z}_2]}{\text{DEN}U} \tag{9.30}$$

The voltage at node 2 is given by

$$\mathbf{V}_2|_{\mathbf{E}_U=0} = \mathbf{V}_1 - \mathbf{Z}_P\mathbf{I}_P = \frac{\mathbf{E}_S\mathbf{Z}_P\mathbf{Z}_1\mathbf{Z}_2}{\text{DEN}U} \tag{9.31}$$

The current through the parallel branch is given by

$$\mathbf{I}_P|_{\mathbf{E}_U=0} = \frac{\mathbf{V}_1 - \mathbf{V}_2}{\mathbf{Z}_P} = \frac{\mathbf{E}_S\mathbf{Z}_1(\mathbf{Z}_U + \mathbf{Z}_2)}{\text{DEN}U} \tag{9.32}$$

The relay current is computed from (9.24).

$$\begin{aligned}\mathbf{I}_R|_{\mathbf{E}_U=0} &= \mathbf{I}_1 + \mathbf{I}_L = \frac{\mathbf{V}_1}{\mathbf{Z}_1} + \frac{\mathbf{Z}_P}{\mathbf{Z}_L}\mathbf{I}_P \\ &= \frac{\mathbf{E}_S\{[\mathbf{Z}_P(\mathbf{Z}_U + \mathbf{Z}_2) + \mathbf{Z}_U\mathbf{Z}_2] + \mathbf{Z}_1(\mathbf{Z}_P/\mathbf{Z}_L)(\mathbf{Z}_U + \mathbf{Z}_2)\}}{\text{DEN}U}\end{aligned} \tag{9.33}$$

The above results are sufficient to compute two **M** matrix parameters.

$$\mathbf{m}_1 = \left.\frac{\mathbf{V}_R}{\mathbf{E}_S}\right|_{\mathbf{E}_U=0} = \frac{\mathbf{Z}_1[\mathbf{Z}_P(\mathbf{Z}_U + \mathbf{Z}_2) + \mathbf{Z}_U\mathbf{Z}_2]}{\text{DEN}U} \tag{9.34}$$

$$\mathbf{m}_3 = \left.\frac{\mathbf{I}_R}{\mathbf{E}_S}\right|_{\mathbf{E}_U=0} = \frac{[\mathbf{Z}_P(\mathbf{Z}_U + \mathbf{Z}_2) + \mathbf{Z}_U\mathbf{Z}_2] + \mathbf{Z}_1(\mathbf{Z}_P/\mathbf{Z}_L)(\mathbf{Z}_U + \mathbf{Z}_2)}{\text{DEN}U} \tag{9.35}$$

Note that $\mathbf{m}_3$ has the dimensions of admittance, but $\mathbf{m}_1$ is dimensionless.

9.2.2 Network Test with E_S Shorted

The tests with $\mathbf{E}_S$ shorted are carried out in exactly the same manner as the previous test. The results, based on the circuit of Figure 9.5 are given without detailed discussion.

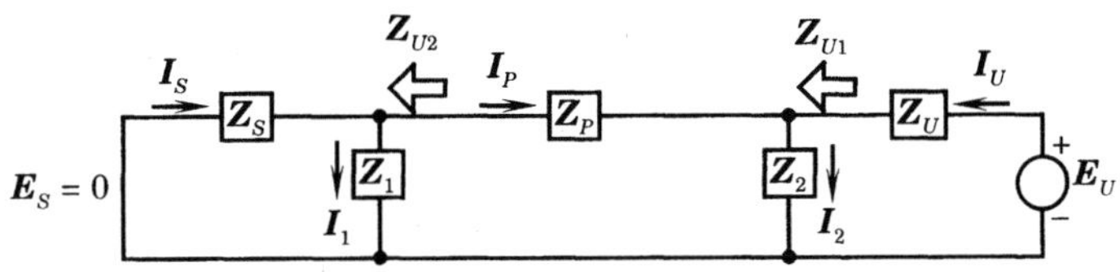

Figure 9.5 Test of the circuit with $\mathbf{E}_S = 0$.

The source current is given by

$$\mathbf{I}_U|_{\mathbf{E}_S=0} = \frac{\mathbf{E}_U}{\mathbf{Z}_{\text{TOTU}}} = \frac{\mathbf{E}_U[(\mathbf{Z}_P + \mathbf{Z}_2)(\mathbf{Z}_S + \mathbf{Z}_1) + \mathbf{Z}_S\mathbf{Z}_1]}{\text{DEN}S} \tag{9.36}$$

where

$$\mathrm{DENS} = \boldsymbol{Z}_U[(\boldsymbol{Z}_P + \boldsymbol{Z}_2)(\boldsymbol{Z}_S + \boldsymbol{Z}_1) + \boldsymbol{Z}_S\boldsymbol{Z}_1] + \boldsymbol{Z}_2[\boldsymbol{Z}_P(\boldsymbol{Z}_S + \boldsymbol{Z}_1) + \boldsymbol{Z}_S\boldsymbol{Z}_1] \tag{9.37}$$

The voltage at node 2 is computed as

$$\boldsymbol{V}_2|_{\boldsymbol{E}_S=0} = \frac{\boldsymbol{E}_U\boldsymbol{Z}_2[\boldsymbol{Z}_P(\boldsymbol{Z}_S + \boldsymbol{Z}_1) + \boldsymbol{Z}_S\boldsymbol{Z}_1]}{\mathrm{DENS}} \tag{9.38}$$

The current in the parallel branch is computed from (9.21).

$$\boldsymbol{I}_P|_{\boldsymbol{E}_S=0} = -\frac{\boldsymbol{E}_U\boldsymbol{Z}_2(\boldsymbol{Z}_S + \boldsymbol{Z}_1)}{\mathrm{DENS}} \tag{9.39}$$

From this result the voltage at node 1, the relay voltage, can be computed.

$$\boldsymbol{V}_1|_{\boldsymbol{E}_S=0} = \boldsymbol{V}_2 + \boldsymbol{Z}_P\boldsymbol{I}_P = \frac{\boldsymbol{E}_U\boldsymbol{Z}_S\boldsymbol{Z}_1\boldsymbol{Z}_2}{\mathrm{DENS}} \tag{9.40}$$

The relay current is computed from (9.24).

$$\boldsymbol{I}_R|_{\boldsymbol{E}_S=0} = \boldsymbol{I}_1 + \boldsymbol{I}_L = \frac{\boldsymbol{E}_U[\boldsymbol{Z}_S\boldsymbol{Z}_2 - \boldsymbol{Z}_2(\boldsymbol{Z}_P/\boldsymbol{Z}_L)(\boldsymbol{Z}_S + \boldsymbol{Z}_1)]}{\mathrm{DENS}} \tag{9.41}$$

Then

$$\boldsymbol{m}_2 = \left.\frac{\boldsymbol{V}_R}{\boldsymbol{E}_U}\right|_{\boldsymbol{E}_S=0} = \frac{\boldsymbol{Z}_S\boldsymbol{Z}_1\boldsymbol{Z}_2}{\mathrm{DENS}} \tag{9.42}$$

$$\boldsymbol{m}_4 = \left.\frac{-\boldsymbol{I}_R}{\boldsymbol{E}_U}\right|_{\boldsymbol{E}_S=0} = \frac{\boldsymbol{Z}_2(\boldsymbol{Z}_P/\boldsymbol{Z}_L)(\boldsymbol{Z}_S + \boldsymbol{Z}_1) - \boldsymbol{Z}_S\boldsymbol{Z}_2}{\mathrm{DENS}} \tag{9.43}$$

Again, we note that $\boldsymbol{m}_4$ is an admittance, but $\boldsymbol{m}_2$ is dimensionless.

9.3 THE CIRCLE LOCI $\boldsymbol{Z} = \boldsymbol{P}/(1 \pm \boldsymbol{Y}_K)$

Before beginning with the construction of $\boldsymbol{Z}_R$ it will be helpful to examine the inverse of the circle $\boldsymbol{Y} = 1 \pm \boldsymbol{Y}_K$ when multiplied by a complex constant $\boldsymbol{P}$, i.e.,

$$\boldsymbol{Z} = \frac{\boldsymbol{P}}{1 \pm \boldsymbol{Y}_K} \tag{9.44}$$

where

$$\boldsymbol{Y}_K = Y_K e^{j\psi}$$

and where ψ is a variable. Then (9.44) may be written as

$$\boldsymbol{Z} = \frac{\boldsymbol{P}}{1 \pm Y_K e^{j\psi}} \tag{9.45}$$

This is recognized to be the form of the second term of $\boldsymbol{Z}_R$ in (9.14) and (9.17). For convenience we define

$$\boldsymbol{P} = c + jd = |\boldsymbol{P}|e^{j\gamma} = \sqrt{c^2 + d^2}e^{j\gamma} = pe^{j\gamma} \tag{9.46}$$

where we define p to be the magnitude of $\boldsymbol{P}$. Then (9.44) can be solved for $\boldsymbol{Y}_K$ with the result

$$\pm\boldsymbol{Y}_K = \frac{\boldsymbol{P} - \boldsymbol{Z}}{\boldsymbol{Z}} \tag{9.47}$$

or

$$\pm\boldsymbol{Y}_K = \pm Y_K e^{j\psi} = \frac{(c - R) + j(d - X)}{R + jX} \tag{9.48}$$

This expression has magnitude-squared

$$Y_K^2 = \frac{(c-R)^2 + (d-X)^2}{R^2 + X^2} \tag{9.49}$$

which clearly is a circle in the Z plane. Result (9.49) can be rearranged as

$$\left(R - \frac{c}{1-Y_K^2}\right)^2 + \left(X - \frac{d}{1-Y_K^2}\right)^2 = \left(\frac{Y_K p}{1-Y_K^2}\right)^2 = S^2 \tag{9.50}$$

Equation (9.50) describes a circle with parameters

$$\begin{aligned} &\text{Center:} \quad \left(\frac{c}{1-Y_K^2}, \frac{d}{1-Y_K^2}\right) \\ &\text{Radius:} \quad S = \frac{pY_K}{1-Y_K^2} \end{aligned} \tag{9.51}$$

Also, from (9.50), we observe that the center is located at

$$\boldsymbol{Z}_o = \frac{c}{1-Y_K^2} + j\frac{d}{1-Y_K^2} = \frac{\boldsymbol{P}}{1-Y_K^2} \tag{9.52}$$

9.4 $\boldsymbol{Z_R}$ LOCI CONSTRUCTION

To construct $\boldsymbol{Z}_R$ loci in the complex impedance plane we assume that the matrix $\mathbf{M}$ is known, i.e., $\boldsymbol{m}_1$, $\boldsymbol{m}_2$, $\boldsymbol{m}_3$, and $\boldsymbol{m}_4$ are known. We then construct the $\boldsymbol{Z}_R$ loci under two conditions:

$$(1) \qquad E = \text{a specified parameter} \tag{9.53}$$

or

$$(2) \qquad \psi = \alpha_n + \theta = \text{a specified parameter} \tag{9.54}$$

From the results of Chapter 8, we would expect these two families of $\boldsymbol{Z}_R$ loci to be orthogonal circles in the Z plane.

We begin the construction by noting the following. When $\boldsymbol{E}_S = 0$, then

$$A \equiv \boldsymbol{Z}_R|_{\boldsymbol{E}_S=0} = \boldsymbol{Z}_Z - \boldsymbol{Z}_Y = -\boldsymbol{Z}_X = Z_X\angle(\alpha_x + \pi) \tag{9.55}$$

Similarly, when $\boldsymbol{E}_U = 0$, then

$$B \equiv \boldsymbol{Z}_R|_{\boldsymbol{E}_U=0} = -\boldsymbol{Z}_X + \boldsymbol{Z}_Y = \boldsymbol{Z}_Z = Z_Z\angle\alpha_z \tag{9.56}$$

These conditions describe two distinct points in the Z plane, which we designate A and B, respectively, as shown in Figure 9.6. From (9.13) we have

$$\boldsymbol{Z}_Y = \boldsymbol{Z}_Z + \boldsymbol{Z}_X = \overline{AB} \tag{9.57}$$

or $\boldsymbol{Z}_Y$ is the line that connects A to B in the direction A-B. Now, when $|\boldsymbol{ne}| = 1$ and $\psi = \theta + \alpha_n = \pi$ we compute, from (9.14),

$$\boldsymbol{Z}_R = \boldsymbol{Z}_Z - \frac{\boldsymbol{Z}_Y}{1 - 1\angle\pi} = \boldsymbol{Z}_Z - \frac{1}{2}\boldsymbol{Z}_Y \tag{9.58}$$

which can be written as

$$\boldsymbol{Z}_R = \left(\boldsymbol{Z}_Z - \frac{1}{2}\boldsymbol{Z}_Y\right) = \frac{1}{2}(\boldsymbol{Z}_R|_{\boldsymbol{E}_S=0} + \boldsymbol{Z}_R|_{\boldsymbol{E}_U=0}) \tag{9.59}$$

or this point is halfway between A and B. We shall call this point $\mathbf{Z}_M$ or we define the point in the Z plane

$$\mathbf{Z}_M = \mathbf{Z}_Z - \frac{1}{2}\mathbf{Z}_Y = -\mathbf{Z}_X + \frac{1}{2}\mathbf{Z}_Y \tag{9.60}$$

and construct the line segment 0-$\mathbf{Z}_M$ shown in Figure 9.7. Clearly, the point $\mathbf{Z}_M$ bisects the line AB into segments of equal length, where each line segment represents one-half of the impedance $\mathbf{Z}_Y$.

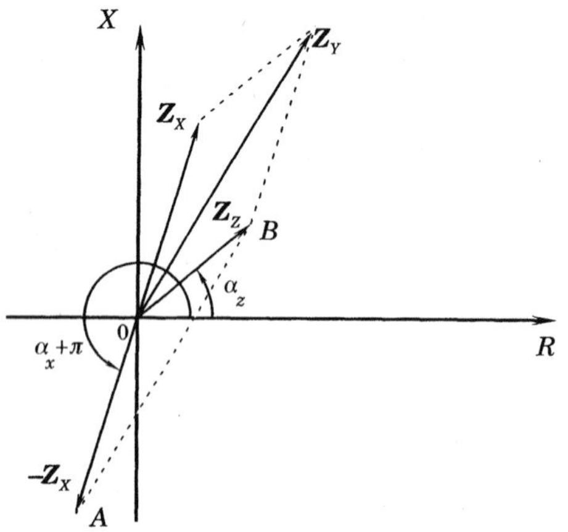

Figure 9.6 Construction of points A, B, and $\mathbf{Z}_y = \overline{AB}$.

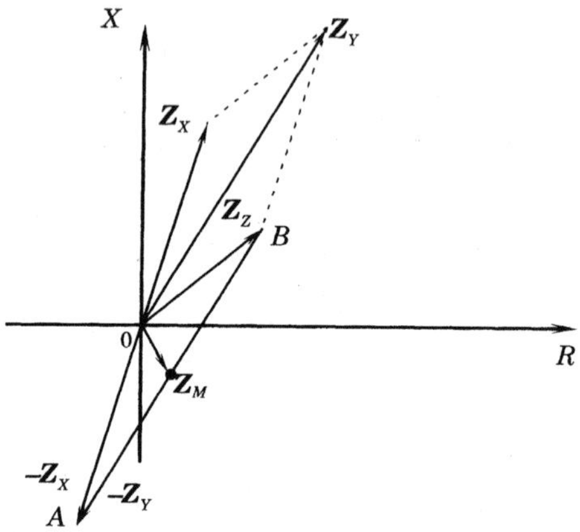

Figure 9.7 Construction of 0-$\mathbf{Z}_M$.

9.4.1 *k* Circles

We now superimpose a family of circles on the construction of Figure 9.7 which correspond to the conditions

$$E = \left|\frac{\mathbf{E}_S}{\mathbf{E}_U}\right| = \text{a parameter} \tag{9.61}$$

For convenience we define a dimensionless parameter as

$$k = \nu E = |\mathbf{n}|\left|\frac{\mathbf{E}_S}{\mathbf{E}_U}\right| \tag{9.62}$$

Then, from (9.14) and the above definitions, we can write

$$\mathbf{Z}_R = \mathbf{Z}_Z - \frac{\mathbf{Z}_Y}{1 - ke^{j\psi}} \tag{9.63}$$

The second term is recognized immediately, from (9.45), to be a circle described by (9.50) and with a radius given by (9.51). We can simplify the result by defining the second term of (9.63) as

$$\mathbf{Z} = R + jX = -\frac{\mathbf{Z}_Y}{1 - ke^{j\psi}} \tag{9.64}$$

which is a circle in the Z plane. Using (9.64) we can compute

$$ke^{j\psi} = \frac{\mathbf{Z} + \mathbf{Z}_Y}{\mathbf{Z}} = \frac{(R + R_Y) + j(X + X_Y)}{R + jX} \tag{9.65}$$

The k circle in the Z plane can be determined by writing the magnitude-squared of (9.65).

$$k^2 = \frac{(R + R_Y)^2 + (X + X_Y)^2}{R^2 + X^2} \tag{9.66}$$

This can be rearranged to the more convenient quadratic form

$$\left(R + \frac{R_Y}{1 - k^2}\right)^2 + \left(X + \frac{X_Y}{1 - k^2}\right)^2 = \left(\frac{kZ_Y}{1 - k^2}\right)^2 \tag{9.67}$$

But, from (9.63) and (9.64) we have

$$\mathbf{Z}_R = \mathbf{Z}_Z + \mathbf{Z} \tag{9.68}$$

which defines a new circle, in the Z_R plane. From (9.68), we can write

$$\begin{aligned} R_R &= R_Z + R \\ X_R &= X_Z + X \end{aligned} \tag{9.69}$$

Making this substitution into (9.67), we get the relay characteristic

$$(R_R - R_k)^2 + (X_R - X_k)^2 = S_k^2 \tag{9.70}$$

where we identify the coordinates of the center $\mathbf{Z}_k$ of the circle to be

$$\mathbf{Z}_k = R_k + jX_k = \left(R_Z - \frac{R_Y}{1 - k^2}\right) + j\left(X_Z - \frac{X_Y}{1 - k^2}\right) \tag{9.71}$$

We also find the radius of the circle from (9.67) to be

$$S_k = \frac{kZ_Y}{1 - k^2} \tag{9.72}$$

A typical k circle is illustrated in Figure 9.8. Expressions can be found to measure the distance to the center from either the point $\mathbf{Z}_M$ or from the point B. The distance from $\mathbf{Z}_k$ to $\mathbf{Z}_M$ represents a limiting value for the radius of the circle, and all circles will have a radius smaller than this critical distance.

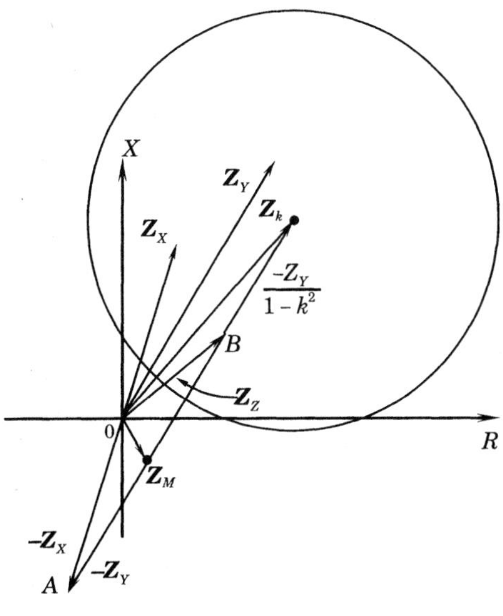

Figure 9.8 Center of E or k circles for $k > 1$.

The following relationships can be computed to illustrate the effect of the parameter k on the location of the center of the family of circles, all of which lie along extensions of the line AB in both directions.

$$\begin{aligned}\frac{\mathbf{Z}_k - \mathbf{Z}_M}{\mathbf{Z}_Y} &< 0 \qquad 0 < k < 1\\ \frac{\mathbf{Z}_k - \mathbf{Z}_M}{\mathbf{Z}_Y} &> 0 \qquad 1 < k < \infty\end{aligned} \tag{9.73}$$

The absolute value of the left-hand side is always greater than $1/2$. Note that when $k = \infty$ and $k = 0$ the circles regenerate to the points A and B, respectively. Moreover, since the radius is always slightly less than $\overline{Z_M Z_k}$ the circles never encircle the point $\mathbf{Z}_M$ except when $k = 1$, which is the circle of infinite radius passing through $\mathbf{Z}_M$ and is itself a perpendicular bisector of the line AB. The locations of the centers of the k circles are given in Figure 9.9.

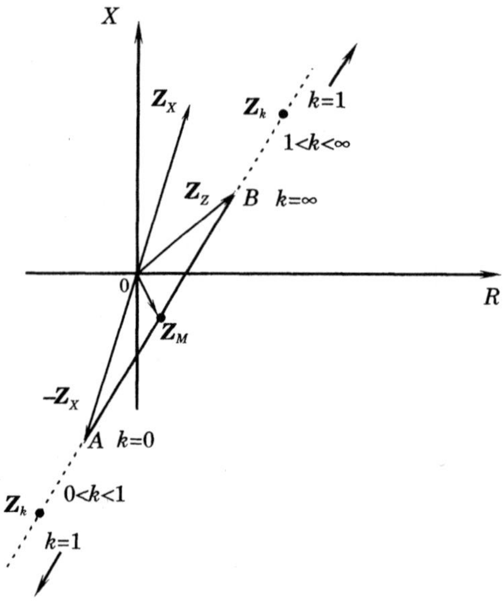

Figure 9.9 Locations of centers C_o of k circles for various values of k.

9.4.2 ψ Circles

Now construct the circles where

$$\psi = \alpha_n + \theta = \text{a parameter} \tag{9.74}$$

where again we write

$$\mathbf{Z}_R = \mathbf{Z}_Z - \frac{\mathbf{Z}_Y}{1 - ke^{j\psi}} \tag{9.75}$$

This is also a family of circles that is orthogonal to the k circle family. To determine the equations for the ψ circles, we write (9.65) as

$$ke^{j\psi} = k\cos\psi + jk\sin\psi = \frac{(R + R_Y) + j(X + X_Y)}{R + jX} \tag{9.76}$$

From this expression, we can solve for the real and imaginary parts to write

$$\begin{aligned}k\cos\psi &= \frac{R(R + R_Y) + X(X + X_Y)}{R^2 + X^2}\\ k\sin\psi &= \frac{R(X + X_Y) - X(R + R_Y)}{R^2 + X^2}\end{aligned} \tag{9.77}$$

We now find the cotangent of the angle ψ to be

$$\cot\psi = \frac{R(R+R_Y)+X(X+X_Y)}{R(X+X_Y)-X(R+R_Y)} \tag{9.78}$$

This can be rearranged to a more convenient quadratic form, as follows.

$$\left(R+\frac{R_Y - X_Y\cot\psi}{2}\right)^2 + \left(X+\frac{X_Y+R_Y\cot\psi}{2}\right)^2 = \left(\frac{Z_Y}{2\sin\psi}\right)^2 \tag{9.79}$$

which is clearly a circle in the Z plane. Now, we use (9.69) to relate this equation to that representing the relay characteristic. The result of this substitution is as follows.

$$(R_R - R_\psi)^2 + (X_R - X_\psi)^2 = S_\psi^2 \tag{9.80}$$

where the center of the circles are located at coordinates

$$\begin{aligned} R_\psi &= R_Z - \frac{R_Y - X_Y\cot\psi}{2} \\ X_\psi &= X_Z - \frac{X_Y + R_Y\cot\psi}{2} \end{aligned} \tag{9.81}$$

and the radius of the circles is given by

$$S_\psi = \frac{Z_Y}{2\sin\psi} \tag{9.82}$$

It can be shown that this family of circles passes through the points A and B. If we identify the center in the Z_R plane as $\mathbf{Z}_\psi$, then we can write

$$\mathbf{Z}_\psi = \mathbf{Z}_Z - \mathbf{Z}_Y\left(\frac{1}{2}+j\frac{1}{2}\cot\psi\right) = \mathbf{Z}_M - j\frac{1}{2}\mathbf{Z}_Y\cot\psi \tag{9.83}$$

where the complex number in parentheses is the center of the Z plane circle prior to its multiplication by $\mathbf{Z}_Y$, which is known from (8.92).

The location of $\mathbf{Z}_\psi$ is shown in Figure 9.10 and is easily constructed once the angle ψ is known. The center of the circle, $\mathbf{Z}_\psi$, is always on the line GH, which is the perpendicular bisector of AB. We may write the following inequalities:

$$\begin{aligned} &\cot\psi > 0, \quad && 0<\psi<\pi/2,\ \pi<\psi<3\pi/2 \\ &\cot\psi < 0, && \pi/2<\psi<\pi,\ 3\pi/2<\psi<2\pi \\ &\cot\psi = 0, && \psi = \pi/2,\ 3\pi/2 \end{aligned}$$

For ψ in the first and third quadrants, $\mathbf{Z}_\psi$ will lie in the H direction from $\mathbf{Z}_M$.

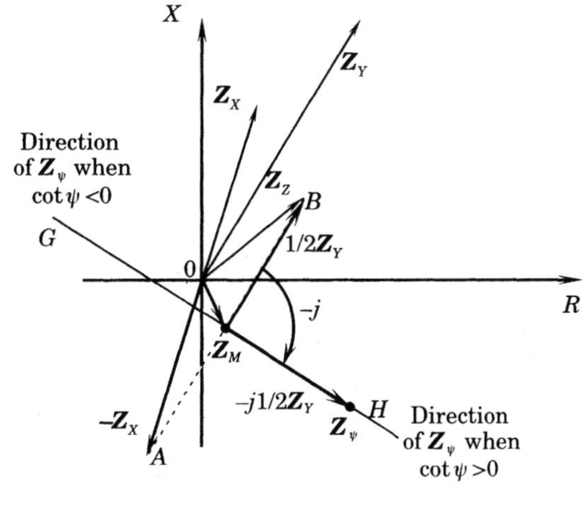

Figure 9.10 Location of the center $\mathbf{Z}_\psi$ for ψ circles.

EXAMPLE 9.1

Construct the families of k and ψ circles for a system with the following data given:

$$\mathbf{Z}_X = 20 + j60 = 63.246\angle 71.565^\circ$$
$$\mathbf{Z}_Z = 40 + j30 = 50.000\angle 36.870^\circ$$
$$\nu = 1.30$$

Solution

First we compute the needed constants. For $\mathbf{Z}_Y$ in ohms, we have

$$\mathbf{Z}_Y = \mathbf{Z}_X + \mathbf{Z}_Z = 60 + j90 = 108.167\angle 56.310^\circ$$
$$k = \nu E = 1.30E$$

k circles

The k circles are defined by (9.70)–(9.72). Values of the parameter k, which corresponds to the ratio of the equivalent system voltages, results in a family of circles, where the impedance $\mathbf{Z}_Y$ is an important constant. For the family of circles to be computed, we are given values of k and each circle corresponds to a particular value of the ratio E. The construction parameters for the k circles are given in Table 9.1. The last column of Table 9.1 gives the distance of the center of each member of the family of circles, located with respect to the point $\mathbf{Z}_M$ as shown in Figure 9.8 and is always greater than the radius S. The value corresponding to $k = 1$ is the perpendicular bisector that bisects the line between A and B, which corresponds to a circle of infinite radius.

TABLE 9.1 Computation of k Circle Coordinates

E	k	k^2	$1-k^2$	kZ_Y	S	$Z_k - Z_M$
0.4	0.52	0.2704	0.7296	56.2466	77.0924	94.1713
0.5	0.65	0.4225	0.5775	70.3082	121.7459	133.2181
0.6	0.78	0.6084	0.3916	84.3699	215.4492	222.1336
0.9862	1.2821	1.6437	−0.6437	138.6750	215.4492	222.1336
1.1824	1.5385	2.5385	−1.3669	166.4101	121.7459	133.2181
1.4793	1.9231	3.6982	−2.6982	208.0126	77.0924	94.1713

ψ Circles

Equations for the ψ circles are given by (9.80)–(9.82). Selected values of the circle construction parameters are given in Table 9.2. Both the k and ψ circles are plotted in Figure 9.11.

TABLE 9.2 Computation of ψ Circle Parameters

ψ	$\cot\psi$	$0.5Z_Y\cot\psi$	$\sin\psi$	S
−150°	1.7321	93.6750	−0.5000	108.1665
−120°	0.5774	31.2250	−0.8660	62.4500
−90°	0.0000	0.0000	−1.0000	54.0833
−60°	−0.5774	−31.2250	−0.8660	62.4500
−30°	−1.7321	−93.6750	−0.5000	108.1665
0°	∞	∞	0.0000	∞
30°	1.7321	93.6750	0.5000	108.1665
60°	0.5774	31.2250	0.8660	62.4500
90°	0.0000	0.0000	1.0000	54.0833
120°	−0.5774	−31.2250	0.8660	62.4500
150°	−1.7321	−93.6750	0.5000	108.1665

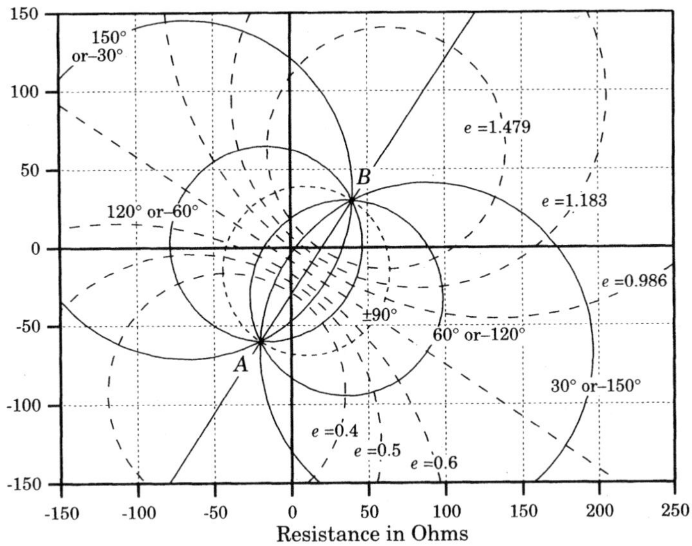

Figure 9.11 Circles of k and ψ circles for Example 9.1. ■

9.5 RELAY APPARENT IMPEDANCE

Suppose we replace the system of Figure 9.1 by the equivalent system of Figure 9.12 where the entire protection equivalent of Figure 9.1, including source impedances, has been included inside the equivalent two port. We solve the two port of Figure 9.12 using the familiar ***ABCD*** constants.

$$\begin{bmatrix} \boldsymbol{E}_S \\ \boldsymbol{I}_S \end{bmatrix} = \begin{bmatrix} \boldsymbol{A} & \boldsymbol{B} \\ \boldsymbol{C} & \boldsymbol{D} \end{bmatrix} \begin{bmatrix} \boldsymbol{E}_U \\ -\boldsymbol{I}_U \end{bmatrix} \tag{9.84}$$

The basic two-port equation (9.84) can be solved for $\boldsymbol{I}_S$ to get

$$\boldsymbol{I}_S = \boldsymbol{C}\boldsymbol{E}_U - \boldsymbol{D}\boldsymbol{I}_U \tag{9.85}$$

But from (9.84),

$$\boldsymbol{E}_S = \boldsymbol{A}\boldsymbol{E}_U - \boldsymbol{B}\boldsymbol{I}_U \tag{9.86}$$

or, solving for the current injected from source #2,

$$\boldsymbol{I}_U = \frac{\boldsymbol{A}\boldsymbol{E}_U - \boldsymbol{E}_S}{\boldsymbol{B}} \tag{9.87}$$

which we substitute into (9.85) to compute

$$\boldsymbol{I}_S = \frac{\boldsymbol{D}}{\boldsymbol{B}}\boldsymbol{E}_S + \frac{\boldsymbol{B}\boldsymbol{C} - \boldsymbol{A}\boldsymbol{D}}{\boldsymbol{B}}\boldsymbol{E}_U \tag{9.88}$$

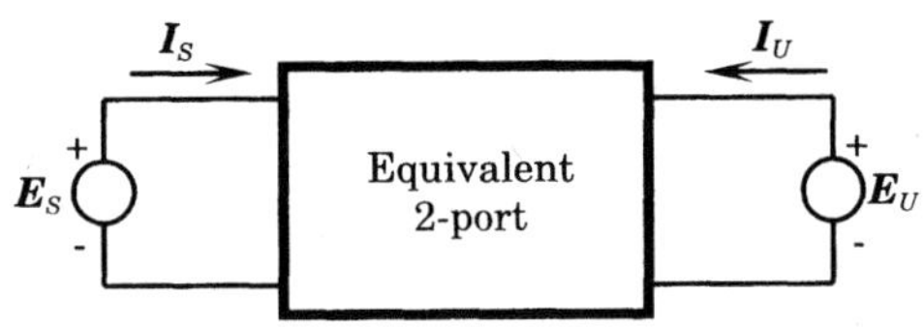

Figure 9.12 Equivalent two-port network.

From two-port network theory we know that, in a reciprocal two port

$$\boldsymbol{AD} - \boldsymbol{BC} = 1 \tag{9.89}$$

Therefore, (9.88) reduces to

$$\boldsymbol{I}_S = \frac{\boldsymbol{DE}_S - \boldsymbol{E}_U}{\boldsymbol{B}} \tag{9.90}$$

This equation is easily solved for $\boldsymbol{E}_S/\boldsymbol{I}_S$ with the result

$$\frac{\boldsymbol{E}_S}{\boldsymbol{I}_S} = \frac{\boldsymbol{B}/\boldsymbol{D}}{1 - \dfrac{\boldsymbol{E}_U}{\boldsymbol{DE}_S}} = \frac{1}{\boldsymbol{Y}_B - \boldsymbol{Y}_V} \tag{9.91}$$

where $\boldsymbol{Y}_B = \boldsymbol{D}/\boldsymbol{B}$

$$\boldsymbol{Y}_V = \frac{\boldsymbol{E}_U}{\boldsymbol{BE}_S} = \frac{e^{j\psi}}{BE} \tag{9.92}$$

and (9.91) is recognized from (8.112) to be a circle in the Z plane.

9.5.1 The Unfaulted System

Consider the unfaulted system shown in Figure 9.13. The equations developed above describe the impedance seen by the relay for the unfaulted case, as well as for situations with faults on the protected line.

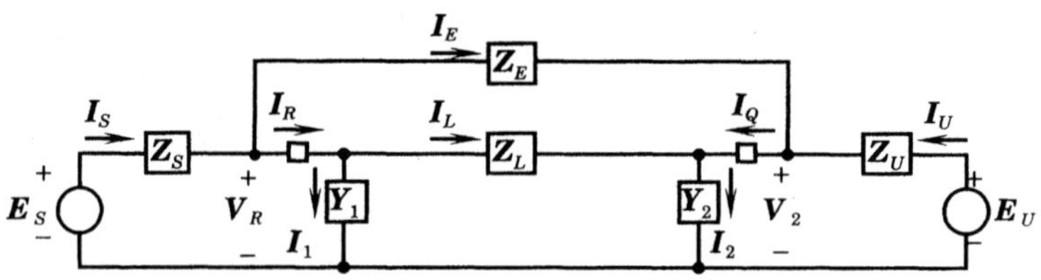

Figure 9.13 Equivalent two-port for a transmission line.

For the protected line in Figure 9.13, the impedance seen by the relay looking into the line is given by

$$\boldsymbol{Z}_R = \frac{\boldsymbol{V}_R}{\boldsymbol{I}_R} \tag{9.93}$$

but, by inspection,

$$\boldsymbol{V}_R = \boldsymbol{E}_S - \boldsymbol{Z}_S\boldsymbol{I}_S \tag{9.94}$$

or

$$\boldsymbol{Z}_R = \frac{\boldsymbol{E}_S - \boldsymbol{Z}_S\boldsymbol{I}_S}{\boldsymbol{I}_R} \tag{9.95}$$

To simplify this expression, it is necessary to resolve the difference between the current $\boldsymbol{I}_R$ and $\boldsymbol{I}_S$ since these currents are not usually equal. From Figure 9.13, we write the following current equation, summing currents at node 1.

$$\boldsymbol{I}_R = \boldsymbol{I}_S - \boldsymbol{I}_E \tag{9.96}$$

But, since the two branches at the top of the figure are in parallel, we can write a constraint between the currents in these branches.

$$\boldsymbol{I}_E = \frac{\boldsymbol{Z}_L}{\boldsymbol{Z}_E}\boldsymbol{I}_L = \frac{\boldsymbol{Z}_L}{\boldsymbol{Z}_E}(\boldsymbol{I}_R - \boldsymbol{I}_1) = \frac{\boldsymbol{Z}_L}{\boldsymbol{Z}_E}(\boldsymbol{I}_R - \boldsymbol{Y}_1\boldsymbol{V}_R) \tag{9.97}$$

Then (9.96) can be written as

$$\boldsymbol{I}_R = \boldsymbol{F}\boldsymbol{I}_S\left(1 - \frac{\boldsymbol{Z}_L\boldsymbol{Z}_S\boldsymbol{Y}_1}{\boldsymbol{Z}_E}\right) + \frac{\boldsymbol{F}\boldsymbol{Z}_L\boldsymbol{Y}_1}{\boldsymbol{Z}_E}\boldsymbol{E}_S \tag{9.98}$$

where $\boldsymbol{F}$ is the complex constant defined by the expression

$$\frac{1}{\boldsymbol{F}} = 1 + \frac{\boldsymbol{Z}_L}{\boldsymbol{Z}_E} \tag{9.99}$$

If we substitute (9.98) into (9.95), we get

$$\begin{aligned} \boldsymbol{Z}_R &= \frac{\boldsymbol{E}_S - \boldsymbol{Z}_S\boldsymbol{I}_S}{\boldsymbol{I}_R} \\ &= \frac{\boldsymbol{E}_S/\boldsymbol{I}_S - \boldsymbol{Z}_S}{\boldsymbol{Y}_D\boldsymbol{E}_S/\boldsymbol{I}_S + \boldsymbol{F} - \boldsymbol{Z}_S\boldsymbol{Y}_D} \end{aligned} \tag{9.100}$$

where we have defined

$$\boldsymbol{Y}_D = \frac{1}{\boldsymbol{Z}_D} = \frac{\boldsymbol{F}\boldsymbol{Z}_L\boldsymbol{Y}_1}{\boldsymbol{Z}_E} \tag{9.101}$$

The result shown in (9.100) can be rewritten in the following form:

$$\boldsymbol{Z}_R = \boldsymbol{Z}_D - \frac{\boldsymbol{F}\boldsymbol{Z}_D^2}{\boldsymbol{E}_S/\boldsymbol{I}_S - \boldsymbol{Z}_S + \boldsymbol{F}\boldsymbol{Z}_D} \tag{9.102}$$

Then, using (9.91) we can write (9.102) as

$$\boldsymbol{Z}_R - \boldsymbol{Z}_D = \frac{\boldsymbol{F}\boldsymbol{Z}_D^2}{\boldsymbol{Z}_1 - 1/(\boldsymbol{Y}_B - \boldsymbol{Y}_V)} = \frac{\boldsymbol{F}\boldsymbol{Z}_D^2(\boldsymbol{Y}_B - \boldsymbol{Y}_V)}{\boldsymbol{Z}_1(\boldsymbol{Y}_B - \boldsymbol{Y}_V) - 1} \tag{9.103}$$

This can be rearranged to write

$$\boldsymbol{Z} = \boldsymbol{Z}_R - \boldsymbol{Z}_D = \frac{1 - ke^{j\psi}}{\boldsymbol{Y}_A - \boldsymbol{Y}_K e^{j\psi}} \tag{9.104}$$

$$\begin{aligned} \text{where} \quad \boldsymbol{Y}_A &= \frac{\boldsymbol{Y}_D^2}{\boldsymbol{F}}(\boldsymbol{Z}_1 - \boldsymbol{B}/\boldsymbol{D}) \\ \boldsymbol{Y}_K &= \frac{k\boldsymbol{Y}_D^2\boldsymbol{Z}_1}{\boldsymbol{F}} \\ ke^{j\psi} &= \frac{\boldsymbol{Y}_V}{\boldsymbol{Y}_B} = \frac{\boldsymbol{Y}_V\boldsymbol{B}}{\boldsymbol{D}} \end{aligned} \tag{9.105}$$

It can be shown that (9.104) is a circle in the Z plane. This proof is left as an exercise. The foregoing results are complex, but can be greatly simplified if we recognize that $\boldsymbol{Y}_1$ is usually small. In this case (9.100) reduces to

$$\boldsymbol{Z}_R = \frac{1}{\boldsymbol{F}}\left(\frac{\boldsymbol{E}_S}{\boldsymbol{I}_S} - \boldsymbol{Z}_S\right) = \frac{1}{\boldsymbol{F}}\left(\frac{\boldsymbol{B}/\boldsymbol{D}}{1 - \boldsymbol{E}_U/\boldsymbol{D}\boldsymbol{E}_S} - \boldsymbol{Z}_S\right) \tag{9.106}$$

which clearly is a circle in the Z_R plane.

EXAMPLE 9.2

Find the ***ABCD*** matrix for the protection equivalent of Figure 9.13, where the protected component is a transmission line with lumped shunt admittances at each end of the line. Find the ***ABCD*** parameters for this system and determine how these parameters are changed if the shunt admittances are either (i) equal, or (ii) zero.

Solution

It is helpful to recognize that the ***ABCD*** parameters are designed for convenience in combining cascaded two-port networks. Therefore, we write the ***ABCD*** matrix as the product of five separate matrices, as follows.

$$\begin{aligned}\begin{bmatrix}\boldsymbol{A} & \boldsymbol{B}\\ \boldsymbol{C} & \boldsymbol{D}\end{bmatrix} &= \begin{bmatrix}\boldsymbol{A}_S & \boldsymbol{B}_S\\ \boldsymbol{C}_S & \boldsymbol{D}_S\end{bmatrix}\begin{bmatrix}\boldsymbol{A}_{YS} & \boldsymbol{B}_{YS}\\ \boldsymbol{C}_{YS} & \boldsymbol{D}_{YS}\end{bmatrix}\begin{bmatrix}\boldsymbol{A}_P & \boldsymbol{B}_P\\ \boldsymbol{C}_P & \boldsymbol{D}_P\end{bmatrix}\begin{bmatrix}\boldsymbol{A}_{YU} & \boldsymbol{B}_{YU}\\ \boldsymbol{C}_{YU} & \boldsymbol{D}_{YU}\end{bmatrix}\begin{bmatrix}\boldsymbol{A}_U & \boldsymbol{B}_U\\ \boldsymbol{C}_U & \boldsymbol{D}_U\end{bmatrix}\\ &= \begin{bmatrix}1 & \boldsymbol{Z}_S\\ 0 & 1\end{bmatrix}\begin{bmatrix}1 & 0\\ \boldsymbol{Y}_1 & 1\end{bmatrix}\begin{bmatrix}1 & \boldsymbol{Z}_P\\ 0 & 1\end{bmatrix}\begin{bmatrix}1 & 0\\ \boldsymbol{Y}_2 & 1\end{bmatrix}\begin{bmatrix}1 & \boldsymbol{Z}_U\\ 0 & 1\end{bmatrix}\end{aligned} \tag{9.107}$$

Note that the parallel elements $\boldsymbol{Z}_E$ and $\boldsymbol{Z}_L$ have been combined for simplicity of notation, but this does not sacrifice accuracy in the analysis. The matrix multiplication gives the following result.

$$\begin{aligned}\boldsymbol{A} &= 1 + \boldsymbol{Z}_S(\boldsymbol{Y}_1 + \boldsymbol{Y}_2) + \boldsymbol{Z}_P(\boldsymbol{Y}_2 + \boldsymbol{Z}_S\boldsymbol{Y}_1\boldsymbol{Y}_2)\\ \boldsymbol{B} &= \boldsymbol{Z}_S + \boldsymbol{Z}_P + \boldsymbol{Z}_U + \boldsymbol{Z}_P(\boldsymbol{Z}_S\boldsymbol{Y}_1 + \boldsymbol{Z}_U\boldsymbol{Y}_2) + \boldsymbol{Z}_S\boldsymbol{Z}_U(\boldsymbol{Y}_1 + \boldsymbol{Y}_2 + \boldsymbol{Z}_P\boldsymbol{Y}_1\boldsymbol{Y}_2)\\ \boldsymbol{C} &= \boldsymbol{Y}_1 + \boldsymbol{Y}_2 + \boldsymbol{Z}_P\boldsymbol{Y}_1\boldsymbol{Y}_2\\ \boldsymbol{D} &= 1 + \boldsymbol{Z}_U(\boldsymbol{Y}_1 + \boldsymbol{Y}_2) + \boldsymbol{Z}_P(\boldsymbol{Y}_1 + \boldsymbol{Z}_U\boldsymbol{Y}_1\boldsymbol{Y}_2)\end{aligned} \tag{9.108}$$

(i) For the special case where $\boldsymbol{Y}_1 = \boldsymbol{Y}_2 = \boldsymbol{Y}$, the result (9.108) simplifies to

$$\begin{aligned}\boldsymbol{A} &= 1 + \boldsymbol{Y}\boldsymbol{Z}_P + \boldsymbol{Y}\boldsymbol{Z}_S(2 + \boldsymbol{Y}\boldsymbol{Z}_P)\\ \boldsymbol{B} &= \boldsymbol{Z}_S + \boldsymbol{Z}_P + \boldsymbol{Z}_U + \boldsymbol{Y}\boldsymbol{Z}_P(\boldsymbol{Z}_S + \boldsymbol{Z}_U) + \boldsymbol{Y}\boldsymbol{Z}_S\boldsymbol{Z}_U(2 + \boldsymbol{Y}\boldsymbol{Z}_P)\\ \boldsymbol{C} &= \boldsymbol{Y}(2 + \boldsymbol{Y}\boldsymbol{Z}_P)\\ \boldsymbol{D} &= 1 + \boldsymbol{Y}\boldsymbol{Z}_P + \boldsymbol{Y}\boldsymbol{Z}_U(2 + \boldsymbol{Y}\boldsymbol{Z}_P)\end{aligned} \tag{9.109}$$

(ii) When $\boldsymbol{Y} = 0$, we have a simple short transmission line in parallel with the system equivalent impedance.

$$\begin{aligned}\boldsymbol{A} &= 1\\ \boldsymbol{B} &= \boldsymbol{Z}_S + \boldsymbol{Z}_P + \boldsymbol{Z}_U\\ \boldsymbol{C} &= 0\\ \boldsymbol{D} &= 1\end{aligned} \tag{9.110}$$

This latter case is of interest in system protection, where it is often possible to ignore shunt susceptances because of the low voltage in the region close to a shunt fault and because the admittance of the line shunt susceptance is much smaller that of the fault. ■

In the special case of a short transmission line with negligible shunt capacitance we have the two-port parameters derived in (9.110). Then, for this case, the relay impedance is written from (9.94) as

$$\boldsymbol{Z}_R = \frac{1}{\boldsymbol{F}}\left(\frac{\boldsymbol{Z}_S + \boldsymbol{Z}_P + \boldsymbol{Z}_U}{1 - \boldsymbol{E}_U/\boldsymbol{E}_S} - \boldsymbol{Z}_S\right) \tag{9.111}$$

where the parameter $\boldsymbol{F}$ is given by (9.99).

9.5.2 *ABCD* Parameters for a Faulted System

When the protected system includes a shunt fault, the equations become more complex. Consider the system shown in Figure 9.14 where the protected line is faulted at a fraction h of the line length from the protective equipment at A. The impedance $\boldsymbol{Z}_K$ represents the total fault impedance, which may include an arcing fault resistance as well as the impedance of the negative and zero sequence networks, if required for correct fault representation. This total impedance depends on the type of fault as well as the impedance between phases or from phase to ground at the fault point.

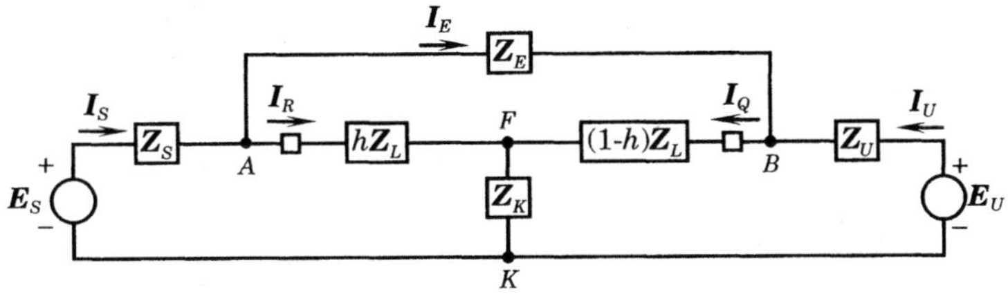

Figure 9.14 Protected component with a shunt fault on line Z_L.

To simplify the computation of system quantities, the wye-connected impedances terminated by points A, B, and K in Figure 9.14 may be replaced by a delta connection of impedances, as shown in Figure 9.15, where the delta impedances consist of elements $\mathbf{Z}_M$, $\mathbf{Z}_G$, and $\mathbf{Z}_H$. In order to define these delta elements, we first define the wye-connected impedance according to their terminal connection. Thus we have, by inspection of Figure 9.14,

$$\begin{aligned} \mathbf{Z}_A &= h\mathbf{Z}_L \\ \mathbf{Z}_B &= (1-h)\mathbf{Z}_L \\ \mathbf{Z}_K &= \text{total fault impedance} \end{aligned} \tag{9.112}$$

Then the delta impedances are computed as follows.

$$\begin{aligned} \mathbf{Z}_M &= \mathbf{Z}_A\mathbf{Z}_B\mathbf{Y}_X = \frac{\mathbf{Z}_N^2}{\mathbf{Z}_K} \\ \mathbf{Z}_G &= \mathbf{Z}_A\mathbf{Z}_K\mathbf{Y}_X = \frac{\mathbf{Z}_N^2}{\mathbf{Z}_B} \\ \mathbf{Z}_H &= \mathbf{Z}_B\mathbf{Z}_K\mathbf{Y}_X = \frac{\mathbf{Z}_N^2}{\mathbf{Z}_A} \end{aligned} \tag{9.113}$$

where, for convenience, we have defined the quantity

$$\mathbf{Y}_X = \frac{\mathbf{Z}_A\mathbf{Z}_B + \mathbf{Z}_B\mathbf{Z}_K + \mathbf{Z}_A\mathbf{Z}_K}{\mathbf{Z}_A\mathbf{Z}_B\mathbf{Z}_K} \equiv \frac{\mathbf{Z}_N^2}{\mathbf{Z}_A\mathbf{Z}_B\mathbf{Z}_K} \tag{9.114}$$

Note that the impedances represented in Figure 9.15 are not the same as those shown in Figures 9.2 to 9.5, because those figures represent the unfaulted case. Figure 9.15 defines a new impedance, $\mathbf{Z}_D$, which is the parallel combination of the two impedances circled in the

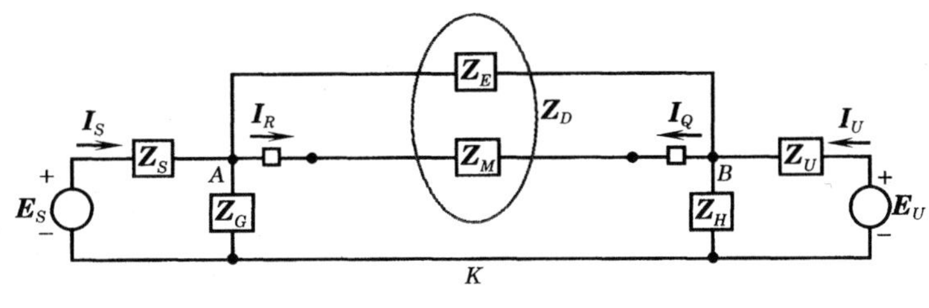

Figure 9.15 Protected component with delta fault equivalent.

figure. Defining this quantity will simplify the computations to follow. Thus, we write

$$Z_D = \frac{Z_E Z_M}{Z_E + Z_M} \tag{9.115}$$

The procedure for computing the ***ABCD*** parameters is straightforward and is based on the defining equation (9.84). The parameters $\boldsymbol{A}$ and $\boldsymbol{C}$ are defined with the #2 port current equal to zero, which means that this port is open, as shown in Figure 9.16.

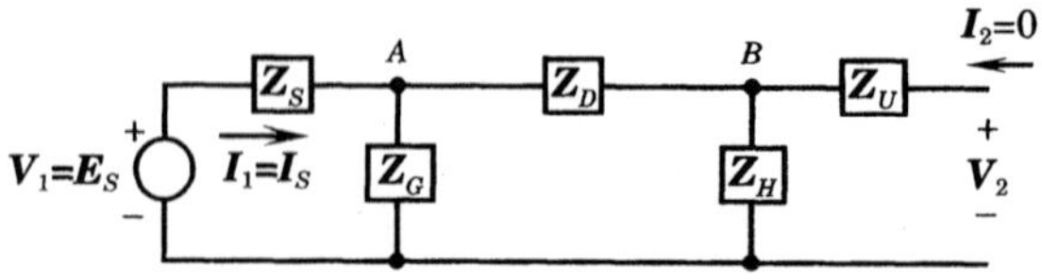

Figure 9.16 System test with the #2 port open.

Under this test condition, we can compute the parameter $\boldsymbol{A}$ as follows.

$$\boldsymbol{A} = \left.\frac{\boldsymbol{V}_1}{\boldsymbol{V}_2}\right]_{I_2=0} \tag{9.116}$$

It is helpful to first compute the driving current from the source $\boldsymbol{E}_S$.

$$\boldsymbol{I}_1 = \frac{\boldsymbol{V}_1}{\boldsymbol{Z}} = \frac{\boldsymbol{E}_S(\boldsymbol{Z}_G + \boldsymbol{Z}_D + \boldsymbol{Z}_H)}{\boldsymbol{Z}_S(\boldsymbol{Z}_G + \boldsymbol{Z}_D + \boldsymbol{Z}_H) + \boldsymbol{Z}_G(\boldsymbol{Z}_D + \boldsymbol{Z}_H)} \tag{9.117}$$

The voltage at node A is computed using Kirchhoff's laws.

$$\begin{aligned}
\boldsymbol{V}_A &= \boldsymbol{V}_1 - \boldsymbol{Z}_S \boldsymbol{I}_1 \\
&= \boldsymbol{E}_S - \frac{\boldsymbol{E}_S \boldsymbol{Z}_S(\boldsymbol{Z}_G + \boldsymbol{Z}_D + \boldsymbol{Z}_H)}{\boldsymbol{Z}_S(\boldsymbol{Z}_G + \boldsymbol{Z}_D + \boldsymbol{Z}_H) + \boldsymbol{Z}_G(\boldsymbol{Z}_D + \boldsymbol{Z}_H)} \\
&= \boldsymbol{E}_S \left(\frac{\boldsymbol{Z}_G(\boldsymbol{Z}_P + \boldsymbol{Z}_H)}{\boldsymbol{Z}_S(\boldsymbol{Z}_G + \boldsymbol{Z}_D + \boldsymbol{Z}_H) + \boldsymbol{Z}_G(\boldsymbol{Z}_D + \boldsymbol{Z}_H)} \right)
\end{aligned} \tag{9.118}$$

Using this result, we compute the open-circuit voltage at the #2 port.

$$\boldsymbol{V}_2 = \frac{\boldsymbol{Z}_H}{\boldsymbol{Z}_D + \boldsymbol{Z}_H} \boldsymbol{V}_A = \boldsymbol{V}_1 \left(\frac{\boldsymbol{Z}_G \boldsymbol{Z}_H}{\boldsymbol{Z}_S(\boldsymbol{Z}_G + \boldsymbol{Z}_D + \boldsymbol{Z}_H) + \boldsymbol{Z}_G(\boldsymbol{Z}_D + \boldsymbol{Z}_H)} \right) \tag{9.119}$$

Then

$$\begin{aligned}
\boldsymbol{A} &= \frac{\boldsymbol{V}_1}{\boldsymbol{V}_2} = \frac{\boldsymbol{Z}_S(\boldsymbol{Z}_G + \boldsymbol{Z}_D + \boldsymbol{Z}_H) + \boldsymbol{Z}_G(\boldsymbol{Z}_D + \boldsymbol{Z}_H)}{\boldsymbol{Z}_G \boldsymbol{Z}_H} \\
&= 1 + \frac{\boldsymbol{Z}_S(\boldsymbol{Z}_G + \boldsymbol{Z}_D + \boldsymbol{Z}_H) + \boldsymbol{Z}_G \boldsymbol{Z}_D}{\boldsymbol{Z}_G \boldsymbol{Z}_H}
\end{aligned} \tag{9.120}$$

We may also compute the parameter $\boldsymbol{C}$ from this same circuit condition. Using (9.116) and (9.119) we write

$$\boldsymbol{C} = \left.\frac{\boldsymbol{I}_1}{\boldsymbol{V}_2}\right]_{I_2=0} = \left(\frac{\boldsymbol{I}_1}{\boldsymbol{V}_1}\right)\left(\frac{\boldsymbol{V}_1}{\boldsymbol{V}_2}\right) \tag{9.121}$$

We can expand this computation as follows.

$$\begin{aligned}
\boldsymbol{C} &= \frac{\boldsymbol{Z}_G + \boldsymbol{Z}_D + \boldsymbol{Z}_H}{\boldsymbol{Z}_S(\boldsymbol{Z}_G + \boldsymbol{Z}_D + \boldsymbol{Z}_H) + \boldsymbol{Z}_G(\boldsymbol{Z}_D + \boldsymbol{Z}_H)} \times \frac{\boldsymbol{Z}_S(\boldsymbol{Z}_G + \boldsymbol{Z}_D + \boldsymbol{Z}_H) + \boldsymbol{Z}_G(\boldsymbol{Z}_D + \boldsymbol{Z}_H)}{\boldsymbol{Z}_G \boldsymbol{Z}_H} \\
&= \frac{\boldsymbol{Z}_G + \boldsymbol{Z}_D + \boldsymbol{Z}_G}{\boldsymbol{Z}_G \boldsymbol{Z}_H}
\end{aligned} \tag{9.122}$$

The second test condition of the two-port network requires that the #2 port be shorted, as shown in Figure 9.17. This network can be tested to determine the parameters $\boldsymbol{B}$ and $\boldsymbol{D}$.

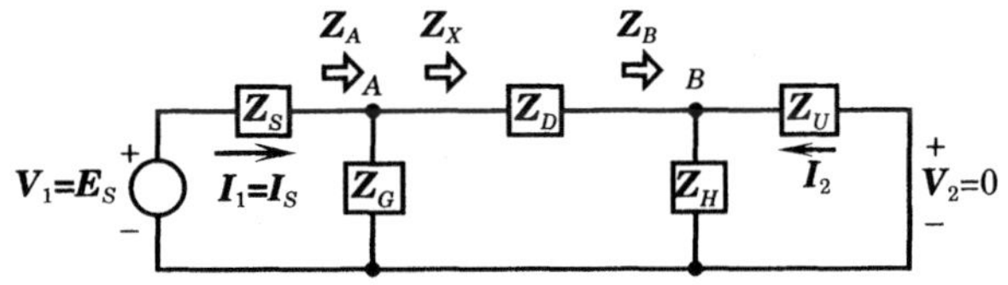

Figure 9.17 System test with the #2 port shorted.

Parameter $\boldsymbol{B}$ is defined from (9.84) as follows.

$$\boldsymbol{B} = \left.\frac{\boldsymbol{V}_1}{-\boldsymbol{I}_2}\right]_{\boldsymbol{V}_2=0} \tag{9.123}$$

Therefore, in order to find $\boldsymbol{B}$, we must solve for the current flowing through the short circuit at the #2 terminal. This is tedious, but straightforward. First, we define the impedance $\boldsymbol{Z}_B$, shown in Figure 9.17.

$$\boldsymbol{Z}_B = \frac{\boldsymbol{Z}_U\boldsymbol{Z}_H}{\boldsymbol{Z}_U+\boldsymbol{Z}_H} \tag{9.124}$$

Then we may write the impedance $\boldsymbol{Z}_X$ as

$$\boldsymbol{Z}_X = \boldsymbol{Z}_D + \boldsymbol{Z}_B = \frac{\boldsymbol{Z}_D(\boldsymbol{Z}_U+\boldsymbol{Z}_H)+\boldsymbol{Z}_U\boldsymbol{Z}_H}{\boldsymbol{Z}_U+\boldsymbol{Z}_H} \tag{9.125}$$

Finally, the impedance $\boldsymbol{Z}_A$ defined in Figure 9.17 may be written as

$$\boldsymbol{Z}_A = \frac{\boldsymbol{Z}_G\boldsymbol{Z}_X}{\boldsymbol{Z}_G+\boldsymbol{Z}_X} = \frac{\boldsymbol{Z}_G[\boldsymbol{Z}_D(\boldsymbol{Z}_U+\boldsymbol{Z}_H)+\boldsymbol{Z}_U\boldsymbol{Z}_H]}{\boldsymbol{Z}_U(\boldsymbol{Z}_G+\boldsymbol{Z}_D+\boldsymbol{Z}_H)+\boldsymbol{Z}_H(\boldsymbol{Z}_D+\boldsymbol{Z}_G)} \tag{9.126}$$

Then, the total impedance seen looking into the #1 port is the sum of (9.126), and the source impedance and this total impedance is equal to the ratio of the input voltage to the input current.

$$\begin{aligned}\frac{\boldsymbol{V}_1}{\boldsymbol{I}_1} &= \boldsymbol{Z}_S+\boldsymbol{Z}_A \\ &= \frac{\boldsymbol{Z}_S[\boldsymbol{Z}_U(\boldsymbol{Z}_G+\boldsymbol{Z}_D+\boldsymbol{Z}_H)+\boldsymbol{Z}_H(\boldsymbol{Z}_D+\boldsymbol{Z}_G)]+\boldsymbol{Z}_G[\boldsymbol{Z}_D(\boldsymbol{Z}_U+\boldsymbol{Z}_H)+\boldsymbol{Z}_U\boldsymbol{Z}_H]}{\boldsymbol{Z}_U(\boldsymbol{Z}_G+\boldsymbol{Z}_D+\boldsymbol{Z}_H)+\boldsymbol{Z}_H(\boldsymbol{Z}_D+\boldsymbol{Z}_G)}\end{aligned} \tag{9.127}$$

Now the currents at the two ports can be related to the current flowing through the impedance $\boldsymbol{Z}_D$, which we shall call the $\boldsymbol{I}_X$. By inspection of Figure 9.17, we write

$$-\boldsymbol{I}_2 = \frac{\boldsymbol{Z}_H}{\boldsymbol{Z}_U+\boldsymbol{Z}_H}\boldsymbol{I}_X \tag{9.128}$$

and

$$\boldsymbol{I}_X = \frac{\boldsymbol{Z}_G}{\boldsymbol{Z}_X+\boldsymbol{Z}_G}\boldsymbol{I}_1 \tag{9.129}$$

Combining and simplifying, we have the ratio

$$\frac{-\boldsymbol{I}_2}{\boldsymbol{I}_1} = \frac{\boldsymbol{Z}_G\boldsymbol{Z}_H}{\boldsymbol{Z}_U(\boldsymbol{Z}_G+\boldsymbol{Z}_D+\boldsymbol{Z}_H)+\boldsymbol{Z}_H(\boldsymbol{Z}_D+\boldsymbol{Z}_G)} \tag{9.130}$$

We may solve for the parameter $\boldsymbol{B}$ as follows.

$$\begin{aligned}\boldsymbol{B} &= \frac{\boldsymbol{V}_1}{-\boldsymbol{I}_2} = \left(\frac{\boldsymbol{V}_1}{\boldsymbol{I}_1}\right)\left(\frac{\boldsymbol{I}_1}{-\boldsymbol{I}_2}\right) \\ &= \frac{\boldsymbol{Z}_S[\boldsymbol{Z}_U(\boldsymbol{Z}_G+\boldsymbol{Z}_D+\boldsymbol{Z}_H)+\boldsymbol{Z}_H(\boldsymbol{Z}_D+\boldsymbol{Z}_G)]+\boldsymbol{Z}_G[\boldsymbol{Z}_D(\boldsymbol{Z}_U+\boldsymbol{Z}_H)+\boldsymbol{Z}_U\boldsymbol{Z}_H]}{\boldsymbol{Z}_G\boldsymbol{Z}_H}\end{aligned} \tag{9.131}$$

Then $\boldsymbol{D}$ may be determined directly from (9.130) by the definition

$$\begin{aligned} \boldsymbol{D} = \left.\frac{\boldsymbol{I}_1}{-\boldsymbol{I}_2}\right]_{V_2=0} &= \frac{\boldsymbol{Z}_U(\boldsymbol{Z}_G+\boldsymbol{Z}_D+\boldsymbol{Z}_H)+\boldsymbol{Z}_H(\boldsymbol{Z}_D+\boldsymbol{Z}_G)}{\boldsymbol{Z}_G\boldsymbol{Z}_H} \\ &= 1+\frac{\boldsymbol{Z}_U(\boldsymbol{Z}_G+\boldsymbol{Z}_D+\boldsymbol{Z}_H)+\boldsymbol{Z}_D\boldsymbol{Z}_H}{\boldsymbol{Z}_G\boldsymbol{Z}_H} \end{aligned} \tag{9.132}$$

This completes the determination of the ***ABCD*** parameters for any system with a shunt fault. The parameters of the protection equivalent must first be determined as described in Chapter 5. Then the fault location must be specified as a fraction h of the total impedance in the protected line. From these specifications, the impedances used in the derivations may be found from (9.112) to (9.115) and substituted into the equations for $\boldsymbol{A}$, $\boldsymbol{B}$, $\boldsymbol{C}$, and $\boldsymbol{D}$. Once the ***ABCD*** parameters are determined, the equations for plotting the relay environment can be carried out using (9.84).

9.6 RELAY IMPEDANCE FOR A SPECIAL CASE

Another special case of (9.106) is of interest, where the only simplification is that we set

$$\left|\frac{\boldsymbol{E}_U}{\boldsymbol{D}\boldsymbol{E}_S}\right| = 1 \tag{9.133}$$

which is often a practical assumption in power system protection studies. Suppose we also define the following parameters.

$$\begin{aligned} \boldsymbol{D} &= de^{j\delta} \\ \frac{\boldsymbol{E}_U}{\boldsymbol{E}_S} &= \frac{1}{E}e^{-j\theta} \\ \psi &= \delta+\theta \end{aligned} \tag{9.134}$$

Then (9.106) becomes

$$\boldsymbol{Z}_R = \frac{1}{\boldsymbol{F}}\left(\frac{\boldsymbol{B}/\boldsymbol{D}}{1-e^{-j\psi}}-\boldsymbol{Z}_S\right) \tag{9.135}$$

where $\boldsymbol{F}$ is defined by (9.99).

This equation may be rearranged to write

$$\begin{aligned} \boldsymbol{Z}_R &= \frac{\boldsymbol{B}}{2\boldsymbol{F}\boldsymbol{D}}\left(1-j\cot\frac{\psi}{2}\right)-\frac{\boldsymbol{Z}_S}{\boldsymbol{F}} \\ &= \left(\frac{\boldsymbol{B}}{2\boldsymbol{F}\boldsymbol{D}}\right)\left[1-j\cot\left(\frac{\psi}{2}\right)-\frac{2\boldsymbol{D}\boldsymbol{Z}_S}{\boldsymbol{B}}\right] \\ &= \left(\frac{\boldsymbol{B}}{2\boldsymbol{F}\boldsymbol{D}}\right)\left[1-j\cot\left(\frac{\psi}{2}\right)-\boldsymbol{H}_S\right] \end{aligned} \tag{9.136}$$

where we have defined the dimensionless, complex quantity $\boldsymbol{H}_S$.

It is helpful to visualize the right side of this equation graphically as shown in Figure 9.18. Note that the ψ function is a scalar and simply changes the distance from the point P to

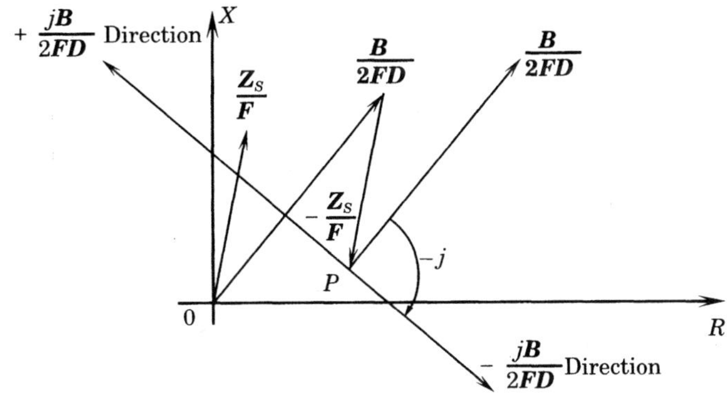

Figure 9.18 Z_R components in the complex Z plane.

Z_R. Examining Figure 9.18, we can make the following observations.

$$\frac{B}{2FD} = \text{a vector operator, variable in magnitude and angle}$$

$$\frac{Z_S}{F} = \text{a translation in the } Z \text{ plane}$$

It is convenient to define

$$\begin{aligned} Z_B &= \frac{B}{2FD} = Z_{BM}e^{j\xi} \\ Z'_S &= \frac{Z_S}{F} = Z'_{SM}e^{j\zeta} \end{aligned} \tag{9.137}$$

Then we can write H_S, which is the dimensionless ratio of these quantities, defined in (9.136).

$$H_S = \frac{Z'_S}{Z_B} = \frac{2DZ_S}{B} = \frac{Z_{BM}}{Z'_{SM}}e^{j(\zeta-\xi)} = he^{j(\zeta-\xi)} \tag{9.138}$$

Using (9.138), we can rewrite the relay apparent impedance (9.136) as

$$Z_R = Z_B\left\{[1 - h\cos(\zeta-\xi)] - j\left[\cot\frac{\psi}{2} + h\sin(\zeta-\xi)\right]\right\} \equiv Z_B H_R \tag{9.139}$$

where we define the complex number in braces to be H_R. In many systems both ζ and ξ will be nearly equal in argument as both are related to transmission impedances. If these angles are equal, then (9.139) becomes

$$H_R = (1-h) - j\cot(\psi/2) \tag{9.140}$$

The quantity H_R is a complex number. The multiplier Z_B in (9.139) expands or contracts Z_R, and its angle ζ rotates the entire locus counterclockwise. We now examine the nature of the approximate equation (9.140) in detail.

The real part H_R from (9.140) is conveniently plotted with in multiples of h as shown in Figure 9.19(a). The imaginary part of H_R is a function only of ψ as noted in Figure 9.19(b).

The function H_R is plotted in Figure 9.20 for selected values of the parameters h and ψ. The line $h = 0$ corresponds to zero source impedance, or an infinite bus behind the relay. The line $Z = h$ corresponds to $Z'_{SM} = Z_{BM}$.

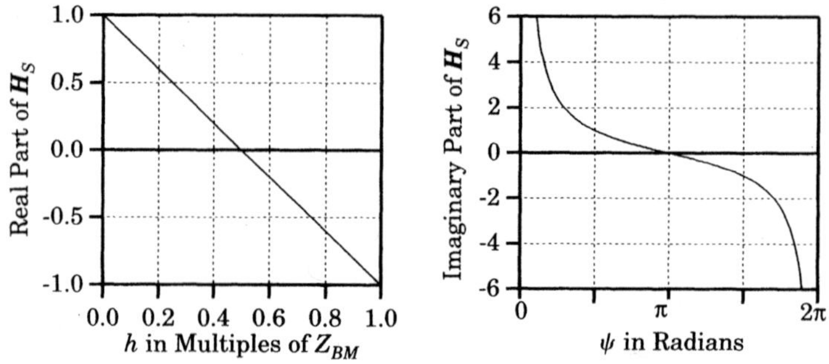

Figure 9.19 Real and imaginary parts of H_R when $\zeta = \xi$.

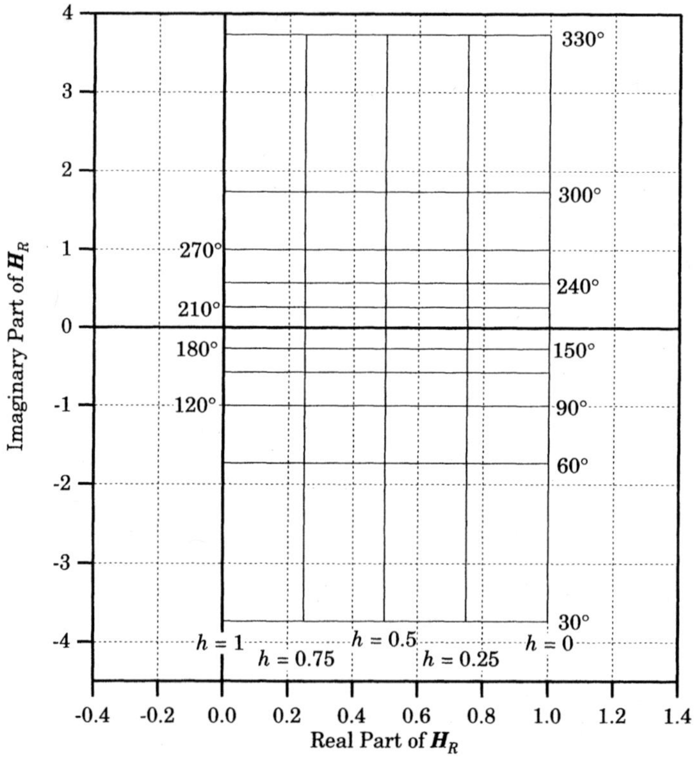

Figure 9.20 H_R for specific values of h and ψ.

From (9.139) the only difference between H_R and Z_R is a scale factor Z_{BM} and an angular rotation ξ. For the purpose of illustration, suppose we let

$$Z_B = Z_{BM}e^{j\xi} = 1e^{j\pi/3} \tag{9.141}$$

Then H_R is scaled by a factor of 1.0 and rotated 60° counterclockwise to give Z_R as shown in Figure 9.21. Admittedly this is a special case where the approximation (9.140) applies. However, the loci shapes are typical.

The grid-like structure of Z_R as a function of h and ψ provides a framework for plotting the relay impedance that can be superimposed on plots such as Figure 9.11. The grid rotation corresponds to the angle of ξ or, in practical transmission systems, to the angle of B/F, where

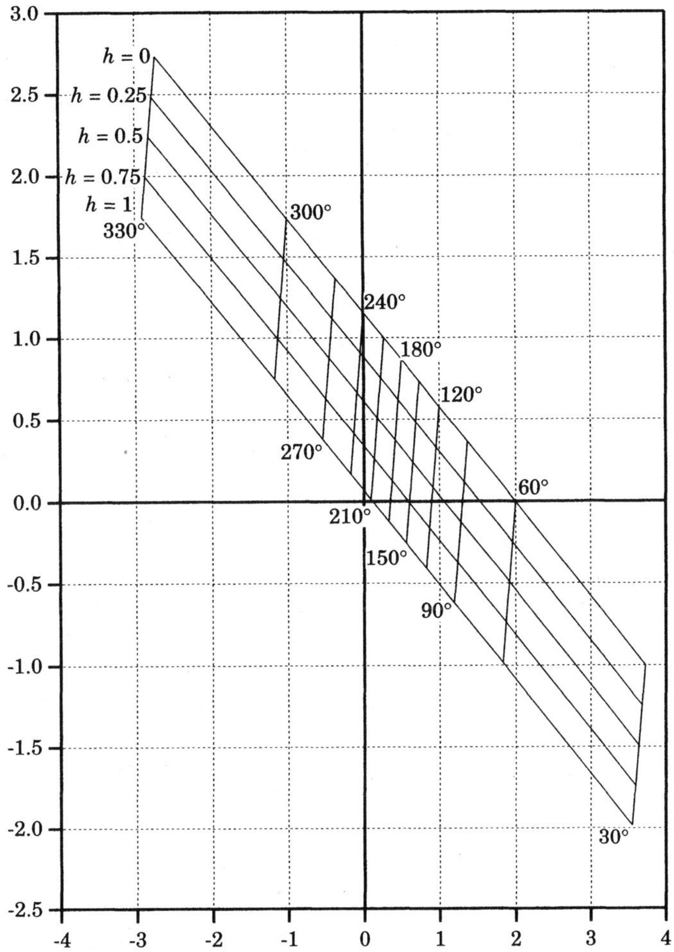

Figure 9.21 Plot of Z_R for specific values of h and ψ.

B is the total impedance between the equivalent voltage sources. The angle ψ is the angle difference between these voltage sources.

The angle of rotation of the grid is the angle of B/F. This is exactly the same as the angle of Z_Y, which is the angle of the line AB in Figure 9.8. Superimposing the two kinds of plots provides information on the impedance seen by the relay from different frames of reference.

EXAMPLE 9.3

Consider the protection equivalent shown in Figure 9.22. Write the expression for the impedance seen by relay R by computing the parameters of (9.135). Assume that the voltage magnitudes of the equivalent circuit are equal, that is, ignore load current.

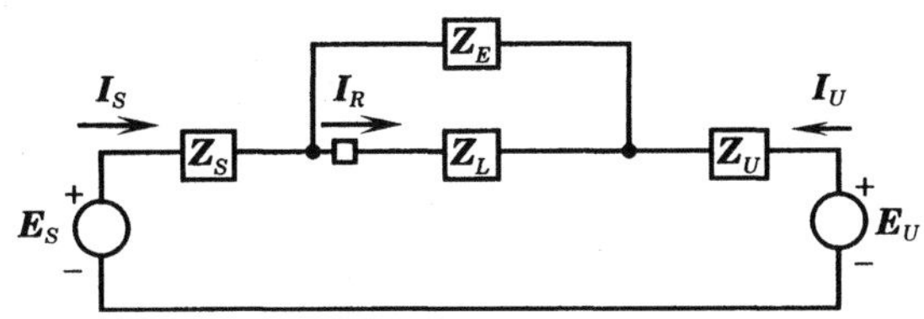

Figure 9.22 Protection equivalent for a transmission line.

Solution

We have derived the ***ABCD*** parameters for this system in example 9.2. From (9.110) we have

$$\begin{aligned} \boldsymbol{B} &= \boldsymbol{Z}_S + \boldsymbol{Z}_P + \boldsymbol{Z}_U \\ \boldsymbol{D} &= 1 \end{aligned} \tag{9.142}$$

We also note that, with a voltage ratio of unity,

$$\frac{\boldsymbol{E}_U}{\boldsymbol{D}\boldsymbol{E}_S} = \frac{\boldsymbol{E}_U}{\boldsymbol{E}_S} = e^{-j\theta} \qquad \text{when } |\boldsymbol{E}_S| = |\boldsymbol{E}_U| \tag{9.143}$$

Then we may write the impedance seen by the relay R as

$$\boldsymbol{Z}_R = \frac{\boldsymbol{Z}_L}{\boldsymbol{Z}_P}\left(\frac{\boldsymbol{Z}_P + \boldsymbol{Z}_U - \boldsymbol{Z}_S}{2} - j\frac{\boldsymbol{Z}_S + \boldsymbol{Z}_P + \boldsymbol{Z}_U}{2}\cot\frac{\theta}{2}\right) \tag{9.144}$$

The multiplier in front of the complex expression is nothing but the complex constant $\boldsymbol{b}$ which is the fraction of the total source current that flows through the relay R. ■

9.7 CONSTRUCTION OF M CIRCLES

The actual impedance seen by the relays is determined by the computation of the parameters of the matrix **M**, described in sections 9.1 through 9.3. We now derive the equations for the **M** parameters, based on the equivalent circuit representing a shunt fault.

Consider the system shown in Figure 9.23, where the impedance seen looking into the protected line at A is desired.

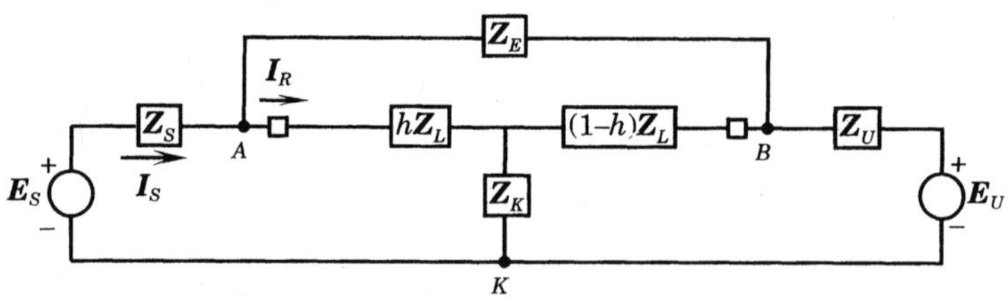

Figure 9.23 Protection equivalent for developing **M** parameters.

Figure 9.23 has a delta equivalent for the faulted line, shown in Figure 9.15. The basic equation relating the matrix **M** parameters to the voltage sources of the protection equivalent are given in (9.1), but repeated here for convenience as Figure 9.24.

$$\begin{bmatrix} \boldsymbol{V}_R \\ \boldsymbol{I}_R \end{bmatrix} = \begin{bmatrix} \boldsymbol{m}_1 & \boldsymbol{m}_2 \\ \boldsymbol{m}_3 & -\boldsymbol{m}_4 \end{bmatrix} \begin{bmatrix} \boldsymbol{E}_S \\ \boldsymbol{E}_U \end{bmatrix} \tag{9.145}$$

The parameters of the **M** matrix are found using (9.3) and (9.4), which require basic short-circuit tests of the system.

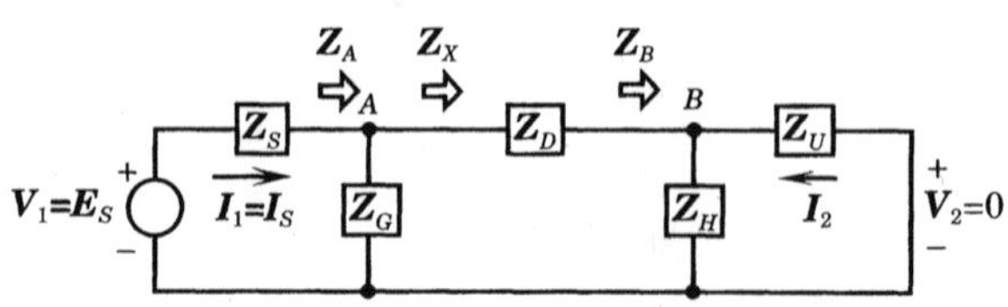

Figure 9.24 Short-circuit test with $\boldsymbol{E}_U = \boldsymbol{V}_2 = 0$.

9.7.1 Short Circuit Test with E_U Shorted

The first test shorts the source $\boldsymbol{E}_U$, which gives the necessary information for finding $\boldsymbol{m}_1$ and $\boldsymbol{m}_3$. It is helpful to sketch the circuit diagram for this condition, with the result shown in Figure 9.17, which is repeated here as Figure 9.24 for convenience.

Under this condition, the driving point impedance seen looking into the network from the source $\boldsymbol{E}_S$ is given by the impedance $\boldsymbol{Z}_A$ in (9.122). From that result we can write the source current as

$$\boldsymbol{I}_S = \frac{\boldsymbol{E}_S}{\boldsymbol{Z}_S + \boldsymbol{Z}_A} = \frac{\boldsymbol{E}_S[(\boldsymbol{Z}_D + \boldsymbol{Z}_G)(\boldsymbol{Z}_U + \boldsymbol{Z}_H) + \boldsymbol{Z}_U\boldsymbol{Z}_H}{\text{DEN}U} \tag{9.146}$$

where

$$\begin{aligned}\text{DEN}U = \boldsymbol{Z}_S[(\boldsymbol{Z}_D + \boldsymbol{Z}_G)(\boldsymbol{Z}_U + \boldsymbol{Z}_H) + \boldsymbol{Z}_U\boldsymbol{Z}_H \\ + \boldsymbol{Z}_G[\boldsymbol{Z}_D(\boldsymbol{Z}_U + \boldsymbol{Z}_H) + \boldsymbol{Z}_U\boldsymbol{Z}_H]\end{aligned} \tag{9.147}$$

We may also compute the relay voltage as follows.

$$\begin{aligned}\boldsymbol{V}_R = \boldsymbol{V}_A &= \boldsymbol{E}_S - \boldsymbol{Z}_S\boldsymbol{I}_S \\ &= \frac{\boldsymbol{E}_S\boldsymbol{Z}_G[\boldsymbol{Z}_D(\boldsymbol{Z}_U + \boldsymbol{Z}_H) + \boldsymbol{Z}_U\boldsymbol{Z}_H]}{\text{DEN}U}\end{aligned} \tag{9.148}$$

Then we can compute the first **M** parameter.

$$\boldsymbol{m}_1 = \left.\frac{\boldsymbol{V}_R}{\boldsymbol{E}_S}\right|_{\boldsymbol{E}_U=0} = \frac{\boldsymbol{Z}_G[\boldsymbol{Z}_D(\boldsymbol{Z}_U + \boldsymbol{Z}_H) + \boldsymbol{Z}_U\boldsymbol{Z}_H]}{\text{DEN}U} \tag{9.149}$$

Now, turning to node B, and using Figure 9.24, we can write

$$\boldsymbol{V}_B = \boldsymbol{V}_A - \boldsymbol{Z}_D\boldsymbol{I}_D = \frac{\boldsymbol{E}_S(\boldsymbol{Z}_G - \boldsymbol{Z}_D)\boldsymbol{Z}_U\boldsymbol{Z}_H}{\text{DEN}U} \tag{9.150}$$

We now determine the current in the impedances $\boldsymbol{Z}_E$ and $\boldsymbol{Z}_M$ of Figure 9.15.

$$\boldsymbol{I}_E = \frac{\boldsymbol{V}_A - \boldsymbol{V}_B}{\boldsymbol{Z}_E} = \frac{\boldsymbol{Z}_D\boldsymbol{I}_D}{\boldsymbol{Z}_E} = \frac{\boldsymbol{E}_S\boldsymbol{Z}_D[\boldsymbol{Z}_G(\boldsymbol{Z}_U + \boldsymbol{Z}_H) + \boldsymbol{Z}_U\boldsymbol{Z}_H]}{\boldsymbol{Z}_E\text{DEN}U} \tag{9.151}$$

$$\boldsymbol{I}_M = \boldsymbol{I}_D - \boldsymbol{I}_E = \frac{\boldsymbol{E}_S(\boldsymbol{Z}_E - \boldsymbol{Z}_D)[\boldsymbol{Z}_G(\boldsymbol{Z}_U + \boldsymbol{Z}_H) + \boldsymbol{Z}_U\boldsymbol{Z}_H]}{\boldsymbol{Z}_E\text{DEN}U} \tag{9.152}$$

Then the relay current is given by

$$\begin{aligned}\boldsymbol{I}_R &= \boldsymbol{I}_M + \boldsymbol{I}_G \\ &= \frac{\boldsymbol{E}_S\{[\boldsymbol{Z}_E(\boldsymbol{Z}_D + \boldsymbol{Z}_G) - \boldsymbol{Z}_D\boldsymbol{Z}_G](\boldsymbol{Z}_U + \boldsymbol{Z}_H) + (2\boldsymbol{Z}_E - \boldsymbol{Z}_D)\boldsymbol{Z}_U\boldsymbol{Z}_H\}}{\boldsymbol{Z}_E\text{DEN}U}\end{aligned} \tag{9.153}$$

Finally, we can compute the **M** parameter

$$\boldsymbol{m}_3 = \left.\frac{\boldsymbol{I}_R}{\boldsymbol{E}_S}\right|_{\boldsymbol{E}_U=0} = \frac{[\boldsymbol{Z}_E(\boldsymbol{Z}_D + \boldsymbol{Z}_G) - \boldsymbol{Z}_D\boldsymbol{Z}_G](\boldsymbol{Z}_U + \boldsymbol{Z}_H) + (2\boldsymbol{Z}_E - \boldsymbol{Z}_D)\boldsymbol{Z}_U\boldsymbol{Z}_H}{\boldsymbol{Z}_E\text{DEN}U} \tag{9.154}$$

9.7.2 Short Circuit Test with E_S Shorted

The short-circuit test with the #1 terminal shorted as shown in Figure 9.25.

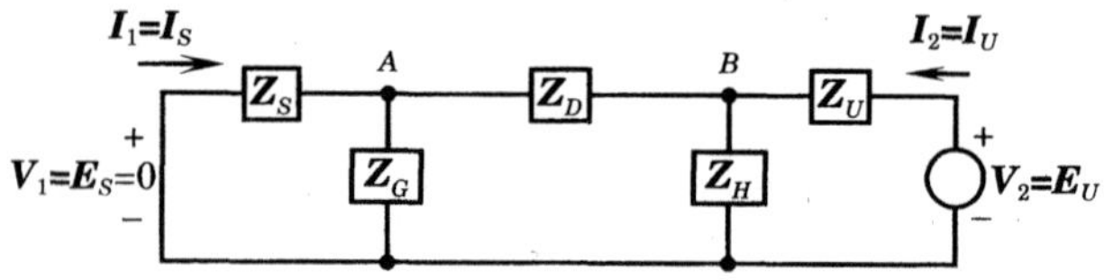

Figure 9.25 Short-circuit test with $E_S = V_1 = 0$.

The impedance seen looking to the left from point B is computed as

$$Z_X = Z_D + \frac{Z_S Z_G}{Z_S + Z_G} = \frac{Z_D(Z_S + Z_G) + Z_S Z_G}{Z_S + Z_G} \tag{9.155}$$

Then, the total impedance seen looking into the network from the source E_U is described, using the notation defined in Section 9.5.1.

$$\begin{aligned} Z &= Z_U + \frac{Z_X Z_H}{Z_X + Z_H} \\ &= \frac{Z_U[(Z_D + Z_H)(Z_S + Z_G) + Z_S Z_G] + Z_H[Z_D(Z_S + Z_G) + Z_S Z_G]}{(Z_D + Z_H)(Z_S + Z_G) + Z_S Z_G} \end{aligned} \tag{9.156}$$

Then, the current entering the network from the voltage source is given by

$$I_U = \frac{E_U}{Z} = \frac{E_U[(Z_D + Z_H)(Z_S + Z_G) + Z_S Z_G]}{\text{DEN}S} \tag{9.157}$$

where

$$\begin{aligned} \text{DEN}S = Z_U[(Z_D + Z_H)(Z_S + Z_G) + Z_S Z_G] \\ + Z_H[Z_D(Z_S + Z_G) + Z_S Z_G] \end{aligned} \tag{9.158}$$

Note the symmetry between (9.158) and (9.147). The voltage at point B can be computed for this short-circuit test condition as

$$V_B = E_U - Z_U I_U = \frac{E_U Z_H[Z_D(Z_S + Z_G) + Z_S Z_G]}{\text{DEN}S} \tag{9.159}$$

Again, note the symmetry between (9.159) and (9.148).

Now, the voltage at node A is computed as

$$V_A = V_B - Z_D I_D \tag{9.160}$$

where the current I_D is given by

$$I_D = I_H - I_U = \frac{-E_U[Z_H(Z_S + Z_G) + Z_S Z_G]}{\text{DEN}S} \tag{9.161}$$

Then

$$V_A = \frac{E_U(Z_H - Z_D)Z_S Z_G}{\text{DEN}S} \tag{9.162}$$

where we again note the symmetry of the results to the previous short-circuit test, this time comparing (9.162) against (9.150).

The external equivalent current may now be computed.

$$\mathbf{I}_E = \frac{\mathbf{V}_A - \mathbf{V}_B}{\mathbf{Z}_E} = \frac{\mathbf{Z}_D\mathbf{I}_D}{\mathbf{Z}_E} = \frac{-\mathbf{E}_U\mathbf{Z}_D[\mathbf{Z}_H(\mathbf{Z}_S + \mathbf{Z}_G) + \mathbf{Z}_S\mathbf{Z}_G]}{\mathbf{Z}_E\text{DENS}} \tag{9.163}$$

Then

$$\mathbf{I}_M = \mathbf{I}_D - \mathbf{I}_E = \frac{-\mathbf{E}_U\mathbf{Z}_D(\mathbf{Z}_E - \mathbf{Z}_D)[\mathbf{Z}_H(\mathbf{Z}_S + \mathbf{Z}_G) + \mathbf{Z}_S\mathbf{Z}_G]}{\mathbf{Z}_E\text{DENS}} \tag{9.164}$$

Finally, the relay current computed as

$$\mathbf{I}_R = \mathbf{I}_M + \mathbf{I}_G = \frac{-\mathbf{E}_U\{(\mathbf{Z}_E - \mathbf{Z}_D)[\mathbf{Z}_H(\mathbf{Z}_S + \mathbf{Z}_G) + \mathbf{Z}_S\mathbf{Z}_G]\}}{\mathbf{Z}_E\text{DENS}} \tag{9.165}$$

We now have the required currents and voltages to determine the remaining short-circuit parameters.

$$\mathbf{m}_2 = \left.\frac{\mathbf{V}_R}{\mathbf{E}_U}\right|_{\mathbf{E}_S=0} = \frac{(\mathbf{Z}_H - \mathbf{Z}_D)\mathbf{Z}_S\mathbf{Z}_G}{\text{DENS}} \tag{9.166}$$

$$\mathbf{m}_4 = \left.\frac{-\mathbf{I}_R}{\mathbf{E}_U}\right|_{\mathbf{E}_S=0} = \frac{(\mathbf{Z}_E - \mathbf{Z}_D)[\mathbf{Z}_H(\mathbf{Z}_S + \mathbf{Z}_G) + \mathbf{Z}_S\mathbf{Z}_G]}{\mathbf{Z}_E\text{DENS}} \tag{9.167}$$

9.7.3 Summary of Short-Circuit Test Results

In summary, the short-circuit test results for the four desired matrix parameters is as follows.

$$\mathbf{m}_1 = \left.\frac{\mathbf{V}_R}{\mathbf{E}_S}\right|_{\mathbf{E}_U=0} = \frac{\mathbf{Z}_G[\mathbf{Z}_D(\mathbf{Z}_U + \mathbf{Z}_H) + \mathbf{Z}_U\mathbf{Z}_H]}{\text{DENU}} \tag{9.168}$$

$$\mathbf{m}_2 = \left.\frac{\mathbf{V}_R}{\mathbf{E}_U}\right|_{\mathbf{E}_S=0} = \frac{(\mathbf{Z}_H - \mathbf{Z}_D)\mathbf{Z}_S\mathbf{Z}_G}{\text{DENS}} \tag{9.169}$$

$$\mathbf{m}_3 = \left.\frac{\mathbf{I}_R}{\mathbf{E}_S}\right|_{\mathbf{E}_U=0} = \frac{[\mathbf{Z}_E(\mathbf{Z}_D + \mathbf{Z}_G) - \mathbf{Z}_D\mathbf{Z}_G](\mathbf{Z}_U + \mathbf{Z}_H) + (2\mathbf{Z}_E - \mathbf{Z}_D)\mathbf{Z}_U\mathbf{Z}_H}{\mathbf{Z}_E\text{DENU}} \tag{9.170}$$

$$\mathbf{m}_4 = \left.\frac{-\mathbf{I}_R}{\mathbf{E}_U}\right|_{\mathbf{E}_S=0} = \frac{(\mathbf{Z}_E - \mathbf{Z}_D)[\mathbf{Z}_H(\mathbf{Z}_S + \mathbf{Z}_G) + \mathbf{Z}_S\mathbf{Z}_G]}{\mathbf{Z}_E\text{DENS}} \tag{9.171}$$

where

$$\begin{aligned}\text{DENU} &= \mathbf{Z}_S[(\mathbf{Z}_D + \mathbf{Z}_G)(\mathbf{Z}_U + \mathbf{Z}_H) + \mathbf{Z}_U\mathbf{Z}_H] \\ &\quad + \mathbf{Z}_G[\mathbf{Z}_D(\mathbf{Z}_U + \mathbf{Z}_H) + \mathbf{Z}_U\mathbf{Z}_H]\end{aligned} \tag{9.172}$$

$$\begin{aligned}\text{DENS} &= \mathbf{Z}_U[(\mathbf{Z}_D + \mathbf{Z}_H)(\mathbf{Z}_S + \mathbf{Z}_G) + \mathbf{Z}_S\mathbf{Z}_G] \\ &\quad + \mathbf{Z}_H[\mathbf{Z}_D(\mathbf{Z}_S + \mathbf{Z}_G) + \mathbf{Z}_S\mathbf{Z}_G]\end{aligned} \tag{9.173}$$

Note that the first two parameters are dimensionless, but the third and fourth parameters have the dimension of admittance.

Computing the **M** parameters for a system with a shunt fault is a tedious, but straightforward, process. The resulting equations are quite general, however, and can be used for any fault location along the protected line, or for any initial conditions of interest. The protective relay current, voltage, and impedance computed using this process in not an approximation, but is exact. The impedance in the protection equivalent can be combined, as required, to substitute directly into (9.168)–(9.173), to give the desired relationships for plotting the relay impedance in the complex plane.

9.8 PHASE COMPARISON APPARENT IMPEDANCE

One of the important applications for the conceptual viewing of impedance seen by the relay is in the application of phase comparison protection. This has been thoroughly studied by several noted investigators, who have usually explained the results of their research in terms of regions in the complex impedance plane. This is particularly convenient for the protection engineer, since it provides a simple means of understanding how a complex relay system works, and identifies clearly the threshold of relay pickup as a region of the complex plane [5–9].

Figure 9.26 illustrates one way of viewing the physical structure of a phase comparison relay scheme. The measured voltage is modified by complex operators $\boldsymbol{k}_1$ and $\boldsymbol{k}_2$, and the current signal is modified by complex constants that can be thought of as impedances $\boldsymbol{Z}_{k1}$ and $\boldsymbol{Z}_{k2}$. This gives the complex variables

$$\begin{aligned} \boldsymbol{S}_1 &= \boldsymbol{k}_1 \boldsymbol{V}_R - \boldsymbol{Z}_{k1}\boldsymbol{I} \\ \boldsymbol{S}_2 &= \boldsymbol{k}_2 \boldsymbol{V}_R - \boldsymbol{Z}_{k2}\boldsymbol{I}_R \end{aligned} \tag{9.174}$$

that are fed to a comparator to determine if the observed quantities fall within the trip zone of the relay. In a phase comparison scheme, it is the phase angle of the quantities $\boldsymbol{S}_1$ and $\boldsymbol{S}_2$ that are used for discrimination. Clearly these quantities have the dimensions of voltage. Usually, it is more convenient to think in terms of impedance, which suggests modifying (9.174) to identify terms with the dimension of impedance. This modification of (9.174) gives

$$\begin{aligned} \boldsymbol{S}_1 &= \boldsymbol{k}_1 \boldsymbol{I}_R(\boldsymbol{Z}_R - \boldsymbol{Z}_{k1}/\boldsymbol{k}_1) \\ \boldsymbol{S}_2 &= \boldsymbol{k}_2 \boldsymbol{I}_R(\boldsymbol{Z}_R - \boldsymbol{Z}_{k2}/\boldsymbol{k}_2) = \left(\frac{\boldsymbol{k}_2}{\boldsymbol{k}_1}\right) \boldsymbol{k}_1 \boldsymbol{I}_R(\boldsymbol{Z}_R - \boldsymbol{Z}_{k2}/\boldsymbol{k}_2) \end{aligned} \tag{9.175}$$

where

$$\boldsymbol{Z}_R = \frac{\boldsymbol{V}_R}{\boldsymbol{I}_R} \tag{9.176}$$

is the apparent impedance seen by the relay. Now, the phase difference between these two quantities is not changed if both equations are divided by the same complex quantity, $\boldsymbol{k}_1 \boldsymbol{I}_R$,

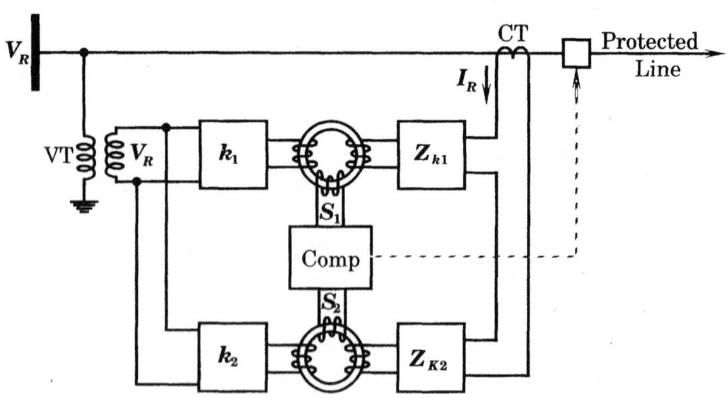

Figure 9.26 Phase comparison relay structure.

which gives the result

$$\begin{aligned} \boldsymbol{S}_A &= \frac{\boldsymbol{S}_1}{k_1\boldsymbol{I}_R} = \boldsymbol{Z}_R - \boldsymbol{Z}_{R1} \\ \boldsymbol{S}_B &= \frac{\boldsymbol{S}_2}{k_1\boldsymbol{I}_R} = \boldsymbol{k}\boldsymbol{Z}_R - \boldsymbol{Z}_{R2} \end{aligned} \tag{9.177}$$

where

$$\boldsymbol{k} = \frac{\boldsymbol{k}_2}{\boldsymbol{k}_1} \tag{9.178}$$

The phase difference between the quantities (9.177) is determined by examining their quotient

$$\frac{\boldsymbol{S}_A}{\boldsymbol{S}_B} = \frac{\boldsymbol{Z}_R - \boldsymbol{Z}_{R1}}{\boldsymbol{k}\boldsymbol{Z}_R - \boldsymbol{Z}_{R2}} \tag{9.179}$$

which can be rearranged to the form

$$\boldsymbol{Z}_R = \frac{(\boldsymbol{S}_A/\boldsymbol{S}_B)\boldsymbol{Z}_{R2} - \boldsymbol{Z}_{R1}}{\boldsymbol{k}(\boldsymbol{S}_A/\boldsymbol{S}_B) - 1} \tag{9.180}$$

This represents a circle in the complex Z plane that is dependent on the complex parameter $\boldsymbol{S_A}/\boldsymbol{S_B}$ as well as the constant impedances $\boldsymbol{Z}_{R1}$ and $\boldsymbol{Z}_{R2}$. Now, define

$$\begin{aligned} \boldsymbol{Z}_R &= R + jX \\ \boldsymbol{Z}_{R1} &= R_1 + jX_1 \\ \boldsymbol{Z}_{R2} &= R_2 + jX_2 \\ \boldsymbol{k} &= e + jf \end{aligned} \tag{9.181}$$

where we recognize that $R + jX$ can be any point in the Z plane. Substituting into (9.177), we have

$$\begin{aligned} \boldsymbol{S}_A &= (R - R_1) + j(X - X_1) \\ \boldsymbol{S}_B &= (eR - fX) + j(fR + eX) - (R_2 + jX_2) \\ &= (eR - fX - R_2) + j(fR + eX - X_2) \end{aligned} \tag{9.182}$$

The phase comparator picks up when

$$\frac{\pi}{2} \le (\arg \boldsymbol{S}_A - \arg \boldsymbol{S}_B) \le \frac{3\pi}{2} \tag{9.183}$$

or when

$$\text{Re}\left(\frac{\boldsymbol{S}_A}{\boldsymbol{S}_B}\right) \le 0 \tag{9.184}$$

We may write the quotient as

$$\frac{\boldsymbol{S}_A}{\boldsymbol{S}_B} = \frac{(R - R_1) + j(X - X_1)}{(eR - fX - R_2) + j(fR + eX - X_2)} \tag{9.185}$$

Then, multiplying numerator and denominator by the complex conjugate of the term in the denominator, we compute (9.184) to be

$$\text{Re}\left(\frac{\boldsymbol{S}_A}{\boldsymbol{S}_B}\right) = \frac{(R - R_1)(eR - fX - R_2) + (X - X_1)(fR + eX - X_2)}{(eR - fX - R_2)^2 + (fR + eX - X_2)^2} \le 0 \tag{9.186}$$

In (9.186) an equality identifies the relay threshold and the trip zone corresponds to values obeying the inequality. Since both sides of (9.186) can be multiplied by the term in the denominator, the numerator determines the relay pick-up zone. Expanding the numerator we have

$$R^2 - \left(\frac{eR_1 + fX_1 + R_2}{e}\right) R + X^2 - \left(\frac{eX_1 - fR_1 + X_2}{e}\right) X \le -\left(\frac{R_1R_2 + X_1X_2}{e}\right) \tag{9.187}$$

Now, we complete the square of the two quadratics on the left side of (9.187) to write

$$\left[R - \left(\frac{eR_1 + fX_1 + R_2}{2e}\right)\right]^2 + \left[X - \left(\frac{eX_1 - fR_1 + X_2}{2e}\right)\right]^2 \le S^2 \tag{9.188}$$

where

$$S^2 = \left(\frac{eR_1 + fX_1 + R_2}{2e}\right)^2 + \left(\frac{eX_1 - fR_1 + X_2}{2e}\right)^2 - \left(\frac{R_1R_2 + X_1X_2}{e}\right) \tag{9.189}$$

This is a circle with center located at

$$\begin{aligned} R_C &= \frac{eR_1 - fX_1 + R_2}{2e} \\ X_C &= \frac{eX_1 - fR_1 + X_2}{2e} \end{aligned} \tag{9.190}$$

and with radius S. Two special cases of the boundary circle (9.188) will now be investigated, where the boundary is defined with an equality replacing the inequality in (9.188).

Case 1: $R = R_1$ *and* $X = X_1$

For this case, one can substitute $R = R_1$ and $X = X_1$ into (9.188). This results in an equality, which means that this point is a solution, that is, this point lies on the boundary circle for any value of the parameters R_1, X_1, R_2, X_2, e, and f.

Case 2: $f = 0$

Another special point is identified where the imaginary part of the complex constant $\mathbf{k} = e + jf$ is zero, i.e., this constant is real. For this special condition, the right-hand side of (9.188) can be factored. Setting $f = 0$ in (9.188) we have

$$\text{RHS} = \left(\frac{eR_1 - R_2}{2e}\right)^2 + \left(\frac{eX_1 - X_2}{2e}\right)^2 = R_O^2 + X_O^2 \tag{9.191}$$

Moreover, the left-hand side can be simplified to the form

$$\text{LHS} = (R - R_O^2)^2 + (X - X_O^2)^2 \tag{9.192}$$

Equating (9.191) and (9.192), and canceling identical terms on both sides gives the result

$$R(R - 2R_O) + X(X - 2X_O) = 0 \tag{9.193}$$

This equation has two solutions:

$$\begin{aligned} &(1) \qquad R = X = 0 \\ &(2) \qquad R = 2R_O,\ X = 2X_O \end{aligned} \tag{9.194}$$

The first solution means that the origin is a solution, or the circle always passes through the origin. The second solution requires that the diameter of the circle, drawn from the origin to twice the distance from the origin to the center, is also a solution. This second solution may or may not be the same as that for case 1, as will now be shown by example.

EXAMPLE 9.4

In order to illustrate the way in which the circle parameters affect the construction and location of the circle, four cases will be investigated, as follows:

Solution

Case 1 in Table 9.3 describes a circle where the impedance $\mathbf{Z}_{R2}$ is zero. This results in the relay threshold being a circle that passes through the origin, as noted in (9.194). Also, the complex constant $\boldsymbol{k}$ is real, and has the value (1, 0). We can think of this circle as being a "base case" and can compare the other three cases against this base case. This is the circle plotted as the solid line in Figure 9.27, and identified in the figure as circle 1. The center of the circle is labeled C_1. Point A is special, since this is the point through which all solutions must pass. Note that

$$\overline{OA} = \mathbf{Z}_{R1} \tag{9.195}$$

The distance OA is also the diameter of Circle 1, and the midpoint of this line locates the center of the circle.

TABLE 9.3 Illustrative Cases for Study

Case	R_1	X_1	R_2	X_2	e	f	R_O	X_O	S_Z
1	4	16	0	0	1	0	2	8	8.246
2	4	16	8	−4	1	0	6	6	10.198
3	4	16	0	0	1	1	10	6	11.661
4	4	16	8	−4	1	1	14	4	15.620

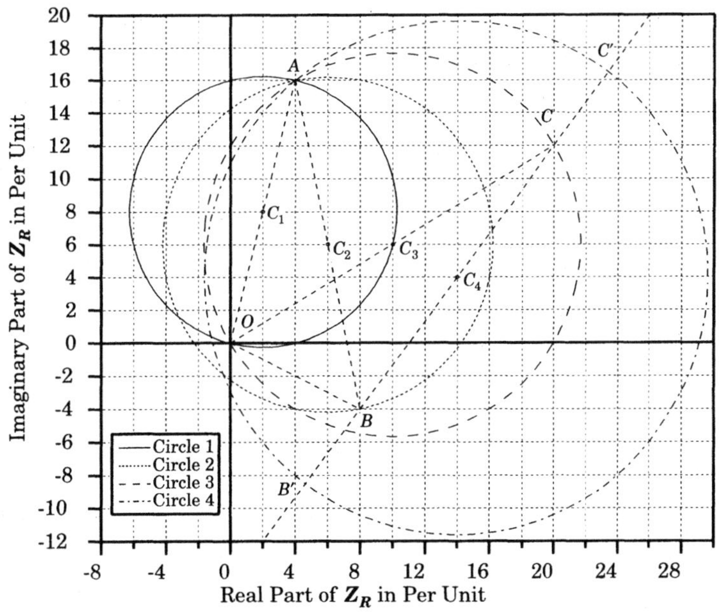

Figure 9.27 Plot of the four cases specified in Table 9.3.

Circle 2 is the same as circle 1 except the impedance $\mathbf{Z}_{R2}$ is not zero, and has been assigned the value $8 - j4$. This circle is plotted using a finely dotted line. Note that circle 2 passes through point A, but does not pass through the origin. Instead of passing through the origin, circle 2 passes through point

B, where it is noted that

$$\overline{OB} = \mathbf{Z}_{R2} \tag{9.196}$$

Thus, in order to expand the circle 1 to include the origin, it is necessary only to define the value (9.196) to fall in the third or fourth quadrant. The center of circle 2 is the point C_2, which is located at the midpoint of the line AB, the circle diameter.

Circle 3 is similar to circle 1 except that the constant $\boldsymbol{k}$ is complex, being assigned the value $1 + j1$, as noted in Table 9.3. This moves the circle to the right, although it still passes through both the origin and point A. The center, however, is located with respect to a new point, denoted as C in Figure 9.27. The center of circle 3 is at point C_3, which is the midpoint of the line OC, the diameter of circle 3. It can be shown that point C can be written as the complex quantity

$$\overline{BC} = \mathbf{Z}_C - \mathbf{Z}_{R2} = |\overline{BC}|\angle\theta = 12 + j16 = 20\angle 53.1^\circ \tag{9.197}$$

The first term is twice the radius of circle 3, measured from the origin, less the value of $\mathbf{Z}_{R2}$. Point C is important since it defines the diameter of circle 3.

Circle 4 includes both the offset impedance $\mathbf{Z}_{R2}$ as well as a the complex constant $\boldsymbol{k}$. This circle is similar to circle 3, but is larger and with its center, C_4, lying at the midpoint of the line BC. Note that circle 4 circumscribes the origin and passes through point A.

The line $B'C'$ has the same angle as BC, but is not the same length. We can write expressions for the points representing B' and C' in terms of the center, which is defined by (9.190).

$$\begin{aligned} \mathbf{Z}_{B'} &= R_{O4} + jX_{O4} - 0.5(\overline{B'C'}) = 4.628 \pm j8.496 \\ \mathbf{Z}_{C'} &= R_{O4} + jX_{O4} + 0.5(\overline{B'C'}) = 23.372 + j16.496 \end{aligned} \tag{9.198}$$

The diameter of circle 4 is defined by the points labeled B' and C'. We can compute the diameter as the magnitude of the difference between the two impedances.

$$\mathbf{Z}_{C'} - \mathbf{Z}_{B'} = \overline{B'C'} = 18.744 + j24.992 = 31.24\angle 53.13^\circ \tag{9.199}$$

Thus, by using the equations of the parameters in the complex plane, we can completely define the parameters associated with the circles shown in Figure 9.27. ■

By proper selection of the parameters that define the phase comparison logic, the circular relay characteristic can be varied, as desired, although all solutions will pass through point A of Figure 9.27.

Referring back to (9.174), we recall that the parameters in the two complex quantities to be compared by the phase comparator are the sum of the relay voltage and a voltage drop caused by the relay current flowing through an impedance. One of these impedances is usually the total impedance of the protected transmission line, or some desired fraction of the total impedance. This is often called the "reach" of the relay. In our modification of the basic equation (9.174), we changed these voltages into impedances by dividing by the relay current, which does not change the phase relationship between the quantities being compared. The impedance called $\mathbf{Z}_{R1}$ can still be thought of as the intended reach of the relay, as it looks into the protected transmission line.

Another important feature of the circular characteristic is that every point on the circle defines a right angle, measured with respect to the apex and either the origin or point B. This is demonstrated by example 9.5.

EXAMPLE 9.5

Show that every point on every circle defines a right angle with respect to the points previously defined.

Solution

In order to demonstrate this point, a slightly different example is selected. All of the parameters for this example are the same as those specified in Table 9.3 except that we select $\mathbf{Z}_{R2} = 8 - j4$ in order to more clearly emphasize the angle construction. The results are shown in Figure 9.28, where four circles are plotted with the same parameters as those of Figure 9.27, except for the difference in $\mathbf{Z}_{R2}$.

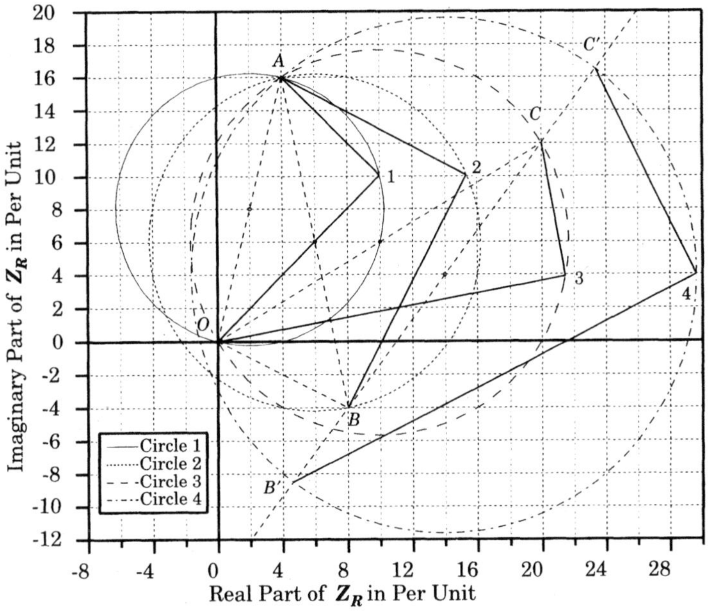

Figure 9.28 The right angle construction defining the circles.

The point of this exercise is to note that lines drawn from the diameter of a circle to any point on that circle to a common point on the circumference always form a right angle. For circle 1, the diameter is the line OA, and selecting point 1 at random, the angle at the point labeled "1" forms a right angle. For circle 2, the diameter is the line AB, and constructing lines to an arbitrary point 2 again forms a right angle. Circle 3 passes through the origin and has diameter OC, so that lines from these points meeting at an arbitrary point 3 form a right angle.

Circle 4 is different since the diameter is defined by (9.189). This diameter can then be used to construct a right angle at any point on the circumference of the circle. This construction is left as an exercise. ■

REFERENCES

[1] Atabekov, G. I., *The Relay Protection of High Voltage Networks*, Pergamon Press, NY, 1960.

[2] Anderson, P. M., *Analysis of Faulted Power Systems*, IEEE Press, Piscataway, NJ, 1995.

[3] Guillemin, E. A., *Communication Networks, Vol. II*, John Wiley & Sons, Inc., New York, 1935.

[4] Nilsson, J. W., *Electric Circuits*, Addison-Wesley Publishing Co., Reading, MA, 1983.

[5] Wedepohl, L. M., "Polarised Mho Distance Relay," *Proc. IEE*, 112, March 1965, pp. 525–535.

[6] Humpage, W. D., and S. Kandil, "Discriminative Performance of Distance Protection Under Fault Operating Conditions," *Proc. IEE*, 115 (1), January 1968, pp. 141-152.

[7] Warrington, A. R. van C., *Protective Relays: Their Theory and Practice*, Vols. I and II, Chapman and Hall, 1969.

[8] Rushton, J., and W. D. Humpage, "Power System Studies for the Determination of Distance Protection Performance," *Proc. IEE.*, 119 (6), June 1972, pp. 677–688.

[9] Cook, V., *Analysis of Distance Protection*, Research Studies Press, Ltd., Letchworth, Hertfordshire, England, and John Wiley & Sons, Inc., New York, 1985.

[10] GEC Measurements, *Protective Relays Application Guide*, The General Electric Company p.l.c. of England, 2nd Edition, 1975.

PROBLEMS

9.1 Derive the result (9.8), based on the definition (9.5) given in Section 9.1 of this chapter.

9.2 Derive result (9.9) based on definition (9.5).

9.3 Derive result (9.9) starting from the result given by (9.8).

9.4 Develop the two-port Thevenin equivalent from network theory, with a special case for a transmission line connected between two of the network nodes. (Hint: see [2]). The general description of the network is shown in Figure P9.4.

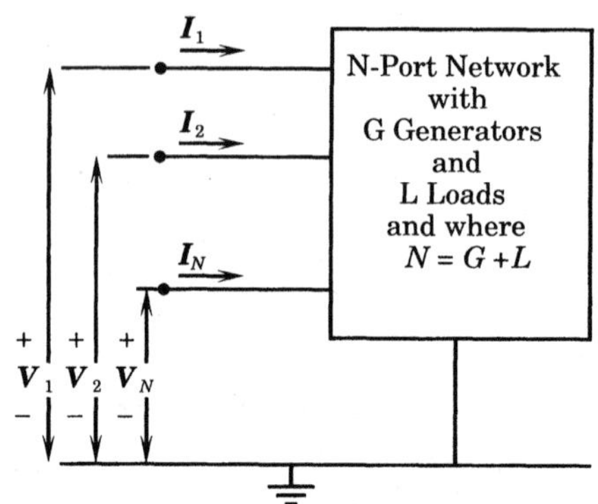

Figure P9.4 A general network description.

9.5 Consider the system shown in Figure P9.5 where the protected system is a short transmission line with impedance $\mathbf{Z}_L$. Determine the matrix **M** where the system data are given as follows.

$$\mathbf{Z}_S = j0.1 \text{ pu}$$
$$\mathbf{Z}_U = j0.2 \text{ pu}$$
$$\mathbf{Z}_L = 0.2 + j0.3 \text{ pu}$$

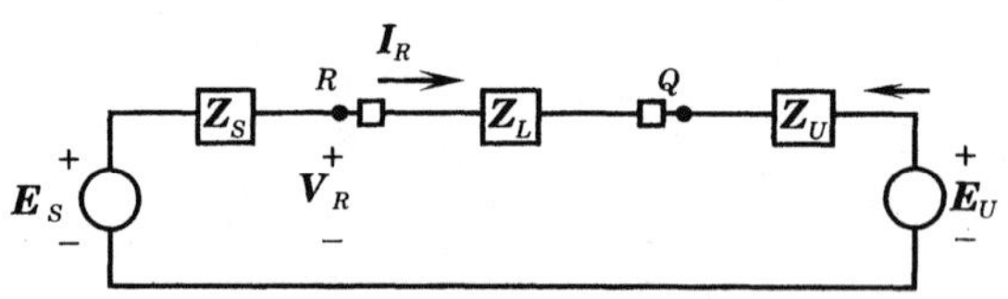

Figure P9.5 System described for problem 9.5.

9.6 Repeat problem 9.5 with the transmission impedance approximated as $\mathbf{Z}_L = j0.3$ pu.

9.7 A 330 kV transmission line 100 km long has the following given parameters:

$R = 4.5\Omega$

$X = 49.2\Omega$

$B_c = 233\mu$ mho

Suppose this line is represented in Figure P9.5 as a nominal pi line in per unit on a 100 MVA base. Let the source impedances be

$$\mathbf{Z}_S = 2 + j5\Omega$$
$$\mathbf{Z}_U = 1 + j6\Omega$$

Compute the matrix **M**.

9.8 Consider the **M** parameters derived for the protection equivalent in Section 9.2, but let the shunt impedances $\mathbf{Z}_1$ and $\mathbf{Z}_2$ be equal. How does this change the results?

9.9 Consider the **M** parameters derived for the protection equivalent in Section 9.2. Determine the **M** parameters if the shunt impedances are removed. Do this by finding the limit as $\mathbf{Z}_1$ and $\mathbf{Z}_2$ increase without bound. What can you conclude regarding the independence of the four parameters?

9.10 Compute the impedances $\boldsymbol{n}$, $\mathbf{Z}_X$, $\mathbf{Z}_Y$, and $\mathbf{Z}_Z$ for the system described in problem 9.5.

9.11 Compute the impedances $\boldsymbol{n}$, $\mathbf{Z}_X$, $\mathbf{Z}_Y$, and $\mathbf{Z}_Z$ for the system described in problem 9.6.

9.12 Compute the impedances $\boldsymbol{n}$, $\mathbf{Z}_X$, $\mathbf{Z}_Y$, and $\mathbf{Z}_Z$ for the system described in problem 9.7.

9.13 Compute the $\mathbf{Z}_R$ loci parameters for the system:
(a) of problems 9.5 and 9.10
(b) of problems 9.6 and 9.11
(c) of problems 9.7 and 9.12

9.14 In constructing k circles, it is desirable to plot a family circles of equal radius, whether k is greater or less than unity. Determine a rule for ensuring that two circles will have equal radii, when one of them is plotted for $k < 1$ and the other for $k > 1$.

9.15 Write expressions that locate the k circle centers measured from (see Figure 9.7)
(a) the Z plane origin
(b) the point B
(c) the point $\mathbf{Z}_M$

9.16 Show that, when $k \to \infty$ and $k = 0$, the k circles degenerate into points A and B, respectively.

9.17 Show that the k circle in the limit as $k \to 1$ is a straight line passing through $\mathbf{Z}_M$.

9.18 Prove that all ψ circles pass through both A and B.

9.19 Construct $\mathbf{Z}_R$ loci for a system where

$$\mathbf{Z}_X = 20 + j70 \text{ ohms}$$
$$\mathbf{Z}_Z = 50 + j30 \text{ ohms}$$
$$\boldsymbol{n} = 1.4\angle 30^\circ$$

9.20 Compute the two-port $\boldsymbol{ABCD}$ parameters for the system
(a) of problem 9.5
(b) of problem 9.6
(c) of problem 9.7

9.21 Verify the development of (9.91).

9.22 Verify (9.102) and show that this is a circle in the Z_R plane.

9.23 Verify the final form of the Z plane circle given by (9.104).

9.24 In Section 9.5.2, the analysis of Figure 9.14 is carried out with an arbitrary impedance $\mathbf{Z}_K$ represented for the shunt fault. Construct a table for the following shunt faults, showing the value that must be used for the impedance $\mathbf{Z}_K$ if the fault is
(a) Three-phase with fault impedance in each phase of $\mathbf{Z}_F$.
(b) One-line-to-ground with fault impedance of $3\mathbf{Z}_F$.
(c) Line-to-line fault with impedance between faulted phases of $\mathbf{Z}_F$.
(d) Two-line-to-ground with phase impedances of $\mathbf{Z}_F$. and a ground impedance of $\mathbf{Z}_G$

9.25 Verify (9.136).

9.26 Construct a spreadsheet to compute the $\boldsymbol{ABCD}$ parameters for an unfaulted system.

9.27 Compute the two-port $\boldsymbol{ABCD}$ parameters for the system:
(a) problem 9.5
(b) problem 9.6
(c) problem 9.7
(d) problem 9.7, but with $\mathbf{Z}_E = 3 + j20\Omega$

(e) same as part (d), but with a shunt reactor at the Q end with a total susceptance of $Y_2 = 10 + j100\mu$mho

9.28 Compute $\boldsymbol{Z}_R$ for the system of Figure 9.13 and using the data of
(a) problem 9.5
(b) problem 9.6
(c) problem 9.7
(d) problem 9.7, but with $\boldsymbol{Z}_E = 3 + j20\Omega$
(e) same as part (d), but with a shunt reactor at the Q end with a total susceptance of $Y_2 = 10 + j100\mu$mho

9.29 Compute $\boldsymbol{H}_S$ as defined by (9.138) for the system of Figure 9.13 and using the data of
(a) problem 9.5
(b) problem 9.6
(c) problem 9.7
(d) problem 9.7, but with $\boldsymbol{Z}_E = 3 + j20\Omega$
(e) same as part (d), but with a shunt reactor at the Q end with a total susceptance of $Y_2 = 10 + j100\mu$mho

9.30 Compute $\boldsymbol{H}_R$ as defined by (9.140) for the system of Figure 9.13 and using the data of
(a) problem 9.5
(b) problem 9.6
(c) problem 9.7
(d) problem 9.7, but with $\boldsymbol{Z}_E = 3 + j20\Omega$
(e) same as part (d), but with a shunt reactor at the Q end with a total susceptance of $Y_2 = 10 + j100\mu$mho

9.31 The circuit of Figure 9.14 has a shunt fault at F, but the type of fault is not identified. Assume that the fault is a one-line-to-ground fault with fault impedance $\boldsymbol{Z}_F$.
(a) Sketch the three sequence networks under the assumption that the power system is a two-port network, with the port currents defined as entering points A and B in Figure 9.14.
(b) Write the two-port Z matrix equations for the sequence networks.
(c) Connect the sequence networks to represent the one-line-to-ground fault. Is this connection a series connection of two-port networks?

9.32 Repeat problem 9.31(c) for a line-to-line fault.

9.33 Consider the figure accompanying problem 9.1. It is claimed in [10] that the impedance seen by the relay for this system is given by

$$\boldsymbol{Z}_R = -\boldsymbol{Z}_S + (\boldsymbol{Z}_S + \boldsymbol{Z}_L + \boldsymbol{Z}_U)\left(\frac{1 - j\cot(\theta/2}{2}\right)$$

Confirm that this is the correct solution for the relay impedance.

9.34 Consider a one-line-to-ground fault at F in Figure 9.14. Construct the sequence network connections for this fault, including the fault impedance.
(a) Assume that the source voltages $\boldsymbol{E}_S$ and $\boldsymbol{E}_U$ are equal, which is an assumption often made in fault studies.
(b) Assume that the source voltages $\boldsymbol{E}_S$ and $\boldsymbol{E}_U$ are not equal. Describe how one can solve the system for the contribution to the total fault current supplied from source $\boldsymbol{E}_S$.

9.35 Verify the ***ABCD*** results computed in Section 9.5.2.

9.36 Verify the **M** parameters given in Section 9.7.3.

9.37 Verify the results given by (9.188) and (9.189).

9.38 Prove that all circles defined by (9.188) pass through the point $R_1 + jX_1$ in the complex Z plane.

9.39 Prove that the origin of the Z plane is a solution of (9.188) when $f = 0$.

9.40 Construct the circle defined by the following parameters:

$$R_1 = 4.0 \qquad R_2 = 2.0 \qquad e = 1.0$$
$$X_1 = 16.0 \qquad X_2 = -4.0 \qquad f = -1.0$$

9.41 Construct the circle defined by the following parameters:

$$R_1 = 4.0 \qquad R_2 = 2.0 \qquad e = 1.0$$
$$X_1 = 16.0 \qquad X_2 = -4.0 \qquad f = -2.0$$

9.42 Experiment with a copy of Figure P9.4 and satisfy yourself that lines drawn to each end of the diameter from any point on the circle forms a right angle at the selected point.

9.43 Consider the general equivalent of a protected component as shown in Figure P9.43, where the ***ABCD*** two-port parameters are known based on viewing the system from the external sources, as noted in the figure. Determine the **M** matrix parameters for the protected component, defined by (9.1), in terms of the ***ABCD*** parameters, defined by (9.84).

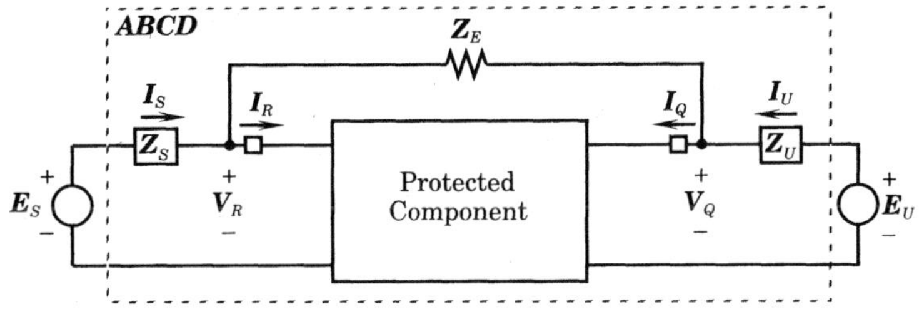

Figure P9.43 General equivalent of a protected component.

10

Admittance at the Relay

For some protective relay problems, there are advantages to viewing the relay characteristic in terms of the apparent impedance seen at the relay. Chapter 9 explores this concept in some detail. Other problems are more easily solved by computing the apparent admittance seen at the relay. This chapter is devoted to a development of the admittance characteristics and its relationship to the impedance characteristic. An emphasis on graphical interpretation is used to clarify these concepts.

10.1 ADMITTANCE DIAGRAMS

From Chapter 9 we have the equations of the relay voltage and current in terms of the Thevenin equivalent voltages, i.e.,

$$\begin{bmatrix} \boldsymbol{V}_R \\ \boldsymbol{I}_R \end{bmatrix} = \begin{bmatrix} \boldsymbol{m}_1 & \boldsymbol{m}_2 \\ \boldsymbol{m}_3 & -\boldsymbol{m}_4 \end{bmatrix} \begin{bmatrix} \boldsymbol{E}_S \\ \boldsymbol{E}_U \end{bmatrix} \tag{10.1}$$

where the variables in (10.1) are defined in Figure 10.1.

If we solve for the admittance $\boldsymbol{Y}_R$ seen by the relay, we have

$$\boldsymbol{Y}_R = \frac{\boldsymbol{I}_R}{\boldsymbol{V}_R} = \frac{\boldsymbol{m}_3\boldsymbol{E}_S - \boldsymbol{m}_4\boldsymbol{E}_U}{\boldsymbol{m}_1\boldsymbol{E}_S + \boldsymbol{m}_2\boldsymbol{E}_U} = \frac{\boldsymbol{m}_3 - \boldsymbol{m}_4\dfrac{\boldsymbol{E}_U}{\boldsymbol{E}_S}}{\boldsymbol{m}_1 + \boldsymbol{m}_2\dfrac{\boldsymbol{E}_U}{\boldsymbol{E}_S}} \tag{10.2}$$

which is a circle for variations in $(\boldsymbol{E}_U/\boldsymbol{E}_S)$. This admittance is the inverse of the impedance $\boldsymbol{Z}_R$.

It is also convenient to define the "input admittance" $\boldsymbol{Y}_I$ seen by $\boldsymbol{I}_S$ at the Thevenin source voltage $\boldsymbol{E}_S$. Thus, by definition,

$$\boldsymbol{Y}_I = \frac{\boldsymbol{I}_S}{\boldsymbol{E}_S} \tag{10.3}$$

This admittance is the inverse of the impedance $\boldsymbol{Z}_I$ shown in Figure 10.1. It can be shown that this function is also a circle in the admittance plane.

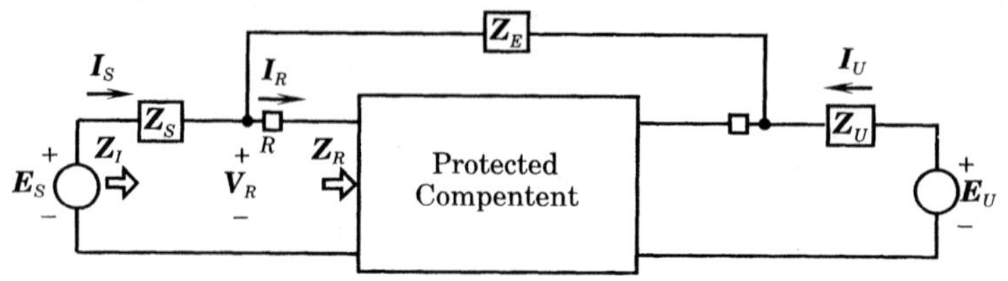

Figure 10.1 Measurements at the relay R.

The admittances $\boldsymbol{Y}_R$ and $\boldsymbol{Y}_I$ correspond to the admittance seen by $\boldsymbol{I}_R$ and $\boldsymbol{I}_S$, respectively, at different points, as shown in Figure 10.1. These admittances correspond to impedances $\boldsymbol{Z}_R$ and $\boldsymbol{Z}_I$, respectively. From Figure 10.1, we write

$$\boldsymbol{E}_S = \boldsymbol{V}_R + \boldsymbol{Z}_S\boldsymbol{I}_S \tag{10.4}$$

which we divide by $\boldsymbol{I}_S$ to compute

$$\boldsymbol{Z}_I = \frac{\boldsymbol{E}_S}{\boldsymbol{I}_S} = \boldsymbol{Z}_S + \frac{\boldsymbol{I}_R}{\boldsymbol{I}_S}\boldsymbol{Z}_R \tag{10.5}$$

or

$$\boldsymbol{Z}_I = \boldsymbol{Z}_S + \boldsymbol{k}\boldsymbol{Z}_R \tag{10.6}$$

where we have defined the complex constant

$$\boldsymbol{k} = \frac{\boldsymbol{I}_R}{\boldsymbol{I}_S} \tag{10.7}$$

Then the input admittance may be written as

$$\boldsymbol{Y}_I = \frac{1}{\boldsymbol{Z}_I} = \frac{1}{\boldsymbol{Z}_R + \boldsymbol{k}\boldsymbol{Z}_S} \tag{10.8}$$

This result is a convenient measure of system conditions or of the relay characteristics as viewed from the input or source terminals.

10.2 INPUT ADMITTANCE LOCI

Consider the symmetrical system viewed from the Thevenin voltages as shown in Figure 10.1. For the Thevenin two port we write the following.

$$\begin{bmatrix} \boldsymbol{E}_S \\ \boldsymbol{I}_S \end{bmatrix} = \begin{bmatrix} \boldsymbol{A} & \boldsymbol{B} \\ \boldsymbol{C} & \boldsymbol{D} \end{bmatrix} \begin{bmatrix} \boldsymbol{E}_U \\ -\boldsymbol{I}_U \end{bmatrix} \tag{10.9}$$

This system has the solution (9.90) or

$$\boldsymbol{I}_S = \frac{\boldsymbol{D}}{\boldsymbol{B}}\boldsymbol{E}_S - \frac{1}{\boldsymbol{B}}\boldsymbol{E}_U \tag{10.10}$$

Then, by definition, the system condition is described in terms of the source voltages by the expression

$$\boldsymbol{Y}_I = \frac{\boldsymbol{I}_S}{\boldsymbol{E}_S} = \frac{\boldsymbol{D}}{\boldsymbol{B}}\left[1 - \frac{\boldsymbol{E}_U}{\boldsymbol{D}\boldsymbol{E}_S}\right] \tag{10.11}$$

which we can write as

$$\boldsymbol{Y}_I = \frac{\boldsymbol{D}}{\boldsymbol{B}}(1 - \boldsymbol{m}e^{-j\psi}) \tag{10.12}$$

where we can define, from (9.134),

$$\begin{aligned} \boldsymbol{D} &= de^{j\delta} \\ \frac{\boldsymbol{E}_U}{\boldsymbol{E}_S} &= \frac{1}{E}e^{-j\theta} \\ \psi &= \delta + \theta \end{aligned} \tag{10.13}$$

so that

$$\boldsymbol{m}e^{-j\psi} = \frac{\boldsymbol{E}_U}{\boldsymbol{D}\boldsymbol{E}_S} = \frac{e^{-j(\delta+\theta)}}{dE} \tag{10.14}$$

or

$$m = \frac{1}{dE} \tag{10.15}$$

Typically for a transmission line $\boldsymbol{D}$ has a magnitude of approximately unity and a very small angle so $1/d$ is also about unity. Therefore, there is little error if we consider $\boldsymbol{m}$ to be a real constant that is approximately unity when $\boldsymbol{E}_S$ is equal to $\boldsymbol{E}_U$.

10.2.1 Y_I Loci For Constant m

We now examine the case where m is a constant and ψ a variable. Suppose we define the dimensionless parameter $\boldsymbol{F}$ such that

$$\boldsymbol{F} = (1 - me^{-j\psi}) = 1 - me^{-j\delta}e^{-j\theta} \tag{10.16}$$

which is plotted in Figure 10.2. Since d is a constant, different values of m correspond to changes in the Thevenin voltage ratio E. Also, since δ is a constant, changes in ψ correspond to changes in the angle θ by which source voltage $\boldsymbol{E}_S$ leads $\boldsymbol{E}_U$.

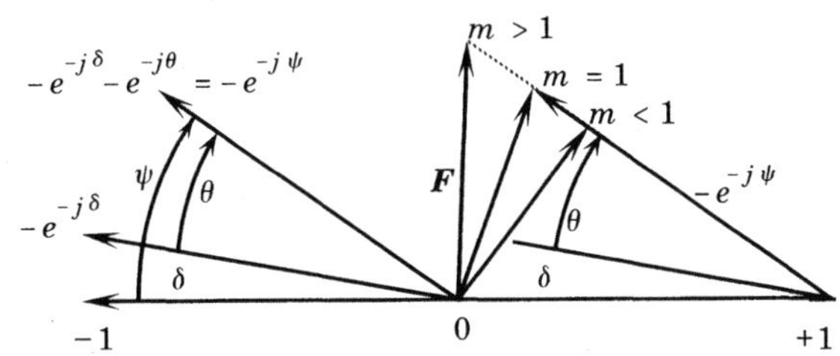

Figure 10.2 Plot of $\boldsymbol{F} = 1me^{j\psi}$ for fixed m and variable θ.

The quantity $\boldsymbol{D}/\boldsymbol{B}$ has dimensions of siemens (or 1/ohms) and, for a transmission line, is approximately equal to the series admittance. Multiplying $\boldsymbol{F}$ by $\boldsymbol{D}/\boldsymbol{B}$ rotates the plot of Figure 10.2 by the phase angle of $\boldsymbol{D}/\boldsymbol{B}$ and scales $\boldsymbol{F}$ by the magnitude of $\boldsymbol{D}/\boldsymbol{B}$. Typically, the angle of rotation will be negative and in the range of $-60°$ to $-90°$.

Suppose we write

$$\frac{\boldsymbol{D}}{\boldsymbol{B}} = ye^{j\phi} \tag{10.17}$$

then ϕ would normally be large, say $270° < \phi < 300°$. Since $\boldsymbol{Y}_I = (\boldsymbol{D}/\boldsymbol{B})\boldsymbol{F}$, this will result in a plot similar to that of Figure 10.3, which is plotted for three different values of the parameter m and for a fixed value of θ. It is apparent from this figure that we can easily find the variation of $\boldsymbol{Y}_I$ as a function of ψ in the Y plane for any fixed m. When $m = 1$, the circular locus for the input admittances passes through the origin. Smaller values of m give smaller radii, and

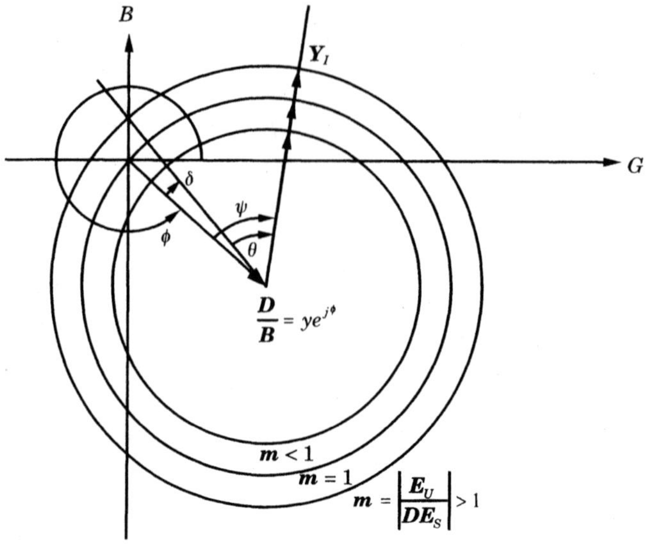

Figure 10.3 Locus of Y_I for constant m and variable θ.

larger values larger radii. The angle θ increases in the clockwise direction, which corresponds to increasing the angle by which the source voltage E_S on the left leads E_U (see Figure 10.1).

10.2.2 Y_I Loci For Constant ψ

When ψ is a constant and m is variable, the locus $Y_I = (D/B)F$ is a line. From (10.12) and (10.17) we write

$$\begin{aligned} Y_I &= \left(\frac{D}{B}\right)F = (ye^{j\phi})(1 - me^{-j\psi}) \\ &= ye^{j\phi} - yme^{j(\phi-\psi)} \end{aligned} \tag{10.18}$$

The first term in (10.18) is the constant $|D/B|$ that locates the center of the circles in Figure 10.3. The line (10.18) goes through this point, making an angle $(\phi - \psi)$ with the horizontal as shown in Figure 10.4. Using this construction it is easy to select values of Y_I corresponding

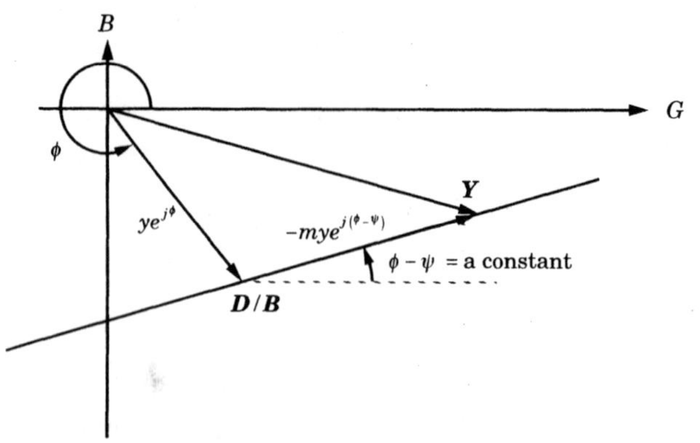

Figure 10.4 Locus of Y_i for constant ψ and variable m.

to variations in m for any fixed ψ. In the second line of (10.18), the first term is a constant and the second term varies with m.

10.3 THE RELAY ADMITTANCE CHARACTERISTIC

We now seek to relate the relay tripping characteristic $\hat{\boldsymbol{Y}}_I$ to $\boldsymbol{Y}_I$. First, however, we must establish a clear notation for the various impedances and admittances. We have previously defined the apparent impedance *seen* at the relay location to be $\boldsymbol{Z}_R$. Note that this is not the relay tripping characteristic, but is a point in the complex Z plane that corresponds to the particular voltage and current that exist at the relay. The *relay threshold characteristic* is a locus of points in the complex Z plane, which we shall designate as $\hat{\boldsymbol{Z}}_R$, or its inverse $\hat{\boldsymbol{Y}}_R$ in the complex Y plane, that defines the region in the Z or Y plane for which the relay will pick up. Since $\boldsymbol{Y}_I$ is convenient for plotting system operating conditions in the complex Y plane and $\hat{\boldsymbol{Z}}_R$ is used to determine relay tripping conditions in the complex Z plane, the intersection of these curves would give us the operating conditions that will cause tripping. For convenience, it is desirable to plot the operating condition and the trip characteristic in the same complex plane.

One way to find the desired intersection would be to plot the image of the $\boldsymbol{Y}_I$ circles in the Z plane for a range of values of m. Then values of m and θ could be read from the plot to determine threshold values. However, this method is inconvenient since the direct reading of m and θ is difficult in the Z plane.

A better method is to transfer the relay threshold characteristic, $\hat{\boldsymbol{Z}}_R$, to the admittance plane and compare this with the system operating characteristics given by $\boldsymbol{Y}_I$. We must be careful, however, to compare system quantities at the same location. Thus, we should not compare the admittance $\hat{\boldsymbol{Y}}_R = 1/\hat{\boldsymbol{Z}}_R$ with $\boldsymbol{Y}_I$ as these admittances refer to measurements made at different locations in the circuit. Instead, we define a new quantity that is the relay characteristic transferred to the input terminals, i.e., from (10.5), define

$$\hat{\boldsymbol{Z}}_I = \boldsymbol{Z}_S + \boldsymbol{k}\hat{\boldsymbol{Z}}_R \tag{10.19}$$

where the circumflex (^) signifies the relay characteristic. We then graphically display the relay characteristic, referred to the input terminals, invert this to obtain a corresponding admittance and compare this relay admittance with $\boldsymbol{Y}_I$, which represents the system operating condition. This is summarized in Figure 10.5, which is not drawn to any particular scale. The impedance $\boldsymbol{Z}_O$ represents the center of the $\boldsymbol{k}\hat{\boldsymbol{Z}}_R$ circle.

The construction of Figure 10.5 begins with the relay characteristic $\hat{\boldsymbol{Z}}_R$. However, this is not exactly the characteristic desired. According to (10.19) we need to multiply this characteristic by the complex constant $\boldsymbol{k}$, which is defined by (10.7). In Figure 10.5, this constant has a value of 0.91 with an angle of 0°. Thus the constant $\boldsymbol{k}$ shrinks the characteristic to get the new circle, $\boldsymbol{k}\hat{\boldsymbol{Z}}_R$, plotted using the gray line. The source impedance is added to the gray circle to obtain the relay characteristic curve as viewed from the input $\hat{\boldsymbol{Z}}_I$ which is the hatched circle displaced upward in the figure. The inverse of this circle gives the admittance circle in the fourth quadrant. The intersection of $\hat{\boldsymbol{Y}}_I$ with any circular m characteristic locates the relay threshold for that value of m. This threshold is identified in Figure 10.5 for $m = 1$, which occurs at a θ value of about 55°.

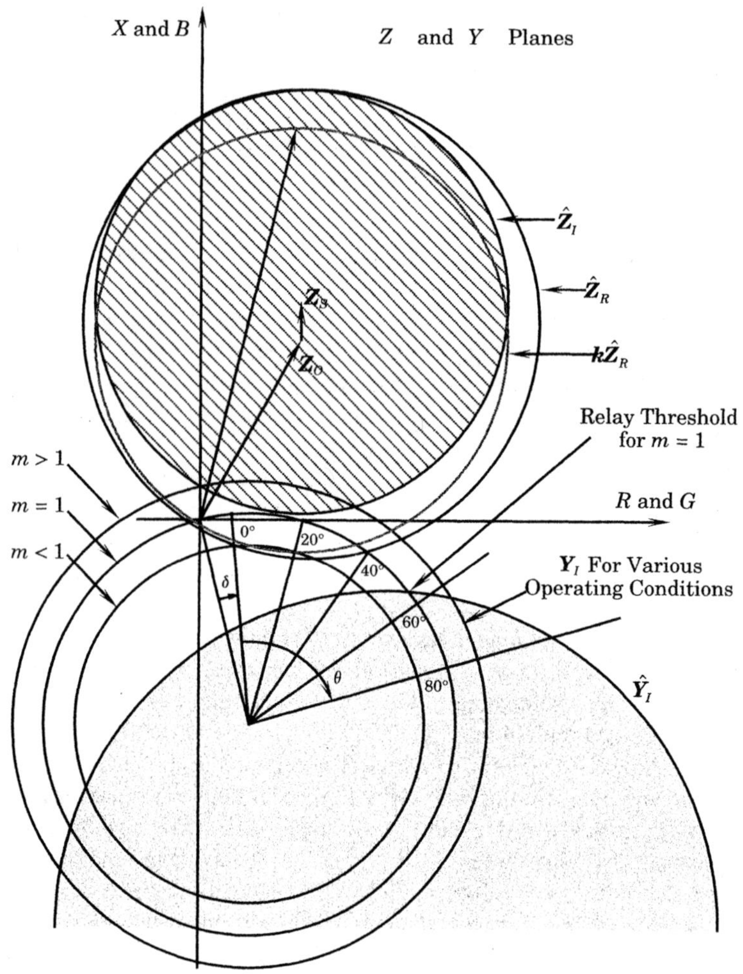

Figure 10.5 Graphical construction comparing Y_I and $\hat{Y}_I$.

EXAMPLE 10.1

A long transmission line connects two systems. The ***ABCD*** parameters for the line and source impedances are given as

$$A = 0.8589\angle 7.75^\circ$$
$$B = 280.0\angle 81^\circ \Omega$$
$$C = 1300\angle 91^\circ \mu S$$
$$D = 0.76\angle 3^\circ$$

For this example, the impedance Z_E of Figure 10.1 is infinite.

The source impedance behind the relay is

$$Z_S = 0 + j40\ \Omega$$

The relay itself is set as a circular (mho) characteristic through the origin, as in Figure 10.6, with Z_o set at an angle of 60° and adjusted so that the maximum relay impedance for a three-phase fault at E_U ($E_U = 0$) is exactly on the relay threshold. Find the limiting phase angle between E_S and E_U for $m = 0.9$, 1.0, and 1.1.

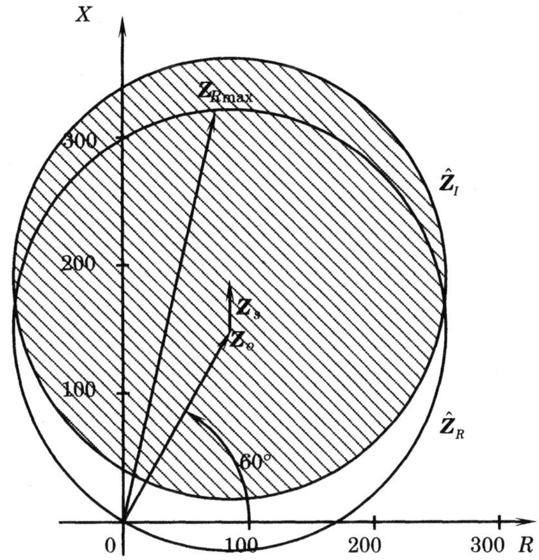

Figure 10.6 The relay and input terminal characteristics.

Solution

First, we determine the relay characteristic We are given that $\mathbf{Z}_{R\max}$ is on the threshold circle, or

$$\mathbf{Z}_{R\max} = \left[\frac{\mathbf{B}}{\mathbf{D}}\left(\frac{1}{1-\dfrac{\mathbf{E}_U}{\mathbf{D}\mathbf{E}_S}}\right) - \mathbf{Z}_S\right]_{E_U=0} = \frac{\mathbf{B}}{\mathbf{D}} - \mathbf{Z}_S$$

$$= 368.42\angle 78^\circ - 40.0\angle 90^\circ$$

$$= 76.599 + j360.370 - j40.0 = 329.400\angle 76.55^\circ$$

Since $\mathbf{Z}_{R\max}$ lies on the threshold, the relay characteristic is that shown in Figure 10.6, where the center $\mathbf{Z}_o$ is still unknown. To find $\mathbf{Z}_o$ we write the equation in the general quadratic form.

$$a(R^2 + X^2) + bR + cX + d = 0 \tag{10.20}$$

Since $R = X = 0$ is a solution, then $d = 0$ and (10.20) may be written as

$$\left(R + \frac{b}{2a}\right)^2 + \left(X + \frac{c}{2a}\right)^2 = r_z^2 \tag{10.21}$$

where

$$r_z = \frac{1}{2a}\sqrt{b^2 + c^2} \tag{10.22}$$

But

$$\mathbf{Z}_R = \mathbf{Z}_{R\max} = 329.400\angle 76.55^\circ = 76.599 + j320.370$$

is also a solution. These values can be substituted into (10.21), as follows.

$$\left(76.599 + \frac{b}{2a}\right)^2 + \left(320.370 + \frac{c}{2a}\right)^2 = r_z^2 \tag{10.23}$$

We also know that $\mathbf{Z}_o$ is along a 60° ray from the origin such that

$$\tan 60^\circ = \sqrt{3} = \frac{c/2a}{b/2a} = \frac{c}{b} \tag{10.24}$$

or

$$c^2 = 3b^2 \tag{10.25}$$

Now let

$$\mathbf{Z}_o = R_o + jX_o \tag{10.26}$$

then

$$R_o = \frac{-b}{2a}, \quad X_o = \frac{-c}{2a}, \quad X_o^2 = 3R_o^2 \tag{10.27}$$

Now, (10.23) becomes

$$(76.599 - R_o)^2 + (320.370 - X_o)^2 = R_0^2 + X_o^2$$

or

$$(76.599 - R_o)^2 + (320.370 - 3R_o)^2 = 4R_o^2$$

Expanding, we compute

$$\begin{aligned}5867.407 - 153.198R_o + R_o^2 \\ +102{,}636.937 - 1109.794R_o + 3R_o^2 = 4R_o^2\end{aligned}$$

and

$$1262.992R_o = 108{,}504.344$$

or

$$R_o = 85.911 = \frac{-b}{2a}$$

$$X_o = 148.80 = \frac{-c}{2a}$$

and

$$r_z = (R_o^2 + X_o^2)^{1/2} = \frac{\sqrt{b^2+c^2}}{2a} = 171.821 \tag{10.28}$$

Now, from (10.19), the relay threshold as viewed from the input terminals is

$$\hat{\mathbf{Z}}_I = \mathbf{Z}_S + k\hat{\mathbf{Z}}_R = \mathbf{Z}_S + \hat{\mathbf{Z}}_R$$

since $\boldsymbol{k} = 1$ for this special case. The relay threshold characteristic has its center at

$$R_o + \mathrm{Re}\mathbf{Z}_S = 85.911 + 0 = 85.911$$
$$X_o + \mathrm{Im}\mathbf{Z}_S = 148.801 + 40.0 = 188.801$$

and has radius $r_z = 171.821$, exactly the same as the relay threshold.

We now compute the center and radius of $\hat{\mathbf{Y}}_I$, the inverse of $\hat{\mathbf{Z}}_I$. This image circle has the Y plane equation

$$d(\hat{G}^2 + \hat{B}^2) + b\hat{G} - c\hat{B} + a = 0$$

or

$$\hat{G}^2 + \hat{B}^2 + \frac{b}{d}\hat{G} - \frac{c}{d}\hat{B} = -\frac{a}{d} \tag{10.29}$$

Since $\hat{\mathbf{Z}}_I$ does not intersect the origin, $d \neq 0$ for this circle. From (10.20) we complete the square to find the square of the radius.

$$\frac{b^2+c^2}{4a^2} - \frac{d}{a} = (171.821)^2$$

$$(85.911)^2 + (188.801)^2 - \frac{d}{a} = (171.821)^2$$

$$\frac{d}{a} = 13{,}504.140$$

and this value may be substituted into (10.29). We also compute

$$\begin{aligned}\frac{b/2a}{d/a} &= \frac{b}{2d} = \frac{-85.911}{13504.140} = -6.362 \times 10^{-3}\\ \frac{c/2a}{d/a} &= \frac{c}{2d} = \frac{-188.821}{13504.140} = -13.981 \times 10^{-3}\\ \frac{b^2+d^2}{4d^2} - \frac{a}{d} &= (40.473 + 195.469 - 74.652) \times 10^{-6}\\ &= 161.890 \times 10^{-6}\end{aligned} \tag{10.30}$$

Now, from (10.29), we complete the square to write

$$\left(\hat{G} + \frac{b}{2d}\right)^2 + \left(\hat{B} - \frac{c}{2d}\right)^2 = \frac{b^2+c^2}{4d^2} - \frac{a}{d} = S_y^2$$

or

$$(\hat{G} - G_o)^2 + (\hat{B} - B_o)^2 = S_y^2 \tag{10.31}$$

Then, from (10.30) and (10.31) the center of admittance threshold is located at

$$G_o = \frac{-b}{2d} = 6.362 \times 10^{-3}$$

$$B_o = \frac{+c}{2d} = -13.981 \times 10^{-3}$$

and from (10.30) the radius is

$$S_y = \sqrt{161.8895} = 12.723 \times 10^{-3}$$

We now compute the operating characteristic $\boldsymbol{Y}_I$ so that it can be plotted together with the input admittance threshold characteristic. From (10.12),

$$\boldsymbol{Y}_I = \frac{\boldsymbol{D}}{\boldsymbol{B}}(1 - \boldsymbol{m}e^{-j\psi})$$

For this problem we have

$$\frac{\boldsymbol{D}}{\boldsymbol{B}} = \frac{1}{368.42\angle 78^\circ} = 2.714 \times 10^{-3}\angle -78^\circ\text{S}$$

$$m = \frac{|\boldsymbol{E}_U|}{|\boldsymbol{D}||\boldsymbol{E}_S|} = \frac{1}{0.76E} = \frac{1.316}{E}$$

We are given values of $E = 0.75$, 1.00, and 1.25. These correspond to

$$\begin{aligned} m &= 1.754 & E &= 0.75\\ m &= 1.316 & E &= 1.00\\ m &= 1.053 & E &= 1.25\end{aligned}$$

These values must be multiplied by $|\boldsymbol{D}/\boldsymbol{B}|$ before plotting in the Y plane, in which case they become 4.762×10^{-3}, 3.572×10^{-3}, and 2.857×10^{-3}, respectively.

We also have

$$\psi = \theta + \delta = \theta + 3^\circ$$

The plot of both the $\boldsymbol{Y}_I$ and $\hat{\boldsymbol{Y}}_I$ circles is shown in Figure 10.7. From the figure we read the following phase angle limits (approximately):

$$\begin{aligned} E &= 1.25 & \theta &= 82^\circ\\ E &= 1.00 & \theta &= 83^\circ\\ E &= 0.75 & \theta &= 86^\circ\end{aligned}$$

Note that θ does not change very fast for changes in m and that $\theta \cong 80°$ is approximately correct for a range of voltages in the Y plane neighborhood of the voltage ratio $E = 1.0$.

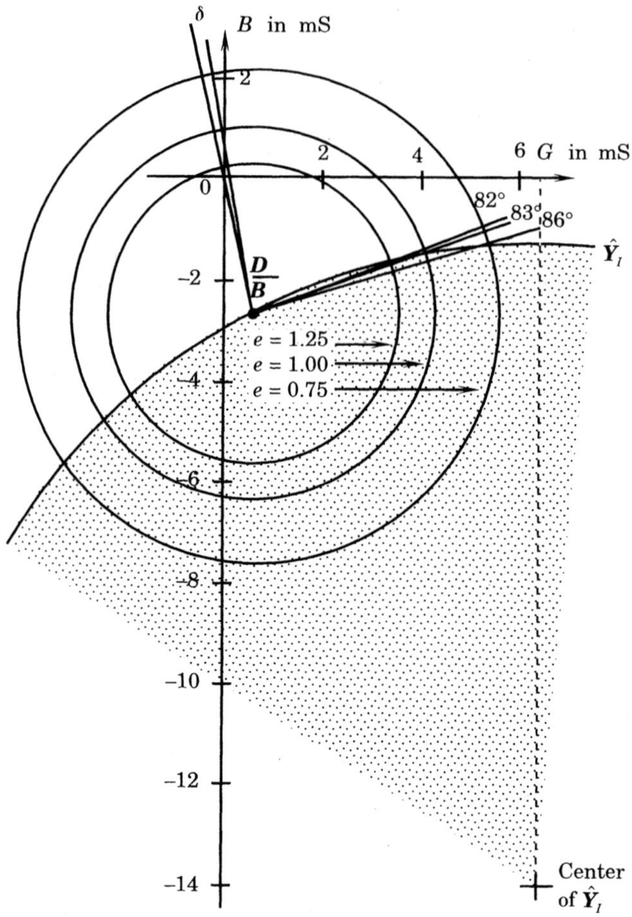

Figure 10.7 Solution of example 10.1 in the Y plane. ■

10.4 PARALLEL TRANSMISSION LINES

Consider a situation where the protected line under consideration is in parallel with $n-1$ other lines as shown in Figure 10.8. Here we let $\mathbf{Z}_1, \mathbf{Z}_2, \ldots, \mathbf{Z}_n$ be the line impedances of lines $1, 2, \ldots, n$ and we assume that these impedances may all be different. Then the impedance seen by relay k is

$$\mathbf{Z}_{Rk} = \frac{\mathbf{V}_R}{\mathbf{I}_{Rk}} \tag{10.32}$$

If we let the source current be $\mathbf{I}_S$, then we can write

$$\mathbf{V}_R = \mathbf{E}_S - \mathbf{Z}_S \mathbf{I}_S \tag{10.33}$$

to get the relay voltage in terms of input quantities.

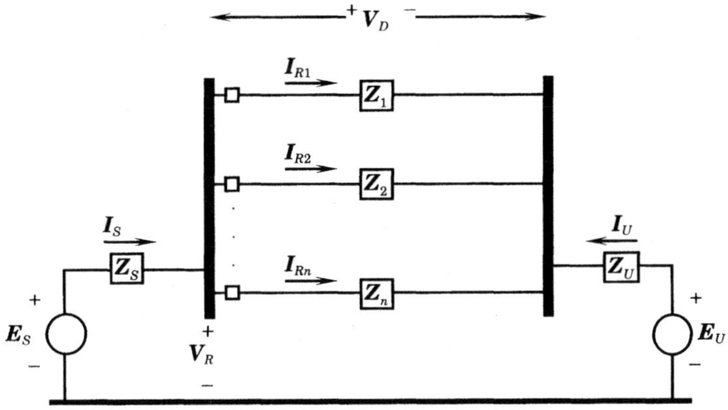

Figure 10.8 A transmission system with n parallel lines.

Since the lines are all in parallel, the voltage drop across all lines is the same. If we call this voltage drop $\boldsymbol{V}_D$, then the value of this voltage is

$$\boldsymbol{V}_D = \boldsymbol{Z}_1\boldsymbol{I}_{R1} = \boldsymbol{Z}_2\boldsymbol{I}_{R2} = \ldots = \boldsymbol{Z}_n\boldsymbol{I}_{Rn} \tag{10.34}$$

We also know that Kirchhoff's law must be satisfied at the relay junction, or

$$\begin{aligned}\boldsymbol{I}_S &= \boldsymbol{I}_{R1} + \boldsymbol{I}_{R2} + \ldots + \boldsymbol{I}_{Rn} \\ &= \boldsymbol{Y}_1\boldsymbol{V}_D + \boldsymbol{Y}_2\boldsymbol{V}_D + \ldots + \boldsymbol{Y}_n\boldsymbol{V}_D = \sum_{i=1}^{n}\boldsymbol{Y}_i\boldsymbol{V}_D\end{aligned} \tag{10.35}$$

where $\boldsymbol{Y}_1, \ldots, \boldsymbol{Y}_n$ are the individual line admittances, and are equal to the reciprocal of the line impedances. From (10.35), we compute

$$\boldsymbol{V}_D = \frac{\boldsymbol{I}_S}{\sum_{i=1}^{n}\boldsymbol{Y}_i} \tag{10.36}$$

which may be substituted into (10.34) to write, at relay k,

$$\boldsymbol{Z}_k\boldsymbol{I}_{Rk} = \frac{\boldsymbol{I}_S}{\sum_{i=1}^{n}\boldsymbol{Y}_i} \tag{10.37}$$

or

$$\boldsymbol{I}_{Rk} = \frac{\boldsymbol{Y}_k\boldsymbol{I}_S}{\sum_{i=1}^{n}\boldsymbol{Y}_i} \tag{10.38}$$

Then the impedance seen by relay k is, from (10.32)

$$\boldsymbol{Z}_{Rk} = \frac{\sum_{i=1}^{n}\boldsymbol{Y}_i\boldsymbol{V}_R}{\boldsymbol{Y}_k\boldsymbol{I}_S} \tag{10.39}$$

which gives the relay impedance in terms of the input current. But from (10.32) we may also compute

$$\boldsymbol{Z}_{Rk} = \frac{\boldsymbol{V}_R}{\boldsymbol{I}_{Rk}} = \frac{\boldsymbol{E}_S - \boldsymbol{Z}_S\boldsymbol{I}_S}{\dfrac{\boldsymbol{Y}_k\boldsymbol{I}_S}{\sum_{i=1}^{n}\boldsymbol{Y}_i}} = \frac{\sum_{i=1}^{n}\boldsymbol{Y}_i}{\boldsymbol{Y}_k}(\boldsymbol{Z}_I - \boldsymbol{Z}_S) \tag{10.40}$$

This expression reduces to $\boldsymbol{Z}_I - \boldsymbol{Z}_S$ when there is only one line, which agrees with (10.6) for the case where $\boldsymbol{k} = 1$.

For the special case of two lines, (10.40) reduces to

$$\boldsymbol{Z}_{R1} = \frac{\boldsymbol{Z}_1 + \boldsymbol{Z}_2}{\boldsymbol{Z}_2}(\boldsymbol{Z}_I - \boldsymbol{Z}_S) \tag{10.41}$$

for relay #1. From this expression we can also compute the relay impedance $\boldsymbol{Z}_{R1}$ as seen at the input terminals, in which case $\boldsymbol{Z}_I = \hat{\boldsymbol{Z}}_{I1}$. From (10.41), for line #1,

$$\hat{\boldsymbol{Z}}_{I1} = \frac{\boldsymbol{Z}_2}{\boldsymbol{Z}_1 + \boldsymbol{Z}_2}\hat{\boldsymbol{Z}}_{R1} + \boldsymbol{Z}_S \tag{10.42}$$

This quantity is plotted in Figure 10.9 for an arbitrary choice of system constants. A similar expression can be developed for line #2, but this is left as an exercise.

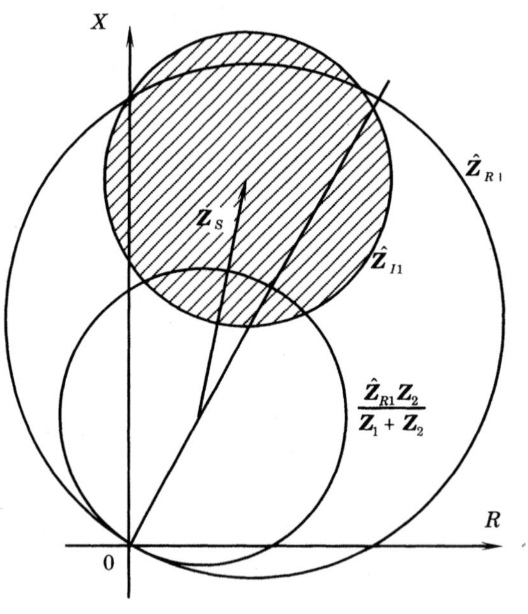

Figure 10.9 Development of $\hat{\boldsymbol{Z}}_{I1}$ from $\boldsymbol{Z}_R$.

Suppose we concentrate on a given relay R of Figure 10.8, located in a line with impedance $\boldsymbol{Z}_L$ with all parallel lines lumped into a single parallel impedance $\boldsymbol{Z}_E$, as shown in Figure 10.10. Then we write the system equations that follow. The relay impedance is

$$\boldsymbol{Z}_R = \frac{\boldsymbol{V}_R}{\boldsymbol{I}_R} \tag{10.43}$$

but

$$\boldsymbol{I}_R = \frac{\boldsymbol{Z}_E}{\boldsymbol{Z}_E + \boldsymbol{Z}_L}\boldsymbol{I}_S = \frac{\boldsymbol{Z}_P}{\boldsymbol{Z}_L}\boldsymbol{I}_S \tag{10.44}$$

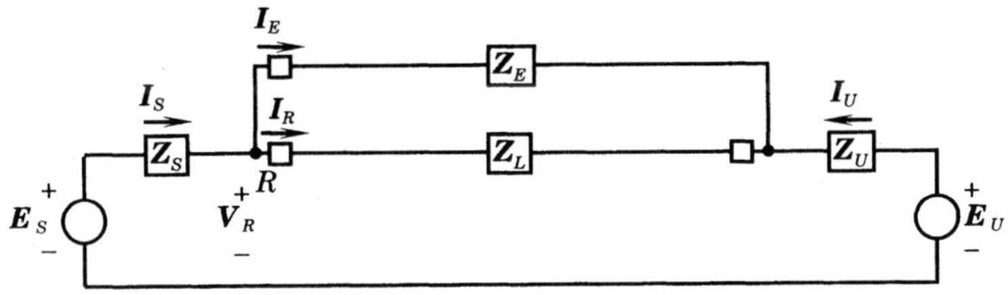

Figure 10.10 Relay impedance in a parallel circuit.

where $\boldsymbol{Z}_P$ is the parallel combination of $\boldsymbol{Z}_L$ and $\boldsymbol{Z}_E$, or

$$\boldsymbol{Z}_P = \frac{\boldsymbol{Z}_E\boldsymbol{Z}_L}{\boldsymbol{Z}_E + \boldsymbol{Z}_L} \tag{10.45}$$

Then

$$\boldsymbol{V}_R = \boldsymbol{E}_S - \boldsymbol{Z}_S\boldsymbol{I}_S = \boldsymbol{E}_S - \frac{\boldsymbol{Z}_S\boldsymbol{Z}_L}{\boldsymbol{Z}_P}\boldsymbol{I}_R \tag{10.46}$$

Now, dividing (10.46) by $\boldsymbol{I}_R$ we get

$$\boldsymbol{Z}_R = \frac{\boldsymbol{E}_S}{\boldsymbol{I}_R} - \frac{\boldsymbol{Z}_S\boldsymbol{Z}_L}{\boldsymbol{Z}_P} \tag{10.47}$$

But we can also write the equations of the system of Figure 10.10 in terms of the two-port transmission parameters as

$$\begin{bmatrix} \boldsymbol{E}_S \\ \boldsymbol{I}_S \end{bmatrix} = \begin{bmatrix} \boldsymbol{A} & \boldsymbol{B} \\ \boldsymbol{C} & \boldsymbol{D} \end{bmatrix} \begin{bmatrix} \boldsymbol{E}_U \\ -\boldsymbol{I}_U \end{bmatrix} \tag{10.48}$$

where the $\boldsymbol{ABCD}$ parameters include the source impedances $\boldsymbol{Z}_S$ and $\boldsymbol{Z}_U$. Then

$$\boldsymbol{I}_U = \frac{\boldsymbol{A}\boldsymbol{E}_U - \boldsymbol{E}_S}{\boldsymbol{B}} \tag{10.49}$$

and

$$\boldsymbol{I}_S = \boldsymbol{C}\boldsymbol{E}_U - \boldsymbol{D}\left(\frac{\boldsymbol{A}\boldsymbol{E}_U - \boldsymbol{E}_S}{\boldsymbol{B}}\right) \tag{10.50}$$

which can be simplified, since $\boldsymbol{AD} - \boldsymbol{BC} = 1$, to

$$\boldsymbol{I}_S = \frac{\boldsymbol{D}}{\boldsymbol{B}}\left(\boldsymbol{E}_S - \frac{\boldsymbol{E}_U}{\boldsymbol{D}}\right) \tag{10.51}$$

If we divide (10.51) into $\boldsymbol{E}_s$, we get the equation of a circle, i.e.,

$$\frac{\boldsymbol{E}_S}{\boldsymbol{I}_S} = \frac{\boldsymbol{E}_S}{\dfrac{\boldsymbol{D}}{\boldsymbol{B}}\left(\boldsymbol{E}_S - \dfrac{\boldsymbol{E}_U}{\boldsymbol{D}}\right)} = \frac{\boldsymbol{B}/\boldsymbol{D}}{1 - \dfrac{\boldsymbol{E}_U}{\boldsymbol{D}\boldsymbol{E}_S}} \tag{10.52}$$

Utilizing (10.44) in conjunction with (10.52) we compute

$$\frac{\boldsymbol{E}_S\boldsymbol{Z}_P}{\boldsymbol{Z}_L\boldsymbol{I}_R} = \frac{\boldsymbol{B}/\boldsymbol{D}}{\left(1 - \dfrac{\boldsymbol{E}_U}{\boldsymbol{D}\boldsymbol{E}_S}\right)}$$

or

$$\frac{\mathbf{E}_S}{\mathbf{I}_R} = \frac{\left(\frac{\mathbf{B}}{\mathbf{D}}\right)\left(\frac{\mathbf{Z}_L}{\mathbf{Z}_P}\right)}{1 - \frac{\mathbf{E}_U}{\mathbf{D}\mathbf{E}_S}} \tag{10.53}$$

From (10.47) we know $\mathbf{Z}_R$ in terms of $\mathbf{E}_S/\mathbf{I}_R$ so we write

$$\mathbf{Z}_R = \left(\frac{\mathbf{Z}_L}{\mathbf{Z}_P}\right)\left(\frac{\mathbf{B}/\mathbf{D}}{1 - \mathbf{E}_U/\mathbf{D}\mathbf{E}_S} - \mathbf{Z}_S\right) \tag{10.54}$$

which agrees with (9.106).

The admittance plane construction for the case of parallel lines can also be developed. The input admittance for relay k is defined as

$$\mathbf{Y}_I = \frac{\mathbf{I}_{Rk}}{\mathbf{E}_S} \tag{10.55}$$

But the current at the relay can be stated in terms of the source current by (10.38). This allows us to write the relay current in terms of the Thevenin source voltages as follows.

$$\mathbf{I}_{Rk} = \frac{\mathbf{Y}_k}{\sum_{i=1}^{n} \mathbf{Y}_i}\left(\frac{\mathbf{D}}{\mathbf{B}}\mathbf{E}_S + \frac{1}{\mathbf{B}}\mathbf{E}_U\right) \tag{10.56}$$

This is substituted into (10.55) to compute

$$\mathbf{Y}_I = \frac{\mathbf{Y}_k}{\sum_{i=1}^{n} \mathbf{Y}_i}\frac{\mathbf{D}}{\mathbf{B}}\left(1 + \frac{\mathbf{E}_U}{\mathbf{D}\mathbf{E}_S}\right) \tag{10.57}$$

This is a circle in the admittance plane similar to (10.11), but scaled by the ratio of the kth line admittance to the sum of all line admittances. If the lines all have the same admittance angles, this scaling factor is a real constant less than unity. It should be noted that this is not an admittance in the usual sense because $\mathbf{E}_S$ and $\mathbf{I}_{Rk}$ are measured at different points in the network. It is, in fact, a scaling of the true input admittance.

10.5 TYPICAL ADMITTANCE PLANE CHARACTERISTICS

It was noted in Chapter 8 that certain impedance plane characteristics can be evaluated more simply in the admittance plane. We examine this concept further by studying the familiar mho characteristic as well as derivatives of this basic relay type. The general equation of a line or circle in the Z and Y planes are as follows.

$$\alpha(R^2 + X^2) + bR + cX + d = 0 \tag{10.58}$$

If $\alpha \neq 0$, this equation can be rearranged to the form

$$\left(R + \frac{b}{2a}\right)^2 + \left(X + \frac{c}{2a}\right)^2 = \frac{b^2 + c^2 - 4ad}{4a^2} = S_Z^2 \tag{10.59}$$

or

$$(R - R_O)^2 + (X - X_O)^2 = S_Z^2 \tag{10.60}$$

This circle has the following properties:

$$\text{Center:} \qquad (R_O, X_O) = \left(\frac{-b}{2a}, \frac{-c}{2a}\right)$$
$$\text{Radius:} \qquad S_Z = \sqrt{\frac{b^2 + c^2 - 4ad}{4a^2}} \tag{10.61}$$

The $1/Z$ transformation takes this circle to the admittance plane, with the general equation

$$d(G^2 + B^2) + bG - cB + a = 0 \tag{10.62}$$

If $d \neq 0$, this equation can be rearranged to write

$$(G - G_O)^2 + (B - B_O)^2 = S_Y^2 \tag{10.63}$$

where, we can derive the following properties of the Y plane properties:

$$\text{Center:} \qquad (G_O, B_O) = \left(\frac{-b}{2d}, \frac{c}{2d}\right)$$
$$\text{Radius:} \qquad S_Y = \sqrt{\frac{b^2 + c^2 - 4ad}{4d^2}} \tag{10.64}$$

If $d = 0$, then (10.62) reduces to the equation of the straight line

$$B = \frac{b}{c}G + \frac{a}{c} \tag{10.65}$$

Two examples are considered, one where the Z circle passes through the origin, and a second example where the Z circle does not pass through the origin. From Chapter 8, we know that the first case results in a straight line in the Y plane, whereas the second case gives a circle in the Y plane.

EXAMPLE 10.2

Consider case 1 from example 9.4, where a Z plane circle goes through the origin, forming the familiar mho relay characteristic. In many cases, it is more convenient to think of this relay characteristic in the admittance plane, where it becomes a straight line. From example 9.4, we are given the data shown here, for convenience, in Table 10.1 Consider only the first case, where the Z plane circle has been shown to pass through the origin, so we expect that the inverse of this circle will be a straight line in the admittance plane.

TABLE 10.1 Data from Four Cases of Example 9.4

Case	R_1	X_1	R_2	X_2	e	f	R_0	X_0	S_Z
1	4	16	0	0	1	0	2	8	8.246
2	4	16	8	–4	1	0	6	6	10.198
3	4	16	0	0	1	1	10	6	11.661
4	4	16	8	–4	1	1	14	4	15.620

The data of Table 10.1 can be expanded, using (9.187)–(9.190) to compute the parameters listed in Table 10.2.

For this example, we concentrate only on case 1 from Tables 10.1 and 10.2. For this case, the parameter $d = 0$, so the Y plane equivalent of the Z plane circle is given by (10.65). The slope and B intercept of this line are computed in Table 10.2. The results are plotted in Figure 10.11, where the admittance values have been multiplied by 30 to force the admittance values to have magnitudes that are closer to those of the impedance values.

TABLE 10.2 Parameters of the Four Cases of Table 10.1

Case	*a*	*b*	*c*	*d*	*b/c*	*a/c*	G_0	B_O	S_Y
1	1	−4	−16	0	0.2500	−.0625	—	—	—
2	1	−12	−12	−32	—	—	−.1875	0.1875	0.3168
3	1	−20	−12	0	1.6667	−.0833	—	—	—
4	1	−28	−8	−32	—	—	−.4375	0.1250	0.4359

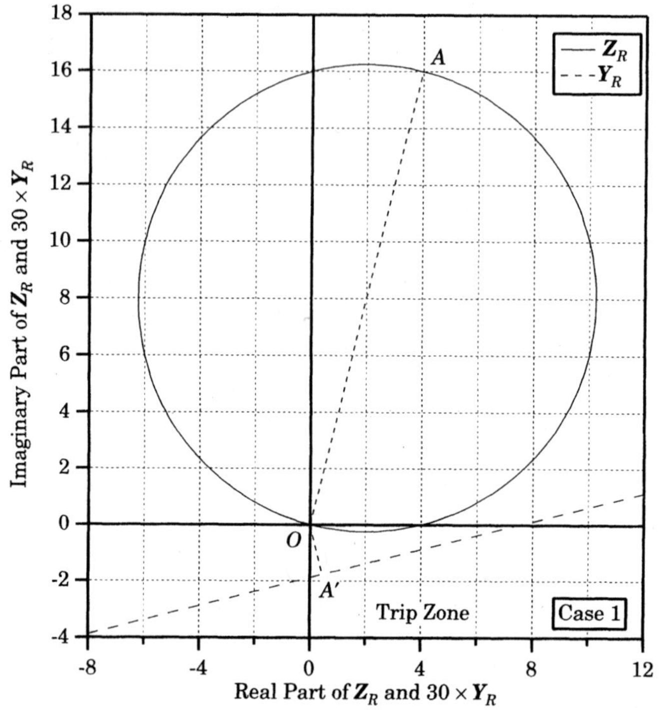

Figure 10.11 Plotted mho characteristics of example 10.2.

The Y plane image of the point A is shown as the Y plane point A' in the figure. This point lies at an angle that is the negative of the angle to the point A in the Z plane. A perpendicular from the relay Y plane characteristic to the origin can be computed from the fact that the perpendicular must have the negative inverse of the slope of the relay characteristic line. Therefore, we can write the equation of the perpendicular line as

$$B = -\frac{c}{b}G \tag{10.66}$$

This equation can be solved simultaneously with (10.65) to determine the point A'. Thus, we write

$$B = -\frac{c}{b}G = \frac{b}{c}G + \frac{a}{c} \tag{10.67}$$

or

$$G = \frac{-ab}{b^2 + c^2} \tag{10.68}$$

This can be substituted into (10.65) to give the B value

$$B = \frac{ac}{b^2 + c^2} \tag{10.69}$$

For this example, the point A' is located at (0.0147, −0.0588). In order to expand the scale for plotting, all Y values are multiplied by 30 in Figure 10.11, so this point is located at (0.441, −1.764).

Now, the relay trip zone consists of the area inside the Z plane circle. Think of this as the area closest to the origin. Then, the corresponding area in the Y plane is that area farthest from the origin, or all of the area below and to the right of the straight line in Figure 10.11, which is labeled "Trip Zone" in the figure. This region corresponds to high admittances, resulting from the high currents due to faults. ■

EXAMPLE 10.3

The second example to be considered from the data of Tables 10.1 and 10.2 is for case 2, where the parameter d is not zero. This results in a Y plane circle, the parameters for which are given in (10.64). Numerical values of these quantities are shown in Table 10.2. The resulting Y plane circle is shown in Figure 10.12, where all admittances have again been multiplied by 30. If this were not done, the Y plane circle would be very small on the scale of the plots in the figure.

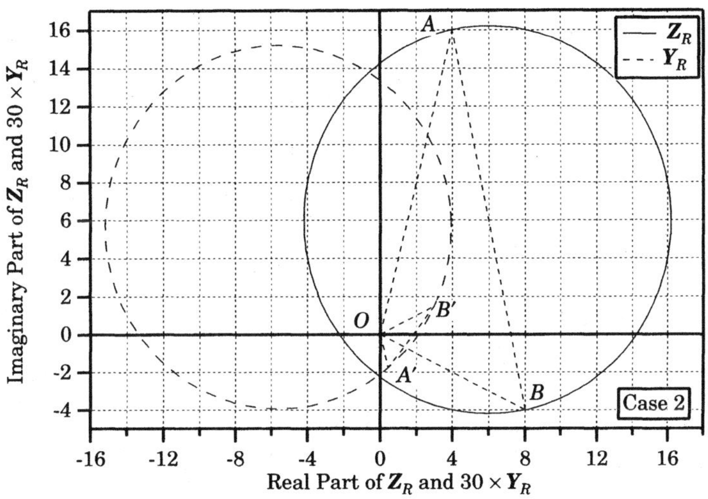

Figure 10.12 Plotted results of example 10.3.

The Z plane points A and B changed to A' and B', respectively, when converted to the admittance plane. Note that the angles to A and A' are equal in magnitude but of opposite sign. Moreover, A is farther from the origin in the Z plane than B, but A' is closer to the origin than B'. Note also that the circles are orthogonal.

Consider the relay trip zone in the impedance plane, which is the area inside the circular mho characteristic. In the admittance plane, however, this is the area for which tripping is prohibited and the trip zone consists of all area outside the Y plane circle. ■

The $1/Z$ transformation of cases 3 and 4 are left as exercises.

10.6 SUMMARY OF ADMITTANCE CHARACTERISTICS

The examples studied in this chapter show that the admittance plane characteristics are very attractive for the study of variations in the relay operation as a function of source voltage magnitudes and angles. Variations in voltage magnitudes can be examined quickly by plotting concentric circles. Since the circles have a common center point, these plots are much easier to construct than the impedance plane plots, where the circle centers for variations of either voltage magnitude or angle are always moving and are somewhat tedious to locate. Moreover, the examination of variations in the angle between the sources is easily constructed, as all of these angles are measured from a common reference.

Certain relay characteristics are easier to represent in the admittance plane than in the impedance plane. This is particularly true of the familiar mho relay, which is simply a straight line in the admittance plane.

REFERENCES

[1] Atabekov, G. I., *The Relay Protection of High Voltage Networks*, Pergamon Press, New York, 1960.

[2] Cook, V., *Analysis of Distance Protection*, Research Studies Press, Ltd., Letchworth, Hertfordshire, England, 1985.

[3] Huelsman, L. P., *Circuits, Matrices, and Linear Vector Spaces*, McGraw-Hill Book Company, New York, 1963.

PROBLEMS

10.1 Show that the admittance given by (10.3) is a circle in the Y plane.

10.2 Derive the equations for plotting the relay apparent admittance for variations in both the angle ψ and the voltage ratio constant k where we define

$$k = \nu E$$

and

$$E = \left| \frac{\boldsymbol{E}_S}{\boldsymbol{E}_U} \right|.$$

10.3 Plot the results of problem 10.2 in the Y_R plane using the data of example 9.1.

10.4 Derive the equations for plotting the relay apparent admittance for variations in both the angle ψ and the voltage ratio constant k where we define

$$k = \nu E$$

and

$$E = \left| \frac{\boldsymbol{E}_S}{\boldsymbol{E}_U} \right|.$$

10.5 Consider the system shown in Figure P10.5, where we describe the protected component by its two-port network parameters, which can be in terms of $\boldsymbol{Z}$, $\boldsymbol{Y}$, $\boldsymbol{G}$, $\boldsymbol{H}$, or $\boldsymbol{ABCD}$ matrix equations [3]. Moreover, if the two-port network meets certain constraints, which is usually true in power systems, and if one of the two-port network descriptions is known, the others can be derived. Suppose, then, that we are given one of the two-port descriptions for the protected component, derive the system two-port description, which views the protected component from the equivalent sources $\boldsymbol{E}_S$ and $\boldsymbol{E}_U$, as shown in Figure P10.5.

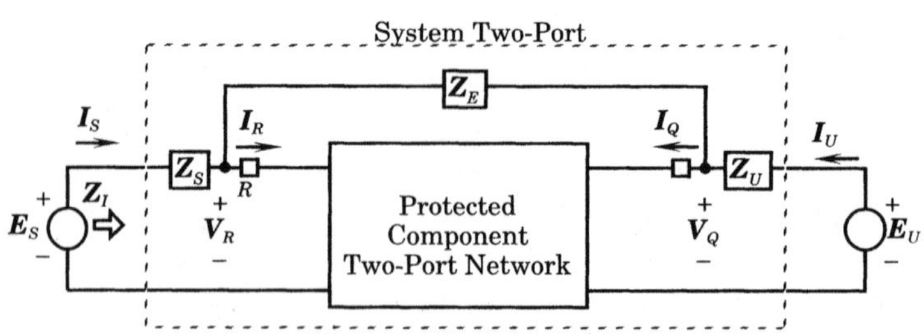

Figure P10.5 A system description of two-port networks.

10.6 Consider the system of Figure 10.1 where the protected system is a short transmission line with impedance $\boldsymbol{Z}_L = 0.3 + j0.8$ per unit. Moreover, let the source impedances be given as $\boldsymbol{Z}_S = 0.1 + j0.3$ and $\boldsymbol{Z}_U = 0.1 + j0.5$ per unit.
(a) Compute $\boldsymbol{m}_1$, $\boldsymbol{m}_2$, $\boldsymbol{m}$, and $\boldsymbol{m}_4$.
(b) Plot $\boldsymbol{Y}_R$ of equation (10.2) in the Y plane.

10.7 A transmission line has an impedance $\boldsymbol{Z}_R = 0.1 + j0.4$ per unit. The source impedance behind the relay is $\boldsymbol{Z}_S = 0.0 + j0.1$ per unit. The line is protected by a mho distance relay with major diameter set to a magnitude of 0.4 per unit at an angle of 60° from the R axis. Compute the following:
(a) The equation for the relay characteristic in the Z plane.
(b) The system input impedance $\boldsymbol{Z}_I$ assuming that $\boldsymbol{k} = 1$.
(c) The equation for the input admittance characteristic $\boldsymbol{Y}_I$.

10.8 Let the $\boldsymbol{ABCD}$ constants of a system be given as

$$\boldsymbol{A} = 0.8\angle 4^\circ$$
$$\boldsymbol{B} = 0.6\angle 70^\circ$$
$$\boldsymbol{C} = 0.3\angle 55^\circ$$
$$\boldsymbol{D} = d\angle \delta^\circ$$

(a) Solve for $\boldsymbol{D}$ using the relation $\boldsymbol{AD} - \boldsymbol{BC} = 1$.
(b) Compute $\boldsymbol{D}/\boldsymbol{B}$. What are the dimensions of this quantity in mks units?
(c) Plot $\boldsymbol{Y}_I$ loci for values of $\boldsymbol{m}$ fixed at 0.8, 1.0, and 1.2.

10.9 Given a system where

$$\boldsymbol{A} = 0.7\angle 0^\circ \qquad \boldsymbol{D} = d\angle \delta^\circ$$
$$\boldsymbol{B} = 300\angle 70^\circ \ \ \Omega \qquad \boldsymbol{Z}_S = 10 + j30 \ \ \Omega$$
$$\boldsymbol{C} = 1200\angle 90^\circ \ \ \mu\text{S}$$

The relay characteristic is the lens characteristic shown in Figure P10.9 where the point $\boldsymbol{Z}_m$ is defined as

$$\boldsymbol{Z}_m = 400\angle 60^\circ \ \Omega$$

Plot the $\boldsymbol{Y}_I$ and $\hat{\boldsymbol{Y}}_I$ characteristics in the admittance plane and determine critical tripping angles θ for $\boldsymbol{m} = 0.8$ and 1.2.

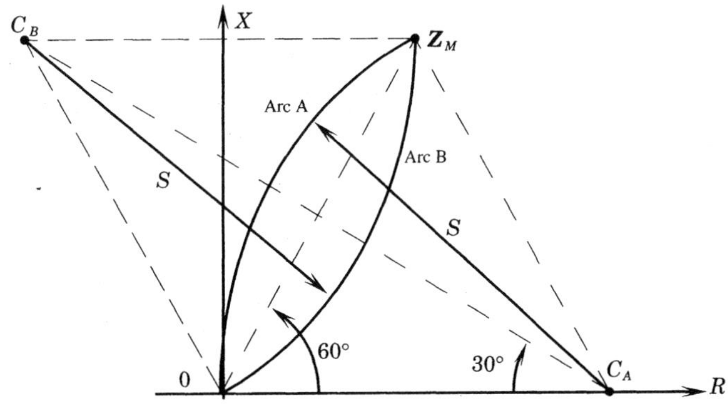

Figure P10.9 A lens relay characteristic.

10.10 A general expression for a circular relay threshold characteristic is given by (8.58), which is the polar form of the characteristic in the Z plane.
(a) Write the rectangular form of this relay threshold characteristic.
(b) Derive the rectangular form of this same relay threshold characteristic, but in the Y plane.

(c) Write the polar form of the Y plane threshold characteristic.
(d) Are there any restrictions to the above equations?

10.11 Some of the theory and derived results in both Chapters 9 and 10 are postulated as true in a reciprocal network.
(a) What is a reciprocal network? Does reciprocity imply passivity?
(b) How can a two-port network be tested for reciprocity?
(c) How is reciprocity recognized in the **Z** and **Y** two-port matrix representations.
(d) Suppose that the impedance matrix for a two-port network is known. Use the Z matrix parameters to derive the reciprocity requirement for the ***ABCD*** parameters.

10.12 A symmetric two-port has been described by: $\mathbf{Z}_{11} = \mathbf{Z}_{22}$ or $\mathbf{Y}_{11} = \mathbf{Y}_{22}$.
(a) How is the symmetric property realized in the ***ABCD*** representation of the two-port?
(b) Can this result determine a rule for the number of independent parameters needed to define a reciprocal two-port network?

10.13 Verify the result given by (10.40).

10.14 Extend the result from problem 10.13 to the following special cases:
(a) Only one transmission line.
(b) Only two transmission lines.
(c) Find the input impedance for case (b).

10.15 Equation (10.42) gives an expression for the input impedance threshold characteristic for relay 1 in a system with two parallel transmission lines. Develop an expression for the input threshold impedance of relay 2.

10.16 Compute the two-port admittance matrix as seen by the source currents of Figure 9.13, but with the line represented as an equivalent pi. Perform this calculation using the ***ABCD*** parameters found in Chapter 9 and use those parameters to find the admittance matrix.
(a) Write the expressions for the admittance parameters in terms of the ***ABCD*** parameters.
(b) Check to make sure that reciprocity is preserved.

10.17 A relay has the tripping characteristic defined by the impedance plane plot of Figure P10.17. Note that three straight lines are required to define the trip zone in the Z plane. Find the corresponding trip zone in the Y plane.

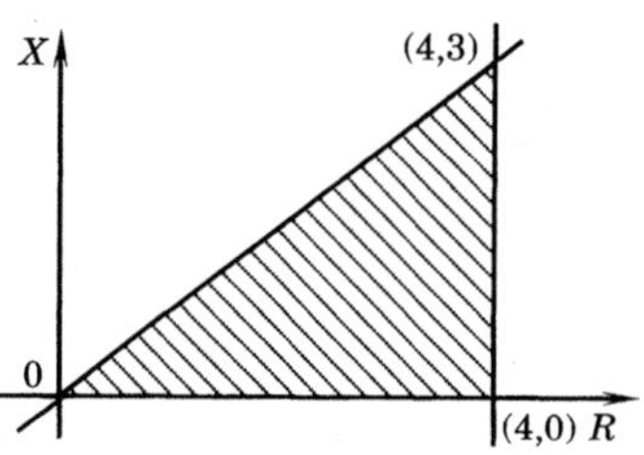

Figure P10.17 A special relay trip zone.

10.18 The system shown in Figure P10.18 has the following parameters:

$$\mathbf{Z}_S = 50 + j150 \quad \Omega$$
$$\mathbf{Z}_L = 250 + j500 \quad \Omega$$
$$\mathbf{Z}_U = 0 + j50 \quad \Omega$$

The relay at the $\mathbf{V}_R$ location is a lens type with lens segments described by the following equations:

Segment A: $(R - 700)^2 + (X + 100)^2 = 750^2$
Segment B: $(R + 500)^2 + (X - 500)^2 = 750^2$

(a) Plot the relay characteristics in the Z plane.
(b) Superimpose typical system operating conditions on the plot.
(c) Estimate graphically the critical trip angles and voltage ratios.

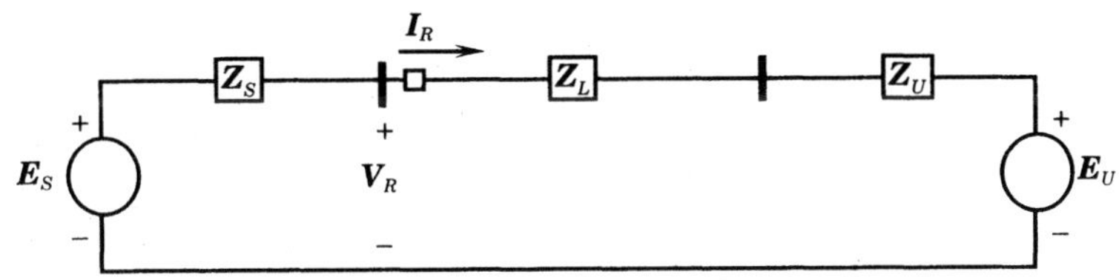

Figure P10.18 System equivalent for problem P10.18

10.19 For the system of problem 10.18, make the following Y plane computations:
(a) The relay characteristics.
(b) The system input characteristic.
(c) The relay threshold characteristics transferred to the input.
(d) Critical trip angles and magnitudes.

10.20 Consider case 1 of Tables 10.1 and 10.2. Assume that the application of this mho characteristic is such that heavy line loading may intrude into the right portion of the mho circle. To combat this problem, a "blinder" is prescribed. The blinder characteristic is a straight line in the Z plane with adjustable slope equal to that of OA of Figure 10.11 but 4 ohms to the right of that line, with the trip zone is to the left of the blinder. Determine the characteristic of the blinder in the Y plane and plot this characteristic together with that of the mho relay.

10.21 Repeat problem 10.20 using data from Tables 10.1 and 10.2.
(a) Case 3
(b) Case 4

PART III
TRANSMISSION PROTECTION

11

Analysis of Distance Protection

11.1 INTRODUCTION

Distance protection of transmission lines is a reliable and selective form of protection for lines where the line terminals are relatively far apart. The most accurate form of comparison for relay quantities is to compare the current entering a circuit with the current leaving that circuit. For many transmission circuits, because of the line length, it is costly to compare the currents at the two line terminals since this requires communications circuits that are at least as long as the transmission line. An alternative method compares the voltage and the current at each line terminal. Their ratio is the apparent impedance seen at the relay location, looking into the line, and therefore measures the distance to a fault along the line. This may not be as accurate as directly comparing the currents entering the circuit, since both the voltage and current undergo violent changes during a fault, but the method has advantages that outweigh the disadvantages.

One problem that will be recognized immediately is that of determining the correct distance measurement for faults of different types. For example, if the impedance from the relay point to a bolted three-phase fault is **Z**, then the impedance to other types of fault will be greater than **Z**, and may also involve the impedance of the ground return path. Moreover, the impedance seen by the individual relays will depend on the connections of the instrument transformers and the resulting secondary currents and voltages presented to the relays. Careful analysis is required to determine precisely the quantities observed by the relays for all possible conditions.

We begin with a basic assumption that there will be three relays for phase fault protection and three different relays for ground fault protection. The first consideration, then, will be to examine the various fault configurations to determine exactly the impedance seen by the six relays under different fault conditions. We perform this analysis using the method of symmetrical components. This requires the analysis of all fault types and the development of relay input quantities for these conditions, for both phase faults and ground faults. Having these fundamental considerations resolved, we will then be able to determine the relay quantities and the impedance seen by the relays under the different conditions.

11.2 ANALYSIS OF TRANSMISSION LINE FAULTS

The system to be analyzed is that shown in Figure 11.1, where we use the two-port Thevenin equivalent to represent the power system to the left and to the right of the line, with the line and the fault location shown in detail. A shunt fault is assumed to occur at a point F, which is defined to be located at a fractional distance h of the total line length, measured from the relay position R. The impedance $\mathbf{Z}_F$ shown in Figure 11.1 represents the arcing fault resistance plus the impedance of the negative- and zero-sequence networks, connected as required for a particular fault representation.

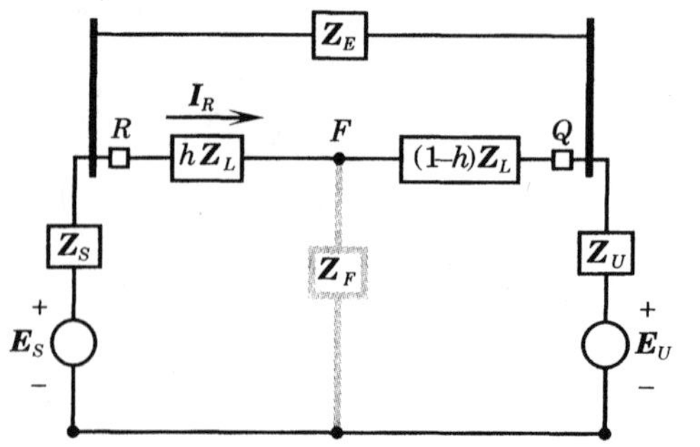

Figure 11.1 Representation of a fault at F as seen by a relay at R.

For this configuration, we would expect the relay at R to see the impedance $h\mathbf{Z}_L$ or some quantity directly related to this value. This must be true no matter what kind of fault, either phase or ground, and involving any number of phases.

Since we are interested primarily in the fault currents, we may redraw Figure 11.1 in the form shown in Figure 11.2, where $\mathbf{E}_F$ is the prefault voltage at F. In Figure 11.2, all network impedances except $\mathbf{Z}_F$ are positive sequence impedances and are therefore subscripted appropriately.

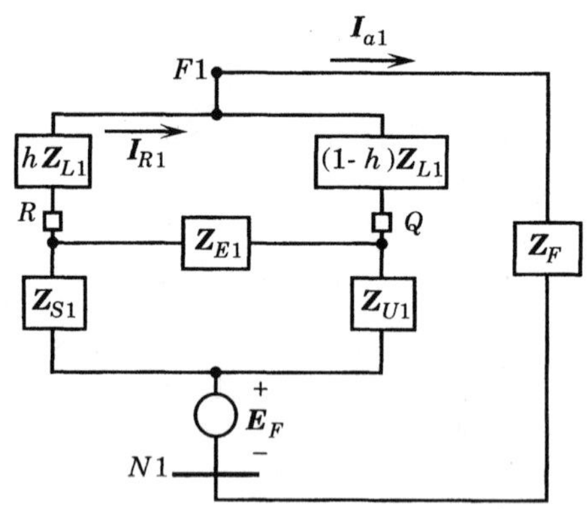

Figure 11.2 Equivalent circuit of the faulted network for a fault at F.

The analysis of the system of Figure 11.2 will be performed in two steps; first for the analysis of phase faults, followed by the analysis of ground faults. For each fault analysis it is necessary to represent the system of Figure 11.2 in terms of its sequence networks. These networks are shown in Figure 11.3. Note that we carry along the identity of both the fault point F as well as the relay point R in each sequence network. The line impedance between R and F, $h\mathbf{Z}_{L1}$, is identical in the positive- and negative-sequence networks, but is different in the zero-sequence network, where it is designated as $h\mathbf{Z}_{L0}$. The source impedances are

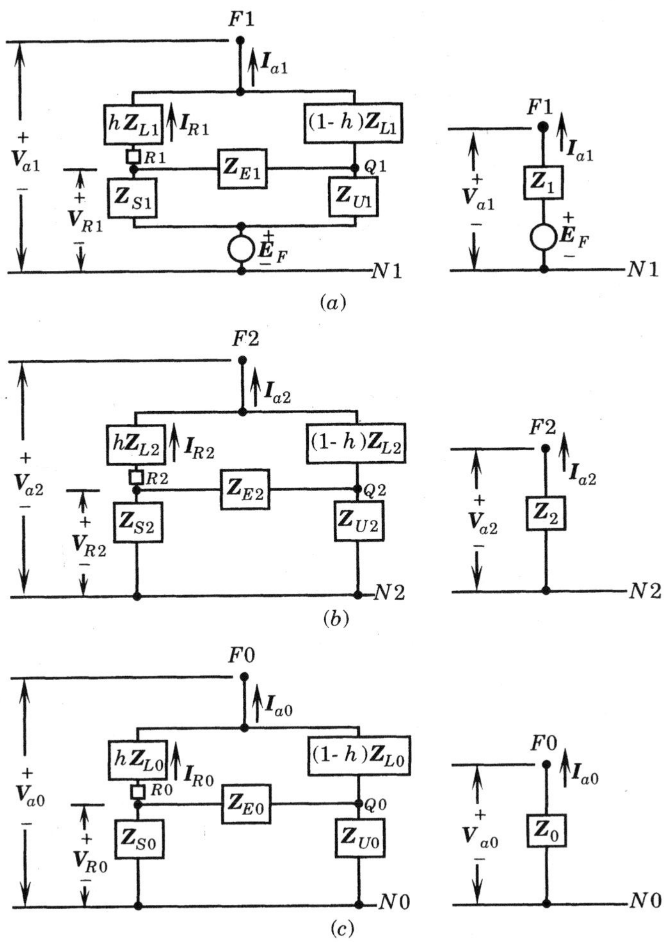

Figure 11.3 Sequence networks for a fault at F and relays at R. (a) Positive-sequence network (left) and positive-sequence equivalent. (b) Negative-sequence network (left) and negative-sequence equivalent. (c) Zero-sequence network (left) and zero-sequence equivalent.

shown as being different for each sequence network, but in practice the difference between the positive and negative sequences is small, and is often neglected. Finally, we note the *equivalent* sequence networks on the right of Figure 11.3. The total sequence impedances, $\mathbf{Z}_0$, $\mathbf{Z}_1$, and $\mathbf{Z}_2$, can be found from Figure 11.3 for any fault location.

11.2.1 Sequence Network Reduction

The network of Figure 11.4 is a general network representation that applies to all sequence networks. This network topology can be analyzed in a general way and applied to the positive-, negative-, and zero-sequence networks in the solution of all types of faults. In order to obtain a general solution, consider the network of Figure 11.4.

This figure represents the basic connections of the fault equivalent in all sequence networks, although some branches may be open depending on the sequence impedances. In all

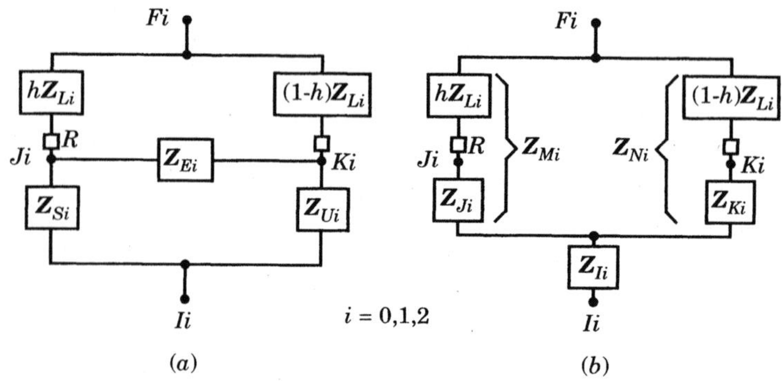

Figure 11.4 Sequence network Δ-Y reduction for $i = 0, 1, 2$. (a) The basic sequence network. (b) The reduced sequence network.

cases, the network of Figure 11.4(a) can be used to represent the impedance of all sequence networks. This network can be reduced to a more convenient form by a delta-wye transformation, of the lower delta, labeled i-j-k, which results in the network of Figure 11.4(b). The transformation is given by the equation

$$\begin{bmatrix} \mathbf{Z}_{Ii} \\ \mathbf{Z}_{Ji} \\ \mathbf{Z}_{Ki} \end{bmatrix} = \frac{1}{\mathbf{Z}_{Si} + \mathbf{Z}_{Ei} + \mathbf{Z}_{Ui}} \begin{bmatrix} \mathbf{Z}_{Ui}\mathbf{Z}_{Si} \\ \mathbf{Z}_{Si}\mathbf{Z}_{Ei} \\ \mathbf{Z}_{Ei}\mathbf{Z}_{Ui} \end{bmatrix} \tag{11.1}$$

which defines new wye impedances in terms of the original delta impedances. Note that this transformation preserves the identity of the protected line and the system voltage and current at the relay location. Since the protection equivalent is an exact representation of the system in all three sequence networks, it is possible to compute the sequence currents and voltage at the relay position for any type of sequence network condition, representative of any balanced or unbalanced shunt fault condition. The basic sequence network fault equivalents are shown on the left side of Figure 11.3. In all cases, these sequence networks can be reduced to find the sequence impedances shown on the right side of Figure 11.3, where the sequence networks are defined in the usual manner [1]. We now consider the application of symmetrical component theory to the analysis of different types of shunt faults.

11.2.2 Phase Faults at F

Shunt phase faults can be either three-phase faults, with balanced three-phase fault impedances to neutral, $\mathbf{Z}_F$, or phase-to-phase faults, with fault impedance $\mathbf{Z}_F$ between phases. We present a detailed analysis for the three-phase faults to illustrate the analytical process.

11.2.2.1 Three-Phase Faults. The three-phase fault is balanced; therefore, only the positive-sequence network is required for the analysis. The fault point connection is shown in Figure 11.5, where we identify the fault impedance $\mathbf{Z}_F$ in each phase for a three-phase fault at the point F.

The positive-sequence network and the fault connection are shown in Figure 11.6. We treat the fault impedance as consisting of both R and X, although in practice this impedance is usually considered to be purely resistive, representing an arcing fault resistance. Since the parallel combination of impedances on the left of Figure 11.6 is simply the positive sequence

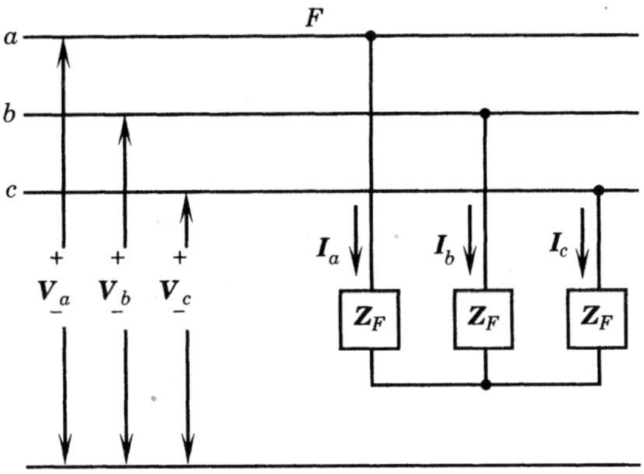

Figure 11.5 Fault connection for the three-phase fault at F.

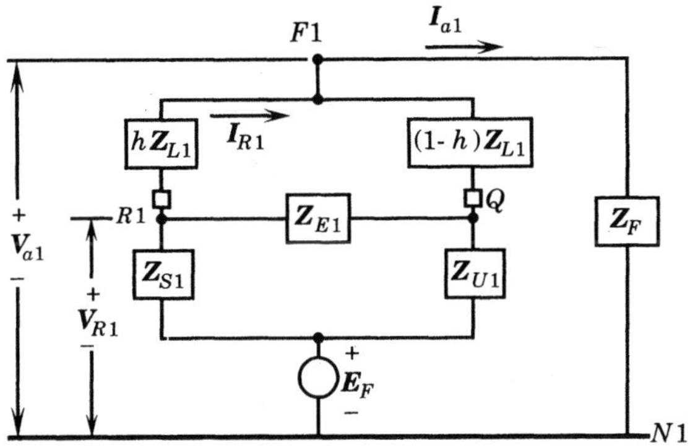

Figure 11.6 Sequence network connection for a three-phase fault at F.

impedance $\mathbf{Z}_1$, we may write the fault current in phase a as

$$\mathbf{I}_{a1} = \frac{\mathbf{E}_F}{\mathbf{Z}_1 + \mathbf{Z}_F} = \frac{1}{\mathbf{K}} \tag{11.2}$$

or

$$\mathbf{K}\mathbf{I}_{a1} = 1 \tag{11.3}$$

where (11.2) defines a complex constant $\mathbf{K}$, an inverse current. We may also write the impedance in the left and right paths of the positive-sequence network as follows. From (11.1) and Figure 11.3, we may write

$$\begin{bmatrix} \mathbf{Z}_{I1} \\ \mathbf{Z}_{J1} \\ \mathbf{Z}_{K1} \end{bmatrix} = \frac{1}{\mathbf{Z}_{S1} + \mathbf{Z}_{E1} + \mathbf{Z}_{U1}} \begin{bmatrix} \mathbf{Z}_{U1}\mathbf{Z}_{S1} \\ \mathbf{Z}_{S1}\mathbf{Z}_{E1} \\ \mathbf{Z}_{E1}\mathbf{Z}_{U1} \end{bmatrix} \tag{11.4}$$

First, for the left path, we define the total impedance from Figure 11.4(b) as

$$\mathbf{Z}_{M1} = \mathbf{Z}_{J1} + h\mathbf{Z}_{L1} \tag{11.5}$$

and for the right path we have

$$\mathbf{Z}_{N1} = \mathbf{Z}_{K1} + (1-h)\mathbf{Z}_{L1} \tag{11.6}$$

Then, from Figures 11.3 and 11.4, we compute the positive-sequence impedance

$$\mathbf{Z}_1 = \mathbf{Z}_{I1} + \frac{\mathbf{Z}_{M1}\mathbf{Z}_{N1}}{\mathbf{Z}_{M1} + \mathbf{Z}_{N1}} \tag{11.7}$$

Now, the current at the relay location can always be computed as a fraction of the total positive-sequence current, that is,

$$\boldsymbol{I}_{R1} = \boldsymbol{C}_1 \boldsymbol{I}_{a1} \tag{11.8}$$

where, in this case,

$$\boldsymbol{C}_1 = \frac{\boldsymbol{Z}_{N1}}{\boldsymbol{Z}_{M1} + \boldsymbol{Z}_{N1}} \tag{11.9}$$

and where $\boldsymbol{Z}_{N1}$ and $\boldsymbol{Z}_{M1}$ are defined in Figure 11.4. Then we may write, from (11.2) and (11.8),

$$\boldsymbol{I}_{R1} = \frac{\boldsymbol{C}_1 \boldsymbol{E}_F}{\boldsymbol{Z}_1 + \boldsymbol{Z}_F} = \frac{\boldsymbol{C}_1}{\boldsymbol{K}} \tag{11.10}$$

or

$$\boldsymbol{K}\boldsymbol{I}_{R1} = \boldsymbol{C}_1 \tag{11.11}$$

The voltage at the fault may be written by inspection of Figure 11.6 as

$$\boldsymbol{V}_{a1} = \boldsymbol{Z}_F \boldsymbol{I}_{a1} = \frac{\boldsymbol{Z}_F}{\boldsymbol{K}} \tag{11.12}$$

where we have incorporated the constant $\boldsymbol{K}$ defined by (11.2), or

$$\boldsymbol{K}\boldsymbol{V}_{a1} = \boldsymbol{Z}_F \tag{11.13}$$

The voltage at the relay may be computed from the voltage at F, given by (11.12), and the voltage across the line impedance between R and F. Thus,

$$\boldsymbol{V}_{R1} = \boldsymbol{V}_{a1} + h\boldsymbol{Z}_{L1}\boldsymbol{I}_{R1} = \frac{\boldsymbol{Z}_F + h\boldsymbol{Z}_{L1}\boldsymbol{C}_1}{\boldsymbol{K}} \tag{11.14}$$

or

$$\boldsymbol{K}\boldsymbol{V}_{R1} = \boldsymbol{Z}_F + h\boldsymbol{Z}_{L1}\boldsymbol{C}_1 \tag{11.15}$$

This completes the calculation of the sequence voltages and currents at locations F and R. Since the fault is balanced, only positive-sequence quantities are needed. Note that there is some merit in computing not just the sequence quantities, but $\boldsymbol{K}$ times these sequence quantities, as this simplifies the result. In every case, $\boldsymbol{K}$ is defined as

$$\boldsymbol{K} = \frac{\boldsymbol{Z}_1 + \boldsymbol{Z}_F}{\boldsymbol{E}_F} \tag{11.16}$$

and has the dimensions 1/current. Therefore, the products $\boldsymbol{KI}$ are dimensionless and the products $\boldsymbol{KV}$ have the dimensions of impedance. These products will be shown to provide a very convenient way of describing all of the faulted conditions as seen by the relay.

We now compute the phase quantities from the sequence quantities. We may always do this by recalling the matrix equation [1]

$$\mathbf{I}_{\mathrm{abc}} = \mathbf{A}\mathbf{I}_{012} \tag{11.17}$$

or

$$\begin{bmatrix} \boldsymbol{I}_a \\ \boldsymbol{I}_b \\ \boldsymbol{I}_c \end{bmatrix} = \frac{1}{k} \begin{bmatrix} 1 & 1 & 1 \\ 1 & \boldsymbol{a}^2 & \boldsymbol{a} \\ 1 & \boldsymbol{a} & \boldsymbol{a}^2 \end{bmatrix} \begin{bmatrix} \boldsymbol{I}_{a0} \\ \boldsymbol{I}_{a1} \\ \boldsymbol{I}_{a2} \end{bmatrix} \tag{11.18}$$

where, from [1],

$$k = \begin{cases} 1 & \text{for the Fortescue transformation} \\ \sqrt{3} & \text{for the power invariant transformation} \end{cases} \tag{11.19}$$

It is common practice to use the value $k = 1$.

For the three-phase fault, the computation of the phase currents is simple and gives the following results. From (11.18) and (11.3) we compute the line currents, as follows. Line currents at F:

$$\begin{aligned} \boldsymbol{KI}_a &= 1 \\ \boldsymbol{KI}_b &= \boldsymbol{a}^2 \\ \boldsymbol{KI}_c &= \boldsymbol{a} \end{aligned} \tag{11.20}$$

We use (11.8) and (11.18) to compute the relay currents. Line currents at R:

$$\begin{aligned} \boldsymbol{KI}_{aR} &= \boldsymbol{C}_1 \\ \boldsymbol{KI}_{bR} &= \boldsymbol{a}^2\boldsymbol{C}_1 \\ \boldsymbol{KI}_{cR} &= \boldsymbol{a}\boldsymbol{C}_1 \end{aligned} \tag{11.21}$$

Line currents at R for delta connected CT's:

$$\begin{aligned} \boldsymbol{K}(\boldsymbol{I}_{aR} - \boldsymbol{I}_{bR}) &= (1 - \boldsymbol{a}^2)\boldsymbol{C}_1 = \sqrt{3}e^{j\pi/6}\boldsymbol{C}_1 \\ \boldsymbol{K}(\boldsymbol{I}_{bR} - \boldsymbol{I}_{cR}) &= (\boldsymbol{a}^2 - \boldsymbol{a})\boldsymbol{C}_1 = \sqrt{3}e^{-j\pi/2}\boldsymbol{C}_1 \\ \boldsymbol{K}(\boldsymbol{I}_{cR} - \boldsymbol{I}_{aR}) &= (\boldsymbol{a} - 1)\boldsymbol{C}_1 = \sqrt{3}e^{j5\pi/6}\boldsymbol{C}_1 \end{aligned} \tag{11.22}$$

We compute the voltages in a similar way, using the same matrix equation with currents replaced by voltages of the identical subscripts given in (11.18), with the following results.

Phase-to-neutral voltages at F:

$$\begin{aligned} \boldsymbol{KV}_a &= \boldsymbol{Z}_F \\ \boldsymbol{KV}_b &= \boldsymbol{a}^2\boldsymbol{Z}_F \\ \boldsymbol{KV}_c &= \boldsymbol{a}\boldsymbol{Z}_F \end{aligned} \tag{11.23}$$

Note that all of the $\boldsymbol{KV}$ products have the dimensions of impedance and are all complex quantities that represent balanced three-phase quantities.

Line-to-line voltages at F:

$$\begin{aligned} \boldsymbol{KV}_{ab} &= (1 - \boldsymbol{a}^2)\boldsymbol{Z}_F \\ \boldsymbol{KV}_{bc} &= (\boldsymbol{a}^2 - \boldsymbol{a})\boldsymbol{Z}_F \\ \boldsymbol{KV}_{ca} &= (\boldsymbol{a} - 1)\boldsymbol{Z}_F \end{aligned} \tag{11.24}$$

Phase-to-neutral voltages at R:

$$\begin{aligned} \boldsymbol{KV}_{aR} &= \boldsymbol{Z}_F + h\boldsymbol{Z}_{L1}\boldsymbol{C}_1 \\ \boldsymbol{KV}_{bR} &= \boldsymbol{a}^2(\boldsymbol{Z}_F + h\boldsymbol{Z}_{L1}\boldsymbol{C}_1) \\ \boldsymbol{KV}_{cR} &= \boldsymbol{a}(\boldsymbol{Z}_F + h\boldsymbol{Z}_{L1}\boldsymbol{C}_1) \end{aligned} \tag{11.25}$$

Line-to-line voltages at R:

$$\begin{aligned} \boldsymbol{KV}_{abR} &= (1 - \boldsymbol{a}^2)(\boldsymbol{Z}_F + h\boldsymbol{Z}_{L1}\boldsymbol{C}_1) \\ \boldsymbol{KV}_{bcR} &= (\boldsymbol{a}^2 - \boldsymbol{a})(\boldsymbol{Z}_F + h\boldsymbol{Z}_{L1}\boldsymbol{C}_1) \\ \boldsymbol{KV}_{caR} &= (\boldsymbol{a} - 1)(\boldsymbol{Z}_F + h\boldsymbol{Z}_{L1}\boldsymbol{C}_1) \end{aligned} \tag{11.26}$$

These computations for the three-phase fault are summarized in Table 11.1, where the quantity $\boldsymbol{Z}_{F1}$ is defined for convenience. This clarifies the difference between the voltages at the fault and those at the relay since, in every case, the equations are exactly the same but the impedance $\boldsymbol{Z}_F$ at the fault point is replaced by $\boldsymbol{Z}_{F1}$ to obtain voltages at the relay point. This type of pattern, but with other defined replacements, will be possible for any type of fault.

TABLE 11.1 Three-Phase Fault Quantities for a Fault at F*

Quantity	Value at F	Value at R
KI_{a0}	0	0
KI_{a1}	1	C_1
KI_{a2}	0	0
KI_a	1	C_1
KI_b	a^2	a^2C_1
KI_c	a	aC_1
$K(I_a - I_b)$	$1 - a^2$	$(1 - a^2)C_1$
$K(I_b - I_c)$	$a^2 - a$	$(a^2 - a)C_1$
$K(I_c - I_a)$	$a - 1$	$(a - 1)C_1$
KV_{a0}	0	0
KV_{a1}	Z_F	Z_{F1}
KV_{a2}	0	0
KV_a	Z_F	Z_{F1}
KV_b	a^2Z_F	a^2Z_{F1}
KV_c	aZ_F	aZ_{F1}
KV_{ab}	$(1 - a^2)Z_F$	$(1 - a^2)Z_{F1}$
KV_{bc}	$(a^2 - a)Z_F$	$(a^2 - a)Z_{F1}$
KV_{ca}	$(a - 1)Z_F$	$(a - 1)Z_{F1}$
Z_{ab}	Z_F	Z_{F1}/C_1
Z_{bc}	Z_F	Z_{F1}/C_1
Z_{ca}	Z_F	Z_{F1}/C_1

*For the 3-Phase Fault, We Define:

$$K = \frac{Z_1 + Z_F}{E_F} \qquad Z_{F1} = Z_F + hZ_{L1}C_1$$
$$Z_{M1} = Z_{J1} + hZ_{L1}$$
$$C_1 = \frac{Z_{N1}}{Z_{M1} + Z_{N1}} \qquad Z_{N1} = Z_{K1} + (1 - h)Z_{L1}$$

11.2.2.2 Phase-to-Phase Faults. Figure 11.7 shows the phase-to-phase fault, where the fault is assumed to occur between phases b and c with a fault impedance of Z_F between the phase conductors. For this condition, the connection of the sequence networks is shown in Figure 11.8. For the network of Figure 11.8, we compute the positive-sequence current at F as follows.

$$I_{a1} = -I_{a2} = \frac{E_F}{Z_1 + Z_2 + Z_F} = \frac{1}{K} \tag{11.27}$$

from which we compute

$$KI_{a1} = 1 \tag{11.28}$$

and

$$K = \frac{Z_1 + Z_2 + Z_F}{E_F} \tag{11.29}$$

Note that the quantity K again has the dimensions of 1/current, but it is defined differently than the three-phase fault case. We also compute

$$V_{a1} = E_F - Z_1I_{a1} = (Z_F + Z_2)I_{a1} = \frac{Z_F + Z_2}{K} \tag{11.30}$$

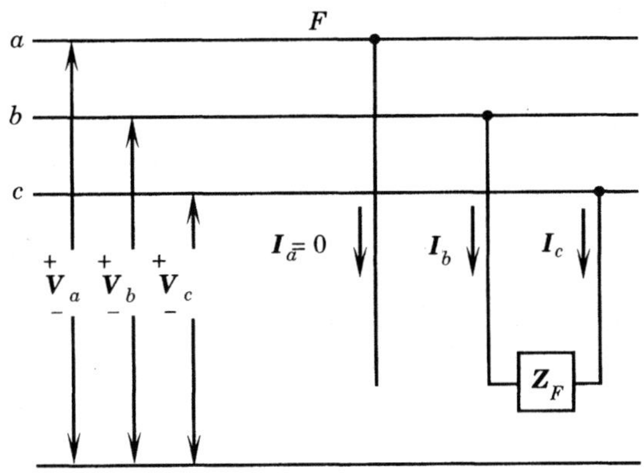

Figure 11.7 Fault connection for a phase-to-phase fault at F.

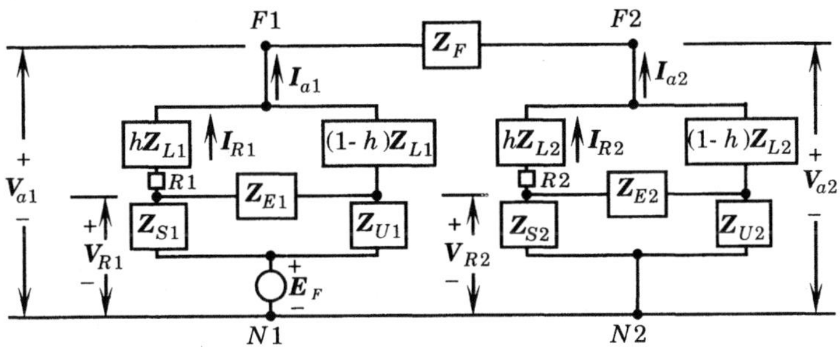

Figure 11.8 Sequence network connections for a phase-to-phase fault at F.

or

$$\mathbf{K}\mathbf{V}_{a1} = \mathbf{Z}_F + \mathbf{Z}_2 \tag{11.31}$$

We also observe that

$$\begin{aligned} \mathbf{K}\mathbf{I}_{a2} &= -1 \\ \mathbf{K}\mathbf{V}_{a2} &= \mathbf{Z}_2 \end{aligned} \tag{11.32}$$

From Figure 11.8, we can solve for the relay current using the wye-delta conversion of Figure 11.4 for the positive-sequence network, and an identical network configuration for the negative sequence except for a change in the subscript notation from "$a1$" to "$a2$." Then we can define

$$\mathbf{Z}_1 = \mathbf{Z}_{I1} + \frac{\mathbf{Z}_{M1}\mathbf{Z}_{N1}}{\mathbf{Z}_{M1} + \mathbf{Z}_{N1}} \tag{11.33}$$

and

$$\mathbf{Z}_2 = \mathbf{Z}_{I2} + \frac{\mathbf{Z}_{M2}\mathbf{Z}_{N2}}{\mathbf{Z}_{M2} + \mathbf{Z}_{N2}} \tag{11.34}$$

where

$$\begin{aligned} \mathbf{Z}_{M2} &= \mathbf{Z}_{J2} + h\mathbf{Z}_{L2} \\ \mathbf{Z}_{N2} &= \mathbf{Z}_{K2} + (1-h)\mathbf{Z}_{L2} \end{aligned} \tag{11.35}$$

where similar positive-sequence impedances are given by (11.5) and (11.6)

Finally, we write the current at the relay location R as a fraction of the current at the fault point F in each of the sequence networks, as follows:

$$\begin{aligned} \boldsymbol{I}_{R1} &= \boldsymbol{C}_1 \boldsymbol{I}_{a1} \\ \boldsymbol{I}_{R2} &= \boldsymbol{C}_2 \boldsymbol{I}_{a2} \end{aligned} \tag{11.36}$$

where $\boldsymbol{C}_1$ is defined by (11.9) and $\boldsymbol{C}_2$ is defined in terms of the negative-sequence network impedances, as follows.

$$\boldsymbol{C}_2 = \frac{\boldsymbol{Z}_{N2}}{\boldsymbol{Z}_{M2} + \boldsymbol{Z}_{N2}} \tag{11.37}$$

From these initial observations, and noting that the zero-sequence voltage and current are both zero, we may again synthesize the phase quantities using the matrix equation (11.18) to compute the values shown in Table 11.2.

TABLE 11.2 Phase-to-Phase Fault Quantities for a Fault at F*

Quantity	Value at F	Value at R
$\boldsymbol{KI}_{a0}$	0	0
$\boldsymbol{KI}_{a1}$	1	$\boldsymbol{C}_1$
$\boldsymbol{KI}_{a2}$	-1	$-\boldsymbol{C}_2$
$\boldsymbol{KI}_a$	0	$\boldsymbol{C}_1 - \boldsymbol{C}_2$
$\boldsymbol{KI}_b$	$\boldsymbol{a}^2 - \boldsymbol{a}$	$\boldsymbol{a}^2\boldsymbol{C}_1 - \boldsymbol{a}\boldsymbol{C}_2$
$\boldsymbol{KI}_c$	$\boldsymbol{a} - \boldsymbol{a}^2$	$\boldsymbol{a}\boldsymbol{C}_1 - \boldsymbol{a}^2\boldsymbol{C}_2$
$\boldsymbol{K}(\boldsymbol{I}_a - \boldsymbol{I}_b)$	$\boldsymbol{a} - \boldsymbol{a}^2$	$(1 - \boldsymbol{a}^2)\boldsymbol{C}_1 + (\boldsymbol{a} - 1)\boldsymbol{C}_2$
$\boldsymbol{K}(\boldsymbol{I}_b - \boldsymbol{I}_c)$	$2(\boldsymbol{a}^2 - \boldsymbol{a})$	$(\boldsymbol{a}^2 - \boldsymbol{a})(\boldsymbol{C}_1 + \boldsymbol{C}_2)$
$\boldsymbol{K}(\boldsymbol{I}_c - \boldsymbol{I}_a)$	$\boldsymbol{a} - \boldsymbol{a}^2$	$(\boldsymbol{a} - 1)\boldsymbol{C}_1 + (1 - \boldsymbol{a}^2)\boldsymbol{C}_2$
$\boldsymbol{KV}_{a0}$	0	0
$\boldsymbol{KV}_{a1}$	$\boldsymbol{Z}_F + \boldsymbol{Z}_2$	$\boldsymbol{Z}_{F1} + \boldsymbol{Z}_2$
$\boldsymbol{KV}_{a2}$	$\boldsymbol{Z}_2$	$\boldsymbol{Z}_{2C}$
$\boldsymbol{KV}_a$	$\boldsymbol{Z}_F + 2\boldsymbol{Z}_2$	$\boldsymbol{Z}_{F1} + \boldsymbol{Z}_2 + \boldsymbol{Z}_{2C}$
$\boldsymbol{KV}_b$	$\boldsymbol{a}^2\boldsymbol{Z}_F - \boldsymbol{Z}_2$	$\boldsymbol{a}^2(\boldsymbol{Z}_{F1} + \boldsymbol{Z}_2) + \boldsymbol{a}\boldsymbol{Z}_{2C}$
$\boldsymbol{KV}_c$	$\boldsymbol{a}\boldsymbol{Z}_F - \boldsymbol{Z}_2$	$\boldsymbol{a}(\boldsymbol{Z}_{F1} + \boldsymbol{Z}_2) + \boldsymbol{a}^2\boldsymbol{Z}_{2C}$
$\boldsymbol{KV}_{ab}$	$(1 - \boldsymbol{a}^2)\boldsymbol{Z}_F + 3\boldsymbol{Z}_2$	$(1 - \boldsymbol{a}^2)(\boldsymbol{Z}_{F1} + \boldsymbol{Z}_2) + (1 - \boldsymbol{a})\boldsymbol{Z}_{2C}$
$\boldsymbol{KV}_{bc}$	$(\boldsymbol{a}^2 - \boldsymbol{a})\boldsymbol{Z}_F$	$(\boldsymbol{a}^2 - \boldsymbol{a})(\boldsymbol{Z}_{F1} + \boldsymbol{Z}_2 - \boldsymbol{Z}_{2C})$
$\boldsymbol{KV}_{ca}$	$(\boldsymbol{a} - 1)\boldsymbol{Z}_F - 3\boldsymbol{Z}_2$	$(\boldsymbol{a} - 1)(\boldsymbol{Z}_{F1} + \boldsymbol{Z}_2) + (\boldsymbol{a}^2 - 1)\boldsymbol{Z}_{2C}$

*For the Phase-to-Phase Fault, We Define:

$$\boldsymbol{K} = \frac{\boldsymbol{Z}_1 + \boldsymbol{Z}_2 + \boldsymbol{Z}_F}{\boldsymbol{E}_F} \qquad \begin{aligned} \boldsymbol{Z}_{F1} &= \boldsymbol{Z}_F + \boldsymbol{C}_1 h \boldsymbol{Z}_{L1} \\ \boldsymbol{Z}_{2C} &= \boldsymbol{Z}_2 - \boldsymbol{C}_2 h \boldsymbol{Z}_{L2} \end{aligned}$$

Again, we note the unique symmetry of the tabulated values, with the definition of new impedances necessary to maintain the symmetry, in this case requiring definitions for $\boldsymbol{Z}_{F1}$ and $\boldsymbol{Z}_{2C}$. The definition of impedance $\boldsymbol{Z}_{2C}$ is stated in terms of $\boldsymbol{Z}_{L2}$, the negative sequence line impedance, but we recognize that $\boldsymbol{Z}_{L1} = \boldsymbol{Z}_{L2}$ for transmission lines.

In reviewing Tables 11.1 and 11.2, we see several interesting patterns:

1. All of the $\boldsymbol{KI}$ product terms are dimensionless, with the currents at R differing from those at F by the complex fraction $\boldsymbol{C}$ for each sequence network.

2. The $\boldsymbol{KV}$ product terms have the dimensions of impedance and are proportional to the voltages at the fault point and at the relay location.
3. The voltage terms at the relay location always include the impedance function representing the voltage at the fault with another term added to represent the voltage rise along the transmission line from the fault point to the relay, where this voltage rise is proportional to the impedance seen by the relay.

Finally, we note that the constants $\boldsymbol{C}_1$ and $\boldsymbol{C}_2$ are expected to be almost equal and are often assumed to be identical. Compare (11.9) and (11.37). If we can approximate these constants to be equal, this simplifies some of the results for the phase voltages and the line-to-line voltages.

11.2.3 Ground Faults at *F*

We now consider ground faults at the point F, using exactly the same analytical technique as for the phase faults. Two types will be considered: the one-line-to-ground fault and the two-line-to-ground fault.

11.2.3.1 The One-Line-to-Ground Fault at F. The system condition for the one-line-to-ground fault is shown in Figure 11.9. The one-line-to-ground fault analysis requires the use of all three sequence networks, arranged as shown in Figure 11.10. For this sequence network connection, we may write

$$\boldsymbol{I}_{a0} = \boldsymbol{I}_{a1} = \boldsymbol{I}_{a2} = \frac{\boldsymbol{E}_F}{\boldsymbol{Z}_0 + \boldsymbol{Z}_1 + \boldsymbol{Z}_2 + 3\boldsymbol{Z}_F} = \frac{1}{\boldsymbol{K}} \tag{11.38}$$

This gives us the following fundamental relations:

$$\boldsymbol{KI}_{a0} = \boldsymbol{KI}_{a1} = \boldsymbol{KI}_{a2} = 1 \tag{11.39}$$

and

$$\boldsymbol{K} = \frac{\boldsymbol{Z}_0 + \boldsymbol{Z}_1 + \boldsymbol{Z}_2 + 3\boldsymbol{Z}_F}{\boldsymbol{E}_F} \tag{11.40}$$

The following voltage relationships are computed at the fault point F from an analysis of the sequence networks of Figure 11.10. First, we write the sequence voltages of the three

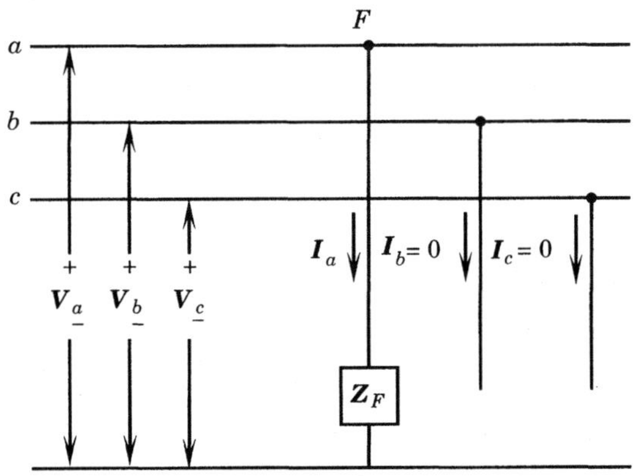

Figure 11.9 Fault connection the single line-to-ground fault at F.

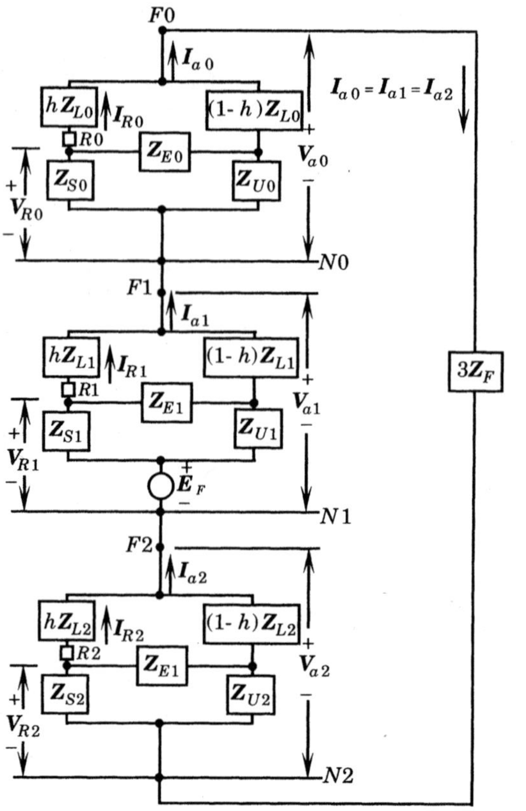

Figure 11.10 Sequence network connections for a one-line-to-ground fault.

networks.

$$
\begin{aligned}
\boldsymbol{V}_{a1} &= (3\boldsymbol{Z}_F + \mathbf{Z}_0 + \mathbf{Z}_2)\boldsymbol{I}_{a1} = \frac{3\boldsymbol{Z}_F + \mathbf{Z}_0 + \mathbf{Z}_2)}{\boldsymbol{K}} \\
\boldsymbol{V}_{a2} &= -\mathbf{Z}_2\boldsymbol{I}_{a1} = \frac{-\mathbf{Z}_2}{\boldsymbol{K}} \\
\boldsymbol{V}_{a0} &= -\mathbf{Z}_0\boldsymbol{I}_{a1} = \frac{-\mathbf{Z}_0}{\boldsymbol{K}}
\end{aligned}
\tag{11.41}
$$

where

$$
\mathbf{Z}_0 = \boldsymbol{Z}_{I0} + \frac{\boldsymbol{Z}_{M0}\boldsymbol{Z}_{N0}}{\boldsymbol{Z}_{M0} + \boldsymbol{Z}_{N0}} \tag{11.42}
$$

These equations define the ***KV*** product terms for the sequence voltages. From (11.39) and (11.41) we have all the information required to synthesize the phase currents and voltages at the fault point F, using the matrix equation (11.18). These will again be computed as ***KI*** and ***KV*** products.

The sequence quantities at the relay point are proportional to those at the fault point, as was found true for phase faults. The proportionality factors are different, however. These factors will be computed and tabulated.

We begin with the computation of the sequence currents at the relay. Let

$$\begin{aligned} KI_{R0} &= C_0 \\ KI_{R1} &= C_1 \\ KI_{R2} &= C_2 \end{aligned} \tag{11.43}$$

where we have defined a new current proportionality constant C_0

$$C_0 = \frac{Z_{N0}}{Z_{M0} + Z_{N0}} \tag{11.44}$$

The constants C_1 and C_2 are defined by (11.9) and (11.37), respectively.

The sequence voltages at the relay are found by inspection of Figure 11.10.

$$\begin{aligned} KV_{R0} &= KV_{a0} + hZ_{L0}KI_{R0} = -Z_0 + C_0hZ_{L0} \\ KV_{R1} &= KV_{a1} + hZ_{L1}KI_{R1} = 3Z_F + Z_0 + Z_2 + C_1hZ_{L1} \\ KV_{R2} &= KV_{a2} + hZ_{L2}KI_{R2} = -Z_2 + C_2hZ_{L1} \end{aligned} \tag{11.45}$$

Having computed all of the sequence quantities at both the fault point and the relay point, we may compute the phase quantities. The results are given in Table 11.3. Again, we note the symmetry of the tabulated values at the fault point and at the relay. The quantities have exactly the same form, but certain impedance quantities must be defined to maintain the symmetry. These quantities are shown at the bottom of Table 11.3.

TABLE 11.3 Single Line-to-Ground Fault Quantities for a Fault at F*

Quantity	Value at F	Value at R
KI_{a0}	1	C_0
KI_{a1}	1	C_1
KI_{a2}	1	C_2
KI_a	3	$C_0 + C_1 + C_2$
KI_b	0	$C_0 + a^2C_1 + aC_2$
KI_c	0	$C_0 + aC_1 + a^2C_2$
$K(I_a - I_b)$	3	$d_0C_1 - d_1C_2$
$K(I_b - I_c)$	0	$d_2(C_1 - C_2)$
$K(I_c - I_a)$	-3	$d_1C_1 - d_0C_2$
KV_{a0}	$-Z_0$	$-Z_{0C}$
KV_{a1}	$3Z_F + Z_0 + Z_2$	$(Z_{F0} + Z_{F1} + Z_{F2}) + Z_{0C} + Z_{2C}$
KV_{a2}	$-Z_2$	$-Z_{2C}$
KV_a	$3Z_F$	$(Z_{F0} + Z_{F1} + Z_{F2})$
KV_b	$3a^2Z_F - d_0Z_0 + d_2Z_2$	$a^2(Z_{F0} + Z_{F1} + Z_{F2}) - d_0Z_{0C} + d_2Z_{2C}$
KV_c	$3aZ_F + d_1Z_0 - d_2Z_2$	$a(Z_{F0} + Z_{F1} + Z_{F2}) + d_1Z_{0C} - d_2Z_{2C}$
KV_{ab}	$3d_0Z_F + d_0Z_0 - d_2Z_2$	$d_0(Z_{F0} + Z_{F1} + Z_{F2}) + d_0Z_{0C} - d_2Z_{2C}$
KV_{bc}	$d_2(3Z_F + Z_0 + 2Z_2)$	$d_2[(Z_{F0} + Z_{F1} + Z_{F2}) + Z_{0C} + 2Z_{2C}]$
KV_{ca}	$3d_1Z_F + d_1Z_0 - d_2Z_2$	$d_1(Z_{F0} + Z_{F1} + Z_{F2}) + d_1Z_{0C} - d_2Z_{2C}$

*For the One Line-to-Ground Fault, We Define the Following:

$K = \dfrac{Z_0 + Z_1 + Z_2 + 3Z_F}{E_F}$ $\quad Z_{F0} = Z_F + C_0hZ_{L0}$ $\quad Z_{0C} = Z_0 - C_0hZ_{L0}$

$Z_{F1} = Z_F + C_1hZ_{L1}$ $\quad Z_{2C} = Z_2 - C_2hZ_{L2}$

$Z_{F2} = Z_F + C_2hZ_{L2}$

Table 11.3 has defined for the first time the quantities $\boldsymbol{d}_0$, $\boldsymbol{d}_1$, and $\boldsymbol{d}_2$. These quantities occur repeatedly in the fault calculations. They represent a balanced set of phasors, shown in Figure 11.11, where the phasor operators are defined in terms of the complex constant $\boldsymbol{b}$, as follows.

$$\begin{aligned} \boldsymbol{d}_0 &= \sqrt{3}\boldsymbol{b} = 1 - \boldsymbol{a}^2 = \sqrt{3}e^{j\pi/6} = \sqrt{3}\angle 30^\circ \\ \boldsymbol{d}_1 &= \sqrt{3}\boldsymbol{ab} = \boldsymbol{a} - 1 = \sqrt{3}e^{j5\pi/6} = \sqrt{3}\angle 150^\circ \\ \boldsymbol{d}_2 &= \sqrt{3}\boldsymbol{a}^2\boldsymbol{b} = \boldsymbol{a}^2 - \boldsymbol{a} = \sqrt{3}e^{-j\pi/2} = \sqrt{3}\angle - 90 \end{aligned} \tag{11.46}$$

where

$$\begin{aligned} \boldsymbol{b} &= e^{j\pi/6} = 1\angle 30^\circ \\ \boldsymbol{d}_0 + \boldsymbol{d}_1 + \boldsymbol{d}_2 &= 0 \end{aligned} \tag{11.47}$$

Using these defined quantities, based on the complex constant $\boldsymbol{b}$, simplifies the notation. It is possible to compute a set of identities associated with the phasor constants defined in (11.46) and (11.47). Some useful identities are given in Table 11.4.

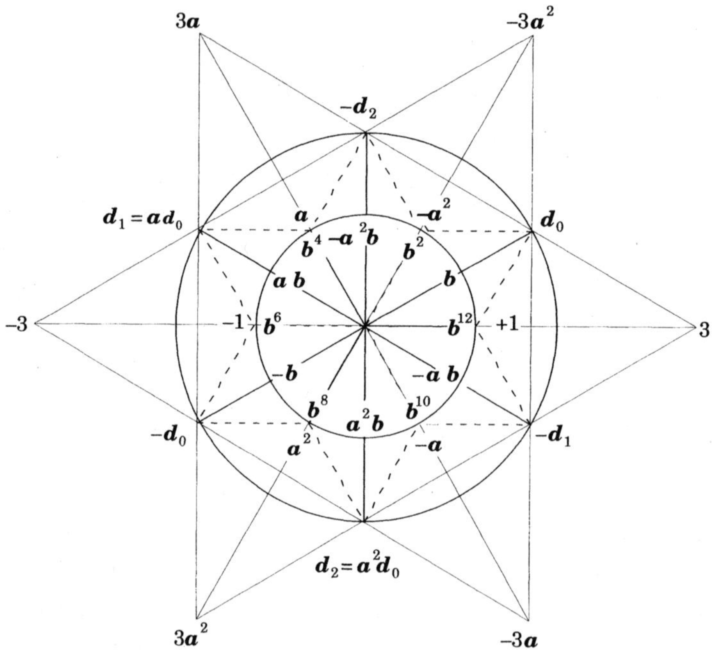

Figure 11.11 Definition of phasor operators $\boldsymbol{d}_0$, $\boldsymbol{d}_1$, and $\boldsymbol{d}_2$.

11.2.3.2 The Two-Line-to-Ground Fault at F. The two-line-to-ground fault is assumed to have impedances $\boldsymbol{Z}_F$ from each line to their connection point and an impedance $\boldsymbol{Z}_G$ to ground, as shown in Figure 11.12. The sequence network connections for this condition are shown in Figure 11.13, where the construction within each sequence network is identical to Figure 11.3.

This fault type is more difficult to analyze algebraically, and the results are complex. As for the other fault types, however, we again discover a striking symmetry between currents and

TABLE 11.4 Identities Involving Phasor Operators d_0, d_1, and d_2

$d_0 - d_1 = 3$	$d_0 + d_1 = -d_2$
$d_1 - d_2 = 3a$	$d_1 + d_2 = -d_0$
$d_2 - d_0 = 3a^2$	$d_2 + d_0 = -d_1$
$\frac{d_0}{d_1} = \frac{d_1}{d_2} = \frac{d_2}{d_0} = a^2$	$\frac{d_0}{d_2} = \frac{d_1}{d_0} = \frac{d_2}{d_1} = a$
$d_0 = ad_2 = a^2 d_1$	$d_0 d_1 = d_1 - d_0 = d_2^2 = -3$
$d_1 = ad_0 = a^2 d_2$	$d_0 d_2 = d_2 - d_1 = d_1^2 = -3a$
$d_2 = ad_1 = a^2 d_0$	$d_1 d_2 = d_0 - d_2 = d_0^2 = -3a^2$

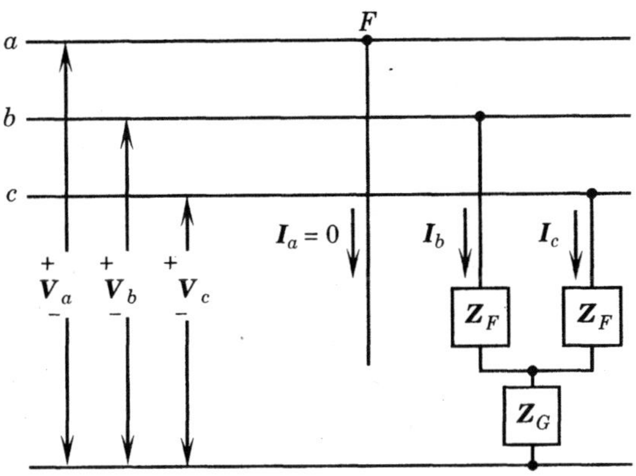

Figure 11.12 Fault connections for the two-line-to-ground fault.

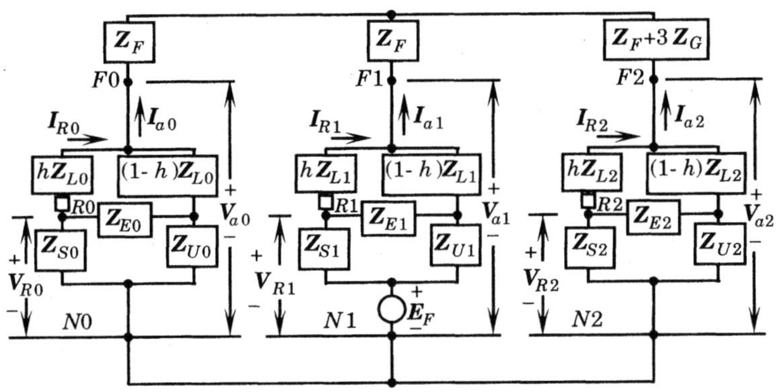

Figure 11.13 Sequence network connections for a two-line-to-ground fault.

voltages at the fault point F and at the relay location R. The results are shown in Table 11.5, where impedances are defined to clarify the tabulation and comparison of results. The results at F and R always have the same form. The currents at R are modified by the constants C_0, C_1, and C_2, since only a fraction of the total fault current flows through the relay current transformers. The voltages are modified by changing two of the impedances, which is necessary since the total fault voltage does not appear at the relay point, even though the form of the voltage equations is exactly the same for both locations.

TABLE 11.5 Two-Line-to-Ground Fault Quantities for a Fault at F*

Quantity	Value at F	Value at R
KI_{a0}	$-Z_{2X}$	$-C_0Z_{2X}$
KI_{a1}	$Z_{2X}+Z_{0X}$	$C_1(Z_{2X}+Z_{0X})$
KI_{a2}	$-Z_{0X}$	$-C_2Z_{0X}$
KI_a	0	$(C_1-C_0)Z_{2X}+(C_1-C_2)Z_{0X}$
KI_b	$-d_0Z_{2X}+d_2Z_{0X}$	$(a^2C_1-C_0)Z_{2X}+(a^2C_1-aC_2)Z_{0X}$
KI_c	$d_1Z_{2X}-d_2Z_{0X}$	$(aC_1-C_0)Z_{2X}+(aC_1-a^2C_2)Z_{0X}$
$K(I_a-I_b)$	$d_0Z_{2X}-d_2Z_{0X}$	$d_0C_1Z_{2X}+(d_0C_1+d_1C_2)Z_{0X}$
$K(I_b-I_c)$	$d_2(Z_{2X}+2Z_{0X})$	$d_2[C_1Z_{2X}+(C_1+C_2)Z_{0X}]$
$K(I_c-I_a)$	$d_1Z_{2X}-d_2Z_{0X}$	$d_1C_1Z_{2X}+(d_1C_1+d_0C_2)Z_{0X}$
KV_{a0}	Z_0Z_{2X}	$Z_{0C}Z_{2X}$
KV_{a1}	$Z_F(Z_{2X}+Z_{0X})+Z_{2X}Z_{0X}$	$Z_{F1}(Z_{2X}+Z_{0X})+Z_{2X}Z_{0X}$
KV_{a2}	Z_2Z_{0X}	$Z_{2C}Z_{0X}$
KV_a	$(Z_F+Z_0)Z_{2X}+(Z_F+Z_2)Z_{0X}+Z_{2X}Z_{0X}$	§See Note
KV_b	$(a^2Z_F+Z_0)Z_{2X}+(a^2Z_F+aZ_2)Z_{0X}+a^2Z_{2X}Z_{0X}$	
KV_c	$(aZ_F+Z_0)Z_{2X}+(aZ_F+a^2Z_2)Z_{0X}+aZ_{2X}Z_{0X}$	
KV_{ab}	$d_0[Z_F(Z_{2X}+Z_{0X})+Z_2Z_{0X}]-d_1Z_2Z_{0X}$	§See Note
KV_{bc}	$d_2[Z_F(Z_{2X}+Z_{0X})+Z_2Z_{0X}]-d_2Z_2Z_{0X}$	
KV_{ca}	$d_1[Z_F(Z_{2X}+Z_{0X})+Z_2Z_{0X}]-d_0Z_2Z_{0X}$	

*For a Two-Line-to-Ground Fault, We Define:

$Z_{2X}=Z_F+Z_2$ $\quad Z_{F1}=Z_F+C_1hZ_{L1}$

$Z_{0X}=Z_F+Z_0+3Z_G$ $\quad Z_{2C}=Z_2-C_2hZ_{L2}$

$Z_{0C}=Z_0-C_0hZ_{L0}$

§For Voltages at R:
Replace Z_F by Z_{F1}
Replace Z_2 by Z_{2C}
Replace Z_0 by Z_{0C}

11.3 IMPEDANCE AT THE RELAY

Using the values of currents and voltages shown in Tables 11.1 through 11.5, we can evaluate the apparent impedance seen by the relay at R for several possible connections of the current and voltage transducers. Some connections that might be explored include:

1. Wye-connected VT's and wye-connected CT's
2. Delta-connected VT's and wye-connected CT's
3. Wye-connected VT's and delta-connected CT's
4. Delta-connected VT's and delta-connected CT's

The choices for instrument transformer connection have been exhaustively treated in the literature [2–4]. From these studies, it is clear that the fourth choice is preferred for distance relays since this connection provides the relay with a ratio of voltages to currents that most nearly match the actual line impedance from the relay to the fault. This is not at all clear from examining the tables of the last section without some detailed algebraic investigations, most of which are left as problems at the end of the chapter.

11.3.1 Relay Impedances for Phase Faults with $C_1 \neq C_2$

Assume that the instrument connections are the delta connections shown in Figure 11.14, which have been shown to be the preferred connection for distance relays [2]. Note that the

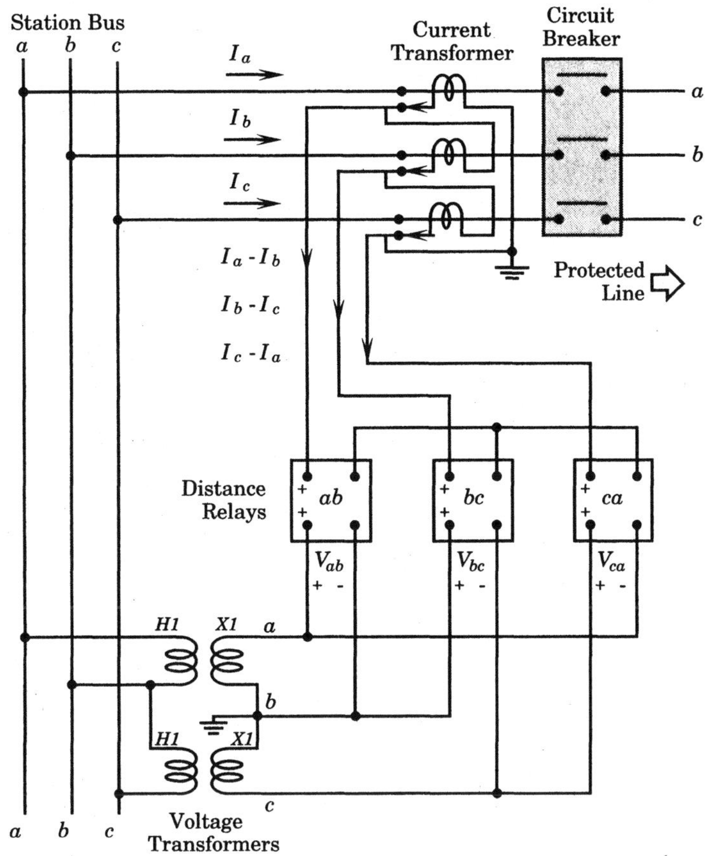

Figure 11.14 Delta current and voltage transformer connections for distance relaying.

current transformers in Figure 11.14 are connected in delta, and the voltage transformers are in an open-delta connection, which is considered appropriate for this application [3]. We now examine the impedances seen by each of the relays for phase faults.

11.3.1.1 Relay Impedances for Three-Phase Faults. For three-phase faults and with delta-connected current and voltage transformers, we compute the following impedances at the relay location. From Table 11.1, we write

$$\begin{aligned} \mathbf{Z}_{ab} &= \frac{\mathbf{K}\mathbf{V}_{ab}}{\mathbf{K}(\mathbf{I}_a - \mathbf{I}_b)} = \frac{\mathbf{Z}_{F1}}{\mathbf{C}_1} = h\mathbf{Z}_{L1} + \frac{\mathbf{Z}_F}{\mathbf{C}_1} \\ \mathbf{Z}_{bc} &= \frac{\mathbf{K}\mathbf{V}_{bc}}{\mathbf{K}(\mathbf{I}_b - \mathbf{I}_c)} = \frac{\mathbf{Z}_{F1}}{\mathbf{C}_1} = h\mathbf{Z}_{L1} + \frac{\mathbf{Z}_F}{\mathbf{C}_1} \\ \mathbf{Z}_{ca} &= \frac{\mathbf{K}\mathbf{V}_{ca}}{\mathbf{K}(\mathbf{I}_c - \mathbf{I}_a)} = \frac{\mathbf{Z}_{F1}}{\mathbf{C}_1} = h\mathbf{Z}_{L1} + \frac{\mathbf{Z}_F}{\mathbf{C}_1} \end{aligned} \tag{11.48}$$

Clearly, the impedance seen by each relay is directly related to the total line impedance from the relay location to the fault location, plus a term depending on the fault impedance. We examine this term more carefully, particularly the effect of the constant $\mathbf{C}_1$ on this term. From

(11.7) we write

$$C_1 = \frac{Z_{N1}}{Z_{N1} + Z_{M1}} = \frac{1}{1 + \dfrac{Z_{M1}}{Z_{N1}}} \tag{11.49}$$

where, from Figure 11.2(*b*) we note that

$$\begin{aligned} Z_{M1} &= \text{impedance in the relay branch} \\ Z_{N1} &= \text{impedance in the parallel branch} \end{aligned}$$

If the fault is very close to the relay location R, then

$$Z_{N1} >> Z_{M1} \quad \text{and} \quad C_1 \to 1.0 \tag{11.50}$$

Similarly, if the fault is very close to the far end of the line, i.e., near Q in Figure 11.1, then

$$Z_{N1} << Z_{M1} \quad \text{and} \quad C_1 \to 0 \tag{11.51}$$

In many cases, except for very short lines, the source impedances are small with respect to the line impedances, such that Z_{M1} is dominated by the line impedance. In this case, we may approximate the value of C_1 by

$$C_1 = \frac{Z_{N1}}{Z_{N1} + Z_{M1}} \cong \frac{(1-h)Z_{L1}}{hZ_{L1} + (1-h)Z_{L1}} = 1 - h \tag{11.52}$$

In this case, the practical limiting minimum value for C_1 for condition (11.51) is about 0.1, which represents placing the relay reach or balance point at 90% of the line length. Note that for $C_1 = 0.1$, the fault impedance in (11.48) is effectively multiplied by a factor of 10, which considerably distorts the impedance seen by the relay. This could be a problem for the relay selectivity, especially on short lines. For long lines, the total impedance in (11.48) will be dominated by hZ_{L1} and the fault impedance term will have less effect. Finally, we note that the effect just described makes the relay see an impedance that is larger than the actual impedance to the fault point, which causes the relay to "underreach" its design balance point.

11.3.1.2 Relay Impedances for Phase-to-Phase Faults. For line-to-line faults, the delta voltages and currents are given in Table 11.2. From these values, we compute the following.

$$\begin{aligned} Z_{ab} &= hZ_{L1} + \frac{Z_F}{C_1 + aC_2} + \frac{3Z_2}{(1-a^2)C_1 + (a-1)C_2} \\ Z_{bc} &= hZ_{L1} + \frac{Z_F}{C_1 + C_2} \\ Z_{ca} &= hZ_{L1} + \frac{Z_F}{C_1 + a^2C_2} - \frac{3Z_2}{(a-1)C_1 + (1-a^2)C_2} \end{aligned} \tag{11.53}$$

For this case, the *b-c* relay sees exactly the line impedance plus a multiple of the fault impedance. The other two relays see impedances that are related to the impedance from the relay to the fault, plus other impedances that are complex and shifted in phase.

Note that the complex constants, C_1 and C_2, are nearly equal in most cases and this simplifies the expressions. Even with this simplification, the *b-c* relay is the only one that will reliably detect the fault. This is as it should be since the line-to-line fault is between lines *b* and *c*, by definition, in the symmetrical component representation.

11.3.1.3 Summary for Phase Faults. The relay impedances seen by the three delta-connected relays for phase faults are summarized in Table 11.6. The quantities displayed in this table are for the general case where $C_1 \neq C_2$. In practical systems these constants are nearly equal, and are often assumed to be equal to simplify the calculations [3], [4]. This special case is discussed in section 11.3.3. Note that the expressions for impedance seen by relays *a-b* and *c-a* for the phase-to-phase fault involve the impedance of the negative sequence network in addition to the fault impedance. This further distorts the impedances seen by these relays and makes fault detection unreliable except for the *b-c* relay, which sees only the line impedance to the fault and the fault impedance.

TABLE 11.6 Relay Impedances for Phase Faults where $C_1 \neq C_2$

Relay	3-Phase Fault	Phase-to-Phase (*b-c*) Fault
a-b	$hZ_{L1} + \frac{Z_F}{C_1}$	$hZ_{L1} + \frac{Z_F}{C_1 + aC_2} + \frac{3Z_2}{d_0C_1 + d_1C_2}$
b-c	$hZ_{L1} + \frac{Z_F}{C_1}$	$hZ_{L1} + \frac{Z_F}{C_1 + C_2}$
c-a	$hZ_{L1} + \frac{Z_F}{C_1}$	$hZ_{L1} + \frac{Z_F}{C_1 + a^2C_2} - \frac{3Z_2}{d_1C_1 + d_0C_2}$

11.3.2 Relay Impedances for Ground Faults with $C_1 \neq C_2$

Using the data for voltages and currents from Tables 11.3 and 11.4, we may compute the impedance seen by the relays for the one-line-to-ground and two-line-to-ground faults. These quantities are generally more complex than those for phase faults, but the procedure is straightforward. The results of these computations are shown in Table 11.7.

TABLE 11.7 Relay Impedances for Ground Faults where $C_1 \neq C_2$

Relay	One-Line-to-Ground Fault	Two-Line-to-Ground Fault
a-b	$hZ_{L1} + \frac{3Z_F + Z_0 - a^2Z_2}{C_1 - aC_2}$	$hZ_{L1} + \frac{d_0Z_F(Z_{2X} + 2Z_{0X}) + 3Z_2Z_{0X}}{d_0C_1Z_{2X} + (d_0C_1 + d_1C_2)Z_{0X}}$
b-c	$hZ_{L1} + \frac{3Z_F + Z_0 + 2Z_2}{C_1 - C_2}$	$hZ_{L1} + \frac{Z_F(Z_{2X} + 2Z_{0X})}{C_1Z_{2X} + (C_1 + C_2)Z_{0X}}$
c-a	$hZ_{L1} + \frac{3Z_F + Z_0 - aZ_2}{C_1 - a^2C_2}$	$hZ_{L1} + \frac{d_1Z_F(Z_{2X} + 2Z_{0X}) - 3Z_2Z_{0X}}{d_1C_1Z_{2X} + (d_1C_1 + d_0C_2)Z_{0X}}$

Note that these impedances generally involve both the zero-sequence and the negative-sequence impedances, in addition to the fault impedance. Also note the impedance of the *b-c* relay for the one-line-to-ground fault on phase a, as the constants C_1 and C_2 approach equality. This relay sees the actual fault impedance plus a very large impedance, which means that this relay will not pick up for this type of fault. For the two-line-to-ground fault, the impedances seen by all three relays is directly related to the actual impedance to the fault point, plus a complex function of the fault impedances. It is also clear that when C_1 and C_2 are equal the second term is changed considerably in magnitude. This will be investigated in the next section.

11.3.3 Relay Impedances when $C_1 = C_2$

For the special case when $C_1 = C_2$, the impedance expressions computed for the three relays can be simplified. These impedances are shown in Tables 11.8 and 11.9. The impedances for the three-phase fault are unchanged, but for other types of faults, the impedance values are not as complex as before. Since the constants C_1 and C_2 are often nearly equal, the values in Tables 11.8 and 11.9 are commonly used for computations.

TABLE 11.8 Relay Impedances for Phase Faults where $C_1 = C_2$

Relay	Three-Phase Faults	Phase-to-Phase Fault
a-b	$h\mathbf{Z}_{L1} + \dfrac{\mathbf{Z}_F}{\mathbf{C}_1}$	$h\mathbf{Z}_{L1} - \dfrac{a\mathbf{Z}_F}{\mathbf{C}_1} - j\dfrac{\sqrt{3}\mathbf{Z}_2}{\mathbf{C}_1}$
b-c	$h\mathbf{Z}_{L1} + \dfrac{\mathbf{Z}_F}{\mathbf{C}_1}$	$h\mathbf{Z}_{L1} + \dfrac{\mathbf{Z}_F}{2\mathbf{C}_1}$
c-a	$h\mathbf{Z}_{L1} + \dfrac{\mathbf{Z}_F}{\mathbf{C}_1}$	$h\mathbf{Z}_{L1} - \dfrac{a^2\mathbf{Z}_F}{\mathbf{C}_1} + j\dfrac{\sqrt{3}\mathbf{Z}_2}{\mathbf{C}_1}$

TABLE 11.9 Relay Impedances for Ground Faults where $C_1 = C_2$

Relay	One-Line-to-Ground Fault	Two-Line-to-Ground Fault
a-b	$h\mathbf{Z}_{L1} - \dfrac{3\mathbf{Z}_F + \mathbf{Z}_0}{\mathbf{d}_1\mathbf{C}_1} + j\dfrac{\mathbf{Z}_2}{\sqrt{3}\mathbf{C}_1}$	$h\mathbf{Z}_{L1} + \dfrac{\mathbf{d}_0\mathbf{Z}_F(\mathbf{Z}_{2X} + 2\mathbf{Z}_{0X}) + 3\mathbf{Z}_2\mathbf{Z}_{0X}}{\mathbf{C}_1(\mathbf{d}_0\mathbf{Z}_{2X} - \mathbf{d}_2\mathbf{Z}_{0X})}$
b-c	∞	$h\mathbf{Z}_{L1} + \dfrac{\mathbf{Z}_F}{\mathbf{C}_1}$
c-a	$h\mathbf{Z}_{L1} + \dfrac{3\mathbf{Z}_F + \mathbf{Z}_0}{\mathbf{d}_0\mathbf{C}_1} - j\dfrac{\mathbf{Z}_2}{\sqrt{3}\mathbf{C}_1}$	$h\mathbf{Z}_{L1} + \dfrac{\mathbf{d}_1\mathbf{Z}_F(\mathbf{Z}_{2X} + 2\mathbf{Z}_{0X}) - 3\mathbf{Z}_2\mathbf{Z}_{0X}}{\mathbf{C}_1(\mathbf{d}_1\mathbf{Z}_{2X} - \mathbf{d}_2\mathbf{Z}_{0X})}$

EXAMPLE 11.1

A 161 kV transmission line is 100 miles long and is protected by distance relays with delta current and voltage connections. The purpose of this example is to apply the concepts described in this chapter for a typical transmission line, including the complex constants, the line impedances, and the sequence network impedances.

The positive- and zero-sequence line impedances are given as follows:

$$\begin{aligned} z_1 &= 0.2799 + j0.6671\ \Omega/\text{mile} \\ z_0 &= 0.5657 + j2.4850\ \Omega/\text{mile} \end{aligned} \tag{11.54}$$

A line-to-line fault between phases b and c occurs at a point 70 miles from the relay location R. At the time of the fault, the source impedances are as follows:

$$\begin{aligned} \mathbf{Z}_{S1} &= \mathbf{Z}_{S2} = 0.002 + j0.005 \text{ pu} \\ \mathbf{Z}_{S0} &= 0.004 + j0.010 \text{ pu} \end{aligned} \tag{11.55}$$

$$\begin{aligned} \mathbf{Z}_{U1} &= \mathbf{Z}_{U2} = 0.003 + j0.006 \text{ pu} \\ \mathbf{Z}_{U0} &= 0.006 + j0.012 \text{ pu} \end{aligned} \tag{11.56}$$

$$\begin{aligned} \mathbf{Z}_{E1} &= \mathbf{Z}_{E2} = 0.20 + j0.50 \text{ pu} \\ \mathbf{Z}_{E0} &= 0.40 + j2.0 \text{ pu} \end{aligned} \tag{11.57}$$

where all values are in per unit with base values of

$$\begin{aligned} S_B &= 100 \text{ MVA} \\ Z_B &= 259.21\ \Omega \end{aligned} \tag{11.58}$$

The fault impedance, $\mathbf{Z}_F$ is assumed to be 0.04 per unit on this same base and is an arcing resistance fault.

Compute the following:

1. The per unit impedance from the relay point to the fault, $h\mathbf{Z}_{L1}$
2. The impedances $\mathbf{Z}_{M1}$ and $\mathbf{Z}_{N1}$
3. The complex constants $\mathbf{C}_1$ and $\mathbf{C}_2$
4. The impedances seen by relays ab, bc, and ca.

Plot the relay impedances in the complex plane, including the components that make up the total impedance for each relay.

Solution

From the given fault location, we know that $h = 0.7$. From this and the given line data, we compute the following:

$$\begin{aligned} \mathbf{Z}_{L1} = \mathbf{Z}_{L2} &= \frac{(100)(0.2799 + j0.6671)}{259.21} \\ &= 0.1080 + j0.2574 \text{ pu} \end{aligned} \tag{11.59}$$

Then

$$\begin{aligned} h\mathbf{Z}_{L1} &= (0.7)(0.1080 + j0.2574) \\ &= 0.0756 + j0.1802 \text{ pu} \end{aligned} \tag{11.60}$$

and

$$\mathbf{Z}_F = 0.04 + j0 \text{ pu} \tag{11.61}$$

From Figures 11.3 and 11.7, we perform the delta-wye conversion for the positive- and negative-sequence networks, with the following results.

$$\mathbf{Z}_{I1} = \frac{\mathbf{Z}_{U1}\mathbf{Z}_{S1}}{\mathbf{Z}_{S1} + \mathbf{Z}_{E1} + \mathbf{Z}_{U1}} = 0.00017\angle 68.1986^\circ \tag{11.62}$$

$$\mathbf{Z}_{J1} = \frac{\mathbf{Z}_{S1}\mathbf{Z}_{E1}}{\mathbf{Z}_{S1} + \mathbf{Z}_{E1} + \mathbf{Z}_{U1}} = 0.00542\angle 68.1986^\circ \tag{11.63}$$

$$\mathbf{Z}_{K1} = \frac{\mathbf{Z}_{E1}\mathbf{Z}_{U1}}{\mathbf{Z}_{S1} + \mathbf{Z}_{E1} + \mathbf{Z}_{U1}} = 0.00388\angle 68.1986^\circ \tag{11.64}$$

Then we may compute the following impedances, defined by Figure 11.3.

$$\begin{aligned} \mathbf{Z}_{M1} &= h\mathbf{Z}_{L1} + \mathbf{Z}_{J1} = 0.0776 + j0.1852 \text{ pu} \\ \mathbf{Z}_{N1} &= (1-h)\mathbf{Z}_{L1} + \mathbf{Z}_{K1} = 0.0363 + j0.0869 \text{ pu} \\ \mathbf{Z}_{M1} + \mathbf{Z}_{N1} &= 0.1139 + j0.2721 \text{ pu} \end{aligned} \tag{11.65}$$

Finally, from (11.7) and (11.9) we compute the total sequence impedance and the constant $\mathbf{C}_1$.

$$\begin{aligned} \mathbf{Z}_1 &= \frac{\mathbf{Z}_{M1}\mathbf{Z}_{N1}}{\mathbf{Z}_{M1} + \mathbf{Z}_{N1}} + \mathbf{Z}_{I1} = 0.0248 + j0.0592 \text{ pu} \\ \mathbf{C}_1 &= \frac{\mathbf{Z}_{N1}}{\mathbf{Z}_{M1} + \mathbf{Z}_{N1}} = 0.3192 + j0.00028 \text{ pu} \end{aligned} \tag{11.66}$$

Since the positive- and negative-sequence networks are identical, we write

$$\mathbf{Z}_2 = \mathbf{Z}_1 \qquad \mathbf{C}_2 = \mathbf{C}_1 \tag{11.67}$$

For the line-to-line fault case, we use the values of Table 11.8 to compute the relay impedances for all three relays. Using these tabulated results, we compute the following.

$$\begin{aligned} \mathbf{Z}_{ab} &= h\mathbf{Z}_{L1} - \frac{\boldsymbol{a}}{\mathbf{C}_1}\mathbf{Z}_F - j\frac{\sqrt{3}}{\mathbf{C}_1}\mathbf{Z}_2 \\ &= 0.1954\angle 67.24^\circ + 0.1253\angle -59.95^\circ + 0.3481\angle -22.78^\circ \end{aligned} \tag{11.68}$$

$$\begin{aligned} \mathbf{Z}_{bc} &= h\mathbf{Z}_{L1} + \frac{1}{2\mathbf{C}_1}\mathbf{Z}_F \\ &= 0.1954\angle 67.24^\circ + 0.0627\angle -0.05^\circ \end{aligned} \tag{11.69}$$

$$\begin{aligned} \mathbf{Z}_{ca} &= h\mathbf{Z}_{L1} - \frac{\boldsymbol{a}^2}{\mathbf{C}_1}\mathbf{Z}_F + j\frac{\sqrt{3}}{\mathbf{C}_1}\mathbf{Z}_2 \\ &= 0.1954\angle 67.24^\circ + 0.1288\angle 60.22^\circ + 0.3475\angle 157.27^\circ \end{aligned} \tag{11.70}$$

The three relay impedances are plotted in Figure 11.15, beginning with the impedance from the relay to the fault point, $h\mathbf{Z}_{L1}$. The various components are added to this quantity to locate the total relay impedance for all three relays. The relay at R is assumed to be a mho type relay, with a circle diameter of exactly $\mathbf{Z}_{L1}$. Clearly, the b-c relay will trip, but the others will see this fault as well outside their trip zone.

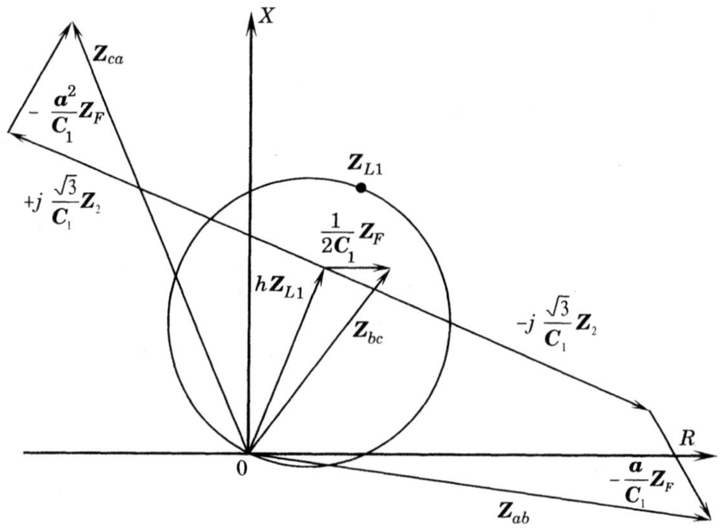

Figure 11.15 Relay impedances for a phase-to-phase fault on phases b-c. ■

The result computed in Example 11.1 and displayed in Figure 11.15 is typical for the line-to-line fault detection using distance relays. Usually two of the relays see the fault as outside the trip zone and only the b-c relay picks up for this type of fault. The three-phase fault, however, will be picked up by all three relays.

11.3.4 Apparent Relay Impedance Plots

A convenient way of viewing the computed complex apparent impedance seen by the relay at R is to plot that impedance in the complex plane as the fault location parameter h is varied from 0 to 1, thereby moving the fault location from the relay terminals to the far end of the transmission line. The impedance seen by the relay is a function of all the parameters of the protection equivalent circuit, shown in Figure 11.1. This concept is studied by means of numerical examples.

EXAMPLE 11.2

The first example studies the effect of the length of an EHV transmission line on the shape of the apparent impedance. The basic system parameters are as follows:

$$\begin{aligned} \text{Base voltage} &= 500 \text{ kV} \\ \text{Base MVA} &= 1000 \text{ MVA} \end{aligned} \tag{11.71}$$

$$\begin{aligned} z &= 0.0007456 + j0.0014920 \text{ pu/km} \\ y &= 0.0 + j0.0011184 \text{ pu/km} \end{aligned} \tag{11.72}$$

$$\begin{aligned} \mathbf{E}_S = \mathbf{E}_U &= 1.0 + j0 \text{ pu} \\ \mathbf{Z}_S = \mathbf{Z}_U &= 0.0 + j0.1 \text{ pu} \\ \mathbf{Z}_E &= 0.002 + j0.04 \text{ pu} \\ \mathbf{Z}_F &= 0.01 + j0.00 \text{ pu} \end{aligned} \tag{11.73}$$

Both the length of the transmission and the effect of the line susceptance will be variables in this example. The transmission line is divided into two sections, separated at the point of the fault application. Both segments of the transmission line are represented as long line equivalent pi sections. Figure 11.16 shows the general view of the system under study.

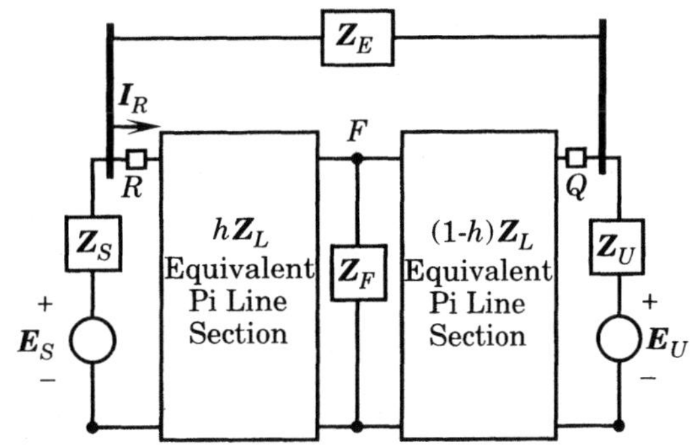

Figure 11.16 Protection equivalent for an EHV line.

Solution

We solve the transmission line be varying only the line length to determine the effect of this parameter on the apparent impedance of the line as seen by the relay. The results are shown in Figure 11.17.

Changing any of the system parameters changes the shape of the resulting system variables. At the relay, the variable of greatest interest are the relay voltage, current, and apparent impedance. This example shows that the apparent impedance extends farther in the positive R direction of the complex plane for the longer lines. An additional parameter that is sometimes neglected in protection studies is the shunt susceptance of the transmission lines. For this 500 kV line the shunt susceptance is rather large. If the susceptance is omitted from the study, the results are considerably in error. The apparent impedance loci have somewhat different shapes, and do not reach nearly as far to the right as for the cases with susceptance included. We conclude that long lines cause the apparent impedance locus to bend toward the right as the fault nears the remote end of the line. Moreover, the apparent impedance at the relay is in error if the shunt susceptance is not included in the analysis.

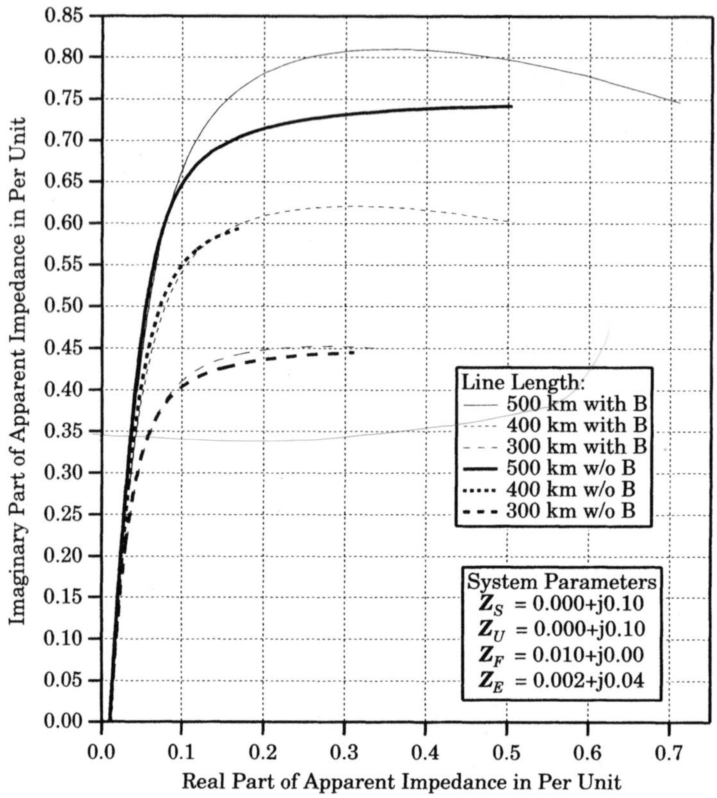

Figure 11.17 Apparent impedance seen by relay R. ■

Other examples will now be pursued to investigate the effect of varying other variables, either for the line or for the connected power system.

EXAMPLE 11.3

A 500 kV transmission line is 400 km long and has the following basic parameters:

$$\begin{aligned} \mathbf{E}_S &= \mathbf{E}_U = 1.0 + j0 \text{ pu} \\ \mathbf{Z}_S &= \mathbf{Z}_U = 0.0 + j0.1 \text{ pu} \\ \mathbf{Z}_F &= 0.001 + j0 \text{ pu} \\ \mathbf{Z}_E &= \text{variable} \end{aligned} \tag{11.74}$$

The external impedance $\mathbf{Z}_E$ is left as a variable so that we can investigate the effect of the external network impedance on the apparent impedance seen by the relay at R. Other parameters are as defined in (11.71)–(11.73). Note that the assumed fault impedance is quite small.

Solution

First, we examine the case where $\mathbf{Z}_E$ is smaller than the equivalent pi impedance for the entire line. The value arbitrarily chosen for study is

$$\mathbf{Z}_E = 0.002 + j0.040 \text{ pu} \tag{11.75}$$

where the X/R ratio is 20:1, the same ratio as the transmission line. This small value of $\mathbf{Z}_E$ represents a highly developed EHV network in parallel with the protected line under study. This impedance is then varied over several decades, multiplying by 10 and 1000.

The apparent impedance seen by the relay at R is shown in Figure 11.18. A mho type of distance measuring relay with a maximum torque angle (MTA) of 60 degrees is assumed to reach 90% of the equivalent pi distance, and is represented by the circle in the figure. The three values plotted represent the base impedance, given by (11.75) and also values 10 and 1000 times greater than this base value. Even higher values produce apparent impedance loci that deviate only slightly from the highest of these values.

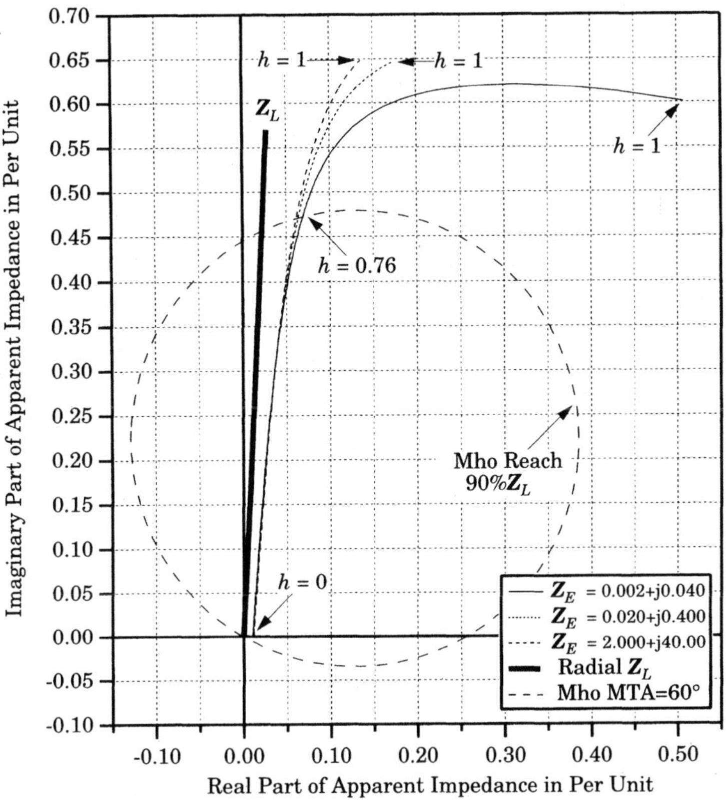

Figure 11.18 Apparent impedance with increasing values of $\mathbf{Z}_E$.

Note that the apparent impedance is not collinear with the radial line impedance, shown as $\mathbf{Z}_L$. This is because the portion of the fault current supplied by the opposite end of the line becomes great as the value of h approaches unity. This causes the locus of the apparent impedance to bend to the right and to extend farther in the $+R$ direction. We conclude that the value of the external impedance is very important in determining the location and the shape of the apparent impedance, plotted in the complex Z plane. ■

EXAMPLE 11.4

Example 11.3 shows the apparent impedance seen by the relay at R as a three-phase fault is moved along the transmission line. Extend this study of the same 500 kV line to compute the complex current and voltage as a function of h, the fault position along the line.

Solution

The circuit of Figure 11.16 is solved for all values of h from 0 to 1, and the complex relay current and voltage are computed for each fault location. The results are plotted as complex phasor loci in Figures 11.19 and 11.20 for the relay phasor voltage and current, respectively.

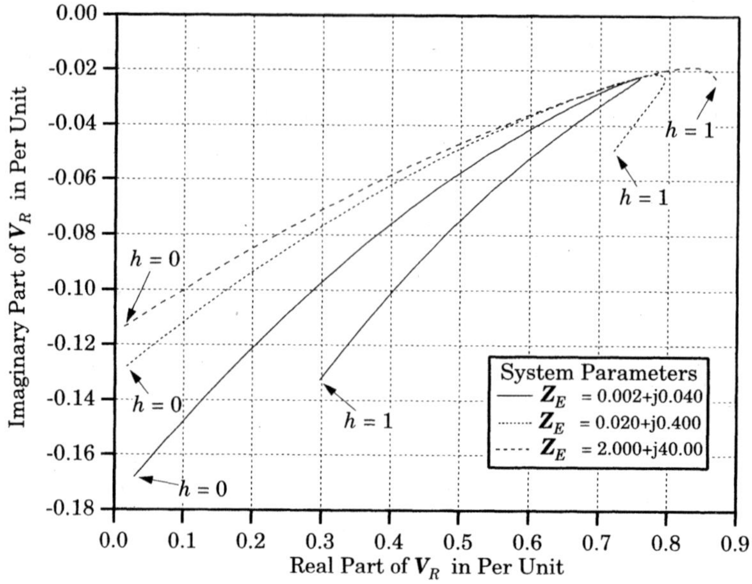

Figure 11.19 Phasor loci for the relay voltage at R.

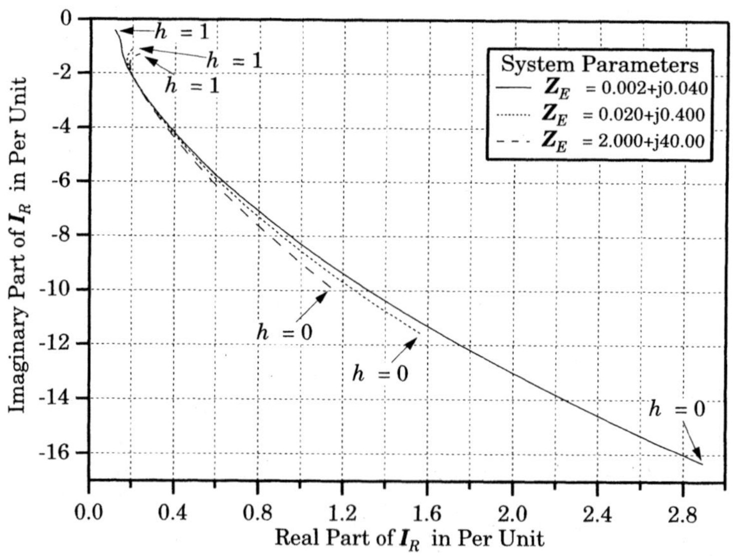

Figure 11.20 Phasor loci for the relay current at R.

Note that the horizontal and vertical axes are not scaled equally in these figures, as is usually done in plotting phasor loci. This has been done in order to show more clearly the curvature of the loci. The voltage loci lie close to the real axis, with imaginary parts that are only slightly negative, as would be expected. For a fault at the relay, the voltage is almost zero, but it increases as the fault moves farther away. For small and reasonably large values of Z_E the voltages reaches a maximum in its real part and then doubles back toward the real axis.

The relay current, shown in Figure 11.20, is largest in magnitude when the external impedance is small, which reflects the fact that a portion of the total relay current is contributed by a current component flowing through the external system. For high values of external impedance, the current locus hooks back as the fault nears the remote end of the line. ■

The preceding examples have demonstrated the effect of different external system impedances on the apparent impedance seen by the relay and the effect on the relay voltage and current. We now explore the effect of the source impedances on the relay quantities.

EXAMPLE 11.5

Plot the relay apparent impedance, voltage, and current for widely different values of source impedances, while leaving the external system impedance unchanged. The line is 400 km long.

Solution

The relay apparent impedances are plotted in Figure 11.21 for three values of source impedances. The first value, plotted as the solid line, is the same as that plotted in Figures 11.17 and 11.18. Two additional curves are plotted, one with the source impedance $\mathbf{Z}_S$ *reduced* by a factor of five, and with source impedance $\mathbf{Z}_U$ *increased* by a factor of five. The second new plot is for the source impedance $\mathbf{Z}_S$ *increased* by a factor of five, and with source impedance $\mathbf{Z}_U$ *reduced* by a factor of five. These are identified in the figures as the "Lo-Hi" and "Hi-Lo" conditions, respectively, where the first term refers to source S and the second to source U impedances. For the smaller source impedance behind the relay position R, the apparent impedance has a smaller resistive component for faults at the extreme end of the line, but is otherwise quite similar to the original impedance. However, for a large source impedance behind the relay, the apparent impedance locus flares far to the right as the fault location approaches the remote end of the line. However, this portion of the locus is well outside the relay characteristic circle. In fact, all three loci cross the relay characteristic circle at approximately the same value of $h = 0.75$.

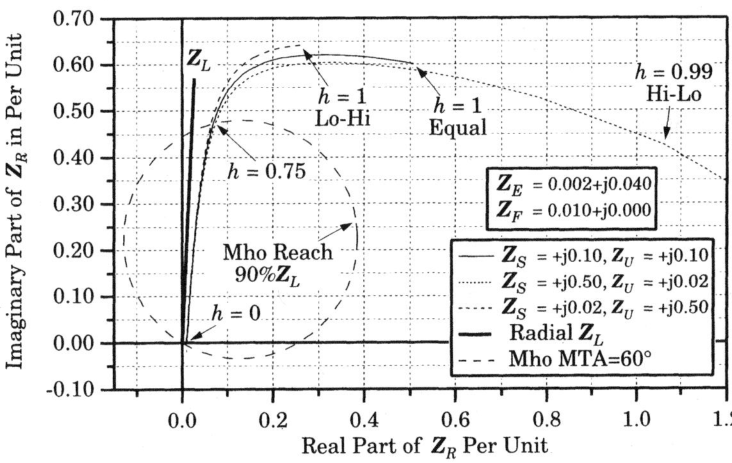

Figure 11.21 Relay apparent impedance for varying source impedances.

Figures 11.22 and 11.23 show the loci of phasor relay voltage and current, respectively, for the same variation in source impedances. Clearly, the source impedances are very important factors in determining the system conditions observed at the relay location.

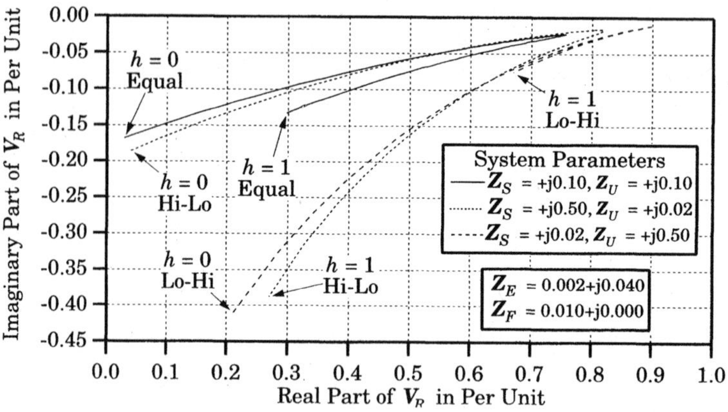

Figure 11.22 Relay voltage for varying source impedances.

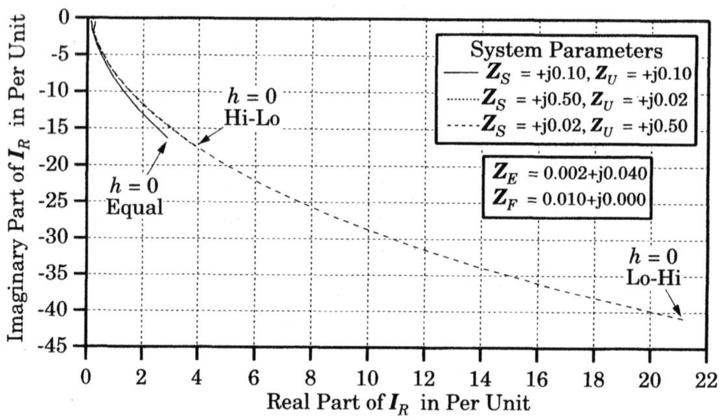

Figure 11.23 Relay current for varying source impedances. ■

11.4 DISTANCE RELAY SETTINGS

The computation of distance relay settings is dependent on the type of relay and the manufacturer's specifications for making the correct computation. The best way to illustrate this process is by example.

EXAMPLE 11.6

A 138 kV transmission line has a nominal load rating of 130 MVA. The line connects buses TB and CP in the 138 kV network, as shown in Figure 11.24. The relay setting is to be determined for the relay R, located at substation TB in the 138 kV network. Most of the current for faults on the protected line will come from the line end opposite that of relay R, due to the network configuration. The line is to be protected by three-zone GCX distance relays. Determine the relay settings.

Solution

Power flow studies under $N-2$ condition, i.e., with any two lines out of service, show that the maximum load on the protected line is 90 MVA. This gives a load current value of

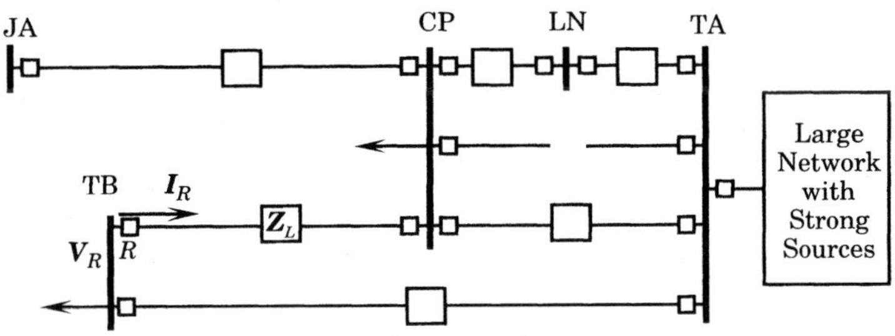

Figure 11.24 Network connections for the protected line and relay R.

$$I_{\text{Load}} = \frac{S}{\sqrt{3}V_{LL}} = \frac{90{,}000 \text{ kVA}}{\sqrt{3}(138\text{kV})} \cong 377A \tag{11.76}$$

The base quantities for the study are as follows:

$$\begin{aligned} S_B &= 200 \text{ MVA} \\ V_B &= 138 \text{ kV (line to line)} \\ Z_B &= 95.22\ \Omega \\ I_B &= 836.74 \text{ A} \end{aligned} \tag{11.77}$$

The instrument transformer ratios are as follows.

$$\begin{aligned} \text{CT ratio} &= 600/5 = 120/1 \\ \text{PT ratio} &= 80{,}500/67.08 = 1200/1 \end{aligned} \tag{11.78}$$

Then the ratio of CT to PT ratios can be computed as

$$\begin{aligned} n_{SP} &= \frac{Z_{\text{sec}}}{Z_{\text{pri}}} = \frac{V_S \text{ (V)}}{I_S \text{ (A)}} \frac{I_P \text{ (A)}}{V_P \text{ (V)}} \\ &= \frac{\text{CT ratio}}{\text{PT ratio}} = \frac{120}{1200} = 0.1 \text{ s } \Omega/\text{pri } \Omega \end{aligned} \tag{11.79}$$

The protected transmission line impedance, in per unit on the given base, is given as

$$\mathbf{Z}_L = 0.01084 + j0.02905 = 0.031\angle 69.5^\circ \text{ pu} \tag{11.80}$$

The Protective Relay. The relay to be used in protecting the line is assumed to be a zoned distance relay where the first and second zones utilize reactance characteristics to measure the distance to the fault, and the third zone uses a directional mho unit. The reactance characteristics of the first two zones can be supervised by the mho unit to restrict the characteristic to the region inside the mho circle. The combined relay characteristic is shown in Figure 11.25.

In Figure 11.25, the line $0L$ represents the impedance of the protected line and the distance $0T$ represents the reach of the first zone. The line $0S$ represents the source impedance behind the relay. Thus the total impedance to the end of zone 1 is represented by the distance ST. For a phase b-c fault, with zero fault impedance, the relay will correctly read a distance $0T$. However, the a-b unit will see an impedance originating at 0 and terminating somewhere along the line TP, assuming zero fault impedance, with the exact termination depending on the system Z_2 as given by (11.68). The general case, including fault resistance, is shown in Figure 11.25. If the impedance seen by the relay is inside the mho unit characteristic, the a-b unit will trip falsely, since the fault is not a phase a-b fault, but if outside the mho unit characteristic, no tripping of the a-b unit will occur. The possibility exists for an overreaching of the a-b phase unit. The overreach problem can be corrected by adjustment of the zone III mho unit characteristic such that the impedance seen by the phase units other than the correct one, b-c, can be made to fall outside the mho unit circle.

Zones 1 and 2 are the reactance characteristics. This type of distance characteristic is particularly well suited for circuits where arc resistance may be large. Since the arc resistance is independent of line length, it may be a significant component in the total impedance seen by the relay, extending rather far

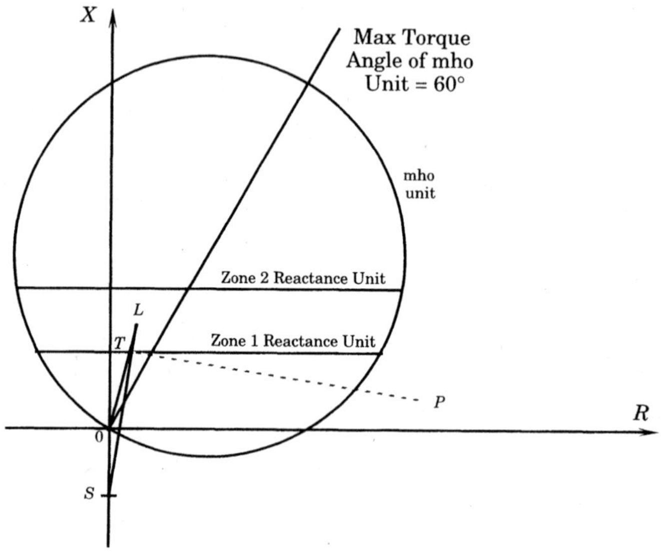

Figure 11.25 Characteristic of a typical three-zone distance relay.

to the right of the line characteristic, which is shown as the line $0L$. This type of relay is well suited, therefore, for short transmission lines, where arc resistance may be large in proportion to the protected line impedance.

Zone I Relay Settings. The relay settings for zone I will be set to reach 85% of the line length. This is in accordance with the protection department policy for this type of line, and is in general agreement with industry practice. The total reactance of zone I is computed in per unit to be

$$X_I = 0.85(0.02905) = 0.0247 \text{ pu} \tag{11.81}$$

Then the impedance seen on the secondary side of the instrument transformers is this per unit value multiplied by the transformation ratio computed above, or

$$\begin{aligned} X_{I(s)} &= 0.0247 n_{SP} = 0.0247(0.1) \\ &= 0.00247 \text{ pu} = 0.235\ \Omega(\text{sec}) \end{aligned} \tag{11.82}$$

Now, the actual primary ohmic value at the zone I threshold is computed as

$$\begin{aligned} X_{I(\text{pri}-\Omega)} &= (0.85)\mathbf{Z}_L Z_B \\ &= (0.85)(0.02905)(95.22) = 2.35\ \Omega(\text{pri}) \end{aligned} \tag{11.83}$$

Therefore, the relay tap setting must be chosen from the available taps that are assumed to be 0.1, 0.2, or 0.3 ohms for this relay. Suppose we select a tap of 0.2 ohms. Then the tap setting T is computed from the following formula, as defined by the manufacturer.

$$\begin{aligned} T_I &= \frac{X_{\min} \times 100\%}{X_{I(\text{pri})}} \\ &= \frac{(0.2)(100)}{0.235} = 85.1\% \end{aligned} \tag{11.84}$$

The relay has separate settings only for the tap decades and units. Since the computed tap is not an integer, it must be rounded to the nearest integral percentage. Some recommend that the result be rounded *up* to 86%. If this is done, the actual reach of the relay will be

$$\text{Actual reach}|_{\text{Rd Up}} = \frac{0.2}{0.86} = 0.2326\ \Omega \tag{11.85}$$

For the case at hand, being only 0.1% removed from the 85% tap, little error is introduced by rounding *down*, rather than up. To check the reach if the result is rounded down, we compute

$$\text{Actual reach}|_{\text{Rd Dn}} = \frac{0.2}{0.85} = 0.2353\ \Omega \tag{11.86}$$

This does not make the relay overreach an unacceptable amount, so we accept this as the final setting of zone I.

Zone II Relay Settings. Zone II should be set to reach through the end of the protected line and a short distance into any lines connected to the far end of the protected line. Referring to Figure 11.24, there are four lines, other then the protected line, connected to the station CP, and the line to LN is the shortest of these lines. The objective we set for zone II is that the relay should not reach more than 20% into the shortest line, namely, the line from CP to LN, which has a reactance of only 0.02303. This makes the reach of zone II

$$\begin{aligned} X_{\text{II}} &= X_{\text{Line}} + (0.2)X_{CP} \\ &= 0.02905 + (0.20)(0.02303) = 0.03366 \text{ pu} \end{aligned} \tag{11.87}$$

In ohms, this value is

$$X_{\text{II(pri)}} = 0.03366(95.22) = 3.205\ \Omega\text{(pri)} \tag{11.88}$$

The secondary value of this reactance is computed as

$$\begin{aligned} X_{\text{II(sec)}} &= 3.205 n_{SP} \\ &= 3.205(0.1) = 0.3205\ \Omega\text{(sec)} \end{aligned} \tag{11.89}$$

To estimate the tap setting, we compute

$$T_{\text{II}} = \frac{(0.2)(100\%)}{0.3205} = 62.402\% \tag{11.90}$$

Since this value is in between integer tap settings, we test the sensitivity of the setting. If 62% is used, then,

$$X_{\Omega\text{(sec)}} = \frac{0.2}{62\%}100 = 0.322\ \Omega\text{(sec)} \tag{11.91}$$

If 63% is used, then,

$$X_{\Omega\text{(sec)}} = \frac{0.2}{63\%}100 = 0.317\ \Omega\text{(sec)} \tag{11.92}$$

Since the 62% value is closest to the target value of 0.32, we shall use this tap setting. Note that the overreach in selecting this setting is not great.

Zone III Mho Unit. The system impedance at the TB bus is known from system studies to be

$$Z_{\text{Sys}} = 0.036 + j0.158 \text{ pu} = 0.162\angle 77.164^\circ \tag{11.93}$$

$$X_{\text{Sys(pri)}} = 0.158(95.22) = 15.045\ \Omega\text{(pri)} \tag{11.94}$$

$$X_{\text{Sys(sec)}} = 15.045 n_{SP} = 1.5045\ \Omega\text{(sec)} \tag{11.95}$$

One method of setting the mho unit is to apply the formula [9]

$$T_{\text{III}} = \frac{Z_{\min} \times 100\%}{Z_{LR}} \cos(60^\circ - \phi) \tag{11.96}$$

where $Z_{\min}$ = min mho unit nameplate reach in Ω(sec)

Z_{LR} = Desired reach in phase-to-neutral Ω(sec)

ϕ = angle of Z_{LR} = angle of line impedance

60° = angle of maximum torque of mho unit

Suppose that the nameplate minimum reach for the mho unit is 1 ohm. Then the magnitude of the desired reach is 0.162 per unit or 1.542Ω(sec) and the line angle is known from (11.80) to be 69.5°. Substituting into (11.96), we get a first estimate of the zone III tap to be

$$T_{\text{III}} = \frac{1.0 \times 100\%}{1.542} \cos(60^\circ - 69.5^\circ) = 63.96\% \tag{11.97}$$

or the third zone tap is set to 64%.

These calculations can be converted into physical settings that are dependent on the relay design. The relay manufacturer must provide information that will enable the protection engineer to convert the foregoing calculations into actual settings. This procedure is simply a matter of following instructions, which are provided by the relay literature. ■

A comment is in order regarding the setting of relays. The procedure for setting a relay depends on the type of relay. The relay is designed to provide a given type of characteristic and the relay designer achieves this objective using his experience and ingenuity. The relay setting is described by characteristic curves, tables, or procedures that the designer develops and which, if correctly followed, ensures the desired relay performance.

11.5 GROUND DISTANCE PROTECTION

Section 11.3 provides the results of impedance calculations as seen by the relay for various ground faults when the relays are supplied with delta currents and voltages. Ground distance relays may also be arranged with phase-to-neutral voltages and currents that are 'compensated' line currents. This scheme is often referred to as "residual compensation" [6].

Consider a single-line-to-ground fault on phase a, for which the sequence network connections are shown in Figure 11.10. Summing the sequence voltages at the fault point F, we write the following.

$$\boldsymbol{V}_{a0} + \boldsymbol{V}_{a1} + \boldsymbol{V}_{a2} = 3\boldsymbol{Z}_F\boldsymbol{I}_{a0} \tag{11.98}$$

or, in terms of the relay voltages and currents,

$$(\boldsymbol{V}_{R0} - h\boldsymbol{Z}_{L0}\boldsymbol{I}_{R0}) + (\boldsymbol{V}_{R1} - h\boldsymbol{Z}_{L1}\boldsymbol{I}_{R1}) + (\boldsymbol{V}_{R2} - h\boldsymbol{Z}_{L1}\boldsymbol{I}_{R2}) = 3\boldsymbol{Z}_F\boldsymbol{I}_{a0} \tag{11.99}$$

From Table 11.3, we know the ratio of currents at the relay point to the total fault currents for the one-line-to-ground fault.

$$\begin{aligned} \boldsymbol{I}_{R0} &= \boldsymbol{C}_0\boldsymbol{I}_{a0} \\ \boldsymbol{I}_{R1} &= \boldsymbol{C}_1\boldsymbol{I}_{a1} \\ \boldsymbol{I}_{R2} &= \boldsymbol{C}_2\boldsymbol{I}_{a2} \end{aligned} \tag{11.100}$$

Using (11.100), we write (11.98) in terms of the total fault current as

$$(\boldsymbol{V}_{R0} - \boldsymbol{C}_0h\boldsymbol{Z}_{L0}\boldsymbol{I}_{a0}) + (\boldsymbol{V}_{R1} - \boldsymbol{C}_1h\boldsymbol{Z}_{L1}\boldsymbol{I}_{a1}) + (\boldsymbol{V}_{R2} - \boldsymbol{C}_2h\boldsymbol{Z}_{L1}\boldsymbol{I}_{a2}) = 3\boldsymbol{Z}_F\boldsymbol{I}_{a0} \tag{11.101}$$

From this result, we solve for the voltage on the faulted phase, phase a, as

$$\begin{aligned} \boldsymbol{V}_{aR} &= \boldsymbol{V}_{R0} + \boldsymbol{V}_{R1} + \boldsymbol{V}_{R2} \\ &= \boldsymbol{C}_0h\boldsymbol{Z}_{L0}\boldsymbol{I}_{a0} + \boldsymbol{C}_1h\boldsymbol{Z}_{L1}\boldsymbol{I}_{a1} + \boldsymbol{C}_2h\boldsymbol{Z}_{L1}\boldsymbol{I}_{a2} + 3\boldsymbol{Z}_F\boldsymbol{I}_{a0} \end{aligned} \tag{11.102}$$

We now add and subtract the term $\boldsymbol{C}_0h\boldsymbol{Z}_{L1}\boldsymbol{I}_{a0}$ to write

$$\begin{aligned} \boldsymbol{V}_{aR} = h\boldsymbol{Z}_{L1}&(\boldsymbol{C}_0\boldsymbol{I}_{a0} + \boldsymbol{C}_1\boldsymbol{I}_{a1} + \boldsymbol{C}_2\boldsymbol{I}_{a2}) \\ &+ (\boldsymbol{C}_0h\boldsymbol{Z}_{L0} - \boldsymbol{C}_0h\boldsymbol{Z}_{L1} + 3\boldsymbol{Z}_F)\boldsymbol{I}_{a0} \end{aligned} \tag{11.103}$$

or

$$\boldsymbol{V}_{aR} = h\boldsymbol{Z}_{L1}\boldsymbol{I}_{aR} + \boldsymbol{C}_0\boldsymbol{I}_{a0}(h\boldsymbol{Z}_{L0} - h\boldsymbol{Z}_{L1}) + 3\boldsymbol{Z}_F\boldsymbol{I}_{a0} \tag{11.104}$$

For the special case where $\boldsymbol{Z}_F = 0$, we write (11.103) as

$$\boldsymbol{V}_{aR} = h\boldsymbol{Z}_{L1}\left[\boldsymbol{I}_{aR} + \boldsymbol{C}_0\boldsymbol{I}_{a0}\left(\frac{\boldsymbol{Z}_{L0}}{\boldsymbol{Z}_{L1}} - 1\right)\right] \tag{11.105}$$

From this expression we see that, should the ground relay be provided with the phase a-to-neutral voltage and the phase a current, the resulting impedance computed will be in error by the second term in the brackets. If, however, we can compensate the current presented to the relay such that the current is equal to the quantity in the brackets, then the relay will indeed measure the impedance $h\boldsymbol{Z}_{L1}$ as required, that is, the compensated relay will see

$$h\boldsymbol{Z}_{L1} = \frac{\boldsymbol{V}_{aR}}{\left[\boldsymbol{I}_{aR} + \boldsymbol{C}_0\boldsymbol{I}_{a0}\left(\frac{\boldsymbol{Z}_{L0}}{\boldsymbol{Z}_{L1}} - 1\right)\right]} \tag{11.106}$$

The desired compensation is accomplished by the current transformer secondary circuit shown in Figure 11.26.

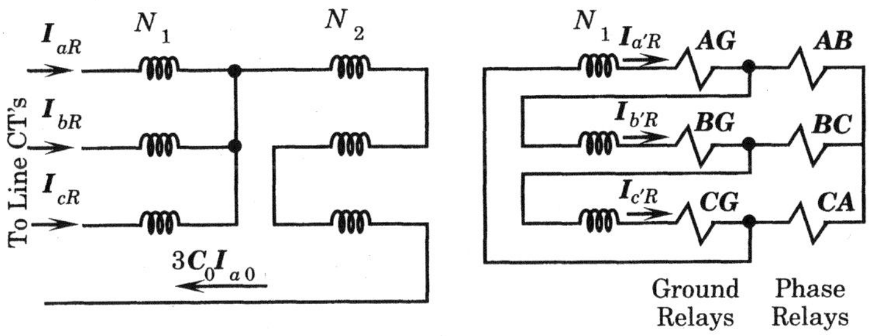

Figure 11.26 Current transformer connections for ground fault compensation in ground distance relaying.

In Figure 11.26, the input currents are the relay currents, which are added together to compute

$$\boldsymbol{I}_{aR} + \boldsymbol{I}_{bR} + \boldsymbol{I}_{cR} = 3\boldsymbol{C}_0\boldsymbol{I}_{a0} \tag{11.107}$$

The windings are arranged such that the currents on the third set of windings are the sum of the currents on the two left sets of windings. Then we may write

$$\begin{aligned} \boldsymbol{I}_{a'R} &= \boldsymbol{I}_{aR} + 3\boldsymbol{C}_0\boldsymbol{I}_{a0}(N_2/N_1) \\ \boldsymbol{I}_{b'R} &= \boldsymbol{I}_{bR} + 3\boldsymbol{C}_0\boldsymbol{I}_{a0}(N_2/N_1) \\ \boldsymbol{I}_{cR} &= \boldsymbol{I}_{cR} + 3\boldsymbol{C}_0\boldsymbol{I}_{a0}(N_2/N_1) \end{aligned} \tag{11.108}$$

which we write as

$$\begin{aligned} \boldsymbol{I}_{a'R} &= \boldsymbol{I}_{aR} + \boldsymbol{I}_{\text{comp}} \\ \boldsymbol{I}_{b'R} &= \boldsymbol{I}_{bR} + \boldsymbol{I}_{\text{comp}} \\ \boldsymbol{I}_{c'R} &= \boldsymbol{I}_{cR} + \boldsymbol{I}_{\text{comp}} \end{aligned} \tag{11.109}$$

where

$$\boldsymbol{I}_{\text{comp}} = 3\boldsymbol{C}_0\boldsymbol{I}_{a0}\frac{N_2}{N_1} \tag{11.110}$$

If we let the turns ratio of the transformation be equal to

$$\frac{N_2}{N_1} = \frac{1}{3}\left(\frac{\boldsymbol{Z}_{L0}}{\boldsymbol{Z}_{L1}} - 1\right) \tag{11.111}$$

then the relay will provide an accurate distance measurement to the fault point and the compensation will be exactly right for zero fault impedance. The relay connection shown in Figure 11.26 provides the ground relays with exactly this compensated current. The phase relays have the usual delta currents discussed in the Section 11.3. For later reference, we write (11.106) as

$$hZ_{L1} = \frac{V_{aR}}{I_{aR} + 3K_C C_0 I_{a0}} = \frac{V_{aR}}{I_{aR} + K_C I_{GR}} \tag{11.112}$$

where we define the compensation constant

$$K_C = \frac{Z_{L0} - Z_{L1}}{3Z_{L1}} \tag{11.113}$$

and where I_{GR} is the total residual current at R.

$$I_{GR} = 3C_0 I_{a0} \tag{11.114}$$

The previous analysis tacitly assumes that the transmission line on both sides of the fault point is completely transposed. This is implied by the use of symmetrical component methods. Other ground fault compensation schemes have been devised to overcome this limitation, which may be important for lines that are not transposed. The fault compensation requirements for untransposed lines may be determined by writing the voltage drop equation across a given line, from R to F, written in the phase frame of reference [6].

$$\begin{bmatrix} V_{aRF} \\ V_{bRF} \\ V_{cRF} \end{bmatrix} = \begin{bmatrix} hZ_{aa} & hZ_{ab} & hZ_{ac} \\ hZ_{ba} & hZ_{bb} & hZ_{bc} \\ hZ_{ca} & hZ_{cb} & hZ_{cc} \end{bmatrix} \begin{bmatrix} I_a \\ I_b \\ I_c \end{bmatrix} \tag{11.115}$$

If the voltage at the fault is zero, then the voltage applied to the phase a ground relay is computed as

$$\begin{aligned} V_a &= hZ_{aa}I_a + hZ_{ab}I_b + hZ_{ac}I_c \\ &= hZ_{aa}\left(I_a + \frac{Z_{ab}}{Z_{aa}}I_b + \frac{Z_{ac}}{Z_{aa}}I_c\right) \end{aligned} \tag{11.116}$$

where the quantity in parentheses represents the compensated current that must be computed. This system has been exhaustively treated in the literature to show that the error in treating untransposed lines by the method based on symmetrical components, summarized in (11.102), results in an error of less than 10% [5], [7], [8]. This type of distance compensation is sometimes referred to as *sound phase compensation* [6], [7].

11.6 DISTANCE RELAY COORDINATION

The coordination of distance relays is relatively simple, but there are certain principles that may be stated. Consider a portion of a transmission system shown in Figure 11.27. We are interested in the relay at R, which is a distance relay that must be coordinated with other distance relays downstream from its position. The relay at R, as well as the downstream relays, are coordinated by having their zone 1 protection operate with no intentional time delay, but with delayed clearing for faults detected in zones 2 and 3. In all cases the relays are assumed to be directional, or to have directional elements, as shown in Figure 11.28.

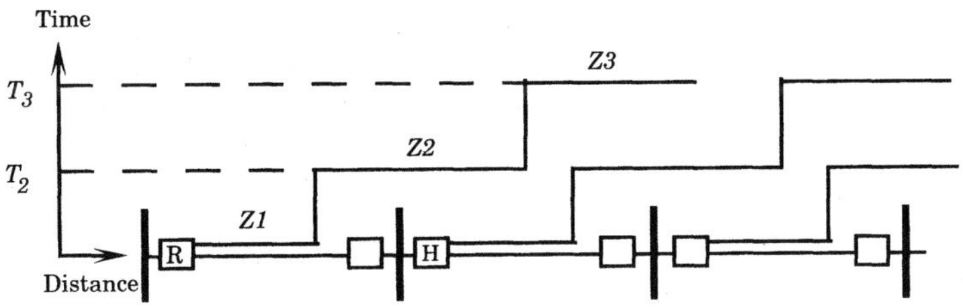

Figure 11.27 Part of a transmission system protected by distance relays.

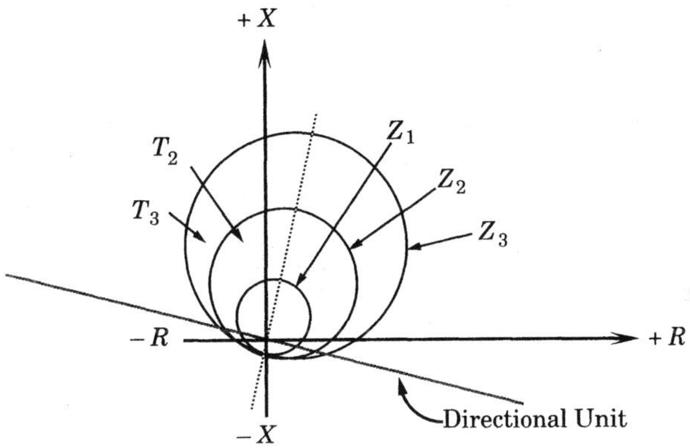

Figure 11.28 Offset mho relays at location R.

The principles of directional distance relay coordination may be stated as follows.

Zone 1 Zone 1 protection is set to trip with no intentional delay. Clearly, this zone must underreach the remote end of the line, since it is not possible to distinguish the exact location of faults near the remote bus. Zone 1 is usually set to reach 85 to 90% of the total line length for phase relays and about 75% for ground relays.

Zone 2 The primary purpose of zone 2 protection is to cover the remote end of the line that is not covered by the zone 1 protection. To do this without error, this zone must reach well beyond the remote bus. This requires a time delay in order to coordinate with the zone 1 protection of the adjacent line relay at H, as noted in Figure 11.29. About 0.25 seconds plus the adjacent breaker opening time is usually recommended to assure this coordination. If the remote relay is a time overcurrent relay rather than a distance

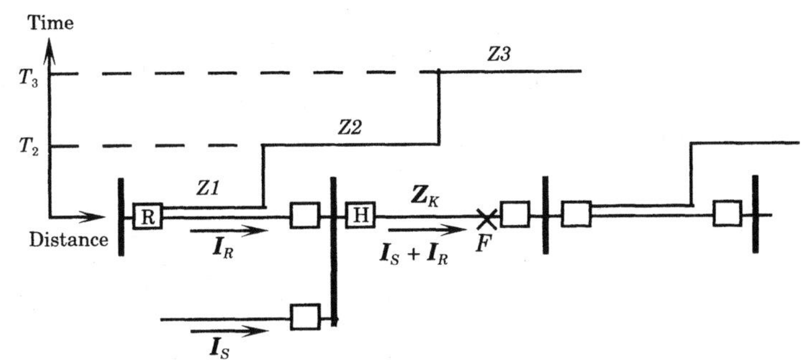

Figure 11.29 Faulted system with infeed current.

relay, a longer coordination time setting is recommended. The impedance setting of zone 2 protection should be at least 120% of the protected line impedance. Zone 2 of the relay R protection should not overreach the zone 1 of the adjacent downstream relay. Figure 11.29 shows a suitable timing arrangement. The coordination of the two line relays must be separated in both the time and impedance coordinates.

Zone 3 The primary purpose of zone 3 protection is to back up the failure of the breaker at H. Should this breaker fail to open for a fault on its protected line, the breaker at R should be caused to open as a backup protection. The reach of this zone can be set to a large value, and should reach at least through the second downstream line. The time delay for zone 3 protection should be large, and is usually set at 1 to 2 seconds.

Occasionally, the zone 3 protection is arranged to look backward rather than forward. This is important for directional comparison relaying, which is discussed in the next chapter.

The system described above assumes a configuration similar to that shown in Figure 11.29. This is not always the case, since other lines may terminate at the downstream bus, and fault currents will enter through these other lines. This affects the impedance seen by the distance relay at R for faults beyond the next bus. Consider, for example, the network arrangement shown in Figure 11.29.

For the situation depicted in the figure, the relay at R does not see the total fault current and this distorts the total impedance "measured" by the relay at R for faults in the adjacent line. For the case depicted in Figure 11.29, it can be shown that the impedance seen by the relay at R is given by

$$\mathbf{Z}_R = (\mathbf{Z}_L + \mathbf{Z}_K) + \frac{\mathbf{I}_S}{\mathbf{I}_R}\mathbf{Z}_K \tag{11.117}$$

The impedance in parentheses is the actual impedance from the relay to the fault point. The additional impedance that the relay "sees" is due to the infeed current $\mathbf{I}_S$ and the error is directly proportional to the value of this infeed current. This makes the fault seem farther away than it actually is, extending the reach of the relay at R. This must be taken into account in setting the zone 2 reach of the relay. As noted above, this reach must not exceed the zone 1 reach of the relay at H. Fault analysis studies of the system will provide data to compute the possible error.

REFERENCES

[1] Anderson, P. M., *Analysis of Faulted Power Systems*, IEEE Press, Piscataway, NJ, 1995.

[2] Lewis, W. A., and L. S. Trippett, "Fundamental Basis for Distance Relaying on 3-Phase Systems," *AIEE Transactions*, 66, pp. 694–708, 1947, also reproduced in the IEEE Book, *Protective Relaying for Power Systems*, S. H. Horowitz, Ed., IEEE Press, 1980.

[3] Mason, C. R., *The Art and Science of Protective Relaying*, John Wiley & Sons, New York, 1956.

[4] Warrington, A. R. van C., *Protective Relays, Their Theory and Practice, Vol. One*, John Wiley & Sons, New York, 1962.

[5] Cook, V., *Analysis of Distance Protection*, Research Studies Press Ltd., John Wiley and Sons, Inc., New York, 1985.

[6] Davison, E.B., and A. Wright, "Some Factors Affecting the Accuracy of Distance Type Protective Equipment Under Earth Fault Conditions," *Proc. IEE*, 110, 1963, pp. 1678–1688.

[7] Adamson, C., and A. Turelli, "Errors of Sound-Phase-Compensation and Residual-Compensation Systems in Earth-Fault Distance Relaying," *Proc. IEE*, 112, 1965, pp. 1369–1382.

[8] Westinghouse Electric Corporation, *Applied Protective Relaying*, Relay-Instrument Division, Newark, NJ. 1957.

[9] General Electric Company, Instructions, Directional Distance (Reactance Relay), Types GCX51A and GCX51B, GEI-98328E, 1980.

PROBLEMS

11.1 Perform the algebraic analysis to verify the entries for the three-phase fault in Table 11.1.

11.2 For the three-phase fault, determine the following:
(a) The ratio of $\boldsymbol{I}_{R1}$ to $\boldsymbol{I}_{a1}$.
(b) The ratio of $\boldsymbol{I}_{S1}$ to $\boldsymbol{I}_{a1}$.
(c) The ratio of $\boldsymbol{I}_{R1}$ to $\boldsymbol{I}_{S1}$.
Make these computations using the original fault equivalent as well as the reduced network shown in Figure 11.4(*a*) as well as 11.4(*b*).

11.3 Perform the algebraic analysis to verify the entries for the phase-to-phase fault in Table 11.2.

11.4 Construct a new Table 11.2 where the complex constants $\boldsymbol{C}_1$ and $\boldsymbol{C}_2$ are equal.

11.5 Perform the algebraic analysis to verify the entries for the one-line-to-ground fault in Table 11.3.

11.6 Use the parameters $\boldsymbol{d}_0$, $\boldsymbol{d}_1$, and $\boldsymbol{d}_2$ to simplify Tables 11.1 and 11.2.

11.7 Perform the algebraic analysis to verify the entries for the two-line-to-ground fault in Table 11.5.

11.8 Construct new tables similar to Tables 11.3 and 11.5, but assuming that $\boldsymbol{C}_1$ and $\boldsymbol{C}_2$ are equal.

11.9 Consider a three-phase fault with delta-connected current and voltage transformers, as shown in Figure 11.14. Examine the limiting conditions on $\boldsymbol{C}_1$, i.e., with $h = 0$ or $h = 1$, and determine the constraints on the system parameters that will justify an assumption that $\boldsymbol{C}_1 = 1$ or $\boldsymbol{C}_1 = 0$.

11.10 Consider a transmission line with distance relays for phase faults and with the relays set to reach 90% of the line length. Both current and voltage transformers are delta connected. An arcing three-phase fault occurs at $h = 0.85$ with a fault resistance of 1.0 ohm. The constant $\boldsymbol{C}_1 = 0.1$. Determine the relay response if
(a) $\boldsymbol{Z}_L = 0 + j10\ \Omega$
(b) $\boldsymbol{Z}_L = 0 + j100\ \Omega$

11.11 Consider a phase-to-phase fault with delta connected current and voltage transformers. Verify the relay apparent impedance shown in Tables 11.6 and 11.8 for the three relays for this type of fault.

11.12 It is proposed that, for the case of three-phase faults and using delta-connected current and voltage transformers, it might be satisfactory to use the approximation $\boldsymbol{C}_1 = 1 - h$. Examine this proposition and determine under what conditions this approximation might be valid.

11.13 Verify Tables 11.7 and 11.9 entries for the one-line-to-ground fault.

11.14 Verify Tables 11.7 and 11.9 entries for the two-line-to-ground fault.

11.15 Compute the impedance seen by the relay at *R* for all types of faults if the relay voltage transformers are connected in wye and the current transformers are also connected in wye.

11.16 Compute the impedance seen by the relay at *R* for three-phase and phase-to-phase faults if the relay voltage transformers are connected in delta, but the current transformers are connected in wye. Use a 90° connection, i.e., use phase *c* current with the line-to-line voltage between phases *a* and *b*, etc.

11.17 Consider a set of three-phase distance relays that are supplied with delta currents and line-to-line voltages, exactly as in Figure 11.13. Verify the impedances computed by the relays when $\mathbf{C}_1 = \mathbf{C}_2$ as given in Table 11.5 for
(a) Three-phase faults
(b) Line-to-line faults

11.18 Analyze three-phase and line-to-line faults where the relays supplied with delta currents and phase-to-neutral voltages.

11.19 Solve example 11.1 by computing the parameters given in Table 11.2 instead of using the derived formulas given in Tables 11.6 and 11.8.

11.20 Consider the distance relaying system described in example 11.1. For this system, compute the impedances seen by the three relays for a one-line-to-ground fault on phase a, located at the same distance from the relay as described in the example. The following zero-sequence system parameters are given:

$$
\begin{aligned}
R_{L0} &= 0.218240 & R_{U0} &= 0.006\\
X_{L0} &= 0.958621 & X_{U0} &= 0.012\\
R_{S0} &= 0.004 & R_{E0} &= 0.40\\
X_{S0} &= 0.010 & X_{E0} &= 2.0
\end{aligned}
$$

11.21 Compute relay apparent impedances for a line-to-line fault on a system with the following data. Assume the relay connections use delta currents and line voltages. Make the computation by solving for the entries in right-hand column of Table 11.3 and using the following system data:

$$
\begin{aligned}
\mathbf{Z}_{S1} &= 0.020 + j0.10 & \mathbf{Z}_{U1} &= 0.010 + j0.10\\
\mathbf{Z}_{S2} &= 0.025 + j0.18 & \mathbf{Z}_{U2} &= 0.015 + j0.12\\
\mathbf{Z}_{S0} &= 0.100 + j0.40 & \mathbf{Z}_{U0} &= 0.030 + j0.30\\
\mathbf{Z}_{E1} &= 0.05 + j0.55 & \mathbf{Z}_{F} &= 0.05 + j0.1\\
\mathbf{Z}_{E2} &= 0.05 + j0.55 & \mathbf{Z}_{L1} &= 0.022 + j0.45 = \mathbf{Z}_{L2}\\
\mathbf{Z}_{E0} &= 0.15 + j1.50 & \mathbf{Z}_{L0} &= 0.50 + j1.39
\end{aligned}
$$

11.22 Use the data from problem 11.21 to find the relay apparent impedance using delta currents and line voltages for a one-line-ground fault.

11.23 Consider a 345 kV transmission line 100 km long with the following basic parameters:

$$
\begin{aligned}
r &= 0.034\ \Omega/\text{km}\\
x &= 0.386\ \Omega/\text{km}
\end{aligned}
$$

The source impedances at the two ends of the line are as follows:

$$
\begin{aligned}
\mathbf{Z}_S &= 0.3 + j0.8\ \Omega\\
\mathbf{Z}_U &= 0.2 + j0.6\ \Omega
\end{aligned}
$$

Compute the following for a three-phase fault located 90 km from the relay
(a) The impedances $\mathbf{Z}_{m1}$ and $\mathbf{Z}_{n1}$
(b) The complex constant $\mathbf{C}_1$
(c) The impedance computed by the relay for a fault resistance of 5 Ω.
(d) Find the relay operation if the distance characteristic is
1. A mho type relay characteristic with diameter equal to the line impedance and inclined at the line impedance angle.
2. A mho type relay characteristic with diameter equal to the line impedance and inclined at an angle of 60 degrees.
3. An impedance relay characteristic with a radius equal to the line impedance magnitude.

11.24 Repeat problem 11.23 for the following conditions:

(a) The three-phase fault is located at $h = 0.9$.

(b) Estimate the maximum fault resistance that can be tolerated and still fall within the relay threshold the two different types of mho characteristics of problem 11.23.

11.25 Consider the distance protection of a transmission line, where the total line impedance is given as

$$\mathbf{Z}_L = 5 + j60\ \Omega$$

The maximum likely fault resistance for close-in faults is computed to be 2 ohms, and for end-of-line faults this resistance is 5 ohms. This gives the protection region in the form of a quadrilateral, similar to that shown in Figure 7.19. Each corner of this impedance quadrilateral represents a point in the Z plane that bounds the estimated values of impedance seen from the relay location. However, depending on the type of fault and the relay viewing that fault, these four points appear to the different relays to lie in different places in the Z plane seen by that relay.

Consider a relay connection that uses delta currents and line-to-line voltages, as shown in Figure 11.14. Compute the location of the quadrilateral seen by all three phase relays for a three phase fault at distances of $h = 0.1$ and $h = 0.9$. Assume that the source impedances are both equal to $0.0 + j0.1$ ohms.

11.26 Repeat problem 11.25 if the fault is at the center of the line and with the fault resistance falling between 0 and 3 ohms. Find the impedance seen by relays *a-b*, *b-c*, and *c-a* for three-phase faults and for a line-to-line fault between phases *b* and *c*. Assume that the negative-sequence source impedance is equal to that of the positive-sequence network.

11.27 Derive a method for finding the relay impedance of the compensated ground distance relay in terms of the parameters used for the phase distance relay, which are expressed in terms of the parameter $\boldsymbol{K}$.

11.28 Calculate the apparent impedance by the phase a relay for a 1LG fault where we specify the following data:

$\mathbf{Z}_F = 0 + j0$	$\mathbf{Z}_{L1} = 2.3409 + j23.523$	$\mathbf{Z}_{L0} = 13.775 + j89.417$
$\mathbf{Z}_{S1} = 0.15 + j1.4$	$\mathbf{Z}_{S2} = 0.22 + j2.0$	$\mathbf{Z}_{S0} = 0.75 + j5.30$
$\mathbf{Z}_{U1} = 0.10 + j1.0$	$\mathbf{Z}_{U2} = 0.15 + j1.5$	$\mathbf{Z}_{U0} = 1.20 + j8.00$
$\mathbf{Z}_{E1} = 0.20 + j2.0$	$\mathbf{Z}_{E2} = 0.20 + j2.0$	$\mathbf{Z}_{E0} = 2.00 + j14$

12

Transmission Line Mutual Induction

12.1 INTRODUCTION

The mutual induction between parallel current-carrying wires complicates the network analysis and the setting of line protective relays, particularly for ground-fault relays. The subject has been explored to considerable lengths in the technical literature, but still remains somewhat a mystery to some engineers. This chapter is devoted to a discussion and explanation of the phenomena surrounding mutual induction, and how it affects the operation of protective systems.

The basic concept of mutual induction is well known and appears in many elementary texts in electrical engineering [1], [2], in books on electric power transmission and power system analysis [3–9], and in most books on power system protection [10–13]. The basic configuration leading to mutual coupling of lines occurs when current-carrying wires are in parallel for significant distances. The current-carrying wire produces a field of flux that links parallel conductors, thereby inducing a voltage in those nearby circuits. We can write the basic equation of flux linkages from field theory, for the affect on circuit 2 due to a current in circuit 1, as follows.

$$\lambda_{21} = \oint_{c_2} \boldsymbol{A} \cdot \boldsymbol{ds}_2 > 0 \quad \text{Wb turn} \tag{12.1}$$

where $\boldsymbol{A}$ = magnetic vector potential $= \dfrac{\mu_0 I_1}{4\pi}\displaystyle\int_{c_1} \frac{1}{r} ds_1$ Wb/m

ds_1 = element of length along circuit 1

ds_2 = element of length along circuit 2

From the mutual flux linkages, we can write the mutual inductance between the two circuits as

$$M = \frac{\lambda_{21}}{I_2} \quad H \tag{12.2}$$

It is convenient to think of the mutual coupling between two current-carrying wires as a one-turn air-core transformer, as shown in Figure 12.1. The currents are defined in relation to the marked terminals of the individual wires, and this marking determines the positive direction of current flow. If we apply a voltage $V_{aa'}$ to wire a-a' of polarity shown, the resulting current in wire a-a' will enter the marked terminal, thereby setting up a flux ϕ_{ba} (the flux linking coil b due to current I_a) that is increasing in the direction shown. Then, by Lenz's law, a flux ϕ_{ab} will be established to oppose ϕ_{ba}, which requires that a current I_b be induced to flow out of the b terminal. Therefore, the b terminal is the marked terminal for circuit b.

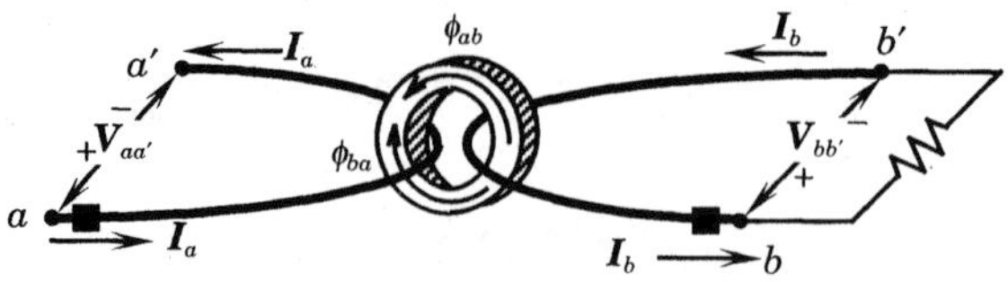

Figure 12.1 An air-core transformer equivalent of coupled wires.

We can also visualize the two wires from a circuit point of view, as shown in Figure 12.2. From Figure 12.2, we can write the circuit voltage equations as follows.

$$\begin{bmatrix} V_{aa'} \\ V_{bb'} \end{bmatrix} = \begin{bmatrix} Z_a & Z_m \\ Z_m & Z_b \end{bmatrix} \begin{bmatrix} I_a \\ I_b \end{bmatrix} \tag{12.3}$$

These equations express the voltage drop along the wires in the defined direction of current. Note that the mutual impedance is equal in the two equations. Figures 12.1 and 12.2 provide two different ways of viewing the mutual coupling, one from the viewpoint of the magnetic fields and the other from the viewpoint of the electric circuit.

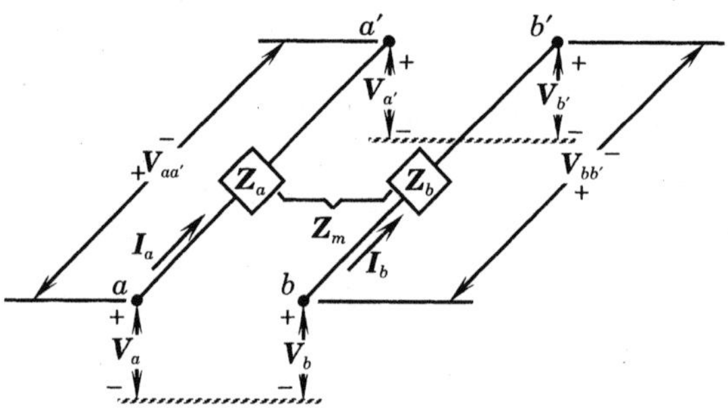

Figure 12.2 Two circuits with mutual coupling.

The same concept can be extended to any number of wires in parallel, with mutual impedances defined between each pair of wires. The computation of these impedances is the subject of the next section.

12.2 LINE IMPEDANCES

This section reviews the method of computing the phase impedances for a transmission line.

12.2.1 Self- and Mutual Impedance

Consider a three-phase transmission line a specified distance above the earth and above the ground-return path, which is specified according to the method of Carson [14]. The circuit configuration is shown in Figure 12.3.

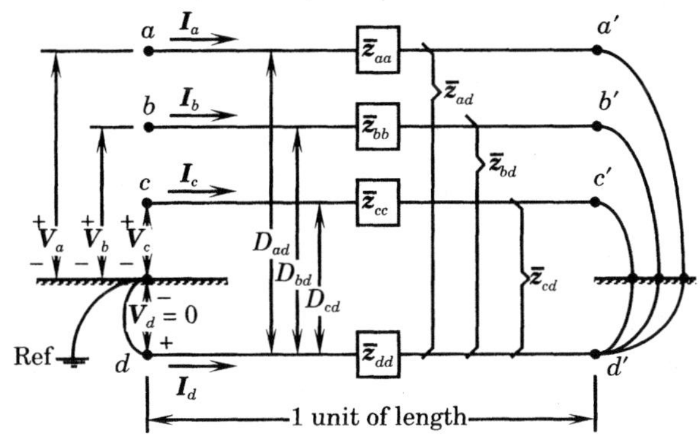

Figure 12.3 Specification for the measurement of impedances.

The impedances shown in Figure 12.3 are called primitive impedances and are combined to determine the total phase impedances of the line [9]. We can write the voltage equation of the three-phase line of Figure 12.3, as follows.

$$\begin{bmatrix} V_a \\ V_b \\ V_c \end{bmatrix} = \begin{bmatrix} z_{aa} & z_{ab} & z_{ac} \\ z_{ba} & z_{bb} & z_{bc} \\ z_{ca} & z_{cb} & z_{cc} \end{bmatrix} \begin{bmatrix} I_a \\ I_b \\ I_c \end{bmatrix} \tag{12.4}$$

where

$$\begin{aligned} z_{ik} &= (r_a + r_d) + j\omega k \ln \frac{D_e}{D_s} \quad \Omega/\text{unit length}, \ i = k \\ &= r_d + j\omega k \ln \frac{D_e}{D_{ik}} \quad \Omega/\text{unit length}, \ i \neq k \end{aligned} \tag{12.5}$$

These expressions can be derived in terms of the primitive impedances pictured in Figure 12.3 and defined in [9]. This review is left as an exercise.

The parameter multiplying the imaginary part of the impedance is a function of the system of units used in the calculations. Table 12.1 gives the numerical value of this coefficient for common system frequencies, and for both natural and base 10 logarithms.

TABLE 12.1 Inductance Multiplying Constants [8]

Constant ωk	Unit of length	Natural Logarithm (ln)	Base 10 Logarithm ($\log_{10}$)
$f = 50$ Hz	km	0.06283	0.1446
	miles	0.10111	0.2328
$f = 60$ Hz	km	0.07539	0.1736
	miles	0.12134	0.2794

Note that the current in each wire of the three-phase system returns to the source end through a fictitious conductor *d* below the surface of the earth. This conductor is assumed to have a geometric mean radius of unity and to lie at a depth that depends on the system frequency and the resistivity of the earth [6], [7], [14]. The value of the fictitious conductor resistance is a function of frequency and is given by

$$r_d = \begin{cases} 9.869 \times 10^{-3} f & \Omega/\text{km} \\ 1.588 \times 10^{-4} f & \Omega/\text{mile} \end{cases} \tag{12.6}$$

Solving (12.6), the ground conductor resistance is computed to be

$$\begin{aligned} r_d &= \left.\begin{cases} 0.04935 & \Omega/\text{km} \\ 0.07940 & \Omega/\text{mile} \end{cases}\right\} \text{ at 50 Hz} \\ r_d &= \left.\begin{cases} 0.05921 & \Omega/\text{km} \\ 0.09528 & \Omega/\text{mile} \end{cases}\right\} \text{ at 60 Hz} \end{aligned} \tag{12.7}$$

Carson defined a distance parameter D_e as a function of the earth resistivity and the system frequency.

$$D_e = k_D \sqrt{\frac{\rho}{f}} \quad \text{units of length} \tag{12.8}$$

where the constant k_D is approximately 2160 or 660 for units of length of feet or meters, respectively.

The value of D_e depends on ρ, the resistivity of the soil. Table 12.2 gives a range of values. When actual earth resistivity data is unavailable, it is not uncommon to assume the earth resistivity of 100 Ω-m, which corresponds to the values in italics in Table 12.2. Estimates of resistivities in various parts of North America are given in [6].

TABLE 12.2 D_e for Various Resistivities at 60 hertz

Return Earth Condition	Resistivity $\rho, \Omega - \text{m}$	D_e in ft @ 50 Hz	D_e in ft @ 60 Hz
Sea water	0.01–1.0	9.3–93.05	27.9–279
Swampy ground	10–100	294–930.5	882–2790
Average damp earth	*100*	*931*	*2790*
Dry earth	1000	2943	8820
Pure slate	10^7	294,300	882,000
Sandstone	10^9	2,943,000	8,820,000

If the transmission line has ground wires, the impedance matrix of (12.4) is expanded to accommodate the additional wires. However, since ground wires are grounded at every tower, the applied voltage of these wires is always zero, so the system of equations can always be reduced to the form shown in (12.4). In a similar fashion, should the three-phase line have bundled conductors for each phase, these can also be combined and reduced to exactly the form of (12.4). The method of performing these reductions is documented in the literature [9].

The impedance matrix of (12.4) is not symmetric since the distances of the different phase wires from the return ground conductor are not equal, and the distances between adjacent phase conductors are also unequal in the general case. Note that it is not possible to have both equal phase spacing and equal height of all phases simultaneously. Therefore, a nonsymmetric impedance matrix will always result. The only way that the impedance matrix can be made symmetric is through transpositions.

Consider a transmission line of length s and having three transpositions, that divide the total line length into fractions f_1, f_2, and f_3. Then we can write the voltage equation as follows.

$$\begin{bmatrix} \boldsymbol{V}_a \\ \boldsymbol{V}_b \\ \boldsymbol{V}_c \end{bmatrix} = \begin{bmatrix} z_s & z_{k1} & z_{k2} \\ z_{k1} & z_s & z_{k3} \\ z_{k2} & z_{k3} & z_s \end{bmatrix} \begin{bmatrix} \boldsymbol{I}_a \\ \boldsymbol{I}_b \\ \boldsymbol{I}_c \end{bmatrix} \tag{12.9}$$

where

$$z_s = (r_a + r_d)s + j\omega k \ln \frac{D_e}{D_s} \quad \Omega \tag{12.10}$$

and

$$\begin{aligned} z_{k1} &= r_d s + j\omega k s \left(f_1 \ln \frac{D_e}{D_{12}} + f_2 \ln \frac{D_e}{D_{23}} + f_3 \ln \frac{D_e}{D_{13}} \right) \quad \Omega \\ z_{k2} &= r_d s + j\omega k s \left(f_1 \ln \frac{D_e}{D_{13}} + f_2 \, ln \frac{D_e}{D_{12}} + f_3 \ln \frac{D_e}{D_{23}} \right) \quad \Omega \\ z_{k3} &= r_d s + j\omega k s \left(f_1 \ln \frac{D_e}{D_{23}} + f_2 \ln \frac{D_e}{D_{13}} + f_3 \ln \frac{D_e}{D_{12}} \right) \quad \Omega \end{aligned} \tag{12.11}$$

The foregoing assumes that wires a, b, and c occupy wire positions 1, 2, and 3, respectively, in the three fractional lengths. If the three fractions are all equal to one-third, then the off-diagonal terms of (12.11) are all equal. Otherwise, the impedance matrix is a nonsymmetric matrix. The degree of unbalance can be improved somewhat by having transpositions that are unequally spaced or by having only one transposition.

12.2.2 Estimation of Mutually Coupled Voltages

We can estimate the magnitude of mutually coupled voltages due only to inductive effects by a direct application of the induction equations [2]. Consider the simple single-phase system shown in Figure 12.4, where a fault current is flowing in line a and we wish to compute the voltage induced in parallel line b [15].

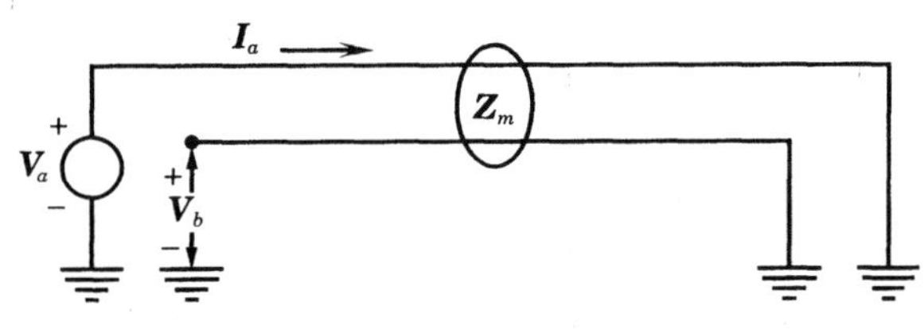

Figure 12.4 Mutual coupling of parallel single-phase lines.

We may write the voltage induced in line b as the product of the current flowing in line a and the mutual impedance between the conductors.

$$\boldsymbol{V}_b = \boldsymbol{Z}_m \boldsymbol{I}_a \tag{12.12}$$

where the voltage induced is due to the mutual inductance of the entire length of the lines. The mutual impedance between two conductors with common earth return is computed as [9]

$$\boldsymbol{Z}_m = r_d + j\omega k \ln \frac{D_e}{D_{ab}} \quad \Omega/\text{unit length} \tag{12.13}$$

Using a typical earth resistivity of 100 Ω-*m* and a system frequency of 60 hertz, we have

$$\begin{aligned} Z_m &= 0.0952 + j0.1213 \ln \frac{2790}{D_{ab}} \quad \Omega/\text{mi} \quad (D_{ab}\ \text{ft}) \\ &= 0.0592 + j0.0628 \ln \frac{931}{D_{ab}} \quad \Omega/\text{km} \quad (D_{ab}\ \text{m}) \end{aligned} \tag{12.14}$$

For example, if the distance between conductors is 20 meters, the mutual impedance is $0.059 + j0.24 \Omega/\text{km}$. Therefore, a fault current of 1000 A will induce a voltage of 250 V/km and if the line is 200 km long, the total induced voltage will be 50 kV.

The example computed above is for a single-phase line. A three-phase line has coupling among all conductors in both circuits, however, if the distances between conductors of the two lines are considered equal, the induction of the different conductors cancels and the mutual coupling is zero. However, there is always some asymmetry, even in transposed lines. From a practical perspective, the mutual coupling of the positive and negative sequences is negligible, but the zero-sequence induction can not be ignored.

Consider the system shown in Figure 12.5, where two three-phase lines are mutually coupled. Zero-sequence currents are equal and in phase in the three conductors. Therefore, we can think of replacing the three conductors by a single conductor for each of the parallel three-phase lines, which reduces the zero-sequence problem to one exactly like that of Figure 12.5. The voltage induced in line b is now three times that of Figure 12.4, since there are three currents contributing to this voltage.

$$V_b = 3 I_{a0} Z_m \tag{12.15}$$

Therefore, we define the zero-sequence mutual impedance as

$$Z_{0m} = \frac{V_{a0}}{I_{a0}} = 3 Z_m \tag{12.16}$$

where Z_m is defined by (12.14). The zero-sequence mutual impedance may be as high as 70% of the zero-sequence self-impedance when the parallel lines are on the same tower. Therefore, zero-sequence mutual coupling has a considerable effect on ground faults and on ground fault relaying.

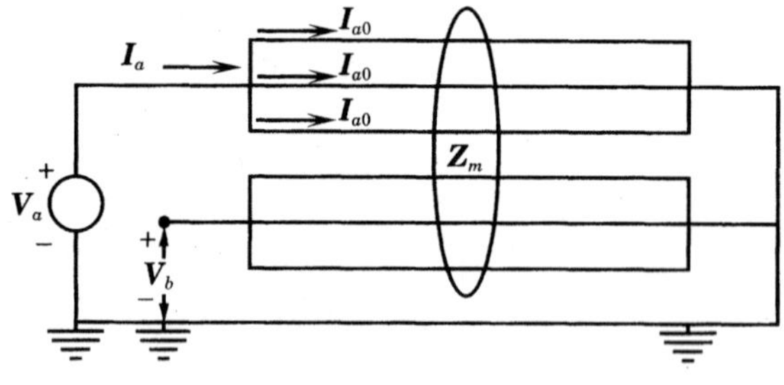

Figure 12.5 Two mutually coupled three-phase lines.

12.2.3 Example of Transmission Line Impedances

Consider two identical 500 kV transmission lines on the same right of way separated by a center-to-center horizontal distance D, as shown in Figure 12.6. Assume that the tower design for each of the lines is identical to that described in Appendix E.

The array of line conductors totals eight wires for each line, six phase wires, and two static wires. The phase wires are bundled in a horizontal arrangement. The description of the

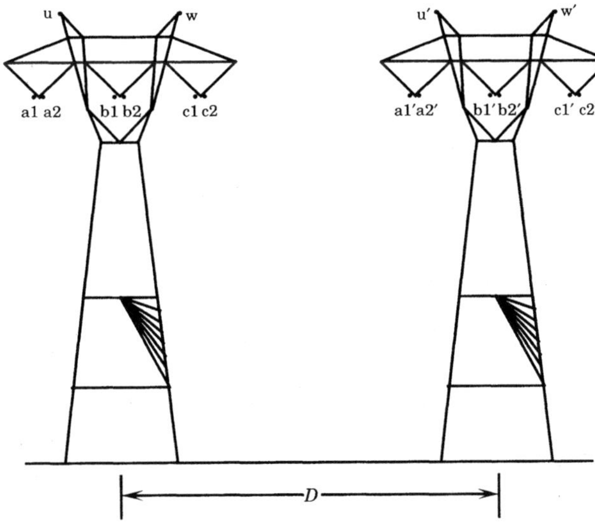

Figure 12.6 Transmission lines on a common right of way.

two lines requires a 16×16 matrix of self- and mutual inductances in an array similar to that for a single line, but twice as large and including both self- and mutual submatrices. Constructing this matrix requires knowledge of the distance from each wire to all the other 15 wires.

The same arrangement is true for the self- and mutual admittances, except that both the above-ground wires and their below-ground images must be represented in a matrix of Maxwell coefficients similar to those illustrated in Appendix E for a three-phase line, but is twice as large for the parallel line case. Construction of this matrix requires computation of the distance from each wire to all the other 15 wires, and also to all 16 image wires.

As the two lines are separated by a greater distance D, the mutual inductances and susceptances become lower in magnitude until, for large separations, the mutual coupling can be neglected.

12.2.3.1 Self- and Mutual Impedances. The self- and mutual impedances of the two parallel lines are computed in a manner similar to that for a single line, except that there are currents flowing in additional lines and all of these currents induce voltages in all the other conductors.

The unbalance in the phase impedances due to differences in spacings between pairs of wires can be minimized by transposing the lines. As desirable as transpositions may be, the fact is that they are seldom used on the higher voltage lines due to the high cost and physical difficulty of rotating the conductors while maintaining the needed interphase spacing and ground clearance. Therefore, transpositions may be regarded as a desirable goal, but one that is often not achieved.

EXAMPLE 12.1

Compute the self- and mutual line impedances for the circuits shown in Figure 12.6 and show that the difference in the phase-to-mutual impedances for the untransposed line can be balanced by transposing the conductors. Distances are provided in problem 12.6. Assume that the distance D between the center line of adjacent towers is 200 feet (about 61 m). It is also assumed that the static wires are grounded at every tower and the voltage of these wires is zero.

Solution

Table 12.3 shows the computed impedances per unit of line length for the untransposed double circuit line. It is helpful to examine the impedance matrix by concentrating on the 3×3 submatrices outlined in Table 12.3 by the lines. The 3×3 group in the upper left hand corner represents the self impedance of the line *abc*. The diagonal terms are unequal because the geometry of the conductor arrangement. The self impedances of the conductors are equal for the same locations. For example, self-impedance $\mathbf{Z}_{aa}$, which occupies the outside position in the line on the left of Figure 12.6, is equal to the self-impedance $\mathbf{Z}_{c'c'}$, which is the outside conductor of the line on the right. The matrix is symmetric because the matrix displays the impedances in the same order that they appear in the physical model depicted in Figure 12.6.

TABLE 12.3 Self- and Mutual Impedances of Untransposed Lines, Ω/mile

Phase	a	*b*	*c*	*a′*	*b′*	*c′*
a	0.18601 *j*0.93694	0.16682 *j*0.36168	0.16228 *j*0.28087	0.14276 *j*0.15541	0.14036 *j*0.13849	0.13599 *j*0.13038
b	0.16682 *j*0.36168	0.19553 *j*0.92099	0.16878 *j*0.35609	0.14770 *j*0.16853	0.14513 *j*0.14854	0.14036 *j*0.13849
c	0.16228 *j*0.28087	0.16878 *j*0.35609	0.19020 *j*0.92568	0.14977 *j*0.19334	0.14770 *j*0.16853	0.14276 *j*0.15541
a′	0.14276 *j*0.15541	0.14770 +*j*0.16853	0.14977 *j*0.19334	0.19020 *j*0.92568	0.16878 *j*0.35609	0.16228 *j*0.28087
b′	0.14036 *j*0.13849	0.14513 *j*0.14854	0.14770 *j*0.16853	0.16878 *j*0.35609	0.19553 *j*0.92099	0.16682 *j*0.36168
c′	0.13599 *j*0.13038	0.14036 *j*0.13849	0.14276 *j*0.15541	0.16228 *j*0.28087	0.16682 *j*0.36168	0.18601 *j*0.93694

If we assume that the two transmission lines are completely transposed, then the impedance matrix is changed to that shown in Table 12.4. The effect of the transpositions is evident in both the self- and mutual submatrices.

TABLE 12.4 Self- and Mutual Impedances of Transposed Lines, Ω/mile

Phase	a	*b*	*c*	*a′*	*b′*	*c′*
a	0.18964 *j*0.92843	0.16644 *j*0.33262	0.16644 *j*0.33262	0.14362 *j*0.15307	0.14362 *j*0.15666	0.14362 *j*0.15595
b	0.16644 *j*0.33262	0.18964 *j*0.92843	0.16644 *j*0.33262	0.14362 *j*0.15595	0.14362 *j*0.15307	0.14362 *j*0.15666
c	0.16644 *j*0.33262	0.16644 *j*0.33262	0.18964 *j*0.92843	0.14362 *j*0.15666	0.14362 *j*0.15595	0.14362 *j*0.15307
a′	0.14362 *j*0.15307	0.14362 *j*0.15595	0.14362 *j*0.15666	0.18964 *j*0.92843	0.16642 *j*0.33262	0.16644 *j*0.33262
b′	0.14362 *j*0.15666	0.14362 *j*0.15307	0.14362 *j*0.15595	0.16644 *j*0.33262	0.18964 *j*0.92843	0.16644 *j*0.33262
c′	0.14362 *j*0.15595	0.14362 *j*0.15666	0.14362 *j*0.15307	0.16644 *j*0.33262	0.16644 *j*0.33262	0.18964 *j*0.92843

Note that the phase impedance matrix has self-impedance partitions that are symmetric about the major diagonal, but the mutual impedance partitions exhibit a different type of symmetry, which is described in Table 12.5, where the symbols α, β, and γ are used to describe the complex impedance quantities in each matrix cell. The diagonal elements are equal, but the off diagonal elements have this unique triangular symmetry. This symmetry can be confirmed by noting the various wire positions in each section of the transposition cycle. It is also noted that the mutual impedance partitions between the primed and unprimed lines are the transpose of those between the unprimed and primed lines. In other words, the upper right partition of the impedance matrix is the transpose and the lower left partition. Finally, we note that all resistances are equal in each of the mutual impedance submatrices.

TABLE 12.5
Mutual Symmetry

α	β	γ
γ	α	β
β	γ	α

The sequence impedances are computed from the phase impedance matrices shown in Tables 12.3 and 12.4 by similarity transformation, as follows. First, we can write the equations in the *a-b-c* frame of reference as

$$\begin{bmatrix} \mathbf{V}_{abc} \\ \mathbf{V}_{a'b'c'} \end{bmatrix} = \begin{bmatrix} \mathbf{Z}_{aa} & \mathbf{Z}_{aa'} \\ \mathbf{Z}_{a'a} & \mathbf{Z}_{a'a'} \end{bmatrix} \begin{bmatrix} \mathbf{I}_{abc} \\ \mathbf{I}_{a'b'c'} \end{bmatrix} \tag{12.17}$$

Then, transforming to the symmetrical component frame of reference, we write

$$\begin{aligned} \begin{bmatrix} \mathbf{V}_{012} \\ \mathbf{V}_{0'1'2'} \end{bmatrix} &= \begin{bmatrix} \mathbf{A}^{-1} & 0 \\ 0 & \mathbf{A}^{-1} \end{bmatrix} \begin{bmatrix} \mathbf{V}_{abc} \\ \mathbf{V}_{a'b'c'} \end{bmatrix} \\ &= \begin{bmatrix} \mathbf{A}^{-1} & 0 \\ 0 & \mathbf{A}^{-1} \end{bmatrix} \begin{bmatrix} \mathbf{Z}_{aa} & \mathbf{Z}_{aa'} \\ \mathbf{Z}_{a'a} & \mathbf{Z}_{a'a'} \end{bmatrix} \begin{bmatrix} \mathbf{I}_{abc} \\ \mathbf{I}_{a'b'c'} \end{bmatrix} \\ &= \begin{bmatrix} \mathbf{A}^{-1}\mathbf{Z}_{aa}\mathbf{A} & \mathbf{A}^{-1}\mathbf{Z}_{aa'}\mathbf{A} \\ \mathbf{A}^{-1}\mathbf{Z}_{a'a}\mathbf{A} & \mathbf{A}^{-1}\mathbf{Z}_{a'a'}\mathbf{A} \end{bmatrix} \begin{bmatrix} \mathbf{I}_{012} \\ \mathbf{I}_{0'1'2'} \end{bmatrix} \end{aligned} \tag{12.18}$$

Carrying out the indicated computation for the untransposed coupled lines described in Table 12.3, we have the results shown in Table 12.6.

TABLE 12.6 Sequence Impedances of Untransposed Lines, Ω/mile

Sequence	0	1	2	0′	1′	2′
0	0.52250 *j*1.59363	0.01828 –*j*0.00992	–0.02567 –*j*0.00422	0.43084 *j*0.46570	0.01372 *j*0.02172	–0.00434 *j*0.02985
1	–0.02567 –*j*0.00422	0.02462 *j*0.59499	–0.04748 –*j*0.02883	0.01195 –*j*0.02275	0.00073 –*j*0.00292	0.00283 –*j*0.00166
2	0.01828 –*j*0.00992	0.04855 *j*0.02666	0.02462 *j*0.59499	–0.02368 –*j*0.01868	–0.00285 –*j*0.00162	–0.00093 –*j*0.00343
0′	0.43084 *j*0.46570	–0.02368 –*j*0.01868	0.01195 –*j*0.02275	0.52250 *j*1.59363	0.01649 –*j*0.02012	–0.01773 –*j*0.01088
1′	–0.00434 *j*0.02985	–0.00093 –*j*0.00343	0.00283 –*j*0.00166	–0.01773 –*j*0.01088	0.02462 *j*0.59499	–0.04736 *j*0.02871
2′	0.01372 *j*0.02172	–0.00285 –*j*0.00162	0.00073 –*j*0.00292	0.01649 –*j*0.02012	0.04871 *j*0.02671	0.02462 *j*0.59499

It is important to note that the positive- and negative-sequence mutual impedances are both very small compared to the self-impedances. For example, the positive-sequence mutual impedance is 0.00073 −j 0.00292, compared to the self impedance of 0.02462 +j 0.59499, or the mutual impedance is about one-half of 1% of the self-impedance. The mutual impedances in both the positive and negative sequences are usually ignored. The zero-sequence mutual impedance, on the other hand, is about one-third of the zero-sequence self-impedance and should not be ignored. This means that the mutual coupling between lines under ground fault conditions will require special consideration to make sure that the protective relays operate correctly in the presence of mutual coupling.

The mutual impedances between sequence networks is another concern in the untransposed case. These couplings carry both positive and negative signs, and some of them are greater in magnitude than the coupling between like sequences. Fortunately, their magnitudes are small enough that there is little error in omitting these couplings between sequence networks. This is fortunate, since there is little advantage to symmetrical component representation of the system if these couplings must be considered.

The sequence impedances for the transposed line are shown in Table 12.7. The transposition of the lines removes all mutual coupling between sequence networks. This is important, since the use of symmetrical components is greatly complicated by mutual coupling. The results shown in Table 12.7 shows coupling between the two lines, but only of the same sequence. In other words, all off diagonal elements of the matrix 3×3 partitions are zero, both for the self- and the mutual partitions.

TABLE 12.7 Sequence Impedance of the Transposed Lines, Ω/mile

Sequence	**0**	**1**	**2**	**0′**	**1′**	**2′**
0	0.52251 j1.59367	0.0 j0.0	0.0 j0.0	0.43085 j0.46568	0.0 +j0.0	0.0 j0.0
1	0.0 j0.0	0.02320 j0.59581	0.0 j0.0	0.0 j0.0	0.00062 −j0.00324	0.0 j0.0
2	0.0 0.0	0.0 j0.0	0.02320 j0.59581	0.0 j0.0	0.0 j0.0	−0.00062 −j0.00324
0′	0.43085 j0.46568	0.0 +j0.0	0.0 j0.0	0.52251 j1.59367	0.0 j0.0	0.0 j0.0
1′	0.0 j0.0	−0.00062 −j0.00324	0.0 j0.0	0.0 j0.0	0.02320 j0.59581	0.0 j0.0
2′	0.0 j0.0	0.0 j0.0	0.00062 −j0.00324	0.0 j0.0	0.0 j0.0	0.02320 j0.59581

■

We conclude that the mutual coupling between the positive and negative sequences for the two lines is very small and can be neglected without introducing significant errors. In contrast, the zero-sequence mutual impedance is about 100 times greater and should not be ignored. In fact, the zero-sequence mutual terms are similar in magnitude to the positive and negative self-impedance terms. It is concluded, from this study of the numerical values of a typical parallel line configuration, that it is safe to ignore mutual coupling of the positive and negative sequences, but mutual coupling of the zero-sequence network should be preserved in all computations. This is in agreement with the observations and conclusions of others [15].

12.2.3.2 Self-and Mutual Admittances. The self-and mutual admittances for the parallel lines of Figure 12.6 can also be computed in a straightforward manner, using the method of images [9].

EXAMPLE 12.2

Compute the self- and mutual admittances for the transmission lines described in Figure 12.6 using the numerical data given in problem 12.6.

Solution

The results are shown in Tables 12.8 and 12.9 for the phase and sequence admittances, respectively, where both of the lines are without transpositions. In both cases, the small real part of the admittance per unit length of the lines is ignored. The phase susceptance matrix of Table 12.8 exhibits the same symmetry as the phase impedance matrix, having symmetric self-admittance partitions and mutual-admittance partitions with symmetry exactly like that shown in Table 12.5.

TABLE 12.8 Phase Admittances of Untransposed Lines, μS/mile

Phase	a	b	c	a′	b′	c′
a	+j6.09766	−j1.21880	−j0.43820	−j0.06770	−j0.03512	−j0.02985
b	−j1.21880	+j6.42580	−j1.20887	−j0.10060	−j0.04708	−j0.03512
c	−j0.43820	−j1.20887	+j6.12635	−j0.23125	−j0.10060	−j0.06770
a′	−j0.06770	−j0.10060	−j0.23125	+j6.12635	−j1.20887	−j0.43820
b′	−j0.03512	−j0.04708	−j0.10060	−j1.20887	+j6.42580	−j1.21880
c′	−j0.02985	−j0.03512	−j0.06770	−j0.43820	−j1.21880	+j6.09766

TABLE 12.9 Sequence Admittances of Untransposed Lines, μS/mile

Sequence	0	1	2	0′	1′	2′
0	0.0	−0.13889	0.13889	0.0	−0.01447	0.01447
	j4.30602	j0.06732	j0.06732	−j0.23834	−j0.08061	−j0.08061
1	0.13889	0.0	0.53712	−0.06257	−0.02033	−0.03024
	j0.06732	j7.17190	−j0.31305	j0.05283	−j0.02793	j0.01746
2	−0.13889	−0.53712	0.0	0.06257	0.03024	0.02033
	j0.06732	−j0.31305	j7.17190	j0.05283	j0.01746	−j0.02793
0′	0.0	0.06257	−0.06257	0.0	−0.12775	0.12775
	−j0.23834	+j0.05283	j0.05283	j4.30602	j0.08663	j0.08663
1′	0.01447	0.02033	−0.03024	0.12775	0.0	0.53967
	−j0.08061	−j0.02793	j0.01746	j0.08663	j7.17190	−j0.30864
2′	−0.014470	0.03024	−0.02033	−0.12775	−0.53967	0.0
	−j0.08061	j0.01746	−j0.02793	j0.08663	j0.30864	j7.17190

Performing a similarity transformation on the phase admittance matrix gives the sequence admittance matrix of the untransposed line, shown in Table 12.9. The symmetry of the sequence admittances is the same as that shown for the sequence impedances, however, the signs are not the same.

The mutual impedances create positive voltage drops in response to currents flowing in the positive direction, which is the result of positive mutual impedance terms in the mutual impedance partitions. In a similar manner, positive sequence linear charge densities on one line produce positive voltages on the neighboring line. However, this is not true for the zero sequence mutual coupling, which carries a negative sign. This means that a positive linear charge density on one line creates a negative zero-sequence voltage on the neighboring line. Moreover, the zero-sequence mutual admittance terms are much greater in magnitude than the positive and negative sequence terms. This agrees with conclusions that the positive and negative sequence coupling can be neglected, but zero-sequence coupling should be retained in computations [15]. Similar calculations with both of the transmission lines being transposed give the results shown in Tables 12.10 and 12.11.

TABLE 12.10 Phase Admittances of Transposed Lines, μS/mile

Phase	a	b	c	a′	b′	c′
a	+j6.13901	−j0.92062	−j0.92062	−j0.05202	−j0.09721	−j0.08638
b	−j0.92062	+j6.13901	−j0.92062	−j0.08638	−j0.05202	−j0.09721
c	−j0.92062	−j0.92062	+j6.13901	−j0.09721	−j0.08638	−j0.05202
a′	−j0.05202	−j0.08638	−j0.09721	+j6.13901	−j0.92062	−j0.92062
b′	−j0.09721	−j0.05202	−j0.08638	−j0.92062	+j6.13901	−j0.92062
c′	−j0.08638	−j0.09721	−j0.05202	−j0.92062	−j0.92062	+j6.13901

TABLE 12.11 Sequence Susceptances of Transposed Lines, μS/mile

Sequence	0	1	2	0′	1′	2′
0	0.0 j4.29778	0.0 j0.0	0.0 j0.0	0.0 −j0.23560	0.0 −j0.0	0.0 −j0.0
1	0.0 j0.0	0.0 j7.05963	0.0 −j0.0	0.0 j0.0	−0.00938 −j0.03977	0.0 j0.0
2	0.0 j0.0	0.0 −j0.0	0.0 j7.05963	0.0 j0.0	0.0 j0.0	0.00938 −j0.03977
0′	0.0 −j0.23560	0.0 +j0.0	0.0 +j0.0	0.0 j4.29778	0.0 j0.0	0.0 +j0.0
1′	0.0 −j0.0	0.00938 −j0.03977	0.0 j0.0	0.0 j0.0	0.0 j7.05962	0.0 −j0.0
2′	0.0 −j0.0	0.0 j0.0	−0.00938 −j0.03977	0.0 j0.0	0.0 j0.0	0.0 j7.05962

■

The effect of the transpositions is to remove the coupling between unlike sequence networks. As for the coupling between like sequence networks, only the zero-sequence networks have mutual coupling large enough to warrant consideration.

The sequence impedance and admittance matrices of Tables 12.6 through 12.11 provide the raw data required for computation of faults on the mutually coupled transmission lines. EHV lines, such as those described here, have large positive-sequence self-susceptances that should be retained in computation. Both the self- and mutual susceptance of the zero sequence should be retained. Note that these computed values assume that the lines are completely transposed. If this is not true, the diagonal character of both the self- and mutual impedance and admittance submatrices is destroyed. This means that the sequence networks are mutually coupled, which introduces complexities that one would like to avoid. Therefore, complete transposition of the lines is often assumed, even when this condition is not exactly true.

12.3 EFFECT OF MUTUAL COUPLING

One of the most difficult problems in transmission line protection is the problem posed by mutual coupling. This usually occurs where lines are on multiple circuit towers, or are on the same right of way. The lines need not be of the same voltage in order to experience mutual coupling, and some coupling exists even for lines that are separated by 100 meters or more.

When mutual coupling exists, the normal polarization of ground currents for the detection of ground faults may be in error. The same is true for ground distance relays, which may sense the direction and distance to the fault incorrectly. The analytical treatment of this subject is discussed in Chapter 14, but we present the concept of mutual coupling here in order to introduce the problems associated with ground directional or ground distance relaying.

The problems associated with correct sensing of the ground fault in mutually coupled lines are caused by zero-sequence voltage inversion. Some protection engineers have discovered, to their dismay, that many of their transmission lines are subject to this problem. Mutual coupling can cause errors in either current polarized protection or in ground distance protection. In some systems, the problem can be solved by various remedial measures, but in other cases, the only recourse is to accept occasional false tripping and plan a strategy of prompt restoration of the tripped line.

12.3.1 Selecting a Reference Phasor

The purpose of this section is to introduce the subject of zero-sequence voltage inversion, using simple system configurations of Figure 12.7 to explain the phenomenon.[1] First, we establish a standard method of measuring the zero-sequence voltage along a transmission line [16]. This can be done using the simple radial system shown in Figure 12.7. Part (*a*) of the figure shows the physical system, which is a radial line connected to a voltage source, with a one-line-to-ground fault at the point F. Part (*b*) shows the sequence network connections for a one-line-to-ground fault at F. Part (*c*) graphically depicts the zero-sequence voltage along the line, as a function of distance from the source voltage. We compute the zero-sequence voltage at any point along the line as

$$\boldsymbol{V}_{a0} = -\boldsymbol{Z}_0 \boldsymbol{I}_{a0} \tag{12.19}$$

The zero-sequence impedance is usually highly reactive, which gives the complex impedance an angle of almost 90 degrees. If we select the current phasor as the reference, this means that the product term $+\boldsymbol{Z}_0\boldsymbol{I}_{a0}$ leads the current by almost 90 degrees, falling in the second quadrant and, because of the minus sign in (12.19), the zero-sequence voltage at the fault will lie in the third quadrant. If the impedance is a pure reactance, as shown in Figure 12.7(*d*), the zero-sequence voltage points straight down. The fault current flows through the entire transmission line impedance, from the source to the fault. At the source, the zero-sequence voltage is zero, and it grows linearly with distance to the fault, as shown in Figure 12.7(*c*), but is always negative. Therefore, with the zero-sequence current as the reference, the voltage is always negative. If this voltage can be shown to be positive for some special system condition, this is not the normal voltage–current relationship and indicates that the fault directional sensing equipment will probably fail. In the discussion that follows, we will assume that the zero-sequence impedance is a pure reactance in order to simplify the discussion.

12.3.2 Transmission System without Mutual Coupling

We now expand the simple example of the previous section to a more practical arrangement, where the power system to which the protected line is connected to sources at both ends. The new system is shown in Figure 12.8. The system represented has four transmission lines, and a one-line-to-ground fault is assumed on one of the lines. This system has two positive-sequence voltage sources, but there are four sources of zero-sequence ground current.

[1]The approach used here is provided by K. H. Engelhardt of Vancouver, B.C., Canada. The author is grateful to Mr. Engelhardt for sharing his experience on the subject.

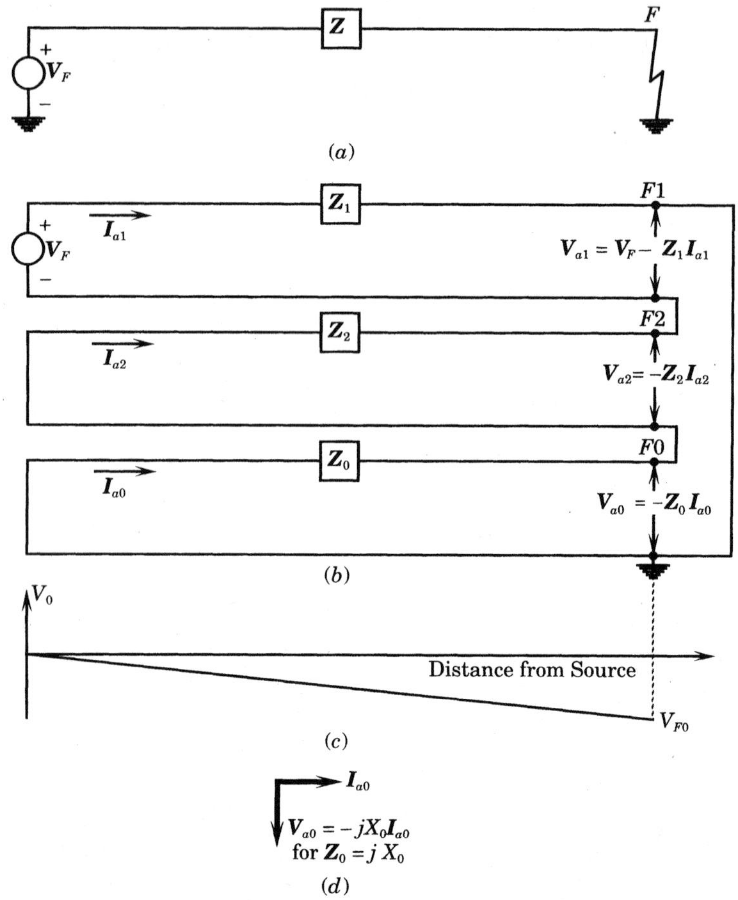

Figure 12.7 Zero-sequence voltage profile on a faulted radial line. (*a*) Radial system with 1LG Fault at *F*. (*b*) Sequence networks for 1LG fault at *F*. (*c*) Zero-sequence voltage profile along the line. (*d*) Zero-sequence voltage using I_{a0} as reference.

As noted previously for the radial system, the zero-sequence voltages are negative at all buses and lag the fault current by 90 degrees. The magnitude of the zero-sequence voltage is greatest at the fault point. Note that the rate of change of voltage with distance is proportional to the fault current, which is not the same in all transmission lines due to the ground sources at buses B and C. The most important conclusion from this example is that all of the zero-sequence voltages are negative and that the zero-sequence currents all flow from the zero-potential bus toward the fault. This means that currents flow *up* through the grounded-wye transformer neutrals.

The zero-sequence voltages and currents at all relay locations are also shown in Figure 12.8. If we assume voltage polarized ground relays, only line 3 has the correct polarization at both ends to cause tripping. The correct polarization requires that the measured fault current lead the voltage by about 90 degrees (abbreviated as "Op" in Figure 12.8). All other lines have a blocking condition at one end or the other. The fault in this example is correctly cleared, and the faulted line will be removed from service.

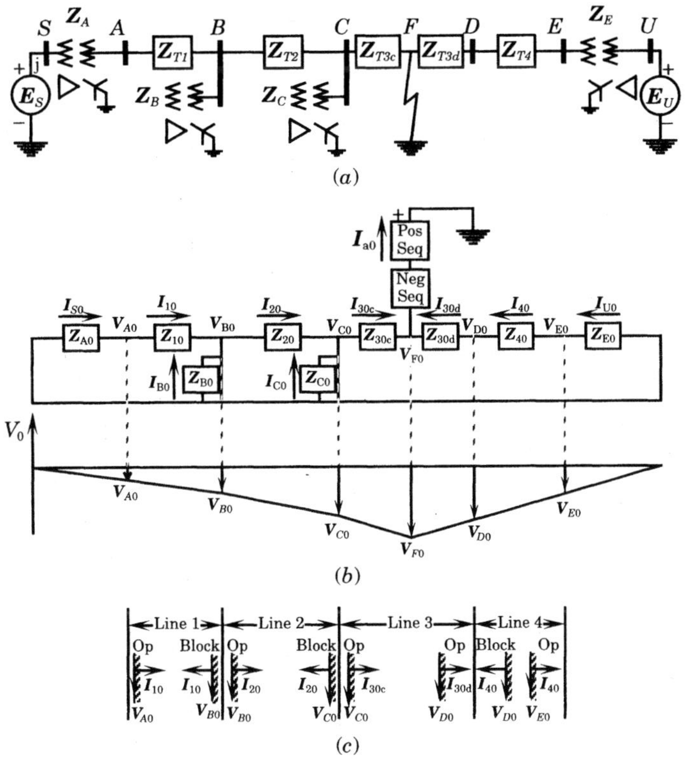

Figure 12.8 Zero-sequence voltage profile with no mutual coupling. (*a*) Typical power system with 1LG fault at F. (*b*) Voltage profile. (*c*) Voltage and current phasors for zero-sequence directional line relays.

12.3.3 Transmission System with Mutual Coupling

We now consider a system that is similar to that of Figure 12.8, but with one line that is mutually coupled to a line that shares the same right of way for a portion of its length. A one-line-to-ground fault will be assumed on the coupled line and the effect of this fault on the system will be examined.

The study system for this new condition is shown in Figure 12.9. Note that the transmission system is exactly the same as that of Figure 12.8, but a new transmission line in a separate, but nearby, system is added. The added line is not connected to the study system except through the mutual coupling of line 3 to a line of the external system.

One way of showing the zero-sequence mutual coupling in the circuit diagram is by use of the ideal transformer, connected as shown between the coupled lines [16]. The phase conductors of both lines are considered to be transposed and mutual coupling exists only in the zero-sequence network. This mutual impedance is shown as $\mathbf{Z}_{M0}$ in the circuit diagram. Observing the polarity marks of the ideal transformer, we can see that the fault current flows into the polarity mark on the faulted line and out of the marked terminal in transmission line 3. This causes the current to flow *up* from the zero-potential plane sources on the right side of line 3 and *down* to the zero-potential plane on the left. The zero-sequence voltage is zero at ground source U and becomes more negative as the current flows toward D. At that point a

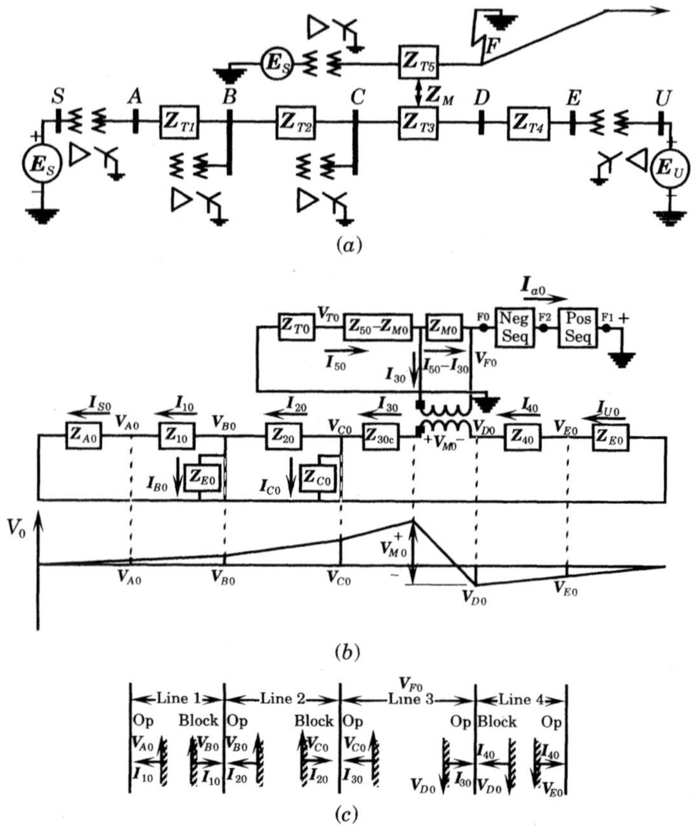

Figure 12.9 Voltage profile with zero-sequence mutual coupling. (*a*) Power system with 1LG fault on a mutually coupled line. (*b*) Voltage profile. (*c*) Voltage and current phasors for zero-sequence directional line relays.

voltage rise is introduced due to the mutual coupling, resulting in a voltage at C that is positive, rather than negative. For the voltage rise as pictured, the voltages at B and A are also positive, but the voltage again goes to zero at source A on the left.

Now examine the voltages and currents seen by the relays at each end of the four lines. Line 3 is of particular interest, since this is the line that is mutually coupled to the neighboring faulted line. At bus D the zero-sequence current flows into the line and the zero-sequence voltage is negative, causing this relay to pick up. At bus C, however, the zero-sequence voltage is positive and the current is flowing *out* of the line, which results in a current leading the voltage by 90°, the correct relationship for relay pickup. Therefore, line 3 will experience a false trip. Note that the other stations to the left of line 3, the current–voltage relationship is such that one terminal sees an operate condition, but the other terminal blocks in every case. Therefore, lines 1 and 2 will be restrained from tripping. It is important to note that only the line that is subjected to the induced voltage from the coupled line will experience false directional tripping. This is because the zero-sequence voltage is negative at the terminal with the entering current and is positive at the terminal with the outgoing current.

Another important observation can be made from this simple example. At bus C of line 3, the voltage is inverted, but the zero sequence source current at that bus flows *down* through all of the transformers on the left to the zero potential bus, including transformer C.

Therefore, had the ground relaying scheme been based on current polarization rather than voltage polarization, the same incorrect tripping of line 3 would have occurred.

12.3.4 Other Examples of Mutual Coupling

Figure 12.10 shows two examples of zero-sequence mutual coupling of transmission lines of the same system, but of different voltages. The system of Figure 12.10(*a*) shows lines of different voltages that are mutually coupled due to lines in the same right-of-way corridor leaving Station C.

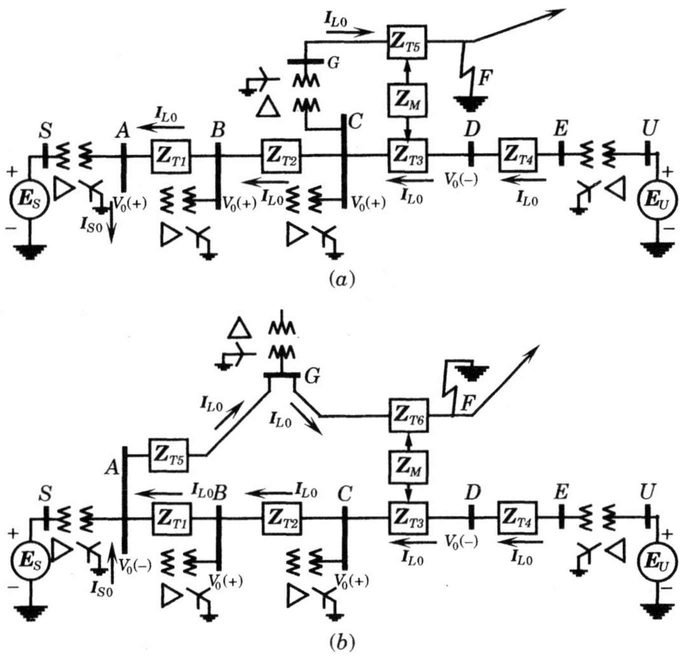

Figure 12.10 Zero-Sequence mutual coupling on lines of same system. (*a*) Power system with a mutually coupled line of different voltage. (*b*) Power system with a mutually coupled line of same voltage.

The figure shows a fault on one of the lines, which could be a subtransmission or distribution line. The positive- and negative-sequence systems for this condition will have two sources and lines in a T configuration to the fault point. The zero-sequence network, however, is exactly like that of Figure 12.9, where the primary loop is isolated from the secondary loop with only the mutual coupling connecting the two loops together. This will cause the zero-sequence voltage to reverse from a negative voltage at bus D to a positive voltage at buses C, B, and A, as shown in Figure 12.10(*a*). The transmission line from C to D may experience a false trip, depending on the magnitude of the induced voltage due to mutual coupling between the two circuits. Note that currents and voltages shown in the figure represent relative direction or polarity only, not magnitudes of the quantities.

Figure 12.10(*b*) shows a different situation, where the fault is on a transmission line of the same system that shares the right-of-way corridor with line $T3$. A fault on the line beyond point G will cause fault current to flow through the mutually coupled line segment, which will cause voltage inversion in Line CD. Therefore, depending on the coupling factor between the two lines, Line CD may experience a false directional comparison and a false trip. Note that the zero-sequence voltage will be positive at C and B, but not at A, where the ground source current will flow up through the transformer neutral and toward the fault.

These conditions are quite common in many power systems and may require careful study by the protection engineer to determine if false tripping might occur. This requires a fault study that represents all zero-sequence couplings in the power system to determine any possible voltage reversal conditions.

Another common condition is the circuit shown in Figure 12.11. This often occurs where a new source or load is tapped off an existing transmission line, with the new bus served through two tapped lines on common towers or sharing the same right-of-way corridor. This situation is sometimes referred to as a "pi-loop" by protection engineers, where the two legs of the pi are mutually coupled.

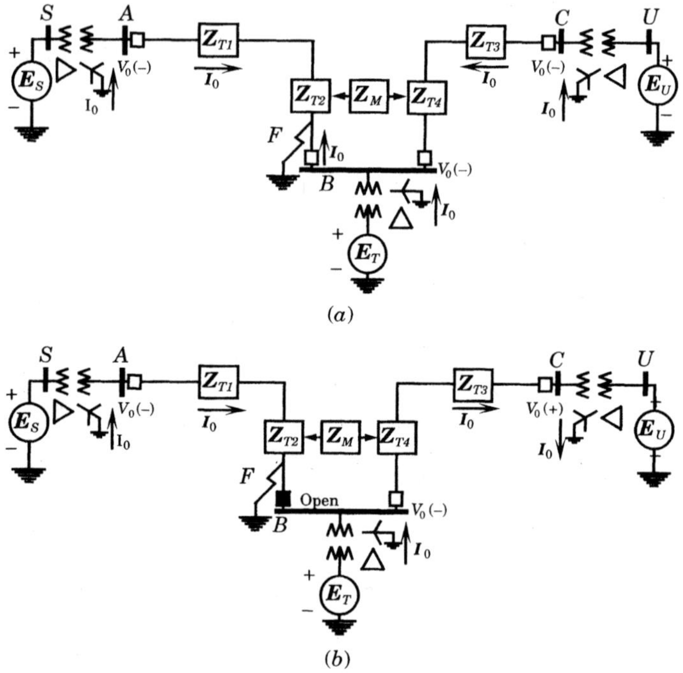

Figure 12.11 Mutual coupling of a tapped transmission line. (*a*) Fault on a mutually coupled tap off a transmission line. (*b*) Fault on a mutually coupled tap after one breaker opens.

Consider a one-line-to-ground fault at F in Figure 12.11(a), where the fault is close to Station B on line AB. Immediately after the fault occurs, all of the zero-sequence voltages will be negative, as shown in Figure 12.11(a). The fault current from Station B is greater that than from Station A due to the close proximity of a source at B. Therefore, there is a high probability that the line terminal at Station B may be cleared before that at Station A. Should this happen, the resulting situation is shown in Figure 12.11(b), where the breaker shown at B is now open and there is no fault current flowing from the source at B to the fault. There is current flowing up through the ground source at B, however, and this current continues along the line to Station C, where it flows *down* through the transformer neutral. Also, due to the mutual coupling, the zero-sequence voltage at Station C will be reversed. The conditions that occur after the opening of the breaker at B are exactly like that shown in Figure 12.9, with the two circuits completely isolated and with mutual coupling between them. The ground relays on Line BC will false trip irrespective to the method used to polarize the ground relays.

There are many other situations that can lead to false directional sensing of ground relays on mutually coupled lines. This suggests that every situation where mutual coupling exists should be carefully analyzed. This type of problem is, unfortunately, becoming more common. Transmission line rights of way are becoming increasingly difficult to obtain and more lines must sometimes be crowded into existing corridors, or multiple circuit towers erected to accommodate the needed circuits. This is often true in the vicinity of large cities, where space is at a premium. Such conditions often lead to situations where improper ground relaying may occur due to the presence of mutually coupled lines.

12.4 SHORT TRANSMISSION LINE EQUIVALENTS

The previous section has introduced three configurations that are often found in power systems. Other configurations are certainly possible, for example, three or more lines in the same right of way, and which may or may not be bused at either end. These more complex arrangements will not be pursued in detail here, although the approach used would be similar to that presented here.

The equivalent circuits discussed here are limited in their application to short transmission lines, and lines where shunt susceptance can be ignored for fault calculations. Long lines are treated in Section 12.5.

12.4.1 General Network Equivalents for Short Lines

General network equivalents of mutually coupled lines are shown in Figure 12.12. Part (a) of the figure shows a general view of the mutual coupling, but the mathematical representation may not be clear from this type of circuit diagram. The exact meaning of this circuit is given as follows.

$$\begin{aligned} \boldsymbol{V}_{12} &= \boldsymbol{Z}_A \boldsymbol{I}_A + \boldsymbol{Z}_M \boldsymbol{I}_B \\ \boldsymbol{V}_{34} &= \boldsymbol{Z}_B \boldsymbol{I}_B + \boldsymbol{Z}_M \boldsymbol{I}_A \end{aligned} \tag{12.20}$$

We can modify these equations by adding and subtracting a voltage proportional to the mutual

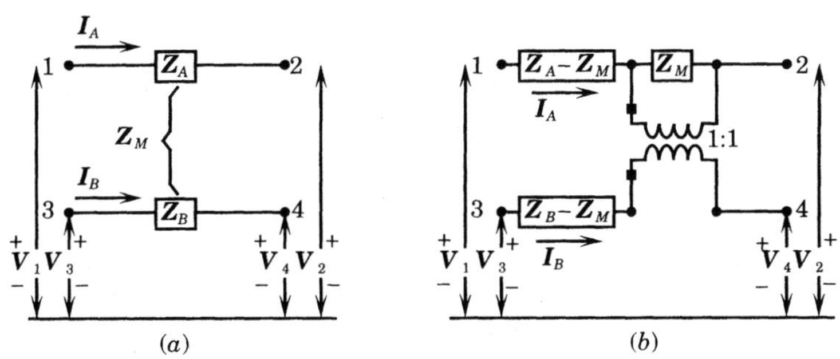

Figure 12.12 Equivalent circuits for two mutually coupled lines. (a) Mutually coupled transmission lines. (b) Equivalent using a 1:1 transformer.

impedance and the line current for each line. The result can be written as

$$\begin{aligned} V_{12} &= (Z_A - Z_M)I_A + Z_M(I_A + I_B) \\ V_{34} &= (Z_B - Z_M)I_B + Z_M(I_A + I_B) \end{aligned} \tag{12.21}$$

These equations describe the circuit of Figure 12.12(b), where a 1:1 ideal transformer is used to inject a voltage drop in each line that is proportional to the mutual impedance. However, since the current flowing through this mutual impedance is the sum of the two currents, the self-impedance of each line must be modified as shown. Figure 12.12(b) provides a useful way of visualizing the effect of mutual coupling. This equivalent is useful in laboratory work since all components of the equivalent are physically realizable. It is also useful conceptually in visualizing the effect of mutual induction, as noted in Section 12.3.

Another method of viewing the mutual coupling between two lines is derived by writing the admittance matrix for the four-terminal network. If we solve (12.10) for the currents, we get

$$\begin{bmatrix} I_A \\ I_B \end{bmatrix} = \frac{1}{\det \mathbf{Z}} \begin{bmatrix} Z_B & -Z_M \\ -Z_M & Z_A \end{bmatrix} \begin{bmatrix} V_1 - V_2 \\ V_3 - V_4 \end{bmatrix} \tag{12.22}$$

Now, if we define the four currents entering the two lines, we can write the admittance matrix for the mutually coupled system, as follows.

$$\begin{bmatrix} I_1 \\ I_2 \\ I_3 \\ I_4 \end{bmatrix} = \begin{bmatrix} I_A \\ -I_A \\ I_B \\ -I_B \end{bmatrix} = \frac{1}{\det \mathbf{Z}} \begin{bmatrix} Z_B & -Z_B & -Z_M & Z_M \\ -Z_B & Z_B & Z_M & -Z_M \\ -Z_M & Z_M & Z_A & -Z_A \\ Z_M & -Z_M & -Z_A & Z_A \end{bmatrix} \begin{bmatrix} V_1 \\ V_2 \\ V_3 \\ V_4 \end{bmatrix} \tag{12.23}$$

We can construct an equivalent circuit that corresponds to (12.23) by recalling that the off-diagonal terms of the admittance matrix are equal to the negative of the admittances between the corresponding nodes of the network. Computing these impedances leads to the lattice network shown in Figure 12.13. Note that this equivalent introduces negative impedances, which are not physically realizable. However, the lattice equivalent is useful for computation and represents the self- and mutual impedances without the need for representing a transformer. Indeed, the lattice circuit is the equivalent circuit for a transformer. The impedances values can be expressed as functions of the original mutually coupled lines as follows.

$$Z_{AX} = Z_A - \frac{Z_M^2}{Z_B} = \frac{Z_A Z_B - Z_M^2}{Z_B} \tag{12.24}$$

$$Z_{BX} = Z_B - \frac{Z_M^2}{Z_A} = \frac{Z_A Z_B - Z_M^2}{Z_A} \tag{12.25}$$

$$Z_{MX} = \frac{Z_A Z_B}{Z_M} - Z_M = \frac{Z_A Z_B - Z_M^2}{Z_M} \tag{12.26}$$

Another interesting extension of the lattice network is to take the limit as the line self-impedances approach zero. This causes the branches labeled Z_{AX} and Z_{BX} to become infinite impedances and leads to the equivalent circuit of an ideal transformation shown in Figure 12.14(a). Note that the negative impedance legs of the lattice have changed position from that found in Figure 12.13.

Now, the transformer self-impedances can be added externally to give the circuit shown in Figure 12.14(b), which is often found in the literature [7], [8].

The equivalents shown in Figures 12.12–12.14 are restricted to applications involving short transmission lines where susceptances can be neglected.

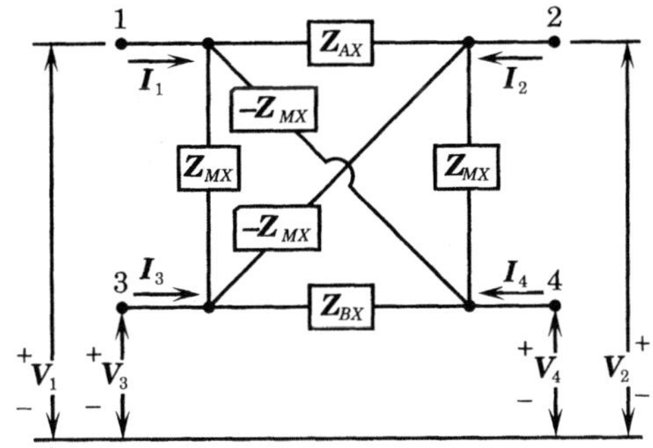

Figure 12.13 Lattice network equivalent of the mutually coupled lines.

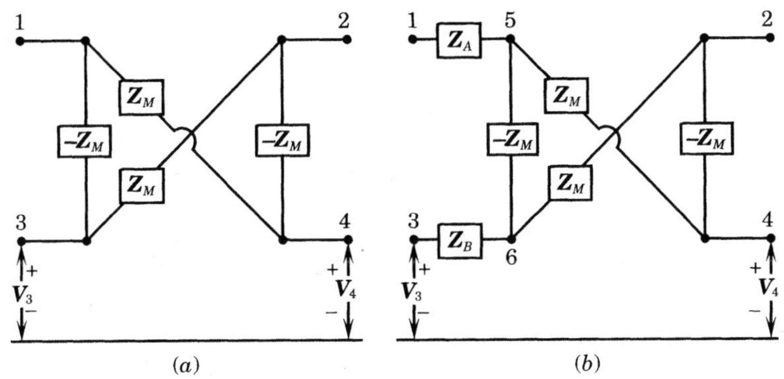

Figure 12.14 Equivalent circuits for a transformer. (*a*) Ideal transformer equivalent. (*b*) Transformers with self impedances.

We now consider the development of equivalent circuits for several network configurations that are commonly found in power systems. The equivalent circuits of the transmission lines require network branches connecting all pairs of terminating nodes of the transmission system, in addition to branches from each node to the reference. We can determine the number N of equivalent circuit branches from the equation

$$N = {}_nC_2 + n = \frac{n(n-1)}{2} + n = \frac{n(n+1)}{2} \tag{12.27}$$

where n is the number of terminating nodes. This relationship is perfectly general, but in practice we will usually have need to apply it for only small values of n. For two transmission lines, the largest value of n occurs when the lines are not bused at either end, in which case $n = 4$, resulting in an equivalent circuit with 10 branches. In many cases the lines are bused at either end, or at both ends, which results in simpler equivalent circuits.

12.4.2 Type 1 Networks

Type 1 networks are defined as those networks where parallel transmission lines terminate at common nodes at both ends of the lines. The equivalent circuits of Figure 12.12 can be used directly for Type 1 networks, i.e., networks in which the lines are bused at both ends. This situation is shown in Figure 12.15.

For Type 1 networks the voltages at the two ends of the lines are equal and the voltage difference across the circuit, from P to Q, are also equal. We can write the voltage equations

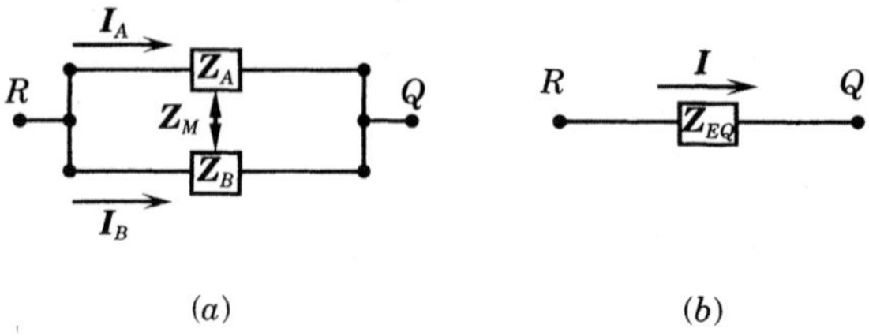

Figure 12.15 Type 1 network equivalent. (*a*) Two circuits bused at both ends. (*b*) Equivalent circuit for (*a*).

of the mutually coupled system as follows.

$$\begin{bmatrix} \boldsymbol{V}_R - \boldsymbol{V}_Q \\ \boldsymbol{V}_R - \boldsymbol{V}_Q \end{bmatrix} = \begin{bmatrix} \boldsymbol{V} \\ \boldsymbol{V} \end{bmatrix} = \begin{bmatrix} \boldsymbol{Z}_A & \boldsymbol{Z}_M \\ \boldsymbol{Z}_M & \boldsymbol{Z}_B \end{bmatrix} \begin{bmatrix} \boldsymbol{I}_A \\ \boldsymbol{I}_B \end{bmatrix} \tag{12.28}$$

Equating the two voltage equations, we can write one current in terms of the other.

$$\boldsymbol{I}_B = \boldsymbol{I}_A \frac{\boldsymbol{Z}_A - \boldsymbol{Z}_M}{\boldsymbol{Z}_B - \boldsymbol{Z}_M} \tag{12.29}$$

Equation (12.29) can be substituted into (12.28) to compute the voltage as a function of the current $\boldsymbol{I}_A$ with the following result.

$$\boldsymbol{V} = \boldsymbol{I}_A \frac{\boldsymbol{Z}_A \boldsymbol{Z}_B - \boldsymbol{Z}_M^2}{\boldsymbol{Z}_B - \boldsymbol{Z}_M} \tag{12.30}$$

This equation can be solved for the current $\boldsymbol{I}_A$, which can be substituted into (12.29) to get the current $\boldsymbol{I}_B$, with both currents expressed as a function of $\boldsymbol{V}$.

$$\begin{aligned} \boldsymbol{I}_A &= \boldsymbol{V} \frac{\boldsymbol{Z}_B - \boldsymbol{Z}_M}{\boldsymbol{Z}_A \boldsymbol{Z}_B - \boldsymbol{Z}_M^2} \\ \boldsymbol{I}_B &= \boldsymbol{V} \frac{\boldsymbol{Z}_A - \boldsymbol{Z}_M}{\boldsymbol{Z}_A \boldsymbol{Z}_B - \boldsymbol{Z}_M^2} \end{aligned} \tag{12.31}$$

The sum of these currents is the total current $\boldsymbol{I}$ flowing in the equivalent. Thus, we have

$$\boldsymbol{I} = \boldsymbol{V} \frac{\boldsymbol{Z}_A + \boldsymbol{Z}_B - 2\boldsymbol{Z}_M}{\boldsymbol{Z}_A \boldsymbol{Z}_B - \boldsymbol{Z}_M^2} \tag{12.32}$$

Now, the ratio of voltage to current gives the impedance of the equivalent.

$$\boldsymbol{Z}_{EQ} = \frac{\boldsymbol{Z}_A \boldsymbol{Z}_B - \boldsymbol{Z}_M^2}{\boldsymbol{Z}_A + \boldsymbol{Z}_B - 2\boldsymbol{Z}_M} \tag{12.33}$$

If the two circuits have equal self-impedance $\boldsymbol{Z}_A$, then the above results simplifies to

$$\boldsymbol{Z}_{EQ} = \frac{\boldsymbol{Z}_A + \boldsymbol{Z}_M}{2} \tag{12.34}$$

This is the equivalent impedance from P to Q shown in Figure 12.15(*b*).

The circuit of Figure 12.15 shows the unfaulted Type 1 network. If a fault is applied to one of the coupled transmission lines, the equivalent circuit must account for the coupling of the line sections on both sides of the fault. A step-by-step development of the equivalent circuit for this condition is shown in Figure 12.16. For the faulted case, the equivalent circuit parameters are a function of h, the fractional distance along the line where the fault occurs. These figures apply during the fault duration, but do not apply following the opening of one of the breakers on the faulted line.

It often occurs that one of the circuit breakers on the faulted line may open before the breakers in the parallel line. This leaves the fault radial from one end of the line. The circuit representing this condition is shown in Figure 12.17.

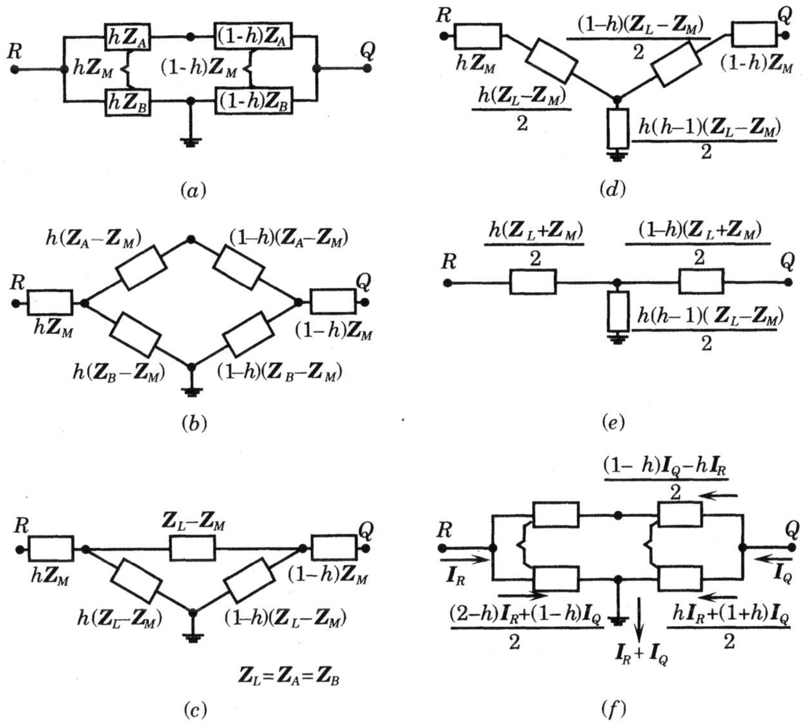

Figure 12.16 Type 1 equivalent with one line faulted [6]. (*a*) Mutually coupled lines *A* and *B*. (*b*) General equivalent circuit. (*c*) Equivalent with $\mathbf{Z}_L = \mathbf{Z}_A = \mathbf{Z}_B$. (*d*) Wye equivalent of (*c*). (*e*) Simplification of equivalent (*d*). (*f*) Current distributions.

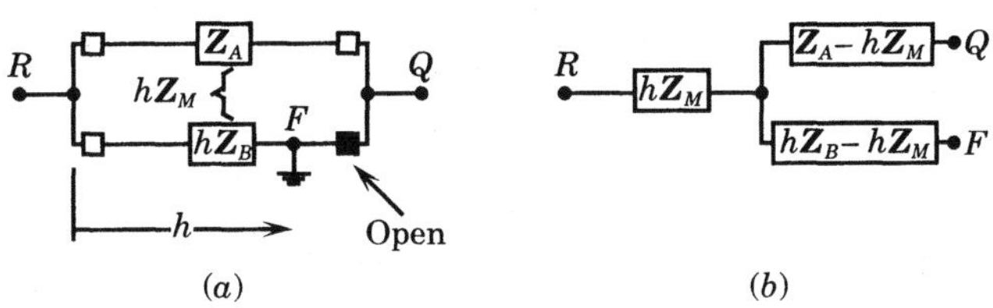

Figure 12.17 Type 1 equivalent following breaker opening at one end. (*a*) Faulted type 1 network with one breaker open. (*b*) Equivalent with one line open at one end.

For this condition, the fault *F* is radial on one line, and that line is mutually coupled to the unfaulted line. The equivalent circuit is similar in topology to a Type 2 circuit, but has the important difference that one of the three terminals is the fault point so that the total zero-sequence current flows through that leg of the equivalent. Note that any contribution to the fault from bus *Q* creates a condition similar to that of Figure 12.9, where the currents in two mutually coupled lines are flowing in opposite directions. This has been noted to cause zero-sequence voltage inversion, and may lead to incorrect relay pickup. When the fractional distance to the fault is nearly unity, corresponding to a fault very close to the breaker at *Q*. There always exists the likelihood that the breaker nearest the fault may open first on zone 1 tripping, thereby setting up the condition described above.

12.4.3 Type 2 Networks

Type 2 networks are defined as mutually coupled lines that are bused at only one end. The general equivalent circuit for this class of networks is shown in Figure 12.18. This circuit is applicable to an unfaulted system. When one of the lines of the Type 2 network is faulted, the equivalent circuit must be modified to reflect the fault condition. Let us assume that the fault occurs on the line RQ at a fractional distance h of that line length measured from bus R. The equivalent at the R end of the line will be similar to that of Figure 12.18, but the equivalent beyond the fault must use one of the constructions shown in Figures 12.12–12.14. One possible result is shown in Figure 12.19.

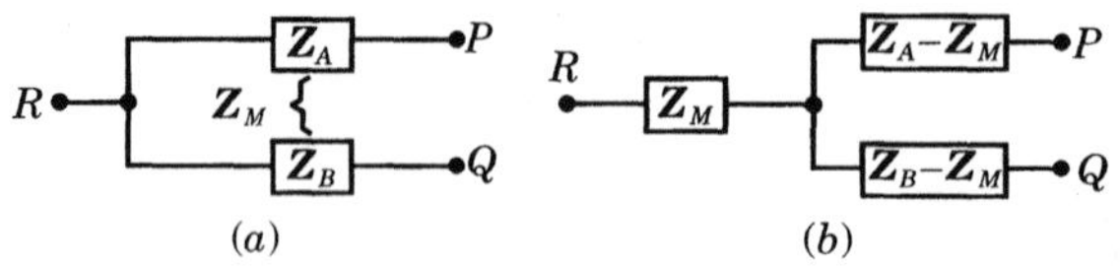

Figure 12.18 Type 2 network and its equivalent. (*a*) General type 2 network. (*b*) Equivalent type 2 network.

The portion of the equivalent to the left of the fault point is exactly the same as Figure 12.17(*b*), but modified for the length of line between bus R and the fault. The portion of the two lines between the fault and buses P and Q, however, must be represented by an equivalent that does not require that the terminals be bused together—hence the insertion of the one-to-one transformer equivalent.

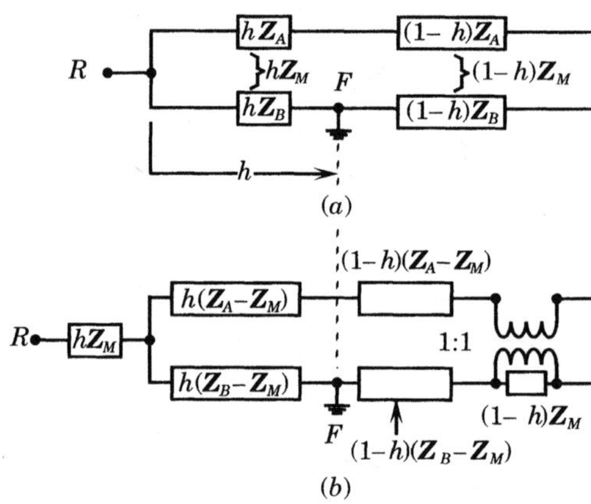

Figure 12.19 Faulted type 2 network equivalent. (*a*) Faulted type 2 network. (*b*) Equivalent of faulted type 2 network.

12.4.4 Type 3 Networks

Type 3 networks are those in which the mutually coupled lines do not terminate at common buses at either end. For this condition, the circuits of Figure 12.12–12.14 are appropriate. These circuits may be used without modification for the conditions before a fault is applied. When a fault is applied to either of the mutually coupled lines, the equivalent circuit must be modified to the form shown in Figure 12.20.

It is important to note that the Type 3 network equivalent is perfectly general and can be converted to other terminations. For example, connecting nodes R and S of Figure 12.20, gives the equivalent circuit of Figure 12.19, with appropriate changes in notation. In a similar way, the equivalent of Figure 12.16 can also be derived, beginning with the network of Figure 12.20. This suggests that there is no need for any equivalent other than that of Figure 12.20, where the four line terminals are distinct.

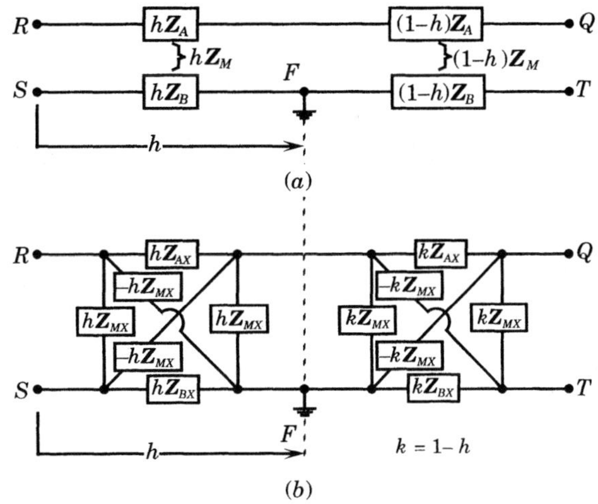

Figure 12.20 Equivalent of type 3 mutually coupled lines. (*a*) Faulted type 3 network. (*b*) Equivalent of faulted type 3 network.

Equivalents similar to Figure 12.20 can be constructed for the positive-, negative-, and zero-sequence networks. The mutual coupling in the positive- and negative-sequence networks is weak, and may be neglected with little error. This results in the two lines being represented only by their self-impedances on each side of the fault point. The mutual coupling between lines is often important in the zero-sequence network, however, and should not be neglected without first making sure that the coupling is negligible. The equivalent circuit of Figure 12.20 is easily determined and represents a very small computational burden. Note that no new nodes are added to the network in constructing the equivalent and that all nodes are identified with physical points along the transmission lines. This is not true of the equivalent construction of Figure 12.14(*b*), which adds fictitious nodes. In solving the network completely, the voltages at these fictitious nodes can be found, but these voltages have no physical significance.

12.4.5 Lines with Appreciable Susceptance

In the foregoing analysis, it has been assumed that the lines are short and that the admittance between lines and from line to ground can be neglected. This is a convenient assumption even when it is not entirely correct, since it greatly simplifies the computations. Moreover, during fault conditions, the line voltages are greatly depressed, which suggests that neglecting the line susceptance usually leads to acceptable results.

For cases where the susceptance is appreciable on short lines, the nominal pi representation of the line can be used. Figure 12.21 shows two ways in which the representation can be carried out.

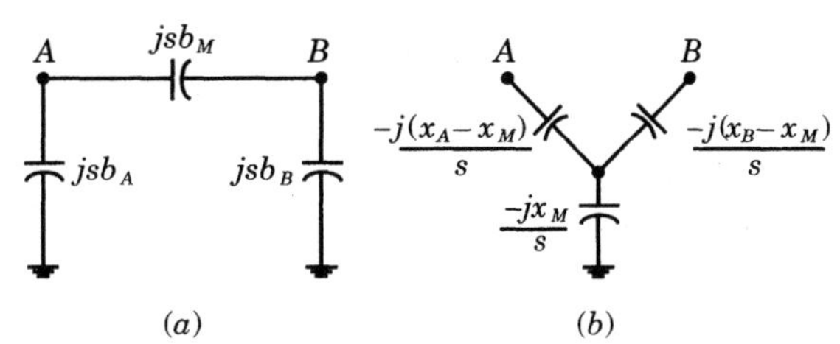

Figure 12.21 Nominal capacitances of two transmission lines [7]. (*a*) Delta admittances. (*b*) Wye impedances.

The nominal susceptances of two transmission lines can be expressed either as admittances or impedances. Figure 12.21(a) shows the nominal admittances expressed as the product of the susceptance b per unit of length multiplied by the line length s. This delta arrangement can be converted to a wye, shown in part (b) of the figure, but with the values expressed as impedance matrix elements per unit length divided by the line length. Either approach is acceptable if used consistently.

For the general case with lines that are not bused at either end, the equivalent circuit including both mutual inductance and susceptance is shown in Figure 12.22. In the figure, the total capacitive reactances for the entire line are shown, rather than the reactance per unit of length. The circuit of Figure 12.22 represents one way of representing the capacitive coupling. The wye connection at each end can be replaced by a delta or pi connection, if this is preferred. This is left as an exercise.

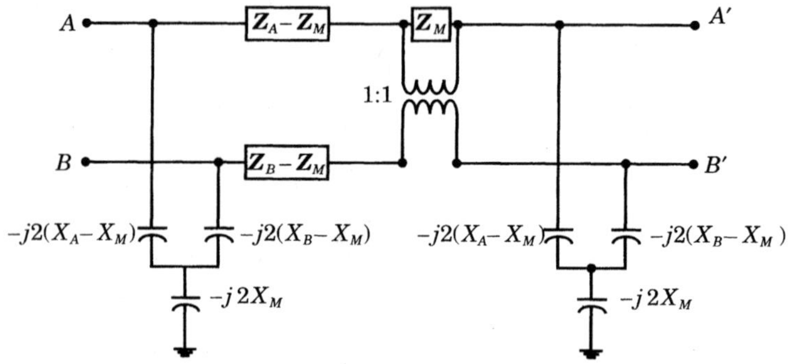

Figure 12.22 Nominal pi circuit for mutual impedance and admittance [7].

It should be emphasized that the representation of capacitance shown in Figure 12.22 is not a long line representation. Long transmission lines require a mathematical representation of the distributed capacitance as well as series impedance. Long transmission lines are discussed in Section 12.5.

12.4.6 Other Network Equivalents

There are other types of networks for which equivalents can be developed—for example, three mutually coupled lines terminating at a common bus, as shown in Figure 12.23. These special cases can be treated in the same manner as the three types described above. Several variations of these special equivalents are treated in the literature [6–8]. These special types of system equivalents are left as exercises.

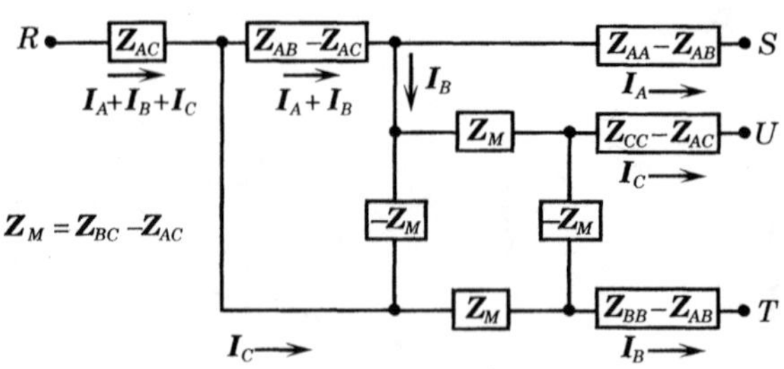

Figure 12.23 Three lines bused at one end and with unequal mutual impedances [7].

12.5 LONG TRANSMISSION LINES

The long transmission presents a more difficult problem to the protection engineer because of the need to model the distributed effects of the line series impedance and shunt admittance. We begin by reviewing briefly the characteristics of a single line, and then extend the equations to consider two mutually coupled, long lines.

12.5.1 The Isolated Long Transmission Line

The isolated long transmission line is defined in terms of the incremental series impedance and shunt admittance, both expressed per unit of length. It is assumed that the line is completely transposed such that the mutual coupling between phases is exactly balanced and a per-phase representation of the circuit is permitted. A sketch of the per-phase line representation of length S is shown in Figure 12.24.

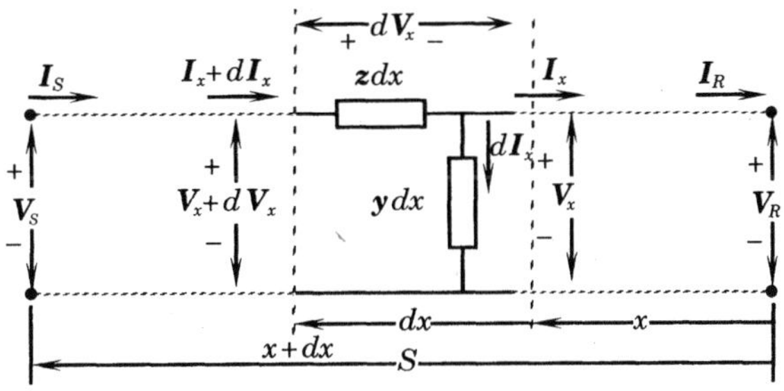

Figure 12.24 Differential currents and voltages of a long line.

We can write the equations for both the differential change in voltage and current associated with the differential length of transmission line. In this derivation, the distance x along the line is arbitrarily measured from the receiving end.

$$\begin{aligned} d\boldsymbol{V}_x &= (\boldsymbol{V}_x + d\boldsymbol{V}_x) - \boldsymbol{V}_x \\ &= \boldsymbol{z}(\boldsymbol{I}_x + d\boldsymbol{I}_x)dx \cong \boldsymbol{z}\boldsymbol{I}_x dx \\ d\boldsymbol{I}_x &= (\boldsymbol{I}_x + d\boldsymbol{I}_x) - \boldsymbol{I}_x \\ &= \boldsymbol{y}\boldsymbol{V}_x dx \end{aligned} \tag{12.35}$$

Rearranging these equations we have

$$\begin{aligned} \frac{d\boldsymbol{V}_x}{dx} &= \boldsymbol{z}\boldsymbol{I}_x \\ \frac{d\boldsymbol{I}_x}{dx} &= \boldsymbol{y}\boldsymbol{V}_x \end{aligned} \tag{12.36}$$

Differentiating with respect to x results in the following form of the equations.

$$\begin{aligned} \frac{d^2\boldsymbol{V}_x}{dx^2} &= \boldsymbol{z}\boldsymbol{y}\boldsymbol{V}_x \\ \frac{d^2\boldsymbol{I}_x}{dx^2} &= \boldsymbol{y}\boldsymbol{z}\boldsymbol{I}_x \end{aligned} \tag{12.37}$$

Taking the Laplace transform of the first equation, we write

$$s^2\boldsymbol{V}_x(s) - s\boldsymbol{V}_x(0) - \boldsymbol{V}'_x(0) = \gamma^2\boldsymbol{V}_x(s) \tag{12.38}$$

where we define the complex constant

$$\gamma^2 = \boldsymbol{zy} \tag{12.39}$$

But we know that

$$\begin{aligned} \boldsymbol{V}_x(0) &= \boldsymbol{V}_R \\ \boldsymbol{V}_x'(0) &= \boldsymbol{zI}_R \end{aligned} \tag{12.40}$$

Substituting (12.40) into (12.38), and solving for the voltage at x, we have

$$\boldsymbol{V}_x(s) = \frac{s\boldsymbol{V}_R + \boldsymbol{zI}_R}{(s^2 - \gamma^2)} \tag{12.41}$$

Then, the solution in the x domain is given by

$$\boldsymbol{V}_x(x) = \boldsymbol{V}_R \cosh \gamma x + \boldsymbol{Z}_c \boldsymbol{I}_R \sinh \gamma x \tag{12.42}$$

for any x. We are usually interested in the voltage at the sending end of the line, which is determined by setting $x = S$ to write

$$\begin{aligned} \boldsymbol{V}_S = \boldsymbol{V}_x(S) &= \boldsymbol{V}_R \cosh \gamma S + \boldsymbol{Z}_C \boldsymbol{I}_R \sinh \gamma S \\ &= \boldsymbol{AV}_R + \boldsymbol{BI}_R \end{aligned} \tag{12.43}$$

where we have defined a new transmission parameter called the "characteristic impedance" as

$$\boldsymbol{Z}_c = \frac{\boldsymbol{z}}{\gamma} = \frac{\boldsymbol{z}}{\sqrt{\boldsymbol{zy}}} = \sqrt{\frac{\boldsymbol{z}}{\boldsymbol{y}}} = \frac{1}{\boldsymbol{Y}_c} \tag{12.44}$$

The parameter γ is called the "propagation constant." The equation (12.42) also defines two of the two-port ***ABCD*** parameters, i.e.,

$$\boldsymbol{V}_S = \boldsymbol{AV}_R + \boldsymbol{BI}_R \tag{12.45}$$

or

$$\begin{aligned} \boldsymbol{A} &= \cosh \gamma S \\ \boldsymbol{B} &= \boldsymbol{Z}_c \sinh \gamma S \end{aligned} \tag{12.46}$$

Equations (12.42) and (12.43) are the classical "long transmission line equations" often used in both communications and power system analysis. Note that these derivations say nothing about a three-phase power transmission line, the tacit assumption being that the line is completely transposed and that the equation applies to only one sequence network, but it can be either the positive or the zero sequence.

The Laplace transform of the current equation may be written as

$$s^2\boldsymbol{I}_x(s) - s\boldsymbol{I}_x(0) - \boldsymbol{I}_x'(0) = \gamma^2\boldsymbol{I}_x(s) \tag{12.47}$$

and with the initial conditions

$$\begin{aligned} \boldsymbol{I}_x(0) &= \boldsymbol{I}_R \\ \boldsymbol{I}_x'(0) &= \boldsymbol{yV}_x(0) = \boldsymbol{yV}_R \end{aligned} \tag{12.48}$$

we compute

$$\boldsymbol{I}_x(s) = \frac{s\boldsymbol{I}_R + \boldsymbol{yV}_R}{(s^2 - \gamma^2)} \tag{12.49}$$

or, in the x domain with $x = S$, the total line length

$$\begin{aligned} \boldsymbol{I}_S = \boldsymbol{I}_x(S) &= \boldsymbol{Y}_c \boldsymbol{V}_R \sinh \gamma S + \boldsymbol{I}_R \cosh \gamma S \\ &= \boldsymbol{CV}_R + \boldsymbol{DI}_R \end{aligned} \tag{12.50}$$

With this result we have defined the following parameters.

$$\begin{aligned} Y_c &= \frac{y}{\gamma} = \frac{y}{\sqrt{zy}} = \sqrt{\frac{y}{z}} = \frac{1}{Z_c} \\ C &= Y_c V_R \sinh \gamma S \\ D &= A = \cosh \gamma S \end{aligned} \tag{12.51}$$

It is convenient to write the long line description in matrix form, as follows.

$$\begin{bmatrix} V_S \\ I_S \end{bmatrix} = \begin{bmatrix} A & B \\ C & D \end{bmatrix} \begin{bmatrix} V_R \\ I_R \end{bmatrix} \tag{12.52}$$

Equations (12.43) and (12.50) are universally used to describe the long power transmission line, where it is assumed that the line is transposed.

12.5.2 Mutually Coupled Long Transmission Lines

The derivation of the previous section shows that the long transmission line requires consideration of the distributed nature of the line inductive and capacitive reactances. Where two long lines are in close proximity, this requirement still holds, but the system of equations is more complex. Incremental parameters of both lines and their mutual impedance and admittance coupling must be considered. Figure 12.25 shows a general view of this distributed parameter system.

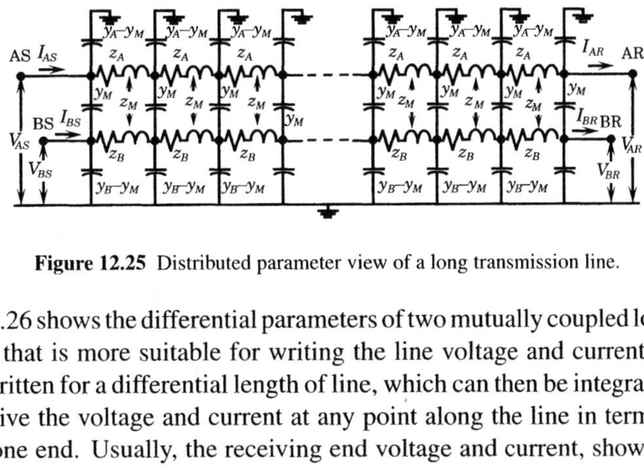

Figure 12.25 Distributed parameter view of a long transmission line.

Figure 12.26 shows the differential parameters of two mutually coupled long transmission lines in a form that is more suitable for writing the line voltage and current equations. The equations are written for a differential length of line, which can then be integrated over the total line length to give the voltage and current at any point along the line in terms of the voltage and current at one end. Usually, the receiving end voltage and current, shown on the right in Figures 12.25 and 12.26 are taken as the reference and the solution derived in terms of these boundary conditions.

The solution of the system of Figure 12.26 will be determined for two cases, first with the line parameters different for the two lines, and second, with equal parameters for the two lines.

12.5.2.1 Long Lines with Distinct Parameters. This system is described by the coupled line equations. First, for line A, which is the top line in Figure 12.26.

$$\begin{aligned} dV_{Ax} &= z_A I_{Ax} dx + z_M I_{Bx} dx = (z_A I_{Ax} + z_M I_{Bx}) dx \\ dI_{Ax} &= (y_A - y_M) V_{Ax} dx + y_M (V_{Ax} - V_{Bx}) dx = (y_A V_{Ax} - y_M V_{Bx}) dx \end{aligned} \tag{12.53}$$

Similarly, for line B, we have

$$\begin{aligned} dV_{Bx} &= z_B I_{Bx} dx + z_M I_{Ax} dx = (z_B I_{Bx} + z_M I_{Ax}) dx \\ dI_{Bx} &= (y_B - y_M) V_{Bx} dx + y_M (V_{Bx} - V_{Ax}) dx = (y_B V_{Bx} - y_M V_{Ax}) dx \end{aligned} \tag{12.54}$$

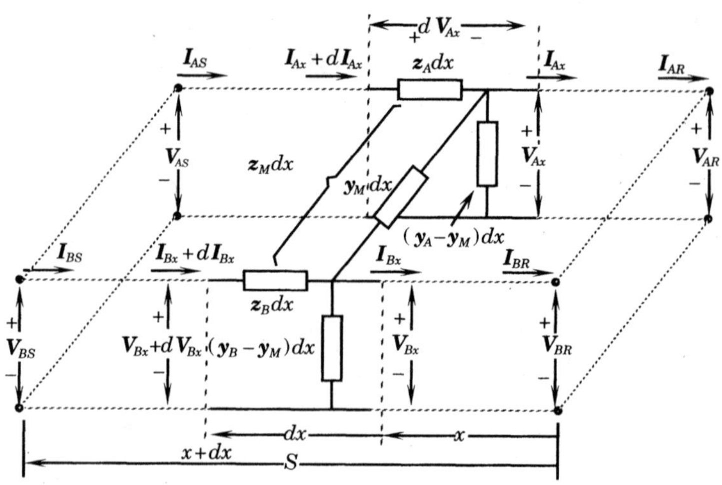

Figure 12.26 Differential quantities for mutually coupled long lines,

where the two lines are coupled by an incremental mutual impedance z_M and by an incremental mutual admittance y_M. These equations can be divided through by dx, and then the derivative of the result taken with respect to x to obtain the following form.

$$\begin{bmatrix} \dfrac{d^2\mathbf{V}_{Ax}}{dx^2} \\ \dfrac{d^2\mathbf{V}_{Bx}}{dx^2} \end{bmatrix} = \begin{bmatrix} \gamma_{AA}^2 & -\gamma_{AB}^2 \\ -\gamma_{BA}^2 & \gamma_{BB}^2 \end{bmatrix} \begin{bmatrix} \mathbf{V}_{Ax} \\ \mathbf{V}_{Bx} \end{bmatrix} \tag{12.55}$$

where we have defined the following system parameters that are analogous to the square of propagation constants of the single-line case.

$$\begin{aligned} \gamma_{AA}^2 &= z_A y_A - z_M y_M \\ \gamma_{BB}^2 &= z_B y_B - z_M y_M \\ \gamma_{AB}^2 &= z_A y_M - z_M y_B \\ \gamma_{BA}^2 &= z_B y_M - z_M y_A \end{aligned} \tag{12.56}$$

Taking the Laplace transform and using the initial conditions, we can write (12.55) as the following matrix equation.

$$\mathbf{M}_V \begin{bmatrix} \mathbf{V}_{Ax}(s) \\ \mathbf{V}_{Bx}(s) \end{bmatrix} = s \begin{bmatrix} \mathbf{V}_{AR} \\ \mathbf{V}_{BR} \end{bmatrix} + \begin{bmatrix} z_A & z_M \\ z_M & z_B \end{bmatrix} \begin{bmatrix} \mathbf{I}_{AR} \\ \mathbf{I}_{BR} \end{bmatrix} \tag{12.57}$$

where we have defined the matrix

$$\mathbf{M}_V = \begin{bmatrix} s^2 - \gamma_{AA}^2 & -\gamma_{AB}^2 \\ -\gamma_{BA}^2 & s^2 - \gamma_{BB}^2 \end{bmatrix} \tag{12.58}$$

The solution of (12.57) requires taking the inverse of the matrix $\mathbf{M}_V$, and this requires a non-zero determinant of that matrix. The determinant is easily computed to be

$$\det \mathbf{M}_V = (s^2 - \gamma_{AA}^2)(s^2 - \gamma_{BB}^2) - \gamma_{AB}^2 \gamma_{BA}^2 \tag{12.59}$$

The solution of (12.59) can be found using the quadratic equation applied to the fourth-order equation in s. We can write this determinant in the form

$$\det \mathbf{M}_V = (s^2 - \gamma_1^2)(s^2 - \gamma_2^2) \tag{12.60}$$

where we recognize that there are two solutions to (12.59). The propagation constants γ_1 and γ_2 can be written in terms of the parameters defined in (12.56). Then we can write the solution to (12.57) in the s domain as

$$\begin{aligned}\begin{bmatrix}\boldsymbol{V}_{Ax}(s)\\ \boldsymbol{V}_{Bx}(s)\end{bmatrix} &= \frac{s}{(s^2-\gamma_1^2)(s^2-\gamma_2^2)}\begin{bmatrix}s^2-\gamma_{AA}^2 & \gamma_{AB}^2\\ \gamma_{BA}^2 & s^2-\gamma_{BB}^2\end{bmatrix}\begin{bmatrix}\boldsymbol{V}_{AR}\\ \boldsymbol{V}_{BR}\end{bmatrix}\\ &+\frac{1}{(s^2-\gamma_1^2)(s^2-\gamma_2^2)}\begin{bmatrix}s^2-\gamma_{AA}^2 & \gamma_{AB}^2\\ \gamma_{BA}^2 & s^2-\gamma_{BB}^2\end{bmatrix}\begin{bmatrix}z_A & z_M\\ z_M & z_B\end{bmatrix}\begin{bmatrix}\boldsymbol{I}_{AR}\\ \boldsymbol{I}_{BR}\end{bmatrix}\end{aligned} \tag{12.61}$$

Solving (12.60) using the quadratic equation readily yields the two complex roots identified in (12.60) as

$$\gamma_1^2 = \frac{1}{2}\left(\begin{matrix}(z_A y_A + z_B y_B - 2z_M y_M)\\ +\sqrt{(z_A y_A + z_B y_B - 2z_M y_M)^2 - 4(z_A z_B - z_M^2)(y_A y_B - y_M^2)}\end{matrix}\right) \tag{12.62}$$

$$\gamma_2^2 = \frac{1}{2}\left(\begin{matrix}(z_A y_A + z_B y_B - 2z_M y_M)\\ -\sqrt{(z_A y_A + z_B y_B - 2z_M y_M)^2 - 4(z_A z_B - z_M^2)(y_A y_B - y_M^2)}\end{matrix}\right) \tag{12.63}$$

The voltages of (12.61) may be written in a matrix form as follows.

$$\begin{bmatrix}\boldsymbol{V}_{Ax}(s)\\ \boldsymbol{V}_{Bx}(s)\end{bmatrix} = \begin{bmatrix}\boldsymbol{A}_{AA}(s) & \boldsymbol{A}_{AB}(s)\\ \boldsymbol{A}_{BA}(s) & \boldsymbol{A}_{BB}(s)\end{bmatrix}\begin{bmatrix}\boldsymbol{V}_{AR}\\ \boldsymbol{V}_{BR}\end{bmatrix} + \begin{bmatrix}\boldsymbol{B}_{AA}(s) & \boldsymbol{B}_{AB}(s)\\ \boldsymbol{B}_{BA}(s) & \boldsymbol{B}_{BB}(s)\end{bmatrix}\begin{bmatrix}\boldsymbol{I}_{AR}\\ \boldsymbol{I}_{BR}\end{bmatrix} \tag{12.64}$$

where

$$\begin{aligned}\boldsymbol{A}_{AA}(s) &= \frac{s(s^2-\gamma_{BB}^2)}{(s^2-\gamma_1^2)(s^2-\gamma_2^2)} & \boldsymbol{A}_{AB}(s) &= \frac{\gamma_{AB}^2 s}{(s^2-\gamma_1^2)(s^2-\gamma_2^2)}\\ \boldsymbol{A}_{BA}(s) &= \frac{\gamma_{BA}^2 s}{(s^2-\gamma_1^2)(s^2-\gamma_2^2)} & \boldsymbol{A}_{BB}(s) &= \frac{s(s^2-\gamma_{AA}^2)}{(s^2-\gamma_1^2)(s^2-\gamma_2^2)}\end{aligned} \tag{12.65}$$

and

$$\begin{aligned}\boldsymbol{B}_{AA}(s) &= \frac{(s^2-\gamma_{BB}^2)z_A + \gamma_{AB}^2 z_M}{(s^2-\gamma_1^2)(s^2-\gamma_2^2)} & \boldsymbol{B}_{AB}(s) &= \frac{(s^2-\gamma_{BB}^2)z_M + \gamma_{AB}^2 z_B}{(s^2-\gamma_1^2)(s^2-\gamma_2^2)}\\ \boldsymbol{B}_{BA}(s) &= \frac{(s^2-\gamma_{AA}^2)z_M + \gamma_{BA}^2 z_A}{(s^2-\gamma_1^2)(s^2-\gamma_2^2)} & \boldsymbol{B}_{BB}(s) &= \frac{(s^2-\gamma_{AA}^2)z_B + \gamma_{BA}^2 z_M}{(s^2-\gamma_1^2)(s^2-\gamma_2^2)}\end{aligned} \tag{12.66}$$

The solution to (12.64) may be found by computing the partial fraction expansion of (12.65) and (12.66). The results can then be written as follows.

$$\begin{aligned}\boldsymbol{A}_{AA} &= \frac{(\gamma_1^2-\gamma_{BB}^2)\cosh\gamma_1 x - (\gamma_2^2-\gamma_{BB}^2)\cosh\gamma_2 x}{\gamma_1^2-\gamma_2^2}\\ \boldsymbol{A}_{AB} &= \frac{\gamma_{AB}^2(\cosh\gamma_1 x - \cosh\gamma_2 x)}{\gamma_1^2-\gamma_2^2}\\ \boldsymbol{A}_{AB} &= \frac{\gamma_{BA}^2(\cosh\gamma_1 x - \cosh\gamma_2 x)}{\gamma_1^2-\gamma_2^2}\\ \boldsymbol{A}_{BB} &= \frac{(\gamma_2^2-\gamma_{AA}^2)\cosh\gamma_1 x - (\gamma_2^2-\gamma_{AA}^2)\cosh\gamma_2 x}{\gamma_1^2-\gamma_2^2}\end{aligned} \tag{12.67}$$

and

$$\begin{aligned}
\boldsymbol{B}_{AA} &= \frac{\gamma_2[(\gamma_1^2 - \gamma_{BB}^2)z_A - \gamma_{AB}^2 z_M] \sinh \gamma_1 x - \gamma_1[(\gamma_2^2 - \gamma_{BB}^2)z_A - \gamma_{AB}^2 z_M] \cosh \gamma_2 x}{\gamma_1\gamma_2(\gamma_1^2 - \gamma_2^2)} \\
\boldsymbol{B}_{AB} &= \frac{\gamma_2[(\gamma_1^2 - \gamma_{BB}^2)z_M - \gamma_{AB}^2 z_B] \sinh \gamma_1 x - \gamma_1[(\gamma_2^2 - \gamma_{BB}^2)z_M - \gamma_{AB}^2 z_B] \cosh \gamma_2 x}{\gamma_1\gamma_2(\gamma_1^2 - \gamma_2^2)} \\
\boldsymbol{B}_{BA} &= \frac{\gamma_2[(\gamma_1^2 - \gamma_{AA}^2)z_M - \gamma_{BA}^2 z_A] \sinh \gamma_1 x - \gamma_1[(\gamma_2^2 - \gamma_{AA}^2)z_M - \gamma_{BA}^2 z_A] \cosh \gamma_2 x}{\gamma_1\gamma_2(\gamma_1^2 - \gamma_2^2)} \\
\boldsymbol{B}_{BB} &= \frac{\gamma_2[(\gamma_1^2 - \gamma_{AA}^2)z_B - \gamma_{BA}^2 z_M] \sinh \gamma_1 x - \gamma_1[(\gamma_2^2 - \gamma_{AA}^2)z_B - \gamma_{BA}^2 z_M] \cosh \gamma_2 x}{\gamma_1\gamma_2(\gamma_1^2 - \gamma_2^2)}
\end{aligned} \tag{12.68}$$

where the variable x can take on any value from 0 to S, the total line length. In many problems, it is the sending-end voltages and currents that are of greatest interest, in which case we let $x = S$.

The currents along the transmission line can be solved in a similar manner, beginning with (12.47) and (12.48) and solving for the second derivatives. The result of that process can be written as follows.

$$\begin{bmatrix} \boldsymbol{I}_{Ax}(s) \\ \boldsymbol{I}_{Bx}(s) \end{bmatrix} = \begin{bmatrix} \boldsymbol{D}_{AA}(s) & \boldsymbol{D}_{AB}(s) \\ \boldsymbol{D}_{BA}(s) & \boldsymbol{D}_{BB}(s) \end{bmatrix} \begin{bmatrix} \boldsymbol{I}_{AR} \\ \boldsymbol{I}_{BR} \end{bmatrix} + \begin{bmatrix} \boldsymbol{C}_{AA}(s) & \boldsymbol{C}_{AB}(s) \\ \boldsymbol{C}_{BA}(s) & \boldsymbol{C}_{BB}(s) \end{bmatrix} \begin{bmatrix} \boldsymbol{V}_{AR} \\ \boldsymbol{V}_{BR} \end{bmatrix} \tag{12.69}$$

where

$$\begin{aligned}
\boldsymbol{D}_{AA} &= \frac{(\gamma_1^2 - \gamma_{BB}^2) \cosh \gamma_1 x - (\gamma_2^2 - \gamma_{BB}^2) \cosh \gamma_2 x}{\gamma_1^2 - \gamma_2^2} \\
\boldsymbol{D}_{AB} &= \frac{\gamma_{BA}^2 (\cosh \gamma_1 x - \cosh \gamma_2 x)}{\gamma_1^2 - \gamma_2^2} = \boldsymbol{A}_{BA} \\
\boldsymbol{D}_{BA} &= \frac{\gamma_{AB}^2 (\cosh \gamma_1 x - \cosh \gamma_2 x)}{\gamma_1^2 - \gamma_2^2} = \boldsymbol{A}_{AB} \\
\boldsymbol{D}_{BB} &= \frac{(\gamma_1^2 - \gamma_{AA}^2) \cosh \gamma_1 x - (\gamma_2^2 - \gamma_{AA}^2) \cosh \gamma_2 x}{\gamma_1^2 - \gamma_2^2}
\end{aligned} \tag{12.70}$$

and

$$\begin{aligned}
\boldsymbol{C}_{AA} &= \frac{\gamma_2[(\gamma_1^2 - \gamma_{BB}^2)y_A - \gamma_{BA}^2 y_M] \sinh \gamma_1 x - \gamma_1[(\gamma_2^2 - \gamma_{BB}^2)y_A - \gamma_{BA}^2 y_M] \sinh \gamma_2 x}{\gamma_1\gamma_2(\gamma_1^2 - \gamma_2^2)} \\
\boldsymbol{C}_{AB} &= -\frac{\gamma_2[(\gamma_1^2 - \gamma_{BB}^2)y_M - \gamma_{BA}^2 y_B] \sinh \gamma_1 x - \gamma_1[(\gamma_2^2 - \gamma_{BB}^2)y_M - \gamma_{BA}^2 y_B] \sinh \gamma_2 x}{\gamma_1\gamma_2(\gamma_1^2 - \gamma_2^2)} \\
\boldsymbol{C}_{BA} &= -\frac{\gamma_2[(\gamma_1^2 - \gamma_{AA}^2)y_M - \gamma_{AB}^2 y_A] \sinh \gamma_1 x - \gamma_1[(\gamma_2^2 - \gamma_{AA}^2)y_M - \gamma_{AB}^2 y_A] \sinh \gamma_2 x}{\gamma_1\gamma_2(\gamma_1^2 - \gamma_2^2)} \\
\boldsymbol{C}_{BB} &= \frac{\gamma_2[(\gamma_1^2 - \gamma_{AA}^2)y_B - \gamma_{AB}^2 y_M] \sinh \gamma_1 x - \gamma_1[(\gamma_2^2 - \gamma_{AA}^2)y_B - \gamma_{AB}^2 y_M] \sinh \gamma_2 x}{\gamma_1\gamma_2(\gamma_1^2 - \gamma_2^2)}
\end{aligned} \tag{12.71}$$

When S, the total line length, is substituted for x in the foregoing equations, the voltage and current at the sending end of the mutually coupled lines are determined.

We can summarize the foregoing equations in matrix form. Writing the sending-end voltages and currents in terms of the receiving end quantities, we have the following.

$$\begin{bmatrix} \boldsymbol{V}_{AS} \\ \boldsymbol{V}_{BS} \\ \hline \boldsymbol{I}_{AS} \\ \boldsymbol{I}_{BS} \end{bmatrix} = \left[\begin{array}{cc|cc} \boldsymbol{A}_{AA} & \boldsymbol{A}_{AB} & \boldsymbol{B}_{AA} & \boldsymbol{B}_{AB} \\ \boldsymbol{A}_{BA} & \boldsymbol{A}_{BB} & \boldsymbol{B}_{BA} & \boldsymbol{B}_{BB} \\ \hline \boldsymbol{C}_{AA} & \boldsymbol{C}_{AB} & \boldsymbol{D}_{AA} & \boldsymbol{D}_{AB} \\ \boldsymbol{C}_{BA} & \boldsymbol{C}_{BB} & \boldsymbol{D}_{BA} & \boldsymbol{D}_{BB} \end{array}\right] \begin{bmatrix} \boldsymbol{V}_{AR} \\ \boldsymbol{V}_{BR} \\ \hline \boldsymbol{I}_{AR} \\ \boldsymbol{I}_{BR} \end{bmatrix} \tag{12.72}$$

An alternate form of the equations may also be of interest.

$$\begin{bmatrix} \boldsymbol{V}_{AS} \\ \boldsymbol{I}_{AS} \\ \hline \boldsymbol{V}_{BS} \\ \boldsymbol{I}_{BS} \end{bmatrix} = \left[\begin{array}{cc|cc} \boldsymbol{A}_{AA} & \boldsymbol{B}_{AA} & \boldsymbol{A}_{AB} & \boldsymbol{B}_{AB} \\ \boldsymbol{C}_{AA} & \boldsymbol{D}_{AA} & \boldsymbol{C}_{AB} & \boldsymbol{D}_{AB} \\ \hline \boldsymbol{A}_{BA} & \boldsymbol{B}_{BA} & \boldsymbol{A}_{BB} & \boldsymbol{B}_{BB} \\ \boldsymbol{C}_{BA} & \boldsymbol{D}_{BA} & \boldsymbol{C}_{BB} & \boldsymbol{D}_{BB} \end{array}\right] \begin{bmatrix} \boldsymbol{V}_{AR} \\ \boldsymbol{I}_{AR} \\ \hline \boldsymbol{V}_{BR} \\ \boldsymbol{I}_{BR} \end{bmatrix} \tag{12.73}$$

The similarity between the mutually coupled line equations and the single line described by its ***ABCD*** parameters is evident in both (12.72) and (12.73).

Comparing the solution of the long line equations for mutually coupled lines with that of a single transmission line, we see that mutual coupling introduces an additional propagation constant into the solution. The solution takes on somewhat the same form as that of the single line, being functions of hyperbolic sines and cosines, but with the added complexity of two propagation parameters, each of which is a function of the parameters of both lines. The propagation constants are complex numbers, where the real part is called the *attenuation constant*, and the imaginary part is called the *phase constant*. This means that there are two attenuation and phase constants for mutually coupled lines, which adds new complexities to any detailed analysis of the transmission system.

12.5.2.2 Long Lines with Identical Parameters. If we assume that the two transmission lines are identical, but still mutually coupled, the system parameters can be simplified. In particular, we assume the following line parameters.

$$\begin{aligned} z_A &= z_B \equiv z_S \\ y_A &= y_B \equiv y_S \end{aligned} \tag{12.74}$$

Then, the following simplifications can be determined.

$$\begin{aligned} \gamma_{AA}^2 &= \gamma_{BB}^2 = z_S y_S - z_M y_M \equiv \gamma_S^2 \\ \gamma_{AB}^2 &= \gamma_{BA}^2 = z_S y_M - z_M y_S \equiv \gamma_M^2 \end{aligned} \tag{12.75}$$

From (12.66) and (12.67), we can determine the following.

$$\begin{aligned} \gamma_1^2 &= \gamma_S^2 + \gamma_M^2 = (z_S - z_M)(y_S + y_M) \\ \gamma_2^2 &= \gamma_S^2 - \gamma_M^2 = (z_S + z_M)(y_S - y_M) \end{aligned} \tag{12.76}$$

$$\begin{aligned} \gamma_1^2 - \gamma_{AA}^2 &= \gamma_1^2 - \gamma_{BB}^2 = \gamma_M^2 \\ \gamma_2^2 - \gamma_{AA}^2 &= \gamma_2^2 - \gamma_{BB}^2 = -\gamma_M^2 \end{aligned} \tag{12.77}$$

Then, the voltage equations are still given by

$$\begin{bmatrix} \boldsymbol{V}_{Ax}(s) \\ \boldsymbol{V}_{Bx}(s) \end{bmatrix} = \begin{bmatrix} \boldsymbol{A}_{AA}(s) & \boldsymbol{A}_{AB}(s) \\ \boldsymbol{A}_{BA}(s) & \boldsymbol{A}_{BB}(s) \end{bmatrix} \begin{bmatrix} \boldsymbol{V}_{AR} \\ \boldsymbol{V}_{BR} \end{bmatrix} + \begin{bmatrix} \boldsymbol{B}_{AA}(s) & \boldsymbol{B}_{AB}(s) \\ \boldsymbol{B}_{BA}(s) & \boldsymbol{B}_{BB}(s) \end{bmatrix} \begin{bmatrix} \boldsymbol{I}_{AR} \\ \boldsymbol{I}_{BR} \end{bmatrix} \tag{12.78}$$

but the matrix elements are simpler for this special case and are given by

$$\begin{aligned} \boldsymbol{A}_{AA} = \boldsymbol{A}_{BB} &= \frac{\cosh \gamma_1 x + \cosh \gamma_2 x}{2} \equiv \boldsymbol{A}_S \\ \boldsymbol{A}_{AB} = \boldsymbol{A}_{BA} &= \frac{(\cosh \gamma_1 x - \cosh \gamma_2 x)}{2} \equiv \boldsymbol{A}_M \end{aligned} \tag{12.79}$$

and

$$\begin{aligned} \boldsymbol{B}_{AA} = \boldsymbol{B}_{BB} &= \left(\frac{z_S + z_M}{2\gamma_1}\right) \sinh \gamma_1 x + \left(\frac{z_S - z_M}{2\gamma_2}\right) \sinh \gamma_2 x \equiv \boldsymbol{B}_S \\ \boldsymbol{B}_{AB} = \boldsymbol{B}_{BA} &= \left(\frac{z_S + z_M}{2\gamma_1}\right) \sinh \gamma_1 x - \left(\frac{z_S - z_M}{2\gamma_2}\right) \sinh \gamma_2 x \equiv \boldsymbol{B}_M \end{aligned} \tag{12.80}$$

In a similar manner, the currents equations are found for this case to be

$$\begin{bmatrix} \boldsymbol{I}_{Ax}(s) \\ \boldsymbol{I}_{Bx}(s) \end{bmatrix} = \begin{bmatrix} \boldsymbol{D}_{AA}(s) & \boldsymbol{D}_{AB}(s) \\ \boldsymbol{D}_{BA}(s) & \boldsymbol{D}_{BB}(s) \end{bmatrix} \begin{bmatrix} \boldsymbol{I}_{AR} \\ \boldsymbol{I}_{BR} \end{bmatrix} + \begin{bmatrix} \boldsymbol{C}_{AA}(s) & \boldsymbol{C}_{AB}(s) \\ \boldsymbol{C}_{BA}(s) & \boldsymbol{C}_{BB}(s) \end{bmatrix} \begin{bmatrix} \boldsymbol{V}_{AR} \\ \boldsymbol{V}_{BR} \end{bmatrix} \tag{12.81}$$

where the first matrix on the right-hand side has been changed and is now identical to the $\boldsymbol{A}$ matrix of the voltage equations given by (12.78). The admittance matrix elements are given by

$$\begin{aligned} \boldsymbol{C}_{AA} = \boldsymbol{C}_{BB} &= \left(\frac{y_S - y_M}{2\gamma_1}\right) \sinh \gamma_1 x + \left(\frac{y_S + y_M}{2\gamma_2}\right) \sinh \gamma_2 x \equiv \boldsymbol{C}_S \\ \boldsymbol{C}_{AB} = \boldsymbol{C}_{BA} &= \left(\frac{y_S - y_M}{2\gamma_1}\right) \sinh \gamma_1 x - \left(\frac{y_S + y_M}{2\gamma_2}\right) \sinh \gamma_2 x \equiv \boldsymbol{C}_M \end{aligned} \tag{12.82}$$

When the transmission lines are physically identical and are described by identical physical parameters, the solutions are simplified. However, there are still two distinct propagation constants for the coupled line system. Using the newly defined constants for the identical line case, we can write the system matrix for the sending end quantities as follows.

$$\begin{bmatrix} \boldsymbol{V}_{AS} \\ \boldsymbol{V}_{BS} \\ \hline \boldsymbol{I}_{AS} \\ \boldsymbol{I}_{BS} \end{bmatrix} = \left[\begin{array}{cc|cc} \boldsymbol{A}_S & \boldsymbol{A}_M & \boldsymbol{B}_S & \boldsymbol{B}_M \\ \boldsymbol{A}_M & \boldsymbol{A}_S & \boldsymbol{B}_M & \boldsymbol{B}_S \\ \hline \boldsymbol{C}_S & \boldsymbol{C}_M & \boldsymbol{D}_S & \boldsymbol{D}_M \\ \boldsymbol{C}_M & \boldsymbol{C}_S & \boldsymbol{D}_M & \boldsymbol{D}_S \end{array}\right] \begin{bmatrix} \boldsymbol{V}_{AR} \\ \boldsymbol{V}_{BR} \\ \hline \boldsymbol{I}_{AR} \\ \boldsymbol{I}_{BR} \end{bmatrix} \tag{12.83}$$

or

$$\begin{bmatrix} \boldsymbol{V}_{AS} \\ \boldsymbol{I}_{AS} \\ \hline \boldsymbol{V}_{BS} \\ \boldsymbol{I}_{BS} \end{bmatrix} = \left[\begin{array}{cc|cc} \boldsymbol{A}_S & \boldsymbol{B}_S & \boldsymbol{A}_M & \boldsymbol{B}_M \\ \boldsymbol{C}_S & \boldsymbol{D}_S & \boldsymbol{C}_M & \boldsymbol{D}_M \\ \hline \boldsymbol{A}_M & \boldsymbol{B}_M & \boldsymbol{A}_S & \boldsymbol{B}_S \\ \boldsymbol{C}_M & \boldsymbol{D}_M & \boldsymbol{C}_S & \boldsymbol{D}_S \end{array}\right] \begin{bmatrix} \boldsymbol{V}_{AR} \\ \boldsymbol{I}_{AR} \\ \hline \boldsymbol{V}_{BR} \\ \boldsymbol{I}_{BR} \end{bmatrix} \tag{12.84}$$

Note that the $\boldsymbol{D}$ matrix is identical to the $\boldsymbol{A}$ matrix, but is traditional to identify the system parameters in this manner.

12.5.2.3 Representation of the Faulted Long Line. In cases where line protection is the primary focus of the analysis, it is convenient to identify the relay location at one end of the coupled lines and the fault point at the opposite end, thereby setting up a practical protection situation. Such is the system described in Figure 12.27. We arbitrarily assume that the fault is on line A and is located at point AF. Fault currents flow to point AF from both ends of the line as shown in the figure. The corresponding fault point on line B is called BF, and currents are defined as flowing to that point from both ends of the line, although these currents obviously sum to zero at the point BF.

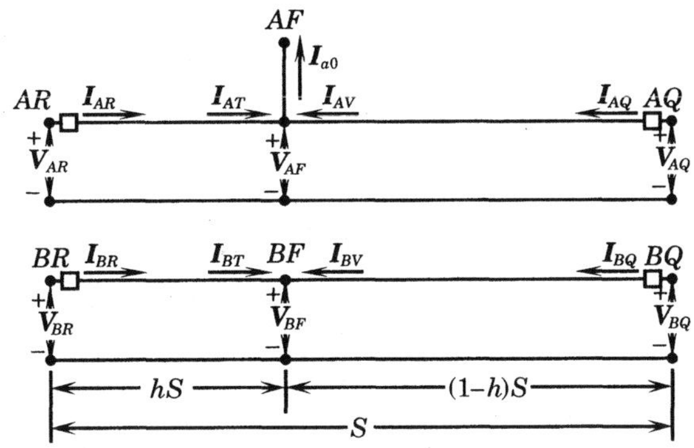

Figure 12.27 Representation of a fault on mutually coupled lines.

Since the lines are long, distributed-parameter systems, the currents at the two ends of each line section must be identified separately. Circuit breakers are assumed at each end of the lines, represented by square boxes in the figure. Note that the line length is designated as S, and that the fault occurs at a fractional distance h from the relays at AR.

12.6 LONG TRANSMISSION LINE EQUIVALENTS

We can use the equations developed in the previous section to determine coupled-line equivalent circuits for various types of line connections to the external power system. However, before equivalents can be developed, it is important to determine the reciprocity of the four-port network. If the network is not reciprocal, equivalents circuits using R-L-C elements are not possible. Therefore, a test for reciprocity is necessary.

12.6.1 Reciprocity and the Admittance Matrix

Before considering any detailed calculations for transmission line equivalents, we examine the reciprocity of the system equation (12.72) or (12.73). Reciprocity is demonstrated by subjecting the network to a series of tests that retain constant terminating impedances at all four network terminals. One way of doing this is to apply an ideal voltage source at one port and measuring the response at the other ports using ideal ammeters, as shown in Figure 12.28. Since both ideal voltage sources and ideal ammeters have zero internal impedance, the voltage sources and ammeters can be interchanged without altering the impedances of the network terminations. An exhaustive test of the network is required, applying the voltage source at each of the four terminals, in turn, and observing (or computing) the resulting ammeter readings. The reciprocity theorem states that if an ideal voltage source $\boldsymbol{E}$ at one point in a network produces a current $\boldsymbol{I}$ at a second point in the network, then the same voltage source $\boldsymbol{E}$ acting at the second point will produce the same current $\boldsymbol{I}$ at the first point [17].

A series of reciprocity tests can be designed, whereby an arbitrary ideal voltage source is applied to each port and currents measured at all other ports as shown in Figure 12.28. However, the information resulting from this series of tests is exactly the same as that determined by

Figure 12.28 The first test of network reciprocity.

constructing the admittance matrix of the four-port network, and examining the symmetry of the resulting admittance matrix, since reciprocal networks are ensured of having a symmetric admittance or impedance matrix representation. Therefore, as an alternative to a classical reciprocity test, we construct the admittance matrix directly from the ***ABCD*** equations.

We begin with the ***ABCD*** equations described in (12.72), which are repeated here for convenience.

$$\begin{bmatrix} V_{AS} \\ V_{BS} \\ \hline I_{AS} \\ I_{BS} \end{bmatrix} = \left[\begin{array}{cc|cc} A_{AA} & A_{AB} & B_{AA} & B_{AB} \\ A_{BA} & A_{BB} & B_{BA} & B_{BB} \\ \hline C_{AA} & C_{AB} & D_{AA} & D_{AB} \\ C_{BA} & C_{BB} & D_{BA} & D_{BB} \end{array}\right] \begin{bmatrix} V_{AR} \\ V_{BR} \\ \hline I_{AR} \\ I_{BR} \end{bmatrix} \tag{12.85}$$

This equation can be written in a more compact matrix notation as follows, with the matrix described in terms of the partitions of (12.85). Thus we have

$$\begin{bmatrix} \mathbf{V}_S \\ \mathbf{I}_S \end{bmatrix} = \begin{bmatrix} \mathbf{A} & \mathbf{B} \\ \mathbf{C} & \mathbf{D} \end{bmatrix} \begin{bmatrix} \mathbf{V}_R \\ \mathbf{I}_R \end{bmatrix} \tag{12.86}$$

where

$$\mathbf{V}_S = \begin{bmatrix} V_{AS} \\ V_{BS} \end{bmatrix} \qquad \mathbf{I}_S = \begin{bmatrix} I_{AS} \\ I_{BS} \end{bmatrix} \tag{12.87}$$

$$\mathbf{V}_R = \begin{bmatrix} V_{AR} \\ V_{BR} \end{bmatrix} \qquad I_R = \begin{bmatrix} I_{AR} \\ I_{BR} \end{bmatrix} \tag{12.88}$$

and the **ABCD** elements of (12.86) are the corresponding submatrices of (12.85). From the $\mathbf{V}_S$ equation of (12.86), we may write

$$\mathbf{B}\mathbf{I}_R = -\mathbf{A}\mathbf{V}_R + \mathbf{V}_S \tag{12.89}$$

or

$$\mathbf{I}_R = -\mathbf{B}^{-1}\mathbf{A}\mathbf{V}_R + \mathbf{B}^{-1}\mathbf{V}_S \tag{12.90}$$

Substituting (12.90) into the current equation of (12.86), we compute

$$\mathbf{I}_S = (\mathbf{C} - \mathbf{D}\mathbf{B}^{-1}\mathbf{A})\mathbf{V}_R + \mathbf{D}\mathbf{B}^{-1}\mathbf{V}_S \tag{12.91}$$

or, in matrix notation, we may write

$$\begin{bmatrix} \mathbf{I}_S \\ -\mathbf{I}_R \end{bmatrix} = \begin{bmatrix} \mathbf{D}\mathbf{B}^{-1} & (\mathbf{C} - \mathbf{D}\mathbf{B}^{-1}\mathbf{A}) \\ -\mathbf{B}^{-1} & \mathbf{B}^{-1}\mathbf{A} \end{bmatrix} \begin{bmatrix} \mathbf{V}_S \\ \mathbf{V}_R \end{bmatrix} \tag{12.92}$$

This is the admittance matrix of the four-port network, expressed in terms of the **ABCD** network parameter submatrices.

We now expand the current vectors to identify the individual currents entering the network. For the sending end, we may write from (12.91),

$$\begin{aligned}\begin{bmatrix} \boldsymbol{I}_{AS} \\ \boldsymbol{I}_{BS} \end{bmatrix} &= \begin{bmatrix} \boldsymbol{D}_{AA} & \boldsymbol{D}_{AB} \\ \boldsymbol{D}_{BA} & \boldsymbol{D}_{BB} \end{bmatrix} \frac{1}{\det \boldsymbol{B}} \begin{bmatrix} \boldsymbol{B}_{BB} & -\boldsymbol{B}_{AB} \\ -\boldsymbol{B}_{BA} & \boldsymbol{B}_{AA} \end{bmatrix} \begin{bmatrix} \boldsymbol{V}_{AS} \\ \boldsymbol{V}_{BS} \end{bmatrix} + \begin{bmatrix} \boldsymbol{C}_{AA} & \boldsymbol{C}_{AB} \\ \boldsymbol{C}_{BA} & \boldsymbol{C}_{BB} \end{bmatrix} \begin{bmatrix} \boldsymbol{V}_{AR} \\ \boldsymbol{V}_{BR} \end{bmatrix} \\ &\quad - \left\{ \begin{bmatrix} \boldsymbol{D}_{AA} & \boldsymbol{D}_{AB} \\ \boldsymbol{D}_{BA} & \boldsymbol{D}_{BB} \end{bmatrix} \frac{1}{\det \boldsymbol{B}} \begin{bmatrix} \boldsymbol{B}_{BB} & -\boldsymbol{B}_{AB} \\ -\boldsymbol{B}_{BA} & \boldsymbol{B}_{AA} \end{bmatrix} \begin{bmatrix} \boldsymbol{A}_{AA} & \boldsymbol{A}_{AB} \\ \boldsymbol{A}_{BA} & \boldsymbol{A}_{BB} \end{bmatrix} \right\} \begin{bmatrix} \boldsymbol{V}_{AR} \\ \boldsymbol{V}_{BR} \end{bmatrix}\end{aligned} \tag{12.93}$$

This equation can be simplified to the form

$$\begin{bmatrix} \boldsymbol{I}_{AS} \\ \boldsymbol{I}_{BS} \end{bmatrix} = \begin{bmatrix} \boldsymbol{Y}_{SS1} & \boldsymbol{Y}_{SS2} \\ \boldsymbol{Y}_{SS3} & \boldsymbol{Y}_{SS4} \end{bmatrix} \begin{bmatrix} \boldsymbol{V}_{AS} \\ \boldsymbol{V}_{BS} \end{bmatrix} + \begin{bmatrix} \boldsymbol{Y}_{SR1} & \boldsymbol{Y}_{SR2} \\ \boldsymbol{Y}_{SR3} & \boldsymbol{Y}_{SR4} \end{bmatrix} \begin{bmatrix} \boldsymbol{V}_{AR} \\ \boldsymbol{V}_{BR} \end{bmatrix} \tag{12.94}$$

where we define the following elements of (12.94).

$$\begin{aligned} \boldsymbol{Y}_{SS1} &= \frac{\boldsymbol{D}_{AA}\boldsymbol{B}_{BB} - \boldsymbol{D}_{AB}\boldsymbol{B}_{BA}}{\det \boldsymbol{B}} & \qquad \boldsymbol{Y}_{SR1} &= \boldsymbol{C}_{AA} - \frac{\boldsymbol{D}_{AA}\boldsymbol{Z}_1 + \boldsymbol{D}_{AB}\boldsymbol{Z}_3}{\det \boldsymbol{B}} \\ \boldsymbol{Y}_{SS2} &= \frac{\boldsymbol{D}_{AB}\boldsymbol{B}_{AA} - \boldsymbol{D}_{AA}\boldsymbol{B}_{AB}}{\det \boldsymbol{B}} & \qquad \boldsymbol{Y}_{SR2} &= \boldsymbol{C}_{AB} - \frac{\boldsymbol{D}_{AA}\boldsymbol{Z}_2 + \boldsymbol{D}_{AB}\boldsymbol{Z}_4}{\det \boldsymbol{B}} \\ & \qquad\qquad \text{and} \\ \boldsymbol{Y}_{SS3} &= \frac{\boldsymbol{D}_{BA}\boldsymbol{B}_{BB} - \boldsymbol{D}_{BB}\boldsymbol{B}_{BA}}{\det \boldsymbol{B}} & \qquad \boldsymbol{Y}_{SR3} &= \boldsymbol{C}_{BA} - \frac{\boldsymbol{D}_{BA}\boldsymbol{Z}_1 + \boldsymbol{D}_{BB}\boldsymbol{Z}_3}{\det \boldsymbol{B}} \\ \boldsymbol{Y}_{SS4} &= \frac{\boldsymbol{D}_{BB}\boldsymbol{B}_{AA} - \boldsymbol{D}_{BA}\boldsymbol{B}_{AB}}{\det \boldsymbol{B}} & \qquad \boldsymbol{Y}_{SR4} &= \boldsymbol{C}_{BB} - \frac{\boldsymbol{D}_{BA}\boldsymbol{Z}_2 + \boldsymbol{D}_{BB}\boldsymbol{Z}_4}{\det \boldsymbol{B}} \end{aligned} \tag{12.95}$$

In (12.95) we have defined for convenience the following $\boldsymbol{Z}$ parameters, each having the dimension of impedance.

$$\begin{aligned} \boldsymbol{Z}_1 &= \boldsymbol{A}_{AA}\boldsymbol{B}_{BB} - \boldsymbol{A}_{BA}\boldsymbol{B}_{AB} \\ \boldsymbol{Z}_2 &= \boldsymbol{A}_{AB}\boldsymbol{B}_{BB} - \boldsymbol{A}_{BB}\boldsymbol{B}_{AB} \\ \boldsymbol{Z}_3 &= \boldsymbol{A}_{BA}\boldsymbol{B}_{AA} - \boldsymbol{A}_{AA}\boldsymbol{B}_{BA} \\ \boldsymbol{Z}_4 &= \boldsymbol{A}_{BB}\boldsymbol{B}_{AA} - \boldsymbol{A}_{AB}\boldsymbol{B}_{BA} \end{aligned} \tag{12.96}$$

The receiving-end currents are also defined by (12.92), which can be expanded as follows.

$$\begin{aligned} \begin{bmatrix} \boldsymbol{I}_{AR} \\ \boldsymbol{I}_{BR} \end{bmatrix} &= \frac{1}{\det \boldsymbol{B}} \begin{bmatrix} \boldsymbol{B}_{BB} & -\boldsymbol{B}_{AB} \\ -\boldsymbol{B}_{BA} & \boldsymbol{B}_{AA} \end{bmatrix} \begin{bmatrix} \boldsymbol{V}_{AS} \\ \boldsymbol{V}_{BS} \end{bmatrix} - \frac{1}{\det \boldsymbol{B}} \begin{bmatrix} \boldsymbol{B}_{BB} & -\boldsymbol{B}_{AB} \\ -\boldsymbol{B}_{BA} & \boldsymbol{B}_{AA} \end{bmatrix} \begin{bmatrix} \boldsymbol{A}_{AA} & \boldsymbol{A}_{AB} \\ \boldsymbol{A}_{BA} & \boldsymbol{A}_{BB} \end{bmatrix} \begin{bmatrix} \boldsymbol{V}_{AR} \\ \boldsymbol{V}_{BR} \end{bmatrix} \\ &= - \begin{bmatrix} \boldsymbol{Y}_{RS1} & \boldsymbol{Y}_{RS2} \\ \boldsymbol{Y}_{RS3} & \boldsymbol{Y}_{RS4} \end{bmatrix} \begin{bmatrix} \boldsymbol{V}_{AS} \\ \boldsymbol{V}_{BS} \end{bmatrix} - \begin{bmatrix} \boldsymbol{Y}_{RR1} & \boldsymbol{Y}_{RR2} \\ \boldsymbol{Y}_{RR3} & \boldsymbol{Y}_{RR4} \end{bmatrix} \begin{bmatrix} \boldsymbol{V}_{AR} \\ \boldsymbol{V}_{BR} \end{bmatrix} \end{aligned} \tag{12.97}$$

The defining matrix elements in this case are stated as

$$\begin{aligned} \boldsymbol{Y}_{RR1} &= \frac{\boldsymbol{Z}_1}{\det \boldsymbol{B}} & \qquad \boldsymbol{Y}_{RS1} &= \frac{-\boldsymbol{B}_{BB}}{\det \boldsymbol{B}} \\ \boldsymbol{Y}_{RR2} &= \frac{\boldsymbol{Z}_2}{\det \boldsymbol{B}} & \qquad \boldsymbol{Y}_{RS2} &= \frac{\boldsymbol{B}_{AB}}{\det \boldsymbol{B}} \\ & \qquad \text{and} \\ \boldsymbol{Y}_{RR3} &= \frac{\boldsymbol{Z}_3}{\det \boldsymbol{B}} & \qquad \boldsymbol{Y}_{RS3} &= \frac{\boldsymbol{B}_{BA}}{\det \boldsymbol{B}} \\ \boldsymbol{Y}_{RR4} &= \frac{\boldsymbol{Z}_4}{\det \boldsymbol{B}} & \qquad \boldsymbol{Y}_{RS4} &= \frac{-\boldsymbol{B}_{AA}}{\det \boldsymbol{B}} \end{aligned} \tag{12.98}$$

It is important to note that the receiving-end currents desired for the admittance matrix are the negative of those given by (12.97). Equations (12.94) and (12.97) express the port currents in terms of the port voltages, which defines the admittance matrix. Rewriting in matrix form, we have

$$\begin{bmatrix} I_{AS} \\ I_{BS} \\ \hline -I_{AR} \\ -I_{BR} \end{bmatrix} = \left[\begin{array}{cc|cc} Y_{SS1} & Y_{SS2} & Y_{SR1} & Y_{SR2} \\ Y_{SS3} & Y_{SS4} & Y_{SR3} & Y_{SR4} \\ \hline Y_{RS1} & Y_{RS2} & Y_{RR1} & Y_{RR2} \\ Y_{RS3} & Y_{RS4} & Y_{RR3} & Y_{RR4} \end{array}\right] \begin{bmatrix} V_{AS} \\ V_{BS} \\ \hline V_{AR} \\ V_{BR} \end{bmatrix} \tag{12.99}$$

The off-diagonal terms of the admittance matrix are exactly the same terms that would be obtained from the reciprocity tests. The equality of $\boldsymbol{Y}_{ik}$ and $\boldsymbol{Y}_{ki}$ for all values of i and k is not at all clear from the terms defined in (12.95) and (12.98). A rigorous proof would be difficult, but a proof can be obtained using computer methods that permit the evaluation of symbolic equations (see problem 12.26). Numerical evaluation of these complex quantities verifies that reciprocity is satisfied, as shown in the following examples.

EXAMPLE 12.3

Compute the ***ABCD*** parameters for two parallel EHV transmission lines that are 200 miles long. The basic line parameters per mile of length, expressed in per unit are as follows:

$$\begin{aligned} z_A &= 0.542331 + j1.65684 \\ z_B &= 0.521516 + j1.59493 \\ z_M &= 0.429016 + j0.448080 \\ y_A &= 0 + j4.19196 \\ y_B &= 0 + j4.07216 \\ y_M &= 0 - j0.209593 \end{aligned}$$

Solution

The given values provide the raw data required to compute the ***ABCD*** parameters for the mutually coupled lines. The results of that computation for the zero-sequence system, using the equations developed above, are given in matrix form, as follows:

$$\begin{bmatrix} \mathbf{A} & \mathbf{B} \\ \mathbf{C} & \mathbf{D} \end{bmatrix} = \left[\begin{array}{cc|cc} \boldsymbol{A}_{AA} & \boldsymbol{A}_{AB} & \boldsymbol{B}_{AA} & \boldsymbol{B}_{AB} \\ \boldsymbol{A}_{BA} & \boldsymbol{A}_{BB} & \boldsymbol{B}_{BA} & \boldsymbol{B}_{BB} \\ \hline \boldsymbol{C}_{AA} & \boldsymbol{C}_{AB} & \boldsymbol{D}_{AA} & \boldsymbol{D}_{AB} \\ \boldsymbol{C}_{BA} & \boldsymbol{C}_{BB} & \boldsymbol{D}_{BA} & \boldsymbol{D}_{BB} \end{array}\right]$$

$$= \left[\begin{array}{cc|cc} 0.8622 & -0.0420 & 0.3845 & 0.3007 \\ +j0.0445 & +j0.0349 & +j1.2697 & +j0.3339 \\ -0.0428 & 0.8709 & 0.3007 & 0.3716 \\ +j0.0358 & +j0.0418 & +j0.3339 & +j1.2255 \\ \hline -0.0033 & -0.0027 & 0.8622 & -0.0428 \\ +j0.1997 & +j0.0070 & +j0.0445 & +j0.0358 \\ -0.0027 & -0.0030 & -0.0420 & 0.8709 \\ +j0.0070 & +j0.1946 & +j0.0349 & +j0.0418 \end{array}\right]$$

The numerical values are expressed here to only a few decimals and are not of sufficient accuracy for precise computation of the admittance matrix. Experience has shown that the basic parameters should be expressed to at least 10 significant figures, and this degree of precision must be maintained throughout the computation of the admittance matrix. Note the equality of the diagonal terms, which proves reciprocity. ■

The foregoing example shows typical values of **ABCD** matrix parameters for mutually coupled lines. The next step in the process of deriving a network equivalent is the conversion of the transmission parameters to the admittance matrix, using the method described above.

EXAMPLE 12.4

Compute the elements of the admittance matrix using the numerical results of Example 12.3. Then extend the result to determine the parameters of the equivalent circuit.

Solution

Using the data given in Example 12.3, but expressed to 10 significant figures of accuracy, we compute the following admittance matrix.

$$\left[\begin{array}{cc|cc} Y_{11} & Y_{12} & Y_{13} & Y_{14} \\ Y_{21} & Y_{22} & Y_{23} & Y_{24} \\ \hline Y_{31} & Y_{32} & Y_{33} & Y_{34} \\ Y_{41} & Y_{42} & Y_{43} & Y_{44} \end{array}\right] = \left[\begin{array}{cc|cc} 0.1585 & 0.0694 & -0.1576 & -0.0686 \\ -j0.6889 & +j0.2826 & +j0.7963 & -j0.2765 \\ 0.0694 & 0.1644 & -0.0686 & -0.1635 \\ +j0.2826 & -j0.7207 & -j0.2765 & +j0.8249 \\ \hline -0.1576 & -0.0686 & 0.1585 & 0.0694 \\ j0.7963 & -j0.2765 & -j0.6889 & +j0.2826 \\ -0.0686 & -0.1635 & 0.0694 & +0.1644 \\ j0.2765 & +j0.8249 & +j0.2826 & -j0.7207 \end{array}\right]$$

Note the unique symmetry of the resulting admittance matrix. First, the matrix is symmetric. However, there are other symmetry patterns that are striking. This is better viewed symbolically, as shown in the pictorial representation of Figure 12.29, using simple geometric shapes to indicate cells of equal value. The symmetry for the case under study, with nonidentical lines, is shown on the left. If the lines have identical parameters, the symmetry is even greater, as shown by the pictorial on the right. This interesting form of symmetry will always result when admittance matrix is derived from **ABCD** parameters.

○	×	△	■
×	●	■	▽
△	■	○	×
■	▽	×	●

Non-Identical Lines

●	×	△	■
×	●	■	△
△	■	●	×
■	△	×	●

Identical Lines

Figure 12.29 Symmetry patterns of the admittance matrix. ■

12.6.2 The Long Line Type 3 Network Equivalent

It was noted in the study of short-line equivalent circuits, that the four-terminal equivalent is the only one required, as the special cases can be found from the general four-terminal equivalent. Therefore, we begin our study of long-line equivalents by first developing the Type 3 equivalent, i.e., the equivalent with all four terminals distinct, as shown in Figure 12.30.

12.6.2.1 Type 3 Network Configuration. Since there are four distinct terminals the network will require 10 elements, according to (12.27), with the elements arranged as shown

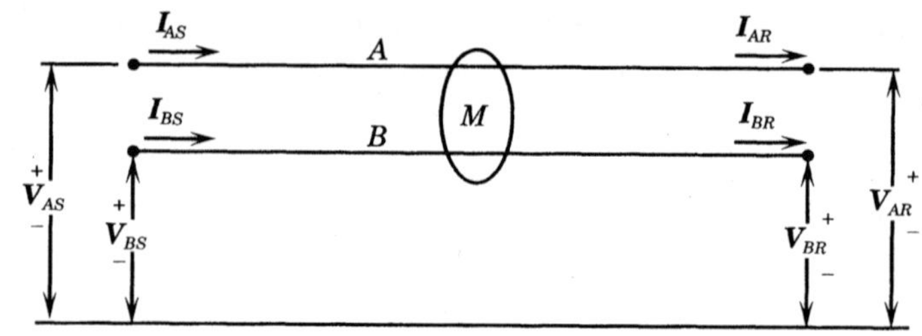

Figure 12.30 Long transmission lines not bused at either end.

in Figure 12.31. The notation used takes advantage of the fact that certain elements are expected to be identical. The circuit of Figure 12.31 is general in the sense that the lines are not required to have identical parameters.

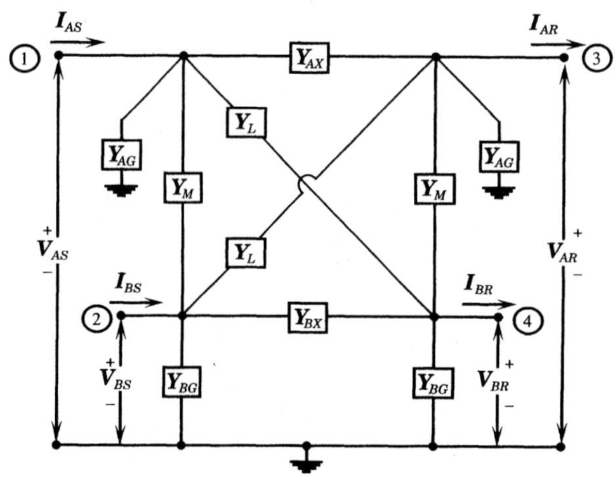

Figure 12.31 Equivalent circuit for type 3 networks.

Since the network obeys the reciprocity theorem, we can easily construct the network of Figure 12.31 from the admittance matrix, which we may write in the following form.

$$\begin{bmatrix} I_1 \\ I_2 \\ \hline I_3 \\ I_4 \end{bmatrix} = \left[\begin{array}{cc|cc} Y_{11} & Y_{12} & Y_{13} & Y_{14} \\ Y_{21} & Y_{22} & Y_{23} & Y_{24} \\ \hline Y_{31} & Y_{32} & Y_{33} & Y_{34} \\ Y_{41} & Y_{42} & Y_{43} & Y_{44} \end{array}\right] \begin{bmatrix} V_1 \\ V_2 \\ \hline V_3 \\ V_4 \end{bmatrix} \tag{12.100}$$

Then, based on the manner of constructing the admittance matrix for Figure 12.31, we compute the following parameters.

$$\begin{aligned} Y_M &= -Y_{12} = -Y_{21} = -Y_{34} = -Y_{43} \\ Y_L &= -Y_{14} = -Y_{41} = -Y_{23} = -Y_{32} \end{aligned} \tag{12.101}$$

$$\begin{aligned} Y_{AX} &= -Y_{13} = -Y_{31} \\ Y_{BX} &= -Y_{24} = -Y_{42} \end{aligned} \tag{12.102}$$

$$\begin{aligned} Y_{AG} &= Y_{11} + Y_{12} + Y_{13} + Y_{14} = Y_{31} + Y_{32} + Y_{33} + Y_{34} \\ Y_{BG} &= Y_{21} + Y_{22} + Y_{23} + Y_{24} = Y_{41} + Y_{42} + Y_{43} + Y_{44} \end{aligned} \tag{12.103}$$

The network equivalent of Figure 12.31 will usually be inserted into a network that represents either the detailed power system or an equivalent of that power system.

12.6.2.2 Type 3 Network in System Analysis. The use of the Type 3 equivalent circuit in fault analysis requires that the line equivalent be incorporated into a larger network representing the entire power system. In most cases, the mutual coupling will be analyzed for only the zero-sequence network, although a similar network could be computed for the positive and negative sequences if the small amount of mutual coupling of those networks is of interest.

Figure 12.32 shows an example of the use of the Type 3 equivalent in fault analysis. Only the zero-sequence network is shown in the figure, which illustrates an example where the Type 3 equivalent is required because of the end-of-line series capacitors on the parallel transmission lines. Because of the series capacitors, the lines do not terminate at a common bus, but terminate at the series capacitors. Since the currents in the two lines will not be equal, the voltages on the line side of the capacitors will not be equal.

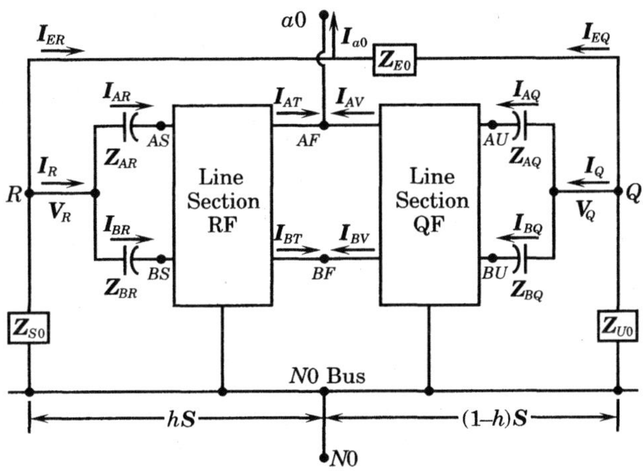

Figure 12.32 Use of the type 3 equivalent in fault analysis.

For the study of faults along the transmission line, two equivalent circuits similar to Figure 12.31 must be used, one for line section RF and another for line section QF. These can be represented as long-line equivalents similar to Figure 12.31, but of proper length between the series capacitors and the fault point. Figure 12.32 shows the fault located at a fractional distance h of the total line length S from bus R.

12.6.3 Long-Line Type 1 Network Equivalents

Type 1 networks are those in which the transmission lines are bused at both ends of the lines. The first type of equivalent is the normal (unfaulted) system, shown in Figure 12.33.

The parameters for this equivalent are readily determined as a special case of the Type 3 equivalent of Figure 12.31. The exact configuration of the resulting equivalent depends on the desired use of the circuit. Figure 12.34 shows two different arrangements that can be used, depending on whether it is necessary to retain the identity of the individual line currents.

A different equivalent is sometimes required when there is a fault along one of the parallel lines. In this case, we require two equivalents, one for each side of the fault, as shown in Fig-

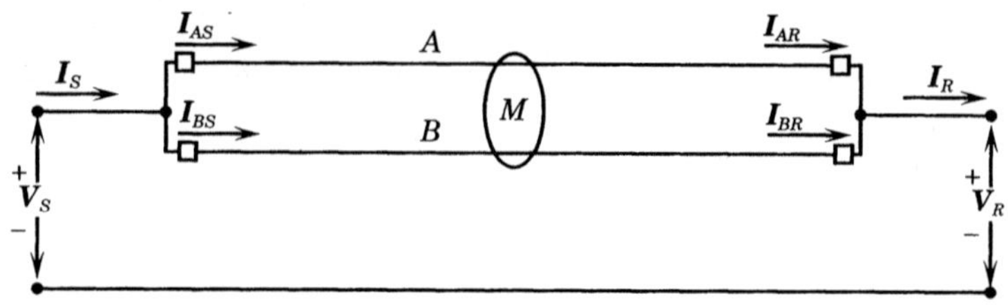

Figure 12.33 Equivalent circuit for identical long, unfaulted, mutually coupled transmission lines.

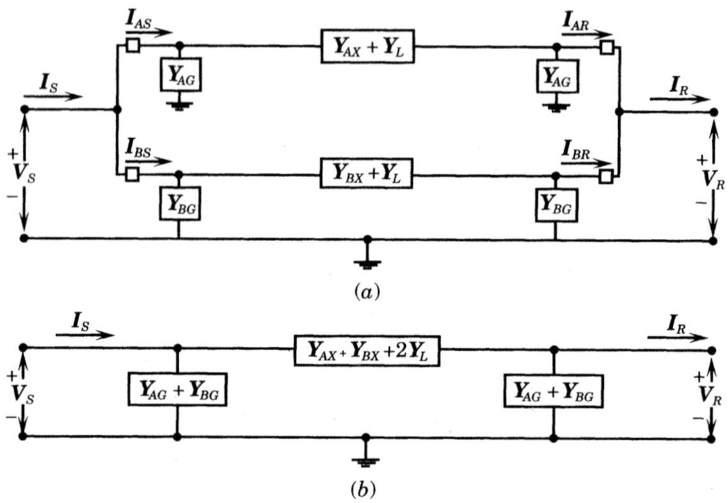

Figure 12.34 Type 1 long-line equivalent networks. (*a*) Individual currents identified. (*b*) Individual currents not identified.

ure 12.35. However, this is recognized as a Type 2 network equivalent since the lines terminate at one bus at one end of each line section, but terminate at two different nodes at the other end.

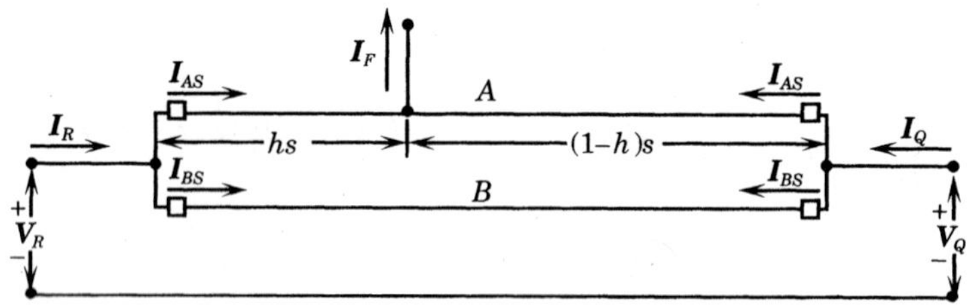

Figure 12.35 Equivalent circuit for the faulted type 1 network.

12.6.4 Long-Line Type 2 Network Equivalents

For this network condition, we construct two equivalents, one for each end of the line, as shown in Figure 12.36. The line currents are named exactly as before, where we identify a sending-end and receiving-end current for each line, using the usual subscript notation.

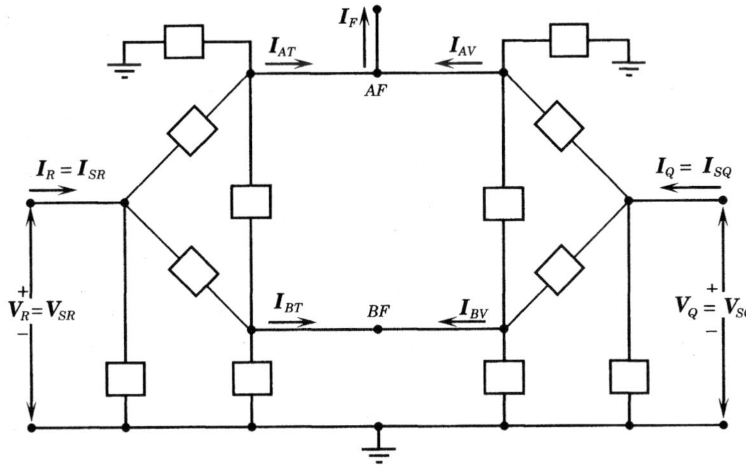

Figure 12.36 Equivalent circuits for the faulted type 1 network.

The Type 2 equivalent for each side of the fault point can be determined from the Type 3 equivalent by connecting the nodes together in the manner prescribed by Figure 12.36. The resulting equivalent for a circuit at the left end of the diagram is shown in Figure 12.37.

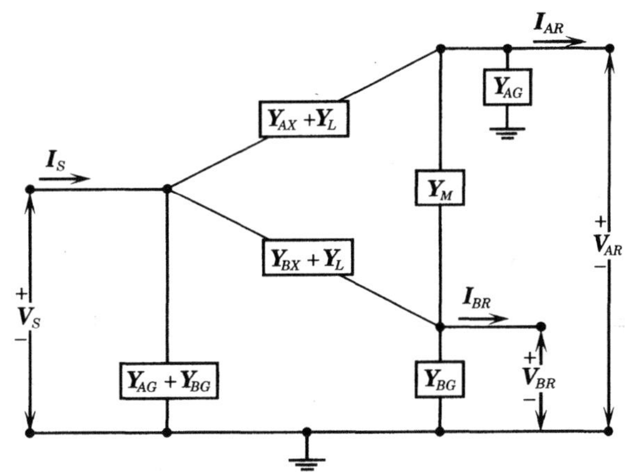

Figure 12.37 Type 2 boundary conditions applied to the left network.

The length of the line on either side of the fault point must be taken into account in finding the **ABCD** parameters for the lines. The line length will be reflected in the admittance matrix and in the circuit parameters computed from that matrix.

12.7 SOLUTION OF THE LONG-LINE CASE

We now develop the solution for fault currents in a power system that includes two mutually coupled transmission lines with nonidentical parameters. A general system configuration is assumed, as shown in Figure 12.38, where a fault is assumed on one of the mutually coupled transmission lines. A general system equivalent, consisting of Thevenin equivalent sources and an external power system, are assumed.

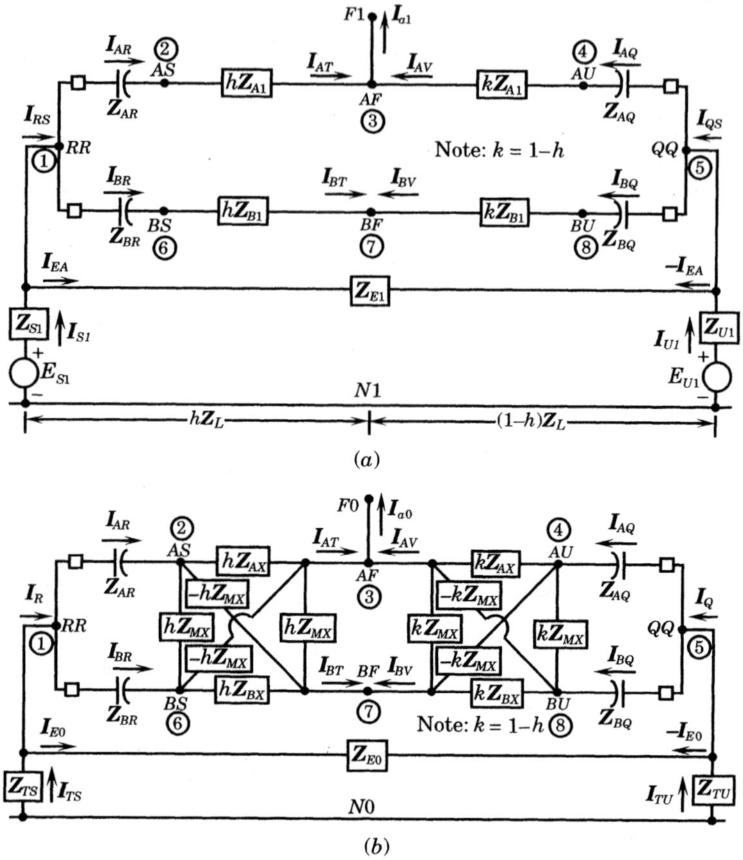

Figure 12.38 Positive and zero-sequence networks of a faulted system. (*a*) Positive sequence network. (*b*) Zero-sequence network.

The cases of greatest interest, insofar as mutual coupling is concerned, are those that involve ground faults. Therefore, we assume that a single or double line-to-ground fault is to be solved. The fault occurs on line A of the mutually coupled pair of lines, and line B is assumed to be in service. Obviously, the system must be changed if line B is not in service, or is out of service and grounded. These situations are considered in Chapter 14. Our task at this point is to describe a method of solution that will serve in any situation of interest that involves the two mutually coupled lines.

In the system selected for solution, series capacitors are shown at the ends of the lines for two reasons. First, some lines do have series capacitors at their ends. If series capacitors are so located, then a Type 3 network equivalent is required, since the voltages at the bus ends of the lines are not equal. Second, the series capacitor impedances can be replaced by equivalent impedances that represent the sequence impedance characteristics of the interconnected power system. Therefore, although the presence of series capacitors complicates the mathematics, the benefits of this representation outweigh the added complexity added by the capacitors.

Also note that we have modified the notation from that used in the previous sections of this chapter to one that is used throughout this book to represent a faulted line, with relay positions at R on the left and at Q on the right. The fault is located at a fractional distance h

from the bus at R. If there are no series capacitors at the ends of the lines, the voltages V_{AS} and V_{BS} are equal to the voltage V_R at node R. In that special case a Type 2 equivalent can be used, or we can continue with the Type 3 equivalents shown, but assign a low value to the series capacitor impedances.

Note that the sequence network nodes have been numbered for convenience and we assume that the admittance matrix for all three sequence networks can be constructed for any fault location.

12.7.1 Determination of the Sequence Impedances

To solve for the fault current it is necessary to know the sequence impedance of the three sequence networks. The solution of the sequence impedances will be developed for the positive-sequence network, but the method can be extended easily to the other sequence networks. The zero-sequence network presents no problem whatsoever, as it is represented by a linear, bilateral network and can be solved using the same technique as prescribed for the positive sequence network.

Consider, then, the positive-sequence network with all sources properly removed. This requires shorting the voltage sources shown in Figure 12.38. The admittance matrix for the eight nodes can be written by inspection, as

$$\begin{bmatrix} I_1 \\ I_2 \\ I_3 \\ \cdots \\ I_8 \end{bmatrix} = \begin{bmatrix} Y_{11} & Y_{12} & Y_{13} & \cdots & Y_{18} \\ Y_{21} & Y_{22} & Y_{23} & \cdots & Y_{28} \\ Y_{31} & Y_{32} & Y_{33} & \cdots & Y_{38} \\ \cdots & \cdots & \cdots & \cdots & \cdots \\ Y_{81} & Y_{82} & Y_{83} & \cdots & Y_{88} \end{bmatrix} \begin{bmatrix} V_1 \\ V_2 \\ V_3 \\ \cdots \\ V_8 \end{bmatrix} \tag{12.104}$$

Examining the circuit carefully, we recognize that the current injected is zero at all nodes except for the current at node 3. Therefore, it is useful to reorder the equations, so that the current and voltage for node 3 are both in row 1 of the matrix equation. Making this adjustment, which is straightforward, we write

$$\begin{bmatrix} I_3 \\ \hline I_1 = 0 \\ \cdots \\ I_8 = 0 \end{bmatrix} = \left[\begin{array}{c|cccc} Y_{33} & Y_{31} & \cdots & Y_{38} \\ \hline Y_{13} & Y_{11} & \cdots & Y_{18} \\ \cdots & \cdots & \cdots & \cdots \\ Y_{83} & Y_{81} & \cdots & Y_{88} \end{array}\right] \begin{bmatrix} V_3 \\ \hline V_1 \\ \cdots \\ V_8 \end{bmatrix} \tag{12.105}$$

We can simplify the notation of (12.105) by assigning fictitious names to the various partitions of the matrix quantities. Thus, we write

$$\begin{bmatrix} I_3 \\ \mathbf{0} \end{bmatrix} = \begin{bmatrix} Y_{33} & \mathbf{Y}_J \\ \mathbf{Y}_K & \mathbf{Y}_L \end{bmatrix} \begin{bmatrix} V_3 \\ \mathbf{V}_X \end{bmatrix} \tag{12.106}$$

The notation used in (12.106) is important. The bold, italic quantities are phasors, but the bold, Roman quantities are matrices. All matrices are defined by the partitions of (12.105). The equation is now in a form that can be readily solved. The second equation can be solved for the unknown voltage, a 7×1 matrix, which can be written as

$$\mathbf{V}_X = -\mathbf{Y}_L^{-1}\mathbf{Y}_K \mathbf{V}_3 \tag{12.107}$$

This voltage can be substituted into the first equation to write the current at node 3 in terms of only the voltage at that node.

$$I_3 = (Y_{33} - \mathbf{Y}_J \mathbf{Y}_L^{-1} \mathbf{Y}_K) V_3 = Y_1 V_3 \tag{12.108}$$

where Y_1 is the total admittance of the positive sequence network. The operation described in (12.108) is often called a Kron reduction. Then the positive-sequence network impedance is given by

$$Z_1 = Y_1^{-1} \tag{12.109}$$

This is the desired result. It represents the total impedance of the positive-sequence network from the fault point $a1$ to the zero-potential bus $N1$. Since the positive and negative sequences are usually almost identical, this result can also be used to represent the negative-sequence network in most cases with very little error. If not, the same procedure can be used to find the total sequence impedance for the negative-sequence network.

Exactly the same procedure is used to determine Z_0, the total impedance of the zero-sequence network. The mutual coupling is represented by the equivalent circuits shown in Figure 12.38 and add nothing to the complexity of the solution.

Knowing the three sequence impedances, the sequence currents of all sequence networks can be found by the proper connection of the networks to represent the type of fault that is to be studied. For example, for the one-line-to-ground fault, the three impedances are placed in series, and an impedance representing the fault impedance, if any, is added [8]. Thus, for the one-line-to-ground fault, we may write

$$I_{a0} = I_{a1} = I_{a2} = \frac{V_F}{Z_0 + Z_1 + Z_2 + 3Z_F} \tag{12.110}$$

where Z_F is the fault impedance. The sequence currents for other types of faults can be found by using the appropriate sequence network connections for that fault type.

12.7.2 Computation of Sequence Voltages and Currents

The computation of the voltages and currents throughout each of the sequence networks is accomplished using the admittance matrices similar to (12.104) that have been computed for each sequence network.

The first step is to compute the inverse of the three sequence admittance matrices to find the sequence network impedance matrices. We can write the form of the impedance matrix for any sequence network as follows, based on the eight node representation of Figure 12.38.

$$\begin{bmatrix} V_1 \\ V_2 \\ V_3 \\ \cdots \\ V_8 \end{bmatrix} = \begin{bmatrix} Z_{11} & Z_{12} & Z_{13} & \cdots & Z_{18} \\ Z_{21} & Z_{22} & Z_{23} & \cdots & Z_{28} \\ Z_{31} & Z_{32} & Z_{33} & \cdots & Z_{38} \\ \cdots & \cdots & \cdots & \cdots & \cdots \\ Z_{81} & Z_{82} & Z_{83} & \cdots & Z_{88} \end{bmatrix} \begin{bmatrix} I_1 = 0 \\ I_2 = 0 \\ I_3 \\ \cdots \\ I_8 = 0 \end{bmatrix} = \begin{bmatrix} Z_{13} \\ Z_{23} \\ Z_{33} \\ \cdots \\ Z_{83} \end{bmatrix} I_3 \tag{12.111}$$

Since the injected currents are all zero except for I_3, the voltages at all nodes depend only on the third column of the impedance matrix. For example, if we inject 1.0 per unit current into node 3, the voltages at all nodes will be exactly equal to the third column of the Z matrix. In an actual fault situation, the current is not 1.0 per unit, but is the sequence current found from the sequence network connection. For example, consider a one-line-to-ground fault, in which case we have, from (12.110)

$$I_{a0} = I_{a1} = I_{a2} \tag{12.112}$$

Now, consider the zero-sequence network. Since the current injected into node 3 of the zero sequence network is the negative of I_{a0}, we can multiply the third column of the Z matrix by $-I_{a0}$ to determine the voltages at all nodes in the zero-sequence network. Knowing the voltages at all nodes, we can compute the currents flowing on all branches of the network. Obviously, a similar procedure gives all the node voltages and branch currents of the positive- and negative-sequence networks as well.

This process requires considerable computation if the power system to be solved is large. However, if the system is represented by a system equivalent similar to that shown in Figure 12.38, the computational burden is small and the complete solution is found rapidly, even on small computers. The result is that the sequence currents and the complete solution of the sequence networks is readily determined by the matrix analysis of the networks. This provides not only the total fault current of each sequence, but the voltages of each node throughout the networks, which leads to the solution of all branch currents, including the currents at the relay location. Knowing the relay voltage and current, we can determine the ground relay polarization and apparent impedance. This is exactly the information needed to evaluate the relay operation for any given fault location along the mutually coupled line.

REFERENCES

[1] Kerchner, R. M., and G. F. Corcoran, *Alternating-Current Circuits*, Third Edition, John Wiley & Sons, Inc., New York, 1951.

[2] Nilsson, J. W., *Electric Circuits*, Addison-Wesley Publishing Co., Reading, MA, 1983.

[3] Westinghouse engineers, *Electrical Transmission and Distribution Reference Book*, Westinghouse Electric Corporation, 1950.

[4] Rothe, F., *An Introduction to Power System Analysis*, John Wiley & Sons, Inc., 1953.

[5] Zaborszky, J., and J. W. Rittenhouse, *Electric Power Transmission, The Power System in the Steady State*, The Ronald Press Company, New York, 1954.

[6] Wagner, C. F., and R. D. Evans, *Symmetrical Components*, McGraw-Hill Book Company, New York, 1933.

[7] Clarke, E., *Circuit Analysis of A-C Power Systems, Volume I, Symmetrical and Related Components*, John Wiley & Sons, Inc., New York, 1943.

[8] Calabrese, G. O., *Symmetrical Components*, The Ronald Press Co., New York, 1959.

[9] Anderson, P. M., *Analysis of Faulted Power Systems*, IEEE Press, Piscataway, NJ, 1995.

[10] Mason, C. R., *The Art and Science of Protective Relaying*, John Wiley & Sons, Inc., New York, 1956.

[11] Warrington, A. R. van C., *Protective Relays, Their Theory and Practice, Volume One*, John Wiley & Sons, Inc., New York, 1962.

[12] Cook, V., *Analysis of Distance Protection*, Second Edition, Research Studies Press, Letchworth, Hertfordshire, England, 1985.

[13] Horowitz, S. H., and A. G. Phadke, *Power System Relaying*, Research Studies Press, Letchworth, Hertfordshire, England, 1992.

[14] Carson, J. R., "Wave Propagation in Overhead Wires with Ground Return," *Bell System Tech. Journal*, 5, October 1926, pp. 539–555.

[15] CIGRÉ Study Committee 34, "Application Guide on Protection of Complex Transmission Network Configurations," CIGRÉ SC34-WG04, Publication E50400-U0048-U211-A3-7600, May 1991.

[16] Englehardt, K. H., "V_0 - Reversal on Mutually Coupled Lines - Relaying Analysis, Experience, and Remedies," a paper presented at the Western Protective Relay Conference, Spokane, Washington, October 28–30, 1974.

[17] IEEE Std 100-1992, *IEEE Dictionary of Electrical and Electronics Terms*, IEEE, New York, 1992.

PROBLEMS

12.1 Investigate the meaning of the primitive impedances shown in Figure 12.2 and show how the phase impedances can be defined in terms of these primitive impedances. Hint: review [9].

12.2 Add two ground wires, u and w, to the phase wires of Figure 12.2 and show how the resulting equation can be reduced to exactly the form of (12.4).

12.3 Assume that one of the phase conductors, say phase a, consists of two identical "bundled" wires. Show that these can be combined into a new phase a composite conductor and that the impedance matrix can be reduced to exactly the form of (12.4).

12.4 Verify (12.10) and (12.11).

12.5 Show that the matrix pattern shown in Table 12.5 for mutually coupled transmission lines is correct. Hint: Tabulate the wire positions for the transposed lines that would be used in making the calculation. It is not necessary to actually compute the numerical values of the matrix to determine the equality of certain terms.

12.6 Verify the calculation of self- and mutual impedances and admittances for the parallel 500 kV lines shown in Figure 12.6. The spacing between conductors is shown in Figure P12.6.

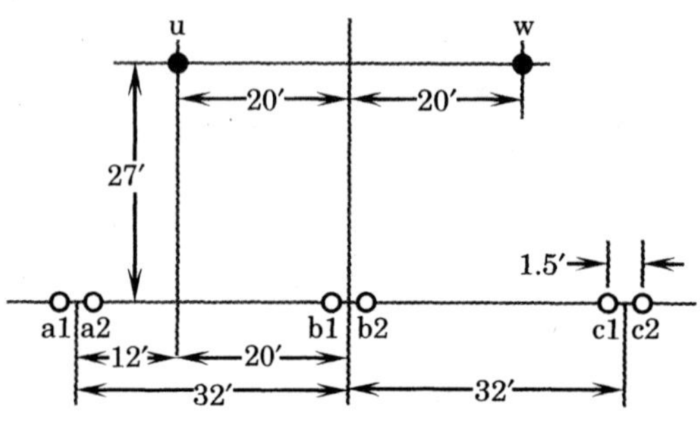

Figure P12.6 Conductor arrangement for a 500 kV line.

The conductor specifications are given in Table P12.6a. The conductor spacings and the distances between conductors and their images are given in Table P12.6b. This is based on a center-line-to-center-line distance D between towers of 200 feet and a phase conductor height of 125 feet above the ground.

(a) Construct a spreadsheet to compute the distances between all pairs of conductors and inductive reactance in ohms per mile (Ω/mile) between each pair of wires at 60 hertz.

(b) Based on the distances determined in part (a), construct a second spreadsheet to compute the heights of all conductors from all image conductors, and from this result, compute the Maxwell potential coefficients in MF^{-1}mile between all pairs of conductors, including image conductors.

(c) Write a computer program to compute the phase impedance per mile of the double circuit line, including a reduction to eliminate the ground wires from the resulting impedances.

(d) Extend the computer program of (c) to add transpositions to the line.

(e) Perform similarity transformation of the result of (d) to find the sequence impedances of the double circuit transmission line.

(f) Repeat (a) through (e) to compute the sequence admittances.

12.7 Determine an equivalent circuit for three parallel transmission lines. Note that there are three different mutual impedances; one for the mutual between lines A and B, another for the mutual between lines B and C, and a third for the mutual impedance between lines C and A. Hint: See [7].

TABLE P12.6a Conductor Characteristics of the 500 kV Line

	Phase Wires	Ground Wires
Conductor Type and Size	**2156kCM ACSR Bluebird**	**7#8 Alumoweld**
Resistance in Ω at 25° C, 60 Hz	0.464 Ω/mile	2.440 Ω/mile (small current)
D_s = Self GMD of Wire	0.0588 ft	0.002085 ft
Radius of Wire	0.07342 ft	0.03208 ft

TABLE P12.6b Conductor-to-Conductor Distances (in feet)

Distances between Phase and Ground Wires and their Images	Distances between Phase Wires and Ground Wires
$H_{a1a1'} = 250$	$D_{a1a2} = D_{b1b2} = D_{c1c2} = 1.5$
$H_{wa1'} = H_{wc2'} = \sqrt{277^2 + 12.75^2} = 277.2932$	$D_{ua1} = D_{wc2} = \sqrt{27^2 + 12.75^2} = 29.8590$
$H_{ub1'} = H_{wb2'} = \sqrt{277^2 + 19.25^2} = 277.6681$	$D_{ub1} = D_{wb2} = \sqrt{27^2 + 19.25^2} = 33.1597$
$H_{uc1'} = H_{wa2'} = \sqrt{277^2 + 51.25^2} = 281.7012$	$D_{uc1} = D_{wa2} = \sqrt{27^2 + 51.25^2} = 57.9272$
$H_{ua2'} = H_{wc1'} = \sqrt{277^2 + 11.25^2} = 277.2284$	$D_{ua2} = D_{wc1} = \sqrt{27^2 + 11.25^2} = 29.2500$
$H_{ub2'} = H_{wb1'} = \sqrt{277^2 + 20.75^2} = 277.7761$	$D_{ub2} = D_{wb1} = \sqrt{27^2 + 20.75^2} = 34.0524$
$H_{uc2'} = H_{wa1'} = \sqrt{277^2 + 52.75^2} = 281.9780$	$D_{uc2} = D_{wa1} = \sqrt{27^2 + 52.75^2} = 59.2584$

12.8 Consider a circuit consisting of three transmission lines all connected at one end to a common bus. The lines are of different lengths and therefore have different self-impedances. However, the lines are physically in parallel such that the two outside lines have the same mutual coupling to the center line, but the mutual coupling between the outside pair of lines is different. Determine the equivalent circuit for the three lines. Hint: See [7].

12.9 Convert the wye equivalent capacitive reactances of Figure 12.22 to an equivalent delta at each end of the line.

12.10 Verify the Type 2 equivalent shown in Figure 12.18.

12.11 Revise Figure 12.19 to one that employs the lattice transformer equivalent of Figure 12.13. Can this circuit be constructed in the laboratory? Explain.

12.12 Verify the short-line equivalent shown in Figure 12.23 for three short transmission lines bused at one end.

12.13 Verify (12.55) and (12.56).

12.14 Verify (12.64).

12.15 Derive the definitions of the parameters γ_1^2 and γ_2^2.

12.16 Check at least one each of the $\boldsymbol{A}$ and $\boldsymbol{B}$ parameters of (12.67) and (12.68) by performing the partial fraction expansion of (12.65) and (12.66), respectively.

12.17 Write the differential equations for the currents in the mutually coupled transmission lines and show that (12.69) represents these currents and their mutual coupling.

12.18 Determine at least one each of the $\boldsymbol{C}$ and $\boldsymbol{D}$ parameters as a function of the Laplace variable s.

12.19 Verify (12.92).

12.20 Perform the reciprocity tests described in Section 12.6.1. This should result in tests of all ammeter readings as a function of the applied voltage applied at all terminals. Compare these test results with the admittance matrix parameters defined in (12.99).

12.21 Prepare a step-by-step procedure for finding the self- and mutual impedances of two bundled-conductor transmission lines that are parallel and in close proximity for their entire length.

12.22 Determine the ***ABCD*** parameters for two mutually coupled 500 kV lines with the following basic parameters. Give the results in per unit on a 1000 MVA base for a line 300 km long.

$$z_A = 0.0232 + j0.5488\ \Omega/\text{mi} \qquad y_A = +j7.6631\ \mu S/\text{mi}$$
$$z_B = 0.0232 + j0.5958\ \Omega/\text{mi} \qquad y_B = +j7.0466\ \mu S/\text{mi}$$
$$z_M = 0.000 + j0.0012\ \Omega/\text{mi} \qquad y_M = -j0.0159\ \mu S/\text{mi}$$

12.23 Using the results of problem 12.22, compute the admittance matrix for the mutually coupled lines. Assume a line length of 300 km.

12.24 Using the results of problem 12.23, compute the long-line Type 3 equivalent circuit for the mutually coupled lines.

12.25 Using the results of problem 12.22, determine the Type 1 equivalent circuit for a fault located 200 km from the relay R, using the long-line equations.

12.26 Use a computer program such as Mathematica or Maple to prove that the admittance matrix of (12.99) is symmetric using symbolic variable representation and evaluation.

13

Pilot Protection Systems

Much of the treatment of transmission line protection in previous chapters has considered each relay to be independent and to operate only on the basis of intelligence gathered from measurements taken at the relay location. This has obvious advantages in simplicity and is often adequate to ensure a high quality of protection and rapid response. In some cases, however, this simple scheme is not adequate. This is the case, for example, where the time delay to clear faults beyond the zone 1 reach for distance relays may be considered unacceptable. This is often true for EHV systems, where the lines carry large power transfers and delayed tripping of faults may cause severe network or stability problems. In such cases, more complex transmission protection schemes are required. Some of these more complex schemes are presented in this chapter, with examples of several different options for providing additional fault clearing intelligence and greater speed in clearing of all faults.

Three important concepts are basic to the design of any protective system; selectivity, reliability, and security. *Selectivity* requires that the protection system must be dependable in identifying faults in its zones of protection. This is achieved by the relay designer, who uses sound logic for fault detection and line tripping. *Reliability* requires that the protection system be operable, that is, that the overall design will ensure appropriate protective action even if some portion of the protective apparatus may have failed. This is achieved by using equipment of high quality, by performing routine testing to ensure that the equipment remains operable, and by designing a protective system that has redundancy. Some engineers insist that the backup system be of a different design or operate on a different principle than the primary system. In many cases, the terms "primary" and "backup" have little real meaning since both systems are of equal quality, but are simply separate systems and may be of different design. For such systems, the two systems are simply called "redundant" or local backup systems.

System *security* refers to the capability of the protection system to refrain from operating when it should *not* operate. This is especially important in EHV circuits, because of the considerable system upset that occurs when a heavily loaded line is opened. For the protection system, this means that the relaying system must be selective and also that precautions are taken to ensure that no operation be initiated, either by the relay logic or other means, that would cause the tripping of important lines or other facilities when not absolutely necessary [1–7].

The schemes discussed in this chapter address all of these concerns and do so in different ways.

13.1 INTRODUCTION

It is instructive to compare protective systems used for EHV transmission lines, all of which share some common characteristics. The systems to be studied are all "pilot" relaying systems, that is, these systems utilize a communications path to send signals from the relaying system at one end of the line to that at the other end. This concept provides the opportunity for the protection decisions made by the relays to be more intelligent, since information from both ends of the line are available for processing. Figure 13.1 shows a generic view of this type of protection system.

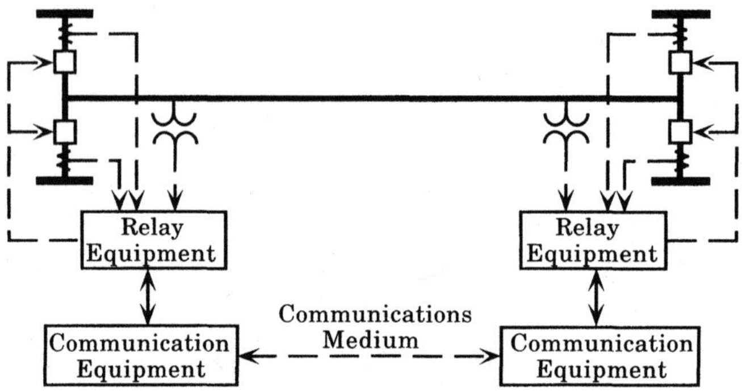

Figure 13.1 General view of a pilot protection system.

The relay systems at each end of the transmission line monitor the local currents and voltages. These signals, or a derived response, are sent to the local relay equipment only, where trip signals may be generated and sent to the circuit breakers at the local relay location. (Signal paths in Figure 13.1 are shown by dashed lines.) Additional equipment is provided that permits each relay to send signals to the relay equipment at the remote end of the line. This provides each relay location with important new information regarding the need for tripping, namely, the view of the disturbance seen from both ends of the line. Both relays can now operate on the basis of the condition observed from both relay locations.

The need for additional intelligence in the protection of some transmission lines arises due to the inadequacy of distance protection for reliable fault detection and selectivity. This occurs due to a number of reasons [8]:

- Intermediate infeeds in multiterminal lines
- Zero-sequence mutual coupling in parallel lines
- Discontinuities due to series capacitors

These difficulties result in zone reach measurements that are variable and are not suitable for correct fault detection and selectivity. An effective solution to these problems is the additional sophistication in the relay logic that incorporates intelligence from all line terminals.

The use of pilot relaying has grown over the years and is now in common use throughout the world [8–10].

As an example, let us assume that the relay equipment at both ends includes distance measuring logic. Now suppose that a fault occurs very close to the relays at the left end of the line. The relays at that end will recognize this fault as a zone 1 fault and will send a trip signal to the local line breakers. This same fault will be seen by the relays at the right end as a zone 2 fault, but these relays are not able to determine if a fault near the left bus is on one side or the other of that bus. This will cause a timer at the right-end relay equipment to be started, which will result in delayed tripping. Adding the pilot channel from the left end to the right end, however, provides a means of eliminating the time delay. A pilot signal generated at the left-end relay informs the right-end relay that the fault is on the protected line, and that tripping should be initiated without delay. This is an example of what is called "transfer trip" logic, although other types of logic can accomplish the desired result.

The timing of events is important in pilot protection. The total time for fault clearing is summarized in Table 13.1. Typical modern relays operate in 8–10 ms, and circuit breakers will clear the fault in 30–50 ms. However, if rapid total fault clearing is required for faults at any location along the transmission line, a means needs to be used that avoids the time delays that accompany zone 2 operation. Rather than wait for zone 2 to time out for a fault near one of the terminals, a pilot signal can be sent between the terminals to verify the need for tripping in an additional 10–30 ms. This is more positive and is faster. Even with telecommunications time, the fault can usually be cleared in 80–90 ms. Note that the times shown in Table 13.1 are added since the various devices depend on the completion of the previous device operation.

TABLE 13.1 Approximate Operating Time of Protective Equipment [1]

Protection Subsystem	Operating Time in ms	Operating Time Cycles, 50 Hz	Operating Time Cycles, 60 Hz
Protective relays	8–10	0.4–0.5	0.5–0.6
Circuit breaker	30–50	1.5–2.5	1.8–3.0
Total relay & CB	38–60	1.9–3.0	2.3–3.6
Pilot signals	10–30	0.5–1.5	0.6–1.9
Total	48–90	2.4–4.5	2.9–5.4

The simple example given, on close examination, has certain possible flaws that will require further study. Suppose, for example, that a *false* trip signal is sent from one end to the other. This may be due to a relaying error to communications noise, or to any other source. This will cause a *false trip* of the transmission line since the pilot channel is not necessarily *secure*, that is, the system described does not have protection against security failures. Procedures are available to guard against such errors, and some of these procedures will be examined in this chapter.

The system of Figure 13.1 is also seen to be vulnerable to tripping failure or delayed tripping unless the communications channel is made redundant. This would suggest that a second communications link be established between the two ends, preferably using a different communications method, a different signal path, or both.

Clearly, the addition of a pilot communications channel gives each protective system additional information to process. For a transmission line, the added information from the remote end is very important. This permits the relay systems to maintain a kind of electronic dialog to reach a common decision as to whether the line should or should not be tripped for the condition being observed. Different kinds of pilot systems and relay decision-making schemes are available to perform these functions.

The pilot protection concept expands the protection system for a given transmission line to include the relay systems at both ends of the line. This will usually include more than one relay system at each end, with the systems communicating through the pilot channels in end-to-end pairs. This expands the concept of protection to "decision/trip centers" at the two ends of the line, with each decision/trip center being provided with information from both points of observation and communicating any resulting decision to both the local circuit breakers and to the remote end of the line. This concept could conceivably be expanded further to provide a given decision/trip center with more global information, which might include data communicated from adjacent lines, from parallel lines, or from other locations. Such an expansion provides opportunities as well as problems. At the present time, the pilot protection schemes use only data transmitted between the two ends of a given line. This is, after all, the most important data and the scheme is limited to the minimum essential data. The future may see more global protection designs, which will require high-bandwidth communications in order to achieve the required speed.

The balance of this chapter presents the various pilot communications methods in common use for transmission protection, and will present the protective system design concepts that employ pilot systems.

13.2 PHYSICAL SYSTEMS FOR PILOT PROTECTION

This section presents several different physical methods of signaling between the two ends of a transmission line for protective relaying purposes. These signaling systems are called *pilot systems*. The term *pilot* is also used to clarify the function to be performed as part of the system protection. For example, the term *pilot relaying* signifies that the relays at the two ends of a transmission line are in communication with each other. The term pilot is also used to clarify other protection functions, such as transfer trip pilot, permissive pilot, or blocking pilot. The term also signifies a reference to the communications medium, e.g., wire pilot, microwave pilot, and power-line carrier pilot. These latter terms designate the physical, rather than the functional, identification of the equipment used in performing the pilot functions. This physical identification is the subject of this section. The functional systems are discussed in the next section.

An overview of the equipment used in pilot protection is shown in Figure 13.2 [1]. The innermost link identifies the physical equipment used for signaling between the line terminals. This can be any form of signaling link, such as a telephone line, a microwave link, a radio link, or a fiber-optic communications system. There are certain items of terminal equipment associated with this telecommunications system, irrespective of its use for protective relaying. This equipment might be used for voice, data, image, or text transmission instead of, or in addition to, protective relaying data. For protective relaying, this will usually be a dedicated telecommunications system, since it is not acceptable that the relay might receive a "busy" signal when trying to communicate.

The next level in Figure 13.2 is the teleprotection equipment level. This includes the physical link and its terminal equipment, and also includes hardware, and perhaps software, at the terminals that are required to take the relay output data and inject it into the communications equipment. This might include coupling equipment, modems, or other devices required to make this translation from the protective equipment to the telecommunications equipment.

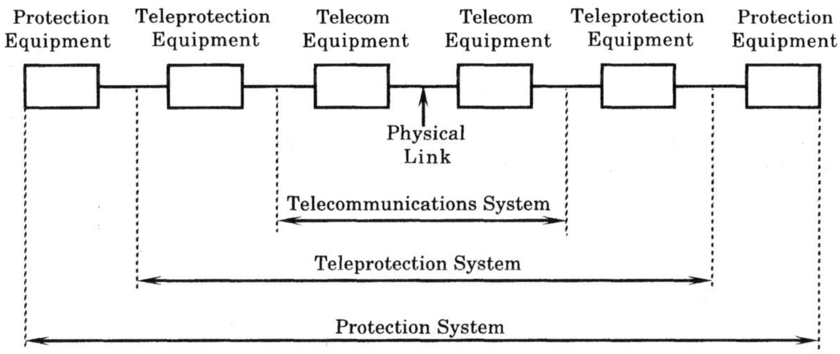

Figure 13.2 Equipment classifications in pilot protection [1].

Finally, the terminal equipment of the pilot protection system is the protective devices. This includes transducers required to gather data from the power system, a logic system to process that data, and determine if a certain signal should be transmitted to the relay system at the opposite end.

13.2.1 General Concepts of Pilot Communications

Pilot protection is a form of line protection that uses communication channels as a means if comparing electrical conditions at the terminals of the line [3]. This is accomplished by the transmission of pilot signals that are used to provide the protective devices with information that can be used to determine the need for tripping the line. The exact method of utilizing the pilot signals depends on several factors. The relay manufacturers have devised many different types of pilot systems. Some of the concepts used in pilot protection are investigated in this chapter.

The pilot communications can take several forms in terms of the technology used for the transfer of information.

13.2.1.1 Signal Form.

Analog signal—A signal in which the information can take on an infinite number of values between its maximum and minimum levels. Frequency, amplitude, phase shift, and pulse width can all be varied. For long distance transmission, the signal may be amplified at certain intermediate locations by amplifiers called *repeaters*.

Digital signal—A signal in which the information is composed of a limited number of levels or states, as opposed to an analog signal which can take on an infinite number of states. In digital signaling, all information is converted into digits before being transmitted. Digital communication offers high channel density with programmability to connect to different types of devices and serving different purposes.

Most power line protection signaling has traditionally been analog, but the trend is to favor digital signaling.

13.2.1.2 Signal Frequency. The signal frequency used in power system protection encompasses a wide bandwidth from dc to the visible spectrum. In particular, the following frequencies have been used for protection signaling.

Direct current—Some types of wire pilot utilize direct current as the form of pilot signal transmission. This form of signal transmission has certain advantages, but is largely obsolete, except for special applications [2].

Power frequency—Some wire pilot systems operate at the power system fundamental frequency by making connections between current transformer secondaries at both ends of the protected circuit through a pair of pilot wires that travel from one end of the line to the other. This is sometimes called "ac wire pilot relaying" [2].

Audio frequencies—The audio frequencies are those to which the human ear is sensitive, or approximately 20 to 20,000 hertz [3]. Since much of the telephone industry equipment is optimized for this frequency band, it is often used for control signaling in power system communications. This technology utilizes tone generators and receivers to transmit and receive the voice frequency carriers, which are often converted to a direct current at the receiver. This type of transmission is used for utility owned, as well as leased, telephone lines between ends of the protected circuit.

Power line carrier frequencies—Power-line carrier (PLC) is a communications system that couples high-frequency carrier signals onto the high-voltage power line conductors through a coupling capacitor (and other equipment) [3], [4]. The frequency used for this type of transmission is in the range of 30 to 600 kHz. This type of system can be used for any type of telecommunications, such as voice, data transfer, or protection signaling applications. The primary motivation for its installation is usually protection.

Radio frequencies—Radio frequencies are those frequencies in the portion of the electromagnetic spectrum that are between the audio-frequency portion and the infrared portion, with practical limits of roughly 10 kHz to 100,000 MHz [3]. Electromagnetic radiation can be used for power system relaying, but this medium is seldom selected because of the possibility of interference, intentional or otherwise, and the requirement for licensing a portion of the radio spectrum for this type of usage.

Microwave frequencies—The microwave radio-frequency band is a term that is loosely applied to radio waves from about 1000 MHz and upward [3], [6]. This type of information transmission is characterized by line of sight transmission between antennas, at which points the signal is passed through repeater equipment and retransmitted. The system may be used for many different applications, such as voice, data, control, and protection.

Visible light frequencies (fiber optics)—In a strict sense, the visible spectrum is that portion of the electromagnetic spectrum that can be perceived by human vision, or nominally covering the wavelength range of about 0.4 μm to 0.7 μm [6]. In the optical communications field, custom and practice have extended the usage of the term to include a much broader portion of the spectrum, or nominally about 0.3 μm through 30 μm. This type of information transmission is becoming more and more prominent in all areas of telecommunications, and may eventually predominate power system communications. This is because of the ability of conductor manufacturers to imbed fiber-optic strands in the core of power conductors, either those used for high voltage or those used for shield wires, thereby forming a wide band communications link between the ends of a transmission line.

13.2.1.3 Signal Transmission Media. The transmission media used for the transmission of protection signals can be summarized as follows.

- *Wire pilot protection*—A pilot protection system in which an auxiliary metallic circuit is used as the communication mechanism between the relays at the circuit terminals [2–4]. Also, "pilot-wire protection."
- *Power line carrier pilot protection*—A form of pilot protection in which the communication mechanism between relays is a pilot signal that is superimposed on the power transmission conductor(s).
- *Microwave pilot protection*—A form of pilot protection in which the communication mechanism between relays is a beamed microwave-radio channel [3–5]. The relay pilot channel is multiplexed with other channels for voice or data transmission.
- *Fiber-optic pilot protection*—A form of pilot protection in which the communication mechanism between relays utilizes a digital fiber-optic light pulse form of transmission.

The foregoing classifications of transmission media separate the pilot systems into significantly different physical systems. These physical systems are apparent from their names. The methods of signal conditioning and transmission, however, may be any number of different types. In power system protection communications, the signaling applications have traditionally been analog transmission in any of these media, although digital systems may predominate in the future as digital systems become available at reasonable costs. It is not clear whether future protection signaling will ever be performed using the *integrated services digital networks* (*ISDN*), which are planned telecommunications systems that utilize digital transmission and switching technologies to support voice, data, text, and image transmission. Such future systems may offer higher reliability for protection functions, because of their inherent multiple redundancies, but at the expense of not having *dedicated* facilities, which are defined as facilities for only a particular function. It would not be satisfactory, for example, for a protective system to receive a "busy signal" indicating that all available channels are in use. The ISDN system may, however, offer a very reliable redundant channel for relaying purposes.

13.2.2 Wire Pilot Systems

Wire pilot relaying systems have been used for many years to protect transmission lines, but these schemes are limited in their application. First, this technique is limited to lines that are short enough to justify the cost of the installation or leasing of separate pilot circuits for relaying. Second, the pilot wires are themselves a transmission line, with the usual series resistance and inductance and shunt capacitance per unit of length. As the pilot transmission lines become long, the pilot line requires special "tuning" to optimize signal transmission, and this is usually feasible only at the two terminals rather than being distributed along the circuit. This also limits the use of wire pilot to relatively short transmission lines.

Wire pilot schemes are divided into two basic types, with variations of each type in practical devices. The first type, shown in Figure 13.3, is called a *circulating current* scheme for pilot protection. Note that a current balance relay is used at each end of the line. The currents shown in the figure represent the normal condition, with current entering one end of the line and leaving the other end. Under this condition, the CT secondary currents flow in the restraint windings of each relay and no current flows through the operate windings. If a fault occurs on the line, the currents at the right-hand end of the line will reverse, which will cause currents to flow in the operate windings of both relays. Note that the pilot wires are an exact low-voltage replica of the transmission line.

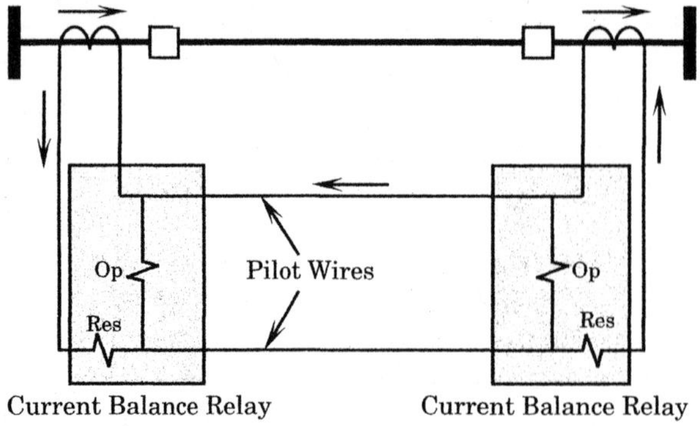

Figure 13.3 Schematic of a circulating-current wire pilot system [2].

The voltage balance system, or "opposed-voltage system" is shown in Figure 13.4. In this system, the pilot wires are crossed and no current flows in the pilot wires under the normal condition shown in the figure, which forces the currents to flow through the restraint windings of each relay. Note that the role of the two windings has been reversed over that illustrated in the circulating current scheme.

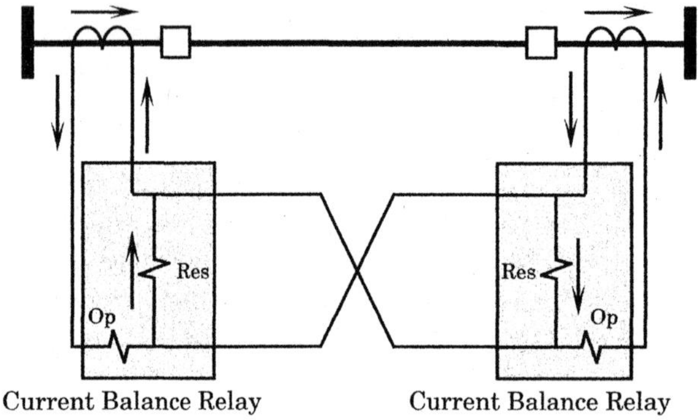

Figure 13.4 Schematic of a voltage balance wire pilot system [2].

For both of the wire pilot schemes illustrated, only the currents of one phase are shown in the diagrams. If this concept were to be adopted, it would require three such systems, one for each phase, which increases the cost and complexity. In practice, the three-phase currents are combined in various ways to provide correct relay action for all types of faults. Several schemes have been devised by the relay manufacturers. One method of combining the phase currents into one signal is shown in Figure 13.5, where the transformer is a kind of summing system for the three-phase currents. The secondary current is connected to the pilot system, as shown in Figures 13.3 and 13.4.

There are problems with wire pilot protection that are not addressed in this brief discussion. In many cases, the pilot channels are leased from a common carrier telephone company, with restrictions on the voltages that can be tolerated on the telephone circuits. Even if the pilot system is owned by the electric utility, there may be problems with induced voltage from a parallel power transmission line, especially under faulted conditions. This requires care in the installation of the pilot circuit. The pilot wires must also be protected from lightning, which may cause considerable damage to the terminal equipment. Moreover, the metallic connection between remote high-voltage stations may introduce problems due to the possibly large difference in ground potential at the two ends of the line during fault conditions. All of

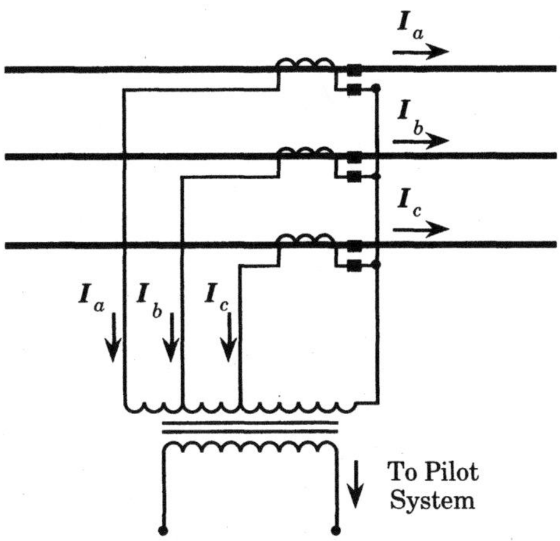

Figure 13.5 A sequence filter used for a pilot wire relay system.

these concerns make wire pilot protection a poor choice for long distance high-voltage lines, where other options have distinct advantages.

13.2.3 Power Line Carrier Pilot Systems

A simple form of power line carrier pilot relaying is shown in Figure 13.6. In this type of pilot relaying, the signal from one relay to the other is transmitted over the power transmission line at power line carrier frequencies. The coupling to the power line is by means of a line tuner and a coupling capacitor voltage transformer (CVT), with impedance matching performed by a hybrid (HYB), which is a transformer arrangement that permits the transmitted and received signals to be separated and also directs the output of the transmitter to the line rather than entering the receiver. In practice a circuit like that of Figure 13.6 might work as follows. The relays at each end of the line are assumed to have distance measuring capability. A close-in fault at either end would be positively detected by the relay at that end, but the relay at the opposite end would have difficulty discriminating between faults on the protected line and those beyond the protected line. However, if the relay close to the fault sends confirmation that the fault is indeed on the protected line, then the opposite-end relay may be permitted to trip without delay. This is termed *permissive tripping*. It has the capability of preventing long delays in tripping faults on the protected line that might otherwise be necessary. A wave trap is

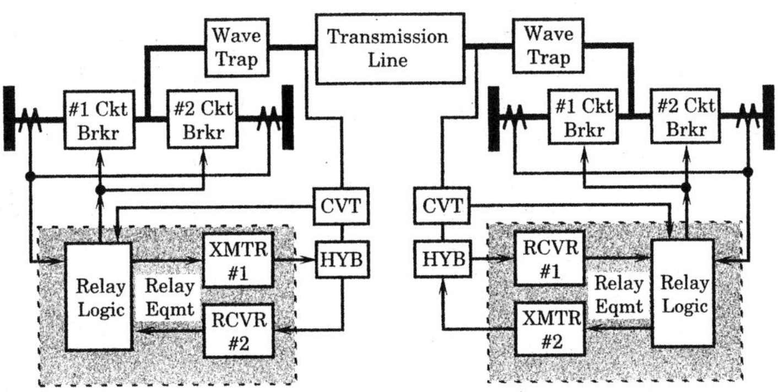

Figure 13.6 Transmission relaying using power line carrier pilot.

used at both ends of the protected line to prevent the carrier frequency signals from spreading to other circuits and possibly causing false tripping of those circuits.

13.2.4 Microwave Pilot Systems

The signaling for system protection may also be provided using microwave communications between the two ends of the transmission line, with equipment arranged as shown in Figure 13.7. In this arrangement, audio tone signals pass through the microwave modulator-demodulator equipment before being transmitted at microwave frequencies to the other end of the line. Usually this arrangement is called a *permissive* scheme since the audio-tone signals "permit" the line relays at the opposite end of the line to trip when a fault is detected.

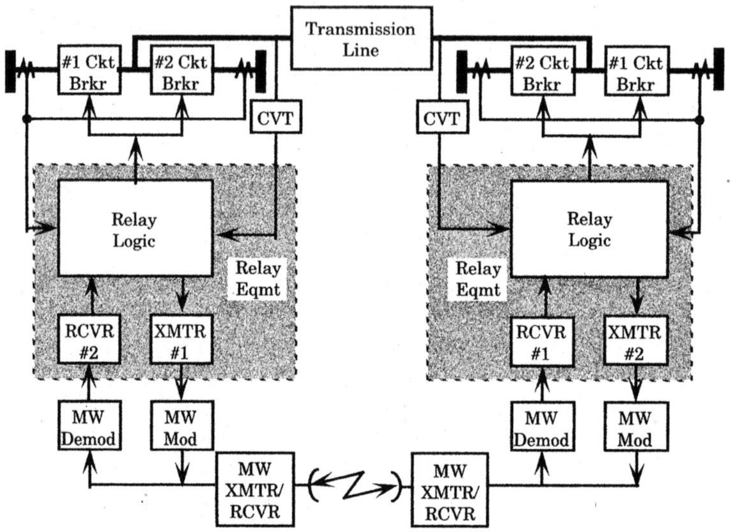

Figure 13.7 Transmission relaying using microwave pilot protection.

The microwave path between the two ends of the transmission line of Figure 13.7 may be quite different from the physical path of the high-voltage circuit. Microwave systems utilize repeater stations that are almost always located at points of the highest terrain. This permits longer "hops" between repeaters, and lower overall cost of installation. In many cases, the intermediate stations are located on mountain tops, or even in high-rise buildings. This may mean that the route of the microwave signal travels much farther than the route of the transmission line being protected. This greater length of communications travel will usually cause no difficulty in the relaying, as long as the time delay of the signal is known and is constant.

Some microwave systems are operated as a loop telecommunications system, with the ability of switching from a normal, say, clockwise direction of signal direction around the loop to an emergency counterclockwise direction of signal transmission. This type of system can cause difficulty in relaying if the resulting distance in the communications path between the two ends of the power transmission line changes due to the microwave switching. This will cause different time delays in the signals between the two ends of the line, which may cause failure of certain kinds of relay logic. This is true, for example, for phase comparison schemes, which rely on knowing the signal delay in order to establish the correct phase comparison.

Directional comparison schemes can be used in microwave pilot systems with little concern for differences in propagation delay.

Microwave systems have a unique characteristic that can be a significant problem in applications such as protection that require a very high system reliability. This is the failure of transmission due to atmospheric "fading" of the signal, which occurs during certain atmospheric conditions, such as periods of high atmospheric water vapor. Signal loss due to fading is greatly improved by the use of "space diversity," where two receiving antennas are used, with the antennas separated vertically on the receiving tower by 10 meters or so. This greatly improves the probability of correct reception of the transmitted signal. Other methods are also used to provide reliable transmission and reception of microwave signals [5]. These special techniques may require careful investigation by the relay engineer, to ensure proper reliability of the pilot protection system.

13.2.5 Fiber-Optic Pilot Systems

Figure 13.8 shows the configuration of a protection system that utilizes fiber-optic signaling, where the fiber optics are assumed to be embedded in the shield wire(s) of the transmission line. The system would be no different if the optical fibers were part of a separate communications system, or embedded in the high-voltage transmission conductors.

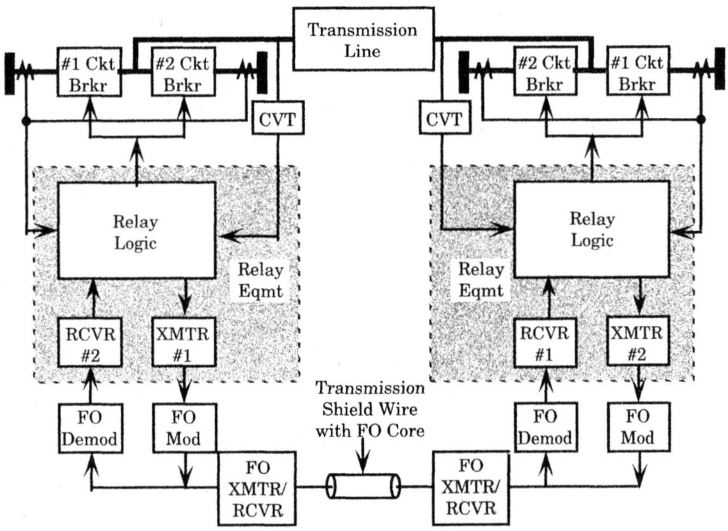

Figure 13.8 Transmission relaying using fiber-optic pilot protection.

The last three pilot systems described, power line carrier current, microwave, and fiber optics, all share the characteristic of providing a communications path between the relay sets at the two ends of the transmission line. The physical apparatus and signaling equipment vary according to the communications technology used to transmit the information. Many relays are not affected by the method of communications and can be connected to any of the three types of systems. The choice of pilot system used will depend on several factors, such as the availability of fiber-optic or microwave paths, cost, reliability, and the type of relay scheme. There are many different types of relay logic that may be used with the pilot systems. This is the subject of the next section.

13.2.6 Guidelines for Pilot Communications Selection

There are no rigid rules that dictate the choice of communications method for pilot relaying. There are some characteristics that can be stated as general guidelines [7–9].

Wire pilot Wire pilot systems are generally applicable under the following conditions:

1. A pair of metallic communication wires is available.
2. The wire pilot pair is less than 9 miles (14.5 km) long.
3. The protected line is two terminal and likely to remain so, or if three terminal each leg is less than 2.3 miles (3.7 km) long and the total line length is less than 6.9 miles (11.1 km).

Power line carrier Power line carrier is generally applicable under the following conditions:

1. The transmission line is too long to consider wire pilot.
2. Fiber-optic systems are unavailable or too expensive.
3. Data communication is not required.

Microwave Microwave communication is used in the protection of long transmission lines where power line carrier does not offer enough channel capacity, or as a second communications path in addition to power line carrier. The following general conditions apply:

1. There is inadequate carrier spectrum available for PLC.
2. Continuous monitoring is required.
3. Voice or data communication is required.

Fiber optics Fiber-optic communication is used in the protection of short transmission lines where PLC is not suitable, or as an alternate communications path in addition to other communications. The following general conditions apply:

1. The transmission line is relatively short since fiber-optic repeaters must be installed every 50–100 km.
2. There is inadequate carrier spectrum available for PLC.
3. Continuous monitoring is required.
4. Voice or data communication is required.
5. A fiber-optic cable is available for relay communications.
6. Ground mat potential differences are not a problem.
7. Economics for using fiber optics is favorable.

13.2.7 Pilot Communications Problems

The use of pilot communications protection is not without problems, and these should be acknowledged. There are two kinds of problems: regulatory problems and operating problems.

Communications are regulated by national and international restrictions on frequency allocations for various purposes. This is necessary in order to avoid conflicts among communications systems and to allocate usage in an orderly manner. Power system communications are assigned certain frequencies for power line carrier and microwave, for example, and usage is restricted to these frequencies. Even given a certain frequency band for this purpose, the engineer needs to make sure that the system being designed does not create internal interference problems.

The problem with the regulation of communications is the restriction of the available frequencies, which can make the protection engineer's job very difficult. Moreover, there has been a trend for the regulatory agencies to allocate power line carrier frequencies for other uses, making a difficult situation even worse. The frequencies are a finite resource, and there are increasing demands on communications from many different sources.

The other problem with communications is the history of difficulties that arise in their operation. The power system is not a friendly environment for communications. Many faults occur due to natural causes, and these causes affect the communications as well as the power system. This means that the pilot channel may not be available when needed, or may not work as designed under hazardous conditions. Power line carrier has been very reliable, but as transmission voltages increase, corona and disconnect arcing present problems of transient suppression for the transmitter and receiver systems. Some have observed that many of the problems associated with high-voltage protection have been due to communications.

13.2.8 Pilot Protection Classifications

Pilot protection has been broadly classified by a CIGRÉ working group into two categories; *unit schemes* and *nonunit schemes*. These schemes are defined in greater detail in Figure 13.9 [10]. This classification will be used here as a convenient grouping of the different schemes. Unit schemes treat the transmission line as a protected unit, and the protection is similar in some respects to differential protection, that is, measurements are taken at all terminals and compared to determine the need for tripping. Non-unit schemes take measurements at all terminals and communicate with distant terminals, but direct comparison of measurements is not performed. These designations are commonly used in Europe, but have not been widely adopted in North America, where the pilot schemes are often classified in terms of the type logic used.

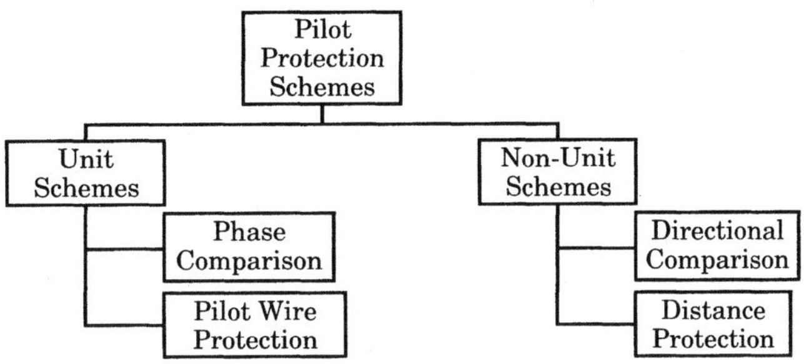

Figure 13.9 Common forms of pilot protection schemes [10].

Many pilot relaying schemes are often referred to as *permissive* schemes, which are defined as follows:

> **Permissive (as applied to a relay system)** A general term indicating that functional cooperation of two or more relays is required before control action can become effective [3–4].

The purpose of the following two sections is to discuss some of the protective schemes that are used for pilot protection of high-voltage transmission lines. The schemes discussed do not constitute an exhaustive list, but the functional descriptions are believed to be exhaustive of those *methods* employed in modern pilot protection systems.

13.3 NON-UNIT PILOT PROTECTION SCHEMES

Two important non-unit pilot protection schemes may be classified as either *transfer trip* or *blocking schemes*. In the *transfer trip* scheme, the relay at each end of the line, recognizing a fault within a designated protection zone, will send a trip signal to the relay at the remote end of the line [4]. The *blocking* system does just the opposite, that is, it sends a *blocking* signal continuously that prevents the remote relay from tripping [4]. The remote relay can operate only when the blocking signal is removed. The two concepts are described below by reference to an example that might be considered typical.

Within the general classifications of *transfer trip* and *blocking*, there are many specific schemes that have fundamental conceptual differences. There are also *underreaching* schemes and *overreaching* schemes, both of which imply the integration of distance measuring equipment within the logic. There are also *direct tripping* and *permissive tripping* schemes.

13.3.1 Directional Comparison Schemes

Directional comparison pilot relaying rely only on the detection of the direction to the fault from the relay location. An example of this type of protection is directional comparison ground relaying, where we assume the presence of ground sources at each end of the faulted line. The fault currents in this case will flow from the ends of the line toward the fault. Either negative or zero-sequence directional relays can be used. The zero-sequence polarizing quantity can be either voltage or current. Even parallel, mutually coupled lines can be protected using directional comparison techniques [10].

Phase fault protection using traveling wave relays is another scheme that is inherently directional. These relays employ the sudden changes in voltage and current that occur due to the fault to determine the direction to the fault from the relay. Comparing directional information between the two ends of the transmission line can be used to positively identify the fault location.

13.3.2 Distance Schemes

The other non-unit pilot scheme that is widely used employs distance protection at both ends of the protected line. Distance protection is inherently directional, and with a measure of distance incorporated in the measurement as well. The distance measurement is usually used with some sort of permissive signaling to provide selectivity and security in its application. Of course, zoned distance protection can be used without pilot signaling, but we assume here that second zone time delays are not acceptable and that faster fault clearing is needed.

In order to appreciate the need for transfer trip or blocking logic, we first consider an example of a transmission line that is protected by distance relays with transfer trip. The relay characteristics for the three zones of protection are shown in Figure 13.10 for the transmission system shown in Figure 13.11. Consider relay R at bus H. Zone 1 is set to reach 80% of the line length and zone 2 reaches about 120% of the line length. Zone 3 reaches forward past the bus beyond the remote end of the protected line. In some cases, zone 3 is set to reach backward. Zone 1 is set for no intentional time delay, but zone 2 has a given time delay (often 0.3–0.4 s) and zone 3 has an even greater time delay, as shown in Figure 13.11.

The control circuit for this distance protection is shown in Figure 13.12. This type of protection is adequate for many transmission lines and is often used at the subtransmission voltage levels. This system is not adequate for many high-voltage lines because of the delayed

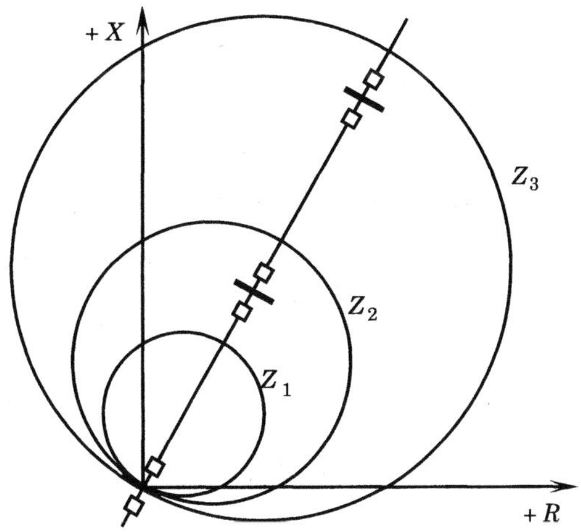

Figure 13.10 Distance relaying elements for transfer trip pilot protection.

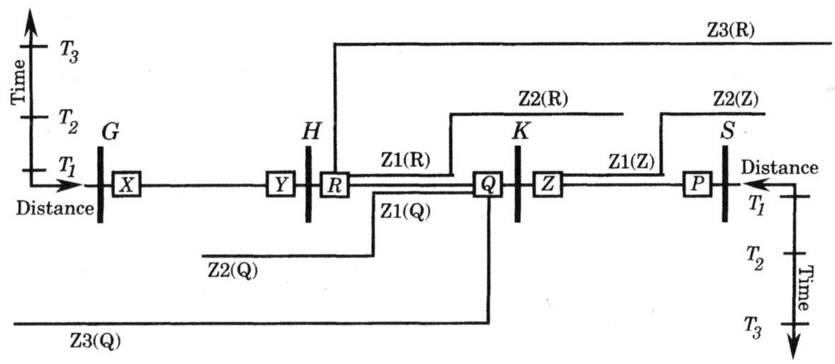

Figure 13.11 Distance relay reach and zoned time delays for transfer trip pilot protection.

tripping for faults occurring on 40% of the line length, which may lead to stability or other system problems on some systems due to the delay in clearing these faults.

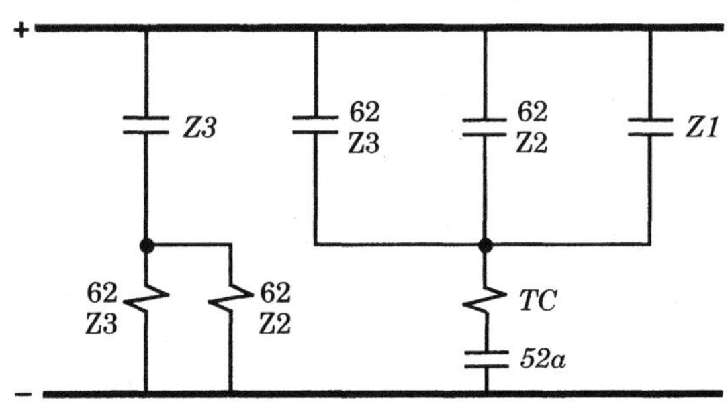

Figure 13.12 Relay contact logic for distance relay protection [8].

Pilot signaling is used to improve the performance of the distance protection system by initiating fast clearing on detection of faults at any portion of the protected line.

The system described above has certain features that are desirable. It offers a method of positive fault detection on the protected line and also for backup protection on the adjacent line. The only major drawback of the scheme is that, without pilot signaling, the fault tripping is delayed about 40% of the time, assuming that faults are equally probable throughout the total line length. This is not acceptable for many high-voltage lines.

13.3.3 Transfer Trip Pilot Protection

Transfer trip pilot protection is designed to provide high-speed tripping of line faults throughout the entire length of the line, and also to provide the benefits of backup protection for adjacent line sections [7–9]. We assume in this discussion that the relay characteristics are the same as that of Figures 13.10 and 13.11, but the system is changed to provide for communications between the relay systems at the two ends of the line.

The design of transfer trip schemes can be direct, permissive, or redundant. Direct schemes send a signal from one relay system that indicates a trip is necessary to the remote end, where the trip is initiated without further checking or delay. Permissive systems "supervise" the transfer trip signal, for example, by monitoring the line condition to see if a fault is detected, even in zone 2 or zone 3. Redundant systems offer a form of security protection by requiring that two transfer trip signals be received, thereby providing assurance that a true signal was sent and not just a noise or false signal. In some systems, transfer trip can also be initiated on detection of breaker failure at either end of the line.

13.3.3.1 Direct Underreaching Transfer Trip. The basic concept in direct underreaching transfer trip is that a zone 1 fault detection at either end will send a trip signal to the relay at the opposite end, in addition to sending a trip signal to the local circuit breakers. The basic control scheme is shown in Figure 13.13. This scheme is called a direct or non-permissive underreaching transfer trip scheme since the zone 1 relays at both ends are set to reach only the original 80% of the total line length. Suppose that a zone 1 fault is detected by the relay system at R in Figure 13.11. This immediately picks up relays $Z1$ and $Z3$ of Figure 13.13. The $Z3$ relay pickup starts timers for zones 2 and 3 time delays, which provide backup protection for adjacent lines. The $Z1$ relay immediately sends a trip signal to the local R breaker(s) and simultaneously transmits a trip signal to the relay system Q at bus K. Figure 13.13 shows the trip signal being transmitted through a communications channel to the opposite end of the line and simultaneously picking up the local trip coil (TC) by the zone 1 contacts. At the Q relay system, the tripping signal is received by circuits that direct the closure of the transfer trip receiving relay (UT). This system provides for positive protection of zone 1 faults at both ends and, with the pilot channel, extends this positive protection to the entire line and completes the line trip before T2 times out.

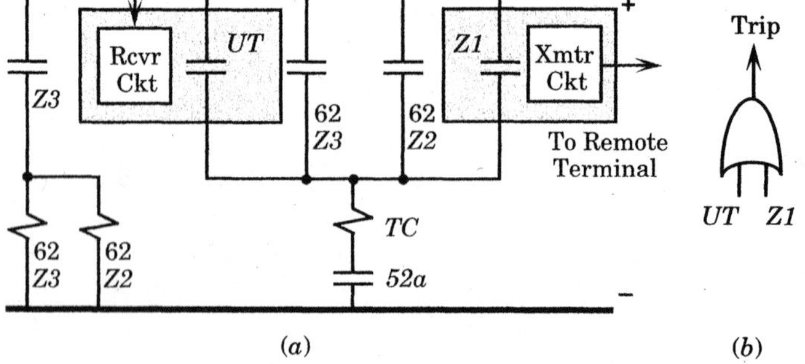

Figure 13.13 Relay contact and zone 1 solid-state logic for a direct underreaching transfer trip scheme [9], [12]. (*a*) Contact logic. (*b*) Zone 1 solid-state logic.

It is important to examine the performance of this or any system due to pilot channel failure. For this direct, underreaching transfer trip scheme, the faulted line will be tripped at both ends by the local relays. For faults beyond the zone 1 reach, tripping will occur with zone 2 time delay. Thus, the *security* of the system is not dependent on the pilot channels, but the *speed* is dependent on successful communications.

On closer examination, the system of Figure 13.13 is flawed, since it is possible that a noisy communications channel will cause the transfer trip receiving circuit to detect a signal, when none has been transmitted. This system is too easily excited by the communications link, which is usually long and may be vulnerable to erroneous excitation by storms, mutual induction, or other means. Thus, we might reject the control of Figure 13.13 as being too risky for some applications.

The system of Figure 13.13 can be made more secure by "supervising" the transfer trip receiving relay with a set of contacts from the zone 3 detector, as shown in Figure 13.14. The zone 3 relay logic is forgiving, but it does require an actual fault to pick up, thereby providing a measure of security.

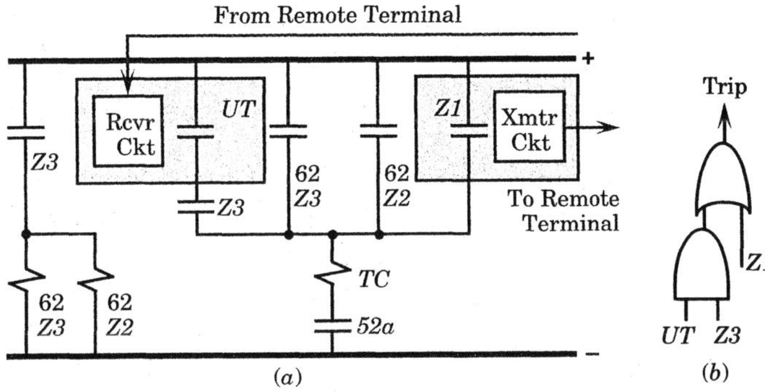

Figure 13.14 Relay contact logic and zone 1 solid-state logic for the supervised direct underreaching transfer trip scheme [9], [12]. (*a*) Contact logic. (*b*) Zone 1 solid-state logic.

Another form of underreaching transfer trip is shown in Figure 13.15, where we now see a normally closed guard relay (GD) inserted into the trip circuit. The transmitter in this system

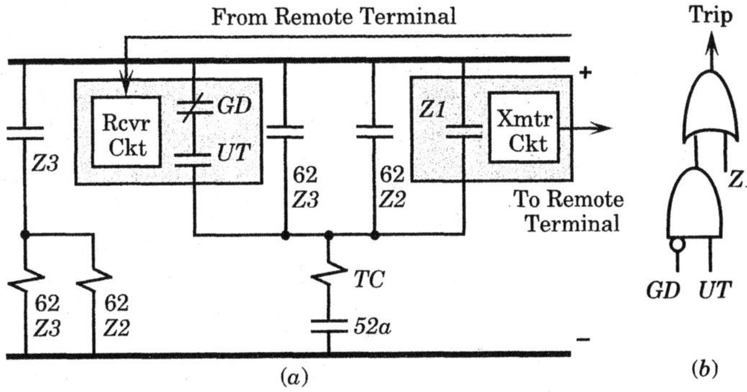

Figure 13.15 Relay contact and zone 1 solid-state logic for the direct underreaching transfer trip with guard frequency security. (*a*) Contact logic. (*b*) Zone 1 solid-state logic.

sends a tone to the remote terminal that picks up the guard receiver and contact GD. When a fault is detected, the transmitter shifts its transmission frequency from the guard frequency to the trip frequency, thereby allowing the GD contacts to close and picking up the UT contacts. Tripping is allowed only if both conditions are satisfied, which is accomplished by a simple shift in the transmitted frequency.

For direct underreaching transfer trip, it is common practice to use directional distance relays for phase faults. The ground fault protection can be either overcurrent or distance protection, depending on the ground fault current sources and magnitudes.

13.3.3.2 Permissive Underreaching Transfer Trip. Another form of underreaching transfer trip is called permissive since two types of relays are used cooperatively to achieve the desired result. The zones of the relay devices are similar to those represented in Figures 13.10 and 13.11 except that a second set of instantaneous elements are employed, which are set to overreach the far end of the transmission line. This is shown in Figure 13.16 as instantaneous overreaching elements OR(R) and OR(Q).

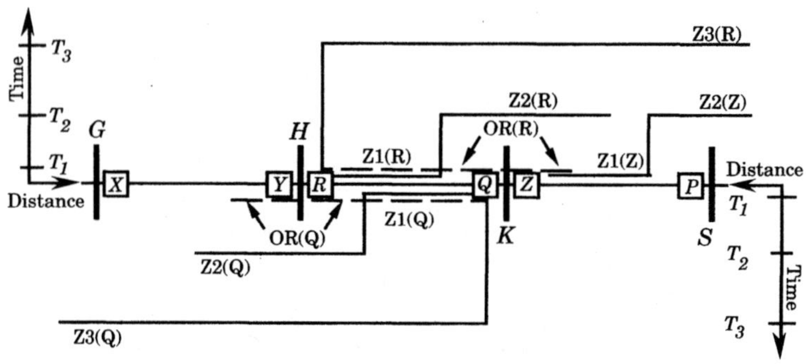

Figure 13.16 Permissive underreaching transfer trip.

The zone 1 device is an underreaching transfer trip device. A fault detection in zone 1 keys a transfer trip frequency in the transmitter, to initiate tripping at the remote end of the line. The OR relays are overreaching, and are basically fault detectors. These overreaching relays are the permissive devices and must operate to permit a received transfer trip signal to trip the circuit. The control circuits for this type of protection is shown in Figure 13.17.

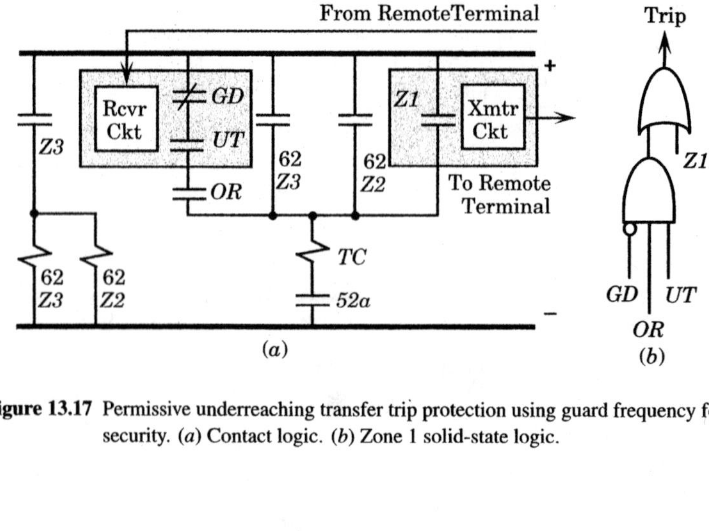

Figure 13.17 Permissive underreaching transfer trip protection using guard frequency for security. (*a*) Contact logic. (*b*) Zone 1 solid-state logic.

The permissive underreaching transfer trip system works as follows. Consider a close-in fault at the R end of the line. The local R zone 1 element sees the fault in its trip zone and immediately sends a trip signal to the R breakers. It also instructs its transmitter to shift from guard to trip frequencies, thereby dropping out the GD contacts and picking up the UT contacts at the Q end. The Q end OR relays, having an overreach setting, will also see the fault near the R end and close the OR contacts at Q, thereby completing the circuit and tripping the breakers at the end remote from the zone 1 fault.

For a fault in the overlapping trip zone, both ends normally trip their local circuit breakers and also send transfer trip signals to the remote ends, thereby assuring positive tripping since the overreaching elements OR will also recognize these faults. This type of protection normally employs distance relays for phase faults and either overcurrent or distance protection for ground faults [9].

13.3.3.3 Direct Overreaching Transfer Trip. Another transfer trip scheme is conceptually an overreaching rather than underreaching scheme. The typical reach and timing of this scheme are shown in Figure 13.18. Here, zone 1 is the same as before and zone 2 overreaches the bus at the far end of the line in both directions. The control is different, however, with the overreaching system using the directional relay of zone 2 to supervise the transfer trip receiving relay, OT, as shown in Figure 13.19 [8].

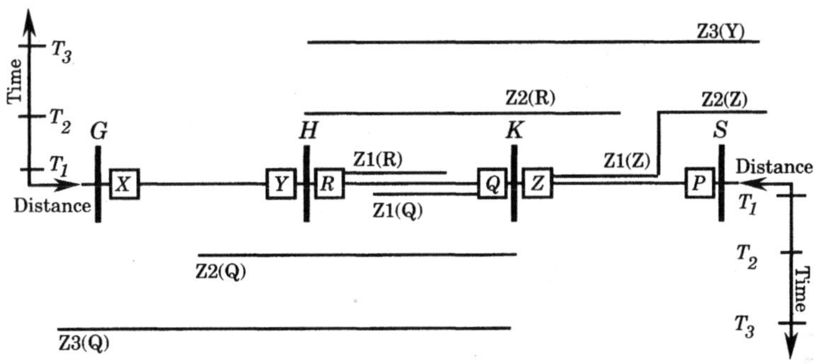

Figure 13.18 Reach and timing for direct transfer trip overreaching scheme.

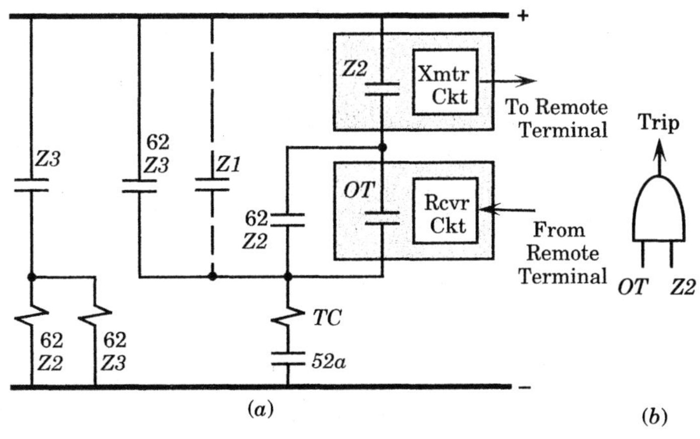

Figure 13.19 Relay control circuit for direct transfer trip overreaching scheme. (*a*) Contact logic. (*b*) Zone 2 solid-state logic.

Consider a close-in fault near relay R at bus H in Figure 13.18. This fault will be cleared at bus H by the zone 1 relays R and at bus K by the Q-OT relays because of the transfer trip receiving relay that is triggered by the R system zone 2 element. If the fault had been on the line GH, the zone 2 relays of relay Q would still pick up, but the transfer trip receiving relay at Q would not be energized since zone 2 at R would correctly refuse to recognize the fault.

A disadvantage of this scheme is the absence of an independent zone 1 trip, making a high-speed trip entirely dependent on the pilot channel. A modification of the scheme can be devised whereby a direct zone 1 trip is inserted, where the zone 1 reach is set for 80–90% of the protected line length (see dashed $Z1$ contact connections in Figure 13.19).

For these systems, as for all transfer trip systems, the signaling of transfer trip information requires that the signal be sent during a fault condition. If the transfer trip signaling is performed using power line carrier, this may fail because the signal transmission takes place over a faulted transmission line. The signal may get through, but the probability of correct transmission is certainly reduced. This is generally true of PLC transfer trip schemes. In some systems, the strength of the pilot signal is increased on fault detection to improve the probability of successful signal transmission through the faulted line.

13.3.3.4 Permissive Overreaching Transfer Trip. An alternate design of an overreaching transfer trip system is the "permissive" system shown in Figure 13.20. This scheme uses high-speed distance elements RO that are set to overreach the remote terminal, similar to the OR element of Figure 13.16. The RO elements serve dual functions. First, they act as transfer trip devices and, on fault detection, initiate the dropping of the guard frequency and keying of the trip frequency. They are also the permissive device, because they must operate in order to permit the transfer trip signal to operate the breakers when a transfer trip signal is received.

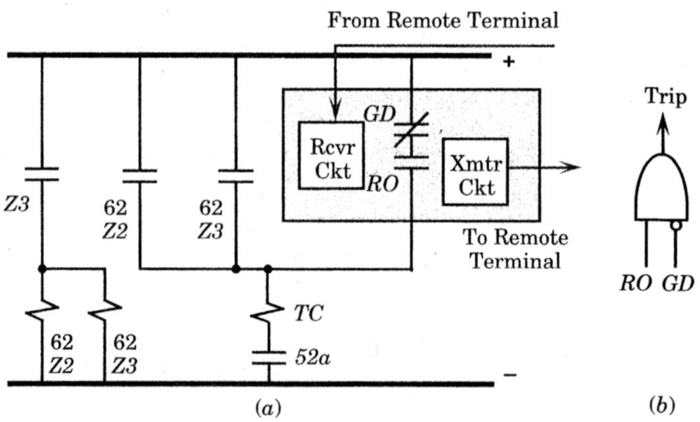

Figure 13.20 Permissive overreaching transfer trip protection. (*a*) Contact logic. (*b*) Zone 1 solid-state logic.

The permissive overreaching transfer trip system normally uses distance relays for phase fault protection and instantaneous overcurrent relays for ground fault protection [9]. All overreaching transfer trip schemes are permissive.

13.3.3.5 Summary of Transfer Trip Schemes. All of the transfer trip schemes employ one transmitter and one receiver at each relay location for the protection of a given line. In some cases, guard signals are also employed to increase the security of the protection and avoid false trips due to communication system noise. In practice, the several frequencies used are selected so that there is no interaction between the various channels. Moreover, the transmitters and receivers are paired, that is, a given transmitted frequency is received only at a receiver at the opposite end of the line, and by no other receiver in that or any nearby system.

All transfer trip schemes share the difficulty of requiring the transmission of a trip signal during a fault. This may sometimes be difficult if power line carrier is used as the signal transmission medium, and the signal would have to pass through the fault to reach the remote end. In practice, it has been observed that the transfer trip signal often does pass through a faulted line section, but this is almost impossible to ensure. This is not a problem for microwave or fiber-optic systems, unless the fiber-optic cable is carried in the faulted phase of the high-voltage line and that conductor becomes severed due to the fault.

13.3.4 Blocking and Unblocking Pilot Protection

Blocking protection is the inverse of transfer trip protection. In a transfer trip scheme, a signal is sent from one relay to the other to order a trip of the remote breakers on sensing a fault that the sending relay verifies to be in a predetermined trip zone. Transfer trip has the disadvantage that any failure of the communications will cause end zone faults to be tripped at a lower speed than desired. Moreover, some transfer trip schemes depend on pilot signaling for proper zone 1 operation.

A blocking scheme does the reverse of the transfer trip scheme. In a blocking scheme, the sending relay monitors the protected line as usual, but it also monitors the region *behind* the sending relay. Any faults detected *behind* the relay are known positively to be faults for which tripping of the protected line is *not* desired. This triggers the transmission of a *blocking* signal from the sending relay to the receiving relay to prevent tripping, by opening the trip circuit of the remote relay.

13.3.4.1 Direct Blocking Scheme. In the direct blocking scheme, the relays for the blocking system will normally employ distance elements in the forward-looking direction, but also with reverse-looking blocking elements BL having no intentional time delay. This scheme, illustrated in Figure 13.21, operates as follows. Consider a fault between buses G and H in Figure 13.21, which obviously is not within the protection zone of line HK. The reverse-looking $ZR(R)$ elements will pick up this fault as positively one for which the line HK should *not* be tripped. The relay R therefore sends relay Q a blocking signal, which positively blocks tripping at Q. Conceptually, this is attractive since the signal transmission takes place over a sound line. This makes the scheme attractive for power line carrier as well as for other forms of pilot signaling.

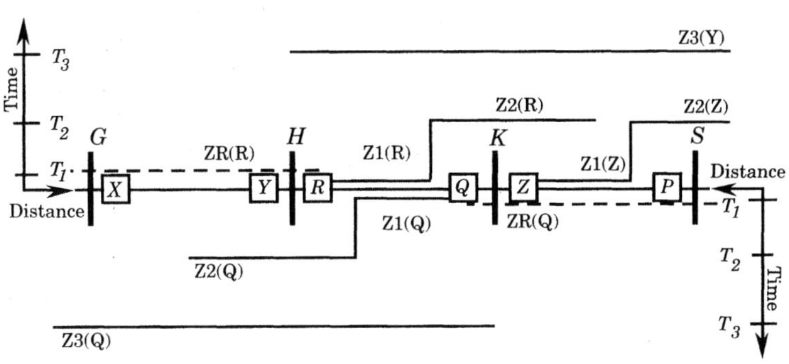

Figure 13.21 Distance-time characteristics for a blocking scheme.

A typical control circuit for a blocking scheme is shown in Figure 13.22 [8]. In the event of a fault *behind* the relay R at H, the remote relay Q zones 2 and 3 will pick up. These relays are prevented from tripping by reception of a blocking signal from the reverse-looking relay ZR at R to open the normally closed contacts BL at the remote end (Q). Clearly, the reverse-looking relays must be faster than the forward-looking *Z2* elements, and the signal channel must also be fast. Coordination is often made more positive by the insertion of a timer in the *Z2-Z3-BL* control path.

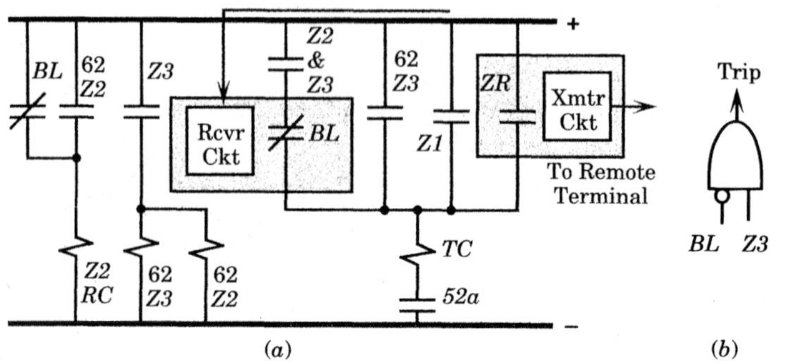

Figure 13.22 Control scheme for direct blocking at Q. (*a*) Contact logic. (*b*) Zone 3 solid-state logic.

Referring to Figure 13.21, consider a fault on the section the protected line HK where the zone 1 protections at the two ends overlap. Such a fault is seen as zone 1 by relays at both ends, and cleared instantaneously. No signal transmission takes place since there is no fault in the reverse-looking direction for either relay. Therefore, the contacts BL remain closed.

Now consider a fault on line HK that is close to the relays R at bus H. Such a fault is tripped in zone 1 by the R relays. The fault is also picked up by the zone 2 and zone 3 elements at Q, which will trip the internal fault after the required time delay. To overcome this time delay, methods have been devised to change the time delay on the zone 2 relay to make it as fast as zone 1, thereby extending the range of zone 1 [8]. Such a method is illustrated in Figure 13.22. For no blocking signal transmitted, the BL contacts remain closed, thereby picking up the "reach change" coil, *Z2RC*, which changes the reach of zone 2 to that of zone 1 for internal faults.

For a fault behind the relay at R, on line section GH, the forward-looking relays at R do not pick up, but those at Q pick up on zones 2 and 3. However, the reverse-looking element at R sends a blocking signal to Q to prevent these overreaching zones from picking up, and thereby ensuring correct protective system operation.

13.3.4.2 Directional Comparison Blocking Scheme. The directional comparison blocking scheme is described in reference to Figures 13.23 and 13.24. Both the phase and ground fault units at R and Q, designated here as $P(R)$ and $P(Q)$, are set to overreach the remote terminals so that they will pick up for all internal faults. Usually these units employ distance measuring modules that are set to overreach by 120–150% of the line length. Special *starting units*, designated S(R) and S(Q) in Figure 13.23, are set with different reach than the overreaching protective relays as shown in the figure. The protective relays and starting units must be relays of the same type, for both phase and ground faults. For example, if the phase relays are distance relays, then the starting units must be distance relays, and similarly for the ground relays.

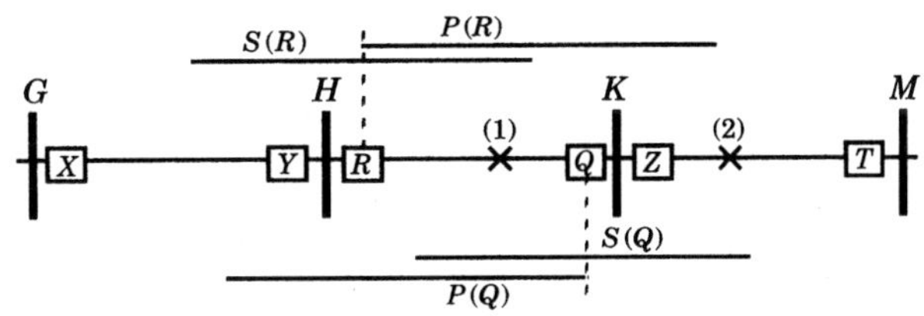

Figure 13.23 A directional comparison blocking scheme [12].

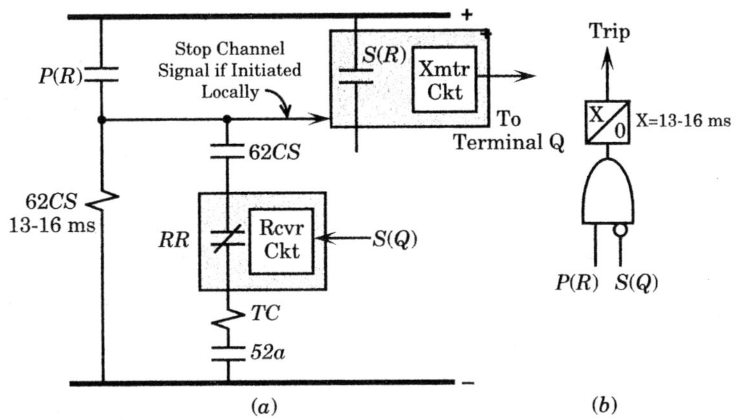

Figure 13.24 Trip logic for the directional comparison blocking scheme. (*a*) Contact logic. (*b*) Solid-state logic.

The contact logic and solid-state logic diagrams are shown in Figure 13.24. The pilot signaling often used for this type of logic is a simple on-off PLC carrier. No signal is transmitted under normal conditions since the *S* units operate only under a fault condition. Table 13.2 summarizes the action of the system for the internal fault (1) and the external fault (2), both of which are illustrated in Figure 13.23.

TABLE 13.2 Summary of Protective Actions at Relay Location *R*

Type of Fault	Events at Station *H*	Events at Station *K*
(1) Internal	*P*(*R*) operates; *S*(*R*) may or may not operate, but *P*(*R*) operation prevents sending of blocking signal. Breaker *R* tripped.	*P*(*Q*) operates; *S*(*Q*) may or may not operate, but *S*(*Q*) operation prevents sending of blocking signal. Breaker *Q* tripped.
(2) External	*P*(*R*) operates; *S*(*R*) does not see fault. Blocking signal received at *R* from Station *K*. *RR* contacts open. No trip.	*S*(*Q*) operates; operates to key XMTR. Blocking signal sent to station *H*. *P*(*Q*) does not see fault. No trip.

The directional comparison blocking scheme is widely used because of its flexibility, reliability, and simplicity of communications. Note that the communications channel is not required for tripping, so faults that may cause problems with correct signaling are eliminated. Overtripping can occur if the signal channel fails, or if the communications fails to operate for external faults within the reach of the starting units. Note that the signal channel is normally off, so there is no means of testing its operation under normal conditions, except for a planned maintenance visit to the site.

13.3.4.3 Directional Comparison Unblocking. The directional comparison unblocking scheme, shown in Figure 13.25, transmits a continuous blocking signal, except during internal faults. Starting units are not required for this scheme. The communications usually employs a frequency shift PLC signal, where a blocking or guard frequency is transmitted continuously during normal conditions. On detection of an internal fault, the signal frequency is shifted to the unblock (or trip) frequency. In some cases, the transmitted power is increased, say from 1 to ten watts, during the unblocking signal period.

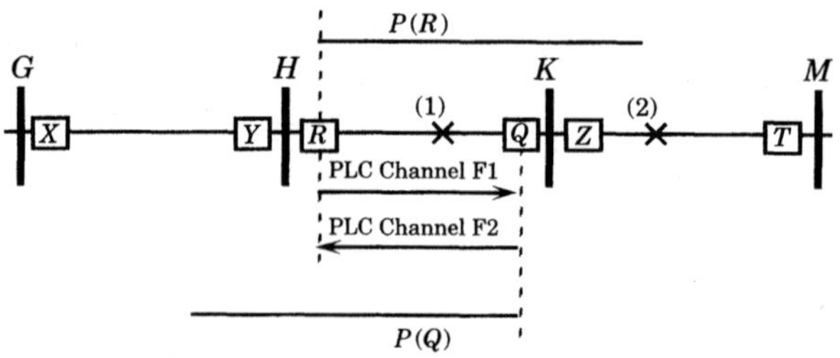

Figure 13.25 A directional comparison unblocking scheme [12].

The contact logic for directional comparison unblocking is shown in Figure 13.26. The frequency-shift channel is monitored continuously so that loss of channel can be detected and alarmed to prevent long channel outages.

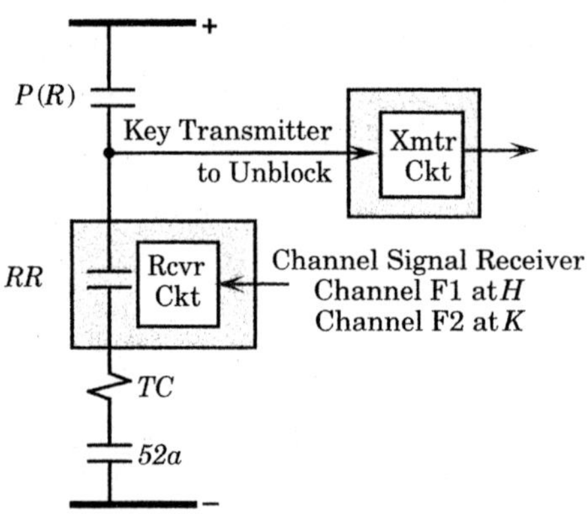

Figure 13.26 Contact logic for directional comparison unblocking [12].

The solid-state logic diagram for this type of system is shown in Figure 13.27. Normally, a block frequency is transmitted and OR-1 has no output. Therefore, both AND-1 and AND-2 are unsatisfied, which means that OR-2 has no output. The block frequency is removed for an internal fault, which means that OR-2 will be satisfied whether the unblock signal is operable or not. This is important, since it is possible that the unblock signal will be shorted out by the fault. When this occurs, OR-1 gives an input to AND-2, which satisfies this gate for 150 ms. Then AND-2 picks up OR-2 to operate relay RR or to provide an input to AND-3. Without this unblock signal, 150 ms is provided for tripping. After 150 ms, lockout is initiated since one of the inputs to AND-2 is removed. This resets RR or removes the input to AND-3.

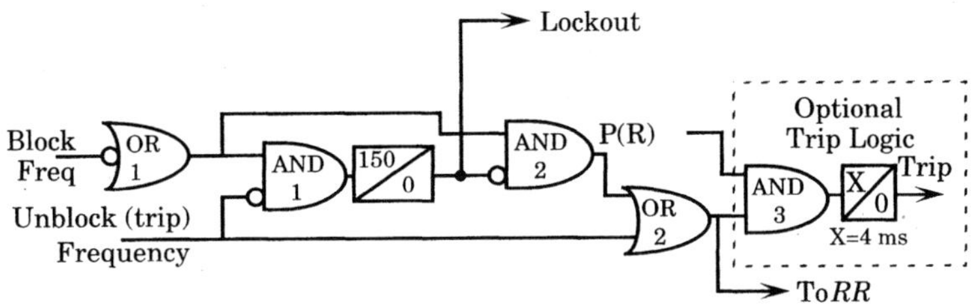

Figure 13.27 Simplified unblocking receiver logic at relay location *R*.

If the unblock signal is received, this results in an input to OR-2 to directly pick up *RR* or to provide input to AND-3, which will result in a trip. The unblock signal also removes an input to AND-1 to stop the timer. Table 13.3 describes the operation of the directional comparison unblocking scheme, in connection with faults (1) and (2) shown in Figure 13.24.

TABLE 13.3 Summary of Protective Actions at Relay Location *R*

Type of Fault	Events at Station *H*	Events at Station *K*
(1) Internal	*P*(*R*) operates. F1 channel shifts to unblock. Loss of block and/or receipt of unblock (F2) operates *RR* or inputs to AND-3. Trip.	*P*(*R*) operates. F2 channel shifts to unblock. Loss of block and/or receipt of unblock (F1) operates *RR* or inputs to AND-3. Trip.
(2) External	*P*(*R*) operates. F1 channel shifts from block to unblock. F2 channel continues to block. No trip.	*P*(*Q*) does not see fault. Loss of block and/or receipt of unblock F1 operates *RR* or inputs to AND-3. No trip.

For this scheme, both the phase and ground fault detectors must operate for all internal faults, which means that they must overreach the remote bus. The reliability and security of directional comparison unblocking schemes make them one of the most attractive schemes for transmission lines using PLC pilot signaling. The only way failure can occur for an external fault is to experience channel failure within 150 ms following the fault occurrence. The scheme is also applicable to multiterminal lines, using separate channels between each terminal and the remote terminals.

13.3.5 Selectivity in Directional Comparison Systems

Directional comparison schemes all require some means of ensuring selectivity in their operation. This is true whether the scheme is based on transfer trip, blocking, unblocking, or some other principle. Embedded in many of these schemes is some form of distance measurement. This is usually described in terms of a particular shape of apparent impedance, plotted in the Z plane. In the modern digital systems, these characteristics can take on a wide variety of shapes that can be varied by a particular selection of settings. Thus, the settings can describe an underreaching, overreaching, or other special shapes. A few examples of these systems are shown in Figure 13.28.

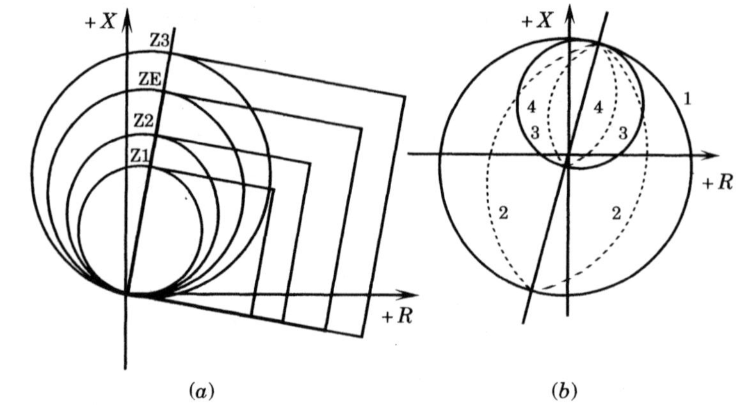

Figure 13.28 A few of the measurement shapes available for selection. (*a*) Characteristics for phase and ground faults. (*b*) Some starting characteristics for three-phase faults.

Figure 13.28(*a*) illustrates a family of characteristics offered by one relay manufacturer [13]. The circles represent three zones of protection, with an additional zone E that is independent of the other three and may be used as a backup zone. The quadrilateral zones are used for ground faults, which often exhibit high resistance. This type of characteristic has proven to be superior to the simple mho characteristic for ground faults. All of these zoned distance measurements are performed using phase comparison techniques, as discussed in Section 9.8.

The characteristics shown in Figure 13.28(*b*) represent a group of starting settings used for different purposes [14]. This type of relay offers a variety of circular and lens shaped characteristics. Many other characteristics are offered as options.

These characteristics may be used with a variety of permissive over- and underreaching transfer trip, as well as blocking schemes.

13.3.6 Other Features of Directional Comparison

Directional comparison schemes often have a number of additional features that can be added to the basic relay system at little additional cost. A few of these added features are described below.

13.3.6.1 High-Speed Reclosing. The use of *high-speed reclosing* is an important concept that, if employed, must be taken into account in the choice of pilot protection schemes. High-speed reclosing is used at many relay locations on the theory that nearly all faults are temporary, and that the immediate reclosing of the circuits will almost always be successful. If the reclosing is not successful, the relays are usually blocked from successive trials at reclosing.

Consider the system shown in Figure 13.29, where line faults are examined at both points 1 and 2, and the protected line is considered to be that from G to H. Suppose first that a temporary fault occurs at 1, on the line adjacent to the protected line. Suppose first that

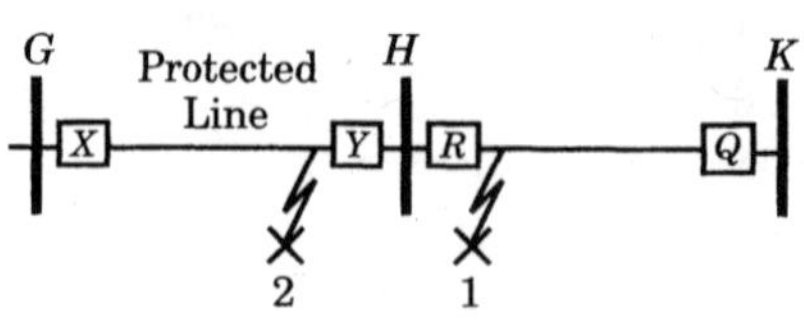

Figure 13.29 Faults on adjacent transmission lines.

line GH is protected by a blocking scheme and that reclosing is used on all lines. If the pilot channel should fail for this situation, blocking will not take place and relays at both X and R will trip and reclose, thus restoring the system to normal.

Now, consider the same temporary fault at 1 with underreaching transfer trip on line GH and with a failed pilot system. In this case, only the breaker at R would trip and would then reclose. Thus, for this example involving fault 1, the blocking scheme causes an unnecessary trip and reclosure of a sound line, but the line outage is temporary and is brief.

Now consider a temporary fault at 2, and again consider the possibility of a failed pilot signal transmission. First, consider a blocking system. Since the fault is internal, there is no need for signal transmission. Breakers at both X and Y will trip and reclose, restoring the system to normal. In fact, this operation does not even reveal the presence of a failed pilot system.

Now consider the same temporary fault #2, but with transfer trip pilot protection. For this case, with the fault close to the relays at Y, the Y relays will clear the local Y breakers in zone 1 time. Having a failed pilot system, there is no way to send the transfer trip signal to X, which will trip with zone 2 time delay, during which time Y will reclose. Successful reclosing is not possible at Y, however, because the fault is maintained at X. Thus a reclosure at Y will still see the fault, and Y will trip and lock out. This example assumes that the reclosure at Y occurs before zone 2 at X times out.

The major advantage of transfer trip is speed. Blocking schemes introduce intentional time delays to ensure that the pilot signals have been received when blocking is necessary. However, blocking has advantages in certain cases—the example cited above being one of those.

13.3.6.2 Power Swing Blocking. Power swing blocking units detect the presence of power swings or out-of-step conditions and block operation of the relay for these events. In some cases the blocking unit can be disabled by zero sequence current, allowing the relay to operate for ground faults even when power swings have been detected.

13.3.6.3 Ground Fault Protection. Some distance relays also have provision for adding directional ground fault protection, which can also operate in either a permissive or blocking mode. Distance measurement can be performed using either zero or negative sequence quantities [14].

13.3.6.4 Switch-Onto-Fault Function. A switch-onto-fault unit provides instantaneous tripping of the line if the circuit breaker is closed into a fault. Various types of logic can be provided to implement this type of function.

13.3.7 Hybrid Schemes

When directional comparison protection schemes were first introduced, the schemes used with power line carrier (PLC) were usually blocking schemes, using an ON-OFF channel, and those used with *microwave* were tripping schemes, using a frequency shift channel. The characteristics of a blocking scheme are attractive for some purposes, and tripping for others. As frequency-shift keyed channels became available on PLC it was possible to offer either tripping or blocking schemes on using either mode of signal transmission. Because of this flexibility, it is necessary to use two terms to classify a particular scheme, one for the channel type and another for the mode of operation.

Several common types of transmission protection pilot schemes are shown in Table 13.4. The first column gives a generic name to the protective scheme. The second column notes the type of pilot channel communications, i.e., whether frequency shift keying (FSK) or an on-off type of signaling. The third column notes the type of permissive that is used, i.e., whether the scheme transfers a trip signal, removes a blocking signal (unblock), or neither. The fourth column notes the relay action, whether to trip or to block tripping. The sixth column shows the type of standby signal transmitted, which is usually a guard or a trip signal. Finally, the seventh column denotes those schemes that normally repeat the signal from the remote source back to that source. It is common practice to identify the pilot scheme that is used for any type of permissive relaying. This will be done for the relay descriptions presented in this book.

TABLE 13.4 Classification of Pilot Relay Schemes [11]

Type of Scheme	Pilot Channel	Trip Permissive	Function		Standby Signal	Repeat Keying
			Trip	Block		
Tripping	FSK	Trip	X	X	Guard	
	FSK	Unblock	X	X	Guard	
Blocking	On-Off	None	X	X	None	
	FSK	Trip	X	X	Guard	
	FSK	Unblock	X	X	Trip	
Hybrid	FSK	Trip	X	X	Guard	X
	FSK	Unblock	X	X	Guard	X

One widely used hybrid relay system uses two principle relay subsystems to recognize and clear faults. The first section is a positive sequence distance measuring relay that is used primarily to clear three-phase faults. This unit has both a static measuring unit for normal applications and a dynamic unit for use with series compensated lines. The second relay section is a negative sequence overcurrent relay, which uses negative sequence measuring and directional characteristics for the detection of unbalanced faults. This has the advantage of being immune from zero-sequence mutual coupling that is often a problem on high-voltage lines.

The hybrid scheme requires a local trip signal as well as a permissive from the remote terminal to initiate a trip. A directional comparison logic is used to determine if the fault is internal or external to the protected line. If the fault is internal, one or more of the tripping functions at each terminal will detect the fault and key the transfer trip transmitter to the trip frequency. The receipt of the trip frequency and operation of one or more of the tripping functions is sufficient to initiate tripping at each terminal. Thus, under normal conditions, the hybrid scheme acts exactly like a permissive overreaching transfer trip scheme.

Now, consider a line terminal with a weak infeed. If the infeed is not sufficient to operate the tripping functions at that terminal, a permissive signal will not be keyed and tripping will not occur because it is necessary for the fault to be detected at all terminals of the line. The hybrid scheme overcomes this deficiency by means of a blocking function and some added logic. The trip signal received from the strong terminal is echoed, or repeated, at the weak terminal, thereby permitting the strong terminal to be tripped. The weak terminal is tripped by additional logic. The signal at the weak terminal is permitted to be repeated only if blocking functions have not operated, or the signal will operate only for internal faults. For external faults, blocking functions will operate to block the echoing of the signal.

A simplified view of a hybrid logic is shown in Figure 13.30. The directional logic for this scheme is similar to that shown in Figure 4.25. In Figure 13.30, the following nomenclature applies:

$$\begin{aligned} M_1T &= \text{positive-sequence tripping} \\ M_1B &= \text{positive-sequence local blocking} \\ D_2T &= \text{negative-sequence directional tripping} \\ D_2B &= \text{negative-sequence blocking} \\ I_1 &= \text{positive-sequence current} \\ I_2 &= \text{negative-sequence current} \end{aligned} \tag{13.1}$$

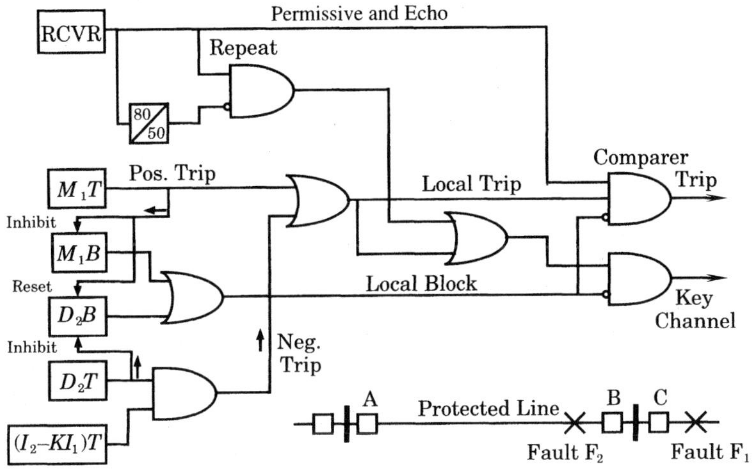

Figure 13.30 Simplified logic of a hybrid trip scheme [15].

The positive-sequence trip function is an overreaching function. Either a positive- or negative-sequence function can result in a local trip function to the comparer gate. Similarly, a local blocking function can come from either a positive or negative sequence pickup. A trip occurs if there is a received channel signal plus a local trip signal and no block signal. There are also inhibit and reset functions that are derived from phase comparator circuits using timers to measure the coincidence between two signals. The inhibit function blocks the input to a timer and a reset blocks the timer if it has not operated, or causes it to reset if it has operated. Thus, the positive sequence function M_1T inhibits M_1B and resets D_2B, and D_2T inhibits D_2B. The 80/50 timer and the AND logic makes up the repeat keying function. If a permissive signal is received at a terminal, this signal keys the channel to send the permissive in the opposite direction if the blocking functions have not operated.

It has been observed that directional comparison relaying schemes can be used on series compensated lines only if the fault current is high enough to cause the capacitor bypass to operate. This is because of the voltage inversion that occurs for a fault just beyond the capacitor, which can defeat the distance measuring device, making the fault appear to be shifted to a different quadrant in the Z plane. The hybrid scheme described here is an improvement on that basic principle, which relies on a transient characteristic of the mho element, which is described below. The positive sequence unit employs a phase comparison principle to describe a polarized mho directional characteristic in the complex IR-IX plane, as shown in

Figure 13.31. The mho characteristic in Figure 13.31 is polarized by using a prefault voltage V_P rather than the actual voltage that is measured when the fault occurs. The prefault voltage is remembered by the relay for a short time, and this memory action prevents collapse of the mho circle when the fault is close to the relay.

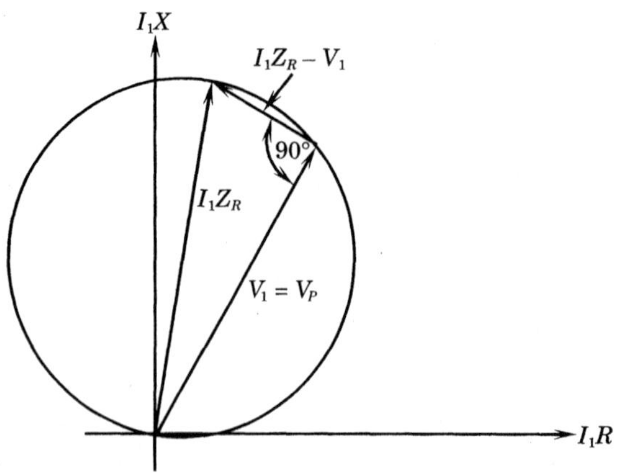

Figure 13.31 The steady-state polarized mho characteristic.

The trip threshold is the 90 degree angle on the circle between the polarizing voltage V_P and the voltage $I_1 Z_R - V_1$ and where V_1 is the positive-sequence restraint voltage and I_1 is the positive-sequence current. When this angle is greater than 90 degrees, tripping occurs. The quantity $I_1 Z_R - V_1$ describes a circle in the complex plane if the angle between this quantity and the polarizing voltage is a right angle. If this angle is more than 90 degrees, the point lies inside the circle and tripping occurs. If the angle is less than 90 degrees, the point is outside the circle and no tripping occurs.

When a voltage inversion occurs, both V_1 and V_P will be inverted, but the polarizing voltage will remain in phase with the source voltage, at least briefly. During this transient, the mho circle changes to the condition shown in Figure 13.32. In this figure, the negative reactance region represents the fault impedance seen by the relay for a forward-direction capacitive reactance fault location. Tripping for faults in this region is, therefore, correct.

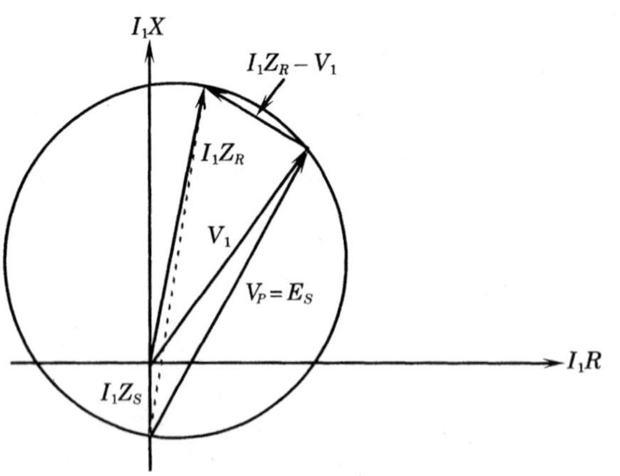

Figure 13.32 The transient polarized mho characteristic.

It can also be shown that, for faults behind the relay, i.e., reverse direction faults, the transient characteristic is well outside the region of fault impedances for the normal reach setting of the line. Therefore, the transient polarized mho characteristic provides a workable relay logic for all three-phase faults, even those for which the relay voltage goes to zero.

The detection of all unbalanced faults in this relay is performed using transient compensated phase comparison, but based on negative sequence, rather than positive sequence, quantities. The reasons for using negative sequence quantities are the following [16]:

1. Negative-sequence measurements are immune to the effects of zero sequence mutual coupling of closely coupled parallel lines.
2. Input currents and voltage can be obtained from the same instrument transformers that supply inputs to the positive-sequence measurements.
3. Negative-sequence current is larger in magnitude than zero-sequence current for one-line-to-ground faults at the far end of the line, thereby providing greater sensitivity.
4. Negative-sequence measuring equipment is easier to field test by opening or shorting the appropriate circuits of the relay to simulate fault conditions.

The polarizing voltage for the negative-sequence measurements is given by

$$V_P = V_2 + kI_2Z_R \tag{13.2}$$

The reference voltage V_2 is supplemented by the adjustable compensating quantity I_2Z_R, which ensures correct polarization for cases when the relay is near a strong source where V_2 may be very small or when the fault is near the end of a series compensated line and the fault current is below the capacitor protection level. A gain k is also provided to adjust the compensating quantity. The compensated polarizing voltage (13.2) and the operating quantity, which is proportional to the negative-sequence current, are compared in phase on each half cycle to produce both tripping and blocking functions.

The hybrid scheme has been used for many years on series compensated lines with excellent results. One of the advantages of this system is the use of negative-sequence quantities for all faults involving the ground. This makes the protection immune to problems of mutually coupled lines. A separate relay module is used to detect three-phase faults, thereby offering complete coverage for all types of faults. This type of logic has been implemented in solid-state systems [15] as well as systems utilizing programmable logic and microprocessor monitoring to detect abnormal system operation and record tripping operations [17].

13.4 UNIT PROTECTION PILOT SCHEMES

The preceding section presents several types of "non-unit" pilot protection schemes, and those schemes are the most common pilot systems. Unit protection schemes measure the current entering the protected component at its several terminals and bases the trip decision on a comparison of these currents.

13.4.1 Phase Comparison Schemes

By far the most common type of unit protection for transmission lines is phase comparison. There are four types of phase comparison schemes:

1. Single-phase-comparison blocking.
2. Dual-phase-comparison blocking.

3. Dual-phase-comparison transfer trip.
4. Segregated phase comparison.

The fourth system, segregated phase comparison, is a current-only system, which has advantages in certain types of transmission lines, particularly lines that include series capacitors. All four types of phase comparison can be current-only, or may be distance supervised systems.

Phase comparison is a concept in protective relaying that has been used for high-voltage transmission lines for many years [18]. The basic concept of phase comparison relaying is to note the exact time of the current passing through zero, or through either a positive or negative threshold value. Records of these crossing patterns are transmitted to the opposite ends of the line, where they are compared. If it is determined that the currents are both entering the line, clearly a line fault exists and the relays at both ends can cause the line to be tripped. The comparison of current waveforms is accomplished by constructing square waves of the unfiltered current waveforms, as shown in Figure 13.33. The square wave labeled I_{SW} (read "SW" as "square wave") is derived from the threshold crossings of the current wave. Pulse trains are also generated as the current wave exceeds both a positive and a negative threshold, shown by the horizontal dashed lines. These square wave trains are labeled I_{SW-P} and I_{SW-N} in Figure 13.33 to identify the positive and negative square waves, respectively.

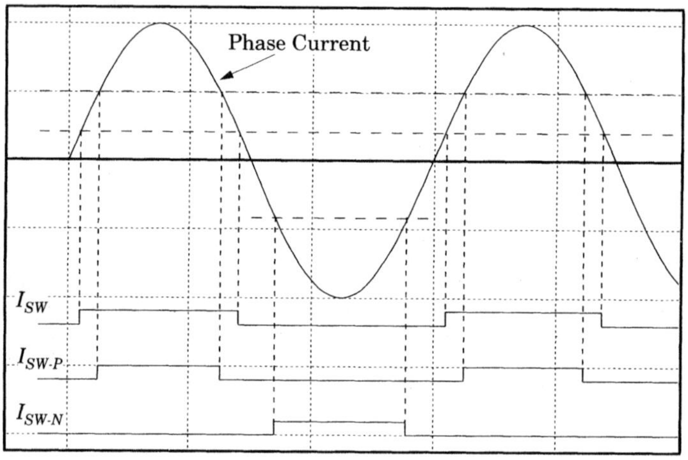

Figure 13.33 Square wave processing from a phase current waveform.

In some types of phase comparison schemes, the measured currents are processed through a sequence network filter, as shown in Figure 13.34, to give a waveform that is converted into a square wave. A similar square wave is generated at the opposite end of the transmission line, and this remote square wave is transmitted by pilot channel to the local relay so it can be compared with the locally generated square wave. A delay circuit is provided to account for the time delay of data transmission. If the two square waves are in phase, this indicates that the currents are in phase and are flowing through the line. This would be the normal condition, or the condition observed for a fault external to the line. If the two waves are out of phase, this indicates that the currents at both ends are flowing into the line, which is the condition for an internal fault. This basic concept is followed in all types of phase comparison schemes, although there are variations that are used in the four types of phase comparison systems. A unit scheme that uses a measuring quantity that is derived as a combination of the phase currents is sometimes called a *composite* type of unit scheme. This concept can

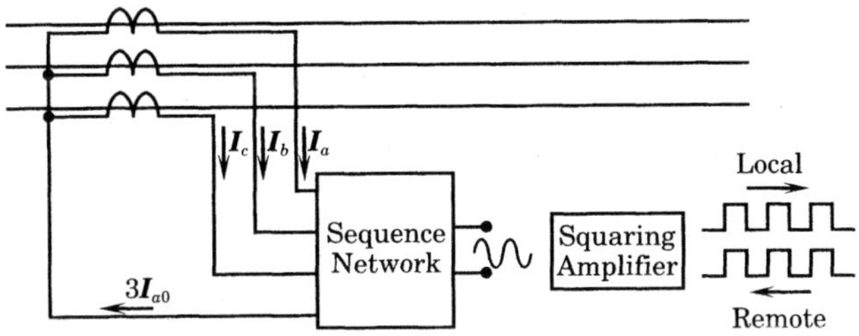

Figure 13.34 A sequence network for combining three-phase currents.

be expanded to consider a network that derives a measuring quantity by combining sequence currents rather than phase currents [18]. Either the phase or sequence currents can be weighted in some manner to produce the final measuring quantity.

When a double fault occurs on a parallel line that involves different phases of the two circuits, the relay measurement of the fault condition may not reliably trip the circuits, since the trip decision may depend on the weighting of the phase currents and on the total fault current distribution. In some cases, the protection will not operate until one of the lines is cleared, reducing the fault condition to a single fault on the remaining line.

These deficiencies are not present for segregated phase comparison scheme, which is discussed below.

13.4.1.1 Single-Phase-Comparison Blocking. A current-only comparison is a simple form of phase-comparison blocking that requires only an on-off PLC channel. Two fault detectors, FD1 and FD2, are used, both having a logic similar to that shown in Figure 13.35. FD1 is called the "carrier start" unit. It is set more sensitively than the FD2 and is arranged to permit the local square wave to key the on-off carrier transmitter. FD2 is set with a higher pickup than FD1, and is used to arm the system for tripping. For transmission lines up to about 100 miles long, FD2 is set at about 125% of FD1, and for longer lines this setting is increased to about 200%. Phase comparison can begin once FD2 operates. Using the two fault detectors permits the coordination of the keying of the carrier square wave with the comparison of the local and remote square waves.

The operation of the system can be understood by studying Figure 13.35. For an internal fault [1], the overcurrent fault detectors at both ends of the faulted line will operate, each having sensed currents above pickup. A flip-flop is energized if the inputs to the AND continue for 4 ms, which provides a continuous output to the trip circuit that is supervised by the FD2 operation. For this trip condition, the square wave inputs from the local input and the receiver output to the AND gate are in phase, and tripping is performed. This system will trip if the currents at the two terminals are up to 90 degrees out of phase. Note that the receipt of a signal from the channel prevents tripping; therefore, this is a blocking system. This also means that the carrier signal does not have to be transmitted through the internal fault.

Note that the local square wave turns the carrier on and off to create the square wave that is received at the remote terminal. For an external fault, blocking is continuous due to the phase of the receiver output being out of phase with the local square wave. Usually the through currents to an external fault are approximately in phase; however, this system will block tripping if these currents are up to 90 degrees out of phase.

Single-phase-comparison blocking can also be devised that uses distance supervision rather than current only. This scheme uses the same on-off channel, but the fault detection and arming techniques are different.

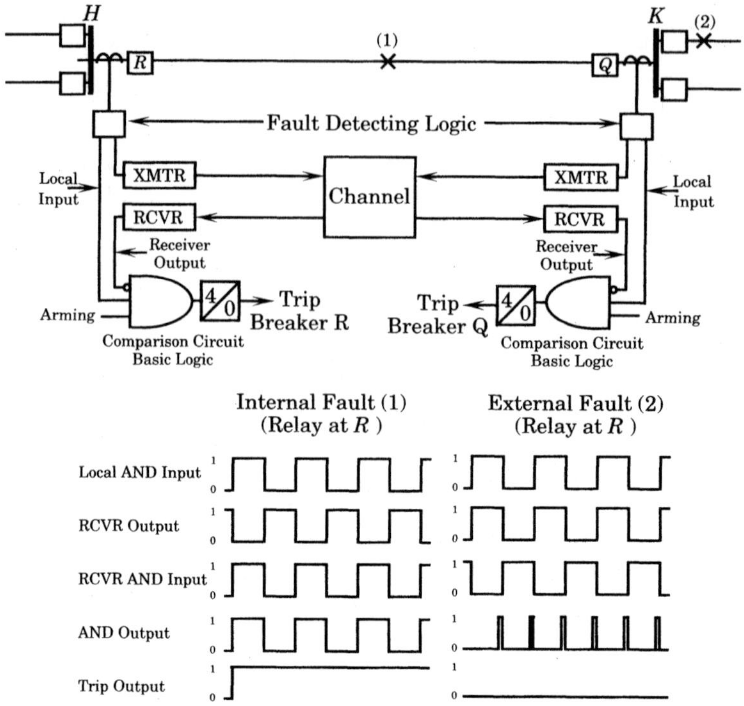

Figure 13.35 The single phase-comparison blocking scheme [18].

13.4.1.2 Dual Phase-Comparison Unblocking. The dual phase-comparison scheme makes phase comparisons on both half cycles of the current wave, which provides faster tripping than the single current comparison scheme. The dual phase-comparison scheme requires a duplex channel with one frequency for each line terminal.

An example of a dual phase-comparison scheme is shown in Figure 13.36 [18]. This scheme uses frequency-shift equipment at each line terminal. One frequency is called "mark" and the other frequency is called "space." The scheme provides continuous channel monitoring since either a mark or space carrier signal is transmitted continuously. This eliminates the need for a carrier start fault detector. The transmitter is keyed to its mark frequency when the square wave goes positive, and is keyed to its space frequency when the square wave is zero.

The operation of the dual phase-comparison scheme is described as follows. For internal faults (1), the single-phase outputs of the sequence-current networks are in phase, even though the physical currents are 180 degrees out of phase. The network output goes through a squaring amplifier that keys the frequency shift transmitter. An adjustable delay circuit is tuned to delay the local square wave by a time equal to the channel delay time. The network output then develops two complementary square waves, called the "local positive" and "local negative" waves. The positive wave has a positive state during the positive half cycle of the sequence current and is compared with the receiver's "mark" output in gate AND 1. The local negative wave is positive during the negative half cycle of the sequence current and is compared to the receiver's "space" output in gate AND 2. For internal faults the local positive and the receiver "mark" wave are in phase, giving an output from the AND 1 gate. Similarly, the local negative and the receiver's "space" wave are in phase, giving an output from the AND 2 gate. If an arming signal is received from either

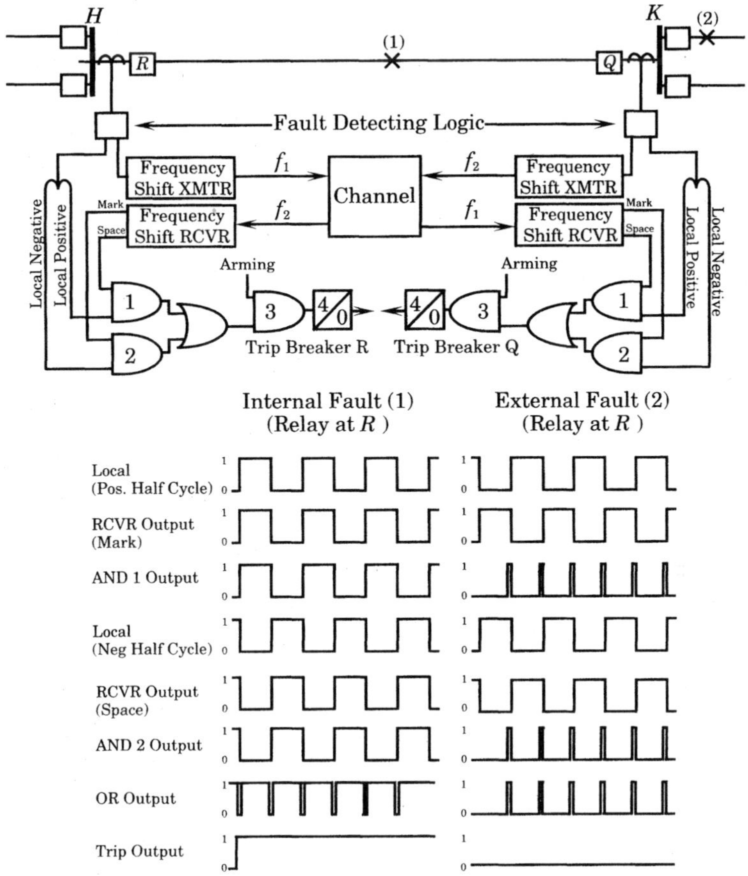

Figure 13.36 A dual phase-comparison unblocking scheme [18].

the fault detector or the protective relay, AND 3 has an output. If this signal persists for 4 ms, the output is sent to a flip-flop that initiates breaker operation. Note that tripping requires that both the mark and space signals be transmitted through an internal fault by means of power line carrier. If both frequencies fail to be received, a loss of channel detector enables both comparison circuits, permitting the system to trip on the basis of local square waves only.

For external faults, such as (2) in Figure 13.36, the 180 degree reversal of one of the currents shifts the square waves 180 degrees, thereby preventing either AND gate from having an output.

Another variation of the dual-channel system uses distance supervision. This type of system is used on long, heavily loaded transmission lines with limited three-phase fault capability.

13.4.1.3 Dual-Phase-Comparison Transfer Trip. Phase-comparison transfer-trip schemes can also be devised. These schemes operate in exactly the same way as the unblocking system described above, except for the channel. Either a current only or a distance supervised system can be designed.

In transfer trip systems, audio tones are usually used over a non-power line carrier channel. In order to trigger tripping for internal faults, the correct mark and space signals must be received. If channel signals are not received, the receiver mark and space signals are clamped to zero, which locks out the system and actuates an alarm.

13.4.1.4 Segregated Phase Comparison. Segregated phase comparison has been developed primarily to solve problems of relaying series-compensated lines and other applications that are plagued with severe waveform distortion. In such systems, the sequence current networks that are designed to operate at the system fundamental frequency, may not provide reliable current representation.

The current measurement system used for segregated phase comparison is shown in Figure 13.37. This should be compared with Figure 13.34, the sequence network method of current measurement. Since each current is measured individually, there is no need to assume that the measured currents are of fundamental frequency. Both sub- and supersynchronous harmonics are measured exactly as they occur on the power system. Thus, even if the measurements are corrupted by harmonics, both measurements are similarly corrupted and the phase comparison will still operate correctly. Experience has shown that even severe waveform distortion is likely to be noted in the current signals at both ends of the line, and a direct comparison of the currents of each phase provides a reliable method of fault detection and selectivity.

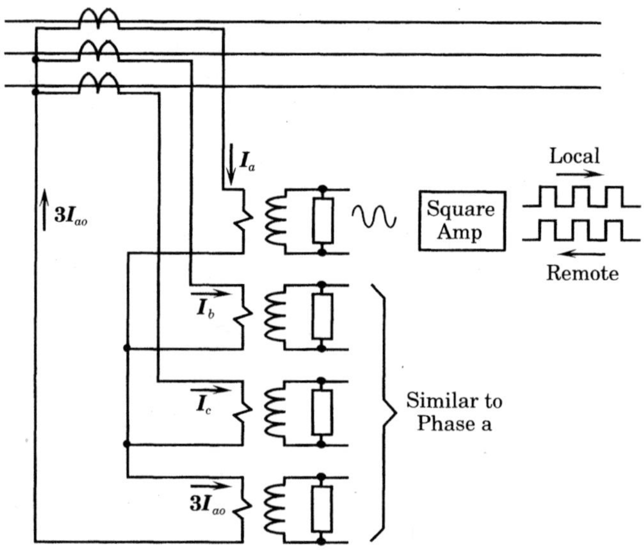

Figure 13.37 The segregated phase comparison measurement system.

Modern versions of segregated phase comparison relaying systems are fully digital systems, which, although primarily a current only protective system, includes optional distance backup protection [19]. The system has been designed to operate correctly and selectively on series compensated lines, where several types of problems have been noted that can cause relaying errors:

- Voltage reversals caused by negative reactance due to series compensation
- Phase unbalance due to bypass system operation and reinsertion
- Abnormal frequencies of 20–400 hertz during the fault and postfault period

Moreover, these systems are designed to operate correctly irrespective of the series capacitor location or the degree of compensation. They can also be used on lines that employ single-pole tripping of circuit breakers.

The operation of the phase comparison system can be described by reference to Figure 13.38, which illustrates an internal fault condition on a transmission line.

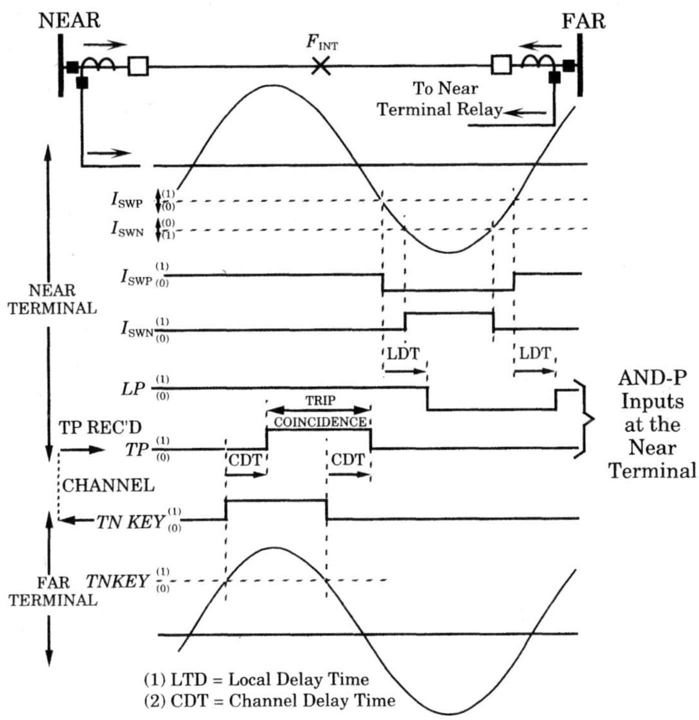

Figure 13.38 Relay logic signals for an internal fault [18].

The signals illustrated in the diagram are defined as follows, moving through the figure from top to bottom. Note that all are digital signals, i.e., they are represented either as a logical "1" or a "0" in the trip logic.

TP Trip positive signal determined by comparing the positive half cycle of the current wave against a keying level at the remote terminal.

TN Trip negative signal determined by comparing the negative half cycle of the current wave against a keying level at the remote terminal.

I_{SWP} Square wave positive. This signal is set to detect a negative current value. Its output is 1 if it is more positive than a negative current setting, otherwise the output is 0.

I_{SWN} Square wave negative. This signal is set to detect a negative current value. Its output is 1 if it is more negative than a negative current setting, otherwise the output is 0.

These signals are shown in Figure 13.38. Two other signals that are not shown in the figure are the following:

I_L A low set overcurrent function that is used as a fault detector to supervise the phase comparison logic.

I_{CD} Rate of change current detector used to initiate keying the other terminal.

The operation of the phase comparison system is described with reference to Figure 13.38, where an internal fault is present on the protected line. At the remote terminal, the trip positive signal is set to a logical 1 when the current level exceeds the TP keying level. This causes a signal to be transmitted to the local terminal, where it is received after a channel time delay CDT, and causes the local TP to be set to a logical 1 where it is compared against the local LP signal, which is also a logical 1, which is the normal state for that signal. Therefore, a trip coincidence time persists until the TP signal is reset to a 0 and this is received at the local terminal after a time delay CDT.

Note that the local signal LP remains high until the current local current wave becomes more negative than I_{SWP}, which occurs during the negative half cycle. This signal is delayed by the delay time LDT to create the local signal LP, as shown in the figure. Had the fault been external to the line, the TP signal would still arrive and would have been delayed by an amount CDT such that the inputs to the AND-P gate would consist of a 1 and a 0, which would not trip the line. Stated another way, the trip coincidence pulse would be shifted to the right, and would appear directly below the $LP = 0$ pulse. By adjusting the two timers LDT and CDT, the relay can be adjusted to correctly trip for internal faults and refrain from tripping due to external faults.

The foregoing applies to the P logic. Another logic, the N logic, uses the I_{SWN} signal in comparison with the TN signal from the remote end to provide a completely separate trip logic. Either the P or the N system is capable of tripping the line. The total system trip logic is shown in Figure 13.39. Note that tripping is supervised by the low current fault detector, I_L, and by the current rate of change. An additional "good channel" signal is also required to ensure the use of a working communications channel. A similar logic is used for each phase and for ground. Any of these separate systems is capable of tripping the faulted transmission line.

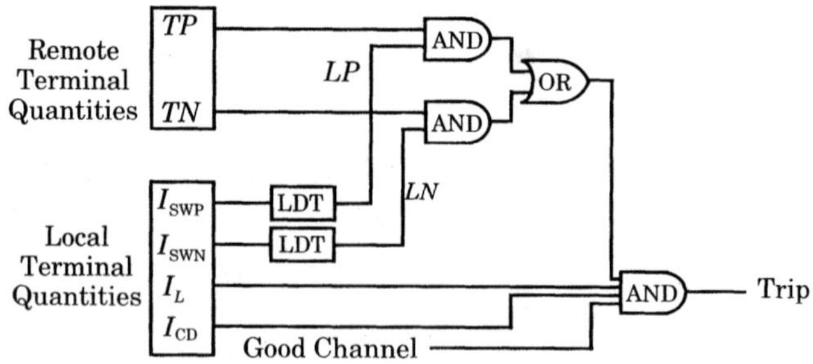

Figure 13.39 Segregated phase comparison tripping logic.

The disadvantage of the segregated phase comparison scheme is that it requires four pilot signals per terminal. One rectangular-wave comparison signal requires about 1 kHz bandwidth, so necessary pilot signals can be transmitted using a standard 4 kHz voice-frequency channel. Where microwave links or fiber-optic cables are available, several high-bandwidth signal channels are available, in which case segregated phase comparison may be the favored line protection scheme [18]. Unit schemes, in general, have a disadvantage in comparison to distance protection schemes in that they require greater bandwidth in the signal channel and their reliability is entirely dependent on the signal reception. Moreover, unit protection does not provide backup for faults that lie outside the protection zone, which is defined as the CT locations at both ends of the protected line. Therefore, distance relays of some type must always be used in conjunction with phase comparison schemes in order to provide backup protection.

However, to balance this extra requirement in data communications, there are several advantages, which can be summarized as follows [18]:

1. The segregated phase comparison protection approach overcomes three of the problems associated with series compensated lines, namely, abnormal frequencies in the current waveform, phase-impedance imbalance, and voltage reversals.
2. High-speed operation, due to angle diversity between phases and to the elimination of filters required in the system conventional phase comparison circuits.
3. The redundancy provided by having four independent logical devices, any of which is designed to detect a fault and trip the line.
4. The method provides the advantages of conventional phase comparison:
   - not responsive to system power swings
   - immune to errors introduced by mutual induction
   - not affected by loss of potential
   - operates correctly for zero-voltage three-phase faults
   - not affected by potential transients associated with CVT's
5. Phase isolation makes it possible to extend the independent-pole concept to include the relaying as well as the circuit breaker, providing advantages where system stability is a problem requiring independent pole operation.
6. Inherent phase selectivity for all types of faults, which provides flexibility in arranging relay and circuit breaker trip circuits to obtain any desired degree of pole-tripping selectivity.

Segregated phase comparison can also be used for the protection of parallel transmission lines and this type of protection provides certain advantages:

- Instantaneous clearing of double faults involving both lines
- Immunity from mutual coupling effects
- Phase selective for all types of single and multiple faults

When applied in this situation, the backup protection should also be phase selective or should be time-delayed for ground faults to avoid a probable canceling of the correct phase selection.

Segregated phase comparisons schemes have been implemented both in solid state logic [18] as well as digital logic [19].

13.4.2 Longitudinal Differential Schemes

Another type of unit protection is referred to, especially in Europe, as "longitudinal protection." Longitudinal protection was formerly restricted to pilot wire systems, and this restricted the applications to lines of about 20 km or less. Some modern longitudinal schemes employ pilot channels to measure line terminal quantities for comparison and decision making.

13.4.2.1 Wire Pilot Schemes. Wire pilot schemes are the most basic form of longitudinal protection, as noted in Section 13.2.2. Using communications by pilot wires provides access to both the amplitude and phase angle of the primary currents in the protected transmission lines. Secondary quantities can be either currents or voltages that are proportional to these basic primary quantities. This forms the basis for differential current balance or voltage balance relaying schemes. In most systems, the measured three-phase quantities are converted into single-phase ac currents for relaying purposes. However, it is possible to construct a "segregated" system where separate circuits are used for each phase. Such a segregated system is required for circuits that employ single pole reclosing.

13.4.2.2 Longitudinal Pilot Schemes. For long lines, pilot wires are not practical due to the requirement of adding a separate circuit for the entire length of the protected transmission line. However, modulation techniques can be used for longer transmission lines that essentially extend the applicability of differential protection to these longer lines. Usually, the communications path will be fiber optics or microwave systems, which replace the pilot wires used on shorter lines.

In longitudinal pilot systems, instantaneous values of current are transmitted to the other terminals by means of the pilot channel. Two methods of modulation are used to transmit these instantaneous current values, *frequency modulation* (*FM*) and *pulse code modulation* (*PCM*).

In FM systems, the instantaneous current values are transmitted as analog quantities to the other terminals using a voice frequency band with frequency modulation. The basic concept is illustrated in Figure 13.40, where the signal wave, shown as the lower trace in the figure, modulates the carrier frequency, with higher signal values resulting in a higher frequency modulated carrier; therefore, the frequency of the carrier follows the signal magnitude. Current differential principles are used at the receiving terminals to determine the need for line tripping. A typical system configuration is shown in Figure 13.41. The instantaneous current values are modulated to FM signals in the range of 0.3–0.4 kHz, and are suitable for transmission on voice grade circuits.

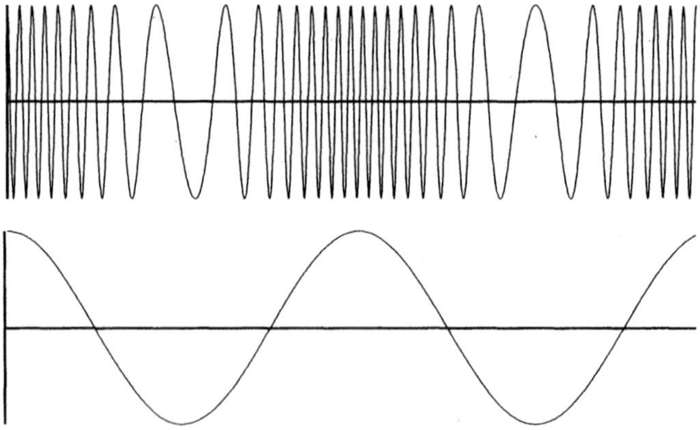

Figure 13.40 The frequency modulation principle.

In the PCM scheme, the instantaneous current values at each terminal are transmitted to the other terminals by pulse coded modulation of the transmitted signal. Figure 13.42 shows the basic circuit for this system.

The output of the mixing current transformer provides a quantity that is proportional to the primary current value that can be used for current comparison. This signal is fed to the data

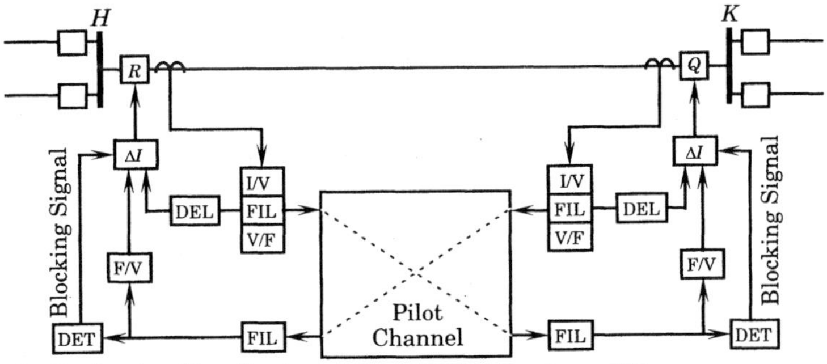

ΔI = Evaluation Circuit
I/V = Current to Voltage Transformer
V/F = Voltage to Frequency Converter
FIL = Bandpass Filter
DEL = Delay Compensation Circuit
DET = Noise Detector

Figure 13.41 An FM current differential protection scheme [1].

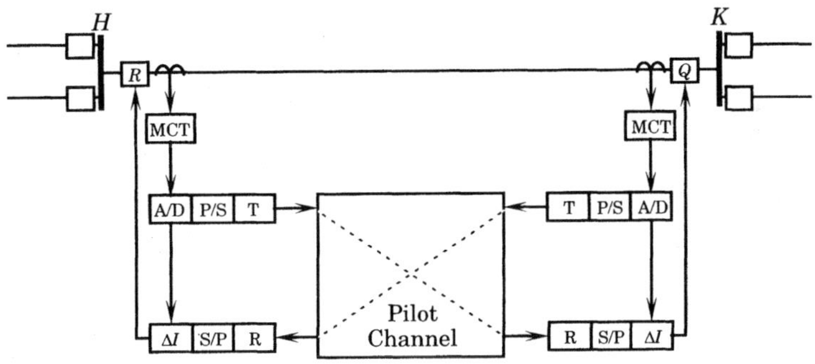

Δ I = Evaluation Circuit
R = Receiver T = Transmitter
MCT = Mixing Current Transformer
S/P = Serial/Parallel Converter
P/S = Parallel/Serial Converter
A/D = Analog/Digital Converter

Figure 13.42 Example of a PCM current differential scheme [1].

terminal, where analog-to-digital conversion is made and the signal transmitted in serial form to the other terminal. At the receiving station, the signals are converted back to parallel data and evaluated using a microprocessor to compare the currents received with those measured locally.

Note that this form of protection is completely dependent on the pilot channel and must be backed up by other types of protection that are not so dependent on communications.

13.5 AN EXAMPLE OF EXTRA HIGH VOLTAGE LINE PROTECTION

Extra high voltage (*EHV*) transmission lines require special treatment in terms of their protective system requirements. This section will address some of the problems associated with EHV transmission and the protection of these important lines.

13.5.1 Considerations in EHV Protection

The classification of "EHV" lines is usually applied to those lines rated at 345 kV or above. In the North America, this means lines rated at 345, 500, 735, and 765 kV. In many other parts of the world, 400 kV is the preferred EHV nominal voltage. These transmission circuits are considered to be the "bulk transmission system," which is usually identified as the system that connects entire utility systems together in a large, long distance, integrated network. The lines are of higher voltage than the "local" transmission lines since they are often needed to transfer large blocks of power over great distances. The need for such transmission may come about due to the availability of energy at the remote site that is favorably priced, or due to the need to restrict local generation due to environmental requirements.

Because of the importance associated with the EHV transmission system, the protective systems at these voltage levels have received special consideration. Power line carrier and microwave pilot relaying is the common choice for these circuits because the lines are too long for wire pilot protection. The first requirement for fault clearing on these systems is high speed, and pilot protection is usually considered to be essential. Generally, the protective systems are static systems and are applied in redundant pairs in order to assure high reliability. The failure to trip a fault on the EHV system has such serious effects that it is relatively easy to justify the cost of redundant pilot relay systems. Moreover, added precautions are also taken to ensure that the relay system itself is secure, and that false or unnecessary trips are reduced to an absolute minimum.

13.5.2 Description of the EHV Pilot Protection

The type of protective system that is often used on EHV transmission lines will be illustrated by means of an example. This will illustrate the types of equipment, their arrangement, and the philosophy of their application.

Consider the protection requirements for a 500 kV line that is part of the backbone EHV bulk power system that ties two large utilities together. The line is relatively long, say 300 km (about 200 miles) or so in length, connecting two major switching stations.

The following design criteria are specified by the protection engineers:

1. Dual redundant relaying systems are required for reliability.
   - The two systems must be of different logical design
   - The two systems must use different communications
   - The two systems should be from different manufacturers
2. The redundant relay systems must be completely independent.
   - Redundant instrument transformers
   - Redundant breaker trip coils
   - Redundant battery systems
3. The two systems must operate using entirely different fault detection principles.
   One system will use directional comparison
   One system will use phase comparison
4. The relaying must be very fast.
   - Static relays will be used
   - Transfer trip will be used as the remote trip logic
5. The relaying must be very secure.
   - False trips must be avoided
   - Transfer trip must be redundant using an "AND" logic

The telecommunications choices for the EHV line are power line carrier, microwave, and fiber optics. Figure 13.43 is a map showing the approximate transmission line route and that of a nearby utility-owned microwave system. The transmission line follows an approximate east–west routing from A to B. Nearby, to the north of the line route, is a utility-owned microwave system connecting communications facilities at C and D, with repeaters arbitrarily designated a, b, ..., v, which are located on high hills or mountain tops along the microwave path. The microwave system is operated as a loop with the loop switching at facility C. The normal microwave path is the southern route, utilizing repeaters a, b, ..., m or 12 hops total. In the event of a repeater failure, the switching logic at C automatically switches to the northern route, utilizing repeaters a, v, u, ..., m or 10 hops total. The microwave path distances between C and D are not the same for the two paths.

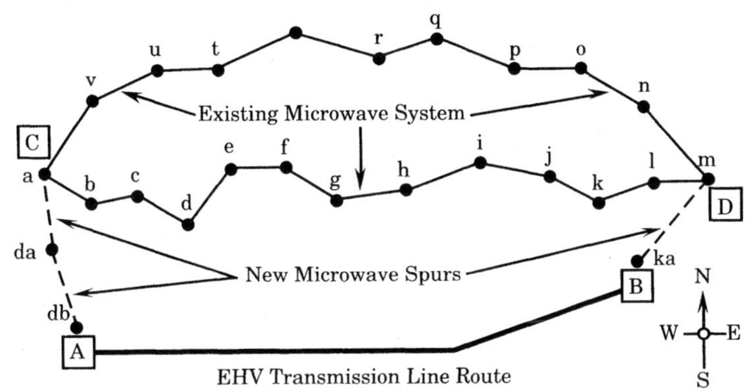

Figure 13.43 Map of the EHV line route and nearby microwave system.

The transmission line terminals are reasonably close to microwave repeaters a and m, on the west and east, respectively, such that microwave spurs can easily be constructed to the EHV switching stations at A and B, as shown in Figure 13.43. This makes the total microwave path between the ends of the line going from db on the west to ka on the east, and utilizing either the south (normal) or north (alternate) path. Due to the existence of the microwave path, this is chosen as one of the relaying communications media, with power line carrier as the other choice. This satisfies the criterion that the two systems be completely independent. Moreover, should PLC communications be questionable during a line fault, the microwave system should not be affected and should operate normally. Should the microwave be affected by atmospheric fading, the power line carrier would not be affected and would not be similarly affected.

Another alternative for the pilot system is fiber optics. It may be possible to justify the incremental expense of adding fiber-optic cores to the transmission line static or phase conductors. This depends on many factors, such as the need for the additional communications capability offered by the optical system, its cost, and its planned utilization. For the present example, we assume that this option is not competitive in price.

The relay logic has been selected to utilize phase comparison and directional comparison relays, with one being dedicated to the PLC system and the other to the microwave. Since the microwave system has a total path length that is variable, this is not a good choice for phase comparison, which requires precise timing of the transmitted signals. Therefore, the phase comparison system will be applied to the power line carrier, leaving the microwave system to utilize directional comparison. Dual static systems will be installed for each system on both the microwave and PLC.

13.5.2.1 The PLC Pilot Protection System. The PLC protection system design is illustrated in Figure 13.44. This system utilizes phase comparison relay logic. The basic concept of the PLC system provides a distance measuring facility to quickly identify faults that are clearly within the primary zone of protection. The protection logic can be summarized as follows:

1. Any condition that causes the local relay to trip the local breakers will simultaneously send a transfer trip signal to the remote end, thereby initiating a rapid trip at that end as well. The transfer trip uses channels #3 and #5 or #4 and #6, depending on the fault location. Note that tripping does not occur at the receiving end unless *both* of the designated channel signals are received, which is the means of providing security against false trips due to noise on one of the channels.
2. Close-in faults are identified in the primary zone of protection and this triggers an immediate trip to the local breakers from the local relay logic and, since a local trip signal is generated, a transfer trip to the remote end is also launched. This provides for fast tripping for end zones faults at both ends of the line. The transfer trip uses two paths. One path uses the PLC TT logic and PLC channels #3 and #5. The second path uses the microwave "cross-trip" path, which simultaneously sends a transfer trip signal using microwave communications. This duplication of transfer trip signalling is referred to as a *cross-tripping* action. The microwave TT logic is described in the next section.
3. Faults in the center portion of the line are also in the primary zone but tripping will not occur unless confirmed by the relays at both ends, using transmitter/receiver channels #1 and #2. Once this confirming relay logic is received at a given end, the breakers at that end are ordered to trip. This trip signal, as noted above, also sends a transfer trip signal to the opposite end to ensure tripping there as well.

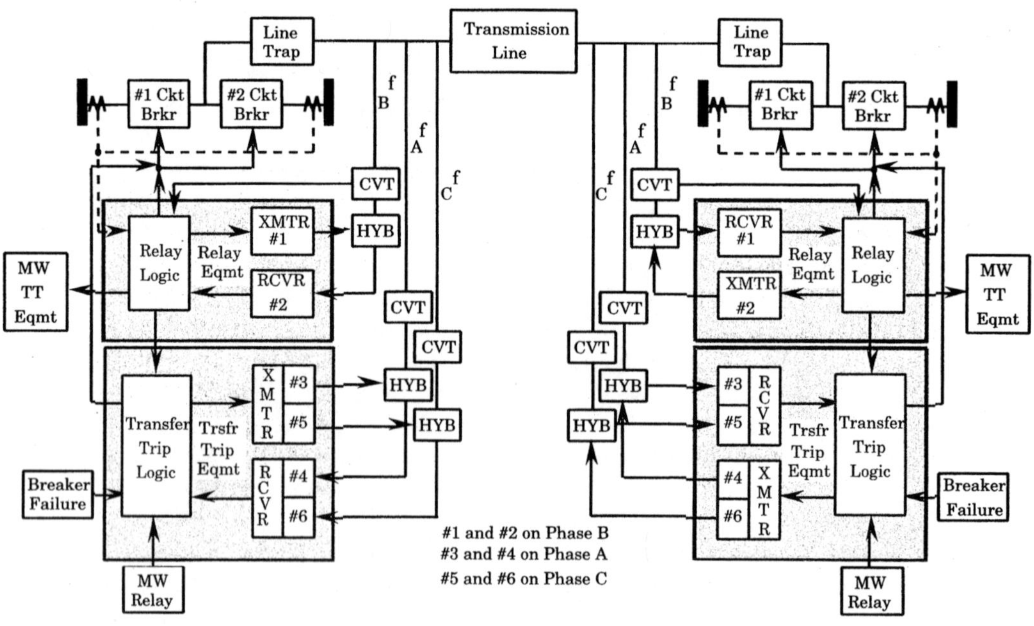

Figure 13.44 Power line carrier protection system.

Note that the relay logic channel uses phase b of the transmission line. Transfer trip channels #3 and #4 use phase A, while channels #5 and #6 use phase C. This provides diversity in the channel utilization and means that any transfer trip signal must receive a signal that uses both phases A and C.

13.5.2.2 The Microwave Pilot Protection System. The microwave protection system for the same transmission line has a similar design, but utilizes directional comparison protection along with distance measuring equipment. The equipment arrangement is shown in Figure 13.45. The trip logic is similar to that of the PLC system, except it is provided using completely different equipment that is supplied from a different manufacturer. This means that the logic design is different and the means of implementing this logic is unique to the manufacturer's approach. This gives a high level of assurance that any fault detected is correctly identified by independent systems, and that the response will be correct. Like the PLC, the microwave system uses dual transfer trip channels, both of which must be received to initiate a trip at the receiving end.

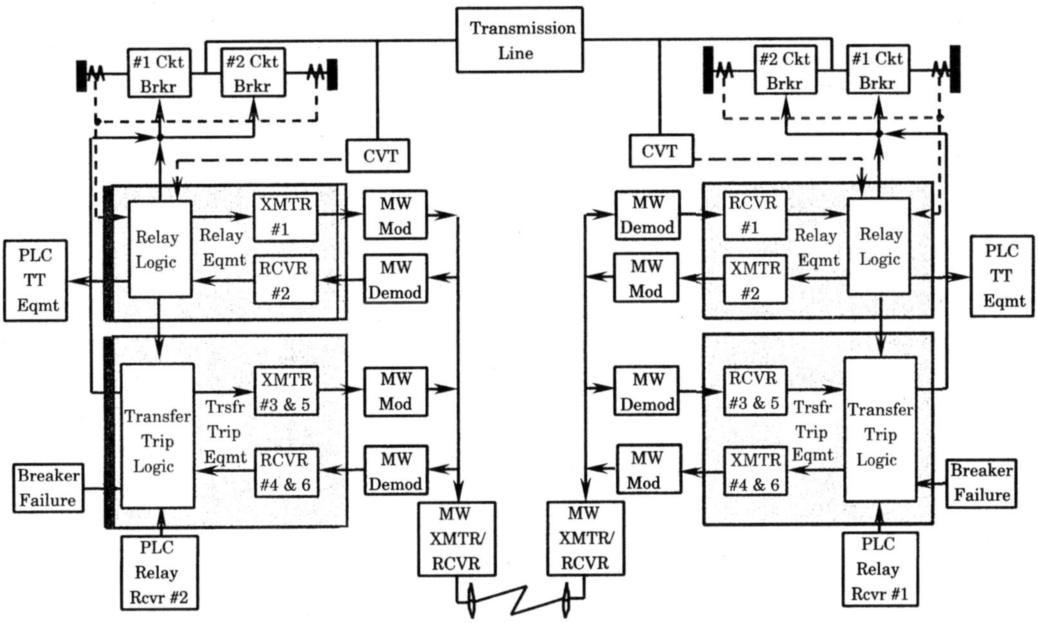

Figure 13.45 Microwave pilot protection system.

The transfer trip function of the two systems are also interconnected. For example, if the microwave relay logic sends a trip signal to its local breakers, it initiates a transfer trip signal in the transfer trip equipment logic, and also sends a transfer trip signal to the PLC transfer trip equipment. This provides additional assurance that the transfer trip signal will indeed be received at the remote end. This dual transfer trip signal initiation is also provided by the PLC relay logic, as shown in Figure 13.44.

The general concepts of the redundant systems described above are typical of EHV protection systems. There is no "standard" protection system for EHV transmission lines. All of these lines are important network elements and their protections are designed with great care. The example cited, however, will provide a sense of the care that is placed on redundant design and security of correct operation.

One feature in the foregoing example has not yet been discussed: the presence of "breaker failure" protection in both ends of both systems. If a circuit breaker fails to open on command, the fault will remain connected to the bus at that breaker location. In most EHV stations there will be two circuit breakers at each end of the protected line, with each breaker connecting the line to an adjacent bus section, as shown in Figure 13.45. If one breaker is inoperative, the fault is permanently connected to the bus section through this stuck breaker. This means that all circuit breakers connected to the now faulted bus section must be cleared. In some cases, this will require the clearing of an unfaulted transmission line, but this depends on the substation arrangement. Also, see Chapter 5 for a description of different station arrangements.

13.5.2.3 Protection Equipment and Controls. For the purpose of this example, we assume the arrangement of equipment at each end of the protected line to be as shown in Figure 13.46. The two ends of the line are designated as Left and Right for this example. There are two circuit breakers at each end of the line that must be tripped for any line fault. This is typical for any ring bus or better switching station design. There are two independent relay systems at each end as well, and these are designated as relays #1 and #2 at their respective line terminals. The relay contacts are designated C1x and C2x for relays 1 and 2, respectively, with x carrying the bus name, L or R, as shown in Figure 13.46. Seal-in relays are also provided, and these are denoted by the letter S with appropriate added characters to identify both the relay number and the appropriate bus name. Each circuit breaker has two trip coils, with each trip coil associated with a different breaker. Separate current transformers are arranged to measure the sum of currents flowing into the line. A single voltage measurement is supplied to both relays. Figure 13.46 provides a distinctive numbering system such that any device can be addressed without ambiguity. Relays designated #1 use power line carrier as their pilot signal and relays designated #2 use microwave communications. There is one battery providing dc power at each terminal. These battery systems are not grounded since ground potential can vary significantly during fault conditions. However, all dc circuits terminate at a common negative bus, which is shown in the diagram by white diamond characters. However, the common negative dc buses at the two line ends are not connected together.

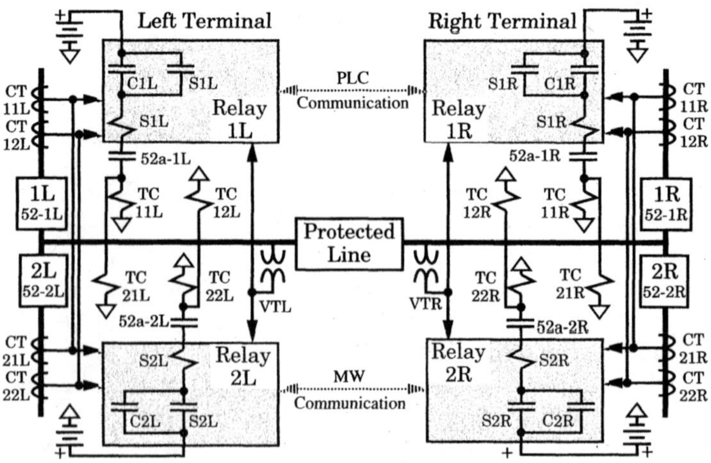

Figure 13.46 Equipment arrangement for the protected line.

The transmission protection is an underreaching distance scheme using the two pilot systems, PLC and microwave (MW). The control circuit arrangement for one terminal is shown in Figure 13.47. A guard signal is continuously transmitted to the opposite terminal. When a zone 1 fault is detected, the detecting relay sends a trip signal to its local circuit breakers by means of the *Z1* contact, which is then sealed in. The relay also orders the transmitted signal to shift from guard to trip at the opposite terminal on both pilot systems, but transfer tripping at that terminal does not occur unless both transfer trip signals are received, hence the series logic of the *TT* contacts, but either the PLC or MW signals can complete the trip independently of the other pilot system. Note that breaker failure also triggers the transfer trip operation. Figures 13.46 and 13.47 must be used in conjunction with Figures 13.44 and 13.45 for complete understanding of the system.

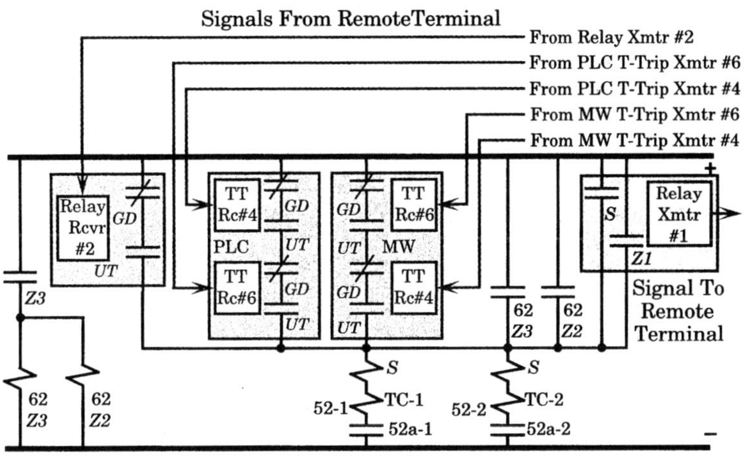

Figure 13.47 The dc control circuits for the left terminal.

13.6 PILOT PROTECTION SETTINGS

As an example of the relay setting procedure for complex transmission line relaying, we consider the determination of settings for a protective scheme utilizing *permissive overreaching transfer trip* (POTT) [20].[1] The computation of relay settings for complex protective schemes is explained in detail in manufacturers' literature for the relay equipment specified for the application These special manufacturer instructions are always recommended for study by the protection engineer, as they provide detailed information regarding the equipment and its application. The relays used for the scheme to be applied in this case are digital relays, which have certain features that are not available for analog relays. In particular, we consider a relay system that includes mho distance elements as the primary protective system plus elements for overcurrent backup for both phase and ground faults. The coordination of the various elements are of particular interest.

POTT schemes are subject to incorrect relay tripping under the following conditions:

1. Current reversals.
2. Weak infeed conditions at one of the line terminals.

[1]Portions of this section are adapted from [20] with permission.

3. Breaker open at one terminal.
4. Switch-onto-fault conditions.

Some relays are provided with special logic to overcome these difficulties. We shall assume that such logic is available.

Another problem occurs when the protection suffers from a loss of communications between the terminals. Some relays incorporate time-delayed backup protection, such as time-overcurrent tripping, that provides assurance of relay action when communications are lost.

13.6.1 Instrument Transformer Settings

The *current transformer ratio* (CTR) should be selected such that the relay has good signal strength during maximum fault conditions, say from 50 to 100 secondary amperes. This helps avoid problems of CT saturation and ensures reliable current signal strength to the relay during the fault. Normal load current is assumed to be in the neighborhood of 5 amperes.

The *voltage transformer ratio* (VTR) should match the primary line-to-neutral voltage to 66.4 volts secondary, line to neutral, or 115 volts line to line.

13.6.2 Maximum Torque Angle

Digital relays usually allow the protection engineer to specify the maximum torque angle of the relay. This angle is usually selected to be equal to or less than the positive sequence line impedance angle.

13.6.3 Distance Element Reach and Time Delay

The reach of the distance element is selected according to the practice of the protection engineer. Figure 13.48 shows a rough measure of the type of reach that is usually considered appropriate. The length of the reach arrows in the figure should be interpreted in proportion to the length of the protected line.

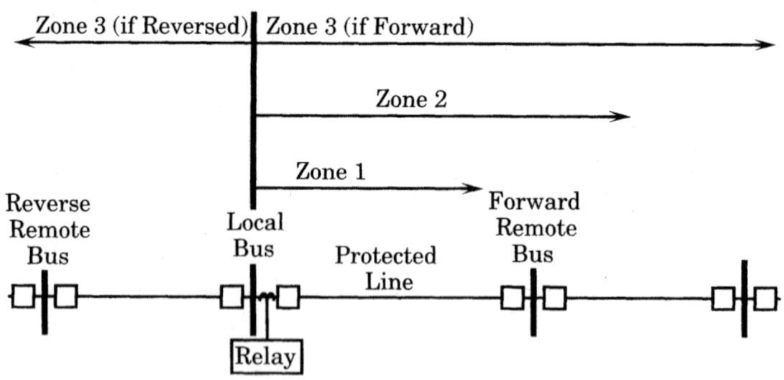

Figure 13.48 Example of distance element reach.

13.6.3.1 Zone 1 Reach. The zone 1 reach is usually set to not overreach the forward line terminal. Zone 1 elements should provide instantaneous protection for three-phase and line-to-line faults inside the zone 1 reach, which is usually set at 80–90% of the line length, independent of communications. The remainder of the line not covered by the zone 1 reach will be protected by zone 2 elements of the relay. It is not practical to extend the zone 1 reach to 100% of the line length due to instrument transformer errors, line data errors, fault study data errors, and small errors in manufacturing tolerance in the relay.

13.6.3.2 Zone 2 Reach. Zone 2 elements provide protection for the portion of the protected line beyond the zone 1 reach. Zone 2 elements also serve as backup protection for close-in faults on the forward adjacent lines. For the POTT system under study here, the zone 2 elements at both terminals must detect internal faults and initiate permissive trip keying. In some cases, parallel source infeeds may increase the apparent impedance to the fault from the relay location. Therefore, the zone 2 reach is usually set at 50% of the shortest forward, adjacent line section. Typical settings will be in the neighborhood of 120–130% of the protected line length. Settings of 150% or more raise a question regarding current reversal for sequentially cleared faults on parallel lines, which require special treatment.

13.6.3.3 Zone 3 Reach. The instantaneous zone 3 elements are used in several ways in the POTT scheme. These include guarding against current reversal and assisting with weak infeed logic. These schemes require that the zone 3 element reach be reversed and must detect all faults in the reverse direction that are also detected by the zone 2 elements at the remote terminal of the protected line. A time-delayed backup is also provided by zone 3 elements for faults behind the relay location.

The zone 3 reach setting must be long enough to detect any fault sensed by the overreaching zone 2 element at the remote terminal. Usually, the reach setting is set to match the remote relay zone 2 reach.

In some cases, the zone 3 elements may be set for forward reach. In this case, the zone 3 elements should act as remote backup for faults at remote buses that are two-line sections from the local relay terminal.

13.6.3.4 Zone Element Time Delays. The zone 1 elements are usually set with no intentional time delay so that tripping of faults within zone 1 will be as fast as possible.

The time delay of zone 2 and zone 3 elements should be set to coordinate with time-step protection at both the remote and local buses. A typical zone 2 delay setting will be 20–30 cycles. This allows time for the remote zone 1 element to pick up, plus breaker operating time. Settings for zone 3 residual overcurrent element, if present, depends on the strength of the zero sequence current source behind the relay.

The zone 3 time delay must coordinate with the zone 2 protection at the remote bus when zone 3 is reversed, as well as the remote zone 2 elements when zone 3 is reversed. A typical phase distance time delay setting for zone 3 is about 60 cycles.

13.6.4 Phase Overcurrent Element Settings

Assume that the digital relay protecting the transmission line has phase overcurrent elements in addition to the distance elements. Let us assume three-phase-overcurrent element levels: high-, medium-, and low-set elements.

13.6.4.1 Low-Set Phase Overcurrent Elements. The 50L element provides fault detector supervision of the mho distance elements and must pick up for all faults for which the distance element is expected to operate. The best setting for the 50L elements is above load current level, but below minimum fault level.

For phase faults, at least one current magnitude must exceed the 50L pickup threshold before the line-to-line distance elements can operate. For three-phase faults, the current magnitude in all three phases must exceed the 50L pickup threshold before the three-phase distance elements can operate.

The 50L elements are also used in the trip unlatch logic. Before the trip output can open, the trip condition must vanish and the current magnitude in all phases must drop below the 50L element thresholds. This ensures that the trip output contact does not attempt to interrupt full trip coil current for the full duration of the faults.

13.6.4.2 Medium-Set Phase Overcurrent Elements. The 50M element provides an important input to the *loss-of-potential* logic (LOP) for cleared VT fuses and also provides nondirectional phase overcurrent protection under those conditions when a loss-of-potential condition is diagnosed. The recognition of an LOP condition blocks the mho distance elements from operation. The line protection under these conditions is provided by a nondirectional time-delayed phase overcurrent element.

It is important to ensure that three-phase LOP conditions are not confused with fault conditions. Once an LOP condition is declared, it remains latched until balanced three-phase voltages are restored. The 50M element setting must be above load current, but below minimum line-to-line or three-phase fault current levels. If the maximum load current is greater than the minimum fault current, the 50M setting must still be below the fault values. This prevents pickup of LOP during a fault. In most cases, when a close-in fault lowers the measured voltage to such a degree that an LOP condition might be suspected, there is an accompanying increase in fault current as viewed from the local terminal.

The 50M element must be time-coordinated with surrounding line terminals. To provide this coordination time, the 50M pickup following LOP starts a definite timer. When the timer expires, the line can be tripped.

13.6.4.3 High-Set Phase Overcurrent Elements. The 50H element function provides a high-set nondirectional phase overcurrent detector in the switch-onto-fault logic. When closing a line breaker into a close-in, three-phase bolted fault, it is likely that the three-phase distance elements will not see the required level of polarizing voltage for pickup. If this occurs, the distance elements are blocked from pickup. The 50H element provides backup protection for this condition. The 50H element measures the current magnitude in each phase and is independent of polarizing voltage. Typical settings for the 50H element is from one-half to one-third the three-phase fault duty at the local line terminal.

13.6.5 Residual Overcurrent Element Settings

The settings of the residual overcurrent elements are determined based on the results of a short-circuit study of the power system. The residual time-overcurrent elements provide current dependent, time-delayed clearing of faults on the protected line and provide backup protection for remote terminals. The magnitude of residual current varies with line switching conditions, fault location, and fault resistance. The system fault study must be used to determine minimum pickup settings and the appropriate time dial setting.

In some cases, three separate residual overcurrent elements are used, providing zoned directional protection for ground faults. Zone 1 is always instantaneous and forward looking, with pickup setting that is greater than the maximum end-of-line ground fault level. Zone 2 is always forward looking and covers that portion of the line not protected by zone 1 and time-delayed backup protection in the adjacent forward line section. A typical setting is 50–60% of the minimum end-of-line SLG fault magnitude. Zone 3 may be either forward or backward looking, with time delay to provide backup protection for local and remote buses. The pickup setting depends on the selected direction of protection. If reversed, the element must pick up for all faults seen by the reverse remote zone 2 relay element. If zone 3 is forward looking, the element must pick up for a SLG fault at the farthest remote bus that the element is expected to protect.

13.6.6 Switch-Onto-Fault Logic

The switch-onto-fault logic permits sensitive overreaching elements to be enabled for a short time after the line breaker is closed. The logic permits instantaneous tripping of the line breaker for end-of-line faults that would normally be cleared with zone 2 time delay. For the POTT scheme, the time required to echo the permissive signal to the remote terminal and back is greater than the time required to issue a trip by means of the switch-onto-fault logic. One relay system provides four different tripping masks, as follows [20]:

MTU	Trip unconditional (no logic qualifiers)
MPT	Trip with permissive trip input and no reverse block
MTB	Trip with the block trip input not asserted
MTO	Trip while the 52BT element is asserted

The MTO mask elements are typically non-time delay overreaching elements. The 52BT time delay setting determines the time during which the switch-onto-fault logic is enabled. The 52BT element may be thought of as an inverted time delayed follower of the 52A input. When the 52A relay input changes from breaker closed to breaker open, the 52BT element changes its logic state from a '0' to '1' and enables the switch-onto-fault logic. When the breaker is closed from an open state, a test for line faults is made, and the 52BT remains high for the 52BT time setting. During this time interval following breaker closure, the assertion of any element selected in the MTO logic mask closes the trip output contacts, thereby enabling the switch-onto-fault protection. The timing of the elements is shown in Figure 13.49.

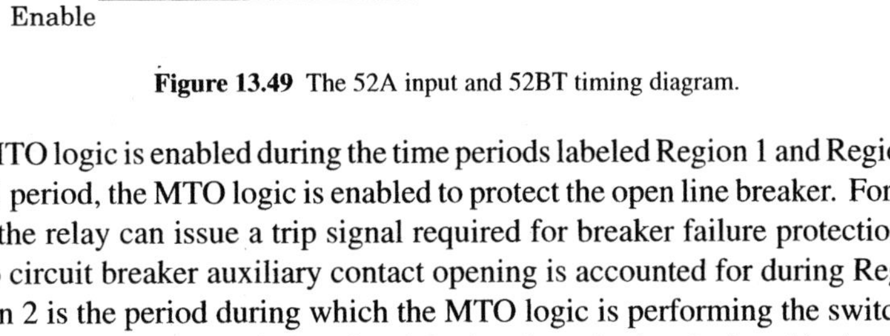

Figure 13.49 The 52A input and 52BT timing diagram.

The MTO logic is enabled during the time periods labeled Region 1 and Region 2. During the Region 1 period, the MTO logic is enabled to protect the open line breaker. For line breaker tank faults, the relay can issue a trip signal required for breaker failure protection. Any time delay due to circuit breaker auxiliary contact opening is accounted for during Region 1.

Region 2 is the period during which the MTO logic is performing the switch-onto-fault testing. The 52BT gives a permissive signal for breaker tripping during this time period.

13.6.7 Current Reversal Logic and Timers

Double-circuit lines create the potential for current reversal due to sequential tripping of a fault near the end of one of the lines. The situation is shown in Figure 13.50. A fault near the end of one of the lines is tripped by instantaneous protection that is independent of the communications scheme. On fault inception, shown in part (*a*) of Figure 13.50, breaker 1 is cleared instantaneously. Parts (*a*) and (*b*) of the figure show the zoned mho element detection at each relay before and after the instantaneous clearing of breaker 1. When the zone 1 element at breaker 1 detects the fault, the zone 2 element picks up and sends a permissive signal to the relay at breaker 2. That breaker detects a zone 2 fault, but must wait for the permissive signal from the relay at breaker 1 before executing a permissive qualified trip.

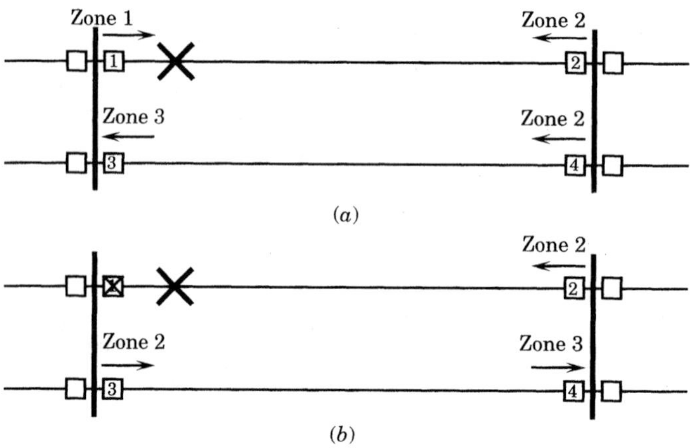

Figure 13.50 Example of current reversal on parallel lines. (*a*) Fault inception, before breaker 1 operation. (*b*) Faulted system, immediately after breaker 1 operation.

Also in part (*a*) of the figure, we note that the relay at 4 also sees the fault as zone 2 and issues a permissive signal to the relay at breaker 3, where the relay has its zone 3 element picked up, indicating the fault is in the reverse direction from that relay.

Immediately following the opening of breaker 1, the situation is shown in part (*b*) of the figure. The currents are now redistributed causing the zone 3 element at 3 and the zone 2 element at 4 to begin dropping out. If the zone 2 element at 3 picks up before the permissive signal from 4 is reset, then breaker 3 will be tripped due to current reversal.

The possible false trip of the unfaulted line can be prevented by a special relay logic to detect the potential for current reversal. The key to the logic is the initial presence of a zone 3 pickup on the unfaulted line, which clearly indicates that the fault is not on that line. This is used to initiate a "reverse block timer" that provides the required time for the permissive signal to be reset as the relay at 3 is changing its state from zone 3 to zone 2. The factors that determine the timer setting are the channel pickup delay, channel reset delay, and the maximum expected clearing time of the fault on the parallel line. For example, if the channel delays are one-half cycle each, the breaker time is three cycles, and the relay time is one cycle, then the timer setting should be five cycles. An interesting extension of this concept is when the faulted line is energized or recloses into the fault. This question is left as an exercise.

13.6.8 Echo Keying

The permissive overreaching transfer trip scheme requires permission from both terminals to realize accelerated tripping for an internal fault along the entire line. When one terminal is open, the relays at the open end are unable to detect an internal fault and are, therefore, unable to issue a permissive to the remote terminal. This usually requires that end-of-line faults are cleared by zone 2 elements, with the associated time delay.

Fast clearing of end-of-line faults can be achieved by the use of "echo keying." Under this arrangement, the relay at the open terminal includes a logic to "echo" the received permissive signal back to the remote sending terminal. When received at the remote end relay, that end will be cleared, thereby achieving the rapid clearing of the entire line, even when one terminal is open.

13.6.9 Weak Infeed Logic and Settings

Occasionally a situation exists in the power system where, with all sources in service, one terminal of a line may not contribute enough fault current to operate the protective elements. If the fault is within the zone 1 reach of the strong terminal, the fault currents may redistribute following the opening of that terminal to permit the weak terminal to then trip the line. However, if the redistribution of currents is insufficient to cause tripping of the weak terminal, it is still desirable that the weak terminal somehow be tripped. This prevents extensive damage that may occur from a continually arcing fault and allows successful autoreclosure from the strong terminal. When the fault is close to the weak terminal, the strong terminal zone 1 will not pick up and the fault will not be cleared until zone 2 operates. It is noted, however, that even though the weak terminal does not contribute a large fault current, its voltages are depressed due to the fault.

Rapid tripping of both terminals of a weak-infeed line can be achieved by application of the proper logic. The strong terminal is permitted to trip by a permissive signal that is echoed back from the weak terminal. The weak terminal is tripped by converting the echoed permissive signal to a trip signal after the following conditions are met:

1. No reverse looking elements are picked up.
2. At least one phase-to-phase undervoltage or residual overvoltage element operates.
3. The line terminal breaker is closed.
4. A permissive trip signal is received for a specified time period.

13.6.10 Loss of Potential Logic

Potential transformers are sometimes protected by fuses or low voltage circuit breakers. A failure of one or more of these devices results in a loss of potential to the polarizing inputs to the relay. The loss of one or more phase voltages disable the relay, making it impossible to discriminate fault direction correctly.

The loss of potential condition is unavoidable, but the existence of this condition can be detected and prevented from causing improper relaying. Once the loss of potential is detected, the protective relays can be disabled and an alarm issued. The problem is to discriminate between loss of potential and a fault condition that reduces the voltage to nearly zero. Correct detection can be achieved by noting the presence of zero sequence voltage in the absence of zero sequence current, plus the absence of positive sequence voltage and overcurrent.

13.6.11 Conclusions Regarding Pilot Protection Settings

The modern transmission line protective system utilizing pilot signaling has developed into a comprehensive system of overlapping protective elements, each with their unique characteristics. Moreover, units of this nature at both ends of the transmission line must be coordinated to work together with the objective of providing primary protection to the line and backup protection to adjacent circuits. Thus, there are many factors to take into account in establishing settings of these complex systems. The best source of information for completing this task is the manufacturer's technical literature. This material usually provides a technical review of the relay system design as well as detailed information for determining settings.

13.7 TRAVELING WAVE RELAYS

The protective relaying concepts presented in the foregoing sections of this chapter represent what might be called conventional transmission protection systems. They are conventional in the sense that the relay logic operates on the basis of the observed behavior of the fundamental frequency voltage and current, or on variables that are functions of these quantities such as apparent impedance. Due to the relatively long period of the fundamental power system frequency, any observation based on these fundamental frequency variables will require that the variables be observed for a long enough time to assure that the faulted condition is truly a case for which relay action is required. This is a basic requirement of any relay scheme based on fundamental frequency variables.

A newer concept in transmission protection is one that is sometimes called *ultra-high speed* protection [21–42]. These schemes are based on the extraction of fault information from a broad bandwidth of frequencies that accompany a fault and the processing of the data digitally to determine the fault location. This type of protection provides a significant advantage for lines with series capacitors, or for the protection of lines adjacent to those with series capacitors.

Most of these schemes are double ended, that is, they use a directional comparison arrangement that is accomplished by linking the two line terminals by a communications channel [21], [26]. Single-ended schemes have also been proposed, where the relaying information is determined from the transients observed at only one terminal of the line [23]. These schemes use correlation techniques that measure the reflection time of a wave component from the time that wave leaves the relay until it is reflected back.

Traveling wave relays utilize the concept of directional wave detection, together with an impedance measurement principal. The directional wave detection utilizes the fact that, when a fault occurs, both the voltage and current disturbances are transmitted throughout the system at nearly the speed of light. The basic concept of traveling wave protective schemes measure the change in voltage and current at the relay R location, i.e.,

$$\begin{aligned} \Delta \boldsymbol{V}_R &= \boldsymbol{V}_R(t=0^-) - \boldsymbol{V}_R(t=0^+) \\ \Delta \boldsymbol{I}_R &= \boldsymbol{I}_R(t=0^-) - \boldsymbol{I}_R(t=0^+) \end{aligned} \tag{13.3}$$

These "delta" quantities are then used to construct a directional comparison scheme that uses signaling to the remote end of the line in the usual manner. The relays are very fast, usually completing their measurement and decision making in about one-quarter cycle, such that the total relay time can be completed in well below one cycle as long as fast signal transmission is available. The direction to the fault is detected by means of ordinary VT's or CT's connected to the line side of the circuit breaker, where the change in voltage and current are measured. Assuming the usual CT and VT polarities, a fault in the forward direction causes a voltage change having the opposite sign to that of the current change. The reverse is true of a fault in the reverse direction, which gives voltage and current changes having the same sign.

The basic concept of traveling wave protection can be illustrated qualitatively by reference to Figure 13.51. The threshold settings of ΔV_R and ΔI_R determine the reach of the protection, and the comparison of the signs of the two quantities determine the direction to the fault. The protection is set in an overreaching mode, as is commonly done in directional comparison schemes. If the signs of the two delta quantities are equal, the fault is in the reverse direction, but if the signs are different, the fault is in the forward direction. Table 13.5 summarizes the possible relay actions. The black arrows in the right-hand column of the table represent the direction of the current wave and the gray arrows represent the voltage wave. The fault is represented as a voltage source according to the compensation theorem. Note that the positive wave direction is defined as moving from A or B toward the fault point F.

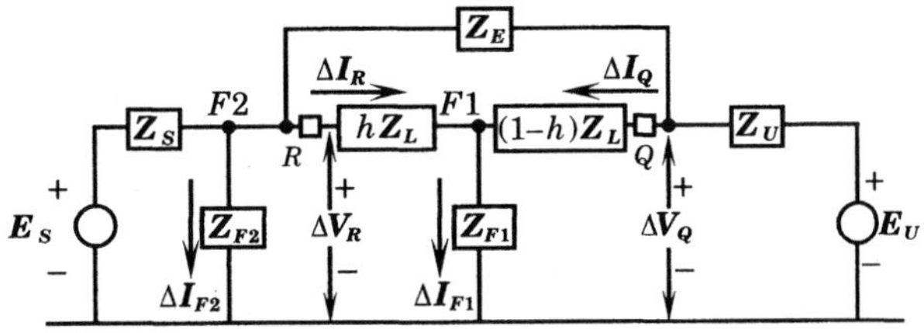

Figure 13.51 A qualitative description of traveling wave protection.

TABLE 13.5 Traveling Wave Relay Actions for Various Fault Locations

		A		***B***		**Fault Specification**
Fault Type	**$v_F(t)$ Polarity**	Δi	Δv	Δi	Δv	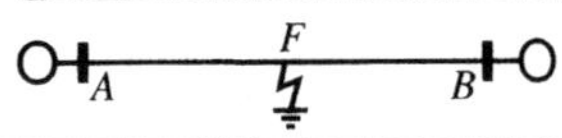
Internal fault	Positive	+	−	+	−	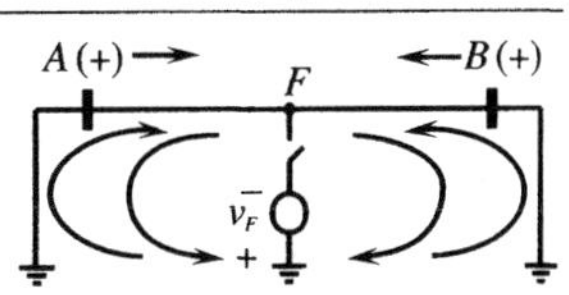
	Negative	−	+	−	+	
External to A	Positive	−	−	+	−	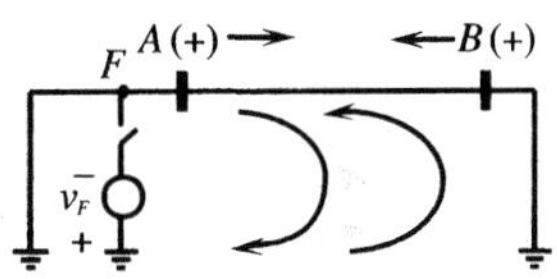
	Negative	+	+	−	+	
External to B	Positive	+	−	−	−	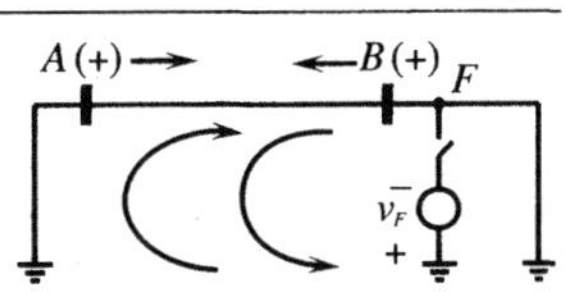
	Negative	−	+	+	+	

First consider the case for an internal fault, as shown in the second row of Table 13.5. If the prefault voltage has a positive polarity, the current waves at both A and B are positive, but the voltage waves are both negative. Exactly the reverse is true if the prefault voltage polarity is negative.

For external faults, when the prefault voltage has a positive polarity, the current nearest the fault has a negative polarity, but the current measured at the opposite terminal has a positive polarity. Both voltages have a negative polarity.

From the principles outlined in Table 13.5 it can be concluded that, in the case of an internal fault, both lines ends will experience a change in current and in voltage that are opposite in sign. For an external fault, one end will always have changes that are equal in sign.

The traveling wave system can be configured to protect 100% of the line by use of pilot signaling as a directional comparison scheme. This makes the system dependent on the communications channel. It is also important that the communications have a short channel delay if the scheme is to have fast response. A block diagram showing the major components of one traveling wave system is shown in Figure 13.52 [29].

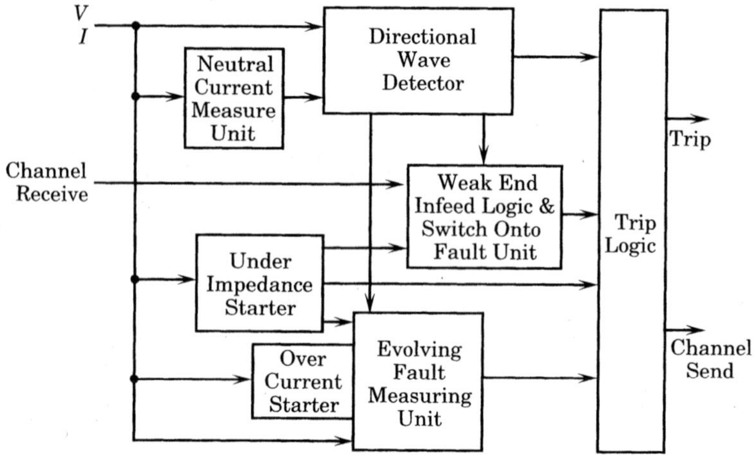

Figure 13.52 Block diagram of a traveling wave protection system.

The directional wave detector is the fault detector, which is based on the principle of directional wave detection, described above. The neutral current improves the detection of high-impedance ground faults. The weak-end infeed unit provides an alternative method of tripping when there is insufficient current infeed at one terminal to generate a channel signal. The evolving fault unit is disabled under normal conditions, but it is picked up by the directional wave detector together with an input from either an under-impedance unit or an overcurrent starter unit. The evolving fault unit then imposes a 35 ms time delay before issuing an output.

Faults are initially detected by the directional wave detector section, where the direction to the fault is detected by noting the polarity of the changes in current and voltage. This requires that changes due to faults be discriminated against changes due to other phenomena, such as line switching, lightning surges, and other events. Lightning surges that do not lead to a fault cause the propagation of high-frequency energy on the system. The directional wave detector damps these high-frequency waves heavily, thereby ensuring that any detected signal should last a preset minimum time before it can be identified as a fault. Faults not due to lightning are similarly detected due to their fundamental frequency behavior. Transients caused by switching surges can be detected by placing the measurement of current and voltage on the line side of the protected line terminating breakers, so that all such disturbances will always be outside the protection zone of the line.

Some traveling wave relays provide multiple protection modes to clear close-in faults at ultra-high speeds independent of the communications channel, and also provide slower clearing of other faults using pilot communications. Other functions, such as ground fault detection are also incorporated into the relay package. One advantage of the traveling wave detection, in addition to high speed, is that this type of detection is not affected by power swings or cleared fuses on voltage transformers, which cause failure in many types of distance protection. Nondirectional backup zones are also provided in some relays.

The logic diagrams for the permissive overreaching directional comparison scheme is shown in Figure 13.53.

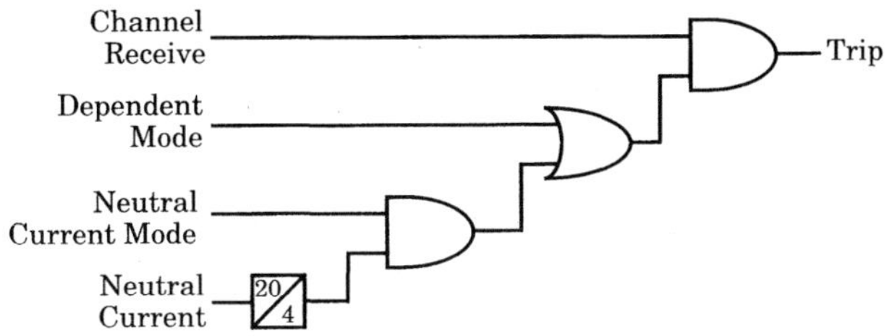

Figure 13.53 Tripping logic for the dependent mode [29].

Note that the tripping logic is designed to utilize neutral current as one of the logical inputs. The object of this feature is to improve the detection of high-impedance ground faults within the protected line. When a line fault occurs this will cause the relay to issue a channel send signal. After a time delay of 20–35 ms a trip signal is generated if the current measuring unit shown in Figure 13.52 continuously senses a neutral current, and a channel receive signal continues to be received.

The principle of traveling wave protection can also be used on multicircuit lines where mutual induction is present, but the evaluation of more complex schemes requires the consideration of the induction of traveling waves on parallel lines due to mutual coupling, which may require that the pickup be desensitized. Traveling wave protection senses the first change in the faulted line and then seals in. Therefore, these schemes cannot detect simultaneous faults or sequential faults, such as faults that involve both circuits of the double-circuit line. This arrangement would require independent main or back-up protection

The traveling wave relay requires no special consideration for communications channels, or for either voltage or current transducer characteristics. It is required, however, that CCVT's be located on the line side of series capacitors, when the series capacitor is located at the end of the line. Several traveling wave relay systems are available for application on series compensated transmission lines [36–42]. Traveling wave protection is normally set in an overreaching mode. In principle, traveling wave protection is superior to distance protection as it is not affected by problems of over- or underreaching, or the problems associated with current reversals. It is subject to problems of mutual induction on parallel lines. Also, since the protection is only active for a short time and then seals in, there may be problems with the detection of slowly evolving faults. It must also be carefully designed to prevent pick up due to lightning arrester operation.

13.8 MONITORING OF PILOT PERFORMANCE

There are many factors that affect pilot protection performance, and many of these factors are external to the relay terminal. These external factors should be considered for monitoring in *sequence-of-events recorders* (*SER*) or *fault recorders* (*FR*) in order to gather data regarding the system performance.

An IEEE committee has examined the various points in the system on which data are often available for analysis to assist in analyzing the performance of pilot relaying systems. These recommendations are summarized in Table 13.6. As noted in the table, SERs usually

TABLE 13.6 Recommended Digital and Analog Monitoring [11]

	Sequence of Events or Fault Recorder Monitored Points	**Type of Pilot System†**						
		1	**2**	**3**	**4**	**5**	**6**	**7**
Digital Inputs	A. Pilot channel							
	1. Start control	x	—	—	—	—	—	—
	2. Stop control[2]	x	x	x	x	—	—	—
	3. Transmitter output	x[6]	—	—	—	x[5]	—	—
	4. Receiver output[1]	x[5]	x	x	x	x	—	—
	5. Guard RCVR output	—	x	x	x	—	—	—
	6. Loss of channel[3]	—	—	x	x	—	x	x*
	B. Fault Detector							
	Operation	x*	—	—	x*	x	—	—
	1. Low-set current	x*	x*	x*	x*	x	x	x*
	2. High-set current							
	3. Distance:	x*	x*	x*	x*	x	x**	—
	(a) Phase	x*	x*	x*	x*	—	—	—
	(b) Ground							
	C. Pilot Trip Output	x	x	x	x	x	x	x
Analog Inputs	D. FR relay voltages							
	1. Phase 1	x	x	x	x	x**	x**	x
	2. Phase 2	x	x	x	x	x**	x**	x
	3. Phase 3	x	x	x	x	x**	x**	x
	E. FR relay currents							
	1. Phase 1	x	x	x	x	x	x	x
	2. Phase 2	x	x	x	x	x	x	x
	3. Phase 3	x	x	x	x	x	x	x
	4.Residual Current	x*	x*	x*	x*	x	x	x
	F. Polarizing quantities							
	1. Current	x*	x*	x*	x*	—	—	—
	2. Voltage	x*	x*	x*	x*	—	—	—
	G. PLC coax signal	x[6]	—	—	—	x[5]	—	x[4]
	H. RCVR output current	x[5]	—	—	—	—	—	—

Notes:
1. Receiver unblock in an unblocking scheme and trip received in transfer trip scheme.
2. Transmitter keyed to trip frequency in an unblocking scheme and transfer trip scheme.
3. Channel supervision in pilot wire scheme.
4. Operate coil Voltage.
5. Use either input labeled 5. Both not necessary.
6. Use either input labeled 6. Both not necessary.
* As applicable
** If supervised by distance relay.

†Pilot System Type:
1. Directional comparison blocking with On-Off PLC
2. Directional comparison unblocking & frequency shift PLC
3. Overreaching transfer trip (tones)
4. Permissive underreaching transfer trip
5. Single phase comparison blocking PLC
6. Dual phase comparison unblocking PLC
7. Pilot wire

monitor the performance of channel control devices, the channel performance, and the trip output. The output of these devices is usually alphanumeric. *Fault recorders* (*FR*) generally monitor system voltages and currents applied to the fault sensing relays. Capturing both types of information provides a good review of pilot channel performance.

IEEE committees often provide summaries of industry experience that can be helpful to the protection engineer [43].

REFERENCES

[1] Lohage, L., et al., "Protection Systems Using Telecommunications," CIGRÉ WG 34/35-05, 1987.

[2] Mason, C. R., *The Art and Science of Protective Relaying*, John Wiley and Sons, Inc., New York, 1956.

[3] IEEE Std 100, *IEEE Standard Dictionary of Electrical and Electronics Terms*, Fifth Edition, IEEE, New York, 1992.

[4] IEEE Std 313-1971, "IEEE Standard for Relays and Relay Systems Associated with Electric Power Apparatus," ANSI C37.90-1971, IEEE, New York, 1971.

[5] Jordan, E. C., Ed., *Reference Data for Engineers: Radio, Electronics, Computer, and Communications*, Seventh Edition, Howard Sams & Co., Indianapolis, 1985.

[6] Murphy, J. J., Chm., IEEE Working Group Report, "Fiber Optic Channels for Protective Relaying," IEEE Paper 88 WM 119-0, presented at the IEEE PES Winter Meeting, New York, January 1988.

[7] Elmore, W. A., "Transmission Line Protection Alternatives," *Proc. American Power Conference*, Chicago, 1971.

[8] GEC Measurements, *Protective Relays Application Guide*, General Electric Company p.l.c. of England, 1975.

[9] General Electric Company, *The Art of Protective Relaying, Transmission and Subtransmission Lines*, Publication GET-7206A, Switchgear Division, Philadelphia. 1971.

[10] Ziegler, G., Ed., "Application Guide on Protection of Complex Transmission Network Configurations," CIGRÉ SC34-WG04, May 1991.

[11] IEEE Committee Report, "Pilot Relaying Performance Analysis," *IEEE Trans.*, PWRD-5, January 1990, pp. 85–102.

[12] Blackburn, J. L., Ed., *Applied Protective Relaying*, Westinghouse Electric Corporation, Pittsburgh, 1976.

[13] Asea Brown Boveri, "Type REL 150/REZ1 Distance Relay," Publication 1MDB06009-EN, March 1992.

[14] Asea Brown Boveri, "Type LZ96a Distance Relay," Publication 1MDB06010-EN, August 1990.

[15] General Electric Company, "Type SLYP-SLCN Static Comparison Relaying: Description and Application," Publication GET-6456A, Switchgear Division, Philadelphia. 1980.

[16] Andrichak, J. G., and R. C. Patterson, "Transmission Line Protection with Positive and Negative Sequence Relays," a paper presented to the Pennsylvania Electric Association, October 2–3, 1975.

[17] General Electric Company, "PLS Hybrid Scheme, Single FSK Channel for Series Compensated Lines," Publication GEK-90671A, Meter and Control Business Department, Malvern, PA, 1991.

[18] Westinghouse Electric Corporation, "SPCU-1A Segregated Phase Comparison Relaying Systems," Application Data Manual 40-201, October 1987.

[19] Asea Brown Boveri, "MSPC (REL 350) Numerical Segregated Phase Comparison System," I.L. 40-201.6, Application Manual, May 1992.

[20] Schweitzer Engineering Laboratories, "SEL-121H, Phase Distance Relay, Ground Directional Overcurrent Relay, Fault Locator," Instruction Manual, Pullman, WA, 1990.

[21] Esztergalyos, J., M. T. Yee, M. Chamia, and S. Liberman, "The Development and Operation of an Ultra High Speed Relaying System for EHV Transmission Lines," CIGRÉ Paper 34-04, presented at the 1978 CIGRÉ Conference, August 30–September 7, 1978.

[22] Chamia, M., and S. Liberman, "Ultra-high-speed Relay for EHV/UHV Transmission Lines – Development Design and Application," *IEEE Trans.*, PAS-97, 1978, pp. 2104-2016.

[23] Hicks, K. L., and W. H. Butt, "Feasibility and Economics of Ultra-High-Speed Fault Clearing," *IEEE Trans.*, PAS-99, 1980, pp. 2138–2145.

[24] Crossley, P. A., and P. G. McLaren, "Distance Protection Based on Traveling Waves," *IEEE Trans.*, PAS-102, 1983, pp. 2971–2983.

[25] Cabeza-Resendez, L. Z., A. N. Greenwood, and T. S. Lauber, "Evaluation of Ultra-high-speed Relay Algorithms," EPRI Report, EL-3996, 1985.

[26] Ramirez, A. R., and R. A. Achilles, "Ultra-High-Speed Relaying Protection: A Setting Methodology," IEEE Paper 86 WM 118-4, presented at the IEEE PES Winter Meeting, February 2–7, 1986, New York.

[27] Mansour, M. M., and G. W. Swift, "Design and Testing of a Multi-Microprocessor Travelling Wave Relay," IEEE Paper 86 WM 115-0, presented at the IEEE PES Winter Meeting, February 2–7, 1986, New York.

[28] Johns, A. T., and E. P. Walker, "Cooperative Research into the Engineering and Design of a New Digital Directional Comparison Scheme," *IEE Proc.*, 135, 1988, pp. 334–368.

[29] Christopoulos, C., D. W. P. Thomas, and A. Wright, "Scheme Based on Traveling-Waves for the Protection of Major Transmission Lines," *Proc. IEE*, 135, Pt. C, 1988, pp. 63–83.

[30] Ermolenko, V. M., V. F. Lachugin, D. R. Lyubarsky, I. N. Popov, and G. V. Sokolova, "High-Speed Wave Directional Relay Protection of UHV Lines," CIGRÉ Paper 34-11, presented at the 1988 CIGRÉ Conference, August 28–September 3, 1988, Paris.

[31] Thomas, D. W. P., and C. Christopoulos, "Ultra-High-Speed Protection of Series Compensated Lines," *IEEE Trans. on Power Delivery*, PWRD-7 1, January 1992, pp. 139–145.

[32] Asea Brown Boveri, "Combined Directional-Wave and Impedance Backup Relay, Types RALZA and RALZB," ABB Publication 1MDB06012-EN, August 1990.

[33] Asea Brown Boveri, "Ultra High-Speed Line Protection Type RALZA, Application Manual:," ABB Publication RF619-001E, Edition 4, RFA, November 1988.

[34] Asea Brown Boveri, "Setting Calculations for Ultra High-Speed Line Protection Type RALZA," ABB Publication RF619-002E, RFA, March 1985.

[35] Asea Brown Boveri, "Commissioning Instructions for Ultra High-Speed Line Protection Type RALZA," ABB Publication RF619-002E, RFA, March 1985.

[36] Asea Brown Boveri, "Ultra High-Speed Line Protection Type RALZB, Application Manual:," ABB Publication RF619-005E, Edition 3, RFA, November 1988.

[37] Asea Brown Boveri, "Setting Calculations for Ultra High-Speed Line Protection Type RALZB," ABB Publication RF619-006E, RFA, March 1985.

[38] Asea Brown Boveri, "Commissioning Instructions for Ultra High-Speed Line Protection Type RALZA," ABB Publication RF619-002E, RFA, March 1985.

[39] Asea Brown Boveri, "Directional-Wave Relay, Type LR91," ABB Publication 1MDB06011-EN, August 1990.

[40] Asea Brown Boveri, "LR91 - An Ultra High-Speed Directional Comparison Relay for Protection of High-Voltage Transmission Lines," ABB Publication CH-ES 23-85.10E, 1985.

[41] Asea Brown Boveri, "Ultra High-Speed Directional Relay Type LR91–Instructions for Installation and Operation," ABB Publication CH-ES 83-85.11E, February 1985.

[42] Jancke, G., and K. F. Akerström, "The Series Capacitor with Special Reference to its Application in the Swedish 220 kV Network," CIGRÉ paper 332, presented at the 1950 session, Paris, 1950.

[43] Report by the IEEE Power Line Carrier Working Group, Power System Relaying Committee, "Power Line Carrier Practices and Experience," IEEE paper 94 SM 428-3 PWRD, 1994.

PROBLEMS

13.1 Define the term *pilot protection*. Hint: Look this term up in the *IEEE Dictionary of Electrical and Electronics Terms*.

13.2 Define the terms *selectivity*, *reliability*, and *security* as applied to power system protection. Check the IEEE definitions of the terms *reliability*, *availability*, and *dependability*, also in connection with their use in system protection.

13.3 A transmission line protective system trips a transmission line in response to a fault on an adjacent line. Is this protective system reliable? Does it exhibit selectivity? Does it exhibit security? Is it dependable?

13.4 A fault occurs in the center of a transmission line, but the protective system at one end does not recognize the fault as a zone 1 fault, but does trip with zone 2 time delay. Is this protective system reliable? Does it exhibit selectivity? Does it exhibit security? Is it dependable?

13.5 Review the terms *pilot*, *pilot channel*, and *pilot protection*. List some systems that might serve as pilot transmission systems.

13.6 Investigate the methods used to construct a fiber-optic communication system. Learn how the signals are transmitted over the fiber lines and why this method of communications is favored for many applications. In particular, find answers to the following questions:

(a) If optical signal transmission is superior, why hasn't this method of transmission always been used?

(b) How is the signal transmitted along the optical fiber and how can these fibers be spliced and repaired, if necessary?

(c) How is the optical signal transmitted and how is it received?

Hint: see the *McGraw-Hill Encyclopedia of Science and Technology*.

13.7 Investigate the development of the integrated system digital network and describe how this might be used in power system protection.

13.8 Prepare a detailed description of the operation of the Direct Underreaching Transfer Trip Scheme shown in Figures 13.10 and 13.13. List the events timewise and describe the relay actions at each event time. Comment on the overall strategy and any problems foreseen in its operation.

13.9 Prepare a detailed description of the operation of the Supervised Direct Underreaching Transfer Trip Scheme shown in Figures 13.10 and 13.14. List the events timewise and describe the relay actions at each event time. Comment on the overall strategy and any problems foreseen in its operation.

13.10 Prepare a detailed description of the operation of the Direct Underreaching Transfer Trip Scheme with Guard Frequency shown in Figures 13.10 and 13.15. List the events timewise and describe the relay actions at each event time. Comment on the overall strategy and any problems foreseen in its operation.

13.11 Prepare a detailed description of the operation of the Permissive Underreaching Transfer Trip Scheme shown in Figures 13.16 and 13.17. List the events timewise and describe the relay actions at each event time. Comment on the overall strategy and any problems foreseen in its operation.

13.12 Prepare a detailed description of the operation of the Direct Overreaching Transfer Trip Scheme shown in Figures 13.18 and 13.19. List the events timewise and describe the relay actions at each event time. Comment on the overall strategy and any problems foreseen in its operation.

13.13 Prepare a detailed description of the operation of the Permissive Overreaching Transfer Trip Scheme with Guard Frequency, as shown in Figure 13.20. List the events timewise and describe the relay actions at each event time. Comment on the overall strategy and any problems foreseen in its operation.

13.14 Prepare a detailed description of the operation of the Direct Blocking Scheme shown in Figures 13.21 and 13.22. List the events timewise and describe the relay actions at each event time. Comment on the overall strategy and any problems foreseen in its operation.

13.15 Prepare a detailed description of the operation of the Directional Comparison Blocking Scheme shown in Figures 13.23 and 13.24. List the events timewise and describe the relay actions at each event time. Comment on the overall strategy and any problems foreseen in its operation.

13.16 Prepare a detailed description of the operation of the Directional Comparison Unblocking Scheme shown in Figures 13.25 and 13.26. List the events timewise and describe the relay actions at each event time. Comment on the overall strategy and any problems foreseen in its operation.

13.17 Prepare a detailed description of the operation of the Hybrid Scheme described in Figures 13.30, 13.31, and 13.32. List the events timewise and describe the relay actions at each event time. Comment on the overall strategy and any problems foreseen in its operation.

13.18 Describe the advantages of the Single-Current-Phase Comparison Blocking Scheme shown in Figure 13.35.

13.19 Describe the advantages of the Dual-Current-Phase Comparison Blocking Scheme shown in Figure 13.36.

13.20 Describe the advantages and disadvantages of the Segregated Phase Comparison Blocking Scheme shown in Figures 13.37, 13.38, and 13.39.

13.21 Prepare out a description of the design of the FM Current Differential Protection Scheme shown in Figure 13.40. List the events timewise and describe the relay actions at each event time. Comment on the overall strategy and any problems foreseen in its operation.

13.22 Prepare out a description of the design of the PCM Current Differential Protection Scheme shown in Figure 13.41. List the events timewise and describe the relay actions at each event time. Comment on the overall strategy and any problems foreseen in its operation.

13.23 The EHV line protection system described in Section 13.5 utilizes two different types of protection, directional comparison and phase comparison, and employs two different means of communication, microwave and PLC. How should the communication systems be assigned to the different protection schemes? Defend your answer.

13.24 Analyze the use of current reversal logic and timers to solve the problems associated with a line closing or reclosing into a fault.

14

Complex Transmission Protection

14.1 INTRODUCTION

There are several aspects of transmission protection that are challenging and present complex problems for the protection engineer. These problems are often due to network configurations that present a protection problem that is not simply routine, but require special analysis. One example is multiterminal transmission lines. Other problems are due to special restrictions on the system operation, such as single-phase switching, which may be dictated by concerns for power system stability. This chapter examines some of these special topics and investigates the application of protective relaying in resolving some difficult problems.

14.2 SINGLE-PHASE SWITCHING OF EHV LINES

EHV transmission lines are important links in the bulk power system, and the loss of these lines is often critical for system stability, power transfer capability, and voltage control. Because of their importance, engineers have devised a number of methods of reducing the outage probability of these lines. One method often used is to automatically reclose the line following any fault related outage, thereby minimizing the outage duration to the few cycles required to deionize the arc and reclose the breakers. This technique is used by some utilities even though there is some hazard that nearby generators may suffer loss of shaft life due to the sudden change in the shaft torque due to the switching [1–4]. In some cases reclosing is not used at switching stations near generating units.

Another important method of gaining many of the same benefits of fast reclosing is to switch only the faulted phases to clear a short circuit. It is well known that nearly all faults are one-line-to-ground faults. This suggests that single phase switching of only the faulted phase will provide significant benefits for the majority of fault conditions. The major benefit is that the two unfaulted phases are left in service, thereby allowing a significant power transfer to continue during the fault detection and clearing operation. This will always have a beneficial effect, especially for system stability. Moreover, single-phase switching can be achieved by making modest changes in the circuit breaker configuration so that each pole is independently operated.

There is, however, a serious disadvantage that accompanies single-phase switching. As the faulted phase is isolated by opening the breakers for the faulted phase at both ends of the line, the faulted phase conductor is still energized due to electromagnetic and electrostatic coupling to the parallel energized phases, as noted in Chapter 12. This means that a voltage persists at the fault point that may be adequate to sustain the fault current, although at a lower current level. This is known as *secondary arc current* (*SAC*) and is a current that usually follows the path of the power arc after disconnecting the faulted phase. As the stored energy in the line is drained, the current rapidly falls under normal switching conditions, and after the current is extinguished, the dielectric of the arc path is quickly reestablished. When single-phase switching is used, the coupling between the deenergized line and the energized lines provides sufficient energy to sustain the SAC for prolonged periods [5].

The SAC is an alternating current and passes through zero twice in each cycle of the applied system voltage. This provides the opportunity for current extinction, provided the dielectric of the arc path can be re-established quickly enough. Successful extinction depends on the outcome of a race between the rate of rise of the line recovery voltage and the withstand voltage across the arc path. The arc path behavior depends on many factors, such as the arc voltage and current, the initial fault current, the arc location, wind velocity, and other meteorological factors. Of all controlling factors, the design engineer can control the recovery voltage, and can thereby influence or control the success of SAC extinction.

The faulted phase is coupled both electromagnetically and electrostatically to the sound phases. Of these two forms of coupling, the capacitive coupling is the more important [6], [7]. The SAC is proportional to the voltage, the interphase capacitance, and the length of the line. The interphase and line-to-ground capacitance can be effectively neutralized by the addition of shunt reactors, both between phases, and from phase to ground. The reactor configuration depends on the capacitance to be neutralized. This capacitance is affected by line transposition, which, if performed exactly, equalizes the capacitances [8].

14.2.1 Control of Secondary Arcs in Transposed Lines

In the analysis of shunt capacitance and the coupling between phases and between phases and ground, the longitudinal currents flowing along the line will be neglected [7]. If we view the total capacitance of the line, we may construct the equivalent circuit of Figure 14.1. A straightforward calculation of the total capacitance of the line will give a value for the capacitance between phases, and that from each phase to ground, as shown in Figure 14.1(*a*). This circuit is valid for all cases, whether the capacitances are balanced by transposition or not, but the equal values of susceptance shown in the figure are valid only for the balanced case. Note the primed values of susceptance are used to indicate *capacitive* susceptance. The delta arrangement of capacitances can always be changed to the equivalent wye, as shown in Figure 14.1(*b*), and this circuit is always correct for any line, but the values shown are for the balanced case. Finally, for the balanced susceptance case illustrated, the equivalent circuit of Figure 14.1(*c*) may be derived, which has the advantage of requiring only four capacitors rather than six.

The capacitive susceptances shown in Figure 14.1 are all given in terms of the positive and zero sequence values, which are determined directly from the phase susceptance (or capacitance) values by means of a similarity transformation of the matrix of phase capacitances [8].

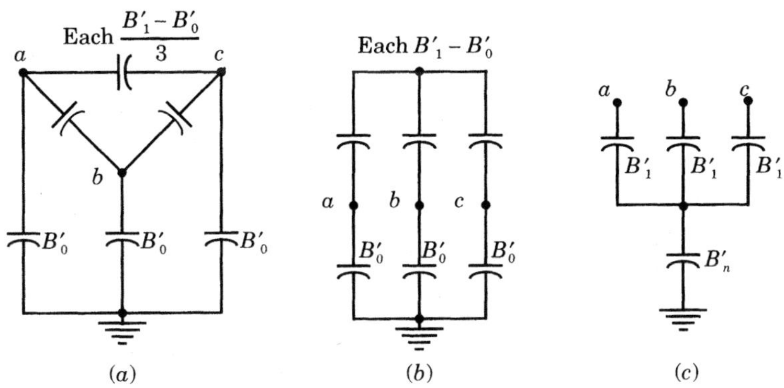

Figure 14.1 Equivalent circuits of the shunt capacitances of a three-phase transmission line. (*a*) Basic line capacitances. (*b*) Delta-wye conversion. (*c*) Balanced equivalent.

Note that, if zero sequence voltages are applied from the *a-b-c* conductors to ground, charging current flows only in the grounded capacitors. If positive sequence voltages are applied, the wye-grounded and wye-ungrounded sets are in parallel, giving a net capacitance of only the positive sequence values.

Banks of inductive reactances may be connected in exactly the same configuration as the capacitances in order to neutralize the capacitive coupling. Such a connection of inductances is shown in Figure 14.2.

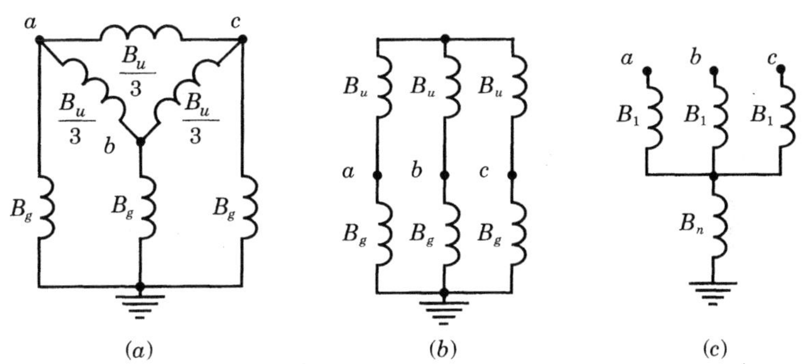

Figure 14.2 Connections of shunt reactors for compensation of capacitive charging current. (*a*) Basic inductive compensation. (*b*) Delta-wye conversion. (*c*) Balanced equivalent.

Other arrangements are possible. The four reactor arrangement of Figure 14.2(*c*) has certain advantages, and this is an arrangement that is often used because of its simplicity. Note that, under balanced conditions, no current flows in the grounded reactance and current will flow in this reactance only under unbalanced conditions, such as during an unbalanced fault.

We analyze the system consisting of wye-connected capacitors and inductors, as shown in Figures 14.1(*b*) and 14.2(*b*). The circuit diagram for this compensation system is shown in Figure 14.3.

When the one-line-to-ground fault occurs on phase *a*, the switch at the fault location is closed and the circuit breaker connecting this line section to the source voltage is opened.

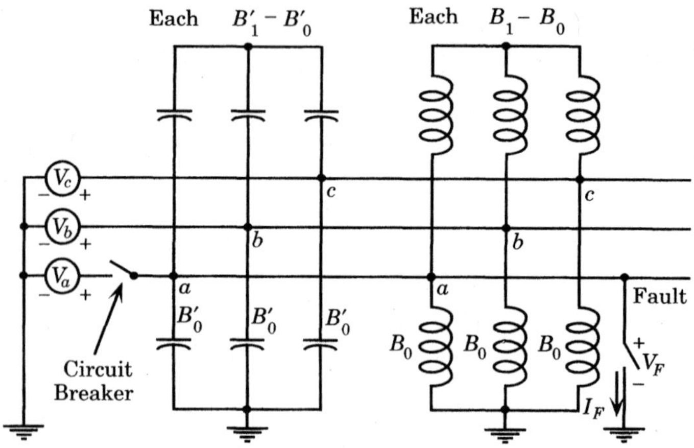

Figure 14.3 Equivalent circuit of a source and a faulted line section with single-phase switching and a one-line-to-ground fault [7].

Let the source voltage be specified as follows:

$$\begin{aligned} \boldsymbol{V}_a &= V + j0 \\ \boldsymbol{V}_b &= -0.5V - j0.866V \\ \boldsymbol{V}_c &= -0.5V + j0.866V \end{aligned} \tag{14.1}$$

Following Kimbark [7], we resolve the applied emfs into two sets; the real components and the imaginary components, which we apply separately by linear superposition.

First set:

$$\begin{aligned} \boldsymbol{V}_a^{(1)} &= V \\ \boldsymbol{V}_b^{(1)} &= -0.5V \\ \boldsymbol{V}_c^{(1)} &= -0.5V \end{aligned} \tag{14.2}$$

Second set:

$$\begin{aligned} \boldsymbol{V}_a^{(2)} &= 0 \\ \boldsymbol{V}_b^{(2)} &= -j0.866V \\ \boldsymbol{V}_c^{(2)} &= +j0.866V \end{aligned} \tag{14.3}$$

We can simplify the circuit shown in Figure 14.3 to show only the system that will be involved with the set of voltages applied, according to (14.2) and (14.3). This gives the simplified circuits shown in Figure 14.4. Figure 14.4(b) has a special symmetry with respect to phase a, where it is noted that this phase is at exactly ground potential. This means that, for the second set of applied emfs, there is no voltage across the fault path when the path is open, and no current when the fault path is closed. Thus, this second set of emfs contributes nothing to the fault current or to the recovery voltage, and it may therefore be ignored. For the first set, the two branches on the left may be ignored, since they do not influence either the voltage or current at the fault point.

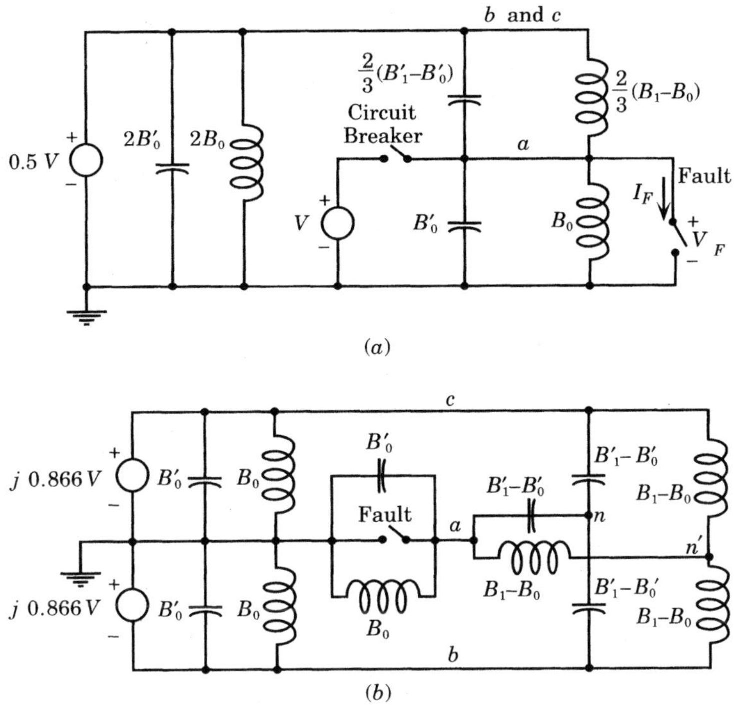

Figure 14.4 Simplified equivalent circuits of the system shown in Figure 14.2 [7]. (*a*) Equivalent circuit for 1st set of emf's. (*b*) Equivalent circuit for 2nd set of emf's.

The recovery voltage across the fault path is the voltage V_F when the fault switch is open. This gives the voltage

$$\boldsymbol{V}_F = -\frac{V[(B_1' - B_0') - (B_1 - B_0)]}{(2B_1' + B_0') - (2B_1 + B_0)} \tag{14.4}$$

This voltage is zero if the two terms in the numerator are equal, or if the capacitance is exactly neutralized by the ungrounded reactors.

If the interphase reactors are omitted, the voltage across the fault path is computed as

$$\boldsymbol{V}_F = \frac{V(B_1' - B_0')}{B_0 - (2B_1' + B_0')} \tag{14.5}$$

with

$$B_1 - B_0 = 0$$

where B_0 is provided only by the grounded reactors.

If both the interphase and grounded reactors are omitted, then the fault path recovery voltage is computed as

$$\boldsymbol{V}_F = -V\frac{B_1' - B_0'}{2B_1' + B_0'} \tag{14.6}$$

Comparing (14.5) and (14.6), it is seen that compensation using only the grounded reactors gives a *higher* recovery voltage than having no compensation at all.

The secondary arc current is computed as the current through the fault switch in Figure 14.4(*a*). This current is computed to be

$$\mathbf{I}_F = -j\frac{V}{3}[(B_1' - B_0') - (B_1 - B_0)] \tag{14.7}$$

Clearly, full compensation makes this fault current equal to zero in exactly the same way that the recovery voltage is caused to be zero.

If there is no compensation at all, the fault current becomes

$$\mathbf{I}_F = -j\frac{V}{3}(B_1' - B_0') \tag{14.8}$$

Note that this current is not affected by the grounded reactors.

For the six reactor scheme shown in Figure 14.2, we conclude that the following reactances are required to completely neutralize the interphase capacitance

$$B_1 - B_0 = B_1' - B_0' \tag{14.9}$$

It is possible, however, that the system requires some shunt compensation irrespective of the fault conditions, for the purpose of absorbing the excess charging current of the EHV line section. Let F be the fraction of the total shunt compensation required. Then we have

$$B_1 = FB_1' \tag{14.10}$$

Then, solving (14.9) and (14.10)

$$B_0 = B_0' - (1 - F)B_1'$$

Returning to the configuration of Figure 14.2(*b*), we compute

$$\begin{aligned} B_u &= B_1' - B_0' \\ B_g &= B_0' - (1 - F)B_1' = FB_1' - B_u \end{aligned} \tag{14.11}$$

The first equation shows that the ungrounded reactors must resonate with the interphase capacitance. This provides at least the degree of shunt compensation given by

$$F_{\min} = \frac{B_1' - B_0'}{B_1'} = 1 - \frac{B_0'}{B_1'} \tag{14.12}$$

For a typical EHV line, this is about 30% compensation. If more shunt compensation is required, this can be supplied by grounded reactors, having the value given by the second part of (14.11).

14.2.2 Secondary Arcs in Untransposed EHV Lines

It is the practice of some utilities to construct EHV lines without transpositions. For untransposed transmission lines the capacitance between phases is not uniform, as assumed in the previous section. For lines with flat conductor configuration, the mid-to-outer phase capacitance is usually over three times that of the outer-to-outer phase capacitance. This means that the analysis performed above is not accurate for the untransposed case.

One solution to the untransposed line case has used a variation of the four-legged reactor arrangement shown in Figure 14.2(*c*), with switches added to create the modified four-legged reactor scheme shown in Figure 14.5. The four switches are closed or opened, depending on which phase is faulted. This simple switching arrangement provides the necessary compensation for the capacitance between the outer phases, and also provides the additional compensation for mid-to-outer phase capacitances [5], [9–12].

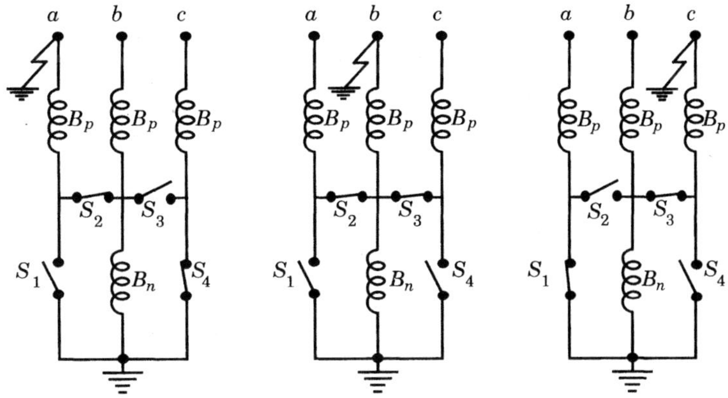

Figure 14.5 Modified four-legged reactor bank for untransposed transmission lines.

The susceptance matrices for the three arrangements shown in Figure 14.5 are given as follows, where we use the notation from the four-legged reactances of Figure 14.2(c).

For a one-line-to-ground fault on the center phase, phase "b,"

$$\mathbf{B} = \begin{bmatrix} B_1 & -B_2 & -B_2 \\ -B_2 & B_1 & -B_2 \\ -B_2 & -B_2 & B_1 \end{bmatrix} \tag{14.13}$$

where we define, for simplicity in writing the result, the susceptances

$$B_1 = \frac{B_p(2B_p + B_n)}{3B_p + B_n} \tag{14.14}$$

$$B_2 = \frac{B_p^2}{3B_p + B_n} \tag{14.15}$$

For a one-line-to-ground fault on phase "a,"

$$\mathbf{B} = \begin{bmatrix} B_3 & -B_4 & 0 \\ -B_4 & B_3 & 0 \\ 0 & 0 & B_p \end{bmatrix} \tag{14.16}$$

where we define the new susceptance quantities

$$\begin{aligned} B_3 &= \frac{B_p(B_p + B_n)}{2B_p + B_n} \\ B_4 &= \frac{B_p^2}{B_p + B_n} \end{aligned} \tag{14.17}$$

Finally, for a one-line-to-ground fault on phase "c,"

$$\mathbf{B} = \begin{bmatrix} B_P & 0 & 0 \\ 0 & B_3 & -B_4 \\ 0 & -B_4 & B_3 \end{bmatrix} \tag{14.18}$$

and where the elements are defined in (14.17).

In practice, the modified four-legged reactor scheme shown in Figure 14.5 has been applied to only one end of the transmission line, with the simple four-legged scheme of Figure 14.35(c) applied to the other end, as shown in Figure 14.6.

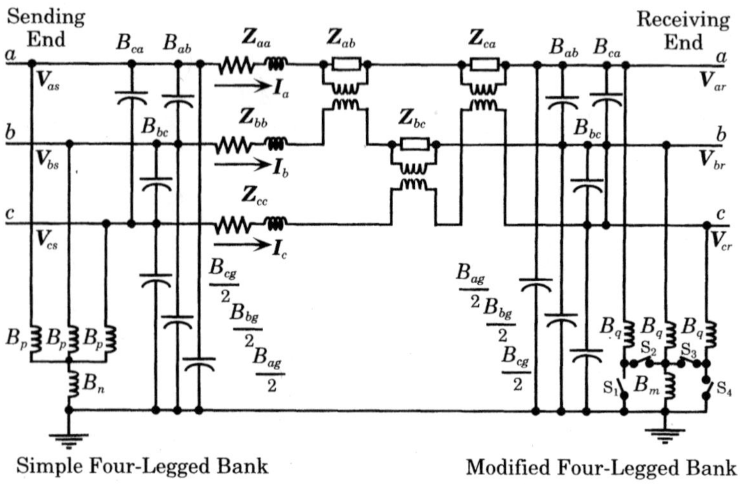

Figure 14.6 Equivalent circuit of a transmission line with both simple and modified four-legged reactor banks.

The above arrangement has been applied in at least one system with good results. Using the circuit of Figure 14.6 and the susceptance matrices computed for various fault conditions, it can be shown that near optimum values of the modified four-legged bank parameters can be found for a given transmission line.

The circuit of Figure 14.6 is complicated by the unbalance of the total line capacitances. Moreover, when one line is faulted and subsequently opened by its breakers, that line is coupled both capacitively and inductively to the two sound lines, as shown in the figure. Thus, for a fault on line x, the total current flowing in the fault is written as

$$\boldsymbol{I}_{Fx} = \boldsymbol{I}_{Cx} + \boldsymbol{I}_{Lx} \tag{14.19}$$

where $\boldsymbol{I}_{Cx}$ = component due to capacitive coupling
$\boldsymbol{I}_{Lx}$ = component due to inductive coupling

The capacitive component is dependent on the voltages of the lines, and the inductive component is dependent on the load current in the unfaulted lines. We can write the two components symbolically as follows.

First for the capacitive component, we write the following using h and k to signify the sound phases

$$\boldsymbol{I}_{Cx} = B_{exh}\boldsymbol{V}_{mh} + B_{exk}\boldsymbol{V}_{mk} \tag{14.20}$$

where $\boldsymbol{V}_{mh}$ = mean or average voltage of phase h
B_{exh} = equivalent interphase susceptance between x and h

and similarly for phase k. The susceptance terms can be found from the matrix expression for a given fault condition. The coupling depends strongly on whether the faulted phase is an outer phase or the center (inner) phase. The interphase susceptance always has three components, one due to the simple four-legged bank, one to the modified four-legged bank, and a third due to the capacitive coupling between phases. These components can all be clearly identified in Figure 14.6.

The inductive component is coupled to the faulted phase by inductive coupling, which causes a series voltage to be induced along the faulted phase. The inductive component of current that flows can be written as

$$\boldsymbol{I}_{Lx} = -B_{xg}(X_{xh}\boldsymbol{I}_h + X_{xk}\boldsymbol{I}_k) \tag{14.21}$$

where B_{xg} = total susceptance of current path to ground
X_{xh} = inductive coupling from phase x to phase h
I_h = current in sound phase h

The total susceptance of the current path depends on the location of the fault and the phase that is faulted (inner or outer). For example, if the fault is at the sending end of the line, it effectively shorts out the simple four-legged bank as well as a fraction (about one-half) of the line-to-ground capacitance. Faults at the receiving end short the modified bank and half of the line-to-ground capacitance. In all cases, the total susceptance is found by knowing the fault location, faulted phase, and the values of the susceptance matrices.

The optimization process to select the best values for the modified four-legged bank is complex and requires a great deal of computation [9–13]. This is considered a reasonable technique despite the effort required since the results are effective.

14.3 PROTECTION OF MULTITERMINAL LINES

One of the most difficult problems in transmission protection is that of applying the protection system to multiterminal lines [14–15]. There are a number of reasons why a high-voltage transmission line might be tapped without incurring the expense of constructing a full switching station and, thereby, dividing the line into three separate line segments. Some of the reasons can be stated as follows [16]:

1. The primary reason is often, simply, to save money. The cost of a high-voltage switching station may convince the planning and design engineers that the system requirements are best served by simply tapping the high-voltage line, at least for a limited period of time, perhaps until the tapped load grows to the point where the investment in circuit breakers is justified.
2. Tapping a high-voltage line is often a good interim solution in the long-term plan for serving a given area. This may permit greater efficiency in utilizing the available resources, while providing improved service to a local area.
3. New rights of way for transmission lines are difficult to obtain. In some cases a three-terminal line may be the only reasonable solution to a given problem.

There are several types of multiterminal configurations that are of interest to the protection engineer. The following terminology is preferred in some countries [17]:

1. Multiterminal lines are lines with three or more terminals and with sources of generation behind each terminal.
2. Tapped lines are lines having one or more terminals with substantial generation behind them and taps at other locations along the line that serve only load, wherein the taps do not have sufficient current injection capability to operate relays.

3. Tapped lines that terminate at a wye-grounded transformer provide a source of zero sequence current and may often be treated in the zero sequence network as a multi-terminal line.
4. In some systems, for example, a tapped load with a part-time cogeneration facility, may have active generation only part of the time, which leads to a situation where the system is described as a multiterminal line when the generation is operating and as a tapped line at other times.

Figures 14.7 and 14.8 show some common configurations of multiterminal lines and tapped lines. Some of the configurations shown in Figures 14.7 and 14.8 might be considered undesirable, and may even be temporary. Such systems are constructed for any number of reasons, and the protection engineer must design a workable protective system. The systems shown in Figures 14.7 and 14.8 might be considered typical, but other configurations can be devised and may be constructed for a particular reason. Mutually coupled lines are discussed in Section 14.4.

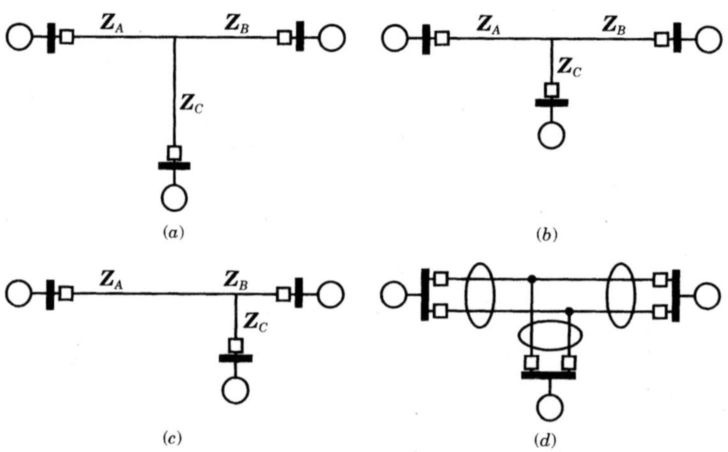

Figure 14.7 Common multiterminal line configurations [17]. (*a*) Equal legs. (*b*) One short leg. (*c*) Two short legs. (*d*) Mutual coupling.

In many cases a line is tapped to serve a load or to provide greater support to a growing load area. This creates a three-terminal line. It is possible that other taps may be made to create four or more terminals. The discussion here is for only the three-terminal case. An extension to more than three terminals, although somewhat difficult, is straightforward.

The difficulty of coordinating relays for a three-terminal line depends on several factors, such as

(a) The relative length (impedance) in the three legs of the line
(b) The Thevenin impedance at each source
(c) The type of relaying used
(d) The transfer impedance between terminals

The principles of three-terminal line protection will be examined by means of an example. Consider the three-terminal line, A-B-C, shown in Figure 14.9. Note that, although a bus is shown at the tap point D, no circuit breakers are installed at this location. This means that the line must be protected from locations A, B, and C for any type of fault and for any possible

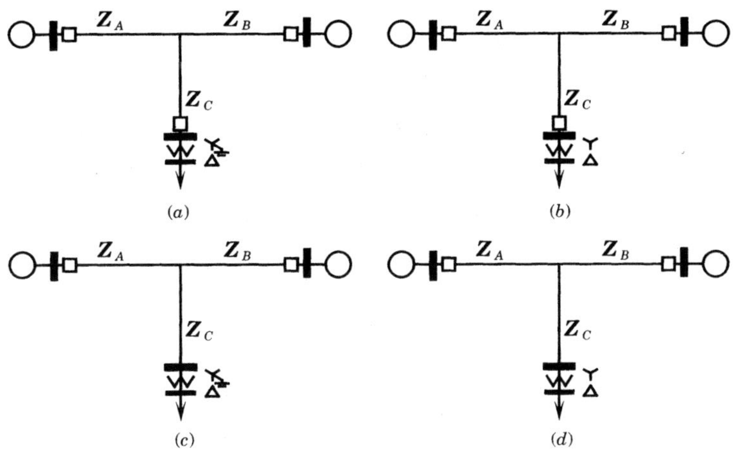

Figure 14.8 Common tapped line configurations [17]. (*a*) Tap with grounded transformer. (*b*) Tap with ungrounded transformer. (*c*) Tap with grounded transformer but no circuit breaker. (*d*) Tap with ungrounded transformer and no circuit breaker.

system condition. We assume that line AB was the original line and that, for some good reason, the line was tapped at D and a tap line constructed to C, where the line is connected into the ac network. Thus, we have sources of fault current at all three points, but not necessarily of the same strength. To be perfectly general, one must consider that the original line is tapped at any point from A to B, that the tap DC might be of any length, and that the three Thevenin sources are of arbitrary strength.

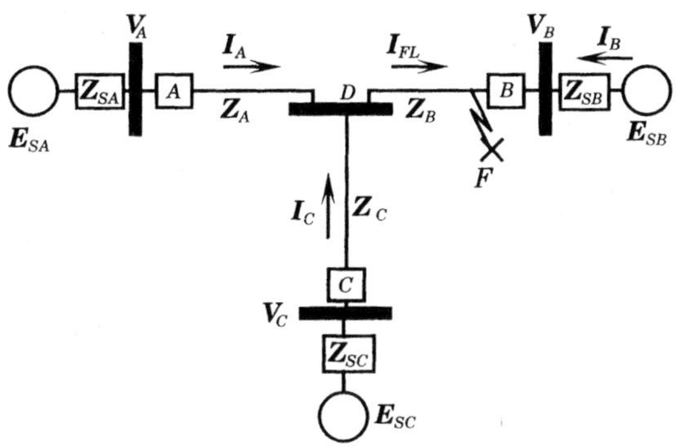

Figure 14.9 A three-terminal transmission line.

To pursue this problem further it is helpful to introduce the concept of the *reach* of a distance relay.

> The *reach* of a relay is the extent of the protection afforded by a relay in terms of the impedance or circuit length as measured from the relay location [18].[1]

Clearly, if the original line is tapped at the center, and the tap CD is exactly the same length as AD and BD, the relaying problem is not difficult. For example, a distance relay, in trying to clear a fault at F that is just inside the zone 1 reach at A, can also be cleared by zone 1 relays

[1] The measurement is usually to the point of fault, but excessive loading or system swings may also come within the reach or operating range of the relay [27].

at C, assuming that the source impedances are also equal. As one departs from this ideal, the coordination of distance relays becomes more difficult and, in some cases, fault clearing will be delayed for zone 2 clearing.

To examine the problem more critically, we first examine the effect on distance relays, and then consider pilot relays and their application in this situation.

14.3.1 Distance Protection for a Three-Terminal Line

To study the problems of distance relays in a three-terminal line, we shall examine the problem presented in Figure 14.9 numerically, to see how the impedance changes for different system and line length conditions. We shall arbitrarily use the following per unit impedances to illustrate the problems in distance protection [14]:

Strong sources	0.1 per unit impedance
Weak sources	1.0 per unit impedance
Short lines	1.0 per unit impedance
Long lines	5.0 per unit impedance

We shall also assume that load currents are negligible, but will comment later on the problems of circuits with large load currents.

For the source voltages equal and in phase, we may construct an equivalent circuit for the system of Figure 14.9, assuming a three-phase fault at point F. This equivalent is shown in Figure 14.10.

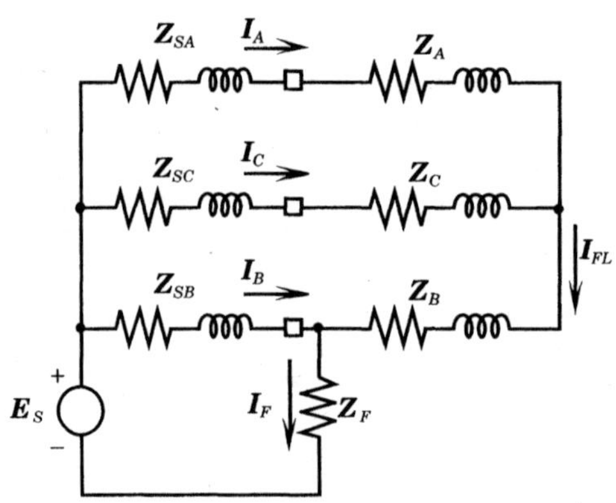

Figure 14.10 Rearrangement of the system of Figure 14.9 for the solution of relay impedances.

The fault impedance is usually small compared to the line impedances and is usually resistive. The fault impedance will be neglected to simplify the calculations. The total fault current is made up of two parts, as follows:

$$\begin{aligned} \boldsymbol{I}_F &= \boldsymbol{I}_B + (\boldsymbol{I}_A + \boldsymbol{I}_C) \\ &= \boldsymbol{I}_B + \boldsymbol{I}_{FL} \end{aligned} \tag{14.22}$$

where we define the fault current flowing toward the fault from the left as I_{FL}, as noted in Figure 14.10. We may now completely solve the network. First we compute the fault current from the left as

$$\boldsymbol{I}_{FL} = \frac{\boldsymbol{E}_S}{\boldsymbol{Z}_B + \dfrac{(\boldsymbol{Z}_{SA} + \boldsymbol{Z}_A)(\boldsymbol{Z}_{SC} + \boldsymbol{Z}_C)}{\boldsymbol{Z}_{SA} + \boldsymbol{Z}_A + \boldsymbol{Z}_{SC} + \boldsymbol{Z}_C}} \tag{14.23}$$

Then we compute

$$\boldsymbol{I}_A = \frac{1}{1 + \dfrac{\boldsymbol{Z}_{SA} + \boldsymbol{Z}_A}{\boldsymbol{Z}_{SC} + \boldsymbol{Z}_C}} \boldsymbol{I}_{FL} \tag{14.24}$$

$$\boldsymbol{I}_C = \frac{1}{1 + \dfrac{\boldsymbol{Z}_{SC} + \boldsymbol{Z}_C}{\boldsymbol{Z}_{SA} + \boldsymbol{Z}_A}} \boldsymbol{I}_{FL} \tag{14.25}$$

and

$$\boldsymbol{I}_B = \left(\frac{(\boldsymbol{Z}_A + \boldsymbol{Z}_{SA})(\boldsymbol{Z}_C + \boldsymbol{Z}_{SC})}{\boldsymbol{Z}_A + \boldsymbol{Z}_{SA} + \boldsymbol{Z}_C + \boldsymbol{Z}_{SC}} + \boldsymbol{Z}_B \right) \frac{\boldsymbol{I}_{FL}}{\boldsymbol{Z}_{SB}} \tag{14.26}$$

The relay voltages are computed as

$$\begin{aligned} \boldsymbol{V}_A &= \boldsymbol{E}_S - \boldsymbol{Z}_{SA}\boldsymbol{I}_A \\ \boldsymbol{V}_B &= \boldsymbol{E}_S - \boldsymbol{Z}_{SB}\boldsymbol{I}_B \\ \boldsymbol{V}_C &= \boldsymbol{E}_S - \boldsymbol{Z}_{SC}\boldsymbol{I}_C \end{aligned} \tag{14.27}$$

Finally, the impedances seen by the relays are computed as follows.

$$\boldsymbol{Z}_{Rk} = \frac{\boldsymbol{V}_k}{\boldsymbol{I}_k} \qquad k = A, B, C \tag{14.28}$$

In writing the equations, we note several striking facts:

1. The example is perfectly general for the three-phase fault and solves the case for an end-of-line fault on any leg of the three-terminal line by simply relabeling the solution quantities.
2. The currents are dependent on the ratio of the total impedances in lines A and C, which includes both the line and source impedances. Moreover, these quantities always appear as their sum, hence varying either line or source impedance will have exactly the same effect.
3. From 2, we learn that the impedance seen by the relays at A and C also depend on the ratio of the two line and source impedances.

EXAMPLE 14.1

Determine the effect on distance relays of varying the length of the tap line C-D. To do this, solve the system of Figure 14.10 for a range of values for the impedance of Z_C while holding all other quantities fixed with the following values:

$$\begin{aligned} \boldsymbol{Z}_A &= \boldsymbol{Z}_B = 1.0 \text{ per unit} \\ \boldsymbol{E}_S &= 1.0 \text{ per unit} \end{aligned}$$

and use two different values for the system impedances, one very strong and one very weak, as follows:

Strong system:

$$\boldsymbol{Z}_{SA} = \boldsymbol{Z}_{SB} = \boldsymbol{Z}_{SC} = 0.1 \text{ per unit}$$

Weak system:

$$Z_{SA} = Z_{SB} = Z_{SC} = 1.0 \text{ per unit}$$

Solution

The numerical work in obtaining the solution is straightforward and is easily obtained using a computer spreadsheet. The results are plotted as follows. First, the currents are solved for the two cases, with strong and weak system ties, with the result shown in Figure 14.11. The effect of the system strength is readily perceived by comparing the top plot against the bottom plot for each current. Clearly, the currents vary much faster with the length of the tap when the system is strong. As the system becomes weaker, the rate of current variation is reduced. It is also clear that, in either case, the change in current magnitude is nonlinear as tap length increases linearly.

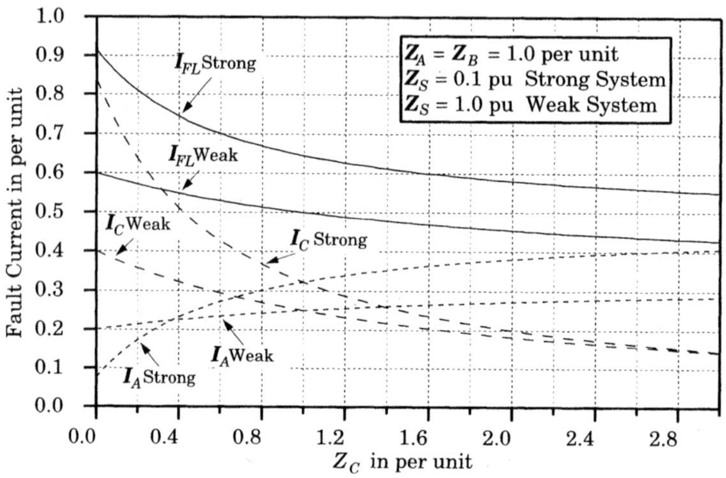

Figure 14.11 Line and total fault currents with varying Z_C.

The impedance to the fault seen by distance relays at A and C are shown in Figure 14.12. The impedance seen by the relay at C varies *linearly* with the tap line length, even though the current is quite nonlinear in its variation. This is because the relay voltage also varies in a nonlinear manner.

The apparent impedance seen by the relay at A, however, varies in a nonlinear manner. For the case of a strong system, this impedance changes very fast as the tap length increases. This is because the infeed from the tap is changing rapidly, making the impedance seen by the relay at A to be much greater than the actual 2.0 per unit impedance in the line AB. When the sources are weak, the source impedance tends to lessen this sensitivity and make the impedance at A more nearly equal to just the line impedance from A to B.

For line C having an impedance below 1.0 per unit the infeed of fault current from line C makes the apparent impedance from relay A to the fault much greater than the actual line impedance. Should relay A zone 1 be set at 80% of the actual line impedance from A to B or 1.8 per unit, for example, the impedance seen by this relay would be very large because of line C infeed.

This is about the setting that would be required, however, if line C should be removed from service for any reason. The problem is the most difficult for low source impedances, which makes the fault appear to be very far away from relay A. Note that there is not such a difficult problem setting the reach of relay C.

If the impedance of line C is more than 1.0 per unit, the problems just stated are approximately reversed. For this condition, the line C relays measure too large an impedance and line A relays could probably be set with reasonable reach. These observations are made only for an end-of-line fault. Other fault locations and fault types also need to be checked.

We conclude that it is very difficult to set distance relays on a tapped line without resorting to delayed clearing for at least some faults.

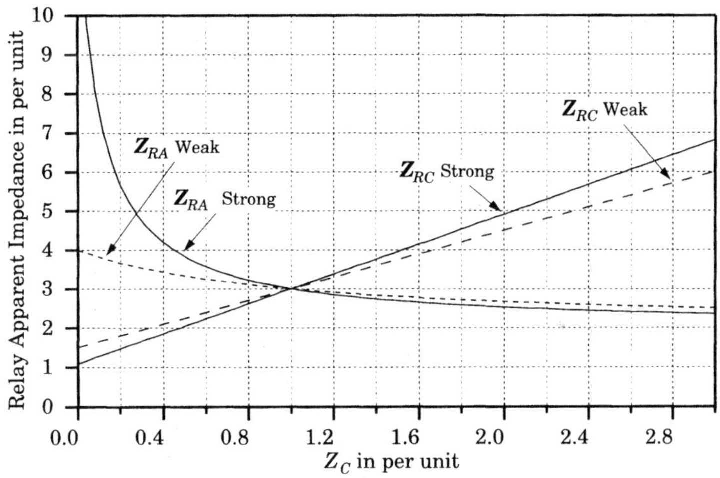

Figure 14.12 Impedances seen by distance relays at A and C as the length of tap CD is increased. ■

The solution in the foregoing example is not a general solution, but rather represents a specific case. It does illustrate the fact that the apparent impedances seen by line distance relays are very much dependent on the impedances of sources and lines in the three-terminal case. For a specific case, the actual system conditions must be solved. Moreover, when any segment of the three-terminal line is taken out of service, the fault current and apparent impedances change dramatically.

14.3.2 Pilot Protection for a Three-Terminal Line

Pilot protection for a three-terminal line may utilize communications of any type, such as wire, microwave, PLC, or fiber optics. Some types of wire pilot schemes are applicable to three-terminal lines, but this is only possible on very short lines. We therefore concentrate on the other types of pilot systems that utilize PLC, microwave, or fiber optics and that might be used on longer transmission circuits.

Pilot systems can be either blocking or tripping schemes, as noted in Chapter 13. Both schemes have certain problems, when applied to three-terminal lines. The nature of these problems will be explored in this section.

14.3.2.1 Blocking Pilot Schemes. Three types of blocking schemes are available: directional comparison, phase comparison, and directional comparison unblocking [14].

DIRECTIONAL COMPARISON. If directional comparison is used, each relay must send a blocking signal when it detects a fault that is "behind" its own location, that is, the relay detects a fault that is outside the zone of protection and it then sends a blocking signal to prevent the other relays on the three-terminal line from tripping. At the same time, all relays must be able to see faults that are within the protection zone so that they can trip if no blocking signal is received. The fault detection is often performed using some sort of distance measurement at all locations. Because of infeed to the fault, as noted before, this means that the relays may have to be set to reach very far. In some cases, it is not possible to achieve good settings for all relays using directional comparison.

PHASE COMPARISON. Phase comparison relays send pulses to the other terminals each half cycle of measured current. The received pulses are compared at each relay to see if they are in phase with locally measured currents or not. The system is designed such that, if the pulses are in phase at a terminal, that terminal will be tripped. If the fault is external to any terminal, the pulses will be out of phase and tripping will be blocked. There are problems with three-terminal lines in setting the detector levels for phase comparison, and sensitivity may be difficult to achieve.

UNBLOCKING. This scheme sends blocking signals continuously from all terminals. If one relay detects a fault in the proper direction, it changes the transmitted signal from blocking to tripping. To trip, a relay must detect a fault and receive either a trip signal or no blocking signal. If distance fault detectors are used, the reach may need to be set to very high values, as noted previously.

14.3.2.2 Transfer Trip Pilot Schemes. There are three basic types of tripping pilot schemes that may be used for three-terminal lines: direct underreaching, permissive underreaching, and permissive overreaching schemes [14].

DIRECT UNDERREACHING. In this scheme the zone 1 relays are set to underreach the remote terminals. If a fault is detected within this reach the terminal is tripped and a trip signal is sent to the other terminals. The problem is to set the zone 1 reach properly for all system conditions, including the tapped portion of the line being temporarily out of service. As noted previously, this may not be possible.

PERMISSIVE UNDERREACHING. In this scheme, two zones are used. Zone 1 is set to underreach the remote terminals, trip the local breaker, and key the transfer trip transmitter. The zone 2 overreaches the remote terminals; however, tripping is permitted only when a trip signal is received from a remote terminal. If one of the terminals is open, the permissive signal is keyed continuously to ensure fast clearing of any internal line faults.

PERMISSIVE OVERREACHING. Here, the transmitter is keyed by a distance relay that is set to overreach the remote terminals, where tripping is permitted only if an internal fault is also detected. With this scheme, the pilot signal is very important, which often eliminates PLC from consideration due to the uncertainty of proper signaling during a fault.

14.3.2.3 Summary of Pilot Relaying Schemes. There are cases where pilot relaying can be arranged for fast clearing of three terminal lines. Some of these schemes use distance devices to measure the proper direction to a fault condition. There are problems with distance measurement that may make it difficult to cover the required reach with distance measuring equipment, although the directional aspect may be acceptable. Clearly, pilot schemes offer a better option than distance relays without the benefit of pilot information. It is not clear that any one scheme is superior to all others, however.

The computation of relay apparent impedances in Section 14.2.2 indicates that these impedances, viewed from one of the terminals, can change dramatically, depending on the impedances of the different line sections. These impedance variations are also important for power line carrier, and this affects the selection of PLC frequencies [19], [20]. Successful PLC performance depends on having an adequate signal-to-noise ratio at the receivers, and this is dependent on the path attenuation. Lines that are short at power frequencies may be many wavelengths long at PLC frequencies, and the PLC frequency line losses may be high. This may make PLC inferior to other types of pilot protection for tapped lines.

14.4 PROTECTION OF MUTUALLY COUPLED LINES

The discussion of transmission protection up to this point has been concerned primarily with the different methods that are used for the protection of a single, two-terminal line. We now consider the protection of mutually coupled transmission lines [17]. These are lines that share the same right of way or have multicircuit towers for all or a portion of the line length. Such lines have significant mutual coupling, especially for ground faults, which complicates the analysis. Figure 14.13 shows three different classes of line configurations, each of which includes lines with mutual coupling.

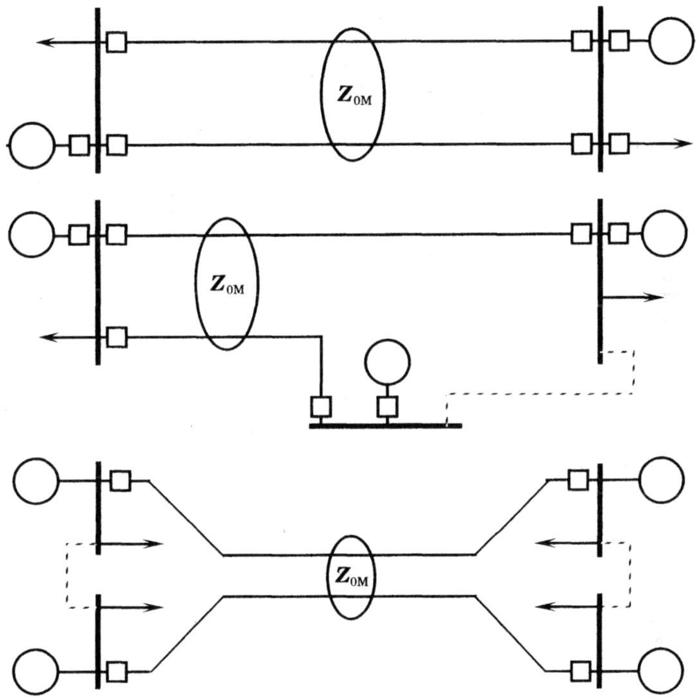

Figure 14.13 Transmission with line segments with mutual coupling Z_{0M}.

The situations shown in Figure 14.13 expose all circuits to conditions where faults on one line affect the mutually-coupled line. Statistics of faults involving both circuits vary considerably, but are usually in the range of 10 to 20 percent [17]. Faults between coupled circuits of different voltage are rare, but can cause considerable damage to the lower voltage circuit. These coupled systems present a challenge for the protection engineer.

The first case shown in Figure 14.13 is the simplest of the three since both lines are connected to the same terminals and share the same source capabilities at both ends of the lines. In this case it is possible to apply compensation methods to counteract the mutual coupling.

14.4.1 Mutual Coupling of Parallel Lines

The computation of mutual coupling of parallel lines is described in Chapter 12. Here, we apply the basic ideas of Chapter 12 to practical problems in transmission systems. The following classifications are of interest [17]:

1. Type 1 networks — Parallel circuits with common positive- and zero-sequence sources.
2. Type 2 networks — Parallel circuits with common positive but isolated zero-sequence sources.
3. Type 3 networks — Parallel circuits with isolated positive- and zero-sequence sources.

Each of the above classes will be illustrated further in order to clarify both the classification and the treatment of the different classes of networks.

The transmission lines that distinguish the three classes of interest cannot be considered in isolation. These lines are part of an interconnected system, in most cases, and the characteristics of that system play an important role in the zero-sequence voltages and currents that result from faults along the mutually coupled lines. However, it can be shown that the worst case, in terms of relay measurement error, occurs when the system Thevenin equivalent at the bus remote from the relay is weak. This may be due to the loss of other lines near the remote bus or to limited network development, either of which are represented by a high source impedance in the Thevenin equivalent. Clearly, then, the worst case to consider is that represented when the remote equivalent is removed. This is fortunate in the sense that it simplifies the analysis, which can be performed simply with only the one source present. We shall take advantage of this fact in the analysis that follows.

14.4.2 Ground Distance Protection of Type 1 Networks

Type 1 networks are defined as networks with common positive- and zero-sequence sources as shown in Figure 14.14. In this case both lines terminate at a common bus at both ends of the lines and there are common sources of ground current for both lines.

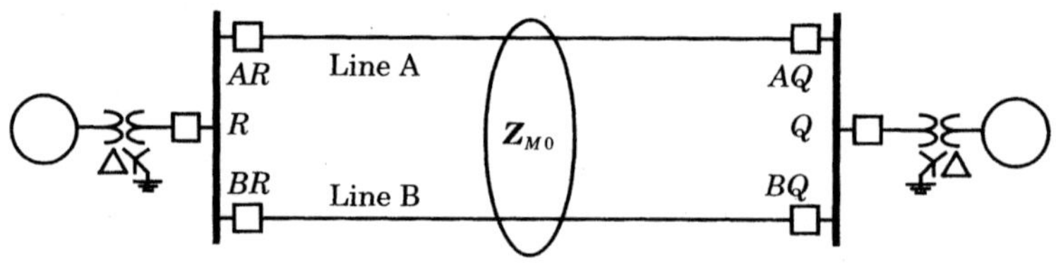

Figure 14.14 A Type 1 parallel circuit with common sources for both positive- and zero-sequence networks [17].

First, consider ground distance protection of the Type 1 circuit of Figure 14.14. The distance protection of phase faults is not affected by the mutual coupling, but the protection of ground faults suffers from a measurement error [21–30].[2] The measurement error occurs because the ground current, $3I_{a0}$, in the faulted line induces a voltage in the unfaulted line.

For this condition, a ground distance relay for a single line A with residual compensation will measure the impedance computed in (11.105), which we may write in the form

$$hZ_{L1} = \frac{V_{AR}}{I_{AR} + C_0 I_{a0}\left(\dfrac{Z_{A0}}{Z_{A1}} - 1\right)} = \frac{V_{AR}}{I_{AR} + k_{CA} I_{GA0}} \tag{14.29}$$

[2]Many of the references on this subject are to papers presented at protective relay conferences that are not indexed for retrieval. Some of those cited are typical examples. These papers provide excellent discussions and are recommended reading.

where

$$k_{CA} = \frac{Z_{A0} - Z_{A1}}{3Z_{A1}} \tag{14.30}$$

and

$$\begin{aligned} I_{AR} &= C_0 I_{a0} + C_1 I_{a1} + C_2 I_{a2} \\ I_{GA0} &= 3C_0 I_{a0} \end{aligned} \tag{14.31}$$

The current I_{GA0} is the residual current in line A and k_{CA} is a complex constant. These quantities were defined in Chapter 11 for a single line. For the mutually coupled case, we will examine other definitions of the constant k_{CA} that are calculated to provide a more effective compensation.

Now consider the zero sequence network for mutually coupled lines as shown in Figure 14.15, where the parallel lines are designated A and B. The lines need not be identical in their self-impedances, but will have a common mutual impedance, as shown in the figure. The relay at AR is assumed to be the relay of interest.

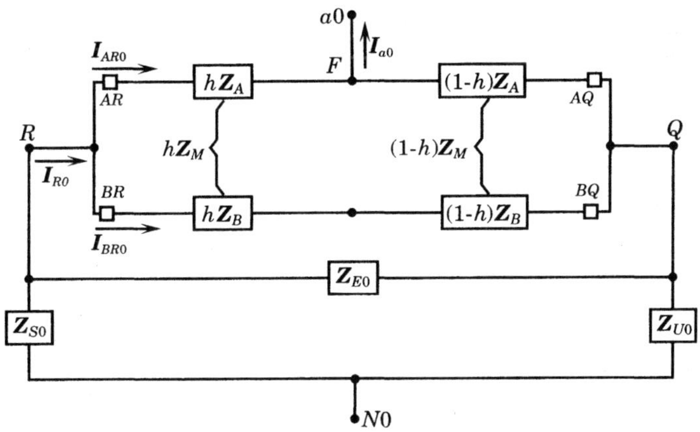

Figure 14.15 Zero sequence network of the coupled lines.

Consider the case of the one-line-to-ground fault with zero fault impedance, where we can write

$$V_a = V_{a0} + V_{a1} + V_{a2} = 0 \tag{14.32}$$

But, we can also write

$$\begin{aligned} V_{a0} &= V_{R0} - hZ_{A0} I_{AR0} - hZ_{M0} I_{BR0} \\ V_{a1} &= V_{R1} - hZ_{A1} I_{AR1} \\ V_{a2} &= V_{R2} - hZ_{A2} I_{AR2} \end{aligned} \tag{14.33}$$

where V_{ai} and V_{Ri} $(i = 0, 1, 2)$ are measured at different points in the three sequence networks.

Substituting (14.33) into (14.32) we get

$$V_R = V_{R0} + V_{R1} + V_{R2} = hZ_{A0} I_{AR0} + hZ_{M0} I_{BR0} + hZ_{A1} I_{AR1} + hZ_{A2} I_{AR2} \tag{14.34}$$

Now we define the fraction of sequence current flowing at the relay locations by the complex constants

$$
\begin{aligned}
\boldsymbol{C}_{A0} &= \frac{\boldsymbol{I}_{AR0}}{\boldsymbol{I}_{a0}} \qquad & \boldsymbol{C}_{A1} &= \frac{\boldsymbol{I}_{AR1}}{\boldsymbol{I}_{a1}} \\
\boldsymbol{C}_{B0} &= \frac{\boldsymbol{I}_{BR0}}{\boldsymbol{I}_{a0}} \qquad & \boldsymbol{C}_{A2} &= \frac{\boldsymbol{I}_{AR2}}{\boldsymbol{I}_{a2}}
\end{aligned}
\tag{14.35}
$$

Using these constants, (14.34) can be written as

$$
\begin{aligned}
\boldsymbol{V}_R = {} & h\boldsymbol{Z}_{A1}(\boldsymbol{C}_{A0}\boldsymbol{I}_{A0} + \boldsymbol{C}_{A1}\boldsymbol{I}_{A1} + \boldsymbol{C}_{A2}\boldsymbol{I}_{A2}) \\
& + h(\boldsymbol{Z}_{A0} - \boldsymbol{Z}_{A1})\boldsymbol{C}_{A0}\boldsymbol{I}_{a0} + h\boldsymbol{Z}_{M0}\boldsymbol{C}_{B0}\boldsymbol{I}_{B0}
\end{aligned}
\tag{14.36}
$$

We recognize that the current in the first line of (14.36) is the total current flowing through relay AR, so we modify the equation to reflect this fact.

$$
\boldsymbol{V}_{AR} = h\boldsymbol{Z}_{A1}\boldsymbol{I}_{AR} + h(\boldsymbol{Z}_{A0} - \boldsymbol{Z}_{A1})\boldsymbol{C}_{A0}\boldsymbol{I}_{a0} + h\boldsymbol{Z}_{M0}\boldsymbol{C}_{B0}\boldsymbol{I}_{a0} \tag{14.37}
$$

Rearranging (14.37), we may write

$$
\boldsymbol{V}_{AR} = h\boldsymbol{Z}_{L1}\left(\boldsymbol{I}_{AR} + \frac{\boldsymbol{Z}_{A0} - \boldsymbol{Z}_{A1}}{3\boldsymbol{Z}_{A1}}\boldsymbol{I}_{GA0} + \frac{\boldsymbol{Z}_{M0}}{3\boldsymbol{Z}_{A1}}\boldsymbol{I}_{GB0}\right) \tag{14.38}
$$

where we have defined the residual currents

$$
\begin{aligned}
\boldsymbol{I}_{GA0} &= 3\boldsymbol{C}_{A0}\boldsymbol{I}_{a0} = \text{residual current in line } A \text{ at relay } AR \\
\boldsymbol{I}_{GB0} &= 3\boldsymbol{C}_{B0}\boldsymbol{I}_{a0} = \text{residual current in line } B \text{ at relay } BR
\end{aligned}
\tag{14.39}
$$

Then, from (11.112) we divide (14.38) by the compensated current to compute the apparent impedance seen by the relay as

$$
\boldsymbol{Z}_{AR} = h\boldsymbol{Z}_{L1}\left(\frac{\boldsymbol{I}_{AR} + \boldsymbol{k}_{SA}\boldsymbol{I}_{GA0} + \boldsymbol{k}_{MA}\boldsymbol{I}_{GB0}}{\boldsymbol{I}_{AR} + \boldsymbol{k}_{CA}\boldsymbol{I}_{GA0}}\right) = h\boldsymbol{Z}_{L1}\left(\frac{\boldsymbol{I}_{\text{ARM}}}{\boldsymbol{I}_{\text{ARC}}}\right) \tag{14.40}
$$

where we have defined the following system constants for line A

$$
\boldsymbol{k}_{SA} = \frac{\boldsymbol{Z}_{A0} - \boldsymbol{Z}_{A1}}{3\boldsymbol{Z}_{A1}} \qquad \boldsymbol{k}_{MA} = \frac{\boldsymbol{Z}_{M0}}{3\boldsymbol{Z}_{A1}} \tag{14.41}
$$

Similar constants can be defined for line B by replacing all A subscripts by the letter B. This will result in different constants if the lines have different physical parameters. We have also defined the current ratio

$$
\text{Measurement index} = 1 - \text{measurement error} = \frac{\boldsymbol{I}_{\text{ARM}}}{\boldsymbol{I}_{\text{ARC}}} \equiv \boldsymbol{k}_e \tag{14.42}
$$

For accurate distance measurement, the measurement index $\boldsymbol{k}_e$ must be as close to unity as possible. Judicious adjustment of the parameter $\boldsymbol{k}_{CA}$ may be used to minimize the measurement error.

The parameter $\boldsymbol{k}_{SA}$ is the compensation constant for a single line. It should be noted that the way $\boldsymbol{k}_{SA}$ and $\boldsymbol{k}_{CA}$ are defined, based on (14.29) and (14.40), the two constants are equal in (14.40). This need not always be the case, as we may change $\boldsymbol{k}_{CA}$ for the case of mutually coupled lines rather than leave it at the value determined for a single line.

Now, consider the measurement index. If this term can be forced to a value of unity, the distance measurement will be exactly right. However, since the denominator may not equal the numerator, there may be an error in the distance measurement. The error is a function of the constants $\boldsymbol{k}_{SA}$ and $\boldsymbol{k}_{CA}$ as well as the parameters associated with the mutual coupling of the nearby line, namely, $\boldsymbol{k}_{MA}$ and $\boldsymbol{I}_{GB0}$. We can, however, make a few general observations about the error, which can be summarized as follows:

- The error is proportional to the mutual impedance.
- The error is proportional to the residual current in line B.
- The relay underreaches when $\boldsymbol{I}_{GB0}$ is in phase with $\boldsymbol{I}_{AR} + \boldsymbol{I}_{GA0}$.
- The relay overreaches when $\boldsymbol{I}_{GB0}$ is out of phase with $\boldsymbol{I}_{AR} + \boldsymbol{I}_{GA0}$.

It has been shown that underreaching as much as 25% may occur [17]. Large overreach errors will occur under certain switching conditions, and these will be examined below.

14.4.2.1 Type 1 Distance Zones for Parallel Lines. Consider a special case of two parallel, mutually coupled lines as shown in Figure 14.16 where line A becomes faulted, and where the source at bus Q is very weak. The distance zones will be computed for the relay located at AR considering the response due to only the source at R. This arrangement is chosen for study as it creates the worst distance measurement error. Adding a source at bus Q will provide infeed from the Q end that will reduce the distance measurement error. Only the one-line-to-ground fault is considered here, but the procedure is the same for other ground faults. Three types of system conditions are of interest, defined here as Types 1.1, 1.2, and 1.3.

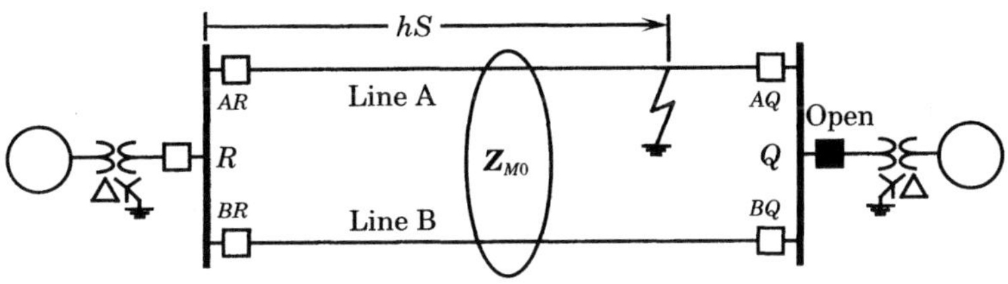

Figure 14.16 Type 1: both lines in service with a fault on line A.

TYPE 1.1 THE NORMAL CONDITION—BOTH LINES IN SERVICE. The case, shown in Figure 14.16, represents the normal system condition for two mutually coupled parallel transmission lines, with a weak system equivalent at the Q end of the coupled lines. It is assumed that both lines are in service and the fault is located on line A at a fractional distance h, of the total line length S, from the relay at AR.

For a one-line-to-ground fault we recognize that the total line current is equal to the total ground or residual current, i.e.,

$$\boldsymbol{I}_{AR} = \boldsymbol{I}_{GA0} \tag{14.43}$$

Moreover, we can show that

$$\boldsymbol{I}_{GB0} = \frac{h}{1 - h + \boldsymbol{k}_{AB}} \boldsymbol{I}_{GA0} \tag{14.44}$$

where we define the complex constant

$$\boldsymbol{k}_{AB} = \frac{\boldsymbol{Z}_{B0} - \boldsymbol{Z}_{M0}}{\boldsymbol{Z}_{A0} - \boldsymbol{Z}_{M0}} \tag{14.45}$$

Substituting (14.43) and (14.44) into (14.40) permits us to express the relay apparent impedance only in terms of the system constants and h, the fault distance parameter. Thus, we compute

$$\boldsymbol{Z}_{AR} = h_1 \boldsymbol{Z}_{A1} \left[\frac{1 + \boldsymbol{k}_{SA} + h\boldsymbol{k}_{MA}/(1 - h + \boldsymbol{k}_{AB})}{1 + \boldsymbol{k}_{CA}} \right] = h_1 \boldsymbol{k}_{ei} \boldsymbol{Z}_{A1} \tag{14.46}$$

The result is an expression that is not an explicit function of the currents. The measurement index is the term in brackets and is given as a function of only the system constants and the fault location. It must be remembered that this is a special case, with no source at the remote end of the line. The impedance expression given by (14.40) is the more general expression, which applies to any situation.

TYPE 1.2 LINE B SWITCHED OFF AND OPEN AT BOTH ENDS. In this case line B is switched out of service, but either is not grounded at all, or is grounded at only one point, such that the current in line B is zero. This condition is pictured in Figure 14.17, where the black squares represent open circuit breakers.

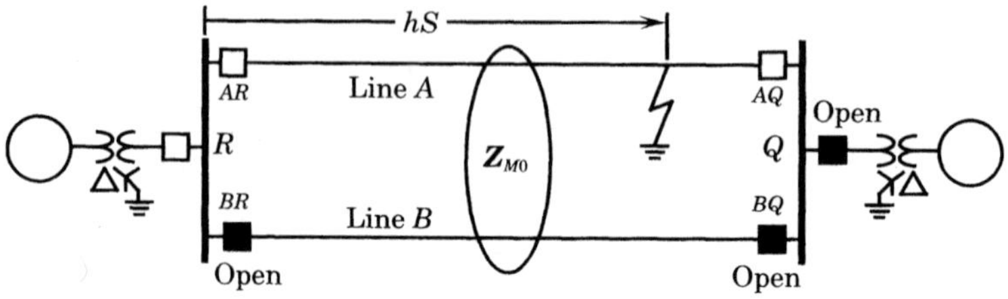

Figure 14.17 Case 2: parallel line B switched off, but not grounded.

For this condition, the current relation (14.43) holds but line current in line B is zero, giving the apparent impedance seen by the relay AR as

$$\mathbf{Z}_{AR} = h_2\mathbf{Z}_{A1}\left[\frac{1+\mathbf{k}_{SA}}{1+\mathbf{k}_{CA}}\right] = h_2\mathbf{k}_{ei}\mathbf{Z}_{A1} \tag{14.47}$$

The measurement index is again the quantity in brackets. The index is smaller than that for the case with both lines in service, given by (14.46). This means that the measured impedance for any fault location will be reduced when the parallel line is out of service, compared to the normal case with both lines in service. The smaller measured impedance results in greater reach, which could lead to overreaching the end of the line for large values of h_2.

TYPE 1.3 PARALLEL LINE B SWITCHED OFF AND GROUNDED AT BOTH ENDS. This case is illustrated in Figure 14.18, where we again note the open circuit breakers in line B, but with the new condition of line grounds attached at both ends of the line. The fault is located a fractional distance h from the relay at AR, as before.

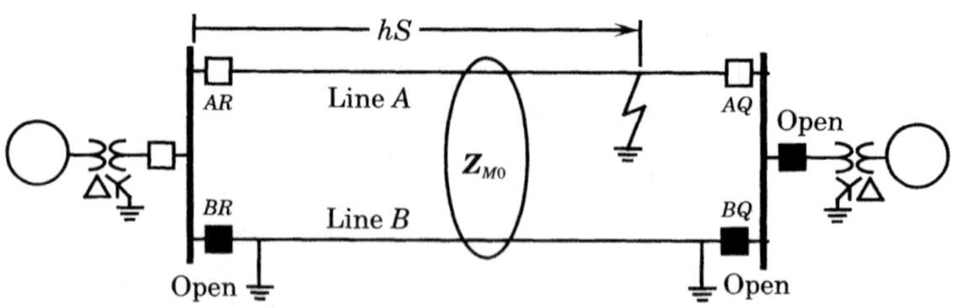

Figure 14.18 Case 3: line B switched off, and grounded at both ends.

A schematic diagram of the zero-sequence network is shown in Figure 14.19, where the mutual induction is represented by one-to-one ideal transformers. This network may be solved for the node voltages as part of the total solution of the sequence networks for any type of fault.

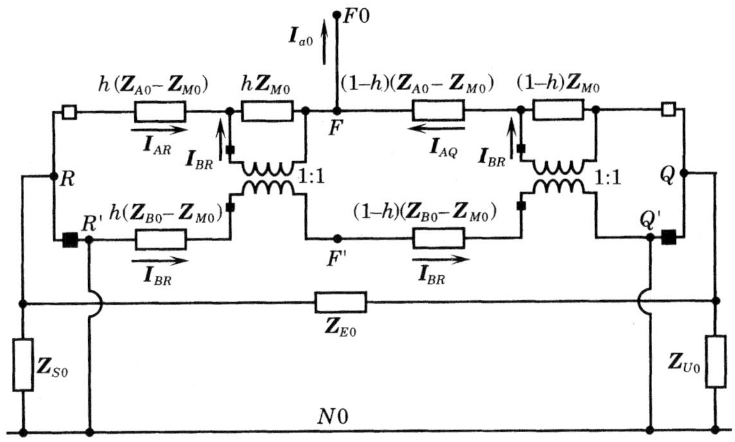

Figure 14.19 Schematic diagram of the zero-sequence network.

Assuming that the node voltages are known, the voltage differences between the terminal nodes can be written as follows:

$$\begin{bmatrix} \boldsymbol{V}_{RF} \\ \boldsymbol{V}_{QF} \\ \boldsymbol{V}_{R'Q'} \end{bmatrix} = \begin{bmatrix} h\boldsymbol{Z}_{A0} & 0 & h\boldsymbol{Z}_{M0} \\ 0 & (1-h)\boldsymbol{Z}_{A0} & -(1-h)\boldsymbol{Z}_{M0} \\ h\boldsymbol{Z}_{M0} & -(1-h)\boldsymbol{Z}_{M0} & \boldsymbol{Z}_{B0} \end{bmatrix} \begin{bmatrix} \boldsymbol{I}_{AR} \\ \boldsymbol{I}_{AQ} \\ \boldsymbol{I}_{BR} \end{bmatrix} \tag{14.48}$$

where we have defined the voltages

$$\begin{bmatrix} \boldsymbol{V}_{RF} \\ \boldsymbol{V}_{QF} \\ \boldsymbol{V}_{R'Q'} \end{bmatrix} \equiv \begin{bmatrix} \boldsymbol{V}_R - \boldsymbol{V}_F \\ \boldsymbol{V}_Q - \boldsymbol{V}_F \\ \boldsymbol{V}_{R'} - \boldsymbol{V}_{Q'} \end{bmatrix} = \begin{bmatrix} \boldsymbol{V}_R - \boldsymbol{V}_F \\ \boldsymbol{V}_Q - \boldsymbol{V}_F \\ 0 \end{bmatrix} \tag{14.49}$$

Since the third equation of (14.48) is equal to zero, we can solve this equation to find a relationship between the current in the mutually coupled line and the faulted line. This gives the result

$$\boldsymbol{I}_{BR} = \frac{1}{\boldsymbol{Z}_{B0}}[-h\boldsymbol{Z}_{M0}\boldsymbol{I}_{AR} + (1-h)\boldsymbol{Z}_{M0}\boldsymbol{I}_{AQ}] \tag{14.50}$$

From (14.50), we note that the current in line B flows in the negative direction ($\boldsymbol{I}_{BR} < 0$) to that defined in Figure 14.19 when

$$h\boldsymbol{I}_{AR} > (1-h)\boldsymbol{I}_{AQ} \tag{14.51}$$

For the special radial case where the source at Q and the system equivalent are both very weak, then (14.51) can be simplified to write

$$\boldsymbol{I}_{BR} = \frac{-h\boldsymbol{Z}_{M0}\boldsymbol{I}_{AR}}{\boldsymbol{Z}_{B0}} \tag{14.52}$$

Now, for a one-line-to-ground fault, we have

$$\begin{aligned} \boldsymbol{I}_{AR} &= \boldsymbol{I}_{GA0} \\ \boldsymbol{I}_{BR} &= \boldsymbol{I}_{GB0} = \frac{-h\boldsymbol{Z}_{M0}\boldsymbol{I}_{GA0}}{\boldsymbol{Z}_{B0}} \end{aligned} \tag{14.53}$$

Then, for the special radial case of interest here, we can write the relay apparent impedance from (14.40) as

$$\begin{aligned} Z_{AR} &= h_3 Z_{A1}\left[\frac{I_{GA0} + k_{SA} I_{GA0} + k_{MA} I_{GB0}}{I_{GA0} + k_{CA} I_{GA0}}\right] \\ &= h_3 Z_{A1}\left[\frac{1 + k_{SA} - k_{MA}(h_3 Z_{M0}/Z_{B0})}{1 + k_{CA}}\right] = h_3 k_{e3} Z_{A1} \end{aligned} \tag{14.54}$$

where the measurement index k_{ei} is the quantity in brackets.

The three radial types are compared in Table 14.1. Comparing the three cases, we see that Type 1.3, with the parallel line out of service and grounded at both ends, gives the lowest impedance measurement. This results in the greatest reach. Type 1.1 gives the highest impedance measurement, which results in the shortest reach. Therefore, to provide compensation that corrects the worst case, we should set k_{CA} to give the correct reach using Type 1.3, since the reach is always smaller for Types 1.1 and 1.2. Should we set the desired reach using Type 1.1 or 1.2, then Type 1.3 would be in danger of overreaching the end of the line. It should be emphasized that the results given in Table 14.1 are for the radial system and represent the worst case. Where current infeed is present at the Q end of the faulted line, the reach variations are not as extreme as this case.

TABLE 14.1 Radial System Type 1 Measurement Indices

Type	Symbol	Measurement Index
1.1	k_{e1}	$\frac{1}{1+k_{CA}}\left(1 + k_{SA} + \frac{hk_{MA}}{1 - h + k_{AB}}\right)$
1.2	k_{e2}	$\frac{1}{1+k_{CA}}(1 + k_{SA})$
1.3	k_{e3}	$\frac{1}{1+k_{CA}}\left(1 + k_{SA} - \frac{hk_{MA} Z_{M0}}{Z_{B0}}\right)$

14.4.2.2 Reach of the Relay at AR. All protective relays measure prescribed system variables and compare those measurements against threshold settings specified by the protection engineer. For distance relays, the system variable of interest is the apparent impedance seen by the relay at the point of measurement. The threshold relay setting is an ohmic distance in the downstream direction from the relay location. For measured impedances less than this threshold setting, the relay should pick up.

Distance relays often have two or more zones of protection that can have different reach settings. Some reach settings are called *underreaching* settings where the reach threshold is set to an ohmic value that is less then the total positive-sequence impedance of the protected line. Reach settings can also be *overreaching* settings, which are commonly used for permissive overreaching pilot relay systems. Overreaching settings are also used for backup settings for distance relays on adjacent downstream lines.

Let h_S be the reach setting of the relay AR. Then the ohmic reach setting of the relay can be written as

$$Z_{ARS} = h_S Z_{A1} \tag{14.55}$$

where $h_S < 1$ underreaching setting
$h_S > 1$ overreaching setting

Note that the reach setting (14.55) is based on the actual ohmic distance to the reach threshold, and is not a function of the system configuration.

Consider the reach of the distance measuring equipment for a fault on line A taking into account the system configuration, particularly the effect of mutually coupled line B. We can examine this by writing a general expression for the three system configurations described above. The impedance seen by relay AR can be written as follows:

$$\mathbf{Z}_{ARi} = h_i \mathbf{k}_{ei} \mathbf{Z}_{A1} \qquad i = 1, 2, 3 \tag{14.56}$$

where the index i refers to the three cases. Examining (14.46), (14.47), and (14.54), or Table 14.1, we can see that the term in brackets always represents the measurement index, defined as $\mathbf{k}_{ei}$ in (14.56), the magnitude of which can be compared for the three cases. It is apparent from the defining equations for each case that

$$\mathbf{k}_{e1} > \mathbf{k}_{e2} > \mathbf{k}_{e3} \tag{14.57}$$

We note that these error constants are complex numbers in general, but may be approximated as real numbers in high-voltage transmission lines due to the high X/R ratio. Also note that the reach for Types 1.1 and 1.3 is a function of the fault location parameter, h, but is constant for Type 1.2. Actually, Type 1.2 is of little interest since it falls between the two extremes, represented by Types 1.1 and 1.3.

Now consider a fault at a fixed distance from the relay. Combining (14.56) and (14.57) we may write

$$\mathbf{Z}_{AR1} > \mathbf{Z}_{AR2} > \mathbf{Z}_{AR3} \tag{14.58}$$

This means that the impedance measured by the relay is the smallest for Type 1.3 and is the greatest for Type 1.1.

The reach threshold of the relay refers to the distance or impedance measurement when a fault occurs at a reach corresponding to the reach setting. For this condition we equate (14.55) and (14.56) to write

$$\mathbf{Z}_{ARS} = \mathbf{Z}_{ARi} \tag{14.59}$$

or

$$h_i = \frac{h_S}{\mathbf{k}_{ei}} \qquad i = 1, 2, 3 \tag{14.60}$$

Using (14.60), we may write the following inequality regarding the reach for the three system configurations for a fault at the threshold.

$$h_1 < h_2 < h_3 \tag{14.61}$$

For any given reach setting, Type 1.3 measures the smallest impedance and has greatest reach, and Type 1.1 measures the highest impedance and has the smallest reach. Knowing these two extremes permits us to arrive at some reasonable compromises regarding the relay settings, such that the relay will perform correctly when viewing a fault under any of the three conditions.

EXAMPLE 14.2

Compute the apparent impedance seen by the relay for the three cases described above. The system impedances, shown in Figures 14.14 and 14.15 are given as follows.

$$\begin{aligned}
\mathbf{Z}_{A1} &= 0 + j0.303 & \mathbf{Z}_{A0} &= 0 + j0.880 \\
\mathbf{Z}_{B1} &= 0 + j0.303 & \mathbf{Z}_{B0} &= 0 + j0.880 \\
\mathbf{Z}_{S1} &= 0 + j0.100 & \mathbf{Z}_{M0} &= 0 + j0.523 \\
& & \mathbf{Z}_{S0} &= 0 + j0.100
\end{aligned}$$

Compute the relay apparent impedance for the three Type 1 conditions at fractional distances beginning at 0.10 in increments of 0.10. Assume that the relay compensation factor $\boldsymbol{k}_{CA}$ is computed as equal to $\boldsymbol{k}_{SA}$ given by (14.41), that is, the compensation is equal to that used for a single transmission line, which is discussed in Chapter 11.

Solution

The solution is plotted in Figure 14.20. The horizontal axis is the true fractional distance from the relay location AR to the fault. The solid black bar is the true impedance to the fault. The next three bars correspond to the apparent impedance measured by the relay for Types 1.1, 1.2, and 1.3, respectively, with each measurement compensated using $\boldsymbol{k}_{CA}$ equal to $\boldsymbol{k}_{SA}$, as given in the problem statement. For Type 1.1, the compensated relay apparent impedance is shown by the bar with /// hatching. This impedance is always greater than the actual impedance to the fault point and the percent error increases as the distance to the fault increases. This is due to the increasing length of line subject to mutual induction, since there is no fault current contribution from the remote end of the line. At $h = 1.0$, the Type 1.1 error is about +35%, which causes a significant underreach of the relay.

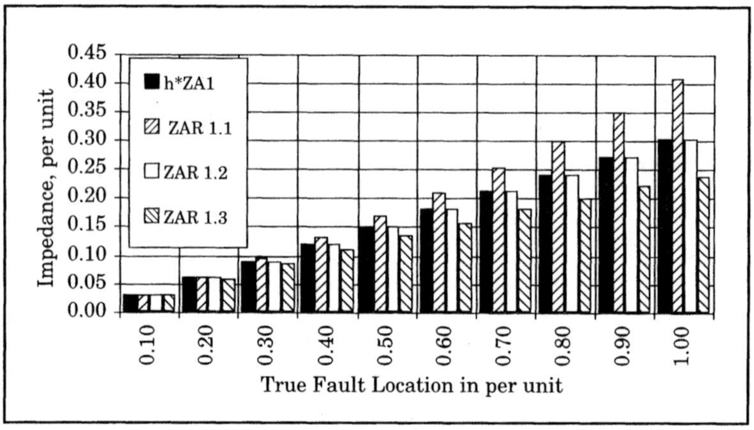

Figure 14.20 Line and relay apparent impedances versus h.

For the transmission line parameters specified, we may compute the system parameters, as follows.

$$\boldsymbol{k}_{SA} = \frac{\boldsymbol{Z}_{A0} - \boldsymbol{Z}_{A1}}{3\boldsymbol{Z}_{A1}} = 0.635 \qquad \boldsymbol{k}_{MA} = \frac{\boldsymbol{Z}_{M0}}{3\boldsymbol{Z}_{A1}} = 0.575$$

The relay apparent impedance for Type 1.2 is exactly right. This makes sense since line B is open for Type 1.2, which makes this case the same as a single line with no mutual coupling and the compensation is calculated for exactly that condition. The relay apparent impedance for Type 1.3 is smaller than the other cases, and the percent error increases with h to a value of −20.9% for a fault at the remote end of the line. An error of this magnitude may cause the relay to reach beyond the remote line terminal, a condition that cannot be tolerated.

For example, for Type 1.3 we may compute the measurement index, using the formula from Table 14.1. For a fault at $h = 0.8$, we have

$$\boldsymbol{k}_{e3} = 0.833$$

If the reach setting $h_S = 0.8$, then from (14.55)

$$h_3 = \frac{h_S}{\boldsymbol{k}_{e3}} = \frac{0.8}{0.833} = 0.96$$

and the relay reaches dangerously close the end of the line for a fault located at the reach threshold, given the Type 1.3 circuit arrangement. Obviously, a smaller reach setting should be considered, which will reduce the fraction of the line that will have zone 1 protection.

For the Type 1.1 system condition, with $h_S = 0.8$, we may compute

$$\boldsymbol{k}_{e1} = 1.235$$

Then, from (14.53)

$$h_1 = \frac{h_S}{\boldsymbol{k}_{e1}} = \frac{0.8}{1.235} = 0.65$$

giving only a 35% overlap for zone 1 coverage for the normal system configuration with both lines in service. ■

In summary, the effect of Type 1.3 is to cause the relay to overreach, which can lead to significant problems. The Type 1.1 error will not lead to incorrect tripping, but it does reduce the zone 1 distance coverage of the relay.

14.4.2.3 Guidelines for the Underreaching Zone. The underreaching zone setting should comply with the following general guidelines [17]:

1. The reach setting should *ensure selectivity*. This means that it should avoid overreach beyond the remote line terminal for the Type 1.3 system configuration.
2. The reach setting should *ensure coverage* of as much of the line as possible. This means that at least 50% of the line length, plus a safety margin, in the most unfavorable situation, which is the Type 1.1 system configuration.

One strategy is to set the zone 1 to about 80% of the line length for Type 1.3 by adjusting $\boldsymbol{k}_{CA}$, but check to ensure sufficient reach for Type 1.1. The following comments apply to the underreaching zone setting [17]:

- If the setting is determined for the worst case, i.e., for Type 1.3, with the line B switched off and grounded at both ends, this assures proper selectivity.
- In many cases, the reach determined using Type 1.1, with line B in service, results in a small overlap with the protection at AQ. Suppose, for example, that the reach for Type 1.1 is 0.6. Since both relays AR and AQ have this same reach, this results in a 20% overlap at the center of the line, where both ends will clear the fault in zone 1. This is not a very large overlap, but the mutual coupling is reduced by infeed in most cases, which will usually improve the overlap.

Other strategies that might be helpful in determining the underreaching zone 1 settings are [17]:

1. Avoid grounding both ends of a line that is out of service. If this practice is strictly followed, the settings can be improved as Type 1.3 is then eliminated from consideration.
2. Grounding both ends of the opened line may be acceptable in some cases, since the overreach is reduced due to infeed at Q, but this should be carefully checked.
3. If autoreclosing is used following faults, the effect of overreach is reduced since a line subjected to a false trip will be quickly and automatically restored.
4. In every case, any possible overreach problem can be solved by pilot protection schemes.

It is noted that #1 is not under control of the protection engineer; therefore, it may not be a wise practice to assume that Type 1.3 can be eliminated with certainty. In fact, multiple grounding of a line that is taken out of service may be dictated by safety considerations for the personnel who are working on the outaged line.

14.4.2.4 Setting the Zone 1 Underreaching Relay. The reach is often measured in ohms but, for a line with uniform impedance per unit length, may also be measured in terms of distance or per unit distance. Per unit distance measurement is often preferred by protection engineers.

The zone 1 relays for transmission line A must be set to reach as far as possible without creating a risk for overreaching past the remote end of the line. Type 1.3 is the critical case with regard to selectivity, so we adjust the relay compensation to minimize the measurement error for this case. From (14.54) we have

$$\mathbf{Z}_{AR3} = h_3\mathbf{Z}_{A1}\left[\frac{1+\mathbf{k}_{SA} - h_3\mathbf{k}_{MA}\mathbf{Z}_{M0}/\mathbf{Z}_{B0}}{1+\mathbf{k}_{CA}}\right] \tag{14.62}$$

where the impedance subscript $AR3$ refers to the impedance at relay AR for system configuration Type 1.3.

We are free to adjust the compensation constant $\mathbf{k}_{CA}$ to minimize the error. Thus, we set $\mathbf{k}_{CA}$ for the correct measurement at the reach threshold.

$$\mathbf{k}_{CA} = \mathbf{k}_{SA} - \frac{h_S\mathbf{k}_{MA}\mathbf{Z}_{M0}}{\mathbf{Z}_{B0}} \tag{14.63}$$

The reach threshold is determined by the reach setting. Equating that setting in ohms to the relay apparent impedance, we have

$$\mathbf{Z}_{AR3} = h_3\mathbf{Z}_{A1}\left[\frac{1+\mathbf{k}_{SA} - h_3\mathbf{k}_{MA}\mathbf{Z}_{M0}/\mathbf{Z}_{B0}}{1+\mathbf{k}_{CA}}\right] = h_S\mathbf{Z}_{A1} \tag{14.64}$$

This gives the equation

$$\frac{\mathbf{k}_{MA}\mathbf{Z}_{M0}}{\mathbf{Z}_{B0}}h_3^2 - (1+\mathbf{k}_{SA})h_3 + \left(1+\mathbf{k}_{SA} - \frac{h_S\mathbf{k}_{MA}\mathbf{Z}_{M0}}{\mathbf{Z}_{B0}}\right) = 0 \tag{14.65}$$

which can be solved for the reach using the quadratic formula.

Now select a reach setting that is reasonable, such as 0.8 or 0.9. Using (14.63) to specify $\mathbf{k}_{CA}$ we can check the resulting reach for system configurations Types 1.1 and 1.2. Using (14.42) for Type 1.2, we write the relay impedance at the reach threshold for this configuration as

$$\mathbf{Z}_{AR2} = h_S\mathbf{Z}_{A1} = h_2\mathbf{Z}_{A1}\left(\frac{1+\mathbf{k}_{SA}}{1+\mathbf{k}_{CA}}\right) \tag{14.66}$$

or

$$h_2 = \frac{h_S(1+\mathbf{k}_{CA})}{1+\mathbf{k}_{SA}} \tag{14.67}$$

Finally, for Type 1.1, we write (14.46) at the relay threshold as follows.

$$\mathbf{Z}_{AR1} = h_S\mathbf{Z}_{A1} = \frac{h_1\mathbf{Z}_{A1}(1+\mathbf{k}_{SA}+h_1\mathbf{k}_{MA}/(1-h_1+\mathbf{k}_{AB}))}{1+\mathbf{k}_{CA}} \tag{14.68}$$

This gives a quadratic equation in the reach h_1 to be solved. The equation may be written in a more convenient form as follows.

$$(1+\mathbf{k}_{SA}-\mathbf{k}_{MA})h_1^2 - [(1+\mathbf{k}_{SA})(1+\mathbf{k}_{AB}) + h_S(1+\mathbf{k}_{CA})]h_1 + h_S(1+\mathbf{k}_{CA})(1+\mathbf{k}_{AB}) = 0 \tag{14.69}$$

This equation can be solved using the quadratic formula.

The results of using this strategy ensures correct measurement for Type 1.3 at the reach threshold, but gives a reach for Type 1.1 and Type 1.2 that are both short of ideal, but still reasonable in most cases. Examining (14.67), we see that the reach for Type 1.2 will always be less than the reach setting. This is also true for Type 1.1, although it is not obvious due to the quadratic form of (14.69). An example will illustrate typical numerical values.

EXAMPLE 14.3

Repeat Example 14.2 but compute the value of $\boldsymbol{k}_{CA}$ for the Type 1.3 configuration.

Solution

First, we compute the value of $\boldsymbol{k}_{CA}$ to minimize the error for Type 1.3. This gives

$$\boldsymbol{k}_{CA} = \boldsymbol{k}_{SA} - \frac{h_S \boldsymbol{k}_{MA} \boldsymbol{Z}_{M0}}{\boldsymbol{Z}_{B0}}$$

or

$$\boldsymbol{k}_{CA} = 0.635 - 0.3417 h_S$$

Suppose we select a reach setting, h_S of 0.8, at which point the Type 1.3 reach measurement will be exactly correct. This gives

$$\boldsymbol{k}_{CA} = 0.635 - 0.3417(0.8) = 0.3616$$

We can verify the result by computing the actual reach using (14.65). The coefficients of the quadratic equation are computed as

$$a = \frac{\boldsymbol{k}_{MA} \boldsymbol{Z}_{M0}}{\boldsymbol{Z}_{B0}} = \frac{(0.575)(j0.523)}{j0.88} = 0.3417$$
$$b = -(1 + \boldsymbol{k}_{SA}) = -1.635$$
$$c = h_S(1 + \boldsymbol{k}_{CA}) = h_S(1.635 - 0.3417 h_S) = 1.0893$$

From these coefficients, we verify the actual reach at the threshold from (14.64). This gives

$$h_3 = \frac{-b \pm \sqrt{b^2 - 4ac}}{2a} = \frac{1.635 \pm \sqrt{1.635^2 - 4(0.3417)(1.0893)}}{2(0.3417)} = 0.8000$$

and the computed reach is exactly correct for the Type 1.3 configuration.

The results of this type of mutual compensation are shown in Figure 14.21. The distance measurement for Type 1.3 is now exactly right when $h = 0.8$. Types 1.1 and 1.2 both underreach. Type 1.1 is the critical case for two reasons: (1) because the underreach is greatest, and (2) because this is the normal system condition and the system has a high probability of existing in this configuration .

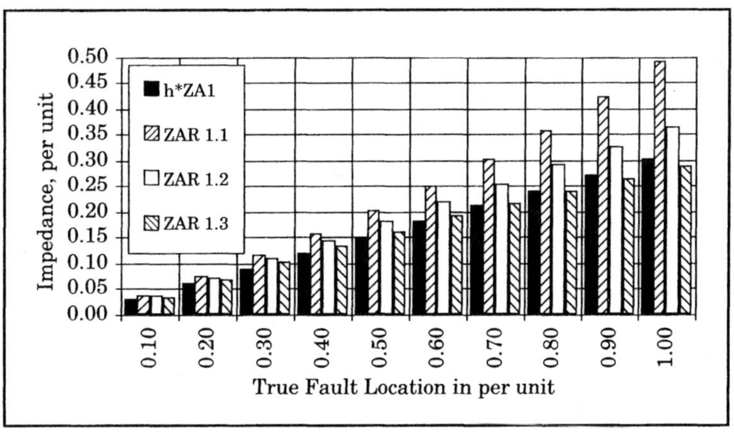

Figure 14.21 Line and relay impedances with mutual compensation.

The numerical values for Figure 14.21 show that Type 1.1 measurement error varies from 22.3% at $h = 0.1$ to 62.4% at $h = 1.0$. The Type 1.2 error is a constant 20.1%, which can be ignored due to the much larger Type 1.1 error. The Type 1.3 error is +17.6% at $h = 0.1$, reduces to zero at $h_3 = h_S = 0.8$, and becomes negative for values of h greater than the reach setting. This is not significant as the relay will not pick up for these values. At $h = 1$, the error is only −5%.

For Type 1.1, the reach must be computed using the quadratic (14.69), which gives the result $h = 0.58$. Thus, the reach is greatly reduced under normal conditions, extending only about 0.08 per unit beyond the center of the line. This results in an overlapping coverage from the two ends of only 16%. This means that 84% of the faults will be cleared by zone 2 protection, which results in delayed clearing most of the time. Thus, another type of ground relaying or other form of protection would be preferred for this system. ■

The reach calculations given in this section are determined from the radial system, with no infeed from the remote end of the faulted line. These approximate solutions are of interest as they represent the worst case. The correct values of reach should be computed using the actual relay currents, modified by self- and mutual compensation. The general expression may be written for a fault at the relay threshold as follows:

$$h_i = h_S \left(\frac{\boldsymbol{I}_{\text{ARC}}}{\boldsymbol{I}_{\text{ARM}}} \right) \tag{14.70}$$

where the two currents are defined in (14.40). The index i refers to the three system configuration types.

The reach of the ground distance relay depends on the transmission system configuration and on the type of current compensation used. For mutually coupled lines, two system configurations define the limiting conditions:

1. Both lines in service, identified here as Type 1.1, results in the greatest underreach and must be checked to ensure adequate coverage. Since this is the normal system condition, the coverage provided by this condition has a high probability of existence.
2. One line out of service and grounded at both ends, identified as Type 1.3, results in the greatest overreach and must be checked to ensure selectivity. For underreaching distance protection, the possibility of the relay reaching beyond the remote end of the protected line must not be permitted. Hence, this condition should be used as the basis for current compensation unless it results in inadequate coverage. If it is not possible to obtain both the required selectivity and adequate coverage, then some type of pilot protection should be used.

The underreaching protection strategy developed here is based largely on the radial system, where the remote source equivalent is infinitely weak. This results in measurement indices, $\boldsymbol{k}_{ei}$, given in Table 14.1, that represent only this worst case. The effect of current infeed at the remote end of the line is to moderate the conclusions of this worst case. The true value of the measurement index for any system is given by (14.40), and this value should be used in checking the true performance of any system.

14.4.2.5 Guidelines for the Overreaching Zone. In setting the overreaching zone 2 for relay AR, we will assume the configuration shown in Figure 14.22, where lines A and B are mutually coupled and the zone 2 ground distance protection from relay AR must reach through a predetermined fraction of line C to provide backup distance protection for relay CQ on line C.

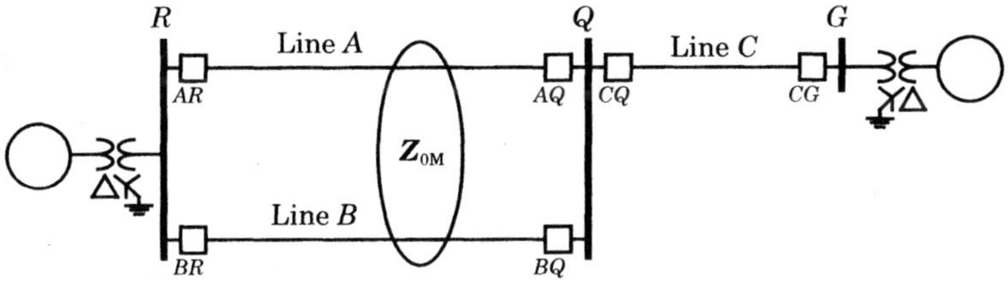

Figure 14.22 Setting the overreaching zone 2 for relay AR.

Overreaching protection schemes should have an overreach safety margin of at least 20% and this margin must be guaranteed for configuration Type 1.1, as this case has the lowest reach of the three cases and is therefore the most unfavorable case for the relay at AR. The other problem that must be examined is the situation line B is out of service and grounded at both ends. This results in the greatest reach and the settings must ensure that this longer reach does not overlap the zone 2 reach of relay CQ. These limitations are illustrated in Figure 14.23.

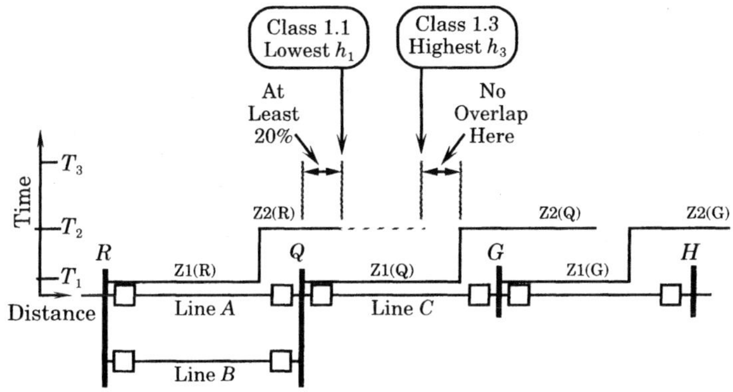

Figure 14.23 Overreaching zone 2 limitations for relay AR.

The difference between Type 1.1 and Type 1.3 reach at AR is sometimes called the *reach extension* [17]. This reach extension occurs following the transition from the normal system to the Type 1.3 configuration. Suppose we write the equations for the impedance for the two cases, but for the special condition that $h = 1$. Then we compute the reach extension

$$\frac{\Delta \boldsymbol{Z}}{\boldsymbol{Z}_{A1}} = \frac{\boldsymbol{Z}_{AR}^{(1.1)} - \boldsymbol{Z}_{AR}^{(1.3)}}{\boldsymbol{Z}_{A1}} = \frac{\boldsymbol{k}_{MA}}{1 + \boldsymbol{k}_{CA}} \left(\frac{1}{\boldsymbol{k}_{AB}} + \frac{\boldsymbol{Z}_{M0}}{\boldsymbol{Z}_{B0}} \right) \tag{14.71}$$

This is the difference in measuring the distance to a fault at Q between Types 1.1 and 1.3, or the reach extension in changing from a normal configuration to a condition with line B out of service and grounded. This quantity is also the per unit reach beyond bus Q for a fault at Q. The reach extension is reduced by source infeed at Q, but this will be ignored here in order to focus attention on the worst case.

First, we derive some general equations that describe the relay distance measurement at AR. The general expression for this measurement is given by (14.40), which is repeated here for convenience.

$$\boldsymbol{Z}_{AR} = h\boldsymbol{Z}_{A1} \left(\frac{\boldsymbol{I}_{AR} + \boldsymbol{k}_{SA}\boldsymbol{I}_{GA0} + \boldsymbol{k}_{MA}\boldsymbol{I}_{GB0}}{\boldsymbol{I}_{AR} + \boldsymbol{k}_{CA}\boldsymbol{I}_{GA0}} \right) \tag{14.72}$$

For one-line-to-ground faults under the condition with both lines A and B in service, we recall the relationship between the residual currents in the lines is given by (14.43) and (14.44).

$$\boldsymbol{I}_{GB0} = \frac{h\boldsymbol{I}_{GA0}}{1 - h + \boldsymbol{k}_{AB}} \tag{14.73}$$

where we can interpret the constant in the denominator as

$$\boldsymbol{k}_{AB} = \left.\frac{\boldsymbol{I}_{GA0}}{\boldsymbol{I}_{GB0}}\right]_{h=1} \tag{14.74}$$

Then, for a fault at $h = 1$, we may substitute for the currents to write (14.72) as

$$\boldsymbol{Z}_{AR} = \boldsymbol{Z}_{A1}\left(\frac{1 + \boldsymbol{k}_{SA} + \boldsymbol{k}_{MA}/\boldsymbol{k}_{AB}}{1 + \boldsymbol{k}_{CA}}\right) \tag{14.75}$$

For setting the zone 2 overreaching relay at AR, the normal condition is that where both lines A and B are in service. Therefore, the zone 2 relay should be compensated by assigning the constant in the denominator of (14.75) the following value.

$$\boldsymbol{k}_{CA} = \boldsymbol{k}_{SA} + \boldsymbol{k}_{MA}/\boldsymbol{k}_{AB} \tag{14.76}$$

This will provide exactly the correct reach to bus Q from bus R by relay AR when both lines A and B are in service. Since the probability of line outage is usually very small, this compensation factor will have a high probability of being correct.

14.4.2.6 Zone 2 Impedance—Type 1.1 Configuration. The zone 2 relay at AR views a fault on line C as having two components, $\boldsymbol{Z}_{RQ}$ and $\boldsymbol{Z}_{QF}$, as shown in Figure 14.24.

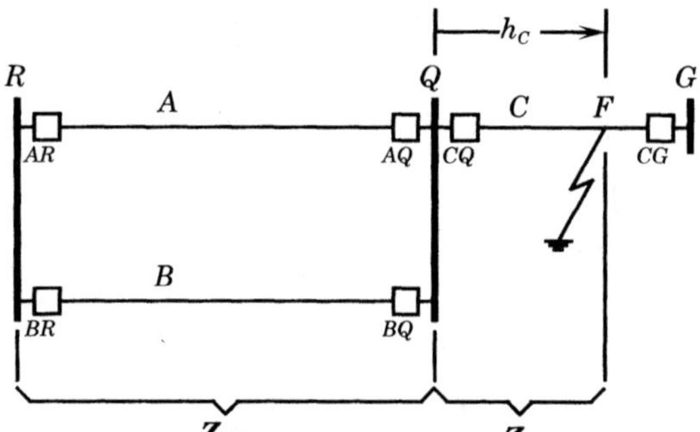

Figure 14.24 Components of impedance measurement at AR.

Thus, we may write an expression for the zone 2 measurement at AR as

$$\boldsymbol{Z}_{AR(2)}^{(1.1)} = \boldsymbol{Z}_{RQ}^{(1.1)} + \boldsymbol{Z}_{QF}^{(1.1)} \tag{14.77}$$

The first term in (14.77) is the normal measured impedance at AR that represents the two lines, A and B, operating in parallel, which is given by (14.75). The second term is a function of the fault location, which can be written in terms of the distance from bus Q, which is identified as the fractional distance h_C in Figure 14.24. We may write the voltage at bus Q in terms of the fault current on line C, in a manner analogous to that used in (11.112) for radial (worst) case.

$$\boldsymbol{V}_Q = h_C\boldsymbol{Z}_{C1}(\boldsymbol{I}_{QF} + \boldsymbol{k}_{SC}\boldsymbol{I}_{GC0}) \tag{14.78}$$

where, for a one-line-to-ground fault on line C we have

$$\boldsymbol{I}_{QF} = \boldsymbol{I}_{AR} + \boldsymbol{I}_{BR} = \boldsymbol{I}_{GA0} + \boldsymbol{I}_{GB0} \equiv \boldsymbol{I}_{GC0} \tag{14.79}$$

and the constant is defined as in (14.41), but using data for line C we get

$$\boldsymbol{k}_{SC} = \frac{\boldsymbol{Z}_{C0} - \boldsymbol{Z}_{C1}}{3\boldsymbol{Z}_{C1}} \tag{14.80}$$

Then, using (14.74), we may write

$$\boldsymbol{I}_{QF} = (1 + 1/\boldsymbol{k}_{AB})\boldsymbol{I}_{GA0} \tag{14.81}$$

Substituting these values into (14.78) we get

$$\boldsymbol{V}_Q = h_C \boldsymbol{Z}_{C1}(1 + 1/\boldsymbol{k}_{AB})(1 + \boldsymbol{k}_{SC})\boldsymbol{I}_{GA0} \tag{14.82}$$

We divide this relay voltage by the compensated relay current at AR, defined by (14.40), to obtain the impedance $\boldsymbol{Z}_{QF}$, as follows.

$$\boldsymbol{Z}_{QF}^{(1.1)} = \frac{h_C \boldsymbol{Z}_{C1}(1 + \boldsymbol{k}_{AB})(1 + \boldsymbol{k}_{SC})}{(1 + \boldsymbol{k}_{CA})\boldsymbol{k}_{AB}} \tag{14.83}$$

Then the total zone 2 impedance measured by the relay at AR may be computed by adding (14.75) and (14.83).

$$\boldsymbol{Z}_{AR(2)}^{(1.1)} = \frac{\boldsymbol{Z}_{A1}}{1 + \boldsymbol{k}_{CA}}(1 + \boldsymbol{k}_{SA} + \boldsymbol{k}_{MA}/\boldsymbol{k}_{AB}) + \frac{h_C \boldsymbol{Z}_{C1}(1 + \boldsymbol{k}_{AB})(1 + \boldsymbol{k}_{SC})}{(1 + \boldsymbol{k}_{CA})\boldsymbol{k}_{AB}} \tag{14.84}$$

In many cases the impedance of line C will be of similar design to that of line A. Therefore, we can simplify (14.84) by defining the ratio

$$\boldsymbol{k}_{XA} = \frac{\boldsymbol{Z}_{C1}}{\boldsymbol{Z}_{A1}} \tag{14.85}$$

Then, we may write (14.84) as

$$\boldsymbol{Z}_{AR(2)}^{(1.1)} = \frac{\boldsymbol{Z}_{A1}}{1 + \boldsymbol{k}_{CA}}\left(1 + \boldsymbol{k}_{SA} + \frac{\boldsymbol{k}_{MA}}{\boldsymbol{k}_{AB}} + \frac{h_C \boldsymbol{k}_{XA}(1 + \boldsymbol{k}_{AB})(1 + \boldsymbol{k}_{SC})}{\boldsymbol{k}_{AB}}\right) \tag{14.86}$$

Now, since the normal condition for lines A and B are to be in service, it is reasonable to define

$$\boldsymbol{k}_{CA} = \boldsymbol{k}_{SA} + \boldsymbol{k}_{MA}/\boldsymbol{k}_{AB} \tag{14.87}$$

so that the distance measurement for $\boldsymbol{Z}_{RQ}^{(1.1)}$ is exactly right. Then (14.86) becomes

$$\boldsymbol{Z}_{AR(2)}^{(1.1)} = \boldsymbol{Z}_{A1}\left[1 + \frac{h_C \boldsymbol{k}_{XA}(1 + \boldsymbol{k}_{AB})(1 + \boldsymbol{k}_{SC})}{\boldsymbol{k}_{AB}(1 + \boldsymbol{k}_{CA})}\right] \tag{14.88}$$

This is the zone 2 impedance measured by the relays at AR when the mutually coupled lines are both in service.

14.4.2.7 Zone 2 Impedance—Type 1.3 Configuration. When line B is out of service and grounded at both ends, the zone 2 distance measurement at AR is changed. For this condition, the impedance $\boldsymbol{Z}_{RQ}$ can be determined from (14.54) with $h = 1$, or

$$\boldsymbol{Z}_{RQ}^{(1.3)} = \boldsymbol{Z}_{A1}\left[\frac{1 + \boldsymbol{k}_{SA} - \boldsymbol{k}_{MA}\boldsymbol{Z}_{M0}/\boldsymbol{Z}_{B0}}{1 + \boldsymbol{k}_{CA}}\right] \tag{14.89}$$

Also, for the Type 1.3 system condition, we may compute $\boldsymbol{Z}_{QF}$ using the QF voltage from (14.78) and recognizing that

$$\boldsymbol{I}_{QF} = \boldsymbol{I}_{AR} = \boldsymbol{I}_{GA0} \tag{14.90}$$

for a one-line-to-ground fault, so that

$$\boldsymbol{V}_Q = h_C \boldsymbol{Z}_{C1}(1 + \boldsymbol{k}_{SC})\boldsymbol{I}_{GA0} \tag{14.91}$$

and

$$\mathbf{Z}_{QF}^{(1.3)} = \frac{\mathbf{V}_{QF}}{(1+\mathbf{k}_{CA})\mathbf{I}_{GA0}} = \frac{h_C\mathbf{Z}_{C1}(1+\mathbf{k}_{SC})}{1+\mathbf{k}_{CA}} \tag{14.92}$$

Therefore, the relay AR impedance measurement for this condition can be expressed as

$$\begin{aligned}\mathbf{Z}_{AR(2)}^{(1.3)} &= \mathbf{Z}_{RQ}^{(1.3)} + \mathbf{Z}_{QF}^{(1.3)} \\ &= \mathbf{Z}_{A1}\left[\frac{1+\mathbf{k}_{SA} - \mathbf{k}_{MA}\mathbf{Z}_{M0}/\mathbf{Z}_{B0} + h_C\mathbf{k}_{XA}(1+\mathbf{k}_{SC})}{1+\mathbf{k}_{CA}}\right]\end{aligned} \tag{14.93}$$

The reach extension due to the line B outage and grounding is given by the difference between (14.88) and (14.93), which gives the result

$$\frac{\Delta\mathbf{Z}_{AR(2)}}{\mathbf{Z}_{A1}} = \frac{\mathbf{k}_{MA}(1/\mathbf{k}_{AB} + \mathbf{Z}_{M0}/\mathbf{Z}_{B0}) + h_C\mathbf{k}_{XA}(1+\mathbf{k}_{SC})/\mathbf{k}_{AB}}{1+\mathbf{k}_{CA}} \tag{14.94}$$

14.4.2.8 Computation of Relay AR Zone 2 Reach. The *reach setting* for relay AR zone 2 can be written as

$$h_{S(2)} = \frac{\mathbf{Z}_{AR(2)}}{\mathbf{Z}_{A1}} \tag{14.95}$$

In practice, this setting would be specified by the protection engineer. Once specified, the actual reach on line C can be computed from (14.95) by solving for the reach h_C for any given value of reach setting. This should be done for both the normal Type 1.1 condition, as well as the Type 1.3 condition, given by (14.84) and (14.93), respectively. This gives two results, one for each configuration, which can be written in a single statement as

$$h_C^{(1.x)} = \frac{1+\mathbf{k}_{CA}}{\mathbf{k}_{XA}\mathbf{k}_{LT}(1+\mathbf{k}_{SC})}\left(h_{S(2)} - \frac{1+\mathbf{k}_{SA}+\mathbf{k}_{MA}\mathbf{k}_{MT}}{1+\mathbf{k}_{CA}}\right) \tag{14.96}$$

where we have defined two type-sensitive parameters, designated $\mathbf{k}_{LT}$ and $\mathbf{k}_{MT}$, which are shown in Table 14.2.

TABLE 14.2 Parameter Variations According to System Type

System Type	k_{LT}	k_{MT}
Type 1.1	$1 + 1/\mathbf{k}_{AB}$	$1/\mathbf{k}_{AB}$
Type 1.3	1	$-\mathbf{Z}_{M0}/\mathbf{Z}_{B0}$

Examining (14.96) and the values in Table 14.2 we can see that, in moving from the normal Type 1.1 system to the Type 1.3 configuration, $\mathbf{k}_{LT}$ becomes smaller (increases h_C) and $\mathbf{k}_{MT}$ changes from positive to negative (increases h_C). This process is explored further in an example.

EXAMPLE 14.4

Use the data of Examples 14.2 and 14.3 to check the zone 2 reach of relay AR if line C has exactly the same parameters as line A. The zone 2 reach setting is 1.5.

Solution

The system constants are all computed in previous examples and are repeated here for convenience.

$$k_{SA} = 0.635$$
$$k_{MA} = 0.575$$
$$k_{AB} = 1.000$$

We are given that

$$h_{S2} = 1.5$$
$$\mathbf{Z}_{C1} = \mathbf{Z}_{A1}$$
$$\mathbf{Z}_{C0} = \mathbf{Z}_{A0}$$

Then, we can compute

$$k_{XA} = \frac{\mathbf{Z}_{C1}}{\mathbf{Z}_{A1}} = 1.0 \qquad k_{LT} = \begin{cases} 2.0 & \text{Type 1.1} \\ 1.0 & \text{Type 1.3} \end{cases}$$

$$k_{SC} = \frac{\mathbf{Z}_{C0} - \mathbf{Z}_{C1}}{3\mathbf{Z}_{C1}} = 0.635 \qquad k_{MT} = \begin{cases} 1.0 & \text{Type 1.1} \\ -0.594 & \text{Type 1.3} \end{cases}$$

The relay apparent impedance for the system configurations are shown in Figure 14.25. The system is compensated for Type 1.1, and the impedance measurement is exactly correct. However, when Type 1.3 is imposed due to outage of line B, the apparent impedance is greatly reduced, raising the possibility the relay overreaching the end of line C.

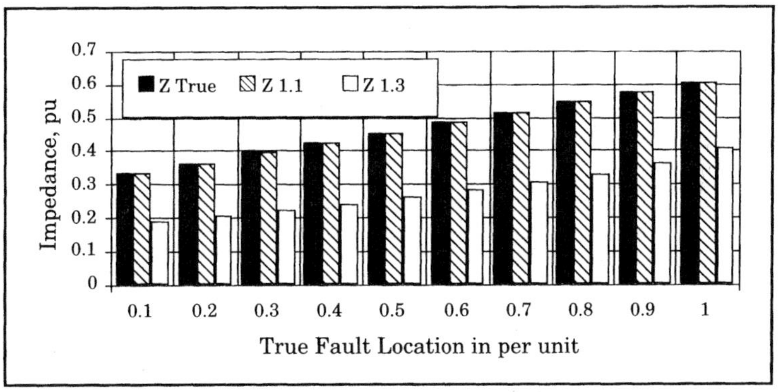

Figure 14.25 Relay apparent impedances for mutual types 1.1 and 1.3.

The relay overreach can be checked by applying (14.96).

Solving the normal case for the reach h_C, we have

$$h_C^{(1.1)} = \frac{1 + k_{CA}}{k_{XA}k_{LC}(1 + k_{SC})}\left(h_{S(2)} - \frac{1 + k_{SA} + k_{MA}/k_{AB}}{1 + k_{CA}}\right)$$

or

$$h_C^{(1.1)} = \frac{2.21}{(1.0)(2.0)(1.635)}\left(1.5 - \frac{1.635 + 0.575}{2.21}\right) = 0.338$$

Thus, the zone 2 reach is short of the desired value of 1.5, but it is an acceptable reach. It also meets the requirement for an overreach safety margin of at least 0.20.

For the case with line B open and grounded, we compute, from (14.96),

$$h_C^{(1.3)} = \frac{1+k_{CA}}{k_{XA}k_{LC}(1+k_{SA})}\left(h_{S(2)} - \frac{1+k_{SA}-k_{MA}Z_{M0}/Z_{B0}}{1+k_{CA}}\right)$$

$$= \frac{2.21}{(1.0)(1.0)(1.635)}\left(1.5 - \frac{1.635-0.342}{2.21}\right) = 1.236$$

This result is unacceptable, as the reach extends beyond the end of line C. Clearly, something must be changed to correct this situation. For example, the zone 2 reach setting could be reduced somewhat, say to 1.3. This new setting must be checked for both overreach requirements. ■

There are several principles that can be summarized regarding the setting of overreaching protection zones [17].

- The reach must extend beyond the remote end of the line plus a safety factor of 20% or so.
- The ground fault compensation factor may require adjustment in order to optimize the relay settings.
- It is important to check the reach extension that might result from the outage and grounding of the mutually coupled line, as this can significantly extend the reach of zone 2.

It should be recognized from the way in which (14.96) is written that a more general form can be written in terms of the self- and mutually compensated currents. The general expression is given by

$$h_C = \frac{1+k_{CA}}{k_{XA}k_{LC}(1+k_{SA})}\left(h_{S(2)} - \frac{I_{ARM}}{I_{ARC}}\right) \tag{14.97}$$

This form of the reach equation should be used to check system configurations with infeed at both ends of line C. Referring to Figure 14.24 it is clear that infeed will usually be provided at both bus Q and at bus G.

14.4.2.9 Distance Measurement on line B. It is tempting to see if an appropriate rule can be devised for residual compensation of mutually compensated lines. For line A we may write the impedance measurement at relay AR as

$$Z_{AR} = \frac{V_{AR}}{I_{ARC}} = \frac{hZ_{A1}(I_{AR}+k_{SA}I_{GA0}+k_{MA}I_{GB0})}{I_{ARC}} \tag{14.98}$$

The optimum compensated current for the relay AR would be

$$I_{ARC} = I_{AR} + k_{SA}I_{GA0} + k_{MA}I_{GB0} \tag{14.99}$$

Then the relay at AR would measure exactly the correct impedance.

If the compensation (14.99) were to be used, however, the relay at BR would measure

$$Z_{BR} = \frac{V_{BR}}{I_{BRC}} = \frac{V_{AR}}{I_{BRC}} \tag{14.100}$$

where we assume that the two relays are connected to the same bus at R. Then

$$Z_{BR} = \frac{V_{AR}}{I_{BR}+k_{SB}I_{GB0}+k_{MB}I_{GA0}} \tag{14.101}$$

This measurement is likely to be smaller than would be desired due to the compensation current, which includes the fault current $\boldsymbol{I}_{GA0}$ in line A. That current is much larger than the current in line B. Therefore, the measured $\boldsymbol{Z}_{BR}$ will be too small and the relay on the unfaulted line will overreach. Note that the constant $\boldsymbol{k}_{MB}$ is defined in exactly the same way as $\boldsymbol{k}_{MA}$, but using the self-impedance of line B rather than line A.

The ratio of the measurement for the two relays can be written as

$$\frac{\boldsymbol{Z}_{BR}}{\boldsymbol{Z}_{AR}} = \frac{\boldsymbol{I}_{AR} + \boldsymbol{k}_{SA}\boldsymbol{I}_{GA0} + \boldsymbol{k}_{MA}\boldsymbol{I}_{GB0}}{\boldsymbol{I}_{BR} + \boldsymbol{k}_{SB}\boldsymbol{I}_{GB0} + \boldsymbol{k}_{MB}\boldsymbol{I}_{GA0}} \tag{14.102}$$

The currents are not equal, but we can write the following equalities for the currents flowing in each line for the one-line-to-ground fault.

$$\begin{aligned} \boldsymbol{I}_{AR} &= \boldsymbol{I}_{GA0} \\ \boldsymbol{I}_{BR} &= \boldsymbol{I}_{GB0} \end{aligned} \tag{14.103}$$

Moreover, if we know that the fault is located a fractional distance h from the relay on line A, we can write the ratio of currents in the two lines as

$$\frac{\boldsymbol{I}_{GB0}}{\boldsymbol{I}_{GA0}} = \frac{h}{1 + \boldsymbol{k}_{AB} - h} \tag{14.104}$$

This allows us to write the ratio of the apparent impedances seen by the relays on the two lines as

$$\frac{\boldsymbol{Z}_{BR}}{\boldsymbol{Z}_{AR}} = \frac{(1 + \boldsymbol{k}_{SA})(1 + \boldsymbol{k}_{AB} - h) + h\boldsymbol{k}_{MA}}{h(1 + \boldsymbol{k}_{SB}) + \boldsymbol{k}_{MB}(1 + \boldsymbol{k}_{AB} - h)} = F(h) \tag{14.105}$$

Now, if the faulted line is properly compensated, $\boldsymbol{Z}_{AR}$ will be measured without error, but $\boldsymbol{Z}_{BR}$ will have an error that is a function of h. Stated mathematically, we have

$$\boldsymbol{Z}_{AR} = h\boldsymbol{Z}_{A1} \tag{14.106}$$

$$\boldsymbol{Z}_{BR} = \boldsymbol{Z}_{AR}\boldsymbol{F}(h) = h\boldsymbol{Z}_{A1}\boldsymbol{F}(h) \tag{14.107}$$

We can compute the error by solving for the value of h, but first we must know the reach setting for both relays to determine the extent of the overreach problem. Let us define h_{SB} to be the reach setting of line B. Then, $\boldsymbol{Z}_{BR}$ will be nonselective if

$$\boldsymbol{Z}_{BR} \le h_{SB}\boldsymbol{Z}_{B1} \tag{14.108}$$

From (14.107) and (14.108), after simplifying, we have the inequality

$$h_{SB} \ge h\boldsymbol{k}_{BA1}\left(\frac{(1 + \boldsymbol{k}_{SA})(1 + \boldsymbol{k}_{AB} - h) + h\boldsymbol{k}_{MA}}{h(1 + \boldsymbol{k}_{SB}) + \boldsymbol{k}_{MB}(1 + \boldsymbol{k}_{AB} - h)}\right) \tag{14.109}$$

where, for convenience in notation, we have used the constant

$$\boldsymbol{k}_{BA1} = \frac{\boldsymbol{Z}_{A1}}{\boldsymbol{Z}_{B1}} \tag{14.110}$$

The inequality (14.109) is a quadratic equation in h, which can be written as

$$ah^2 - bh + c \ge 0 \tag{14.111}$$

$$\begin{aligned} \text{where } a &= \boldsymbol{k}_{BA1}(1 + \boldsymbol{k}_{SA} - \boldsymbol{k}_{MA}) \\ b &= \boldsymbol{k}_{BA1}(1 + \boldsymbol{k}_{SA})(1 + \boldsymbol{k}_{AB}) - h_{SB}(1 + \boldsymbol{k}_{SB} - \boldsymbol{k}_{MB}) \\ c &= h_{SB}\boldsymbol{k}_{MB}(1 + \boldsymbol{k}_{AB}) \end{aligned} \tag{14.112}$$

Given the reach setting for relay BR, (14.112) can be solved to find the value of h that identifies the boundary for selectivity. The relay at BR will be non-selective for values of h less than this boundary value.

EXAMPLE 14.5

The following normalized line impedance values are given.

$$\mathbf{Z}_{A1} = \mathbf{Z}_{B1} = j0.3 \text{ per unit}$$
$$\mathbf{Z}_{A0} = \mathbf{Z}_{B0} = j1.0 \text{ per unit}$$
$$\mathbf{Z}_{M0} = j0.5 \text{ per unit}$$

Solve for the impedance measurements at AR and BR as a function of h. Find the limiting value of h if the reach setting of line B is 0.8 per unit.

Solution

The applicable equations are (14.106) and (14.107). Solving these equations we get the two functions pictured in Figure 14.26.

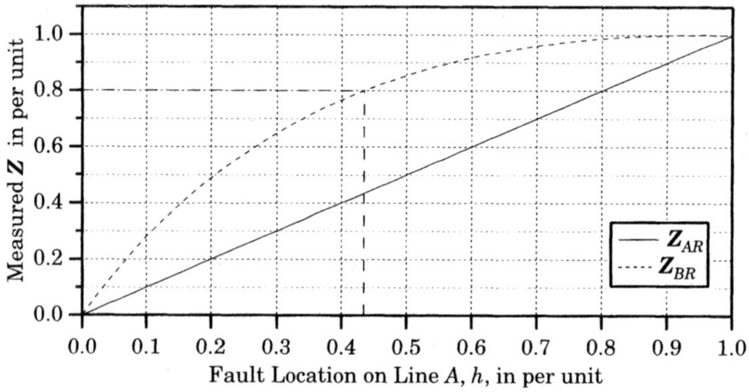

Figure 14.26 Apparent impedance seen by relays at AR and BR.

The relay at BR measures an impedance equal to its threshold setting of 0.8 when the fault on line A is at a distance of 0.434 from the relay. Any fault between 0 and 0.366 on line A will cause false pickup of the relay at BR.

This solution can also be obtained by solving the quadratic equation (14.111), which results in two solutions: 1.675 and 0.434. Only the smaller of the two solutions has any physical meaning, and is the correct answer. ■

It is important to take some kind of countermeasure to prevent the type of false tripping of unfaulted line B described above. One method that has been used is to compare the magnitude of the zero-sequence currents in the two lines and to compensate only the line having the greater current, which will be the faulted line.

14.4.3 Distance Protection of Type 2 Networks

Type 2 networks are defined as networks with common positive-sequence sources, but with different zero-sequence sources as shown in Figure 14.27. In this case both lines terminate at a common bus at both ends of the lines and there are common sources of positive-sequence current for both lines.

For circuits of this type, the system analysis is similar to that of the Type 1.3 network with one line out of service and grounded at both ends. This makes distance ground relay protection difficult. It also presents problems for directional ground overcurrent relays in line B.

Consider, for example, the small system shown in Figure 14.28, where a fault occurs on a line external to the mutually coupled lines A and B. The sequence network connections for

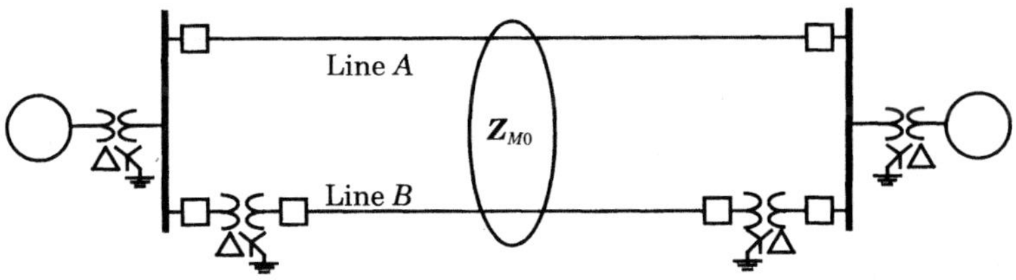

Figure 14.27 A type 2 parallel circuit with common positive but isolated zero-sequence source impedances [17].

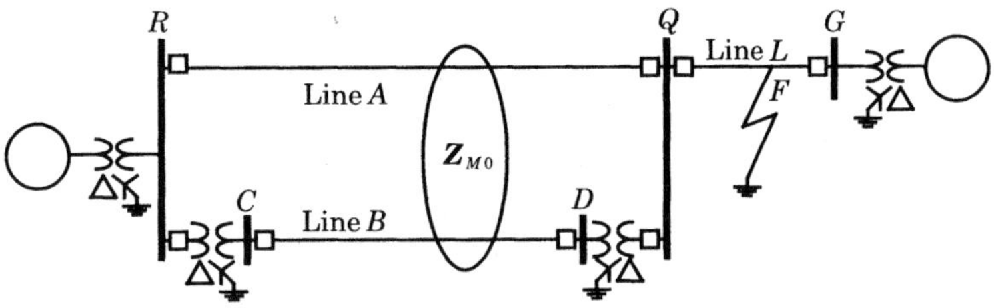

Figure 14.28 A Type 2 network with external fault.

a one-line-to-ground fault at point F are shown in Figure 14.29. The positive- and negative-sequence networks can be represented by the total sequence impedances, but the zero-sequence network requires further analysis because of the mutually coupled lines.

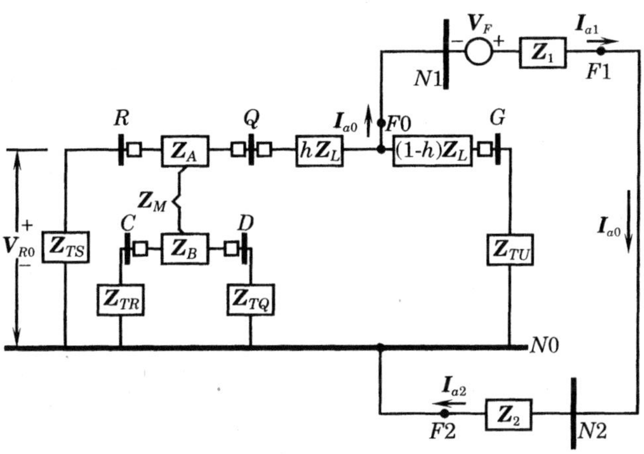

Figure 14.29 Sequence network connections for a SLG fault.

The comments in regard to Type 1 networks also apply here. Ground distance protection will present difficult problems and may not be suitable for this type of system. Moreover, ground fault protection utilizing zero-sequence current or voltage polarization may also cause false tripping. This type of network is further complicated since the terminations of the parallel lines are not necessarily in the same substations, which effectively rules out schemes that compare line currents.

The conclusion is that, for Type 2 networks, phase-comparison or negative-sequence relays should be considered for ground fault protection.

14.4.4 Distance Protection of Type 3 Networks

Type 3 networks are defined as networks that have separate positive- and zero-sequence sources, as shown in Figure 14.30. In this case both lines terminate at different buses at both ends of the lines and there are different sources of ground current for both lines. This type of system is not usually found, as most modern power systems tend to be interconnected. This type of system may be found where the two lines are of different voltage and the connections between the voltage levels are of relative high impedance. Obviously, this type of network could occur where the two power systems are intentionally not interconnected, even for lines of the same voltage.

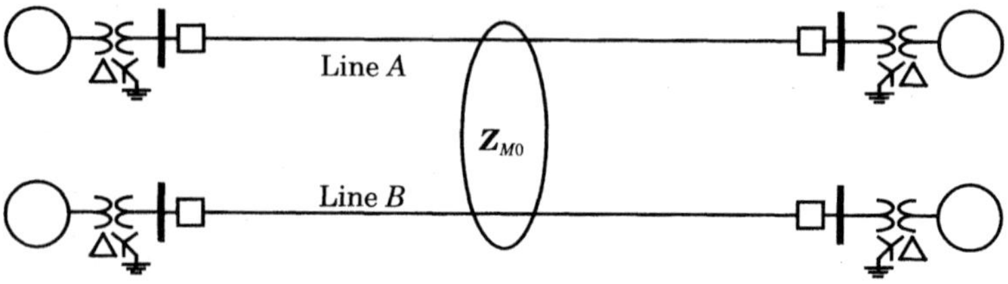

Figure 14.30 The type 3 parallel circuit with isolated positive- and zero-sequence sources impedances [17].

The comments for Type 1 and 2 networks also apply for this network arrangement. Comparison of ground currents in the two lines may not be possible, due to the separation of terminating substations.

As noted in Type 1.3 and Type 2 networks, this type of network presents the real possibility of a grounded zero-sequence line in parallel with the faulted line. This will cause the faulted line relays to overreach.

Compensation methods for this type of network are usually not applicable. Ground distance relays must be applied with caution and will not always provide adequate and secure protection. Polarization of ground overcurrent relays may also fail to provide correct sensing. We conclude that the best protection for Type 3 networks is a system that is not subject to these mutual induction problems, such as current phase comparison.

REFERENCES

[1] Batchelor, J. W., D. L. Whitehead, and J. S. Williams, "Transient Shaft Torques in Turbine Generators Produced by Transmission-Line Reclosing," *Protective Relaying for Power Systems*, Stanley H. Horowitz, Ed., IEEE Press, 1980, p. 513.

[2] Joyce, J. S., T. Kulig, and D. Lambrecht, "Torsional Fatigue of Turbine-Generator Shafts Caused by Different Electrical System Faults and Switching Operations," *Protective Relaying for Power Systems*, Stanley H. Horowitz, Ed., IEEE Press, 1980, p. 519.

[3] Jackson, M. C., S. D. Umans, R. D. Dunlop, S. H. Horowitz, and A. C. Parikh, "Turbine-Generator Shaft Torques and Fatigue: Part I-Simulation Methods and Fatigue Analysis," *Protective Relaying for Power Systems*, Stanley H. Horowitz, Ed., IEEE Press, 1980, p. 828.

[4] Dunlop, R. D., S. H. Horowitz, A. C. Parikh, M. C. Jackson, and S. D. Umans, "Turbine-Generator Shaft Torques and Fatigue: Part II-Impact of System Disturbances and High Speed Reclosure," *Protective Relaying for Power Systems*, Stanley H. Horowitz, Ed., IEEE Press, 1980, p. 537.

[5] Wolff, R. F., "765 kV Faults Test Single-Phase Switching," Electrical World, September 1, 1980, pp. 38–41.

[6] Knudsen, N., "Single-Phase Switching of Transmission Lines Using Reactors for Extinction of the Secondary Arc," CIGRÉ paper 310, Paris, 1962.

[7] Kimbark, E. W., "Suppression of Ground-Fault Arcs on Single-Pole-Switched EHV Lines by Shunt Reactors," *Trans. IEEE*, PAS-83, March 1984, pp. 285–290.

[8] Anderson, P. M., *Analysis of Faulted Power Systems*, IEEE Press, Piscataway, NJ, 1995.

[9] Shperling, B. R., A. Fakheri, and B. J. Ware, "Compensation Scheme for Single-Pole Switching on Untransposed Transmission Lines," *Trans. IEEE*, PAS-97m (4), July/August 1978, pp. 1421–1429.

[10] Shperling, B. R., and A. Fakheri, "Single-Phase Switching Parameters for Untransposed EHV Transmission Lines," *Trans. IEEE*, PAS-98 (2), March/April 1979, pp. 643–648.

[11] Fakheri, A. J., J. Grzan, B. R. Shperling, and B. J. Ware, "The Use of Reactor Switches in Single Phase Switching," CIGRÉ paper 13-06, Paris, 1980.

[12] Geszti, P. O., G. Bán, A. Dán, I. Horváth, J. Kisvölcsey, I. Benkó, S. Csida, Ju I. Lyskov, and N. P. Antonova, "Problems of Single-Pole Reclosing on Long EHV Transmission Lines," CIGRÉ paper 33-10, Paris, 1982.

[13] Perry, D. E., R. M. Hasibar, J. W. Chadwick, B. J. Ware, A. J. Fakheri, and R. S. Bayless, "Investigation and Evaluation of Single Phase Switching on EHV Networks in the United States," CIGRÉ paper 39-08, Paris, 1984.

[14] AIEE Committee Report, "Protection of Multiterminal and Tapped Lines," *Trans. AIEE*, 80, April 1961, pp. 55–65, reprinted in *Protective Relaying for Power Systems*, Stanley H. Horowitz, Ed., IEEE Press, 1980, p. 157.

[15] IEEE Committee Report, "Protection Aspects of Multi-Terminal Lines," IEEE Pub. TH0056-2-PWR, 1979, pp. 1–18, reprinted in *Protective Relaying for Power Systems*, Stanley H. Horowitz, Ed., IEEE Press, 1980, p. 513.

[16] Esztergalyos, J., and E. Einarsson, "Ultra High Speed Protection of Three-Terminal Lines," CIGRÉ paper 34-06, Paris, 1982.

[17] Ziegler, G., et al., "Application Guide on Protection of Complex Transmission Network Configurations," CIGRÉ, SC34-WG04, CIGRÉ Publication E50400-U0048-U211-A3-7600, Paris, May 1991.

[18] IEEE Std 100-1992, *IEEE Dictionary of Electrical and Electronics Terms*, Christopher J. Booth, Ed., IEEE, New York, 1992.

[19] Durbak, D. W., and J. R. Stewart, "PLC Signal Attenuation in Branched Networks," *IEEE Trans.*, PWRD-5, April 1990, pp. 878–883.

[20] Davison, E. B., and A. Wright, "Some Factors Affecting the Accuracy of Distance-Type Protective Equipment Under Earth-Fault Conditions," *Proc. IEE*, 110, (9), 1963, pp. 1678–1688.

[21] Adamsson, C., and A. Türeli, "Errors of Sound-Phase-Compensation and Residual-Compensation Systems in Earth-Fault Distance Relaying," *Proc. IEE.*, 112, (7), 1965, pp. 1369–1382.

[22] Wheeler, S. A., "Influence of Mutual Coupling Between Parallel Circuits on the Setting of Distance Protection," *Proc. IEE.*, 117, (2), 1970, pp. 439–444.

[23] Newbould, A, and J. B. Royle, "A New Method of Mutual Compensation for Distance Relays," *Developments in Power System Protection*, London, 1980, pp. 201–205.

[24] Rockefeller, G. D., "The Influence of Mutual Impedance on Ground Distance Relays," Annual Protective Relaying Conference, Georgia Institute of Technology, Atlanta, GA, April 1968.

[25] Ungrad, H., and V. Narayan, "Behavior of Distance Relays Under Earth Fault Conditions on Double-Circuit Lines," *Brown Boveri Review*, 1969, pp. 494–501.

[26] Adams, M. G. and J. H. Fish, "Coupling Effects Related to Grounding of Multicircuit Transmission Lines on Common Towers," Third International Conference on Live-Line Maintenance, Atlanta, GA, June 6–9, 1983.

[27] Humpage, W. D., et al, "Distance-Protection Performance Under Conditions of Single-Circuit Working in Double Circuit Transmission Lines," *Proc. IEE*, 117, (4), 1970, pp. 766–770.

[28] Engelhardt, K. H., "V_0-Reversal on Mutually Coupled Lines–Relaying Analysis, Experience, and Remedies," Western Protective Relay Conference, Spokane, Washington, October 28–30, 1974.

[29] Elmore, W. A., "Zero Sequence Mutual Effects on Ground Distance Relays and Fault Locators," Forty-Sixth Annual Protective Relaying Conference, Georgia Institute of Technology, Atlanta, GA, April 29-May 1, 1992.

[30] American National Standard, "Definitions for Power Switchgear," ANSI/IEEE Standard C37.100-1981, IEEE Pub. SHO8375.

PROBLEMS

14.1 Consider an untransposed transmission line consisting of phase wires a, b, and c.

(a) Write the linear charge, $Q = CV$, equations for an untransposed transmission line. Identify the three-phase charge densities, in c/m, due to the three phase-to-neutral voltages and the line capacitances.

(b) Compute the capacitances from phase wire to ground, per unit of length from the phase capacitances.

14.2 Extend the solution of problem 14.1(a) to the case of a completely transposed line. Devise simplifying names for the terms that result due to the total line averaging of individual matrix parameters.

14.3 Consider the equivalent circuit of the transmission line susceptances shown in Figure 14.1(b).

(a) Apply balanced positive-sequence voltages at the connection points labeled a, b, and c, and determine the total positive-sequence current that will result from these applied voltages. Compute the currents as functions of the phase voltages and the network susceptances.

(b) Apply zero-sequence voltages to the connection points labeled a, b, and c, and determine the total zero-sequence currents that flow into these three terminals in terms of the applied voltages and network susceptances.

(c) Use the results of parts (a) and (b) to analyze the equivalent circuit shown in Figure 14.1(c). Determine the parameters of that circuit in terms of the parameters of Figure 14.1(b).

14.4 Given the instantaneous current flowing in a capacitor as a function of capacitor voltage

$$i = \frac{dq}{dt} = C\frac{dv}{dt}$$

(a) Write the given equation in the Laplace domain.

(b) Convert the results of (a) into the phasor domain.

(c) Expand (c) to apply to a three-phase circuit.

(d) Convert the result of (c) into symmetrical components and define the circuit positive- and zero-sequence susceptances.

14.5 Evaluate the results of the previous problem for a completely transposed line.

14.6 Consider the single-line-to-ground fault as a two-port network, with voltage sources for phases b and c as the two-port voltage sources.

(a) Develop this two-port equivalent of the system of Figure 14.3, where phase a is the faulted phase. This may require a step-by-step network modification to arrive at the two-port equivalent.

(b) Analyze the two-port network equivalent to determine the fault current as a function of the network parameters.

14.7 Verify the circuit connections shown in Figure 14.4.

14.8 Verify the recovery voltage across the fault path, given by (14.4).

14.9 Verify the equation for the secondary arc current for the transposed line, given by (14.7).

14.10 Consider the wye-connected, four-reactor system shown in Figure 14.5(c). Derive the equation that expresses the three-phase currents flowing into the reactor system as a function of the three reactor phase voltages. Express the result as a susceptance matrix as in (14.13). Compare your computed values of susceptance in that matrix with (14.14) and (14.15).

14.11 Consider the unbalanced system of Figure 14.5(a), where a one-line-to-ground fault is applied to phase b. Determine the susceptance matrix values for this condition. Solve this system of equations to verify the results shown in (14.13).

14.12 Consider the unbalanced system of Figure 14.5(a), where a one-line-to-ground fault is applied to phase a. Determine the susceptance matrix values for this condition.

14.13 Consider the unbalanced system of Figure 14.5(c), where a one-line-to-ground fault is applied to phase c. Determine the susceptance matrix values for this condition.

14.14 Consider the multiterminal system of Figure P14.14. Compute the apparent impedances measured by relays at a, b, and c for a three-phase, zero-impedance fault at point F on the line from B to P. Also, quantify the measurement error for all three relays.

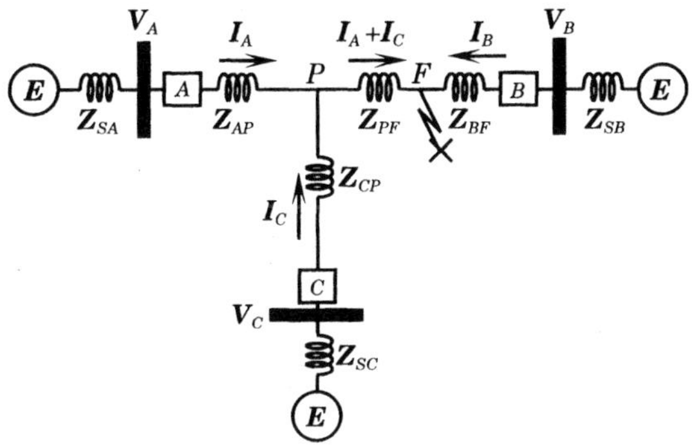

Figure P14.14 A multiterminal line.

14.15 Compute the total current of the three sources for the system of Figure P14.14 with a zero-impedance fault at F.

14.16 Evaluate qualitatively the worst case in terms of the relative size of impedance in the system of Figure P14.14.

14.17 Develop a relay threshold setting strategy for the relays of Figure P14.14, where we assume that the following impedance inequalities hold.

$$Z_{BP} > Z_{AP} > Z_{CP}$$

14.18 Consider the system with relay settings as described in the previous problem. Let the total impedances of branches AP and CP be equal and let the impedance of branch BP be three times that of the others.

(a) Compute the measured impedance and fault currents of relays A and C.
(b) Compare these results with the threshold setting computed from the previous problem.
(c) Compare the results with the zone 1 setting where $k = 0.85$.
(d) Compute the fault current seen by relay A or C.

14.19 Consider the setting of the overreaching zone for the multiterminal system of Figure P14.14. Assume that the impedance of the faulted branch B is greater than that of A or C. Determine the settings of the overreaching zone settings for relays A and C if an overreach of 20% is specified.

14.20 Derive (14.38) and (14.40).

14.21 Derive (14.44).

14.22 Verify (14.46) for the Type 1.1 system condition.

14.23 Consider a system of identical parallel transmission lines, A and B, having the following parameters.

$$\begin{aligned} Z_{L1} &= 0.30 \quad \Omega/\text{km} \\ Z_{L0} &= 1.00 \quad \Omega/\text{km} \\ Z_{M0} &= 0.50 \quad \Omega/\text{km} \end{aligned}$$

(a) Determine the compensation constant for relays in line A that will minimize the zone 1 error for Type 1.3, i.e., with line B open and grounded at both ends.

(b) For the conditions found for (a), determine the reach for Type 1.2, i.e., the case with line B open.

(c) For the same conditions, determine the reach for Type 1.1, i.e., the case with line B in service and unfaulted.

14.24 Verify (14.47) for the Type 1.2 system condition.

14.25 Verify (14.54) for the Type 1.3 system condition with a very weak source at Q.

14.26 Construct a spreadsheet to calculate the apparent impedance quantities given in examples 14.2 and 14.3. Plot the results and compare with the impedance plots shown in Figures 14.20 and 14.21.

14.27 Calculate the reach extension for the system of Figure 14.22, where the extension of reach is caused by having the mutually coupled lines A and B change from a Type 1.1 to Type 1.3 condition.

14.28 Derive (14.88) for the overreaching relay measurement for the Type 1.1 system condition.

14.29 Derive (14.93) for the overreaching relay measurement for the Type 1.3 system condition.

14.30 Derive (14.96) for the zone 2 reach of the relay AR as a function of the relay setting.

14.31 Repeat the computations of example 14.4 using a zone 2 reach setting of 1.3. Are the resulting values of reach for Types 1.1 and 1.3 system configurations acceptable?

15

Series Compensated Line Protection

15.1 INTRODUCTION

Series-compensated transmission lines utilize series capacitors to cancel a portion of the inductive reactance of the line, and thereby improving the power transmission capability of the line. Series compensation has been applied mostly to long transmission lines, such as those found in the Western United States, in South America, and other locations where the transmission distances are great and where large power transfers over these distances are required.

There are several reasons for favoring series compensation of long EHV transmission lines.

1. The lower line impedance improves stability.
2. The lower line impedance improves voltage regulation.
3. Adding series capacitance provides a method of controlling the division of load among several lines.
4. Increasing the loading of a line improves the utilization of the transmission system and, therefore, the return on the capital investment.

For the reasons cited, series-compensated transmission lines have become rather common in locations where the distances between load centers is great and large transmission investments are required. Even though the series compensation has been known to create problems in system protection and subsynchronous resonance, the return is usually considered worth the extra engineering effort required to properly design and operate these more complex transmission facilities. Moreover, these problems have been shown to have technically feasible solutions, so that there is no longer the fear that series compensation should be avoided. In fact, the installations of series compensation is growing, both in the number of installations and in the total MVA of capacitors installed.

The addition of series capacitors in the transmission circuit makes the design of the protection more complex [1–6]. The degree of complexity depends on the size of the series capacitor and its location along the transmission line. Series compensation is usually stated as a percent of the line inductive reactance, which is referred to as the "degree of compensation." Thus, we speak of a line that is 50% or 70% compensated, meaning that 50% or 70% of the inductive reactance, respectively, is being installed as series capacitive reactance. The higher degrees of compensation make the relaying more difficult. However, as noted above, the relay manufacturers have developed relays for series compensated lines that overcome the problems experienced in the past.

The location of the series capacitors is also important in the design of the protective system. Distance relays measure the distance between the relay and the fault by effectively computing the impedance seen looking into the line at the relay location. If a large part of the inductive reactance is canceled by capacitive reactance, the relay apparent impedance measurement is directly affected. In many cases, the series capacitors are located at the ends of the transmission line. Locating the series capacitance at the ends of the line is often the least expensive alternative, since no midline station construction is required. However, this alternative may cause a distance relay to measure a negative reactance for a close-in fault, which the distance element interprets to be a fault *behind* the relay. This creates a problem for the relay designer, but this problem has been overcome in modern protection designs.

The protective strategy is also affected by the method of series capacitor bypass. The capacitor banks are protected from overvoltage by bypass arrangements that shunt the fault current around the capacitor. Some bypass designs accomplished this using spark gaps, some of which are triggered gaps the firing of which can be controlled. A newer concept is the use of metal oxide varistors instead of spark gaps, which can be designed to more precisely bypass the capacitor, and reinsert the capacitor immediately when the fault current disappears. The varistor protection is faster and more precise in both bypassing and reinserting the capacitor. This is important in its effect on the line protection, since the apparent impedance seen by the relay changes greatly from the unbypassed state to the bypassed state, and then back to the normal, unbypassed state.

In order to understand the problems associated with the protection of series compensated lines, it is necessary to examine the network aspects of a faulted line with a series capacitor. First, we shall consider the protected line viewed by the relay before capacitor bypass occurs. Then we will look at the same line with the capacitor bypassed. In both cases, we will examine the relay current, voltage, and apparent impedance in order to understand the complexity of events viewed by the relay during a fault. We shall do this for varying fault locations, for different series capacitor locations, and for varying degrees of compensation.

15.1.1 The Degree of Compensation

A convenient method of specifying the amount of series compensation on a transmission line is by means of a quantity called the *degree of compensation* [7]. This quantity, usually designated as k, is defined as

$$k = \frac{\operatorname{Im} \Delta \boldsymbol{B}}{\operatorname{Im} \boldsymbol{B}} \tag{15.1}$$

where $\boldsymbol{B}$ is the complex quantity associated with the familiar $\boldsymbol{ABCD}$ representation of the transmission line. The numerator of (15.1) represents the total amount of series compensation to be added, in ohms, and the ratio defined as k expresses this compensation as a per unit quantity, based on the reactive part of the $\boldsymbol{B}$ parameter. Actually, it is probably more common to express the degree of compensation in percent, by multiplying (15.1) by 100. Degrees of compensation in the range of 50–70% are common.

Note that the degree of compensation is specified in terms of the ***B*** parameter, since this is equal to the physical series impedance between the line terminals. This total series impedance is often referred to as the equivalent pi impedance. The degree of compensation is specified in terms of the imaginary part of this impedance.

15.1.2 Voltage Profile on Series Compensated Lines

The voltage profile of the transmission line is the magnitude of the voltage plotted as a function of distance measured along the length of the line. Transmission lines exhibit a voltage profile that is represented as a smooth curve of voltage versus distance. The actual shape of this curve depends on the line loading. The terminal voltages depend on the type of voltage control used, if any, at the two ends of the line and the total solution of the network at any given load level.

Series compensated lines have a voltage profile that is a smooth function of distance, but exhibit step changes in voltage at the series capacitors. The capacitor voltage depends on the current flowing in the line at that point. For a long transmission line the current is a continuous function of distance, but varies along the length of the line. Therefore, on such a long line with more than one series capacitor, the voltage change across the two series capacitors is not necessarily the same.

Figure 15.1 shows the voltage profile of a 500 kilometer, 500 kilovolt transmission line. The line is loaded to twice its surge impedance loading, which leads to a voltage profile that is concave upward.[1] Lightly loaded lines have voltage profiles that are concave downward, and lines loaded to exactly surge impedance loading have linear voltage profiles. The line illustrated in Figure 15.1 has two series capacitors that are placed at distances that divide the line into three segments of equal length. Note that the voltages across the two series capacitors are not equal, since the current varies along the length of the line. Also note that the larger series capacitors may have a rather high capacitor voltage due to the high currents at high load levels.

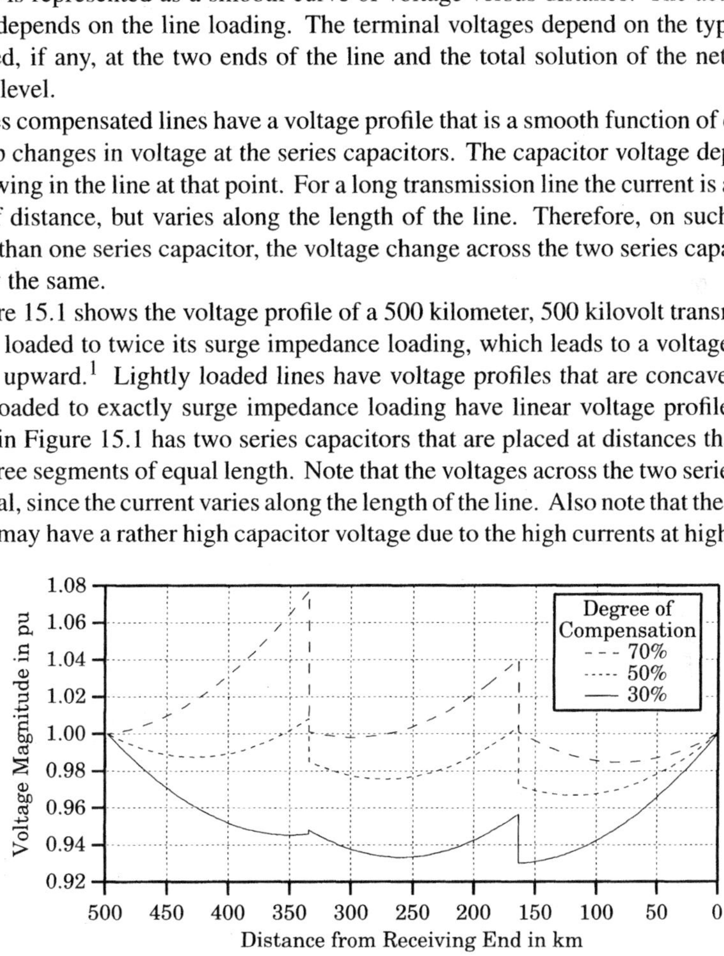

Figure 15.1 500 kV line voltage profile with heavy line loading.

If the line is faulted, the currents are much larger and the voltage across the series capacitors can become extremely large. Under these conditions, the series capacitors are bypassed by some means in order to protect the capacitor bank from permanent damage.

[1]A curve is *concave upward* on an interval if the curve on that interval lies above the straight line tangent to the curve and represents the slope of the line on that interval. If the curve lies below the tangent line, it is said to be *concave downward*. [7]

Several methods of capacitor bypass have been used, and this is the primary method of series capacitor protection. Series capacitor protection is not explored in detail in this book. This function is usually performed by the capacitor bank manufacturer, who is in a position to fully understand the protective requirements of the capacitor bank equipment. Series capacitor bank bypass is discussed further in Section 15.3.

15.2 FAULTS WITH UNBYPASSED SERIES CAPACITORS

A convenient way to visualize the problems of protecting a series compensated line is to examine the fundamental frequency voltage, current, and apparent impedance for varying system and fault conditions.[2] The apparent impedance is conveniently plotted in the complex RX plane. The use of RX plots should not be viewed as an assumption that distance relays are used for the line protection, but the plots simply provide a method of comparing different system conditions.

To perform this type of comparison, it is necessary to solve for the system voltages and currents for the different system and fault conditions of interest. For this purpose, a protection equivalent, similar to that of Figure 15.2, is solved. This is a perfectly general method of viewing the protection for any system, under any shunt fault condition. The resulting currents and voltages are functions of the capacitor location. The capacitor locations to be studied here are: (a) capacitors of equal size at both ends of the line, and (b) one capacitor at the center of the line. In all cases, the total degree of compensation will be the same. The following sections will analyze the voltage, current, and apparent impedance signals provided by various series capacitor and system configurations.

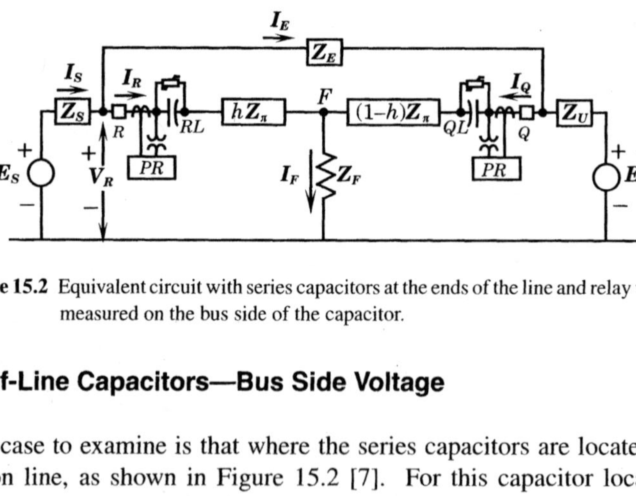

Figure 15.2 Equivalent circuit with series capacitors at the ends of the line and relay voltage measured on the bus side of the capacitor.

15.2.1 End-of-Line Capacitors—Bus Side Voltage

The first case to examine is that where the series capacitors are located at the ends of the transmission line, as shown in Figure 15.2 [7]. For this capacitor location, there is a choice of voltage measurements for the protective relays. In many cases, the relay voltage is measured at the bus, or on the bus side of the series capacitor. This is the arrangement shown in Figure 15.2, where the protective relays (PR) are supplied with the line current and a voltage that is measured on the bus side of the series capacitor.

For many high-voltage systems under faulted conditions, the system voltages are low enough that currents flowing in the distributed system susceptances may be neglected for

[2]Portions of this and following sections are taken from the book *Series Compensation of Power Systems*, ©1996 by P. M. Anderson and R. G. Farmer, [7].

practical purposes. This is not true of EHV lines, where the total shunt susceptance is high. In the solution of Figure 15.2 for a long 500 kilovolt line, for example, each line section should be represented in the solution as an equivalent pi line section in order to exactly account for the shunt susceptances of the line. These equivalent pi sections would replace the rectangular boxes in Figure 15.2 that represent transmission line sections.

The solution is assumed to be a fundamental frequency, phasor solution. Since there are only three nodes in the network, at R, Q, and F, the voltages and currents in the equivalent circuit are easily found. The relay voltage is assumed to be measured on the bus side of the capacitor, that is, at point R in Figure 15.2. The case where the voltage is measured on the line side of the capacitor is taken up in Section 15.2.2.

We are particularly interested in the voltage and current measured at the relay locations. The relays at R measure the current at R flowing toward the fault, and the voltage measured either at R or at RL (see Figure 15.2). As the fault F is moved along the line, the currents and voltages observed by the relay R will change. In addition to the current and voltage, we are also interested in the apparent impedance seen by the relay, since many transmission protective schemes employ distance measuring equipment, or use phase or amplitude comparison techniques that are equivalent to distance measurements. The quantity to be examined is the impedance seen by the relay, or

$$\boldsymbol{Z}_R = \frac{\boldsymbol{V}_R}{\boldsymbol{I}_R} = \frac{\boldsymbol{E}_S - \boldsymbol{Z}_S \boldsymbol{I}_S}{\boldsymbol{I}_S - \boldsymbol{I}_E} \tag{15.2}$$

The relay impedance can be viewed under many different system and fault conditions by changing the parameters in the equivalent circuit and solving the equations. Of particular interest is the variation in impedance seen by the relay as the fault is moved from one end of the line to the other. This will be done by varying the parameter h from 0 to 1 in small steps. We are particularly interested in viewing these results for different degrees of series compensation.

The effect of series compensation on the relay voltage and current is illustrated by an example. The parameters for the equivalent circuit of Figure 15.2 are given in (15.3), where all values are in per unit on the base values given in (15.4).

$$\begin{aligned}
\boldsymbol{E}_S = \boldsymbol{E}_U &= 1.0 + j0.0 \text{ pu} \\
\boldsymbol{Z}_S = \boldsymbol{Z}_U &= 0.0 + j0.1 \text{ pu} \\
\boldsymbol{Z}_F &= 0.01 + j0.00 \text{ pu} \\
\boldsymbol{Z}_E &= 0.002 + j0.04 \text{ pu} \\
z &= 0.00012 + j0.0024 \text{ pu/m} \\
y &= 0.000 + j0.0018 \text{ pu/m} \\
s &= 200 \text{ miles}
\end{aligned} \tag{15.3}$$

$$\begin{aligned}
S_B &= 1000 \text{ MVA} \\
V_B &= 500 \text{ kV} \\
Z_B &= 250\ \Omega \\
Y_B &= 4000\ \mu\text{S}
\end{aligned} \tag{15.4}$$

The equivalent pi parameters for the transmission line are completely determined from the impedance and admittance per unit length and the total line length. The resulting relay current magnitude is plotted as a function of h, the fault location, in Figure 15.3. The fault current magnitude for a fault at point R, on the line side of the circuit breaker (on the bus side of the capacitor) at R, is approximately 16 per unit for all cases. The capacitor bypass circuits shown in Figure 15.2 are assumed to be nonoperative for this initial study.

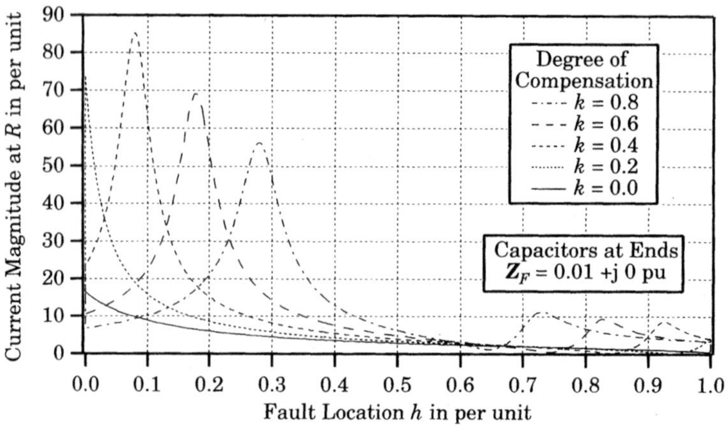

Figure 15.3 Relay R current magnitude as a function of fault position h.

At $h = 0+$, with the fault at point RL, just to the right if the capacitor at the R end of the line, the current jumps to a new value, depending on the size of the capacitor. For $k = 0.2$ the current reaches a magnitude of almost 75 per unit, assuming that there is no protective capacitor bypass to limit the current. This is due to a near resonant condition for the fault at this point when the degree of compensation, k, is small. Since the source impedance behind the relay is an inductive reactance of 0.1 per unit, the total reactance from the sources to the fault is capacitive. In this case, with the degree of compensation set to 20%, the capacitor has a value of about 0.05 per unit capacitive reactance, which greatly reduces the total impedance between source S and the fault. Moreover, for the case simulated, the current flowing through the external impedance Z_E supplies about 40% of the total fault current. For the $k = 0.2$ case, the current reaches its maximum value at $h = 0$ due to this near resonant condition, but this is not true in all cases.

For higher degrees of compensation, the resonant condition occurs at different fault locations. For $k = 0.4$, the circuit resonant condition occurs at about $h = 0.08$. As the degrees of compensation increases, the resonant point moves further down the transmission line, the magnitude of the peak current becomes smaller, and the sharpness of resonance is reduced due to line resistance. Note that, in all cases, the currents plotted assume that the capacitor protective bypass system does not operate, such that the currents computed are the maximum that can occur. Later, these same studies will include the effect of capacitor bypass protective systems, which greatly limits the current.

The magnitude of the current at R, as the fault is moved beyond the resonant point, falls to a low value until the fault nears the Q end of the line, where new resonances occur due to the high current entering the line from the source at U. The current contribution from source U is an exact mirror image of the current at S because of the symmetrical source impedances selected here for study. As the remote source current increases, it causes the voltage to rise at the fault point, thereby increasing the current at R, as noted in Figure 15.3.

Figure 15.4 shows a plot of the voltage magnitude as a function of fault location. As with the current magnitude, the voltage reaches resonant peaks that are very high, with the fault location at which resonance occurs depending on the degree of compensation, and therefore the size of the capacitor. Since the source impedance for the case simulated in Figure 15.4 is small, the resonance for small degrees of compensation occurs near the bus. For larger capacitor sizes, the resonances occur farther away from the bus. There are rather high resonances noted at the far end of the line, where the resonance is with the source at the Q end of the line. This causes peaking of the fault current contribution from the far end, and this current is the major contribution to faults located near the Q bus. The resonances at the far end of the line are also due, in part, to the high shunt susceptance of this EHV line. These high remote-end resonances are not as great in lower voltage lines.

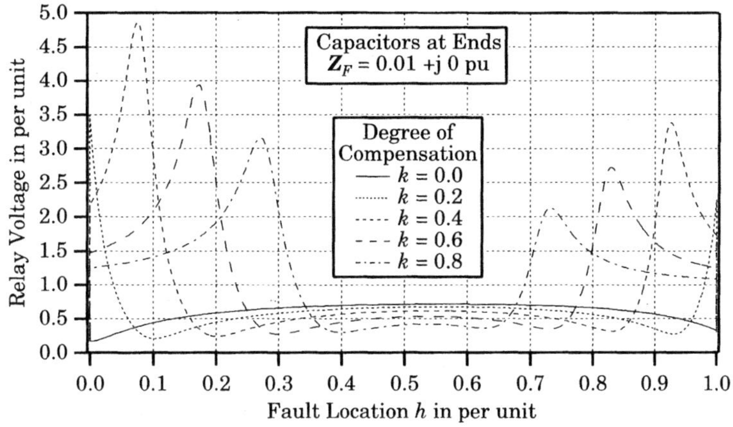

Figure 15.4 Voltage magnitude as a function of fault position h.

It is important to examine more than the current and voltage magnitudes. The fundamental frequency current is a phasor, and the angle as well as the magnitude is important. Figure 15.5 shows a plot of the locus of the phasor current and voltage at R, plotted in the complex plane for the uncompensated line. Note that the scales of the axis have been changed to show the current at one-tenth of its actual per unit value, in order to better compare the current and voltage phasors. The locus of the current phasor for a fault at $R(h = 0)$ starts at $2.88 - j\,16.26$ per unit and reduces in magnitude as the fault moves along the line. The voltage phasor starts at a low value of $0.03 - j\,0.17$ per unit and increases in magnitude as the fault is moved. As the current at the far end of the line begins to dominate the fault current, the voltage phasor reverses direction, creating a hook shape in the voltage phasor locus.

When series compensation is added to the line, the relay R phasor loci change dramatically, as shown in Figure 15.6 for the case of 60% compensation. Here, the notation $h = 0-$ refers to the fault location on the bus side of the series capacitor, i.e., at R. The location $h = 0+$ is for the fault on the line side of the series capacitor, i.e., at RL, which is the same distance from bus R, $h = 0$, but on the line side of the capacitor bank. Note that the per unit current is divided by ten for improved scaling of the two plots. First, examine the locus corresponding to the relay current.

The current phasor locus forms a large circle as the current goes through its resonant values, reaching a maximum of 85.2 per unit at an angle of about 2 degrees or nearly unity power factor. Resonance occurs at $h = 0.17$, which is quite close to the capacitor bank. This is shown clearly in Figure 15.3. The circular locus continues as h increases with the current finally returning almost to the origin, at which point another resonance is experienced, causing the locus to trace a very small circular path in the left-half plane. The voltage phasor locus

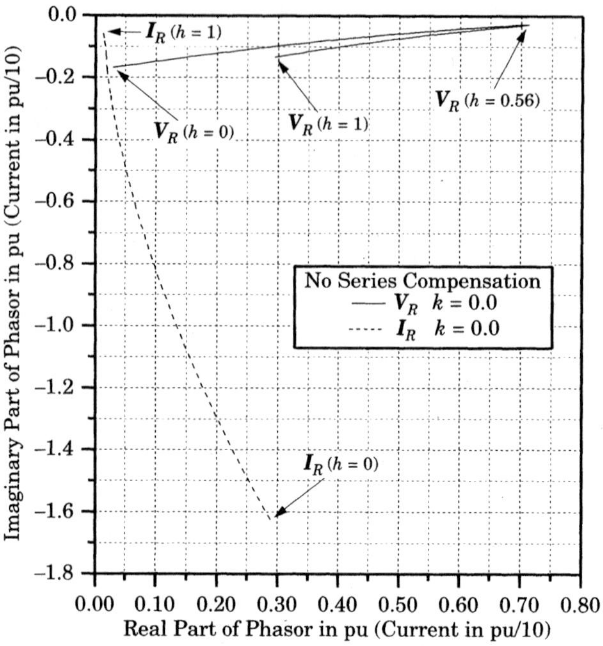

Figure 15.5 Phasor loci for the uncompensated line as a function of h.

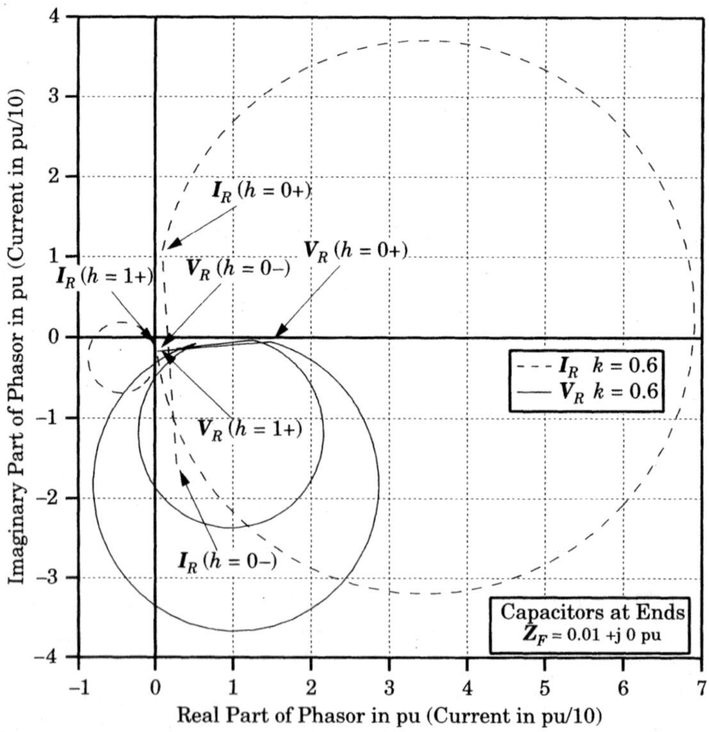

Figure 15.6 Phasor loci for $k = 0.6$ for varying h.

traces a similar type of locus, making a large circular path at the first resonance and ending with a small circular path in the opposite direction for the second resonance. At $h = 1+$, with the fault on the far side of the remote capacitor, i.e., at Q, the voltage drops to a very low value because the fault is on the bus at the remote end of the external impedance.

The relay must somehow observe all of this complexity and make a correct decision regarding the need to trip. One way that this can be accomplished is by measuring the apparent impedance seen from the relay location. Apparent impedance is a phasor quantity equal to the relay voltage divided by the relay current. The apparent impedance is plotted for different degrees of compensation in Figure 15.7. The heavy line represents the equivalent pi series impedance of the transmission line, with one end at the origin and the opposite end labeled $\mathbf{Z}_\pi$. A circle representing a mho characteristic is superimposed on the line impedance, with the major diagonal of the mho characteristic equal to the total line impedance. The mho circle is used to represent a typical relay trip zone, and is not intended to imply that this is the exact characteristic of any particular protective relay. It is simply a rough frame of reference to graphically display a typical relay trip zone.

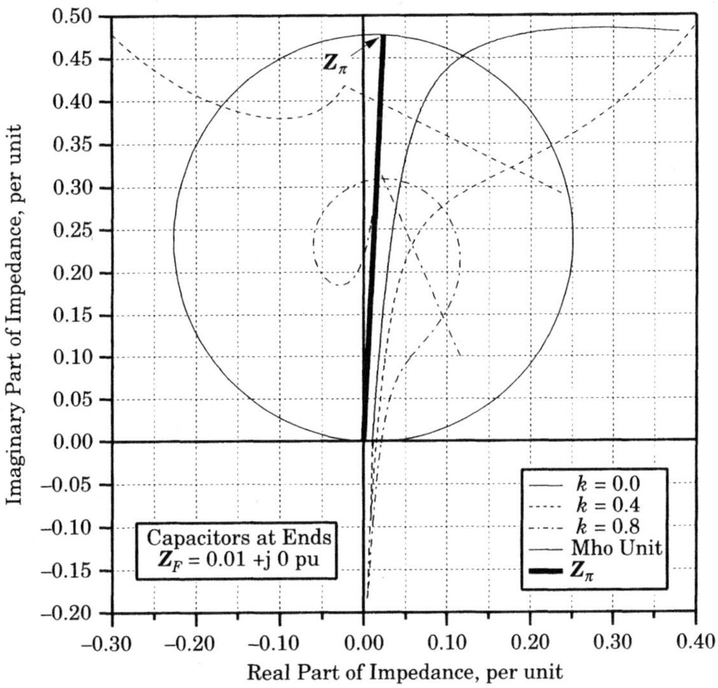

Figure 15.7 Apparent impedance seen by relay R as a function of h.

Figure 15.7 also shows typical plots of impedance seen by the relay for degrees of compensation ranging from 0 to 80%. All the plots of Figure 15.7 begin at position RL, which is the line side of the series capacitor at R. In the Z plane, this is a point displaced in the positive real direction by the value of the fault resistance. See Figure 15.2. This point corresponds to the point $h = 0+$ in our notation, to indicate that it is at the bus, but on the line side of the capacitor. When the capacitive reactance is nonzero, this point always lies in the fourth quadrant, with the exact position of the point moving farther in the negative X direction as the degree of compensation is increased. It is assumed that the relay current and voltage are both measured at the bus, i.e., the point R of Figure 15.2.

The first plot of interest is the solid curve just to the right of the transmission line impedance. This solid line represents the locus of the impedance seen at the relay location for the uncompensated line as the fault is moved from R to Q. The curve is not a straight line because of the way in which the currents in Figure 15.3 combine to form the total fault current. As the fault moves close to the far end of the line, the contribution of the total fault current from the Q end increases and the current contribution flowing clockwise through $\mathbf{Z}_E$ increases as well.

For the impedance loci representing series compensated lines, the impedance locus takes on a nearly circular shape as the current and voltage go through a resonance. For degrees of compensation less than 0.6, these loci are completely outside the plotted portion of the Z plane shown in Figure 15.7. As the fault is moved past the capacitor, with the fault at the point Q in Figure 15.2, the impedance takes an abrupt jump to its final point. This final point is usually, but not always, within the circular mho characteristic used as the protective zone reference in Figure 15.7. The QL end of the distributed parameter line can be located by noting the end of the smooth, curved portion of the locus. The locus then moves sharply to the right, representing the jump in apparent impedance as the fault is changed from QL to Q.

In order to get a more accurate picture of the impedance loci that fall outside the boundaries of Figure 15.7, a wider view of the complex impedance plane is shown in Figure 15.8. Here, we see that the impedance loci for $k = 0.2$ and 0.4 both have circular shapes and undergo significant movement for very small changes in the distance parameter h. The large circular locus, computed for $k = 0.2$, is plotted in Figure 15.8 with a Δh resolution of 0.002. Note that the impedance moves large ohmic distances in the complex plane for these very small changes in fault location. This occurs because of the sharpness of the resonance for this value of compensation. The same is true for $h = 0.4$, but to a much lesser extent.

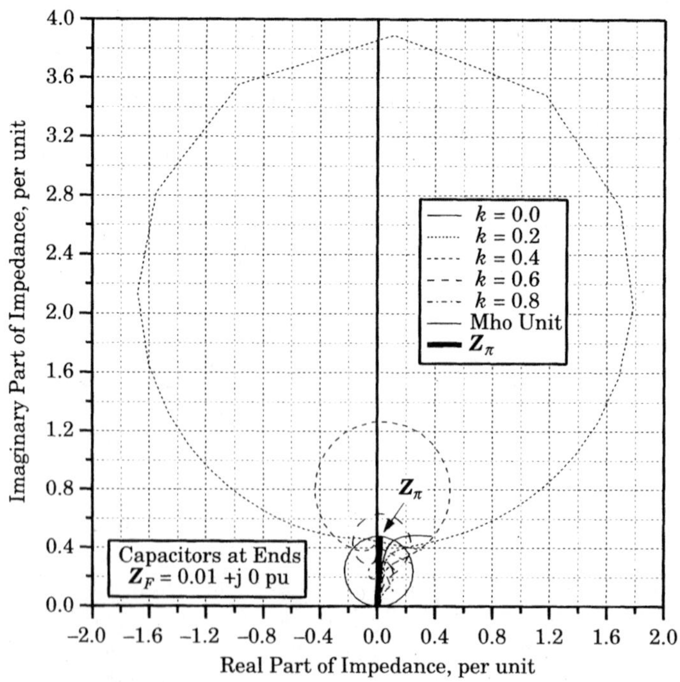

Figure 15.8 Expanded view of impedance plane plots as a function of h.

In summary, it is seen that the voltages, currents, and apparent impedances seen by the relay at R undergo significant changes that are very much dependent on both the fault location and the degree of compensation. These observations hold for the condition that the capacitor has the capability of carrying the large resonant currents that would naturally occur for a resonant circuit. In practice, we know that the capacitor is protected against such large currents by bypass devices. The effect of varistor protection on the relay quantities is explored later in this chapter.

15.2.2 End-of-Line Capacitors—Line Side Voltage

Another method of obtaining the relay quantities, where the series capacitors are at the ends of the lines, is to measure the voltage on the line side of the capacitors. In Figure 15.7, it is noted that close-in faults are viewed by the local relay as falling in the fourth quadrant. This presents a problem for the distance-measuring elements of the relay. The problem is not without solution, but is still something that requires special treatment.

If the relay voltage is measured on the line side of the capacitor, however, this problem is largely eliminated. The metering connections are shown in Figure 15.9. Using this method of measurement, the currents are exactly the same as before, but the voltage and impedance seen by the relay are quite different. As before, the capacitor bypass devices are assumed to be nonoperative for this test.

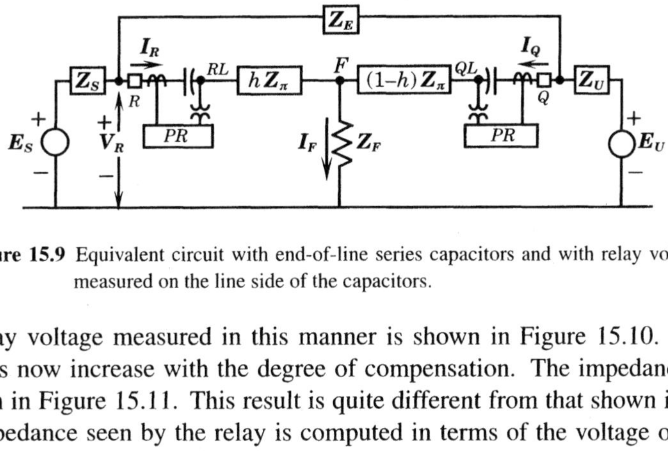

Figure 15.9 Equivalent circuit with end-of-line series capacitors and with relay voltage measured on the line side of the capacitors.

The relay voltage measured in this manner is shown in Figure 15.10. Note that the resonant peaks now increase with the degree of compensation. The impedance seen by the relay is shown in Figure 15.11. This result is quite different from that shown in Figure 15.7, where the impedance seen by the relay is computed in terms of the voltage on the bus side of the series capacitor. In Figure 15.7, the impedance is seen as negative for all degrees of compensation when the fault is close to the relay R, but this is not the case in Figure 15.11. That is an important difference. A relay with fault locator logic that assumes the line to be inductive, would interpret a negative reactance as a fault *behind* the relay. Relays often have a trip zone similar to the nominal mho characteristic shown in Figure 15.11. Such relays are usually not practical for series compensated lines. However, if the relay voltage is measured on the line side of the capacitor, there is no negative reactance observed by the relay. The observed resonances for faults at the far end of the line are noted in both cases, with the impedance excursion being somewhat greater in Figure 15.11. Taking the voltage measurement on the line side of the series capacitor is an improvement, as it eliminates the serious problem noted in Figure 15.7 of impedance measurements outside of the normal trip zone of a mho distance element.

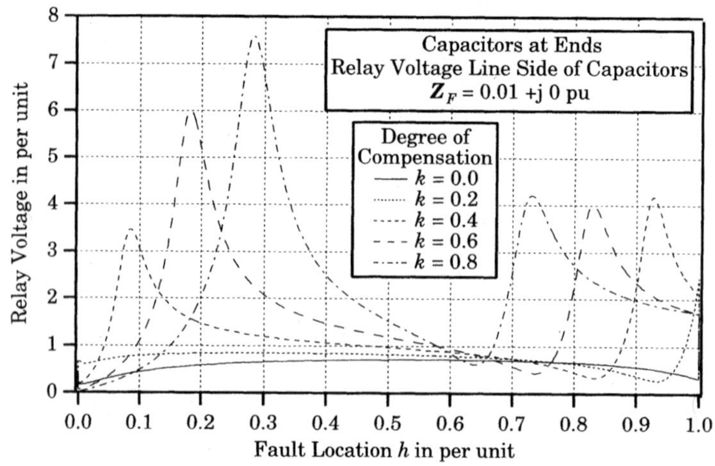

Figure 15.10 Relay voltage magnitude as a function of fault position h with voltage measured on the line side of the series capacitors.

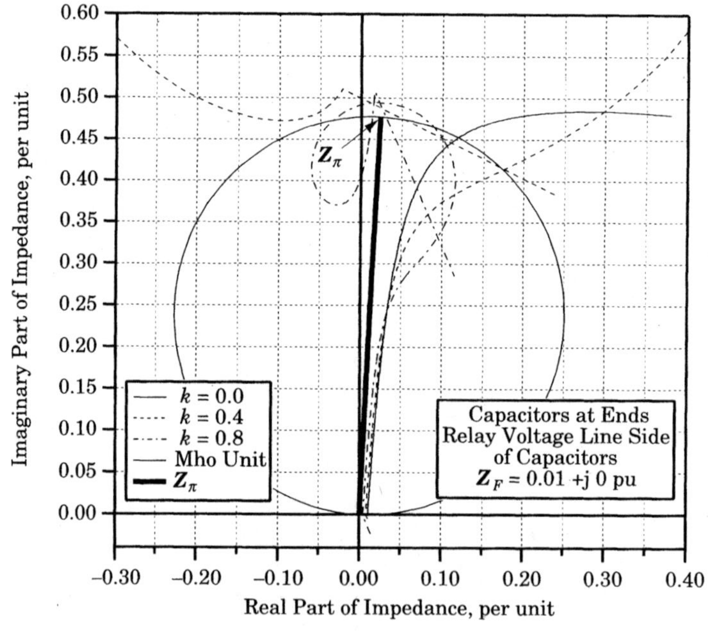

Figure 15.11 Relay apparent impedances with relay voltage measured on the line side of the series capacitors.

15.2.3 Capacitors at the Center of the Line

The effectiveness of series compensation is increased if the series capacitors are placed at the center of the line. This is especially important for very long lines, as the effectiveness is a function of line length. For lines of 300 km or less in length, the gain in effectiveness is not great. For a line with a capacitor at the center of the line, as in Figure 15.12, the currents, voltages, and impedances seen by the relay are examined.

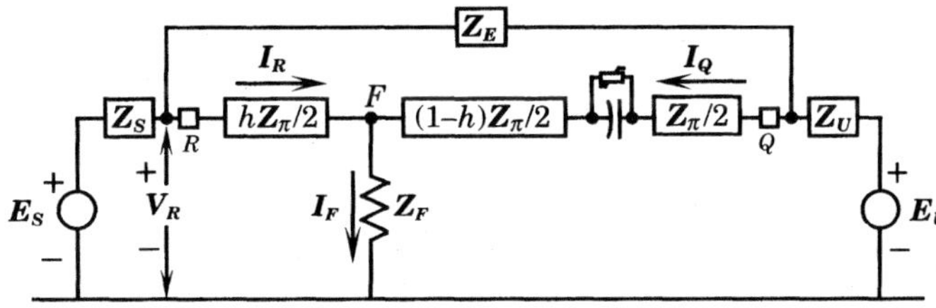

Figure 15.12 Series capacitor at the center of the line with fault at fractional distance h on the left side of capacitor.

Figure 15.13 shows the current magnitude seen at relay R as a function of fault location. For small degrees of compensation, the current magnitude for faults to the left of the capacitor location are nearly the same as the uncompensated current magnitude. For higher degrees of compensation, this is no longer true, although the difference is not great. For faults located beyond the capacitor bank, typical resonances are observed for the higher degrees of compensation, but the resonant peaks are not nearly as great as for the case with capacitors at the ends of the line.

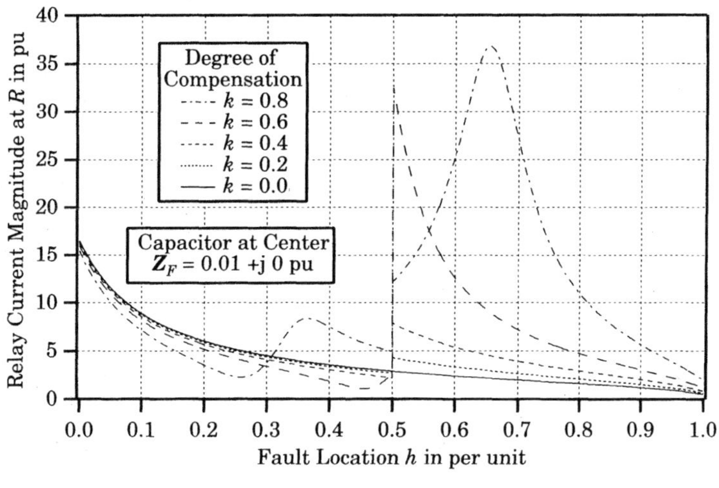

Figure 15.13 Relay current magnitude as a function of fault location h.

The relay voltage as a function of fault location is shown in Figure 15.14, where the voltage resonance at high degrees of compensation are shown. Although the voltage peaks at R are much higher than the uncompensated line, they are not as great as for the case with capacitors at the ends of the line. See Figure 15.4 for comparison.

Figures 15.13 and 15.14 display only the magnitude of the phasor current and voltage for varying fault location. When the degree of compensation is 0.6, a resonant condition begins to develop, but it is incomplete. Both the current and voltage increase in magnitude, but there is no smooth resonant peak in either variable. Instead, the series capacitor introduces a discontinuity in the resonance.

The complex phasor current and voltage for 60% series compensation is shown in Figure 15.15, which may be compared to Figure 15.6 for the case with capacitors at the ends of the line. As the fault location is moved along the line, a resonance in both current and voltage

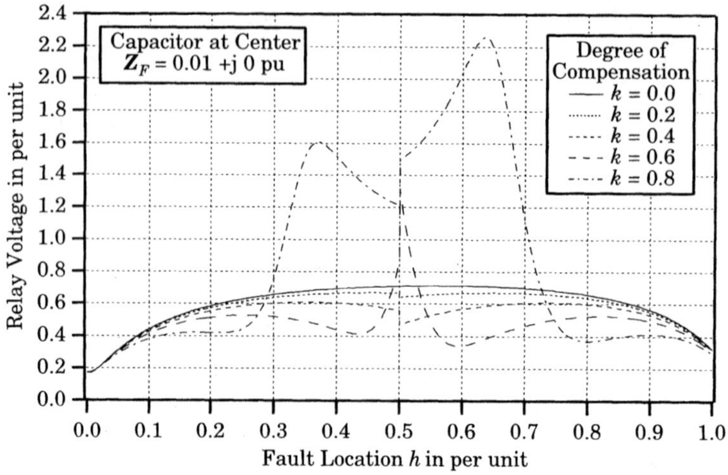

Figure 15.14 Relay voltage as a function of fault location h.

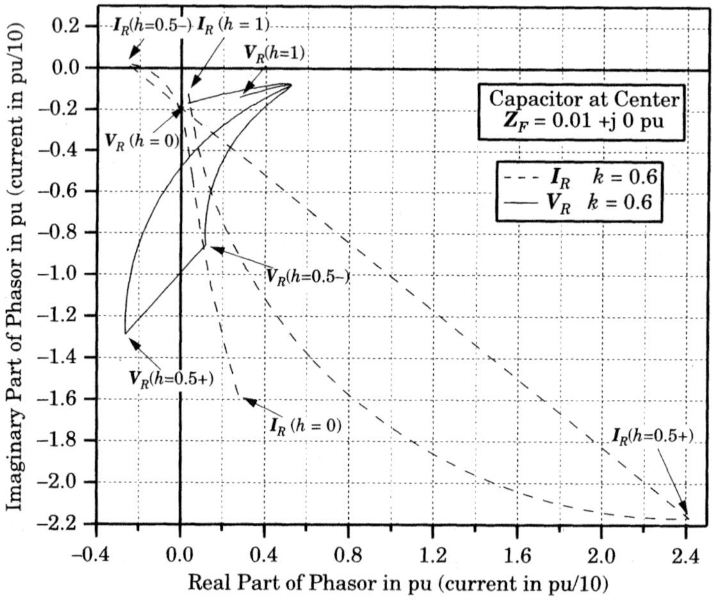

Figure 15.15 Phasor loci of current and voltage versus fault location, $k = 0.6$.

begins to develop but this is interrupted by the series capacitor, which causes a large change, especially in the current. The voltage change is especially interesting, since it moves into the third quadrant. Note that, at $h = 0.5+$, the current leads the voltage, as viewed from the relay at R. This is a special condition called a *voltage inversion*, which is explored in greater detail later. This condition exists for faults on the far side of the capacitor at locations up to a value of $h = 0.55$ before the real part of the voltage again becomes positive.

When the degree of compensation is increased to 0.8, the loci of currents and voltages is entirely different, as shown in Figure 15.16. For this higher compensation degree, the resonances are fully developed, both in the current and in the voltage. Both variables experience a small resonance on the near side of the capacitor, but have a very large resonance for faults beyond the capacitor. Note that the voltage does not invert, as it did for 60% compensation. Moreover, the current reaches very high values and becomes leading. This behavior is due to the high value of capacitive reactance. The impedance seen by the relay at *R* is quite complex, as shown in Figure 15.17.

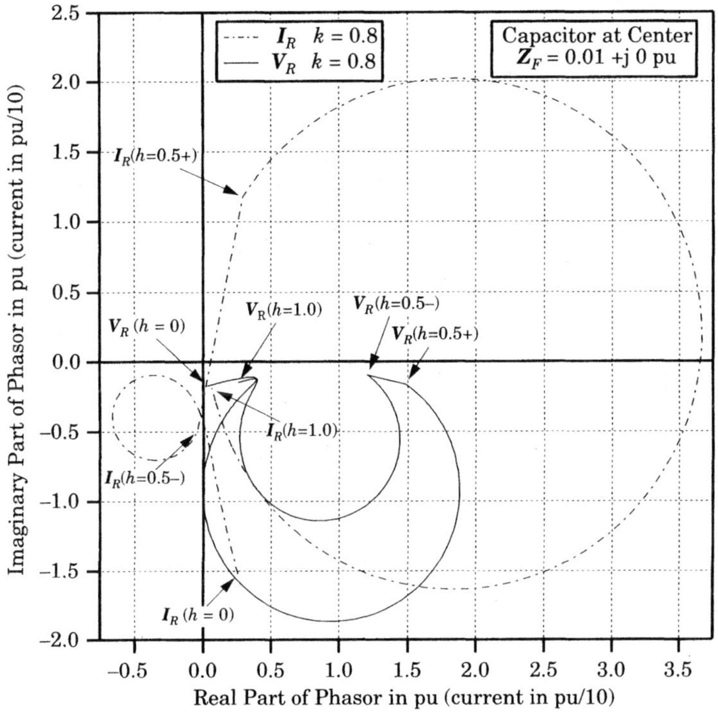

Figure 15.16 Phasor loci of current and voltage versus fault location, $k = 0.8$.

The impedance is mostly within the nominal mho circle, which indicates that any normal line protective scheme will have no problem with correctly detecting the fault. The impedance loci for degrees of compensation greater than 0.4, however, undergo rapid and unusual changes as the fault moves through the resonant conditions just beyond the center of the line. The impedance loops toward the origin as the resonance occurs, which indicates that the relay greatly overreaches the actual fault position. Just beyond the resonant point, the impedance becomes negative, moving outside the directional mho trip zone, but finally returns to the trip zone as the fault reaches the far end of the line. This indicates that a simple distance relay would not provide correct discrimination for all faults.

15.2.4 Conclusions on Series Compensation Effects

The unbypassed series capacitors in a transmission line greatly alter the currents, voltages, and apparent impedance seen by the relays. The relay variables are distorted more severely when the capacitors are at the ends of the lines than when at the center. The distortion is sufficient to complicate the distance measurement to the fault and suggests that special

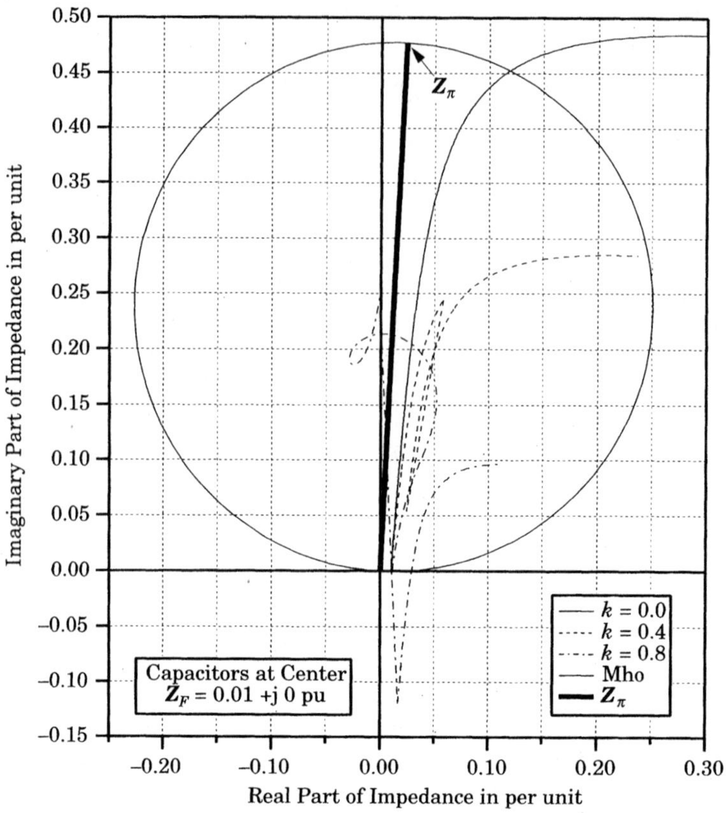

Figure 15.17 Locus of impedances seen by the relay at R as a function of h.

relays will be required that are designed to overcome these problems. Note, however, that the foregoing conclusions are for unbypassed series capacitors, which cause significant changes in the system variables.

When the series capacitor is at the center of the line, the distortion in the relay variables is not as great as for end-of-line capacitors. In some cases, depending on the degree of compensation, conventional line relays may work quite well for this type of compensation. However, this should be carefully checked for each application. Even for capacitors at the center of the line, when the degree of compensation is high the current, voltage, and impedance loci are very complex and take large resonant variations. This suggests that simple rules of thumb are not adequate for the analysis of these problems and that rigorous analytical methods are required to fully understand the relaying quantities.

15.3 SERIES CAPACITOR BANK PROTECTION

Series capacitor banks are made up of series and parallel combinations of capacitor elements as required in order to achieve the desired bank rating. Figure 15.18 shows a typical arrangement of individual capacitor units arranged in parallel and series arrays as required for a given bank rating.

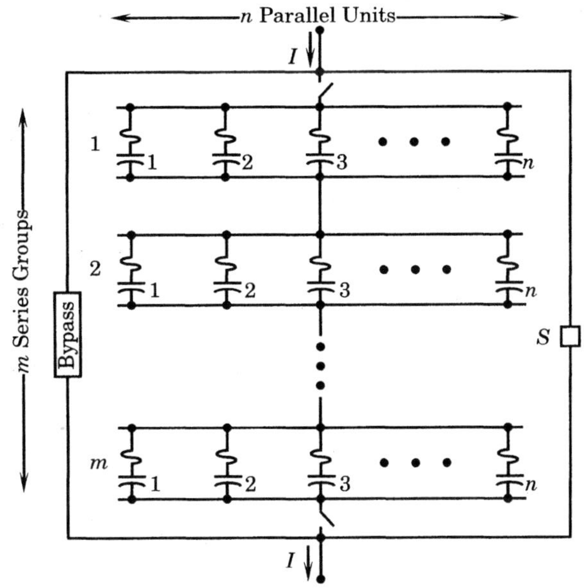

Figure 15.18 Electrical structure of a series capacitor bank.

Each individual capacitor is individually fused, either internally, i.e., within the capacitor unit enclosure, or externally, as shown in Figure 15.18. A bypass switch or breaker S is provided to bypass the entire bank for maintenance. In some cases, each series group has a bypass switch or breaker, making it possible to remove only selected groups from service, and thereby allowing the reactance rating of the bank to be modified by switching. A bypass device is also provided to shunt damaging fault currents around the capacitor bank, operating independently of the bypass breaker.

The capacitor bank is designed to carry a given rated current. Capacitor units are placed in parallel in sufficient number to carry the highest continuous current anticipated. This parallel arrangement is called a *group*. The groups shown in Figure 15.18 are labeled $1 \ldots m$. Groups are connected in series to provide the required ohmic value of compensation.

For the structure of Figure 15.18, let X_U be the reactance of a single capacitor unit. If we assume that all units are identical, then the reactance of a complete group of n_P capacitor units in parallel is given by

$$X_G = \frac{X_U}{n_P} \tag{15.5}$$

Connecting n_S identical groups in series to provide the required reactance gives the capacitor bank reactance of one phase,

$$X_P = n_S X_G = \frac{n_S X_U}{n_P} \tag{15.6}$$

The current rating of the bank is also of interest. Let

$$I_U = \text{rated current of a capacitor unit} \tag{15.7}$$

Then, the rated current of the capacitor bank, assuming all groups to have the same number of units, is given by

$$I_N = n_P I_U \tag{15.8}$$

Capacitor units must be connected in parallel in order to satisfy the desired current rating. Then, groups must be added in series in order to achieve the ohmic value corresponding to the desired degree of compensation.

15.3.1 Series Capacitor Bypass Systems

Series capacitors are subject to high stress during fault conditions, where the fault currents through the bank can create capacitor voltages that exceed the capacitor ratings [8], [9]. The capacitors are protected from overvoltage by the installation of bypass systems in parallel with the capacitors. During fault conditions, the bypass systems conduct most of the fault current through the parallel protective device, thereby ensuring that the capacitor is unharmed by the fault current. When the fault current ceases to flow and the voltage across the series capacitor returns to normal, the bypass impedance is increased to its prefault value, thereby allowing the current to again flow through the capacitor bank.

The types of capacitor bypass protection are varied and depend on the approach taken by various utilities and equipment vendors. They can, however, be conveniently grouped as follows:

(a) Bypass gaps
(b) Nonlinear resistors
(c) Bypass gaps supplemented by nonlinear resistors

These three types of capacitor segment protection are shown in Figure 15.19.

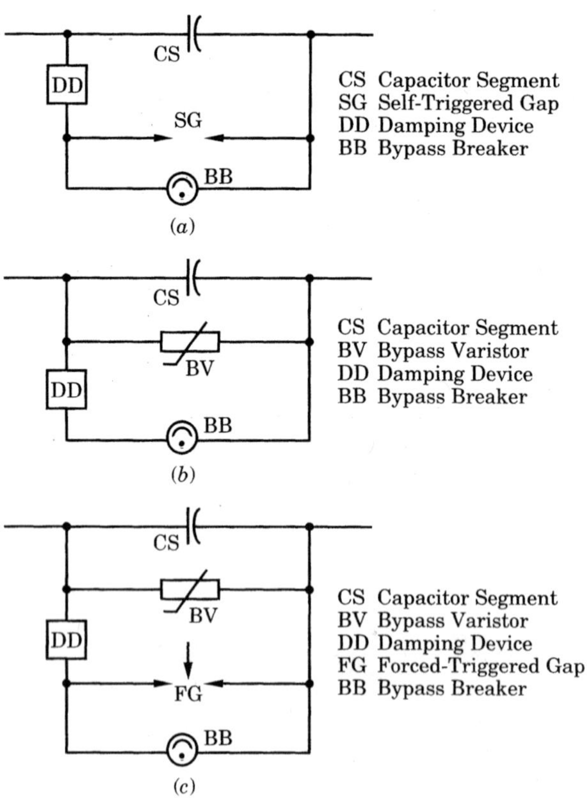

Figure 15.19 Series capacitors protected by parallel bypass. (a) Single gap overvoltage protector for one capacitor segment. (b) Gapless varistor overvoltage protector for one capacitor segment. (c) Varistor and triggered gap protection for one capacitor segment.

The entire capacitor bank is built up in a modular fashion, using components such as those shown in the figure, with the modules mounted on an insulated platform. It should be emphasized that the capacitor protection must be designed for a particular capacitor segment. In some series capacitor installations, series capacitors may be arranged in separate *segments*, which are arranged electrically in series, often on the same platform. The segments are not necessarily designed to have the same reactance—in fact, there are certain advantages to having segments of different sizes. If this is the case, the capacitor protections will also have different ratings.

The series capacitor segments shown in Figure 15.19 include a "damping device"[3] that is a current limiting circuit, usually consisting of an inductor and resistor in series. When the gap fires, or the breaker closes, this device causes a fast ringdown of the current at a frequency that depends on the inductance and capacitance of the circuit, usually in the range of 300 to 1000 hertz. The damping is such that the ringdown decays the current to a negligible value in less than one cycle of the fundamental frequency. Capacitors must be protected from the overvoltages that would occur due to the high currents that flow on the transmission line in response to line faults. Different protective responses are often designed for faults on the line that includes the series capacitor, which are designated as *internal* faults. Faults on other nearby lines are called *external* faults, and these more remote faults may require a different type of capacitor protection.

15.3.1.1 Bypass Gaps. The simplest type of installation is the parallel self-triggered bypass gap, an example of which is shown in Figure 15.19(*a*). Capacitor overvoltage protection for each segment is provided by a spark gap, installed in parallel with the capacitor, which is designed to protect the capacitor from the overvoltage that results from fault currents flowing on the transmission line. The gap is set to flash over at a given voltage, usually 2.0 to 3.5 per unit (where 1.0 per unit is equal to crest voltage at rated current). The gap flashes in a few microseconds following a fault, which bypasses the capacitors completely. The gap conduction is interrupted when the transmission line breakers operate to isolate an internal fault. For external faults, some means must be provided for extinguishing the gap conduction. This is sometimes accomplished by injecting compressed air through the gap to extinguish the arc. The injection of air is controlled by monitoring the current through the gap. A forced-triggered gap may also be used. This type of gap ensures consistent flashover at a given voltage. The bypass breaker is used by an operator to remove the capacitor bank from service for maintenance, and for reinserting the bank into service following these intentional removals. The bypass breaker is usually ordered to close when the gaps are fired, which provides a backup protection for the capacitor.

Bypass systems similar to that shown in Figure 15.19(*a*) have been in use since the 1950s in the western United States and have achieved a credible service record [10–11]. The major drawback to this concept is the complexity of the equipment and the fact that there may be some doubt regarding the successful reinsertion, which may be very important for system stability and for restoring power transfer quickly. These early systems often employed line relaying that depended on the capacitor being bypassed in order to obtain correct distance measurement for the fault condition. Thus, any error or delay in the bypass function increases the probability of delayed line relaying as well as capacitor damage. Despite these reservations, these systems have worked satisfactorily for years.

[3]The correct name for this device is the "discharge current limiting device" according to IEEE Standard 824-1994.

15.3.1.2 Metal Oxide Varistor Protection. The second type of capacitor bypass system is one that employs a nonlinear resistor, called a *varistor*, in the bypass circuit, as shown in Figure 15.19(*b*). The varistor voltage–current relationship can be described by the equation

$$v = ki^{\beta} \tag{15.9}$$

where v = resistor voltage
i = resistor current
β = a constant < 1.0

The solution to (15.9) for various values of current may be found by taking the logarithm to write

$$\alpha = \ln v = \ln k + \beta \ln i \tag{15.10}$$

Then

$$v = e^{\alpha} = e^{(\ln k + \beta \ln i)} \tag{15.11}$$

The earliest types of nonlinear resistors were made of silicon carbide, for which the constant β has a value of about 0.2–0.5. Figure 15.20 shows a plot of (15.9) for values of per unit current up to 10, and using a constant k set arbitrarily to limit the voltage to about 1.5 per unit for this range of current.

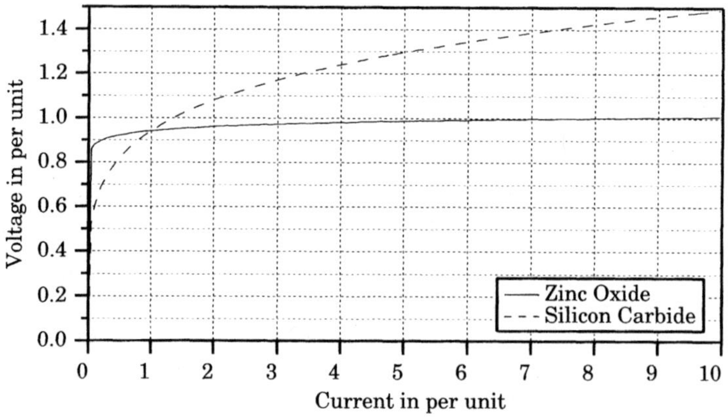

Figure 15.20 Voltage–current characteristics of two typical varistors.

The dashed curve in Figure 15.20 is typical for silicon carbide varistors. Note that, as the voltage approaches the limiting value this causes large currents to flow through the varistor, thereby effectively limiting the voltage across the capacitor. Because there is substantial current flowing even at normal voltage, however, it is not feasible to leave this type of resistor permanently in parallel with the capacitor, as it would increase the losses. Therefore, silicon carbide varistors are usually used in conjunction with a spark gap in series with the silicon carbide varistor. As the fault current raises the voltage across the capacitors, the gap will flash causing current to flow through the varistor. This helps to limit the fault current and limits the voltage across the capacitors. After the fault is cleared, the capacitors are reinserted by causing compressed air to extinguish the arc across the gap. Note that the voltage across the capacitor increases significantly as the fault current increases, as shown by the upper curve in

Figure 15.20. The varistor is not nonlinear enough to effectively "clamp" the voltage to an exact value. Silicon carbide varistors have been replaced in modern series capacitor installations with metal-oxide varistors.

Metal oxide varistors are characterized by a value of β that is an order of magnitude smaller than that for silicon carbide. The lower plot in Figure 15.20 is typical for zinc oxide varistors. Note that this varistor has two excellent characteristics for this application. First, the current flowing at normal voltages is negligible, which means that the resistor can be permanently installed in parallel with the capacitor without the need for a series gap. Second, this varistor effectively clamps the capacitor voltage to a given design level more effectively than silicon carbide. Figure 15.19(*b*) shows a typical arrangement for capacitor protection utilizing metal oxide varistors [12–15].

It should be noted that the arrangement of Figure 15.19(*b*) introduces the concept of the value of resistance that will be effective, and in the rating of that resistor. The previous system, with only a gap, is sensitive only to voltage. The varistor system requires an energy rating for the varistor to assure that both the capacitor and the varistor will survive the highest fault current magnitude for the required length of exposure. If the varistor is specified for too great an energy capacity, the cost will be excessive. The varistor bypass system also requires some method of measuring or estimating the accumulated energy absorbed by the varistor during a fault condition. If there is danger of exceeding the varistor rating, the bypass breaker must be ordered to close, thereby protecting both the capacitor and the varistor. The bypass breaker can be reopened by the system operator.

The metal oxide varistor (sometimes abbreviated as MOV), because of its greater nonlinearity, is superior in its ability to limit the capacitor voltage [12–15]. The normal leakage current through the varistor is on the order of milliamperes, which is considered acceptable. Detailed studies and field tests have shown the effectiveness of the metal oxide varistors in aiding transient stability of the system [14]. The MOV devices provide several benefits, including instantaneous reinsertion, lower capacitor protective levels, high reliability, and low maintenance costs.

15.3.1.3 Bypass Gap and Nonlinear Resistor. The arrangement shown in Figure 15.19(*b*) is sometimes altered to provide a gap in parallel with the varistor for faults of unusually high magnitude and as a safety measure, such as the system of Figure 15.19(*c*). This system provides three levels of protection. For external faults, the varistor will conduct as soon as the voltage exceeds the design value. However, if the energy accumulated by the varistor reaches a level close to the varistor rating, the forced-triggered gap is ordered to flash, which can usually be completed in a few milliseconds. On ordering the gap to flash, the bypass breaker is also ordered to close, which will usually require 40 to 50 milliseconds, thereby bringing the third level of protection into service.

Figure 15.21 shows a typical plot of line current in the faulted phase of a transmission line. The current has two parts, one part that flows through the capacitor and another component that flows through the varistor. When the instantaneous fault current reaches a high value, the capacitor voltage, which lags the current by 90 degrees, is also much higher than normal. When the voltage reaches the design value, the varistor conducts. This is seen to occur in short pulses, alternating positive and negative, as shown in Figure 15.21(*a*). These short pulses of current cause power to be dissipated in the varistor, and the accumulation of these power pulses results in a buildup of energy. Since the pulses occur rapidly, essentially all of the energy is contained within the varistor and causes its temperature to rise.

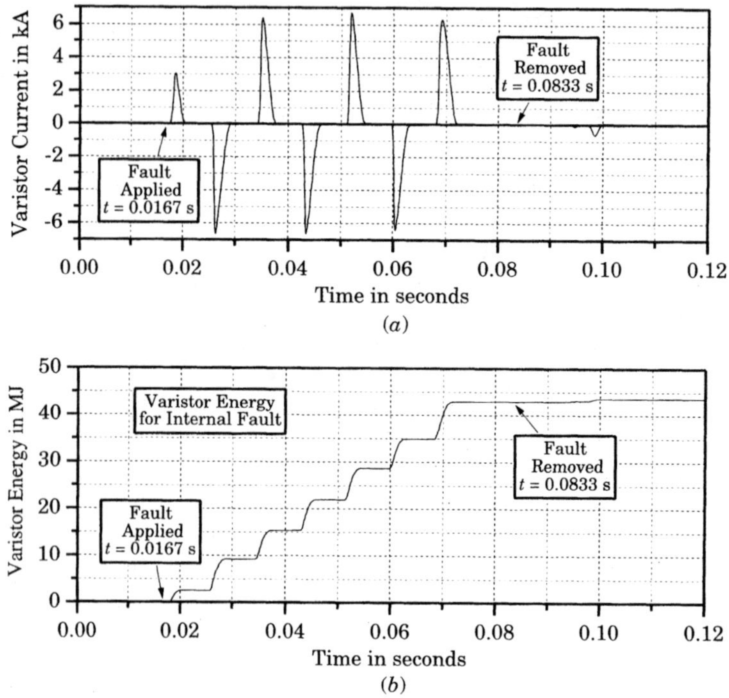

Figure 15.21 Plot of faulted phase current and varistor energy buildup. (a) Faulted phase current through capacitor and varistor. (b) Varistor energy buildup for a four cycle internal fault.

The total line current in the faulted phase is the sum of the capacitor current and the varistor current, although the two current components do not flow simultaneously. These currents are shown in Figure 15.22 for the first three cycles of the fault. The sum of the two components is a fundamental frequency current that grows in amplitude to the full fault value.

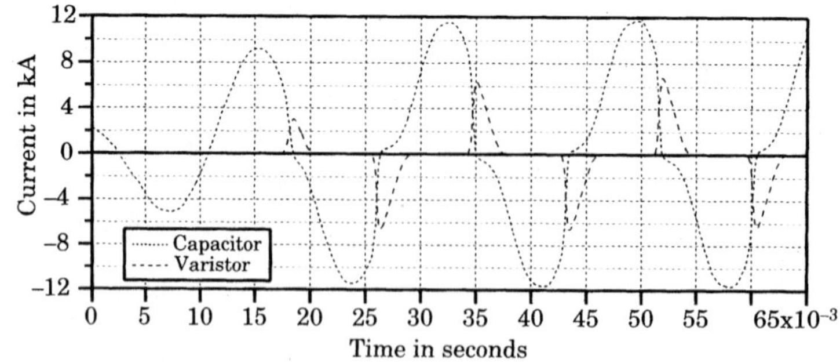

Figure 15.22 Total line current in the faulted phase.

Figure 15.23 shows a typical series compensation system that is designed to provide the three levels of capacitor protection described above. The equipment for two segments of compensation are arranged on a common high-voltage platform, where each segment has its

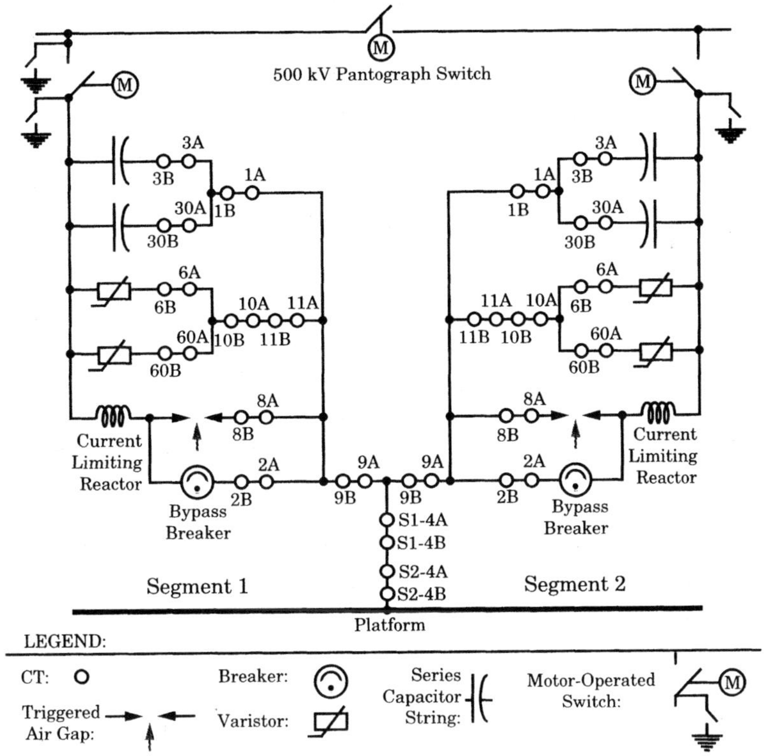

Figure 15.23 A typical series capacitor system with two segments.[4]

own measurements and series capacitor protection. Note that the equipment is arranged such that all current measurements are taken at platform potential. The varistor is always in the circuit and conducts immediately in response to a capacitor overvoltage and this is the primary protection for external faults. The triggered gap is designed to spark over when excessive current or varistor energy level is detected. The bypass breaker is programmed to close any time that the triggered gap is commanded to spark over.

Not shown in the figure are two platform control cabinets, one for each segment. Each platform cabinet contains redundant controls, two platform fiber-optic interfaces, two platform power supplies, and two varistor analog units and pulsers to trigger gap firing. The platform fiber-optic interface is a system of electronic circuits that has the function of transforming the platform analog current measurements into digital signals for transmission to the ground controllers, which are housed in ground-level control cabinets. The platform fiber-optic interfaces are powered by platform power supplies, which are powered by current transformers 9A and 9B, provided for this purpose. The varistor analog pulsers are electronic components that use analog signals to monitor and detect varistor energy, based on measured varistor currents. The output of the pulser is a pulse that is used to trigger the sparkover of the gap. All of these items are redundant, for reliability. The gap igniter consists of an expulsion tube that injects a stream of ions into the gap, which causes a reliable breakdown of the gap voltage near the design protective level, thus bypassing the capacitors.

[4]The author gratefully acknowledges the Mead-Adelanto Project in Southern Nevada and the project management who granted permission to use this drawing.

Note that all current measurements on the platform are redundant, with systems A and B operating independently. Failures of individual capacitor or varistor units can be detected by measuring the imbalance in current that occurs when one unit in a group fails. Capacitor units are individually fused. The master control of the series capacitor bank is performed by digital control computers, located in the ground control cabinets. Platform-to-ground communications is by dual paths of fiber-optic columns, arranged so that the failure of one column does not cause failure of either control system A or B. The system shown in Figure 15.23 is only for one phase.

15.3.2 A Fundamental Frequency Varistor Model

The development of a fundamental frequency model of the nonlinear varistor has been developed by detailed simulation to a series R-X model of the varistor that closely fits experimental data [13]. Figure 15.24 indicates the process, where the nonlinear device is to be replaced by a series R-X model, where the exact values of R_C and X_C become functions of the current flowing through the varistor. This is indicated in Figure 15.24 by the instantaneous current i_L, which is shown in the figure as controlling both R_C and X_C.

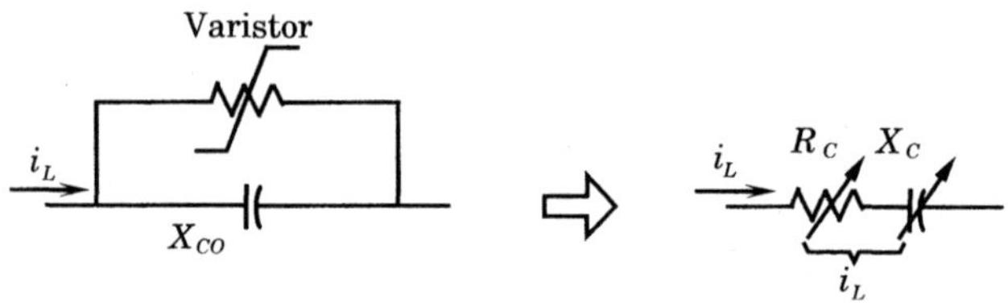

Figure 15.24 Series R_C-X_C model of the series capacitor and varistor.

A model of the series impedance is shown in Figure 15.25, where it is clear that both the resistance and capacitive reactances are nonlinear, and are functions of the normalized capacitor bank current I_{LN}. The values plotted are expressed in per unit, where one per unit I_L

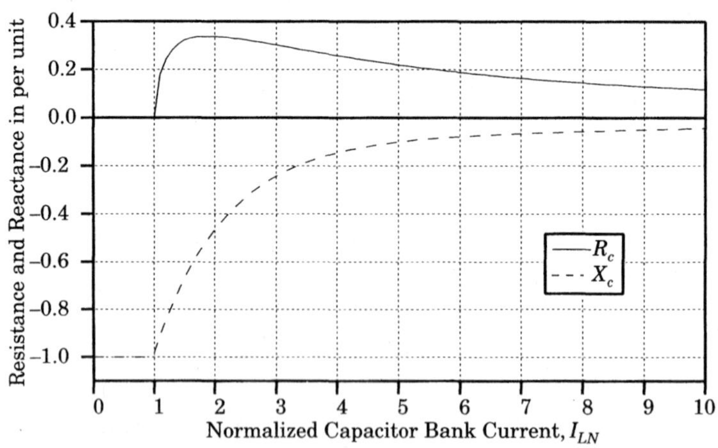

Figure 15.25 Effective resistance and reactance of the varistor-protected series capacitor bank as a function of normalized bank current, I_{LN}.

is the capacitor bank current rating at which the varistor begins to conduct. For bank currents below this value, the series circuit is a constant capacitive reactance equal to the series capacitor rating. When the current exceeds the varistor conducting threshold, the current in the varistor increases rapidly, as shown in Figure 15.20, which decreases the effective series impedance of the circuit and diverts current from the capacitor. As larger values of current flow through the varistor, the circuit becomes predominately resistive, as shown in Figure 15.25. For large current values, the capacitive reactance is less than 5% of its normal value, but there is still a small capacitive reactance component.

Solving a circuit containing the nonlinear R-X combination of Figure 15.24 requires an iterative procedure. An initial value of current is estimated, based on the system condition, and then the circuit with the nonlinear impedance is solved using the values corresponding to this estimated current. This solution usually shows an error in the original estimate, so that a new estimate is made and the circuit solved a second time. This is repeated until the estimated current and the solution agree to the desired precision. The resulting solution is a fundamental frequency equivalent system solution of the parallel circuit containing the capacitor and varistor bypass system.

Another way of viewing the R-X characteristic of the varistor is to plot the data of Figure 15.25 in the complex impedance plane. This result is shown in Figure 15.26, where the normalized values of R_C and X_C are both in per unit, with the base impedance value being the capacitor reactance rating in ohms.

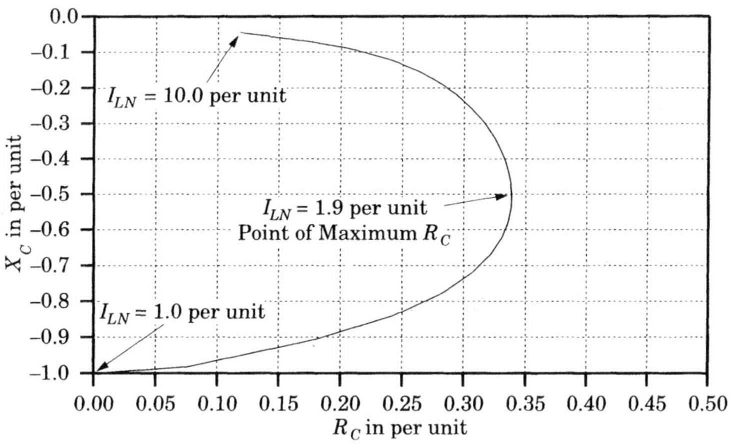

Figure 15.26 Varistor-protected capacitor bank effective resistance and reactance in the complex impedance plane.

To analyze the varistor equivalent circuit, we begin by defining the following currents.

$$\begin{aligned} I_N &= \text{Rated capacitor bank current, (A rms)} \\ I_B &= \text{Base current} = \frac{S_{B-3\phi}}{\sqrt{3}V_{B(L-L)}}, \text{ (A rms)} \\ I_L &= \text{Line or capacitor current, (A rms)} \\ i_{TH} &= \text{Capacitor threshold current rating} = \sqrt{2}I_{TH}, \text{(A)} \end{aligned} \tag{15.12}$$

The capacitor rated current, I_N, is the rms value of current that the capacitor is designed to carry continuously. The varistor conducting threshold i_{TH} is the instantaneous value of capacitor bank current at which the varistor is designed to begin conducting. This is related to the rms value of the threshold current, I_{TH}, as noted in (15.12). The line current, I_L, is the total current flowing in the line, whether it flows through the capacitor, the varistor, or both. We normalize the line current and identify the result by appending the per unit subscript "u" for clarity.

$$I_{Lu} = \frac{I_L}{I_B} \text{ per unit on the system base} \tag{15.13}$$

This is the current through the capacitor when the varistor is not conducting, expressed in per unit, and where I_B is the base current in amperes on the system study base.

The varistor conducts for that part of each half cycle where

$$I_L > I_{TH} \text{ A} \tag{15.14}$$

or

$$I_{Lu} > \frac{I_{TH}}{I_B} \text{ per unit on system base} \tag{15.15}$$

Now, let X_C be the capacitive reactance of the protected capacitor bank in ohms. Then, from [13] we write the following general equation for the equivalent series resistance and reactance in ohms.

$$\begin{aligned} R_c &= X_N(0.0745 + 0.49e^{-0.243I_{LN}} - 35.0e^{-5.01I_{LN}} - 0.6e^{-1.4I_{LN}})\Omega \\ X_c &= X_N(0.1010 - 0.0057I_{LN} + 2.088e^{-0.8566I_{LN}})\Omega \end{aligned} \tag{15.16}$$

$$\text{where } I_{LN} = \frac{I_L}{I_{TH}} \text{ per unit on the } I_{TH} \text{ base} \tag{15.17}$$

Note that the series equivalent parameters given by (15.16) are defined in terms of the rated series capacitor bank reactance, X_N, which is the rated value of the capacitive reactance. The equivalent parameters are also defined in terms of the per unit current I_{LN} which is in per unit on the capacitor threshold rating, not on a system base.

For convenience, we define

$$I_{TH} = k_P I_N \tag{15.18}$$

In many cases, the value of k_P is in the range of 1.7 to 2.0. The line current at which the varistor begins to conduct is given by (15.15). The per unit line current with the varistor conducting may be determined by a change of base.

$$I_{LN} = \frac{I_L}{I_{TH}} = \frac{I_L}{k_P I_N} = \left(\frac{I_B}{k_P I_N}\right)\left(\frac{I_L}{I_B}\right) = K_{bc} I_{Lu} \tag{15.19}$$

where we have defined the parameter

$$K_{bc} = \frac{I_B}{k_P I_N} \tag{15.20}$$

EXAMPLE 15.1

Determine the magnitude of line current in per unit for which the varistor will begin conducting under the following defined conditions:

$$S_B = 1000 \text{ MVA} \qquad I_N = 1000 \text{ A}$$

$$V_B = 500 \text{ kV} \qquad k_P = 2.0$$

$$I_B = 1154.701 \text{ A} \qquad K_{bc} = \frac{1154.701}{2000} = 0.577$$

$$Z_B = 250\ \Omega$$

The varistor begins conducting when $I_L = I_C = 1.0$ per unit on I_{TH} base, i.e.,

$$I_{Lu} = \frac{I_{LN}}{K_{bc}} = \frac{1.0}{K_{bc}} = \frac{2000}{1154.7} = 1.732 \text{ per unit on system base} \tag{15.21}$$

In order to make the equivalent impedance equations more useful, a change of base is necessary, which requires the substitution of (15.19) into (15.16). This gives the following result.

$$\begin{aligned} R_c &= X_N(0.0745 + 0.49e^{-0.243K_{bc}I_{Lu}} - 35.0e^{-5.0K_{bc}I_{Lu}} - 0.6e^{-1.4K_{bc}I_{Lu}})\ \Omega \\ &= X_N f_1(I_{Lu}) \\ X_c &= X_N(0.101 - 0.005749K_{bc}I_{Lu} + 2.088e^{-0.8566K_{bc}I_{Lu}})\ \Omega \\ &= X_N f_2(I_{Lu}) \end{aligned} \tag{15.22}$$

The equations (15.22) give the value of the series equivalent R and X parameters as functions of the total line current in per unit when the varistor is conducting. ■

Note that the foregoing equations give the varistor parameters in ohms. To normalize to the system base, both equations must be divided by the system base impedance.

$$\begin{aligned} R_c &= \frac{X_N f_1(I_{Lu})}{Z_B} \quad \text{per unit} \\ X_c &= \frac{X_N f_2(I_{Lu})}{Z_B} \quad \text{per unit} \end{aligned} \tag{15.23}$$

15.3.3 Relay Quantities Including Varistor Bypass

To determine the effect of the series capacitor bypass on the relay voltage and current, the system must be solved for the given series capacitor location(s) and with the capacitor in parallel with a varistor of appropriate rating. In order to estimate the rating, a test case is considered, where the total delivered power in two parallel 500 kilovolt lines is taken to be nominally 2000 megawatt. When one line is faulted and removed from service, the remaining line must carry the entire 2000 megawatt, so we let this be the series capacitor rating. This corresponds to about 2300 amperes. Therefore, let us set the rating at 2500 amperes, since we do not want the series capacitor to bypass during normal operation in the neighborhood of 2000 megawatt.

The system configuration for varistor protected series capacitors at the ends of the line is shown in Figure 15.27, where the voltage measurement is assumed to be on the bus side of the capacitor at both ends of the line. The varistor will bypass the capacitor according to the rules described in Section 15.3.2. The fundamental frequency relay quantities are computed according to these rules, with the results given below.

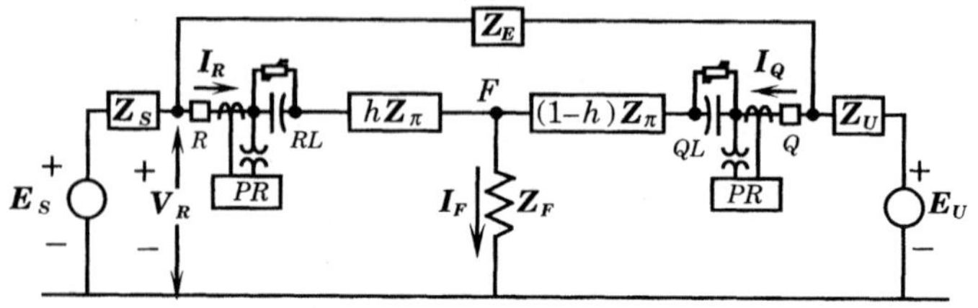

Figure 15.27 System configuration with varistor protected series capacitors at the ends of the line, with bus side voltage measurements.

The current magnitude for varying fault location is shown in Figure 15.28. This figure should be compared with the currents shown in Figure 15.3, which plots the same case but with no bypass of the capacitors. The very high resonant current peaks have been eliminated by the varistor protection and the current magnitude becomes similar to the uncompensated line, but with slightly higher magnitude due to two factors: (1) the small capacitive reactance in the bypassed compensation system, and (2) the fact that one of the capacitors remains unbypassed for certain fault locations.

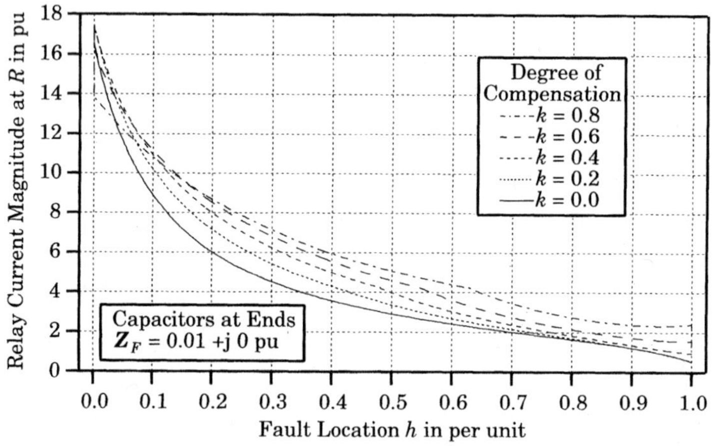

Figure 15.28 Relay current magnitude as a function of fault location h.

The point at which bypassing occurs is shown in Figure 15.29, which shows that the varistor is either conducting (UP) or not conducting (DOWN). This view of the bypass situation shows that, for small degrees of compensation, there is a region near the center of the line where neither of the capacitors is bypassed. However, for high degrees of compensation there is a region near the center of the line when both capacitors are bypassed simultaneously. The fault locations where switching of the varistor occurs can be seen on some of the plots of current, voltage, and impedance.

The voltage seen by the relay at R is shown in Figure 15.30. As with the current magnitudes, the resonant peaks of voltage noted in Figure 15.4 have been eliminated by the varistor and the relay voltage is at a more reasonable level.

A plot of the relay voltage and current in the complex plane, for a degree of compensation of 60%, is shown in Figure 15.31. Having removed the resonances in both the current and voltage, the voltage and current phasors in the complex plane are more restricted in their

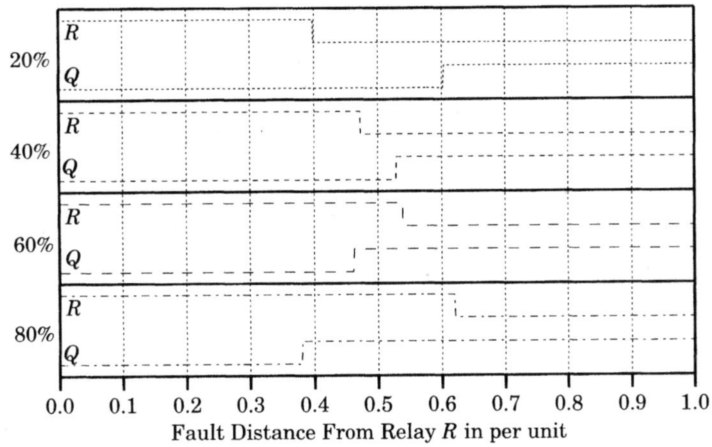

Figure 15.29 Varistor firing for capacitors at line ends.

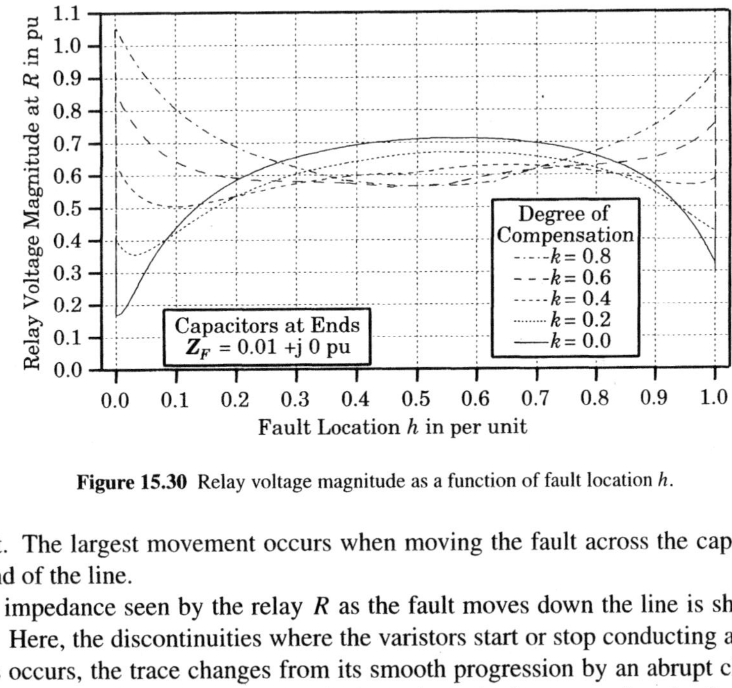

Figure 15.30 Relay voltage magnitude as a function of fault location h.

movement. The largest movement occurs when moving the fault across the capacitor at the relay R end of the line.

The impedance seen by the relay R as the fault moves down the line is shown in Figure 15.32. Here, the discontinuities where the varistors start or stop conducting are apparent. When this occurs, the trace changes from its smooth progression by an abrupt change in direction. At the remote end of the line, the large jump is due to moving the fault from one side of the capacitor to the other. One item of interest is whether the relay experiences over- or underreaching, since the traces all travel outside the nominal mho circle. These crossing points are summarized in Table 15.1. This shows a slight tendency to overreach[5] for the higher degrees of compensation. The mho characteristic used for the measurements in Table 15.1 is shown in Figure 15.32, where the diameter of the mho circle is arbitrarily set as being exactly

[5]The *reach* of a relay is defined as: "The extent of protection afforded by a relay in terms of the impedance or circuit length as measured from the relay," ANSI/IEEE Std C37.100-1981.

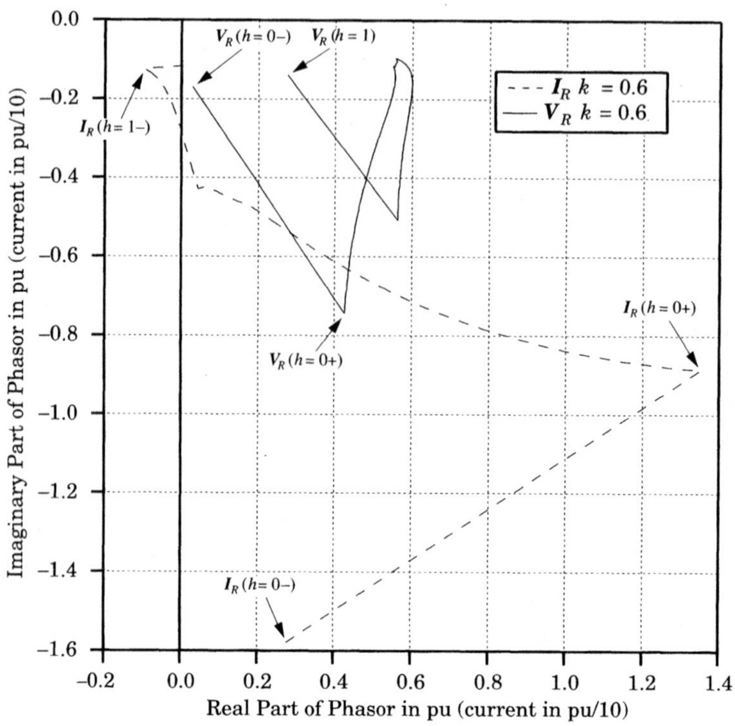

Figure 15.31 Current and voltage phasor loci with varistor protection.

the total line impedance in the complex plane. The reach measurements will vary depending on the mho characteristic used.

TABLE 15.1 Reach of Relay R for Various Degrees of Compensation

$k = 0.2$	$k = 0.4$	$k = 0.6$	$k = 0.8$
$h = 0.892$	$h = 0.906$	$h = 0.924$	$h = 0.984$

The relay impedance plot of Figure 15.32 should be compared with the unbypassed case, shown in Figure 15.7. The loci are more restricted, due to the lack of resonance and there is not such a large excursion initially into the fourth quadrant, in Figure 15.32. Still, there is a problem for the fault detector for close-in faults, especially at high degrees of compensation, as the apparent impedance moves into the fourth quadrant for close-in faults.

15.3.4 Effect of System Parameters

The foregoing examples of protective relay current, voltage, and impedance measurement have been computed with certain assumptions regarding the power system. These assumptions include the following:

1. The Thevenin equivalent sources at each end of the line are exactly equal, both in their Thevenin equivalent impedance, but also in their voltage magnitudes and angles.

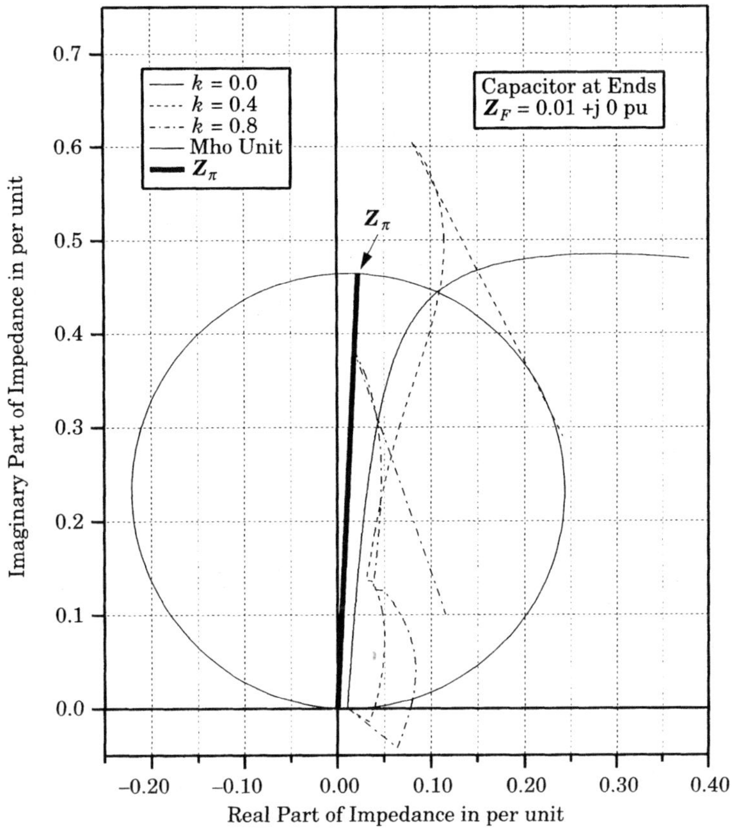

Figure 15.32 Loci of impedance seen by relay R as a function of fault location h.

2. The external impedance has been modeled as if a strong integrated system exists in parallel with the protected line, resulting in a small external impedance.
3. The fault impedance has been modeled as a pure arcing resistance, but of small magnitude.
4. Only three-phase faults have been considered.
5. The varistor can be accurately modeled as a fundamental frequency device using the model derived in [13].

Assumption 1 creates a condition where the fault currents flowing from the two sources are symmetrical. A plot of the current magnitude I_R and I_Q with respect to fault location along the line will show that these currents are exact mirror images of each other. The same is true of the other currents and voltages in the system. This assumption ignores any load current flowing in the system. It also assumes that both sources are of equal strength, in terms of their ability to supply fault current. Assumption 2 is arbitrarily adopted, but will not be valid for some systems, especially for the case of a developing system that has few high voltage transmission lines. However, these cases are readily accommodated by simply increasing the external impedance to the correct value, which can be determined by a short-circuit study. In the limit, the external impedance can be assumed infinite, which would seldom be the case for a developed, integrated power system.

Assumption 3 assumes a very low value of arc resistance, which might be correct if maximum currents are of interest. However, high arc resistance is also of interest in many cases, for example, where the line may be contacted by tree branches. Also, the impedance used here to represent the fault might actually represent both the fault impedance and the impedance of the negative and zero sequence networks, if an unbalanced fault is to be considered. In that case, the fault impedance will also have a reactance as well as a resistance, and this reactance must also consider the location and size of the series capacitor bank. Assumption 4 has been made simply for convenience. However, unbalanced faults of any type are readily solved using the method of symmetrical components, where the fault impedance is adjusted to include the appropriate sequence network impedances. This will not be pursued in detail here since the techniques are well known.

Assumption 5 is simply stated, but is more difficult to justify. The series capacitor current flows on each half cycle up to the value where the varistor begins to conduct, and the varistor current flows for a portion of that half cycle. It is assumed that these two current segments sum to an equivalent fundamental frequency current. This assumption is discussed in greater detail in [7].

We now examine the effect of the above assumptions briefly, again using an example for comparison of computed currents, voltages, and impedance at the relay location. This will be done only for the unbypassed capacitor cases. For convenience, the previous unbypassed case will be referred to as the "base case." Only the case with capacitors at the ends of the line are examined. Table 15.2 shows the parametric changes to be studied.

TABLE 15.2 Parametric Impedance Changes Selected for Study

Z Change	Z_S	Z_U	Z_E	Z_F
Base Case	$0.0 + j0.1$	$0.0 + j0.1$	$0.002 + j0.04$	$0.01 + j0.0$
$\mathbf{Z}_E$	$0.0 + j0.1$	$0.0 + j0.1$	$0.04 + j0.80$	$0.01 + j0.0$
$\mathbf{Z}_S$ and $\mathbf{Z}_U$	$0.0 + j2.0$	$0.0 + j0.005$	$0.002 + j0.04$	$0.01 + j0.0$
$\mathbf{Z}_F$	$0.0 + j0.1$	$0.0 + j0.1$	$0.002 + j0.04$	$0.05 + j0.0$

15.3.4.1 Effect of Increased External Impedance. The effect of increased external impedance is readily studied by varying this impedance. The original external impedance, given in (15.3), is $0.002 + j0.040$ per unit. This impedance is increased by a factor of 20 to a new value of

$$\text{New } \mathbf{Z}_E = 0.04 + j0.80 \text{ per unit} \tag{15.24}$$

The result of this change is shown in Figures 15.33 through 15.36.

Figure 15.33 shows the fault current magnitudes for varying fault location along the line. This figure should be compared to the base case shown in Figure 15.3.

The peaks of current magnitude for 40, 60, and 80% compensation are considerably higher for this case than for the base case, but the 20% compensation case is much lower. The resonances at the remote end of the line are almost completely eliminated by the higher external impedance, however. The same comments apply to the voltage magnitude plots, shown in Figure 15.34, where the resonant peaks are very high for the higher degrees of compensation, compared to Figure 15.4. Again, the higher external impedance has almost eliminated the resonant voltage peaks at the remote end of the line, which were great for the base case, as noted in Figure 15.4.

The loci of the current and voltage phasors in the complex plane are shown in Figure 15.35 for 60% compensation. This is the familiar circular resonance plot, both for faults close to the relay and again for remote faults, but the magnitude of the resonant loci are greater for the

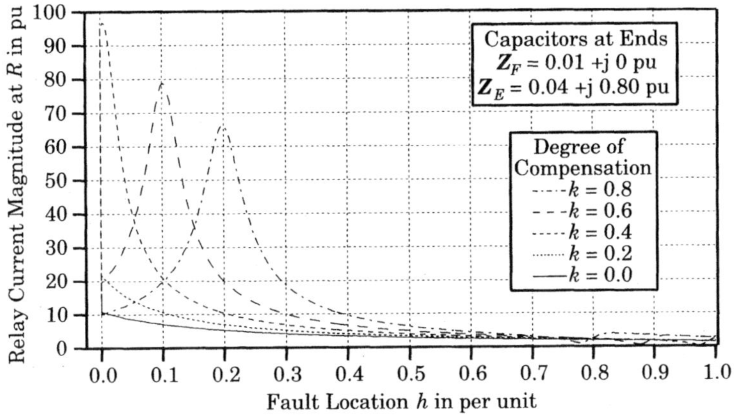

Figure 15.33 Relay current magnitude as a function of fault location h.

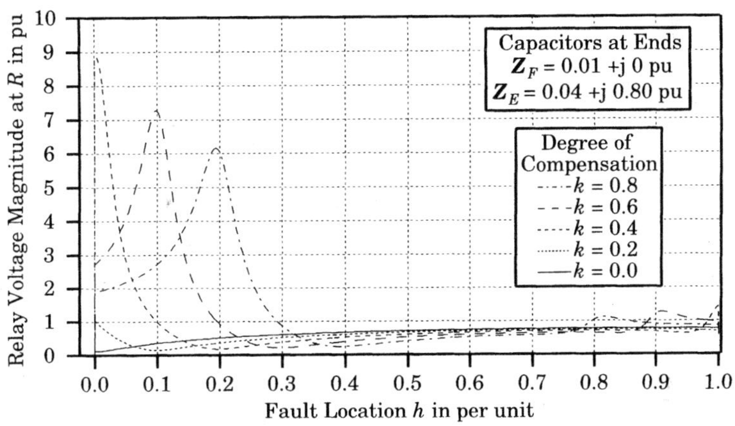

Figure 15.34 Relay voltage magnitude as a function of fault location h.

case with high external impedance, shown in Figure 15.35. Compare with Figure 15.6.

The impedance seen by the relay R is shown in Figure 15.36, which should be compared to Figures 15.7 for the base case. In Figure 15.7, large resonant excursions of the loci are evident for all degrees of compensation except 80%, which is well contained within the mho circle. For the case of high external impedance, however, all of the loci have much larger resonant circular excursions, and leave the mho circle at a lower value of h. An indicator of reach is given for different h values in Table 15.3.

TABLE 15.3 Reach of Relay R for Various Degrees of Compensation

Case	$k = 0$	$k = 0.4$	$k = 0.8$
Base Case	$h = 0.910$	$h = 0.806$	$h = 1.0$
High Z_E	$h = 0.915$	$h = 0.913$	$h = 0.758$

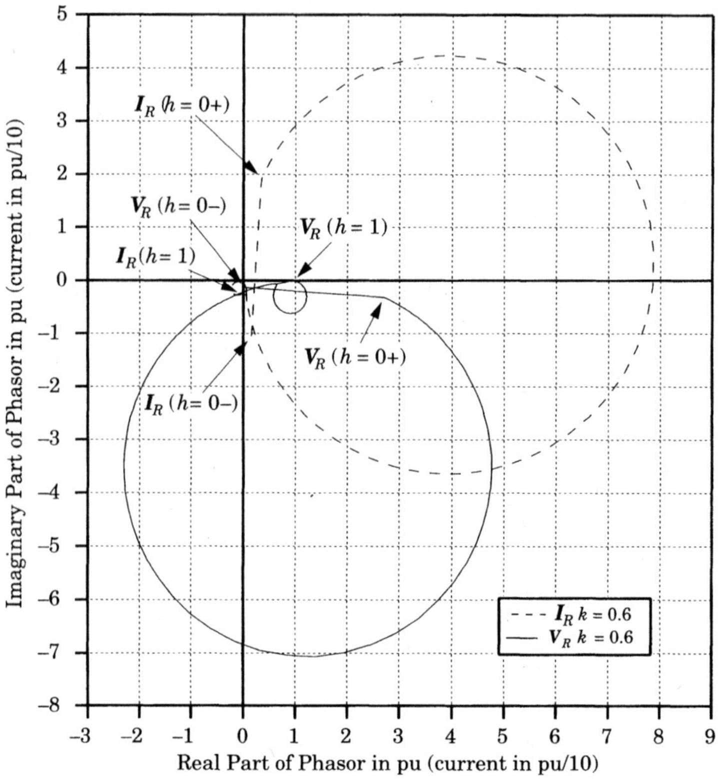

Figure 15.35 Current and voltage phasor loci with high external impedance.

In Figure 15.36 both impedance loci extend well outside the nominal mho circle for this case with high external impedance. This is due to the reduction of infeed at the remote end of the protected line.

15.3.4.2 Effect of Increased Source Impedance. To understand the effect of different source impedances, the source impedance both $\mathbf{Z}_S$ and $\mathbf{Z}_U$ are changed by a factor of 20, with $\mathbf{Z}_S$ being increased and $\mathbf{Z}_U$ being decreased. The new values of these parameters are

$$\begin{aligned}\text{New } Z_S &= 0 + j2.000 \text{ per unit}\\ \text{New } Z_U &= 0 + j0.005 \text{ per unit}\end{aligned} \tag{15.25}$$

A comparison of the computed results are described as follows. The resonant current peaks of this example, shown in Figure 15.37, are interesting since they are higher than those of the base case of Figure 15.3, even though the current source behind the relay is much weaker. The source of the increased current is the voltage source at the remote end of the line, providing increased current through the external impedance. The voltage profile peaks, as shown in Figure 15.38, are somewhat smaller than the base case shown in Figure 15.4.

15.3.4.3 Effect of Increased Fault Impedance. The effect of increased fault resistance has an effect that is most noticeable in the impedance measured by the relay. As the fault resistance is increased, the locus of impedances seen by the relay move straight to the right, reflecting directly the larger resistance in the circuit. Clearly, if the resistance is large enough,

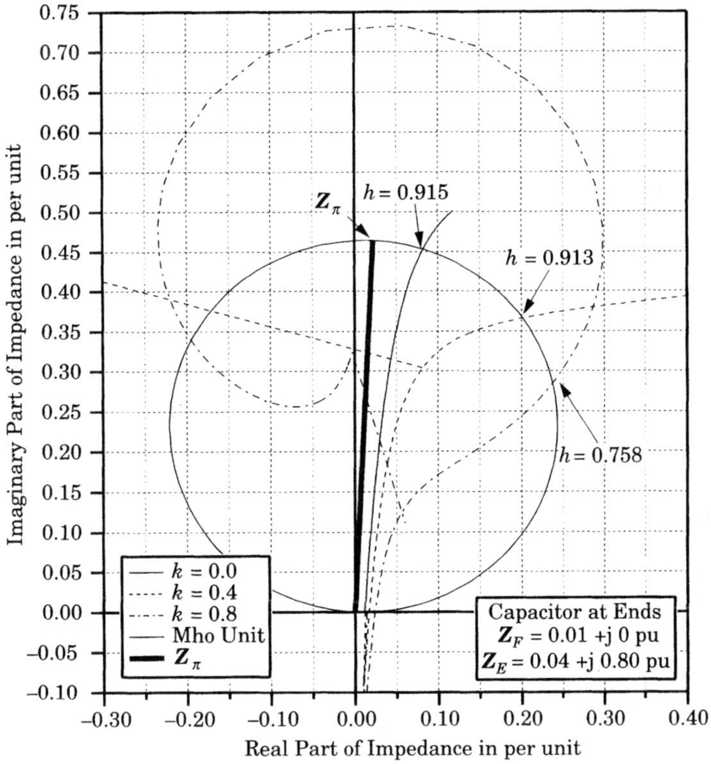

Figure 15.36 Loci of impedance seen by relay R as a function of fault location h.

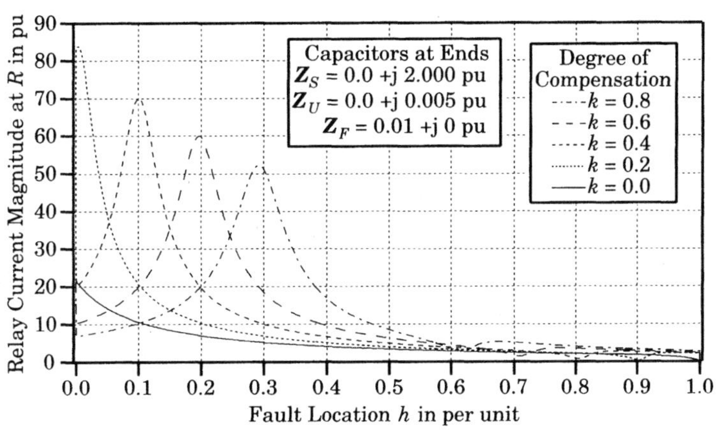

Figure 15.37 Relay current magnitude as a function of fault location h.

there exists the possibility that the fault impedance seen by the relay will fall outside the designated protection zone. In any event, higher fault resistance causes the impedance loci to move to the boundary of the protection zone for lower values of h, that is, for faults that are closer to the relay. This is an effective overreach of the relay, which views this close-in fault as lying outside the zone of protection.

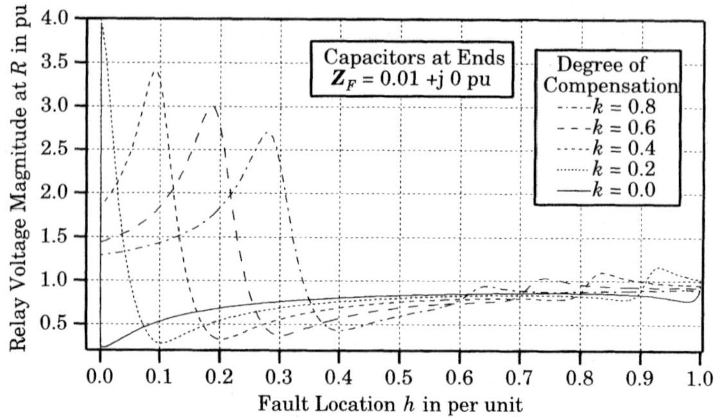

Figure 15.38 Relay voltage magnitude as a function of fault location h.

In order to study the effect of the fault impedance more closely, an increase from the base case value is made. The new fault impedance is given as

$$\text{New } \mathbf{Z}_F = 0.05 + j0.0 \text{ per unit} \tag{15.26}$$

This would be a fault resistance of 12.5 ohms, which is not unreasonable. The impedance seen by the relay is shown in Figure 15.39, where the effect of the higher fault resistance is immediately noted by the shift of all loci to the right. The reach of the relay is summarized in Table 15.4.

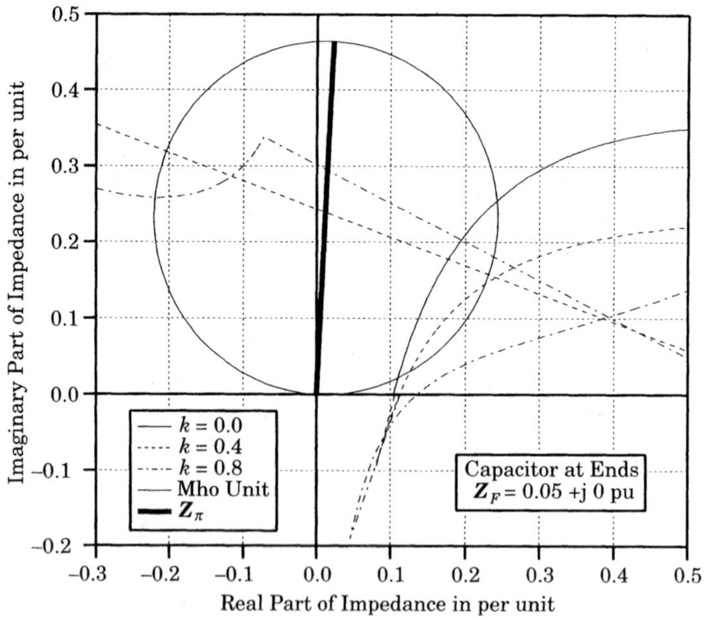

Figure 15.39 Loci of impedance seen by relay R as a function of fault location h.

It is clear that as the degree of compensation increases, the overreach of the relay also increases. However, the more serious problem is the loss of fault detection entirely. This will

TABLE 15.4 Reach of Relay R for Varying k and High Fault Impedance

$k = 0$	$k = 0.2$	$k = 0.4$	$k = 0.6$	$k = 0.8$
$h = 0.544$	$h = 0.536$	$h = 0.508$	$h = 0.0$	$h = 0.0$

be most likely at the higher degrees of compensation, where the impedance loci may be forced completely out of the nominal mho circle.

15.4 RELAY PROBLEMS DUE TO COMPENSATION

This section examines some of the problems that occur in the protection of transmission lines that are due to the presence of series capacitors. It has been shown that the series capacitors may change the observable characteristic of the transmission line significantly, and problems have been noted that may cause problems for conventional transmission line protective relaying schemes. These special problems are examined here in detail.

The descriptions in Sections 15.2 and 15.3 show that the relay protecting a series compensated transmission line is presented with changing information following the incidence of a fault. Initially, the line relays will see fundamental frequency currents and voltages of the line and series capacitor. Depending on the fault location, the relays may be viewing a fundamental frequency resonant condition, with very high currents or voltages. Moreover, depending on the fault location and the degree of compensation, it has been shown that the impedance seen by the line protection may interpret the line fault as being outside of the designated protective zone.

15.4.1 The Effect of Transient Phenomena

When a fault occurs on the transmission line, an abrupt change in the power system occurs and this change is accompanied by a transient response. Immediately following the application of the fault, high-frequency transients (100–1000 hertz) are excited as the system inductances and capacitances respond to the change in the network. Later, when the capacitor bypass system operates, assuming that the fault current is great enough to cause bypass operation, the relays then see currents and voltages that are much different. Moreover, in evolving from the unfaulted line, to the faulted and unbypassed, to faulted and bypassed, the system undergoes rapid transient changes. The relay must be designed to make the correct decision under rapidly changing system conditions, which might include resonances and transients of both super- and subsynchronous frequencies. Moreover, it must do so quickly and reliably.

The transient response of the system due to faults are of two types, *high-frequency transients* and *low-frequency transients*. The high-frequency transients are due to the natural frequencies of the series and shunt inductances and capacitances of the system. Any change in the network, such as a fault or series capacitor bypass, requires the readjustment in stored energy between the line inductances and the distributed shunt capacitances, which causes high-frequency currents to flow. The typical characteristics of EHV lines for both phase and ground faults is shown in Table 15.5 [16].

The initial magnitude of transient current associated with a phase or ground fault can be estimated by assuming that the voltage traveling wave from the fault consists of equal magnitude waves traveling in each direction. Then, we may compute the initial rms current

TABLE 15.5 Characteristics of Fault Conditions on EHV Transmission Lines

	Phase Faults	Ground Faults
Characteristic impedance, Z_C	300 Ω	600 Ω
Velocity of propagation	3×10^5 km/s	2×10^5 km/s
Time constant	$\tau_1 = \dfrac{L_1}{R_1} = 35\text{-}75$ ms	$\tau_0 = \dfrac{L_0}{R_0} \leq 15$ ms

magnitude as

$$I = 0.5\frac{\sqrt{2}V_{LL}}{\sqrt{3}Z_C}\ \text{A} \tag{15.27}$$

where the numerator gives the peak value of voltage in volts, and the factor 0.5 is required to find the magnitude of the wave traveling in one direction. For example, for a 750 kilovolt line, the initial magnitude of transient current for phase faults is about 1000 amperes.

The initial transients are made up of periodic components with frequencies of oscillation that are a function of the distance to the fault and the source impedance. As a general order of magnitude, these frequencies can be characterized as shown in Table 15.6 [16].

TABLE 15.6 Transient Frequencies of Oscillation

Line Length	Frequency
150 km	500–1000 Hz
300 km	250–500 Hz
600 km	125–250 Hz

Frequencies up to 1000 hertz may have magnitudes of up to 15% of the fundamental [17]. The high-frequency currents damp out quickly, as noted by the time constants in Table 15.5, but they will persist for a few cycles, which is longer than the usual relay time of high-speed protective systems. Most relays are designed to interpret the measured quantities in terms of their fundamental frequency behavior, and must filter out the high-frequency transient responses using low-pass filters. On long transmission lines, these high-frequency transients may be quite different at the two ends of the line, and identical relays at the two ends may see quite different transient voltages and currents. Moreover, since the fault may occur at any point in the network, the spectrum of high-frequency transients is great, and it is difficult to predict with precision [18].

The low-frequency transients are due to the resonant condition created between the series capacitors and the network series inductances, which are always transients of subsynchronous frequency. When a fault occurs, the series-compensated system experiences a transition from the prefault to the faulted state. Viewed in the complex Z plane, the apparent impedance locus moves from its prefault value, which is usually remote from the origin, to a new location much closer to the origin and along the line representing the line impedance. The shape of this transition has sometimes been described as a logarithmic spiral. This transition can cause distance measuring elements to either over or under reach, and can cause false directional sensing as well. The reach error can be overcome by low-pass filtering of the measured quantities. The directional sensing error can be corrected by relay polarization using unfaulted phase voltages or by memory polarization. The low-frequency transient behaves much the same as dc offset, which can cause saturation in current transformers [18]. However, because of the series capacitor, there is no dc offset in series-compensated lines. One study indicates that if the CT performance is satisfactory for dc offset, it is adequate for offset due to the presence of subsynchronous frequencies in the waveform [19]. As a general rule, the line relays must be designed to operate correctly even in the presence of these low-frequency transients, since it is difficult to provide effective filtering for the low frequencies. These low-frequency transients may also cause subsynchronous resonance problems.

15.4.2 The Effect of Phase Impedance Unbalance

Phase impedance unbalance occurs as the result of series capacitor bypass systems operating to bypass and reinsert the series capacitors unsymmetrically. This problem is apparent whether the bypass system is based on spark gaps or metal oxide varistor devices. Spark gaps flash when the voltage across a series capacitor, in any given phase, reaches a predetermined level. This occurs at different times in each phase, resulting in a large phase unbalance for a short time. Varistor devices conduct during a portion of each half cycle, with the capacitor carrying the total current for the remaining portion of the half cycle. In either case, these abrupt changes occur at the same time that the line protective relays are making their decision regarding the necessity for tripping the line. The relays see a transmission line that is undergoing rapid change, from normal to faulted, to severe unbalanced conditions, all in rapid sequence. In systems with spark gap capacitor protection, there is also the possibility that the bypass system will fail to operate in one or more phases, further complicating the conditions observed from the relay location.

Several approaches have been taken to overcome these problems. One approach simply requires that the bypass of capacitors be complete before correct relaying is assumed to occur. This causes delay in fault clearing that may not be acceptable in some cases. Another approach is to provide completely separate relays on each phase, rather than having a single line relay with measured data from all three phases. This may lead to an acceptable solution with the relays of any one phase causing the line to be tripped, or the breaker of the faulted phase to be tripped if independent pole tripping is used. This solution is somewhat more expensive, but is often used on EHV circuits to overcome phase impedance unbalance.

15.4.3 Subsynchronous Resonance Effects

Another problem that affects the protection of series compensated lines is the phenomenon of subsynchronous resonance. This is a resonant condition that occurs for any series LC circuit, where the electrical resonant frequency may be computed as

$$f_{er} = \frac{1}{2\pi\sqrt{LC}} = \frac{\omega_0}{2\pi}\sqrt{\frac{X_C}{X_L}} = f_0\sqrt{\frac{X_C}{X_L}} = f_0\sqrt{k} \tag{15.28}$$

In (15.28) f_{er} is the resonant frequency of the electrical system, f_0 is the system fundamental frequency, X_L is the line inductive reactance at the fundamental frequency, X_C is the line capacitive reactance at fundamental frequency due to the series compensation, and k is the degree of compensation. The frequency of oscillation is noted to be below the fundamental frequency, since the capacitive reactance is always less than the total line inductive reactance. In practice, the subsynchronous resonance frequency is in the range of about 15 to 90% of the system fundamental frequency.

The subsynchronous frequency response superimposes a subsynchronous component on the current and voltage that are observed by the transmission line protective relays. Because the frequency is so low, it is difficult to filter the subsynchronous frequency component rapidly enough to provide a satisfactory relay speed of response. For this reason, the line protective relays for series-compensated lines are often designed to operate correctly in the presence of currents and voltages with subsynchronous components. The subsynchronous frequency component can be relatively large and the current peaks may exceed the current peaks for the uncompensated line due to the addition of the synchronous and subsynchronous values. These higher currents may cause capacitor overvoltage devices to bypass the capacitor [20].

The subject of subsynchronous resonance is treated in greater detail in Chapter 23. The phenomenon is mentioned here simply because it is one of the problems affecting the line protective relaying that must be considered in the relay design.

15.4.4 Voltage and Current Inversions

It was noted in Section 15.2 that the apparent impedance measured by the relay is greatly affected by series compensation. The result depends on the degree of compensation and the location of the capacitor on the transmission line. The apparent impedance also depends on the type of capacitor bypass system, and whether bypass operates for a given fault. We now consider a different type of problem that occurs because the network resonances cause the phasor currents and voltages to occupy different positions in the complex plane than would be found in uncompensated systems. These unusual phase characteristics of the currents and voltages are often referred to in the literature as *reversals* or *inversions* of the phasor quantities.[6]

***15.4.4.1 Midline Series Compensation*.** Consider a 500 kilovolt transmission line 500 kilometers long, with 70% series compensation located at the center of the line. The apparent impedance measured by the line relay is to be examined under two conditions: (1) under a "weak" system condition, with source impedances of 0.2 per unit; and (2) under a "strong" system condition with source impedances of 0.1 per unit. The Z-plane plots of apparent impedance for the two conditions are shown in Figures 15.40 and 15.41, respectively. In both plots, the heavy straight line from the origin to the point labeled $\mathbf{Z}_\pi$ represents the total equivalent pi impedance of the 500 km line if that line were served radially from the relay terminal. For that special case, the impedance at various points along the line is linear with distance from the relay measuring equipment. The plotted mho circle has a maximum torque angle of 60°.

First, consider the apparent impedance of the line with center-line series capacitors and with the higher source impedance, represented by the dashed line in Figure 15.40, for h ranging from zero to one. The interconnected system is represented by an external impedance. The apparent impedance measured by the relay arcs upward and to the right, with a notch showing the change in moving the fault past the series capacitor. In order to reach 80% of the line length, the mho circle with the series capacitor and the higher source impedance must be considerably smaller than that of the uncompensated line, which is represented by the larger circle. That would be an appropriate relay characteristic for the uncompensated line. For the illustrated setting of reach, a fault on the near side of the series capacitor would fall outside the

[6] The term *current reversal* also indicates a change in direction of current, which is usually due to circuit breaker operation. The meaning of the term is usually clear from the context of the usage, but in this book the term *reversal* is used for directional change only, as described in Section 13.6.7.

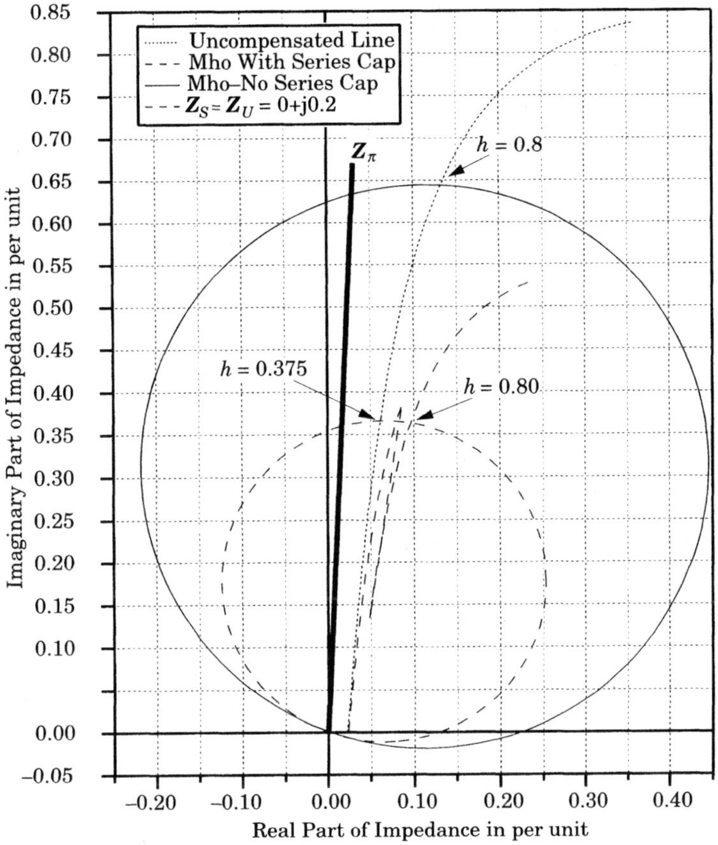

Figure 15.40 Line with 70% center series capacitors and "weak" sources.

trip zone. A higher value of reach, or larger circle, would correct this problem. If the series capacitor were bypassed, however, the smaller mho circle is not at all adequate, as the reach on the uncompensated line would be restricted to only 37.5% of the line length, which would leave the middle 25% of the line unprotected by the zone 1 relays. This situation requires special treatment, as a compromise setting that will be suitable for both the compensated and uncompensated line is not possible. One solution might be to delay clearing until the capacitor is bypassed, giving a delay that may not be acceptable. Another possibility is to employ a permissive overreaching scheme, which is discussed in the next section.

Now, consider the apparent impedance of the same line with the same series capacitor at the line center, but with a smaller source impedance. This might represent a maximum generation condition, for example. This situation is illustrated in Figure 15.41. Now, the apparent impedance is at a near-resonant condition and undergoes both voltage and current inversions. The impedance locus makes a large resonant excursion as the fault location is moved toward the series capacitor, then plunges to the fourth quadrant as the fault moves past the series capacitor. The relays see a quite different impedance for the two system conditions.

The reasons for the changes noted above are not clear from the Z-plane observations and requires additional study. Plots of the relay voltage and current are helpful in understanding the process, and these are shown in Figures 15.42 and 15.43 for degrees of compensation of

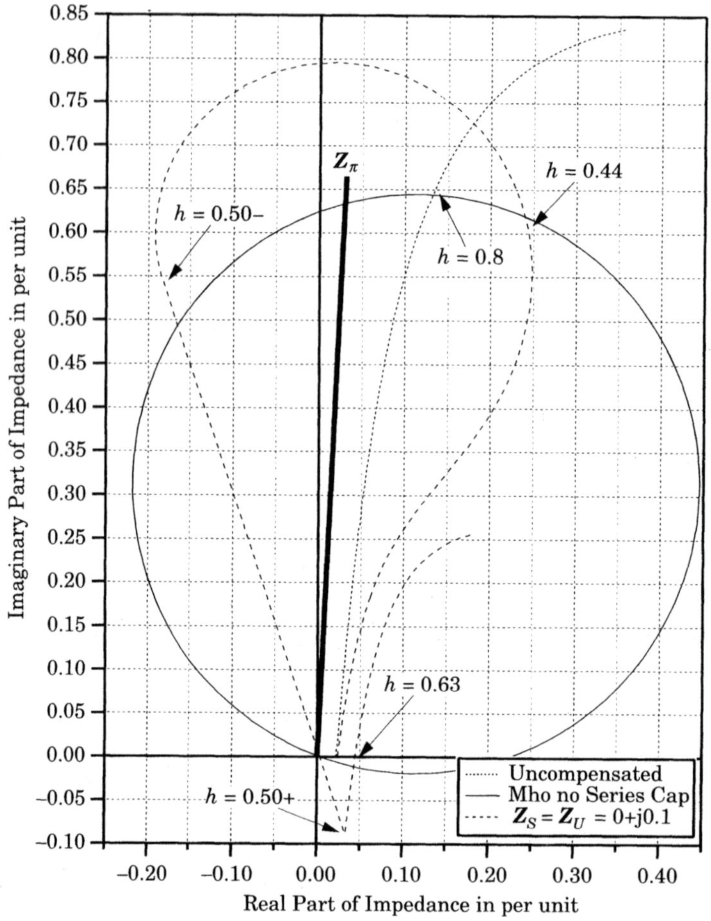

Figure 15.41 Line with 70% center series capacitors with "strong" sources.

60, 70, and 80%. The plotted curves are determined using the following parameters for the fault equivalent circuit:

$$\begin{aligned} S_B &= 1000 \text{ MVA} & \mathbf{Z}_E &= 0.015 + j0.30 \text{ per unit} \\ V_B &= 500 \text{ kV} & \mathbf{Z}_S &= 0 + j0.10 \text{ per unit} \\ Z_B &= 250\ \Omega & \mathbf{Z}_U &= 0 + j0.10 \text{ per unit} \end{aligned} \tag{15.29}$$

Note that an external system impedance is used to represent the interconnected system that is electrically in parallel with the protected line.

It is helpful to view Figures 15.42 and 15.43 together. First, we note the three loci all start at the same point and end at very nearly the same point. The 70% compensation case is a near resonant condition, as both the voltage and current at 70% are greater than for the 80% compensation case. Although interesting, this does not explain the behavior very clearly. It is more helpful to look for regions of voltage and current inversion.

Normal voltages are expected to have magnitudes of about 1.0 per unit, usually with a small lagging phase angle depending on the system condition. When a fault occurs, the voltage will be reduced in magnitude, but the voltage phasor remains in the fourth quadrant,

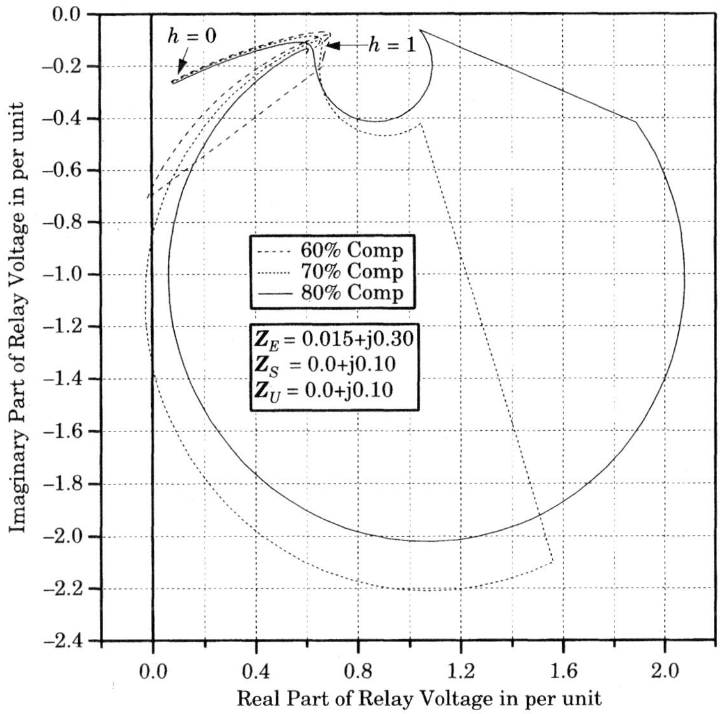

Figure 15.42 Relay voltage phasor loci for varying fault location.

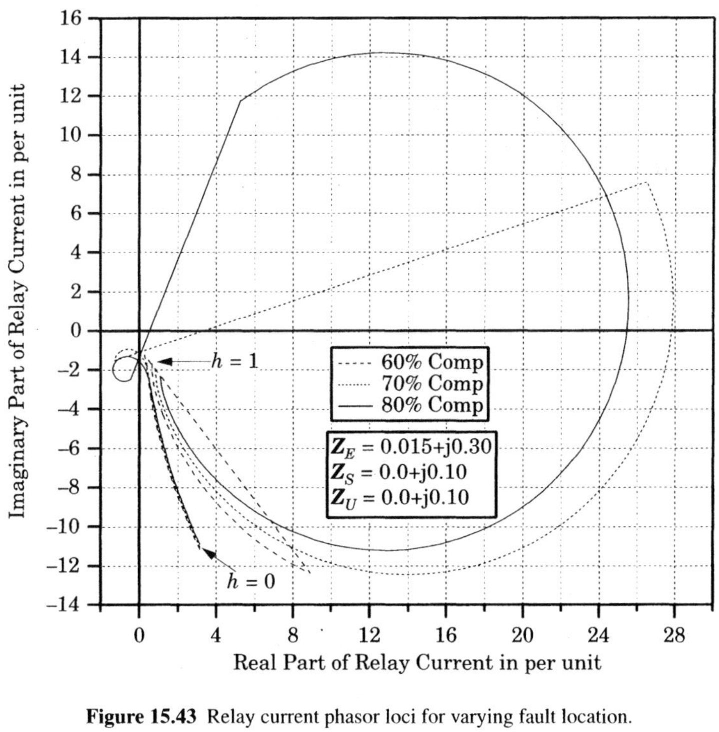

Figure 15.43 Relay current phasor loci for varying fault location.

even for close-in faults that cause the voltage magnitude to become very low. This is observed in Figure 15.42, where the voltage magnitude for a fault just beyond the bus is small, but still in the fourth quadrant. As the fault location is moved farther away from the bus, the voltage magnitude moves in the positive-real direction until the center-line series capacitor is reached. As the fault is moved past the series capacitor, the voltage jumps to a new value. As the fault is moved still farther from the relay the voltage becomes more lagging. If the capacitive reactance is large enough the voltage may even move into the third quadrant, as it does at 60% compensation. At 70% compensation, resonant conditions cause the voltage locus to circle about, but for faults beyond the capacitor the voltage moves into the third quadrant. We call any voltage movement into the third quadrant a *voltage inversion*. This occurs for the case under study with both 60 and 70% compensation, but it does not occur at 80% compensation.

The fault current phasor loci, shown in Figure 15.43, always start at the same point and remain essentially the same for all degrees of compensation until the fault is moved beyond the series capacitor. Fault currents are normally lagging currents, as the transmission system is largely inductive. As the fault moves away from the relay location, the fault current reduces in magnitude, but the currents remain lagging with a large lagging phase angle, which is sometimes assumed to be -90 degrees. As the loci approach the center of the line, resonances occur at the higher degrees of compensation that cause the currents to loop into the third quadrant. This is important in understanding the apparent impedance behavior, as will be shown later. Now, as the fault is moved past the series capacitor, the current takes a large jump in magnitude and, for the higher degrees of compensation, becomes leading. We refer to currents in the first quadrant as a *current inversion* as they are in the opposite sense of the normal currents in an inductive circuit, which are lagging. For the case under study, current inversion occurs for both 70 and 80% compensation.

A summary of the findings from this case, and verified by many other cases, is shown in Figure 15.44. Part (*a*) of the figure shows the typical phasor currents and voltage observed from the previous figures. The following notation is used in Figure 15.44.

$$
\begin{aligned}
\text{where } \boldsymbol{V}_{NOR} &= \text{normal voltage phasor} \\
\boldsymbol{I}_{NOR} &= \text{normal current phasor} \\
\boldsymbol{V}_{INV} &= \text{inverted voltage phasor} \\
\boldsymbol{I}_{IR} &= \text{inverted current phasor in right-half plane} \\
\boldsymbol{I}_{IL+} &= \text{inverted current phasor in second quadrant} \\
\boldsymbol{I}_{IL-} &= \text{inverted current phasor in third quadrant}
\end{aligned} \tag{15.30}
$$

We find it convenient to identify three types of current inversions, as noted above.

As noted from Figure 15.42, the voltage inversion causes the voltage locus to move into the third quadrant, as shown also in Figure 15.44(*a*). The case illustrated in Figure 15.42 does not amplify this movement, but other cases will show this voltage moving far into the third quadrant, for example, when the source impedances are larger.

The current inversion notation shows the inverted current in the first quadrant, but there is also an "inverted" current that moves into the second and third quadrants, which are referred to here with the subscripts IL with a sign added to indicate the second (+) or third (−) quadrant. Figure 15.44(*a*) illustrates all the variations representing many cases that have been examined. Note that no voltages venture into the second quadrant for the cases illustrated, although the current IL inversion may do so for cases with different system parameters.

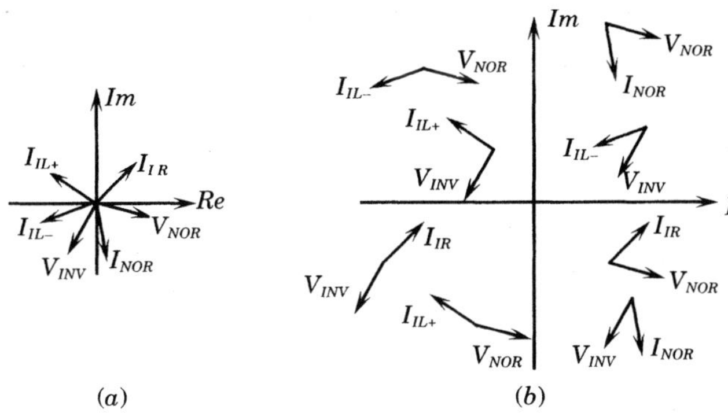

Figure 15.44 Typical phasor representation of fault currents and voltages. (*a*) Typical voltage and current phasor positions. (*b*) Resultant Z-plane apparent impedances.

Now, if we take combinations of the currents and voltages from Figure 15.44(*a*) and combine them, we can determine the location of the impedance represented by dividing the voltage by the current, and noting the resulting phase angle. These combinations are all plotted in Figure 15.44(*b*). This shows how apparent impedances can fall into any quadrant as the result of current and voltage inversions. The following observations can be made in summary:

1. Normal, first quadrant, impedances result from normal current and normal voltage, but can also result from an inversion of both variables.
2. Second $(IL+)$ and third $(IL-)$ quadrant impedances result from an inversion of current.
3. Third quadrant impedances may result from simultaneous inversion of voltage and current (IR), or from normal voltages with $(IL+)$ currents.
4. Impedances in the fourth quadrant result from inversion of either current of voltage, but not both.
5. Current inversions of type I_{IR} always result in negative reactance, i.e., the lower half of the Z plane.
6. Voltage inversions are always in the third quadrant if the applied voltages are close to the reference (0°) axis.

It should be noted that the phasor values represented in Figure 15.44 are typical, but are certainly not constant phasors. The phasor currents and voltage change rapidly with fault location. The actual system presents a gradual movement from one condition to the other as the fault position changes. Figure 15.44 is helpful, however, in understanding the mechanism of apparent impedance dependency on voltage and current inversion. These conclusions would be possible if voltage and current inversion are considered assuming a purely reactive system, as this will provide logic to locate the impedance only in either the top or bottom half of the Z plane.

15.4.4.2 End-of-Line Series Compensation. When the series capacitors are located at both ends of the line, the situation is different, but still complex. End-of-line series capacitors are more likely to create the conditions for voltage and current inversions, since there is no line impedance between the relay location and the series capacitor. In order to understand this situation, we examine the same system as described in the previous section, with the same 500 kilovolt transmission line that is 500 km long. The only thing that is changed is the series capacitor location. It is assumed that the series compensation is equally divided between the two ends.

Figures 15.45 and 15.46 show the relay voltage and current for varying degrees of compensation as a function of the fractional line distance to the fault for capacitors located at the ends of the line. As the fault is moved from the bus to the line side of the series capacitor, a current reversal occurs immediately and continues until the fault is moved far enough that the line reactance exceeds the series capacitor reactance. This distance, therefore, depends on the degree of compensation, but is between 10 and 20% of the line length for the values plotted. Up to this distance, the voltage and current combine to give in an apparent impedance in the fourth quadrant. Moving beyond this point, the current crosses the real axis and is no longer reversed, but voltage inversion now occurs as the voltage moves into the third quadrant, hence the impedance remains in the fourth quadrant of the Z plane. Moving the fault still farther along the line eventually reaches a point where both the current and voltage are in their normal domains, and the impedance enters the first quadrant.

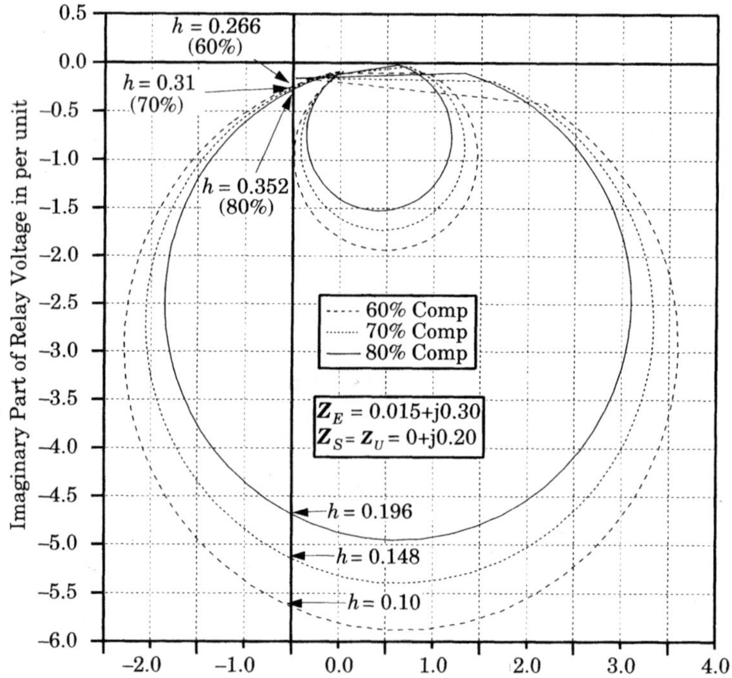

Figure 15.45 Relay voltage for varying degrees of series compensation.

Now, consider the relay currents, shown in Figure 15.46, for the same set of conditions. For faults just beyond the capacitor, the current is inverted and lies in the first quadrant, or an I_{IR} type of inversion. This condition persists for increasing values of h until the loci cross the real axis. The current is then "normal" until the fault approaches the remote terminal, after which a current inversion of type I_{IL} occurs. These inversions are interesting as the current moves through both the second and third quadrants in traversing the resonant condition. The current inversion forces the apparent impedance into the second quadrant of the complex impedance plane.

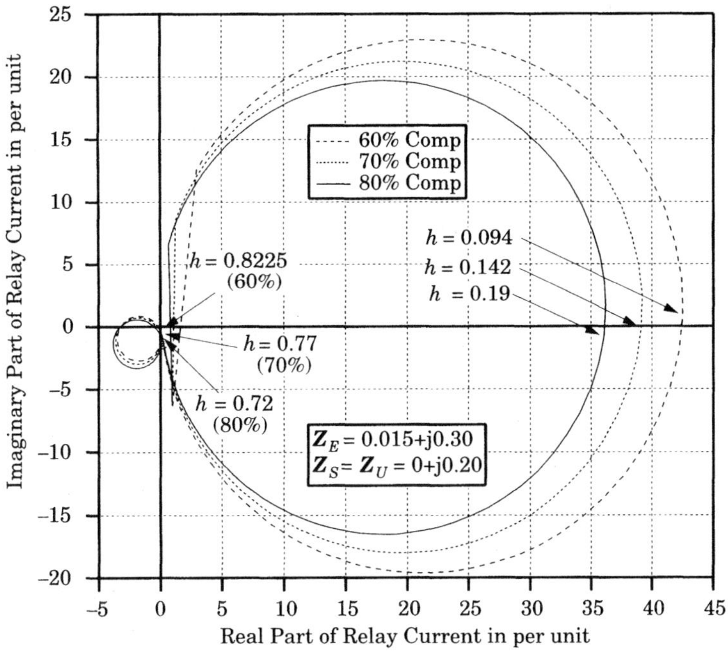

Figure 15.46 Relay current phasor loci for varying degrees of series compensation.

A review of the voltage and current reversals in the Z plane is shown in Figure 15.47 for the case of 70% compensation. When plotted in the impedance plane, the initial current reversal of type $\boldsymbol{I}_{IR}$ is confined to a very small portion of the total locus in the fourth quadrant. This condition quickly changes to one caused by a voltage inversion, still in the fourth quadrant. Both current and voltage are then "normal" from about 0.32 to 0.79, or for about the middle one-half of the transmission line. At distance 0.79, the remote-end resonance begins to unfold, forcing a type IL current inversion, which moves the Z plane locus into the second quadrant, where it remains until the remote capacitor is reached.

15.4.4.3 Apparent Impedance Observations. Series capacitors cause two types of problems that challenge the relay logic in positively identifying faults on the transmission line. First, for close-in faults, current or voltage inversions will cause the apparent impedance to fall in the fourth quadrant. Under this condition, the directional sensing of the relay logic is not affected, but the fault falls outside the normal trip zone. The second challenge comes from current inversions that force the apparent impedance into the second quadrant of the impedance plane. This results from a resonant condition that often moves the relay apparent impedance beyond the normal limits of the trip zone, but also causes the directional measurement to fail. Various schemes have been devised to provide correct fault detection in spite of these difficulties. Some of these schemes are discussed in Section 15.5.

The effects noted regarding the effect of series compensation on apparent impedance are all dependent on the system parameters. Table 15.7 summarizes observations from parametric studies of different system conditions.

Clearly, voltage and current inversions are more likely at high degrees of compensation and are more severe with end-of-line capacitors. For capacitors located at the center of the

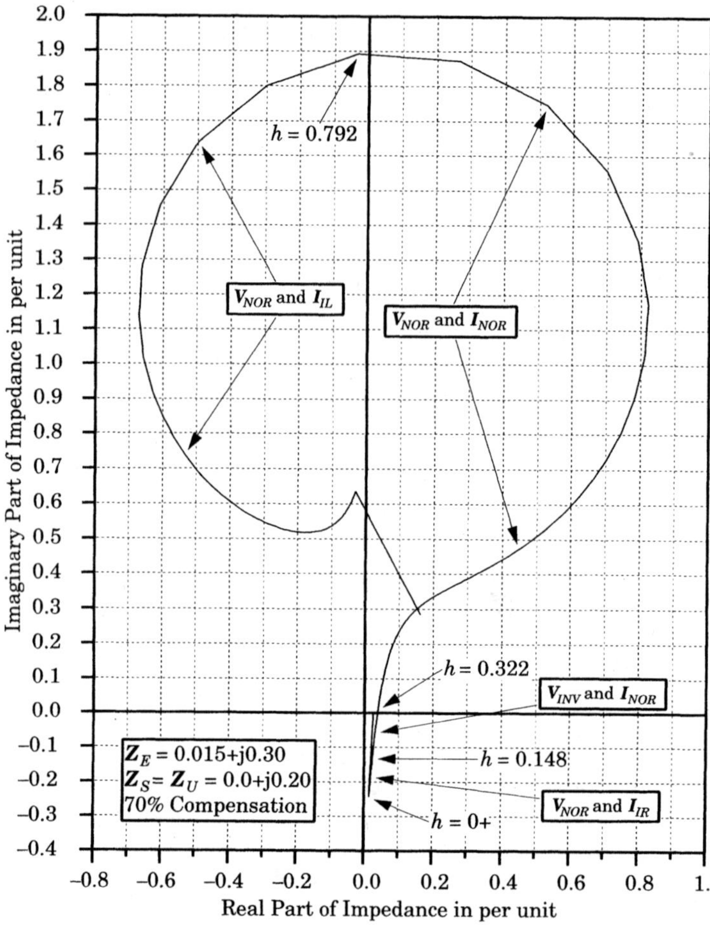

Figure 15.47 Relay apparent impedance with depiction of reversal conditions.

TABLE 15.7 Degrees of Compensation Causing Current or Voltage Inversions

Center Capacitor	**Z_E = 0.015 + j0.30 per unit**		**Z_E = 0.05 + j1.00 per unit**	
Capacitor	$Z_S = Z_U$ = j0.10	$Z_S = Z_U$ = j0.20	$Z_S = Z_U$ = j0.10	$Z_S = Z_U$ = j0.20
Invert *V*	60, 70%	80%	60, 70, 80%	60, 70, 80%
Invert *I*	70, 80%	80%	80%	None
End Capacitors	**Z_E = 0.015 + j0.30 per unit**		**Z_E = 0.05 + j1.00 per unit**	
Capacitors	$Z_S = Z_U$ = j0.10	$Z_S = Z_U$ = j0.20	$Z_S = Z_U$ = j0.10	$Z_S = Z_U$ = j0.20
Invert *V*	60, 70, 80%	60, 70, 80%	60, 70, 80%	60, 70, 80%
Invert *I*	60, 70, 80%	60, 70, 80%	60, 70, 80%	60, 70, 80%

line, high external system impedance and high source impedance tends to eliminate the current inversions. Studies using more widely varying conditions may lead to different results, which should be checked by the protection engineer.

15.4.5 Problems Due to Voltage Inversions

A voltage inversion has been shown to occur when the total reactance between the fault and the relay is capacitive. This will always occur when the series capacitors are located at the ends of the line, and for faults on the far side, but close to the series capacitor. The apparent impedance seen by the relay is shown in Figure 15.48, assuming that the relay voltage is measured at the bus. For faults immediately beyond the series capacitor, i.e., at ($h = 0+$), the apparent impedance is in the fourth quadrant until the capacitor bypass is complete. For capacitors with spark gap protection, the bypassing requires a short time to complete and bypassing may not occur at all for high impedance faults. The relay apparent impedance remains in the fourth quadrant for all faults in approximately the first one-third of the line length for the case illustrated. At a fractional distance of 0.75 the impedance leaves the nominal mho circle. If the entire series compensation is placed at the relay end of the line, the distortion is worse. Note that the impedance locus moves off the plotted area and completes a circular resonance before returning to the vicinity of the mho circle as the fault nears the remote terminal ($h = 1-$). Finally, the impedance for a fault at the remote bus is within the trip zone ($h = 1+$).

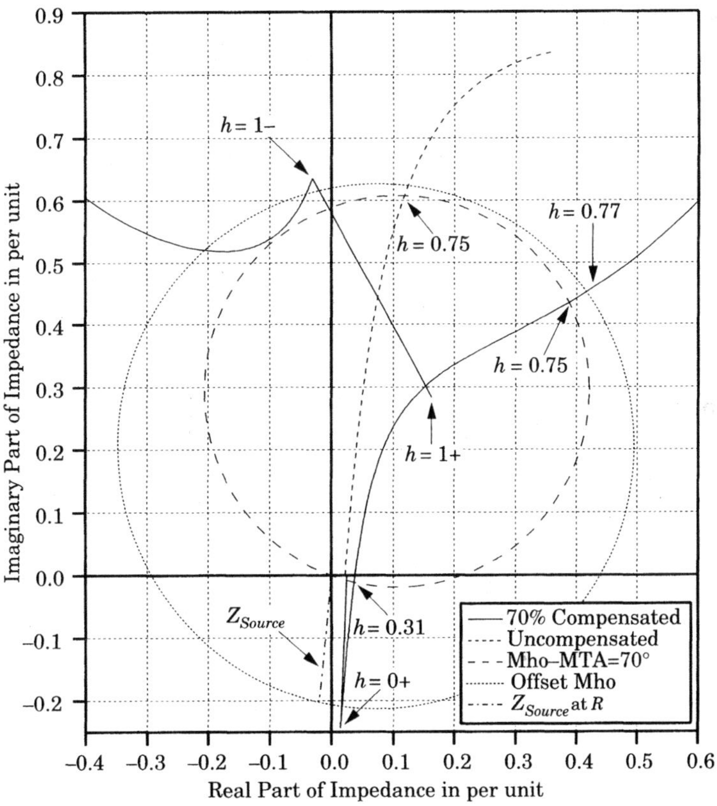

Figure 15.48 Addition of memory to the relay enlarges the trip zone.

For those close-in faults that result in apparent impedance in the fourth quadrant, the relay voltage is usually in the third quadrant, but returns to the fourth quadrant as the impedance again enters the trip zone depicted by the mho characteristic. The relay current for close-in faults is leading the voltage by almost 90 degrees, but the angle rotates clockwise as the fault moves farther from the bus. The relay current always remains in the right half of the complex plane, but the voltage moves into the left-half plane for faults in this critical region. This means that the directional measurement for faults in this region is correct, but impedance moves outside the trip zone for faults that should be correctly detected. The problem associated with the negative impedance for close-in faults can be corrected by adding a memory circuit to the relay in order to provide correct fault directional sensing. This gives a trip locus similar to the offset mho characteristic shown in Figure 15.48. This will ensure correct fault detection even for faults close to the series capacitor, or just beyond the bus. The relay must also correctly clear the fault after the bypassing of the capacitor and the normal mho characteristic is adequate for this task.

The negative impedance noted for close-in faults in Figure 15.48 will also affect distance relays on lines feeding into the bus. The tendency is to amplify the negative impedance, which may cause false tripping of these adjacent lines unless this condition is taken into account in the protection design of those lines. One solution might be to delay the first zone tripping until the series capacitors are bypassed; but if this is not acceptable, then some form of directional comparison should be added to the protective schemes of the adjacent lines.

The foregoing indicates that distance-measuring equipment on underreaching zones must be set very short in order to avoid overreach. Moreover, faults near the relay may appear outside the trip zone to distance relay equipment. These problems occur before the series capacitors are bypassed, which takes a finite amount of time for completion. Some line protection systems are designed to wait until the series capacitor bypass occurs before determining the need to trip. However, in many cases where lines are series compensated, the time delay required to ensure bypass may not be acceptable. In such cases, relaying schemes must be used that have the capability of operating correctly irrespective of the state of bypass equipment, presence of subsynchronous frequencies in the measured quantities, voltage or current inversions, and transient changes that occur due to fault inception and the eventual operation of bypass equipment.

The conclusions regarding apparent impedance problems due to series capacitors can be stated as follows. It may be possible to use distance relays, but each case will require careful studies of the application. Distance relays have been used and are sometimes preferred by protection engineers. However, the distance relays used for this application must have special design features, such as memory circuits, in order to be acceptable. To overcome the changes in measured impedance caused by the fault, gap flashing, and varistor conduction, underreaching zone settings must be kept shorter than the settings computed for a line with unbypassed series capacitor. Moreover, the reach setting of an overreaching zone must be greater than that computed for an uncompensated line. These requirements usually rule out the use of direct or underreaching transfer trip schemes. Permissive overreaching or blocking schemes are usually preferred for series compensated transmission lines.

15.4.6 Problems Due to Mutual Induction

Parallel transmission lines are subject to mutual coupling, both inductive and capacitive. The lines may be of the same or different voltage, and may be bused at one end, at both ends, or not bused at either end. The length of line with relatively close lateral distance between circuits is the important variable in determining the magnitude of the mutual coupling. Mutual coupling is much greater in the zero-sequence network than in the positive and negative sequence networks. For example, the positive-sequence mutual impedance is less than 1% of the self-impedance and is usually ignored. Zero sequence mutual reactance, however, is on the order of one-third the zero sequence self-impedance, and remains relatively high even for lines separated by 100 meters or more.

In series compensated transmission lines, the series capacitor compensates the zero sequence self-impedance, but the mutual impedance remains the same as the uncompensated line. Therefore, the relative effect of series compensation is worse than that observed in the uncompensated line. This fact, coupled with possible voltage and current inversions, greatly complicates the problems for some types of protection. Distance protection, for example, becomes very difficult with mutual coupling unless special design features are incorporated to overcome the difficulty.

We can determine the effect of mutual coupling in the zero-sequence network by observing the equivalent circuit in Figure 15.49. Viewed in this manner, it is clear that the reactance of the series capacitor affects only the self-impedance of the two lines and that the mutual impedance is unchanged from the uncompensated case.

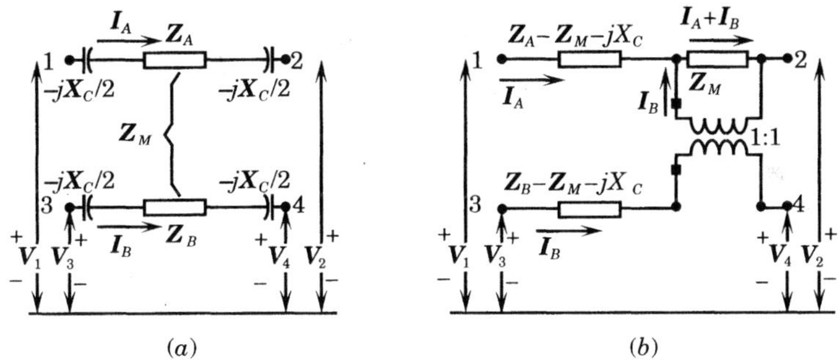

Figure 15.49 Equivalent circuit of zero sequence mutual coupling. (*a*) Mutually coupled transmission lines. (*b*) Equivalent using a 1:1 transformer.

The result is that higher fault currents will flow, at least until the series capacitor is bypassed, and these higher currents will induce higher voltages in the mutually coupled line.

The solution of problems related to mutual induction in series compensated lines is not greatly different from uncompensated lines. The possible solutions to this type of protection problem are beyond the scope of this book, but methods of analyzing mutual coupling are treated in the literature [21], [22].

15.4.7 Problems in Reach Measurement

Directional comparison schemes require some measure of distance. For underreaching schemes, the relay threshold is purposely less than the total line length. Overreaching schemes extend their reach beyond the remote end of the protected line. These distance measurements are affected by the series capacitors and can cause significant errors in the relay measurements. This phenomenon is investigated briefly in the following sections.

15.4.7.1 Underreaching Schemes. Underreaching protection for transmission lines requires that the reach of the zone 1 relays be set short of the remote end of the line, with settings of 80 to 90% of the total line length being common. The error in distance measurement depends on the series capacitor arrangement for that line and on the assumptions regarding capacitor bypass made in setting the reach of the relays. Figure 15.50 shows the three most common arrangements for the series capacitors and an indication of the reach setting, represented by the quantity h_S, where this parameter is defined as

$$h_S = \frac{X_S}{X_L} \tag{15.31}$$

where X_S is the ohmic distance representing the reach setting and X_L is the total inductive reactance of the line in ohms. The total impedance could be used, making the reach a complex number, but for high-voltage lines there is little error in using only the reactance.

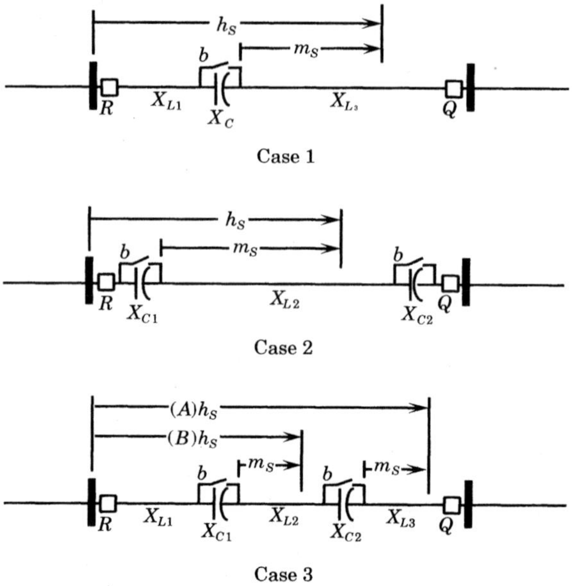

Figure 15.50 Cases of interest for underreaching protection. Case 1: One mid-line series capacitor. Case 2: One or two end-of-line series capacitors. Case 3: One or two mid-line series capacitors.

The reach setting depends on the fraction of the total line impedance represented by X_S, but it also depends strongly on the presence of one or more series capacitors and whether these capacitors are assumed to be in service or bypassed. To account for these variations, we can write a functional expression for the distance setting as follows.

$$X_S = f_S(X_{Li}, X_{Ck}, m_S, b) \tag{15.32}$$

where X_{Li} = inductive reactances i spanned by X_S
X_{Ck} = capacitive reactances k spanned by X_S
m_S = per unit distance in line section of zone end
b = binary switching function of capacitor bypass

The definition (15.32) recognizes that there may be more than one line section, and these sections are referenced by the index i. Likewise, there may be in X_S more than one series capacitor, so these are referenced by the index k. The parameter m_S is a convenient per unit distance measure from the last capacitor to the reach threshold. This measure is not affected by capacitor bypass. Finally, the parameter b is a binary switching parameter indicating whether any given series capacitor is bypassed or not. This function can be written mathematically as

$$X_C = bX_{CO} = \begin{cases} X_{CO} & \text{if } b = 0 \text{ (bypass switch open)} \\ 0 & \text{if } b = 1 \text{ (bypass switch closed)} \end{cases} \tag{15.33}$$

and where X_{CO} is the normal capacitive reactance of a particular capacitor bank. Thus, (15.32) or its per unit equivalent (15.31) provides a short-hand notation that functionally describes the variables that define the reach setting.

Now, the reach setting must not be confused with the measured reach of the relay R. Consider the reach measured by the relay for a fault located at the reach threshold. This reach can also be described by a function similar to (15.32), which we can write as follows:

$$X_R = f_R(X_{Li}, X_{Ck}, m, b) \tag{15.34}$$

where all the functions have the same meaning as in (15.32). Note, however, the per unit distance m is not necessarily the same as that given for the reach setting. This is because one assumption regarding series capacitor bypass might be made in computing the reach setting and that assumption does not correctly describe the actual reach of the relay due to unforeseen capacitor bypass actions. For example, the reach settings might be computed assuming that all of the series capacitors are bypassed, but the reach measured by the relay will be different from the setting if that bypass assumption is not true. If the assumptions regarding the capacitor switching are identical for (15.32) and (15.34), then the two values of reach distance are equal and the values of the parameter m are also equal. The problem with setting distance relays on series compensated lines is that one can never be absolutely sure of the assumptions regarding the state of switching parameter b.

One can construct a mathematical test, however, to determine the range of errors that might occur. The test is simply to equate (15.32) and (15.34).

$$f_S(X_{Li}, X_{Ck}, m_S, b_S) = f_R(X_{Li}, X_{Ck}, m, b_R) \tag{15.35}$$

This equation can be solved for the parameter m measured by the relay to determine if this value differs from the value of m_S used in determining the reach setting. Note that, in either case, the m-parameter is related to the total reach by a constant and is, therefore, a convenient measure of reach since it is not affected by series capacitor bypass.

For any given physical transmission line, we can write the reach setting as follows:

$$X_S = m_{Si} \sum_{i=1}^{N_X} X_{Li} - \sum_{k=1}^{N_C} X_{Ck} \tag{15.36}$$

where all the parameters depend on the physical arrangement of components in the transmission line. For underreaching protection, the three cases shown in Figure 15.50 are of interest. The reach setting shown in the figure, h_S, is defined in (15.31).

For example, consider the system condition shown in Figure 15.50 case 1, with a single midline series capacitor. For this case, we may write the reach setting as follows, depending on the assumption regarding series capacitor bypass.

(S1) Set reach assuming the series capacitors are bypassed ($b = 1$):

$$h_{S1} = \frac{X_{L1}}{X_L} + m_S \frac{X_{L3}}{X_L} \tag{15.37}$$

(S2) Set reach assuming the series capacitors are not bypassed ($b = 0$):

$$h_{S2} = \frac{X_{L1} - X_C}{X_L} + m_S \frac{X_{L3}}{X_L} = \frac{X_{L1}}{X_L} - k + m_S \frac{X_{L3}}{X_L} \tag{15.38}$$

where k is the degree of compensation.

Now, for a fault at the reach threshold ($m = m_S$), we can also write the actual reach of the relay in two ways, depending on the bypass action. Then we can compare the relay reach against the reach setting using (15.35).

(b_1, s_1) Setting s_1, fault at $m = m_S$ and bypass actually occurs ($b = 1$):

$$h_R = \frac{X_{L1}}{X_L} + m_S \frac{X_{L3}}{X_L} \qquad h_{S1} = \frac{X_{L1}}{X_L} + m_S \frac{X_{L3}}{X_L} \tag{15.39}$$

or

$$h_R = h_{S1} \tag{15.40}$$

(b_1, s_2) Setting s_2, fault at $m = m_S$ and bypass actually occurs ($b = 1$):

$$h_R = \frac{X_{L1}}{X_L} + m_S \frac{X_{L3}}{X_L} \qquad h_{S2} = \frac{X_{L1}}{X_L} - k + m_S \frac{X_{L3}}{X_L} \tag{15.41}$$

or

$$h_R = h_{S2} + k \tag{15.42}$$

We can also write out the solution for the two cases where bypass does not occur following the fault.

(b_0, s_1) Setting s_1, fault at $m = m_S$ and no bypass occurs ($b = 0$):

$$h_R = \frac{X_{L1}}{X_L} - k + m_S \frac{X_{L3}}{X_L} \qquad h_{S1} = \frac{X_{L1}}{X_L} + m_S \frac{X_{L3}}{X_L} \tag{15.43}$$

or

$$h_R = h_{S1} - k \tag{15.44}$$

(b_0, s_2) Setting s_2, fault at $m = m_S$ and no bypass occurs ($b = 0$):

$$h_R = \frac{X_{L1}}{X_L} - k + m_S \frac{X_{L3}}{X_L} \qquad h_{S2} = \frac{X_{L1}}{X_L} - k + m_S \frac{X_{L3}}{X_L} \tag{15.45}$$

or

$$h_R = h_{S2} \tag{15.46}$$

Thus, we get an array of solutions for h_R, which are summarized in Table 15.8.

TABLE 15.8 Computed m for Reach Threshold Fault for Case (a)

Bypass Assumed for Relay Settings?	Computed h_R for Fault at Threshold: Bypassed	Not Bypassed
Yes: s_1	$h_R = h_{s1}$	$h_R = h_{s1} - k$
No: s_2	$h_R = h_{s2} + k$	$h_R = h_{s2}$

From Table 15.8, we see that the measured reach at the relay agrees with the setting when the assumptions regarding the capacitor bypass switching is correct, but the reach will be in error when the assumptions disagree with the measured reach. If the degree of compensation is high, the error in measured reach can be large. The case where $h_R = h_{S1} - k$ represents an overreach condition. Moreover, if the settings are computed assuming no bypass ($b = 0$), the relay might reach well beyond the protected line, a situation that cannot be tolerated. Therefore, for underreaching protection it is better to set the relay reach for zone 1 assuming bypass of the series capacitors will occur. Should the capacitors not be bypassed during a fault, in contradiction to the setting assumption, the reach will be greater than desired. This may result in a gap near the center of the line that will be cleared with zone 2 time delay, but false tripping will be avoided. However, the possible magnitude of the error must be examined. It may be so large that underreaching distance protection will be positively eliminated from consideration.

The foregoing has concentrated on case 1 of Figure 15.50. However, the same technique can be applied to the other configurations as well. These questions are left as exercises.

It is important to consider the likelihood of series capacitor bypass in setting the relay reach. It is noted in Figure 15.29 that capacitor bypass depends on the fault location as well as the degree of compensation. A calculation similar to that of Figure 15.29 may be helpful in estimating the probability of bypass for faults at the reach threshold, as well as other faults along the line. There may also be design considerations that should be examined. For faults at certain locations, one capacitor may be bypassed and the other not bypassed. This depends on the degree of compensation, on the location of the capacitors on the transmission line, and on the magnitude of currents feeding the fault.

The foregoing illustrates the difficulty inherent in applying distance protection to series compensated lines. Other forms of protection are usually preferred because of the problems described above.

15.4.7.2 Overreaching Schemes. Overreaching schemes are designed such that zone 1 extends well beyond the remote end of the series compensated transmission line [31]. An example of an overreaching scheme is shown in Figure 15.51. It is intended that the overreach will always span line L entirely, reaching into line N by 20–50% of the impedance of that line. The reach calculation depends on the size and location of any series capacitors on line N, as well as on line L, and whether these series capacitors fall within the reach. Both the reach setting and the actual measured reach depend on the bypassing of series capacitors on both lines. The overreach must be great enough that it still overreaches bus Q when the capacitors are bypassed for any reason. The uncertainty in settings is greater than the underreaching case, as there are more capacitors to be considered, as well as the probability of the bypass of capacitors on line L for faults on adjacent lines, which is a function of the system strength behind line L.

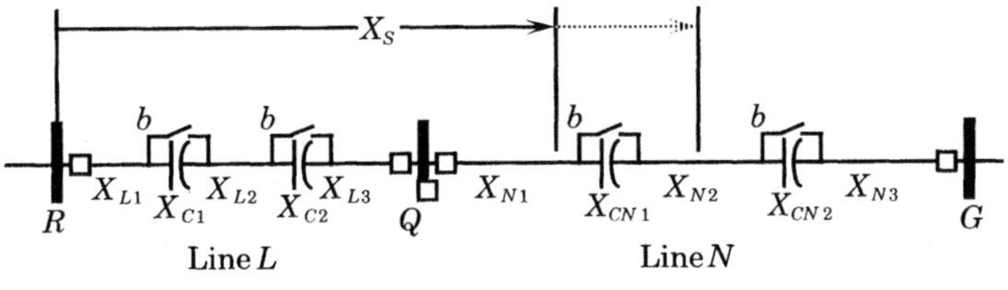

Figure 15.51 Permissive overreaching distance pilot protection.

We can write a general expression for the per unit reach for the overreaching case, as follows:

$$h_S = 1 - k_L + \sum_{i=1}^{N_{XN}} m_i \frac{X_{Ni}}{X_L} - \sum_{k=1}^{N_{CN}} \frac{X_{CNk}}{X_L} \tag{15.47}$$

where k_L is the degree of compensation of line L. As in the underreaching case, there are three cases of interest, depending on the number and location of series capacitors on line N, as shown in Figure 15.52.

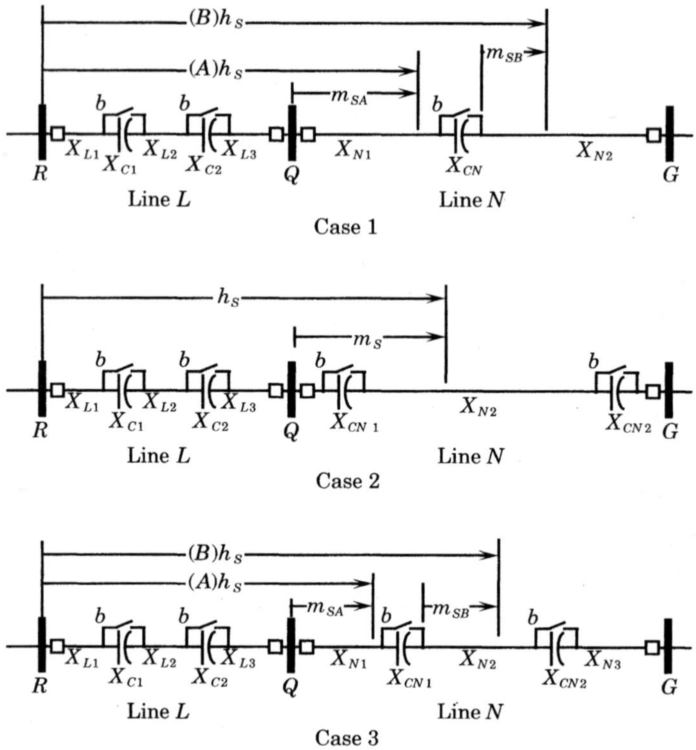

Figure 15.52 Overreaching protection cases of interest. Case 1: One midline series capacitor on line N. Case 2: One or two end-of-line series capacitors on line N. Case 3: One or two midline series capacitors on line N.

Let us examine case 1 of Figure 15.52 as an example of computing the overreach settings. For case 1, the general per unit reach may be written as follows.

$$h_S = 1 - k_L + \frac{X_{N1}}{X_L} - \frac{k_N X_N}{X_L} + m_S \frac{X_{N1}}{X_L} \tag{15.48}$$

where k_L is the degree of compensation for line L and k_N is the degree of compensation for line N. Two variations of case 1 can be considered: $1(A)$ the reach setting falls short of the series capacitor on line N, or $1(B)$ the reach setting lies beyond that capacitor. For (A) the capacitor can be simply ignored in (15.47), and for case $1(B)$ the equation is correct as written.

For each assumption made regarding series capacitor bypass on line N, we may compute two different per unit reach settings depending on the assumed bypass state of the capacitors on line L. This results in two different variations of (15.47), one with k at its normal value and another case with k set equal to zero. A third possibility exists, where one of the series capacitors on line L is assumed to be bypassed, and the other not bypassed. Although quite possible, this last case is not considered further here.

These cases require an examination of three different reach settings, corresponding to the following three capacitor bypass conditions:

S1. Bypass capacitors on both transmission lines
S2. Bypass capacitors only on line N
S3. No capacitor bypass on either line.

One could also consider capacitor bypass only on line L, but that is not likely since the fault is on line N. The above three conditions must be examined to determine the measured reach of the relay when measuring the distance to a fault at the reach threshold. This gives nine conditions to be examined, resulting in a 3×3 matrix of results. For case $1(B)$ the results are shown in Table 15.9.

TABLE 15.9 Case $1(B)$ Reach Calculation for Three Setting Assumptions

Overreach Setting	Measured Reach at Relay R for Threshold Faults = h_{RB}		
	$b_L = 1, b_N = 1$	$b_L = 0, b_N = 1$	$b_L = 0, b_N = 0$
h_{SB1}	h_{SB1}	$h_{SB1} - k_L$	$h_{SB1} - k_L - \dfrac{k_N X_N}{X_L}$
h_{SB2}	$h_{SB2} + k_L$	h_{SB2}	$h_{SB2} - \dfrac{k_N X_N}{X_L}$
h_{SB3}	$h_{SB3} + k_L + \dfrac{k_N X_N}{X_L}$	$h_{SB3} + \dfrac{k_N X_N}{X_L}$	h_{SB3}

The diagonal terms in Table 15.9 represent conditions where the measure of threshold faults is exactly right. This occurs for terms in which the actual bypass condition agrees with the assumptions for the reach settings. The off diagonal terms with positive quantities are subtracted from h_{SBi} and represent conditions where one or more capacitors are in service, but were assumed to be bypassed. This causes the relay to overreach. Likewise, positive added terms represent one or more capacitors that were assumed to be in service in making the reach setting, but are bypassed for the condition represented. These cells represent underreaching conditions.

Similar tabulations can be derived for case 2 and case 3. These cases are left as exercises.

For the overreaching case, it is usually considered essential that the reach setting be determined with the capacitors bypassed. This may lead to large reach extensions when the capacitors are not bypassed, which may be considered an unacceptable security risk, as the reach may extend well beyond the remote end of line N.

These results are not exhaustive. Where there are two series capacitors in line L, they are assumed to be both bypassed or neither bypassed in the analysis described above. As noted in Figure 15.29 this is not likely to be true, depending on the fault location and on the degree of compensation.

One must conclude that setting the reach for distance relays on series compensated lines is difficult and involves several compromises. However, the compromises may be acceptable in some cases. For example, a situation may arise where it can be shown that the series capacitor bypass on a given line will not operate for faults on neighboring lines. This limits the chances for error by eliminating a row and column from the matrices used to describe the results, thereby limiting the hazards. Each case must be judged on its own merits and on the physical factors of the system under study. However, it is clear that determining reach settings for distance measuring devices on series compensated lines is fraught with difficulties, and there is always a probability for error in the relay operation, irrespective of the rules selected for computing the settings.

15.5 PROTECTION OF SERIES COMPENSATED LINES

The protection of series compensated lines has been a concern of protection engineers since the earliest applications of series capacitors to transmission lines. In some of the earliest applications of series capacitors, the protection was accommodated by limiting the degree of compensation and using the same line protective relays that were used for uncompensated lines, although it was recognized that this practice could not be continued indefinitely [23]. As series compensation became more common, the relay manufacturers began to provide new devices, or modifications of existing relays, that were especially designed for series compensated lines. As a result, many relays are now available that are candidates for use on series compensated lines.

15.5.1 Current Phase Comparison

The segregated current phase comparison principle, described in Section 13.4, has been widely used on series compensated lines. This scheme is not affected by the series capacitors at all due to the fundamental way in which it discriminates between an internal and an external fault. The primary disadvantage of this principle is its complete dependence on communications. Although this has been noted as the cause of failure on some occasions, these systems have provided good service and continue to be specified for new transmission lines. It is also noted that nearly all permissive schemes at the higher voltages are dependent to some degree on communications.

Some protection engineers favor the use of current phase comparison for one system and use a different type of scheme for a second type of line protection. This approach has the advantage of not relying entirely on one form of relay logic, by using two types of devices based on different principles. For those who use this approach, current phase comparison is often one of the two types selected for series compensated lines because this provides one method of fault identification that is not affected by the series capacitors.

15.5.2 Directional Comparison Schemes

The second category of relays that has been widely used for series compensated lines falls into the general classification of directional comparison. There are several good choices of available directional comparison schemes that are either recommended for series compensated lines or have special units that can be added to the basic system to make the relays suitable for this application.

15.5.2.1 Hybrid Schemes. The hybrid scheme described in Chapter 13 has been rather widely used in North America and with excellent success. The earlier versions of this relay have been replaced with digital systems that offer the same basic coverage as the original devices, and many added features as well.

The traveling wave schemes, also described in Chapter 13, are usually directional comparison schemes but operate using a different principle than other systems. These relays take advantage of the fact that it takes a finite amount of time for the stored energy in the series capacitor to change in response to a fault current. During this brief period as the capacitor state begins to change, the traveling wave relay completes its measurement. This means that the presence of the series capacitor does not affect the performance of this type of relay.

15.5.2.2 Distance Schemes. A number of the newer directional comparison relays using basic distance measuring principles have been adapted for use on series compensated lines. These relays use a combination of distance measurement and memory to defeat the voltage inversions that have been a source of difficulty in measurements on series compensated lines.

As noted in Section 15.4.4 and 15.4.5, series capacitors often cause voltage and current inversions, which cause errors in distance measurement, and occasionally in directional measurement as well. Some relays, however, have been designed to defeat these problems. Other relays, such as current phase comparison and traveling wave relays, are immune from these problems because of their inherent measurement techniques.

One of the difficulties of voltage and current inversions is that the impedance measurement may fall outside the first quadrant in the Z plane. This is likely to occur before the capacitor is bypassed, or at the instant of fault initiation. Since it is not usually desirable to purposely delay the relay until bypass action can be completed, relays have been devised to overcome these difficulties.

One method of defeating voltage and current inversions is to apply the concept of *memory* in the relay measurements. Suppose that a voltage inversion takes place at fault initiation. However, if the relay can remember the voltage condition just prior to the fault, the inversion effect can be overcome. This technique is used in the hybrid relay described in Chapter 13.

One relay that accommodates all of the foregoing features is based on a *generalized fault criterion* (GFC) that will work for both three-phase and ground faults [32–34]. The basis for the GFC is based on two phasor voltage measurements:

$$\begin{aligned} \boldsymbol{V}_{KZ1} &= \boldsymbol{V}_R - \boldsymbol{Z}_K\boldsymbol{I}_R \\ \boldsymbol{V}_{KZ2} &= \boldsymbol{V}_R - (-\boldsymbol{Z}_K\boldsymbol{I}_R) \end{aligned} \tag{15.49}$$

where the phasor quantities are shown in Figure 15.53. A phase comparison of these voltages is carried out with the threshold condition given by

$$90^\circ \leq \arg(\boldsymbol{V}_{KZ1}) - \arg(\boldsymbol{V}_{KZ2}) \leq 270^\circ \tag{15.50}$$

These quantities are compared in a phase comparison scheme to provide the circular GFC relay threshold shown in Figure 15.53. The memory action brings the voltage V_{kz2} into action temporarily, extending the active trip zone to cover all four quadrants for a brief time. The time the memory voltage can be used is limited to about 100 ms. Coupled with permissive or blocking logic, this can be made into a workable measurement system to positively identify a fault in the transmission line, and to avoid tripping for faults on adjacent lines.

The impedance characteristic of Figure 15.53 may be modified by the addition of resistance relay blinders, which restrict the fault zone to any desired resistance value, both positive and negative.

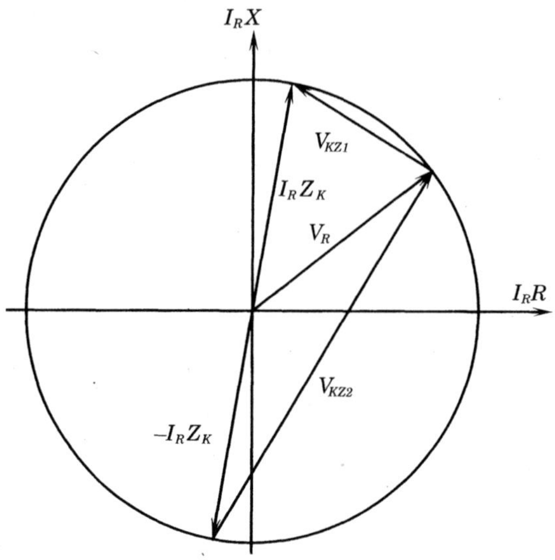

Figure 15.53 Measuring quantities for the generalized fault criteria.

The polarizing voltages for the three phases are given as

$$\begin{aligned} \boldsymbol{V}_{PA} &= \boldsymbol{V}_{R1} + 0.05(\boldsymbol{V}_B - \boldsymbol{V}_C)e^{j\pi/2} \\ \boldsymbol{V}_{PB} &= \boldsymbol{V}_{R1}e^{-j2\pi/3} + 0.05(\boldsymbol{V}_C - \boldsymbol{V}_A)e^{j\pi/2} \\ \boldsymbol{V}_{PC} &= \boldsymbol{V}_{R1}e^{j2\pi/3} + 0.05(\boldsymbol{V}_A - \boldsymbol{V}_B)e^{j\pi/2} \end{aligned} \tag{15.51}$$

where $\boldsymbol{V}_{R1}$ is a positive sequence voltage derived by filtering the raw phase voltages. This voltage is memorized for a brief period and is available as a polarizing voltage to provide correct fault direction sensing should an unbalanced fault develop into a three-phase fault. Memory is especially important for three-phase faults since all voltages experience inversion for this type of fault and the only way to achieve correct directional determination is through the use of memorized prefault polarizing voltage.

Faults on series compensated lines are accompanied with two types of transients that must also be considered. Low-frequency transients due to subsynchronous currents sometimes appear in the impedance plane as a logarithmic spiral, which will sometimes cross the vertical axis during the transient from the unfaulted to the faulted state.

In conclusion, it can be stated that distance relaying can be applied to series compensated transmission lines, using some of the modern distance systems.

Most of these systems are modular, and the relays can be used by replacing only those modules that are required to enhance relay measurement and decision making for series compensated lines. This type of protection may be particularly attractive for a utility that has such a modular distance protection system in use on lines that have no series compensation, and where there may be interest in using the same type of relay on series compensated lines as well. Using the same system reduces the inventory of spare parts and reduces the training requirements of personnel in the setting, monitoring, and maintenance of these relay systems.

15.5.2.3 Traveling Wave Protection. Traveling wave relays, described in Chapter 13, are not affected by the presence of series capacitors. These schemes are based on the extraction of fault information from a broad bandwidth of frequencies that accompany a fault and the processing of the data digitally to determine the fault location. The directional discrimination is accomplished in 2–4 ms, which is short enough that the voltage across the series capacitor bank is usually changed very little from the prefault condition.[7] This is because the change in capacitor voltage must be accompanied by a change in stored energy in the capacitor, which requires a finite amount of time due to the largely inductive circuits in the power system. This means that the capacitor will not influence the short time measurement. This provides a significant advantage for lines with series capacitors, or for the protection of lines adjacent to those with series capacitors.

Most of these schemes are double-ended schemes; that is, they use directional comparison arrangement that is accomplished by linking the two line terminals by a communications channel [38], [43]. Single-ended schemes have also been proposed, where the relaying information is determined from the transients observed at only one terminal of the line [40]. These schemes use correlation techniques that measure the reflection time of a wave component from the time that wave leaves the relay until it is reflected back.

In principle, traveling wave protection is superior to distance protection as it is not affected by problems of over- or underreaching, or the problems associated with current inversions. It is subject to problems of mutual induction on parallel lines. Also, since the protection is only active for a short time and then seals in, there may be problems with the detection of slowly evolving faults. It also must be carefully designed to prevent pickup due to lightning arrester operation or on transients due to switching.

15.5.3 Directional Overcurrent Ground Protection

Protection against ground faults can often be provided quite adequately using directional residual overcurrent protection. In this form of protection, the directional relay is fed the zero-sequence voltage and current measurements and a trip decision is made based only on these quantities. In most cases, the protection is set in an overreaching mode. Since the zero-sequence current and voltage are nearly zero under nonfault conditions, and only appear during the ground fault, this type of relaying has certain features in common with traveling wave protection, which also sees trip characteristics only during a fault. One difference is that overcurrent protection operates only for ground vaults. One disadvantage of this form of ground protection is that the reach is a function of the source impedance, which may vary significantly for some systems, for example, where the difference between maximum and minimum generation is great. However, this form of ground fault protection is preferred by some for its relative simplicity.

Negative sequence overcurrent protection is also preferred in some cases, especially where parallel transmission lines are subject to mutual induction in the zero sequence network.

Both residual- and negative-sequence overcurrent protection have a disadvantage where single-pole reclosing is required because the relay detects a fault during the dead time when one pole is opened.

15.6 LINE PROTECTION EXPERIENCE

Clearly, the protection of series compensated transmission lines presents a challenge to the relay designer and to the engineers that apply these relays. In the early applications of series compensation, the protection was accommodated by limiting the degree of compensation

[7] Gaps often flash in about one-fourth cycle, although varistor protection is faster.

and using the same line protective relays used for uncompensated circuits, although it was recognized that this practice could not be continued for higher degrees of compensation [55]. As series compensation became more common, relay manufacturers began to provide devices that were especially designed for series compensated lines. Examples of these devices are described in the preceding sections.

The natural question arises, then, as to the experience with these line protective relays when applied to series compensated lines. In seeking an answer to this question, the author conducted interviews with experienced protection engineers, who have designed and operated the systems associated with the 500 kV transmission system of the western United States. Their experience is assumed to be similar to that of other utilities around the globe. These engineers were asked questions regarding their experience with the problems noted in Section 15.4.2, and their responses are summarized below.

15.6.1 The Effect of Transient Phenomena on Protection

Most of the relays that have been designed for the protection of series compensated lines include special filtering that rejects the high frequencies that occur at fault inception and when capacitors are bypassed. These phenomena have not been a problem and have not caused either false trips or failure to detect line faults.

Gap firing has been noted to affect power line carrier performance, but no relaying problems have been cited that are believed to be due to this phenomenon.

Low frequencies are noted to be of higher energy level and can affect the reach of mho elements. It is acknowledged that these lower frequencies are difficult to filter, but it is not clear that the subsynchronous frequencies have caused transmission line relay failure.

15.6.2 The Effect of Phase Impedance Unbalance

It is acknowledged that phase unbalance on untransposed lines is increased with series compensation, but this has apparently caused no problems in line protection. One utility has many series compensated lines, most of which are untransposed, with no reported problems due to this feature. Most of the utilities questioned have their longer lines transposed, but report no problems on short lines, which usually are not transposed.

15.6.3 The Effect of Voltage and Current Inversions

Problems with voltage and current inversions have been solved by better relay design. The use of memory and cross-polarization for fault location logic has been very effective. Traveling wave relays do not see these inversions, and have provided excellent operating experience.

15.6.4 The Effect of Fault Locator Error

Relays that use mho or offset mho characteristics are acknowledged to have problems under certain conditions, depending on the degree of compensation and the capacitor location.

However, these problems have not been noted as causing relay failure in the newer relay types that have been developed for series compensated lines.

15.6.5 The Effect of Transducer Error

The utilities contacted report no problems due to CT saturation causing failure of the protection system. One utility specifies CT's with large cores in order to avoid this problem. Another utility reports that problems may occur if CT saturation occurs before gap firing. There are no reported problems due to high frequency on the accuracy of the current transformers, however.

There are also errors due to CCVT reproduction of the step function accompanying a fault, which will appear at the relay as an exponential decay rather than a step function. This will make a mho element underreach in the first half cycle, and then overreach in the second half cycle. One relaying problem has been identified that was probably caused by this phenomenon.

When series capacitors are located at the ends of the line, the relay engineer has a choice for the location of the voltage and current measurements, either on the line side of the capacitor or on the bus side. In the past, most utilities placed both transducers on the bus side of the capacitor, but there seems to be a trend toward placing the voltage measurement on the line side of the capacitor.

15.6.6 Autoreclosing of Transmission Lines

Of the utilities questioned, some do not employ autoreclosing as a matter of design, which is predicated on the problem of maintaining stability following unsuccessful reclosure. Others use single-pole tripping and reclosing when required to improve stability performance. Reclosing can be a problem if the fault current is too small to bypass the capacitor, so the line recloses on a capacitor with a trapped charge. One utility has reported a relay problem due to a trapped charge on the capacitor.

15.6.7 Requirements for Protection System Studies

The utilities contacted often use the relay manufacturer's *model power system* facilities to examine relay performance for series compensated lines. Some of the utilities also perform *electromagnetic transients program* (*EMTP*) studies, and manufacturers also seem to be performing more digital studies of this type. Studies of the relay application on a particular line are judged to be important and are recommended.

15.6.8 General Experience with Line Protection

All utilities contacted noted satisfactory performance with most of the relays described in this chapter. Directional comparison relays have provided excellent service, and several utilities reported that no failures have been experienced by these relays. A few problems have been experienced with current comparison techniques, but these problems have been due to the pilot communications and not to the relays themselves. The newer designs have provided excellent performance records. None of the utilities questioned have experience with the distance relays on series compensated lines. Experience with traveling wave relays is excellent, even on high impedance faults.

REFERENCES

[1] Harder, E. L., J. E. Barkle, and R. W. Ferguson, "Series Capacitors During Faults and Reclosing," *AIEE Trans.*, 70, 1951, pp. 1627–1641.

[2] Berdy, John, "Protection of Circuits with Series Capacitors," *AIEE Trans.*, 82, February 1963, pp. 929–935.

[3] Blackburn, J. L., Ed., *Applied Protective Relaying*, Westinghouse Electric Corp., Newark, NJ, 1976.

[4] Mason, C. Russell, *The Art and Science of Protective Relaying*, John Wiley & Sons, New York, 1956.

[5] Warrington, A. R. van C., *Protective Relays: Their Theory and Practice, Vol. 1*, John Wiley and Sons, New York, 1962.

[6] Warrington, A. R. van C., *Protective Relays: Their Theory and Practice, Vol. 2*, John Wiley and Sons, New York, 1969.

[7] Anderson, P. M., and R. G. Farmer, *Series Compensation of Power Systems*, PBLSH! Inc., Encinitas, CA, 1996.

[8] IEEE Std 824-1994, IEEE Standard for Series Capacitors in Power Systems, draft of 1994 standard, IEEE, New York, 1994.

[9] International Electrotechnical Commission, Technical Committee No. 33, Power Capacitors, 33 (Secretariat) 103, Paris, August 1987.

[10] Courts, A. L., and E. C. Starr, "Experience with 500 kV Series Capacitor Installations and New Protection Schemes on the BPA System," CIGRÉ paper 31-09, presented at the CIGRÉ 1980 session, August 27–September 4, 1980, Paris.

[11] Hartley, R. H., R. G. Farmer, L. A. Kilgore, D. G. Ramey, and E. R. Taylor, "EHV Series Capacitor Applications Considering Subsynchronous Oscillations," CIGRÉ paper 31-06, presented at the CIGRÉ 1980 session, August 27–September 4, 1980, Paris.

[12] Hamann, J. R., "A Zinc Oxide Varistor Protective System for Series Capacitors," *IEEE Trans.*, PAS-100, March 1981, pp. 929–937.

[13] Goldsworthy, Daniel L., "A Linearized Model for MOV-Protected Series Capacitors," *IEEE Trans.*, PWRS-2 (4), November 1987, pp. 953–958.

[14] Iravani, M. R., "Effect of Energy and Voltage Ratings on ZnO Protective Systems of Series Capacitors on Transient Stability and Transient Torques," *IEEE Trans.*, PWRS-3 (4), 1988, pp. 1662–1669.

[15] Hamann, J. R., S. A. Miske, Jr., I. B. Johnson, and A. L. Courts, "A Zinc Oxide Varistor Protective System for Series Capacitors," *IEEE Trans.*, PAS-100, 1981, pp. 929–937.

[16] Souillard, M., "Protection of EHV Long Distance Transmission Lines—Distance Measurement and Directional Functions; Special Case of Series Condenser Compensated Lines," CIGRÉ paper 34-09, presented at the 1978 Session, Paris, 1978.

[17] Lewis, W. P., "Effects of Transients on EHV Protection," *Electrical Times*, January/February 1967.

[18] IEEE Committee Report, "EHV Protection Problems," *IEEE Trans.*, PAS-100 (5), May 1981, pp. 2399–2406.

[19] Batho, J. L., J. E. Hardy, and N. Tolmunen, "Series Capacitor Installations in the B.C. Hydro 500 kV System," *IEEE Trans.*, PAS-96 (6), November/December 1977, pp. 1767–1776.

[20] Alexander, G. E., J. G. Andrichak, S. D. Rowe, and S. B. Wilkinson, "Series Compensated Line Protection: A Practical Evaluation," a paper presented to the Pennsylvania Electric Association, Pittsburgh, PA., January 26–27, 1989.

[21] Wagner, C. F., and R. D. Evans, *Symmetrical Components*, McGraw-Hill Book Company, New York, 1933.

[22] Clarke, Edith, *Circuit Analysis of A-C Power Systems, Volume I, Symmetrical and Related Components*, John Wiley & Sons, Inc., New York, 1943.

[23] Hinman, W. L., "Segregated Phase Comparison Relaying Facilitates Independent-Pole Protection," Westinghouse Engineer, January 1974, pp. 20–27.

[24] Westinghouse Electric Corporation, "SPCU-1A Segregated Phase Comparison Relaying Systems," Application Data Manual 40-201, October 1987.

[25] Asea Brown Boveri, "MSPC (REL 350) Numerical Segregated Phase Comparison System," I.L.40-201.6, Application Manual, May 1992.

[26] Andrichak, J. G., and R. C. Patterson, "Transmission Line Protection with Positive and Negative Sequence Relays," a paper presented to the Pennsylvania Electric Association, October 2–3, 1975.

[27] General Electric Company, "Type SLYP-SLCN Static Directional Comparison Relaying, Description and Application, General Electric Publication GET-6456A, 1980.

[28] Alexander, G. E., W. Z. Tyska, and S. B. Wilkinson, "Improvements in Polyphase Line Protection," a paper presented at the Western Protective Relay Conference, Spokane, WA, October 19–22, 1987.

[29] Alexander, G. E., J. G. Andrichak, and W. Z. Tyska, "Series Compensated Line Protection: Practical Solutions," a paper presented at the Western Protective Relay Conference, Spokane, WA, October 22–25, 1990.

[30] General Electric Company, "PLS Hybrid Scheme Single FSK Channel for Series Compensated Lines," General Electric Publication GEK-90671A, 1991.

[31] Ziegler, G., Ed. "Application Guide on Protection of Complex Transmission Network Configurations," CIGRÉ SC34-WG04, May 1991.

[32] Asea Brown Boveri, "Distance Relay, Type 150/REZ 1," ABB Publication 1MDB06009-EN, March, 1992.

[33] Asea Brown Boveri, "Distance Protection REZ1, Application Manual:," ABB Publication UG03-7018E, Part 1A, Edition 5, November 1990.

[34] Asea Brown Boveri, "Distance Relay, Type REZ1, Application Manual: Application in Series Compensated Networks," ABB Publication UG03-7018E, Supplement 1, 1991.

[35] Asea Brown Boveri, "Distance Relay, Type LZ96a, ABB Publication 1MDB06010-EN, August 1990.

[36] Asea Brown Boveri, "Distance Relay, Type LZ96a, Application Manual," ABB Publication CH-ES 25-96.11E, November 1990.

[37] Asea Brown Boveri, "Distance Relay, Type LZ96a, Application Manual, Supplement for Series Compensated Lines (SCL)," Undated bulletin.

[38] Chamia, M., and S. Liberman, "Ultra-high-speed Relay for EHV/UHV Transmission Lines—Development Design and Application," *IEEE Trans.*, PAS-97, 1978, pp. 2104–2116.

[39] Hicks, K. L., and W. H. Butt, "Feasibility and Economics of Ultra-High-Speed Fault Clearing," *IEEE Trans.*, PAS-99, 1980, pp. 2138–2145.

[40] Crossley, P. A., and P. G. McLaren, "Distance Protection Based on Traveling Waves," *IEEE Trans.*, PAS-102, 1983, pp. 2971–2183.

[41] Cabeza-Resendez, L. Z., A. N. Greenwood, and T. S. Lauber, "Evaluation of Ultra-high-speed Relay Algorithms," EPRI Report, EL-3996, 1985.

[42] Johns, A. T., and E. P. Walker, "Cooperative Research into the Engineering and Design of a New Digital Directional Comparison Scheme," *IEE Proc.*, 135, 1988, pp. 334–368.

[43] Christopoulos, C., D. W. P. Thomas, and A. Wright, "Scheme Based on Traveling-Waves for the Protection of Major Transmission Lines," *IEE Proc.*, 135, Pt. C, 1988, pp. 63–83.

[44] Thomas, D. W. P., and C. Christopoulos, "Ultra-High-Speed Protection of Series Compensated Lines," *IEEE Trans. on Power Delivery*, PWRD-7 (1), January 1992, pp. 139–145.

[45] Asea Brown Boveri, "Combined Directional-Wave and Impedance Backup Relay, Types RALZA and RALZB," ABB Publication 1MDB06012-EN, August 1990.

[46] Asea Brown Boveri, "Ultra High-Speed Line Protection Type RALZA, Application Manual:," ABB Publication RF619-001E, Edition 4, RFA, November 1988.

[47] Asea Brown Boveri, "Setting Calculations for Ultra High-Speed Line Protection Type RALZA," ABB Publication RF619-002E, RFA, March 1985.

[48] Asea Brown Boveri, "Commissioning Instructions for Ultra High-Speed Line Protection Type RALZA," ABB Publication RF619-002E, RFA, March 1985.

[49] Asea Brown Boveri, "Ultra High-Speed Line Protection Type RALZB, Application Manual:," ABB Publication RF619-005E, Edition 3, RFA, November 1988.

[50] Asea Brown Boveri, "Setting Calculations for Ultra High-Speed Line Protection Type RALZB," ABB Publication RF619-006E, RFA, March 1985.

[51] Asea Brown Boveri, "Commissioning Instructions for Ultra High-Speed Line Protection Type RALZA," ABB Publication RF619-002E, RFA, March 1985.

[52] Asea Brown Boveri, "Directional-Wave Relay, Type LR91," ABB Publication 1MDB06011-EN, August 1990.

[53] Asea Brown Boveri, "LR91—An Ultra High-Speed Directional Comparison Relay for Protection of High-Voltage Transmission Lines," ABB Publication CH-ES 23-85.10E, 1985.

[54] Asea Brown Boveri, "Ultra High-Speed Directional Relay Type LR91—Instructions for Installation and Operation," ABB Publication CH-ES 83-85.11E, February 1985.

[55] Jancke, G., and K. F. Akerström, "The Series Capacitor with Special Reference to its Application in the Swedish 220 kV Network," CIGRÉ paper 332, presented at the 1950 session, Paris, 1950.

PROBLEMS

15.1 Consider the following short transmission line:

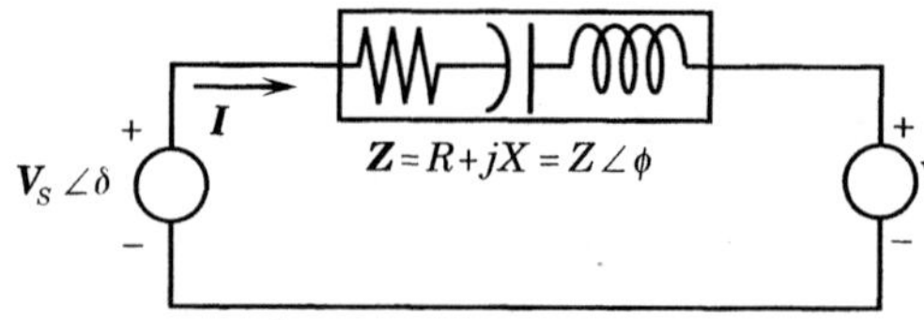

Figure P15.1 Short transmission line.

Determine the power received at the receiving end of the line as a function of the line impedance, the end-of-line voltages, and the voltage angle δ. Also compute the sending-end voltage required to sustain this received power.

15.2 Construct a spreadsheet to compute the sending-end voltage magnitude and angle a short transmission line with the following parameters:
$V_2 = 1.0$ per unit magnitude at 0 degrees
$R + jX_L = 0.002 + j0.050$ per unit
$P_2 = 0$ to 10 per unit at 80% PF lag and 80% PF lead
Degrees of compensation of 0.0, 0.3, 0.5, and 0.7
Perform the calculations and plot the results. What can you conclude about the effect of series compensation on voltage regulation and angle spread across the line?

15.3 For the system of problem 15.1 determine the optimum received power with respect to the voltage angle δ and comment on this result.

15.4 The solution found in problem 15.3 is suboptimal, since the line reactance can also be controlled to an extent by adding series compensation. Determine the optimum power delivered to the receiving end of the line if we optimize the line reactance. Plot the results of the calculations.

15.5 Using the value of reactance calculated in problem 15.4, compute the receiving end power with this computed constraint.

15.6 Consider a 500 kV line with total $R = 6$ and $X_L = 120$ ohms. Perform the following calculations, using a spreadsheet.
(a) Normalize the line impedance on a 100 MVA base and calculate the net line reactance for degrees of compensation between 0 and 1 in small steps for plotting the result.
(b) Calculate the voltage ratio, assuming no restrictions, for the values of k used in (a).
(c) Calculate the net line impedance magnitude and the received active power for all values of k. Plot the results.

15.7 Extend the calculations of the previous problem, but for this new calculation limit the voltage ratio to a magnitude of not more than 1.0. Using this restricted value of voltage ratio, compute the following:
(a) The received active and reactive powers.
(b) The sending active and reactive powers.
(c) Plot the results.

15.8 Consider the simple equivalent circuit of a power system shown in Figure 15.2. Write a computer program to determine the voltage and current at relay R as a function of the fault location parameter h. Consider only the fundamental frequency component of currents and voltages and ignore the line susceptances.

15.9 Discuss the advantages and disadvantages of the following locations for series capacitors on a transmission line:
(a) Two equal series capacitors located at the ends of the line.
(b) One series capacitor located at the center of the line.
(c) Two equal series capacitors located at a distance of one-third the length of the line from each end.
(d) One series capacitor located at one end of the line.
(e) One series capacitor located at an arbitrary distance from one end of the line.
Assume that the degree of compensation is the same for all cases.

15.10 Show that the effectiveness of series capacitors is a function of the series capacitor location. For simplicity, assume that only one series capacitor is to be installed, and that it can be located anywhere along the line.
(a) Determine analytically the optimum location for the series capacitor and plot the effectiveness of compensation. Effectiveness of compensation is defined as the ratio of the net reduction in transfer reactance to the capacitive reactance of the series compensation. Assume that the line is long, and that a simple series impedance model of the line is not adequate.
(b) Plot the effectiveness in per unit as a function of the capacitor location, for lines of length 400, 600, 800, and 1000 km.
(c) Tabulate the effectiveness for end-of-line locations and center locations of the series capacitor and compute the percent differences.

15.11 Consider a series capacitor bank that is protected by a metal-oxide varistor of the type described in Section 15.3.2. The line variables are to be normalized on a 1000 MVA base and a 345 kV base. The rated current for the capacitor bank is to be 1500 A rms. The value taken for k_P is 1.7. Determine the per unit line current at which the varistor begins to conduct and write the equations for the series equivalent R-X model of the varistor and capacitor, as given by the Goldsworthy model.

15.12 Consider the two-segment series compensation scheme shown in Figure 15.23, which is to be installed in a 500 kV transmission line. The series capacitors have a design rating of 1800 A and are to be built up of individual capacitor units rated 45 A and 7.5 kV. Show how the desired current rating can be achieved in parallel groups, and how such groups can be stacked to achieve the desired ohmic rating.

15.13 Determine the total ratings of the two segments of series capacitors in Figure 15.23, using the construction developed in problem 15.12.

15.14 It is specified that the fiber-optic communications from platform to ground of the Figure 15.23 series compensation system must be arranged such that the loss of either of the redundant signal columns will not result in loss of either control system A or B. Prepare a sketch showing how this can be accomplished.

15.15 Consider Figure 15.50, showing three cases of interest for determining reach settings for underreaching transfer trip protection. Can you think of any other cases that might be of interest? Explain.

15.16 Consider the underreaching relay configurations of Figure 15.50(*a*). Verify the reach calculations shown in Table 15.8.

15.17 Calculate and tabulate the underreaching relay configurations of Figure 15.50(*b*).

15.18 Calculate and tabulate the underreaching relay configurations of Figure 15.50(*c*).

15.19 Calculate the relay reach calculations for the overreaching condition for case 1(A) shown in Figure 15.52.

15.20 Calculate the relay reach calculations for the overreaching condition for case 1(B) shown in Figure 15.52.

15.21 Calculate the relay reach calculations for the overreaching condition for case 2 shown in Figure 15.52.

15.22 Calculate the relay reach calculations for the overreaching condition for case 3(B) shown in Figure 15.52.

15.23 Consider a series compensated transmission line with permissive underreaching distance protection, as shown in Figure 15.50, case 1. The degree of compensation is 0.75. Let the relay reach setting be 80% of the line reactance with the series compensation bypassed. The total line reactance, X_L, is 1.2 per unit.

(a) Compute the reach h_R with the series capacitor in service.

(b) Compute the reach h_R with the series capacitor out of service.

(c) Comment on these reach calculations. Would you recommend these settings? Why, or why not?

15.24 Consider a permissive overreaching transfer trip scheme on a series compensated line with 40% compensation. Determine a suitable overreaching relay setting if the adjacent lines are the same length as the protected line.

15.25 Consider a typical series capacitor installation such as Figure 15.23. Make a list of the protections you can think of that are necessary for the system.

PART IV
APPARATUS PROTECTION

16

Bus Protection

16.1 INTRODUCTION

The protection of electric apparatus differs in a fundamental way from the protection of transmission lines or distribution feeders. Transmission line protection is complicated by having the relaying equipment located at the terminals of the line, often separated by great distances, making it rather difficult to compare observations of the protective equipment at the line terminals. Electric apparatus, on the other hand, is always located at a given place. This makes it possible to compare the observations at several different ports of the device as a part of the decision process that leads to a tripping action. Since the protective equipment is at the same location as the apparatus, the various measurements are relatively easy to connect in any desired manner to achieve a given purpose.

Electric apparatus, however, is usually much more complex than transmission or distribution lines, the latter being nothing more than passive wires strung from one place to another.[1] Some types of apparatus, such as synchronous machines, are very expensive and complex devices. This tends to make their protection more complex. Moreover, there are many things that can go wrong with complex equipment such as generators, and each hazard usually requires a different type of protective equipment. This is often considered a good investment since the equipment is very costly to repair or replace and, when damaged, it may be out of service for a very long time and at great financial loss to the owner. Because of this high concentration of investment in complex equipment, the protective devices are designed to protect every conceivable type of major hazard that might cause the equipment to require costly repair.

There is a similarity in many of the types of equipment protections, whether applied to machines or passive devices. For example, a fundamental concept is the use of differential protection, which measures the sum of all currents entering a device. In normal operation, this sum should be zero, whether the device is a generator, a transformer, or a bus section.

[1]There is considerable discussion regarding the insertion into transmission lines of electronic devices that can be used for control purposes, which will complicate transmission line protection even more.

If there is a net inward flow of current, there is obviously a fault within the device terminals and differential relays are designed to make this measurement and take appropriate action. Differential protection is much easier to apply for a piece of equipment than for a transmission line, as noted earlier, since there is no need for complicated communications between the different measurement points. However, there are other complications and some of these will be discussed in this chapter. See Chapter 3 for a general description of differential relays.

Bus protection is sometimes called "bus-zone protection," because the bus protection includes all of the apparatus connected to the bus [1]. Although this terminology is more descriptive, it is not widely used and most engineers refer to this type of protection as simply "bus protection." This common term will be used throughout this chapter.

16.2 BUS FAULTS

The simplest type of apparatus protection in many respects is the relaying for faults on bus sections. The bus section itself is not physically complex, but there are interesting problems to be explored in the design of a good protective system for bus sections. This chapter begins with a description of the types of faults that have been observed in bus sections and a description of the general requirements of bus protection. This is followed by a description of the types of bus protection that are found in practice and a discussion of their relative merits.

Bus faults have been observed to be relatively rare compared with line faults [1], [2]. Bus faults account for only 6–7% of all faults, whereas line faults account for over 60% [4]. Statistics on bus faults are not widely published, but one set of bus fault statistics shows the relative frequency of different types of faults, as shown in Table 16.1 [3]. Note that most of the bus faults are ground faults in every cause category. It is also interesting to note that a relatively large number of bus faults are caused by human error, due to leaving safety grounds connected to the bus after repair or routine maintenance work, or due to operators opening loaded disconnects. The largest number of failures, however, are due to flashovers and insulation failures, which are often initiated by inclement weather. The statistics are presumed to be for outdoor bus construction, not for metal-clad or gas-insulated substations.

TABLE 16.1 Bus Fault Statistics [3]

Reported Cause of Fault	Type and Number of Faults					Tot. No.	Tot. %
	1LG	2LG	3LG	3PH	?		
Flashover	20	6	1	—	—	27	21.0
Breaker failure	16	2	2	—	—	20	15.5
Switchgear insulation failure	19	2	—	—	1	22	17.0
Other insulation failure	4	1	1	3	—	9	7.0
Current Transformer failure	3	—	—	—	—	3	2.3
Disc. opened or grounded	8	1	5	1	—	15	11.6
Safety grounds left On	6	1	8	—	—	15	11.6
Accidental contact	5	—	2	—	—	7	5.4
Falling debris	4	1	—	1	—	6	4.7
Miscellaneous/unknown	2	1	—	1	1	5	3.9
Total for each fault type	87	15	19	6	2	129	
Percentage for each type	67.4	11.6	14.7	4.7	1.6		100.0

16.3 BUS PROTECTION REQUIREMENTS

Considerable thought has been given by protection engineers regarding the need for bus protection. If dedicated bus protection is not used, clearing of bus faults will be performed by the backup protective zones of the lines terminating at the bus. This may be satisfactory in terms of the reliability of fault clearing, but it will generally be rather slow and the fault will be allowed to do more damage to the bus structure and surrounding equipment than would be the case with fast clearing. This is certainly true if the lines employ distance protection, since the bus will always be in zone 2. It will also be true for overcurrent protection, since faults on the remote reaches of the line will produce smaller fault currents and will therefore clear more slowly. Slow clearing of bus faults will often cause transient stability problems, since the bus is usually common to several terminating lines and the bus outage may result in multiple line outages. Moreover, slow clearing of faults may endanger personnel who work in the station. This suggests that dedicated, fast bus protection may be considered necessary, even though bus faults occur infrequently.

The other side of the argument is that dedicated bus protection presents the possibility of tripping the bus unnecessarily, thereby causing the outage of connected load and possibly creating system stability problems. Such unnecessary trips, which are often called "security failures,"[2] may be due to relay setting error, to instrument transformer saturation, or to human error in working on the relays or the nearby station relay equipment. This possibility, coupled with the fact that the frequency of bus faults is small and is viewed by some engineers as adequate reason for not using special bus protection. This is reported to be the case in France, for example [1].

These arguments are countered by the fact that bus faults, although rare, are likely to cause extensive damage and may possibly destroy an entire station. The effect of a bus fault could result in an extended outage of an important point of power transfer in the power system, which could be costly both in terms of repair and also in the forced modification of system operation. These arguments tended to prevail in the 1930s. Today, the need for bus protection in major stations is often required by protection engineers [5], [6]. In the final analysis, both system and bus configurations are factors to be considered.

Assuming that it has been decided that bus protection has benefits that are sufficient to merit its implementation, what are the general requirements for this type of protection? The requirements can be stated quite simply: speed and security.

High speed is necessary to limit the damage that might be caused by a bus fault. This is particularly true of high-voltage buses that have the capability of supporting very large fault currents. Bus fault clearing should always be faster than the connected line backup clearing times. Indeed, if the backup clearing is faster, there is no need for special bus protection. Most high-voltage transmission bus relays should provide relay times on the order of one cycle. Fast bus fault clearing is also required to improve the safety of the substation environment, where workers are often present.

Security is very important for bus relays since many circuits can be tripped by tripping a single bus, with the number of possible line outages depending on the bus arrangement. There are several prominent causes of security failures in bus relays. These include faults in the relay circuits, lack of proper selectivity, incorrect relay settings, mechanical shock to the relay

[2]The practice in the United Kingdom is to call this type of failure "instability." This term is rejected because of its quite different meaning in North America. The word *security*, meaning "freedom from risk," is more appropriate for this situation.

panel (for electromechanical relays), and maintenance personnel errors. Early experience in bus protection was not entirely favorable, with a relatively large percentage of failures [3]. To eliminate these problems, bus relays are often installed as two independent systems, and placed in the relay house in separate cabinets that are located at least two meters apart. Tripping, then, occurs only if both systems order the tripping action, using an "and" logic. This has the disadvantage that the failure of either relay system causes complete failure of the bus relaying function. A more comprehensive solution is to install three relay systems and use a two-out-of-three voting system for tripping. This might be considered for important substations, since the cost of the protection is small compared to the cost of the failure to clear bus faults.

Improved security can also be improved by appropriate station design wherein the bus arrangement, and therefore the bus protection, can be separated into multiple protection zones for smaller bus sections. When a bus fault occurs, the resulting bus outage will affect a smaller portion of the entire station and would usually result in a smaller system disturbance. Similarly, if a bus protection security failure should occur, the effect on the power system will be less severe if the protected bus connects fewer system components. This suggests a useful linkage between station design and protection that should be investigated on any important station where bus failure has the potential for system-wide impact. Indeed, it might be argued that there is little need for bus protection with some station designs.

16.4 BUS PROTECTION BY BACKUP LINE RELAYS

Backup protection is almost always provided for a bus by the line relays of lines that are connected to that bus. Consider, for example, bus H of Figure 16.1, which shows a part of a large system serving a subtransmission network. Should a fault occur at bus H the directional relays shown will not provide fast clearing of the fault, since the bus at H is not in the primary protection zone of any line relays. The bus fault will be cleared by the breakers shown in dark color, however, after these relays time out to operate their second zones. This would be typical for distance relays, for example, which would not reach all the way from their respective locations to bus H in their primary zone, but probably only about 90% of that distance. The second zone, however, would reach beyond bus H and would clear the bus fault.

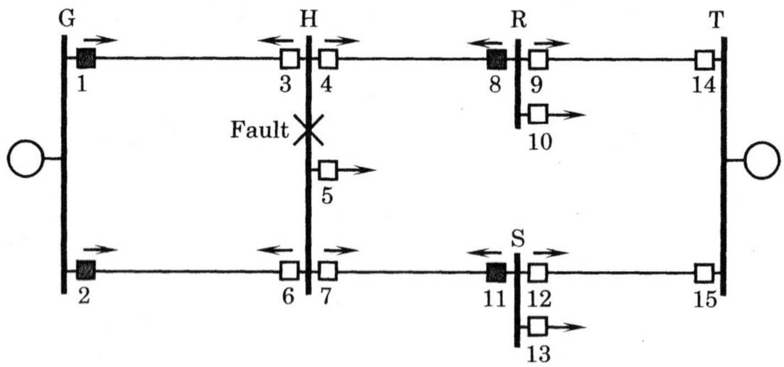

Figure 16.1 Relay clearing for a fault at bus H.

The backup protection illustrated by Figure 16.1 may also be implemented using over-current relays, where faults more distant from a given relay will pick up, but due to the magnitude of the current, may trip slowly. This is a function of the line impedance and the source impedance *behind* the relay. Overcurrent relays may be very difficult to coordinate on the system of Figure 16.1, but may work well if the subtransmission system is fed radially from only one end.

This type of protection is regarded as adequate for lower voltage stations or those that are minor switching points in the network. For high-voltage stations that are considered part of the bulk power system, a dedicated bus protection system is often preferred, since the damage caused by a slowly cleared fault may be quite serious.

16.5 BUS DIFFERENTIAL PROTECTION

Differential protection is by far the most common method of providing bus protection. There are, however, several different methods of using the differential concepts to protect a bus. The differential concepts and problems inherent in bus protection are presented first, followed by a brief description of the most common methods of bus differential protection.

16.5.1 Current Transformers for Bus Protection

The major problem of bus differential protection has historically been the unequal saturation of the current transformers used for the current measurements around the bus. This is due to the large differences in fault current magnitude measured at the different bus connection points and to the residual magnetism in the current transformers [6].

16.5.1.1 Bushing Current Transformers. Bushing current transformers are the most widely used types, primarily because of their lower installed cost. A bushing current transformer has a tapped secondary winding on an annular magnetic core. The core encircles the high-voltage conductor in the bushings of circuit breakers, transformers, generators, or other electrical equipment. This conductor forms a one-turn primary winding of the bushing current transformer (CT). The secondary turns are distributed completely around the annular core to minimize leakage reactance. Because of the tapped secondary, bushing current transformers are usually called multiratio bushing current transformers.

16.5.1.2 Window-Type Current Transformers. Window-type current transformers have a magnetic core with a center opening through which the high-voltage conductor is passed, forming the one-turn primary of the transformer. The secondary winding is wound on the core in a distributed manner. Usually, the transformer is encased in an insulating material.

16.5.1.3 Wound-Type Current Transformers. The wound-type current transformer is similar in design to a power transformer, with both primary and secondary wound on a common core.

16.5.1.4 Auxiliary Current Transformers. Auxiliary current transformers are sometimes required in the secondary circuit in order to match the ratios of the several current transformers connected to the bus terminals.

16.5.1.5 Current Transformer Accuracy. Current transformers are classified according to their accuracy, and these classifications are described in the standards [7]. The standards define the minimum steady-state performance of current transformers at high overcurrent levels. The performance is described by an identification letter and a number selected from the following: (C,T) (10, 20, 50, 100, 200, 400, 800). The letter classification described the transformer in terms of its construction characteristics.

The C classification refers to bushing current transformers with uniformly distributed windings and any other type of current transformer construction that results in a leakage flux that has a negligible effect on the ratio error. This permits the ratio correction to be calculated at any current level, knowing the burden and the excitation characteristics of the transformer.

The T classification refers to wound current transformers or any other current transformer where the leakage flux has an appreciable effect on the ratio. For this type of current transformer, ratio correction must be determined by test.

The numerical part of the classification is the secondary terminal voltage rating. It specifies the secondary voltage that can be delivered by the full winding at 20 times rated secondary current, without exceeding a 10% ratio error. Moreover, the ratio correction is limited to 10% for any current from 10 to 20 times rated current and at any lesser burden. For example, relay accuracy class C100 prescribed that the ratio correction can be calculated such that it will not exceed 10% at any current from 1 to 20 times rated secondary current if the burden does not exceed 1.0 ohm (1.0 Ω x 5 A x 20 x rated current = 100 V).

16.5.1.6 Current Transformer Problems. Current transformers are not linear transducers because of the iron core. Thus, depending on the magnitude of fault current being measured, various levels of saturation occur in the various current transformers. This complicates the problem of bus protection. One of the reasons for saturation is that some of the measured currents may be much higher than the others, which causes the measuring current transformers on those terminals to saturate.

Another important cause of saturation is due to the direct current component of fault current. This is especially important on bus locations that are electrically close to synchronous generators. The time constant of the dc component of fault current is governed by the L/R ratio of the system impedance and can vary from 0.01 to 0.3 seconds or more near generating plants [6].

Another contributor to saturation is total burden of the secondary circuit, including the CT winding resistance as well as the resistance of the leads. In some substations, these leads may be quite long, which may require the use of larger than normal wire sizes or the use of parallel secondary wiring.

16.5.2 Differential Protection Concepts and Problems

Differential protection is based on Kirchhoff's laws. Most of the systems used are current differential systems and are based on the "current law" that requires the sum of all currents entering the bus to sum to zero. Should an internal fault occur, however, the sum of currents measured at the current transformer locations will not be zero, and tripping should occur. This type of bus protection is arranged as shown in Figure 16.2, which shows the connections for one phase. The current transformer secondaries are added together to give the sum of the currents in all four lines, and the sum is sent to the differential relay (device type 87). Theoretically, should a fault occur that is external to the current transformer connections, say at the point labeled 1 in the figure, the total current flowing to that fault will be exactly equal to the total current entering the bus on lines 2, 3, and 4, and no current will flow to the differential relay. However, should a fault occur on the bus, either between phases or from phase to ground,

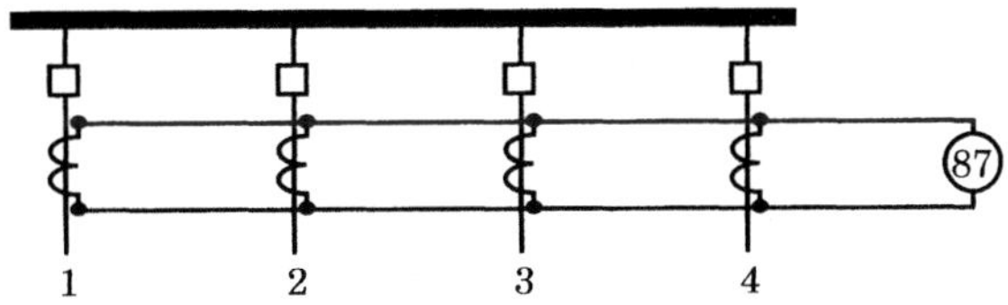

Figure 16.2 Connections for bus differential protection based on Kirchhoff's current law.

the sum of the line currents will equal the total bus fault current and the relay will correctly measure this quantity.

The foregoing discussion assumes that the current transformers are perfect and that the secondary currents are exact replicas of the line currents. Unfortunately, in a practical situation this is not the case. Consider, for example, the case of the line fault on line #1. For iron-cored current transformers, the high current flowing to the fault on line #1 is likely to saturate the current transformer in that line, whereas the unfaulted lines will carry smaller currents and their current transformers will experience either no saturation or varying degrees of saturation, depending on the total current flowing in the lines. The current measurement is further complicated by the dc component of fault current, which gives the primary current two components, a symmetrical ac component and a dc component. The dc component starts at a maximum value and decays exponentially. We can write the total primary current as follows:

$$i_P = I_{\max}[\sin(\omega t + \psi - \phi) + e^{-Rt/L}\sin(\psi - \phi)] \tag{16.1}$$

where $\phi = \tan^{-1}\left(\frac{\omega L}{R}\right)$

ψ = Phase angle after voltage zero of fault occurrence

The first term in (16.1) is the symmetrical ac current, and the second term is the dc component. The dc component starts at a maximum value and decays exponentially with a time constant of L/R, where L and R are the inductance and resistance of the primary circuit between the bus and the fault. The dc offset is an important part of the primary current if the fault should occur close to the moment the voltage is crossing zero, as this leads to the maximum dc offset. The dc offset persists longest in high-voltage circuits, where the time constant is larger than on lower voltage circuits, and is specially troublesome at locations close to synchronous generators. The dc offset causes the magnetizing flux of the transformer core to have both ac and dc components, with the dc component being large for high-voltage circuits. The dc component of flux is time varying and induces a component of current in the secondary winding, which leads to potentially large errors on CT accuracy that is also dependent on the secondary burden. Figure 16.3 shows an oscillogram of a fully offset primary current and the resulting secondary current. Note that immediately after the fault occurrence, the secondary current is small and increases slowly toward a normal magnitude. This may delay tripping for an internal fault, but can also cause false tripping of an external fault as the external fault will result in an imbalance in the readings of the current transformers of Figure 16.2.

To analyze this problem more carefully, consider the simpler arrangement shown in Figure 16.4, where there are only two lines connected to the bus. This may be considered an equivalent of the connection of Figure 16.2, with all of the lines that are not saturated represented by an equivalent line #1 in Figure 16.4 and with line #2 representing the faulted line with saturated current transformer.

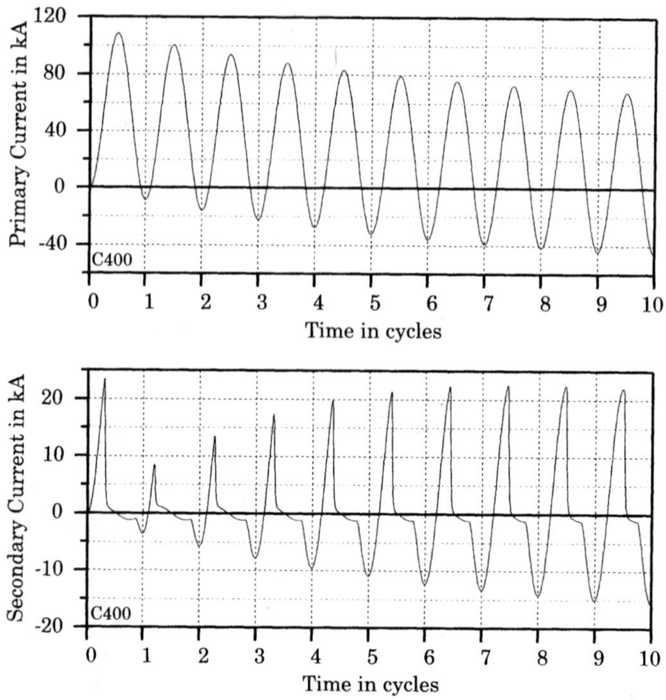

Figure 16.3 Primary current in line #1 with saturated CT.

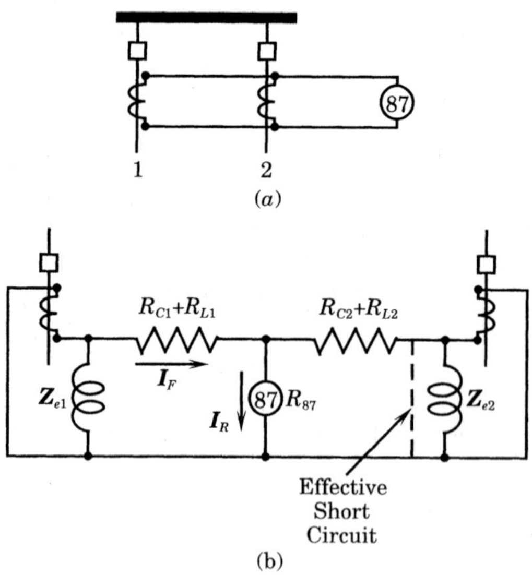

Figure 16.4 Example of the effect of CT saturation. (*a*) Physical circuit. (*b*) Equivalent circuit.

From the equivalent circuit note that the magnetizing impedance for current transformer #2 is effectively a short circuit. Then no current flows in the secondary of CT #2. The fault current flowing in CT #1, labeled I_F, divides inversely as the parallel impedances to give a relay current of

$$I_R = \frac{R_{C2} + R_{L2}}{R_{87} + R_{C2} + R_{L2}} I_F \cong \frac{R_{C2} + R_{L2}}{R_{87}} I_F \tag{16.2}$$

where R_{C2} = CT# 2 secondary resistance
R_{L2} = CT# 2 secondary lead resistance
R_{87} = Differential relay resistance

with the approximate expression being true if the relay impedance is large.

Thus, increasing the relay impedance reduces the current through the relay when one CT is saturated. One way of controlling the effect of saturation, therefore, is to add resistance to the relay branch. The relay voltage is computed as

$$V_R = R_{87} I_R = (R_{C2} + R_{L2}) I_F \tag{16.3}$$

Some relays are designed with high impedance, and these devices would work well in this situation.

In more practical situations, with several lines terminating on the protected bus section, the currents will usually all be different and the CT's may not have equal burdens. This makes it difficult to predict the effect of saturation for all possible external fault conditions. Moreover, since fault currents are high in modern transmission networks, it is not practical to provide current transformers with cores that are large enough to ensure freedom from core saturation. Early applications of current differential protection solved these problems with time delay, but this is usually considered an inadequate solution in modern power systems. Another early solution was a form of biased percentage differential protection, which also suffered from the need for current transformers with very large cores. Because of these difficulties, the current differential form of bus protection is seldom used.

16.5.3 Differential Protection with Overcurrent Relays

Bus differential protection can be performed using overcurrent relays in a connection similar to that of Figure 16.2. This arrangement experiences all of the problems with CT saturation, discussed above. Since the overcurrent relay may receive a rather high current for an external fault, it will be necessary to set the relay pickup to a high value in order to avoid false tripping.

The most troublesome aspect of current transformer saturation is due to the dc component of the fault current. Following a fault on a line near the bus of Figure 16.2, the fault current on the CT primary may be expressed as

$$i_1 = I_{\max}[Ke^{-t/\tau} + \sin(\omega t + \phi)] \tag{16.4}$$

where

$$\tau = \frac{X_2}{\omega r_a} = \text{dc offset time constant, s}$$

The dc offset time constant varies widely, depending on the type of equipment connected to the bus on the faulted line. For example, typical values of τ are given by [7], [8]:

Two-pole turbine generators	0.09 s
Four-pole turbine generators	0.20
Salient-pole generators with dampers	0.15
Salient-pole generators without dampers	0.30
Transformers	0.04
Lines	0.005–0.03

Therefore, it makes a great difference in the duration of the dc component if the fault current is limited by a generator impedance rather than a line impedance. If a generator is present, the dc component will persist for a relatively long time and CT saturation will be much more serious. This suggests that it would be impractical to overcome the effects of saturation merely by delaying tripping of the overcurrent relay by selecting a larger time dial setting, which might be practical for a situation involving only lines.

Mason gives a formula for estimating the maximum flux density in the current transformer in terms of the symmetrical components of the CT primary current [8].

$$B_{\max} = \frac{\sqrt{2} R_2 I_1 N_1 \tau}{A N_2^2} \quad \text{T} \tag{16.5}$$

where I_1 = primary current in A

$R_2 = R_{C2} + R_{L2}$ = total secondary resistance (Ω)

τ = primary circuit dc time constant (s)

A = cross-sectional area of CT core (m^2)

N_1 = number of primary turns (= 1 for bushing CT)

N_2 = number of secondary turns

Values of $B_{\max}$ greater than about 1.5 Tesla will cause saturation of the current transformer and values as high as 1.8 Tesla will cause very high saturation. Older current transformers will saturate at lower values. If dc saturation occurs it is very difficult to determine the error characteristics of the CT's, particularly if instantaneous relay pickup is desired. Methods of analysis have been presented in [3].

As noted previously, the effect of saturation of one current transformer can be counteracted by adding resistance to the relay circuit. The resistance added should not be so high as to cause serious overvoltage in the relay. Mason suggests that the resistance added should not be so high that the CT's cannot supply at least 1.5 times pickup current under minimum bus fault current conditions [8].

Mason also offers the following practical rules for applying overcurrent relays to bus differential protection [8]:

1. Locate the junction point of the CT's at a central point with respect to the CT locations and use heavy wire in order to hold down the resistance of the leads.
2. Choose CT ratios so that the maximum magnitude of external fault current is less than about 20 times the CT rating.
3. Set the relay pickup at least twice the load current of the most heavily loaded circuit.
4. Use inverse-time overcurrent relays to provide some time delay and account for the dc component of current

16.5.4 Bus Protection with Percent Differential Relays

Using percent differential relays instead of overcurrent relays in bus differential protection is a great improvement. For this type of relay, the problem is one of providing adequate restraint in the relay to make up for the inadequacy of the CT's as they tend to saturate with large fault currents. One solution is the use of a "multi-restraint" relay that consists of three groups of restraint units and one operating unit, as shown in Figure 16.5. The restraint windings are unidirectional, that is, current flowing through the restraint winding in either direction causes restraint, or flowing through the operating winding provides closing bias. The restraint windings are paired such that one group sees the difference of line currents and another sees the sum [9].

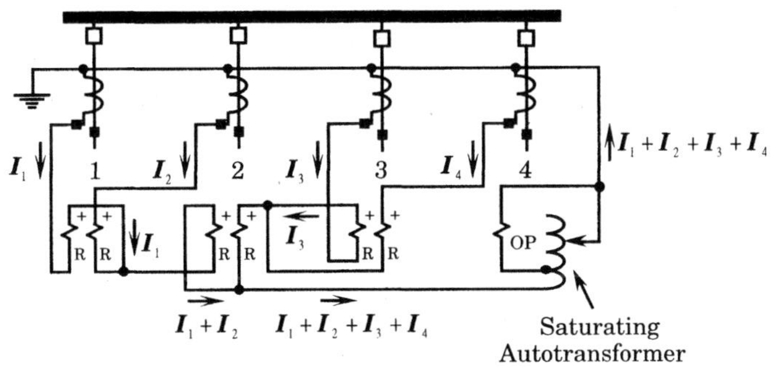

Figure 16.5 Multirestraint percentage differential relay [9].

In addition to the multiple restraint windings, this relay has a "variable percentage" characteristic, provided through a saturating auto transformer, which also provides a shunt for the dc component, thereby improving the sensitivity of the unit.

16.5.5 Bus Differential Protection with Linear Couplers

The problems associated with current transformer saturation make bus differential relaying very difficult because it is hard to predict the degree of saturation and to design appropriate countermeasures. One effective solution to this problem is to eliminate CT saturation by eliminating the iron from the current transducer. This is accomplished by a device called a "linear coupler," which is an air-cored mutual reactor on a nonmagnetic toroidal core [9], [10]. The advantages of this device are as follows:

1. Elimination of saturation in current measurements
2. High-speed performance with negligible transient response
3. Very reliable design
4. Easy to set and maintain
5. Can be operated without damage with secondaries open

The only disadvantage is that this solution requires the purchase of all new CT's since all circuits on the bus must be equipped with linear couplers if they are to be used at all. This may be costly in cases where there are already CT's in place that the engineer would like to use.

The linear coupler has a large number of secondary turns with a linear characteristic of about 5 volts per 1000 primary ampere turns. Because of their unique characteristic, linear coupler differential relaying uses a voltage differential scheme, as shown in Figure 16.6 for a typical four circuit bus.

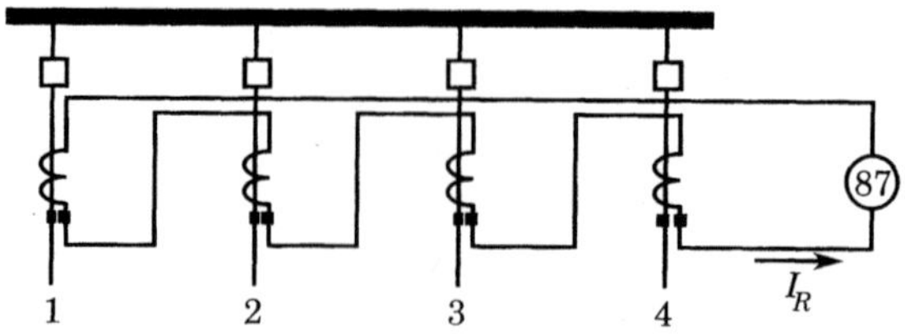

Figure 16.6 A linear coupler bus differential protection scheme [9].

The linear coupler output is a secondary voltage V_s that is proportional to the primary current I_p, or expressed in a phasor equation

$$\boldsymbol{V}_s = jX_M\boldsymbol{I}_p \tag{16.6}$$

where X_M is the mutual inductive reactance of the device. Then the relay current may be computed as

$$\boldsymbol{I}_R = \frac{\sum_i \boldsymbol{V}_{2i}}{\boldsymbol{Z}_R + \sum_i \boldsymbol{Z}_{Ci}} \tag{16.7}$$

where $\boldsymbol{Z}_C$ = self-impedance of linear coupler
$\boldsymbol{Z}_R$ = Impedance of relay

For this type of protection, the impedance of the leads can be neglected compared to the 30–80 ohms of impedance for the relay and 2–20 ohms for each linear coupler in the circuit.

For a bus fault, all of the line currents are directed toward the faulted bus and the secondary voltages of the linear couplers add together, with the total voltage impressed across the relay. This provides high-speed protection that is not complicated by estimating the errors due to saturation. The only problem in this type of protection is the possible error due to linear couplers with different characteristics, which may cause a differential voltage to the relay during certain external faults. Since linear couplers have an accuracy of about 1%, the worst case is to have all the nonfaulted lines with an error of +1% and the faulted line with an error of −1%, for a total of 2% error. If the relay is set to operate for a minimum internal fault of x amperes, then it would also operate for a maximum external fault of $50x$ amperes. This suggests that a useful performance criterion is stated by the ratio [9]

$$\frac{\text{Max external}}{\text{Min internal}} = 50 \tag{16.8}$$

If a safety factor of 2/1 is desired, this ratio should be set at 25/1. In many cases the maximum external fault will be a three-phase fault, while the minimum internal fault will be a one-line-to-ground fault through an impedance.

The linear coupler bus protection is effective in solving the problems due to current transformer saturation. The system has two disadvantages, however. First, the sensitivity of the linear coupler system is limited by the maximum external fault conditions, as noted above. Second, this system requires the installation of linear couplers that are dedicated only to bus differential protection.

16.5.6 High Impedance Bus Differential Protection

Another type of bus differential relaying is the high impedance bus differential scheme, which is a current differential relaying scheme using overvoltage relays. This system is also designed to defeat the problems of CT saturation [11]. The basic structure of this scheme is shown in Figure 16.7.

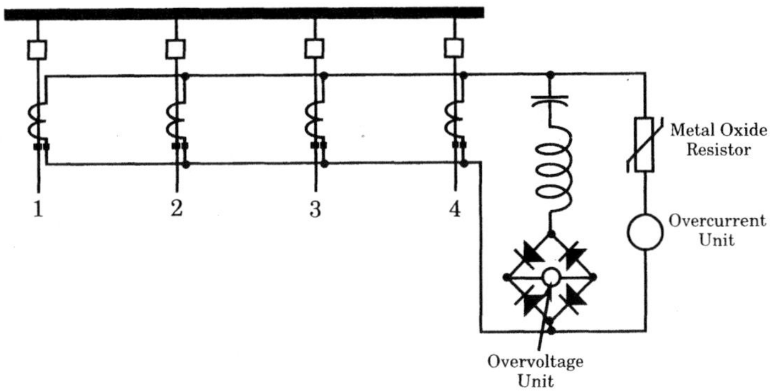

Figure 16.7 Bus differential protection using high impedance voltage relays.

In this scheme, conventional CT's are used such as bushing CT's or other types that have low impedance secondary windings. This concept is a good example of reaping the benefits of adding resistance to the relay circuit in order to control the effects of saturation. This is true since the full-wave bridge rectifier adds substantial resistance to that leg of the circuit. The series L-C circuit is tuned to the system fundamental frequency in order to make the relay responsive only to the fundamental component of current, thereby improving selectivity. The disadvantage of this idea is that it slows the response to a degree, but this is countered somewhat by the addition of a high-speed overcurrent relay in series with a voltage-limiting, metal-oxide resistor, which is placed in the circuit to limit the voltage that is produced when faults occur, the voltage being quite high because of the high resistance of the bridge rectifier (about 3000 ohms). Thus, the overcurrent unit provides the tripping for high-magnitude bus faults, and its minimum pickup can be conveniently set quite high to permit it to ignore external faults.

This type of bus protection is particularly well suited for situations where a large number of circuits are connected to the bus. The essential requirements for proper application of this principle are the following [12]:

(a) Equal current transformer ratios on all connections.

(b) Low current transformer secondary winding resistance.

(c) Adequate knee-point voltage output from the current transformers.

(d) Low burden from connecting leads.

The knee-point referred to in (c) is usually taken to be equal to or greater than twice the relay voltage setting required to ensure security against tripping for external faults just outside the bus protection zone, and under maximum fault conditions. Correcting auxiliary current transformers can be used to correct for unequal CT ratios.

EXAMPLE 16.1

As an example of bus differential protection, consider the protection of the north and south buses of the breaker-and-a-half station shown in Figure 16.8. All current transformers are 2000/5 except for breaker positions 3N and 3T, which are 3000/5. Therefore, these two positions require 3/2 auxiliary current transformers, as shown in the figure. The relays used for this application are differential voltage relays [13]. These relays are high impedance voltage sensing relays that operate from the voltage produced by differentially connected current transformers. This relay is also provided with an optional instantaneous overcurrent unit, which is set to operate for high current internal faults that occur within the protected zone. Because the voltage sensing unit is subject to damage due to overvoltage during severe internal faults, these units are protected by nonlinear resistance devices that are connected in parallel with the voltage sensing relays. The overcurrent relays are connected in series with the voltage protective devices such that positive fault detection is ensured under all conditions.

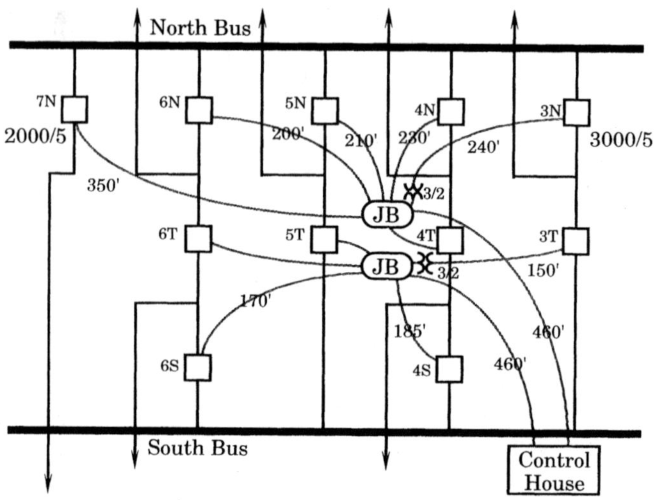

Figure 16.8 A breaker-and-a-half station requiring bus protection.

A typical connection of the relays is shown in Figure 16.9, which depicts a conventional differential scheme. In this type of connection all of the CT secondaries are connected in wye. The voltage sensing relay is designated 87L, which is protected by a nonlinear resistor against overvoltage. The optional overcurrent unit is device 87H, which is in series with the protective device. A lockout relay, device 86, is used to prevent reclosing, once a bus fault is detected. The lockout contacts also bypass the protective

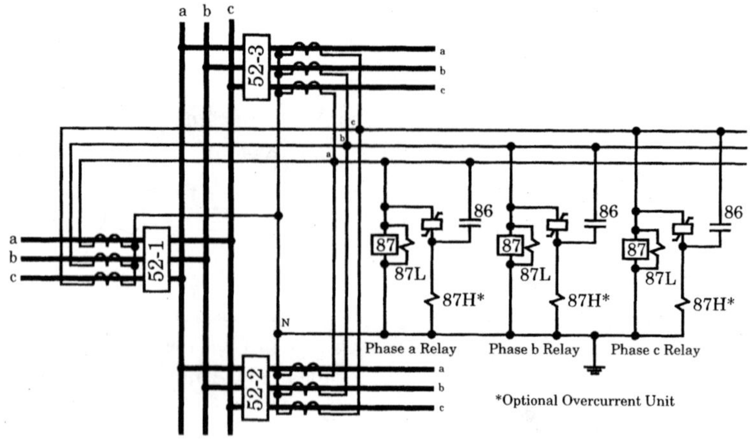

Figure 16.9 External ac connections for the bus differential relays [13].

device and thereby limit the thermal energy dissipation in the protective device. Figure 16.9 shows only three lines connected to the protected bus, but a larger number of lines can be connected to the summing junctions. Note that one relay is required for each phase.

This type of relay is applicable for bus protection where all current transformers have negligible leakage reactance. This is usually true of current transformers with toroidal cores having distributed windings, such as most bushing current transformers. It is preferable that all CT's in the differential circuit have the same ratio, although it is usually possible to accommodate a CT with a different ratio, if necessary; but this is considered a case that may require special treatment.

The external dc connections for the relay are shown in Figure 16.10. A timer, 62X, is used to initiate breaker failure protection following a suitable time delay. A seal-in unit (TSI) is added to ensure continued pickup of the relay.

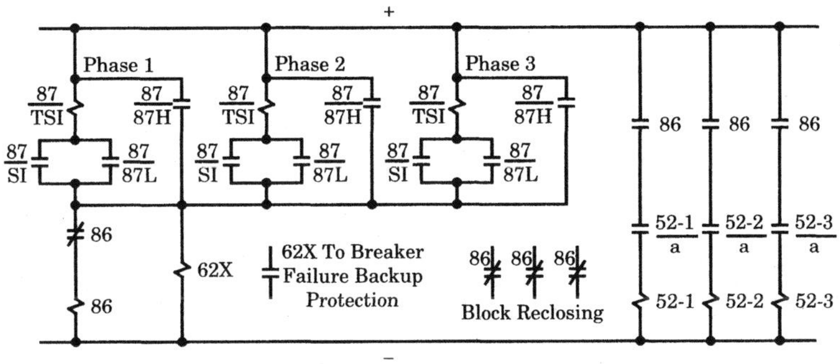

Figure 16.10 External dc connections for the bus differential relay [13].

Referring again to Figure 16.8, the problem is to provide bus protection to the north and south buses of the station. One of the problems in a large station is the resistance of the leads from each circuit breaker location to the junction box, labeled JB in the figure, as well as the distance to the control house where the relays are located. During a fault, the voltage developed at the secondary of the current transformers provides the emf to drive each line current contribution to the junction box, where all currents are summed and sent to the differential relay. The relay must trip for any fault on the bus, but must not trip for a fault just outside the bus CT's. Figure 16.11 shows the case of a one-line-to-ground fault on phase #1, just outside the bus CT's of one line. Only the faulted phase is shown in the figure. In analyzing this circuit it is important to remember that the relay is a high impedance device, so the impedance seen looking to the right in the diagram is high. Any appreciable voltage at the summing junction will trip the relay. The worst case is for the maximum fault on the line with the highest lead resistance, or the longest leads from the CT to the summing junction. If all CT's are ideal, there will be a negligible voltage V_R due to the external fault and the relay will not trip, since the current flowing

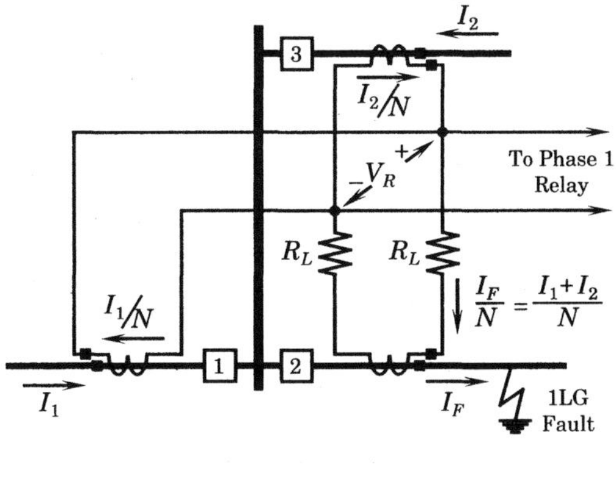

Figure 16.11 A single-line-to-ground fault outside the protected bus.

to the summing junction is equal to the current leaving that junction. The worst condition occurs when the external fault is large enough to saturate the current transformer that carries the fault current, which is the CT on the lower right of Figure 16.11. Under this condition, the rate of change of flux linkages in the transformer is unable to create any secondary induced voltage and the secondary winding is seen as simply its winding resistance, assuming the secondary leakage reactance is negligible (a fundamental assumption in the application of this type of relay). However, the other circuits connected to the summing junction do not have saturated CT's and all contribute currents that flow in the circuit consisting of the lead resistance and the CT secondary winding resistance. This creates a voltage at the summing junction that may be computed as

$$V_R = \frac{(R_S + 2R_L)I_F}{N} \tag{16.9}$$

where V_R = voltage across the relay
R_S = CT secondary winding resistance
R_L = one-way cable resistance from junction to CT
I_F = fault current, rms primary amperes
N = CT ratio

The factor of two is required to account for the fact that the current must flow through both outgoing and return cable lines for the single-line-to-ground fault. For a three-phase fault, there will be no current in the return line, since all currents add to zero in the return connection. Therefore, (16.9) can be modified to account for this by multiplying the cable resistance term by a factor, as shown in (16.10), where the factor k is given the value 1 or 2, depending on whether the fault is three phase or one-line-to-ground, respectively.

$$V_R = \frac{(R_S + kR_L)I_F}{N} \tag{16.10}$$

The voltage computed in (16.10) is the maximum that can be developed across the differential relay. It represents an extreme condition, since the CT on the faulted line will probably not lose all of its ability to induce a secondary voltage.

For an internal fault, that is a fault on the protected bus, all CT's operate into the high impedance of the relay, and, perhaps, with one or more idle CT's as well. In this case, the voltage at the summing junction will be approximately the open-circuit secondary voltage of the current transformers, which is very high. Even for a high impedance bus fault, the relay voltage will be much greater than that calculated by (16.10). The relay can be set with a pickup value computed by (16.10), which ensures that there will be no pickup for external faults, and providing excellent pickup for a fault on the protected bus. In order to provide a margin over and above the worst external fault condition, the pickup setting is computed as

$$V_R = \frac{1.6K(R_S + kR_L)I_F}{N} \tag{16.11}$$

where

$$K = \text{CT performance factor}$$

and the value 1.6 is a margin factor to provide a factor of safety in the minimum pickup setting. The current transformer performance factor is shown in Figure 16.12 [12]. This factor is computed as a function of the relay voltage (16.9) divided by the knee-point voltage E_S for the poorest CT in the circuit. This is explained in more detail below.

The minimum internal fault current that will cause the 87L unit to operate is computed as

$$I_{\min} = \left(\sum_{i=1}^{n} I_i + I_R + I_P\right) N \tag{16.12}$$

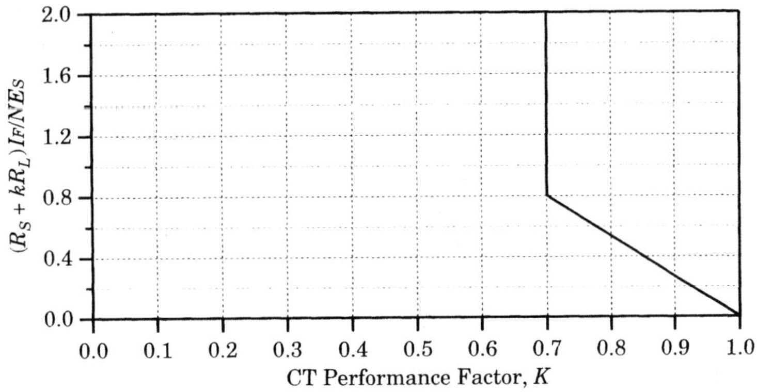

Figure 16.12 CT performance factor as a function of fault conditions [13].

where I_{min} = minimum internal fault current to trip 87L
I_i = secondary excitation current of CT i at pickup voltage
I_R = current in 87L at pickup voltage
I_P = current in protective resistor at pickup voltage
n = total number of breakers connected to the bus
N = CT turns ratio

The current in the relay can be determined from the relay voltage and the known resistance of the relay, which is given by the manufacturer as 1700 ohms, or

$$I_R = \frac{V_R}{1700} A \tag{16.13}$$

Solution

The actual calculation of the relay settings is performed according to methods provided by the manufacturer, which are direct and easy to follow. Two methods are provided. The first is based on using (16.11) for the greatest fault current on any circuit connected to the protected bus, either for three-phase or one-line-to-ground faults. A simplified method uses simply the maximum interrupting rating of the connected circuit breakers rather than the computed maximum fault current. With either method, the resistance R_L is based on the one-way distance from the junction point to the most distant CT.

For the station configuration shown in Figure 16.8, the maximum distances from the junction point to the most remote breakers are shown, with the distances in feet noted on the diagram. The cable used is #9 (19/22) wire, which has a resistance 20°C of 0.8233 ohms/1000 feet. If we consider that the maximum working condition is to be 75°C, then we compute the cable resistance to be

$$R(@75°C) = 0.8233\left(\frac{234.5+75}{234.5+20}\right) = 1.0012\ \Omega/\text{kft} \tag{16.14}$$

or about 1 ohm per 1000 feet.

The secondary resistance of the current transformers is computed from an estimate of the resistance per turn and the number of turns. For the station in Figure 16.8, the current transformers have the following characteristics:

$$\text{Winding resistance} = 0.0013\ \Omega/\text{turn}$$
$$\text{Lead resistance} = 0.04\ \Omega$$

For the two ratings of current transformers used in this station, we compute the following secondary resistance values.

$$\begin{aligned} R_S &= 0.0013 \times 400 + 0.04 = 0.56\ \Omega \text{ for 2000/5 CT} \\ R_S &= 0.0013 \times 600 + 0.04 = 0.82\ \Omega \text{ for 3000/5 CT} \end{aligned} \tag{16.15}$$

The two longest runs from the junction boxes to the circuit breakers are noted from Figure 16.8 to be 350 feet (2000/5 CT) and 240 feet (3000/5 CT). Therefore, the values of R_L for these two runs is computed as

$$\begin{aligned} R_L &= \frac{350 \times 1.001}{1000} = 0.35\ \Omega \text{ for the 2000/5 CT} \\ R_L &= \frac{240 \times 1.001}{1000} = 0.24\ \Omega \text{ for the 3000/5 CT} \end{aligned} \tag{16.16}$$

The maximum interrupting rating of the circuit breakers for the station of Figure 16.8 is known to be 40 kA. We can use this value in conjunction with the simplified method to compute the relay settings. The value of E_S must be taken from the knee of the saturation curve of the current transformers. This value is known to be 400 volts for the 2000/5 current transformers (but not shown in this example).

The current transformer performance factor is taken from Figure 16.12. We can compute the ordinate of Figure 16.12, which is a function of the CT saturation voltage, as follows. Since we are using the breaker rating, the fault is assumed to be a one-line-to-ground fault.

$$\frac{(R_S + kR_L)I_F}{NE_S} = \frac{(0.56 + 2 \times 0.35)40{,}000}{(2000/5)400} = 0.315 \tag{16.17}$$

Referring to Figure 16.12, the CT performance factor is approximately 0.88. Substituting into (16.11), we compute, for the 2000/5 CT's that are located at a distance of 350 feet from the junction box:

$$\begin{aligned} V_R &= \frac{1.6K(R_S + kR_L)I_F}{N} \\ &= (1.6)(0.88)(0.56 + 2 \times 0.35) \times \frac{40{,}000}{400} \times \frac{3}{2} = 266\ \text{V} \end{aligned} \tag{16.18}$$

where the 3/2 factor is the ratio of the auxiliary current transformer.

As a check, we also compute the relay voltage for the 3000/5 CT's that are located 240 feet from the junction box. These current transformers have a knee voltage of about 600 volts. Therefore, the ordinate of Figure 16.12 is

$$\frac{(R_S + kR_L)I_F}{NE_S} = \frac{(0.82 + 2 \times 0.24)40{,}000}{(3000/5)600} = 0.144 \tag{16.19}$$

and the performance factor for this current transformer is about 0.95. Then the relay voltage based on the 3000/5 as the worst CT is computed to be

$$\begin{aligned} V_R &= \frac{1.6K(R_S + kR_L)I_F}{N} \\ &= \frac{(1.6)(0.95)(0.82 + 2 \times 0.24)(40{,}000)}{600} \cong 132\ \text{V} \end{aligned} \tag{16.20}$$

Clearly, the 2000/5 CT is the worst case. The 266 volt worst-case value can also be checked to see how it compares with the voltage at 10 amperes excitation on the CT saturation curve, which corresponds to about 900 volts. The relay manufacturer suggests that the relay voltage should not exceed 67% of this 10 ampere value, and 67% of 900 volts is over 600 volts. Hence, the 266 volt relay threshold is satisfactory from that point of view.

The instantaneous overcurrent 87H setting of the relay is based on the setting of the 87L unit. To determine this setting the manufacturer provides a curve for this purpose, which is shown in Figure 16.13. Since the 87L setting is 266 volts, the setting of the 87H is read from the plot to be 2.0 amperes.

If the actual fault currents are used rather than the circuit breaker ratings, a more accurate and sensitive setting can be determined, if this is considered necessary. Using the circuit breaker rating gives a setting that provides an excellent margin of safety against false trips, but results in reduced sensitivity.

The foregoing is a simplified calculation of the relay setting, and there are other checks on the relay setting that can be made. The characteristics of the nonlinear protective device should be checked, as this is a factor in determining the minimum pickup current, as noted in (16.13).

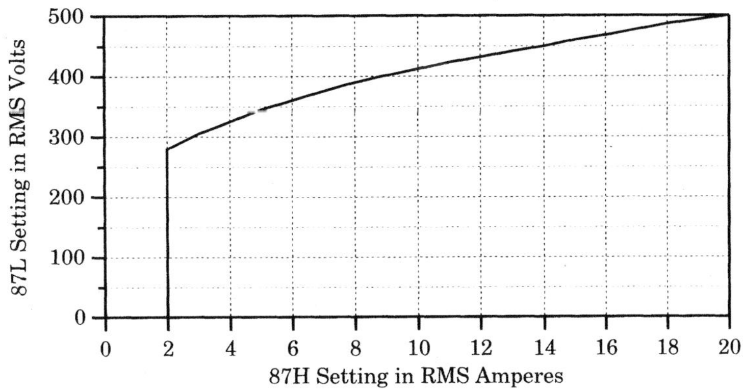

Figure 16.13 Pickup setting of the 87H versus settings of the 87L relay.

Determining the relay setting requires that information about the current transformers be available, as well as information on the station dimensions, CT ratios, and the circuit breaker ratings or calculated fault currents. Once all of this information is available, however, the computation of the settings is not difficult. ■

The success of this method of bus protection depends on the proper selection of current transformers [12]. This requires optimum-ratio, low-reactance CT's. It has been noted that some CT's with low ratios sometimes have inadequate knee-point voltage. In such cases, an alternative solution might be the use of biased differential relays and in some very difficult cases, both methods may be necessary, which is sometimes referred to as a moderately high impedance relay [14].

16.6 OTHER TYPES OF BUS PROTECTION

It is probably fair to say that bus differential protection is the most common type of bus protection. However, there are other forms of bus protection that should be mentioned, even though they represent somewhat special cases. The first to be discussed is a form of protection called "fault bus" protection that is applicable to certain switching station designs. The second type is a combined protective system where a bus and a transformer can be protected as a unit.

16.6.1 Fault Bus Protection

Fault bus protection is used in cases where an insulating bus support structure can be built that is completely isolated from ground as in the case of metal-clad switchgear or other stations that can be completely housed in a metal enclosure. The basic structure is shown in Figure 16.14, where all equipment in the enclosure is isolated from ground except for the single ground connection that contains the protective relays.

The overcurrent relay controls an auxiliary relay that trips all circuit breakers connected to the bus. It is important that all framework, circuit breaker tanks, supporting steel structure, as well as the high-voltage bus conductors, be isolated from ground by supporting insulators. This type of protection is excellent for isolated phase construction, in which case all faults are ground faults. The scheme is probably used most often for indoor switchgear.

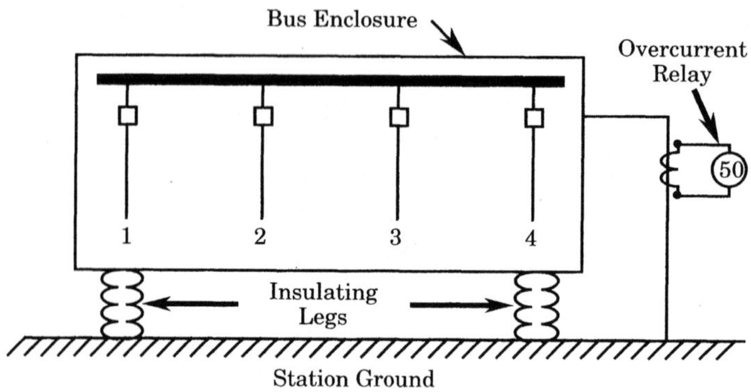

Figure 16.14 Diagram of the bus fault method protection scheme.

16.6.2 Combined Bus and Transformer Protection

Another type of bus protection is sometimes employed when a transformer and a bus can logically be protected as a unit, as shown in Figure 16.15. This particular type of combined protection uses a different restraint winding in the differential relay for each circuit. Note that there is no circuit breaker between the bus side of the transformer and the transformer itself, which is a rather common situation. Lacking the bus side circuit breaker, bus differential protection is not possible, so the transformer and bus are included in the same protection zone.

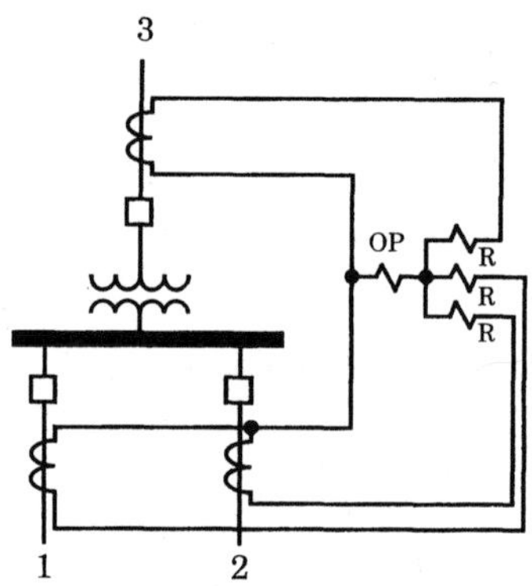

Figure 16.15 Combined bus and transformer differential protection.

The arrangement shown in Figure 16.15 is only one of several arrangements for combined bus-transformer protection that can be devised [9]. Another arrangement provides separate differential protection for the bus and for the transformer, but with both protective schemes tripping all breakers to de-energize the entire bus-transformer combination.

16.6.3 Bus Protection Using Auxiliary CT's

Most of the differential bus protection schemes suffer from two types of problems: (1) current transformer saturation, and (2) careful current transformer selection, matching, and design restrictions on the burden presented to each transformer. Current transformer saturation is particularly difficult for buses that are near synchronous generators, because of the high dc component of fault currents supplied by the generators.

One type of bus differential protection that solves these problems is the "half-cycle" protection scheme that uses auxiliary current transformers for proper impedance matching and detects the fault condition before the current transformers reach saturation [15]. The design is based on the fact that a conventional CT will deliver current for about 2.0 ms, even under fault conditions, before saturation occurs. The half-cycle relay is designed to detect the fault and initiate tripping before saturation occurs.

The impedance of the half-cycle relay is moderate, usually in the range of 100 to 250 ohms. This is much less than the usual high impedance system that has several thousand ohms impedance. However, a few hundred ohms is great enough that the resistance of leads and other incidental impedances is of little concern. The half-cycle relay includes a restraint system that permits high resistance in the CT secondary circuits.

A schematic of the half-cycle relay system is shown in Figure 16.16. The main CT's can be of any type and of different ratios that may be required for other functions for which these CT's are used. Main CT's may have CT ratios that differ by factors as large as 10 to 1. The auxiliary CT's are part of the half-cycle protective system and are adjusted to effectively cause each circuit to present the same turns ratio.

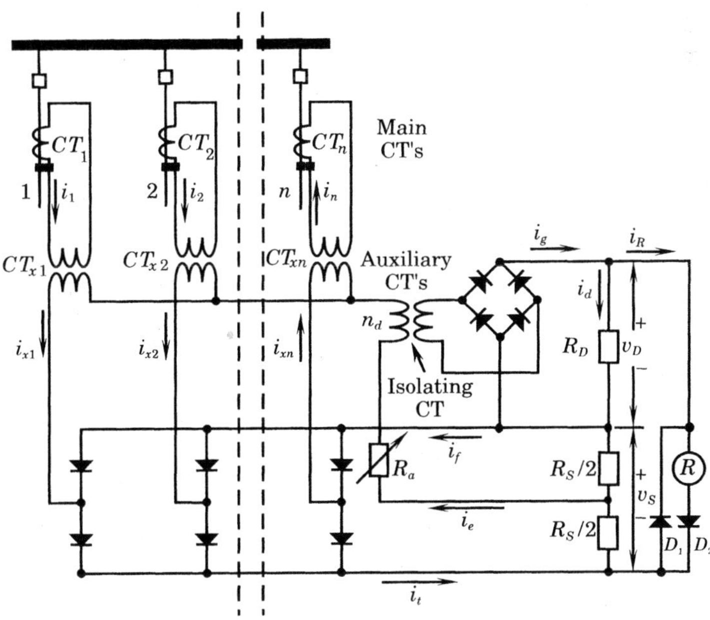

Figure 16.16 Schematic of the half-cycle bus differential system [15].

The relay utilizes a full wave rectifier in the secondary of each circuit, only one phase of which is illustrated in Figure 16.16. This gives a unidirectional pulsating current, the sum of which is the total incoming current, i_t, shown in the figure. This current flows through two resistances, R_S and R_D, which causes the measured voltage drops shown in the figure. Should an internal fault occur, or current flow due to CT ratio error, this will result in a differential error current i_e flowing from the center tap of R_S through an adjusting resistor and the primary of the isolating CT. The error current output of the isolating transformer is passed through a full wave rectifier to produce i_g, which flows through R_D to produce voltage drop v_D. This voltage is compared to the voltage v_S across R_S in a comparator circuit to develop the relay threshold condition, with tripping of relay R occurring when

$$v_D > v_S \tag{16.21}$$

When this condition is satisfied, a current i_R will flow through the relay R in the direction shown in the figure. The relay R is a dry reed relay with an operating time of less than 1 millisecond, which facilitates tripping before CT saturation occurs.

We may rewrite the threshold equation (16.21) as follows:

$$\begin{aligned} v_D - v_S &> 0 \\ v_D - v_S &= (R_R + R_{D2})i_R \cong R_R i_R > 0 \end{aligned} \tag{16.22}$$

where we assume that the diode resistance is small compared to the relay resistance. Now, rearranging (16.22) we write the threshold inequality as

$$v_D - v_S - R_R i_R > 0 \tag{16.23}$$

The voltages may be written by inspection of the circuit of Figure 16.10 as follows.

$$\begin{aligned} v_D &= R_D i_D = R_D(i_g - i_R) = R_D(n_d i_e - i_R) \\ v_S &= -\frac{R_S}{2}(i_t + i_R) - \frac{R_S}{2}(i_t + i_R - i_e) \\ &= -R_S i_t - R_S i_R + \frac{R_S}{2} i_e \end{aligned} \tag{16.24}$$

Substituting these voltages into (16.23) and rearranging, we compute the threshold inequality

$$i_e > s i_t + k i_R \tag{16.25}$$

where

$$\begin{aligned} s &= \frac{R_S}{n_d R_D + R_S/2} \\ k &= \frac{Z_R + R_D + R_S}{n_d R_D + R_S/2} \end{aligned} \tag{16.26}$$

The dimensionless quantity s is the slope of the relay threshold characteristic and k is a dimensionless characteristic of the comparator circuit. The equation (16.25) gives the conditions under which the relay will trip. Replacing the inequality with an equal sign gives an equation for the relay threshold, which is usually plotted in the i_e versus i_t plane.

The largest possible error current that will ensure no relay operation is given by setting $i_R = 0$ in (16.25). Then the boundary of secure operation is given by

$$i_e = s i_t \tag{16.27}$$

The literature sometimes refers to this boundary as the "stability" limit, a somewhat confusing term. Security limit is a more precise term. This limit is represented by a straight line through the origin.

We may also compute the minimum possible error current. This occurs when all of the total current enters the relay differential circuit, or

$$i_{e(\min)} = i_t \tag{16.28}$$

where the subscript (min) is attached for emphasis. Substituting this into (16.25) we compute the minimum differential operating current in terms of the relay current.

$$i_{e(\min)} = \frac{k}{1-s} i_R \tag{16.29}$$

We now examine the principles of operation of the half-cycle relay system.

16.6.3.1 Normal Conditions. With correct CT ratios, the total current entering the relay, i_t, will be exactly equal to the current leaving, i_f, and this current will circulate in the loop containing the full wave rectifiers. The current i_e will be zero. Thus, we write these conditions as follows:

$$\begin{aligned} i_t &= i_f & i_R &= i_d = 0 \\ i_e &= i_g = 0 & v_S &= v_D = \frac{R_S R_D}{2(R_S + R_D)} i_t \end{aligned} \tag{16.30}$$

The voltage expression assumes that the resistance of diode D2 is small compared to the resistance R_D.

Minor CT ratio errors will result in a small error current i_e flowing from the center tap of R_s. This current will be very small due to the high impedance of the isolating transformer. This is because the isolating transformer output current i_g will not flow unless the condition of (16.21) is satisfied.

16.6.3.2 External Fault with No Saturation. For small external fault currents that do not saturate the CT's, the condition will be similar to that for normal conditions, given above.

16.6.3.3 External Fault with CT Saturation. For large external faults that cause CT saturation, the conditions are more complex; however, before saturation occurs the conditions described above will prevail. The magnitude of error current flowing will depend on the impedance of the relay circuit and that of the saturated CT, that is,

$$i_e = \frac{\text{Relay Ckt Z}}{\text{Sat CT Sec Z}} i_t = \frac{R_{et}}{R_{CT}} i_t \tag{16.31}$$

The current i_f is the current in the faulted CT and is equal to

$$i_f = i_t - i_e \tag{16.32}$$

or, at the threshold condition given by (16.29)

$$i_f = (1 - s) i_t \tag{16.33}$$

Now, let R_{fx} be the maximum effective saturated CT secondary resistance. At the threshold condition we may write

$$R_{et} i_e = \left(R_{fx} + \frac{R_s}{2} \right) i_f \cong R_{fx} i_f \tag{16.34}$$

where we define the total impedance seen by the differential current as

$$R_{et} = R_a + n_d^2 R_D + Z_{md} \tag{16.35}$$

and where Z_{md} is the short-circuit impedance of the isolating CT. From (16.33) and (16.34) we compute

$$R_{fx} = \frac{s}{1 - s} R_{et} \tag{16.36}$$

This is the maximum permissible effective CT secondary resistance as seen from the relay. Note that this resistance depends on $s < 1$ so the effective impedance will usually be larger than the differential circuit resistance and will typically be 100 times or so greater than the maximum permissible resistance of other differential systems [15].

16.6.3.4 Internal Faults. For internal faults, the secondary current flows in the differential circuit and the relay operates. The impedance of this circuit is R_{et} which is on the order of a few hundred ohms. The relay burden is high, and saturation is likely to occur. The relay should operate on the initial rise of current and before saturation occurs.

16.6.3.5 Operating Characteristics. The operating characteristic for external faults is shown in Figure 16.17, which is plotted for a typical value of $s = 0.8$. The upper line in the figure represents the case of (16.28), where all of the relay input current flows through the differential circuit. Operation above this line is impossible.

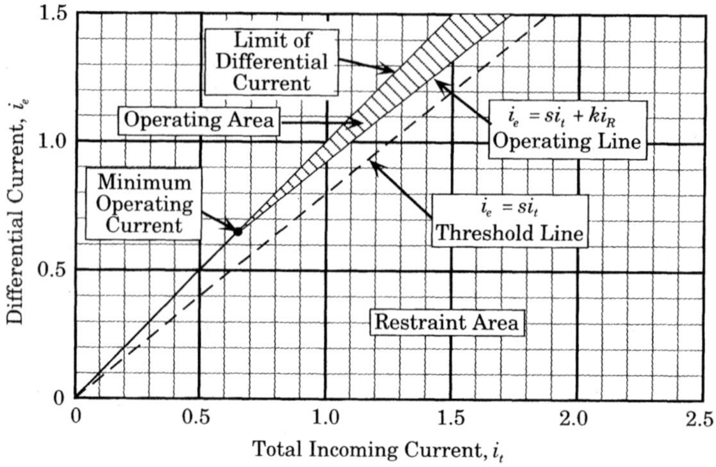

Figure 16.17 Relay characteristic for an external fault [15].

The threshold line is described by (16.27), and operation below this line will always be restrained operation. The operating line is given by (16.25) and the differential current must be large enough to fall above this line in order to cause tripping. This restricts the region of possible false tripping for external faults to the shaded triangular region shown.

For internal faults where all CT's contribute fault current the same operating characteristic as shown in Figure 16.17 will apply. For this case, the differential relay current will lie on the line

$$i_e = i_t \tag{16.37}$$

which is the top line in the figure, labeled "limit of differential current." Clearly, this is an operating condition.

For internal faults with parallel, but idle CT's, a different characteristic is required, as shown in Figure 16.18. This characteristic is more sensitive than that for external faults as evidenced by the much larger operating region. This characteristic results from the fact that, at the beginning of each half cycle, adequate current flows in the relay differential circuit to cause tripping before currents begin to flow in the parallel idle CT's.

The half-cycle differential relay principle is based on the fact that CT saturation cannot occur instantaneously, and in practical circuits requires 1–2 milliseconds to occur (about 0.1 to 0.2 cycle at 60 hertz). During this short interval the relay evaluates the need for action and initiates tripping in less than one-half cycle. An advantage of this design is that the relay requires no special or dedicated current transformers and can tolerate added burdens on the CT's that might be used for other purposes than protection.

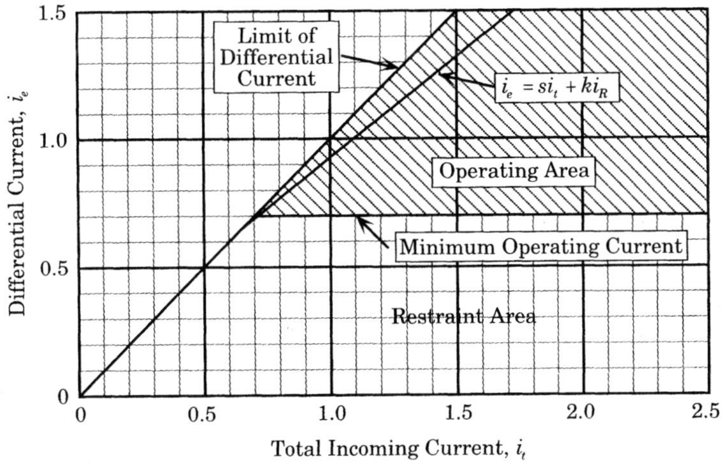

Figure 16.18 Relay operating characteristic for internal faults.

16.6.4 Directional Comparison Bus Protection

The situation may arise in which bus protection needs to be added to an existing station, but the expense of changing all the existing current transformers is excessive. In such cases, it may be possible to use the existing line CT's for bus fault protection [14], [16]. This requires directional relays, fault detectors, and a timer.

The directional relays are installed on each circuit connected to the bus requiring protection, adjusted so that they reach beyond the bus and into the connected circuit, and all are connected in series. In addition, fault detectors, usually instantaneous overcurrent relays, are placed on several circuits, but not necessarily all circuits to make sure that a fault exists. For phase faults, the fault detectors are connected to those circuits with the strongest sources and for ground faults they are connected to the neutrals of power transformers. A timer is required to provide coordination, which is needed since it is necessary for all the series-connected contacts to close for a bus fault. Since only the direction and not the magnitude of currents are compared, current transformer saturation is usually not a problem. Moreover, it is not necessary to have matched current transformer ratios and multiple use of the transformers is acceptable. Ground fault detectors are often located in transformer neutrals in one or more of the connected circuits.

The saturation of current transformers is usually not a problem when the comparison is made between current direction rather than current magnitude. The several connected circuits can have current transformers of different ratios and can also be used for other forms of relaying or for metering. This type of system is more complex due to the larger number of relay contacts, suggesting that maintenance may be more demanding.

16.7 AUXILIARY TRIPPING RELAYS

In many bus protection installations, the bus protection energizes an auxiliary tripping relay, which has individual tripping contacts for each breaker connected to the protected bus [6].

16.7.1 Lockout Relays

Lockout relays are often used to prevent a re-energization of the bus until an inspection can be performed. In some cases, each individual circuit breaker may have its own separate lockout contacts, even if the breaker is tripped by the lockout relay tripping contacts. Some of these lockout relays are reset manually, but provision can be made for remote resetting by means of a motor-driven reset mechanism. Other relays use an electric seal-in coil to lock the relay in the operated position, with resetting being performed by a push-button switch.

16.7.2 Non-Lockout Relays

Some bus protection applications use non-lockout auxiliary tripping relays. This arrangement permits automatic re-energization of the bus. Some relays automatically reset once the relay coil is de-energized, or when the fault is cleared in the case of a bus fault. This type of device is appropriate for unattended stations with open bus work and no remote control facilities. If successful, automatic re-energization of the bus will reduce the outage time.

16.8 SUMMARY

In earlier times, there were persuasive arguments against using any form of bus protection. This was due to the poor performance record of bus protective installations, to the low incidence of bus faults, and because of the significant problems caused by inadvertent tripping of bus protective relays. These early concerns are now largely overcome with newer relays that are designed to take into account the possibility of current transformer saturation and impedance mismatch, which were often the cause of the early poor performance. Another factor in this changing attitude is the larger power flows in modern high-voltage systems, which makes the delayed clearing of bus faults a very serious problem and one that cannot be tolerated.

The result has been the invention of better methods of providing bus protection that address the problems noted above. As a result, bus protection is becoming commonplace, and the experience with bus protection is now considered quite satisfactory.

REFERENCES

[1] Warrington, A. R. van C., *Protective Relays: Their Theory and Practice, Volume One*, John Wiley & Sons, Inc., New York, 1962.

[2] AIEE-EEI Committee Report, "Bus Protection," *AIEE Trans.*, 58, May 1939, pp. 206–222.

[3] GEC Measurements, *Protective Relays Application Guide*, General Electric Company p.l.c. of England, 1975.

[4] The Electricity Council, *Power System Protection, Vol. 1, Principles and Components*, Peter Peregrinus, Ltd., London, 1981.

[5] Horowitz, S. H., Ed., *Protective Relaying for Power Systems*, Selected Reprint Series, IEEE Press, New York, 1980, p. 301.

[6] ANSI/IEEE Std C37.97-1979, "IEEE Guide for Protective Relay Applications to Power System Buses," IEEE, New York, 1979.

[7] ANSI/IEEE Std C57.13-1978, "Current Transformer Accuracy Classification for Relaying," IEEE, New York, 1978.

[8] Mason, C. R., *The Art and Science of Protective Relaying*, John Wiley & Sons, New York, 1956.

[9] Blackburn, J. L., Ed., *Applied Protective Relaying*, Westinghouse Electric Corporation, Relay Division, Newark, NJ, 1976.

[10] Harder, E. L., E. H. Klemmer, W. K. Sonnemann, and E. C. Wentz, "Linear Couplers for Bus Protection," *AIEE Trans.*, 61, 1942, pp. 241–248.

[11] Seeley, H. T., and F. von Roeschlaub, "Instantaneous Bus-Differential Protection Using Bushing Current Transformers," *AIEE Trans.*, 67, 1948, pp. 1709–1718.

[12] Warrington, A. R. van C., *Protective Relays: Their Theory and Practice, Volume Two*, John Wiley & Sons, Inc., New York, 1969.

[13] Instruction Manual, "Differential Voltage Relays, Type PVD," GEK-45405, General Electric Company, Philadelphia, 1977.

[14] Horowitz, S. H., and A. G. Phadke, *Power System Relaying*, Research Studies Press, Ltd., Somerset, England, 1992.

[15] Forford, T., and J. R. Linders, "A Half Cycle Bus Differential Relay and Its Application," *IEEE Transactions*, PAS-93, July/August 1974, pp. 1110–1117.

[16] Haug, H., and M. Foster, "Electronic Bus Zone Protection," CIGRÉ paper 31-11, 1968 session, Paris, June 1968.

PROBLEMS

16.1 Review the following station arrangements presented in Section 5.2 and determine the number of line (or other termination) outages due to a bus fault. Perform this analysis for the following types, and use the number of connections pictured in Section 5.2 for determining the number of outages. What conclusions can be drawn from this study?

(a) Single bus, single breaker
(b) Main and transfer
(c) Double bus, single breaker
(d) Double bus, double breaker
(e) Ring bus
(f) Breaker and a half

16.2 Repeat question 1, but this time determine the number of circuits on outage due to a false trip of a bus protection system. What conclusions can be drawn from this study?

16.3 Review the concept of providing bus protection by means of line backup relays, described in connection with Figure 16.1. Assume that the bus configuration of bus H is a ring bus, shown in Figure P16.3. Compare the bus protection for these two systems, assuming that in both cases the bus protection is provided by zone 2 from the connected line relays.

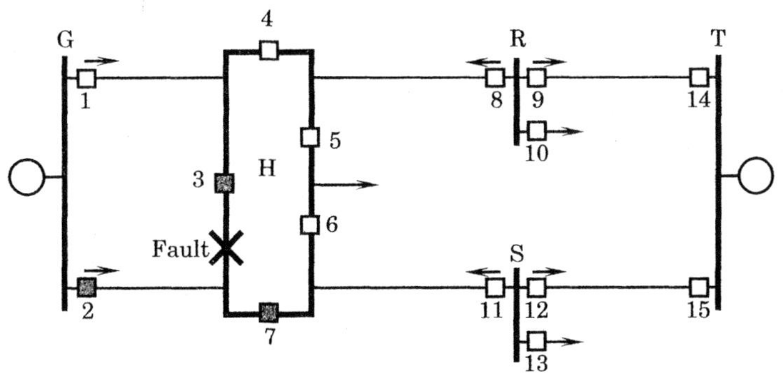

Figure P16.3 Ring bus sections with line overcurrent protection.

16.4 Consider a station similar to the one shown if Figure 16.8, but with shorter leads to the various breaker positions. Assume that the maximum lead length is 200 feet to positions 7N and 3N. Compute the relay settings for this configuration.

16.5 Derive an expression from which one can determine the amount of resistance that should be added to the relay inherent resistance in order to assure selectivity of a differential relay with one current transformer saturated, as shown in Figure 16.4.

16.6 Sketch the bus differential connections for the station arrangements listed in problem 16.1.

16.7 In designing a bus protection scheme, the engineer has three options regarding the re-energization of the bus following a bus protection operation. These are:

1. Manually reclose breakers after inspection and repair, if required.
2. Reclose breakers by supervisory control, provided no lockout system exists or given that lockout relays can be remotely reset.
3. Reclose the breakers by automatic reclosing relays, provided that no lockout of the breaker close circuit exists.

Discuss the issues that might dictate the use of these options. In particular, consider the three options for buses of the following types:

(a) Buses in metal clad switchgear, or high-voltage SF_6 buses that are essentially totally enclosed in a protected environment.

(b) Buses that use air as the primary insulation, which are subject to faults from foreign objects, lightning, small animals or birds, dirty insulator flashover, etc.

16.8 Provide arguments against any form of bus protection in a power system. Also, record arguments in support of bus protection.

17

Transformer and Reactor Protection

17.1 INTRODUCTION

This chapter describes the protection of a class of shunt-connected devices that are important components in any power system. The principal member of this family of devices is the power transformer, and most of this chapter is devoted to transformer protection. In addition to transformers, this chapter also describes the protection of regulating transformers, shunt reactors, and static var compensators.

The power transformer is one of the most important classes of hardware in the electric power system, and transformer protection is an essential part of the overall system protection strategy. Transformers are used in a wide variety of applications, from small distribution transformers serving one or more users to very large units that are an integral part of the bulk power system. Moreover, transformers have a wide variety of features, including tap changers, phase shifters, and multiple windings, that require special consideration in the protective system design. Some transformers, such as zig-zag transformers, have very special design features.

An important consideration in transformer protection is the high cost of the transformer and the relative long outage time that occurs when a large transformer fails. The proper type of protection can often detect incipient faults before they become major, and thereby prevent major physical damage and long outage times.

Transformers suffer from several different types of stresses, due to overheating, short circuits, and open circuits. Most transformers, especially the smaller units, are not protected from overheating, although there are protective devices that can be used for this purpose. Open circuits are not detected at all, in most cases, since they do not present a particular hazard. This leaves short circuits as the major focus of transformer protection. Surge protection is also provided for transformers, but surge phenomena are beyond the scope of this book.

Short-circuit protection includes internal short circuits, such as turn-to-turn faults, and turn-to-ground faults. It also includes external short circuits, such as bushing flashovers, that are also within the protection zone of the transformer relays.

The most common form of transformer protection is differential relaying, which treats the transformer as a unit, making measurements at all of the transformer terminals. This is very convenient since, unlike transmission lines, the transformer terminals are all located at the same station. There are complications, as will be noted later, that must be overcome in order to make this application practical. There are several types of faults, other than short circuits, that a large transformer should be protected against. These are described below, and the appropriate protection strategies are outlined.

Transformer protection also presents certain problems in the application of current transformers. CT's are available in many different types, ratios, and characteristics. Moreover, the CT should be matched to the ratings of the different transformer windings and this matching may not be exact. Many transformers are equipped with bushing CT's.

Transformers also may be connected to provide phase shift, such as the common Δ-Y connection with its 30° phase shift. This means that the CT connections must be arranged to correct for this phase error if the relaying is to operate correctly. Transformers may be designed with special features, such as tap changers or phase shifters, that present problems in the protection design. Grounding transformers also require special consideration for protection. An overview of transformer protection is provided in [1].

Finally, this chapter introduces the protection required for a static var compensator or SVC. The SVC includes a power transformer as an integral part of the system, but also includes inductances, capacitors, or both connected to the secondary side of the transformer bank where the secondary currents are controlled by solid-state switching. This permits the SVC to serve as a very fast voltage controller, which can provide significant benefits in power system operation, voltage stability, and angle stability.

These problems are all discussed below, beginning with a consideration of transformer faults.

17.2 TRANSFORMER FAULTS

For convenience in analysis, we divide transformer faults into two classes: external faults and internal faults.

17.2.1 External Faults

External faults are those faults or hazards that occur outside the transformer. These hazards present stresses on the transformer that may be of concern and may shorten the transformer life. These faults include the following [2]:

- *Overloads* Overloads cause the transformer to overheat and have the potential to cause permanent damage or loss of life to the unit. The time constant for overheating is long, however, and it may take many hours of exposure for the condition to become serious. In most cases, no protection is provided for overload, but an alarm will often be used to warn operating personnel of the condition. One cause of overload may be due to unequal load sharing of parallel transformers or unbalanced loading of three-phase banks.
- *Overvoltage* Overvoltage can be either due to short-term transient conditions or long-term power-frequency conditions. *Transient overvoltages* cause end-turn stresses and possible breakdown. These transients are protected against by surge protective devices that are designed for this purpose. *Power frequency overvoltages* occur due to an emergency operating condition, such as a sudden loss of load on an isolated portion

the system that causes the voltage to rise. This condition causes overfluxing of the transformer and an increase in stress on the winding insulation. Overfluxing increases iron losses and may result in a large increase in exciting current. Such conditions result in rapid heating of the iron circuits of the transformer, with possible damage to core lamination insulation and even to winding insulation.

- *Underfrequency* Underfrequency also is caused by a major system disturbance that causes an imbalance between generation and load. The condition is similar to overvoltage in that exciting current is greatly increased at low frequencies, causing overfluxing of the transformer iron circuits. The transformer may be able to continue operation at either high voltage or underfrequency, but the two conditions experienced at the same time could be very serious. Usually, the ratio of voltage to frequency should not be allowed to exceed 1.1 per unit, which is usually called a "volts per hertz" limit.
- *External system short circuits* System faults that are external to the transformer protection zone, but cause high transformer currents, can cause transformer winding damage. Large external fault currents cause high mechanical stress in the transformer windings, with the maximum stress occurring during the first cycle. This short time frame makes it almost impossible to protect the transformer from experiencing these stresses. The protection strategy for these events is, therefore, a matter of transformer design.

Most of the foregoing conditions are often ignored in specifying transformer relay protection, depending on the criticality of the transformer and its importance in the system. The exception is protection against overfluxing, which may be provided by devices called "volts per hertz" relays, which detect either high voltage or underfrequency, or both, and will disconnect the transformer if this quantity exceed a given limit, which is usually 1.1 per unit.

17.2.2 Internal Faults

Internal faults are faults that occur within the transformer protection zone. This classification includes not only faults within the transformer enclosure but also external faults that occur inside the transformer CT locations. Transformer internal faults are divided into two classifications for discussion; incipient faults and active faults. Incipient faults are faults that develop slowly, but that may develop into major faults if the cause is not detected and corrected. Active faults are caused by the breakdown in insulation or other components that create a sudden stress situation that requires prompt action to limit the damage and prevent further destructive action.

17.2.2.1 Incipient Faults. Incipient faults are of three kinds: transformer overheating, overfluxing, or overpressure.

- *Overheating* Overheating may be due to several different internal transformer conditions, as follows:
  1. Poor internal connections, in either the electric or the magnetic circuits.
  2. Loss of coolant due to leakage.
  3. Blockage of coolant flow.
  4. Loss of fans or pumps that are designed to provide cooling.
- *Overfluxing* Overfluxing was discussed above under externally caused faults. It is mentioned again here since continued periods of overfluxing may gradually lead to

insulation breakdown of the magnetic circuit insulating materials or the electric circuit insulation.

- *Overpressure* Overpressure in the transformer tank occurs due to the release of gases or products that accompany the localized heating due to any cause. For example, a turn-to-turn fault may burn slowly, releasing gases in the process, or local heating of insulation may give off gases. These gases accumulate in the enclosed transformer tank as an increase in pressure, which may develop suddenly or slowly over a long period of time.

The foregoing transformer faults are called incipient faults since they usually develop slowly, often in the form of a gradual deterioration of insulation due to some cause. This deterioration may eventually become serious enough to cause a major arcing fault that will be detected by protective relays. If the condition can be detected before major damage occurs, the needed repairs can often be made more quickly and the unit placed back into service without a prolonged outage. Major damage may require shipping the unit to a manufacturing site for extensive repair, which results in an extended outage period.

17.2.2.2 Active Faults. Active faults are faults that occur suddenly and that usually require fast action by protective relays to disconnect the transformer from the power system and limit the damage to the unit. For the most part, these faults are short circuits in the transformer, but other difficulties can also be cited that require prompt action of some kind.

The following classifications of active faults are considered:

1. Short circuits in wye-connected windings
   Grounded through a resistance
   Solidly grounded
   Ungrounded
2. Short circuits in delta-connected windings
3. Phase-to-phase short circuits (in 3 phase transformers)
4. Turn-to-turn short circuits
5. Core faults
6. Tank faults

Each of the above will be considered in turn.

1. **Short circuits in wye-connected windings.** The short-circuit conditions examined here are in wye connected windings of wye-delta connected transformers. It can be shown that the current flowing in such a fault can be quite different on the delta side of the transformer than the wye side [2], [3]. Consider the three-phase transformer bank shown in Figure 17.1, where there is a source on the delta side of the bank, but no source on the wye side. A fraction h of the phase a winding is faulted to ground. We assume that the grounding resistance R_N is much greater than the transformer leakage reactance. Therefore, we write the fault current as

$$I_F = 3I_{a0} = \frac{hV_a}{R_N} \tag{17.1}$$

Because of the fault, the effective turns ratio of the transformer is computed as

$$a_F = \frac{a}{h} = \frac{n_\Delta}{h\,n_Y} \tag{17.2}$$

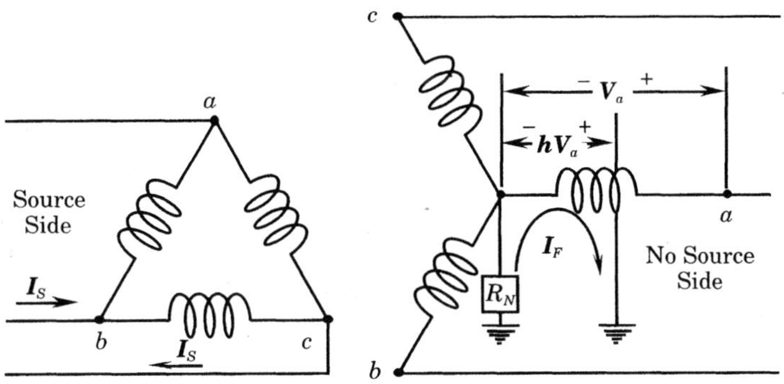

Figure 17.1 Ground fault in a wye-connected transformer winding.

where a is the design turns ratio. This means that the effective current transformation is changed by the factor h, or

$$I_\Delta = \frac{h}{a} I_Y \tag{17.3}$$

Therefore the fault current will be transformed on the delta side to currents flowing in lines B and C, with magnitude

$$I_S = \frac{hI_F}{\sqrt{3}} = \frac{h^2 V_a}{\sqrt{3}\, R_N} \tag{17.4}$$

Thus, the fault current on the wye side varies in direct proportion to the fraction h of the winding faulted, measured from the neutral, but the fault current flowing on the delta side varies as the square of this fraction. This makes fault detection very difficult on the delta side because of the low magnitude of current, which may be even less than the full load current. It is convenient to express both currents in terms of the maximum fault current, computed with $h = 1$.

$$I_{F\max} = \frac{V_a}{R_F} \tag{17.5}$$

Then

$$I_F = hI_{F\max} \tag{17.6}$$

and

$$I_S = \frac{h^2 I_{F\max}}{\sqrt{3}} \tag{17.7}$$

The fault currents on the two sides of the transformer are plotted in Figure 17.2, which illustrates the low value of current for faults near the neutral, especially the currents on the delta side.

If the transformer is solidly grounded, i.e., with zero grounding resistance, the only impedance seen by the fault current is the transformer leakage impedance. This impedance varies as the square of the number of turns. Moreover, the voltage at the fault point is not proportional to the number of turns for faults near the neutral because of the increased leakage. Therefore, the impedance function becomes very complex and the current on the wye side has a minimum at about 40% of the total winding faulted and increases as the fault point approaches the neutral, then dropping quickly to zero at the neutral [2], [3]. This is illustrated in Figure 17.3 [2]. For ungrounded

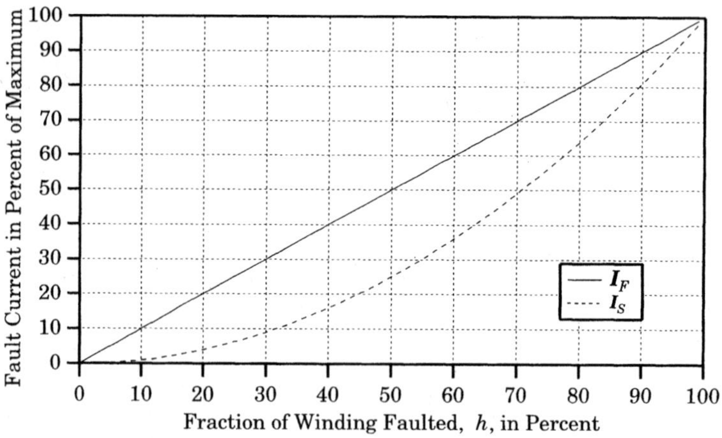

Figure 17.2 Wye winding ground fault currents with wye connection grounded through an impedance [2], [3].

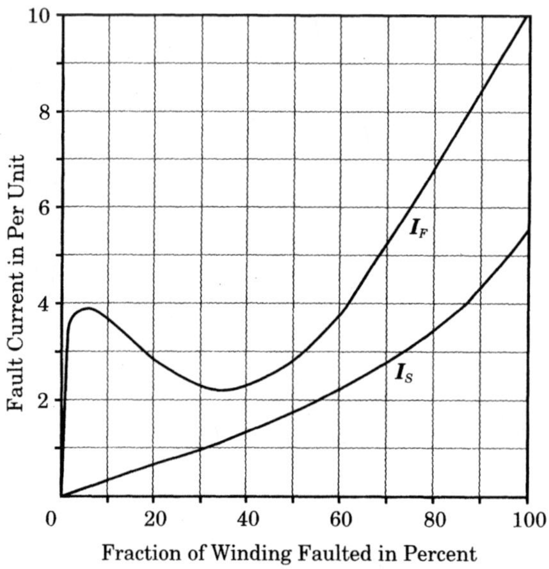

Figure 17.3 Fault current and delta line current for wye winding faults with a solidly grounded neutral.

wye banks, in the configuration discussed here with a source only on the delta side, there is no fault current at all.

2. **Short circuits in delta-connected windings.** The computation of fault current in the delta connected windings is an even more complex function of fault position. The fault current magnitude is much less than for the wye case. The impedance will vary between about 25% and 50%, based on the transformer rating. The minimum fault current occurs when the fault is at the center of one of the delta windings, in which case the fault current may be no more than the rated current. Note that the fault current is supplied by two of the phases in some proportion, each of which will be relatively small.
3. **Phase-to-phase short circuits in three-phase transformers.** Phase-to-phase faults between transformer windings can only occur in three-phase units, and their occur-

rence is of low probability. When such faults do occur, the currents are large, and are comparable in their relative magnitude to the phase-to-ground currents computed above.

4. **Turn-to-turn short circuits in transformer windings.** Inter-turn faults are unlikely in low-voltage transformers unless the windings have been damaged mechanically by large through currents due to external faults, which can crack the insulation.

   For high-voltage transformers connected to a high-voltage system, the unit is likely to be damaged by steep front traveling waves or impulses that can be much higher than the rated transformer voltage. The risk of turn-to-turn flashover is greatest in the end turns of the winding which are prone to failure for this type of event. Shorting a few turns will cause large fault currents to flow in the shorted section, but the terminal currents will not be greatly affected, making detection difficult. It is claimed that up to 80% of all high-voltage transformer failures are due to this cause [2].

5. **Core Faults.** The transformer core laminations are carefully insulated from each other to prevent eddy currents from crossing the gap between adjacent laminations. Even the bolts that clamp the laminations together are insulated from each of the laminations to prevent the bolts from causing a magnetic short across the laminations. Any overheating or overfluxing of the transformer provides the possibility of causing a magnetic short of this kind due to the deterioration of the insulation between laminations or around the bolts.

   Such a shorted path will allow eddy currents to flow and will greatly increase the core losses and cause localized heating to occur. This condition does not greatly affect the terminal currents of the transformer, making this type of fault difficult to detect by electrical relays connected to the terminals. This type of damage will often be detected by gas relays, once the damage becomes great enough to cause breakdown of insulation materials, which is accompanied by emission of gases and increasing the tank pressure above the oil.

6. **Tank Faults.** For an oil-immersed transformer, the primary coolant is the flow of oil around the core and coils of the transformer. Should the enclosing tank develop a leak, this fault may cause a dangerous overheating and reduction of the insulation. A similar effect could be caused by any blockage of the oil flow in cooling ducts or pipes. Although quite different from an electrical fault, this type of fault can be quite serious in its effect on the unit. Some transformers are also cooled by external fans that circulate air through radiators that are intended to cool the circulating oil. Failure of these fans can also be the cause of excess heating.

7. **Bushing Flashovers.** The transformer bushings are within the protection zone of electrical relays, if external current transformers are connected external to the bushings on both sides of the unit. Bushing flashovers may occur due to lightning or other surge phenomena, resulting in a shorted path to the grounded transformer tank. It is important to note that these faults, or faults on external transformer connections, are detected by some, but not all, types of transformer protective schemes.

Summaries of transformer failures over an extended period are shown in Tables 17.1 and 17.2. Table 17.1 gives the various causes of transformer failures and Table 17.2 gives the percentages of the various categories, measured over two different time periods. Most of the transformer failures are due to winding and tap changer failures.

TABLE 17.1 Causes of Transformer Failure [1]

1. Winding Failures
a. Turn-to-turn insulation failure
b. Surges due to lightning, switching, etc.
c. Moisture
d. External faults (insulation failure)
e. Overheating
f. Open winding
g. Deterioration
h. Improper blocking of turns
i. Grounds
j. Phase-to-phase failures
k. Mechanical failures

2. Tap Changer Failures
a. Mechanical
b. Electrical
c. Contacts
d. Leads
e. Tracking
f. Overheating
g. Short circuits
h. Oil leak
i. External fault

3. Bushing Failures
A. Aging, contamination, and cracking
b. Flashover due to animals
c. Flashover due to surges
d. Moisture
e. Low oil

4. Terminal Board Failures
a. Loose connections
b. Leads (opened)
c. Links
d. Moisture
e. Insufficient insulation
f. Tracking
g. Short circuit

5. Core Failure
a. Core insulation failure
b. Ground strap burned away
c. Shortened laminations
d. Loose clamps, bolts, wedges

6. Miscellaneous Failures
a. Bushing current transformer failure
b. Metal particles in oil
c. Damage in shipment
d. External faults
e. Bushing flange grounding
f. Poor tank weld
g. Auxiliary system failures
h. Overvoltage
i. Overloads
j. Unidentified problems

TABLE 17.2 Transformer Failure Statistics 1955–1982 [1]

Failure Cause Classification	Failures 1955–1965%	Failures 1975–1982%
1. Winding failures	51	51
2. Tap changer failures	49	19
3. Busing failures	41	9
4. Terminal board failures	19	6
5. Core failures	7	2
6. Miscellaneous failures	12	13
Total	100	100

17.2.3 Fault Protection Philosophy

The basic philosophy of protective devices is different for incipient faults than for active faults. We can summarize the objectives of transformer fault protection as follows:

Active fault protection The protection must provide fast isolation of the faulted transformer in order to remove the faulty equipment from the power system, thereby minimizing the effect of the disturbance, and also to minimize the damage to the transformer. Faults that are not cleared promptly may cause substantial transformer damage, requiring long and expensive repairs, and may even present a hazard to personnel who are in the vicinity.

Incipient fault protection Incipient faults do not require fast detection and equipment isolation. These faults develop slowly and there is time for careful observation and testing. Moreover, these faults are usually not detected by the same protective devices as used for the detection of active faults. This suggests that supplementary protective systems be used to detect incipient faults.

It is important to recognize that the two types of protection are usually considered to be mutually exclusive and complimentary. Neither is considered to be adequate for both types of faults. The only exception to this philosophy is the use of Buchholz protection in Europe, which has sometimes been claimed to be the only protection required; a philosophy that has not been widely accepted in North America. This is discussed further below.

17.3 MAGNETIZING INRUSH

When a transformer is first energized, there is a transient inrush of current that is required to establish the magnetic field of the transformer. This is not a fault condition and should not cause protective relays to operate. However, under certain conditions, depending on the residual flux in the transformer core, the magnitude of inrush current can be as great as eight to ten times normal full load current and can be the cause of false tripping of protective relays. This is rather serious, since it is not clear that the transformer is not internally faulted. The sensible response is, therefore, to thoroughly test the transformer before making any further attempts at energization. This will be expensive and frustrating, especially if the tests show that the transformer is perfectly normal. Since this is such an important concept, it will be examined in some detail in order to understand the reason for high inrush current and to learn what steps can be used in protective relays to prevent their tripping due to magnetizing inrush.

There are several factors that control the magnitude and duration of the magnetizing inrush current [4], [5], [6]:

1. Size of the transformer bank
2. Strength of the power system to which the bank is connected
3. Resistance in the system from the equivalent source to the bank
4. Type of iron used in the transformer core
5. Prior history of the bank and the existence of residual flux
6. Conditions surrounding the energization of the bank, e.g.,
   - Initial energization
   - Recovery energization from protective action
   - Sympathetic inrush in parallel transformers

17.3.1 Magnetizing Current Magnitude

Consider a transformer that is to be energized from a bus voltage that is sinusoidal. Then the steady-state flux is the integral of the voltage, or

$$\phi = \frac{1}{N}\int \sin\omega t\, dt = \frac{-1}{\omega N}\cos\omega t \tag{17.8}$$

as shown in Figure 17.4. Note that the flux lags the voltage by 90 degrees. If the circuit is energized as the voltage wave is passing through zero, and there is zero residual flux, the resulting flux wave must start at zero rather than its negative maximum shown in the Figure 17.4, and will change by $2\phi_{\max}$ over the next half cycle.

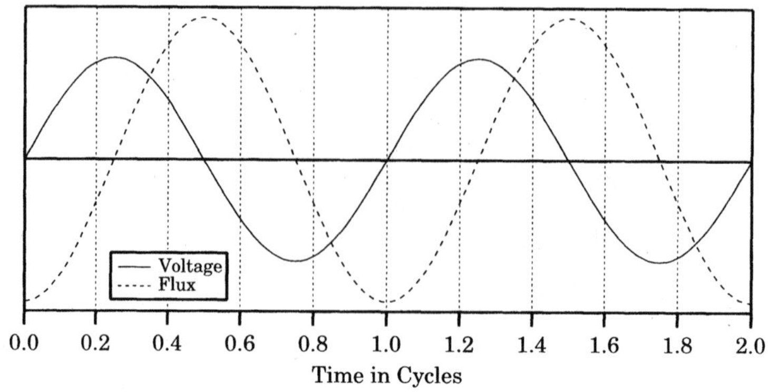

Figure 17.4 Steady-state voltage and flux waveforms.

As the flux builds, the exciting current grows with the flux. If the winding inductance were linear, the current would have exactly the same waveform as the flux, that is,

$$i = \frac{1}{L}\int v\,dt = -\frac{1}{\omega L}\cos\omega t \tag{17.9}$$

However, the inductance is not linear and saturation can be expected to occur since transformers are usually designed to operate near the knee of the saturation curve under normal conditions. Taking the flux to twice its normal maximum will cause hard saturation, requiring very large exciting currents. Even this is not the worst case. Suppose the transformer is energized at the zero point on the voltage wave with a residual flux of $\phi_{\max}$. In this case the saturation will be even greater. Exciting currents as great as 500 times normal are not unusual for such a condition [5]. Moreover, the current wave will be fully offset from the time axis.

The way in which saturation causes severe exciting current buildup is illustrated in Figure 17.5. The saturation curve on the left shows the exciting current required in order to

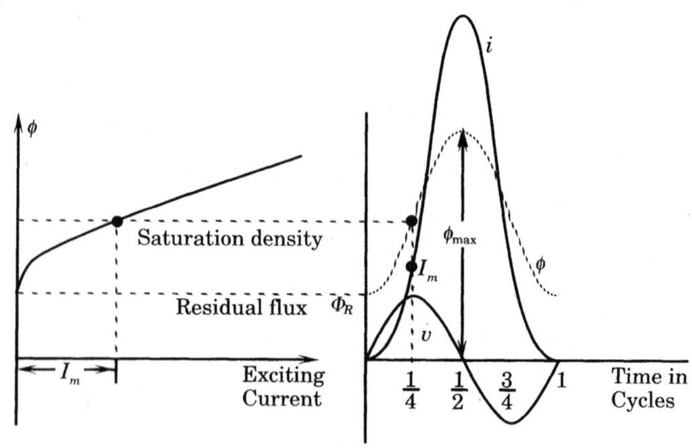

Figure 17.5 Derivation of the inrush current wave from the excitation saturation curve [2].

provide a given level of flux. For each point on the flux wave, starting at the residual flux value ϕ_R, a value of current may be found from the saturation curve and plotted on the time axes. This is illustrated for one value of current, labeled I_m. Plotting many different points gives the fully offset current pulse shown. Note that the current waveform is not sinusoidal, but is a sharp pulse, with the peak occurring at maximum flux.

The decay of the excitation current is rapid for the first few cycles, but then decays very slowly. Usually several seconds are required for the current to reach normal levels. The time constant governing this decay is not a constant L/R, since the inductance is varying due to saturation. Thus, the time constant is small at first, then increases as saturation is reduced. Moreover, the time constant is a function of the transformer size and may vary from 10 cycles for small transformers to 1 minute for large sizes [5]. The decay of exciting current also depends on the resistance seen looking into the power system. If the transformer is close to a generator, this resistance will be very small and the exciting current will damp very slowly. Moreover, the current will still be distorted in its waveform for an extended period, as much as 30 minutes, after initial energization.

For three-phase transformer banks, each phase will have a different excitation current, since the point on the voltage wave at which excitation begins is different for all three phases. For example, if the bank is energized when the phase a voltage is at its peak, and with no residual flux, then phase a may not saturate at all, but phase b will probably saturate with a positive pulse not unlike that of Figure 17.5 and phase c will experience a negative pulse of about the same magnitude as phase b.

Inrush currents also depend on the type of transformer core design, the type of three-phase connection, and the type of steel [2]. The type of steel may be very important, since it has been shown that the magnetizing ampere-turns required for cold-rolled steel of modern units are much greater than for older hot-rolled steel cores. The type of transformer connection is also important, with wye and delta windings giving different excitation results.

17.3.2 Magnetizing Inrush Current Harmonics

The magnetizing inrush current contains all orders of harmonics, but the second and third harmonics are by far the greatest in magnitude. The dc offset of the current is also significant, as noted from the waveform of Figure 17.5, which represents a totally offset wave. The most important harmonics are the following [2]:

1. *DC or offset component* A dc component is always present in the inrush current of a three-phase transformer, with different offsets in each of the three phases. If the residual flux happens to equal the normal required steady state flux for that phase at the instant of switching, then that phase will have no dc component in the magnetizing inrush current, but large dc offsets will occur in both of the other phases.
2. *Second harmonic* The second harmonic current is present in all inrush waveforms of all three phases. The proportion of second harmonic current varies with the degree of saturation, but is always present as long as the dc offset is present in the core flux. The minimum second harmonic magnitude has been shown to be about 20% of the excess magnetizing current (over its steady-state value). It is important to note that normal fault currents do not contain second harmonic components or any other even harmonics. Moreover, saturation of iron-cored devices may cause distortion in the currents, but these distorted currents contain only odd harmonics. Similarly, saturated current transformers will contain only odd harmonics.

3. *Third harmonic* The inrush current also contains large amounts of third harmonic current, in about the same proportion as the second harmonic. In three-phase transformers, the third harmonic currents in the three phases are all in phase, and may not appear in the line current of delta connected banks. It is also important to note that third harmonics currents are likely to flow as the result of CT saturation.
4. *Higher harmonics* Many higher harmonics are present in the inrush current, but their proportion is much smaller than those discussed above. For this reason these higher harmonics are not of great interest, although there has been some interest in detecting the fifth harmonic.

The most important observation regarding the harmonic content of the magnetizing inrush current is the unique presence of the second harmonic current. This harmonic is not present in either normal or fault currents on a three phase system and provides an excellent method of testing the current to determine if the condition seen by the protective system is a fault or merely a normal inrush condition. More will be said on this subject in later sections.

17.3.3 Sympathetic Inrush in Parallel Banks

A unique condition exists where transformer banks are connected in parallel. Consider the situation where one bank is in service and the second, parallel bank is switched into service. In this case, the second bank can have what is called a "sympathetic" inrush. This due to the fact that the inrush current flowing in the newly switched bank can find a parallel path in the previously energized bank. In fact, the dc component flowing may saturate the core of the energized bank, causing this bank to experience an apparent inrush. This apparent inrush will not be as large as an initial inrush. The magnitude depends on the size of the unit and on the strength of the power system.

17.4 PROTECTION AGAINST INCIPIENT FAULTS

Incipient faults are not detectable at the transformer terminals, which means that the normal methods of applying relay protection will not work for this type of fault. Indeed, in many cases there is no protection provided for these types of faults since incipient fault conditions are not necessarily critical. There are methods that can be employed, however, and these may be quite sensible for some transformers. This is especially true of very important transformers, where long outages are to be avoided.

17.4.1 Protection Against External Incipient Faults

External faults are those that affect the transformer, but are due to a condition that exists external to the transformer unit. The most common of these conditions are overheating, overfluxing, and circulating current.

OVERHEATING. A transformer that is part of an interconnected network may experience temporary overheating that will lead to loss of life if not corrected. The transformer rating is based on the maximum temperature rise above an assumed maximum ambient temperature and under this condition no overload is permissible. At a lower ambient temperature, some overload may be sustained. The amount and length of overload that is permissible depends on the recent history of loading, which determines the operating temperature of the unit. No definite rule can be stated except that the windings must not overheat. Violation of this rule can subject the winding to drastic redu ction of useful life. A temperature measurement at the hottest location in the winding is required for a direct determination of possible winding damage, but this is not practical. Protection, therefore, is usually based on somehow modeling the temperature performance of the unit.

One method of providing overheating protection is the method of thermal imaging [3], [7]. A thermal sensor is placed in a small compartment near the top of the oil, with a heating element in the compartment that is supplied a current proportional to the transformer loading. This produces a temperature in the compartment that is proportional to the temperature rise of the windings. A heat-sensitive resistor can be calibrated, as one leg of a bridge, to provide an indication of the overheating condition.

A more refined method is the temperature-time integrator, which integrates the total period of overheating in the transformer life, thereby providing an overheat history of the unit [3]. This can be used as an indication of the probable loss of life of the transformer.

The aging of the transformer, as a result of repeated periods of overheating, takes place over a long period of time. In many cases, special protection is not provided for this hazard, although external temperature indicating instruments may be available for observation and recording by operating personnel. The need for this type of protection must also be balanced against the possible false trip of the protection, which would cause a potentially long outage and inspection that may be unnecessary.

OVERFLUXING. Transformers are usually designed to operate near the knee of the iron core saturation curve. This means that any overvoltage or underfrequency will cause higher than normal flux in the core material, and may result in substantial increases in core temperature. As a general statement of the condition, we may write the flux as

$$\phi = k\left(\frac{V}{f}\right) \tag{17.10}$$

This function can be measured approximately by connecting an R-C load to a voltage transformer, as shown in Figure 17.6. From the figure, we write

$$\begin{aligned} \boldsymbol{V}_2 &= \frac{1}{n}\boldsymbol{V}_1 \\ \boldsymbol{I}_2 &= \frac{\boldsymbol{V}_2}{\boldsymbol{Z}} = \frac{\boldsymbol{V}_1}{n\boldsymbol{Z}} = \frac{V_1}{n\sqrt{R^2 + X_C^2}} \end{aligned} \tag{17.11}$$

Then

$$V_0 = I_2 X_C = \frac{V_1 X_C}{n\sqrt{R^2 + X_C^2}} \tag{17.12}$$

If the circuit components are chosen such that

$$R^2 \gg X_C^2 \tag{17.13}$$

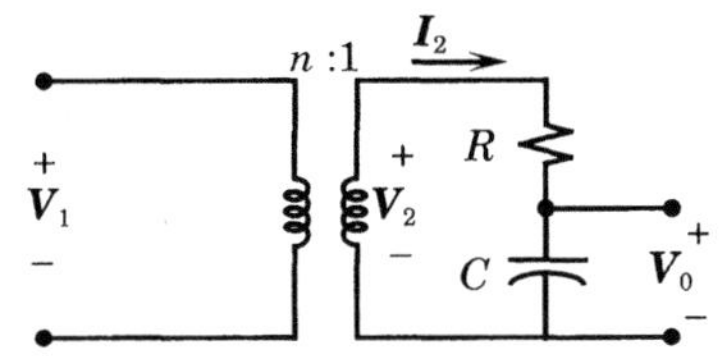

Figure 17.6 Measurement of the overfluxing condition.

then

$$V_0 \cong \frac{V_1 X_C}{nR} = \frac{V_1}{2\pi f n R C} = k\frac{V_1}{f} \tag{17.14}$$

or the output voltage of the circuit is proportional to the ratio of voltage over frequency.

An overfluxing condition does not require high speed tripping, in fact, this would be undesirable since overvoltage may occur as a transient condition for which tripping should be avoided. This type of protection is sometimes recommended for generator step-up transformers, where the risk of overfluxing may be high.

OTHER EXTERNAL INCIPIENT FAULT CONDITIONS. Another external condition that may be troublesome is the occurrence of circulating currents in parallel transformer banks. This can occur if the parallel banks have different tap settings, for example, which causes a continuous circulation of reactive power. This will cause unnecessary heating of both units that may become a problem. The condition will appear as an external disturbance to both banks. Protection is not usually provided for this specific condition, but the overheating protection discussed above would suffice.

17.4.2 Protection Against Internal Incipient Faults

Internal incipient faults in windings may be due to the heating associated with a poor electrical connection, limited arcing of a turn-to-turn fault, or other localized heating. It could also be due to a core fault, such as a lamination breakdown, or shorting of core bolts, that causes excess heating due to eddy currents. Another cause could be ineffective cooling due to coolant flow failure, clogging of coolant ducts, low oil content, failure of oil pumps, or other cooling aids. The purpose of providing protection against these failures is to limit the damage such that the transformer can be repaired without an extended outage. The relays used for this type of internal fault are usually associated with measuring the gas given off due to the fault or the tank pressure that results from the condition.

One type of relay that has found high popularity in Europe, and to a limited extent in North America, is the Buchholz relay, named after its inventor [8]. The relay is applied to transformers of the "conservator" type, that is, transformers with an external tank used as an expansion reservoir for cooling oil. The relay is placed in the pipe connecting the main transformer tank to the conservator tank and is designed to detect the flow or presence of gases at this location or a sudden rush of oil through the pipe. Small faults will give off gases that will rise to the top of the oil and pass through the pipe, where they can be detected and tested, if necessary. Severe faults will cause a rush of oil toward the conservator and can cause an immediate trip. This type of relay is capable of detecting the following types of faults [3]:

1. Hot spots on the core due to a short in lamination insulation.
2. Core bolt insulation failure.
3. Faulty joints.
4. Inter-turn faults or other winding faults.
5. Loss of oil due to leakage.
6. Major winding faults, either between windings or to ground.

Because of the excellent coverage of fault conditions, this type of relay is sometimes used as a main protection. It should be noted, however, that this relay does not protect against bushing faults, which may be inside the protection zone of differential relays, but will not be detected by the Buchholz relay. This relay design is also not applicable for transformers that have no conservator tank.

For transformers that have a gas cushion at the top of the tank rather than a conservator, the sudden pressure relay is often used. This relay measures the rate-of-change of pressure in the oil or in a small bellows that is immersed in the oil.

Problems with the gas actuated relays can arise if the settings are too sensitive, such that the relay is tripped due to shock, vibration, or earth tremors. These relays are generally rather slow to operate unless the fault is quite severe. Many relay engineers use the gas actuated relay as a backup to electrical relays, which are usually faster for internal faults.

17.5 PROTECTION AGAINST ACTIVE FAULTS

The main protection provided for active faults is almost always unit protection, using percentage differential relays. Other types of protection can be used as backup or supplemental protection. Transformer protection requires a careful consideration of current transformer connections and accuracy analysis. The connections require study because of the inherent 30 degrees phase shift in delta-wye banks, which are very common. The ratio of the CT's is also important, since the CT ratio may not exactly match the transformer turns ratio. This section will examine current transformer applications to transformer protection and will discuss the common methods of providing transformer protection against active faults.

17.5.1 Connections for Differential Protection

There are two requirements that dictate the current transformer connections for power transformer protection [4]:

1. The relay *must not* operate for normal load or for external faults.
2. The relay *must* operate for internal faults of a given severity.

A rule of thumb often applied to the connection of current transformers for power transformer protection is as follows [4]:

CT's on a wye-connected winding should be connected in delta.

CT's on a delta-connected winding should be connected in wye.

Making the connections in this way ensures that, for external faults, the CT secondary currents are equal and the differential protection will not trip the transformer, thereby satisfying the first requirement.

17.5.1.1 Delta-Wye Bank CT Connections. A typical connection for a delta-wye-connected power transformer is shown in Figure 17.7. Note that the wye-connected windings of the power transformer, which carry line currents, have delta-connected CT's, and therefore have secondary currents that are proportional to the *difference* in the line currents. The delta-connected side of the power transformer has line currents that are the *difference* of the individual delta phase currents, hence this side has wye-connected CT's, which also carry these difference currents. Therefore, the differential relays see difference currents on both sides of the transformer when the fault is external, and no tripping is initiated. For example, the left-most relay has current $i_a - i_c$ entering from the top and this identical current leaving on the bottom of the diagram. It can be shown that it does not matter how the CT connections are

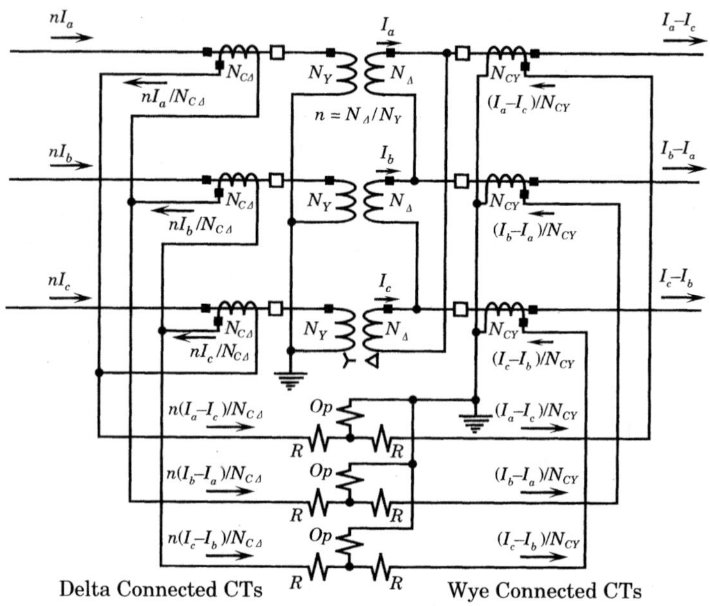

Figure 17.7 CT connections for a Δ-Y transformer bank.

made as long as they have their wye connection on the delta side and vice versa. We assume only that the three line currents add to zero on each side of the transformer.

The CT connections of Figure 17.7 are verified by first checking to make sure that the differential relay will not trip for normal loading conditions or for external faults. To do this, one first assumes current directions, in this case entering the wye-connected winding, on the left in the figure. This fixes current directions in the delta-connected windings and, hence, the difference or line currents leaving the circuit breakers on the delta side. We assume that the windings have a $1{:}n$ turns ratio and that exciting currents are negligible. With these through line currents flowing, the relay must not pick up. If the CT's are connected as shown, then the currents in the restraint windings (R) of each relay must be equal, such that there is no current in the operate winding (Op) of any relay. Equating the currents in the restraint windings we get the restraint on the various system ratios as

$$n = \frac{N_\Delta}{N_Y} = \frac{N_{C\Delta}}{N_{CY}} = \frac{\sqrt{3}I_{LY}}{I_{L\Delta}} = \frac{\sqrt{3}V_{LL\Delta}}{V_{LLY}} \tag{17.15}$$

It is noted that there are no zero-sequence currents leaving the delta connection of the power transformer; therefore, there are no zero-sequence currents flowing in the CT secondaries on that side either. However, this is also true in the restraint coils on the wye side, where the CT's are connected in delta. This means that the relays will see no zero-sequence currents at all, even for ground faults. However, the relays will correctly trip the transformer for an internal ground fault, operating with only positive- and negative-sequence currents.

***17.5.1.2 Current Transformer Ratios*.** The previous example assumed that the power transformer ratio was $1{:}n$, such that the currents in each winding of the power transformer have this ratio. We now examine the relationship between the line currents and the CT secondary currents. To establish a mathematical notation, refer to the wye-delta transformer connection shown in Figure 17.8.

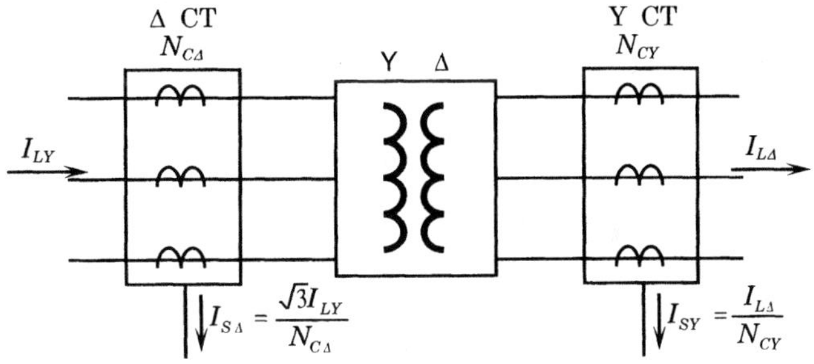

Figure 17.8 Current notation in the delta-wye connection.

The CT primary current magnitudes are labeled $I_{L\Delta}$ and I_{LY} on the delta and wye sides of the transformer, respectively, with the subscripts also noting the power transformer bank connection. The CT secondary currents are labeled $I_{S\Delta}$ and I_{SY} where these subscripts refer to the type of CT connection, being, respectively, delta and wye, as shown in the figure. The CT ratios, labeled N in Figure 17.8, are also subscripted according to the CT secondary connection shown. We are interested in the ratio

$$\frac{I_{S\Delta}}{I_{SY}} = \sqrt{3}\frac{N_{CY}}{N_{C\Delta}}\frac{I_{LY}}{I_{L\Delta}} \cong 1.0 \tag{17.16}$$

Note that the Δ and Y subscripts of the N terms in (17.16) refer to the connection of the current transformers, not the power transformers, and the C prefix is intended to serve as a reminder of this fact. We would like this ratio to be as near to unity as possible in order to avoid tripping the relay for an external fault, in which case we would like the two secondary currents to be nearly equal. The ratio of the primary or line currents, on the right side of (17.16), depends only on the power transformer turns ratio, as shown in Figure 17.7. The ratios of the current transformer banks, also on the right side of (17.16), are chosen so that the currents flowing in the restraint coils of the relay under full load conditions do not exceed the restraint coil ratings, with the ratios usually chosen to be as small as possible for maximum sensitivity. The designer will often select one of the CT's to have multiple ratios, so that the equation can be very nearly satisfied using one of the available ratios. An example will illustrate the process.

EXAMPLE 17.1

The delta-wye transformer in Figure 17.8 is rated as follows.

$$50 \text{ MVA} \quad V_\Delta = 115 \text{ kV} \quad V_Y = 69 \text{ kV} \quad N_{CY} = 600/5$$

Find a suitable ratio $N_{C\Delta}$ for the delta connected CTs on the 69 kV side.

Solution

First we compute the transformer full-load line currents.

$$I_{L\Delta} = \frac{50 \times 10^6}{\sqrt{3}(115{,}000)} = 251.02 \text{ A}$$

$$I_{LY} = \frac{50 \times 10^6}{\sqrt{3}(69{,}000)} = 418.37 \text{ A}$$

With full-load line current flowing, the restraint coil in the secondary circuit of the wye-connected CT's will see only 251.02/120 or about 2 amperes, which is well within its rating. We note that the ratio of the full load currents is 1.667, which agrees with the rated voltage ratio.

From (17.16) we write

$$\frac{I_{S\Delta}}{I_{SY}} = \sqrt{3}\frac{(418.37)(600/5)}{(251.02)N_{C\Delta}} \cong 1.0$$

from which we compute

$$N_{C\Delta} \cong 346$$

This presents an interesting problem since a ratio of 350 would be a CT ratio of 1750/5, which is not a standard rating. Hence, the desired ratio is about halfway between two standard ratings: 300 and 400. It is best to choose the higher value, which gives higher voltage on the secondary side of the CT. This tends to minimize the effects of secondary lead resistance, which is analogous to our using high-voltage transmission to minimize losses. Therefore, we set

$$N_{C\Delta} = 400 = 2000/5$$

We may compute the differential current that will flow under full load conditions as follows.

$$I_{S\Delta} = \sqrt{3}\frac{I_{LY}}{N_{C\Delta}} = \sqrt{3}\frac{418.37}{2000/5} = 1.81 \text{ A}$$

$$I_{SY} = \frac{I_{L\Delta}}{N_{CY}} = \frac{215.02}{600/5} = 2.09 \text{ A}$$

$$I_{SY} - I_{S\Delta} = 0.28 = 13\% \text{ of } I_{SY}$$

If we select a percentage differential relay with a 50% threshold, we will have an excellent margin of security. ■

17.5.2 Differential Protection of Transformers

The most common method of transformer protection utilizes the percentage differential relay as the primary protection, especially where speed of fault clearing is considered important [9]. It was noted earlier that Buchholz relays will clear all but bushing faults, but these relays are relatively slow and are designed to perform a different function than high-speed, active fault clearing. It is probably true that most active faults involve arcing to ground, and can probably be cleared by ground relays. Still, the differential relays predominate the fault detection for power transformers, and this type of protection is recommended by most transformer manufacturers, especially for the larger sized banks [4], [5]. Moreover, the trend in standards for reduced fault-withstand time in power transformers [11] requires that fast clearing of transformer faults be emphasized.

17.5.2.1 Percent Slope of Differential Relays. The percentage differential relays usually recommended by transformer manufacturers are often available with different percent slopes, and a slope can be selected that is adequate for the type of unbalance that is inherent due to differences in transformer ratios and CT accuracy.

There are three sources of error that tend to cause unbalances in the CT secondary currents during external faults [4]:

1. Tap changing in the power transformer
2. Mismatch between CT currents and relay tap ratings
3. Differences in accuracy of the CT's on either side of the transformer bank

Moreover, we can make the following observations regarding current transformer error in their application to power transformers:

1. Taps are often available on power transformers in $\pm k\%$ change in the transformation ratio, thus providing a range of $2k\%$. In this case, the practice is to choose CT ratios and relay taps to balance the currents at the midpoint of the tap range. This means that the maximum error that can occur is $k\%$.

2. The maximum mismatch between CT currents and relay tap ratings is one-half the difference between two relay tap ratings, expressed in percent.
3. The percent difference between CT errors is found by analyzing the maximum external fault that produces the greatest error, computed as a steady-state quantity.

The usual practice is to add all three errors together, expressed as a percent, and then add 5% or so to this value to determine the minimum slope that the relay characteristic should have.

EXAMPLE 17.2

Consider a delta-wye-grounded transformer similar to that shown in Figure 17.7. The transformer is rated 42 MVA, 69 kV delta-12.5 kV wye. The transformer has load tap changing of ±10%. The transformer is to be protected by a percentage differential relay, which is an electromagnetic induction disk relay with two restraint coils and one operating coil [5]. A one-line diagram of the relay and its connection to the transformer is shown in Figure 17.9. The relay has taps of 5.0, 5.5, 6, 6.6, 7.3, 8.0, 9.0, and 10.0. The relay is nominally a 50% percentage differential relay, which means that the operating current must be at least 50% of the restraint current to cause tripping. Determine the current transformer ratios and select an appropriate tap setting for this relay.

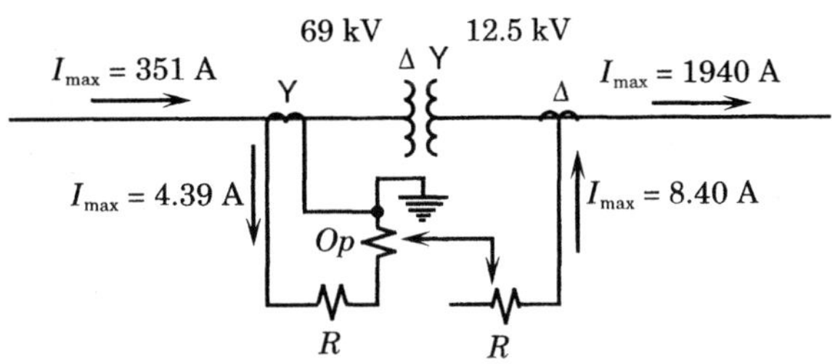

Figure 17.9 One line diagram of the transformer protection.

Solution

First we need to calculate the maximum or rated current at both voltage levels. First, at the 69 kV terminals under full load, we have

$$I_{L69} = \frac{42{,}000}{\sqrt{3}(69)} = 351 \text{ A}$$

A 400/5 CT ratio will be about right for this side. Then the full load relay current on the 69 kV side will be

$$I_{R69} = 351\frac{5}{400} = 4.39 \text{ A}$$

For the 12.5 kV side, the full load current is computed as

$$I_{L12} = \frac{42{,}000}{\sqrt{3}(12.5)} = 1940 \text{ A}$$

A 2000/5 CT ratio will be about right for this side. Then the full load relay current for the 12.5 kV side will be

$$I_{R69} = 1940\frac{5\sqrt{3}}{2000} = 8.40 \text{ A}$$

These computed currents are shown in Figure 17.9.

The relay is provided with taps so that the ratio of the two relay currents will balance the relay as closely as possible. Thus, the ratio of the relay currents determines the relay tap ratio. Let's assume that the 5 tap will be used for the 4.36 A side. Then we can compute the ratio as follows.

$$\frac{4.39}{8.40} = \frac{5}{x}$$

and

$$x = 9.57$$

There is no relay tap of 9.57, but a tap of 10 is close. If we use this tap for the 12.5 kV side, the relay mismatch under normal load conditions is computed as

$$\% \text{ Error} = \frac{10.0 - 9.57}{9.57} \times 100 = 4.49\%$$

This is the error due only to the tap mismatch. There are also errors due to the fact that the transformer has load tap changing and a CT error of about five percent. To be conservative, we assume that all errors apply at their maximum amount at the same time. Thus we compute

LTC error	10.0%
CT error	5.0%
Relay error	4.5%
Total	19.5%

Then the sensitivity margin is

$$\text{Margin} = 50\% - 19.5\% = 30.5\%$$

This is a safe margin and the foregoing calculations are acceptable. ■

17.5.2.2 Magnetizing Inrush Suppression. There are several methods that have been used to prevent the tripping of a sound transformer due to large inrush currents that accompany initial energization of the unit. The common methods used are the following [12], [13]:

1. Desensitize the relay during startup
2. Supervise the relay with voltage relays
3. Add time delay
4. Detect magnetizing inrush by observing the current harmonics

These methods can be further described, as follows:

1. Methods have been devised to desensitize the differential relay and prevent tripping during startup. One method parallels the operating coil with a resistor, with the resistor circuit being closed by an undervoltage relay *b* contact. When the transformer bank is de-energized, the undervoltage relay resets, thereby closing the resistor bypass circuit. On startup, the operating coil is bypassed until the undervoltage relay picks up, which is delayed for a suitable time [4].

   Another method uses a fuse to parallel the differential relay operating coil. The fuse size is selected to withstand normal startup currents, but internal fault currents are sufficient to blow the fuse and divert all current to the operating coil.
2. The voltage supervised relay measures the three-phase voltage as a means of differentiating between inrush current and a fault condition, a fault being detected by the depression in one of the three-phase voltages [4], [5]. This concept is usable for either fast or slow relays, and constitutes an improvement over the techniques of method (1).
3. Simply adding time delay to the differential relays during energization of the transformer is effective, but it must be accompanied by some method of overriding the

time delay if an actual fault occurs during startup. Usually, time delay is used in conjunction with other relay intelligence.

4. Harmonic current restraint is another method that is used. It was noted earlier that the second harmonic of the total current is almost ideal for determining whether a large inrush of current is due to initial energization or to a sudden fault. Most differential relays use filters to detect the second, and sometimes the fifth, harmonic current and restrain tripping when this current is present [12], [13].

The minimum second harmonic current found in a large number of actual transformer energizing tests was about 23% [5]. Using this as a basis, some differential relays are designed to restrain operation as long as the second harmonic exceeds 15% of the fundamental. For internal faults there is still sufficient energy in the fundamental and other harmonics to cause tripping.

Magnetizing inrush detection continues to be a subject of interest [13–19]. Many of the newer techniques employ digital logic and one report investigates the use of artificial neural network training to discriminate between magnetizing inrush and fault currents.

17.5.2.3 Three Winding Transformer Protection. The three-winding transformer can be protected by differential relays with the connection depending on the system connections of the transformer. The most common situation is probably one in which the transformer is connected to the system at both its high- and low-voltage terminals. In this case, all three windings of the transformer must have their own restraint coil in the differential relay scheme, as shown in Figure 17.10.

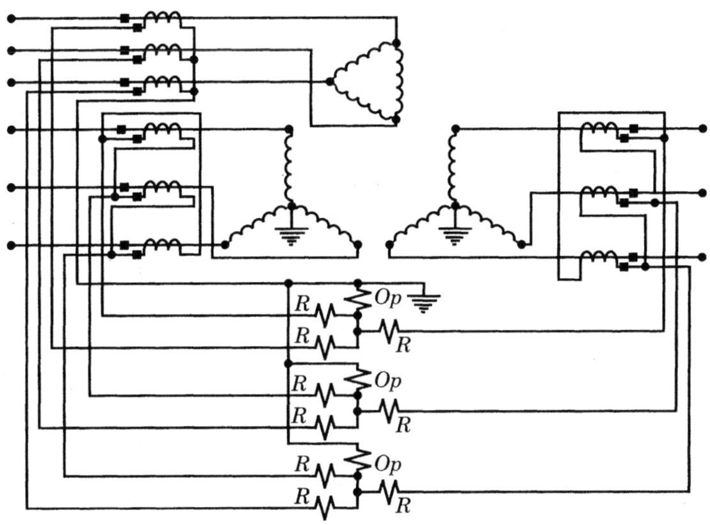

Figure 17.10 Differential protection of a three-winding transformer.

If the three-winding transformer has only one winding connected to the source of supply, then the two load windings can have their CT outputs summed into the same restraint coil [2]. For example, in Figure 17.10, suppose the delta-connected winding is energized by the power system and both the wye windings serve loads at different voltages. Then the CT's of these wye-connected windings can be connected in parallel, which forms the sum of these currents.

This cannot be done if two or more of the three windings are connected to the power system source voltages.

17.5.2.4 Parallel Transformer Banks. There is a tendency to parallel transformer banks without separate breakers for each bank, which results in considerable savings in breaker costs. The resulting configuration is shown in Figure 17.11, where it is noted that separate differential protection is required for each bank, in addition to differential protection for all equipment inside the protection zone, defined by the two circuit breakers. Should the individual bank protections be omitted, the result is transformer protection with only one-half the sensitivity of that for a single transformer, since the CT's must be rated at least twice that of a single bank. We assume that both banks have the same rating. If one bank is smaller, then the imbalance in sensitivity is even worse.

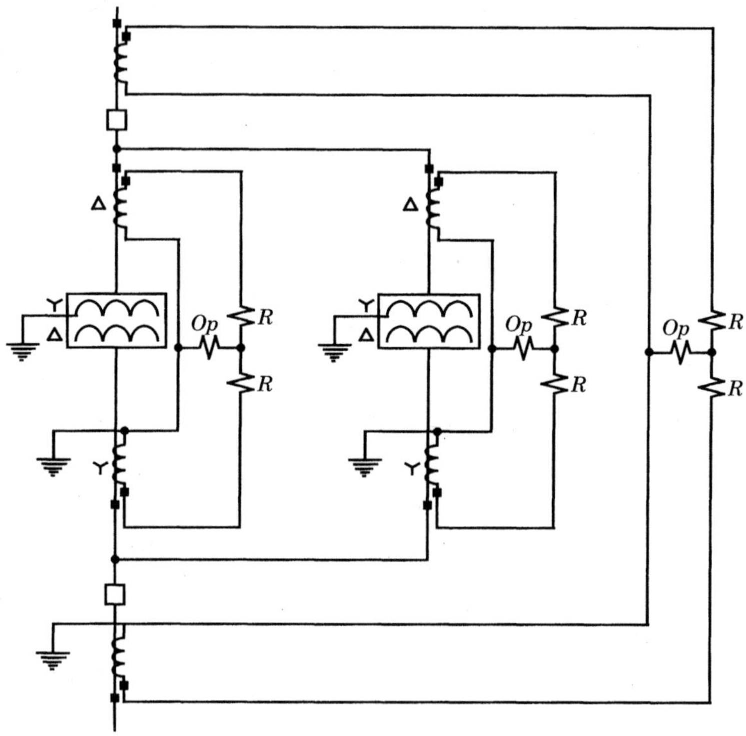

Figure 17.11 Protection of parallel transformers with common breakers.

17.5.2.5 Autotransformer Protection. Autotransformers can also be protected using differential protection schemes as described above. As an alternative, autotransformers can be protected using a scheme that is based on Kirchhoff's current law. This scheme, shown in Figure 17.12, requires that the neutral connections of the autotransformer must be available externally for CT connection. The relays used in this scheme are high impedance relays [2].

This type of protection has good selectivity and high speed. Moreover, it is not affected by tap changing of the autotransformer or by magnetizing inrush currents. This protection is not responsive to inter-turn faults, which must be detected by another means, such as Buchholz relays. If the autotransformer has a tertiary winding, faults in that winding are not detected by this scheme.

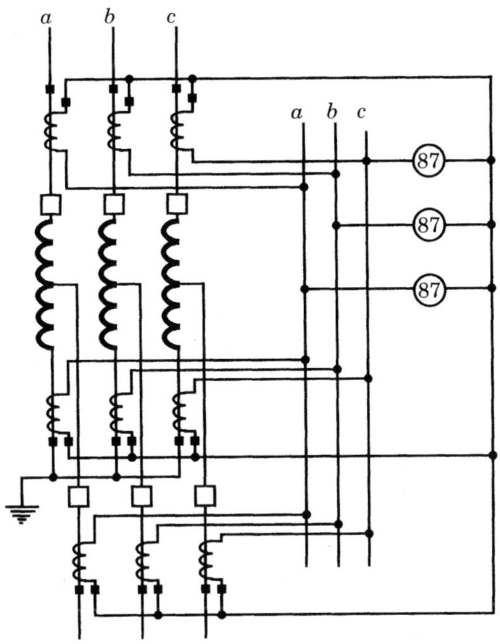

Figure 17.12 A protection scheme for autotransformer units [2].

17.5.2.6 Problems with Differential Relays. In applying differential protection it is important to use current transformers of similar characteristics on both sides of the power transformer. If the two sets of CT's are of different characteristics, any current flowing in the operating coil of the relay tends to add to the burden of the more accurate CT and reduce the burden of the less accurate one. If this is the case, it is sometimes recommended that a shunt burden, having saturation characteristics similar to the less accurate CT, be added across the terminals of the more accurate CT, thereby making the two sets equally poor, but still better balanced.

If only one set of CT's have poor accuracy, there is also the hazard of "locking in" for internal faults. This means that the less accurate CT is unable to sustain any secondary induced emf or its secondary winding is effectively shorted. Thus the better CT's secondary currents are shunted around the operating coil and tripping is blocked.

17.5.3 Overcurrent Protection of Transformers

Overcurrent protection is often used for small transformers, especially for installations that are connected to the source at only one winding and provide service to a load. An extremely inverse relay characteristic is recommended, with an instantaneous unit for severe faults. A very inverse relay in the ground connection of a grounded wye will provide ground fault protection [2].

Overcurrent protection is usually not recommended by transformer manufacturers, except as backup protection. One utility is known to have used overcurrent protection on transformers for years, and apparently with good success [4]. This application uses very inverse relays, set at 40% of rated transformer current, and with a time dial setting of 0.5 or 1.0. Instantaneous units are set at 200%–300% of rated current. This application requires repeated field tests to make sure that the unit will not trip on magnetizing inrush with the 40% setting. It is noted that, if phase current balancing autotransformers are required in this scheme, it is almost the same cost as some of the differential protection schemes [5].

An interesting variation of overcurrent protection is to connect overcurrent relays in a "rough balance" differential arrangement, as shown in Figure 17.13 [2]. A definite minimum time overcurrent relay is used in this scheme, but the ratios of the current transformers are selected to purposely not have an exact balance. A typical unbalance might be about 1 ampere with the relay set at 25% of 5 amperes, or 1.25 amperes. The relay will protect against overloads of more than 25% and simultaneously provide differential protection against internal faults.

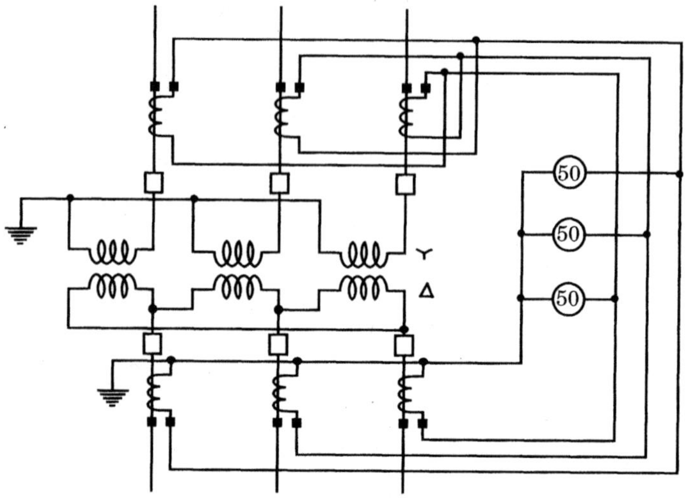

Figure 17.13 Rough balance transformer protection.

17.5.4 Ground Fault Protection of Transformers

Ground fault protection of transformers can be achieved using either differential relays or overcurrent relays, in various arrangements. The exact relay selection and connection depends on the type of transformer bank. A few examples are explored below.

Figure 17.14 shows an example of ground fault protection of a delta-wye bank using a combination of overcurrent and differential relays. If the delta side has a source of ground current, relay 51G will detect ground faults in the delta winding or in the phase windings between the current transformers and the power transformer windings, assuming that an external source of zero-sequence current is available. The wye-connected winding has differential protection. For the external fault illustrated, it is shown that the differential relay will not operate. However, if the ground fault is internal to the current transformers and zero-sequence currents are supplied from the power system, the relay will pick up. The auxiliary transformer is necessary if the current transformers in the phase and neutral terminals of the wye-connected windings have different ratios.

When the transformer wye connection is grounded through an impedance, differential protection may not have sufficient sensitivity to operate correctly. In many cases, this problem can be solved by using a sensitive time-overcurrent relay in the transformer impedance-grounded neutral. Another option is to replace the differential relay with a high-impedance ground detector relay (device type 64). Although the fault current level falls as the fault posi-

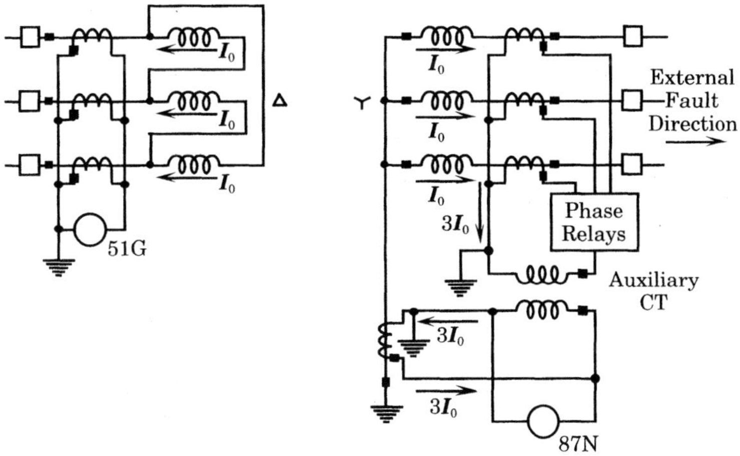

Figure 17.14 Overcurrent and differential ground fault protection for a delta-wye transformer bank [1].

tion nears the neutral, this method can still provide reasonable coverage of the entire winding [1], [3].

17.6 COMBINED LINE AND TRANSFORMER SCHEMES

Occasionally, where a transmission line terminates at a transformer, the possibility exists for omitting the line side circuit breaker and protecting the line and transformer together as a unit. The motivation for doing this is to save the cost of the line side circuit breaker, and is not based on any superior protection concepts.

Many different physical arrangements are possible, some of which are shown in Figure 17.15. In each case, the protection zone is between the CT locations, spanning both line and transformer. The type of transformer connection is important, especially in considering ground relaying requirements. In every case, the saving in one circuit breaker position must

Figure 17.15 Line and transformer unit protection examples.

be balanced by the increased relaying complexity. One example is the requirement for transfer tripping on detecting an internal transformer fault by a pressure relay, which must trip the local breaker and also send a signal to the remote end of the line to initiate tripping there. This requires communications that are not otherwise needed for transformer tripping.

17.6.1 Non-Unit Protection Schemes

First, we consider the line and transformer system where the protection consists of separate relays for the line and for the transformer.

17.6.1.1 Line Phase Fault Protection. One method of providing high-speed protection for phase faults is to utilize distance relays at the end of the line remote from the transformer. In Figure 17.15(*a*) and (*b*), for example, the distance relays are installed at Bus H and are set to reach 100% of the line length and well into the transformer impedance. This provides fast clearing of faults located anywhere on the line, but the protection for transformer faults is questionable. The advantage of this arrangement is that there is little problem of overreaching the transformer impedance. The nominal rule is to extend the first zone halfway through the transformer. This type of line protection is not affected by varying system conditions.

If the system Thevenin impedance is nearly constant, overcurrent relays could be used rather than distance relays, since the total fault current can be readily computed. If both ends of the line are connected to system sources voltages, the overcurrent relays must be directional. Instantaneous relays can be used as well, if the system impedance is nearly constant.

17.6.1.2 Line Ground Fault Protection. The transmission line ground relaying depends on the availability of a ground source of fault current at the transformer. For the condition shown in Figure 17.15(*a*), the line side of the transformer is wye-grounded. In this case, high speed ground fault protection can consist of a directional ground fault relay in the transformer neutral.

For the system of Figure 17.15(*b*), the line side of the transformer is delta connected, and the only source of ground current is from the system behind bus H. In this case, wye-connected and grounded CT's at bus H can be used as a source of polarizing current for a ground fault on the line. Figure 17.15(*c*) presents a similar problem, where a source of ground current for line protection is available at only one end of the line.

17.6.2 Line and Transformer Unit Protection

Protecting the line and transformer together using some combination of line and transformer relays presents problems in providing good sensitivity in the transformer protection, since there are no CT's on the line side of the transformer. If line side CT's are used, then the transformer protection can be provided in the usual manner, but with a means of transfer tripping the line at bus H in the event of a transformer fault.

One scheme that has been used is to provide tapped transformers and differential relays at both ends of the line, as shown in Figure 17.16. If an overall differential system is to be considered, the method must take into account the fact that the zero-sequence current on one side of the transformer can not be reproduced on the other side. This means that summation windings, such as those described in Section 13.2.2 for wire pilot schemes, are not applicable.

The protection method of Figure 17.16 can utilize any type of pilot signaling. The concept is not without problems, however. Phase protection is acceptable, but the ground protection is not adequate [3].

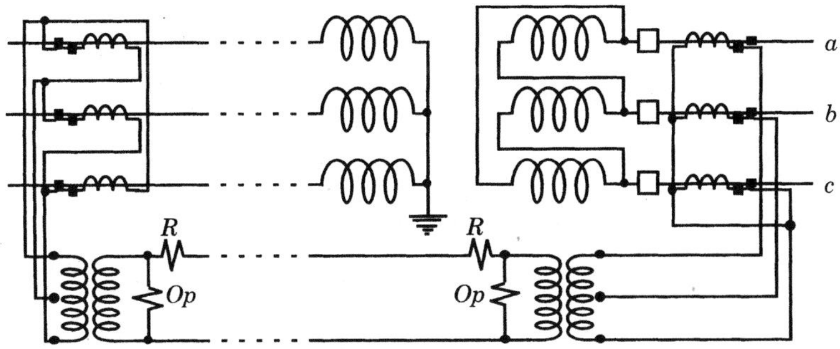

Figure 17.16 One method of line-transformer unit protection [2], [3].

17.7 REGULATING TRANSFORMER PROTECTION

Regulating transformers are designed to provide a controllable change in the power system voltage angle, voltage magnitude, or both. Some designs are capable of only a change in phase angle, and some can change only the voltage magnitude. Regulating transformers are constructed using many different winding configurations and connections, depending on the rating, the amount of phase shift, and whether magnitude regulation is also provided. Phase angle regulators can be designed for either discrete or continuous angle change. Continuous phase angle regulation requires a load tap changer, thereby providing several tap positions for a range of angle changes.

Providing protection for the exciting winding of a regulating transformer presents a rather difficult problem since ordinary power transformer differential relays are not sensitive enough to detect faults in this high-impedance winding. The system shown in Figure 17.17 represents an in-phase regulating transformer, i.e., a transformer that does not intentionally introduce phase shift, but only a magnitude change in the regulated voltage.

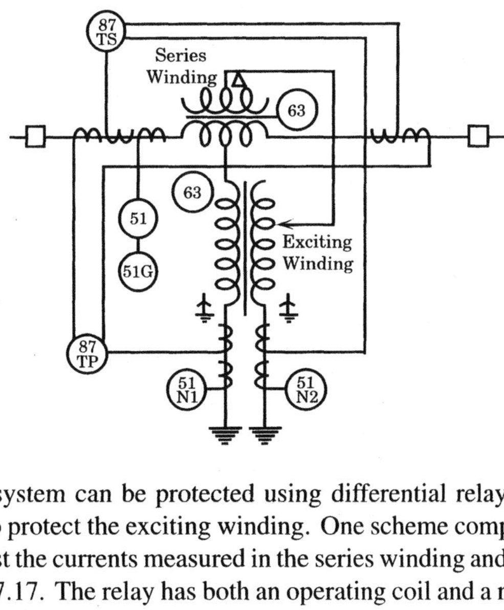

Figure 17.17 Protection of a regulating transformer [21].

The total transformer system can be protected using differential relaying, but special purpose relays are required to protect the exciting winding. One scheme compares the current in the exciting winding against the currents measured in the series winding and the transformer neutral, as shown in Figure 17.17. The relay has both an operating coil and a restraint coil and is set to pick up with a given percentage of unbalance greater than the maximum unbalance due to regulation. The transformer manufacturers often provide protection of special design for regulating transformers.

Regulating transformers have shunt connected exciting windings and series connected voltage regulating windings, which may be either in-phase or phase-shifting types. Since the series winding must carry full rated current, this winding has a very small impedance and is properly protected by percentage differential relays. The exciting winding, however, has a very high impedance and is not protected by the differential relays. This means that special protection must be added to protect the excitation windings [21].

One way of doing this is to provide a current balance protection circuit that compares the currents flowing in each winding. Normally, the current in the exciting winding will be about 10% of that in the series winding, with the exact ratio depending on the tap setting. If this ratio exceeds about 15%, there is a high likelihood of a fault in the exciting winding or the tap changing equipment. The current transformer connections are important for this type of protection. This is discussed in detail in the literature [4], [21].

Regulating transformers should also have backup protection using Buchholz or other gas accumulator protective devices. If the regulating transformer also incorporates phase shifting of more than about 10 degrees, then special types of relaying must be used [4]. Such modified differential protection to accommodate phase shifting tends to be complex.

Phase shifting transformers cannot be protected with the scheme shown in Figure 17.17 because of the phase shift introduced by the series winding. This is sometimes accomplished by inducing a voltage in the series winding from one of the other two phases. As a result, an external fault may produce a current that is larger on one side of the differential relay than the other side, causing an unwanted pickup of the relay. This type of transformer requires special protection, usually requiring current transformers inside the transformer enclosure, as well as bushing CT's. In some cases, the sudden pressure relay may be the primary protection.

17.8 SHUNT REACTOR PROTECTION

Shunt reactors are designed for connection to the ends of high-voltage transmission lines or to high-voltage pipe-type cables for the purpose of controlling the line voltage by absorbing reactive power. In many cases, shunt reactors are not switched, but are continuously in service. However, at some installations the reactors are switched off during periods of high circuit loading. Therefore, any protective device may clear reactor faults by switching either the reactor breaker, if one exists, or by switching the entire circuit on which the reactor is installed. It is common for reactors to be installed at both ends of EHV lines, and sized to prevent the line voltage from exceeding design values when energized from one end. Since there is usually some uncertainty as to which end of a line may be energized (or de-energized) first, reactors are usually installed at both ends of the line.

There are two general types of shunt reactors [22]. One is a dry-type reactor of an air core or core-less design. These are installed open to the atmosphere and are cooled by natural convection of air circulating between core layers and turns. Dry-type reactors are limited to voltages up to about 34.5 kV and are often installed on the tertiary of a transformer, which is connected to the high-voltage line being compensated. One problem with dry-type reactors is the lack of any metallic housing or shield. As a result, a high-intensity external magnetic field is generated when the reactor is energized. This means that adequate clearance must be maintained to surrounding metal structure that might form a closed loop for currents. In some cases, shielding is required to prevent such unwanted currents in adjacent structural or other metal objects. Dry type reactors are single-phase units. The advantage of the dry type reactor is lower initial and operating costs, lower weight, and lower losses. The disadvantages are the limitation in voltage and kVA rating and the high-intensity external magnetic field. Also, since there is no iron core, there is no magnetizing inrush current when the reactor is energized.

The second type is an oil-immersed shunt reactor, which is designed as either core-less or with gapped iron cores. The iron-cored reactors are subject to severe inrush currents, more so than the core-less type. The core-less oil-immersed reactors are surrounded by a magnetic shield, which contains the magnetic flux within the reactor tank. The magnetic circuit of the shunt reactor is similar to that of a transformer, except that small air gaps are introduced in the iron core to improve the linearity of inductance and to reduce residual or remnant flux. Oil-immersed reactors can be either single phase or three phase units. They can be either self-cooled or forced cooled.

17.8.1 Dry Type Reactors

Dry-type shunt reactors are usually connected to the tertiary of a transformer bank as shown in Figure 17.18. The reactor can be switched on the supply side, as shown in Figure 17.18(*a*) or on the neutral side as shown in part (*b*) of that figure. The grounding transformer has a grounded wye-connected primary and a broken-delta-connected secondary with a grounding resistor. This arrangement provides a limited amount of ground current. This is a high-resistance method of detecting a fault on the tertiary circuits.

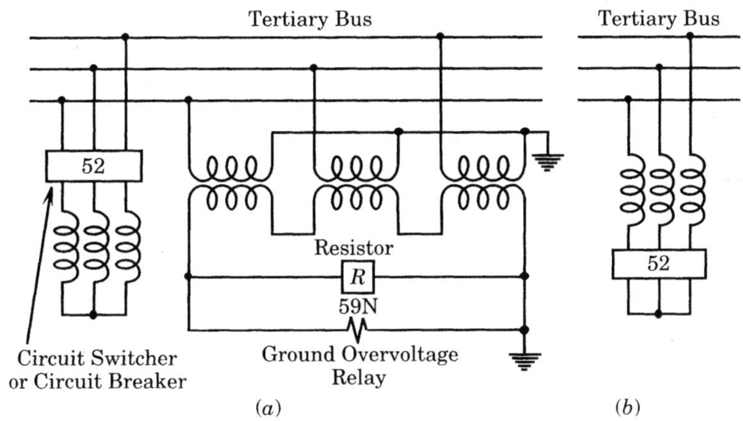

Figure 17.18 Typical installations for dry-type shunt reactors [22]. (*a*) Reactor with 3-pole supply side switching and with grounding transformer. (*b*) Reactor with 2 or 3-pole neutral side switching.

Dry-type shunt reactors are subject to three types of failure [22]:

1. Phase-to-phase faults on the tertiary bus, resulting in high-magnitude phase currents.
2. Phase-to-ground faults on the tertiary bus, resulting in a low-magnitude ground current that depends on the size of the grounding transformer and resistor.

3. Turn-to-turn faults within the reactor, resulting in a very small change in the reactor phase current.

Phase-to-phase faults are not likely in the reactors themselves, since they are single-phase units with considerable separation between units. Faults between the windings and ground are also unlikely since the reactors are mounted on insulated supports with standard clearances to ground.

It should be noted that transmission system is usually not affected by a faulted dry-type reactor since even a shorted phase of the reactor will have only a minor effect on the phase current. Such a fault could evolve into a more serious fault, such as a phase-to- phase fault, in which case the reactor would have to be isolated by switching it out of service. If the reactor is not equipped with a switching device, then the transformer bank must be tripped. The loss of the reactor may cause an unacceptable voltage rise in the high-voltage circuit, and this possibility must be checked.

Dry-type reactors are protected against phase-to-phase faults by overcurrent, differential, or negative sequence relays, or a combination of these schemes [22]. Phase-to-ground protection is illustrated in Figure 17.18(a). The broken delta output of the grounding transformer is monitored by an overvoltage relay. The relay is equipped with a harmonic filter to reject any third harmonic, thereby improving the relay sensitivity. A phase-to-ground fault is not very serious, and it is common practice to alarm this condition rather than tripping. Note that the relay is not able to distinguish between a fault on the reactor or any other part of the tertiary circuit.

Turn-to-turn faults are very difficult to detect and the current change for such faults is small and of the same order of magnitude as encountered in normal operation. A voltage unbalance scheme has been used for this purpose [22].

17.8.2 Oil-Immersed Reactors

Oil-immersed reactors are usually connected to one or both ends of long transmission lines, as shown in Figure 17.19. The reactors are needed to control the voltage of the line, especially when one line circuit breaker is open. This condition causes the voltage to rise to unacceptable levels. The shunt reactors are wye-connected with a solidly grounded neutral connection. In some cases the reactors are not switched, but are permanently connected.

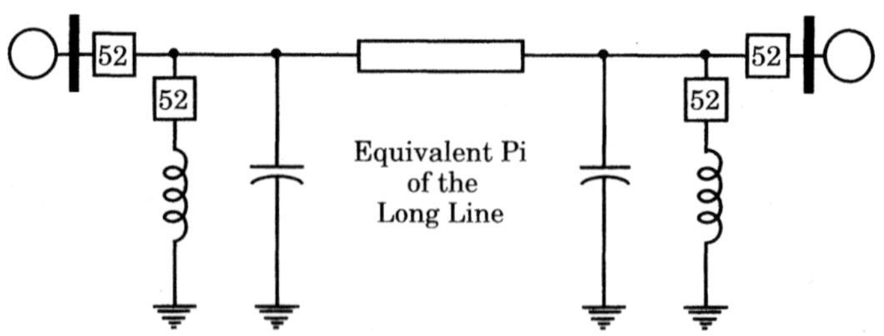

Figure 17.19 One-line diagram of line-connected switched shunt reactors.

In some installations, shunt reactors are connected to the substation bus and are generally solidly grounded and wye connected.

17.8.2.1 Failure Modes of Shunt Reactors. The failure modes of both line or bus connected, oil-immersed reactors are as follows:

1. Equipment failure, such as bushing failure, insulation failure, etc., which results in large changes in the magnitude of phase currents. Since oil-immersed reactors have their windings in close proximity to a grounded tank, grounding failures can occur, with the resulting fault current being dependent on the location of the ground in the reactor winding.
2. Turn-to-turn faults within the reactor winding. These faults result in small changes in current. These faults may result in a change in reactor impedance, increased operating temperature, internal tank pressure, and gas accumulation. They may evolve into major faults if not detected.
3. Miscellaneous failures, such as low oil and loss of cooling.

The usual practice for reactor fault clearing is to trip the local line breaker and transfer trip the remote end line breaker. If the reactor is switched, the sequence is to trip the line breakers, open the reactor switch, then reclose the line breakers. Some switching devices have fault interrupting capability and are capable of clearing faults, if within the switcher rating. Making the best use of this capability requires a coordination between the switcher and the line circuit breaker.

The distributed shunt susceptance of the transmission line forms a parallel circuit with the shunt reactor that can become resonant, usually with a resonant frequency close to the 60 hertz. Such a resonance can be troublesome and should be avoided.

Another problem situation occurs when two transmission lines are close enough to have substantial mutual induction, it is possible for higher than rated voltage to develop on a shunt reactor of a line that is out of service. This problem may be prevented by switching out the shunt reactors after switching the transmission line.

If the transmission line is series compensated, there may be a problem when the line is de-energized. A parallel resonant circuit due to the reactor and the line shunt susceptance will induce a subsynchronous current in the de-energized line that starts at about rated voltage and damps out slowly at the subsynchronous frequency. This can cause false operation of distance relays used to protect the shunt reactors unless the relays are designed to counteract this problem.

17.8.2.2 Protection Practices for Shunt Reactors. The detection of large magnitude faults is generally performed by overcurrent, differential, or distance relays. One of the problems with these applications is false tripping of the reactor during switching the device on or off. Switching causes dc offset with long-time constants and low-frequency current components.

Turn-to-turn faults are difficult to detect with confidence. Overcurrent relays are generally agreed to be inadequate for this purpose. Distance or ground relays offer some improvement, but are not without problems. Generally, the most reliable method of detecting turn-to-turn faults is with sudden-pressure relays or gas-accumulator relays.

Some oil-immersed reactors are designed with forced cooling, and the continued operation of the cooling motors is critical. Loss of cooling can be detected by monitoring oil flow using flow indicators, monitoring the ac voltage supply to the fans and pumps, or by monitoring the temperature.

Overvoltage can occur on transmission circuits, but disconnecting the shunt reactors under such a condition will make the voltage even higher. This problem must be solved by opening the line circuit breakers.

In many installations, line reactors are switched in and out as system requirements dictate. The switchers are usually single pole devices that are not mechanically linked, and with each pole independently operated. This can result in the three phases being energized or de-energized at different times, creating large imbalances in the system voltages. Two methods have been used to detect pole disagreement. The first method utilizes auxiliary contacts on the pole operators to ensure that if all three poles do not open or close at the same time, all poles will be tripped or the backup breaker tripped. A second method uses a pole disagreement relay to compare the currents in the three phases.

17.9 STATIC VAR COMPENSATOR PROTECTION

The static var compensator is a special type of power system voltage control that utilizes thyristor switching of inductors or capacitors to achieve fast effective voltage control at the point of connection [24]. The connection of one phase of such a device is shown in Figure 17.20.

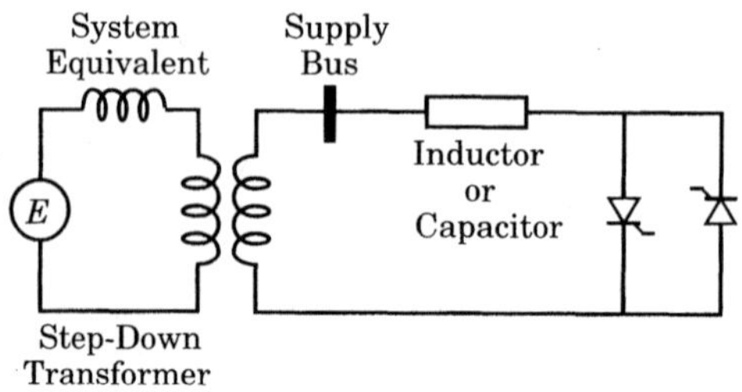

Figure 17.20 One phase of a thyristor switched inductor or capacitor.

First, consider a *thyristor controlled reactor* (*TCR*). If the thyristor gate pulses are always turned on, the thyristors effectively become diodes, which results in a sinusoidal current through the inductor. By proper control of the thyristor delay angle, the current can be caused to flow only in pulses, thereby reducing the effective value of the current. The resulting total current is not a pure sine wave, but consists of odd harmonics whose amplitudes depend on the delay angle.

For a three-phase device, it is common to connect the controlled inductor and controlling thyristors in a delta connection, which permits the third harmonic to circulate in the delta. Higher harmonics are filtered out using passive series tuned filters [24]. In some installations, mechanically switched capacitors are connected in parallel with the TCR to provide a wider range of voltage control. Switching a capacitor will cause the voltage to rise a small amount, and the TCR is used to lower that voltage in a controlled fashion. When the system voltage tends to be high, all mechanically switched capacitors will be switched out of service.

Capacitors are not as easily controlled as inductors. When the controlled element is a capacitor the switching is performed as integral half-cycle control, where the capacitor is either fully in or out of the circuit. This mode of operation is called a *thyristor switched capacitor* (*TSC*). The switching is accomplished by blocking the gate pulses to both thyristors, which causes the current flow to stop at the instant of zero crossing.

It is not uncommon for static var compensators to employ both TCR's and TSC's, which can be used to provide a suitable range of controlled voltage levels under varying system conditions. Static var compensators are fundamentally voltage controllers, but are also effective in improving transient stability and in damping objectionable system oscillations that occur following disturbances. Since high-voltage thyristors are expensive, it is common to incorporate a step-down transformer to provide an SVC voltage to less than 20 kV.

17.9.1 A Typical SVC System

An example of a typical static var compensation system is shown in Figure 17.21. This system is a ±350 MVA system, including two 175 MVA thyristor controlled reactors, two 150 MVA thyristor switched capacitors, and a set of three filters for the 5th, 7th, and 13th harmonics which are capacitive at the fundamental frequency and provide a total of 50 MVA reactive power to the power system. The SVC system is connected to a 230 kV bus through a circuit breaker and step-down transformer, which provides a 20 kV bus for the SVC system. Both the TCR's and the TSC's are connected in delta, thereby permitting third harmonic currents to circulate in the delta, but not enter the power system. The TSC branches are equipped with reactors to limit the current stresses on the thyristors. The small circles in Figure 17.21 show where current measurements are made for use by protective relays.

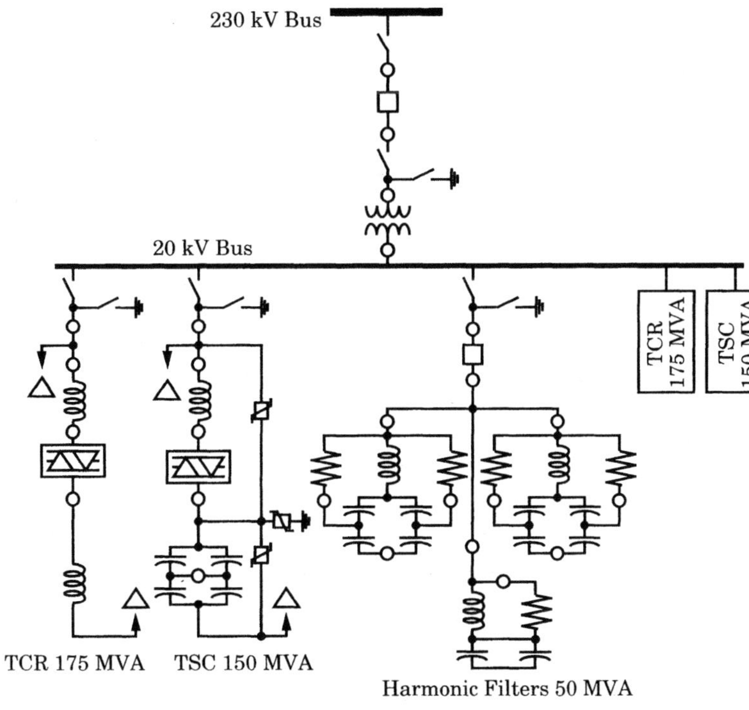

Figure 17.21 A typical static var compensation system [25].

17.9.2 SVC Protection Requirements

All of the equipment in a static var compensator must be protected from damage due to all types of system hazards. This includes overcurrent, overvoltage, excess reactive power loading, unbalance due to component failure, phase-to-phase faults, phase-to-ground faults, failure of cooling equipment, control malfunctions, and any other type of failure that can place undue stress on the SVC components.

Both the TCR and TSC branches of the SVC shown in Figure 17.21 are connected in delta, but this is not always the case. The equipment configuration and voltage ratings are optimized to best utilize the available thyristor voltage rating and current-carrying capability, and to minimize the number of thyristors that must be connected in series. This may result in a wye connection for some branches and delta connection in other branches [26]. It is also noted that some SVC designs require only a TCR and fixed capacitor (or filters), whereas others may require only TSC branches [26].

It is common to require that the protection be provided as two independent systems, which are sometimes called the main and backup protections, or simply systems A and B, described in Table 17.3(a) and (b). The two systems are required to be different in their design and principle of operation, and use different measurement transducers where possible. It must be possible to remove either protective system for maintenance or repair, with the second system continuing to function normally. Every effort must also be made to secure the

TABLE 17.3(a) Summary of SVC "A" Protections

230 kV bus protection A	
87B differential current	51 ground fault
High-voltage transformer A	
50/51 instantaneous overcurrent	50/51N instantaneous ground I
87T transformer differential	87N high impedance ground I
63 gas pressure	49 thermal protection
Bucholz protection	
20 kV bus A	
87B differential current	50/51N instantaneous ground I
59F overvoltage	
TSC 1 A	
87/50N current differential	49 overload protection
37 undercurrent	60 voltage unbalance
59N unbalance compensation	
TCR1 A	
87/50N current differential	46 phase balance
49V overload protection	87 delayed step differential
Filter branch A	
87 current differential	50BF breaker failure
5th harmonic filter A	
50/51 delayed step overcurrent	50/51N instantaneous overcurrent
49 thermal protection	37 undercurrent
60 voltage unbalance	49R resistor overload protection
7th harmonic filter A	
50/51 delayed step overcurrent	50/51N instantaneous overcurrent
49 thermal protection	37 undercurrent
60 voltage unbalance	59N voltage unbalance
49R resistor overload protection	
13th harmonic filter A	
50/51 delayed step overcurrent	50/51N instantaneous overcurrent
49 thermal protection	37 undercurrent
60 voltage unbalance	59N voltage unbalance
49R resistor overload protection	

protective systems against false tripping, which can be due to any cause such as geomagnetic induced currents, high system harmonics, or the presence of spurious signals of various kinds.

TABLE 17.3(b) Summary of SVC "B" Protections

230 kV bus protection B	
59 overvoltage	27 undervoltage
High-voltage transformer B	
50/51 instantaneous overcurrent	51N instantaneous ground fault
87T transformer differential	87N high impedance ground I
63 gas pressure	49 thermal protection
Bucholz protection	
20 kV bus B	
51N ground fault	59N open delta ground fault
59F overvoltage	
TSC 1 B	
87 current differential	50/51 instantaneous overcurrent
60 voltage unbalance	49V overload protection
TCR1 B	
87 current differential, delayed	50/51N instantaneous overcurrent
49 overload, 5th harmonic	49V overload protection
Filter branch B	
50/51 overcurrent protection	50/51 N instantaneous overcurrent
5th harmonic filter B	
50/51 delayed step overcurrent	50/51N instantaneous overcurrent
49 thermal protection	37 undercurrent
60 voltage unbalance	59N voltage unbalance
49R resistor overload protection	
7th harmonic filter B	
50/51 delayed step overcurrent	50/51N instantaneous overcurrent
49 thermal protection	37 undercurrent
60 voltage unbalance	59R voltage unbalanace
49R resistor overload protection	
13th harmonic filter B	
50/51 delayed step overcurrent	50/51N instantaneous overcurrent
49 thermal protection	37 undercurrent
60 voltage unbalance	59N voltage unbalance
49R resistor overload protection	

In some cases, the descriptions of the two system may appear to be identical, but the equipment used to meet the protection requirement should always be different in the two systems, using different hardware design or different tripping logic.

The capacitors in the TSC are usually connected in parallel arrays as shown in Figure 17.21. This provides a method of determining failure of individual capacitor units by measuring the current in the bridge connection between the parallel arrays. Under normal conditions, the two branches are balanced and the current in the bridge is zero. If one capacitor unit fails, a bridge current flows that can be computed based on knowledge of the capacitor array structure. Failure of a second unit in the same branch will cause an even greater unbalance current. A criterion can be established, such as 10% overvoltage, beyond which additional failures will cause the SVC to be shut down for repair, which is accomplished by blocking the thyristors. Once the TSC is blocked, the disconnect switch can be operated to isolate and ground the TSC for repair.

Thyristor protection is aided by a monitoring system that will detect any failed thyristors. This monitoring system has the capability of tripping the SVC and isolating the branch containing the failure thyristors.

The SVC system requires a source of auxiliary power, which is not shown in Figure 17.21. Auxiliary power is usually supplied by two independent power sources and designed so that no combination of failed components will cause total loss of auxiliary power.

Many faults in the SVC branches can be interrupted by control action of the thyristors, which can quickly block the flow of current in the defective branch while allowing the unfaulted branches to continue in operation. The tripping of the circuit breakers serves as backup for fault removal. Clearly, there must be coordination between thyristor blocking and breaker operation. Breaker operation must be limited to only those faults that require complete removal of the entire SVC from the power system.

In some cases, the primary protection for a branch fault may be fast control of the thyristor valves. This can be used for overvoltage or overcurrent faults, or for capacitor failure that results in current unbalance. The backup protection for these functions may be conventional relays, which may need to be slightly delayed to assure that the control completes its blocking action before the backup relay picks up.

REFERENCES

[1] IEEE Std. C37.91-1985, "IEEE Guide for Protective Relay Applications to Power Transformers," IEEE, New York, 1985.

[2] Warrington, A. R. van C., *Protective Relays: Their Theory and Practice*, 1, John Wiley & Sons, Inc., New York, 1962.

[3] GEC Measurements, *Protective Relays Application Guide*, General Electric Company p.l.c. of England, 1975.

[4] Mason, C. R., *The Art and Science of Protective Relaying*, John Wiley & Sons, Inc., New York, 1956.

[5] Blackburn, J. L., Ed., *Applied Protective Relaying*, Westinghouse Electric Corp., Pittsburgh, 1976.

[6] Rockefeller, G. D., W. K. Sonnemann, and C. L. Wagner, "Magnetizing Inrush Phenomena in Transformer Banks," *AIEE Trans.*, 77 (pt. III), October 1958, pp. 884–892.

[7] Horowitz, S. H., Ed., *Protective Relaying for Power Systems*, IEEE Press Reprint Series, New York, 1980.

[8] Baleriaux, H., and F. Brohet, "Buchholz Relays," CIGRÉ paper 321, presented at the Paris session, 1960.

[9] Warrington, A. R. van C., *Protective Relays: Their Theory and Practice*, 2, John Wiley & Sons, Inc., New York, 1969.

[10] Blackburn, J. L., *Protective Relaying, Principles and Applications*, Marcel Dekker, Inc., New York, 1987.

[11] Brown, R., and S. Khan, "Planning and Protection Implications of Reduced Fault-Withstand Time for Power Transformers," *IEEE Trans.*, PWRD-1 (4), October 1986, pp. 60–67.

[12] Hayward, C. D., "Harmonic-Current Relays for Transformer Differential Protection," *AIEE Trans.*, 60, 1941, pp. 377–382.

[13] Sharp, R. L., and W. E. Glassburn, "A Transformer Differential Relay with Second Harmonic Restraint," *AIEE Trans.*, 77, December 1958, pp. 913–918.

[14] Schweitzer, E. O., R. R. Larson, and A. J. Flechsig, "An Efficient Inrush Detection Algorithm for Digital Computer Relay Protection of Transformers," *IEEE Trans.*, PAS-97 (2), March–April 1978, p. 323, 1A14.

[15] Horowitz, S. H., and A. G. Phadke, *Power System Relaying*, Research Studies Press, Ltd., Taunton, Somerset, England, 1992.

[16] Sidhu, T. S., M. Sachdev, H. C. Wood, and M. Nagpal, "Design, Implementation and Testing of a Microprocessor-Based High-Speed Relay for Detecting Transformer Winding Faults," *IEEE Trans.*, PWRD-7 (1), January 1992, pp. 108–117.

[17] Liu, P., O. P. Malik, D. Chen, G. Hope, and Y. Gao, "Improved Operation of Differential Protection of Power Transformers for Internal Faults," *IEEE Trans.*, PWRD-7 (4), October 1992, pp. 1912–1919.

[18] Sidhu, T. S., and M. S. Sachdev, "On Line Identification of Magnetizing Inrush and Internal Faults in Three-Phase Transformers," *IEEE Trans.*, PWRD-7 (4), October 1992, pp. 108–117.

[19] Phadke, A., and J. Thorp, "A New Computer Based, Flux Restrained, Current Differential Relay for Power Transformer Protection," *IEEE Trans.*, PAS-102 (11), November 1983, pp. 3624–3629.

[20] Perez, L. G., A. J. Flechsig, J. L. Meador, and Z. Obradovic, "Training an Artificial Neural Network to Discriminate Between Magnetizing Inrush and Internal Faults." *IEEE Trans.*, PWRD-9 (1), January 1994, pp. 434–441.

[21] Sen, P. K., and B. R. Craig, "Application and Protection Considerations of Large Phase-Shifting Transformers," American Power Conference, Chicago, April 24–26, 1989.

[22] ANSI/IEEE Std. C37.109-1988, IEEE Guide for the Protection of Shunt Reactors," IEEE Standards, New York, 1988.

[23] Central Station Engineers, *Electrical Transmission and Distribution Reference Book*, Westinghouse Electric Corporation, East Pittsburgh, PA., Fourth Edition, 1955.

[24] Mohan, N., T. M. Undeland, and W. P. Robbins, *Power Electronics, Converters, Applications, and Design* Second Edition, John Wiley & Sons, Inc., New York, 1989.

[25] Private Communication, Keeler and Maple Valley SVC Plants, Bonneville Power Administration, Portland, OR, 1992.

[26] Private Communication, Mead Adelanto SVC Project, Los Angeles Department of Water and Power, Project Managers, 1996.

PROBLEMS

17.1 Derive (17.7). References 2 and 3 may be helpful.

17.2 A new distribution substation serving an urban area is to have installed a 69-12.47 kV (Δ-Y grounded) three-phase transformer rated 12/16/20 MVA, where the 12 MVA is the self-cooled rating and the higher ratings are for two levels of forced cooling. The transformer is to be protected by percentage differential relays with 15, 25, and 40% slope. Bushing type CT's are to be used. Transformer percentage differential relays, Type BDD, manufactured by the General Electric Company are to be used. These relays have available relay tap ratios of 2.9, 3.2, 3.5, 3.8, 4.2, 4.6, 5.0, and 8.7. The utility has set the following general rules for transformer protection:

(a) Connect CT's in wye on the transformer delta side and in delta on the transformer wye side.

(b) Do not use less than 40 turns (200/5) on bushing CT's.

(c) Choose the high and low side CT ratios such that each is the closest available ratio above the force-cooled rating.

(d) Choose the high and low side relay taps such that the ratio of low side tap to high side tap is as close as possible to the ratio of the low side relay current to the high side relay current.
(e) Check that the high and low side relay current corresponding to the transformer self-cooled rating does not exceed that relay tap setting.
(f) Check that the high and low side relay current corresponding to the transformer force-cooled rating does not exceed two times that relay tap setting.
(g) Check that the maximum internal fault will not thermally damage the relay.
(h) Check that the relay tap mismatch does not exceed 15%.

The fault currents at the transformer location are computed by digital computer, with the following results:

One-line-to-ground fault:	1879 A
Three-phase fault	3356 A

The relay is capable of 220 amperes for up to 1.0 second.

Check items (a) through (h). Assume that Figure 17.7 gives the transformer connection.

17.3 Ground fault protection is to be provided for a delta-wye transformer bank using overcurrent relays. Sketch a possible relay connection diagram for this type of protection

17.4 Devise a protection scheme for a grounding transformer. Consider the following types of grounding transformers:
(a) A zig-zag transformer
(b) A wye-delta transformer

17.5 A power plant start-up four-winding transformer is has the winding configuration shown in Figure P17.5.

The transformer has the following ratings:

525-13.8-13.8 kV
84/112/140-42/56/70-42/56/70 MVA
OA/FOA/FOA

The transformer manufacturer has provided the following positive and zero-sequence impedance data:

$$Z_{H-Y} = 4.4\% \text{ at } 42 \text{ MVA}$$
$$Z_{H-X} = 11.5\% \text{ at } 29.4 \text{ MVA}$$
$$Z_{H-Z} = 6.6\% \text{ at } 42 \text{ MVA}$$
$$Z_{Y-X} = 10.7\% \text{ at } 29.4 \text{ MVA}$$
$$Z_{Y-Z} = 14.4\% \text{ at } 42 \text{ MVA}$$
$$Z_{X-Z} = 16.8\% \text{ at } 29.4 \text{ MVA}$$

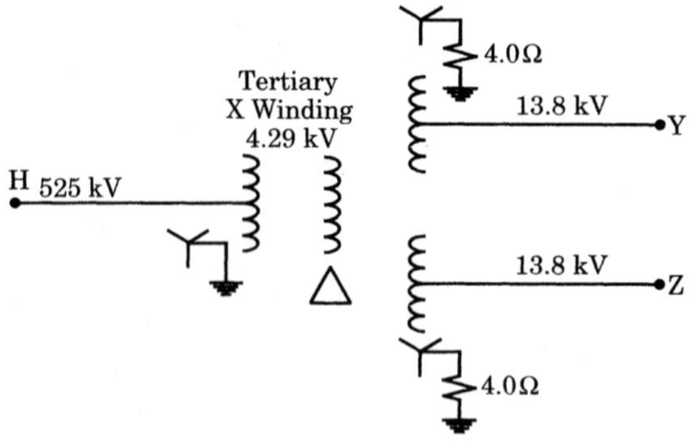

Figure P17.5 A plant startup transformer.

The available fault current is provided by a fault study:

1. At the 525 kV terminals:
   Three phase: 27416 A
   Line to ground: 30947 A
2. At the 13.8 kV terminals:
   Three phase: 38437 A
   Line to ground: 1897 A

For a line-to-ground fault at the 13.8 kV terminals of the transformer, what fault current will be measured by a relay in the 525 kV neutral ground connection?

Hint: It will be necessary to construct an equivalent circuit for the four-winding transformer. Suitable equivalents can be found in the literature [23].

17.6 Prepare a discussion of the need for backup protection for transformers.

17.7 Some conditions provide currents that can contribute to a relay pickup when there is no fault present. Describe some of these currents and comment on what might be done to desensitize the relay to these types of currents.

17.8 Examine the problems of providing transformer protection using fuses. Can the fuses provide overload as well as fault protection? List some of the things to consider in choosing a fuse for transformer protection.

17.9 Consider the causes that lead to the overheating of a transformer. List as many of these as you can. Next, consider the possible results of the overheating.

17.10 A transformer that is connected to a power system is tripped by its protective relays. This particular transformer is a very important component in the bulk power system and communication facilities are available to permit the system operator to re-energize the transformer by sending close signals to the station circuit breakers. Prepare a discussion of the pros and cons of permitting trial re-energization of a transformer when there is no way the system operator can be absolutely sure that a transformer fault exists. Include in your discussion the possible use of automatic reclosing of the transformer breakers, a practice that is often employed for transmission lines.

17.11 Discuss the possible damage that can be caused to a transformer by through fault currents.

17.12 A three-phase thyristor-controlled capacitor bank is wye connected and is protected by differential relays. Sketch the overcurrent and current differential relay connections to the three legs of the wye-connected system.

17.13 Repeat problem 17.12 if the capacitor bank is connected in delta rather than wye. Draw a phasor diagram of the currents.

18

Generator Protection

18.1 INTRODUCTION

Electric rotating machinery represents a class of equipment that is very complex and, therefore, subject to many different types of failure. The ac windings are subject to the same types of failures as transformers, and protection must be provided to remove faulted machines to preserve system integrity as well as to limit the damage to the equipment. Machines are also subject to faults in the magnetic lamination insulation that can lead to heating and eventual significant damage. Machines are also subject to the problems associated with overheating and the loss of life of insulation that accumulates due to overheating. Therefore, the cooling media that control the heating of current-carrying windings and the magnetic steel circuits are possible causes of machine failure that must be considered.

Rotating machines, like transformers, are also subject to failure due to overfluxing. This can be caused by overvoltage or underfrequency, exactly as in transformers. Synchronous machines are subject to overvoltage due to self excitation, which can be imposed on the machine due to purely external loading conditions.

Rotating machines are also complicated because of the fact that important parts of the machine are in mechanical motion. This gives rise to problems related to vibration, bearing failure, mechanical resonance, and other purely mechanical failure modes that can, nonetheless, cause the machine to be removed from service. Moreover, mechanical parts are subject to aging and wear that require periodic monitoring in order to preserve the expected life of the machine and maintain it in good working condition.

Protective devices are also placed on the electrical side of a turbine-generator unit to provide purely mechanical protection. Examples include low or high frequency protection for the turbine blading and overspeed protection due to sudden loss of load. In some cases, the basic disturbance is due to some event or condition on the electric system that can readily be detected by monitoring electrical quantities, which can then be used to remove the machine before any mechanical damage can result.

This chapter will concentrate on synchronous generator protection. This recognizes that the majority of generators are of the synchronous type, although induction generators are also in use, usually in smaller sizes. Protection of induction generators is similar to the protection of induction motors, which is discussed in the next chapter.

18.2 TYPES OF GENERATOR PROTECTION

There are many different types of faults that synchronous generators may experience and, therefore, many different types of protection. All generators will not have the same level of protection, however. As a general rule, the larger, more expensive machines will have the greatest number of different protective systems simply because serious damage of these units is very costly, both in terms of the repair and also the cost due to the unavailability of the unit. Nearly all generators will have basic protection against stator short circuits, but not all generators will have all of the other protective measures described in this chapter. For example, there are some differences in the protection provided to machines located in attended stations and those in unattended locations. Generally, those machines in unattended locations must have protective devices to shut down the unit for precautionary conditions where an operator might be able to keep the unit on line by changing its operating conditions.

Table 18.1 provides an overview of synchronous generator protections, as discussed in six different references.[1] Although not an exhaustive list, it is clear from this tabulation that there are a relatively large number of conditions for which generators are provided special protection. None of the cited references covers all of the items, which reflects the generator owner preference for the importance of the different subjects. There are still other protective systems that are used occasionally but are not shown at all in Table 18.1, and these will be discussed briefly at the end of this section.

All synchronous generators must have stator protection, and most units have several types of protection, each of which guards against different types of problems. There is little disagreement regarding the need for basic stator short-circuit protection, and most engineers agree that this is best done using differential protection. Beyond this basic protection the opinions vary as to the need for special relaying. Some argue that too many protective systems increase the preventive maintenance and testing time, resulting in a reduction of protective system reliability, and therefore tend to favor simpler systems and frequent testing. Others prefer more elaborate protective systems on the theory that generator outages are very costly and even moderate limitations on the amount of damage will more than pay for the additional relaying and testing. The final decision rests on the benefit derived from the added complexity in terms of improved generator availability. This is not easy to measure accurately since major outages are rare events.

This chapter does not discuss the general physical structure of synchronous machines. The reader is referred to one of the many excellent books on electric machinery for this type of information.

[1] See the list of references at the end of the chapter

TABLE 18.1 References on Generator Protection Topics

Problem Location	Type of Problem	Reference Number [1]	[2]	[3]	[4]	[5]	[6]
Stator winding	Phase fault	T	T	T	T	T	T
	Ground fault	T	T	T	T	T	T
	Turn-to-turn fault	N	N	D	D	—	T
	Open circuit	D	N	—	—	—	—
	Overheating	A	U	A	A	D	A
	Overvoltage	R	T	R	R	R	T
	Unbalanced current	T	T	T	T	R	T
	Generator fault backup	—	—	—	T	—	T
	External fault backup	T	T	—	—	T	T
	Generator VT fuse blown	D	T	—	—	—	—
	Single phasing	D	—	—	—	—	—
Field winding	Shorted winding	N	D	D	D	—	T
	Grounded winding	D	T	T	T	D	T
	Open field winding	D	—	D	D	—	—
	Overheating	D	—	—	A	—	A
Other systems	Overspeed	—	D, O	T, O	T, O	D, O	T
	Vibration	D	T, A	—	D	—	T
	Motoring	D	D	D	D	D	T
	Bearing overheating	D	U, A	U, A	D	—	—
	Overheated coolant	D	D	—	D	—	—
	Fire	D	—	—	—	—	—
	Startup protection	—	—	—	—	D	T
	Loss of excitation	T	T	T	D	D	—
	Loss of synchronism	D	O	O	D	—	T
	Voltage regulator failure	—	—	D	—	—	T
	Plant auxiliaries	—	—	D	—	—	T

Note: T = Generator trip recommended; A = Alarm only recommended; D = Described; R = Protection recommended in some cases; N = Not usually provided; U = Needed for unattended stations; O = Provided by other protective systems

18.3 STATOR PROTECTION

There are many different types of stator protection. First, we examine the protection against short circuits in the stator, and then consider other types of stator problems that might be considered for special protective devices.

Stator faults result from insulation breakdown that causes an arc to develop, either from phase to phase or from the phase conductor to the grounded magnetic steel laminations of the stator. The cause of the insulation breakdown may be due to overvoltage, overheating, or mechanical damage of the winding insulation due to faults. The overvoltage that may cause an insulation failure might be due to lightning or switching surges, which are usually protected against by surge protective devices. Overheating may be due to prolonged unbalanced loading or to the loss of cooling, either of which may cause insulation deterioration over a period of time.

18.3.1 Phase Fault Protection

Phase faults in generators are rare, but they can occur and must be protected against. Phase faults can develop in the winding end turns, where all three phase windings are in close proximity, or in slots if there are two coils in the same slot. Phase faults often change to ground faults, but they must be detected in either event.

The standard method of protection against phase faults is the differential relay, and usually the percentage differential type of relay. Most of the manufacturers recommend this type of protection for all units larger than about 1 MVA. The CT connections for a wye-connected generator with a unit step-up transformer are shown in Figure 18.1. If there is no unit step-up transformer, the transformer and transformer relay are simply deleted from the figure. Note that the neutral current transformers are used by both the generator and the step up transformer relays.

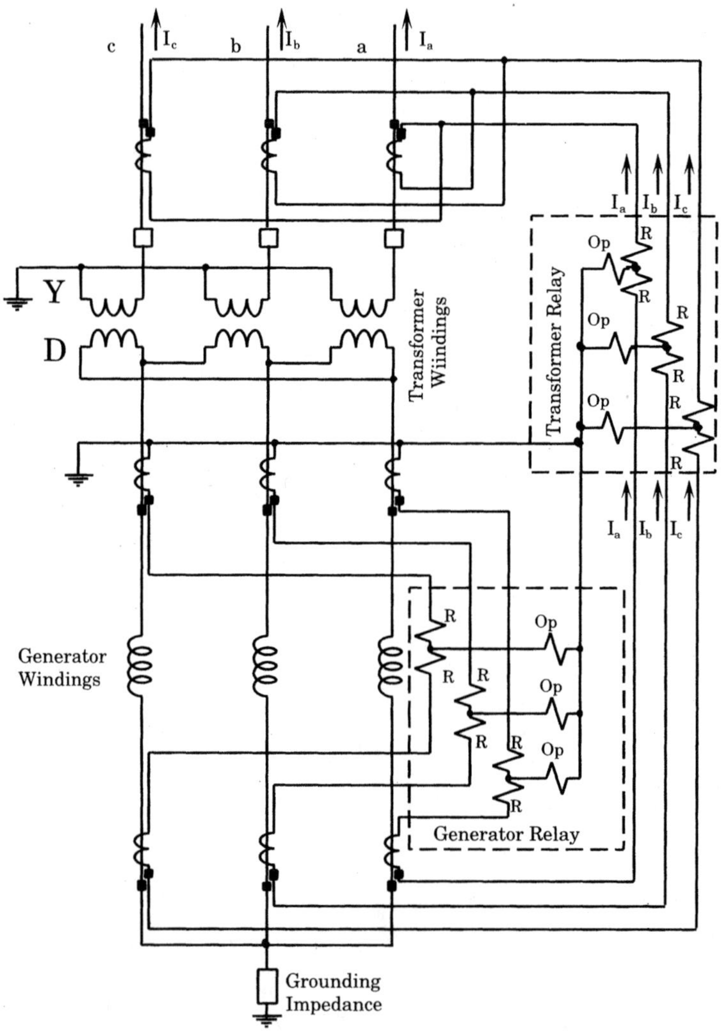

Figure 18.1 Percentage differential relay connections for a unit generator and transformer with phase sequence a-b-c [2].

The generator CT's are often shared by relaying and switchboard metering so these CT's are usually rated at about 25% more than rated full load current so that the meter deflections will be less than full scale. This will not usually be possible if the transformer connects to a double breaker arrangement, such as that found in a ring bus, breaker-and-a-half, or double bus-double breaker arrangement. In these cases, the CT's may have to be rated according to the normal current through the two breakers, which may be greater than the generator rated current.

The generator CT's are subject to CT errors due to manufacturing tolerances and differences in secondary burden. It is recommended that identical units be used in the generator relaying, if possible, to minimize CT ratio errors.

The arrangement shown in Figure 18.1 is common for large generators, where the generator and generator step-up (GSU) transformer operate as a unit and are protected as a unit. The CT's used for the transformer protection should be chosen using the methods discussed in Chapter 17. The question arises regarding the effect of magnetizing inrush current on this transformer protection and the need for harmonic or other restraint. In normal operation, the transformer is brought up to full voltage gradually, as the generator is brought from turning gear to full speed. Therefore, at startup, the generator step-up transformer is never switched on suddenly.

There is the possibility of magnetizing inrush in the GSU transformer following a system fault near the generator that is cleared by opening and then reclosing line breakers. This is analyzed by considering the three-terminal equivalent of the transformer, as shown in Figure 18.2.

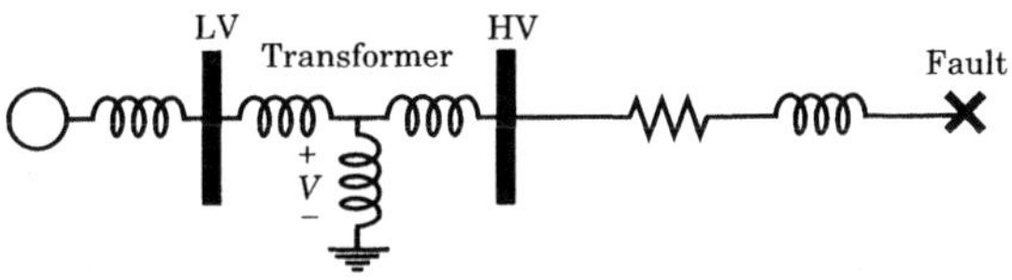

Figure 18.2 Equivalent of the generator step-up transformer with a system fault.

The voltage V across the magnetizing branch of the transformer during the fault is reduced only by 20% or so during the fault, so the switching that occurs does not create a large transformer transient. Moreover, the flux at the moment of switching is in the right direction, not in the opposing direction as we usually assume for worst-case system transformer switching. Another important factor is that the fault current interruption is achieved at the current zero, at which time the voltage is maximum and the flux is zero, thereby giving no exciting current transient.

If there are large system transformers connected to the high-voltage bus at the generator switching station, switching of these transformers may be of concern. Such switching should not affect the generator differential protection, but the dc offset associated with the transient current may saturate the generator CT's. This reinforces the concept that the CT's be of identical design so they will saturate together and uniformly. We conclude that no harmonic restraint is required in the GSU transformers, but some engineers still provide this feature as a safeguard against unnecessary tripping of the generator.

18.3.2 Ground Fault Protection

Most of the generator stator winding faults are phase-to-ground faults. This is true because the windings are always in close contact with steel slots that are at ground potential and, in some designs, are not close to other phase conductors except for the end turns. Phase-to-phase faults are less severe, in a sense, since the damage due to these faults can sometimes be repaired by carefully taping the damaged insulation. This is not true for ground faults, which often melt the steel laminations and may cause so much steel damage that the core laminations will have to be rebuilt, resulting in a very long outage.

Note that the generator in Figure 18.1 is grounded through an impedance, which is installed to limit the ground fault current. It is the usual practice to ground the neutral in some manner, so the generator and transformer low-voltage winding are not just floating above ground. In many cases, resistance grounding is achieved using a distribution transformer, as shown in Figure 18.3. As long as the neutral is to be grounded, adding impedance (usually resistance) has the advantage of limiting the fault current of phase-to-ground faults, and thereby limiting the damage done by these faults. If the neutral impedance is high enough, the ground fault current may be limited to rated generator current, or even less. There is no general agreement as to the best value for this current-limiting impedance. If the impedance is too high, the fault current is low and the phase relaying lacks sensitivity. For example, a current lower than rated current makes it difficult for the differential relays to operate for ground faults, and suggests that sensitive ground fault relays be used in addition to differential protection.

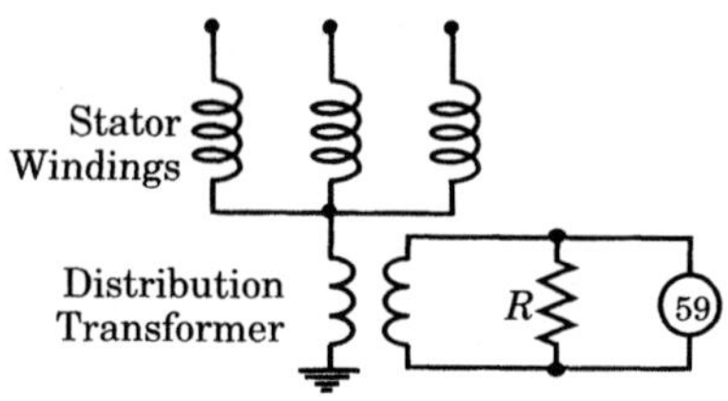

Figure 18.3 Grounding a generator neutral through a distribution transformer.

If the neutral resistance is too high, there is also the risk of creating a transient ferro-resonance circuit with the capacitance to ground of the stator windings and connected lines [3]. The usual restriction is to compute a maximum neutral resistance based on the total capacitive susceptance of the three stator phases. If we let C be the capacitance of each phase winding, the leads, and the capacitive surge arrester, then

$$X_C = \frac{1}{3\omega C} \quad \Omega \tag{18.1}$$

and

$$R \leq \frac{1}{3\omega C} \quad \Omega \tag{18.2}$$

Since C is small, perhaps a few tenths of a microfarad, the maximum resistance will typically be a few thousand ohms. This resistance is often achieved by placing a small resistance in the secondary of a distribution transformer, which is reflected to its primary side by the square of the turns ratio. The power rating of the resistor and the rating of the transformer need to be checked to make sure both are adequate [4].

For low values of grounding resistance, the fault current will be high and the damage due to ground faults may be great. As the value of the grounding resistance decreases, however, the sensitivity of the ground relays decreases, since less of the available voltage will appear across the zero-sequence network, as shown in Figure 18.4. The ground relays respond to the voltage across the grounding resistance, and this voltage must be of a reasonable value for good relay sensitivity. Some engineers consider this to be an advantage, in a way, since the ground relays are less likely to operate for faults on the low voltage side of the GSU transformer. Others

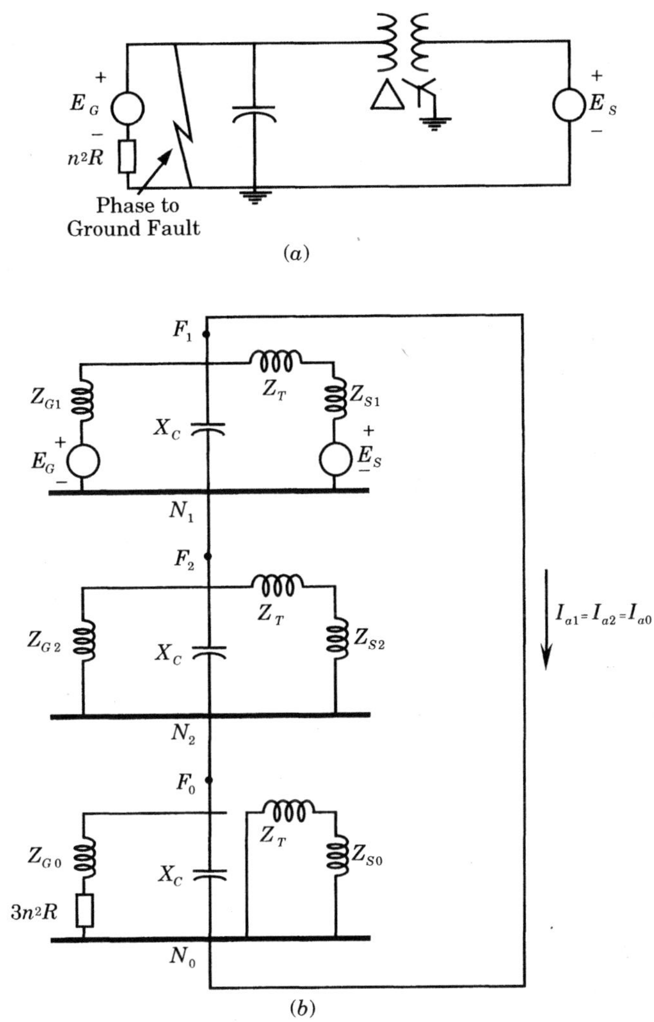

Figure 18.4 Sequence network connections for a stator ground fault. (*a*) One line diagram. (*b*) Sequence network connections.

simply set the goal of limiting the ground fault current to 15 amperes [2]. If this criterion is used, then the value of the grounding resistance in the secondary of the distribution transformer is given as

$$R = \frac{V_L}{15\sqrt{3}n^2} \quad \Omega \tag{18.3}$$

where V_L is the generator rated line-to-line voltage in volts and n is the distribution transformer turns ratio. It is also important to check the volt-ampere rating of the distribution transformer and of the resistor. These ratings depend on the value of the resistance and the action to be taken by the protective device. If the protection is only to sound an alarm, the ratings should be the continuous rating of both the transformer and the resistor. If the generator is to be tripped, then a short time rating of both can be used [2]. However, the transformer should not be too small, since very small units have higher leakage inductance.

The overvoltage relay (device 59 in Figure 18.3) used for ground faults should be sensitive to fundamental frequency currents and insensitive to third harmonic currents and voltages, which may be 15% or so of the fundamental due to unbalanced loading on the generator. Some relays that are used for this type of ground protection have harmonic restraint, which filters out the third harmonic and allows the relay to respond only to the fundamental [5].

EXAMPLE 18.1

The computation of the grounding of a generator by a resistor will illustrate some of the considerations involved and will provide a rough idea as to the magnitude of system quantities in a typical situation [4]. Let the generator and GSU transformer have the following physical parameters:

Generator	Rated volt-amperes	100 MVA
	Rated voltage	13.8 kV
	Maximum overvoltage	18.0 kV
	Winding capacitance	0.22 μF/phase
Transformer	Rated voltage	13.8Δ – 230 Y kV
	Primary capacitance	0.006 μF/phase
Connections	Capacitance	0.001 μF/phase

Determine the size and rating of the grounding resistance and the ratings of the distribution transformer.

Solution

First, we compute the total capacitance per phase, including the generator, the transformer primary, and the connections.

Generator	0.220
Connections	0.001
Transformer	0.006
Total	0.227 μF/phase

The capacitance of the connections depends very much on whether cable is used between the generator terminals and the transformer, since cable has high capacitance per unit length. The total capacitance is not greatly affected, however [4].

The residual capacitive reactance for the three phases is computed as

$$X_C = \frac{1}{3\omega C} = \frac{10^6}{3(120\pi)(0.227)} = 3895.13 \quad \Omega$$

If the total grounding impedance is made equal to the above reactance, then the total resistive current for a fault at the generator terminals will be

$$I_R = \frac{13800}{\sqrt{3}(3895.13)} = 2.045 \quad \text{A}$$

and, since the capacitive current is equal to this amount, the total fault current is computed to be

$$I_F = I_R + jI_C = 2.045 + j2.045 = 2.892\angle 45^\circ \quad \text{A}$$

The grounding transformer primary knee-point voltage should be not less than 1.3×13.8 or 18 kV. During a fault at the terminals, the normal line-to-neutral voltage is 7.97 kV, which may be increased by fast field forcing. Suppose the field response increases this line-to-neutral voltage by 30%, or to 10.36 kV. The distribution transformer should have at least this rated voltage. Let us assume that the transformer has a standard 12 kV rated voltage.

Under field forcing, the current will also be increased by this same 30% or

$$I_{F\,\max} = 1.3(2.892) = 3.76 \quad \text{A}$$

Then, the maximum transformer continuous loading is computed as

$$S = 3.761(10.36) = 38.954 \quad \text{kVA}$$

A 30-second maximum duty is more than adequate for the short time the fault current will flow, and about *six times* the transformer rating can flow for this shorter period without transformer damage. The transformer rating, however, should be based on the maximum output current and the maximum voltage that the transformer can produce. Using the knee point voltage for this maximum, we compute

$$S = \frac{(3.761)(18.0)}{6.0} = 11.28 \quad \text{kVA}$$

This value is rounded up to the nearest standard rating, or 15 kVA.

The transformer may sustain a continuous loading due to third harmonic currents flowing through the neutral, the machine, and the capacitance. The capacitive reactance is much smaller for third harmonic currents. Another source of continuous loading is due to currents below the ground relay threshold. Suppose that these factors combine to cause an effective current of 25% rated (a very pessimistic assumption), then this gives a loading of

$$P = \frac{(0.25 \times 2.892)^2(12000)}{\sqrt{3}} = 1.25 \quad \text{kW}$$

This value is very small compared to the transformer rating.

There is also a choice of transformer rated secondary voltage, which can be chosen to give a suitable secondary current. The usual distribution transformer has a rating of 240 volts. Using this as a possibility, we compute the size of the resistor that would be required in the secondary circuit to give 3895 ohms in the primary. The total secondary resistance should be

$$R_{\text{tot}} = (3895.13)\left(\frac{240}{12000}\right)^2 = 1.558 \quad \Omega$$

This total is made up of the winding resistance and the added external resistance. The winding resistance can be computed from the total copper loss of the 15 kVA, 12 kV transformer, which is given as 2.3% or 345 W [7]. Then, since the rated secondary current is 15000/240 or 62.5 A, we have

$$R_{\text{wdg}} = \frac{345}{(62.5)^2} = 0.088 \quad \Omega$$

and the added external resistance is the difference, or

$$R_{\text{ext}} = 1.558 - 0.088 \cong 1.50 \quad \Omega$$

This resistance should be able to carry a current of

$$I_R = \frac{(1.3)(2.893)(12000)}{240} = 188 \quad \text{A for 30s}$$

This gives a resistor short time rating of

$$P = \frac{(188)^2(1.50)}{1000} = 53 \quad \text{kW}$$

The above procedure determines both the resistor size and its rating. ■

The method described above for ground fault protection has been widely used for many years. However, this scheme has two major disadvantages. First, it will not detect faults near the generator neutral and, second, it is not self-monitoring. An open or short circuit in the relay or its current transformer may not be detected until a ground fault occurs. Other schemes have been suggested that are to be installed in addition to the basic scheme of Figure 18.3 and which are completely independent from that scheme. If properly designed, the supplementary scheme will not have the same disadvantages, and will have the capability of continuously monitoring the generator grounding system [8]. One supplemental scheme has been proposed that injects low-frequency 15 hertz currents in the neutral and then compares the injected current with that leaving the stator, with any significant difference indicating a ground somewhere on the stator winding. Other schemes monitor the third harmonic current flowing in the neutral and compares the observed current against the stator terminal current. Reference 8 provides a detailed discussion of three different schemes that have been tested.

18.3.2.1 Grounding Methods. Although the distribution transformer method of generator grounding is quite common, it is certainly not the only method in use. The general practices used for the grounding of generators throughout the world are summarized in [6] and [9], with the most common connections shown in Figure 18.5. The connection shown in Figure 18.5(*a*) grounds the machine neutral directly, in which case it is called "high resistance" grounding. In this case the resistor is typically sized to limit ground fault current to 25 amperes or less [6]. In some cases a voltage transformer is connected across the grounding resistor for metering and relaying, with the voltage rating selected exactly as in the distribution transformer grounding case.

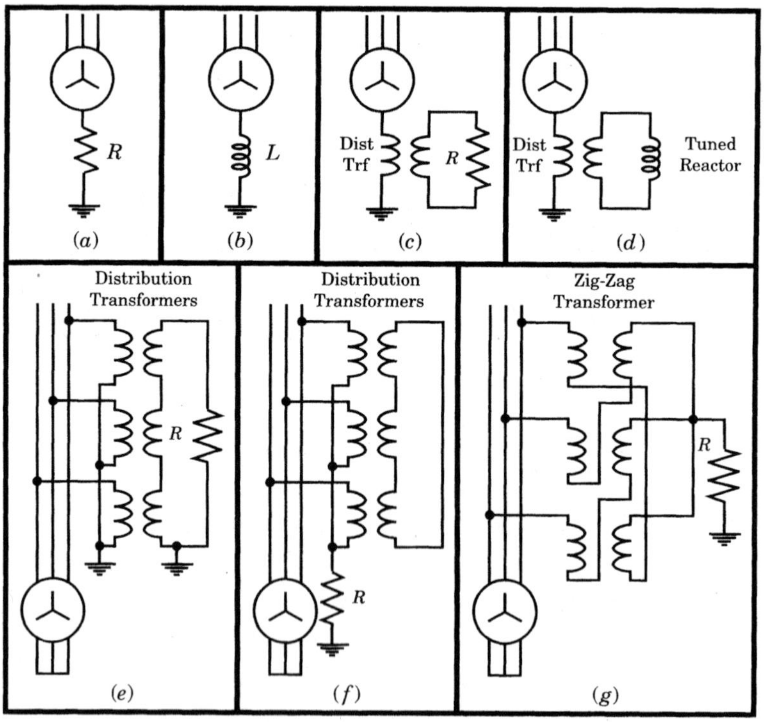

Figure 18.5 Methods used for generator grounding.

Another variation of the direct grounding through a resistor, as shown in Figure 18.5(*a*), is referred to as "low resistance" grounding, which permits much higher fault currents. Some utilities permit fault currents as high as 1500 amperes or even more.

Figure 18.5(b) shows the neutral grounded directly through an inductor, which is usually one with low inductive reactance. With this type of grounding, the allowable fault currents are much higher than in the resistance grounding case, with typical limiting values of 25–100% of the generator three-phase fault current.

Figure 18.5(c) is the common distribution transformer grounding discussed above. The transformer primary voltage rating is selected with a rating equal to or greater than the generator voltage rating. Secondary voltage ratings are usually 120 volts or 240 volts. The transformer voltage rating should be high enough that the transformer does not saturate on phase to ground faults when the machine is operating at a terminal voltage of 1.05 per unit. The resistance is selected using the technique outlined above, which is considered to be "high resistance" grounding and is often selected to limit a single line-to-ground fault to 25 primary amperes or less [6].

Figure 18.5(d) shows the neutral grounded through a distribution transformer, with an inductive reactance in the secondary. The ohmic value of this reactor is chosen such that its value viewed from the primary side is equal to one-third the zero-sequence capacitive reactance of the generator, leads, and GSU primary. This type of grounding limits the one-line-to-ground fault current to no more than one ampere, a condition in which an arc will not be maintained. This method is not applicable if the circuit capacitance changes for different system conditions.

Figure 18.5(e) shows one method of adding ground fault resistance to a generator that has no neutral ground. Note that the distribution transformers are connected in wye on the primary side and delta on the secondary side, allowing zero sequence current to circulate through the resistor. The resistor is sized with a high ohmic value and rated in a similar manner to that used for neutral grounding through a distribution transformer.

Figure 18.5(f) shows still another way of accomplishing the same result as Figure 18.5(e), except this is usually a "medium resistance" method [6]. In this case the zero-sequence current flows through the grounding resistor.

Figure 18.5(g) is also considered a "medium resistance" grounding method and differs from the previous method in the transformer connection. The zig-zag transformer connection is a common grounding method, with zero sequence current flowing to ground through the resistor.

A thorough review of generator grounding methods is provided in [9], which is highly recommended for further study.

18.3.2.2 Ground Fault Current Magnitude. Stator winding ground faults near the neutral of the generator do not produce very large fault currents and may be difficult to detect, even with sensitive ground relays. Let us assume that a fraction h of the winding closest to the neutral is unprotected. For a fault at this critical point, the fault current is computed as follows [3]:

$$I_F = \frac{hV_a}{R_n} = \frac{hV_L}{\sqrt{3}\,R_n} \quad \text{A} \tag{18.4}$$

where V_L = rated line-to-line voltage in (V)
R_n = neutral resistance in (Ω)

Now define the ground relay pickup current for this critical fault location as follows:

$$I_{PU} = k \cdot CTR \quad \text{A} \tag{18.5}$$

where CTR = neutral CT ratio

k = fraction of rated secondary current that gives pickup

Equating (18.4) and (18.5) we compute the unprotected fraction of the winding h to be

$$h = \frac{\sqrt{3}\,kI_{PR}R_n}{V_L} \tag{18.6}$$

EXAMPLE 18.2

A 13,800 kV generator has a ground relay pickup current of one ampere, a neutral resistor of 300 ohms, and a neutral CT rated 200/5.

Solution

For this combination, we compute the following. First, the fraction k of primary rated current that corresponds to the critical point is computed as follows. Therefore

$$I_{pu} = \frac{1}{5} \times \frac{200}{5} = 8 \text{ A}$$

and

$$k = \frac{1}{5} = 0.2$$

so that

$$h = \frac{\sqrt{3}(0.2)(40)(300)}{13800} = 0.3012$$

Therefore, about 30% of the winding will be unprotected for the combination of values chosen. ■

The result computed in the example is a bit pessimistic. Generator ground fault protection using overcurrent relays with carefully selected pickup and time dial setting have been estimated to protect as much as 96% of the total winding [7].

18.3.3 Turn-to-Turn Fault Protection

Differential stator winding protection will not detect turn-to-turn faults until they become shorted to ground. Turn-to-turn faults are rare, but they can have high currents and produce significant core damage. The greatest danger is fire, which is aided by forced air cooling that supplies ample amounts of oxygen to the fire. Hydrogen cooled machines present less of a fire hazard since hydrogen will not support combustion. Since the probability of turn-to-turn faults is low, protection is usually not provided. There is always the back-up protection provided by ground fault relaying, since turn-to-turn faults will eventually develop into ground faults.

The only type of turn-to-turn differential fault protection that is feasible is for those generators with multiturn coils in their windings, as shown in Figure 18.6. This is common for hydro generators but not for large steam turbine generators, which usually have only one turn windings. This type of protection is called "split-phase" relaying in the United States and "transverse differential" protection in Europe [2], [4]. It uses percentage differential relays, similar to those used for other stator winding relaying.

18.3.4 Stator Open Circuit Protection

Stator open circuit protection is usually not provided, since no permanent damage is likely although it is possible that damage will accumulate due to a high resistance connection. An open circuit in one of the stator windings will cause single phasing of the generator and

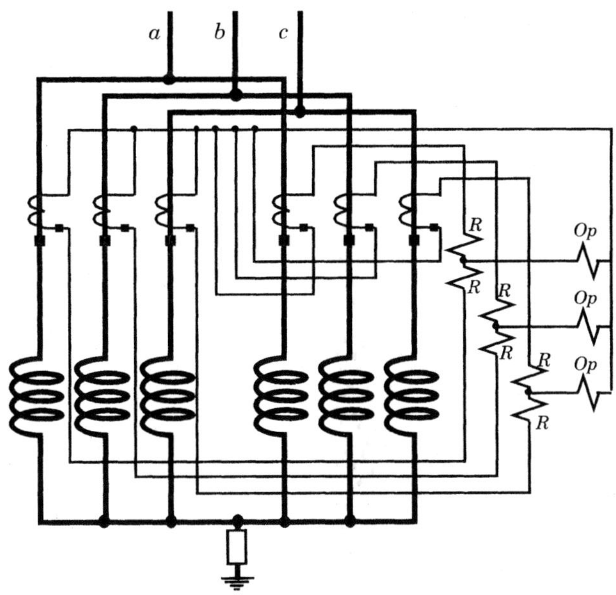

Figure 18.6 Split-phase protection for parallel winding generators.

high negative-sequence currents. This will cause an alarm to alert the operator of the condition. Open circuits are highly unlikely in a generator, and any possible damage will occur slowly; hence there is no need for prompt tripping of the unit.

18.3.5 Overheating Protection

Overheating of a synchronous generator may occur due to one of the following causes:

1. Overload
2. Failure of the ventilation or hydrogen cooling system
3. Shorted laminations in the stator iron
4. Core bolt insulation failures in the stator iron

Excessive overload is not likely since the prime mover rating is usually not much greater than the generator rating. There is the possibility of overload due to high active power load coupled with high excitation. If the power factor is below rating, this will give an alarm for high excitation. Failure of the cooling system is also likely to be detected by operator alarms. The other failure, involving core failures and heating will develop slowly and must be detected by temperature measurements of some kind.

Temperature detection is often accomplished using embedded thermocouples in the stator winding slots, placing the thermocouples throughout the windings in several locations. Another measurement technique is to record the input and output cooling medium to note any marked changes in the readings. Smaller generators are often provided with "replica" type temperature estimating devices that use stator current in a heat storage enclosure to estimate actual machine temperature.

All of these devices are used to alarm the operator of possible serious problems. At unattended stations, the output of the temperature indicator may be used to shut down the unit.

18.3.6 Overvoltage Protection

One type of overvoltage in a generator is that due to transient surges caused by lightning or switching surges. These transients are protected by surge protective devices that are designed for this purpose.

Power frequency overvoltages are possible if the generator controls are defective or have inadequate transient response. A defective voltage regulator, for example, can cause the exciter to ramp to its ceiling voltage. If the voltage control is performed manually, a sudden change in load will result in an increase in voltage. The loss of load may cause high voltage on units that are remotely located in the system. This is particularly true of remote hydro units since it may not be possible for the governor to close the wicket gates of large hydro units fast enough to prevent an overvoltage due to loss of load. The result is overspeed, which is associated with overvoltage. This type of overvoltage is not likely on a steam unit, since they have tighter control against overspeed and are designed to limit overspeed to low values.

Steam turbine generator units are not always provided with overspeed or overvoltage protection, but this type of protection is often recommended for hydro units or combustion turbine units. In many cases, the desired protection is provided by the voltage regulating equipment. If not, it can be provided by overvoltage relays or overfrequency relays. Overvoltage relays should have a time delay and a pickup of about 110% of rated voltage. An instantaneous unit is sometimes provided with a pickup of 130%–150% of rated voltage. Some types are compensated for varying frequency and should be supplied from a voltage transformer that is different from that supplying the voltage regulator. Some of these relays are used to simply insert a large resistance in the exciter field circuit.

18.3.7 Unbalanced Current Protection

Unbalanced loading of a synchronous generator causes negative-sequence currents to flow in the stator windings. These currents are reflected to the rotor as double-frequency currents in the rotor iron, which can cause severe rotor heating and may soften or weaken slot edges and retaining rings. These harmful conditions may be caused by

1. One open phase of a line
2. One open pole of a circuit breaker
3. A close-in unbalanced fault that is not promptly cleared
4. A stator winding fault

The first three of these causes are for failures external to the machine. Unbalanced faults near a generator cause more severe heating in synchronous machines than balanced faults. If the fault persists for very long, the rotor metal will melt. Since the damage is done mostly to the rotor, the withstand criteria is usually stated in terms of the rotor withstand time, which is determined from the expression

$$K = I_2^2 t \tag{18.7}$$

where K is a constant. The relative magnitude of the parameter K is illustrated for several different operating conditions in Figure 18.7, where the identification of the five operating conditions are given in Table 18.2.

Clearly, the value of K can vary over a rather wide range, but for most practical systems a voltage regulator will be operative and the connecting system will not be infinite. This suggests values of K closer to 45, and being lower if the voltage regulator is inoperative. ANSI standards

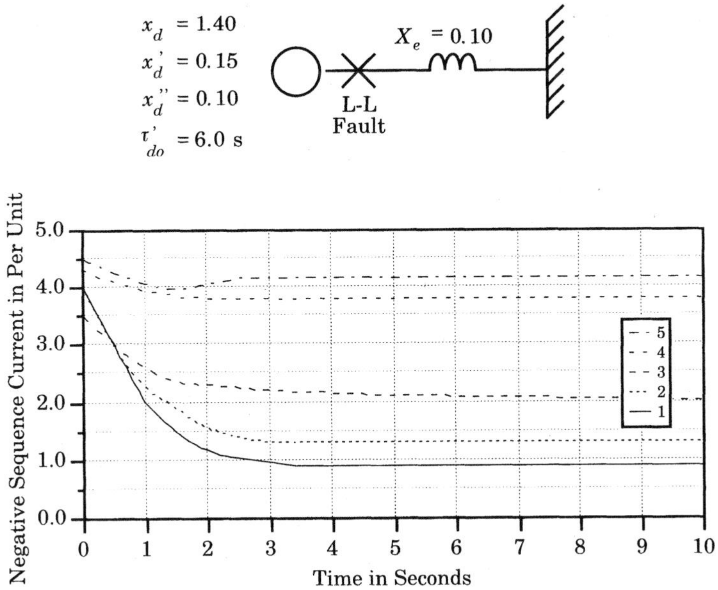

Figure 18.7 Relative magnitude of negative sequence currents for line-to-line faults on a typical machine under different conditions [10].

TABLE 18.2 Identification of Operating Conditions for the Curves of Figure 18.7

Curve	Operating Condition	I_2^2 at 10 s
1	No load without voltage regulator	18
2	Full load without voltage regulator	21
3	Full load with voltage regulator	45
4	Full load with voltage regulator, infinite bus	152
5	No load with voltage regulator, infinite bus	168

set the permissible magnitude of K for "indirectly cooled" steam turbine generators at 30, and the standard for hydro machines or engine-driven generators at 40. The standard limit for "directly cooled" machines up to 800 MVA rating is set at 10, and some very large machines have a limit of only 5. The standards recommend early inspection for machines subjected to faults above the recommended limit and up to twice that limit. Above twice the standard limit, serious damage can be expected.

Figure 18.8 shows typical characteristics of negative sequence relays that are available commercially [10]. Two values of K are illustrated. The upper set of curves are plotted for $K = 90$ and just below the constant K line is the relay characteristic for that K factor. Note that the relay characteristic parallels the K factor line for all values above 1.0 per unit. The lower curves in Figure 18.8 are for $K = 30$ and a similar relay characteristic. The relays can be selected to protect a generator having K values of anywhere from 5 to 90 for negative sequence currents values from 1 to 5 per unit. Large machines usually require the protection of a lower value of K. This K is due to the trend toward an increase in machine ratings while the physical size has remained about the same. This means that the larger, newer units have the capability of overheating more quickly, since larger currents are induced in the rotors of these units.

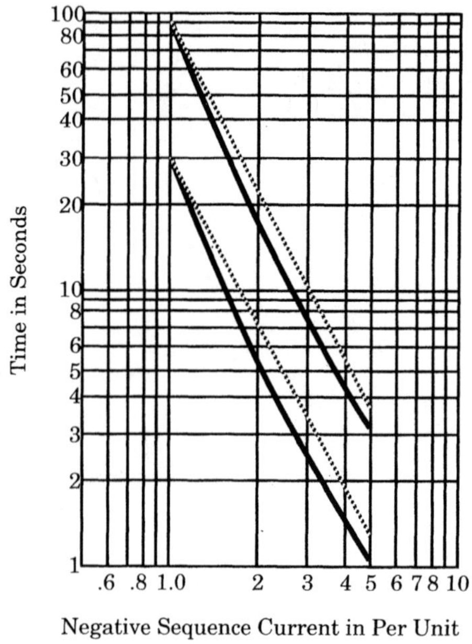

Figure 18.8 Time current characteristic for negative sequence relays.

18.3.8 Backup Protection

There are two types of backup protection that might be applied to a generator: backup of relays protecting the generator protection zone and backup of relays protecting external zones. Some types of backup protection may be graded to coordinate with both internal and external protective devices.

The negative-sequence relay discussed in the previous section might be considered a form of backup protection, since most faults should be cleared by the stator differential protection with the negative sequence relay acting as backup.

Balanced faults that are not cleared promptly can also cause considerable damage to a generator and backup protection is warranted. One type of such protection is to provide a distance relay that is supplied with current from a CT in the generator neutral and voltage from the generator terminals. Such a relay can recognize balanced faults both internal and external to the generator [10]. The connection makes the relay directional from the neutral, but gives it reach in both directions from the voltage transformer location, thereby sensing both generator and GSU transformer faults. This type of relay can also be set to reach through the transformer, making it operable for phase faults on the high voltage side of the transformer. Coordination is achieved through time delay.

18.4 ROTOR PROTECTION

There are several different types of rotor protection, each type guarding the rotor from a particular type of fault. From this viewpoint, the protection against unbalanced loading, using negative sequence relays, can be considered a type of rotor protection since the effect of negative sequence currents is likely to result in rotor damage. In this section, we consider some other types of rotor problems and the protective systems used to avoid or limit rotor damage.

18.4.1 Shorted Field Winding Protection

Shorted turns in the generator field winding have the potential for distorting the field across the air gap, as illustrated in Figure 18.9. This is due to the unsymmetrical ampere turns of mmf in different parts of the field winding. If the air gap flux is badly distorted, there can be very distorted forces acting on the rotor, since the forces vary as the square of the flux density. Once there are unequal forces on opposite sides of the rotor, there is tendency for the rotor to warp. The unbalanced force can be very large, as much as 50 to 100 tons, tending to warp the rotor [4]. In some cases the rotor may be displaced enough to contact the stator iron core.

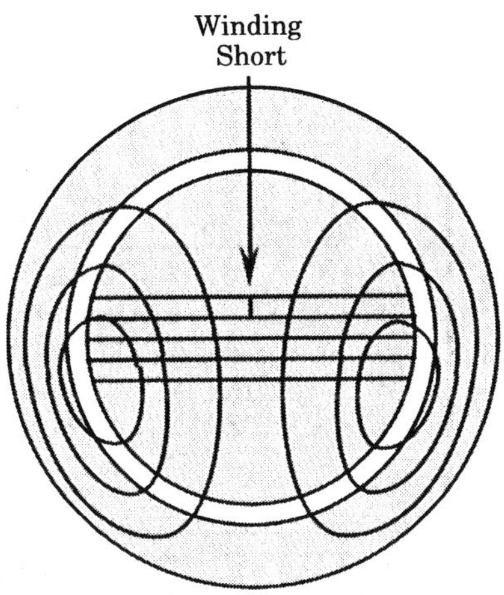

Figure 18.9 Field flux pattern with shorted field turns.

Another effect of the unbalanced forces on the rotor is severe vibration, which may cause damage to bearings. The machine can be spared from serious damage by vibration detectors, which can alarm the operator or trip the unit.

The mechanism that causes the shorted winding is often due to grounding of the winding at two different places.

18.4.2 Grounded Field Winding

The field winding of a synchronous machine is usually floating with respect to ground. A single ground fault, therefore, does not draw any fault current, although it does stress the insulation in portions of the winding. The real danger is a second ground, which can set up significant forces, as discussed above. Dual grounds can also draw very large currents and may cause extensive damage to the field conductor and rotor steel. The best way to prevent this from occurring is to detect the first ground, thereby preventing a more serious chain of events. The generator main and field breakers should be tripped on the occurrence of the first ground. The exception is in attended stations, where the unit trip may be delayed until a more convenient time to schedule the repair. During this period, any unusual vibration should immediately trip the unit. It is best to not trip the turbine following a rotor ground fault and tripping of the generator and field. The rotor has better cooling when running at rated speed, and the excess heat is carried away as the generator continues to run for several minutes at rated speed. Allowing the unit to coast down also introduces the danger of aggravated vibration due to natural modal frequencies of vibration, which may be worse due to the rotor distortion.

There are several methods of detecting a rotor circuit ground. The methods, shown in Figure 18.10, are summarized as

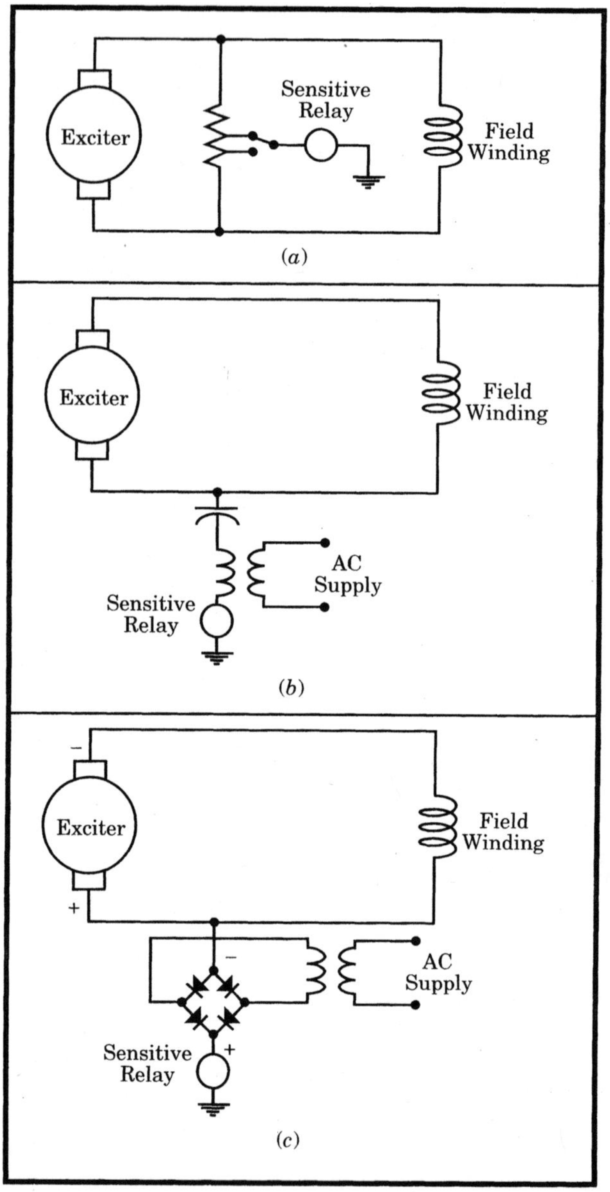

Figure 18.10 Methods of detecting field winding grounds. (*a*) Potentiometer method. (*b*) AC injection method. (*c*) DC injection method.

1. Potentiometer method
2. AC injection method
3. DC injection method

The potentiometer method measures the voltage to ground of a center tapped resistor, connected across the exciter output voltage. If some point in the winding becomes grounded, there will be a potential between that point and the point to which the voltage relay is connected.

The only problem is that, should a point very close to the center of the winding become grounded, the center tapped potentiometer would not detect it. To check that this has not happened, a manual switch is arranged to move the test point from the center to some other point along the resistor. The operator can check this periodically to ensure that the system is sound.

A better method is the ac injection method, which connects an ac voltage to the field winding through a capacitor. Should any point on the field winding become grounded, the circuit will be complete and the relay will trip. This system has no blind point. There is a disadvantage, however, in that some current will flow through the capacitance from the field winding to the rotor body, through the rotor body, the bearings, and to ground. This has the potential of causing erosion of the metal in the bearings.

A still better method is the dc injection method. The dc output of the transformer-rectifier unit is connected to bias the positive side of the field circuit to a negative voltage relative to ground. A ground at any point on the field winding will complete the circuit to the grounded side of the relay. The relay is a sensitive current relay in this case, but must not be so sensitive that it will trip due to normal insulation leakage current. The current through the relay is limited by the high impedance of the rectifier.

A special problem is presented in the case of the brushless exciter. This exciter is an alternator-rectifier exciter that is physically on the rotor and rotating at synchronous speed. The field for the alternator is stationary. The basis of the design is that there is no need for brushes between the field winding and the casing of the machine. In this type of exciter, there is usually no access to the field circuit; therefore, the previous types of ground detectors will not work. The only portion of the excitation system that is accessible to the outside is the field of the alternator exciter. Any severe fault on the field winding will require excess field current, hence excess excitation for the alternator rectifier [4]. The backup is a vibration monitor that can take action if a severe field winding fault occurs.

18.4.3 Open Field Winding

Field winding open circuits are rare, but prompt action is required when an open occurs because it will be accompanied with arcing that can do great damage to the rotor iron. An open circuit that does not involve ground will cause a sudden drop in field current that can be detected by a loss-of-field relay. This type of protection is discussed in the next section.

18.4.4 Overheating of the Field Winding

The temperature of the field winding can be monitored by an ohmmeter type of detector that measures the winding resistance of the field [4]. Such an instrument is often calibrated in temperature, rather than ohms, for a direct estimation of the winding temperature.

18.5 LOSS OF EXCITATION PROTECTION

In this section we consider the problems associated with the loss of excitation in a synchronous generator. First we discuss the conditions that surround the loss of excitation, and then we examine the types of devices that are used to protect both the system and the generator from this disturbance.

18.5.1 Induction Generator Operation

Under normal operation of a synchronous generator, the field created by the rotor windings locks in with revolving mmf of the stator windings and the rotor moves at synchronous speed, at least in the steady state. When excitation is lost, the rotor field suddenly loses its mmf and the rotor begins to move away from synchronism, having lost its strong magnetic coupling with the stator mmf. During this time, the governor is still set to deliver a given amount of power to the generator, so the generator will accelerate, inducing large slip frequency currents in the rotor in order to maintain the power output as an induction generator. Actually, the power requirement will be reduced as the slip increases due to the governor characteristic, the increase in stator current, and possibly the lowering of terminal voltage, but the total power is still quite large. Also, since the excitation has collapsed, the generator begins to absorb reactive power from the system in very large amounts, which depresses the voltage. This could lead to voltage collapse if the system is weak. The large increase in reactive power, at leading power factor, creates large stator currents that may reach two to four times rated current and the rotor begins to overheat.

The degree of rotor heating depends on several factors including the initial generator loading, the conditions causing the loss in excitation, and the way the generator is connected to the system. In a cylindrical rotor generator, the rotor currents will flow through the rotor body and through the field winding, if that winding has been shorted or is connected through a field discharge resistor, and will also flow through the rotor coil wedges. These currents oscillate at slip frequencies and with magnitudes that are proportional to the generated power. These large rotor currents will cause very high and possibly dangerous temperatures in the rotor in a very short time. In most cases, the time required for these currents to cause serious damage is only a few seconds if the generator has suffered a complete loss of excitation

For hydro machines, there are almost always amortisseur windings that are designed to carry slip currents, so these generators can continue to operate in the induction generator mode without damage. It is estimated that small solid rotor machines, of 50 MVA or so, can withstand induction generator operation for 3 to 5 minutes without damage but large machines of 500 MVA or more must be tripped in 20 seconds or less. This means that there is no time for an operator to evaluate the problem and determine the proper corrective action. The protective system must be fast in order to prevent severe damage.

There is also a potential for considerable stress to the power system during the transition described above. The generator that suffered the loss of excitation will draw its excitation from the power system as reactive power. This will depress the voltage at the generator terminals and possibly in the surrounding area if the system is relatively weak. This could cause problems in customer loads and could lead to voltage collapse.

Clearly, it is essential that large steam turbine generators that suffer a loss of excitation be tripped automatically and soon after the loss of excitation.

18.5.2 Loss of Field Protection

Loss of field protection is often provided using a mho type of distance relay. The reason for this choice of device requires explanation. Generator operation is confined to a region in the P-Q plane, as shown in Figure 18.11.

Four broken curves are illustrated in the figure, showing the generator capabilities for operation as various power factors and for different cooling conditions, represented for four different hydrogen pressures labeled H_1 through H_4. The rated power factor is assumed to be

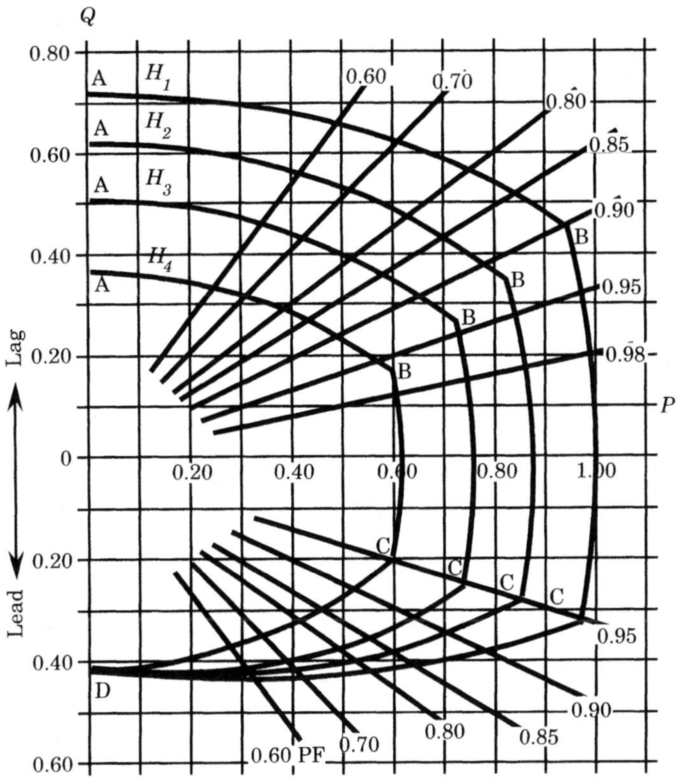

Figure 18.11 Synchronous generator capability curves in the P-Q plane.

0.90 lagging, and the outer curve at point B represents this rated condition. The arcs from B to C correspond to the generator rated volt-ampere conditions for power factors near unity and for the different hydrogen pressures. The limiting condition in this B-C region is the generator stator current.

As the power factor falls below the rated value in the lagging direction, the machine is generating reactive power in excess of its rating, requiring large field currents. This corresponds to the region labeled A-B, and the limiting condition in this region is overheating of the field due to high field current. In the leading power factor region, represented as the region labeled C-D, the limiting condition is overheating of the ends of the armature core and the end structure of the machine stator. This heating is due to eddy currents that are set up by armature reaction leakage flux, which rotates at synchronous speed. This leading power factor region may also be limited by instability, since the coupling between rotor and stator fields is weak when the field excitation is low. The limits here are sometimes shown as straight lines rather than the circular arcs shown in Figure 18.11.

Usually the generator operation will be in the range given in the figure as B-C, with the power factor in the neighborhood of unity. In this region we can write the equation of the limiting circles B-C as

$$P^2 + Q^2 = h_i^2 \tag{18.8}$$

where h = radius of the circular arc at any hydrogen pressure
i = hydrogen pressure index

The phasor current of the generator at a given voltage can be expressed in terms of the real and reactive power as

$$\boldsymbol{I} = \frac{P - jQ}{\boldsymbol{V}^*} \tag{18.9}$$

Dividing the phasor voltage $\boldsymbol{V}$ by the current gives the complex impedance seen looking toward the power system from the generator. This impedance is given by

$$\boldsymbol{Z} = \frac{V^2(P + jQ)}{h_i^2} \tag{18.10}$$

Now, at any power factor, we can relate the power factor to the radius of the limiting circle in the P-Q plane as follows.

$$\begin{aligned} F_P &= \cos\theta = \frac{P}{\sqrt{P^2 + Q^2}} = \frac{P}{h_i} \\ F_Q &= \sin\theta = \frac{Q}{\sqrt{P^2 + Q^2}} = \frac{Q}{h_i} \end{aligned} \tag{18.11}$$

or

$$\begin{aligned} P &= h_i \cos\theta \\ Q &= h_i \sin\theta \end{aligned} \tag{18.12}$$

Then the limiting circles in the P-Q plane can be converted to limiting circles in the R-X plane by substituting (18.12) into (18.10) with the result

$$\boldsymbol{Z} = \frac{V^2}{h_i}(\cos\theta + j\sin\theta) = \frac{V^2}{h_i}e^{j\theta} \tag{18.13}$$

This equation represents a family of concentric circles in the Z plane with the center at the origin, as shown in Figure 18.12, with each circle corresponding to a given hydrogen pressure. Note, however, that the circle in the Z plane corresponding to the highest hydrogen pressure is the circle with the smallest radius. Thus, higher generator outputs, represented in the P-Q plane by points far from the origin, are represented in the R-X plane by points closer to the origin. Faults are represented by points very close to the Z origin compared to these normal operating conditions, for which the currents are much lower than for faulted conditions. Still, the limiting conditions are circles in both planes.

Operation of the generator under normal conditions, say at lagging power factor, will be represented by points in the first quadrant of the Z plane that are quite far from the origin, but moving closer to the origin as the loading is increased. Operation in the fourth quadrant of the Z plane may be limited by instability.

If a mho relay is installed at the generator stator terminals, looking into the network, the relay would see a picture similar to that of Figure 18.12 under normal conditions. If excitation is lost, the operating point in the Z plane will begin to move as the slip increases. The relay impedance is computed as noted in Section 9.4, where we find that the impedance seen by the relay may be written in terms of the $ABCD$ parameters as

$$\boldsymbol{Z}_R = \frac{\boldsymbol{B}/\boldsymbol{D}}{1 - \dfrac{\boldsymbol{E}_U}{\boldsymbol{D}\boldsymbol{E}_S}} - \boldsymbol{Z}_S \tag{18.14}$$

We simplify this using the assumption that the source voltages both behind and in front of the

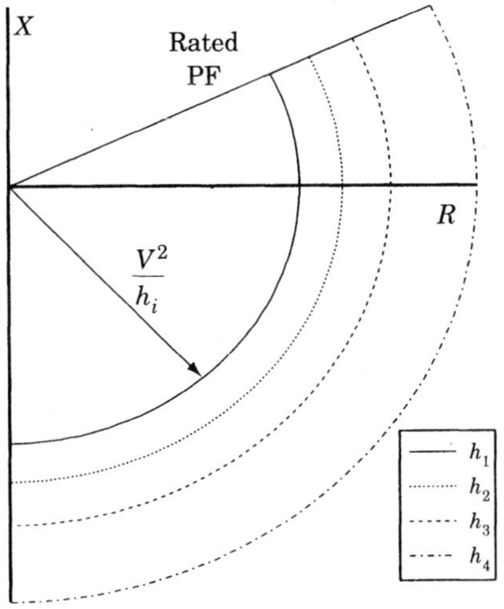

Figure 18.12 Stator current limited operating circles represented in the Z plane.

relay are approximately equal (see Section 9.5). Then we write

$$\boldsymbol{Z}_R = \frac{\boldsymbol{B}}{2\boldsymbol{D}}\left(1 - j\cot\frac{\theta}{2}\right) - \boldsymbol{Z}_S \tag{18.15}$$

where θ is the angle by which $\boldsymbol{E}_S$ leads $\boldsymbol{E}_U$ as described in Chapter 9. The parameter $\boldsymbol{B}$ is the total impedance of the system between the two equivalent source voltages, as shown in Figure 18.13.

In Figure 18.13, the relay is located at the origin. Behind the relay is the equivalent impedance of the generator, which varies from its synchronous impedance at steady state to the subtransient impedance at the onset of faults. In the forward direction, the relay sees the network impedance and the remote source impedance. The operating point in the Z plane depends on the phase angle between the two source impedances, which is given by θ. It is constructed by first plotting the line ab, with length $\boldsymbol{B}/\boldsymbol{D}$. According to (18.15), the locus lies on a line at right angles to the midpoint of the line ab, and at a distance that depends on the angle θ. Figure 18.13 shows two possible operating points, labeled "1" and "2," the first point corresponding to a large angle θ and the second point to a smaller angle. The operating point moves to the right along the bisector line and approaches infinity as the angle θ approaches zero, which corresponds to no load on the generator. When excitation is lost, the voltage behind the relay, $\boldsymbol{E}_S$, goes to zero and the relay sees an impedance that corresponds to point a in Figure 18.13. This point is in the third or fourth quadrant at a distance from the origin that varies with time. When the excitation voltage collapses, the flux dies away slowly and during this period the ratio $\boldsymbol{E}_S/\boldsymbol{E}_U$ decreases as the rotor angle increases. The location of the operating point in the Z plane moves from its starting point as shown in Figure 18.13 and moves to the left as the angle increases (see Figure 9.11). If the generator internal voltage could go to zero, the impedance would be exactly $-\boldsymbol{Z}_S$, which is denoted as point a in the figure. This point is never reached because of induction, which provides the generator with an effective internal emf. If this emf is constant, giving a constant ratio of the generator to the remote internal voltages, the locus of points will follow a circular path around point a in a clockwise direction. It should be noted that the foregoing is a very approximate analysis. The important thing to note is that the impedance seen from the generator terminals moves toward the fourth quadrant and the loss of excitation

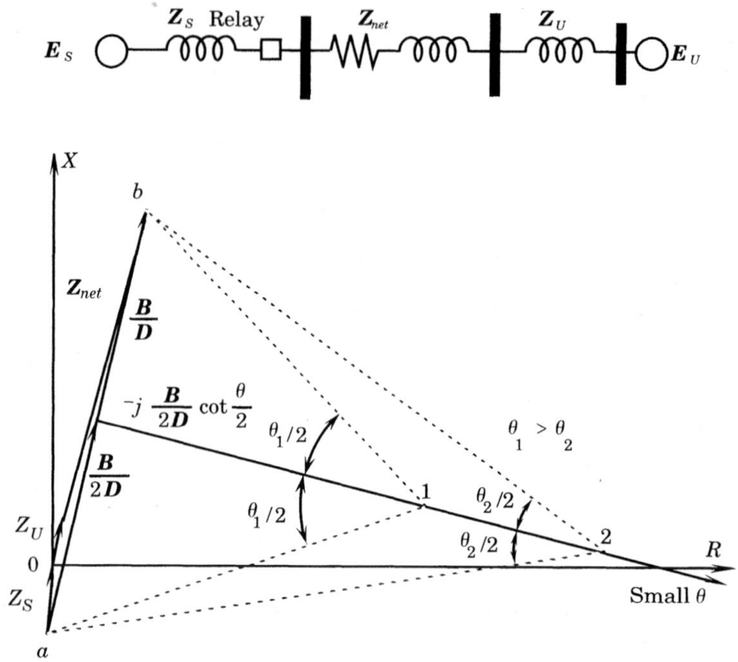

Figure 18.13 Graphical construction of the Z plane locus of $\mathbf{Z}_R$.

could be detected by a mho type element that is centered on the $-X$ axis. Computer studies of the path traced in the Z plane verify the approximate description given above [1], [4], [11].

A typical loss of excitation relay characteristic is shown in Figure 18.14. This relay employs a mho unit with the center of the offset circle located on the negative X axis. This

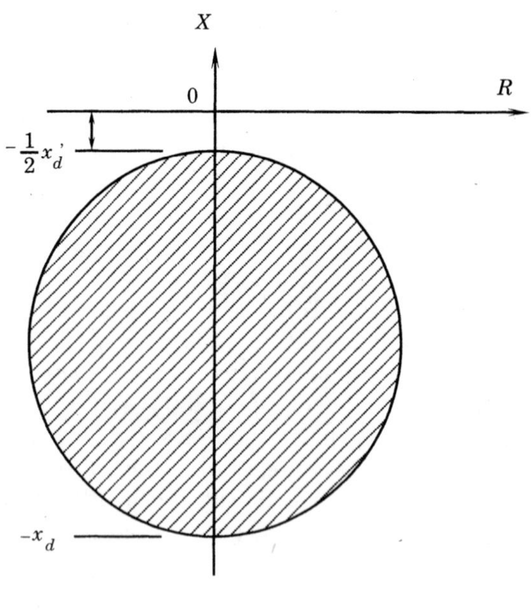

Figure 18.14 Typical loss of excitation relay characteristic.

characteristic has been noted to not clear loss-of-excitation conditions if the loss is only partial, since the relay circle is purposely kept as small as possible to avoid false trips. Other loss of excitation relay characteristics have been devised that can be defended as entirely acceptable, but employ a trip zone with a different shape than a circle [5].

18.6 OTHER GENERATOR PROTECTION SYSTEMS

The protective systems considered above are designed to protect the stator and rotor windings. There are many other protective systems that are required for synchronous generators, some of which are considered below.

18.6.1 Overspeed Protection

It has been noted above that overspeed causes overvoltage, which may be protected against using overvoltage relays. Basically, however, overspeed control is part of the turbine control system. Most large steam turbine controls have two or three separate speed control units, with one of these being strictly a mechanical centrifugal device that will close the turbine control valves even if electrical power is lost to the controls that require electric input.

Overspeed protection must be selective and must not shut the unit down due to a temporary loss of load, even if the cause is serious, for example, a short circuit. Short circuits anywhere near a generator will collapse the voltage and the generator experiences a loss of load. Since the turbine power is unchanged, the turbine-generator unit will overspeed until the governor throttles the turbine input back. Faults are usually temporary, however, and there is no need to shut the unit down unless the fault is on the generator or GSU transformer. Other nearby faults have exactly the same effect on the generator, however, and the protections must be designed to discriminate. Faults in the generator or step-up transformer will trip the unit before overspeed can become a problem. Nearby transmission faults should not trip the unit, however.

Most large steam turbine units have protective devices that are designed to distinguish between a load rejection and a fault. Both cause overspeed and a sudden loss of generated power, but the generator current behavior is quite different. Faults cause an increase in current, while load rejection causes a decrease in current. The generator should not be tripped for the fault condition, assuming that it is not a generator or transformer fault, but the generator will have to be tripped if the load is lost permanently. Turbine controls are designed to make this distinction, to run back in the case of load rejection and to do nothing in the case of a network fault except the usual speed governing reaction.

These turbine controls are outside the scope of this book, but there are cases where there needs to be coordination between the protective actions taken on the electrical side with those taken in the power plant. In the case of overspeed, backup protection can be interlocked with the turbine controls that will monitor bus frequency and order a generator trip if this frequency becomes too high. Obviously, this must be coordinated with the turbine controls

18.6.2 Generator Motoring Protection

Motoring protection is provided for the steam or hydraulic turbine, or for the power system, and not for the generator. Motoring is not harmful to the generator in any way. The protection is usually considered as part of the generator protective system since it uses electrical quantities, usually in the form of sensitive power relays.

Steam turbines have the tendency to overheat when the steam supply is cut off for some reason and the turbine generator system begins motoring. The overheating is due to the loss of steam, which normally keeps the turbine blading at a stable temperature. When the steam flow is interrupted, the blade temperature rises. Most modern steam turbines will overheat when the steam flow is less than about 10% of full load. The time constant is large and varies from 30 seconds to 30 minutes, depending on the type of turbine and its design. The practice varies regarding the need for protection and the turbine manufacturer should be consulted regarding any requirements. When protection is required, a reverse power relay is often used and is set to trip at about 0.5% reverse power.

Hydraulic turbines sometimes lose their supply of water, for example, when trash blocks the input gates to the penstocks. When this happens, the low flow can cause cavitation and possible turbine runner damage and will often result in motoring of the generator. To protect against these possibilities, hydro-units are often provided with power relays that are set to trip the unit for low power output or for reverse power of a few percent of normal load.

Diesel engine driven generators almost always are provided with antimotoring protection when the engine fails. There are two reasons for this requirement. First, the failed engine is a relatively large load, and the generator will draw 15% or so of its rated power from the system. This may be more than the system can supply. Second, there is the danger of explosion or fire from unburned fuel in the Diesel engine if driven by the electrical system.

Combustion turbines that are directly connected to their generator will require from 10% to 50% power if driven as a mechanical load by the motoring generator. Such units are usually protected from generator motoring in order to avoid loading the system with this unnecessary load, even though the motoring is not harmful to the generator or to the turbine.

18.6.3 Vibration Protection

Vibration protection is also a part of the turbine protection system and is not normally supplied or specified by the electrical protection engineer. Still, as noted above, vibration detection is the backup for faults that cause unbalanced rotor heating due to rotor faults or negative-sequence rotor heating. It is important that there be coordination between the electrical protections and the mechanical backup.

Vibration detection equipment should always be in service on a turbine-generator system. It is not adequate, for example, to manually add the vibration detection recorder *after* having detected a field ground. The reason is that the second field ground may occur quite soon after the first one and before any manual action is possible. The vibration detector should at least provide an alarm, but preferably should issue a time-delayed trip of the generator and field breakers. As noted previously, it is preferable that the vibration detector not trip the turbine, but only the generator and field breakers.

This type of redundancy in protection is important. The electrical detection equipment is designed on the basis of the *cause* of problems that lead to rotor heating and possible unbalance. Vibration detectors, on the other hand, are based on the *effect* of the unbalance. Failure of one type of detector is not likely to cause failure of the other; hence, the protection system that includes both has good reliability.

18.6.4 Bearing Failure Protection

Bearing temperature can be measured by a temperature sensitive device placed in a hole in the bearing. The temperature measurement can actuate an alarm to alert the operator or, for unattended stations, can be used to trip the unit.

On some units the bearings are lubricated by oil that is circulated through the bearing under pressure. On these units, the temperature of the oil can be monitored. Provision can also be made to cause an alarm or a trip if the lubricating oil stops flowing for any reason.

18.6.5 Coolant Failure Protection

Some of the larger generators have stator windings that are cooled by water circulating through the windings. If the flow of cooling water is impeded or stopped for some reason, the winding will quickly overheat. The cooling water is circulated by water pumps, which are backed up by standby pumps.

Flow detectors placed in the water line monitor the adequacy of the coolant flow. If the flow is reduced for any reason, these detectors can start the standby pump. If this restores the normal flow, no further action is taken, but if not restored within 1 minute, the turbine and generator are tripped sequentially, starting with the turbine and then the generator [4].

18.6.6 Fire Protection

Special fire protection equipment is sometimes used to quickly extinguish fires within the generator housing [1]. One method that has been used releases water into the generator enclosure on detection of a fire. A better system injects carbon dioxide into the generator rather than water. This is safer and does not add to the damage caused by the fire.

Many large machines operate in a hydrogen cooled environment. Since hydrogen will not support combustion, additional fire protection is often not considered necessary for these units. Instead, a method is required to monitor the purity of the hydrogen to make sure no air is mixed with the hydrogen to form an explosive mixture. Most generator shells are designed to withstand such an explosion of the maximum intensity possible from a hydrogen and air mixture.

18.6.7 Generator Voltage Transformer Fuse Blowing

Some of the relays used for generator protection require voltage as one of their inputs. An example is the distance backup scheme discussed in the previous section. Should the voltage transformer supplying these relays lose its potential supply, the relays lose their restraint and the impedance seen from the relay appears to be zero. This means that the loss of a fuse, which causes no damage whatever, will result in the trip of the generating unit. Even the loss of one fuse in a three-phase voltage supply will change the magnitude and phase of the supplied voltages to cause the undesired trip. One solution to this problem is to supervise the voltage sensitive relays with an voltage balance relay, which compares the magnitude of voltage from two sources and energizes an alarm when the two sources are not nearly equal. This and other schemes are described in [2].

18.6.8 Protection of Power Plant Auxiliaries

In the operation of steam power plants, it is desirable to automatically transfer essential power plant auxiliaries from their main supply to an auxiliary supply when the main supply fails. The essential loads that are transferred are such essential loads as boiler feed pumps, induced draft or forced draft fans, and other loads that are necessary for maintaining boiler output. It would be ideal to transfer all auxiliary load, but if the alternate supply does not have adequate capacity to carry all of this load, the essential loads should be isolated for switching to alternate service.

The basic premise in this protection is that the two sources should not be paralleled since this raises the interrupting duty and the possibility of circulating currents. Therefore, the transfer scheme must first drop the main supply and then quickly close the breaker for the alternate supply. Before doing anything, however, the alternate supply must be checked to ensure adequate voltage. In addition, the motor loads in the power plant must be checked to make sure that there are no faulted circuits. The logic shown in Figure 18.15 is suggested for this type of protection [5].

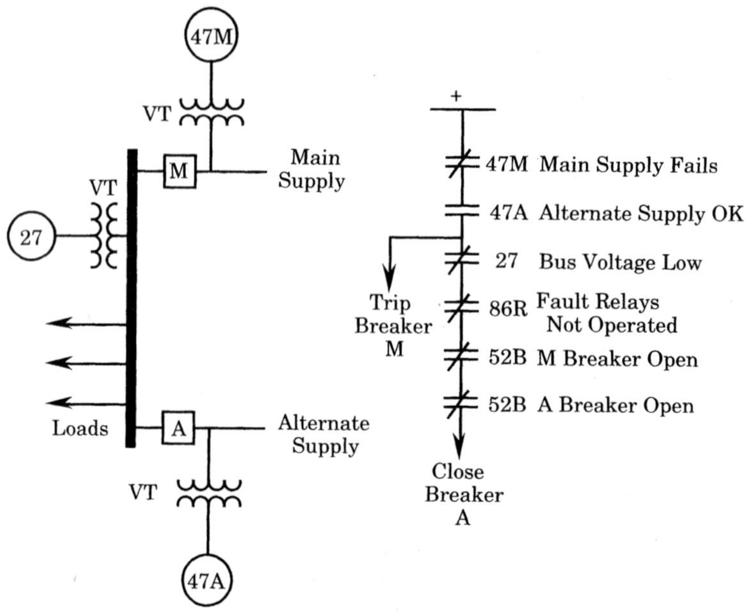

Figure 18.15 Bus transfer scheme using polyphase voltage relays [5].

18.7 SUMMARY OF GENERATOR PROTECTION

This section presents a summary of genérator protection as applied to a unit generator- transformer and discusses the different modes of tripping that are often recommended by the manufacturers.

18.7.1 Unit Generator-Transformer Protection

An overview of the protection that would be expected for a unit generator-transformer system is summarized in Figure 18.16 and Table 18.3. The protective relays shown in Figure 18.16 and described in Table 18.3 are the minimum recommended protection for steam turbine generators, hydro generators, and combustion turbine generators where the generator uses a unit transformer for stepping up to transmission voltages. In most cases, the protections shown apply equally to attended or unattended generating stations and for almost all ratings [7]. It is assumed that high impedance grounding of the generator is used, with the grounding achieved

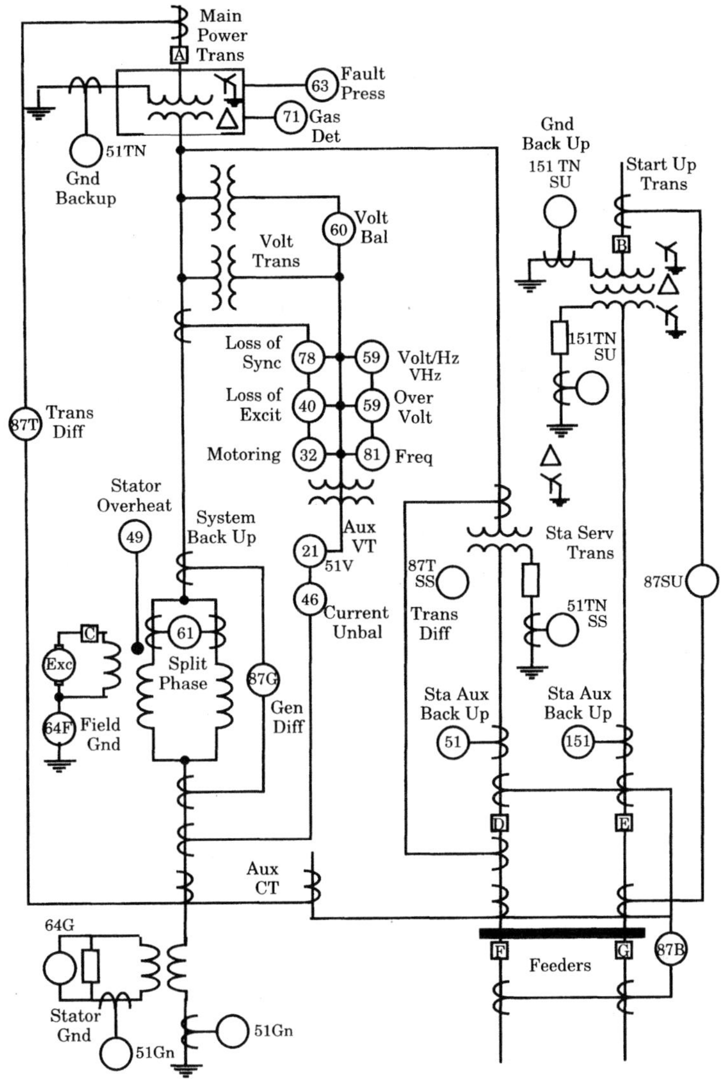

Figure 18.16 Unit generator and transformer protection.

through use of a distribution transformer and the grounding resistance chosen to limit the current to 10 primary amperes or less.

Figure 18.16 presents a typical generator protective plan. There are many variations of generator protection, some of which are documented in the literature [12–16].

Table 18.3 lists all of the protections illustrated, giving the generic type of relay and the action taken by each device. The Type 21 distance relays are installed as backup protection for the transmission lines. The overcurrent (Types 51 and 151) devices are also serve as backup protection. Most of the devices in Figure 18.16 are labeled as to their function. Table 18.3 gives, in addition to the relay function, the device that is controlled or the breaker(s) tripped by each relay. This is discussed in greater detail below.

TABLE 18.3 Description of Generator Protective Devices

Device	Function	Relay Controlled of Breaker Tripped
21	Distance relay	21X
21X	Auxiliary device	86G
40	Field relay	94G1
46	Current unbalance	94G1
49	Thermal relay	Alarm
51	Time overcurrent relay	86G
51/GN	Generator overcurrent	86G
51/TN	Transformer overcurrent	86G
51TN/SS	Station service transformer OC	86G
51V	System backup overcurrent	86G
59	Overvoltage relay	94G1
59 V/Hz	Volts/hertz relay	94G1
60	Voltage balance relay	Alarm
61	Split phase (hydrogenerators)	86G,87X
63	Pressure relay	86G
64F	Field ground relay	86G
64G	Generator ground relay	86G, 87X
71	Gas detector relay	Alarm
78	Loss of synchronism relay	94G2
81	Frequency relay	94G2
86B	Bus lockout relay	D, E, F, G
86G	Generator lockout relay	A, B, D, H, BFI
86T/SU	Startup Transformer lockout	B, E
87B	Bus Differential relay	86B
87G	Generator differential relay	86G, 87X
87T	Unit transformer differential	86G
87T/SS	Station service transformer diff.	86G
87T/SU	Startup transformer differential	86T/SU
87X	Differential-air-cooled machines	CO2
94G1	Generator zone trip relay	A, C, H, BFI
94G2	Generator zone trip relay	A, BFI
151	Backup station auxiliary OC	86T/SU
151TN	Startup transformer ground backup	86T/SU
151TN/SU	startup transformer ground	86T/SU
86G	Tripping hand-reset lockout relay	
94G1	Tripping self-reset auxiliary relay	
94G2	Tripping self-reset auxiliary relay	

18.7.2 Unit Generator-Transformer Trip Modes

In order to fully understand the tripping action that takes place for each of the generating system protections, it is necessary to know the configuration of the trip circuits. The relays in the generator protective systems do not usually energize the circuit breaker trip coils directly, but pick up a tripping device. For the system described here, there are three trip relays used [7]:

The reason for having three different trip modes is that it provides a degree of flexibility that would not be possible otherwise. Each different mode is designed to carry out different types of tripping operations. In some cases, for example, the prime mover is not tripped, giving the operator time to check the disturbance and resynchronize the generator, thereby preventing long and unnecessary outages.

The 86G device trips the main and field circuit breakers, the turbine, and the boiler. This trip mode is used for all generator or transformer faults and for all backup operations. This trip mode requires human intervention to hand reset the trip relay. Station auxiliaries are transferred to the standby source of power.

Trip relay 94G1 is an automatically reset auxiliary relay that initiates tripping of only the main breakers and the field breakers. This trip mode is used in cases where the problem that causes the tripping may be quickly corrected, thereby permitting reconnection of the generator after a short period of time. Station auxiliaries are transferred to the standby source of power.

Trip relay 94G2 is an automatically reset auxiliary relay that initiates tripping of only the main generator breakers. This trip mode is used on detection of disturbances that might allow fast reconnection to the system, hence the station auxiliaries are left connected to the generator terminals. An example of this type of tripping is a loss of synchronism or low system frequency due to a system disturbance. For such disturbances the generator and its auxiliaries are not faulted and are ready to be reconnected after a short time delay. For this type of condition, it is not advisable to switch the plant auxiliaries to the standby source, which may be out of phase with the generator or may not be available at all. In this tripping mode the generator carries its own auxiliaries while the operators attempt to restore the system to a condition where reconnection is possible.

The foregoing are all electrical protections for the generator and unit step up transformer. There are other protective systems for the turbine and boiler that are often used to cross trip the generator. Turbine protective systems often provide for *sequential* tripping. In this case, the turbine valves are tripped first and a set of auxiliary contacts on these valves are used to initiate tripping of the generator main and field breakers. This is backed up by reverse power relays that sense the beginning of motoring to initiate tripping of the 94G1 trip relay. The sequential trip is the normal method of tripping a unit when there is no emergency detected.

Another trip mode that is sometimes used is a *simultaneous* trip of the turbine valves and the generator main and field breakers, sometimes with a time delay in the breaker trip command.

18.7.3 Breaker Failure Protection of the Generator

Generator manufacturers usually recommend the use of breaker failure protection as an integral part of the overall generator protection [7]. A simplified form of breaker failure protection is shown in Figure 18.17. This scheme requires the use of fault detectors (50BF) and timers (62BF). Separate fault detectors and times must be used for each breaker, such as breakers 2 and 3 in Figure 18.17. The primary and backup protective systems, on detecting a fault, will attempt to trip the circuit breakers and, at the same time, will start the breaker failure timers (62BF2 and 62BF3). If the breaker does not clear the fault before the timer completes its cycle, the timer will trip the other breakers that are required to isolate the fault. Thus, if breaker 3 fails, all breakers on bus N must be tripped. If breaker 2 fails, breaker 1 and the remote end of line A must be tripped. The line should be tripped by a transfer trip signal since the remote end relays may not detect the internal generator fault.

The scheme shown in Figure 18.17 is unusual since it uses the breaker auxiliary "a" contacts in the trip circuit to control the backup timer in case the fault current level is below the pickup of the fault detector relays (50BF/2 and 50BF/3), which might be the case for a low level transformer fault or low level unbalanced currents. This is a bit unusual since the

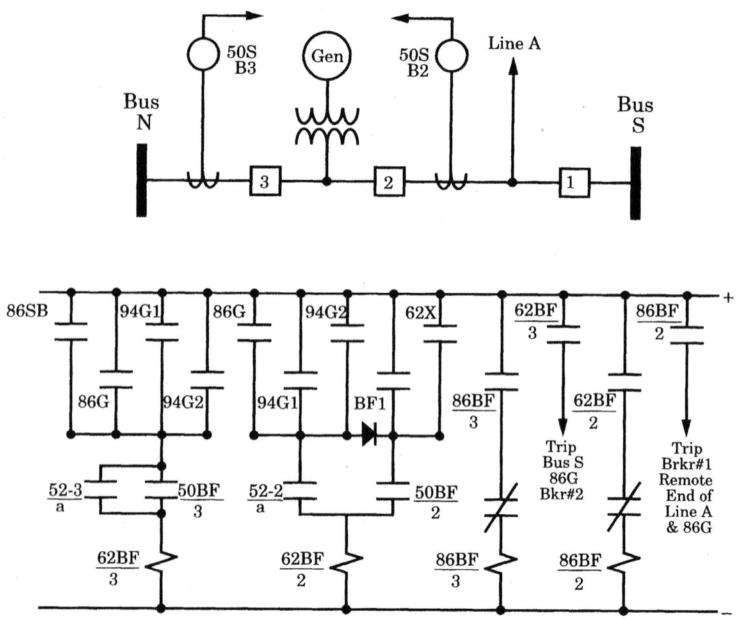

Figure 18.17 Tripping logic for generator protection [13].

breaker "a" contacts may have been partly the cause of the breaker failure. In this case they are required for low-level faults.

The schemes shown in Figures 18.16 and 18.17 imply the use of only one lockout (86G) relay. Some engineers prefer to split the generator zone protection into groups with a separate lockout relay for each group. For example, the primary relays could be one group and the backup relays another. This would eliminate protection failure due to the malfunction of the lockout relay.

If there are to be two lockout relays, the dual system is operated as a one-out-of-two energize to trip logic. Any system trip that requires the opening of the generator breakers will trip one of the lockout relays, which in turn will trip the unit trip relays and also the turbine trip logic. One possible grouping of generator lockout relay trip functions might be specified as follows:

Inputs to Generator Lockout Relay 86G1

1. Generator transformer lockout relay trip
2. Generator loss of field voltage
3. Generator voltage regulator trip
4. Generator negative-sequence relay trip
5. Accidental energization of the generator
6. Turbine valves closed and reverse power detected
7. Reverse power for 30 seconds detected

Inputs to Generator Lockout Relay 86G2

1. Generator stator differential relay trip
2. Generator neutral ground relay trip

3. Distance backup relay trip
4. Generator loss of field voltage
5. Generator excitation system trip
6. Subsynchronous relay trip
7. Turbine valves closed and reverse power detected
8. Reverse power for 30 seconds detected

Obviously, other schemes for dividing the relay functions between two or more lockout relays are possible. Note that certain critical functions are included in both of the above systems.

The generator step-up transformer lockout relay might have the following functions as inputs:

Generator Step-Up Transformer Lockout Relay

1. Transformer differential trip
2. Transformer internal fault or gas detection
3. Transformer primary winding neutral overcurrent
4. Transformer fire detected
5. Transformer volts/hertz trip
6. High-voltage bus differential trip
7. High-voltage breaker failure
8. Low-voltage ground fault
9. Low-voltage breaker failure (if any)
10. Station service transformer primary phase overcurrent
11. Station service transformer differential trip
12. Station service transformer ground differential
13. Station service transformer ground overcurrent
14. Station service transformer fire detected
15. Station service transformer internal fault or gas detection

There are many other protective systems in a complex power plant that can cause the trip of the generator. In a modern steam power plant the system of protection extends to the boiler, boiler auxiliaries, and turbine control systems. Loss of important components that disable the plant will also send an input to one of the generator lockout relays.

REFERENCES

[1] Horowitz, S. H., Ed., *Protective Relaying for Power Systems*, IEEE Press Book, New York, 1980.

[2] Mason, C. R., *The Art and Science of Protective Relaying*, John Wiley & Sons, New York, 1956.

[3] Warrington, A. R. van C., *Protective Relays: Their Theory and Practice, Volume One*, John Wiley & Sons, Inc. New York, 1962.

[4] GEC Measurements, *Protective Relays Application Guide*, General Electric Company p.l.c. of England, 1975.

[5] Blackburn, J. L., Ed., *Applied Protective Relaying*, Westinghouse Electric Corporation, 1976.

[6] Basilesco, J., and J. Taylor, "Report on Methods for Earthing of Generator Step-Up Transformer and Generator Winding Neutrals as Practiced Throughout the World," Electra, 121, December 1988, pp. 86–101.

[7] Power Systems Management Business Dept., "Protective Relaying Guide," A guide for short circuit protection and protection for abnormal operating conditions for generating stations, General Electric Company, Philadelphia, PA, 1980.

[8] Pope, J. W., "A Comparison of 100% Stator Ground Fault Protection Schemes for Generator Stator Windings," *IEEE Trans.*, PAS-103, April 1984, pp. 832–840.

[9] ANSI/IEEE C37.101-1985, "IEEE Guide for Generator Ground Protection," IEEE, New York, 1985, *IEEE Guides and Standards for Protective Relaying Systems - 1989 Edition*, IEEE, New York, 1989.

[10] Barkle, J. E., and W. E. Glassburn, "Protection of Generators Against Unbalanced Currents," *AIEE Trans.*, 72 (III), 1953, pp. 282–285.

[11] General Electric Company, "Use of R-X Diagram in Relay Work," Publication GET-2230B, General Electric Switchgear Dept., Philadelphia, 1967.

[12] Landoll, L. E., et al., IEEE AC Generator Protection Guide Working Group Report, Summary of the 'Guide for ac Generator Protection,' ANSI/IEEE C37.102-1987, IEEE Trans., v. PWRD-4, April 1989, pp. 957–964.

[13] Berdy, J., "Protective Relaying Guide," General Electric Company Publication, Schenectady, NY, May 1977.

[14] Baldwin, M. S., "Generator Protection Seminar Notes," Westinghouse Electric Corp. Publication, Pittsburgh, PA, February 1978.

[15] Ungrad, H., V. Narayan, M. Fiorentzis, and M. Ilar, "Coordinated Protection of a Large Power Plant and the Connected Network," CIGRÉ paper 34-05, presented at the 1978 session, Paris.

[16] Benmouyal, G., "Design of a Universal Protection Relay for Synchronous Generators," CIGRÉ paper 34.09, presented at the 1988 session, Paris.

PROBLEMS

18.1 A synchronous generator is connected to a power system through a step-up transformer and circuit breaker. The generator is wye connected and grounded through a resistance. A line-to-ground fault develops in one phase of the generator stator winding. The stator protection trips the generator breaker, removing the unit from the transmission system and preventing any fault current being contributed from the power system. However, the fault persists since the stator winding continues to produce an emf, which sustains the fault current. How can the fault be removed?

18.2 Sketch the sequence networks for the representation of winding faults on a synchronous generator. Assume a linear winding impedance distribution throughout each phase winding.

18.3 For the fault condition described in problem 18.2, write out the expressions for the sequence impedances on the neutral side and on the line side of the fault point F.

18.4 Consider the system described in problems 18.2 and 18.3. Compute the fault current, if the fault is assumed to be

(a) a three-phase fault

(b) a phase-to-phase fault

(c) a one-line-to-ground fault

(d) a one-line-to-ground fault with high neutral impedance

18.5 A delta-connected generator is to be protected by current percentage differential relays. Sketch the CT connections for one of the three phases that will provide the desired differential protection.

18.6 A 974 MVA, 22 kV generator is unit connected to a 345 kV transmission bus and grounded through a distribution transformer as shown in Figure P18.6.

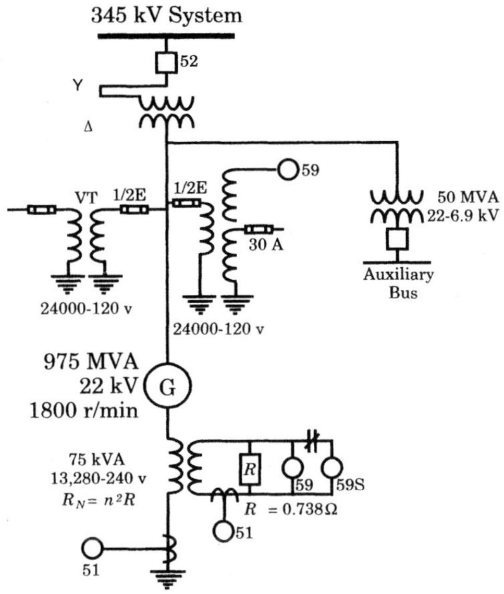

Figure P18.6 A 974 MVA generating unit.

The phase-to-ground capacitive reactance of the generator, transformers, leads, and associated equipment is 6780 ohms per phase. The distribution transformer is rated 13,280-240 volts. The secondary resistor is 0.738 ohms.

(a) Compute the grounding resistance reflected to the primary of the distribution transformer.

(b) Sketch the sequence networks and their connection to compute the fault current for a one-line-to-ground fault. Solve this system for the zero-sequence current I_0.

(c) Compute the following quantities:
The zero-sequence current flowing in the neutral resistor
The zero-sequence current in the distributed capacitance
The total fault current flowing in the generator neutral
The fault current contribution from the distributed capacitance

(d) Determine the current flowing in the distribution transformer secondary circuit and the voltage across the secondary resistor for a fault at the generator terminals.

(e) Let us assume that the 59 relay is set at 5.4 volts. What percentage of the generator winding is covered for ground faults?

(f) Relay 59S is a startup overvoltage relay that is set at 7 volts. What percentage of the generator winding is protected during startup?

18.7 An 810 MVA generator is connected to a 230 kV transmission network through a unit step up transformer. The following data are given for the generating unit:

Generator:	Rated voltamperes	810 MVA
	Rated voltage	24 kV
	Max overvoltage	31 kV
	Winding capacitance	0.43 μF/phase
GSU transformer:	Rated voltage	24 Δ-230 Y kV
	Primary capacitance	0.00975 μF/phase
Connections:	Capacitance	0.00245 μF/phase

Generator ground fault protection is to be provided by grounding the neutral by a resistor on the secondary of a neutral connected distribution transformer. Determine the size and rating of the grounding resistance and the ratings of the distribution transformer.

18.8 A synchronous generator is rated 991 MVA, 26 kV, 0.9 pf lag. The underexcitation protection limit is specified as a circular arc passing through point D (0, −410), C (940, −310), and an intermediate point E (500, −420). Find the equation for the limiting circular arc, the location of the center, and radius of the circle.

18.9 Unbalanced generator loading can lead to overheating. One way to detect unbalanced loading is to measure negative-sequence currents in the generator terminals and use these measurements as an unbalance relay quantity. This can be done using a bridge connection, as shown in Figure P18.9. Use phasor diagrams of the positive and negative currents to show that negative sequence currents flow through the bridge branch.

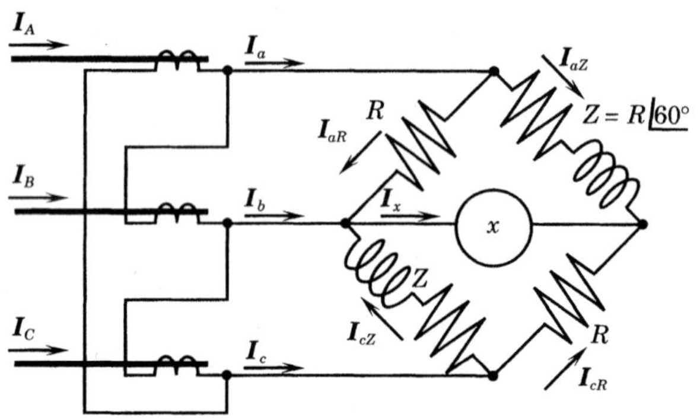

Figure P18.9 Generator negative-sequence bridge protection.

18.10 The statement is made in Section 18.4.1 that the air gap distortion due to shorted field turns can warp the rotor because of unbalanced forces that vary as the square of the flux density. Show that this is true, based on fundamentals of electromagnetic field theory.

18.11 A description of several different types of generator ground protections is given in [9]. As a research problem, investigate the theory of one of the following schemes described in that reference and report on your findings to the class:

1. Ground Overvoltage, Complete Shutdown
2. Ground Overvoltage, Permissive Shutdown
3. Ground Overvoltage, Alarm and Time-Delay Shutdown
4. Ground Overvoltage, Alarm
5. Startup Ground Overvoltage, Complete Shutdown
6. Ground Fault Neutralizer Overvoltage, Alarm and Time-Delay Orderly Shutdown
7. Wye-Broken Delta Voltage Transformer Ground Overvoltage, Complete Shutdown
8. Startup Wye-Broken Delta Voltage Transformer Ground Overvoltage, Complete Shutdown
9. Secondary Connected Current Transformer, Time-Delay Ground Overcurrent, Complete Shutdown
10. Primary Connected Current Transformer, Time-Delay Ground Overvoltage, Complete Shutdown.
11. Instantaneous Ground Overcurrent, Alarm or Complete Shutdown
12. Generator Leads Ground Overcurrent, Complete Shutdown
13. 3-Wire Generator Leads Window Current Transformer Instantaneous Ground Overcurrent, Complete Shutdown

14. 4-Wire Generator Leads Window Current Transformer Instantaneous Ground Overcurrent, Complete Shutdown
15. Generator Percentage Differential, Complete Shutdown
16. Generator Percentage Differential and Polarized Neutral Overcurrent, Complete Shutdown
17. Delta-Connected Generator, Generator Percentage Differential, Complete Shutdown

19

Motor Protection

19.1 INTRODUCTION

Motor protection is similar to generator protection in many ways, but there are some characteristics of motors that require special consideration. Motor physical characteristics are important in the design of the protective system, and this probably applies to motors more than other types of apparatus. Thus, we must take into consideration the starting current versus time, the locked rotor current, and the thermal capability of the motor in different types of applications and operating duty cycles. Conditions that cause rotor heating are particularly difficult for the protection system designer to solve effectively, since direct measurements of the rotor are not possible in many types of motors.

The protection of small motors is specified in the National Electric Code [1] and will not be discussed in this chapter. There are also NEMA standards [2] for motor protection. Small motors are usually provided with built-in protection of some type that is usually quite adequate, considering the low cost of the small motors. The cost and degree of protection must be balanced against the risk of hazards that can cause loss of the motor. The larger and higher voltage motors represent a large investment, and it is worthwhile to protect that investment by special protective systems. Therefore, our concentration is on the protection of large, high-voltage induction motors, such as those found in the drive applications in electric generating stations and other heavy industrial drives. Protection of such motors is also described in the technical literature [3–11].

There are several types of industrial motors that could be discussed. The protection of large synchronous motors and synchronous condensers is similar to that of synchronous generators, although the protection may not be as elaborate as that of generators since the motors are usually smaller and do not warrant the investment of elaborate protective designs. For the most part, we are concerned with induction motors, both squirrel cage and wound rotor types. In a few special cases, motor protections that apply only to synchronous motors are presented.

19.2 INDUCTION MOTOR ANALYSIS

The analysis of three-phase induction motors is presented in many textbooks on electric machinery [12], [13]. The analytical methods will not be explored in detail here, but the basic equations and circuit models will be presented and analyzed. For further information, any of the textbooks on electric machinery can be consulted. The emphasis here is on the three-phase induction motor.

19.2.1 Normalization of the Basic Equations

One of the induction motor problems of considerable interest to the protection engineer is the motor behavior when starting. The starting characteristics depend on the motor design and on the shaft load that must be accelerated from rest to normal running speed. During startup, the current drawn by the motor is several times the normal full-load current and there is always the danger of protective devices tripping the motor before it reaches normal running speed and normal running current magnitude.

19.2.1.1 The Swing Equation. Starting is greatly affected by the inertia of the driven load, and loads with a high moment of inertia are more likely to cause problems for the protective system. Therefore, it is important to understand the dynamics of the motor performance in accelerating a shaft load. The basic equation for acceleration is given by Newton's law for a rotational system, which can be written as

$$J\frac{d\omega}{dt} = T_a \quad \text{Nm} \tag{19.1}$$

where J = moment of inertia (kg· m^2)
T_a = net accelerating torque (N· m)
ω = shaft angular velocity (rad/s)
t = time (s)

The equation (19.1) is usually called the *swing equation.* It is customary to normalize the swing equation, as with all power system equations. To do this, we select the following base quantities.

$$\begin{aligned} V_B &= V_{LL} = \text{base (rated line-to-line) voltage (v)} \\ S_{B3} &= \text{base three-phase (volt-amperes)} \end{aligned} \tag{19.2}$$

Then

$$\begin{aligned} I_B &= \frac{S_{B3}}{\sqrt{3}\,V_B} = \text{base current (A)} \\ Z_B &= \frac{V_B^2}{S_{B3}} = \text{base impedance } (\Omega) \end{aligned} \tag{19.3}$$

Note that, once the base voltage and three-phase volt-amperes (19.2) are selected, the other base quantities (19.3) can be derived [14]. For induction motors, it is convenient to select the base voltage and base current from the motor nameplate values at full load, in which case the base volt-amperes would be computed as well as the base impedance.

19.2.1.2 Normalization of the Swing Equation. The time in (19.1) has the base value that is constrained by the system frequency, i.e.,

$$t_B = \frac{1}{\omega_B} = \frac{1}{2\pi f_B} \quad \text{s} \tag{19.4}$$

We can relate the three-phase torque to three-phase power by the fundamental relation

$$P_{3\phi} = T_{3\phi}\omega \quad \text{W} \tag{19.5}$$

If the three-phase power is equal to the base volt-amperes, we can write the following relationship among the base quantities.

$$P_{3\phi} = S_{B3} = T_{B3}\omega_B \tag{19.6}$$

Therefore the three-phase base torque is given by

$$T_{B3} = \frac{S_{B3}}{\omega_B} \quad \text{N}\cdot\text{m} \tag{19.7}$$

The moment of inertia term in (19.1) can be written dimensionally as

$$J_B = T_{B3}t_B^2 = \frac{T_{B3}}{\omega_B^2} = \frac{S_{B3}}{\omega_B^3} \tag{19.8}$$

Now, it is customary in problems involving the swing equation to define a normalized "inertia constant" that is defined as a function of the energy of the rotating mass when rotating at rated angular velocity [15], [16]. Thus we define

$$H = \frac{W_k}{S_{B3}} \quad \text{MJ/MVA or s} \tag{19.9}$$

where the dimension of H is noted to be seconds. This is convenient since values of H can often be estimated with reasonable accuracy if exact values are not known. At least errors of an order of magnitude would usually not occur. A shaft load with an H of 10 or more would represent a large moment of inertia. It has been shown that H can be computed from a moment of inertia given in the English units in lbm-ft^2 by the equation

$$W_k = 2.311525 \times 10^{-10}\left(WR^2\right)n_R^2 \quad \text{MJ} \tag{19.10}$$

where WR^2 = moment of inertia (lbm· ft^2)
n_R = shaft rated synchronous speed (rev/min)

From (19.9), we can write

$$H = \frac{J\omega_B^2}{2S_{B3}} \tag{19.11}$$

or, solving for J,

$$J = \frac{2HS_{B3}}{\omega_B^2} = (2H\omega_B)\left(\frac{S_{B3}}{\omega_B^3}\right) = (2H\omega_B)\,J_B \equiv J_u J_B \tag{19.12}$$

where a subscript "u" has been added to emphasize that this variable is a per unit quantity. Then, the swing equation (19.1) can be written as

$$J_u J_B \frac{d\omega}{dt} = T_{au}T_{B3} \quad \text{N}\cdot\text{m} \tag{19.13}$$

Substituting from (19.12) and (19.7), we have the normalized swing equation

$$\frac{2H}{\omega_B}\frac{d\omega}{dt} = T_{au} \quad \text{pu} \tag{19.14}$$

where H and T_{au} are normalized quantities, ω is in radians per second, and the time t is in seconds. One further simplification is possible if we recognize

$$\omega_u = \frac{\omega}{\omega_B} \quad \text{pu} \tag{19.15}$$

to write

$$2H\frac{d\omega_u}{dt} = T_{au} \quad \text{pu} \tag{19.16}$$

where all quantities are in per unit except time and the inertia constant, which are in seconds. Note that the dimensionality of the equation is preserved since the normalized inertia constant H has units of seconds.

The accelerating torque is the made up of the net torque acting to accelerate the shaft of the machine. This can be written as

$$T_{au} = T_{Mu} - T_{Lu} \tag{19.17}$$

where T_{Mu} = net developed mechanical torque of the motor
T_{Lu} = net retarding torque of the load

The per unit motor mechanical torque has two components, which can be written as

$$T_{Mu} = T_{M1} + T_{M2} \tag{19.18}$$

where T_{M1} = positive-sequence mechanical torque
T_{M2} = negative-sequence mechanical torque (19.19)

If balanced voltages are applied to the motor, the negative-sequence torque is zero. However, if unbalanced voltages are applied, the negative-sequence torque is a negative quantity, thereby reducing the total torque available to accelerate the shaft load.

19.2.1.3 Symmetrical Component Transformation. The mechanical torque equation requires that the symmetrical components of motor torque be computed. This can be accomplished by writing the total three-phase apparent power in phase coordinates and transforming that result into symmetrical component coordinates. The three-phase apparent power in per unit can be written as

$$S_{3\phi} = P_{3\phi} + jQ_{3\phi} = \mathbf{V}_{abc}^T\mathbf{I}_{abc}^* \quad \text{VA} \tag{19.20}$$

where
$$\begin{bmatrix} \boldsymbol{V}_a \\ \boldsymbol{V}_b \\ \boldsymbol{V}_c \end{bmatrix} = \begin{bmatrix} 1 & 1 & 1 \\ 1 & \boldsymbol{a}^2 & \boldsymbol{a} \\ 1 & \boldsymbol{a} & \boldsymbol{a}^2 \end{bmatrix} \begin{bmatrix} \boldsymbol{V}_{a0} \\ \boldsymbol{V}_{a1} \\ \boldsymbol{V}_{a2} \end{bmatrix} \tag{19.21}$$

are the three phase voltages. A similar definition applies to the currents, with the result in amperes [14]. The product (19.20) gives the result

$$S_{3\phi} = P_{3\phi} + jQ_{3\phi} = \boldsymbol{V}_a\boldsymbol{I}_a^* + \boldsymbol{V}_b\boldsymbol{I}_b^* + \boldsymbol{V}_c\boldsymbol{I}_c^* \quad \text{VA} \tag{19.22}$$

This result assumes that there is negligible mutual coupling such that each phase current causes a voltage drop only in its own phase.

It is convenient to write (19.21) in the following matrix form.

$$\mathbf{V}_{abc} = \mathbf{A}\mathbf{V}_{012} \tag{19.23}$$

where all of the voltages are phasors and the complex constant $\boldsymbol{a}$ is defined as

$$\boldsymbol{a} = -\frac{1}{2} + j\frac{\sqrt{3}}{2} \tag{19.24}$$

We can show that [14]

$$\mathbf{V}_{abc}^{T} = (\mathbf{A}\mathbf{V}_{012})^{T} = \mathbf{V}_{012}^{T}\mathbf{A}^{T} \tag{19.25}$$

and

$$\mathbf{I}_{abc}^{*} = \mathbf{A}^{*}\mathbf{I}_{012}^{*} \tag{19.26}$$

Therefore, (19.22) becomes

$$S_{3\phi} = \mathbf{V}_{abc}^{T}\mathbf{I}_{abc}^{*} = \mathbf{V}_{012}^{T}\mathbf{A}\mathbf{A}^{*}\mathbf{I}_{012}^{*} = 3\mathbf{V}_{012}^{T}\mathbf{I}_{012}^{*} \quad \text{VA} \tag{19.27}$$

The proof of this result is left as an exercise.

We can normalize (19.27) by dividing both sides of the equation by the base three-phase voltage amperes, S_{B3}, to get the following result.

$$S_{3\phi u} = \frac{S_{3\phi}}{S_{B3}} = \frac{3\mathbf{V}_{012}^{T}\mathbf{I}_{012}^{*}}{S_{B3}} = \frac{3\left(\boldsymbol{V}_{a0}\boldsymbol{I}_{a0}^{*} + \boldsymbol{V}_{a1}\boldsymbol{I}_{a1}^{*} + \boldsymbol{V}_{a2}\boldsymbol{I}_{a2}^{*}\right)}{\sqrt{3}\,V_B I_B} \tag{19.28}$$

Now, we recognize the relationship between line-to-line and line-to-neutral voltages by defining a base line-to-neutral voltage

$$V_{B\phi} = \frac{V_B}{\sqrt{3}} \quad V \tag{19.29}$$

Substituting (19.29) into (19.28) we have

$$\begin{aligned} S_{3\phi u} &= \frac{3\left(\boldsymbol{V}_{a0}\boldsymbol{I}_{a0}^{*} + \boldsymbol{V}_{a1}\boldsymbol{I}_{a1}^{*} + \boldsymbol{V}_{a2}\boldsymbol{I}_{a2}^{*}\right)}{3\,V_{B\phi} I_B} \\ &= \boldsymbol{V}_{a0u}\boldsymbol{I}_{a0u}^{*} + \boldsymbol{V}_{a1u}\boldsymbol{I}_{a1u}^{*} + \boldsymbol{V}_{a2u}\boldsymbol{I}_{a2u}^{*} \quad \text{pu} \end{aligned} \tag{19.30}$$

where the "u" subscript has been added to emphasize the per unit quantities in the result.

Now, induction motors are usually wye connected with the common point of the wye being ungrounded. Therefore, zero-sequence currents cannot flow in the motor and we are left with a simplified result for the motor.

$$S_{3\phi u} = \boldsymbol{V}_{a1u}\boldsymbol{I}_{a1u}^{*} + \boldsymbol{V}_{a2u}\boldsymbol{I}_{a2u}^{*} \quad \text{pu} \tag{19.31}$$

The per unit three-phase active power can be written as

$$P_{3\phi u} = \Re\left(\boldsymbol{V}_{a1u}\boldsymbol{I}_{a1u}^{*} + \boldsymbol{V}_{a2u}\boldsymbol{I}_{a2u}^{*}\right) \quad \text{pu} \tag{19.32}$$

Then the per unit torque is given by

$$T_M = T_{3\phi u} = \frac{P_{3\phi u}}{\omega_u} = \frac{1}{\omega_u}\Re\left(\boldsymbol{V}_{a1u}\boldsymbol{I}_{a1u}^{*} + \boldsymbol{V}_{a2u}\boldsymbol{I}_{a2u}^{*}\right) \quad \text{pu} \tag{19.33}$$

or

$$T_M = T_{M1} + T_{M2} \quad \text{pu} \tag{19.34}$$

The problem can now be stated as follows. Given a set of three-phase voltages, balanced or unbalanced, we can determine the positive- and negative-sequence voltages. These sequence voltages are applied to the positive- and negative-sequence induction motor equivalent circuits to determine the developed torque for each sequence. We also must write an expression for the load torque, which is usually a function of rotor angular velocity. These torque equations are substituted into the swing equation (19.18) and (19.17) to give the swing equation (19.16) in a form ready for solution. The resulting equations are all in per unit based on the motor nameplate rated voltage and rated full-load current.

19.2.2 Induction Motor Equivalent Circuits

The induction motor can be analyzed by means of convenient circuit models that are usually attributed to Steinmetz. The circuit model for the positive and negative sequences are shown in Figure 19.1 [12], [13]. The resistance and reactance on the far left of the figure, R_T and X_T, represent the Thevenin equivalent impedance parameters of the power system supplying the motor. The input voltages are $\boldsymbol{V}_{a1}$ and $\boldsymbol{V}_{a2}$ for the positive- and negative-sequence networks, respectively. The voltages $\boldsymbol{V}_{M1}$ and $\boldsymbol{V}_{M2}$ represent the motor terminal voltages applied to the two sequence networks.

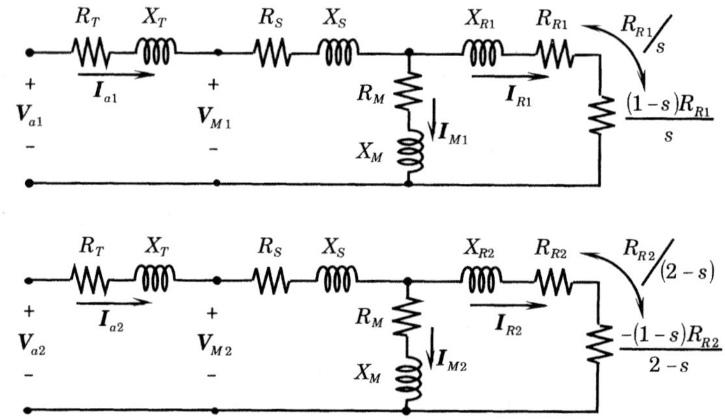

Figure 19.1 Induction motor positive- and negative-sequence equivalents.

The resistance and reactance R_S and X_S on the immediate left side of the mutual impedances represent the motor stator impedance in both the positive- and negative-sequence circuits. The mutual inductive reactance represents the motor excitation and the mutual resistance represents the losses due to hysteresis and eddy currents that result from exciting the core of the machine. The total resistance to the right represents the rotor of the machine. In the positive sequence, the rotor resistance is R_{R1}/s, and this is usually divided into two parts, the rotor positive-sequence resistance R_{R1}, which represents the rotor I^2R losses, and resistance on the extreme right in the figure, which represents the power transferred across the air gap that is available to drive the shaft load. In the negative-sequence network, the slip is changed from the positive-sequence value of s to the new value of $(2 - s)$.

In many types of problems, the rotor resistance and reactance to both positive- and negative-sequence currents can be considered constant. However, in the analysis of motor starting, these parameters should be modeled as a function of the slip [9], [11]. It has been shown that it is reasonably accurate to consider both the resistance and reactance as linear functions of slip or rotor speed, as shown in Figure 19.2 [4], [9].

If we assume the linear characteristic shown in the figure, we can write the following equations for the positive-sequence circuit.

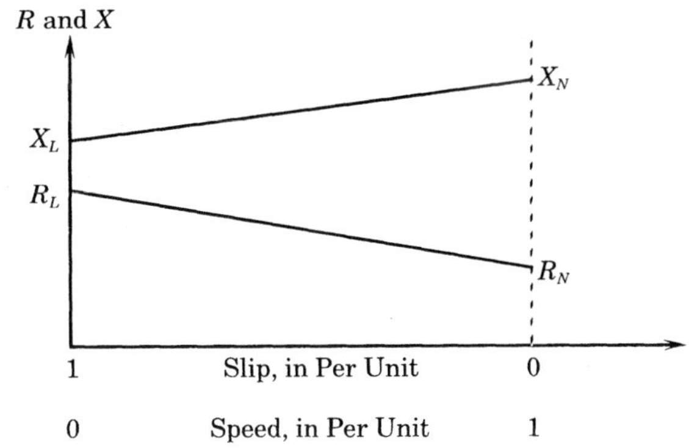

Figure 19.2 Variation of rotor resistance and reactance with slip.

$$\begin{aligned} R_{R1} &= R_N + (R_L - R_N)s \\ X_{R1} &= X_N - (X_N - X_L)s \end{aligned} \tag{19.35}$$

$$\begin{aligned} \text{where } R_N, X_N &= \text{rotor resistance, reactance running} \\ R_L, X_L &= \text{locked-rotor resistance, reactance} \\ s &= \text{slip} \end{aligned} \tag{19.36}$$

A similar relationship holds for the negative-sequence network, but with s replaced by $2 - s$.

$$\begin{aligned} R_{R2} &= R_N + (R_L - R_N)(2 - s) \\ X_{R2} &= X_N - (X_N - X_L)(2 - s) \end{aligned} \tag{19.37}$$

These equations are important when studying the starting characteristics of the motor, a condition that is of considerable interest to the protection engineer.

19.2.2.1 The Positive-Sequence Equivalent. The positive-sequence network equivalent is solved by first analyzing the impedance seen looking into the parallel impedances shown in Figure 19.3.

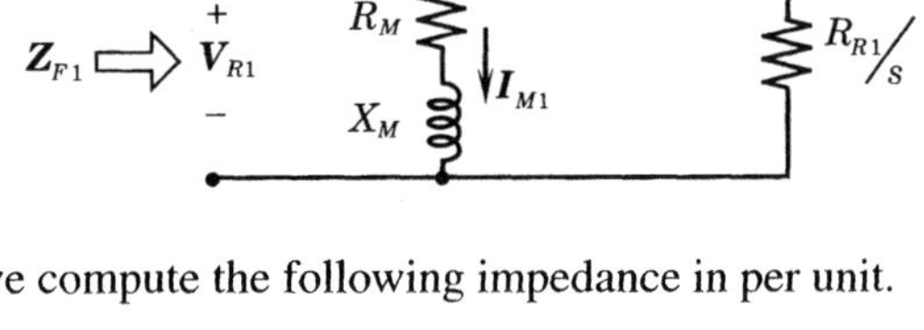

Figure 19.3 Analysis of the impedance $\mathbf{Z}_{F1}$.

From Figure 19.3, we compute the following impedance in per unit.

$$\mathbf{Z}_{F1} \equiv R_{F1} + jX_{F1} \tag{19.38}$$

$$\begin{aligned} \text{where } R_{F1} &= \frac{(R_{R1}/s + R_M)(R_{R1}R_M/s - X_{R1}R_M) + (X_{R1} + X_M)(R_{R1}X_M/s + X_{R1}R_M)}{(R_{R1}/s + R_M)^2 + (X_{R1} + X_M)^2} \\ X_{F1} &= \frac{(R_{R1}/s + R_M)(R_{R1}X_M/s + X_{R1}R_M) - (X_{R1} + X_M)(R_{R1}R_M/s - X_{R1}X_M)}{(R_{R1}/s + R_M)^2 + (X_{R1} + X_M)^2} \end{aligned} \tag{19.39}$$

Then we may combine (19.38) with the stator and source impedances to write the total positive-sequence impedance of the motor as

$$\begin{aligned} \mathbf{Z}_1 &\equiv R_1 + jX_1 = Z_T + Z_S + Z_{F1} \\ &= (R_T + R_S + R_{F1}) + j\,(X_T + X_S + X_{F1}) \quad \text{pu} \end{aligned} \tag{19.40}$$

The total positive-sequence input current is computed as follows.

$$\mathbf{I}_{a1} = \frac{\mathbf{V}_{a1}}{\mathbf{Z}_1} = \frac{\mathbf{V}_{a1}}{R_1 + jX_1} \quad \text{pu} \tag{19.41}$$

It is helpful in the algebraic analysis to define the following simplifications of the voltage and current. Let

$$\begin{aligned} \mathbf{V}_{a1} &\equiv E_1 + jF_1 \text{ pu} \\ \mathbf{I}_{a1} &\equiv A_1 + jB_1 \text{ pu} \end{aligned} \tag{19.42}$$

Then, from (19.41), we may write the per unit current as

$$\mathbf{I}_{a1} = \frac{(E_1 + jF)(R_1 - jX_1)}{R_1^2 + X_1^2} = \frac{(E_1R_1 + F_1X_1) + j\,(F_1R_1 - E_1X_1)}{R_1^2 + X_1^2} \quad \text{pu} \tag{19.43}$$

Using the normalization technique of Section 19.2.1.2, the input apparent power can be written in vector form as

$$P_{\text{IN}} + jQ_{\text{IN}} = 3\mathbf{V}_{012}\mathbf{I}_{012}^* \quad \text{VA} = \mathbf{V}_{012}\mathbf{I}_{012}^* \quad \text{pu} \tag{19.44}$$

Then

$$\begin{aligned} P_{\text{IN}} + jQ_{\text{IN}} &= \mathbf{V}_{a1}\mathbf{I}_{a1}^* + \mathbf{V}_{a2}\mathbf{I}_{a2}^* \\ &= (P_1 + jQ_1) + (P_2 + jQ_2) \text{ pu} \end{aligned} \tag{19.45}$$

where we ignore the zero-sequence product since we assume that the motor is wye connected, but with an ungrounded neutral. Therefore, no zero-sequence currents can flow. With the result from (19.42) and (19.44), we can write the positive-sequence contribution to the input apparent power.

$$P_1 + jQ_1 = (E_1A_1 + F_1B_1) + j\,(F_1A_1 - E_1B_1) \quad \text{pu} \tag{19.46}$$

Now, the rotor voltage of Figure 19.3 can be computed as follows.

$$\mathbf{V}_{R1} = \mathbf{V}_{a1} - \mathbf{I}_{a1}\mathbf{Z}_1 = \mathbf{I}_{a1}\mathbf{Z}_{F1} \quad \text{pu} \tag{19.47}$$

Having computed (19.47), the positive-sequence excitation current can be computed as

$$\mathbf{I}_{M1} = \frac{\mathbf{V}_{R1}}{R_M + jX_M} \quad \text{pu} \tag{19.48}$$

and the positive-sequence rotor current can then be computed.

$$\begin{aligned} \mathbf{I}_{R1} &= \mathbf{I}_{a1} - \mathbf{I}_{M1} = \frac{\mathbf{Z}_M}{\mathbf{Z}_{R1} + \mathbf{Z}_M}\mathbf{I}_{a1} \\ &= \left\{ \begin{array}{l} \dfrac{[R_M\,(R_{R1}/s + R_M) + X_M\,(X_{R1} + X_M)]}{(R_{R1}/s + R_M)^2 + (X_{R1} + X_M)^2} \\ + j\dfrac{[X_M\,(R_{R1}/s + R_M) - R_M\,(X_{R1} + X_M)]}{(R_{R1}/s + R_M)^2 + (X_{R1} + X_M)^2} \end{array} \right\} \mathbf{I}_{a1} \quad \text{pu} \end{aligned} \tag{19.49}$$

This can be simplified by defining the coefficient in (19.49) as a complex constant.

$$\begin{aligned} \mathbf{I}_{R1} &\equiv A_{R1} + jB_{R1} = (K_{R1} + jK_{I1})\mathbf{I}_{a1} \\ &= (K_{R1}A_1 - K_{I1}B_1) + j\,(K_{R1}B_1 + K_{I1}A_1) \end{aligned} \tag{19.50}$$

where the notation for the complex coefficient to the current in (19.49) is simplified. Knowing the rotor voltage and the positive-sequence input current, we can compute both the stator copper losses and the core losses, as

$$P_{SC1} = R_S I_{a1}^2 \quad \text{pu} \tag{19.51}$$

$$P_{CL1} = R_M I_{M1}^2 \quad \text{pu} \tag{19.52}$$

Then the positive-sequence air-gap power can be written as

$$P_{AG1} = P_1 - P_{SC1} - P_{CL1} = I_{R1}^2 \frac{R_{R1}}{s} \quad \text{pu} \tag{19.53}$$

Now, the rotor copper losses are given in per unit by

$$P_{RC1} = R_{R1} I_{R1}^2 \quad \text{pu} \tag{19.54}$$

It should be emphasized that these results are in per unit. If the same equations are written in mks units, as is done in many textbooks [12], [13], the results will have the factor "3" on each of these computed values of power, which recognizes that the equivalent circuit is a per-phase equivalent and there are three identical phases. This factor of "3" is not required in the normalized equations since x per unit of power is the same whether x per unit per phase or x per unit three phase.

Motors also have friction, windage, and stray load losses that are functions of the rotor angular velocity. We can model these rotational losses approximately as

$$P_{\text{ROT}} = P_{\text{CON}}\omega = P_{\text{CON}}(1 - s) \quad \text{pu} \tag{19.55}$$

where the constant P_{CON} can be assumed to be a few percent of the motor rating, say in the range of 0.05–0.08 per unit.

The positive sequence developed torque of the induction motor is computed as the air-gap power less the rotor copper losses.

$$P_{D1} = P_{AG1} - P_{RC1} = R_{R1} I_{R1}^2 \frac{1-s}{s} = P_{AG1}(1 - s) \quad \text{pu} \tag{19.56}$$

The shaft output power is the developed power less the rotor core losses. Then, the shaft mechanical torque is determined from the developed power.

$$\begin{aligned} T_{M1} &= \frac{P_{S1}}{\omega} = \frac{(P_{AG1} - P_{\text{CON}})(1-s)}{\omega_S(1-s)} = \frac{(P_{AG1} - P_{\text{CON}})}{\omega_S} \\ &= R_{R1} I_{R1}^2 \frac{1-s}{s} - P_{\text{CON}} \quad \text{pu} \end{aligned} \tag{19.57}$$

where we recognize that the synchronous angular velocity in per unit is equal to one per unit. In many cases it is not important to include the rotational loss term, as it is always small. Equation (19.57) is the desired result, as it permits the solution positive-sequence torque term in the startup equation. The only required input variable is the positive-sequence voltage at the source Thevenin equivalent. Note that we could also compute the motor terminal voltage as simply the voltage drop across the Thevenin equivalent impedance. However, this voltage is not required for the computation of the positive-sequence torque.

There are no assumptions made in the foregoing derivation regarding the applied voltage. If the applied voltage is not a balanced three-phase set of voltages, the unbalanced voltages can be resolved into their symmetrical components. Only the positive-sequence component is used in the foregoing development. We now repeat the same type of development for the negative-sequence network of Figure 19.1.

19.2.2.2 The Negative-Sequence Equivalent. The procedure for computing the negative-sequence torque is similar to that for the positive-sequence torque, the major difference being the function of slip in the negative-sequence motor equivalent circuit.

For the negative-sequence equivalent of Figure 19.1, we assume that the negative-sequence voltage for the Thevenin equivalent has been computed. Then, we solve the negative-sequence network for the currents. The first step is to find the solution for that part of the network shown in Figure 19.4 by computing the impedance seen looking into this portion of the network.

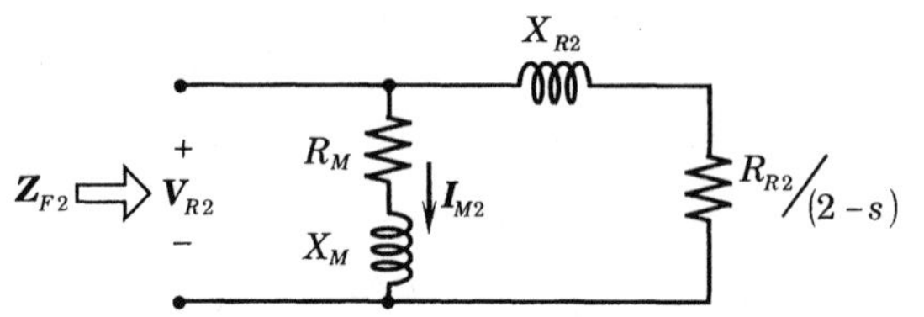

Figure 19.4 Analysis of the impedance $\boldsymbol{Z}_{F2}$.

The solution of Figure 19.4 gives the following results.

$$\begin{aligned} R_{F2} &= \frac{\left(\frac{R_{R2}}{2-s} + R_M\right)\left(\frac{R_{R2}R_M}{2-s} - X_{R2}X_M\right) + (X_{R2} + X_M)\left(\frac{R_{R2}X_M}{2-s} + X_{R2}R_M\right)}{\left(\frac{R_{R2}}{2-s} + R_M\right)^2 + (X_{R2} + X_M)^2} \\ X_{F2} &= \frac{\left(\frac{R_{R2}}{2-s} + R_M\right)\left(\frac{R_{R2}X_M}{2-s} + X_{R2}R_M\right) - (X_{R2} + X_M)\left(\frac{R_{R2}R_M}{2-s} - X_{R2}X_M\right)}{\left(\frac{R_{R2}}{2-s} + R_M\right)^2 + (X_{R2} + X_M)^2} \end{aligned} \tag{19.58}$$

Then

$$\boldsymbol{Z}_2 = \boldsymbol{Z}_T + \boldsymbol{Z}_S + \boldsymbol{Z}_{F2} = (R_T + R_S + R_{F2}) + j\,(X_T + X_S + X_{F2}) \tag{19.59}$$

The negative-sequence motor current can be computed as

$$\boldsymbol{I}_{a2} = \frac{\boldsymbol{V}_{a2}}{\boldsymbol{Z}_2} = \frac{E_2 + jF_2}{R_2 + jX_2} = \frac{(E_2R_2 + F_2X_2) + j\,(F_2R_2 - E_2X_2)}{R_2^2 + X_2^2} \equiv A_2 + jB_2 \tag{19.60}$$

The negative-sequence apparent power is given by

$$P_2 + jQ_2 = \boldsymbol{V}_{a2}\boldsymbol{I}_{a2}^* = (E_2A_2 + F_2B_2) + j\,(F_2A_2 - E_2B_2) \quad \text{pu} \tag{19.61}$$

The negative-sequence air-gap power is given by

$$P_{ag2} = P_2 - P_{SC2} - P_{CL2} = I_{R2}^2\frac{R_{R2}}{2-s} \tag{19.62}$$

The negative-sequence rotor current can be computed as

$$\boldsymbol{I}_{R2} = \frac{\boldsymbol{Z}_M}{\boldsymbol{Z}_{R2} + \boldsymbol{Z}_M}\boldsymbol{I}_{a2} \equiv (K_{R2} + jK_{I2})\boldsymbol{I}_{a2} \equiv A_{R2} + jB_{R2} \tag{19.63}$$

where the complex constants can be computed from the circuit parameters.

The magnetizing current is computed as

$$\boldsymbol{I}_{M2} = \boldsymbol{I}_{a2} - \boldsymbol{I}_{R2} \tag{19.64}$$

The negative-sequence core losses are given by

$$P_{CL2} = R_M I_M^2 \quad \text{pu} \tag{19.65}$$

The rotor copper losses are given by

$$P_{RC2} = R_{R2} I_{R2}^2 \quad \text{pu} \tag{19.66}$$

Then the negative-sequence developed mechanical power is computed as the difference between the air-gap power and the rotor copper losses.

$$P_{D2} = P_{AG2} - P_{RC2} = -R_{R1} I_{R2}^2 \frac{1-s}{2-s} \quad \text{pu} \tag{19.67}$$

and we note that this developed power is always negative.

The negative-sequence mechanical torque is computed as

$$T_{M2} = \frac{P_{D2}}{\omega} = \frac{P_{D2}}{\omega_S (1-s)} = -\frac{I_{R2}^2 R_{R2}}{2-s} \quad \text{pu} \tag{19.68}$$

This completes the analysis of the positive- and negative-sequence torques.

19.2.3 The Net Accelerating Torque

The net accelerating torque consists of two parts, the mechanical torque developed by the motor and the retarding or load torque presented by the driven load.

19.2.3.1 The Mechanical Torque. The total net torque is the total mechanical torque developed by the motor less the load torque presented by the shaft load. The total mechanical torque is the sum of the two sequence torques, given by (19.57) and (19.68).

$$T_M = T_{M1} + T_{M2} = \frac{R_{R1} I_{R1}^2 (1-s)}{s} - \frac{R_{R2} I_{R2}^2}{2-s} - P_{\text{CON}} \quad \text{pu} \tag{19.69}$$

Note that this torque can be computed for any value of slip, given the sequence voltages. If the voltages are balanced, the negative-sequence torque is zero.

19.2.3.2 The Load Torque. The load torque depends on the type of load, with fans having a different torque-speed characteristic than pumps or other rotating equipment. The equation for the load torque as a function of speed can be written in several different ways.

One model of load torque is to write the total load as a function of the initial and final torques and the angular velocity. If we let T_I be the initial value and T_F be the final value, we can write the following general equation.

$$T_{L1} = T_I + (T_F - T_I)\,\omega^n \quad \text{pu} \tag{19.70}$$

A second load torque model also uses the initial and final values of torque, but in a different manner.

$$T_{L2} = T_I (1-\omega)^m + T_F \omega^n \quad \text{pu} \tag{19.71}$$

Examples of these two different models of load torque are shown in Figure 19.5 using typical values of the model parameters. Load torque model 1 causes the torque to increase monotonically from its initial to its final value. Load torque model 2 causes the net torque to fall after the initial starting value and finally rise sharply to its final value [17]. This type of behavior might be used to model loads that are difficult to start, making the torque at low speeds relatively high compared to the torque that is required once the load begins moving.

19.2.3.3 The Accelerating Torque. The net torque acting to accelerate the shaft is the difference between the motor developed torque and the load torque, or

$$T_A = T_M - T_L \quad \text{pu} \tag{19.72}$$

This value of torque is substituted into the swing equation to solve for the rotor speed as a function of time.

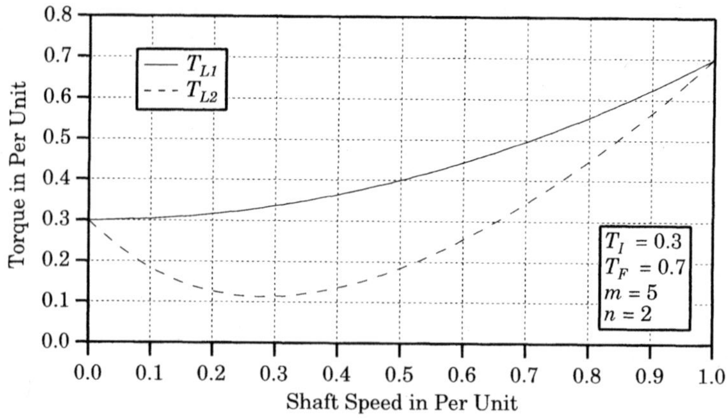

Figure 19.5 Load torque using two different models.

19.2.3.4 The Swing Equation. The swing equation can be written as

$$\frac{d\omega}{dt} = \frac{T_M - T_L}{2H} = F(\omega, t) \tag{19.73}$$

with the initial conditions

$$\omega = 0 \quad \text{at} \quad t = 0 \tag{19.74}$$

The differential equation (19.73) is a nonlinear equation that is not easily integrated, but can be solved by numerical methods. The integration begins with the condition (19.74) and computes the motor currents and torque contributions for both the positive- and negative-sequence networks for any given set of applied three-phase voltages, balanced or unbalanced. The numerical integration can be terminated as the values of developed mechanical torque and load torque become equal, which forces the accelerating torque (19.72) to zero. As the accelerating torque approaches zero, the time derivative of ω approaches zero and the speed approaches a constant that results in a small value of slip. The final speed depends on the torque required by the load, which is given by T_F in (19.70) or (19.71).

19.2.3.5 Integration of the Swing Equation. The swing equation can be easily integrated numerically to compute the shaft speed as a function of time and, in so doing, compute all of the motor currents, voltages, and other electrical parameters as well. To do this, we let $t_0, t_1, \ldots, t_n$ be equally spaced values of time such that

$$t_{n+1} - t_n = h \quad n = 0, 1, 2, \ldots \tag{19.75}$$

Using the notation,

$$\omega_n = f(t_n) \tag{19.76}$$

we can write the derivative of speed as

$$\frac{d\omega}{dt} = \lim_{t_{n+1} \to t_n} \left[\frac{\omega_{n+1} - \omega_n}{t_{n+1} - t_n} \right] = \lim_{h \to 0} \left[\frac{\omega_{n+1} - \omega_n}{h} \right] \tag{19.77}$$

Now from (19.73), we can write the difference equation

$$\frac{\omega_{n+1} - \omega_n}{h} = F(\omega_n, t_n) \tag{19.78}$$

or

$$\omega_{n+1} = \omega_n + hF(\omega_n, t_n) \tag{19.79}$$

We begin the process by computing the initial values of (19.79), i.e., for the value $n = 0$, which corresponds to a value of $\omega_0 = 0$. Solving (19.79) for this value gives speed for the next time step. The process terminates when

$$\begin{aligned} n &= k \\ \omega_k - \omega_{k-1} &= 0 \end{aligned} \tag{19.80}$$

This is a very simple numerical integration algorithm called Euler's rule. The only control on the accuracy of the method is the size of the time step, h. Choosing relatively small values of h ensures reasonable accuracy. More complex numerical integration algorithms are available that employ more powerful error control measures.

19.2.4 Motor Electrical and Mechanical Performance

Using the foregoing equations for the motor performance, the numerical values of developed torque, load torque, current, and other variables of interest can be computed as a function of time. An example will be used to illustrate the performance of an induction motor starting.

EXAMPLE 19.1

Consider a 10 HP, 220 volt, 6 pole, 60 Hz, three-phase induction motor, for which the following motor data are available [12]:

$$\begin{aligned} R_S &= 0.294\ \Omega & X_S &= 0.503\ \Omega \\ R_N &= 0.144\ \Omega & X_N &= 0.209\ \Omega \\ R_M &= \text{not given} & X_M &= 13.25\ \Omega \end{aligned}$$

This is rather typical of the type and amount of data that the protection engineer will likely find for most motors. The no-load rotational loss and stray load loss are estimated to be about 400 W at rated voltage and frequency. These losses may be considered to be constant and independent of load [12]. Determine the normalized motor equivalent circuit.

Solution

The data for the motor is not entirely adequate for a complete solution. However, we know that most induction motors have the capability of driving rated shaft load at a small value of slip. The nameplate full load current is 18.8 amperes. In order to solve the equations, we need to know the full load slip. This can be solved in two ways. An algebraic solution of the motor electrical and mechanical equations can be performed, iterating on the value of slip until the accelerating torque is zero, to within a given allowed error. The second method is to solve the differential equations of the system for a time period long enough for the accelerating torque to reach zero to within a prescribed error allowance.

The base quantities for the motor can be summarized as follows.

$$V_B = 220\ V \qquad I_B = 18.8\ \text{A}$$

Then

$$S_B = \sqrt{3}\, V_B I_B = 7163.76\ \text{VA}$$

$$Z_B = \frac{V_B^2}{S_B} = 6.756\ \Omega$$

Then, the per unit impedance parameters can be summarized as follows.

$$\begin{aligned} R_S &= 0.0435\ \text{pu} & X_S &= 0.745\ \text{pu} \\ R_N &= 0.0213\ \text{pu} & X_N &= 0.0309\ \text{pu} \\ R_M &= \text{not known} & X_M &= 1.961\ \text{pu} \end{aligned}$$

We know that the no-load losses of the motor are about 400 watts. This value can be used to solve for the value R_M and provides one approach. Some analysts ignore the resistive branch and solve the motor equations with only the shunt reactance X_M in the circuit. We will use another technique, whereby we estimate that

$$\frac{X_M}{R_M} = 7.7$$

This gives a value for R_M of 0.2547 per unit. After solving the motor equivalent circuit, we can check to see if this results in no-load losses in the neighborhood of 400 watts.

Our solution method will be to compute the air-gap power in two different ways. One computation will be based on the estimated value of slip using (19.53). The other method is to compute the air-gap power as the positive-sequence input power less the stator winding and core losses. The two values will usually differ, but can be made equal by iterating on the slip until the two values are equal.

Now, if rated full load current flows in the stator windings, the voltage across the positive-sequence mutual branch of Figure 19.1 can be computed assuming rated voltage is applied to the motor and finding the impedance $\mathbf{Z}_{F1}$ from (19.39). The result of this calculation gives

$$\mathbf{Z}_{F1} = 0.803 + j0.425 \quad \text{pu}$$

Adding the stator impedance gives the total positive-sequence impedance.

$$\mathbf{Z}_1 = 0.846 + j0.499 \quad \text{pu}$$

Then the positive-sequence stator current can be computed as

$$\mathbf{I}_{a1} = \frac{\mathbf{V}_{a1}}{\mathbf{Z}_1} = 0.876 - j0.547 \quad \text{pu}$$

and

$$P_1 + jQ_1 = 0.876 + j0.517 \quad \text{pu}$$

The positive-sequence stator losses can be computed as

$$P_{SL1} = R_S I_{a1}^2 = 0.045 \quad \text{pu}$$

The positive-sequence rotor current can be computed from (19.49) to give

$$\mathbf{I}_{R1} = 0.835 - j0.057 \quad \text{pu}$$

Knowing both the stator and rotor currents, we can compute the current in the mutual branch by subtracting the two currents.

$$\mathbf{I}_{M1} = 0.041 - j0.460 \quad \text{pu}$$

Then the positive-sequence core losses can be computed from

$$P_{CL1} = R_M I_{M1}^2 = 0.118\,pu = 389\,\text{W}$$

This result is a reasonable one, so we conclude that our estimate of R_M is acceptable. In some cases, it may be necessary to try different values of this parameter in order to find a value for these losses that is acceptable.

Now the positive-sequence air-gap power can be computed from the above results.

$$P_{AG1} = P_1 - P_{SC1} - P_{CL1} = 0.777 \quad \text{pu}$$

As a check on this results, we can also compute

$$P_{AG1} = R_{R1}\frac{I_{R1}^2}{s} = 0.764 \quad \text{pu}$$

This does not check exactly since the assumed slip is not exactly right, but it is close enough to use as an initial estimate. One solution for the air-gap torque is based on an estimated value of slip, taken as 0.02, which results in a final value of accelerating torque that is not exactly zero. The correct solution can be found algebraically by iterating on the slip. From the air-gap power mismatch, it is noted that the slip should be somewhat less than the assumed value of 0.02. The correct value of slip is actually 0.01939.

Figure 19.6 shows the motor-developed torque, the load torque and the accelerating torque as functions of slip. The solution is that value of slip where the accelerating torque is zero.

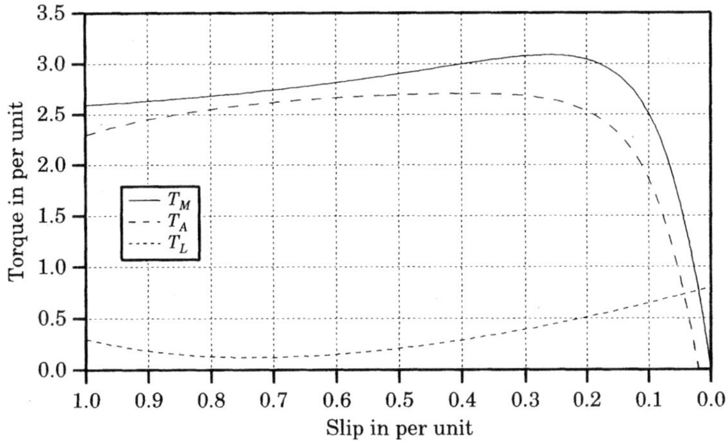

Figure 19.6 Motor, accelerating, and load torques as a function of slip.

Thus, the motor behavior is completely determined for this case of balanced applied voltages and assumed initial conditions. A similar process can be followed if unbalanced voltages are applied. This is left as an exercise. ■

If the motor equations are solved algebraically, iterating on the slip until the accelerating torque is zero, the process must be carried out to several decimals of accuracy because the torque-speed curve of the motor varies rapidly in the neighborhood of the solution, as shown in Figure 19.6. It is also noted that this type of solution gives an inaccurate understanding of the machine behavior during motor starting, as it ignores the smooth transition from the transient to the steady-state condition as the acceleration approaches zero. This will be illustrated by an example.

EXAMPLE 19.2

Use the equivalent circuit of Example 19.1 to determine the motor performance for values $H = 4$, 8, and 12 s. Let the load torque be represented by (19.71) with the initial value given as 0.3 per unit and the final value as 0.9 per unit. Thus, we can write the load torque as

$$T_L = 0.3\,(1 - \omega)^5 + 0.8\omega^2 \quad \text{pu}$$

where we have let $m = 5$ and $n = 2$, exactly as in Figure 19.5. Assume that the applied voltages are perfectly balanced rated voltages.

Solution

Motor startup performance can be computed from the equivalent circuit and the given load torque characteristic. First, we compute the motor speed in accelerating the load with three different values of the inertia constant. The results are shown in Figure 19.7. For any given H, the acceleration is almost constant until the motor approaches a speed where the developed torque and the load torque are nearly equal. As the accelerating torque approaches zero, the acceleration of the shaft approaches zero until the speed approaches a constant value at a small value of slip.

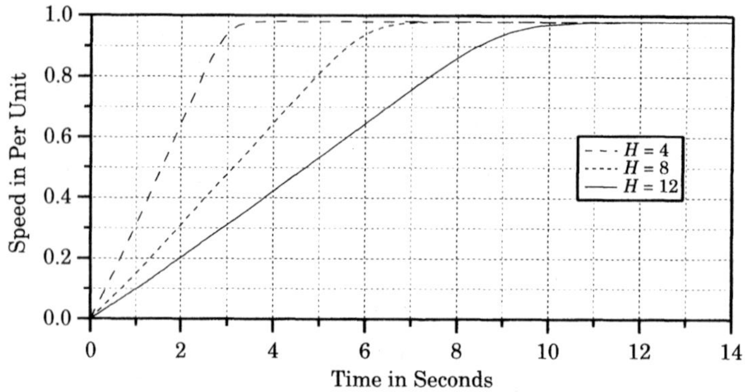

Figure 19.7 Motor speed in per unit at startup.

The three values of torque represent a typical "normal" moment of inertia (for $H = 4$), a moderately high value ($H = 8$), and a very high inertia ($H = 12$). Note that, for the higher values of inertia constant, the motor requires a relatively long time to reach normal operating speed.

The torques for the same values of inertia constant are shown in Figure 19.8.

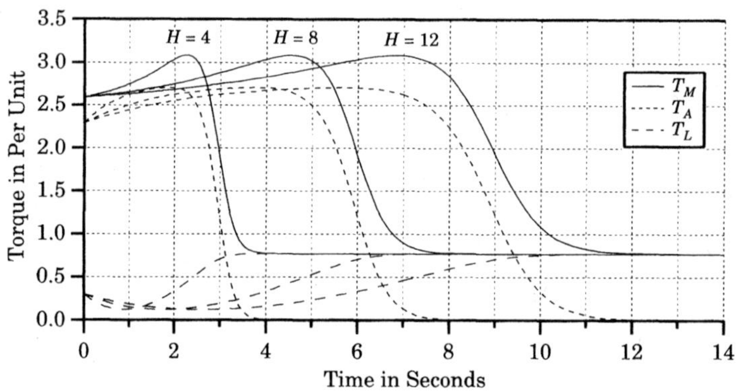

Figure 19.8 Motor developed, accelerating, and load torques in per unit.

The motor developed torque starts at the same value for each of the loads, since the initial load torque is the same for all three conditions. For the case illustrated, there is a substantial value of accelerating torque initially. Note the increase in developed torque as the motor accelerates, with this torque reaching a maximum value before falling rapidly until the motor and load torques become equal as the accelerating torque goes to zero.

Finally, we examine the motor currents at startup. Current behavior is important since motor overcurrent relays are often used to detect motor internal faults, and it is essential that these overcurrent relays not trip on startup. For the three inertias used in this example, the starting currents are shown in Figure 19.9.

The initial current is about 7.5 per unit, and the current remains above 5 per unit for over 7 seconds for the case with the largest inertia. Note that the initial and final values of current are identical for all three values of inertia constant. This is because the load torque is the same for all three cases and the applied voltage is also the same.

Another way of viewing the current is in the complex plane, where we plot the phasor locus, or the imaginary part of the current against the real part, as shown in Figure 19.10. Here, we see that the phasor

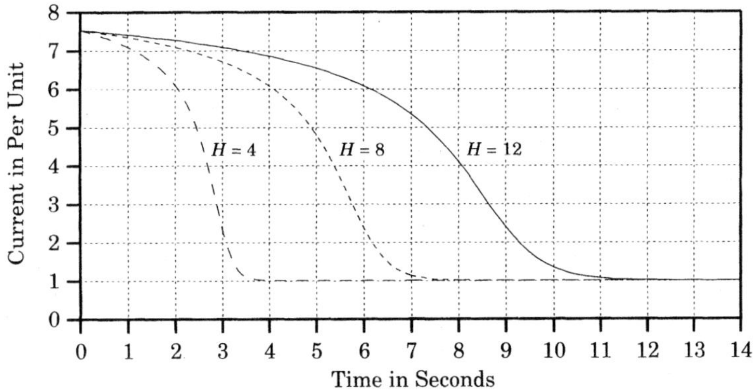

Figure 19.9 Motor current magnitude during starting in per unit.

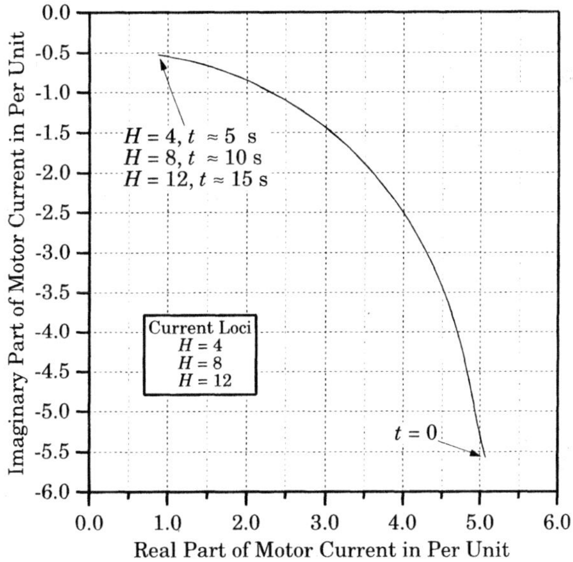

Figure 19.10 Complex plane plot of motor starting current.

motor current is identical for all three cases. Only the time required to complete the movement along the loci from start to finish is different. The approximate time required to achieve the final steady-state value is noted in the figure.

This example shows that the time required to accelerate a motor to a steady-state running condition depends on the torque requirements of the load and the total inertia of motor and load. The current drawn by the motor during startup can be several times greater than the normal full-load motor current. This high current is sustained for several seconds, and this may pose a problem for the protective relaying. ■

The previous example illustrates the behavior of a relatively small, integral horsepower motor with perfectly balanced phase voltages. In this case, the negative-sequence currents and torque are zero. We now examine the effect caused by applying unbalanced voltages to the motor, which cause negative-sequence currents to flow.

EXAMPLE 19.3

Compute the motor starting characteristics for the case of unbalanced applied voltages. For this example, the motor applied voltages are arbitrarily set as follows:

$$V_a = 1.0\angle 0^\circ$$
$$V_b = 0.8\angle -120^\circ$$
$$V_c = 0.9\angle +120^\circ$$

Solution

For these values of applied voltage a negative-sequence component of voltage will exist and negative-sequence currents will flow, creating a negative-sequence torque that will have a negative magnitude. This will affect the motor acceleration and, since the load torque is a function of speed, the load torque will also differ from the balanced case. A comparison of these torques for the case with $H = 8$ seconds is shown in Figure 19.11.

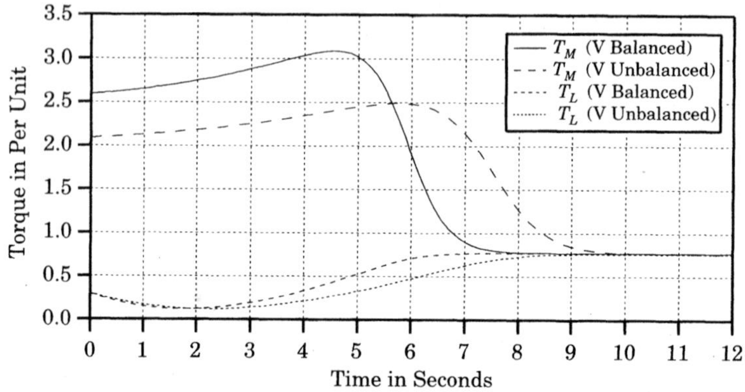

Figure 19.11 Comparison of torques for the balanced and unbalanced applied voltages, $H = 8$ seconds.

The developed motor torque is significantly lower than normal due to the unbalanced applied voltages. Because of the lower developed torque, the rotating inertia accelerates more slowly to its final value.

The effect of the unbalanced voltages on rotor speed shows clearly the reduction in acceleration due to the unbalanced condition. This is illustrated in Figure 19.12.

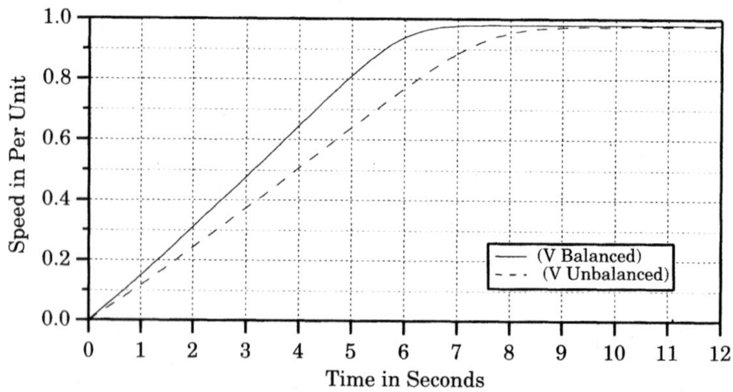

Figure 19.12 Comparison rotor angular velocity for balanced and unbalanced applied voltages, $H = 8$ seconds.

Finally, we examine the effect of the unbalanced supply voltages on the motor currents. This result is shown in Figure 19.13. Due to the unbalanced supply voltages, the initial current is somewhat smaller than that for the normal case. However, because of the slower acceleration of the inertia, the current persists at a high level for a longer period of time. Figure 19.13 shows three current magnitudes for the unbalanced voltage case, the total current (long dashes), the positive-sequence current (short dashes), and the negative-sequence current (alternating dashes and dots). The magnitude of the negative-sequence current is not great, but its effect is considerable.

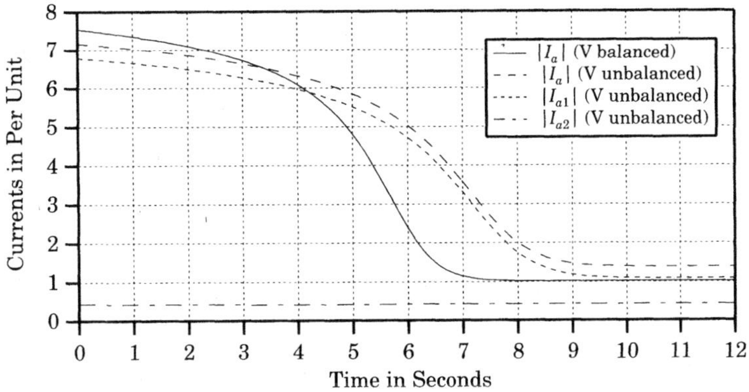

Figure 19.13 Comparison of motor currents for balanced and unbalanced applied voltages, $H = 8$ seconds. ■

19.3 INDUCTION MOTOR HEATING

One of the most difficult problems in motor protection is to provide protection against overheating of the motor. This is especially difficult for large motors that drive high-inertia loads, because the high value of motor starting current has a high probability of heating the motor beyond established thermal limits. In order to understand this problem, we first review briefly the physical fundamentals of heat transfer.

19.3.1 Heat Transfer Fundamentals

Heat can be transferred from one medium to another by three methods:

- Conduction
- Convection
- Radiation

All three of these concepts will be examined with the view toward understanding the important mechanisms that apply to induction motors.

19.3.1.1 Heat Transfer by Conduction. Conduction can take place in a body only when different parts of the body are at different temperatures [18], [19]. The direction of heat flow is always from the points of higher temperature to the points of lower temperature.

Consider the slab of material shown in Figure 19.14. The body is a slab of material L units thick and having an area of A. The temperature on the left surface is at a higher temperature than the right surface. The heat flow q is from the higher temperature surface to the lower temperature surface. We can write an equation that describes the heat flow as follows:

$$q = \frac{kA}{L}\Delta\theta = \frac{1}{\boldsymbol{R}}(\theta_1 - \theta_2) \quad \text{W} \tag{19.81}$$

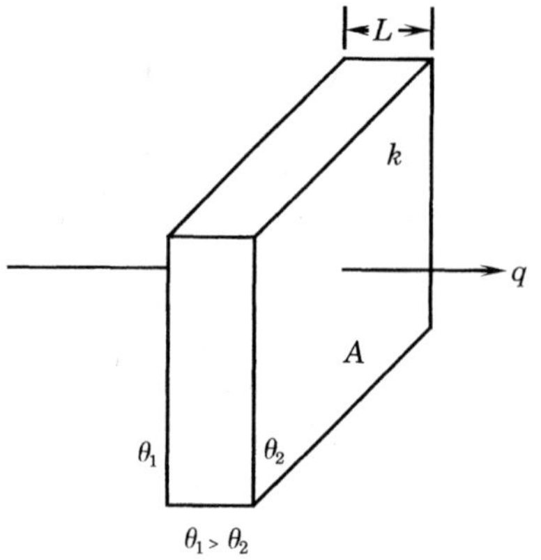

Figure 19.14 Heat flow by conduction through a body.

where q = heat flow (W)
θ = temperature (K)
A = area (m^2)
L = length of path (m)
k = thermal conductivity of material (W/m · K)
$\boldsymbol{R}$ = thermal resistance (K/W)

The thermal resistance is defined as[1]

$$\mathcal{R} = \frac{L}{kA} \quad \text{K/W} \tag{19.82}$$

which is similar to the definition of electrical resistance. In fact, it is common practice to construct electric analogs of thermal heat transfer problems, which often simplifies the analysis of multiple series, parallel, or other material configurations [19–22].

The thermal conductivity of the material is a property that can be found in various references [18], [19]. A few examples are shown in Table 19.1. Note that materials that have high electrical conductivity also have high thermal conductivity.

The previous discussion considers the flow of heat through a material, from a high temperature region to a lower temperature region. However, another important concept is the retention of heat within a material. The rate at which a material can receive heat is given by

$$q = \frac{dU}{dt} \tag{19.83}$$

where the variable U is called the internal energy and is specified by the following relation.

$$dU = mc_p\, d\theta \tag{19.84}$$

[1] A caligraphic $\mathcal{R}$ is used for thermal resistance to distinguish it from electrical resistance.

TABLE 19.1 Thermal Conductivity of Common Materials [19]

k (W/mK)	Material
0.166	Asbestos
36	Carbon steel (1.5%)
54	Carbon steel (0.5%)
204	Aluminum
386	Copper (at 20°C)

where m = mass (kg)
c_p = specific heat (J/kg · K)
θ = temperature (K)

The specific heat at constant pressure of materials is a parameter that is published in many references. Table 19.2 gives a few typical values.

TABLE 19.2 Specific Heat of Common Materials at 100° K [23]

c_p in J/kg K	Material
481	Aluminum
254	Copper
108	Gold
216	Iron

Therefore, we can write

$$q = mc_p\frac{d\theta}{dt} = \mathcal{C}\frac{d\theta}{dt} \quad \text{W} \tag{19.85}$$

where we define the *thermal capacitance*[2]

$$\mathcal{C} = mc_p \quad \text{J/K} \tag{19.86}$$

Note the similarity between (19.85) for the thermal capacitance and the equation for the current flowing in a capacitor in an electric network

$$i = C\frac{dV}{dt} \tag{19.87}$$

where V is the voltage across the capacitor.

If we combine the concepts of thermal resistance and capacitance, we have a practical lumped-parameter model for the thermal system. For example, if we have an insulating material in contact with a metal, such as copper, we can treat the insulator as purely resistive and the metal as purely capacitive. Then, for a given heat flow through the materials in response to suddenly subjecting the insulation to a high temperature θ_1 we can write

$$q = \frac{\theta_1 - \theta}{\mathcal{R}} = \mathcal{C}\frac{d\theta}{dt} \tag{19.88}$$

[2]A caligraphic $\mathcal{C}$ is used for thermal capacitance to distinguish it from electrical capacitance.

Solving for the temperature rise in the metal as the heat flows through the insulator, we get

$$\theta = \theta_1 - (\theta_1 - \theta_0)\, e^{-t/RC} \tag{19.89}$$

where θ_0 is the initial temperature of the metal. This is exactly the same solution as the voltage V_C across the capacitor in an electric RC network due to a suddenly applied dc voltage, as shown in Figure 19.15, where the initial voltage on the capacitor is V_0 and the switch is closed at time zero.

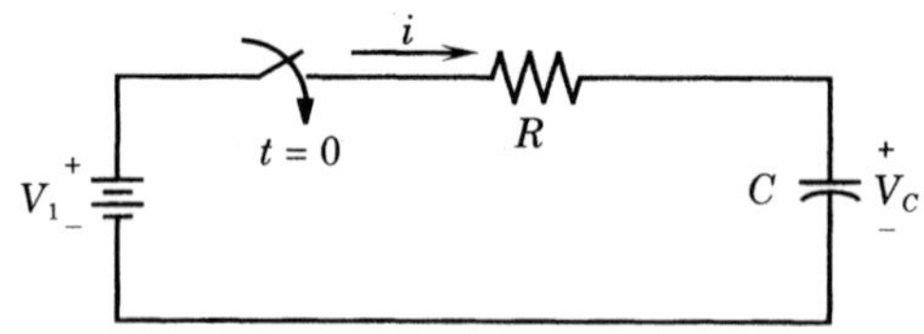

Figure 19.15 Electric analog of (19.89).

For an electric motor, the source of heat is the I^2R losses in the stator windings, the core losses in the stator iron, and the I^2R losses in the rotor. The motor windings and core have thermal capacitance that is capable of storing heat and heat will flow from the sources through the surrounding materials. In the steady-state, the heat is removed from the motor by convection, which is discussed in the next section. During the starting of the motor, however, it can be assumed that all of the heat is stored in the motor metals and the starting current causes an increase in the temperature of these metals. The purpose of a thermal model is to determine the temperature rise in the metals and to assure that this rise is not excessive.

19.3.1.2 Heat Transfer by Convection. Convection is the transfer of heat from one place to another by the actual motion of hot fluid. Convection can be either *forced* or *natural* (free). In convection, heat is lost or gained by a surface at one temperature being in contact with a fluid at another temperature. This process depends on the following:

(a) The shape of the surface
(b) The orientation of the surface (e.g., vertical versus horizontal)
(c) The type of fluid (gas or liquid)
(d) The density, viscosity, specific heat, and thermal conductivity of the fluid
(e) The velocity of the fluid
(f) Whether evaporation, condensation, or formation of scale can or is likely to take place.

In thermodynamics, convection is described by the transportation of internal energy of the fluid, which serves as a carrier of stored thermal energy. As the fluid passes over the hot surface, energy is transferred to the fluid by conduction. This energy is then carried or convected downstream by the fluid and is diffused throughout by conduction within the fluid.

An approximate lumped-parameter model of convection can be constructed by writing

$$q = hA\,\Delta\theta \quad \text{W} \tag{19.90}$$

where h is the convection heat transfer coefficient.

For the motor starting problem, we will assume that convection can be ignored for the brief period of interest. This would not be an acceptable assumption for the steady-state problem, but the motor starting is such a brief period of time that only a small amount of heat will be convected to ambient during this transient period.

19.3.1.3 Heat Transfer by Radiation. All warm bodies radiate heat. Radiation energy from a hot surface is defined by the *emissive power* E as follows.

$$E = \sigma\theta^4 \quad \text{W/m}^2 \tag{19.91}$$

The outgoing radiation to the space surrounding the hot object is computed as

$$q = \varepsilon\sigma A\theta^4 \quad \text{W} \tag{19.92}$$

where

$$\varepsilon = \text{emissivity of the material}$$

Emissivities for various materials are published in references.

Radiation is an important type of heat transfer for some problems, but the importance depends on many factors. For example, the radiative heat loss from transmission line conductors may account for about 30% of the total heat loss in still air, but accounts for less than 5% when the wind speed exceeds 5 m/s, thereby increasing the forced convection heat loss [24]. When the motor is starting at room temperature, radiation is not an important heat transfer mechanism because the conductor termperatures are too low. As the rotor accelerates and the conductors reach an elevated temperature, conduction and convection become the effective mechanisms. We will ignore radiation heat transfer for the transient motor starting problem, as this is not likely to be a major factor in determining the heat buildup over the short time under consideration.

19.3.1.4 Heat Transfer Summary. Heat transfer can be analyzed to determine the flow of heat from a hot region to surrounding cooler regions. In many cases, an electrical analog of the heat transfer process provides a convenient and effective way of defining the problem on a lumped-parameter basis. For the motor starting problem, we assume an adiabatic process.[3] This is like building an adiabatic wall around the motor such that no heat can either enter or leave the confines of that wall during the period of interest. This means that all heat generated due to the various sources of losses will transfer through the motor materials and will be stored in those materials, raising the material temperatures in the process. We would like to know how these temperatures behave during starting and if there is a danger of tripping the motor during this time period.

Motors make use of convection heat transfer to exhaust heat to the surrounding atmosphere during steady-state operation. In many motor designs, this convection is forced, by blowing air through the air gap of the machine. Although this is an important method of cooling, our purpose here is to model only the rapid increase in motor temperatures during the startup process to make sure that motor thermal protection will not trip the motor and prevent it from reaching normal operating conditions.

19.3.2 A Motor Thermal Model

The purpose of this section is to derive a lumped-parameter thermal model for the induction motor that can be used to compute the transient temperature rise that occurs in the motor during startup. The concepts of the previous section will be used in constructing an electric analog of the thermal heat transfer problem.

The structure of an induction motor is pictured in Figure 19.16. The stator is the outer portion of the figure, with slots to accommodate the stator windings. Currents flow in the

[3] An *adiabatic process* takes place in such a way that no heat flows into or out of the system.

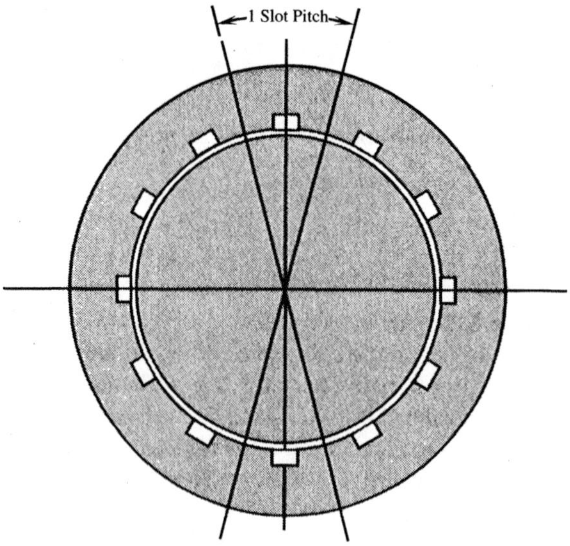

Figure 19.16 End cross-sectional view of an induction motor.

three-phase stator windings to develop a rotating magnetic field that revolves around the stator at a given fraction of synchronous speed, which is dependent on the number of poles. The motor pictured is a two-pole machine. The rotor is the inner portion of the cross-sectional view and is separated from the stator by an air gap. The rotor has imbedded short-circuited bars through which rotor current can flow, thereby permitting the motor to develop torque. Several references provide excellent descriptions of the general design concepts for rotating machines [12], [13].

It should be noted that the structure of the machine shown in Figure 19.16 is repetitive, with each pole pitch having an identical structure. If we slice the machine longitudinally along the pole pitch boundaries, we have the wedge structure shown in Figure 19.17. This structure is repeated around the periphery of the machine and all repetitions are exactly the same. Moreover, we can assume that all carry the same currents in both stator and rotor. Due to symmetry, the walls of the machine sectors pictured are always at the same temperature, so there is no heat transfer between sectors.

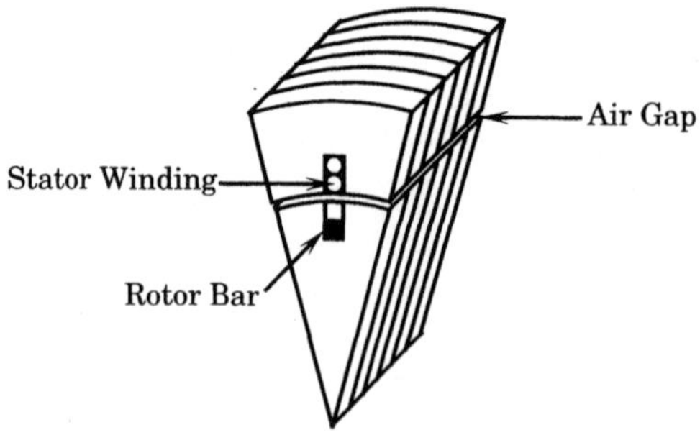

Figure 19.17 One-pole pitch of an induction motor.

The basic heat conduction equation for a cylindrical geometry is provided by the thick-wall pipe equation, which can be written as

$$q = \frac{2\pi kd}{\ln(r_o/r_i)}(\theta_i - \theta_o) \tag{19.93}$$

$$\begin{aligned} \text{where} \quad k &= \text{thermal conductivity} \quad (\text{W/m-K}) \\ d &= \text{length of pipe} \quad (\text{m}) \\ r_i &= \text{inside pipe radius} \quad (\text{m}) \\ r_o &= \text{outside pipe radius} \quad (\text{m}) \\ \theta_i &= \text{inside pipe temperature} \quad (\text{K}) \\ \theta_o &= \text{outside pipe temperature} \quad (\text{K}) \end{aligned} \tag{19.94}$$

This gives the total heat in watts conducted from the inside of the pipe to the outside, for a given thermal conductivity and pipe dimension [25]. But, as noted from Figure 19.17, we can divide the motor into sectors, with each sector having exactly the same radial heat flow characteristics. Therefore, for an n sector system, where n is the number of slots on the machine stator, we can solve for only $1/n$ of the total heat transferred, which we can define as q_n to write

$$q_n = \frac{\theta_i - \theta_o}{\mathcal{R}_n} \tag{19.95}$$

where we define the thermal resistance of the nth sector as

$$\mathcal{R}_n = \frac{n \ln (r_o/r_i)}{2\pi k d} \tag{19.96}$$

For the case under study here, the parameter d is the thickness of the iron laminations. Note that n is always an even integer that is divisible by 3 (6, 12, 18, ...) for a three-phase machine.

Now, (19.95) is the formula for heat transfer when the entire inside diameter has a uniform temperature of θ_i. In the wedge sector of Figure 19.17, the heat sources are the stator or rotor conductors, which are finite and do not cover the entire inside surface. However, for a machine with many slots, (19.95) is a good approximation.

19.3.2.1 A Lumped-Parameter Model of the Motor. We can narrow the finite element model to only one element, sliced out of the center of the machine. Since the faces of adjacent wedges are at the same temperature, no heat will flow in the longitudinal direction except at the ends of the machine. These end effects will be ignored. All heat will flow from the heat sources, the winding and cores losses, toward the material nodes where that heat energy can be stored. Note that we are ignoring convective heat flow to the ambient. We use the ambient temperature as a reference or initial condition for the motor. If the motor is restarting from a hot condition, the initial temperature would be the hot condition of the materials, not the temperature of the air surrounding the motor.

An elementary electric analog of the motor thermal nodes is given in Figure 19.18. There are two sources of heat in the model, which represent the copper losses in the stator windings and rotor bars and the induced eddy currents in the iron laminations. The stator and rotor iron represent heat sinks, represented by the large black dots in the figure the capacity of which can be estimated from the mass of the material and its specific heat. Capacities of the stator and rotor windings can be similarly estimated. The temperatures in the model are θ_S for the stator and θ_R for the rotor copper, θ_{SC} for the stator core, θ_{RC} for the rotor core, and θ_A for the ambient temperature both outside the stator and in the air gap. Heat sources due to winding conductor losses are shown as current sources. Thermal resistances represent the paths of heat flow from points of higher to lower temperatures.

An electrical analog of the thermal system is shown in Figure 19.19. This is simply a redrawing of the lumped-parameter model pictured in Figure 19.18 except that the thermal resistances and capacitances from the stator core to ambient have been combined into a single parameter representing the dual paths from the core center of mass and the ambient temperature

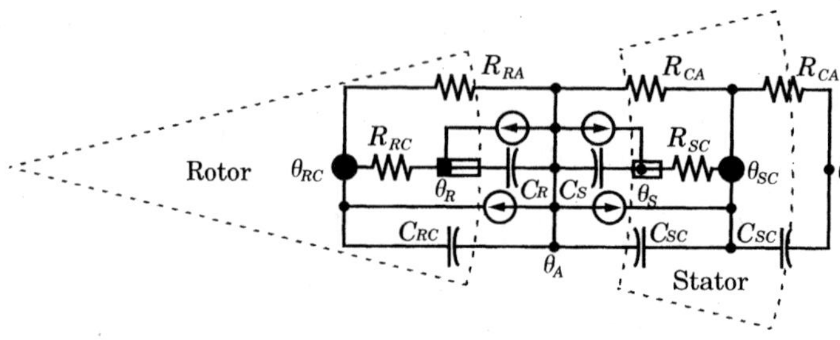

Figure 19.18 Elementary thermal model of the motor.

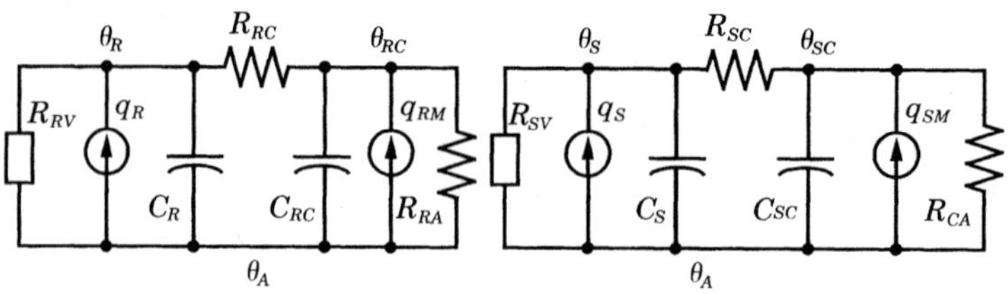

Figure 19.19 Electric analog of the motor thermal system.

of the air gap and the air outside the motor. The other added feature in Figure 19.19 is the resistances shown as rectangles labeled R_{RV} and R_{SV}. These resistances represent convection heat flow to the air-gap ambient due to forced convection. These parameters can be ignored during startup, but would represent important paths for heat flow during steady-state operation. The heat sources due to induced currents in the iron structure have been ignored, as these heat sources are small compared to the stator and rotor heat sources.

A simplified electrical analog has also been presented in the literature, where the rotor core temperature is not specifically identified [9]. This reduced model is shown in Figure 19.20. The heat source due to induced current in the stator iron circuits is sometimes ignored. This simplified model requires only three temperatures and there are fewer components to determine, making the model easier to support for calculations. Study results using this model have been shown to be reasonable [11]. Note that the rotor temperature is completely decoupled from the stator winding and stator core temperatures in this model.

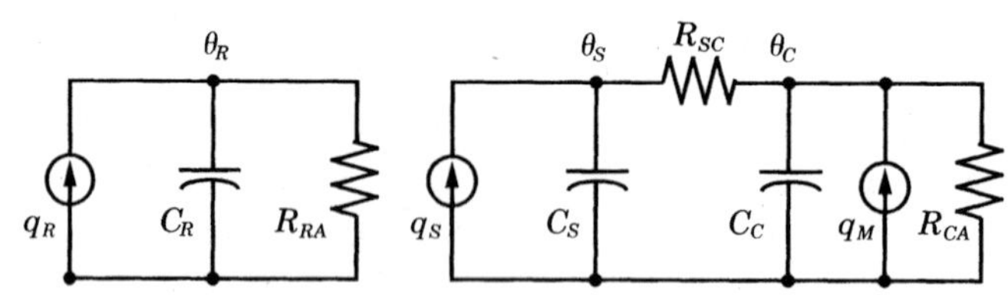

Figure 19.20 Simplified electric analog of the motor thermal system [9].

We can write the differential equations of the analog, solving for the voltages of the stator, core, and rotor, as follows.

$$\frac{d\theta_S}{dt} = a_S P_{SL} + b_{SS}\theta_C = F_S(q_S, \theta_S, \theta_C, t) \tag{19.97}$$

$$\frac{d\theta_C}{dt} = a_C P_{CL} - b_C\theta_C + b_{CC}\theta_S = F_C(q_M, \theta_S, \theta_C, t)$$

$$\frac{d\theta_R}{dt} = a_R P_{RL} - b_R \theta_R = F_R(q_R, \theta_R, t) \tag{19.98}$$

where

$$\begin{aligned} a_S &= \frac{1}{\mathcal{C}_S} \quad b_{SS} = \frac{1}{\mathcal{R}_{SC}\mathcal{C}_S} \\ a_C &= \frac{1}{\mathcal{C}_C} \quad b_C = \frac{1}{\mathcal{R}_{CA}\mathcal{C}_C} + \frac{1}{\mathcal{R}_{SC}\mathcal{C}_C} \qquad b_{CC} = \frac{1}{\mathcal{R}_{SC}\mathcal{C}_C} \\ a_R &= \frac{1}{\mathcal{C}_R} \quad b_R = \frac{1}{\mathcal{R}_{RA}\mathcal{C}_R} \end{aligned} \tag{19.99}$$

Note that all of the b coefficients of the equations are inverse time constants. Also note that (19.98), the rotor equation, is uncoupled from the two stator equations (19.97). Comparing the temperature notation in Figures 19.19 and 19.20, we recognize that the temperatures shown in Figure 19.20 are, in fact, temperature *differences* between the motor parts and ambient.

The analog solution relates the electrical quantities to thermal quantities. In particular, the current sources must be related to the physical losses in the machine, since that is where the heat is generated. The voltages are related to the temperature of the three nodes. We can summarize these relationships as follows.

$$\begin{aligned} P_{SL} &= I_S^2 R_S = q_S \quad \text{W} \\ P_{CL} &= I_M^2 R_M = q_C \quad \text{W} \\ P_{RL} &= I_R^2 R_R = q_R \quad \text{W} \end{aligned} \tag{19.100}$$

In (19.97) and (19.98) the temperatures are actually incremental changes from the ambient temperature, although the notation does not show this explicitly.

The loss terms are computed from the electrical model of the motor. Using Euler's rule, we can convert these differential equations to difference equations with the following result.

$$\begin{aligned} \theta_{S(n+1)} &= \theta_{S(n)} + hF_{S(n)} \\ \theta_{C(n+1)} &= \theta_{C(n)} + hF_{C(n)} \\ \theta_{R(n+1)} &= \theta_{R(n)} + hF_{R(n)} \end{aligned} \tag{19.101}$$

19.3.2.2 Thermal Model Parameters. The model parameters consist of thermal resistances and capacitances. The thermal resistances are computed from (19.96) using values corresponding to the stator and rotor dimensions and the thermal conductivities, given in Table 19.1. This computation will give values that correspond to motor running conditions. Starting values are estimated to be three times the running values [9]. Thermal capacitances are computed from (19.86). This requires a computation or estimate of the mass of the material in the segment under study.

19.3.2.3 Thermal Model Performance. The performance of the thermal model must be integrated together with the induction motor electrical equations. The electrical equations give the values of copper and core losses, which become the sources of thermal energy in the equivalent circuit of Figure 19.19. The values of thermal resistance and capacitance can be computed. Therefore, all of the coefficients in (19.99) are known or can be computed for the lumped parameter model.

EXAMPLE 19.4

Compute the temperatures of the stator windings, rotor bars, and core for the motor described in Example 19.2. Estimate the values of motor thermal parameters by reference to published data of [9].

Solution

Typical computed temperatures are shown if Figure 19.21, where 1.0 per unit corresponds to 100° C. The parameters used for the data plotted in Figure 19.20 are given in Table 19.3. The computed results are very sensitive to the thermal capacitance, but not nearly so sensitive to variations in thermal resistance. This is fortunate since precise calculations of resistance are not possible, whereas the mass of iron and copper could be determined with reasonable accuracy for many motors.

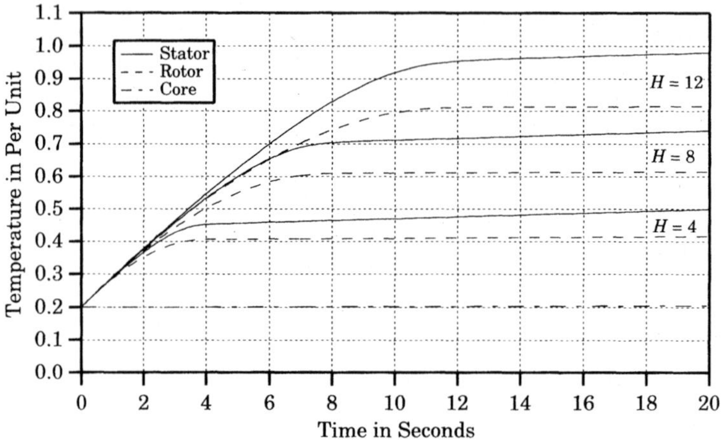

Figure 19.21 Stator, rotor, and core temperatures.

TABLE 19.3 Per Unit Values for the Thermal Model

Stator		Rotor	
Parameter	**Per Unit Value**	**Parameter**	**Per Unit Value**
R_{SC}	20.0	R_{RN}	43.9
R_{CN}	6.67	R_{RL}	130.0
R_{CL}	20.0	C_R	22.0
C_R	20.0		
C_C	100.0		
$\theta_{S\max}$	100°C	$\theta_{S\max}$	80°C

It is common practice to set the temperature limit to 100 and 80°C for the stator and rotor windings, respectively [9]. If these limits are used for this motor, then the rotor exceeds its limiting values for the case of starting the highest inertia load and the stator winding is close to its limit. The lower inertia loads do not cause the temperatures to approach the limiting values due to the shorter acceleration time. ■

The temperature plots of Figure 19.21 are typical of the temperature rises one would expect. The model is useful for determining the relative importance of the different parameters. For example, the values of thermal resistance have no effect on the initial rise and curvature of the temperature plots. This shape is determined exclusively by the capacitance. The thermal resistance comes into play later and may affect the possibility of overheating 20 or 30 seconds after startup. However, for these longer time periods, the model is no longer correct, as convection heat transfer begins to have an effect in cooling the motor to reach a steady-state temperature distribution.

Computations of the thermal effects described here have not been commonly performed in the past, and the analog parameters required for the computation are not usually known. Therefore, we may not have high confidence in the data used, nor do we know that all motors have parameters that are similar in magnitude within a small margin of error, even after normalization. In the calculation of motor or generator electrical parameters, we have considerable experience and can estimate the machine parameters with high confidence. This is not yet true of the motor thermal model parameters, although our confidence in this process should grow as we gain experience.

It is also noted that some of the induction motor parameters are not known and are sometimes difficult to estimate. This is true of the mutual terms of the motor equivalent circuit, both resistance and reactance.

19.3.2.4 Modeling Thermal Limits. A motor is designed with the capability to absorb a given amount of thermal energy that originates from I^2R losses in the motor windings and losses due to induced currents in the iron parts. Consider the transfer of heat from losses to the thermal capacitance of the motor. We can write this heat flow as

$$q = C\frac{d\theta}{dt} = 3I^2R \tag{19.102}$$

where the parameter R in (19.102) is an electrical, not a thermal, resistance. The factor of 3 is necessary since the motor equivalent is for only one phase and we assume a three-phase motor. We can solve this equation for the temperature rise by integration.

$$\theta = \frac{3R}{C}\int_0^t I^2 dt \tag{19.103}$$

where RI^2 is in watts and C is in J/K. Now suppose that we let I be the limiting value of current, that is, I is a given constant. Then the integration gives

$$\theta = \frac{3RI^2t}{C} \tag{19.104}$$

Now, let us normalize this equation. We choose the following base values.

$$\begin{aligned} \theta_B &= \text{the maximum allowable temperature} \quad \text{(K)} \\ V_B &= \text{the motor rated voltage} \quad \text{(V)} \\ I_B &= \text{the motor nameplate full load current} \quad \text{(A)} \end{aligned} \tag{19.105}$$

Then, we can derive other base quantities from (19.105).

$$\begin{aligned} S_B &= \sqrt{3}V_B I_B \quad \text{VA} \\ C_B &= \frac{S_B t_B}{\theta_B} \quad \text{J/K} \end{aligned} \tag{19.106}$$

Now, dividing (19.104) by the base temperature, we can write, in per unit

$$\theta = \left(\frac{3V_B^2 I_B^2}{S_B^2}\right) RI^2t \quad \text{s} \tag{19.107}$$

We can show that the quantity in parentheses is equal to unity. Thus, with all parameters in per unit on the given bases, we have

$$\theta = RI^2t \quad \text{s} \tag{19.108}$$

or

$$I^2t = \frac{\theta}{R} = K_{\text{LIM}} \tag{19.109}$$

represents the thermal energy limit [17]. Thus one per unit temperature per unit of resistance becomes the limiting normalized heat energy. Taking the logarithm of this equation, we have

$$\ln t = \ln K_{\text{LIM}} - 2 \ln I \tag{19.110}$$

Thus, the thermal limit is a straight line when plotted on log-log paper, and has a slope of −2.

An example of the application of the thermal limit in motor starting is shown in Figure 19.22, where the thermal limit is set arbitrarily at 350. It would be very difficult to fit a fuse characteristic between the starting current and the thermal limit. Note that the coordination margin is most critical if the motor is started with unbalanced applied voltages.

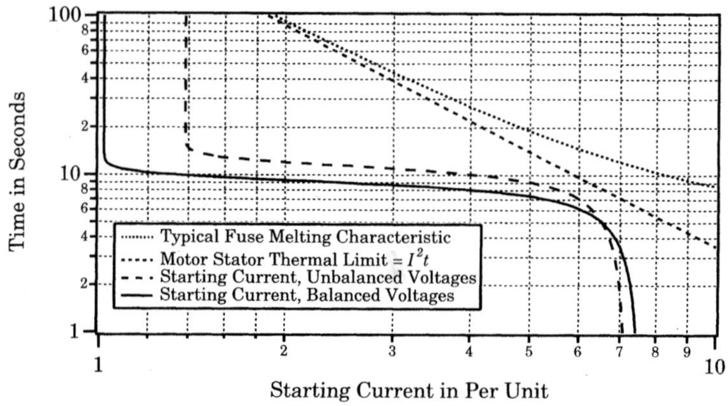

Figure 19.22 Motor starting current and the thermal limit.

The rotor heating is usually the most critical part of the motor. Therefore, a thermal model of the rotor is a useful concept. We now investigate the physical realization of this concept.

19.3.2.5 Thermal Relay Realization. One commercial digital motor protection relay expands the concepts of the previous sections to derive a simple model for protection of the rotor, which is the motor component most likely to overheat due to starting high inertia loads. This model reasons that the heat input can be written as

$$q_{\text{IN}} = 3I_R^2\boldsymbol{R}_R = C_R\frac{d\theta_R}{dt} \quad \text{W} \tag{19.111}$$

where I_R = rotor current (A)
R_R = rotor resistance (Ω)
C_R = rotor thermal capacitance (J/K)
θ_R = rotor temperature (K)

This equation can be normalized on the three-phase base volt-amperes at rated voltage to write

$$R_R I_R^2 = \frac{C_R}{\omega_B}\frac{d\theta_R}{dt} = \boldsymbol{H}_C \frac{d\theta_R}{dt} \tag{19.112}$$

where I_R = rotor current (pu)
R_R = rotor resistance (pu)
C_R = rotor thermal capacitance (pu)
θ_R = rotor temperature (pu)
ω_B = base radian frequency (rad/s)
t = time (s)

and where we have defined a new parameter

$$\boldsymbol{H}_C = \frac{C_R}{\omega_B} \quad \text{s} \tag{19.113}$$

which is the thermal inertia constant in seconds. This is analogous to the mechanical inertia constant described in (19.9) and (19.16). The units of "seconds" is necessary since the time variable in (19.112) is in seconds.

We can write the solution of (19.112) for the rotor temperature, with the result

$$\theta_R = \frac{1}{\boldsymbol{H}_C}\int_0^t R_R I_R^2 dt \tag{19.114}$$

If a limiting value of current, such as locked-rotor current, is permitted to flow in the rotor, a limiting value of temperature will be reached in a value of time determined by the equation. Since the locked-rotor current is a given constant, we have

$$\theta_R = \frac{R_R I_R^2 t}{\boldsymbol{H}_C} \tag{19.115}$$

or, in the limiting case of locked rotor current, we can write

$$\theta_{\text{LIM}} = \frac{R_R I_{LR}^2 t_{\text{LIM}}}{\boldsymbol{H}_C} \tag{19.116}$$

This means that locked rotor current will cause the rotor to reach its limiting temperature in t_{LIM} seconds. Clearly, if locked rotor current is permitted to flow longer than the limiting time, the rotor will become overheated.

This concept has been used to construct a simple rotor model based on (19.115) that is used in commercial protective relays for large induction motors [17]. Recalling the definition of the thermal inertia, we can compute

$$C_R = \frac{R_R I_{LR}^2 (t_{\text{LIM}}\omega_B)}{\theta_{\text{LIM}}} = \frac{R_R I_{LR}^2 t_u}{\theta_{\text{LIM}}} \quad \text{pu} \tag{19.117}$$

where t_u is the per unit time. Now, if we construct the analog on the basis of

$$I_{LR}^2 t_u = \theta_{\text{LIM}} \tag{19.118}$$

then

$$C_R = R_R \quad \text{pu} \tag{19.119}$$

We also recall that, for a thermal resistance we can write

$$\Delta\theta = \mathcal{R}q = \theta_R - \theta_A \tag{19.120}$$

where the boldface $\boldsymbol{R}$ is a thermal resistance, not an electrical resistance. Then, using the analogy (19.118), we can write

$$\boldsymbol{R}q = I^2\,(t_R - t_A) \tag{19.121}$$

in reference to ambient temperature, t_A. If the motor is started at some operating temperature, greater than ambient, then

$$\boldsymbol{R}q = I^2\,(t_R - t_O) \tag{19.122}$$

where we replace the ambient temperature time with an operating temperature time, t_o, such that the time difference in parentheses will be smaller. The rotor thermal model is shown in Figure 19.23. This thermal model has been used successfully in commercial motor protective relays.

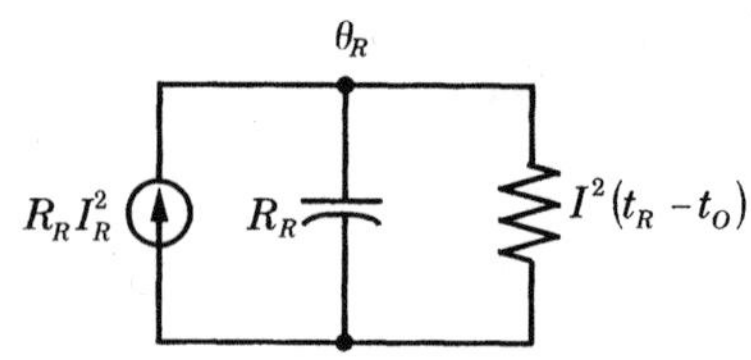

Figure 19.23 The Zocholl rotor thermal model [17].

19.4 MOTOR PROBLEMS

Motor problems can be classified in two broad categories, with the classification depending on the origin of the hazard:

- Problems due to internal hazards or failures within the machine
- Problems due to external, or system imposed, hazards

Each of these categories will be discussed briefly in order to present the types of hazards for which motor protection might be appropriate.

19.4.1 Motor Problems Due To Internal Hazards

Internal hazards are those that occur due to some type of failure within the motor terminals, where the failure may be either electrical or mechanical. This includes winding short circuits and bearing failures as well as incipient hazards, such as overheating of windings or bearings, that may lead to eventual failure if the cause of the problem is not identified and corrected.

Winding short circuits may be either phase-to-phase or phase-to-ground, but all short circuits are likely to develop into faults involving ground since the windings are in close proximity to grounded stator iron. This suggests that both phase and ground protection may be required.

A problem in applying short-circuit protection is being able to distinguish between a shorted winding and the normal starting current of the motor, which will usually be several times larger than normal running current. Figure 19.24 shows a typical plot of motor current, torque, and power as a function of slip. Note that the motor current remains at over 75% of the initial starting value until the motor is nearing running speed. Moreover, since the motor

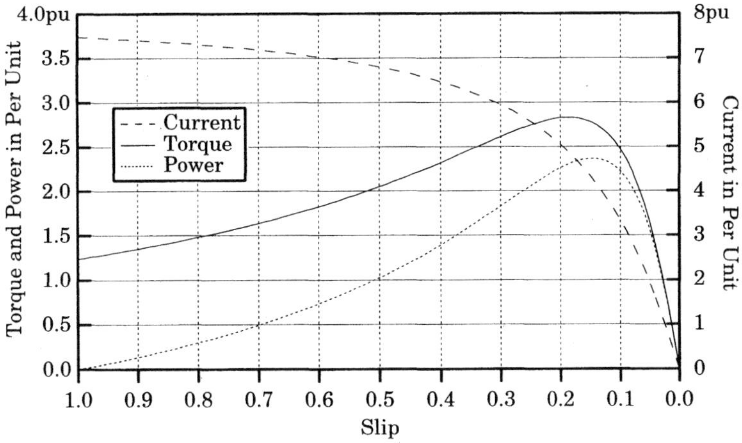

Figure 19.24 Motor current, power, and torque as a function of slip.

is accelerating during this time, the starting current is at high values for 80–90% of the total starting time. This means that any motor overcurrent protection must be set to permit currents of starting current magnitude to exist for the entire starting time without tripping.

The duration of the high starting current depends largely on the driven mechanical load. Large fans, such as those used for forced draft or induced draft fans in steam plants, are very slow to accelerate to running speed. This means that the motor may draw five to eight times normal current on starting, and these currents will persist for several seconds.

Another problem is due to the nature of the driven load, which may be capable of stalling the motor and causing high currents similar to starting currents. The stalled motor condition must be cleared, either by changing the mechanical load or by tripping the motor before excessive heating causes serious deterioration of insulation. This presents a problem in protective system design that will allow maximum necessary use of the motor and still guard against serious conditions, even temporary "normal" loadings, that may be damaging to the motor.

Another internal motor hazard is the failure of the motor bearings. Motors up to about 500 hp often have ball or roller bearings. When these bearings fail, the result is usually catastrophic and occurs very quickly, causing immediate shutdown and possible mechanical damage to the motor. There is little chance of detecting an incipient failure by relaying before actual failure occurs. Relays may act to quickly disconnect the motor, however.

Larger motors usually have sleeve bearings that are lubricated by pressurized oil supply systems. For this type of system, incipient failure may often be detected by monitoring the bearing oil temperature and sounding an alarm if the oil temperature exceeds a given value. Should lubricating oil supply be lost entirely, the bearing will seize in just a few minutes. This is not detectable prior to occurrence, but the motor current will jump rapidly upon shaft seizure. Motor thermal overload measurements on the stator will not protect against bearing failure or seizure.

19.4.2 Motor Problems Due To External Hazards

There are several different kinds of external hazards that affect motor performance, some of which can be monitored or controlled by protective relays. The following external hazards are of interest:

1. Unbalanced supply voltage
2. Single phasing of supply voltage
3. Low supply voltage
4. Low supply frequency
5. Reverse phase sequence in the supply voltage
6. Motor stalling due to excessive load
7. Loss of synchronism (synchronous motors only)
8. Loss of excitation (synchronous motors only)

None of the above conditions can be blamed on the motor or result from any failure within the motor, but they represent sources of either direct or incipient motor failure that require detection and possible protective system action. Each of the above will be briefly discussed.

19.4.2.1 Unbalanced Supply Voltage. Unbalanced supply voltage causes negative-sequence currents to circulate in the motor, which increases the stator and rotor heating. Unbalanced supply voltage may be due to the presence of unbalanced single-phase loads in the vicinity of the motor or lack of adequate transpositions in the supply lines. The worst case of unbalanced supply voltage is "single phasing" of the motor, where one phase is completely lost. This special case is discussed in the next section.

To analyze the effect of unbalanced applied voltages on an induction motor we refer to the equivalent circuit of the motor for both the positive and negative sequences, shown in Figure 19.1. From the diagram, we can compute the total impedance seen by positive- and negative-sequence currents when the motor is running with slip s as follows.

$$\begin{aligned} Z_1 &= \sqrt{\left(R_s + \frac{R_r}{s}\right)^2 + (X_s + X_r)^2} \\ Z_2 &= \sqrt{\left(R_s + \frac{R_r}{2-s}\right)^2 + (X_s + X_r)^2} \end{aligned} \tag{19.123}$$

We note that the two impedances with $s = 1$ (stalled rotor) are equal, but when running they are not equal. The shunt impedance term has been omitted for simplicity in (19.123). For balanced conditions, this allows us to write the ratio

$$\frac{I_{\text{Start}}}{I_{\text{Run}}} \cong \frac{\sqrt{\left(R_s + \frac{R_r}{s}\right)^2 + (X_s + X_r)^2}}{\sqrt{(R_s + R_r)^2 + (X_s + X_r)^2}} \tag{19.124}$$

The only difference between the impedances of (19.123) and (19.124) is the rotor resistance terms. But for a given slip the total impedance is not influenced very greatly by the resistance terms due to the relatively large magnitude of the reactance terms. Therefore, the approximation is often made that

$$\frac{I_{\text{Start}}}{I_{\text{Run}}} \cong \frac{Z_1}{Z_2} \tag{19.125}$$

and this ratio for a typical induction motor is in the range of 5 to 8.

We also know that

$$I_{a1} = \frac{V_{a1}}{Z_1}$$
$$I_{a2} = \frac{V_{a2}}{Z_2} \tag{19.126}$$

Taking the ratio of I_{a2} to I_{a1} we compute

$$\frac{I_{a2}}{I_{a1}} = \left(\frac{V_{a2}}{V_{a1}}\right)\left(\frac{Z_1}{Z_2}\right) = \left(\frac{V_{a2}}{V_{a1}}\right)\left(\frac{I_{\text{Start}}}{I_{\text{Run}}}\right) \tag{19.127}$$

Suppose that the ratio of starting to running current is 6.0 and the ratio of negative- to positive-sequence applied voltage is 0.05 (5% negative sequence), then the ratio of negative- to positive-sequence currents will be 30%. This is a typical result. The imbalance in the applied voltage is amplified by the ratio of sequence impedances by about the ratio given by (19.125).

The total torque of the induction motor is that due to the sum of the positive- and negative-sequence currents. This sum is given by

$$T_m = T_{m1} + T_{m2} = \frac{R_r}{\omega_s}\left[\frac{I_{r1}^2}{s} - \frac{I_{r2}^2}{2-s}\right] \tag{19.128}$$

The negative-sequence current usually produces very little torque, especially if the unbalance is small, which implies a small negative-sequence current. Its major effect is to increase the losses, primarily the stator I^2R losses. The winding carrying the largest current will overheat, but in time the excess heat is distributed throughout the machine more or less uniformly [26]. This may cause the machine to be derated, with the derating being highly dependent on the ratio of sequence impedances given by (19.125). For example, for a machine having a ratio of starting-to-running currents of 8.0, a 10% unbalance in applied voltage causes a derating to about 55% of rated output [26]. This suggests that, in some cases, it may be necessary to monitor the phase balance of the applied voltage to prevent continuous motor overheating. Figure 19.11 provides an example of motor torque for severely unbalanced voltages.

19.4.2.2 Single Phasing of the Supply Voltage. Single phasing of the supply voltage occurs when a fuse or switching device in one of the three supply lines is opened. This is the most serious case of unbalanced supply voltage, since the voltage of one of the three phases is not just reduced, but becomes zero. This condition is discussed and analyzed in many references [29–31]. The solution is based on the connection of the sequence networks shown in Figure 19.25.

The connection of Figure 19.25 shows that, for one line open, the positive- and negative-sequence networks are connected in series, with the two sequence currents being equal in magnitude, but of opposite sign. We may compute the sequence voltages and currents for line a open, with the following result [14], [27].

$$I_{a1} = -I_{a2} = j\frac{I_b}{\sqrt{3}}$$
$$V_{a1} - V_{a2} = j\frac{V_{bc}}{\sqrt{3}} \tag{19.129}$$

This corresponds to 100% unbalance in the sequence currents. If the motor is running when the open-circuit condition occurs, it will still develop positive torque and will continue to operate if the shaft load is not too great. The unbalance will cause increased heating that

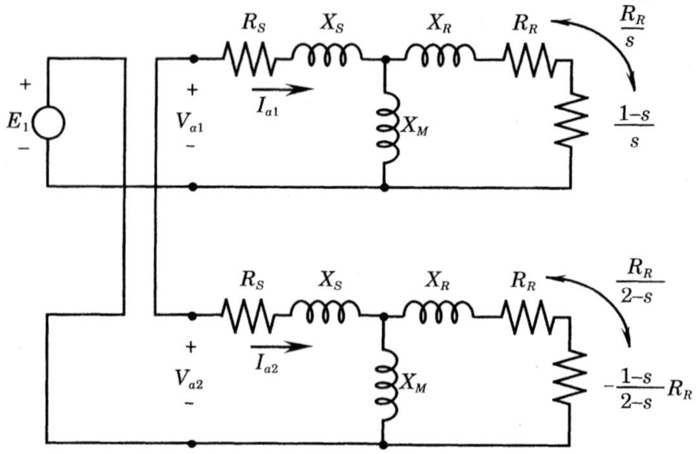

Figure 19.25 Sequence network connections with one open supply line.

may eventually trip the motor, depending on the type of protection used. Starting torque under this condition is zero.

19.4.2.3 Low Supply Voltage. When the supply voltage to a motor is low, the motor slows down and draws more current for the same load than under normal voltage conditions. The effect of low voltage is graphically illustrated in Figure 19.26, where the average torque is plotted for four different values of applied voltage.

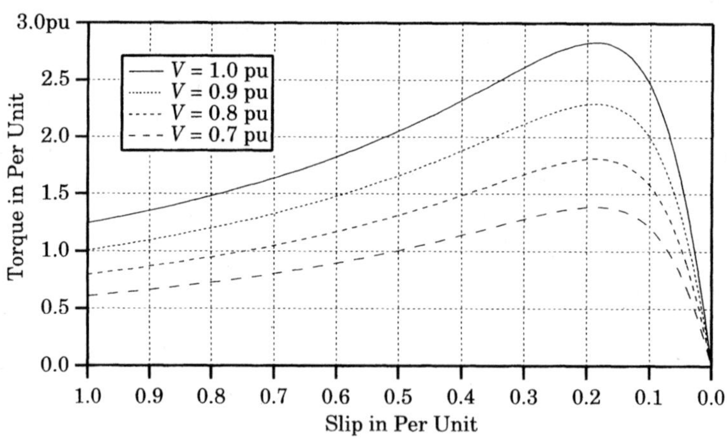

Figure 19.26 Effect of low voltage on the induction motor torque.

Suppose that the motor was originally operating at a 3% slip, which corresponds to a load torque of about 1.09 per unit in Figure 19.26. When the voltage is reduced to 90% of normal the developed torque moves to the lower curve and, in order to develop the 1.09 per unit torque, the slip must increase to about 3.9%. Thus, the motor slows down and draws more current from the supply in order to drive the shaft load.

In some cases, a large drop in voltage may cause the motor to stall. This suggests that low voltage is serious enough to warrant protection in certain cases, with at least an alarm of the condition being a possible form of protection. In unattended locations, the low-voltage protection may be used to trip the unit.

19.4.2.4 Low System Frequency. Low system frequency causes an induction motor to run at a reduced speed. If the speed reduction is small, the effect on the motor will not be great but there will be reduced cooling since the shaft rotation usually is designed to assist the circulation of air through the air gap and provide cooling to the metal parts of the motor. This means that any reduction in speed from normal will increase the motor heating. However, reduction in speed will also result in a reduction of shaft load in most cases. This means that the current drawn by the motor will be very much dependent on the type of shaft load. The power consumption of the motor will be reduced, but the reactive power consumption is often increased [29]. It has also been noted that the reduction in output for fans is less than for pumps, especially pumps working into a high head, such as boiler feed pumps. It is also noted that some protective relays have different characteristics at reduced frequency than at normal frequency.

If the load driven by the motor falls off substantially due to the low frequency, there may be need to trip the motor if it is unable to provide the required output to its mechanical load. Reduced frequency is especially difficult to assess in the case of power plant auxiliaries, where there is a combination of fans and pumps, all of which are required to maintain plant output. In many cases, however, there are multiple fans or pumps with total rating that is greater than the normal full load requirement, which indicates that it may not be necessary to reduce plant output with reduced frequency. When underfrequency occurs on the power system, all power plants are overloaded and the problem would be made much more serious if plant motors are caused to trip, resulting in lowering the plant outputs [28]. These conditions are discussed more fully in the next chapter.

In the case of the individual motor, the greatest problem caused by underfrequency would seem to be overheating due to reduced cooling. The reduction in frequency, and the resulting lowering of speed, does not necessarily damage the motor, but may cause problems for the driven load.

If the supply frequency is *higher* than normal, air circulation is increased and cooling improved. Shaft mechanical load, however, is increased for this condition with most types of shaft load.

19.4.2.5 Supply Voltage Reverse Phase Sequence. A reversal in the phase sequence of the supply voltage causes the motor to reverse direction of rotation. For most motors, this is not damaging to the motor, but it does subject the motor to a difficult restart. It may be a cause for serious concern for the driven load, and this may conceivably be dangerous to nearby personnel. Because of this, it may be necessary to provide protection for the reverse phase sequence condition.

19.4.2.6 Motor Stalling Due to Excessive Load. All induction motors will stall if the load torque exceeds the maximum torque that the motor is capable of developing. Stalling may occur on startup, if the load is too great at low speeds where the developed motor torque is not great. Stalling may occur when the motor is running, if a large load is applied. This may occur, for example, in such loads as a mill or pulverizer, where the shaft load may be quite variable due to the inflow or size of materials to be ground. Another cause of stalling is due to single phasing of the supply voltage, which decreases the motor developed torque.

When stalling occurs, the motor current changes from its normal run value to the stall or startup value, which will usually be from five to eight times greater than normal full load run current. The motor is not designed to carry this much current continuously and protection is required to ensure that the motor is disconnected before permanent damage occurs.

19.4.2.7 Synchronous Motor Loss of Synchronism. Loss of synchronism protection is often applied to synchronous motors that drive loads where sudden overloads may occur. This is important where the overload may exceed the pull-out torque of the motor and cause it to fall out of step. In this case, the motor may stall and draw heavy stator currents and will cause large slip frequency currents to flow in the rotor. It is preferable to disconnect the motor in such cases rather than allow it to run as an induction motor, with high rotor currents. In some applications, it may be possible to temporarily disconnect the load and allow the motor to resynchronize.

19.4.2.8 Synchronous Motor Loss of Excitation. Synchronous motors are also subject to a loss of excitation while running under load. This will cause the motor to run as an induction motor, with higher than normal rotor currents and rotor heating. In some cases, the motor is adequately protected by overheating and loss of synchronism protection.

19.5 CLASSIFICATIONS OF MOTORS

There are several ways to classify motors, and, in many cases, the protection differs within the various classifications. The following are arbitrary classifications that illustrate the point:

The motor is classified by:

- Size or horsepower rating
- Service requirement:
  (a) essential service motors
  (b) nonessential service motors
- Type:
  induction
  synchronous
- Location:
  attended locations
  unattended locations

It has been noted previously that we are concerned in this book with large, high-voltage motors, and mostly with induction motors.

19.5.1 Motors Classified by Service

Motor classification by service is important in considering protection because the protective system design may be quite different depending on the motor service.

19.5.1.1 Essential Service Motors. Motors designated as being an essential service are required to maintain service to the motor load under all reasonable conditions. In some cases, this may require that the motor be switched from one source of supply to another in order to remain in service. It may also mean that the motor is kept in operation for all but the most severe disturbances, even at the cost or risk of accelerated loss of life. For example, protection may be designed to trip the motor for short circuits and for no other reason [29].

Warrington [30] separates motors that serve power plant auxiliaries into essential and nonessential classifications. The essential classification includes the following motors [30]:

Essential Service Power Plant Auxiliary Motors

Boiler feed pumps	Stokers
Condensate pumps	Circulating water pumps
Forced draft fans	Pulverizer feeders
Induced draft fans	Pulverizers—unit type
Primary air fans	Excitation drive motors

Clearly, the plant will not operate without these auxiliary motors, at least not at full output. In many cases, there are multiple fans, pumps, and other drives such that output is maintained, perhaps at a reduced level, when one motor is tripped. In any case, it is desirable to keep all of the above motors in service unless there is serious trouble that requires prompt and decisive action by the protective system.

Essential service motors are usually designed for full voltage starting and for fast restart after interruption. In some cases, essential service motors may be backed up with dc motors, or steam turbine drives.

19.5.1.2 Nonessential Service Motors. Motors in nonessential service may be out of service and not affect the output of the total plant. For the case of a generating station, typical nonessential motors are the following [30]:

Nonessential Service Power Plant Auxiliary Motors

Coal handling equipment	Coal crushers
Central coal pulverizers	Conveyors
Clinker grinders	Vent fans
Air compressors	Service pumps

Clearly, the power plant can operate normally for a period of time without these motors, although not indefinitely.

19.5.2 Motors Classified by Location

Motors in unattended locations must be protected against all possible types of abnormal conditions, which may require an extensive protective system design, especially if the motor is in essential service. Motors in attended stations are monitored by the operators, who often are able to judge the need to trip the motor for a given condition. This may permit the use of alarms rather than automatic tripping of the motor for abnormal conditions that are temporary, or that can be tolerated in times of urgent need for continuity of service.

Examples include the motor overheating or the lubricating oil overheating. If the condition is not excessive and the operating environment such that a trip can be delayed, the operator may elect to keep the motor in service until the urgency of service is past. In many cases, the operator has information that the protective system does not have, such as the variation in service or load requirements, that may make continued service an acceptable risk unless the motor condition is extremely serious.

19.5.3 Summary of Motor Classifications

The foregoing suggests that the application of motor protection must take into account a number of conditions, many of which are external to the motor. The internal concerns are still present, such as the desire to remove faulted equipment promptly and to prevent unnecessary internal damage due to abnormal operation. However, conditions external to the motor may modify the protective system design to permit better utilization of the motor, even if this causes some additional risk or sacrifice of life in the unit.

In the protective systems described below, these considerations will be noted from time to time.

19.6 STATOR PROTECTION

Stator protection of induction motors takes many different forms, with the most elaborate protection being necessary for large motors at unattended locations where all types of problems must be detected and cleared. This section presents the following different types of stator protections:

1. Phase fault protection
2. Ground fault protection
3. Locked rotor protection
4. Overload protection
5. Undervoltage protection
6. Reverse phase rotation protection
7. Unbalanced supply voltage protection
8. Loss of synchronism protection (synchronous motors)
9. Loss of excitation protection (synchronous motors)
10. Sudden supply restoration protection

There are many ways to provide these various protective functions. We shall investigate only the most common of these.

19.6.1 Phase Fault Protection

Phase faults seldom occur in motors, since almost all faults quickly become ground faults and can be cleared by ground relays. It is common practice, however, to provide protection against phase faults. This is often in the form of overcurrent relays or, on larger motors, phase winding differential relays, which are much faster and more sensitive than overcurrent relays. Moreover, differential relays will not operate during starting, which could be a problem with overcurrent relays.

The pickup setting of phase overcurrent relays should be about four times rated current, but with adequate time delay so that they do not operate during starting. Instantaneous phase overcurrent relays must obviously be set well above locked rotor current. Overcurrent relays are often omitted for essential service motors to ensure that the motor is tripped for short circuits and for no other reason.

Figure 19.22 shows a comparison of a typical fuse characteristic on the same time-current axes as the starting current. These curves must be viewed with caution, since the characteristics are not the same types of plots. The starting current curve is sometimes called a "trace," and it depicts a simulation of the transient value of rms current versus time, with time plotted on the vertical axis in the figure [31]. The fuse characteristic plot is not a transient current trace. The fuse characteristic shows the time for the fuse to melt at various values of current, with times greater than that plotted indicating a trip condition. By the time the starting current trace reaches about 90% of locked rotor, the fuse may be close to melting. A similar comparison would apply to an overcurrent relay. Coordination is very tight. It is almost impossible to select a relay or fuse characteristic that closely matches the motor starting current trace. Obviously, the fuse characteristic shown in Figure 19.22 will not protect the motor due to overload.

The setting for an instantaneous overcurrent relay must be greater than locked rotor current. This setting must still be below the available phase-to-phase fault current. The setting for phase differential relays, however, can be to the left of the starting current trace, say at about 0.1 per unit on the horizontal scale.

One source recommends that instantaneous phase relays be set at 1.6 times locked rotor current, but less that one-third the three-phase fault current available from the system [31]. Moreover, the ratio of three-phase fault current to locked rotor current should be greater than about 5.0 for good protection.

It is possible that adequate phase fault protection might be provided by using overcurrent and instantaneous relays in two phases of the motor, with the option of using thermal overload relays in the third phase [31].

19.6.2 Ground Fault Protection

Almost all motor faults quickly develop into ground faults, hence it is essential that ground fault protection be provided. Induction motors are almost always wye connected with ungrounded neutral. Ground faults are easily detected using instantaneous relays in the neutral of the wye-connected supply transformers, and set at about 20% of full load current with a time dial of 1.0. There may be a problem with false tripping due to unequal saturation of the current transformer on motor starting. This can be helped by using a lower relay tap, forcing all current transformers to saturate uniformly, or by placing a resistor in series with the ground relay [32], [33]. Another effective technique is to use a "donut" or window type CT with the donut core surrounding all three-phase conductors. This solves any problems with CT saturation on starting and is effective in detecting ground fault currents. Instantaneous ground relays are also used by some protection engineers.

19.6.3 Locked Rotor Protection

Locked rotor protection is required to prevent excessive overheating if the motor stalls while running or if it is unable to accelerate large mechanical loads. A typical induction motor can carry locked rotor current safely for about 20 seconds [26]. If the motor starting is fully accomplished in about 10 seconds, this leaves a good margin for time discrimination in setting an overcurrent relay delay time. However, if the load inertia is great, such as a large fan, the motor starting time may be close to the thermal capability of the motor and other means must be used to detect the need for tripping.

One method of detecting a locked rotor condition is to use a bi-metal thermal relay. The motor heats according to an I^2t function and the bi-metallic element matches this characteristic well. These relays have a relatively large thermal time constant, which amounts to a built-in delay. This is sometimes termed an "over-run" and may be as high as 45% [26]. Thus, if locked rotor current can be sustained for only 20 seconds, the relay over-run will be 9 seconds, requiring a setting of 11 seconds for adequate locked rotor protection.

Locked rotor protection on starting is difficult to accomplish using overcurrent relays. As noted in Figure 19.22, the lower end of the heating curve represents the acceptable locked rotor heating time. It may be tempting to try and fit an inverse overcurrent characteristic between the starting trace and the motor capability curve, but there is little coordination time in that region. In fact, by the time the starting trace makes the first bend away from locked rotor current, the relay contacts are already very nearly closed and the current does not drop below pickup until the current is about 0.5 locked rotor. Contact closure may occur before reaching that lower current in accelerating a high inertia load, even though the relay characteristic is always above the starting current trace.

An alternative to the bimetal relay is to use a distance relay and timer for locked rotor protection. The impedance characteristic at locked rotor is inside the relay trip zone, as shown in Figure 19.27. The relay is positioned to measure the impedance seen looking into the motor terminals. If the motor does not accelerate well in starting, the impedance is almost a constant, with low power factor, which can easily be detected by a mho type distance element. If the motor does accelerate, the impedance rapidly increases and the power factor improves, thereby moving the impedance outside of the mho element trip zone.

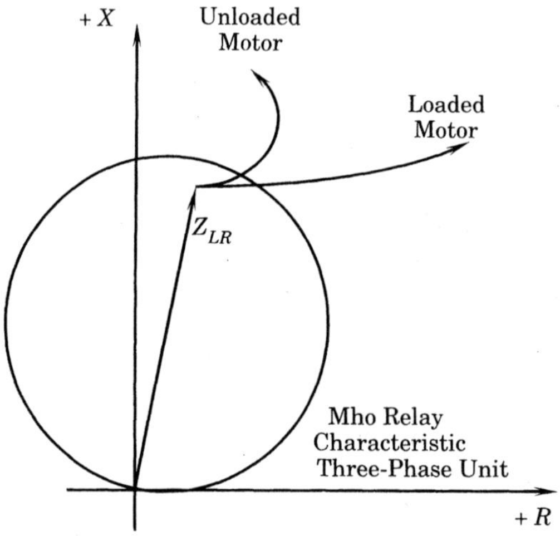

Figure 19.27 Distance relay characteristic for locked rotor protection [33].

19.6.4 Overload Protection

Overload protection is designed to prevent the motor from overheating beyond design limits. This type of protection is often omitted on essential service motors, but may be considered for other applications where it is considered prudent to protect the motor from unusual loading conditions. Replica type thermal overload relays are often used for overload protection, but this solution is not entirely satisfactory. One problem in providing this type of protection is that the motor thermal capability is often not known accurately and is usually not precise. Moreover, it is difficult to define damage, loss of life, and safe zones of operation. Figure 19.28 illustrates the problem of matching the thermal relay characteristic with a typical motor capability curve.

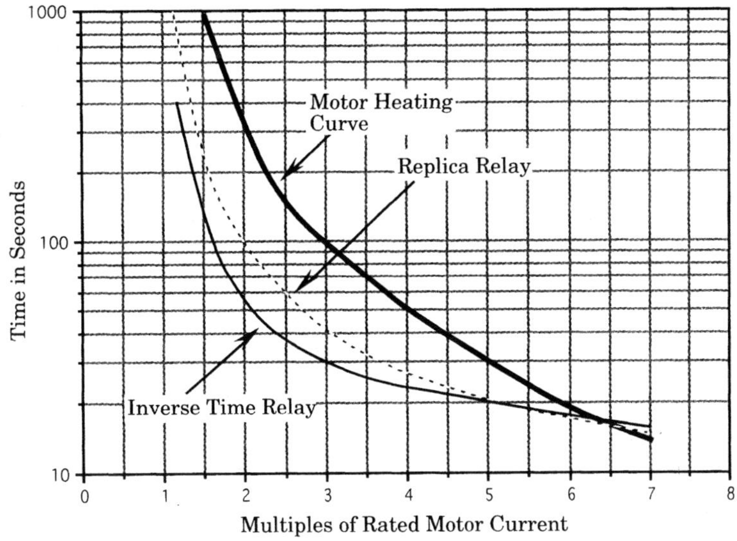

Figure 19.28 Comparison of a typical motor thermal characteristic against replica and overcurrent relay characteristics [29].

The replica-type relay characteristic protects the motor against long-term overloads, but coordination is questionable for short-term overloads. This suggests that the replica relay be supplemented in this range using a long-time-overcurrent relay.

The above solution is still ineffective in cases of extreme variations in load, such as jogging motors. This is due to the different cooling characteristics between the relay and the motor. Variations in load may "ratchet" the relay into premature tripping. One solution to this difficult problem is to measure the stator coil temperature directly using embedded exploring coils placed in the stator slots. This type of thermal relay uses exploring coils that are connected in one leg of a Wheatstone Bridge circuit. Both ac and dc types are available. This solution is also effective in protecting the motor against blocked ventilation, against which current operated relays are not effective.

19.6.5 Undervoltage Protection

Undervoltage usually causes a drop in speed, as shown in Figure 19.26, and an increase in current. One exception to this rule is for a motor driving a fan, where the load falls off rapidly as speed is reduced. Undervoltage protection is usually recommended on all motors that are not essential service motors [29]. The protection should be applied on at least one phase of small motors and on all three phases of motors over 1500 hp. The protective relay should have some type of inverse time characteristic. Undervoltage protection is a standard feature of many ac motor controllers.

Protection may be in the form of inverse undervoltage relays, although overcurrent and overload relays also provide a measure of protection. If the undervoltage condition is severe, the motor should be quickly disconnected.

Low voltage may prevent a motor from starting properly due to the reduced torque at low voltage. This will cause higher than normal starting currents that should not be permitted to flow for a long period of time since the motor may never come up to speed. Overcurrent and thermal relays provide some protection, but undervoltage relays are preferred.

19.6.6 Reverse Phase Rotation Protection

Reverse phase rotation is not required to protect the motor, but to protect the load and personnel working in the vicinity due to the reverse direction of rotation of the motor and load. If reverse phase rotation presents a hazard, protection may be provided by a reverse phase relay. These relays also will operate when phases are badly unbalanced but are not designed as protection against single phasing of the supply voltages.

19.6.7 Unbalanced Supply Voltage Protection

The conditions that exist due to unbalanced supply voltage are discussed in Section 19.4.2.1. The result of unbalanced supply voltage is to cause an even greater unbalance in the motor currents. The worst case of unbalanced supply voltage is single phasing of the supply, which also needs to be considered in a protective system design.

Unbalanced supply voltage protection may be provided by a three-phase induction disk voltage relay. This device can be used to prevent the motor from being started unless all three phases are present and also in the correct sequence. It does not protect the motor from overheating when the unbalance occurs during running, however.

Another type of protection uses bimetallic devices in each of the three phases with contacts arranged so that all three must move together or the contacts meet and trip the motor. These same bimetallic spirals also protect the motor against overload [30]. Another type of protection uses a negative-sequence element to provide greater sensitivity. These relays detect even small amounts of negative-sequence currents, which is a direct measure of the unbalance. Still another type of protection measures negative-sequence voltage on the motor supply bus. This is attractive in some installations, since one relay can protect a group of motors connected to the same bus.

A special case of unbalanced supply voltage occurs when one phase becomes opened. This condition is discussed in detail in [31]. Single phasing can be protected by several different types of relays [32].

1. Phase balance relays
2. Reverse phase voltage relay
3. Negative-sequence overcurrent relay with time delay
4. Instantaneous negative-sequence overcurrent relay
5. Negative-sequence voltage relay
6. Thermal relays:
   embedded temperature detectors
   replica type
7. Overcurrent relays
8. Phase failure relays

Each of the above is described in detail in [31], where the conclusion is reached that types 1, 4, and 8 are the most effective for detecting single phasing. When one phase opens, currents of about 120 hertz are induced in the rotor squirrel cage (or the field winding of a synchronous motor). The high frequency causes skin effect that forces the currents to the surface of the rotor bars. Extreme heating can result with possible rotor damage occurring before the stator begins to overheat. Thus, it is sometimes said that unbalanced supply voltages create more of a rotor problem than a stator problem. For this reason, this subject is discussed more fully in Section 19.7.

19.6.8 Loss of Synchronism in Synchronous Motors

Synchronous motors that are required to start under load should have loss of synchronism protection. If the motor does not start under load, the use of undervoltage and loss of excitation protection should provide adequate protection against loss of synchronism.

Loss of synchronism or "pull out" protection is often provided for motors that may experience large voltage dips or sudden increases in load that exceed the pull out torque of the motor. If the excitation remains in service with the motor out of step, the motor draws large currents and is often unable to resynchronize. Disconnecting the excitation is not recommended, since this will cause the motor to run as an induction motor, which will not solve the problem and simply leads to additional rotor heating. A relay that is sensitive to motor power factor is a good candidate for loss of synchronism protection, since the motor power factor is very low during out-of-step operation [26].

Another type of protection is a relay that counts successive power reversals in a specified time interval. The out-of-step motor experiences a pulsating power demand on the system. These pulsations can be detected and counted for a given time interval to detect the condition and trip the motor [31], [33].

19.6.9 Loss of Excitation in Synchronous Motors

Loss of excitation protection is indirectly provided on motors that have protection against stator overheating, rotor overheating, and loss of synchronism. Motors that do not have these protections may require special relays for loss of excitation protection.

Loss of excitation protection may be provided by an undercurrent relay with time delay in the field circuit. Some types of these relays detect even partial loss of field and may also operate on loss of synchronism [31].

Both loss of excitation and loss of synchronism may be detected by a wattmeter type relay that is biased to have maximum output when the motor is operating at its normal power factor. On loss of excitation, the power factor drops rapidly and is quickly detected by the wattmeter element.

19.6.10 Sudden Supply Restoration Protection

When the supply voltage to a synchronous motor is lost, the motor continues to rotate and generate a terminal voltage, driven by its connected load. If the supply voltage is then restored, the phase difference may be great enough to cause a large surge in the motor current. Therefore, it is often considered necessary to trip a synchronous motor following the loss of supply voltage so that a normal startup procedure can be followed after the supply is restored.

The type and arrangement of the protection will depend on the specific connection, and on the presence of other load at the motor bus. If there is no other connected load near the motor, then an under-power relay can be used to detect loss of supply. If there is other local load, this load will cause the power to reverse at the motor terminals, when its operation changes from motoring to generating. In this case, a reverse power relay might be used if the other load is always present. If there is no other load near the motor, overvoltage or underfrequency relays will provide the desired protection.

19.7 ROTOR PROTECTION

It was noted earlier that unbalanced supply voltage creates unbalanced currents in the stator, but also causes currents in the rotor at about twice the supply frequency. These currents cause rapid heating of the rotor that may be damaging, and the rotor should be protected against this hazard. Because of the high frequency of rotor currents, the resistance seen by these currents is also about five times greater than the dc resistance seen by positive-sequence currents [26]. Therefore, the heating effect of one per unit negative-sequence current is greater than that due to one per unit positive-sequence current.

19.7.1 Rotor Heating

The total heating of the motor may be described in terms of the total I^2R loss, which may be written as

$$\begin{aligned} P_{\text{Loss}} &= \text{positive-sequence losses} + \text{negative-sequence losses} \\ &= 3\left(I_{a1s}^2 R_{1s} + I_{a1r}^2 R_{1r}\right) + 3\left(I_{a2s}^2 R_{2s} + I_{a2r}^2 R_{2r}\right) \end{aligned} \quad (19.130)$$

For the stator, the resistance to positive- and negative-sequence currents is approximately equal, i.e.,

$$R_{1s} = R_{2s} = R_s \quad (19.131)$$

Suppose we write, for the rotor,

$$R_{2r} = kR_{1r} \quad k > 1 \quad (19.132)$$

then (19.8) may be written as

$$P_{\text{Loss}} = 3R_s\left(I_{a1s}^2 + I_{a2s}^2\right) + 3R_{1r}\left(I_{a1r}^2 + kI_{a2r}^2\right) \quad (19.133)$$

For the rotor, however, the resistances to the two sequence currents, reflected to the rotor, are quite different, with the negative-sequence resistance being the larger by a considerable amount. Since this fact is known, measuring the sequence currents in the stator forms a sensible basis for estimating rotor heating and this is done in some protective systems.

19.7.2 Rotor Protection Problems

The problem in providing good rotor protection for induction machines is that there is no way to directly measure currents or temperatures on the rotor itself. Therefore, the effects seen in the stator currents or temperatures must be used to provide protection that will also shield the rotor from excessive heating. The criterion for protection should be that prolonged unbalanced conditions should not be allowed, due to the excessive rotor heating, but unnecessary disconnection due to brief unbalances should be avoided. One way to protect the motor from the prolonged periods of overheating is to derate the motor when the voltages are unbalanced, or in situations where the motor is likely to see unbalanced voltages. Since no direct rotor measurements are possible, the stator protection must be relied on to give rotor protection as well.

The protection of wound rotor motors is more difficult than for squirrel cage induction motors, and the stator protections described above may not be adequate for wound rotor designs [29]. It is recommended that the protection engineer consult the motor manufacturer regarding protection of wound rotor motors.

Synchronous motors should have protection against field winding faults and against grounding of the field winding. The protection described for generators is generally applicable to synchronous motors.

19.8 OTHER MOTOR PROTECTIONS

This section presents some diverse topics that don't fit well within the scope of the preceding sections. The first is special protective systems for the motor bearings. Following that, we discuss some of the modern concepts of digital protection that claim to protect the entire motor against most hazards.

19.8.1 Bearing Protection

Bearings can be protected by using a relay to measure the metal temperature of the bearings or of the lubricating oil. In unattended stations the increase of bearing or oil temperature to a set threshold limit can be used to trip the motor. In attended stations, an alarm would usually be used to alert the operator to investigate the cause of the overheating, which may be due to low oil level or faulty detector. It would be especially important not to shut down essential service motors unless the incipient failure has a very high probability of causing extensive damage.

19.8.2 Complete Motor Protection

Some of the newer motor protective relays combine several functions in one relay. One device separates the line currents into positive and negative components and provides a trip characteristic that is proportional to the quantity

$$I_{a1}^2 + kI_{a2}^2 \tag{19.134}$$

This is approximately equal to the heating characteristic of the motor. The relay is also temperature compensated allowing the relay characteristic to approximate the heating of the motor. This prevents shutdown on overload unless absolutely necessary [26].

Other recent types of motor protection schemes are microprocessor based and incorporate several different protective functions in the same device. One manufacturer offers a motor protector that provides 12 different functions in a single relay case, including the following protections [34]:

1. Stator overheating
2. Rotor overheating
3. Blocked rotor
4. Ground overcurrent
5. Instantaneous overcurrent
6. Load loss (undercurrent)
7. Phase current unbalance

8. Instantaneous phase reversal
9. Motor bearing overheating
10. Load bearing overheating
11. Load case overheating
12. Load jamming

The overheating protection is based on the computation of slip dependent rotor resistance. Embedded temperature devices can be utilized to optimize the thermal model. Not only is the protection complete but, being a digital device, it also has the capability of self-diagnosis, event recording, and communications of information to a computer or to the protection engineer's office.

19.9 SUMMARY OF LARGE MOTOR PROTECTIONS

Figure 19.29 summarizes the types of protection that are often employed for large induction motors. The system shown is assumed to have two motors fed from a common supply that is grounded on the load side. The load bus has relays to monitor the bus voltage and protect the motors against unbalance (device 47) and undervoltage (device 27). The remainder of the relays use motor currents or temperatures and are unique to each protected motor. All of the relays shown in the figure are described further in Table 19.4, except for relay 55. This is a power factor relay that might be used for synchronous motor protection against loss of field. The other relays are applicable to either induction or synchronous motors.

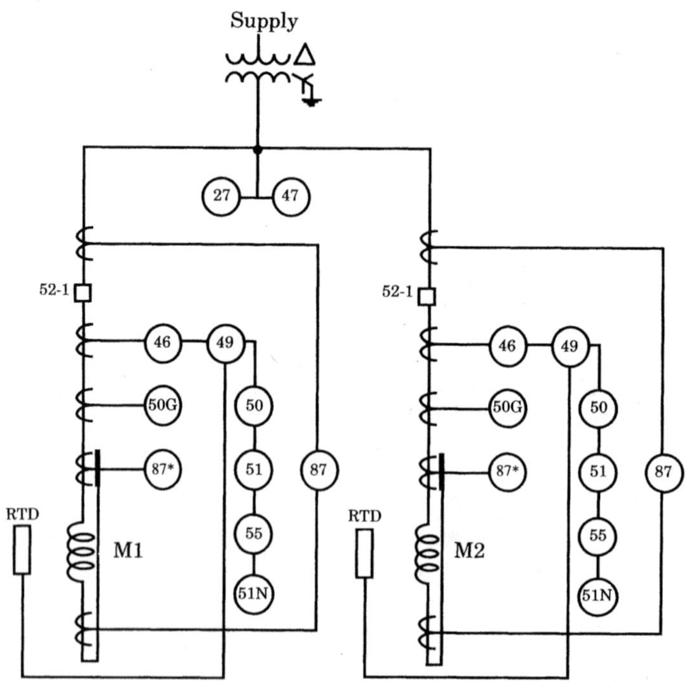

Figure 19.29 Typical protection for large induction motors [31].

TABLE 19.4 Typical Protection Systems for Large Motors [31]

Device	Description	Typical Setting	Remarks
49	Thermal relay	Set for motor safe operating temperature	Overload protection, blocked ventilation, high ambient temperature
51	Overcurrent	Current set at 0.5 locked rotor	LR protection when starting time > 30 s
		Time delay to make operate time < start time	LR protection when starting time < 30 s
50*	Instantaneous overcurrent	Set at twice locked rotor	Fault protection
50G	Instantaneous overcurrent	0.25A	Ground fault protection with donut type CTs
46	Reverse phase Phase balance	Set 2A, $I > 3A$ Set 1A, $I < 3A$	Unbalanced current protection
47	Phase sequence voltage relay	Negative sequence = 5%	Unbalanced voltage protection
27	Undervoltage relay	Low voltage 75% to 80%	Undervoltage protection
87ϕ*	Winding differential	0.25 A	Phase and ground protection
87	Winding differential	10%	Phase fault protection alternative, use where 3 ph fault I available less than $5 \times$ LR current
51N 50N	Overcurrent Instantaneous overcurrent	Pickup 0.5 A time 0.1 s at inst. setting, inst. set $4 \times$ FL current	Use where 50G is not applicable

*Use where minimum available 3ϕ fault current is less than five times motor starting amount and 87ϕ cannot be used.

The protective systems shown in Figure 19.29 are conventional relays that could be electromechanical, static, or digital. The degree of protection in all cases should be weighed against the cost of the motor and the cost of having the motor out of service for prolonged periods.

REFERENCES

[1] ANSI/NFPA 70-1987, National Electrical Code, American National Standards Institute, New York, 1987.

[2] NEMA MG1-1993, Motors and Generators, National Electrical Manufacturers Association, New York, 1993.

[3] ANSI/IEEE 37.96-1988, "Guide for Motor Protection," IEEE, New York, 1988.

[4] Gafford, B. N., W. C. Duesterhoeft, Jr., and C. C. Mosher, III, "Heating of Induction Motors on Unbalanced Voltages," *AIEE Trans.*, 78, 1959, pp. 282–287.

[5] Martiny, W. J., R. M. McCoy, and H. B. Margolis, "Thermal Relationships in an Induction Motor Under Normal and Abnormal Operation," *AIEE Trans.*, 79, 1960.

[6] ANSI/IEEE Std 620, "IEEE Trial Use Guide for Construction and Interpretation of Thermal Limit Curves for Squirrel-Cage Motors over 500 HP," IEEE, New York, 1981.

[7] Eliasen, A. N., "The Protection of High-Inertia Drive Motors During Abnormal Starting Conditions," *IEEE Trans.*, PAS-99 (4), 1980.

[8] Eliasen, A. N., "High-Inertia Drive Motors and Their Starting Characteristics," *IEEE Trans.*, PAS-99 (4), 1980, pp. 1472–1482.

[9] Zocholl, S. E., E. O. Schweitzer, III, and A. Aliaga-Zegerra, "Thermal Protection of Induction Motors Enhanced by Interactive Electrical and Thermal Models," *IEEE Trans.*, PAS-103 (7), 1984, pp. 1749–1755.

[10] Dymond, J. H., "Stall Time, Acceleration Time, Frequency of Starting: The Myths and the Facts," *IEEE Trans. on Industrial Applications.*, 29 (1), January/February 1993.

[11] Schweitzer, E. O., III, and S. E. Zocholl, "Aspects of Overcurrent Protection for Feeders and Motors," a paper presented at the IEEE I&CPS Technical Conference, Irvine, CA, May 1–5, 1994.

[12] Fitzgerald, A. E., C. Kingsley, Jr., and A. Kusko, *Electric Machinery; The Processes, Devices, and Systems of Electromechanical Energy Conversion*, McGraw-Hill Book Co., New York, 1971.

[13] McPherson, G., *An Introduction to Electrical Machines and Transformers*, John Wiley & Sons, Inc., New York, 1981.

[14] Anderson, P. M., *Analysis of Faulted Power Systems*, IEEE Press, Piscataway, NJ, 1995.

[15] Kimbark, E. W., *Power System Stability*, IEEE Press, Piscataway, NJ, 1995.

[16] Anderson, P. M., and A. A. Fouad, *Power System Control and Stability*, Revised Printing, IEEE Press, 1993.

[17] Zocholl, S. E., "Motor Analysis and Thermal Protection," *IEEE Trans.*, PWRD-5 (3), July 1990, pp. 1275–1280.

[18] Richards, J. A., F. W. Sears, M. R. Wehr, and M. W. Zermanshy, *Modern University Physics*, Addison-Wesley Publishing Co., Inc., Reading, MA, 1960.

[19] Reynolds, W. C., and H. C. Perkins, *Engineering Thermodynamics*, McGraw-Hill Book Co., New York, 1970.

[20] Takahashi, Y., M. J. Rabins, and D. M. Auslander, *Control and Dynamic Systems*, Addison-Wesley Publishing Co., Inc., Reading, MA, 1970.

[21] Raven, F. H., *Automatic Control Engineering*, McGraw-Hill Book Co., New York, 1968.

[22] Cochin, I., *Analysis and Design of Dynamic Systems*, Harper & Row, Publishers, New York, 1980.

[23] Weast, R. C., Ed., *Handbook of Chemistry and Physics*, 45th Ed., The Chemical Rubber Co., Cleveland, 1964.

[24] Morgan, V. T., *Thermal Behavior of Electrical Conductors: Steady, Dynamic and Fault-Current Ratings*, Research Studies Press Ltd., Taunton, Somerset, England, 1991.

[25] Todd, J. P., and H. B. Ellis, *Applied Heat Transfer*, Harper & Row, Publishers, New York, 1982.

[26] GEC Measurements, *Protective Relays Application Guide*, General Electric Company p.l.c. of England, 1975.

[27] Wagner, C. F., and R. D. Evans, *Symmetrical Components*, McGraw-Hill Book Company, New York, 1933.

[28] Bauman, H. A., G. R. Hahn, and C. N. Metcalf, "The Effect of Frequency Reduction on Plant Capacity and on System Operation," *Trans. AIEE*, 74, 1955, pp. 1632–1637.

[29] Mason, C. R., *The Art and Science of Protective Relaying*, John Wiley & Sons, New York, 1956.

[30] Warrington, A. R. van C., *Protective Relays: Their Theory and Practice, Volume One*, John Wiley & Sons, Inc. New York, 1962.

[31] Westinghouse Relay-Instrument Division, *Applied Protective Relaying*, Westinghouse Electric Corporation, Newark, NJ, 1976.

[32] Horowitz, S. H., Ed., *Protective Relaying for Power Systems*, IEEE Press Selected Reprint Volume, New York, 1980.

[33] Blackburn, J. L., Ed., *Applied Protective Relaying*, Westinghouse Electric Corporation, Newark, NJ, 1973.

[34] Product Bulletin, Pro-Star Motor Protection System, ABB Protective Relay Division, 35 N. Snowdrift Road, Allentown, PA, 1989.

PROBLEMS

19.1 Verify (19.16) by working through the entire derivation to satisfy yourself that this is correct, and that you understand what every term in the equation means.

19.2 What is the purpose of normalizing the induction motor equations? Why not just solve the system in mks units?

19.3 Review (19.21), the synthesis equation. What is the origin of this equation and what does it mean?

19.4 Suppose that a given passive load has physical parameters such that it can be modeled as shown in Figure P19.4.
Derive the equation for the complex three-phase volt-amperes of this circuit, where it is recognized that the wye-connected impedances are mutually coupled.

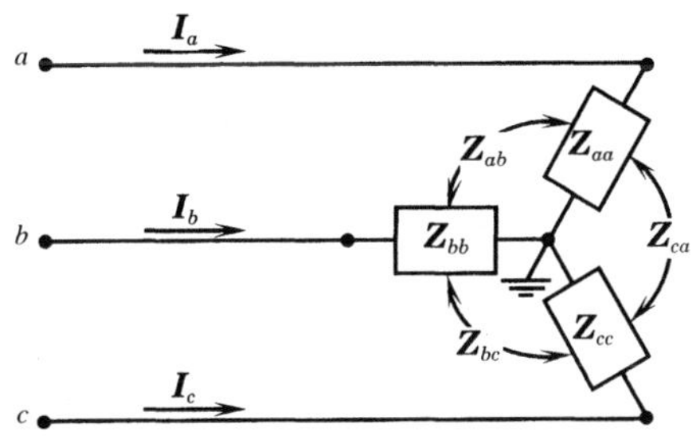

Figure P19.4 A Wye-connected load with mutual phase impedances.

19.5 Verify (19.25) and (19.26).

19.6 Verify (19.27).

19.7 Prepare a spreadsheet to perform the numerical integration to compute the induction motor starting solution. This can be done in a straightforward manner by first entering the applied voltages and the motor data for the equivalent circuit. The initial value of slip is known to be 1.0. After working through the various motor equations, the value of $F(\omega, t)$ can be computed. Any numerical integration method can be used to compute the value of ω for the next time step. The number of time steps required is a function of the time step h.

19.8 Example 19.1 states that the motor current for the 10 HP motor is 20 amperes. Verify this result by computing the motor current, assuming balanced applied rated voltage, given that the load torque is specified by (19.71) with an initial torque of 0.3 and a final torque of 0.8, and with exponents of $m = 5$ and $n = 2$.

19.9 Use the spreadsheet of problem 19.8 to verify the behavior of the motor when applied with unbalanced voltages, as described in example 19.3.

19.10 Two different materials are connected along a common boundary as shown in Figure P19.10. The outer surfaces of the combined materials have different temperatures. Construct an electric analog for the solution of heat transfer through the two materials in the direction shown in the figure.

19.11 Two different materials are connected along a common boundary as shown in Figure P19.11. The outer surfaces of the combined materials have different temperatures. Construct an electric analog for the solution of heat transfer through the two materials in the direction shown in the figure.

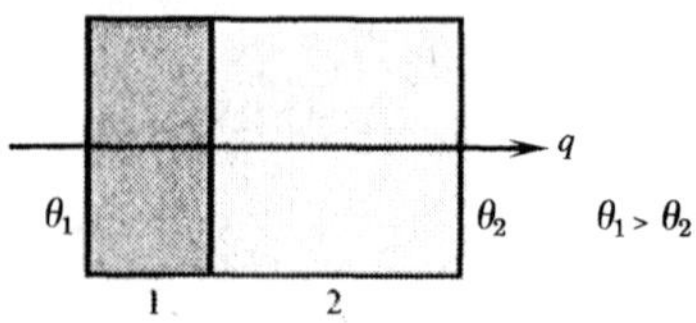

Figure P19.10 Two materials joined at one surface.

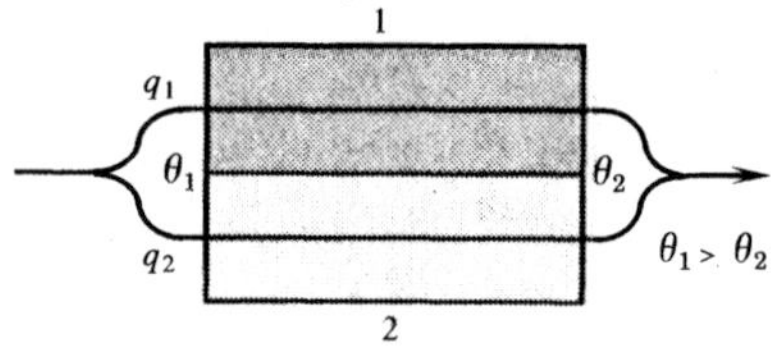

Figure P19.11 Two materials joined at one surface.

19.12 Three materials are joined and bolted together, as shown in Figure P19.12. Construct an electric analog for the solution of heat transfer through the combined materials from left to right. Note that the heat transfer can take place through both the joined materials and the bolt.

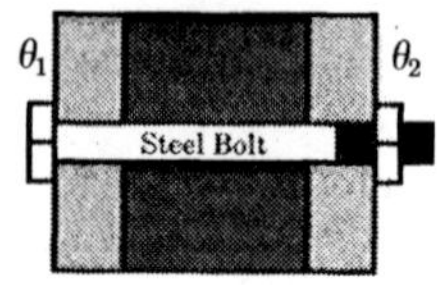

Figure P19.12 Three materials bolted together.

19.13 Make a dimensional check of (19.81).

19.14 Solve the network of Figure 19.15 and show that the capacitor voltage as defined in the figure results in exactly the form of (19.89).

19.15 Verify (19.97) and (19.98).

19.16 Add the thermal equations of the induction motor to the spreadsheet model developed in problem 19.8. Use this new model to verify the results of example 19.4, using the data of Table 19.3.

19.17 It has been suggested that the thermal capacitance for stator and rotor can be determined by the product of the mass of the metal and the specific heat of the material. Suppose that this lumped-parameter model is located at the center of mass of rotor and stator. Write the equations for such a model.

19.18 Verify the normalization of (19.111) as given in (19.112).

19.19 Prepare a list of potential internal motor hazards. Separate your list into two types of hazards: (1) those that you would expect to detect by electrical measurements and (2) those that you would not expect to detect by electrical measurements.

19.20 One method of preventing a motor overcurrent trip during startup would be to install two separate overcurrent relays, one for normal running conditions and the other for starting conditions. On starting, the starting protection is in service and, as the motor comes up to speed, a centrifugal switch replaces the starting protection by the running protection. Comment on the feasibility of this plan. Can you identify any weakness in the scheme?

19.21 Section 19.4.2 provides a list of motor problems that are due to hazards external to the motor. List these hazards and determine the type of protection that might be used for each.

19.22 Verify (19.123) and (19.124). Assume that the excitation branch of the motor equivalent circuit can be neglected for the purpose of this calculation.

19.23 It is proposed that the positive-sequence induction motor equivalent circuit be solved by Thevenin's theorem. This can be accomplished by replacing the left portion of the positive-sequence network of Figure 19.1 by the following equivalent (neglecting the mutual resistance), as shown in Figure P19.23.

(a) Compute the open circuit voltage and the short circuit current for the circuit to the left of the terminals $a - b$. Then determine the Thevenin impedance for that circuit. Call the resulting Thevenin impedance $R_1 + jX_1$.

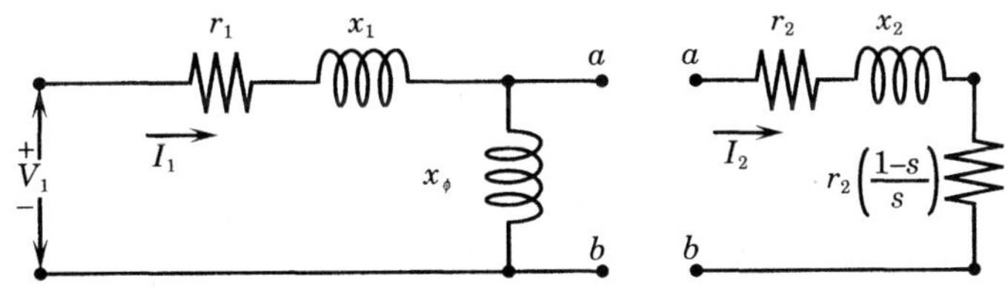

Figure P19.23 A Thevenin equivalent circuit for an induction motor.

(b) Solve for the rotor current in terms of the Thevenin equivalent circuit voltage and impedance.

(c) Write the expression for the following motor parameters:
1. The positive-sequence air-gap power
2. The developed power
3. The developed torque

19.24 Develop a spreadsheet to solve for the rotor current by solving the Thevenin Equivalent system derived in problem 19.23. Use the following induction motor parameters for the calculations.

$$r_1 = 0.294\ \Omega/\text{ph} \qquad x_1 = 0.503\ \Omega/\text{ph}$$

$$r_2 = 0.144\ \Omega/\text{ph} \qquad x_2 = 0.209\ \Omega/\text{ph}$$

$$V_1 = \frac{220}{\sqrt{3}}\ V \qquad x_\phi = 13.25\ \Omega/\text{ph}$$

19.25 Figure P19.25 shows the positive-sequence equivalent circuit for an induction motor. The goal is to determine the positive-sequence current entering the motor for a given applied voltage and slip. One way of doing this is to first form an equivalent of the system seen looking into the air gap, identified by the points FA and FB in the figure. Call this equivalent impedance $\mathbf{Z}_{F1}$ and determine its parameters as a function of the circuit parameters. Then

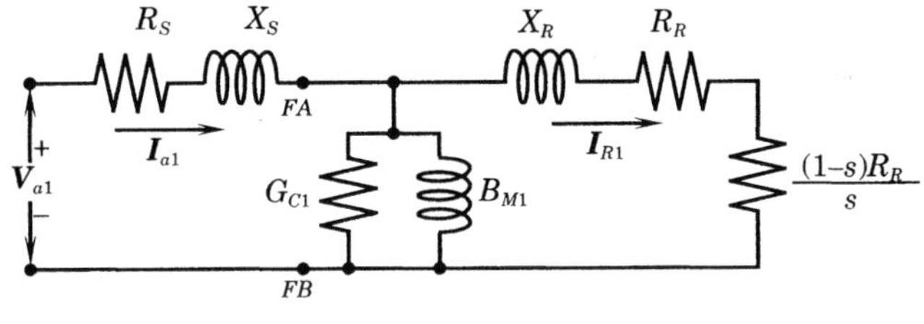

Figure P19.25 Positive-sequence induction motor equivalent.

add the derived equivalent to the stator impedance to give the total positive-sequence motor equivalent.

(a) Based on the new equivalent circuit, compute the current entering the motor as a function of slip.

(b) Knowing both the applied positive-sequence voltage and the current, find the positive-sequence stator losses.

19.26 Repeat problem 19.25 for the stator losses for the negative-sequence network. Recall that the negative-sequence equivalent for the induction motor is different from the positive-sequence network.

PART V
SYSTEM ASPECTS OF PROTECTION

20

Protection Against Abnormal System Frequency

The protections that have been discussed in the last three chapters have considered faults or other problems that originate in a generator, exciter, transformer, or motor that cause the breakers at a station to be tripped. In this chapter, we consider another category of problems that originate in the power transmission system, rather than in the power plant or substation components. Protective systems are devised to monitor power system behavior, and to make sure that the life of the generating unit is not sacrificed unnecessarily due to an emergency system condition for which detection and unit protection is feasible. Several different conditions are discussed in this broad category in this and the following chapters. This chapter presents the need for abnormal frequency protection and describes protective measures that are commonly used. The next three chapters address other types of system disturbances for which special protections are required.

20.1 ABNORMAL FREQUENCY OPERATION

Both the generator and the generator step-up transformer are protected by volts per hertz relays to prevent damage due to emergency low-frequency conditions. There are other abnormal frequency conditions that are of concern and that are not within the generator protection zone. Severe system disturbances on the network sometimes cause heavily loaded lines to be tripped, which may bring about the cascaded tripping of other lines. This sometimes results in the separation of the interconnected system into islands. Assuming that the system splits into two islands, which is often the case, one island will invariably have a surplus of generation and the other a surplus of load, since the separation usually does not result in islands whose loads and generation are in equilibrium. This results in high frequency in one island and low frequency in the other due to the imbalance between load and generation in each of the islands. There are two aspects of off-normal frequency operation of steam-turbine generating units: one pertaining to the generator and one pertaining to the turbine [1–3]. In the discussion that follows, we assume that a system is split into two islands, one with overfrequency and one with underfrequency and will consider the effect on both generator and turbine in each island.

20.2 EFFECTS OF FREQUENCY ON THE GENERATOR

First, consider the effects of sustained operation of a generator that is in an emergency island, created as a result of some system disturbance, which produces the condition that either over- or underfrequency exists in the island.

20.2.1 Overfrequency Effects

In an islanded operation where load has been lost for any reason, all generators will see this disturbance as a loss of load, which is often termed a *load rejection* as far as the generating unit is concerned. This will result in an increase in speed, which should be controlled rapidly by the prime mover speed governors. Assuming that the load reference settings of the speed governors are not changed, the governor "droop" setting (about 5% in North America) will determine the final change in generator speed or frequency per unit change in load. Should the overspeed condition persist, the generator is usually not in a hazardous condition since generator loading is lower than normal and cooling is improved due to the increased speed. Moreover, since load has been lost, voltages are likely to be high, resulting in reduced excitation. If the excitation is greatly reduced, the generator might be tripped by loss-of-excitation protection, depending on the sensitivity and settings of these relays. It is also possible that a unit could be tripped due to high voltage, should a voltage regulator be out of service. Generator tripping may not necessarily be a serious matter since there is already too much generation in the overfrequency island, and such a trip is likely to be an isolated event.

20.2.2 Underfrequency Effects

In an underfrequency island the reverse of the above conditions will prevail. Here, the generators are all overloaded and the speed, and hence cooling, are below normal. Because of the overload conditions, system voltages are likely to be low, causing generator excitations to be increased, perhaps to their limits. This raises the possibility of thermal overload of both the stator and the rotor. The possibility exists for unit trip due to stator overheating, rotor overheating, overexcitation, and underfrequency (volts/hertz). Moreover, the entire island is short of generation and a trip of any unit could start a cascading of unit trips and a rapid deterioration of the island to a complete blackout condition. Therefore, it is very important that the protection not be overly sensitive and initiate unit trips unless absolutely necessary [1–5].

The ANSI standards provide guidelines regarding the short-time thermal capability required of steam turbine generators, as shown in Table 20.1. These values are based on a constant increment of heat added and can be extended to other values of time by the expression [3], [6]

$$\left(x^2 - 1\right) t = K \tag{20.1}$$

where x = per unit value of stator current or field voltage
K = a constant
≈ 41 for the stator
≈ 33 for the rotor

It is important that some protection be provided to protect the generator from exceeding the limits imposed by Table 20.1. Most modern voltage regulators provide overexcitation limiting

TABLE 20.1 Generator Short Time Thermal Capability [6]

Permissible Time in Seconds	Stator Limit in Per Unit of Rated Current	Stator Limit in Per Unit of Rated Field Voltage
120	1.16	1.12
60	1.30	1.25
30	1.54	1.46
10	2.26*	2.08

that will reduce the excitation voltage to a safe value after an appropriate time delay. This voltage reduction may also reduce the stator current, but the stator current overload is system dependent and voltage reduction cannot be relied on for stator protection.

For generator continuous operating times exceeding those shown in the table some limitation must be applied, either as a reduction in MVA to reduce the current or a reduction in voltage to reduce the volts/hertz and counteract the excess flux due to low frequency. Appropriate limitations are shown in Figure 20.1.

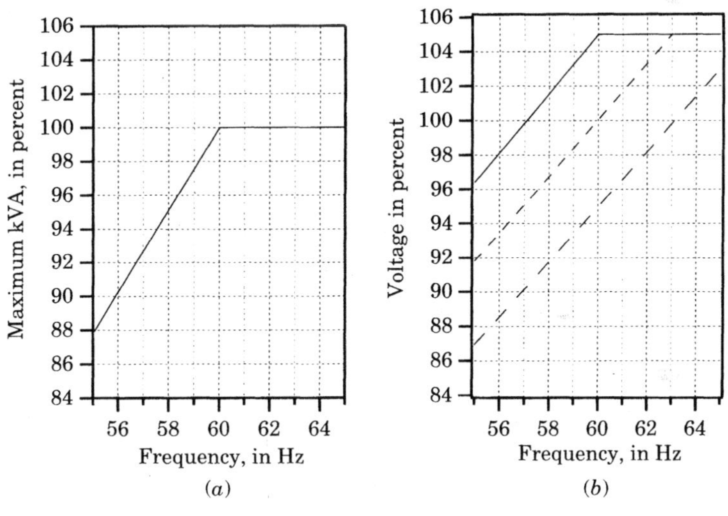

Figure 20.1 Generator underfrequency operation limitations [3]: (*a*) Maximum kVA. (*b*) Voltage.

To satisfy the volts per hertz limitations, the voltage of the generator can be reduced in the same proportion as the frequency reduction to return to normal flux conditions. This may also reduce the volt-ampere loading to acceptable values, but there is no assurance that this will be the case. Some manufacturers publish a time limitation for the volts per hertz criterion, as shown in Figure 20.2 [3]. The protection trip time must be adjusted to honor these time restrictions.

The protection of the generator against over- and underfrequency conditions does not require any specific additions to the protective systems discussed previously. The primary difference here is one of protecting the generator from unnecessary loss of life rather than the removal of a faulted generator from the system. This same philosophy extends to the protection of the turbine, which is the subject of the next section.

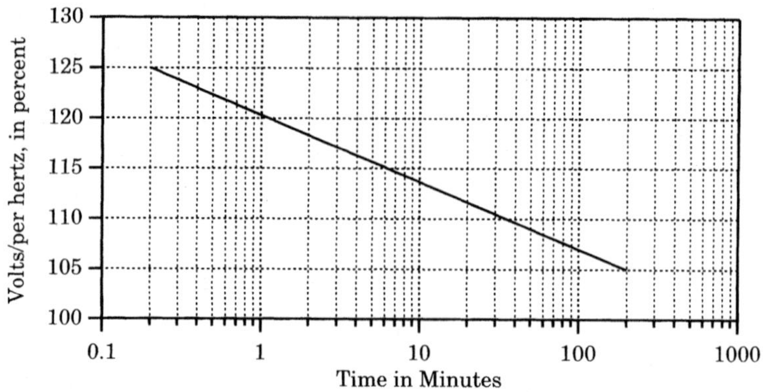

Figure 20.2 Time limit for volts/hertz protection [3].

20.3 FREQUENCY EFFECTS ON THE TURBINE

Before considering specific turbine protection for over- and underfrequency operation, we consider some features of steam turbine design that are important from the standpoint of off-normal frequency operation. This discussion deals only with steam turbines. Hydro turbines are not affected by over- or underfrequency to any marked degree.

Steam turbines are constructed with many stages of blades or buckets, from very short blades in the high pressure section to blades several feet long in the lower pressure sections. Each blade is like a steel beam that is anchored at its root and therefore can display several natural bending modes of oscillation. The natural frequencies of oscillation for turbine blades consist of two parts: one that depends on the natural frequencies of oscillation when the blade is stationary and the other that depends on the speed of oscillation and other factors, such as the radius of the shaft and the length of the blade [3].

The effect of these natural frequencies of blade oscillation for different rotor speeds is illustrated by a set of curves called a "Campbell diagram," an example of which is shown in Figure 20.3 for a particular stage of long, tuned turbine blades. The heavy, nearly horizontal lines represent the natural resonant frequencies for this stage of blading. The diagonal lines represent the harmonics of a given shaft speed, e.g., at rated speed of 60 rev/s the second harmonic is 120 hertz, the third is 180 hertz, etc. The turbines are designed such that the natural blade resonant frequencies fall between these integral harmonics of the driven speed. Note, however, that should the shaft speed be either above or below normal (60 rev/s), this would constitute a harmonic driving force to the blades that may be very close to the natural frequency of vibration. As with many mechanical oscillators, the blade response is very tightly tuned, as shown in Figure 20.4, i.e., it is a high Q system with very high gain at or near resonance but very low gain at frequencies just slightly removed from resonance. The peak of the amplification factor curve is inversely proportional to the damping, which is always very small, hence the peak is very high. For turbines, the stress due to this magnification is not great at low load levels, although there is some danger of high stress during start up. At rated load, however, it has been found that it is not reasonable to design blades to survive the stress of a resonant condition. The failure mechanism is related to the endurance of the material when subjected to many cycles of high stress vibrations, as shown in Figure 20.5. The left portion of the figure shows the stress that results from off nominal frequencies and the right side plots

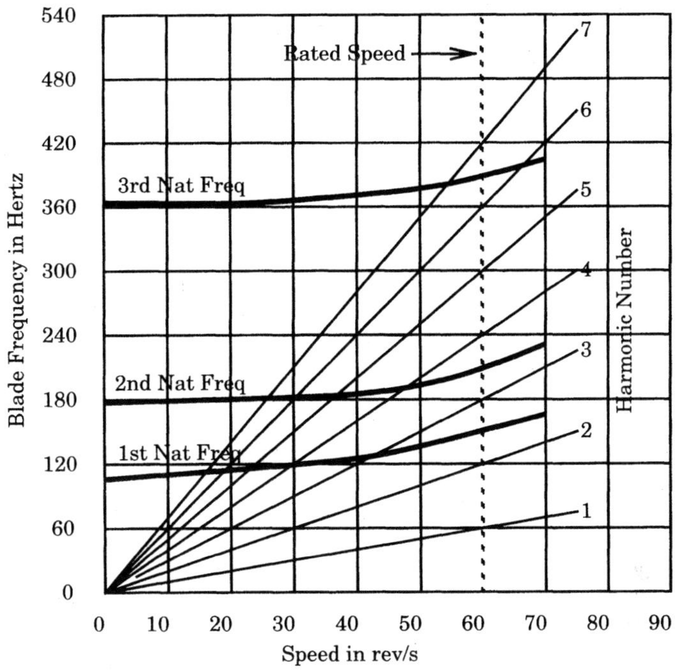

Figure 20.3 Campbell diagram for one stage of tuned turbine blades [3], [7].

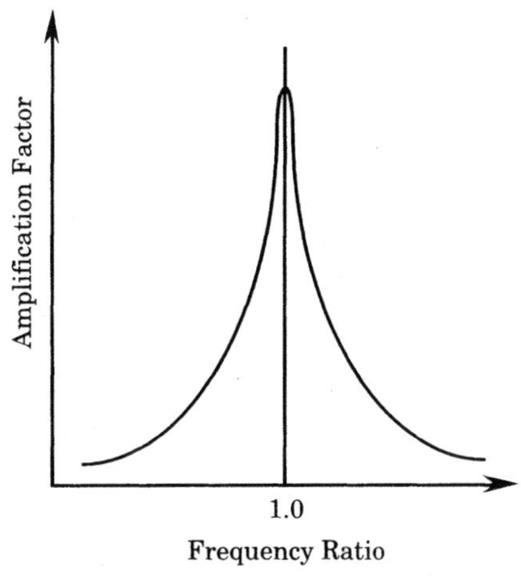

Figure 20.4 Amplification factor for forced blade oscillation [3].

this stress against the endurance limit of the material. If the stress is small, as shown by "a" in the figure, then the material can endure a very large number of vibration cycles. This means that the turbine could run almost continually at this stress level (speed) without inflicting blade damage.

As the stress (speed deviation) increases, the number of cycles of vibration that can be endured is much lower (note the logarithmic scale). The life of the turbine may see many excursions into these high stress regions. It is the *cumulative* experience that measures the total loss of life of the machine.

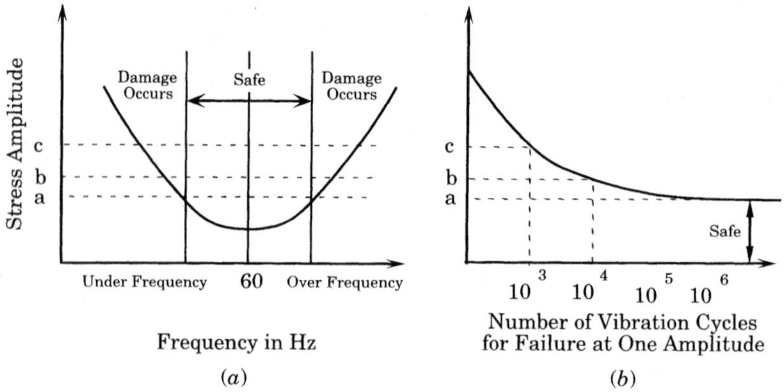

Figure 20.5 Stress amplitude versus frequency and endurance [7]. (*a*) Stress amplitude. (*b*) Number of vibration cycles.

The turbine manufacturers have developed guidelines that show the total accumulated time that a turbine can be operated at various off nominal speeds. Figure 20.6 is a composite of these curves for turbines of two different manufacturers [3], [7].

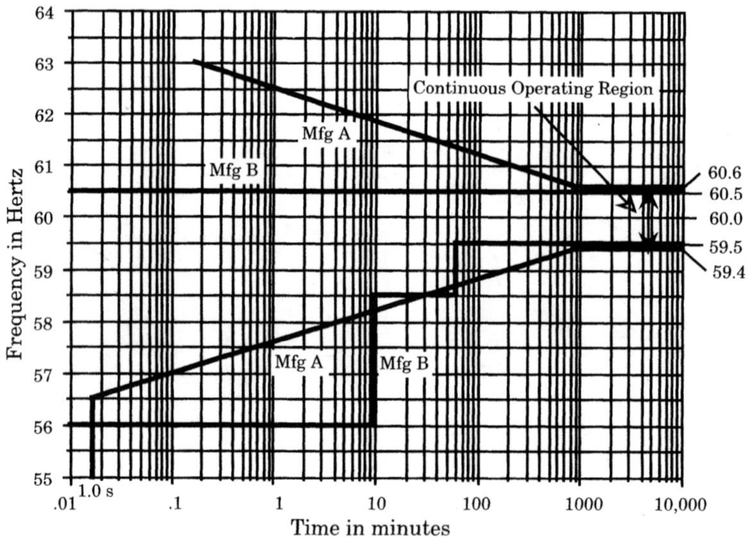

Figure 20.6 Turbine off frequency lifetime [3], [7].

Manufacturer A depicts the turbine off-normal frequency lifetime as two horizontal lines at 60.6 and 59.4 hertz, between which the unit can operate continuously without loss of life due to turbine blade limitations. As the frequency deviation increases either up or down, the lifetime of the unit is shown by slanting lines. The total lifetime below 56.5 hertz is shown as 1.0 second.

Manufacturer B shows continuous operation between 60.5 and 59.5 hertz. Here, it is assumed that there will be no significant operation above 60.5 due to prompt governor action. At below normal frequencies, the unit lifetime drops in steps with the region between 58.5 and 59.5 being limited to 60 minutes and that between 56.0 and 58.5 being limited to 10 minutes.

It is important to note that the effects of off-normal frequency operation are cumulative. For example, using Manufacturer A's curves, a total of 2 minutes of operation at 58 hertz leaves the unit with only about 2 minutes' additional operating lifetime at that frequency.

20.3.1 Overfrequency Effects

Overfrequency operation always follows a load rejection due to some cause. Assuming a 5% droop characteristic in the governor, a 50% load rejection will result in a 2.5% rise in speed, to a frequency of 61.5 hertz. Using Curve A of Figure 20.6, the turbine has a lifetime of about 30 minutes at this frequency. Thus, even if the governor is not responding quickly, the operator has time to take manual control action to reduce the governor load reference setting before a substantial loss of life occurs. For higher frequency excursions, an overfrequency relay can be used to initiate runback of the governor load reference motor, which readjusts the desired turbine power output.

20.3.2 Underfrequency Effects

Underfrequency operation is usually more critical than overfrequency and, as noted previously, it is important that the unit not be tripped if it can possibly be kept in operation. It is also important that the turbine be protected by underfrequency relaying to prevent lengthy excursions that expend large amounts of turbine life. A 1978 survey by the North American Electric Reliability Council showed that 38% of the units in North America utilized underfrequency protection, and that 26% of all underfrequency schemes were designed specifically for protection of the turbine [8]. This same survey also showed that, at the time of the survey, there was no general consensus regarding the need for underfrequency protection for turbine blading, even though most manufacturers recommend such protection.

Turbine underfrequency protection is complicated by the fact that the frequency response of the system depends on the size of the disturbance which, in this case, is the loss of generation to support the load, and also on the parameters that govern the system dynamic performance. Another very important element in the equation is the design and response of underfrequency load shedding protection. Underfrequency load shedding is usually installed by mutual agreement of all operating utilities in an interconnected region. If it is assumed that these relays are all installed and working properly, an estimate can be made of the dynamic frequency dip that follows a predefined system separation, including the shedding of load as the frequency falls. Knowing this characteristic, the turbine underfrequency protection can be designed.

Another basic assumption that must be made is the size of load change disturbance that the system is designed to withstand. As the magnitude of the load disturbance is increased, the depth of frequency decay increases, and the amount of load to be shed also increases. These basic conditions must be known in order to estimate the actual frequency decay following the loss of generation. Once this expected rate of frequency performance is known, the underfrequency protection for the turbine can be designed.

20.4 A SYSTEM FREQUENCY RESPONSE MODEL

The dynamic frequency behavior of a power system can be estimated by modeling the speed control of the total system as it responds to changes in load. Since every power system is an island of some discrete size, we model the system under the assumption that the equations are normalized to the size of the total island, that is, the system base volt ampere magnitude is the sum of the ratings of all of the generating units [9]. The system model, then, is similar to the model of any individual generating unit, except that it is scaled in size to represent the total system. We also assume that the system generation is composed primarily of steam turbine units, which is surely the predominate type of generation in most modern systems, at least in North America and Western Europe. This model may not represent systems that are served primarily by hydrogeneration.

The concept of a uniform frequency model has been explored by numerous investigators dating back 50 years or more. Our approach is similar to that of Rudenberg [9], who provides a number of references on the subject as well as a mathematical derivation of the basic concept. Similar and related approaches have been pursued more recently through work on energy functions [10], [11]. The basic ideas are also important in the work on system area control simulators [12], [13], as well as the work on long-term dynamics [14], [15]. In addition to these resources, certain ideas have also been adopted from the work on coherency based dynamic equivalents [16], [17], as well as the work on transient energy stability analysis [18]. A related, but quite different approach has been taken in the work on emergency control [19], but that model is more complex than believed necessary and is more difficult to use than the method presented here. The analysis and results found here are similar to that of [20] and [21], but our model is simpler. References 22–24 provide still other methods of analyzing the problem of frequency behavior, in varying degrees of complexity. The approach pursued here is to provide the minimum order model that retains the essential average frequency shape of a system with typical time constants and active speed governing [25], [26].

The basic system model is shown in Figure 20.7, where we assume that every generating unit has a speed governor and a turbine that produces mechanical power, P_m. Under normal steady-state conditions, the mechanical power is balanced by the electrical power output, P_e, of the generators. Any imbalance between mechanical and electrical power produces accelerating power that acts through the rotating inertia to accelerate the shaft and thereby create an incremental change, $\Delta\omega$, in frequency. Clearly, if every unit is of this type, we may represent the entire system using this simple model by aggregating all units together as if they were on the same bus. In actual systems, the generating units of various different types, but are predominantly driven by reheat steam turbine prime movers. This is an important assumption, as we shall see.

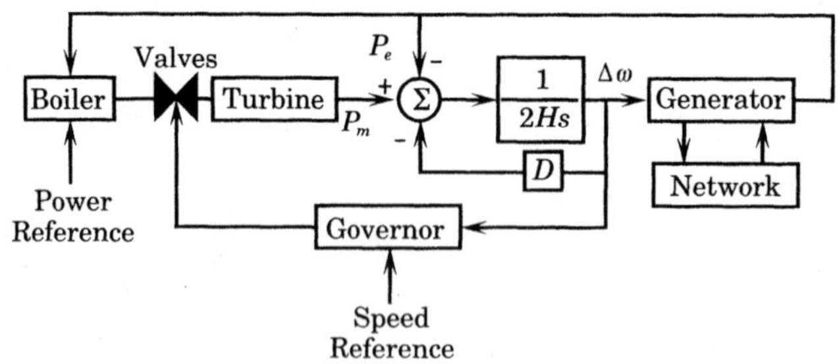

Figure 20.7 Block diagram of the system frequency model.

Let us examine the components of the system shown in Figure 20.7. In particular, we seek ways to simplify the system description without losing the essential dynamic characteristics. Simplification of the mathematics is justified since the model is only approximate and may not represent the exact behavior at any particular location in the island, but is rather an estimate of the weighted average frequency of the entire island.

We begin with the swing equation, shown by the block diagram of Figure 20.8, which computes the speed deviation as a function of accelerating power. This is simply an expression of Newton's law. The inertia constant in this equation, H, is rather large, with typical values in the range of 3–5 seconds. When multiplied by 2, this is a very important parameter that will play a dominant role in the frequency behavior. The damping constant, D, is a simplified representation of the damping inherent in the steam turbine and generator to changes in speed, as well as the frequency dependence of the system loads.

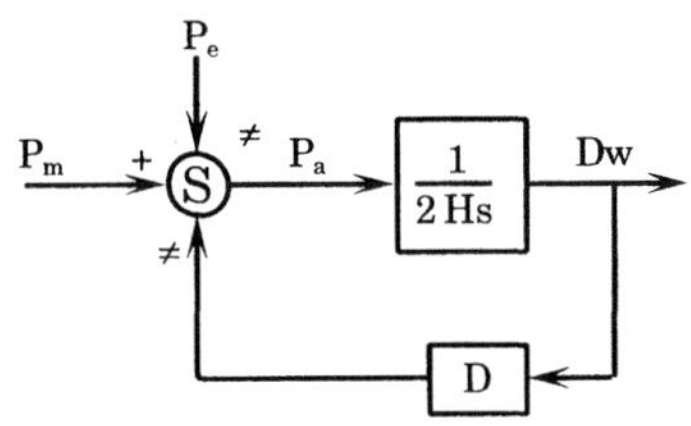

Figure 20.8 Block diagram for the swing equation.

The turbine is assumed to be a single reheat turbine, which is common in North America and in many other parts of the world. A typical model for a reheat steam turbine is shown in Figure 20.9. There is one predominate time constant in this model, the reheat time constant, T_R. This constant is typically in the range from 7 to 11 seconds [25]. The time constants associated with the steam chest and the crossover are very small and will be neglected in our model. The fractions of total mechanical power produced by the high-pressure turbine, F_H, is about 0.2–0.3 per unit, with the remainder produced by the intermediate and low-pressure turbines. Note that the total of the steam turbine fractions must always equal unity.

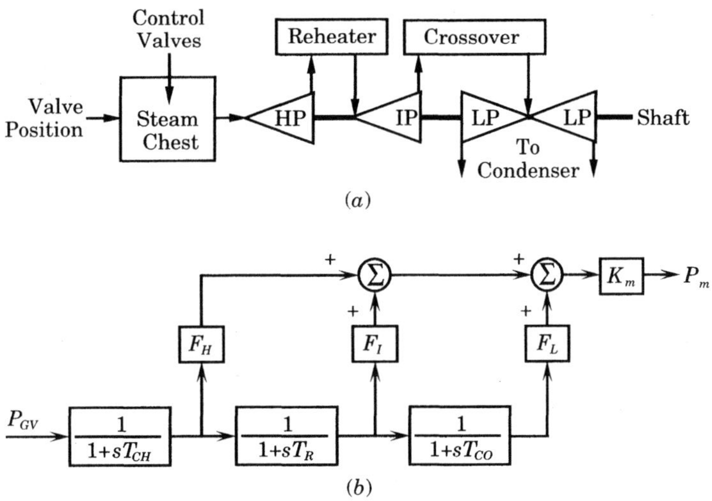

Figure 20.9 Block diagram for a single reheat steam turbine [25]. (*a*) Typical reheat turbine configuration. (*b*) Mathematical model of the reheat turbine.

The constant K_m is an overall gain constant that will permit further tuning of the turbine model output power. Initially, we will set this constant to a value near to unity, but later will revise it to properly account for the load power factor and the dispatch of generating units in the island. The input to the turbine model is the speed governor power demand, P_{GV}, which is proportional to the throttle valve area or the governor valve position, and which is controlled by the speed governor.

The speed governor model may be of many different types, but the model shown in Figure 20.10 is typical. The time constants in the governor model are all quite small compared to the reheat time constant and will be neglected, which makes the entire forward loop gain equal to unity. This means that the governor is assumed to act very fast compared to the change in speed (frequency). We also neglect the governor nonlinearities on the assumption that operation is in the linear range.

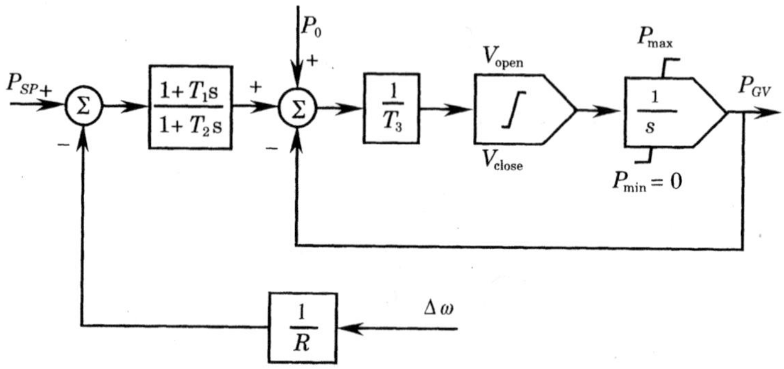

Figure 20.10 Block diagram of a typical speed governor.

The final constant to be considered is the speed droop or regulation of the governor which is the constant R shown in the feedback path of Figure 20.10. In North America, R is set to about 0.05, which makes its inverse a constant feedback gain of 20. This is a significant gain constant and must be retained in the model.

Making the foregoing assumptions to simplify the model, we arrive at the *reduced order system frequency response* (*SFR*) model shown in Figure 20.11, where all parameters are in per unit on an MVA base equal to the total rating of all generating units in the island [27]. Note that only six constants describe the model behavior:

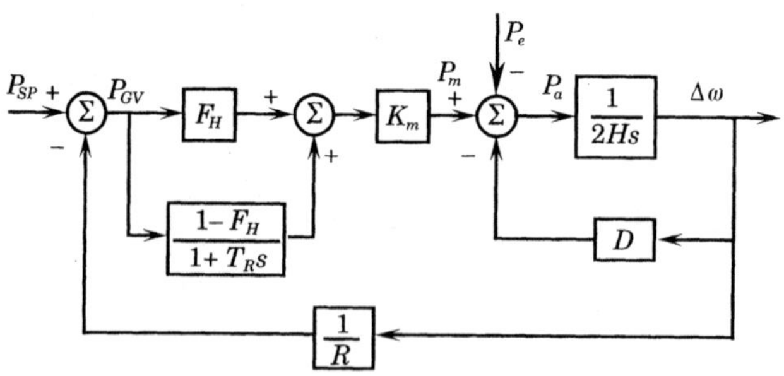

Figure 20.11 The reduced order SFR model.

The gain factor, K_m

The damping factor, D

The inertia constant, H

The reheat time constant, T_R

The high pressure power fraction, F_H

The speed droop or regulation, R

The predominate physical constants are the inertia and the reheat steam constant. The governor droop and turbine steam fractions are common design parameters that are also important in the system behavior. All parameters are readily estimated based on common knowledge of typical system designs. Note that the system model is only of second order.

The notation used in Figure 20.11 is described as follows:

P_{SP} = incremental power set point, per unit
P_m = turbine mechanical power, per unit
P_e = generator electrical load power, per unit
$P_a = P_m - P_e$ = accelerating power, per unit
$\Delta\omega$ = incremental speed, per unit
F_H = fraction of total power generated by the HP turbine
T_R = reheat time constant, seconds
H = inertia constant, seconds
D = damping factor
K_m = mechanical power gain factor

The system block diagram of Figure 20.11 can be easily reduced to a forward transfer function and a feedback transfer function. The closed loop transfer function of the system frequency can then be written in the following form:

$$\Delta\omega = \left(\frac{R\omega_n^2}{DR + K_m}\right)\left(\frac{K_m\,(1 + F_H T_R s)\,P_{SP} - (1 + T_R s)\,P_e}{s^2 + 2\varsigma\omega_n s + \omega_n^2}\right) \tag{20.2}$$

where we compute the system undamped natural frequency and damping factor to be given by

$$\omega_n^2 = \frac{DR + K_m}{2HRT_R}$$
$$\varsigma = \left(\frac{2HR + (DR + K_m F_H)\,T_R}{2\,(DR + K_m)}\right)\omega_n \tag{20.3}$$

The SFR model is linear; therefore the variables may be either the full value of the variables or incremental values. The model is the same in either case. We are interested in small changes in frequency from normal, hence we shall consider all variables to be incremental variables.

We also note that there are two inputs to the system, the governor power set point and the generated electric power or system load power. The governor power set point is adjusted by signals from the energy control center, which schedules the generation of all units as required to meet the load demand. Since we are interested in behavior over a period of a few seconds, the power set point reference may be considered constant (or its incremental value zero). Therefore, we may rearrange the system to show only the incremental electric power as the input. Let us also redefine this input as a "disturbance" power, P_d. We select the sign of the disturbance power to be the opposite of the electrical generated power, since we are usually interested in a power unbalance where the load is greater than the generation. Therefore, we define

$$P_d = -P_e \tag{20.4}$$

where a negative value of P_d corresponds to the step load increase, such as an island that forms with excess load, therefore requiring a negative step change in disturbance power. With this sign convention, a positive disturbance is one that will cause the system frequency to increase.

With this minor change, we may redraw the SFR model to that shown in Figure 20.12, where the incremental disturbance power is the input and the per unit incremental speed or frequency is the output.

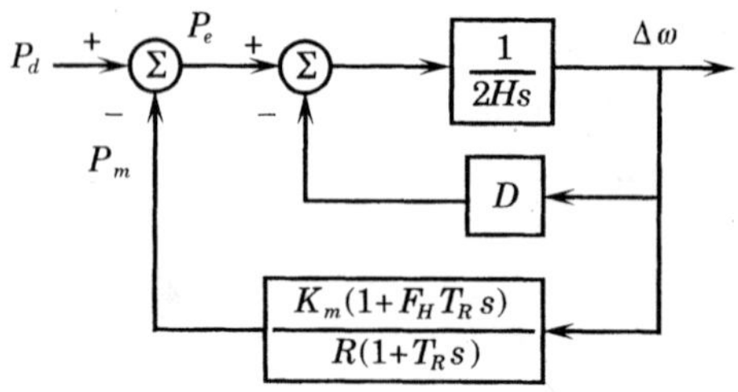

Figure 20.12 The simplified SFR model with disturbance power as the input.

For this system model we compute the frequency response in per unit to be

$$\Delta\omega = \left(\frac{R\omega_n^2}{DR + K_m}\right)\left(\frac{(1 + T_R s)\, P_d}{s^2 + 2\varsigma\omega_n s + \omega_n^2}\right) \tag{20.5}$$

and the per unit speed or frequency can be computed for any P_d.

For sudden disturbances, large or small, we are usually interested in the disturbance power in the form of a step function, i.e.,

$$P_d(t) = P_{\text{step}}\, u(t) \tag{20.6}$$

where P_{step} is the disturbance magnitude in per unit based on the system base MVA, S_{SB}, (defined later) and $u(t)$ is the unit step function. In the Laplace domain, we write

$$P_d(s) = \frac{P_{\text{step}}}{s} \tag{20.7}$$

and this expression can be substituted into (20.5) with the result

$$\Delta\omega = \left(\frac{R\omega_n^2}{DR + K_m}\right)\left(\frac{(1 + T_R s)\, P_{\text{step}}}{s\left(s^2 + 2\varsigma\omega_n s + \omega_n^2\right)}\right) \tag{20.8}$$

This equation can be solved directly to write, in the time domain,

$$\Delta\omega(t) = \frac{R P_{\text{step}}}{DR + K_m}\left[1 + a\, e^{-\varsigma\omega_n t}\sin\left(\omega_r t + \phi\right)\right] \tag{20.9}$$

where

$$a = \sqrt{\frac{1 - 2T_R\varsigma\omega_n + T_R^2\omega_n^2}{1 - \varsigma^2}}$$
$$\omega_r = \omega_n\sqrt{1 - \varsigma^2} \tag{20.10}$$

and

$$\phi = \phi_1 - \phi_2$$

$$\phi_1 = \tan^{-1}\left(\frac{\omega_r T_R}{1 - \varsigma\omega_n T_R}\right) \qquad \phi_2 = \tan^{-1}\left(\frac{\sqrt{1 - \varsigma^2}}{-\varsigma}\right) \tag{20.11}$$

The time response is a damped sinusoidal frequency offset, as shown in Figure 20.13. Note that the result has been multiplied by 60 to give the results in hertz.

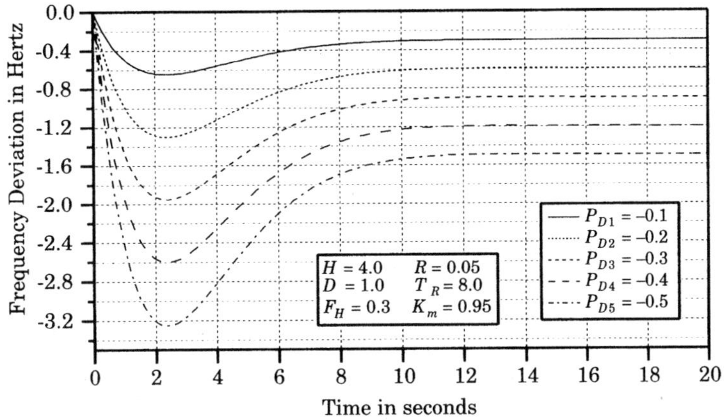

Figure 20.13 Time response for islanding under excess load conditions.

20.4.1 Effect of Disturbance Size, P_{step}

As an example of the computation of the SFR consider a system with the following typical parameters.

For load disturbances from -0.1 to -0.5 per unit in increments of -0.1

$$R = 0.05 \qquad H = 4.0\ s \qquad K_m = 0.95$$
$$F_H = 0.3 \qquad T_R = 8.0\ s \qquad D = 1.0$$

Then we compute

$$\omega_n = 0.559 \qquad \varsigma\omega_n = 0.481 \qquad \phi_1 = 141.396^\circ$$
$$\varsigma = 0.861 \qquad \omega_r = 0.284 \qquad \phi_2 = 149.416^\circ$$
$$a = 7.168 \qquad \sqrt{1-\varsigma^2} = 0.509 \qquad \phi = -8.020^\circ$$

The results, for values of P_{step} from -0.1 to -0.5 in increments of -0.1 are shown in Figure 20.13.

Note that the response is underdamped ($\zeta < 1$) and that the damping exponential $\zeta\omega_n$ is fairly large, resulting in almost negligible oscillation beyond the first swing. Also note that the maximum frequency deviation occurs at exactly the same frequency for all cases, which indicates that the time of maximum deviation is not a function of P_{step}.

20.4.2 Normalization

The equations shown above are typical of any reheat unit in the system, with all equations assumed to be in per unit on some base. Let us assume that all are in per unit on a common system volt-ampere base, S_B. We now combine all units into a single large unit that represents all generating units in the entire system. This can be done by adding the power equations, as follows:

$$\sum_i (2H_i s + D_i)\ \Delta\omega = \sum_i P_{mi} - \sum_i P_{ei} \tag{20.12}$$

$$\sum_i P_{GVi} = \sum_i P_{SPi} - \sum_i \left(\frac{1}{R_i}\right)\Delta\omega \tag{20.13}$$

$$(1 + T_R s)\sum_i P_{mi} = K_m\,(1 + F_H T_R s)\sum_i P_{GVi} \tag{20.14}$$

where we assume that all equations are on a common system base S_B. We now re-normalize (20.12) to (20.14) to the total system base S_{SB} which is equal to the sum of the ratings of all generating units in the system.

$$S_{SB} = \sum_{i=1}^{n} S_{Bi} \tag{20.15}$$

This change of base multiplies (20.12–14) by the ratio of the two bases with the result

$$\begin{aligned}\frac{S_B}{S_{SB}}\sum_i 2H_i s\,\Delta\omega &= \frac{K_m\;(1 + F_H T_R s)}{(1 + T_R s)}\\ &\times\left[\frac{S_B}{S_{SB}}\sum_i P_{SPi} - \frac{S_B}{S_{SB}}\sum_i\left(\frac{1}{R_i}\right)\Delta\omega\right] - \frac{S_B}{S_{SB}}\sum_i P_{ei}\end{aligned} \tag{20.16}$$

which defines the equivalent generator parameters.

$$\begin{aligned}H &= \frac{S_B}{S_{SB}}\sum_i H_i\\ \frac{1}{R} &= \frac{S_B}{S_{SB}}\sum_i \frac{1}{R_i}\end{aligned} \tag{20.17}$$

The normalized value of these parameters on the total system base will be typical of those for a single unit on its own base.

20.4.3 Slope of the Frequency Response

It is instructive to examine the effect on the transient response of the several important parameters in the system. Since the solution is easily written in closed form, by (20.9), we easily compute the slope of the response.

$$\frac{d\,\Delta\omega}{dt} = \frac{a\omega_n R P_{\text{step}}}{DR + K_m} e^{-\varsigma\omega_n t}\sin(\omega_r t + \phi_1) \tag{20.18}$$

where all parameters are previously defined. We are particularly interested in the value of (20.18) at two points: first, when $t = 0$, which corresponds to the maximum rate of change of slope, and second, the time at which the slope is zero, which corresponds to the maximum frequency deviation.

1. $t = 0$

$$\left.\frac{d\omega}{dt}\right|_{t=0} = \frac{aR\omega_n P_{\text{step}}}{DR + K_m}\sin\phi_1 = \frac{P_{\text{step}}}{2H} \tag{20.19}$$

2. $\dfrac{d\omega}{dt} = 0$

$$0 = \frac{aR\omega_n P_{\text{step}}}{DR + K_m} e^{-\varsigma\omega_n t}\sin(\omega_r t + \phi_1) \tag{20.20}$$

The condition in (20.20) is satisfied when $\omega_r t + \phi_1 = n\pi$ for n an integer, including zero. If we call this time t_z, we compute

$$t_z = \frac{n\pi - \phi_1}{\omega_r} = \frac{1}{\omega_r} \tan^{-1}\left(\frac{\omega_r T_R}{\varsigma\omega_n T_R - 1}\right) \tag{20.21}$$

These slope parameters are clearly observed in Figure 20.13. The initial slope depends only on P_{step} and H, hence it changes for each run plotted in the figure. However, t_z is not a function of P_{step}, hence the maximum frequency deviation occurs at exactly the same time (about 2.4 s) for all disturbances computed. Note also that Figure 20.13 is plotted in hertz, so the frequency equation must be multiplied by 60 to check the results.

Another parameter that can be readily checked is the regulation. Governors are set with a regulation or droop R to give a steady-state speed versus power relationship of

$$\Delta\omega_{ss} = \frac{RP_{\text{step}}}{DR + K_m} \tag{20.22}$$

where both ω_{ss} and P_{step} are incremental, normalized quantities. Note that P_{step} must take the sign of P_d, which would be negative for an increase in load or loss of generation. Thus, in Figure 20.13, when $P_{\text{step}} = -0.3$ and for $R = 0.05$ we compute, for $(DR + K_m)$ of unity,

$$\Delta\omega_{ss} = -(0.05)(0.3) = -0.015 \text{ per unit } = -0.9 \text{ Hz}$$

which is clearly observed in Figure 20.13. This result can be also be verified mathematically by the final value theorem

$$\lim_{t\to\infty} \Delta\omega(t) = \lim_{s\to 0} s\,\Delta\omega(s) = \frac{RP_{\text{step}}}{DR + K_m} \tag{20.23}$$

It is difficult to visualize the effect of each physical parameter of the SFR model without plotting the results. Therefore, each parameter is now varied in turn and the results plotted to illustrate the effect of that parameter.

20.4.4 The Effect of Governor Droop, *R*

To show the effect of governor droop, the value of R is varied from 0.05 to 0.10 in increments of 0.01 per unit. The results are shown in Figure 20.14.

Actual observed system responses in islanded situations have sometimes shown the net system regulation to be in the neighborhood of 0.1. This indicates that some generating units are operating with governor valves blocked, giving a regulation of 1.0 per unit for these units. This increases the net regulation for the system in proportion to the size of the units with blocked valves. Note in Figure 20.14 that the steady-state regulation is exactly as given by (20.22). The simulation shown in Figure 20.14 must be extended for a rather long time to accurately observe the final value.

It is also important to note that the assumed droop setting R has absolutely no effect on the initial rate of frequency decline. This is important. Even if all governors are at the extreme valve-closing end of their individual backlash limits, as in following a gradual load decrease, a sudden loss of load would require a rapid change to a valve open condition. While traversing this backlash region, the system is operating essentially open loop and the natural 100% turbine regulation prevails. However, the initial rate of change is the same as for a tightly tuned system with no backlash. This effect can be noted in Figure 20.12, where the regulation term is seen to be affected by the lag of the feedback term, which is a "lag-lead" function. The difference the regulation makes is in the recovery time, the maximum offset, and the final value. It is

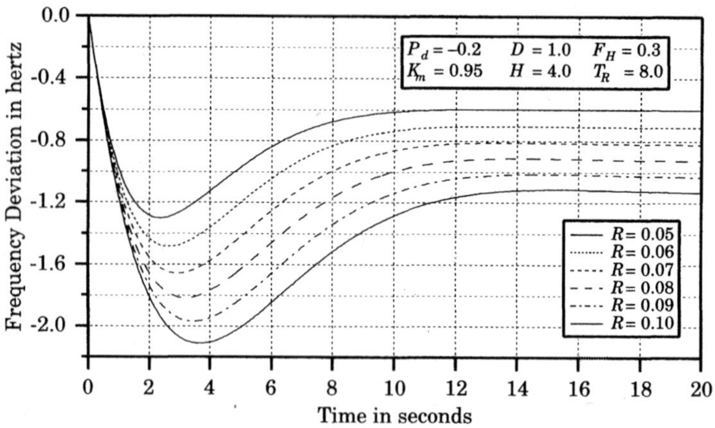

Figure 20.14 Frequency response for varying values of R from 0.05 to 0.10 in steps of 0.01.

important that this recovery be fast in order to limit the time of exposure at frequencies below about 57 to 58 hertz.

The value of R is probably bounded between about 0.05 and 0.10 for most systems. This effectively bounds the maximum frequency deviation for a given disturbance to the values shown in the shaded portion of Figure 20.15. Here, P_{step} is the disturbance step magnitude. Figure 20.15 is plotted with nominal values ($H = 4.0$ s, $T_R = 8.0$ s, $F_H = 0.3$) of the other parameters.

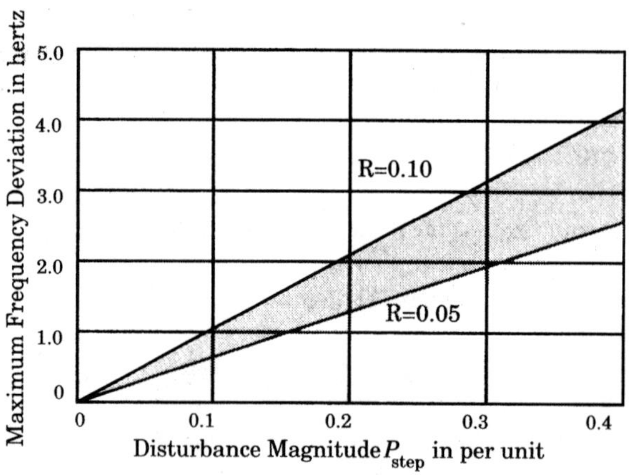

Figure 20.15 Range of maximum frequency deviation versus disturbance size.

20.4.5 The Effect of Inertia, *H*

The effect of inertia is examined by using typical values for all parameters and varying H, with the result plotted in Figure 20.16. The value of H has a direct effect on the initial slope, as noted earlier, and this is clearly shown in Figure 20.16.

The value of H affects all the measures of frequency response, including ω_n, and ζ, as given by (20.3), and thereby all other computed parameters as given by (20.10) and (20.11), including the initial slope and the time t_z of maximum frequency deviation as given by (20.19)

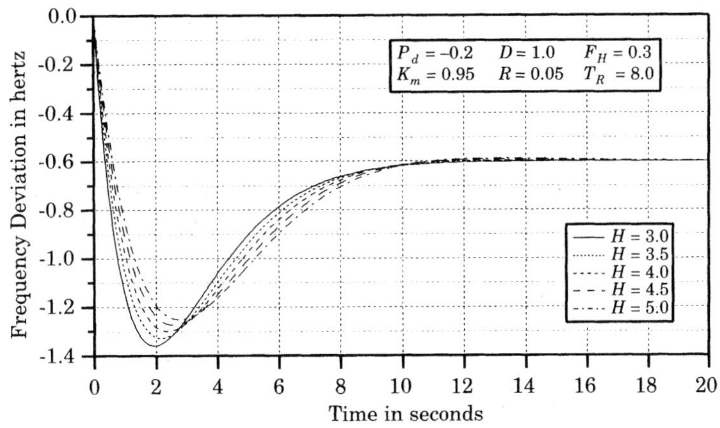

Figure 20.16 Frequency response for varying values of H from 3.0 to 5.0 seconds in increments of 0.5 seconds.

and (20.20). It is also interesting to observe that $(1/2H)$ is the forward loop gain in Figure 20.12 and that H is involved only in the feed-forward loop, while all other parameters are feedback parameters.

The most pronounced effect of high values of H is to reduce the initial rate of frequency decline, and to delay and reduce in the maximum deviation. Note that H does not affect the final steady-state value of frequency. The family of inertia responses are interesting due to the overlap of response curves. This characteristic is not exhibited by the other response plots. The higher inertia values result in a slower drop in frequency, which is logical, and also a slower recovery. Since the response is slower for the higher inertias, the governor has more time to respond and therefore limits the maximum frequency deviation to smaller values.

20.4.6 The Effect of Reheat Time Constant, T_R

The reheat time constant is an important system parameter. Usually 70% or more of the turbine output is delayed by the reheat time constant and the variation of this time lag has a pronounced effect on frequency performance. Figure 20.17 shows a range that is typical of large generating units. T_R has an effect on the damping ratio ζ and the natural undamped frequency ω_n, but does not effect the initial slope or the final value.

The major effect of the reheat time constant is to produce a lag in the response of the frequency following its initial dip. This lag also increases the maximum frequency deviation and delays the peak value of this maximum in proportion to the size of the reheat lag. This means that the reheat constant will not affect underfrequency relay settings, which are set to be responsive to the initial drop in frequency, but this parameter has an effect on the response and the total time of exposure to low frequencies. This may be important in the effect on turbine blading and the loss of life due to low frequency fatigue.

20.4.7 The Effect of High-Pressure Fraction, F_H

The constant F_H measures the fraction of shaft power developed by the high-pressure turbine on a single reheat system. This is the fraction of shaft power that is not delayed by reheating. Figure 20.18 shows the effect of varying only F_H, with other parameters at their

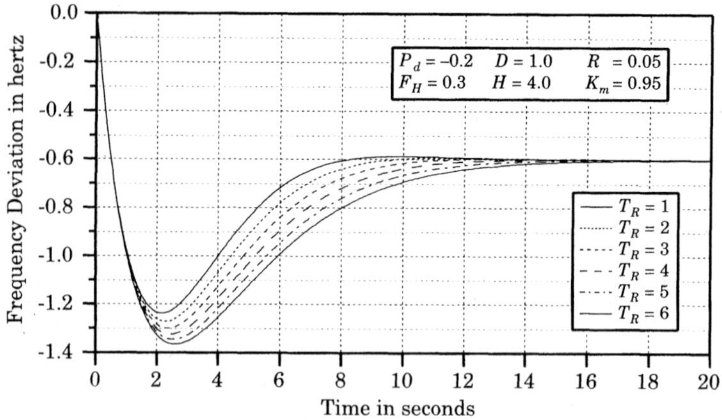

Figure 20.17 Frequency response for varying values of T_R.

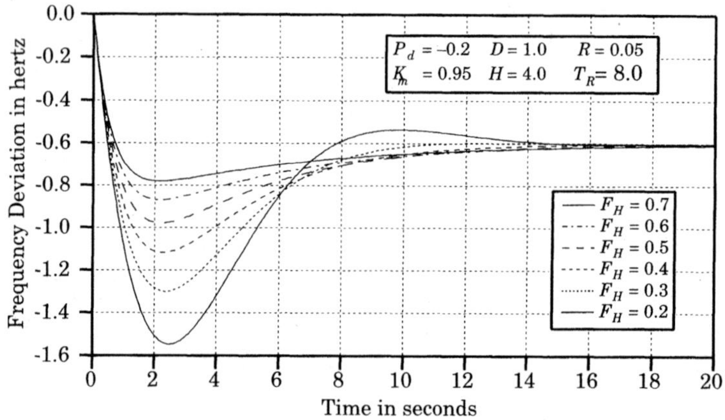

Figure 20.18 Frequency response for varying values of F_H.

nominal values. Large values of F_H have a pronounced effect on ζ and can make the system overdamped ($\zeta > 1$). When this occurs, the frequency response is not given by (20.9) but is a combination of exponentials. The frequency response equation for this condition is computed from (20.8) by factoring the quadratic to write

$$\Delta\omega(s) = \frac{T_1 T_2 R \omega_n^2 P_{\text{step}}}{DR + K_m}\left(\frac{1 + T_R s}{s(1 + T_1 s)(1 + T_2 s)}\right) \tag{20.24}$$

Then

$$\Delta\omega(t) = \frac{T_1 T_2 R \omega_n^2 P_{\text{step}}}{DR + K_m}\left(1 + \frac{T_1 - T_R}{T_2 - T_1} e^{-t/T_1} - \frac{T_2 - T_R}{T_2 - T_1} e^{-t/T_2}\right) u(t) \tag{20.25}$$

In the numerical example plotted in Figure 20.18, when $F_H < 0.4$, the system is underdamped. These are the two curves with the largest frequency displacement in the figure. The underdamped cases exhibit overshoot of the final value. The curves representing $F_H > 0.4$ are overdamped.

20.4.8 The Effect of Damping, D

Power system loads are known to be sensitive to system frequency. One way of characterizing this dependence is to model the load as having a constant component as well as a frequency dependent component.

$$P_L = P_{L0}\left(1 + k_f \Delta\omega\right) \tag{20.26}$$

Then the incremental change in load is a function of the incremental change in frequency. But this effect is already included in the model in the form of the damping constant D, where we may write

$$\Delta P_L = D\,\Delta\omega \tag{20.27}$$

where all values are in per unit. The damping constant is also interpreted as representing the viscous damping of the turbine and generator to oscillations. From Figure 20.12, however, we note that the product (20.27) is of the same sign as the electrical power, which is exactly the incremental load of the system. Thus, the model given by Figure 20.12 has both components of load represented, the constant portion and the frequency dependent portion.

From (20.3), we see that the system undamped natural frequency and damping are functions of D, but this dependence always appears as the product DR. Since R is small, nominally about 0.05 per unit, the product DR will also be small and the effect of D is therefore diminished. We can illustrate the effect of D on the frequency response by plotting for various values of this parameter, with the results seen in Figure 20.19.

Note the similarity between Figure 20.14, which shows variation in R, and Figure 20.19 that illustrates variation in D. The effect of varying the two parameters is much the same, but R plays a much more important role than D in the results. We conclude that, even though the load has a frequency dependent component, it is not nearly as important as the droop settings of the governors in the resulting frequency response. It also shows the importance of keeping the governor droop settings closely tuned.

20.4.9 System Performance Analysis

Performance analysis for a given system using the SFR model is relatively easy, providing data from an actual system disturbance is available to determine the parameters. Actual system disturbance records are necessary since some system parameters are almost impossible to determine exactly due to unknown operating conditions. Regulation or droop (R), for example, depends on turbine droop adjustment and the effect of units with blocked governor valves, such as large nuclear units. However, this effect is clearly revealed in the frequency record of an actual disturbance by observing the steady-state frequency error. This steady-state error can be computed from (20.22) to be

$$f_{ss} = \frac{60\,R\,P_{\text{step}}}{DR + K_m} \quad \text{Hz} \tag{20.28}$$

We can begin by making rough estimates of D, since this parameter will not make any significant difference. The value of K_m is usually known, at least approximately, by the system control center, or it can be estimated based on the operating and spinning reserve policy of the islanded utility. There remains only R and P_{step} to be found. If the disturbance magnitude is known, and it usually is easily estimated for a known system disturbance, then R can be computed using (20.28). As noted previously, a value of about 0.1 is not uncommon, even though governors are nominally set to 0.05.

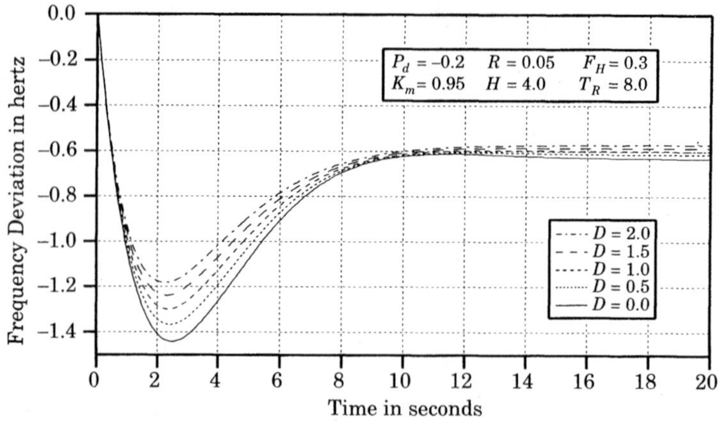

Figure 20.19 Frequency response for various values of the parameter D.

20.4.10 Use of the SFR Model

Data for an actual system separation is used to validate the SFR model. The physical system is assumed to have separated and islands created, one with excess generation and one with excess load. The simulation presented is for the island with excess load. The result is shown in Figure 20.20, where the irregular line represents the measured system frequency and the smooth line is the simulation of the average island frequency.

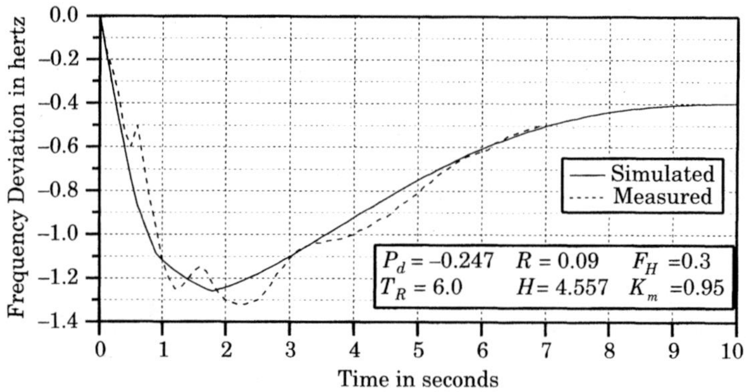

Figure 20.20 Model validation for an actual power system disturbance.

Note that the data from the physical system include the effect of the oscillation of local machines with respect to other machines in the island. This is usually the case, and measurements from different parts of the island will show these local oscillations. The local machines oscillate about the average system frequency, which is computed by the SFR model.

For the case illustrated in Figure 20.20, the power system underfrequency load shedding relays caused load to be shed at various times during the initial frequency decline. These load shedding events were computed in per unit, based on the estimated total island generation, as shown in Table 20.2.

A total of almost 16% of the island load was shed in this event, and the system frequency response was well controlled. Note that the simulation, using approximate values of

TABLE 20.2 Load Shed in Underfrequency Island

Time in Seconds	Increment Load Shed in Percent	Total Load Shed in Percent
0.2	0.2070	0.2070
0.5	0.1980	0.4050
0.6	6.4880	6.8930
0.9	6.2100	13.103
1.8	2.7120	15.815
Total	15.815	15.815

the system generation parameters, provides a reasonably good estimate of the actual system underfrequency event.

20.4.11 Refinements in the SFR Model

The model used in the previous sections is convenient for its simplicity, but it uses assumptions that are not entirely necessary. We will discuss two refinements that may be desirable for some studies where the greater accuracy may be warranted. One refinement adjusts the mechanical power, and the other corrects the electrical power.

20.4.11.1 Mechanical Power. The mechanical power in the previous model neglects the fact that all systems normally allocate a given fraction of all generation as "spinning reserve," to provide a generation margin for exactly the type of emergency being simulated [28]. This means that not all of the rated system capability is scheduled, and that only a specified fraction of that capability is available to respond to an islanding event. We may define the mechanical power of all turbines in service as follows:

$$K_m = \frac{P_{m0}\ in\ MW}{S_{SB}} = \frac{1}{S_{SB}}\sum_i S_{Bi}F_{Pi}(1 - f_{SR}) = F_P(1 - f_{SR}) \tag{20.29}$$

where S_{Bi} = MVA rating of unit i
S_{SB} = system MVA rating
F_{Pi} = power factor of unit $i = F_P$
f_{SR} = fraction of S_{SB} on spinning reserve

This model of the mechanical power takes into account the spinning reserve of the system. This means that not all of the system generation is available immediately, since the governor references are set to use only a fraction of the unit ratings. The spinning reserve fraction used here is the average for the entire system.

This gives an estimate for the value of the gain K_m used in the system equations. It assumes that all operating generators are at the same power factor.

20.4.11.2 Electrical Power. The model of the electric power in the frequency response model may be limited due to the inability of the generators to deliver an amount of power equal to the size of the initial disturbance. Let us assume a simple generator model, with constant voltage behind transient reactance, a generator step-up transformer, and an equivalent system impedance between the generator and load, as shown in Figure 20.21. The total system impedance seen by the equivalent generator is designated as X.

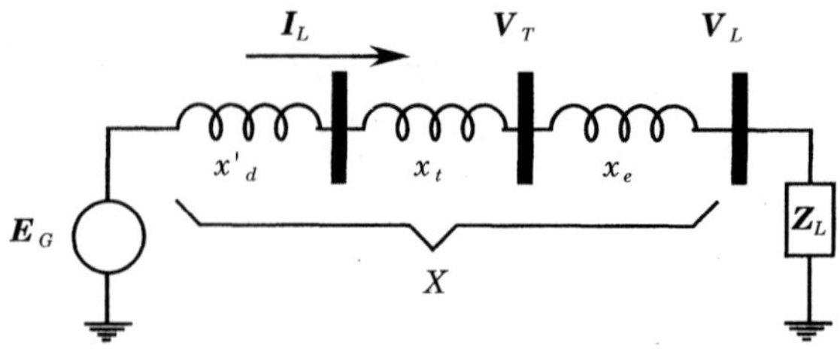

Figure 20.21 Equivalent system following the disturbance.

We may draw a phasor diagram of the system voltages both before and after the disturbance as shown in Figure 20.22, assuming constant voltage E_G behind the transient reactance. Note that the power factor of load currents is assumed to be unchanged by the disturbance, i.e., the angle θ is a constant angle. The constant generator internal voltage describes an arc with center at the origin. This does not mean that there is no exciter, but that the excitation is fast enough to effectively hold this internal voltage constant.

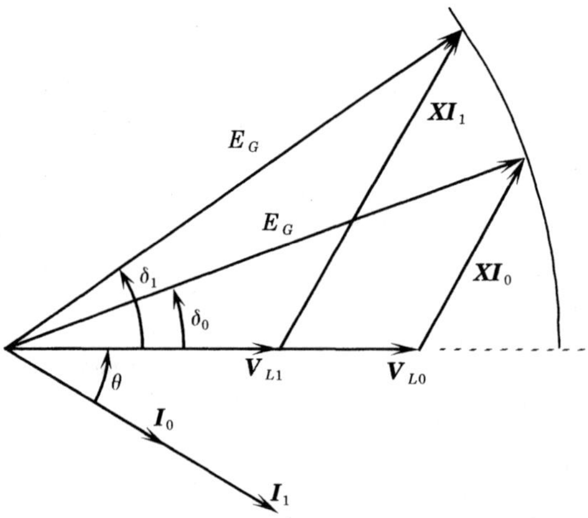

Figure 20.22 Phasor diagram of conditions before and after disturbance.

From the phasor diagram, we may write the following voltage equations.

$$\begin{aligned} V_L &= E_G \cos\delta - XI\sin\theta \\ XI\cos\theta &= E_G \sin\delta \end{aligned} \tag{20.30}$$

Then we may compute the load voltage

$$V_L = E_G \cos\delta - E_G \sin\delta \tan\theta \tag{20.31}$$

and the load power

$$P_L = \frac{E_G^2}{X}\left(\sin\delta\ \cos\delta - \sin^2\delta\ \tan\theta\right) \tag{20.32}$$

Recall that θ is a constant power factor angle. We may now compute the maximum load power that may be delivered in terms of the angle δ by taking the derivative of (20.32) with respect to δ. Performing this operation, we compute the following constraint between the two angles.

$$\tan 2\delta = \frac{1}{\tan\theta} \tag{20.33}$$

Using this relationship, we may write the following equation for the maximum load power, that is, when the derivative of (20.32) is zero. This gives the maximum deliverable load power to be

$$P_{L\max} = \frac{E_G^2}{2X}\left(\frac{1-\sin\theta}{\cos\theta}\right) = \frac{E_G^2}{2X}\tan\delta \tag{20.34}$$

Proof of this equation is left as an exercise. The important thing to note is that there is a definite maximum load power that can be delivered by the system generation. This maximum power is a function of the system reactance seen looking into the network, the excitation voltage of the equivalent generator, and the generator torque angle.

From the basic swing equation, we may write

$$\frac{d\,\Delta\omega}{dt} = \frac{1}{2H}\,(P_m - P_e) \quad \text{pu} \tag{20.35}$$

Then, at maximum power delivered to the load, this equation defines the maximum slope of the frequency curve.

$$\left.\frac{d\,\Delta\omega}{dt}\right]_{\max} = \frac{1}{2H}\,(P_m - P_{e\max}) \quad \text{pu} \tag{20.36}$$

If we also incorporate the modified mechanical power, we may write the revised equation as follows.

$$\left.\frac{d\,\Delta\omega}{dt}\right]_{\max} = \frac{1}{2H}\left[F_P\,(1-f_{SR}) - \frac{E_G^2}{2X}\left(\frac{1-F_R}{F_P}\right)\right] \quad \text{pu} \tag{20.37}$$

where $F_P = \cos\theta =$ power factor

$F_R = \sin\theta =$ reactive factor

The limitation on electrical power can be used simply as a maximum value that can be substituted for the disturbance power in the frequency response model.

20.4.12 Other Frequency Response Models

One application for frequency response models is for the setting of underfrequency load shedding relays, where relays are used to shed portions of the load and thereby preserve a reasonable load-generation balance. One method of estimating frequency behavior for this purpose is to model the frequency decline following the disturbance using a model that reflects the load and generator behavior. This model results in a first-order differential equation that has the solution [29]

$$\Delta\omega(t) = \frac{P_{\text{step}}}{D}\left(1 - e^{-(D/2H)t}\right) \tag{20.38}$$

This model, which is attractive for its simplicity, describes the frequency trajectory as it falls from its initial value in an exponential fashion. The rate of decline is faster than that predicted by (20.9) since it does not model the governor behavior. The derivation of this model is an interesting one and is left as an exercise. The slope of the frequency function using this model is given by

$$\frac{d\,\Delta\omega}{dt} = \frac{P_{\text{step}}}{2H}e^{-\frac{D}{2H}t} \tag{20.39}$$

This model is often used for determining the time at which load shedding relays should be employed. The initial slope computed by (20.39) is exactly the same as that computed by (20.18). Note that only the inertia and damping are required to find the frequency for a disturbance of any size, since the governor behavior is ignored.

A comparison of the two methods of computing the frequency as a function of time is shown in Figure 20.23 and the results are summarized for two disturbance values in Table 20.3.

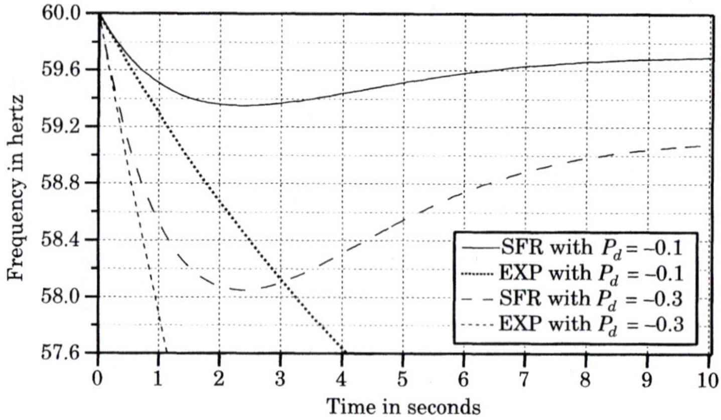

Figure 20.23 Comparison of two methods or computing the frequency response to a step type load disturbance.

TABLE 20.3 Frequency Deviations Computed by the Two Methods

Time Seconds	$P_{step} = -0.1$ SFR	EXP	% Error	$P_{step} = -0.3$ SFR	EXP	% Error
1.0	–0.491	–0.705	43.6	–1.474	–2.115	43.5
2.0	–0.642	–1.327	112.9	–1.925	–3.982	106.9

The system parameters used in the System Frequency Response (SFR) model for Figure 20.23 are as follows.

$$H = 4.0 \qquad R = 0.05$$
$$D = 1.0 \qquad T_R = 8.0$$
$$F_H = 0.3 \qquad K_m = 0.95$$

The exponential model is reasonably accurate for the first one-half second, but after that the error builds rapidly.

20.4.13 Conclusions Regarding Frequency Behavior

The SFR model of system frequency behavior following an islanding event or a large disturbance on the interconnected system is a greatly simplified model of system behavior. The model developed for this purpose omits many details and ignores small time constants in an effort to provide a model that may be useful in approximating the system performance, including the essential behavior of speed governing and turbine response. However, in spite of the model simplicity, the comparison with actual system disturbances and detailed stability simulations are encouraging. Moreover, the model provides an understanding of the way in which important system parameters affect the frequency response. Such understanding is difficult to achieve in high-order models, where performance is a very complex function of many state variables.

From Figure 20.23, it is clear that, although the exponential model gives the correct initial rate of change of frequency, an error begins to accumulate at the start of the disturbance and by the end of 1 second this error is quite large. Table 20.3 shows the percent error, which is observed to be about 40% after 1 second and over 100% after 2 seconds. Clearly, the error accumulates very fast.

20.5 OFF NORMAL FREQUENCY PROTECTION

The frequency deviation that accompanies system islanding disturbances is caused by the imbalance between load and generation. This effect is most serious in the island that has an excess of load, since speed governing is usually effective in reducing the generation in islands that have an excess of generation. Therefore, most of the concern is for the island with an excess of load. Since there is no direct control of utility load, the primary method of restoring frequency is to shed load in appropriate amounts. This must be done with considerable planning, since there is no merit in shedding excessive amounts of load and thereby creating a need for reducing generation.

The problem of load shedding has been studied for many years in North America with some of the early investigations dating to the mid 1950s [30–33]. It has also been noted that the power plants themselves are subject to failure at low frequencies since plant auxiliaries are unable to maintain normal output, with the critical frequency being about 10–15% below normal (6–9 hertz on a 60 hertz system) [34]. The most logical correction for a load-rich area is the switching off of a portion of the load. Some loads, such as ore crushers and rolling mills, for example, should not be allowed to return to service during the underfrequency period since the material being worked in the milling process may jam the machine and promote stalling.

Several scholarly treatments of underfrequency load shedding have been presented in the literature over the years [35–51]. A 1978 industry survey recorded the practice of utilities in underfrequency and undervoltage relay applications [51]. This survey showed that over one-third of the generating units have underfrequency protection, with most of these schemes being applied for turbine blade protection. It was also revealed that, in most cases, no effort was made to coordinate the turbine protection with the system underfrequency load shedding relays. No cases of unit tripping due to underfrequency were reported, which indicates the rare occurrence of this type of disturbance.

Recently, a standard has been prepared that provides a guide for abnormal frequency protection of power generating plants [52], [53]. This guide states two major objectives in connection with the operation of a steam generating station at abnormal frequency:

1. Protect equipment from damage that could result from operation at abnormal frequency.
2. Prevent cascaded tripping that may lead to a complete plant shutdown if limiting conditions are not reached during the abnormal frequency condition.

The plant systems that are most affected by abnormal frequency conditions are the following:

Generator
Unit step-up transformer

Turbine

Station auxiliaries

Limitations in the generator, step-up transformer, and turbine have been discussed in this and the preceding chapters. For the station auxiliaries, those that are most limiting are boiler feed pumps, circulating water pumps, and condensate pumps. This is due to the fact that each percent reduction in speed, brought about by low frequency, causes a large percent loss in pumping capacity. The frequency at which the loss of pump capacity becomes critical will vary from one plant to another. Tests have shown that the plant capability will begin to decrease at 57 hertz and that frequencies below 55 hertz are critical for continued plant operation due to the reduction in pumping rate. These limitations are not covered in the ANSI standard.

20.6 STEAM TURBINE FREQUENCY PROTECTION

The primary underfrequency protection for steam turbines is the automatic load shedding program that is designed to maintain a reasonable load-to-generation balance. This protection is not located in the power plant, but is distributed throughout the power system in order to protect each region equally. In some cases, this protection will provide a load-generation balance that will make it unnecessary for special turbine underfrequency protection to operate. However, it must be recognized that system underfrequency load shedding makes certain assumptions as to how the system will break up into distinct islands, and that the assumed island boundaries may not always occur as planned. Therefore, it is still necessary for the turbine to have its own protection to cover those rare events when unusual frequency excursions occur due to the method of system breakup or the failure of the load shedding program. Turbine underfrequency protection, then, is the last line of protection and its operation may lead to blackout of the underfrequency island.

The following design criteria are suggested by the ANSI guide [52]:

1. Establish trip points and time delays based on the manufacturer's turbine abnormal frequency limits (similar to Figure 20.6).
2. Coordinate the turbine generator underfrequency tripping relays with the system automatic load shedding program.
3. Failure of a single underfrequency relay should not cause an unnecessary trip of a generator.
4. Failure of a single underfrequency relay to operate during an underfrequency condition should not jeopardize the overall protection scheme.
5. Static relays should be considered, as their accuracy, speed of operation, and reset capability are superior to the electro-mechanical relays.
6. The turbine underfrequency protection system should be in service whenever the unit is synchronized to the system, or while separated from the system but supplying auxiliary load.
7. Provide separate alarms to alert the operator for each of the following conditions:
   - A situation of less than the nominal system frequency band on the electric system
   - An underfrequency level detector output indicating a possible impending trip of the unit.
   - An individual relay failure.

It is also recognized that unnecessary generator trips during frequency excursions, from which the system should recover, must be avoided. Moreover, it is important that the scheme be designed to minimize the stress on the turbine. These two criteria suggest that the underfrequency operation be broken up into small bands with separate protection and timing in each band, since the turbine loss of life is greatly increased as the frequency falls. At least five or six bands are suggested. Since the turbine damage is cumulative, the operating experience within each band should be preserved in a nonvolatile memory.

The ANSI Guide [52] provides two suggested protection schemes, only one of which is presented here. This scheme is a multi-setpoint scheme with frequency logic and accumulating counters, as shown in Figure 20.24. It is designed to protect a turbine with generalized abnormal frequency operating limits similar to those shown in Figure 20.6. The settings are based on the following criteria [52]:

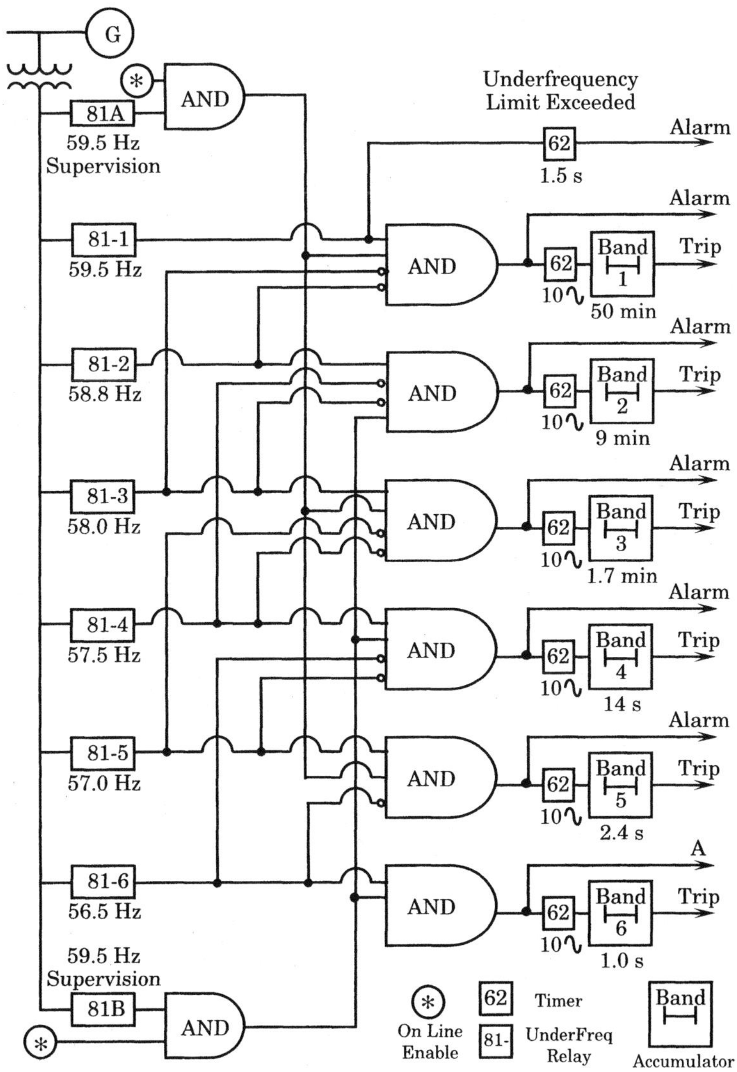

Figure 20.24 Block diagram of a turbine protection scheme [51].

1. The frequency set points are set equal to or slightly higher than the steps of the turbine abnormal frequency operating limits.
2. The time delay setpoints are set equal to or slightly lower than the steps of the turbine blade time restricted operating limits.
3. The settings are modified in some cases to provide coordination with the load shedding scheme.

20.7 UNDERFREQUENCY PROTECTION

The underfrequency turbine protection scheme should be coordinated with the other underfrequency protective systems. Most large systems employ a load shedding scheme that uses underfrequency relays to drop a portion of the system load when the frequency is below normal. These protective schemes reduce the magnitude of the load imbalance in the island and make it possible for the system to recover to an operating condition from which the operators can begin the process of restoring the system to normal. As noted earlier, the unnecessary tripping of any turbines during this emergency period makes the load imbalance worse, and may lead to complete collapse.

Therefore, the load shedding scheme should drop load before the turbine protection starts dropping generating units.

20.7.1 A Typical Turbine Protection Characteristic

The following information is required in order to provide the necessary coordination [52]:

1. The system frequency response characteristic for possible load imbalance conditions, including the effect of the load shedding scheme.
2. A time-frequency characteristic of the proposed turbine underfrequency protection scheme.

We assume that the frequency response characteristic derived in Section 20.4 is adequate to satisfy the first requirement. This must be compared with the turbine underfrequency limits and the turbine underfrequency protection characteristics. Such a comparison is shown in Figure 20.25, where we assume a turbine protection similar to that shown in Figure 20.24.

The characteristics shown in Figure 20.25 must be used with caution. The curves may not be simply compared by their placement in the time versus frequency coordinates since the total damage to the turbine blading is cumulative and the required coordination time is between the system frequency response and the *remaining life* of the blading. The accumulators in the protection scheme add the time in each interval during each underfrequency incident and accumulate the total for each interval. The turbine protection will not trip the unit until its underfrequency experience exceeds that allowed, which should agree with the limits set in the accumulators. A simple comparison between the underfrequency response and the turbine protection will ensure that the unit is not tripped due to exceeding the turbine underfrequency limits on the one underfrequency excursion. Usually, a rather wide coordination margin would be advisable for Figure 20.25. For example, it would be wise to limit any single underfrequency excursion to less than 10% of the turbine trip limits, which limits the loss of life experienced by the turbine due to any single incident. No such limits are shown in Figure 20.25, but can be achieved by proper design of the load shedding characteristics.

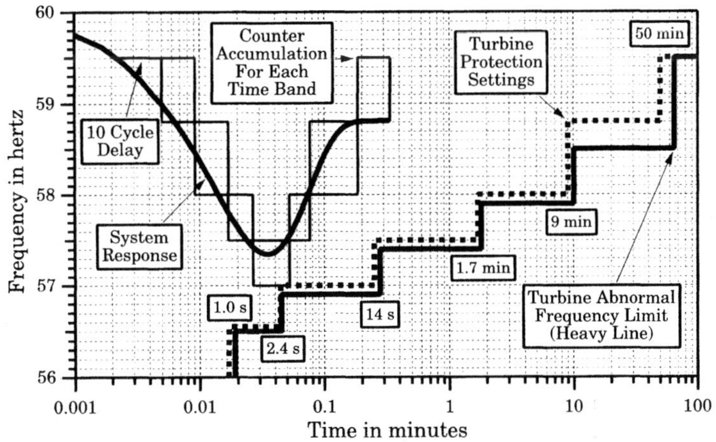

Figure 20.25 Comparison of system underfrequency response and turbine underfrequency protection.

20.7.2 Load Shedding Relay Characteristics

The creation of a load imbalance in a power system may cause such an excess of load over generation that there is no alternative but to shed some of the load. This is accomplished using underfrequency relays, with timers, to drop specified amounts of load at predetermined times.

In many cases, the underfrequency condition arises due to the breakup of a large system into two or more islands. The boundaries of the islands are seldom known prior to the disturbance that causes the breakup. The exception to this is the case where certain anticipated types of disturbances are detected by special protective schemes, which are preprogrammed to cause the creation of islands at specified boundaries. This concept is discussed in Chapter 21. In most cases, however, it is necessary to install load shedding relays throughout the power system so that any possible island configuration will be protected against underfrequency operation. In most cases, a large number of individual utilities are interconnected to form a large system, which may be forced into islands with many different boundaries. Thus, it is necessary for all of the utilities that make up the interconnected system to come to an agreement as to the amount and timing of load shedding, so that all portions of the system behave in approximately the same manner when load shedding is required, irrespective of the exact cut set that defines the separation.

20.7.2.1 Load Shedding Criteria. In order to assure proper coordination of the load shedding relays and the turbine underfrequency limits, coordination criteria are required. From Figure 20.6 it is clear that operation below 56 hertz should never occur and even 56.5 hertz is limiting for some turbines. A margin above these values should be added for safety, say to 57 hertz. The underfrequency limits at 57 hertz are shown to be about 0.1 minutes (6 s), but this is the accumulated time to failure. Only a prudent fraction of this time should be allowed on any one underfrequency excursion, say 10% or about 0.6 seconds. This will permit ten significant load upsets before the cumulative damage would predict the replacement of turbine blading. Since the allowable time below 57 hertz is so small, a better approach is to shed adequate amounts of load so that system operation below 57 hertz is avoided entirely.

One set of criteria for optimizing the amount of load shedding and its coordination with turbine underfrequency protection has been proposed as follows [45]:

(a) The load shedding program should prevent underfrequency excursions below 57.0 hertz for longer than 30 cycles (on a 60 hertz base) and the system frequency should be able to recover following the overload disturbance.

(b) The system frequency should recover fast enough to provide adequate margin with the turbine protection schemes.

(c) Frequency overshoot due to overshedding of load should be limited to less than 61 hertz to prevent conflict with turbine over-frequency limitations.

20.7.2.2 Definition of the Initial Load Imbalance. A distinction should be made regarding the cause of the load imbalance, since this affects the computed frequency response. Several situations may be encountered, with the following descriptions being typical.

1. The utility is part of a large interconnection, with portions of the load served from purchased power that is imported over tie lines from neighboring utilities. A large disturbance causes all or a portion of the utility system to island due to line tripping. The island may occur at preplanned boundaries, or it may be random and dependent on the nature and location of the initiating disturbance. In either case, imported power is cut off and the island is left with load that exceeds the island generation.
2. The utility is operated as an islanded system, either by design or due to a temporary switching of connections to the neighboring systems. A load imbalance may be caused by the loss of one or more generating units, which results in an excess of load over the remaining generation.
3. The utility is part of a large interconnection, as in 1, and suffers the loss of one or more large generating units. This causes a local incremental drop in frequency, which causes power to flow toward the generation deficient region over interconnecting transmission lines. This may cause significant overload of one or more of these lines, which are tripped (either by protective relays or by operator action) as a result of the overload, thereby causing even more critical loading of the remaining lines. Eventually several lines trip, isolating the generation deficient region, which suffers an under frequency response.

Other scenarios can be conceived that will result in a significant load imbalance. The point to note is that some disturbances cause a change in the remaining island *generation* and some do not. The net disturbance power, however, is expressed in per unit based on the total *remaining generation* in the island following disturbance. We define the following quantities.

P_L = island load in MW
P_{Go} = generation at time of islanding in MW
P_{Gx} = generation tripped in MW

Prior to the islanding action, the load and generation (plus imported power) in the island area are equal, but after islanding they are unequal.

Then for Case 1, we have

$$P_{d1} = \frac{P_L - P_{Go}}{P_{Go}} = \frac{P_d}{P_{Go}} \text{ per unit} \tag{20.40}$$

For Case 2, we compute

$$P_{d2} = \frac{P_{Gx}}{P_{Go} - P_{Gx}} = \frac{P_d}{P_{Go} - P_{Gx}} \text{ per unit} \tag{20.41}$$

In the first case, the size of the disturbance is the difference between load and generation. In the second case, the size of the disturbance is the amount of generation tripped. In both cases, the numerator represents the size of the disturbance in MW, but the denominator for the two cases is different. The difference in the per unit disturbance size can be quite large. Case 2 starts out with a loss of generation, but when the separation occurs it can be modeled as a variation of Case 1.

EXAMPLE 20.1

Consider a system with a total generation of 10,000 MW. Compute the per unit disturbance for Cases (1) and (2).

Solution

The results of the calculation are shown in Table 20.4. The point of this exercise is to illustrate the importance of computing the disturbance power with the correct base generation for the island. This calculation should carefully account for cases in which the total amount of generation is changed at the time of separation. The difference is more pronounced for higher per unit values of disturbance power. Note, for example, that a 0.40 per unit disturbance power amounts to a net 0.67 per unit disturbance for loss of generation is compared to only a 0.40 per unit disturbance for a loss of load. The point is that the base MVA is computed by totaling the remaining generation in the island. It is this total that determines the stored energy (H) and other island generation parameters.

TABLE 20.4 Normalized Disturbance Power for Different Conditions of Remaining Generation with Initial Generation of 10,000 MW

P_d	P_{d1}	P_{d2}
500	0.05	0.0526
1000	0.10	0.1111
1500	0.15	0.1765
2000	0.20	0.2500
2500	0.25	0.3333
3000	0.30	0.4286
3500	0.35	0.5385
4000	0.40	0.6667
4500	0.45	0.8182
5000	0.50	1.0000
5500	0.55	1.2222
6000	0.60	1.5000
6500	0.65	1.8571
7000	0.70	2.3333
7500	0.75	3.0000
8000	0.80	4.0000
8500	0.85	5.6667
9000	0.90	9.0000
9500	0.95	19.0000
10000	1.00	Infinite

■

If we select a goal of 57 hertz as the minimum allowable operating frequency, we can estimate the maximum disturbance for a typical system from the plots of Figure 20.26, which show the frequency deviation without any load shedding. This indicates that, for load disturbances greater than about 0.4 per unit, based on (20.40), the frequency deviation is very likely to drop below 57 hertz.

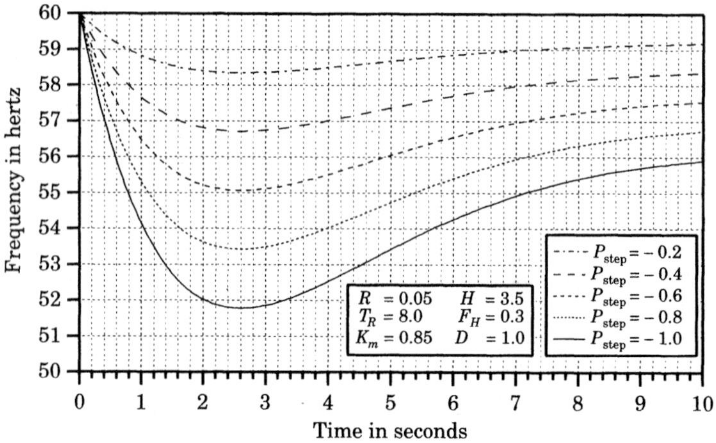

Figure 20.26 Typical system frequency responses to load disturbances.

It is clear that load must be shed for disturbances greater than 0.40 per unit. The amount of load shed and the timing of these shedding events must be determined in some logical manner.

Some utilities set the first step of load shedding at 59.5 hertz. This is chosen for several reasons [46]. One reason is that the large turbine-generator sets are not rated for continuous operation below about 59.5 hertz, with the exact value depending on the manufacturer as noted in Figure 20.6. Setting the initial load shed frequency at a relatively high value, such as 59.5 hertz, also tends to limit the maximum frequency deviation. Values higher than 59.5 hertz are not advised because of the need for coordination with automatic underfrequency loading of some hydro and other peaking units.

There is also an argument for setting this first load shedding step at about 59.0 hertz. Above this frequency setting the turbine can operate for 3.3 hours up to continuous operation at 59.4 or 59.5 hertz. If the frequency should level off in this range, as shown in the −0.2 per unit curve of Figure 20.26, the operator will have time to remedy the situation by starting rapid-start generating units or manually tripping nonessential load.

The second variable is the number of steps at which load will be shed and the amount to be shed at each frequency step. Utilities have experimented with load shedding steps in the range of three to six. The larger number of steps requires the use of more relays, but the steps are all smaller, both in the frequency spread between steps and in the amount of load to be shed at each step.

We illustrate the method of computing the settings of underfrequency relays by the following example.

EXAMPLE 20.2

An underfrequency turbine protection scheme is to be designed for a system where the largest disturbance is assumed to be a sudden load increase of 0.6 per unit. The average system parameters for this example are as follows:

$$K_m = 0.95 \qquad F_H = 0.3 \qquad D = 1.0$$
$$T_R = 8.0 \qquad H = 3.5 \qquad R = 0.06$$

Two protection plans are to be designed and compared. The first plan is shown in Table 20.5. It consists of four equal load shedding steps of 0.0625 per unit, for a total load shed of 0.25 per unit.

TABLE 20.5 Frequency Settings of a Four Step Protection Plan

Relay	Frequency	Load Trip	Delay
f1	59.5 Hz	0.0625 per unit	0.1 s
f2	59.2 Hz	0.0625 per unit	0.1 s
f3	58.9 Hz	0.0625 per unit	0.1 s
f4	58.6 Hz	0.0625 per unit	0.1 s

The second plan is shown in Table 20.6. It consists of six load shedding steps of unequal size, and taken at steps that are closer together than the first plan, but with the same total load shed. Note that the larger steps are taken first, followed by smaller steps of load shedding.

TABLE 20.6 Frequency Settings for a Six Step Protection Plan

Relay	Frequency	Load Trip	Delay
f1	59.5 Hz	0.048 per unit	0.1 s
f2	59.3 Hz	0.048 per unit	0.1 s
f3	59.1 Hz	0.042 per unit	0.1 s
f4	58.8 Hz	0.040 per unit	0.1 s
f5	58.5 Hz	0.036 per unit	0.1 s
f6	58.2 Hz	0.030 per unit	0.1 s

Both load shedding plans use a time delay of 0.1 s, or six cycles, for each load shedding step. A comparison of the two plans is shown in Figure 20.27, which is a close-up view of the time frame when the load shedding is being initiated. Note that the two schemes do not present very different results, which indicates that the total amount of load shed is more important than the exact time of shedding. It is important to detect the need for shedding very quickly, however, and to accomplish all load shedding before the frequency can decay to a hazardous level.

The result of the two load shedding schemes is better viewed over a longer time period, as shown in Figure 20.28. Clearly, the two schemes are so nearly the same, having shed the same total load, that they are hardly distinguishable over the longer time frame. The net effect is about right, since the frequency does not fall below 57 hertz, although it does come close. Note that we can't wait until 1 second after the disturbance to initiate load shedding, as this is clearly too late. In this example, the load shedding is initiated at about 0.1 seconds after the disturbance, with the first blocks of load to be shed after a 0.1 second time delay, or at about 0.2 seconds. This is about the timing that we should aim for in the design of a system. ■

20.7.2.3 Load Shedding Protection Design. We now seek to develop a logical methodology for the design of a load shedding scheme that will ensure that our objectives are met. First, we record some general observations that will guide our development load shedding scheme:

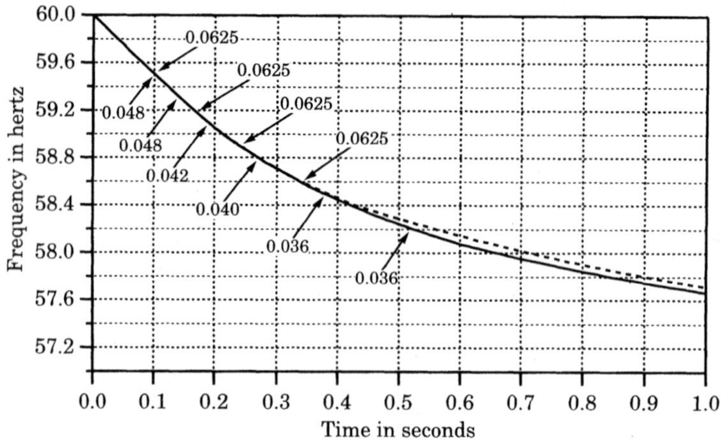

Figure 20.27 Comparison of the two load shedding schemes.

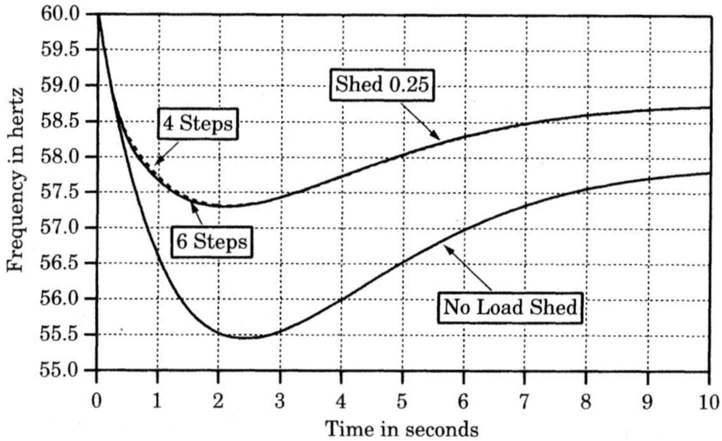

Figure 20.28 Result of the two different shedding schemes.

1. We must be prepared for the worst possible system separation disturbance, since there is no way of knowing a priori exactly how the system will separate or the magnitude of the load imbalance.
2. It is important to determine very quickly how much load shedding is required for a given disturbance.
3. At the moment of system separation, the initial slope is the only clue as to the magnitude of the disturbance.
4. It is better to shed too much load than not enough, and the load shedding must begin without excessive delay.

The worst possible disturbance is subject to debate, but the load shedding plan will inherently be limited based on this assumption. It has been noted that the larger system upsets are less probable than small upsets, but large disturbances still have a finite probability [54]. To shed adequate load for very large step disturbances, either large amounts of load must be shed at certain steps or many steps of shedding must be used. Even this strategy may not work, since an adequate amount may not be shed soon enough to arrest the frequency decline. A better strategy would be to shed larger amounts of load at the first step or two, based on the observed slope of the frequency decay. This suggests an adaptive strategy that can adjust the amount of load shed, based on observed system frequency decay.

If the load shedding is delayed too long, there is danger of exceeding the nominal 57 hertz minimum frequency. This is especially true of the larger disturbances, which cause the frequency to decay faster than the smaller load upsets (see Figure 20.26). Therefore, as the size of the disturbance increases, there is greater need for rapid action. One way of accomplishing this is to trigger the first load shedding step at a given frequency, say 59.5 hertz, that is high enough to ensure quick removal of the first step of load.

Clearly, the only observed quantity that gives any clue as to the size of the disturbance is the initial slope of frequency decline. As noted previously, this slope is equal to

$$m_o = \left.\frac{d\,\Delta\omega}{dt}\right|_{t=0} = \frac{60P_{\text{step}}}{2H} \text{ Hz/s} \tag{20.42}$$

where m_o is defined to be the observed slope in hertz per second. Suppose that this quantity is observed continuously, so that load shedding can be triggered when the slope exceeds some critical value. Knowing the slope, we can determine the size of the disturbance if the inertia constant is known. Now the inertia constant is normalized on the base of the total generation in the island. We know its normalized value is approximately 3 to 5 seconds. This permits us to quickly estimate the size of the disturbance to be

$$P_{\text{step}} = \frac{2Hm_o}{60} \text{ per unit} \tag{20.43}$$

We don't need to know this value in MW, only in per unit, in order to estimate the percent of load that must be shed.

The final guideline addresses the question of the total percent of load that should be shed. If too much is shed, the operator will reconnect the load in a few minutes and the inconvenience to customers will not be great. However, if too little load is shed, permanent damage will accumulate on all turbines, with possible long-term effects that are very expensive to the utility and, ultimately, to the customers. Therefore, if an error is to be made, it should be made in the direction of shedding a little too much load, rather than too little.

If we use the initial slope to estimate the magnitude of the island disturbance, every substation in the island will observe a slightly different slopes, and will therefore shed load based on different estimates of the disturbance. This is illustrated by Figure 20.29, which shows the frequencies observed at several different locations in a system following loss of a large nuclear generating unit and subsequent islanding [45]. Note that one location shows a faster frequency decline than the average, which is computed by the SFR model and shown by the dashed line. This part of the system will shed more load than necessary. Other buses, however, see a decline that is smaller than the average and will result in load shedding that is less than the desired amount. On the average, however, all buses will observe frequency rates of change that are about right since the trend in frequency is to follow the system average, as computed by the SFR model. It should also be noted that, in the area where the generator was tripped, which suffers the more rapid decay, the load shed will be greater than the average, but this is entirely proper since it will help balance the load to the remaining generation in every locality of the system and thereby minimize the transmission flows within the island. These intrasystem flows could themselves be a source of trouble, particularly if the unbalance is great enough to cause lines to begin tripping.

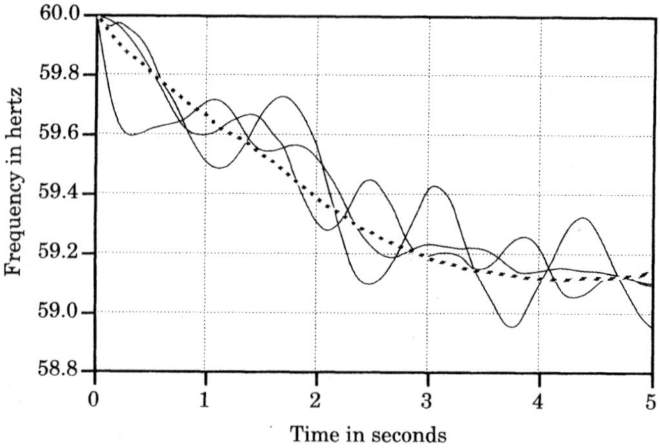

Figure 20.29 Simulated generating unit frequencies and the SFR computed average frequency after islanding [44].

We conclude that, although a few stations may measure frequency rates of change that are higher or lower than the true average rate of change, the system as a whole will probably shed the correct amount of load, since the average frequency behavior will prevail. This assumes, of course, that the frequency measurements and load shedding protective systems are located at many points throughout the system. Clearly, if the measurement and protection is at only a few locations, a rather large error might result.

Having set down the above guidelines, we must make a choice regarding the maximum disturbance for which protection will be provided. We arbitrarily set this maximum disturbance at 100% of the total generation in the islanded portion of the system. The curves of disturbances up to and including this 1.0 per unit disturbance are shown in Figure 20.26. We also observe that no load shedding is required for disturbances smaller than about 0.4 per unit as these disturbances do not cause the frequency to fall below the 57 hertz threshold.

First, we assume the following system parameters are typical of any large power system.

$$H = 3.5 \text{ s} \qquad F_H = 0.3$$
$$T_R = 8.0 \text{ s} \qquad K_m = 0.85$$
$$R = 0.06 \qquad D = 1.0$$

From these average system parameters, we may compute the undamped natural frequency and damping factors to be

$$\omega_n^2 = \frac{DR + K_m}{2HT_R} = 0.27083$$

$$\omega_n = 0.5204$$

$$\varsigma = \omega_n \left(\frac{2HR + (DR + K_m F_H) T_R}{2(DR + K_m)} \right)$$

$$= 1.615\omega_n = 0.84$$

Table 20.7 shows the computed initial slope and maximum frequency deviation for the typical system condition:

TABLE 20.7 Initial Slope and Maximum Deviation versus Upset

P_{step}	m_o		$\Delta\omega_{max}$	f_{min}
pu	pu/s	Hz/s	Hz	Hz
–0.2000	–0.0286	–1.7143	–1.6438	58.358
–0.3648	–0.0521	–3.1260	–3.0000	57.000
–0.4000	–0.0571	–3.4268	–3.2876	56.712
–0.6000	–0.0857	–5.1429	–4.9313	55.069
–0.8000	–0.1143	–6.8571	–6.5751	53.425
–1.0000	–0.1429	–8.5714	–8.2189	51.781

The second row of Table 20.7 represents the maximum step change of load that can be permitted if the frequency is not to decline below 57 hertz. We set this as the limiting value. Therefore, when the magnitude of the observed initial negative slope is greater than 3.126 hertz/s, load shedding must be triggered. Note that the initial slope for the −0.3648 per unit step disturbance is −0.0521 pu/s. It is proposed that we estimate the load that must be shed by subtracting 0.0521 per unit/s from the computed slope for any disturbance. We can do this since the system is linear. Therefore, the incremental load shed should be equal to the following.

$$\frac{P_{shed}}{2H} = \frac{|P_{step}|}{2H} - 0.0521 \quad \text{per unit/s} \tag{20.44}$$

Solving for P_{shed} and substituting (20.43) for the step change, we compute

$$P_{shed} = |P_{step}| - 0.1042H = H\left(\frac{|m_o|}{30} - 0.1042\right) \quad \text{per unit} \tag{20.45}$$

where m_o is defined to be the initial slope in Hz/s. From this equation, we may compute the incremental step function of load to be shed for various sizes of the initial disturbance. The incremental amount of load shed is therefore a linear function of the initial disturbance, and is therefore a linear function of the initial slope.

There is a fundamental deficiency in computing the amount of load shed by (20.45). As noted previously, it is prudent to begin load shedding at a frequency of 59.0–59.5 hertz. Moreover, the relays have a finite time delay so that the actual time at which the load is shed will be delayed. The computation (20.45) would be just barely adequate if all load were shed at time $t = 0$. Clearly, the computed amount is the minimum, which we can consider to be a "static load shed amount" that ignores the dynamics of the system frequency decay up to the actual shedding instant. Moreover, it is not prudent to shed all the load in one step, since system conditions may change and the required amount of load shedding will also change. It is better to shed load in several steps, with appropriate delay for each step.

The suggested total load shed is shown in Figure 20.30. The straight line labeled "Static Load Shed" is computed by (20.45). To account for the delay in implementing the load shedding, we add 5% to the Static Load Shed line to obtain a new line called the "Dynamic Load Shed," which is estimated to be adequate to account for the delay in initiating the various steps of load shedding. Actually, this precaution may not be necessary, since we have previously observed that the *amount* of load shed is more important than the exact time of load shedding. However, the additional 5% provides a factor of safety that may be considered prudent.

There is another factor that is very important, and that is the size of the various steps of load shedding. The strategy is as follows. Let the first step of load shedding occur when the

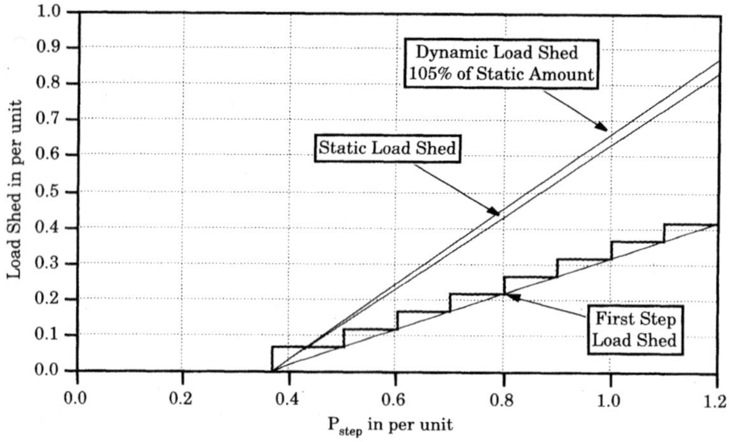

Figure 20.30 A load shedding strategy based on initial slope.

frequency reaches 59.5 hertz, and let the size of this first step be one-half of the total static load shed value for any given disturbance size. Making this first step as large as possible will make it possible to arrest the declining frequency, even for very large disturbances, well before the 57 hertz threshold can be reached. This is important since large disturbances cause the frequency to decline very rapidly, and there is little time to act if the frequency trajectory is to be reversed before it is too late.

Therefore, we use the knowledge gained by observing the initial slope to shed half of the required amount in the first step. This gives the first step of load shedding a linear relationship with the disturbance size and also with initial slope, as shown in Figure 20.30. All steps after the first one can be any convenient size, such as 0.1 per unit. These additional steps must sum to the total required load and should be scheduled to occur without excessive delay. Steps of 0.1 per unit, scheduled at each 0.5 hertz of frequency deviation, seem to work nicely.

The stairstep curve for the first step of load shedding, shown in Figure 20.30, is a practical alternative. This stairstep characteristic can be easily obtained by using fixed amounts between the minimum and maximum limits of the slope. The logic can be easily achieved using standard protection hardware. An example will illustrate the effectiveness of the design.

EXAMPLE 20.3

The effectiveness of the protective design described above is tested by a step load disturbance of -1.0 per unit. Show that the design concept is sound for this large disturbance and that the frequency decline is arrested before the critical 57.0 hertz level is reached.

Solution

We use Figure 20.30 to determine the amount of load to be shed at the design load shedding frequencies. This results in the table of load shed values and frequencies shown in Table 20.8.

Using the load shedding steps shown in the table, we compute the frequency response using the SFR model. The result is shown in Figure 20.31. Clearly, the frequency excursion has been prevented from exceeding the 57 hertz limit. This is the maximum design load disturbance. All other disturbances must be checked, however, to ensure compliance with the performance criteria. If the stairstep design is used, the extreme right side of each step must be checked for compliance. ■

20.7.2.4 Turbine Protective Margin. The load shedding protection described in Section 20.7.2.3 should be checked against the turbine protection described in Section 20.6 and

TABLE 20.8 Load Shed Schedule for 1.0 Per Unit Step

Frequency of Load Shedding, Hz	Amount of Load Shed Per Unit
59.5	0.317
59.2	0.10
58.9	0.10
58.6	0.10
58.3	0.05
Total Load Shed	0.635 × 1.05 = 0.667

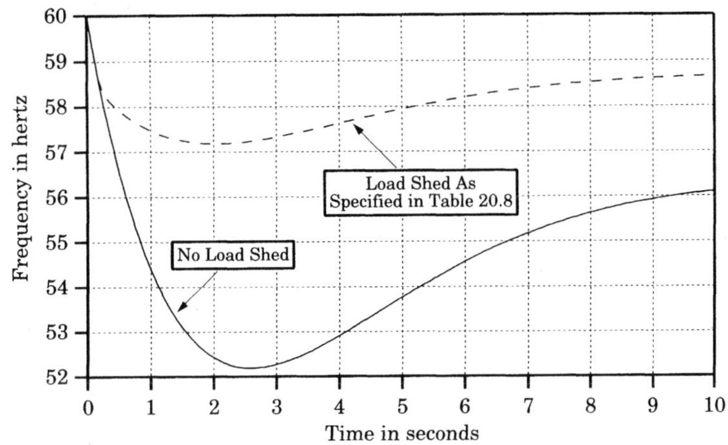

Figure 20.31 Result of load shedding according to the adaptive schedule.

[52]. The SFR model can be plotted against the turbine cumulative damage curve. Such a comparison is plotted in Figure 20.32.

The result shown in Figure 20.32 must be interpreted with caution. First, the time scale here is in minutes rather than seconds. This is more convenient for the turbine frequency limitation plot. We also note that the turbine cumulative damage curve represents the total life of the turbine, not the amount of damage that can be sustained in a single event, whereas the frequency response is for only one such event.

From the SFR calculation for the case shown in Figure 20.32, the frequency is in the 57.0–57.5 hertz band for 1.55 seconds, whereas the turbine protective relay is set for tripping the unit after 16 seconds has accumulated in this band. This means that this single excursion has used approximately 9.7% of the total turbine life. After about 10 such events, the turbine will be tripped by its protective relays and must be carefully examined for damage.

The foregoing discussion emphasizes the importance of developing a method of estimating the probability and frequency of occurrence of a load imbalance condition. Suppose, for example, that we can determine from statistical records that the disturbance plotted in Figure 20.32 will occur with a frequency of 0.01 events per year. This means that this type of event will occur, on the average, once every 100 years. This is far more than the intended lifetime of the turbine, so these events would not be of great concern. However, if the frequency should turn out to be 0.1 events per year, such an event would occur once every 10 years. This is much

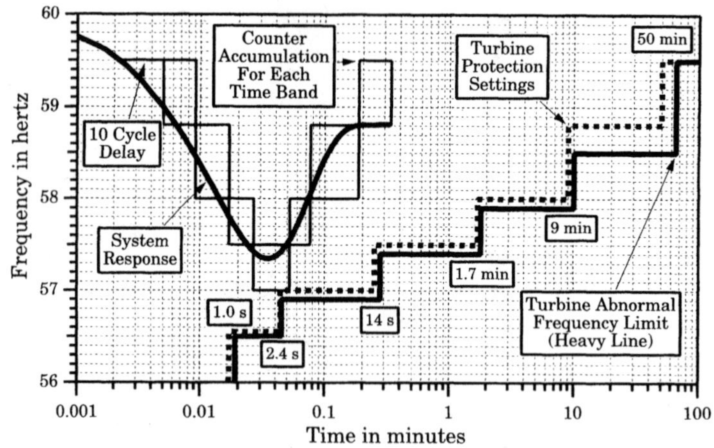

Figure 20.32 Comparison of system frequency response with load shedding and the turbine cumulative damage and protection curves.

more serious, since the turbine will probably be expected to have a useful life of about 40 years. The turbine lifetime is, therefore, threatened by disturbances in this frequency band, which may require a redesign of the underfrequency load shedding to avoid this band altogether.

The next higher turbine life expenditure is shown in Figure 20.32 as 1.7 minutes or 102 seconds. The frequency excursion plotted remained in this band for 0.5784 seconds on the way down and 1.445 seconds on the way back up, for a total accumulated time of 2.023 seconds. This represents 1.98% of the turbine lifetime of 102 seconds. Even if such events should occur at a frequency of 0.1 per year, this is not a serious threat to the unit.

The coordination of the underfrequency protection for all levels of initiating disturbances must be checked against the turbine cumulative damage curves and the turbine protection. If sufficient data is available to compute probabilities and frequencies of occurrence of these events, this would be valuable information in determining the adequacy of the protection.

20.7.3 Load Shedding Relay Connections

The connections of the load shedding relays are important since the failure to trip as designed may lead to turbine blade damage, long outage times, and excessive financial loss. It is important to design a system where a false trip of an underfrequency relay will not cause the unnecessary trip of a generating unit. Consider the scheme shown in Figure 20.33 [27].

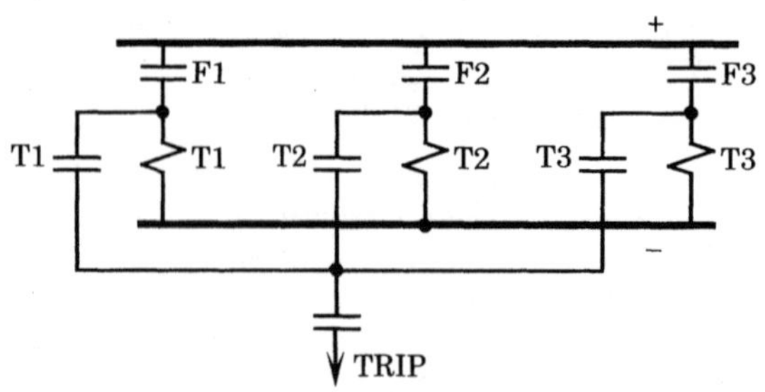

Figure 20.33 Three-step underfrequency protection scheme [27].

The scheme in Figure 20.33 can be directly extended to any desired number of frequency steps, each with its independent trip timer. Although simple to construct, this scheme lacks reliability, since failure of any relay fails the protection for that step. Moreover, the false trip of any step can cause the unnecessary trip of a unit.

The weaknesses of the scheme shown in Figure 20.33 are addressed in the scheme shown in Figure 20.34, where three relays are used for each step in a two out of three voting scheme. In this case, a single relay failure will not cause the system to fail, whether an operational failure or a false trip failure. However, it could be argued that this refinement is not necessary when the load shed at any one station is a small fraction of the total load to be shed. Many other schemes can be devised [50–53].

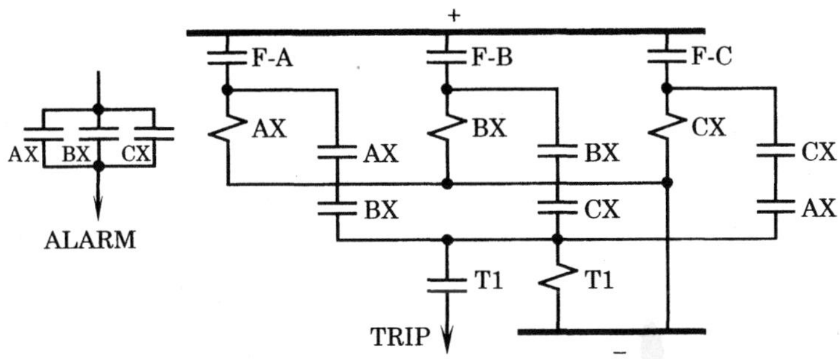

Figure 20.34 Two out of three voting scheme using three relays and one timer per step [27].

REFERENCES

[1] IEEE Working Group on Methods of System Preservation During Under-Frequency Conditions, "A Status Report on Methods Used For System Preservation During Underfrequency Conditions," Power System Relaying Committee Report, *IEEE Transactions*, PAS-94 (2), March/April 1975, pp. 360–366.

[2] Baldwin, M. S., and H. S. Schenkel, "Determination of Frequency Decay Rates During Periods of Generation Deficiency," *IEEE Transactions*, PAS-95 (1), January/February 1976, pp. 26–36.

[3] Warner, R. E., T. L. Dillman, and M. S. Baldwin, "Off Frequency Turbine-Generator Unit Operation," *Proc. Am. Power Conf.*, 38, 1976, pp. 570–580.

[4] Ghoneim, M. Z., M. H. Saleh, M. A. N. Askourah and S. A. Abdel Rahman, "Estimation of Optimum Load Shedding in Transmission Systems Using Mathematical Programming Techniques," *IEEE Trans.*, PAS-98 (1), March/April 1978, pp. 587–593.

[5] Khan, M. A., and K. Kuppusamy, "Optimum Load Curtailment Under Emergency Conditions Using Constant Matrices," IEEE paper A 79-113-2, presented at the IEEE PES Winter Meeting, New York, February 1979.

[6] American National Standards, "Requirements for Cylindrical-Rotor Synchronous Generator," ANSI C50.13-1977, American National Standards Institute, 1430 Broadway, New York, 10018

[7] Berdy, J., P. G. Brown and L. E. Goff, "Protection of Steam Turbine Generators During Abnormal Frequency Conditions," A paper presented at the Missouri Valley Electric Association Engineering Conference, Kansas City, MO, April 1978.

[8] Task Force on Underfrequency and Undervoltage Relaying, "Underfrequency and Undervoltage Relay Applications," a report by the National Electric Reliability Council, Princeton, NJ, 1978.

[9] Rudenberg, R., *Transient Performance of Electric Power Systems; Phenomena in Lumped Networks*, MIT Press, Cambridge, MA, 1967.

[10] Tavora, C., and O. J. M. Smith, "Stability Analysis of Power Systems," Report ERL-70-5, College of Engineering, University of California, Berkeley, 1970.

[11] Stanton, K. N., "Dynamic Energy Balance Studies for Simulation of Power Frequency Transients," Proc. PICA Conference, IEEE PES, C26-PWR, 1971.

[12] Virmani, S., S. Kim, R. Podmore, T. Athay, and D. Ross, "Development and Implementation of Advanced Automatic Generation Control, Final Report of Task 1, Modeling and Analysis of the WEPCO System," and "App. A, AGC Simulation Program User's Guide," SCI Proj 5215, DOE Contract EC-77-01-2118, Palo Alto, 1979.

[13] Smith, L. M., and J. H. Spare, "Area Control Simulator Program; v 1, Technical Manual," EPRI EL-1648, Palo Alto, 1980.

[14] Schulz, R. P., A. E. Turner, and D. N. Ewart, "Long Term Power System Dynamics," EPRI Report 90-70-0, Palo Alto, 1974.

[15] Schulz, R. P., and A. E. Turner, "Long Term Power System Dynamics, Phase II," EPRI EL-367, Palo Alto, 1977.

[16] deMello, R. W., R. Podmore, and K. N. Stanton, "Coherency Based Dynamic Equivalents for Transient Stability Studies," EPRI Report 904, Palo Alto, 1975.

[17] Podmore, R., and A. Germond, "Development of Dynamic Equivalents for Transient Stability Studies," EPRI EL-456, Palo Alto, 1977.

[18] Athay, T., V. R. Sherkat, R. Podmore, S. Virmani, and C. Puech, "Transient Energy Stability Analysis," SCI Proj 5158, DOE Contract EX-76-C-01-2076, Palo Alto, 1979.

[19] MIT Staff Report, "Emergency State Control of Slow Speed Dynamics," Final Report, DOE Contract E-49-18-2075, Cambridge, 1979.

[20] Kirchmayer, L. K., *Economic Control of Interconnected Systems*, John Wiley and Sons, New York, 1959.

[21] Concordia, C., F. P. deMello, L. K. Kirchmayer, and R. P. Schulz, "Effect of Prime-Mover Response and Governing Characteristics on System Dynamic Performance," Proc. Am. Power Conf, XXVIII, 1966, pp. 1074–1085.

[22] Chan, M. L., R. D. Dunlop, and F. Schweppe, "Dynamic Equivalents for Average System Frequency Behavior Following Major Disturbances," *IEEE Trans.*, PAS-91 (6), Jul/Aug 1972, pp. 1637–1642.

[23] Crevier, D., and F. C. Schweppe, "The Use of Laplace Transforms in the Simulation of Power System Frequency Transients," *IEEE Trans.*, PAS-94 (2), March/April 1975.

[24] Manohar, V. N., and M. K. Sinha, "Choice of Power System Model for Study of Nuclear Plant Dynamics," CIGRÉ paper 32-12, Paris, 1976.

[25] IEEE Committee Report, "Dynamic Models for Steam and Hydro Turbines in Power System Studies," *IEEE Transactions*, PAS-92, July/December 1973, pp. 1904–1915.

[26] Weber, G. A., M. C. Winter, S. M. Follen, and P. M. Anderson, "Nuclear Plant Response to Grid Electrical Disturbances," EPRI Final Report NP-2849, February 1983.

[27] Anderson, P. M., and M. Mirheydar, "A Low-Order System Frequency Response Model," *IEEE Transactions on Power Systems*, 5 (3), 1990, pp. 720–729.

[28] Baldwin, M. S., M. M. Merrian, H. S. Schenkel, and D. J. VandeWalle, "An Evaluation of Loss of Flow Accidents Caused by Power System Frequency Transients in Westinghouse PWR's," Westinghouse Report WCAP-8424, May 1975.

[29] New, W. C., Ed., "Load Shedding, Load Restoration and Generator Protection Using Solid-State and Electromechanical Underfrequency Relays," General Electric Company Publication GET-6449, 1977.

[30] AIEE Committee Report, "Automatic Load Shedding," *AIEE Trans.*, 74, December 1955, pp. 1143–1146.

[31] Bauman, H. A., G. R. Hahn, and C. N. Metcalf, "The Effect of Frequency Reduction on Plant Capacity and on System Operation," *AIEE Trans.*, 73, 1954 (February 1955 section), pp. 1632–1637.

[32] Holgate, R., "The Effect of Frequency and Voltage," *AIEE Trans.*, 73, 1954, pp. 1637–1646.

[33] Squire, P. J., "Operation at Low Frequency in Great Britain," *AIEE*, 73, 1954, pp. 1647–1650.

[34] Dalziel, C. F., and E. W. Steinback, "Underfrequency Protection of Power Systems for System Relief," *AIEE Trans.*, 78, December 1959, pp. 1227–1238.

[35] IEEE Committee Report, "The Effect of Frequency and Voltage on Power System Load," presented at the IEEE PES Winter Meeting, New York, January/February 1966.

[36] IEEE Committee Report, "Survey of Underfrequency Relay Tripping of Load Under Emergency Conditions," *IEEE Trans.*, PAS-87, March 1968, pp. 1362–1366.

[37] Lokay, H. E., and V. Burtnyk, "Application of Underfrequency Relays for Automatic Load Shedding," *IEEE Trans.*, PAS-87 (3), March 1968, pp. 776–783.

[38] Horowitz, S. H., A. Politis, and A. F. Gabrielle, "Frequency Actuated Load Shedding and Restoration, Part 2–Implementation," *IEEE Trans.*, PAS-90, 1971, p. 1460.

[39] Lokay, H. E., and P. O. Thoits, "Effects of Future Turbine Generator Characteristics on Transient Stability, *IEEE Trans.*, PAS-90, November/December 1971, pp. 2427–2431.

[40] Maliszewski, R. M., R. D. Dunlop, and G. L. Wilson, "Frequency Actuated Load Shedding and Restoration, Part 1–Philosophy," *IEEE Trans.*, PAS-90, 1971, p. 1452.

[41] Hohn, A., and P. Novacek, "Last-Stage Blades of Large Steam Turbines," *Brown Boveri Review*, 59, January 1972, pp. 42–53.

[42] Merrian, M. M., and D. J. Vandewalle, "The Effect of Grid Frequency Decay Transients on Pressurized Water Reactors," *IEEE Trans.*, PAS-95, January/February 1976, pp. 269–274.

[43] Somm, E., and Z. S. Stys, "The Development of Last-Stage Blades for Large Steam Turbines," *Proc. American Power Conf.*, 38, 1976, pp. 581–589.

[44] Hahn, R. S., S. Disgupta, E. Baytch, and R. D. Willoughby, "Maximum Frequency Decay Rate for Reactor Coolant Pump Motors," *IEEE Trans.*, NS-26, February 1977, pp. 863–870.

[45] Darlington, A. N., "Response of Underfrequency Relays on the Peninsular Florida Electric System for Loss of Generation," A paper presented at the Georgia Tech Relay Conference, May 4, 1978.

[46] Smaha, D. W., C. R. Rowland, and J. W. Pope, "Coordination of Load Conservation with Turbine-Generator Underfrequency Protection," *IEEE Trans.*, PAS-99, May/June 1980, pp. 1137–1150.

[47] Baldwin, M. S., W. A. Elmore, and J. J. Bonk, "Improve Turbine-Generator Protection for Increased Plant Reliability," *IEEE Trans.*, PAS-99, May/June 1980, pp. 982–989.

[48] Narayan, V., "Monitoring Turbogenerators in the Underfrequency Range," *Brown Boveri Review*, 67, September 1980, pp. 530–534.

[49] Narayan, V., H. Schindler, C. Pendinger, and D. Carreau, "Frequency Excursion Monitoring of Large Turbo-Generators," *IEE* (London) Conference on Developments in Power System Protection, Publication No. 185, June 10–12, 1980, pp. 45–48.

[50] Hicks, K. L., "Hybrid Load Shedding is Frequency Based," *IEEE Spectrum*, February 1983, pp. 52–56.

[51] Report of the Task Force on Underfrequency and Undervoltage Relaying," Underfrequency and Undervoltage Relay Applications to Large Turbine-Generators," National Electric Reliability Council, Research Park, Princeton, NJ, July 1978.

[52] American National Standard, "Guide for Abnormal Frequency Protection for Power Generating Plants," ANSI/IEEE C37.106-1987, New York.

[53] IEEE Working Group Report, "Summary of the 'Guide for Abnormal Frequency Protection for Power Generating Plants,' ANSI/IEEE C37.106-198X," *IEEE Trans. on Power Delivery*, 3, January 1988, pp. 153–158.

[54] Ewart, D. N., "Whys and Wherefores of Power System Blackouts," *IEEE Spectrum*, April 1978, pp. 36–41.

PROBLEMS

20.1 Compute the maximum permissible time that the stator current of a synchronous generator exceeds its rated value by the following stated percentages:
(a) 10%
(b) 25%
(c) 75%
(d) 100%
(e) 150%

20.2 Compute the maximum permissible time that the field voltage of a synchronous generator exceeds its rated value by the following stated percentages:
(a) 10%
(b) 25%
(c) 75%
(d) 100%
(e) 150%

20.3 An underfrequency event causes the frequency of an island to fall to 58 hertz for 2 minutes, recover to 59 hertz for 1 hour, and finally return to normal (60 hertz). Determine the total percent loss of life for a turbine of manufacturer A, using the lifetime curves of Figure 20.6.

20.4 An underfrequency event on a power system results in a frequency of 57 hertz for about 1 minute. Determine the loss of life accumulated by steam turbines in the underfrequency island.

20.5 Write the differential equations that are described in block diagram form in Figure 20.8, and explain the meaning of the equations. Why is the inertia constant multiplied by 2?

20.6 What is a typical value of the inertial constant, H, in Figure 20.8,
(a) for an individual machine, on the machine base?
(b) for an entire island, on the base of all island machines?

20.7 Write the s-domain equations for the turbine model of Figure 20.9 and simplify by retaining only the largest time constants. Draw a simplified block diagram for the turbine.

20.8 Write the s-domain equations for the speed governor system of Figure 20.10 and simplify to obtain a linear model suitable for small signal analysis. Draw a simplified block diagram for the speed governor.

20.9 Derive (20.2) and (20.3) from the block diagram of Figure 20.11.

20.10 Verify the block diagram of the *SFR* model, given in Figure 20.12, beginning with the block diagram of Figure 20.11.

20.11 Verify (20.5).

20.12 Solve (20.8) in order to verify the time domain solution given by (20.9), (20.10), and (20.11). Hint: Use a good table of Laplace transforms to save time in finding the transformation.

20.13 Compute the final value of the frequency deviation for the case of a step change in disturbance power of -0.3. See (20.23). Use data from Figure 20.13.

20.14 Write a small computer program to compute the system frequency response given by (20.9). Using this program will provide a way of verifying many of the results shown in this chapter.

20.15 Compute the final value of the frequency deviation in hertz, for the values of the system parameters given in Figure 20.14. Your computation should match the observed values given in the figure.

20.16 Compute the initial rate of change and the maximum frequency deviation for each of the parametric values of H given in Figure 20.16. Explain why the final value is always the same for these cases.

20.17 Compute the maximum frequency deviation for each of the parametric values of T_R given in Figure 20.17. Explain why the initial slope and the final value is always the same for these cases.

20.18 Compute the maximum frequency deviation for each of the parametric values of F_H given in Figure 20.18. Explain why the initial slope and the final value is always the same for these cases. Also, provide a physical explanation as to why the frequency dip is smaller as this parameter is increased. Why is one of the values more oscillatory than the others, and why is this feature not noticed in any other plots?

20.19 Compute the initial rate of change, the maximum frequency deviation, and the final value for the five cases illustrated in Figure 20.19. What can you conclude regarding the effect of changes in the parameter D, as compared to the parameter R?

20.20 How does the introduction of spinning reserve, as defined in (20.29) change the solution of the system of Figure 20.11?

20.21 Derive (20.34).

20.22 Show that we can write the power delivered to the load in Figure 20.21 by the expression

$$P_L = K\frac{E_G^2}{X}$$

Define K as a function of θ and δ.

20.23 Write the constant K of problem 20.22 as a function only of θ.

20.24 Write the constant K of problem 20.22 as a function only of δ.

20.25 Derive (20.38).

20.26 Compute the size of the disturbance power for cases (1) and (2), making a table similar to Table 20.4. The system total generation is 5000 MW.

21

Protective Schemes for Stability Enhancement

21.1 INTRODUCTION

A growing number of protective schemes are primarily designed for improving power system stability or enhancing system security [1]. These schemes differ from others, such as line or apparatus protection schemes, in that they are not necessarily applied to a particular device or protective zone, but are safeguards designed to alter or preserve the system structure, security, or connectivity. Other stability enhancement schemes are protections identified with individual generating units that are designed to prevent the tripping of units under conditions where the generator protection may see a condition that looks like a fault, but is actually just the normal dynamic oscillation of the power system that occurs as the result of a disturbance.

Since these schemes are designed to enhance stability, we begin this chapter with a brief review of some stability fundamentals. This is followed by a discussion of some stability enhancement protective schemes.

21.2 REVIEW OF STABILITY FUNDAMENTALS

Power system stability is a complex subject that is beyond the scope of this book. Our discussion here is somewhat superficial, but it deals with the fundamental concept of the stability of dynamical systems and introduces an elementary concept that will help in understanding the importance of protection in maintaining the stability of the power system.

21.2.1 Definition of Stability

The concept of stability is central to several disciplines in electrical engineering, physics, and mathematics. Control systems that do not exhibit stable performance would be almost worthless and stability theory is central to control system design and analysis. For example, the controls for an elevator should move the passenger compartment to the correct floor and align it to the floor level for safe entry and exit of passengers. The system would be of little value if this objective were not achieved.

Power systems have a number of controls and protective devices to ensure that the response to any change in system conditions is controlled, thereby giving service to customers that can be relied on to have reasonably constant and dependable characteristics. The system must be robust in its design and should be able to withstand naturally occurring disturbances, such as electrical storms, high winds, or grass fires, as well as customer-related disturbances, such as the starting of large motors. The power system is designed to withstand events that have a reasonable probability of occurrence and should continue to operate in a normal manner in response to such events, except under extremely severe conditions.

What exactly do we mean by the term *stability*? Several definitions have been presented in the literature [2] and a few of these will be reviewed.

> *Stability* is the ability of a power system to remain in synchronous equilibrium under steady operating conditions, and to regain a state of equilibrium after a disturbance has occurred [2].

The disturbance can be of any type whatsoever but, in many cases, short circuits are of particular interest as these conditions have a reasonable expectation, are relatively severe, and have the potential of causing instability if not removed promptly. Extremely large disturbances are possible, and may result in cascading outages due to large power surges that result in system separations. The power system is not normally designed to withstand these rare and unusually large disturbances, but the anticipation of such events may lead to the design of appropriate controlled response schemes. Underfrequency load shedding is an excellent example of such a scheme.

Stability is also defined in a mathematical context, in which case the stability under consideration is the stability of *solutions* of the differential equations that describe the power system.

> *Stability* refers to the behavior of the solution of the nonlinear differential equations that are used to represent the physical power system.

In most of our work related to the stability of power systems and its reliance on system protection, we are most concerned with the behavior of the physical power system rather than the behavior of the system equations. However, it is recognized that analytical methods are important in providing the capability of performing system studies in which the system response to a given disturbance can be examined. Indeed, it is through computer simulations that power systems design objectives are achieved.

21.2.2 Power Flow Through an Impedance

The basic equation for power flow through an impedance is central to the concept of stability, since it illustrates the importance of impedance or transfer admittance in limiting the power flow through a network.

Consider the simple system shown in Figure 21.1, where we write the equation of power flow in the direction of defined current flow.

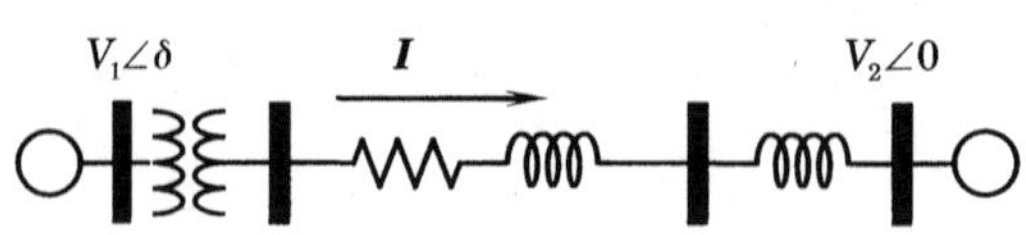

Figure 21.1 Power flow through an impedance.

We arbitrarily take the voltage on the right as the phasor reference to compute

$$\boldsymbol{I} = \frac{\boldsymbol{V}_1 - \boldsymbol{V}_2}{\boldsymbol{Z}} = \frac{V_1\angle\delta - V_2\angle 0}{\boldsymbol{Z}} \tag{21.1}$$

where

$$\boldsymbol{Z} = R + jX \tag{21.2}$$

$\boldsymbol{Z}$ is the total series impedance between the two voltage sources, and may include several lines, transformers, or other series impedances.[1] We then compute the complex power leaving the voltage source on the left as

$$P_1 + jQ_1 = \boldsymbol{V}_1\boldsymbol{I}_1^* \tag{21.3}$$

or

$$\begin{aligned} P_1 + jQ_1 = {} & \frac{V_1^2}{Z^2}(R + jX) \\ & - \frac{V_1 V_2}{Z^2}[(R\cos\delta - X\sin\delta) + j(X\cos\delta + R\sin\delta)] \end{aligned} \tag{21.4}$$

We are most interested in the active power injected at node 1, which we can write from (21.4) as

$$P_1 = \frac{V_1^2 R}{Z^2} - \frac{V_1 V_2}{Z^2}(R\cos\delta - X\sin\delta) \tag{21.5}$$

In most power systems, particularly at the higher voltages, the resistance is much smaller than the reactance, so we often write (21.5) neglecting the resistance terms as follows.

$$P_1 = \frac{V_1 V_2}{X}\sin\delta = P_{\max}\sin\delta \tag{21.6}$$

Thus, we note that the maximum power that can be transferred between the two sources is dependent on the product of the voltage magnitudes and inversely proportional to the total reactance between the sources. We will make use of this important equation in establishing a fundamental stability concept. First, however, since the case presented above is an unusually simple system, we now consider a more general system description.

21.2.3 Two-Port Network Representation

Figure 21.2 shows the same Generator 1 connected to a more general network. We like to think of this individual generator responding to any type of disturbance within the network by changing its rotor angle in an effort to continue to transmit the power P_1 if at all possible. The generator labeled 2 on the right may be thought of as the inertial center of the entire system to which Generator 1 is connected. We will assume that this equivalent generator looks very large compared to Generator 1. It has a huge moment of inertia and a very large volt-ampere rating. Hence, the injection of power by Generator 1 has little effect on this large generator, which we usually consider to be an "infinite bus," since it is fully capable of holding the voltage V_2 at exactly the same value at all times, no matter how severe the network disturbance. We now derive the Generator 1 power equation in general terms, based on the two-port network shown, with currents injected as noted in Figure 21.2. For the network, we write the matrix

[1] Note carefully the notation. We use the bold italic typeface to indicate phasors, such as current of voltage phasors, and complex numbers, such as impedances. Bold Roman type is used to indicate matrices.

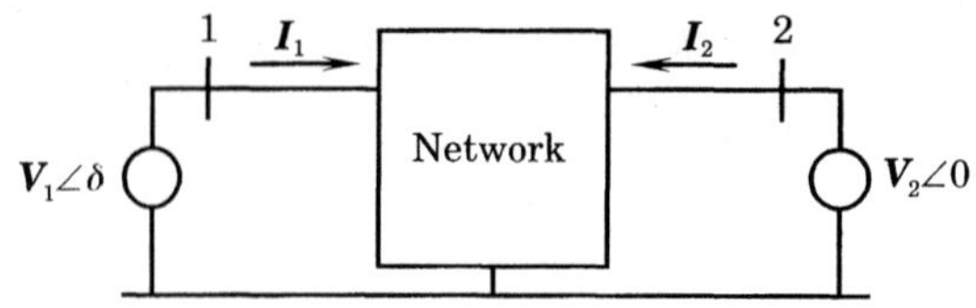

Figure 21.2 Generator 1 connected to a general network.

two-port equation

$$\begin{bmatrix} I_1 \\ I_2 \end{bmatrix} = \begin{bmatrix} Y_{11} & Y_{12} \\ Y_{21} & Y_{22} \end{bmatrix} \begin{bmatrix} V_1 \\ V_2 \end{bmatrix} \tag{21.7}$$

where all matrix quantities are complex. In particular, we may write the admittances in the first equation as

$$\begin{aligned} Y_{11} &= G_{11} + jB_{11} \\ Y_{12} &= G_{12} + jB_{12} \end{aligned} \tag{21.8}$$

The two voltages are defined exactly as before. Since the current leaving Generator 1 is defined by the admittance equation (21.7), we may again compute the power injected by Generator 1 by solving (21.3), with the result

$$\begin{aligned} P_1 + jQ_1 = {} & V_1^2(G_{11} - jB_{11}) \\ & + V_1V_2[(G_{12}\cos\delta + B_{12}\sin\delta) + j(G_{12}\sin\delta - B_{12}\cos\delta)] \end{aligned} \tag{21.9}$$

From this equation we readily compute the power injected by Generator 1 to be

$$P_1 = V_1^2 G_{11} + V_1V_2(G_{12}\cos\delta + B_{12}\sin\delta) \tag{21.10}$$

This is the same as (21.5), except using admittance rather than impedance quantities. It is more general than (21.5), however, since the two-port admittance description can always be constructed from a large network, once the inertial center machine has been identified in that network and the system impedances are known. This is not always easy in an actual system, but the concept is valid. In many cases, the infinite bus is represented using an arbitrarily large machine that is connected to the actual machine buses through ideal phase shifters [2], thereby constructing a dynamic equivalent of the system.

Since the network seen by Generator 1 is only a two port, it can be represented immediately by the circuit shown in Figure 21.3, using only three admittances. Any network connecting the two generators can be simplified to this form, although it may require Y- Δ transformations to do so. With the system described in this way, we may write the matrix admittances as

$$\begin{aligned} Y_{11} &= y_{10} + y_{12} \\ Y_{12} &= Y_{21} = -y_{12} \\ Y_{22} &= y_{20} + y_{12} \end{aligned} \tag{21.11}$$

where the lowercase values represent actual admittances shown in Figure 21.3 and the uppercase characters represent matrix elements of (21.7).

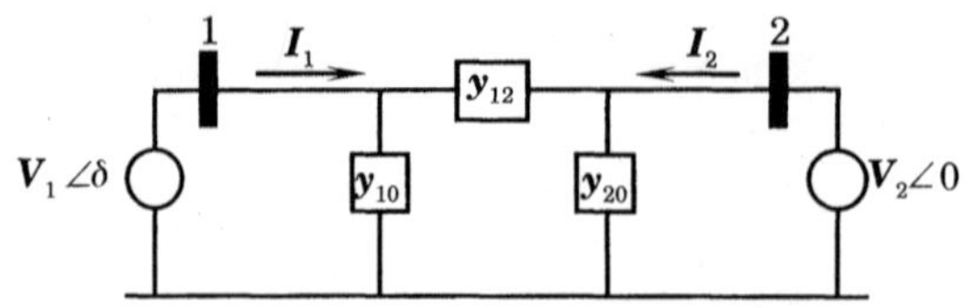

Figure 21.3 The equivalent two port seen by Generator 1.

EXAMPLE 21.1

Consider the network shown in Figure 21.4, where we represent Generator 1 by a constant voltage behind its transient reactance. The impedance $\mathbf{Z}_L$ represents a local load connected to the generator bus. This bus is connected to parallel transmission lines $\mathbf{Z}_U$ and $\mathbf{Z}_V$ by the equivalent impedance $\mathbf{Z}_E$. The infinite bus is connected to the same parallel lines by the source impedance $\mathbf{Z}_S$. A fault with impedance $\mathbf{Z}_F$ is connected to one of the transmission lines.

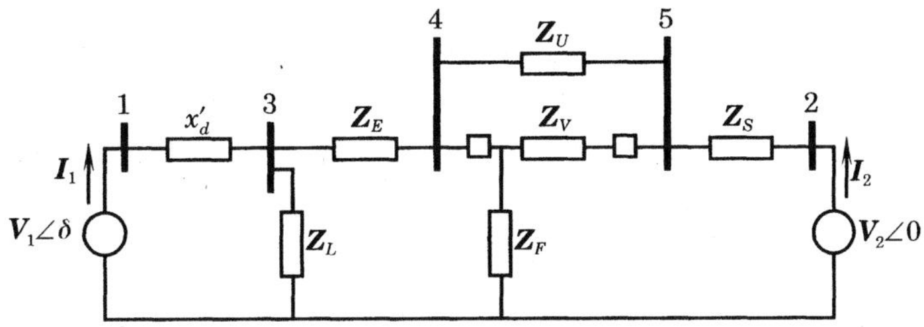

Figure 21.4 A network with faulted line.

This network has five nodes, numbered as shown in Figure 21.4, with the two generator nodes numbered 1 and 2, as before. To get the equation into the form of (21.10), we perform a network reduction. Actually, there will be three network reductions required:

1. Before the fault occurs
2. During the fault
3. After the fault has been cleared and the faulted line opened

The process is the same, however, for each case. We begin by writing the admittance matrix equations for the five nodes, only two of which have nonzero injection currents.

$$\begin{bmatrix} \mathbf{I}_1 \\ \mathbf{I}_2 \\ \hline \mathbf{I}_3 = 0 \\ \mathbf{I}_4 = 0 \\ \mathbf{I}_5 = 0 \end{bmatrix} = \left[\begin{array}{cc|ccc} \mathbf{Y}_{11} & \mathbf{Y}_{12} & \mathbf{Y}_{13} & \mathbf{Y}_{14} & \mathbf{Y}_{15} \\ \mathbf{Y}_{21} & \mathbf{Y}_{22} & \mathbf{Y}_{23} & \mathbf{Y}_{24} & \mathbf{Y}_{25} \\ \hline \mathbf{Y}_{31} & \mathbf{Y}_{32} & \mathbf{Y}_{33} & \mathbf{Y}_{34} & \mathbf{Y}_{35} \\ \mathbf{Y}_{41} & \mathbf{Y}_{42} & \mathbf{Y}_{43} & \mathbf{Y}_{44} & \mathbf{Y}_{45} \\ \mathbf{Y}_{51} & \mathbf{Y}_{52} & \mathbf{Y}_{53} & \mathbf{Y}_{54} & \mathbf{Y}_{55} \end{array}\right] \begin{bmatrix} \mathbf{V}_1 \\ \mathbf{V}_2 \\ \hline \mathbf{V}_3 \\ \mathbf{V}_4 \\ \mathbf{V}_5 \end{bmatrix} \qquad (21.12)$$

Here, it is noted that only injection currents 1 and 2 are nonzero. Because of the zero injections, this equation may be reduced by Kron reduction to give the result

$$\begin{bmatrix} \mathbf{I}_1 \\ \mathbf{I}_2 \end{bmatrix} = \left\{ \begin{bmatrix} \mathbf{Y}_{11} & \mathbf{Y}_{12} \\ \mathbf{Y}_{21} & \mathbf{Y}_{22} \end{bmatrix} - \begin{bmatrix} \mathbf{Y}_{13} & \mathbf{Y}_{14} & \mathbf{Y}_{15} \\ \mathbf{Y}_{23} & \mathbf{Y}_{24} & \mathbf{Y}_{25} \end{bmatrix} \begin{bmatrix} \mathbf{Y}_{33} & \mathbf{Y}_{34} & \mathbf{Y}_{35} \\ \mathbf{Y}_{43} & \mathbf{Y}_{44} & \mathbf{Y}_{45} \\ \mathbf{Y}_{53} & \mathbf{Y}_{54} & \mathbf{Y}_{55} \end{bmatrix} \begin{bmatrix} \mathbf{Y}_{31} & \mathbf{Y}_{32} \\ \mathbf{Y}_{41} & \mathbf{Y}_{42} \\ \mathbf{Y}_{51} & \mathbf{Y}_{52} \end{bmatrix} \right\} \begin{bmatrix} \mathbf{V}_1 \\ \mathbf{V}_2 \end{bmatrix}$$
$$= \begin{bmatrix} \mathbf{Y}'_{11} & \mathbf{Y}'_{12} \\ \mathbf{Y}'_{21} & \mathbf{Y}'_{22} \end{bmatrix} \begin{bmatrix} \mathbf{V}_1 \\ \mathbf{V}_2 \end{bmatrix} \qquad (21.13)$$

where the new matrix values are indicated by primes. In this way, any network can be so reduced to a two-port network, with only the Generator 1 and the infinite bus represented. Obviously, the resulting matrix depends on the condition of the faulted line, as noted previously. ■

21.2.4 The Swing Equation

The equation that determines the dynamic behavior of Generator 1 for a network disturbance, such as a fault, is called the *swing equation*. This equation is nothing more than Newton's second law, expressed in terms of the power system quantities. The basic system to be considered is the generator and its prime mover, which can be described by the second order differential equation

$$J\frac{d^2\theta}{dt^2} = T_a \qquad \text{Nm} \tag{21.14}$$

$$\begin{aligned} \text{where}\quad J &= \text{moment of inertia (kg - m}^2\text{)} \\ \theta &= \text{physical angle of shaft (rad)} \\ t &= \text{time (s)} \\ T_a &= \text{accelerating torque (Nm)} \end{aligned} \tag{21.15}$$

We can write the angle of the shaft in general terms as follows. Imagine a mark on the outer radius of the shaft that identifies the shaft angle with respect to a fixed point on the shaft supporting structure. Then the total shaft angle can be written as

$$\theta = \omega_R t + \delta_m + \alpha \qquad \text{rad} \tag{21.16}$$

$$\begin{aligned} \text{where}\quad \omega_R &= \text{shaft rated angular velocity (rad/s)} \\ \delta_m &= \text{torque angle (rad)} \\ \alpha &= \text{arbitrary angle to a fixed reference (rad)} \end{aligned} \tag{21.17}$$

The first term in (21.16) gives the angular position of a synchronously rotating reference. If there were no disturbance to the shaft, the total angle θ would advance linearly with time and the angle δ_m with respect to the synchronously rotating reference would be constant. Any disturbance would cause the torque angle δ_m to vary, usually in a damped oscillatory manner for a small disturbance, but it may increase or decrease monotonically without oscillation for a large disturbance.

The derivative of (21.16) is the shaft mechanical angular velocity, which we can write as

$$\omega_m = \frac{d\theta}{dt} = \omega_R + \frac{d\delta_m}{dt} \qquad \text{rad/s} \tag{21.18}$$

The subscript m is used here to indicate that this is the *mechanical* angular velocity of the shaft, not the radian frequency of the electric system. Taking the derivative again gives the angular acceleration of the shaft.

$$\frac{d\omega_m}{dt} = \frac{d^2\theta}{dt^2} = \frac{d^2\delta_m}{dt^2} \qquad \text{rad/s}^2 \tag{21.19}$$

Thus, we see that the angular acceleration of the swing equation can be written in terms of the acceleration of the torque angle, or as the derivative of the angular velocity. Thus, substituting (21.19) into (21.14), we can write

$$J\frac{d^2\delta_m}{dt^2} = T_a \qquad \text{Nm} \tag{21.20}$$

There is a unique relationship between the shaft angular velocity and the electric system radian frequency that depends entirely on the number of poles in the generator. That relationship can be written in terms of angular velocities as

$$\omega = \frac{p}{2}\omega_m \qquad \text{electrical rad/s} \tag{21.21}$$

where p = number of poles
ω = radian frequency of generated voltage (rad/s)
ω_m = mechanical shaft speed of generator (rad/s)

We may also write (21.21) in terms of angles, as follows.

$$\omega = \frac{d\delta}{dt} = \frac{p}{2}\frac{d\delta_m}{dt} = \frac{p}{2}\omega_m \tag{21.22}$$

The base values for the angular velocities can be written from (21.21) as

$$\omega_B = \frac{p}{2}\omega_R \tag{21.23}$$

where ω_B = electric system base radian frequency
ω_R = generating unit rated shaft angular velocity (21.24)

Now, taking the derivative of (21.22), dividing by (21.23) term by term, and rearranging the result, we can compute

$$\frac{d^2\delta_m}{dt^2} = \frac{\omega_R}{\omega_B}\frac{d^2\delta}{dt^2} = \frac{2}{p}\frac{d^2\delta}{dt^2} \tag{21.25}$$

This result can be substituted into the swing equation (21.20) to express the swing equation in terms of the electrical angle δ.

$$J\frac{d^2\delta_m}{dt^2} = \frac{2J}{p}\frac{d^2\delta}{dt^2} = T_a \qquad \text{Nm} \tag{21.26}$$

In computations involving power systems it is convenient to normalize all equations with respect to a common system volt-ampere base. Normalizing the swing equation requires that we find an expression for the base torque. Torque is related to power and speed by the expression

$$\text{Torque} = \frac{\text{power}}{\text{speed}} \tag{21.27}$$

when expressed in appropriate units. As a base quantity, it is more convenient to refer to the base three-phase voltamperes and the rated shaft angular velocity. Thus we can write the base torque as

$$T_B = \frac{S_{B3}}{\omega_R} = \frac{pS_{B3}}{2\omega_B} \tag{21.28}$$

where S_{B3} is the system three-phase voltampere base.

We can also write the moment of inertia in terms of the rotating energy at rated speed.

$$J = \frac{2W_k}{\omega_R^2} = \frac{p^2 W_k}{2\omega_B^2} \tag{21.29}$$

Then the normalized swing equation can be written as

$$\frac{2W_k}{\omega_B S_{B3}}\frac{d^2\delta}{dt^2} = T_{au} \qquad \text{pu} \tag{21.30}$$

This is a convenient form of the equation since δ is the electrical angle of the generated voltage of this generator. It has become common practice to define an "inertia constant" as the ratio

$$H = \frac{W_k}{S_{B3}} \quad \text{s} \tag{21.31}$$

This quantity is not literally a constant, but its numerical value is usually in the range of 2–4 seconds for most generating units. Making this substitution into (21.30) we can write

$$\frac{2H}{\omega_B}\frac{d^2\delta}{dt^2} = T_{au} = T_{mu} - T_{eu} - \frac{D}{\omega_B}\frac{d\delta}{dt} \quad \text{pu} \tag{21.32}$$

where we have introduced the components of the accelerating torque that are further defined as follows.

$$\begin{aligned} T_{mu} &= \text{prime mover mechanical torque} && \text{pu} \\ T_{eu} &= \text{generator shaft torque} && \text{pu} \\ \frac{D}{\omega_B}\frac{d\delta}{dt} &= \text{viscous damping torque} && \text{pu} \end{aligned} \tag{21.33}$$

Rearranging the terms, we can write

$$\frac{2H}{\omega_B}\frac{d^2\delta}{dt^2} + \frac{D}{\omega_B}\frac{d\delta}{dt} = T_{mu} - T_{eu} \quad \text{pu} \tag{21.34}$$

This form of the equation is convenient, but can be made even more useful for our purpose here if we convert each torque term into a term with dimensions of per unit power. This can be done using (21.27), which requires that we multiply each term of the equation by the per unit angular velocity ω. Thus we have

$$\frac{2H\omega_u}{\omega_B}\frac{d^2\delta}{dt^2} + \frac{D\omega_u}{\omega_B}\frac{d\delta}{dt} = P_{mu} - P_{eu} \quad \text{pu} \tag{21.35}$$

This is a convenient form of the equation since the electric generated power can be written in terms of the angle δ, resulting in a differential equation that can be solved for the torque angle δ. In some cases, the equation is simplified by assuming that the mechanical power is constant, which is usually true for a short time following the initiation of a disturbance, since it takes a finite amount of time for the speed governor to respond, change the prime mover valve position, and for the prime mover power to change.

Returning to the system of Figure 21.4, if we neglect resistance and losses, we can write the swing equation as

$$\frac{2H_1}{\omega_B}\frac{d^2\delta_1}{dt^2} + \frac{D_1}{\omega_B}\frac{d\delta_1}{dt} + V_1V_2B_{12}\sin\delta_1 = P_{m1} \tag{21.36}$$

Even in this simplified form, the equation is nonlinear, because of the sine function and the nature of the disturbance. The disturbance in the system is usually provided by the term B_{12}, which changes abruptly when the network is changed by the disturbance. In our approximate analysis here, we will consider the mechanical power of the turbine to be a constant. The major force that causes the generating unit to respond to the disturbance is the rapid change in accelerating power, which is approximately the difference between the nearly constant mechanical power and the rapidly changing electrical power. This is often plotted as shown in Figure 21.5, where we have plotted the mechanical power as a constant of 0.8 per unit and the electrical power for the three conditions, pre-fault, faulted, and post-fault.

The pre-fault condition, which is the upper sinusoidal curve, shows that the system is in equilibrium when the torque angle is δ_o. For this condition, there is no acceleration and the

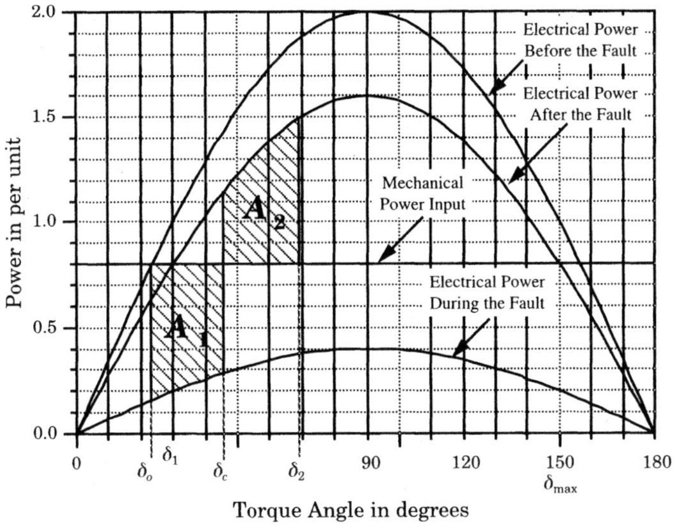

Figure 21.5 Power versus torque angle.

system is at rest. When the fault occurs, the operating angle is still δ_o since that angle cannot change instantaneously, but the electric power is now given by the lower curve, representing the power characteristic during the fault. The shaft begins to accelerate along this P-δ characteristic until the fault is cleared at angle δ_c, the clearing angle. Now, the electrical power suddenly jumps to the post-fault curve and the angle continues to accelerate to its maximum value, δ_2, at which point it reverses direction. The angle continues to oscillate about the mechanical power until it finally settles at a new equilibrium point at a new value of δ_1, which is given by the intersection of the post-fault electrical power curve and the mechanical power line. Since the total susceptance of the network is lower due to the line outage, it requires a larger torque angle to maintain this value of generated power.

If we assume the speed is nearly constant at 1.0 per unit, we can simplify (21.35) to write

$$2H\frac{d\omega}{dt} + D\omega = P_{mu} - P_{eu} \qquad \text{pu} \tag{21.37}$$

We now consider a special case where the damping coefficient is zero. If we neglect the damping term in (21.37) we may write the following expression.

$$2H\omega\frac{d\omega}{dt} = \omega(P_m - P_e) = \frac{1}{\omega_B}\frac{d\delta}{dt}(P_m - P_e) \qquad \text{pu} \tag{21.38}$$

or

$$H\frac{d\omega^2}{dt} = \frac{1}{\omega_B}(P_m - P_e)\frac{d\delta}{dt} \tag{21.39}$$

where we note that the angular velocity variable is in per unit. We now integrate both sides of (21.39) and multiply by dt with the result

$$H\omega_B\int_{\omega_1}^{\omega_2} d\omega^2 = \int_{\delta_1}^{\delta_2}(P_m - P_e)\,d\delta \tag{21.40}$$

The left side can be evaluated directly to write

$$H\omega_B(\omega_2^2 - \omega_1^2) = \int_{\delta_1}^{\delta_2}(P_m - P_e)\,d\delta \tag{21.41}$$

For a stable system, the left-hand side of (21.41) must be zero since the shaft speed must return to its original value; otherwise, the generator would accelerate or decelerate until it separates from the system. Setting the left-hand side of (21.41) to zero gives a criterion for stability, which we write as

$$\int_{\delta_1}^{\delta_2} (P_m - P_e)\, d\delta = 0 \tag{21.42}$$

where the subscripts 1 and 2 refer to the limits of the integration. These limits are at points where the shaft speed is synchronous, such that the left side of (21.41) is zero. This is true at the starting point δ_o and also at the upper limit of angle travel δ_2

We break up the integration of (21.42) in two steps, integrating first from δ_o to δ_c and then from δ_c to δ_2 as shown in Figure 21.5. However, we also recognize that the integral changes sign at δ_c to write

$$\int_{\delta_o}^{\delta_c} (P_m - P_e) d\delta = \int_{\delta_c}^{\delta_2} (P_e - P_m)\, d\delta \tag{21.43}$$

But these integrals are both areas in the P-δ plane and are exactly the areas labeled A_1 and A_2 in Figure 21.5. Thus, for stability, these areas must be equal. This effectively places a limit on the clearing angle δ_c, which must occur soon enough to limit the area A_1 to an amount that is equal to or less than the maximum area available for A_2, which is limited by the area between the post-fault curve and the mechanical power level, designated as δ_2. In Figure 21.5, δ_{max} is approximately 150 degrees, so the case illustrated in Figure 21.5 is well below the stability limit.

This simple concept has some important implications for the protective system, including the following:

1. It is important to clear the fault as quickly as possible, thereby limiting the area A_1 to as small a region as possible.
2. It would be beneficial if the faulted line could be reclosed, perhaps after a brief delay needed to allow the arc to deionize. This would permit area A_2 to extend all the way to the upper curve in Figure 21.5, after reclosing, thereby giving a larger margin for stability.

In the remainder of this chapter, we consider some system protective schemes that are intended to aid system stability in different ways.

21.3 SYSTEM TRANSIENT BEHAVIOR

The behavior of the power system when confronted by large disturbances is usually studied using stability computer programs designed for this purpose. The stability programs solve the differential equations of the entire system, including the swing equations for all generating units and of all transient control responses. The actions of generator excitation systems, speed governors, SVCs, and HVDC converters are especially important, since these devices help control the power flows throughout the system. In this section we review some of the results of disturbances and the resulting control actions that are important from the viewpoint of the protection engineer.

21.3.1 Stability Test System

In order to illustrate the essentials of power system stability, we concentrate on a small test system. This provides a way of illustrating typical system behavior without introducing the complexity and data requirements of a large, interconnected power system. The test system of interest is shown in Figure 21.6. This system has four generating units, which results in three natural modes of oscillation. Areas 2 and 3 represent large power systems consisting of many generators, loads, and transmission lines. These systems are represented by an equivalent generator and load to represent those large, remote areas. Area 1 represents the system of interest for detailed study. This system has two generating plants and three major load centers. All the transmission lines in Area 1 are series-compensated 500 kV lines, with lines labeled A, B, and C having 70% and the remaining lines 60% series compensation. The largest load is located at bus 7 and the voltage at this bus is controlled by a static var compensator, or SVC, which is rated at 1500 MVA, either positive or negative. Power flows are scheduled among the three areas to represent purchases and sales of power among the areas. The points labeled "M" indicate the points of measurement between areas. The underlying subtransmission and distribution of power within Area 1 are not shown, but are assumed to be present. However, the load centers are far enough apart that there is no assumption of lower voltage interconnection between these regions. The system is tested by applying a fault, as shown in the figure.

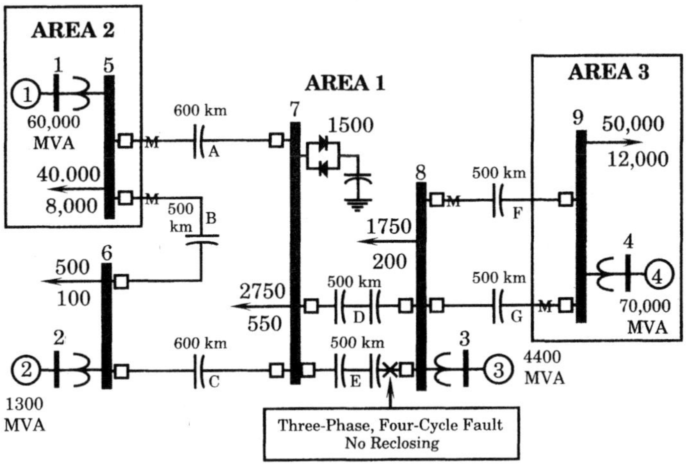

Figure 21.6 A four-machine stability test system.

For illustration of system behavior, we assume a scheduled purchase of power from Area 3 with delivery to both Area 1 and Area 2. The condition of immediate interest is a purchase of 700 MW by Area 1 and 700 MW or more by Area 2. Note that all of the Area 2 purchase must flow through Area 1. A three-phase, four-cycle fault is applied near bus 8, as shown in Figure 21.6. The system behavior in response to this disturbance is of interest. It is assumed that no reclosing of the faulted line is attempted, since the system design criteria include a requirement that the system must be stable for any fault, even if reclosing should fail to operate. This results in a higher design standard than one that assumes correct reclosing under all conditions.

21.3.2 Effect of Power Transfer

The effect on clearing of the fault on line E is first examined for different levels of power transfer. The power transfer is assumed to be scheduled from Area 3, in the eastern part of the system, to Area 2 in the west. In all cases, 700 MW is transferred from Area 3 to Area 1 as a means of ensuring adequate spinning reserve to Area 1. Additional scheduled transfers are also in effect from Area 3 to Area 2, which increases the east to west line loadings.

The dynamic performance of the system is illustrated in Figures 21.7 through 21.9 for Generators 1, 2, and 3, respectively. Similar results would be obtained for the transmission line system variables. The fault on line E causes an acceleration of the shafts of all generators, with the angles of the shafts advancing with respect to the reference generator, which is taken

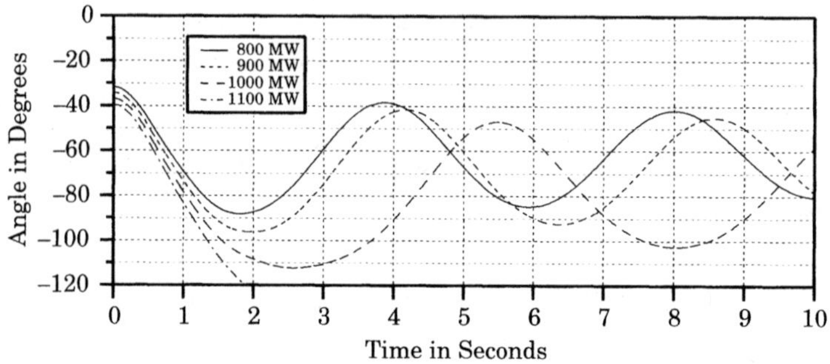

Figure 21.7 Generator 1 angle with respect to Generator 4.

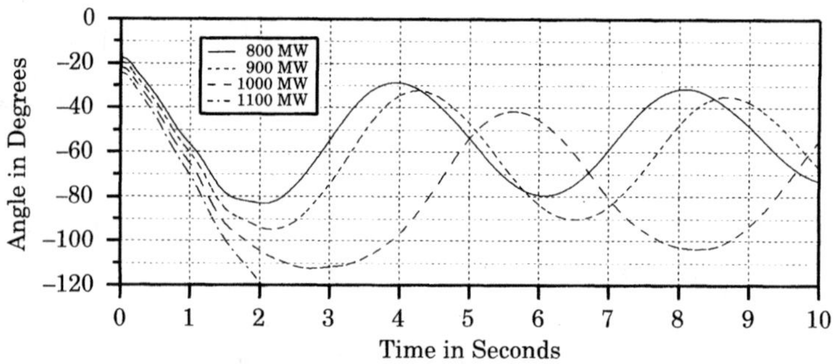

Figure 21.8 Generator 2 angle with respect to Generator 4.

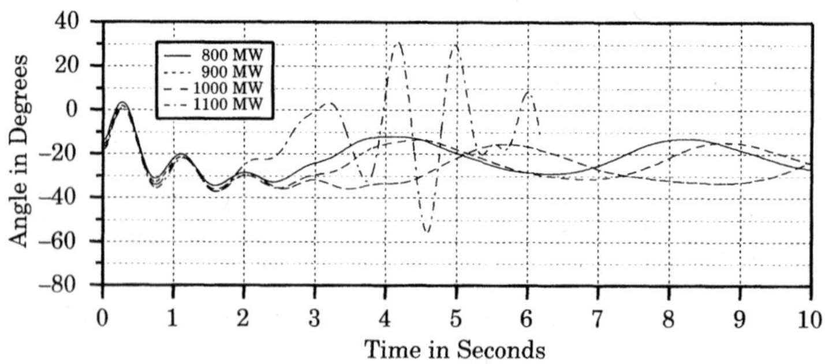

Figure 21.9 Generator 3 angle with respect to Generator 4.

to be the large inertia of Area 3, at Generator 4. Since power is flowing from east to west, the angles of all generators to the east of the reference machine are negative and accelerate to larger negative values following the fault. As the pre-fault power transfer level is increased, the angular separation increases and its rate of change also increases. Clearly, there is a limit to the amount of power that can be transferred through the system, and this limit is determined, in this case, by stability. The 1000 MW level shown in the figure is approaching the stability limit. The amount of angular change in the oscillations and in the total angle offset increases with power transfer. The power transfer shown in the figures is increased in 100 MW increments from 800 MW to a total of 1100 MW delivered from Area 3 to Area 2. Note that the first swing for the 1000 MW case is very large, indicating that this is close to the stability limit. Indeed, if the transfer is increased from 1000 to 1100 MW, instability occurs and the angle of Generator 1 grows monotonically in the negative direction. Note that the Generator 2 angle follows closely the behavior of the Generator 1 equivalent system. Generator 3, however, shows a different type of behavior, becoming highly oscillatory but not growing, at least until the end of simulation at just over 6 seconds. This generator will remain in step with the reference Generator 4 as the system separates.

The condition illustrated in Figures 21.7 to 21.9 represents a special type of instability called a "system separation." Note that none of the four generators goes out of step due to the disturbance, but the system tends to be split in two, with the break occurring between bus 7 and bus 8. The mechanism causing this split is not yet clear, and will be investigated further.

The problem with this system is that the power transfer requirement is too great for the one remaining transmission line between nodes 7 and 8. At the 1100 MW level, the total flow on the two lines from bus 8 to bus 7 is 3171.2 MW. This is well within the capability of only one line, as it represents just over 3.5 times the surge impedance loading of these lines. However, the change from two lines to one carrying that much power does not occur gradually, but suddenly, due to the fault. Moreover, carrying the larger power on one line requires that increased var support be supplied to hold the voltage, and we are not sure that this var support is available. We now investigate further what is happening on this line.

The active power flowing along the unfaulted transmission line from bus 8 to bus 7 is shown in Figure 21.10. Note that the power has difficulty rising above about 3500 MW, although this should be an adequate amount of power to satisfy system requirements. As the scheduled transfer is increased to 1100 MW, the actual transfer falls rather than increasing and the machines at the two ends of the system begin slipping poles. The power transfer across this line is effectively lost at this level of transfer. However, the reason for the collapse of power transfer is not evident from Figure 21.10.

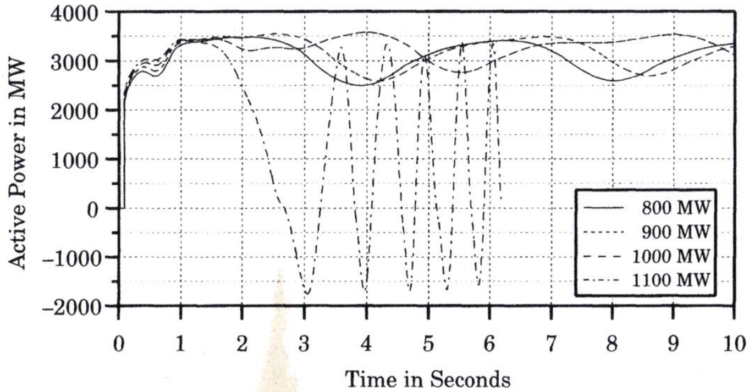

Figure 21.10 Active power flow from bus 8 to bus 7 in MW.

Figure 21.11 shows the reason for the loss of power transfer. The voltage at bus 7 collapses about 1 second after fault clearing when the scheduled transfer is 1100 MW. As the transfer is increased beyond 800 MW, the voltage falls to rather low levels, and at the 1100 MW level there is no hope of sustaining the required power transfer. The 1500 MVA static var controller at bus 7 is not capable of controlling the voltage with the larger transfers. This is evident to some degree in all levels of transfer in Figure 21.11, but becomes increasingly serious above the 900 MW level.

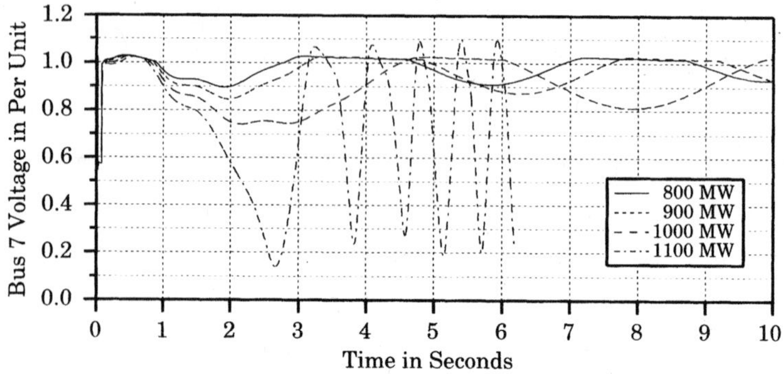

Figure 21.11 Bus 7 voltage in per unit.

Figure 21.12 shows the susceptance of the SVC at each level of scheduled transfer. The susceptance jumps to its maximum value immediately when the fault is applied and remains at this maximum output much of the time. When the voltage tends to rise above the desired value of 1.02 per unit, the susceptance is reduced to maintain the desired voltage. Much of the time the SVC is at a limiting value of 1.0 per unit (on a 1500 MVA base) and the amount of time at this limiting value increases as the scheduled transfer increases. Note that the dips in the susceptance in Figure 21.12 correlate with the leveling of the voltage in Figure 21.11. However, the voltage is not adequately controlled and collapses completely at the 1100 MW level. The SVC does not have an adequate rating to permit this level of transfer.

For normal conditions, with all lines in service, the reactive power requirement to hold the bus 7 voltage to 1.02 per unit is shown in Table 21.1. The SVC at bus 7 is rated at 1500 MW, which provides a rather generous rating above the normally required reactive power requirements under steady-state system conditions. However, when one of the transmission lines

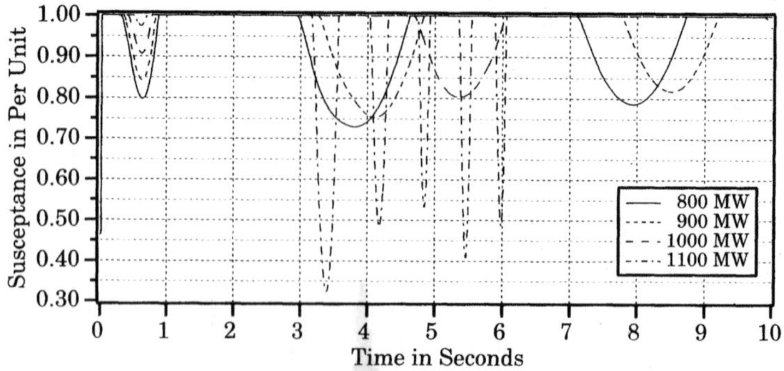

Figure 21.12 Susceptance of the bus 7 static var controller.

is lost under heavy transfer conditions, the SVC is not capable of maintaining the desired voltage.

TABLE 21.1 Steady-State SVC Reactive Power Requirements at Bus 7

Scheduled Power Transfer (MW)	800	900	1000	1100
Bus 7 SVC Reactive Power (MVAR)	721.9	735.9	787.5	843.2

As the scheduled transfer is increased above 800 MW, the SVC is required to operate at its maximum level more of the time. At the 1000 MW transfer level and above, the SVC is operating at maximum almost all of the time after the fault. Control of the voltage at these higher levels of transfer would require a much larger SVC, which may not be an economical solution. Increasing the degree of series compensation in the lines between bus 7 and bus 8 might be more economical and could be more effective. The cost of vars is lower for series capacitors than for SVCs.

21.3.3 Effect of Circuit Breaker Speed

The effect of circuit breaker speed can be examined by assuming that one or more of the circuit breakers on the faulted line from bus 7 to bus 8 are slow to operate. In the initial runs described in the previous section, a fault clearing time of four cycles is assumed. Suppose that the breaker operating time is increased to 12 or 14 cycles. This would represent a very long time for a 500 kV circuit breaker, but might be typical of a condition where the breaker suffers a mechanical malfunction.

Under this delayed clearing condition, the initial effect is greatest on Generator 3, since the fault is electrically on the high-voltage bus of this generator. Delaying the fault clearing effectively unloads this generator, as it is not able to deliver its generated power past the nominally zero voltage at the bus. This causes a high imbalance between the prime mover power output and the generator power, with this difference acting to accelerate the shaft. This is a very large power plant, with a rating of 4400 MVA. For example, this could represent five identical 880 MVA generators. The combined inertia of these generators is great, and they will not accelerate rapidly. However, if the fault is not cleared for several cycles, the combined plant will accelerate to the point where synchronism is lost and the generator will be tripped by its protective relays due to overspeed, overvoltage, overfrequency, overcurrent, or other protective means.

The effect of delayed clearing is shown in Figures 21.13 for generator angles and Figure 21.14 for generator angular velocity. Only Generators 1 and 3 are plotted since Generator 2 tends to follow Generator 1. With 12 cycles of breaker delay time, a system separation occurs and Generator 1 angle decreases monotonically. Generator 3 is stable and remains in synchronism with the Area 3 equivalent. At 14 cycles breaker delay time, Generator 3 angle advances to the point where this generator becomes unstable. The eastern and western parts of the system separate.

Until the fault is cleared, the transmission voltage at Generator 3 is zero, so this generator accelerates in a linear manner. At the greater breaker delay time, this acceleration has reached the point where generator stability is lost. This illustrates that breaker clearing time is an important parameter in maintaining stability. Circuit breakers are usually subjected to periodic maintenance to ensure their continued performance in rated operating time.

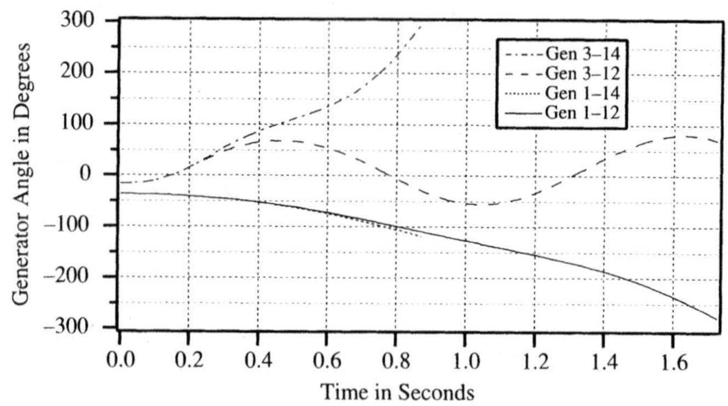

Figure 21.13 The effect of delayed clearing on generator angles.

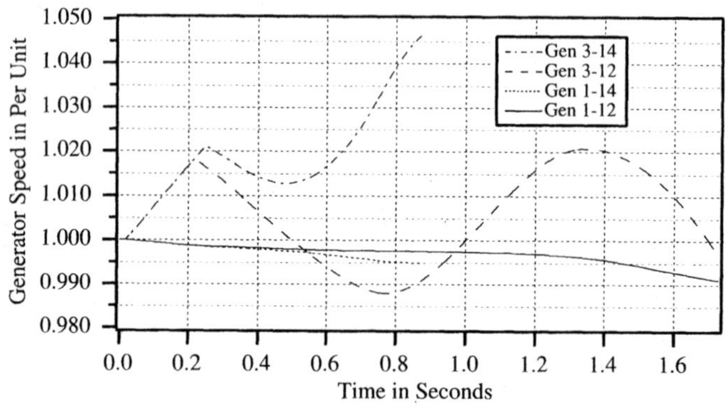

Figure 21.14 Effect of delayed clearing on generator angular velocity.

21.3.4 Effect of Reclosing

Fast reclosing following fault detection and clearing is a method sometimes used for improving stability. If there is a reasonable probability that the fault is temporary and can be removed by simply opening and immediately reclosing the faulted line, then automatic reclosing can be very effective as a means of improving transient stability. However, the price to be paid by unsuccessful reclosing must always be considered. In many systems that are stability limited, automatic reclosing following fault detection and clearing is not permitted, since the consequences of reclosing into a permanent fault may lead to widespread outages and possible blackout. Therefore this option must be studied very carefully before making a decision to employ automatic reclosing for stability enhancement. Automatic reclosing is discussed in greater detail in Section 21.4.

21.3.5 Relay Measurements During Transients

The measurement of transmission line performance under fault conditions, such as those described above, are of considerable interest to the protection engineer. The apparent impedance seen by the relay is important since so many types of transmission relays utilize some form of distance measurement. As an illustration of the apparent impedance seen by relays of the test system, the case with 800 MW of scheduled interchange is examined.

Figure 21.15 shows the apparent impedance for the relays at bus 7 on the unfaulted transmission line connecting bus 7 and bus 8. Just before the fault occurs, this bus 7 relay sees an apparent impedance in the second quadrant since power is flowing toward this relay. After the fault is applied, the apparent impedance is less than half the total line impedance. This line is series compensated at 60%, so we would expect the impedance to be 40% of the uncompensated line impedance. This would not be true of the faulted line as the high currents in that line would cause the series capacitors to be bypassed. There is little movement of the apparent impedance during the fault, in this case, but after the fault is cleared, the impedance jumps to a new location in the second quadrant, which is closer to the origin than the pre-fault point due to the increased load on the line. After fault clearing, the system continues to oscillate with each oscillation intercepting the outer relay characteristic and possibly also intercepting the inner mho characteristic. The mho relay characteristic illustrated has a diameter equal to 90% of the total line impedance, with the circle diameter inclined at a maximum torque angle of 75° from the real axis, noted in the figure as MTA75.

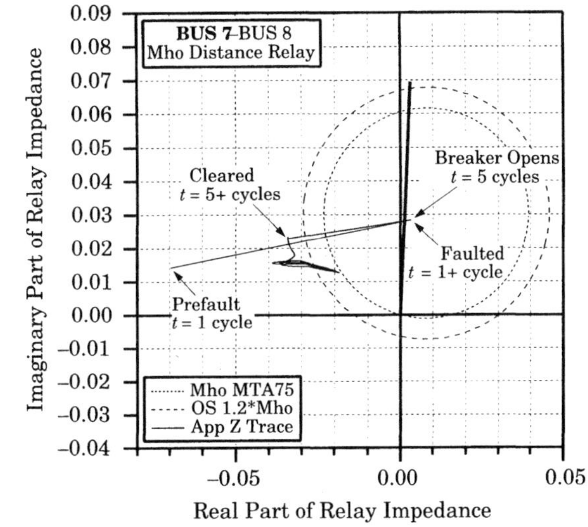

Figure 21.15 Apparent impedance, relay 7, line 7-8, 800 MW transfer.

The outer circle is a typical out-of-step relay characteristic, which is discussed in detail in Section 21.5. It is important to note that in moving from pre-fault to faulted, to post-fault conditions, the apparent impedance jumps to the inside of the mho characteristic with absolutely no time delay. This is very important as transit time provides the key to detecting out-of-step conditions.

The system oscillations that occur following fault clearing may last for several seconds. The simulation shown in Figure 21.15 shows only 10 seconds of time and therefore fails to show the exact final impedance.

Figure 21.16 shows the apparent impedance observed by the relay at the bus 8 end of line 7-8, also for the 800 MW level power transfer. Note that the initial impedance for this relay is in the first quadrant since the power flow is positive in the 8–7 direction. Since the fault is effectively at bus 8, this impedance is approximately at the origin of the Z plane. However, for a fault on bus 8 the fault is *behind* the bus 8 relay, so the apparent impedance is actually in the third or fourth quadrant and is clearly outside of the mho circle. The final oscillation of the apparent impedance, in this case, crosses both relay characteristics repeatedly, as the system adjusts to the new loading condition. Since this loading is very heavy, the final impedance might be very close to falling within the mho circle for this relay. If power transfers of this magnitude are to

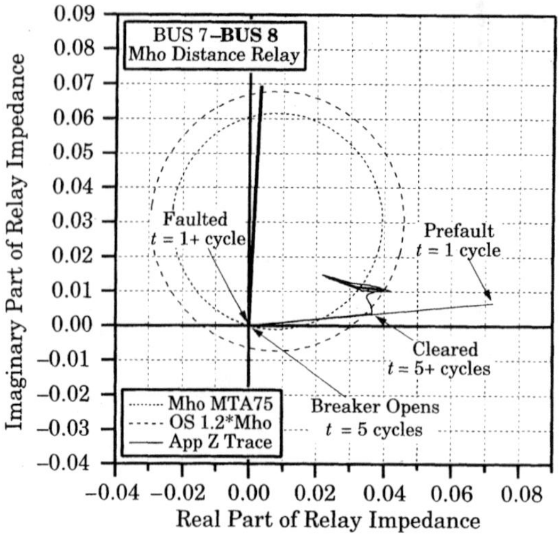

Figure 21.16 Apparent impedance, relay 8, line 7-8, 800 MW transfer.

be anticipated, a more restricted Z plane relay characteristic, or one using blinders, would be appropriate.

It is also interesting to determine the apparent impedance seen by line relays one bus away from the fault. The relays at bus 5 and bus 6 looking in the direction of bus 7 will both be examined, both for the 800 MW transfer level.

The relay apparent impedance on line 5–7 at bus 5 is shown in Figure 21.17 with the transfer level at 800 MW. The pre-fault impedance is in the second quadrant, well to the left of the origin. When the fault is applied, however, the impedance jumps into the mho circle, causing this relay to pick up. Note that after fault clearing, the apparent impedance locus, which is off scale in the figure, oscillates in the second and third quadrants. A small portion of this oscillation is noted on the left side of the figure. The relay at the bus 7 end of line 5–7 does not pick up for this fault. Therefore, there is no danger of tripping this line if we assume that some type of permissive relaying scheme is used, which would be typical of the protection on 500 kV transmission lines. The final relay of interest is at bus 6 on line 6–7, shown in Figure 21.18. This figure shows the entire relay locus from prefault to the end of the 10 second

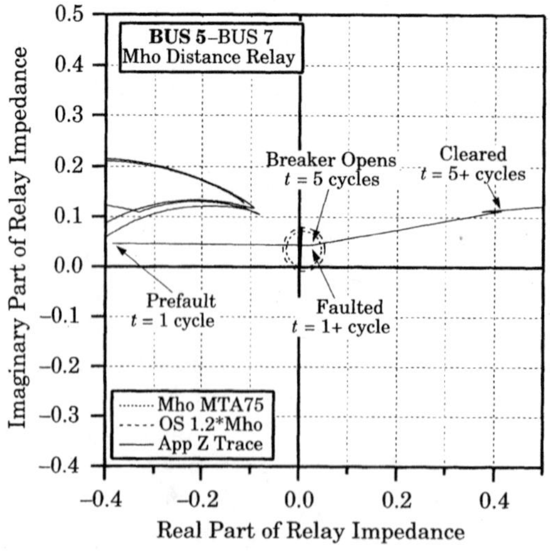

Figure 21.17 Apparent impedance, relay 5, line 5–7, 800 MW transfer.

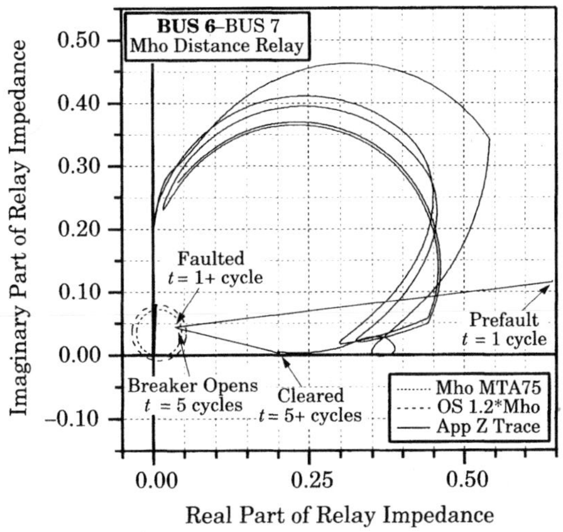

Figure 21.18 Apparent impedance, relay 6, line 6–7, 800 MW transfer.

simulation. This type of oscillatory behavior is not uncommon, and in some cases the loci may move large distances in the Z plane in a very short time. This bus 6 relay also picks up when the fault is applied because the mho relay overreaches due to the series compensation in the line. However, the relay at bus 7 of the line does not pick up, so the permissive scheme will prevent a false trip of the line.

The foregoing Z plane loci are for the 800 MW level, which is a stable operating condition. We now examine the 1100 MW transfer level, which results in an unstable condition following the fault on line E. First, we examine the apparent impedance seen by the relay at bus 7 on line D, which is shown in Figure 21.19. The impedance locus following fault clearing moves in circular excursions, with each oscillation passing through the combined relay characteristics for line D. Only five oscillations are shown in Figure 21.19, which is limited to the 10 seconds of the simulation. Some of these oscillations show the limited resolution of the simulation, which specified a data output increment of 0.002 seconds or 0.12 cycles on a 60 hertz basis. In this small step size, it is noted in the larger circles that the apparent impedance takes very large steps. It is interesting that the oscillations are almost perfectly circular in shape.

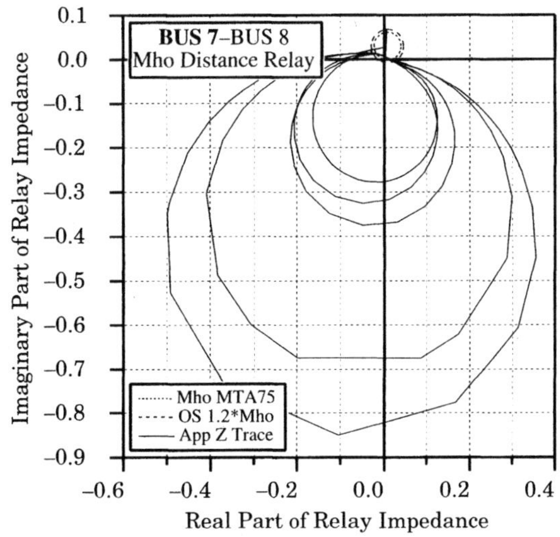

Figure 21.19 Apparent impedance, relay 7, line D, 1100 MW transfer.

The oscillatory behavior observed at the bus 8 relay on line D is similar to that shown in Figure 21.19, except that the circular characteristics extend both in the upper as well as the lower half plane. The apparent impedance excursions away from the origin are even greater for the bus 8 relay than those shown in Figure 21.19, and the movement around the circular path is faster. Otherwise, the two characteristics are similar in their circular shape and in passing through both of the relay characteristics. Thus, the relays at both ends of the transmission line will observe multiple passes through their characteristic circles. The question of relay pickup becomes one of the timing of the loci movements. This will be investigated for the first pass through the relay circles.

To determine the performance of the protection, we focus our attention on the relay characteristics. All of the impedance oscillations travel through both the mho and OS relay circles, as shown in Figure 21.19. Protection for out-of-step conditions is often dependent on timing, and on the fact that the apparent impedance requires a finite time to travel from the pre-fault to the faulted position in the Z plane. These time delays can be used to determine the type of system condition observed. The initial application and removal of a fault causes the impedance to jump into and out of the mho characteristic instantaneously. However, subsequent traverses of the mho and OS characteristics are slower, and these timings can be used to advantage.

A close-up view of the two relay characteristics is shown in Figures 21.20 and 21.21 for the relays at bus 7 and bus 8, respectively. First, Figure 21.20 shows that the direction of travel is clockwise and the times of crossing the relay characteristics are noted. For example, the locus crosses the OS circle at 0.801 seconds and the mho circle at 1.074, giving a total time of 0.273 seconds between characteristics. The total time inside the mho characteristic is 2.850 – 1.074 or 1.776 seconds. Finally, the time to cross the OS circle is 2.984 – 2.850 or 0.134 seconds.

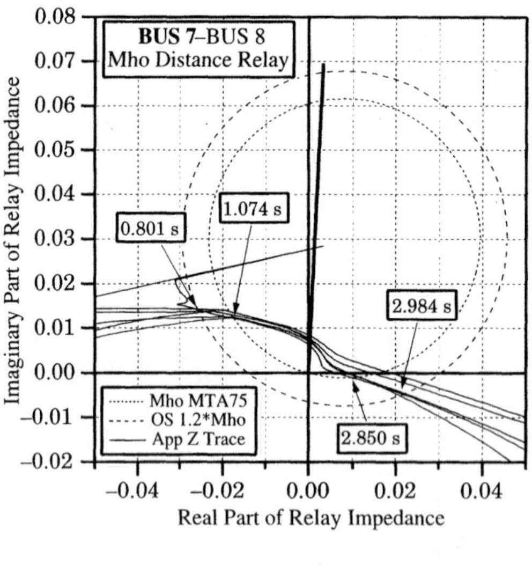

Figure 21.20 Apparent impedance, relay 7, line D, 1100 MW transfer.

For the relay at bus 8, shown in Figure 21.21, the post-fault position is between the two relay characteristics. The movement is counterclockwise. The time spent within the mho characteristic is 3.012 – 0.661 or 2.351 seconds and the time between characteristics is 3.042 – 3.012 or 0.03 seconds, which is only 1.8 cycles. These noted times are for the first

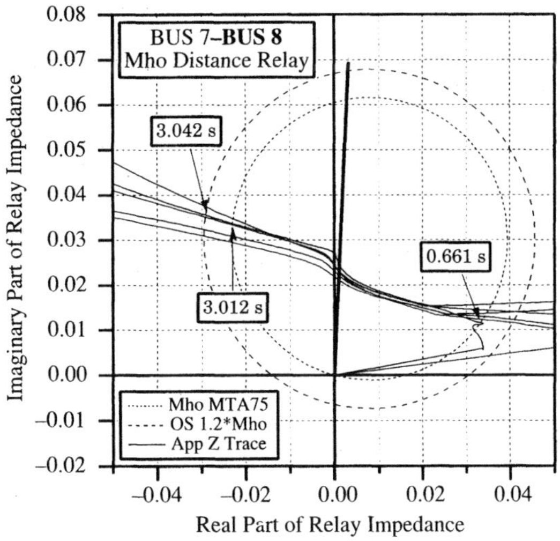

Figure 21.21 Apparent impedance, relay 8, line D, 1100 MW transfer.

pass through the relay characteristics. Other passes are to follow, so the relays have several opportunities to evaluate the need for tripping. Since this is a true out-of-step condition, the relay logic should be adjusted to ensure tripping.

Protective relays for out-of-step detection use of the time delays observed in the apparent impedance locus. Every passage past a relay characteristic circle causes that relay to either pick up or reset. Measuring the time between relay pickup of the two types of relays, and the time between pickup and reset of each relay, can provide information that is useful in determining the type of system condition being observed. This subject is explored in greater detail in Section 21.5.

Impedance plots such as those illustrated can be readily drawn using the data computed in stability simulations. These plots help the protection engineer visualize the transient system behavior from any given relay location.

21.4 AUTOMATIC RECLOSING

Automatic reclosing, or autoreclosing, is a control scheme for quickly reclosing breakers after clearing a fault in order to restore the system to normal as quickly as possible. Sufficient outage time must be allowed for the fault path to deionize if the scheme is to succeed. This usually requires at least 10–30 cycles on a 60 hertz system, with the deionizing time being highly dependent on the line voltage. See Section 21.4.3.4 for a discussion of deionization time.

Automatic reclosing of circuit breakers, after a time delay to account for deionization of the arc, can be a very effective method of preserving stability. This can be illustrated by considering the case introduced in Section 21.3.2, where 1100 MW is scheduled to be delivered from Area 3 to Area 2, with an additional 700 MW delivered to Area 1. For a fault at bus 8 on one of the lines from bus 8 to bus 7, the simulation plots of Section 21.3.2 show that the system is unstable. Let us reconsider this same fault condition, but with reclosing of the line permitted on the assumption that the fault is temporary and, once reclosed, the system will be restored to its normal condition.

The results of successful reclosing are shown in Figure 21.22. where it is clear that the system with successful reclosing is stable and reasonably well damped. This illustrates the benefit of automatic reclosing following fault clearing. In this case, reclosing was completed 30 cycles after fault clearing. The scheduled transfer is 1100 MW.

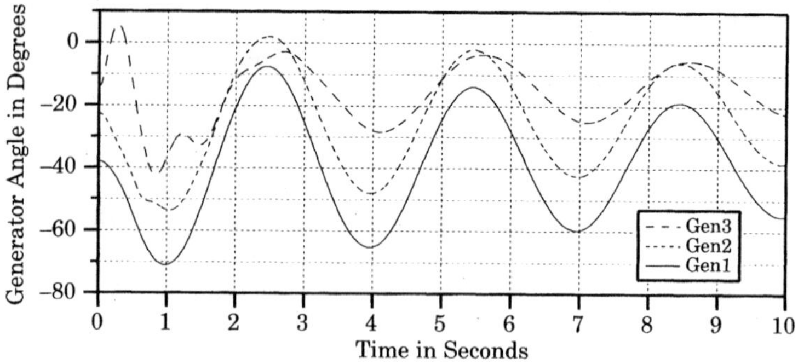

Figure 21.22 The effect of successful reclosing of the faulted line E.

Unfortunately, there is no way to guarantee that reclosing will be successful, even though statistics show that a high percentage of faults are temporary and are successfully cleared by opening the line and then reclosing after a time delay for deionization of the arcing fault.

To illustrate the potential hazard of automatic reclosing, the same system condition is simulated with the scheduled transfer reduced from 1100 MW to only 800 MW, and the fault is now a permanent three-phase fault. The results are shown in Figure 21.23, where it is evident that Generators 1 and 2 are going to separate from the system. How the system splits into islands depends on the response of many transmission line relays. Load is likely to be lost and possible widespread blackout is possible. This is called an unsuccessful reclosing. The fault is applied not once, but twice, which gives the system a much more severe disturbance to overcome. Finally, the faulted line is lost. The result is a system separation.

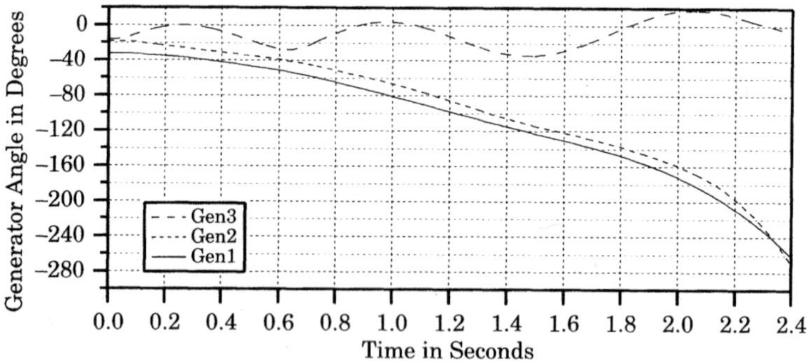

Figure 21.23 The effect of reclosing into a permanent fault with 800 MW of scheduled power transfer.

Additional studies of this same system show that, with only 700 MW of scheduled transfer, the system remains connected but suffers large voltage dips that might be considered unacceptable. However, even if the voltage dips are allowed, the scheduled transfer limit must be reduced from 1100 MW to 700 MW if unsuccessful reclosing is to be tolerated. However, if reclosing is not used, up to 1000 MW can be scheduled and stability maintained.

21.4.1 The Need for Fast Reclosing

Power system stability requires the synchronous operation of all the generating units in the power system. The units do not need to have the same phase angle, but they must have the same average speed. Units tend to oscillate following a disturbance, and sometimes these oscillations are poorly damped. A stable system is one in which oscillations are positively damped, such that the unit oscillations settle in a relatively short time to a new steady-state operating condition in which all units have the same speed or frequency. In order to maintain stability following a severe disturbance, it is necessary for the generators to have the capability of exchanging "synchronizing power" through the transmission system. When a line or other network branch is faulted and removed, it is necessary for the system to seek a new operating condition, with new voltage phase angles at all generators. The generating units move toward these new phase angles by adjusting their generated power, with the movement of each individual unit being dependent initially on a quantity called the *synchronizing power coefficient*, a parameter that determines how much the output power of the unit changes as the voltage angle changes. In order to make the required readjustment, the generators must be able to change their power output to achieve the new required balance. After an initial angle adjustment due to the synchronizing power coefficient, the speed governors take over and these devices determine the final outcome of the power readjustment [3]. The rescheduling of generated power at the various units is performed by the energy control center by ordering small changes in the governor settings, but this is a relatively slow process.

Some faults, even permanent faults with sustained line outages, do not cause system instability. If the line that is tripped is not essential to maintaining an important path for the flow of synchronizing power among the machines, then its outage following a fault will probably not cause instability. Adding autoreclosing to the protective relays of these circuits may be justified simply to restore the system to normal following temporary faults without unnecessary delay, but reclosing is not needed for stability.

Important lines, especially tie lines that connect important generating stations, often *require* autoreclosing in order to maintain system stability for a given desired operating condition. This means that one or more generators will become unstable unless the system is restored to normal in a short time, often just a few cycles. Since 90% or more of all line faults are temporary in nature, this means that autoreclosing will usually be successful in restoring these essential circuits. As a general rule, autoreclosing is used on many high-voltage lines of the bulk power system. This practice subjects the system to the occasional double exposure to a permanent fault, but this occurs so seldom that it may be a risk worth taking, particularly if autoreclosing is essential to maintaining stability on the occurrence of temporary faults. However, even if autoreclosing is used on all lines, the system design criteria may require that the system be stable when confronted by reclosing into a fault. System design criteria may be different from one utility to another, depending on the system structure and operating experience. In North America, regional reliability councils reach agreement on general reliability criteria that should be employed, taking into account the general structure and sensitivity of the interconnected system.

21.4.2 Disturbance Considerations in Reclosing

The disturbance considerations that enter into the design and installation of auto-reclosing deal with the voltage levels of the application and the types of disturbances that are likely at each voltage level.

21.4.2.1 Voltage Levels. At the lower voltage levels that are typical of distribution circuits, say up to about 34.5 kV, most of the circuits are radial and serve a limited number of users. Autoreclosing at this level is useful in order to limit the outage time of the customers. It may also be helpful in coordinating line switching with lateral circuit fusing, as noted in Chapter 6. There is no consideration for stability at this level, since these circuits either have no generation or very small generators that may be customer-owned units. If reclosing is used in this case, it may have to be delayed to give these small generating units time to switch off prior to reclosing. These units may become unstable when the system is faulted, but this is not likely to be of great concern to the rest of the interconnected system.

At the subtransmission level, autoreclosing may be very helpful in keeping the system intact without undue delay. Practices vary regarding autoreclosing at this voltage level. Some utilities use autoreclosing on all lines and other utilities use autoreclosing only on lines that are considered tie lines, or are in parallel with higher voltage tie lines. There may be a problem in applying autoreclosing at the subtransmission voltage level since much of the line relaying is distance relaying, which permits time-delayed clearing in zone 2 and fast clearing in zone 1. This means that 75 or 80% of the faults are cleared almost immediately, but end-of-line faults are tripped sequentially. This does not lend itself to accurate autoreclosing, since the total dead time would be different for the two ends, depending on the fault location.

At the EHV transmission level, autoreclosing is often used. At this voltage level, nearly all of the primary line relaying is by pilot schemes, in which all line faults are tripped in a few cycles. In this case, a timed reclosure works very well and is successful a high percentage of the time.

The design criteria used for power systems is based on judgment and on the probability of system disturbances. Power systems that cover large geographical areas often have long transmission lines connecting load centers to remote generation facilities. Such systems tend to be stability limited. If the lines are restricted to small power transfers, stability might be ensured. But if larger transfers are required, stability following a disturbance, such as a fault, may be at risk. Engineers test the system for stability using computer stability simulations of faults throughout the system. The three-phase fault with small fault impedance is the most severe disturbance, since no power can be transmitted through a zero-impedance, three-phase fault. This provides a severe test of the system. If the system can survive a three-phase fault, applied to the system for a specified duration, such as three to five cycles, this provides confidence that the system will survive most naturally occurring disturbances, which are likely to be of lesser severity. Sometimes a line-to-ground fault, with delayed clearing, is also checked.

Some engineers argue that the system design criteria should not assume post-fault line reclosing. This is common for systems that are transient stability limited. The logic of this argument is that the line rated capacity would have to be reduced to the point where the system could survive reclosing into a permanent fault. A higher line rating can often be used if post-fault reclosing is not employed.

The design criteria used by a given utility is a function of the strength of the interconnected power system, which may involve many different utilities, and the relative dependence of the system security on the transmission lines. In North America, regional reliability councils recommend design criteria that are adequate for most system conditions. Individual utilities, however, may have somewhat different criteria to account for special conditions unique to their system topology, operating conditions, and protection requirements.

21.4.2.2 Fault Types. The type of fault encountered on the system also enters into the application of autoreclosing. One way to classify the fault types is as follows [3]:

Transient faults are faults, such as insulator flashovers, that are quickly cleared by switching the line and do not recur if the line is quickly reclosed. The most common cause for this type of fault is lightning, but transient faults can occur due to swinging wires or temporary contact with foreign objects, such as trees or blowing debris.

Semipermanent faults are faults that might clear themselves if left to burn for a short time. Examples are contacts with tree branches that sometimes occur during stormy weather. If the contacting branch is burned away, the fault may clear itself.

Permanent faults are faults that do not clear themselves, but that must be repaired, such as a broken conductor. Underground cable faults are almost always permanent faults. For example, damage to the cable due to intrusion by a drag line or other excavation machinery creates a permanent fault.

Most of the faults on EHV transmission lines are transient faults. At this level, tree clearance is usually maintained with considerable care and semipermanent faults are not common. Permanent faults can occur, but are rare.

Subtransmission lines experience more semipermanent faults, but also have a high percentage of transient faults. The application of autoreclosing at this voltage level may depend on factors other than fault type. Distribution lines have more tree contact than the higher voltages and will experience more semipermanent faults. As noted above, however, stability is not usually an issue for distribution lines, but reclosing may be employed for improved service to the loads served.

21.4.3 Reclosing Considerations

There are several items that require consideration that relate to the reclosing operations themselves.

21.4.3.1 Number of Reclosures. At the EHV level the first reclosure is almost always successful and single reclosures are easily justified. Multiple reclosures, however, are not warranted at these voltage levels and are seldom used.

At the subtransmission voltage level, practices vary. It is not unusual to find two or three reclosures on radial lines, but only one or two on tie lines [3]. If the reclosures are delayed, it may be necessary to check synchronism prior to reclosing.

At the distribution level, multiple reclosures are almost always warranted and a high percentage have been shown to be beneficial [3].

21.4.3.2 Reclosing Success. The success of reclosing depends a great deal on the speed of tripping. Fast clearing of the fault ensures less damage to lines and equipment, and also limits the ionization of the fault path. Fast fault clearing also improves system stability and limits the shock to the system. Therefore, breaker speed is a very important factor in the success of stability. Circuits with very high speed relays and circuit breakers will have a higher probability of successful reclosures. This is usually the case at the highest transmission levels, where breakers are usually fast and the relaying is high speed pilot relaying. This gives the arcing fault little time to become well established and improves the probability of successful deionization and reclosing.

21.4.3.3 Definitions. Following are some definitions of the terms used to describe autoreclosing. These definitions are taken from [4], [5], and [6], unless otherwise indicated.

Antipump (pump free) Device (power switchgear). A device that prevents reclosing after an opening operation as long as the device initiating closing is maintained in the closing position [6].

Antipumping. A feature incorporated in the circuit breaker or reclosing scheme whereby, in the event of a permanent fault, repeated operations of the circuit breaker are prevented when the closing impulse lasts longer than the sum of the protective relay and circuit breaker operating times [4].

Arcing time The time interval between the instant of the first initiation of the arc and the instant of final arc extinction in all poles. For switching devices that use switching resistors, a distinction must be made between the arcing time up to the instant of the extinction of the main arc, and the arcing time up to the instant of the breaking of the resistor current [6]. A similar distinction must be made between the time instant of separation of the circuit breaker contacts and the instant of extinction of the fault arc [4].

Closing impulse time. The time during which the closing contacts of the autoreclose relay are made [4].

Closing time. The interval of time between the initiation of the closing operation and the instant when metallic continuity is established in all poles. Notes:

1. Closing time includes the operating time of any auxiliary equipment necessary to close the switching device, and that form an integral part of the switching device.
2. For switching devices that embody switching resistors, a distinction should be made between the closing time up to the instant of establishing a circuit at the secondary arcing contacts, and the closing time up to the establishment of a circuit at the main or primary arcing contacts, or both [6]. Also, the time for the energizing of the circuit breaker closing circuit to the making of the circuit breaker contacts [4].

Counting relay. A relay, often of the electromagnetic type, with a ratchet mechanism that is driven forward one step each time its coil is energized. A contact is operated after a chosen number of steps and the mechanism may be manually or electrically reset [4].

Dead time (autoreclose relay). The time between the autoreclose scheme being energized and the operation of the contacts which energize the circuit breaker closing circuit. On all but instantaneous or very high speed reclosing schemes, this time is virtually the same as the circuit breaker dead time [4].

Dead time (of a circuit breaker on a reclosing operation). The interval between the interruption in all poles on the opening stroke and re-establishment of the circuit on the reclosing stroke. Notes:

(A) In breakers using arc-shunting resistors, the following intervals are recognized and the one referred to should be stated:

1. Dead time from interruption on the primary arcing contacts to re-establishment through the primary arcing contacts.
2. Dead time from interruption on the primary arcing contacts to re-establishment through the secondary arcing contacts.
3. Dead time from interruption on the secondary arcing contacts to re-establishment through the primary arcing contacts.
4. Dead time from interruption on the secondary arcing contacts to re-establishment through the secondary arcing contacts.

(B) The dead time of an arcing fault on a reclosing operation is not necessarily the same as the dead time of the circuit breakers involved, since the dead time of the fault is the interval during which the faulted conductor is de-energized from all terminals [6].

Deionizing time. The time following the extinction of an overhead line fault arc necessary to ensure dispersion of ionized air so that the arc will not re-strike when the line is re-energized [4]. See Section 21.4.3.4.

High-speed reclosing scheme. A scheme whereby a circuit breaker is automatically reclosed within one second after a fault trip operation [4].

Lockout. A feature of an auto-reclose scheme which, after tripping of the circuit breaker, prevents further automatic reclosing [4]. Also, an opening operation followed by the number of closing and opening operations that the mechanism will permit before locking the contacts in the open position (for an automatic circuit recloser) [6].

Low speed or delayed reclosing scheme. A scheme whereby the automatic reclosing of a circuit breaker following a fault trip operation is delayed for a time in excess of one second [4].

Multishot reclosing. An operating sequence providing more than one reclosing operation on a given fault before lock-out of the circuit breaker occurs [4].

Operation counter. A counter, usually of the electromagnetic cyclometer type, arranged to indicate the number of automatic operations, either closing or tripping, performed by a circuit breaker since its commissioning [4].

Opening time (of a mechanical switching device). The time interval between the time when the actuating quantity of the release circuit reaches the operating value, and the instant when the primary arcing contacts have parted. Any time delay device forming an integral part of the switching device is adjusted to its minimum setting or, if possible, is cut out entirely for the determination of opening time. Note: The opening time includes the operating time of an auxiliary relay in the release circuit when such a relay is required and supplied as part of the switching device [6]. Also, the time between the energizing of the circuit breaker trip coil and the instant of separation of the contacts [4].

Operating time (circuit breaker). The time from the energizing of the trip coil until the fault arc is extinguished [4].

Operating time (protection). The time from the inception of the fault to the closing of the tripping contacts. Where a separate tripping relay is used, its operating time is included [4]

Reclaim time. The time following a successful closing operation, measured from the instant the auto-reclose relay closing contacts make, which must elapse before the auto-reclose relay will initiate a reclosing sequence in the event of a further fault incident [4].

System disturbance time. The time between the inception of the fault and the circuit breaker contacts making on successful reclosing [4].

Spring windup time. On motor-operated spring-closed breakers, this is the time required for the motor to charge the spring fully after a closing operation [4].

Single shot reclosing. An operating sequence providing only one reclosing operation, lock-out of the circuit breaker occurring on subsequent tripping [4].

Trip-free (release free) as applied to a mechanical switching device. A descriptive term that indicates that the opening operation can prevail over the closing operation during specified parts of the closing operation [6].

Many of the above definitions are shown in Figures 21.25 and 21.26, which show the operating sequence of a single-shot autoreclosing scheme for transient and permanent faults, respectively. Some of the foregoing definitions are also discussed in Chapter 3.

21.4.3.4 Arc Deionization. An important aspect of reclosing has to do with the arc deionization and the time required for safe reclosure of the circuit breaker. Unfortunately, arc deionization is not a constant, but varies with voltage and other factors. Most references on the subject define a quantity called the "minimum dead time," which is the most optimistic estimate of the time the circuit should be dead prior to attempting a reclosure. It is imperative that enough time be allowed for the arc to extinguish and the arc path cool sufficiently for the reclosure to be successful.

Figure 21.24 shows estimates of the minimum dead time as given by different sources [3–5]. The simplest formula is the linear equation [3]

$$T_{\min} = 10.5 + \frac{V}{34.5} \quad \text{cycles} \tag{21.44}$$

where V is the voltage in kV. Clearly, this time becomes more and more important at the higher voltages as it extends the total operating time.

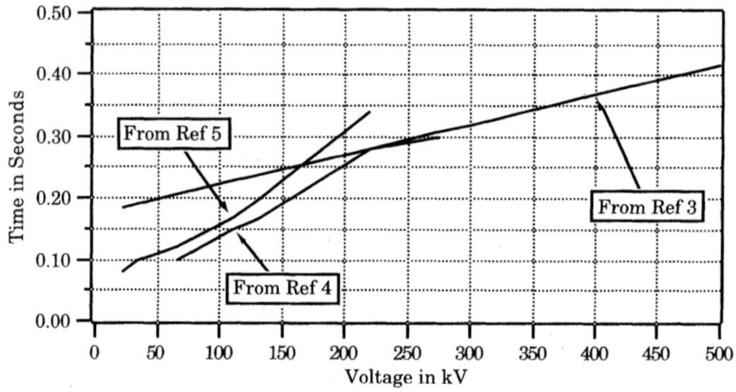

Figure 21.24 Estimates of arc deionization time [3–5].

Figures 21.25 and 21.26 give timing diagrams for a single-shot reclosing scheme with transient and permanent faults, respectively [3].

Reference 3 provides data regarding the probability of successful reclosure in a given time, as shown in Table 21.2. The values in the table are interpreted as follows. At 115 kV, for example, the probability of a transient fault being successfully cleared in 8.5 cycles is 95%, but the probability of successful clearing in only 6 cycles is only 75%.

TABLE 21.2 Probability of Successful Transient Fault Clearing at Specified Minimum Dead Times, in cycles on a 60 hertz Basis [3]

Rated Line Voltage in kV	Minimum Dead Time 95% Probability	Minimum Dead Time 75% Probability
23	4.0	—
46	5.0	3.5
69	6.0	4.0
115	8.5	6.0
138	10.0	7.5
161	13.0	10.0
230	18.0	14.0

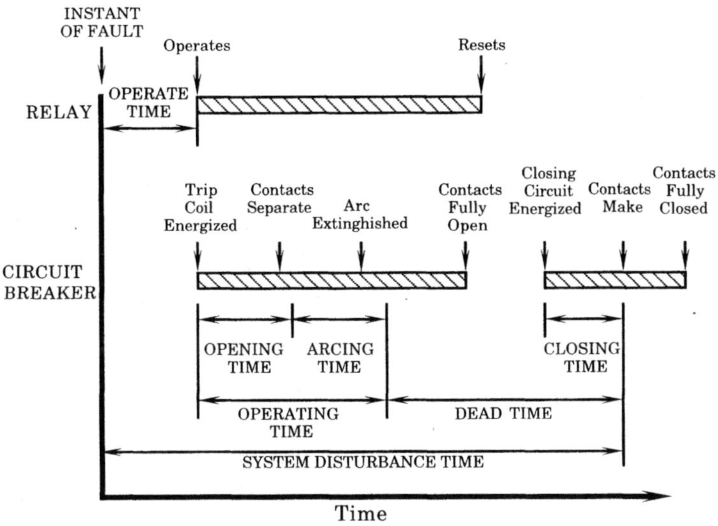

Figure 21.25 Single-shot autoreclose operation for transient faults.

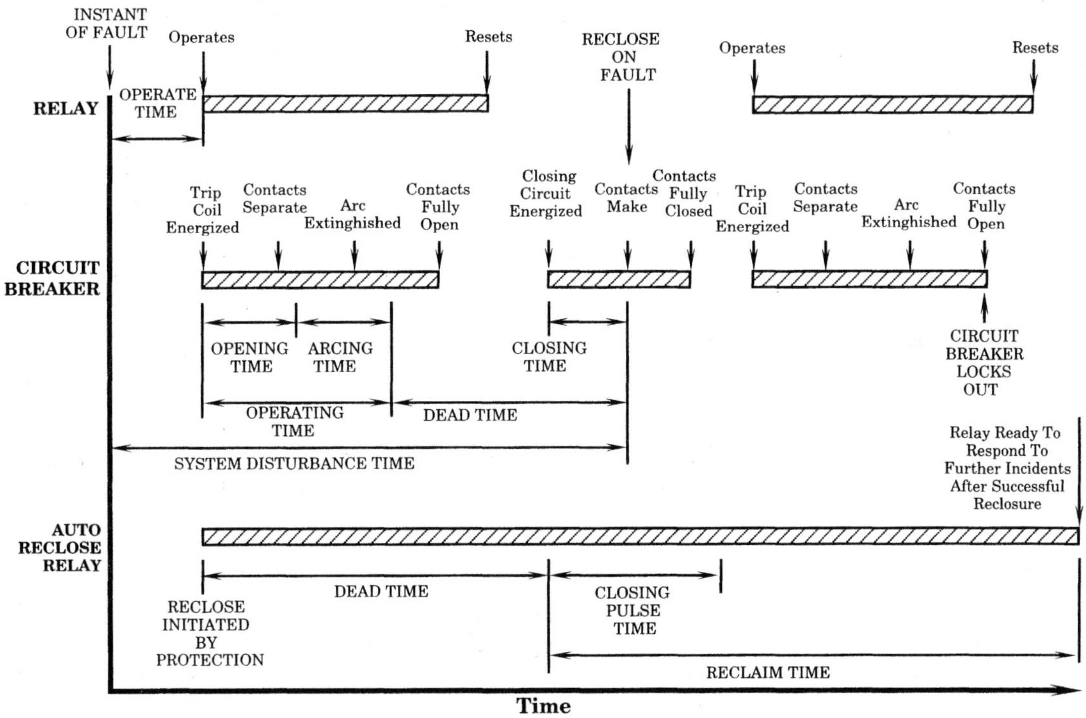

Figure 21.26 Operation of a single-shot autoreclose scheme for a permanent fault.

Comparing this table with Figure 21.24, the 95% probability values align almost exactly with the lowest curve in the figure [4]. This would suggest that the estimates of dead time that are longer than those represented by the lower curve are very conservative, and have a very high probability of successful realization.

21.4.4 Reclosing Relays

There are several types of relays that should be included under the subject of reclosing. These include single-shot and multi-shot reclosing relays, synchro-check relays, and the different types of line relays.

21.4.4.1 Breaker Operation. Before discussing reclosing relays, we first consider the mechanical operation of a circuit breaker and the sequence of operation for the main and auxiliary contacts. A typical time sequence of events in the reclosing cycle is shown in Figure 21.27. The auxiliary contacts 52a and 52b are both actuated directly by the main breaker contact travel or by the operating mechanism of the main contacts. Auxiliary contact operation occurs within a small bandwidth, shown by the hatched regions in the figure, with these regions represent the uncertainty of the exact time of opening or closing of the contacts. The same is true of the arc interruption by the main contacts, which depends on several factors as discussed previously. A similar uncertainty governs restrike of the arc, should this occur. The important time in the reclosing cycle is the relative time of operation of the "a" and "b" contacts. Usually, the timing of these contacts is not overlapped, but separated by a small time interval, as shown in Figure 21.27.

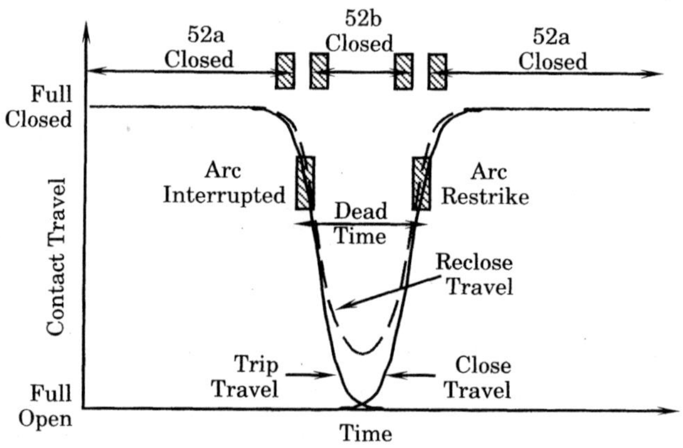

Figure 21.27 Typical circuit breaker reclosing cycle [3].

On EHV systems with high levels of fault current, the reclosing duty on the circuit breaker can be very severe. By its very nature, reclosing requires that the breaker be able to interrupt a fault not once, but twice, and with very little time between the two interruptions. Therefore, the breaker must be designed for this longer, and more severe, sequence of operations. Most of the circuit breakers at the EHV level are oil breakers or air breakers, and these types have significantly different interruption performance.

Oil Circuit Breakers. Oil circuit breakers are commonly used for voltages up to about 300 kV. There are two types of oil breakers; one roughly classified as "bulk oil" and the other as "small oil volume," where the second type has been designed to limit the amount of oil in order to reduce the fire hazard.

The operating mechanism is almost always the "trip free" design, which means that a collapsible link is arranged so that the breaker can be tripped even when the normal closing action is initiated. The reclosing cycle of these breakers must allow sufficient time for the collapsible links to reset after the first trip before applying the closing command. Various ingenious devices have been invented to limit this reset time. The mechanisms are usually solenoid or pneumatically operated, with the latter being more common at the higher voltages. Typical operating times for pneumatically operated circuit breakers of the 138 kV class [4] are shown in Figure 21.28(*a*).

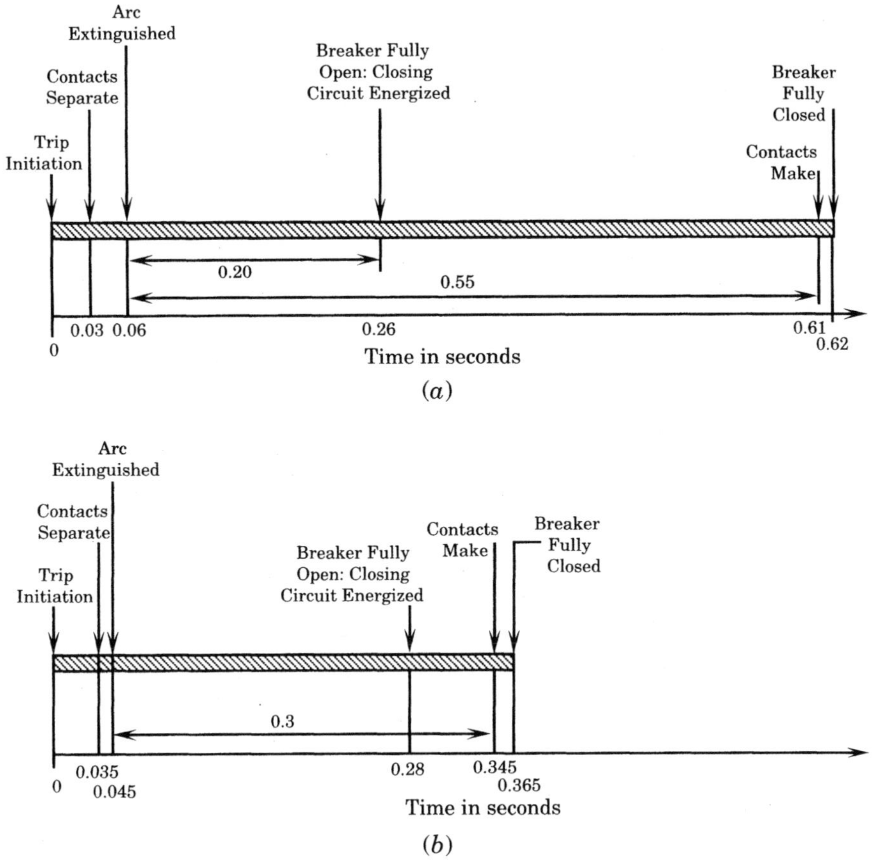

Figure 21.28 Typical trip close operation for circuit breakers [4]. (*a*) Pneumatically operated bulk oil circuit breaker. (*b*) 400 kV air blast circuit breaker.

AIR BLAST CIRCUIT BREAKERS. Air blast breakers were developed for the higher transmission voltages. These breakers are of two types: pressurized head breakers and nonpressurized head breakers [4].

Pressurized head breakers maintain compressed air in a chamber surrounding the main contacts. The trip signal also triggers an auxiliary compressed air supply that blasts a stream of air past the parting contacts to the atmosphere. After the arc is cleared, compressed air is again allowed to enter the chamber. If air pressure is lost, there is the possibility of the arc restriking. For this reason, a series of isolators are arranged to isolate the main contacts after tripping. These isolators must be inhibited when reclosing is used.

Nonpressurized head air blast breakers have air at atmospheric pressure surrounding the main contacts. When a trip signal is given, high pressure air from a compressed air supply is introduced from a separate tank. This slows the action somewhat, and this type of breaker is being phased out. Typical operating times for air blast breakers are shown in Figure 21.28(*b*).

Comparing the air blast breaker operating time with the deionization time at the higher transmission voltages shows that the deionization time tends to predominate the dead time for reclosing.

21.4.4.2 Single-Shot Reclosing Relays. It was noted previously that single-shot reclosing is often used in the higher voltage transmission applications since the probability of success is very high on EHV lines. Single-shot reclosing relays may be either of electromechanical or static design and the control of the reclosing operation is different for the two different types. Figure 21.29 shows a typical electromechanical control for a reclosing relay. The operation is described as follows [3]:

> When the breaker is tripped by protective relays, 101SC will be closed so that as 52bb closes, reclosing is initiated. This energizes the 52X coil through the pre-closed contacts 79X, 101SC, 43, 52LC, 52LPC, and 52Y. The 52X make contact energizes the 52CC breaker close coil to close the breaker. At the same time, another 52X make contact "seals" the 52X coil, and a third 52X make contact energizes 79X-O to toggle the 79X latching unit. This opens the 79X contact in the closing circuit and closes the 79X contact in the reset motor circuit, closing the breaker and the 52a contact energizes 52Y to de-energize 52X. The 52X/52Y scheme prevents pumping when the system is closed manually into a permanent fault.
>
> If the breaker remains closed, 52a stays closed and the motor is energized. After a preset interval, timer motor contact 79M energizes the 79X-R reset coil to toggle the 79 unit back and restore the reclosing relay to its normal condition, ready for a subsequent breaker trip.
>
> If the breaker does not remain closed, 52a re-opens to de-energize the motor 79M, and the relay is locked out. The relay remains in this state and can produce no further reclosing action until the breaker is closed manually and the motor times out to reset 79X.

Other schemes may be used for electromechanical reclosing relays, but the one shown in Figure 21.29 is typical.

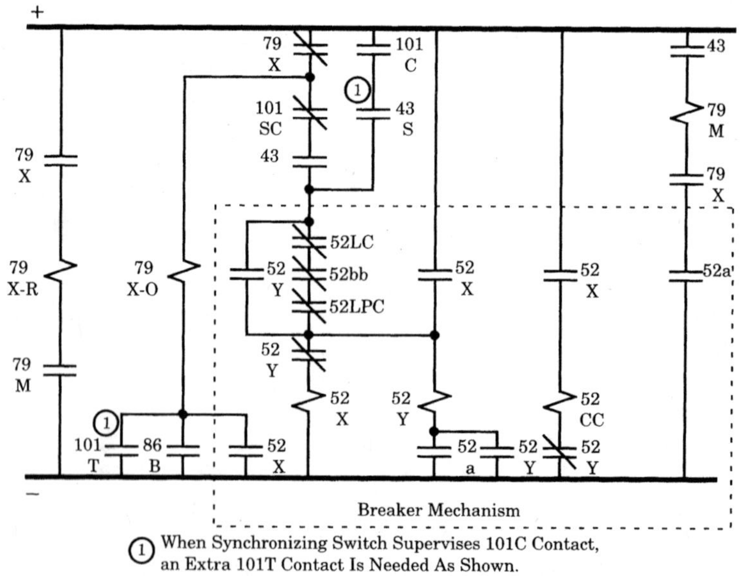

Figure 21.29 External schematic for an electromechanical single-shot reclosing relay [3].

Many of the current generation of reclosing relays are of static, solid-state, or digital design. These designs incorporate the breaker control by means of solid-state device logic. An example, also from [3], is shown in Figure 21.30.

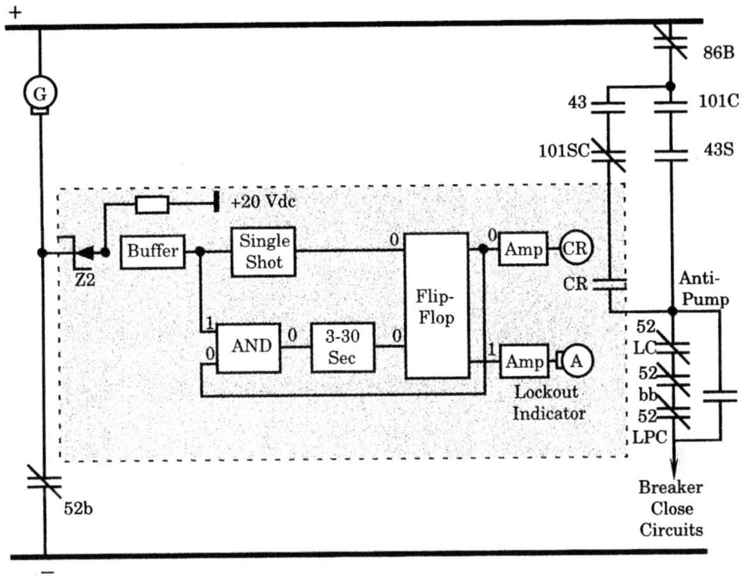

Figure 21.30 Logic diagram for a typical static reclosing relay [3].

The description of the devices numbers is given in Table 21.3. The relay is functionally identical to the electromechanical device shown in Figure 21.29. The static relay logic is described as follows [3]:

> The single-shot function provides an output during the closing stroke of the breaker, similar to the action of a 52X contact. It has a short output immediately following a "1" output. It then reverts to a "0", regardless of whether the input "1" is short or continuous. Thus, with the relay reset and the breaker closed, the 52b switch is open, and a continuous "1" exists at the single-shot input with a steady-state "0" output. For this condition, the flip-flop outputs are as shown. With the negated input to the upper amplifier, relay CR is continuously energized, providing a closed contact CR in the breaker close circuit.
>
> As the breaker is tripped by the protective relays, the only open contact in this breaker close circuit, 52bb, closes. Shortly after, 52b closes to produce a "0" input to the single shot. Then, as the breaker recloses, 52b opens, putting a "1" on the single shot input. A short "1" output follows to operate the flip-flop. The upper output of the flip-flop changes to a "1", putting a "1" on the negated amplifier input and de-energizing the CR relay. This opens contact CR in the close circuit. The lower output of the flip-flop changes to a "0", operating the amber lamp to indicate a lockout.
>
> If the breaker stays closed (52b open), the two "1" inputs to AND permit an output to reset the timer. If this condition continues for the reset interval (adjustable from 3 to 30 s), the lower input to the flip-flop is energized. This resets the flip-flop, turns off the amber light, and closes CR ready for the next automatic reclose operation.
>
> If the breaker re-trips before the reset timer times out, the closing of 52b removes the "1" from the AND to stop the timer and prevent the reset. Further action is blocked until the breaker is closed manually.

TABLE 21.3 Device Descriptions for Figure 21.29

Device Number	Device Description
43	Authomatic operation cutout switch
43S	Synchronizing switch
52a	Circuit breaker auxiliary switch, open when breaker is open
52b	Circuit breaker auxiliary switch, closed when breaker is open
52aa	Circuit breaker auxiliary switch, open when breaker is open
52bb	Circuit breaker auxiliary switch, closed when breaker is open
52LC	Latch check switch
52LPC	Breaker low pressure switch
52CC	Circuit breaker closing coil
52X	Breaker control relay
52Y	Breaker cutoff relay
79	Automatic reclosing relay
79M	Timer unit of automatic reclosing relay
79X	Toggle unit of automatic reclosing relay
	O-operator coil
	R-reset coil
	Contacts shown after reset coil energized (reset position)
86B	Bus lockout relay
101	Manual control switch
A	Amber lockout indicator lamp

Other logic circuits are possible, some of which include additional functions, such as intentional time delay. Some relays also have the capability of equalizing the duty of the reclosing breakers on ring bus or breaker-and-a-half bus arrangements so that the initial reclosing is not always assigned to the same breaker.

21.4.4.3 Multishot Reclosing Relays. There are applications where multishot reclosing is desirable and multishot reclosing relays are available for these locations. Two shot reclosing, with one instantaneous and one-time delayed reclosure is also available. Other reclosing relays have three or more reclosures available. Still others are designed to coordinate with synchronism check relays, for locations where system separations are likely to occur and the two sides of the breaker are not in synchronism.

Multishot relays are not as widely used for EHV transmission protection. Hence, these relays are not as important in stability enhancement as the single shot reclosing relays. As noted earlier, any delay in the reclosing scheme is detrimental to stability.

21.4.4.4 Synchro-Check Relays. Synchro-check relays are not designed for automatic synchronization of two systems that are not synchronized. The synchro-check relay simply verifies that the voltages on both sides of a breaker are approximately equal both in magnitude and phase angle. The relays supervise automatic reclosing and prevent the large disturbance that can occur from reclosing when the two sides are out of synchronism. Figure 21.31 shows the connection of a synchro-check relay. The components are defined in Table 21.4.

Synchronism check relays of the type shown in Figure 21.31 are slow and are not capable of performing automatic synchronism. The relay has no provision for energizing the closing coil of a breaker. The function of the relay is to ensure that the two monitored voltages are

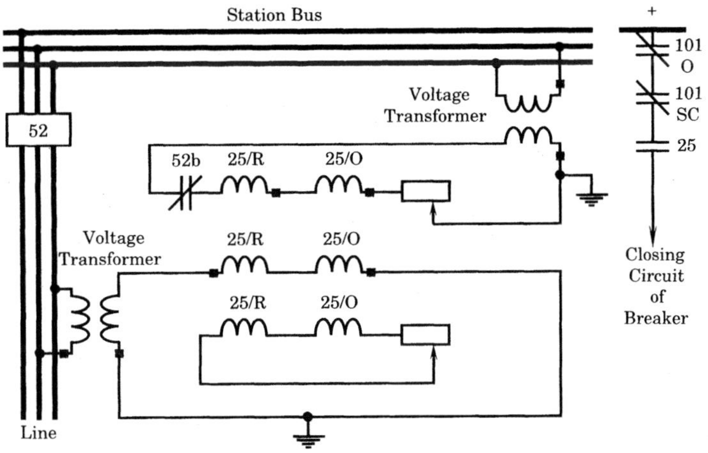

Figure 21.31 Synchronism check relay schematic [3].

TABLE 21.4 Device Descriptions for Figure 21.31

Device Number	Device Description
25	Synchro verifier relay
25/O	Operating electromagnet
25/R	Resitraint electromagnet
52	Power circuit breaker
52b	Breaker auxiliary contact, closed when breaker is open
101	Circuit breaker control switch
101-O	Contact as shown in the off position
101-SC	Contact closed during and after a close operation

nearly equal, in which case the relay contacts (25) are closed. The 52b contact in Figure 21.31 ensures that the synchro verifier contact is open immediately after tripping.

The typical characteristic of the synchro-verifier relay is shown in Figure 21.32. One of the voltages serves as the reference voltage. If the other voltage falls within the shaded area, the relay contacts are closed and reclosing is permitted. Both the angle and the magnitude are adjustable. There is also a requirement that the two voltages remain within the circle for a period of time, and not just pass through this region. The beat frequency between the two voltages is monitored, and this frequency must fall below a given threshold, at a given angle setting, to permit closing.

The connections for a synchro-verifier relay are shown on the right side of Figure 21.32. Note that both V_1 and V_2 must be closed before the relay picks up. Device 43 is the automatic reclose lockout relay and device 79 is the reclosing relay.

21.4.4.5 Digital Reclosing and Synchronism Check Relay. Electromechanical synchronism check relays similar to the one shown in Figure 21.32 have been used for many years. However, there are a number of newer devices available that make use of solid-state and digital technologies to expand the capability of these systems. In some cases, these modern devices provide not only synchronism checking, but a number of additional protective functions, such

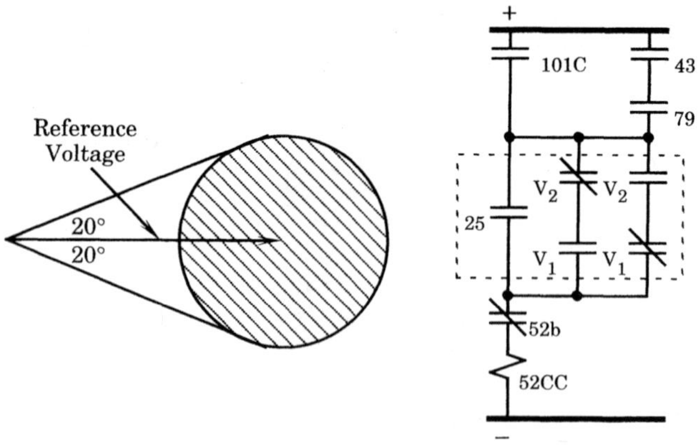

Figure 21.32 Typical synchro-verifier relay closing characteristic and relay connections [3].

as time-overcurrent, voltage and frequency measurement, loss-of-potential identification, as well as tripping and reclosing logic of different kinds. Event reporting and sequence-of-event reporting are also provided in some devices [7]. In distribution and subtransmission systems, a typical application would find one such relay is dedicated to a given circuit breaker.

21.4.5 Reclosing Switching Options

Another important consideration in automatic reclosing is the choice of switching options. Several options are possible, depending on the system and the anticipated conditions. A few of these options are discussed below.

21.4.5.1 Single-Phase Switching. In most cases, all three phases are opened by the circuit breaker following a fault and all three are automatically reclosed at the same time. There are significant advantages, however, to switching only the faulted phase (see Chapter 14). Since most faults involve only one or two phases, the fault can usually be cleared by switching only the faulted phases. Leaving the unfaulted phases in service permits the continued transfer of synchronizing power on these unfaulted phases during the switching operation, which can be a significant factor in maintaining the stability of the system.

In order to implement single-phase switching it is necessary to separate the three poles of the circuit breaker and equip each pole with its own tripping and closing mechanisms. This is normal with most air blast circuit breakers and for many types of oil circuit breakers.

The major disadvantage to single-phase switching is the longer deionization time. This is due to the capacitive and inductive coupling to the faulted phase from the unfaulted phases, which tends to sustain the voltage at the fault. This complicates the relaying and may lead to communications interference. Also, see Chapter 14 for a discussion of this phase coupling.

Where single-phase switching is used, distance relaying works very well since both phase- and ground-fault measuring units are provided for each separate phase and these units readily determine the faulted phase. However, on EHV transmission systems, the lines are often protected using unit system protection employing some type of pilot signal. These pilot systems simply send a trip signal to the remote end, without any identification of the phase on which the fault is observed. If single-phase switching is to be used with pilot protection, clearly some sort of phase identification is required. One type of phase selection relay is described in [4].

21.4.5.2 Live Line, Dead Bus or Dead Line, Live Bus. *Live line, dead bus/dead line, live bus (LLDB/DLLB)* control is sometimes used in a reclosing scheme for a transmission or subtransmission circuit. The concept employed here is that the circuit breaker is not reclosed unless one side or the other has essentially zero voltage. This idea is the complement of synchronism verification, which can only be performed when both sides of the breaker are live and in synchronism. There is no need to check synchronism when one side of the breaker sees a dead system.

21.4.5.3 Bus Protection versus Line Protection. For the most part, reclosing is used only for lines. The question arises as to its application on bus faults as well as line faults. There is no technical reason why reclosing should not be used for bus faults, but it is seldom done. One of the reasons is that buses are better shielded than lines and, therefore, suffer fewer faults. Little would be gained by adding reclosing for the relatively rare bus fault.

On high-voltage systems, bus arrangements are used that permit one bus to be faulted with no loss of power transfer capability through the station except, of course, during the fault. This is true of arrangements such as double bus-double breaker, or breaker-and-a-half bus arrangements. Since stability depends primarily on maintaining transfer power, there is little to be gained by adding bus reclosing. Should an important circuit have a single bus arrangement, then bus reclosing may be quite logical and would help maintain stability.

The ring bus is sometimes used, even on some EHV systems. These stations are vulnerable when one breaker is out-of-service for maintenance, since the bus path redundancy is lost during this period. Bus fault reclosing may help maintain stability for this situation.

21.4.5.4 Delayed Autoreclosing. On transmission systems that are highly interconnected, and where the loss of a single line seldom leads to stability problems, delayed reclosing may offer advantages. In most reclosing schemes, the designer tries to achieve the absolute minimum timing in order to restore the faulted line as quickly as possible in order to improve stability. However, the fast reclosing action increases the probability of unsuccessful reclosing for some faults.

One way of improving the probability of successful reclosing is to delay the reclosing long enough that there is a virtual certainty that full deionization has occurred. One utility has reported a 10% improvement in successful reclosures by delaying the reclosing for up to 60 seconds [4]. This solution is highly dependent on the network topology and is not recommended for every system.

21.4.6 Reclosing at Generator Buses

Although reclosing is generally a valuable aid to stability, there is concern regarding the reclosing into a permanent fault at a generator bus. This subject was thoroughly reviewed in the 1970s, and in response to these investigations the manufacturers issued recommended practices regarding reclosing at generator buses and the effect of such reclosing on units of their manufacture. A good review of the subject is provided in a report published by IEEE [8]. The concern is mostly with steam turbine-generators. These units have long shafts with complex modes of oscillation, which could place the generating unit at risk when reclosing into a close-in fault (also see Chapter 23).

The shaft torsional fatigue problem is complex because there are so many different frequencies, any one of which may become excited due to a fault clearing and reclosing event. Figure 21.33 illustrates a typical steam turbine-generator shaft, where the major masses on the shaft are shown, with the masses connected by shaft segments that act as torsional springs. The unit pictured has four turbines, a high pressure, intermediate pressure, and two low pressure units, plus a generator and exciter all on the same shaft. This would not be unusual for a tandem compound unit. Different unit designs will have different numbers of masses and springs. This type of model is called a *spring-mass model* and is reasonably accurate for studying the shaft behavior. A shaft with n masses will have $n-1$ natural frequencies of oscillation. These frequencies will usually fall between 10 and 50 hertz for a 60 hertz machine.

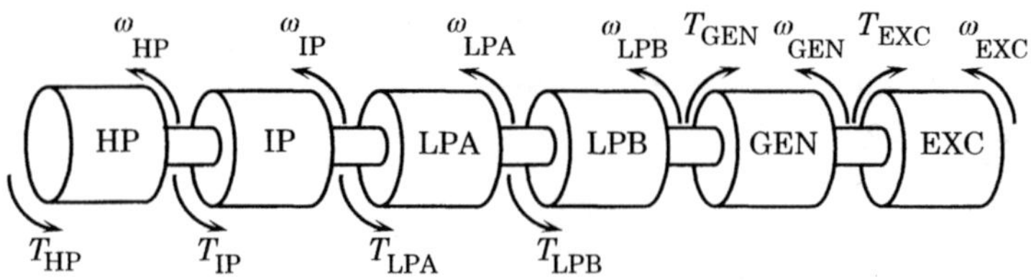

Figure 21.33 A typical turbine generator shaft arrangement.

A fault and switching operation causes a sudden change in generator power that will stimulate the shaft natural torsional frequencies. Should the time between the fault application and its clearing, which represents a pulse of power to the generator, have a period that is related to one of the shaft natural frequencies, that frequency will be excited and torque oscillations of this frequency will occur that are often poorly damped. While this oscillation persists, the autoreclosing into a permanent fault may cause either amplification of this frequency or a partial cancellation, depending on the timing of the second pulse of generator power. Examples have been published showing that the results may be quite different, based on this timing, and that a change in timing of as little as 0.05 seconds can make the difference between torque amplification and cancellation at the lower shaft frequencies [18]. Even smaller timing changes would similarly affect the higher shaft frequencies. This would suggest that very careful timing might be used to cause cancellation of torque, but this is not the case since any change in timing that causes torque cancellation in one of the natural frequencies may cause torque amplification in another frequency. Indeed, a detailed analysis of the shaft behavior is required to determine exactly what will happen. Usually, these studies can be performed only by the machine manufacturer, since they have detailed shaft models and the best model data for the simulations.

The conflict is one of balancing the risk of equipment damage or failure against reliability and the possible loss of customer service. On the one hand, it is argued that system stability is well served by the practice of fast reclosing and that no generating units are known to have been permanently damaged by this practice. On the other hand, the practice could lead to shaft cumulative damage that is greater than necessary, and expensive repairs might be required during the normal lifetime of the equipment.

Both the utilities and the equipment manufacturers have made detailed studies of this problem, beginning in the mid-1970s [9–25]. As a result of the intensive industry study and evaluation of this problem, some utilities have stopped the practice of high speed reclosing on lines near generating stations. Others have evaluated the practice and decided to continue reclosing on all lines because of the considerable benefit derived in terms of system stability and reliability of service. One such utility reported that, based on their statistics, only 0.2% of high speed reclosures would result in reclosure on a permanent three-phase fault, and an even smaller percentage would be considered "close in" to the generating plant bus [14]. Moreover, this same utility reported no evidence of shaft damage due to high speed reclosing practice over a period of about 40 years.

One solution that has been suggested is to permit high speed autoreclosing at a generating plant only for single line-to-ground faults. This could be full three-phase reclosing or single-phase switched reclosing, since the magnitude of the torque disturbance is about the same either way, and is much lower than the torque disturbance for a three-phase fault. The added complexity of equipment, controls, and testing needs to be considered when proposing schemes that are unusually complex, however. It may be difficult to ensure proper discrimination and operation of the more complex scheme.

The types of disturbances that can occur have been described as falling into one of the following three classes [22]:

Class 1	Single torsional excitation (faults, synchronizing out of phase, load rejection, and planned line switching)
Class 2	Double torsional excitation (fault clearing)
Class 3	Multiple torsional excitation (automatic reclosing of a faulted line)

Each additional torque transient following the initial transient superimposes additional torsional shocks to the shaft system. These added shocks can lead to attenuation, at best, or severe amplification, at worst. Moreover, there is considerable uncertainty as to the exact instant the torque transient will occur, making the result highly unpredictable. The simple clearing of a fault can have highly unfavorable timing and result in considerable torque amplification.

Another important factor is the fault location, with the maximum shaft torque being reduced as the fault is removed farther from the generator bus. Table 21.5 provides a rough estimate of the torsional fatigue that may accompany different fault and switching conditions.

Because of the many variables that determine the shaft stresses and resulting torsional fatigue, several investigators have suggested a probabilistic approach to the problem [23]. Studies have been conducted on a number of generating units in service and the results analyzed using probabilistic methods. In analyzing these results, the following definitions have been used (quoted directly from [23]):

> *Unrestricted high speed reclosing* refers to the practice of reclosing transmission line circuit breakers at both ends of the line as rapidly as possible following a fault tripout, regardless of the fault type. Successful reclosing cannot occur until the fault arc has extinguished and the dielectric strength has recovered. This may require more than a second for lines having a large amount of shunt capacitance. For short transmission lines, reclosing times may be as short as 15 cycles.
>
> *Delayed reclosing* refers to automatic reclosing of the transmission line breakers where additional time delay is used beyond the minimum required for arc extinguishing and dielectric recovery. A ten second time delay was used with this type of reclosing in the study. The main reasons for adding time delay include the following:
>
> 1. To improve the percentage of successful reclosings.
> 2. To reduce the probability of system transient instability following an unsuccessful reclosure back into a major fault. Time delays of one-half to several seconds are typically used.
> 3. To avoid the risk of major shaft fatigue damage associated with unsuccessful reclosing into multiphase faults. Delay times of at least 10 seconds have been recommended for this purpose.
>
> *Sequential reclosing* refers to the practice of reclosing the line first from the end which is remote from the plant, after which, on the basis of check relays (usually voltage and phase

TABLE 21.5 Impact of Different Fault Conditions and Switching Operations on the Torsional Fatigue of Steam Turbine-Generator Shafts [22]

Class	Type of Disturbance	Location or Condition		Percent Fatigue Per Incident (Negligible → Severe)
1 Single Torsional Excitation	L-L fault	HV bus		
		Gen bus		
	3Ph fault	HV bus		
		Gen bus		
	Sync out of phase			
	Full load rejection			
	Normal line switching	dP<0.5 (1)		
		dP>0.5		
2 Double Torsional Excitation	L-L fault	V = 0.2 (2)		
		V = 0.0		
	3Ph fault	V = 0.2		
		V = 0.0		
3 Multiple Torsional Excitation	L-G fault	3 Pole Sw	S	
		(3)	U	
		1 Pole Sw	S	
			U	
	L-L fault	3 Pole Sw	S	
			U	
	3 Ph fault	3 Pole Sw	S	
			U	

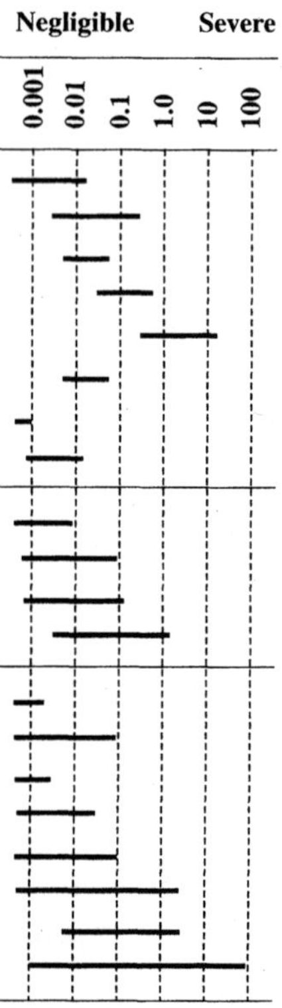

(1) dP = magnitude of disturbance power
(2) V = magnitude of voltage at the fault
(3) S = successful reclosure
U = unsuccessful reclosure

angle) to ensure that the fault no longer exists, the plant end breaker is automatically closed. One to several seconds time delay is inherent in this type of reclosing. This type of reclosing is therefore particularly applicable where a short delay is desired anyway for reducing the risk of transient instability.

The effectiveness of sequential reclosing in avoiding the compounding of fatigue damage when reclosing into permanent faults is, of course, a function of the network. Sequential reclosing would be of little value in tight networks where the "remote" line terminal is electrical close to the plant, or if another generating plant is located close to the "remote" end of the line.

Selective reclosing is the practice of distinguishing the type of fault and permitting high speed reclosing only for single-line-to-ground and line-to-line faults. Selective reclosing might also be applied on some other basis, such as distance from the plant, to screen out the more severe, close-in faults.

The computations assumed the following percentages of the different types of faults:

Three-phase faults	3%
Double-line-to-ground faults	3%
Line-to-ground and line-to-line	94%

It was also assumed that 30% of the shaft fatigue life is reserved for other possible events such as bus faults, out-of-phase synchronization accidents, and remote line faults. This leaves 70% of the shaft fatigue life that might be considered prudently "available" for responding to faults near the generating unit. Finally, it was assumed that the frequency of faults has a mean value of four faults per 100 miles per year. With these boundary conditions and assumptions, the probabilities shown in Table 21.6 are provided.

TABLE 21.6 Probabilities of Expending 70% of the Shaft Life in 40 Years of Exposure Due to Reclosing Practices on Lines Connected to the Plant [23]

Type of Reclosing	3% Three-Phase Four faults/ 100 mi/yr	3% Three-Phase Eight faults/ 100 mi/yr	4% Three-Phase Four faults/ 100 mi/yr
1. Unrestricted high-speed reclosing	0.21	0.70	0.30
2. Selective high-speed reclosing	0.009	0.20	0.014
3. Sequential high-speed reclosing	0.001	0.04	0.002
4. Delayed reclosing (10 s)	0.0003	0.015	0.0009
5. No reclosing	<0.0001	0.001	<0.0001

These results indicate that the frequency of faults is a very important parameter in determining the probability of loss of life. Areas that do not have very much lightning, for example, will have a better experience than those with heavy lightning or other hazards that lead to faults. The type of faults experienced is also important but not as important as the frequency of faults. It is also quite clear that unrestricted high speed reclosing subjects the units to a much greater hazard than more conservative methods. This would indicate that unrestricted high speed clearing should be used with considerable caution and only in cases where there are significant benefits due to this practice.

Studies have been conducted to try and establish a "screening guide" that can be used by a utility to determine if dangerous shaft torques can result from autoreclosing at a given location for a given generating unit. One manufacturer conducted studies of many of the large turbine-generators of their design and computed the total magnitude of change in power ΔP that would stress the shaft to the allowable short-circuit shear stress level for the material [11]. From these studies, a safe allowable value of ΔP was suggested that would limit the actual stress to one-half this maximum value. If the conditions at a particular unit indicate that higher values might be possible, the owner should consult the manufacturer for more detailed calculations.

An IEEE Committee has published a screening guide that is available for protection engineers to estimate if a particular switching event is severe enough to require greater study, perhaps in consultation with the turbine-generator manufacturer [25]. This guide is based on the screening quantity ΔP, which is a function of both the switching angle and the system impedance. Moreover, this quantity can be computed using conventional transient stability programs. This screening guide, however, is limited to steady-state switching and is clearly not intended for emergency line switching, faulty synchronization, full load rejection, fault clearing, or reclosing. The guide concludes that any switching event with a value of ΔP that is less than 0.5 per unit is acceptable, and no further investigation is required. If the value of ΔP does exceed 0.5 per unit then further study is needed, including detailed computation of transient electrical and mechanical torques.

21.5 LOSS OF SYNCHRONISM PROTECTION

Another type of protection that is designed as an aid to system stability is termed *loss of synchronism* or *out of synchronism* protection. This type of protection is designed to detect the condition following a large disturbance that may cause the generators in one part of the system to accelerate while generators in another part of the system decelerate, thereby creating a condition where the two parts of the system are likely to separate.

21.5.1 System Out-of-Step Performance

A simple example of this situation is shown in Figure 21.34, where a transmission line is represented by its impedance and the power system by the Thevenin equivalent impedances to the right and left of the line. The equivalent voltage sources are represented as phasors of assumed equal magnitude in the upper part of the figure. The source $S2$ is assumed to vary in phase from voltage $S1$. As $S2$ rotates clockwise, lagging more and more behind $S1$, the locus for the voltage creates a circle in the complex plane. The voltage at the midpoint of the transmission line is also described by a circle and we note that, when the two voltages are 180 degrees apart, the voltage at the midpoint is exactly zero in all three phases. This would be viewed by the transmission line relays as a three-phase fault at the center of the transmission line and the relays would pick up, clearing the line. If the voltages have swung that far apart, separating the system may be the correct thing to do, but location of the separation can be very important. It would be desirable to separate at a point where the remaining generation and load at both sides would be approximately equal. This ideal is difficult to achieve in general.

21.5.1.1 Representation in the Z Plane. A more detailed description of the behavior of the two systems on either side of the transmission line is examined in detail in Chapter 9.

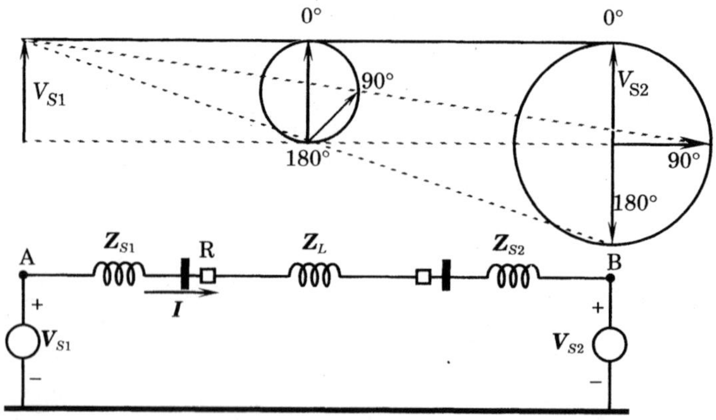

Figure 21.34 Two equivalent systems going out of step.

From that description, we know that the total impedance between A and B is plotted in the Z plane as shown in Figure 21.35, with a perpendicular bisector through the point Q. This line represents the locus of points for any angle θ by which $S1$ leads $S2$, when the voltage magnitudes are exactly equal. This locus bisects the triangle APB. The point Q is the electrical center of the system and is the point at which the voltage collapses to zero, since this is the point where the angle θ equals 180 degrees.

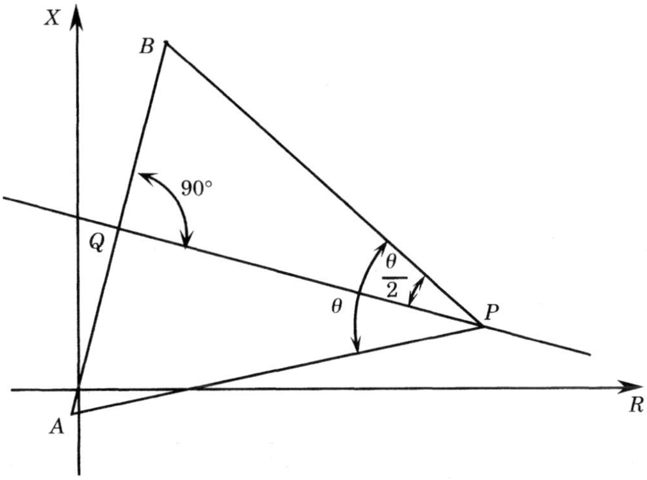

Figure 21.35 Locus of variable angle θ by which A leads B.

For this simplified description of the system, we can write the impedance seen by a relay at R of Figure 21.34 by the equation

$$\mathbf{Z}_R = \frac{\mathbf{Z}_T}{2}\left(1 - j\cot\frac{\theta}{2}\right) - \mathbf{Z}_{S1} \tag{21.45}$$

where $\mathbf{Z}_T$ = total impedance between A and B

As the angle θ advances, the point P moves to the left. This simple result is due to the voltages of the sources being equal. For the more general case, we define

$$\frac{\mathbf{V}_{S1}}{\mathbf{V}_{S2}} = ne^{j\theta} \tag{21.46}$$

When the two voltage magnitudes are not equal, the locus of points becomes the arc of a circle with the center of the circle lying along an extension of the line AB in the Z plane. A sample of the resulting loci is shown in Figure 21.36, where we have plotted one locus for $n < 1$ and one for $n > 1$. The points on the three loci that correspond to the same angle θ also fall on a circle that passes through points A and B, as shown in Figure 21.36. Thus we may compute, for example,

$$\frac{\overline{P'A}}{\overline{P'B}} = \frac{|\mathbf{V}_{S1}|}{|\mathbf{V}_{S2}|} = n \tag{21.47}$$

Under normal conditions, the operating point P would be expected to lie quite far to the right or left, depending on the direction of power flow. For a fault on the transmission line, the impedance will be quite close to the origin along a line determined by the impedance from relay to fault, including the fault resistance. After clearing the fault, the operating point P

again moves far to the right or left, but then moves along the appropriate locus as the angle between the source voltages increase. This locus may enter a relay protective zone, causing that relay to trip. This may or may not be a desired reaction, and *out-of-synchronism* (*OS*) relays are designed to supervise this transient behavior.

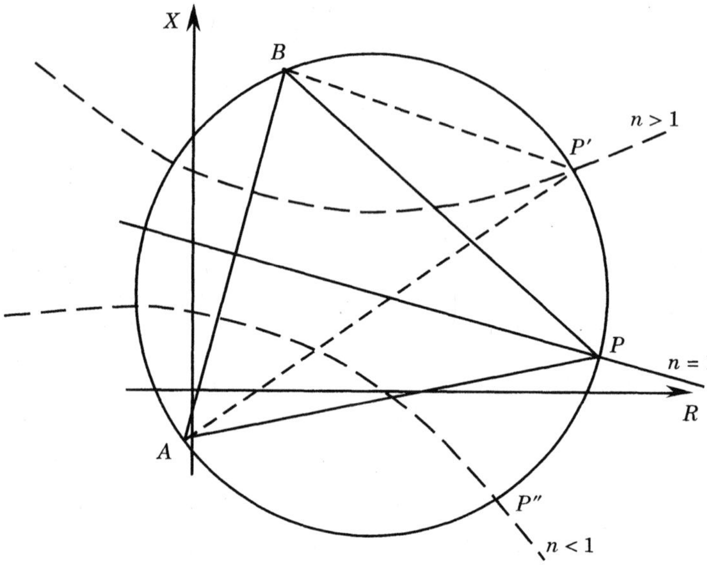

Figure 21.36 General loci of the loss of synchronism characteristic.

21.5.1.2 Protection Requirements. The fundamental objectives for OS protection may be summarized by the following statements [3]:

1. Block tripping at all locations when the swinging out-of-step condition is expected to be stable.
2. Trip lines to separate the system for a true out-of-step condition.
3. Control the separation to preserve a reasonable balance between the load and generation on each side of the point of separation and avoid shedding any load unnecessarily.
4. Block tripping or automatic reclosing at one end of the line at which separation takes place.
5. Initiate tripping only when systems are less than 120 degrees apart and when the angle is in the closing direction, in order to minimize breaker stress.
6. Minimize any possible OS conditions by the following precautions:
   (a) Use high-speed line relays
   (b) Use high-speed excitation systems on generators
   (c) Use loss-of-field relays on all generators
   (d) Provide adequate transmission capacity
   (e) Trip generators on the loss of critical lines
   (f) Apply braking resistors or insert series capacitors for critical fault conditions
   (g) Use fast valving of turbines to control overspeed
   (h) Use independent pole tripping to increase power flow through the fault point and minimize separation during the fault

These objectives represent an ideal that may never be reached on any system for many reasons, not the least of which is cost. System planners and designers have the difficult task of doing the best they can with limited resources. It is up to the protective system designer to provide system protection for the system as it exists, which often falls short of the above objectives.

The requirements for OS protection are difficult to achieve. First, it is necessary to detect a true OS condition, and clearly distinguish this from a fault for which fast clearing is always required. Knowing that a swinging of generation sources is under way, it is necessary to determine if tripping is required. Therefore, some protective systems are designed for OS Blocking and others for OS Tripping.

21.5.2 Out-of-Step Detection

The swinging of generation sources following a disturbance is often a slow oscillation of the voltage angles between the two sources. For stable swings, these voltages will vary in rather poorly damped oscillations with a frequency that depends on the inertia of the two sources and the impedance between them. Usually this oscillation will be at less than 1 hertz, with oscillations of 0.5–0.8 hertz being common. This gives a period of oscillation greater than 1 second, which is a rather long time for observation of the phenomenon. If the disturbance is large enough, however, there will be no oscillation, but a monotonic increase in angle until the voltage collapses at some point, as noted in Figure 21.11, and relay operation takes place. The problem is that the fault relays are not designed to determine the out-of-step condition and may trip incorrectly at the wrong place, or in the wrong sequence.

Since the movement of the voltage angle is often rather slow, time is one system parameter that is useful to help distinguish between a slowly changing angle separation and the jump change due to a fault. Relays designed for this type of detection are usually Z plane devices that can measure the apparent impedance seen from the relay location and note the sequence by which this impedance moves past certain guideposts set up for this purpose.

One common type of detector is the "blinder." Blinders are ohm units that have a straight line characteristic in the Z plane, such as the lines A and B shown in Figure 21.37. The trip zone for blinder A is to its left and for B to its right, the unhatched area in each case. For a line fault the impedance seen by the relay jumps immediately into the fault zone, thereby

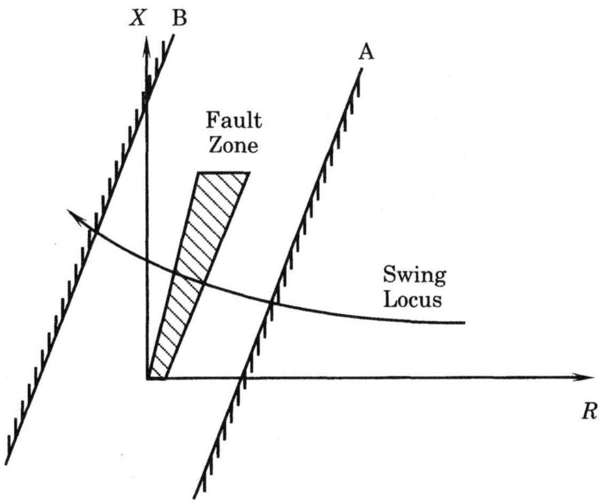

Figure 21.37 Typical blinder characteristics in the Z plane.

picking up both the fault relay and the blinders at the same time. For a swinging out-of-step condition, however, the angle moves slowly along the locus, which picks up first the blinder A (or B), followed by the relay at an instant later. This time discrimination can be used to tell the difference between a fault that should be cleared quickly and a possible swinging out-of-step condition. The fault relay characteristic is not shown in Figure 21.37, but its characteristic usually covers the entire fault zone and a surrounding area as well.

Another type of detection, which also relies on timing for selectivity is to provide a mho-type unit as shown in Figure 21.38. This figure shows a mho type fault relay, with its characteristic completely surrounded by another mho type unit, which acts as the OS detector. As before, timing is used to discriminate a fault from an out-of-step condition.

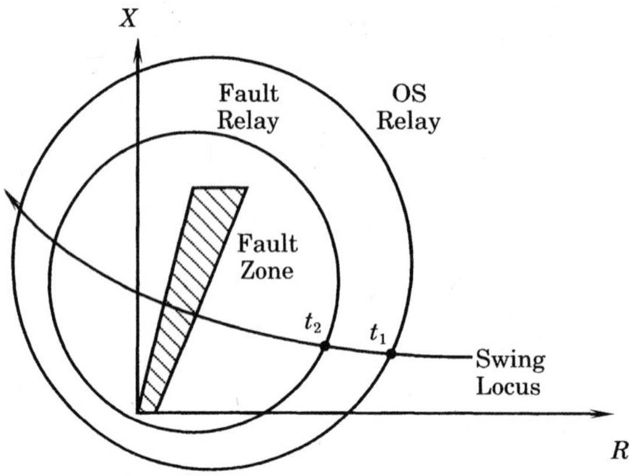

Figure 21.38 A mho-type out-of-step detector.

21.5.3 Out-of-Step Blocking and Tripping

There are situations where it is desirable to block out-of-step tripping. For example, it is necessary to block at locations where tripping will separate the system into greatly unbalanced islands with large differences between load and generation. This would be true in situations where two systems are interconnected by only a few lines, and the line being called upon to trip for a swinging condition may be the last line holding the system together. Blocking is also required when there is no fault on the protected line, but only a transient penetration of the trip zone due to the oscillatory condition. For example, see Section 21.3.5.

There are other situations where tripping is necessary in response to a system disturbance that is known to lead to instability. This may require that the system be separated into islands, which can be stabilized and later reconnected. This action will temporarily block scheduled power transfers that are too large to be sustained following a critical disturbance.

It is not possible to determine if tripping or blocking is desirable using fault studies. In most cases, stability studies of the oscillatory condition following all kinds of disturbances will be required to determine where OS tripping is desirable and where OS blocking should be used. If these studies show that the system should recover from a given disturbance, then blocking should be used.

Timing is important in the detection scheme. For a fault condition, the impedance seen by the relays jumps instantaneously from a normal load value to a small impedance in the fault zone, as shown in Figures 21.15, 21.16, and 21.38. For a swinging out-of-step condition, however, the impedance locus usually moves rather slowly along the swing locus path. If an OS relay is used, the swing locus enters the OS zone at time t_1 and does not enter the fault relay protection zone until time t_2. This time difference permits the OS relay to pick up first and take action to prevent tripping by the fault relay.

One way of arranging the controls to accomplish blocking is shown in Figure 21.39. Here, the fault relay is a distance type relay (21), the blocking relay is device 68, and an auxiliary relay is shown as device Y. When a fault occurs, the 21 contacts pick up. One set of contacts, the one on the left in the figure, shorts the auxiliary relay coil Y to prevent blocking and the other set of contacts trips the circuit breaker. If the disturbance is not a fault, but the impedance enters the OS relay trip zone, relay 68 picks up first, which in turn picks up the auxiliary relay. This relay opens the trip coil circuit, which blocks tripping. Since a fault picks up both 21 and 68, the blocking relay is usually given a small time delay to permit tripping to be completed for a true fault condition.

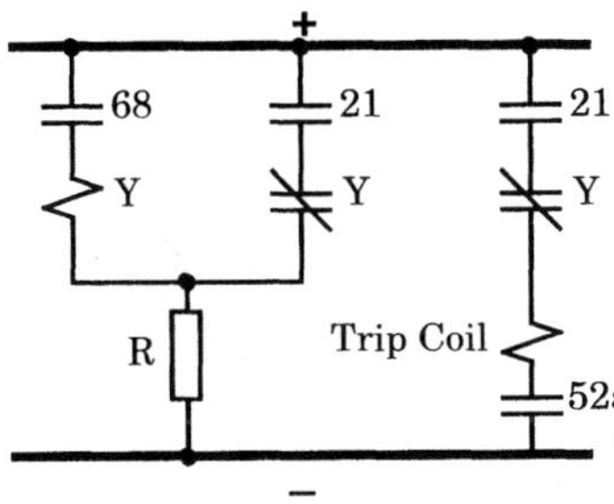

Figure 21.39 Control circuit for OS blocking [26].

The blocking scheme described is applicable for blocking either tripping or reclosing. It is not ideal, however. For faults near the edge of the mho characteristic, the relay responds slowly because of its low energy in this marginal region. However, the 68 relay has high energy and responds quickly for this fault condition. Sometimes an additional surrounding OS relay is added, and timing is measured as the impedance seen by the relay moves from one characteristic to the other, thereby giving better discrimination of the out-of-step condition.

Three terminal line relaying also presents a problem because of infeed from the extra terminal, which shortens the reach of the distance relays. This can cause pickup of the 68 relay, but not the 21 relay, thus blocking a true fault condition. Out-of-step blocking should not be used for three-terminal transmission lines.

Another problem with the distance characteristic shown in Figure 21.38 is that the region of the Z plane enclosed by the fault relay is large and the OS relay characteristic is even larger. Blinders are sometimes used to reduce the OS pickup zone to a smaller area, immediately surrounding the fault relay. Even the fault relay itself can be reduced by selecting a more discriminating relay with a characteristic that is closer to the true fault zone, for example, having an elliptical rather than a circular shape.

Another blocking and tripping scheme with greater selectivity is sometimes required. One example is illustrated in Figure 21.40, where three loci, A, B, and C are shown in a situation where distance relays are protecting a line. Zone 2 protection (21-2) will overreach the remote line terminal and will be tripped with a time delay, whereas zone 1 (21-1) will trip without time delay.

Swing locus A penetrates the OS distance characteristic, but only briefly and then leaves the zone, allowing the relay to reset. Locus B penetrates both the OS and zone 2 characteristics, and then returns to a normal condition far to the right in the Z plane. Locus C represents a probable out-of-step condition for which tripping is desired. The question is how to distinguish between conditions B and C, one being a stable condition and the other unstable. One difference is that the time spent in traversing the various zones is greater for locus C than for B.

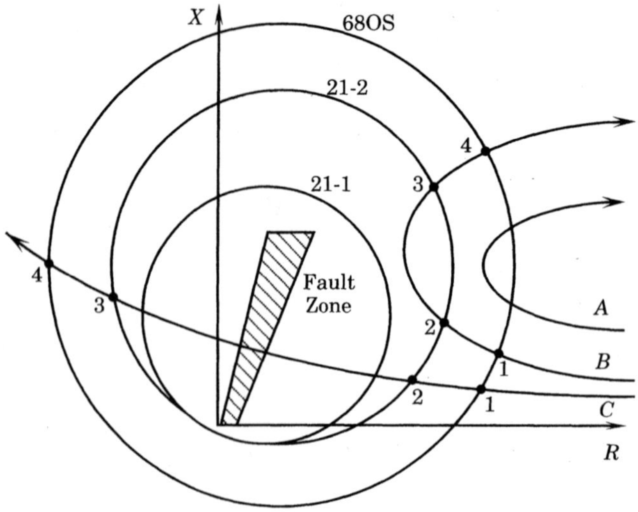

Figure 21.40 OS blocking and tripping schemes.

One method of providing the OS trip logic for these conditions is shown in Figure 21.41. Zone 2 of the distance relay is used in conjunction with the distance characteristic of the OS relay and several timers to provide the necessary tripping logic for a swing locus such as Locus C in Figure 21.40.

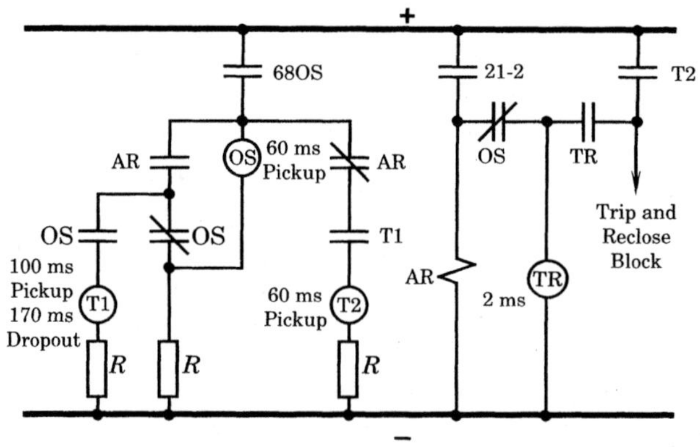

Figure 21.41 Out-of-step tripping and blocking scheme [3].

If a fault occurs, all relays are picked up almost simultaneously. Contacts 21-2 pick up relay AR, which shorts the OS coil and timer and picks up the TR timer, which in turn times out and trips the breakers. Should a locus enter the 68OS zone and a fault occur before the 60 ms timer times out, the same result is achieved since the OS coil becomes shorted, due to the 21-2 pickup, and does not operate.

On a swinging no-fault condition the locus enters the 68OS characteristic, then the 60 ms OS timer times out. This separates the 21-2 relay contacts from the TR timer so the only way the breaker can be tripped is through the telephone relay T2. Also, when the OS timer times out, its contacts hold its coil in a picked-up condition as long as 68OS remains closed by separating from the AR contacts. A third set of OS contacts close to provide for the pick up of the T1 timer if the 21-2 relay picks up. Suppose the locus does not enter the 21-2 zone. Then the T1 timer is never started. This would correspond to locus A in Figure 21.39. If, however, the swing is similar to locus B or C, upon entering the 21-2 zone, relay AR picks up, which starts the T1 timer.

After 100 ms, if the locus is still inside 21-2, the T1 contacts are closed, but the T2 timer cannot be started because the AR back contacts are still open. If 21-2 resets before the 60 ms time for T2, the line will be tripped as the locus leaves the 21-2 zone, which corresponds to locus C. If, however, the locus quickly leaves the 68OS zone, as in locus B, then tripping is blocked due to the reset of 68OS.

The logic is controlled by timers. The setting of these timers is critical and depends on the speed of movement in the Z plane for various types of disturbances, and whether the system has restoring forces or is being torn apart by the disturbance. This can only be determined by stability studies.

The foregoing analysis is based on electromechanical relays, but similar logic can be provided for solid-state relays using digital logic.

21.5.4 Circuit Breaker Considerations

The protective system design during out-of-step conditions should take into account the fact that the system may be separated at the last circuit breaker connecting two large systems. Moreover, the separation may occur when the two systems are swinging close to 180° apart as the breaker contacts begin to separate. This means that the peak-to-peak voltages may be as much as four times normal across the opening contacts of the breaker. Re-ignition is possible under these conditions. Tripping may therefore have to be delayed until the systems drift closer together again.

21.5.5 Pilot Relaying Considerations

Much of the foregoing has considered only distance relays, since the measurement of distance is a necessary aid to determining out-of-step conditions. We now examine the performance of pilot relaying systems during an out-of-step condition.

21.5.5.1 Phase Comparison Pilot. Phase comparison systems respond only to the current flowing in the line. Since the swing condition produces a through current, rather than current components at each end directed toward an internal fault, these relays see no unusual condition and are therefore not affected by the out-of-step oscillations.

21.5.5.2 Transfer Trip. Transfer trip relaying may be either under-reaching or over-reaching transfer trip. For under-reaching transfer trip schemes, swings entering the circle of either zone 1 relay will cause a trip of both terminals. For over-reaching transfer trip, the swings must enter both zone 2 characteristics simultaneously to cause a transfer trip operation.

21.5.5.3 Directional Comparison. Swings entering a protected zone between the terminals of the line will cause simultaneous high speed tripping. Swings external to the line protection zone will block tripping.

21.5.6 Out-of-step Relaying Practice

The relaying practice of utilities is usually somewhat different than the ideal objectives outlined earlier. A review of utility practice in this type of relaying may be summarized as follows [26]:

1. Initiate out-of-step relaying by line relays
2. Allow generator loss-of-field relays to trip generators during out-of-step conditions
3. Restrict relay trip sensitivity at higher power factors
4. Block tripping in selected locations
5. Block reclosing following out-of-step tripping
6. Initiate out-of-step tripping using special OS relays

Clearly, there are methods available to the protection engineer who has need for OS tripping or blocking. The need for this type of protection must be verified by stability studies and the conditions under which the action is required can only be determined by these methods. Once the conditions are known, the protective system can be designed.

21.6 SPECIAL PROTECTION SCHEMES

Another type of protective scheme that is designed to preserve system stability is the special protection scheme or SPS [27]. In certain power systems, studies of system performance may show that large disturbances on important lines or facilities can cause violent and often disastrous effects. This may occur in systems that are interconnected by long or weak tie lines, which may be heavily loaded. When this occurs, the system may break apart in a way that is not predictable and with the possible creation of power system islands having large generation-to-load imbalances. These poorly balanced islands may not be able to survive, resulting in total blackout. One way to prevent this disastrous result is to separate the interconnected system in a controlled fashion, such that the resulting islands can be assured of having a reasonable balance between load and generation. This increases the probability of survival of the islands.

Another condition that requires special treatment is the occurrence of a disturbance by one utility that, because of its nature or size, causes serious consequences to the transmission system of a neighboring utility. In the interconnected systems of North America, the interconnection is designed and operated such that this type of condition should not occur. In other words, it is a design requirement, either expressed or implied, that one should not design a system that creates a serious problem for a neighboring utility. This requires that prudent measures be taken during the design process that will assure neighboring utilities that a new facility will not cause them operating difficulties, even during unusual events such as faults or other unplanned disturbances. An example of this second type of disturbance might be a bipole block of a large HVDC converter station that is interconnected on the ac side to one or more neighboring utilities. Since the HVDC system may be carrying a large load prior to its interruption, its outage will cause the tie lines between the utilities to absorb the sudden change, with possible tripping of sound transmission lines.

The foregoing are examples of situations that can be determined, through stability studies, to be serious enough to call for preplanned control action. These actions constitute a form of dynamic security assurance, where the actions to be taken in response to a given condition are preplanned.

21.6.1 SPS Characteristics

The type of control scheme described above is known by various names, in addition to SPS, depending on the originator, for example [27]:

Special stability controls
Dynamic security controls
Contingency arming schemes
Remedial action schemes
Adaptive protection schemes
Corrective action schemes
Security enhancement schemes

These schemes provide different types of control actions, depending on the problem created when the disturbance occurs. Some schemes trip generators, others intentionally open transmission lines, and some create islanded systems at predetermined locations. The schemes have several traits in common:

1. All are dynamic security control systems and are designed to control power system stability in cases where the uncontrolled response is likely to be more damaging than the controlled response.
2. All are devised by off-line analysis, as opposed to on-line real-time control. The reasons for this is that the power system response is too fast to allow time for the usual sequential control system logic, which might be summarized as
   (a) make the observations in real time,
   (b) determine the scope of the disturbance,
   (c) decide what action is required, and then
   (d) take the needed action.

   In particular, item (c) may take a relatively long time, even if it is possible to create a logic that will always make the correct decision.
3. Many of these schemes are armed or disarmed, as required, in order to meet the needs of the system at a particular time. In other words, the special control logic may not be required under certain operating conditions, in which case the SPS is disarmed.
4. All of the schemes provide a particular type of remedial action that is designed to alleviate a certain observed system condition, or to take a predetermined action when a certain event occurs whose resulting effects are calculated to be too serious to ignore.

21.6.2 Disturbance Events

Special protection schemes are designed to take special action in response to certain event disturbances [28–35]. Typical event disturbances are:

- Transmission faults
- Cascading outages of lines
- Generation outages
- Sudden, large load changes
- HVDC pole or bipole blocking
- Combinations of the above

When a prescribed event disturbance occurs, the SPS will take a predetermined action. This predetermined action is designed on the basis of computer studies of the system behavior, where it is noted that, without special action, the resulting system response creates serious problems. Based on these studies, carefully designed fast control actions are also studied that will provide more satisfactory system conditions than those that result without the special actions. This may require drastic actions, such as shedding of load, or dropping of generation, but even these heroic actions are much better than risking total system collapse.

21.6.3 SPS Design Procedure

The design of an SPS follows a logical procedure. It is necessary first to understand the system responses to disturbances. Some disturbances are more serious than others, depending on the type, location, complexity, and duration of the disturbance. Some may be found to be very serious, and may be such that the protective devices normally used for system protection are inadequate. This may dictate the installation of several devices, some of which may be of special design. The following is a description of a typical design procedure.

21.6.3.1 Definition of Critical Conditions. The critical disturbances are those expected to have a devastating effect on the power system under a particular operating condition. Their identification will probably require many stability studies, for different operating conditions and different disturbance scenarios. In some cases, the engineer seeks guidance in this search from actual disturbances that have occurred and resulted in serious consequences such as loss of stability, islanding, loss of load, or blackout. This procedure should result in a clear definition as to the disturbances of interest, how these disturbances can be detected, and their effect. It is also possible to devise corrective measures using computer simulation as a tool, including the action to be taken, the required speed of that action, and the results of the control actions.

21.6.3.2 Definition of Recognition Triggers. Recognition triggers are devices that are used to identify the need for SPS response. Usually the triggers are relays of various types and may be used in combinations. For example, line relays on two different lines, perhaps at different locations, can be used as input triggers in an "and" or "or" logic to take a particular action. In some schemes that have been designed, there are dozens of such input triggers that consist of

- AC system faults in a given location or within a given distance
- Low voltage of a given magnitude at a certain location
- Blocking one or both poles of an HVDC converter
- Close-in faults that may cause HVDC commutation failure
- Multiphase faults on a given circuit
- Apparatus faults, such as transformer faults

In some cases, these input triggers are combined in hardwired or digital logic to form super triggers, each of which may dictate different controlled responses, such as

1. Tripping of one generator
2. Tripping of two or more generators
3. Tripping of one or more ac transmission lines
4. Tripping of HVDC transmission systems

In a few cases, transfer trip signals are sent hundreds of miles to remote facilities to order the desired control action.

21.6.3.3 Operator Control of SPS. SPS are often designed for operator review. The power system operator is given guidelines, such as nomographs or computer displays, to provide information concerning the need for "arming" a particular scheme. Operator control is necessary because the SPS is not always required. When line loadings are light, for example, any disturbance that causes the lines to trip may not cause large system stresses. However, when loadings exceed some critical value, the loss of a line may require that the SPS perform some rapid control action, such as the tripping of a generator. This means that the SPS is armed and disarmed by the system operator in response to system conditions. A typical operator nomograph is shown in Figure 21.42.

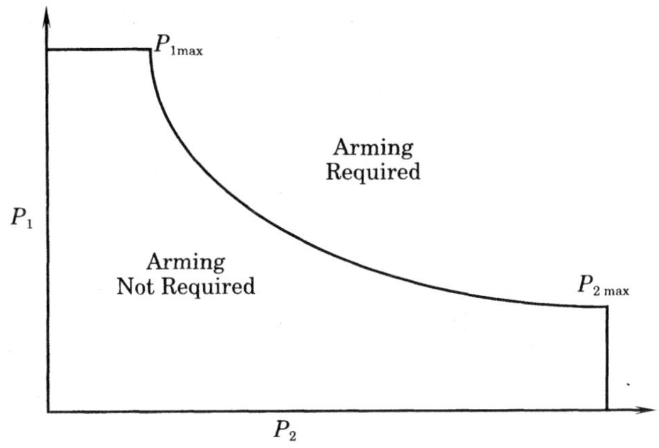

Figure 21.42 Operator nomograph for contingency arming.

The axes in Figure 21.42 may represent the power flowing through key facilities on the system, perhaps through a transformer at one location and transmission line or lines at another location. The two readings are combined to locate a point on the plane that either requires arming or not. The operator can then take the appropriate action to activate (arm) or deactivate (disarm) the SPS. In some cases these nomographs are simplifications that are simply expedient, but still give the operator a reasonable tool to use for removing unnecessary controls from service when not required. One reason for the disarming is that the arming scheme might take the programmed action when it should not do so, which represents either a security failure of the equipment or a design error. The arming schemes are carefully designed to prevent this type of failure, but a probability still exists that the action could be triggered, perhaps with serious or costly results.

21.6.4 Example of a Special Protection Scheme

A good example of an SPS is the system designed for the Intermountain Power Project (IPP) in the western United States [35]. The IPP is a project that consists of the Intermountain Generating Station (IGS), located in the State of Utah; the Intermountain Converter Station (ICS), also located in Utah; the Adelanto Converter Station (ACS), located in Southern California; and a 784 km ± 500 kV HVDC transmission line connecting ICS and ACS. The geographic layout of IPP is shown in Figure 21.43. In addition to the above facilities, there are interconnections to the existing transmission system in Utah, Nevada, and Southern California. The IGS consists of two 827 MW generating units, and the HVDC system has a design rating of nominally 1600 MW, so that the entire output of the power plant can be delivered to Southern California, or portions of the output can be delivered to the local transmission system.

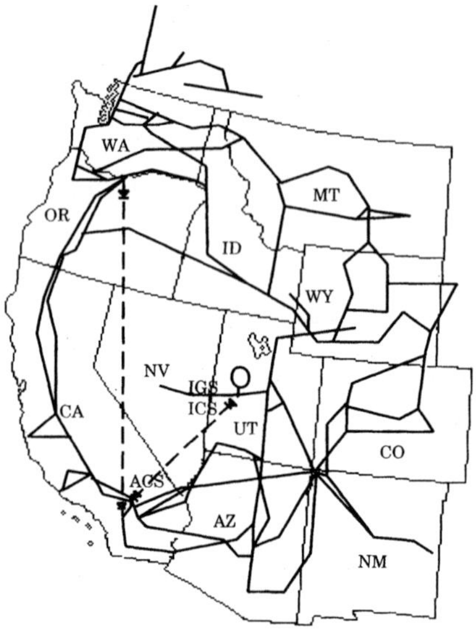

Figure 21.43 Location of the intermountain power project [35].

In the studies that were made in the design process, it was learned that, under certain loading conditions of the bulk transmission system, a pole or bipole outage of the HVDC system would immediately force the generating plant output into the Utah and Nevada transmission networks, with the possibility of creating serious voltage depression in those networks and possible line tripping or loss of load. The SPS was designed to trip IGS units, based on observations leading to outages of the HVDC systems near the generating station.

The SPS is required to be operative under heavily loaded system conditions, which requires observation of several key transmission corridors in the western U.S. system. In particular, the Pacific Intertie, linking Southern California to the Pacific Northwest, and the Arizona to Southern California transmission corridors are very important, since these systems are normally loaded with power flowing toward Southern California. The need for SPS arming is determined at the Energy Control Center (ECC) of the Los Angeles Department of Water and Power, by observing these transmission loadings.

The SPS itself consists of measurements taken at both ICS, ACS, and HVDC converter stations at both Intermountain and Adelanto. The following input triggers are defined:

Fault type
Fault location
Voltage level
One ac line outage
Two ac line outages
Pole block
Bipole block

There are 17 of these input triggers, located at the two ends of the HVDC system. These input trigger states are transmitted by microwave and fiber-optic signal communications to a computer logic unit located at generating station. There the input triggers are combined to identify 13 single and six double (or multiple) contingency disturbances, which are called contingency triggers. The logic is very complex and was determined on the basis of hundreds of stability studies [35]. Finally, these contingency triggers, for both normal and breaker failure conditions, are combined by OR logic to form five "supertriggers" that are used to trip the generators or take other measures to prevent the propagation of the disturbance into the surrounding transmission system. The system is arranged so that other remedials can be incorporated, if required in the future. This might include tripping lines, applying dynamic brakes, fast ramping of the HVDC system, or other measures that are appropriate for the observed disturbance and system condition. The entire SPS is completely redundant, so that it is possible to remove one system for maintenance and still have a fully operational system in service. The SPS is not designed to require agreement in the trip commands of the two systems.

The arming of the system, which is performed at the Energy Control Center (ECC) is also automatic, but has manual backup. The entire system is illustrated in Figure 21.44. The arming is based on charts that have been prepared with the aid of stability studies. These charts are organized in *chart sets* with each set belonging to a three-dimensional array on the dispatch computer at the ECC. The three dimensions of the array are

Regional power flow
Configuration of the system
Supertrigger pickup

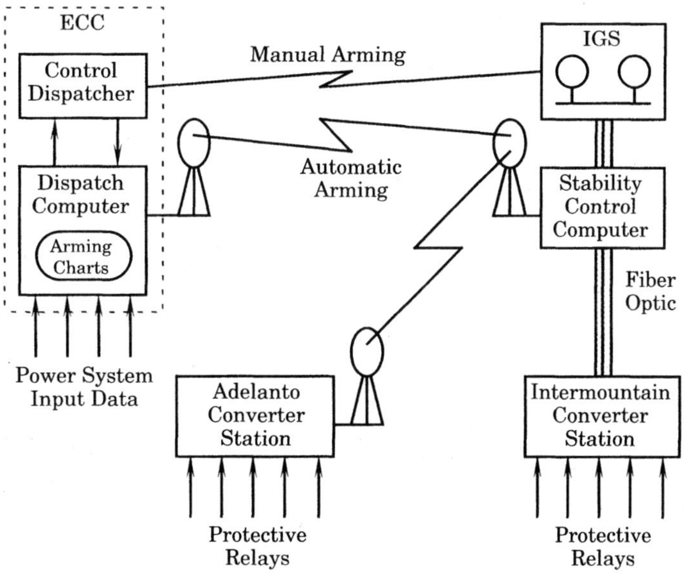

Figure 21.44 Diagram of the arming and data acquisition system [35].

The computer first searches along the regional power flow axis to obtain a column of charts corresponding to the power flowing on the Pacific Intertie, the Arizona-California Corridor, and the Utah Transmission Corridor. A chart set is then selected that corresponds to the system configuration, which is particularly sensitive to the following:

Number of IPP generating units in service
HVDC poles in operation
Major Utah lines out of service
Intermountain ac lines out of service
Adelanto ac lines out of service

Finally, each chart set contains five individual charts, one for each super trigger. A sample chart for supertrigger 5 is shown in Figure 21.45 and is typical of the appearance of all charts. This chart tells the operator whether he should arm the system to trip 0, 1, or 2 generating units, depending on the dc line loading and the regional transfers for which the chart is constructed. Based on this information, the dispatch computer will send the arming information to the stability control computer to set the arming condition. Then, when the appropriate input trigger combinations arrive at the computer, the correct number of generating units will be tripped automatically. This is important, since there is not time to compute the system condition once the triggers are picked up, as stability requires prompt action, usually within a few cycles.

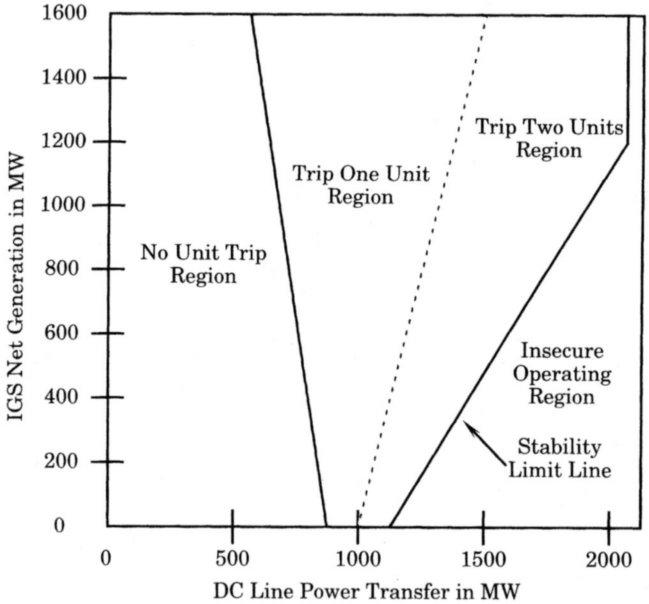

Figure 21.45 Typical contingency arming chart for supertrigger 5 and for a given regional transfer condition.

If both the stability control computer and the dispatch computer are unavailable for some reason, the system operator can look at the charts in a notebook and telephone the power plant to order manual arming by the plant operator, who enters the correct arming into a push-button panel at the plant. The computer systems are completely redundant, but the manual arming procedure provides an additional level of redundancy, and hence improves the probability of correct operation.

The procedure for constructing a contingency arming system such as this is complex and involves the cooperation of many engineers, from system planning, protection, design, communications, and operations. It illustrates the lengths to which system designers must go in order to ensure adequacy and security in the operation of an interconnected power system. It also illustrates a different use for the information contained in the protective systems that are located throughout the power system. These devices, taken collectively, have a great store of information that is useful in many ways. We usually think of protective devices as being dedicated to a single task, and this is often the case. However, the information regarding the operation of a particular protection can be important in ordering preplanned actions to literally save the system from degradation or even total collapse. The future will probably see more of this type of mixed use of protective device information, particularly as the relays themselves become entirely digital, which will simplify somewhat the monitoring, storage, and communications of their states. There also seems to be a trend to delay or cancel the construction of new transmission facilities, which may dictate the design and implementation of additional special protection schemes, similar to the one described here.

21.7 SUMMARY

Some power systems are stability limited, and the desired operating conditions may require that protective systems be provided in order to operate the system in a secure manner. To meet these requirements, a number of protective schemes have been developed that have power system stability enhancement as their primary purpose. The implementation of these schemes requires a knowledge of the dynamic system performance, which can best be determined by stability studies. This often requires a cooperative effort between the engineers responsibility for stability analysis and the protection engineers. Through this type of cooperation, a protective strategy can be developed that will ensure the maintenance of system security in an environment that could otherwise lead to system breakup and possible loss of loads.

REFERENCES

[1] Anderson, P. M., and A. A. Fouad, *Power System Control and Stability*, IEEE Press, Piscataway, NJ, 1995.

[2] Podmore, R., and A. Germond, "Development of Dynamic Equivalents for Transient Stability Studies," EPRI EL-456, Palo Alto, 1977.

[3] Westinghouse Relay-Instrument Division, *Applied Protective Relaying*, Westinghouse Electric Corporation, Newark, NJ, 1976.

[4] GEC Measurements, *Protective Relays Application Guide*, General Electric Company p.l.c. of England, 1975.

[5] Warrington, A. R. van C., *Protective Relays: Their Theory and Practice*, Volume One, John Wiley & Sons, Inc. New York, 1962.

[6] Jay, Frank, Editor in Chief, *IEEE Standard Dictionary of Electrical and Electronics Terms*, Fourth Edition, ANSI/IEEE Std 100-1988, The Institute of Electrical and Electronics Engineers, New York, 1988.

[7] SEL-279 Instruction Manual, Reclosing Relay, Voltage Relay, Synchronism Check Relay, Schweitzer Engineering Laboratories, Inc., Pullman, WA, March 18, 1994.

[8] ASME/IEEE Committee Report, "State of the Art Symposium - Turbine Generator Shaft Torsionals," Prepared by the ASME/IEEE Joint Working Group on Power Plant/Electric System Interaction, IEEE publication 79 TH0059-6-PWR, IEEE, New York, 1979.

[9] ANSI Standard C 50.13-1965, "Requirements for Cylindrical Rotor Synchronous Generators," American National Standards Institute, New York, 1965.

[10] Abolins, A., D. Lambrecht, J. S. Joyce, and L. T. Rosenberg, "Effect of Clearing Short Circuits and Automatic Reclosing on Torsional Stress and Life Expenditure of Turbine-Generator Shafts," *IEEE Trans.*, PAS-95, January-February 1975, pp. 14–25.

[11] Baldwin, M. S., and R. H. Daugherty, "Transmission Line Switching and its Effect on Generators," a paper presented at the Missouri Valley Electric Assn. meeting, Kansas City, April 1977.

[12] Fujikura, T., T. Komukai, and M. Udo, "Statistical Approach to Analysis of Turbine Generator Shaft Torques at High Speed Reclosing, *IEEE Trans.*, PAS-96, July-August 1977, p. 1070.

[13] Jackson, M. C., S. D. Umans, R. D. Dunlop, S. H. Horowitz, and A. C. Parikh, "Turbine-Generator Shaft Torques and Fatigue: Part I - Simulation Methods and Fatigue Analysis," *IEEE Trans.*, PAS-98 (6), 1979, pp. 2299–2307.

[14] Dunlop, R. D., S. H. Horowitz, A. C. Parikh, M. C. Jackson, and S. D. Umans, "Turbine-Generator Shaft Torques and Fatigue: Part II - Impact of System Disturbances and High Speed Reclosure," *IEEE Trans.*, PAS-98, (6), 1979, pp. 2308–2328.

[15] Undrill, J. M., and L. N. Hannett, "Turbine Generator Impact Torques in Routine and Fault Operations," *IEEE Trans.*, PAS-98 (7), 1979, pp. 618–628.

[16] Lokay, H. E., D. G. Ramey, and W. R. Brose, "Turbine-Generator Shaft Loss-of-Life Concepts for Power System Disturbances," Proceedings of the American Power Conference, Chicago, 1978.

[17] Martinich, T. G., H. W. Dommel, and P. J. Fenwick, "The Effect of Electrical Disturbances on the Resultant Shaft Torques of Turbogenerator Units," *IEEE Trans.*, PAS-98 (4), 1979, p. 1144.

[18] Ramey, D. G., A. C. Sismour, and G. C. Kung, "Important Parameters in Considering Transient Torques on Turbine-Generator Shaft Systems," *IEEE Trans.*, PAS-99, January-February 1980, pp. 311–317.

[19] Rusche, P. A., "Turbine Generator Shaft Stresses Due To Network Disturbances - A Bibliography with Abstracts," *IEEE Trans.*, PAS-99, July-August 1980, p. 1329.

[20] Hammons, T. J., "Stressing of Large Turbine-Generators at Shaft Couplings and LP Turbine Final-Stage Blade Roots Following Clearance of Grid System Faults and Faulty Synchronization," *IEEE Trans.*, PAS-99, July-August 1980, pp. 1652–1667.

[21] Canay, I. M., H. J. Rohrer, and K. E. Schnirel, "Effect of Electrical Disturbances, Grid Recovery Voltage and Generator Inertia on Maximization of Mechanical Torques in Large Turbogenerator Sets," *IEEE Trans.*, PAS-99, July-August 1980, pp. 1357–1370.

[22] Dunlop, R. D., S. H. Horwitz, J. S. Joyce, and D. Lambrecht, "Torsional Oscillations and Fatigue of Steam Turbine-Generator Shafts Caused by System Disturbances and Switching Events," CIGRÉ paper 11-06, presented at the 1980 Session, Paris, August-September 1980.

[23] Bowler, C. E. J., P. G. Brown, and D. N. Walker, "Evaluation of the Effect of Power Circuit Breaker Reclosing Practices on Turbine-Generator Shafts," *IEEE Trans.*, PAS-99, September-October 1980, pp. 1764–1779.

[24] Hammons, T. J., "Shaft Torsional Phenomena in Governing Large Turbine-Generators with Non-Linear Steam Valve Stroking Constraints," *IEEE Trans.*, PAS-99, July-August 1980, p. 1320.

[25] IEEE Committee Report, "Effects of Switching Network Disturbances on Turbine-Generator Shaft Systems," *IEEE Trans.*, PAS-101, September 1982, pp. 3151–3157.

[26] Mason, C. R., *The Art and Science of Protective Relaying*, John Wiley & Sons, New York, 1956.

[27] Anderson, P. M., and B. K. LeReverend, "Industry Experience with Special Protection Schemes," IEEE/CIGRÉ Committee Report, IEEE Trans., PWRS (11), August 1996, pp. 1166–1179.

[28] Klinger, M., W. A. Mittelstadt, and C. W. Taylor, "Transient Stability Controls Used by Bonneville Power Administration To Mitigate Delays of Planned Facilities," CIGRÉ paper 32-01, presented at the 1982 Session, Paris, September 1982.

[29] Taylor, C. W., "Hydro Generator Tripping in the Pacific Northwest for Stability Enhancement," a paper presented for the Generator Tripping Panel, IEEE PES Summer Meeting, Los Angeles, July 1983.

[30] Taylor, C. W., J. M. Hauer, L. A. Hill, W. A. Mittelstadt, and R. L. Cresap, "A New Out-of-Step Relay with Rate of Apparent Resistance Augmentation," *IEEE Trans.*, PAS-102, pp. 631–639, March 1983.

[31] Lubkeman, D. L., and G. T. Heydt, "The Application of Dynamic Programming in a Discrete Supplementary Control for Transient Stability Enhancement of Multimachine Power Systems," *IEEE Trans.*, PAS-104, September 1985, pp. 2342–2348.

[32] Alexander, G. E., and T. U. Patel, "Severe Line Outage Detector (SLOD) Generation Reduction Scheme for a Generating Station," *IEEE Trans.*, PWRD-1, January 1986, pp. 118–124.

[33] Ohura, Y., K. Matsuzawa, H. Ohtsuka, N. Nagai, T. Gouda, H. Oshida, S. Takeda, and S. Nishida, "Development of a Generator Tripping System for Transient Stability Augmentation Based on the Energy Function Method," *IEEE Trans.*, PWRD-1, July 1986, pp. 68–77.

[34] Doudna, J. H., "Application and Implementation of Fast Valving and Generator Tripping Schemes at Gerald Gentleman Station," *IEEE Trans.*, PWRS-3, August 1988, pp. 1155–1166.

[35] Beshir, M. J., J. H. Gee, and R. L. Lee, "Contingency Arming System Implementation for the Intermountain Power Project HVDC Transmission System," *IEEE Trans.*, PWRS-4, May 1989, pp. 434–442.

PROBLEMS

21.1 Review definitions of stability in several textbooks, both from mathematics and Electrical Engineering.

21.2 Consider a 500 kV transmission line with the conductor and ground wire spacings shown in Figure P21.2.

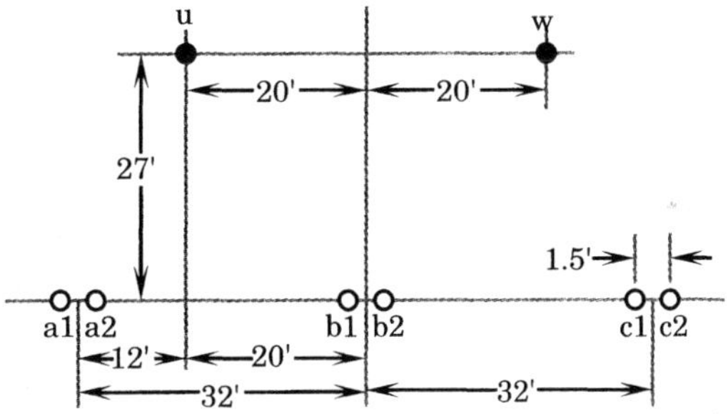

Figure P21.2 A 500 kV transmission line configuration.

The conductor data for the line of Figure 21.46 is given in Table 21.7.

TABLE P21.2 Conductor Characteristics of the 500 kV Line

	Phase Conductors	Ground Conductors
Conductor size and type	2156 kCM ACSR Bluebird	7#8 Alumoweld
Resistance, at 25° C, 60 Hz	0.0464 Ω/mile	2.440 Ω/mile small current
D_s = self GMD of wire	0.0588 ft	0.002085 ft
Radius of wire	0.07342 ft	0.01604 ft

The statement is made that the resistance term can often be neglected. Consider a 500 kV transmission line of nominal design and determine the resistance and inductive reactance per unit of length of the line under study. Would you say that it is reasonable to neglect the resistance term?

21.3 Verify (21.4) and (21.5).

21.4 Write out a complete derivation of the equal area criterion in order to develop a clear understanding of this important concept.

21.5 Write the swing equation (21.36) as a differential equation in terms of angular velocity, rather than shaft angle.

21.6 The data for a turbine-generator unit is given by the manufacturer as follows: 525 MVA, 24 kV, 3600 rpm, moment of inertia = 444,000 lb-ft^2. What is the value of H, the inertia constant?

21.7 It is postulated that the fault location pictured in Figure 21.6 is the most severe location. Can you support this assumption?

21.8 It is noted in Figure 21.11 that the SVC is not capable of maintaining the voltage at bus 7, although it has adequate rating to support this bus under steady-state conditions. Why is this SVC inadequate?

21.9 Suppose that the decision is made to reinforce the system described in Section 21.3.2 to be able to sustain a power transfer of 1100 MW, taking into account that faults may occur on one of the lines connecting bus 7 to bus 8. What could be done to reinforce the system to meet this objective?

21.10 Figures 21.15 and 21.16 show the apparent impedance seen by the relays at buses 7 and 8, respectively, for a case where there is no instability or system separation. Is there any danger of relay false tripping for this case?

21.11 Can you suggest a way in which the out-of-step condition pictured in Figures 21.20 and 21.21 could be detected?

21.12 The utility management has decided that reclosing line D or E into a permanent fault would be a reasonable risk if this would not happen more than once every 10 years. Data for these and similar lines indicate that a line permanent fault occurs approximately every m years and that when this happens the line repairs require about r hours. Using this experience data, will we be able to show the management that we will meet the criterion of not reclosing into a fault more than once in 10 years? A typical value for the annual failure rate for a 500 kV transmission line is provided by the company statistician to be 0.0001 outages per km and the mean repair time is 10 hours.

22

HVDC Protection

22.1 INTRODUCTION

High-voltage direct current (*HVDC*) systems are becoming more and more common in modern power systems and their protection must be coordinated with other protective systems of the ac network. The purpose of this chapter is to provide an introduction to the subject of HVDC protections. The HVDC protective systems are different than those on the rest of the power system. One major difference stems from the fact that the HVDC converter stations are purchased as a complete system, including the protections. In some parts of the converter system, it is difficult to clearly distinguish between a control and a protection, since both functions are performed by the same or similar devices. In fact, certain HVDC protections are provided by control actions, using the same equipment that controls the converter in its normal operation.

The emphasis of this chapter is on the two-terminal HVDC system, consisting of two converter stations, a rectifier and an inverter, connected by a dc transmission line. In some cases the line is omitted, and the rectifier and inverter are simply connected "back-to-back." Many of the protective systems are the same in the back-to-back system except that the line protection is omitted or greatly simplified. Future HVDC systems will undoubtedly include several converter stations, interconnected by a dc transmission network. Such multiterminal systems will probably include protections that are not described here, but they will be extensions of the functions discussed below. The dc transmission network will require dc circuit breakers, which are difficult to design. However, there are proposed designs for dc circuit breakers and dc transmission networks will eventually be constructed.

22.2 DC CONVERSION FUNDAMENTALS

The HVDC converters designed and constructed since the early 1970s are solid-state power electronic designs, using high-voltage, high-power thyristors for the switching elements. The older designs used mercury-arc devices, and several of these systems are still in service. The principles of operation are the same in either case, since the electronic device used is basically a switch. These are often referred to as "valves" in the literature.

22.2.1 Rectifier Operation

It has been shown that the most practical converter is the three-phase bridge, which is sometimes called a *Graetz* bridge. A typical arrangement is shown in Figure 22.1, which shows a three-phase bridge supplied by an equivalent circuit on the left, representing the supply transformer and its leakage reactance, L_c. The supply transformer bank can be connected in either a delta or wye arrangement, with the wye connection having a distinct advantage, as will be shown. The dc load is not shown, but is assumed to be to the right in the figure. If the currents are flowing out from the converter, it is operating as a rectifier, and if the current is flowing in, the operation is that of an inverter. A dc transmission system must always have at least one rectifier, but there can be multiple inverters delivering power to different points. This assumes a dc transmission network of more than two terminals. The most common arrangement is to have one rectifier and one inverter, with a single dc transmission line connecting the two converter stations. In some cases, the line is just a few feet long, connecting two asynchronous ac systems, in which case it is called a "back-to-back" dc converter system.

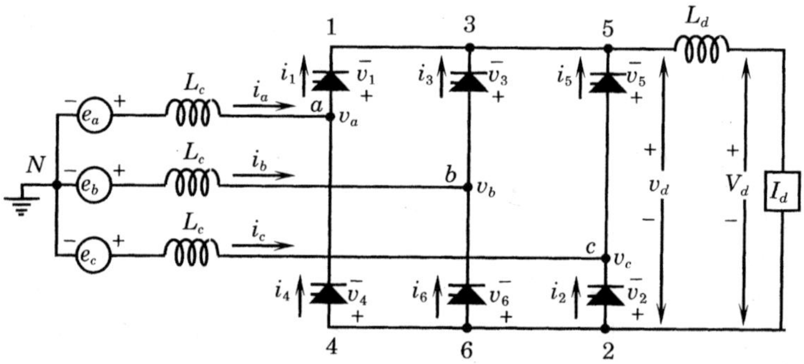

Figure 22.1 A three-phase or Graetz bridge converter.

22.2.1.1 Uncontrolled Six-Pulse Rectifier Operation. For the bridge converter, there are always two valves conducting in series. Assume, for example, that the phase a voltage is the most positive voltage, such as during the period from zero to 60 degrees in Figure 22.2. Then valve 1 will conduct current to the load and the return current will flow through valve 2, since phase c voltage is the most negative voltage during this period. As time moves to the next 60 degree period, from 60 degrees to 120 degrees, phase b is the most positive, and phase c is the most negative, so the current leaves the converter through valve 3 and returns through valve 2. From 120 degrees to 180 degrees, phase b is still the most positive, but now phase a is the most negative, so the current leaves through valve 3 and returns through valve 4 (see the shaded region of Figure 22.2)

Consider the period from 120 degrees to 180 degrees, represented by the shaded area in Figure 22.2. During this period, the voltage e_b is the most positive, which will cause valve 3 to conduct. Then the commutating voltage e_b is given by

$$e_b = E_m \cos(\omega t - 120°) \tag{22.1}$$

During this same period, the minimum voltage across the rectifier is that of phase a, which is

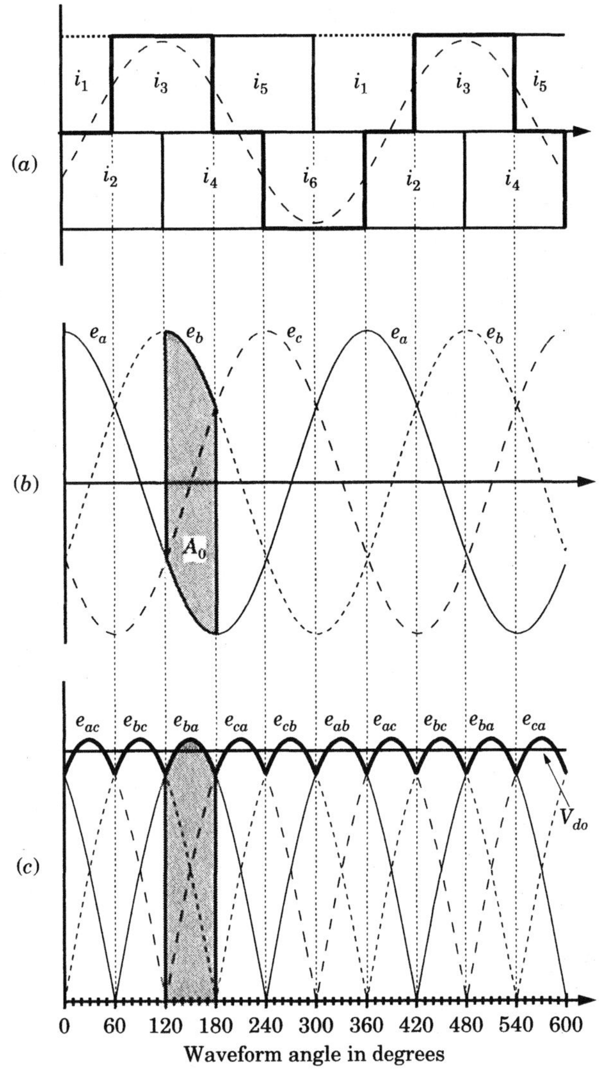

Figure 22.2 Uncontrolled rectifier waveforms. (*a*) Secondary current. (*b*) Voltage waveforms. (*c*) Rectified voltages.

given by

$$e_a = E_m \cos \omega t \tag{22.2}$$

Then, the total voltage across the entire rectifier during this time period is

$$\begin{aligned} e_{ba} &= e_b - e_a = E_m[\cos(\omega t - 2\pi/3) - \cos \omega t] \\ &= -\sqrt{3}E_m \cos(\omega t + \pi/6) = \sqrt{3}E_m \cos(\omega t - 5\pi/6) \end{aligned} \tag{22.3}$$

During this period, the average value of the rectified voltage can be determined by integrating the instantaneous rectifier voltage over any 60 degree period similar to that pictured in Figure 22.2. For the area pictured, we have

$$V_{do} = \frac{3}{\pi} A_o = \frac{3}{\pi} \int_{2\pi/3}^{\pi} \sqrt{3}E_m \cos(\omega t - 5\pi/6) d\omega t = \frac{3\sqrt{3}}{\pi} E_m \tag{22.4}$$

where A_0 corresponds to the shaded area and is equal to the integral of (22.3) over the 60 degree interval. The actual rectified voltage is a series of pulses, shown by the heavy line on Figure 2.22(c). The average value, V_{do}, is shown as the straight line in Figure 22.2(c) and represents the value determined by (22.4). This voltage is called the *ideal no-load direct voltage*. Note that the rectified waveform results in six voltage pulses every 360 degrees. Therefore, this type of circuit is referred to as a six-pulse bridge rectifier.

We usually find it convenient to work with rms line-to-line voltages. For the converter transformer secondary, we can write

$$E_m = \sqrt{2}V_{LN} = \frac{\sqrt{2}V_{LL}}{\sqrt{3}} \tag{22.5}$$

where V_{LL} is the rms value of the line-to-line voltage. Then (22.4) becomes

$$V_{do} = \frac{3\sqrt{2}V_{LL}}{\pi} \tag{22.6}$$

The converter transformer secondary current, shown in Figure 22.2(a), is made up of pulses of current that are created by the switching of the valves. The valves are numbered such that commutation occurs from valve 1 to valve 3, then from 2 to 4, then from 3 to 5, then from 4 to 6, then from 5 to 1, then from 6 to 2. The secondary current is made up of these square wave pulses, which roughly approximates a sine wave. For example, for the phase b transformer secondary current during the period shown in Figure 2.22, we can write

$$i_b = i_3 \text{ or } i_6 \tag{22.7}$$

where we write the equation with the qualifying *or* since i_3 and i_6 do not flow at the same time. In Figure 22.2(a), the phase b current is shown both as a dashed sine wave and also in terms of these pulses that occur when valves 3 and 6 are conducting. The total secondary current consists of large pulses, shown by the heavy lines in Figure 22.2(a). Figure 22.1 shows that i_3 and i_6 are connected to phase b. The load current is always carried by two valves in series. In this segment of the time domain, the current carrying valves are numbers 3 and 6, which results in current pulses in phase b.

The *peak inverse voltage* (PIV) of the rectifier is the maximum instantaneous voltage across the diode in the reverse direction. This can be computed by summing around the loop consisting of the supply circuit and any conducting valve, and solving for the peak voltage across any nonconducting valve. For example, during the period highlighted in Figure 22.2, we can sum voltages around the loop b-3-1-a-N-b as follows.

$$-v_3 + v_1 - e_a + e_b = 0 \tag{22.8}$$

However, the voltage across valve 3 is zero, since that valve is conducting. Therefore, we can solve (22.8) for the instantaneous voltage across valve 1.

$$v_1 = e_a - e_b = e_{ab} = \sqrt{3}E_m \cos(\omega t + \pi/6) \tag{22.9}$$

The peak value of this voltage is the *peak inverse voltage* for this valve.

$$\text{PIV} = \sqrt{3}E_m = \frac{\pi V_{do}}{3} = 1.047\ V_{do} \tag{22.10}$$

This value can be verified by summing voltages around any loop in a similar manner during any portion of the operating cycle of the bridge.

The peak-to-peak voltage ripple can be determined by solving for the difference in the instantaneous voltage at peak and at its minimum value.

$$\begin{aligned}\text{PPR} &= v_d(\pi/6) - v_d(0) = \sqrt{3}E_m\left(\cos 0 - \cos\frac{\pi}{6}\right)\\ &= \sqrt{3}E_m\left(1 - \frac{\sqrt{3}}{2}\right) = \left(\sqrt{3} - \frac{3}{2}\right)E_m = 0.23205E_m\end{aligned} \tag{22.11}$$

The average value of direct current can be found by integrating the current pulse of magnitude I_d over its time duration.

$$I_{\text{ave}} = \frac{1}{2\pi}\int_{-\pi/3}^{5\pi/3} I_d d\omega t = \frac{I_d}{3} \tag{22.12}$$

22.2.1.2 The Controlled Rectifier. When thyristor valves are used, it is possible to delay the valve conducting turn-on time. This delay angle is usually called *alpha*, which modifies (22.4) to the following form.

$$\begin{aligned}V_d &= \frac{1}{\pi/3}\int_0^{\pi/3+\alpha}\sqrt{3}E_m\sin\left(\omega t + \frac{\pi}{3}\right)d\omega t = \frac{3\sqrt{3}}{\pi}E_m\cos\alpha\\ &= V_{do}\cos\alpha\end{aligned} \tag{22.13}$$

This condition is shown in Figure 22.3, where the delay angle of firing of the valves is shown as 15 degrees. This causes the square wave pulses of current to also be delayed by this same

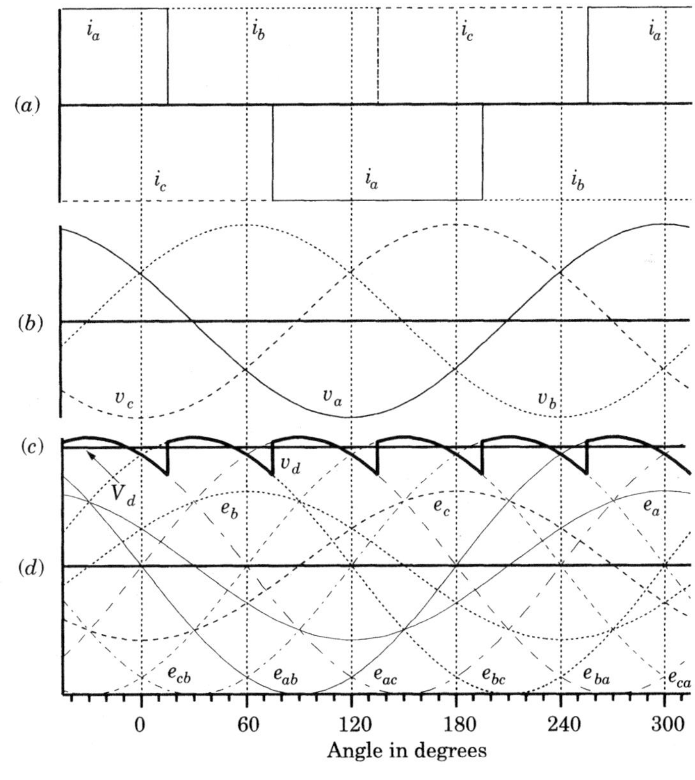

Figure 22.3 Controlled rectifier waveforms of current and phase voltage illustrated for a delay angle of 15° (*a*) Line currents. (*b*) Line-to-neutral voltages, (*c*) Direct voltage. (*d*) Applied line-to-line voltages.

angle. Varying the delay angle provides a control over the direct voltage and, therefore, over the power transferred over the dc line.

22.2.1.3 Rectifier with Overlap Less Than 60 degrees. The waves shown in Figure 22.3 are an idealized representation of the valve firing. In physical circuits, the commutation inductance L_c causes an additional delay. This delay is necessary since it is not possible to switch currents in an inductive circuit instantaneously. The converter transformer has inherent inductance that makes this additional delay unavoidable. The angle of this delay is called the *overlap angle*, and it is represented here by the symbol μ. Therefore, the overlap time is μ/ω seconds. In normal operation the overlap angle is less than 60 degrees, with typical full load values being in the range of 20–25 degrees.

Figure 22.4 shows the effect of overlap on the number of valves conducting at any instant of time. When the overlap angle is zero, only two valves are conducting at any instant. As the overlap angle increases, three valves are conducting for brief periods, but between commutations only two valves conduct. The angle interval during which two valves conduct is $60° - \mu$. When the overlap angle increases beyond 60 degrees, four valves are conducting for brief periods, as noted for the 90 degree overlap condition in Figure 22.4. This is an abnormal mode of operation that is usually avoided.

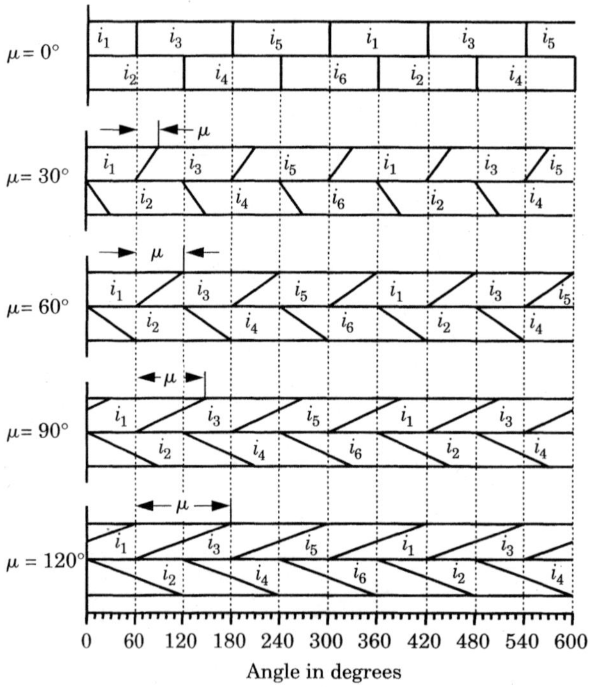

Figure 22.4 Effect of overlap on the number of valves conducting.

The normal mode of operation is for the overlap angle to be less than 60 degrees. In this situation, two valves will conduct for a period, followed by three valves conducting. Two valves conducting has already been analyzed, with the resulting average value of direct voltage given by (22.9). The sum of the delay angle and the overlap angle is called the *extinction angle*, and is written in this book as[1]

$$\delta = \alpha + \mu \tag{22.14}$$

[1]This notation comes from Kimbark [1] and others. Some refer simply to the sum as the commutating margin angle.

We can analyze the operation of the system using an example, where we examine the period during which both valves 1 and 3 are conducting, and where valve 2 is also conducting. Beginning at $\omega t = a$, we can write

$$\begin{aligned} i_1 &= I_d \\ i_3 &= 0 \end{aligned} \tag{22.15}$$

At the end of the interval $\omega t = a + \mu = \delta$, we have

$$\begin{aligned} i_1 &= 0 \\ i_3 &= I_d \end{aligned} \tag{22.16}$$

Referring to Figure 22.1, we can write the voltage equation around the loop N-3-1-N as follows.

$$e_b - e_a = \sqrt{3}E_m \sin\omega t = L_c\frac{di_3}{dt} - L_c\frac{di_1}{dt} \tag{22.17}$$

But

$$i_1 + i_3 = I_d \tag{22.18}$$

so that

$$\frac{di_1}{dt} + \frac{di_3}{dt} = 0 \tag{22.19}$$

Therefore,

$$\sqrt{3}E_m \sin\omega t = 2L_c\frac{di_3}{dt} \tag{22.20}$$

We now divide this equation by $2L_c$ and integrate with respect to time, setting the lower limit of integration according to (22.15) and the upper limit by (22.16).

$$\frac{\sqrt{3}E_m}{2L_c}\int_{\alpha/\omega}^{t} \sin\omega t\, dt = \int_0^{i_3} di_3 \tag{22.21}$$

This gives the result

$$I_{s2}(\cos\alpha - \cos\omega t) = i_3 = I_d - i_1 \tag{22.22}$$

where we have defined the current

$$I_{s2} = \frac{\sqrt{3}E_m}{2\omega L_c} \tag{22.23}$$

From (22.22), we see that the current i_3 has two components, a dc term and a sinusoidal component. At the end of the commutation period, the direct current is given by (22.16), or

$$I_d = I_{s2}(\cos\alpha - \cos\delta) \tag{22.24}$$

which gives the current in terms of both the ignition and extinction angles.

We now develop a similar equation for the direct voltage with the aid of Figure 22.5. From (22.2) we have determined that the shaded area between the voltage curves is

$$A_0 = \frac{\pi V_{do}}{3} \tag{22.25}$$

The integration to determine the average voltage also determines the area. In a like manner, we can write, in reference to Figure 22.5

$$A_1 = \frac{\pi \Delta V_d}{3} \tag{22.26}$$

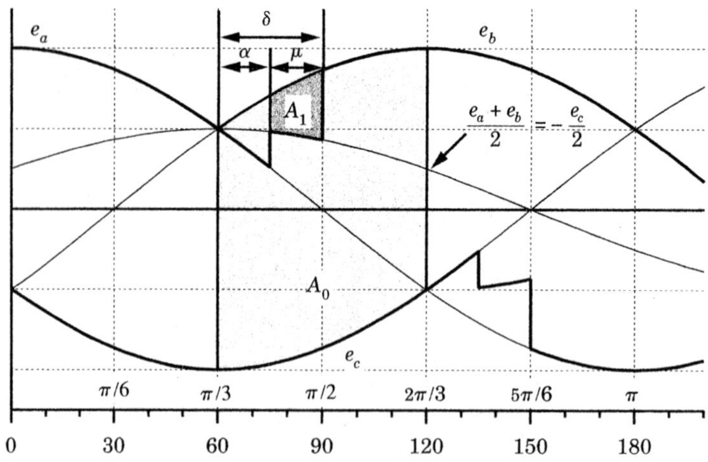

Figure 22.5 Reduction in direct voltage caused by overlap.

where A_1 is the area lost due to overlap, which is related directly to the reduction in the average voltage. From Figure 22.5, we can compute the area A_1 as follows.

$$A_1 = \int_\alpha^\delta \left(e_b - \frac{e_a + e_b}{2} \right) d\theta = \int_\alpha^\delta \left(\frac{e_a - e_b}{2} \right) d\theta$$
$$= \frac{\sqrt{3}E_m}{2} \int_\alpha^\delta \sin\theta d\theta = \frac{\sqrt{3}E_m}{2}(\cos\alpha - \cos\delta) \tag{22.27}$$

Then the reduction in the average voltage is given by

$$\Delta V_d = \frac{3}{\pi} A_1 = \frac{3\sqrt{3}E_m}{\pi}(\cos\alpha - \cos\delta) \tag{22.28}$$

The direct voltage with overlap is given by

$$V_d = V_{do}\cos\alpha - \Delta V_d = \frac{V_{do}(\cos\alpha + \cos\delta)}{2} \tag{22.29}$$

Comparing (22.28) with the current equation (22.24) we can find a relationship between the voltages and currents, which we can write as

$$\frac{\Delta V_d}{V_{do}} = \frac{I_d}{2I_{s2}} \tag{22.30}$$

This result can be used in (22.29) to write the average direct voltage as

$$V_d = V_{do}\left(\cos\alpha - \frac{I_d}{2I_{s2}} \right) = V_{do}\cos\alpha - R_c I_d \tag{22.31}$$

where R_c is called the *equivalent commutating resistance*. This quantity can be further defined using (22.31) and (22.22).

$$R_c = \frac{V_{do}}{2I_{s2}} = 6fL_c = \frac{3X_c}{\pi} \tag{22.32}$$

The current and voltage waveforms for the six-pulse bridge rectifier are shown in Figure 22.6. An equivalent circuit for the bridge rectifier is shown in Figure 22.7.

22.2.1.4 Rectifier with Overlap More Than 60 degrees. If the overlap angle is greater than 60 degrees, the rectifier operation is abnormal. Reference to Figure 22.4 shows that at

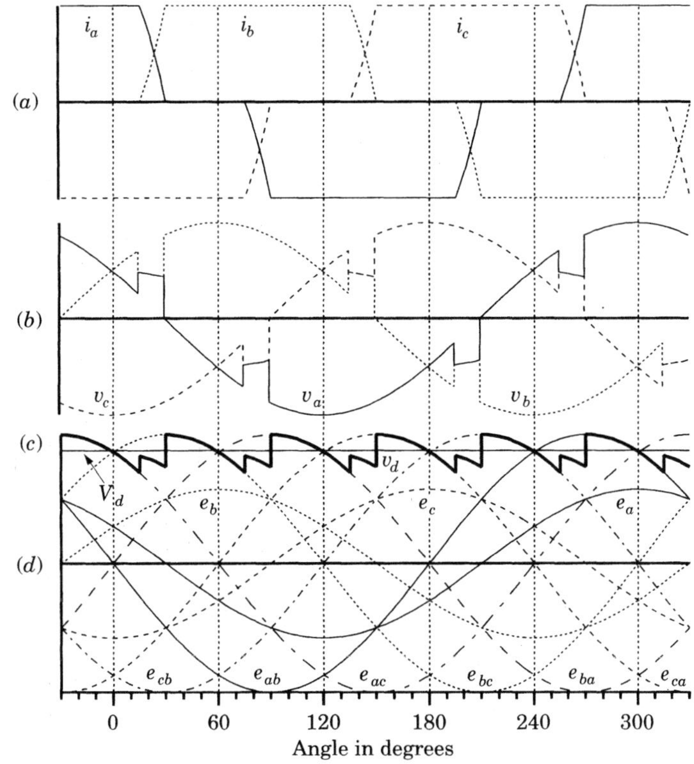

Figure 22.6 Currents and voltages of the three-phase bridge. (*a*) Line currents. (*b*) Line-to-neutral voltages. (*c*) Direct voltage. (*d*) Line-to-line voltages.

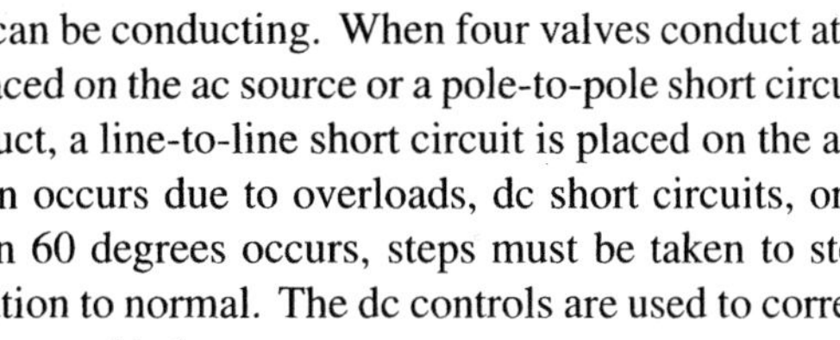

Figure 22.7 Equivalent circuit for the six-pulse bridge rectifier.

times as many as four valves can be conducting. When four valves conduct at the same time, a three-phase short circuit is placed on the ac source or a pole-to-pole short circuit on the dc side. When only three valves conduct, a line-to-line short circuit is placed on the ac side for a short time. This mode of operation occurs due to overloads, dc short circuits, or low ac voltage. When an overlap greater than 60 degrees occurs, steps must be taken to stop the operation promptly and return the operation to normal. The dc controls are used to correct this abnormal condition, and this will be discussed below.

22.2.2 Inverter Operation

For a complete HVDC system to operate it is also necessary to design an inverter, to deliver the dc power back to the ac network at the receiving end of the dc transmission line. The valves are unidirectional devices, so it is not possible to reverse the current direction.

Therefore, the reversal of power flow through the converter is accomplished by reversing the polarity of the direct voltage. This can be done if the delay angle is greater than 90 degrees and less than 180 degrees. In this case, the angle of interest is called the *extinction angle* and is measured in the reverse direction from the delay angle.

From the previous discussion for the rectifier, it is noted that when the delay angle exceeds 90 degrees the average value of the resulting direct voltage will be negative. Mathematically, we can write the conditions as

$$\begin{aligned} 0^\circ < \alpha < 90^\circ & \qquad \text{rectification occurs} \\ 90^\circ < \alpha < 180^\circ & \qquad \text{inversion occurs} \end{aligned} \tag{22.33}$$

In all practical devices, overlap is always present, so the foregoing relations depend on overlap as well as delay angle. The value of delay angle at which inversion begins is called the *transition value* of α, and this can be written as

$$\alpha_t = \pi - \delta = \frac{\pi - \mu}{2} < 90^\circ \tag{22.34}$$

An additional small margin may be allowed for thyristor switching, which further restricts the range of the delay angle. Figure 22.8 shows the commutation of current from valve 1 to valve 3. The current must be completely commutated to valve 3 before the angle reaches π radians, at which time the voltage reverses direction. Failure to complete commutation before reaching this limit is called a *commutation failure*. The commutation voltage for HVDC converters are usually furnished by synchronous generators.

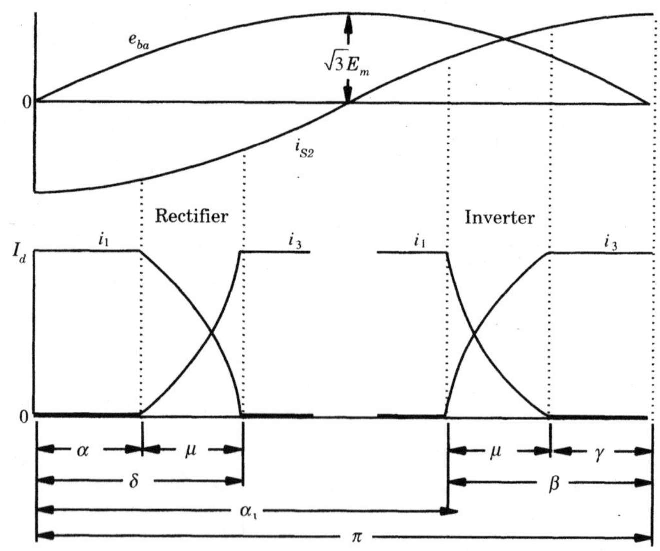

Figure 22.8 Relationship among angles of rectifier and inverter operation.

The ignition angle has been defined as the angle by which ignition must be delayed from the instant ($\omega t = 0$ for valve 3) at which the commutating voltage (e_{ba} for valve 3) becomes zero and is changing in the positive direction. The *extinction angle* δ is measured by the delay from that same angle reference ($\omega t = 0$ for valve 1). For the inverter, we define a new angle β as the ignition angle and γ for the extinction angle, as noted in Figure 22.8. In a sense, the angle β is the mirror image of the angle δ. For the inverter, the angle α is subscripted with i to clarify that this is an angle greater than 90 degrees as required by (22.33).

Based on the defined angles in Figure 22.8, we can write the following angular relationships.

$$\beta = \pi - \alpha_i \tag{22.35}$$

$$\gamma = \pi - \delta = \pi - \mu - \alpha_i \tag{22.36}$$

$$\mu = \delta - \alpha = \beta - \gamma \tag{22.37}$$

It has been noted above that, for inverter operation, the inverter dc voltage must be negative. However, when considering only inverter operation, it is common practice to write the voltage as a positive quantity, with the understanding that it is negative with respect to the rectifier voltage. This avoids complications due to carrying along a minus sign in all calculations.

The inverter equations can be developed in a similar manner as that used for the rectifier equations. This is accomplished by changing the sign of v_d and making the substitutions $\cos\alpha = -\cos\beta$ and $\cos\delta = -\cos\gamma$. Then the inverter equations can be written as follows.

$$I_d = I_{s2}(\cos\gamma - \cos\beta) \tag{22.38}$$

$$V_d = \frac{V_{do}(\cos\gamma + \cos\beta)}{2} \tag{22.39}$$

For constant ignition angle β, we can write

$$V_d = V_{do}\cos\beta + R_c I_d \tag{22.40}$$

Inverters are often controlled to operate at constant extinction angle γ, in which case we can write

$$V_d = V_{do}\cos\gamma - R_c I_d \tag{22.41}$$

An equivalent circuit for the inverter is shown in Figure 22.9, based on (22.40).

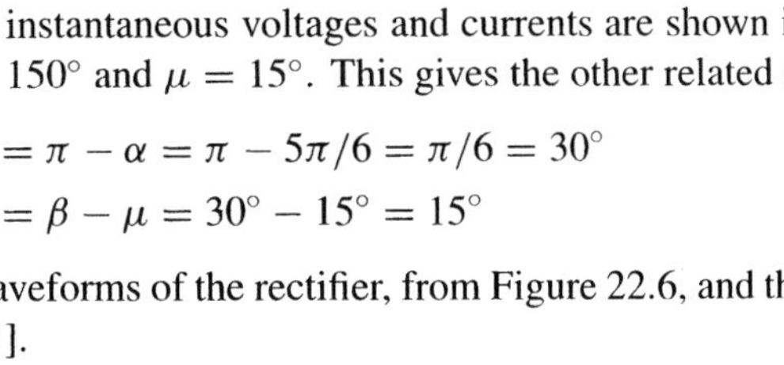

Figure 22.9 Equivalent circuit for the inverter.

The waveforms for the instantaneous voltages and currents are shown in Figure 22.10, which is constructed for $\alpha = 150°$ and $\mu = 15°$. This gives the other related angles as

$$\beta = \pi - \alpha = \pi - 5\pi/6 = \pi/6 = 30° \tag{22.42}$$

$$\gamma = \beta - \mu = 30° - 15° = 15° \tag{22.43}$$

A comparison of the waveforms of the rectifier, from Figure 22.6, and the inverter, from Figure 22.10, is instructive [1].

1. The average voltage across a rectifier valve is negative and is called an *inverse voltage*, whereas the average voltage across the inverter valve is positive.
2. In both modes of operation, the voltage across a valve is positive just before conduction begins, but in the rectifier it is positive for a shorter time period that in the inverter.
3. In both modes of operation, the voltage across a valve is negative immediately after thyristor blocking, but in the inverter it is negative for a much shorter time period than in the rectifier. This time is called the *commutation margin*.

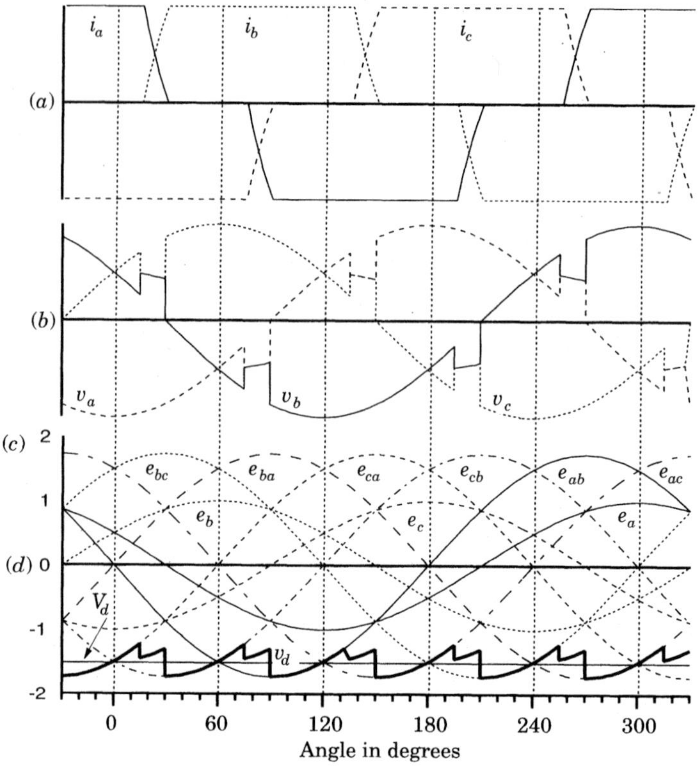

Figure 22.10 Currents and voltages of the three-phase bridge inverter. (*a*) Line currents. (*b*) Line-to-neutral voltages. (*c*) Direct voltage. (*d*) Line-to-line voltages.

4. Changes in voltage across the valves occur at ignition and extinction, but the voltage jump is greater in the inverter than in the rectifier. The formulas for the major voltage jumps are given in Table 22.1.

TABLE 22.1 Major Voltage Jumps in Converter Operation

Operation	Rectifier	Inverter
Ignition	$\sqrt{3}E_m \sin\alpha$	$\sqrt{3}E_m \sin\beta$
Extinction	$\sqrt{3}E_m \sin\delta$	$\sqrt{3}E_m \sin\gamma$

22.2.3 Multibridge Converters

In older designs of three-phase bridge converters that used mercury-arc valves, there was a voltage limit for the system that depended on the maximum allowable peak inverse voltage of the valves. This is no longer true with modern thyristor valve bridges since valves can be connected in series to provide any desired voltage. Moreover, current ratings can be increased to the desired level by paralleling thyristors, although this is usually not necessary in modern designs since thyristors of adequate current rating are usually available. The optimum system design depends on the desired power to be transmitted, the cost of the major components, and the anticipated operating conditions of the entire system.

The most common arrangement for cascading three-phase bridges is the system shown in Figure 22.11, where one three-phase bridge is supplied by a wye-wye connected transformer and the other by a wye-delta transformation. This gives a 12-pulse bridge, which provides advantages in both cost and space requirements. The 12-pulse design doubles the voltage over that obtained with only a single bridge and eliminates the lower harmonics of the six-pulse design, such that the lowest harmonics requiring filtering are the 11th and 13th on the ac side and the 12th harmonic on the dc side. The converter is usually fed from a single three-winding transformer rather than two two-winding banks, which saves one set of transformer circuit breakers.

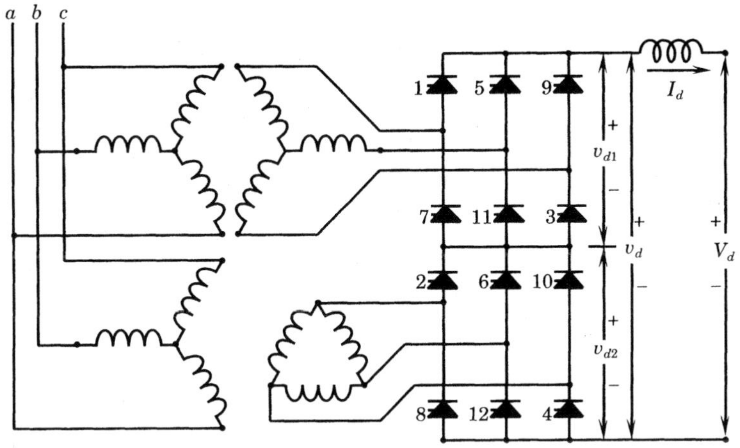

Figure 22.11 A 12-pulse converter station arrangement.

The system shown in Figure 22.11 is described as a monopolar system. The monopolar system has the advantage that the return current can be carried in the earth, thereby simplifying the transmission line. However, a bipolar system has several advantages over monopolar designs. Figure 22.12 shows the basic differences between the two system designs.

The two poles of the bipolar design are nearly independent of each other, which provides several advantages. First, there are very few equipment items that are common to both poles, which increases reliability due to the minimization of common mode failures. When one pole

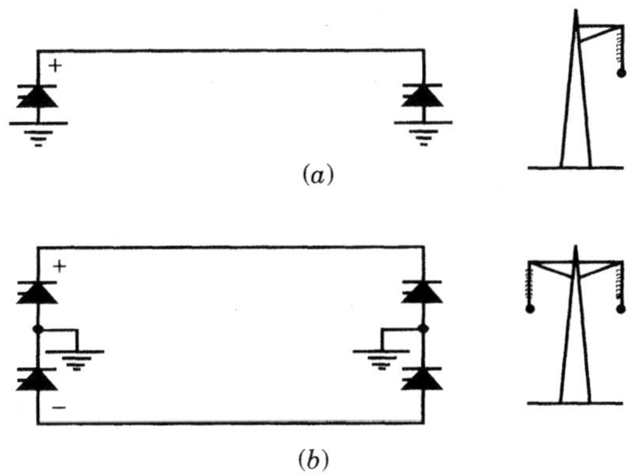

Figure 22.12 Monopolar and bipolar HVDC system designs. (*a*) Monopolar system. (*b*) Bipolar system.

of the bipolar system is out of service for any reason, the system can operate in the monopolar mode, using either the ground for the return current path or, by special switching arrangements, use the overhead line conductor of the pole that is out of service. Note that the monopolar system shown in Figure 22.12(a) uses ground return for the current path. In some cases this may not be permitted, in which case a metallic return, similar to the bipolar system, is necessary. Ground return currents may cause erosion of parallel pipelines, for example, leading to the need for expensive cathodic protection of the pipe. Ground return is used on several systems, for example, systems serving power to islands, such as found in New Zealand and in Sweden.

Harmonics are a factor that must be considered in HVDC systems and harmonics are reduced with the 12-pulse converters stations. The harmonics generated can be express by the formula

$$h = pq \pm 1 \tag{22.44}$$

where h = harmonic number
p = pulse number
q = an integer

Therefore, the harmonics generated on both the ac and dc sides for common 6- and 12-pulse bridge systems can be enumerated, as shown in Table 22.2.

TABLE 22.2 Harmonics Generated by HVDC Converters

Pulse No.	DC Side	AC Side
6	0, 6, 12, 18, 24, ...	1, 5, 7, 11, 13, 17, 19, 23, 25, ...
12	0, 12, 24, ...	1, 11, 13, 23, 25, ...

The amplitudes of the harmonics decrease with increasing order, so it is most important to filter the lower harmonics. Harmonic filters are almost always required on the ac side of the converter and may be required on the dc side as well. Fortunately, the filter appears as a capacitive reactance at the fundamental frequency of the power system, so there is some benefit to the power system in placing filters in service, as they supply a portion of the reactive power required by the converter station. The amount of filtering is a function of the loading of the converter, so filters are often constructed in relatively small sizes, such that individual filter units can be added as the load is increased and removed as load is decreased.

Harmonics can be further reduced by increasing the pulse number, say, from 12 to 18 or 24. However, these higher pulse numbers increase the complexity of the converter transformers. It is simpler and less costly to construct 12-pulse converter transformers and to provide the necessary ac filtering for the resulting harmonics.

There are many details that must be considered in the specification and design of a dc system. These concepts are beyond the scope of this book, but have been treated in many technical papers and in several excellent textbooks [1–5].

22.2.4 Basic HVDC Control

The basic concept of HVDC control is to establish the desired current in the line from rectifier to inverter by varying the dc voltage at the converters. Then the desired power can be caused to flow according to ohms law. This can be visualized by expanding the equivalent circuits developed for rectifier and inverter into the more complete equivalent circuit shown in Figure 22.13, where the equivalent circuit has been extended to include transformer tap changing on both converter stations and the commutating resistances are separated into rectifier and inverter resistances.

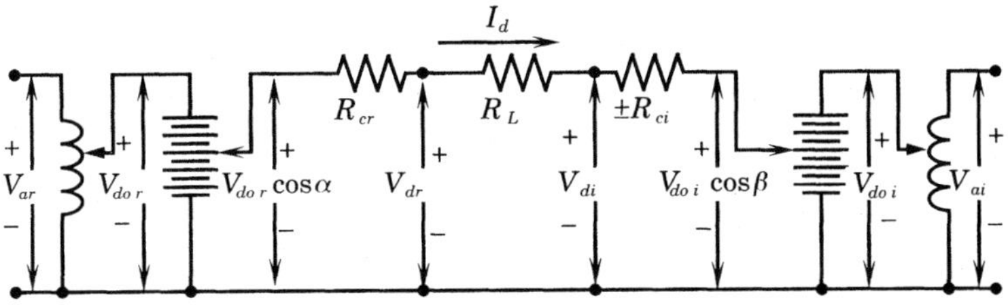

Figure 22.13 Equivalent circuit of dc line and converter stations.

Referring to Figure 22.13, the line current can be computed as

$$I_d = \frac{V_{dor}\cos\alpha - V_{doi}\cos(\beta \text{ or } \gamma)}{R_{cr} + R_L \pm R_{ci}} \tag{22.45}$$

where the $+R_{ci}$ in the denominator is used with $\cos\beta$ in the numerator and $-R_{ci}$ in the denominator is used with $\cos\gamma$ in the numerator. Let us assume that beta is the controlled angle at the inverter since this angle can be directly controlled, whereas gamma must be controlled indirectly through the control of beta. The current flowing in the line depends only on the terminal voltages, since the total circuit resistance is constant for any given mode of operation. Therefore, current and power transmitted are controlled by varying the terminal voltages.

The internal voltages can be controlled by either of two methods: thyristor control or control of the ac bus voltages. Using thyristor control, the ignition angle can be varied, thereby reducing the internal voltage, or if a given delay is present, the internal voltage can be increased by reducing the delay angle, up to the ideal rectifier voltage, V_{dor}. A similar observation can be made with respect to the inverter. The ideal rectifier voltage is directly proportional to the applied ac voltage, which is usually controlled by changing the tap position of the converter transformer. Thyristor control is very fast, but tap changing control requires 5 to 6 seconds for each step change.

Manual control of the HVDC system has practical limits due to the slow human response to sudden changes in the network. For example, a sudden short circuit in the ac network may depress the voltage in the vicinity of the converter station, which would result is a sudden change in dc power flow. It would be preferred that the dc line flow be immune from these disturbances. This can be accomplished by providing automatic control of the converters.

The desirable features of the dc control system can be summarized as follows [1]:

1. Limit the maximum current to avoid damage to the valves and other current-carrying equipment.
2. Limit the fluctuation in current due to fluctuation of ac voltages.
3. Maintain as high a power factor as possible.
4. Prevent commutation failures in the inverter.
5. Provide adequate anode voltage prior to ignition.
6. Maintain nearly constant, rated voltage at the sending end of the line to minimize losses for a given transmitted power.
7. Control the delivered power or, in some cases, the frequency at one end of the line.

The control characteristics of a typical rectifier and inverter are shown in Figure 22.14, which illustrates the direct voltage and current at some point in the dc system, such as the rectifier end of the system. Several different operating conditions are illustrated in the diagram. The solid lines represent normal operating characteristics. The upper half plane represents positive direct voltage, the condition where Converter #1 is the rectifier and Converter #2 is the inverter, with power flowing from #1 to #2. The bottom half plane is for negative direct voltage, with Converter #2 as the rectifier and Converter #1 as the inverter, with power flowing from #2 to #1. The solid line *ABHM* is the normal rectifier operating characteristic with power flowing from #1 to #2. The operating point *P* gives the value of direct voltage and current at the point represented by the diagram, where Converter #1 is operating in the *constant current* (*CC*) mode and Converter #2 operating in the *constant extinction angle* (*CEA*) mode. This would be a normal operating condition to transfer power from Converter #1 to Converter #2.

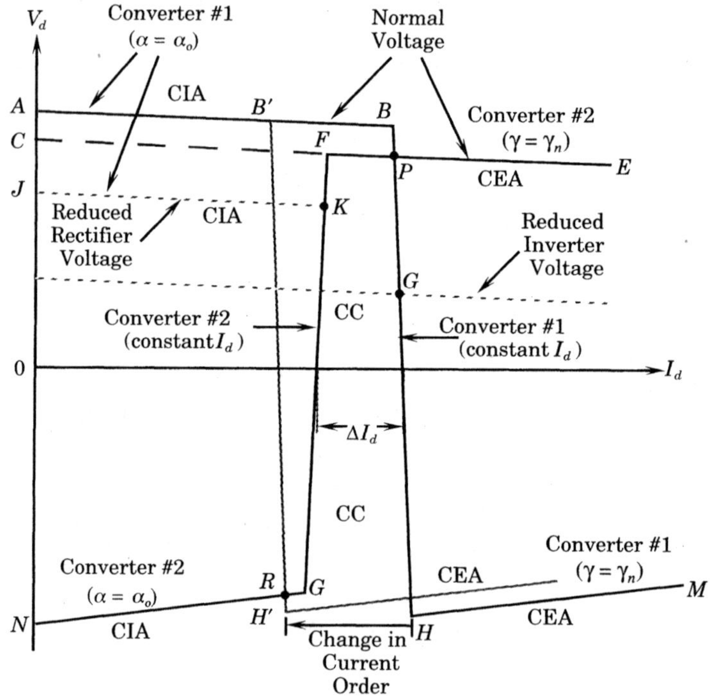

Figure 22.14 Typical HVDC control characteristics.

The rectifier CC regulator mode is described by the straight nearly vertical line *BH* in Figure 22.14. The rectifier current regulator also includes a minimum value of alpha, usually called α_o, which results in the horizontal constant portion *AB* of the rectifier characteristic. Thus, the complete rectifier characteristic is the line *ABHM* in Figure 22.14.

For power transfer from #1 to #2, the inverter employs a *constant extinction angle* (*CEA*) regulator, which has a characteristic similar to the line *CFE* extending from left to right at some fixed distance along the voltage axis. This line usually has a small negative slope, as illustrated. Since there can be only one value of voltage and of current, that value is given by the intersection of the rectifier and inverter characteristics, represented by the point *P* in the figure. The rectifier current controller measures the actual current. If this current is smaller than the desired value, the current regulator advances the firing time by decreasing the delay angle. This increases the rectifier internal voltage in proportion to $\cos\alpha$, thereby increasing I_d and moving the current, and hence the line BH, to the right and closer to the desired value.

This process continues until the measured current agrees with the current controller set point. The inverter characteristic can be moved vertically in the $V_d - I_d$ plane by changing the inverter transformer tap position, thereby varying the ac voltage at the inverter. As the voltage is changed, the CEA regulator restores the desired value of γ. This changes the internal direct voltage at the inverter in proportion to the ac voltage since $\cos\gamma$ is constant, and this tends to change the direct current. The current is quickly restored to the set value by the current regulator at the rectifier

Now consider a voltage reduction at the rectifier, which shifts the rectifier characteristic downward from AB to JK in the figure, giving a new operating point at K. The system operates at lower voltage, and with lower current than before, so that the power is considerably reduced in proportion to the voltage reduction. The inverter now changes to constant current regulation and the rectifier to constant ignition angle. If the rectifier voltage is restored to normal, the controls quickly resume the normal system operation in response. However, if the voltage remains at low values, the inverter tap changer will operate to slowly move the operating point back toward its normal value.

A voltage reduction at the inverter can be analyzed by considering a lower level of the line *CFE*, which moves the operating point down to the G in Figure 22.14. Under this condition, the line operates at a lower voltage, but with approximately the same current as before such that the power is reduced in proportion to the reduction in voltage. Should the voltage recover in a short time, the operation is returned to normal. However, if the voltage remains at the low level, the inverter tap changer will operate to either restore the voltage to normal or to the maximum level permitted by the tap-changer range.

Another type of system change is caused by a change in the current order of only Converter #1, which results in the vertical segment moving from BH to $B'H'$, that is, a movement toward a lower current order. The solution for the direct voltage and current now rests at the point R, which results in a negative voltage. This means that the power suddenly flows in the opposite direction, which is called a *power reversal*. Such a reversal might constitute a very large disturbance to the power system, occurring at two different places at the same time. This is an unwanted type of event and must be avoided. This requires that the current orders at the two converters must be moved together. Power reversals can be eliminated by maintaining a minimum α_i above $100°$ for the inverter.

The foregoing describes only one type of converter control, which is called the constant current mode of operation. Other control options, such as constant power control, are described in the following section.

22.3 CONVERTER STATION DESIGN

A general description of a converter station design is described in order to provide an understanding of the types of devices necessary to control and protect the dc line and terminal equipment.

22.3.1 A Typical Converter Station

A typical bipolar converter station arrangement is shown in Figure 22.15. This station consists of two 12-pulse bipolar converters that are independently fed from identical converter transformers. Note that the secondaries of these transformers are connected differently, one secondary being connected in delta and the other in wye, in order to provide the 12-pulse rectification or inversion. Note also that the two poles making up this bipole system are completely independent, so that one pole can continue to operate when the other pole is out of service for any reason. This type of operation requires that the return current flow either through the ground, using the ground electrode system, or through the unused dc line conductor, which can be switched to provide "metallic return" operation when one pole is out of service. A network of switches is provided for this purpose. Clearly, one pole can be isolated for maintenance while the other pole is working in a monopole mode of operation.

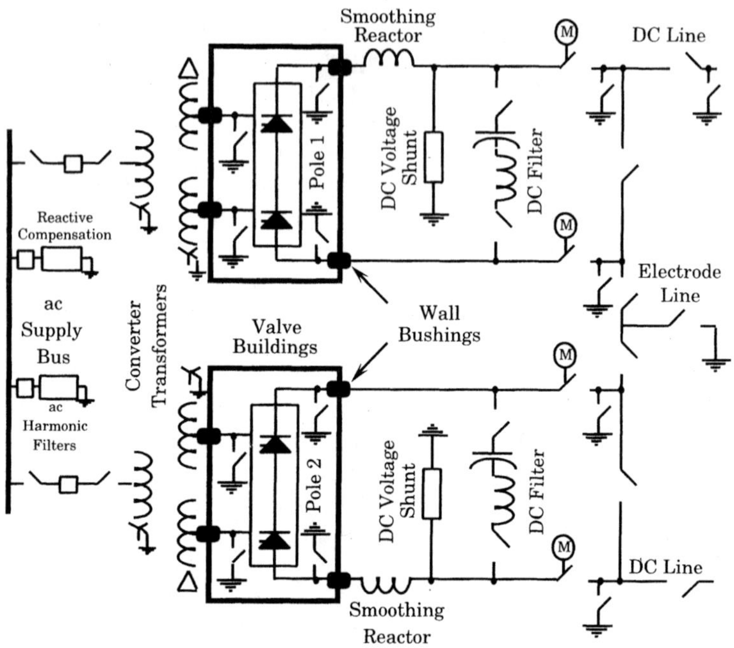

Figure 22.15 Typical configuration of a bipole converter station [1].

Note the wall bushings on both the ac and dc sides of the converter, where the electric conductors penetrate the wall of the building. These bushings insulate the high-voltage conductor from the wall, and also incorporate transducers for measurement of current passing through the valve building walls. The wall bushings also identify the protective zone for the converters and other equipment located inside the valve buildings.

EXAMPLE 22.1 Example of a Typical HVDC Converter Rating

The ratings of a typical HVDC converter station are given to provide a sense of the ratings available. The Intermountain HVDC System described is located in the western United States [7], [8] and has the following ratings:

Power rating, dc	1600 MW	Rectifier nominal firing angle	15°
Direct Voltage	± 500 kV	Inverter nominal firing angle	17°
Direct Current	1600 A		
Normal control modes:			
Rectifier		Constant current	
Inverter		Constant extinction angle	

Pole Overload Power Ratings:	
One second	2.0 per unit
Continuous	1.5 per unit

Note that this system is designed for a high 1 second overload. ■

22.3.2 HVDC Control Hierarchical Structure

The controls of a modern HVDC system are hierarchical in structure, as shown in Figure 22.16. The design is structured for high reliability and availability, hence many of the controls are duplicated. The higher level controls, at the station level and bipole level include only those functions that are absolutely necessary at those levels. Most critical functions are implemented at as low a level as possible in order to minimize the effect of a control failure. The system depicted in the figure is for a dual bipole system, but the control arrangement for a single bipole system is similar.

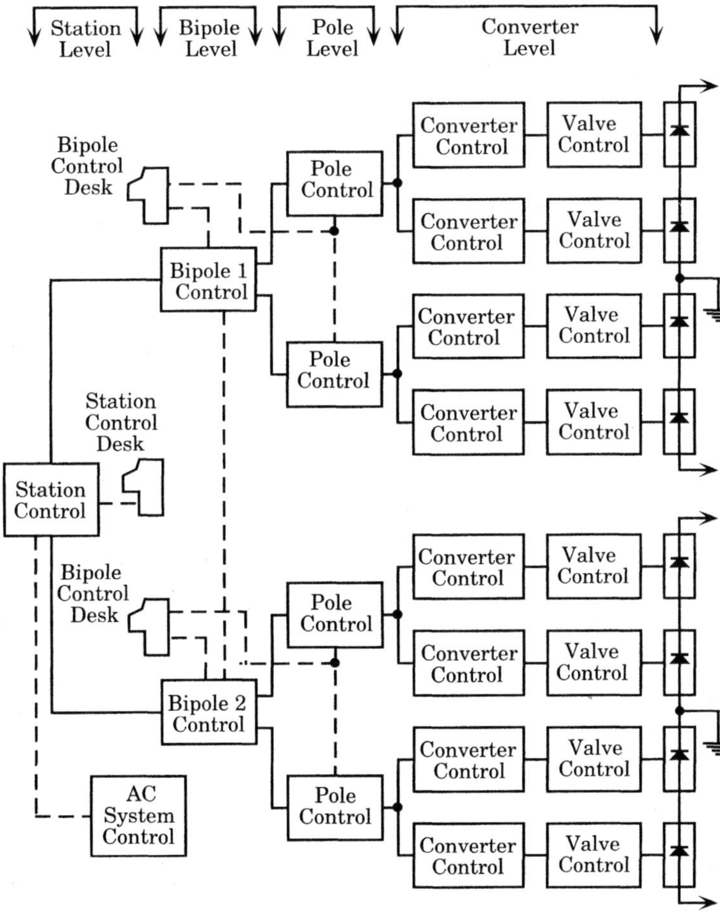

Figure 22.16 The hierarchy of typical HVDC controls with a dual bipole system illustrated [6].

The *station level* of control provides control for the ac circuit breakers and disconnect switches. It also controls the switching of ac filters and reactive compensation that are common to both bipoles. This level also controls the total dc power order (level), and this information is downloaded to the bipole level for implementation.

The HVDC transmission system equipment below the station level and including the different levels of equipment at each terminal is shown in Figure 22.17.

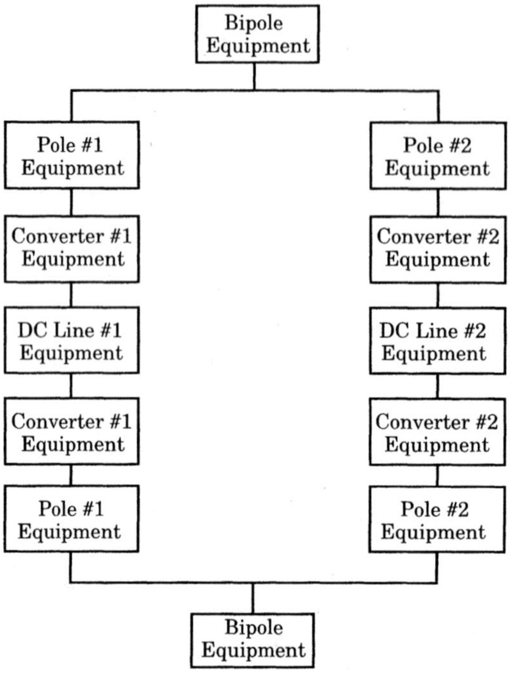

Figure 22.17 Physical connection diagram of HVDC equipment.

The *bipole control level* allocates the bipole power order between the two poles. The actual power control is performed at the *pole control level.* The pole level also performs the communications between terminals. The current control is usually at the pole level. Other pole level controls are the frequency control (for islanded operation), power runback control, overload control, active power modulation control, and controls for synchronizing with the ac system. The pole level also manages overloads of the poles.

With careful design, a relatively small amount of equipment is involved at the bipole level, which tends to make the two poles quite independent. This has the advantage that bipole failures are rare and the individual poles can operate independent of failures in the other pole. Typical equipment that will usually be included at the bipole level are shown in Table 22.3. The pole level common equipment is where most of the control is executed and is where some of the dc protections are found.

The converter level controls include the converter firing control, the control pulse generator, and converter sequence controls. Reactive power modulation, minimum firing voltage, and subsynchronous damping (see Section 22.6.3) are managed at the converter level. The HVDC protective systems are at the converter level, for the most part, but protections also exist at the pole level for some functions. The equipment that will typically be included in the converter level are listed in Table 22.3.

The various controls need to be managed in some logical manner, and this function is performed by the converter sequence controller. The location of the sequence controller

TABLE 22.3 Subsystem Identification for HVDC Equipment Groups

Equipment Group	Subsystems Included in Group
Bipole common equipment	Bipole controls
	Ground electrode system
	AC filters
	AC shunt capacitor banks
	AC shunt reactors
	Reactive power balance equipment
	Auxiliary power systems for bipole equipment
	Metallic return transfer system
Pole common equipment	Pole controls
	DC filters
	DC pole line switching equipment
	Power line carrier noise filters
	Pole bus and neutral arresters
	Neutral bus switching equipment
	Pole measuring equipment
	Auxiliary power systms for pole equipment
Converter common equipment	Converter controls
	Converter AC connection
	Converter transformers
	Wall bushings
	Converter arresters
	Valves
	Valve cooling equipment
	Smoothing reactor
	Short circuit current measuring system
	Valve hall grounding switches

hierarchy is shown in Figure 22.18. The converter sequence controller receives manual orders for required changes in the operation of the system or the maneuvers required to reach a new control objective. The sequence controller includes telemetered communications between the terminals and the execution of line switching operations. Examples of sequence controls include the following:

- *Rectifier/inverter control* Change in the direction of transmitted power is usually ordered manually, but can also be ordered automatically under specified conditions.
- *Power control/current control* Switching of current control modes can be ordered automatically on failure of the telecommunication channel or of other control equipment.
- *Trip/reclose* A trip order may be given manually or by the system protection to stop operation of the converter. Reclose is given only by manual order and includes energizing the converter transformer.
- *Start/stop* A start order unblocks the control pulses in the released converter. A stop order blocks the thyristor control pulses, but the converter remains energized.

The control functions structure for a typical system is shown in Figure 22.18, where the protection is shown with outputs directed to the converter firing control as well as to tripping of breakers as required. The protective function receives measured information from both

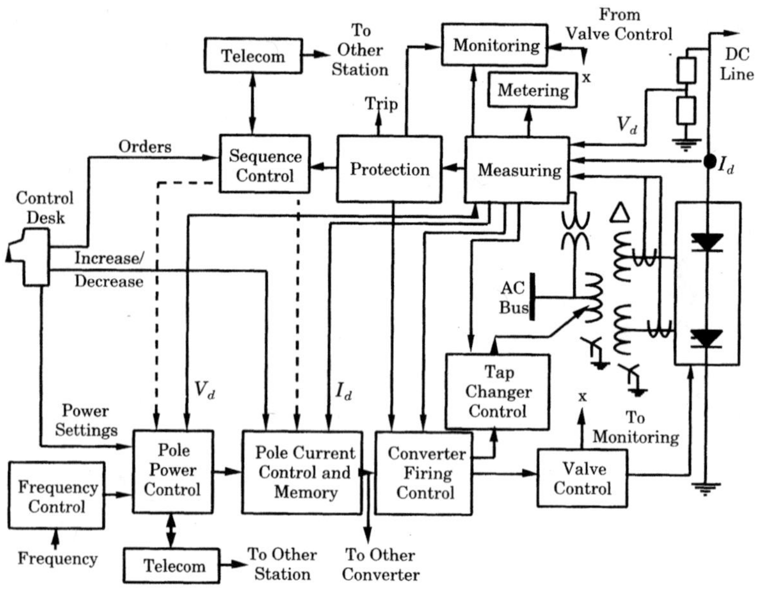

Figure 22.18 The structure of HVDC control showing the interaction of system protection with other controls [6].

the ac and dc side. The instrument transformers are shared with the control and monitoring functions. Note that the protection function has outputs that affect both the sequence control and the converter firing control functions. The control paths to the ac side are not shown in Figure 22.18, but the protection system is connected to the ac circuit breakers as well as to the dc controls.

One function that might be considered part of the protection is the frequency control function. This is provided in order to control the frequency on the ac side in the event of an ac system separation, in which there may not be adequate generation for effective ac frequency control. The frequency can be controlled within the HVDC system rating by modulating the power flowing on the dc line, and thereby helping to balance the load and generation on the ac side.

22.3.3 General Philosophy of HVDC Protection

The philosophy of HVDC protection is similar to that for ac system protection. This philosophy can be summarized as follows [7]:

- Security against unwanted or unnecessary trips
- Dependability in actual fault clearing
- Selectivity; separation of the minimum system around the fault
- Backup protection to guarantee fault clearing in the event of failure of the primary protective system
- Redundancy to increase reliability

A modern HVDC protective scheme will usually be designed to meet some or all of the following requirements:

1. Any abnormal conditions, including faults, that expose the equipment to a hazard or that present an unacceptable operating condition must be detected by the protective system and the stressed equipment removed from service or otherwise relieved of the stress imposed by the abnormality. Moreover, this action must be controlled, such that the system will continue to operate in the best possible manner under the emergency condition.
2. All protections must be fully redundant and, where possible, based on different designs or operating principles.
3. It is essential that the removal of equipment from service, in response to a disturbance, be limited to only those items that absolutely must be removed from service and that no sound equipment be involved in this action, insofar as possible.
4. Precautions should be taken to make sure that no protective systems operate unless there exists a genuine disturbance that requires protective action. In other words, false tripping of sound items of equipment must be avoided at all times.
5. All protective systems should have dual and completely independent communication lines to both the ac circuit breakers or to the converter valves, such that any breaker opening or valve blocking actions can take place using either of the two redundant communications paths.
6. The protective system should be designed with overlapping protective zones. Moreover, every fault should be detected by both primary and backup protections, which should be based on different measuring principles and may be designed to operate at different speeds, with the primary system being faster than the backup system.
7. All dc protective systems must be coordinated with all nearby ac protections in order to ensure the best performance of both systems, including the optimum recovery following the clearing of a disturbance.
8. All protective systems must be arranged such that testing of the protective devices can be permitted without affecting the operation of the HVDC system.

Many of the foregoing principles are restatements of design objectives for any protective system. The HVDC system is complicated, however, by the extensive array of equipment and the complexity of the controls. Moreover, many HVDC systems have a very high power rating and can present a severe shock to the ac system if the dc system should be suddenly removed from service. It is especially important that the bipole outages be kept to an absolute minimum, due to the severity of these outages on the system external to the HVDC system itself [8–10]. Therefore, it is important that the HVDC system protections be very secure from improper actions and that no more equipment be removed from service than necessary for any disturbance.

Figure 22.19 shows a typical block diagram of measuring points and protections for one pole of an HVDC system. The dc current is measured by dc current transducers (DCCTs) or zero flux current transformers. Faults are always cleared by redundant protections. One fast protection is provided by the sequence system, which provides fast blocking of the valves as well as sending a signal to trip the circuit breakers. Another source of protection is by means of individual protections for the various system components.

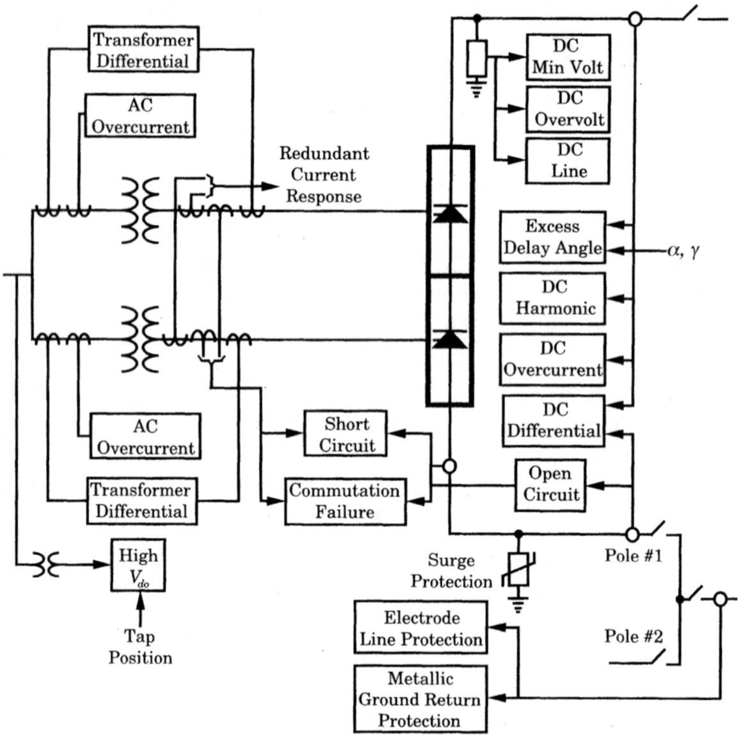

Figure 22.19 Typical HVDC protective systems for one pole [10].

22.3.4 General Categories of HVDC Protection

The protection for the HVDC converter station can be separated into the following protection zones [1]:

1. AC Side Protection
2. DC Side Protection
   - Valve Protection
   - Other DC Protections

Each of these will be discussed in turn.

22.4 AC SIDE PROTECTION

The major ac side protective zones are shown in Figure 22.20, which shows the main protective zones associated with only one of the two poles. The other pole will have identical ac protective systems.

22.4.1 AC Line Protection

The ac supply for the pole may be provided by a short ac line, which will require protection. Such a feeder line would be on the left of Figure 22.20. The type of line protection is not shown, but it must be high-speed line protection for both phase and ground faults. In

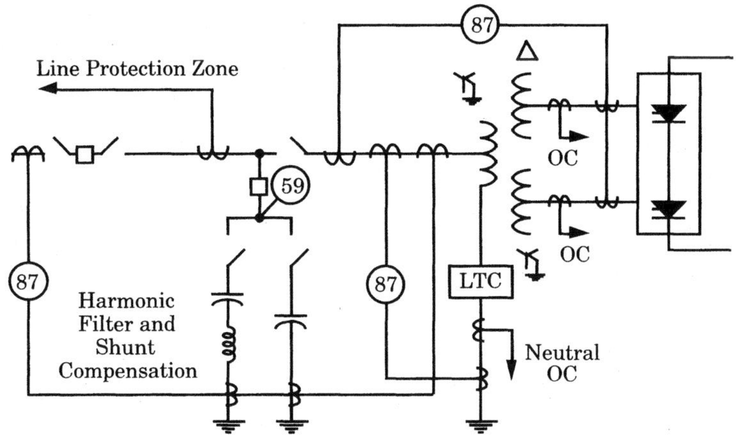

Figure 22.20 One-line diagram of the AC side protection for one pole [13].

many cases, this protection will be some form of pilot relaying. The ac line breaker must be tripped for all major pole and line faults. In some cases, there may be a short length of line on the ac source side, in which case the breaker separates the bus protective zone from the harmonic filter and capacitor zone. Inputs to the protection are from CT's located at the converter terminal boundary, with current and voltage polarization for zero-sequence ground faults from the converter transformer neutral and the bus voltage transformers.

22.4.2 AC Bus Protection

A rigid bus will normally be used to tie the supply breaker to the converter transformer. This bus also supplies the harmonic filters and the reactive support for this pole of the converter. The main breaker must be tripped for all bus faults.

The protected zone for the ac bus lies between the current transformers on the source side of the breaker to the converter transformer high-voltage winding. This zone includes the harmonic filters and reactive support, using the neutral end CT's of each shunt connection. Bus differential relays would normally be used for this protection.

22.4.3 Converter Transformer Protection

The converter transformer is usually supplied in the form of single-phase units, with each transformer having two secondary windings that are connected in delta for one valve group and in wye for the other group. A fourth transformer is often provided as a spare, but is not connected unless one of the in-service units should fail.

Each transformer is usually provided with pressure, transformer winding temperature, oil temperature, gas accumulation, and oil level alarms. The transformer protection will usually be provided by transformer differential relays connected as shown in Figure 22.20. Second harmonic restraint is commonly provided to suppress tripping when the transformer is energized.

High-speed ground fault protection is also provided across the high-voltage bushing to ground. Overcurrent protection (not shown) is usually provided as backup for the transformer differential protection. The phase currents are measured on the line side of the converter transformer.

22.4.4 Filters and Reactive Support Protection

It is common to find both harmonic filters and shunt capacitors for reactive support on the bus supporting each pole of the converter. The filters are required for the 11th and 13th harmonics [14]. Since the converter station absorbs reactive power in proportion to the active power load on the pole, large capacitor banks are usually required for reactive support. The filters are largely capacitive in nature at the power system fundamental frequency and may supply about half of the needed reactive support.

The shunt banks and harmonic filters must have overvoltage protection. For this purpose, the same protection can be used for both types of shunt devices. This requires a voltage measurement on the converter bus and an overvoltage relay. The capacitor banks have a voltage distribution that is dependent on the number and location of failed individual capacitor units. This is detected and an alarm signal generated to warn of serious unbalances in the structure.

Both fundamental and harmonic frequency currents are measured in the filter branches. Inverse-time overcurrent protection is often used.

A scheme for detecting faulty capacitor units is also recommended. This is accomplished by measuring and comparing fundamental frequency currents in two capacitor chains, either in the same branch or in two different branches. Unbalances exceeding a limiting value will cause an alarm.

22.5 DC SIDE PROTECTION OVERVIEW

The dc side protection includes equipment at the bipole, pole, and converter levels. There are many different types of equipment and many different failure modes that require detection and protective action. In some cases, the required protective action is to order a change in control or operating conditions, and in other cases the protective action will block the pole or bipole, depending on the nature of the disturbance.

Since the protections are an integral part of the control systems, the control system itself must be redundant to guard against failure of the controller. One way this can be accomplished is to provide two identical controllers and require that both controllers agree that a given control action is required. This can be accomplished by designating one of the controllers as the active controller, with the other as backup. Certain disturbances are predefined that require confirmation by both controllers before ordering any protective action. The active controller will monitor the usual control quantities and, having observed a predefined situation that requires protective action and does not require fast tripping action, control will be transferred to the redundant controller. The redundant controller then takes the required action based on its independent measurement of the system. This prevents an error or failure in the first controller from creating a false trip, or failing to trip in a fault condition. An even more secure method of accomplishing this purpose would be to have three controllers, with two-out-of-three agreement being required for any control action. It is common to require redundant controllers at the bipole level, but not necessarily at the pole level.

In order to meet the requirements set forth in Section 22.3.3, particularly items 2 and 6, the dc side protections for both of the redundant control systems are divided into two or more protection blocks, with each block having its own power supply. The two blocks are entirely independent with a minimum of equipment in common. Any fault that does not require fast tripping will be processed by a transfer to the redundant control system.

22.5.1 Valve Protection

The converter valves are solid-state thyristor valves in all modern converter designs. These thyristors are the heart of the converter system and have physical limits that must not be exceeded if valve damage is to be avoided.

22.5.1.1 General Description of the Valves. The valves are connected to form a 12-pulse bridge circuit for each pole of the converter station. The valve protection is provided by microprocessor control, but the concept is similar to an analog "zone of protection" system, similar to that found on ac systems. A diagram of the major items of equipment on the dc side of the converter station is shown in Figure 22.15. All of the valves are connected as components in a 12-pulse circuit, as shown in Figure 22.11. This valve group for each pole forms a protection zone. Several different types of protections are provided, depending on the hazard being controlled.

22.5.1.2 Valve Short-Circuit Protection. The protection zone for valve short circuits is the 12-pulse bridge circuit from the wall bushings on the ac side to the dc wall bushings at the neutral bus and at the smoothing reactor terminal. Similar protections are normally provided for each pole. Short circuits across valves are phase-to-phase short circuits that occur within the valve short-circuit protection zone. Short circuits are cleared by blocking the converter and by tripping the ac side circuit breakers.

One type of detection scheme compares the highest current amplitude in the ac currents at the wall bushings and the amplitude of the dc currents at their wall bushings in a form of differential current protection. If the valve operation is normal, these magnitudes are equal. Any short circuit within the valve group results in excess current on the ac side, which is seen as a phase-to-phase fault. A valve short circuit can occur when a sound valve is fired in conjunction with the faulty commutation of another valve. The fault currents are maximum for the rectifier.

The protection system should not operate for ac ground faults between the converter transformer and the valve bridge. The protection used is a form of differential protection that compares the maximum ac current with the dc currents at both the high-voltage and low-voltage sides of the converter pole. This permits the protection to avoid operation for ground faults on the ac conductors in the converter. The protective action is to block the faulted pole and trip the ac circuit breakers at the ac supply bus. The protection is enabled when the following condition is satisfied:

$$I_{\text{ac}} > \left[0.5 + 0.2\left(\frac{I_d}{I_{dn}}\right)\right] I_{dn} \tag{22.46}$$

where I_{ac} = current on the ac side of the pole
I_d = dc current at the dc high voltage line bushing
I_{dn} = dc current at the dc neutral bushing

A small time delay of approximately 1 millisecond is usually provided. It is necessary that fault detection be made before the third valve in a faulty three-phase group is fired. Coordination with other protections is not considered necessary. This protection is provided in dual redundant protection systems.

22.5.1.3 Converter Overcurrent Protection. The protection zone for converter overcurrent protection is the 12-pulse bridge systems of each pole. The objective of the protection is to detect overcurrents that may cause unusual stress in the converter equipment, particularly the thyristor valves.

The overcurrent protection is provided in two parts. The first part detects dc overcurrents and gives a transfer trip after a preset delay or, in the case of high overcurrents, an instantaneous trip.

The second part is a thermal overload protection that is a provided by a computer model of the valve heating and cooling. The valve losses are computed by the formula

$$P_{th} = k_1 I_d^2 + k_2 I_d \tag{22.47}$$

The losses are computed and the result supplied to a network representing the thyristor cooling properties. The resulting calculation is the thyristor temperature rise. This is added to the cooling water temperature and the result compared to a temperature threshold. The overcurrent protection gives a trip order if the computed temperature exceeds the threshold.

The protective actions taken on pickup of the overcurrent protection are as follows:

1. Transfer converter control to the redundant system.
2. Block the pole and simultaneously fire bypass pairs.
3. Trip the ac circuit breakers.
4. Isolate the pole and line at both ends.
5. Out-of-range alarm for cooling water temperature.

Time delays are also used. The fast part of the protection has a time delay of about 4 ms. The slow part requires an inverse time characteristic to match the cooling characteristic of the valves. Valve bypassing is the intentional firing of valves in parallel with a bridge valve that experiences a temporary failure [2].

The overcurrent protection is provided in identical redundant systems.

22.5.1.4 Commutation Failure Protection. Commutation failure is the inability to transfer current to the valve next in line to conduct. This may occur either because of low voltage or distortion of the voltage waveform. Commutation failures cause no damage to equipment and are usually temporary. Repeated commutation failures can occur, however, and this will require either temporary or permanent pole blocking [17], [21]. A commutation failure is not a fault, but is an indication of the failure of the valve control pulses, or of faults on the ac side.

The protection zone for commutation failures is the 12-pulse valve bridge. The objectives of the protection is to detect commutation failures and, if the failure persists, to remove the bridge from service.

Commutation failures are detected on a six-pulse bridge basis. In the six-pulse bridge, there are intervals when all ac phase currents are zero but a direct current is flowing through two opposed valves in the bridge. These intervals are detected. Persistent faults may be due to valve misfire, e.g., no control pulses or continuous control pulses. Intermittent faults are usually due to ac system disturbances. These conditions are used to distinguish between the two conditions.

The protective actions taken in response to commutation failure are the following:

1. Instantaneous advancing of the firing angle in the faulty converter to improve recovery.
2. Transfer to the redundant control system.
3. Blocking and simultaneous bypass of the faulty converter for the slow part of the protection and tripping of the ac breakers feeding the converter.

4. Blocking of the converter when the commutation failure is detected in only one of the six-pulse groups.

The commutation failure protection must be coordinated with the following other systems:

(a) AC protections, where time delays must coordinate with the longest clearing time for ac side faults.
(b) Valve misfire protection.
(c) DC harmonic protection.

Backup protection is provided by (b) and (c).

22.5.1.5 Valve Misfire Protection. In valve misfire protection, the protective zone is the thyristor valve. The objectives of valve misfire protection are as follows:

- To detect the failure of a valve to conduct when a control pulse is applied.
- To detect unintentional valve firing.
- To prevent a valve that fails to conduct from being selected into a bypass pair.
- To select into a bypass pair a valve that is firing unintentionally.
- To supervise the system operation.

Valve misfiring is detected by monitoring the performance of the valve and comparing this performance with the control signal arriving at the valve in the form of a control pulse, whose duration corresponds to the required conduction time. The control pulse duration is compared to the actual valve firing duration to determine correct firing. Valve firing outside the desired interval is also detected to determine improper firing. The protection has a fast part that has low sensitivity, and a slower part that is more sensitive.

The protective switching actions that result on pickup of valve misfiring are the following:

1. Transfer to redundant control.
2. Block the converter (and insert bypass pairs if an inverter).
3. Trip the ac side circuit breakers.

A time delay of 60-80 ms is used in order to coordinate with the commutation failure protection, which should be slower than the valve misfire protection.

The valve misfire protection should coordinate with the commutation failure protection and the dc harmonic protection and should be interlocked with the ac fault protection by a "low commutation voltage" signal.

The backup protections for valve misfire are the commutation failure and dc harmonic protections.

22.5.1.6 Voltage Stress Protection. The protective zone for voltage stress protection is all converter equipment exposed to ac voltage. This includes the thyristor valves and the converter transformer.

The objectives of voltage stress protection are twofold. First, it is designed to detect high commutation voltages and to prevent further increase of voltage by interlocking the converter transformer tap changers. Second, it takes the faulty converter out of service in the case of persistent ac overvoltage.

The principle of voltage stress detection is determined by observing the ac voltage as a function of time and by noting the converter transformer tap settings. Based on these observations, an ideal value of dc voltage is computed by the protection and compared with a preset reference value. When this computed value exceeds the reference, pickup is initiated. When the voltage is too high, the tap changers are blocked from further operation and an alarm is given. At a higher voltage level, the converter is taken out of service after a preset time delay. Another protection feature compares the positions of the transformer taps and sounds an alarm if these tap positions exceed a preset difference.

The protective actions that are taken following pickup are as follows:

1. On detection of a small overvoltage, the tap changers are inhibited from further raising the voltage.
2. A transfer in made to the redundant control system.
3. On detection of a high overvoltage, the converter is blocked and the ac circuit breakers are tripped.
4. If the tap changers are out of range or out of step positions, an alarm is sounded.

The voltage stress protection must coordinate to avoid unnecessary alarms or trips in the case of permanent ac network voltage changes due to normal switching in the network. Moreover, this protection should not operate when the converter transformer tap changers are operating normally.

This system is a redundant protective system. Some backup is provided by the converter transformer tap changer control equipment. Additional backup is provided by excessive delay angle protections, which are described in the next section.

22.5.1.7 Excessive Delay Angle Protection. The protection zone for this protection is the thyristor valves. The objective of the protection is to protect the resistors in the valve damping circuits against overload during operation at too large a delay angle (α) and extinction angle (γ). It serves as a backup to the dc minimum voltage protection. Moreover, this protection detects large delay angle operation due to faults in the tap changer controls or faults in the converter firing control system.

Detection of excessive delay angles is based on the computed losses in the valve damping circuits. When operating with excessive delay or extinction angles, the losses in these circuits increase over normal operation due to the higher amplitudes of breakdown and recovery voltages across the valve when they are fired or extinguished. The voltage across the valve is derived from the capacitive taps in the wall bushings. This voltage is applied to an RC circuit with the same time constant as the valve damping circuit. The resulting voltage is smoothed to obtain a voltage proportional to the damping resistor losses. When this voltage exceeds a preset threshold, the protection picks up.

The protective actions taken on pickup are the following:

1. Interlock the tap changer from stepping when abnormally large delay or extinction angles exist.
2. Transfer to the redundant control system.
3. Block the converter at sustained operation with abnormal angles.

The backup protections to this protection are the voltage stress protection, if the condition is the result of high commutation voltage, and the dc minimum voltage protection.

22.5.1.8 DC Harmonics Protection. The protective zone for the dc harmonics protection is the converter. The objective of this protection is to detect abnormal harmonics in the converter current. These harmonics are often generated by valve disturbances, by the ac network disturbances, and by control equipment malfunctions. The dc harmonic protection provides backup for the commutation failure and the valve misfire protections.

The detection principle is based on having prior knowledge concerning the filtering of the dc current harmonics, especially with respect to the fundamental and second harmonics. When the observed harmonics exceed these preset values, the protection is picked up. For high amplitudes of the harmonics the pole is removed from service. This protection is designed to have an inverse time characteristic.

The protective action of this system is as follows:

1. Transfer control to the redundant system.
2. Block the converter.

The dc harmonics protection is designed to coordinate with the following other protective systems:

- Commutation failure protection.
- Valve misfire protection.
- AC protections, particularly ac line distance protections that are expected to determine the maximum duration of an ac fault.

The settings of the dc harmonics protection depend on several factors [14], [15]. The protection should not pick up for normal harmonics that occur with ac network asymmetries. Any unsymmetrical ac network condition results in second harmonic currents on the dc side. For severe ac faults, the dc harmonics protection should not trip before the ac protective systems.

However, the dc harmonics protection should detect current harmonics that are generated due to persistent valve misfiring in the rectifier or persistent commutation failure in the inverter. Also, failure of the converter firing control to provide proper firing pulses should be detected.

The settings of the dc harmonics protection are based on known values of expected normal harmonics levels. Time delays of 100 ms–3 seconds are applied, with the longer time delay being applied prior to a converter block.

22.5.2 Other DC Side Protective Functions

DC side protection includes all protective functions on the dc side except for those defined above as valve protections.

22.5.2.1 General Description. The dc side protection includes the dc line and terminal equipment. These protections are different from those normally encountered on the ac network, where we usually have specific protective schemes for each protective zone. The dc side protections are a part of the dc control system and, in some cases, the distinction between control and protection is somewhat obscure. In modern systems, much of the protection and control is provided in the form of microcomputer devices that are programmed to provide the desired function, whether a control function or a protective function. Some of the protective systems take the form of differential measurements, while others are computer models of the physical system. Most are redundant systems, and some employ voting logic for reliability and security of control action.

22.5.2.2 Converter DC Differential Protection. The protective zone for the converter dc differential protection is the dc side of the converter between the transducer in the dc wall bushing on the low-voltage terminal and the transducer in the dc reactor on the line side (see Figure 22.15). The objective of this protection is to detect ground faults within the protective zone and clear the fault by initiating the proper switching actions.

The detection system monitors the dc current in the neutral bus and in the high-voltage dc bus. These measurements are compared in the protective system by a differential scheme with a significant difference in these currents being the indicator of a ground fault in the protective zone. The protection has one fast part with low selectivity and a slow part with higher selectivity.

The switching actions that are taken on pickup are as follows:

1. Transfer to the redundant control system.
2. Block the converter.
3. Trip the ac side circuit breakers.
4. Isolate the pole and line at both ends.

As noted before, the protection has both a fast and a slow part. The fast part of the protection compares the current differential to the following function.

$$I_{\mathrm{REF1}} = \left[0.4 + 0.2\frac{I_d}{I_{dn}}\right] I_{dn} \tag{22.48}$$

If the current differential exceeds this reference value, the system switching is ordered following a 3 ms delay. See (22.46) for a definition of the currents.

The slow part of the protection uses two different reference signals. The first signal is computed as

$$I_{\mathrm{REF2}} = \left[0.05 + 0.2\frac{I_d}{I_{dn}}\right] I_{dn} \tag{22.49}$$

Any current in excess of this reference is delayed 25 ms to ensure proper fault detection, and then is delayed an additional 80 ms before pickup. The second slow reference signal is computed as

$$I_{\mathrm{REF3}} = \left[0.04 + 0.15\frac{I_d}{I_{dn}}\right] I_{dn} \tag{22.50}$$

Any current in excess of this reference is delayed 25 ms and is then delayed an additional 50 ms before pickup.

Coordination is required between the converter dc differential protection and the pole dc differential protection and the dc line protection.

The backup protections for the converter dc differential protection are the pole dc differential protection, and the converter transformer differential and bus differential protection for faults on the outdoor ac system.

22.5.2.3 DC Line Protection. The protective zone of the dc line protection is the dc transmission line. The objective of this protection is to detect ground faults on the dc line and, by means of control action, extinguish the fault current. If the fault is not permanent, the control action should restore power transmission after a suitable time delay to allow for fault deionization.

The detection of ground faults is based on observations of the direct voltage at the line terminal, using a dc voltage divider circuit. A dc ground fault is characterized by a voltage collapse to a low-voltage level at a high rate of voltage change. The protection uses both the voltage level and the rate of change to detect the ground fault on the dc line.

The derivative of the direct voltage is determined by a differentiator circuit and compared against two different reference signals, designated as Ref1 and Ref2. Ref2 is the lower of the two signals and is used to open a window of time during which the derivative must exceed Ref1 and, in addition, the magnitude signal must exceed its reference, Ref3. If these conditions are all met the control is enabled to take appropriate action. The control is further complicated to avoid false trips of actions outside of this protection zone, such as inverter faults or traveling waves on the line, which will also create large voltage rates of change.

The protective action that is taken is active only at the rectifier. The redundant protection is slaved to the active protection during this period. Following a fault detection, the converters at the rectifier are forced to a full inversion operation, thereby preventing the rectifier from supplying any current to the fault. The inverter control attempts to maintain ordered current by increasing its margin of commutation to about 100 degrees. When the line discharge reaches a very low value, the inverter is brought back to minimum extinction angle and the line will remain de-energized with neither the rectifier nor inverter feeding current to the fault until a preset deionization time has elapsed. Following this delay, the rectifier stop order is removed and a restart attempt is made. If this attempt fails, control is transferred to the redundant system and this system takes over further attempts to restore the line to normal operation. Up to three full voltage restart attempts can be made, with an adjustable deionization time between each attempt. If all attempts fail, the pole is blocked and both the pole and the line are isolated at both ends. The restart attempts are blocked when the telecommunications channel between rectifier and inverter is out of service or when the system is already operating at a reduced voltage.

DC line fault protection must be coordinated with the ac system protections. In particular, care must be taken that the dc line protection does not operate for ac faults at the rectifier or inverter, when starting or stopping a pole, or during commutation failures in the inverter [15], [17].

The back-up protections for dc line faults are the dc minimum voltage protection and the excessive delay angle protection.

***22.5.2.4 DC Minimum Voltage Protection*.** The protective zone of the dc minimum voltage protection is the dc line and all equipment connected to the dc line, including the thyristor valves and the bypass pairs. The objective of this protection is to detect ground faults on the dc line and system, thereby serving as a backup to the primary dc line protection. This protection also backs up the voltage dependent current order limit (VDCOL) protective system.

Two different principles are involved in the detection scheme used for this protection. The first detection system is simply to compare the observed dc voltage against a preset reference value, which is similar to the dc line protection. The second type of detection is to pick up the protection when the firing angle a is greater than about 80 degrees and the dc current is greater than the highest allowed continuous bypass pair current.

The protective actions that are taken on pickup are described as follows:

1. Transfer control to the redundant control system.
2. Block the converter.
3. Isolate the pole and dc line by prearranged switching.

This protective system must be coordinated with the dc line protection and the excessive delay angle protection. This system serves as the backup protection for the dc line protection and the VDCOL.

22.5.2.5 DC Overvoltage Protection. The protective zone for dc overvoltage protection is all quipment exposed to the dc line voltage. The objective of this protection is to detect overvoltage on the dc line and equipment when starting a pole against an open-ended dc line [14], [15].

Overvoltages on the dc side may occur for a number of different reasons and may be of long duration, especially on peak rectification when a pole is started against an open dc line, in which case the voltage is limited to 1.15 per unit by the overvoltage limiter in the pole equipment. The dc current will be very low during this open-circuit condition. These conditions can be combined to provide effective detection of the condition and to initiate proper control action.

The protective actions ordered for these conditions are as follows:

1. Transfer control to the redundant control system.
2. Block the converter.
3. Isolate the pole and line at both ends of the line.

This protection must be coordinated with the overvoltage limiter. It serves as the backup protection for the dc overvoltage protection at the other end of the line and for the overvoltage limiter.

22.5.2.6 Pole DC Differential Protection. The protective zone for the pole dc differential protection is the dc side of the converter, including the dc filters. See Figure 22.15. The objective of this protection is to detect ground faults within the protective zone and to remove the faulted converter pole from service.

The detection principle is current differential protection between the dc current at the neutral bus and at the dc line terminal, with the two current measurements being performed outside of the filters. Measures are taken to inhibit the protection during operation of arresters or other transient grounding equipment. This permits the protection to ignore normal arrester operation, but it will operate for arrester failure. This protection is designed to include a fast and relatively insensitive detector, and a second detector that is slower, but more sensitive.

The pickup protective actions taken by the pole dc differential protection are as follows:

1. Transfer control to the redundant control system.
2. Block the converter.
3. Trip the ac side circuit breakers feeding the converter.
4. Isolate the pole and line at both ends.

The pole differential protection must be coordinated with the dc line protection and the converter dc differential protection. When a ground fault occurs on the dc bus at the rectifier, both the dc line and pole differential protections will pick up. In this case, the fast protection module timing should be selected to trip before the dc line protection, which will try and extinguish the fault current by reduced voltage operation and restart the system. Since the pole differential knows that the fault is not on the line, its logic should prevail and block the converter. If the dc line fault current is low, and the fault is not picked up by the fast module of line protection, then the line protection will attempt a restart after a preset delay. In this case, the slow module of the pole differential protection will order a trip during this restart attempt.

This protection is backed up by the dc line protection for ground faults that cause a significant drop in dc voltage.

22.5.2.7 Electrode Open-Circuit Protection. The protective zone of the electrode line protection is the neutral bus equipment. See Figure 22.15. The objective of the protective system is to detect an open circuit of the electrode line and to relieve the neutral bus equipment from any overvoltage that may accompany an open electrode line.

The detection principle is to measure the voltage to ground of the pole neutral bus, since overvoltage at this bus is an indicator of an open electrode line.

On detection of an open electrode line, the following actions are taken:

1. Transfer control to the redundant control system, using a voltage level at the neutral bus that is slightly greater than that used for metallic return.
2. Check the voltages at both neutral buses and determine if they are both rising over the metallic return level. If this is found to be the case, close the switching device between the neutral buses and ground. This will cause the differential current to flow to ground as the bipole continues to operate, unless an overcurrent in the ground switch protection is detected.
3. If the neutral bus voltage rises in monopolar operation, the ground switch will be closed and the pole blocked with appropriate isolation.

This protection must be coordinated with the bus arrester design and the withstand capabilities of the neutral bus equipment.

This protection backs up the electrode line open-circuit protection in the other pole when in bipole operation. When in monopole operation, this protection backs up the pole dc differential protection since, on opening the electrode line, the current will pass through the neutral bus arrester, which is in the protective zone of the pole dc differential protection.

22.5.2.8 DC Filter Protection. The protective zone of this protection is the dc filter. The objective of this protection is to detect overload on the dc filter components and to relieve the filter from being overstressed by blocking the pole. It also interlocks the operation of the dc filter switches if the filter current becomes too high.

The detection principle used for dc filter protection is to measure the current through the filter bank and compare this magnitude against a preset reference. Tripping is delayed sufficiently to avoid operation of the protection during transient overloads that the filters are designed to withstand.

The protective actions taken on pickup are as follows:

1. Transfer control to the redundant control system.
2. Block the converter.

This protection must be coordinated with the excessive delay angle protection when operation is at high firing angles.

22.5.2.9 Voltage-Dependent Current-Order Limit. The voltage-dependent current order limiter (VDCOL) provides a current limitation that is dependent on the magnitude of the measured dc voltage. Its characteristic is similar to that shown in Figure 22.21 [19]. When the relative strength of the ac system is low, the speed of dc recovery following faults must be slowed in order to ensure successful recovery. The reason for this precaution is that weak ac systems experience large voltage fluctuations with changes in dc power level. This means that the rapid power control of the dc system following a fault will cause rapid and large changes in the reactive power drawn from the ac network and large ac voltage changes. The result may be unsuccessful fault recovery in the dc system due to commutation failures.

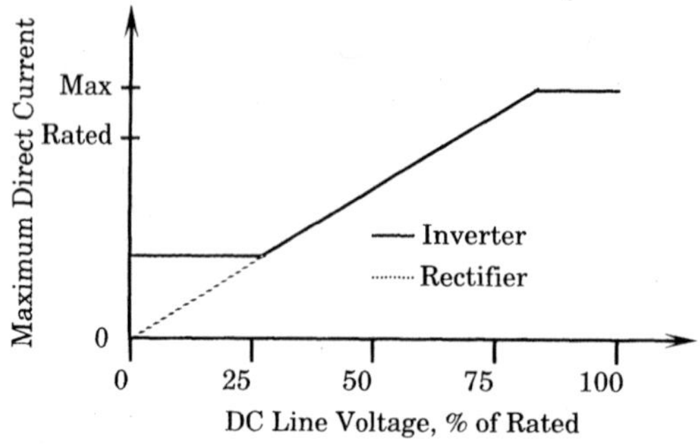

Figure 22.21 Voltage-dependent current order limiter characteristic [19].

The protection zone of the VDCOL is the converter and the nearby ac network. The objective of the VDCOL is to prevent damage to the converter equipment from long periods of operation at large currents and low dc voltage, and to prevent the dc system from causing voltage collapse in the ac system because of excessive consumption of reactive power by the converter stations [16].

The VDCOL is an effective means of controlling the rate of recovery following a dc disturbance, thereby limiting the amount of reactive power drawn from the ac system. This, in effect, limits the rate of change in the reactive power drawn by the converters, and thereby assists in the system recovery following the disturbance.

22.6 SPECIAL HVDC PROTECTIONS

There are other dc protections that are often applied to HVDC systems, primarily as a means of coordination with the ac system and to protect the ac system against certain unusual operating conditions that are attributed to the dc system.

22.6.1 General Description

Many dc systems are designed to transport large blocks of power over long distances. This suggests that such dc systems will have a high power rating and that rapid changes in the converter operation, such as pole or bipole blocks, might be viewed by the ac system as large disturbances. This is further aggravated by the extremely fast controls available for the dc system, which have the capability of ordering changes of hundreds of megawatts of power in only a matter of microseconds. No other power controls on the ac system have this high rate of response and the traditional ac controls and protections were designed to view such rapid changes as emergency conditions. This means that the integration of an HVDC system into an ac system requires considerable planning, and that the system response to various dc control actions should be carefully examined to see how these actions might affect the surrounding ac system.

Another unique thing about HVDC systems is that many large dc disturbances inject a large change in the ac system at two locations, not just one. This further enlarges the regions that might be affected by a dc disturbance and widens the possible candidates of ac equipment that might interact with the dc controls. Some of these broad system questions are discussed in this section.

22.6.2 Reverse Power Protection

One of the unique control features of a dc system lies in its ability to reverse the direction of power flow very quickly. This can be done purposely, or it can occur as the result of a failure within the dc controls, (see Section 22.2.4).

The protective zone of reverse power protection is the ac system in the neighborhood of the converter stations at both ends of the dc line, or at several terminals for a multiterminal HVDC system. Any generating plants near a dc terminal may be affected by a reverse of power and any heavily loaded transmission lines may also be affected. The objective of reverse power protection is to detect a reversal in the direction of power delivered by the dc system.

The reverse power protection measures the dc current and dc voltage, from which a calculation of the dc power magnitude and direction is obtained. If the calculated power direction is not in accord with the desired direction, the pole carrying the reverse power is blocked.

The protective actions that take place are as follows:

1. Transfer control to the redundant control system.
2. Block the converter.
3. Isolate the pole and line at both ends.

This protection backs up the reverse power protection at the other end of the line.

Power reversal of a dc system of high rating is a very serious disturbance, not unlike simultaneous three-phase faults at two points in the ac network. When this occurs, very fast action is required as the immediate remedial action, and further action may be required within the ac system, such as tripping of generators at the sending end of the dc line and shedding load at the receiving end. The detailed actions that are required depend on the size of the disturbance and whether this initial disturbance propagates to cause ac line outages and possible islanding or break up of the system.

The HVDC system is capable of very fast reaction to this type of disturbance. Nearby ac generators will be affected by the power reversal, but the these generators and their prime movers have considerable inertia, and are often able to sustain large disturbances without becoming unstable. Therefore, it is not a foregone conclusion that instability will result following a power reversal, but this possibility must be carefully checked by computer simulation.

22.6.3 Torsional Interaction Protection

In Chapter 23, it is noted that large turbine-generator shafts can be described by a spring-mass model, with $n-1$ natural frequencies of oscillation for a shaft with n masses. Most of these frequencies fall in the range of 10 to 50 hertz, on a 60 hertz basis, and are referred to as "subsynchronous" modes of oscillation.

The steel shaft and associated masses forms a highly tuned oscillatory system. This means that a very small amount of energy injected into this system at exactly one of the natural frequencies of oscillation will cause the spring-mass system to oscillate, with possibly large amplitude. Continued operation with such an oscillation is very likely to cause permanent fatigue life expenditure.[2] See Chapter 23 for additional information on shaft fatigue.

The controls of the HVDC converter are high-order control systems and have their own natural frequencies of oscillation, just as any higher order system. Clearly, there exists the

[2]Fatigue life expenditure is the gradual loss of strength of a metal due to prolonged and repeated stressing, such as bending or twisting. The endurance limit of the material occurs with the expenditure of100% of the fatigue life, resulting in a crack in the metal.

possibility of these controls having a natural mode of oscillation that is equal to, or an exact multiple of, one of the shaft natural modes. Should this be the case, it is important that torsional oscillation of the shaft be prevented in order to prevent shaft fatigue damage. At least one HVDC system has exhibited the potential for torsional interaction of the type described [14], [16], [18], [19]. Since that occurrence, all HVDC systems have been designed with special controls to prevent this type of interaction.

The usual solution of the torsional interaction problem is to tune the controls to avoid this type of problem. Thus, there is no special protective system for torsional interaction, but the controls are designed to prevent this phenomenon from occurring. It has been shown that careful tuning of the current controls of the HVDC system will eliminate the response at critical frequencies, giving improved torsional stability [17]. Most HVDC system controllers include a device for modulating the dc line power as a system control to provide damping of ac system oscillations. It has been shown that special damping filters may be required in this modulation controller to improve damping at subsynchronous frequencies of interest. Using this technique, the subsynchronous frequencies are not necessarily eliminated, but any interaction that results can be caused to be well damped.

22.6.4 Self-Excitation Protection

When a synchronous generator is operating at a lagging power factor, the excitation required is divided into two parts, one to produce the internally generated emf and another part to overcome the effect of armature reaction. If the generator operates at a leading power factor, however, the armature reaction is magnetizing rather than demagnetizing; hence, less field current is required to hold a given terminal voltage. If the generator operates into a system that is made more and more capacitive, a point is reached at which the field current is zero, and the terminal voltage is still held at its normal terminal voltage by the magnetizing armature reaction. This is a limiting condition, however, since a further increase in capacitive loading produces an unstable condition and the voltage of the generator can become very high.

A self-exciting condition can be reached for generators located near one of the converters of an HVDC system [21]. The filters and reactive power correction required for the dc system operation consists of large capacitor banks. As the dc system loading increases to its rated value, more and more of these reactive support and harmonic suppressing banks must be placed into service to provide the reactive power and filtering required by the converter. Should the dc system encounter a problem that requires the converters to be blocked suddenly, leaving the large capacitor banks connected, any nearby generator could go into self-excitation because of the largely capacitive load seen from the generator terminals. Moreover, the rise in voltage is rapid and can quickly reach damaging levels.

One way of preventing self-excitation for machines near a converter terminal is through proper system management. This means that operating restrictions be placed on the valve groups, filters, and machines. One type of protection monitors the field current of the generator, and if it goes below a predetermined level an order is given to trip filters. This, of course, depends on the number and type of machines and filters. The maximum allowable time from detection of the condition to filter tripping is on the order of 200 ms [6]. It is also necessary to make sure that, with the rising voltage, the filter circuit breakers will not restrike.

22.6.5 Dynamic Overvoltage Protection

The preceding section presents some of the factors relating to the large capacitor banks and filters that are part of a large HVDC installation. The size of each switchable capacitor bank is limited by the strength of the ac network and is given by the equation

$$\Delta V = \frac{\text{Capacitor bank size}}{\text{SCR} \times P_{dc}} \tag{22.51}$$

where ΔV = the tolerable ac voltage variation for each switching
SCR = the minimum short-circuit ratio
P_{dc} = the rating of the HVDC system

The minimum short-circuit ratio is defined as

$$\text{SCR} = \frac{\text{Minimum short-circuit MVA}}{P_{dc}} \tag{22.52}$$

is a quantity known for HVDC installations as it is an important design parameter. The voltage change is, therefore, simply the capacitor bank size divided by the minimum short-circuit MVA.

The concept of *dynamic overvoltage* (DOV) is a fundamental frequency phenomenon. The DOV that will be experienced at each switching of a capacitor bank is computed as

$$\begin{aligned} \text{DOV} &= 1 + \frac{\text{Capacitor bank size}}{\text{Minimum available short-circuit MVA}} \\ &= 1 + \frac{\text{Capacitor bank size}}{\text{SCR} \times P_{dc}} \quad \text{pu} \end{aligned} \tag{22.53}$$

If the ac system is weak, the minimum available short-circuit MVA will be small, leading to high values of DOV. In such cases, the converter can be controlled to act as a static reactive voltage controller and can control the ac voltage dynamically [10], [12], [14–16].

Typical values of DOV that might be permitted at the ac bus are usually in the range of 5%. The specification may also include the amount of dc power variation that may be permitted following a capacitor bank switching event with maximum DOV.

22.7 HVDC PROTECTION SETTINGS

HVDC systems are usually delivered by an equipment manufacturer as a complete or *turnkey* system, including the protective systems for both the ac and dc side of the converter stations. This means that the supplier will design all protections and provide recommended settings that are based on the specifications. The protection engineer at the utility must ensure coordination with the local ac system protective systems with those supplied with the HVDC engineers. Since this interface is at an ac bus, this coordination is not unusual.

The important difference between this activity and that of normal ac system protection is the preparation of the specification, since this is the only information that the supplier will have concerning the ac system in which the HVDC system must operate. This must include information about the strength of the ac system at both converter locations and a statement of the utility requirements to ensure proper reliability throughout the protective system. For example, this would include requirements for redundant protective devices, measurement systems, processing systems, and tripping controls. Since the controls are often digital, this should include requirements for power supplies, digital hardware, software, and construction standards that will assure separation of redundant systems.

Table 22.4 provides a summary of some of the HVDC protections, showing both the main (M) and backup (B) systems.

TABLE 22.4 A Summary of HVDC Protective Systems*

Protected Item	Overcurrent	Overvoltage	Undervoltage	Differential	Ground Fault	Line
AC bus protection				M, B		
AC filters and capacitor banks	B	B		M	B	
Converter Transformers	B			M		
Valve groups	B		B	M		
DC line and filters	B					M

Note: * M = main protection, B = backup.

22.8 SUMMARY

HVDC protections are unique to the equipment requirements, their design, and ratings. They may vary from one manufacturer to another. However, the examples cited in this chapter provide a general description of the types of protection required, the protective zones, and the coordination of the many different devices. It is a complex protection system. A partial summary of HVDC protections is given in Table 22.4.

HVDC protections must be coordinated on the ac side with the existing system protection and system capabilities. This requires a coordinated plan by the utility protection engineer and the HVDC manufacturer to ensure proper coordination and satisfactory system operation.

REFERENCES

[1] Kimbark, E. W, *Direct Current Transmission, Vol. I*, Wiley-Interscience, New York, 1971.

[2] Arrillaga, J., *High Voltage Direct Current Transmission*, IEE Power Engineering Series, No. 6, 1988.

[3] Weedy, B. M., *Electric Power Systems*, Second Edition, John Wiley & Sons, New York, 1972.

[4] El-Hawary, M. E., *Electrical Power Systems, Design and Analysis*, IEEE Press, Piscataway, NJ, 1995.

[5] Wadhwa, C. L., *Electrical Power Systems*, Second Edition, John Wiley & Sons, New York, 1991.

[6] Bahrman, M. P. "Control System Design," a panel presentation on HVDC Converter Station Design and Procurement Methods, IEEE Power Engineering Society Summer Meeting, Los Angeles, July 1983.

[7] Melvold, D. J., and T. Sumi, "The Proposed Southern California-Intermountain Power Project HVDC Transmission System," Symposium Record, "Incorporating HVDC Power Transmission Into System Planning," U. S. Department of Energy, Phoenix, AZ, March 24–27, 1980, pp. 237–253.

[8] Beshir, M. J., J. H. Gee, and R. L. Lee, "Contingency Arming System Implementation for the Intermountain Power Project HVDC Transmission System," *IEEE Trans.*, PWRS-4, May 1989, pp. 434–442.

[9] Povh, D., "Protection System Design," a panel presentation on HVDC Converter Station Design and Procurement Methods, IEEE Power Engineering Society Summer Meeting, Los Angeles, July 1983.

[10] Mortensen, K. N., "Performance Aspects of HVDC Control and Protection: CU HVDC Operating Performance," a panel presentation on Performance of HVDC Control and Protection Systems, IEEE Power Engineering Society Summer Meeting, Los Angeles, July 1983.

[11] Lee, R. L., M. J. Beshir, and J. H. Gee, "Planning Considerations for the Intermountain HVDC Transmission System," *IEEE Trans.*, PWRD-1 (1), January 1986, pp. 225–231.

[12] Bahrman, M. P., "Protection Systems," HVDC Seminar Notes, Phoenix, AZ, ASEA, February 1984.

[13] Molnar, A. J., "Protection of HVDC Terminals," a paper from General Electric HVDC Systems, Malvern, PA, presented to the Pennsylvania Electric Association Relay Committee, Washington, PA, January 22–23, 1987.

[14] Wu, C. T., "AC/DC System Interaction," a panel presentation on HVDC Converter Station Design and Procurement Methods, IEEE Power Engineering Society Summer Meeting, Los Angeles, July 1983.

[15] Melvold, D. J., "A Potpourri of Procurement Pitfalls and Considerations for Future HVDC Systems," a panel presentation on Performance of HVDC Control and Protection Systems, IEEE Power Engineering Society Summer Meeting, Los Angeles, July 1983.

[16] Weaver, T. L., "Miles City DC Tie Requirements," a panel presentation on Performance of HVDC Control and Protection Systems, IEEE Power Engineering Society Summer Meeting, Los Angeles, July 1983.

[17] Shockley, P. R., "Performance of the Pacific HVDC Intertie Control System Under Faulted Conditions," a panel presentation on Performance of HVDC Control and Protection Systems, IEEE Power Engineering Society Summer Meeting, Los Angeles, July 1983.

[18] Bahrman, M. P., E. V. Larsen, H. S. Patel, and R. J. Piwko, "Experience with HVDC-Turbine-Generator Torsional Interaction at Square Butte," *IEEE Trans.*, PAS-99, May/June 1980, pp. 966–975.

[19] Hingorani, N., S. Nilsson, M. Bahrman, J. Reeve, E. V. Larsen, and R. J. Piwko, "Subsynchronous Frequency Stability Studies of Energy Systems Which Include HVDC Transmission," Symposium Record, "Incorporating HVDC Power Transmission Into System Planning," U. S. Department of Energy, Phoenix, AZ, March 24–27, 1980, pp. 389–398.

[20] Grund, C. E., "Square Butte HVDC System Design Performance," a panel presentation on Performance of HVDC Control and Protection Systems, IEEE Power Engineering Society Summer Meeting, Los Angeles, July 1983.

[21] Rashwan, M. M., and C. V. Thio, "Control, Protection, and Operating Performance of the Nelson River HVDC System," a panel presentation on Performance of HVDC Control and Protection Systems, sponsored by the Transmission and Distribution Committee, presented at the IEEE Power Engineering Society Summer Meeting, Los Angeles, July 1983.

PROBLEMS

22.1 As a research project, perform a brief study of high voltage direct current transmission systems. References 1 and 2 are recommended for study.

22.2 In Figure 22.1, the supply voltage is shown as having a balanced set of wye-connected voltages, designated e_a, e_b, and e_c with all phase voltages having a peak magnitude of E_m. This representation is an equivalent for the converter transformer, which has a line-to-neutral rms voltage that depends on the transformer tap ratio, a. Thus, we can write

$$aV_{LN} = \frac{E_m}{\sqrt{2}}$$

Derive an expression for the ideal no-load direct voltage of a six-pulse rectifier in terms of the converter transformer primary line-to-line voltage and the turns ratio a.

22.3 A six-pulse bridge rectifier is supplied from a 230 kV ac transmission system. The converter transformer is rated 230-115 kV. Determine the no-load dc rectifier voltage if the ac supply voltage is 115 kV and delay angle is (a) 15°, (b), 30°, (c) 45°.

22.4 Consider the three-phase bridge circuit shown in Figure 22.1 during the time interval when valves 1 and 2 are conducting. If the voltage of phase a is given by $E_m \cos(\omega t + 60°)$, find the following voltages:

(a) e_b

(b) e_c

(c) i_a as a function of the total dc current, I_d

(d) i_b

(e) v_1 through v_6 (the voltages across the six valves), when v_1 is conducting.

22.5 For the uncontrolled three-phase rectifier shown in Figure 22.2, compute the ratio of the applied rms ac voltage to the average value of the rectified dc voltage.

22.6 Find an expression for the instantaneous voltage across the rectifier from Figure 22.2. Then find the peak inverse voltage from this result.

22.7 What is the ratio of the ideal no-load direct voltage to the applied peak phase voltage in the three-phase bridge rectifier?

22.8 Verify (22.4) by performing the indicated integration.

22.9 Prepare a spreadsheet to determine the plotted quantities shown in Figure 22.3 for the controlled rectifier with no overlap. This process is described fully in several references, with Kimbark [1] being a recommended resource.

22.10 Prepare a spreadsheet to determine the plotted quantities shown for the controlled rectifier with overlap, shown in Figure 22.6.

22.11 The transformer line-to-line secondary voltage of a three-phase bridge converter rectifier is 170 kV. Compute the direct voltage of the converter when the overlap angle is 15 degrees and delay angle is (a) 0°, (b) 15°, (c) 30°, (d) 45°.

22.12 A three-phase bridge rectifier is delivering 200 MW with a direct voltage of 200 kV. (a) If we can assume that $\alpha = 30°$ and $\mu = 15°$, what is the ideal no-load direct voltage of the rectifier? (b) Given the conditions of (a), what is the equivalent commutating reactance of the rectifier?

22.13 Consider the 12-pulse converter consisting of two three-phase bridge rectifiers in cascade, as shown in Figure 22.11. The applied ac voltage of the lower bridge is assumed to lag that of the upper bridge by 30 degrees. Thus, we can write

$$v_{d1} = \sqrt{3}E_m \cos(\omega t - \pi/6) \qquad 0 \le \omega t \le \pi/3$$

$$v_{d2} = \sqrt{3}E_m \cos(\omega t - \pi/3) \qquad \pi/6 \le \omega t \le \pi/2$$

(a) Compute the total voltage dc voltage

$$v_d = v_{d1} + v_{d2}$$

(b) Compute the average value of the rectified voltage.

22.14 Prepare a spreadsheet to determine the plotted quantities shown for the controlled rectifier with overlap, shown in Figure 22.10.

23

SSR Protection

23.1 INTRODUCTION

This chapter continues with a consideration of protective systems that are designed to protect equipment against disturbances that arise on the network rather than within the equipment. The topic of concern in this chapter is *subsynchronous resonance*, which is usually abbreviated simply as *SSR*. SSR is not a common phenomenon and it is not of concern in some power systems. Because of its special nature, SSR is not well known to many power system engineers. Therefore, we begin this chapter with a brief introduction to the subject. This introduction is followed with a description of the types of protection that may be necessary on power systems where SSR can be shown to be a potential hazard.

23.2 SSR OVERVIEW

Subsynchronous resonance is a condition that can exist on a power system where the network has natural frequencies that fall below the fundamental frequency of the generated voltages. Transient currents flowing in the ac network have two components; one component at the frequency of the driving voltages and another component at a frequency that depends entirely on the elements of the network. For a network with only series resistance and inductance, an isolated transient, such as switching a load, will consist of a fundamental component and a dc component that decays with a time constant that depends on the L/R ratio of the equivalent impedance between source and load. Since loads are frequently switched on and off, the transient currents usually appear as random noise, superimposed on the fundamental frequency currents. The addition of shunt capacitors to the network result in new natural frequencies of oscillation that are always greater than the fundamental frequency. In networks containing series capacitors, the currents will include oscillatory components with frequencies that depend on the relative magnitude of the transmission line L and C elements, but have frequencies that are below the system fundamental frequency.

A general equivalent circuit seen looking into a large network is shown in Figure 23.1. The network seen from the generator includes transmission lines that have series compensation to reduce the total line reactance. This accounts for the presence of the capacitor in the series equivalent.

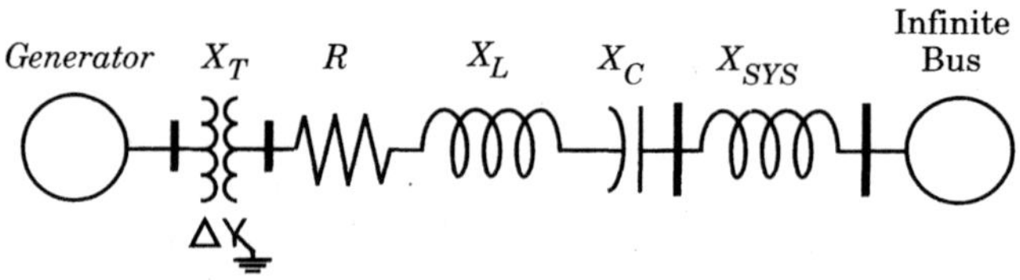

Figure 23.1 The system equivalent seen from a generator.

We can write a general expression for the current in a simple radial R-L-C network as

$$i(t) = K[A\sin(\omega_1 t + \psi_1) + Be^{-\varsigma\omega_2 t}\sin(\omega_2 t + \psi_2)] \tag{23.1}$$

where all of the parameters in the equation are functions of the network parameters except ω_1, which is the frequency of the driving voltages of all the generators. Note the presence of another frequency, labeled ω_2 in (23.1), where this new frequency is a function of the network elements.

To understand the dual frequency condition, consider the simplified system shown in Figure 23.2, where a switch is closed to start the flow of current. This figure could represent the initiation of a fault on any power system with a series compensated line. A trapped charge is assumed to exist on the capacitor. It can be shown that the total current that flows under these conditions can be written as [1]

$$i(t) = i_s(t) + i_t(t) \tag{23.2}$$

where $$i_s(t) = \frac{E_m}{Z}\sin(\omega t + \phi - \theta) = \text{steady-state component} \tag{23.3}$$

$$\begin{aligned} i_t(t) &= \frac{E_d}{\beta L}e^{-at}\sin\beta t - \frac{E_m}{Z}\sin(\phi - \theta)e^{-at}\cos\beta t \\ &= \text{transient component} \end{aligned} \tag{23.4}$$

The parameters used in these equations are as follows.

E_m = peak value of the supply voltage

$$E_d = E_m\sin\phi - \frac{Q_o}{C} - \frac{E_m\omega L}{Z}\cos(\phi - \theta) - \frac{E_m R}{2Z}\sin(\phi - \theta) \tag{23.5}$$

$$Z = \sqrt{R^2 + \left(\omega L - \frac{1}{\omega C}\right)^2} = \text{impedance of the circuit}$$

$$a = \frac{R}{2L} = \text{damping factor}$$

$$\beta = \sqrt{\frac{1}{LC} - \frac{R^2}{4L^2}} \quad \text{where} \quad \frac{R^2}{4L^2} < \frac{1}{LC} \tag{23.6}$$

$$\theta = \tan^{-1}\left(\frac{\omega L - 1/\omega C}{R}\right)$$

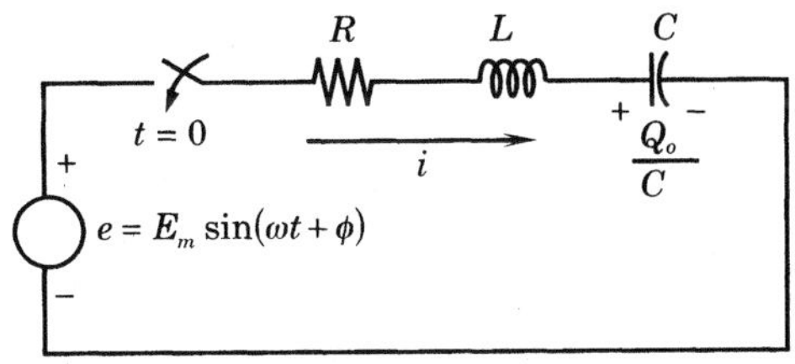

Figure 23.2 A series R-L-C network with sinusoidal supply voltage.

Note that the transient component of current has a radian frequency β that depends on the network parameters. This transient component, given by (23.4), has two parts, and both parts decay exponentially.

In power transmission lines that are series compensated, the ratio of the capacitive reactance of the series capacitors to the total equivalent pi inductive reactance of the line is called the *degree of compensation*. This is often written as

$$k = \frac{X_C}{X_L} = \frac{1}{\omega^2 LC} \tag{23.7}$$

where k is the degree of compensation. Typical values for the degree of compensation range from about 0.2 to 0.8. There is a practical upper limit to the degree of compensation and values greater than about 0.7 require careful study [2]. However, as a theoretical network of interest in the simple circuit of Figure 23.2, the degree of compensation can have any value, with the capacitive reactance being either smaller than the inductive reactance ($k < 1$) or larger than the inductive reactance ($k > 1$). In power systems, only degrees of compensation less than unity are considered.

EXAMPLE 23.1

Consider the simple circuit shown in Figure 23.2, where an R-L-C network is energized at $t = 0$ by closing the switch. The supply voltage is sinusoidal with a frequency of 60 hertz. The total impedance, exclusive of the capacitor, is the equivalent pi resistance and reactance of a 500 kV transmission line that is 200 miles long. Compute the current that flows after the switch is closed.

The equivalent pi impedance in the transmission line is given as

$$Z_\pi = 5.6299 + j115.9443 \quad \Omega$$

Calculate the current as a function of time in per unit on a 100 MVA base. Assume that the supply system has a Thevenin equivalent impedance of $0.001 + j0.002$ per unit.

Solution

The total transmission line impedance in per unit is computed by dividing the ohmic impedance by the base impedance for 500 kV on a 100 MVA base.

$$Z_\pi = \frac{5.6299 + j115.9443}{2500} = 0.003226 + j0.069502 \text{ pu}$$

Let the degree of compensation be set at $k = 0.4$. Then the total line reactance is reduced to 0.04170 per unit. Now, as a practical matter let us assume a system source impedance of

$$Z_{\text{sys}} = 0.001 + j0.002 \text{ pu}$$

This gives a total network impedance of

$$Z = 0.004226 + j0.071502 \text{ pu}$$

The applied voltage has a phase angle that can be any desired value. Let this phase angle, ϕ, be 25 degrees, an arbitrary choice. The impedance phase angle of the total impedance, θ, is computed from

(23.6) to be

$$\theta = \tan^{-1}\left(\frac{0.071502}{0.004226}\right) = 84.476^\circ$$

Given the stated values of all parameters, the steady-state current, the transient current, and the total current are computed, with the results shown in Figure 23.3. The steady-state current is represented by the solid line, and it is displaced from the integral cycle markers by the arbitrarily selected angle $\phi - \theta$. The transient current, shown by the line with short dashes, has an initial value that is exactly the negative of the steady-state current such that the total current will start at zero. Note that the transient current has a lower frequency than the steady-state current. For the case plotted, this frequency is 37.9 hertz. The transient component is lightly damped. As the degree of compensation is increased, the transient component frequency increases and would reach a value equal to the fundamental frequency of the applied voltage for $k = 1$, since this would result in a purely resistive network. For degrees of compensation with $k > 1$, the transient current component has a frequency that is greater than that of the applied voltage. The total current is shown in Figure 23.3 as a line with long dashes, and is the sum of the two components. Note that the total current reaches maximum values that exceeds the steady-state peak values, as would be expected.

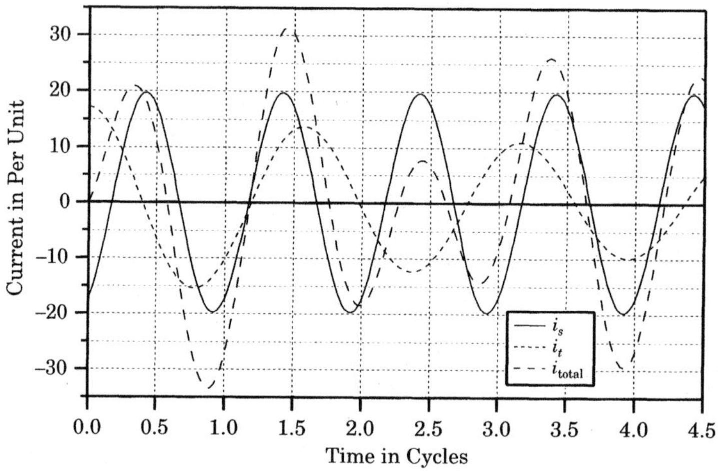

Figure 23.3 Network steady-state, transient, and total currents. ■

Currents similar to (23.1) flow in the stator windings of the generators and are mathematically reflected in the generator rotor according to Park's transformation. This transformation makes the 60 hertz component appear to the rotor as a dc current, in the steady state, but the currents of frequency ω_2 are modulated into currents of frequencies containing the sum $(\omega_1 + \omega_2)$ and difference $(\omega_1 - \omega_2)$ of the two network frequencies. The difference frequencies are called *subsynchronous* frequencies. The subsynchronous currents may induce high shaft torques on the turbine-generator rotor. The forced frequencies of oscillation of the shaft are seen to be the fundamental complement of the network subsynchronous resonant frequency.

The presence of subsynchronous torques on the generator is of interest because the turbine-generator shaft itself has natural modes of oscillation. A typical lumped spring-mass model of a steam turbine-generator shaft is shown in Figure 23.4, which represents a shaft of a typical steam generating unit with a high pressure, intermediate pressure, and two low pressure turbines, the latter designated LPA and LPB. Other shaft components are the generator and a shaft connected exciter. Like any spring-mass system, this shaft system will oscillate with different "modes" of oscillation. For an n mass system, there exist $n-1$ distinct modal frequencies of oscillation and most of these frequencies for a turbine generator shaft are subsynchronous. In practical turbine-generator shafts, these natural frequencies are usually between 10 and 50 hertz for a 60 hertz generating unit. As described in Chapter 20, the mechanical system is subject to torque amplification if driven by oscillatory torques at frequencies that are close to one of the natural frequencies of oscillation.

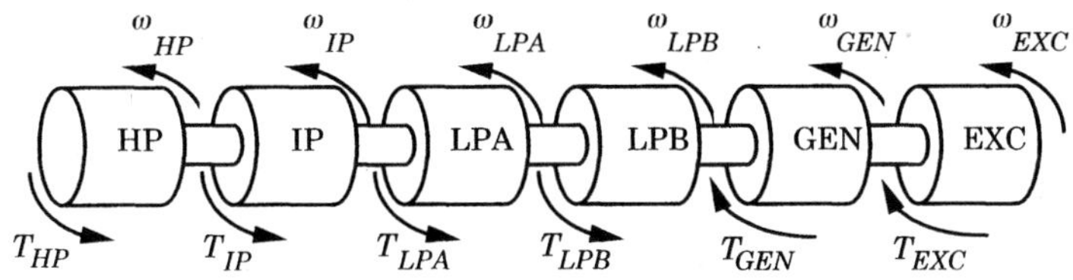

Figure 23.4 Spring-mass model of a steam turbine-generator shaft.

Should the forced subsynchronous torques coincide with one of the shaft natural modes of oscillation, the forced oscillations of the shaft at this natural frequency will occur, sometimes with high amplitude. This condition is called *subsynchronous resonance*, which can cause shaft fatigue life expenditure and possible shaft damage or even failure.

EXAMPLE 23.2

A typical generating unit has a synchronous generator driven by a steam-turbine prime mover consisting of a high-pressure and low-pressure turbine sections. A diagram of the physical system is shown in Figure 23.5. The equations of motion for the shaft system can be written in matrix form as follows:

$$\mathbf{J}\ddot{\theta} + \mathbf{D}\dot{\theta} + \mathbf{K}\theta = \mathbf{T}$$

where $\mathbf{J}$ = a 3×3 diagonal matrix of mass moments of inertia
$\mathbf{D}$ = a 3×3 diagonal matrix of damping constants
$\mathbf{K}$ = a 3×3 nondiagonal matrix of spring constants
$\mathbf{T}$ = a 3×1 vector of applied torques

The data for the equations of motion are summarized as follows:

$$J_1 = 1216 \text{ lbf - ft - s}^2 \qquad K_{12} = 35.28 \times 10^6 \text{ lbf - ft}$$
$$J_2 = 6975 \text{ lbf - ft - s}^2 \qquad K_{23} = 70.40 \times 10^6 \text{lbf - ft}$$
$$J_3 = 4060 \text{ lbf - ft - s}^2$$

The damping constants are unknown and are usually found by an iterative process [3]. Find the modal frequencies for the shaft system.

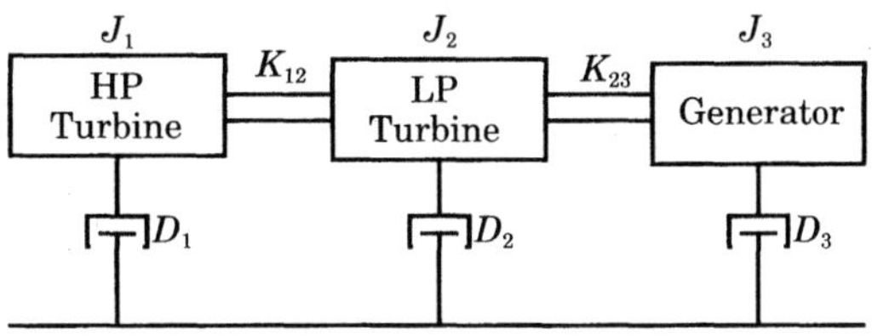

Figure 23.5 A three-mass shaft system [3].

Solution

We can write the system equations, given the numerical data, as follows.

$$\begin{bmatrix} 1216.0 & & \\ & 6975.0 & \\ & & 4060.0 \end{bmatrix} \begin{bmatrix} \ddot{\theta}_1 \\ \ddot{\theta}_2 \\ \ddot{\theta}_3 \end{bmatrix} + \begin{bmatrix} D_1 & & \\ & D_2 & \\ & & D_3 \end{bmatrix} \begin{bmatrix} \dot{\theta}_1 \\ \dot{\theta}_2 \\ \dot{\theta}_3 \end{bmatrix}$$

$$+10^6 \begin{bmatrix} 35.28 & -35.28 & 0 \\ -35.28 & 105.70 & -70.40 \\ 0 & -70.40 & 70.40 \end{bmatrix} \begin{bmatrix} \theta_1 \\ \theta_2 \\ \theta_3 \end{bmatrix} = \begin{bmatrix} T_1 \\ T_2 \\ T_3 \end{bmatrix}$$

The three equations are coupled through the spring constant matrix. The damping constants are unknown, but these constants are relatively small and can be neglected with little error in the computation of modal frequencies. This is done by computing the eigenvalues of the system, which are found to have the following values for the modal frequencies.

$$\begin{aligned} \omega_1 &= 196.53 \quad \text{rad/s} & f_1 &= 31.29 \quad \text{Hz} \\ \omega_2 &= 151.24 \quad \text{rad/s} & f_2 &= 24.08 \quad \text{Hz} \\ \omega_3 &= 0.00 \end{aligned}$$

Each modal frequency corresponds to a particular eigenvalue. The third mode is the shaft rigid-body mode, where all masses on the shaft move exactly together as one solid body. Note that there are only two modal frequencies for the shaft, although there are three masses. The modal frequency numbers do not correspond to the inertia numbers, but are simply a means of identifying the different modal frequencies. The modal frequencies are subsynchronous, which is typical for a turbine-generator shaft system. ■

23.2.1 Types of SSR Interactions

There are several ways in which the system and the generator may interact with subsynchronous effects. A few of these interactions are basic in concept and have been given special names. We mention three of these that are of particular interest [3]:

- Induction generator effect
- Torsional interaction
- Transient torque

Each of the above effects will be discussed briefly.

23.2.1.1 Induction Generator Effect. Induction generator effect (IGE) is caused by self-excitation of the electrical network. The resistance of the generator to subsynchronous current, viewed looking into the generator at the armature terminals, is a negative resistance over much of the subsynchronous frequency range. This is typical of any voltage source in any electric network. The network also presents a resistance to these same currents that is a positive resistance. However, if the negative resistance of the generator is greater in magnitude than the positive resistance of the network at one of the network natural frequencies, growing subsynchronous currents can be expected. This is the condition known as the induction generator effect. Should this condition occur, the generator may experience subsynchronous torques at or near a natural shaft frequency, which may cause large and sustained oscillations that could be damaging to the shaft.

23.2.1.2 Torsional Interaction. Torsional interaction occurs when a generator is connected to a series compensated network, which has one or more natural frequencies that are synchronous frequency complements of one or more of the torsional natural modes of the turbine-generator shaft. When this happens, generator rotor oscillations will build up and this motion will induce armature voltage components at both subsynchronous and supersynchronous frequencies. Moreover, the induced subsynchronous frequency voltage is phased to sustain the subsynchronous torque. If this torque equals or exceeds the inherent mechanical damping of the rotating system, the system will become self-excited. This phenomenon is called *torsional interaction* (TI).

The network may be capable of many different subsynchronous natural frequencies, depending on the number of lines with series compensation and the degree of compensation installed on each line. Moreover, switching of the network lines can cause these natural frequencies, as viewed from the generator, to change. The engineer must evaluate the network frequencies under all possible switching conditions to determine all possible conditions that may be threatening to the generators. Another condition that can greatly increase the number of discrete network subsynchronous frequencies is the outage of series capacitor segments. The series compensation in high-voltage systems usually consists of several capacitor segments that are connected in series, with each series segment consisting of parallel capacitors as required to carry the line current. This permits individual segments to be removed from service for maintenance and still permit nearly normal loading of the lines. However, individual segments can fail, thereby changing the network natural frequencies and greatly increasing the number of possible frequencies that can be observed from an individual generator. This increases the work required to document and analyze the network frequencies as seen by each generating station.

Another possible source of subsynchronous currents is the presence in the network of HVDC converter stations. The controls of these converters are very fast in their control of dc power, but the controls can have other modes of oscillation that may be close to a natural mode of oscillation of a nearby generator. Systems that include HVDC converters also must be carefully checked to see if these controls might induce subsynchronous currents in the generator stators, leading to torsional interaction. See Chapter 22.

23.2.1.3 Transient Torques. Transient torques are torques that result from large system disturbances, such as faults. System disturbances cause sudden changes in the network, resulting in sudden changes in currents with components that oscillate at the natural frequencies of the network. In a transmission system without series capacitors, these transients are always dc transients, which decay to zero with a time constant that depends on the ratio of inductance to resistance. For networks that contain series capacitors, the transient currents will be of a form similar to (23.1), and will contain one or more oscillatory frequencies that depend on the network capacitance as well as the inductance and resistance. In a simple radial R-L-C system, there will be only one such natural frequency, which is exactly the situation for (23.1). If any of these frequencies coincide with the complement of one of the natural modes of shaft oscillation, there can be peak torques that are quite large and these torques are directly proportional to the magnitude of the oscillating current. Currents due to short circuits, therefore, can produce very large shaft torques both when the fault is applied and also when it is cleared. In a real power system there may be many different subsynchronous frequencies involved and the analysis is quite complex.

Of the three different types of interactions described above, the first two, IGE and TI, may be considered as small disturbance conditions, at least initially. The third type, transient torque, is definitely not a small disturbance and nonlinearities of the system also enter into the analysis. From the viewpoint of analysis, it is important to note that the induction generator and torsional interaction effects may be analyzed using linear methods. Eigenvalue analysis is appropriate for the study of these problems and the results of eigenvalue studies give both the frequencies of oscillation and also the damping of each oscillatory mode. The other method used for linear analysis is called the *frequency scan* method, where the network seen by the generator is also modeled as a function of frequency and the frequency is varied over a wide range of subsynchronous values. This requires that the generator be represented as a tabulation of generator impedance as a function of subsynchronous frequency, which must be provided by the generator manufacturer. This is considered the best model of the generator performance at subsynchronous frequencies, and is often the preferred method of analysis, with eigenvalue analysis used as a complementary check on the frequency scan results.

23.2.2 A Brief History of SSR Phenomena

SSR was apparently identified for the first time in 1971 in the United States. Two separate shaft failures occurred at the Mohave Generating Station in Southern Nevada. The first failure occurred in late 1970 and the second in late 1971 [4], [5]. These failures were at first thought to be due to generator self-excitation due to induction generator effect. Both failures occurred when one unit at the generating station was operating radially from the network, at the end of a single series-compensated transmission line, and with exactly seven out of eight of the series capacitor segments in service. Observations of line currents and the resulting vibration were similar in the two cases, with a 30.5 hertz component noted in the generator current. After the second failure, however, it was noted that the induction generator hypothesis could not be correct, since having all eight series capacitors in service would provide even larger negative resistance than that experienced with only seven segments in service, and would therefore sustain induction generator action. However, the unit at Mohave had operated in the identical radial configuration on several occasions, with all eight capacitor segments in service, and no subsynchronous oscillations were noted for that operating condition. This realization led to a more careful analysis of the Mohave system with the result that the condition was identified as a resonance between the natural frequencies of the network and the generator rotor, the first case of subsynchronous resonance in torsional interaction.

Following this early experience, the industry initiated several projects to better understand and prevent this type of hazard. The work was largely carried out by engineers from the major electric equipment manufacturers and from utilities in the western United States, where series compensated transmission lines are widely used. One major focus for organization of the work was the IEEE Task Force on Subsynchronous Resonance.[1] This group held symposia at major IEEE meetings to inform engineers as to the causes of SSR and the types of analysis required to determine the possible hazard to a generating unit. They also published a report giving valuable information on the phenomenon [4]. Later, they established a set of common terms and definitions for SSR so that precise meaning could be conveyed by a given terminology [6–8]. They have also published helpful bibliographies of technical papers dealing with the subject [9–12]. In 1981, the IEEE Working Group published a summary of known types of countermeasures that can be used to control SSR [13], [14]. Finally, they published two benchmark models to be used to verify computer programs used for the analysis of the SSR phenomenon [15], [16]. The work of this group helped a great deal in providing understand-

[1]This group later became the IEEE Working Group on Subsynchronous Resonance and still later the Subcommittee on Torsional Effects, under the System Dynamic Performance Committee.

ing of the problems associated with subsynchronous oscillations and in learning how to deal effectively with these problems.

The early emphasis on subsynchronous resonance was in connection with series compensated transmission systems. This was appropriate since the only known turbine-generator failure occurred because of resonance in such a transmission system. In 1977, however, an incident occurred in North Dakota where a turbine-generator torsional vibration occurred on a generating unit that is close to an HVDC converter station. Tests were in progress at the Square Butte HVDC converter to analyze the effect of a power modulation controller. These tests showed that the controller destabilized the first torsional mode of a turbine generator at the nearby Milton Young Generating Station [17]. This led to a special investigation to more fully understand this problem and to provide suitable corrections to the HVDC controls [18].

The western United States is not the only region that has reported interest in SSR control and prevention. Several countries that have reported studies on the subject including Argentina [19], Brazil [20], Canada [21], [22], Italy [23], Japan [24], South Africa [25], Sweden [26], and Turkey [27]. In the United States the series compensation and HVDC systems were first introduced in the western states. HVDC is becoming much more common as new systems are designed and installed. Series compensation has been added to certain high voltage lines in the eastern interconnected system of the United States [28]. As series compensation and HVDC transmission become more common, more will be learned about SSR and additional countermeasures will probably be devised. Recent transmission developments include fast solid-state controls of series compensation, phase shifters, and other devices [29]. These new devices will replace the discrete frequencies of older series compensation systems by a band of frequencies that may cause problems in many different turbine-generator shafts, but the control of the series compensation also provides hope of a new control of this phenomenon. Therefore, it appears that subsynchronous resonance analysis will become a necessary activity in the design of the protective systems for many steam turbine-generator units. Note that it is only the steam turbine-generator shafts that are subject to damage by SSR. Hydrogenerator shafts have been shown to be immune to this type of damage [30].

Another potential cause for subsynchronous torsional interactions has been identified with *static reactive compensators*, or *static var compensators* (*SVCs*) [31], [32]. Studies have been made in conjunction with the extensive HVDC interconnections between the northeastern United States and the Canadian Province of Quebec. Developments in this region involve both HVDC converter stations and a large SVC installation. It has been shown that, if the SVC has a large control variation that is comparable in magnitude to nearby generating units, there can be torsional interactions between the SVC and the generating unit under certain conditions. Special filters for the SVC controls can be designed to mitigate the effects. This is an example of still another potential problem that can arise, and that may require special study and countermeasures to provide a fully satisfactory solution. SVCs are also becoming more common, which emphasizes the need for careful analysis of possible SSR effects in many different systems.

23.3 SSR SYSTEM COUNTERMEASURES

The protection provided for turbine-generator units against the possibility of SSR damage is a class of devices known as *SSR countermeasures* [13], [14]. There are several types of SSR countermeasures, which are presented here under two classifications: system countermeasures and unit countermeasures. The system class of devices is discussed in this section. Unit countermeasures is presented in the following section. System countermeasures are controls, filters, and operating strategies that are applied to the power system, rather than being located to effect only one particular generating unit. This means that these system countermeasures may affect the operation of more than one turbine-generator unit, although the primary intent may be to provide protection to a certain nearby unit.

23.3.1 Network and Generator Controls

Since SSR is caused by interactions between the network and the generator, one countermeasure that may be feasible is to simply alter the network in some way when potentially damaging SSR currents flow. Another way of protecting a generating unit from SSR induced damage is simply to remove the unit from service. These countermeasures are described below.

23.3.1.1 System Switching. There are many different ways in which the network can be altered by switching various network elements and some of these options may be feasible and cost effective as countermeasures. Some switching actions may simply change the SSR alert to an emergency of another nature, which suggests that any network altering scheme must be carefully examined under a wide range of system conditions.

One type of system switching that is feasible in some cases is to bypass the series capacitors under certain generating plant loading conditions. It is well known that turbine-generator modal damping increases with plant loading. In some cases, torsional interaction may only be a problem when the plant loading is below a certain minimum level and damping is poor. If this concept is applicable, a very inexpensive countermeasure is simply to bypass the transmission line series capacitors when the plant loading is reduced below this critical level. This is feasible if the series compensation is not required at the lower loading. Removing the series capacitors, or even a portion of the total series compensation, may not be a workable solution in all cases, but it is a good solution when the conditions warrant its use [33].

23.3.1.2 Series Capacitor Voltage Control. Another form of network switching is a controlled series capacitor bypass resistor that is timed to suppress subsynchronous components in the transmission line current. This device is referred to in the literature as the NGH scheme[2] [34–37]. The concept behind this scheme is relatively simple and may be explained with reference to Figure 23.6.

In Figure 23.6(a) a fundamental voltage wave, represented by a solid line, is combined with a dc voltage, illustrated as a dashed line. The sum of the two waves, represented by the short-dash line, consists of half cycles that are alternating longer and shorter in duration than the fundamental . In Figure 23.6(b), the fundamental voltage is combined with a subsynchronous voltage. Again, it is noted that some half cycles are shorter and some are longer than those of the fundamental. Clearly, if there were no dc or subsynchronous components at all, then all half cycles would be exactly the same length. An effective method of determining the presence of a subsynchronous current in a series capacitor, then, is simply to measure the time between zero crossings of the voltage across the capacitor. The current through a parallel resistor bypass circuit is controlled by monitoring the time between zero crossings. On each half cycle that exceeds the normal duration, the zero crossing can be caused to arrive sooner by forcing a reduction in the capacitor voltage thereby forcing the voltage to more nearly approximate a pure fundamental wave.

The control circuit for this scheme is shown in Figure 23.7. A linear resistor is arranged in series with back-to-back thyristors, with the series combination connected across the series

[2]The scheme was named using the initials of its inventor, Narain G. Hingorani.

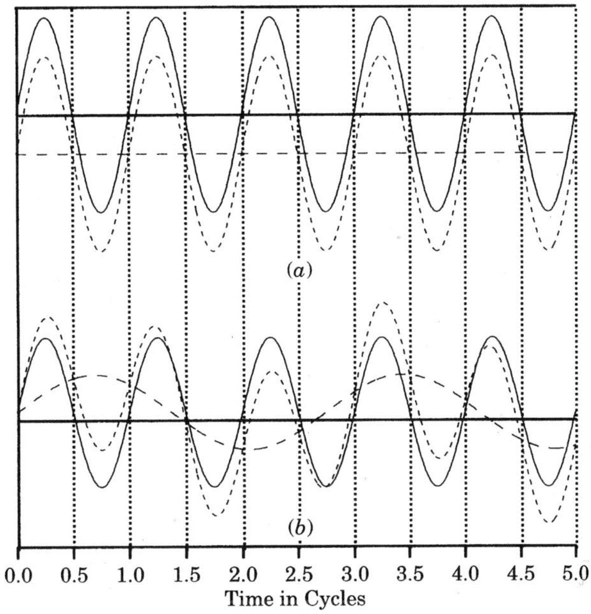

Figure 23.6 Combining a fundamental voltage with a dc voltage or a subsynchronous voltage.

compensation of the transmission line. At each zero crossing of the capacitor voltage, the timer is set to begin counting the time until the next zero crossing is found. If the elapsed time exceeds the one-half cycle period, the thyristor is fired to discharge the capacitor through the resistor, thereby bringing the capacitor voltage zero crossing sooner. The thyristor stops conducting when the capacitor voltage, and the resistor current, reaches zero. A new cycle of counting and control action follows immediately.

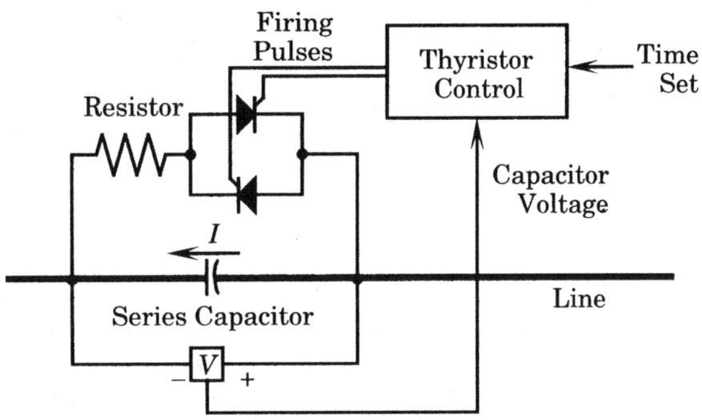

Figure 23.7 Control circuit for the NGH scheme.

A test installation of the NGH scheme has been installed in Southern California, near the Mohave Generating Station and its operation is being evaluated [38].

23.3.1.3 Thyristor Controlled Series Capacitors. The thyristor controlled series capacitor is similar in concept to the NGH scheme, but uses thyristor control of a inductor bypass circuit rather than a resistor. Several different types of systems have been under development and others are likely to emerge as the technology becomes more mature [39–43]. One typical arrangement is shown in Figure 23.8.

The thyristors for this duty must be sized to meet the high fault current bypass requirements and the high di/dt during turn-on for protective action during faults. The currents flowing

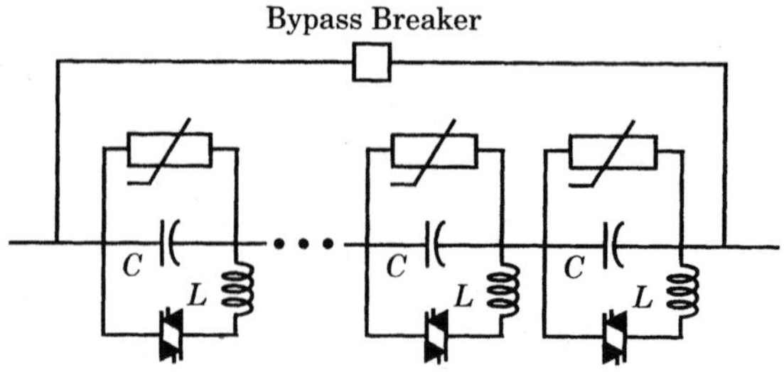

Figure 23.8 Thyristor controlled series capacitors [39].

in the thyristors are caused to flow in controlled pulses, as shown in Figure 23.9. If the current flowing in the transmission line has a subsynchronous component, the capacitor voltage will also have a subsynchronous component. This will cause the zero crossing of the voltage to be delayed past its normal crossing time. However, if the thyristors are pulsed at the right time, the pulse circulates in the L-C loop, forcing the voltage to zero. This causes the capacitor current to carry the pulse in addition to its normal current. These effects are shown clearly in Figure 23.9, which is a recording of a field test of a thyristor controlled series capacitor installed on the Bonneville Power Administration system. Note the shape of the voltage wave, which would appear to have a subsynchronous component that would lead to a delayed zero crossing. However, by timing and sizing the thyristor pulse correctly, the capacitor current is also pulsed and the capacitor voltage goes through zero as the current pulse reaches its maximum value. The capacitor current pulse is rich in higher harmonics, but with a synchronous component as well. The high-frequency harmonics see the capacitor as a very low impedance path, which explains why the current pulse tends to circulate through the capacitor. The line current shows no evidence of the current pulse. Careful field measurements have shown that the harmonic content of the line current is changed very little from its value with the capacitors completely bypassed. This tends to support the observation that the pulse of high-frequency current is largely restricted to the inductance and capacitance loop.

Thyristor controlled series capacitors have been installed in a few locations as of the mid-1990s, and most of these installations are experimental [39–43].

One way to control the thyristor controlled series capacitor to force the required zero crossing of the capacitor voltage is to use that voltage as the input to a controller, as shown in Figure 23.10. At higher line loadings with greater current, the capacitor voltage will increase, which requires larger current pulses to force the required zero crossing.

In order to qualify as an SSR countermeasure, the protective system must have been installed in the field and tested to assure compliance with one or more of the objectives of countermeasures:

1. To respond to system conditions that are capable of causing a subsynchronous torsional response by the protected generating unit,
2. To provide positive damping of a turbine-generator torsional mode, and,
3. To ensure that the countermeasure is capable of limiting generating unit damage to the protective levels required.

The thyristor controlled series capacitor is not yet considered an SSR countermeasure as the technology is new and largely untested, although the technology shows considerable promise. It must be realized, however, that this may not provide adequate protection for all generating units unless all compensated transmission lines were equipped with thyristor controlled series capacitors. This would be expensive and is not likely to be the best solution.

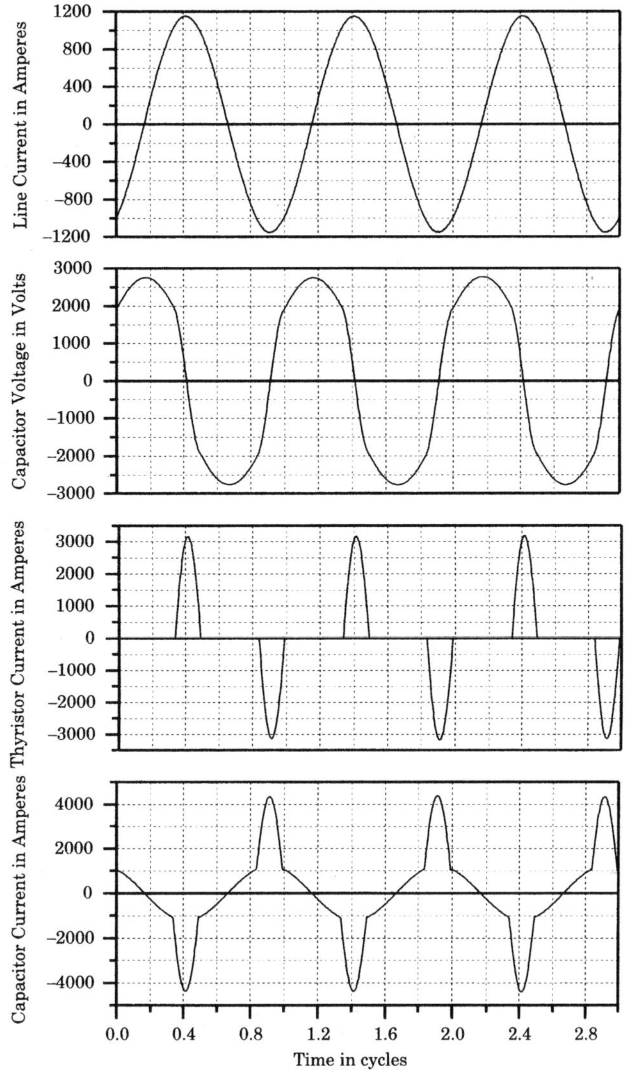

Figure 23.9 Test results of a thyristor controlled series capacitor.

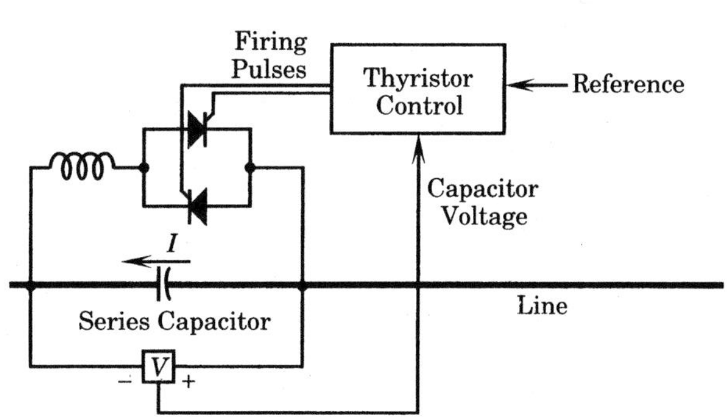

Figure 23.10 A control for the thyristor controlled series capacitor.

However, if only a few lines are identified as posing a threat to a given generator, adding thyristor control to the capacitors in those lines may provide a reasonable solution.

23.3.1.4 Unit Tripping. Tripping a generating unit that experiences SSR oscillations is a measure that will not usually be taken except as the last resort. The generating unit is surely needed for its capacity and voltage support or it would not be in service in the first place. Removing a steam generating unit from service to solve a temporary problem should only be done if the unit is under high stress and in danger of loss of life. Removal under these conditions may be controlled by SSR relaying, which is discussed in Section 23.4.2.

One reason for not tripping a steam generating unit is that it may be several hours before the unit can be restarted. It is often impractical to suddenly ramp a fossil steam unit back to a low load level without the unit tripping as a result of the steam generator upset. One reason for this is that the steam generator is not as stable in its operation at light load. Nuclear units are usually capable of bypassing the steam past the turbine and directly into the condenser, but this is not true of most fossil steam units in North America.[3] Experience with large fossil steam units shows that there are many possible delays in restarting the boiler and fuel systems, as well as the many plant auxiliaries, in attempting to get the unit back to its normal loading after a unit trip. Delays of 10 to 12 hours are not unusual. The loss of the unit for such a long time can have serious economic penalties, especially if the lost energy must be supplied by emergency purchase from neighboring utilities.

In the final analysis, it is clear that damage to a generating unit can be prevented by tripping the unit. This is not always an economic solution and must be carefully examined for its possible costs and other consequences. Indeed, the sudden tripping of the unit also contributes to the loss of life of the unit because of the sudden stresses that this action imposes. However, unit tripping schemes have been devised by utilities as one of the measures taken to prevent unit damage due to SSR [38].

23.3.2 Generator and System Modifications

Another method of preventing SSR problems is to make modifications in the system or generating unit design. These countermeasures may be taken in addition to other measures and may provide additional benefits at a modest cost, if initiated early enough in the design process.

23.3.2.1 Turbine-Generator Modifications. It is not possible to design the turbine-generator so that the SSR hazard is completely eliminated. However, certain design features may be taken that will provide improved unit behavior. One hazard that can often be solved by design improvements is the induction generator effect.

Induction generator effect occurs when the net system resistance, including the generator, is negative at frequencies corresponding to system series resonance. In some cases, calculations showing several ohms of negative resistance have been documented [44]. This effect can often be improved by the addition of pole face amortisseurs to the generator rotor. These low resistance bars can reduce the effective rotor resistance by a significant amount and the reduction may be sufficient to completely eliminate induction generator effect as a problem. This must be determined early enough in the design so that appropriate studies can be made and changes in the manufacturing process initiated in a timely way.

[3]It should be noted that bypass valving is often used in Europe, where once-through steam generators are common and bypass valving is part of the control strategy for these units.

Another change that may be required is not in the generator, but in the step-up transformer. An excellent location for certain filters or other SSR countermeasures is at the neutral of the generator step-up transformer. The neutral is a good location since the voltage to ground is low at this point and any added devices will not require high-voltage insulation. The basic insulation level (BIL) of the transformer neutral may have to be increased, however, to account for additional impedances connected between the neutral and ground. This change also must be determined early in the design phase, and may be considered a good investment even if the requirement for neutral connected countermeasures are not to be pursued initially, but may be needed later in the life of the unit.

23.3.2.2 Generator Circuit Series Reactance. One form of generator circuit series reactance modification is described above in Section 23.3.1, where the series compensation is removed under certain system conditions. This is likely to be applicable only in cases where the unit is radially connected to the bulk power system.

Another way of modifying the generator series reactance is in the form of a dynamic filter, which varies the voltage across a series transformer that is placed in the generator leads. This type of counter measure is discussed below under the "Dynamic Filter" category.

It is not practical to increase the generator stator reactance to limit the flow of subsynchronous currents, because any such increase will have greater effect on the fundamental frequency current than to the subsynchronous currents. This means that any modification of the generator series circuit reactance must be done by some sort of filtering, such that only the subsynchronous currents are affected. To date, no schemes for doing this have been proposed and this option for controlling subsynchronous current has not been used as a countermeasure.

A variation of series reactance modification is the insertion of filters in series with the windings of the generator step-up transformer. This type of countermeasure has been used and is described in the next section.

23.4 SSR UNIT COUNTERMEASURES

The second classification of countermeasures are devices that are installed in connection with a particular generating unit. In a sense, one may think of these countermeasures as a form of turbine-generator protection. Their purpose, however, is not to remove a generating unit that has suffered from internal damage, as most protective devices are designed to do. Instead, the purpose of these countermeasures is to protect the generating unit from damaging interactions with the power system, and to do so before a substantial fraction of the turbine-generator life has been lost due to the interaction. The reason for this philosophy is that turbine-generators are very expensive to repair if shaft damage is sustained, and any necessary repair will force the unit to be out of service for a very long time. It is simply not acceptable to allow conditions on the network that will damage the turbine-generator units in view of this costly and time-consuming repair process. If such conditions should appear, for any cause whatever, the source of the damaging interaction must be found and controlled. However, when such an interaction first appears, it is very important to protect the affected generating units from permanent damage.

There are two general classifications of this type of unit protection countermeasure. The first type is a family of filters and controls that are designed to shield the generating unit from those frequencies that would cause torsional interaction or self-excitation. The second type is a family of relays that are the last line of defense for the unit, and will trip the unit if any form of subsynchronous interaction should occur.

23.4.1 Filtering and Damping

Several types of filtering and damping countermeasures have been described in the technical literature, but only of few of them have actually been installed at generating stations. Some of the promising ideas and practical implementations of filters and dampers are described below.

23.4.1.1 Static Blocking Filters. Static blocking filters have been designed for SSR protection of a generating unit. These filters are connected at the generator step-up transformer neutral or on the high-voltage side of the transformer. The filter consists of L-C filter segments that are tuned to each of the shaft torsional frequencies. This arrangement essentially isolates the machine from the system insofar as the critical subsynchronous frequencies are concerned. It does this by combining the filter characteristics with the system impedance to produce a parallel resonance at the rotor complementary frequencies. The result is a control over torsional interaction and transient torque effects. Moreover, since the filter is tuned to the natural frequencies of the generating unit, it is not impacted by system changes or by future system development [13].

One such static blocking filter has been installed at the Navajo Generating Station in Northern Arizona since 1976 [45–47]. The Navajo generating station has three identical turbine-generators with a total plant rating of 2250 MW. Each unit has five natural torsional modes of oscillation. The electric system frequencies corresponding to these natural torsional modes are computed as ($f_e = 60 - f_m$). For the Navajo units, these electric system frequencies are shown in Table 23.1.

TABLE 23.1 Electrical Frequencies Corresponding to the Navajo Natural Frequencies of Oscillation [39]

Mode Number	Natural Frequency (Hz)	Electrical Frequency (Hz)
1	15.8	44.2
2	20.2	39.8
3	26.0	34.0
4	33.2	26.8
5	51.5	8.5

Studies of the Navajo system showed that the subsynchronous frequencies of the system can shift for a change in the number of units on line, for lines out of service, and for changes in the amount of series compensation in the lines. For this reason, it was decided that any electrical system frequency from 10 hertz to 45 hertz could be possible, which means that the system could be tuned to any of the first four natural modes of the units. Therefore, without some form of countermeasure, the units were at risk of torsional interaction and possible damage.

The solution to this problem was to construct a static blocking filter with the configuration shown in Figure 23.11. The filter for each phase consists of four separate filters in series, with the assembly connected to each phase of the neutral of the step-up transformer. Each filter section has a high Q parallel resonant circuit that is tuned to block the electric currents at the frequency corresponding to the first four torsional modes, but to present negligible resistance to 60 hertz currents.

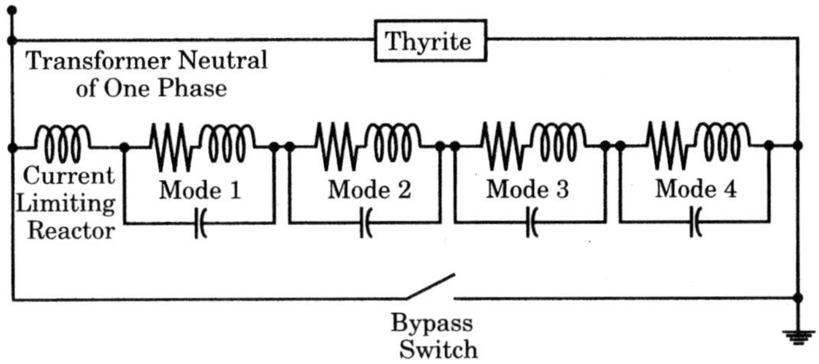

Figure 23.11 The Navajo SSR blocking filter [46], [47].

Passive filters, such as the one described above, are subject to detuning as capacitor elements fail and as the ambient temperature changes. If the filter segments are sharply tuned, calculations should be made of the effect of the detuning to make sure that sufficient resistance is still available at the modal frequencies when the filter tuning is not optimum. For the higher subsynchronous frequencies and requirements for high apparent resistance, the quality factor, Q, of the filter must be quite high, generally greater than 150 [48]. Such a filter will be very sharply tuned and may require automatic tuning if the device is to be effective. If the filter becomes detuned, it may be necessary to add an additional countermeasure to overcome this deficiency.

Another proposed type of static filter consists of a shunt series or parallel passive filter connected from the generator leads to ground, as shown in Figure 23.12. The filter is simply a series resonant circuit connected from line to ground at the generator terminals. The filter shown is a series filter, although a parallel-tuned connection will work as well [49]. Such a device is capable of adding damping to a critical shaft frequency. Tuning is computed according to the phase modulation theory [50], [51]. Although effective for damping a critical frequency, these devices are probably not practical because of the relatively high power rating required.

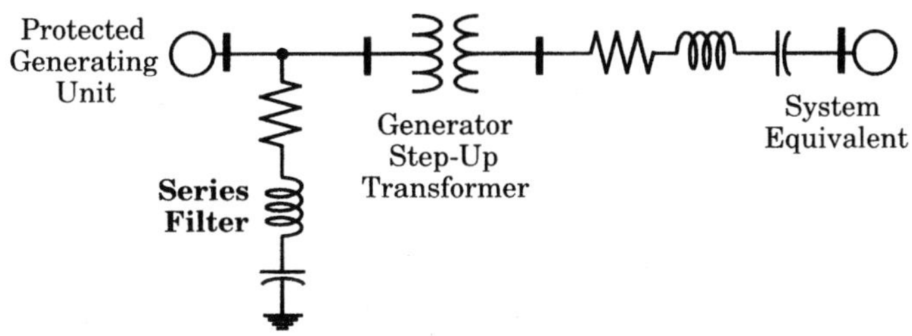

Figure 23.12 A passive shunt countermeasure [49].

23.4.1.2 Line Filters. Filters can also be connected to the transmission lines themselves. One scheme provides a shunt across the series capacitors in each phase, which is designed to act as an inductive and resistive bypass path to the flow of subsynchronous currents [52]. This type of arrangement is particularly effective in counteracting induction generator effect because it introduces substantial resistance to the flow of subsynchronous currents for frequencies up to about 90% of the fundamental.

The filter arrangement consists of a damping resistor in series with the parallel combination of a reactor and a capacitor that are tuned to the system fundamental frequency. Since the filter is a parallel tuned circuit, it has high impedance and carries little current at the fundamental frequency. However, at other frequencies the parallel tuned circuit has reduced impedance and carries higher current through the resistor, thereby increasing the resistance seen by a generator that might otherwise experience induction generator effect. The scheme requires rather expensive components due to the high-voltage level of the installed equipment, but it has been employed at several locations. If only one torsional frequency is critical, on a single circuit, the bypass damping filter is also effective for torsional interaction as well as induction generator effect.

Another line filter scheme is designed to reduce the amplitude of the subsynchronous mmf in the machine [53]. The basic concept is based on the fact that the rotating mmf wave in a synchronous machine is purely a circular wave when created by balanced stator currents. If the stator currents are not balanced, the mmf wave created is of lower amplitude. If the subsynchronous currents are unbalanced currents, the mmf wave created by these currents will produce subsynchronous torques of lower magnitude. The line filter, then, must be designed to present balanced impedances to the fundamental frequency currents, but unbalanced impedances to the subsynchronous currents.

The basic concept can be realized by either series or parallel resonance compensation systems, as shown in Figure 23.13. Part (*a*) of the figure shows the normal balanced series compensation impedances. Part (*b*) shows a series compensation that creates an unbalance by series connected components. Part (*c*) accomplishes the unbalance by using a parallel connection of components. These two schemes are called the series resonant and parallel resonant circuits, respectively [53].

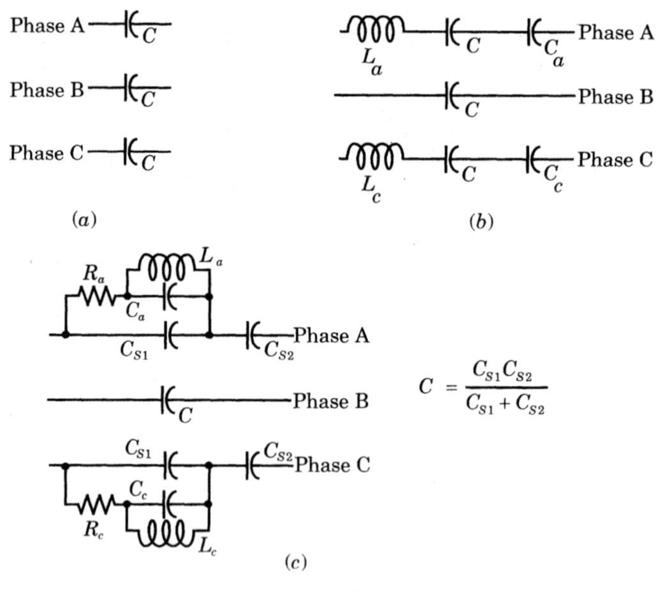

Figure 23.13 Series and parallel resonant compensation schemes [51]. (*a*) Conventional series compensation. (*b*) Series resonant compensation. (*c*) Parallel resonant compensation.

SERIES RESONANT SCHEME. For the series resonant scheme, phases a and c are modified, but phase b is unchanged. The modified phases are changed by connecting a resonant L-C circuit in series with the normal series capacitor. The parameters of the resonant circuits are governed by the equation

$$\omega_o = \sqrt{\frac{1}{L_a C_a}} = \sqrt{\frac{1}{L_c C_c}} \tag{23.8}$$

where ω_o = natural resonant frequency of the series network

Therefore, each resonant frequency has only one independent variable. The ratios C/C_a and C/C_c represent the degrees of asymmetry and also determine the amount of decoupling of the electrical and mechanical systems at the subsynchronous frequencies.

The resonant frequencies of the modified phases with respect to the unmodified phase resonant frequency ω_e is given by

$$\frac{\omega_{ek}}{\omega_e} = \sqrt{\frac{1 + (C/C_k)}{1 + (\omega_e/\omega_o)^2 (C/C_k)}} \qquad k = a, c \tag{23.9}$$

PARALLEL RESONANT SCHEME. For this scheme, shown in Figure 23.13(*c*), two of the phases are modified by shunting a portion of the compensating capacitance by parallel resonant circuits. The inductance and capacitance of the resonant circuits are still governed by (23.8), which ensures that the three phases are kept symmetrical at the fundamental frequency. At any other frequency, however, the three phases will have unsymmetrical phase reactances. The degree of asymmetry is controlled by tuning the resonant circuits.

The advantage of these types of countermeasures is that they consist entirely of passive elements. There is no requirement for any monitoring or control of the compensating elements, hence no control equipment to tune or maintain. It is also possible that the parameters can be optimized for a given installation to provide the maximum effect.

23.4.1.3 Dynamic Filters. A "dynamic filter," as the term has been used in the literature, is an active series filtering device that is used to control subsynchronous currents [13], [14]. One device was proposed soon after the first incidence of SSR in the western United States [54], [55]. The basic structure of this dynamic filter is shown in Figure 23.14. The filter is placed in series with the generator and is controlled to null subsynchronous voltages generated by rotor oscillation.

Subsynchronous voltages produce armature currents that always tend to enhance the rotor oscillations. The dynamic filter is arranged to detect these voltages and, with proper

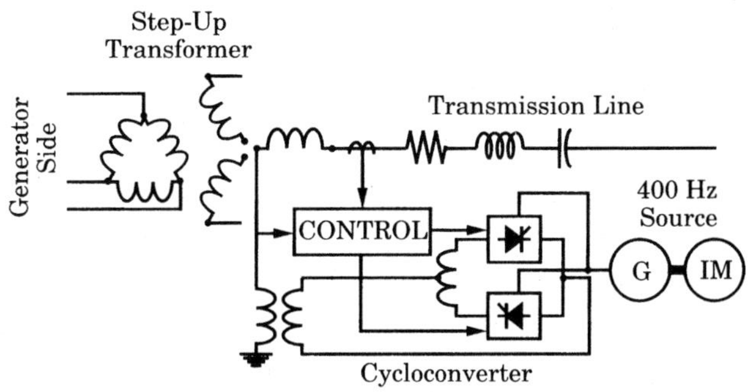

Figure 23.14 One phase of a dynamic filter using a cycloconverter.

control, introduce a voltage that is in phase opposition and of sufficient magnitude to cancel the subsynchronous rotor induced voltage. If the dynamic filter can overpower the voltage due to rotor oscillation, it will provide positive net damping.

The dynamic filter is effective in preventing self-excitation of the generator due to torsional interaction. Moreover, it is not affected by off-normal system frequency operation and is largely immune to system changes, including the number of series capacitors in service at any given time. It is not capable of providing protection for transient torques without a very large expenditure for a powerful countermeasure rating. It also requires a rather complex control system and a separate power supply. Dynamic filters have been demonstrated in the laboratory, but no devices of this type have been placed in service [13].

A dynamic filter similar to that described above has also been described and tested in the laboratory by other investigators [56]. They also provide stability analysis to show that the dynamic filter provides protection for both the induction generator effect as well as torsional interaction.

Another type of dynamic filter has been proposed in the form of a thyristor controlled, static phase shifter [57]. The phase shifter is connected in series with the generator step-up transformer on the high-voltage winding. It provides discrete steps of phase change to modify the active power transferred. The controller uses generator speed as the input signal and employs a bang-bang control strategy. The designers conclude that four discrete steps of 0.6% each are adequate to provide damping of all unstable shaft torsional modes of the IEEE First Benchmark Model [15].

23.4.1.4 Dynamic Stabilizers. The "dynamic stabilizer," as the term has been used in the literature [14], is an active shunt device that is designed to control subsynchronous frequency currents at the generator terminals and prevent these frequencies from entering the stator windings. Several different types have been proposed and one system has been installed and successfully operated for several years. A few different types of devices that can be classified as dynamic stabilizers are discussed below.

THYRISTOR-CONTROLLED REACTOR. The first dynamic stabilizer description appeared in the early 1980s [13], [14]. The device is connected as a shunt at the generator terminals, as shown in Figure 23.15 [13]. It consists of a *thyristor-controlled reactor* (*TCR*) that is connected to the generator leads. Control is achieved by modulating the thyristor firing angles in response to measured turbine-generator speed oscillations. In the absence of rotor oscillations, the TCR presents a constant reactive load to the generator. The unit is sized to provide subsynchronous current at the generator terminals that cancels the current from the transmission system.

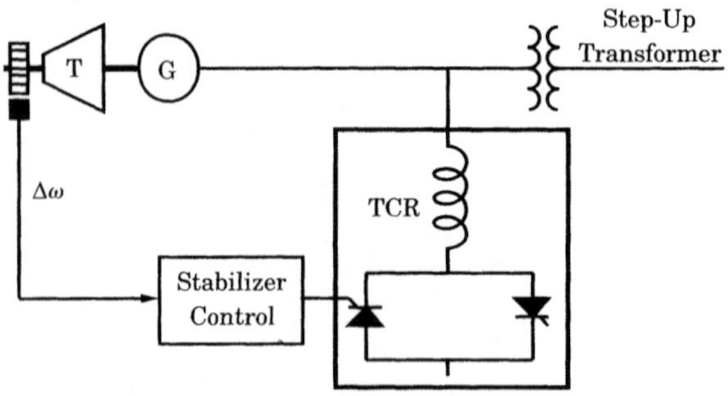

Figure 23.15 Schematic of a dynamic stabilizer arrangement [14].

The theory of the dynamic stabilizer has been proven by extensive derivation and field tests [58], [61]. The advantages of this type of countermeasure are as follows [14]:

1. Connection as a shunt device eliminates the requirement for carrying continuous generator current, so the device can have a much smaller rating than a series-connected scheme.
2. The TCR provides damping for shaft oscillations initiated by any means, making it relatively easy to check system performance.
3. The dynamic stabilizer is not sensitive to variations in the system fundamental frequency and is unaffected by ambient temperature.
4. It provides protection for all modes of torsional oscillation within the bandwidth of the control circuit.
5. The system is relatively easy to maintain and can be readily disconnected from the generator for testing.

On the negative side, the dynamic stabilizer does not provide protection against induction generator effect or for transient torque problems. It will provide damping of oscillations under these conditions, however. The connection to the generator leads is rather difficult, and care must be taken to minimize the possibility of faults in this part of the network. The device also generates harmonics that must be controlled.

Another type of dynamic filter using thyristor-controlled reactors is shown in Figure 23.16 [62], [63]. Computer studies have shown that it is possible to damp modal oscillations using turbine and generator speeds as input signals with a compensating filter to control the inductor-converter unit.

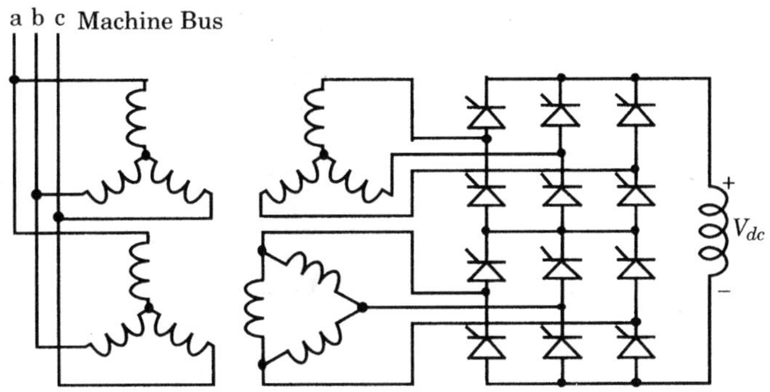

Figure 23.16 A line commutated inductor-converter countermeasure [58].

Still another type of reactive power controller uses a TCR system, similar to the static var controller that is used for transmission voltage control [64]. This type of countermeasure is shown in Figure 23.17.

One of the advantages claimed for this type of dynamic filter is that it has the capability of damping all oscillatory modes at all compensation levels. It is also possible that this type of control may be effective in preventing generator self-excitation that may occur at high levels of series compensation.

DYNAMICALLY CONTROLLED RESISTOR BANK. Another type of dynamic filter is described as a "dynamically controlled resistor bank," which uses a line commutated converter

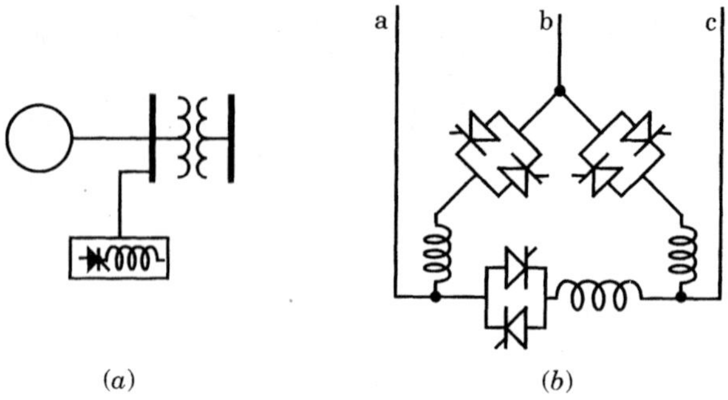

Figure 23.17 Application of a TCR for damping SSR oscillations [57]. (*a*) System connection. (*b*) Schematic of TCR system.

connected to the generator terminals or step-up transformer [65]. The system arrangement is shown in Figure 23.18. In operation, the converter power is modulated according to a derived control law to add damping to the torsional modes of shaft oscillation. The basic idea of this scheme is that power dissipation in the resistor bank is increased whenever the incremental generator speed or the incremental speed of a torsional mode exceeds zero. Computer studies used a resistor bank rated at 5.6% of the generator rating. Clearly, damping can be added by this means, and would appear to be effective for both torsional interaction and transient torques. A somewhat similar system is described in [56].

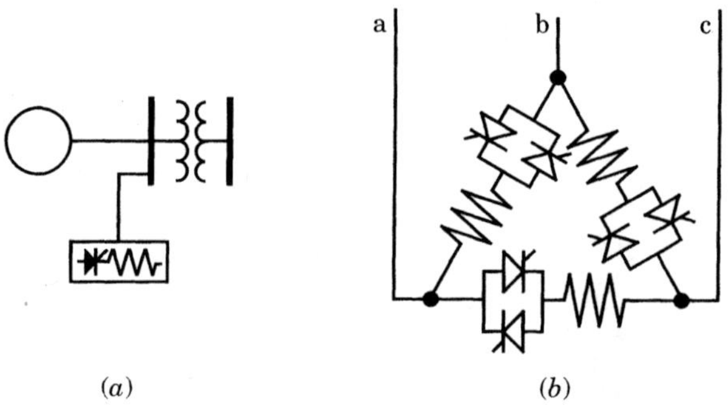

Figure 23.18 Dynamically controlled resistor bank. (*a*) System connection. (*b*) Schematic of resistor control.

The dynamic filters described above are but examples of the types of controls that can be designed using modern solid-state technology. This is an active area of investigation, and other schemes will almost certainly emerge that will merit consideration.

23.4.1.5 Excitation System Dampers. The voltage regulator amplifiers that are part of the generator excitation system are readily accessible, and it is easy to inject new supplementary control signals at the voltage regulator. Therefore, it is not surprising that a number of efforts have been made to use the excitation system as a means of damping subsynchronous oscillations. In these schemes, the excitation system is modulated in response to torsional oscillations of the turbine-generator shaft. If these modulations are properly phased and of adequate magnitude, it is possible to improve the damping of shaft oscillations. In some cases, excitation system dampers are used as supplementary damping enhancements to assist other methods [44].

Early studies of excitation damping schemes were published in the mid- to late 1970s [66–71]. One method that shows promise is based on linear optimal control theory [72–74]. The optimal controller is designed to shift the eigenvalues of the shaft modes as far as possible to the left in the complex plane. The result is a very good improvement in the damping of the torsional modes. It is noted that the studies cited have assumed a very fast static excitation system with very small time constants.

This method of control is not practical for all units due to the relatively large time constants associated with the exciter. For rotating exciters utilizing either dc or ac rotating machines, the time constants are not small enough to provide effective control in the subsynchronous frequency range of interest. Stated another way, the machine transfer function seen looking into the voltage regulator is a low-pass filter with its magnitude falling off at frequencies below 1 hertz. Machines with these exciters are not good candidates for excitation system dampers.

Excitation system dampers are effective in reducing self-excitation due to torsional interaction with low-level oscillations. To be effective, however, the excitation system must have a high initial speed of response [13]. The effectiveness is also dependent on the following factors:

1. The relative position of the exciter mass on the shaft.
2. The capability of the exciter to induce subsynchronous voltages in the armature circuits.
3. The power rating of the excitation system.

Excitation dampers are not fast enough to prevent unit damage due to transient torques from large disturbances and the damping provided is not significant when the oscillations are large in magnitude. Clearly, the excitation system damper is a supplemental type of control that is usually used together with another countermeasure.

Excitation system dampers have been in service since 1976 [75]. Moreover, field tests of the devices have shown that the excitation system damper is very effective in controlling torsional interaction [76], [77].

23.4.2 Unit Relaying and Monitoring

The last line of defense against subsynchronous resonance is a protective relay that can detect the problem and trip the unit before shaft fatigue has accumulated to a significant level. Most of the other countermeasures are designed to make sure that there is good damping of subsynchronous oscillations and to shield the turbine-generator unit from experiencing these oscillations, insofar as possible. Despite all precautions, there may be occasions when a generating unit will be exposed to subsynchronous oscillations. There is always the possibility that conditions are not ideal. The countermeasures in service may not be working correctly, or may be disabled. Should such an unforeseen event occur, it is not an acceptable risk that the generating unit should be caused to sustain long or growing subsynchronous oscillations. Therefore, SSR relays are usually installed to make sure that the unit life is preserved, even under conditions that are extremely rare. In some cases, where the risk of SSR damage is limited or the dangerous conditions are rare, the relay may be the only SSR countermeasure required.

***23.4.2.1* SSR *Protective Relays*.** Three different types of SSR relays have been provided to meet the requirements of detection of potential damaging oscillations and removal of the unit. Two of the relays are based primarily on the monitoring of the frequency content of the generator currents. The third relay models the mechanical behavior of the turbine-generator shaft. Each of these relays will be briefly described.

ARMATURE CURRENT SSR RELAY. The first SSR protective relay was designed by protection engineers of the Southern California Edison Company soon after the first SSR incident at the Mohave Generating Station [78]. Edison protection engineers noted that, on the two Mohave SSR incidents, negative-sequence relays protecting the generating units issued alarms, indicting the presence of negative-sequence currents. This indicated that the negative-sequence relay design might be sensitive to subsynchronous frequencies. Laboratory tests of a similar relay confirmed this characteristic, in fact, the negative-sequence relay produced a signal at 30 hertz, even without the presence of any negative-sequence current. This concept became the basis for the design of a new SSR relay, that was later designated the TEX relay, by the addition of suitable band-pass and blocking filters. The TEX relay has the following design features [78]:

1. Detects positive-sequence currents in the 20–40 hertz range,
2. Provides two subsynchronous current level detectors that are separately adjustable,
3. Is relatively insensitive to low system frequency operation,
4. Is relatively insensitive to generator negative-sequence currents, and
5. Has sufficient time delay to override SSR currents associated with normal system operations, such as system faults and series capacitor switching.

A schematic diagram of the TEX relay is shown in Figure 23.19. The three-phase currents enter the relay at terminals 3, 5, and 7 and leave at terminals 4, 6, and 8. A voltage is developed between terminals 15 and 17 that is dependent on the phase current frequency, magnitude, and sequence of phase rotation. Current level detectors, D_1 and D_2, are contained in the full wave rectifier, which is connected across terminals 16–17. The relay has three filters, designated F_1, F_2, and F_3. F_1 is a dual tuned filter that is designed to bypass the 60 hertz components of positive- and negative-sequence currents and also to block any subsynchronous current components, thereby forcing the subsynchronous currents through the level detector circuits. F_3 is a synchronous rejection filter, which works with F_1 to ensure high subsynchronous current detection sensitivity and low response to both positive- and negative-sequence currents at or near synchronous frequency. F_2 is a harmonic band-pass filter that is added to ensure that higher harmonics will not interfere with the relay operation.

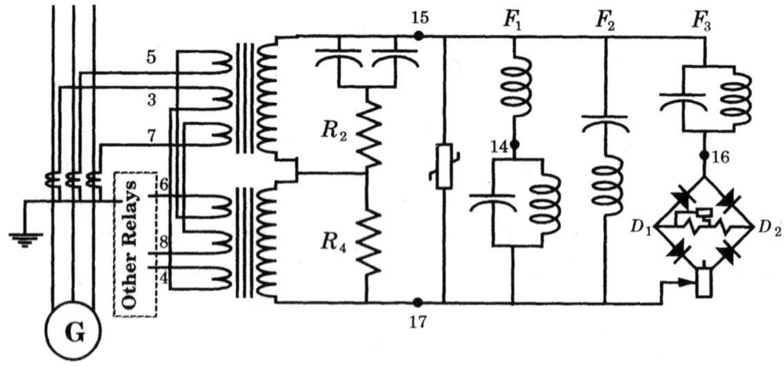

Figure 23.19 Schematic diagram of the TEX relay [78].

The minimum pickup current is set at 0.16 amperes, but this setting can be increased by increasing the size of resistor R_4. Detector D_2 can be further adjusted by varying its bypass resistor. The contacts corresponding to detectors D_1 and D_2 can be used to trip the generator, a transmission line, or a transformer, as required.

The shape of the current pickup curves is adjustable and, if required, the shape at a given subsynchronous frequency can be changed by a minor modification of the filter network. It is important that precautions be taken so the relay will not pick up during system underfrequency conditions that may persist following a system separation, and that may persist for several minutes.

The relay time response varies between three and 10 cycles, on a 60 hertz base. External timers may be required for a particular application in order to provide coordination with other protective devices. The exact settings to be used will usually depend on the application. The manufacturer of the generator should be consulted to learn the allowable subsynchronous generator current that can be permitted. Subsynchronous currents due to faults, line switching, and series capacitor switching should be studied to determine the limits of subsynchronous currents under a variety of operating conditions. All of these factors have a bearing on the proper settings of the relays.

TEX relays have been installed at a number of locations and were the primary protection at some locations at the time they were first designed. Since that time, newer designs have been developed and these newer devices have begun to replace the TEX installations.

TORSIONAL MOTION RELAY. The torsional motion relay was developed in the late 1970s and was first applied at the Navajo Generating Station in northern Arizona. The relay is designed to detect any of the following conditions [79–82]:

- generator self-excitation
- torsional interaction
- torque amplification

These three processes are quite different and require unique approaches to their detection. The principal variables required for detection are shown in Figure 23.20.

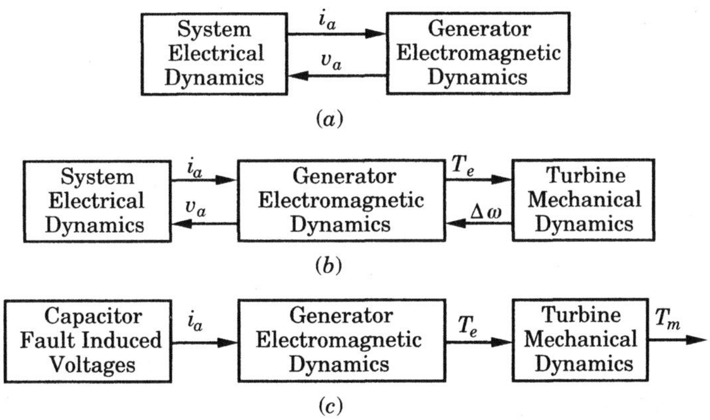

Figure 23.20 Principal variables for the SMF SSR relay application [75]. (*a*) Electrical self-excitation. (*b*) Torsional Interaction. (*c*) Mechanical torque amplification.

When electrical self-excitation occurs, the method of detection requires the use of only electrical quantities, as noted in Figure 23.20(*a*). Each self-excitation event will generate a component of generator voltage and current at a frequency below 60 hertz that is superimposed on the normal 60 hertz quantities, and that increases with time. The TEX relay recognizes self-excitation, but the device is frequency dependent in the frequency range of interest (usually

30–60 hertz) and the sensitivity deteriorates as the frequency approaches 60 hertz. This is the opposite to the required sensitivity under self-excitation conditions.

Torsional interaction is a phenomenon that requires the use of both electrical and mechanical quantities for its detection. As with self-excitation, the generator appears as a negative resistance and oscillations are enhanced by the mechanical system. Detection requires finely tuned pickups at the natural shaft resonant frequencies and the monitoring of shaft mechanical variables to detect the torsional behavior. This monitoring must be sensitive to the shaft stress at the critical frequencies of oscillation in order to provide adequate protection of the unit. This can be accomplished by filtering the speed signal at the critical frequencies to develop a signal proportional to shaft stress [79].

Torque amplification measurement is an open-loop process, as shown in Figure 23.20(*c*). Torque amplification occurs when the turbine-generator mechanical system responds to the discharge of energy from the series capacitors in the faulted transmission lines, with the maximum response occurring one-half second or so following the fault. The relay must be designed to recognize that the torque is growing and will cause damage. In some cases this requires very fast tripping of the unit, perhaps within the first half cycle of torsional response [82]. The torque response must be determined from speed measurements on the shaft, possibly at both ends, and this information must be processed very quickly to provide timely relay action.

The torsional motion relay is constructed with modules for analog processing and a digital tripping logic. Diagrams of these sections are shown in Figures 23.21 and 23.22, respectively [79]. The modal speed sensing consists of three functions: the speed sensor, the differential speed transducer, and the modal speed filters, as shown in Figure 23.21. The speed sensor consists of a rotating toothed wheel with magnetic pickup. Speed sensors are commonly located at both ends of the turbine-generator shaft. The pickup coil voltage for constant rotor speed is a constant amplitude sinusoid with frequency proportional to the generator speed times the number of teeth in the pickup wheel. When the speed includes torsional shaft motion, the torsional frequencies modulate the fundamental output voltage. This voltage is processed by the differential speed transducer to produce an output that consists of a dc voltage proportional to the fundamental speed and ac components due to the differential torsional shaft motion. This signal is processed by the modal speed filters, which are band-pass Butterworth filters that are tuned to each of the turbine-generator shaft natural frequencies. The output of each filter represents the separated modal speed components present at that toothed wheel location. Measuring the speed at different locations provides adequate data to compute the shaft displacement at shaft locations other than the point of measurement.

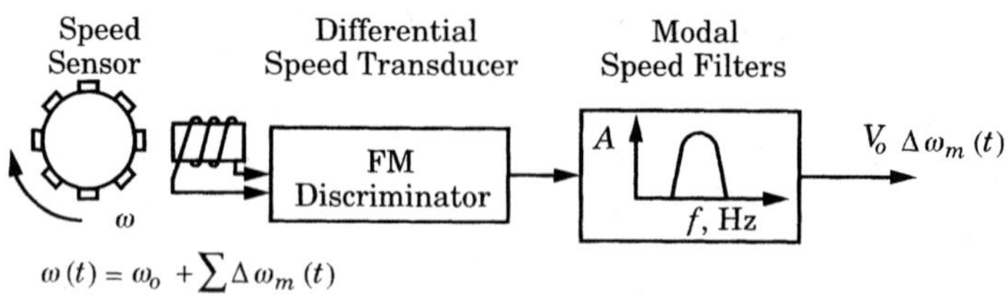

Figure 23.21 Torsional motion relay analog of speed sensing and analysis.

The digital tripping logic, shown in block diagram form in Figure 23.22, includes four features: the detection of single modal amplitude response, the modal stability, the dual mode response, and the transient response. The modal tripping is directed by 16 level detectors, four for each modal filter, which drive all but the transient protection logic.

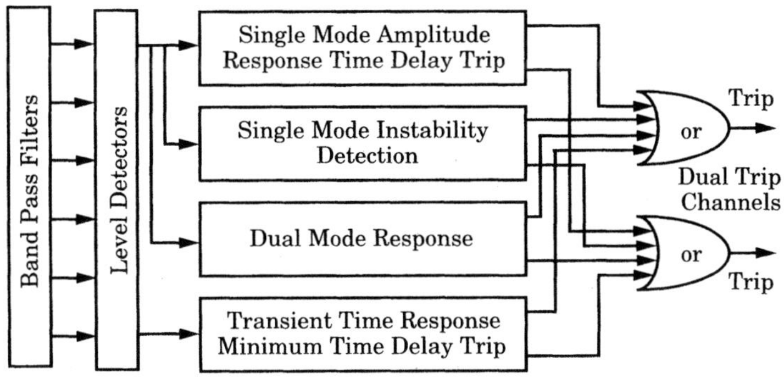

Figure 23.22 Torsional motion relay digital tripping logic.

Tripping can be directed by any one of four functions, as shown in Figure 23.22, namely:

- Single mode trip logic
- Steady-state instability trip logic
- Dual mode trip logic
- Transient mode trip logic

Each of these logical trip modes is described briefly.

Single mode trip logic. The single mode trip logic provides protection for unstable and growing responses, for limit cycle oscillations, and for stable transient responses. It accomplishes this function by using four different level detection circuits to detect oscillations at four discrete levels, thereby approximating an inverse time characteristic, as shown in Figure 23.23.

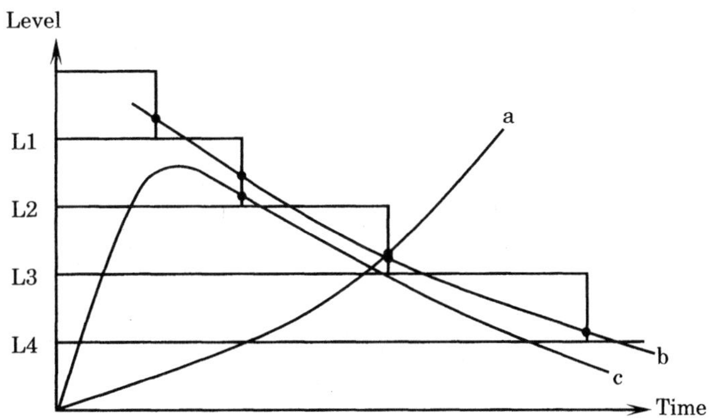

Figure 23.23 Single mode trip pickup levels.

This provides protection for (a) unstable and growing responses, (b) transient start limit-cycle responses, and (c) stable transient responses. This logic provides protection for the three types of response shown in Figure 23.23, but is not able to distinguish between them. This function is provided by the next logic, the steady-state instability trip logic.

Steady-state instability trip logic. The steady-state instability trip logic provides fast tripping of the generator for a transient-free unstable oscillation, thereby allowing generator tripping before substantial damage can accrue. Detection is provided by timing the interval of transient response growth between level 4 and level 3. The logic is based on the fact that a stable fault induced transient response usually achieves its maximum value in less than 1 second. Therefore, if the transient time between level 4 and level 3 is greater than or equal to 1 second, the response is unstable and tripping can be ordered without further testing.

Dual mode trip logic. The dual mode trip logic is designed to recognize the possibility that more than one mode of transient response may occur during a single incident. During a single mode response, one particular shaft section will be undergoing greater damage than other sections. The limit on a given shaft section may not be the same for all modes. Under dual mode response using only single mode trip logic may result in extracting more than a desired life, say 10 percent, from this shaft section for a single event. The dual mode trip scheme looks at all possible pairs of modes, taken from the output of Figure 23.21, and provides a trip when any two modes pick up the level 2 detectors simultaneously and remain picked up for a time greater than one-half the time limit for either single mode trip. This helps to minimize the total machine damage for any multimodal responses that may occur.

Transient mode trip logic. The transient mode protection logic provides basic protection for low level transients that decay more slowly than a preset rate, or that are unstable. In many cases, the basic transient mode protection is provided by other equipment, such as a static or dynamic filter. The transient protection mode provides backup for the primary protection equipment, which may fail during a significant transient event that is of greater magnitude than the equipment was designed to withstand. The transient protection mode also provides protection to the unit when other protection equipment or filters are out of service.

The transient protection logic must detect the loss of primary protection in less than 0.5 second. The philosophy is to detect total shaft torque on the principal limiting shaft sections and trip with minimum time delay when this torque exceeds a preset value. In many cases there are known to be limiting shaft sections that will be stressed heavily during transients. These shaft sections can be associated with oscillatory modes that are the controlling factors for the known stresses. The modal responses of interest are detected at the toothed wheel by a wide pass-band filter, which rejects unwanted modes as well as 60 hertz energy. Transient protection of these filtered signals can be triggered if the signal exceeds the normal level of signal that would occur if the primary protection is in service. Usually, the critical signals without primary protection will be an order of magnitude higher than normal if that protection is out of service, making the detection for this mode of transient protection easy to set with confidence.

Torsional motion relay experience. Relays have been installed at many locations throughout North America, where there is possible hazard to turbine-generator shafts due to either series compensated transmission lines or to nearby HVDC converter stations. Over a decade of experience has been accumulated with these relays and a number of unit trips have resulted due to stresses on unit shafts that would have caused excessive damage [80], [82].

SUBSYNCHRONOUS CURRENT RELAY. The subsynchronous current relay utilizes a special technique to detect very low values of subsynchronous currents and employs a special logic to determine if these currents represent a potential danger to the turbine generator [83–87]. A schematic diagram of the relay detection system is shown in Figure 23.24.

It is important to note that the subsynchronous frequency components of armature current are directly related to both the magnitude and frequency of the electrical torques and the resultant mechanical stresses in the turbine-generator shafts [83]. The stator currents of frequency f_{er} interact with the machine flux to produce air-gap torques that cause the generator

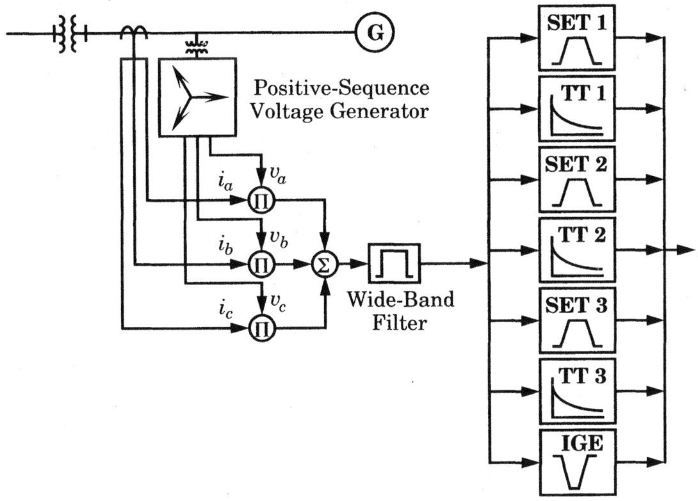

Figure 23.24 Schematic diagram of the subsynchronous current relay [83].

rotor to oscillate at a frequency $(f_o - f_{er})$. These air-gap torques decay at the same rate as the transient currents. If, however, the frequency of the air-gap torques should correspond to a torsional natural frequency f_n of the turbine-generator shaft, growing oscillations will result.

The system shown in Figure 23.24 is the signal extraction circuit. Phase current measurements are modulated by positive-sequence phase voltages for phase locking. The output of these multipliers will typically include a dc component plus other components that oscillate at frequencies of $(f_o + f_{er})$ and $(f_o - f_{er})$, where f_{er} is the positive-sequence subsynchronous-frequency component of the phase current and f_o is the system fundamental frequency. This doubling of the frequency components provides an excellent opportunity for filtering out any dc component as well as the supersynchronous components, leaving only the required subsynchronous components. This filtering is performed by the wide band filter, shown in Figure 23.24. These filtered signals are directed to three different types of circuits:

The self-excited trip detection (SET) circuits
The transient trip detection (TT) circuits
The induction generator effect detection (IGE) circuits

Two of these special circuits are shown in greater detail in Figure 23.25. The relay is capable of detecting either a growing or decaying subsynchronous signal. A decaying signal is detected by the *transient trip module* (TT) shown in Figure 23.25. However, a signal whose frequency is growing or constant is detected by the *self-excited trip* (SET) module. If the signal is constant or growing, and with frequency anywhere in the subsynchronous range, this signal is detected by the *induction generator effect* module (IGE), which is identified as IGE in Figure 23.24.

Referring to Figure 23.25, the transient trip module includes two different detection schemes. The first is an *instantaneous transient trip* (ITT) function, which will initiate a trip if the subsynchronous signal is greater than a prescribed threshold value. The input to the TT module, which includes the MRTT, described below, is the output of the wide-band filter. This signal is essentially a representation of the electrical torque, T_e and will usually be a signal that is decaying exponentially, as shown in Figure 23.26.

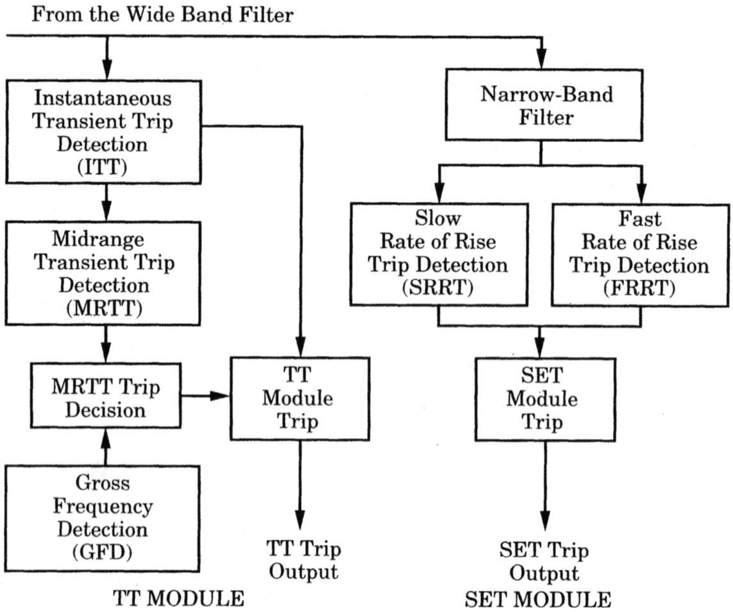

Figure 23.25 Subsynchronous current relay signal detection [78].

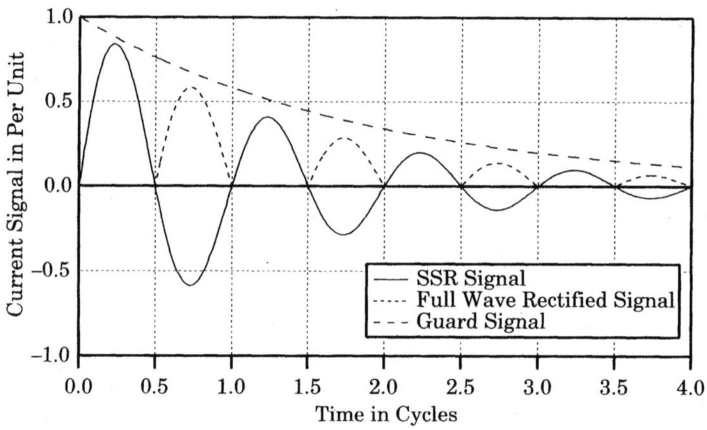

Figure 23.26 The MMRT trip characteristic.

THE TT MODULE. Shaft damage is likely to occur if the frequency of the TT signal is close to one of the shaft torsional modes and if the magnitude is greater than a threshold level. Therefore, the MRTT trip is based on meeting the following two conditions:

1. The frequency of T_e is within ± 3 hertz of a modal frequency.
2. Two peaks of the rectified T_e exceed the exponentially decaying guard signal.

The relay time must also be limited to about 0.05 second (3 cycles at 60 hertz) to protect the shaft in worst-case situations. This requires that the frequency of oscillation and magnitude of signal be checked simultaneously. The frequency is checked by counting the time between zero crossings. The magnitude of two full half cycles, corresponding to three zero crossings, is checked. If two half cycles are above the guard signal magnitude and the period of three consecutive zero crossings is within ± 3 hertz of a torsional, the relay initiates a unit trip.

In addition to the MRTT section, the TT module has an ITT section. This section initiates a relay trip instantaneously whenever $T_e \leq$ ITT, irrespective of frequency. The quantity ITT is the relay setting for the instantaneous trip section.

THE SET MODULE. The SET module consists of a narrow-band filter tuned to a torsional mode, a full wave rectifier, and logic components to detect either a *slow rate-of-rise* (*SRRT*) or a relatively *fast rate-of-rise* (*FRRT*) of the oscillating torque signal. The SRRT is intended to protect the unit against subsynchronous currents whose magnitudes are either constant or growing slowly with time. The FRRT section is intended to provide protection against currents with rapidly growing magnitudes that cannot be detected by SRRT. Since the SET module gets its input from a narrow-band filter, it should only respond to subsynchronous signals that correspond exactly to the shaft natural torsional frequency to which the narrow-band filter is tuned.

The SRRT monitors the subsynchronous oscillation over a period of time given by

$$T = T_0 + T_1 \tag{23.10}$$

where $T =$ time to initiate trip
$T_0 =$ fixed time delay

The parameter T_1 can be determined for a given subsynchronous current I_S by solving the integral equation

$$K = \int_0^{T_1} I_S dt \tag{23.11}$$

where K is a preset constant and I_S is the time-varying magnitude of the peaks of the subsynchronous current.

For the special case where the subsynchronous current has constant magnitude

$$I_S = I_{SC}$$

then

$$T = T_0 + \frac{K}{I_{SC}} \tag{23.12}$$

The SRRT circuit will generate a trip signal when the subsynchronous current satisfies the following criteria:

1. The current I_S must be greater than a threshold, designated I_{SS} to start a timer that measures the trip time T,
2. Each succeeding peak of I_S, following initiation of the timer, must be equal to or greater than the peak immediately preceding it, or

$$I_{p(n+1)} \geq I_{p(n)} \tag{23.13}$$

where $I_{p(n)} =$ peak value of I_S at the nth peak
$I_{p(n+1)} =$ peak value of I_S at the $(n+1)$th peak

3. At time $t = T$, I_S must be greater than a threshold level I_{SF}

The FRRT detection is similar to the SRRT, but is designed to give a shorter trip time for a subsynchronous current with a rapid growth rate. The trip time, in this case, is designated

T_2 and the requirements for detection are:

1. The current I_S must be greater than a threshold, designated I_{SFR}
2. Each succeeding peak of I_S, following initiation of the timer, must be greater than the preceding peak by a selected ratio $(1+\delta)$, i.e.,

$$I_{P(n+1)} \geq (1+\delta) I_{p(n)} \tag{23.14}$$

3. After completing the time T_2, one more rising peak must be encountered before permitting a trip to be generated.

It should be noted that FRRT and SRRT do not preclude each other. A subsynchronous signal could be first detected by SRRT and then FRRT, and if the rate of growth has slowed so as to not pick up FRRT, SRRT could eventually trip the unit. However, for any occurrence, the maximum trip time, as given by (23.10) is T, and the minimum trip time is T_2. The trip time could fall in between the two limits, as in the case where a signal spends part of its growing time as an SRRT case and the remainder as an FRRT case.

MULTIFREQUENCY RESPONSE. Some events will generate multiple frequency responses in the electrical torque signal, and these present a more difficult challenge to the detector circuits. To analyze this situation, consider a subsynchronous current with components at two frequencies, ω_1 and ω_2, i.e.,

$$I(t) = I_1 \sin \omega_1 t + I_2 \sin \omega_2 t \tag{23.15}$$

We assume that the narrow-band filter is set to pass only the frequency ω_1. There are several different conditions that are of interest to investigate, as follows [77–78]:

$$\begin{aligned} &\text{Case 1:} && I_1 = I_2 \text{ and } \omega_1 \gg \omega_2 \\ &\text{Case 2:} && I_1 = I_2 \text{ and } \omega_1 > \omega_2 \\ &\text{Case 3:} && I_1 \neq I_2 \text{ and } \omega_1 \gg \omega_2 \\ &\text{Case 4:} && I_1 \neq I_2 \text{ and } \omega_1 > \omega_2 \end{aligned} \tag{23.16}$$

Typical waveforms for cases 1 and 3 are shown in Figure 23.27 and for cases 2 and 4 in Figure 23.28. For cases 1 and 3, shown in Figure 23.27, the frequencies are widely separated and the filter will successfully reject the signal component of frequency ω_2, at least when those signals are of constant or slowly growing magnitude. If the unwanted signals are growing rapidly, the transient detection scheme should detect them on the basis of their growing magnitude. Figure 23.27 shows that the zero crossings can be quite different depending on the magnitudes of the two components, but the *gross frequency detection* (*GFD*) function should still be able to successfully identify the frequency of interest. See Figure 23.25.

When the two frequencies are close together, as shown in Figure 23.28, the narrow-band filter may not be able to distinguish the frequencies. This situation can be analyzed further as follows. We use a following trigonometric identity to write in terms of the sum and difference of the two frequencies. This leads to the following new equations.

$$\sin x \pm \sin y = 2 \sin\left(\frac{x \pm y}{2}\right) \cos\left(\frac{x \mp y}{2}\right) \tag{23.17}$$

For case 2, when $I_2 = I_1$, we can rewrite (23.15) as

$$\begin{aligned} I &= I_1[\sin(\omega_1 t) + \sin(\omega_2 t)] \\ &= 2I_1 \sin\left(\frac{(\omega_1+\omega_2)t}{2}\right) \cos\left(\frac{(\omega_1-\omega_2)t}{2}\right) \end{aligned} \tag{23.18}$$

This is similar to an amplitude modulated signal with carrier suppressed.

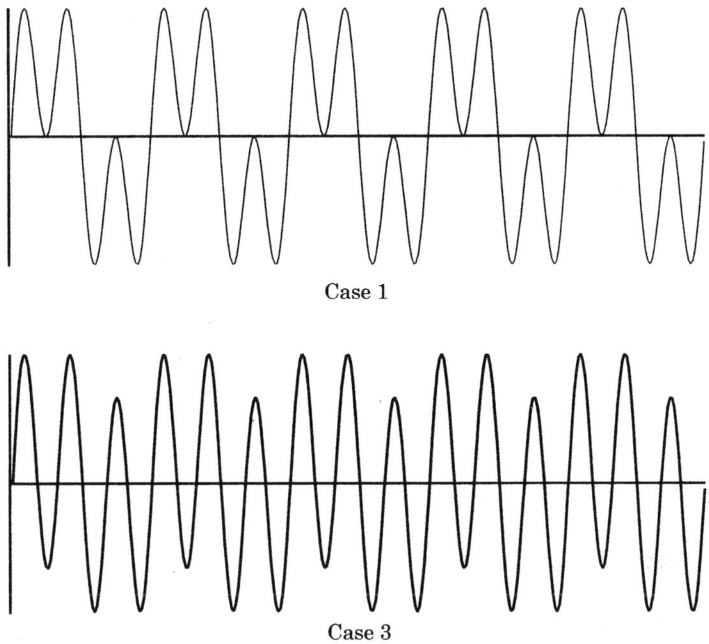

Figure 23.27 Multimodal oscillations for cases 1 and 3. Case 1: $I(t) = I_1 \sin 3\omega t + I_1 \sin \omega t$. Case 3: $I(t) = 4I_1 \sin 3\omega t + I_1 \sin \omega t$.

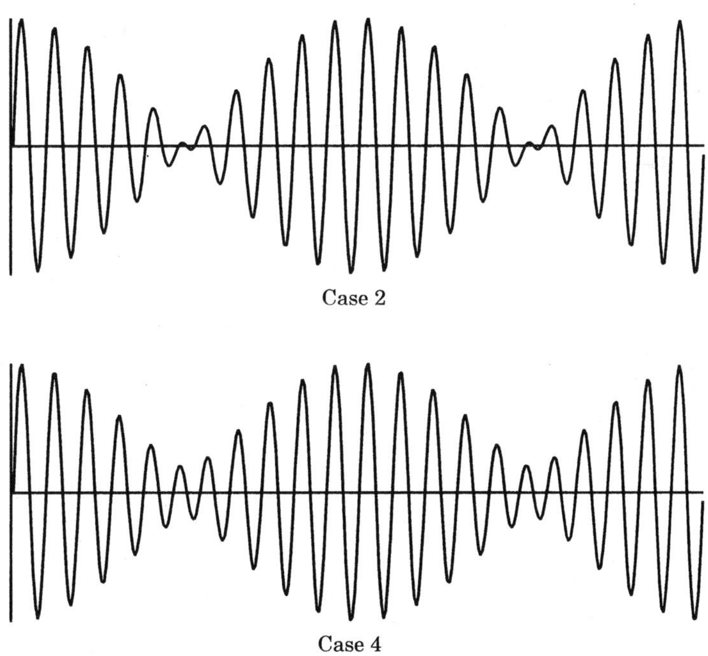

Figure 23.28 Multimodal oscillations for cases 2 and 4. Case 2: $I(t) = I_1 \sin(1.1\omega t) + I_1 \sin \omega t$. Case 4: $I(t) = 1.5I_1 \sin(1.1\omega t) + I_1 \sin \omega t$.

For case 4, when $I_2 \neq I_1$, we can rewrite (23.15) as

$$\begin{aligned} I_2 &= kI_1 \\ I &= I_1[\sin(\omega_1 t) + k\sin(\omega_2 t)] \\ &\quad + I_1\sin(\omega_2 t) - I_1\sin(\omega_2 t) \\ &= 2I_1\sin\left(\frac{(\omega_1+\omega_2)t}{2}\right)\cos\left(\frac{(\omega_1-\omega_2)t}{2}\right) \\ &\quad -(I_1 - I_2)\sin\omega_2 t \end{aligned} \tag{23.19}$$

Equation (23.19) is also an amplitude modulated waveform, but this equation has a term with the carrier present.

When the two frequency components are constant or growing slowly with time, there will be magnitude peaks and valleys that may cause variations in the trip time. When the period of the slow component is too small for the SRRT module to complete a trip, the FRRT may eventually generate a trip, as shown in Figure 23.29, which illustrates a waveform at the output of the narrow-band filter.

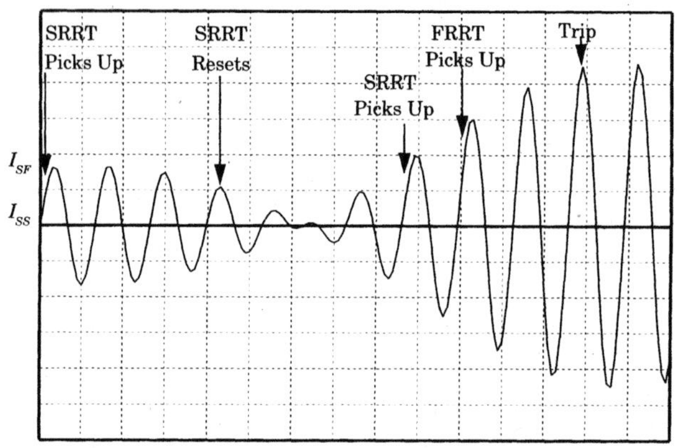

Figure 23.29 A multimodal self-excited case.

SUBSYNCHRONOUS CURRENT RELAY SETTINGS. The relay settings of the subsynchronous current relay are based entirely on a knowledge of the turbine-generator modal frequencies, and the electrical and mechanical characteristics at these frequencies. The basic philosophy of the method is to limit the loss-of-life of the shaft at one frequency, f_n, to less than 1% of shaft life per incident. The loss-of-life is estimated based on a shaft fatigue curve similar to that shown in Figure 23.30, which shows the general shape of the stress versus number of cycles for a generating unit. They are based on the equation

$$N_j = A\left(\frac{I_O}{I_S}\right)^B \tag{23.20}$$

where I_O = subsynchronous current that produces shaft torques equalling the endurance limit of the shaft, pu

I_S = subsynchronous current, pu

A = a constant, typically about 10^6

B = a constant (shown in Figure 23.30 as 6.64)

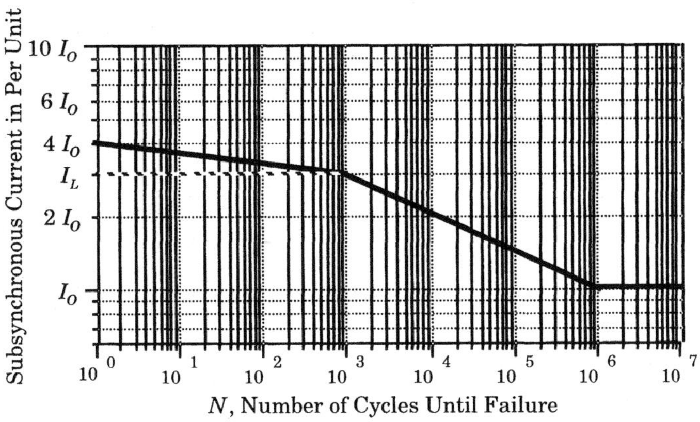

Figure 23.30 Fatigue life curve for a turbine-generator unit [87].

Then, it is required to maintain N according to the constraint

$$\sum_{j=1}^{n} \frac{1}{N_j} < 0.01 \tag{23.21}$$

where $n = f_n t_c$

f_n = natural torsional frequency of a turbine-generator unit, Hz

t_c = time for the percent loss-of-life to reach I_L when current grows exponentially from I_{SF}

I_L = subsynchronous current that will produce shaft torques equalling the elastic limit of a critical shaft section, pu

In the plot of Figure 23.30, it is presumed that

$$I_L < I_S < I_O \tag{23.22}$$

The following assumptions are required for the relay setting [87]:

1. Shaft shear stress can be converted to torque based on using linear elastic formulas.
2. The turbine-generator unit is oscillating at a torsional natural frequency so that shaft torque can be directly related to generator oscillation by use of a mode shape and the shaft spring constant.
3. As the oscillation grows, the relation between the electrical torque of the generator and the shaft torque is the same as it would be for a steady-state oscillating torque applied at the generator. This means that the instantaneous generator oscillating velocity is equal to the applied torque divided by the modal damping factor.
4. The generator is operating at 1.0 per unit flux level, so that the oscillating current and electrical torque are equal in per unit.
5. The per unit loss-of-life per cycle in the shaft for any level of subsynchronous current is equal to the reciprocal of the corresponding number of cycles to failure.

The subsynchronous current corresponding to the endurance limit is defined as I_O. This means that, when the oscillating current is below this value, there should be no loss-of-life to the turbine-generator shaft due to oscillations at the torsional frequency f_n. Therefore, the relay setting of the SET and IGE modules should be

$$I_{SF} = I_O$$
$$I_{SS} < I_{SF}, \text{ typically } \frac{I_{SF}}{2} \tag{23.23}$$

The value of K in (23.11) and (23.12) should be chosen such that a constant subsynchronous current of magnitude of $2I_O$ will cause less than one 1% loss-of-life before the unit is tripped, including relay and circuit breaker time. This gives the value of K to be

$$K = 2I_O \left[\frac{0.01A}{2^B f_n} - t_b \right] \tag{23.24}$$

where t_b = sum of relay and circuit breaker operating times, s

The settings for the FRRT are chosen so the relay can provide protection against oscillations that are growing too fast for detection by the SRRT. This must trip the unit before permanent damage can occur. This setting is based on a parameter δ that is defined by the following equation.

$$\delta = e^{\sigma_c/2f_n} - 1 \tag{23.25}$$

where σ_c = critical value of undamping, s^{-1}

Then the critical value of undamping must be solved from the following equation [87].

$$t_b + \frac{\ln(1 + K\sigma_c/I_{ss})}{\sigma_c} = \frac{\ln(1 + 0.01AB\sigma_c/f_n)}{B\sigma_c} + \frac{\ln(I_{sf}/I_{ss})}{\sigma_c} \tag{23.26}$$

where all parameters have been previously defined. The relay pickup value should be set to a value between I_{SS} and I_{SF} such as

$$I_{SFR} = 0.9I_{SF} \tag{23.27}$$

The transient trip module assumes that the transient currents are in the form

$$I_S = I_F e^{-\sigma_e t} \sin \omega_n t \tag{23.28}$$

where I_F = initial value of the decaying sinusoid
ω_n = a torsional natural frequency

The shaft torque oscillation will have this same frequency and the envelope of the oscillatory torque wave will have a form:

$$T_s \approx \frac{I_f K_1}{\sigma_e}(1 - e^{-\sigma_e t}) \tag{23.29}$$

where the constant K_1 is determined by the mechanical design of the shaft system. Examples of these settings are given in the literature [87], [88]. Other methods of computing the settings of the subsynchronous current relay have been developed [85] and are recommended for study by the protection engineer.

23.4.2.2 **SSR *Monitors*.** The foregoing SSR countermeasures and protective relays are designed to detect and protect the turbine-generator from permanent damage that may occur for any reason. These devices, however, do not have the capability of recording the total life history of the turbine-generator shaft oscillations such that the remaining life, following many different incidents, may be estimated. Since shaft fatigue damage is cumulative, it is necessary to record the nature of all severe oscillatory events over a long period of time in order to make a remaining life estimate. The shaft monitor is designed to perform this task. This rounds out the total unit protection by providing records of all types of oscillations that occur. These records are valuable for assessing the impact of each event, and their cumulative effect. However, they are also useful in providing a check on SSR computer simulations that are made to estimate the types of events that might occur. This provides a useful surveillance role on the unit life experience, including the results of unusual system switching configurations that have not been studied or simulated and that might cause unexpected damage.

The SSR monitor design should include at least the following features.

1. Shaft speed deviation detectors
2. Other metering required to construct a shaft analog model
3. Methods of processing the continuous data and to decide if an event of significant magnitude has occurred and should be recorded and analyzed
4. Suitable recording equipment, alarm functions, and signal processing required to analyze the recorded data

A block diagram of a typical monitor is shown in Figure 23.31. Other types of monitors are discussed in the literature [92–99]. The monitor design shown in Figure 23.31 uses speed signals at both ends of the turbine-generator shaft, plus the generator power output, to derive an electrical analog of the mechanical shaft system. This mathematical model provides estimates of the oscillations that occur at each of the shaft sections, and these signals are directed to detector, recording, and alarm circuits.

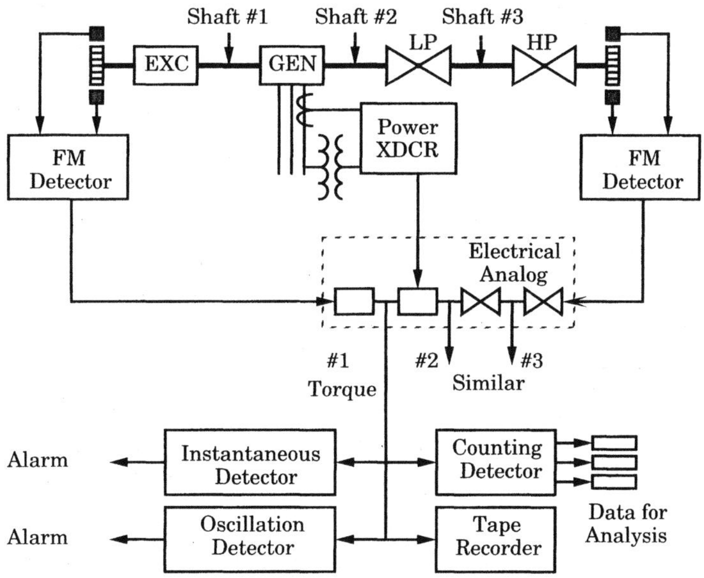

Figure 23.31 Block diagram of a typical torsional monitor [86].

The speed signal for the monitor is derived from a toothed wheel at each end of the shaft with magnetic pickups to sense the passing of each tooth. The output is a frequency modulated signal, with carrier frequency of 3600 hertz for 3600 rpm machines, and with velocity deviation information contained in the modulation. A phased-locked loop demodulator provides the required velocity deviation signal. Note that two pickups are used for each measurement, with these devices located 180 degrees apart. The velocity deviation signals of the two are added together to cancel noise introduced by lateral vibration.

The output signals from the FM decoders are directed to the electrical analog of the mechanical system. The output of this process is a signal proportional to the shaft torque oscillation at each shaft section. Data on the torsional stress history of the shaft can be stored by the monitor for later retrieval and analysis. The recorded data gives the number of torque oscillation cycles at various torque levels that are experienced by each shaft section. These torque waveforms are often very complex and appropriate rules for counting the material damage are required. The equipment counts the reversals of torque according to a defined methodology that has been derived by metallurgical engineers and used for fatigue studies.

The historical data provides a cumulative record that may be used with stress-life curves, similar to the one shown in Figure 23.30, to determine possible loss-of-life. These curves are modified to include the effects of shaft size, notch stress concentration, surface finish, environmental effects, and average loading.

Alarms are generated from the detectors to alert the operators of potentially damaging experiences of the shaft. Two different types of alarms are provided. For very high level torques, an instantaneous alarm will be triggered. The level of this alarm can be set to correspond to the desired portion of the stress curve that has been exceeded. A second alarm is to alert the operator of sustained oscillatory torques. These alarms are set to sound before the oscillations grow to a large enough value to cause high fatigue damage.

Other torsional monitor designs place speed detectors at each shaft section, thereby avoiding the necessity of mathematically modeling the shaft system [94]. This requires the addition to the shaft of a metallic band with holes in the band that are accurately spaced. The difference in reluctance can then be used in a magnetic pickup to provide a signal for the speed at that shaft section. Other methods have been devised to add shaft speed measurements to a shaft that has no previous provision for such measurement [100].

23.4.2.3 Comments on SSR Relays. The early solution to providing a protective relay that can respond to the presence of subsynchronous current was the TEX relay, which was improvised from a negative-sequence relay. This was not considered an adequate nor permanent solution to the need for generating unit protection against unforeseen subsynchronous currents that could be damaging to a unit. Having recognized that need, the industry has developed two relays that are specifically designed to recognize and respond to potentially damaging subsynchronous currents. The two relays, the torsional motion relay and the subsynchronous current relay, operate on different principles to provide the kind of protection that is considered necessary. Setting the relays requires an extra effort by the protection engineer, as the settings introduce the need for knowledge of the turbine-generator torsional behaviour. These relays are now applied in many locations where there is a perceived threat of SSR.

23.5 SUMMARY

Subsynchronous resonance is a complex phenomenon that requires serious study by the protection engineer. In many cases, this will be a team effort, involving engineers who specialize in steam turbine-generator systems, in system measurements, and in the analysis of power system dynamics. The extra study is well worth the effort, however, since the benefits of series compensation are often great. It is also rewarding to study and master the complications of this intriguing subject.

REFERENCES

[1] Kerchner, R. M., and G. F. Corcoran, *Alternating-Current Circuits*, Third Edition, John Wiley & Sons, Inc., New York, 1951.

[2] Breuer, G. D., H. M. Rustebakke, R. A. Gibley, and H. O. Simmons, "The Use of Series Capacitors to Obtain Maximum EHV Transmission Capability," *IEEE Trans.*, PAS-84, November 1964.

[3] Anderson, P. M., B. L. Agrawal, and J. E. Van Ness, *Subsynchronous Resonance in Power Systems*, IEEE Press, New York, 1990.

[4] IEEE Committee Report, *Analysis and Control of Subsynchronous Resonance*, IEEE Pub 76 CH1066-0-PWR, IEEE, New York, 1976.

[5] Hall, M. C., and D. A. Hodges, "Experience with 500 kV Subsynchronous Resonance and Resulting Turbine Generator Shaft Damage at Mohave Generating Station," IEEE Pub 76 CH1066-0-PWR, IEEE, New York, 1976.

[6] IEEE Committee Report, "Proposed Terms and Definitions for Subsynchronous Resonance in Series Compensated Transmission Systems," IEEE Pub 76 CH1066-0-PWR, IEEE, New York, 1976.

[7] IEEE Committee Report, "Proposed Terms and Definitions for Subsynchronous Oscillations," *IEEE Trans.*, PAS-99 (2), March/April 1980, pp. 506–511.

[8] IEEE Committee Report, "Terms, Definitions and Symbols for Subsynchronous Oscillations, *IEEE Trans.*, PAS-104, June 1985, pp.1326–1334.

[9] IEEE Committee Report, "A Bibliography for the Study of Subsynchronous Resonance Between Rotating Machines and Power Systems," *IEEE Trans.*, PAS-95 (1), January/February 1976, pp. 216–218.

[10] IEEE Committee Report, "First Supplement to A Bibliography for the Study of Subsynchronous Resonance Between Rotating Machines and Power Systems," *IEEE Trans.*, PAS-98 (6), November/December 1979, pp. 1872–1875.

[11] IEEE Committee Report, "Second Supplement to A Bibliography for the Study of Subsynchronous Resonance Between Rotating Machines and Power Systems," *IEEE Trans.*, PAS-104 (2), February 1985, pp. 321–327.

[12] IEEE Committee Report, "Third Supplement to A Bibliography for the Study of Subsynchronous Resonance Between Rotating Machines and Power Systems," *IEEE Trans.*, PWRS-6 (2), May 1991, pp. 830–834.

[13] IEEE Committee Report, "Symposium on Countermeasures for Subsynchronous Resonance, IEEE Pub 81 TH0086-9-PWR, 1981.

[14] IEEE Committee Report, "Countermeasures to Subsynchronous Resonance Problems," *IEEE Trans.*, PAS-99 (5), September/October 1980, pp. 1810–1818.

[15] IEEE Committee Report, "First Benchmark Model for Computer Simulation of Subsynchronous Resonance," *IEEE Trans.*, PAS-96 (5), September/October 1977, pp. 1565–1570.

[16] IEEE Committee Report, "Second Benchmark Model for Computer Simulation of Subsynchronous Resonance," *IEEE Trans.*, PAS-104, May 1985, pp. 1057–1066.

[17] Bahrman, M. P., E. V. Larsen, R. J. Piwko, and H. S. Patel, "Experience with HVDC-Turbine-Generator Torsional Interaction at Square Butte," *IEEE Trans.*, PAS-99, May/June 1980, pp. 966–975.

[18] Piwko , R. J., and E. V. Larsen, "HVDC System Control for Damping of Subsynchronous Oscillations," EPRI Report EL-2708, 1982.

[19] Achilles, R. A., and A. R. Ramirez, "Silicon Carbide Varistor Competitiveness in EHV Series Capacitor Reinsertion," IEEE paper 86 WM 037-6, presented at the IEEE PES Winter Meeting, February 2–7, 1986, New York.

[20] de Franco, N., J. E. Neville, and J. G. René, "Feasibility Study of the Itaipu (Sete Quedaw) Transmission," CIGRÉ paper 32-04, presented at the 1976 CIGRÉ Conference, Paris.

[21] Mansour, Y., T. G. Martinich, and J. E. Drakos, "B. C. Hydro Series Capacitor Bank Staged Fault Test," IEEE paper 83 WM 028-8, presented at the IEEE PES Winter Meeting, January 30–Febraury 4, 1983, New York.

[22] Hughes, M. B., R. W. Leonard, and T. G. Martinich, "Measurement of Power System Subsynchronous Driving Point Impedance and Comparison with Computer Simulations," *IEEE Trans.*, v. PAS-103, March 1984, pp. 619–630.

[23] Paris, L., G. Zini, M. Valtorta, G. Manzoni, A. Invernizzi, N. de Franco, and A. Vian, "Present Limits of Very Long Distance Transmission Systems," CIGRÉ paper 37-12, presented at the 1984 CIGRÉ Conference, Paris.

[24] Ichihara, Y., T. Kotake, K. Watanabe, S. Takeda, K. Katsuke, H. Suzuki, and R. Fujiwara, "Laboratory Test and Feasibility Study of Two Countermeasures to Subsynchronous Resonance," IEEE paper 82 SM 333-3, presented at the IEEE PES Summer Meeting, July 18–23, 1982, San Francisco.

[25] Jenings, G. D., R. G. Harley, and D. C. Levy, "Sensitivity of Subsynchronous Resonance Predictions to Turbo-Generator Modal Parameter Values and to Omitting Certain Active Subsynchronous Modes," IEEE paper 86 SM 483-2, presented at the IEEE PES Summer Meeting, July 20–25, 1986, Mexico City.

[26] Ahlgren, L., K. E. Johansson, and A. Gadhammar, "Estimated Life Expenditure of Turbine-Generator Shafts at Network Faults and Risk for Subsynchronous Resonance in the Swedish 400 kV System," *IEEE Trans.*, v. PAS-97 (6), November/December 1978, pp. 2005–2018.

[27] Iliceto, F., E. Cinieri, F. Gatta, and A. Erkan, "Optimal Use of Reactive Power Resources for Voltage Control in Long Distance EHV Transmission - Applications to the Turkish 420 kV System," CIGRÉ paper 38-13, presented at the 1988 CIGRÉ Conference, Paris.

[28] Maliszewski, R. M., B. M. Pasternak, H. N. Scherer, Jr., M. Chamia, H. Frank, and L. Paulsson, "Power Flow Control in a Highly Integrated Transmission Network," CIGRÉ paper 37-303, presented at the 1990 CIGRÉ Conference, Paris.

[29] IEEE Report, "Current Activity in Flexible AC Transmission Systems," IEEE Pub. 92 TH 0465-5 PWR, IEEE, New York, April 1992.

[30] Andersson, G., R. Atmuri, R. Rosenqvist, and S. Torseng, "Influence of Hydro Units' Generator to Turbine Inertia-Ratio on Damping of Subsynchronous Oscillations," *IEEE Trans.*, PAS-103, August 1984, pp. 2352–2361.

[31] Rostamkolai, N., R. J. Piwko, E. V. Larsen, D. A. Fisher, M. A. Mobarak, and A. E. Poitras, "Subsynchronous Torsional Interactions with Static Var Compensators - Concepts and Practical Implications," *IEEE Trans.*, PWRS-5, November 1989, pp. 1324–1332.

[32] Rostamkolai, N., R. J. Piwko, E. V. Larsen, D. A. Fisher, M. A. Mobarak, and A. E. Poltras, "Subsynchronous Torsional Interactions with Static Var Compensators - Influence of HVDC," *IEEE Trans.*, v. PWRS-5 (4), November 1990, pp. 1324–1332.

[33] IEEE Committee Report, "Series Capacitor Controls and Settings as Countermeasures to Subsynchronous Resonance," *IEEE Trans.*, PAS-101 (6), June 1982, pp. 1281–1287.

[34] Hingorani, N. G., "A New Scheme for Subsynchronous Resonance Damping of Torsional Oscillations and Transient Torque - Part I," *IEEE Trans.*, PAS-100 (4), April 1981, pp. 1852–1855.

[35] Hedin, R. A., K. B. Stump, and N. G. Hingorani, "A New Scheme for Subsynchronous Resonance Damping of Torsional Oscillations and Transient Torque - Part II, Performance," *IEEE Trans.*, PAS-100 (4), April 1981, pp. 1856–1863.

[36] Hingorani, N. G., R. A. Hedin, K. B. Stump, and B. Bhargava, "Evaluation of "NGH" Damping Scheme Applied to Mohave Generator," IEEE Pub. 81TH0086-9-PWR, 1981, pp. 70–80.

[37] Hedin, R. A., and K. B. Stump, "Evaluation of the NGH Subsynchronous Resonance-Damping Scheme," EPRI Report EL-2327, March 1982.

[38] Perez, A. J., "Mohave Project Subsynchronous Resonance Unit Tripping Scheme," IEEE Pub. 81TH0086-9-PWR, 1981, pp. 20–22.

[39] Urbanek, J., R. J. Piwko, D. McDonald, and N. Martinez, "Thyristor-Controlled Series Compensation–Equipment Design for the Slatt 500 kV Installation," EPRI, Proc. Flexible AC Transmission Systems Conf., Boston, 1992.

[40] Larsen, E. V., K. Clark, C. Wegner, S. Nyati, J. K. Hooker, R. W. Delmerico, and D. H. Baker, "Thristor Controlled Series Compensation–Control Design and Dynamic Performance," EPRI, Proc. Flexible AC Transmission Systems Conf., Boston, 1992.

[41] Juette, G., P. Luetzelberger, A. Schultz, S. M. McKenna, and D. R. Torgerson, "Advanced Series Compensation (ACS) Main Circuit and Related Components," EPRI, Proc. Flexible AC Transmission Systems Conf., Boston, 1997.

[42] Agrawal, B. L., R. A. Hedin, R. K. Johnson, A. H. Montoya, and B. A. Vossler, "Advanced Series Compensation (ASC) Steady-State, Transient Stability and Subsynchronous Resonance Studies," EPRI, Proc. Flexible AC Transmission Systems Conf., Boston, 1997.

[43] Hedin, R. A., V. Henn, S. Weiss, A. H. Montoya, and D. R. Torgerson, "Advanced Series Compensation (ACS) Transient Network Analyzer Studies Compared with Digital Simulation Studies," EPRI, Proc. Flexible AC Transmission Systems Conf., Boston, 1997.

[44] Farmer, R. G., A. L. Schwalb, and Eli Katz, "Navajo Project Report on Subsynchronous Resonance Analysis and Solutions," from the IEEE Symposium Publication, *Analysis and Control of Subsynchronous Resonance*, IEEE Publication, 76 CH1066-0-PWR, New York, 1976.

[45] Tice, J. B., and C. E. J. Bowler, "Control of the Phenomenon of Subsynchronous Resonance," *Proc. Am. Power Conf.*, 37, 1975, pp. 916–922.

[46] Limebeer, D. J. N., R. G. Harley, and M. A. Lahoud, "Suppressing Subsynchronous Resonance with Static Filters," *IEE Proc. C*, Gen., Trans., & Dist., 128 (1), 1981, pp. 33–44.

[47] Tang, J. F., and J. A. Young, "Operating Experience of Navajo Static Blocking Filter," IEEE Pub. 81TH0086-9-PWR, 1981, pp. 23–26.

[48] Kilgore, L. A., D. G. Ramey, and W. H. South, "The Dynamic Filter and Other Solutions to the Subsynchronous Resonance Problem," *Proc. Am. Power Conf.*, 37, 1975, pp. 923–929.

[49] Ooi, B.-T., and G. Joos, "A Passive Shunt Device to Suppress Torsional Interactions in Synchronous Generators," IEEE paper 85 SM 353-8, presented at the IEEE PES Summer Meeting, July 14–19, 1985, Vancouver.

[50] Kilgore, L. A., D. G. Ramey, and M. C. Hall, "Simplified Transmission and Generation System Analysis Procedures for Subsynchronous Resonance Problems," *IEEE Trans.*, PAS-96 (6), November/December 1977, pp. 1840–1845.

[51] Ooi, B.-T., "Phase Modulation Theory of Electromechanical Damping in Synchronous Generators," *IEEE Trans.*, PAS-100 (5), May 1981, pp. 2211–2218.

[52] Undrill, J. M., and F. P. deMello, "Subsynchronous Oscillations, Part II," *IEEE Trans.*, PAS-95, 1976, pp. 1456–1464.

[53] Erdis, A.-A., "Series Compensation Schemes Reducing the Potential of Subsynchronous Resonance," IEEE paper 89 SM 664-4 PWRS, presented at the IEEE PES Summer Meeting, July 9–14, 1989, Long Beach.

[54] Kilgore, L. A., E. R. Taylor, D. G. Ramey, R. G. Farmer, E. Katz, and A. L. Schwalb, "Solutions to the Problems of Subsynchronous Resonance in Power Systems with Series Capacitors," *Proc. Am. Power Conf.*, 35, 1973, pp. 1120–1128.

[55] Kilgore, L. A., D. G. Ramey, and W. H. South, "The Dynamic Filter and Other Solutions to the Subsynchronous Resonance Problem," *Proc. Am. Power Conf.*, 37, 1975, pp. 923–929.

[56] Ichihara, Y., T. Kotake, K. Watanabe, S. Takeda, K. Katsuki, and R. Fujiwara, "Laboratory Test and Feasibility Study of Two Countermeasures to Subsynchronous Resonance," *IEEE Trans.*, PAS-102 (2), February 1983, pp. 300–311.

[57] Irvani, M. R., and R. M. Mathur, "Damping Subsynchronous Oscillations in Power Systems Using a Static Phase Shifter," *IEEE Trans.*, PWRS-1 (2), May 1986, pp. 76–83.

[58] Ramey, D. G., I. A. White, J. H. Dorney, and F. H. Kroening, "Application of the Dynamic Stabilizer to Solve an SSR Problem," *Proc. Am. Power Conf.*, 43, 1981, pp. 605–609.

[59] Ramey, R. G., D. S. Kimmel, J. H. Dorney, and F. H. Kroening, "Dynamic Stabilizer Verification Tests at the San Juan Station," *IEEE Trans.*, PAS-100 (12), December 1981, pp. 5011–5019.

[60] Putman, T. H., and D. G. Ramey, "Theory of the Modulated Reactance Solution for Subsynchronous Resonance," *IEEE Trans.*, PAS-101 (6), June 1982, pp. 1527–1535.

[61] Kimmel, D. S., M. P. Carter, J. N. Bednarek, and W. H. Jones, "Dynamic Stabilizer On-Line Experience," *IEEE Trans.*, PAS-103 (1), January 1984, pp. 72–75.

[62] Wasynczuk, O., "Damping Subsynchronous Resonance Using Reactive Power Control," *IEEE Trans.*, PAS-100 (3), March 1981, pp. 1096–1104.

[63] Wasynczuk, O., "Damping Subsynchronous Resonance Using Energy Storage," *IEEE Trans.*, PAS-101 (4), April 1982, pp. 905–914.

[64] Hammad, A. E., and M. El-Sadek, "Application of a Thyristor Controlled Var Compensator for Damping Subsynchronous Oscillations in Power Systems," *IEEE Trans.*, PAS-103 (1), January 1984, pp. 198–212.

[65] Wasynczuk, O., "Damping Shaft Torsional Oscillations Using a Dynamically Controlled Resistor Bank," *IEEE Trans.*, PAS-100 (7), July 1981, pp. 2240–2249.

[66] Saito, O., H. Mukae, and K. Murotani, "Suppression of Self-Excited Oscillations in Series Compensated Transmission Lines by Excitation Control of Synchronous Machines," *IEEE Trans.*, PAS-94, September/October 1975, pp. 1777–1788.

[67] Raina, V. M., W. J. Wilson, and J. H. Anderson, "The Control of Rotor Torsional Oscillations Excited by Supplementary Exciter Stabilization," IEEE paper A76-457-2, presented at the IEEE PES Summer Meeting, July 18–23, 1976, Portland.

[68] Ooi, B.-T., and M. M. Sartawi, "Concepts of Field Excitation Control of SSR in Synchronous Machines," IEEE paper F77 771-5, presented at the IEEE PES Summer Meeting, July 17–22, 1977, Mexico City.

[69] Hurry, S., J. Nanda, and C. Kasshkari, "Effect of Excitation Control on Self-Excited Oscillations of Dual Excited Synchronous Machines," *Proc. 1977 Power Industry Computer Applications Conf.*, pp. 473–480.

[70] Hamden, H. M. A., and F. M. Hughes, "Excitation Controller Design for the Damping of Self Excited Oscillations in Series Compensated Lines," IEEE Pub. 78 CH1361-5-PWR, IEEE, NY, 1978.

[71] El-Serafi, A. M., and A. A. Shaltout, "Control of Subsynchronous Resonance Oscillations by Multi-Loop Excitation Controller," IEEE Pub. 79CH1417-5-PWR, 1979.

[72] Yan, A., M. D. Wvong, and Y. Yu, "Excitation Control of Torsional Oscillations," IEEE paper A 79 505-9, presented at the IEEE PES Summer Meeting, July 15–19, 1979, Vancouver.

[73] Yan, A., M. D. Wvong, and Y. Yu, "Excitation Control of Torsional Oscillations," IEEE Pub. 79CH1458-9-PWR, New York, 1979.

[74] Yan, A., and Y. Yu, "Multi-Mode Stabilization of Torsional Oscillations Using Output Feedback Excitation Control," *IEEE Trans.*, PAS-101 (5), May 1982, pp. 1245–1253.

[75] Bowler, C. E. J., and D. H. Baker, "Concepts of Supplemental Torsional Damping by Excitation Modulation," IEEE Pub. 81TH0086-9-PWR, New York, 1981, pp. 64–69.

[76] Bowler, C. E. J., and R. A. Lawson, "Operating Experience with Supplemental Excitation Damping Controls," IEEE Pub. 81TH0086-9-PWR, New York, 1981, pp. 27–33.

[77] Bowler, C. E. J., D. H. Baker, N. A. Mincer, and P. R. Vandiveer, "Operation and Test of the Navajo SSR Protective Equipment," *IEEE Trans.*, PAS-97 (4), July/August 1978, pp. 1030–1035.

[78] Ning, T. S., "Subsynchronous Overcurrent Relay for Generators Connected to Series Compensated Power System," IEEE Special Pub. 76 CH 1066-0-PWR, New York, 1977, pp. 30–36.

[79] Bowler, C. E. J., J. A. Demcko, L. Mankoff, W. C. Kotheimer, and D. Cordray, "The Navajo SMF Type Subsynchronous Resonance Relay," *IEEE Trans.*, PAS-97 (5), September/October 1978, pp. 1489–1495.

[80] Baker, D. H., C. E. J. Bowler, F. Rose, N. A. Mincer, and P. R. Vandiveer, "Test Performance and Operating Experience with the Navajo SSR Protective Equipment," Proc. Am. Power Conf., 39, Chicago, 1977, pp. 1144–1154.

[81] Bowler, C. E. J., D. H. Baker, N. A. Mincer, and P. R. Vandiveer, "Operation and Test of the Navajo SSR Protective Equipment," *IEEE Trans.*, PAS-97, July/August 1978, pp. 1030–1035.

[82] Bowler, C. E. J., and L. Mankoff, "Experience with SMF Type Subsynchronous Relays," IEEE Special Pub. 81TH0086-9-PWR, New York, 1981, pp. 43–46.

[83] Sun, S. C., S. Salowe, and C. R. Mummert, "Protective Relay Application for Turbine-Generator Unit SSR Protection," *Proc. Am. Power Conf.*, 43, Chicago, 1981, pp 610–615.

[84] Farmer, R. G., and B. L. Agrawal, "Application of Subsynchronous Oscillation Relay - Type SSO," *IEEE Trans.*, PAS-100 (5), May 1981, pp. 2442–2451.

[85] Agrawal, B. L., and R. G. Farmer, "Test Results of Subsynchronous Oscillation Relay - Type SSO," IEEE Special Pub. 81TH0086-9-PWR, New York, 1981, pp. 54–63.

[86] Sun, S. C., S. Salowe, E. R. Taylor, and C. R. Mummert, "A Relay for Protection Against Subsynchronous Oscillations," IEEE Special Pub. 81TH0086-9-PWR, New York, 1981, pp. 47–53.

[87] Sun, S. C., S. Salowe, E. R. Taylor, and C. R. Mummert, "A Subsynchronous Oscillation Relay - Type SSO," *IEEE Trans.*, PAS-100 (7), July 1981, pp. 3580–3589.

[88] ABB Relay Division, SSO Relay Application Data, Publication 40-174, ABB, Orlando, December 1984.

[89] Schwalb, A. L., "Navajo Project Subsynchronous Resonance Monitor and Relaying Equipment," IEEE Special Pub., 76 CH 1066-0-PWR, New York, 1977, pp. 77-80.

[90] Ramey, D. G., R. G. Farmer, B. L. Agrawal, and J. A. Demcko, "Subsynchronous Resonance Tests and Torsional Monitoring System Verification at the Cholla Station," *IEEE Trans.*, v. PAS-99 (6), September/October 1980, pp. 1900–1907.

[91] Raczkowski C., and G. C. Kung, "Turbine-Generator Torsional Frequencies - Field Reliability and Testing," a paper presented at the 40th annual American Power Conference, Chicago, 1978.

[92] Hurley, J. D., and W. H. South, "Torsional Monitor Equipment for Turbine-Generator Units," *Proc. Am. Power Conf.*, 41, Chicago, 1979, pp. 1163–1169.

[93] Stein, J., and H. Fick, "The Torsional Stress Analyzer for Continuously Monitoring Turbine Generators," *IEEE Trans.*, PAS-99 (2), March/April 1980, pp. 703–710.

[94] Ahlgren, L., K. Walve, N. Fahlén, and S. Karlsson, "Countermeasures Against Oscillatory Torque Stresses in Large Turbogenerators," CIGRÉ paper 31-07, presented at the 1982 CIGRÉ session, Paris.

[95] Horne, B. P., and D. N. Walker, "Torsional Vibration Monitoring," EPRI Workshop on Generator Monitoring and Surveillance, October 19–21, 1983, Dallas, Texas.

[96] White, J. C., D. N. Walker, and A. J. Perez, "Torsional Monitoring of Large Steam Turbine-Generators," ASME paper 83-JPGC-PWR-2, ASME, 1983, New York.

[97] Edmonds, J. S., and A. J. Perez, "A Torsional Vibration Monitoring Program for Large Steam Turbine-Generators," *Proc. Am. Power Conf.*, Chicago, 1983.

[98] Lambrecht, D., T. S. Kulig, W. Berchtold, J. Van Hoorn, and H. Fick, "Evaluation of the Torsional Impact of Accumulated Failure Combinations on Turbine Generator Shafts as a Basis of Design Guidelines," CIGRÉ paper 11-06, presented at the 1984 CIGRÉ Conference, Paris.

[99] Walker, D. N., R. J. Placek, C. E. J. Bowler, J. C. White, and J. S. Edmonds, "Turbine-Generator Shaft Torsional Fatigue and Monitoring," CIGRÉ paper 11-07, presented at the 1984 CIGRÉ Conference, Paris.

[100] Demcko, J. A., and J. P. Christy, "Self Calibrating Power Angle Instrument: Vol. 1, A Functional Overview, and Vol. 2, Theory, Operation, and Maintenance," Electric Power Research Institute Report GS-6475, Palo Alto, 1989.

PROBLEMS

23.1 Verify (23.4) for the time domain solution of transient current in an R-L-C circuit.

23.2 Write the differential equation for the simple series R-L-C network shown in Figure 23.2. Solve this equation using Laplace Transform methods.

23.3 Write the differential equation for the fault current flowing in the left loop of the system shown in Figure P23.3. Include the effect of the pre-fault power flowing from the source on the left to the source on the right.

Solve the differential equations for this circuit after the switch closes at $t = 0$. The circuit contains two storage elements, an inductor and a capacitor and the initial inductor current and capacitor voltage can be obtained from the pre-fault system solution. Let the external equivalant voltage be taken as the voltage reference. The pre-fault Thevenin source voltage leads the reference by ϕ radians or, in the time domain we have the source voltages

$$e = E_m \sin(\omega t + \phi)$$
$$e_o = E_{mo} \sin(\omega t)$$

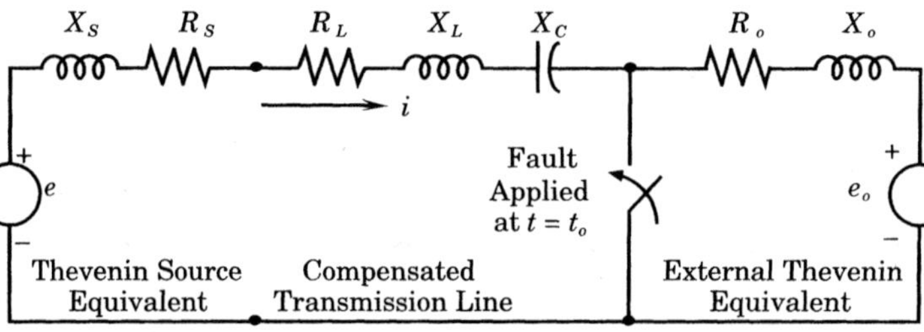

Figure P23.3 Fault applied to a small power system.

23.4 Develop a spreadsheet to provide a numerical solution for problem 23.3 using the numerical data given below. Theory tells us that the solution of the differential equations will exhibit two frequencies, one due to the frequency of the driving voltages and another that depends only on the parameters of the network. This latter frequency is a subsynchronous frequency as long as the degree of compensation is less than unity, which is always the case in power transmission systems.

$R_o = 0.01$ pu	external ckt resistance
$X_o = 0.10$ pu	external ckt reactance
$R = 0.003226$ pu	(200 mile 500 kv line)
$X_L = 0.0695016$ pu	(200 mile 500 kv line)
$k = 0.4$	degree of compensation
$R_S = 0.001$ pu	source resistance
$X_S = 0.005$ pu	source reactance
$E_{mo} = 1.500$ pu	external source Emax
$E_m = 1.500$ pu	source Emax
$\phi = 30°$	source voltages angle
$t_o = 0.015$ s	switching time of fault

23.5 A model for a turbine-generator shaft is shown in Figure P23.5. The turbine masses are arbitrarily numbered from one to n, and adjacent turbines are connected by a spring, with spring constant K, and a dashpot of damping D. These parameters represent the material behavior of the shaft section connecting the masses. Another set of dashpots from each mass to the reference represent the damping due to the relative motion of the turbine blades.

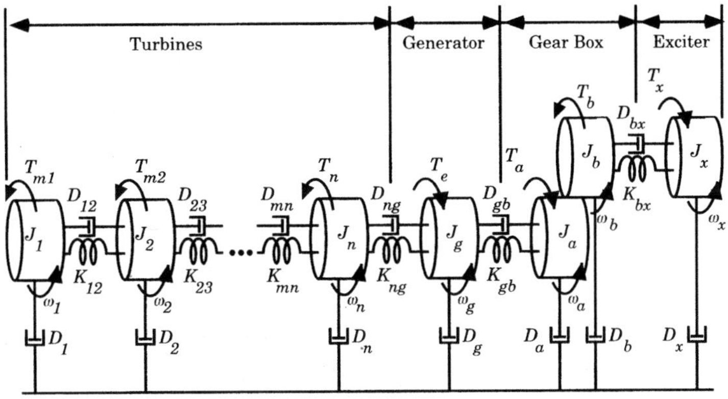

Figure P23.5 Lumped spring-mass model of a turbine-generator shaft.

Each turbine provides an accelerating torque T, shown as positive in the counterclockwise direction viewed looking at the shaft from the left end. A positive acceleration due to this torque results in an angular acceleration in the counterclockwise direction and in an instantaneous angular velocity, as shown in the figure. The generator is denoted by the subscript g in the figure with its torque T_e shown as positive in the direction opposite to the driving torque of the turbines. This means that the electromagnetic torque of the generator represents a load or retarding torque to the turbines. The generator angular velocity is measured as a positive quantity in the same direction as the turbines, namely, counterclockwise.

A shaft mounted exciter is also represented at the far right end of the shaft, with its parameters denoted by the subscript x. The exciter torque is also represented as a load torque. A gear box is also represented between the generator and the exciter, having moments of inertia J_a and J_b. No load torque is represented for the gear box as any losses in the gears should be small compared to the rated torques under consideration. The gear ratio is given as

$$R_G = \frac{\omega_a}{\omega_b} > 1$$

(a) Make a table of all quantities under consideration, including the name of the quantity, its symbol, its units in the mks system, and its dimensions in the [VIT] (voltage, current, time) system.

(b) Write the differential equations of the shaft system.

23.6 In the shaft system of the previous problem, we are usually concerned about the power, or the ability to do work. Convert the torque equations to power equations.

23.7 Referring to problem 23.6, normalize the power equations by dividing by the base voltamperes of the system, which is taken as the rating of the generator.

23.8 We now extend the shaft problem even further. We are only interested in very small deviations of the problem variables. Suppose that we define all variables in the form

$$x = x_o + \Delta x$$

where x is a generic representation of any variable. Convert the derived shaft equations to incremental equations in the incremental variables Δx. Justify in words why this is possible.

23.9 Construct a block diagram of the Laplace equations for the shaft system. Use an ellipsis "..." to indicate that the number of turbine masses is unspecified.

PART VI
RELIABILITY OF PROTECTIVE SYSTEMS

24

Basic Reliability Concepts

24.1 INTRODUCTION

The reliability of protective systems is extremely important for the secure operation of a power system. Random failures are certain to occur from time to time, especially when extremes in weather or other causes present hazards that the power system was not designed to withstand. Even during these extreme conditions, it is not acceptable that the power system be permitted to collapse and cease operating. Therefore, protective systems are designed to accurately determine the cause and location of a failed component and to selectively disconnect these failed circuit elements in a controlled manner.

Reliability in protective systems has always been provided by careful design based on sound judgment of experienced engineers. These methods have served remarkably well and the high availability of the power systems shows that protective system design has been carefully planned and implemented. For the most part, these systems have not been designed using reliability mathematics or models, but are mostly the result of experience and sound judgment. Indeed, reliability mathematics is not a substitute for sound judgment, but is a useful tool that provides the protective system designer additional information regarding a design. This information, and the methods described in this and the following chapters, give the protection engineer valuable concepts for design and powerful methods for analyzing the design reliability. Reliability methods also provide important analytical tools that can be used to evaluate and compare alternative designs of both the power system and the protective systems.

This chapter presents the necessary background mathematics from probability and reliability theory. These are concepts that will be familiar to most engineers and, if this is the case, this chapter will serve as a reference and will document the notation used in this book. For the engineer who has not made a formal study of probability and reliability, this chapter should serve as an introduction to the subject. It should be recognized, however, that this brief introduction is far from a complete treatment of probability and reliability. The interested reader will find many excellent texts on probability and reliability for further reading (e.g., see references [1–7] at the end of the chapter).

24.2 PROBABILITY FUNDAMENTALS

There are several ways in which probability theory can be developed, but one of the simplest and most satisfying is the axiomatic approach [1], [2].

24.2.1 The Probability Axioms

We assign to each event A a number $\Pr(A)$ which we call the probability of the event A. This number is chosen to satisfy the following three conditions:

$$\begin{aligned}&1.\ \Pr(A) \geq 0\\&2.\ \Pr(S) = 1\\&3.\ \text{If } AB = \{0\}, \text{ then } \Pr(A + B) = \Pr(A) + \Pr(B)\end{aligned} \tag{24.1}$$

where the space S is the certain event.[1] Elements of the probability space are often called experimental outcomes. The subsets of the space S are called events. The axioms (24.1) are the basis of the theory of probability, and all conclusions are based either directly or indirectly on these axioms. The following properties can be developed from the axioms.

1. The probability of the impossible event is zero. This is stated mathematically as

$$\begin{aligned}&\Pr\{\emptyset\} = 0\\&\{\emptyset\} = \text{The empty set or impossible event}\end{aligned} \tag{24.2}$$

2. For any event A

$$\Pr(A) = 1 - \Pr(\bar{A}) \leq 1 \tag{24.3}$$

where $\bar{A}$ = the complement of A
= the set consisting of all elements of S not in A

3. For any A and B

$$\Pr(A + B) = \Pr(A) + \Pr(B) - \Pr(AB) \leq \Pr(A) + \Pr(B) \tag{24.4}$$

FREQUENCY INTERPRETATION. In engineering work we must deal with real-world problems that are defined in terms of a physical system. The description of a real world experiment must be compatible with the axioms. The physical experiment can usually be defined in the following way.

$$\Pr(A) \cong \frac{n_A}{n} \tag{24.5}$$

where n_A = total number of outcomes favorable to A
n = total number of times the experiment is repeated

The reader should be able to determine that this definition of probability is not in conflict with the axioms.

24.2.2 Events and Experiments

The mathematical model of probability requires a clear definition of an experiment and the set of all possible outcomes of the experiment. In elementary games of chance, such as

[1] In this book, we use the common notation "Pr" to indicate probability. This notation will avoid confusion with the letter P, which is used as the symbol for power.

tossing a die, the outcomes are clearly defined and may be enumerated. In more complex physical problems, a careful definition of the experiment is required.

Let us identify the elements ξ as the elementary outcomes of an experiment E. Therefore, in a given experimental trial we observe the single outcome, ξ_i. Now, define the event A as

$$A = \{\xi_i\} \tag{24.6}$$

or we say that a trial that has outcome ξ_i is the event A. Probability theory is concerned with the probabilities of events, and the concept of a well-defined experiment E helps us determine the events of interest.

The set S, sometimes called the "basic set," is the set that contains all possible outcomes of the experiment. In many cases we are interested in events that contain one or more elements ξ_i and a certain experimental outcome selects a particular set of elements comprising a subset of the basic set.

Events are also defined in terms of some given proposition and a particular event is said to occur whenever that proposition is true. If we let π_A represent a proposition concerning elements ξ in the basic set S, we write

$$A = \{\xi : \pi_A(\xi)\} \tag{24.7}$$

which we read as "A is the set of ξ for which the proposition $\pi_A(\xi)$ is true," and we say that the *event* A occurs iff (if and only if) ξ is a member of the set A. Here, we distinguish between the elementary outcome ξ and the elementary event $\{\xi\}$. The certain event S occurs at every trial of an experiment since S contains all possible elements. Also, the impossible event $\emptyset$ never occurs. The event $A + B$ occurs whenever A or B or both occur since $A + B$ contains all elements in A or in B or in both A and B.

24.2.3 Venn Diagrams

A Venn diagram is a pictorial representation of a sample space, where the sets or experimental outcomes are represented by an area within the diagrammed space. We usually call these sets events. Examples of Venn diagrams are shown in Figure 24.1, which illustrate two different situations pertaining to the events A_1 and A_2. In Figure 24.1(a) the two sets share the common subset C_3. For the sets shown in Figure 24.1(b) there is no common subset, in which case we say that these events are disjoint.

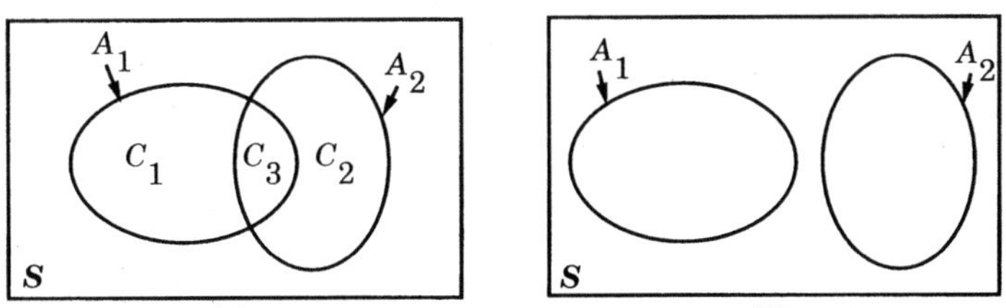

Figure 24.1 Venn diagrams illustrating two situations for events A_1 and A_2. (a) Two sets with common subset. (b) Two disjoint sets.

In some experiments it is possible to define the events in such a way that they completely fill the sample space S. Examples of this are shown in Figure 24.2, where the sets are the individual cards in a common deck of playing cards and the sample space is the collection of all 52 cards that might be used in an experiment, or in a particular card game.

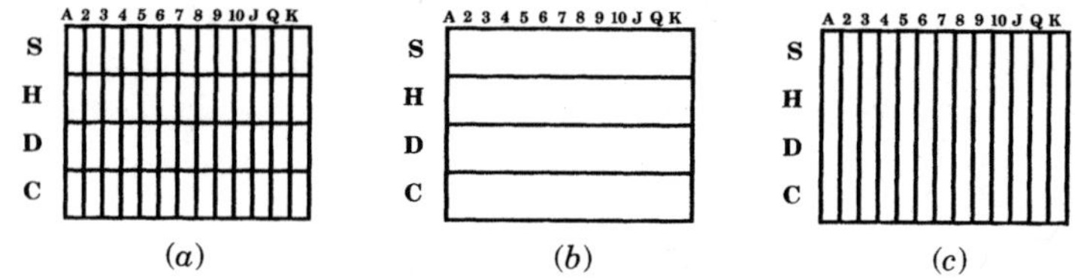

Figure 24.2 Venn diagram showing events for a deck of playing cards. (*a*) Sample space representing cards in a deck. (*b*) Events for the suits; spades, hearts, diamonds, clubs. (*c*) Events for the card face values; 2–10, J, Q, K, A.

The events shown in Figure 24.2 are examples of the sample space being *partitioned* into nonoverlapping sets that completely fill the sample space.

24.2.3.1 Union of Events. The *union* of two events A_1 and A_2 is that portion of the sample space that contains all elements of A_1 and all elements of A_2 or of both A_1 and A_2. Figure 24.1(*a*) provides a good example of the union of the events A_1 and A_2. Here the two events are depicted in terms of the disjoint subsets C_1, C_2, and C_3, where C_1 is that portion completely within A_1, C_2 is that portion completely within A_2 , and C_3 is common to both A_1 and A_2. Thus, we may write the union of A_1 and A_2 as the new event B, where

$$B = A_1 \cup A_2 = A_1 + A_2 = C_1 + C_2 + C_3 \tag{24.8}$$

Note that we often write the union as a sum, using the plus sign rather than the union symbol. The union of the events A_1 and A_2 of Figure 24.1(*b*) is defined to be the event D and is given by

$$D = A_1 \cup A_2 = A_1 + A_2 \tag{24.9}$$

The result (24.9) appears to be the same as (24.8), but referring to the Venn diagram we see that this is not the case. In many problems, it is convenient to think of the areas of the Venn diagram as having mass that corresponds to the fraction of the total area of the sample space, that is, the mass of S is unity. Using this concept, we see that (24.9) is clearly greater than (24.8) because of the overlapping mass represented by C_3 and which is only counted once in (24.8) but is counted twice in (24.9).

24.2.3.2 Intersection of Events. The *intersection* of two events A_1 and A_2 is that portion of the sample space that is contained in both A_1 and A_2. Referring to Figure 24.1(*a*), we see that the intersection of the events A_1 and A_2 is exactly the event C_3 or

$$C_3 = A_1 \cap A_2 = A_1 A_2 \tag{24.10}$$

This result also shows that we often write the intersection as a product in engineering work. When two events have no common elements, such as the two events shown in Figure 24.1(*b*), we say that these events are *disjoint* or *mutually exclusive*.

24.2.4 Classes and Partitions

A class of events is a set of events. For example, define a class of events $\tilde{A}$ by

$$\tilde{A} = \{A_1, A_2, A_3, A_4\} = \{A_i : i \in J\},\ J = \{1, 2, 3, 4\} \tag{24.11}$$

where the given class has four member events. We define the union and intersection of events in $\tilde{A}$ in a straightforward way. Thus we write the union and intersection of the member events in (24.11) as

$$A_1 \cup A_2 \cup A_3 \cup A_4 = \bigcup_{i \in J} A_i \tag{24.12}$$

$$A_1 \cap A_2 \cap A_3 \cap A_4 = \bigcap_{i \in J} A_i \tag{24.13}$$

The union and intersection are defined more formally in the next section. The class $\hat{A}$ is said to form a complete system of events if at least one of them is sure to occur, e.g., for a class of n member events

$$\bigcup_{i=1}^{n} A_i = S \tag{24.14}$$

The mutually exclusive set of member events is of particular interest and is given the name *partition*. A class of events is said to form a *partition* if one and only one of the member events must occur on each trial. A partition divides the Venn diagram into non-overlapping sets. (See Figure 24.2.)

24.2.5 Rules for Combining Probabilities

There are several rules that occur so commonly in practice that they should be stated for emphasis [3], [8].

24.2.5.1 Rule 1 Independent Events. Two events are said to be *independent* if the occurrence of one event does not affect the probability of occurrence of the other event. If two events A and B are independent, then we may write

$$\Pr(AB) = \Pr(A)\Pr(B) \tag{24.15}$$

EXAMPLE 24.1

Tossing a pair of dice is a classical example of two independent events. The face of the first die that shows upward does not affect, in any way, the face that will occur on the toss of the second die. ■

24.2.5.2 Rule 2 Mutually Exclusive Events. Two events are said to be mutually exclusive, or disjoint, if they cannot occur at the same time. Thus, when event A occurs, event B cannot possibly occur. For mutually exclusive events, their intersection is zero.

$$AB = \emptyset \tag{24.16}$$

or

$$\Pr(AB) = 0 \tag{24.17}$$

EXAMPLE 24.2

Consider tossing a coin. If heads appears in any toss, then tails cannot possibly appear. These events, heads and tails, are disjoint and their intersection is the null set. ■

24.2.5.3 Rule 3 Complementary Events. Two events are said to be complementary if, when one outcome does not occur, the other must occur. If the events A and B are complementary, then we may write the following.

$$\Pr(A) + \Pr(B) = 1 \tag{24.18}$$

EXAMPLE 24.3

In tossing a coin, the events *head* and *tail* are complementary since, if heads does not occur then tails must occur, and vice versa. Also consider the deck of playing cards whose sample space is shown in

Figure 24.2. Clearly, the sum of all events, in any of the three cases shown, completely fills the sample space and the probability of all events is unity. This is true no matter how the sample space is divided, i.e., whether by individual card faces, by suits, or by face values ■

24.2.5.4 Rule 4 Conditional Events. Conditional events are events that occur on the condition of the occurrence of another event or events. Consider two events A and B and the probability of event A occurring only on the condition that event B has occurred. This is depicted mathematically by the following equation.

$$\Pr(A|B) = \frac{\Pr(A \cap B)}{\Pr(B)} \tag{24.19}$$

In words, we say that the probability of A, given that B has occurred (or A given B), is the probability of the intersection of A and B, divided by the probability of the event B. Refer again to the deck of cards shown by the Venn diagram in Figure 24.2. What is the probability of drawing the 5 of hearts from a full shuffled deck? Clearly, the answer is 1/52. However, if we reword the problem to ask the probability of drawing the 5 of hearts from the full deck, *given that a heart is drawn*, the answer is 1/13. The answer is quite different because the problem was restated to add a condition that was not present in the original problem statement. Note that conditional probabilities can be zero, for example, the probability of drawing the 5 of hearts, given that a spade is drawn.

We can now define the probability of intersection $\Pr(A_1A_2)$ in terms of conditional probabilities. We write

$$\begin{aligned}\Pr(A_1A_2) &= \Pr(A_1)\Pr(A_2|A_1)\\ \Pr(A_1A_2) &= \Pr(A_2)\Pr(A_1|A_2)\end{aligned} \tag{24.20}$$

We interpret these equations from the Venn diagram of Figure 24.1(a). First, we see by inspection of that figure that $\Pr(A_1A_2) = \Pr(C_3)$ since the intersection refers to the region where the two events overlap, or C_3. Now consider the concept of uniform mass density with the total mass of the sample space being unity. Then we write, from the first form of (24.20),

$$\Pr(A_1A_2) = (M_{C1} + M_{C3})\left(\frac{M_{C3}}{M_{C1} + M_{C3}}\right) = M_{C3} = \Pr(C_3) \tag{24.21}$$

Clearly, for the probability $\Pr(A_2|A_1)$ the probability space is not S but is only $C_1 + C_3$, which gives the second mass ratio term in (24.20).

24.2.5.5 Rule 5 Simultaneous Events. The simultaneous occurrence of two events A and B implies the occurrence of both A and B. There are two cases of interest.

A and B independent If the events are independent, then the probability of occurrence of one event is not influenced by the occurrence of the other. Then we write

$$\Pr(AB) = \Pr(A)\Pr(B) \tag{24.22}$$

A and B dependent If A and B are not independent, then the probability of occurrence of one is affected by the occurrence of the other and (24.20) applies.

24.2.5.6 Rule 6 Occurrence of One of Two Events. The occurrence of at least one of two events A and B is the occurrence of A or B or both. In terms of the Venn diagram of Figure 24.1(a) we write

$$A_1 + A_2 = C_1 + C_2 + C_3 \tag{24.23}$$

There are three special cases to consider: events that are independent but not disjoint, events that are independent and disjoint, and events that are not independent. Disjoint events are events that have no common elements.

(a) **Events that are independent but not disjoint** First we write the probability and reduce the resulting expression.

$$\begin{aligned}\Pr(A_1 \cup A_2) &= \Pr\{A_1 \text{ OR } A_2 \text{ OR BOTH } A_1 \text{ AND } A_2\}\\ &= 1 - \Pr\{\bar{A}_1 \cap \bar{A}_2\}\\ &= 1 - \Pr(\bar{A}_1)P(\bar{A}_2)\\ &= 1 - (1 - \Pr(A_1))(1 - \Pr(A_2))\\ &= \Pr(A_1) + \Pr(A_2) - \Pr(A_1)\Pr(A_2)\end{aligned} \tag{24.24}$$

This result can be more easily found by reference to the Venn diagram of Figure 24.1(a). Considering the uniform mass distribution, we see clearly that the probability of the union is not just the sum of the probabilities, since this sum would cause the intersection to be counted twice. Hence the probability of the intersection must be subtracted from the sum.

(b) **Events that are independent and disjoint** If the events are disjoint, their intersection has probability zero, as shown in Figure 24.1(b). In this case we write

$$\Pr(A_1 \cup A_2) = \Pr(A_1) + \Pr(A_2) \tag{24.25}$$

This is always true when the two areas of the Venn diagram do not overlap.

(c) **Events that are not independent** Dependent events form the third special case. For this case we may write

$$\begin{aligned}\Pr(A_1 \cup A_2) &= \Pr(A_1) + \Pr(A_2) - \Pr(A_1 \cap A_2)\\ &= \Pr(A_1) + \Pr(A_2) - \Pr(A_2|A_1)\Pr(A_1)\\ &= \Pr(A_1) + \Pr(A_2) - \Pr(A_1|A_2)\Pr(A_2)\end{aligned} \tag{24.26}$$

EXAMPLE 24.4

What is the probability that a card drawn from a full deck is a red card, a face card, or both? To solve this problem, define the following events.

$$\begin{aligned}A &= \{\text{red}\}\\ B &= \{\text{face card}\}\end{aligned}$$

Then

$$\begin{aligned}\Pr(A) &= 26/52 \qquad \Pr(A|B) = 6/12\\ \Pr(B) &= 12/52 \qquad \Pr(B|A) = 6/26\end{aligned}$$

Then, using the first form of (24.26), we compute

$$\begin{aligned}\Pr(A \cup B) &= (26/52 + 12/52) - (6/26 \times 26/52) = 32/52\\ \Pr(A \cap B) &= \Pr(A|B)\Pr(B) = \Pr(B|A)\Pr(A)\\ &= 6/12 \times 12/52 = 6/26 \times 26/52 = 6/52\end{aligned}$$

The reader can repeat the computation using the second form of (24.26). ■

24.2.5.7 Rule 7 Conditional Probability. Consider an event A that is dependent on a number n of disjoint events B_i. From (24.20) we write

$$\Pr(AB) = \Pr(A|B)\Pr(B) \tag{24.27}$$

If we write this equation for each of the events B_i and sum the results over all n disjoint events, we have the following result.

$$\sum_{i=1}^{n} \Pr(AB_i) = \sum_{i=1}^{n} \Pr(A|B_i)\Pr(B_i) \tag{24.28}$$

But the right side of (24.28) must include all contributions to the probability of the event A, or this is just $\Pr(A)$. Thus, we conclude that

$$\Pr(A) = \sum_{i=1}^{n} \Pr(AB_i) = \sum_{i=1}^{n} \Pr(A|B_i)\Pr(B_i) \tag{24.29}$$

The reader can readily justify this equation by constructing an elementary Venn diagram that includes a partitioned space with an event that intersects all partitions.

24.3 RANDOM VARIABLES

An important concept in probability theory, and in its application to protective systems, is the concept of a *random variable*. We now consider this concept and will introduce definitions and nomenclature that will prove useful in applying this concept to practical problems.

24.3.1 Definition of a Random Variable

A *random variable* or *rv* is a real function of the elements of the sample space S. In this book, we represent the random variable by capital italic letters (e.g., X, Y, or T) and any particular value of that random variable by the lower case italic letters (x, y, or t). Consider an experiment conducted in the sample space S, where we assign to every outcome ζ of that experiment a number $X(\zeta)$, according to some rule. We then call $X(\zeta)$ a random variable (rv). One can think of this process as mapping all elements of the sample space into points on the real line.[2]

EXAMPLE 24.5

The experiment consists of tossing a die and flipping a coin. If a head shows on the coin, the random variable X is given the value shown by the top face of the die, however, if a tail shows on the coin, then X is given the negative of the die face value. Thus we map all outcomes into 12 points on the real line, six of which are positive and the rest are negative. ■

Note that it is not necessary that all points in the sample space be mapped into a unique point on the real line. For example, in the foregoing experiment we could map the die values all to the same point every time the coin face is heads (or tails). The rules of the experiment determine these mappings. The random variable can be almost any type of function, but we require that it not be multivalued.

We now present a formal definition of a random variable [2].

Definition. A random variable X is a process of assigning a number $X(\zeta)$ to every outcome ζ of an experiment. The resulting function must satisfy the following two conditions but is otherwise arbitrary:

[2] Random variables can also be complex, but we will limit our discussion to real random variables. Complex random variables are discussed in the literature [2], [8].

1. The set $\{X \leq x\}$ is an event for every x.
2. The probabilities of the events $\{X = \infty\}$ and $\{X = -\infty\}$ equal zero, i.e.,

$$\Pr(X = \infty) = 0 \qquad \Pr(X = -\infty) = 0$$

The second condition states that, although infinity values are allowed for some outcomes of the experiment, we require that the probabilities associated with these outcomes constitute a set with zero probability.

A discrete random variable is an rv having only discrete values. The example cited above is a discrete random variable. There should be no confusion regarding the sample space and the values assigned to the random variable. Sample spaces may be either discrete or continuous, and random variables can be assigned values that are either discrete or continuous from either type of sample space. Moreover, some rv may be mixed, that is, they may be assigned some values that are discrete and some that are continuous.

24.3.2 The Distribution Function

The mathematics of random variables is based on the outcome of an experiment that is an event. For the event $\{X \leq x\}$ we assign the probability $\Pr\{X \leq x\}$ of that event. We call this probability by a special name $F_X(x)$, which we define as the *cumulative probability distribution function* of the random variable X.

$$F_X(x) = \Pr\{X \leq x\} \tag{24.30}$$

In many references this function is called the cdf, or simply the "distribution function." The argument x can take on any value in the range of $-\infty$ to $+\infty$.

The distribution function has the following properties, which are derived from the fact that this function is a probability. These properties are [8]:

$$\begin{aligned} &1.\ F_X(-\infty) = 0 \\ &2.\ F_X(\infty) = 1 \\ &3.\ 0 \leq F_X(x) \leq 1 \\ &4.\ F_X(x_1) \leq F_X(x_2) \text{ if } x_1 < x_2 \\ &5.\ \Pr\{x_1 < X \leq x_2\} = F_X(x_2) - F_X(x_1) \\ &6.\ F_X(x^+) = F_X(x) \end{aligned} \tag{24.31}$$

The distribution defined by (24.30) is a nondecreasing function of the random variable X that always starts at zero, on the extreme left, and moves toward unity, on the extreme right.

24.3.3 The Density Function

The probability density function is defined as the derivative of the distribution function. Probability density functions are usually denoted by $f_X(x)$ and are computed as follows.

$$f_X(x) = \frac{dF_X(x)}{dx} \tag{24.32}$$

The probability density function is often called the pdf or simply the "density" of the random variable X. The density function exists for continuously defined random variables, but there may be values of x where the density does not exist for discrete random variables. At these points, the density may have abrupt spikes, and may be zero at all other points. The properties of a proper density function can be stated as follows.

$$1.\ 0 \le f_x(x) \qquad \text{for all } x$$

$$2.\ \int_{-\infty}^{\infty} f_x(x)\,dx = 1 \tag{24.33a}$$

$$3.\ F_x(x) = \int_{-\infty}^{x} f_x(u)\,du$$

$$4.\ \Pr(x_1 < X < x_2) = \int_{x_1}^{x_2} f_x(x)\,dx \tag{24.33b}$$

24.3.4 Discrete Distributions

When X is a discrete random variable, which is the case of interest for many power system application, the distribution will have a stairstep form, which is given mathematically by the equation

$$F_X(x) = \sum_{i=1}^{N} \Pr\{X = x_i\}u(x - x_i) \tag{24.34}$$

where u is the unit step function and N is the number of discrete states. The magnitude of each step in this stairstep function will be the probability of that value of X where the step occurs.

The discrete density function is the derivative of (24.34) and is given by

$$f_X(x) = \sum_{i=1}^{N} \Pr(x_i)\delta(x - x_i) \tag{24.35}$$

where δ is the unit impulse function and N is the number of discrete states.

The joint distribution of the two variables X and Y may be written as

$$F_{XY}(x, y) = \Pr\{X \le x, Y \le y\} \tag{24.36}$$

The special case where the two random variables are independent is often applicable. In this special case, (24.36) becomes

$$F_{XY}(x, y) = F_X(x)F_Y(y) = \Pr\{X \le x\}\Pr\{Y \le y\} \tag{24.37}$$

The quantity (24.37) still obeys all the rules for a distribution function, but instead of mapping to the real line it maps to a two-dimensional area in the x-y plane.

EXAMPLE 24.6

The experiment is the rolling of a single die and the outcome of the experiment is the value of the top face of the die after it comes to rest. The sample space in this experiment has six elements, ξ_1, ξ_2, ξ_3, ξ_4, ξ_5, and ξ_6 where ξ_i implies that face i was the outcome of that toss. For this simple experiment, there is a rather obvious way in which we can define a random variable, namely, by letting the value of the rv be equal to the value on the upward face of the die. Note that we have established a rule that assigns to each outcome, which is not a number, a real number. The probability of each outcome of this experiment is 1/6, assuming that the die is fair. We can now evaluate the distribution function for this rv.

For $x < 1$, we have

$$F_X(x) = \Pr\{X \le x\} = \Pr\{\emptyset\} = 0 \qquad x < 1$$

For $1 \le x < 2$,

$$F_X(x) = \Pr\{X \le x\} = \Pr\{\xi_1\} = 1/6 \qquad 1 \le x < 2$$

For $2 \le x < 3$

$$F_X(x) = \Pr\{X \le x\} = \Pr\{\xi_1, \xi_2\} = 1/3 \qquad 2 \le x < 3$$

and so forth. A summary of all possible outcomes is as follows:

$$F_X(x) = \Pr\{X \le x\} = \begin{cases} 0 & x < 1 \\ 1/6 & 1 \le x < 2 \\ 1/3 & 2 \le x < 3 \\ 1/2 & 3 \le x < 4 \\ 2/3 & 4 \le x < 5 \\ 5/6 & 5 \le x < 6 \\ 1 & 6 \le x \end{cases} \tag{24.38}$$

This distribution function is plotted in Figure 24.3. Note that this function obeys all of the requirements for a distribution function given by (24.31). We can write this function in a more compact notation using the unit step function. From (24.33), we write for this example function

$$F_X(x) = \frac{1}{6}\sum_{i=1}^{6} u(x-i) \tag{24.39}$$

The distribution function shown in Figure 24.3 is typical of distributions for discrete random variables. These functions may have jump discontinuities depending on the definition of the random variable.

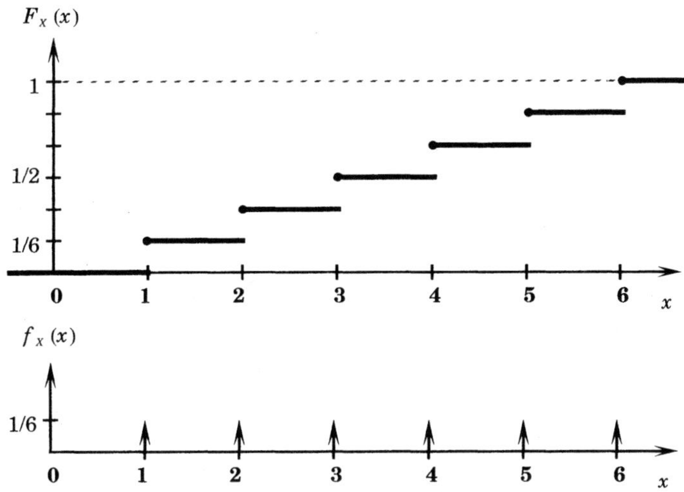

Figure 24.3 Graph of the example distribution and density functions.

The density function for this random variable is a series of delta functions each of strength 1/6 and occurring at $x = 1, 2, \ldots, 6$. ■

24.3.5 Continuous Distributions

Some distributions are described by continuous random variables rather than discrete random variables. An example of a continuous rv is given by the following example.

EXAMPLE 24.7

Business calls at an office are received at random during the interval (0,1), which represents the time of the normal business day. The experiment is to monitor the received business calls during the day and assign a number to each call that is the exact time of the call arrival, measured from the beginning of the business day, and normalized such that the total time of the day (8 hours) is equal to unity. In this case, both the experiment outcome and the random variable are represented by the same number. The probability that the time that a call is received falls between t_1 and t_2 is given by

$$\Pr\{t_1 \le x \le t_2\} = t_2 - t_1$$

Since calls cannot be received before the office opens in the morning, we note that

$$F_X(x) = \Pr\{X \le x\} = \Pr\{\emptyset\} = 0 \qquad x < 0$$

During the day the distribution function is given by

$$F_X(x) = \Pr\{X \le x\} = \Pr\{0 \le t \le x\} \qquad 0 \le x \le 1$$

After the office closes for the day, all calls have been received and we write

$$F_X(x) = \Pr\{X \le x\} = \Pr\{0 \le t \le 1\} = 1 \qquad x < 1$$

This distribution is continuous, although it has points where the slope changes abruptly. The distribution and density are plotted in Figure 24.4.

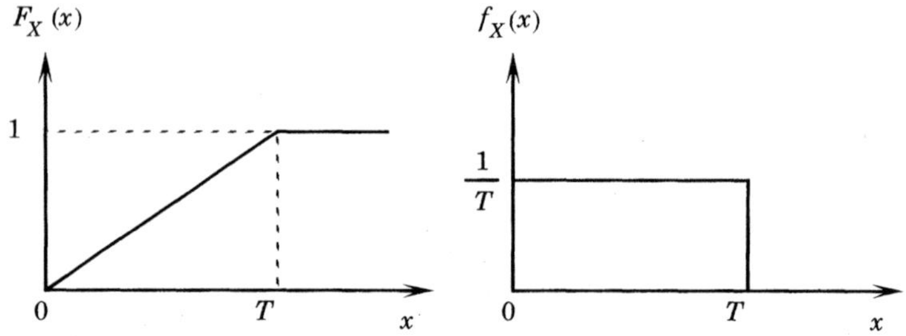

Figure 24.4 Example of continuous distribution and density functions. ■

24.3.6 Moments

Moments are important in the study of random variables. The nth moment is defined by the equation

$$m_n = E\{X^n\} = \int_{-\infty}^{\infty} x^n f_X(x)\,dx \tag{24.40}$$

The first moment is called the *mean* or *expected value* of the random variable and is computed with the parameter $n = 1$. The mean is often given a special symbol, which we shall call η.

The *variance* or dispersion of an rv is defined as the second moment. [2]

$$\sigma^2 = \int_{-\infty}^{\infty} x^2 f_X(x)\,dx \tag{24.41}$$

The central moment is defined by the integral equation [2]

$$\mu_n = E\{(X - \mu)^n\} = \int_{-\infty}^{\infty} (x - \mu)^n f_X(x)\,dx \tag{24.42}$$

where

$$\mu = E(X)$$

24.3.7 Common Distribution Functions

Several distribution functions occur so often in practical problems that they are given special names and their characteristics are well known. A few of these distributions that are useful in engineering work are described below.

24.3.7.1 Discrete Distributions. There are several discrete distributions that are frequently used in practical probability problems. The most common of these are the following.

THE DISCRETE UNIFORM DISTRIBUTION. The discrete uniform distribution is valuable in describing experiments such as those dealing with coin tossing, dice rolling, and other games. In these games, the discrete random variable x has a uniform distribution if

$$f_X(x) = \frac{1}{n}\sum_{i=1}^{n}\delta(x - x_i) \tag{24.43}$$

This distribution is limited to values of X taken from the n values $x_1, x_2, \ldots, x_n$, each with probability $1/n$. The distribution function for this case is found by integrating (24.43) to get

$$F_X(x) = \frac{1}{n}\sum_{i=1}^{n}u(x - x_i) \tag{24.44}$$

THE BINOMIAL DISTRIBUTION. The binomial distribution has many applications in engineering. This distribution is useful in cases where a large number of experiments, n, result in two possible outcomes, which are often designated as success p or failure q. The binomial distribution can be represented by the general polynomial expression $(p + q)^n$, where p is the probability of success, q is the probability of failure, and n is the number of trials. To be applicable, four conditions are required:

1. There must be a fixed number of trials so that n is known,
2. Each trial must result in either a success, p, or a failure, q, i.e., only these two outcomes are possible and $p + q = 1$;
3. All trials must have identical probabilities of success and failure; i.e., the values of p and q are constant; and
4. All trials must be independent.

Such trials are called Bernoulli trials.

The expression $(p + q)^n$ can be expanded in polynomial form to give the following result.

$$\begin{aligned}(p+q)^n &= p^n + np^{n-1}q + \frac{n(n-1)}{2!}p^{n-2}q^2 + \ldots \\ &+ \frac{n(n-1)\ldots(n-r-1)}{r!}p^{n-r}q^r + \ldots + q^n\end{aligned} \tag{24.45}$$

In this polynomial expansion, the coefficient of the $(r + 1)$th term represents the number of ways or combination that exactly r failures and, therefore, a total of $(n - r)$ successes can occur in n trials. This coefficient is the familiar quantity designating the number of combinations of n things taken r at a time, usually written as ${}_nC_r$ where

$${}_nC_r = \frac{n!}{r!(n-r)!} \tag{24.46}$$

Then the probability of exactly r successes or $(n-r)$ failures in n trials is computed from the expression

$$\Pr\{\text{exactly } r \text{ successes}\} = \frac{n!}{r!(n-r)!} p^r q^{n-r} = {}_nC_r p^r q^{n-r} \tag{24.47}$$

The density function for the binomial distribution is given by the expression

$$f_X(x) = \sum_{k=0}^{n} {}_nC_k p^k (1-p)^{n-k} \delta(x-k) \tag{24.48}$$

The binomial distribution function is the integral of the density function and may be written as follows.

$$F_X(x) = \sum_{k=0}^{n} {}_nC_k p^k (1-p)^{n-k} u(x-k) \tag{24.49}$$

The mean or expected value of the binomial distribution is given by

$$E(X) = \sum_{k=0}^{n} k\, {}_nC_k p^k (1-p)^{n-k} = np \tag{24.50}$$

Finally, the variance of the binomial distribution can be shown to be equal to

$$E(X^2) - E^2(X) = np(1-p) \tag{24.51}$$

Examples of the binomial density function for $n = 9$ and varying p are shown in Figure 24.5.

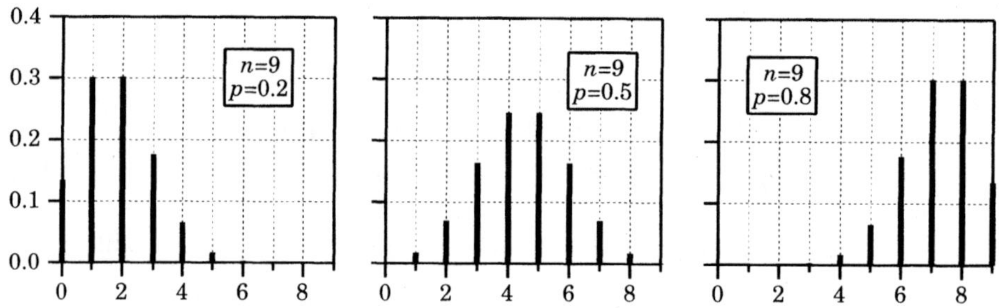

Figure 24.5 The binomial density function for $n = 9$ and variable p.

The binomial distribution function is the integral of the density. Examples corresponding to the densities of Figure 24.5 are shown in Figure 24.6.

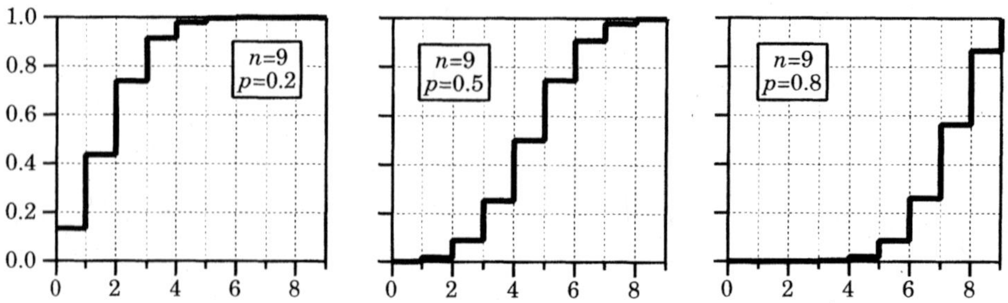

Figure 24.6 The binomial distribution for $n = 9$ and variable p.

The peaks of the density function occur near the expected value np and fall off on each side of this value. When $p = 0.5$, the falloff is symmetrical, but it is asymmetrical for p either larger or smaller that this value.

When p is fixed and n allowed to vary, the density function behaves as shown in Figure 24.7. Again, the peak occurs near the mean and falls off on either side, and the peak becomes smaller for larger values of n.

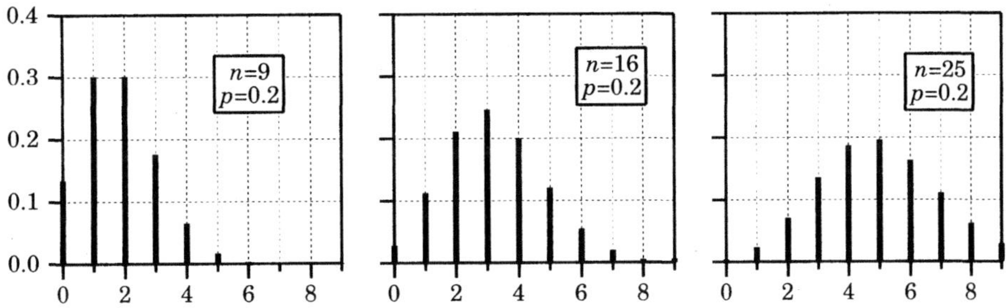

Figure 24.7 The binomial density for $p = 0.2$ and variable n.

Since many statistical studies involve repeated Bernoulli trials, this distribution is of great practical importance. In power system reliability analysis, the binomial distribution is widely used in computing the reliability of large systems of generators or other system components that can be considered as either the up (success) state, with probability p, or in the down (failed) state, with probability q. Because of its wide usage, the binomial distribution has been tabulated in several sources [9].

THE POISSON DISTRIBUTION. The Poisson distribution, named for the French mathematician Siméon Denis Poisson, has the following density and distribution functions.

$$f_X(x) = e^{-b} \sum_{k=0}^{\infty} \frac{b^k}{k!} \delta(x - k) \tag{24.52}$$

$$F_X(x) = e^{-b} \sum_{k=0}^{\infty} \frac{b^k}{k!} u(x - k) \tag{24.53}$$

where $b > 0$ is a real constant.

Plots of the Poisson density function with various values of the parameter b are shown in Figure 24.8.

Note that the peak of the density function is near the value of b and with falloff on either side. Symmetry begins to develop as b becomes larger. Values of the Poisson distribution are often read from tables.

The Poisson distribution applies to many counting type problems, such as the number of telephone calls made or received during a period of time or the number of faults occurring on the power system in a given period [10], [11].

EXAMPLE 24.8

We compare the binomial and Poisson distributions by the following problem (from [12]). Assume that a large static var system is protected by digital protective devices, which have the capability of checking their cpu's to determine their ability to function properly. Three such devices are specified for the protective system under study and two implementation schemes have been suggested. The first scheme places the three protections in parallel, with any one of the three then being able to identify a

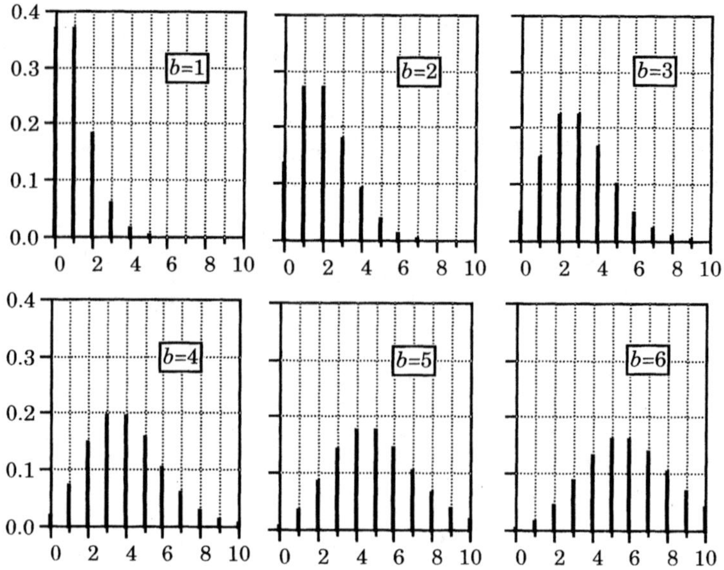

Figure 24.8 The Poisson density function for various values of b.

fault and take appropriate action. The second scheme places only one device in service, with the other two switched off on the theory that the nonenergized device will have a lower failure rate. One of the standby devices will be switched into service if the active device fails its self-check test. We would like to know which arrangement has the higher probability of failure.

Solution

The first arrangement is described by a binomial distribution since it has all of the features required for this distribution. For the three devices in service, the probability of success is given by adding all the terms in the binomial expansion that include at least one p. This gives

$$\begin{aligned}\Pr\{\text{success parallel}\} &= p^3 + 3p^2q + 3pq^2 \\ &= p^3 - 3p^2 + 3p\end{aligned} \tag{24.54}$$

where p is the probability of successful operation of one device.

When arranged in the cold standby configuration, the binomial distribution does not apply since the probability of success depends on the failure of the preceding unit. Here, the Poisson distribution applies and we compute

$$\Pr\{\text{success redundant}\} = e^{-b} + be^{-b} + \frac{b^2}{2!}e^{-b} \tag{24.55}$$

To compare the two results. we assume that the probability of a single unit in the parallel case is equal to the probability of success of a single standby unit, or

$$\begin{aligned}p &= e^{-b} = 0.9 \\ b &= 0.10536\end{aligned}$$

and

$$\begin{aligned}\Pr\{\text{success parallel}\} &= 0.99900 \\ \Pr\{\text{success standby}\} &= 0.99982\end{aligned}$$

It is easier to compare the probabilities of failure.

$$\begin{aligned}\Pr\{\text{failure parallel}\} &= 0.00100 \\ \Pr\{\text{failure standby}\} &= 0.00018\end{aligned}$$

Thus, we see that the probability of failure of the parallel system is about five times greater than that of the standby system. ■

24.3.7.2 Continuous Distributions. Another important class of distributions is represented by continuous functions of the random variable. Several of these are very important in all areas of science and engineering.

THE NORMAL (GAUSSIAN) DISTRIBUTION. The normal distribution is often called Gaussian, after the German mathematician.[3] A random variable is called *Gaussian* or *normal* if its density function has the form

$$f_X(x) = \frac{1}{\sqrt{2\pi\sigma_X^2}} e^{-(x-\mu_X)^2/2\sigma_X^2} \tag{24.56}$$

where $\sigma_X > 0$ and $-\infty < \mu_X < \infty$ are real constants. The shape of this curve is the well-known bell-shaped density function of Figure 24.9. The factor μ is called the *mean* and σ is the *standard deviation* of the normal distribution. The bell-shaped curve is always symmetrical about the mean and the standard deviation determines the shape of the density, with small values of σ leading to a density with most of its weight concentrated close to the mean.

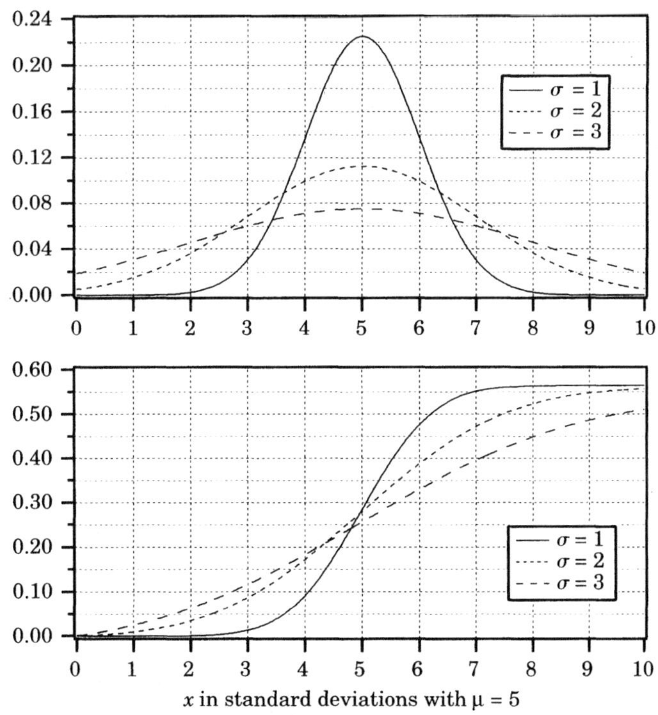

Figure 24.9 The normal or Gaussian density function (top) and distribution function (bottom), plotted for $\mu = 5$.

[3] The normal distribution is said to have been first published by De Moivre in 1733 and was known to Laplace no later than 1744, but through historical error has been attributed to Gauss, whose earliest publication on the subject occurred in 1809.

The distribution function is given by

$$F_X(x) = \frac{1}{\sqrt{2\pi\sigma_X^2}} \int_{-\infty}^{x} e^{-(u-\mu_X)^2/2\sigma_X^2}\, du \tag{24.57}$$

This integral has no known closed form solution and must be evaluated using numerical methods. The distribution is shown in the lower part of Figure 24.9, which is found by integrating the density traces of the upper part of the figure.

The numerical evaluation of the normal distribution would always be required for problem solving except for the fact that published tables are widely available. These tabulations are usually prepared for the *unit normal distribution*, which is computed by substituting u for x in (24.57), where u is defined as

$$u = \frac{x - \mu}{\sigma} \tag{24.58}$$

This reduces (24.57) to the unit normal form

$$F(x) = \frac{1}{2\pi} \int_{-\infty}^{x} e^{-u^2/2}\, du \tag{24.59}$$

which has a standard deviation of unity and a mean of zero. A tabulation of the ordinates and several areas of interest are given in Table 24.1.

The areas given in Table 24.1 are designated by the shaded regions in the plots above each column. Thus, if we designate ξ as the random value of a unit normal variate, with $x \geq 0$, then we can find the following relationships.

$$\begin{aligned}
&\Pr(\xi \leq x) = F(x) && (a)\\
&\Pr(\xi \geq x) = 1 - F(x) = R(x) && (b)\\
&\Pr(\xi \leq -x) = F(-x) = 1 - F(x) = R(x) && (c)\\
&\Pr(|\xi| \geq x) = F(-x) - R(x) = 2R(x) && (d)\\
&\Pr(-x \leq \xi \leq x) = W(x) && (e)
\end{aligned} \tag{24.60}$$

Proof of these relationships is left as an exercise.

THE UNIFORM DISTRIBUTION. The uniform density and distribution are defined by the following functions.

$$f_X(x) = \begin{cases} 1/(b-a) & a \leq x \leq b \\ 0 & \text{elsewhere} \end{cases} \tag{24.61}$$

$$F_X(x) = \begin{cases} 0 & x < a \\ (x-a)/(b-a) & a \leq x < b \\ 1 & x \geq b \end{cases} \tag{24.62}$$

These functions are plotted in Figure 24.10. The uniform distribution has many practical uses in engineering.

THE WEIBULL DISTRIBUTION. The Weibull distribution, developed by W. Weibull of Sweden, has the important property that it has no definite shape and can take on a number of different shapes depending on the values of its parameters. The density and distribution functions are as follows.

$$f_X(x) = \left\{ \frac{\beta x^{\beta-1}}{\alpha^\beta} \exp\left[-\left(\frac{x}{\alpha}\right)^\beta\right] \right\} u(x) \tag{24.63}$$

$$F_X(x) = \left\{ 1 - \exp\left[-\left(\frac{x}{\alpha}\right)^\beta\right] \right\} u(x) \tag{24.64}$$

TABLE 24.1 The Unit Normal Distribution and Areas of Integration [13], [14]

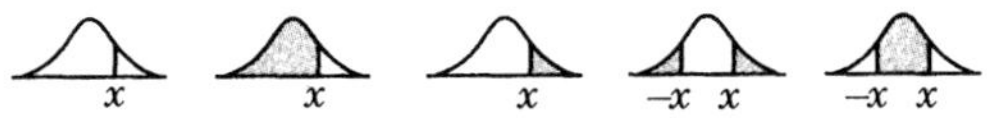

x	$f(x)$	$F(x)$	$R(x)$	$2R(x)$	$W(x)$
0.0	0.3989	0.5000	0.5000	1.0000	0.0000
0.1	0.3970	0.5398	0.4602	0.9203	0.0797
0.2	0.3910	0.5793	0.4207	0.8415	0.1585
0.3	0.3814	0.6179	0.3821	0.7642	0.2358
0.4	0.3683	0.6554	0.3446	0.6892	0.3108
0.5	0.3521	0.6915	0.3085	0.6171	0.3829
0.6	0.3332	0.7257	0.2743	0.5485	0.4515
0.7	0.3123	0.7580	0.2420	0.4839	0.5161
0.8	0.2897	0.7881	0.2119	0.4237	0.5763
0.9	0.2661	0.8159	0.1841	0.3681	0.6319
1.0	0.2420	0.8413	0.1587	0.3173	0.6827
1.1	0.2179	0.8643	0.1357	0.2713	0.7287
1.2	0.1942	0.8849	0.1151	0.2301	0.7699
1.3	0.1714	0.9032	0.0968	0.1936	0.8064
1.4	0.1497	0.9192	0.0808	0.1615	0.8385
1.5	0.1295	0.9332	0.0668	0.1336	0.8664
1.6	0.1109	0.9452	0.0548	0.1096	0.8904
1.7	0.0940	0.9554	0.0446	0.0891	0.9109
1.8	0.0790	0.9641	0.0359	0.0719	0.9281
1.9	0.0656	0.9713	0.0287	0.0574	0.9426
2.0	0.0540	0.9772	0.0228	0.0455	0.9545
2.1	0.0440	0.9821	0.0179	0.0357	0.9643
2.2	0.0355	0.9861	0.0139	0.0278	0.9722
2.3	0.0283	0.9893	0.0107	0.0214	0.9786
2.4	0.0224	0.9918	0.0082	0.0164	0.9836
2.5	0.0175	0.9938	0.0062	0.0124	0.9876
2.6	0.0136	0.9953	0.0047	0.0093	0.9907
2.7	0.0104	0.9965	0.0035	0.0069	0.9931
2.8	0.0079	0.9974	0.0026	0.0051	0.9949
2.9	0.0060	0.9981	0.0019	0.0037	0.9963
3.0	0.0044	0.9987	0.0013	0.0027	0.9973

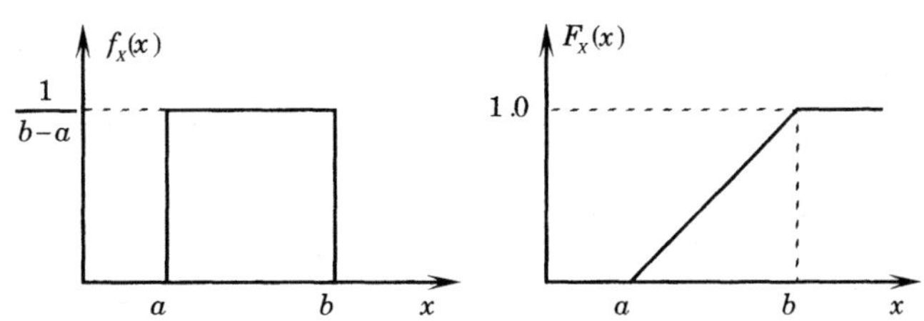

Figure 24.10 The uniform density and distribution.

where

$$x \geq 0, \qquad \alpha > 0, \qquad \beta > 0$$

The Weibull distribution, because of its variable shape, is often used in the analysis of statistical data. Special graph paper, called Weibull paper, is available for plotting data and the paper is designed so that the parameters can be easily determined. Plots of the Weibull density and distribution are shown in Figure 24.11. The parameter β is called the "shape parameter" for this distribution.

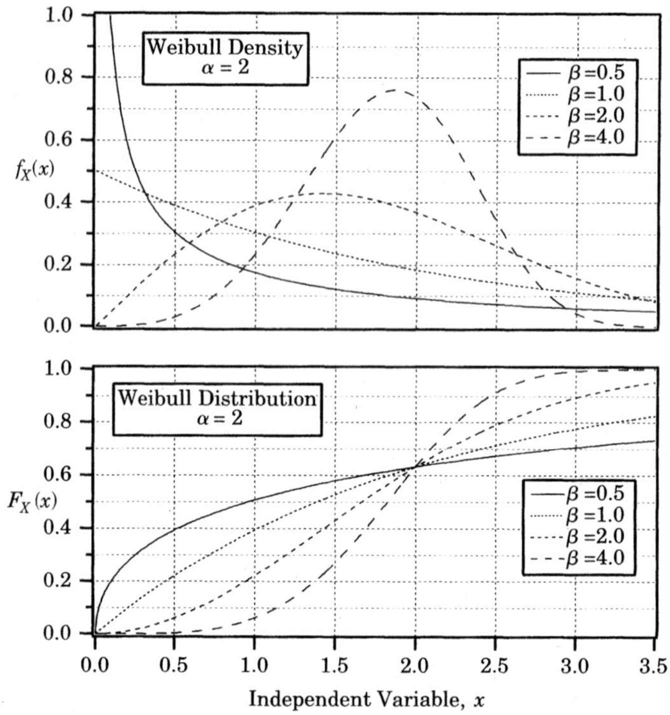

Figure 24.11 The Weibull density function (top) and distribution (bottom) for various Weibull parameter values.

Some of the characteristics of the Weibull Distribution can be summarized as follows [15]:

1. For values of $0 < \beta < 1$, Weibull distribution is suitable for representing early-type failures where the failure rate decreases with age.
2. For $\beta = 1$ the function is suitable for chance failures and the useful-life period failure rate of items.
3. For $\beta > 1$, the function is suitable for wear-out types of failure.
4. The value of α is often called the characteristic life parameter, or sometimes the scale parameter.

THE EXPONENTIAL DISTRIBUTION. The exponential distribution is one of the most important distributions in the reliability of engineering systems. It is a special case of the Weibull distribution where $\beta = 1$ and is very important in many physical systems. The exponential density and distribution functions are stated as follows.

$$f_X(x) = \begin{cases} \dfrac{1}{b} e^{-(x-a)/b} & x > a \\ 0 & x < a \end{cases} \tag{24.65}$$

$$F_X(x) = \begin{cases} 1 - e^{-(x-a)/b} & x > a \\ 0 & x < a \end{cases} \tag{24.66}$$

where

$$-\infty < a < \infty \qquad b > 0$$

The exponential distribution is illustrated in Figure 24.12.

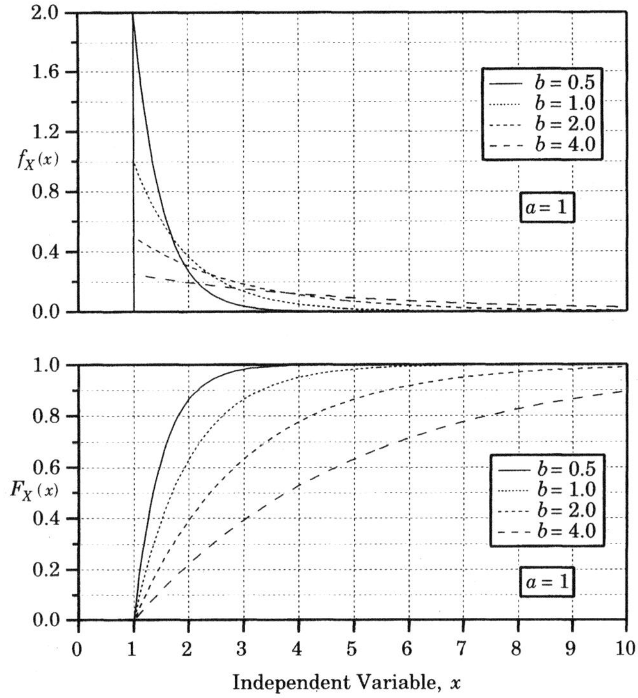

Figure 24.12 Exponential density (top) and distribution (bottom).

We will investigate the exponential distribution more carefully when we consider reliability theory in Section 24.5.

24.3.8 Random Vectors

In some experiments it is possible to define more than one random variable for a given problem. In such problems it is convenient to use the concept of a random vector $\mathbf{x}(\xi)$ defined in an n-dimensional Euclidean space R^n. Thus $\mathbf{x}$ maps S into R^n where we define

$$\mathbf{x} = \begin{bmatrix} x_1 \\ x_2 \\ \cdots \\ x_n \end{bmatrix} \tag{24.67}$$

Moreover, if we let $\mathbf{r} \in R^n$ then we can define an event as

$$\{\xi : \mathbf{x}(\xi) \in \mathbf{r}\} \tag{24.68}$$

As in the scalar case, we define $\mathbf{x}$ such that we may associate a probability with each ξ in S, and this may be done if the event $\{\xi\}$ is such that for every $\mathbf{r}$ in R^n such that $\mathbf{x}(\xi) \leq \mathbf{r}$. We may then state the following [12].

Definition. A real valued *random vector* is any vector function $\mathbf{x}$ that maps S into R^n such that

1. $\{\xi : \mathbf{x}(\xi) \leq \mathbf{r}\}$ is an event for every $\mathbf{r}$ in R^n, and
2. $\Pr\{\xi : \mathbf{x}(\xi) = -\infty\} = \Pr\{\xi : \mathbf{x}(\xi) = +\infty\} = 0$

This random vector definition may be particularly interesting in power system applications since it is often appropriate to develop an n-dimensional vector representation of an n-bus system. This provides an orderly method of identifying or noting events at different buses.

24.3.9 Stochastic Processes

Stochastic processes are important in the study of probability and reliability, and certain processes are important in power system protection.

Definition. A stochastic process $\mathbf{x}(t)$ is a family of random vectors $\{\mathbf{x}(t), t \in T\}$ indexed by a parameter t belonging to an index set T.

It is clear from the definition that when $t = t_i$ is specified, $\mathbf{x}(t_i)$ is a random-variable vector, and $\{\mathbf{x}(t_i) \leq \mathbf{a}\}$ is used to represent the set of outcomes for which $\mathbf{x}$ at $t = t_i$ is less than or equal to $\mathbf{a}$ so that

$$\{\mathbf{x}(t_i) \leq \mathbf{a}\} = \{\xi : \mathbf{x}(t_i, \xi) \leq \mathbf{a}\} \tag{24.69}$$

is an event for every $t_i \in T$ and $\mathbf{a} \in R^n$.

Definition. The probability distribution function of the stochastic process $\mathbf{x}(t)$ is defined as

$$\begin{gathered} \mathbf{F_X}(\mathbf{a}, t) = \Pr\{\mathbf{x}(t) \leq \mathbf{a}\} \\ \text{for all } \mathbf{a} \in R^n \text{ and } t \in T. \end{gathered} \tag{24.70}$$

Because the Poisson distribution is central to the analysis of power system protection, we investigate this interesting distribution more thoroughly.

In order to characterize the probabilistic nature of power system disturbances, it is helpful to introduce the counting process $\{N(t), t \geq 0\}$, which counts the number of points in an interval, the points having been distributed by some stochastic process. The process begins at $t = 0$, the time when the observations are begun, and are observed on the interval $(0, t)$, during which time the value of $N(t)$ is observed. A typical graph of $N(t)$ is shown in Figure 24.13.

Since the counting process $N(t)$ changes as a function of time in a random manner, the observed value is a random variable and the family of random variables $\{N(t), t \geq 0\}$ is an integer-valued stochastic process.

If we let $t_1, t_2, \ldots$ represent the times at which the value of $N(t)$ changes, then we can compute the random variables

$$\begin{aligned} T_1 &= t_1 \\ T_2 &= t_2 - t_1 \\ &\cdots \\ T_n &= t_n - t_{n-1} \end{aligned} \tag{24.71}$$

which are called the "inter-arrival times."

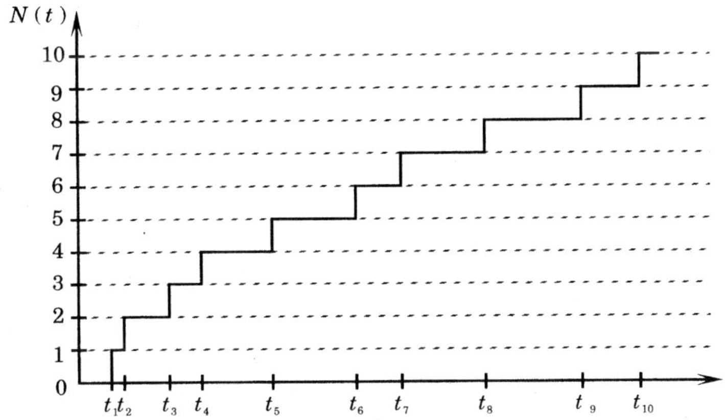

Figure 24.13 The counting function for a typical Poisson process.

The *Poisson Process* is an important integer-valued process if the following assumptions on $\{N(t), t \geq 0\}$ are made:

1. $N(t)$ has stationary independent increments.
2. For any two instants of time, designated t_m and t_n and with $t_m < t_n$ the number $N(t_n) - N(t_m)$ of counts in the interval $t_n - t_m$ is Poisson distributed with mean $\lambda(t_n - t_m)$.

Then for $k = 0, 1, 2, \ldots$ the following may be computed [10].

$$\begin{aligned} \Pr\{N(t_n) - N(t_m) = k\} &= \frac{[\nu(t_n - t_m)]^k}{k!} e^{-\nu(t_n - t_m)} \\ E[N(t_n) - N(t_m)] &= \nu(t_n - t_m) \\ \mathrm{Var}[N(t_n) - N(t_m)] &= \nu(t_n - t_m) \end{aligned} \tag{24.72}$$

The Poisson process is characterized by stationary independent integer-valued unit jumps. Examples include the following:

Accidents, traffic counts, and misprints
Demands for service, maintenance, sales, and admissions
Counts of particles suspended in liquids
Shot noise in vacuum tubes
Number of calls per minute at a telephone exchange
Number of accidents per year under stable conditions
Number of disturbances on a power system in a given time

This process is a natural choice of stochastic process to characterize the occurrence of disturbances on a power system. Therefore, we examine the axioms that define the process as well as a test that can be applied to data to verify our assumption regarding power system disturbances.

Consider events occurring in time on the interval 0 to ∞. For $t > 0$, let $N(t)$ be the number of events that have occurred on the interval $(0, t]$. Then $N(t) \geq 0$ and $N(t + k) - N(t) > 0$ for any $k > 0$. We now state the following axioms [10]:

Axiom 1 Since we begin counting at $t = 0$, we define $N(0) = 0$.

Axiom 2 The process $\{N(t), t \geq 0\}$ has independent increments.

Axiom 3 For any $t > 0$, $0 < \Pr\{N(t) > 0\} < 1$, or in words, in any interval, no matter how small, there is a positive probability that an event will occur, but it is not certain that an event will occur.

Axiom 4 For any $t \geq 0$,

$$\lim_{h \to 0} \frac{\Pr\{N(t+h) - N(t) \geq 2\}}{\Pr\{N(t+h) - N(t) = 1\}} = 0 \tag{24.73}$$

In words we interpret this to say that, in sufficiently small intervals, at most one event can occur, that is, it is not possible for events to occur simultaneously.

Axiom 5 The counting process $N(t)$ has stationary increments, that is, for any two points $t_n > t_m > 0$ and for any $h > 0$, the random variables $N(t_n) - N(t_m)$ and $N(t_n + h) - N(t_m + h)$ are identically distributed.

There are various modifications of the definition of a Poisson process that have been found useful.

(a) If axiom 4 is dropped, we have a "generalized Poisson process."

(b) If axiom 5 is dropped, we have a "nonhomogeneous Poisson process."

By definition, the counting process $\{N(t), t \geq 0\}$ that satisfies axioms 1 to 5 is the Poisson process defined by (24.72).

24.3.10 Power System Disturbances

We presume that the counting of power system disturbances begins at some arbitrary time $t = 0$ and that axiom 1 is satisfied. One could argue that power system disturbances may occur simultaneously, but the probability of exact simultaneity is zero. In this regard, power system disturbances are no different from the processing of incoming telephone calls or the answering of signals from competing customers for an elevator in a building. These events have in common the fact that some type of hardware must receive the request for service (or counting) and that this hardware can process only one request at a time. Therefore, even if two disturbances occur so nearly together that we would have difficulty measuring any time difference, our counting of the events would record the one disturbance before the other, just as in the other examples cited. We conclude that, although axiom 4 may not be strictly obeyed, we are unable to detect any violation.

Axiom 5 deals with stationarity. Certain power system disturbances, such as the failure and disconnection of generating units are probably stationary since there is no evidence to the contrary. Line faults, however, are probably not stationary because so many faults are caused by inclement weather, which is seasonal. In transmission reliability, this is often solved by keeping two different sets of line outage statistics, one for fair weather and one for stormy weather. Stationarity can be logically assumed as long as one deals with only one of these types of statistical data.

24.4 FAILURE DEFINITIONS AND FAILURE MODES

We now consider the concept of failure and define more precisely what is meant or implied by this term.

24.4.1 Failure Definitions

First we introduce several definitions that are taken from the U.S. military standards for reliability [16].

Failure The event, or inoperable state, in which any item or part of an item does not, or would not, perform as previously specified.

Failure, catastrophic A failure that can cause item loss.

Failure, critical A failure, or combination of failures, that prevents an item from performing a specified mission.

Failure, dependent Failure which is caused by the failure of an associated item(s). Not independent.

Failure, independent Failure which occurs without being caused by the failure of any other item. Not dependent.

Failure mode The consequence of the mechanism through which the failure occurs, i.e., short, open, fracture, excessive wear.

Failure rate The total number of failures within an item population, divided by the total number of life units expended by that population, during a particular measurement interval under stated conditions.

These are applicable definitions to power systems. In addition to the above, we find the following failure definition in the IEEE standards [17].

Failure The termination of the ability of an item or equipment to perform its required function. Note: Failures may be unannounced and not detected until the next test or demand (unannounced failure), or they may be announced and detected by any number of methods at the instant of occurrence (announced failure).

Failure rate The mean number of failures of a component per unit exposure time. Usually exposure time is expressed in years and failure rate is given in failures per year.

The IEEE definition of failure is particularly important in protective systems, where failures are often *unannounced* and not detected until the next inspection. This may also be true of protection-related components such as voltage and current transducers, which are installed as part of the protective system. For example, current or voltage transformers that are used for normal metering of electrical quantities, where these quantities are transmitted to a central control center, will have *announced* failures since the loss of the required metering will be detected. However, similar transducers dedicated exclusively to protective systems may have only *unannounced* failures. The reliability modeling for the two conditions is quite different. This will be explored later.

24.4.2 Modes of Failure

The foregoing discussion about the reliability of an item has tacitly assumed that the item is either working or failed, or, in other words, the item may exist in either of two states. Some items, however, exhibit more than two states. A common example is an item, such as a resistor or a capacitor, that can fail shorted or fail open. Whether this causes failure of the system in which the component is installed depends on the circuit configuration and the designed purpose of that circuit. Some complex components have many different failure modes. A good example in system protection is a circuit breaker, which can fail to open on command, to close on command, to make a current, to break a current, etc.

As an example of a component with two modes of failure, consider a diode, which can fail either shorted or open. Figure 24.14 shows two different diode connections that might be of interest. For the diodes, we define the following states, using the overbar to indicate a failed state.

$$\begin{aligned} x_n &= \text{normal} \\ \bar{x}_s &= \text{failed shorted} \\ \bar{x}_o &= \text{failed open} \end{aligned} \tag{24.74}$$

Since these states are disjoint and are the only possible states, we may write

$$\Pr(x_n + \bar{x}_s + \bar{x}_o) = \Pr(x_n) + \Pr(\bar{x}_s) + \Pr(\bar{x}_o) = 1 \tag{24.75}$$

The reliability R of a single diode is defined as the probability that the diode is neither open nor shorted.

$$R = \Pr(x_n + \bar{x}_s + \bar{x}_o) = \Pr(x_n) + \Pr(\bar{x}_s) + \Pr(\bar{x}_o) = \Pr(x_n) \tag{24.76}$$

Figure 24.14 shows two diodes in either a series or a parallel arrangement. System success is defined as the existence of a unidirectional path from left to right. For the series arrangement, the system is failed if either diode fails open or if both fail shorted. Then

$$R_S = \Pr\{\text{success(series)}\} = \Pr(x_{n1}x_{n2} + x_{n1}\bar{x}_{s2} + x_{n2}\bar{x}_{s1}) \tag{24.77}$$

For the parallel arrangement, the system is failed if either diode fails shorted or if both fail open.

$$R_P = \Pr\{\text{success(parallel)}\} = \Pr(x_{n1}x_{n2} + x_{n1}\bar{x}_{o2} + x_{n2}\bar{x}_{o1}) \tag{24.78}$$

This concept is very important in system protection. Relays are often installed in redundant pairs, just like the diodes in parallel in Figure 24.14. Moreover, relays always have contacts that can fail either shorted or open, exactly like the diodes. Therefore, the previous equations apply directly to two relay contacts that are installed in parallel.

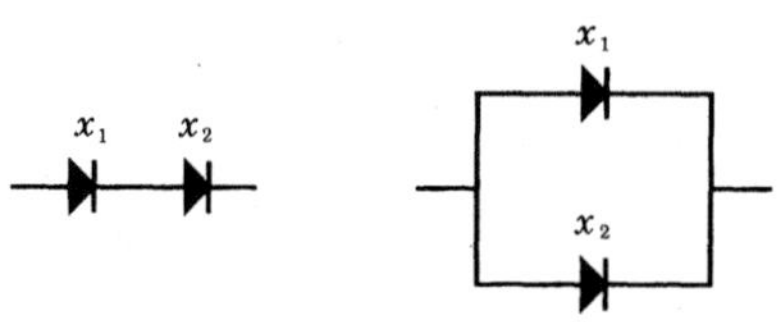

Figure 24.14 Diodes in series and parallel.

When relay contacts fail shorted, we often refer to this event as a "security failure," since this causes an unnecessary tripping of the protected circuit. When relays fail open, they are unable to trip when needed and that is the reason relays are often installed in parallel. This subject is discussed in greater depth in Chapter 26.

24.5 RELIABILITY MODELS

This section presents some basic definitions and concepts of reliability that will be found useful in this and the following chapters. We begin with some basic definitions and show how reliability is based on probability theory. Then we present several concepts and terms that are an essential part of reliability assessment and prediction.

24.5.1 Definition of Reliability

There are many different definitions of reliability and many of them have become IEEE or ANSI standards. This is because many different technical disciplines find it necessary to define the term "reliability" as it applies to that particular discipline. Textbooks on reliability also define the term, often in somewhat different ways [4], [6], [7], [12]. The following definitions are all appropriate.

Reliability (IEEE standard) The ability of an item to perform a required function under stated conditions for a stated period of time [17].

Reliability (Military standard) (1) The duration or probability of failure-free performance under stated conditions. (2) The probability that an item can perform its intended function for a specified interval under stated conditions. (For nonredundant items, this is equivalent to definition (1). For redundant items, this is equivalent to the definition of mission reliability) [16].

Reliability (textbook) The probability of remaining in the operating state as a function of time, given that the system started in the operating state at time $t = 0$.

Other definitions are found in the literature. Note that the definition includes a requirement for operation over a specified period of time. This is an important requirement in the definition, since no systems are capable of operating indefinitely without failure.

24.5.1.1 The Failure Process. The definitions all embrace the concept of time as an important element in reliability. Mathematically, we define the random variable T as the failure time of an item.[4] The probability of failure, then, is given by the equation

$$\Pr(T \le t) = F_T(t) \tag{24.79}$$

where T, the failure time, is a random variable. This equation relates the probability of the failure event to the distribution function. We define reliability as the probability of success in terms of the failure distribution as follows.

$$R(t) = \Pr\{\text{success}\} = 1 - F_F(t) = \Pr(T > t) \tag{24.80}$$

where $R(t)$ is called the reliability of the item. The failure density function is the derivative of the failure distribution

$$f_T(t) = \frac{dF_T(t)}{dt} \tag{24.81}$$

Consider a large population of N items all of which have the same failure density. Over time, the items fail independently all with a probability of failure given by (24.79). Let $X(t)$ be a random variable that represents the items surviving at time t, then X has a binomial distribution with $p = R(t)$. Then we may write, from (24.47)

$$\Pr\{X(t) = n\} = \frac{N!}{n!(N-n)!}[R(t)]^n[1 - R(t)]^{N-n} \tag{24.82}$$

where

$$n = 0, 1, \ldots N$$

At any time t, the number of items still working is a random variable, but we can compute the expected number of the survivors by finding the expected value of $X(t)$, that is

$$n(t) = E[X(t)] = NR(t) \tag{24.83}$$

Solving for $R(t)$, we have

$$R(t) = \frac{n(t)}{N} \tag{24.84}$$

Stated in words, the reliability at time t is the average fraction of survivors at that time. We can also relate this result to the distribution function. From (24.80)

$$F_T(t) = 1 - R(t) = 1 - \frac{n(t)}{N} = \frac{N - n(t)}{N} \tag{24.85}$$

[4]The word "item" is a nonspecific term used to denote any product, including systems, materials, parts, subassemblies, sets, accessories, etc.

The density is computed from (24.81).

$$f_T(t) = \frac{dF_T(t)}{dt} = -\frac{1}{N}\frac{dn(t)}{dt} = \lim_{\Delta t \to 0} \frac{n(t) - n(t + \Delta t)}{N\,\Delta t} \tag{24.86}$$

Equation (24.85) expresses the average fraction of units failed, and (24.86) is the slope of this average fraction. Note that the failure density is *normalized* in terms of the size of the population N. The failure density function is described as follows.

> The function $f(t)dt$ is the probability that the component failure occurs during the interval $[t, t + dt)$, given that the component was new and working or was repaired to as good as new at time zero.

24.5.1.2 The Hazard Rate. We have seen that the failure density is a normalized quantity, where normalization is based on the total population of items. In many cases, it is more important to normalize the result in terms of the number of survivors $n(t)$. Using (24.86), we write the hazard function as

$$h(t) = \lim_{\Delta t \to 0} \frac{n(t) - n(t + \Delta t)}{n(t)\,\Delta t} \tag{24.87}$$

where the quantity $h(t)$ is called the *hazard function or hazard rate*. We can relate the hazard function to the density function by observing from (24.86) and (24.87) that

$$h(t) = \frac{Nf_T(t)}{n(t)} = \frac{f_T(t)}{R(t)} = \frac{f_T(t)}{1 - F_T(t)} \tag{24.88}$$

The hazard rate is further defined as follows.

> The hazard or failure rate is the probability that the component experiences a failure per unit of time at time t, given that the component was new or was just repaired at time zero and has survived to time t.

The quantity $h(t)dt$ is the probability that the component fails during the interval $[t, t + dt)$ given that the component age is t, where the term age means that the component was working at time zero and has survived to time t. This assumes the continuation of the normal state to time t, i.e., no failure has occurred in the interval $[0, t]$. Clearly, the hazard rate is a conditional probability, which is conditioned on the fact that the component remains normal to time t.

Now, from the definition of the reliability function, we write

$$R(t) = 1 - F_T(t) = 1 - \int_0^t f_T(\xi)\,d\xi \tag{24.89}$$

where ξ is a dummy variable of integration. We may also write the hazard function in terms of the survivor function as follows.

$$\begin{aligned} h(t) &= [f(t)]\left[\frac{1}{R(t)}\right] = \left[-\frac{1}{N}\frac{dn(t)}{dt}\right]\left[\frac{N}{n(t)}\right] \\ &= -\frac{d}{dt}\ln n(t) \end{aligned} \tag{24.90}$$

or

$$\ln n(t) = -\int_0^t h(\xi)d\xi + c \tag{24.91}$$

where ξ is a dummy variable and c is a constant of integration. Then

$$n(t) = e^c \exp\left[-\int_0^t h(\xi)\,d\xi\right] \tag{24.92}$$

Inserting the initial conditions

$$n(0) = N = e^c \tag{24.93}$$

gives

$$n(t) = N \exp\left[-\int_0^t h(\xi)\, d\xi\right] \tag{24.94}$$

and substituting into (24.83)

$$R(t) = \exp\left[-\int_0^t h(\xi)\, d\xi\right] \tag{24.95}$$

This completes the relationships among the variables reliability, hazard, and failure density. It is particularly important to note that, if the hazard function is constant, we have the following.

$$\begin{aligned} h(t) &= \lambda = \text{a constant} \\ \int_0^t h(\xi)\, d\xi &= \int_0^t \lambda d\xi = \lambda t \end{aligned} \tag{24.96}$$

$$\begin{aligned} R(t) &= e^{-\lambda t} = 1 - F_T(t) \\ f_T(t) &= \lambda e^{-\lambda t} \end{aligned} \tag{24.97}$$

We will see later that this is a very important special case.

The hazard function varies with the operating lifetime of the item, with the form of that variation depending on the type of equipment and the modes of failure. One description that is often used is described as a "bathtub curve" because of the function shape. Examples of such curves are shown in Figure 24.15. Figure 24.15(a) is typical of electrical and electronic systems where, following system start-up and debugging, the hazard rate is nearly constant for much of the total lifetime until the components begin to fail due to aging. Mechanical systems, as shown in Figure 24.15(b), tend to wear out constantly, although the lifetime can be extended by preventive maintenance programs where components are replaced as soon as excessive wear is noted. Preventive maintenance returns the item to nearly a condition approaching the lowest hazard that can be achieved. If this is repeated throughout the system lifetime, as shown by the "life extension" line on the right of Figure 24.15(b), then the hazard rate can be held at a nearly constant value.

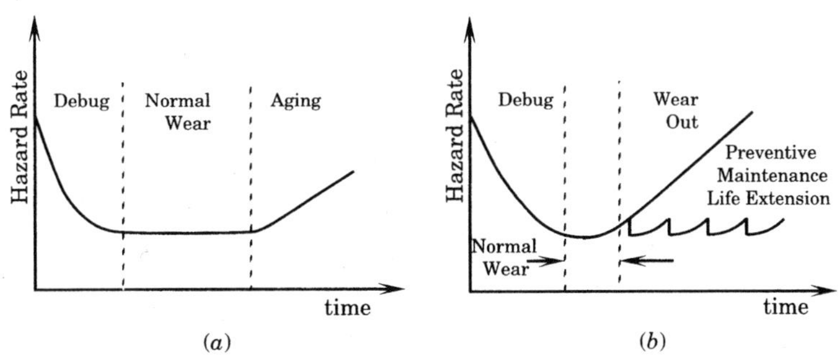

Figure 24.15 Typical hazard function variation with age. (a) Electrical item. (b) Mechanical Item.

The foregoing analysis is a probabilistic description of a failure process, which may be described as a birth to death process [7]. A component at birth is in a new condition, with a reliability of unity. As the component ages, the reliability $R(t)$ declines, but at all times represents the probability of survival up to and including the time t. We have noted that this is equal to the number of survivors of a population of components, all placed in service at the same time, to the total population of components. The unreliability $F(t)$, on the other hand, is the probability of failure (death) up to, but not including, age t, and is equal to the number of failures divided by the starting population.

24.5.1.3 The Mean Time To Failure. Another parameter that is important in characterizing the failure process is the time to failure, which is designated as the lifetime, T. The time to failure is the time elapsed from time zero, at which time the component jumped to a reliability of unity, to the first failure of that component. Since T is a random variable, it is not possible to predict exactly when the failure will occur. However, we can compute the expected value or mean of this quantity from the first moment. It is customary to call this result the *mean time to failure*, or *MTTF*, or simply the variable m.

$$MTTF = m = \int_0^\infty t f_T(t)dt \tag{24.98}$$

This is an important parameter for the failure process of a component and is information that is often available from component manufacturers.

24.5.2 The Repair Process

The preceding analysis is based on the concept that the items, whose reliability is described by a time-to-failure distribution, are not repairable or the repair of the failed item is not considered. Nonrepairable items are defined as those that cannot be repaired, where repair is not economical, or where the item lifetime up to the time of catastrophic failure is the point of interest. This is appropriate for many items, such as light bulbs, fuses, and similar low-cost items where repair is not even considered.

In power systems we are often dealing with very expensive items that cannot economically be replaced in the event of a failure. Such items are repaired routinely. We also recognize two different types of restoration activities. The first is called preventive maintenance, where the item is taken out of service on a regular schedule for inspection, cleaning, and replacement of questionable parts. This practice tends to keep the item in a condition that is nearly "as good as new," at least on the average. If preventive maintenance is performed often enough, serious failures may be reduced considerably. Preventive maintenance is performed regularly on many different types of power system equipment, such as generators, turbines, boilers, circuit breakers, and protective devices of all kinds.

The second type of repair takes place following the detection of failure. In this case the failed item is inoperable and must be restored to working condition. This activity is often referred to simply as "repair."

24.5.2.1 Ideal Repair. One model of repair is called ideal repair, where the repair time is assumed to be zero. One example of ideal repair is the case where failed components are replaced immediately. An example of this is so-called two- or three-shot fuses, which are sometimes used on distribution circuits. Since most faults are temporary, a fault will cause failure of the first fuse, which is immediately replaced with a new fuse element by the fuse holding equipment. If the fault was temporary, service is restored with only a very short interruption and with no service call.

From the definition of the time-to-failure density, we may write, for the item called A,

$$\begin{aligned} F_T(t) &= \Pr\{A \text{ will fail prior to } t\} = \Pr\{T \leq t\} \\ f_T(t)dt &= \Pr\{A \text{ will fail in the interval } [t, t+dt)\} \end{aligned} \tag{24.99}$$

where T is the random variable corresponding to the time to failure of the item. Stated in words, we say that $F_T(t)$ is the probability that the item fails during the interval $[0, t)$, given that it is new or was repaired at time zero. Now define the following.

$$\begin{aligned} f_{T1}(t) &= \text{density function to the first failure} \\ f_{T2}(t) &= \text{density function to the second failure} \end{aligned} \tag{24.100}$$

Assume that the first failure occurs at $\tau < t$, so that we have the second lifetime as $t - \tau$. Then the probability that the second failure will occur in the interval $[t, t + \Delta t)$ for a given τ is given by

$$f_{T2}(t)\Delta t = f_{T1}(\tau)\Delta\tau \cdot f_{T1}(t - \tau) \tag{24.101}$$

Then for any $t > \tau$, the density of the second failure is computed by the convolution integral

$$f_{T2}(t) = \int_0^t f_{T1}(\tau)\Delta\tau \cdot f_{T1}(t - \tau)d\tau \tag{24.102}$$

This process can be repeated over and over again, if necessary. This is not a problem of great interest in system protection, but there are applications as noted above.

24.5.2.2 Repair and Preventive Maintenance. Let us define "repair" to mean that the item has failed and repair personnel are dispatched immediately to determine the cause of the failure and to correct the problem. For power systems, which often span large regions, this may include relatively large time periods for travel, for locating spare parts, and for making the actual repair to correct the failure. All of these times, for all similar types of item failure, must be averaged to determine the mean repair time. Since all items fail eventually, the item lifetime may be a series of UP and DOWN states representing periods of service and of repair.

Preventive maintenance requires that time be scheduled for inspection and repair of an item. This is usually done periodically. The repair time now becomes a random variable and the element lifetime becomes a series of UP and DOWN states, with the item sequentially in service and in preventive maintenance, in addition to periods of repair to correct failures.

24.5.2.3 Probabilistic Repair Parameters. The repair process can be described in terms of probabilistic parameters in the same way the failure process is described in terms of the failure distribution, failure density, hazard rate, and *MTTF*. For the repair process it is appropriate to define a repair distribution, which is defined as follows [7].

$$G(t) = \text{repair probability at time } t \tag{24.103}$$

The repair distribution is the probability that the repair is completed before time t, given that the component failed at time zero. This is a cumulative distribution of the same type as the failure distribution $F(t)$; that is, it is a monotonically increasing function starting at zero at time zero and increasing to unity as time approaches infinity. A component that is nonrepairable has a repair distribution of zero.

It is also appropriate to define a repair density function, as follows.

$$g(t) = \frac{dG(t)}{dt} \tag{24.104}$$

The repair density is described more precisely as follows.

The quantity $g(t)dt$ is the probability that the component repair is completed during the interval $[t, t + dt)$, given that the component failed at time zero.

We may also define the repair rate $n(t)$, as follows.

$$n(t) = \text{repair rate} \tag{24.105}$$

The repair rate is the probability that the component is repaired per unit time at time t given that the component failed at time zero and has been failed to time t.

When the repair rate is not a function of time, it is simply designated as n. In this case, the component has the same probability of being repaired irrespective of the time of failure. A nonrepairable component has a repair rate of zero.

It is also appropriate to consider the repair time T as a random variable. Since it is not possible to predict this time, we compute the mean time to repair as the first moment of the repair density.

$$MTTR = r = \int_0^\infty t g(t) dt \tag{24.106}$$

This parameter is often referred to in the literature as the *MTTR*, or simply as r.

24.5.3 The Whole Process

The whole process consists of repeated cycles of failure and repair; that is, the repair-to-failure process and the failure-to-repair process are repeated indefinitely. This concept is introduced in [7] as means of presenting the total system process in an orderly manner. We begin by assuming that a component jumps to the normal state at time $t = 0$, or it is as good as new initially. As time progresses, different components fail and are repaired repeatedly. At any given time, or during any given interval, this will usually find some components working and others failed, and undergoing repair. This means that, in order to characterize the whole process, a new treatment of the meaning of time is required, since the previous treatment of the two processes both considered the time to start at time zero.

For the whole process it is convenient to think in terms of the *availability* of the system, $A(t)$, where availability is defined as follows.

The availability is defined as the probability that the component is normal at time t, given that it was as good as new at time zero.

Availability is the ensemble average of those components that are working at time t. It is, therefore, greater than the reliability, which is unity immediately after repair, but monotonically approaches zero from that point onward. Thus, we can write, for repairable components,

$$A(t) \geq R(t) \tag{24.107}$$

For nonrepairable components, the two quantities are equal.

Another quantity that is useful in characterizing the whole process is the unavailability $U(t)$, which is defined as follows.

The unavailability $U(t)$ is the probability that a component is in the failed state at time t, given that it jumped into the normal state at time zero.

The unavailability is the ensemble average of those components that are failed at time t. Now, we know that the reliability and failure distribution are related mathematically. Since the component is either available or unavailable, the probability of one or the other is unity.

$$A(t) + U(t) = 1 \tag{24.108}$$

We also know from (24.84) that

$$R(t) + F(t) = 1 \tag{24.109}$$

Combining (24.102) through (24.104), we find that, for repairable systems

$$F(t) \geq U(t) \tag{24.110}$$

For nonrepairable components, the equality holds.

24.5.3.1 The Conditional Failure Intensity. We now define an important parameter, which is called the conditional failure intensity of the whole process [7].

$$\lambda(t) = \text{conditional failure intensity} \tag{24.111}$$

> The conditional failure intensity is the probability that the component fails per unit time at time t, given that it is in the normal state at time zero and remains normal until time t.

This parameter is further described as follows. The quantity $\lambda(t)dt$ is the probability that the component fails during the interval $[t, t + dt)$, given that the component was as good as new at time zero and normal at time t. Note that this is very similar to the quantity $h(t)dt$, but the two are not identical. The term $\lambda(t)dt$ requires the continuation of the normal state to time t, i.e., there may be no failure in the interval $[0, t]$.

The following special cases are of interest. For the *general case* we must conclude that

$$\lambda(t) \neq h(t) \qquad \text{For the general case} \tag{24.112}$$

For a nonrepairable component,

$$\lambda(t) = h(t) \qquad \text{Nonrepairable component} \tag{24.113}$$

For components with a constant failure rate λ we may write

$$\lambda(t) = \lambda \qquad \text{Constant failure rate} \tag{24.114}$$

24.5.3.2 The Unconditional Failure Intensity. The conditional failure intensity is conditional on the health of the component at time t. If this condition is not considered, we can define the unconditional failure intensity, as follows [7].

$$w(t) = \text{unconditional failure intensity} \tag{24.115}$$

> The unconditional failure intensity is the probability that a component fails per unit time at time t, given that it jumped into the normal state at time zero.

Stated another way, the quantity $w(t)dt$ is the probability that the component fails during the interval $[t, t+dt)$, given that the component was as good as new at time zero. If the component is not repairable, the unconditional and conditional failure intensities are equal.

Note that both the unconditional and conditional failure intensities are failures per unit time, but they assume different populations. Both assume that the component is as good as new at time zero, but the conditional intensity adds the qualification that the component still be normal at time t. An example of the difference in definitions is shown in Figure 24.16.

From Figure 24.16, we may write the following.

$$\begin{aligned} \lambda(t)dt &= \frac{xdt}{M} \\ w(t)dt &= \frac{xdt}{N} \end{aligned} \tag{24.116}$$

The unconditional failure intensity is often referred to as the *frequency* of failure [4]. This concept is discussed further in the next chapter.

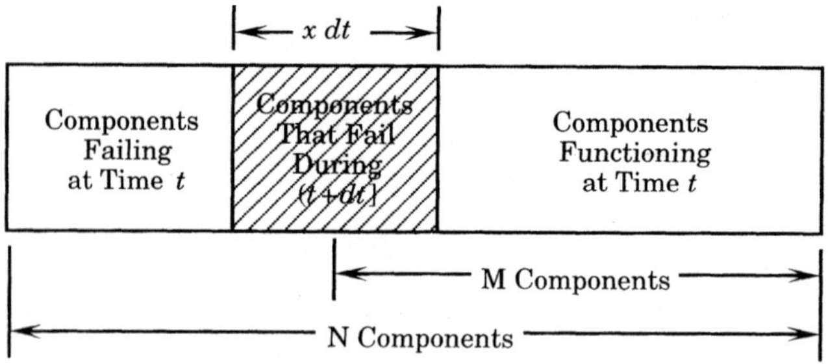

Figure 24.16 Definition of conditional failure intensity $\lambda(t)$ and unconditional failure intensity $w(t)$ [7].

24.5.3.3 The Expected Number of Failures (ENF). The expected number of failures is denoted as follows.

$$W(t, t + dt) = \text{expected number of failures} \tag{24.117}$$

$W(t)$ is the expected number of failures during $[t, t + dt)$, given that the component jumped into the normal state at time zero.

This parameter is most useful if integrated over an interval. Thus, the integral

$$W(t_1, t_2) = \int_{t_1}^{t_2} w(t)dt \tag{24.118}$$

gives the number of failures on the interval $[t_1, t_2)$. Two values of interest are the following.

$$\begin{aligned} W(0, t) &= \int_0^t w(t)dt \\ &= \begin{cases} F(t) & \text{nonrepairable component} \\ \to \infty & \text{large } t\text{, repairable component} \end{cases} \end{aligned} \tag{24.119}$$

24.5.3.4 The Conditional Repair Intensity. The conditional repair intensity is defined as follows.

$$\mu(t) = \text{conditional repair intensity} \tag{24.120}$$

The conditional repair intensity is the probability that a component is repaired per unit time at time t, given that it jumped into the normal state at time zero and is failed at time t.

The conditional repair intensity $\mu(t)$ is not the same as the repair rate $n(t)$. Their relationship is similar to the relationship between $\lambda(t)$ and $h(t)$. In particular, we can write

$$\begin{aligned} \mu(t) &= n(t) = 0 && \text{nonrepairable components} \\ \mu(t) &= \mu = h && \text{constant repair rate } h \end{aligned} \tag{24.121}$$

24.5.3.5 The Unconditional Repair Intensity. The unconditional repair intensity is defined as follows.

$$v(t) = \text{unconditional repair intensity} \tag{24.122}$$

The unconditional repair intensity is the probability that the component is repaired per unit time at time t, given that it jumped into the normal state at time zero.

The conditional and unconditional repair intensities are different in exactly the same sense that the conditional and unconditional failure intensities differ, i.e., they involve different populations.

24.5.3.6 The Expected Number of Repairs. The expected number of repairs is defined as follows.

$$V(t, t+dt) = \text{expected number of repairs} \tag{24.123}$$

The expected number of repairs during $[t, t+dt)$, given that the component jumped into the normal state at time zero.

The integral over a given interval gives the number of repairs over that interval.

$$V(t_1, t_2) = \int_{t_1}^{t_2} v(t)dt \tag{24.124}$$

The following special cases are of interest.

$$\begin{aligned} V(0, t) &= \int_0^t v(t)dt \\ &= \begin{cases} 0 & \text{for nonrepairable components} \\ \infty & \text{for large } t \end{cases} \end{aligned} \tag{24.125}$$

24.5.3.7 The Mean Time Between Failures. The mean time between failures is the sum of the MTTF and MTTR.

$$\text{MTBF} = \text{MTTF} + \text{MTTR} = m + r \tag{24.126}$$

The MTBF is sometimes called the "cycle time," which implies that the component cycles up and down between working and nonworking states, with an average total cycle time of MTBF. The inverse of this cycle time is the period.

24.5.3.8 Summary of Whole Process Variables. Table 24.2 summarizes the relations among the variables for the repair-to-failure process.

TABLE 24.2 Repair-to-Failure Process Probabilistic Parameters Relations [7]

Failure rate $h(t)$	General $h(t)$	(1) $R(t) + F(t) = 1$ (2) $R(0) = 1,\ R(\infty) = 0$ (3) $F(0) = 0,\ F(\infty) = 1$ (4) $f(t) = \dfrac{dF(t)}{dt}$ (5) $f(t)dt = F(t + dt) - f(t)$ (6) $F(t) = \int_0^t f(u)du$ (7) $R(t) = \int_0^\infty F(u)du$	(8) $\text{MTTF} = \int_0^\infty tf(t)dt$ (9) $h(t) = \dfrac{f(t)}{1 - F(t)}$ (10) $R(t) = \exp\left[-\int_0^t h(u)du\right]$ (11) $F(t) = 1 - \exp\left[-\int_0^t h(u)du\right]$ (12) $f(t) = h(t)\exp\left[-\int_0^t h(u)du\right]$
	Const. $h(t) = \lambda$	(13) $R(t) = e^{-\lambda t}$ (15) $f(t) = \lambda e^{-\lambda t}$	(14) $F(t) = 1 - e^{-\lambda t}$ (16) $\text{MTTF} = m = {}^1/_\lambda$

Table 24.3 summarizes the relationship among the variables for the failure-to-repair process.

Table 24.4 summarizes the relationship among the variables for the whole process.

24.5.4 Constant Failure and Repair Rate Model

An important special case that is often applicable in the analysis of physical system is the case that considers the failure and repair rates to be constant parameters. This special case

TABLE 24.3 Failure-to-Repair Process Probabilistic Parameters [7]

Failure rate $h(t)$	General $h(t)$	(1) $G(t) = g(t) = n(t) = 0$ For nonrepairable item (2) $G(0) = 0, G(\infty) = 1$ (3) $g(t) = \frac{dG(t)}{dt}$ (4) $g(t) = G(t + dt) - G(t)$ (5) $G(t) = \int_0^t g(u)du$	(6) $G(t_2) - G(t_1) = \int_{t_1}^{t_2} g(u)du$ (7) $\text{MTTR} = \int_0^\infty tg(t)dt$ (8) $n(t) = \frac{g(t)}{1 - G(t)}$ (9) $G(t) = 1 - \exp\left[-\int_0^t m(u)du\right]$ (10) $g(t) = n(t)\exp\left[-\int_0^t m(u)du\right]$
	Const. $h(t) = \mu$	(11) $G(t) = 1 - e^{-\mu t}$ (13) $\text{MTTR} = r = 1/\mu$	(12) $g(t) = \mu e^{-\mu t}$

TABLE 24.4 Probabilistic Parameter Relations for the Whole Process [7]

	Repairable	Non-repairable
Fundamental relations	(1) $A(t) + U(t) = 1$ (2) $A(t) > R(t)$ (3) $U(t) < F(t)$ (4) $w(t) = f(t) + \int_0^t f(t-u)v(u)du$ (5) $v(t) = \int_0^t g(t-u)w(u)du$ (6) $W(t, t+dt) = w(t)dt$ (7) $V(t, t+dt) = v(t)dt$ (8) $W(t_1, t_2) = \int_{t_1}^{t_2} w(u)du$ (9) $W(t_1, t_2) = \int_{t_1}^{t_2} v(u)du$ (10) $U(t) = W(0, t) - V(0, t)$ (11) $\lambda(t) = \frac{w(t)}{1 - U(t)}$ (12) $\mu(t) = \frac{v(t)}{U(t)}$	(1) $A(t) + U(t) = 1$ (2) $A(t) = R(t)$ (3) $U(t) = F(t)$ (4) $w(t) = f(t)$ (5) $v(t) = 0$ (6) $W(t, t+dt) = w(t)dt$ (7) $V(t, t+dt) = 0$ (8) $W(t_1, t_2) = \int_{t_1}^{t_2} w(u)du = F(t_2) - F(t_1)$ (9) $V(t_1, t_2) = 0$ (10) $U(t) = W(0, t) = F(t)$ (11) $\lambda(t) = \frac{w(t)}{1 - U(t)}$ (12) $\mu(t) = 0$
Stationary values	(13) MTBF = MTTF + MTTR (14) $0 < A(\infty) < 1, 0 < U(\infty) < 1$ (15) $0 < w(\infty) < 1, 0 < v(\infty) < 1$ (16) $w(\infty) = v(\infty)$ (17) $W(0, \infty) = \infty, V(0, \infty) = \infty$	(13) $\text{MTBF} = \infty$ (14) $A(\infty) = 0, U(\infty) = 0$ (15) $w(\infty) = 0, v(\infty) = 0$ (16) $w(\infty) = v(\infty) = 0$ (17) $W(0, \infty) = 1, V(0, \infty) = 0$
Remarks	(18) $w(t) \neq \lambda(t), v(t) \neq \mu(t)$ (19) $\lambda(t) \neq h(t), \mu(t) \neq n(t)$ (20) $w(t) \neq f(t), v(t) \neq g(t)$	(18) $w(t) \neq \lambda(t), v(t) = \mu(t) = 0$ (19) $\lambda(t) = h(t), \mu(t) = n(t) = 0$ (20) $w(t) = f(t), v(t) = g(t) = 0$

assumes the following constant parameters.

$$\begin{aligned} h(t) &= \lambda \\ n(t) &= \mu \end{aligned} \tag{24.127}$$

If we substitute these constant values into (10-12) of Table 24.1, we compute the following.

$$\begin{aligned} F(t) &= 1 - e^{-\lambda t} \\ R(t) &= e^{-\lambda t} \\ f(t) &= \lambda e^{-\lambda t} \end{aligned} \tag{24.128}$$

Then the MTTF is computed from (8) of Table 24.2 to be

$$\text{MTTF} = m = \int_0^{\infty} t\lambda e^{-\lambda t} dt = \frac{1}{\lambda} \tag{24.129}$$

In a similar way we solve for the repair-to-failure parameters, with the following result.

$$\begin{aligned} G(t) &= 1 - e^{-\mu t} \\ g(t) &= \mu e^{-\mu t} \\ \text{MTTR} &= r = \frac{1}{\mu} \end{aligned} \tag{24.130}$$

Knowing these probabilistic variables of both failure and repair processes, we can solve for the unconditional failure intensities and the expected number of failures and repairs. The results of this analysis for a constant failure and repair rates are summarized in Table 24.5.

TABLE 24.5 Results for the Constant Failure and Repair Rate Model [7]

	Repairable	**Non-repairable**
Repair to failure	$h(t) = \lambda$ $R(t) = e^{-\lambda t}$ $F(t) = 1 - e^{-\lambda t}$ $f(t) = \lambda e^{-\lambda t}$ $\text{MTTF} = 1/\lambda$	$h(t) = \lambda$ $R(t) = e^{-\lambda t}$ $F(t) = 1 - e^{-\lambda t}$ $f(t) = \lambda e^{-\lambda t}$ $\text{MTTF} = 1/\lambda$
Failure to repair	$n(t) = \mu$ $G(t) = 1 - e^{-\mu t}$ $g(t) = \mu e^{-\mu t}$ $\text{MTTR} = 1/\mu$	$n(t) = \mu$ $G(t) = 1 - e^{-\mu t}$ $g(t) = \mu e^{-\mu t}$ $\text{MTTR} = 1/\mu$
Dynamic behavior of the whole process	$U(t) = \frac{\lambda}{\lambda + \mu}\left[1 - e^{-(\lambda + \mu)t}\right]$ $A(t) = \frac{\mu}{\lambda + \mu} + \frac{\lambda}{\lambda + \mu} e^{-(\lambda + \mu)t}$ $w(t) = \frac{\lambda\mu}{\lambda + \mu} + \frac{\lambda^2}{\lambda + \mu} e^{-(\lambda + \mu)t}$ $v(t) = \frac{\lambda\mu}{\lambda + \mu}\left[1 - e^{-(\lambda + \mu)t}\right]$ $W(0, t) = \frac{\lambda\mu}{\lambda + \mu} t + \frac{\lambda^2}{(\lambda + \mu)^2}\left[1 - e^{-(\lambda + \mu)t}\right]$ $V(0, t) = \frac{\lambda\mu}{\lambda + \mu} t - \frac{\lambda\mu}{(\lambda + \mu)^2}\left[1 - e^{-(\lambda + \mu)t}\right]$ $\frac{dU(t)}{dt} = -(\lambda + \mu)U(t) + \lambda,\ U(0) = 0$	$U(t) = 1 - e^{-\lambda t} = F(t)$ $A(t) = e^{-\lambda t} = R(t)$ $w(t) = \lambda e^{-\lambda t}$ $v(t) = 0$ $W(0, t) = 1 - e^{-\lambda t} = F(t)$ $V(0, t) = 0$ $\frac{dU(t)}{dt} = -(\lambda + \mu)U(t) + \lambda,$ $U(0) = 0$
Stationary value of the whole process	$U(\infty) = \frac{\lambda}{\lambda + \mu} = \frac{r}{m + r}$ $A(\infty) = \frac{\mu}{\lambda + \mu} = \frac{m}{m + r}$ $w(\infty) = \frac{\lambda\mu}{\lambda + \mu} = \frac{1}{m + r}$ = frequency $v(\infty) = \frac{\lambda\mu}{\lambda + \mu} = w(\infty)$ $\frac{U(t)}{U(\infty)} = 0.632$ for $t = \frac{1}{\lambda + \mu}$ $0 = -(\lambda + \mu)U(\infty) + \lambda$	$U(\infty) = 1$ $A(\infty) = 0$ $w(\infty) = 0$ $v(\infty) = 0 = w(\infty)$ $\frac{U(t)}{U(\infty)} = 0.632$ for $t = \frac{1}{\lambda}$ $0 = -\lambda U(\infty) + \lambda$

REFERENCES

[1] Pfeiffer, P. E., *Concepts of Probability Theory*, McGraw-Hill Book Co., New York, 1965.

[2] Papoulis, A., *Probability, Random Variables, and Stochastic Processes*, Second Edition, McGraw-Hill Book Co., New York, 1984.

[3] Von Alven, W. H., Ed., *Reliability Engineering*, Prentice-Hall, Englewood Cliffs, NJ, 1964.

[4] Billinton, R., and R. N. Allan, *Reliability Analysis of Engineering Systems: Concepts and Techniques*, Plenum Press, New York, 1983.

[5] Billinton, R., and R. N. Allan, *Reliability Analysis of Power Systems*, Plenum Press, New York, 1984.

[6] Dhillon, B. S., and C. Singh, *Engineering Reliability, New Techniques and Applications*, John Wiley & Sons, New York, 1981.

[7] Kumamoto, H., and E. J. Henley, *Probabilistic Risk Assessment and Management for Engineers and Scientists*, Second Edition, IEEE Press, Piscataway, NJ, 1996.

[8] Peebles, P. Z., Jr., *Probability, Random Variables, and Random Signal Principles*, Second Edition, McGraw-Hill Book Co., New York, 1987.

[9] *Tables of the Cumulative Binomial Probabilities*, Ordnance Corps Pamphlet ORDP-1, U. S. Department of Commerce, Office of Technical Services, Washington, DC, 1952.

[10] Parzen, E., *Stochastic Processes*, Holden-Day, San Francisco, 1962.

[11] Shooman, M. L, *Probabilistic Reliability: An Engineering Approach*, McGraw-Hill Book Co., 1968.

[12] Davenport, W. B., Jr., *Random Processes*, McGraw-Hill Book Co., New York, 1970.

[13] Wadsworth, G. P., *Introduction to Probability and Random Variables*, McGraw-Hill Book Co., New York, 1960

[14] Bolz, R. E., and G. L. Tuve, Editors, *Handbook of Tables for Applied Engineering*, Chemical Rubber Co., Cleveland, OH, 1970.

[15] Tobias, P. A., and D. Trindade, *Applied Reliability*, Van Nostrand Reinhold Company, New York, 1986.

[16] MIL-STD-721C, "Military Standard Definitions of Terms for Reliability and Maintainability," June 12, 1981, U.S. Government Printing Office: 1981-023-4380.

[17] IEEE Std 100-1988, *IEEE Standard Dictionary of Electrical and Electronics Terms*, John Wiley & Sons, New York, 1988.

[18] Kececioglu, D., *Reliability Engineering Handbook, Volumes 1 and 2*, Prentice-Hall, Englewood Cliffs, NJ, 1991.

PROBLEMS

24.1 An experiment consists of tossing a coin three times and recording the resulting head or tail by the notation H or T, respectively.

(a) How many elements are there in the probability space S?

(b) What is the probability of tossing a tail on the second toss?

(c) What is the probability of tossing a head on the first toss?

24.2 Distinguish between an elementary *event* and an elementary *outcome* of an experiment.

24.3 Consider a toss of a pair of dice, with the outcome consisting of a 2 and a 4, giving a total face value of 6. Distinguish between the event and outcome related to this experimental result.

24.4 What is the probability of tossing an even number with one die?

24.5 How many possible outcomes exist in the tossing of a pair of dice?

24.6 What is the probability of tossing a 7 when tossing a pair of dice?

24.7 An electrical manufacturer purchases components from three different suppliers, which can be considered as providing products that can be considered independent. The manufacturer has prepared specifications for the acceptability of the components that can be determined by testing. Let A be the event that a component meets specification, and let B_k be the event that the component comes from supplier number $k (k = 1, 2, \text{or } 3)$. Write out an expression for the probability that a given component meets the specifications.

24.8 What is the probability of drawing a four or a club from a deck of cards in a random draw from a well-shuffled deck?

24.9 What is the probability of drawing a 4, given that a club has been drawn? Write your solution as a conditional probability.

24.10 A university department seeks to hire two new faculty members. One applicant is an experienced teacher who is a leader of research in his field. The second is a young lady who is just completing her graduate studies and who is perceived as wanting very much to work with the more experienced applicant. It is estimated that the older applicant has a 50–50 chance of accepting an offer. If he accepts the offer, the younger person has a 90% chance of accepting, but if he does not, then this chance drops to an estimated 40%. (a) What is the probability that the young lady will accept an offer, and (b) what is the probability that both will accept?

24.11 A protective device in a power plant is designed such that it has a high probability of operating successfully when a given dangerous plant failure occurs and it also has a high probability of not operating when the given dangerous plant failure does not occur. Let F be the failure event and let W be the event where the protective device works. Also let $\Pr(F/W) = p$ and $\Pr(\bar{F}/\bar{W}) = q$. It is desired to make p and q as near to unity as possible. These conditional probabilities are based on a given knowledge of the event of the dangerous plant failure. For brevity, let $\Pr(F) = p_F$, which is assumed to be small.
(a) If $p = q = 0.98$ and $p_F = 0.001$ compute $\Pr(F/W) = p$.
(b) Compute $\Pr(F\bar{W})$.
(c) What conditions must be satisfied to make $\Pr(F/W)$ greater?

24.12 Statistics show that, when thunderstorms are present, the probability of a transmission line faulting is 75%, but this probability falls to only 10% during fair weather. Moreover, weather statistics show that the probability of thunderstorms is 25%. (These are fictitious statistics.) What is the probability that a thunderstorm is present, given that the transmission line is faulted?

24.13 Test the function shown in Figure P24.13 to see if it can be classified as a valid density function.

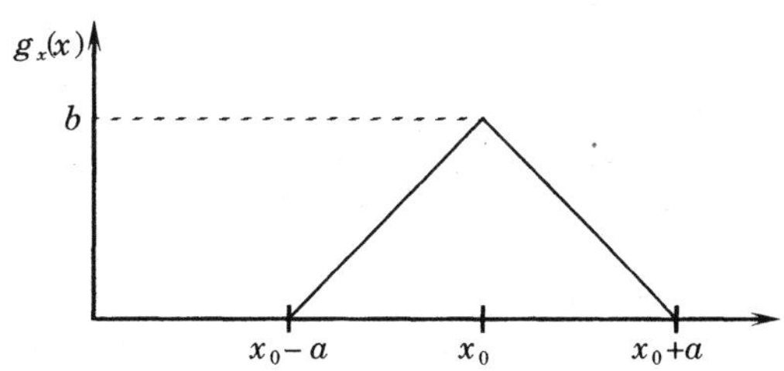

Figure P24.13 A possible density function.

24.14 Pascal's Triangle has been devised to simplify the determination of the combination of n things taken r at a time, which is written as

$$_nC_r = \binom{n}{r} = \frac{n!}{r!(n-r)!}$$

Investigate Pascal's Triangle, construct a sample triangle, and explain its usage.

24.15 A small island is served by a power system consisting of only three generators, two 30 MW units and one 50 MW unit, giving a total capacity of 110 MW when all units are in service. Use the binomial distribution to determine the probability of the following.

(a) For the 30 MW units, find the probability of zero, 1, and 2 units out of service if their probability of outage is $q = 0.02$.

(b) Make the same calculation for the 50 MW unit, with $q = 0.02$.

(c) Combine the calculations of (a) and (b) to construct a probability outage table with stepwise increasing values of outage and their associated probabilities.

24.16 Consider the problem of cars arriving at a service station for gasoline and assume that their arrivals are Poisson distributed. The arrivals are estimated to occur at a rate of 50/hr and the station has only one pump. If all cars require 1 minute for refueling, what is the probability that a waiting line will occur at the pump?

24.17 A small factory has installed 10 identical machines that are used in a manufacturing process. The machines are loaded by workers and, once ready, are turned on to process the product. Each machine runs on the average of 12 minutes per hour. If the machines are placed into service randomly, we can model their operation as a binomial process, in particular with the probability of exactly k machines being in service at the same time express as a binomial distribution. What is the probability of seven or more machines requiring power service at the same time? If the supply transformer has a rating that will accommodate only six machines operating simultaneously, what is the probability of overloading the supply transformer?

24.18 Verify the result given by (24.58).

24.19 Verify the results given by (24.60).

24.20 A resistor manufacturer produces 100 ohm resistors with a tolerance of $\pm 10\%$. Assuming that resistance values are normally distributed with a standard deviation of 5 ohms, what is the probability of any particular resistor being out of tolerance?

24.21 The pickup time for a digital relay A is determined by thousands of tests to be 27 ms with a standard deviation of 5 ms. Similar tests of relay B give a pickup time of 30 ms with a standard deviation of 2 ms. Which is the more reliable relay for situations where a clearing time of 30 ms is required?

24.22 Repeat problem 24.21 if the required pickup time is relaxed from 30 ms to 34 ms. Comment on the difference in the results of this and the previous problem.

24.23 The conductor specified by the design department of a utility for a new transmission line has a published tensile strength of 1100 pounds with a standard deviation of 120 pounds.

(a) What portion of wire samples will survive a tension of 1300 pounds?

(b) What is the probability of a conductor breaking if the average tension is 800 pounds?

(c) What portion of conductors will survive a tension of 850 but not 1320 pounds?

24.24 A population of capacitors is known to fail according to a Weibull distribution with characteristic life $\alpha = 20{,}000$ power-on hours. Evaluate the probability that a new capacitor will fail by 100, 1000, 20,000, and 30,000 hours, for the cases where the shape parameter β is 0.5, 1.0, or 2.0 [16]. Comment on the computed results.

24.25 The probability distribution has been described by the expression

$$e^b e^{-b} = 1$$

Use this expression to compute the probability of exactly k failures will occur in a system.

24.26 Verify (24.77) and (24.78). Use the state enumeration approach.

24.27 For the capacitors described in problem 24.24, calculate the failure rates at these times for the three shape parameters specified.

24.28 Verify (24.110).

25

Reliability Analysis

Chapter 24 provides the basic mathematical foundations of reliability. This chapter presents several techniques that are useful in the analysis of physical systems. The first section treats simple methods that are intuitive but still useful in many engineering applications. This is followed by methods that are more powerful, providing the capability of analyzing complex systems, such as protective systems.

25.1 RELIABILITY BLOCK DIAGRAMS

In all physical systems, the reliability is dependent on the system configuration, that is, on the arrangement and connection of the various components that make up the system design. The physical arrangement leads to an evaluation of the logic that describes a given failure mode. This logic can usually be described in general ways, and equations can be written to compute the reliability of that logic. It should be emphasized that it is the logic that is being analyzed rather than the physical arrangement.

Reliability block diagrams are graphical representations of a physical system, the operation of which can often be represented as a network of objects with each object representing a certain operation or function. For example, we might represent a digital controller as a series connection of two objects, one representing the digital hardware and the other the software. Since the failure of either the hardware or the software would result in an inoperable system, we would conclude that these objects are logically in series. The analogy to a series electrical network is rather obvious, since two electrical components in series must both be *working* for the series path to work. Note that we are attempting to represent the logic of the system, not necessarily its physical connection of hardware (or software) items. Such graphs that are intended to represent the reliability of a system are called network diagrams or, more commonly, *reliability block diagrams* (*RBD*).

A group of items connecting two points in a reliability block diagram is called a *path*. A path represents one means of accomplishing a given task. Some paths are redundant, in which case the failure of one path does not result in the failure of the system. If a path is redundant, it has at least one functional duplicate such that the failure of a path does not cause system failure.

The basic items in a path are called *elements*, which may be components, circuits, items of equipment, or even entire systems or subsystems. If an element is redundant, it has at least one functional duplicate.

The notation used in reliability block diagrams is directed toward finding the reliability of the system represented by the diagram. The following terminology may be helpful.

$$\begin{aligned} R &= \text{reliability, or probability of system success} \\ Q &= \text{unreliability, or probability of system failure} \end{aligned} \tag{25.1}$$

$$\begin{aligned} p &= \text{reliability of an element} \\ q &= \text{failure of an element} \end{aligned} \tag{25.2}$$

In some cases, it is convenient to identify elements by letters of the alphabet, such as A, B, C, etc. In this case, we can write

$$\begin{aligned} \Pr(A) &= \text{the event: success of } A = \Pr\{A \text{ works}\} \\ \Pr(\bar{A}) &= \text{the event: failure of } A = \Pr\{A \text{ fails}\} \end{aligned} \tag{25.3}$$

In thinking of the *RBD* as an analog of an electric network, the failure of interest, in some cases, is to determine if a path becomes *open*, with an open circuit or removal of the path being interpreted as a failure. Clearly, there are other modes of failure for an electric network. It could fail shorted to ground or reference, in which case a signal is not transmitted at all, similar to an open path. However, in the *RBD*, failure is represented as the *removal* of the element, which is representative of an *open* failure in an electric network analogy. Reliability block diagrams have been largely replaced by fault trees, because of the greater versatility of the fault tree in analysis. However, the *RBD* is an effective way of viewing simple hardware configurations and of analyzing their reliabilities.

25.1.1 Series Systems

Consider a system of elements $A_1, A_2, \ldots, A_n$, as shown in Figure 25.1. We call this system a *series system*, not because the elements are physically connected in series, although they might be, but because all of the elements must be working for the entire system to work. Diagrams such as Figure 25.1 are often called *network diagrams* or *reliability block diagrams*. It is important to emphasize that this is not necessarily the physical connection of elements, but a pictorial description of the system operating logic.

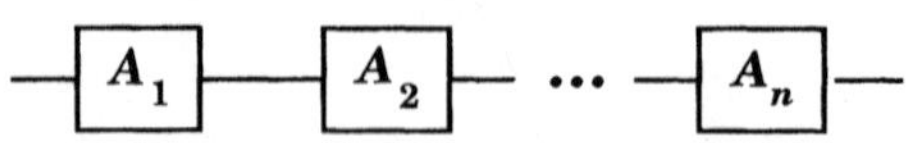

Figure 25.1 A series system of n element.

If we assume that all n elements of the series system are independent, then the entire system operates successfully only if all n elements are in the success state, i.e., all elements are working. Let A_i indicate the success of the ith unit.

Then the reliability of the series system of independent units is given a

$$\begin{aligned} R_S &= \Pr(A_1 A_2 \ldots A_n) = \Pr(A_1)\Pr(A_2)\cdots\Pr(A_n) \\ &= R_1 R_2 \cdots R_n = \prod_{i=1}^{n} R_i \end{aligned} \tag{25.4}$$

Clearly, for the series system, the failure of any one element causes the failure of the entire system.

EXAMPLE 25.1

Consider the simple electric circuit shown in Figure 25.2.

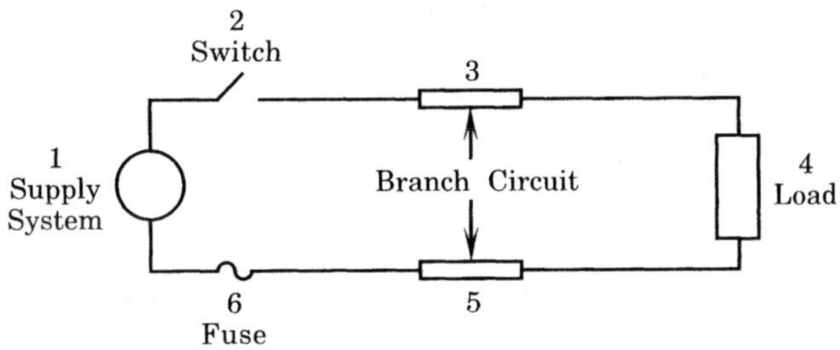

Figure 25.2 A single-phase circuit supplying a load.

The load is served by a single-phase circuit with a switch for controlling the load and a fuse for the circuit protection. Consider only failure modes that cause the circuit to fail open, that is, ignore short circuits and consider only open circuits.

Solution

The components in the circuit can be readily enumerated:

1. Supply system
2. Switch
3. Branch circuits
4. Load
5. Fuse

An open circuit in any of these components will block the flow of current and cause service to the load to fail. We define the failure events as follows.

$$E_i = \{\text{Component } i \text{ open}\} \tag{25.5}$$

Then

$$\Pr\{E_i\} = \Pr\{\text{Component } i \text{ Open}\} \tag{25.6}$$

and the reliability of any component can be expressed as

$$R_i = 1 - \Pr\{E_i\} \tag{25.7}$$

We assume that the component failures are independent. Clearly, there may be dependent failure modes, such as overheating of the circuit wiring causing eventual melting of the fuse. We shall ignore this type of failure with the assumption of independence of the five components. Then the reliability of the circuit is computed as follows.

$$R = \prod_{1}^{6} R_i \tag{25.8}$$

This circuit has several different ways it can fail, with one failure mode associated with each component. If one of the components has a high probability of failing open, the reliability will be low because the reliability of that one element is low. The reliability of the series system can be no higher than the reliability of the least reliable component. This example could be expanded to consider short-circuit failure modes as well. ■

25.1.2 Parallel Systems

For some systems, certain elements are in parallel logic. As a general case, consider the system shown in Figure 25.3, where n elements have a parallel logic. System success occurs if any one of the n elements works.

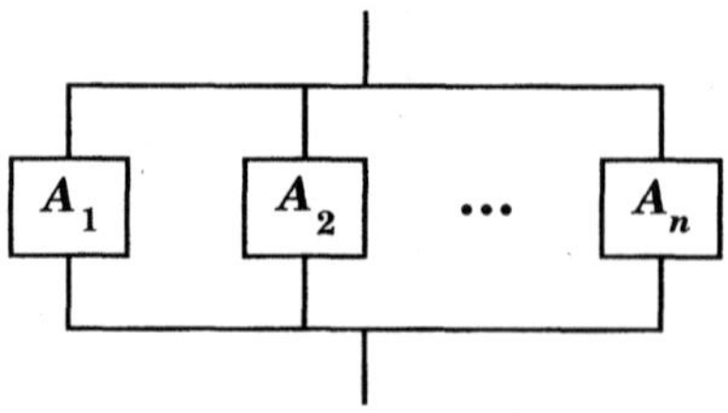

Figure 25.3 A parallel system of n elements.

The probability of success of the parallel system is given by the probability of the union of the n successful events

$$\begin{aligned} R_P &= \Pr(A_1 + A_2 + \cdots + A_n) \\ &= [\Pr(A_1) + \Pr(A_2) + \cdots + \Pr(A_n)] \\ &\quad - \left[\Pr(A_1 A_2) + \Pr(A_2 A_3) + \cdots + \Pr\left(A_i A_j\right)_{i \neq j}\right] \\ &\quad + \cdots + (-1)^n \Pr(A_1 A_2 \ldots A_n) \end{aligned} \tag{25.9}$$

For the parallel logic, it is easier to compute the probability of failure rather than the probability of success. Stated in words, we can say that *all* elements must be failed for the system to be in the failed state.

$$R_P = 1 - \Pr\left(\bar{A}_1 \bar{A}_2 \ldots \bar{A}_n\right) = 1 - Q_P = 1 - \prod_{i=1}^{n} Q_i \tag{25.10}$$

Clearly, if any one of the parallel elements is working, the system works.

Many complex systems can be modeled as an interconnection of series and parallel logic subsystems. One can derive formulas for series-parallel, parallel-series, and other combinations.

EXAMPLE 25.2

Consider a microwave communications system used in conjunction with a protective system. Since the reliability of the transmission of information is very important, the transmitters at a particular site are redundant. If each transmitter has a probability of failure given by $\Pr(F_i)$, write an expression for the reliability of the system.

Solution

Using (25.10), we write the following expression.

$$R = 1 - \Pr(F_1 F_2) \tag{25.11}$$

This expression says that the system must experience failure of both transmitters to cause the system to fail. If the transmitters are completely independent of each other, with independent power supplies and with separate antennas and other auxiliary equipment, then the reliability expression can be written as follows.

$$R = 1 - \Pr(F_1)\Pr(F_2) \tag{25.12}$$

The designer of the system should consider ways to avoid overlapping failures of the two transmitters. For example, using transmitters of different design, supplied by different manufacturers, will avoid simultaneous failures due to a flaw in the design. Also note that, when one of the transmitters is down for inspection, the redundancy is lost. ■

25.1.3 Series-Parallel and Parallel-Series Systems

Systems can be organized in such a way that several parallel systems are connected in series, as shown in Figure 25.4. This connection is called a series-parallel system. Each parallel system is analyzed using (25.10) to compute the success probability of that parallel connection. These probabilities are then considered as a series connection and analyzed using (25.4).

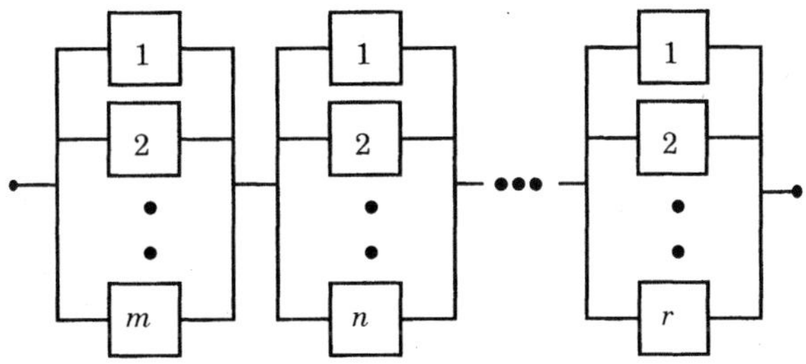

Figure 25.4 A series-parallel connection.

A special case of the series-parallel connection can be defined to consist of exactly m identical elements of reliability p in each parallel group and with n such groups connected in series. For this special case, the reliability is given as follows [1].

$$R = \left[1 - (1 - p)^m\right]^n \tag{25.13}$$

For constant n and p, the reliability R increases as m increases, but when m and p are fixed, R decreases as n increases. It has been noted that reliability improves slowly for $p > 0.8$ as m increases beyond 3 [1].

Another variation is to have several series connections connected in parallel, to form a "parallel-series" connection, as shown in Figure 25.5. This is also analyzed in two steps, first for the individual series connections, and finally considering the parallel connection of the interim results for the series combinations.

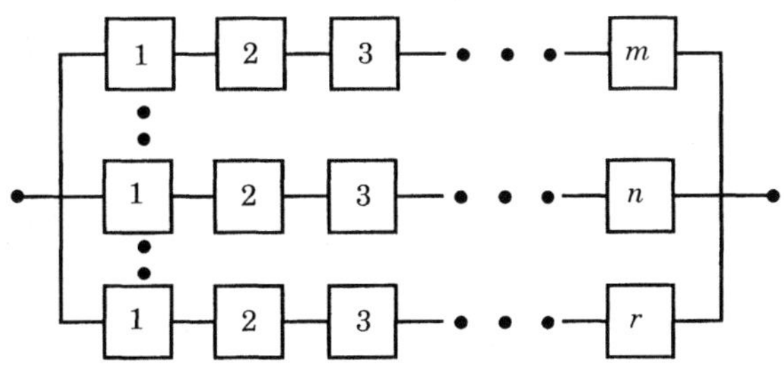

Figure 25.5 A parallel-series connection.

A special case of this arrangement is to have n identical elements of reliability p in each series group and with m such groups in parallel. For this arrangement, the reliability is computed as follows.

$$R = 1 - \left(1 - p^n\right)^m \tag{25.14}$$

With n and p fixed in this arrangement, R increases as the number of parallel paths m increases, but for m and p fixed, R decreases as n increases. For $p < 0.5$, however, increasing m results in negligible increase in R.

25.1.4 Standby Systems

Figure 25.5 shows a parallel arrangement of components, which is sometimes called a redundant system since the system works as long as any one of the parallel paths is working. A variation of this basic parallel arrangement is to consider one of the parallel elements to be in standby, with the additional requirement that a switch be arranged to connect the standby element when the working element fails. This arrangement is shown in Figure 25.6, where the normally working element is called the "hot unit" and the standby element the "cold unit." These terms are used to indicate that, in many cases, the standby unit is not energized. This would be advantageous if the standby unit has a lower failure rate when not energized. It is assumed that the switch is controlled by a logic system that detects the failure of A and orders the switch to operate.

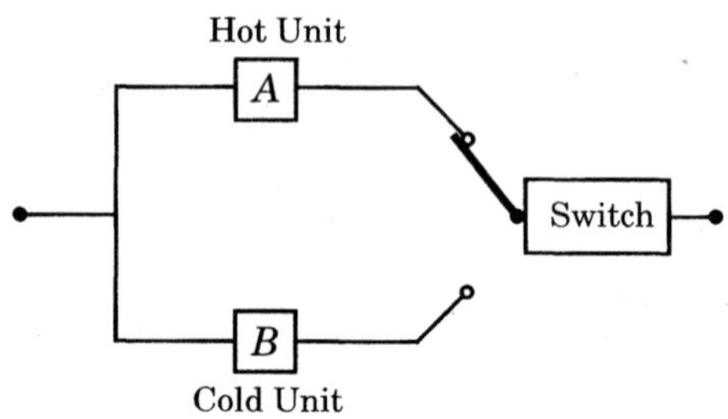

Figure 25.6 Standby redundancy.

There are two aspects to consider in this switched system. One approach is to ignore time, and work with the average long-time probabilities of failure of the components A, B, and the switch. This can take into account the different modes of switch failure, e.g., the switch operates when not required and the switch fails to operate when required.

A second approach, which is not pursued here, is to take the time of failure into account. With this more detailed approach, the failure of each element is represented by a failure density function. The standby unit can be represented by different failure densities when operating on standby. The different failure modes of the switch can have different densities. This approach is left as an exercise, but is acknowledged to be more rigorous [1].

The switch shown in Figure 25.6 may be considered to be both a decision and switching device, i.e., the switch (or some switching logic not illustrated) must determine correctly when to switch and must perform the switching operation without failure. The analysis of standby redundant system requires knowledge of the probability of correct operation of the switching element. For the general non-time-varying case, we can analyze the reliability using the conditional probability approach.

$$\begin{aligned} \Pr\{\text{failure}\} &= \Pr\{\text{failure given switching works}\} \cdot \Pr\{\text{switching works}\} \\ &\quad + \Pr\{\text{failure given switching fails}\} \cdot \Pr\{\text{switching fails}\} \end{aligned} \tag{25.15}$$

If we let the letter Q designate the probability of failure, then (25.15) may be written as

$$Q = Q_A Q_B P_S + Q_A (1 - P_S) \tag{25.16}$$

where $P_S = \Pr\{\text{switching works}\}$

EXAMPLE 25.3

Consider that the switch in Figure 25.6 is perfectly reliable and compute the reliability of the standby redundant system if the two components are identical.

Solution

The system described is the same as an equivalent single unit, which is allowed to fail once [1]. If A fails, B is switched into service so the system does not fail. The system fails only when the second failure occurs, that is, when B fails following the failure of A, where we assume that A is nonrepairable. This situation can be described by a Poisson distribution, where we write

$$\Pr\{x \text{ components fail in time } t\} = \frac{(\lambda t)^x e^{-\lambda t}}{x!} \tag{25.17}$$

Expanding, we have

$$\Pr\{\text{no components fail}\} = e^{-\lambda t}$$
$$\Pr\{\text{one component fails}\} = \lambda t e^{-\lambda t}$$

Therefore,

$$R(t) = e^{-\lambda t} + \lambda t e^{-\lambda t} = (1 + \lambda t)\, e^{-\lambda t} \tag{25.18}$$

This approach is readily extended to a standby system with two or more redundant elements. ■

25.1.5 Bridge Networks

Another type of network that sometimes occurs is the bridge network, shown in Figure 25.7. This network can also be analyzed using the *conditional probability approach* by writing the following expression. Let R_S be the reliability of the system and R_E the reliability of element E. Then we write

$$R_S = R_S\{\text{given } E \text{ is good}\}\, R_E + R_S\{\text{given } E \text{ is bad}\}\,(1 - R_E) \tag{25.19}$$

The statement that E is "good" implies that E is working and the component E can be replaced by a short circuit. However, if E is "bad," this implies that the component is failed (open) and there is no connection through the component labeled E. It is helpful to sketch the network for the two parts, as shown in Figure 25.8. The sketch for the part where E is good will show E being replaced by a short circuit and the sketch for the part where E is bad will show E replaced by an open circuit [1–3].

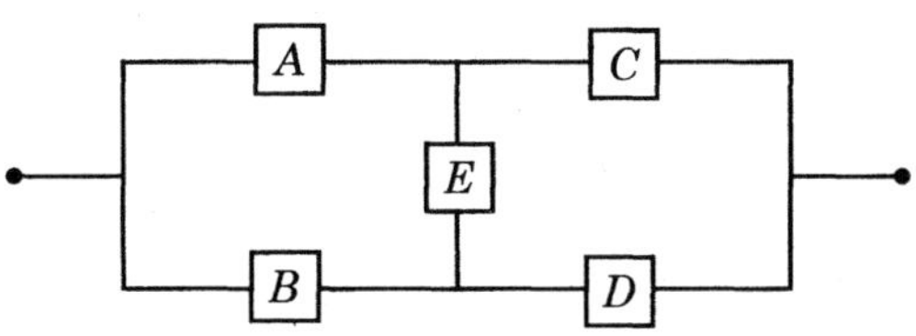

Figure 25.7 The bridge network.

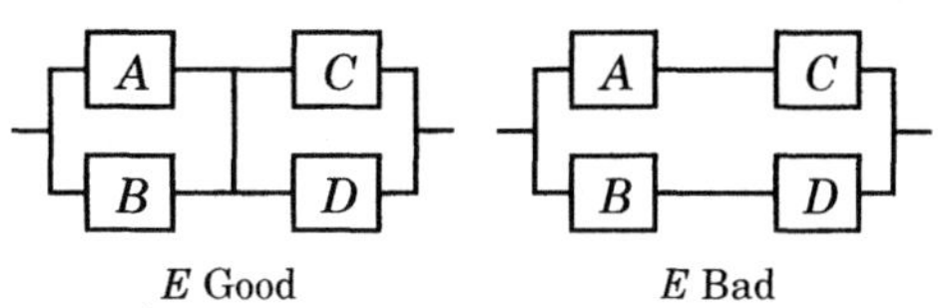

Figure 25.8 Effect of the two conditions for element E.

The solution for the first part of (25.19), when E is good, requires that we write an expression for two parallel systems. This is most easily done by writing the equation in terms of the failure of the components.

$$R_S\,\{\text{given } E \text{ is good}\} = (1 - Q_A Q_B)\,(1 - Q_C Q_D) \tag{25.20}$$

where Q_i = failure of component i

The solution for the second part of (25.19), when E is bad, requires the analysis of a parallel arrangement of two series circuits. Series circuits are most easily analyzed in terms of their success, or reliability R. Note that the system will fail only if *both* parallel paths fail. Therefore, we may write the following equation for the reliability with E bad.

$$R_S\,\{\text{given } E \text{ is bad}\} = 1 - (1 - R_A R_C)\,(1 - R_B R_D) \tag{25.21}$$

The system reliability is given by (25.19), which is the sum of (25.20) and (25.21). If the reliability of all components are identical and are equal to R, this sum reduces to the following.

$$R_S = 2R^2 + 2R^3 - 5R^4 + 2R^5 \tag{25.22}$$

There are other techniques that can be used, such as the cut set method, described in the next section.

EXAMPLE 25.4

Consider the breaker-and-a-half switching station shown in Figure 25.9. The station designer is particularly interested in the reliability of a particular power transmission through the station, which normally enters the station on line *1* and exits on line *5*. Since the power transfers are usually large, a high financial penalty accompanies loss of this path. Construct a reliability block diagram of the station components between terminals *1* and *5*.

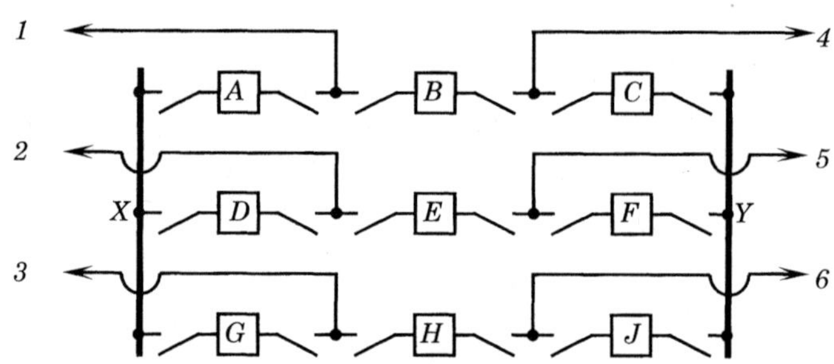

Figure 25.9 A breaker-and-a-half switching station layout.

Solution

The construction of the *1-5* path is shown in Figure 25.10. Six different pairs of station terminal nodes result in a reliability block diagram with the same general structure, and these similar arrangements are noted in Figure 25.10. Clearly, this is a bridge network, exactly like the one in Figure 25.7, but this is not the only connection topology and it is not the only bridge connection. The others are left as exercises.

■

25.1.6 Cut Sets

The cut set method provides a convenient tool for complex network analysis. A cut set is defined as follows.

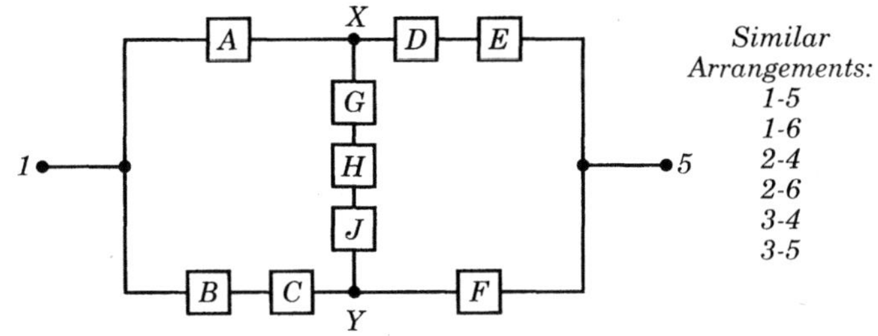

Figure 25.10 Reliability block diagram of Figure 25.9 from line 1 to line 5.

Definition. A cut set is a set of components whose failure:

(a) Results in system failure, or

(b) Results in loss of continuity between the input and the output of the reliability block diagram.

Cut sets are usually easy to describe, especially where a network diagram is available.

EXAMPLE 25.5

Consider the bridge network of Figure 25.7. We can define the cut sets by inspection, as follows:

AB, CD

AED, BEC, ABE, ABC, ABD, ACD, BCD, CED

ABED, ABCD, ABCE, ACDE, BCDE

ABCDE ■

A *minimal cut set* is a set of components whose failure results in system failure and the removal of any of its components results in a set that is no longer a cut set. All components of a minimal cut set must be failed to cause system failure.

EXAMPLE 25.6

For the bridge network of Figure 25.7, the following are the minimal cut sets.

AB, CD, AED, BEC

The bridge network can be analyzed by placing these four parallel systems in series and analyzing as a parallel-series system, as shown in Figure 25.11.

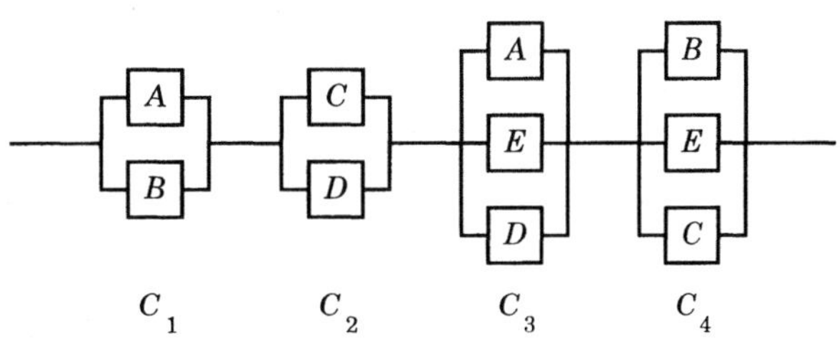

Figure 25.11 Minimal cut sets for the bridge network.

If we let Q_S be the probability of system failure, we can write this probability in terms of the cut sets. Since the same components appear in more than one cut set, the probability of failure is not just the series system probability.

The probability can, however, be stated as the probability of the union of all cut sets. In writing this union, we take advantage of the fact that some cut sets are disjoint. In the present example, this is true of C_1 and C_2 only. Therefore, we write the failure probability as follows.

$$\begin{aligned} Q_S &= \Pr(C_1 \cup C_2 \cup C_3 \cup C_4) \\ &= \Pr(C_1) + \Pr(C_2) + \Pr(C_3) + \Pr(C_4) \\ &\quad - \Pr(C_1 \cap C_3) - \Pr(C_1 \cap C_4) - \Pr(C_2 \cap C_3) - \Pr(C_2 \cap C_4) \\ &\quad - \Pr(C_3 \cap C_4) + \Pr(C_1 \cap C_2 \cap C_3) + \Pr(C_1 \cap C_2 \cap C_4) \\ &\quad + \Pr(C_3 \cap C_3 \cap C_4) - \Pr(C_1 \cap C_2 \cap C_3 \cap C_4) \end{aligned} \tag{25.23}$$

This is the general expression for the probability of failure of the bridge network. ■

EXAMPLE 25.7

In many physical systems, the probability of joint failures of two or more components may be very small. When this is true, it is possible to approximate the probability of failure to simply the sum of the failure probabilities of the cut sets.

$$\begin{aligned} Q_S &= \sum_{i=1}^{N} \Pr(C_i) = \Pr(C_1) + \Pr(C_2) + \Pr(C_3) + \Pr(C_4) \\ &= Q_A Q_B + Q_C Q_D + Q_A Q_E Q_D + Q_B Q_E Q_C \end{aligned} \tag{25.24}$$

This approximation may be necessary for large and very complex systems, where the number of cut sets may number in the thousands. This will be discussed further in the section on fault trees, where computer programs have been prepared to determine the cut sets and to evaluate the probability of system failure.

For the special case where the components of the bridge are identical, the probability of failure reduces to

$$Q_S = 2Q^2 + 2Q^3 = 2Q^2(1 + Q) \tag{25.25}$$

Note that this special equation incorporates the assumption underlying (25.24) and the relative magnitude of joint failure probabilities. ■

EXAMPLE 25.8

Consider the bridge network for the breaker-and-a-half switching station, shown in Figure 25.10. Find the minimal cut sets for this bridge network.

Solution

The minimal cut sets for the connection from 1 to 5 are as follows: (AB), (AC), (DF), (EF), (AGF), (AHF), (AJF), (DGB), (DGC), (DHB), (DHC), (DJB), (DJC), (EGB), (EGC), (EHB), (EHC), (EJB), (EJC). There are 19 cuts in this minimal set. This result should be compared with the solution to Example 25.5 to see if there is agreement. The difference is that some of the branches of the system of Figure 25.10 have more than one component. ■

25.2 FAULT TREES

Fault trees provide a very flexible and powerful method for the reliability analysis of complex systems. The technique was first proposed by H. A. Watson of Bell Telephone Laboratories in 1961 to analyze the Minuteman Launch Control System. The concept has been expanded and refined by others. The technique is now widely used and has proven valuable in many different types of analysis [2–4]. For example, this method of analysis is often used for the analysis of protective system designs for nuclear power plants [3], [4].

25.2.1 Fault Tree Conventions

The construction of a fault tree provides a method of defining the events leading to a particular failure event of interest, with this failure event being the top of the tree or the "top event." The branches of the tree consist of fault events that lead to the top event, and these lower events are, in turn, described in terms of still more detailed fault events. Finally, at the end of each branch of the tree are the basic failures attributed to components.

Fault trees provide a convenient method of graphically visualizing causal relations between events. This is done by constructing building blocks that are displayed as either *gate symbols* or *event symbols*. The gate symbols employed are summarized in Table 25.1. These gate symbols are used to connect events according to their causal relationships. A gate can have two or more input events, but only one output event. Table 25.1 describes the causal relation that must exist among the input events in order for the output event to occur. The causal relation expressed by the AND gate or the OR gate is deterministic since the occurrence of the output is controlled completely by the input events.

TABLE 25.1 Fault Tree Gate Symbols

Gate Symbol	Gate Name	Causal Relation
	AND gate	Output event occurs if all input events occur simultaneously
	OR gate	Output event occurs if any one of the input events occurs
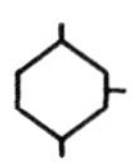	Inhibit gate	Input produces output when conditional event occurs
	Priority AND gate	Output event occurs if all input events occur in the order from left to right
	Exclusive OR gate	Output event occurs if one, but not both, of the input events occur
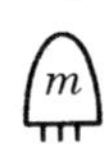	m out of n gate	Output even occurs if m out of n input events occur

The inhibit gate, however, represents a probabilistic causal event. The event at the bottom is the input event and that at the side is the conditional event, which is conditioned by the input event. For this gate, the input event causes the output event with the probability of occurrence of the conditional event.

The priority AND gate is the logical equivalent to the AND gate, with the restriction that the inputs must occur in a definite order. If these inputs should occur, but in the wrong order, there will be no output event.

The exclusive OR gate has an output when either one, but not both, of the input events occur.

The m out of n voting gate must have at least m of its inputs occur for an output to occur, but the order is of no importance.

In many problems, the entire fault tree can be constructed using only AND and OR gates. The other gates are required for more complex problems.

The other graphical aid used in the construction of fault trees is the event symbol. These symbols are shown in Table 25.2. Used together, the gate symbols and event symbols permit the engineer to completely describe how the occurrence of a particular failure mode of a given system can occur.

TABLE 25.2 Fault Tree Event Symbols

Symbol	Name	Meaning of Symbol
	Circle	Basic event with sufficient data
	Diamond	Undeveloped event
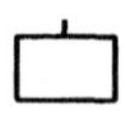	Rectangle	Event represented by a gate
	Oval	Conditional event used with the inhibit gate
	House	House event. Either occurring or not occurring
	Triangles	Transfer symbol

25.2.2 System Analysis Methods

In order to put fault trees in perspective as one of the methods of reliability analysis, it is helpful to review the basic concepts that have evolved for the analysis of large systems. Much of the analytical work can be summarized as being of two basic types: forward analysis and backward analysis [2], [6], [7].

Forward analysis, often called "bottom-up" analysis, begins with basic failure events of components and determines the consequences of these failure events, combining results to reach the top event representing a particular mode of system failure. Analytical methods that utilize forward analysis include event tree analysis, failure mode and effects analysis, criticality analysis, and preliminary hazards analysis methods [2]. These are examples of *inductive analysis*, where the analysis begins with the individual events and works toward a generalization.

Deductive analysis, on the other hand, constitutes reasoning from the general to the specific. This "backward analysis," or "top-down" analysis, begins with the top event that represents system failure mode of interest and identifies the causal relations leading to that event. Fault tree analysis is a deductive process. Using this concept, the analyst works downward through the fault tree to the basic failure events. Note that fault tree analysis does not identify the top event, nor does it provide a method of identifying all possible hazard events for the system. The top event is presumed to be given or postulated prior to the analysis. Clearly, some systems can have different top events, with large or complex systems having a very large number of system hazards that may be of interest for particular studies. Although fault tree analysis is often considered a backward analytical method, computer programs have been constructed that perform the analysis using both forward and backward methods [2].

25.2.3 System Components

A *system* is composed of components that are arranged to permit the achievement of a particular objective [2]. The system may also require materials (feed stocks or fuels), information (measurements or computed controls), and personnel (operators and repair or maintenance personnel). All of these things may have an impact on the hazard defined as the top event. The fault tree must clearly show the interaction of components, through the system topography, to produce failure events. The same components may have different characteristics in different physical systems. Usually, the interrelation among the components can be determined through piping, wiring, or other diagrams as well as drawings that show the location and connection of the components.

The environment, plant personnel, and aging of components provide the mechanisms that may lead to failure, and it is only through the components that these external forces provide causal means of failure. This is shown graphically in Figure 25.12.

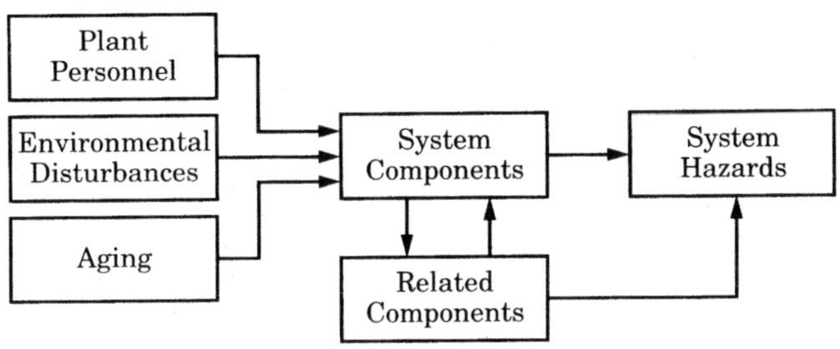

Figure 25.12 Causal means of failure [2].

For example, a key parameter of a component may drift away from its design value due to aging, and this may cause a system hazard or may cause the failure of a related component that leads to the top event. Components located in the same vicinity may be destroyed by a common mode failure, such as a fire, leading to the system hazard under study.

25.2.4 Component Failures

Component failures are classified as primary failures, secondary failures, or command faults [2], [4].

A *primary failure* is defined as the component being in a nonworking state for which that component is held accountable. A primary failure requires repair action on the component to restore it to a working state. Primary failures occur when the component is working within its design envelope and is hence "accountable" for working under these conditions. Aging or random environmental stress is often the cause of primary failure. For example, a transmission line conductor is struck by lightning, is thereby faulted to a tower, and burns down. A diode is capable of passing factory acceptance tests, but later its parameters drift a bit and failure occurs when carrying rated current (within the design envelope).

A *secondary failure* is the same as a primary failure, but the component is not held responsible for the failure. This type of failure is due to excessive stress being placed on the component caused by operating conditions that are out of design tolerance due to environmental conditions, by related components, or by human operator actions (or inactions). For example, a power transformer has voltage ratings that are calculated to provide long life, but excessive overvoltage causes insulation failure due to operation outside of the apparatus rating (outside the design envelope). An operator may be the cause of a secondary failure, if the operator has a means of relieving an overload condition, for example, but fails to take the required action to provide the needed relief in a timely manner.

A *command fault* is defined as a condition where the component is in a nonworking condition because of incorrect control signals or noise. Repair is not usually required to correct command faults, which can usually be corrected by control or operator action to return the component to a working state. For example, an operator causes power to an important load to be lost because the wrong breaker is opened during routine switching for repair of a neighboring circuit. Service is restored by correcting the switching error.

These component failure characteristics are further described in Figure 25.13, where the possible causes of the failures are shown in the outside circle.

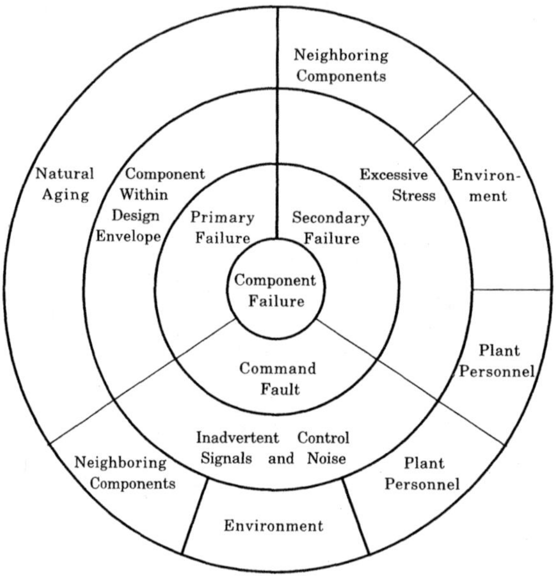

Figure 25.13 Component failure characteristics [2].

A term that is widely used in fault tree analysis is the term *fault*. The events pictured in the fault tree are often referred to as "faults." A distinction must be made between a fault and a failure. A *failure* is a basic abnormal occurrence for a component. It is an inoperable state, in which a component is not able to perform its intended function [6]. A *fault* is the immediate cause of failure [6]. It is a general type of occurrence, which may not refer to the complete loss of an item, but just an improper action or operation of that item. For example, a relay may close without command, due to physical or seismic shock, but this does not imply failure of the relay, since the component is not damaged by the action, although the event may be unwanted. Thus, we say that all failures are faults, but not all faults are failures [4]. This distinction is important to observe in analyzing power system protection, where it is common to refer to short circuits or other events requiring protective action as faults. More precisely, a temporary short circuit should be termed a *fault*, but a permanent short circuit is a *failure*. The general usage of the term fault will usually be clear from the context.

25.2.5 Basic Fault Tree Construction

Fault tree construction has evolved to the point where definitive rules of construction can be provided. These rules are stated as follows [4]:

1. Enter statements in the event boxes as faults; state precisely what fault has occurred and when it occurs.
2. Examine each boxed event statement and ask the question, "Can this fault consist of a component failure?"
   (a) If the answer to the question is "Yes," classify the event as a state-of-component fault, terminate the event in an OR gate and look for primary, secondary, and command modes of failure.
   (b) If the answer to the question is "No," classify the event as a state-of-system fault, then terminate the event in an appropriate gate, as required, and look for the minimum necessary and sufficient immediate cause or causes of the fault.
3. Observe the "No Miracles Rule." If the normal functioning of a component propagates a fault sequence, then assume that the component functions normally.
4. Observe the "Complete the Gate Rule." All inputs to a particular gate should be completely defined before further analysis of any one of them is undertaken.
5. Observe the "No Gate-to-Gate Rule." Gate inputs should be properly defined fault events, and gates should never be directly connected to other gates.

The first rule is sometimes called a "what-when" rule. The "what" condition describes the faulted state of the component or subsystem, and the "when" condition describes a limiting system condition that makes this particular state a fault. Sometimes, it is necessary to write rather long statements in order to convey precisely the fault condition. Abbreviate the words, but not the concepts.

The second rule helps the fault tree builder organize the construction work. If the stated fault can be caused by a component, this event is near the end of the tree branch, which will be terminated as a primary failure, secondary failure, or command mode fault of that component. This always requires an OR gate, as shown in Figure 25.14.

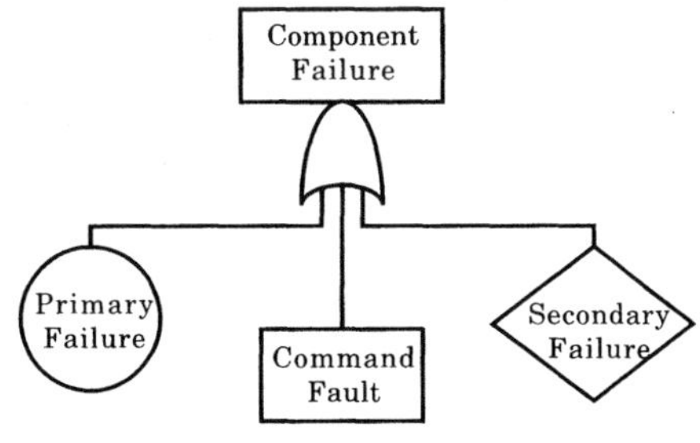

Figure 25.14 Fault tree of a component failure.

A *primary failure* is any failure of the component that occurs in an environment for which that component is qualified, i.e., the operation takes place within the design envelope of the component. For example, a circuit breaker fails to interrupt a current that is lower in magnitude than the interrupting rating of the breaker. The root cause of the failure may be due to different basic failure mechanisms, but the breaker has suffered a failure due to an occurrence of some kind that caused fault interruption to fail.

A *secondary failure* is any failure of a component that occurs in an environment for which that component has *not* been qualified. For example, a circuit breaker fails to interrupt a current that is greater than the interrupting rating of the breaker. The breaker is not designed to interrupt this magnitude of current, hence its failure rate in attempting to do so will be considerably higher than a primary current interruption failure rate. Such secondary faults may be of interest in certain types of analysis, but are often omitted from the fault tree in order to restrict the scope of investigation.

A command mode fault implies a proper operation of the component, but at the wrong time or the wrong place. For example, a relay closes without a command signal from the relay logic, perhaps because the case is bumped by a person working near the relay cabinet. Another example is the tripping of a transmission line for a fault outside the primary protection zone, which is due to improper setting of the relay reach. In this case, the relay operates correctly, but the setting is incorrect. In both cases, the failures noted are not random occurrences of a basic failure mechanism. These faults are grouped together under the name "command mode faults." No repair of the component is required to correct command mode faults, but other actions may be required to ensure that these faults do not persist or reoccur.

In some fault trees the secondary failure or command fault may be of little significance, of very low probability, or outside the scope of study, in which case these events may be omitted. The primary failure is always present, however. Note that the secondary failure shown in Figure 25.14 is an undeveloped event (hence, the diamond icon). It may be necessary to add several additional branches to the fault tree to complete the analysis of this event. Command faults, if present, are shown as events (square boxes), and this implies an extension of the fault tree below this level to other primary basic events.

EXAMPLE 25.9

Consider the network shown in Figure 25.15, showing an electric motor and its supply circuit [2], [3], [8]. To construct a fault tree for this system, the first thing to define is the top event. In this case, the basic system hazard is that the motor fails to start on command.

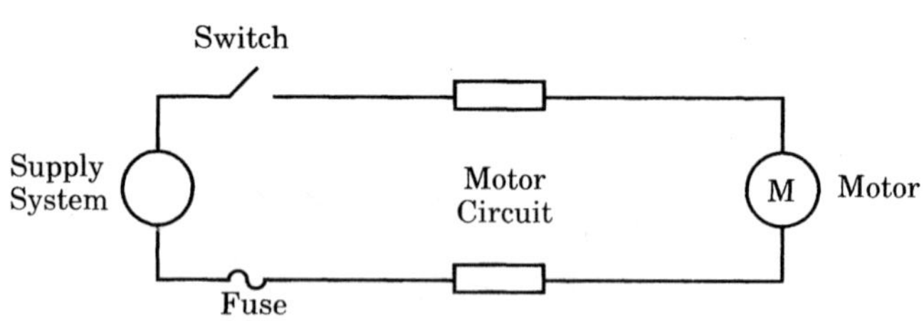

Figure 25.15 An electric motor and its supply system.

The fault tree for this system is shown in Figure 25.16 [2], including all primary, secondary, and command fault events.

The primary or basic failures are all shown on the left side of the tree and the secondary failures, which are undeveloped, are on the right. The primary failures are random failures due to normal aging or deterioration of the components. The secondary failures are due to environmental conditions (overheating, etc.); out-of-tolerance conditions (vibration, thermal stress, etc.); or improper lubrication or maintenance. The command faults are due to inadvertent control signals, which in the first case is motor power failure.

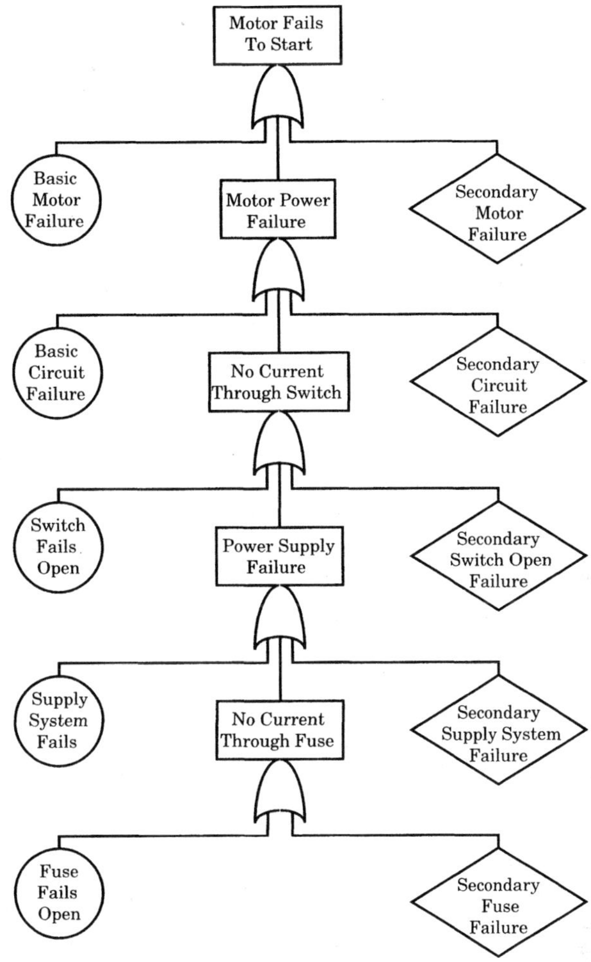

Figure 25.16 Fault tree for the system of Figure 25.15.

These failures are not due to the motor but to the failure of support systems to control the power supply to the motor. ■

EXAMPLE 25.10

Consider the system shown in Figure 25.17 [4]. The function of the control is to regulate the pumping of water from a reservoir to a pressure tank. It takes an average of 10 minutes to completely fill the tank to the desired pressure, and a timer T is set to trip the pump should it run longer than 10 minutes. As a backup protection, a pressure switch is used to monitor the tank pressure and open the motor circuit on overpressure. The pressure switch contacts are closed when the tank is empty and remain closed until the critical tank pressure is reached.

When the tank is empty, the pressure contacts P are closed. The operator closes momentary contact SW to fill the tank. This picks up relays K1 and K2, which starts the motor. Should the motor run longer than 10 minutes, the timer times out, opening contact T, which drops out K1 and stops the motor. Also, if the pressure exceeds the setting of the pressure switch P, this opens contact P, which drops out K2 and stops the motor. The fault tree for this system, with no secondary failures illustrated, is shown in Figure 25.18. The top event of interest is the rupture of the pressure tank. ■

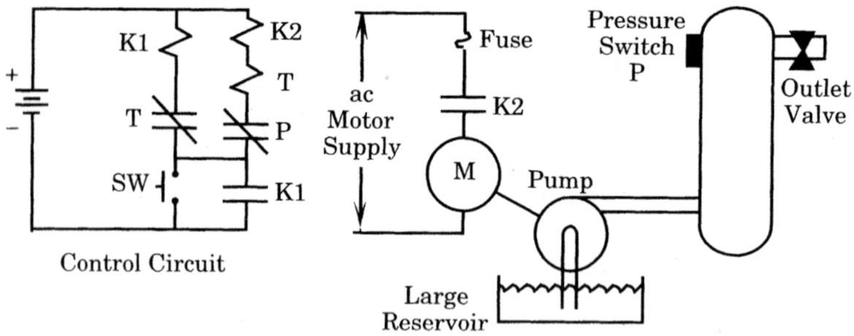

Figure 25.17 Pressure tank control and protection system.

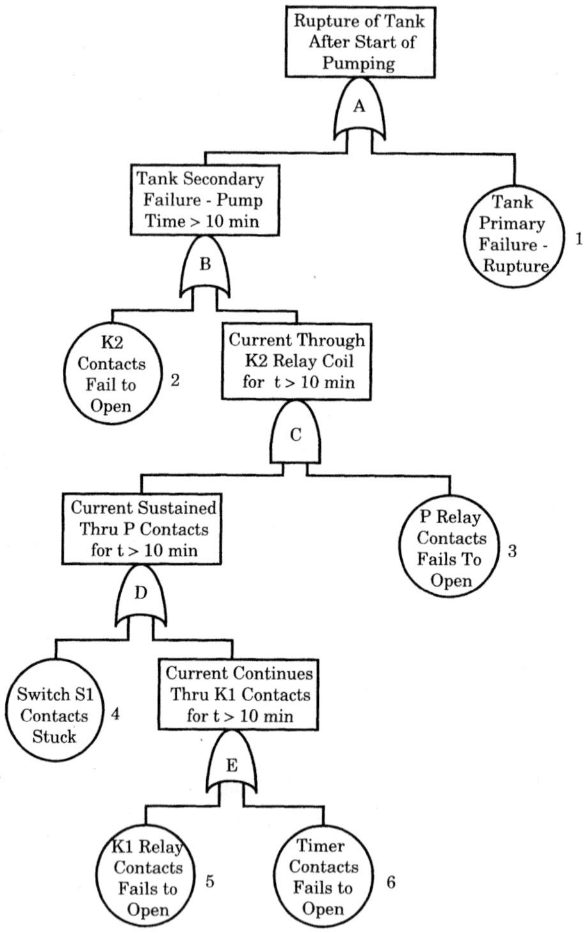

Figure 25.18 Fault tree for example 25.10.

25.2.6 Decision Tables

The fault tree construction method described in the previous section is heuristic and requires judgment on the part of the engineer to construct a correct and useful tree. This can be a difficult and tedious process for complex systems, and it may lead to errors. Methods have been developed to make the fault tree construction more rigorous and less likely to result in costly errors. One such method is to use decision tables.

The decision table method requires that sufficient information regarding the process be known, and that a library of component models exists. If these conditions are met, the tree construction process can be fast and systematic. A computer program is even available that will automate the tree construction [9], [10]. The process is described as follows:

1. A set of events, called output events, is assigned to each output from each component, which gives the possible states of the component. For example, an electromechanical relay may be characterized as having contact states normal, failed shorted, and failed open. Any number of states that are required to completely specify the output of that component can be used.
2. A set of input events is also assigned to describe the states of the input of each component. For example, the relay input may be described as voltage applied or no voltage applied to the relay coil.
3. Inputs from the system environment are considered basic input events. Inputs from other components are state-of-system or state-of-component events.
4. Each component is modeled by a decision table, which is similar to a truth table. This table gives the rules as to how an output event occurs as the result of certain input events. A component may have several input events, but it has only one output event.
5. The connection of thc various components defines the system. This is done by connecting the inputs to a component from the outputs that drive that component. The top event output is the hazard of interest.

The decision method of fault tree construction is demonstrated by example.

EXAMPLE 25.11

Consider the relay control circuit shown in Figure 25.19. A single overcurrent relay is set to pick up for faults on a radial line and the relay is connected in a control circuit to open the line circuit breaker. To analyze this system using decision tables, we construct input and output tables for four components; the breaker, the trip coil, the relay contacts, and the relay.

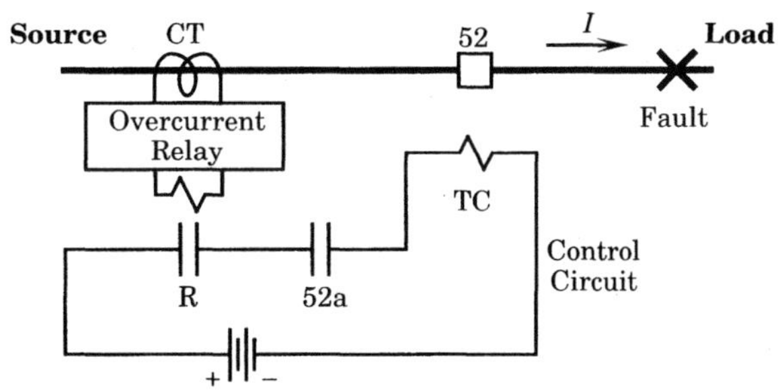

Figure 25.19 Relay system for decision table analysis.

The top event for this system is the failure to trip when there is a fault on the line. With this system hazard defined, the four decision tables are examined to determine those basic events that lead to this top event, with the results shown in Table 25.3.

1. The Circuit Breaker decision table shows that the top event, No Trip, occurs under three different input conditions, the last two of which are basic events. The event "Inadequate Force" on the trip mechanism is due to the low output of the Trip Coil.

TABLE 25.3 Decision Tables for Four Components of the System

1. Circuit Breaker		**2. Circuit Breaker Trip Coil**	
Input	Output	Input	Output
Adequate force	Trips	Current OK	Adequate force
Inadequate force	No trip	Current low	Inadequate force
52a open		Trip coil faulted	
Battery failed		Trip coil open	
3. Relay Contact		**4. Relay**	
Input	Output	Input	Output
Relay pickup	Current OK	Overcurrent	Relay pickup
No relay pickup	Current too low	No Overcurremt	No relay pickup
Contact bent		Relay failed	
		Wrong setting	
		CT failed	

2. The Trip Coil decision table, under Inadequate Force, identifies three input events that will lead to this result. The last two of these are basic events, but the event "Current Low" is due to the control circuit components.
3. The Relay Contact decision table shows an output that gives low current. This event is caused by either of two input events, the second of which is a basic event. The event "No Relay pickup" is due to the relay component.
4. The Relay component decision table shows the output No Relay pickup. This event is due to any of four input events, the last three of which are basic failure events. The first event, No Overcurrent, is of no interest, since there is no need to trip the overcurrent relay when this event occurs. Therefore, this event is ignored.

The fault tree resulting from this analysis is shown in Figure 25.20. The words along the left side identify the component fault corresponding to the event described in the square boxes. ■

25.2.7 Signal Flow Graphs

Signal flow graphs are a useful aid in the construction of certain fault trees. One application of signal flow graphs that is particularly appealing is in the analysis of feedback control systems, and certain protective systems fall into this category. For example, in Chapter 22 it was noted that the HVDC controls and protections are implemented using the same equipment. Signal flow graphs may be helpful in analyzing such protections. There are many examples of protective control systems in power plants and other industrial systems. For example, a pump is used to maintain the proper level of a fluid in a tank, and is designed to prevent overflow of the tank. Clearly, this control system may be also viewed as a protective system.

Signal flow graphs were introduced by S. J. Mason as an alternative to control system block diagram construction [10]. The method provides a very general way of writing the system equations, which may be integrodifferential equations, Laplace domain algebraic equations, or sampled data equations. In all cases, the form of the equations is

$$\text{Effect at } j = \sum (\text{transmission from } k \text{ to } j)(\text{cause at } k) \tag{25.26}$$

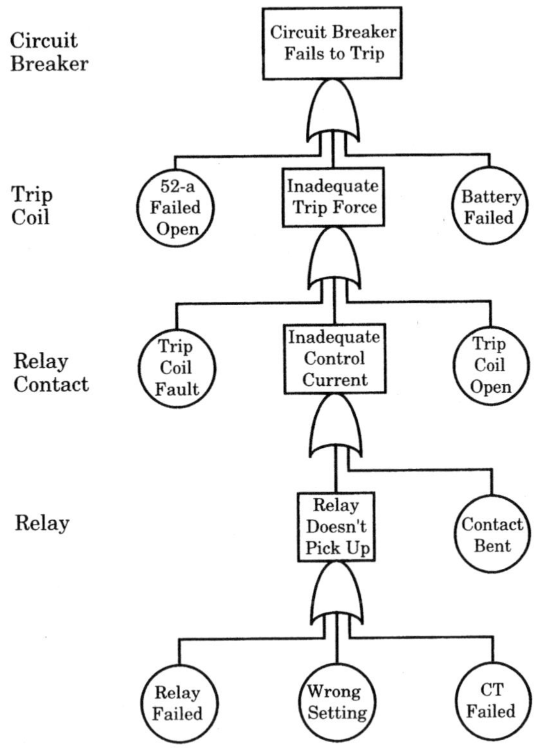

Figure 25.20 Fault tree for the relay system of example 25.11.

Mathematically, the equations will be written as

$$x_j = \sum_{k=1}^{N} t_{kj} x_k \qquad j = 1, 2, \ldots N \tag{25.27}$$

where t_{kj} = transmission function from k to j

The transmission function represents the contribution of the variable x_k to the value of the variable x_j. In the signal flow graph, the variables are represented as nodes and the transmission functions as directed branches. The construction of the signal flow graph is a matter of following the cause and effect relations throughout the system and relating each variable to itself and to the other variables.

The solution for the functional relationship between any input and the output variables is found using the general gain formula of Mason [11], which is stated as follows [11-14].

$$M = \sum_{k} \frac{M_k D_k}{D} \tag{25.28}$$

where M_k = gain of the kth forward path

$$D = 1 - \sum_{m} P_{m1} + \sum_{m} P_{m2} - \sum_{m} P_{m3} + \cdots \tag{25.29}$$

and where P_{mr} = gain product of the mth possible combination of r nontouching loops

D_k = value of D for that part of the graph not touching the kth forward path

The technique in constructing fault trees is to introduce new variables, representing the failure of the system variables. One can then compute the total system gain with any variable failed. Moreover, a fault tree can be constructed to represent failure of variables or transmission functions. This technique will not be pursued here in detail. The interested reader is referred to the treatment in [2].

25.3 RELIABILITY EVALUATION

The previous section provides several methods of constructing logic diagrams and trees to show how failure may occur. These graphical methods are very useful in providing fundamental information regarding the characteristic of a system failure. This section presents methods of evaluating these constructions to determine the adequacy of the system design. Two types of evaluation are presented: qualitative analysis and quantitative analysis.

25.3.1 Qualitative Analysis

In many systems, a qualitative analysis is very useful in evaluating the effect of various component failures on the top event. An engineer with experience and good judgment can sometimes determine whether design changes are required without performing numerical analysis, once the basic failure events and effects are structured in a fault tree. This graphical structuring of the failure events provides useful insight into the system performance that is sometimes difficult to achieve by writing equations. This section presents several techniques that are often used by engineers to evaluate qualitatively the design of a system relative to its likelihood of failure.

25.3.1.1 Cut Sets. Cut sets were introduced in Section 25.1.6, where the application of the cut set concept was illustrated using network diagrams. Cut sets are also useful in fault tree analysis both for qualitative and quantitative evaluation. Computer programs have been prepared to determine the cut sets of complex fault trees [2], [3], [15]. It is very important to note that only the minimum cut sets are of interest. This is because of the very large number of cut sets that are typically generated from a fault tree. It has been observed that a system with a few hundred components may have billions of cut sets. The minimal cut set is one in which, should any basic event be removed from the set, the remaining events no longer constitute a cut set.

Several methods have been used for this type of evaluation. The methods are usually characterized as either "bottom-up" or "top-down" techniques. One top-down method is described in [2] and is presented here to illustrate the concept of qualitative analysis. The method described is applicable as long as there are no mutually exclusive events in the fault tree.

It has been observed that OR gates increase the number of cut sets, but AND gates enlarge the size of the cut sets. A top-down procedure for finding the minimal cut sets is described as follows [1]:

1. Alphabetize the gates.
2. Number each basic failure event.
3. Locate the uppermost gate in the first row and first column of a matrix.
4. Iterate either of the fundamental permutations (a) or (b) in a top-down fashion.
   (a) Replace OR gates by a vertical arrangement of the input to the gates, and increase the cut sets.

(b) Replace AND gates by a horizontal arrangement of the input to the gates, and enlarge the size of the cut sets.

5. When all gates are replaced by basic events, obtain the minimal cut sets by removing supersets. A superset is a cut set that includes some other cut sets.

This technique is illustrated using an example.

EXAMPLE 25.12

Determine the minimal cut sets for the fault tree developed in Example 25.10. This fault tree already has the gates alphabetized and the basic failure events numbered, as shown in Figure 25.18. Therefore, the only remaining tasks are those identified above as 3, 4, and 5.

We begin with the upper left corner of the matrix, where we place the top event, i.e.,

A

The gate below A is an OR gate, so we replace A with a vertical arrangement of the inputs to gate A, which gives the following array:

1
B

Note that we always place the basic failure events on top. Now, gate B is also an OR gate, with inputs 2 and C, so the matrix now has three rows, but still just one column, as follows.

1
2
C

Gate C is an AND gate, so we expand the matrix to include a new column. The inputs to C are 3 and D, which replaced C in the above array.

1	
2	
3	*D*

Now gate D is an OR gate with inputs 4 and E. This expands the array vertically, with the result

1	
2	
3	4
3	*E*

Finally, gate E is an OR gate with basic event inputs 5 and 6. Expanding vertically again, we have the following.

1	
2	
3	4
3	5
3	6

Then the minimal cut sets for this fault tree are the following.

$$\{1\} \quad \{2\} \quad \{3, 4\} \quad \{3, 5\} \quad \{3, 6\}$$

Note that there are no supersets, so this is the final result. ■

For large fault trees, this task is much more difficult. One of the available computer programs may be used for these large problems [2].

25.3.1.2 Qualtitative Importance. Suppose that the minimal cut sets of a fault tree have been determined. An understanding of the relative importance of the various cut sets can be determined by first listing the cut sets in order, beginning with the single component cut sets and moving down through the multiple event cut sets. Such an ordering was obtained in Example 25.12.

Now make a rough estimate of the *order of magnitude* of the various component failures. Suppose, for example, that most of the failures are of the order 10^{-3}. Then each single component cut set will have a probability on the order of 10^{-3}, two component cut sets will have a probability on the order of 10^{-6}, etc. One may often conclude from this kind of estimation that the cut sets of only the single components are significant. This focuses the designer's attention on these single points of failure. If these can be eliminated by design changes, the total system probability of failure will be improved.

25.3.1.3 Common Mode Failure Analysis. The primary failures in a fault tree are not necessarily independent. In many cases there are single basic causes of failure that may affect more than one component. This common cause of failure may be environmental, such as fire, flood, or earthquake. It may be due to a common design flaw that affects identical components of the same design. The common cause may be due to improper maintenance or adjustment of components by personnel who are following improper instructions or practices.

In some cases, we can identify the common cause of failure and can examine this cause to estimate its effect. For example, in the system of Example 25.10, there are four relays and the effect of failure modes having a bearing on the top event were analyzed in that example. However, all of these relays are supplied by the same battery power supply. Therefore, failure of the single battery must be considered, even though the probability of battery failure may be quite low.

To identify the minimal cut sets that are susceptible to common cause failures, we must first define the common cause categories that can cause component failure dependence. Table 25.4 lists several common cause categories that might be of interest for evaluation.

TABLE 25.4 List of Common Cause Categories for Evaluation [3]

Manufacturer
Location
Seismic susceptibility
Flood susceptibility
Temperature
Humidity
Radiation
Wear-out susceptibility
Test degradation
Maintenance degradation
Operator interactions
Energy sources
Dirt or contamination

For each common cause category, we identify the components that are susceptible to failure under that category, for example, all components of a particular manufacturer, or all that are located in a flood plain. Some analysts prefer to code the component names, using a code that identifies those categories that should be considered for each, and entering this code on the fault tree near that component basic failure event. Having the codes established, one can identify the potentially susceptible minimal cut sets that are potential problems. For example, this method will identify those cut sets whose primary failures all have the same element of a given category. Finally, a screening should be made to determine those cut sets that may require further analysis. The screening can be based on a rough numerical analysis or on engineering judgment. If one needs to be absolutely sure of the analysis, as in a safety investigation, then every possibility should be carefully investigated.

25.3.1.4 Other Qualitative Methods. There are other methods of qualitative analysis that are used for special types of systems. Fault trees that contain exclusive OR gates, for example, are called noncoherent fault trees and they require special methods of analysis. One such method is called *prime implicants*. The method is based on a technique called *consensus*, which is based on the use of decision tables [2].

Systems with control loops may require special study, since such systems are not readily described by simple Boolean functions [2]. In such systems, the order in which the component failures occur may be important and the analysis is much more complex.

25.3.2 Quantitative Analysis

Quantitative analysis is often required before a given design can be approved as adequate, or capable of meeting a design specification. This section presents several methods of analysis that are often used. The emphasis is on fault tree analysis.

25.3.2.1 Top Event Analysis. One of the most important types of quantitative analysis, especially for complex systems, is the quantitative analysis of the top event of a fault tree. The definition of system failure can be expressed as the top event, which is determined by combining all failure modes or hazards in an OR gate, as shown in Figure 25.21. System success, then, is defined as the nonoccurrence of all failure modes. In some studies, only one particular failure mode is of interest, in which case that top event is the only one for which a fault tree is constructed.

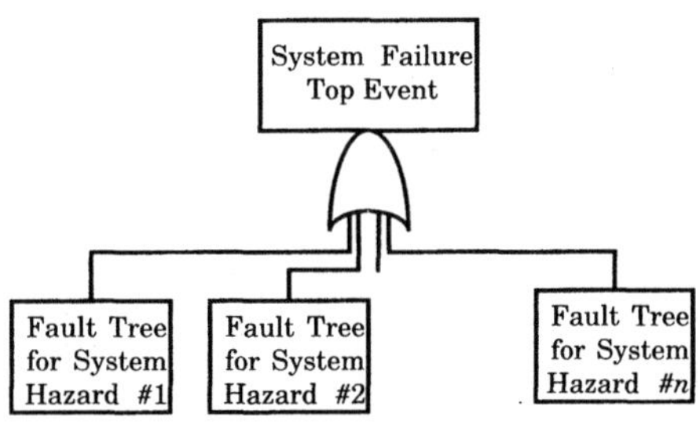

Figure 25.21 A fault tree for system failure by different failure modes.

The probabilistic parameters, developed in Chapter 24, can be used to describe the system failure, whether of the top event for the system or for a particular hazard or failure mode of interest [from 2].

System availability $A_S(t)$ The system availability is the probability that the top event does not exist in time t. This is the probability of the system operating successfully when the top event is an OR combination of all system hazards. It is the probability of the nonoccurrence of a particular hazard when the top event is that of a single system hazard.

System unavailability $U_S(t)$ The system unavailability is the probability that the top event exists at time t. This is either the probability of system failure or the probability of a particular system hazard at time t, depending on the definition of the top event. The system availability and unavailability are complimentary, and obey the equation

$$A_S(t) + U_S(t) = 1 \tag{25.30}$$

System reliability $R_S(t)$ The system reliability is the probability that the top event does not occur over the time interval $(0, t]$. The definition of system reliability requires the continuation of the nonoccurrence of the top event and obeys the following inequality.

$$R_S(t) \le A_S(t) \tag{25.31}$$

The reliability is often used to describe the top event for catastrophic or non-repairable system failures.

System unreliability $F_S(t)$ The system unreliability is the probability that the top event occurs before time t. This is the complement of the system reliability, or the following identity holds.

$$R_S(t) + F_S(t) = 1 \tag{25.32}$$

The system unreliability is equal to or greater than the unavailability.

$$F_S(t) \ge U_S(t) \tag{25.33}$$

System failure density $f_S(t)$ The system failure density is the first derivative of the system failure distribution.

$$f_S(t) = \frac{dF_S(t)}{dt} \tag{25.34}$$

$f_S(t)\,dt$ is the probability that the top event occurs during $[t, t + dt)$, given that it does not occur before time t.

System conditional failure intensity $h_S(t)$ The system conditional failure intensity is the probability that the top event occurs per unit time at time t, given that it does not exist at time t. A large value of the conditional failure intensity means that the system is near failure.

System unconditional failure intensity $w_S(t)$ The unconditional failure intensity of the system is the probability that the top event occurs per unit time at time t. The quantity $w_S(t)\,dt$ is the probability that the top event occurs during $[t, t + dt)$.

Expected number of top events $W_S(t, t+dt)$ The expected number of top events during the interval $(t, t + dt]$ is given by the relation

$$W_S(t, t + dt) = w_S(t)dt \tag{25.35}$$

The expected number of top events during the interval $[t1, t2)$ is given by the integral

$$W_S(t_1, t_2) = \int_{t_1}^{t_2} w_S(t)dt \tag{25.36}$$

Mean time to failure $MTTF_S$ The mean time to first failure is the expected length of time to the first occurrence of the top event. This corresponds to the average human lifetime and is a suitable parameter for catastrophic system hazards. This parameter is given by the following equation.

$$MTTF_S = \int_0^\infty t\, f_S(t)dt \tag{25.37}$$

We now apply the foregoing definitions to fault trees. Unless otherwise stated, it is assumed that all primary events are mutually independent. First, however, it is necessary to review the basic rules of Boolean algebra.

25.3.2.2 Boolean Algebra. Fault trees are usually easy to construct, and this exercise forces the analyst to think clearly about the failure modes of the system under study. Once constructed, the logic of the fault tree can be analyzed using Boolean algebra. This is an invaluable tool, both for preparing computer programs to analyze fault trees, and to evaluate complex protective systems. (See Table 25.5).

TABLE 25.5 Boolean Algebra Rules*

Boolean Basic Laws	Expressions For	
	Intersections	**Unions**
1. Commutative	$X \cdot Y = Y \cdot X$	$X + Y = Y + X$
2. Associative		$X + (Y + Z) = (X + Y) + Z$
3. Distributive	$X \cdot (Y + X) = (X \cdot Y) + (X \cdot Z)$	$X + (Y \cdot Z) = (X + Y) \cdot (X + Z)$
4. Idempotent	$X \cdot X = X$	$X + X = X$
5. Absorption	$X \cdot (X + Y) = X$	$X + (X \cdot Y) = X$
6. Complementation		
7. de Morgan's law	$\overline{(X \cdot Y)} = \bar{X} + \bar{Y}$	$\overline{(X + Y)} = \bar{X} \cdot \bar{Y}$
8. Operations with ϕ and Ω	$\phi \cdot X = \phi$	$\phi + X = X$
	$\Omega \cdot X = X$	$\Omega + X = \Omega$
	$\bar{\phi} = \Omega$	$\bar{\Omega} = \phi$
9. Reduction	$X + (\bar{X} \cdot Y) = X + Y$	$\bar{X} \cdot (X + \bar{Y}) = \bar{X} \cdot \bar{Y} = \overline{(X + Y)}$

*The symbol Ω indicates the universal set and ϕ indicates the empty set. In engineering notation, Ω is often replaced by 1 and ϕ is replaced by 0 (zero).

25.3.2.3 Availability and Unavailability. The following development pertains to fault trees with independent basic events. This assumption is applicable to many physical systems.

INDEPENDENT BASIC EVENTS. Events are said to be independent if the occurrence of one event does not in any way influence the occurrence of the other events. Consider the events $B_1, B_2, \ldots B_n$, which are independent events. Then we may write the following expression for the intersection of the events.

$$\Pr(B_1 \cap B_2 \cap \cdots B_n) = \Pr(B_1)\Pr(B_2)\cdots\Pr(B_n) \tag{25.38}$$

SYSTEM WITH ONE AND GATE. Consider a system with one AND gate, as shown in Figure 25.22. For this system we write the unavailability as follows.

$$U_S(t) = \Pr(B_1 \cap B_2 \cap \cdots B_n) = \Pr(B_1)\Pr(B_2)\cdots\Pr(B_n) \tag{25.39}$$

Clearly, the assumption of independence is essential for the AND gated fault tree. The availability can be computed as follows.

$$A_S(t) = 1 - \Pr(B_1)\Pr(B_2)\cdots\Pr(B_n) \tag{25.40}$$

SYSTEM WITH ONE OR GATE. Consider a system with one OR gate, as shown in Figure 25.23. For this system, we may write the following expression for the system unavailability.

$$U_S(t) = \Pr(B_1 \cup B_2 \cup \cdots B_n) \tag{25.41}$$

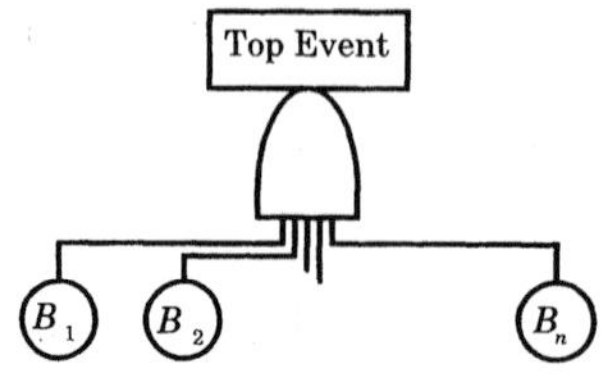

Figure 25.22 Gated AND fault tree.

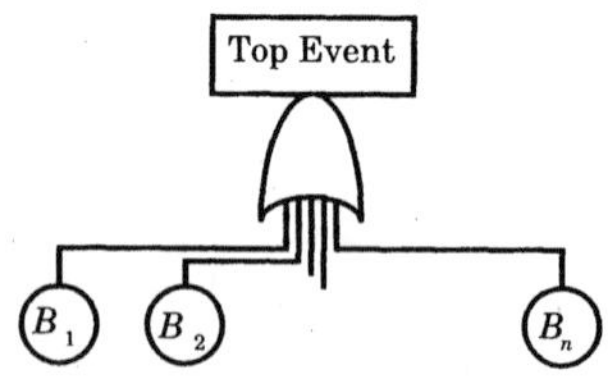

Figure 25.23 Gated OR fault tree.

The system availability is computed by taking the compliment of the unavailability, with the following result.

$$\begin{aligned} A_S(t) &= \Pr\left(\bar{B}_1 \cap \bar{B}_2 \cap \cdots \bar{B}_n\right) \\ &= \Pr\left(\bar{B}_1\right)\Pr\left(\bar{B}_2\right)\cdots\Pr\left(\bar{B}_n\right) \\ &= [1-\Pr(B_1)]\,[1-\Pr(B_2)]\cdots[1-\Pr(B_n)] \end{aligned} \tag{25.42}$$

Then the unavailability can be evaluated as follows.

$$\begin{aligned} U_S(t) &= 1 - A_S(t) \\ &= 1 - [1-\Pr(B_1)]\,[1-\Pr(B_2)]\cdots[1-\Pr(B_n)] \end{aligned} \tag{25.43}$$

The evaluation of the unavailability requires the expansion of (25.43). For $n = 2$, the result is readily computed as follows.

$$\begin{aligned} U_S(t) &= \Pr(B_1 \cup B_2) \\ &= 1 - [1-\Pr(B_1)]\,[1-\Pr(B_2)] \\ &= \Pr(B_1) + \Pr(B_2) - \Pr(B_1)\Pr(B_2) \end{aligned} \tag{25.44}$$

The reason for subtracting the intersection of the two probabilities is readily explained using a Venn diagram. This is left as an exercise.

SYSTEM WITH m OUT OF n VOTING GATE. The fault tree for a system with m out of n voting, shown in Figure 25.24, produces an output if more than m components are detected as failed. This type of system is often used in protective systems where it is essential that a spurious signal or operation of any one of the components will not lead to system failure.

As an example of this type of system, consider a two out of three system. In this case, we can construct the fault tree using ordinary AND and OR gates, with the result shown in Figure 25.25.

The m out of n system can always be decomposed into a system of equivalent AND and OR gates. A more direct approach for a system of identical components, however, is to use the binomial distribution. Let Q be the probability of failure of any one of the components. Then, we may write the probability of failure as the probability of m outcomes favorable to failure from n trials, summed over all successful occurrences of failure [2].

$$U_S(t) = \Pr(m \le k \le n) = \sum_{k=m}^{n} {}_nC_k Q^k (1-Q)^{n-k} \tag{25.45}$$

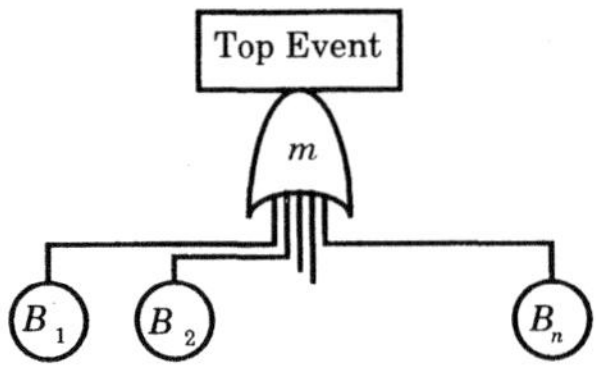

Figure 25.24 An m out of n voting system.

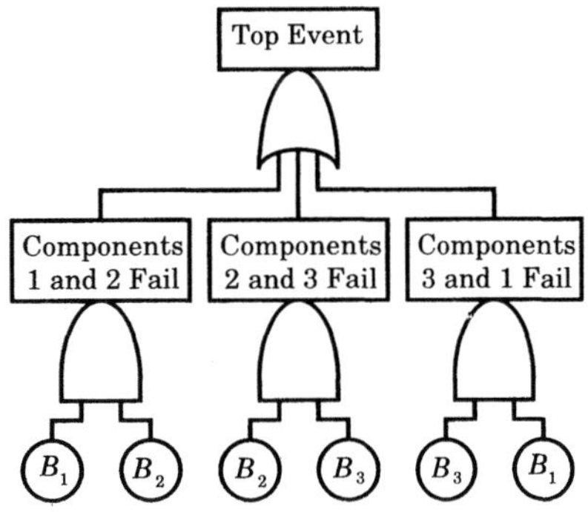

Figure 25.25 A two out of three voting system.

25.3.2.4 Quantitative Importance. The concept of the *importance* of a component or a cut set to the occurrence of the top event has received quite a lot of attention. The analysis of importance is somewhat like a sensitivity analysis and is often useful in considering system design and optimization. Several importance measures have been devised and are described in the literature [19]. Two importance measures that are often used are the Birnbaum basic event importance and the Fussell-Vesely (F-V) importance.

The Birnbaum evaluates the structural importance, which is quantified as the change in top event unavailability with a given component failed. The F-V importance determines the probability of a given component contributing to the cut set failure. This topic is not pursued here in detail, but the interested reader is referred to the literature for a complete analysis of the subject [19].

25.3.2.5 Top Event Prevention. In many complex systems it is important to identify a collection of design elements that is necessary and sufficient for achieving a level of performance that is deemed satisfactory. This may involve a required level of service to the public as well as the optimization of the investment in the system design. This idea has led to the concept of a *prevention set* of components, which specifies combinations of basic events that can prevent the occurrence of the top event of the fault tree. A technique called "Top Event Prevention Analysis" has been developed for finding these prevention sets. This technique has been applied to nuclear power plant design and similar problems [20], [21].

The basic idea of the prevention set is to find a set of components that, if prevented from failure, would provide a specific level of protection against the occurrence of the top event. This can be extended to find a minimal prevention set, which is a set of events having the property of being collectively both necessary and sufficient to preclude the occurrence of the top event at the desired design level.

25.4 OTHER ANALYTICAL METHODS

The foregoing discussion of quantifying the availability or unavailability is based on a given fault tree. Quantification can also be performed using system descriptions other than fault trees. A few of these alternate methods of analysis are mentioned below.

25.4.1 Reliability Block Diagrams

The reliability block diagram (RBD) or network diagram provides the same information of the system as a fault tree, at least for simple systems. In many problems, the reliability block diagram is easy to construct, provides excellent understanding of the system, and is relatively easy to analyze. The construction and use of the reliability block diagram is best explained by an example.

EXAMPLE 25.13

Consider the radial circuit feeding a motor that is described in example 25.10. The hazard of interest is the failure of the motor to start. The reliability block diagram depicts the compliment of that failure, viz., the successful starting of the motor. The components shown in the block diagram are those that must work correctly in order to ensure successful motor starting. The RBD is shown in Figure 25.26.

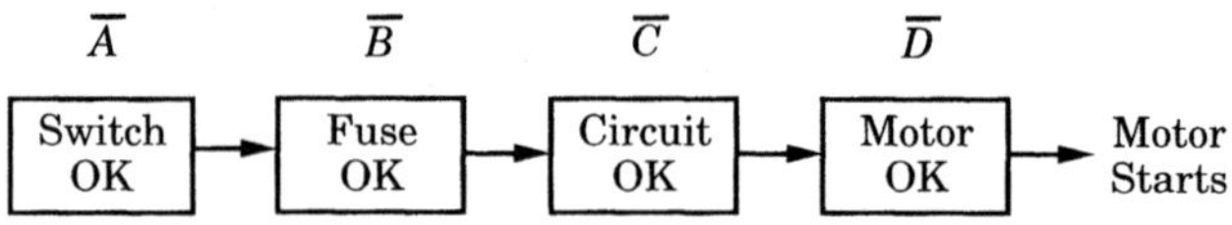

Figure 25.26 RBD for the system of example 25.9. ■

The system of the preceding example is very simple, but it illustrates several points of difference between RBDs and fault trees. First, the RBD depicts the success of the system, rather than the system failure. Second, only the primary basic component success states are implied by the blocks of the network, whereas the fault tree explicitly shows both the primary and secondary failures. Third, command faults are not explicitly shown in the RBD, for example, the inadvertent closing of the switch is not considered.

Suppose we define the following basic failure events.

$A =$ basic failure of switch to close on command

$B =$ basic failure of fuse to carry current

$C =$ basic failure of circuit to carry current

$D =$ basic failure of motor

If we let S be the successful start of the motor when the operator closes the switch, then the top event of the fault tree can be written as the failure Q of the system due to the failure of any of the series-connected components.

$$Q = A \cup B \cup C \cup D \tag{25.46}$$

The RBD expresses the compliment of (25.46), or the success of the system.

$$S = \bar{A} \cap \bar{B} \cap \bar{C} \cap \bar{D} \tag{25.47}$$

This is intuitively correct, since it is clear that all of the components must be free of failure for this simple system to work.

EXAMPLE 25.14

It is proposed that the system of the preceding example might be improved if the circuit feeding the motor is made redundant, as shown in Figure 25.27. Construct the RBD for the new system and write the expression for system successful motor starting.

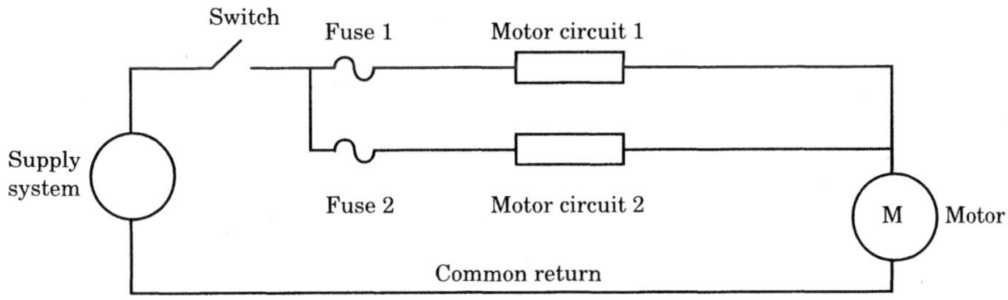

Figure 25.27 Motor circuit with redundant fused supply circuits.

Since there are no new types of components, the nomenclature described previously is still adequate, although it is necessary to distinguish between the items in circuit 1 and those in circuit 2. The RBD can be drawn by inspection, with the result shown in Figure 25.28.

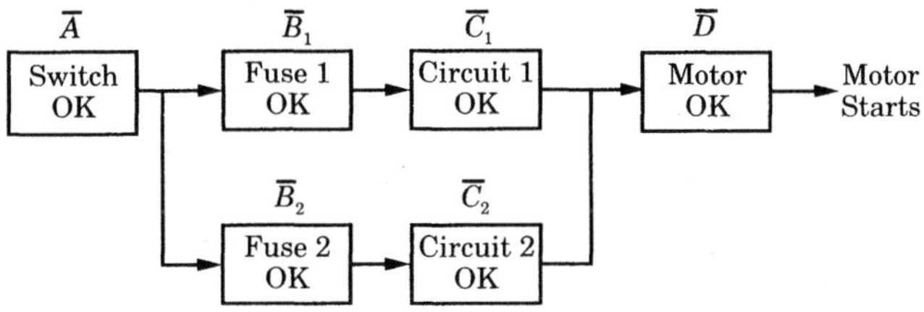

Figure 25.28 RBD for the revised motor system.

The failure of this system will not occur unless both of the circuits are failed, which is often called "overlapping" failures. We assume that the motor common return is perfectly reliable. Then the failure is written as follows.

$$Q = A \cup [(B_1 \cup C_1) \cap (B_2 \cup C_2)] \cup D \tag{25.48}$$

The RBD expresses the success of the system, which can be written in terms of the basic failure events as follows.

$$S = \bar{A} \cap \left[\left(\bar{B}_1 \cap \bar{C}_1\right) \cup \left(\bar{B}_2 \cap \bar{C}_2\right)\right] \cap \bar{D} \tag{25.49}$$

Note that the AND and OR operations are simply reversed between the two expressions. Stated in words, system success occurs if the switch works and either circuit 1 or circuit 2 works and the motor works. Each circuit works if both the fuse works and the circuit works, i.e., each is capable of carrying the current. ■

In many cases, the engineer can write the statement in words and then convert that statement into a mathematical expression.

25.4.2 Success Trees

A success tree is the mathematical dual of a fault tree. In a success tree, the top event is the successful working of the system, rather than its failure. This is sometimes helpful, especially for complex systems where it is easier to define success than failure. This difference is conceptual, as there is no difference in the amount of detail or the amount of work involved. The choice of which logic tree to use, in this case, is a matter of personal preference. In most cases, a success tree can be changed to a fault tree by changing all OR gates to AND gates, and changing all AND gates to OR gates. More complex gates require careful examination.

In the foregoing example, it was noted that the reliability block diagram is a network description of success of the system. Success trees can be constructed that convey exactly the same information. This is left as an exercise.

25.4.3 Truth Tables

A truth table is a tabular expression of all states of basic events in a system that lead to either the occurrence or the nonoccurrence of the top event and the corresponding probabilities of these events. The sum of the failure probabilities from the table gives the probability of failure of the system.

Consider the two gates shown in Figure 25.29. Figure 25.29(a) is an AND gate with two basic events B_1 and B_2. The truth table for this gate is shown in Table 25.6. If we let U_S be the system unavailability, this is given by the first row of Table 25.6, or we may write the unavailability as follows.

$$U_S(t) = \Pr(B_1)\Pr(B_2) \tag{25.50}$$

Figure 25.29(b) is an OR gate with two basic failure events. The truth table for this gate is shown in Table 25.7.

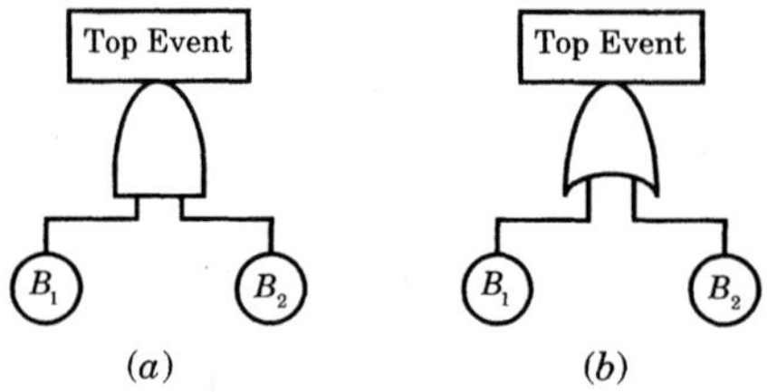

Figure 25.29 Gated (a) AND and (b) OR fault trees.

TABLE 25.6 Truth Table for an AND Gate with Two Basic Events

State No.	Basic Event B_1	Basic Event B_2	Top Event	State Probability
1	Exists	Exists	Exists	$\Pr(B_1)\Pr(B_2)$
2	Exists	Not exist	Not exist	$\Pr(B_1)\Pr(\overline{B}_2)$
3	Not exist	Exists	Not exist	$\Pr(\overline{B}_1)\Pr(B_2)$
4	Not exist	Not exist	Not exist	$\Pr(\overline{B}_1)\Pr(\overline{B}_2)$

TABLE 25.7 Truth Table for an OR Gate with Two Basic Events

State No.	Basic Event B_1	Basic Event B_2	Top Event	State Probability
1	Exists	Exists	Exists	$\Pr(B_1)\Pr(B_2)$
2	Exists	Not exist	Exists	$\Pr(B_1)\Pr(\overline{B}_2)$
3	Not exist	Exists	Exists	$\Pr(\overline{B}_1)\Pr(B_2)$
4	Not exist	Not exist	Not exist	$\Pr(\overline{B}_1)\Pr(\overline{B}_2)$

From Table 25.7, the system unavailability is given by states 1, 2, and 3. This is written mathematically as follows.

$$\begin{aligned} U_S(t) &= \Pr(B_1)\Pr(B_2) + \Pr(B_1)\Pr(\bar{B}_2) + \Pr(\bar{B}_1)\Pr(B_2) \\ &= \Pr(B_1)\Pr(B_2) + \Pr(B_1)[1 - \Pr(B_2)] + [1 - \Pr(B_1)]\Pr(B_2) \\ &= \Pr(B_1) + \Pr(B_2) - \Pr(B_1)\Pr(B_2) \end{aligned} \tag{25.51}$$

which agrees with (25.44).

EXAMPLE 25.15

Consider the motor supply circuit of Figure 25.27, where each circuit consists of a fuse and the circuit wiring to the motor. Develop a truth table for the states for only the fuse and supply circuit components of the motor supply system.

Solution

Since there are four components, and each component is assumed to have two states, working or failed, there are 2^4 unique states. The truth table for these states are given in Table 25.8, where we use the notation W for *working* and F for *failed*.

TABLE 25.8 Truth Table for the Motor Fuse and Circuit System

State Number	Fuse $\bar{B}_1$	Curcuit $\bar{C}_1$	Fuse $\bar{B}_2$	Curcuit $\bar{C}_2$	System State
1	W	W	W	W	W
2	F	W	W	W	W
3	W	F	W	W	W
4	F	F	W	W	W
5	W	W	F	W	W
6	F	W	F	W	F
7	W	F	F	W	F
8	F	F	F	W	F
9	W	W	W	F	W
10	F	W	W	F	F
11	W	F	W	F	F
12	F	F	W	F	F
13	W	W	F	F	W
14	F	W	F	F	F
15	W	F	F	F	F
16	F	F	F	F	F

If the probability of being in the working and failed states are known, then the probability of the system for each state is known. The probability of the system being found in the working or failed states can be determined by summing the contributions to each state from the last column. The availability of the system can be written as follows.

$$\begin{aligned} A_S(t) &= \sum \Pr(\text{Rows } 1, 2, 3, 4, 5, 9, 13) \\ &= \Pr(\bar{B}_1)\Pr(\bar{C}_1)\Pr(\bar{B}_2)\Pr(\bar{C}_2) + \Pr(B_1)\Pr(\bar{C}_1)\Pr(\bar{B}_2)\Pr(\bar{C}_2) \\ &\quad + \Pr(\bar{B}_1)\Pr(C_1)\Pr(\bar{B}_2)\Pr(\bar{C}_2) + \Pr(B_1)\Pr(C_1)\Pr(\bar{B}_2)\Pr(\bar{C}_2) \\ &\quad + \Pr(\bar{B}_1)\Pr(\bar{C}_1)\Pr(B_2)\Pr(\bar{C}_2) + \Pr(\bar{B}_1)\Pr(\bar{C}_1)\Pr(\bar{B}_2)\Pr(C_2) \\ &\quad + \Pr(\bar{B}_1)\Pr(\bar{C}_1)\Pr(B_2)\Pr(C_2) \end{aligned} \tag{25.52}$$

The unavailability can be computed as the complement of (25.52). ■

The truth table method requires the construction of an exhaustive state enumeration of the system, and this is often a tedious process. However, it is a reliable method of computation that forces the analyst to clearly develop each state of the system. For systems with n components, there are 2^n states, and this makes the truth table very large for systems of even moderate size.

25.4.4 Structure Functions

The structure function method describes the state of each basic event and of the system by a *binary indicator variable*. Suppose that we assign the binary indicator variable Y_i to the basic event i, then we may write [2]

$$Y_i = \begin{cases} 1, & \text{when the basic event } i \text{ occurs} \\ 0, & \text{when the event } i \text{ does not occur} \end{cases} \tag{25.53}$$

The top event is associated with the binary indicator variable $\psi(\mathbf{Y})$, which is related to the state of the system by the *structure function*, ψ, where

$$\psi(\mathbf{Y}) = \begin{cases} 1, & \text{when the top event is occurring} \\ 0, & \text{when the top event is not occurring} \end{cases} \tag{25.54}$$

is the structure function for the top event and where

$$\begin{aligned} \mathbf{Y} &= [Y_1, Y_2, \ldots, Y_n] \\ &= \text{vector of basic event states} \end{aligned} \tag{25.55}$$

The event unions and intersections correspond to the Boolean OR and AND operators, and to algebraic operations, as shown in Table 25.9.

TABLE 25.9 Event, Boolean, and Algebraic Operations [2]

Event	Boolean	Algebraic	Notes*
B_i	$Y_i = 1$	$Y_i = 1$	Event i exists
$\bar{B}_i$	$Y_i = 0$	$Y_i = 0$	Event i does not exist
$B_i \cap B_j$	$Y_i \wedge Y_j = 1$	$Y_i Y_j = 1$	$\Pr(B_i \cap B_j) = E(Y_i \wedge Y_j)$
$B_i \cup B_j$	$Y_i \vee Y_j = 1$	$1-(1-Y_i)(1-Y_j) = 1$	$\Pr(B_i \cup B_j) = E(Y_i \vee Y_j)$
$B_1 \cap \ldots \cap B_n$	$Y_1 \wedge \ldots \wedge Y_n = 1$	$Y_1 \times \ldots \times Y_n = 1$	$\Pr(B_1 \cap \ldots \cap B_n) = E(Y_1 \wedge \ldots \wedge Y_n)$
$B_1 \cup \ldots \cup B_n$	$Y_1 \vee \ldots \vee Y_n = 1$	$1-(1-Y_1) \times \ldots \times (1-Y_n) = 1$	$\Pr(B_1 \cup \ldots \cup B_n) = E(Y_1 \vee \ldots \vee Y_n)$

Note: $E(x)$ = expected value of x, a probability.

The Boolean operators can be manipulated following the rules reviewed in Table 25.10. Using these rules, the top event can be represented in terms of the structure function.

For the AND gate fault tree shown in Figure 25.22, the top event exists if and only if all input basic events exist. This may be written in terms of the system structure function as follows.

$$\psi(\mathbf{Y}) = \psi(Y_1, Y_2, \ldots Y_n) = Y_1 \wedge Y_2 \wedge \ldots \wedge Y_n \tag{25.56}$$

In this expression, Y_i is the indicator variable for the basic event B_i. Expressed in terms of algebraic operators, we may write (25.56) as follows.

$$\psi(\mathbf{Y}) = \prod_{i=1}^{n} Y_i = Y_1 Y_2 \ldots Y_n \tag{25.57}$$

TABLE 25.10 Rules for Boolean Manipulations [2]

Laws	Algebraic Interpretation & Remark
Identities:	
$B \vee B = B$	$1-(1-Y)(1-Y)=Y$
$B \wedge B = B$	$YY = Y$
Commutative laws:	
$B_1 \vee B_2 = B_2 \vee B_1$	$1-(1-Y_1)(1-Y_2)=1-(1-Y_2)(1-Y_1)$
$B_1 \wedge B_2 = B_2 \vee B_1$	$Y_1Y_2 = Y_2Y_1$
Associative laws:	
$B_1 \vee (B_2 \vee B_3) = (B_1 \vee B_2) \vee B_3$	Can be written as $B_1 \vee B_2 \vee B_3$
$B_1 \wedge (B_2 \wedge B_3) = (B_1 \wedge B_2) \wedge B_3$	Can be written as $B_1 \wedge B_2 \wedge B_3$
Distributive laws:	
$B_1 \wedge (B_2 \vee B_3) = (B_1 \wedge B_2) \vee (B_1 \wedge B_3)$	
$B_1 \vee (B_2 \wedge B_3) = (B_1 \vee B_2) \wedge (B_1 \vee B_3)$	
Absorptive laws:	
$B_1 \wedge (B_1 \wedge B_2) = B_1 \wedge B_2$	$Y_1Y_1Y_2 = Y_1Y_2$
$B_1 \vee (B_1 \wedge B_2) = B_1$	$1-(1-Y_1)(1-Y_1Y_2)=Y_1$
Set definitions:	
$B \wedge 1 = B$	$Y \cdot 1 = Y$
$B \wedge 0 = 0$	$Y \cdot 0 = 0$
$B \vee 0 = B$	$1-(1-Y)(1-0)=Y$
$B \vee 1 = 1$	$1-(1-Y)(1-1)=1$

The gated OR tree of Figure 25.23 indicates that the top event, system failure, occurs if any of the input basic events occur. In this case, the structure function is written as

$$\psi(\mathbf{Y}) = Y_1 \vee Y_2 \vee \ldots \vee Y_n \tag{25.58}$$

or, in algebraic form

$$\begin{aligned}\psi(\mathbf{Y}) &= 1 - \prod_{i=1}^{n}(1-Y_i)\\ &= 1-(1-Y_1)(1-Y_2)\ldots(1-Y_n)\end{aligned} \tag{25.59}$$

EXAMPLE 25.16

Consider the fault tree shown in Figure 25.25 for a two out of three voting system [2]. Assume that the probability of all of the basic events occurring are equal to 0.6, which we may write in terms of binary indicator variables as follows.

$$E(Y_1) = E(Y_2) = E(Y_3) = 0.6 \tag{25.60}$$

Compute the probability of the top event for the two out of three system and compare this result with another system that is simply a two component parallel connection of components.

Solution

For the two out of three system, we may write the probability of the top event as follows.

$$\begin{aligned}\psi(\mathbf{Y}) &= (Y_1 \wedge Y_2) \vee (Y_2 \wedge Y_3) \vee (Y_3 \wedge Y_1)\\ &= 1-[1-(Y_1 \wedge Y_2)][1-(Y_2 \wedge Y_3)][1-(Y_3 \wedge Y_1)]\\ &= 1-[1-Y_1Y_2][1-Y_2Y_3][1-Y_3Y_1]\end{aligned} \tag{25.61}$$

This can be expanded and simplified using the absorption laws as follows.

$$\psi(\mathbf{Y}) = 1 - \begin{bmatrix} 1 - Y_1Y_2 - Y_2Y_3 - Y_3Y_1 + Y_1Y_2Y_2Y_3 + Y_2Y_3Y_3Y_1 \\ +Y_3Y_1Y_1Y_2 - Y_1Y_2Y_2Y_3Y_3Y_1 \end{bmatrix}$$
$$= Y_1Y_2 + Y_2Y_3 + Y_3Y_1 - 2Y_1Y_2Y_3 \tag{25.62}$$

The probability of the top event is then expressed as the expected value of the binary indicator variable ψ.

$$\begin{aligned} Q_S(t) &= E(\psi(\mathbf{Y})) = E(Y_1Y_2) + E(Y_2Y_3) + E(Y_3Y_1) - 2E(Y_1Y_2Y_3) \\ &= E(Y_1)E(Y_2) + E(Y_2)E(Y_3) + E(Y_3)E(Y_1) \\ &\quad - 2E(Y_1)E(Y_2)E(Y_3) \\ &= 3 \times (0.6)^2 - 2 \times (0.6)^3 = 0.648 \end{aligned} \tag{25.63}$$

This can be compared with the parallel system of only two components by computing the expected value of that system top event.

$$Q_S(t) = E(\psi(\mathbf{Y})) = E(Y_1) + E(Y_2) - E(Y_1)E(Y_2) \tag{25.64}$$

If the components have the same probability of failure as above, the result is computed to be

$$Q_S(t) = 2 \times 0.6 - (0.6)^2 = 0.84 \tag{25.65}$$

Thus, it is seen that the parallel system has an 84% chance of experiencing the top event (system failure), but the two out of three system has only a 64.8% of system failure. ■

25.4.5 Minimal Cut Sets

The system unavailability or top event can also be computed using minimal cut sets. Consider a system that has m minimal cut sets, defined as follows.

$$\begin{aligned} C_1 &= \{B_{1,1}, B_{2,1}, \ldots, B_{n_1,1}\} \\ &\ldots \\ C_j &= \{B_{1,j}, B_{2,j}, \ldots, B_{n_j,j}\} \\ &\ldots \\ C_m &= \{B_{1,m}, B_{2,m}, \ldots, B_{n_m,m}\} \end{aligned} \tag{25.66}$$

Now, denote the event B_{ij} by the indicator variable Y_{ij}. The top event occurs only when all of the basic events in any of the cut sets occur simultaneously. Moreover, any cut set causes the top event to occur. This can be expressed using the indicator variable as follows.

$$\psi(\mathbf{Y}) = \bigvee_{j=1}^{m} \left[\bigwedge_{i=1}^{n_j} Y_{ij} \right] \tag{25.67}$$

In algebraic form, we may write (25.67) as

$$\psi(\mathbf{Y}) = \bigvee_{j=1}^{m} \left[\prod_{i=1}^{n_j} Y_{ij} \right] = 1 - \prod_{j=1}^{m} \left[1 - \prod_{i=1}^{n_j} Y_{ij} \right] \tag{25.68}$$

This equation can be expressed as a fault tree, but its construction is left as an exercise.

25.4.6 Minimal Path Sets

A minimal path set is the dual of a minimal cut set. A minimal path set is a collection of basic events. If none of these events occur, then the top event will not occur. Let us represent the minimal path sets by the notation

$$\begin{aligned} P_1 &= \{B_{1,1}, B_{2,1}, \ldots, B_{n_1,1}\} \\ &\ldots \\ P_j &= \{B_{1,j}, B_{2,j}, \ldots, B_{n_j,j}\} \\ &\ldots \\ P_m &= \{B_{1,m}, B_{2,m}, \ldots, B_{n_m,m}\} \end{aligned} \tag{25.69}$$

where the B_{ij} represent the basic events for the path set. These events can also be represented by indicator variables. The top event occurs if at least one basic event in all minimal path sets occur. Thus, we write the structure function as follows.

$$\begin{aligned} \psi(\mathbf{Y}) &= \bigwedge_{j=1}^{m} \left[\bigvee_{i=1}^{n_j} Y_{ij} \right] \\ &= \bigwedge_{j=1}^{m} \left[1 - \prod_{i=1}^{n_j} [1 - Y_{ij}] \right] = \prod_{j=1}^{m} \left[1 - \prod_{i=1}^{n_j} [1 - Y_{ij}] \right] \end{aligned} \tag{25.70}$$

The minimal path set solution can also be represented as a fault tree. This is left as an exercise.

25.5 STATE SPACE AND MARKOV PROCESSES

The foregoing treatment of the reliability of a protective system has omitted any discussion as to how the repair of failed components enters into the reliability modeling. This will now be considered, and a model will be developed for protective system performance including repair.

25.5.1 The Markov Process

Markov models are functions of two variables, the state of the system and the time of observation. This leads to four kinds of model formulations, where either variable can be considered to be either discrete or continuous. The Markov process describes the transition in time from one state of a system to another state during a small time interval Δt and the probability of a transition occurring during this time interval. The occurrences are discrete and time is continuous; therefore, we describe this process as a discrete-state, continuous-time process. The following basic assumptions are necessary in deriving the Markov Process:

1. The probability that a transition occurs in time Δt is $h(t)\,\Delta t$, where $h(t)$ is the hazard function associated with the two states in question. When $h(t) = \lambda$ is a constant, the process is called *homogeneous*. If the hazard functions are time dependent, then the process in *nonhomogeneous*.
2. The probability of more than one transition in time Δt are infinitesimals of a higher order and can be neglected.

The Markov model is defined by a set of probabilities p_{ij} that define the probability of transition from any state i to another state j. The transition depends only on the states i and j and is completely independent of all past states except the last one, state i.

The Markov process can be specified by a set of differential equations and the initial conditions for these equations. Since the transitions depend only on the last state, the differential equations are always first order. The constants are determined from a transition probability matrix, where the rows represent the probability of being in a state i at time t and the columns represent the probability of being in state j at time $t + \Delta t$. A typical transition matrix for a system with $n + 1$ states is given in Table 25.11.

TABLE 25.11 A Typical Transition Matrix

	$s_0(t+\Delta t)$	$s_1(t+\Delta t)$	$s_2(t+\Delta t)$	...	$s_n(t+\Delta t)$
$s_0(t)$	p_{00}	p_{01}	p_{02}	...	p_{0n}
$s_1(t)$	p_{10}	p_{11}	p_{12}	...	p_{1n}
$s_0(t)$	p_{20}	p_{21}	p_{22}	...	p_{2n}
...	...	...	...	...	...
$s_n(t)$	p_{n0}	p_{n1}	p_{n2}	...	p_{nn}

The elements in the transition matrix represent the probability that in time Δt the system will undergo a transition from initial state i to final state j. The terms on the diagonal represent the probability that the system will remain in its initial state. The case of greatest interest is that in which the hazard rates are all constants. This means that the failure and repair rates are described by exponential distributions. (See Section 25.5.4.)

For example, consider a system with only two states, as shown in Figure 25.30. This system has two states, labeled 1 and 2. The probability of making a transition from one state to the other in time Δt is shown as the transition rate multiplied by the time, as noted on the diagram. The probability of remaining in either state 1 or 2 is the probability of not making a transition away from that state. The process depicted is a repairable process. State 1 is the UP or working state and state 2 is the DOWN or failed state. The transition rate from UP to DOWN is the failure rate λ and the transition rate from DOWN to UP is the repair rate μ.

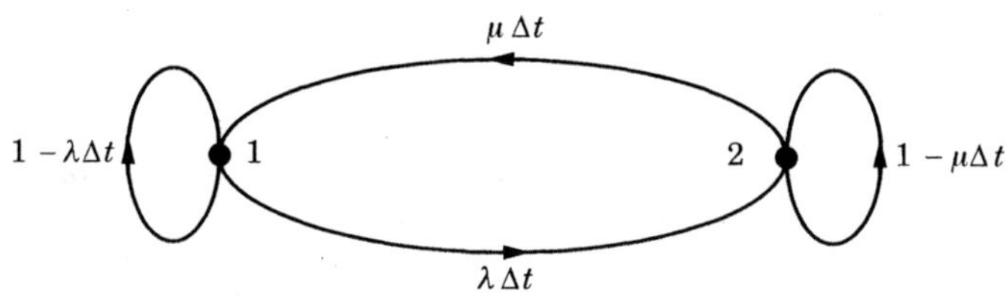

Figure 25.30 Markov graph for a system with two states.

By inspection of Figure 25.30, and using conditional probability concepts, we may write the following equations:

$$\begin{aligned} P_1(t+\Delta t) &= P_1(t)(1-\lambda\,\Delta t) + P_2(t)(\mu\,\Delta t) \\ P_2(t+\Delta t) &= P_2(t)(1-\mu\,\Delta t) + P_1(t)(\lambda\,\Delta t) \end{aligned} \tag{25.71}$$

where P_k is defined as the probability of residing in state k. Rearranging, we have the following differential equations in the limit as Δt approaches zero.

$$\begin{aligned} \frac{P_1(t+\Delta t) - P_1(t)}{\Delta t} &= \mu P_2(t) - \lambda P_1(t) = \frac{dP_1(t)}{dt} \\ \frac{P_2(t+\Delta t) - P_2(t)}{\Delta t} &= \lambda P_1(t) - \mu P_2(t) = \frac{dP_2(t)}{dt} \end{aligned} \tag{25.72}$$

These equations may be solved for the two state probabilities as functions of time, with the following result.

$$P_1(t) = \frac{\mu}{\lambda+\mu} + \frac{e^{-(\lambda+\mu)t}}{\lambda+\mu}\left[\lambda P_1(0) - \mu P_2(0)\right]$$
$$P_2(t) = \frac{\lambda}{\lambda+\mu} + \frac{e^{-(\lambda+\mu)t}}{\lambda+\mu}\left[\mu P_2(0) - \lambda P_1(0)\right] \tag{25.73}$$

We must make some assumption regarding the initial probabilities of the two states in order to get a complete solution. It is common to assume that the system is in the working state initially. Therefore, the probability of being in state 1 initially is unity, which makes being in state 0 an impossibility.

$$P_1(0) = 1$$
$$P_2(0) = 0 \tag{25.74}$$

Then, we compute

$$P_1(t) = \frac{\mu}{\lambda+\mu} + \frac{\lambda e^{-(\lambda+\mu)t}}{\lambda+\mu}$$
$$P_2(t) = \frac{\lambda}{\lambda+\mu} - \frac{\lambda e^{-(\lambda+\mu)t}}{\lambda+\mu} \tag{25.75}$$

A plot of these functions is shown in Figure 25.31. The probability of state 1 is called the availability, $A(t)$, and that of state 2 the unavailability, $U(t)$. Note that the availability falls off from its initial (new) value and asymptotically approaches a constant value.

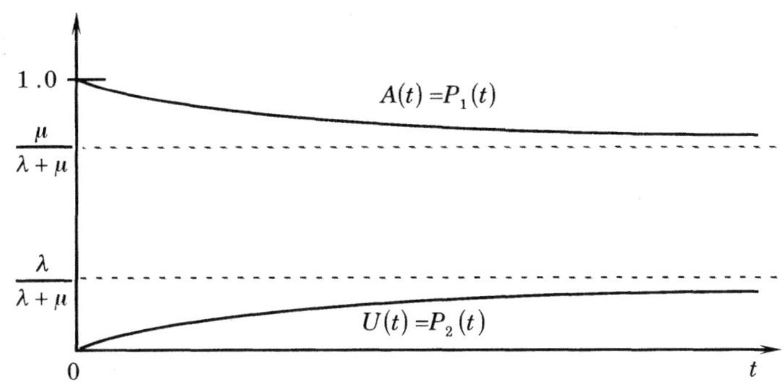

Figure 25.31 State probabilities as a function of time.

25.5.2 Stationary State Probabilities

The limiting state or stationary probabilities can be found from (25.75) by letting the time parameter approach infinity, in which case we compute

$$P_1 = P_1(\infty) = \frac{\mu}{\lambda+\mu}$$
$$P_2 = P_2(\infty) = \frac{\lambda}{\lambda+\mu} \tag{25.76}$$

For the exponential distribution, we know that the failure rate is the inverse of the mean time to failure. We can also compute the mean time to repair as the inverse of the repair rate. Thus we write

$$MTTF = m = \frac{1}{\lambda}$$
$$MTTR = r = \frac{1}{\mu} \tag{25.77}$$

It is common practice to refer to the probability P_1 as the availability of the system and the probability P_2 as the unavailability. It is important to clearly distinguish between the availability, as defined above, and the reliability, which is given by

$$R(t) = e^{-\lambda t} \tag{25.78}$$

The availability is the probability of *being found* in the operating state, whereas the reliability is the probability of *remaining* in the operating state. Note that the sum of the probabilities P_1 and P_2 of the two states is equal to unity.

25.5.3 General Algorithm for Markov Analysis

The determination of the differential equations that represent the state transition process represented by the Markov model can be determined in a general form that is applicable to systems of any size and complexity. The one practical limitation is that the transition intensities must be constants. If the transition intensities are not constant there are methods that can be used, but the process becomes more difficult [17]. In the following development, we consider these transition intensities to be constant failure and repair rates.

The state probabilities of the proposed Markov model can be computed using matrix differential equations, which can be constructed using the following rule [2].

$$\begin{aligned}\frac{dp_i(t)}{dt} &= \dot{p}_i(t) = (\text{inflow to state } i) - (\text{outflow from state } i)\\ &= \sum_{j \neq i} (\text{rate of transition to state } i \text{ from state } j) \times p_j\\ &\quad - \sum_{j \neq i} (\text{rate of transition from state } i \text{ to state } j) \times p_i\end{aligned} \tag{25.79}$$

where $p_i(t)$ = probability of system state i at time t

We can write (25.79) in matrix form as

$$\dot{\mathbf{p}} = \mathbf{Tp} \tag{25.80}$$

where $\mathbf{T}$ = matrix of state transitions

The state transitions are the constant failure and repair rates associated with the transitions noted on the Markov graph and can be determined by inspection of that graph.

The solution of the vector differential equation (25.80) can be written as [18]

$$\mathbf{p}(t) = \mathbf{p}_0 e^{\mathbf{A}t} \tag{25.81}$$

where $\mathbf{p}_0$ = a vector of initial conditions of all states

The exponential function

$$e^{\mathbf{A}t} = \mathbf{I} + \mathbf{A}t + \mathbf{A}^2\frac{t^2}{2!} + \ldots = \mathbf{I} + \sum_{k=1}^{\infty} \frac{\mathbf{A}^k t^k}{k!} \tag{25.82}$$

converges absolutely and uniformly in any finite interval of the time axis [2].

The initial conditions may be chosen such that the system is in some state with a certain probability and other state probablities having a probability of one minus the chosen state probability. It is common practice to assume initial conditions as that state where all components are UP with a probability of unity all other states have probability zero. The differential equations can be integrated step by step until convergence to a final value is achieved, which is determined when all variations in the state probabilities become arbitrarily small. The result is a timewise record of the transient solution of the state probabilities over time to the steady-state solution. The initial conditions have no influence on the final result.

In many cases, we are most interested in the final values of the state probabilities. This corresponds to the condition where the derivatives of (25.80) vanish, or we have the algebraic system of equations

$$\mathbf{0} = \mathbf{Tp} \tag{25.83}$$

Usually, the transient solution is of no particular interest as the final value is the condition sought. This may not be true for protective systems where self-testing is used, but this has not been explored in any detail.

Now, it can be shown that all columns of **T** sum to zero, so the determinant of **T** is zero, which means that the equations are not linearly independent. However, we can discard one of the equations and substitute the equation

$$\sum_{i=1}^{n} p_i = 1 \tag{25.84}$$

since we know that the sum of the state probabilities is a certainty. We can clarify the resulting equation as follows. Suppose we write the transition matrix as

$$\mathbf{T} = \begin{bmatrix} t_{11} & t_{12} & \dots & t_{1n} \\ t_{21} & t_{22} & \dots & t_{2n} \\ \dots & \dots & \dots & \dots \\ t_{n1} & t_{n2} & \dots & t_{nn} \end{bmatrix} \tag{25.85}$$

The off-diagonal elements of **T** are the failure and repair rates that represent the transitions between states of the system; the diagonal elements are the transitions out of each state (but with a negative sign).

Making the substitution of (25.84) for the nth row of the **T** matrix, we get a new transition matrix that we identify with the subscript n to distinguish it from (25.85) to write

$$\mathbf{T}_n = \begin{bmatrix} t_{11} & t_{12} & \dots & t_{1n} \\ \dots & \dots & \dots & \dots \\ t_{(n-1)1} & t_{(n-1)2} & \dots & t_{(n-1)n} \\ 1 & 1 & \dots & 1 \end{bmatrix} \tag{25.86}$$

Then, we can write the new steady-state equation as

$$\begin{bmatrix} t_{11} & t_{12} & \dots & t_{1n} \\ \dots & \dots & \dots & \dots \\ t_{(n-1)1} & t_{(n-1)2} & \dots & t_{(n-1)n} \\ 1 & 1 & \dots & 1 \end{bmatrix} \begin{bmatrix} p_1 \\ \dots \\ p_{n-1} \\ p_n \end{bmatrix} = \begin{bmatrix} 0 \\ \dots \\ 0 \\ 1 \end{bmatrix} = \mathbf{b} \tag{25.87}$$

where we recognize that the right-hand side, now designated **b**, is no longer zero, but is changed due to the insertion of (25.84) as the nth equation. Then the final solution of the steady-state condition can be written

$$\mathbf{p} = \mathbf{T}_n^{-1}\mathbf{b} \tag{25.88}$$

to give the probability of every state in the system. The inverse of the transition matrix can be found using any desired method. Modern software solutions are readily available to compute numerical results.

If the system is not too large, it is often possible to obtain a Laplace transform solution of (25.80) in closed form, which gives an expression for the probability of each state in the system as a function of time [17]. However, this is not practical on systems where the dimensions of the matrix **T** are large. Systems that can be described by a set of linear, constant coefficient differential equations can be solved using modern computer software.

The Markov model is a very useful tool for the analysis of even rather large systems. Since the probability of each state in the system is determined, the solution can be used in many different ways, such as the computation of the system reliability, availability, mean time to failure, and other measures of performance. These computations usually require that the system logic be considered, for example, whether the elements are in a series, parallel, or other type of logical arrangement. For example, if a series logic is assumed, then all components must be in the operating state for system success, which will be expressed in the probability of only one system state. Parallel or other more complex system logical descriptions will provide the information to determine the definition of reliability or other factors of interest. However, the basic information to compute these measures of system performance is the state probabilities, which is provided by the Markov model.

25.5.4 Model of Two Repairable Components

Consider a system that consists of two repairable components, where the components are not identical and have their own unique failure and repair rates. For this case the state space diagram is shown in Figure 25.32.

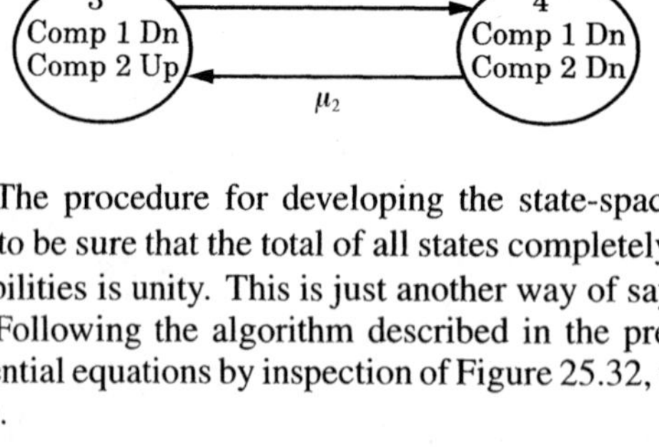

Figure 25.32 State Space diagram for two repairable components.

The procedure for developing the state-space diagrams is relatively simple, but one needs to be sure that the total of all states completely defines the system and the total of all the probabilities is unity. This is just another way of saying that there are no undefined states.

Following the algorithm described in the previous section, we can write the Markov differential equations by inspection of Figure 25.32, with the following result (also see problem 25.35).

$$\begin{bmatrix} \dot{p}_1(t) \\ \dot{p}_2(t) \\ \dot{p}_3(t) \\ \dot{p}_4(t) \end{bmatrix} = \begin{bmatrix} -(\lambda_1+\lambda_2) & \mu_2 & \mu_1 & 0 \\ \lambda_2 & -(\lambda_1+\mu_2) & 0 & \mu_1 \\ \lambda_1 & 0 & -(\lambda_2+\mu_1) & \mu_2 \\ 0 & \lambda_1 & \lambda_2 & -(\mu_1+\mu_2) \end{bmatrix} \begin{bmatrix} p_1(t) \\ p_2(t) \\ p_3(t) \\ p_4(t) \end{bmatrix} \tag{25.89}$$

The definition of *availability* for the system of two repairable items depends on their connection. If connected in a series logic, the failure of either one of the items constitutes a total system failure; therefore, states 2, 3, and 4 are all failure states and only state 1 is an operating state. On the other hand, if the items are connected in a parallel logic, both items must be failed simultaneously for system failure to occur, which means that only state 4 is the failure state and the other three states are all operating states.

25.5.5 Markov Models with Special Failure Modes

The Markov modeling process is readily extended to special cases; for example, one component may have a partial outage state and a full outage state. This is not likely to occur for a single component, but may well occur for a system or a subsystem. The state space diagram is shown in Figure 25.33, which is the diagram for the failure modes of a transformer that has oil pumps and air circulating fans that are required in order to obtain full rated output of the transformer. If only the fans or pumps fail, the output is derated, but the transformer still operates at this lower rating.

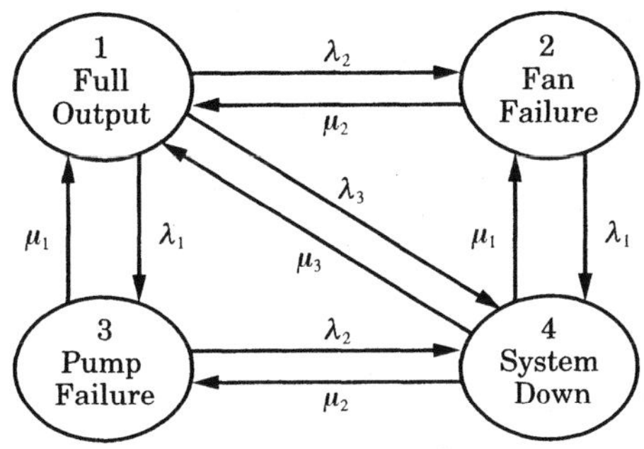

Figure 25.33 Markov model of a system with partial failure.

Note that this example requires a third failure rate in addition to the failure rates for fans and pumps. This failure rate represents all types of failure that cause the transformer to be placed on full outage without making the transition through a partial outage state, for example, a transformer core or winding failure. We also observe that there may be several fans and several pumps, in which case there may be multiple derated states for each item and the Markov graph becomes more complex. The procedure for constructing the graph is still straightforward, however.

Some examples of systems of interest to the power engineer that exhibit partial outage states are boilers, generators, and HVDC converters. For system protection, most of the systems and subsystems are either up or down, with no intermediate states. There can be outages of redundant items, but most of the items in the protective system are not subject to a partial outage. A possible exception may be in a digital protective device where a partial memory failure may disable some protective functions, but may leave other functions operable.

25.5.6 Failure Frequency and Duration

Consider the case of a single component and its state space diagram, shown in Figure 25.30. If we should observe the history of this repairable component over a long period of time, we would see a sequence of up and down cycles, similar to that shown in Figure 25.34.

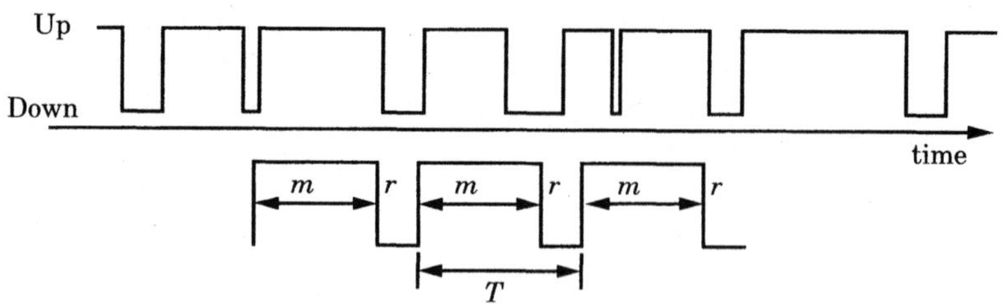

Figure 25.34 One component failure and repair history.

The top portion of the component history shows the actual cycles of up and down times. The mean time to failure, m, is the mean or expected value of the up times. The mean time to repair, r, is the mean of the down times. These mean values are shown approximately in the lower diagram, which also shows the period T of the average of the up-down cycle, which is often called the *mean time between failures* (*MTBF*). Thus, we can write

$$\begin{aligned} m &= \text{MTTF} = \frac{1}{\lambda} \\ r &= \text{MTTR} = \frac{1}{\mu} \\ T &= \text{MTBF} = m + r = \frac{1}{w} \end{aligned} \tag{25.90}$$

where w is the frequency of failure (and also the frequency of repair). We can now write the probabilities of the up and down states in terms of the frequency of failure, as follows.

$$\begin{aligned} P_0 &= \frac{m}{m+r} = \frac{m}{T} = \frac{1}{\lambda T} = \frac{w}{\lambda} \\ P_1 &= \frac{r}{m+r} = \frac{r}{T} = \frac{1}{\mu T} = \frac{w}{\mu} \end{aligned} \tag{25.91}$$

Then we may write the frequency as

$$w = P_0\lambda = P_1\mu \tag{25.92}$$

This result, although derived from a very simple example, is perfectly general. Stated in words, *the frequency of encountering any given state is computed as the probability of residing in that state times the rate of departure from that state.* In some systems, there may be several rates of departure, in which case they must be added together. For example, the two-state system of Figure 25.32 has the rates of entry and departure shown in the following Table 25.12 [15].

TABLE 25.12 Rates of Departure and Entry for a Two Component System

State Number	Component 1	Component 2	Rate of Departure	Rate of Entry
1	up	up	$\lambda_1 + \lambda_2$	$\mu_1 + \mu_2$
2	down	up	$\lambda_2 + \mu_1$	$\lambda_1 + \mu_2$
3	up	down	$\lambda_1 + \mu_2$	$\lambda_2 + \mu_1$
4	down	down	$\mu_1 + \mu_2$	$\lambda_1 + \lambda_2$

For this same system, the steady-state probabilities of the four states and the frequencies of encounter are given in Table 25.13 [15]. Again, it can be verified that frequency is the product of probability and departure rates.

TABLE 25.13 State Probabilities and Frequencies*

State Number	State Probability	Frequency of Encournter
1	$\mu_1\mu_2/D$	$\mu_1\mu_2(\lambda_1+\lambda_2)/D$
2	$\lambda_1\mu_2/D$	$\lambda_1\mu_2(\mu_1+\lambda_2)/D$
3	$\mu_1\lambda_2/D$	$\mu_1\lambda_2(\lambda_1+\mu_2)/D$
4	$\lambda_1\lambda_2/D$	$\lambda_1\lambda_2(\mu_1+\mu_2)/D$

*Where $D = (\lambda_1+\mu_1)(\lambda_2+\mu_2)$

Frequencies and durations are important measures of a system's reliability performance. Probabilities are decimal numbers and, in many cases, the probability of success is a decimal that begins with several nines (0.9999xxx). With such numbers, it is difficult to understand the effect of a given system change or to compare two different system changes.

Frequencies and durations of failure are more easily understood. We get a better judgment of the importance of a failure if we compute that it is expected to occur several times a year, for example, or only once in 10 years or once in 100 years. Such numbers help one decide if it is worthwhile to budget large expenditures to improve the system. Failure durations are especially important to the user of a piece of equipment, or of a source of power. For example, if failure durations are long, it may be economical to invest in an alternate or standby source of supply.

Note that the frequency is the limiting value of the independent failure intensity defined in Section 24.5.3.2.

REFERENCES

[1] Von Alven, W. H., Ed. *Reliability Engineering*, Prentice-Hall, Inc., Englewood Cliffs, NJ, 1964.

[2] Kumamoto, H., and E. J. Henley, *Probabilistic Risk Assessment and Management for Engineers and Scientists*, 2nd Ed., IEEE Press, Piscataway, NJ, 1996.

[3] McCormick, N. J., *Reliability and Risk Analysis*, Academic Press, New York, 1981.

[4] Roberts, N. H., W. E. Vesely, D. F. Haasl, and F. F. Goldberg, *Fault Tree Handbook*, NUREG-0942, U. S. Nuclear Regulatory Commission, Washington, DC, 1981.

[5] Rayes, L. G., and J. E. Riley, "WAM-E User's Manual," Report NP-4460-CCM, Electric Power Research Institute, Palo Alto, 1986.

[6] MIL-STD-721C, Military Standard, Definitions of Terms for Reliability and Maintainability," Philadelphia, PA, June 12, 1981.

[7] Fussell, J. B., *Generic Techniques in Systems Reliability Assessment*, E. J. Lynn, and J. W. Lynn, Eds., Noordhoff, Leyden, 1976.

[8] Salem, S. L., G. E. Apostolakis, and D. Okrent, "A New Methodology for the Computer Aided Construction of Fault Trees, *Annuls of Nuclear Energy*, 4, 1977, pp. 417–433.

[9] Apostolakis, G. E., S. L. Salem, and J. S. Wu, "CAT: A Computer Code for the Automatic Construction of Fault Trees," Report NP-705, Electric Power Research Institute, Palo Alto, 1978.

[10] Mason, S. J., "Feedback Theory - Some Properties of Signal Flow Graphs," *Proc. IRE*, 41 (9), September 1953, pp. 1144–1156.

[11] Mason, S. J., Feedback Theory - Further Properties of Signal Flow Graphs," *Proc. IRE*, 44 (7), July 1956, pp. 920–926.

[12] Kuo, B. C., *Automatic Control Systems*, Prentice-Hall, Inc., Englewood Cliffs, NJ, 1962.

[13] Mason, S. J., Feedback Theory - Further Properties of Signal Flow Graphs," *Proc. IRE*, 44 (7), July 1956, pp. 920–926.

[14] Fussell, J. B., E. B. Henry, and N. H. Marshall, "MOCUS—A Computer Program to Obtain Minimal Cut Sets from Fault Trees," Aerojet Nuclear Company, ANCR-1166, Idaho Falls, ID, 1974.

[15] Billinton, R., and R. N. Allan, *Reliability Analysis of Engineering Systems: Concepts and Techniques*, Plenum Press, New York, 1983.

[16] Dai, S.-H., and M.-O Wang, *Reliability Analysis in Engineering Applications.*, Van Nostrand Reinhold, New York, 1992.

[17] Shooman, M., L., *Probabilistic Reliability: An Engineering Approach*, McGraw-Hill Book Company, New York, 1968.

[18] Ogata, K., *State Space Analysis of Control Systems*, Prentice-Hall, Englewood Cliffs, NJ, 1967.

[19] Henley, E. J., and H. Kumamoto, *Reliability Engineering and Risk Assessment*, Prentice-Hall, Inc., Englewood Cliffs, NJ, 1981.

[20] Youngblood, R. W., and R. B. Worrell, "Top Event Prevention in Complex Systems," ASME/JSME Pressure Vessels & Piping Conference, PVP-vol. 296, July 1995.

[21] Worrell, R. B., and D. P. Blanchard, "'Top Event Prevention Analysis' to Eliminate Requirements Marginal to Safety," ASME/JSME Pressure Vessels & Piping Conference, PVP-vol. 296, July 1995.

PROBLEMS

25.1 Consider the small system shown in Figure 25.2. Let

$$Ei = \{\text{component faulted to ground}\}$$

Compute the reliability of the system assuming all failures are ground fault failures.

25.2 Assume that, for the simple system of Figure 25.2, the only failure modes of interest are components failing open and failing faulted. With this assumption, compute the reliability of each component against both kinds of failure, and compute the reliability of the system against both kinds of hazard.

25.3 Let p_i be the reliability of component i of a series system. Then we can write the unreliability of the component as

$$q_1 = 1 - p_i$$

Derive an expression for the reliability of the series system that uses the unreliability of each item rather than the reliability. Also find the reliability for the special case when all components are identical.

25.4 Consider a series system with six identical components, each of which has a reliability of 0.95. Compute the system reliability

(a) Using the exact formula

(b) Using the approximate formula derived in problem 25.3.

25.5 Repeat problem 25.4 if the reliability of all components is 0.98.

25.6 A parallel system consists of three identical components. Compute the reliability of the system under the following conditions.

(a) The element reliabilities $p_i = 0.95$.

(b) The element reliabilities $p_i = 0.99$.

25.7 Derive a general formula for the reliability of a series-parallel system where each parallel subsystem has m elements in parallel and there are n of these parallel subsystems connected in series.

25.8 Develop an approximate formula for the series-parallel system reliability of identical elements, each having a small value of q.

25.9 Derive a general formula for the reliability of a parallel-series system where each series system has n elements and there are m such series groups connected in parallel.

25.10 Develop an approximate formula for the parallel-series system reliability of identical elements, each series string having a reliability of r.

25.11 Three components of a system are connected in parallel. However, the system is considered successful if two or more of the components are working. Determine the reliability of this system if the components have different reliabilities.

25.12 Determine the reliability of the system described in problem 25.11 if all three components are identical.

25.13 For the system of three identical components in parallel, let the reliability of a component be $p = 0.95$. Compute the reliability for the following conditions:
(a) All three elements are required for successful operation.
(b) Only two out of three elements are required for successful operation.
(c) Only one element is required for successful operation.

25.14 Consider a system with standby redundancy, with elements A and B as shown in Figure 25.6. We define the following types of switching failure:
1. The switch works when required, with probability p_w.
2. The switch switches without command, with probability q_s.
3. The switch refrains from switching unless required, p_r.
4. On switching the contacts make successfully, probability p_c.

Determine the reliability of the system.

25.15 Refer to the breaker-and-a-half switching station pictured in Figure 25.9. A power transfer through the station from line 1 to line 6 is of interest. Prepare a reliability block diagram of the station between these two terminals.

25.16 Distinguish between a *fault* and a *failure*.

25.17 Distinguish among the following: a primary fault, a secondary fault, and a command fault.

25.18 Define clearly what is meant by inductive analysis as opposed to deductive analysis.

25.19 Is fault tree analysis an inductive or deductive approach to system analysis?

25.20 Distinguish among the following: a failure mechanism, a failure mode, and a failure effect.

25.21 Complete the following table that describes the controls for the flow of fuel to an engine. The subsystem of interest consists of a valve and the valve actuator. Classify the events as viewed from the system, subsystem, valve, or actuator level [6]. The entries in each of the right-hand columns should be "mechanism," "mode," or "effect," or no entry at all.

TABLE P25.21 Controls for the Flow of Fuel to an Engine

Event	System	Subsystem	Valve	Actuator
No flow from subsystem when required				
Valve unable to open				
Binding of actuator stem				
Corrosion of actuator stem				

25.22 A positioning system, shown in Figure P25.22(*a*), is powered by a dc motor, which is controlled by the operator depressing a toggle switch to move the motor shaft load to the desired angular position.

The basic fault tree construction rules are stated in Section 25.2.5. Apply rule #2 to this system. Identify as many system faults as you can and classify each fault as being either state of component or state of system. Do this under two different sets of system environments: (1) system operating, and (2) system in standby.

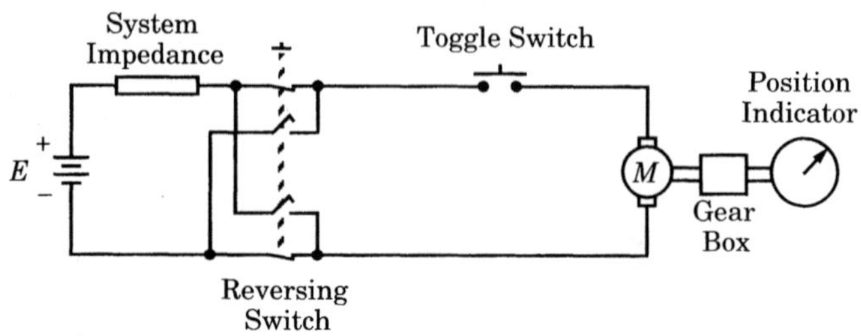

Figure P25.22 (*a*) A motor-driven positioning system.

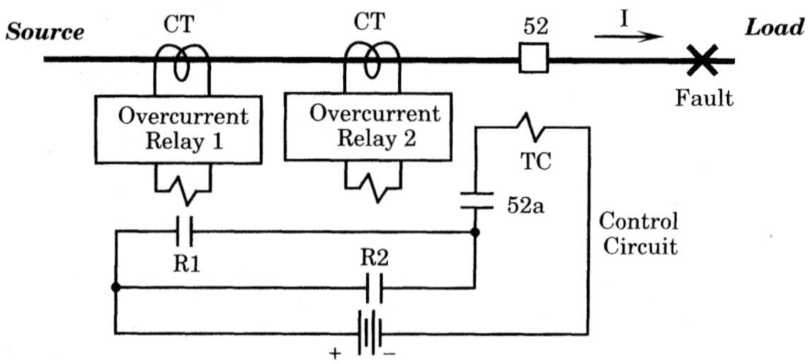

Figure P25.22 (*b*) Overcurrent protection with redundant relays.

25.23 Reconsider the overcurrent relay system described in example 25.11, but modify the system to have redundant relays as shown in Figure P25.22(*b*). (a) Construct decision tables for this system. (b) Construct the fault tree.

25.24 Determine the minimum cut sets for the fault tree constructed from problem 25.23.

25.25 Sketch a Venn diagram of two independent events, B_1 and B_2, where the areas of the two events, as pictured in the Venn diagram, overlap. Show that this confirms (25.44). What is the significance of the overlap of the two events?

25.26 Evaluate the top event for the fault tree shown in Figure P25.26 using the method outlined in Section 25.3.1. Determine the minimal cut sets for this system.

25.27 Develop a success tree for the system described by the fault tree in problem 25.26.

25.28 Tabulate for each state in the fault tree of problem 25.26 the binary indicator variables, Y_i, and the structure function, $\psi(\mathbf{Y})$. There are exactly 2^n unique states for n basic events, or 32 states in this case.

The reader can complete Table P25.28. Note that this task is usually performed by a computer program.

25.29 Develop the fault tree that is the equivalent of (25.68).

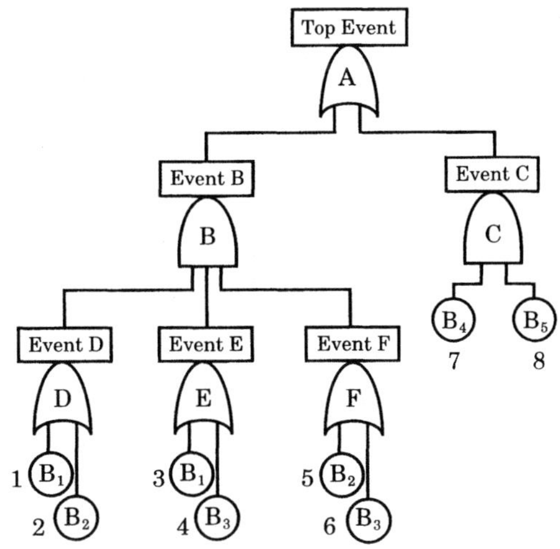

Figure P25.26 A fault tree of a physical system.

TABLE P25.28 Fault Tree States for the Fault Tree of Problem 25.26

State	Y_1	Y_2	Y_3	Y_4	Y_5	$\psi(\mathrm{Y})$
1	0	0	0	0	0	0
2	0	0	0	0	1	0
3	0	0	0	1	0	0
4	0	0	0	1	1	1

25.30 Develop the fault tree that is the equivalent of (25.70).

25.31 Sketch a Markov graph for a system consisting of two identical components, each with failure rate λ and repair rate μ. Develop the transition probability matrix for this system and evaluate the limiting state probabilities.

25.32 Continue with the system described in problem 25.31 and examine the special cases described below. In each case determine the availability and the unavailability.

(a) The two identical components are connected in series.

(b) The two identical components are connected in parallel.

25.33 Develop a Markov graph for a system consisting of three nonidentical components.

25.34 A Markov model of a system can be constructed using the following set of rules [2]:

1. The probability of a transition in time Δt from one state to another is given by $h(t)\Delta t$, where $h(t)$ is the hazard function or transition intensity associated with the two states in question. If all the $h(t)$'s are constant, $h(t) = \lambda$, the model is called *homogeneous*. If any of the transitions are functions of time, the model is *nonhomogeneous*.
2. The probability of more than one transition in time Δt is an infinitesimal of a higher order and can be neglected.

Use this basic model to write the state transition equations for the system shown in Figure P25.34 and show that the result is a system of differential equations in the limit as Δt approaches zero.

25.35 Repeat the procedure of problem 25.34 for the system described by the Markov graph of Figure 25.32.

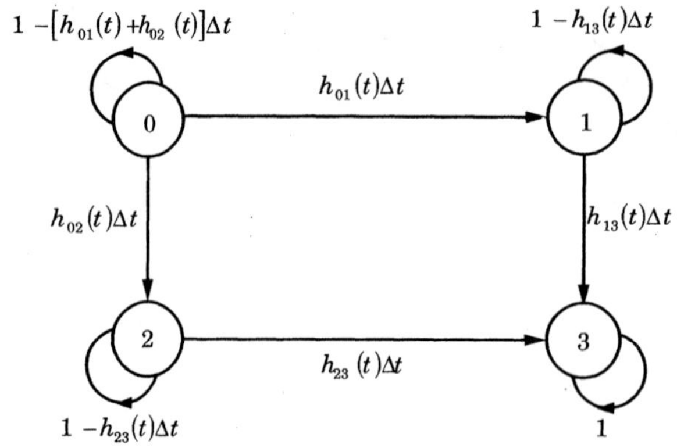

Figure P25.34 Markov graph for two distinct nonrepairable elements.

25.36 Modify the Markov equation given by (25.89) to correctly model the transformer described by the Markov graph of Figure 25.33.

26

Reliability Concepts in System Protection

26.1 INTRODUCTION

This chapter presents some fundamental concepts of reliability as they apply to power system protection. We begin by considering disturbances that occur on the power system and treat the occurrence of these disturbances as random variables with certain probability distributions. We also examine typical protective system configurations and analyze the basic failure modes of these systems. Then we examine the sequential operation that occurs following the pickup of a protective device through the tripping of the circuit breakers, and describe these sequential operations as a stochastic process.

26.2 SYSTEM DISTURBANCE MODELS

Protective systems are designed to recognize certain types of power system disturbances and to isolate those parts of the system on which the disturbance occurs. The first task in developing a reliability model is to understand the power system disturbance as an *event*. This concept will help us develop a probabilistic model of disturbances and to think in terms of the probability of occurrence of a certain event, such as a fault on a transmission line, a generator overheating, or a system underfrequency event.

26.2.1 A Probabilistic Disturbance Model

Power system disturbances are caused by component failures [1]. This process can be used to generate random variables since disturbances are events and with every event we can associate a certain probability. In order to formalize the process, we will find it helpful to define certain quantities.

It is noted in Section 24.3 that a random variable is a function X that assigns to each outcome ξ of an experiment a real number $X(\xi) = r$. It is convenient to think of this process as a functional mapping of events in an event space S to numbers r on the real line R. Thus, the element r, a real number, is called the value of X at ξ. However, it is not sufficient to know the value of r. In order to qualify as a random variable, we must also know something about the event described by the element ξ. Probabilities are defined for events in terms of elements ξ in the event space S. This mapping procedure can be visualized as the process depicted in Figure 26.1.

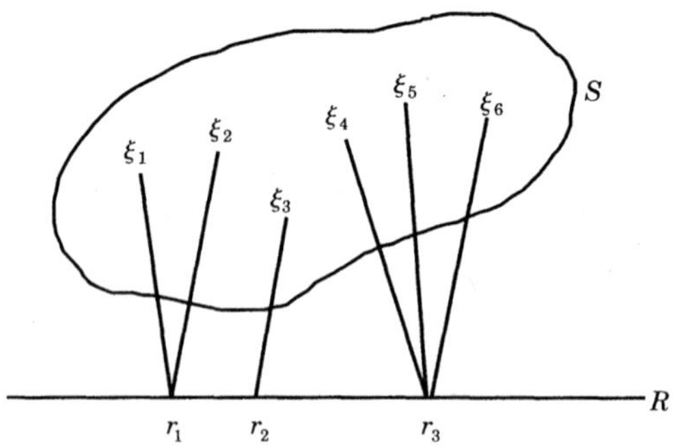

Figure 26.1 Mapping outcomes ξ_i in S to elements r_i in R by $X(\xi_i) = r_i$.

To each element r there corresponds a set of elements ξ in S that are called the inverse image of r. For example, the inverse image of r_1 is given by

$$X^{-1}(r_1) = \{\xi_1, \xi_2\} \tag{26.1}$$

It is also useful to define the *indicator function* for an event. Suppose that we let E be any event, or any subset of S, then we define the indicator function $I_E(\xi)$ for E as follows.

$$I_E(\xi) = \begin{cases} 0 \text{ if } \xi \notin E \\ 1 \text{ if } \xi \in E \end{cases} \tag{26.2}$$

This mapping is shown in Figure 26.2, where we illustrate the event space E and it's complement E^C.

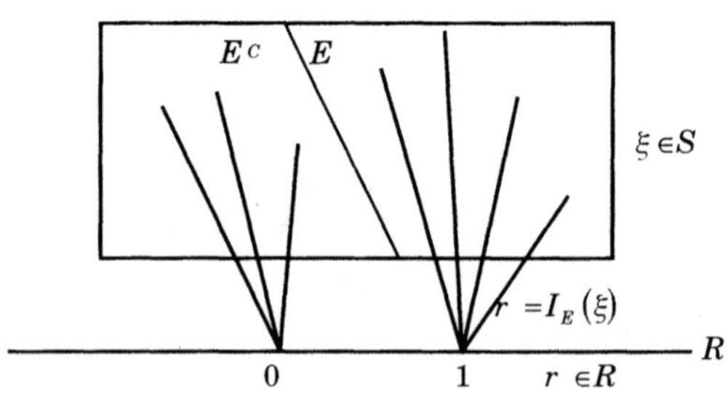

Figure 26.2 Indicator function mapping.

These concepts are applied to power system disturbances or failures by using the failure to generate random variables. This is possible since disturbances are events with which we can associate a probability. For example, suppose we let the event space S consist of all shunt faults that might occur at any bus. Since the buses and bus faults are distinct and mutually exclusive, the event space contains subspaces A_i ($i = 1$ to n) that partition[1] the space S. Thus,

[1] A class of events is said to form a *partition* if one and only one of the member events must occur on each trial. A partition divides the Venn diagram into nonoverlapping sets.

we can let the random variable X be defined as

$$X(\xi_i) = \sum_{i \in J} r_i I_{Ai}(\xi_i) \tag{26.3}$$

and each event ξ_i maps through $X(\xi_i)$ into a single point r_i on R. This is depicted graphically in Figure 26.3.

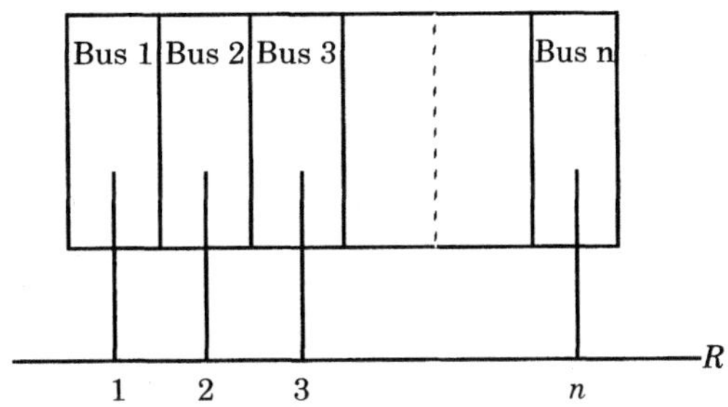

Figure 26.3 Mapping bus events to the real line by the indicator function.

It is clear that the condition described by (26.3) is a random variable since all the elements of the definition of a random variable are satisfied. We also note that the probability of an event $\xi \in A_i$ is given by

$$\Pr(A_i) = \Pr(X = r_i) = p_i \tag{26.4}$$

This idea can be extended to include in the description a separate accounting for the different kinds of shunt faults at any bus. Suppose that we consider the following types only:

1. One-line-to-ground fault
2. Two-line-to-ground fault
3. Line-to-line fault
4. Three-phase fault

Then the event space is further partitioned, as shown in Figure 26.4. This new mapping can be performed by a random variable similar to (26.3) but with A_i replaced by A_{ik} where

$$A_{ik} = \{\xi_k : X(\xi_k) = r_i\} \qquad k = 1, 2, 3, 4 \tag{26.5}$$

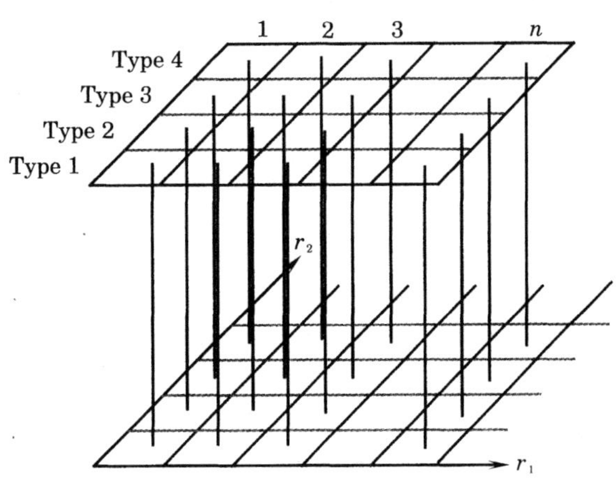

Figure 26.4 Mapping of bus faults to the real plane R^2 of bus shunt fault location and fault type.

The above definition of a random variable for shunt faults assumes that the event space S consists only of this type of disturbance, i.e., that the union of all A_{ik} is the entire space S. Since shunt faults represent an important type of power system disturbance, this choice of events is logical. However, it should be noted that this concept for generating a random variable can be extended to other disturbances. For example, suppose that we define the following types of disturbances that result from system component failures of different types:

1. Shunt faults (short circuits)
2. Longitudinal faults (open lines, series compensation failures)
3. Step changes in bus power injection (generator loss)
4. Step changes in load (load or tie line loss)

One can readily envision the expansion of the event space to include all of these additional types of disturbances. The event space structure is well defined, and probabilities are definable on this space. As long as we avoid any consideration of simultaneous faults, the problem is well posed. Simultaneous faults have a very low probability, so this does not present a serious restriction to the basic concept. It is also clear that we can add transmission lines to the disturbance space, and defining probabilities for various kind of shunt and series disturbances for lines.

A standard definition for a disturbance is [2]

> A *disturbance* in a power system is a sudden change or a sequence of changes in one or more of the parameters of the system, or in one or more of the operating quantities.

When disturbances occur, the protective system must determine, based on the observed parameters that it monitors and the protective device logic, whether this event has the precise characteristics for which protective action is required. The protective system, then, has two choices:

1. Take the predefined protective action for which it was designed, or
2. Refrain from any action, i.e., ignore the event.

In some cases, the action (1) may require concurrence of other protective system observers prior to initiating any action. Since system disturbances, either large or small, are occurring almost all the time, the protective systems must make the above decision continually. In most cases, the correct decision is (2), but when (1) is the desired action, it must be performed according to design, which usually requires prompt action.

26.2.2 Disturbance Distribution

Disturbances occur on the power system all of the time. Loads are switched on and off constantly, and faults occur as items fail due to any cause. These events are all discrete random events that must be modeled using discrete distributions. As noted in Chapter 24, this type of discrete event is described by the Poisson distribution. This distribution describes the probability of isolated events occurring a specified number of times in a given time interval when the rate of occurrence, or hazard rate, is constant. Only the occurrence of the event is counted, the nonoccurrence is not counted.

We write the Poisson distribution with time as the independent variable as

$$F_n = \frac{(\lambda t)^n e^{-\lambda t}}{n!} \tag{26.6}$$

where n is the number of failures in the time interval of interest, F_n is the probability of exactly n failures in time t, and λ is the failure rate.

The expected value of the Poisson distribution is computed as follows where, for simplicity, we let $b = \lambda t$.

$$\begin{aligned} E(n) &= \sum_{n=0}^{\infty} nF_n = \sum_{n=0}^{\infty} \frac{nb^n e^{-b}}{n!} \\ &= \sum_{n=1}^{\infty} \frac{nb^n e^{-b}}{n!} \quad \text{since the } n = 0 \text{ term is zero} \\ &= b\sum_{n=1}^{\infty} \frac{b^{n-1} e^{-b}}{(n-1)!} = b \end{aligned} \tag{26.7}$$

where the last term is unity since the summation of probabilities for all n must be unity. Since the hazard rate is constant, the expected value of failure in an interval t may be written as

$$b = \lambda t \tag{26.8}$$

and the probability of n failures is given by (26.6).

EXAMPLE 26.1 *Transmission Line Failures*

A certain type of transmission line has a constant failure rate of 0.002 failures per kilometer-year. What is the probability of a given number of failures on lines of various lengths for a period of 1 year?

Solution

We arbitrarily take line lengths that are multiples of 50 km. Then we compute the hazard rate λ in failures per year for the entire line by multiplying the length L in km by 0.002 failures/km-yr. Then

$$b = \lambda t = \lambda_{km} L \text{ (1 yr)} = 0.002L \tag{26.9}$$

where L = length of line in km

The annual results are shown in Figure 26.5.

We observe that, for short lines, there is a very high probability of zero failures, but this probability becomes smaller as the length of the line increases. For a line length of 450 km, the probability of one fault is almost as high as that for zero faults annually and the probability for as many as four faults is emerging as having a probability large enough to consider. ■

From Example 26.1 we see that, as the line length increases, the expected value of failure increases, the probability of zero failures decreases and that for one or more annual failures increases. Common sense would arrive at this same general conclusion, but the Poisson distribution allows the engineer to make numerical estimates that are consistent with the laws of probability. Also note that, as the time interval increases, the number of failures would be expected to increase. This can be easily verified by comparing Figure 26.5 with Figure 24.8, which is plotted for larger values of the b parameter.

26.2.3 Disturbance Classifications

One way in which disturbances can be described is by a "topological classification," as shown in Table 26.1 [1].

Series disturbances are caused by changes in the system longitudinal structure or control. This may occur due to open conductors, to changes in transformer taps, phase shifter controls, or HVDC converter controls.

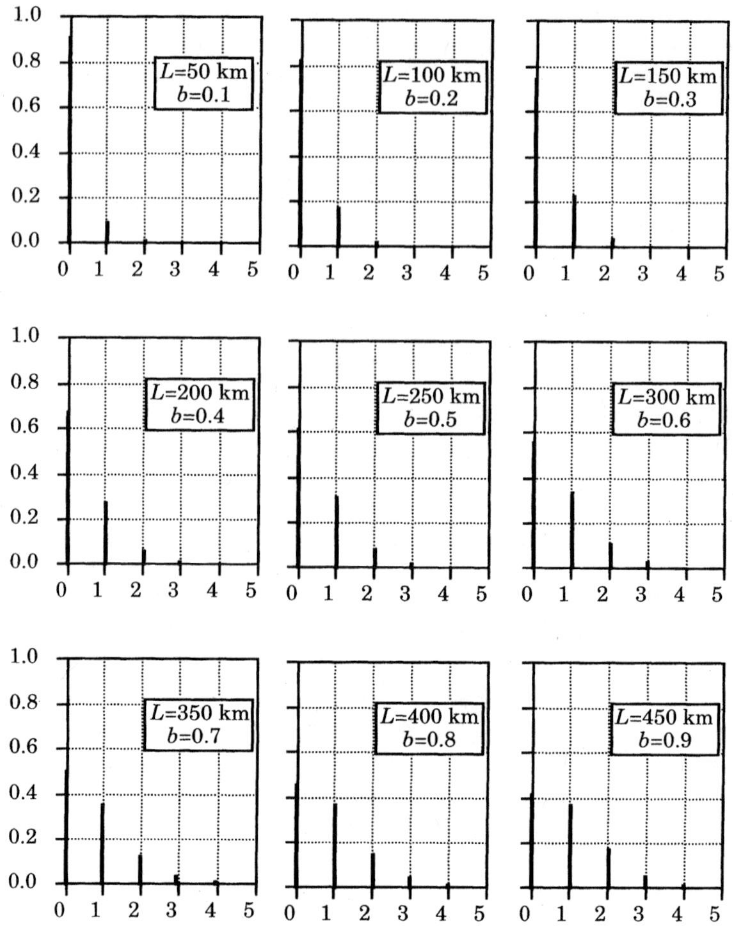

Figure 26.5 Poisson distributions for transmission lines annual failures.

TABLE 26.1 Topological Classification of Power System Disturbances

Series Disturbances	Shunt Disturbances
One line open	One-line-to-ground fault
Two lines open	Two-line-to-ground fault
Three lines open	Line-to-line fault
Change in impedance	Three-phase fault
Change in controlled line flow	Change in nodal injection

Shunt disturbances affect the network physical relationship between phases or between phases and ground at a given point in the transmission system structure. The most severe types of shunt disturbances are short circuits, but changes in injection, such as a change in generation or load, is also a shunt disturbance as this event changes the effective shunt impedance between a node and the reference (considering a generator to be a Thevenin equivalent impedance with negative resistance). Disturbances in HVDC systems or in series compensation systems are series disturbances, which also require protective system response.

It is possible that both types of disturbances occur at the same time, for example, if two lines become entangled with the result that one line is burned open and one or both are faulted to ground at the same time.

Another classification of disturbance is to consider the disturbance location as either affecting a network branch or a node. Such a classification is shown in Table 26.2 [2].

TABLE 26.2 Disturbance Location Classifications

Branch Disturbances	Node Disturbances
Open lines	Open bus or breaker poles
Shunt line faults	Shunt bus faults
Series impedance changes	Shunt injection changes

The various classifications are presented to instill the sense of these disturbances as *events*, with each event having a given probability of occurrence. These probabilities or the failure rate associated with a given type of disturbance can be determined by observing the system over a long period of time. Disturbance events are considered to be *random failures* of the power system, each with a certain probability of occurrence. It is due to the random nature of these events that reliability modeling and prediction are possible.

26.2.4 Probabilistic Model of Disturbances

We now develop a probabilistic model of system disturbances. We describe the disturbances as random variables and develop the associated probability functions.

The different types of disturbances have different probabilities of occurrence, and the disturbances occur at random locations and at random times. Therefore, a given disturbance is described by its type and location as a joint probability. We illustrate the technique for line faults, but it could be repeated for bus faults, generator faults, or failures of other types of apparatus. The line fault is described by the type of fault and the location, or the line on which the fault occurs. These events are described in terms of their probability density functions (*pdf*) and cumulative probability distribution functions (*cdf*).

Suppose we describe fault locations by the random variable X, and fault types by the random variable Y. Combining the distributions for fault locations and fault types, we may compute the joint distribution of the random variables X and Y, which are defined on a three-dimensional probability space.

Since these two discrete random variables are independent, we may write the joint probability distribution as

$$F_{XY}(x, y) = F_X(x)F_Y(y) = \Pr\{X \leq x\}\Pr\{Y \leq y\} \tag{26.10}$$

The quantity (26.10) obeys all the rules for a distribution function, but instead of mapping to the real line, it maps to a two-dimensional region in the x-y plane, as shown in Figure 26.4. The joint density function can also be found for the two random variables. This is left as an exercise. The example given has been restricted to two random independent variables, but the same concept can be extended directly to define random variables for other network parts, such as buses, and to other disturbances, such as series faults.

EXAMPLE 26.2 *Line Fault Type and Location*

Consider the three-node power system shown in Figure 26.6. Our objective is to describe line faults that occur during a particular year on the four lines of that system.

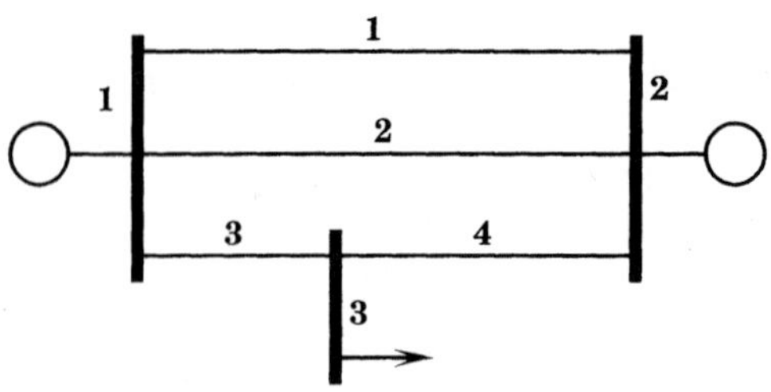

Figure 26.6 Example system with three nodes and four branches.

We assume that the four lines are not exactly alike and have different probabilities of failure, due to differences in line length and line design. Let us also assume that the line fault probabilities for one or more faults over a certain period of observation are given in Table 26.3.

TABLE 26.3 Failure Probabilities for Transmission Lines in Figure 26.6

Line Number	Probability	Line Number	Probability
Line 1	0.300	Line 3	0.175
Line 2	0.300	Line 4	0.225

Since these four lines represent the entire system and the period of observations is long enough so that faults are observed on all lines, the probabilities must add to unity as these probabilities define the entire sample space. The density function for line faults is shown in Figure 26.7. In this figure, the vertical jumps at each of the discontinuities represent the probability of that line being faulted in the system consisting of four lines, with these probabilities having the values given above.

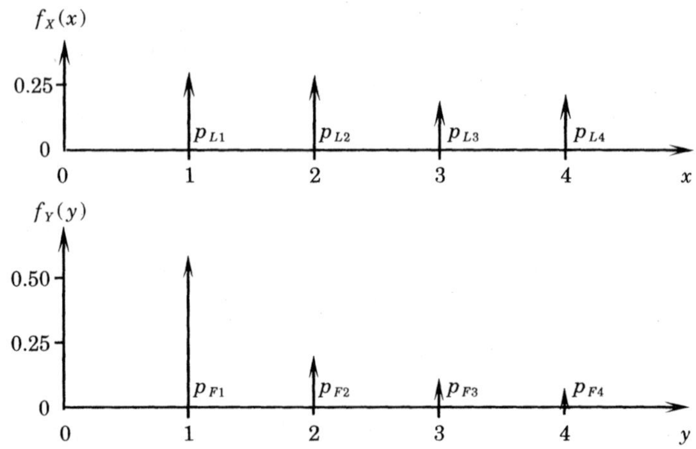

Figure 26.7 Probability density functions for random variables X and Y.

We can also describe the type of fault that occurs by a distribution function. Let us assume that only four types of faults are possible, with those types and their probabilities defined as shown in Table 26.4

(The probabilities are fictitious and are chosen for ease in plotting rather than to describe any real situation. Usually the probability of the one-line-to-ground fault is greater than 0.6.) The density functions for the two random variables are also shown in Figure 26.7, and the cumulative distribution functions are plotted in Figure 26.8, but to a different scale. These distributions are very similar in form, but the numerical data for the two are different.

TABLE 26.4 Fault Type Probabilities for Transmission Lines in Figure 26.6

Fault Type	Probability	Fault Type	Probability
One line to ground	0.600	Two line to ground	0.150
Line to line	0.200	Three phase	0.050

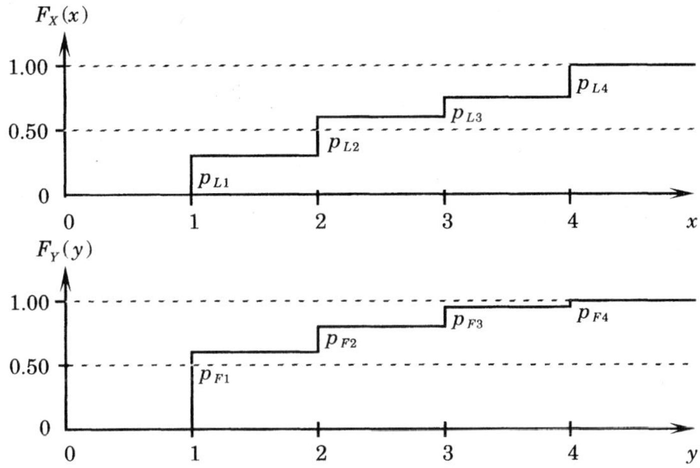

Figure 26.8 Cumulative distribution functions for X and Y.

Both distributions obey the rules stated above for distributions of any random variable. As in the previous case, the size of the jump in the distribution is equal to the probability of that type of fault, considering that only the four types of faults listed above are possible occurrences. Obviously, this concept could be extended to consider all possible types of faults, either series or shunt, as described in Tables 26.1 and 26.2

One could also define a third random variable that defines the location of a fault along the length of any transmission line, as an alternative the fault location could be modeled as a uniform distribution, making faults at all locations of equal probability. This third random variable gives us three experiments for which the set of all possible outcomes are the sample spaces S_L, S_F, and S_K. Each fault represents the simultaneous occurrence of one outcome in each of these experiments and is, in effect, an event in the sample space formed by the Cartesian product of S_L, S_F, and S_K. With this sample space given as

$$S = S_L \times S_F \times S_K \tag{26.11}$$

and the fault events are given by

$$D_m = A_i \times B_j \times C_k \qquad m = (i, j, k) \tag{26.12}$$

The variable D represents the occurrence on all lines i of all type j faults in all line sections k. To complete the specification of this joint experiment, probabilities must be defined for these events. Suppose we defined the probabilities $\Pr(A_i)$, $\Pr(B_j)$, and $\Pr(C_k)$ as the probabilities of events A_i, B_j and C_k. We recognize that the probability of occurrence of the joint event $\Pr(A_i \times B_j \times C_k)$ cannot, in general, be obtained from the individual probabilities unless the events are independent. For our experiment, the events are independent, so we can write

$$\Pr\left(A_i \times B_j \times C_k\right) = \Pr\left(A_i\right)\Pr\left(B_j\right)\Pr\left(C_k\right) \tag{26.13}$$

This concept can be extended to obtain the joint distribution and density functions from the relationships

$$F_{X_L Y_L Z_L}(r, s, v) = F_{X_L}(r)\, F_{Y_L}(s)\, F_{Z_L}(v) \tag{26.14}$$

and similarly for the density function. Since Z_L is also discrete, the joint density and distributions can be expressed as triple summations.

If we assume independence exists between the random variables X_L and Y_L, then assuming that the line identification is independent of its type, the joint distribution and density functions can be obtained by multiplication of the individual functions. In this case, the density and distribution functions can be written as follows.

$$\begin{aligned} f_{X_L Y_L Z_L}(r, s) = {} & p_{L1} p_{F1} \delta(r-1)\,\delta(s-1) + p_{L1} p_{F2} \delta(r-1)\,\delta(s-2) \\ & + p_{L1} p_{F3} \delta(r-1)\,\delta(s-3) + p_{L1} p_{F4} \delta(r-1)\,\delta(s-4) \\ & + p_{L2} p_{F1} \delta(r-2)\,\delta(s-1) + p_{L2} p_{F2} \delta(r-2)\,\delta(s-2) \\ & + p_{L2} p_{F3} \delta(r-2)\,\delta(s-3) + p_{L2} p_{F4} \delta(r-2)\,\delta(s-4) \\ & + p_{L3} p_{F1} \delta(r-3)\,\delta(s-1) + p_{L3} p_{F2} \delta(r-3)\,\delta(s-2) \\ & + p_{L3} p_{F3} \delta(r-3)\,\delta(s-3) + p_{L3} p_{F4} \delta(r-3)\,\delta(s-4) \end{aligned} \tag{26.15}$$

The joint distribution of the two variables X and Y defined previously may be written as

$$\begin{aligned} F_{XY}(x, y) &= F_X(x) F_Y(y) = \Pr\{X \le x\} \Pr\{Y \le y\} \\ &= \sum_{n=1}^{N} \sum_{m=1}^{M} \Pr(x_n) \Pr(y_m)\, u(x - x_n)\, u(y - y_m) \end{aligned} \tag{26.16}$$

Expanding for values of M and N equal to four, we have

$$\begin{aligned} F_{X_L Y_L Z_L}(r, s) = {} & p_{L1} p_{F1} u(r-1)\, u(s-1) + p_{L1} p_{F2} u(r-1)\, u(s-2) \\ & + p_{L1} p_{F3} u(r-1)\, u(s-3) + p_{L1} p_{F4} u(r-1)\, u(s-4) \\ & + p_{L2} p_{F1} u(r-2)\, u(s-1) + p_{L2} p_{F2} u(r-2)\, u(s-2) \\ & + p_{L2} p_{F3} u(r-2)\, u(s-3) + p_{L2} p_{F4} u(r-2)\, u(s-4) \\ & + p_{L3} p_{F1} u(r-3)\, u(s-1) + p_{L3} p_{F2} u(r-3)\, u(s-2) \\ & + p_{L3} p_{F3} u(r-3)\, u(s-3) + p_{L3} p_{F4} u(r-3)\, u(s-4) \end{aligned} \tag{26.17}$$

The magnitude of each pulse of the density and each step of the distribution is the product of probabilities, which is characteristic of joint probabilities. In this discussion, we ignore the third random variable in order to simplify the problem and to be able to plot the results in three-dimensional space, but recognize that the third random variable could be included in the analysis. The results of the joint density and distribution are shown in Figures 26.9 and 26.10, respectively. Note that the joint probabilities for line number and fault type occur as products in (26.15) and (26.16). These products can be computed from the given probability data of Tables 26.3 and 26.4. The joint probabilities are shown in Table 26.5.

Combining the distributions for fault locations and fault types, we may compute the joint distribution of the random variables X and Y, which must be defined in the three-dimensional probability space shown in Figures 26.9 and 26.10. The numerical values shown in Table 26.5 can be seen in these figures, at least approximately.

This joint distribution is plotted as a three-dimensional distribution in Figures 26.9 and 26.10. The quantity (26.16) still obeys all the rules for a distribution function, but instead of mapping to the real line it maps to a two-dimensional area in the x-y plane.

The joint density function of two random variables is given by the second derivative function.

$$f_{X,Y}(x, y) = \frac{\partial^2 F_{X,Y}(x, y)}{\partial x\, \partial y} \tag{26.18}$$

This relationship is evident from the two-dimensional plots of Figures 26.9 and 26.10.

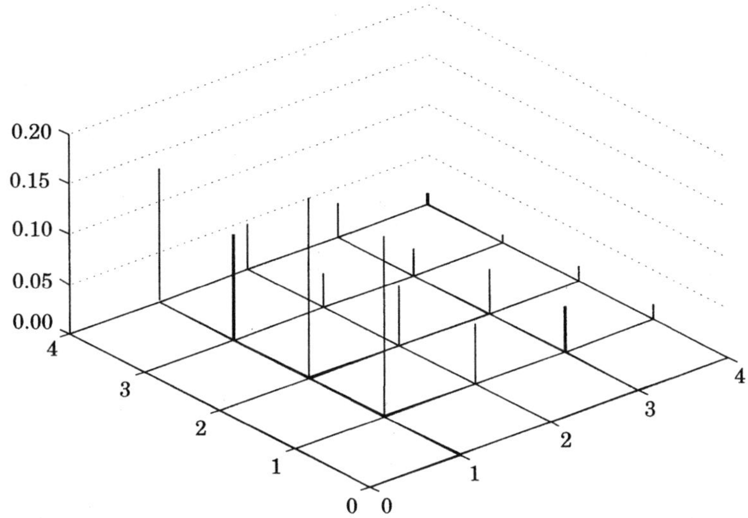

Figure 26.9 The density for random variables of Figures 26.7 and 26.8.

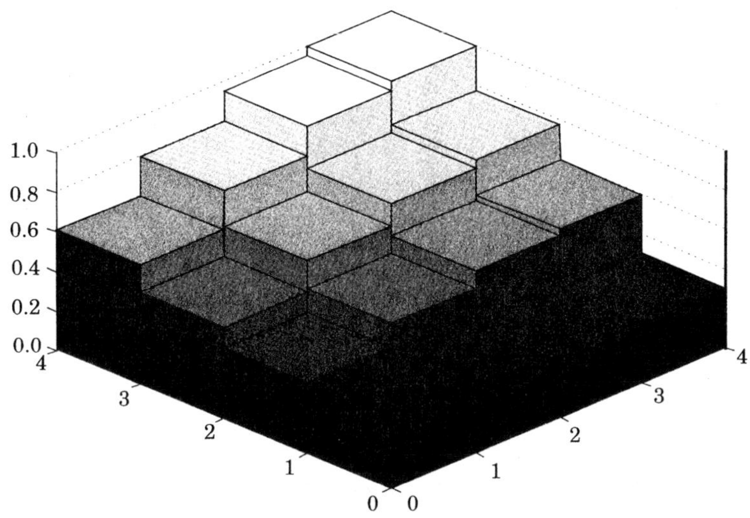

Figure 26.10 The distribution random variables of Figures 26.7 and 26.8.

TABLE 26.5 Joint Probabilities for Faults of All Types on All Lines

Line/Fault	Fault 1	Fault 2	Fault 3	Fault 4
Line 1	0.18000	0.06000	0.04500	0.01500
Line 2	0.18000	0.06000	0.04500	0.01500
Line 3	0.10500	0.03500	0.02625	0.00875
Line 4	0.13500	0.04500	0.03375	0.01125

■

26.2.5 Disturbance Joint Probability Density

Consider a power system where a family of different types of disturbances requiring protective system recognition are to be analyzed. The various density functions derived above are meaningful only for the particular sample space defined for the random variables involved. Integration of any of these over a particular region of the appropriate sample space gives the absolute probability of those events represented in that region. However, if the class sample spaces are combined to form an overall sample space representing all reasonably possible disturbances, the probabilities as calculated above are actually conditioned upon the probability of which class of disturbance occurs. As such, these must be multiplied by this probability in order to obtain the overall absolute probability. This is accomplished by generalizing and combining the above densities to form the total density.

As a first step in this process, the random variable W is defined to distinguish between the various classes of disturbances. In keeping with the previously used method of defining a random variable, the sample space of disturbances is defined, somewhat arbitrarily, as

$$S_C = \{\varepsilon_1, \varepsilon_2, \ldots, \varepsilon_6\} \tag{26.19}$$

where we define

$$\begin{aligned} \varepsilon_1 &= \text{faulted line} \\ \varepsilon_2 &= \text{open line} \\ \varepsilon_3 &= \text{faulted bus} \\ \varepsilon_4 &= \text{open bus} \\ \varepsilon_5 &= \text{faulted load} \\ \varepsilon_6 &= \text{faulted generator} \end{aligned} \tag{26.20}$$

Obviously, other definitions could be proposed. The events of interest are of the form

$$C_{L_m} = \{\varepsilon_m\} \tag{26.21}$$

which occur with probability $p_{C_{Lm}}$. The random variable W is therefore defined as

$$W(\varepsilon_m) = w_m = m \tag{26.22}$$

associating with each class of disturbance its class number. The density function

$$f_W(w) = \sum_{m=1}^{6} p_{C_{Lm}} \delta(w - w_m) \tag{26.23}$$

can be integrated to obtain the probability of any particular group of system disturbances. In terms of the random variable W, the previously derived density functions can now be written as absolute densities in the proper form as conditional densities. For example, for line faults, we can write

$$f_{X_L Y_L Z_L | W(r,s,v)|w=1} = \sum_{i=1}^{L} \sum_{j=1}^{4} \sum_{k=1}^{M} p_{Li} p_{Fj} p_{Kk} \delta(r - i)\, \delta(s - j)\, \delta(v - k) \tag{26.24}$$

where the indices i, j, and k refer to faults on line i of type j in line section k. Similar conditional probability densities can be written for other types of disturbances.

26.3 TIME-INDEPENDENT RELIABILITY MODELS

The analysis of certain aspects of the reliability of a system can be performed independently of time. The arrangement of the components in the system and the interaction of the components are often independent of time. In such cases, a time independent analysis of the system structure is very helpful in understanding the system reliability. This section presents some basic concepts of this type of analysis for protective systems.

Consider the basic protection for a transmission line shown in Figure 26.11. This protection system has redundant items of hardware for all functions except for the circuit breaker itself. The instrument transformers, relays, trip coils, circuit breaker *a* contacts, and even the batteries are redundant. This is about the maximum practical redundancy, since duplicating the circuit breaker, by placing two breakers in series, would be quite expensive.[2] We will use this protective system to study the technique for developing reliability models for protective systems.

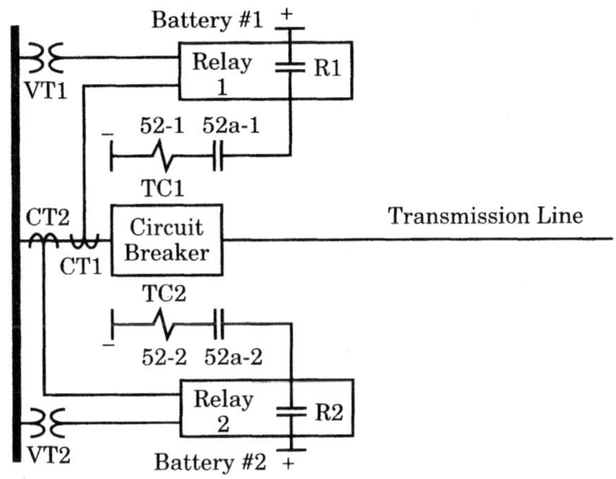

Figure 26.11 Transmission line protection with redundant components.

Clearly, either subsystem 1 or 2 can operate the circuit breaker and clear any fault detected by the relay logic. Note that there is no way in which relay 1 can operate trip coil 2, or vice versa. The two systems are wired in such a way that each system is complete and separate.

26.3.1 The Protection and The Protected Component

Consider a power system component, such as a transmission line or transformer, represented by item C in Figure 26.12. The component C is completely surrounded by circuit breakers that can isolate C when it becomes faulted. The circuit breakers are controlled by

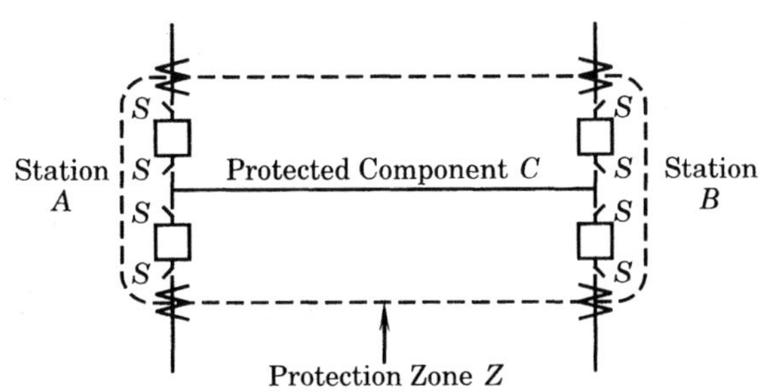

Figure 26.12 Protected component C and protection zone Z.

[2]A series connection of two circuit breakers is sometimes required. One example is for the switching of large shunt capacitor banks, where the possibility of voltage doubling may require series breakers to prevent restrike. Another example is in connection with a non-utility-owned generator, where both the generator owner and the utility require their separate and independent circuit breakers for safety and operational reasons.

protective relays located at the component terminals. These relays sense the component condition by means of input sensors or transducers, such as current and voltage transformers, located at the component terminals. Some protective systems utilize direct connections or communications between protective devices at the terminals to assist in accurately detecting a fault condition in C. The protective system consists of the input transducers, the relays, the communication system, the circuit breaker trip coils, and the physical contacts of the circuit breakers, all of which are referred to here as simply as the "protection," or by the letter P.

The protected component C is surrounded by a protective zone, Z, shown as the dashed line in Figure 26.12. The boundaries of the protective zone are defined by the location of current transducers just outside the circuit breakers that are used to isolate the C. Any ancillary equipment inside of Z and connected to C are observed by the protection and are therefore included as part of C. This includes the circuit breaker enclosures as well as other equipment such as voltage transformers, lightning arresters, wave traps, or other devices needed for measurements or other purposes. Disconnect switches, S, are provided to isolate the circuit breakers for maintenance. These switches are a part of C since they are inside zone Z. In some cases, the current transformers that define Z may be circuit breaker bushing current transformers.

Note that the circuit breaker is included in both C and in P. The circuit breaker case and the high-voltage bushings of the breaker may be involved in faults that will be detected by P, resulting in isolation of C (and perhaps of adjacent components). The circuit breaker mechanism is a necessary part of P, since it is only through this mechanism that isolation of C can be achieved. It is conceivable that the breaker mechanism could be faulted to ground, making this device a part of both C and P.

When P operates as designed, any C failure that causes abnormal currents and voltages will be detected by P. This will result in command signals being sent to the circuit breakers to open and isolate C. The isolated C may then be inspected, repaired, and returned to service.

The question naturally arises as to what exactly is included in the failure rates λ_C and λ_P for the C and P, respectively. All apparatus inside the breaker positions around C are clearly included and are sometimes identified in data bases as "terminal related" faults (as opposed to line-related faults). Bushing faults of the circuit breakers would be included in this category. However, failures of the breaker mechanism due to any reason, except faulting to ground, should correctly be included in the computation of the failure rate of the protection. We conclude, therefore, that parts of the circuit breaker are identified with the protective system and parts of it are identified with the protected components on either side of the breaker. Circuit breaker failures that cause pickup of C will be included in the statistics that determine λ_C and circuit breaker failures that contribute to the failure of P will be included in the statistics that determine the failure rate λ_P. The circuit breaker is unique in that it "belongs" to both C and to P.

26.3.2 System Reliability Concepts

We now review briefly some important concepts of reliability theory that will form the basis for analyzing systems such as that shown in Figure 26.11. Clearly, the concept of redundancy is important. It takes no special study or knowledge of reliability theory to see that the redundant system of Figure 26.11 is more reliable for tripping the circuit breaker than a system that has no protection redundancy. Redundancy provides more than one way of successfully accomplishing the required task.

The following notation is often used in reliability engineering [3–6]. Let

$$\begin{aligned} R &= \Pr\{\text{success of a given system}\} = \text{reliability} \\ F &= \Pr\{\text{failure of a given system}\} = 1 - R = \overline{R} \end{aligned} \tag{26.25}$$

where the superior bar over a probability is interpreted as the *not* or complement of that probability, in this case, the unreliability of the system.

We also adopt the following notation to represent the behavior of a given item or element of a system.

$$\begin{aligned} p_i &= \Pr\{\text{success of element } i\} = r_i \\ q_i &= \Pr\{\text{failure of element } i\} = \overline{r}_i \end{aligned} \tag{26.26}$$

It is also convenient to use the following notation in considering the outcome of the event A.

$$\begin{aligned} \Pr(A) &= \Pr\{A \text{ works}\} = p_A \\ \Pr(\overline{A}) &= \Pr\{A \text{ fails}\} = q_A \end{aligned} \tag{26.27}$$

We shall now utilize these mathematical symbols for developing a mathematical model for redundancy. Before analyzing any system, however, it will be helpful to first develop some general expressions for common element configurations that are found in almost all hardware systems.

EXAMPLE 26.3 *Analysis of a Redundant Relay System*

We apply the series and parallel reliability concepts to analyze the protective system control diagram of Figure 26.11, which permits us to sketch the *RDB* of Figure 26.13. First, we note that subsystem 1 is a series logic consisting of battery *BA*1, voltage transformer *VT*1, current transformer *CT*1, relay *R*1, trip coil *TC*1, and breaker front contact 52*a*1. Failure of any one of these items will cause failure of relay system 1; therefore, this is a series system logic. A similar conclusion can be drawn for relay system 2. The circuit breaker main contacts, denoted as *CB*, are not redundant and must be treated separately from the relay systems. Compute the reliability of the total system.

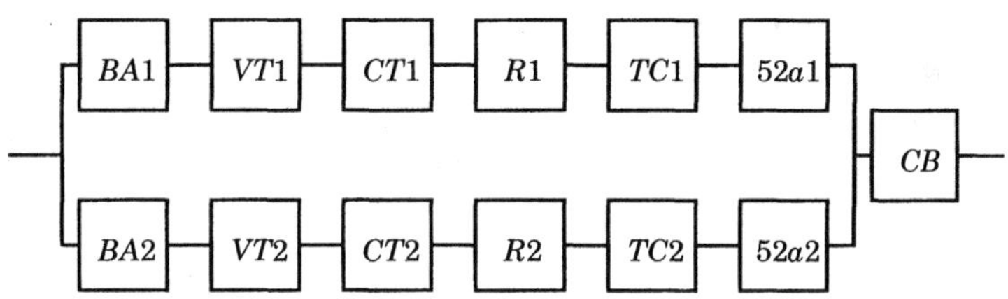

Figure 26.13 Reliability block diagram for the system of Figure 26.11.

Solution

We adopt the following notation introduced earlier to indicate the success or failure of an item. Therefore, for subsystem 1, since the separate hardware items are independent, we compute the reliability as the product of the individual items.

$$\begin{aligned} R_1 &= \Pr\{BA1 \text{ works}\} \Pr\{VT1 \text{ works}\} \Pr\{CT1 \text{ works}\} \\ &\quad \times \Pr\{R1 \text{ works}\} \Pr\{TC1\text{works}\} \Pr\{52a1 \text{ works}\} \\ &= p_{BA1} p_{VT1} p_{CT1} p_{R1} p_{TC1} p_{52a1} \end{aligned} \tag{26.28}$$

In a similar fashion, we may write for subsystem 2,

$$\begin{aligned} R_2 &= \Pr\{BA2 \text{ works}\} \Pr\{VT2 \text{ works}\} \Pr\{CT2 \text{ works}\} \\ &\quad \times \Pr\{R2 \text{ works}\} \Pr\{TC2\text{works}\} \Pr\{52a2 \text{ works}\} \\ &= p_{BA2} p_{VT2} p_{CT2} p_{R2} p_{TC2} p_{52a2} \end{aligned} \tag{26.29}$$

Since subsystems 1 and 2 are in parallel logic, the total system fails only when these two systems fail simultaneously, which is the product of the failure probabilities of the two parallel items. Thus, we write

$$F = (1 - R_1)(1 - R_2) = 1 - R_1 - R_2 + R_1 R_2 \tag{26.30}$$

Since system success and failure are complementary events, the system reliability is given as the complement of the unreliability.

$$R_{12} = 1 - F = R_1 + R_2 - R_1 R_2 \tag{26.31}$$

We can also write a special equation for the case where all elements in each of the two systems are identical. If we assume this to be the case we can simplify the above result to write

$$R_{12} = 2R_1 - R_1^2 \tag{26.32}$$

The failure of the circuit breaker causes failure of the system; therefore, this is a series logic and the reliability of the total system, R_S, is the product of the reliabilities of the redundant protection components and that of the circuit breaker.

$$R_S = \left(2R_1 - R_1^2\right) R_{CB} \tag{26.33}$$

Clearly, a low reliability of the circuit breaker will result in a low reliability for the system. ■

26.3.2.1 Dual Failure Modes of Protective Systems. We now consider a feature of protective systems that is not true of all physical systems—that of dual failure modes. A Markov graph for a system with dual failure modes is shown in Figure 26.14. The transition rates for the two failure modes are not necessarily the same.

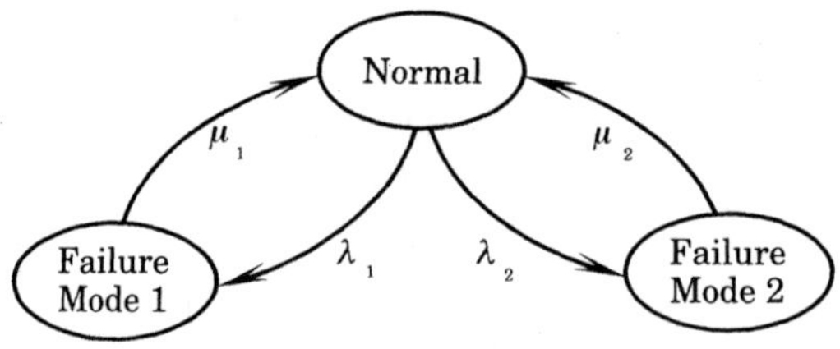

Figure 26.14 Markov graph for a system with two failure modes.

For protective systems, these two modes of failure are defined as follows [7].

Failure mode 1 The system **fails to operate** when a fault occurs in the protected zone.

Failure mode 2 The system **operates unnecessarily**, either when there is no fault, or for a fault outside of the protected zone.

For simplicity, we call failure mode 1 an *operational failure*, and failure mode 2 a *security failure* [8]. Clearly, connecting two protective systems in a parallel logic, similar to that of Figure 26.11, greatly increases the probability of a security failure, since *either* relay can cause an unnecessary trip even though the other relay system maintains the proper restraint and selectivity. We also note that a security failure might create a serious disturbance for the power system, especially if the line loadings are high, or when some components are out of service for maintenance. In this case, any disruption may place the entire system at considerable hazard. It has become the practice in the reliability literature to identify these two failure modes as the *Fail Dangerous* (*FD*) and *Fail Safe* (*FS*) modes of failure [9], [10]. Fail Dangerous is the same as the operational failure and Fail Safe is the same as the security failure. These terms have their origin in the protection of nuclear reactors and chemical processes, where an operational failure may indeed lead to a dangerous condition, but where a security failure results only in a safe shutdown of the facility. In this book, we will use both names for the two failure modes.

These two failure modes are sometimes required to analyze electronic components that may be placed in series or parallel in order to increase the rating or otherwise enhance the design. Solid-state diodes, for example, can fail either open (similar to an operational failure) or shorted (similar to a security failure). Placing two such diodes in parallel in a circuit design is similar in concept to having dual redundant relays in a protective system. In general, any device that performs switching can experience these two, quite different, modes of failure. This is true of diodes, thyristors, relays, and circuit breakers.

26.3.2.2 Operational Failure (Fail Dangerous). Consider a protective system similar to that shown in Figure 26.11. This system has been analyzed for *operational failures* in example 26.3. If we use the subscript o to indicate *operational failure*, we may write the probability of failure of the dual redundant system as

$$F_o = q_{o1} q_{o2} = q_o^2 \tag{26.34}$$

where q_{o1} = Pr{failure of system # 1}
q_{o2} = Pr{failure of system # 2}
$q_o = q_{o1} = q_{o2}$ for identical systems

For a more general system, consider n nonidentical relay systems in a parallel connection, where the different systems all have their separate trip coils and breaker front contacts (a straightforward extension of Figure 26.11). We also assume that each system is designed to perform the same type of protective function. In this more general case, the probability of operational failure may be written as follows:

$$F_o = \prod_{i=1}^{n} q_{oi} = \prod_{i=1}^{n} (1 - p_{oi}) \tag{26.35}$$

where p_{oi} = Pr {success of system i in avoiding operational failures} (26.36)

Also,

$$R_o = \text{Pr}\,\{\text{system operational success}\} = 1 - F_o \tag{26.37}$$

Thus, we observe that parallel redundancy improves the system reliability in terms of lowering the probability of operational failure. This is apparent due to the product of failure probabilities in (26.35), each of which is a number less than unity, such that the product of terms is an even lower number and is always less than the failure probability of the least reliable member. With failure a very small number, success will be closer to unity, as shown in (26.37).

26.3.2.3 Security Failure (Fail Safe). Again considering the redundant relay system of Figure 26.11, we note that *either* relay may cause a *security failure*, independent of the other relay system. The parallel physical arrangement creates a series logic, which we can describe by a series network. That is, the security failure of either relay causes the security failure of the entire system. We use the subscript s to represent the unnecessary operation due to security failure. Then we define

$$p_{si} = \text{Pr}\,\{\text{success of system } i \text{ in avoiding security failures}\} \tag{26.38}$$

and, using the series logic, we can write

$$R_s = \text{Pr}\,\{\text{no security failures}\} = \prod_{i=1}^{n} p_{si} \tag{26.39}$$

and the probability of failure is given by

$$F_s = 1 - R_s = 1 - \prod_{i=1}^{n} p_{si} = 1 - \prod_{i=1}^{n} (1 - q_{si}) \tag{26.40}$$

Obviously, the larger the number of redundant relay systems, the greater the probability of security failure.

26.3.2.4 Optimization [5]. Since operational and security failures are mutually exclusive, we may write the probability of failure considering either failure mode by adding the two failure probabilities. The total probability of failure then is given by

$$F = 1 - \prod_{i=1}^{n} (1 - q_{si}) + \prod_{i=1}^{n} q_{oi} \tag{26.41}$$

or the reliability is

$$R = \prod_{i=1}^{n} (1 - q_{si}) - \prod_{i=1}^{n} q_{oi} \tag{26.42}$$

If the elements of the system are identical, (26.42) simplifies to

$$R = (1 - q_s)^n - q_o^n \tag{26.43}$$

This reliability can be optimized by taking the derivative or R with respect to n and solving for n to determine the optimum number of parallel connected elements. If we call this optimum value $\hat{n}$ the result is as follows.

$$\hat{n} = \frac{\ln\left[\dfrac{\ln q_o}{\ln (1 - q_s)}\right]}{\ln\left(\dfrac{1 - q_s}{q_o}\right)} \tag{26.44}$$

A system designer who has general knowledge of the range of values the different failure probabilities can use this formula to gain understanding of the role of the different failure modes. One important thing that this does not tell the protection engineer is the severity of the failure caused by the different failure modes. Failure to clear a fault, for example, might be judged more serious than a security failure. On the other hand, there are usually backup devices to provide assurance of clearing the fault, whereas there is often no hardware provided to prevent the security failure.

26.3.2.5 Dual Redundant Systems. Since duplicate relaying systems are quite common, we write the formulas for the probabilities of both operational and security failure for this special arrangement.

For operational failure of a dual redundant system, we write

$$\begin{aligned} F_o &= q_{o1} q_{o2} \\ R_o &= 1 - F_o = 1 - q_{o1} q_{o2} \end{aligned} \tag{26.45}$$

For security failure or unnecessary operation, of the dual redundant system, we write

$$\begin{aligned} F_s &= 1 - p_{s1} p_{s2} = 1 - (1 - q_{s1})(1 - q_{s2}) \\ R_s &= 1 - F_s = (1 - q_{s1})(1 - q_{s2}) \end{aligned} \tag{26.46}$$

It is important to note that the network diagrams and fault trees for operational and security failures are entirely different. Even the elements in the network diagrams are often different since some elements are not capable of causing a security failure.

EXAMPLE 26.4 *Probabilities of Operational and Security Failures*

Consider the dual redundant protective system shown in Figure 26.11 and the protection zone of Figure 26.12. Add to these items a model of the circuit breaker. Analyze the entire system for the probabilities of both operational and security failures.

Solution

OPERATIONAL FAILURES. First, we create a model of the system for operational failures. The system has three elements;

1. Relay system 1
2. Relay system 2
3. Circuit breaker (item 3)

Due to relay redundancy, the relaying will fail only when the two relay systems experience overlapping failures. Failure of the circuit breaker to open on command will also fail the system. Therefore, the network diagram for the operational failure of this system, including the circuit breaker, is shown in Figure 26.15.

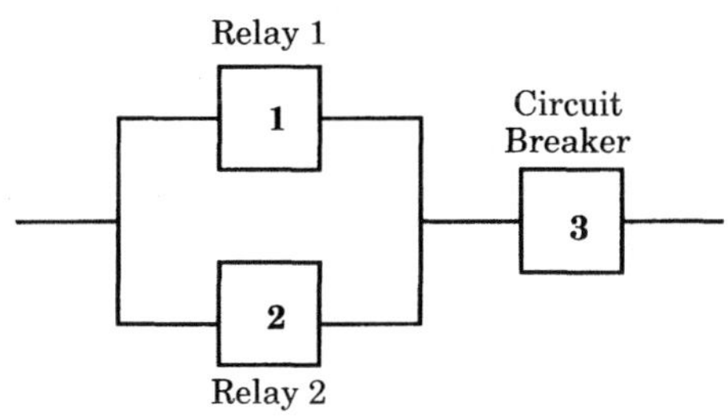

Figure 26.15 Network diagram for operational failure.

For operational failure, the system is in a failed state if 3 fails or if 1 and 2 have overlapping failures. We can clarify the mathematics by use of a Venn diagram to depict the various failure probabilities graphically [3], as shown in Figure 26.16. First, we represent the failure events of the three systems in a probability space S as the three areas in Figure 26.16. The simultaneous failure of relay systems q_{o1} and q_{o2} is represented by the *intersection* of the ovals labeled 1 and 2, and this corresponds to the region shown with hatching (\\\\\\). This intersection represents the simultaneous or overlapping failure of the

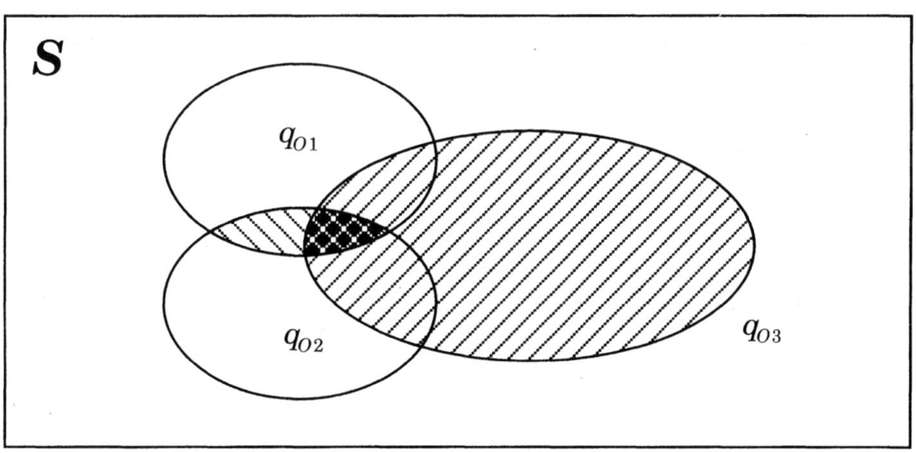

Figure 26.16 Venn diagram for operational failures.

two relay systems and is given by

$$F_{oR} = q_{o1}q_{o2} \tag{26.47}$$

The failure of the circuit breaker is shown by the large hatched oval with hatching (///). This failure is represented by q_{o3}. The system failure is represented by the *union* of the two hatched regions and is given by

$$F_o = q_{o1}q_{o2} + q_{o3} - q_{o1}q_{o2}q_{o3} \tag{26.48}$$

The first term represents the first hatched region. The second term represents the large circuit breaker failure region. The final term, representing the double-hatched area, must be subtracted to avoid counting the crosshatched region twice.

SECURITY FAILURES. For security failure, the system will fail if either of the relays fails or if the circuit breaker fails in a security failure mode. The network diagram for this mode of failure is shown in Figure 26.17.

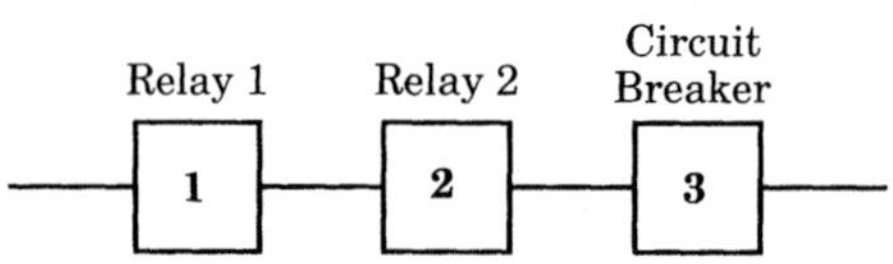

Figure 26.17 Network diagram for security failure.

This is a simple series logic or we have the result for security failure as follows. First, we compute the probability of success of the system, that is, the probability of no security failure. This can only occur if all elements are free of this type of failure.

$$R_s = p_{s1}p_{s2}p_{s3} \tag{26.49}$$

or the probability of security failure is given by

$$\begin{aligned} F_s &= q_{s1} + q_{s2} + q_{s3} \\ &\quad - q_{s1}q_{s2} - q_{s2}q_{s3} - q_{s3}q_{s1} \\ &\quad + q_{s1}q_{s2}q_{s3} \end{aligned} \tag{26.50}$$

This result may be clearly explained by the use of a Venn diagram, and this task is left as an exercise.

Now let us assume some numerical values for the failure probabilities and examine the results for sensitivities. Arbitrarily, we assume the two relay systems are identical and set the following probabilities of failure.

$$q_{o1} = q_{o2} = 0.01$$

$$q_{o3} = \begin{cases} 0.01 \\ 0.10 \end{cases}$$

$$q_{s1} = q_{s2} = 0.05$$

$$q_{s3} = \begin{cases} 0.01 \\ 0.10 \end{cases}$$

Note that we will choose two different values of failure for the circuit breaker to see how much difference this makes in the final result. Using the formulas from (26.47) and (26.49), we compute

$$F_o = 0.0001 + 0.9999q_{o3} = \begin{cases} 0.0101 & q_{o3} = 0.01 \\ 0.1001 & q_{o3} = 0.10 \end{cases}$$

$$F_s = 0.0975 + 0.9925q_{s3} = \begin{cases} 0.1074 & q_{s3} = 0.01 \\ 0.1969 & q_{s3} = 0.10 \end{cases}$$

For operational failures, improving the breaker operational failure rate by a factor of 10 improves the total system failure rate by about that same factor. However, for security failures, improving the breaker security failure rate by a factor of 10 improves the system failure rate by about a factor of 2. This is because the total security failure rate is largely determined by the relays, which are configured in such a way as to dominate this mode of failure. ■

In the foregoing example, both the relay systems and the circuit breaker system are assumed to undergo both operational and security failures. One might question how the circuit breaker can fail in the security failure mode. Since the breaker has moving contacts that can be forced to open or close by some means other than current flowing in the trip or close coils, we must conclude that breakers are indeed subject to security failures. Not all elements are capable of this failure mode, however, and each component in a system must be examined to determine the physical possibility of security failure. All components are subject to operational failure.

26.3.2.6 Fault Tree Analysis. As an example of fault tree construction, Figure 26.18 shows a fault tree for failure of the protective system shown in Figure 26.11. Almost the entire fault tree is constructed of OR gates, which indicates that there are many different items that can cause the failure of a protective system. The only requirement for the AND gate is in connection with the relays themselves, which are fully redundant. Note that the event *breaker mechanism fails* is not fully developed in this fault tree because information relevant to the event is not available. The nature of this type of failure may be different for breakers of different designs. A second undeveloped event is shown for the event *#2 System Fails*. This event would be a duplicate of the #1 system failure event and must be evaluated, but is omitted to simplify the figure.

26.3.3 Coherent Protection Logic

The protective system having dual failure modes can be described more rigorously by using coherent logic [10]. For a system to be described as coherent, the following three properties must hold.

1. *Causality*: The protection should not pick up if none of the relays picks up. This should be true whether the relays are connected in series, parallel, or some other logical connection.
2. *Nonnegative contribution*: If the protective system is picked up due to the pickup of one of the relays in that system, the system should not drop out if another relay picks up.
3. *Relevance*: Each relay should have a chance to trigger pickup of the protective system.

A protective system logic is coherent if and only if the following conditions are satisfied:

(a) The system can be represented as a logic diagram, including only AND or OR gates, and

(b) Each relay output is connected to at least one gate of the system fault tree.

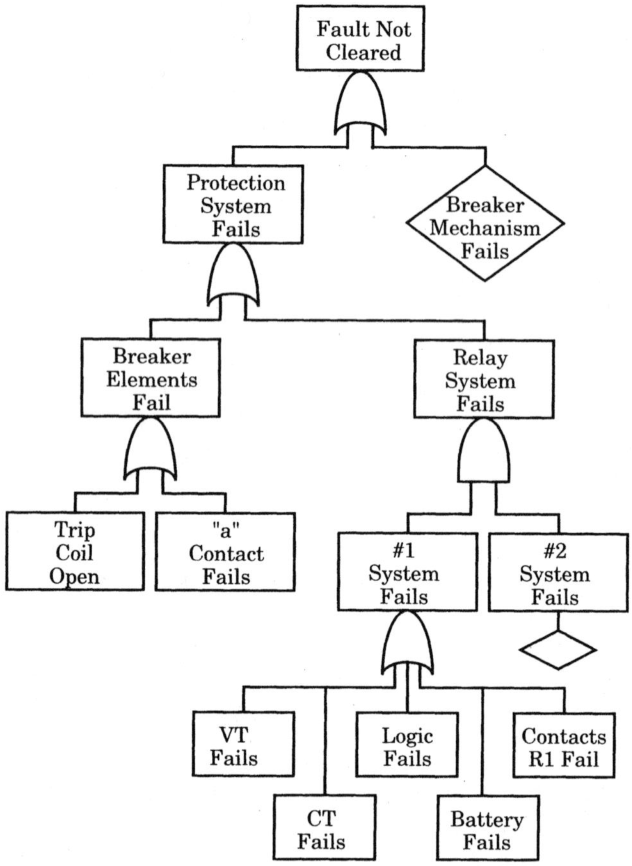

Figure 26.18 Fault tree for one of the protective systems of Figure 26.7.

Coherent logic is applicable for protective systems having two or more relays arranged in some type of logical configuration to cause system pickup under the target hazard condition.

26.3.3.1 Two-Relay Systems. First we consider a protective system that utilizes two relays. We are interested in defining the conditions under which the protective system picks up under different system conditions and with different relay logic arrangements. Figure 26.19 shows the two types of logic for systems of two relays. Since there are only two relays, they can be connected in two ways, either is (a) series or (b) parallel. The series logic requires two-out-of-two for system pickup, whereas the parallel system requires only one-out-of-two

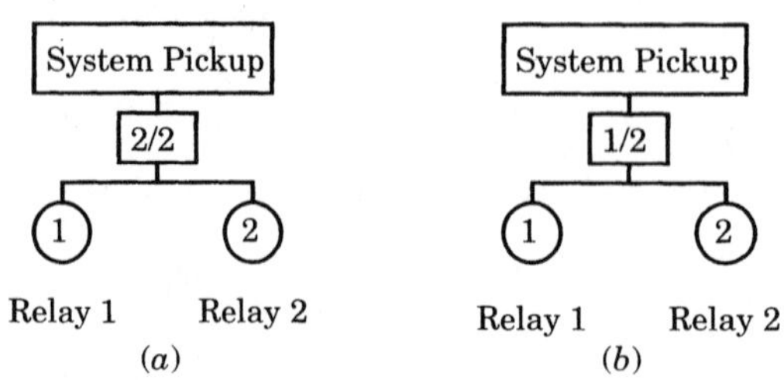

Figure 26.19 Protective systems with two relays. (*a*) Two-relay series system. (*b*) Two-relay parallel system.

for system pickup. In these logic diagrams, we read x/y to mean "x out of y" required for the top event to occur.

Figure 26.19 does not indicate the environment in which the relays are found. The system can be Fail Save (FS) only in an environment where no hazard (fault) is present and can be Fail Dangerous (FD) only in an environment where a hazard exists. Therefore, a complete system description should indicate the existence of a hazard.

FS FAILURE. The protective system can fail safe only in an environment where no hazard is present. Consider the series connection of relays, as shown in Figure 26.19(a). The FS condition for a two-relay series logic in an environment with no hazard present requires that both relays FS simultaneously, requiring a pickup logic using an AND gate, as shown in Figure 26.20(a). For the series system to FS, the minimal cut set is

$$\{1, 2\} \tag{26.51}$$

For the series connection, the system is FS if *both* relays are FS, that is, tripping cannot occur unless both relays pick up.

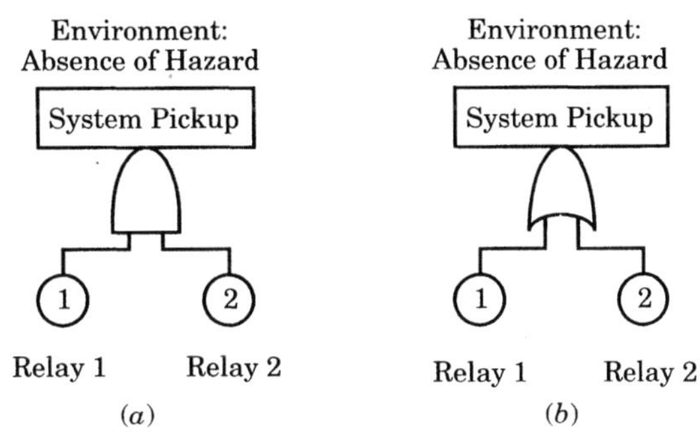

Figure 26.20 Fail Save fault trees for two-relay systems in an environment with the absence of a hazard. (a) Two-relay series system. (b) Two-relay parallel system.

For a parallel connection of the two relays in the absence of a hazard, the pickup logic requires the use of an OR gate, as shown in Figure 26.20(b). In this case, there are two minimal cut sets.

$$\{1\}, \{2\} \tag{26.52}$$

For the parallel connection, the system is FS if *either* of the relays is FS.

FD FAILURE. The two-relay protective system can fail dangerous only in an environment where a hazard is present. The series relay connection for this case is shown in Figure 26.21(a), for which the top event is system pickup. However, the system is FD if and only if it *fails* to pick up in the presence of a hazard. Thus, the top event that we need to consider is the complement of the top event shown in Figure 26.21(a), which is obviously "no system pickup." Therefore, the fault tree for the series connection in the hazardous environment is shown in Figure 26.21(b).

A comparison of the fault trees representing the FS condition, Figure 26.20(a), and the FD condition, Figure 26.21(b), brings out the following observations:

1. The environment is complemented.
2. The top event is complemented.
3. The AND gate is replaced by an OR gate.
4. The basic events are complemented.

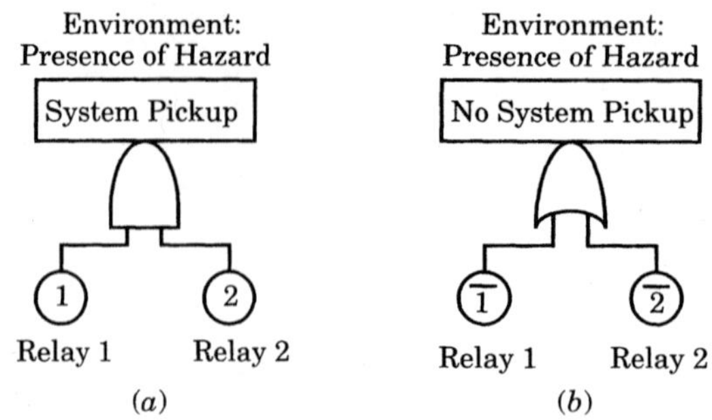

Figure 26.21 Series system FD fault trees in the presence of a hazard. (*a*) Series system, pickup. (*b*) Series system, no pickup.

The first comparison is due to the difference in the definition of the environment for the two types of failure. The remaining three differences are a statement of De Morgan's law, which can be stated in equation form as follows.

$$\overline{1 \text{ AND } 2} = \bar{1} \text{ OR } \bar{2} \tag{26.53}$$

From Figure 26.21(*b*), we can determine that the minimal cut sets for series system FD failure are

$$\{\bar{1}\}, \{\bar{2}\} \tag{26.54}$$

The series system is FD if a relay is FD, i.e., if *either* relay fails to pick up in the presence of a hazard.

Next, consider the parallel connection of two relays in the presence of a hazard. The pickup logic is shown in Figure 26.22(*a*). As before, it is necessary to complement the top event to obtain a top event that represents failure to pick up, as shown in Figure 26.22(*b*). Note that we have again complemented the environment and used De Morgan's law to complete the fault tree (*b*) from (*a*).

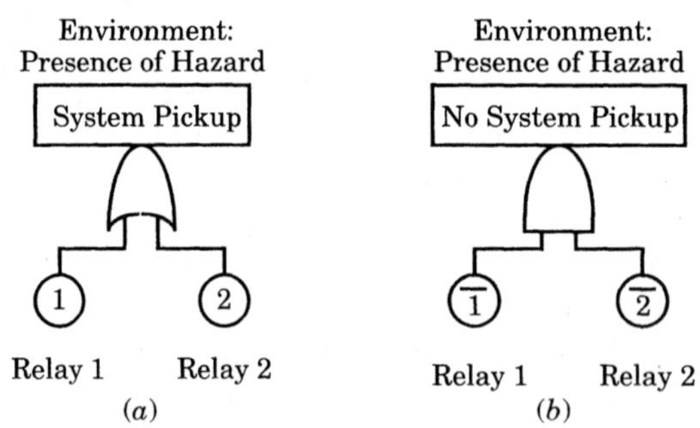

Figure 26.22 Parallel system FD fault trees in the presence of a hazard. (*a*) Parallel system, pickup. (*b*) Parallel system, no pickup.

The FD fault tree of Figure 26.18(*b*) gives a single minimal cut set.

$$\{\bar{1}, \bar{2}\} \tag{26.55}$$

Thus, the parallel system fails dangerous if *both* relays are FD.

Table 26.6 gives the minimal cut sets for a two-relay system. The series system has fewer FS failures than the parallel system, but the parallel system has fewer FD failures.

26.3.3.2 Three-Relay Systems. Two-relay systems are commonly used in power systems for applications such as line or transformer protection. However, some systems require a higher level of reliability and security and are usually protected by more complex systems

TABLE 26.6 Minimal Cut Sets for a Two-Relay Protective System

Failure Mode	Series System	Parallel System
FS	$\{1, 2\}$	$\{1\}, \{2\}$
FD	$\{\bar{1}\}, \{\bar{2}\}$	$\{\bar{1}, \bar{2}\}$

using three or more independent protective devices, connected in an appropriate logical manner. Protecting a nuclear reactor from a scram[3] shutdown is an example of a system that must have high resistance in both FS and FD conditions. Certain important transmission lines may also require this more failure resistant protective system design.

Three-relay systems can exist in several different configurations, for example,

1. Series system
2. AND-OR system
3. 2-out-of-3:G system
4. OR-AND system
5. Parallel system

The notation 2-out-of-3:G means 2 out of 3 good, which distinguishes this from 2-out-of-3 failed. These five protective logics are shown in Figure 26.23.

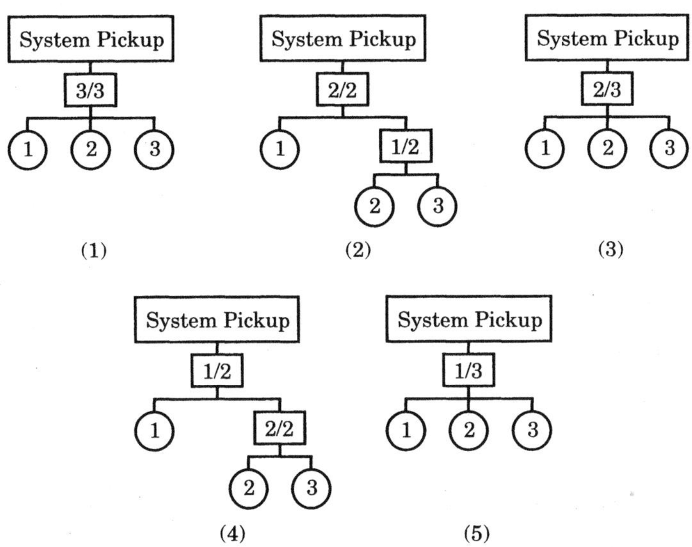

Figure 26.23 Coherent protective systems using three relays [10]. (1) Series system. (2) AND/OR system. (3) 2-out-of-3: G system. (4) OR/AND system. (5) Parallel system.

The analysis of the three-relay systems for both FS and FD modes of failure is carried out in the same manner as for the two-relay systems. The results are summarized below.

[3]A *scram* is the rapid shutdown of a nuclear reactor.

The FS fault trees are summarized in Figure 26.24 and the FD fault trees are summarized in Figure 26.25. Both results are summarized in Table 26.7.

For the three-relay system there are two extremes, represented by the series and parallel systems. The series system is not FS unless all three relays are FS and this is the best system insofar as FS events are concerned. The parallel system, on the other hand, requires that all three relays be FD before the system is FD. The other three systems lie in between these two extremes. The order of ranking for FS events is (1), (2), (3), (4), and (5), with (1) best for FS failures. For FD failures, the order is reversed, with the parallel system being best and the series system worst.

26.3.3.3 Analysis of Coherent Systems. It has been shown that fault trees can be constructed for relay systems containing two and three relays. The same concepts can be used to extend the analysis to systems of four or more relays. In every case, the top events can be computed if the basic events can be evaluated quantitatively. This requires the introduction of quantitative parameters at the basic event or relay level, and eventually extending the analysis to the system level.

RELAY LEVEL PARAMETERS. At the relay level, there are three states: normal, FS, and FD. Moreover, the FS state is conditioned by the absence of a hazard, but the FD state is conditioned by the presence of a hazard. The normal state is conditioned by either the presence or the absence of a hazard.

We introduce the following probabilistic parameters for the three states [10]:

Conditional FS probability, a_i The conditional probability that relay i picks up, given an environment without a hazard.

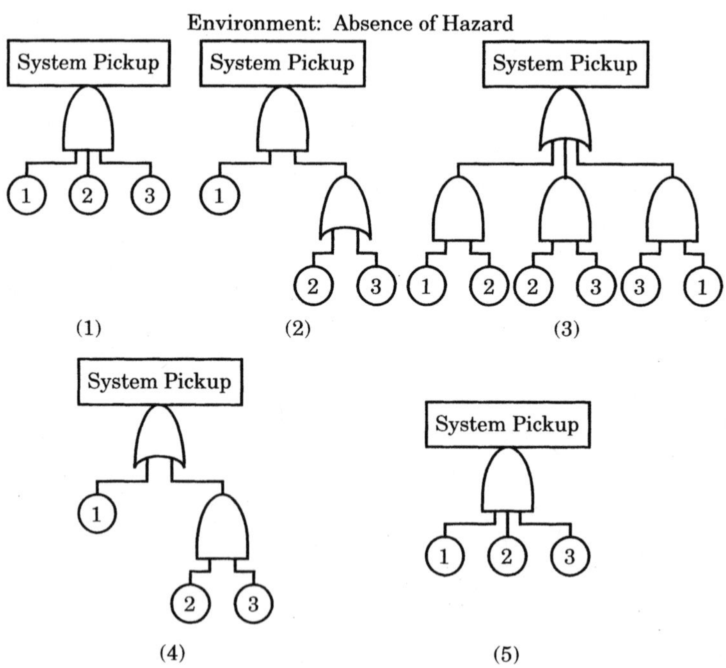

Figure 26.24 FS fault trees for three-relay systems [10]. (1) Series system. (2) AND/OR system. (3) 2-out-of-3: G system. (4) OR/AND system. (5) Parallel system.

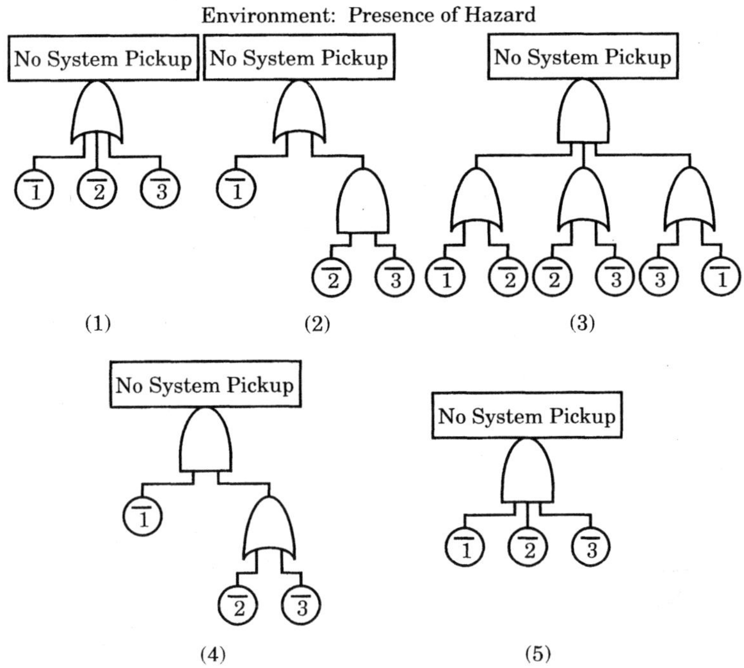

Figure 26.25 FD fault trees for the three-relay system [10]. (1) Series system. (2) AND/OR system. (3) 2-out-of-3: G system. (4) OR/AND system. (5) Parallel system.

TABLE 26.7 Minimal Cut Sets for the Three-Relay System

Mode	Series (1)	AND-OR (2)	2-out-of-3G (3)	OR-AND (4)	Parallel (5)
FS	{1, 2, 3}	{1, 2}	{1, 2}	{1}	{1}
		{1, 3}	{2, 3}	{2, 3}	{2}
			{3, 1}		{3}
FD	$\{\bar{1}\}$	$\{\bar{1}\}$	$\{\bar{1}, \bar{2}\}$	$\{\bar{1}, \bar{2}\}$	$\{\bar{1}, \bar{2}, \bar{3}\}$
	$\{\bar{2}\}$	$\{\bar{2}, \bar{3}\}$	$\{\bar{2}, \bar{3}\}$	$\{\bar{1}, \bar{3}\}$	
	$\{\bar{3}\}$		$\{\bar{3}, \bar{1}\}$		

Conditional FD probability, b_i The conditional probability that relay i does not pick up in an environment with a hazard.

Demand probability, p The probability of occurrence of a hazard, i.e., the probability of a fault occurring for which the protection should operate.

We also define the following additional parameters.

Unconditional FS probability, a_i^* The unconditional probability that relay i is FS when it is placed in an unknown environment. The event occurs (1) if the environment happens to contain no hazard, and (2) if the relay falsely picks up and causes a system trip. Thus

$$a_i^* = a_i\,(1 - p) \tag{26.56}$$

Unconditional FD probability, b_i^* The unconditional probability that relay i is FD when placed in an unknown environment. The event occurs (1) if the environment happens to contain a hazard, and (2) if the relay fails to pick up and clear the hazard. Thus

$$b_i^* = b_i p \tag{26.57}$$

Reliability, R_i The unconditional probability that relay i responds correctly when placed in an unknown environment. The correct response occurs in two cases: (1) the environment happens to contain no hazard, and the relay is not FS; or (2) the environment happens to contain a hazard and the relay is not FD. Thus

$$\begin{aligned} R_i &= (1 - a_i)(1 - p) + (1 - b_i)p \\ &= 1 - a_i^* - b_i^* \end{aligned} \tag{26.58}$$

EXAMPLE 26.5

A set of 200 relays are purchased and tested in a controlled test where all relays are placed, for a given time period, in environments either with or without a hazard present. Under these test conditions, 10 relays were found to be FS and five relays were found to be FD. Then the conditional probabilities for these relays are computed as follows.

$$a_i = \frac{10}{200} = 0.050$$

$$b_i = \frac{5}{200} = 0.025$$

Note that the environment for the two tests is different, so the two probabilities do not necessarily sum to unity and may have a sum greater than unity. ■

EXAMPLE 26.6

Compute the unconditional probabilities and the reliability of the relays described in Example 26.5.

Solution

The data required for the solution are known. Therefore, we can substitute into (26.35)–(26.37) to obtain the reliability parameters.

$$\begin{aligned} a_i^* &= a_i(1 - p) = (0.05)(1 - 0.0001) = 0.049{,}995 \\ b_i^* &= b_i p = (0.025)(0.0001) = 0.000{,}0025 \\ R_i &= 1 - a_i^* - b_i^* \\ &= 1 - 0.049995 - 0.0000025 \\ &= 0.9500025 \end{aligned}$$

The numbers used in these examples are not necessarily typical of actual hardware, but are selected to make a sensible example. ■

System Level Parameters. The relay level parameters can be extended to the system level, as follows. Let the subscript "S" stand for "system."

Conditional FS probability, a_S The conditional probability that the relay system picks up, given an environment without hazard.

Conditional FD probability, b_S The conditional probability that the relay system does not pick up, given an environment with a hazard.

Unconditional FS probability, a_S^* The unconditional probability that the relay system is FS when placed in an unknown environment . The event occurs (1) if the environment happens to contain no hazard, and (2) if the relay system incorrectly picks up in this environment. Thus

$$a_S^* = a_S(1 - p) \tag{26.59}$$

Unconditional FD probability, b_S^* The unconditional probability that the relay system is FD when placed in an unknown environment. Thus

$$b_S^* = b_S p \tag{26.60}$$

System reliability, R_S The unconditional probability that the relay system responds correctly when placed in an unknown environment. The correct system response occurs exclusively in cases where (1) the environment happens to contain no hazard and the relay system is not FS, or (2) the environment happens to contain a hazard and the relay system is not FD. Thus

$$\begin{aligned} R_S &= (1 - a_S)(1 - p) + (1 - b_S)p \\ &= 1 - a_S^* - b_S^* \end{aligned} \tag{26.61}$$

Unreliability, Q_S The complement of reliability.

$$\begin{aligned} F_S &= 1 - R_S = a_S(1 - p) + b_S p \\ &= a_S^* + b_S^* \end{aligned} \tag{26.62}$$

EXAMPLE 26.7

Compute the system conditional reliability parameters of a two-relay system with relays having parameters computed in example 26.6. Make the computation for two conditions: (a) a series system, and (b) a parallel system. Compute the unconditional system parameters and the system reliability if the demand probability is estimated to be 0.0001.

Solution

(a) *The Series system.* The minimal cut sets for the series system is given in Table 26.3 to be $\{1, 2\}$ for the FS failure and $\{\bar{1}\}$ and $\{\bar{2}\}$ for the FD failure. Therefore, we compute the conditional system probabilities.

$$\begin{aligned} a_S &= a_1 a_2 \\ &= (0.05)(0.05) = 0.0025 \\ b_S &= b_1 + b_2 - b_1 b_2 \\ &= 0.025 + 0.025 - (0.025)^2 \\ &= 0.049375 \end{aligned}$$

From the conditional probabilities and the demand probability, we can compute the unconditional system probabilities for the series system.

$$\begin{aligned} a_S^* &= a_S(1 - p) \\ &= (0.0025)(1 - 0.0001) \\ &= 0.00249975 \\ b_S^* &= b_S p \\ &= (0.049375)(0.0001) \\ &= 0.00000494 \end{aligned}$$

Finally, we compute the reliability and unreliability for the series system.

$$\begin{aligned} R_S &= 1 - a_S^* - b_S^* \\ &= 1 - 0.00249975 - 0.00000494 \\ &= 0.997495 \\ F_S &= 1 - R_S \\ &= 1 - 0.997495 \\ &= 0.002505 \end{aligned}$$

(b) *The parallel system.* The minimal cut sets for the parallel system are given in Table 26.3 to be {1} and {2} for the FS failure and $\{\bar{1}, \bar{2}\}$ for the FD failure. Thus, we write the conditional probabilities as follows.

$$\begin{aligned} a_S &= a_1 + a_2 - a_1 a_2 \\ &= 0.05 + 0.05 - (0.05)^2 \\ &= 0.0975 \\ b_S &= b_1 b_2 \\ &= (0.025)^2 \\ &= 0.000625 \end{aligned}$$

From the conditional probabilities, we compute the unconditional probabilities for the parallel system.

$$\begin{aligned} a_S^* &= a_S\,(1 - p) \\ &= (0.0975)\,(1 - 0.0001) \\ &= 0.09749025 \\ b_S^* &= b_S p \\ &= (0.000625)\,(0.0001) \\ &= 6.25 \times 10^{-8} \end{aligned}$$

The reliability and unreliability of the parallel system are computed as follows.

$$\begin{aligned} R_S &= 1 - a_S^* - b_S^* \\ &= 1 - 0.09749025 - 6.26 \times 10^{-8} \\ &= 0.90251 \\ F_S &= 1 - R_S \\ &= 1 - 0.90251 \\ &= 0.09749 \end{aligned}$$

The parallel system is noted to have a lower reliability than the series system for the data used in this example. ■

It is observed that the reliability is not an adequate parameter to use in evaluating the performance of protective systems since it treats the FS and FD failures with equal weight. Actually, FD failures have the potential of being much more costly than FS failures, since FS failures may cause only an inconvenience or a small cost, depending on the system conditions. A more realistic parameter, called the *expected loss*, has been proposed [16]. This parameter is defined as follows:

FS loss C_a: The loss caused by one FS failure of the protective system

FD loss C_b: The loss caused by one FD failure of the protective system

Expected loss I_S: The expected value of the sum of the FS and FD losses, computed as

$$\begin{aligned} I_S &= C_a a_S^* + C_b b_S^* \\ &= C_a a_S (1 - p) + C_b b_S p \end{aligned} \tag{26.63}$$

This concept permits a more realistic evaluation of the cost of failure. For example, consider a protective system for a synchronous generator. A FS failure will trip the generator unnecessarily, and it may be several hours before the machine is inspected and put back in service, with the lost generation replaced by emergency purchase of energy at a high cost. The total loss could be many thousands of dollars. Compare this with a FD failure, where a permanent loss of generator life is experienced due to the FD failure of the protection. This results in the generator being rebuilt years ahead of schedule, at a loss of millions of dollars.

EXAMPLE 26.8

Compute the expected loss for the previous example if the following loss factors are given.

$$\begin{aligned} C_a &= \$100 \\ C_b &= \$10,000,000 \end{aligned}$$

Solution

Substituting into (26.63), we compute the following expected loss for the series and parallel systems.

For the series system,

$$\begin{aligned} I_S &= C_a a_S^* + C_b b_S^* \\ &= (100)\,(0.00249975) + (10{,}000{,}000)\,(0.00000494) \\ &= 0.2499 + 49.375 \\ &= 49.63 \end{aligned}$$

For the parallel system,

$$\begin{aligned} I_S &= C_a a_S^* + C_b b_S^* \\ &= (100)\,(0.09749025) + (10{,}000{,}000)\left(6.25 \times 10^{-8}\right) \\ &= 9.749 + 0.625 \\ &= 10.374 \end{aligned}$$

The reliability calculation shows the series system to be the most reliable by a wide margin. However, depending on the relative costs of FS and FD failures, this may not be the best choice. For the costs used in this example, the expected loss for the parallel system is about one-fifth that of the series system. ■

Table 26.8 summarizes the reliability parameters for a two-relay system.

TABLE 26.8 Probabilistic Parameters for a Two-Relay System

Parameter	Series System (1)	Parallel System (2)
a_S	$a_1 a_2$	$a_1 + a_2 - a_1 a_2$
b_S	$b_1 + b_2 - b_1 b_2$	$b_1 b_2$
R_S	$1 - a_S(1-p) - b_S p$	$1 - a_S(1-p) - b_S p$
Q_S	$a_S(1-p) + b_S p$	$a_S(1-p) + b_S p$
I_S	$C_a a_S(1-p) + C_b b_S p$	$C_a a_S(1-p) + C_b b_S p$

A comparison of the series and parallel arrangements for the two-relay system is an interesting exercise. If the relays are identical and have identical probabilistic parameters, one can show the following:

1. The series system is superior to the parallel system if

$$I_{S\ \text{series}} < I_{S\ \text{parallel}} \tag{26.64}$$

We can show that (26.64) hold approximately when

$$\frac{C_b p}{C_a(1-p)} < \frac{a}{b} \tag{26.65}$$

2. The parallel system is superior to the series system if

$$I_{S\ \text{series}} > I_{S\ \text{parallel}} \tag{26.66}$$

We can show that this relationship holds approximately if

$$\frac{C_b p}{C_a(1-p)} > \frac{a}{b} \tag{26.67}$$

Proof of these expressions is left as an exercise.

EXAMPLE 26.9

Plot the unreliability as a function of the probability of system hazard for various values of the parameters a and b. A typical plot is shown in Figure 26.26, where the parameters have the following values:

$$a = 0.05$$
$$b = 0.01, 0.001, \text{and } 0.0001$$

The plots labeled Q_S are the system unreliability for the series connection, and those labeled Q_P are for the parallel connection. In each case, three values of the b parameters are used in the calculation.

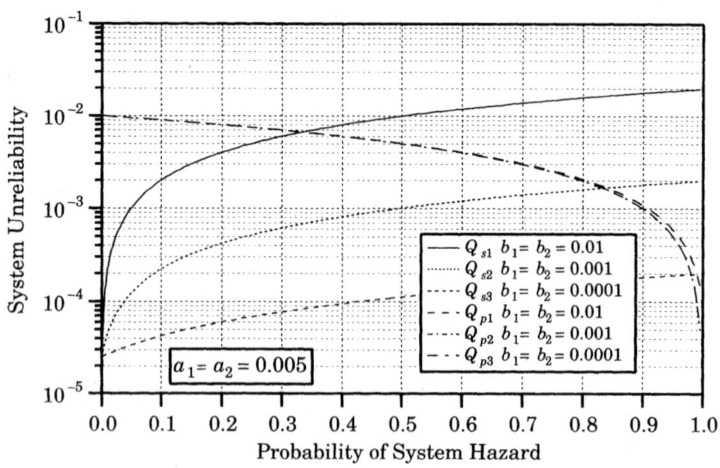

Figure 26.26 Plot of system unreliability versus probability of hazard.

All the plots in Figure 26.26 are linear with the probability of system hazard, but the results cover so many decades of values that semilogarithmic plots show the variations more clearly. For the values of the parameters chosen, the three curves for the parallel connection are almost identical, with the only difference being at values of p very close to unity, which is a part of the curve that is of little interest since

p is usually close to zero. Note that, for values of p less than about 0.2, the parallel connection is the higher unreliability for all values of the parameters selected for study. For $p = 0$, the series connection has an unreliability of a^2, which is equal to 0.0025 for the cases plotted. We conclude, therefore, that the series connection has the higher reliability and should be chosen if this is the only criterion governing the selection. ■

EXAMPLE 26.10

Plot the Expected Loss, I_S, for the same parametric values used in example 26.9 and using the following values of the Loss Parameters.

$$C_a = 1.0$$
$$C_b = 10,000$$

These values are chosen to reflect the fact that the FD failure is much more costly than suffering a FS type of failure. The result is shown in Figure 26.27. The cost of the series-connected systems is greater than that of the parallel connected protective system. This is true for all values of the parameter b, which confirms that the commonly used parallel connection for protective relays is probably the best choice in most situations, even though it may be true that the series system has a higher reliability considering both FS and FD types of failure. For any given application, the results should be checked, however, to be sure that the parallel connection is the best choice.

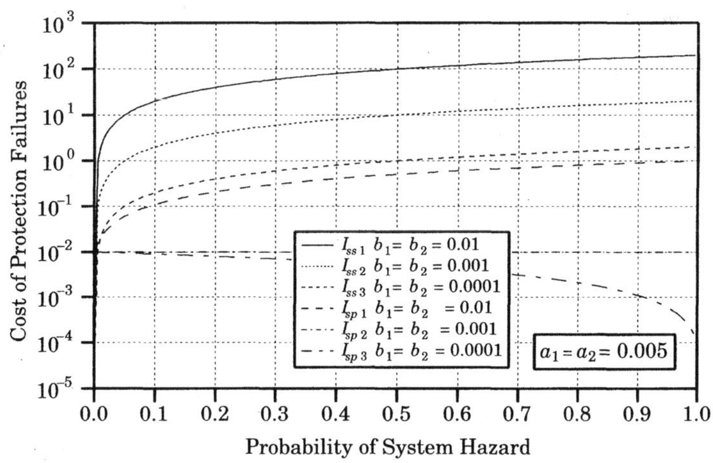

Figure 26.27 Cost of protection failure versus probability of system hazard. ■

26.3.4 Protective System Analysis

We now examine a variety of common protective system arrangements to determine their reliability block diagrams. We will develop logic or network diagrams for these typical systems so that the reliability of different configurations can be observed and computed.

26.3.4.1 Protective System Configurations. We examine common protective system configurations and define their functions. In this analysis, we define three systems that will be considered in the analysis: the main protective system, the local backup system, and the remote backup system.

MAIN PROTECTIVE SYSTEM (MPS). Three different protection control configurations are examined and compared—each having different degrees of redundancy. These systems are shown in Figure 26.28. Each control configuration assumes only a single battery, a single breaker mechanism, and redundant relays. Different degrees of redundancy are assumed for the other components.

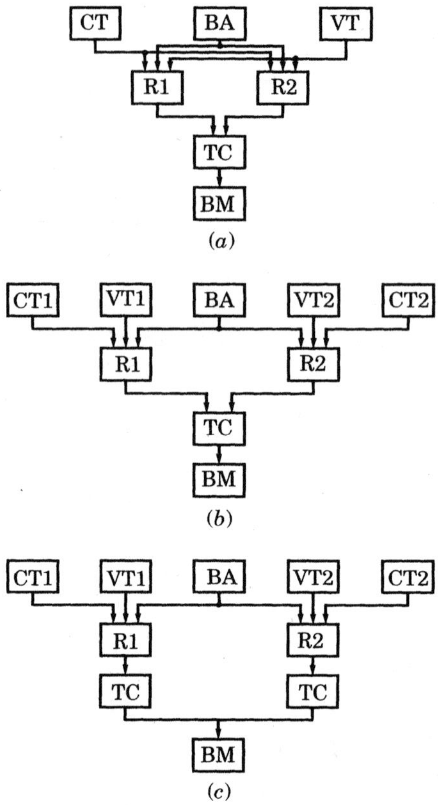

Figure 26.28 Block diagrams of three typical power system relay and circuit breaker control configurations. (1) Redundant relays. (2) Redundant instrument transformers and relays. (3) Redundant instrument transformers, relays, and circuit breaker trip coils.

Control configuration (1) assumes redundancy only in the relays. Configuration (2) has redundant relays and instrument transformers. Configuration (3) adds redundant trip coils. These configurations are typical of industry practice. The degree of redundancy depends on many factors, with a tendency to offer higher degrees of redundancy for the more important circuits and the higher voltage levels, which often are designed to carry very large power transfers.

The other important factor in the protective system layout is the station arrangement. Three different arrangements will be considered, and these are shown in Figure 26.29. In our analysis, we will consider the clearing of the fault **F** on line HR at Station H. For the single bus arrangement of Figure 26.29(a), this requires the tripping of breaker 2. However, for both the ring bus of Figure 26.29(b) or the breaker-and-a-half scheme of Figure 26.29(c), fault clearing requires the tripping of both breakers 2 and 3. We assume that breaker 8 at Station R always operates correctly.

Note that the proper fault clearing for station arrangements of Figure 26.29 (b) and (c) are also affected by the choice of control configurations, since they require the successful operation of two relay-breaker controls for the successful clearing of the fault. This will require a modification of the control schematic, and these modifications will depend on the station arrangement.

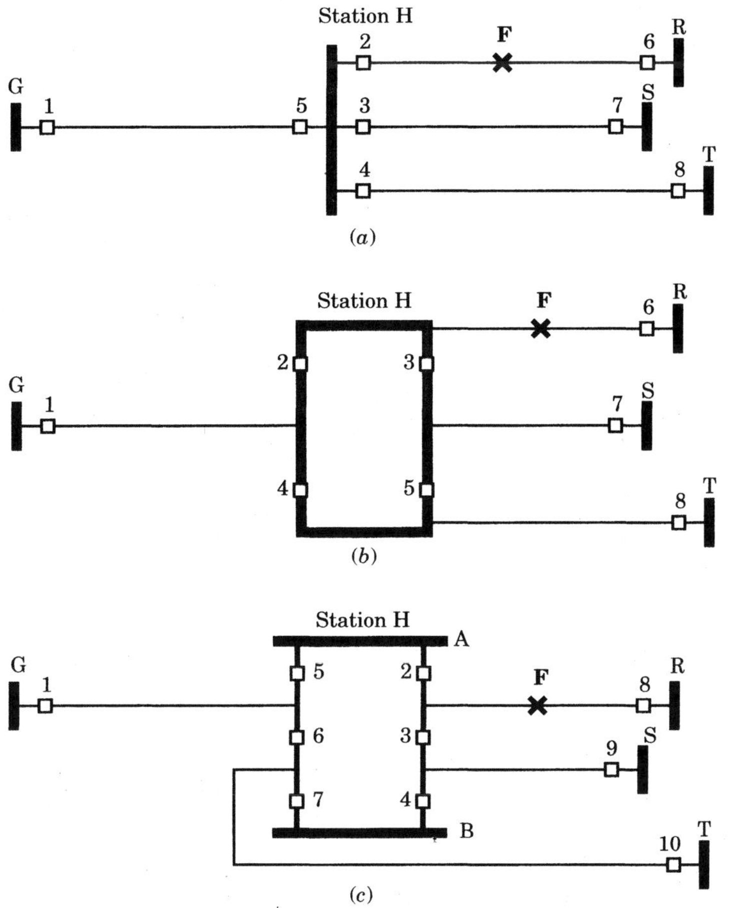

Figure 26.29 Typical transmission switching station arrangements. (*a*) Single bus at station H. (*b*) Ring bus at station H. (*c*) Breaker-and-a-half at station H.

The dual breaker arrangements also mean that, in the logic diagram for successful fault clearing (or in the fault tree), the two breaker trips will be in series logic since failure of either will cause failure of the *main protective system* (*MPS*).

All control configurations shown in Figure 26.28 provide redundancy in fault detection, but not in the circuit breaker mechanism. This is common in the industry due to the high cost of the circuit breakers. Should the breaker mechanism fail, or should all relay operations fail, we rely on backup protective systems for fault clearing. Two types of backup are usually employed, a local backup system (LBS) and a remote backup system (RBS).

LOCAL BACKUP SYSTEM (LBS). The application of a LBS will be illustrated by means of an example [11]. Consider the switching stations of Figure 26.29 with a fault F on line HR and assume that breaker 2 fails to operate, that is, it remains permanently closed. This is sometimes called a "stuck breaker" condition or failure to open on command [14].

For the single bus arrangement of Figure 26.29(*a*), a failure of breaker 2 requires the local backup system to trip all breakers (3, 4, and 5) connected to bus H. This splits the system at bus H, and may have serious consequences for the integrity of the power system, but it does clear the fault at F.

For the ring bus arrangement of Figure 26.29(*b*), the control logic requires the tripping of breakers 2 and 3 to clear fault F. Since breaker 2 is inoperative, the LBS must trip breaker 4. This leaves the fault connected to line GH and requires the delayed tripping of breaker 1 at G by remote backup. Note, however, that lines HS and HT are still connected together at H and the system is not completely severed at bus H.

Now consider the breaker-and-a-half scheme of Figure 26.29(*c*) and the control logic for tripping breakers 2 and 3. Since breaker 2 has failed, the LBS must trip breaker 5 and any other breakers connected to rack bus A. This clears the fault and leaves all lines in service except the faulted line HR. Moreover, all other lines connected to bus H remain in service.

The foregoing example illustrates the improvement in performance that results from more complex switching arrangements. This is achieved at greater cost. It also imposes an interesting requirement, namely, that every fault at bus H must be successfully cleared by not one, but two circuit breakers. Since a faulted condition places a high stress on the circuit breaker, the probability of failure under these conditions may be high compared to the switching of normal load currents. Requiring two successful switchings to take place for every fault may lead to more breaker failure events than simpler switching schemes. However, the more complex switching schemes are effective in maintaining maximum system connectivity, which is very important to system stability and security.

Now consider the breaker-and-a-half station with breaker 3 stuck and the same fault illustrated in Figure 26.29 (*c*). Breaker 3 is the center breaker in the breaker-and-a-half connection. This causes the local backup system to trip breaker 4, thereby disconnecting the sound line from H to S. Failure of the center breaker of a breaker-and-a-half scheme is similar to a breaker failure for a Ring Bus, in that it will always result in the loss of a sound circuit element.

This illustrates the importance of the local backup system, in combination with the station switching arrangement. It also illustrates some of the reliability effects that must be considered in protective system evaluation.

A simplified Local Backup System is shown in Figure 26.30. The *main protective system* (*MPR*) relay contacts close upon fault recognition and this picks up either contacts 62X or 62Y or both, in addition to the trip coil (52T). Relay 50 is a multiple- contact instantaneous overcurrent relay, whose contacts remain closed as long as the fault is uncleared, thus allowing 62X or 62Y to energize the timer 62. Device 94 is an auxiliary tripping device. When the timer completes its timing cycle, contacts 62 close, energizing breaker failure relay 86. Relay 86 has multiple contacts that initiate tripping of all breakers adjacent to the failed breaker. The timing of the various devices is shown in Figure 26.31.

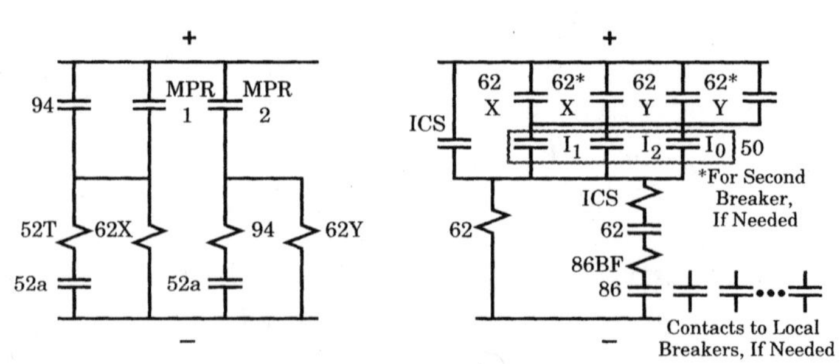

Figure 26.30 Schematic for breaker failure and local backup protection [11].

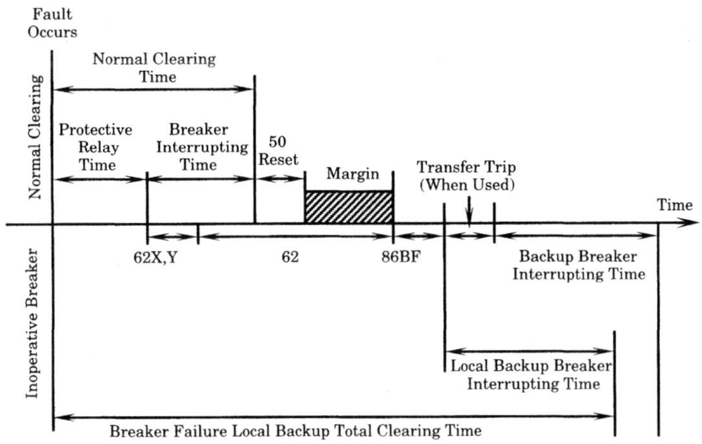

Figure 26.31 Breaker failure local backup timing.

REMOTE BACKUP SYSTEM. In addition to the MPS and the LBS, a complete protective system strategy must include a *remote backup system* (*RBS*). This is necessary, in many cases, to completely isolate the fault. For example, consider the configuration of Figure 26.29(*b*), with the ring bus and with the failure of breaker 2. The LBS acts to open local breakers 3 and 4, leaving the fault radial from bus G (again, assuming that breaker 6 operates correctly). Remote backup clearing of breaker 1 at bus G is required to complete the fault isolation. The fault is well beyond the primary protective zone of the line relays at G, but delayed clearing will take place to clear the fault. The coordination of the RBS relays with the MPS and LBS relays may be achieved by time coordination using delayed clearing of the remote breakers. Examples of this type of coordination are shown in Figure 26.32. For the cases illustrated, the coordination is achieved by sequential tripping, with a suitable *coordination time interval* (*CTI*) to ensure proper relay operation and coordination. This means that the fault clearing will be somewhat slower than normal,[4] but this is not a high price to pay for a double contingency consisting of a faulted line and a failed breaker. If these events are independent, the probability of the combined event, which is the product of the individual probabilities, will be extremely low. However, if these two events can be considered common cause, for example, the high fault current causes the breaker failure, the probability will be much higher. (This will be the case if the breaker does not have adequate interrupting rating.)

26.3.4.2 Total System Operational Failure. Consider a protective system strategy that incorporates all three relay systems, the main protective system, the local backup system, and the remote backup system. Moreover, we assume that these systems fail only by FD or operational failures, which we designate here as the "failure mode," as opposed to the security mode. For this section, we ignore security failures, but these failures can be evaluated using a similar technique.

The three protective systems under consideration will usually have a control configuration similar to those shown in Figure 26.28. Backup systems may be simpler due to the omission of redundant elements. In any case, we note the series logic of these systems. The logical pattern is always as follows:

[4] It is possible that the remote backup can be made faster by means of transfer trip using pilot channel signaling.

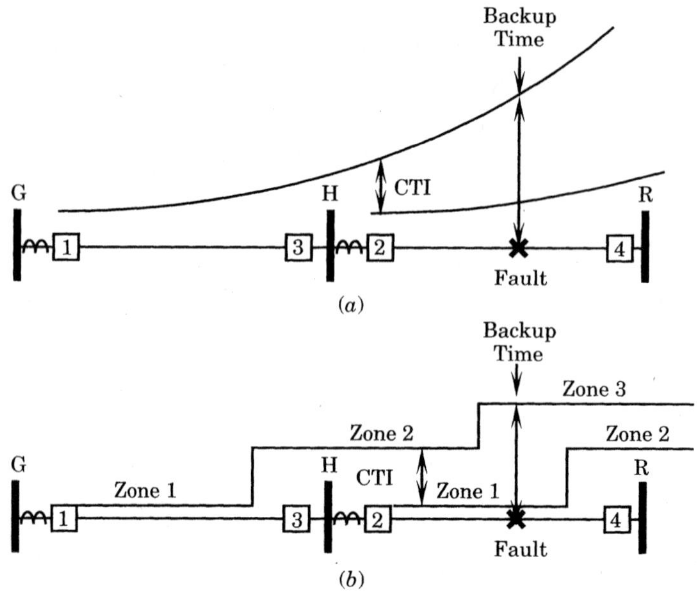

Figure 26.32 Remote backup at bus G breaker 1 for a fault F on line HR. (*a*) overcurrent relays. (*b*) Distance relays.

1. Instrument transformers and battery
2. Relays
3. Trip coil(s)
4. Breaker mechanism

For analysis, we group these components into two groups, a protective relay (electrical) group and a circuit breaker (mechanical) group. Somewhat arbitrarily, we include (1), (2), and (3) in the relay subsystem, and (4) in the circuit breaker subsystem. This simplifies the notation for computing reliabilities. The subsystems thus defined are shown in Table 26.9. Note that any of these subsystems may consist of more than one element, depending on the control configuration (Figure 26.28) and the bus arrangement (Figure 26.29).

We now define the event

$$\{\text{SFC}\} = \{\text{successful fault clearing}\} \tag{26.68}$$

TABLE 26.9 Protective Systems and Defined Subsystems

Subsystem	MPS	LBS	RBS
Relay subsystem	main protective relays (MPR)	local backup relays (LBS)	remote backup relays (RBS)
Breaker subsystem	main circuit breakers (MCB)	local circuit breakers (LCB)	remote circuit breakers (RCB)

and compute the protective system reliability as

$$R = \Pr\{\text{SFC}\} \tag{26.69}$$

To compute this probability we first compute the reliability of each subsystem and then combine the result. In order to simplify the notation, we write

$$\begin{aligned} R_{\text{MPS}} &= \Pr\{\text{MPS works}\} \\ &= \Pr\{\text{MPR works and MCB works}\} \end{aligned} \tag{26.70}$$

Since the MPR and MCB are independent systems, we simplify (26.70) to write

$$\begin{aligned} R_{\text{MPS}} &= \Pr\{\text{MPR works and MCB works}\} \\ &= \Pr\{\text{MPR works}\}\Pr\{\text{MCB works}\} \\ &= R_{\text{MPR}}R_{\text{MCB}} \end{aligned} \tag{26.71}$$

We also define

$$\begin{aligned} F_{\text{MPS}} &= \Pr\{\text{MPS fails}\} = 1 - R_{\text{MPS}} \\ &= F_{\text{MPR}} + F_{\text{MCB}} - F_{\text{MPR}}F_{\text{MCB}} \end{aligned} \tag{26.72}$$

The LBS and RBS have definitions similar to (26.71) and (26.72), with only the subscripts changed.

To compute the reliability (26.69) we evaluate the probabilities associated with both the local systems and the remote systems. Clearly, one or the other must work correctly for successful fault clearing. If we let LPS designate the "local protective systems" and RBS the "remote backup systems," we may write

$$\begin{aligned} R = &\Pr\{\text{LPS works}\} + \Pr\{\text{RBS works}\} \\ &- \Pr\{\text{LPS works and RBS works}\} \end{aligned} \tag{26.73}$$

In compact notation, and taking advantage of independence, we write

$$R = R_{\text{LPS}} + R_{\text{RBS}} - R_{\text{LPS}}R_{\text{RBS}} \tag{26.74}$$

Now the LPS consists of both the main protective system and the local backup system, with control configuration assumed similar to Figure 26.33. Note that the local backup system depends on successful primary relay operation. This gives the logic diagram of Figure 26.33, for which we compute

$$R_{\text{LPS}} = R_{\text{MPR}}\left(R_{\text{MCB}} + R_{\text{LBS}} - R_{\text{MCB}}R_{\text{LBS}}\right) \tag{26.75}$$

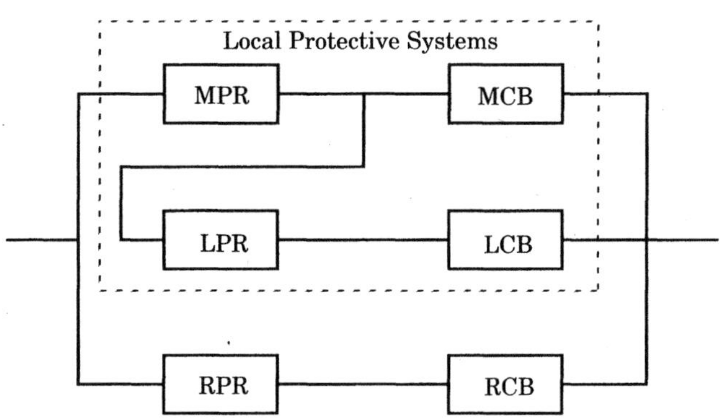

Figure 26.33 Network diagram for successful fault clearing.

where

$$R_{\mathrm{LBS}} = R_{\mathrm{LPR}} R_{\mathrm{LCB}} \tag{26.76}$$

Combining (26.75) and (26.76) we compute, after some algebraic manipulation

$$\begin{aligned} R = {} & R_{\mathrm{MPR}} R_{\mathrm{MCB}} + R_{\mathrm{MPR}} F_{\mathrm{MCB}} R_{\mathrm{LBS}} \\ & + R_{\mathrm{MPR}} F_{\mathrm{MCB}} F_{\mathrm{LBS}} R_{\mathrm{RBS}} + F_{\mathrm{MPR}} R_{\mathrm{RBS}} \end{aligned} \tag{26.77}$$

which agrees with the success tree result derived by others [12].

26.3.4.3 Block Diagrams of Operational Failure. Protective systems are designed in different equipment configurations, and these different arrangements have different reliabilities. The objective here is to develop a method for comparing different protective system arrangements by construction of a network diagram, from which an analysis of the configuration can be made. Network diagrams are constructed for the three different switching station arrangements shown in Figure 26.29 and for the three control configurations shown in Figure 26.28. Other arrangements can be evaluated using the same techniques, but these nine unique arrangements will illustrate the method. To simplify the analysis, we examine only the local protective system of Figure 26.33, but the remote backup system can be added in a straightforward way.

As an example of the technique, refer to control configuration 1 of Figure 26.28 and the single bus, single breaker station arrangement shown in Figure 26.29. The local protective system has redundancy only in the relays, and the entire system network diagram is shown in Figure 26.34.

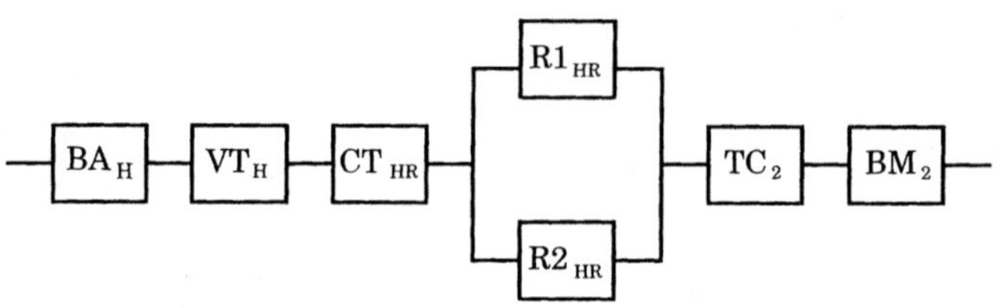

Figure 26.34 Network diagram of control configuration 1 for the successful clearing of faults at breaker 2 of a single breaker system.

This simple diagram clearly shows the redundancy of the relays, with their parallel logic, and the series logic of the separate subsystems. This example also illustrates the need for a clear numbering system to denote which relays and which breakers are being depicted. This is accomplished using subscripts as noted in Table 26.10.

For the system of Figure 26.34 all of the subscripts are the same since the relays control only one breaker, breaker 2. Backup relaying systems sometimes control different breakers that those controlled by the primary relays, hence the need to distinguish between the two. The subscripts show the locations of measurements or control actions.

For Figure 26.34, we can write the probability of successful operation as follows.

$$\begin{aligned} R &= \Pr\{\text{successful primary fault clearing}\} \\ &= \Pr\{BA \cap VT \cap CT \cap (R1 \cup R2) \cap TC \cap BM\} \\ &= R_{\mathrm{BA}} R_{\mathrm{VT}} R_{\mathrm{CT}} (R_{\mathrm{R1}} + R_{\mathrm{R2}} - R_{\mathrm{R1}} R_{\mathrm{R2}}) R_{\mathrm{TC}} R_{\mathrm{BM}} \end{aligned} \tag{26.78}$$

Most relay schemes employ local backup to ensure fault removal in the event that the local

TABLE 26.10 Abbreviated Apparatus Designations

Abbreviation	Subsystem
BA_H	Battery at station H
VT_x	Voltage measurement at bus x
CT_x	Current measurement on line HR
$R1_{HR}$	Primary relay 1 for protective zone HR
$R2_{HR}$	Primary relay 2 for protective zone HR
TC_x	Trip coil for protective on breaker x
BM_x	Breaker mechanism of breaker x
Subscript *x*	Location of measurement or control action
Subscripts H, R	Identification of relay protective zone

breaker(s) do not open for any reason. Following the same example analyzed above, the breaker failure trip logic of Figure 26.30 is used to trip all breakers adjacent to the failed breaker 2. The logic of Figure 26.30 gives the network diagram shown in Figure 26.35. The network diagram shows clearly the redundant nature of the local backup clearing for breaker failure. In normal operation, the trip coil of breaker 2 will be energized and breaker mechanism 2 will open the faulted line. However, should this system be inoperative, the timer 62 will time out and, if the fault persists (50 remains picked up), the 86 contacts will close to trip breakers 3, 4, and 5. This type of control logic is an example of "standby redundancy," since the backup system in not switched into service unless the primary system fails, with the timer used to detect failure of the primary system.

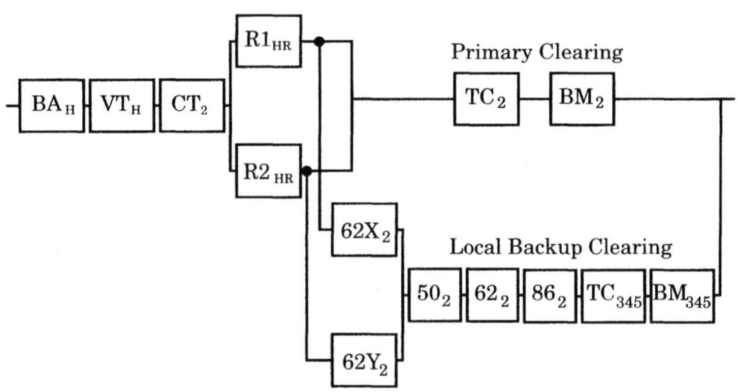

Figure 26.35 Cold standby arrangement of primary and local backup protection systems using control configuration 1.

Figures 26.36, 26.37, and 26.38 show network diagrams of the three control configurations applied to the single bus–single breaker, ring bus, and breaker-and-a-half schemes, respectively. Writing out the reliability expressions, similar to (26.78), is a straightforward extension of previous work. Analysis of operational failures can also be performed by other methods such as event trees [13]. This is left as an exercise.

The foregoing procedure provides a method of analyzing the reliability of a given protective system configuration. The reliability analysis requires that the probability of successful

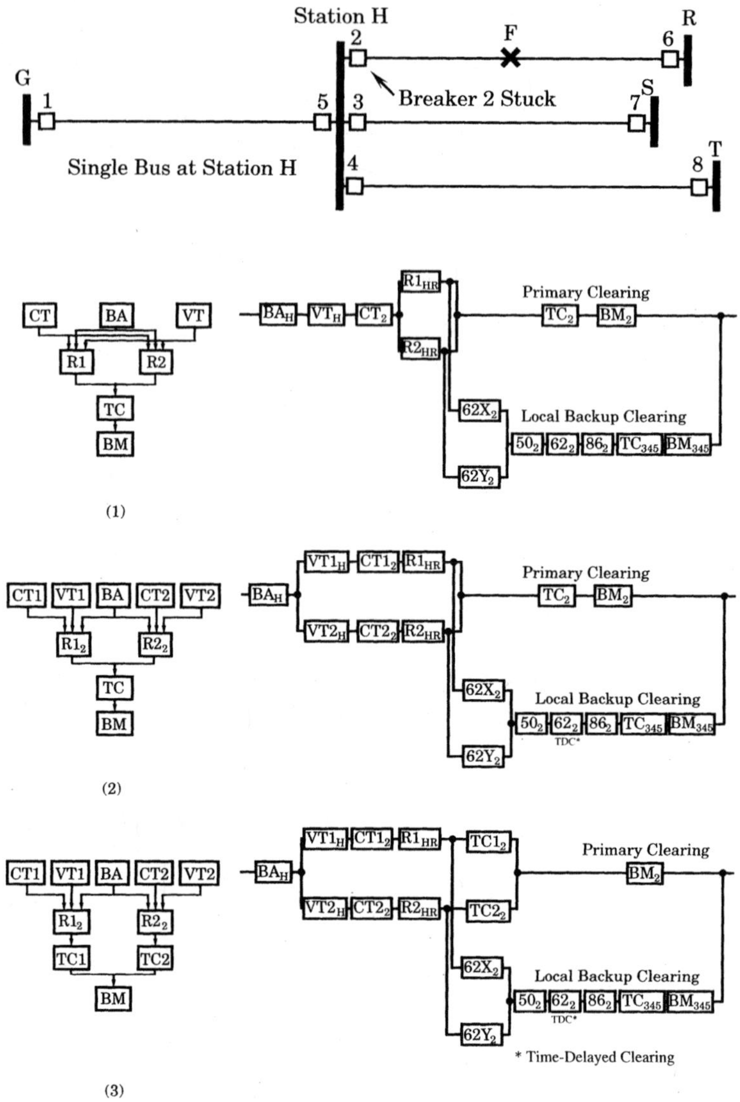

Figure 26.36 Network diagrams for the protection of line HR at Station H using a single bus–single breaker switching arrangement. (1) Redundant relays, (2) redundant instrument transformers and relays, (3) redundant instrument transformers, relays, and circuit breaker trip coils.

operations be known for all subsystems. This requires knowledge of each subsystem, a reliability model for that subsystem, and data to support the reliability model. Appropriate models for this type of equipment are discussed in Chapter 24.

The foregoing analysis conveniently ignores the element of time. Reliability is a time-dependent quantity, and the analysis of systems often requires that time be included in the system analysis. This is the subject of Section 26.4.

26.3.4.4 Block Diagrams of Security Failure. In considering security failure of a protective system, it is necessary to examine the security of the primary system, the local

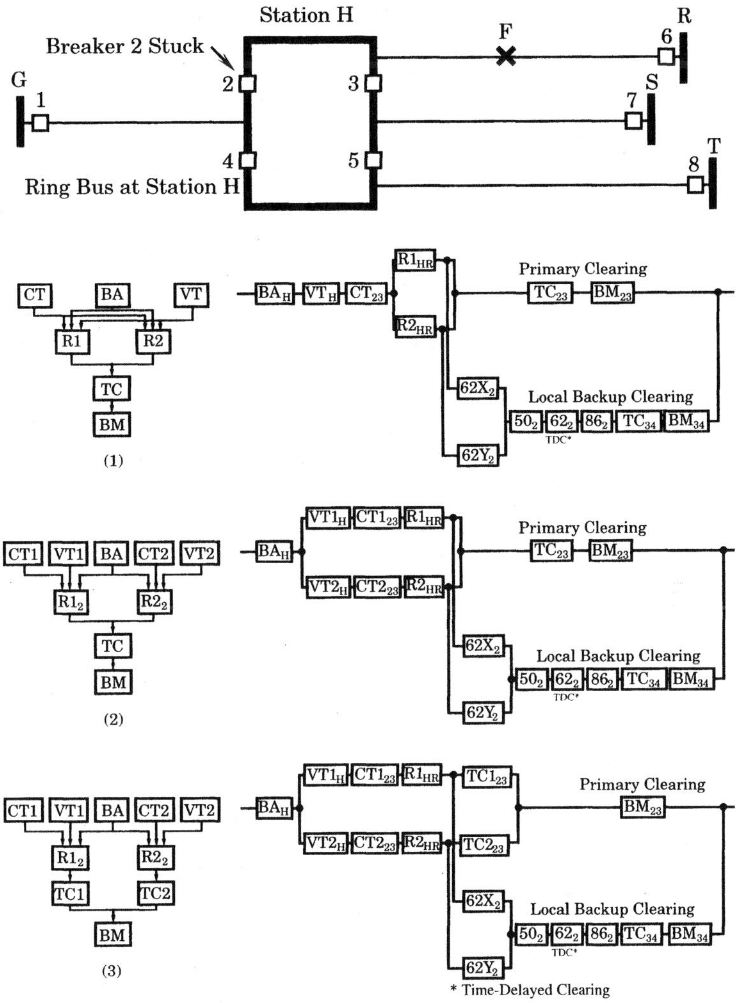

Figure 26.37 Network diagrams for the protection of line HR at Station H using a ring bus switching arrangement. (1) Redundant relays, (2) redundant instrument transformers and relays, (3) redundant instrument transformers, relays, and circuit breaker trip coils.

backup system, and the remote backup system. It is also necessary to clearly define what is meant by failure. In this case, it is appropriate to define failure as follows.

> *Security failure* is the unnecessary or fail safe disconnection of the primary protected component due any cause or causes in the absence of a hazard.

The focus should be on the primary protected component, such as the protected line, transformer, or bus section. This immediately removes remote backup systems from consideration, since these remote protections trip circuit elements other than the primary protected element. This may be cause for concern, should the remote backup system suffer a false trip, but it has no direct impact on the primary protected element. This narrows the focus of attention to the primary and local backup systems.

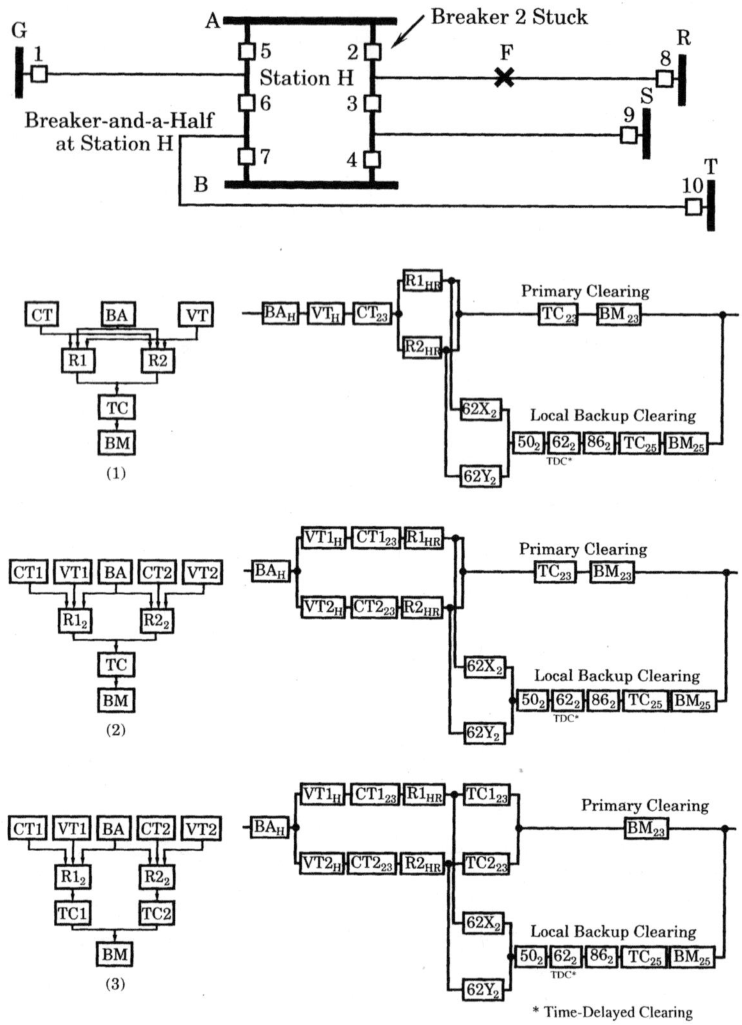

Figure 26.38 Network diagrams for the protection of line HR at Station H using a breaker-and-a-half switching arrangement. (1) Redundant relays, (2) redundant instrument transformers and relays, (3) redundant instrument transformers, relays, and circuit breaker trip coils.

As an example of security failure, consider the control configurations of Figure 26.28. Clearly, any of these control configurations can be the cause of an unnecessary trip of the primary breaker(s) at one terminal of a protected element, thereby qualifying as a security failure. Not all of the components in the three different control configurations are capable of contributing to a security failure, however. Table 26.11 reviews the possibilities of security failures for each component.

The current transformers, battery, and circuit breaker trip coil are not security failure risks. It is difficult to conceive that current will somehow be forced through a trip coil, of sufficient magnitude to trip the breaker, unless one or more of the relay contacts are closed. It is *possible* that a fault in the control circuitry could cause this to occur, but this would have to be considered rare. Failure of the battery could cause operational failure, but not security failure.

TABLE 26.11 Security Failure Capability of Control Components

Component	Security Failure Comments
Current transformer	Not a security failure risk
Voltage transformer	Security failure risk for distance protection
Battery	Not a security failure risk
Relay 1	Security failure risk
Relay 2	Security failure risk
Trip coil	Not a security failure risk
Breaker mechanism	Security failure risk

Failure of the current transformer will usually cause loss of current sensing, not increased current output.

Of the elements that are candidates for security failure risk, the relays and the breaker mechanism are the most likely, due to their moving contacts that can be forced into a change from their normal open/close position. This could occur due to a severe mechanical jolt, for example. The voltage transformer is a possible security failure risk in the case of distance protection. Failure of the voltage transformer will usually result in zero voltage input to the relay logic, which would cause distance protection logic to equate this to a close-in fault. This would not affect overcurrent protection or many other types of primary protection. In the final analysis, the engineer must examine the various protective device components and determine their possible security failure risks in view of the type of protective system installed. The RDB for the protective system components is shown in Figure 26.39, since any of the four components can be the cause of security failure.

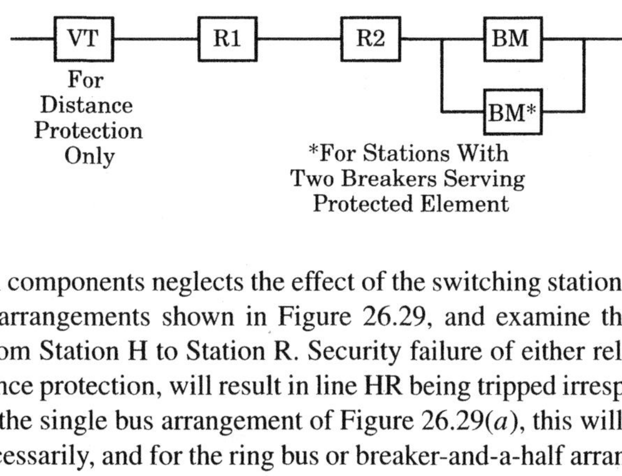

Figure 26.39 RDB of control components for security failure.

This analysis of control components neglects the effect of the switching station arrangement. Consider the station arrangements shown in Figure 26.29, and examine the risk of security failure on the line from Station H to Station R. Security failure of either relay or the voltage transformer, for distance protection, will result in line HR being tripped irrespective of the station arrangement. For the single bus arrangement of Figure 26.29(*a*), this will result in breaker 2 being tripped unnecessarily, and for the ring bus or breaker-and-a-half arrangements of Figure 26.29(*b*) and (*c*), the security failure will trip both breakers 2 and 3. This is not true for the breaker mechanism security failure. When there is only one breaker, as in the single bus arrangement, the line will be opened at bus H. For stations with two breakers serving the protected line, however, both would have to suffer security failure simultaneously in order to open the line at bus H. This conclusion leads to the RDB for the protective system components shown in Figure 26.39, since any of the four components can be the cause of security failure.

From Figure 26.39, it is observed that the voltage transformer or either relay can be the sole cause of a security failure. If the protected line is served by only one circuit breaker, the security failure of that breaker will result in an open line and a security failure of that line. If the protected line is served by more than one breaker, all local breakers for the protected line would have to suffer overlapping security failures to cause unnecessary opening.

We conclude that security failure analysis requires a complete examination of the protective system components and the station arrangement, taking into consideration the type of protective scheme employed. In most cases, there are several components for which security failure is either impossible or highly unlikely. An examination of typical failure data shows that the relays are the most likely cause of security failure, and human intervention is often the cause of security failure.

26.3.5 Specifications for Transmission Protection

As an example of the application of reliability concepts to a practical system encountered in power system protection, consider a typical EHV power transmission system. For circuits of the EHV voltage levels, the transmission lines carry large blocks of power. Loss of these lines can have serious financial consequences; therefore, measures must be taken to ensure that security failures, and the accompanying unnecessary line openings, be kept at a minimum. On the other hand, faults on these circuits cause very large upsets to the system, and the danger of instability and possible cascading of line outages might occur if faults are not removed promptly. Therefore, these protective systems must have the very highest possible speed and security.

26.3.5.1 Relay Specifications. In order to provide high probability of clearing all faults on the lines, it is necessary to employ completely independent protective systems. The relay systems usually employed on EHV transmission lines are advanced permissive systems that require the concurrence of a tripping decision from the two ends of the line in order to avoid a false trip. This requires the use of fast communications between the two terminals of the protected line. Moreover, to be completely independent the communications systems must employ *different* communications media, not just different channels of the same medium, and should also employ different detection logic to assure reliable operation. In summary, the principles on which these systems are based are the following:

1. At least two independent relaying systems
   - Independent relays
   - Independent voltage transformers
   - Independent current transformers
   - Independent circuit breaker trip coils
2. Independent and different communications for each system
3. Different types of detection and analysis methods for each system

26.3.5.2 Switching Station Specifications. It was noted previously that the network diagrams for the various switching station arrangements are quite different. The different station arrangements result in a different selection of breakers that must be tripped in the case of a fault. Figures 5.11 and 5.12 give one-line diagrams for a variety of different switching station arrangements. These different station arrangements lead to different failure modes.

Station failure can have serious consequences for the power system. This suggests that the design of the station cannot be adequately specified without considering the effect of station failures on the power system as a whole. For example, consider the effect of a stuck breaker. We have noted that different station designs result in different system switchings, depending on which breaker fails. It is also clear that some station designs are less likely to cause a system separation than others. The point is that the station reliability is only one factor in system reliability and the inherent reliability of the station and the reliability of the transmission system must be studied together. This may require power flow, reliability, and system stability studies, for example, to determine the effect of a station failure. If the results are hazardous, a more robust station design may be warranted, even though such an alternative may be more expensive. Knowing the frequency of a given type of failure and the cost of each failure occurrence, the engineer can compute the effectiveness of the various alternatives.

Station failure is also dependent on the failure modes of the circuit breakers and other devices in the station. Circuit breakers are critical items and have many different modes of failure.

CIRCUIT BREAKER FAILURE MODES. One of the unique features of stations is the many different failure modes, especially of the circuit breakers. IEEE Standard 500 [14] lists the 12 different failure modes of circuit breakers, interrupters and relays. These are listed, together with their failure rates in failures per million hours and in failures per million operations, in Table 26.12.

TABLE 26.12 Average Failure Rates for Circuit Breakers [14]

	Failure Rate*	
Failure Mode	**Fail/Mhr**	**Fail/Mops**
All modes	4.83	882.
Catastrophic failure	1.59	379.
1. Does not close on command	0.76	243.
2. Does not open on command	0.35	118.
3. Closes without command	0.03	
4. Opens without command	0.16	
5. Does not make current	0.03	7.58
6. Does not break current	0.03	11.4
7. Fails to carry current	0.05	
8. Breakdown to earth (internal)	0.02	
9. Breakdown to earth (external)	0.02	
10. Breakdown between poles	0.03	
11. Breakdown across open pole (internal)	0.09	
12. Breakdown across open pole (external)	0.02	
Degraded failure	3.24	503.

*Failure rates are given in failures per million hours and in failures per million operations

Since stations have a large number of switching devices that can fail in any of these modes, the total number of possible failures is very large.

The IEEE standard gives values for many of the different failure modes. Failure rates are given in the standard based on time of service as well as the number of operations.[5] Table 26.9 gives the recommended failure rates of circuit breakers from all voltage classes to use in system studies. The average repair rate is given by the standard as 12.1 hours for all modes of failure.

[5]IEEE Std 500 gives three values for each item, which are called the low, recommended, and high values. Table 26.9 gives only the recommended values.

The failure rate of all modes for circuit breakers is highly dependent on the voltage class of the breaker, with the higher voltage classes having higher failure rates. Table 26.13 shows a breakdown of failure rates by voltage class. Note that almost all modes experience a higher failure rate as the voltage class increases.

TABLE 26.13 Circuit Breaker Failure Rates by Voltage Class [14]

Failure Mode*	Failure Rate, failures/Mhr				
	63–100 kV	100–200 kV	200–300 kV	300–500 kV	Over 500 kV
All modes	2.4	6.62	10.28	24.0	17.6
Catastrophic	0.46	1.83	2.97	5.25	12.0
1. No close on command	0.46	0.97	1.31	2.31	4.36
2. No open on command	0.08	0.30	0.72	1.36	3.27
3. Closes w/o command	0.02	0.02	0.11	0.10	0.00
4. Opens w/o command	0.03	0.11	0.30	0.34	2.18
5. No make current	0.01	0.02	0.17	0.10	0.00
6. No break current	0.01	0.05	0.04	0.27	0.00
7. Fail to carry current	0.01	0.08	0.05	0.34	0.00
8. Internal ground	0.01	0.04	0.05	0.07	0.00
9. External ground	0.01	0.02	0.05	0.13	0.00
10. Fault between poles	0.00	0.02	0.00	0.00	1.09
11. Internal fault across pole	0.02	0.15	0.15	0.13	1.09
12. External fault across pole	0.00	0.05	0.02	0.10	0.00
Degraded	1.94	4.79	7.31	18.7	5.59

*See Table 26.9 for a complete description of each failure mode

The tabulations of failure rates for circuit breakers show that failure to close on command is usually the most prominent failure mode, and this is true for all voltage classes. This is somewhat surprising, since there is little stress on the circuit breaker for a closing operation. At the EHV levels, station arrangements are often used that serve each circuit with two circuit breakers at both terminals, so a complete failure to close either end of the circuit would require the overlapping failure to close on command of both circuit breakers at one end. This makes the failure to successfully connect the circuit a rather low probability event. When such failures occur, repair personnel would be dispatched to determine the cause of the failure and take corrective action, but the power system will not be adversely affected due to the failure of one breaker to close on command in stations with two breakers serving each line.

Failure to open on command is usually second in the ranking of failure rates. When the circuit breaker fails to open, this may be during a fault condition when the circuit breaker is operating under high stress, perhaps near its interrupting rating. Failure to operate leaves the fault connected to the station, and backup protection must operate after a suitable time delay. This leaves the fault connected longer, with possible destabilizing results. Moreover, if the fault magnitude is close to the circuit breaker rating, the damage to the breaker may be substantial, which will require a more extensive repair.

STATION FAILURE MODES. Consider the effect of a fault on one of the circuits connected to a given station. When a fault occurs, the protective systems are designed to recognize the fault and initiate its removal as quickly as possible. This will require the successful operation of all circuit breakers that are connected to the faulted component. This, in turn, requires that all the support systems within the station are working as designed.

Simple station arrangements have only one breaker connected to the faulted circuit, but more complex arrangements have two or more breakers on each end of each circuit, all of which experience the stress of fault interruption. Because of this stress, there is a greater probability of breaker failure on the more complex stations since more than one breaker must operate successfully at each terminal. This is, therefore, a series logic since the failure of any one of the breakers results in the failure to trip the faulted circuit. It is common to have four breakers on each circuit, two at each end. If all have the same failure rate, the rate of failing to clear the circuit is four times that of a single breaker. Some station designs, such as the ring tripod or the ring bridge, have three breakers on each circuit, which raises the failure rate to as much as six times that of a single breaker (see Chapter 5 to review these station arrangements). As noted in Section 26.3.4, the more complex stations are favored for the higher voltage circuits since they have a higher probability of leaving the unfaulted circuits connected following the loss of a faulted circuit.

This complexity is not without problems, however, since the larger number of breakers tends to increase the failure modes of the station. Data on the effect of terminal-related transmission failures has been accumulated for many years by a group of North American utilities [15]. An analysis of the data for a six-year period is summarized in Table 26.14.

TABLE 26.14 Terminal-Related Failures Rates for Stations [15]

Voltage>	**230 kV**		**345 kV**	
Station Configuration	**Failure Rate (per term – yr)**	**Duration (hr/term – yr)**	**Failure Rate (per term – yr)**	**Duration (hr/term – yr)**
Single bus, single breaker	0.084	1.052	0.020	0.282
Ring bus	0.130	0.996	0.045	0.160
Breaker and a half	0.178	0.074	0.192	0.068

These data indicate that the terminal-related failure rates on the more complex station configurations are greater than on the simpler station arrangements, which is a direct result of the increased complexity required of the control and protection systems on the more complex designs. In contrast, however, the average duration of outage per terminal- year has the opposite effect, with the more complex station designs having the lower outage durations.

The stuck breaker, or failure to open on command, is the most serious breaker failure mode and is the only one for which special protective systems are designed. It is common, at the EHV level of transmission, to require reliability at the level delivered by at least the breaker-and-a-half arrangement, and important stations may require double breaker–double bus switching to minimize the hazard of system breakup following stuck breaker failures. These EHV stations usually supply each outgoing feeder by two or more circuit breakers. This provides high reliability in the sense that the line can be fed normally with one of the circuit breakers out of service for either planned or forced outage. It also gives greater reliability against failure of the circuit breakers to close on command, but clearly this type of failure is often of little concern.

Some station arrangements are subject to serious system breakup due to a stuck breaker failure. This is true of the ring bus and, to a lesser extent, of the breaker-and-a-half scheme.

26.3.5.3 Communications Specifications. EHV transmission lines usually employ protective systems that utilize high-speed communications between the relay equipment at the two ends of the line. These systems are usually designed to provide high levels of reliability against both operational (FD) and security (FS) failures, and the communications between the two ends of the line are an important contributor to this reliability. In many cases, these protections will utilize two different types of communication, for example, one using *microwave* (*MW*) and the other using *power line carrier* (*PLC*), to provide both redundancy and independence for the communications media. Fiber-optic communications are sometimes available on either the phase conductors or the static wires of transmission lines, and are available for protective system communications. The older systems of the 230 kV or lower voltage classes often use only PLC protection systems.

The specifications for communications should include the following requirements:

1. Independent transmission media
   - microwave
   - fiber optics
   - power line carrier
   - telephone circuit
2. Independent power supplies
3. Redundant equipment for
   - transmitters
   - receivers
   - communications channels
   - antennas (to minimize microwave fading)

In addition to the requirement for independence and redundancy, there is often a requirement for cross-wiring of communications. This means that the protective system utilizing communications medium 1, will send its messages on both medium 1 and medium 2 at the same time. This provides additional protection against a communications failure or fading of signal strength.

26.4 TIME-DEPENDENT RELIABILITY MODELS

In many reliability problems, time is an important variable. This is true of protective system where timing of relay operations is controlled to provide correct coordination of devices. It is also important in a given protective strategy that calls for the sequential operation of several different devices.

The equations that describe the reliability of a system may be written to include the time variable. For example, consider a parallel system of two elements, for which we may write

$$\begin{aligned} R_P(t) &= 1 - q_1(t)\, q_2(t) \\ &= p_1(t) + p_2(t) - p_1(t)\, p_2(t) \end{aligned} \tag{26.79}$$

Equations for other configurations may also be written to recognize that time is an important variable. We now develop some of the basic reliability mathematics that describe the time behavior of systems.

26.4.1 Failure Distributions of Random Variables

Let T_A be a random variable (rv) corresponding to the time to failure of a given system A. Then we define

$$\begin{aligned} T_A &= \text{time from the moment } A \text{ was placed} \\ &\quad \text{in service until it fails} \\ F_A(t) &= \Pr\{A \text{ will fail prior to } t\} = \Pr\{T_A \le t\} \\ f_A(t) &= \frac{dF_A(t)}{dt} \end{aligned} \tag{26.80}$$

These distributions have the mathematical properties usually associated with probability distributions (see Chapter 24).

Now let C represent a composite system formed by combinations of other elements, A and B, and with joint distributions defined as follows. If we define a random variable T_B corresponding to the time to failure time of element B, we can define similar distributions for this element exactly as in (26.80). We may then define the joint distribution

$$F_{AB}(t_A, t_B) = \Pr\{T_A \le t_A, T_B \le t_B\} \tag{26.81}$$

Clearly, we must keep track of the two time measurements separately. We now examine some elementary composite systems that are of interest in system protection.

26.4.1.1 Series Connection of A and B. If two elements are connected in a series logic, such that both must work in order that the series system work, the composite system will fail if either A or B fails. Consider the first element

$$\begin{aligned} P_A(t) &= \Pr\{A \text{ works}\} \\ &= 1 - \Pr\{A \text{ fails}\} \\ &= 1 - \Pr\{T_A \le t\} \\ &= \Pr\{T_A > t\} \\ &= 1 - F_A(t) \end{aligned} \tag{26.82}$$

and similarly for element B. Similar equations may also be written for the composite system C.

$$\begin{aligned} \Pr\{C \text{ works}\} &= \Pr\{\text{both } A \text{ and } B \text{ work}\} \\ &= \Pr\{T_A > t \text{ and } T_B > t\} \\ &= \Pr\{T_A > t\}\, P\{T_B > t\} \end{aligned} \tag{26.83}$$

where the last line of (26.83) can be written if the two systems are statistically independent.

Combining the results, we may write

$$(1 - F_C) = (1 - F_A)(1 - F_B) \tag{26.84}$$

or

$$F_C = F_A + F_B - F_A F_B \tag{26.85}$$

Obviously, from (26.83)

$$T_C = \min(T_A, T_B) \tag{26.86}$$

which is intuitively correct.

The failure density of the series system is computed by taking the derivative of the failure distribution (26.85).

$$\begin{aligned} f_C(t) &= f_A(t) + f_B(t) - F_A(t)\, f_B(t) - F_B(t)\, f_A(t) \\ &= f_A(t)\,[1 - F_B(t)] + f_B(t)\,[1 - F_A(t)] \end{aligned} \tag{26.87}$$

EXAMPLE 26.11 *Series System*

Prepare a sketch of the region in the t_A, t_B plane representing the failure time of a series connection of components A and B.

Solution

We construct the times to failure as the axes of a two-dimensional space as shown in Figure 26.40. The failure region is the shaded region below, which corresponds to the value of T_A, and is shown in this example as the smaller of the two failure times.

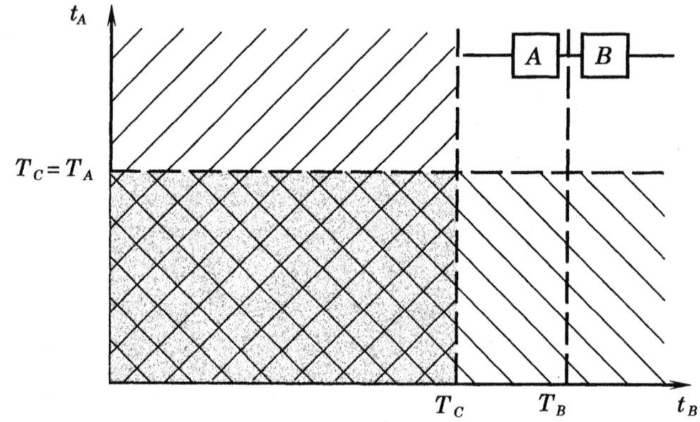

Figure 26.40 Failure region for a series-connected system of A and B. ■

26.4.1.2 Parallel Connection of A and B. If two elements are connected in a parallel logic, such that the system works when either A or B works, the composite system C will fail only when both A and B fail. For this situation, we may write

$$\begin{aligned} \Pr\{C \text{ fails}\} &= \Pr\{T_C \le t\} \\ &= \Pr\{A \text{ fails and } B \text{ fails}\} \\ &= \Pr\{T_A \le t \text{ and } T_B \le t\} \end{aligned} \tag{26.88}$$

If A and B are independent, then

$$\Pr\{C \text{ fails}\} = \Pr\{T_A \le t\}\Pr\{T_B \le t\} \tag{26.89}$$

or

$$F_C(t) = F_A(t)\, F_B(t) \tag{26.90}$$

This requires that

$$T_C = \max(T_A, T_B) \tag{26.91}$$

The failure density function for the parallel connection is computed as the derivative of (26.67), with the following result.

$$f_C(t) = f_A(t)\, F_B(t) + f_B(t)\, F_A(t) \tag{26.92}$$

EXAMPLE 26.12 ***Parallel System***

Sketch the region in the t_A, t_B plane representing the failure time of a parallel connection of items A and B.

Solution

We construct the times to failure as the axes of a two-dimensional space, as shown in Figure 26.41. The failure time of the A system is the maximum of the two failure times. The shaded area is the region of failed operation of the parallel system.

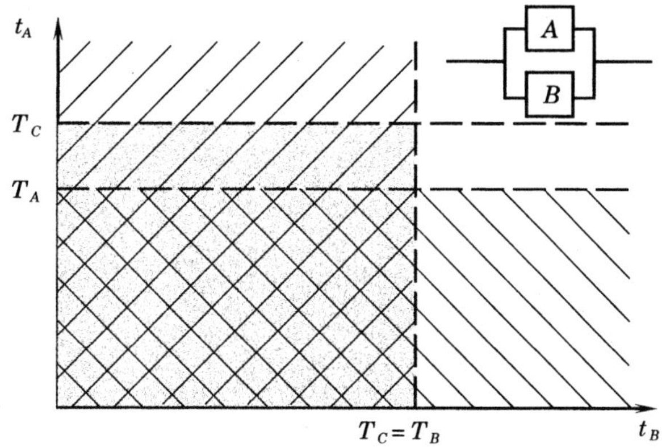

Figure 26.41 Failure region for a parallel connection of A and B. ■

The preceding examples help to visualize the failure distribution for the series and parallel systems. The failure distribution of the total system, in each case, is the mass of the shaded region in the plane.

26.4.1.3 Standby Redundancy. For some composite systems, it is important to note that certain redundant elements are on standby and that no operating time accumulates for these elements until the primary element fails. This situation is shown in Figure 26.42. We assume that the switching is perfect.

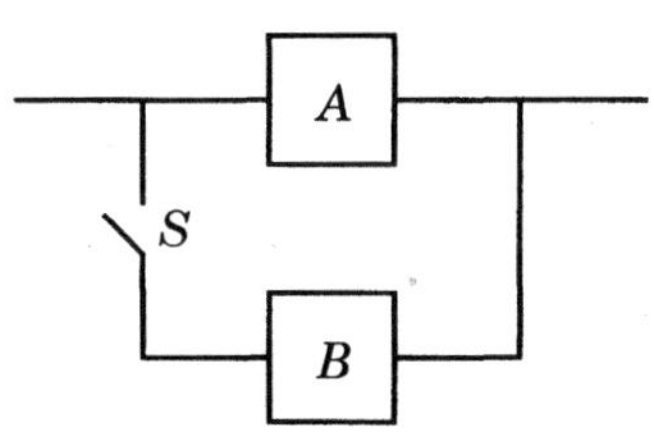

Figure 26.42 Standby redundancy with element B on standby.

Backup relaying is a form of standby redundancy. The backup system B is not required to operate unless a predesigned condition occurs, at which time this protective system is placed in operation. The switch, in this case, comprises contacts that are closed by a timer, or other device.

In terms of the time of operation, we view the redundant system as shown in Figure 26.43. We assume that element B does not fail when de-energized, an assumption that may not always be acceptable. If this is a valid assumption, then the system will be successfully operating at time t if either of the following conditions are met:

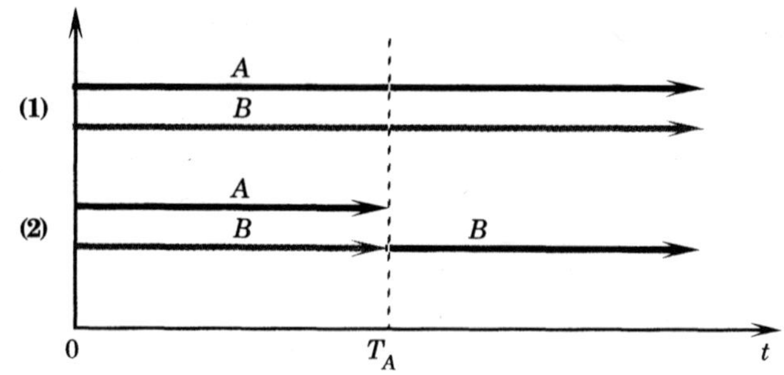

Figure 26.43 Success options for a standby system.

1. A succeeds up to time t.
2. A fails at time $T_A < t$ and B operates successfully from T_A to t.

These conditions are depicted graphically in Figure 26.43. Since we assume that the switch is perfectly reliable, we may write

$$T_C = T_A + T_B \tag{26.93}$$

This equation represents the sum of two random variables. It can be shown that the failure distribution in this case is given by the convolution integral, i.e.,

$$F_C = F_{A+B} = \int_0^t F_A(t-u)\, f_B(u)\, du \tag{26.94}$$

where we convolve the distribution of A with the density of B. This concept has applications in power system protection if the stated assumptions are correct for the system under consideration.

EXAMPLE 26.13 *Standby Redundant System*

Sketch the region in the t_A, t_B plane representing the failure time of a standby redundant connection of components A and B. Perfect switching logic and operation are assumed.

Solution

We construct the times-to-failure variables as the axes of a two-dimensional space, as shown in Figure 26.44. The failure time of the total system is the sum of the two failure times. The shaded area is the

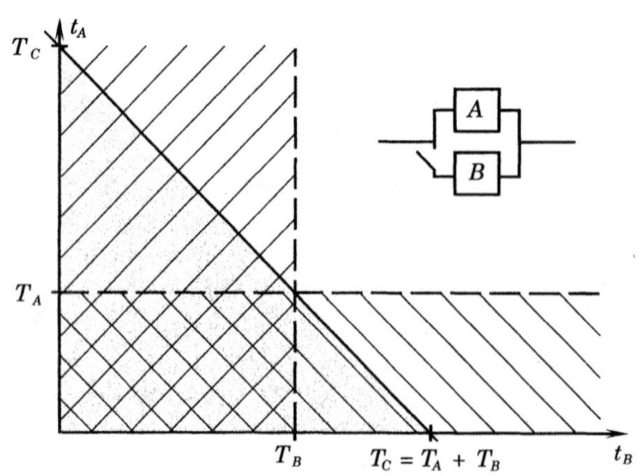

Figure 26.44 Failure region for a standby redundant system.

region of failed operation of the standby system for the condition that system B does not fail when it is on standby. ■

26.4.1.4 Sequential Operation. Another time-dependent situation of interest in system protection is a sequential operation. The hardware configuration is shown in Figure 26.45. Here, we assume the following conditions apply:

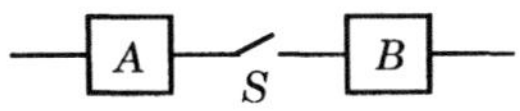

Figure 26.45 System arrangement for a sequential operation.

1. A is placed in operation at $t = 0$; B is idle.
2. A logical decision in A determines if switch S should be closed (ideal switching assumed).
3. B is placed in operation at $t = T_X$, where T_X is a random variable.

The composite system C fails if either of two conditions, illustrated in Figure 26.45, are satisfied:

(i) Failure occurs at $t < T_X$ due to failure of A, i.e., $T_A < T_X$.
(ii) Failure occurs at $T_X > t$, due to
 (a) failure of A; $T_A \leq t$
 (b) failure of B; $T_B + T_X - t$.

The success of either condition is shown by the shaded regions in Figure 26.46. Each of the above possibilities must be examined.

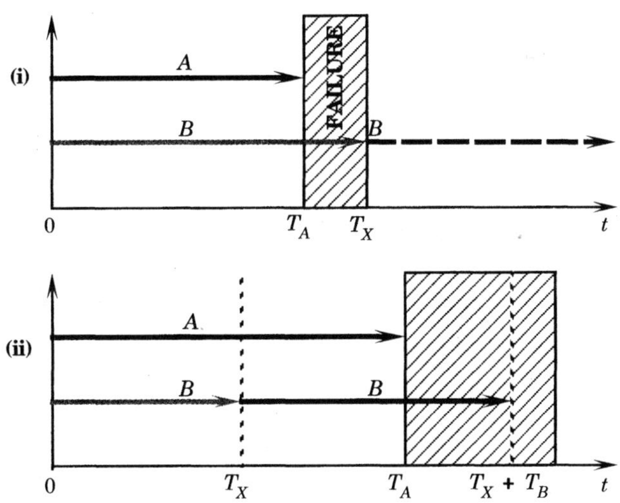

Figure 26.46 Time-to-failure possibilities of sequential operation.

(i) Failure occurs for $T_X > t$. For this condition, designated as C, the system C fails in mode 1, i.e.,

$$\Pr\{C_1\} = \Pr\{T_A \leq t,\ T_X > t\} \tag{26.95}$$

Assuming that T_A and T_X are independent, then

$$\Pr\{C_1\} = \Pr\{T_A \le t\} \Pr\{T_X > t\} \tag{26.96}$$

or

$$F_{C1}(t) = F_A(t)[1 - F_X(t)] \tag{26.97}$$

(ii) Failure occurs for $T_B + T_X \le t$. We compute the success mode 2 as

$$\Pr\{\text{success}\} = 1 - \Pr\{\text{failure}\} = 1 - F_{C2}(t) \tag{26.98}$$

But

$$\begin{aligned} \Pr\{\text{success}\} &= \Pr\{A \text{ succeeds and } B \text{ succeeds and } T_X \le t\} \\ &= \Pr\{T_A > t\} \Pr\{T_X + T_B > t\} \Pr\{T_X \le t\} \\ 1 - F_{C2} &= (1 - F_A)(1 - F_{XB}) F_X \end{aligned} \tag{26.99}$$

Now, since the operations (i) and (ii) are disjoint, we may add the two probabilities, with the following result.

$$\begin{aligned} F_C &= F_{C1} + F_{C2} \\ &= F_A(1 - F_X) + 1 - (1 - F_A)(1 - F_{BX}) F_X \\ &= 1 + F_A - F_X(1 - F_{XB} + F_A F_{XB}) \end{aligned} \tag{26.100}$$

where

$$F_{XB} = F_{T_X + T_B} = \int_0^t f_X(u) F_B(t - u)\, du \tag{26.101}$$

Protective systems have elements and subsystems that operate sequentially, making the above analysis of direct interest. The operating times and the waiting times are not exact measures, but are random variables. For example, the *main protective relay* (*MPR*) subsystem and the *main circuit breaker* (*MCB*) subsystem operate sequentially as elements A and B in the foregoing analysis.

26.4.2 Composite Protection System

We apply the concepts of the preceding section to a practical protection system at one terminal of a transmission line. The network diagram for this system is shown in Figure 26.47. The system illustrated includes the main or primary protection (M), the local backup protection (L), and the remote backup protection (R). We add switches to the diagram to serve as a reminder that the system operates sequentially. In some cases, these are literally switches

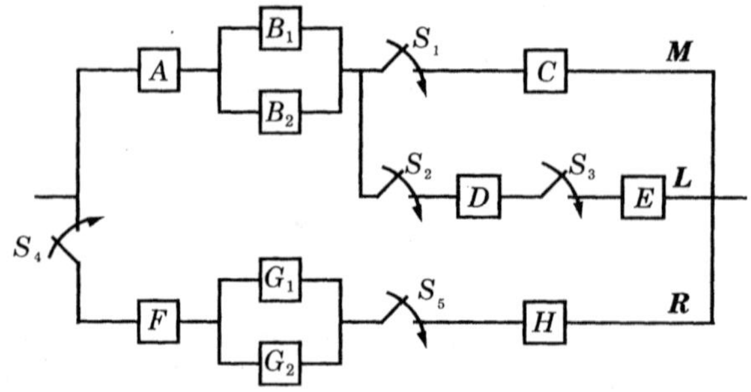

Figure 26.47 Network diagram for fault clearing at one-line terminal.

or relay contacts, in other cases they may be solid-state devices or logical controls. We must also associate a random operating time with each of the switches. The identification of the components in the boxes may be determined from Figure 26.33 and the station arrangement. Here these functions are assigned letters for simplicity of notation, so that single subscripts can be used for each type of device.

26.4.2.1 The Main Protection System. As the first step in the analysis of Figure 26.47, we analyze the main protective system (M), which consists of the topmost line of items labeled A, B, and C, plus the switch S_1. The M system consists of a MPR system and a MCB, as noted in Figure 26.33.

MPR. For this subsystem let the random operating time be assigned the random variable T_{MPR}, which is the random time to failure of the MPR.

The MPR system consists of two parts, labeled A and B in Figure 26.47. The A part consists of all elements in series logic, such as the battery and instrument transformers. The B parts are the relays, which are connected in a parallel logic.

For the protective relays of the parallel B elements, we know that the random operating time is given by

$$T_B = \max(T_{B1}, T_{B2}) \tag{26.102}$$

and

$$\begin{aligned} F_B &= \Pr\{B \text{ fails}\} \\ &= \Pr\{B_1 \text{ fails and } B_2 \text{ fails}\} \\ &= \Pr\{T_{B1} \le t\}\, P\{T_{B2} \le t\} \\ &= F_{B1}(t)\, F_{B2}(t) \end{aligned} \tag{26.103}$$

The last line of (26.103) is true since the two relays are independent. For elements A and B in series, we write

$$F_{\text{MPR}}(t) = F_A(t) + F_B(t) - F_A(t)\, F_B(t) \tag{26.104}$$

These two systems are also observed to be independent systems and the random variables T_A and T_B are statistically independent.

MPS. The MPS consists of all elements A, B, and C, plus the switch S_1. The total operating time of this system will also depend on the random operating time of the switch, which we designate as T_1. Now, the actual time switching time T_1 depends on the time of the fault T_X and the relay time T_{MPR}, which are both random numbers, that is,

$$T_1 = T_X + T_{\text{MPR}} \tag{26.105}$$

where the time of the switch itself is considered negligible. Actually, the switch depicted in Figure 26.47 is a part of the MPR, the last action of which closes relay contacts to permit current to flow through the circuit breaker trip coil. It would be possible to treat the contact operating time as a third random switching time, but it is simpler to consider this time to be a part of the total MPR time. Since T_X is measured in days and T_{MPR}, in cycles, for practical purposes (26.105) reduces to only the time of the fault, or

$$T_1 \cong T_X \tag{26.106}$$

FAILURE MODES. The sequential system has two failure modes, which are summarized in Figure 26.46. We now examine these two failure modes.

1. *Failure mode* 1, $T_X > t$.
   First, we define the quantity

$$\{M_1\} = \{\text{failure for } T_X > t\} \tag{26.107}$$

Then,

$$\begin{aligned}\Pr(M_1) &= \Pr\{T_{\text{MPR}} \le t \text{ and } T_X > t\} \\ &= \Pr\{T_{\text{MPR}} \le t\}\, P\{T_X > t\} \\ &= F_{\text{MPR}}(t)\,[1 - F_X(t)] = F_{M1}(t)\end{aligned} \tag{26.108}$$

where we again take advantage of statistical independence. Finally, then

$$T_{M1} = T_{\text{MPR}} \le t < T_X \tag{26.109}$$

2. *Failure mode 2*, $T_X \le t$.
   Here, we define

$$\{M_2\} = \{\text{failure at } T_X \le t\} \tag{26.110}$$

Then

$$\Pr\{\text{success}\} = 1 - \Pr(M_2) = 1 - F_{M2}(t) \tag{26.111}$$

But

$$\begin{aligned}\Pr\{\text{success}\} &= \Pr\{\text{MPR works and MCB works and } T_X \le t\} \\ &= \Pr\{T_{\text{MPR}} > t\}\Pr\{T_{\text{MPR}} + T_{\text{MCB}} > t\}\Pr\{T_X \le t\}\end{aligned} \tag{26.112}$$

Now define

$$F_{XC} = \int_0^t f_{\text{MPR}}(u)\, F_{\text{MCB}}(t-u)\, du \tag{26.113}$$

We may write this equation as

$$1 - F_{M2}(t) = [1 - F_{\text{MPR}}(t)]\,[1 - F_{XC}(t)]\, F_X(t) \tag{26.114}$$

Then

$$F_{M2} = 1 - F_X(1 - F_{\text{MPR}})(1 - F_{XC}) \tag{26.115}$$

and

$$T_{M2} = \min(T_{\text{MPR}},\ T_X + T_{\text{MCB}}) \tag{26.116}$$

where

$$T_{\text{MPR}} > T_X$$

Combining the two failure modes and noting that they are disjoint, we compute

$$\begin{aligned}F_{\text{MPS}}(t) &= F_{M1}(t) + F_{M2}(t) \\ &= 1 + F_{\text{MPR}} - F_X(1 - F_{XC} + F_{\text{MPR}}F_{XC})\end{aligned} \tag{26.117}$$

The random time of system failure is given by

$$\begin{aligned}T_{\text{MPS}} &= \min(T_{M1},\ T_{M2}) \\ &= \min[T_{\text{MPR}},\ \min(T_{\text{MPR}},\ T_X + T_{\text{MCB}})] \\ &= \min(T_{\text{MPR}},\ T_X + T_{\text{MCB}})\end{aligned} \tag{26.118}$$

The other subsystems in Figure 26.47 can be evaluated using similar techniques. These are left as exercises.

26.4.2.2 Random Time Evaluation. In a Monte Carlo simulation of the operation of a power system, the simulation moves along the time axis in discrete steps, evaluating the system performance at each step. An operating history accumulates for all system components and average failure rates can be computed. For any given component, the probability of failure increases with time. Consider a given component A, whose reliability can be described by an exponential distribution. Then the probability of failure of that component is given by

$$\begin{aligned}\Pr\{\text{failure before time } t\} &= \Pr\{T_A \le t\} \\ &= F_{T_A}(t) \\ &= 1 - R_A(t) \\ &= 1 - e^{-\lambda_A t}\end{aligned} \tag{26.119}$$

where λ is the constant failure rate of that component. For the exponential distribution, the mean lifetime of the component is the inverse of λ.

For power system components, we would expect to use equipment that has long mean lifetimes, say at least a few years, and we perform periodic maintenance to keep the component operational and almost as good as new for many years. However, we assume that maintenance is scheduled for important subsystems on a regular, but not on a daily, weekly, or even monthly basis. In some cases, even annual maintenance may not be achieved, which means that certain components always exist for which failure has a relatively high probability. A Monte Carlo simulation can be used to simulate these random failures, inspections, and repairs.

As the simulation steps through time, certain disturbances will be scheduled, and some components will be found to have failed due to random events. For protective systems, there is always the probability that some portion of the protective hardware has failed and any fault will have to be cleared by either local backup or remote backup equipment.

Let us assume that a given disturbance type and location are known and the identity of any random equipment failure has been determined. We now direct our attention to the disturbance clearing and the time sequence of the protective system operation. This clearing sequence will then be the determining factor for the power system performance.

On the time scale of the disturbance occurrence and component failure, which may be measured in hours, days, or even years, the elapsed time of the disturbance clearing is very small and is usually measured in cycles. We arbitrarily let the disturbance clearing time consist of three distinct parts, which are shown in Figure 26.48. These parts are defined by the equation

$$T_C = T_R + T_D + T_B \quad \text{cycles} \tag{26.120}$$

where T_C = total clearing time
T_R = relay time
T_B = breaker time
T_D = delay time

All are assumed to measured in the same units, such as cycles. Note that all of these times are random variables. For primary protection, the intentional delay time is often zero, but for either local or remote backup protection there will be intentional delay for proper coordination.

Figure 26.48 shows typical measures of the clearing time for both normal clearing and local backup clearing. Remote backup must also include the transfer trip or other communication signaling time.

It is important to note that the random time variables in (26.120) are statistically independent associated with independent items of equipment.

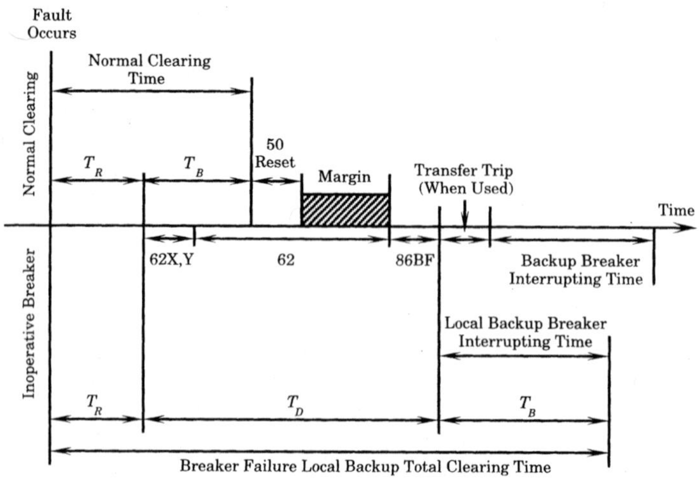

Figure 26.48 Components of the fault clearing time.

It would be reasonable to consider these variable to be Gaussian random variables, each with mean μ and standard deviation σ. Since the distribution of the sum of normal random variables is also normal we can compute the sum and its statistical properties [16], [17]. One way of describing a normal random variable is to write the density as follows.

$$f_n\,(t;\,\mu,\sigma) = \frac{1}{\sigma\sqrt{2\pi}}\exp\left[\frac{-\,(t-\mu)^2}{2\sigma^2}\right] \tag{26.121}$$

Then, using appropriate subscripts defined in (26.120), we can write the distribution of the sum T_C as

$$f_C\,(t) = n\,(t:\mu_C,\ \sigma_C) \tag{26.122}$$

$$\text{where}\quad \begin{aligned} \mu_C &= \mu_R + \mu_B && \text{for primary relaying} \\ &= \mu_R + \mu_B + \mu_D && \text{for backup relaying} \end{aligned} \tag{26.123}$$

and

$$\begin{aligned} \sigma_C^2 &= \sigma_R^2 + \sigma_B^2 && \text{for primary relaying} \\ &= \sigma_R^2 + \sigma_B^2 + \sigma_D^2 && \text{for backup relaying} \end{aligned} \tag{26.124}$$

This means that the total clearing time will always be the sum of the mean operating times of the protective devices involved in the clearing. It also means that the variance of the total clearing is increased substantially, due to the law relating to Gaussian random variables.

REFERENCES

[1] Anderson, P. M., A. Bose, K. J. Timko, and F. E. Villaseca, "A Probabilistic Approach to Stability Analysis," Final Report EL-2797, Electric Power Research Institute, Palo Alto, 1983.

[2] IEEE Committee Report, "Terms and Definitions for Power System Stability," *IEEE Trans.*, PAS-101, (7), July 1982, pp. 1894–1898.

[3] Billinton, R., and R. N. Allan, *Reliability Analysis of Engineering Systems: Concepts and Techniques*, Plenum Press, New York, 1983.

[4] Billinton, R., and R. N. Allan, *Reliability Analysis of Power Systems*, Plenum Press, New York, 1984.

[5] Dhillon, B. S., and C. Singh, *Engineering Reliability, New Techniques and Applications*, John Wiley & Sons, New York, 1981.

[6] Henley, E. J., and H. Kumamoto, *Reliability Engineering and Risk Assessment*, Prentice-Hall, Inc., Englewood Cliffs, NJ. 1981.

[7] Grimes, J. D., "On Determining the Reliability of Protective Relay Systems," *IEEE Trans. on Reliability*, R-19 (3), August 1970, pp. 82–85.

[8] Anderson, P. M., "Reliability Modeling of Protective Systems," *IEEE Trans.*, PAS-103, August 1984, pp. 2207–2214.

[9] Kontoleon, J. M., "Analysis of a Dynamic Redundant System," *IEEE Trans. on Reliability*, R-27 (2), June 1978.

[10] Henley, E. J., and H. Kumamoto, *Designing for Reliability and Safety Control*, Prentice-Hall, 1985.

[11] Westinghouse Relay Division, *Applied Protective Relaying*, Westinghouse Electric Corporation, Newark, 1976.

[12] Kuruganty, P. R. S., and R. Billinton, "Protection System Modeling in a Probabilistic Assessment of Transient Stability," *IEEE Trans.*, PAS-100, May 1981, pp. 2163-2170.

[13] Allan, R. N., and A. N. Adraktas, "Terminal Effects and Protection System Failures in Composite System Reliability Evaluation," *IEEE Trans.*, PAS-101, December 1982, pp. 4557–4562.

[14] IEEE Std 500-1984, *IEEE Guide to the Collection and Presentation of Electrical, Electronic, Sensing Component, and Mechanical Equipment Reliability Data for Nuclear-Power Generating Stations*, IEEE, New York, 1984.

[15] Lauby, M. G., D. D. Klempel, C. G. Dahl, R. O. Gunderson, E. P. Weber, and P. J. Lehman, "The Effects of Terminal Complexity and Redundancy on the Frequency and Duration of Forced Outages," *IEEE Trans.*, PWRS-2, November 1987, pp. 856–863.

[16] Wadsworth, G. P., and J. G. Bryan, *Introduction to Probability and Random Variables*, McGraw-Hill Book Co., New York, 1960.

[17] Papoulis, A., *Probability, Random Variables, and Stochastic Processes*, Mc-Graw-Hill Book Company, New York, 1984.

PROBLEMS

26.1 The Poisson process is of critical importance in modeling power system disturbances. Review the axiomatic definition of the Poisson process. Hint: see any good text on random variables and stochastic processes. What kind of special Poisson process does the power system fit?

26.2 The argument might be logically made that power system disturbances should be restricted in their distribution to a "generalized" Poisson process. What argument can you put forward to the effect that this restriction is not necessary.

26.3 A random variable defined by the notation X_L is hypothesized for faults on transmission lines. Write out a description of the sample space, the events of interest and any restrictions placed on the definition of X_L as a random variable.

26.4 Consider further the proposition of the previous problem that faults on a transmission line can be represented by random variables. For this to be possible it must be possible to define a probabilities for the line faults. Devise a method for computing the probability of line faults.

26.5 Having defined a random variable for line faults and having shown that line faults can be represented as a random variable X_L, determine the probability distribution of transmission line faults.

26.6 Based on the result of problem 26.5, determine also the density function for the random variable X_L.

26.7 Develop expressions for the distribution and density of fault types occurring on a transmission line. For simplicity, assume that only four types of faults occur:

1. One-line-to-ground
2. Two-lines to-ground
3. Line-to-line
4. Three-phase

26.8 Based on the construction derived in problem 26.7, how do we interpret $F_{Y_L}(3.5)$?

26.9 Equations (26.11) and (26.12) describe a random variable that locates the fault along a transmission line. Develop this description further and write the failure density and distribution for a transmission line that we arbitrarily divide into M equal segments.

26.10 Consider the disturbances arising from the inadvertent opening of lines in a network. This could be due to a failure mode of circuit breakers in which the breaker opens without a normal open command, or it could be due to false trips of protective relays. Let S_L be the set of all lines in the system and let R_L be a random variable associated with the probability of line opening. Write the corresponding density and distribution functions associated with these probabilities. Assume that there are three types of open lines: one phase open, two phases open, and three phases open.

26.11 Buses in the power system are also subject to the four types of faults defined in the text for line faults. A random variable is required describing the location of the faults, according to the bus number identification. Let the random variables X_B and Y_B be defined on appropriate sample spaces to model the faulted bus number and the fault type, respectively. Write expressions for the probability density and distribution for both random variables and also write equations for the joint density and distribution of a particular type of fault on a particular bus.

26.12 Determine the conditional density for faulted buses.

26.13 Sketch and explain a Venn diagram to depict the probability of security failure in a system having three elements connected in series, as shown in Figure 26.17 The equation for the probability of security failure is given by (26.50).

26.14 Consider the 3-out-of-4:G system shown in Figure P26.14. Develop the FS and FD fault trees for this system. Hint: develop the FS fault tree first and modify it appropriately to find the FD fault tree.

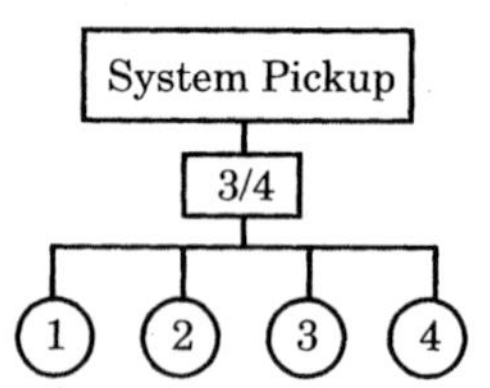

Figure P26.14 A 3-Out-of-4:G system.

26.15 A set of 1000 fire alarms is tested in the laboratory under both FS and FD conditions, with the following results:

(a) Under FS conditions, 50 units created a false alarm.
(b) Under FD conditions, 10 units failed to alarm.

Determine the conditional FS and FD probabilities and their sum. Does the sum add to unity?

26.16 Repeat problem 15, but assume that under FS conditions 500 units created a false alarm and under FD conditions 600 units failed to alarm. Answer the same questions as in problem 15.

26.17 Given that the demand probability of a hazardous condition is 0.001, compute the unconditional FS and FD probabilities for the system described in problem 26.15.

26.18 Given a demand probability of 0.001, compute the unconditional probabilities for the system described in problem 26.16.

26.19 Compute the reliability of the system described in problem 26.15.

26.20 Compute the reliability of the system described in problem 26.16.

26.21 A series system consists of two sensors with the conditional probabilities computed in problem 26.15. Determine the minimum cut sets for FS and FD failure and the conditional system probabilities for this system.

26.22 A parallel system consists of two sensors with the conditional probabilities computed in problem 26.15. Determine the minimum cut sets for FS and FD failure and the conditional system probabilities for this system.

26.23 Verify (26.65) and (26.67)

26.24 Consider the system shown in Figure P26.26, where a single overcurrent relay is used to protect a line from short-circuit conditions. The fault tree of this system for a faulted environment is shown in Figure 25.20. Construct a decision table and a fault tree for this system for an environment that contains no short-circuit hazard.

26.25 Construct an event tree for an overcurrent relaying system that includes one breaker mechanism, one trip coil, one set of sensors (CT and VT), and redundant relays. Consider the battery to be perfectly reliable in order to reduce the complexity of the tree.

26.26 Evaluate the following for the local backup system of Figure P26.26.

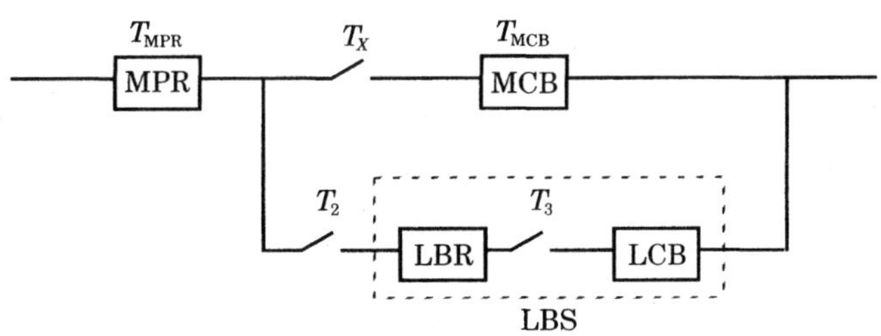

Figure P26.26 Logic diagram for a standby redundant local backup system.

(a) The random time to failure of only the local backup system, T_{LBS}

(b) The random time to failure standby redundant pair, T_{SR}, that includes the main circuit breaker and the local backup system.

(c) The random time to failure of the the main protective system and the standby redundant pair, T_{CLS} that we designate the "combined local and primary system" or CLS.

(d) The failure distribution of the CLS.

26.27 Extend the analysis of the previous problem to include the remote backup system RBS. The system components are shown in Figure P26.27.

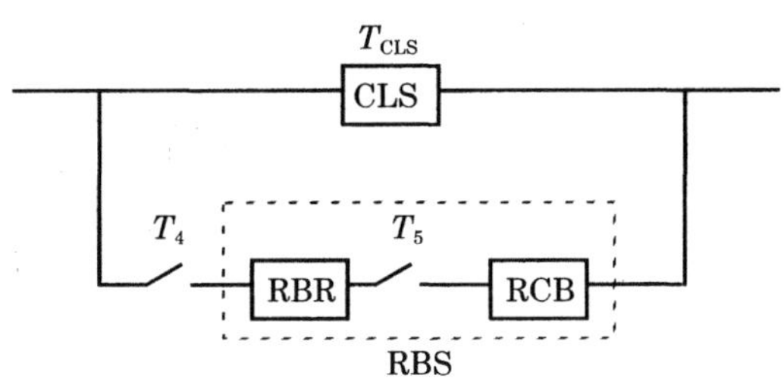

Figure P26.27 Standby redundancy logic for the remote backup system.

27

Fault Tree Analysis of Protective Systems

27.1 INTRODUCTION

A power system is a very complex structure and it is difficult to generalize the protective system requirements. Protection design at the distribution level is quite different than protective systems for subtransmission and transmission systems. Apparatus protection is even more specialized. From a reliability modeling viewpoint, it is not possible to formulate general rules for reliability modeling, since the system has so many different components with quite different protection requirements.

It has been noted that the reliability modeling of a system depends on several factors, including:

1. System component monitoring
2. Repair policy and performance
3. System inspection and maintenance policy

System monitoring is important in reliability modeling since we are dealing with a system of repairable components. However, repair cannot be initiated unless there is knowledge of component failure. This kind of knowledge is usually provided by some kind of monitoring. At the distribution level, circuit outages will usually be promptly reported by the customer and repair crews can be dispatched in response. At the bulk transmission level, all lines, stations, and generators are monitored by a computer-controlled *energy management system*, so that outages are observed and the operator is given appropriate alarms of failures. In both of the foregoing cases, the failures of components are *announced*. At the intermediate voltages, such as subtransmission systems, failures may not be announced because the components are not all monitored and it may be some time before a particular failure is noticed. This may be true of equipment that is used only occasionally, for example, in some types of switched devices such as capacitors or reactors. It may also be true of lines or other components that are not monitored, but are part of an interconnected transmission system such that an outage may go undetected for a time. This introduces the concept of failures that are *unannounced*, which complicates the reliability modeling. Note that protective device failures are usually unannounced.

Repair policy and repair performance are also important in determining the availability of a repairable component. This includes logistics and spare parts policy. It also includes the size of the repair staff and their home base location. Power systems often span many hundreds of kilometers of distance from one boundary of a system to another. The repair crew may be a long way from the site of a particular failure and travel time may be a significant part of the outage time. The stocking and location of spare parts is equally important. Also, the availability of a repair crew of adequate size to respond to the failure may be important, especially when natural disasters such as storms or earthquakes can cause multiple failures spread over large parts of the power system. This means that the MTTR of a component may vary significantly depending on the component location, the availability of spare parts, the location of spare parts, and the size of the repair crews. All of these parameters affect the MTTR and the availability of the system.

System maintenance policy is important since some components on the power system are out of service for significant periods for preventive maintenance. This is especially important for generating units and may also be true of some transmission lines where washing insulators, or similar routine maintenance, is required. This affects the availability of the component that is subject to such maintenance. Routine inspection is important for protective systems since this is often the only way of ensuring that the protective system is in an operable condition. This raises the question as to the optimum timing of inspection and preventive maintenance for protective systems. The time between inspections is an important parameter in the reliability modeling of protective systems. This is investigated in Chapter 28.

All of the foregoing conditions are a necessary part of reliability modeling. In modeling a protective system for reliability analysis, the components are not all the same in terms of their monitoring, repair cycles, and inspection cycles. This means that the analytical technique used for reliability analysis must be robust enough to permit the use of a variety of components having quite different types of reliability models. One technique that has the required capability is the fault tree methodology. For this reason, this chapter is devoted exclusively to fault tree analysis. An example will be chosen for analysis that will use a variety of different reliability modeling techniques.

27.2 FAULT TREE ANALYSIS

Chapter 25 presents the basic rules of fault tree construction and describes several methods of fault tree analysis. For fault trees of large or complex systems, it is desirable to have analytical methods that employ a computer. Most of the methods used for complex systems are computer based methods, and most of these use some form of minimal cut set analysis [1], [2], [3]. Although it is not essential that the user of one of the computer programs understand all of the details of the computation, it is desirable to have a basic understanding of the process. Therefore, this section reviews some of the concepts used in fault tree analysis with the motivation of providing the engineer with an overview.[1] The method described here is based on kinetic tree theory, which uses minimal cut sets to determine the probability of the top event of the tree. This method is limited to fault trees consisting of AND, OR, and INHIBIT gates, and these methods are generally satisfactory for protective systems. Other methods are available that are more general [3] and permit the evaluation of common cause

[1]This development follows closely that of [2], which is recommended for further reading.

failures. This may be important in some types of protective system evaluations, for example, where it is necessary to evaluate the failure of the power supply system that may affect several different components or subsystems of protective hardware.

27.2.1 System Nomenclature

Before developing the equations, it will be useful to establish a system nomenclature that will clearly indicate whether a variable refers to a component, a cut set, or to the system as a whole. The basic guideline used here is that an unsubscripted variable refers to the component, a variable with the subscript "C" refers to a cut set, and a variable with the subscript "S" refers to the entire system. Therefore, it will be possible to discuss a variable, such as unavailability, at the component level, at the cut set level, and at the system level. Several of the variables of interest are summarized in Table 27.1.

TABLE 27.1 Reliability Variables for Fault Tree Analysis

Name	Definition	Component	Cut Set	System
Unavailability	Probability of a failed state at time t	$U(t)$	$U_C(t)$	$U_S(t)$
Unconditional failure intensity	Expected number of failures per unit time at time t	$w(t)$	$w_C(t)$	$w_S(t)$
Unconditional repair intensity	Expected number of repairs per unit time at time t	$v(t)$	$v_C(t)$	$v_S(t)$
Conditional failure intensity	Probability of a failure per unit time at time t, given no failures up to time t	$h(t)$	$h_C(t)$	$h_S(t)$
Conditional repair intensity	Probability of a repair per unit time at time t, given no repairs up to time t	$n(t)$	$n_C(t)$	$n_S(t)$
Expected number of failures in $[0, t)$	Expected number of failures during time interval $[0, t)$	$W(0, t)$	$W_C(0, t)$	$W_S(0, t)$
Expected number of repairs in $[0, t)$	Expected number of repairs during time interval $[0, t)$	$V(0, t)$	$V_C(0, t)$	$V_S(0, t)$
Unreliability	Probabilty of one or more failures in the time interval $[1, t)$	$F(t)$	$F_C(t)$	$F_S(t)$

It is possible to determine exact time-dependent reliability parameters at the level of the component or the cut set. For the system as a whole, however, this is not always possible, in which case upper- and lower bounds are computed that bracket the parameters of interest. This is adequate for most systems and often provides a good estimate of the system parameters. In most cases, it will be assumed that the system component can be described by constant failure rates λ and repair rates μ.

27.2.2 Calculation of Component Parameters

The computation of all fault tree parameters is based on the calculation of unconditional failure intensities from given densities, then the expected numbers W and V, then the availability A and unavailability U, and finally the conditional intensities, h and n. The process is illustrated in Figure 27.1 and assumes that the failure and repair densities are known.

27.2.2.1 Component Unconditional Intensities. From the failure and repair densities, the unconditional intensities may be computed, as follows. It is assumed that the components

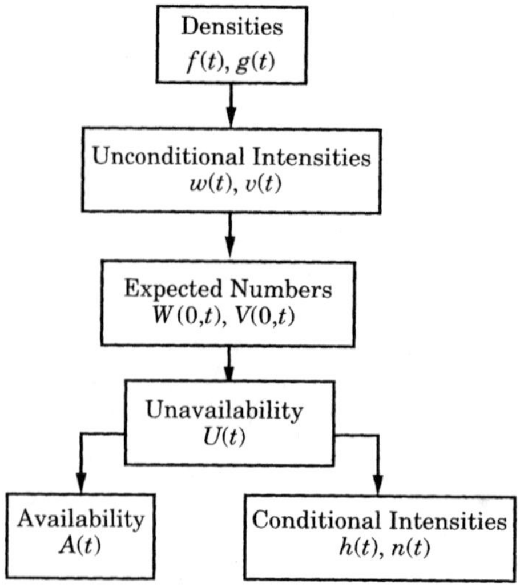

Figure 27.1 Flow chart for computing probabilistic parameters [2].

that fail during the interval $[t, t + dt)$ are of two types, where these two types are illustrated in Figure 27.2.

Type 1 This type of component was repaired during $[u, u + du)$ and has been normal (N) to time t, then fails (F) during $[t, t + dt)$, given that the component jumped to the normal state at $t = 0$.

Type 2 This type of component has been normal to time t and fails during $[t, t + dt)$, given that the component jumped to the normal state at $t = 0$.

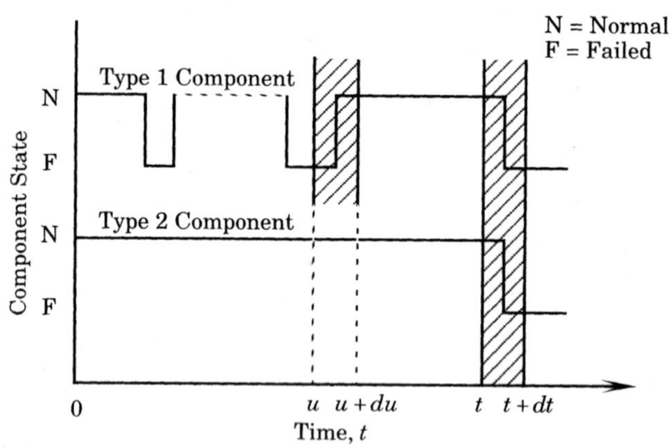

Figure 27.2 Types of components failing during $[t, t + dt)$.

For the Type 1 component, the failure event probability is

$$\Pr\{\text{type } 1\} = v(u)du \cdot f(t - u)dt \tag{27.1}$$

This result is explained as follows. The first part $v(u)du$ is the probability that the component is repaired during $[u, u + du)$, given that it is as good as new at $t = 0$. The second part of (27.1), $f(t - u)dt$, is the probability that the component has been normal to time t, and failed during $[t, t + dt)$, given that it was as good as new at $t = 0$, and was repaired at time u, since the component failure characteristics depend only on the survival age $t - u$ at time t, and do not depend on the history before time u.

The probability for the second type of component is computed as

$$\Pr\{\text{type 2}\} = f(t)dt \tag{27.2}$$

This is determined by definition of what is meant by the relationship (27.2), as given by (24.99), which corresponds to the Type 2 failure event.

The desired result is $w(t)\,dt$, which is the probability that the component fails during $[t, t + dt)$, given that it jumped to the normal state at time zero. This probability is equal to the sum of the two parts. Therefore, we write

$$w(t)dt = f(t)dt + dt \int_0^t v(u)f(t-u)du \tag{27.3}$$

or

$$w(t) = f(t) + \int_0^t v(u)f(t-u)du \tag{27.4}$$

The components that are repaired during $[t, t+dt)$ consist of the following type of components.

Type 3 Components that are failed during $[u, u + du)$, have been failed to time t, and are repaired during $[t, t + dt)$, given that the components jumped to the normal state at time zero.

This type of component is illustrated in Figure 27.3.

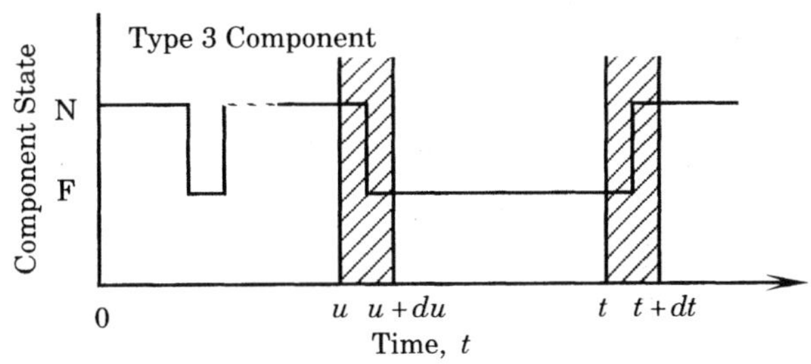

Figure 27.3 Component that is repaired during $[t, t + dt)$.

For this type of component, following the same type of reasoning as given previously, we compute

$$v(t) = \int_0^t w(u)g(t-u)du \tag{27.5}$$

Therefore, (27.4) and (27.5) give the unconditional failure intensity $w(t)$ and unconditional repair intensity $v(t)$ in terms of the respective densities. The convolution integrations can be computed by numerical integration when the densities are given. The Laplace transform can be used to determine these unconditional intensities analytically, if desired. This is left as an exercise.

27.2.2.2 Expected Number of Failures and Repairs. The expected number of failures is computed as the integral of the unconditional failure intensity, as given by (24.118). Thus, for the number of failure during $(0, t]$ we compute

$$W(0,t) = \int_0^t w(t)dt \tag{27.6}$$

Similarly, for the number of repairs during $(0, t)$ we have, from (24.124),

$$V(0,t) = \int_0^t v(t)dt \tag{27.7}$$

Both (27.6) and (27.7), the number of failures and repairs over a given period, are excellent performance measures of a protective system.

27.2.2.3 Component Unavailability. The unavailability is computed as follows. Let $x(t)$ be defined as an indicator variable, defined as follows.

$$x(t) = \begin{cases} 1 & \text{if the component is in a failed state} \\ 0 & \text{if the component is in a normal state} \end{cases} \tag{27.8}$$

$$\begin{aligned} x_{0,1}(t) &= \text{number of failures to time } t \\ x_{1,0}(t) &= \text{number of repairs to time } t \end{aligned} \tag{27.9}$$

Then

$$x(t) = x_{0,1}(t) - x_{1,0}(t) \tag{27.10}$$

Now compute the expected value of (27.10) as follows.

$$\begin{aligned} E(x(t)) &= E(x_{0,1}(t)) - E(x_{1,0}(t)) \\ &= 1 \times \Pr(x(t) = 1) + 0 \times \Pr(x(t) = 0) \\ &= \Pr(x(t) = 1) = U(t) \end{aligned} \tag{27.11}$$

But, by definition

$$\begin{aligned} E(x_{0,1}(t)) &= W(0,t) \\ E(x_{1,0}(t)) &= V(0,t) \end{aligned} \tag{27.12}$$

so that

$$\begin{aligned} E(x(t)) &= E(x_{0,1}(t)) - E(x_{1,0}(t)) \\ U(t) &= W(0,t) - V(0,t) \end{aligned} \tag{27.13}$$

or the unavailability can be computed directly from the number of failures and repairs on the interval (0,t). Also, from definitions (27.6) and (27.7), we compute

$$U(t) = \int_0^t [w(u) - v(u)]du \tag{27.14}$$

Finally, from the definition of unavailability (24.108), we can compute the availability.

$$A(t) = 1 - U(t) \tag{27.15}$$

This completes the computation of all parameters in Figure 27.1 except the conditional intensities.

27.2.2.4 Component Conditional Intensities. The conditional failure intensity $h(t)$ has been described as a conditional probability. From the definition of the conditional probability rule (24.20) we may write.

$$\Pr(A \cap C) = \Pr(C)\Pr(A|C) \tag{27.16}$$

Now, suppose that all events C under consideration have a certain property W which alters the conditional probability rule as follows.

$$\Pr(A \cap C|W) = \Pr(A|C \cap W)\Pr(C|W) \tag{27.17}$$

where we can define the events

$$
\begin{aligned}
C &= \text{component is normal at time } t \\
A &= \text{component fails during } (t, t+dt] \\
W &= \text{component jumped into the normal state at time zero}
\end{aligned}
$$

Now, at most one failure can occur during a sufficiently small interval of time. Also, note that event A implies event C. Hence, (27.17) can be simplified since the simultaneous occurrence of A and C is just the occurrence of A. Therefore, we can write

$$\Pr(A|W) = \Pr(A|C \cap W)\Pr(C|W) \tag{27.18}$$

Now, from the definition of the various probabilistic parameters, we may write

$$
\begin{aligned}
\Pr(A|W) &= w(t)dt \\
\Pr(A|C \cap W) &= h(t)dt \\
\Pr(C|W) &= A(t)
\end{aligned}
\tag{27.19}
$$

Therefore, we compute the conditional failure intensity as

$$h(t) = \frac{w(t)}{A(t)} = \frac{w(t)}{1 - U(t)} \tag{27.20}$$

When the hazard rate is constant and is a known quantity, we can alter the computation as follows. First, we have

$$h(t) = \lambda = \text{a constant} \tag{27.21}$$

From this constant and the unavailability, we can use (27.20) to compute

$$w(t) = \lambda[1 - U(t)] \tag{27.22}$$

which is simpler than the evaluation of the convolution integral (27.4).

The conditional repair intensity is given as

$$n(t) = \frac{v(t)}{U(t)} \tag{27.23}$$

When the conditional repair intensity is known and is a constant μ, we can compute the unconditional repair intensity from (27.23), i.e.,

$$v(t) = \mu U(t) \tag{27.24}$$

If the failure and repair rates are time varying, then (27.22) and (27.24) cannot be used, and the convolution method must be used to determine the unconditional intensities.

27.2.3 Computation of Minimal Cut Set Parameters

A cut set occurs if all of the basic events in the cut set occur. The probability of a cut set occurring at time t is defined as the unavailability $U_C(t)$ and is computed from the intersection of the basic events in the cut set.

27.2.3.1 Cut Set Unavailabilities. The computation of unavailabilities is usually straightforward, given unavailabilities of the components and the definition of the cut sets. For a cut set of n members, we can write the unavailability of the cut set in terms of the unavailability of the cut set members.

$$U_C(t) = \Pr\{B_1 \cap B_2 \cap \ldots \cap B_n\} = \prod_{k=1}^{n} U_k(t) \tag{27.25}$$

where n = number of cut set members
B_i = basic event i for cut set k
$U_k(t)$ = Pr{kth basic event occurring at time t}

The computation of the cut set unavailability must take into account the physical configuration of the components. For example, if the components are in series logic, then each component is a cut set and the cut set unreliabilities are just the component unreliabilities.

$$U_{Ck}(t) = U_k(t) \quad k = 1, \ldots, n \qquad \text{For a series system} \tag{27.26}$$

If the components are all in parallel, a cut through all components is the minimal cut set.

$$U_C(t) = U_1(t)U_2(t)\ldots U_n(t) \qquad \text{For a parallel system} \tag{27.27}$$

27.2.3.2 Conditional and Unconditional Intensities. The computation of conditional and unconditional intensities can be approached as follows. First, consider the conditional failure intensity of the cut set. In words, we can write the following definition.

$$h_C(t) = \Pr\begin{Bmatrix}\text{occurrence of the cut set per unit time at time } t, \\ \text{given that there is no cut set failure at time } t\end{Bmatrix} \tag{27.28}$$

Moreover, we can write

$$h_C(t)dt = \Pr\begin{Bmatrix}\text{cut set occurs during the time interval } [t, t+dt), \\ \text{given that the cut set does not exist at time } t\end{Bmatrix} \tag{27.29}$$

Rewriting in mathematical symbols, we write

$$\begin{aligned} h_C(t)dt &= \Pr(K_C(t, t+dt)|\bar{K}_C(t)) \\ &= \frac{\Pr(K_C(t, t+dt) \cap \bar{K}_C(t))}{\Pr(\bar{K}_C(t))} \\ &= \frac{\Pr(K_C(t, t+dt))}{\Pr(\bar{K}_C(t))} \end{aligned} \tag{27.30}$$

where we define the events

$$K_C(t, t+dt) = \{\text{occurrence of the cut set during } [t, t+dt)\}$$
$$\bar{K}_C(t) = \{\text{nonexistence of the cut set failure at time } t\}$$

The last line of (27.30) is true since the first term in the numerator implies the second term, i.e., if the cut set occurs during the stated interval, then its nonexistence at the beginning of that interval is ensured.

We can analyze the cut set occurrence as follows [2]. Cut set failure occurs if and only if one basic event in the cut set does not exist at t and this basic event occurs from t to $t + dt$, and all the other basic events exist at t.

$$\Pr(K_C(t, t+dt)) = \sum_{j=1}^{n} \Pr\begin{Bmatrix}\text{basic event } j \text{ occurs in } t \text{ to } t+dt \\ \text{and the other basic events exist at } t\end{Bmatrix} \tag{27.31}$$

Since all basic events are disjoint, the above equation can be written as

$$\begin{aligned} \Pr(K_C(t, t+dt)) &= \sum_{j=1}^{n} \Pr\begin{Bmatrix}\text{basic event } j \\ \text{occurs in } t, t+dt\end{Bmatrix} \Pr\begin{Bmatrix}\text{other events} \\ \text{exist at } t\end{Bmatrix} \\ &= \sum_{j=1}^{n} w_j(t)dt \cdot \prod_{\substack{k=1 \\ k\neq j}}^{n} U_k(t) \end{aligned} \tag{27.32}$$

Then (27.30) can be written as

$$h_C(t)dt = \frac{\sum_{j=1}^{n} w_j(t)dt \cdot \prod_{\substack{k=1 \\ k\neq j}}^{n} U_k(t)}{1 - U_C(t)} \tag{27.33}$$

Each term in the summation of the numerator is the probability of the jth basic event occurring during $[t, t + dt)$ with the condition that the other basic events already exist at t. The denominator is the probability of the nonexistence of the cut set failure at time t. Now, using (27.20), but applied to cut sets, we recognize that (27.33) is exactly

$$h_C(t) = \frac{w_C(t)}{1 - U_C(t)} \tag{27.34}$$

where

$$w_C(t) = \sum_{j=1}^{n} w_j(t) \cdot \prod_{\substack{k=1 \\ k\neq j}}^{n} U_k(t) \tag{27.35}$$

In a similar manner, we compute the cut set parameters related to repair, with the following results.

$$n_C(t) = \frac{v_C(t)}{U_C(t)} \tag{27.36}$$

$$v_C(t) = \sum_{j=1}^{n} v_j(t) \cdot \prod_{\substack{k=1 \\ k\neq j}}^{n} [1 - U_k(t)] \tag{27.37}$$

27.2.3.3 Expected Number of Failures and Repairs. The expected numbers of cut set failures and repairs over the interval $(0, t)$ are readily obtained from the differential parameters, with the following results.

$$W_C(0, t) = \int_0^t w_C(u)du \tag{27.38}$$

$$V_C(0, t) = \int_0^t v_C(u)du \tag{27.39}$$

27.2.4 Computation of System Parameters

Having computed the reliability parameters for the cut sets, it is possible to extend the computation to the system level [2].

27.2.4.1 System Unavailability. To compute the system unavailability, we first define an event d_i as follows.

d_i = all basic events in the ith minimal cut set exist at time t.

Then the ith minimal cut set failure exists at time t. The top event S can be expressed in terms of d_i as

$$S = \bigcup_{i=1}^{N_C} d_i \tag{27.40}$$

where N_C = total number of minimal cut sets

Then the unavailability of the system can be expressed in terms of the probability of existence of the minimal cut sets.

$$U_S(t) = \Pr\left(\bigcup_{i=1}^{N_C} d_i\right) = \sum_{i=1}^{N_C} \Pr(d_i) - \sum_{i=2}^{N_C}\sum_{j=1}^{i-1} \Pr(d_i \cap d_j) + \cdots + (-1)^{N_C}\Pr(d_1 \cap d_2 \cap \cdots \cap d_{N_C}) \tag{27.41}$$

The mth term on the right-hand side of (27.41) is the contribution to the unavailability from m out of N_C minimal cut sets being simultaneously failed at time t. The computation involves intersections rather than unions, which makes the work somewhat easier, at least for small systems.

For large systems with complex fault trees, the computation of the exact system unavailability by the method of (27.41) is very time consuming. In these cases, it is easier to compute upper and lower bounds of the unavailability. There are several methods of computing the upper and lower bounds. One method uses the "inclusion-exclusion principle" to obtain the following result [2].

$$\text{Lower bound} \le U_S(t) \le \text{Upper bound}$$

$$\sum_{i=1}^{N_C} \Pr(d_i) - \sum_{i=2}^{N_C}\sum_{j=1}^{i-1} \Pr(d_i \cap d_j) \le U_S(t) \le \sum_{i=1}^{N_C} \Pr(d_i) \tag{27.42}$$

This result can be written in terms of the cut set unavailabilities, as follows.

$$\sum_{i=1}^{N_C} U_{Ci}(t) - \sum_{i=2}^{N_C}\sum_{j=1}^{i-1}\prod_{i,j} U(t) \le U_S(t) \le \sum_{i=1}^{N_C} U_{Ci}(t) \tag{27.43}$$

where $\prod_{i,j}$ is the product of the $U(t)$'s for the basic events that are members of either cut set i or j.

There are other methods employed in the various computer programs used for fault tree analysis. Some of the approximate methods are exact when the cut sets are disjoint sets of basic events.

27.2.4.2 System Unconditional Intensities. Considering the entire system, the unconditional intensity $w_S(t)$ is the expected number of times that the top event occurs per unit time at time t. Therefore, the quantity $w_S(t)dt$ is the expected number of occurrence of the top event in the interval extending from t to $t + dt$.

Define the following event.

$$e_i = \begin{Bmatrix} \text{the event that the } i\text{th cut set failure} \\ \text{occurs in the time from } t \text{ to } t + dt \end{Bmatrix} \tag{27.44}$$

Then

$$\Pr(e_i) = w_{Ci}(t)dt \tag{27.45}$$

If the top event is to occur in the interval $[t, t + dt)$, then none of the cut set failures can exist at time t and, moreover, one or more of the cut set failures must occur during the interval $[t, t + dt)$. Mathematically,

$$w_S(t)dt = \Pr\left(A \cap \bigcup_{i=1}^{N_C} e_i\right) \tag{27.46}$$

where A = {the event that none of the cut set failures exist at t}

$\bigcup_{i=1}^{N_C} e_i$ = {the event that one or more cut set failures occur at t}

We can also write (27.46) in terms of the cut sets that fail to occur at t

$$w_S(t)dt = \Pr\left(\bigcup_{i=1}^{N_C} e_i\right) - \Pr\left(B\bigcup_{i=1}^{N_C} e_i\right) \tag{27.47}$$

$$\text{where } B = a\bigcup_{j=1}^{N_C} d_j = \left\{\begin{matrix}\text{union of events of the } j\text{th cut set} \\ \text{failure existing at time } t\end{matrix}\right\} \tag{27.48}$$

From (27.47), we can think of the system intensity as having two components, which can be identified as follows.

$$w_S(t) = w_S^{(1)}(t) - w_S^{(2)}(t) \tag{27.49}$$

The first term is the contribution from one or more cut sets failing during the period $[t, t+dt)$. The second term represents the contribution from one or more cut sets failing during the interval $[t, t+dt)$ but where other cut sets are already failed and have not been repaired. This second term can be considered a second-order correction, which can be ignored in many cases without sacrificing greatly in accuracy if the events are of low probability.

Expanding the first term of (27.49), we write

$$\begin{aligned} w_S^{(1)}(t)dt &= \Pr\left(\bigcup_{i=1}^{N_C} e_i\right) \\ &= \sum_{i=1}^{N_C} w_{Ci}(t) - \sum_{i=2}^{N_C}\sum_{j=1}^{i-1} \Pr(e_i \cap e_j) + \\ &\quad \ldots + (-1)^{N_C-1}\Pr(e_1 \cap e_2 \cap \ldots \cap e_{N_C}) \end{aligned} \tag{27.50}$$

The first term of the expansion is the enumeration of the contributions from each of the cut sets occurring individually. The second and following terms are the simultaneous occurrence of two or more cut sets, where these events do not exist at time t, and then all simultaneously occur during the interval $[t, t+dt)$. This equation can be used to get useful bounds on the intensity. If only the first term is used, an upper bound is found and if the first two terms are used, a lower bound is computed. This method of bounding the intensity is often adequate and greatly simplifies the computation.

27.2.4.3 Other System Parameters. Once the unconditional intensity and unavailability are known the other parameters of interest can be readily computed. From the basic equation

$$w_S(t)dt = [1 - U_S(t)]h_S(t)dt \tag{27.51}$$

we readily compute the conditional failure intensity of the system.

$$h_S(t) = \frac{w_S(t)}{1 - U_S(t)} \tag{27.52}$$

We can also compute the expected number of system failures from the integral of the unconditional intensity.

$$W_S(0, t) = \int_0^t w_S(t)dt \tag{27.53}$$

In cases where the unconditional intensity is represented by an upper or lower bound, similar bounding is computed for the above parameters. This is often an adequate solution if the range between the two bounds is not too great.

27.2.4.4 Short Cut Methods. Short cut methods have been developed that permit quick estimation of the system parameters [4]. The methods require that the failure and repair rates, λ and μ, of each component be given and the minimal cut sets be known. It assumes exponential distributions of the component failure and repair rates and independence of the component failures. All of the basic equations are given in Table 24.4.

For a nonrepairable component, we compute the unavailability as

$$U_i = 1 - e^{-\lambda_i t} \cong \lambda_i t \tag{27.54}$$

If the component is repairable, the unavailability depends on the repair rate as well as the failure rate.

$$U_i = \frac{\lambda_i}{\lambda_i + \mu_i}[1 - e^{-(\lambda_i + \mu_i)t}] \tag{27.55}$$

Now, if t is large and if the failure rate is much less than the repair rate, which will almost always be the case, we can simplify (27.55) as follows.

$$U_i \cong \frac{\lambda_i}{\lambda_i + \mu_i} \cong \frac{\lambda_i}{\mu_i} \tag{27.56}$$

These simplifications usually predict an unavailability that is a bit larger than the true value, and is therefore a conservative estimate.

To predict the cut set unavailability, we write

$$U_C = \prod_{i=1}^{N_C} U_i \tag{27.57}$$

From (27.35) and (27.22) we may write

$$w_C(t) = \sum_{j=1}^{N_C}[1 - U_j(t)]h_j(t) \prod_{\substack{k=1 \\ k \neq j}}^{N_C} U_k(t) \tag{27.58}$$

Now substituting (27.57) and assuming that the unavailability is very small compared to unity, (27.58) becomes

$$w_{Ci}(t) \cong U_{Ci}(t) \sum_{j=1}^{N_C} \frac{\lambda_j}{U_j} \tag{27.59}$$

The cut set failure rate can be computed as

$$\lambda_{Ci} = \frac{w_{Ci}}{1 - U_{Ci}} \tag{27.60}$$

Finally, the system parameters are computed approximately from the cut set parameters.

$$\begin{aligned} U_S &\cong \sum_{i=1}^{N_C} U_{Ci} \\ \lambda_S &\cong \sum_{i=1}^{N_C} \lambda_{Ci} \\ w_S &\cong \sum_{i=1}^{N_C} w_{Ci} \end{aligned} \tag{27.61}$$

Certain restrictions for the use of the approximate equations apply. These restrictions are stated as follows [2]:

Nonrepairable components:

$$\lambda_i t < 0.1 \quad (27.62)$$

Repairable components:

$$t > 2/\mu_i \quad (27.63)$$

Sample studies indicate that system unavailability can be determined to a reasonable accuracy for certain systems. Clearly, as with all engineering approximations, these equations should be used with care. However, it is often valuable to obtain a rough solution without spending a great amount of time, which would often be required to get an exact solution. Moreover, in some cases, the input data may not justify the larger effort for high precision.

27.3 ANALYSIS OF TRANSMISSION PROTECTION

The application of fault tree analysis in system protection will be illustrated by constructing a detailed fault tree for a transmission line protective system. In constructing the fault tree, all the basic rules of fault tree construction will be carefully followed [1]. The protective system to be analyzed is based on the system described in Section 13.5.2, with a few exceptions. The protection employs redundant relays and utilizes both power line carrier and microwave communications for transfer trip pilot signaling. This system is typical of EHV transmission protection designs that have been installed in North America.

27.3.1 Functional Specification for the Protective System

The basic PLC protective system is shown in Figure 27.4, which is a variation of the configuration of Figure 13.44. The microwave system for this study is exactly that shown in Figure 27.5. The PLC system to be analyzed, shown in Figure 27.4, separates the various signal requirements onto the three phases of the transmission line. In the analysis, it is assumed that the system is designed to operate as follows:

1. The transmission line is divided into three zones:
   One zone covering approximately 90% of the left end of the line
   One zone covering approximately 90% of the right end
   One zone covering the middle 80% of the line
   (Hereafter, the two ends of the transmission line will be referred to as the "left" and "right" ends and the equipment items at each end will be thereby clearly identified).
2. The protection logic is a non-unit system using permissive underreaching distance protection and supplemental transfer trip. This means that the relay logic at each end can accurately pick up only for a zone 1 fault. Faults beyond 90% of the line length away from the relay are cleared with zone 2 time delay. To avoid time delay in clearing these end-of-line faults, the relay logic at the end near the fault will pick up without time delay and this relay will also initiate immediate tripping at the remote end. This is accomplished in two ways, first by sending a relay trip permissive and also by initiating a transfer trip action to the remote relays.

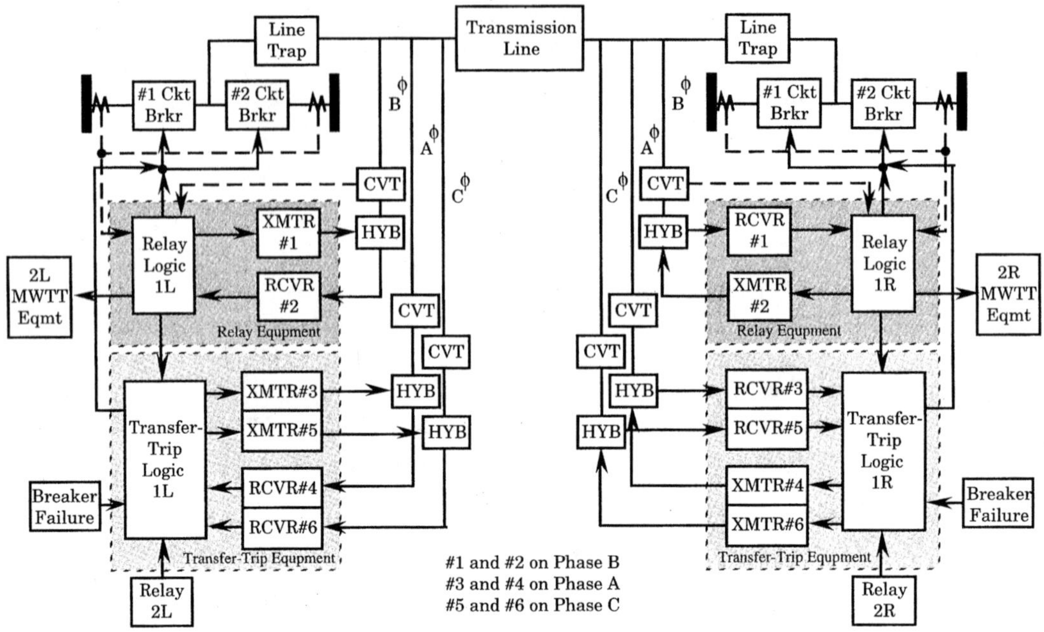

Figure 27.4 Transmission line with power line carrier signaling.

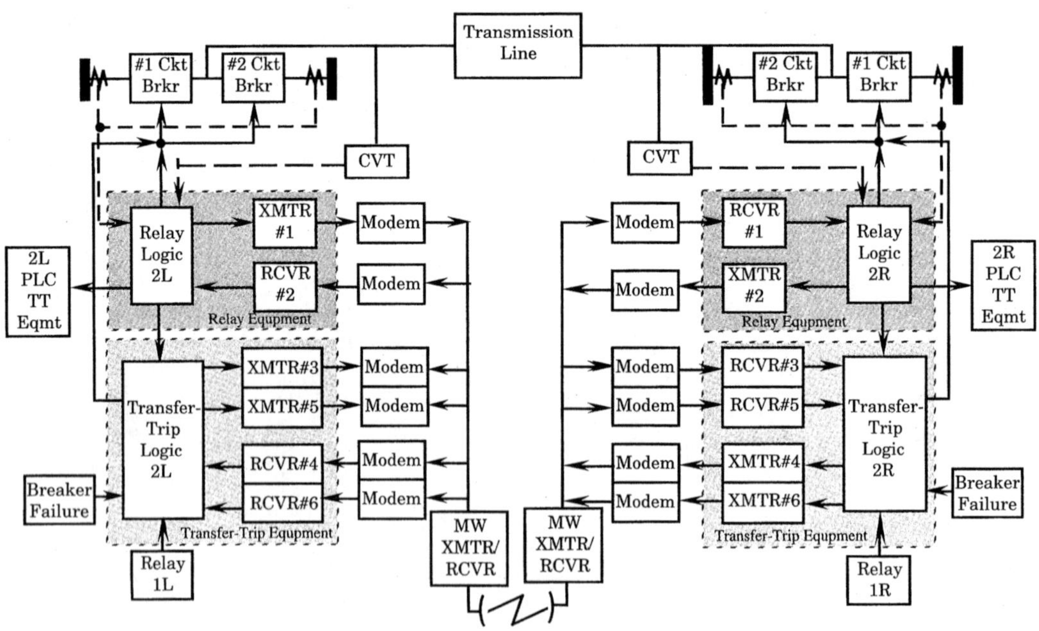

Figure 27.5 Transmission line with microwave pilot system.

3. Faults that are in the middle portion of the line initiate a response by all the line relays at both ends. In this case the relays at both ends of the line will see the fault as a zone 1 fault and all relays should initiate tripping by local breaker, by permissive signaling to the remote relay, as well as by transfer trip signaling.
4. We assume that there are two relay systems; one using PLC communications as shown in Figure 27.4 and the other using microwave as in Figure 27.5. The relay hardware and signaling for the two systems are independent with few common modes of failure. Positive tripping is enhanced by configuring each relay to send a transfer trip to the relay at the other end of the line *every time the relay sends a trip signal to its local breakers*. This concept is extended by having the transfer trip signal sent not only over its normal communications link, but *also* over the redundant relay system's communications link. This cross tripping ensures a higher reliability should either transfer trip path fail. See the cross-trip boxes in Figure 27.4 that show the signal being sent from the relay logic to the microwave transfer trip (2L (or 2R) MWTT Eqmt) and also a transfer trip input from the microwave transfer trip logic (relay 2L (or 2R)).
5. Transfer trip is, by its very nature, somewhat subject to security failures, such as a noise signal that may be interpreted as a trip signal by the receiving relay logic. Security is improved by requiring that transfer trip signals over two separate channels *both be received* before tripping is enabled by the receiving relay. In Figure 27.4, both channels 3 and 5 are required for left to right transfer tripping and both channels 4 and 6 are required for right to left transfer trip.

We assume that Figures 27.4 and 27.5 and the foregoing description adequately define the functional specification of the relay operation for a transmission line fault. The other information that is required is a specification of the relay control circuits. This is shown in Figure 27.6 for the local circuit breaker controls and Figure 27.7 for the relay dc control circuits for the left terminal. The right terminal is identical, except for labeling.

The control circuit shown if Figure 27.6 consists of the battery power supply, the protective relay contacts C1 and C2, the trip coils of the circuit breakers, and the circuit breaker front

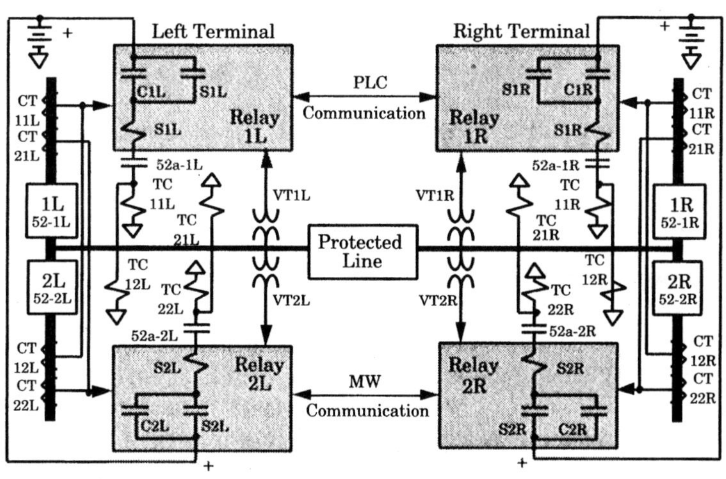

Figure 27.6 Control circuit for the transmission Relay.

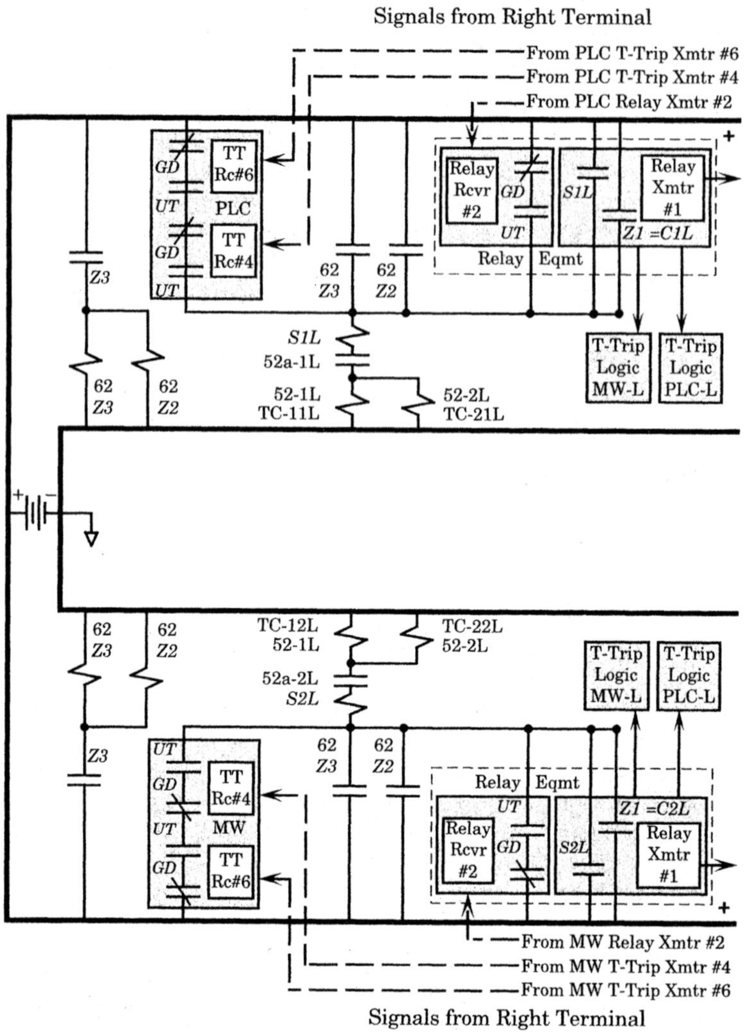

Figure 27.7 Left terminal control circuit showing permissive and transfer trip signaling by both PLC and MW.

contacts 52a. A seal-in relay S is also shown, which provides a means of maintaining current in the trip coil should the C contacts open prematurely for any reason. It is assumed that there are two circuit breakers at each end of the line, as shown in Figure 27.4, and that there are two relay systems, each with its own zone 1 C contact as shown in Figure 27.6. Moreover, the relay systems are completely redundant, with redundant current and voltage transducers, as well as redundant circuit breaker trip coils. For the PLC system, we identify the relay contacts as C1(L or R), with C2(L or R) being the contact for a companion microwave protective system. Note that *either* relay contact will cause the opening of *both* local circuit breakers 1 and 2. A similar control circuit is assumed to be present at each end of the transmission line, and these will be referred to as the left control circuit and right control circuit, respectively. The dc common connection at each end is usually grounded at only one point, but is shown by triangles in Figure 27.6 to avoid complicating the figure.

The dc control circuit for tripping the circuit breakers is shown in Figure 27.7 for the left terminal. Under normal conditions, the right terminal relay transmitter sends a guard tone to the left terminal relay using signal channel 2. This picks up the normally closed guard contacts, thereby preventing unwarranted tripping of relay receiver 2. When a zone 1 right fault is detected by the relay 1R, the zone 1R contacts C1R picks up to order tripping of the local circuit breakers and the relay PLC transmitter 2 shifts its transmitted tone from guard to a permissive tone, picking up the Left *UT* contacts thereby allowing the relay at the left terminal of the line to trip the breakers at that end. Exactly the same action is taken by relay 2R, which picks up contacts C2R, sending trip signals to the local breakers, and shifting the transmitted microwave permissive tone from guard to trip. In addition, both relays 1R and 2R cause tone shifts to be made in the transfer trip channels 4 and 6, which permits both the PLC and MW transfer trip systems to also pick up.

Therefore, for a fault near the right terminal that would normally be a zone 2 fault for left terminal relays, three modes of fast tripping of the left terminal breakers are placed in operation. First, the permissive tone shifts from the relay guard *GD* to trip *UT* frequency. However, the PLC and MW transfer trip circuits also perform a tone shift from guard to trip on channels 4 and 6. The transfer trip will not operate, however, unless the tone shift is received on both channels of either transfer trip system. This logic is shown in Figure 27.7 as mechanical contacts, but in many modern systems the actual switching is performed using solid-state electronic devices. Control circuits similar to Figure 27.7 can be drawn for the right terminal. Development of these control circuits is left as an exercise. There are differences in the receiving end hardware that depend on the form of signaling, the transmitters and receivers used, and different trip coils that are energized.

Figures 27.4 through 27.7 specify the resolution or level of component detail that is to be considered in the fault tree evaluation. The various systems shown include both active and passive components that are considered to be a part of the protective system. The passive wiring that connects the components could be considered in the analysis, but these items will be ignored here. Only the major items identified in these figures are to be considered in the fault tree analysis.

27.3.2 The Top Event

The top event of any fault tree represents the undesired failure event. The description of this event must define the top event precisely including *what* the "fault" is and *when* it occurs. The undesired event, in the case of clearing a transmission line short circuit, is "failure to clear the fault by the primary relays in a specified time." Therefore, it is also necessary to set a specification as to the time required for normal tripping of a fault. The normal tripping time has several components, which are often identified as follows.

$$T_C = T_R + T_B + T_M \text{ cycles} \tag{27.64}$$

where T_C = total fault clearing time
T_R = relay time
T_B = breaker time
T_M = margin for coordination

The time to clear the fault is usually measured in cycles.[2] The relay time is usually small, say one cycle or even less. Breaker time depends on the type of circuit breaker, the type of

[2]See Table 13.1 for typical fault clearing and pilot signaling times.

mechanical design of the breaker, the method of forcing contact separation, and the mechanism used for extinguishing the arc. For EHV lines, the fastest breakers are about one cycle, with two or three cycles being commonly used. The margin time is added to account for various physical problems that may cause the total time to vary from the ideal, such as the contact operating time, the differences in opening time due to varying fault current magnitudes, the ambient temperature effects, humidity effects, and other factors. As a practical consideration, one may consider that any margin time is divided in some logical manner between the relay time and the breaker time, so that only these two times need to be specified as the limiting design specifications for the relay and breaker subsystems, respectively. The important thing to note, insofar as the fault tree construction is concerned, is that the clearing of the transmission line fault will be considered a failure if the time to clear exceeds T_C, where this time limit is specified as follows.

$$T_C = T_{BM} + T_{PM} \tag{27.65}$$

where T_{PM} = protective relay operating time margin
T_{BM} = breaker operating time margin

and where each of these time specifications include a suitable margin. Therefore, we can state the top event of the fault tree as follows.

Failure to Clear Transmission Line Fault in $t < T_c$.

It is assumed that T_C is a specification of the protective system and is a given quantity. Note that the top event must include a "what" and a "when." In this case, the "what" is failure to clear transmission line fault and the "when" is the time restriction $t < T_C$. The top event will be avoided only when all four of the circuit breakers trip in a timely manner in response to correct relay detection and coordinated commands between the protective relays and also between the relays and the circuit breakers at both ends of the line.

The time restriction might be used to purposely disqualify the delayed clearing of the line. For example, in a pilot system similar to that described above, the relay reach might be incorrectly set resulting in relays at both ends of the line viewing the fault as a zone 2 fault. The line will tripped, but not within the specified time, hence violating the time restriction.

Having defined the top event, the next question to ask is "Can this top event consist of a component failure?" For this system, there are several components that can cause the top event to occur, but the fault (i.e., the short circuit) itself is not a protective system component failure. Therefore, we classify the top event as a "state-of-system" fault and look for the minimum necessary and sufficient immediate cause or causes leading to the top event [1], [2]. This gives the fault tree shown in Figure 27.8, where the time limit subscripts *BM* and *PM* refer to the breaker margin and protective margin, respectively.

This directs our attention to two major subsystems that must work in a coordinated way to clear a transmission line fault; the breaker failure subsystem (BF) and the protection subsystem (PS). Each of these subsystems is now considered separately.

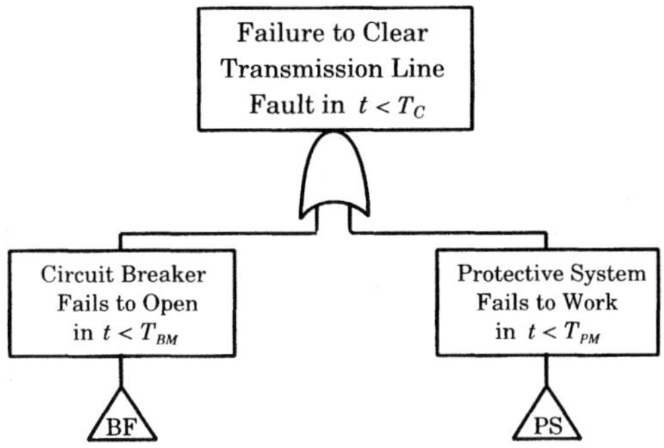

Figure 27.8 Events leading to the top event.

27.3.3 Failure of the Circuit Breakers

The breaker failure branch of the fault tree will now be constructed. It is recognized immediately that the fault event of the breaker failure subsystem (BF) will occur if any of four circuit breakers terminating the transmission line should fail to open or if the battery at either terminal is failed. This gives the fault tree construction shown in Figure 27.9. The items described here as "batteries" are usually systems that include multiple battery cells, battery chargers, and ancillary equipment. For the present analysis we will treat the battery as a simple component with a known basic failure rate. The circuit breakers are rather complex systems that often need more detailed component modeling depending on the purpose of the study. The breaker model is described below.

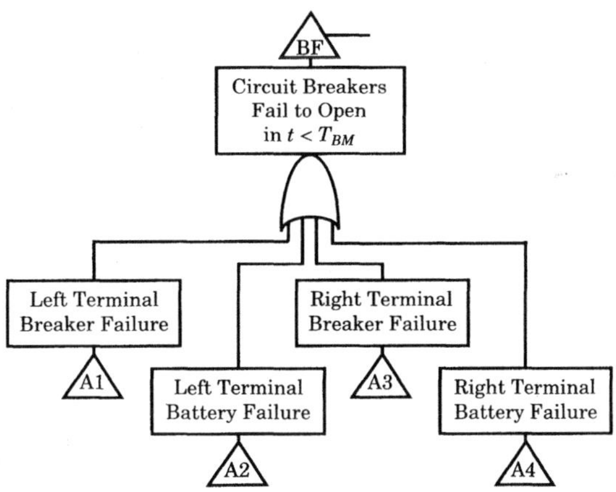

Figure 27.9 Circuit breaker failure tree.

The first failure events of Figure 27.9 to be evaluated are those due to the batteries at the left and right terminals. The fault tree for battery failure at the left terminal is shown in Figure 27.10, where the basic event is tagged Y1. Note that an identical fault tree can be drawn for the right battery failure, which will have its basic event tagged as Y12, as noted by the parentheses. Hereafter, all primary failures for the circuit breaker failure events (BF) will be designated Y1, Y2, etc. The battery fault could be due to secondary failures, such as failure of the battery charger. Command mode failures, such as operating at the wrong time or in the wrong sequence, are not likely for the battery systems. We will not pursue secondary failures in this investigation, but they could easily be added to the fault tree.

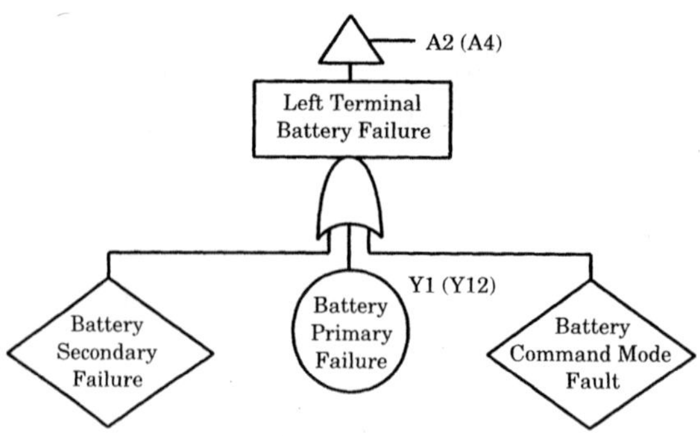

Figure 27.10 Fault tree branch for battery faults at the left terminal.

Each of the breaker failure events of Figure 27.9 is a fault with several types of failures, including basic component failures. It is noted that some of these failures are associated with the breaker itself and others with the type of control used. Therefore, we separate these two functions as we extend the fault tree in Figure 27.11 for left terminal only. The fault tree for the right terminal will have an identical structure.

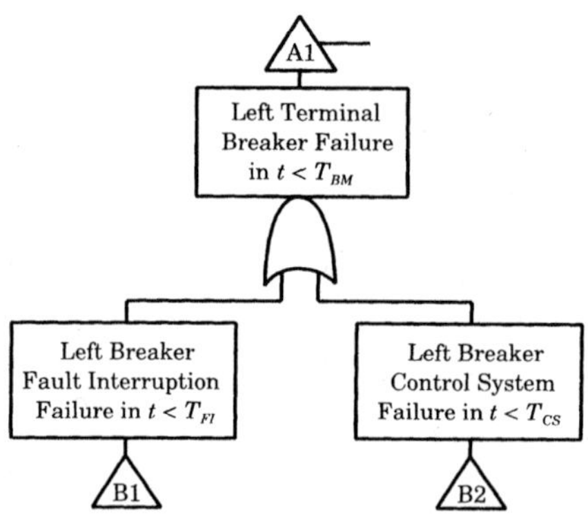

Figure 27.11 State of component fault for the circuit breaker 1L.

The circuit breaker subsystem A1 consists of two subsystems: a fault interruption subsystem and a control subsystem. The fault interruption system consists of the breaker operating mechanism and associated components that are needed to extinguish the arc. The breaker operating mechanism consists of the main current carrying contacts and the mechanism that causes these contacts to move apart to interrupt the circuit. We shall refer to this entire subsystem as the "breaker mechanism" although it must be recognized that this includes both the means of moving the contacts apart as well as that for lengthening, cooling, and eventually extinguishing the arc. There is the tacit assumption that the breaker is initially latched closed and, once triggered by the trip coil, a well-defined action takes place. This action may be complex and interesting, but is beyond the scope of analysis here.

The breaker interruption control consists of the breaker front contacts (52a) and the trip coils. The front contacts are designed to have the same position, open or closed, as the circuit breaker main contacts. Since the protected line is assumed to be in service prior to the line fault, then both the main contacts and the front contacts are initially closed. If the front contacts are closed, current can flow through the trip coils once any of the protective relay or transfer trip contacts close to complete the breaker operating circuit. Some circuit breakers are provided with more than one trip coil, and the breakers under study here will be assumed to be of this type. This is shown clearly in Figures 27.6 and 27.7. The trip coils and the front contacts are part of the control system as shown in the control diagram of Figure 27.6.

In Figure 27.11, the time specification has been separated into two parts, one part for the control system and the other for the fault interruption. These actions are not concurrent, but are additive.

$$T_{BM} = T_{FI} + T_{CS} \tag{27.66}$$

We now expand the fault tree branch B1 relating to circuit breaker fault interruption failure. This type of failure can be caused by a failed component. Therefore, below each of the failure events, following the procedure for fault tree construction, we immediately construct an OR gate and look for the primary, secondary, and command modes of failure for the circuit breaker, as shown in Figure 27.12 [1]. Note that failure to interrupt the fault current in either left terminal circuit breaker constitutes a system failure, since all breakers must perform correctly to avoid failure.

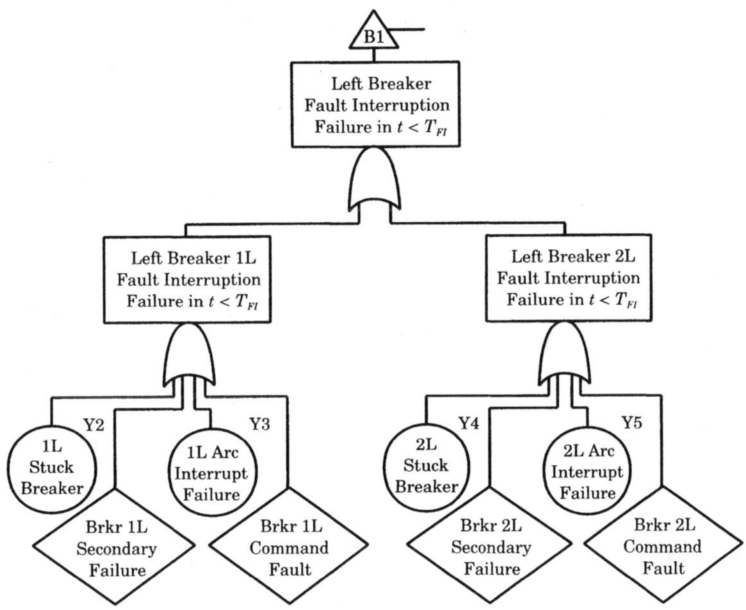

Figure 27.12 Left terminal circuit breaker fault interruption failure.

Clearly, the breaker mechanism can be in a stuck closed condition, which is a primary failure, represented by the circled event, and there may be several different failure mechanisms leading to the stuck breaker condition. These mechanisms are not explored further here, but the failure rates for all such conditions can be added together to obtain the stuck breaker failure rate, which is identified in Figure 27.12 as the primary failure Y2. The breaker arc interruption equipment may also fail, which is designated as primary failure Y3. The breaker may also suffer secondary failures to clear the fault, where the secondary failures are defined for conditions where the breaker is operating outside its design envelope. This might occur, for example, if the fault current to be interrupted is greater than the interrupting rating of the breaker. Secondary failures are represented by diamond boxes and will not be explored further in the present evaluation, but the fault tree could be expanded in a straightforward way to investigate the effect of secondary failures. The third type of fault noted is a command fault. This type of fault involves the proper operation of the breaker, but at the wrong time or in the wrong sequence. For example, the breaker may open without a proper opening command due to mechanical failure. Another example of a command fault could be a false operation due to relay setting error. These failure modes will not be explored further, so the event is shown in a diamond-shaped box. This brings us to the terminus of this branch of the fault tree. The bottom events are basic failures for which failure and repair rate data are known, such that the probability of each basic failure can be computed.

The circuit breaker control systems are defined in Figure 27.6. Consider the trip coils and front contacts (52a) for the breakers 1L and 2L at the left end of the protected line. The trip coils and 52a contacts associated with the two breakers are shown with a different perspective in Figure 27.13. The system is designed such that no single component failure will cause failure of both breakers to be successfully tripped. Failure of this system requires overlapping failure of two or more components. Since there are six control components in this group at each end of the protected line, an exhaustive consideration of overlapping failures requires that we consider the various combinations of multiple component failures that might occur 2, 3, 4, or 5 at a time. There is no need to consider six at a time as that event is assured to cause failure of both breakers 1L and 2L. By considering all of the various outage combinations, we can assemble the results shown in Table 27.2.

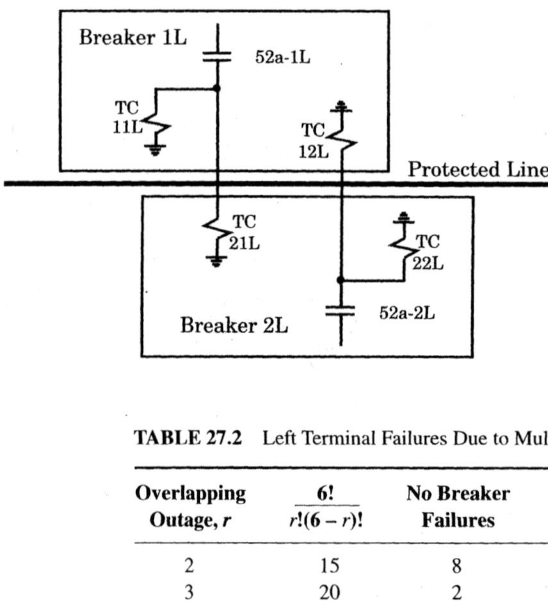

Figure 27.13 Trip coil and 52a contacts for left-end breakers.

TABLE 27.2 Left Terminal Failures Due to Multiple Component Outages

Overlapping Outage, r	$\frac{6!}{r!(6-r)!}$	No Breaker Failures	1L Only Failures	2L Only Failures	1L & 2L Failures
2	15	8	3	3	1
3	20	2	6	6	6
4	15	0	2	2	11
5	6	0	0	0	6

The system design is robust, since eight of 15 ways that two overlapping component outages can occur result in no failures of the circuit breaker operation. As the number of overlapping component outages increases, however, the breaker failures increase rapidly. An exhaustive analysis that considers all of these multiple outages is the correct way to analyze the effect of breaker failures due to random failure of front contacts and trip coils. However, we also recognize that the probabilities associated with multiple failures is small, and the probability of more than two outages at a time is very small and can usually be ignored with little error. We chose this approximate method of analysis where we consider only two overlapping component outages, but recognize that extending the fault tree to provide a complete analysis is entirely possible. Therefore, with this assumption, we construct the fault tree shown for the left terminal breakers in Figure 27.14. A similar fault tree can be drawn for the right terminal breakers.

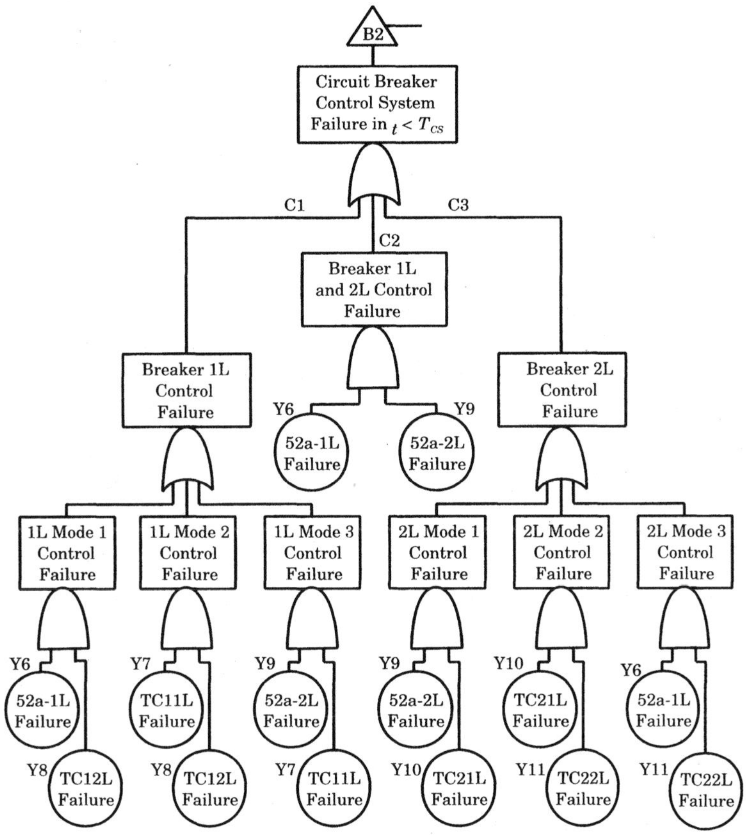

Figure 27.14 Fault tree branch for circuit breaker control system failure.

Each of the fault events of Figure 27.14 identified as C1 and C2 can be caused by component failures. Therefore, following the rules of fault tree construction [1], we can construct an OR gate below each of these events and add events representing primary, secondary, and control mode faults. Note that certain components associated with breaker 2L are involved in the fault tree for the control of breaker 1L, and vice versa. Basic failures are considered two at a time and proceed through AND gates. There is only one way that both breakers can fail due to loss of two of these components, and that is the loss of both front contacts (branch C3).

Since all the fault trees represented by Figures 27.11 through 27.14 all terminate in basic primary and secondary component failures, this terminates these branches of the fault tree. We can simply create these fault trees by ignoring the secondary and command mode failures, since these events are not considered in the present analysis. The analysis of command mode faults may be of interest in some studies. Command mode faults for breaker auxiliary contact can occur due to proper relay contact operation, but at the wrong time or in the wrong sequence. For example, a workman performing repairs near the relay cabinet may jar the relay mounting, causing mechanical relay contacts to operate. Another example of command mode faults is due to the relay logic having an improper reach, and thereby initiating pickup correctly, but for a fault on an adjacent line.

Figure 27.15 summarizes the analysis of the breaker failure branch of the top event for breakers 1L and 2L. This completes the analysis of the fault tree construction, beginning with Figure 27.11, for the left terminal. Since the right terminal is assumed to be identical in its component makeup to the left terminal, the fault trees will be identical, although it is possible that the fault tree basic failure events for the other terminal may require different reliability data. We assume that these fault tree branches are identical to those shown in Figures 27.9 through 27.15.

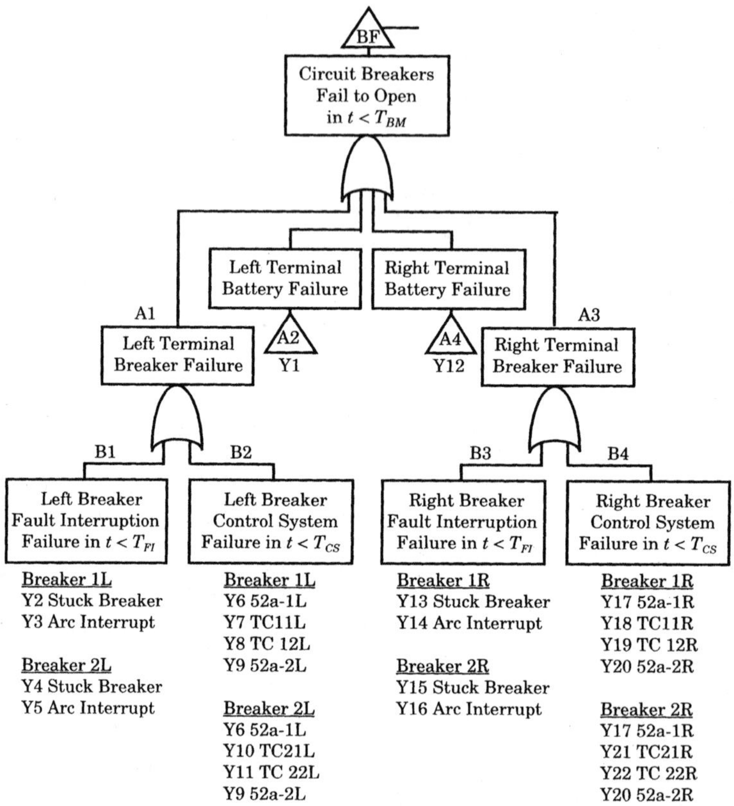

Figure 27.15 Summary of breaker failure data for all circuit breakers.

Rather than construct three additional fault trees, we can summarize these results by a tabulation of the basic event tags for all breaker failure (BF) fault trees. This tabulation is shown in Figure 27.15. Based on the assumed breaker and control configurations, it requires 22 basic failure events to describe the BF portion of the fault tree. Note that it is not possible to separate the breaker controls according to breaker number because breaker trip coils are assigned to different relays and are supplied from different batteries. This makes the control system more reliable.

27.3.4 Protective System Failure

In addition to circuit breaker failure, the failure of the protective relay system represents another way in which the top event can be reached. We now examine the contribution of protective system failures to the top event. This leads to the second fault event shown in

Figure 27.8, viz., the protective system fails to work in the specified time T_{PM}, which includes the protective relay time plus an allowable margin. We abbreviate this system as "PS" in the analysis, as noted in Figure 27.8.

The faults associated with the protective system must be divided according to the transmission line zones of protection, since different components of the protection hardware must operate for transmission faults in these different zones. This constitutes state-of-system events leading to the failure event for system protection. First, however, we evaluate the communications links as these are required for all successful protective system operations.

27.3.4.1 Modes of Pilot Signaling. The protection has two basic modes of signaling, a permissive trip mode and a transfer trip mode. Both modes use the same technique, but use different hardware items to accomplish the objectives. Both of the pilot signaling techniques assume that a guard tone is transmitted from relays at one end of the line to relays at the other end. If a fault is sensed and a time delay, if any, expires, then the tone is shifted from the guard frequency to an operate, or trip, frequency.

The first mode of operation is referred to here as the *permissive* mode. If a fault is sensed by relay 1L at the left terminal, the permissive signal is sent by transmitter 1L over PLC channel 1 (PC1) to relay receiver 1R at the right terminal and to the relay 1L logic, shifting the transmitted tone frequency from guard to trip. If the fault is sensed by relays at the right terminal, the tone shift is initiated by the relays at that terminal but using PLC channel 2 to signal the relay at the left terminal. The signaling channels are noted in Figure 27.4 for PLC only. The permissive mode can be summarized by the hardware items noted in Figure 27.16, which shows the signaling equipment only in the direction from left to right. The lower part of the figure shows relay 2L signaling relay receiver 2R over microwave channel 1. A figure exactly like 27.16 can be drawn for the signal channels between relays at the right terminal and those on the left, which use channel 2 for both the PLC and MW signals, as noted in Figures 27.4 and 27.5.

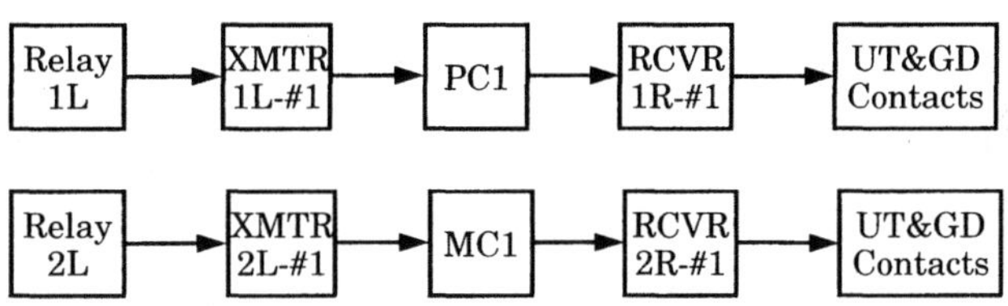

Figure 27.16 The permissive mode of left terminal protection signaling.

The second mode of signaling makes use of the transfer trip capability of the system design. This mode is depicted in Figure 27.17, which again shows only the hardware used for signaling from the left terminal to the right terminal. A similar diagram can be sketched for right-to-left signaling, which uses different hardware items throughout. Note that each left terminal relay, 1L or 2L, signals the remote terminal through both the power line carrier and microwave media. An additional requirement, not evident in the figure, is placed on the received signal transfer trip logic, which requires that *both* channels be received before tripping can take place at the receiving end. See Figure 27.7 for a detailed description of the remote tripping logic.

One further complication of the signaling is that both the actions shown in Figure 27.16 and in Figure 27.17 occur simultaneously, and the reception of the tone shift in either the permissive or the transfer trip systems will initiate tripping of circuit breakers at the right terminal. This means that any one relay, working correctly, will pick up for a zone 1 fault and

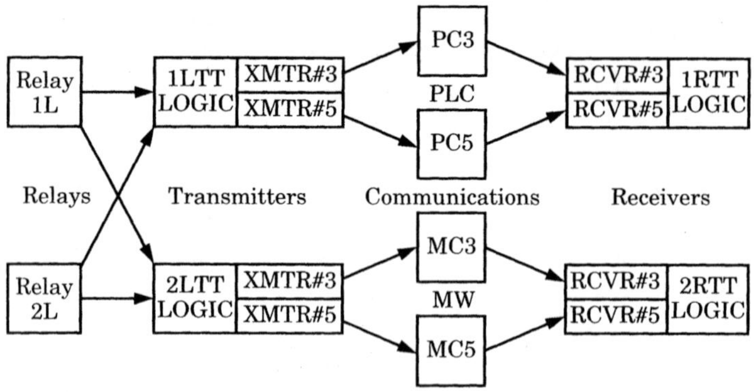

Figure 27.17 The transfer trip mode of left terminal protection signaling.

will trip all four circuit breakers, assuming that all components are working and each relay has both the permissive and transfer trip methods of accomplishing the fault clearing.

27.3.4.2 Transmitter and Receiver Modeling. The transmitters and receivers shown in Figures 27.16 and 27.17, as well as those shown in Figures 27.4 and 27.5, will be modeled as basic events in the reliability analysis. If detailed design information is available, these system models can be expanded to account for the complexity of design, but that will not be pursued further here. If the relays are working correctly but both signaling systems are failed, tripping will still occur, but the time limitation will be violated, which will be considered a failure in this analysis.

27.3.4.3 Communication Links. As noted in Chapters 13 and 14, complex relay systems used for high-voltage transmission protection employ communications between the relays at the two ends of the line to ensure proper fault clearing with the required selectivity. Therefore, the communication links become an essential part of the relay hardware and must be analyzed as part of the protective system. Not only are the communication links essential to system operation, communication failure will cause failure of the protection system. This is due to the need for dropping the guard signal in order to permit tripping in the specified time. Before analyzing the system protection, it will be helpful to first understand the communications links. Different types of communication links may be used in a given transmission protection system and all of these links require analysis. Fortunately, however, many of these links are identical in their hardware configurations, which may simplify the analysis.

Consider the power line carrier protective scheme shown in Figure 27.4. This protective scheme employs six different communication links, labeled 1 through 6 in the diagram. Note that all of these links utilize exactly the same type of equipment, but certain failures will affect more than one link. An example of this common mode failure is the failure of transmission phase A, which will cause the failure of communication links 3 and 4. The CVTs and hybrids are also single points of failure for more than one link.

A general fault tree for any power line carrier communications link is shown in Figure 27.18. Secondary failures and command faults have not been shown in the figure, but could be considered in a straightforward manner.

To simplify the notation in the fault tree, the following abbreviations have been used in the figure:

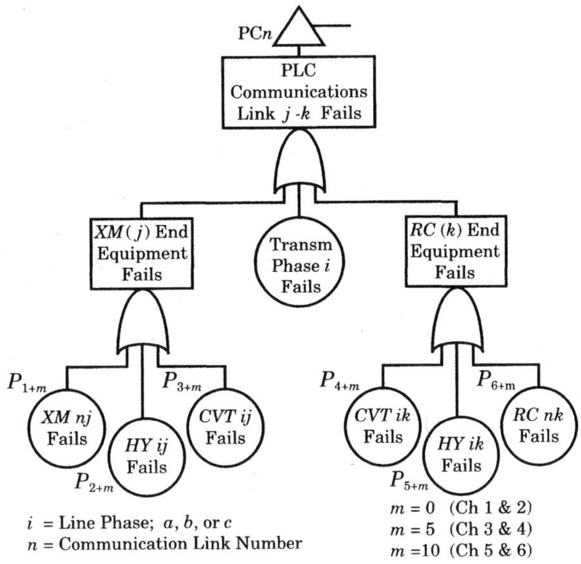

Figure 27.18 Fault tree for a PLC communications link from j to k.

$$\begin{aligned} CVT &= \text{capacitive voltage transformer} \\ HY &= \text{Hybrid} \\ n &= \text{communication link number} \\ i &= \text{transmission phase, } a, b, \text{ or } c \\ j, k &= \text{transmission network node numbers} \end{aligned}$$

Only two types of terminal hardware are employed in this example. The CVT is used to couple the carrier frequency on to the transmission line conductor, where wave traps prevent the transmitted frequency from propagating throughout the power system. The hybrids are impedance matching devices that present a low impedance for transmission to the remote receiver and high impedance to prevent a transmitted signal from being absorbed by the local receiver. The only other component in the system is the transmission conductor. A complete system also has transmitters and receivers, but these are considered in this analysis to be part of the terminal equipment.

Note that all gates in the fault tree are OR gates. This is a series logic, which means that the failure of any one of the components causes the failure of that communications link. It also means that the failure rate of the link is equal to the sum of the failure rates of the components.

In summary, then, we write the failure rate of the power line carrier (PC) communication link from j to k for any phase x as

$$\lambda_{PC} = \lambda_{CVTxj} + \lambda_{HYxj} + \lambda_{TLx} + \lambda_{CVTxk} + \lambda_{HYxk} \tag{27.67}$$

where the subscript TL refers to the "transmission line" conductor over which the carrier current is designated to flow.[3] Four failure rates are due to the equipment at the two ends. The remaining failure rate is due to the transmission line conductor and this failure rate will vary with line length and with the complexity of the line terminations.

[3]It should be noted that the communications signal may be successfully transmitted even if the transmission line fails. This is not at all reliable, however, and should not be depended on.

If the equipment at the two ends are identical, which might often be the case, then the equation can be simplified to the following form.

$$\lambda_{PC} = \lambda_{TLx} + 2\lambda_{CVTx} + 2\lambda_{HYx} \tag{27.68}$$

In this case, there is no requirement to keep track of equipment at each end separately. The important thing to note is that there are five items of equipment that are capable of causing failure of the PLC communications link. Table 27.3 shows the variable labels that will be used for the basic events for the PLC Communication links. Some reliability databases account for transmission line failure rates as either terminal-related or line-related failures. Such data includes line failures that are due to failed terminal equipment. The data needed in (27.68) is the failure rate for just the line itself.

TABLE 27.3 Basic Events for PLC Channel Component Failures

Channel	Phase	L-CVT	L-HYB	LINE	R-CVT	R-HYB
1 and 2	B	P1	P2	P3	P4	P5
3 and 4	A	P6	P7	P8	P9	P10
5 and 6	C	P11	P12	P13	P14	P15

The microwave communications link is quite different than the PLC link, as noted in the descriptions of Section 13.5.2. The microwave system has terminal equipment such as antennas, filter-isolators, mixers, IF amplifiers, FM detectors, baseband amplifiers, and transmitter klystrons [5]. Some of the equipment is installed in duplicate, with one system operating in hot standby mode for high reliability. Between the terminals are a number of microwave "hops" that depend on the number and location of microwave repeaters. Each repeater contains essentially the same type of equipment. In many cases, two receiving antennas are separated vertically by 10 to 15 meters on the repeater tower to utilize what is called "space diversity." The basic concept is that, should one receiver fail to receive a strong signal due to atmospheric fading, the other receiver at a different elevation will likely succeed in reception.

A possible fault tree for the MW communications link is shown in Figure 27.19, which represents the equipment in a single hop between a transmitter and receiver. The representation of the terminal microwave equipment is similar to the PLC scheme. The signal transmission failure representation is quite different, however. Whereas the PLC system has only the one transmission medium, the MW link has a series of hops, with the number being variable depending on the distance and terrain. Each hop, however, consists of a tower-to-tower signal transmission system that consists of a transmitter, the line-of-sight path, and the receiver. The transmitter and receiver at a given tower share a power supply system, which is usually redundant. The failure rate of the MW communications link is computed as follows.

$$\lambda_{MC} = \lambda_{MWXMj} + \lambda_{MDj} + \sum_{i} \lambda_{Hi} + \lambda_{MWRCk} + \lambda_{DMk} \tag{27.69}$$

where the subscript notation is identified as follows.

MWXM = microwave frequency transmitter
MWRC = microwave frequency receiver
MD = input signal modulator
DM = output signal demodulator
H = microwave repeater equipment
i = hop number
j, k = communication link node numbers

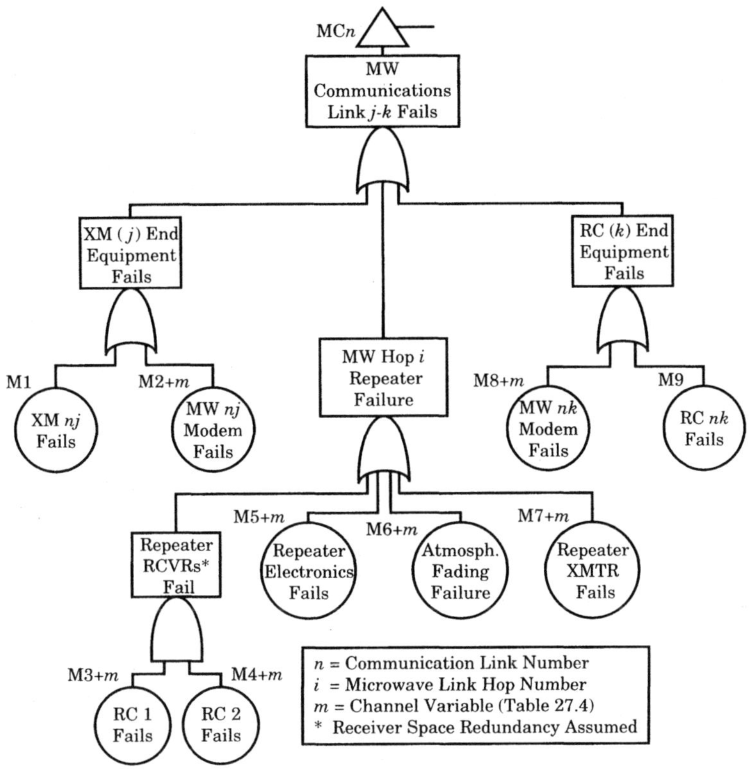

Figure 27.19 Fault tree for a MW communications link from j to k.

Table 27.4 shows the variable labels that will be used for the basic events for the microwave communication links. This is a simple analysis of the microwave system. Analysis of an existing system must include all equipment in order to provide the basis for a detailed fault tree of the system.

TABLE 27.4 Basic Events for Microwave Channel Component Failures

Chan	Xm*j*	Mod*j*	RC1	RC2	Elec	AF	RepX	Mod*k*	Rcvr*k*
1	M1	M2	M3	M4	M5	M6	M7	M8	M9
2	M1	M10	M11	M12	M13	M14	M15	M16	M9
3	M1	M17	M18	M19	M20	M21	M22	M23	M9
4	M1	M24	M25	M26	M27	M28	M29	M30	M9
5	M1	M31	M32	M33	M34	M35	M36	M37	M9
6	M1	M38	M39	M40	M41	M42	M43	M44	M9

Multiply above by number of hops

Clearly, there are multiple failure modes for the communications links used in transmission protection. There is no redundancy in the systems represented here, and redundancy is not always used in practice. The entire protective scheme is redundant, however, if both a PLC system and a MW system are used in a redundant protection configuration. Experience has shown that many protection failures are due to loss of communications. A more detailed model of the microwave repeater stations is recommended if the details of the equipment at each location are known. This equipment varies with the construction date and manufacturer of the systems.

Hereafter, in this chapter, the analysis of the protective system reliability will refer to communication links by number, and the failure of these links will be treated as hardware failure modules with known fault trees. The failure rates for the communication links are based on the fault trees shown in Figures 27.18 and 27.19 and on (27.68) and (27.69). The repair times will always depend on the location of the failure, and will vary across the span of distance from one end of the line to the other, and on the logistics of moving repair personnel and spare parts to the various locations. In some cases, the weather may even play an important part in these logistics—for example, where mountain-top microwave repeaters might be completely snowed in for several months each year. It is assumed that the repair times can be estimated, based on the distance, terrain, and recorded experience in making repairs on a link or one with similar physical characteristics.

Line protection is to be analyzed based on the distance protection scheme shown in Figure 27.20, where the line is terminated at stations called "Left" and "Right," as shown in the figure. Relays at both ends have reach settings of nominally 80–90% of the total line length, with faults within this reach considered as zone 1 faults. This divides the line into three zones, which are designated zone 1 Left, zone 1 Center, and zone 1 Right.

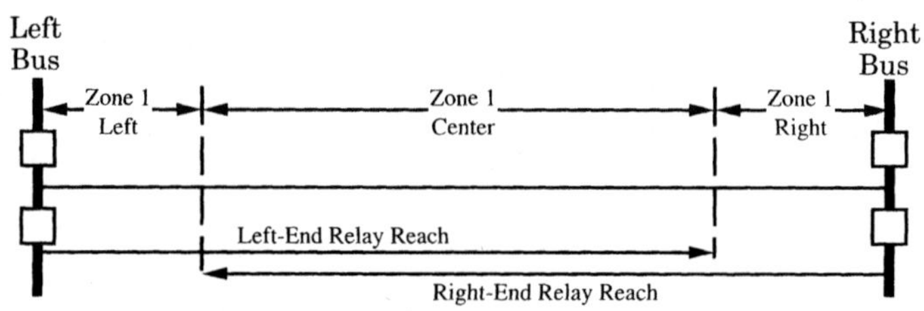

Figure 27.20 Identification of line protection zones.

We now consider how the protective systems can contribute to system failure. In doing so, we note that either relay at either end of the line has the ability, by design, to trip all four line breakers in response to a zone 1 fault. Therefore, failure implies loss of all of these systems, since all line faults must appear to be zone 1 faults to relays at either the left or right terminals. This logic requires an OR gate and leads to the fault tree of Figure 27.21. Relay

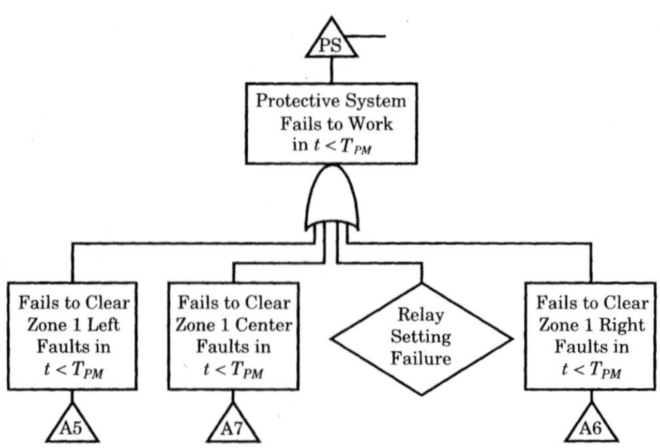

Figure 27.21 First step in the protective system fault tree.

setting errors, shown as secondary failures, are human errors, which have often been ignored due to the difficulties in modeling. Recent advances now show promise for including human errors in fault trees [2]. However, relay setting errors will be ignored in the present fault tree development in order to concentrate on the hardware failures.

The primary failure events depicted in Figure 27.21 are not related to any specific components failures, so these are state-of-system events. Therefore, we immediately look for the minimum necessary and sufficient immediate cause or causes for each of these events. Faults near one end of the line are zone 1 faults for protective equipment at that end, and protective equipment at that end is relied on to clear the fault in the prescribed time limit.

27.3.4.4 Failure to Clear Left End Zone 1 Faults. Referring to Figure 27.21, consider the failure to clear zone 1 left fault. We construct the fault tree segment A5, shown in Figure 27.22 where the events B5 and B6 under the AND gate are state-of-system events. Therefore, we seek the minimum necessary and sufficient immediate cause or causes leading to these events. Basic failure L1 is a failure of left zone 1 of the transmission line, which can be determined a fraction of the total line failure rate proportioned according to the length of line that is not reached by zone 1 relays at the right end of the line. For faults in this zone the system is wholly dependent on the left-end protection system for detection and clearing within the prescribed time specification.

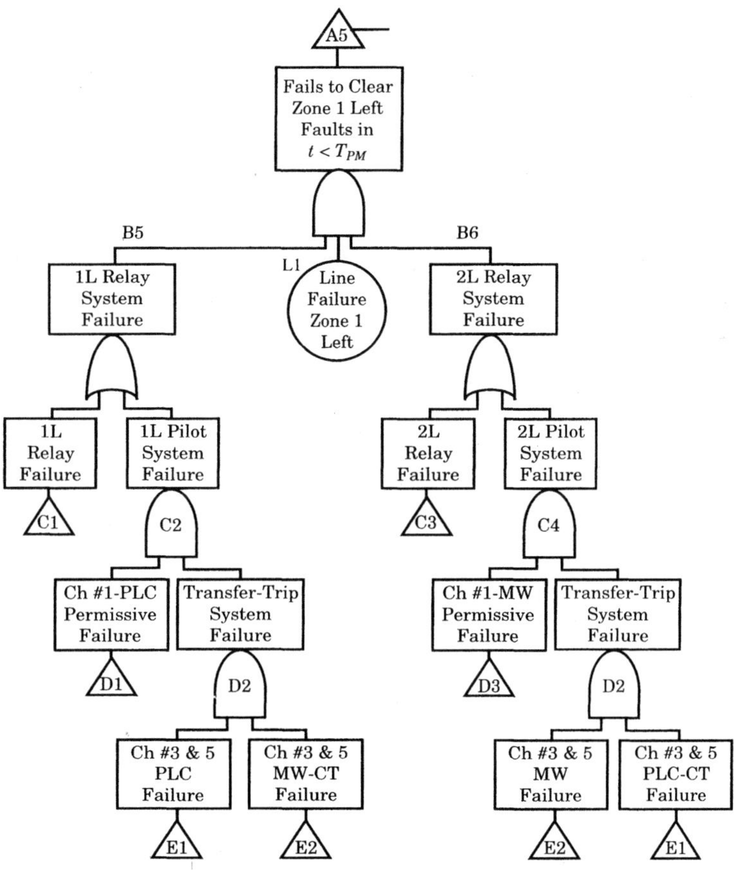

Figure 27.22 Fault tree for failure to clear close-in left end fault.

From Figure 27.22, it is clear that branches B5 and B6 are identical in structure, but model different relays and different permissive signaling equipment. The transfer trip equipment is the same for the two branches, although relay 1L considers the microwave system as cross-trip equipment, labeled CT under gate D2, whereas relay 2L labels the power line carrier system as cross-trip. However, the transfer trip hardware for the two relays are identical in their operation and in their fault tree construction. Note that the two transfer-trip systems are connected through an OR gate, due to the connection of the contacts at the receiving end, as shown in Figure 27.7.

Since either of the two left relays is capable of performing this fault clearing, failure occurs only if both relays fail, hence the AND gate at A5. The fault trees for B5 and B6 are identical in structure but refer to different relays. Relay 1L is a failed system if the relay itself is failed, which is identified as event C1 in Figure 27.22. However, in order to complete faulted line removal in the required time constraint, it is essential that pilot signal transmission be available, either in the form of permissive or transfer trip signaling. If both of the permissive and transfer trip pilot systems are failed, the 1L relay system fails to achieve the require speed of response. For transfer-trip signaling, both the carrier and microwave systems must both be received, so the failure of either mode of signaling causes transfer trip failure.

Figure 27.23 gives the fault tree for failure of relay 1L failure, which is identified in Figure 27.23 as C1. This includes the basic failure events for all components plus secondary and command fault failures of this relay system. The basic events are described by a numbering system that is written as X1, X2, ..., X8 for all basic events. Secondary and command faults are not pursued further in the analysis.

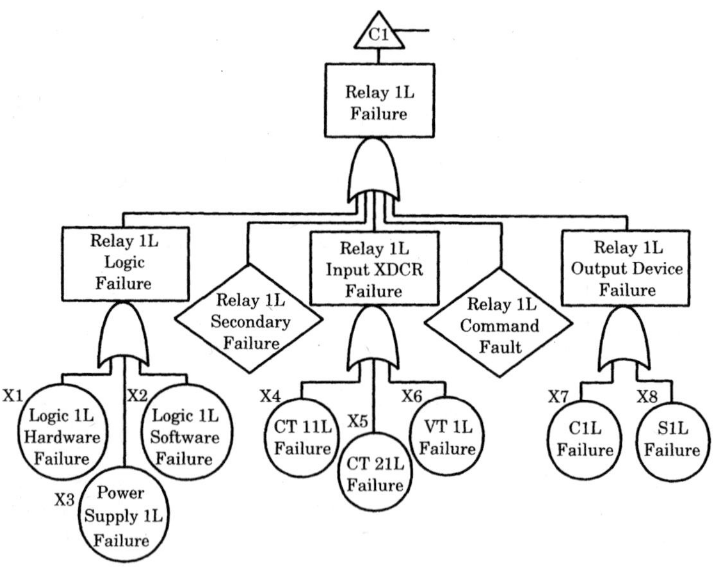

Figure 27.23 Fault tree C1 for relay 1L failure.

An interesting failure event included in Figure 27.23 is the failure of relay software, which is appropriate only if the relay employs digital equipment. Caution must be used in the treatment of digital computer software, because it does not obey the same rules as hardware. For example, redundancy is often used to improve the reliability of hardware items, but redundant identical computer software does not improve reliability if both computer codes contain the same errors. However, if there are redundant digital devices, it would be possible for each device to use different software, perhaps using a different approach to the computation. In this analysis we treat software failures as a basic event.

Figure 27.24 shows the fault tree for relay 1L permissive signaling failure, identified as D1 in Figure 27.22, which terminates in basic failure events, indicating the end of this branch of the fault tree. We identify failures due to loss of communications or failure of equipment items at the receiving relay on the right line terminal. The fault tree for permissive failure of relay 2L is identical to that of Figure 27.24 except that relay 2L uses microwave pilot. That subtree would be labeled D3, will communicate with relay 2R, and will have a communications subtree labeled MC2. However, the basic failure events for the terminal equipment will be the same as those of Figure 27.24.

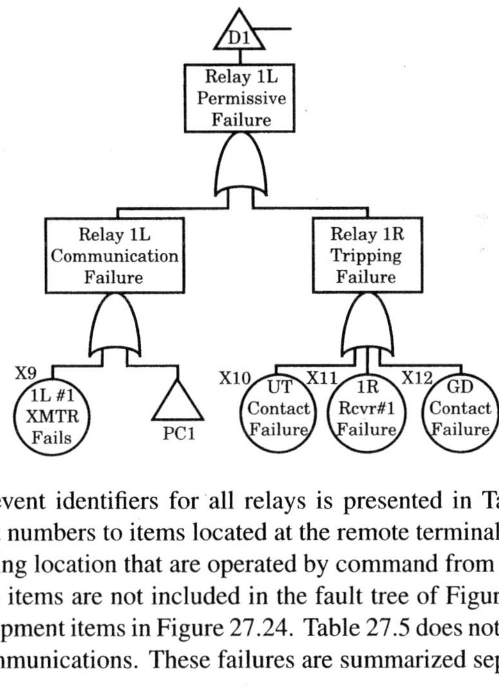

Figure 27.24 Fault tree for relay 1L permissive failure.

A summary of basic event identifiers for all relays is presented in Table 27.5. This table also assigns basic event numbers to items located at the remote terminal, and associated with equipment at the receiving location that are operated by command from the transmitting relay. These four permissive items are not included in the fault tree of Figure 27.23, but are included with other pilot equipment items in Figure 27.24. Table 27.5 does not include failures attributed to transfer-trip communications. These failures are summarized separately below.

TABLE 27.5 Relay Basic and Permissive Failure Item Tags

	Sending Relay Basic Items								Permissive Items				
	HW	**SW**	**PS**	**CT1**	**CT2**	**VT**	**Cx**	**Sx**	**Ch #**	**XM**	**RC**	**GD**	**UT**
1L	X1	X2	X3	X4	X5	X6	X7	X8	PC1	X9	X10	X11	X12
2L	X13	X14	X15	X16	X17	X18	X19	X20	MC1	X21	X22	X23	X24
1R	X25	X26	X27	X28	X29	X30	X31	X32	PC2	X33	X34	X35	X36
2R	X37	X38	X39	X40	X41	X42	X43	X44	MC2	X45	X46	X47	X48

We now address the failures associated with transfer-trip pilot, shown in Figure 27.22 as E1 and E2 for the PLC and MW systems, respectively. The E1 subtree is shown in Figure 27.25.

We group together the events that constitute failure of the transfer trip terminal equipment. These groups are identified as 1LTX for the relay 1L transfer-trip transmitter equipment, which

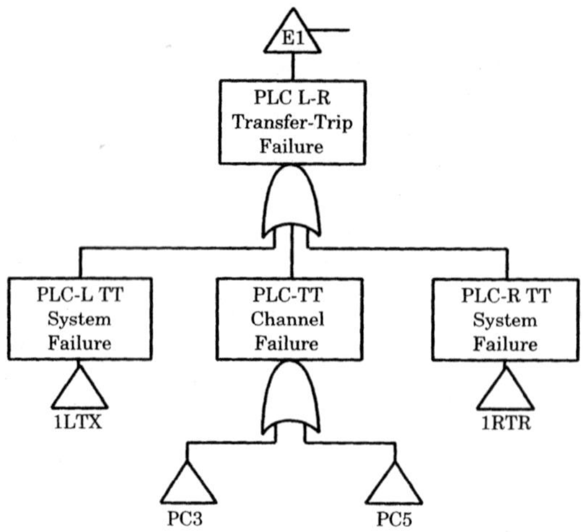

Figure 27.25 Fault tree for relay 1L PLC transfer-trip failure.

is shown in Figure 27.26, and 1RTR for relay 1R transfer-trip receiver equipment, which is shown in Figure 27.27. Secondary faults are not shown in these fault trees but could easily be added, if desired. These two fault trees terminate in basic events, which indicates that this is the termination of this branch of the fault tree.

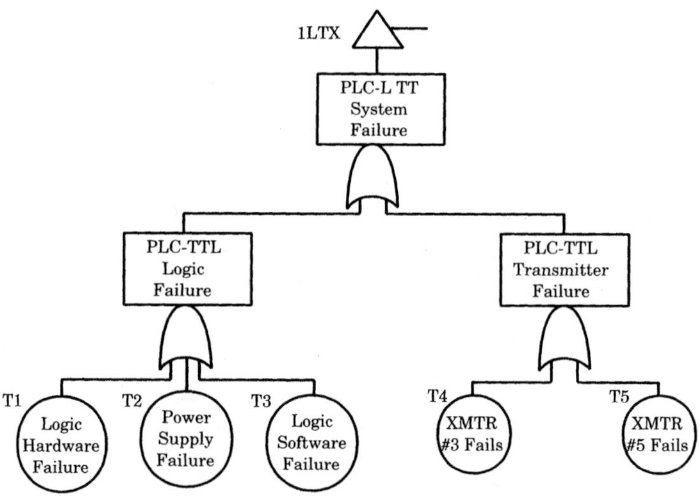

Figure 27.26 Fault tree of the transfer-trip module 1LTX.

Table 27.6 expands the fault trees shown above to identify basic failures for other transfer-trip logic modules. A total of 44 basic failure events are required to represent these transfer-trip systems.

Command mode faults of the transfer-trip system may require special study by the protection engineer. The communications link is often considered the most failure-prone subsystem in pilot relaying. The communications may be corrupted by noise that may make correct reception difficult. In some system studies it may be important to place a high failure rate on communications channels to determine the effect on the protective system. This type of analysis is not investigated here, but it might be an important study in some cases, especially if a particular system has experienced difficulty.

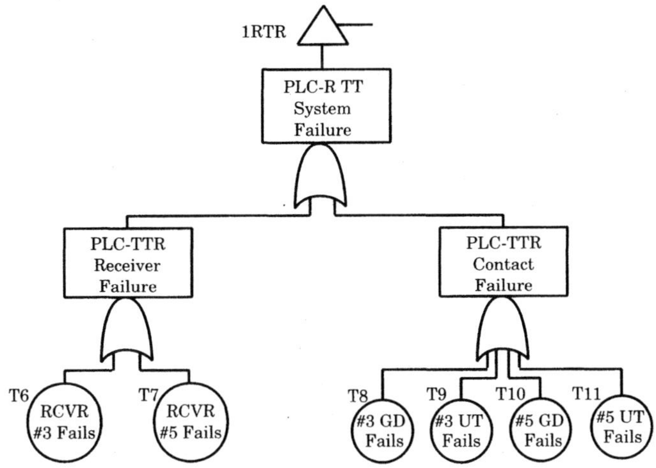

Figure 27.27 Fault tree for L-R transfer-trip receiving system 1RTR.

TABLE 27.6 Basic Events for Transfer-Trip Logic Failure

Sending Terminal			**Receiving Terminal**		
Left–Right	Left Logic	Xmtr 3 & 5	L-R	Rcvr 3&5	Right Trip Contacts
PC 1LTX MC 2LTX	T1 T2 T3 T12 T13 T14	T4 T5 T15 T16	1RTR 2RTR	T6 T7 T17 T18	T8 T9 T10 T11 T19 T20 T21 T22
Right–Left	Right Logic	Xmtr 4 & 6	R-L	Rcvr 4&6	Left Trip Contacts
PC 1RTX MW 2RTX	T23 T24 T25 T34 T35 T36	T26 T27 T37 T38	1LTR 2LTR	T28 T29 T39 T40	T30 T31 T32 T33 T41 T42 T43 T44

27.3.4.5 Failure to Clear Right-End Zone 1 Faults. The fault tree for failure to clear close-in right-end faults, shown in Figure 27.28, is exactly the same as that developed above for close-in left-end faults. The hardware items used in clearing the right-end faults are the same type, but are different hardware items. Therefore, the numbering and identification is different and must be determined with care.

The failure to clear right-end faults is conditioned by the line failure in zone 1 at the right end of the line. This failure rate will usually be determined based on a proportion of the line failure rate to the length of right-end zone 1 as a proportion of the total line length.

27.3.4.6 Failure to Clear Midline Zone 1 Faults. The final event leading to the Protective System (PS) gate is the failure to clear midline zone 1 faults. This type of failure results when a fault occurs in the center of the line where relays at both ends of the line view the fault as a zone 1 fault. Since relays at both ends should recognize and initiate clearing of this type of fault, failure can only occur when protective systems at both ends fail, either due to relay failure or due to pilot system failure, where pilot performance results in failure meet the top event time restriction. The fault tree for midline zone 1 faults is shown in Figure 27.29. Note that this fault tree terminates in basic events for equipment items that are the same as those used in clearing zone 1 faults.

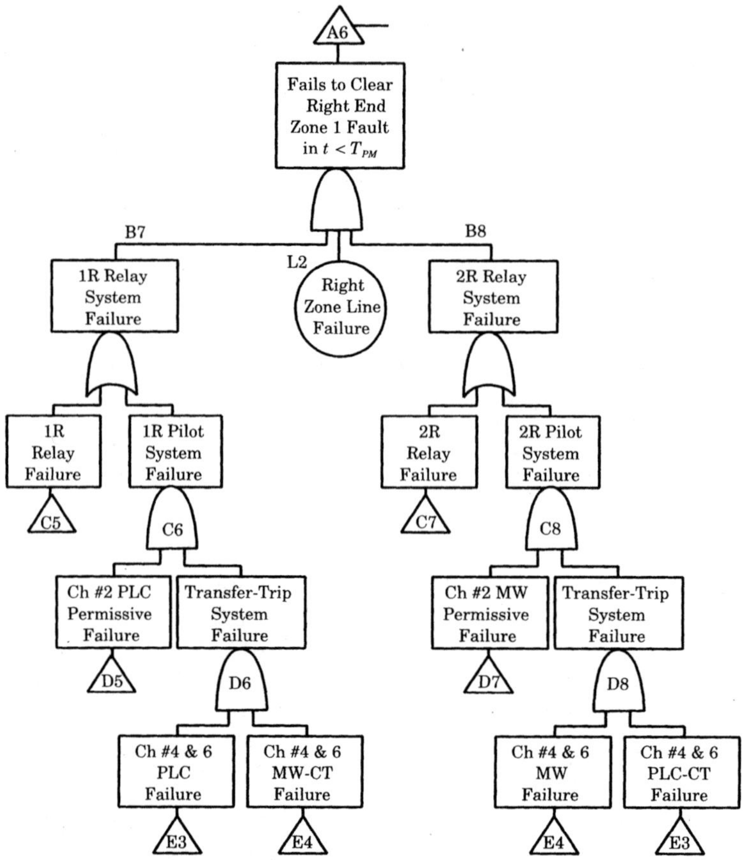

Figure 27.28 Fault tree to clear zone 1 right-end faults.

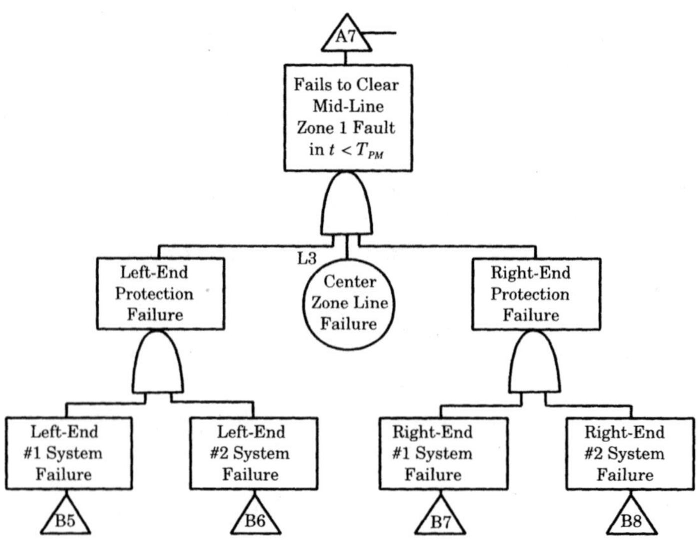

Figure 27.29 Failure to clear midline zone 1 faults.

The foregoing description of a permissive and transfer-trip pilot system is intended as an example as to how fault trees for a complex protection system can be developed. The protection engineer will need to vary this procedure to match the actual equipment in the system under study.

27.4 FAULT TREE EVALUATION

The evaluation of the fault tree is accomplished by evaluating the tree structure using Boolean algebra. The notation is simplified by using the alphanumeric shorthand identifiers for each event. Algebraic notation is used, rather than the Boolean notation. Table 25.9 compares the two types of notation. In engineering work, the algebraic notation is more common.

A top-down evaluation[4] of the fault tree is started by writing the algebraic equation for the Figure 27.7. Referring to the top event simply as TOP, we write

$$\text{TOP} = \text{BF} + \text{PS} \tag{27.70}$$

27.4.1 Breaker Failure Evaluation

The breaker failure event (BF) is evaluated first. This branch of the fault tree is summarized in Figure 27.9 for one of the circuit breakers. From the fault tree, we can write the following equation.

$$\text{BF} = \text{A1} + \text{A2} + \text{A3} + \text{A4} \tag{27.71}$$

for simplicity in the analysis, we assume that the equipment associated with each breaker is of the same type and in the same arrangement. Some of that equipment is not only the same type, it is the same equipment. For example, on each end of the line, there is but one battery. The battery at the remote end is the same type of equipment, but it is a different battery and the failure of that remote battery must be identified as a different event. This means that the primary component failure events must be carefully noted.

An equation can be written for each of the B level events of (27.71). From Figure 27.11 we can write the following

$$\text{A1} = \text{B1} + \text{B2} \tag{27.72}$$

The event B1 focuses attention on the left terminal fault interruption. The fault tree for this event is given by Figure 27.12, which allows us to write

$$\text{B1} = \text{Y2} + \text{Y3} + \text{Y4} + \text{Y5} \tag{27.73}$$

Event B2 is the circuit breaker control system failure event, which has the fault tree shown in Figure 27.14. We can write this event in terms of the basic event data as follows.

$$\begin{aligned} \text{B2} = {} & (\text{Y6} \cdot \text{Y8} + \text{Y7} \cdot \text{Y8} + \text{Y7} \cdot \text{Y9}) + (\text{Y6} \cdot \text{Y9}) \\ & + (\text{Y9} \cdot \text{Y10} + \text{Y10} \cdot \text{Y11} + \text{Y6} \cdot \text{Y11}) \end{aligned} \tag{27.74}$$

Then, A1 can be written in terms of basic events by substitution.

Event A2 is the left terminal battery, which is modeled as a basic event, for which we write

$$\text{A2} = \text{Y1} \tag{27.75}$$

[4]Fault tree evaluation can also be conducted using a bottom-up method.

With this result, we can summarize the events leading to failure of the left terminal as the sum of (27.72) and (27.75).

$$\begin{aligned} A1 + A2 = \sum_{i=1}^{5} Yi &+ (Y6 \cdot Y8 + Y7 \cdot Y8 + Y7 \cdot Y9) + (Y6 \cdot Y9) \\ &+ (Y9 \cdot Y10 + Y10 \cdot Y11 + Y6 \cdot Y11) \end{aligned} \tag{27.76}$$

This completes the A1 and A2 branches of the tree, since all events are given in terms of component primary failure events, in this case, in terms of events Y1 through Y11. Note that our notation is such that all breaker component primary failures events begin with the letter Y and terminate in a unique number.

The foregoing process is continued for branch A3, for which the following events are identified.

$$A3 = B3 + B4 \tag{27.77}$$

Note that event B3 is not the same as B1, the two events refer to breakers at opposite ends of the line. However, since we assume the breakers at the two ends to be identical types, with identical fault trees, the form of events B3 and B4 will be similar to B1 and B2, respectively. Therefore, we immediately write these results as follows, with help from Figure 27.15.

$$B3 = Y13 + Y14 + Y15 + Y16 \tag{27.78}$$

$$\begin{aligned} B4 = (Y17 \cdot Y19 + Y18 \cdot Y19 + Y18 \cdot Y20) + (Y17 \cdot Y20) \\ + (Y20 \cdot Y21 + Y21 \cdot Y22 + Y17 \cdot Y22) \end{aligned} \tag{27.79}$$

We also observe that event A4 is simply the basic failure rate Y2. Using this fact together with the two equations for the right terminal, we can write

$$\begin{aligned} A3 + A4 = \sum_{i=12}^{16} Yi &+ (Y17 \cdot Y19 + Y18 \cdot Y19 + Y18 \cdot Y20) \\ &+ (Y17 \cdot Y20) \\ &+ (Y20 \cdot Y21 + Y21 \cdot Y22 + Y17 \cdot Y22) \end{aligned} \tag{27.80}$$

Then the breaker failure event BF is the sum of (27.76) and (27.80).

$$\begin{aligned} BF &= A1 + A2 + A3 + A4 \\ &= \sum_{i=1}^{5} Yi + (Y6 \cdot Y8 + Y7 \cdot Y8 + Y7 \cdot Y9) + (Y6 \cdot Y9) \\ &\qquad + (Y9 \cdot Y10 + Y10 \cdot Y11 + Y6 \cdot Y11) \\ &\quad + \sum_{i=12}^{16} Yi + (Y17 \cdot Y19 + Y18 \cdot Y19 + Y18 \cdot Y20 \\ &\qquad + (Y17 \cdot Y20) \\ &\qquad + (Y20 \cdot Y21 + Y21 \cdot Y22 + Y17 \cdot Y22) \end{aligned} \tag{27.81}$$

The greatest contribution to the result is the single failure rate terms. The product terms, being products of small numbers, will not likely contribute nearly as much to the event BF. The important terms, therefore, are the breaker mechanism failure rates and the arc interruption failure rates. Faults occurring as a result of control system components are the terms in the result that are second-order terms.

27.4.2 Protective System Failure Evaluation

The protective system failure event is identified in Figure 27.8 as PS. This event can fail in three ways, as shown in Figure 27.21, where the three failure modes refer to the failure to clear left-end, right-end, or midline faults. These events are designated A5, A6, and A7, respectively, in Figure 27.21. From this figure, we write the following Boolean expression.

$$\mathrm{PS} = \mathrm{A5} + \mathrm{A6} + \mathrm{A7} \tag{27.82}$$

Any line fault can be cleared by proper action of any one of the four relays, providing that all circuit breakers, controls, and communications work correctly. Circuit breaker failures are described by (27.81) and need not be repeated here. Therefore, we concentrate on the line protective systems, consisting of the relays, the communications, and the control devices at both line terminals.

27.4.2.1 Failure to Clear Left End Zone 1 Faults. Failure to clear left-end faults are described in the fault tree of Figures 27.22 through 27.27. If only the component primary failures are considered, the results are summarized in Figure 27.22, for which we may write the following Boolean expressions.

$$\mathrm{A5} = \mathrm{L1} \cdot \mathrm{B5} \cdot \mathrm{B6} \tag{27.83}$$

Event L1 refers to the failure of this segment of the transmission line, which results in the need for protective system pickup. Event B5 refers to the relay equipment associated with relay 1L and B6 to equipment associated with relay 2L. Since either relay 1L or 2L are capable of clearing such faults, failure to clear zone 1 faults will occur only during the overlapping failure of the line segment and both relay systems, which is described by an AND gate and by the product of the three events in (27.83). Event B5 refers only to the equipment associated with relay 1L, as described in Figure 27.22. For this system we write

$$\mathrm{B5} = \mathrm{C1} + \mathrm{C2} = \mathrm{C1} + (\mathrm{D1} \cdot \mathrm{D2}) = \mathrm{C1} + \mathrm{D1} \cdot (\mathrm{E1} + \mathrm{E2}) \tag{27.84}$$

where the individual events are defined as

$$\mathrm{C1} = \sum_{i=1}^{8} = \mathrm{Xi} \tag{27.85}$$

$$\mathrm{D1} = \mathrm{PC1} + \sum_{i=9}^{12} \mathrm{Xi} \tag{27.86}$$

Also,

$$\mathrm{E1} = \mathrm{1LTX} + \mathrm{PC3} + \mathrm{PC5} + \mathrm{1RTR} \tag{27.87}$$

where, from Table 27.6

$$\mathrm{1LTX} = \sum_{i=1}^{5} \mathrm{Ti} \tag{27.88}$$

and

$$\mathrm{1RTR} = \sum_{i=6}^{11} \mathrm{Ti} \tag{27.89}$$

Also, from Figure 27.25,

$$\text{E1} = \text{PC3} + \text{PC5} + \sum_{i=1}^{11} \text{Ti} \tag{27.90}$$

and

$$\text{E2} = \text{2LTX} + \text{MC3} + \text{MC5} + \text{2RTR} \tag{27.91}$$

where

$$\text{2LTX} = \sum_{i=1}^{16} \text{Ti} \tag{27.92}$$

$$\text{2RTR} = \sum_{i=17}^{22} \text{Ti} \tag{27.93}$$

Then

$$\text{E2} = \text{MW3} + \text{MW5} + \sum_{i=12}^{22} \text{Ti} \tag{27.94}$$

Combining (27.90) and (27.94), we get

$$\text{D2} = \text{E1} + \text{E2} = \text{PC3} + \text{PC5} + \text{MW3} + \text{MW5} + \sum_{i=1}^{22} \text{Ti} \tag{27.95}$$

In summary, then, for B5 we get

$$\begin{aligned} \text{B5} &= \text{C1} + \text{D1} \cdot (\text{E1} + \text{E2}) \\ &= \sum_{i=1}^{8} \text{Xi} \\ &\quad + \left(\text{PC1} + \sum_{i=9}^{12} \text{Xi}\right)\left(\text{PC3} + \text{PC5} + \text{MW3} + \text{MW5} + \sum_{i=1}^{22} \text{Ti}\right) \end{aligned} \tag{27.96}$$

We follow a similar procedure to compute B6.

$$\text{B6} = \text{C3} + \text{C4} = \text{C3} + \text{D3} \cdot \text{D2} = \text{C3} + \text{D3} \cdot (\text{E1} + \text{E2}) \tag{27.97}$$

where

$$\text{C3} = \sum_{i=13}^{20} \text{Xi} \tag{27.98}$$

and

$$\text{D3} = \text{MC1} + \sum_{i=21}^{24} \text{Xi} \tag{27.99}$$

or

$$\begin{aligned} \text{B6} &= \sum_{i=13}^{20} \text{Xi} \\ &\quad + \left(\text{MC1} + \sum_{i=21}^{24} \text{Xi}\right) \cdot \left(\text{PC3} + \text{PC5} + \text{MW3} + \text{MW5} + \sum_{i=1}^{22} \text{Ti}\right) \end{aligned} \tag{27.100}$$

Finally, then, A5 is computed as the product given by (27.83).

27.4.2.2 Failure to Clear Right End Zone 1 Faults. The event A6 is constructed in the same manner as A5 since it describes an identical system, but installed at the right terminal, with right-to-left pilot signaling. From Figure 27.28 we can write

$$A6 = L2 \cdot B7 \cdot B8 \tag{27.101}$$

where B7 and B8 are found in a manner similar to that followed above for B5 and B6. B7 depends on the basic failures associated with relay 1R given by C5, the permissive signaling items for that relay given as D5, and transfer-trip signaling items given by D6 in Figure 27.28. For B7 we can write

$$B7 = C5 + C6 = C5 + D5 \cdot D6 = C5 + D5 \cdot (E3 + E4) \tag{27.102}$$

where

$$C5 = \sum_{i=25}^{32} Xi \tag{27.103}$$

$$D5 = PC2 + \sum_{i=33}^{36} Xi \tag{27.104}$$

Using the logic of Figures 27.25 to 27.27 and Table 27.6, we can write

$$\begin{aligned} E3 &= 1RTX + PC4 + PC6 + 1LTR \\ &= \sum_{i=23}^{27} Ti + PC4 + PC6 + \sum_{i=28}^{33} Ti = PC4 + PC6 + \sum_{i=23}^{33} Ti \end{aligned} \tag{27.105}$$

and, similarly,

$$E4 = MW4 + MW6 + \sum_{i=34}^{44} Ti \tag{27.106}$$

Then (27.103) through (27.106) can be used to construct B7 in terms of basic events.

In a similar manner, we can write the equations making up B8 as follows.

$$B8 = C7 + C8 = C7 + D7 \cdot (E3 + E4) \tag{27.107}$$

$$C7 = \sum_{i=37}^{44} Xi \tag{27.108}$$

$$D7 = MW2 + \sum_{i=45}^{48} Xi \tag{27.109}$$

and, finally,

$$A6 = L2 \cdot B7 \cdot B8 \tag{27.110}$$

27.4.2.3 Failure to Mid-Line Zone 1 Faults. The failure to clear midline zone 1 fault is described in Figure 27.29 and all of the component failures contributing to this event have already been found. Thus, we can write

$$A7 = L3 \cdot (B5 \cdot B5) \cdot (B7 \cdot B8) \tag{27.111}$$

The fault tree representing PS failure is now complete. This fault tree is, therefore, in a proper form for evaluation of the top event. The fault tree branches representing circuit breaker failure and protective system failure can be combined to give the complete fault tree representing the failure to clear a fault on the transmission line in the allotted time.

27.4.3 Determination of Minimal Cut Sets

Now that the complete fault tree for transmission protection is known, it will be evaluated to determine the minimal cut sets for the system. This can be done following the procedure of Section 25.3. For complex fault trees this process is tedious and prone to human error. Therefore, we shall determine the minimal cut set of the case under study using a computer program.

There are several computer programs that are available for fault tree analysis. We shall use the program SETS, which is available commercially [7]. The program permits entering the representation of the fault tree using a personal computer and, once the data is entered, provides several editing, quantification, and reporting functions that are generally useful in fault tree analysis. For the problem at hand only the cut sets will be determined as a function of the basic event alphanumeric tags.

The analysis of the transmission protection fault trees presented in the previous sections can be used to determine the number of minumum cut sets for the system. These results are shown in Table 27.7.

TABLE 27.7 Minimum Cut Set Results of Fault Tree Analysis*

Gate Name	A5	A6	A7	PS	BF	TOP
No of min. cut sets	12,458	12,458	42,384,802	42,409,698	24	42,409,722

* The author is indebted to Richard Worrell of Logic Analysts, Inc. for computing the number of cut sets in the fault tree using the SETS computer program.

It is apparent, from viewing the results of the analysis, that a great many minimum cutsets are created in the analysis of a problem of this size, and that it would be almost hopeless to perform the evaluation by hand computation. The results for the breaker failure portion of the tree, shown as BF in the table, are not difficult to evaluate because there are 24 basic events and most of the gates are OR gates. However, there is no need for hand calculation or counting, since computer programs are available for this type of tedious evaluation.

The protective system, labeled PS in the table, is much more complex. The PS fault tree contains nine different AND gates. This greatly expands the number of cutsets and minimum cut sets. The event PS occurs when A5, A6, AND A7 occur. These AND gates cause a great many multiple outage events that lead to the TOP event of the system. In order to get some understanding of the A5, A6, and A7 branches, they were evaluated separately with the results shown in Table 27.7. The A5 and A6 results are identical since the structure of these trees are identical. Most of the minimum cut sets occur in gate A7, where many of the gates are AND gates.

Table 27.8 shows the order of minimum cut sets in the fault tree. The BF branch consists of only first- and second-order failure events, and these can be readily enumerated by examining the fault tree. The PS branch has a large number of higher order failures because of the AND gates. These are failures where multiple order overlapping failures of basic events result in system failure. Note that the highest order of failure in both A5 and A6 is only sixth order, but branch A7 has higher order events, up to the 10th order. The futility of hand computation for such a complex system is evident from these quantifications of minimum cut set numbers.

No numerical data is used in this example of fault tree analysis, but if data were to be provided for all basic events, then numerical results could also be determined. Our motivation

TABLE 27.8 Summary of Fault Tree Analytical Results

Event Order	TOP Event	BF Subtree	PS Subtree	A5 Branch	A6 Branch	A7 Branch
1	10	10	0	0	0	0
2	24	14	10	0	0	10
3	648	0	648	138	138	392
4	49,974	0	49,974	10,962	10,962	28,050
5	4,130,820	0	4,130,820	1,320	1,320	4,128,180
6	28,943,550	0	28,943,550	38	38	28,943,474
7	8,332,940	0	8,332,940	0	0	8,332,940
8	907,272	0	907,272	0	0	907,272
9	43,700	0	43,700	0	0	43,700
10	784	0	784	0	0	784

here is to describe the method of fault tree construction for a typical complex protective system, and this objective can be accomplished without numerical computation. In practical problems, the numerical results may be of great importance, especially if comparisons are to be made of different hardware systems used to make up the integrated line protective system. The manufacturers of the various components can provide failure rates for their equipment and the maintenance section of the utility can provide practical estimate of repair times. Used together with the fault tree, all of the measures described in Section 27.2 can be computed.

It is also important to determine the importance measures and especially the vulnerability of the top event to specific types of failure. Work has also been performed to provide strategies for top event prevention. This concept is important as it can provide useful guidance as to the basic events that have the greatest influence on the top event, and investing in more reliable components for these basic events can be given high priority.

27.4.4 Constant Failure Rate-Special Cases

In many cases it is possible to approximate the failure rate of most power system components as having constant failure rates, leading to an exponential failure distribution. This makes it easy to compute the component availability and unavailability, assuming a Markov model for each component. This process has been described in Section 25.5.

One of the strengths of the fault tree method is that the individual components need not all have the same type of mathematical model as long as the probability of the failure event can be computed by some process. We limit our discussion to repairable components, but within that limited classification there are different ways of modeling the unavailability, and these differences are extremely important in protective systems. Two different models are often used that are applicable to the protective system problem. These two cases are: (1) equipment is monitored, such that failures are always announced failures; and (2) failures are not detectable until a periodic inspection or test is performed, or when use of the system is demanded.

Monitored components always provide a means of sending the operator an alarm or signal of some kind to announce that the failure has occurred. In this case the operator will take appropriate action to have the component inspected and, if necessary, replaced or repaired. This process always leads to prompt detection and repair of components. For the power system, this will be the case for most of the power generating and transmission components, at least at the higher voltage levels, where modern control centers monitor all lines and generators by sophisticated on-line computer systems. Some subtransmission and distribution circuits may not be continuously monitored, but this situation varies from one utility to another and will often vary within any given utility since some circuits serve more essential load than others.

Some of the key components in the protective system are not monitored at all, for example, many of the relays and associated detection and control equipment. The circuit breakers are monitored, at least at the higher voltages, but the ability of the breaker to operate is often unknown. The situation may arise where a given breaker has not operated in quite a long time, and its ability to perform is uncertain. Even if the breaker is capable of interrupting normal load currents, this is no assurance that fault current interruption will be successful. We must conclude, therefore, that both the circuit breakers and the protective system will suffer unannounced failures. Such failures will not be detected until the next scheduled inspection for many types of protection equipment. However, modern digital protective devices can be considered continuously monitored.

27.4.4.1 Monitored Systems. For monitored systems, each component is modeled by constant failure and repair rates, resulting in the familiar exponential unavailability distribution. If we define the failure rate as λ and the repair rate as μ, we can write the unavailability of the component from (22.69).

$$U(t) = \frac{\lambda}{\lambda + \mu} - \frac{\lambda e^{-(\lambda+\mu)t}}{\lambda + \mu} \tag{27.112}$$

In most practical cases, this function quickly reaches the asymptotic value, which is often written as follows:

$$U(t) = \frac{\lambda}{\lambda + \mu} = \frac{\lambda T_D}{1 + \lambda T_D} \tag{27.113}$$

where T_D = mean down time = MTTR

In many cases, this equation can be conservatively approximated by

$$U(t) \approx \lambda T_D \tag{27.114}$$

because the product λT_D is small compared to unity. This approximation is within 10% if the product (27.114) is less than 0.1.

For power system protection reliability evaluation, the monitored components are usually the transmission lines and generators or other important items of system equipment. In some cases, the circuit breakers may be considered to be monitored, depending on the utility practice.

27.4.4.2 Periodically Test Systems. For components that are inspected and tested periodically, a different model is required that includes the time between inspections. If we let T be the time between periodic tests, the unavailability is zero (item 100% available) immediately after an inspection, but it increases steadily to the next scheduled test.

$$U(T) = 1 - e^{-\lambda T} \approx \lambda T \tag{27.115}$$

If we approximate the exponential by the linear function, then we can say that the average unavailability between tests is given by

$$U_{\text{ave}} = \frac{\lambda T}{2}, \qquad 0 < t < T \tag{27.116}$$

This average value is applicable to fault tree evaluation if we assume that the demand on the component occurs uniformly at any time during the interval between inspections [1].

When an inspection shows that a component has failed, then that component will remain unavailable until its repair can be completed. This extends the down time of the inspection and repair period and permits us to estimate the total average unavailability as follows.

$$U(t) = \frac{\lambda T}{2} + \lambda T_R \tag{27.117}$$

where T_R = mean repair time

Note that the units must be consistent. If time is given in hours, then the failure rate must be given in failures per hour. Note that the repair time is not the same as the total downtime, which includes both repair plus the undetected downtime from the time of failure to the time of detection. Usually the repair time is small compared to T, in which case the unavailability may be estimated by (27.116).

For repairable components, the unconditional failure intensity $w(t)$ is a complex function of time as noted in (27.4). However, this function approaches λ as time elapses and the constant failure rate is often considered accurate enough for fault tree evaluation. Thus, we often see the approximation

$$w(t) \cong \lambda \tag{27.118}$$

and this applies to both repairable and nonrepairable components [1].

REFERENCES

[1] Roberts, N. H., W. E. Vesely, D. F. Haasl, and F. F. Goldberg, *Fault Tree Handbook*, NUREG-0492, U. S. Nuclear Regulatory Commission, Washington, DC, 1981.

[2] Kumamoto, H,. and E. J. Henley, *Probabilistic Risk Assessment and Management for Engineers and Scientists*, IEEE Press, Piscataway, NJ, 1996.

[3] McCormick, N. J., *Reliability and Risk Analysis; Methods and Nuclear Power Applications*, Academic Press, Inc., New York, 1981.

[4] Fussell, J., "How to Hand-Calculate System Reliability and Safety Characteristics," *IEEE Trans.*, R-24 (3), 1975.

[5] Nelson, F. R., "Power Systems Communications – Microwave Radio, Mobile Radio, and Tone Multiplex Systems," a contribution to the *Standard Handbook for Electrical Engineers*, 10th edition, Donald G. Fink, Editor-in-Chief, McGraw-Hill Book Company, New York, 1975.

[6] Endrenyi, J., *Reliability Modeling in Electric Power Systems*, John Wiley & Sons, Inc., Ltd., London.

[7] Worrell, R. E., "SETS Reference Manual," Report SAND83-2675, Sandia Laboratories, Albuquerque, NM, May 1985.

PROBLEMS

27.1 Determine the Laplace Transforms of unconditional failure intensities given by (27.4) and (27.5).

27.2 Determine the failure and repair density functions, $f(t)$ and $g(t)$ for the case where the failure and repair rates are constants, expressed by the Greek letters λ and μ.

27.3 Find the simultaneous solution of the Laplace equations from problem 27.1, under the conditions of constant failure and repair rates found in problem 27.2.

27.4 Compute the expected number of failures over the interval from $(0, t)$ for the system described by problem 27.3.

27.5 Computed the expected number of repairs over the interval from $(0, t)$ for the system described by problem 27.3.

27.6 Determine the unavailability and the availability of the system described by problem 27.3.

27.7 From Table 24.4, (11), we are given an expression for the conditional failure intensity $\lambda(t)$ in terms of the unconditional failure intensity $w(t)$ and the unavailability $U(t)$. Check this for the case where the failure rate is constant, using the results of problems 27.3 and 27.6.

27.8 Determine the stationary or limiting state probabilities of the system described in problems 27.6 and 27.7. Then construct a Markov graph of the system.

27.9 Consider a Type 1 component, as defined by Figure 27.2, with a failure rate of 0.001 failures/hr. Determine the following reliability parameters as a function of time:

$$R(t),\ F(t),\ f(t),\ U(t),\ A(t),\ w(t),\ v(t),\ W(0,t),\ V(0,t)$$

Do this for times $t = 10$, 100, 1000, and 10,000 hours using a spreadsheet, under the following assumptions:

(a) The component is not repairable.

(b) The component is repairable with MTTR = 10 hr.

27.10 Repeat problem 27.9 for a component having a failure rate of 0.002 failures/hr and an MTTR of 30 hr.

27.11 Repeat problem 27.9 for a component having a failure rate of 0.003 failures/hr and an MTTR of 60 hr.

27.12 Compute the cut set parameters for a three-component series system, where the three components are those from problems 27.9, 27.10, and 27.11.

27.13 Determine the cut set parameters for a two-component parallel system consisting of components 1 and 2 from problems 27.9 and 27.10.

27.14 Determine the cut set parameters for a three-component parallel system consisting of components 1, 2, and 3 from problems 27.9, 27.10, and 27.11, respectively.

27.15 A series system consists of the nonrepairable components described in problems 27.9 and 27.10. Compute the following for the system:

(a) The upper bound for the unavailability at $t = 100$ hr.

(b) The lower bound for the unavailability at $t = 100$ hr.

27.16 Extend problem 27.15 in two directions:

(a) First, repeat the calculation of upper and lower bounds of the series system, assuming nonrepairable components, but calculate for $t = 10$ hr and 1000 hr.

(b) Next, repeat the same calculations for the same components, but assume that they are repairable.

(c) Tabulate the results and comment on any trend observed.

27.17 Let basic events be represented by the capital letters $B_1, B_2, \ldots, B_n$, and their existence at time t. Moreover, let the probability of the event B, given by $\Pr(B)$, be the unavailability of the basic event $U(t)$. Now, a system is described by n basic events connected to a single AND gate. Define the top event as the unavailability of n basic events $B_1, B_2, \ldots, B_n$ that are defined as being independent. Write the expression for the unavailability of the system.

27.18 The n basic events described in the previous problem are connected to a single OR gate. The top event exists at time t if and only if at least one of the basic events occurs at time t. Determine the system unavailability.

27.19 Consider the fault tree shown in Figure P27.19, where the gates are identified by number and the basic events by letter. The top event is T.

(a) Write out the expression for the top events in terms of the basic events, with the understanding that the letter designation is equal to the probability of the basic event.

(b) Compute the probability of the top event if each basic event has a probability of occurrence of 0.1.

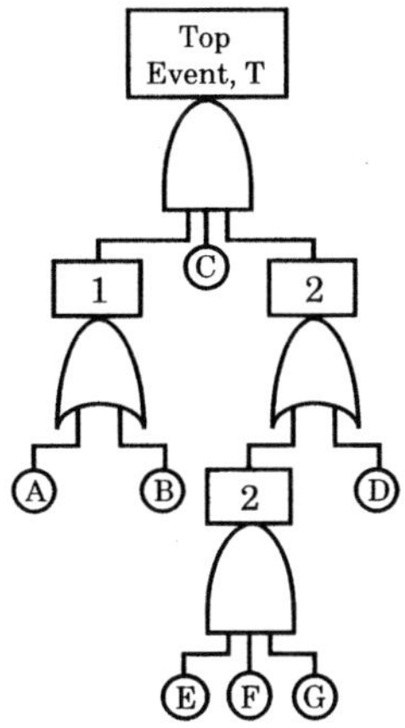

Figure P27.19 A fault tree.

27.20 Repeat problem 27.9 if the basic event D is identical to basic event F.

27.21 Consider the small network shown in Figure P27.21 [6]. Nodes B and C are supplied from A, and A is the only source of supply. The modes of failure of the power system are defined as follows:

1. The system fails if either B or C are not supplied.
2. The system fails, due to line overload, if all the load at B and C must be supplied by only one line.

Assuming the only failures of interest are line failures, enumerate the state space of successes and failure states.

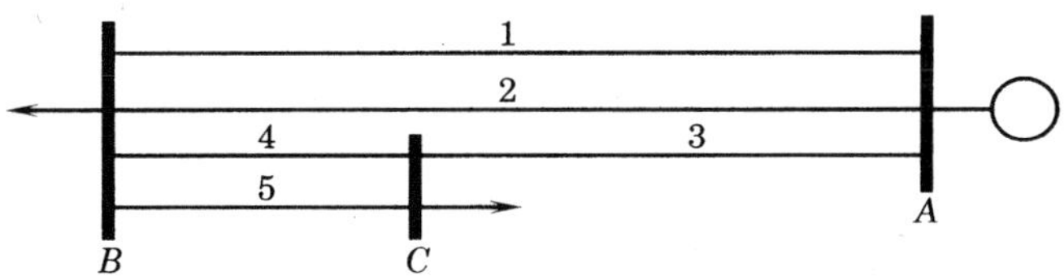

Figure P27.21 A small segment of a transmission system.

27.22 Construct a fault tree for the system of Problem 27.21.

27.23 (a) Evaluate the top event for the fault tree shown in Figure P27.23. Note that the event D is repeated. This suggests that the fault tree can be simplified.
(b) Determine the simplified construction of this fault tree.

27.24 Write out the solution to problem 27.15 for the case where there are only two basic events. Illustrate this result using a Venn diagram.

27.25 A 2-Out-of-3:G system consists of three basic components, B_1, B_2, B_3. Determine the minimal cut sets for this system.

27.26 Write out the solution to problem 27.15 for the case where there are three basic events and illustrate your solution using a Venn diagram.

27.27 Consider an m-out-of-n: gate where the top event is the failure of the system, which consists of basic events $\Pr(B_1) = \Pr(B_2) = \ldots = \Pr(B_n) \equiv Q$ or all basic events are identical. Find the unavailability of top event.

27.28 Compare the computation of unavailability using fault trees and reliability block diagrams. Make the comparison for a series system of two basic components and for a parallel system of two basic components. Hint: note that the *RBD* is a success diagram, whereas the fault tree is a failure diagram.

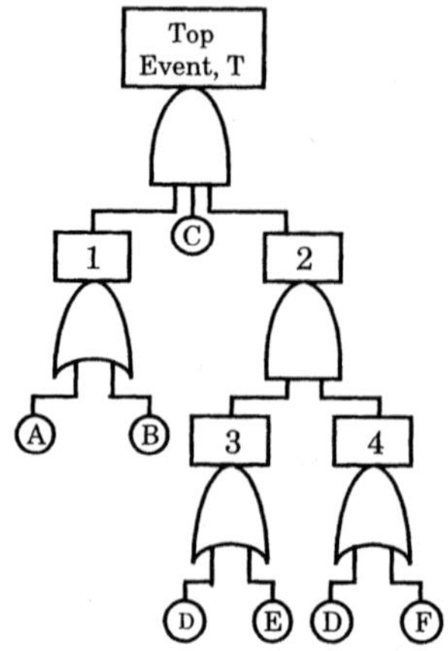

Figure P27.23 A fault tree.

27.29 Figure 27.14 gives a fault tree branch that describes failures associated with relay contacts. A relay contact primary failure is the basic event, which means that the relay contacts fail to close when operating in an environment for which the relay is qualified. Two types of secondary faults are also described: (1) a relay contact secondary failure, and (2) a relay contact command fault. Describe these secondary faults more fully.

27.30 Verify all of the entries in Table 27.2.

27.31 Verify equation (27.81).

27.32 Verify equations (27.94) and (27.98).

28

Markov Modeling of Protective Systems

28.1 INTRODUCTION

The mathematical model of protective systems must consider the system being protected as well as the viability of the protective equipment to work as designed. As time moves forward, the probability of the protected component failing will increase or, in other words, the protected system availability gradually falls from its initial value of unity. This same trend occurs for the protective system itself, and it may suffer a complete failure so that, when the protected component fails, the protection does not operate. Since failures of the protection are often unannounced, it may not be possible to know if a failure has occurred unless an inspection of the protection is conducted. Usually, such inspections are made periodically at predetermined intervals. To simplify the notation between *protected system* and *protective system*, we shall henceforth refer to the protected system as the *plant*, or the protected *component* and the protective system as simply the *protection*.

One way of characterizing the entire system is to compute the expected number of failures of that system, including the protection, over a specified period. The computation is based on the "demand rate" on the protection, which is the frequency at which the plant is expected to fail [1]. Another factor is the availability of the protection to perform its function, which is sometimes called the "fractional dead time" [2]. Then the expected number of failures is computed as

$$W = kD \tag{28.1}$$

where k = fractional dead time
D = demand rate on the protection

The fractional dead time, k, is literally the fraction of the total time during which the protection is in a failed condition. This can be considered the long-term expected value of the protection being unavailable. The demand rate is the rate at which failures occur on the plant, which would require protective action. For example, if the demand rate is given in occurrences per year then the expected number of failures W will be in failures per year. The demand rate should apply only for those types of failures to which the protection is designed to respond.

It should be noted that the protection has a failure rate λ, but the protection can not be linked to the plant by a simple AND gate because the fractional dead time is a function not only of λ, but also of the time at which the demand occurs. Protective systems are usually inspected periodically, at fixed intervals T. Therefore, the fractional dead time is a function of both λ and T, as well as the rate of restoration of the protection following inspection. We usually assume that the protection is restored to as good as new condition at each inspection.

As a general rule, we can make the following assumptions regarding the mathematical modeling of protective systems.

(a) An exponential distribution is usually applicable to the protective system components, so that constant hazard rates can be used.

(b) The Markov process is an applicable foundation for modeling.

(c) The plant is defined by a *protection zone*, shown in Figure 26.12.

Before applying these basic assumptions to protective system models, some observations regarding the testing of systems are presented.

28.2 TESTING OF PROTECTIVE SYSTEMS

Protective systems are unique, in comparison with other types of power system equipment, in that they normally execute no control actions. The decision-making logical elements of the protection are, of course, continually active, but the output is usually restrained from action. It is only on those rare occasions, when faults occur within the protection zone, that the protection should respond, and it should do so dependably and with high reliability. Unfortunately, there is always the possibility that the protective system will already be in a failed condition when the fault occurs. This suggests that frequent testing of the protective system would be beneficial.

Consider a protected power system component that experiences periods of being up, or working, and periods when it is down, or failed, as shown in Figure 28.1, where the record

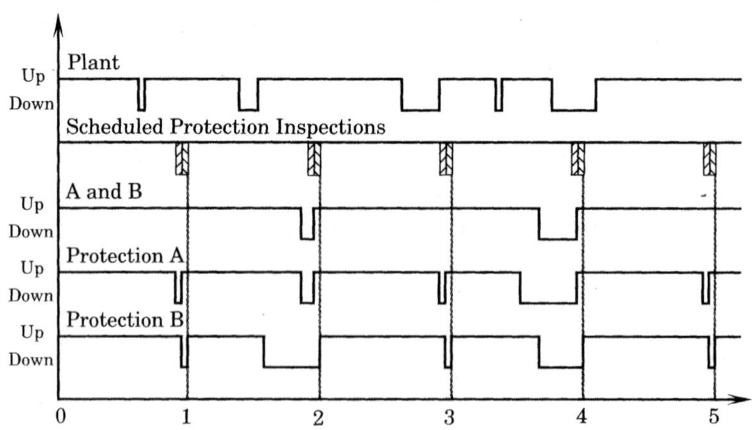

Figure 28.1 Time line of up and down periods for a protected element and the associated redundant protective systems A and B.

of these fluctuations is shown as the top trace. The protective system associated with this component is scheduled for inspection and testing at regular intervals, which are labeled simply 1, 2, ... in the figure. At each scheduled inspection, both protective systems A and B are inspected in a staggered manner. The figure shows A being inspected first (///), followed immediately by the inspection of B (\\\). Since the inspections are staggered, the plant is usually ensured of having adequate protection. Just prior to inspection epoch 2, however, the protective systems fail sequentially, so there exists a brief period when both A and B are down, which is labeled "A and B" in the figure.

There is no system failure, however, since there is no failure of the plant during this overlapping protection failure. Now examine the condition just prior to inspection epoch #4. In this case the plant fails during an overlapping protective system failure. This causes the plant fault to be cleared by backup protection, which results in a more severe disruption to the power system. Since the protective systems are assumed to be independent, the probability of such an overlapping failure is given as

$$\Pr\{A \cap B \text{ down}\} = \left[\left(\frac{\lambda}{\lambda+\mu}\right) - \left(\frac{\lambda}{\lambda+\mu}\right) e^{-(\lambda+\mu)t}\right]^2 \tag{28.2}$$

where the equation assumes that the protective systems are identical, and have failure rate λ and repair rate μ.

The system is in a somewhat exposed condition during the testing of protective systems. Consider a system shown in Figure 28.1. Clearly, only one of the protections is undergoing inspection at any given time. However, during the tests of either A or B, the protective system is in a precarious condition, since the system is without its designed redundancy. In fact, during this brief period, the probability of failure is given by

$$\Pr\{\text{failure}\} = \left(\frac{\lambda}{\lambda+\mu}\right) - \left(\frac{\lambda}{\lambda+\mu}\right) e^{-(\lambda+\mu)t} \tag{28.3}$$

This probability is much greater than that of (28.2).

Testing is also subject to human error due to test personnel manipulating the protective equipment, which is itself a hazard since workers can cause unwanted action that is difficult to predict [3]. Worker error may leave the system in a nonoperable condition due to some flaw in their testing procedure or instructions, due to a human misdeed, judgmental error, bungling, or just a mistake in carrying out the task. As a result of human and other testing hazards, the failure rate during testing is considerably greater than normal. This suggests that it may not be beneficial to test often, as was previously concluded.

Another factor that must be considered in power system protection is the effect of the weather on the protection demand rate. It is well known that transmission lines have a higher failure rate during foul weather. For this reason, testing of the protection of transmission lines is usually postponed if the weather is bad.

It is postulated that there should be an optimum frequency for performing tests of protective equipment and that the tests themselves should conform to an optimum test duration. These parameters, frequency and duration of testing, are related to the test methodology and may be computed by application of reliability theory. It is conceivable that a testing scheme can be developed and calculations can be performed so that the testing can be planned in the most intelligent manner possible.

It should also be recognized that many protective system designs are based on digital technology, and enjoy the advantages of stored program devices. This suggests that tests could conceivably be performed often and that the tests themselves could be very brief. This may require that the testing be automatically performed in order to accommodate a requirement for both high frequency and short duration. We address this possibility and discuss a mechanism for designing tests, whether they be manual or automatic, or some combination of the two.

28.2.1 The Need For Testing

It should not be necessary to defend the need for testing of a protective system, since the failure of the protection may have serious consequences. Briefly, we can state the purpose of testing as follows:

- To ensure that protective system reliability goals are maintained.
- To reduce the exposure time due to failed, but undetected, protection components.

We assume that any protective system has been designed to comply with certain reliability criteria. For example, a typical criterion may require that the protective system not fail with a frequency of more than once in n years. The only way to make sure that this performance is maintained is to test the system regularly. This procedure serves two purposes. First, it ensures that any failed components are replaced or repaired. Second, it provides one method of computing the probability of protective system failure corresponding to a particular testing schedule, assuming that this schedule is constant. If the protective system is found to have a high probability of failure, then the interval between inspections is obviously too great, and appropriate measures can be taken to correct the schedule. These tests must be regarded as "as found" tests, since all components continue to age and to draw nearer to failure each passing day. Thus, testing may not restore the system to "as good as new" condition, but it does ensure that the system is still operational at a particular time. Also, with proper replacement of any worn mechanical parts, the system can be maintained in nearly as good as new condition since electronic components in the protection maintain a nearly constant failure rate.

To examine the testing situation more carefully, consider the time line shown in Figure 28.2. The figure depicts the time line between the kth test and the $(k-1)$st test, where the scheduled time between inspection visits is defined as T. We define the following quantities.

$$\begin{aligned} T_u &= \text{duration of undetected failures} \\ T &= \text{time between inspections} \\ T_F &= \text{time to failure (a random variable)} \\ t &= \text{the time variable} \end{aligned} \tag{28.4}$$

Since failures are random, the time to failure is a random variable and we may write the probability of failure in terms of the failure distribution.

$$P\{T_F \le t\} = F_{T_F}(t) = \text{the failure distribution} \tag{28.5}$$

If we assume an exponential distribution, which would be common for electrical or electronic equipment, we have the following distribution.

$$F_{T_F}(t) = \begin{cases} 1 - e^{-\lambda t} & t > 0 \\ 0 & t < 0 \end{cases} \tag{28.6}$$

where it assumed that the equipment is placed in service at time zero. The mean time to failure is the expected value

$$E(t) = m = \frac{1}{\lambda} \tag{28.7}$$

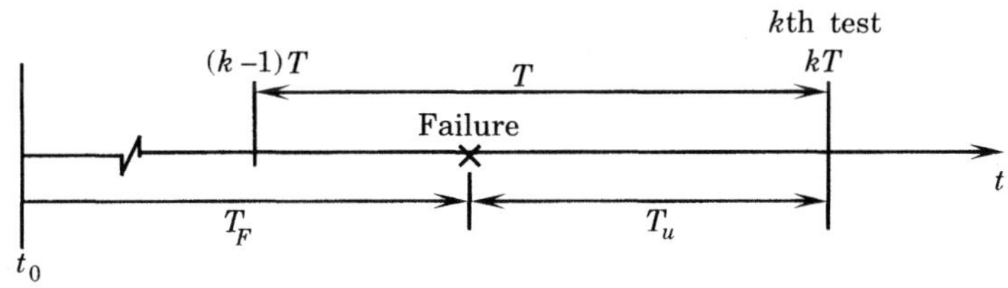

Figure 28.2 Time line of testing and protection failure.

where the expected value is the mean time to failure, m. The duration of undetected failure is of particular interest, since the power system is in some jeopardy during this time and it may be a long time to the next scheduled inspection visit. This time is computed as follows.

$$\begin{aligned} T_u &= kT - T_F \qquad k = 1, 2, \ldots \\ 0 &\le T_u \le T \end{aligned} \tag{28.8}$$

The time duration of undetected failures, T_u, is also a random variable since the time to failure is random. However, this duration is bounded by the time between inspection visits.

Let us assume the following:

1. Inspection is periodic with period T.
2. As a result of inspection, the protective system is restored to as good as new condition at each inspection visit.

This means that we can effectively reset the clock to zero following each inspection visit, and the time line simplifies to that shown in Figure 28.3. Under these conditions, we can rewrite the repair-to-failure distribution as follows from (24.79).

$$\begin{aligned} \Pr\{T_F \le T\} &= F_{T_F}(T) \\ &= \Pr\{\text{failure occurs before } T\} \\ &= 1 - e^{-\lambda T} \cong \lambda T \end{aligned} \tag{28.9}$$

where the final approximation is true if $\lambda T \ll 1$.

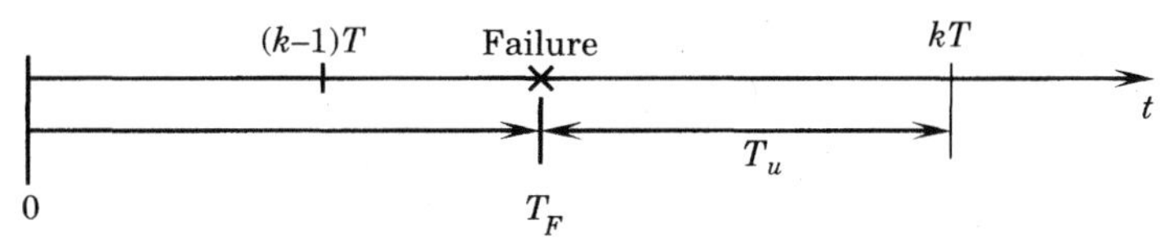

Figure 28.3 Failure time line with clock reset to zero.

The probability that a failure does not occur is called the reliability and is computed for the exponential distributed case as

$$R(t) = e^{-\lambda t} \tag{28.10}$$

The duration of undetected failures is given by

$$T_u = kT - T_F \tag{28.11}$$

We can bound T_u by reducing the time between inspection visits. For example, if T is 1 year, we could have very long periods of undetected failure. Since failures are random, the time of undetected failures could be as large as 1 year, even though the mean time to failure is much larger, usually many years.

For example, if the protective system failure rate is one failure per million hours, this equates to 0.00876 failures per year with a mean time to failure of 114.15 years. Annual visits are, on the average, quite frequent compared to the mean time to failure. However, if the system is defective and the failure rate is one failure per thousand hours, this equates to 8.76 failures per year, or about 0.7 failures per month. This results in a failure every 1.37 months, which would indicate that annual inspection visits are not adequate.

When a protective system fails, one of the following events is likely to occur:

(a) Detection of the failure does not occur until the next scheduled inspection visit because no fault occurs during the intervening period that would require operation of the protection.

(b) Detection of the failure occurs prior to the next scheduled inspection visit due to:
  (i) Catastrophic failure due to a fault and, with the protective system failed, requiring backup protection operation. This is followed by an investigation, which identifies the failure.
  (ii) Routine station inspection (daily, weekly, etc.) shows that, even though faults have occurred, no relay targets appear on the failed equipment.
  (iii) Protection self-testing, which is available on some modern devices, and is able to alarm the fact that a protective test has failed.

(c) Repair is initiated to fix or replace the failed equipment.
  - Normal mean time to repair is the elapsed repair time
  - The clock is reset to zero on this protective system

During periods of good weather, faults may be infrequent and condition (a) will likely occur, but there is no guarantee that this will be the case. It is always possible that a fault, even a momentary fault, will occur and the failed protective system will not work. This emphasizes the importance of checking the relay targets regularly to ensure that all faults are observed by all of the proper relay systems. The exposure to undetected failures depends in some degree on the inspection practice of the utility.

28.2.2 Reliability Modeling of Inspection

Research has been performed to optimize the availability of a protective system, including the routine testing of the system, in order to identify and correct any hidden failures. One of the early reports, by Jacobs [3], shows that it is possible to find an optimum interval between inspections depending only on the failure rate of the protective system and the time required to complete the inspection. That study assumes that the units being inspected are unavailable to the system during the period of inspection. Others [4–10] have elaborated on this basic idea.

Another aspect of the problem is the management of maintenance of the protective system. This has also been addressed by several researchers [5], [7], [11–17] who have added significantly to the literature of this area. Veseley and Goldberg [11] describe a computer program that is used to predict the time-dependent system unavailability, including the effect of different testing schemes. Kontoleon [7], [12], [14], [18], [20] has investigated the use of computer techniques to optimize the availability for a given k-out-of-n:G system, for example, to determine the optimal order of supervision for a given test interval. He was one of the first to introduce the terminology of Fail Safe (FS) and Fail Dangerous (FD) modes of failure for a protective system, and has produced computer algorithms to demonstrate the optimization techniques.

A group of Japanese authors, principally Inagaki, Inoue, Kohda, Kumamoto, and Takami have contributed generously to the technical literature dealing with inspection scheduling of protective system consisting of multiple units with staggered inspection schedules [15], [19], [21–29]. They have shown, for example, that it is possible to improve the availability by staggered testing, dividing the units into subgroups, with staggered subgroup inspections, where the entire process can be optimized using mathematical programming techniques. They also introduced the concept of cost effectiveness as an integral part of the optimization process.

Much of the work described above has been performed to improve and optimize the performance of protective systems for nuclear power plants, where the cost of failure is great and where rather complex multiple unit protective systems are required. The concepts, however, are applicable to any type of protective system and are recommended for study by any engineer charged with improving the performance of a protective system.

Another group of technical papers have addressed the reliability of systems used for protection of transmission, distribution, and conventional power plant apparatus. Singh and Patton [30] present a unique Markov model for a transmission protective system that takes into account the fact that a failed unit may go undetected until the next inspection. They refer to this as the "unreadiness" probability and present a method of computing this quantity.[1] A contribution, by Yip, et al. [31], presents an analysis of a particular distance protection system, and shows how the availability of this system is improved by monitoring and self-checking of the equipment. These techniques are relatively new, but are considered important in the long term, especially as digital protective systems become more common. These concepts have also been discussed by CIGRÉ working groups, who have published summary papers on the subject [32], [33].

It is not possible to present all of these concepts here, but some of the basic ideas are presented in the following sections. The interested reader is encouraged to examine the cited literature for additional information. The subject is very important in many types of protective systems, and the practicing protection engineer can benefit from the study and application of these concepts.

28.3 MODELING OF INSPECTED SYSTEMS

This section presents examples of probabilistic models for protective systems, all of which include the concept that the protective units must be inspected periodically. There are many ways of doing the inspections. For example, consider a protective system that consists of several units that work either independently, or in some logical interconnection, in order to achieve a certain protection objective. The several units could all be tested simultaneously, although this may leave the protected plant exposed. The unit inspections, therefore, are usually staggered in some manner. The scheduling may be simple if the units are identical and have identical failure distributions, but the best schedule is not at all obvious for the general case. It is also recognized that some types of protective units can be inspected while the protection is in operation, although there are obvious hazards in this practice. However, if the units are to be disconnected for inspection, what is the best order of performing the task? If the inspection takes a given amount of time for each type of unit, what is the optimum period between inspections of that unit? Moreover, failure of the protective systems are costly, with some types of failure costing more than others. How can the cost of the various options be taken into account? Some of these questions will be addressed in this section.

[1] Unreadiness probability is examined further in Section 28.5.

28.3.1 Optimal Inspection Interval

For any given protective system it is possible to optimize the system availability by making the proper selection of the inspection interval corresponding to a given inspection duration. As with any analytical procedure, it is necessary to first state the assumptions of the study. The following assumptions are appropriate for the problem at hand.

1. The protective system is composed of individual protective units, which are usually arranged in a redundant configuration. The total number of units depends on the particular system and the protective system design, but the units are not necessarily identical.
2. At the beginning of the study period, all units are assumed to be good, and during the inspection all failed units are assumed to be repaired or replaced by new units.
3. Protective units are independent and are assumed to be either good or failed.
4. Protective unit failures can be described by exponential distributions, such that constant failure rates are known for each unit.
5. Inspection is perfect such that failures are detected without affecting the power system in any way.
6. The inspection duration is short compared to the interval between inspections, and this duration can sometimes be neglected, if necessary, without introducing serious error.
7. All protective units of a given component are inspected and repaired, if necessary, at each inspection visit. The order in which the unlike units are inspected must be specified.
8. Testing is not permitted during periods of high component risk, such as foul weather, which may slightly modify the test schedule.

Assumption (1) describes the protective system in general terms, but in all studies this description must be made more precise. Most protective systems are redundant, often with unlike protective units. This system configuration must be made clear prior to any analytical study. Assumption (2) states that the system is operating perfectly at the start of the study period. Assumption (3) is usually valid, but may not be true if there are common cause failures that can fail all redundant units. Assumption (4) is true for most electronic equipment and may be a reasonable approximation for many types of protective devices. Other distributions can be specified, however, if the failure distributions are known, although this will probably make the analysis more complex. Assumption (5) is not always true, and special studies may be made to determine the effect of inspection error. Assumption (6) may be desirable for some studies, but is not necessary or desirable for others. However, it is important to recognize at the outset whether operator error is to be taken into account. Finally, assumption (7) is a statement that will usually be true. The order in which the units are to be inspected is sometimes important. Where the units are not identical, it is important to specify the exact order of inspection of all units. Assumption (8) states a practical limitation, but this has little effect on the analysis as the disruption to the test schedule is not great.

The concept of optimizing the interval between tests can be described more clearly by computing the average availability of the protective system over time, including the fact that the protective unit is unavailable during the test period [4]. Consider a protective system inspection that removes the entire system for simultaneous inspection. Figure 28.4 shows two examples of the availability of the protective system plotted over multiple test intervals, each

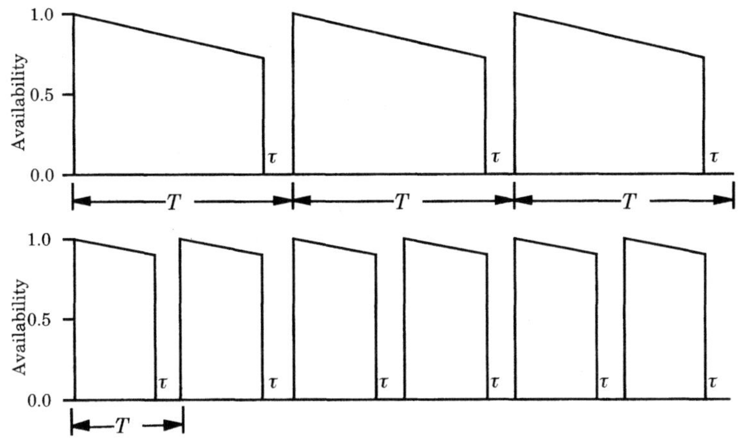

Figure 28.4 Comparison of long and short inspection intervals.

of length T. The example in the top of the figure shows three complete test intervals, each of which consists of an operating period, during which the availability declines exponentially, followed by the inspection period τ, which removes the protection from service for testing and any required repair or replacement. For the entire period shown, the protection is completely unavailable for a total period of 3τ and is available with a certain probability for the rest of the time. Should the test interval T be increased, the reliability becomes lower as described by the exponential availability function.

The lower part of Figure 28.4 shows the same system, with the interval between inspections, T, much shorter than before, but with the same test duration, τ. Comparing the total period with the previous example, the lower plot shows a period of six τ when the protection is completely unavailable due to the inspection, although the availability averages to a somewhat higher value during the operating periods.

In both cases, it is assumed that the failure rate of the system is constant, the inspection time is constant, and that the protection is restored to as good as new condition following each inspection. During the operating period, the availability declines gradually until the test period begins, at which time the availability falls to zero for the duration, τ, of the inspection. If the interval between inspections should be further decreased, the total time the system has zero availability is increased and, in the limit, the protective system would be on inspection all of the time, resulting in zero availability. On the other hand, if the interval between tests is made very long at a constant τ, then the system would have a low average availability. This suggests that there should be an optimum test interval that would maximize the availability over the test interval.

Consider one cycle of the operate and inspection sequence, shown in Figure 28.4. It is assumed that the failure rate of the system is constant, with a value λ and that the availability of the system is described by an exponential function of time. The inspection is accomplished in a period of duration τ, which is defined as follows.

$$\text{Inspection duration} = \tau = d_V + d_E + r \tag{28.12}$$

where d_V = visual inspection duration
d_E = experimental testing duration
r = repair time, if required

The visual inspection duration is the time required to make an inspection of the protective device, which might reveal components that appear to have suffered damage, wear, overheating, or other obvious signs that may make replacement advisable. The experimental test duration is the time required to make mechanical, electrical, or other field tests, using appropriate instruments and recorders, to verify the performance of the protection to stimuli that are characteristic of system faults that the device is designed to detect. Finally, the repair time is the average time required to repair or replace components, or the entire protective device, if the unit fails to respond as required. Note that the total inspection time, τ, does not include logistics time, such as the time required to reach the inspection site. This time may be significant, if the site is far removed from the test personnel offices, but should not be charged against the total inspection time because the protection is assumed to be operating during logistics time. It is noted, however, that the inspection personnel will probably charge their time, including logistics, to the testing of the protection, and these time charges, if used to estimate τ, will distort the true inspection time.

Referring to Figure 28.4, we compute the average availability over the entire period T as follows:

$$A \doteq \frac{1}{T}\int_0^{T-\tau} e^{-\lambda u}\,du = \frac{1-e^{-\lambda(T-\tau)}}{\lambda T} \tag{28.13}$$

The unavailability is more interesting to plot than the availability. The unavailability is easily computed as the complement of the availability.

$$U = 1 - A = \frac{\lambda T + e^{-\lambda(T-\tau)} - 1}{\lambda T} \tag{28.14}$$

This function has a value of unity for an inspection interval, T, equal to the inspection duration, τ, and reduces gradually to a minimum as T is increased. For very large inspection intervals, the unavailability again goes to unity. Plots of the unavailability of a single protective system is shown in Figure 28.5, where all plots are made for a test duration of 10 hours, which is probably typical of the inspection time required for protective systems. As the failure rate of the protective system is reduced from one failure per thousand hours to one per million hours, the optimum time between inspections increases, as shown in the figure.

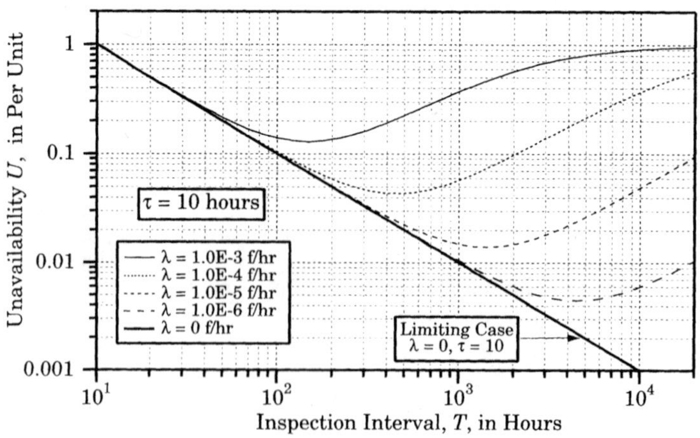

Figure 28.5 Unavailability of a protective system with $\tau = 10$ hours and for variable failure rates.

The limiting case shown in Figure 28.5 occurs for a perfectly reliable protective system, i.e., with $\lambda = 0$. The equation for this line is determined by taking the limit of (28.14) as λ approaches zero, with the following result.

$$\lim_{\lambda \to 0} U(T) = \lim_{\lambda \to 0} \frac{e^{-\lambda(T-\tau)} - 1}{\lambda T} = \lim_{\lambda \to 0} \frac{[1 - \lambda(T - \tau)] - 1}{\lambda T} = \frac{\tau}{T} \tag{28.15}$$

This function is a straight line on the log-log plot of Figure 28.5. Clearly, as the failure rate is reduced, the optimum test interval increases and the unavailability is reduced as the minimum moves downward and to the right.

Figure 28.6 shows a different parametric plot of the unavailability, where the inspection duration is the varied parameter and the failure rate is held constant at 0.0001 failures per hour (a mean time to failure of 10,000 hours). This family of plots shows clearly that shorter inspection durations result in lower unavailability, but more frequent inspections are required if the optimum unavailability is to be achieved. The lowest curve in the family corresponds to a test duration of 0.1 hours or 360 seconds. Such a short duration might be achieved through the use of digital automatic test equipment, where tests of short duration are performed frequently. Clearly, using brief and frequent tests, coupled with the use of equipment with low failure rates, can result in a protective system with a very low unavailability. This approach illustrates how the engineer can alter the test interval for a system of a given failure rate in order to optimize the performance of the system.

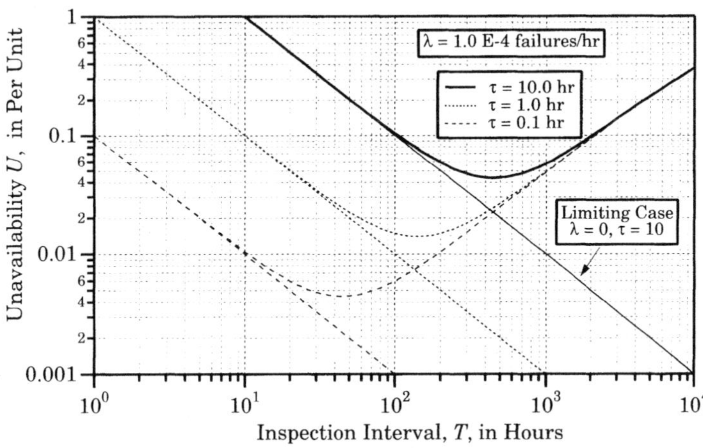

Figure 28.6 Unavailability of a protective system with $\lambda = 0.0001$ failures per hour and for variable inspection durations.

The optimum inspection interval from (28.14) can be found by differentiating that equation with respect to the inspection interval and setting the resulting derivative equal to zero. This results in the following equation for the optimum inspection interval

$$(1 + \lambda T_o)e^{-\lambda T_o} = e^{-\lambda \tau} \tag{28.16}$$

where the inspection interval has been subscripted with the letter "o" to indicate an optimum solution. Unfortunately, this equation cannot be solved explicitly for T_o. However, the equation can be solved explicitly for τ as a function of T_o. This can be done by taking the logarithm of each side of (28.16), with the following result.

$$\lambda\tau = \lambda T_o - \ln(1 + \lambda T_o) \tag{28.17}$$

or

$$\tau = T_o - \left(\frac{1}{\lambda}\right)\ln(1 + \lambda T_o) \tag{28.18}$$

Thus, the optimum inspection duration can be determined directly for a given inspection interval. We can also determine the minimum unavailability corresponding to the optimum condition by rearranging (28.16) to write

$$e^{-\lambda(T_o-\tau)} = \frac{1}{1 + \lambda T_o} \tag{28.19}$$

This result can be substituted into the unavailability equation (28.14) to give the following optimal solution.

$$U_o = \frac{1}{1 + \dfrac{1}{\lambda T_o}} \tag{28.20}$$

An approximate solution to (28.16) can be obtained by assuming that the exponents λT_o and $\lambda\tau$ are small, and replacing the exponential with the first two terms of the exponential series expansion. This results in the approximate formula

$$T_o = \sqrt{\frac{\tau}{\lambda}} \tag{28.21}$$

A comparison between the exact solution (28.16) and the approximate solution (28.21) is shown in Figure 28.7. The result (28.21) always gives a test interval that is smaller than the optimum, but it makes a good starting point if one wishes to use (28.16) to iterate toward a more accurate solution.

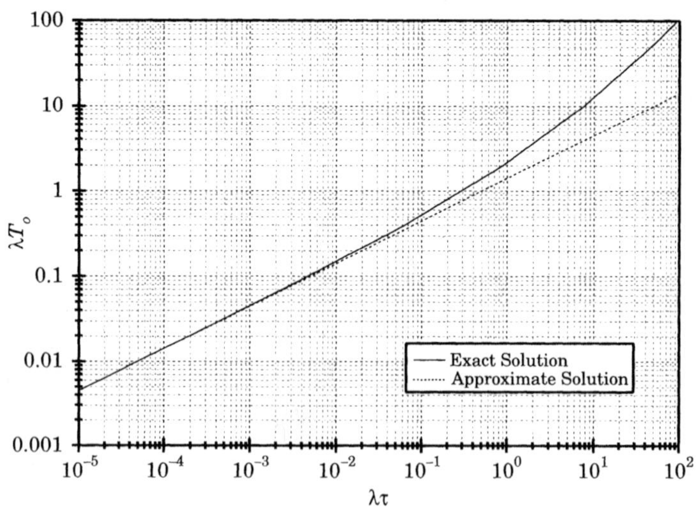

Figure 28.7 Comparison between exact and approximate solutions for T_o.

In many cases, the approximate solution is quite adequate. Usually, the inspection duration is known and, given the failure rate, λ, the quantity $\lambda\tau$ can be computed. Then the value of λT_o can be read from the curve.

EXAMPLE 28.1

Compute the optimum value of the test interval, T_o, for each of the curves shown in Figures 28.5 and 28.6.

Solution

The optimum test interval can be computed by trial and error from (28.16) for any given failure rate to find T_o and this result substituted into (28.20) to find the resulting minimum unavailability. The results for the four curves of Figure 28.6 are shown in Table 28.1, which gives the optimum inspection interval for the four different failure rates. These results can be clearly observed in Figure 28.5.

TABLE 28.1 Optimum Inspection Interval and Minimum Unavailability for Figure 28.6

λ, f/hr	τ, hr	T_o	λT_o	U_{min}
1.0E-3	10.0	148.41	0.14841	0.12899
1.0E-4	10.0	454.86	0.045486	0.04359
1.0E-5	10.0	1422.26	0.0142226	0.01402
1.0E-6	10.0	4491.76	0.0044916	0.00447

Figure 28.6 uses the inspection duration as a parameter, but with fixed failure rate. The computed optimum values of inspection interval and the resulting optimum unavailability are shown in Table 28.2.

TABLE 28.2 Optimum Inspection Interval and Minimum Unavailability for Figure 28.7

λ, hr	τ, f/hr	T_o	λT_o	U_{min}
10.0	1.0E-4	454.86	0.045586	0.04359
1.0	1.0E-4	142.59	0.014259	0.01406
0.1	1.0E-4	44.70	0.004470	0.00445

The result is interesting in that it shows that a change in *either* the failure rate or the inspection duration of two orders of magnitude results in an improvement in the unavailability of about one order of magnitude. ■

28.3.2 Optimization for Redundant Systems

The foregoing analysis is performed for a single protective system that is removed from service once every T hours for inspection and maintenance, with the total period of the inspection requiring τ hours. This shows clearly that an optimum inspection period can be computed for the protective system, based on the failure rate of the system and the duration τ of the inspection.

We now consider a more practical case where redundant protective systems are employed for the protection of a particular power system component. For simplicity, we assume that the redundant protections are identical, but the method is equally applicable if the two systems have different failure rates and different inspection durations.

The timing of the inspections are shown in Figure 28.8. The inspection interval is still T, as in the case for the nonredundant protective system. At each inspection visit for the redundant case, however, both System A and System B are inspected in a staggered manner.

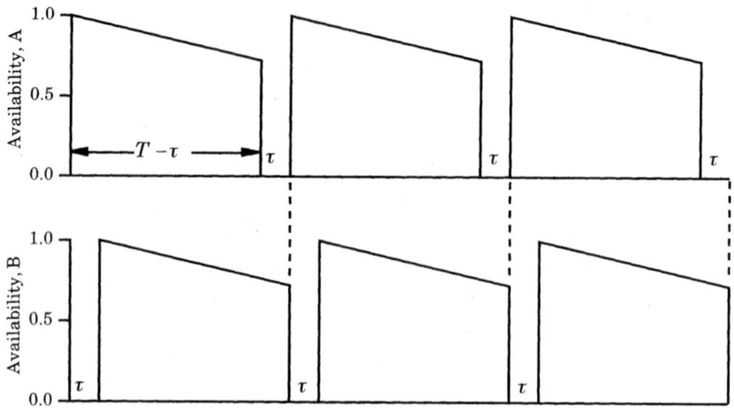

Figure 28.8 Availability of identical redundant protective systems A and B with inspection interval T and inspection duration τ.

For the case shown in Figure 28.8, System A is inspected first, immediately followed by the inspection of System B. Note carefully, however, that the in-service period for each system is still $T-\tau$ and this period is immediately followed by the inspection time τ.

For System A, we compute the average availability over the entire period as follows.

$$\Pr(\mathrm{A}) = \frac{1}{T}\int_0^{T-\tau} e^{-\lambda u}\,du = \frac{1-e^{-\lambda(T-\tau)}}{\lambda T} \tag{28.22}$$

In a similar manner, we may compute the average availability of System B.

$$\Pr(\mathrm{B}) = \frac{1}{T}\int_\tau^{T} e^{-\lambda(u-\tau)}\,du = \frac{1-e^{-\lambda(T-\tau)}}{\lambda T} \tag{28.23}$$

It should not be surprising that these two results are identical, since the period of in-service and out-of-service times are exactly the same. The average availability of the total system, then, is given by the union of the two probabilities.

$$\begin{aligned} A &= \Pr(\text{system available}) \\ &= \Pr(\mathrm{A}) + \Pr(\mathrm{B}) - \Pr(\mathrm{A})\Pr(\mathrm{B}) \end{aligned} \tag{28.24}$$

Solving (28.24) for the average availability, we compute

$$A = \frac{2}{\lambda T}(1-e^{-\lambda(T-\tau)}) - \left(\frac{1-e^{-\lambda(T-\tau)}}{\lambda T}\right)^2 \tag{28.25}$$

Finally, the unavailability is the complement of the availability.

$$U = 1 - A \tag{28.26}$$

This unavailability is plotted in Figure 28.9. Compare the scale of the y-axis with that of Figure 28.5 to see the effect of redundancy in reducing the unavailability.

For small values of the exponent, the exponential can be replaced by the first two terms of the series expansion. Thus, for very small failure rates the unreliability becomes

$$U_{\text{lim}} = \left(\frac{\tau}{T}\right)^2 \tag{28.27}$$

The comparison between this result and that of Figure 28.5 for the single protective system is instructive. First, we note that the slope of the limiting case in the redundant system falls off

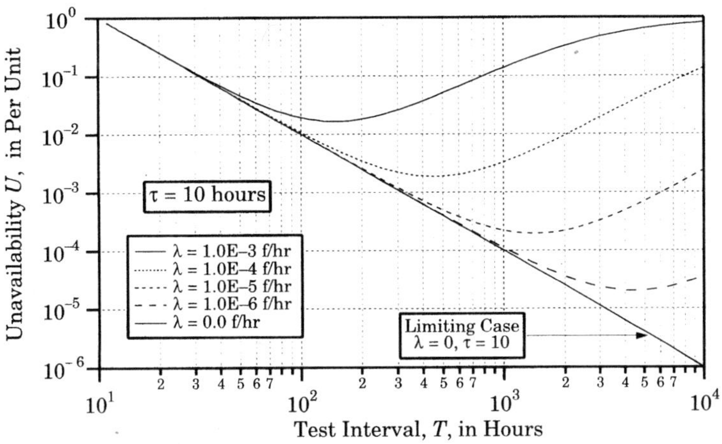

Figure 28.9 Unavailability of redundant protective systems with $t = 10$ hours and for variable failure rates.

at twice the rate, in the log-log plot, of that for the single protective system. This is because the limiting unavailability of the redundant case is the square of that for the single protection. The optimum time intervals for constant inspection duration for each value of failure rate are exactly the same as for the single system. This interesting result is explained as follows. For the single protective system, we can write the unavailability as

$$U_1 = 1 - \Pr(A) \tag{28.28}$$

and the optimum value of the unavailability is

$$\frac{dU_1}{dT} = -\frac{d\Pr(A)}{dT} = 0 \tag{28.29}$$

Similarly, for the dual protective system, we can write

$$U_2 = 1 - [2\Pr(A) - (\Pr(A))^2] = (1 - \Pr(A))^2 = U_1^2 \tag{28.30}$$

Therefore, the optimum of the redundant system may be computed as

$$\frac{dU_2}{dT} = 2U_1\frac{dU_1}{dT} = 0 \tag{28.31}$$

or

$$\frac{dU_1}{dT} = 0 \tag{28.32}$$

which is exactly the same result as (28.29). Therefore, the optimum values of unavailability for the redundant system occur at the same values of T as determined for the single protective system. The numerical values of the unavailability for the redundant system are the square of the unavailability for the single system. This results in a significant improvement in the optimum values, as noted by comparing Figure 28.9 with 28.5.

EXAMPLE 28.2

Compute the optimum values of unavailability for the redundant protective system shown in Figure 28.9.

Solution

The result can be taken directly from the values computed in Example 28.1 and squaring the unavailability. This gives the results shown in Table 28.3.

TABLE 28.3 Optimum Inspection Interval and Minimum Unavailability for the Redundant System of Figure 28.9

λ, f/hr	τ, hr	T_o	λT_o	$U_{\min}$
1.0E-3	10.0	148.41	0.14841	0.01664
1.0E-4	10.0	454.86	0.045486	0.00190
1.0E-5	10.0	1422.26	0.0142226	0.00019
1.0E-6	10.0	4491.76	0.0044916	0.00002

The tabulated values can be confirmed approximately by observing the minima of the four curves of Figure 28.9. ■

28.3.3 Optimal Design of *k*-out-of-*n*:*G* Systems

Many different protective systems can be described as having a k-out-of-n:G operating logic. Therefore, a general solution of the optimal design of such systems has wide applicability, from simple 1-out-of-2 systems that are so common in transmission and distribution protection to more complex multiple-sensor systems used in nuclear power plants and other hazardous industrial applications. This section presents a general optimal design of the k-out-of-n:G system. The optimal configuration is based on an evaluation of the cost of failures, whether fail safe (*FS*) or fail dangerous (FD) types. The optimal values for k, n, and the inspection interval are derived, based on the cost evaluation technique of Takami, et al. [25], where the total cost includes both the initial cost as well as the expected loss due to failure. The method also includes an evaluation of common cause types of failure, which may be a problem in many types of protective systems.

In the following discussion, the protected system will be referred to as the "plant," where the physical system being protected could be any component in the power system.

28.3.3.1 Assumptions. The following assumptions form the basis for the probabilistic evaluation.

1. The k-out-of-n:G system is made up of n protective units, and the plant is tripped if k or more units pick up.
2. FD protection failures are detected by inspection and cannot be detected in any other way.
3. Inspections of the protective system are conducted at regularly scheduled intervals.
4. Protective units are inspected in sequential order, and the unit being inspected is removed from service during inspection. Following the complete inspection, the entire protective system is assumed to be as good as new.
5. *FS* failures are detected on their occurrence, and when this type of failure occurs, repair is initiated without delay. While undergoing repair, the protection being repaired is removed from service.

The notation used in the derivation is as follows:

λ_1 = independent FD failure rate

λ_{C1} = common cause FD failure rate

λ_2 = independent *FS* failure rate

λ_{C2} = common cause FS failure rate
μ = protective unit repair rate
D = plant demand rate
T = inspection interval of the k out of n system
τ = inspection duration of a unit
T_M = mission time of the plant
w_{FD} = expected frequency of FD occurrences in the mission time
w_{FS} = expected frequency of FS shutdowns in the mission time
w^* = upper bound of w
w^*_{FD} = upper bound of w_{FD}
w^*_{FS} = upper bound of w_{FS}
n_{FD} = number of FD failures in mission time T_M
n_{FS} = number of FS failures in mission time T_M
T_1^*, T_2^* = lower and upper bound of T, $T_1^* \leq T \leq T_2^*$, where T_2^* is the plant inspection interval
C_{FD} = loss per FD occurrence
C_{FS} = loss per FS occurrence
C_T = cost per inspection of a unit
C_U = cost of a unit
J = cost functional

28.3.3.2 Probability of FD Failures. An FD type of failure occurs when $n-k+1$ units have failed and a hazard occurs for which the system should operate. The probability that the k-out-of-n system suffering an FD failure at time r is computed as follows.

$$\begin{aligned}\Pr(FD_r) = &\sum_{i=n-k+1}^{n} \binom{n}{i} (e^{-\lambda_i r})^{n-i}(1-e^{-\lambda_1 r})^i + 1 - e^{-\lambda_{c1} r} \\ &- \left[\sum_{i=n-k+1}^{n} \binom{n}{i} (e^{-\lambda_1 r})^{n-1}(1-e^{-\lambda_1 r})^i\right](1-e^{-\lambda_{c1} r})\end{aligned} \tag{28.33}$$

The probability density function of the FD failure at time r is the derivative of the probability (28.33). We call this density $f_1(r)$, where

$$f_1(r) = \frac{d\Pr(FD_r)}{dr} \tag{28.34}$$

The probability density function for the conditional probability that the FD failure occurs at time t under the condition that the k-out-of-n system fails at time r is given by

$$f_2(t|r) = \begin{cases} 0 & t \leq r \\ De^{-D(t-r)} & t > r \end{cases} \tag{28.35}$$

Define the event

$$\{FD_T\} = \{FD \text{ Occurs in time } 0 \rightarrow T\}$$

Then, the probability that the FD event occurs in the interval between inspections, from 0 to

T, is given by

$$\Pr(FD_T) = \int_0^T dt \int_0^t f_2(t|r) f_1(r) dr \tag{28.36}$$

Usually, it is quite accurate to make the following assumptions.

$$\begin{aligned} 1 - e^{-\lambda_1 T} &\cong \lambda_1 T \\ 1 - e^{-\lambda_{C1}} T &\cong \lambda_{C1} T \\ 1 - e^{-DT} &\cong DT \end{aligned} \tag{28.37}$$

Then, we can approximate (28.36) by

$$\Pr(FD_T) \cong \binom{n}{n-k+1} \frac{D\lambda_1^{n-k+1} T^{n-k+2}}{n-k+2} + \frac{\lambda_{C1} DT^2}{2} \tag{28.38}$$

Usually the probability $\Pr(FD_T)$ is very small and is approximately equal to the frequency of occurrence of FD failures. Therefore, we write, approximately, the number of occurrences during a mission time T_M as the fraction T_M/T of the frequency, or the number of FD type failures is given by

$$W_{FD} \cong \binom{n}{n-k+1} \frac{D\lambda_1^{n-k+1} T_M T^{n-k+1}}{n-k+2} + \frac{\lambda_{C1} DT_M T}{2} \tag{28.39}$$

28.3.3.3 Probability of FS Failures. Security failures of type FS occur when any k units all generate trip signals, thereby causing an unnecessary or spurious trip of the plant. The frequency of occurrence of such FS trips is determined by a Markov model of the system [25]. The Markov graph for the k-out-of-n system is shown in Figure 28.10.

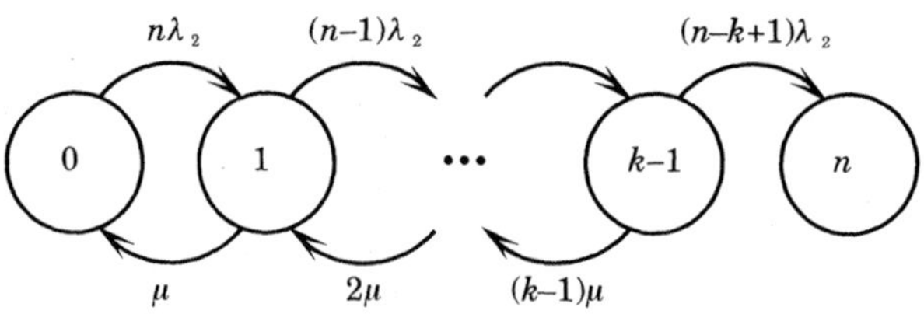

Figure 28.10 Markov graph for FS failures of a k-out-of-n system.

It has been shown that the mean time between failures for FS failures for a k-out-of-n system is given by [26]

$$MTBF = \sum_{j=0}^{k-1} \frac{1}{(n-j)\lambda_2 \rho_j} \sum_{r=0}^{j} \rho_r \tag{28.40}$$

where

$$\rho_j = \binom{n}{j} \left(\frac{\lambda_2}{\mu}\right)^j$$

and

$$\rho_0 = 1$$

For the case where $\mu \gg \lambda_2$ this simplifies to

$$MTBF = \frac{\mu^{k-1}}{\binom{n}{k} k \lambda_2^k} \tag{28.41}$$

The frequency of FS failures caused by independent failures of protective units is the reciprocal of the $MTBF$, or

$$w_{FS} = \binom{n}{k} \frac{k\lambda_2^k}{\mu^{k-1}} \tag{28.42}$$

The number of failures in mission time T_M caused by independent unit failures is computed as

$$W_{\text{ind}} = \binom{n}{k} \frac{k\lambda_2^k T_M}{\mu^{k-1}} \tag{28.43}$$

and the number of common cause failures in the mission time is given by

$$W_{CC} = \lambda_{C2} T_M \tag{28.44}$$

The total number of failures in the mission time is given by the sum of (28.43) and (28.44).

$$W_{FS} = \left[\binom{n}{k} \frac{k\lambda_2^k}{\mu^{k-1}} + \lambda_{C2}\right] T_M \tag{28.45}$$

28.3.3.4 Optimization. To optimize the design of the k-out-of-n protective system, four different kinds of cost are to be considered [25].

1. The expected cost due to FD failures of the k-out-of-n system during the mission time.
2. The expected cost due to FS failures of the k-out-of-n system during the mission time.
3. The cost of inspections during the mission time.
4. The installed cost of the n units that make up the k-out-of-n system.

The cost (1) is proportional to the number of FD events during the mission time and this number is given by (28.39). The cost (2) is proportional to the number of FS events during the mission time and this number is given by (28.45). The cost (3) is inversely proportional to the inspection interval, and this can be thought of as the cost of assuring the reliability of the system. The cost (4) is the initial installed cost of the k-out-of-n system.

The frequency of FD events depends on the inspection interval, so this interval must be optimized in order to determine the best overall design of the k-out-of-n system.

The optimization problem, then, is to determine the values of n, k, and the inspection interval T that minimize the cost functional [25]

$$J = C_{FD} W_{FD} + C_{FS} W_{FS} + C_T \frac{n T_M}{T} + C_S n \tag{28.46}$$

subject to

$$\begin{aligned} &1 \le n \le n^* \\ &1 \le k \le k^* \\ &W_{FD} \le W_{FD}^* \\ &W_{FS} \le W_{FS}^* \\ &T_1^* \le T \le T_2^* \end{aligned} \tag{28.47}$$

where the terms on the right-hand side of (28.47) are judiciously selected. From [25], the optimization can be determined by the following procedure.

Step 1 When the values of n and k are fixed, the solution for the inspection interval can be found that will minimize J. Since the practical upper limit of n is about 5, this task is not unreasonable to perform by exhaustive search.

Step 2 The combination of n, k, and the inspection interval T that will minimize J is the optimal solution.

EXAMPLE 28.3

Use the model described above and compute the cost functional for a variety of system conditions. First, vary the protective unit configuration and reliabilities, as follows.

1. Examine the following configurations, with failure rates as specified in the following table. Three configurations of protective units are to be considered:
   1-out-of-2
   2-out-of-3
   2-out-of-4
   The failure rates of the protective units will be varied over four decades. Common cause failure rates will be set to zero in Case (a), but will be set to a constant 10^{-7} for Case (b). This variation will provide a broad examination of the sensitivity of the cost functional to failure rate. The assumed system data are given in Table 28.4.

TABLE 28.4 Basic Parameters for the k-Out-of-n Example

Case	n	k	$\lambda_1 = \lambda_2$	$\lambda_{C1} = \lambda_{C2}$
(a)	2	1	10^{-6}	0
	3	2	10^{-6}	0
	4	2	10^{-6}	0
(b)	2	1	10^{-6}	10^{-7}
	3	2	10^{-6}	10^{-7}
	4	2	10^{-6}	10^{-7}

2. The rest of the problem variables will be held to constant values, and set as shown in the following table.

TABLE 28.5 Other Parameters for the k-Out-of-n Example

μ y^{-1}	D y^{-1}	T_M y	T^*_1 y	T^*_2 y	C_{FD} pu	C_{FS} pu	C_T pu	C_S pu
1095	2.0	10.0	0.05	1.00	10^6	10^4	1.0	100

The foregoing constants are selected as typical of those found in actual power systems. The costs are selected to place a somewhat high cost on FS failures, which would tend to favor those protective systems with high redundancy.

Solution

Sampled results are plotted in Figures 28.11 and 28.12, where the protection failure rate is one failure per million hours and the inspection duration is 8 hours. For both cases plotted, the 2/3 system has the lowest cost for inspection durations of less than about 0.3 years, but for inspection intervals of greater than this amount, the 2/4 system has the lower cost. If the inspection interval is 1 year, the 2/4 system has much lower costs.

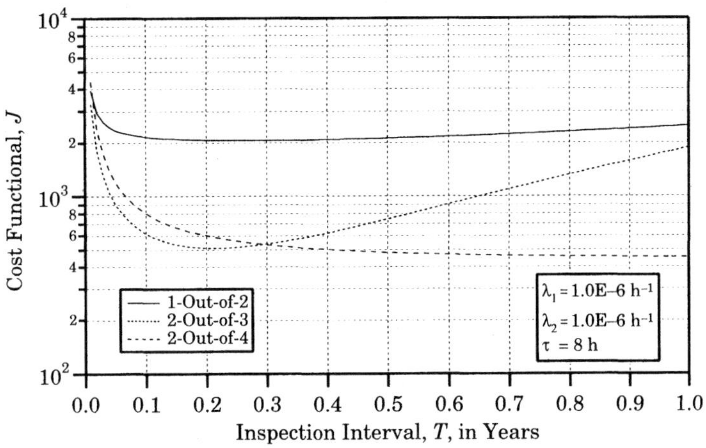

Figure 28.11 Cost functional variation without common cause failures.

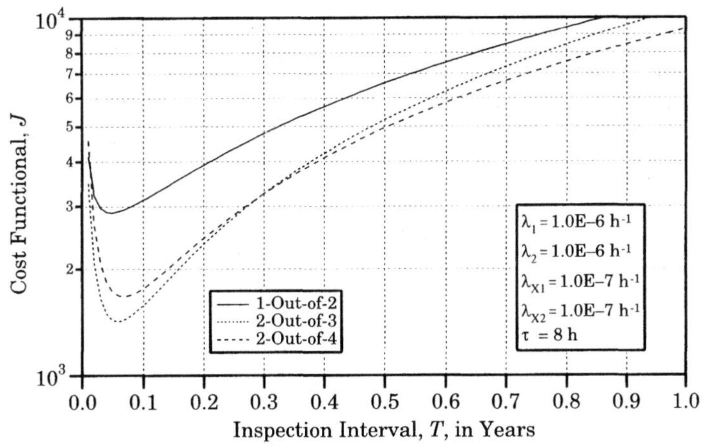

Figure 28.12 Cost functional variation with common cause failures.

For the failure rates shown in the illustration, the 1/2 system always has the highest cost. This is not always the case. For example, if the protection failure rate is one failure every 10,000 hours, then the 1/2 system has the lowest cost if the inspection interval is greater than about 6 months.

There are many variations that can be studied, using the data provided in this example. The reader may wish to try different variations in the data to determine how the cost functional varies.

Clearly, the common cause failure increases the total cost significantly. Also note that, with common cause failures included, the cost rises rapidly as the inspection interval increases. This suggests that, if common cause failures are a significant problem, reducing the inspection interval may be quite effective in reducing the total cost. ■

The foregoing example illustrates that there are many factors that enter into the cost evaluation of a protective system. The important parameters are the failure and repair rates of the protection, as well as the failure rate of the plant and the installed cost of the protective equipment. Reliability evaluation provides a method of analyzing these costs, once the various parameters are known. Even if the parameters are not known precisely, reasonable ranges of the parameters can be examined, and the effect on the total cost functional can be determined.

Moreover, the computation is readily carried out using a spreadsheet. The cost of failure is an important parameter. Surveys have shown the cost of protection failures can be very high [34]. The technique discussed here provides a means of minimizing the total cost. Other examples of protection optimization have appeared in the literature and are recommended for further study [27–29].

28.4 MONITORING AND SELF TESTING

In the foregoing analysis, it has been assumed that the protective equipment is not monitored and that all failures are unannounced. In some cases, certain types of failure may be alarmed so that maintenance personnel can investigate the failure. This is possible for those parts of the protective system that are always active, such as the power supplies. This is not possible for some types of equipment, such as the relay logic, which may suffer noncatastrophic failures that leave the system working, but not correctly. Another limitation to alarms is that they are effective only at locations where personnel are on duty at all times, so that the alarm can be observed. Many protective systems are located at remote places that are unmanned, making the response to an alarm difficult, if not impossible.

Clearly, redundancy of protective equipment is an important concept for improving the performance of the system. This could be carried to a higher level, wherein the components within the protective device are redundant, or where standby circuits are designed to take over if the main circuit does not meet certain standards of performance. This type of redundancy in the internal design is not usually practiced because of the excessive cost and complexity that this concept introduces. However, this concept can be used by the designer if a reliability design objective is provided that warrants the added cost of redundant circuitry.

Another method that is used for improving the reliability of protective equipment is to design built-in monitoring and self-checking facilities. These facilities are designed to detect any failure inside the relay as soon as possible and to cause an alarm to operate, thereby notifying the operating personnel that the relay requires immediate attention. Techniques for designing such built-in monitoring facilities have been described in the literature and the benefits of this approach have been evaluated [31].

28.4.1 Monitoring Techniques

Monitoring techniques are usually designed to operate continuously. They can be used for the following purposes [31]:

- To detect unavailability of a module
- To reduce a tendency for false tripping
- To prevent false tripping

At least two methods of achieving the monitoring function are:

1. Monitor those circuits that should have constant outputs, irrespective of the input conditions to the relay, e.g., a power supply.
2. Detect illogical operation of a module, e.g., in a distance relay, the pick up of zone 1 with no pickup of zone 2.

Monitor systems have a topology similar to that shown in Figure 28.13. Figure 28.13 shows a typical connection of a monitoring system.

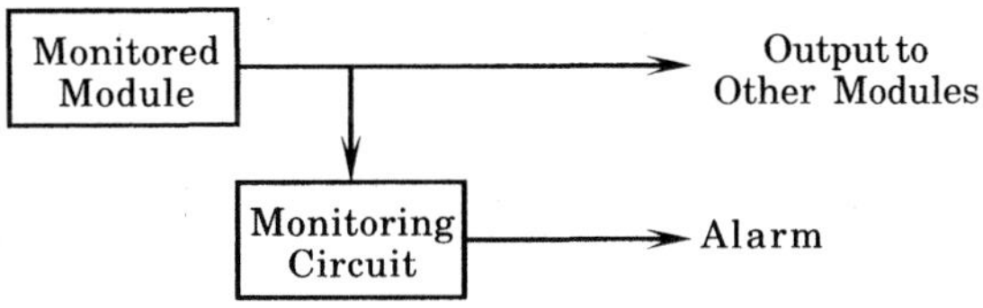

Figure 28.13 Typical structure of a monitoring application.

28.4.2 Self-Checking Techniques

Self-checking systems do not operate continuously, as do monitoring facilities, but operate only periodically and for a very short time. These facilities are designed to determine if the protective system is operative. When the self-testing is being performed, the protective system may be either completely or partially out of service. There are several methods of self-checking that have been discussed [32]:

1. Apply artificial voltages and currents to the protective system inputs to apply a specific condition for which the protection should respond.
2. Apply test signals to only the input stages of the protective system circuits.
3. Force the relay comparators to operate by temporarily changing the relay setting to create a condition for which pick up should occur, e.g., expand the circular characteristic of a distance relay to pick up for normal load conditions.
4. Invert the operating criterion of the comparator so that the measuring element is changed from the normal restraint state to an operate state.

Note that the last two methods described make use of currents and voltages from the power system input transducers. This effectively includes these input transducers in the check of the relay module.

Self-checking systems have a topology similar to that shown in Figure 28.14. This arrangement provides for disabling the module output, performing the desired test, and then restoring the system back to normal operation. Note that the self-checking circuit injects signals into the module under test and then checks the output to determine if the response is correct.

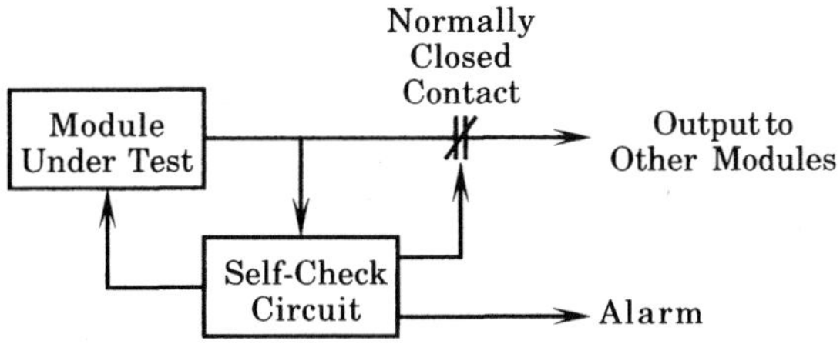

Figure 28.14 Typical structure of a self-checking application.

Both monitoring and self-checking have the ability to reduce the average unavailability of the protective system by reducing the duration and increasing the frequency of testing. Moreover, some of these systems may also reduce the probability of security failures of the protective system. There are some disadvantages, however. It may be very expensive to apply these techniques to every part of the protective system. Also, self-checking removes the protection from operation, which increases the unavailability for the duration of the self-checking activity. Effective use of these concepts requires excellent design of the monitoring and self-check systems.

28.4.3 Monitoring and Self-Checking Systems

The reliability of a protective system that employs monitoring or self-checking facilities can be determined by applying the conditional probability approach[2] [32], [33]. This can be expressed in general terms as follows. Consider a protective system A, which contains a component B. Then we may write

$$\Pr(\bar{A}) = \sum_{i=1}^{n} \Pr(\bar{A}|B_i) \cdot \Pr(B_i) \tag{28.48}$$

where $\bar{A}$ = failure of system A

B_i = the ith mutually exclusive state of component B

n = the total number of mutually exclusive states of B

For monitoring circuits, consider a module A having a failure rate λ_A, where this failure rate is interpreted as the rate of encountering a failure-to-trip state. Also, let T be the time between periodic inspections. Then, we may write the unavailability for an exponentially distributed system as

$$U = \frac{1}{T}\int_0^T (1 - e^{-\lambda_A u})du = 1 - \frac{1 - e^{-\lambda_A T}}{T} \cong \frac{\lambda_A T}{2} \tag{28.49}$$

where the approximation applies when $\lambda_A T \ll 1$.

Consider a protective system A with a monitoring circuit M, which has a failure rate λ_M and let A_M represent the module A together with monitoring system M. Then, using the conditional probability approach, we may write

$$\begin{aligned}\Pr(\bar{A}_M) &= \Pr(\bar{A}_M|M) \cdot \Pr(M) + \Pr(\bar{A}_M|\bar{M}) \cdot \Pr(\bar{M}) \\ &= 0 \cdot \left(1 - \frac{\lambda_M T}{2}\right) + \left(\frac{\lambda_A T}{2}\right) \cdot \left(\frac{\lambda_M T}{2}\right) \\ &= \left(\frac{\lambda_A T}{2}\right) \cdot \left(\frac{\lambda_M T}{2}\right)\end{aligned} \tag{28.50}$$

This assumes that the repair time is negligible compared to the total time that the protective system is in service.

EXAMPLE 28.4

Compute the probability of failure of a monitored protective system that has a failure rate of 0.1 failures per year, a testing interval of 1 year, and a monitoring system that has only 20% the complexity of the total protective system.

Solution

Since the monitor has only 20% the complexity of the total protective system, we approximate its failure rate as $(0.2)(0.1) = 0.02$ failures per year. Then we compute

$$U_A = \frac{\lambda_A T}{2} = \frac{0.1 \times 1.0}{2} = 0.05$$

$$U_{A_M} = \left(\frac{\lambda_A T}{2}\right) \cdot \left(\frac{\lambda_M T}{2}\right) = 0.05 \times \frac{0.02 \times 1.0}{2} = 0.005$$

The system unavailability is improved by a factor of 10 by monitoring. ■

---

[2]This development is from [29], which is recommended for further study.

Now consider a protective system A with a self-checking circuit S that has a single failure mode with failure rate λ_S. Let the self-testing interval be denoted as T_S and let A_S represent the module A together with the self-checking facility S. Then, we may write

$$\begin{aligned}\Pr(\bar{A}_S) &= \Pr(\bar{A}_S|S)\cdot\Pr(S) + \Pr(\bar{A}_S|\bar{S})\cdot\Pr(\bar{S}) \\ &= \left(\frac{\lambda_A T_S}{2}\right)\cdot\left(1-\frac{\lambda_S T}{2}\right) + \left(\frac{\lambda_A T}{2}\right)\cdot\left(\frac{\lambda_S T}{2}\right)\end{aligned} \tag{28.51}$$

This equation is developed based on the assumption that the time required for the self-checking is negligible compared to the time between successive self-checking periods.

EXAMPLE 28.5

Compute the probability of failure of a self-checked system where the protective system has a failure rate of 0.1 failures per year, the testing interval is 1 year, the self-checking interval is 1 week, and the self-checking system failure rate is 0.02 failures per year.

Solution

We apply the conditional probability method to compute the unavailability when self-checking is used. From (28.41), we write

$$\begin{aligned}U_{A_S} &= \left(\frac{\lambda_A T_S}{2}\right)\cdot\left(1-\frac{\lambda_S T}{2}\right) + \left(\frac{\lambda_A T}{2}\right)\cdot\left(\frac{\lambda_S T}{2}\right) \\ &= \left(\frac{0.1\times 1/52}{2}\right)\cdot\left(1-\frac{0.02\times 1}{2}\right) + \left(\frac{0.1\times 1}{2}\right)\cdot\left(\frac{0.02\times 1}{2}\right) \\ &= 0.00145\end{aligned} \tag{28.52}$$

The unavailability without self-checking is computed in Example 28.4 to be 0.05. Therefore, the self-checking system improves the unavailability by a factor of about 35. ■

An example for an actual distance relay is given in [32], which shows that monitoring circuits tend to reduce the unavailability of the protection by 38 to 61%, with the wide range depending on the type of fault. Moreover, adding self-checking in addition to monitoring reduces the unavailability by another 7–12%. This study also shows a substantial reduction in potential false trip failures, which were reduced by 55%. These figures are for only one particular relay design and the application of special monitoring and self-check circuits to that design. However, the results show that these methods can be very effective in improving the performance of protective systems.

28.4.4 Automated Testing

It has been noted that increasing the frequency of testing also increases the chances for human error and a resulting security failure of any protective system. This is especially true if the tests are permitted only during the late night hours, when the workers have been busy for unusually long periods and are therefore not at their peak effectiveness. These conditions reinforce the concept of making inspection and testing as automatic as possible.

Automation of testing also opens the possibility of making the tests shorter, which provides additional advantages, as previously noted. This must be weighed against the following negative factors:

(a) The added cost of automation.
(b) The possibility of failure of the automatic test equipment (ATE).

Fortunately, item (b) can be minimized in at least three ways:

1. Require that the supplier of the ATE comply with acceptance testing requirements defined by the utility.
2. Require that the supplier of the ATE also provide diagnostic procedures that are designed to check the correct operation of the ATE equipment.
3. Cycle the ATE prior to each test in a nonoperational pre-test trial to assure proper operation of this equipment.

If the test automation can be realized at a reasonable cost, it helps to solve several problems and will result in more reliable protective systems. ATE equipment is available from a number of manufacturers, and considerable experience has been accumulated in the application of this type of test equipment in a variety of industrial and military applications. Power industry committees have also been active in the investigation of equipment that can be used for automatic testing, as well as for monitoring and self-checking of protective apparatus [32], [33]. One report has been provided to document the advantages of automated testing of a protective relay [36].

It is difficult to translate the foregoing concepts into mathematical models that include all the desired features. The models described in this chapter represent the state of the art at the time of writing. These models include some, but not all, of the features that might be desired by some protection engineers. Modeling the protection to include all of these features is a difficult task, and more remains to be done in the search for full-featured protective system models.

28.5 THE UNREADINESS PROBABILITY

A unique concept for the analysis of the reliability of protective systems was introduced by the idea of an "unreadiness probability," which was first described in a scholarly paper by Singh and Patton in 1980 [30]. The time of occurrence of an undetected fault is not known, and the protection failure rate and mean duration can not be estimated directly. However, the authors point out that an unreadiness probability can be estimated from field data, where this probability is defined as follows:

Unreadiness Probability The probability that the protective system fails to respond when called upon to operate in the presence of a fault.

It is further asserted that this probability can be estimated as the following limiting ratio, which is given the variable identification C.

$$C = \frac{\text{number of times the protection has failed to respond}}{\text{number of times the protection is called upon to respond}} \tag{28.53}$$

This is really a conditional probability, which might be further defined as follows.

$$C = \Pr(\bar{A}|B)\Pr(B) \tag{28.54}$$

where we define the following events.

$$\bar{A} = \{\text{the protection is not operable}\}$$
$$B = \{\text{a fault occurs}\}$$

Even though the protection is in a nonoperable condition, it is not known to have failed as long as there is no need for it to operate. Indeed, it is not possible to assert that the protection has failed until it is either called upon to operate, or is tested, with the result that it is found to be in a failed state. Thus, the concept of an unreadiness probability is an appropriate measure to apply to protective systems and is more meaningful than the unavailability.

The transition diagram for the protective system and protected plant is shown in Figure 28.15. The five states are defined as follows:

State 1 The plant and the protective system are both UP
State 2 The plant is DOWN but the protection is UP.
State 3 The plant is UP and the protection is DOWN.
State 4 The plant and the protection are both DOWN.
State 5 The plant is UP and the protection is on inspection.

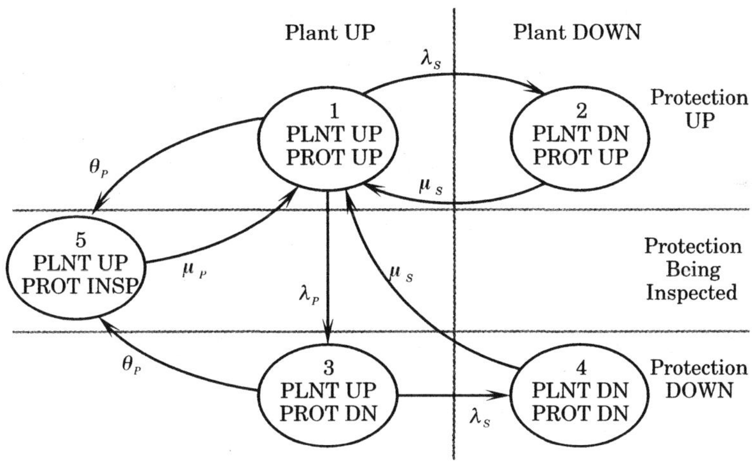

Figure 28.15 State transition diagram of plant and protection states [30].

The following assumptions are considered to be appropriate for the system under study [30].

1. Testing detects the failure of the protective system with certainty.
2. Testing the protective system does not cause plant failure.
3. When the plant is failed, the protection cannot fail.
4. Failures of the protective system are statistically independent of the failures of the plant.

Now, if we let $p_1, \ldots, p_5$ be the steady-state probabilities of the five states, we can compute these steady-state values. The system presented in [30] leaves the type of reliability distribution unspecified, such that a variety of distributions can be studied. Here, we are interested in the exponential distribution. The state probabilities are found to be the following.

$$p_1 = \frac{\mu_s \mu_p(\theta_p + \lambda_s)}{(\theta_p + \lambda_s + \lambda_p)[\theta_p \mu_s + \mu_p(\theta_p + \mu_s)]} \quad (28.55)$$

$$p_2 = \frac{\lambda_S}{\mu_S} p_1 \quad (28.56)$$

$$p_3 = \frac{\lambda_p}{\theta_p + \lambda_s} p_1 \quad (28.57)$$

$$p_4 = \frac{\lambda_s \lambda_p}{\mu_s(\theta_p + \lambda_s)} p_1 \tag{28.58}$$

$$p_5 = \frac{\theta_p(\theta_p + \lambda_s + \lambda_p)}{\mu_p(\theta_p + \lambda_s)} p_1 \tag{28.59}$$

All states are now described as functions of the probability of state 1. One can easily verify that the sum of the state probabilities is equal to unity. This is left as an exercise.

$$\sum_{i=1}^{5} p_i = 1 \tag{28.60}$$

The original publication of this method investigated the exponential, gamma, and uniform distributions. Our interest here will be restricted to the exponential distribution.

The unreadiness probability is defined in terms of the states of the Markov model, described above. The original definition of this probability was given as follows [30].

$$C = \text{unreadiness probability} = \frac{p_3}{p_1 + p_3} \tag{28.61}$$

The probability of state 3 can be written as

$$p_3 = p_1 \left[k_1 \left(\frac{G_S}{G_P} \right) - 1 \right] = p_1 \left(\frac{k_1 G_S - G_P}{G_P} \right) \tag{28.62}$$

where

$$k_1 = \frac{\lambda_s + \lambda_p}{\lambda_s} \tag{28.63}$$

and

$$G_s = \frac{\lambda_s}{\theta_p + \lambda_s} \tag{28.64}$$

$$G_p = \frac{\lambda_s + \lambda_p}{\theta_p + \lambda_s + \lambda_p} \tag{28.65}$$

With the foregoing preparation, we compute the unreadiness probability as follows.

$$C = \frac{p_3}{p_1 + p_3} = \frac{\dfrac{k_1 G_S - G_P}{G_P}}{1 + \dfrac{k_1 G_S - G_P}{G_P}} = \frac{k_1 G_S - G_P}{k_1 G_S} \tag{28.66}$$

or, simplifying,

$$C = 1 - \frac{G_P}{k_1 G_S} = 1 - \left(\frac{\theta_P + \lambda_S}{\theta_P + \lambda_S + \lambda_P} \right) = \frac{\lambda_P}{\theta_P + \lambda_S + \lambda_P} \tag{28.67}$$

The only variable in this result that the engineer has much control over is the inspection rate θ_P, since the other parameters are statistical failure rates of the plant and protection, and are not easily changed. The inspection rate is the inverse of the mean time between inspections, which can be varied if there is reason to make a change in protective system inspection policy.

Figure 28.16 shows a plot of the unreadiness probability plotted as a function of the time interval between inspections for specific values of the failure rates. As the interval between inspections is increased, the probability of the protective system failing to respond to a fault increases monotonically until, at very high intervals, the probability levels off. The limiting value can be computed by setting the value of θ_P in (28.67) to zero, which corresponds to never inspecting the protective system. On the other hand, increasing the frequency of inspections always causes a reduction in the unreadiness probability. Note that, according to the unreadiness probability definition, the time required to complete the inspection is not a factor in the result.

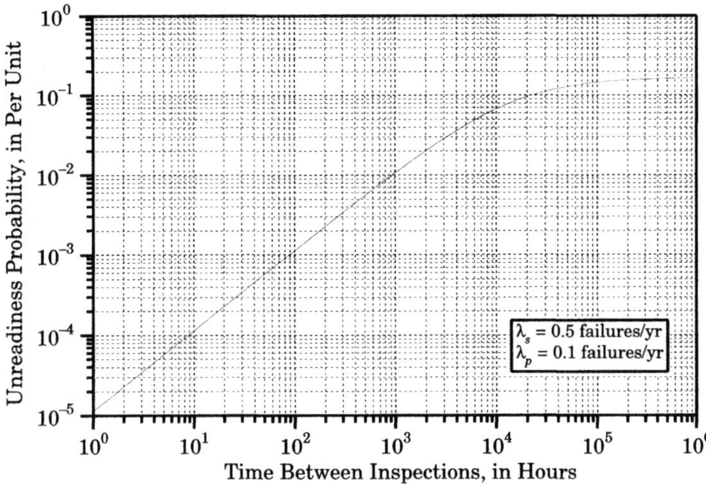

Figure 28.16 Unreadiness probability as a function of inspection interval.

The *mean duration* of an undetected fault is the duration of residing in state 3 of the Markov graph. This duration is computed as follows.

$$d_u = \frac{p_3}{p_1 \lambda_p} = \frac{1}{\lambda_P}\left[\left(\frac{\lambda_S + \lambda_P}{\lambda_S}\right)\left(\frac{G_s}{G_p}\right) - 1\right] = \frac{1}{\theta_P + \lambda_S} \tag{28.68}$$

The duration increases as the mean time between inspection increases, or as the inspection rate θ_P, becomes small. This means that, if the protective system is not inspected very often, there will be long periods during which faults will not be detected by the protection. If plotted, this function will have the same shape as the unreadiness probability, reaching a limiting value that can be determined by setting θ_P equal to zero. This limiting value is the mean time between failures of the plant.

28.6 PROTECTION ABNORMAL UNAVAILABILITY

The *abnormal unavailability* is a probabilistic measure of the inability of a protected system to recognize and respond to a component failure because of protective system unreadiness. This index of performance provides a direct and physically significant definition of unreadiness, and can be computed based on typical system transition rates for an exponentially distributed system. Using this model, it is possible to estimate the optimal value of the protection inspection interval, i.e., the time between inspections of the protection. The model takes into account the operation of backup protection, the removal of the protection for inspection, the failure and repair of the protected component, and the occurrence of common cause failures, in addition to the normal clearing of a faulted component [35], [36]. The physical system is assumed to consist of a protected component and its associated protection zone, as defined in Figure 28.17.

The protection zone shown in Figure 28.17 represents a typical situation for a transmission line, but it could just as well represent a transformer or other type of apparatus. Usually, the clearing of a fault on component C will require current measurements of appropriate system

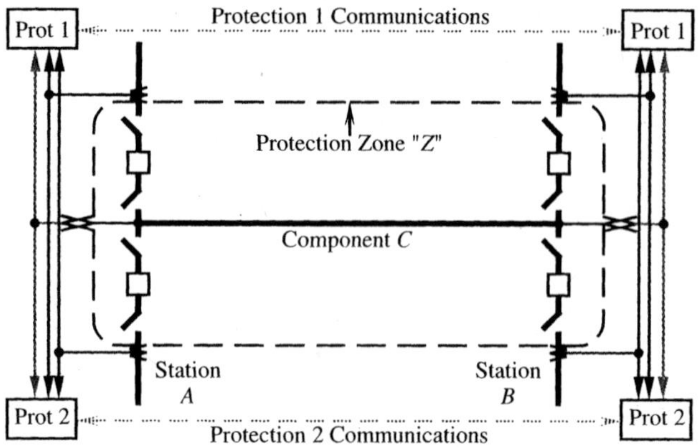

Figure 28.17 A protected component and the protection zone.

variables at the terminals of C, the pickup of one or more protections assigned to C, and the correct operation of all circuit breakers. In most cases, the protection will consist of redundant protective devices, often with redundant measurement transducers, that will command operation of one or more circuit breakers. We assume redundant protective systems, identified as P_1 and P_2. The reliability of the protective systems are to be represented in this analysis by failure and repair rates of the two redundant protective systems.

28.6.1 Assumptions

The system consisting of a single component protected by redundant protective systems is assumed to adhere to the following assumptions, which are stated in terms of a protected component C and two redundant protective devices P_1 and P_2 [35].

1. An inspection of a protective device or a fault on the protected component C must occur in order to detect the failure of the protective device.
2. A protective device must be taken out of service to be inspected.
3. Inspection always detects protective device failure.
4. Repair always restores the protection to as good as new.
5. The time required to test a protective device is equal to the time required to repair or replace a failed device.
6. Common mode failures of C, P_1, and P_2 can occur.
7. Common mode failures and repairs of P_1 and P_2 can occur.
8. Common mode failures of C and P_1 only, or of C and P_2 only cannot occur.
9. When C fails, P_1, P_2, or the backup protection isolates C before any other event can occur. In this analysis, the backup protection is assumed to be perfectly reliable.
10. The repair of C takes much longer than protection inspection and repair.
11. P_1, P_2, or both can fail or be inspected while C is isolated for repair or maintenance.
12. When C is in the UP state, P_1 and P_2 will not be inspected simultaneously.
13. When C is isolated, P_1 and P_2 can be inspected simultaneously using overlapping or common mode inspections.
14. C will not be brought back into service if both P_1 and P_2 are known to be in the DOWN state.

The definitions of the states for which C, P_1, and P_2 can reside are as follows:

UP	The element is operational.
DN	The element is in an announced-failed state. In the case of protective devices, transition to this state occurs when the device is removed for inspection or when C fails but the protection fails to respond because it was in the unannounced-failed state.
DU	The element is in the unannounced-failed state. Only protective devices can fail unannounced. These failures are determined by regular inspection or are discovered when the protective device fails to respond to a demand caused by the failure of C.
ISO	This state represents a condition in which the faulty component C is isolated from the rest of the system so it can be repaired. If P_1 and P_2 fail to isolate C, the backup protection is called on to remove C from service. The operation of backup protection results in the removal of additional components X, which can only be restored to service by manual switching.

The definitions of the transition rates used to construct the Markov model are as follows:

λ_C, μ_C	failure, repair rates of protected component C
λ_{P1}, μ_{P1}	failure, repair rates of protective device P_1
λ_{P2}, μ_{P2}	failure, repair rates of protective device P_2
λ_{CC}	common cause failure of C, P_1, and P_2
λ_{PP}, μ_{PP}	common cause failure, repair rates of P_1 and P_2
θ_{P1}, θ_{P2}	inspection rates of P_1 and P_2
θ_{PP}	common cause inspection rates of P_1 and P_2
ψ_N	normal switching rates initiated by P_1 or P_2
ψ_B	backup switching rates initiated by backup protection
ψ_M	manual switching rate to isolate C and restore X

The Markov model of the system involves four states of the component C and multiple transitions between these four states. One way to view this situation is to identify these four states as layers, with defined transitions between layers. These layers are shown in Figure 28.18. The layers are defined as follows:

Layer 1:	C is UP
Layer 2:	C is DOWN
Layer 3:	C is ISOLATED
Layer 4:	$C + X$ are ISOLATED

When C fails there is a transition from layer 1, thc UP statc, to layer 2, the DOWN state, with a transition rate of λ_C, as shown in Figure 28.18. The other layer transitions are switching rates, as defined above, except for the transition from the C ISOLATED to C UP state, which is the repair rate for C. The status of the protective devices is not shown in the figure, but is discussed below.

When C transits from layer 1 (UP) to layer 2 (DOWN) with rate λ_C, the system states depend on the status of the protections P_1 and P_2. Once in the DOWN state, C will be ISOLATED, i.e., moved to layer 3, using the normal circuit breaker switching rate ψ_N, if at

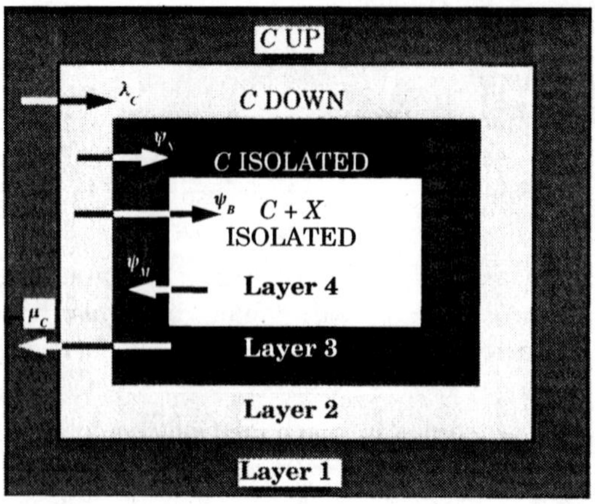

Figure 28.18 Subdivision of the state space into four layers.

least of one protective device is available. However, if both P_1 and P_2 are in a failed state, either announced or unannounced, then $C + X$ will be isolated using the backup protection with a switching rate ψ_B, which results in a transition from layer 2 to layer 4. The system remains in layer 4 until a manual switching, with rate ψ_M, returns the system to layer 3, where X is restored and C remains ISOLATED. This manual rate depends on the system operation. It may be performed by a system operator, using switching commands from a control center, or it may require visual inspection prior to any physical switching or, perhaps, both. This transition may require several minutes to even hours to complete. However, once C is in layer 3, repair can be completed with a transition rate μ_C back to layer 1. It is important to note that the transition of C to layer 1 will not be permitted unless at least one protection is available.

The Markov model for the system is shown in Figures 28.19 to 28.21. The system has 32 states and 111 transitions. Figure 28.19 shows all possible states in layers 1 and 2, as well as the states that interface with these layers from layers 3 and 4. The transitions between layers, defined in Figure 28.18, are all shown in Figure 28.19. Any state not shown in Figure 28.19

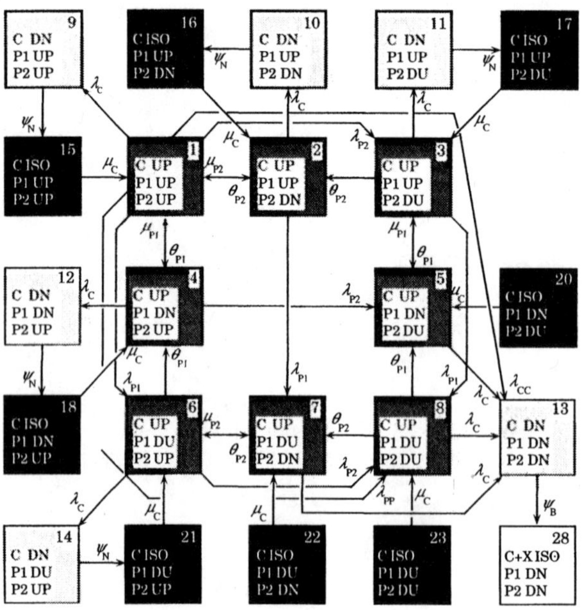

Figure 28.19 Markov model for layers 1 and 2, with interface to 3 and 4.

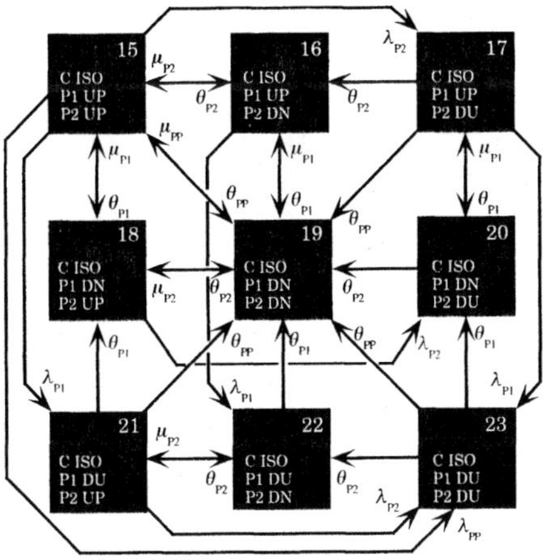

Figure 28.20 Markov model for layer 3 where C is ISO.

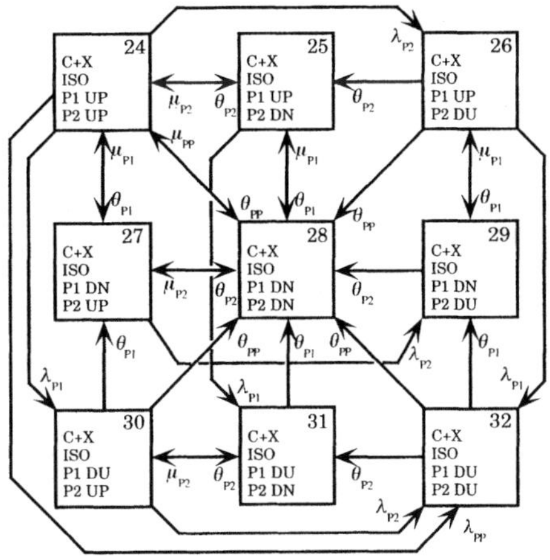

Figure 28.21 Markov model for layer 4 where $C+X$ are ISO.

indicates that the transition from that state to another in layer 1 is not possible, according to the assumptions.

Figures 28.20 and 28.21 show the rest of the Markov model, specifically the transitions between states in layers 3 and 4, respectively. The number of states and transitions between states are identical in each of these figures. The only difference is the fact that $C + X$ is isolated in layer 4, whereas only C is isolated in layer 3. A transition ψ_M occurs between each of the corresponding states in these figures, for example, from state 24 to state 15. These models are the direct result of Assumption 11, which states that the protective devices can fail and/or be inspected while C is isolated. If this assumption is changed, these transitions will be different.

The matrix of state transitions is shown in Figure 28.22, where the diagonal terms have been omitted due to space restrictions, but are computed as specified in Chapter 25 by (25.79). The steady-state probability of residing in each state of the Markov model can be computed using the theory of continuous Markov processes described in Chapter 25. Solving the Markov equation for the steady-state condition, we obtain the probabilities of each of the 32 states of the Markov model. We can then identify those states that result in the abnormal unavailability condition, which we can identify as

$$U_{AB} = P_{13} + P_{19} + P_{20} + P_{22} + P_{23} + P_{28} + P_{29} + P_{31} + P_{32} \tag{28.69}$$

We can also compute the unreadiness probability, P_{UR}, as the quantity

$$P_{UR} = \frac{w_{nr}}{w_{nr} + w_n} \tag{28.70}$$

$$\begin{aligned} \text{where } w_{nr} &= \lambda_{CC} P_1 + \lambda_C (P_5 + P_7 + P_8) \\ &= \text{frequency of protective systems not} \\ &\quad \text{responding when called upon to do so} \end{aligned} \tag{28.71}$$

and

$$\begin{aligned} w_n &= \lambda_C (P_1 + P_2 + P_3 + P_4 + P_6) \\ &= \text{frequency of protective systems} \\ &\quad \text{responding when called upon to do so} \end{aligned} \tag{28.72}$$

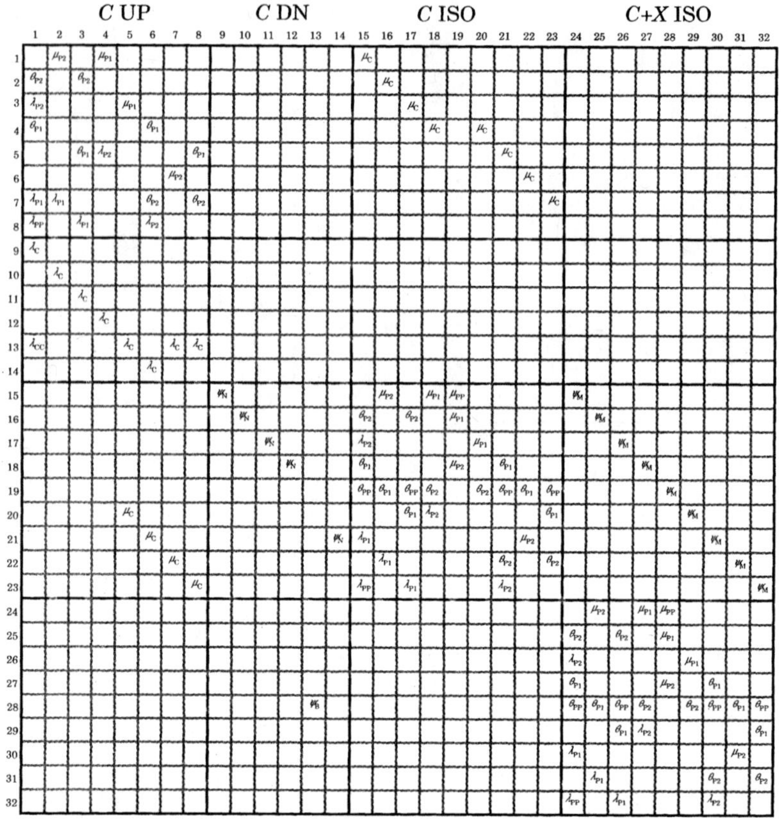

Figure 28.22 State transitional probability matrix for redundant protection.

Then

$$P_{UR} = \frac{(\lambda_{CC}/\lambda_C)P_1 + P_5 + P_7 + P_8}{(1 + \lambda_{CC}/\lambda_C)P_1 + P_2 + P_3 + P_4 + P_5 + P_6 + P_7 + P_8} \tag{28.73}$$

The steady-state probabilities for each system state can be computed from the state transition matrix using typical values of the various transition rates. The numerical solution can also be used to determine the optimum interval between inspections of the protection, which is the inverse of the inspection rate, assuming the inspection rates for the two protective systems to be equal. The other transition rates are assumed to be as shown in Table 28.6.

If the inspection rate is changed, the system can be solved repeatedly to determine the abnormal unavailability as a function of the inspection rate. The result of this calculation, using the given numerical data, is shown in Figure 28.23, where two indices of performance are plotted as a function of the inspection interval.

The results of Figure 28.23 show that the abnormal unavailability provides a method of finding the optimum inspection interval for the system, which corresponds to the minimum value of the abnormal unavailability. If the inspection interval is too short, the protection is on inspection nearly all of the time, forcing the abnormal unavailability to a value of unity. Moreover, if the inspection interval is too large, the result is an abnormal unavailability that is much higher than the optimum value. Fortunately, the minimum is not sharply defined, giving the protection engineer a reasonable latitude in choosing the inspection interval but the index rises very fast if too much testing is set as the normal schedule.

TABLE 28.6 Transition Rates Used in Calculations

Transition	Transition Rate	Mean Time
λ_C	2 failures/year	0.5 years
μ_C	175 repairs/year	50 hours
$\lambda_{P1} = \lambda_{P2}$	0.1 failures/year	10 years
$\mu_{P1} = \mu_{P2}$	876 repairs/year	10 hours
$\lambda_{PP} = \lambda_{CC}$	0.00876 failure/year	1 million hours
μ_{PP}	500 repairs/year	17.5 hours
$\theta_{P1} = \theta_{P2}$	12 inspections/year	730 hours
ψ_N	43200 operations/hour	5 cycles
ψ_B	21600 operations/hour	10 cycles
ψ_M	0.1 operations/hour	10 hours

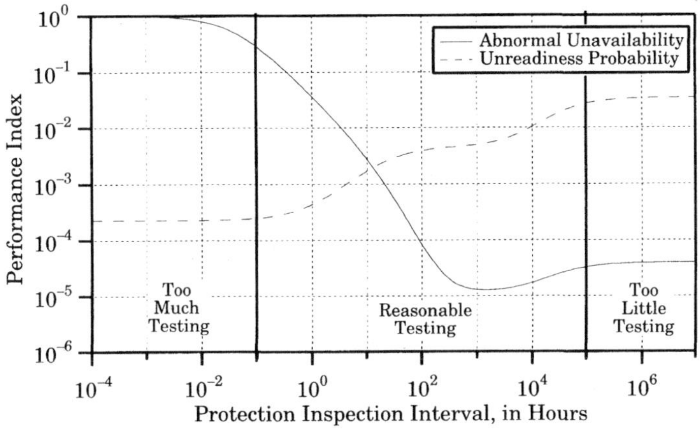

Figure 28.23 Variation of performance indices with inspection interval.

Figure 28.23 also shows the unreadiness probability, which is noted to become smaller as the inspection interval is reduced. This makes sense, as high inspection rates confirm the readiness of the protection often, but having the protection out of service so often is not beneficial if the rate is excessively high. The abnormal unavailability offers a better index of performance as it provides a method of optimizing the inspection interval or its inverse, the inspection rate.

We now examine the sensitivity of the Markov model to variations in the system failure and repair rates. Figure 28.24 shows the abnormal unavailability plotted for three different values of the component failure rate. The variation in abnormal unavailability is not great, with all curves having a similar shape and with all finding the optimum at exactly the same value of inspection interval.

Figure 28.25 shows the effect of changing the repair rate of the component C. The variation of this parameter does affect the inspection interval that results in the optimum value of abnormal unavailability, with the optimum interval being longer for the smaller repair rates (longer MTTR). However, the optimum interval is not sharply defined with respect to the repair rate.

We now examine the sensitivity of abnormal unavailability with respect to variations in the failure and repair rates of the protection. Figure 28.26 shows the result for different values of the inspection failure rate. In this case the optimum value varies significantly with respect to the protection failure rate and with the lowest rate ($MTTF = 100$ yr), the protective system is

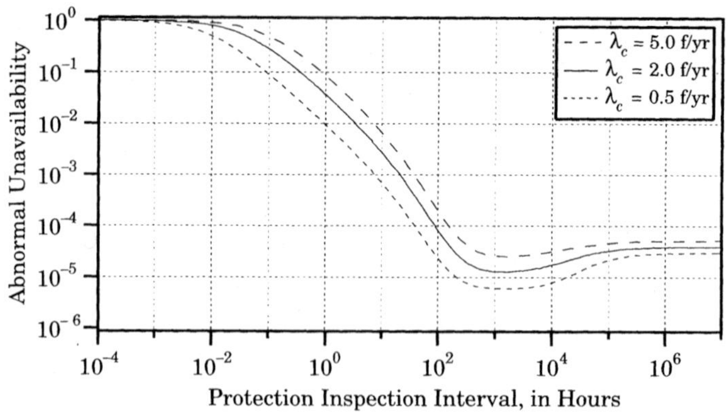

Figure 28.24 Sensitivity to component C failure rate.

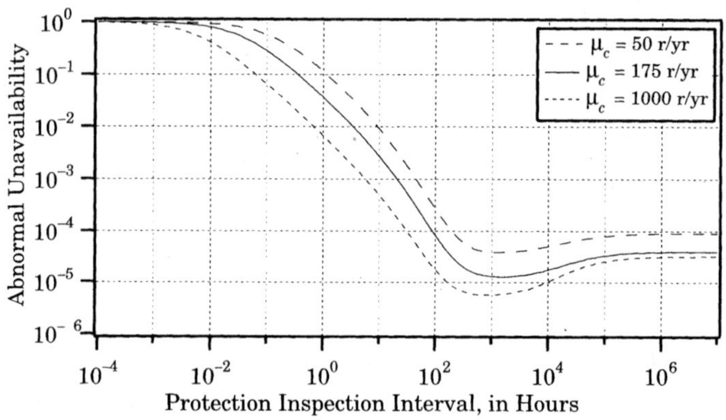

Figure 28.25 Sensitivity to component C repair rate.

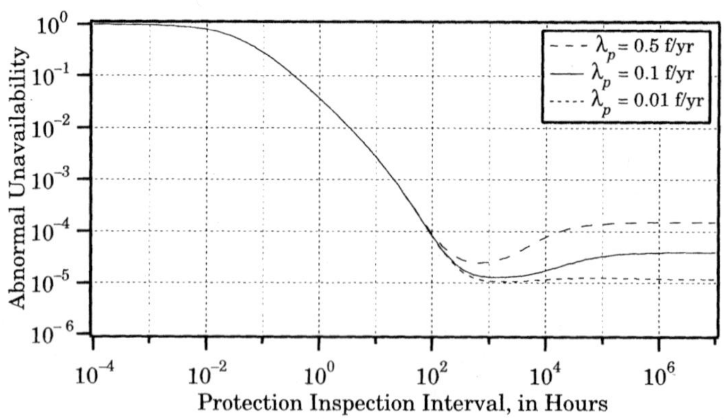

Figure 28.26 Sensitivity to protection failure rate.

so reliable that the inspection interval can be extended almost indefinitely. This is a reasonable result, since the equipment has a very high availability when the failure rate is this low.

The sensitivity of the result to a variation in the protection repair rate is shown in Figure 28.27. Here, the optimum inspection interval is also highly dependent on the protection

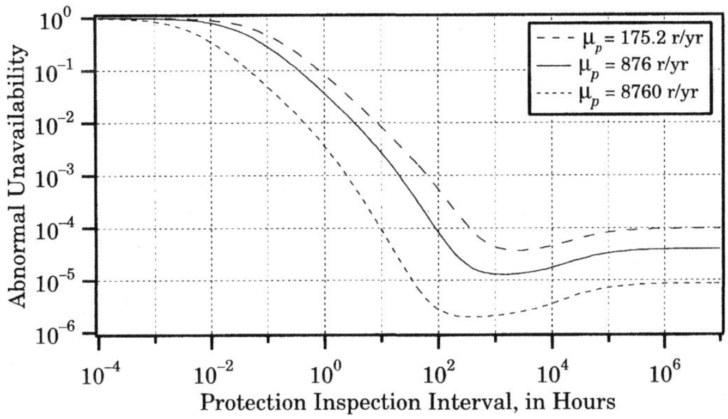

Figure 28.27 Sensitivity to protection repair rate.

repair rate, with very fast inspections favoring much smaller intervals between inspections. This is consistent with the concept of automated testing of modern digital equipment, which can be tested very often, and with each test requiring only a small time interval. The highest rate examined here is one repair per hour, which would not be adequate for the replacement of failed equipment.

Finally, we examine the sensitivity of the common mode failure rate where one event causes failure of the component as well as both protective systems. This result is shown in Figure 28.28 for a fivefold increase in the failure rate.

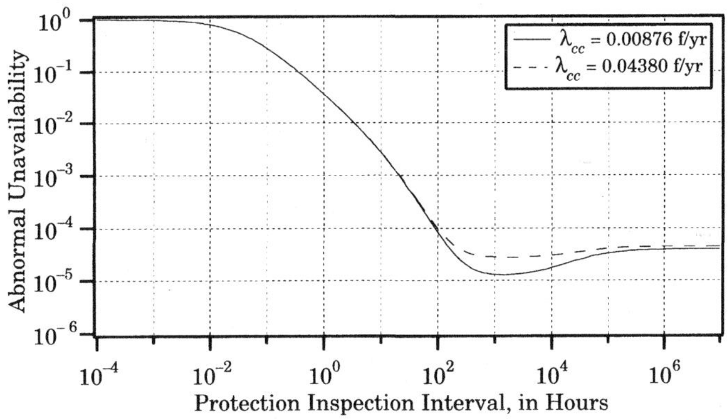

Figure 28.28 Sensitivity to common mode failure of C and P.

Increasing the common mode failure rate causes a rather large increase in the abnormal unavailability, but the optimum inspection interval is not changed. Moreover, the value of the performance index for very large inspection would appear to be merging, although the two never become equal for the range of values used here.

28.7 EVALUATION OF SAFEGUARD SYSTEMS

The protective systems for power system components are designed to respond to failure of a component and to isolate the failed component so that the remainder of the power system can operate in a nearly normal way. Many protective systems can be described as performing this type of responsive action. In most cases, protective system is not able to control the component operating environment or to *prevent* the failure of the component in any way.

Another type of protective system is designed to control the component environment and, therefore, prevent a particular type of catastrophic failure. These systems are sometimes called *safeguards*, and will be so designated in this book. In power systems, and particularly in thermal power plants, there are protective systems that can be described as safeguards, for example,

- Transformer overpressure relays
- SSR countermeasures
- Fire sprinkler systems
- Pressure relief valves
- Nuclear reactor scram systems

These systems are designed to recognize a hazard that would probably cause failure of the plant and to take action to preserve the life of the plant by eliminating the hazard. Some of these systems actually change the component environment to eliminate the hazard, which is the case for fire sprinkler systems and pressure relief valves.

The objective of this section is to describe a method for computing the expected number of failures of safeguard systems. The method provides a probabilistic description of both the protected element, the "plant," as well as the safeguard. This model embraces the concept of protective system periodic testing and allows for redundant protective equipment. It does not include the duration of testing explicitly, but it could be argued that this time is small (hours) compared to the test interval (weeks or months). The model was developed for the purpose of evaluating protective equipment in chemical plants [1], but is equally adaptable to the protection of power system elements.

28.7.1 Definitions and Assumptions

The objective in this development is to derive a probabilistic model for a safeguard system.[3] First, we define the following variables that will be used in the development.

1. L: $L =$ length of life of the plant, a random variable. We write the expression $L = t$ to mean that the plant failure, for which protection is required, occurs at time t.
2. S: $S =$ length of life of the safeguard system, a random variable.
3. T: $T =$ the inspection interval for the safeguard system.
4. W: $W(0, T) =$ the expected number of plant hazards that occur between 0 and T.
5. D: $D =$ the demand rate on the safeguard system.
6. w: $w(t) =$ expected number of plant hazards at time t per unit time.
7. H: $H(t, t+dt] =$ the event wherein the plant hazard occurs in the interval $(t, t+dt]$, or $t < L \leq t + dt$.

Note that the variables 1, 2, and 3 have units of time. Variables 5 and 6 have units of inverse time.

The following assumptions further define the system.

(a) The protective system is redundant with n identical, independent protective devices, all of which have the constant failure rate λ. Therefore, we may write

[3]This derivation closely follows that found in [1], which is recommended for study.

$$\Pr\{S \leq t\} = (1 - e^{-\lambda t})^n \tag{28.74}$$

where n is the number of identical protective devices.

(b) The probability distribution of plant lifetime, L, is dependent on the lifetime of the protective system, S.

$$\Pr\{L \leq t | S = u\} = \begin{cases} 0 & t \leq u \\ 1 - e^{-D(t-u)} & t > u \end{cases} \tag{28.75}$$

The interpretation of (28.75) is as follows. If the safeguard is functioning, then $t \leq u$, and there will be no uncleared hazards on the plant, hence the probability of failure is zero. Note that this would not be true of some types of protective systems, but is true of safeguards, which are designed to remove hazards. The second part of the equation shows that, after failure of the safeguard system ($t > u$), the plant failure probability is exponentially distributed, with failure rate D, which shows that the probability of not suffering from the hazard begins to decline following an exponential distribution, once the safeguard is no longer available to remove the hazard.

28.7.2 The Unconditional Hazard Rate

From the definition of the unconditional failure intensity, we write the probability of the plant hazard occurring in an incremental interval.

$$w(t)dt = \Pr\{H(t, t + dt]\} \tag{28.76}$$

It can be shown that this probability can be written as

$$\Pr\{H(t, t + dt]\} = dt\, Dn\lambda e^{-Dt} \int_0^t e^{(D-\lambda)u}(1 - e^{-\lambda u})^{n-1} du \tag{28.77}$$

Therefore, we can write the unconditional failure intensity as

$$w(t) = Dn\lambda e^{-Dt} \int_0^t e^{(D-\lambda)u}(1 - e^{-\lambda u})^{n-1} du \tag{28.78}$$

28.7.3 Expected Number of Failures

The expected number of failures, $W(0, T)$, is probably the most useful of all reliability parameters and is, therefore, an important parameter to compute. From the definition of $W(0, T)$ we can write the following equation.

$$W(0, T) = \int_0^T w(t)dt = Dn\lambda \int_0^T e^{-Dt} \int_0^t e^{(D-\lambda)u}(1 - e^{-\lambda u})^{n-1} du dt \tag{28.79}$$

The right side of (28.79) is evaluated as follows. We can expand the term in parentheses to write

$$(1 - e^{-\lambda u})^{n-1} = \sum_{i=0}^{n-1} \binom{n-1}{i} (-1)^i e^{-\lambda i u} \tag{28.80}$$

This result can be substituted into (28.79) with the following result.

$$W(0,T) = \sum_{i=0}^{n-1} \binom{n-1}{i} (-1)^i \frac{n}{D-(i+1)\lambda} \left(\lambda e^{-DT} - \frac{De^{-(i+1)\lambda T}}{i+1} \right) + \sum_{i=0}^{n-1} \binom{n-1}{i} (-1)^i \left(\frac{n}{i+1} \right) \tag{28.81}$$

It can be shown that the second summation of (28.81) is equal to unity. Therefore (28.79) can be written as

$$W(0,T) = 1 + \sum_{i=0}^{n-1} \binom{n-1}{i} (-1)^i \left(\frac{n}{D-\lambda(i+1)} \right) \left(\lambda e^{-DT} - \frac{De^{-(i+1)\lambda T}}{i+1} \right) \tag{28.82}$$

Note that this result gives the expected number of system failures, integrated over the period T between inspections of the safeguard.

In some cases, the engineer may be interested in the long term average hazard rate. This can be computed as follows.

$$\bar{w} = \frac{W(0,T)}{T} \tag{28.83}$$

The result $W(0,T)$ has several interesting properties.

1. Let us assume that $\lambda T \ll 1$ and $DT \ll 1$. These assumptions are usually valid because the hazard rates are small. Under these conditions, we make the following approximations:

$$e^{-\lambda t} = 1 - \lambda t + \lambda^2 t^2 - \cdots \cong 1 - \lambda t \tag{28.84}$$

Then

$$1 - e^{-\lambda t} \cong \lambda t \tag{28.85}$$

Substituting these approximations into the integrand of (28.79), we have the following.

$$\begin{aligned} e^{-Dt} e^{(D-\lambda)u} (1-e^{-\lambda u})^{n-1} &\cong (1-Dt)[1+(D-\lambda)u]\lambda^{n-1}u^{n-1} \\ &\cong \lambda^{n-1}u^{n-1} \end{aligned} \tag{28.86}$$

Substituting (28.86) into (28.79) and integrating, we get the following result.

$$W(0,T) \cong \int_0^T \int_0^t Dn\lambda^n u^{n-1} du dt = \frac{D\lambda^n T^{n+1}}{n+1} \tag{28.87}$$

This result provides a simpler solution to evaluate if the approximations regarding λT and DT are satisfied, which will often be the case. The approximate model is quite close for small values of inspection period T, but the error increases for large T.

2. From (28.82) we can compute an approximation of the number of failures for limiting values of D. First, for large values of D, we compute the following limit.

$$\begin{aligned} \lim_{D\to\infty} W(0,T) &= 1 + \sum_{i=0}^{n-1} \binom{n-1}{i} (-1)^i \left(\frac{ne^{-(i+1)\lambda T}}{i+1} \right) \\ &= 1 - \binom{n}{2} e^{-2\lambda T} + \binom{n}{3} e^{-3\lambda T} - \binom{n}{4} e^{-4\lambda T} + \cdots + \binom{n}{n} e^{-n\lambda T} \\ &= (1-e^{-\lambda T})^n \end{aligned} \tag{28.88}$$

We can also compute an approximation for very small D.

$$\lim_{D\to 0} W(0,T) = 1 + \sum_{i=0}^{n-1} \binom{n-1}{i} (-1)^i \left(\frac{n}{i+1}\right) = 0 \tag{28.89}$$

This result shows that the protected plant takes on the characteristic of the safeguard system when DT is large.

3. We can also compute an approximation of the number of failures for limiting values of λ, with the following results.

$$\lim_{\lambda\to\infty} W(0,T) = 1 + \sum_{i=0}^{n-1} \binom{n-1}{i} (-1)^i \left(\frac{n}{i+1}\right) e^{-DT} = 1 - e^{-DT} \tag{28.90}$$

$$\lim_{\lambda\to 0} W(0,T) = 1 + \sum_{i=0}^{n-1} \binom{n-1}{i} (-1)^i \left(\frac{n}{i+1}\right) = 0 \tag{28.91}$$

This result shows that the total system takes on the characteristic of the protected element alone when the protection has high unavailability, which is the case when there is no protection. However, if the protection has extremely high availability, it reduces the number of failures to a very low value.

The computation of the expected number of failures of a protective system is an excellent way to evaluate or compare competing systems. This requires that the unconditional hazard rate (28.78) of the system be known. Integration of this unconditional rate gives the expected number of failures.

REFERENCES

[1] Kumamoto, H., and E. J. Henley, *Probabilistic Risk Assessment and Management for Engineers and Scientists*, 2nd Ed., IEEE Press, Piscataway, NJ, 1996.

[2] Green, A. E., and A. J. Bourne, *Reliability Technology*, Wiley-Interscience, New York, 1972.

[3] Jacobs, I. M., "Reliability of Engineered Safety Features as a Function of Testing Frequency," *Nuclear Safety*, 9 (4), July–August 1968.

[4] Hirsh, H. M., "Setting Test Intervals and Allowable Bypass Times as a Function of Protection System Goals," *IEEE Trans. on Nuclear Sci.*, N-18, February 1971, pp. 488–494.

[5] Bonhomme, N. M., "Evaluation and Optimization of Inspection and Repair Schedules with Applications to Nuclear Power Generating Stations," Ph.D. Dissertation, Carnegie-Mellon University, Pittsburgh, May 1972.

[6] Chay, S. C., W. D. Loftus, and M. Mazumdar, "A Probabilistic Approach to Safety Testing and Maintenance," *ASME Trans.*, J of Engineering for Power, 96, July 1974, New York, pp. 174–180.

[7] Kontoleon, J. M., N. Kontoleon, and N. G. Chrysochoides, "Optimum Active-Inactive Protective Times in Supervised Protective Systems for Nuclear Reactors," *Nuclear Sci. and Engr.*, 55, 1974, pp. 219–224.

[8] Chay, S. C., and M. Mazumdar, "Determination of Test Intervals in Certain Repairable Standby Protective Systems," *IEEE Trans. Rel.*, R-24 (3), August 1975, New York, pp. 201–205.

[9] Dressler, E., and H. Spindler, "The Effect of Test and Repair Strategies on Reactor Safety," *Proc. IAEA Symposium on Reliability of Nuclear Power Plants*, April 14–18, 1975, Innsbruck.

[10] Apostolakis, G. E., and P. P. Bansal, "Effect of Human Error on the Availability of Periodically Inspected Redundant Systems," *IEEE Trans. Rel.*, R-26, August 1977, New York, pp. 220–228.

[11] Vesely, W. E., and F. F. Goldberg, "Time Dependent Unavailability Analyses of Nuclear Safety Systems," *IEEE Trans. Rel.*, R-26, October 1977, New York, pp. 257–260.

[12] Kontoleon, J. M., "Analysis of a Dynamic Redundant System," *IEEE Trans. Rel.*, R-27 (2), June 1978, New York, pp. 116–119.

[13] Kumamoto, H., and E. J. Henley, "Protective Systems Hazard Analysis," *Ind. Engr. Chem.*, 13 (4), 1978, pp. 274–276.

[14] Kontoleon, J. M., "Optimal Supervision Interval and Order of Supervision in Nuclear Reactor Protection Systems," *Nucl. Sci. Engr.*, 66, 1978, pp. 9–13.

[15] Inagaki, T., K. Inoue, and H. Akashi, "Improvement of Supervision Schedules for Protective Systems," *IEEE Trans. Rel.*, R-28 (2), June 1979, New York, pp. 141–144.

[16] Inagaki, T., K. Inoue, and H. Akaski, "Improvement of Supervision Schedules for Protective Systems," *IEEE Trans. Rel.*, R-28, June 1979, New York, pp. 141–144.

[17] Sayers, B., "Safety and Risk in an Chemical Plant (a case study)," *1979 Proc. Ann. Rel. and Maint. Symp.*, 1979, pp. 174–180.

[18] Kontoleon, J. M., "Analysis of a Dynamic Redundant System with Non-Identical Units," *IEEE Trans. Rel.*, R-29, April 1980, New York.

[19] Inagaki, T., K. Inoue, and H. Akashi, "Optimization of Staggered Inspection Schedules for Protective Systems," *IEEE Trans. Rel.*, R-29 (2), June 1980, pp. 170–173.

[20] Kontoleon, J. M., "Fail-to-safe and Fail-to-danger Analysis of Logic Protective Networks," *IEEE Trans. Rel.*, R-29 (5), December 1980, New York, pp. 436–437.

[21] Kumamoto, H., E. J. Henley, and K. Inoue, "Computer Aided Protective System Hazard Analysis," *Computers and Chemical Engineering*, 5 (2), 1981, pp. 93–98.

[22] Kohda, T., H. Kumamoto, K. Inoue, and T. Takami, "Optimal Logical Structure of Sensor Systems Composed of Non-Identical Sensors," *Microelectronic Reliability*, 22 (3), 1981, pp. 445–456.

[23] Inoue, I., T. Kohda, H. Kumamoto, and I. Takami, "Optimal Structure of Sensor Systems with Two Failure Modes," *IEEE Trans. Rel.*, R-31 (1), April 1982, New York, pp. 119–120.

[24] Kumamoto, H., H. Ohtsuka, and K. Inoue, "Expected Number of Failures Caused by Protective Systems," *IEEE Trans. Rel.*, R-31 (2), June 1982, New York, pp. 219–221.

[25] Takami, I., K. Inoue, E. Sakino, and H. Kumamoto, "Optimal Reliability Design of k-out-of-n Systems," *Reliability in Electrical and Electronic Components and Systems*, E. Langer and J. Moltoft, eds., North-Holland Pub. Co., 1982, pp. 201–205.

[26] Kohda, T., K. Inoue, H. Kumamoto, and T. Takami, "Optimal Logical Structure of Safety Monitoring Systems of Different Types,", *Reliability in Electrical and Electronic Components and Systems*, E. Langer and J. Moltoft, eds., North-Holland Pub. Co., pp. 446–450.

[27] Kohda, T., H. Kumamoto, and K. Inoue, "Optimal Shut-Down Logic for Protective Systems," *IEEE Trans. Rel.*, R-32 (1), April 1983, New York, pp. 26–29.

[28] Kohda, T., K. Akinari, H. Kumamoto, and K. Inoue, "Reliability Evaluation of Protective System with Scheduled and Unscheduled Maintenance," *Microelectronic Reliability*, 24, 1984, pp. 957–960.

[29] Kohda, T., H. Kumamoto, K. Inoue, and E. J. Henley, "Optimal Logic for Multi-Channel Protective Systems During On-Line Maintenance," *IEEE Trans. Rel.*, R-36 (1), April 1987, New York, pp. 25–31.

[30] Singh, C., and A. D. Patton, "Protection System Reliability Modeling: Unreadiness Probability and Mean Duration of Undetected Faults," *IEEE Trans. Rel.*, R-29 (4), October 1980, New York.

[31] Yip, H. T., G. C. Weller, and R. N. Allan, "Reliability Evaluation of Protection Devices in Electrical Power Systems," *Proc. Fourth National Reliability Conference*, (1), July 1983, pp.2c/4/1–2c/4/22.

[32] Gantner, J., and S. H. Horowitz, "Use of Equipment Built-In Automatic Testing, Self-Checking and Monitoring with a View to Improving Reliability," summary of a Working Group 34.03 report, *Electra* (111), February 1987, full text of the Working Group report available from CIGRÉ.

[33] Monseu, M., and S. H. Horowitz, "Evaluation of Characteristics and Performance of Power System Protection," summary of a Working Group 34.04 report, *Electra* (111), February 1987, full text of the Working Group report available from CIGRÉ.

[34] Anderson, P. M., and B. K. LeReverend, "Industry Experience with Special Protection Schemes," *IEEE Trans. on Power Systems*, 11 (3), August 1996, pp. 1166–1179.

[35] Anderson, P. M., and S. K. Agarwal, "An Improved Model for Protective-System Reliability," *IEEE Trans. on Reliability*, 41 (3), September 1992, pp. 422–26.

[36] Anderson, P. M., R. F. Ghajar, G. M. Chintaluri, and S. M. Magbuhat, "A New Reliability Model for Redundant Protective Systems - Markov Models," *IEEE Trans. on Power Systems.*, v. 12 (2), 1997, pp. 573–578.

[37] Kumar (Agarwal), S., and R. Billinton, "Graph Theory Concepts in Frequency and Availability Analysis," *IEEE Trans. on Rel.*, R-34 (4), 1985, New York, pp. 290–294.

PROBLEMS

28.1 Verify and explain (28.2).

28.2 A protective device is known to have the following failure and repair rates:

$$\lambda = 0.1 \text{ failures/yr}$$
$$\mu = 0.1 \text{ repairs/hr}$$

Compute the failure rate of the protection system under the following conditions:
(a) The protection is not redundant.
(b) The protection is redundant.
Determine these failure rates at times of 10, 100, 1000, and 10,000 hr. Assume that the protection was good when installed at time $t = 0$. Experiment with doubling the failure rate and the repair rate and note the effect on the probability of failure.

28.3 Show that the expected value of an exponentially distributed component is the inverse of the failure rate, as given in (28.7).

28.4 Verify (28.13).

28.5 Verify (28.16).

28.6 Verify (28.20).

28.7 Verify (28.27).

28.8 Construct a spreadsheet to analyze the unreadiness probability described in Section 28.5. Using your spreadsheet, plot the value of C for a system failure rate of 0.5 failures per year and a protection failure rate of 0.05 failures per year.

28.9 Using only the definitions of the state probabilities for the unreadiness probability problem, prove that (28.68) is correct.

28.10 Derive (28.33).

28.11 Verify the computation of the state probabilities for the system of Figure 28.15.

28.12 A method is given in reference 40 to determine the probabilities of the various states of a Markov graph using the concepts of graph theory. Use this method to determine the state probabilities of the system of Figure 28.15.

28.13 In their paper on unreadiness probabilities, Singh and Patton introduce the parameters E_s and E_p that are defined as follows:

$$E_s = \frac{\theta_p}{\theta_p + \lambda_s}$$
$$E_p = \frac{\theta_p}{\theta_p + \lambda_s + \lambda_p}$$

Show that the following are true.

$$D = \frac{\lambda_s + \lambda_p}{\lambda_s}\left(\frac{1 - E_s}{\lambda_s} + \frac{1 - E_s}{\mu_s} + \frac{E_s}{\mu_p}\right)$$

$$p_1 = \frac{1}{D}$$

28.14 A model of a power system component protected by only one protective system has been investigated in [35]. The Markov graph for that model is shown in Figure P28.14.

This system is small enough that it is possible to write explicitly the probability of all system states. One method for doing this is a graph theory concept, described in [37]. Use this method to write out the probability of states 1–8 and prepare a plot of the abnormal unavailability of the system.

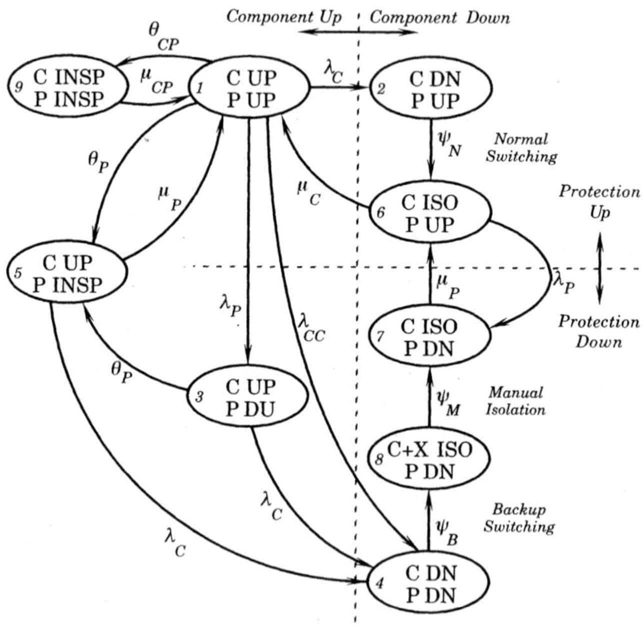

Figure P28.14 Markov graph of a component with only one protection.

28.15 Verify (28.80).

28.16 Use the results of problem 28.15 to evaluate the inside integral of expression (28.79).

28.17 Use the result of problem 28.16 to complete the integration of (28.79).

28.18 Property 1 for the expected number of failures of a safeguard specifies that both $DT \ll 1$ and $\lambda T \ll 1$. Prove the result given by (28.87).

28.19 Property 2 for the expected number of failures of a safeguard specifies that D is very large. Use this assumption to prove (28.88).

A

Protection Terminology

There are many terms and definitions used in power system protection that must be clearly understood prior to any detailed discussion of the subject. This appendix presents some of the terms that are required for a preliminary understanding. Other terms are defined throughout the text.

Terms and definitions related to the general subject of power system protection are presented first. This is followed by terms used in relaying and a classification of relays used in the industry. Finally, circuit breaker terms and definitions are discussed.

A.1 PROTECTION TERMS AND DEFINITIONS

There are many terms used in the publications and discussions associated with power system protection. We present here some of the basic terms and provide formal definitions of each.

We begin with definitions associated generally with power system protection. These definitions have been presented formally in the IEEE standards related to nuclear power generating stations. Several of these general definitions form an excellent basis for power systems in general. These definitions follow:

Protection system The electric and mechanical devices and circuitry, from sensors of the process variable to the actuation device input terminals, involved in generating those signals associated with the protective function [1], [2].

Protective function The sensing of one or more variables associated with a particular generating station condition, the signal processing, and the initiation and completion of the protective action at values of the variables established in the design bases [1], [3], [4].

Protective action The initiation of the operation of a sufficient number of actuators to effect a protective function [1], [2–5].

In some complex process industries, such as electric power generating plants, there are many systems that require protective system surveillance and action. Examples include the plant electric systems, hydraulic systems, pneumatic systems, safety systems, and others.

Electric power transmission and distribution system protection deals almost exclusively with electric system protection. This is also an important aspect of electric power generating stations. This book concentrates on the protection of electric systems, although many of the principles can be applied to other problems.

The protective function, in electric system protection, is provided by *relays*, which are defined as follows:

> *Relay (general)* An electric device that is designed to interpret input conditions in a prescribed manner and, after specified conditions are met, to respond to cause contact operation or similar abrupt change in associated electric control circuits.
>
> *Note A* Inputs are usually electrical, but may be mechanical, thermal, or other quantities. Limit switches and similar simple devices are not relays.
>
> *Note B* A relay may consist of several units, each responsive to specified inputs, the combination providing the desired performance characteristic [1], [6–11].

The actuator that provides the protective action, in an electric system, is usually a device called a circuit breaker, which is defined as follows:

> *Circuit breaker (general)* A mechanical switching device capable of making, carrying, and breaking currents under normal circuit conditions, and also, making, carrying for a specified time, and breaking currents under specified abnormal circuit conditions, such as those of short circuit [1], [12], [13].
>
> *Note 1* A circuit breaker is usually intended to operate infrequently, although some types are suitable for frequent operation.
>
> *Note 2* The medium in which circuit interruption is performed may be designated by a suitable prefix, such as, air-blast circuit breaker, air circuit breaker, compressed-air circuit breaker, gas circuit breaker, oil circuit breaker, vacuum circuit breaker, etc.
>
> *Note 3* Circuit breakers are classified according to their application or characteristics and these classifications are designated by the following modifying words or clauses delineating the several fields of application, or pertinent characteristics:

High-voltage power	Rated 1.0 kV ac and above
Low-voltage power	Rated below 1.0 kV
Other (see ANSI Standards [14])	

Finally, we introduce a general term that is often used in system protection.

> *Switchgear* A general term covering switching and interrupting devices and their combination with associated control, instrumentation, metering, protective, and regulating devices, also assemblies of these devices with associated interconnections, accessories, and supporting structures used primarily in connection with the generating, transmission, distribution, and conversion of electric power [1].

Switchgear is a broad term that covers all of the hardware usually associated with the protective system, including both function and actuation.

A.2 RELAY TERMS AND DEFINITIONS

In addition to the protective system definitions, it is important for the student of system protection to be familiar with definitions used in connection with the relays themselves. These definitions are, for the most part, taken from IEEE Standard 313 [11], which contains many definitions associated with relaying and protective systems in general. We cite here only those that are considered necessary as a beginning. Other definitions are introduced throughout the book as needed.

Auxiliary relay A relay whose function is to assist another relay or control device in performing a general function by supplying supplemental actions.

Relay logic is often designed with more than one relay required to achieve the desired action. Auxiliary relays are often used to provide the necessary supplementary action, such as providing additional current-carrying capacity, providing a seal-in function, adding time delay, or providing some kind of interlock.

Burden The load impedance imposed by a relay on an input circuit, expressed in ohms and phase angle at a specified condition.

The relay represents an electrical load to the input device; for example, an overcurrent relay has its input supplied by a current transformer. The input device has a rating that states its capability of providing the required quantity. The relay burden gives the impedance of the relay, such that the rating of the input device can be determined.

Contact A conducting part that acts with another conducting part to make or break a circuit.

Back contact ("b" contact) A contact that is closed when the relay is reset.

Front contact ("a" contact) A contact that is closed when the relay is picked up.

Contact opening time The time a contact remains closed while in process of opening following a specified change of input.

Contacts are depicted in relay circuit diagrams as shown in Figure A.1. Note that the "a" contact is always shown in drawings with the contact open, even though the contacts may be normally closed during operation. Similarly, the "b" contact is always shown as normally closed, even though its normal operating state may occur with the input quantity having sufficient magnitude to hold the contact open.

Figure A.1 ANSI standard graphic symbols for "a" and "b" contacts [15]. (*a*) "a" contact. (*b*) "b" contact.

We usually think of a contact as one of the precious metals (gold or silver) that is caused to connect or "make" with a second contact to complete a circuit mechanically. Actually, this can also be accomplished electronically, for example, by means of a thyristor.

Dropout A term for contact operation (opening or closing) as a relay just departs from pickup. Also, the maximum value of an input quantity that will allow a relay to depart from pickup.

Input A physical quantity or quantities to which a relay is designed to respond.

Memory action A method of retaining an effect of an input after the input ceases or is greatly reduced so that this (stored) input can still be used to produce the relay response.

Pickup The action of a relay as it makes designated response to progressive increase of input. Also, the minimum value of an input quantity reached by progressive increase that will cause the relay to reach pickup state from reset.

Polarization A term identifying the input that provides a reference for establishing the direction of system phenomena, such as the direction of power flow or direction to a fault or other disturbance.

Relay An electric device designed to respond to input conditions in a prescribed manner and, after specified conditions are met, to respond to cause contact operation or similar change in associated electric control circuits.

Electromagnetic relay An electromechanical relay magnet element, which is energized by the inputs quantity.

Static Relay A relay (or relay unit) in which the designated response is developed by electronic, solid state, magnetic, or other components without mechanical motion. A relay which is composed of both static and electromechanical units in which the designed response is accomplished by static units may be referred to as a static relay.

Reset The action of a relay as it makes designated response to decrease in input. Reset, as a qualifying term, denotes the state of a relay when all response to decrease in input has been completed. Reset is also used to identify the maximum value of an input quantity reached by progressive decreases that will permit the relay to reach the state of complete reset from pickup. (Note: In defining the designated performance of relays having multiple inputs, reset describes the state when all inputs are zero and also when some input circuits are energized, if the resulting state is not altered from the zero-input condition.)

Electrically reset relay A relay that is so constructed that it remains in the picked-up condition even after the input quantity is removed; an independent electric input is required to reset the relay.

Hand reset relay A relay that is so constructed that it remains in the picked up condition even after the input quantity is removed; specific manual reset action is required to reset the relay.

Self-reset relay (*automatically reset relay*) A relay that is so constructed that it returns to its reset position following an operation after the input quantity is removed.

The terms pickup, reset, and dropout are further described with the aid of the following discussion. Consider the two relay contacts shown in Figure A.2 and imagine that these contacts are part of an electromagnetic relay that is actuated by current to close a clapper on an electromagnetic relay circuit. Neglecting saturation, the force f acting on the member that holds the moving contact varies as the square of the current. The moving part is restrained from moving to the right by the spring "S." The contacts shown are "a" contacts since they are normally open when there is no current, hence no external force acting in the closing direction.

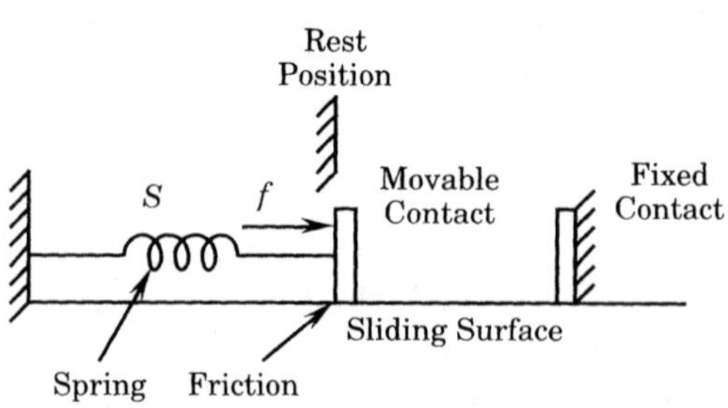

Figure A.2 Diagram of the contacts of a simple electromechanical relay.

Now imagine that an input current flows that is just sufficient to hold the contacts closed. This value of current is called the "pickup" current. A slight reduction in this input current will cause the contacts to part, but they do not return to the rest position since there is still sufficient current and friction to balance the force of the spring. Under this situation, we say that the relay has "dropped out." If the current is not reduced such that the relay movable contact returns to the rest position, opening the contacts, we say that the relay has "reset," and the value of current that just permits this to occur is the reset value. Reset action will occur, after a time delay, once the spring force exceeds the electromagnetic force of the current, plus any restraining forces such as friction, stiction, and other mechanical restraints.

(We note that, in magnetic devices that reduce an air gap when picking up, the current that will hold the relay in a picked-up state may be less than the pick-up current, since the air gap reduces during operation, requiring less current to produce the same force.)

Note that the rest position defined in Figure A.1 is not necessarily the zero current position. For example, it may be part of the relay design that a set of "b" contacts are attached that just open when the current exceeds that corresponding to the rest position. Any current below that which corresponds to the rest position current will cause the "b" contacts to remain closed.

Seal-in relay An auxiliary relay that remains picked up through one of its own contacts, which bypasses the initiating circuit until deenergized by some other device.

Setting (*noun*) The desired characteristic, obtained as a result of having set a device, stated in terms of calibration markings or of actual performance bench marks, such as pickup current and operating time at a given value of input.

Trip-free relay An auxiliary relay whose function is to open the closing circuit of an electrically operated switching device so that the opening operation can prevail over the closing operation.

The foregoing definitions are used repeatedly in the study of relays and protective systems in general. For additional definitions, the reader is referred to IEEE Standard 313 [11] and the IEEE Dictionary [1].

A.3 CLASSIFICATION OF RELAY SYSTEMS

Relays and relay systems may be classified by functions, by input, by operating principle, and by performance characteristics. The following tabulations give examples of each classification. The tabulations are not necessarily exhaustive, but are presented to illustrate the various forms of usage that have evolved in this discipline. We begin by presenting a classification of relays by functions, which are necessarily rather broad classifications. These classifications are presented in Table A.1, on the following page.

A second classification is according to the input quantities to which the relay responds. The terms "over" and "under" are qualifying terms which may be used to prefix the input quantities, thereby classifying the relay on a performance basis. Other qualifying terms are "ground," "residual," and "neutral." Classifications according to input are given in Table A.2.

A third classification is according to operating principle or structure. A list of the members of this classification are given in Table A.3.

Table A.4 classifies relays by the relay performance characteristics. In Table A.4, the general performance prefix "over" indicates that the relay will operate for values of the input in excess of a predetermined threshold, while the prefix "under" indicates operation for input less than a given threshold. In some cases the relay operates for values both greater than and less than a predetermined value and are called over-and-under input relays.

TABLE A.1 Classification of Relays by Function [11]

Function	Examples
Protective relay	See Tables A.2, A.3, A.4
Regulating relay	See Tables A.2, A.3, A.4
Monitoring relay	Alarm, fault detector, network phasing, synchronism check, verification
Programming relay	Accelerating, initiating, network master, phase selection, reclosing, synchronizing
Auxiliary relay	Blocking, closing, control, lockout, receiver, seal-in, timing, trip tree

TABLE A.2 Classification of Relays by Input [11]

Current relay	Pressure relay
Flow relay	Temperature relay
Power relay	Voltage relay

TABLE A.3 Classification of Relays by Operating Principle or Structure* [11]

Balance	Harmonic restraint	Quotient
Current balance	Latching	Replica temperature
Electrically reset	Mechanically reset	Restraint
Electromagnetic	Multirestraint	Self-reset
Electromechanical	Notching	Static
Gas accumulator	Percentage differential	Sudden pressure
Gas pressure	Phase sequence	Thermal
Hand reset	Product	

*Add the word "relay" to each entry

TABLE A.4 Classification of Relays by Performance Characteristics* [11]

Conductance	Mho	Positive sequence
Definite minimum time	Modified impedance	Power
Definite time	Negative sequence	Rate of change
Differential	Neutral	Reactance
Directional overcurrent	Open phase	Residual
Directional power	Overcurrent	Resistance
Directional	Overload	Reverse current
Distance	Overvoltage	Sector impedance
Frequency	Phase balance	Susceptance
Ground	Phase comparison	Time overcurrent
High speed	Phase	Undercurrent
Inverse time	Phase reversal	Undervoltatge
Linear impedance	Phase undervoltage	Zero sequence

*Add the word "relay" to each entry

In addition to the above there are some general classifications that are often useful. These are given in Table A.5.

TABLE A.5 Some General Relay Classifications

Classification	Types
Speed	Slow speed High speed (50 ms or faster)
Intelligence source	Local Local plus remote Global
Computation	Analog Digital
Response designation	Static Mechanical action
Nature of response	Instantaneous Time delayed Definite time Inverse time versus current

A brief study of the various relay classifications gives an indication of the extent of the effort that has been invested in this activity, both in protective device design and in the application of relay devices. This book is more concerned with applications, but it is obvious that some knowledge of available equipment is necessary.

A.4 CIRCUIT BREAKER TERMS AND DEFINITIONS

This is not a book with principal emphasis on circuit interruption or arc extinction. Nonetheless, these topics are important in the science of protection and deserve at least a brief consideration. A few definitions are presented here that will clarify the terminology related to circuit breakers and circuit interruption. Most of these terms are from ANSI Standards [16], [17], although the wording has been changed in some cases to make the definitions more readable in the present setting.

Circuit breaker A circuit breaker is a device for closing, carrying, and interrupting a circuit between separable contacts under both load and fault conditions as prescribed in (C37.4) standards.

Circuit breakers are further defined in terms of the medium in which its contacting members and the circuit closing and interruption occur. This includes oil circuit breakers, air circuit breakers, compressed air circuit breakers, magnetic air circuit breakers, and oilless (other than oil) circuit breakers.

Circuit breakers are also classified in terms of voltage.

Operating voltage The operating voltage of a circuit breaker is the rms line-to-line voltage of the system on which it is operated.

Recovery Voltage The recovery voltage is the voltage that occurs across the terminals of a pole of an ac circuit interrupting device upon interruption of current.

The phenomenon of interruption, or failure of the interruption process, is of considerable interest. Thus we define the following:

Restrike A restrike is a resumption of current between the contacts of a circuit interrupting device during an opening operation after an interval of zero current of 1/4 cycle at normal frequency or longer.

Reignition Reignition is a resumption of current between the contacts of a circuit interrupting device during an opening operation after an interval of zero current of less than 1/4 cycle at normal frequency.

It is well known that, when a fault occurs on a three-phase system, the currents in the three phases immediately following the application of the fault will exhibit dc offset in varying amounts, depending on the exact time of the fault with respect to the fundamental ac current wave. This requires the following definitions relating to circuit breaker currents.

Total short-circuit current (*asymmetrical*) The total short circuit current is the combination of the symmetrical component and the dc component, if any, of the short circuit current.

Symmetrical component (*ac component*) The symmetrical component of the short-circuit current is the normal-frequency ac component of the total current in rms amperes.

DC component The dc component of the normal-frequency short-circuit current is that portion of the total short-circuit current that constitutes the asymmetry.

Degree of asymmetry The degree of asymmetry of a current at any time is the ratio of the dc component to the peak value of the symmetrical component, determined from the envelope of the current wave at that time.

Making current The making current of a circuit breaker, when closed on a short circuit, is the rms value of the total current, which is measured from the envelope of the current wave at the time of its first major peak.

Another group of definitions deals with construction details of the circuit breaker and its auxiliary control circuits. Since the relay system interacts directly with some of these auxiliary control circuits, these definitions are of direct interest.

Auxiliary switch An auxiliary switch is a switch operated by the main switching device for signaling, interlocking, or other purposes.

Auxiliary switches in circuit breakers are classified as "a," "b," "aa," "bb," and "LC" for the purpose of specifying definite contact positions with respect to the main circuit breaker contacts. Thus, we have the following terms:

"a" auxiliary switch An "a" auxiliary switch is one that is open when the main switching device contacts are open.

As an aid to the memory, think of "a" meaning in "agreement" with the main circuit breaker contacts.

"b" auxiliary switch A "b" auxiliary switch is one that is closed when the main switching device contacts are open.

As an aid to the memory, think of "b" as meaning "backwards" from the main circuit breaker contacts.

"aa" auxiliary switch An "aa" auxiliary switch is one that is open when the main switching device operating mechanism is in the de-energized or non-operated position.

"bb"auxiliary switch A "bb" auxiliary switch is one that is closed when the main switching device operating mechanism is in the de-energized or non-operated position.
"LC"auxiliary switch An "LC" auxiliary switch is one that indicates the readiness of the mechanical linkage of the operating mechanism to close the breaker.

The abbreviation "LC" means "latch checking." A latch checking auxiliary switch is closed when the circuit breaker mechanism linkage is relatched after an opening operation of the breaker.

Circuit breaker mechanism A circuit breaker mechanism is the complete assembly of levers and other parts that actuates the movable contacts of the circuit breaker.

The circuit breaker is constructed so that it may be operated electrically to initiate either closing or opening of the circuit breaker switching device. This action is initiated by causing current to flow in coils provided for both closing and tripping.

Closing Coil A closing coil of a circuit breaker is a coil used in the electromagnet that supplies power for closing a circuit breaker.
Trip Coil A trip coil of a circuit breaker is a coil used in the electromagnet that initiates the opening of a circuit breaker.

It is also useful to consider certain functional definitions, such as the following:

Automatic tripping (automatic opening) Automatic tripping of a circuit breaker signifies the tripping of a circuit interrupter under predetermined or other conditions without the intervention of operating personnel. In other words, not *manual* tripping.

There are several types of automatic tripping that are of interest in system protection. The major types are the following:

Series overcurrent tripping Series overcurrent tripping signifies the tripping of a circuit breaker from a trip coil in series with the main circuit, responsive to an increase in the main circuit current above a predetermined value.
Shunt tripping Shunt tripping signifies the tripping of a circuit breaker by a trip coil energized from the same or a separate circuit or source of power; the trip coil circuit being closed through a relay, switch, or other means.
Transformer overcurrent tripping Transformer overcurrent tripping signifies the tripping of a circuit breaker from a trip coil in series with the secondary windings of a current transformer whose primary winding is in series with the main circuit current above a predetermined value.

In addition to automatic tripping, the circuit breaker may be operated manually.

Non-automatic tripping Non-automatic tripping signifies the tripping of a circuit interrupter only in response to an act of operating personnel.

There are also several definitions regarding the term "trip free" as applied to circuit breakers. These definitions include the following:

Trip Free A circuit breaker is trip free when the tripping mechanism can trip the breaker even though the normal closing action is applied.
Mechanically trip free A circuit breaker is trip free when the tripping mechanism can trip even though:

1. In a manually operated circuit breaker, the operating lever is being moved toward the closed position, or
2. In a power-operated circuit breaker, the operating mechanism is being moved toward the closed position either by continued application of closing power or by means of a maintenance closing lever.

Electrically trip free (*anti-pump*) An electrically operated circuit breaker is electrically trip free when the tripping mechanism can trip even though the closing control circuit is energized, and the closing mechanism will not reclose it after tripping until the closing control circuit is opened and again closed. However, the breaker may be held closed by the maintenance closing lever, unless it is also mechanically trip free.

It is important that circuit breakers be trip free for safety reasons. Suppose, for example, a breaker is positioned in a substation and protects a radial line. Suppose further that the line is undergoing maintenance and is permanently grounded by the maintenance personnel. Electrically, the line has a permanent three-phase fault to ground. Now, should an operator attempt to close the circuit breaker by manually energizing the closing coil, a trip-free breaker will trip the breaker even though the operator continues to hold the closing switch handle in the closed position.

REFERENCES

[1] Jay, F., Ed., *IEEE Standard Dictionary of Electrical and Electronics Terms, 2nd Ed.*, IEEE Std 100-1977, Published by IEEE, New York and distributed by John Wiley and Sons, New York.

[2] IEEE Std 380-1975, "Definitions of Terms Used in IEEE Nuclear Power Generating Stations Standards," IEEE, New York, 1975.

[3] IEEE Std-279-1971 (ANSI N42.7-1972), "Criteria for Protection Systems for Nuclear Power Generating Stations," IEEE, New York, 1971.

[4] IEEE Std 379-1972 (ANSI N41.2-1972), "Guide for the Application of the Single Failure Criterion to Nuclear Power Generating Station Protection Systems," IEEE, New York, 1972.

[5] IEEE Std 379-1976, a 1976 revision of reference 4.

[6] ANSI C37.100-1975, "Definitions for Power Switchgear," American National Standards Institute, New York, 1975.

[7] IEEE Std 474-1973, "Specifications and Test Methods for Fixed and Variable Attenuators, DC to 40 GHz," IEEE, New York, 1973.

[8] IEEE Power System Relaying Committee, Power Engineering Society 31.8, IEEE, New York.

[9] IEEE Switchgear Committee, Power Engineering Society 31.11, IEEE, New York.

[10] IEEE Industrial Control Committee, IAS 34.10, IEEE, New York.

[11] IEEE Std 313-1971 (ANSI C37.90-1971), "Relays and Relay Systems Associated with Electric Power Apparatus," IEEE, New York, 1971.

[12] ANSI C2.2-1960, "Safety Rules for the Installation and Maintenance of Electric Supply and Communications Lines," American National Standards Institute, New York, 1960.

[13] IEEE Std 27-1974 (C37.20-1974), "Switchgear Assemblies Including Metal Enclosed Bus," IEEE, New York, 1974.

[14] ANSI C37.100, 1975, "Definitions for Power Switchgear," American National Standards Institute, New York, 1975.

[15] IEEE Std 315-1971 (ANSI Y32.2-1970), "IEEE Standard and American National Standard Graphic Symbols for Electrical and Electronics Diagrams," IEEE, New York, 1971.

[16] ANSI C37.4-1953, "Alternating-Current Power Circuit Breakers," American National Standards Institute, New York, 1953.

[17] ANSI C37.4-1953, "Supplement to and Partial Revision to C37.4-1953, American National Standards Institute, New York, 1964.

B

Protective Device Classification

The devices in switching equipment are referred to by numbers, with appropriate suffix letters when necessary, according to the functions they perform.

These numbers are based on a system adopted as standard for automatic switchgear by IEEE, and incorporated in American Standard C37.2-1970. This system is used in connection diagrams, in instruction books, and in specifications.

B.1 DEVICE DEFINITION AND FUNCTION

1. **Master Element** is the initiating device, such as control switch, voltage relay, float switch, etc., which serves either directly, or through such permissive devices as protective relay system, except as specifically provided by device functions 48, 62, and 79 described later.
2. **Time-delay starting,** or **closing relay** is a device which functions to give a desired amount of time delay before or after any point of operation in a switching sequence or protective relay system, except as specifically provided by device functions 48, 62, and 79 described later.
3. **Checking** or **interlocking relay** is a device which operates in response to the position of a number of other devices, (or to a number of predetermined conditions), in an equipment, to allow an operating sequence to proceed, to stop, or to provide a check of the position of these devices or of these conditions for any purpose.
4. **Master contactor** is a device, generally controlled by device No.1 or equivalent, and the required permissive and protective devices, that serves to make and break the necessary control circuits to place an equipment into operation under the desired conditions and to take it out of operation under other or abnormal conditions.
5. **Stopping device** is a control device used primarily to shut down an equipment and hold it out of operation. [This device may be manually or electrically actuated, but excludes the function of electrical lockout (see device function 86) on abnormal conditions.]

6. **Starting circuit breaker** is a device whose principal function is to connect a machine to its source of starting voltage.
7. **Anode circuit breaker** is one used in the anode circuits of a power rectifier for the primary purpose of interrupting the rectifier circuit if an arc back should occur.
8. **Control power disconnecting device** is a disconnective device such as a knife switch, circuit breaker, or pullout fuse block, used for the purpose of connecting and disconnecting the source of control power to and from the control bus or equipment. **Note:** Control power is considered to include auxiliary power which supplies such apparatus as small motors and heaters.
9. **Reversing device** is used for the purpose of reversing a machine field or for performing any other reversing functions.
10. **Unit sequence switch** is used to change the sequence in which units may be placed in and out of service in multiple-unit equipments.
11. Reserved for future application.
12. **Over-speed device** is usually a direct-connected speed switch which functions on machine over-speed.
13. **Synchronous-speed device**, such as centrifugal-speed switch, a slip-frequency relay, a voltage relay, and undercurrent relay or any type of device, operates at approximately synchronous speed of a machine.
14. **Under-speed device** functions when the speed of a machine falls below a predetermined value.
15. **Speed** or **frequency-matching device** functions to match and hold the speed or the frequency of a machine or of a system equal to, or approximately equal to, that of another machine, source or system.
16. Reserved for future application.
17. **Shunting** or **discharge switch** serves to open or to close a shunting circuit around any piece of apparatus (except a resistor), such as a machine field, a machine armature, a capacitor or a reactor. **Note:** This excludes devices which perform such shunting operations as may be necessary in the process of starting a machine by devices 6 or 42, or their equivalent, and also excludes device 73 function which serves for the switching of resistors.
18. **Accelerating** or **decelerating device** is used to close or to cause the closing of circuits which are used to increase or to decrease the speed of a machine.
19. **Starting-to-running transition contactor** is a device which operates to initiate or cause the automatic transfer of a machine from the starting to the running power connection.
20. **Electrically operated valve** is an electrically operated, controlled or monitored valve in a fluid line. **Note:** The function of the valve may be indicated by the use of the suffixes.
21. **Distance relay** is a device which functions when the circuit admittance, impedance, or reactance increases or decreases beyond predetermined limits.
22. **Equalizer circuit breaker** is a breaker which serves to control or to make and break the equalizer or the current-balancing connections for a machine field, or for regulating equipment, in a multiple-unit installation.
23. **Temperature control device** functions to raise or lower the temperature of a machine or other apparatus, or of any medium, where its temperature falls below, or rises

above, a predetermined value. **Note:** An example is a thermostat which switches on a space heater in a switchgear assembly when the temperature falls to desired value as distinguished from a device which is used to provide automatic temperature regulation between close limits and would be designated as 90T.

24. Reserved for future application.
25. **Synchronizing** or **synchronism-check device** operates when two ac circuits are within the desired limits of frequency, phase angle or voltage, to permit or to cause the paralleling of these two circuits.
26. **Apparatus thermal device** functions when the temperature of the shunt field or the armortisseur winding of a machine, or that of a load limiting or load shifting resistor or of a liquid or other medium exceeds a predetermined value; or if the temperature of the protected apparatus, such as a power rectifier, or of any medium decreases below a predetermined value.
27. **Undervoltage relay** is a device which functions on a given value of undervoltage.
28. **Flame detector** is a device that monitors the presence of the pilot or main flame in such apparatus as a gas turbine or a steam boiler.
29. **Isolating contactor** is used expressly for disconnecting one circuit from another for the purposes of emergency operation, maintenance, or test.
30. **Annunciator relay** is a non-automatically reset device that gives a number of separate visual indications upon the functioning of protective devices, and which may also be arranged to perform a lockout function.
31. **Separate excitation device** connects a circuit such as the shunt field of a synchronous converter, to a source of separate excitation during the starting sequence; or one which energizes the excitation and ignition circuits of a power rectifier.
32. **Directional power relay** is one which functions on a desired value of power flow in a given direction, or upon reverse power resulting from arc back in the anode or cathode circuits of a power rectifier.
33. **Position switch** makes or breaks contact when the main device or piece of apparatus, which has no device function number, reaches a given position.
34. **Master sequence device** is a device such as a motor-operated multicontact switch, or the equivalent, or a programming device, such as a computer, that establishes or determines the operating sequence of the major devices in an equipment during starting and stopping or during other sequential switching operations.
35. **Brush-operating,** or **slip-ring-short-circuiting, device** is used for raising, lowering, or shifting the brushes of a machine, or for short-circuiting its slip rings, or for engaging or disengaging the contacts of a mechanical rectifier.
36. **Polarity** or **polarizing voltage device** operates or permits the operation of another device on a predetermined polarity only or verifies the presence of a polarizing voltage in an equipment.
37. **Undercurrent** or **underpower relay** functions when the current or power flow decreases below a predetermined value.
38. **Bearing protective device** functions when the current or power flow decreases below a predetermined value.
39. **Mechanical condition monitor** is a device that functions upon the occurrence of an abnormal mechanical condition (except that associated with bearings as covered under device function 38), such as excessive vibration, eccentricity, expansion, shock, tilting, or seal failure.

40. **Field relay** functions on a given or abnormally low value or failure of machine field current, or on an excessive value of the reactive component of armature current in an ac machine indicating abnormally low field excitation.
41. **Field circuit breaker** is a device which functions to apply, or to remove, the field excitation of a machine.
42. **Running circuit breaker** is a device whose principal function is to connect a machine to its source of running or operating voltage. This function may also be used for a device, such as a contactor, that is used in series with a circuit breaker or other fault protecting means, primarily for frequent opening and closing of the circuit.
43. **Manual transfer** or **selector device** transfers the control circuits so as to modify the plan of operation of the switching equipment of some of the devices.
44. **Unit-sequence starting relay** is a device which functions to start the next available unit in a multiple-unit equipment on the failure or on the non-availability of the normally preceding unit.
45. **Atmospheric condition monitor** is a device that functions upon the occurrence of an abnormal atmospheric condition, such as damaging fumes, explosive mixtures, smoke, or fire.
46. **Reverse-phase,** or **phase-balance, current relay** is a relay which functions when the polyphase currents are of reverse-phase sequence, or when the polyphase currents are unbalanced or contain negative phase-sequence components above a given amount.
47. **Phase-sequence voltage relay** functions upon a predetermined value of polyphase voltage in the desired phase sequence.
48. **Incomplete sequence relay** is a relay that generally returns the equipment to the normal, or off, position and locks it out if the normal starting, operating, or stopping sequence is not properly completed within a predetermined time. If the device is used for alarm purposes only, it should preferably be designated as 48A (alarm).
49. **Machine,** or **transformer, thermal relay** is a relay that functions when the temperature of a machine armature, or other load carrying winding or element of a machine, or the temperature of a power rectifier or power transformer (including a power rectifier transformer) exceeds a predetermined value.
50. **Instantaneous overcurrent,** or **rate-of-rise relay** is a relay that functions instantaneously on an excessive value of current, or on an excessive rate of current rise, thus indicating a fault in the apparatus or circuit being protected.
51. **Ac time overcurrent relay** is a relay with either a definite or inverse time characteristic that functions when the current in an ac circuit exceeds a predetermined value.
52. **Ac circuit breaker** is a device that is used to close and interrupt an ac power circuit under normal conditions or to interrupt this circuit under fault or emergency conditions.
53. **Exciter** or **dc generator relay** is a relay that forces the dc machine field excitation to build up during starting or which functions when the machine voltage has built up to a given value.
54. Reserved for future application.
55. **Power factor relay** is a relay that operates when the power factor in an ac circuit rises above or below a predetermined value.

56. **Field application relay** is a relay that automatically controls the application of the field excitation to an ac motor at some predetermined point in the slip cycle.
57. **Short-circuiting** or **grounding device** is a primary circuit switching device that functions to short circuit or to ground a circuit in response to automatic or manual means.
58. **Rectification failure relay** is a device that functions if one or more anodes of a power rectifier fail to fire, or to detect an arc-back or on failure of a diode to conduct or block properly.
59. **Overvoltage relay** is a relay that functions on a given value of overvoltage.
60. **Voltage** or **current balance relay** is a relay that operates on a given difference in voltage, or current input or output of two circuits.
61. Reserved for future application.
62. **Time-delay stopping** or **opening relay** is a time-delay relay that serves in conjunction with the device that initiates the shutdown, stopping, or opening operation in an automatic sequence.
63. **Pressure switch** is a switch which operates on given values or on a given rate of change of pressure.
64. **Ground protective relay** is a relay that functions on failure of the insulation of a machine, transformer or of other apparatus to ground, or on flashover of a dc machine to ground. **Note:** This function is assigned only to a relay which detects the flow of current from the frame of a machine or enclosing case or structure of a piece of apparatus to ground, or detects a ground on a normally ungrounded winding or circuit. It is not applied to a device connected in the secondary circuit or secondary neutral of a current transformer, or in the secondary neutral of current transformer connected in the power circuit of a normally grounded system.
65. **Governor** is the assembly of fluid, electrical, or mechanical control equipment used for regulating the flow of water, steam, or other medium to the prime mover for such purposes as starting, holding speed or load, or stopping.
66. **Notching** or **jogging device** functions to allow only a specified number of operations of a given device, or equipment, or specified number of successive operations within a given time of each other. It also functions to energize a circuit periodically or for fractions of specified time intervals, or that is used to permit intermittent acceleration or jogging of a machine at low speeds for mechanical positioning.
67. **Ac directional overcurrent relay** is a relay that functions on a desired value of ac overcurrent flowing in a predetermined direction.
68. **Blocking relay** is a relay that initiates a pilot signal for blocking or tripping on external faults in a transmission line or in other apparatus under predetermined conditions, or cooperates with other devices to block tripping or to block reclosing on an out-of-step condition or on power swings.
69. **Permissive control device** is generally a two-position, manually operated switch that in one position permits the closing of a circuit breaker, or the placing of an equipment into operation, and in the other position prevents the circuit breaker or the equipment from being operated.
70. **Rheostat** is a variable resistance device used in an electric circuit, which is electrically operated or has other electrical accessories, such as auxiliary, position, or limit switches.

71. **Level switch** is a switch which operates on given values, or on a given rate of change, of level.
72. **Dc circuit breaker** is used to close and interrupt a dc power circuit under normal conditions or to interrupt this circuit under fault or emergency conditions.
73. **Load-resistor contactor** is used to shunt or insert a step of load limiting, shifting, or indicating resistance in a power circuit, or to switch a space heater in circuit, or to switch a light, or regenerative load resistor of a power rectifier or other machine in and out of circuit.
74. **Alarm relay** is a device other than an annunciator, as covered under device No. 30, which is used to operate, or to operate in connections with, a visual or audible alarm.
75. **Position changing mechanism** is a mechanism that is used for moving a main device from one position to another in an equipment; as for example, shifting a removable circuit breaker unit to and from the connected, disconnected, and test positions.
76. **Dc overcurrent relay** is a relay that functions when the current in a dc circuit exceeds a given value.
77. **Pulse transmitter** is used to generate and transmit pulses over a telemetering or pilot-wire circuit to the remote indicating or receiving device.
78. **Phase angle measuring,** or **out-of-step protective relay** is a relay that functions at a predetermined phase angle between two voltages or between two currents or between voltage and current.
79. **Ac reclosing relay** is a relay that controls the automatic reclosing and locking out of an ac circuit interrupter.
80. **Flow switch** is a switch which operates on given values, or on a given rate of change, of flow.
81. **Frequency relay** is a relay that responds to the frequency of an alternating electrical input quantity.
82. **DC reclosing relay** is a relay that controls the automatic closing and reclosing of a dc circuit interrupter, generally in response to load circuit conditions.
83. **Automatic selective control** or **transfer relay** is a relay that operates to select automatically between certain sources or conditions in an equipment, or performs a transfer operation automatically.
84. **Operating mechanism** is the complete electrical mechanism or servo-mechanism, including the operating motor, solenoids, position switches, etc., for a tap changer, induction regulator or any similar piece of apparatus which has no device function number.
85. **Carrier** or **pilot-wire receiver relay** is a relay that is operated or restrained by a signal used in connection with carrier-current or dc pilot-wire fault directional relaying.
86. **Locking-out relay** is an electrically operated, hand or electrically reset relay that functions to shut down and hold an equipment out of service on the occurrence of abnormal conditions.
87. **Differential protective relay** is an electrically operated, hand or electrically reset, relay that functions to shut down and hold an equipment out of service on the occurrence of abnormal conditions.
88. **Auxiliary motor** or **motor generator** is one used for operating auxiliary equipment such as pumps, blowers, exciters, rotating magnetic amplifiers, etc.

89. **Line switch** is used an a disconnecting load-interrupter, or isolating switch in an ac or dc power circuit, when this device is electrically operated or has electrical accessories, such as an auxiliary switch, magnetic lock, etc.
90. **Regulating device** functions to regulate a quantity, or quantities, such as voltage, current, power, at a speed, frequency, temperature, and load, at a certain value or between certain (generally close) limits for machines, tie lines or other apparatus.
91. **Voltage directional relay** is a relay that operates when the voltage across an open circuit break or contactor exceeds a given value in a given direction.
92. **Voltage and power directional relay** is a relay that permits or causes the connection of two circuits when the voltage difference between them exceeds a given value in a predetermined direction and causes these two circuits to be disconnected from each other when the power flowing between them exceeds a given value in the opposite direction.
93. **Field changing contactor** functions to increase or decrease in one step the value of field excitation on a machine.
94. **Tripping** or **trip-free relay** functions to trip a circuit breaker, contactor or equipment, or to permit immediate tripping by other devices; or to prevent immediate reclosure of a circuit interrupter, in case it should open automatically even though its closing circuit is maintained closed.

95, 96, 97. Used only for specific applications on individual installations where none of the assigned numbered functions from 1 to 94 is suitable.

B.2 DEVICES PERFORMING MORE THAN ONE FUNCTION

If one device performs two relatively important functions in an equipment so that it is desirable to identify both of these functions, this may be done by using a double function number and name such as

50/51 Instantaneous and Time-Overcurrent Relay

B.2.1 Suffix Numbers

If two or more devices with the same function number and suffix letter (if used) are present in the same equipment, they may be distinguished by numbered suffixes as for example, 52X-1, 52X-2 and 52X-3, when necessary.

B.2.2 Suffix Letters

Suffix letters are used with device function numbers for various purposes. In order to prevent possible conflict each suffix letter should have only one meaning in an individual equipment. All other words should use the abbreviations as contained in the latest revision of ANSI Y1.1, or should use some other distinctive abbreviation, or be written out in full each time they are used. The meaning of each single suffix letter, or combination of letters, should be clearly designated in the legend on the drawings or publications applying to the equipment.

Lowercase (small) suffix letters are used in practically all instances on electrical diagrams for the auxiliary, position, and limit switches. Capital letters are generally used for all other suffix letters.

The letters should generally form part of the device function designation, are usually written directly after the device function number, as for example, 52CS, 71W, or 49D. When it is necessary to use two types of suffix letters in connection with one function number, it is often desirable for clarity to separate them by a slanted line or dash, as for example, 2-D/CS or 20D-CS.

The suffix letters which denote parts of the main device, and those which cannot or need not form part of the device function destination, are generally written directly below the device function number on drawings, as for example,

$$\frac{52}{CC} \quad \text{or} \quad \frac{43}{A}.$$

B.2.3 Separate Auxiliary Devices

X	
Y	Auxiliary relay 1[1]
Z	
R	Raising relay
L	Lowering relay
O	Opening relay or contactor
C	Closing relay or contactor
CS	Control switch
CL	Auxiliary relay, open (energized when main device is in open position)
OP	Auxiliary relay, open (energized when main device in open position)
U	"Up" position-switch relay
D	"Down" position-switch relay
PB	Push button

B.2.4 Actuating Quantities

These letters indicate the **condition** or **electrical** quantity to which the device responds, or the medium in which it is located, such as

A	Air, or amperes or alternating
C	Current
D	Direct or discharge
E	Electrolyte
F	Frequency, or flow or fault
H	Explosive
J	Differential
L	Level, or liquid
P	Power, or pressure
PF	Power factor

[1]In the control of a circuit breaker with so-called X-Y relay control scheme, the X relay is the device whose main contacts are used to energize the closing coil or the device which in some other manner, such as by the release of stored energy, causes the breaker to close. The contacts of the Y relay provide the antipump feature for the circuit breaker.

Q	Oil
S	Speed or suction or smoke
T	Temperature
V	Voltage, volts, or vacuum
VAR	Reactive power
VB	Vibration
W	Water, or watts

B.2.5 Main Devices

These letters denote the **location of the main device in the circuit**, in which the device is used or the type of circuit or apparatus with which it is associated, when this is necessary, such as:

A	Alarm or auxiliary power
AN	Anode
B	Battery, or blower, or bus
BK	Brake
BL	Block (valve)
BP	Bypass
BT	Bus tie
C	Capacitor, or condenser, compensator, or carrier current or case or compressor
CA	Cathode
CH	Check (valve)
D	Discharge (valve)
E	Exciter
F	Feeder, or field, or filament, or filter, or fan
G	Generator, or ground[2]
H	Heater
L	Line or logic
M	Motor, or metering
N	Network
P	Pump or phase comparison
R	Reactor, or rectifier, or room
S	Synchronizing or secondary or strainer or sump or suction (valve)
T	Transformer, or thyratron
TH	Transformer (high-voltage side)
TL	Transformer (low-voltage side)
TM	Telemeter
U	Unit

[2]Suffix "N" is generally used in preference to "G" for devices connected in the secondary neutral of a machine or power transformer, except in the case of transmission line relaying, where the suffix "G" is more commonly used for those relays which operate on ground faults.

B.2.6 Main Device Parts

These letters denote **parts of the main device**, divided into the two following categories:

1. **All parts, except auxiliary contacts, position switches, limit switches, and torque limit switches.**

BK	Brake
C	Coil, or condenser, or capacitor
CC	Closing coil
HC	Holding coil
M	Operating motor
MF	Fly-ball motor
ML	Load-limit motor
MS	Speed adjusting, or synchronizing, motor
S	Solenoid
SI	Seal-in
TC	Trip coil
V	Valve

2. **All auxiliary contacts, position and limit switches** for such devices and equipment as circuit breakers, contactors, valves and rheostats, and contacts of relays. These are designated as follows:

a	Contact that is open when the main device is in that standard reference position, commonly referred to as the non-operated or de-energized position, that closes when the device assumes the opposite position
b	Contact that is closed when the main device is in the standard reference position, commonly referred to as the non-operated or de-energized position, and that opens when the device assumes the opposite position
aa	Contact that is open when the operating mechanism of the main device is in the non-operated position and that closes when the operating mechanism assumes the opposite position
bb	Contact that is closed when the operating mechanism of the main device is in the non-operated position and that opens when the operating mechanism assumes the opposite position

B.3 STANDARD REFERENCE POSITIONS OF SOME TYPICAL DEVICES

Device	Standard Reference Position
Power circuit breaker	Main contacts open
Disconnecting switch	Main contacts open
Load-break switch	Main contacts open
Valve	Closed position

Device	Standard Reference Position
Gate	Closed position
Clutch	Disengaged position
Turning gear	Disengaged position
Power electrodes	Maximum gap position
Rheostat	Maximum resistance position
Adjusting means[1]	Low or down position
Relay[2]	De-energized position
Contactor[2]	De-energized position
Relay (latched-in type)	See 2-9.7.2 (C37.2-1970)
Contactor (latched-in type)	Main contacts open
Temperature relay[3]	Lowest temperature
Level detector[3]	Lowest level
Flow detector[3]	Lowest flow
Speed switch[3]	Lowest speed
Vibration dectector[3]	Minimum vibration
Pressure switch[3]	Lowest pressure
Vacuum switch[3]	Lowest pressure, i.e., highest vacuum

Note: If several similar auxiliary switches are present on the same device, they should be designated numerically, 1, 2, 3, etc., when necessary.

The simple designation "a" or "b" is used in all cases where there is no need to adjust the contacts to change position at any particular point in the travel of the main device or where the part of the travel where the contacts change position is of no significance in the control or operating scheme. Hence the "a" and "b" designations usually are sufficient for circuit breaker auxiliary switches.

B.3.1 Other Switches

These letters cover **all other distinguishing features or characteristics or conditions**, which serve to describe the use of the device or it contacts in the equipment such as:

A	Accelerating, or automatic
B	Blocking, or backup
C	Close, or cold
D	Decelerating, detonate, down, or disengaged
E	Emergency or engaged
F	Failure, or forward

[1]These may be speed, voltage, current, load, or similar adjusting devices comprising rheostats, springs, levers, or other components for the purpose.

[2]These electrically operated devices are of the non-latched-in type, whose contact position is dependent only on the degree of energization of the operating or restraining or holding coil or coils which may or may not be suitable for continuous energization. The de-energized position of the device is that with all coils de-energized.

[3]The energizing influences for these devices are considered to be, respectively, rising temperature, rising level, increasing flow, rising speed, increasing vibration, and increasing pressure.

H	Hot, or high
HR	Hand reset
HS	High speed
L	Left, local, low, lower, or leading
M	Manual
OFF	Off
ON	On
P	Polarizing
R	Right, raise, reclosing, receiving, remote, or reverse
S	Sending, or swing
T	Test, or trip, or trailing
TDC	Time-delay closing
TDO	Time-delay opening
U	Up

B.3.2 Representation of Device Contacts on Electrical Diagrams

On electrical diagrams the "b" contacts of all devices, including those of relays and those with suffix letters or percentage figures, should be shown as closed contacts, and all "a" contacts should be shown as open contacts.

For those devices that have no de-energized or non-operated position, such as manually operated transfer or control switches (including those of the spring-return type) or auxiliary position indicating contacts on the housings or enclosures of a removable circuit breaker unit, the preferred method of representing these contacts is as an "a" switch. Each contact should, however, be identified on the elementary diagram as to when it closes.

In the case of latched-in or hand-reset locking-out relays, which operate from protective devices to perform the shutdown of an equipment and to hold it out of service, the contacts should preferably be shown in the normal non-locking-out position. In general, any devices, such as electrically operated latched-in relays, which have not been specifically covered in the above paragraphs should have their contacts shown in the position most suitable for the ready understanding of the operation of the devices in the equipment, and sufficient description should be present, as necessary, on the elementary diagram to indicate the contact operation.

C

Overhead Line Impedances

This appendix presents impedances that may be used for overhead lines of a wide variety of voltages and physical arrangements.

The positive sequence reactance for a three phase line is computed from the formula [1], [2].

$$\begin{aligned} x_1 &= 0.1213 \ln \frac{D_m}{D_s} \quad \Omega/\text{mile} \\ &= 0.1213 \ln D_m + 0.1213 \ln \frac{1}{D_s} \quad \Omega/\text{mile} \\ &= x_d + x_a \end{aligned} \tag{C.1}$$

where x_a = inductive reactance due to flux linkage out to one foot radius

x_d = inductive reactance due to flux linkages beyond one foot radius

The quantity x_a is tabulated in wire tables, such as C.1 and is often given in ohms per mile. The quantity x_d is a function only of the wire spacing. Clearly, D_m and D_s in (C.1) must be in the same units.

For distribution lines, where shorter line lengths are common, the inductive reactance is often given in ohms per thousand feet, as noted in Table C.1. In this case, it is convenient to compute the wire spacing in inches rather than feet, because of the closer spacing used at distribution voltages. With D_m in inches, we write from (C.1)

$$\begin{aligned} x_d &= 0.1213 \ln \frac{d_m}{12} \quad \Omega/\text{mile} = 0.1213 \ln D_m - 0.3014 \quad \Omega/\text{mile} \\ &= \frac{1}{5.28}(0.1213 \ln D_m - 0.3014) \quad \Omega/\text{kft} = 0.02297 \ln D_m - 0.057 \quad \Omega/\text{kft} \\ &= 0.05289 \log_{10} D_m - 0.057 \quad \Omega/\text{kft} \end{aligned} \tag{C.2}$$

The plot shown in Figure C.1 is convenient for use on problems where the spacing is measured in inches, such as distribution lines. Tables C.2a and C.2b give the resistance characteristics

TABLE C.1 Characteristics of Open Wire Distribution Lines

Wire Size			Diam in	Mass	Approx Amp		Resistance	Reactance X1
AWG	Strands	kCM	inches	lb/kft	Capacity (1)		in Ω/kft (2)	in Ω/kft (3)
				Copper – Hard Drawn				
8	(1)	16.51	.128	50	50	80	.656	.126
6	(1)	26.25	.162	80	70	110	.413	.121
4	(3)	41.74	.254	128	110	161	.263	.114
2	(7)	66.37	.292	205	145	210	.167	.109
1	(7)	83.69	.328	258	170	245	.132	.106
1/0	(7)	105.5	.368	326	200	285	.105	.1035
2/0	(7)	133.3	.414	411	240	335	.083	.101
3/0	(7)	167.8	.464	518	280	390	.066	.098
4/0	(7)	211.6	.522	653	330	450	.053	.095
	(19)	250.0	.574	772	375	510	.045	.092
	(19)	300.0	.629	926	425	575	.037	.090
	(19)	350.0	.679	1081	475	635	.032	.088
	Al/Steel			**ASCR**				
6	6/1	26.25	.198	36.2	55	85	.675	.128
4	6/1	41.74	.250	57.6	75	120	.425	.125
2	6/1	66.37	.316	91.6	110	165	.267	.126
1/0	6/1	105.54	.398	145.6	150	225	.168	.124
2/0	6/1	133.1	.477	183.7	175	260	.134	.122
3/0	6/1	167.8	.502	231.6	210	305	.106	.118
4/0	6/1	211.6	.563	292.1	245	355	.084	.110
	26/7	266.8	.642	366.8	290	410	.066	.088
	26/7	366.4	.721	462.4	340	480	.053	.086
	26/7	397.5	.783	546.4	380	535	.045	.084
	26/7	477.0	.858	655.7	430	605	.037	.082
	26/7	556.5	.927	765.0	480	670	.032	.080
	26/7	795.0	1.108	1093.0	620	850	.022	.076
	(Strands)			**All Aluminum – Hard Drawn**				
4	(7)		.232	39.0	75	115	.424	.114
2	(7)		.292	62.0	105	160	.267	.109
1/0	(7)		.368	98.5	145	215	.168	.103
2/0	(7)		.414	124.3	170	250	.134	.101
3/0	(7)		.464	156.7	200	290	.106	.098
4/0	(7)		.522	197.6	240	340	.084	.095
	(7)	266.8	.586	249.1	280	400	.066	.092
	(19)	336.4	.666	315.7	330	465	.053	.088
	(19)	397.5	.724	373.0	370	520	.045	.086
	(19)	477.0	.793	447.6	425	590	.037	.084
	(19)	556.5	.856	522.0	465	645	.032	.082
	(37)	795.0	1.026	746.0	605	820	.022	.079
Size				**Copperweld – Copper**				
8A			.199	74.3	60	90	.664	.127
6A			.230	101.6	84	120	.418	.123
4A			.290	161.5	115	165	.263	.118
2A			.366	256.8	185	220	.166	.112

(1) Conductor at 80 C, 40 C ambient, emissivity = 0.5 for Cu, 0.2 for A.
Lower current values correspond to still air, higher values to air moving at 2 ft/s

(2) Resistance of conductor in Ω/kft at 60 Hz, 25 C

(3) Reactance of conductor out to one ft in Ω/kft at 60 Hz
Total reactance = $x_a + x_d$ where x_d = external reactance beyond 1 ft obtained from curve in Figure C.1.

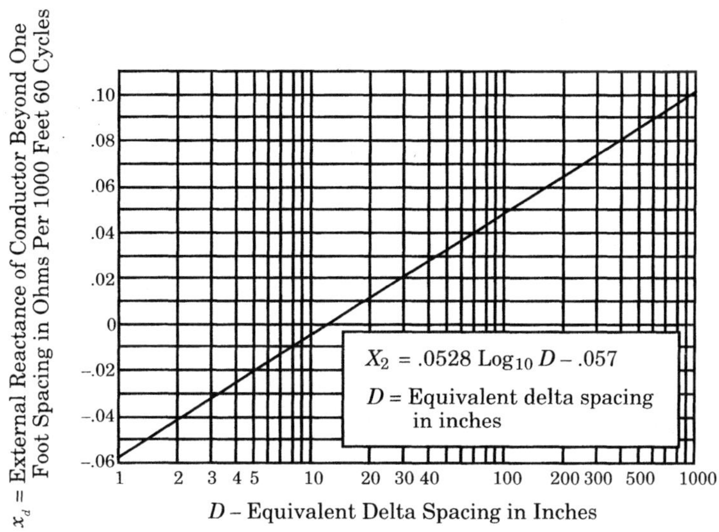

Figure C.1 External reactance as a function of equivalent delta spacing of conductors in a three-phase line.

for a wide range of ACSR conductors, which are often used for high voltage transmission lines. Tables C.3a and C.3b give the inductive reactances x_a for bundled conductors, which are commonly used for high-voltage transmission lines.

In Table C.3 we adopt the following notation for the bundled conductor circuits. x_a is the component of inductive reactance due to the magnetic flux within a 1 foot radius. The remaining component of inductive reactance, x_d is that due to the other phases. The total inductive reactance per phase is the sum of x_a and x_d.

The following formula can be used to calculate additional values of x_a.

$$x_a = 0.2794 \log_{10} \left[\frac{1}{\left[N(\text{GMR})A^{N-1} \right]^{1/N}} \right] \quad \Omega/\text{mile} \tag{C.3}$$

where GMR = geometric mean radius (ft)

N = number of conductors per phase

$A = \dfrac{S}{2\sin(\pi/N)}$ for $N > 1$

S = bundle spacing (ft)

The value for x_d is obtained from the formula

$$x_d = 0.2794 \log_{10}(\text{GMD}) \quad \Omega/\text{mile} \tag{C.4}$$

where GMD is the geometric mean distance between phases in feet. Tables C.3a and C.3b provide reactances x_a for 2 conductor bundles. Table C.3c provides data for 3 conductor bundles. Tables C.4a and C.4b give the inductive reactance spacing factor x_d. Finally, Table 5 gives data for common types of static wires.

TABLE C.2a Resistance of ACSR Conductors (Ω/mile)

Code	Area cmil	Strands AL	Strands ST	Diam. in	DC 25 C	AC – 60Hz 25 C	AC – 60Hz 50 C	AC – 60Hz 75 C	AC – 60Hz 100 C
Expanded	3108000	62/8	19	2.500	0.0294	0.0333	0.0362	0.0389	0.0418
Expanded	2294000	66/6	19	2.320	0.0399	0.0412	0.0453	0.0493	0.0533
Expanded	1414000	58/4	19	1.750	0.0644	0.0663	0.0728	0.0793	0.0859
Expanded	1275000	50/4	19	1.600	0.0736	0.0736	0.0808	0.0881	0.0953
Kiwi	2167000	72	7	1.737	0.0421	0.0473	0.0515	0.0552	0.0593
Bluebird	2156000	84	19	1.762	0.0420	0.0464	0.0507	0.0545	0.0586
Chukar	1780000	84	19	1.602	0.0510	0.0548	0.0599	0.0647	0.0696
Falcon	1590000	54	19	1.545	0.0567	0.0594	0.0653	0.0707	0.0763
Lapwing	1590000	45	7	1.502	0.0571	0.0608	0.0664	0.0719	0.0774
Parrot	1510500	54	19	1.506	0.0597	0.0625	0.0686	0.0744	0.0802
Nuthatch	1510500	45	7	1.466	0.0602	0.0636	0.0697	0.0755	0.0813
Plover	1431000	54	19	1.465	0.0630	0.0657	0.0721	0.0782	0.0843
Bobolink	1431000	45	7	1.427	0.0636	0.0668	0.0733	0.0794	0.0856
Martin	1351500	54	19	1.424	0.0667	0.0692	0.0760	0.0825	0.0890
Dipper	1351500	45	7	1.385	0.0672	0.0705	0.0771	0.0836	0.0901
Pheasant	1272000	54	19	1.382	0.0709	0.0732	0.0805	0.0874	0.0944
Bittern	1272000	45	7	1.345	0.0715	0.0746	0.0817	0.0886	0.0956
Grackle	1192500	54	19	1.333	0.0756	0.0778	0.0855	0.0929	0.1000
Bunting	1192500	45	7	1.302	0.0762	0.0792	0.0867	0.0942	0.1002
Finch	1113000	54	19	1.293	0.0810	0.0832	0.0914	0.0993	0.1080
Bluejay	1113000	45	7	1.259	0.0818	0.0844	0.0926	0.1019	0.1090
Curlew	1033500	54	7	1.246	0.0871	0.0893	0.0979	0.1070	0.1150
Ortolan	1033500	45	7	1.213	0.0881	0.0905	0.0994	0.1080	0.1170
Tanager	1033500	36	1	1.186	0.0885	0.0915	0.1010	0.1090	0.1180
Cardinal	954000	54	7	1.196	0.0944	0.0915	0.1060	0.1150	0.1250
Rail	954000	45	7	1.165	0.0954	0.0963	0.1080	0.1170	0.1260
Catbird	954000	36	1	1.140	0.0959	0.0978	0.1090	0.1180	0.1270
Canary	900000	54	7	1.162	0.1000	0.0987	0.1120	0.1220	0.1320
Ruddy	900000	45	7	1.131	0.1010	0.1020	0.1130	0.1230	0.1340

TABLE C.2b Resistance of ACSR Conductors (Ω/mile)

Code	Area cmil	Strands AL	Strands ST	Diam. in	DC 25 C	AC – 60Hz 25 C	AC – 60Hz 50 C	AC – 60Hz 75 C	AC – 60Hz 100 C
Mallard	795000	30	19	1.140	0.111	0.114	0.125	0.137	0.147
Drake	795000	26	7	1.108	0.112	0.114	0.125	0.137	0.147
Condor	795000	54	7	1.093	0.113	0.115	0.127	0.138	0.149
Cuckoo	795000	24	7	1.092	0.113	0.114	0.127	0.137	0.148
Tern	795000	45	7	1.063	0.114	0.116	0.128	0.139	0.150
Coot	795000	36	1	1.040	0.115	0.117	0.129	0.141	0.152
Redwing	715500	30	19	1.081	0.124	0.126	0.139	0.151	0.164
Starling	715500	26	7	1.051	0.125	0.126	0.139	0.151	0.164
Stilt	715500	24	7	1.036	0.126	0.127	0.141	0.153	0.165
Gannet	666600	26	7	1.014	0.134	0.135	0.149	0.162	0.176
Flamingo	666600	24	7	1.000	0.135	0.137	0.151	0.164	0.177
—	653900	18	3	0.953	0.140	0.142	0.156	0.171	0.184
Egret	636000	30	19	1.019	0.139	0.143	0.157	0.172	0.186
Grosbeak	636000	26	7	0.990	0.140	0.142	0.156	0.170	0.184
Rook	636000	24	7	0.977	0.142	0.143	0.157	0.172	0.186
Kingbird	636000	18	1	0.940	0.143	0.145	0.160	0.174	0.188
Swift	636000	36	1	0.930	0.144	0.146	0.161	0.175	0.189
Teal	605000	30	19	0.994	0.146	0.150	0.165	0.180	0.195
Squab	605000	26	7	0.966	0.147	0.149	0.164	0.179	0.193
Peacock	605000	24	7	0.953	0.149	0.150	0.165	0.180	0.195
Eagle	556500	30	7	0.953	0.158	0.163	0.179	0.196	0.212
Dove	556500	26	7	0.927	0.160	0.162	0.178	0.194	0.211
Parakeet	556500	24	7	0.914	0.162	0.163	0.179	0.196	0.212
Osprey	556500	18	1	0.879	0.163	0.166	0.183	0.199	0.215
Hen	477000	30	7	0.883	0.185	0.190	0.209	0.228	0.247
Hawk	477000	26	7	0.858	0.187	0.188	0.207	0.226	0.245
Flicker	477000	24	7	0.846	0.189	0.190	0.209	0.228	0.247
Pelican	477000	18	1	0.814	0.191	0.193	0.212	0.232	0.250
Lark	397500	30	7	0.806	0.222	0.227	0.250	0.273	0.295
Ibis	397500	26	7	0.783	0.224	0.226	0.249	0.271	0.294
Brant	397500	24	7	0.772	0.226	0.227	0.250	0.273	0.295
Chickade	397500	18	1	0.743	0.229	0.231	0.254	0.277	0.300
Oriole	336400	30	7	0.741	0.262	0.268	0.295	0.322	0.349
Linnet	336400	26	7	0.721	0.265	0.267	0.294	0.321	0.347
Merlin	336400	18	1	0.684	0.270	0.273	0.300	0.328	0.355
Ostrich	300000	26	7	0.680	0.297	0.299	0.329	0.359	0.389

TABLE C.3a Inductive Reactance x_a of ACSR Bundled Conductors in ohms per mile for 1 foot radius at 60 Hz

Code	Area cmil	Strands AL	Strands ST	Diam. in	GMR ft.	Single Cond.	2 – Conductor Spacing (In.) 6	9	12	15	18
Expanded	3108000	62/8	19	2.500	0.0900	0.2922	0.1881	0.1635	0.1461	0.1326	0.1215
Expanded	2294000	66/6	19	2.320	0.0858	0.2980	0.1910	0.1664	0.1490	0.1355	0.1244
Expanded	1414000	58/4	19	1.750	0.0640	0.3336	0.2088	0.1842	0.1668	0.1532	0.1422
Expanded	1275000	50/4	19	1.600	0.0578	0.3459	0.2150	0.1604	0.1730	0.1594	0.1484
Kiwi	2167000	72	7	1.737	0.0571	0.3474	0.2158	0.1912	0.1737	0.1602	0.1491
Bluebird	2156000	84	19	1.762	0.0588	0.3438	0.2140	0.1894	0.1719	0.1584	0.1473
Chukar	1780000	84	19	1.602	0.0536	0.3551	0.2196	0.1950	0.1775	0.1640	0.1529
Falcon	1590000	54	19	1.545	0.0523	0.3580	0.2211	0.1965	0.1790	0.1655	0.1544
Lapwing	1590000	45	7	1.502	0.0498	0.3640	0.2241	0.1995	0.1820	0.1685	0.1574
Parrot	1510500	54	19	1.506	0.0506	0.3621	0.2231	0.1985	0.1810	0.1675	0.1564
Nuthatch	1510500	45	7	1.466	0.0486	0.3670	0.2255	0.2009	0.1835	0.1699	0.1589
Plover	1431000	54	19	1.465	0.0494	0.3650	0.2245	0.1999	0.1825	0.1689	0.1579
Bobolink	1431000	45	7	1.427	0.0470	0.3710	0.2276	0.2030	0.1855	0.1720	0.1609
Martin	1351500	54	19	1.424	0.0482	0.3680	0.2260	0.2014	0.1840	0.1704	0.1594
Dipper	1351500	45	7	1.385	0.0459	0.3739	0.2290	0.2044	0.1869	0.1734	0.1623
Pheasant	1272000	54	19	1.382	0.0466	0.3721	0.2281	0.2035	0.1860	0.1725	0.1614
Bittern	1272000	45	7	1.345	0.0444	0.3779	0.2310	0.2064	0.1890	0.1754	0.1644
Grackle	1192500	54	19	1.333	0.0451	0.3760	0.2301	0.2055	0.1880	0.1745	0.1634
Bunting	1192500	45	7	1.302	0.0429	0.3821	0.2331	0.2085	0.1910	0.1775	0.1664
Finch	1113000	54	19	1.293	0.0436	0.3801	0.2321	0.2075	0.1901	0.1765	0.1655
Bluejay	1113000	45	7	1.259	0.0415	0.3861	0.2351	0.2105	0.1931	0.1795	0.1685
Curlew	1033500	54	7	1.246	0.0420	0.3847	0.2344	0.2098	0.1923	0.1788	0.1677
Ortolan	1033500	45	7	1.213	0.0402	0.3900	0.2370	0.2124	0.1950	0.1815	0.1704
Tanager	1033500	36	1	1.186	0.0384	0.3955	0.2398	0.2152	0.1978	0.1842	0.1732
Cardinal	954000	54	7	1.196	0.0402	0.3900	0.2370	0.2124	0.1950	0.1815	0.1704
Rail	954000	45	7	1.165	0.0386	0.3949	0.2395	0.2149	0.1975	0.1839	0.1729
Catbird	954000	36	1	1.140	0.0370	0.4000	0.2421	0.2175	0.2000	0.1865	0.1754
Canary	900000	54	7	1.162	0.0392	0.3930	0.2386	0.2140	0.1965	0.1830	0.1719
Ruddy	900000	45	7	1.131	0.0374	0.3987	0.2414	0.2168	0.1994	0.1858	0.1748
Mallard	795000	30	19	1.140	0.0392	0.3930	0.2386	0.2140	0.1965	0.1830	0.1719
Drake	795000	26	7	1.108	0.0373	0.3991	0.2416	0.2170	0.1995	0.1860	0.1749
Condor	795000	54	7	1.093	0.0370	0.4000	0.2421	0.2175	0.2000	0.1865	0.1754
Cuckoo	795000	24	7	1.092	0.0366	0.4014	0.2427	0.2181	0.2007	0.1871	0.1761
Tern	795000	45	7	1.063	0.0352	0.4061	0.2451	0.2205	0.2030	0.1895	0.1784
Coot	795000	36	1	1.040	0.0377	0.3978	0.2409	0.2163	0.1989	0.1853	0.1743
Redwing	715500	30	19	1.081	0.0373	0.3991	0.2416	0.2170	0.1995	0.1860	0.1749
Starling	715500	26	7	1.051	0.0355	0.4051	0.2446	0.2200	0.2025	0.1890	0.1779
Stilt	715500	24	7	1.036	0.0347	0.4078	0.2460	0.2214	0.2039	0.1904	0.1793
Gannet	666600	26	7	1.014	0.0343	0.4092	0.2467	0.2221	0.2046	0.1911	0.1800
Flamingo	666600	24	7	1.000	0.0355	0.4121	0.2481	0.2235	0.2061	0.1925	0.1815

TABLE C.3b Inductive Reactance x_a of ACSR Bundled Conductors in ohms per mile for 1 foot radius at 60 Hz

Code	Area cmil	Strands AL	Strands ST	Diam. in	GMR ft.	Single Cond.	2 – Conductor Spacing (In.) 6	9	12	15	18
—	653900	18	3	0.953	0.0308	0.4223	0.2532	0.2286	0.2111	0.1976	0.1865
Egret	636000	30	19	1.019	0.0352	0.4061	0.2451	0.2205	0.2030	0.1895	0.1784
Grosbeak	636000	26	7	0.990	0.0335	0.4121	0.2481	0.2235	0.2061	0.1925	0.1815
Rook	636000	24	7	0.977	0.0327	0.4150	0.2496	0.2250	0.2075	0.1940	0.1829
Kingbird	636000	18	1	0.940	0.0304	0.4239	0.2540	0.2294	0.2119	0.1984	0.1873
Swift	636000	36	1	0.930	0.0301	0.4251	0.2546	0.2300	0.2125	0.1990	0.1879
Teal	605000	30	19	0.994	0.0341	0.4099	0.2470	0.2224	0.2050	0.1914	0.1804
Squab	605000	26	7	0.966	0.0327	0.4150	0.2496	0.2250	0.2075	0.1940	0.1829
Peacock	605000	24	7	0.953	0.0319	0.4180	0.2511	0.2265	0.2090	0.1955	0.1844
Eagle	556500	30	7	0.953	0.0327	0.4150	0.2496	0.2250	0.2075	0.1940	0.1829
Dove	556500	26	7	0.927	0.0314	0.4200	0.2520	0.2274	0.2100	0.1964	0.1854
Parakeet	556500	24	7	0.914	0.0306	0.4231	0.2536	0.2290	0.2115	0.1980	0.1869
Osprey	556500	18	1	0.879	0.0284	0.4321	0.2581	0.2335	0.2161	0.2025	0.1915
Hen	477000	30	7	0.883	0.0304	0.4239	0.2540	0.2294	0.2119	0.1984	0.1873
Hawk	477000	26	7	0.858	0.0289	0.4300	0.2571	0.2325	0.2150	0.2015	0.1904
Flicker	477000	24	7	0.846	0.0284	0.4321	0.2626	0.2335	0.2161	0.2025	0.1915
Pelican	477000	18	1	0.814	0.0264	0.4410	0.2596	0.2380	0.2205	0.2070	0.1959
Lark	397500	30	7	0.806	0.0277	0.4352	0.2626	0.2350	0.2176	0.2040	0.1930
Ibis	397500	26	7	0.783	0.0264	0.4410	0.2639	0.2380	0.2205	0.2070	0.1959
Brant	397500	24	7	0.772	0.0258	0.4438	0.2681	0.2394	0.2219	0.2084	0.1973
Chickadee	397500	18	1	0.743	0.0241	0.4521	0.2647	0.2435	0.2260	0.2125	0.2014
Oriole	336400	30	7	0.741	0.0255	0.4452	0.2676	0.2401	0.2226	0.2091	0.1980
Linnet	336400	26	7	0.721	0.0243	0.4511	0.2731	0.2430	0.2255	0.2120	0.2009
Merlin	336400	18	1	0.684	0.0222	0.4620	0.2712	0.2485	0.2310	0.2175	0.2064
Ostrich	300000	26	7	0.680	0.0229	0.4583	0.2581	0.2466	0.2291	0.2156	0.2045

TABLE C.3c Inductive Reactance x_a of ACSR Bundled Conductors in ohms per mile for 1 foot radius at 60 Hz

Code	Area	Strands		Diam.	GMR	3 – Conductor Spacing (In.)				
	cmil	AL	ST	in	ft.	6	9	12	15	18
Expanded	3108000	62/8	19	2.500	0.0900	0.1535	0.1207	0.0974	0.0793	0.0646
Expanded	2294000	66/6	19	2.320	0.0858	0.1554	0.1226	0.0993	0.0813	0.0665
Expanded	1414000	58/4	19	1.750	0.0640	0.1673	0.1345	0.1112	0.0931	0.0784
Expanded	1275000	50/4	19	1.600	0.0578	0.1714	0.1386	0.1153	0.0973	0.0852
Kiwi	2167000	72	7	1.737	0.0571	0.1719	0.1391	0.1158	0.0977	0.0830
Bluebird	2156000	84	19	1.762	0.0588	0.1707	0.1379	0.1146	0.0966	0.0818
Chukar	1780000	84	19	1.602	0.0536	0.1744	0.1416	0.1184	0.1003	0.0854
Falcon	1590000	54	19	1.545	0.0523	0.1754	0.1426	0.1193	0.1013	0.0866
Lapwing	1590000	45	7	1.502	0.0498	0.1774	0.1446	0.1213	0.1033	0.0885
Parrot	1510500	54	19	1.506	0.0506	0.1768	0.1440	0.1207	0.1026	0.0879
Nuthatch	1510500	45	7	1.466	0.0486	0.1784	0.1456	0.1223	0.1043	0.0895
Plover	1431000	54	19	1.465	0.0494	0.1777	0.1449	0.1217	0.1036	0.0889
Bobolink	1431000	45	7	1.427	0.0470	0.1797	0.1469	0.1237	0.1056	0.0909
Martin	1351500	54	19	1.424	0.0482	0.1787	0.1459	0.1227	0.1046	0.0899
Dipper	1351500	45	7	1.385	0.0459	0.1807	0.1479	0.1246	0.1066	0.0918
Pheasant	1272000	54	19	1.382	0.0466	0.1801	0.1473	0.1240	0.1060	0.0912
Bittern	1272000	45	7	1.345	0.0444	0.1820	0.1492	0.1260	0.1079	0.0932
Grackle	1192500	54	19	1.333	0.0451	0.1814	0.1486	0.1253	0.1073	0.0925
Bunting	1192500	45	7	1.302	0.0429	0.1834	0.1506	0.1271	0.1093	0.0946
Finch	1113000	54	19	1.293	0.0436	0.1828	0.1500	0.1267	0.1087	0.0939
Bluejay	1113000	45	7	1.259	0.0415	0.1848	0.1520	0.1287	0.1107	0.0959
Curlew	1033500	54	7	1.246	0.0420	0.1843	0.1515	0.1282	0.1102	0.0954
Ortolan	1033500	45	7	1.213	0.0402	0.1861	0.1533	0.1300	0.1119	0.0972
Tanager	1033500	36	1	1.186	0.0384	0.1879	0.1551	0.1318	0.1138	0.0990
Cardinal	954000	54	7	1.196	0.0402	0.1861	0.1533	0.1300	0.1119	0.0972
Rail	954000	45	7	1.165	0.0386	0.1877	0.1549	0.1316	0.1136	0.0988
Catbird	954000	36	1	1.140	0.0370	0.1894	0.1566	0.1333	0.1153	0.1005
Canary	900000	54	7	1.162	0.0392	0.1871	0.1543	0.1329	0.1130	0.0982
Ruddy	900000	45	7	1.131	0.0374	0.1890	0.1562	0.1310	0.1149	0.1001
Mallard	795000	30	19	1.140	0.0392	0.1871	0.1543	0.1330	0.1130	0.0982
Drake	795000	26	7	1.108	0.0373	0.1891	0.1563	0.1333	0.1150	0.1002
Condor	795000	54	7	1.093	0.0370	0.1894	0.1566	0.1338	0.1153	0.1005
Cuckoo	795000	24	7	1.092	0.0366	0.1899	0.1571	0.1354	0.1157	0.1010
Tern	795000	45	7	1.063	0.0352	0.1914	0.1586	0.1326	0.1173	0.1026
Coot	795000	36	1	1.040	0.0377	0.1887	0.1559	0.1330	0.1145	0.0998
Redwing	715500	30	19	1.081	0.0373	0.1891	0.1563	0.1350	0.1150	0.1022
Starling	715500	26	7	1.051	0.0355	0.1911	0.1583	0.1310	0.1170	0.1031
Stilt	715500	24	7	1.036	0.0347	0.1920	0.1583	0.1359	0.1179	0.1036
Gannet	666600	26	7	1.014	0.0343	0.1925	0.1597	0.1364	0.1184	0.1046
Flamingo	666600	24	7	1.000	0.0355	0.1934	0.1606	0.1374	0.1193	0.1080

TABLE C.4a Inductive Reactance Spacing Factor, x_d

	0.0	0.1	0.2	0.3	0.4	0.5	0.6	0.7	0.8	0.9
0	—	−0.2794	−0.1953	−0.1461	−0.1112	−0.0841	−0.0620	−0.0433	−0.0271	−0.0128
1	0.0	0.0116	0.0221	0.0318	0.0408	0.0492	0.0570	0.0644	0.0713	0.0779
2	0.0841	0.0900	0.0957	0.1011	0.1062	0.1112	0.1159	0.1205	0.1249	0.1292
3	0.1333	0.1373	0.1411	0.1449	0.1485	0.1520	0.1554	0.1588	0.1620	0.1651
4	0.1682	0.1712	0.1741	0.1770	0.1798	0.1825	0.1852	0.1878	0.1903	0.1928
5	0.1953	0.1977	0.2001	0.2024	0.2046	0.2069	0.2090	0.2112	0.2133	0.2154
6	0.2174	0.2194	0.2214	0.2233	0.2252	0.2271	0.2290	0.2308	0.2326	0.2344
7	0.2361	0.2378	0.2395	0.2412	0.2429	0.2445	0.2461	0.2477	0.2493	0.2508
8	0.2523	0.2538	0.2553	0.2568	0.2582	0.2597	0.2611	0.2625	0.2639	0.2653
9	0.2666	0.2680	0.2693	0.2706	0.2719	0.2732	0.2744	0.2757	0.2769	0.2782
10	0.2794	0.2806	0.2818	0.2830	0.2842	0.2853	0.2865	0.2876	0.2887	0.2899
11	0.2910	0.2921	0.2932	0.2942	0.2953	0.2964	0.2974	0.2985	0.2995	0.3005
12	0.3015	0.3025	0.3035	0.3045	0.3055	0.3065	0.3074	0.3084	0.3094	0.3103
13	0.3112	0.3122	0.3131	0.3140	0.3149	0.3158	0.3167	0.3175	0.3185	0.3194
14	0.3202	0.3211	0.3219	0.3228	0.3236	0.3245	0.3253	0.3261	0.3270	0.3278
15	0.3286	0.3294	0.3302	0.3310	0.3318	0.3326	0.3334	0.3341	0.3349	0.3357
16	0.3364	0.3372	0.3379	0.3387	0.3394	0.3402	0.3409	0.3416	0.3424	0.3431
17	0.3438	0.3445	0.3452	0.3459	0.3466	0.3473	0.3480	0.3487	0.3494	0.3500
18	0.3507	0.3514	0.3521	0.3527	0.3534	0.3540	0.3547	0.3554	0.3560	0.3566
19	0.3573	0.3579	0.3586	0.3592	0.3598	0.3604	0.3611	0.3617	0.3623	0.3629
20	0.3635	0.3641	0.3647	0.3653	0.3659	0.3665	0.3671	0.3677	0.3683	0.3688
21	0.3694	0.3700	0.3706	0.3711	0.3717	0.3723	0.3728	0.3784	0.3740	0.3745
22	0.3751	0.3756	0.3762	0.3767	0.3773	0.3778	0.3783	0.3789	0.3794	0.3799
23	0.3805	0.3810	0.3815	0.3820	0.3826	0.3831	0.3836	0.3841	0.3846	0.3851
24	0.3856	0.3861	0.3866	0.3871	0.3876	0.3881	0.3886	0.3891	0.3896	0.3901
25	0.3906	0.3911	0.3916	0.3920	0.3925	0.3930	0.3935	0.3939	0.3944	0.3949
26	0.3953	0.3958	0.3963	0.3967	0.3972	0.3977	0.3981	0.3986	0.3990	0.3995
27	0.3999	0.4004	0.4008	0.4013	0.4017	0.4021	0.4026	0.4060	0.4035	0.4039
28	0.4043	0.4048	0.4052	0.4056	0.4061	0.4065	0.4069	0.4073	0.4078	0.4082
29	0.4086	0.4090	0.4094	0.4098	0.4103	0.4107	0.4111	0.4115	0.4119	0.4123
30	0.4127	0.4131	0.4135	0.4139	0.4143	0.4147	0.4151	0.4155	0.4159	0.4163
31	0.4167	0.4171	0.4175	0.4179	0.4182	0.4186	0.4190	0.4194	0.4198	0.4202
32	0.4205	0.4209	0.4213	0.4217	0.4220	0.4224	0.4228	0.4232	0.4235	0.4239
33	0.4243	0.4246	0.4250	0.4254	0.4257	0.4261	0.4265	0.4272	0.4272	0.4275
34	0.4279	0.4283	0.4286	0.4290	0.4293	0.4297	0.4300	0.4307	0.4307	0.4311
35	0.4314	0.4318	0.4321	0.4324	0.4328	0.4331	0.4335	0.4342	0.4342	0.4345
36	0.4348	0.4352	0.4355	0.4358	0.4362	0.4365	0.4368	0.4375	0.4375	0.4378
37	0.4382	0.4385	0.4388	0.4391	0.4395	0.4398	0.4401	0.4408	0.4408	0.4411
38	0.4414	0.4417	0.4420	0.4423	0.4427	0.4430	0.4433	0.4439	0.4439	0.4442
39	0.4445	0.4449	0.4452	0.4455	0.4458	0.4461	0.4464	0.4470	0.4470	0.4473
40	0.4476	0.4479	0.4482	0.4485	0.4488	0.4491	0.4494	0.4500	0.4500	0.4503
41	0.4506	0.4509	0.4512	0.4515	0.4518	0.4521	0.4524	0.4530	0.4530	0.4532
42	0.4535	0.4538	0.4541	0.4544	0.4547	0.4550	0.4553	0.4558	0.4558	0.4561
43	0.4564	0.4567	0.4570	0.4572	0.4575	0.4578	0.4581	0.4586	0.4586	0.4589
44	0.4592	0.4595	0.4597	0.4600	0.4603	0.4606	0.4609	0.4614	0.4614	0.4616
45	0.4619	0.4622	0.4624	0.4627	0.4630	0.4632	0.4635	0.4640	0.4640	0.4643
46	0.4646	0.4648	0.4651	0.4654	0.4656	0.4659	0.4661	0.4667	0.4667	0.4669
47	0.4672	0.4674	0.4577	0.4680	0.4682	0.4685	0.4687	0.4692	0.4692	0.4695
48	0.4697	0.4700	0.4702	0.4705	0.4707	0.4710	0.4712	0.4717	0.4717	0.4720
49	0.4722	0.4725	0.4727	0.4730	0.4732	0.4735	0.4737	0.4742	0.4742	0.4744
50	0.4747	0.4749	0.4752	0.4754	0.4757	0.4759	0.4761	0.4766	0.4766	0.4769

TABLE C.4b Inductive Reactance Spacing Factor, x_d

	0.0	0.1	0.2	0.3	0.4	0.5	0.6	0.7	0.8	0.9
51	0.4771	0.4773	0.4776	0.4778	0.4780	0.4783	0.4785	0.4787	0.4790	0.4792
52	0.4795	0.4797	0.4799	0.4801	0.4804	0.4806	0.4808	0.4811	0.4813	0.4815
53	0.4818	0.4820	0.4822	0.4824	0.4827	0.4829	0.4831	0.4834	0.4836	0.4838
54	0.4840	0.4843	0.4845	0.4847	0.4849	0.4851	0.4854	0.4856	0.4858	0.4860
55	0.4863	0.4865	0.4867	0.4869	0.4871	0.4874	0.4876	0.4878	0.4880	0.4882
56	0.4884	0.4887	0.4889	0.4891	0.4893	0.4895	0.4897	0.4900	0.4902	0.4904
57	0.4906	0.4908	0.4910	0.4912	0.4914	0.4917	0.4919	0.4921	0.4923	0.4925
58	0.4927	0.4925	0.4931	0.4933	0.4935	0.4937	0.4940	0.4942	0.4944	0.4946
59	0.4948	0.4950	0.4952	0.4954	0.4956	0.4958	0.4960	0.4962	0.4964	0.4966
60	0.4968	0.4970	0.4972	0.4974	0.4976	0.4978	0.4980	0.4982	0.4984	0.4986
61	0.4988	0.4990	0.4992	0.4994	0.4996	0.4998	0.5000	0.5002	0.5004	0.5006
62	0.5008	0.5010	0.5012	0.5014	0.5016	0.5018	0.5020	0.5022	0.5023	0.5025
63	0.5027	0.5029	0.5031	0.5033	0.5035	0.5037	0.5039	0.5041	0.5043	0.5045
64	0.5046	0.5048	0.5050	0.5052	0.5054	0.5056	0.5058	0.5060	0.5062	0.5063
65	0.5065	0.5067	0.5069	0.5071	0.5073	0.5075	0.5076	0.5078	0.5080	0.5082
66	0.5084	0.5086	0.5087	0.5089	0.5091	0.5093	0.5095	0.5097	0.5098	0.5100
67	0.5102	0.5104	0.5106	0.5107	0.5109	0.5111	0.5113	0.5115	0.5116	0.5118
68	0.5120	0.5122	0.5124	0.5125	0.5127	0.5129	0.5131	0.5132	0.5134	0.5136
69	0.5138	0.5139	0.5141	0.5143	0.5145	0.5147	0.5148	0.5150	0.5152	0.5153
70	0.5155	0.5157	0.5159	0.5160	0.5162	0.5164	0.5166	0.5167	0.5169	0.5171
71	0.5172	0.5174	0.5174	0.5178	0.5179	0.5181	0.5183	0.5184	0.5186	0.5188
72	0.5189	0.5191	0.5193	0.5194	0.5196	0.5198	0.5199	0.5201	0.5203	0.5204
73	0.5206	0.5208	0.5209	0.5211	0.5213	0.5214	0.5216	0.5218	0.5219	0.5221
74	0.5223	0.5224	0.5226	0.5228	0.5229	0.5231	0.5232	0.5234	0.5236	0.5237
75	0.5329	0.5241	0.5242	0.5244	0.5245	0.5247	0.5249	0.5250	0.5252	0.5253
76	0.5255	0.5257	0.5258	0.5260	0.5261	0.5263	0.5265	0.5266	0.5268	0.5269
77	0.5271	0.5272	0.5274	0.5276	0.5277	0.5279	0.5280	0.5282	0.5283	0.5285
78	0.5287	0.5288	0.5290	0.5291	0.5293	0.5294	0.5296	0.5297	0.5299	0.5300
79	0.5302	0.5304	0.5305	0.5307	0.5308	0.5310	0.5311	0.5313	0.5314	0.5316
80	0.5317	0.5319	0.5320	0.5322	0.5323	0.5325	0.5326	0.5328	0.5329	0.5331
81	0.5332	0.5334	0.5335	0.5337	0.5338	0.5340	0.5341	0.5343	0.5344	0.5346
82	0.5347	0.5349	0.5350	0.5352	0.5353	0.5355	0.5356	0.5358	0.5359	0.5360
83	0.5362	0.5363	0.5365	0.5366	0.5368	0.5369	0.5371	0.5372	0.5374	0.5375
84	0.5376	0.4378	0.5379	0.5381	0.5382	0.5384	0.5385	0.5387	0.5388	0.5389
85	0.5391	0.5392	0.5394	0.5395	0.5396	0.5398	0.5399	0.5401	0.5402	0.5404
86	0.5405	0.5406	0.5408	0.5409	0.5411	0.5412	0.5413	0.5415	0.5416	0.5418
87	0.5419	0.5420	0.5422	0.5423	0.5425	0.5426	0.5427	0.5429	0.5430	0.5432
88	0.5433	0.5434	0.5436	0.5437	0.5438	0.5440	0.5441	0.5442	0.5444	0.5445
89	0.5447	0.5448	0.5449	0.5451	0.5452	0.5453	0.5455	0.5456	0.5457	0.5459
90	0.5460	0.5461	0.5463	0.5464	0.5466	0.5467	0.5468	0.5470	0.5471	0.5472
91	0.5474	0.5475	0.5476	0.5478	0.5479	0.5480	0.5482	0.5483	0.5484	0.5486
92	0.5487	0.5488	0.5489	0.5491	0.5492	0.5493	0.5495	0.5496	0.5497	0.5499
93	0.5500	0.5501	0.5503	0.5504	0.5505	0.5506	0.5508	0.5509	0.5510	0.5512
94	0.5513	0.5514	0.5515	0.5517	0.5518	0.5519	0.5521	0.5522	0.5523	0.5524
95	0.5526	0.5527	0.5528	0.5530	0.5531	0.5532	0.5533	0.5535	0.5536	0.5537
96	0.5538	0.5540	0.5541	0.5542	0.5544	0.5545	0.5546	0.5547	0.5549	0.5550
97	0.5551	0.5552	0.5554	0.5555	0.5556	0.5557	0.5559	0.5560	0.5561	0.5562
98	0.5563	0.5565	0.5566	0.5567	0.5568	0.5570	0.5571	0.5572	0.5573	0.5575
99	0.5576	0.5577	0.5578	0.5579	0.5581	0.5582	0.5583	0.5584	0.5586	0.5587
100	0.5588	0.5589	0.5590	0.5592	0.5593	0.5594	0.5595	0.5596	0.5598	0.5599

TABLE C.5 Electrical Characteristics of Overhead Ground Wires[1]

Alumoweld Strand[2]	Resistance (Ohms/Mile) Small Currents dc-25°C	60 Hz-25°	75% of Cap. dc-75°C	60 Hz-75°	60 Hz Reactance @ 1 ft Inductive Ω/mile	Capacitive Ω-miles	60 Hz GMR ft
7 No. 5	1.217	1.240	1.432	1.669	0.707	0.1122	0.002958
7 No. 6	1.507	1.536	1.773	2.010	0.721	0.1157	0.002633
7 No. 7	1.900	1.937	2.240	2.470	0.735	0.1191	0.002345
7 No. 8	2.400	2.440	2.820	3.060	0.749	0.1226	0.002085
7 No. 9	3.020	3.080	3.560	3.800	0.763	0.1260	0.001858
7 No. 10	3.810	3.880	4.480	4.730	0.777	0.1294	0.001658
3 No. 5	2.780	2.780	3.270	3.560	0.707	0.1221	0.002940
3 No. 6	3.510	3.510	4.130	4.410	0.721	0.1255	0.002618
3 No. 7	4.420	4.420	5.210	5.470	0.735	0.1289	0.002333
3 No. 8	5.580	5.580	6.570	6.820	0.749	0.1324	0.002078
3 No. 9	7.040	7.040	8.280	8.520	0.763	0.1358	0.001853
3 No. 10	8.870	8.870	10.440	10.670	0.777	0.1392	0.001650

Single Layer ACSR[3]	DC 25 C	Resistance (Ohms/Mile) 60 Hz I = 0[4]	I = 100	I = 200	60 Hz Reactance @ 1 ft Inductive Ohms/Mile at 75 C I = 0	I = 100	I = 200	Capacitive MΩ-Miles
Brahma	0.394	0.470	0.510	0.565	0.500	0.520	0.545	0.1043
Cochin	0.400	0.480	0.520	0.590	0.505	0.515	0.550	0.1065
Dorking	0.443	0.535	0.575	0.650	0.515	0.530	0.565	0.1079
Dotterel	0.479	0.565	0.620	0.705	0.515	0.530	0.575	0.1091
Guinea	0.531	0.630	0.685	0.780	0.520	0.545	0.590	0.1106
Leghorn	0.630	0.760	0.810	0.930	0.530	0.550	0.605	0.1131
Minorca	0.765	0.915	0.980	1.130	0.540	0.570	0.640	0.1160
Petrel	0.830	1.000	1.065	1.220	0.550	0.580	0.655	0.1172
Grouse	1.080	1.295	1.420	1.520	0.570	0.640	0.675	0.1240

Steel Conductors (7-Strand) Grade[5]	Dia. In.	Resistance (Ohms/Mile) 60 Hz I = 0[3]	I = 30	I = 60	60 Hz Reactance @ 1 ft Inductive Ohms/Mile I = 0	I = 30	I = 60	Capacitive MΩ-Miles
Ordinary	1/4	9.5	11.4	11.3	1.3970	3.7431	3.4379	0.1354
Ordinary	9/32	7.1	9.2	9.0	1.2027	3.0734	2.5146	0.1319
Ordinary	5/16	5.4	7.5	7.8	0.8382	2.5146	2.0409	0.1288
Ordinary	3/8	4.3	6.5	6.6	0.8382	2.2352	1.9687	0.1234
Ordinary	1/2	2.3	4.3	5.0	0.7049	1.6893	1.4236	0.1148
E.B.	1/4	8.0	12.0	10.1	1.2027	4.4704	3.1565	0.1354
E.B.	9/32	6.0	10.0	8.7	1.1305	3.7783	2.6255	0.1319
E.B.	5/16	4.9	8.0	7.0	0.9843	2.9401	2.5146	0.1288
E.B.	3/8	3.7	7.0	6.3	0.8382	2.5997	2.4303	0.1234
E.B.	1/2	2.1	4.9	5.0	0.7049	1.8715	1.7615	0.1148
E.B.B.	1/4	7.0	12.8	10.9	1.6764	5.1401	3.9482	0.1354
E.B.B.	9/32	5.4	10.9	8.7	1.1305	4.4833	3.7783	0.1319
E.B.B.	5/16	4.0	9.0	6.8	0.9843	3.6322	3.0734	0.1288
E.B.B.	3/8	3.5	7.9	6.0	0.8382	3.1168	2.7940	0.1234
E.B.B.	1/2	2.0	5.7	4.7	0.7049	2.3461	2.2352	0.1148

[1]Data compiled from EHV Transmission Line Reference Book.
[2]Data compiled from E.D. 3015 – Copperweld Steel Company.
[3]Data compiled from "Resistance and Reactance of Aluminum Conductors" – ALCOA.
[4]Conductor Current in Amperes.
[5]Data compiled from "Symmetrical Components" Wagner & Evans (Book) McGraw-Hill.

REFERENCES

[1] Anderson, P. M., *Analysis of Faulted Power Systems*, IEEE Press, Piscataway, NJ, 1995.

[2] Westinghouse Engineers, "Electrical Transmission and Distribution Reference Book," Westinghouse Electric Corporation, East Pittsburgh, PA, 1950.

D

Transformer Data

TABLE D.1 Impedances for Two-Winding Power Transformers

High-Voltage Winding Insulation Class kV	Low-Voltage Winding Insulation Class kV	Impedance Limits in Percent: Class OA OW OA/FA* OA/FA/FOA* Min	Max	Class FOA FOW Min	Max
15	15	4.5	7.0	6.75	10.5
25	15	5.5	8.0	8.25	12.0
34.5	15	6.0	8.0	9.0	12.0
	25	6.5	9.0	9.75	13.5
46	25	6.5	9.0	9.75	13.5
	34.5	7.0	10.0	10.5	15.0
69	34.5	7.0	10.0	10.5	15.0
	46	8.0	11.0	12.0	16.5
92	34.5	7.5	10.5	11.25	15.75
	69	8.5	12.5	12.75	18.75
115	34.5	8.0	12.0	12.0	18.0
	69	9.0	14.0	13.5	21.0
	92	10.0	15.0	15.0	23.25
138	34.5	8.5	13.0	12.75	19.5
	69	9.5	15.0	14.25	22.5
	115	10.5	17.0	15.75	25.5
161	46	9.5	15.0	13.5	21.0
	92	10.5	16.0	15.75	24.0
	138	11.5	18.0	17.25	27.0
196	46	10.0	15.0	15.0	22.5
	92	11.5	17.0	17.25	25.5
	161	12.5	19.0	18.75	28.5
230	46	11.0	16.0	16.5	24.0
	92	12.5	18.0	18.75	27.0
	161	14.0	20.0	21.0	30.0

Source: Westinghouse Electric Corp. Used with permission.

* The impedances are expressed in percent on the self-cooled rating of OA/FA and OA/FA/FOA.

Definition of transformer classes:

OA – Oil-immersed, self-cooled OW – Oil immersed, water-cooled.

OA/FA – Oil-immersed, self-cooled/forced-air-cooled/forced-oil-cooled.

FOA – Oil-immersed, forced-oil-cooled wth forced-air cooler.

FOW – Oil-immersed, forced-oil-cooled with water cooler.

Note: The through impedance of a two-winding autotransformer can be estimated knowing rated circuit voltages, by multiplying impedance obtained from this table by the factor (HV– LV/HV).

TABLE D.2 Approximate Distribution Transformer Impedances

	2400/4160 Y to 120/240 Volts 60 Hertz			4800/8320 Y to 120/240 Volts 60 Hertz			7200/12470 Y to 120/240 Volts 60 Hertz			14440/24940 GrdY-120/240 V 60 Hz			34500 GrdY/ 19920† to 120 /240 V 60 Hz		
kVA	%IR	%IX	%IZ	%IR	%IX	% IZ	%IR	%IX	% IZ	%IR	%IX	% IZ	%IR	%IX	% IZ
5	1.9	1.6	2.5	2.2	1.6	2.7	2.2	2.2	3.1	2.5	2.2	3.3			
10	4.4	1.0	1.7	1.4	1.0	1.7	1.4	1.0	1.7	1.6	1.0	1.9	1.4	1.0	1.7
15	1.2	1.2	1.7	1.2	1.2	1.7	1.3	1.2	1.8	1.4	1.7	2.2	1.4	1.7	2.2
25	1.1	1.3	1.7	1.1	1.3	1.8	1.2	1.6	2.0	1.3	1.8	2.2	1.3	1.5	2.0
37.5	0.9	1.4	1.7	1.0	1.4	1.7	1.1	1.4	1.8	1.1	1.6	1.9	1.2	1.7	2.1
50	1.0	1.2	1.6	1.0	1.2	1.6	1.1	1.3	1.7	1.1	1.8	2.1	1.2	1.5	1.9
75	0.9	1.3	1.6	1.0	1.2	1.6	1.0	1.5	1.8	1.1	1.9	2.2	1.0	1.6	1.9
100	0.9	1.6	1.8	0.9	1.4	1.7	0.9	1.4	1.7	1.0	2.0	2.2	1.0	1.5	1.8
167	0.9	1.7	1.9	0.9	1.7	1.9	0.9	1.7	1.9	0.9	2.1	2.3	0.9	1.8	2.0
	240 480			240 480			240 480			240 480			240 480		
250	0.8	2.9	3.0	0.8	2.9	3.0	0.8	2.9	3.0	0.8	2.9	3.0	0.8	2.9	3.0
333	0.8	3.2	3.3	0.8	3.2	3.3	0.8	3.2	3.3	0.8	3.1	3.2	0.8	3.1	3.2
500	0.7	3.2	3.3	0.7	3.2	3.3	0.7	3.2	3.5	0.7	3.3	3.4	0.7	3.3	3.4

When only one of the two low voltage windings is loaded, the %IR and %IX (on kVA base equal to capacity of that winding, or one-half nameplate kVA) are approximately 0.75 and from 0.6 to 1.25 of full winding %IR and full winding %IX, respectively.
† 125 kV (From General Electric Company. Used with permission.)

TABLE D.3 Full Load Currents for Distribution Transformer Single-Phase Circuits

	Circuit Voltage												
kVA	120	240	480	2400	4160	4330	4880	6600	6900	7200	7620	11.5 k	13.2 k
1.5	12.5	6.3	3.1	.63	.36	.35	.31	.23	.22	.21	.20	.13	.11
2.5	20.8	10.4	5.2	1.04	.60	.58	.52	.38	.36	.35	.33	.22	.19
3	25.0	12.5	6.3	1.25	.72	.69	.63	.45	.43	.42	.39	.26	.23
5	41.7	20.8	10.4	2.08	1.20	1.16	1.04	.76	.72	.69	.66	.43	.38
7.5	62.5	31.3	15.6	3.13	1.80	1.73	1.56	1.14	1.09	1.04	.98	.65	.57
10	83.3	41.7	20.8	4.17	2.40	2.31	2.08	1.52	1.45	1.39	1.31	.87	.76
15	125	62.5	31.3	6.25	3.61	3.46	3.13	2.27	2.17	2.08	1.97	1.30	1.14
25	208	104	52.1	10.4	6.01	5.77	5.21	3.79	3.62	3.47	3.28	2.17	1.89
37.5	313	156	78.1	15.6	9.01	8.66	7.81	5.68	5.43	5.21	4.92	3.26	2.84
50	417	208	104	20.8	12.00	11.55	10.4	7.58	7.25	6.94	6.56	4.35	3.79
75	625	313	156	31.3	18.00	17.32	15.6	11.4	10.9	10.4	9.84	6.52	5.68
100	833	417	208	41.7	24.00	23.10	20.8	15.2	14.5	13.9	13.1	8.70	7.58
150	1250	625	313	62.5	36.10	34.64	31.3	22.7	21.7	20.8	19.7	13.0	11.4
200	1667	833	417	83.3	48.10	46.19	41.7	30.3	29.0	27.8	26.2	17.4	15.2
250	2083	1042	521	104	60.10	57.74	52.1	37.9	36.3	34.7	32.8	21.7	18.9
333	2775	1388	694	139	80.00	76.91	69.4	50.5	48.3	46.2	43.7	29.0	25.2
500	4167	2083	1042	208	120.	115.47	104	75.8	72.5	69.4	65.6	43.5	37.9

TABLE D.4 Full Load Currents for Distribution Transformer Three-Phase Circuits

	Circuit Voltage												
kVA	**208**	**240**	**480**	**2400**	**4160**	**4330**	**4800**	**6900**	**7200**	**8320**	**11.5k**	**13.2k**	**33 k**
4.5	12.5	10.8	5.41	1.08	.62	.60	.54	.36	.38	.31	.23	.20	.08
7.5	20.8	18.0	9.02	1.80	1.04	1.00	.90	.61	.63	.52	.38	.33	.13
9	25.0	21.7	10.8	2.17	1.25	1.20	1.08	.73	.75	.62	.45	.39	.16
10	27.8	24.1	12.0	2.41	1.39	1.33	1.20	.80	.84	.69	.50	.44	.17
15	41.6	36.1	18.0	3.61	2.08	2.00	1.80	1.20	1.26	1.04	.75	.66	.26
22.5	62.5	54.1	27.1	5.41	3.12	3.00	2.71	1.80	1.88	1.56	1.13	.98	.39
25	69.4	60.1	30.1	6.01	3.47	3.33	3.01	2.00	2.09	1.73	1.26	1.09	.44
30	83.3	72.2	36.1	7.22	4.16	4.00	3.61	2.41	2.51	2.08	1.51	1.31	.52
37.5	104	90.2	45.1	9.02	5.20	5.00	4.51	3.01	3.14	2.60	1.88	1.64	.66
45	125	108	54.1	10.8	6.25	6.00	5.41	3.60	3.77	3.12	2.26	1.97	.79
50	139	120	60.1	12.0	6.94	6.67	6.01	4.01	4.18	3.47	2.51	2.19	.87
75	208	180	90.2	18.0	10.4	10.00	9.02	6.01	6.28	5.21	3.77	3.28	1.31
100	278	241	120	24.1	13.9	13.33	12.0	8.02	8.37	6.94	5.02	4.37	1.75
112.5	312	271	135	27.1	15.6	15.00	13.5	9.02	9.41	7.81	5.65	4.92	1.97
150	416	361	180	36.1	20.8	20.00	18.0	12.0	12.6	10.4	7.53	6.56	2.62
200	555	481	241	48.1	27.8	26.67	24.1	16.0	16.7	13.9	10.0	8.75	3.50
225	625	541	271	54.1	31.2	30.00	27.1	18.0	18.8	15.6	11.3	9.84	3.94
300	833	722	361	72.2	41.6	40.00	36.1	24.1	25.1	20.8	15.1	13.1	5.25
450	1249	1083	541	108	62.5	60.00	54.1	36.0	37.7	31.2	22.6	19.7	7.87
500	1388	1203	601	120	69.3	66.67	60.1	40.1	41.8	34.7	25.1	21.9	8.74
600	1665	1443	722	144	83.3	80.00	72.2	48.1	50.2	41.6	30.1	26.2	10.5
750	2082	1804	902	180	104	100.01	90.2	60.1	62.8	52.0	37.7	32.8	13.1
1000	2776	2406	1203	241	139	133.34	120	80.0	83.7	69.4	50.2	43.7	17.5
1500	4164	3609	1804	361	208	200.01	180	120	126	104	75.3	65.6	26.2

E

500 kV Transmission Line Data

For the purpose of illustrations and examples used throughout the book, it is convenient to use data that are consistent and realistic. Since many of the examples cited pertain to power transmission lines, it is appropriate to use the same line parameters for all illustrations. This makes the various examples consistent, and ensures the reader that comparisons between examples are not due to the choice of transmission line parameters.

The transmission line parameters chosen for illustration are almost exactly those of lines that exist in the western United States. All frequency-dependent data for the standard line is based on a frequency of 60 hertz.

E.1 TOWER DESIGN

The transmission towers are lattice steel construction with a spacing between conductors as shown in Figure E.1.

The conductors of the standard line are horizontal twin bundled 2156 kCM (1.092 mm^2)[1] ACSR (Bluebird), with a stranding of 84/19. The conductors have an outside diameter of 4.475 cm (1.762 in). The horizontal spacing between adjacent phases is 9.754 meters (32 ft) and the horizontal spacing between conductors of the same phase bundle is 45.7 cm (18 in.). The shield wire is 7#8 alumoweld conductor. The horizontal separation between the shield wires is 12.2 m (40 ft), and the shield wires are centered at an elevation 8.23 m (27 ft) above the phase wires. The height of the phase wires above the ground is nominally 27.4 m (90 ft), but base extension structures are used to raise the tower height, as required, for ground clearance.

The line is constructed with approximately 2 towers per km. For the sake of illustration, the capital cost of the transmission line is taken to be US$350,000 per km.

[1]There are 1974 circular mils in 1 square millimeter [1].

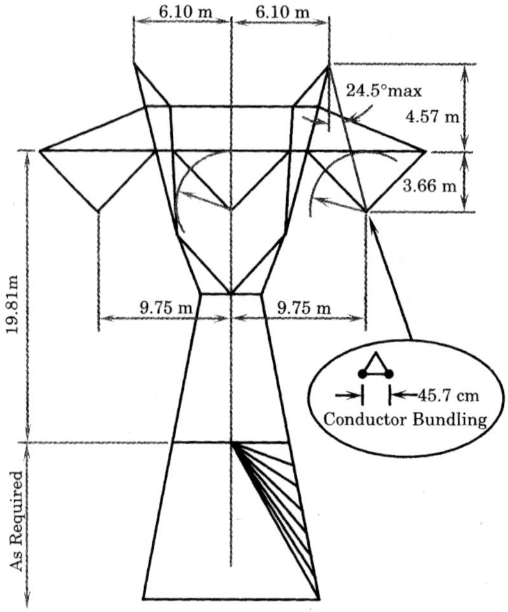

Figure E.1 Tower configuration of the standard line.

E.2 UNIT LENGTH ELECTRICAL CHARACTERISTICS

The following unit length parameters are given for the standard line:

$$\begin{aligned} z &= 0.01864 + j0.3728 = 0.373288\ \Omega/\text{km}\ \angle 0.3728\ \text{rad} \\ &= 0.0300 + j0.6000 = 0.60075\ \Omega/\text{mile}\ \angle 0.3728\ \text{rad} \end{aligned} \tag{E.1}$$

$$\begin{aligned} y &= 0 + j4.4739\ \mu\text{S/km} \\ &= 0 + j7.200\ \mu\text{S/mile} \end{aligned} \tag{E.2}$$

The characteristic impedance and admittance of the line are computed as follows:

$$\begin{aligned} Z_C &= \frac{1}{Y_C} = \sqrt{\frac{z}{y}} = \sqrt{\frac{0.01864 + j0.3728}{j4.4739 \times 10^{-6}}} \\ &= 288.765275 - j7.214626\ \Omega = 288.855388\ \Omega \angle -1.431203^\circ \end{aligned} \tag{E.3}$$

The propagation constant is also computed from unit length parameters.

$$\begin{aligned} \gamma &= \alpha + j\beta = \sqrt{zy} \\ &= 0.00003228\ \text{neper/km} + j0.00129190\ \text{radians/km} \\ &= 0.00005195\ \text{neper/mi} + j0.00207911\ \text{radians/mi} \end{aligned} \tag{E.4}$$

E.3 TOTAL LINE IMPEDANCE AND ADMITTANCE

Let the length of the standard line be set to 320 km (about 200 miles). Then we can compute the following.

$$\begin{aligned} \ell &= 320\ \text{km} \\ \gamma\ell &= 0.0103287\ \text{neper} + j0.4134077\ \text{rad} \end{aligned} \tag{E.5}$$

Then the total nominal line impedance and admittance can be computed as

$$\begin{aligned} \boldsymbol{Z} &= z\ell = 5.965163 + j119.303269\ \Omega \\ \boldsymbol{Y} &= y\ell = 0 + j1431.639227\ \mu\text{S} \end{aligned} \tag{E.6}$$

E.4 NOMINAL PI

The nominal pi is an approximate model for the transmission line, which sets the total line impedance, given in (E.6), for the series impedance between the ends of the line and divides the total susceptance into two equal susceptances of one-half the total amount, with these two susceptances located at the ends of the line. The arrangement is shown in Figure E.2.

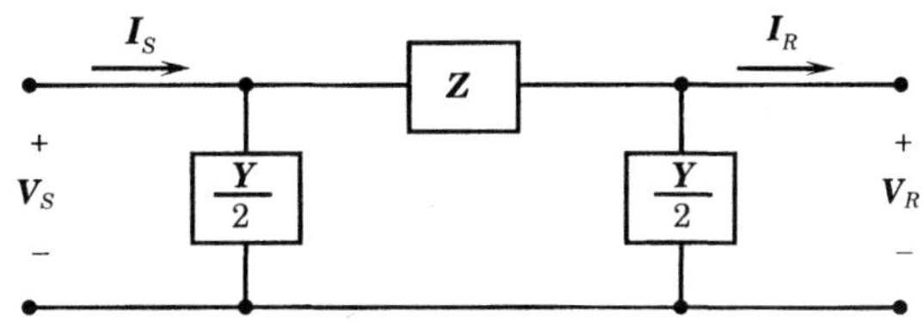

Figure E.2 Nominal pi circuit for the transmission line.

Using the data computed above, the parameters for the nominal pi line are as follows:

$$\begin{aligned} \boldsymbol{Z} &= 5.965163 + j119.303269\ \Omega \\ \boldsymbol{Y}/2 &= +j715.819614\ \mu\text{S} \end{aligned} \tag{E.7}$$

E.5 *ABCD* PARAMETERS

Another convenient way of representing the transmission line of Figure E.2, is by the use of the two-port network equations, particularly the ***ABCD*** equations, which are used for analyzing cascaded two-port networks. These equations are written as

$$\begin{bmatrix} \boldsymbol{V}_s \\ \boldsymbol{I}_s \end{bmatrix} = \begin{bmatrix} \boldsymbol{A} & \boldsymbol{B} \\ \boldsymbol{C} & \boldsymbol{D} \end{bmatrix} \begin{bmatrix} \boldsymbol{V}_R \\ \boldsymbol{I}_R \end{bmatrix} \tag{E.8}$$

For the case where the transmission line is represented as a symmetrical pi, that is, with equal shunt susceptance at each end of the line, we have

$$\begin{aligned} \boldsymbol{A} &= \cosh \gamma\ell & \boldsymbol{B} &= Z_C \sinh \gamma\ell \\ \boldsymbol{C} &= Y_C \sinh \gamma\ell & \boldsymbol{D} &= \boldsymbol{A} \end{aligned} \tag{E.9}$$

The ***ABCD*** parameters for the standard transmission line, with varying lengths, are given in Table E.1 (see page 1290).

E.6 EQUIVALENT PI

For long transmission lines, the equivalent pi is preferred to the nominal pi for transmission representation in system studies. This line is pictured in Figure E.3.

The parameters of the equivalent pi circuit for the 320 km line are computed as follows:

$$\begin{aligned} \boldsymbol{Z}_\pi &= \boldsymbol{Z}_C \sinh \gamma\ell = \boldsymbol{B} \\ &= 5.629872 + j115.944278\ \Omega \end{aligned} \tag{E.10}$$

TABLE E.1 ***ABCD*** **Constants for the Standard Transmission Line with Varying Length**

ℓ km	ReA	ImA	ReB, Ω	ImB, Ω	ReC, μS	ImC, μS	ℓ mi
0	1.0000000	0.0000000	0.0000000	0.000000	0.00000	0.00000	0.0
20	0.9996664	0.0000167	0.3727398	7.455627	−0.00050	89.46750	12.4
40	0.9986659	0.0000667	0.7449823	14.906293	−0.00398	178.87532	24.9
60	0.9969992	0.0001500	1.1162305	22.347039	−0.01342	268.16379	37.3
80	0.9946673	0.0002664	1.4859886	29.772913	−0.03180	357.27337	49.7
100	0.9916718	0.0004158	1.8537623	37.178974	−0.06208	446.14458	62.1
120	0.9880146	0.0005981	2.2190591	44.560293	−0.10720	534.71816	74.6
140	0.9836984	0.0008129	2.5813892	51.911958	−0.17008	622.93499	87.0
160	0.9787258	0.0010599	2.9402655	59.229075	−0.25363	710.73622	99.4
180	0.9731002	0.0013389	3.2952046	66.506777	−0.36071	798.06329	111.8
200	0.9668254	0.0016495	3.6457270	73.740218	−0.49417	884.85791	124.3
220	0.9599055	0.0019912	3.9913575	80.924586	−0.65682	971.06219	136.7
240	0.9523451	0.0023636	4.3316258	88.055098	−0.85142	1056.61860	149.1
260	0.9441493	0.0027662	4.6660673	95.127008	−1.08070	1141.47007	161.6
280	0.9353236	0.0031984	4.9942227	102.135611	−1.34733	1225.55997	174.0
300	0.9258737	0.0036597	5.3156395	109.076241	−1.65394	1308.83219	186.4
320	0.9158060	0.0041495	5.6298717	115.944278	−2.00310	1391.23118	198.8
340	0.9051272	0.0046670	5.9364805	122.735150	−2.39733	1472.70194	211.3
360	0.8938444	0.0052116	6.2350350	129.444339	−2.83909	1553.19011	223.7
380	0.8819651	0.0057825	6.5251120	136.067377	−3.33075	1632.64198	236.1
400	0.8694972	0.0063790	6.8062973	142.599855	−3.87466	1711.00453	248.5
420	0.8564490	0.0070003	7.0781852	149.037426	−4.47305	1788.22546	261.0
440	0.8428291	0.0076454	7.3403798	155.375802	−5.12810	1864.25322	273.4
460	0.8286468	0.0083136	7.5924946	161.610765	−5.84192	1939.03709	285.8
480	0.8139113	0.0090038	7.8341534	167.738163	−6.61652	2012.52713	298.3
500	0.7986325	0.0097152	8.0649908	173.753916	−7.45383	2084.67430	310.7
520	0.7828206	0.0104467	8.2846519	179.654019	−8.35570	2155.43044	323.1
540	0.7664861	0.0111974	8.4927935	185.434542	−9.32389	2224.74831	335.5
560	0.7496399	0.0119662	8.6890839	191.091636	−10.36007	2292.58163	348.0
580	0.7322932	0.0127520	8.8732035	196.621534	−11.46581	2358.88512	360.4
600	0.7144574	0.0135538	9.0448449	202.020552	−12.64259	2423.61450	372.8
620	0.6961446	0.0143703	9.2037135	207.285095	−13.89177	2486.72655	385.3
640	0.6773669	0.0152004	9.3495278	212.411654	−15.21462	2548.17912	397.7
660	0.6581367	0.0160430	9.4820195	217.396816	−16.61232	2607.93117	410.1
680	0.6384669	0.0168969	9.6009339	222.237258	−18.08593	2665.94280	422.5
700	0.6183707	0.0177607	9.7060302	226.929755	−19.63640	2722.17524	435.0
720	0.5978613	0.0186334	9.7970818	231.471181	−21.26457	2776.59094	447.4
740	0.5769524	0.0195135	9.8738765	235.858508	−22.97116	2829.15353	459.8
760	0.5556579	0.0203999	9.9362166	240.088811	−24.75680	2879.82790	472.2
780	0.5339921	0.0212912	9.9839196	244.159272	−26.62197	2928.58016	484.7
800	0.5119693	0.0221862	10.0168177	248.067175	−28.56707	2975.37775	497.1
820	0.4896042	0.0230834	10.0347588	251.809915	−30.59236	3020.18936	509.5
840	0.4669116	0.0239815	10.0376059	255.384995	−32.69798	3062.98505	522.0
860	0.4439068	0.0248792	10.0252379	258.790031	−34.88395	3103.73618	534.4
880	0.4206050	0.0257751	9.9975496	262.022750	−37.15018	3142.41549	546.8
900	0.3970218	0.0266678	9.9544515	265.080995	−39.49644	3178.99711	559.2

$$Y_\pi/2 = \frac{\cosh\gamma\ell - 1}{Z_C \sinh\gamma\ell} = \frac{A-1}{B} = 0.527286 + j726.184687\ \mu\text{S} \tag{E.11}$$

Values of the equivalent pi parameters for lines of different length are given in Table E.2. Note that the nominal pi is in error by only about 1% for a line of 200 km length, but the error is over

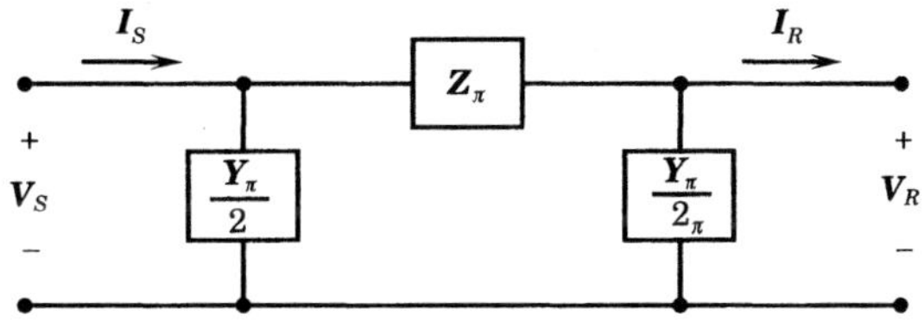

Figure E.3 Equivalent pi circuit for a long transmission line.

TABLE E.2 Standard Line Impedances and Admittances for Varying Length [3]

ℓ km	ReZ_{nom}, Ω	ImZ_{nom}, Ω	ImY_{nom}, μS	ReZ_π, Ω	ImZ_π, Ω	Re$Y_\pi/2$, μS	Im$Y_\pi/2$, μS
0	0.00000	0.00000	0.00000	0.00000	0.00000	0.0000000	0.00000
20	0.37282	7.45645	44.7387	0.37274	7.45563	0.0001244	44.74121
40	0.74565	14.91291	89.4774	0.74498	14.90629	0.0009955	89.49736
60	1.11847	22.36936	134.2161	1.11623	22.34704	0.0033620	134.28338
80	1.49129	29.82582	178.9549	1.48599	29.77291	0.0079768	179.11427
100	1.86411	37.28227	223.6936	1.85376	37.17897	0.0155983	224.00507
120	2.23694	44.73873	268.4323	2.21906	44.56029	0.0269936	268.97093
140	2.60976	52.19518	313.1710	2.58139	51.91196	0.0429395	314.02706
160	2.98258	59.65163	357.9098	2.94027	59.22908	0.0642252	359.18881
180	3.35540	67.10809	402.6485	3.29520	66.50678	0.0916543	404.47169
200	3.72823	74.56454	447.3872	3.64573	73.74022	0.1260473	449.89134
220	4.10105	82.02100	492.1259	3.99136	80.92459	0.1682433	495.46361
240	4.47387	89.47745	536.8647	4.33163	88.05510	0.2191032	541.20456
260	4.84670	96.93391	581.6034	4.66607	95.12701	0.2795114	587.13046
280	5.21952	104.39036	626.3421	4.99422	102.13561	0.3503791	633.25785
300	5.59234	111.84681	671.0808	5.31564	109.07624	0.4326465	679.60355
320	5.96516	119.30327	715.8196	5.62987	115.94428	0.5272859	726.18469
340	6.33799	126.75972	760.5583	5.93648	122.73515	0.6353045	773.01871
360	6.71081	134.21618	805.2970	6.23503	129.44434	0.7577478	820.12343
380	7.08363	141.67263	850.0357	6.52511	136.06738	0.8957029	867.51706
400	7.45645	149.12909	894.7745	6.80630	142.59986	1.0503017	915.21820
420	7.82928	156.58554	939.5132	7.07819	149.03743	1.2227251	963.24592
440	8.20210	164.04199	984.2519	7.34038	155.37580	1.4142066	1011.61978
460	8.57492	171.49845	1028.9906	7.59249	161.61077	1.6260368	1060.35982
480	8.94775	178.95490	1073.7294	7.83415	167.73816	1.8595677	1109.48665
500	9.32057	186.41163	1118.4681	8.06499	173.75392	2.1162177	1159.02147
520	9.69339	193.86781	1163.2068	8.28465	179.65402	2.3974768	1208.98611
540	10.06621	201.32427	1207.9456	8.49279	185.43454	2.7049120	1259.40304
560	10.43904	208.78072	1252.6843	8.68908	191.09164	3.0401731	1310.29547
580	10.81186	216.23717	1297.4230	8.87320	196.62153	3.4049995	1361.68736
600	11.18468	223.69363	1342.1617	9.04484	202.02055	3.8012269	1413.60348
620	11.55750	231.15008	1386.9005	9.20371	207.28509	4.2307943	1466.06943
640	11.93033	238.60654	1431.6392	9.34953	212.41165	4.6957527	1519.11177
660	12.30315	246.06299	1476.3779	9.48202	217.39682	5.1982733	1572.75800
680	12.67597	253.51945	1521.1166	9.60093	222.23726	5.7406567	1627.03667
700	13.04880	260.97590	1565.8554	9.70603	226.92976	6.3253436	1681.97740
720	13.42162	268.43236	1601.5941	9.79708	231.47118	6.9549248	1737.61103
740	13.79444	275.88881	1655.3328	9.87388	235.85851	7.6321542	1793.96960
760	14.16726	283.34526	1700.0715	9.93622	240.08881	8.3599608	1851.08650
780	14.54009	290.80172	1744.8103	9.98392	244.15927	9.1414631	1908.99653
800	14.91291	298.25817	1789.5490	10.01682	248.06718	9.9799844	1967.73599
820	15.28573	305.71463	1834.2877	10.03476	251.80992	10.8790695	2027.34279
840	15.65855	313.17108	1879.0264	10.03761	255.38500	11.8425024	2087.85653
860	16.03138	320.62754	1923.7652	10.02524	258.79003	12.8743269	2149.31865
880	16.40420	328.08399	1968.5039	9.99755	262.02275	13.9788678	2211.77252
900	16.77702	335.54044	2013.2426	9.95445	265.08099	15.1607546	2275.26361

4% for a line of 400 km length. Longer lines have progressively larger errors, if represented as a nominal pi.

From (E.7) it is seen that Z_π is equal to B. However, the other $ABCD$ parameters are not so simply related to the parameters of the equivalent pi circuit. It is easily shown that the following relations apply:

$$\mathbf{A} = \mathbf{D} = 1 + \frac{\mathbf{Z}_\pi \mathbf{Y}_\pi}{2} \tag{E.12}$$

$$\mathbf{C} = \mathbf{Y}\left(1 + \frac{\mathbf{Z}_\pi \mathbf{Y}_\pi}{4}\right) \tag{E.13}$$

E.7 SURGE IMPEDANCE LOADING

The surge impedance loading of a transmission line is described in Chapter 2. For the standard 500 kV line, the surge impedance loading is computed as follows.

$$\text{SIL} = \frac{V_{L-L}^2}{R_C} = \frac{(500)^2}{288.765275} = 865.755 \text{ MW} \tag{E.14}$$

Note that, if the voltage is given as the line-to-line voltage in kV, and the surge impedance is given as a pure resistance in ohms, then the SIL is in MW.

E.8 NORMALIZATION

In many computations, it is convenient to normalize all quantities. Many system analysts prefer to normalize the electric system equations to a 100 MVA base. For the 500 kV line, this gives the following base quantities.

$$\begin{aligned} S_B &= 100 \text{ MVA} & Z_B &= 2500\ \Omega \\ V_B &= 500 \text{ kV} & Y_B &= 400\ \mu\text{mho} \\ I_B &= 115.4701 \text{ A} \end{aligned} \tag{E.15}$$

Then any quantity can be normalized by dividing by the appropriate base quantity. For example, the equivalent pi parameters are given in per unit as follows for a line 320 km long. First, the characteristic impedance in per unit is

$$\mathbf{Z}_C = \frac{288.765275 - j7.214626}{2500} = 0.115506 - j0.0028858 \text{ pu} \tag{E.16}$$

The equivalent pi series impedance is computed in per unit as follows:

$$\begin{aligned} \mathbf{Z}_\pi = \mathbf{Z}_C \sinh \gamma\ell &= \frac{5.62987 + j115.94428\ \Omega}{2500} \\ &= 0.002252 + j0.046378 \text{ pu} \\ &= 0.046432 \text{ pu} \angle 87.220089^\circ \end{aligned} \tag{E.17}$$

The per unit admittance for the equivalent pi line of 320 km length is

$$\begin{aligned} \mathbf{Y}_\pi/2 = \frac{\cosh \gamma\ell - 1}{\mathbf{Z}_C \sinh \gamma\ell} &= 0.52729 + j726.1947 \\ &= 0.0013182 + j1.8154617 \text{ pu} \\ &= 1.8154622 \text{ pu} \angle 89.958397^\circ \end{aligned} \tag{E.18}$$

E.9 LINE RATINGS AND OPERATING LIMITS

Transmission lines are designed to a rated voltage and all apparatus connected to that line are also designed to specified voltage ratings. The ratings are usually given in terms of a design value, plus a maximum operating limit that should not be exceeded. For the nominal 500 kV transmission lines, these ratings are given as follows.

$$\begin{aligned} V &= 500 \text{ kV nominal voltage} \\ &= 1.0 \text{ pu} \end{aligned} \tag{E.19}$$

$$\begin{aligned} V &= 525 \text{ kV operating voltage [2]} \\ &= 1.05 \text{ pu} \end{aligned} \tag{E.20}$$

$$\begin{aligned} V &= 550 \text{ kV maximum continuous operating voltage} \\ &\cong 1.10 \text{ pu} \end{aligned} \tag{E.21}$$

$$\begin{aligned} V &= 576 \text{ kV maximum design voltage [2]} \\ &\cong 1.15 \text{ pu} \end{aligned} \tag{E.22}$$

The current-carrying limit for a transmission line may be based on several factors. The conductor current carrying capability per phase for the line is 3610 amperes based on a maximum conductor temperature of 75°C, ambient temperature of 25°C, and a wind speed of 0.609 m/s (2 ft/s). This corresponds to 3118 MW, which is nearly four times the SIL for the standard line, an unreasonably high level. Since it is not reasonable to operate the line at such a high current level, the physical line design may impose a lower limit based upon conductor sag. Conductor sag must be limited to provide adequate separation between the high-voltage conductors and the earth's surface or objects passing under the line. Since current flow increases conductor temperature, which increases sag, the line current capacity may be sag limited. The line itself may have a high current limit but the line terminal equipment may impose a lower limit. This is particularly true where series capacitors are applied, since series capacitor costs vary as the square of the current magnitude. Therefore, it may not be economical to provide series capacitors with an unusually high rating unless the line is expected to accommodate such high power flows on a regular basis.

REFERENCES

[1] Jay, F., Ed., *IEEE Standard Dictionary of Electrical and Electronics Terms*, ANSI/IEEE Std 100-1988. 4th Ed., IEEE, New York.

[2] ANSI/IEEE Standard C37.100-1981, "IEEE Standard Definitions for Power Switchgear," The Institute of Electrical and Electronics Engineers, New York, 1981.

[3] Anderson, P. M., and R. L. Farmer, "Series Compensation of Power Systems," PBLSH, Inc., Encinitas, CA, 1996.

Index

D

S

About the Author

Dr. P. M. Anderson has been employed his entire career in electric power engineering. He has worked as a utility engineer for Iowa Public Service and as a program manager for the Electric Power Research Institute at Palo Alto, California. He has also been the principal engineer of Power Math Associates, an engineering consulting company.

However, most of his career has been as a university professor of electrical engineering. This included experiences at Iowa State University, Arizona State University, and Washington State University. At Iowa State University he was active in the establishment of a graduate program in Power Engineering. At Arizona State University, Dr. Anderson served as the Electrical Power Systems Professor supported by Arizona Public Service and Salt River Project. Most recently Dr. Anderson was a Schweitzer Visiting Professor of electrical engineering at Washington State University where he taught power system protection and had the opportunity to teach from the manuscript of this book.

Dr. Anderson has authored or co-authored four previous technical books in power system engineering. They are: *Analysis of Faulted Power Systems*, *Power System Control and Stability* (with A. A. Fouad), *Subsynchronous Resonance in Power Systems* (with B. Agrawal and J. E. Van Ness), and *Series Compensation in Power Systems* (with R. G. Farmer).

Dr. Anderson has written many technical papers. He is a Registered Professional Engineer, has been active in many IEEE committees, and is a Life Fellow of the IEEE. He is a member of the IEEE Press Editorial Board and serves as the Editor of the IEEE Press Series in Power Engineering.